THE ROYAL HORTICULTURAL SOCIETY

GARDENERS' ENCYCLOPEDIA *of* PLANTS *and* FLOWERS

Editor-in-chief
CHRISTOPHER BRICKELL

BCA

LONDON   NEW YORK   SYDNEY   TORONTO

A Dorling Kindersley Book

FIRST EDITION

Senior Editor Jane Aspden
Editors Liza Bruml, Joanna Chisholm, Roger Smoothy, Jo Weeks
Additional editorial assistance from Jane Birdsell, Lynn Bresler, Jenny Engelmann, Kate Grant, Shona Grimbly, Susanna Longley, Andrew Mikolajski, Diana Miller, Celia Van Oss, Anthony Whitehorn

Senior Art Editor Ina Stradins
Designer Amanda Lunn

REVISED EDITION

Managing Editor Jane Aspden
Managing Art Editor Bob Gordon
Editors Claire Folkard, Andrew Mikolajski
Editorial Assistants Link Hall, Libby Waldren

Designers Rachel Gibson, Sasha Kennedy
Page make-up Debbie Rhodes
DTP Designer Chris Clark

Photographers Clive Boursnell, Andrew Butler, Eric Crichton, Neil Fletcher, John Glover, Jerry Harpur, Andrew Lawson, Andrew de Lory, with Neil Holmes, Jacqui Hurst
Illustrators Vanessa Luff, Amanda Lunn, Eric Thomas

Advisors and consultants In addition to the contributors listed on the Contents page, Dorling Kindersley would like to thank Barry Ambrose, John Bond, Tony Clements, Steven Davis, Sheila Ecklin, Barbara Ellis, the late Thomas Everett, Jim Gardiner, Ralph Gould, David Kerly and the staff of the Royal Horticultural Society at Vincent Square and Wisley Garden

Copyright © 1989, 1994 Dorling Kindersley Limited, London

First edition published in Great Britain in 1989 by Dorling Kindersley Limited, 9 Henrietta Street, London WC2 8PS

Reprinted and updated 1990, 2/1990, 3/1990, 4/1990, 1991, 2/1991

Revised and expanded edition published in Great Britain in 1994 by Dorling Kindersley Limited

This edition published in 1994 by BCA by arrangement with Dorling Kindersley
Reprinted 1996, 2/1996, 1997

All rights reserved. No part of this book may be reproduced, stored in a retrieval system, or transmitted in any form or by any means, electronic, mechanical, photocopying, recording or otherwise, without the prior permission of the copyright owner.

CN 9651

Text film output by Graphical Innovations, London
Colour reproduction by Colourscan, Singapore
Printed and bound in Italy by A. Mondadori Editore, Verona

Preface

The Royal Horticultural Society Gardeners' Encyclopedia of Plants & Flowers has become a standard reference book with a wide readership in the four years since its publication. The arrangement of plants through the photographic section of the book – by plant type, size, season of interest and colour – has proved immensely popular, while the accuracy of its horticultural information has set high standards.

The publication of this, the first revised edition, demonstrates the Society's commitment to updating the work on a regular basis to ensure that it remains the best possible source of information on plants for amateur gardeners. There is continuous expansion of the range of plants available, which the Editor-in-Chief of this work, Christopher Brickell, highlighted as a chief reason for publishing this encyclopedia. To match this growth the revised edition has 16 new pages in the photographic section, including new feature boxes on a number of genera that have gained in popularity; for example penstemons, daylilies and crocuses. Other feature boxes have been expanded, among these hostas, daffodils and tulips, to take into account changes in the popularity of cultivars.

New pages explain the Society's prestigious Award of Garden Merit (AGM), which recognizes plants of outstanding excellence for garden decoration or use. Throughout the encyclopedia you will notice the symbol ♛ against any plant that has earned the award, and I hope that you will come to associate this symbol with the Society's traditional standards of horticultural excellence.

This new improved 'variety' of *The Royal Horticultural Society Gardeners' Encyclopedia of Plants & Flowers* is even better suited to fulfil its purpose to delight while it informs. I congratulate the Editor-in-Chief, his contributors, and publishers, on the publication of this new revised edition and commend it to you.

Sir Simon Hornby
President, The Royal Horticultural Society
London, Summer, 1994

Contents

Preface 5	**THE PLANT CATALOGUE**	**CLIMBERS** 165
How to Use this Book 8	**TREES**	Clematis 172
Plant Origins and Names 10	Large 38	Ivies 181
	Medium 48	**GRASSES, BAMBOOS, RUSHES and SEDGES** 182
	Magnolias 49	
Creating a Garden 13	Small 59	
	Hollies 72	**FERNS** 186
The Planter's Guide 31	**CONIFERS**	
	Large 75	**PERENNIALS**
	Medium 79	Large 190
	Small 82	Delphiniums 194
	Dwarf conifers 84	Medium 197
	SHRUBS	Irises 198
	Large 86	Peonies 200
	Lilacs 90	Phlox 206
	Medium 97	Pelargoniums 210
	Camellias 98	Penstemons 212
	Rhododendrons and Azaleas 102	Daylilies 221
	Hydrangeas 114	Chrysanthemums 224
	Small 123	Michaelmas daisies 226
	Fuchsias 134	Bromeliads 229
	Heathers 148	Small 231
	ROSES	Primulas 236
	Shrub and Old Garden 150	Carnations and Pinks 244
	Modern 155	Hostas 252
	Miniature 161	Begonias 259
	Climbing 162	Orchids 260
		African violets 266

ANNUALS and BIENNIALS
270

ROCK PLANTS

| Large 294 |
| Small 309 |

BULBS
(including CORMS and TUBERS)

| Large 340 |
| Gladioli 343 |
| Lilies 348 |
| Dahlias 350 |
| Medium 352 |
| Tulips 354 |
| Daffodils 360 |
| Small 370 |
| Crocuses 372 |
| Hyacinths 375 |

WATER PLANTS
386

| Water lilies 390 |

CACTI and other SUCCULENTS

| Large 392 |
| Medium 397 |
| Small 405 |

✽ Tint panels indicate features on separate plant groups or genera ✽

THE AWARD OF GARDEN MERIT
417

THE PLANT DICTIONARY
425

Glossary of Terms
627

Index of Common Names
629

Acknowledgments
640

CONTRIBUTORS

Susyn Andrews *Hollies*
Larry Barlow *Chrysanthemums*
Kenneth A. Beckett *The Planter's Guide, Shrubs, Climbers, Bromeliads*
John Brookes *Creating a Garden*
Eric Catterall *Begonias*
Allen J. Coombes *Plant Origins and Names, Trees, Shrubs, Glossary of Terms*
Philip Damp *Dahlias*
Kate Donald *Peonies, Daffodils*
Kath Dryden *Rock plants*
Raymond Evison *Clematis*
Diana Grenfell *Hostas*
Peter Harkness *Roses*
Terry Hewitt *Cacti and other Succulents*
David Hitchcock *Carnations and Pinks*
Hazel Key *Pelargoniums*
Sidney Linnegar *Irises*
Brian Mathew *Irises, Bulbs*
Victoria Matthews *Climbers, Lilies, Tulips*
David McClintock *Grasses, Bamboos, Rushes and Sedges*
Diana Miller *Perennials, African violets*
John Paton *Perennials*
Charles Puddle *Camellias*
Wilma Rittershausen (with **Sabina Knees**) *Orchids*
Peter Q. Rose *Ivies*
Keith Rushforth *Conifers*
A.D. Schilling *Rhododendrons and Azaleas*
Arthur Smith *Gladioli*
Philip Swindells *Ferns, Primulas* (with **Kath Dryden** and **Jack Wemyss-Cooke**), *Water plants, Water lilies*
John Thirkell *Delphiniums*
Alan Toogood *Annuals and Biennials*
Major General Patrick Turpin *Heathers*
Michael Upward *Perennials*
John Wright *Fuchsias*

How to Use this Book

The Royal Horticultural Society Gardeners' Encyclopedia of Plants & Flowers is the ideal reference when planning a garden, selecting plants or identifying specimens; it provides a wealth of information on the appearance and cultivation of thousands of individual plants.

How to plan your garden
In *Creating a Garden* you will find advice on garden styles, how to position and group plants and the use of colour and texture – all the background information you need before selecting individual specimens. *The Planter's Guide* recommends plants for particular sites or uses. Cross-references to *The Plant Catalogue* and *Plant Dictionary* lead you to the photograph and description of any plant you choose.

How to identify and select plants
If you know a plant but cannot recall its name, have a specimen that you want to identify or simply wish to choose plants for your garden, *The Plant Catalogue* will provide the answer. First turn to the appropriate chapter, e.g. Shrubs, Perennials or Bulbs. Then decide on the size and season of interest you want. In each section you will find plants grouped by colour from which to make your choice. By comparing a specimen with the photograph and its description, a plant may easily be identified. Other plants, similar in type, size, season of interest and colour, are listed at the bottom of each page. Such plants are either illustrated elsewhere (if they are equally appropriate to another section) or described in *The Plant Dictionary*.

How to find a particular plant
The Plant Dictionary lists all plants in the encyclopedia alphabetically by genus, with clear and detailed descriptions of plants not

The Plant Catalogue

Size categories (based on plant heights)

	LARGE	MEDIUM	SMALL
TREES	over 15m (50ft)	10–15m (30–50ft)	up to 10m (30ft)
CONIFERS	over 15m (50ft)	10–15m (30–50ft)	up to 10m (30ft)
SHRUBS	over 3m (10ft)	1.5–3m (5–10ft)	up to 1.5m (5ft)
PERENNIALS	over 1.2m (4ft)	60cm–1.2m (2–4ft)	up to 60cm (2ft)
ROCK PLANTS	over 15cm (6in)	—	up to 15cm (6in)
BULBS (including CORMS and TUBERS)	over 60cm (2ft)	23–60cm (9in–2ft)	up to 23cm (9in)
CACTI and other SUCCULENTS	over 1m (3ft)	23cm–1m (9in–3ft)	up to 23cm (9in)

The colour wheel
Within each section the plants are grouped by the colour of their main feature. They are arranged in the same order: from white through reds, purples and blues to yellows and oranges. Variegated plants are categorized by the colour of their variegation (i.e. white or yellow), succulents by the colour of their flowers, if produced.

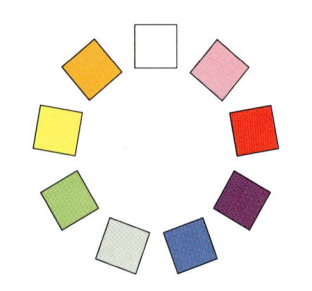

The symbols

- ☼ Prefers sun
- ☼ Prefers partial shade
- ● Tolerates full shade
- pH Needs acid soil
- ○ Prefers well-drained soil
- ◐ Prefers moist soil
- ● Prefers wet soil
- ♚ Award of Garden Merit

❄ Half hardy – can withstand temperatures down to 0°C (32°F)
❄❄ Frost hardy – can withstand temperatures down to -5°C (23°F)
❄❄❄ Fully hardy – can withstand temperatures down to -15°C (5°F)

CHAPTER HEADINGS
Where appropriate each chapter is subdivided into sections, according to the approximate size of the plants (see chart, left) and their main season of interest or type.

COLOUR BOXES
Colour boxes at the top of the page show the colour range of plants featured beneath, e.g.

PLANT PORTRAITS
Colour photographs assist the identification and selection of plants.

CAPTIONS
Captions describe the plants in detail and draw attention to any special uses they may have

SIZE AND SHAPE
For Trees, Conifers and Shrubs a scale drawing shows the size and shape of each plant at maturity. For other plants their approximate height (H) and spread (S) are given at the end of each caption. (The 'height' of a trailing plant is the length of its stems, either hanging or spreading.)

PLANT NAMES
The full botanical name is given for each plant. Synonyms (syn.) and common names are included where appropriate.

AWARD OF GARDEN MERIT
This symbol indicates that the plant has been given the Award of Garden Merit. See p.417.

CULTIVATION AND HARDINESS
Symbols show a plant's preferred growing conditions and hardiness. However, the climatic and soil conditions of your particular site should also be taken into account as they may affect a plant's growth. For frost tender plants the minimum temperature required for its cultivation is given.

Shrubs/medium SPRING INTEREST
WHITE–PINK

Aronia arbutifolia (Red chokeberry)
Deciduous shrub, upright when young, later arching. Clusters of small, white flowers with red anthers appear in late spring, followed by red berries. Dark green foliage turns red in autumn.

Prunus mume 'Omoi-no-m...'
Deciduous, spreading shrub with fragrant, semi-double, occasionally single, pink-flushed, white flowers wreathing young growths in early spring before oval, toothed leaves appear.

Myrtus communis (Common myrtle)
Evergreen, bushy shrub with aromatic, glossy, dark green foliage. Fragrant, white flowers are borne from mid-spring to early summer, followed by purple-black berries.

Viburnum plicatum 'Pink B...'
Deciduous, bushy shrub. Dark g... leaves become reddish-purple in autumn. In late spring and early summer bears white, later pink, blooms, followed by red, then black, fruits.

Chaenomeles speciosa 'Moerloosii'
Vigorous, deciduous, bushy shrub. Has glossy, dark green leaves and pink-flushed, white flowers in early spring, followed by greenish-yellow fruits.

HOW TO USE THIS BOOK

illustrated and cross-references to the photographs, as well as advice on cultivation and planting sites for the whole genus. Synonyms are also listed. If only the common name is known, then turn to the *Index of Common Names*, which will give you the correct botanical name and lead you either to an illustration or to *The Plant Dictionary*.

Plant names

How plants were introduced and a clear explanation of botanical nomenclature are given in *Plant Origins and Names*. The plant names in this book are those considered to be correct as the book goes to press, but as taxonomic work is continually in progress names may be superseded. Many synonyms have been included, with cross-references in *The Plant Dictionary*, so that plants may also be found using alternative or older names.

The Feature Boxes

Plant groups or genera of special interest to the gardener are presented in separate feature boxes within the appropriate chapters.

GROUP CHARACTERISTICS
The introduction outlines the merits of each group of plants and gives guidance on cultivation and planting positions.

FLOWER FORMS
Detailed descriptions of the flower forms and horticultural classifications within a genus are given where appropriate.

LINE DRAWINGS
Clear line drawings show the different flower forms.

PLANT PORTRAITS
Close-up photographs of individual flowers or plants allow quick identification or selection.

PLANT NAMES
Full botanical name is given and the group or classification where appropriate. Plant descriptions appear in *The Plant Dictionary*.

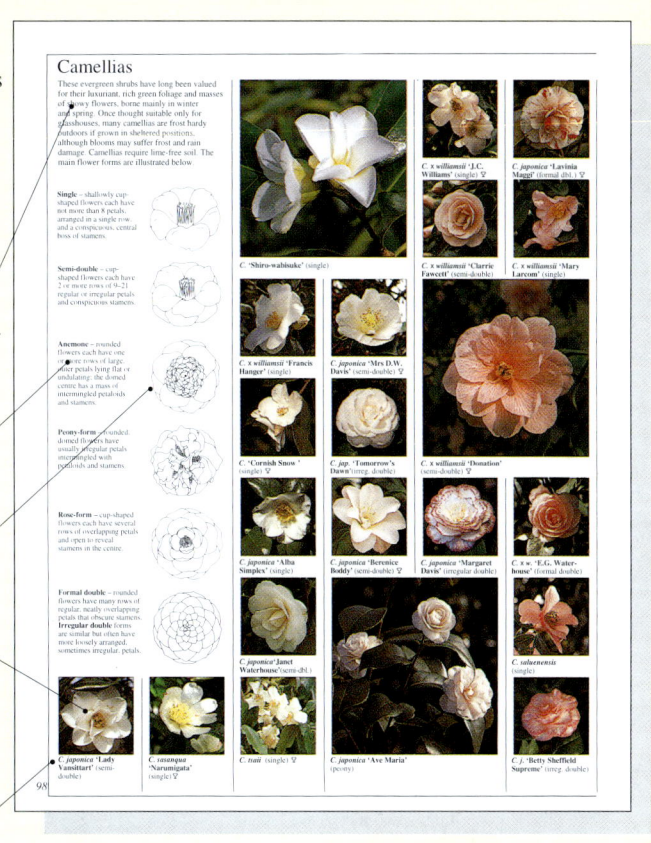

CROSS-REFERENCES
Plants featured elsewhere in the book but similar in type, size, season of interest and colour to those illustrated are listed for consideration.

The Plant Dictionary

The Plant Dictionary contains entries for every genus in the encyclopedia and includes over 4000 recommended plants not described in *The Plant Catalogue*. It also functions as the index to *The Plant Catalogue*.

GENUS NAMES
The genus name is followed by the family name and, where appropriate, any common names.

GENUS ENTRIES
Each genus entry covers its distinctive characteristics, hardiness range and advice on siting plants, their cultivation, propagation, and, if relevant, pruning and pests and diseases.

SYNONYMS
Synonyms cross-refer to the correct botanical name.

PLANT DESCRIPTIONS
Entries on species and cultivars give full botanical names, synonyms and common names. The symbol ♣ indicates that the plant has received the Award of Garden Merit. Hardiness and cultivation needs are included if they are specific to the plant. The *Glossary of Terms* explains all technical terms.

Abbreviations are listed on page 640.

Plant Origins and Names

Common names
Although many have familiar common names, plants are usually listed under their botanical names. Why should this be? There are a number of reasons. Many plants either do not have a common name, or they share a common name with others. More confusingly, the same common name may be used in different regions to describe different plants: in Scotland 'plane' refers to *Acer pseudoplatanus* (sycamore), in England to the London plane (*Platanus* × *hispanica*), while in North America the native plane (*Platanus occidentalis*) may be called plane or sycamore. Unlike botanical names, which bring related plants together by grouping them in separate genera (all true hollies belong to the genus *Ilex*), the same word is often used in common names for quite unrelated plants, as in sea holly (*Eryngium*), hollyhock (*Alcea*) and summer holly (*Arctostaphylos diversifolia*), none of which is related to the true holly. Conversely, one plant may have several common names: 'heartsease', 'love-in-idleness' and 'Johnny-jump-up' are all charming titles for *Viola tricolor*. Even greater confusion arises when a vernacular common name is in Malay, Chinese or Arabic. In botany, as in other scientific disciplines, the use of Latin has been found to be a convenient and precise basis of a universal language for naming plants.

The binomial system
Greek and Roman scholars laid the foundations of our method of naming plants, and their practice of observing and describing nature in detail was continued in the monasteries and universities of Europe, where classical Latin remained the common language. The binomial system in use today, however, was largely due to the influence of the famous eighteenth-century Swedish botanist, Carl Linnaeus (1707–78). In his definitive works *Genera plantarum* and *Species plantarum*, Linnaeus classified each plant by using two words in Latin form, instead of adopting the descriptive phrases that had been in common usage among the botanists and herbalists of his day. The first word was the name of the genus (e.g. *Ilex*) and the second the specific epithet (e.g. *aquifolium*). Together they provided a name by which a particular plant (species) could be universally known (*Ilex aquifolium*, English or common holly). Other species in the same genus were then given different epithets (*Ilex crenata*, *Ilex pernyi*, *Ilex serrata* and so on).

The meaning of plant names
A greater appreciation of botanical names may be gained by knowing something of their meaning. A name may be commemorative: the *Fuchsia* is a tribute to Leonhart Fuchs, a German physician and herbalist. It may tell us where a plant comes from, as with *Parrotia persica* (of Persia, now Iran). A plant may bear the name of the collector who introduced it: *Primula forrestii* was brought into cultivation by George Forrest. Or the name may tell us something about physical character: *Pelargonium* derives from the Greek word *pelargos* (a stork), an appropriate description of the fruits of these plants which resemble storks' bills; *quinquefolia*, the epithet of *Parthenocissus quinquefolia*, means with foliage made up of five leaflets, from the Latin *quinque* ('five') and *folium* ('leaf').

Generic names, like all Latin nouns, are either masculine, feminine or neuter and the specific epithets, usually adjectives, therefore agree in gender with the genus.

Plant introductions
Just as the origins of plant names may be traced to Classical civilization, so can the first plant introductions. At its peak, the Roman Empire covered a vast area, from western Europe to Asia, and as the Romans travelled

DICENTRA SPECTABILIS
Robert Fortune introduced this graceful plant when he returned in 1846 from northern China, where the Dicentra *was already widely cultivated. It has since become one of the most popular of cottage garden plants.*

PLANT ORIGINS AND NAMES

they brought with them plants used for food or ornament, such as the Spanish chestnut, peach, fig and many herbs. Later, North America proved to be a major source of new plants; by the end of the sixteenth century some of these could be found growing in British gardens and in the seventeenth century many more were introduced by private collectors such as the two John Tradescants (father and son), who were gardeners to Charles I and others of the nobility. In 1824 the Royal Horticultural Society employed David Douglas, a Scot, to travel to America's west coast and gather vast quantities of seed from hitherto unknown species including the Douglas fir and Sitka spruce.

The greatest rewards for the plant collector, however, were to be found in the rich and diverse flora of the East. Access to China was restricted until the Treaty of Nanking in 1842, but in the following year Robert Fortune began an expedition, under the aegis of the Royal Horticultural Society, that was to yield many ornamental garden plants. In his wake went the French botanist-missionaries, including Armand David who discovered the lovely *Davidia* tree.

The Golden Age of plant hunting was the early part of this century, when thousands of species were introduced to cultivation. Ernest Henry Wilson, one of the most famous plant hunters, who collected in the East for the Royal Botanic Gardens at Kew and the

MAGNOLIA WILSONII
One of the finest summer-flowering shrubs, Magnolia wilsonii *was discovered in 1904 by Ernest Henry Wilson, who was collecting for Messrs Veitch (nurserymen) in Sichuan, China. Wilson later brought back many thousands of species for the Arnold Arboretum in Boston.*

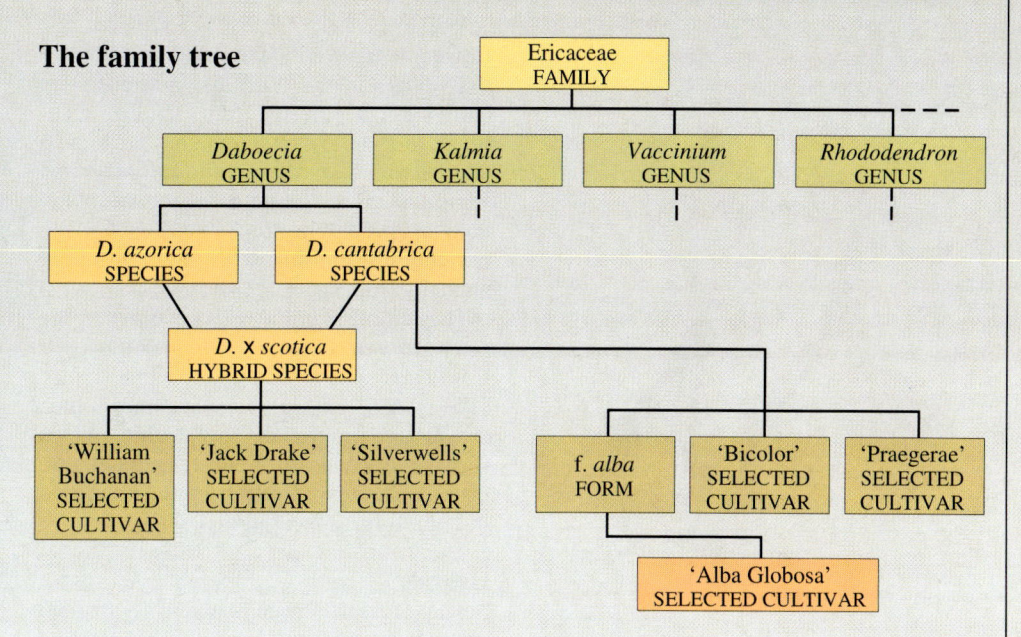

The family tree

```
Ericaceae FAMILY
├── Daboecia GENUS
│   ├── D. azorica SPECIES
│   └── D. cantabrica SPECIES
│       └── D. x scotica HYBRID SPECIES
│           ├── 'William Buchanan' SELECTED CULTIVAR
│           ├── 'Jack Drake' SELECTED CULTIVAR
│           └── 'Silverwells' SELECTED CULTIVAR
├── Kalmia GENUS
├── Vaccinium GENUS
└── Rhododendron GENUS
    ├── f. alba FORM
    │   └── 'Alba Globosa' SELECTED CULTIVAR
    ├── 'Bicolor' SELECTED CULTIVAR
    └── 'Praegerae' SELECTED CULTIVAR
```

Arnold Arboretum in Boston, discovered over 3000 species. In the same period, George Forrest introduced many rhododendrons and other plants from China and Tibet. Frank Kingdon Ward, who had travelled widely in the Himalayas, published several very readable accounts of his experiences in the 1920s, collecting unusual primulas, lilies, rhododendrons and gentians.

Plant-hunting expeditions to areas of the Sino-Himalayan region, closed for many years to westerners, continue to be sponsored today, and new plants are still being introduced, if in less dramatic quantities than at the turn of the century.

International codes
Since its instigation the Linnaean system of plant classification has been developed by scientists so that the entire plant kingdom is divided and subdivided into a multi-branched 'family tree', according to each plant's botanical characteristics. International cooperation has been essential in order that the system be reliable for scientific, commercial and horticultural use. To this end, there are now rules laid down in the *International Code of Nomenclature for Cultivated Plants* (1980) and the *International Code of Botanical Nomenclature* (1988).

The plant kingdom
The plant kingdom may be broadly divided into vascular plants and non-vascular plants. Vascular plants are of most interest to the gardener and have specialized conducting tissue that enables them to grow in a wider range of habitats and reach a larger size than the non-vascular plants such as algae, fungi, mosses and liverworts. Vascular plants are classified into many groups, mainly according to the way they bear their seeds. For example conifers, part of the Gymnosperm group, are distinct as they bear their naked seeds in cone-like fruits. The basic division within these groups is the family.

The family
The botanical division of the family may contain clearly related and specialized plants such as the orchids (Orchidaceae) and the bromeliads (Bromeliaceae), or, like the Rosaceae, embrace plants as diverse in terms of what they offer the gardener as *Alchemilla, Cotoneaster, Crataegus, Geum, Malus, Prunus, Pyracantha, Sorbus* and *Spiraea*.

CARL LINNAEUS
Linnaeus in Lapp costume: engraving from a portrait by M. Hoffman of about 1737.

PLANT ORIGINS AND NAMES

DAVID DOUGLAS
Crayon drawing by Sir Daniel Macnee, 1828

Plants are grouped in particular families according to the structure of their flowers, fruits and other organs.

The genus and its species

A family may contain one genus (*Eucryphia* is the only genus in Eucryphiaceae) or many (the daisy family Compositae has over 1000 genera). Each genus comprises related plants, such as oaks (*Quercus*), maples (*Acer*) and lilies (*Lilium*), with several features in common. A genus may contain one or many species. Thus a reference to a member of the genus *Lilium* could be to any of the lilies, but one to *Lilium candidum* would denote one particular lily (in this case the Madonna lily). Just as some families form closely related groups, so certain genera form separate horticultural groups within families. An example of this would be the heaths or heathers (including *Erica* and *Calluna*) within the family Ericaceae, which also includes *Rhododendron*, *Kalmia* and *Vaccinium*.

Subspecies, varieties and forms

Species are usually variable in the wild and may be split into three botanically recognized but occasionally overlapping subdivisions. The subspecies (subsp.) is a distinct variant, usually because of the plant's geographical distribution; the variety (botanical *varietas*, abbreviated to 'var.') differs slightly in its botanical structure; and the form (*forma*, f.) has only minor variations, such as of habit or colour of leaf, flower or fruit.

Cultivars

Many plants grown in today's gardens may be adequately described by their botanical names, but numerous variants exist in cultivation which differ slightly from the normal form of the species. These forms may be of considerable horticultural interest, for their variegated leaves or variously coloured flowers and so on. They may be found as individuals in the wild and introduced to cultivation, be selected from a batch of seedlings, or occur as a mutation; they are known as cultivars – a contraction of 'cultivated varieties'. To come true to type many cultivars need to be propagated vegetatively (by cuttings, grafting or division) or grown annually from specially selected seed.

The naming of cultivars is governed by the *International Code of Nomenclature for Cultivated Plants*. Cultivars named since 1959 must be given vernacular names, which are printed in roman type within quotes (*Phygelius aequalis* 'Yellow Trumpet'), to distinguish them clearly from wild varieties in Latin form, appearing in italic.

Hybrids

Sexual crosses between botanically distinct species or genera are known as hybrids and are indicated by a multiplication sign. If the cross is between species in different genera, the result is called an intergeneric hybrid and, when two (occasionally three) genera are concerned, the name given is a condensed form of the names of the genera involved: x *Cupressocyparis* covers hybrids between all species of *Chamaecyparis* and *Cupressus*. If more than three genera are involved, then the hybrids are called after a person and given the ending *-ara*. Thus x *Potinara*, covering hybrids of *Brassavola*, *Cattleya*, *Laelia* and *Sophronitis*, commemorates M. Potin of the French orchid society. Most common, however, are hybrids between species in the same genus (interspecific hybrids), which are given a collective name similar to a species name but preceded by a multiplication sign; for example *Epimedium* x *rubrum* covers hybrids between *E. alpinum* and *E. grandiflorum*.

When one plant is grafted onto another, a new plant may occasionally arise at the point of grafting which contains the tissues of both parents. For naming purposes, these graft chimaeras or graft hybrids are treated in the same way as sexual hybrids, but instead of placing a multiplication sign before the name, they are denoted by a plus sign, as in + *Laburnocytisus adamii*, a graft hybrid between species of *Laburnum* and *Cytisus*.

GENTIANA SINO-ORNATA
George Forrest had a special interest in gentians, collecting hundreds of specimens during his trips to China. He first discovered this plant in 1904, during a hazardous expedition high in the mountain ranges of northwestern Yunnan. It is one of his finest introductions, and is admired for its exquisite beauty and late and prolonged flowering during autumn, when few other plants bring colour into the garden.

Cultivars of hybrids should be listed under a botanical name if one is available (*Viburnum* x *bodnantense* 'Dawn') or, if the parentage is complex or obscure, by giving the generic followed solely by the cultivar name (*Rosa* 'Buff Beauty').

Group names

Botanical names of hybrids can provide a convenient way of gathering together cultivars of like parentage, as with the cultivated forms of *Camellia* x *williamsii*. But occasionally, especially with some orchids and many annuals, such names have not been allocated for groups with hybrid parents. For these, group names in modern language and roman type without quotation marks are used. Thus the orchid group *Paphiopedilum* Freckles covers hybrid cultivars with similar parentage, as does the group of annuals *Dahlia*, Disco Series. Distinct members within a group may be recognized as cultivars, for example *Miltoniopsis* Anjou 'St Patrick' or *Viola*, Imperial Series 'Orange Prince'.

Name changes

It is often confusing and frustrating to come across name changes in new publications. Long-established and often well-known names disappear, only to be replaced by new, unfamiliar ones. But there are good reasons for such changes. A plant may originally have been incorrectly identified; research may discover that a plant earlier had a different name (the *International Code of Botanical Nomenclature* states that the valid name for a plant is the earliest correctly published one); a name may be found to apply to two different plants; or new scientific knowledge may mean that a plant's classification, such as the genus to which it belongs, has to be changed. In this Encyclopedia, numerous synonyms have been given in order to minimize the problems of identifying or purchasing plants.

CREATING *a* GARDEN

How to group plants effectively, plan and
structure your garden, and use colour, light
and texture to enhance its natural beauty.

CREATING A GARDEN

Style

Gardens, like homes, may have many different atmospheres or styles. Some are instantly familiar, such as the haphazard abundance of an English cottage garden or the sparse elegance of a Japanese garden, while others, such as a *jardin sauvage*, may prove harder to define.

Design influences

Styles may have particular historical associations, showing for example the influence of the formal, stepped stone terraces of Renaissance Italy, the elegant grandeur of classical French gardens, such as Versailles, or the showy, colourful and exotic plantings of Edwardian and Victorian England.

Different styles have also developed as a result of regional cultures. This is particularly true of the United States where East-coast pioneer and picket gardens, full of wooden decks and open white fences, contrast with glamorous West-coast gardens, with patios and pools, or the sleepy gardens of the South with their magnolias and oaks shading the verandah.

These examples may seem too ambitious for the average gardener, but they can provide the source of his or her inspiration and be adapted to a particular site. It may be impossible to reproduce a classical French layout or an elegant Italian hillside landscape in a small city garden, but a sense of classical formality is within the reach of everyone with the use of just a few stone steps, a small fountain and a handsome stone container filled with annuals, or perhaps an architectural, evergreen shrub or clipped tree. When choosing a style, bear in mind the climate, available space and soil type.

Proportion

To create a style in your own garden, be sensitive to the scale and mood of the surrounding landscape and buildings. A classical-style house, for instance, may well require a garden constructed to a formal layout; the proportions between the windows and doors could suggest the pattern made in the garden by paths, pools, terraces, borders and lawn areas. Care should be taken with the relationships of these elements, to each other and the garden as well as to the house. Whereas broad sweeps of colour and grand paths lend elegance to a large garden with open views, narrow borders of diminutive flowers and little winding paths would be lost in so large a space, although ideally suited to a small, informal cottage garden.

Perspective

You can alter the apparent shape of your garden by creating false perspective in your design, using the edges of different areas (paths, beds or

STYLE

terraces) as 'lines'. Horizontal lines will make the site seem wider while vertical lines will increase its length. To accentuate length even more, make vertical lines converge as they approach the bottom of the garden. Skillful placing of upright or prostrate trees and shrubs will enhance this effect.

Attention may be drawn to a particular spot, within or beyond the garden, by creating a dynamic, or moving, line such as a curving path that leads to a focal point – a pond, arbour, specimen tree or magnificent view. Formal or enclosed gardens with no dominant point of interest are, however, best served by static patterns, such as those created by box hedges bordering geometric beds.

Materials

Paths, paved areas, fencing and garden buildings should harmonize with the house and its surroundings in materials as well as in form. You may have difficulty in deciding what will be appropriate; try determining instead what will not be suitable and continue by a process of elimination. For example, crazy paving will not enhance an eighteenth century town house.

Try to make the garden a continuation of the interior in terms of colour and style. This concept may be taken further, so that the view from within blends harmoniously with the internal décor and does not jolt the senses, whether you are viewing the interior from the garden, or the garden from the interior. Plants are, of course, the most important 'materials' used to create a garden's style. Your choice will be limited by the location, soil type and climate of the area. Bright Mediterranean flowers such as rock roses will not thrive in a shady woodland, nor will heathers and camellias flourish on open, chalky downland. Take advantage of the assets at your disposal and do not try to fight nature. Not only will inappropriate plants never

ITALIAN-STYLE TERRACE
Above: This elegant terrace has been created on a modest scale using terracotta pots and a small, wall-mounted fountain. The strong, horizontal lines of the pergola are softened by climbing roses, honeysuckle and the decorative leaves of Vitis coignetiae.

EXOTIC GARDEN
Left: A Japanese mood has been conjured here by combining a clear stream, cascading down stepped rocks, with a carefully placed, distinctive, dwarf Japanese maple (Acer palmatum 'Dissectum Atropurpureum').

SEMI-FORMAL GARDEN
Right: The classical pavilion sets the formal tone of this country garden. Shrubs and hedges are clipped into strong, carefully related, architectural shapes and lead the eye towards the austere building and to the bonsai tree, on a plinth in front of it, which provides the focal point. The path has been designed to harmonize with the pavilion, in materials as well as design.

CREATING A GARDEN

thrive, but very often when they are grown away from their natural habitat they appear to be irritatingly out of place.

Making a style your own
Most important of all, do not forget your own relationship with the garden – let it reflect your personality. If you are a relaxed person who likes pottering, you might allow herbaceous perennials to overspill their formal boundaries, or leave a few self-sown seedlings to interrupt a well-groomed gravel path. On the other hand, if you like everything around you to be ordered and tidy, you will probably be happier with a neat, symmetrical layout, perhaps blocks of annuals or Hybrid Tea (modern) roses rather than old-fashioned (old garden) varieties.

Do not overlook more practical requirements, either. If you have a young family, there is little point in siting an attractive lily pond in the middle of the garden as you will have to cover it with ugly safety mesh. Instead lay large areas to lawn and plant robust trees and shrubs. If a person using a wheelchair will be spending time in the garden, some stepped levels should be removed. Should you want to introduce a variety of levels, use gently sloping ramps or undulating banks. You might also build some raised beds; these would make gardening easier for the handicapped or for elderly gardeners.

Consider realistically how much time you are willing to spend on maintenance, as some gardens are much more time-consuming than others. If time is limited, concentrate on trees, shrubs and ground cover; perennials and bulbs provide variety of colour, form and texture, but usually demand rather more gardening time. For maximum commitment grow more specialized plants, for example rock plants that may have demanding cultivation requirements, or plant or sow annuals and biennials.

Keeping the planting simple
A common mistake made by gardeners is to include too many varieties of plants in a small area. There is something to be learnt from a specialist's garden – the rose grower's, for example, or that of the enthusiast who cultivates only conifers and heathers. Their gardens look good because the choice of plants is restricted, giving rise to a comparatively simple design.

If you do not want to be quite so limited, you can prevent a mixed planting from becoming too complex and unfocused by designing

COTTAGE-STYLE GARDEN
Below left: A haphazard mixture of mainly perennial plants is here combined with the shrub Piptanthus nepalensis, *which scrambles up an old stone wall, while beneath it* Corydalis ochroleuca, Hypericum 'Hidcote' *and lady's mantle* (Alchemilla mollis) *grow happily through each other. Note the limited colour range of the planting, making it informal rather than chaotic.*

JAPANESE-STYLE GARDEN
Below right: Bamboo, stepping stones and, in the distance, complementary foliage plants evoke a tranquil atmosphere in this Japanese-style garden. Such a style may be adapted to suit sunny or shaded sites and requires very little maintenance.

within a particular colour range and by taking special care when grouping different varieties of plants together. This applies to foliage as well as to flowers, as is demonstrated so effectively in the famous white garden at Sissinghurst in Kent.

Remembering the detail
Once the overall style and layout of your garden have been decided upon and the range of plants within it chosen, make sure that the furnishings do not ruin the effect. A wrought-iron bench, for instance, might be too sophisticated for many gardens, whereas a simple wooden one would rarely jar. An elaborate Italianate urn would probably be too grand for an informal, suburban garden, even if planted with humble daisies. A period garden may be difficult to furnish economically, but on the whole plain containers and ornaments are successful, especially in small, confined areas, and have the additional advantage of suiting a variety of plantings. Once you have decided on the style of furniture for your garden, do not stray from your original concept, as slight deviations will soon create a sense of clutter and all your planning will have been in vain.

ROMANTIC GARDEN
Left: A circular seat is placed in the leafy shade of a country garden made romantic by rough grass and a beautiful, sprawling shrub rose.

INFORMAL GARDEN
Below: Even in this deceptively simple planting of herbaceous perennials, bulbs, annuals and shrubs, each plant has been carefully selected to balance the others in form, colour and texture. The barberry (Berberis) and elegant, arching sprays of Solomon's seal (Polygonatum x hybridum) provide continuity and cohesiveness to the planting scheme.

CREATING A GARDEN

Structure

'Structuring' a garden does not mean physically shifting earth or laying paving slabs, but developing it from the basic ground plan by using elements of the planting together with man-made features, such as paths and gravel areas, to relate a garden to its setting.

Introducing different levels

Once you have decided on the style of garden you wish to create and have sketched a layout, you can start to turn your plan into three-dimensional reality. Consider the differences in height that occur, or could be introduced to the site, and then decide on the plant masses, their relationship to the landscape beyond the garden and to the other elements in the garden, including each other.

Changes in ground level may already exist and can be accentuated by building steps at these points or to a pool, a vantage point or a summerhouse. Fences and walls or structures such as pergolas, dovecotes and arbours all give a vertical emphasis or become useful screens, as can hedges, trees or certain shrubs. Raised beds are invaluable for adding interest to a small, flat plot, and much variety can be introduced by arranging plants of different heights.

> **USING STRUCTURAL PLANTS**
> *An upright tree such as a juniper may dominate too small a group of plants (A), but may help relate that group to the scale of a wall behind it (B). The same tree may block a view (C) or sited to one side of it act as a counterbalance (D). By incorporating another similar group within the field of vision, a more sophisticated, repetitive effect may be created. Movement and progression will result from placing one juniper forward of the other (E), but sited equidistant from the viewpoint the two plantings provide an elegant frame (F).*

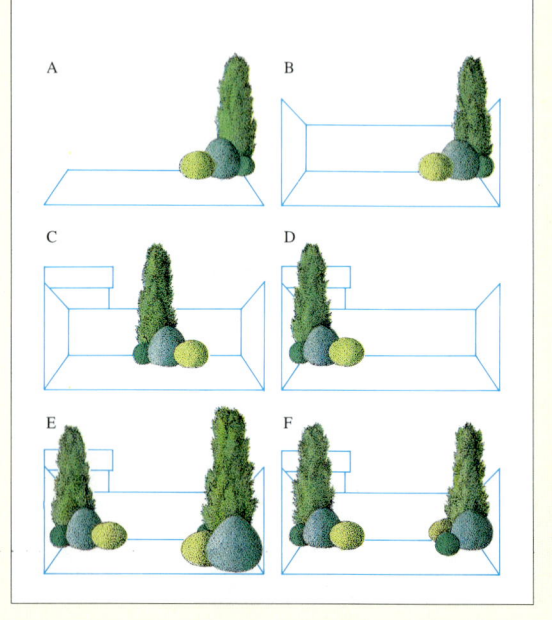

VISUALLY RESHAPING A FLAT SITE
Above: The forms of trees and shrubs may be used to introduce a variety of levels to a site. Prostrate and dwarf conifers contrast with the light, horizontal branches of golden-leaved Abies, *the open-branched locust tree (*Robinia pseudoacacia*) and, in the background, the dark, upright form of* Picea.

CHANGES IN LEVEL
*Left: Steps and a low wall serve to emphasize a change in level. The domed forms of the topiary yews (*Taxus*), flanking a flowering crab apple (*Malus*), highlight the sobriety of the former and the fresh abandon of the latter.*

STRUCTURE

Proportionate plantings

When planning your design, avoid crowding the basic pattern with a 'liquorice allsorts' sprinkling of plants. Think instead in terms of proportionate masses and build up the structure by adding key plants judiciously. There is much more to the look of a garden than the sum total of the individual plants that are growing within it; no matter how special a plant might be, to create a harmonious whole, it must relate to its neighbour. Moreover, each group of plants must then be in proportion to the bed in which it is situated, which in turn should relate to the whole garden.

The three-dimensional shape of the garden is largely determined by the plant masses and their siting. When you next come across a satisfying group of shrubs or perennials, analyse how the individual shapes of the plant masses are working structurally and how together they combine to form a unified whole. You will find that it is the proportionate relationship that they create with each other that makes the colour masses work so satisfactorily.

Using structural plants

No selection of plants is seen in isolation – there is always the backdrop of a garage, wall, neighbour's house or view. The style and planting of a garden should blend rather than jar with this background. By introducing structural, or architectural, plants (those with strong outlines or forms), you can provide the bridge between the garden and what lies beyond it, help to relate the garden to the house or bind together scattered plants. Whether spiky and aggressive like a yucca or soft and romantic like a weeping willow, the distinctive outline will add a focal point to the design. Structural plants help to contribute solidity and scale and, by using key plants of diminishing size, create an illusion of space through false perspective.

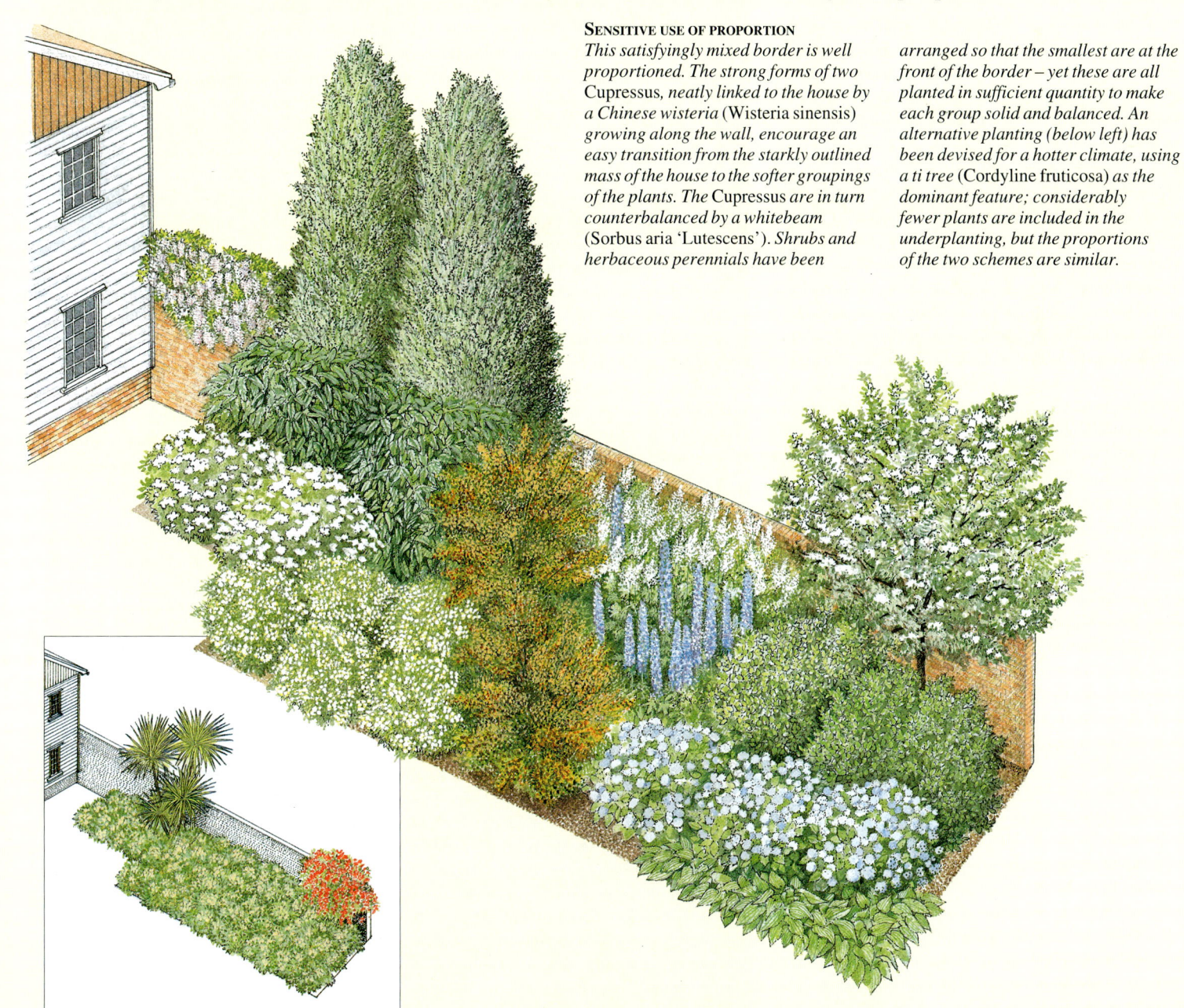

CREATING HARMONY
Purple-leaved elder helps relate the background trees to the foreground perennials, including Salvia officinalis 'Purpurascens'.

SENSITIVE USE OF PROPORTION
This satisfyingly mixed border is well proportioned. The strong forms of two Cupressus, *neatly linked to the house by a Chinese wisteria* (Wisteria sinensis) *growing along the wall, encourage an easy transition from the starkly outlined mass of the house to the softer groupings of the plants. The* Cupressus *are in turn counterbalanced by a whitebeam* (Sorbus aria 'Lutescens'). *Shrubs and herbaceous perennials have been arranged so that the smallest are at the front of the border – yet these are all planted in sufficient quantity to make each group solid and balanced. An alternative planting (below left) has been devised for a hotter climate, using a ti tree* (Cordyline fruticosa) *as the dominant feature; considerably fewer plants are included in the underplanting, but the proportions of the two schemes are similar.*

CREATING A GARDEN

Gardening through the Year

The most successful gardens are those that provide interest throughout the year. Since no individual plant is at its peak for all twelve months, its impact in a group will alter as the year progresses. For example, a tree that is spectacular in spring, when it is covered in blossom, may fade into the background during the rest of the year. The appearance of the group as a whole will therefore change with the seasons.

When selecting plants, remember to take into account seasonal variations in their appearance and thus ensure that each grouping sustains year-round interest. A planting of summer-flowering perennials alone may look dull in spring, autumn and winter. Consider all the merits of each plant – its size, habit, leaf form, colour, bark and texture – and not merely the flower colour. These are the long-term qualities of a plant, which will be on display after its main season of interest has passed.

Ideally, you should provide a succession of 'feature' plants against a relatively unchanging background of shrubs and trees. In this way, when the eye-catching flowers of one plant in the bed are over, another begins to blossom. Alternatively, you might like to plan a succession of feature plants in the garden as a whole, rather than just one border. In this way, the focal point will move from one area of the garden to another.

Spring
In the rock garden, Narcissus cyclamineus *nod cheerfully above* Euphorbia myrsinites.

Mid-summer
Above: In this cottage garden – suggestive of a wild meadow – red poppies, perhaps garish in a more dense planting, look delightful sprinkled among the cooler tones of white bellflowers and a few pale blue delphiniums.

Late autumn
Right: Evergreen and berried trees and shrubs may now be fully appreciated. Upright, dark green conifers contrast attractively with the dazzling, silvery-white leaves of Elaeagnus umbellata*, while the red berries of* Cotoneaster horizontalis *provide a colourful foreground.*

GARDENING THROUGH THE YEAR

Winter to spring

During winter, flower interest in the garden is likely to be minimal, and it is then that the bold shapes and foliage colour provided by evergreens, such as conifers, holly and ivy, come into their own. Less obviously, the colours and textures of twigs, branches, bark and berries of deciduous trees and shrubs also have roles to play in providing visual interest during the winter months. The branches of the elegant willow, for instance, can look quite brilliant in winter sunlight; silver birches are especially attractive, graceful, open trees with lovely, white bark that glistens on even the greyest day. The shapes of deciduous as well as evergreen plants play an important part at this time, whether silhouetted against a clear, winter sky or bearing a cloak of snow.

With the arrival of spring, the garden is soon awash with colour. Some is provided by spring-flowering trees and shrubs, such as flowering cherries and magnolias, and by myriad rock plants, but most of the colour comes from an abundance of flowering bulbs and corms. As these die down, they are followed by camellias, forsythias, rhododendrons and azaleas and the fresh green of young leaves, with scented viburnum and lilac flowers appearing in succession through late spring.

Summer to autumn

In early summer, more colour is added by a profusion of perennials and biennials, often lasting until autumn. Shrub interest diminishes as the season progresses, which is when the annuals make their contribution, brightening up the heavy green face of summer. Annual flowers, although often extremely bright, last for only short periods, so it is necessary to sow seeds and transplant seedlings successively to maintain the display. Self-seeding annuals will emerge year after year in a random way, often providing short explosions of colour in unexpected places. The wealth of interesting plants that bloom in late summer, including clematis, remontant (repeat-flowering) roses, as well as annuals and herbaceous perennials, creates a pitfall of its own: you will have to be selective otherwise the effect in the garden as a whole will be overwhelming.

As summer fades, the greenery starts to flare into the fiery colours of autumn, the intensity of the colours and their duration depending on the weather. In your planting include trees or shrubs with spectacular autumn leaves, such as acers; they can make the garden as handsome in this season as in any other. Late-flowering perennials, such as Michaelmas daisies and chrysanthemums, brighten up the garden until the arrival of the first frosts, when the flowers rot and the browned leaves fall, returning us to the winter landscape.

SEASONAL PLANTING
In this small plot (9 square metres), the interplay between a group of shrubs, perennials and bulbs can be followed throughout the seasons. Although plants will have changing roles to play at different times of year, each has been chosen to relate successfully within the whole as well as to its neighbour.

WINTER
Evergreen rosemary contrasts with fleshy, purple Bergenia leaves, variegated Iris foetidissima foliage and feathery twigs of Acer palmatum 'Senkaki'. Snowdrops and winter aconites (Eranthis hyemalis) will stud the ground by early spring.

SPRING
Bergenia now bears its tight, red flower clusters and the scented blooms of Viburnum opulus open. Fresh green is provided by young Acer leaves and delphinium shoots.

SUMMER
Blue spikes of delphinium dominate early on. Foreground interest moves from the fern-like, grey leaves of the Achillea to its long-lasting, plate-like, yellow flower heads.

AUTUMN
Yellowish Acer and bronze-red Viburnum foliage command most attention, although later the Viburnum berries will also become noteworthy. Iris seed pods open to reveal their orange fruits and late-flowering chrysanthemums brighten the foreground.

CREATING A GARDEN

Planning

A visually satisfying composition of plants comprises several elements. Look at the characteristics of the individual plants, and consider the effects those characteristics have on each other. The shapes and forms of their leaves and flowers, their colour and texture are all important. Think also about the changing size of plants: a small shrub planted now may dominate the garden in five years' time.

Any combination of plants should not be seen in isolation but should be considered in relation to the overall garden plan, the house and its surroundings. For example, a small tree or large shrub backing a small group of iris, ballota and meadow rue would work well, but replace the single tree or shrub with a bank of trees and more of the smaller plants would be needed to prevent them from being 'swamped'. Similarly, two irises, three ballotas and a meadow rue would probably be too few plants when sited against a two-storey house.

Restricting your choice of plants
Most people tend to be shy of buying more than one or two of a particular kind of plant, or, with such a wide range of suitable subjects available, cannot resist the temptation of including yet another species. It is just this magpie instinct that makes overall planting schemes look fractured and detracts from the strong, simple lines of the basic concept. The plant collector and the garden designer are very often a difficult pair to reconcile!

Aim for simple, bold plant masses to begin with; later, if the effect seems too stark, widen the range of plants slightly to add a little light relief. Keep your choice of suitable plants narrow and try to see them growing, by visiting well-established gardens, before you make your final selection.

The magnificent, large-scale plantings within grand borders in the gardens of historic houses that can be admired by the general public are made up of groups of perhaps twelve of this plant and eight of that; to simulate them on a domestic scale calls for an iron discipline. The same applies to planting bulbs – twos and threes tend to look messy; try groups or drifts instead. Remember, your planting is garden furnishing, and you would not cover individual chairs in the same room with contrasting fabrics if you wished to create a calm, comfortable atmosphere to relax in and enjoy.

PLANNING

SHAPE AND SCALE
Above left: Huge-leaved Gunnera manicata, *tall giant lily* (Cardiocrinum giganteum), *bushy ferns and yellow flag irises* (Iris pseudacorus), *although very varied in shape and scale, still form a cohesive planting for a damp site.*

COLOUR RANGES
Above: The transition in colour tone from the golden fronds of Juniperus x media 'Pfitzeriana' *through the orange-yellow clump of coneflowers* (Rudbeckia) *to the reddish-orange montbretia* (Crocosmia) *is smooth and pleasing. More varieties in smaller quantities might be difficult for the eye to absorb within this strong colour range.*

SIMPLE, BOLD PLANT MASSES
Far left: Erect, silvery-blue Arizona cypress (Cupressus glabra) *is balanced by the horizontal, golden foliage of* Acer shirasawanum 'Aureum'; *both trees seem stabilized by the solid-looking plants at their feet.*

STRUCTURED INFORMALITY
Left: Old stone slabs and exposed gravel punctuate the planting of lady's mantle (Alchemilla mollis), *variegated* Cornus, *roses and sweet Williams, which might otherwise appear unruly.*

23

CREATING A GARDEN

Siting your plants

A useful discipline when selecting and siting plants is to start with the largest and work towards the smallest. First decide on the background planting, or 'chorus line'. This could be either trees or shrubs, depending on the scale of the location. Forest trees may provide a link with a wood beyond, but they can become too large for the average town garden, whereas medium-sized trees will provide privacy, shelter and a 'backdrop' for your garden. Climbers may be grown around the trees or trained along garden walls or fencing. Against this background, site the occasional feature plant – a more decorative small tree or a large shrub.

Think in the same way about shrubs. Use a backing group to cover a fence, to provide a screen between your garden and your neighbour's or to camouflage unsightly buildings. Evergreens are obviously more suitable for this than deciduous shrubs. I find many conifers too demanding in both their shape and colour for inclusion in the chorus line: they make better specimen plants. Plant a group of shrubs – two or three in a small town garden, maybe seven or eight in a larger country one.

In front of these shrubs, plant the flowering specimens and among the perennials and biennials introduce the smallest shrubs, including many herbs such as thyme, sage or lavender. At this stage consider carefully the position of any feature plants.

Allowing for growth

It is essential, when planning where to site your plants, to envisage your scaled associations of plant masses not only as they appear now but

CREATING A BACKDROP
Above: Climbing plants such as Clematis 'Nelly Moser', *here grown through* Rosa 'American Pillar', *may be used to soften or screen buildings as well as to act as a foil for other plants.*

VERDANT CHORUS LINE
Left: The large shrubs Chimonanthus praecox *and* Corokia cotoneaster *form a verdant 'chorus line' for smaller, more decorative flowering plants, in this case purple sage, lady's mantle (*Alchemilla mollis*) and* Campanula persicifolia. *The stone wall also makes a handsome, contrasting backdrop.*

PLANNING

also as they will have developed in five years' time. Site the plants to allow enough room for them to expand and merge together over that time span. Beware of selecting trees that will ultimately become far too large for the scale of your garden as a whole, even if they look delightful at the garden centre.

If the planting seems too far apart initially, fill in the gaps between your groupings with bulbs, annuals or short-lived perennials that may easily be moved later. These will then bulk out the planting scheme, providing a temporary source of interest and colour, until the larger plants have reached maturity.

CLASSIC BORDER PLANTING
The striking foliage of Elaeagnus pungens 'Maculata', *the rich red rose and the golden-yellow flowers of* Potentilla fruticosa *provide a classic background for a mass of low-growing perennials and violas.*

PLANTING FOR THE FUTURE
*It is easy to forget how quickly trees and shrubs grow, so they should be well spaced. It may help to mark out 1m sections with string. This bed has been planted for winter interest. Rosebud cherry (*Prunus subhirtella *'Autumnalis') dominates the bed, while two evergreen Mexican orange blossoms (*Choisya ternata*) are contrasted with darker* Sarcococca humilis, *which bears fragrant flowers in late winter. Against these, the variegated* Iris foetidissima *stands out well. White-flowered* Bergenia *'Silberlicht' is placed under the cherry and beyond the Mexican orange blossoms are grouped three* Helleborus argutifolius.

YEAR 1

YEAR 3

CONTINUING CARE AND MAINTENANCE
This bed should reach its peak of glory after about five years. Trimming may be necessary 6–10 years after planting.

YEAR 5

Prunus subhirtella 'Autumnalis'

Helleborus argutifolius

Bergenia 'Silberlicht'

Choisya ternata

Sarcococca humilis

Iris foetidissima

CREATING A GARDEN

Using Colour

Everyone interprets the moods that colours create in different ways, and many people change their minds according to the weather, the time of day or their emotional state. Individuals with perfectly normal colour vision perceive colours differently, and colour blindness is a surprisingly common phenomenon.

Colour choice is first and foremost about personal preferences. Although there are some fundamental design rules on how to use colour, including the avoidance of such obvious clashes as hot orange with delicate pink, they really constitute only a broad guide for the use of flat planes of basic colour. Often they have little to do with the realities of a garden – the textures of living plants, the changing light, the varying hues of a colour range, the season of the year and the location.

The intricacies of colour perception

The key to perceiving colour is light. Soft colours look wonderful in the early morning, in the evening and in dull or damp weather. In the Mediterranean sunshine, these colours would look blanched, and strong colours, which may be overpowering in a soft light, come into their own. Happily nature has ensured that plants with strong-coloured flowers are usually native to sunny regions.

The damp light of a temperate climate is quite unlike humid, tropical light or clear, desert light; sharp, winter light is quite distinct from the hazy light of a summer noon, and both differ from misty, autumnal and dusky, evening lights. All have different effects on colours, altering our perception of them and transforming the mood of the garden. If you are only going to be using your garden at one time of day, for example the evening, bear this in mind when choosing your plants: white flowers can take on a luminous quality in fading light but dark blue ones will be invisible.

Apart from climatic conditions, other factors determine how we notice colour. The sea, for example, can alter colour perception, reflecting light for some distance inland. A background building or fence will also influence colour tones by reflecting or absorbing light: whether shiny or matt, made of glass, stone, brick or timber, it will influence our view of the colours of plants grown in front of it. Colours themselves affect each other too. The many shades of green modify any colours placed in front of them, as do the mousey-brown colours of a wintery landscape. The individual colour

GLOWING PINKS
Right: The pinks and purple-blues of this late summer bed set off each other excellently. The key to this success is the positioning of Artemisia ludoviciana 'Silver Queen' *and* Agapanthus *between the strong pinks of* Aster novae-angliae 'Alma Potschke' *and ice-plant* (Sedum spectabile 'Brilliant').

AUTUMNAL HUES
Far right: Rich bronze-red Japanese maple (Acer palmatum) *and golden-yellow* Fothergilla major – *in classic colours associated with a temperate autumn – glow even in the rather misty light typical of this season.*

USING COLOUR

SUBTLE COLOURS
Far left: The pastel colours of catmint (Nepeta × faassenii) *and* Campanula lactiflora 'Loddon Anna' *are used here to provide a subtle backdrop for the vivid carmine-red of the phlox.*

COLOUR AND TEXTURE
Left: This planting deliberately juxtaposes extremes of colour and texture, producing an almost layered effect. The sword-shaped leaves and reddish-orange flowers of monbretia (Crocosmia) *contrast strongly with the yellow of feathery dill and erect, spiky* Verbascum; *the yellow creates a link between the foreground and the background.*

STRONG COMPOSITIONS
Below: This interesting contrast of a deep yellow daylily (Hemerocallis) *set against the intense purple-blues of* Hydrangea *and* Agapanthus *makes a striking statement.*

CREATING A GARDEN

masses in a group influence each other, as well as affecting the whole planting. A largely white planting with a touch of purple will create one mood; but what about the reverse – a largely purple planting with a touch of white? The mood of one will be bright and sharp, the other sombre and tranquil.

Considering the character of the site
In the days before the introduction of foreign species and mass hybridization, there was an indigenous range of colours in every area. To help design a harmonious garden you might keep to this spirit of planting by analysing the location of the garden; the particular flavour of a place is dictated by its climate and the shapes and colours of its natural plant forms.

The background colour may be a dark, coniferous green with the grey of granite or it may be deciduous green with mellower sandstone or limestone. In chalky areas, indigenous vegetation might include a proportion of grey foliage with the darker greens of yew or box. Although there are some bright-coloured flowers in the temperate zones, the really brilliantly coloured flowers originate in hotter climates, where the intensity of the sun accentuates them. Thus the nearer the equator you go, the brighter the indigenous flowers.

Seasonal colour associations
As well as geographical factors, there is a natural progression of dominant colours through the seasons. Pale spring colours transmute into the blues of early summer; hotter pinks of mid-summer are transformed into yellows and bronzes as autumn approaches, before turning into the browns of winter. When choosing plants for individual plant groups it is wise to avoid the use of certain colours against these seasonal backdrops, lest they introduce a jarring note. For example, refrain from setting soft pink against spring green, using purple in mid-summer or planting bright blue in autumn.

Even within the walls of a town garden with a backdrop of bricks and mortar, the seasons should still influence your colour choices. In temperate zones, for a natural look, try whites and pale lemon in spring; pinks, blues and greys in mid-summer; in late summer a little red but mainly yellows; and in autumn some bronze and purple.

FOLIAGE AND FLOWERS
Below right: Foliage and flower colours may be combined with equal dominance. In this planting, the huge, variegated leaves of Siberian bugloss (Brunnera macrophylla 'Dawson's White') are well set off by the surrounding, pale pink Potentilla nepalensis 'Miss Willmott', pink Astilbe flowers and golden-leaved Lamium maculatum 'Aureum'.

SATISFYING CONTRASTS
Below: Not only does the texture of Bowles' golden sedge (Carex elata 'Aurea') and the two hostas combine excellently but also the colour and form of these plants further complement the design.

USING COLOUR

Colour preferences
There remain, of course, more personal preferences for certain colour combinations. Strong colours are often more appreciated by the young than by the elderly, as they are invigorating. It follows that, in the garden, strong colours may be more appropriate when associated with various kinds of activity – near a tennis court, for instance, or around a pool. Softer colours, including shades of grey and pink, are calm and peaceful. Change pink to salmon, add a touch of purple and terracotta, and a broadly soft range becomes quite solid.

Making your personal mark
Your individual style and feelings about the moods that different colours create will be the final factors conditioning the colours in your garden, just as they are in the decoration of your home. If the garden is small, the result of linking the colours with those in the room adjacent to it will have a dramatic effect on both room and garden. Where room and garden meet, avoid clashes such as pink roses against orange-red brick or curtains. Conversely where the garden is larger, gradually change the colours in it so that, although those of the flowers nearest the house still link with the interior, those at the bottom of the garden work with the backdrop of the fence, hedge or countryside beyond.

Colour may be used to enhance perspective, too: hotter ranges of colour (reds, oranges and pinks) look better closer to the house, while cool colours (blues and whites) introduced at a distance from it will increase the sense of space. Once you begin thinking in colour ranges, you will start to eliminate many of the plant options and this will help further to simplify your plan.

A SENSE OF CALM
Above: Soft colours and forms intermingle harmoniously in this shady site, where the pinkish-mauve of the lacecap hydrangea is echoed in the foreground by the purple Ajuga leaves and the tiny, pink flowers of Lamium maculatum. *The delicate tracery of the ferns blends comfortably with the bolder, bluish leaves of* Hosta sieboldiana var. elegans.

SYMPATHETIC PLANTING
Right: This damp corner in a cottage garden contains a deliberately limited colour scheme, yet with invigorating results. When set against the ferns and grasses, the yellow heads of giant cowslip (Primula florindae) *and* Ligularia przewalskii *seem natural, because they have an affinity with the character and location of the site.*

CREATING A GARDEN

Texture

When considering different plant groupings, avoid thinking only in terms of colour combinations. A really successful design is a blend not only of colour – foliage as well as flowers – but of texture as well.

Texture is the soft, fluffy quality of fennel foliage and the dense, furry surface of grey stachys leaves, the spiky nature of many cacti and the smooth surface of evergreen leaves. It is an important aspect of the garden, contributed more by leaves than flowers (whose main role is to provide colour).

Try to set plants with very different textures next to one another. Group the rough with the smooth, the hairy with the spiky, and so on. It is the proportionate, yet contrasting relationship of purple fennel and grey stachys that we find so appealing. Varying foliage textures influence the quality of light reflected or diffused by the plant and so the effect of a mass of flat, dull leaves will be notably different from a mass of bright, shiny leaves. For a damp spot, ornamental grasses will contrast well with feathery ferns and the leathery leaves of hostas.

Arrange your plants so that you maintain an interesting balance of texture and shape in all seasons. Think, also, about the texture of objects in your garden. A large, smooth stone placed by a composition of plants of bold shapes can create a most pleasing scene.

PLAY ON TEXTURES
Above: Spiky branches of Scotch thistle (Onopordum acanthium) *provide a dramatic foil for glossy-leaved ivy, soft-leaved* Nicotiana *and fine-leaved lilies and marguerites. The paving slabs, the smooth terracotta pot (planted with marguerites) and the carved stone container (with* Helichrysum petiolare 'Aureum') *add further variety to the scene.*

DISTINCTIVE TEXTURES
Far left: The creamy plumes of pampas grass (Cortaderia selloana) *will help to add a textural dimension to any garden, but they must be in proportion to the surrounding trees and shrubs if they are not to look out of place.*

SUCCESSFUL COMBINATIONS
Left: This blend of flower and leaves skilfully exploits the changing textures and colours between Bowles' golden sedge (Carex elata 'Aurea') *in the foreground, the feathery plumes of a deep red* Astilbe *and pink Candelabra primulas through to the enormous, prickly-edged leaves of* Gunnera manicata.

The Planter's Guide

The following lists suggest plants that are suitable for growing in particular situations, or that have special uses or characteristics. For each category the list is broadly subdivided into groups, following the arrangement of the Plant Catalogue, pp.37–416. Individual plants that appear in the Plant Catalogue are followed by page numbers. Refer to the Plant Dictionary if a whole genus or a plant not followed by a page number is recommended.

Plants are not always consistent in their growing habits and much of their success depends on climate, available nutrients in the soil, planting site and weather. If a plant is listed for a particular purpose (coastal sites, shade, etc.) it should thrive there happily in normal growing conditions, even if the situation is not the plant's naturally preferred location.

○ – evergreen or semi-evergreen plants, including those with overwintering rosettes of leaves.

Plants for hedges and windbreaks

△ – moderately to fast growing.

Trees
Alnus cordata, p.40△
Arbutus andrachne ○
♛ *Arbutus unedo*, p.67 ○
♛ *Carpinus betulus*
♛ *Carpinus betulus* 'Fastigiata', p.74
Crataegus monogyna
Fagus sylvatica, p.42
♛ *Ilex aquifolium*, p.72 ○
♛ *Ilex aquifolium* 'Argentea Marginata', p.72 ○
♛ *Laurus nobilis* ○
Melaleuca viridiflora var. *rubriflora* ○△
Metrosideros excelsa, p.57 ○△
Nothofagus dombeyi, p.46 ○△
Nothofagus obliqua, p.42△
♛ *Olea europaea* ○
Populus x *canadensis* 'Robusta', p.39△
Prunus lusitanica ○
Syzygium paniculatum, p.55 ○
Umbellularia californica, p.48 ○
♛ *Zelkova serrata*, p.45

Conifers
♛ *Abies grandis*, p.78 ○△
♛ *Cedrus deodara*
Cephalotaxus harringtonia ○
Chamaecyparis lawsoniana ○△
x *Cupressocyparis leylandii*, p.75 ○△
Cupressus macrocarpa ○△
Juniperus communis ○
♛ *Larix decidua* △
♛ *Picea omorika*, p.77 ○△

♛ *Pinus nigra* ○
♛ *Pinus radiata*, p.78 ○△
Pseudotsuga menziesii subsp. *glauca*, p.76 ○△
♛ *Taxus baccata* ○
Thuja plicata ○△
Tsuga canadensis, p.81 ○△

Shrubs
♛ *Berberis darwinii*, p.87 ○
♛ *Buxus sempervirens* 'Suffruticosa', p.147 ○
♛ *Choisya ternata*, p.97 ○
Codiaeum variegatum, p.147 ○
Cotoneaster simonsii, p.118△
Dodonaea viscosa 'Purpurea', p.121 ○△
Duranta erecta, p.120 ○△
Elaeagnus x *ebbingei* ○△
♛ *Escallonia* 'Langleyensis', p.112 ○
Euonymus japonicus 'Macrophyllus' ○
♛ *Griselinia littoralis* ○
Hibiscus rosa-sinensis ○
Lavandula cvs ○
Leptospermum scoparium cvs ○△
Ligustrum ovalifolium, p.96 ○△
Lonicera nitida ○
Photinia x *fraseri* 'Birmingham', p.87△
♛ *Pittosporum tenuifolium*, p.97 ○△
♛ *Prunus laurocerasus* ○
♛ *Prunus lusitanica* ○
♛ *Pyracantha* x *watereri*, p.106 ○△
Rosmarinus officinalis and cvs ○
Tamarix ramosissima, p.90△

Roses
Rosa californica△
♛ *Rosa* 'Céleste', p.151△
♛ *Rosa* 'Felicia', p.151△
Rosa 'Frühlingsmorgen'△
♛ *Rosa gallica* var. *officinalis*△
♛ *Rosa gallica* 'Versicolor', p.153△
♛ *Rosa* 'Geranium', p.153△
♛ *Rosa glauca*, p.152△
Rosa 'Great Maiden's Blush', p.151△
♛ *Rosa* 'Marguerite Hilling', p.152△
♛ *Rosa* 'Nevada', p.151△
♛ *Rosa* 'Penelope', p.150△
♛ *Rosa* 'Président de Sèze'△
Rosa rugosa, p.153△
♛ *Rosa* 'Tuscany Superb'△

Grasses (including Bamboos)
Arundo donax△
Cortaderia selloana 'Sunningdale Silver', p.182 ○
♛ *Fargesia nitida* ○
♛ *Miscanthus floridulus*△
Phyllostachys bambusoides, p.184 ○△
♛ *Semiarundinaria fastuosa*, p.184 ○
♛ *Stipa gigantea*, p.183 ○

Perennials
Echinops bannaticus, p.192△
Eupatorium purpureum, p.195△
Filipendula camtschatica△
♛ *Helianthus atrorubens* 'Monarch'
♛ *Macleaya microcarpa* 'Coral Plume', p.191△
Phormium tenax ○
♛ *Rudbeckia* 'Goldquelle', p.193△
Silphium laciniatum△
Veronica virginica f. *alba*, p.205

Plants for coastal and exposed sites

Trees
Acer pseudoplatanus and cvs
Agonis flexuosa, p.64 ○
Alnus incana, p.40
♛ *Arbutus unedo*, p.67 ○
♛ *Eucalyptus cocciferа*, p.46 ○
♛ *Eucalyptus globulus* ○
♛ *Eucalpytus gunnii*, p.46 ○
Ficus macrophylla ○
Fraxinus excelsior
Ilex aquifolium cvs, pp.70–71 ○
♛ *Laurus nobilis* ○
Melaleuca viridiflora var. *rubriflora* ○
Salix alba
♛ *Schefflera actinophylla*, p.58 ○
Schinus molle ○
♛ *Sorbus aria* 'Lutescens', p.52
Tabebuia chrysotricha, p.71
Thevetia peruviana, p.67 ○
Tipuana tipu

Conifers
x *Cupressocyparis leylandii*, p.75 ○
Cupressus macrocarpa ○
Cupressus sempervirens, p.79 ○
Pinus nigra var. *nigra*, p.78 ○

Shrubs
Acacia verticillata ○
Bupleurum fruticosum, p.116 ○
Cytisus x *spachianus* ○
Duranta erecta, p.120 ○
♛ *Elaeagnus pungens* 'Maculata', p.97 ○
♛ *Erica cinerea* 'Eden Valley', p.149 ○
♛ *Escallonia rubra* var. *macrantha* 'Crimson Spire' ○
Euonymus japonicus ○
Euphorbia characias subspp., p.126 ○
♛ *Felicia amelloides* 'Santa Anita', p.138
Genista hispanica, p.140
Halimium lasianthum subsp. *formosum*, p.139 ○
♛ *Hebe franciscana* 'Blue Gem' ○
Hibiscus rosa-sinensis ○
♛ *Hippophäe rhamnoides*, p.94
♛ *Lavandula* 'Hidcote', p.137 ○
Malvaviscus arboreus, p.90 ○
Pyracantha coccinea 'Lalandei' ○
Rosmarinus officinalis, p.137 ○
♛ *Senecio* 'Sunshine', p.140 ○
Viburnum tinus, p.119 ○

Climbers
Antigonon leptopus, p.169 ○
♛ *Bougainvillea glabra*, p.174 ○
♛ *Eccremocarpus scaber*, p.177 ○
Ercilla volubilis ○
Euonymus fortunei 'Coloratus' ○
Ficus pumila ○
Hedera canariensis ○
Muehlenbeckia complexa
Pandorea jasminoides, p.168 ○
Pyrostegia venusta, p.178 ○
Schisandra rubriflora, p.171
Solandra maxima, p.166 ○
Tripterygium regelii
♛ *Tropaeolum tuberosum* 'Ken Aslet', p.176
♛ *Wisteria sinensis*, p.175

Perennials
Anaphalis margaritacea, p.203
Anchusa azurea 'Loddon Royalist', p.218
Anthurium andraeanum, p.228 ○
♛ *Artemisia absinthium* 'Lambrook Silver' ○
Centaurea hypoleuca 'John Coutts', p.243
Echinacea purpurea
Erigeron 'Charity', p.242
Eryngium variifolium, p.250 ○
Euphorbia griffithii 'Fireglow', p.222
Geranium sanguineum, p.302
IRISES, pp.198–9, some ○
♛ *Kniphofia caulescens*, p.227 ○
Lupinus 'Thundercloud'
Peperomia obtusifolia 'Variegata', p.269 ○
Phormium tenax ○
♛ *Pilea cadierei*, p.264 ○
Romneya coulteri, p.190
♛ *Salvia argentea*, p.204
Senecio x *hybridus* Series ○
Senecio cineraria 'Silver Dust', p.286 ○
Tradescantia fluminensis ○

Annuals and Biennials
Antirrhinum majus and cvs
Calendula officinalis, Series and cvs
Clarkia amoena and Series
Coreopsis tinctoria, p.288
Cynoglossum amabile 'Firmament', p.286
Dahlia, Coltness Hybrids, p.281
♛ *Eschscholzia californica*, p.290
Gilia capitata, p.285
Helichrysum bracteatum cvs
Helipterum roseum, p.273
Impatiens, Novette Series, pp.273, 279 ○
♛ *Kochia scoparia* f. *trichophylla*, p.287
♛ *Limnanthes douglasii*, p.288
Matthiola
Portulaca grandiflora, Series and cvs
Tagetes

Rock plants
Achillea clavennae, p.321 ○
♛ *Aethionema grandiflorum*, p.300 ○
♛ *Dianthus deltoides* ○
Draba aizoides ○
Epilobium glabellum, p.299 ○
♛ *Iberis sempervirens*, p.294 ○
♛ *Origanum laevigatum*, p.302
♛ *Oxalis enneaphylla*
Phlox subulata 'Marjorie', p.326 ○
♛ *Pulsatilla vulgaris*, p.296
Saxifraga paniculata ○
♛ *Sedum spathulifolium* 'Cape Blanco', p.339 ○
♛ *Sempervivum arachnoideum*, p.337 ○
Thlaspi rotundifolium, p.313
♛ *Viola cornuta*, p.297

Bulbs, Corms and Tubers
Amaryllis belladonna, p.352
Canna x *generalis* cvs
Crinum
Crocus
DAFFODILS, pp.360–62
Eucharis grandiflora, p.355 ○
Freesia
Galtonia candicans, p.341
Hippeastrum
HYACINTHS, p.375

THE PLANTER'S GUIDE

Hymenocallis
Nerine
Scilla
🏆 *Sprekelia formosissima*, p.357
TULIPS, pp.354–6
🏆 *Veltheimia bracteata*, p.369
Zantedeschia aethiopica ○

Shrubs preferring wall protection

🏆 *Abutilon megapotamicum* ○
Acacia pravissima, p.71 ○
Artemisia arborescens, p.145 ○
Buddleja crispa, p.113
Ceanothus impressus, p.115 ○
🏆 *Chaenomeles speciosa* 'Moerloosii', p.100
Chimonanthus praecox
Cytisus × *spachianus* ○
Daphne odora 'Aureo-marginata', p.144 ○
Elsholtzia stauntonii, p.143
🏆 *Escallonia* 'Iveyi', p.88 ○
🏆 *Fabiana imbricata* f. *violacea*, p.115 ○
Feijoa sellowiana, p.113 ○
🏆 *Fremontodendron* 'California Glory', p.93 ○
Garrya elliptica, p.95 ○
Jasminum mesnyi, p.167 ○
Lagerstroemia indica, p.65
🏆 *Leptospermum scoparium* 'Red Damask', p.101 ○
Lonicera fragrantissima
🏆 *Melianthus major* ○
Olearia × *scilloniensis* ○
Pyracantha atalantioides 'Aurea', p.94 ○
🏆 *Robinia hispida*, p.111
🏆 *Rosa banksiae* 'Lutea', p.164 ○
🏆 *Rosa* 'Mermaid', p.164
Rosmarinus officinalis, p.137 ○
Salvia involucrata 'Bethellii', p.195
Solanum crispum 'Glasnevin', p.174 ○
🏆 *Tibouchina urvilleana*, p.92
Viburnum foetens, p.119

Plants for quick cover

Shrubs
🏆 *Ceanothus thyrsiflorus* var. *repens*, p.138 ○
Hypericum calycinum, p.140 ○
Lantana camara ○

Climbers
Hedera helix (small cvs) ○
🏆 *Hydrangea petiolaris*, p.168
Lonicera japonica cvs ○
Pueraria lobata
🏆 *Trachelospermum asiaticum* ○
🏆 *Trachelospermum jasminoides*, p.167 ○

Ferns
Dryopteris dilatata
🏆 *Polystichum aculeatum* ○
Polystichum setiferum cvs, pp.187, 189 ○

Perennials
🏆 *Alchemilla mollis*, p.254
🏆 *Anthemis punctata* subsp. *cupaniana*, p.240 ○
Duchesnea indica ○
🏆 *Euphorbia amygdaloides* var. *robbiae*, p.235 ○
Galeobdolon argentatum ○

Geranium macrorrhizum, p.243 ○
Heterocentron elegans, p.248 ○
Lamium maculatum and cvs ○
🏆 *Osteospermum jucundum*, p.243 ○
Pulmonaria (most), some ○
Stachys byzantina, p.268 ○
🏆 *Symphytum* × *uplandicum* 'Variegatum', p.202

Annuals and Biennials
Lathyrus odoratus 'Bijou', p.276
Portulaca grandiflora, Series and cvs
Sanvitalia procumbens, p.288
Tropaeolum majus, Series and cvs

Rock plants
Acaena anserinifolia ○
Arabis caucasica and cvs ○
Aubrieta ○
Campanula poscharskyana, p.329
Helianthemum ○
Phlox douglasii cvs ○
Phlox subulata ○
Phuopsis stylosa, p.300
Polygonum affine and cvs
🏆 *Polygonum vacciniifolium*, p.335 ○
🏆 *Saxifraga stolonifera* ○
🏆 *Tiarella cordifolia*, p.295 ○
Waldsteinia ○

Ground-cover plants for shade

Shrubs
Daphne laureola subsp. *philippi*, p.126 ○
Epigaea asiatica ○
Euonymus fortunei 'Kewensis' ○
Gaultheria shallon, p.132 ○
🏆 *Leucothöe fontanesiana* ○
Mahonia repens ○
Pachysandra terminalis, p.336 ○
Ruscus hypoglossum, p.146 ○
Sarcococca humilis, p.144 ○
Vinca minor, p.146 ○

Climbers
Asteranthera ovata ○
Berberidopsis corallina, p.171 ○
Epipremnum aureum 'Marble Queen', p.179 ○
Ficus pumila ○
🏆 *Hedera colchica* 'Dentata', p.181 ○
🏆 *Hydrangea petiolaris*, p.168
🏆 *Rhoicissus capensis* ○
Rhoicissus rhomboidea ○
Schizophragma hydrangeoides

Ferns
🏆 *Adiantum venustum*, p.189
Blechnum penna-marina ○
Polypodium vulgare, p.189 ○
Polystichum setiferum cvs, pp.187, 189 ○

Perennials
🏆 *Alchemilla mollis*, p.254
Asarum caudatum ○
Aspidistra elatior ○
🏆 *Bergenia cordifolia* 'Purpurea', p.233 ○
Brunnera macrophylla
Duchesnea indica ○
Epimedium perralderianum ○
🏆 *Euphorbia amygdaloides* var. *robbiae*, p.235 ○
Galeobdolon argentatum ○
Geranium macrorrhizum, p.243 ○
HOSTAS (some), pp.252–3
Luzula sylvatica 'Marginata' ○
Pellionia daveauana, p.267 ○

🏆 *Plectranthus oertendahlii* ○
Pulmonaria saccharata, p.235 ○
Symphytum grandiflorum
Tellima grandiflora 'Rubra', p.266 ○
Tolmiea menziesii ○
🏆 *Tradescantia fluminensis* 'Variegata', p.264 ○
Waldsteinia ternata, p.333 ○

Rock plants
Asarina procumbens, p.332
Cardamine trifolia, p.310
Homogyne alpina ○
Mitchella repens ○
Prunella grandiflora, p.329
🏆 *Saxifraga stolonifera* ○
🏆 *Saxifraga* × *urbium* ○
🏆 *Tiarella cordifolia*, p.295 ○

Ground-cover plants for sun

Conifers
Juniperus communis 'Prostrata' ○
Juniperus conferta ○
Juniperus sabina var. *tamariscifolia*, p.84 ○
🏆 *Juniperus squamata* 'Blue Carpet' ○
🏆 *Microbiota decussata*, p.84 ○
Picea abies 'Inversa' ○

Shrubs
Arctostaphylos uva-ursi, p.336 ○
🏆 *Calluna vulgaris* 'White Lawn' ○
🏆 *Ceanothus thyrsiflorus* var. *repens*, p.138 ○
Cotoneaster cochleatus ○
Cotoneaster 'Skogholm' ○
🏆 *Erica carnea* 'Springwood White', p.148 ○
Gaultheria myrsinoides ○
Hebe pinguifolia 'Pagei', p.299 ○
Hypericum calycinum, p.140 ○
Lantana montevidensis, p.137 ○
Leiophyllum buxifolium ○
🏆 *Rosmarinus officinalis* 'Prostratus' ○
Salix repens, p.126
Stephanandra incisa 'Crispa'

Climbers
Anredera cordifolia ○
Campsis radicans
Clematis armandii, p.172 ○
🏆 *Clematis rehderiana*, p.173
Clematis tangutica, p.173
Decumaria sinensis ○
Hardenbergia comptoniana, p.166 ○
🏆 *Hibbertia scandens* ○
Kennedia rubicunda, p.165 ○
🏆 *Lathyrus latifolius*, p.171
🏆 *Lonicera japonica* 'Halliana', p.175 ○
🏆 *Parthenocissus tricuspidata*, p.178
Pueraria lobata
Pyrostegia venusta, p.178 ○
🏆 *Trachelospermum asiaticum* ○
🏆 *Vitis coignetiae*, p.178
Vitis davidii ○

Perennials
🏆 *Anthemis punctata* subsp. *cupaniana*, p.240 ○
Centaurea montana, p.248
🏆 *Euphorbia polychroma*, p.239
Geranium sanguineum, p.302
Heterocentron elegans, p.248 ○
Lamium maculatum, p.233 ○
Lysimachia punctata, p.217
Nepeta × *faassenii*, p.250
🏆 *Osteospermum jucundum*, p.243 ○

Peltiphyllum peltatum, p.202
🏆 *Phlomis russeliana*, p.219
Rheum palmatum 'Atrosanguineum', p.191
Stachys byzantina, p.268 ○

Annuals and Biennials
Any of spreading habit.

Rock plants
🏆 *Acaena microphylla*, p.337 ○
Arabis caucasica 'Variegata', p.310 ○
🏆 *Armeria maritima* 'Vindictive', p.328 ○
Aubrieta cvs ○
🏆 *Aurinia saxatilis*, p.298 ○
Campanula poscharskyana, p.329
🏆 *Dianthus gratianopolitanus*, p.325 ○
🏆 *Dryas octopetala*, p.323 ○
Helianthemum 'Ben More', p.302 ○
🏆 *Hypericum olympicum*
🏆 *Iberis sempervirens*, p.294 ○
🏆 *Lithodora diffusa* 'Heavenly Blue', p.305 ○
Nierembergia repens, p.322
🏆 *Phlox douglasii* 'Crackerjack', p.327 ○
Phuopsis stylosa, p.300
Polygonum affine and cvs
Thymus praecox ○
Veronica prostrata 'Kapitan', p.305

Plants for paving and wall crevices

Annuals and Biennials (not walls)
Ageratum houstonianum (small cvs)
🏆 *Ionopsidium acaule*
🏆 *Limnanthes douglasii*, p.288 ○
Lobelia erinus ○
Lobularia maritima
Malcolmia maritima, p.275
Nemophila maculata, p.271
Nemophila menziesii, p.285
Portulaca grandiflora, Series and cvs

Rock plants
🏆 *Acaena microphylla*, p.337 ○
🏆 *Aethionema* 'Warley Rose', p.324 ○
Alyssum montanum ○
Aubrieta ○
Campanula poscharskyana, p.329
🏆 *Dianthus deltoides* ○
🏆 *Erinus alpinus*, p.314 ○
Gypsophila repens and cvs ○
Helianthemum ○
🏆 *Hypericum olympicum*
🏆 *Lithodora diffusa* 'Heavenly Blue', p.305 ○
Parahebe lyallii ○
Phlox douglasii cvs ○
🏆 *Ramonda myconi* (wall only), p.330 ○
Saxifraga cotyledon, p.300 ○
🏆 *Sedum spathulifolium* 'Cape Blanco', p.339 ○
Sempervivum montanum, p.338 ○
Thymus praecox and cvs ○

Plants for dry shade

Trees and Conifers
🏆 *Ilex aquifolium*, p.72 ○
Taxus baccata 'Adpressa' ○

Shrubs
🏆 *Buxus sempervirens* ○
Daphne laureola ○

THE PLANTER'S GUIDE

Elaeagnus × *ebbingei* ○
Gaultheria shallon, p.132 ○
Hypericum × *inodorum* 'Elstead', p.140
Lonicera pileata, p.146 ○
Mahonia aquifolium, p.127 ○
Osmanthus decorus ○
Ruscus aculeatus ○
Viburnum rhytidophyllum, p.88 ○
Vinca major ○
Vinca minor, p.146 ○

Climbers
Berberidopsis corallina, p.171 ○
Cissus striata ○
Epipremnum aureum 'Marble Queen', p.179 ○
Hedera canariensis ○
♈ *Lapageria rosea*, p.170 ○
♈ *Lonicera japonica* 'Halliana', p.175 ○
♈ *Philodendron scandens*, p.180 ○

Ferns
♈ *Asplenium scolopendrium*, p.189 ○
Ceterach officinarum, p.187 ○
♈ *Cyrtomium falcatum*, p.187 ○
♈ *Davallia canariensis*
Microlepia strigosa, p.186 ○
Nephrolepis exaltata, p.188 ○
Polypodium vulgare, p.189 ○
♈ *Pteris cretica*, p.187 ○

Perennials
Achimenes
♈ *Alchemilla mollis*, p.254
Chirita ○
♈ *Epimedium pinnatum* subsp. *colchicum* ○
Galeobdolon argentatum ○
Iris foetidissima ○
♈ *Kohleria digitaliflora*, p.208
Luzula sylvatica 'Marginata' ○
Streptocarpus saxorum, p.248 ○
Symphytum grandiflorum
Tellima grandiflora ○
Tolmiea menziesii ○
♈ *Tradescantia zebrina* 'Quadricolor' ○

Bulbs, Corms and Tubers
♈ *Clivia miniata*, p.363 ○
Haemanthus albiflos ○
Hyacinthoides hispanica, p.358
Hyacinthoides non-scripta, p.358

Plants for moist shade

Shrubs
Clethra arborea ○
Crataegus laevigata 'Punicea'
♈ *Kalmia latifolia*, p.111 ○
Lindera benzoin, p.101
Neillia thibetica, p.110
♈ *Paeonia lutea* var. *ludlowii*, p.201
Paeonia suffruticosa 'Rock's Variety', p.200
♈ *Pieris formosa* var. *forrestii* 'Wakehurst', p.112 ○
Pittosporum eugenioides ○
♈ *Prunus laurocerasus* ○
RHODODENDRONS, pp.102–104, most ○
♈ *Salix magnifica*
Sarcococca ruscifolia ○
Skimmia japonica, p.145 ○
♈ *Viburnum* 'Pragense', p.109 ○

Climbers
Akebia quinata, p.166, sometimes ○
Asteranthera ovata ○
Decumaria sinensis ○
Dioscorea discolor, p.179 ○
♈ *Humulus lupulus* 'Aureus', p.166
♈ *Hydrangea petiolaris*, p.168
♈ *Lonicera tragophylla*
Macleania insignis ○
Mikania scandens ○
Passiflora coccinea, p.165 ○
♈ *Pileostegia viburnoides*, p.168 ○
♈ *Schizophragma integrifolium*, p.168
Smilax china
♈ *Thunbergia mysorensis*, p.167 ○
♈ *Trachelospermum jasminoides*, p.167 ○

Ferns
Athyrium nipponicum, p.189
♈ *Blechnum tabulare* ○
Cyathea australis, p.74 ○
Cyathea medullaris ○
♈ *Dicksonia antarctica*, p.186 ○
Dryopteris goldiana
Lunathyrium japonicum ○
Lygodium japonicum ○
♈ *Matteuccia struthiopteris*, p.188
♈ *Onoclea sensibilis*, p.188
Osmunda claytoniana
Polystichum munitum, p.186 ○
♈ *Selaginella martensii*, p.187 ○
♈ *Woodwardia radicans* ○

Perennials
Actaea pachypoda, p.223
Anemone × *hybrida* cvs
Anthurium scherzerianum, p.266 ○
♈ *Aruncus dioicus*, p.190
Begonia rex and hybrids ○
Bergenia ○
♈ *Calathea zebrina*, p.230 ○
Cardamine pentaphyllos, p.234
♈ *Convallaria majalis*, p.232
Deinanthe caerulea
Dichorisandra reginae, p.216 ○
♈ *Digitalis* × *mertonensis* ○
Helleborus orientalis, pp.264, 265, 266 ○
HOSTAS, pp.252–3
♈ *Kirengeshoma palmata*, p.227
♈ *Maranta leuconeura* 'Erythroneura', p.267 ○
Polygonatum × *hybridum*, p.202
PRIMULAS (many), pp.236–7
Ruellia devosiana, p.241 ○
♈ *Trillium grandiflorum*, p.232
♈ *Uvularia grandiflora*, p.238
Vancouveria hexandra, p.295

Bulbs, Corms and Tubers
Arisaema
Arisarum proboscideum
♈ *Arum italicum* 'Pictum', p.378
Camassia leichtlinii, p.341
♈ *Galanthus elwesii*, p.383
Galanthus nivalis and cvs
♈ *Galanthus plicatus* subsp. *plicatus*
Leucojum aestivum, p.340
♈ *Leucojum vernum*, p.370
♈ *Narcissus cyclamineus*, p.362

Plants for sandy soil

Trees
♈ *Acacia dealbata*, p.57 ○
Acer negundo
Agonis flexuosa, p.64 ○
Banksia serrata ○
♈ *Betula pendula* 'Dalecarlica', p.47
♈ *Castanea sativa*
Celtis australis, p.41
♈ *Cercis siliquastrum*, p.62
Eucalyptus ficifolia ○
Gleditsia triacanthos
Melia azedarach, p.50
Nothofagus obliqua, p.42
♈ *Phoenix canariensis* ○
♈ *Quercus ilex* ○
Schinus molle ○

Conifers
♈ *Abies grandis*, p.78 ○
× *Cupressocyparis leylandii* and cvs ○
Cupressus glabra ○
Juniperus ○
♈ *Larix decidua*
♈ *Pinus pinaster*, p.77 ○
♈ *Pinus radiata*, p.78 ○
Pseudotsuga menziesii subsp. *glauca*, p.76 ○
Thuja occidentalis and cvs ○

Shrubs
Berberis empetrifolia, p.127 ○
Calluna vulgaris and cvs ○
Ceanothus thyrsiflorus and forms ○
Cistus ○
Cytisus scoparius forms
♈ *Erica arborea* var. *alpina*, p.148 ○
Erica cinerea and cvs ○
Erica pageana ○
Genista tinctoria, p.127
Hakea lissosperma ○
Lavandula ○
Pernettya mucronata and cvs ○
Physocarpus opulifolius
Rosa pimpinellifolia, p.148
Rosmarinus officinalis and cvs ○
♈ *Spartium junceum*, p.117
Ulex europaeus, p.127
♈ *Yucca gloriosa*, p.107 ○

Climbers
Adlumia fungosa
Anredera cordifolia ○
Bomarea andimarcana ○
♈ *Clianthus puniceus*, p.165 ○
Kennedia rubicunda, p.165 ○
Merremia tuberosa ○
Mutisia oligodon ○
Periploca graeca
Petrea volubilis, p.166 ○
Semele androgyna ○
Solanum wendlandii, p.174 ○
♈ *Streptosolen jamesonii*, p.180 ○
♈ *Tropaeolum tricolorum*, p.165
Vitis vinifera 'Purpurea', p.178

Perennials
♈ *Acanthus spinosus*, p.215
♈ *Aphelandra squarrosa* 'Louisae', p.220 ○
Artemisia ludoviciana var. *albula*, p.230
Asphodeline lutea, p.202
Billbergia nutans, p.229 ○
Centranthus ruber, p.209
♈ *Cryptanthus zonatus* ○
Echinops sphaerocephalus, p.190
♈ *Eryngium* × *tripartitum*, p.217
Foeniculum vulgare 'Purpureum'
Gaillardia × *grandiflora* cvs
Limonium latifolium 'Blue Cloud', p.250
Nepeta × *faassenii*, p.250
♈ *Origanum vulgare* 'Aureum', p.254
Papaver orientale
PELARGONIUMS, pp.210–11 ○
♈ *Romneya coulteri*, p.190
Ruellia devosiana, p.241 ○
♈ *Sansevieria trifasciata* 'Laurentii', p.231 ○
♈ *Strelitzia reginae*, p.231 ○

Annuals and Biennials
Anchusa capensis cvs
Antirrhinum majus and cvs
Brachycome iberidifolia, p.286
Chrysanthemum segetum, p.289
Cleome hassleriana
Coreopsis tinctoria, p.288
Exacum affine, p.283 ○
Helichrysum bracteatum, Monstrosum Series, p.292
Impatiens, Novette Series, pp.273, 279 ○
♈ *Limnanthes douglasii*, p.288
Limonium sinuatum, p.274
Linaria maroccana 'Fairy Lights', p.282
Lobularia maritima
Mentzelia lindleyi, p.288
Papaver rhoeas, Shirley Series, pp.274, 280
Portulaca grandiflora, Series and cvs
Schizanthus
Tagetes
Verbena × *hybrida*, Series and cvs

Rock plants
Acaena caesiiglauca, p.338 ○
Achillea × *kellereri*, p.323 ○
♈ *Aethionema* 'Warley Rose', p.324 ○
♈ *Arabis ferdinandi-coburgii* 'Variegata', p.336 ○
♈ *Arenaria montana*, p.322
♈ *Armeria juniperifolia*, p.313 ○
♈ *Cytisus* × *beanii*, p.297
♈ *Dianthus deltoides* ○
♈ *Gypsophila repens* ○
Helianthemum ○
Iberis saxatilis, p.322 ○
Linum suffruticosum subsp. *salsoloides*
Phlox bifida, p.329 ○
♈ *Saponaria ocymoides*, p.326
Sedum ○
Sempervivum ○

Bulbs, Corms and Tubers
Babiana rubrocyanea, p.375
Brodiaea coronaria
Crocus
Freesia
IRISES (bulbous species), pp.198–9
Ixia
Muscari
Narcissus tazetta and Div.8 hybrids
Ornithogalum
Scilla
Tigridia pavonia, p.367
Zephyranthes

Cacti and other Succulents (all)

Plants for clay soils

□ – tolerates slow-draining soil; others require reasonable drainage.

Trees
Alnus glutinosa □
Castanospermum australe ○
Drimys winteri, p.52 ○
Fraxinus
♈ *Juglans nigra*, p.41
Melaleuca viridiflora, var. *rubriflora* ○ □
Oxydendrum arboreum, p.52 □
Populus □
♈ *Pterocarya fraxinifolia* □
♈ *Quercus palustris*, p.43
♈ *Quercus robur*
♈ *Salix* 'Chrysocoma', p.48 □
♈ *Salix matsudana* 'Tortuosa', p.59 □

Conifers
Cryptomeria ○ □
Metasequoia □
♈ *Taxodium distichum*, p.78 □

THE PLANTER'S GUIDE

Shrubs
Aronia arbutifolia, p.100 □
Calycanthus floridus □
Clethra alnifolia
♛ *Cornus alba* 'Sibirica', p.120 □
♛ *Kalmia latifolia*, p.111 ○
Ledum groenlandicum, p.124 ○ □
Magnolia virginiana ○
Salix caprea □
Salix purpurea □
Sambucus racemosa
♛ *Tetrapanax papyrifer*, p.96 ○
Viburnum lentago
Viburnum opulus

Climbers
Celastrus scandens
♛ *Humulus lupulus* 'Aureus', p.166
♛ *Rosa filipes* 'Kiftsgate', p.162
♛ *Vitis coignetiae*, p.178

Ferns
♛ *Matteuccia struthiopteris*, p.188 □
♛ *Onoclea sensibilis*, p.188 □
♛ *Osmunda regalis*, p.188 □
Polystichum setiferum cvs, pp.185 ○
Thelypteris palustris, p.188 □
♛ *Woodwardia radicans* ○ □
Woodwardia virginica □

Perennials
♛ *Aruncus dioicus*, p.190 □
Cyperus papyrus, p.183 ○ □
Filipendula ulmaria 'Aurea', p.254 □
♛ *Gunnera manicata*, p.192 □
Helonias bullata ○ □
Houttuynia cordata 'Chamaeleon', p.387 □
♛ *Iris laevigata*, p.199 □
Lythrum □
Mimulus guttatus □
Peltiphyllum peltatum, p.202 □
♛ *Primula florindae*, p.237 □
♛ *Primula japonica* □
Scrophularia auriculata 'Variegata' □
Trollius □

Water plants
♛ *Butomus umbellatus*, p.388 □
♛ *Caltha palustris*, p.391 □
♛ *Lysichiton americanus*, p.391 □
♛ *Pontederia cordata*, p.388 □
Ranunculus lingua, p.391 □
Sagittaria latifolia, p.386 □
Thalia dealbata □

Plants for chalk and limestone

Trees
Acer negundo 'Variegatum', p.53
♛ *Cercis siliquastrum*, p.62
Crataegus
♛ *Fagus sylvatica*, p.42
♛ *Fraxinus ornus*, p.50
Ilex aquifolium cvs, pp.72–73 ○
Malus
♛ *Morus nigra*
Phillyrea latifolia ○
♛ *Prunus avium* 'Plena', p.50
♛ *Robinia pseudoacacia* 'Frisia', p.55
Sorbus aria and cvs
Tilia tomentosa

Conifers
♛ *Calocedrus decurrens*, p.80 ○
Cedrus libani, p.77 ○
Chamaecyparis lawsoniana and cvs ○
× *Cupressocyparis leylandii* and cvs ○
Cupressus glabra ○

Juniperus ○
♛ *Picea omorika*, p.77 ○
♛ *Pinus nigra* ○
♛ *Taxus baccata* and cvs ○
Thuja orientalis and cvs ○
Thuja plicata and cvs ○

Shrubs
♛ *Berberis darwinii*, p.87 ○
Buddleja davidii and cvs
Ceanothus impressus, p.115 ○
♛ *Choisya ternata*, p.97 ○
Cistus ○
Cotoneaster, some ○
Deutzia
LILACS, p.90
Malus sargentii, p.86
Malus sieboldii, p.100
Nerium oleander, p.90 ○
Philadelphus
♛ *Phlomis fruticosa*, p.140 ○
Potentilla (all shrubby species)
Rosa rugosa, p.153
Viburnum tinus, p.119 ○
Vitex agnus-castus
Yucca aloifolia, p.123 ○

Climbers
Campsis radicans
♛ *Celastrus orbiculatus*
CLEMATIS, pp.172–73, some ○
♛ *Eccremocarpus scaber*, p.177 ○
IVIES, p.181 ○
Lonicera, some ○
♛ *Passiflora caerulea*, p.174 ○
♛ *Rosa* 'Albéric Barbier', p.162 ○
♛ *Rosa* 'Albertine', p.163 ○
♛ *Rosa banksiae* 'Lutea', p.164 ○
♛ *Trachelospermum jasminoides*, p.167 ○
♛ *Wisteria sinensis*, p.175

Ferns
♛ *Asplenium scolopendrium*, p.189 ○
Asplenium trichomanes, p.187 ○
♛ *Dryopteris filix-mas*, p.186
Polypodium vulgare 'Cornubiense', p.188 ○

Perennials
♛ *Acanthus spinosus*, p.215
♛ *Achillea filipendulina* 'Gold Plate', p.220
Bergenia ○
Doronicum
Eryngium, some ○
Gypsophila paniculata cvs
Helenium
IRISES (most), pp.198–9, some ○
Salvia nemorosa
♛ *Scabiosa caucasica* 'Clive Greaves', p.249
Sidalcea
Verbascum ○
Veronica spicata

Annuals and Biennials
Ageratum houstonianum and cvs
Calendula officinalis and Series and cvs
Callistephus chinensis, Series and cvs
Calomeria amaranthoides, p.282
Cheiranthus cheiri and Series and cvs
Gomphrena globosa, p.283
Lavatera trimestris 'Silver Cup', p.276
Limonium sinuatum, p.274
Lobularia maritima
Matthiola
Salvia viridis, p.283
Tagetes
Ursinia anthemoides, p.290
Xeranthemum annuum
Zinnia

Rock plants
Aethionema ○
Alyssum ○
Campanula (most rock garden species), some ○
Chrysanthemum hosmariense, p.294 ○
Dianthus (most rock garden species) ○
Draba ○
Erysimum helveticum, p.321 ○
♛ *Gypsophila repens* ○
Helianthemum ○
Leontopodium alpinum, p.294
Origanum dictamnus
Papaver burseri (*P. alpinum* group)
♛ *Saponaria ocymoides*, p.326
Saxifraga (most) ○
Thymus caespititius, p.323 ○
Veronica (all rock garden species), some ○

Bulbs, Corms and Tubers
Babiana
Chionodoxa
Colchicum
♛ *Crinum* × *powellii*, p.342
Crocus
♛ *Cyclamen hederifolium*, p.382
DAFFODILS, pp.360–62
GLADIOLI, p.343
Leucocoryne ixioides, p.358
♛ *Lilium regale*, p.348
Muscari
Pancratium illyricum, p.363
Scilla
TULIPS, pp.354–6
Zephyranthes

Plants requiring neutral to acid soil

Trees
♛ *Arbutus menziesii* ○
CAMELLIAS, pp.98–9 ○
Embothrium coccineum, p.67 ○
Stuartia
♛ *Styrax japonica*, p.51

Conifers
Abies ○
Picea (most) ○
Pinus densiflora ○
Pinus pumila ○
♛ *Pseudolarix amabilis*, p.81
Pseudotsuga ○
♛ *Sciadopitys verticillata*, p.80 ○
♛ *Tsuga heterophylla* ○

Shrubs
Arctostaphylos (some) ○
CAMELLIAS, pp.98–9 ○
♛ *Desfontainia spinosa*, p.113 ○
Epacris impressa, p.125 ○
HEATHERS (most), pp.148–9 ○
Leucothoë ○
Pernettya ○
Philesia magellanica ○
Pieris ○
RHODODENDRONS, pp.102–104, most ○
Styrax officinale, p.88
Telopea speciosissima, p.112 ○
Vaccinium, most ○
Zenobia pulverulenta, p.108

Climbers
Agapetes (several) ○
Asteranthera ovata ○
Berberidopsis corallina, p.171 ○
Mitraria coccinea, p.166 ○

Perennials
Cypripedium reginae, p.260
Drosera ○
Nepenthes ○
♛ *Sarracenia flava*, p.254
Trillium
Uvularia

Rock plants
Arctostaphylos ○
Cassiope ○
Corydalis cashmeriana
Cyananthus
Epigaea ○
Galax urceolata, p.299 ○
♛ *Gentiana sino-ornata*, p.335
Leucothoë keiskei
Lithodora diffusa ○
Mitchella repens ○
Ourisia ○
Pernettya ○
Phyllodoce ○
Pieris nana ○
Shortia ○
Vaccinium, most ○

Plants with decorative fruits or seed heads

Trees
Annona reticulata
Arbutus ○
Cornus kousa
Cotoneaster frigidus
Crataegus (most)
HOLLIES (most), pp.72–73, most ○
♛ *Koelreuteria paniculata*, p.66
MAGNOLIAS, p.49, some ○
Malus (most)
Schinus molle ○
Sorbus (most)

Conifers
Abies (some) ○
Cedrus (some) ○
Picea (some) ○
Pinus (some) ○

Shrubs
Aucuba japonica, p.122 ○
Berberis (most), some ○
Callicarpa bodinieri
Cotoneaster (most), some ○
Decaisnea fargesii, p.92
Euonymus (many), some ○
Gaultheria mucronata and cvs ○
♛ *Hippophaë rhamnoides*, p.94
Hypericum × *inodorum* 'Elstead', p.140
Pyracantha ○
ROSES (most), pp.150–64, some ○
Sambucus racemosa
Skimmia (some) ○
Symphoricarpos
Viburnum (several), some ○

Climbers
Actinidia chinensis
Akebia, sometimes ○
Cardiospermum halicacabum
♛ *Celastrus orbiculatus*
Clematis orientalis
Holboellia coriacea ○
ROSES (several), pp.150–64, some ○
Trichosanthes cucmerina var. *anguina*
♛ *Tropaeolum speciosum*, p.170

Perennials
Actaea
Clintonia borealis
Disporum hookeri

34

THE PLANTER'S GUIDE

Duchesnea indica ○
Iris foetidissima ○
Ophiopogon ○
❦ *Physalis alkekengi*
Phytolacca
Podophyllum

Annuals and Biennials
Briza maxima
Capsicum annuum cvs
Coix lacryma-jobi, p.184
Lagurus ovatus, p.182
Lunaria annua, p.277
Martynia annua, p.272
Nicandra physalodes
Nigella damascena and cvs
Zea mays

Rock plants
❦ *Acaena microphylla*, p.337 ○
❦ *Cornus canadensis*, p.322
❦ *Dryas octopetala*, p.323 ○
Gaultheria (most) ○
Maianthemum
Mitchella repens ○
Nertera granadensis, p.335 ○
Pulsatilla (most)

Bulbs, Corms and Tubers
❦ *Allium christophii*, p.365
Arisaema triphyllum, p.366
❦ *Arum italicum* 'Pictum', p.378
❦ *Cardiocrinum giganteum*, p.342

Water plants
Nelumbo
Nuphar lutea, p.391
Thalia dealbata

Plants with aromatic foliage

Trees
Agonis flexuosa, p.64 ○
Eucalyptus ○
❦ *Laurus nobilis* ○
Populus balsamifera
Populus trichocarpa
Sassafras albidum, p.42
Umbellularia californica, p.48 ○

Conifers
❦ *Calocedrus decurrens*, p.80 ○
Chamaecyparis ○
Cupressus ○
Juniperus ○
❦ *Pseudotsuga menziesii* ○
Thuja (most) ○

Shrubs
Aloysia triphylla, p.113
❦ *Artemisia abrotanum*, p.146
❦ *Choisya ternata*, p.97 ○
Elsholtzia stauntonii, p.143
❦ *Helichrysum italicum* ○
Hyssopus officinalis, p.138
Lavandula (most) ○
Lindera
❦ *Myrtus communis*, p.100 ○
PELARGONIUMS (scented-leaved forms), pp.210–11 ○
Prostanthera ○
Rhododendron rubiginosum ○
Rosmarinus officinalis, p.137 ○
Salvia officinalis cvs ○

Perennials
❦ *Artemisia absinthium* 'Lambrook Silver' ○
Chamaemelum nobile ○

Geranium macrorrhizum, p.243 ○
Houttuynia cordata 'Chamaeleon', p.387
Mentha, some ○
Monarda didyma
Myrrhis odorata, p.204
Origanum vulgare
Perovskia atriplicifolia
Tanacetum parthenium, p.271

Rock plants
Mentha requienii ○
❦ *Origanum laevigatum*, p.302
Satureja montana
Thymus ○

Plants with fragrant flowers

Trees
Bauhinia variegata, p.71
Clethra arborea ○
Drimys winteri, p.52 ○
MAGNOLIAS, p.49
❦ *Malus hupehensis*, p.48
Pittosporum tenuifolium, p.97 ○
Pittosporum undulatum ○
❦ *Robinia pseudoacacia*
❦ *Styrax japonica*, p.51
❦ *Tilia* 'Euchlora'
Virgilia oroboides ○

Shrubs
Buddleja davidii and cvs
Chimonanthus praecox
❦ *Choisya ternata*, p.97 ○
❦ *Cytisus battandieri*, p.93 ○
Daphne (many), most ○
❦ *Hamamelis mollis*
LILACS (most), p.90
Lonicera fragrantissima
MAGNOLIAS, p.49
Osmanthus ○
Philadelphus (many)
Pittosporum tobira ○
ROSES (many), pp.150–62, some ○
Sarcococca ○
Viburnum (many), some ○

Climbers
❦ *Clematis montana* 'Elizabeth'
❦ *Hoya carnosa*, p.168 ○
Jasminum (many), most ○
❦ *Lathyrus odoratus*
Lonicera (many), some ○
Mandevilla laxa
ROSES (many), pp.162–4, some ○
❦ *Stephanotis floribunda*, p.165 ○
Trachelospermum ○
Wattakaka sinensis
Wisteria

Perennials
CARNATIONS and PINKS (most), pp.244–5 ○
❦ *Convallaria majalis*, p.232
Cosmos atrosanguineus, p.213
Crambe cordifolia, p.190
❦ *Hedychium gardnerianum*, p.196
HOSTAS (some), pp.252–3
Iris graminea
❦ *Iris unguicularis* ○
Meehania urticifolia
❦ *Nicotiana sylvestris*, p.190
Petasites fragrans ○
❦ *Primula elatior*, p.237
❦ *Primula veris*, p.237
Tulbaghia natalensis
Verbena x *hybrida* 'Defiance', p.280 ○

Annuals and Biennials
Centaurea moschata, p.287
Erysimum cheiri, Series and cvs ○
Exacum affine, p.283 ○
Lathyrus odoratus and cvs
Lobularia maritima
Matthiola incana
Nicotiana alata, p.203
Primula Series
Reseda odorata, p.271
Scabiosa atropurpurea

Rock plants
Alyssum montanum ○
Dianthus (most) ○
Erysimum helveticum, p.321 ○
Papaver nudicaule
Primula auricula
Viola odorata ○

Bulbs, Corms and Tubers
❦ *Arisaema candidissimum*, p.379
Chlidanthus fragrans, p.380
Crinum bulbispermum
❦ *Crocus angustifolius*
Crocus longiflorus
Cyclamen persicum, p.384
Cyclamen repandum
❦ *Eucharis* x *grandiflora*, p.369 ○
Hymenocallis
LILIES (several), pp.348–9
Narcissus jonquilla and Div.7 hybrids
Narcissus tazetta and Div.8 hybrids
Ornithogalum arabicum, p.364
Polianthes tuberosa

Flowers for cutting

Shrubs
Calluna vulgaris (tall cvs) ○
Camellia japonica cvs ○
Erica ○
Forsythia
❦ *Hamamelis mollis*
LILACS, p.90
Lonicera fragrantissima
Philadelphus
ROSES (some), pp.150–64, some ○
Salix caprea
Turraea obtusifolia, p.142 ○

Perennials
Anaphalis
Anchusa azurea
Anemone x *hybrida* cvs
Astrantia major, p.241
CARNATIONS and PINKS, pp.244–5 ○
Cattleya (most)
CHRYSANTHEMUMS, pp.224–5
Cymbidium (most) ○
DELPHINIUMS (most), p.194
❦ *Helleborus niger*, p.265
Phalaenopsis (most) ○
Phlox paniculata cvs
Rudbeckia (most)
❦ *Strelitzia reginae*, p.231 ○

Annuals and Biennials
Amaranthus caudatus, p.278
Callistephus chinensis, Series and cvs
Centaurea cyanus and cvs
Centaurea moschata, p.287
Cosmos, Bright Lights Series
Gaillardia pulchella 'Lollipops', p.290
Gypsophila elegans, p.270
Helipterum roseum, p.273
Lathyrus odoratus and cvs
Limonium sinuatum, p.274
Matthiola cvs

Moluccella laevis, p.287
Xeranthemum annuum
Zinnia elegans group (tall hybrids)

Bulbs, Corms and Tubers
Allium (tall species)
Alstroemeria (tall species and cvs)
DAFFODILS (tall species and cvs), pp.360–62
DAHLIAS, pp.350–51
GLADIOLI (most), p.343
LILIES (some), pp.348–9
❦ *Nerine bowdenii*, p.368
Ornithogalum thyrsoides, p.364
Polianthes tuberosa
TULIPS (tall cvs), pp.354–6
Zantedeschia aethiopica ○

Flowers for drying

Trees
❦ *Acacia dealbata*, p.57 ○
Acacia longifolia ○
Acacia verticillata ○

Shrubs
Acacia (most) ○
Calluna vulgaris and cvs
Cassinia ○
❦ *Fothergilla major*, p.97 ○
Garrya elliptica, p.95 ○
Helichrysum (most) ○
Holodiscus discolor, p.89
Lavandula ○
LILACS, p.90
Rosmarinus officinalis and cvs ○

Perennials
Achillea, some ○
Artemisia (several), some ○
Astilbe (most)
❦ *Catananche caerulea* 'Major', p.249
Echinops
Eryngium, some ○
Eupatorium (some)
❦ *Gypsophila paniculata* 'Bristol Fairy', p.203
Limonium (most)
Lythrum
Rodgersia
Solidago (most)
Typha

Annuals and Biennials
Amaranthus caudatus, p.278
Centaurea cyanus
Gilia capitata, p.285
Gomphrena globosa, p.283
Gypsophila elegans, p.270
Helichrysum bracteatum, Monstrosum Series, p.292
Helipterum
Limonium sinuatum, p.274
Moluccella laevis, p.287
Onopordum acanthium, p.274
Salvia viridis and Series
Scabiosa atropurpurea
Tagetes, Erecta group hybrids
Xeranthemum annuum

Trailing plants for walls or baskets

Conifers
Juniperus conferta ○
Juniperus horizontalis and cvs ○
❦ *Juniperus squamata* 'Blue Carpet' ○
❦ *Microbiota decussata*, p.84 ○

THE PLANTER'S GUIDE

Shrubs
❦ *Ceanothus thyrsiflorus* var. *repens*, p.138 ○
❦ *Cotoneaster microphyllus* ○
Hebe pinguifolia 'Pagei', p.299 ○
❦ *Helichrysum petiolare*, p.145 ○
Leptospermum rupestris, p.130 ○
Salix lindleyana
Salix repens, p.126

Perennials
Alloplectus nummularia ○
❦ *Campanula isophylla* ○
Columnea (most) ○
❦ *Cyanotis kewensis* ○
Episcia cupreata, p.265 ○
❦ *Lotus berthelotii*, p.247 ○
Pelargonium peltatum and cvs ○
Pellionia daveauana, p.267 ○
❦ *Peperomia scandens* ○
Ruellia devosiana, p.241 ○
Tradescantia fluminensis and cvs ○
❦ *Tradescantia zebrina*, p.265 ○
Verbena peruviana ○

Annuals and Biennials
❦ *Limnanthes douglasii*, p.288 ○
Lobelia erinus cvs
Nemophila maculata, p.271 ○
Nolana paradoxa
Petunia, Cascade Series
Petunia, Jamboree Series
Portulaca grandiflora, Series and cvs
Sanvitalia procumbens, p.288 ○
Tropaeolum majus, Series and cvs

Rock plants
Acaena 'Blue Haze' ○
Arabis caucasica ○
Cymbalaria muralis ○
❦ *Cytisus* x *beanii*, p.297 ○
❦ *Euphorbia myrsinites*, p.319 ○
❦ *Gypsophila repens* ○
Lithodora diffusa cvs ○
❦ *Lysimachia nummularia* 'Aurea', p.334 ○
Oenothera macrocarpa, p.333 ○
Othonna cheirifolia, p.306 ○
❦ *Parahebe catarractae*, p.304 ○
❦ *Parochetus communis*, p.332 ○
Phlox subulata ○
❦ *Polygonum vacciniifolium*, p.335 ○
Pterocephalus perennis subsp. *perennis*, p.328 ○
❦ *Saxifraga stolonifera* ○

Plants for containers

Trees
Acer negundo
Cordyline australis and cvs ○
Crataegus laevigata and cvs
Eucalyptus (when young) ○
Ficus (most) ○
Ilex aquifolium and cvs ○
Jacaranda mimosifolia, p.53
❦ *Laurus nobilis* ○
Malus (small species and cvs)
Melia azedarach, p.50
❦ *Olea europaea* ○
❦ *Phoenix canariensis* ○
Prunus (small species and cvs)
Sorbus (small species and cvs)
Washingtonia ○

Conifers
All the small species and cvs of:
Abies ○
Chamaecyparis ○
Juniperus ○
Picea ○
Pinus ○
Thuja ○
❦ *Thujopsis dolabrata* ○

Shrubs
Buxus sempervirens and cvs ○
❦ *Catharanthus roseus*, p.130 ○
Erica ○
FUCHSIAS, pp.134–5
Hebe ○
HYDRANGEAS, p.114
Lavandula ○
❦ *Myrtus communis*, p.100 ○
Pittosporum ○
RHODODENDRONS (most), pp.102–104, most ○
ROSES (most), pp.150–64, some ○
Santolina ○
Senecio (shrubby species) ○
Spiraea
Viburnum tinus, p.119 ○

Climbers
❦ *Cissus antarctica*, p.180 ○
CLEMATIS (small cvs), pp.172–3, some ○
❦ *Cobaea scandens*, p.174 ○
❦ *Eccremocarpus scaber*, p.177 ○
Hedera helix and cvs ○
Ipomoea, some ○
Jasminum (climbing species), some ○
Lathyrus (climbing species)
Lonicera (climbing species), some ○
Mandevilla ○
Passiflora ○
❦ *Stephanotis floribunda*, p.165 ○
Tropaeolum (climbing species)

Ferns
Adiantum (most)
Asplenium scolopendrium 'Marginatum', p.189 ○
Athyrium nipponicum, p.189
Polypodium vulgare 'Cornubiense', p.188 ○
Polystichum setiferum 'Divisilobum', p.187 ○

Perennials
Agapanthus
Bergenia ○
DAYLILIES, p.221
Geranium, some ○
Geum
Heuchera x *brizoides* cvs ○
HOSTAS, pp.252–3
Phormium ○
PRIMULAS (tall species and cvs), pp.236–7
Pulmonaria, some ○
❦ *Rudbeckia fulgida* var. *sullivantii* 'Goldsturm', p.220
Salvia (many)
Stachys, some ○
Verbena, some ○
Veronica spicata

Annuals and Biennials
Ageratum
Browallia speciosa, p.230
Calendula officinalis, Series and cvs
Callistephus chinensis, Series and cvs
Coleus blumei, Series and cvs
Impatiens, Novette Series, p.279 ○
❦ *Kochia scoparia* f. *trichophylla*, p.287
Lobelia erinus cvs
Nemesia strumosa and Series
Petunia
Salpiglossis sinuata, Series and cvs
Tagetes
Viola x *wittrockiana* hybrids

Rock plants
All rock plants are suitable, the following being recommended:
Campanula (many), some ○
Dianthus (many) ○
Geranium (several), some ○
Hebe ○
Helianthemum ○
❦ *Iberis sempervirens*, p.294 ○
Penstemon (many), some ○
Phlox (several) ○
Polygonum affine
Primula auricula and hybrids
❦ *Saponaria ocymoides*, p.326 ○
Saxifraga (many) ○
❦ *Silene schafta*, p.327

Bulbs, Corms and Tubers
All bulbous plants are suitable, the following being recommended:
Begonia x *tuberhybrida* hybrids
Crinum
CROCUSES, pp.372–3
DAFFODILS, pp.360–62
HYACINTHS, p.375
LILIES (most), pp.348–9
TULIPS, pp.354–6
Zantedeschia aethiopica 'Crowborough', p.341 ○

Water plants
Watertight containers are ideal for:
Aponogeton distachyos, p.387
Eichhornia crassipes, p.388 ○
Menyanthes trifoliata, p.386
Nelumbo nucifera and cvs
❦ *Pontederia cordata*, p.388
Thalia dealbata
WATER LILIES (small cvs), p.390

Architectural plants

Trees
Cordyline ○
Dracaena draco, p.74 ○
Eucalyptus (many) ○
Jacaranda mimosifolia, p.53
Kalopanax septemlobus, p.54
MAGNOLIAS (several), p.49, some ○
❦ *Paulownia tomentosa*, p.51
❦ *Phoenix canariensis* ○
Salix (several)
❦ *Trachycarpus fortunei*, p.58 ○
Trochodendron aralioides, p.58 ○
Washingtonia ○

Conifers
Abies ○
Araucaria ○
❦ *Calocedrus decurrens*, p.80 ○
Cedrus ○
❦ *Juniperus* x *media* 'Pfitzeriana', p.85 ○
❦ *Metasequoia glyptostroboides*, p.76
Picea ○
❦ *Pseudolarix amabilis*, p.81
❦ *Sciadopitys verticillata*, p.80 ○
❦ *Sequoia sempervirens* ○
❦ *Sequoiadendron giganteum*, p.76 ○
❦ *Taxodium distichum*, p.78
❦ *Tsuga heterophylla* ○

Shrubs
❦ *Aesculus parviflora*, p.89
Brachyglottis repanda, p.97 ○
❦ *Cycas revoluta*, p.122 ○
Daphniphyllum macropodum, p.87 ○
❦ *Eriobotrya japonica* ○
❦ *Fatsia japonica* and *F.j.* 'Variegata', p.121 ○

Mahonia (most) ○
Parkinsonia aculeata ○
Protea ○
❦ *Rhus hirta* and ❦ *R.h.* 'Laciniata', p.94
Yucca ○

Climbers
Epipremnum aureum 'Marble Queen', p.179 ○
❦ *Hedera colchica* 'Dentata', p.181 ○
❦ *Monstera deliciosa*, p.180 ○
Schizophragma hydrangeoides
❦ *Schizophragma integrifolium*, p.168 ○
❦ *Vitis coignetiae*, p.178

Ferns
Asplenium scolopendrium 'Marginatum', p.189 ○
❦ *Blechnum tabulare* ○
Cyathea australis, p.74 ○
❦ *Dicksonia antarctica*, p.186 ○
❦ *Matteuccia struthiopteris*, p.188
❦ *Platycerium bifurcatum*, p.186 ○
Polystichum munitum, p.186 ○
❦ *Woodwardia radicans* ○
Woodwardia virginica

Perennials
❦ *Acanthus spinosus*, p.215
Angelica archangelica, p.192
Berkheya macrocephala, p.222
Crambe cordifolia, p.190
❦ *Cynara cardunculus*, p.192
Echinops bannaticus, p.192
Ensete ventricosum, p.197 ○
❦ *Gunnera manicata*, p.192
Heliconia ○
Ligularia (most)
Macleaya
Meconopsis (most), several ○
Peltiphyllum peltatum, p.202
Phormium tenax and cvs ○
Rodgersia

Annuals and Biennials
Amaranthus tricolor cvs
Helianthus annuus and cvs
Humea elegans, p.274
Onopordum acanthium, p.274
Silybum marianum, p.274
Verbascum densiflorum ○

Bulbs, Corms and Tubers
Arisaema (most)
Arum creticum, p.359
Begonia rex and hybrids ○
Begonia x *tuberhybrida* hybrids
Canna x *generalis* cvs
Dracunculus vulgaris, p.345
GLADIOLI (most), p.343
Sauromatum venosum, p.357
Zantedeschia aethiopica ○

Water plants
Colocasia esculenta and cvs ○
Eichhornia crassipes, p.388 ○
❦ *Lysichiton americanus*, p.391
Nelumbo nucifera and cvs
Orontium aquaticum, p.391
Sagittaria
Thalia dealbata

Cacti and other Succulents
Most of the larger species, especially:
❦ *Aeonium tabuliforme*, p.414 ○
Aloe (most) ○
Carnegiea gigantea, p.392 ○
Cereus (most) ○
Cyphostemma juttae, p.395
Euphorbia candelabrum ○
Opuntia (most) ○

The PLANT CATALOGUE

A photographic and descriptive guide to over 4250 garden plants, arranged by plant type, size, season of interest and colour.

TREES/large SPRING INTEREST
WHITE–YELLOW

Aesculus hippocastanum
(Horse-chestnut)
Vigorous, deciduous, spreading tree. Has large leaves with 5 or 7 leaflets and spires of white flowers, flushed pink and yellow in centres, in spring. Spiny fruits contain glossy, brown nuts in autumn.

Aesculus × carnea 'Briotii'
Deciduous, round-headed tree. Leaves, consisting of 5 or 7 leaflets, are glossy, dark green. Panicles of red flowers are borne in late spring.

Acer macrophyllum
(Oregon maple)
Deciduous, round-headed tree with large, deeply lobed, dark green leaves that turn yellow and orange in autumn. Yellowish-green flowers in spring are followed by pale green fruits.

OTHER RECOMMENDED PLANTS: MAGNOLIAS, p.49
Acer cappadocicum 'Aureum'
Calodendrum capense
Fraxinus excelsior 'Jaspidea'

Trees/large SUMMER INTEREST
WHITE

Populus maximowiczii
Fast-growing, deciduous, conical tree. Oval, heart-shaped, bright green leaves have green-veined, white undersides and turn yellow in autumn. Bears long, pendent seed heads surrounded by silky, white hairs in late summer.

Aesculus chinensis
(Chinese horse-chestnut)
Slow-growing, deciduous, spreading tree. Leaves are glossy, dark green with 7 leaflets. Slender spires of white flowers are produced in mid-summer.

Populus alba
(Abele, White poplar)
Deciduous, spreading tree with wavy-margined or lobed leaves, dark green above, white beneath, turning yellow in autumn.

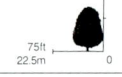

OTHER RECOMMENDED PLANTS:
Acer platanoides 'Drummondii'   *Ceiba pentandra*
Aesculus indica   *Robinia pseudoacacia*
Castanea sativa   *Schima wallichii*

Trees/large SUMMER INTEREST

WHITE–PURPLE

***Castanea sativa* 'Albomarginata'**
Deciduous, spreading tree. Has glossy, white-edged, dark green leaves that turn yellow in autumn. Spikes of creamy-yellow flowers in summer are followed by edible fruits in autumn.

Prunus serotina
(Black cherry, Wild rum cherry)
Deciduous, spreading tree. Spikes of fragrant, white flowers appear in early summer followed by red fruits that turn black in autumn. Glossy, dark green leaves become yellow in autumn.

Liriodendron tulipifera (Tulip tree)
Vigorous, deciduous, spreading tree. Deep green leaves, with a cut-off or notched tip and lobed sides, turn yellow in autumn. Tulip-shaped, orange-marked, greenish-white flowers appear in mid-summer.

Brachychiton acerifolius,
syn. *Sterculia acerifolia*
(Illawarra flame tree)
Deciduous tree with clusters of bright scarlet flowers in late winter, spring or summer before 3–7-lobed, lustrous leaves develop. Min. 7–10°C (45–50°F).

Fagus sylvatica* f. *purpurea, syn. *F.s.* f. *atropunicea*
(Copper beech, Purple beech)
Deciduous, round-headed tree with oval, wavy-margined, purple leaves. In autumn, leaves turn a rich coppery colour.

PURPLE–GREEN

***Acer platanoides* 'Crimson King'**
Vigorous, deciduous, spreading tree. Leaves are large, lobed and deep reddish-purple, turning orange in autumn. Tiny, red-tinged, deep yellow flowers are carried in mid-spring.

Populus* × *canescens
(Grey poplar)
Vigorous, deciduous, spreading tree with slightly lobed leaves, grey when young, glossy, dark green in summer and yellow in autumn. Usually bears greyish-red catkins in spring.

***Populus* × *canadensis* 'Robusta'**
Fast-growing, deciduous, conical tree with upright branches. Broadly oval, bronze, young leaves mature to glossy, dark green. Bears long, red catkins in spring.

Acer platanoides 'Royal Red'
Fagus sylvatica 'Dawyck Purple'
Fagus sylvatica 'Riversii'
Fagus sylvatica 'Rohanii'

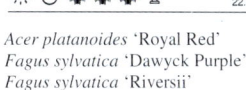

Knightia excelsa
Lagerstroemia speciosa
MAGNOLIAS, p.49
Toona sinensis, p.53

Acer platanoides
Acer platanoides 'Emerald Queen'
Acer saccharinum
Acer velutinum

Ailanthus altissima
MAGNOLIAS, p.49
Salix alba 'Caerulea'
Sorbus thibetica

Trees/large SUMMER INTEREST
■ GREEN

***Populus* x *canadensis* 'Serotina de Selys'**, syn. *P.* x *c.* 'Serotina Erecta'
Fast-growing, deciduous, upright tree. Has broadly oval, grey-green leaves, pale green when young, and red catkins in spring.

Fagus sylvatica* f. *pendula (Weeping beech)
Deciduous, weeping tree with oval, wavy-edged, mid-green leaves that in autumn take on rich hues of yellow and orange-brown.

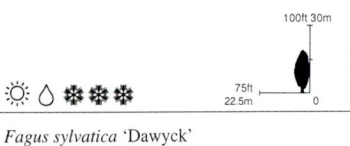

Alnus cordata (Italian alder)
Fast-growing, deciduous, conical tree. Yellow, male catkins appear in late winter and early spring, followed by heart-shaped, glossy, deep green leaves. Has persistent, round, woody fruits in autumn.

Quercus canariensis (Algerian oak, Mirbeck's oak)
Deciduous or semi-evergreen tree, narrow when young, broadening with age. Large, shallowly lobed, rich green leaves become yellowish-brown in autumn, often persisting into late winter.

Quercus macranthera (Caucasian oak)
Deciduous, spreading, stout-branched, handsome tree with large, deeply lobed, dark green leaves.

Acer lobelii (Lobel's maple)
Deciduous tree of narrow, upright habit, well-suited for growing in restricted space. Has wavy-edged, lobed leaves that turn yellow in autumn.

Alnus incana (Grey alder)
Deciduous, conical tree useful for cold, wet areas and poor soils. Yellow-brown catkins are carried in late winter and early spring, followed by oval, dark green leaves.

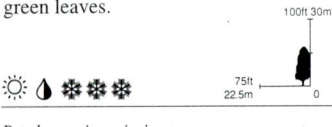

Quercus robur* f. *fastigiata
Deciduous, upright, columnar tree of dense habit carrying lobed, dark green leaves.

***Populus nigra* 'Italica'** (Lombardy poplar)
Very fast-growing, deciduous, narrowly columnar tree with erect branches, diamond-shaped, bright green leaves and red catkins in mid-spring.

Juglans regia (Walnut)
Deciduous tree with a spreading head. Leaves, usually with 5 or 7 leaflets, are aromatic, bronze-purple when young, glossy, mid-green when mature. Produces edible nuts.

Betula maximowicziana
Castanea dentata
Celtis occidentalis
Fagus orientalis

Fagus sylvatica 'Dawyck'
Fagus sylvatica f. *laciniata*
Fraxinus americana
Fraxinus angustifolia

Fraxinus excelsior f. *diversifolia*
Fraxinus oxycarpa 'Raywood'
Fraxinus pennsylvanica
Fraxinus pennsylvanica 'Patmore'

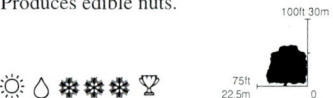

Gleditsia triacanthos 'Shademaster'
Gleditsia triacanthos 'Skyline'
Gymnocladus dioica
Juglans cathayensis

☐ GREEN

Tilia oliveri
Deciduous, spreading, open tree with pointed, heart-shaped leaves, bright green above and silvery-white beneath. Produces small, fragrant, greenish-yellow flowers in summer, followed by winged fruits.

Juglans nigra (Black walnut)
Fast-growing, deciduous, handsome, spreading tree with large, aromatic leaves of many pointed, glossy, dark green leaflets. Produces edible nuts in autumn.

Quercus muehlenbergii
Deciduous, round-headed tree with sharply toothed, bright green leaves.

Celtis australis (Nettle tree)
Deciduous, spreading tree. Has oval, pointed, sharply toothed, dark green leaves and small, purple-black fruits.

Platanus x ***hispanica***, syn.
P. x *acerifolia* (London plane)
Vigorous, deciduous, spreading tree with ornamental, flaking bark. Has large, sharply lobed, bright green leaves. Spherical fruit clusters hang from shoots in autumn.

Nothofagus procera
(Rauli, Southern beech)
Fast-growing, deciduous, conical tree. Leaves, with many impressed veins, are dark green, turning orange and red in autumn.

Juglans cinerea
Platanus orientalis
Populus balsamifera
Populus x *berolinensis*

Populus x *canadensis* 'Eugenei'
Populus x *candicans*
Populus deltoides
Populus lasiocarpa

Populus nigra
Populus szechuanica
Populus trichocarpa
Pterocarya fraxinifolia

Pterocarya stenoptera
Quercus cerris
Quercus imbricaria
Quercus mongolica var. *grosseserrata*

Trees/large SUMMER INTEREST
■ GREEN

Sassafras albidum
Deciduous, upright, later spreading tree. Aromatic, glossy, dark green leaves vary from oval to deeply lobed and turn yellow or red in autumn. Has insignificant, yellowish-green flowers in spring.

Juglans ailantifolia var. cordiformis
Deciduous, spreading tree with large, aromatic leaves consisting of many glossy, bright green leaflets. Long, yellow-green, male catkins are borne in early summer. In autumn has edible nuts.

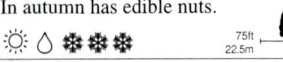

Quercus nigra (Water oak)
Deciduous, spreading tree with glossy, bright green foliage retained until well into winter.

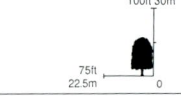

Nothofagus obliqua
(Roblé, Southern beech)
Fast-growing, deciduous, elegant tree with slender, arching branches. Has deep green leaves that turn orange and red in autumn.

Firmiana simplex, syn. *F. platanifolia*, *Sterculia platanifolia* (Chinese parasol tree)
Robust, deciduous tree with large, lobed leaves, small, showy, lemon-yellow flowers and papery, leaf-like fruits. Min. 2°C (36°F).

Tilia 'Petiolaris'
(Pendent silver lime)
Deciduous, spreading tree with pendent branches. Pointed, heart-shaped leaves, dark green above, silver beneath, shimmer in the breeze. Has fragrant, creamy-yellow flowers in late summer.

Quercus petraea 'Columna'
Deciduous, upright, slender tree with large, wavy-edged, leathery, dark green leaves, tinged bronze when young.

Quercus castaneifolia
Deciduous, spreading tree with sharply toothed leaves, glossy, dark green above, grey beneath.

Fagus sylvatica (Common beech)
Deciduous, spreading tree with oval, wavy-edged leaves. These are pale green when young, mid- to dark green when mature, and turn rich yellow and orange-brown in autumn, when nuts are produced.

Populus x canadensis 'Eugenei'
Populus x candicans
Populus deltoides
Populus lasiocarpa

Populus nigra
Populus szechuanica
Populus trichocarpa
Pterocarya fraxinifolia

Pterocarya stenoptera
Quercus cerris
Quercus imbricaria
Quercus mongolica var. grosseserrata

Tilia americana
Tilia cordata
Tilia 'Euchlora'
Tilia mongolica

GREEN

Quercus frainetto (Hungarian oak)
Fast-growing, deciduous, spreading tree with a large, domed head and handsome, large, deeply lobed, dark green leaves.

Quercus palustris (Pin oak)
Fast-growing, deciduous, spreading tree with slender branches, pendulous at the tips. Deeply lobed, glossy, bright green leaves turn scarlet or red-brown in autumn.

Carya ovata (Shag-bark hickory)
Deciduous tree with flaking, grey bark. Has dark green leaves, usually consisting of 5 slender leaflets, that turn golden-yellow in autumn.

Populus alba 'Raket', syn. *P.a.* 'Rocket'
Deciduous, upright, narrow tree. Leaves, often lobed, are dark green with white undersides. In autumn, foliage turns yellow.

Quercus laurifolia
Deciduous, round-headed tree with narrow, glossy, bright green leaves, bronze-tinged when young, that are retained until late in the year.

Tilia platyphyllos
Tilia tomentosa
Ulmus americana
Ulmus carpinifolia

Ulmus glabra
Ulmus 'Hollandica'
Ulmus procera
Ulmus 'Sarniensis'

Ulmus 'Vegeta'
Zelkova carpinifolia

GREEN–YELLOW

Quercus rubra (Red oak)
Fast-growing, deciduous, spreading tree. Attractively lobed leaves, often large, are deep green becoming reddish- or yellowish-brown in autumn.

Liriodendron tulipifera **'Aureomarginatum'**
Vigorous, deciduous tree. Deep green leaves have yellow margins, cut-off or notched tips and lobed sides. Bears cup-shaped, greenish-white flowers, splashed orange, in summer on mature trees.

Pterocarya x *rehderiana*
Very fast-growing, deciduous, spreading tree. Has glossy, bright green leaves consisting of narrow, paired leaflets that turn yellow in autumn and long catkins of winged fruits in late summer and autumn.

Alnus incana 'Aurea'
Fagus sylvatica 'Aurea Pendula'
Populus alba 'Richardii'
Robinia pseudoacacia 'Frisia', p.55

43

Trees/large AUTUMN INTEREST
PINK–RED

Chorisia speciosa (Floss silk tree)
Fast-growing, deciduous tree, the trunk and branches studded with thick, conical thorns. Pink to burgundy flowers appear as indented, light green leaves fall.
Min. 15°C (59°F).

Liquidambar styraciflua
(Sweet gum)
Deciduous, conical to spreading tree. Shoots develop corky ridges. Lobed, glossy, dark green leaves turn brilliant orange, red and purple in autumn.

Quercus ellipsoidalis
Deciduous, spreading tree with deeply lobed, glossy, dark green leaves that turn dark purplish-red, then red in autumn.

Acer pseudoplatanus f. ***erythrocarpum***
Vigorous, deciduous, spreading tree with lobed, deep green leaves. Wings of young autumn fruits are bright red.

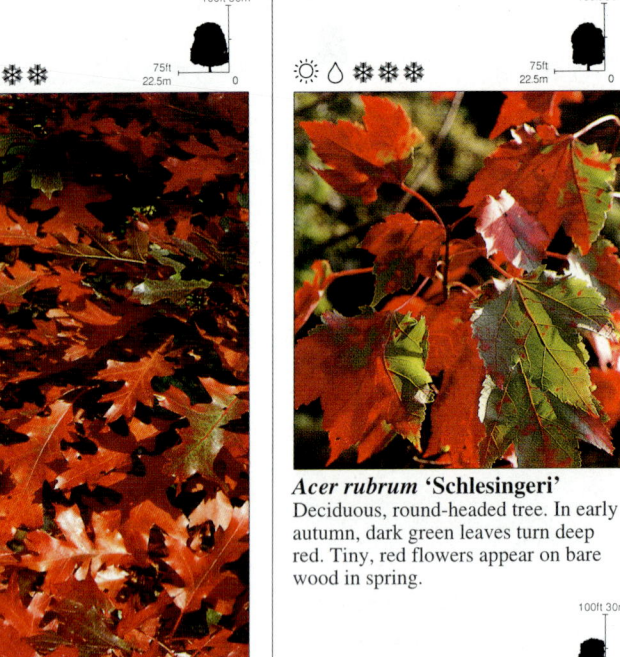

Acer rubrum 'Schlesingeri'
Deciduous, round-headed tree. In early autumn, dark green leaves turn deep red. Tiny, red flowers appear on bare wood in spring.

Acer rubrum 'Scanlon'
Deciduous, upright tree. Has lobed, dark green foliage that in autumn becomes bright red, particularly on acid or neutral soil. Clusters of small, red flowers decorate bare branches in spring.

Quercus coccinea (Scarlet oak)
Deciduous, round-headed tree. Glossy, dark green leaves have deeply cut lobes ending in slender teeth. In autumn, they turn bright red, usually persisting for several weeks on the tree.

Acer rubrum (Red maple)
Deciduous, round-headed tree. Dark green leaves turn bright red in autumn, producing best colour on acid or neutral soil. In spring, bare branches are covered with tiny, red flowers.

OTHER RECOMMENDED PLANTS:
Acer rubrum 'Columnare', p.55
Acer rubrum 'October Glory'
Acer rubrum 'Red Sunset'
Acer saccharum 'Green Mountain'
Lagerstroemia speciosa
Quercus coccinea 'Splendens'

RED–YELLOW

Cercidiphyllum japonicum (Katsura)
Fast-growing, deciduous, spreading tree. Leaves, bronze when young, turn rich green, then yellow to purple in autumn, especially on acid soil. Fallen leaves smell of burnt toffee.

Spathodea campanulata (African tulip tree, Flame-of-the-forest)
Evergreen, showy tree. Leaves have 9–19 deep green leaflets. Clusters of tulip-shaped, scarlet or orange-red flowers appear intermittently. Min. 16–18°C (61–4°F).

Quercus phellos (Willow oak)
Deciduous, spreading tree of elegant habit. Narrow, willow-like, pale green leaves turn yellow then brown in autumn.

Zelkova serrata
Deciduous, spreading tree with sharply toothed, finely pointed, dark green leaves that turn yellow or orange in autumn.

Nyssa sylvatica (Black gum, Tupelo)
Deciduous, broadly conical tree with oval, glossy, dark to mid-green leaves that turn brilliant yellow, orange and red in autumn.

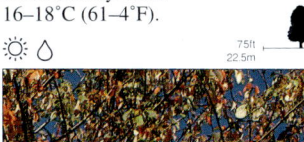

Prunus avium (Gean, Wild cherry)
Deciduous, spreading tree with red-banded bark. Has sprays of white flowers in spring, deep red fruits and dark green leaves that turn red and yellow in autumn.

Quercus alba (American white oak)
Deciduous, spreading tree. Deeply lobed, glossy, dark green leaves turn reddish-purple in autumn.

Acer platanoides 'Lorbergii'
Vigorous, deciduous, spreading tree. Deeply divided, pale green leaves with slender lobes turn yellow or reddish-orange in autumn. Tiny, yellow flowers appear in mid-spring.

Sophora japonica 'Violacea'
Fast-growing, deciduous, round-headed tree. Large sprays of pea-like, white flowers, tinged with lilac-pink, appear in late summer and early autumn.

Acer cappadocicum
Acer platanoides
Betula maximowicziana
Betula szechuanica

Betula utilis
Carya cordiformis
Carya glabra
Fraxinus oxycarpa 'Raywood'

Gleditsia triacanthos
Gymnocladus dioica
Liquidambar styraciflua 'Lane Roberts'
Liriodendron chinense

Populus alba, p.38
Populus tremula
Populus trichocarpa
Quercus velutina

45

Trees/large WINTER/ALL YEAR INTEREST

WHITE–GREEN | GREEN

Betula ermanii
Deciduous, open-branched, elegant tree that has peeling, pinkish-white bark, distinctively marked with large lenticels. Oval, glossy, green leaves give excellent autumn colour.

Ficus benghalensis (Banyan)
Evergreen, wide-spreading tree with trunk-like prop roots. Has oval, leathery leaves, rich green with pale veins, to 20cm (8in) long, and small, fig-like, brown fruits. Min. 15–18°C (59–64°F).

Eucalyptus coccifera (Tasmanian snow gum)
Evergreen tree with peeling, blue-grey and white bark and aromatic, pointed, grey-green leaves. Bears clusters of white flowers, with numerous stamens, in summer.

Eucalyptus dalrympleana (Mountain gum)
Vigorous, evergreen tree. Creamy-white, young bark becomes pinkish-grey, then peels. Leaves are long, narrow and pendent. Clusters of white flowers appear in late summer and autumn.

Eucalyptus gunnii (Cider gum)
Evergreen, conical tree with peeling, cream, pinkish and brown bark. Leaves are silver-blue when young, blue-green when mature. Clusters of white flowers, with numerous stamens, appear in mid-summer.

Betula papyrifera (Canoe birch, Paper birch)
Vigorous, deciduous, open-branched, round-headed tree with peeling, shiny, white bark, yellowish catkins in spring and oval, coarsely serrated leaves that turn clear yellow in autumn.

Ficus elastica 'Doescheri' (Rubber plant)
Strong-growing, evergreen, upright then spreading tree with oblong to oval, leathery, lustrous, deep green leaves, patterned with grey-green, yellow and white. Min. 10°C (50°F).

Archontophoenix alexandrae (Alexandra palm, Northern bungalow palm)
Evergreen palm with feather-shaped, arching leaves. Mature trees bear sprays of small, white or cream flowers. Min. 15°C (59°F).

OTHER RECOMMENDED PLANTS:
Betula pendula
Betula utilis
Betula utilis var. jacquemontii, p.57

Eucalyptus globulus
Ficus elastica 'Decora'
Metrosideros robusta
Tilia platyphyllos 'Princes Street'

Archontophoenix cunninghamiana
Castanospermum australe
Ficus benjamina
Ficus lyrata

Ficus macrophylla
Ficus religiosa
Ficus rubiginosa
Nothofagus menziesii

46

GREEN

Quercus x *turneri*
Semi-evergreen, rounded, dense tree. Lobed, leathery, dark green leaves fall just before new foliage appears in spring.

Nothofagus dombeyi
Evergreen, loosely conical tree of elegant habit with shoots that droop at the tips. Leaves are sharply toothed, glossy and dark green.

Betula pendula 'Tristis'
(Weeping birch)
Deciduous, slender, elegant tree with white bark and lightly pendulous branchlets. Triangular to diamond-shaped leaves provide excellent golden colour in autumn.

Macadamia integrifolia
(Macadamia nut, Queensland nut)
Evergreen, spreading tree with edible, brown nuts in autumn. Has whorls of leathery, semi-glossy leaves and panicles of small, creamy-yellow flowers in spring.
Min. 10–13°C (50–55°F).

Quercus suber (Cork oak)
Evergreen, round-headed tree with thick, corky bark. Oval, leathery leaves are glossy, dark green above and greyish beneath.

Washingtonia robusta
(Thread palm)
Fast-growing, evergreen palm with large, fan-shaped leaves and, in summer, tiny, creamy-white flowers in large, long-stalked sprays. Black berries appear in winter-spring. Min. 10°C (50°F).

Syagrus romanzoffianus, syn. *Arecastrum romanzoffianum*
(Queen palm)
Majestic, evergreen palm. Has feather-shaped leaves with lustrous, green leaflets. Mature trees carry clusters of yellow flowers in summer. Min. 18°C (64°F).

Quercus x *hispanica* **'Lucombeana'** (Lucombe oak)
Semi-evergreen, spreading tree with toothed leaves, glossy, dark green above, grey beneath.

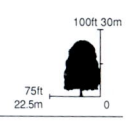

Phoenix canariensis
Quercus ilex
Roystonea regia
Salix matsudana 'Tortuosa', p.59

Tamarindus indica
Terminalia catappa
Washingtonia filifera

47

Trees/large WINTER/ALL YEAR INTEREST

- GREEN
- YELLOW

Betula albosinensis
(White Chinese birch)
Deciduous, open-branched, elegant tree with serrated, oval to lance-shaped, pale green leaves. Peeling bark is honey-coloured or reddish-maroon with a grey bloom.

Umbellularia californica
(Californian laurel)
Evergreen, spreading tree with aromatic, leathery, glossy, dark green leaves and creamy-yellow flowers in late spring. Pungent leaves may cause nausea and headache when crushed.

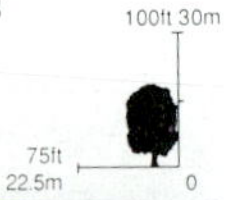

Salix alba var. **vitellina**
(Golden willow)
Deciduous, spreading tree, usually cut back hard to promote growth of strong, young shoots that are bright orange-yellow in winter. Lance-shaped, mid-green leaves appear in spring.

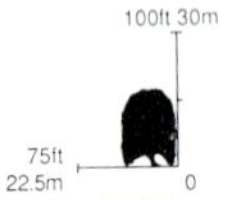

Nothofagus betuloides
Evergreen, columnar tree with dense growth of oval, glossy, dark green leaves on bronze-red shoots.

Salix 'Chrysocoma', syn. *S. alba* 'Tristis', *S.* x *sepulcralis* 'Chrysocoma' (Golden weeping willow)
Deciduous tree with slender, yellow shoots falling to the ground as a curtain. Yellow-green, young leaves mature to mid-green.

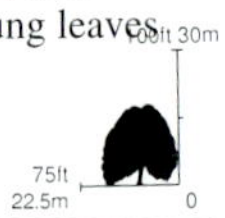

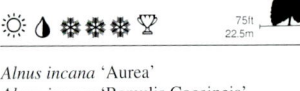

Alnus incana 'Aurea'
Alnus incana 'Ramulis Coccineis'
Betula maximowicziana
Corylus colurna

Trees/medium SPRING INTEREST

- WHITE

Malus hupehensis (Hupeh crab)
Vigorous, deciduous, spreading tree. Has deep green leaves, large, fragrant, white flowers, pink in bud, from mid- to late spring, followed by small, red-tinged, yellow crab apples in late summer and autumn.

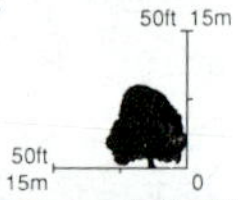

Salix daphnoides (Violet willow)
Fast-growing, deciduous, spreading tree. Has lance-shaped, glossy, dark green leaves, silver, male catkins in spring and purple shoots with bluish-white bloom in winter.

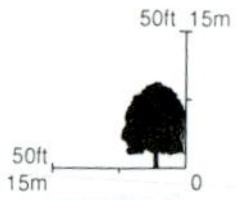

Malus baccata var. **mandschurica**
Vigorous, deciduous, spreading tree with dark green leaves and a profusion of white flowers in clusters in mid-spring, followed by long-lasting, small, red or yellow crab apples.

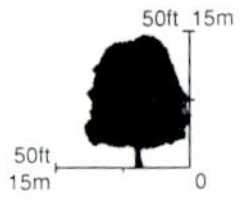

OTHER RECOMMENDED PLANTS:
Dillenia indica
MAGNOLIAS, p.49
Malus baccata
Malus prattii

Magnolias

A mature magnolia in full bloom is one of the most spectacular sights in the spring garden. Most magnolias are elegant in habit and though slow-growing, eventually form imposing trees and shrubs that are valuable as focal points or as single specimens in lawns. Some species, such as *Magnolia stellata*, may be successfully grown in the smallest of gardens.

Magnolia flowers are generally saucer- or goblet-shaped and often have a subtle fragrance. The colour range includes pure white or white flushed or stained with pink or purple, pink and rich wine-purple. The genus also includes some evergreen, summer-flowering species. These, and cultivars that are not fully hardy, make excellent subjects for planting against a sunny wall. Some magnolias prefer acid or neutral soil, but most tolerate any soil provided it is humus-rich. Plenty of organic matter should be dug into the soil before planting.

M. wilsonii ♀

M. grandiflora 'Exmouth' ♀

M. campbellii

M. x *soulangeana* 'Rustica Rubra' ♀

M. stellata 'Water Lily' ♀

M. tripetala

M. campbellii var. *mollicomata*

M. 'Norman Gould'

M. cylindrica ♀

M. hypoleuca ♀

M. x *soulangeana* 'Etienne Soulange-Bodin' ♀

M. liliiflora 'Nigra' ♀

M. 'Wada's Memory' ♀

M. salicifolia ♀

M. denudata ♀

M. sprengeri

M. 'Charles Coates'

M. x *veitchii* 'Peter Veitch'

M. 'Heaven Scent' ♀

M. kobus

M. 'Manchu Fan'

M. campbellii 'Charles Raffill' ♀

M. stellata ♀

M. x *wieseneri*

M. fraseri

M. x *loebneri* 'Leonard Messel' ♀

M. sprengeri 'Wakehurst'

M. campbellii 'Darjeeling'

49

Trees/medium SPRING INTEREST

☐ WHITE
☐ WHITE–PINK

Pyrus calleryana 'Chanticleer', syn. *P. c.* 'Cleveland Select'
Deciduous, conical tree with glossy leaves that turn purplish in autumn. Sprays of small, white flowers appear in spring. Resists fireblight.

Prunus mahaleb
Deciduous, round-headed, bushy tree that bears a profusion of fragrant, cup-shaped, white flowers from mid- to late spring. Rounded, glossy, dark green leaves turn yellow in autumn.

Halesia monticola (Silver bell, Snowdrop tree)
Fast-growing, deciduous, conical or spreading tree. Masses of pendent, bell-shaped, white flowers appear in late spring before leaves, followed by 4-winged fruits in autumn.

Prunus padus (Bird cherry)
Deciduous, spreading tree, conical when young. Bears fragrant, white flowers in pendent spikes during late spring, followed by small, black fruits in late summer. Dark green leaves turn yellow in autumn.

Prunus avium 'Plena'
Deciduous, spreading tree with reddish-brown bark and masses of double, pure white flowers in spring. Dark green foliage turns red in autumn.

Melia azedarach (Bead tree, Persian lilac)
Deciduous, spreading tree. Has dark green leaves with many leaflets and fragrant, star-shaped, pinkish-lilac flowers in spring, followed by pale orange-yellow fruits in autumn.

Cornus nuttallii (Mountain dogwood, Pacific dogwood)
Deciduous, conical tree. Large, white bracts, surrounding tiny flowers, appear in late spring. Has oval, dark green leaves.

Fraxinus ornus (Manna ash)
Deciduous, round-headed tree. Has deep green leaves with 5–9 leaflets. Panicles of scented, creamy-white flowers appear in late spring and early summer.

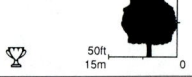

Prunus serrulata var. **spontanea** (Hill cherry)
Deciduous, spreading tree bearing cup-shaped, white or pink flowers from mid- to late spring. Oval leaves, bronze when young, mature to deep green.

Michelia doltsopa, p.57
Prunus padus var. *commutata*
Prunus padus 'Plena'
Prunus padus 'Watereri'
Prunus pensylvanica
Rehderodendron macrocarpum

MAGNOLIAS, p.49
Malus x *robusta*
Prunus 'Spire', p.61
Prunus x *yedoensis*, p.61

50

Trees/medium SUMMER INTEREST

▨▨ PINK–YELLOW

☐ WHITE

Prunus 'Kanzan'
Deciduous, vase-shaped tree. Large, double, pink to purple flowers are borne profusely from mid- to late spring amid bronze, young leaves that mature to dark green.

Paulownia tomentosa, syn. *P. imperialis*
(Foxglove tree, Princess tree)
Deciduous, spreading tree. Has large, lobed, mid-green leaves and terminal sprays of fragrant, foxglove-like, pinkish-lilac flowers in spring.

Styrax japonica
Deciduous, spreading tree bearing in early summer a profusion of pendent, fragrant, bell-shaped, white flowers amid glossy, dark green foliage.

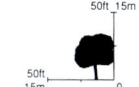

Ostrya virginiana (American hop hornbeam, Ironwood)
Deciduous, conical tree with dark brown bark and deep green leaves, yellow in autumn. Has yellowish catkins in spring, followed by greenish-white fruit clusters.

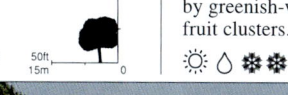

Malus 'Profusion'
Deciduous, spreading tree. Dark green foliage is purple when young. Cup-shaped, deep purplish-pink flowers are freely borne in late spring, followed by small, reddish-purple crab apples in late summer and autumn.

Gleditsia triacanthos 'Sunburst'
Deciduous, spreading tree with fern-like, glossy foliage that is golden-yellow when young, deep green in summer.

Davidia involucrata
(Dove tree, Ghost tree, Pocket handkerchief tree)
Deciduous, conical tree with heart-shaped, vivid green leaves, felted beneath. Large, white bracts appear on mature trees from late spring.

Aesculus x *neglecta* 'Erythroblastos', p.63
Catalpa x *erubescens* 'Purpurea'
Cercis siliquastrum, p.62
MAGNOLIAS, p.49

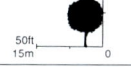

Prunus 'Okame'
Tabebuia rosea

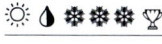

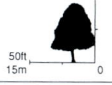

OTHER RECOMMENDED PLANTS:
Acer rufinerve f. *albolimbatum*
Arbutus menziesii
Catalpa ovata

Cladrastis lutea, p.56
Cornus controversa
Cunonia capensis
Eucryphia cordifolia

51

Trees/medium SUMMER INTEREST
☐ WHITE

Oxydendrum arboreum
(Sorrel tree)
Deciduous, spreading tree with glossy, dark green foliage that turns bright red in autumn. Sprays of white flowers appear in late summer and autumn.

Catalpa speciosa
Deciduous, spreading tree. Heads of large, white flowers marked with yellow and purple are borne in mid-summer among glossy, mid-green leaves.

Drimys winteri (Winter's bark)
Evergreen, conical, sometimes shrubby tree with long, glossy, pale or dark green leaves, usually bluish-white beneath. Bears clusters of fragrant, star-shaped, white flowers in early summer.

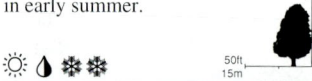

***Quercus cerris* 'Argenteovariegata'**, syn. *Q. c.* 'Variegata'
Deciduous, spreading tree. Strongly toothed or lobed, glossy, dark green leaves are edged with creamy-white.

Sorbus vestita, syn. *S. cuspidata*
Deciduous, broadly conical tree. Has very large, veined, grey-green leaves, white-haired when young. Heads of pink-stamened, white flowers in late spring or early summer are followed by russet or yellowish-red fruits.

Stuartia pseudocamellia
Deciduous, spreading tree with ornamental, peeling bark. Bears white flowers in mid-summer. Foliage is mid-green, turning orange and red in autumn.

Catalpa bignonioides
(Indian bean tree)
Deciduous, spreading tree. Large, light green leaves are purplish when young. White flowers marked with yellow and purple appear in summer, followed by long, cylindrical, pendent pods.

Cornus macrophylla
Deciduous, spreading tree. Clusters of small, creamy-white flowers appear in summer. Glossy, bright green leaves are large, pointed and oval.

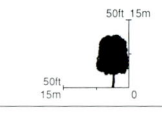

***Sorbus aria* 'Lutescens'**
Deciduous, spreading tree, upright when young. Young foliage is silvery, maturing to grey-green. White flowers in late spring and early summer are followed by orange-red fruits in autumn.

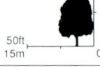

Gevuina avellana
Hoheria populnea
Lyonothamnus floribundus var. *aspleniifolius*
MAGNOLIAS, p.49

Malus trilobata
Pterostyrax hispida
Stuartia monadelpha, p.56
Stuartia sinensis

Styrax obassia
Weinmannia trichosperma

52

WHITE–PINK

Acer pseudoplatanus 'Simon Louis Frères'
Deciduous, spreading tree. Young leaves are marked with creamy-white and pink; older foliage is pale green with white markings.

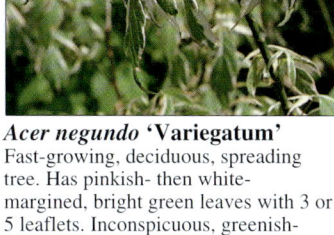

Acer negundo 'Variegatum'
Fast-growing, deciduous, spreading tree. Has pinkish- then white-margined, bright green leaves with 3 or 5 leaflets. Inconspicuous, greenish-yellow flowers appear in late spring.

Aesculus indica 'Sydney Pearce'
Deciduous, spreading tree with glossy, dark green leaves, bronze when young and orange or yellow in autumn. Pinkish-white flowers, marked red and yellow, appear from early to mid-summer.

Albizia julibrissin, p.64
Clusia major
Lagunaria patersonii
Manglietia insignis

Robinia x *ambigua* 'Decaisneana'

PURPLE–GREEN

Jacaranda mimosifolia, syn. *J. acutifolia* of gardens, *J. ovalifolia*
Fast-growing, deciduous, rounded tree with fern-like leaves of many tiny, bright green leaflets. Has trusses of vivid blue to blue-purple flowers in spring and early summer. Min. 7°C (45°F).

Broussonetia papyrifera (Paper mulberry)
Deciduous, round-headed tree. Dull green leaves are large, broadly oval, toothed and sometimes lobed. In early summer, small globes of purple flowers appear on female plants.

Sorbus thibetica 'John Mitchell', syn. *S.* 'Mitchellii'
Strong-growing, deciduous, conical tree with dark green leaves, silvery beneath. Has white flowers in spring and brown fruits in late summer.

Hovenia dulcis (Raisin-tree)
Deciduous, spreading tree with large, glossy, dark green leaves. In summer it may bear small, greenish-yellow flowers, the stalks of which become red, fleshy and edible.

Toona sinensis, syn. *Cedrela sinensis*
Deciduous, spreading tree with shaggy bark when old. Dark green leaves with many leaflets turn yellow in autumn. Bears fragrant, white flowers in mid-summer. Shoots are onion-scented.

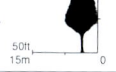

Populus tremula 'Pendula' (Weeping aspen)
Vigorous, deciduous, weeping tree. Leaves, reddish when young, grey-green in summer and yellow in autumn, tremble in the wind. Has purplish catkins in late winter and spring.

Acer negundo var. *violaceum*
Olea europaea
Populus tremula
Pyrus amygdaliformis

Salix alba var. *sericea*

Trees/medium SUMMER INTEREST
■ GREEN

Quercus marilandica (Black Jack oak)
Deciduous, spreading tree. Large leaves, 3-lobed at the apex, are glossy, dark green above, paler beneath, and turn yellow, red or brown in autumn.

Fraxinus velutina (Arizona ash)
Deciduous, spreading tree. Leaves vary but usually consist of 3 or 5 narrow, velvety, grey-green leaflets.

Quercus garryana (Oregon oak)
Slow-growing, deciduous, spreading tree with deeply lobed, glossy, bright green leaves.

***Tilia cordata* 'Rancho'**
Deciduous, conical, dense tree, spreading when young. Has small, oval, glossy, dark green leaves and clusters of small, fragrant, cup-shaped, yellowish flowers are borne in mid-summer.

Meliosma veitchiorum
Deciduous, spreading tree with stout, grey shoots and large, dark green, red-stalked leaves with 9 or 11 leaflets. Small, fragrant, white flowers in late spring are followed by violet fruits in autumn.

Gleditsia japonica
Deciduous, conical tree with a trunk armed with spines. Shoots are purplish when young. Fern-like leaves consist of many small, mid-green leaflets.

Emmenopterys henryi
Deciduous, spreading tree. Large, pointed, dark green leaves are bronze-purple when young. Clusters of white flowers (some bearing a large, white bract) are rarely produced except in hot summers.

Idesia polycarpa
Deciduous, spreading tree with large, heart-shaped, glossy, dark green leaves on long stalks. Small, fragrant, yellow-green flowers in mid-summer are followed in autumn, on female plants, by red fruits hanging in clusters.

Kalopanax septemlobus, syn.
K. pictus, *K. ricinifolius*,
Acanthopanax ricinifolium
Deciduous, spreading tree with spiny stems, large, 5–7-lobed, glossy, dark green leaves and umbels of small, white flowers, followed by black fruits in autumn.

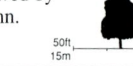

Quercus macrocarpa (Bur oak)
Slow-growing, deciduous, spreading tree. Large, oblong-oval, lobed, glossy, dark green leaves turn yellow or brown in autumn.

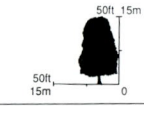

Quercus macrolepis, syn.
Q. ithaburensis subsp. *macrolepis*
Deciduous or semi-evergreen, spreading tree. Has grey-green leaves with angular lobes.

***Quercus rubra* 'Aurea'**
Slow-growing, deciduous, spreading tree. Large, lobed leaves are clear yellow when young, becoming green by mid-summer. Produces best colour in an open but sheltered position.

Acer giraldii
Acer monspessulanum
Aesculus glabra
Alnus glutinosa 'Imperialis'

Eucommia ulmoides
Fagus grandifolia
Meliosma pinnata
Nothofagus antarctica

Ostrya carpinifolia
Phellodendron amurense
Platycarya strobilacea
Pyrus calleryana 'Bradford'

Quercus acutissima
Quercus aliena
Quercus dentata
Ulmus parvifolia

Trees/medium AUTUMN INTEREST

■ GREEN–YELLOW ■ WHITE–RED

Phellodendron chinense
Deciduous, spreading tree. Aromatic leaves, with 7–13 oblong leaflets, are dark green, turning yellow in autumn. Pendent racemes of greenish flowers in early summer are followed on female trees by berry-like, black fruits.

Robinia pseudoacacia 'Frisia'
Deciduous, spreading tree with luxuriant leaves divided into oval leaflets, golden-yellow when young, greenish-yellow in summer and orange-yellow in autumn.

Ulmus 'Dicksonii', syn. *U. carpinifolia* 'Sarniensis Aurea', *U.* 'Wheatleyi Aurea' (Cornish golden elm, Dickson's golden elm)
Slow-growing, deciduous, conical tree of dense habit. Carries small, broadly oval, bright golden-yellow leaves.

Alnus glutinosa 'Aurea'
Catalpa bignonioides 'Aurea'
Quercus robur 'Concordia'

Eucryphia x nymansensis
Evergreen, columnar tree. Some of the leathery, glossy, dark green leaves are simple, others consist of 3 (rarely 5) leaflets. Clusters of large, white flowers open in late summer or early autumn.

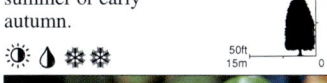

Syzygium paniculatum, syn. *Eugenia australis*, *E. paniculata* (Australian brush cherry)
Evergreen tree with glossy leaves, coppery when young. Has creamy-white flowers, with reddish sepals, and fragrant, rose-purple fruits. Min. 10°C (50°F).

OTHER RECOMMENDED PLANTS:
Chorisia speciosa, p.44
Cornus capitata
MAGNOLIAS p.49

Acer davidii 'Madeline Spitta'
Deciduous tree with upright branches that are striped green and white. Glossy, dark green foliage turns orange in autumn after the appearance of winged, green fruits that ripen reddish-brown.

Sorbus hupehensis 'Rosea', syn. *S. h.* var. *obtusa*
Deciduous, spreading tree with leaves of 4–8 pairs of blue-green leaflets turning orange-red in late autumn. White flowers in spring are followed by long-lasting, pink fruits.

Sorbus commixta, syn. *S. discolor* of gardens
Vigorous, deciduous, spreading tree. Leaves have 6–8 pairs of glossy, deep green leaflets that turn orange and red in autumn. White flowers in spring are followed by bright red fruits.

Rhus potaninii
Rhus verniciflua
Sorbus americana
Sorbus esserteauiana

Sorbus aucuparia
(Mountain ash, Rowan)
Deciduous, spreading tree. Leaves have mid-green leaflets that turn red or yellow in autumn. Bears white flowers in spring and red fruits in autumn.

Acer rubrum 'Columnare'
Deciduous, slender, upright tree with lobed, dark green foliage becoming a fiery column of red and yellow in autumn.

Sorbus hupehensis
Sorbus scalaris

55

Trees/medium AUTUMN INTEREST
RED–YELLOW

Acer rufinerve
(Snake-bark maple)
Deciduous tree with arching branches striped green and white. In autumn, lobed, dark green leaves turn brilliant red and orange.

Aesculus flava, syn. *A. octandra*
(Sweet buckeye, Yellow buckeye)
Deciduous, spreading tree. Glossy, dark green leaves, with 5 or 7 oval leaflets, redden in autumn. Has yellow flowers in late spring and early summer followed by round fruits (chestnuts).

Quercus x heterophylla
(Bartram's oak)
Deciduous, spreading tree with toothed, glossy, bright green leaves that turn orange-red and yellow in autumn.

Stuartia monadelpha
Deciduous, spreading tree with peeling bark and glossy, dark green leaves that turn orange and red in autumn. Small, violet-anthered, white flowers appear in mid-summer, followed by small fruits.

Acer cissifolium subsp. **henryi**, syn. *A. henryi*
Deciduous, spreading tree. Dark green leaves with 3 oval, toothed leaflets turn bright orange and red in autumn.

Nyssa sinensis
Deciduous, spreading tree. Has long, narrow, pointed leaves that are purplish when young, dark green when mature and brilliant scarlet in autumn.

Sorbus 'Joseph Rock'
Deciduous, upright tree. Bright green leaves composed of many leaflets turn orange, red and purple in autumn. White flowers in late spring are followed by large clusters of small, yellow berries in late summer and autumn.

Acer saccharum 'Temple's Upright'
Deciduous, columnar tree. In autumn, large, lobed leaves turn brilliant orange and red.

Parrotia persica (Persian ironwood)
Deciduous, spreading, short-trunked tree with flaking, grey and fawn bark. Rich green leaves turn yellow, orange and red-purple in autumn. Small, red flowers are borne on bare wood in early spring.

Acer capillipes
(Snake-bark maple)
Deciduous, spreading tree. Has lobed, bright green leaves that turn brilliant red and orange in autumn. Older branches are striped green and white.

Cladrastis lutea (Yellow wood)
Deciduous, round-headed tree. Leaves of 7 or 9 rounded-oval leaflets are dark green, turning yellow in autumn. Clusters of fragrant, pea-like, yellow-marked, white flowers appear in early summer.

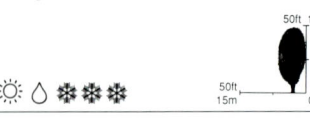

Acer maximowiczianum
Diospyros kaki
Dipteronia sinensis
HOLLIES, pp.72–3

Liquidambar formosana
Maclura pomifera
Malus 'Hopa'
Malus hupehensis, p.48

Malus 'Profusion', p.51
Malus x robusta
Malus x robusta 'Yellow Siberian'
Morus nigra

Oxydendrum arboreum, p.52
Sorbus aucuparia 'Fructu Luteo'
Sorbus decora

Trees/medium WINTER INTEREST
WHITE–YELLOW

Betula utilis var. **jacquemontii**
(West Himalayan birch)
Deciduous, open-branched, elegant tree with bright white bark. Oval, serrated, mid-green leaves turn clear yellow in autumn.

Michelia doltsopa
Evergreen, rounded tree with oval, glossy, dark green leaves, paler beneath. Strongly scented, magnolia-like flowers, with white to pale yellow petals, appear in winter-spring.

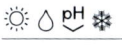

Acacia dealbata
(Mimosa, Silver wattle)
Fast-growing, evergreen, spreading tree. Has feathery, blue-green leaves with many leaflets. Racemes of globular, fragrant, bright yellow flower heads are borne in winter-spring.

OTHER RECOMMENDED PLANTS:
Bauhinia variegata
Salix daphnoides, p.48
Tabebuia chrysotricha, p.71

RED

Arbutus x andrachnoides
Evergreen, bushy, spreading tree with peeling, reddish-brown bark and glossy, dark green foliage. Clusters of small, white flowers in autumn to spring are followed by small, strawberry-like, orange or red fruits.

Metrosideros excelsa (New Zealand Christmas tree, Pohutukawa)
Robust, evergreen, wide-spreading tree. Oval leaves are grey-green above, white felted beneath. Showy tufts of crimson stamens appear in large, terminal clusters in winter. Min. 5°C (41°F).

Trees/medium ALL YEAR INTEREST
WHITE–GREEN

***Ficus benjamina* 'Variegata'**
Evergreen, dense, round-headed, weeping tree, often with aerial roots. Has slender, pointed, lustrous leaves that are rich green with white variegation. Min. 15–18°C (59–64°F).

Trochodendron aralioides
Evergreen, broadly conical tree with glossy, dark green foliage. In late spring and early summer bears clusters of unusual, petal-less, wheel-like, green flowers.

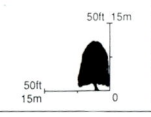

Schefflera actinophylla
(Queensland umbrella tree)
Evergreen, upright tree with large, spreading leaves of 5–16 leaflets. Has large sprays of small, dull red flowers in summer or autumn. Min. 16°C (61°F).

Acer pensylvanicum
(Snake-bark maple)
Deciduous, upright tree. Shoots are boldly striped green and white. Large, lobed, mid-green leaves turn bright yellow in autumn.

Eucalyptus pauciflora
(White Sally)
Evergreen, spreading tree with peeling, white, young bark and red, young shoots. In summer, white flower clusters appear amid glossy, bright grey-green foliage.

Trachycarpus fortunei
(Chusan palm, Windmill palm)
Evergreen palm with unbranched stem and a head of large, deeply divided, fan-like, mid-green leaves. Sprays of fragrant, creamy-yellow flowers appear in early summer.

Eucalyptus niphophila, syn. *E. pauciflora* subsp. *niphophila*
(Snow gum)
Evergreen, spreading tree. Has patchwork-like, flaking bark, red-rimmed, grey-green leaves and white flowers in summer.

Quercus myrsinifolia
Evergreen, rounded tree with narrow, pointed, glossy, dark green leaves, reddish-purple when young.

Prunus maackii
Deciduous, spreading tree with peeling, yellowish-brown bark. Produces spikes of small, white flowers in mid-spring and pointed, dark green leaves that turn yellow in autumn.

OTHER RECOMMENDED PLANTS:
Acer davidii
Betula 'Jermyns'
Cinnamomum camphora

Cordyline australis 'Veitchii'
Cyathea medullaris
Eucalyptus glaucescens
Ficus lyrata

HOLLIES, pp.72–3
Laurelia sempervirens
Laurus nobilis
Lithocarpus densiflorus

MAGNOLIAS, p.49
Salix x *rubens* 'Basfordiana'

■ GREEN–ORANGE

Trees/small SPRING INTEREST
☐ WHITE

Jubaea chilensis, syn. *J. spectabilis* (Chilean wine palm, Coquito)
Slow-growing, evergreen palm with a massive trunk and large, silvery-green leaves. Has small, maroon and yellow flowers in spring and woody, yellow fruits in autumn. Min. 10°C (50°F).

Livistona chinensis (Chinese fan palm, Chinese fountain palm)
Slow-growing, evergreen palm with a stout trunk. Has fan-shaped, glossy leaves, 1–3m (3–10ft) across. Mature trees bear loose clusters of berry-like, black fruits in autumn. Min. 7°C (45°F).

Quercus agrifolia
(Californian live oak)
Evergreen, spreading tree bearing rigid, spiny-toothed, glossy, dark green leaves.

***Salix matsudana* 'Tortuosa'**
(Dragon's-claw willow)
Fast-growing, deciduous, spreading tree with curiously twisted shoots and contorted, narrow, tapering, bright green leaves.

Corynocarpus laevigata
Evergreen, upright tree, spreading with age. Has leathery leaves and clusters of small, greenish flowers in spring-summer. Plum-like, orange fruits appear in winter. Min. 7–10°C (45–50°F).

Mespilus germanica (Medlar)
Deciduous, spreading tree or shrub. Has dark green leaves that turn orange-brown in autumn, white flowers in spring-summer and brown fruits in autumn, edible when half rotten.

Amelanchier laevis
Deciduous, spreading tree or large shrub. Oval, bronze, young leaves turn dark green in summer, red and orange in autumn. Sprays of white flowers in spring are followed by rounded, fleshy, red fruits.

Aesculus californica
(California buckeye)
Deciduous, spreading, sometimes shrubby tree. Dense heads of fragrant, sometimes pink-tinged, white flowers appear in spring and early summer. Small, dark green leaves have 5–7 leaflets.

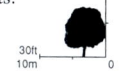

Crataegus laciniata,
syn. *C. orientalis*
Deciduous, spreading tree with deeply lobed, hairy, dark green leaves. A profusion of white flowers in late spring or early summer is followed by red fruits tinged with yellow.

***Cornus florida* 'White Cloud'**
Deciduous, spreading tree. Massed flower heads, comprising large, white bracts around tiny flowers, appear in spring. Oval, pointed, dark green leaves turn red and purple in autumn.

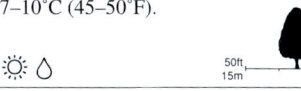

Betula alleghaniensis
HOLLIES, pp.72–3
Prunus serrula
Thespesia populnea

OTHER RECOMMENDED PLANTS:
Agonis flexuosa, p.64
Amelanchier arborea
Amelanchier asiatica

Arbutus andrachne
Bauhinia variegata 'Candida', p.71
Cornus 'Eddie's White Wonder', p.70
Crataegus crus-galli

59

Trees/small SPRING INTEREST
☐ WHITE

Prunus 'Shogetsu', syn. *P.* 'Shimidsu'
Deciduous, round-topped tree. In late spring, pink buds open to large, double, white flowers that hang in clusters from long stalks. Mid-green leaves turn orange and red in autumn.

Cornus 'Porlock'
Deciduous, spreading tree. Creamy-white bracts around tiny flowers turn to deep pink in summer. These are often followed by heavy crops of strawberry-like fruits in autumn.

Prunus incisa (Fuji cherry)
Deciduous, spreading tree. White or pale pink flowers appear in early spring. Sharply toothed, dark green leaves are reddish when young, orange-red in autumn.

Prunus 'Taihaku', syn. *P.* 'Tai Haku' (Great white cherry)
Vigorous, deciduous, spreading tree. Very large, single, pure white flowers are borne in mid-spring among bronze-red, young leaves that mature to dark green.

Prunus 'Ukon'
Vigorous, deciduous, spreading tree. Semi-double, pale greenish-white flowers open from pink buds in mid-spring amid pale bronze, young foliage that later turns dark green.

Prunus 'Shirotae', syn. *P.* 'Mount Fuji'
Deciduous, spreading tree with slightly arching branches. Large, fragrant, single or semi-double, pure white flowers appear in mid-spring. Foliage turns orange-red in autumn.

Crataegus ellwangeriana
Crataegus x lavallei 'Carrierei'
Docynia delavayi
Gordonia axillaris

Halesia tetraptera
MAGNOLIAS, p.49
Malus 'Frettingham's Victoria'
Malus 'Golden Hornet', p.71

Malus 'John Downie', p.69
Malus prunifolia, p.69
Malus toringoides
Malus transitoria

Malus 'Veitch's Scarlet', p.68
Parrotiopsis jacquemontiana
Prunus 'Hally Jolivette'

■ PINK

Prunus x *yedoensis*
(Yoshino cherry)
Deciduous, round-headed tree with spreading, arching branches and dark green foliage. Sprays of pink buds open to white or pale pink flowers in early spring.

Prunus 'Spire',
syn. *P.* x *hillieri* 'Spire'
Deciduous, vase-shaped tree, conical when young. Soft pink flowers appear profusely from early to mid-spring. Dark green leaves, bronze when young, turn brilliant orange-red in autumn.

Prunus 'Hokusai', syn.
P. 'Uzu-zakura'
Deciduous, spreading tree. Oval, bronze, young leaves mature to dark green, then turn orange and red in autumn. Semi-double, pale pink flowers are borne in mid-spring.

Prunus subhirtella 'Stellata',
syn. *P.* 'Pink Star'
Deciduous, spreading tree. Pink flowers with narrow, pointed petals, red in bud, open from early to mid-spring. Dark green leaves turn yellow in autumn.

Malus x *arnoldiana*
Deciduous, low, spreading tree with arching branches. In mid- to late spring red buds open to fragrant, pink flowers that fade to white. Bears small, red-flushed, yellow crab apples in autumn. Leaves are oval.

Prunus 'Shirofugen'
Deciduous, spreading tree with bronzed-red leaves turning orange-red in autumn. Pale pink buds open to fragrant, double, white blooms that turn pink before they fade in late spring.

Prunus 'Pandora'
Deciduous tree, upright when young, later spreading. Massed, pale pink flowers appear in early spring. Leaves are bronze when young, dark green in summer and often orange and red in autumn.

Prunus sargentii (Sargent cherry)
Deciduous, spreading tree. Oval, dark green leaves are red when young, turning brilliant orange-red in early autumn. Clusters of blush-pink flowers appear in mid-spring.

Prunus 'Pink Perfection'
Deciduous, upright tree that bears double, pale pink flowers in late spring. Oval leaves are bronze when young, dark green in summer.

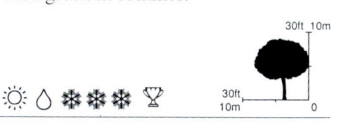

Cercis canadensis
Cercis canadensis 'Forest Pansy', p.65
Cornus florida 'Apple Blossom'
Cydonia oblonga 'Lusitanica'

Cydonia oblonga 'Vranja', p.66
MAGNOLIAS, p.49
Malus coronaria 'Charlottae'
Malus 'Katherine'

Malus 'Marshall Oyama', p.70
Malus 'Professor Sprenger', p.70
Malus 'Red Jade'
Melaleuca viridiflora var. *rubriflora*

Prunus davidiana
Staphylea holocarpa var. *rosea*, p.87

Trees/small SPRING INTEREST
■ PINK

***Prunus* 'Accolade'**
Deciduous, spreading tree with clusters of deep pink buds opening to semi-double, pale pink flowers in early spring. Toothed, mid-green leaves turn orange-red in autumn.

***Prunus subhirtella* 'Pendula Rubra'**
Deciduous, weeping tree that bears deep pink flowers in spring before oval, dark green leaves appear; these turn yellow in autumn.

Cercis siliquastrum (Judas tree)
Deciduous, spreading, bushy tree. Clusters of pea-like, bright pink flowers appear in mid-spring, before or with heart-shaped leaves, followed by long, purplish-red pods in late summer.

***Prunus persica* 'Prince Charming'**
Deciduous, upright, bushy-headed tree with narrow, bright green leaves. Double, deep rose-pink flowers are produced in mid-spring.

***Malus* 'Magdeburgensis'**
Deciduous, spreading tree with dark green foliage. Dense clusters of large, semi-double, deep pink flowers appear in late spring, occasionally followed by small, yellow crab apples in autumn.

***Prunus* 'Kiku-shidare'**, syn. *P.* 'Cheal's Weeping'
Deciduous, weeping tree. Has double, bright pink flowers that cover pendent branches from mid- to late spring.

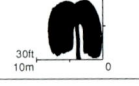

Malus floribunda
Deciduous, spreading, dense-headed tree with pale pink flowers, red in bud, appearing from mid- to late spring, followed by tiny, pea-shaped, yellow crab apples in autumn.

Dombeya* x *cayeuxii (Pink snowball)
Evergreen, bushy tree with rounded, toothed, hairy leaves to 20cm (8in) long. Pink flowers appear in pendent, ball-like clusters in winter or spring. Min. 10–13°C (50–55°F).

***Prunus* 'Yae-murasaki'**, syn. *P.* 'Yae-marasakizakura'
Deciduous, spreading tree with bright green leaves, bronze when young, orange-red in autumn. Semi-double, deep pink flowers are produced in mid-spring.

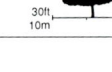

Bauhinia variegata, p.71
Cornus florida f. *rubra*
Crataegus laevigata 'Punicea'
Malus 'Almey'

Malus x *atrosanguinea*
Malus 'Chilko'
Malus 'Cowichan', p.68
Malus 'Dorothea'

Malus 'Neville Copeman'
Malus 'Van Eseltine'
Prunus x *amygdalo-persica* 'Pollardii'
Prunus 'Kursar'

Prunus mume 'Pendula'
Prunus persica 'Klara Meyer'
Pseudocydonia sinensis
Rhodoleia championii

62

RED–YELLOW

Malus 'Royalty'
Deciduous, spreading tree with glossy, purple foliage. Crimson-purple flowers appear from mid- to late spring, followed by dark red crab apples in autumn.

Acer pseudoplatanus 'Brilliantissimum'
Slow-growing, deciduous, spreading tree. Lobed leaves are salmon-pink when young, then turn yellow and finally dark green in summer.

Malus 'Lemoinei'
Deciduous, spreading tree. Oval leaves are deep reddish-purple when young, later becoming tinged with bronze. Wine-red flowers in late spring are followed by dark reddish-purple crab apples in autumn.

Michelia figo
Evergreen tree or rounded shrub. Has oval, glossy, rich green leaves and banana-scented, creamy-yellow flowers, edged maroon, in spring-summer. Min. 5°C (41°F).

Aesculus x neglecta 'Erythroblastos'
Deciduous, spreading tree. Leaves with 5 leaflets emerge bright pink, turn yellow, then dark green, and finally orange and yellow in autumn. May bear panicles of flowers in summer.

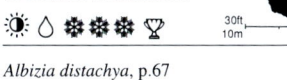

Sophora tetraptera
Semi-evergreen, spreading tree or large shrub with dark green leaves composed of many tiny leaflets. Clusters of golden-yellow flowers appear in late spring.

Albizia distachya, p.67
Cassia fistula
Erythrina americana
Laburnum x watereri 'Vossii', p.66

Malus niedzwetskyana
Sophora microphylla
Thevetia peruviana, p.67
Tipuana tipu

Trees/small SUMMER INTEREST
WHITE

Cornus alternifolia 'Argentea'
Deciduous, spreading tree, grown for its attractive, narrowly oval, white-variegated leaves. Has small heads of white flowers in spring.

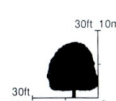

Acer crataegifolium 'Veitchii'
Deciduous, bushy tree with branches streaked with green and white. Small, pointed, dark green leaves, blotched with white and paler green, turn deep pink and reddish-purple in autumn.

Cornus controversa 'Variegata' (Wedding-cake tree)
Deciduous tree with layered branches. Clusters of small, white flowers appear in summer. Leaves are bright green with broad, creamy-white margins and turn yellow in autumn.

OTHER RECOMMENDED PLANTS:
Aralia elata 'Variegata'
Carrierea calycina
Chionanthus virginicus, p.89

Cornus kousa
Cornus kousa var. *chinensis*
Cornus mas 'Variegata', p.89
Cotoneaster frigidus

63

Trees/small SUMMER INTEREST

☐ WHITE

Crataegus flava
(Yellow haw)
Deciduous, spreading tree. Has small, dark green leaves and white flowers in late spring and early summer, followed by greenish-yellow fruits.

Agonis flexuosa
(Peppermint tree, Willow myrtle)
Evergreen, weeping tree. Aromatic, lance-shaped, leathery leaves are bronze-red when young. In spring-summer, mature trees carry an abundance of small, white flowers. Min. 10°C (50°F).

Hoheria angustifolia
Evergreen, columnar tree with narrow, dark green leaves. Shallowly cup-shaped, white flowers are borne from mid- to late summer.

Eucryphia glutinosa
Deciduous, upright or spreading tree. Glossy, dark green leaves, consisting of 3–5 leaflets, turn orange-red in autumn. Large, fragrant, white flowers appear from mid- to late summer.

Hoheria lyallii
Deciduous, spreading tree with deeply toothed, grey-green leaves. Clusters of white flowers are borne in mid-summer.

Eucryphia lucida
Evergreen, upright, bushy tree with narrow, glossy, dark green leaves and fragrant, white flowers in early or mid-summer.

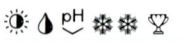

Crataegus prunifolia
Crataegus tanacetifolia
Elaeocarpus cyaneus
Franklinia alatamaha

Fraxinus mariesii
Hoheria 'Glory of Amlwch'
Hoheria sexstylosa
MAGNOLIAS, p.49

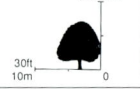

Plumeria alba
Sophora japonica 'Pendula'
Xanthoceras sorbifolium, p.88

☐ WHITE–PINK

Maackia amurensis
Deciduous, spreading tree with deep green leaves consisting of 7–11 leaflets. Dense, upright spikes of white flowers appear from mid- to late summer.

Albizia julibrissin (Silk tree)
Deciduous, spreading tree. Large leaves are light to mid-green and divided into many leaflets. Clusters of brush-like, clear pink flowers appear in late summer or autumn.

Cornus florida 'Spring Song'
Deciduous, spreading tree. Pink bracts, surrounding tiny flowers, appear in spring-summer. Leaves are oval, pointed and dark green, turning red and purple in autumn.

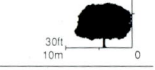

Rothmannia capensis

64

■ PINK–PURPLE

Lagerstroemia indica
(Crape myrtle)
Deciduous, rounded tree or large shrub. Has trusses of flowers with strongly waved, pink, white or purple petals in summer and early autumn.

Malus yunnanensis var. ***veitchii***
Deciduous, upright tree with lobed, heart-shaped leaves, covered with grey down beneath. Bears white, sometimes pink-tinged, flowers in late spring and a mass of small, red-flushed, brown crab apples in late summer and autumn.

***Crataegus laevigata* 'Paul's Scarlet'**
Deciduous, spreading tree. Has toothed, glossy, dark green leaves and a profusion of double, red flowers in late spring and early summer.

***Aesculus pavia* 'Atrosanguinea'**
Deciduous, round-headed, sometimes shrubby tree. In summer, panicles of deep red flowers appear among glossy, dark green leaves, which have 5 narrow leaflets.

***Cercis canadensis* 'Forest Pansy'**
Deciduous, spreading tree or shrub. In mid-spring has flowers that are magenta in bud, opening to pale pink, before heart-shaped, reddish-purple leaves appear.

Acer palmatum f. *atropurpureum*, p.90
Aesculus pavia
Corylus maxima 'Purpurea', p.92
Erythrina americana

Prunus spinosa 'Purpurea', p.90
Punica granatum
Virgilia oroboides

■ PURPLE–GREEN

***Prunus cerasifera* 'Nigra'**
Deciduous, round-headed tree with deep purple leaves, red when young. Pink flowers are borne in profusion from early to mid-spring.

Ehretia dicksonii
Deciduous, spreading tree with stout, ridged branches and large, dark green leaves. Large, flattish heads of small, fragrant, white flowers are borne in mid-summer.

***Pyrus salicifolia* 'Pendula'**
Deciduous, weeping, mound-shaped tree with white flowers in mid-spring and narrow, grey leaves.

Acer platanoides 'Globosum'
Elaeagnus angustifolia, p.92
Fagus sylvatica 'Purpurea Pendula'
Morus alba 'Pendula'

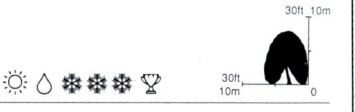

Pyrus elaeagrifolia
Pyrus salicifolia
Salix pentandra
Sinowilsonia henryi

Trees/small SUMMER INTEREST

◻ GREEN

Pseudopanax ferox
Evergreen, upright tree with long, narrow, rigid, sharply toothed leaves that are dark bronze-green overlaid white or grey.

Juglans microcarpa, syn. *J. rupestris* (Little walnut, Texan walnut)
Deciduous, bushy-headed tree with large, aromatic leaves of many narrow, pointed leaflets that turn yellow in autumn.

Cydonia oblonga 'Vranja'
Deciduous, spreading tree. Pale green leaves, grey-felted beneath, mature to dark green and set off large, white or pale pink flowers in late spring and, later, very fragrant, golden-yellow fruits.

Ulmus 'Camperdownii'
Deciduous, strongly weeping tree with sinuous branches. Leaves are very large, rough and dull green.

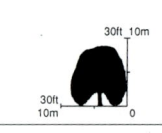

Betula pendula 'Youngii' (Young's weeping birch)
Deciduous, weeping tree forming a mushroom-shaped dome of thread-like branchlets. Has triangular, serrated leaves and smooth, white bark that is fissured black at maturity.

◻ GREEN–YELLOW

Acer carpinifolium (Hornbeam maple)
Deciduous tree of elegant habit, often with several main stems. Prominent-veined, hornbeam-like leaves turn golden-brown in autumn.

Acer shirasawanum 'Aureum'
Deciduous, bushy tree or large shrub. Has rounded, many-lobed, pale yellow leaves.

Morus alba 'Laciniata'
Deciduous, spreading tree. Has rounded, deeply lobed, glossy leaves that turn yellow in autumn and bears edible, pink, red or purple fruits in summer.

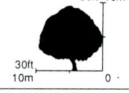

Koelreuteria paniculata (Golden-rain tree, Pride of India)
Deciduous, spreading tree with mid-green leaves, turning yellow in autumn. Bears sprays of yellow flowers in summer, followed by inflated, bronze-pink fruits.

Laburnum × watereri 'Vossii'
Deciduous, spreading tree. Leaves, consisting of 3 leaflets, are glossy, deep green. Pendent chains of large, yellow flowers are borne in late spring and early summer.

Aralia elata
Celtis sinensis
Cornus alternifolia
Neolitsea sericea

Pseudopanax arboreus
Quercus pontica
Tetracentron sinense
Zelkova abelicea

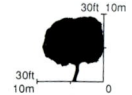

Aralia elata 'Aureo-variegata'
Laburnum anagyroides
Michelia figo, p.63

Trees/small AUTUMN INTEREST

YELLOW–ORANGE

Genista aetnensis
(Mount Etna broom)
Almost leafless, rounded tree with many slender, bright green branches and a profusion of fragrant, pea-like, golden-yellow flowers in mid-summer.

Albizia lophantha, syn. *A. distachya*, *Paraserianthes distachya*
Fast-growing, deciduous, spreading tree. Has fern-like, dark green leaves comprising many leaflets. Creamy-yellow flower spikes appear in spring-summer.

Thevetia peruviana, syn. *T. neriifolia* (Yellow oleander)
Evergreen, erect tree with narrow, lance-shaped, rich green leaves and funnel-shaped, yellow or orange-yellow flowers from winter to summer. Min. 16–18°C (61–4°F).

Laburnum alpinum
(Scotch laburnum)
Deciduous, spreading tree. Leaves consist of 3 leaflets and are glossy, dark green. Long, slender chains of bright yellow flowers appear in late spring or early summer.

Embothrium coccineum
(Chilean firebush)
Evergreen or semi-evergreen, upright, suckering tree with lance-shaped, glossy, deep green leaves. Clusters of brilliant orange-red flowers are borne in late spring and early summer.

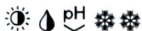

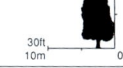

Caesalpinia gilliesii, p.93
Plumeria rubra, p.70
Tecoma stans, p.70

WHITE–RED

Sorbus cashmiriana
Deciduous, spreading tree with leaves consisting of 6–9 pairs of rich green leaflets. Pink-flushed, white flowers in early summer are followed by large, white fruits in autumn.

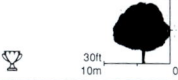

Cornus florida 'Welchii'
Deciduous, spreading tree. Bears white bracts, surrounding tiny flowers, in spring. Dark green leaves, edged with white and pink, turn red and purple in autumn.

Arbutus unedo (Strawberry tree)
Evergreen, spreading tree or shrub with rough, brown bark and glossy, deep green leaves. Pendent, urn-shaped, white flowers appear in autumn-winter as previous season's strawberry-like, red fruits ripen.

OTHER RECOMMENDED PLANTS:
HOLLIES, pp.72–3
Lagerstroemia indica, p.65
Schinus molle
Schinus terebinthifolius
Sorbus insignis
Sorbus prattii

67

Trees/small AUTUMN INTEREST
■ RED

Sorbus vilmorinii
Deciduous, spreading, arching, elegant tree. Leaves of 9–14 pairs of dark green leaflets become orange- or bronze-red in autumn. Has white blooms in late spring and small, deep pink fruits in autumn.

Crataegus macrosperma var. acutiloba
Deciduous, spreading tree with broad, sharply toothed, dark green leaves. White flowers with red anthers in late spring are followed by bright red fruits in autumn.

Photinia davidiana
Evergreen, spreading tree or large shrub with narrow, glossy, dark green leaves, older ones turning red in autumn. Sprays of white flowers in early summer are followed by clusters of bright red fruits in autumn.

Malus 'Cowichan'
Deciduous, spreading tree. Has dark green foliage, reddish-purple when young. Pink flowers appear in mid-spring, followed by reddish-purple crab apples.

Malus 'Veitch's Scarlet'
Deciduous, spreading tree with dark green foliage. Carries white flowers in late spring and crimson-flushed, scarlet crab apples in autumn.

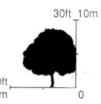

Acer palmatum var. coreanum
Deciduous, bushy-headed tree or large shrub. Leaves are deeply lobed and mid-green, turning brilliant red in autumn. Small, reddish-purple flowers are borne in spring.

Cotoneaster frigidus
Crataegomespilus dardarii 'Jules d'Asnières'
Crataegus crus-galli
Crataegus ellwangeriana
Crataegus laevigata 'Punicea'
Crataegus monogyna
Crataegus phaenopyrum
HOLLIES, pp.72–3
Malus 'Chilko'
Malus x *purpurea*
Malus 'Red Jade'
Rhus typhina
Sorbus sargentiana

■ RED

Acer japonicum **'Aconitifolium'**
Deciduous, bushy tree or large shrub. Deeply divided, mid-green leaves turn red in autumn. Reddish-purple flowers appear in mid-spring.

Crataegus pedicellata
Deciduous, spreading tree with sharply toothed, lobed, dark green leaves that turn orange and red in autumn. White flowers with red anthers in late spring are followed by bright red fruits in autumn.

Malus **'John Downie'**
Deciduous tree, narrow and upright when young, conical when mature. White flowers, borne amid bright green foliage in late spring, are followed by large, edible, red-flushed, orange crab apples in autumn.

Malus prunifolia
Deciduous, spreading tree. Has dark green leaves and fragrant, white flowers in mid-spring. In autumn bears long-lasting, small, red or occasionally yellowish crab apples.

Acer japonicum **'Vitifolium'**
Vigorous, deciduous, bushy tree or large shrub with large, rounded, lobed, mid-green leaves that turn brilliant red, orange and purple in autumn.

Rhus trichocarpa
Deciduous, spreading tree. Large, ash-like leaves with 13–17 leaflets are pinkish when young, dark green in summer and purple-red to orange in autumn. Bears pendent, bristly, yellow fruits.

Acer ginnala (Amur maple)
Deciduous, spreading tree or large shrub. Clusters of fragrant, creamy-white flowers are borne in early summer amid dainty, bright green leaves that turn red in autumn.

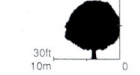

69

Trees/small AUTUMN INTEREST

■ RED

■■ ORANGE–YELLOW

Acer triflorum
Slow-growing, deciduous, spreading tree with peeling, grey-brown bark. Leaves, composed of 3 leaflets, are dark green, turning brilliant orange-red in autumn. Clusters of tiny, yellow-green flowers appear in late spring.

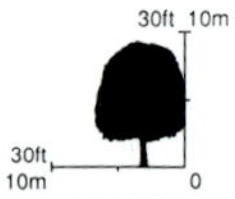

Cornus 'Eddie's White Wonder'
Deciduous, spreading tree or shrub. Large, white bracts, surrounding insignificant flowers, appear in late spring. Oval leaves are mid-green, turning red and purple in autumn.

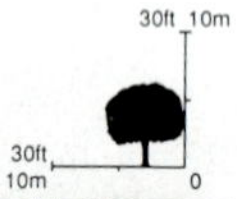

Malus 'Professor Sprenger'
Deciduous, rounded, dense tree. Dark green leaves turn yellow in late autumn. White flowers, pink in bud, open from mid- to late spring and are followed by orange-red crab apples in autumn.

Plumeria rubra (Frangipani)
Deciduous, spreading tree or large shrub, sparingly branched. Has fragrant flowers, in shades of yellow, orange, pink, red and white, in summer-autumn. Min. 13°C (55°F).

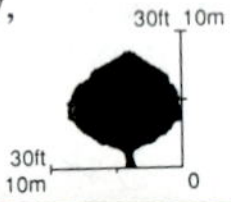

Malus x zumi 'Calocarpa'
Deciduous, spreading tree. Dark green leaves are sometimes deeply lobed. White flowers in late spring are followed by dense clusters of long-lasting, cherry-like, red crab apples in autumn.

Malus 'Marshall Oyama'
Deciduous, upright tree with dark green leaves. Pink-flushed, white flowers borne in late spring are followed by a profusion of large, rounded, crimson and yellow crab apples in autumn.

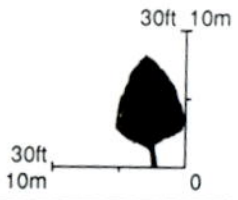

Tecoma stans, syn. *Bignonia stans, Stenolobium stans* (Yellow bells, Yellow elder)
Evergreen, rounded, upright tree or large shrub. Leaves have 5–13 leaflets. Has funnel-shaped, yellow flowers from spring to autumn. Min. 13°C (55°F).

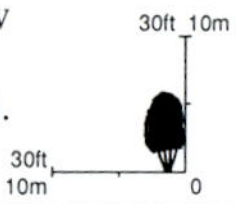

Acer circinatum
Acer palmatum
Cornus florida
Crataegus x *lavallei* 'Carrierei'

Crataegus x *prunifolia*
HOLLIES, pp.72–3
Malus 'Eleyi'
Malus 'Neville Copeman'

Malus niedzwetskyana
Malus 'Royalty', p.63
Photinia nussia
Sorbus aucuparia 'Sheerwater Seedling'

Acer crataegifolium
Acer palmatum 'Senkaki', p.94
Annona reticulata
Malus toringoides

Trees/small WINTER INTEREST

☐ YELLOW
☐ WHITE–PINK
☐ YELLOW

Bauhinia variegata 'Candida'
Deciduous tree, rounded when young, spreading with age. Has broadly oval, deeply notched leaves and fragrant, pure white flowers, 10cm (4in) across, in winter-spring or sometimes later. Min. 15–18°C (59–64°F).

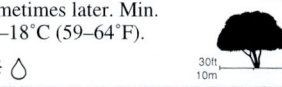

Acacia pravissima (Ovens wattle)
Evergreen, spreading, arching tree or shrub. Has triangular, spine-tipped, silver-grey phyllodes (flat, leaf-like stalks) and small heads of bright yellow flowers in late winter or early spring.

Acacia baileyana (Cootamundra wattle)
Evergreen, spreading, graceful tree with arching branches and finely divided, blue-grey leaves. Clusters of small, golden-yellow flower heads appear in winter-spring.

Picrasma ailanthoides, syn. *P. quassioides* (Quassia)
Deciduous, spreading tree with glossy, bright green leaves, composed of 9–13 leaflets, that turn brilliant yellow, orange and red in autumn.

Malus 'Golden Hornet', syn. *M. × zumi* 'Golden Hornet'
Deciduous, spreading tree with dark green foliage and open cup-shaped, white flowers in late spring. In autumn, branches are weighed down by a profusion of golden-yellow crab apples.

Bauhinia variegata, syn. *B. purpurea* of gardens
Deciduous, rounded tree with broadly oval, deeply notched leaves. Fragrant, magenta to lavender flowers, to 10cm (4in) across, appear in winter-spring, sometimes later. Min. 15–18°C (59–64°F).

Tabebuia chrysotricha (Golden trumpet tree)
Deciduous, round-headed tree with dark green leaves, divided into 3–5 oval leaflets, and rich yellow flowers, 7cm (3in) long, borne in late winter or early spring. Min. 16–18°C (61–4°F).

Crataegus tanacetifolia
Cydonia oblonga 'Lusitanica'
Cydonia oblonga 'Vranja', p.66
HOLLIES, pp.72–3

OTHER RECOMMENDED PLANTS:
Dombeya × cayeuxii, p.62
Prunus incisa 'February Pink'
Prunus subhirtella 'Autumnalis'

Azara microphylla, p.95

Hollies

The common holly, *Ilex aquifolium*, is one of the best-known evergreen trees, but there are many other hollies, including lesser-known *Ilex* cultivars, that make attractive garden plants. In size they range from tall, specimen trees to small shrubs useful in the rock garden or for growing in containers. Hollies respond well to pruning and many may be clipped to form good hedges. Leaves of different species and cultivars may be smooth-edged or spiny and vary considerably in colour, several having gold, yellow, cream, white or grey variegation. Small, often white, male and female flowers, borne on separate plants during summer, are followed by attractive, red, yellow or black berries. In almost all cases hollies are unisexual, that is the berries are borne only on female plants, so to obtain fruits it is usually necessary to grow plants of both sexes. Most plants are fully hardy.

I. fargesii var. *brevifolia*

I. aquifolium 'Pyramidalis' ♀

I. aquifolium 'Argentea Marginata Pendula'

I. × *altaclerensis* 'Camelliifolia' ♀

I. aquifolium 'Argentea Marginata' ♀

I. ciliospinosa

I. pernyi

I. crenata var. *paludosa*

I. fargesii

I. aquifolium 'Scotica'

I. × *altaclerensis* 'Lawsoniana' ♀

I. macrocarpa

I. × *koehneana*

I. crenata 'Helleri'

I. × *altaclerensis* 'Balearica'

I. crenata 'Latifolia'

I. cornuta 'Burfordii'

I. × *altaclerensis* 'N.F. Barnes'

I. verticillata

I. aquifolium ♀

I. opaca

I. × *aquipernyi*

I. crenata 'Convexa' ♀

I. × *altaclerensis* 'Belgica'

I. × *meserveae* 'Blue Princess' ♀

I. aquifolium 'Silver Milkmaid' ♀

Trees/small ALL YEAR INTEREST
WHITE–PURPLE

I. aquifolium 'Silver Queen'

I. aquifolium 'Elegantissima'

I. aquifolium 'Mme Briot'

I. x altaclerensis 'Belgica Aurea'

I. aquifolium 'Crispa Aurea Picta'

I. aquifolium 'Ovata Aurea'

I. aquifolium 'Aurifodina'

I. aquifolium 'Watereriana'

I. serrata f. *leucocarpa*

I. aquifolium 'Pyramidalis Aurea Marginata'

I. x altaclerensis 'Camelliifolia Variegata'

I. chinensis

I. crenata 'Variegata'

I. pedunculosa

Pittosporum crassifolium 'Variegatum'
Evergreen, bushy-headed, dense tree or shrub with grey-green leaves edged with white. Clusters of small, fragrant, deep reddish-purple flowers appear in spring.

Pittosporum eugenioides 'Variegatum'
Evergreen, columnar tree. Wavy-edged, glossy, dark green leaves have white margins. Honey-scented, pale yellow flowers are borne in spring.

Acer pectinatum subsp. laxiflorum
Deciduous, spreading tree with arching branches streaked white and green. In late summer has pale red, winged fruits. Pointed, red-stalked, dark green leaves turn orange in autumn.

Grevillea banksii
Evergreen, loosely branched tree or tall shrub. Has leaves divided into 5–11 slender leaflets, silky-downy beneath. Spider-like, red flowers appear in dense heads intermittently throughout the year. Min. 10°C (50°F).

Dracaena marginata 'Tricolor', syn. *D. cincta* 'Tricolor'
Slow-growing, evergreen, upright tree or shrub with narrow, strap-shaped, cream-striped, rich green leaves, prominently edged with red. Min. 13°C (55°F).

Cordyline australis 'Atropurpurea'
Slow-growing, evergreen tree with purple to purplish-green leaves. Has terminal sprays of white flowers in summer and small, globular, white fruits in autumn. Min. 5°C (41°F).

OTHER RECOMMENDED PLANTS:
Eriobotrya japonica
Kigelia africana
Pittosporum tenuifolium, p.97

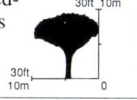

Polyscias guilfoylei 'Victoriae', p.96

73

Trees/small ALL YEAR INTEREST
GREY–GREEN

Leucadendron argenteum
(Silver tree)
Evergreen, conical to columnar tree, spreading with age. Leaves are covered with long, silky, white hairs. Has insignificant flowers set in silvery bracts in autumn–winter. Min. 7°C (45°F).

Butia capitata,
syn. *Cocos capitata* (Jelly palm)
Slow-growing, evergreen palm. Feather-shaped leaves, composed of many leathery leaflets, are strongly arching to recurved, 2m (6ft) or more long. Min. 5°C (41°F).

Carpinus betulus 'Fastigiata'
Deciduous, erect tree, with a very distinctive, flame-like outline, that becomes more open with age. Oval, prominently veined, dark green leaves turn yellow and orange in autumn.

Cyathea australis
(Australian tree fern)
Evergreen, upright tree fern with a robust, almost black trunk. Finely divided leaves, 2–4m (6–12ft) long, are light green, bluish beneath. Min. 13°C (55°F).

Eucalyptus perriniana
(Spinning gum)
Fast-growing, evergreen, spreading tree with rounded, grey-blue, young leaves joined around stems. Leaves on mature trees are long and pendulous. White flowers appear in late summer.

Meryta sinclairii (Puka, Pukanui)
Evergreen, round-headed tree with large, glossy, deep green leaves. Greenish flowers appear sporadically in spring to autumn, followed by berry-like, black fruits. Min. 5°C (41°F).

Pittosporum dallii
Evergreen, rounded, dense tree or shrub. Has purplish stems and sharply toothed, deep green leaves. Clusters of small, fragrant, shallowly cup-shaped, white flowers are borne in summer.

Beaucarnea recurvata, syn. *Nolina recurvata*, *N. tuberculata* (Elephant's foot, Pony-tail)
Slow-growing, evergreen tree or shrub with a sparsely branched stem. Recurving leaves, 1m (3ft) long, persist after turning brown. Min. 7°C (45°F).

Dracaena draco (Dragon tree)
Slow-growing, evergreen tree, eventually with a wide-branched head. Has stiff, lance-shaped, grey- or blue-green leaves. Mature trees bear clusters of orange berries, usually from mid- to late summer. Min. 13°C (55°F).

Lithocarpus henryi
Slow-growing, evergreen, broadly conical tree with glossy, pale green leaves that are long, narrow and pointed.

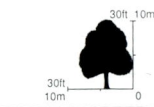

Chrysalidocarpus lutescens,
syn. *Areca lutescens*
(Golden-feather palm, Yellow palm)
Evergreen, suckering palm, forming clumps of robust, cane-like stems. Has long, arching leaves of slender, yellowish-green leaflets. Min. 16°C (61°F).

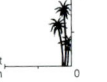

Acer griseum (Paper-bark maple)
Deciduous, spreading tree with striking, peeling, orange-brown bark. Dark green leaves have 3 leaflets and turn red and orange in autumn.

Castanopsis cuspidata
Cyathea arborea
Eucalyptus ficifolia
HOLLIES, pp.72–3

Howea forsteriana
Ligustrum lucidum 'Excelsum Superbum', p.97
Pandanus tectorius

Pisonia umbellifera
Pittosporum crassifolium
Pittosporum tobira
Quercus alnifolia

Quercus coccifera
Ravenala madagascariensis
Sapium sebiferum
Schefflera digitata

CONIFERS/large
BLUE–GREEN

***Abies concolor* 'Argentea'**, syn. *A. c.* 'Candicans'
Conical conifer with silvery foliage that contrasts well with dark grey bark. Oblong to ovoid, pale blue or green cones are 8–12cm (3–5in) long.

Pinus* x *holfordiana (Holford pine)
Broadly conical, open conifer with large cones, brown when ripe. Pendent, glaucous blue-green leaves are held in 5s.

Cupressus cashmeriana (Kashmir cypress)
Handsome, broadly conical conifer, spreading with age, with aromatic foliage borne in pendent, flat, glaucous blue sprays. Bears small, globose, dark brown, mature cones.

x *Cupressocyparis leylandii* (Leyland cypress)
Vigorous, upright, columnar conifer, tapering at the apex. Will grow 1m (3ft) a year, so is a popular screening plant. Dark green or grey-green foliage is held in flat sprays.

Cedrus atlantica* f. *glauca (Blue Atlas cedar)
Conical conifer with silvery-blue foliage that is very bright, especially in spring. Erect, cylindrical cones are produced in autumn. Is widely planted as a specimen tree.

Pinus peuce (Macedonian pine)
Upright conifer, forming a slender pyramid. Has dense, grey-green foliage and cylindrical, green cones with white resin that ripen brown in autumn. Is an attractive tree that grows consistently well in all sites.

OTHER RECOMMENDED PLANTS:
Abies concolor
Abies procera
Abies procera 'Glauca', p.79
Cedrus atlantica
Cedrus atlantica f. *fastigiata*
Cedrus deodara
Cedrus libani, p.77
Chamaecyparis lawsoniana 'Triomf van Boskoop'
Chamaecyparis lawsoniana 'Wisselii'
Chamaecyparis nootkatensis
Cupressus lusitanica
Juniperus virginiana
Larix kaempferi
Picea pungens

Conifers/large
GREEN

Chamaecyparis lawsoniana 'Intertexta'
Elegant, weeping conifer with aromatic, grey-green foliage carried in lax, pendulous sprays. Old trees become columnar with some splayed branches.

Pinus coulteri
(Big-cone pine, Coulter pine)
Fast-growing conifer with large, broadly ovoid, prickly cones, each 1–2kg (2–4½lb). Grey-green leaves in crowded clusters are sparsely set on branches. Grows in all soils, even heavy clays.

Abies veitchii (Veitch fir)
Upright conifer with dark green leaves, silvery beneath, and cylindrical, violet-blue cones.

Pseudotsuga menziesii subsp. **glauca** (Blue Douglas fir)
Fast-growing, conical conifer with thick, grooved, corky, grey-brown bark, aromatic, glaucous blue-green leaves, and sharply pointed buds. Cones have projecting, 3-pronged bracts.

Pinus strobus (Eastern white pine, Weymouth pine)
Conifer with an open, sparse, whorled crown. Has grey-green foliage and cylindrical cones. Smooth, grey bark becomes fissured with age. Does not tolerate pollution.

Metasequoia glyptostroboides (Dawn redwood)
Fast-growing, deciduous, upright conifer with fibrous, reddish bark. Soft, blue-green leaves turn yellow, pink and red in autumn. Cones are globose to ovoid, 2 cm (¾in) long.

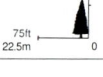

Pinus montezumae
(Montezuma pine, Rough-barked Mexican pine)
Majestic conifer with a rounded crown and grey-green or blue-grey tufts of long, upswept leaves. Bears ovoid cones up to 15cm (6in) long.

Sequoiadendron giganteum
(Big tree, Giant redwood, Wellingtonia)
Very fast-growing, conical conifer. Has thick, fibrous, red-brown bark and sharp, bluish-green leaves. Is one of the world's largest trees when mature.

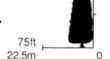

Abies alba
Abies amabilis
Abies cephalonica
Abies homolepis

Abies nordmanniana
Araucaria heterophylla
Chamaecyparis lawsoniana 'Green Pillar', p.81

Chamaecyparis obtusa
Chamaecyparis pisifera
Cryptomeria japonica
x Cupressocyparis leylandii 'Leighton Green'

■ GREEN

Cedrus libani (Cedar of Lebanon)
Spreading conifer, usually with several arching stems. Branches carry flat layers of dark grey-green foliage and oblong to ovoid, greyish-pink cones, 8–15cm (3–6in) long.

Pinus muricata (Bishop pine)
Fast-growing, often flat-topped conifer. Leaves are blue- or grey-green and held in pairs. Ovoid cones, 7–9cm (3–3½in) long, rarely open. Does particularly well in a poor, sandy soil.

Picea omorika (Serbian spruce)
Narrow, conical conifer, resembling a church spire, with dark green leaves that are white below. Branches are pendulous and arch out at tips. Violet-purple cones age to glossy brown. Grows steadily in all soils.

Araucaria araucana (Chile pine, Monkey puzzle)
Open, spreading conifer with grey bark, wrinkled like elephant hide. Has flattened and sharp, glossy, dark green leaves and 15cm (6in) long cones. Makes a fine specimen tree.

Pinus ponderosa (Western yellow pine)
Conical or upright conifer, grown for its distinctive, deeply fissured bark, with smooth, brown plates, and bold greyish-green foliage. Bears ovoid, purplish-brown cones.

Pinus jeffreyi (Black pine, Jeffrey pine)
Upright, narrow-crowned conifer with stout, grey-green leaves, 12–26cm (5–10in) long. Bark is black with fine, deep fissures and shoots have an attractive, greyish bloom.

Pinus wallichiana, syn. *P. chylla*, *P. excelsa*, *P. griffithii* (Bhutan pine, Himalayan pine)
Conical conifer with long, drooping, blue-green leaves in 5s. Has smooth bark, grey-green on young trees, later fissured and dark, and cylindrical cones.

Pinus pinaster (Cluster pine, Maritime pine)
Vigorous, domed conifer with a long, branchless trunk. Has grey-green leaves and whorls of rich brown cones. Purple-brown bark is deeply fissured. Is well-suited to a dry, sandy soil.

Ginkgo biloba (Maidenhair tree)
Long-lived, deciduous conifer, upright when young, spreading with age. Has fan-shaped, 12cm (5in) long, bright green leaves. Bears fruits, with edible kernels, in late summer and autumn, if male and female plants are grown together.

Cupressus macrocarpa
Juniperus chinensis 'Keteleeri', p.79
Larix decidua
Picea orientalis

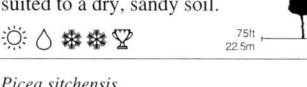

Picea sitchensis
Picea smithiana
Pseudotsuga menziesii
Sequoia sempervirens

Thuja plicata
Thuja plicata 'Atrovirens'
Thujopsis dolabrata
Tsuga heterophylla

Conifers/large

🟩 GREEN   🟩🟧 GREEN–ORANGE

Pinus nigra var. **nigra** (Austrian pine)
Broadly crowned conifer, with well-spaced branches, often with several stems. Paired, dark green leaves are densely tufted. Tolerates an exposed site.

Pinus radiata, syn. *P. insignis* (Monterey pine)
Very fast-growing conifer, conical when young, domed when mature. Black bark contrasts well with soft, bright green leaves. Makes an excellent windbreak.

Pinus heldreichii var. **leucodermis**, syn. *P. leucodermis* (Bosnian pine)
Dense, conical conifer with dark green leaves held in pairs. Ovoid cones, 5–10cm (2–4in) long, are cobalt-blue in their second summer, brown when ripe.

Abies grandis (Giant fir, Grand fir)
Very vigorous, narrow, conical conifer, with a neat habit. Mid-green leaves have an orange aroma when crushed. Cones, 7–8cm (3in) long, ripen red-brown. Makes a useful specimen tree.

Picea abies (Common spruce, Norway spruce)
Fast-growing, pyramidal conifer with dark green leaves. Narrow, pendulous, glossy, brown cones are 10–20cm (4–8in) long. Much used as a Christmas tree but less useful as an ornamental.

x **Cupressocyparis leylandii** '**Harlequin**' (Variegated Leyland cypress)
Fast-growing, columnar conifer with a conical tip. Grey-green foliage, with patches of clear ivory-white, is held in plume-like sprays.

x **Cupressocyparis leylandii** '**Castlewellan**'
Upright, vigorous conifer, slightly slower-growing than the species, grown for its bronze-yellow foliage.

Picea orientalis '**Skylands**'
Dense, upright, graceful conifer that retains the gold coloration of short, glossy leaves throughout the year. Narrowly oblong cones are dark purple, males turning brick-red in spring.

Taxodium distichum (Bald cypress, Swamp cypress)
Deciduous, broadly conical conifer with small, globose to ovoid cones. Yew-like, fresh green leaves turn rich brown in late autumn. Grows in a very wet site, producing special breathing roots.

x *Cupressocyparis leylandii* 'Robinson's Gold'
Picea orientalis 'Aurea'
Thuja plicata 'Aurea'

Conifers/medium
BLUE–GREEN

Abies procera 'Glauca'
Upright conifer with smooth, silvery bark and blue foliage. In drought conditions unsightly cracks will appear in bark and wood. Purplish-brown cones are cylindrical, 15–25cm (6–10in) long.

Picea glauca 'Coerulea'
Dense, upright, conical conifer with needle-like, blue-green to silver leaves and ovoid, light brown cones.

Juniperus chinensis 'Keteleeri'
Dense, regular, slender, columnar conifer with scale-like, aromatic, greyish-green leaves and peeling, brown bark. Makes a reliable, free-fruiting form for formal use.

Picea breweriana
(Brewer's spruce)
Upright conifer with level branches and completely pendulous branchlets, to 2m (6ft) long. Leaves are stout and blue-green. Bears oblong, purplish cones, 6–8cm (2½–3in) long.

Picea pungens 'Koster'
Upright conifer with whorled branches. Has scaly, grey bark and attractive, needle-like, silvery-blue leaves, which fade to green with age. Tends to suffer from aphid attack.

Chamaecyparis lawsoniana 'Pembury Blue'
Magnificent, conical conifer with aromatic, bright blue-grey foliage held in pendulous sprays.

Picea engelmannii (Engelmann spruce, Mountain spruce)
Broadly conical conifer. Leaves encircle shoots and are prickly or soft, lush, glaucous or bluish-green. Bears small, cylindrical cones. Is good for a very poor site.

Cupressus sempervirens (Italian cypress, Mediterranean cypress)
Very narrow, upright conifer, fast-growing when young. Aromatic, grey-green foliage is held in erect sprays. Glossy, grey-brown cones are globose or ovoid.

Pinus parviflora
(Japanese white pine)
Slow-growing, conical or spreading conifer with fine, bluish foliage and purplish-brown bark. Leaves are held in 5s. Bears ovoid cones, 5–10cm (2–4in) long.

OTHER RECOMMENDED PLANTS:
Abies lasiocarpa
Chamaecyparis lawsoniana 'Fletcheri'
Chamaecyparis lawsoniana 'Wisselii'

Chamaecyparis nootkatensis 'Pendula'
Chamaecyparis pisifera 'Squarrosa'
Cupressus cashmeriana, p.75
Cupressus glabra

Juniperus recurva
Picea glauca
Picea likiangensis
Picea mariana

Picea pungens 'Hoopsii'
Pinus armandii
Pinus x *holfordiana*, p.75
Tsuga mertensiana

79

Conifers/medium
■ GREEN

Fitzroya cupressoides, syn.
F. patagonica (Patagonian cypress)
Vase-shaped to sprawling conifer with red-brown bark that peels in long strips. White-lined, dark green leaves are held in open, pendulous, wiry sprays.

Pinus rigida (Northern pitch pine)
Conical conifer, often with sucker shoots from trunk. Twisted, dark green leaves are borne in 3s. Ovoid to globose, red-brown cones, 3–8cm (1¼–3in) long, persist, open, on the tree.

Cunninghamia lanceolata
(Chinese fir)
Upright conifer, mop-headed on a dry site, with distinctive, thick and deeply furrowed, red-brown bark. Glossy, green leaves are sharply pointed and lance-shaped.

Calocedrus decurrens,
syn. ***Libocedrus decurrens***
(Incense cedar)
Upright conifer with short, horizontal branches and flaky, grey bark, brown beneath. Has flat sprays of aromatic, dark green leaves. Resists honey fungus.

Podocarpus salignus
Upright conifer. Leaves are willow-like, 5–11cm (2–4in) long, and glossy above. Attractive, fibrous, red-brown bark peels in strips.

Phyllocladus trichomanoides
Slow-growing conifer, conical when young, developing a more rounded top with age. Leaf-like, deep green, modified shoots, 10–15cm (4–6in) long, have 5–10 lobed segments.

Pinus cembra (Arolla pine)
Dense, conical conifer with dark green or bluish-green leaves grouped in 5s. Ovoid, bluish or purplish cones, 6–8cm (2½–3in) long, ripen brown.

Pinus thunbergii
(Japanese black pine)
Rounded conifer, conical when young, with dark green leaves and grey-brown cones, 4–6cm (1½–2½in) long. Buds are covered with a silky cobweb of white hairs. Tolerates sea spray well.

Austrocedrus chilensis,
syn. ***Libocedrus chilensis***
(Chilean incense cedar)
Conical conifer with flattened, feathery sprays of 4-ranked, small, dark green leaves, white beneath.

Sciadopitys verticillata
(Japanese umbrella pine)
Conical conifer with reddish-brown bark. Deep green leaves, yellowish beneath, are whorled at the ends of shoots, like umbrella spokes. Ovoid cones ripen over 2 years.

Pinus contorta var. ***latifolia***
(Lodgepole pine)
Conical conifer with bright green leaves, 6–9cm (2½–4in) long. Small, oval cones remain closed on the tree. Is suitable for a wet or coastal site.

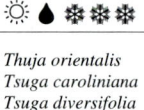

Abies amabilis
Abies delavayi
Abies veitchii, p.76
Chamaecyparis lawsoniana 'Kilmacurragh'

Juniperus chinensis
Picea omorika, p.77
Pinus coulteri, p.76
Pinus heldreichii var. leucodermis, p.78

Pinus wallichiana, p.77
Saxegothaea conspicua
Taxus baccata
Taxus baccata 'Fastigiata'

Thuja orientalis
Tsuga caroliniana
Tsuga diversifolia

■ GREEN

Pseudolarix amabilis, syn. *P. kaempferi* (Golden larch)
Deciduous, open-crowned conifer, slow-growing when young. Has clusters of linear, fresh green leaves, 2.5–6cm (1–2½in) long, which gradually turn bright orange-gold in autumn.

Pinus banksiana (Jack pine)
Slender, conical, scrubby-looking conifer with fresh green leaves in twisted, divergent pairs. Curved cones, 3–6cm (1¼–2½in) long, point forward along shoots.

Picea morrisonicola (Taiwan spruce)
Upright, conical conifer, becoming columnar with age. Needle-like, deep green leaves are pressed down on slender, pale brown shoots. Cones are cylindrical and 5–7cm (2–3in) long.

Torreya californica (California nutmeg)
Upright conifer with very prickly, glossy, dark green leaves, yellowish-green beneath, similar to those of yew. Fruits are olive-like.

Chamaecyparis thyoides (White cypress)
Upright conifer with aromatic, green or blue-grey leaves in rather erratic, fan-shaped sprays on very fine shoots. Cones are small, round and glaucous blue-grey.

Tsuga canadensis (Canada hemlock, Eastern hemlock)
Broadly conical conifer, often with several stems. Grey shoots have 2-ranked, dark green leaves, often inverted to show silver lines beneath. Cones are ovoid and light brown.

Chamaecyparis lawsoniana 'Lanei'
Upright conifer that forms a neat column of aromatic, golden-yellow-tipped foliage.

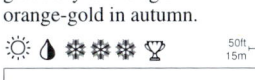

Chamaecyparis lawsoniana 'Green Pillar'
Conical conifer with upright branches. Aromatic foliage is bright green and becomes tinged with gold in spring. Is suitable for hedging as requires little clipping.

Pinus contorta (Beach pine, Shore pine)
Dense, conical or domed conifer. Has paired, bright green leaves and conical to ovoid cones, 3–8cm (1¼–3in) long. Is well-suited to a windy, barren site and tolerates waterlogged ground.

Pinus halepensis (Aleppo pine)
Conical, open-crowned conifer with an open growth of bright green leaves, 6–11cm (2½–4½in) long, and ovoid, glossy, brown cones. Young trees retain glaucous, juvenile needles for several years.

Pinus virginiana (Scrub pine, Virginia pine)
Conifer of untidy habit. Grey- to yellow-green leaves are 4–7cm (1½–3in) long. Young shoots have a pinkish-white bloom. Bears oblong to conical, red-brown cones, 6cm (2½in) long.

Chamaecyparis obtusa 'Crippsii', p.83
Chamaecyparis pisifera 'Filifera Aurea', p.85
Chamaecyparis pisifera 'Plumosa'
Cupressus macrocarpa 'Goldcrest', p.81
Juniperus chinensis 'Aurea'
Juniperus drupacea
Pinus densiflora
Taxus baccata 'Fastigiata Aurea'
Thuja occidentalis
Thuja occidentalis 'Fastigiata'
Thujopsis dolabrata

Conifers/small
■ GREEN

Picea mariana 'Doumetii'
Densely branched, globose or broadly conical conifer with short, needle-like, silvered, dark green leaves and pendulous, ovoid, purplish cones.

Juniperus virginiana 'Robusta Green', syn. *J. chinensis* 'Robusta Green'
Slow-growing, column-like, conical conifer, making only 7–8cm (3in) a year, with aromatic, green foliage and small, grey-green juniper berries.

Pinus bungeana (Lace-bark pine)
Slow-growing, bushy conifer with dark green foliage, planted for its exquisite, grey-green bark that flakes to reveal creamy-yellow patches, darkening to red or purple.

Chamaecyparis lawsoniana 'Columnaris'
Narrow, upright conifer that forms a neat column of aromatic, blue-grey foliage. Will tolerate poor soil and some clipping. Is an effective, small, specimen tree.

Pinus aristata (Bristle-cone pine)
Slow-growing, bushy conifer. Leaves are in bundles of 5, very dense and blue-white to grey-green, flecked with white resin. Ovoid cones, 4–10cm (1½–4in) long, have bristly prickles. Is the oldest-known living plant, over 4000 years old.

Juniperus virginiana 'Burkii'
Slow-growing, dense, upright conifer. Aromatic, blue-greyish-green foliage develops a purplish tinge in winter. Scale- and needle-like leaves occur on same shoot. Has small fruits with a pronounced bloom.

Pinus sylvestris f. fastigiata
Upright conifer with erect branches forming a narrow, obelisk shape. Has flaky, red-brown bark, blue-green foliage and conical cones. Suffers wind-damage in an exposed site.

Juniperus chinensis 'Obelisk'
Slender, irregularly columnar conifer. Has ascending branches and long, prickly, needle-like, aromatic, dark green leaves. Tolerates a wide range of soils and conditions but is particularly suited to a hot, dry site.

OTHER RECOMMENDED PLANTS:
Cephalotaxus harringtonia
Chamaecyparis lawsoniana 'Ellwoodii'
Chamaecyparis lawsoniana 'Fletcheri'
Juniperus communis
Juniperus x *media* 'Hetzii'
Juniperus rigida
Juniperus scopulorum 'Skyrocket', p.84
Juniperus squamata 'Meyeri'
Picea breweriana, p.79
Pinus pumila
Podocarpus alpinus
Podocarpus macrophyllus
Pseudotsuga menziesii 'Fletcheri'
Taxus baccata 'Adpressa'

GREEN–YELLOW

Pinus cembroides
(Mexican stone pine, Pinyon)
Slow-growing, bushy conifer, rarely more than 6–7m (20–22ft) high. Scaly bark is a striking silver-grey or greyish-brown. Leaves, in 2s and 3s, are sparse and dark green to grey-green.

Thujopsis dolabrata 'Variegata'
Slow-growing, broadly conical, bushy conifer. Stout, hatchet-shaped leaves have irregular, creamy patches above and are silvery beneath.

Taxus cuspidata (Japanese yew)
Evergreen, spreading conifer. Leaves are dark green above, yellowish-green beneath, sometimes becoming tinged red-brown in cold weather. Tolerates very dry and shady conditions.

Chamaecyparis obtusa 'Crippsii'
Attractive, small-garden, conical conifer, grown for its flattened sprays of aromatic, bright golden foliage. Bark is stringy and red-brown. Cones are round, 1cm (½in) across, and brown.

Abies koreana (Korean fir)
Broadly conical conifer. Produces cylindrical, violet-blue cones when less than 1m (3ft) tall. Leaves are dark green above, silver beneath.

Cryptomeria japonica 'Pyramidata'
Narrowly columnar or obelisk-shaped conifer. Foliage is blue-green when young, maturing to dark green.

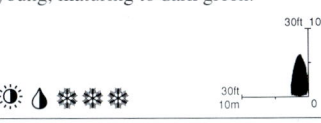

Cedrus deodara 'Aurea'
Slow-growing, upright conifer with pendent branch tips and golden-yellow leaves when young in spring-summer. Foliage matures to yellowish-green. Makes a dramatic, small-garden evergreen.

Cryptomeria japonica 'Cristata'
Conical conifer with twisted, curved shoots and soft, fibrous bark. Foliage is bright green, ageing brown.

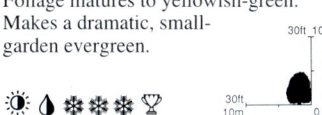

Pinus pinea
(Stone pine, Umbrella pine)
Conifer with a rounded crown on a short trunk. Leaves are dark green, but blue-green, juvenile foliage is retained on young trees. Broadly ovoid cones ripen shiny brown; seeds are edible.

Cupressus macrocarpa 'Goldcrest'
Fast-growing, conical conifer with aromatic, golden-yellow foliage held in plume-like sprays that are useful in flower arrangements. Dislikes clipping.

Chamaecyparis obtusa 'Tetragona Aurea'
Taxus baccata 'Aurea', p.85
Taxus baccata 'Semperaurea'
Thuja koraiensis

Thuja occidentalis 'Rheingold'
Tsuga canadensis 'Aurea', p.85

Dwarf conifers

Dwarf conifers are valuable plants, especially for the small garden, requiring very little attention and providing year-round interest. They can be planted as features in their own right, displaying to advantage their varied shapes, habits and often striking colours, or, in the rock garden for example, to provide scale or act as a foil for other plants such as bulbs. Several species and cultivars are spreading and good for ground cover.

Almost all conifers are suitable for a wide range of growing conditions, although *Cedrus* and *Juniperus* do not tolerate shade and *Juniperus* and *Pinus* are best for dry, freely drained, sandy soils. Most *Abies*, *Taxus*, *Thuja* and *Tsuga* are particularly shade tolerant. Some species may be clipped to form a low hedge but new growth will seldom be made if plants are cut back into wood more than 3 or 4 years old.

Juniperus squamata 'Blue Star' ♀

Juniperus procumbens 'Nana' ♀

Podocarpus nivalis

Juniperus horizontalis 'Douglasii'

Juniperus chinensis 'Stricta'

Juniperus horizontalis 'Turquoise Spreader'

Juniperus sabina 'Cupressifolia'

Juniperus squamata 'Chinese Silver'

Thuja occidentalis 'Caespitosa'

Abies balsamea 'Nana'

Juniperus recurva 'Densa'

Picea pungens 'Montgomery'

Abies concolor 'Compacta' ♀

Microbiota decussata ♀

Abies cephalonica 'Meyer's Dwarf'

Juniperus sabina var. *tamariscifolia*

Abies lasiocarpa var. *arizonica* 'Compacta' ♀

Picea omorika 'Gnom'

Juniperus virginiana 'Grey Owl' ♀

Juniperus procumbens

Pseudotsuga menziesii 'Fretsii'

Picea abies 'Ohlendorffii'

Juniperus squamata 'Holger' ♀

Juniperus × *media* 'Pfitzeriana Glauca'

Abies lasiocarpa 'Roger Watson'

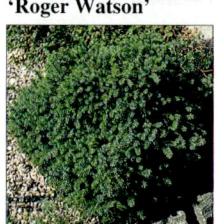

Picea abies 'Reflexa'

Pinus sylvestris 'Doone Valley'

Juniperus scopulorum 'Springbank'

Juniperus sabina 'Mas'

Juniperus scopulorum 'Skyrocket'

Picea mariana 'Nana' ♀

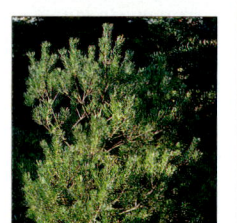

Pinus sylvestris 'Nana'

Pseudotsuga menziesii **'Oudemansii'**

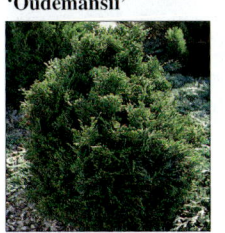

Chamaecyparis lawsoniana **'Gnome'**

Chamaecyparis lawsoniana **'Minima'**

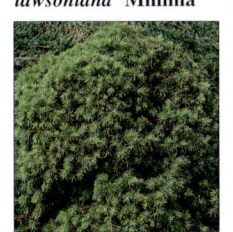

Picea abies **'Gregoryana'**

Chamaecyparis obtusa **'Intermedia'**

Juniperus × *media* **'Pfitzeriana'** ♀

Pinus heldreichii var. *leucodermis* **'Schmidtii'** ♀

Chamaecyparis obtusa **'Nana Pyramidalis'**

Juniperus communis **'Hibernica'** ♀

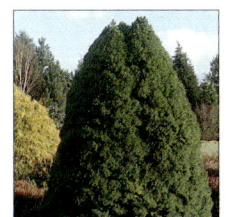

Picea glauca var. *albertiana* **'Conica'** ♀

Cedrus libani **'Sargentii'**

Pinus held. var. *leucodermis* **'Compact Gem'**

Thuja orientalis **'Aurea Nana'** ♀

Thuja orientalis **'Semperaurea'**

Thuja occidentalis **'Filiformis'**

Thuja plicata **'Hillieri'**

Juniperus davurica **'Expansa Variegata'**

Juniperus × *media* **'Pfitzeriana Aurea'**

Juniperus × *media* **'Plumosa Aurea'** ♀

Cryptomeria japonica **'Spiralis'**

Juniperus × *media* **'Blue Gold'**

Cryptomeria japonica **'Sekkan-sugi'**

Taxus baccata **'Dovastonii Aurea'** ♀

Chamaecyparis obtusa **'Nana Aurea'** ♀

Chamaecyparis pisifera **'Filifera Aurea'** ♀

Thuja plicata **'Collyer's Gold'**

Taxus baccata **'Aurea'**

Tsuga canadensis **'Aurea'**

Pinus sylvestris **'Aurea'** ♀

Abies nordmanniana **'Golden Spreader'** ♀

Thuja plicata **'Stoneham Gold'** ♀

Cryptomeria japonica **'Elegans Compacta'** ♀

Pinus sylvestris **'Gold Coin'**

SHRUBS/large SPRING INTEREST
☐ WHITE

Osmanthus delavayi, syn. *Siphonosmanthus delavayi*
Evergreen, rounded, bushy shrub with arching branches. Has small, glossy, dark green leaves and a profusion of very fragrant, tubular, white flowers from mid- to late spring.

Pieris japonica
Evergreen, rounded, bushy, dense shrub with glossy, dark green foliage that is bronze when young. Produces drooping racemes of white flowers during spring.

Amelanchier lamarckii
Deciduous, spreading shrub. Young leaves unfold bronze as abundant sprays of star-shaped, white flowers open from mid- to late spring. Foliage matures to dark green, then turns brilliant red and orange in autumn.

Osmanthus × burkwoodii, syn. × *Osmarea burkwoodii*
Evergreen, rounded, dense shrub. Glossy foliage is dark green and sets off a profusion of small, very fragrant, white flowers from mid- to late spring.

Anopterus glandulosus
Evergreen, bushy shrub or, occasionally, small tree. Has narrow, glossy, dark green leaves, amid which clusters of cup-shaped, white or pink flowers appear from mid- to late spring.

Dipelta yunnanensis
Deciduous, arching shrub with peeling bark and glossy leaves. In late spring produces tubular, creamy-white flowers, marked orange inside.

Malus sargentii
Deciduous, spreading shrub or small tree. A profusion of white flowers in late spring is followed by long-lasting, deep red fruits. Oval, dark green leaves are sometimes lobed.

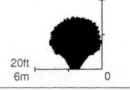

***Viburnum plicatum* 'Mariesii'**
Deciduous, bushy, spreading shrub with tiered branches clothed in dark green leaves, which turn reddish-purple in autumn. Large, rounded heads of flowers with white bracts appear in late spring and early summer.

Staphylea pinnata (Bladder nut)
Deciduous, upright shrub that in late spring carries clusters of white flowers, tinted pink with age, followed by bladder-like, green fruits. Foliage is divided and bright green.

OTHER RECOMMENDED PLANTS:
Amelanchier canadensis
Amelanchier laevis, p.59
CAMELLIAS, pp.98–9
Exochorda racemosa
Halesia tetraptera
HEATHERS, pp.148–9
LILACS, p.90
MAGNOLIAS, p.49
Myrtus bullata
Parrotiopsis jacquemontiana
Pieris formosa 'Henry Price'
Poncirus trifoliata
Prunus laurocerasus
RHODODENDRONS, pp.102–104
Staphylea colchica

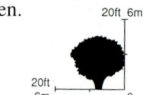

WHITE–RED

Dipelta floribunda
Vigorous, deciduous, upright, tree-like shrub with peeling, pale brown bark. Fragrant, pale pink flowers, marked yellow inside, open in late spring and early summer. Has pointed, mid-green leaves.

Viburnum x carlcephalum
Deciduous, rounded, bushy shrub. In late spring large, rounded heads of pink buds open to fragrant, white flowers. These are borne amid dark green foliage that often turns red in autumn.

Enkianthus campanulatus
Deciduous, bushy, spreading shrub with red shoots and tufts of dull green leaves that turn bright red in autumn. Small, bell-shaped, red-veined, creamy-yellow flowers appear in late spring.

Staphylea holocarpa var. **rosea**, syn. *S. h.* 'Rosea'
Deciduous, upright shrub or spreading, small tree. From mid- to late spring bears pink flowers, followed by bladder-like, pale green fruits. Bronze, young leaves mature to blue-green.

Photinia x fraseri 'Birmingham'
Evergreen, upright, bushy, dense shrub with glossy, dark green leaves that are bright purple-red when young. Broad heads of small, white flowers are carried in late spring.

CAMELLIAS, pp.98–9
Greyia sutherlandii, p.101
HEATHERS, pp.148–9
MAGNOLIAS, p.49

Photinia x *fraseri* 'Red Robin'
Pieris 'Forest Flame'
Prunus triloba 'Multiplex'
RHODODENDRONS, pp.102–104

PURPLE-ORANGE

Daphniphyllum macropodum
Evergreen, bushy, dense shrub with stout shoots and dark green leaves. Small flowers, green on female plants, purplish on male plants, appear in late spring.

Corylopsis glabrescens
Deciduous, open shrub. Oval leaves, with bristle-like teeth along margins, are dark green above, blue-green beneath. Drooping spikes of fragrant, bell-shaped, pale yellow flowers appear in mid-spring on bare branches.

Berberis darwinii
(Darwin's barberry)
Vigorous, evergreen, arching shrub. Has small, glossy, dark green leaves and a profusion of rounded, deep orange-yellow flowers from mid- to late spring followed by bluish berries.

Acacia verticillata
Ceanothus thyrsiflorus
LILACS, p.90
RHODODENDRONS, pp.102–104

Shrubs/large SUMMER INTEREST
☐ WHITE

Escallonia leucantha
Evergreen, upright shrub. Narrow, oval, glossy, dark green leaves set off large racemes of small, shallowly cup-shaped, white flowers in mid-summer.

Olearia virgata
Evergreen, arching, graceful shrub with very narrow, dark grey-green leaves. Produces an abundance of small, star-shaped, white flower heads in early summer, arranged in small clusters along stems.

Viburnum rhytidophyllum
Vigorous, evergreen, open shrub with long, narrow, deep green leaves. Dense heads of small, creamy-white flowers in late spring and early summer are succeeded by red fruits that mature to black.

Styrax officinale
Deciduous, loose to dense shrub or small tree. Fragrant, bell-shaped, white flowers appear in early summer among oval, dark green leaves with greyish-white undersides.

Sparmannia africana (African hemp)
Evergreen, erect shrub or small tree. Has large, shallowly lobed leaves and clusters of white flowers, with yellow and red-purple stamens, in late spring and summer. Min. 7°C (45°F).

Ligustrum sinense
Deciduous or semi-evergreen, bushy, upright shrub with oval, pale green leaves. Large panicles of fragrant, tubular, white flowers are borne in mid-summer, followed by small, purplish-black fruits.

Buddleja davidii 'Peace'
Vigorous, deciduous, arching shrub. Long, pointed, dark green leaves, white-felted beneath, set off long plumes of fragrant, white flowers from mid-summer to autumn.

Escallonia 'Iveyi'
Evergreen, upright shrub. Glossy, dark green foliage sets off large racemes of fragrant, tubular, pure white flowers, with short lobes, borne from mid- to late summer.

Xanthoceras sorbifolium
Deciduous, upright shrub or small tree with bright green leaves divided into many slender leaflets. In late spring and early summer produces spikes of white flowers with red patches inside at the base of the petals.

OTHER RECOMMENDED PLANTS:
Clethra arborea
Clethra barbinervis, p.108
Cotoneaster lacteus, p.95

Cotoneaster 'Rothschildianus'
Cotoneaster x *watereri* 'John Waterer'
Deutzia scabra, p.106
Elaeocarpus cyaneus

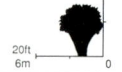

Eucryphia milliganii, p.107
HEATHERS, pp.148–9
Hibiscus syriacus 'Diana'
HYDRANGEAS, p.114

LILACS, p.90
MAGNOLIAS, p.49
Melaleuca armillaris
Myrtus bullata

☐ WHITE

Brugmansia x candida, syn. *Datura x candida*
Semi-evergreen, rounded shrub or small tree. Has downy, oval leaves and strongly scented, pendulous, white flowers, sometimes cream or pinkish, in summer-autumn. Min. 10°C (50°F).

Myrtus luma, syn. *M. apiculata*, *Amomyrtus luma*, *Myrceugenia apiculata*
Strong-growing, evergreen shrub with peeling, golden-brown and grey-white bark and cup-shaped flowers amid aromatic leaves in summer-autumn.

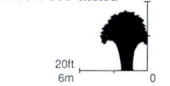

Abutilon vitifolium 'Album'
Fast-growing, deciduous, upright shrub. Large, bowl-shaped, white blooms, pink-tinged when young, are freely borne in late spring and early summer amid deeply lobed, sharply toothed, grey-green leaves.

Abelia triflora
Vigorous, deciduous, upright shrub with pointed, deep green leaves. Small, extremely fragrant, white flowers, tinged pale pink, appear in mid-summer.

Chionanthus virginicus (Fringe tree)
Deciduous, bushy shrub or small tree. Has large, glossy, dark green leaves that turn yellow in autumn. Drooping sprays of fragrant, white flowers appear in early summer.

Cornus mas 'Variegata'
Deciduous, bushy, dense shrub or small tree. Small, star-shaped, yellow flowers appear on bare branches in early spring before white-edged, dark green leaves develop.

Holodiscus discolor
Fast-growing, deciduous, arching shrub. Has lobed, toothed, dark green leaves and large, pendent sprays of small, creamy-white flowers in mid-summer.

Aesculus parviflora (Bottlebrush buckeye)
Deciduous, open shrub. Leaves are bronze when young, dark green in summer and yellow in autumn. Panicles of red-centred, white flowers appear from mid- to late summer.

Olearia macrodonta
Osmanthus fragrans
Philadelphus magdalenae
Prunus lusitanica

Pyracantha atalantioides
Pyracantha 'Mohave'
RHODODENDRONS, pp.102–104
Rubus 'Benenden', p.106

Sambucus canadensis 'Maxima'
Viburnum 'Pragense', p.109

89

Lilacs

The heady scent of the lilac (*Syringa*) epitomizes early summer. The flowers are produced in abundance and are excellent for cutting. Apart from the classic lilacs and mauves, colours include white, pink, cream and rich red-purple; double forms are also available. Most lilacs grown in gardens are vigorous shrubs derived from *S. vulgaris*. They may eventually become tree-like and are therefore best planted at the back of a shrub border or in groups in a wild garden; they may also be used as an informal hedge. Where space is restricted choose from the smaller-growing species, some of which may be grown in containers. Spent flower heads are best removed, but care should be taken not to damage the new shoots that form below the flowers. Otherwise little pruning is required, though older, straggly plants may be rejuvenated by hard pruning in winter.

S. 'Michel Buchner'

S. 'Monge'

S. 'Masséna'

S. *meyeri* 'Palibin' ♡

S. 'Mme Florent Stepman'

S. 'Maréchal Foch'

S. 'Paul Thirion'

S. 'Charles Joly' ♡

S. x *persica* ♡

S. 'Mme F. Morel'

S. 'Decaisne'

S. 'Jan van Tol'

S. 'Mme Lemoine' ♡

S. 'Mrs Edward Harding' ♡

S. 'Clarke's Giant'

S. 'Président Grévy'

S. 'Cora Brandt'

S. x *chinensis* 'Alba'

S. 'Mme Antoine Buchner' ♡

S. 'Esther Staley' ♡

S. *microphylla* 'Superba' ♡

S. *yunnanensis*

S. 'Congo'

S. 'Blue Hyacinth'

S. 'Primrose'

Shrubs/large SUMMER INTEREST

WHITE–PINK

Clethra delavayi
Deciduous, open shrub with lance-shaped, toothed, rich green leaves. Dense, spreading clusters of pink buds opening to scented, white flowers appear in mid-summer.

Tamarix ramosissima, syn. *T. pentandra*
Deciduous, arching, graceful shrub or small tree with tiny, narrow, blue-green leaves. In late summer and early autumn bears large, upright plumes of small, pink flowers.

Abelia x *grandiflora*
Vigorous, semi-evergreen, arching shrub. Has glossy, dark green foliage and an abundance of fragrant, pink-tinged, white flowers from mid-summer to mid-autumn.

Nerium oleander (Oleander)
Evergreen, upright, bushy shrub with leathery, deep green leaves. Clusters of salver-form, pink, white, red, apricot or yellow flowers appear from spring to autumn, often on dark red stalks. Min. 10°C (50°F).

Kolkwitzia amabilis 'Pink Cloud'
Deciduous, arching shrub that bears a mass of bell-shaped, pink flowers amid small, oval, mid-green leaves in late spring and early summer.

Aralia elata 'Variegata'
Cornus alternifolia 'Argentea', p.63
Ligustrum lucidum
LILACS, p.90

MAGNOLIAS, p.49
Rothmannia capensis
Sorbaria kirilowii
Viburnum cinnamomifolium

RED–PURPLE

Buddleja davidii 'Royal Red'
Vigorous, deciduous, arching shrub. Has long, pointed, dark green leaves with white-felted undersides and plumes of fragrant, rich purple-red flowers from mid-summer to autumn.

Acer palmatum f. *atropurpureum*, syn. *A. palmatum* 'Atropurpureum'
Deciduous, bushy-headed shrub or small tree with lobed, reddish-purple foliage that turns brilliant red in autumn. Small, reddish-purple flowers are borne in mid-spring.

Buddleja davidii 'Harlequin'
Vigorous, deciduous, arching shrub. Leaves are long, pointed and dark green with creamy-white margins. Plumes of fragrant, red-purple flowers appear from mid-summer to autumn.

Buddleja colvilei
Deciduous, arching shrub, often tree-like with age. Large, white-centred, deep pink to purplish-red flowers are borne in drooping racemes amid dark green foliage during early summer.

Malvaviscus arboreus (Sleepy mallow)
Vigorous, evergreen, rounded shrub. Serrated, bright green leaves are soft-haired. Has bright red flowers with protruding stamens in summer-autumn. Min. 13–16°C (55–61°F).

Cotinus coggygria 'Notcutt's Variety'
Deciduous, bushy shrub with deep reddish-purple foliage. Long-lasting, purplish-pink plumes of massed, small flowers are produced in late summer.

Buddleja colvilei 'Kewensis'
Callistemon viminalis
Cercis canadensis 'Forest Pansy', p.65
Cotinus coggygria 'Royal Purple'

Erythrina x *bidwillii*, p.113
Erythrina crista-galli, p.113
Feijoa sellowiana, p.113
HEATHERS, pp.148–9

Shrubs/large SUMMER INTEREST

■ RED–PURPLE

■ GREEN–YELLOW

Acer palmatum var. *heptalobum* **'Rubrum'**
Deciduous, bushy-headed shrub or small tree. Large leaves are red when young, bronze in summer and brilliant red, orange or yellow in autumn. Has small, reddish-purple flowers in mid-spring.

Buddleja alternifolia
Deciduous, arching shrub that can be trained as a weeping tree. Has slender, pendent shoots and narrow, grey-green leaves. Neat clusters of fragrant, lilac-purple flowers appear in early summer.

Decaisnea fargesii
Deciduous, upright, semi-arching, open shrub with stout, blue-bloomed shoots and large, deep green leaves of paired leaflets. Racemes of greenish flowers in early summer are followed by pendent, sausage-shaped, bluish fruits.

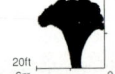

Elaeagnus angustifolia (Oleaster)
Deciduous, bushy shrub or spreading, small tree. Has narrow, silvery-grey leaves and small, fragrant, creamy-yellow flowers, with spreading lobes, in early summer, followed by small, oval, yellow fruits.

Prunus spinosa **'Purpurea'**
Deciduous, dense, spiny shrub or small tree. Bright red, young leaves become deep reddish-purple. Bears saucer-shaped, pale pink flowers from early to mid-spring, followed by blue-bloomed, black fruits.

Tibouchina urvilleana, syn. *T. semidecandra* (Glory bush)
Evergreen, slender-branched shrub. Velvet-haired leaves are prominently veined. Has satiny, blue-purple flowers in clusters from summer to early winter. Min. 7°C (45°F).

Corylus maxima **'Purpurea'**
Vigorous, deciduous, open shrub or small tree with deep purple leaves and purplish catkins, with yellow anthers, that hang from bare branches in late winter. Edible nuts mature in autumn.

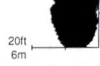

Acer palmatum var. *heptalobum* **'Lutescens'**
Deciduous, bushy-headed shrub or small tree. Large, lobed leaves become clear yellow in autumn. In mid-spring produces small, reddish-purple flowers, followed by winged fruits.

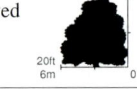

Paliurus spina-christi (Christ's thorn, Jerusalem thorn)
Deciduous, bushy shrub with slender, thorny shoots. Has oval, glossy, bright green leaves, tiny, yellow flowers in summer and curious, woody, winged fruits in autumn.

LILACS, p.90
Melaleuca hypericifolia
RHODODENDRONS, pp.102–104
Sambucus nigra 'Guincho Purple'

Acer shirasawanum 'Aureum', p.66
Aralia elata 'Aureovariegata'
Buddleja x *weyeriana*
Caesalpinia pulcherrima

Cassia didymobotrya, p.117
Colutea arborescens, p.116
Dendromecon rigida, p.117
Fremontodendron californicum

☐ YELLOW

***Brugmansia* 'Grand Marnier'**, syn. *Datura* 'Grand Marnier'
Evergreen, robust shrub with large, oval to elliptic leaves. Pendent, flared, trumpet-shaped, peach-coloured flowers open from an inflated calyx in summer. Min. 7–10°C (45–50°F).

Brugmansia sanguinea, syn. *B. bicolor*, *Datura sanguinea*
Semi-evergreen, erect to rounded shrub or small tree with lobed, young leaves. Has large, trumpet-shaped, yellow and orange-red flowers from late summer to winter. Min. 10°C (50°F).

Crotalaria agatiflora
(Canary-bird bush)
Evergreen, loose, somewhat spreading shrub with grey-green leaves. Racemes of greenish-yellow flowers appear in summer and also intermittently during the year. Min. 15°C (59°F).

Cytisus battandieri, syn. *Argyrocytisus battandieri*
(Moroccan broom, Pineapple broom)
Semi-evergreen, open shrub. Leaves have 3 silver-grey leaflets. Pineapple-scented, yellow flowers appear in summer.

Caesalpinia gilliesii
Deciduous, open shrub or small tree. Has finely divided, dark green leaves and bears short racemes of yellow flowers with long, red stamens from mid- to late summer.

Genista cinerea
Deciduous, arching shrub that produces an abundance of fragrant, pea-like, yellow blooms from early to mid-summer. Has silky, young shoots and narrow, grey-green leaves.

Buddleja globosa
Deciduous or semi-evergreen, open shrub with dark green foliage. Dense, rounded clusters of orange-yellow flowers are carried in early summer.

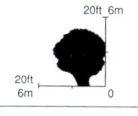

***Fremontodendron* 'California Glory'**
Very vigorous, evergreen or semi-evergreen, upright shrub. Has rounded, lobed, dark green leaves and large, bright yellow flowers from late spring to mid-autumn.

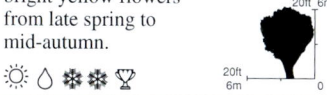

Fremontodendron mexicanum
Ligustrum ovalifolium 'Aureum'
Myrtus luma 'Glanleam Gold'
Osmanthus fragrans f. *aurantiacus*
Sambucus canadensis 'Aurea'
Spartium junceum, p.117
Tecoma stans, p.70

Shrubs/large AUTUMN INTEREST
RED–YELLOW

Cotoneaster 'Cornubia'
Vigorous, semi-evergreen, arching shrub. Clusters of white flowers, produced in early summer amid dark green foliage, are followed by large, pendent clusters of decorative, bright red fruits.

Rhus hirta 'Laciniata', syn. *R. h.* 'Dissecta', *R. typhina* 'Laciniata'
Deciduous, spreading, open shrub or small tree with velvety shoots. Fern-like, dark green leaves turn brilliant orange-red in autumn, when deep red fruit clusters are also borne.

Cotinus coggygria 'Flame'
Deciduous, bushy, tree-like shrub with dark green leaves that turn brilliant orange-red in autumn. From late summer, showy, plume-like, purplish-pink flower heads appear above the foliage.

Hippophäe rhamnoides (Sea buckthorn)
Deciduous, bushy, arching shrub or small tree with narrow, silvery leaves. Tiny, yellow flowers borne in mid-spring are followed in autumn on female plants by bright orange berries.

Hamamelis vernalis 'Sandra'
Deciduous, upright, open shrub. Bears small, fragrant, spidery, deep yellow blooms in late winter and early spring. Oval leaves are purple when young, mid-green in summer, purple, red, orange and yellow in autumn.

Euonymus myrianthus
Evergreen, bushy shrub with pointed, leathery, mid-green leaves. Dense clusters of small, greenish-yellow flowers in summer are followed by yellow fruits that open to show orange-red seeds.

Acer palmatum var. heptalobum
Deciduous, bushy-headed shrub or small tree with large, lobed, mid-green leaves that turn brilliant red, orange or yellow in autumn. Bears small, reddish-purple flowers in mid-spring.

Acer palmatum 'Senkaki', syn. *A.p.* 'Sango Kaku' (Coral-bark maple)
Deciduous, bushy-headed shrub or small tree. Has bright coral-pink, young shoots in winter. Palmate leaves are orange-yellow in spring, green in summer and pink then yellow in autumn.

Pyracantha atalantioides 'Aurea'
Vigorous, evergreen, upright, spiny shrub, arching with age. Has narrowly oval, glossy, dark green leaves and white flowers in early summer, followed by large clusters of small, yellow berries in early autumn.

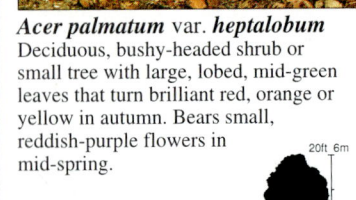

OTHER RECOMMENDED PLANTS:
Acer ginnala, p.69
Acer japonicum 'Vitifolium', p.69
Cotinus coggygria 'Royal Purple'
Cotinus obovatus
Cotoneaster × *watereri* 'John Waterer'
Decaisnea fargesii, p.92
Hibiscus mutabilis
Photinia davidiana, p.81
Pyracantha 'Golden Charmer', p.119
Pyracantha 'Mohave'
Sorbus scopulina
Tibouchina urvilleana, p.92
Viburnum betulifolium, p.118

Shrubs/large WINTER INTEREST

☐ YELLOW

■ RED

☐ GREEN–YELLOW

***Mahonia × media* 'Charity'**
Evergreen, upright, dense shrub with large leaves composed of many spiny, dark green leaflets. Slender, upright, later spreading spikes of fragrant, yellow flowers are borne from early autumn to early spring.

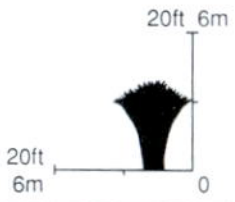

Garrya elliptica (Silk-tassel bush)
Evergreen, bushy, dense shrub with leathery, wavy-edged, dark green leaves. Grey-green catkins, longer on male than female plants, are borne from mid-winter to early spring.

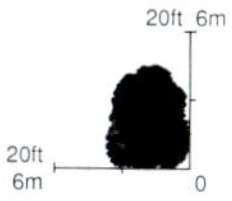

***Mahonia × media* 'Buckland'**
Evergreen, upright, dense shrub. Has large leaves with many spiny, dark green leaflets. Clustered, upright then spreading, long, branched spikes of fragrant, yellow flowers appear from late autumn to early spring.

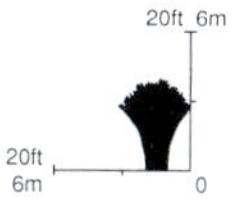

Cotoneaster lacteus
Evergreen, arching shrub suitable for hedging. Oval, dark green leaves set off shallowly cup-shaped, white flowers from early to mid-summer. Long-lasting, red fruits are carried in large clusters in autumn-winter.

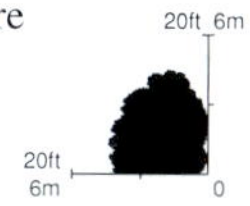

Azara microphylla
Elegant, evergreen shrub or small tree. Has tiny, glossy, dark green leaves and small clusters of vanilla-scented, deep yellow flowers in late winter and early spring.

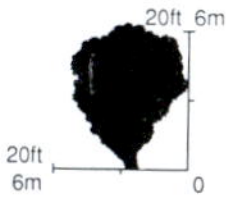

Hamamelis virginiana
(Virginian witch hazel)
Deciduous, open, upright shrub. Small, fragrant, spidery, yellow flowers with 4 narrow petals open in autumn as leaves fall. Broadly oval leaves turn yellow in autumn.

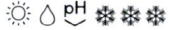

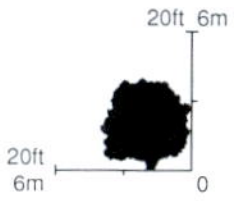

***Hamamelis × intermedia* 'Diane'**
Deciduous, open, spreading shrub that produces fragrant, spidery, deep red flowers on bare branches from mid- to late winter. Broadly oval, mid-green leaves turn yellow and red in autumn.

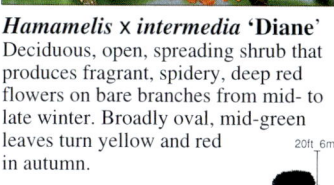

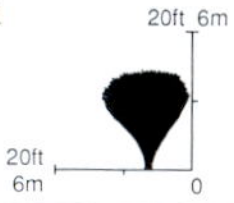

Cyphomandra crassicaulis, syn. *C. betacea* (Tree tomato)
Evergreen, sparingly branched shrub or small tree, upright when young, with large, heart-shaped, rich green leaves. Has edible, tomato-like, red fruits from summer to winter. Min. 10°C (50°F).

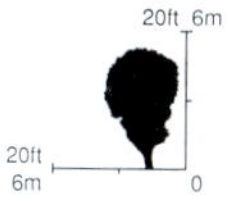

***Corylus avellana* 'Contorta'**
Deciduous, bushy shrub with curiously twisted shoots and broad, sharply toothed, mid-green leaves. In late winter, bare branches are covered with pendent, pale yellow catkins.

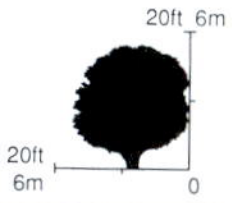

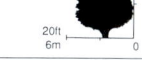

Brugmansia aurea
Cotoneaster 'Rothschildianus'
HOLLIES, pp.72–3
Tecoma stans, p.70

OTHER RECOMMENDED PLANTS:
Euphorbia pulcherrima, p.120
Euphorbia pulcherrima 'Paul Mikkelson'
HEATHERS, pp.148–9

Cornus mas
Hamamelis japonica
Hamamelis mollis 'Pallida'
Mahonia acanthifolia

Shrubs/large WINTER INTEREST

■ YELLOW

Hamamelis × intermedia 'Arnold Promise'
Deciduous, open, spreading shrub. Large, fragrant, spidery, yellow flowers with 4 narrow, crimped petals appear from mid- to late winter. Broadly oval, green leaves turn yellow in autumn.

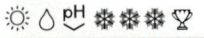

Hamamelis japonica 'Sulphurea'
Deciduous, upright, open shrub. In mid-winter, fragrant, spidery, pale yellow flowers with 4 narrow, crimped petals are borne on leafless branches. Broadly oval, dark green leaves turn yellow in autumn.

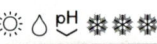

Hamamelis mollis 'Coombe Wood'
Deciduous, open, spreading shrub. From mid- to late winter bears very fragrant, spidery, golden-yellow flowers with 4 narrow petals. Broadly oval, mid-green leaves turn yellow in autumn.

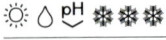

Acacia pravissima, p.71
Buddleja madagascariensis
Jasminum nudiflorum, p.121
Senecio grandifolius

Shrubs/large ALL YEAR INTEREST

■ WHITE–GREEN

Dracaena deremensis 'Warneckii', syn. *D. d.* 'Souvenir de Schriever', *D. fragrans* 'Warneckii'
Slow-growing, evergreen shrub. Erect to arching, lance-shaped leaves are banded grey-green and cream. Min. 15–18°C (59–64°F).

Ligustrum ovalifolium
Vigorous, evergreen or semi-evergreen, upright, dense shrub with glossy, mid-green leaves. Dense racemes of small, rather unpleasantly scented, tubular, white flowers appear in mid-summer, followed by black fruits.

Prunus lusitanica 'Variegata'
Slow-growing, evergreen, bushy shrub with reddish-purple shoots. Has oval, glossy, dark green, white-edged leaves. Fragrant, shallowly cup-shaped, creamy-white flowers in summer are followed by purple fruits.

OTHER RECOMMENDED PLANTS:
Cordyline fruticosa 'Imperialis'
Dracaena marginata 'Tricolor', p.73
Elaeagnus × ebbingei

Prunus lusitanica subsp. **azorica**
Evergreen, bushy shrub with reddish-purple shoots and bright green leaves, red when young. Bears spikes of small, fragrant, white flowers in summer, followed by purple fruits.

Tetrapanax papyrifer, syn. *Fatsia papyrifera* (Rice-paper plant)
Evergreen, upright, suckering shrub. Long-stalked, circular leaves are deeply lobed. Has bold sprays of small, creamy-white flowers in summer and black berries in autumn-winter.

Euonymus japonicus 'Macrophyllus'
Garrya elliptica 'James Roof'
Grevillea banksii, p.73
Griselinia littoralis 'Dixon's Cream'

Griselinia littoralis 'Variegata'
Evergreen, upright shrub of dense, bushy habit. Leathery leaves are grey-green, marked with bright green and creamy-white. Bears inconspicuous, yellow-green flowers in late spring.

Pittosporum 'Garnettii'
Evergreen, columnar or conical shrub of dense, bushy habit. Rounded, grey-green leaves, irregularly edged with creamy-white, become tinged with deep pink in cold areas. May bear small, greenish-purple flowers in spring-summer.

Polyscias guilfoylei 'Victoriae'
Slow-growing, evergreen, rounded shrub or small tree with leaves that are divided into several oval to rounded, serrated, white-margined, deep green leaflets. Min. 15–18°C (59–64°F).

Griselinia lucida
HOLLIES, pp.72–3
Pittosporum crassifolium 'Variegatum', p.73
Synadenium grantii 'Rubrum'

GREEN–YELLOW

Schefflera elegantissima, syn. *Aralia elegantissima* (False aralia)
Evergreen, upright, open shrub. Large leaves have 7–10 coarsely toothed, lustrous, grey-green, sometimes bronze-tinted, leaflets.
Min. 13°C (55°F).

Ligustrum lucidum 'Excelsum Superbum'
Evergreen, upright shrub or small tree. Large, glossy, bright green leaves are marked with pale green and yellow-edged. Small, tubular, white flowers open in late summer and early autumn.

Shrubs/medium SPRING INTEREST
WHITE

Pieris floribunda (Fetterbush, Mountain fetterbush)
Evergreen, bushy, dense, leafy shrub with oval, glossy, dark green leaves. Greenish-white flower buds appear in winter, opening to urn-shaped, white blooms from early to mid-spring.

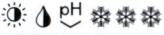

Enkianthus perulatus
Deciduous, bushy, dense shrub. Dark green leaves turn bright red in autumn. A profusion of small, pendent, urn-shaped, white flowers is borne in mid-spring.

Brachyglottis repanda (Pukapuka, Rangiora)
Evergreen, bushy shrub or tree, upright when young, with robust, downy, white stems. Has veined leaves, white beneath, and fragrant, white flower heads in summer.
Min. 3°C (37°F).

Osmanthus heterophyllus 'Aureomarginatus'
Evergreen, upright shrub. Sharply toothed, holly-like, glossy, bright green leaves have yellow margins. Small, fragrant, white flowers are produced in autumn.

Fothergilla major, syn. *F. monticola*
Deciduous, upright shrub with glossy, dark green leaves, slightly bluish-white beneath, that turn red, orange and yellow in autumn. Tufts of fragrant, white flowers appear in late spring.

Pieris japonica 'Scarlett O'Hara'
Evergreen, rounded, bushy, dense shrub. Young foliage and shoots are bronze-red, leaves becoming glossy, dark green. Produces sprays of white flowers in spring.

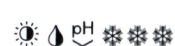

Pittosporum tenuifolium
Evergreen, columnar, later rounded shrub or small tree with purple shoots and wavy-edged, oval, glossy, mid-green leaves. Bears honey-scented, purple flowers in late spring.

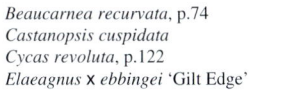

Elaeagnus pungens 'Maculata'
Evergreen, bushy, slightly spiny shrub. Glossy, dark green leaves are marked with a large, central, deep yellow patch. Very fragrant, urn-shaped, creamy-white flowers open from mid- to late autumn.

Choisya ternata (Mexican orange blossom)
Evergreen, rounded, dense shrub with aromatic, glossy, bright green leaves composed of 3 leaflets. Clusters of fragrant, white blooms open in late spring and often occur again in autumn.

Beaucarnea recurvata, p.74
Castanopsis cuspidata
Cycas revoluta, p.122
Elaeagnus x *ebbingei* 'Gilt Edge'

Euonymus japonicus 'Ovatus Aureus'
Pittosporum tenuifolium 'James Stirling'
Schefflera arboricola
Schefflera digitata

OTHER RECOMMENDED PLANTS:
Buddleja asiatica
CAMELLIAS, pp.98–9
Gardenia thunbergia

Osmanthus x *burkwoodii*, p.86
Pieris taiwanensis
RHODODENDRONS, pp.102–104
Viburnum x *burkwoodii* 'Anne Russell'

Camellias

These evergreen shrubs have long been valued for their luxuriant, rich green foliage and masses of showy flowers, borne mainly in winter and spring. Once thought suitable only for glasshouses, many camellias are frost hardy outdoors if grown in sheltered positions, although blooms may suffer frost and rain damage. Camellias require lime-free soil. The main flower forms are illustrated below.

Single – shallowly cup-shaped flowers each have not more than 8 petals, arranged in a single row, and a conspicuous, central boss of stamens.

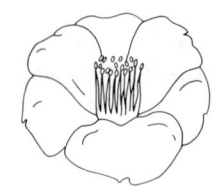

Semi-double – cup-shaped flowers each have 2 or more rows of 9–21 regular or irregular petals and conspicuous stamens.

Anemone – rounded flowers each have one or more rows of large, outer petals lying flat or undulating; the domed centre has a mass of intermingled petaloids and stamens.

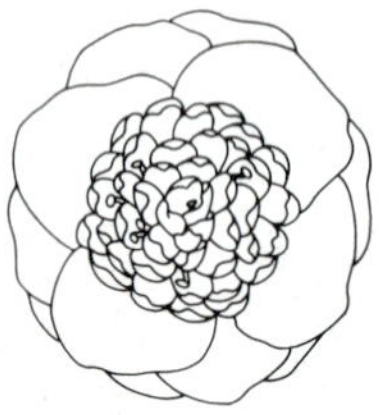

Peony-form – rounded, domed flowers have usually irregular petals intermingled with petaloids and stamens.

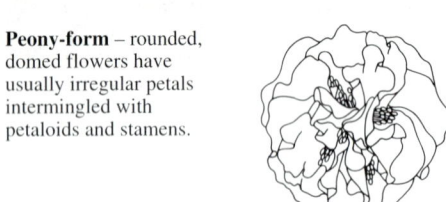

Rose-form – cup-shaped flowers each have several rows of overlapping petals and open to reveal stamens in the centre.

Formal double – rounded flowers have many rows of regular, neatly overlapping petals that obscure stamens. **Irregular double** forms are similar but often have more loosely arranged, sometimes irregular, petals.

C. japonica 'Lady Vansittart' (semi-double)

C. sasanqua 'Narumigata' (single)

C. 'Shiro-wabisuke' (single)

C. x *williamsii* 'Francis Hanger' (single)

C. japonica 'Mrs D.W. Davis' (semi-double)

C. 'Cornish Snow' (single)

C. jap. 'Tomorrow's Dawn' (irreg. double)

C. japonica 'Alba Simplex' (single)

C. japonica 'Berenice Boddy' (semi-double)

C. japonica 'Janet Waterhouse' (semi-dbl.)

C. tsaii (single)

C. japonica 'Ave Maria' (peony)

C. x *williamsii* 'J.C. Williams' (single)

C. japonica 'Lavinia Maggi' (formal dbl.)

C. x *williamsii* 'Clarrie Fawcett' (semi-double)

C. x *williamsii* 'Mary Larcom' (single)

C. x *williamsii* 'Donation' (semi-double)

C. japonica 'Margaret Davis' (irregular double)

C. x *w.* 'E.G. Waterhouse' (formal double)

C. saluenensis (single)

C. j. 'Betty Sheffield Supreme' (irreg. double)

C. × *williamsii* 'Bow Bells' (single)

C. japonica 'Elegans' (anemone) ♀

C. japonica 'Jupiter' (single) ♀

C. 'Anticipation' (peony) ♀

C. japonica 'Adolphe Audusson' (semi-double) ♀

C. × *williamsii* 'Brigadoon' (semi-double) ♀

C. × *williamsii* 'Water Lily' (double) ♀

C. japonica 'Rubescens Major' (formal double) ♀

C. japonica 'Alexander Hunter' (single) ♀

C. reticulata 'Mudan Cha' (variable)

C. reticulata 'Zaotaohong' (variable)

C. 'Dr Clifford Parks' (variable) ♀

C. japonica 'Althaeiflora' (peony)

C. × *williamsii* 'St Ewe' (single) ♀

C. × *williamsii* 'Golden Spangles' (single)

C. reticulata 'Zaomudan' (irregular double)

C. japonica 'Guilio Nuccio' (semi-double) ♀

C. reticulata 'Captain Rawes' (semi-double) ♀

C. 'Inspiration' (semi-double) ♀

C. japonica 'Elegans Supreme' (anemone)

C. japonica 'Jessie Burgess' (semi-double)

C. × *williamsii* 'Caerhays' (variable)

C. japonica 'R.L. Wheeler' (variable) ♀

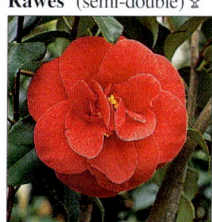

C. japonica 'Apollo' (semi-double)

C. × *williamsii* 'Mary Christian' (single) ♀

C. 'Innovation' (peony)

C. japonica 'Gloire de Nantes' (semi-double) ♀

C. japonica 'Julia Drayton' (variable)

C. × *williamsii* 'George Blandford' (semi-double) ♀

C. 'Leonard Messel' (semi-double) ♀

C. 'William Hertrich' (semi-double)

C. reticulata 'Houye Diechi' (semi-double)

C. japonica 'Mathotiana' (formal double)

Shrubs/medium SPRING INTEREST

WHITE–PINK

Aronia arbutifolia
(Red chokeberry)
Deciduous shrub, upright when young, later arching. Clusters of small, white flowers with red anthers appear in late spring, followed by red berries. Dark green foliage turns red in autumn.

Prunus mume 'Omoi-no-mama'
Deciduous, spreading shrub with fragrant, semi-double, occasionally single, pink-flushed, white flowers wreathing young growths in early spring before oval, toothed leaves appear.

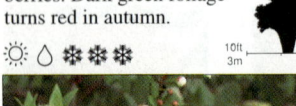

Myrtus communis
(Common myrtle)
Evergreen, bushy shrub with aromatic, glossy, dark green foliage. Fragrant, white flowers are borne from mid-spring to early summer, followed by purple-black berries.

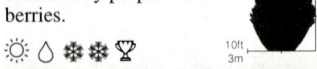

Viburnum plicatum 'Pink Beauty'
Deciduous, bushy shrub. Dark green leaves become reddish-purple in autumn. In late spring and early summer bears white, later pink, blooms, followed by red, then black, fruits.

Malus sieboldii
Deciduous, spreading shrub with arching branches. Bears white or pale to deep pink flowers in mid-spring followed by small, red or yellow fruits. Dark green leaves, often lobed, turn red or yellow in autumn.

Chaenomeles speciosa 'Moerloosii'
Vigorous, deciduous, bushy shrub. Has glossy, dark green leaves and pink-flushed, white flowers in early spring, followed by greenish-yellow fruits.

PINK–RED

Cotoneaster divaricatus
Deciduous, bushy, spreading shrub. Leaves are glossy, dark green, turning red in autumn. Shallowly cup-shaped, pink-flushed, white flowers in late spring and early summer are followed by deep red fruits.

Acer palmatum 'Corallinum'
Very slow-growing, deciduous, bushy-headed shrub or small tree. Lobed, bright reddish-pink, young foliage becomes mid-green, then brilliant red, orange or yellow in autumn. Reddish-purple flowers appear in mid-spring.

Ribes sanguineum 'Pulborough Scarlet'
Deciduous, upright shrub that in spring bears pendent, tubular, deep red flowers amid aromatic, dark green leaves, with 3–5 lobes, sometimes followed by black fruits with a white bloom.

Prunus mume 'Beni-shidori', syn. *P.m.* 'Beni-shidon'
Deciduous, spreading shrub with fragrant, single, carmine flowers in early spring before pointed, dark green leaves appear.

Banksia coccinea
Evergreen, dense shrub with toothed, dark green leaves, grey-green beneath. Flower heads comprising clusters of bright red flowers with prominent styles and stigmas are borne in late winter and spring. Min. 10°C (50°F).

Daphne × *burkwoodii* 'Variegata'
MAGNOLIAS, p.49
RHODODENDRONS, p.102–104
Skimmia japonica, p.145

Spirea 'Arguta'
Spirea prunifolia
Viburnum × *burkwoodii* 'Park Farm Hybrid'
Viburnum carlesii 'Diana'

Callistemon speciosus
CAMELLIAS, pp.98–9
Chaenomeles × *superba* 'Etna'
Cytisus 'Windlesham Ruby'

HEATHERS, pp.148–9
Pieris japonica 'Daisen'
Prunus tenella 'Fire Hill'
RHODODENDRONS, pp.102–104

■■ RED–PURPLE

Enkianthus cernuus f. rubens
Deciduous, bushy shrub with dense clusters of dull green leaves that turn deep reddish-purple in autumn. Small, bell-shaped, deep red flowers appear in late spring.

Greyia sutherlandii
Deciduous or semi-evergreen, rounded shrub. Coarsely serrated, leathery leaves turn red in autumn. Spikes of small, bright red flowers appear in spring with new foliage.
Min. 7–10°C (45–50°F).

Leucospermum reflexum
Evergreen, erect shrub with ascending branchlets. Has small, blue-grey or grey-green leaves. Slender, tubular, crimson flowers with long styles are carried in tight, rounded heads in spring-summer.
Min. 10°C (50°F).

Leptospermum scoparium 'Red Damask'
Evergreen, upright, bushy shrub. Narrow, aromatic, dark green leaves set off sprays of double, dark red flowers in late spring and summer.

Telopea truncata
(Tasmanian waratah)
Evergreen, upright shrub, bushy with age. Has deep green leaves and dense, rounded heads of small, tubular, crimson flowers in late spring and summer.

Berberis thunbergii f. atropurpurea
Deciduous, arching, dense shrub. Reddish-purple foliage turns bright red in autumn. Globose to cup-shaped, red-tinged, pale yellow flowers in mid-spring are followed by red fruits.

Ceanothus 'Delight'
Ceanothus 'Italian Skies'
Ceanothus papillosus
Cestrum 'Newellii'

Daphne genkwa
LILACS, p.90
RHODODENDRONS, pp.102–104
Rosmarinus officinalis, p.137

☐ YELLOW

Corylopsis pauciflora
Deciduous, bushy, dense shrub. Oval, bright green leaves, bronze when young, have bristle-like teeth. Bears fragrant, tubular to bell-shaped, pale yellow flowers from early to mid-spring.

Kerria japonica var. japonica,
syn. *K. j.* 'Pleniflora'
Vigorous, deciduous, graceful shrub. Double, golden-yellow flowers are borne along green shoots from mid- to late-spring. Leaves are narrowly oval, sharply toothed and bright green.

Berberis gagnepainii
Evergreen, bushy, dense shrub. Massed, globose to cup-shaped, yellow flowers appear among long, narrow, pointed, dark green leaves in late spring followed in autumn by blue-bloomed, black berries.

Lindera benzoin
(Benjamin, Spice bush)
Deciduous, bushy shrub with aromatic, bright green leaves that turn yellow in autumn. Tiny, greenish-yellow flowers in mid-spring are followed by red berries on female plants.

Forsythia suspensa
Deciduous, arching, graceful shrub with slender shoots. Nodding, narrowly trumpet-shaped, bright yellow flowers open from early to mid-spring, before mid-green leaves appear.

Berberis aggregata
Berberis sargentiana
Corokia x virgata
Corylopsis spicata

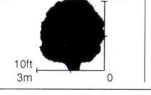

Forsythia x intermedia 'Spring Glory'
Lonicera morrowii
RHODODENDRONS, pp.102–104

Rhododendrons and Azaleas continued

R. 'Narcissiflorum' (azalea) ♀

R. 'Freya' (azalea)

R. 'Hawk Crest' (rhododendron)

R. luteum (azalea) ♀

R. 'Medway' (azalea)

R. 'Curlew' (rhododendron) ♀

R. 'George Reynolds' (azalea)

R. 'Fabia' (rhododendron) ♀

R. 'Yellow Hammer' (rhododendron) ♀

R. 'Glory of Littleworth' (azalea x rhodo.)

R. 'Frome' (azalea)

R. 'Moonshine Crescent' (rhododendron)

R. macabeanum (rhododendron) ♀

R. 'Gloria Mundi' (azalea)

Shrubs/medium SPRING INTEREST

☐ YELLOW

Berberis verruculosa
Slow-growing, evergreen, bushy shrub. Has glossy, dark green leaves with blue-white undersides. Clusters of small, cup-shaped, bright yellow flowers in late spring and early summer are followed by blue-black fruits.

Berberis julianae
Evergreen, bushy, dense shrub. Has glossy, dark green leaves, yellow flowers in late spring and early summer and egg-shaped, blue-black fruits in autumn.

Forsythia x intermedia **'Spectabilis'**
Vigorous, deciduous, spreading shrub with stout growths. A profusion of large, deep yellow flowers is borne from early to mid-spring before sharply toothed, dark green leaves appear.

Hibbertia cuneiformis, p.116
Kerria japonica var. *japonica*, p.101
Mahonia napaulensis
Mahonia 'Undulata'

Ochna serrulata
Ribes odoratum

104

Shrubs/medium SUMMER INTEREST

☐ YELLOW–ORANGE
☐ WHITE

Azara serrata
Evergreen, upright shrub with glossy, bright green foliage and rounded bunches of fragrant, yellow flowers in late spring or early summer.

Berberis x *stenophylla*
Evergreen, arching shrub with slender shoots and narrow, spine-tipped, deep green leaves, blue-grey beneath. Massed, golden-yellow flowers appear from mid- to late spring followed by small, blue-black fruits.

Carissa macrocarpa 'Tuttlei', syn. *C. grandiflora* 'Tuttlei'
Evergreen, compact and spreading shrub with thorny stems and leathery leaves. Has fragrant flowers in spring-summer and edible, plum-like, red fruits in autumn. Min. 13°C (55°F).

Philadelphus 'Beauclerk'
Deciduous, slightly arching shrub. Large, fragrant flowers, white with a small, central, pale purple blotch, are produced from early to mid-summer. Leaves are dark green.

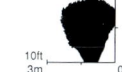

Forsythia x *intermedia* 'Beatrix Farrand'
Vigorous, deciduous, bushy, arching shrub with stout shoots. A profusion of large, deep yellow flowers appears from early to mid-spring before oval, coarsely toothed, mid-green leaves emerge.

Berberis x *lologensis* 'Stapehill'
Vigorous, evergreen, arching shrub. Glossy, dark green foliage sets off profuse racemes of globose to cup-shaped, orange flowers from mid- to late spring.

Acacia podalyriifolia (Mount Morgan wattle, Queensland silver wattle)
Evergreen, arching shrub that produces racemes of bright yellow flowers in spring. Has blue-green phyllodes (flattened, leaf-like stalks).

Berberis linearifolia 'Orange King'
Evergreen, upright, stiff-branched shrub with narrow, rigid, dark green leaves. Bears large, globose to cup-shaped, deep orange flowers in late spring.

Exochorda x *macrantha* 'The Bride'
Deciduous, arching, dense shrub that forms a mound of pendent branches. Large, white flowers are produced in abundance amid dark green foliage in late spring and early summer.

Berberis x *lologensis*
Cytisus 'Firefly'
Cytisus scoparius f. *andreanus*, p.141
Forsythia x *intermedia* 'Arnold Giant'

Forsythia x *intermedia* 'Lynwood'
Forsythia ovata
RHODODENDRONS, pp.102–104

OTHER RECOMMENDED PLANTS:
Acradenia frankliniae
Aronia x *prunifolia*
Brugmansia arborea

Buddleja fallowiana var. *alba*
Cistus x *cyprius*, p.129
Clethra alnifolia
Cleyera japonica

Shrubs/medium SUMMER INTEREST
☐ WHITE

Deutzia scabra
Deciduous, upright shrub with narrowly oval, dark green leaves that from early to mid-summer set off dense, upright clusters of 5-petalled, white blooms.

Philadelphus 'Belle Etoile'
Deciduous, arching shrub. Very fragrant, white flowers each with a pale purple mark at the base are borne profusely among mid-green foliage in late spring and early summer.

Pyracantha × watereri
Evergreen, upright, dense, spiny shrub with glossy, dark green foliage. Shallowly cup-shaped, white flowers in early summer are succeeded by bright red berries in autumn.

Spiraea canescens
Deciduous shrub with upright shoots arching at the top. Small heads of white flowers are borne in profusion amid narrowly oval, grey-green leaves from early to mid-summer.

Deutzia × magnifica 'Staphyleoides'
Vigorous, deciduous, upright shrub. Large, 5-petalled, pure white blooms, borne in dense clusters in early summer, have recurved petals. Leaves are bright green.

Aronia melanocarpa (Black chokeberry)
Deciduous, bushy shrub. White flowers appear in late spring and early summer, followed by black fruits. Has glossy, dark green leaves that turn red in autumn.

Fallugia paradoxa (Apache plume)
Deciduous, bushy shrub that bears white flowers in mid-summer, followed by silky, pink- and red-tinged, green fruits. Dark green leaves are finely cut and feathery.

Rubus 'Benenden', syn. *R.* 'Tridel'
Deciduous, arching, thornless shrub with peeling bark. Large, rose-like, pure white flowers are borne among lobed, deep green leaves in late spring and early summer.

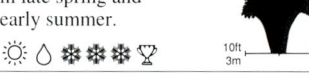

Olearia nummulariifolia
Evergreen, rounded shrub with stiff, upright shoots densely covered with small, very thick, mid- to dark green leaves. Small, fragrant, white flowers appear in mid-summer.

Cotoneaster 'Firebird'
Cotoneaster prostratus
Cotoneaster serotinus
Deutzia 'Joconde'

Deutzia × *magnifica*
Deutzia setchuenensis
FUCHSIAS, pp.134–5
Gardenia augusta 'Fortuniana', p.128

Hakea suaveolens
HEATHERS, pp.148–9
Hebe brachysiphon
Hebe salicifolia

HYDRANGEAS, p.114
Ligustrum japonicum
Ligustrum vulgare
Lyonia ligustrina

☐ WHITE

Sorbaria sorbifolia,
syn. *Spiraea sorbifolia*
Deciduous, upright shrub that forms thickets by suckering. Mid-green leaves consist of many sharply toothed leaflets. Large panicles of small, white flowers appear in summer.

Yucca gloriosa (Spanish dagger)
Evergreen shrub with a stout stem crowned with a tuft of long, pointed, deep green leaves, blue-green when young. Bears very long panicles of bell-shaped, white flowers in summer-autumn.

Philadelphus 'Boule d'Argent'
Deciduous, bushy, arching shrub with dark green foliage that sets off clusters of slightly fragrant, semi-double to double, pure white flowers from early to mid-summer.

Philadelphus 'Dame Blanche'
Deciduous, bushy, compact shrub with dark, peeling bark. Dark green foliage sets off slightly fragrant, semi-double to loosely double, pure white flowers borne in profusion from early to mid-summer.

Prinsepia uniflora
Deciduous, arching, spiny shrub. From late spring to summer bears small, fragrant, white flowers amid narrow, glossy, dark green leaves followed by cherry-like, deep red fruits. Grows best in hot sun.

Eucryphia milliganii
Evergreen, upright, narrow shrub. Has tiny, dark green leaves, bluish-white beneath, and small, white flowers, borne in mid-summer.

Escallonia virgata
Deciduous, spreading, graceful shrub with arching shoots and small, glossy, dark green leaves. Bears racemes of small, open cup-shaped, white flowers from early to mid-summer.

Lyonia ovalifolia
Myrtus communis, p.100
Myrtus communis var. *tarentina*
Nandina domestica

Olearia avicenniifolia
Olearia ilicifolia
Olearia phlogopappa var. *subrepanda*, p.128
Olearia 'Talbot de Malahide'

Philadelphus delavayi
Philadelphus 'Virginal'
Physocarpus opulifolius
Pyracantha angustifolia

Rhaphiolepis indica
RHODODENDRONS, pp.102–104
Rhodotypos scandens, p.128
Sorbaria aitchisonii

Shrubs/medium SUMMER INTEREST
☐ WHITE

Osteomeles schweriniae
Evergreen, arching shrub with long, slender shoots. Leaves, consisting of many small leaflets, are dark green. Clusters of small, white flowers in early summer are followed by red, later blue-black, fruits.

Styrax wilsonii
Deciduous, bushy shrub with slender shoots that produce an abundance of yellow-centred, white flowers in early summer. Leaves are small and deep green.

Ceanothus incanus
Evergreen, bushy shrub. Has spreading, spiny shoots, broad, grey-green leaves and large racemes of white flowers in late spring and early summer.

Philadelphus 'Lemoinei'
Deciduous, upright, slightly arching shrub that produces profuse racemes of small, extremely fragrant, white flowers from early to mid-summer.

Ozothamnus rosmarinifolius, syn. *Helichrysum rosmarinifolium*
Evergreen, upright, dense shrub with woolly, white shoots and narrow, dark green leaves. Clusters of fragrant, white flower heads open in early summer.

Zenobia pulverulenta
Deciduous or semi-evergreen, slightly arching shrub, often with bluish-white-bloomed shoots. Glossy leaves have a bluish-white reverse when young. Bears fragrant, bell-shaped, white flowers from early to mid-summer.

Carpenteria californica
Evergreen, bushy shrub. Glossy, dark green foliage sets off fragrant, yellow-centred, white flowers borne during summer.

Symplocos paniculata (Sapphire berry)
Deciduous, bushy shrub or small tree. Panicles of small, fragrant, white flowers in late spring and early summer are followed by small, metallic-blue berries. Has dark green leaves.

Clethra barbinervis
Deciduous, upright shrub with peeling bark. Has oval, toothed, dark green leaves that turn red and yellow in autumn. Racemes of fragrant, white flowers are borne in late summer and early autumn.

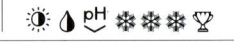

Cornus alba 'Elegantissima'
Vigorous, deciduous shrub. Young shoots are bright red in winter. Has white-edged, grey-green leaves and small, creamy-white flowers in late spring and early summer followed by white fruits.

Spiraea nipponica
Spiraea 'Snow White'
Spiraea veitchii
Stachyurus chinensis 'Magpie'
Viburnum dilatatum
Viburnum plicatum
Viburnum plicatum 'Nanum Semperflorens'
Weigela 'Candida'
Weigela florida 'Variegata', p.130
Weigela praecox 'Variegata'

108

☐ WHITE

Viburnum dilatatum 'Catskill'
Deciduous, low, spreading shrub with sharply toothed, dark green leaves that turn yellow, orange and red in autumn. Flat heads of creamy-white flowers in late spring and early summer are followed by bright red fruits.

Olearia x haastii
Evergreen, bushy, dense shrub, good for hedging. Has small, oval, glossy, dark green leaves and is covered with heads of fragrant, daisy-like, white flowers from mid- to late summer.

Philadelphus coronarius 'Variegatus'
Deciduous, bushy shrub with racemes of very fragrant, creamy-white flowers in late spring and early summer and mid-green leaves broadly edged with white.

Spiraea nipponica 'Snowmound', syn. *S.n.* var. *tosaensis* of gardens
Deciduous, spreading shrub with stout, arching, reddish branches. Small, narrow, dark green leaves set off profuse, dense clusters of small, white flowers in early summer.

Eriogonum giganteum (St Catherine's lace)
Evergreen, rounded shrub with oblong to oval, woolly, white leaves. Small, white flowers are carried in branching clusters to 30cm (12in) or more wide in summer. Min. 5°C (41°F).

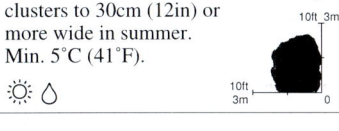

Viburnum 'Pragense', syn. *V.* x *pragense*
Evergreen, rounded, bushy shrub that has dark green foliage and domed heads of white flowers opening from pink buds in late spring and early summer.

Leptospermum polygalifolium, syn. *L. flavescens*
Evergreen, arching, graceful shrub with small, glossy, bright green leaves. Bears an abundance of small, pink-tinged, white flowers in mid-summer.

Philadelphus delavayi f. **melanocalyx**
Deciduous, upright shrub, grown for its extremely fragrant flowers, with pure white petals and deep purple sepals, opening from early to mid-summer. Leaves are dark green.

Hibiscus syriacus 'Red Heart'
Deciduous, upright shrub that bears large, white flowers with conspicuous red centres from late summer to mid-autumn. Oval leaves are lobed and deep green.

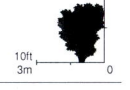

Shrubs/medium SUMMER INTEREST

WHITE–PINK

Lonicera xylosteum
(Fly honeysuckle)
Deciduous, upright, bushy, dense shrub. Creamy-white flowers are produced amid grey-green leaves in late spring and early summer and are followed by red berries.

***Acer palmatum* 'Butterfly'**
Slow-growing, deciduous, mounded shrub or small tree with lobed, grey-green leaves edged with cream and pink. In mid-spring bears small, reddish-purple flowers.

PINK

***Deutzia longifolia* 'Veitchii'**
Deciduous, arching shrub with narrow, pointed leaves and large clusters of 5-petalled, deep pink flowers from early to mid-summer.

Protea neriifolia
Evergreen, bushy, upright shrub with narrow leaves. Flower heads, about 13cm (5in) long, are red, pink or white, the bracts tipped with tufts of black hair, and appear in spring-summer. Min. 5–7°C (41–5°F).

Lonicera tatarica
Deciduous, bushy shrub. Tubular to trumpet-shaped, 5-lobed, white, pink or red flowers cover dark green foliage in late spring and early summer and are succeeded by red berries.

Stephanandra tanakae
Deciduous, arching shrub with orange-brown shoots and sharply toothed, mid-green leaves that turn orange and yellow in autumn. Small, yellow-green buds open to white flowers from early to mid-summer.

***Escallonia* 'Donard Seedling'**
Vigorous, evergreen, arching shrub with small, glossy, dark green leaves. Masses of pink flower buds open to white blooms, flushed with pale pink, from early to mid-summer.

Neillia thibetica
Deciduous, arching shrub. Slender spikes of rose-pink flowers are borne profusely in late spring and early summer. Leaves are sharply toothed.

Deutzia 'Magicien'
Deutzia pulchra
Deutzia scabra 'Plena'
Duvernoia adhatodoides

Justicia adhatoda
Melastoma candidum
Neillia sinensis
RHODODENDRONS, pp.102–104

Abelia x *grandiflora*, p.90
Ceanothus 'Marie Simon'
Deutzia x *elegantissima* 'Fasciculata'
Grevillea rosmarinifolia

HYDRANGEAS, p.114
Kolkwitzia amabilis
LILACS, p.90
Weigela florida

■ PINK

***Escallonia* 'Apple Blossom'**
Evergreen, bushy, dense shrub. From early to mid-summer apple-blossom-pink flowers are borne in profusion amid glossy, dark green leaves.

Robinia hispida (Rose acacia)
Deciduous shrub of loose habit with brittle, bristly stems that carry dark green leaves composed of 7–13 leaflets. Pendent racemes of deep rose-pink blooms open in late spring and early summer.

***Indigofera heterantha*,
syn. *I. gerardiana***
Deciduous, slightly arching shrub. Has greyish-green leaves consisting of many small leaflets and spikes of small, purplish-pink flowers from early summer to early autumn.

***Lavatera olbia* 'Rosea'**, syn. *L.* 'Rosea'
Semi-evergreen, erect shrub that produces abundant clusters of hollyhock-like, deep pink flowers throughout summer. Has lobed, sage-green leaves.

Kalmia latifolia (Calico bush)
Evergreen, bushy, dense shrub. In early summer large clusters of pink flowers open from distinctively crimped buds amid glossy, rich green foliage.

Medinilla magnifica
Evergreen, upright shrub, with sparingly produced, 4-angled, robust stems and boldly veined leaves. Pink to coral-red flowers hang in long trusses beneath large, pink bracts in spring-summer. Min. 16–18°C (61–4°F).

***Hibiscus syriacus* 'Woodbridge'**
Deciduous, upright shrub. From late summer to mid-autumn large, reddish-pink flowers, with deeper-coloured centres, appear amid lobed, dark green leaves.

***Hibiscus rosa-sinensis*
'The President'**
Evergreen, bushy shrub with toothed, oval, glossy, dark green leaves. In summer bears large, magenta-centred, bright pink flowers with prominent, yellow anthers. Min. 15°C (59°F).

Lavatera assurgentiflora
Semi-evergreen shrub with twisted, grey stems. Clusters of hollyhock-like, darkly veined, deep cerise blooms open in mid-summer. Palmate, mid-green leaves are white-haired beneath.

Ceanothus 'Perle Rose', p.133
Escallonia 'Edinensis'
FUCHSIAS, pp.134–5
HEATHERS, pp.148–9

HYDRANGEAS, p.114
Justicia carnea, p.133
Kalmia latifolia 'Ostbo Red'
LILACS, p.90

Podalyria calyptrata
RHODODENDRONS, pp.102–104
Richea scoparia
Robinia kelseyi

Rondeletia amoena
Rubus ulmifolius 'Bellidiflorus'
Spiraea x *billiardii*

111

Shrubs/medium SUMMER INTEREST

PINK–RED

Melaleuca elliptica
(Granite bottlebrush)
Evergreen, rounded shrub with long, leathery, usually greyish-green leaves. Flowers, consisting of a brush of red stamens, are borne in dense, terminal spikes in spring-summer.

Cestrum elegans
Vigorous, evergreen, arching shrub. Nodding shoots carry downy, deep green foliage. Dense racemes of tubular, purplish-red flowers in late spring and summer are followed by deep red fruits.

RED

Pieris formosa var. *forrestii* 'Wakehurst'
Evergreen, bushy, dense shrub. Young leaves are brilliant red in early summer, becoming pink, creamy-yellow and finally dark green. Bears urn-shaped, white flowers in spring-summer.

Lonicera ledebourii
Deciduous, bushy shrub. Red-tinged, orange-yellow flowers are borne amid dark green foliage in late spring and early summer and are followed by black berries. As these ripen, deep red bracts enlarge around them.

Calycanthus occidentalis
(California allspice)
Deciduous, bushy shrub. Leaves are large, aromatic and dark green. Fragrant, purplish-red flowers with many strap-shaped petals appear during summer.

Bauhinia galpinii,
syn. *B. punctata*
Semi-evergreen or evergreen, spreading shrub, occasionally semi-climbing. Has 2-lobed leaves and, in summer, fragrant, bright brick-red flowers. Min. 5°C (41°F).

Escallonia 'Langleyensis'
Evergreen or semi-evergreen, arching shrub with small, glossy, bright green leaves and an abundance of rose-pink flowers from early to mid-summer.

Crinodendron hookerianum,
syn. *Tricuspidaria lanceolata*
(Lantern tree)
Evergreen, stiff-branched shrub. In late spring and early summer, lantern-like, red flowers hang from shoots clothed with narrow, dark green leaves.

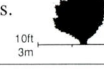

Telopea speciosissima (Waratah)
Evergreen, erect, fairly bushy shrub with coarsely serrated leaves. Has tubular, red flowers in dense, globose heads, surrounded by bright red bracts, in spring-summer.

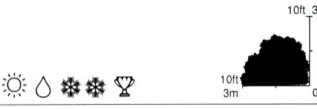

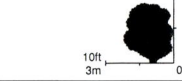

Escallonia 'Donard Beauty'
Hibiscus schizopetalus
Leptospermum scoparium 'Nicholsii'
Lonicera tatarica 'Hack's Red'
Pavonia hastata
Spiraea x *billiardii* 'Triumphans'
Spiraea japonica cvs, p.133
Sutherlandia frutescens, p.135
Abelia floribunda
Abutilon x *hybridum* 'Ashford Red'
Callistemon speciosus
Callistemon subulatus
Cestrum 'Newellii'
Correa 'Mannii'
Escallonia rubra var. *macrantha* 'Crimson Spire'

■ RED

Erythrina crista-galli
(Cockspur coral-tree)
Deciduous, mainly upright shrub or small tree. Leaves have 3 oval leaflets. Has leafy racemes of crimson flowers in summer-autumn. Dies back to ground level in winter in cold areas.

Erythrina × *bidwillii*
Deciduous, upright shrub with pale to mid-green leaves divided into 3 leaflets, up to 10cm (4in) long. Bright red flowers are carried in racemes in late summer or autumn.

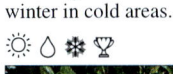

Desfontainia spinosa
Evergreen, bushy, dense shrub with spiny, holly-like, glossy, dark green leaves. Long, tubular, drooping, red flowers tipped with yellow are borne from mid-summer to late autumn.

Callistemon citrinus 'Splendens'
Evergreen, arching shrub with broad, lemon-scented, grey-green leaves that are bronze-red when young. In early summer, bright red flowers are borne in bottlebrush-like spikes.

Callistemon rigidus
Evergreen, bushy, slightly arching shrub with long, narrow, sharply pointed, dark green leaves and dense spikes of deep red flowers in late spring and early summer.

Rhus glabra (Smooth sumach)
Deciduous, bushy shrub with bluish-white-bloomed, reddish-purple stems. Deep blue-green leaves turn red in autumn. Bears panicles of greenish-red flower heads in summer followed by red fruits on female plants.

FUCHSIAS, pp.134–5
Grevillea 'Poorinda Constance'
Hibiscus rosa-sinensis
Kunzea baxteri

Macleania insignis
Phygelius capensis 'Coccineus'
Templetonia retusa
Weigela 'Bristol Ruby'

■■ RED–PURPLE

Feijoa sellowiana
(Pineapple guava)
Evergreen, bushy shrub or tree. Dark green leaves have white undersides. In mid-summer bears large, dark red flowers with white-edged petals, followed by edible, red-tinged, green fruits.

Aloysia triphylla, syn. *Lippia citriodora* (Lemon verbena)
Deciduous, bushy shrub. Leaves are pale green and lemon-scented. Racemes of tiny, lilac-tinged, white flowers appear in early summer.

Acer palmatum 'Bloodgood'
Deciduous, bushy-headed shrub or small tree with deep reddish-purple leaves that turn brilliant red in autumn. Small, reddish-purple flowers in mid-spring are often followed by decorative, winged, red fruits.

Buddleja crispa
Deciduous, upright, bushy shrub that from mid- to late summer bears racemes of small, fragrant, lilac flowers with white eyes. Has woolly, white shoots and oval, greyish-green leaves.

Acalypha wilkesiana
(Copperleaf, Jacob's coat)
Evergreen, bushy shrub. Oval, serrated leaves are 10cm (4in) or more long, rich copper-green, variably splashed with shades of red. Min. 16°C (61°F).

Hibiscus sinosyriacus 'Lilac Queen'
Deciduous, spreading, open shrub. From late summer to mid-autumn produces large, pale lilac flowers with red centres. Broad, lobed leaves are dark green.

Buddleja fallowiana
Buddleja 'Lochinch'
Calycanthus floridus
Carmichaelia australis

Chordospartium stevensonii
Dais cotinifolia
LILACS, p.90
RHODODENDRONS, pp.102–104

113

Hydrangeas

Valued for their late summer flowers, hydrangeas are versatile shrubs that thrive in a variety of situations. Larger-growing species, some of which may become tree-like with age, are best suited to light woodland, while there is a wide range of cultivars, mostly of *H. macrophylla*, that make excellent border plants. Some may also be grown in containers. Colours range from white through pink, red and purple to blue, but the truest blue is obtained only on acid soil. Lacecap hydrangeas have a central corymb of small, fertile flowers surrounded by showy, large, coloured bracts; mopheads (or hortensias) have domed heads of sterile bracts only.
H. paniculata cultivars bear larger though fewer cone-shaped flowerheads if pruned hard in spring. Hydrangea flower heads persist on the plants for many months, and may be dried for winter decoration indoors.

H. paniculata 'Floribunda' ♚

H. aspera subsp. *aspera* ♚

H. macrophylla 'Lanarth White' ♚

H. paniculata 'Unique' ♚

H. heteromalla 'Bretschneideri' ♚

H. involucrata 'Hortensis' ♚

H. macrophylla 'Lilacina'

H. quercifolia ♚

H. paniculata 'Praecox' ♚

H. paniculata 'Brussels Lace'

H. macrophylla 'Générale Vicomtesse de Vibraye' ♚

H. macrophylla 'Blue Wave' ♚

H. arborescens 'Annabelle' ♚

H. macrophylla 'Hamburg'

H. macrophylla 'Veitchii' ♚

H. serrata 'Bluebird'

H. arborescens 'Grandiflora' ♚

H. paniculata 'Pink Diamond'

H. macrophylla 'Altona' ♚

H. serrata

H. macrophylla 'Blue Bonnet'

Shrubs/medium SUMMER INTEREST

■ PURPLE ■■ PURPLE–BLUE ■ GREEN

Prostanthera ovalifolia
Evergreen, bushy, rounded shrub with tiny, sweetly aromatic, oval, thick-textured leaves. Cup-shaped, 2-lipped, purple flowers appear in short, leafy racemes in spring-summer. Min. 5°C (41°F).

Prostanthera rotundifolia
(Round-leaved mint-bush)
Evergreen, bushy, rounded shrub with tiny, sweetly aromatic, deep green leaves and short, leafy racemes of bell-shaped, lavender to purple-blue flowers in late spring or summer.

Fabiana imbricata f. *violacea*
Evergreen, upright shrub with shoots that are densely covered with tiny, heath-like, deep green leaves. Tubular, lilac flowers are borne profusely in early summer.

Eleutherococcus sieboldianus
Deciduous, bushy, elegant shrub. Has glossy, bright green leaves, divided into 5 leaflets, and is armed with spines. Clusters of small, greenish flowers appear in early summer.

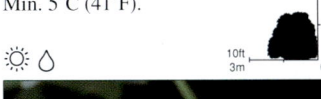

Melaleuca nesophila
(Western tea-myrtle)
Evergreen, bushy shrub or small tree with oval, grey-green leaves. Flowers, consisting of a brush of lavender to rose-pink stamens, are borne in rounded, terminal heads in summer.

Solanum rantonnetii 'Royal Robe'
Evergreen, loosely rounded shrub with smooth, bright green leaves. In summer has clusters of rich purple-blue flowers that open almost flat. Min. 7°C (45°F).

Hibiscus syriacus 'Blue Bird'
Deciduous, upright shrub that carries large, red-centred, lilac-blue flowers from late summer to mid-autumn. Has lobed, deep green leaves.

Zanthoxylum piperitum
(Japan pepper)
Deciduous, bushy, spiny shrub or small tree with aromatic, glossy, dark green leaves composed of many leaflets. Small, red fruits follow tiny, greenish-yellow, spring flowers.

Abutilon x *suntense* 'Violetta'
Fast-growing, deciduous, upright, arching shrub that carries an abundance of large, bowl-shaped, deep violet flowers in late spring and early summer. Vine-like leaves are sharply toothed and dark green.

Sophora davidii, syn. *S. viciifolia*
Deciduous, bushy shrub with arching shoots. Produces short racemes of small, pea-like, purple and white flowers in late spring and early summer. Grey-green leaves have many leaflets.

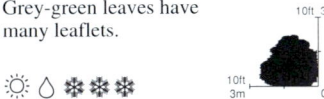

Ceanothus impressus
Evergreen, bushy shrub. Spreading growth is covered with small, crinkled, dark green leaves. Deep blue flowers appear in small clusters from mid-spring to early summer.

Ptelea trifoliata 'Aurea'
Deciduous, bushy, dense shrub or low tree. Leaves, consisting of 3 leaflets, are bright yellow when young, maturing to pale green. Bears racemes of greenish flowers in summer, followed by winged, green fruits.

Ceanothus x *veitchianus*
HYDRANGEAS, p.114
Teucrium fruticans 'Azureum'
Tibouchina urvilleana, p.92

Oplopanax horridus
Rhamnus imeretinus
Ricinus communis, p.287
Sambucus racemosa 'Plumosa'

Shrubs/medium SUMMER INTEREST

GREEN–YELLOW

Itea ilicifolia
Evergreen, bushy shrub with arching shoots and oval, sharply toothed, glossy, dark green leaves. Long, catkin-like racemes of small, greenish flowers appear in late summer and early autumn.

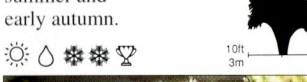

Cornus alba 'Spaethii'
Vigorous, deciduous shrub with bright red, young shoots in winter. Bright green leaves are yellow-edged. Bears small, creamy-white flowers in late spring and early summer, followed by rounded, white fruits.

Piptanthus nepalensis, syn. *P. laburnifolius*
Deciduous or semi-evergreen, open shrub with leaves consisting of 3 large, dark blue-green leaflets. Racemes of pea-like, bright yellow flowers appear in spring-summer.

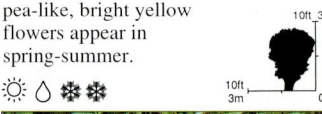

Hibbertia cuneiformis, syn. *Candollea cuneiformis*
Evergreen, upright, bushy shrub with small, oval leaves, serrated at tips. Has small clusters of bright yellow flowers, with spreading petals, in spring-summer. Min. 5–7°C (41–5°F).

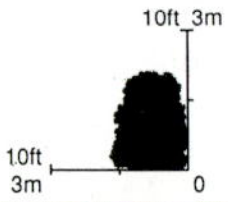

Callistemon pallidus
Evergreen, arching shrub. Grey-green foliage is pink-tinged when young and in early summer is covered with dense spikes of creamy-yellow flowers that resemble bottlebrushes.

Bupleurum fruticosum
(Shrubby hare's ear)
Evergreen, bushy shrub with slender shoots. From mid-summer to early autumn rounded heads of small, yellow flowers are borne amid glossy, dark bluish-green foliage.

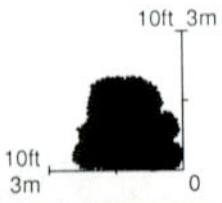

Physocarpus opulifolius 'Dart's Gold'
Deciduous, compact shrub with peeling bark and oval, lobed, golden-yellow leaves. Produces clusters of shallowly cup-shaped, white or pale pink flowers in late spring.

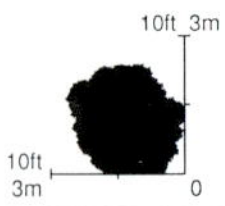

Colutea arborescens
(Bladder senna)
Fast-growing, deciduous, open shrub. Has pale green leaves with many leaflets, pea-like, yellow flowers throughout summer and bladder-like seed pods in late summer and autumn.

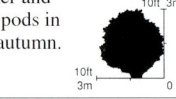

Jasminum humile
(Yellow jasmine)
Evergreen, bushy shrub that bears bright yellow flowers on long, slender, green shoots from early spring to late autumn. Leaves, with 5 or 7 leaflets, are bright green.

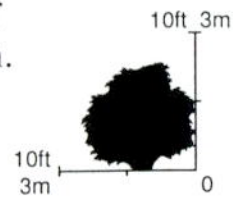

Callistemon sieberi
Cestrum parqui
Choisya ternata 'Sundance'
Cornus alba 'Gouchaultii'

Correa reflexa
Cytisus nigricans, p.140
Genista tenera 'Golden Showers'
Hypericum 'Rowallane'

Jasminum humile f. *wallichianum*
Ligustrum vulgare 'Aureum'
Lonicera morrowii
Medicago arborea

Potentilla 'Friedrichsenii', p.139
Sambucus racemosa 'Plumosa Aurea'
Vestia foetida
Weigela 'Looymansii Aurea'

Shrubs/medium AUTUMN INTEREST

YELLOW–ORANGE

Coluteα × mediα
Vigorous, deciduous, open shrub. Grey-green leaves have many leaflets. Racemes of yellow flowers, tinged with copper-orange, appear in summer, followed by bladder-like, papery, red-tinged seed pods.

Dendromecon rigidα
Vigorous, evergreen, upright shrub, best grown against a wall. Large, fragrant, golden-yellow flowers appear amid grey-green foliage from spring to autumn.

Cαssiα corymbosα
Vigorous, evergreen or semi-evergreen shrub. Leaves have 4–6 oval, bright green leaflets; sprays of bowl-shaped, rich yellow flowers appear in late summer. Min. 7°C (45°F).

Cαssiα didymobotryα
(Golden wonder)
Evergreen, rounded, sometimes spreading shrub with leaves of several leaflets; spikes of rich yellow flowers open from glossy, blackish-brown buds throughout the year. Min. 13°C (55°F).

Spαrtium junceum
(Spanish broom)
Deciduous, almost leafless, upright shrub that arches with age. Fragrant, pea-like, golden-yellow flowers appear from early summer to early autumn on dark green shoots.

Abutilon pictum 'Thompsonii'
Robust, evergreen, upright shrub with 3–5-lobed, serrated, rich green, heavily yellow-mottled leaves. Yellow-orange flowers with crimson veins are borne from summer to autumn. Min. 5–7°C (41–5°F).

Abutilon × *hybridum* 'Golden Fleece'
Banksia baxteri
Cassia × *floribunda*
Cestrum aurantiacum

Colutea orientalis
Juanulloa mexicana, p.141
Justicia spicigera, p.142
Ochna serrulata

WHITE–RED

Colletiα hystrix
Almost leafless, arching, stoutly branched shrub armed with rigid, grey-green spines. Pink flower buds open in late summer to fragrant, tubular, white blooms that last into autumn.

Cαlliαndrα hαemαtocephαlα
[pink form]
Evergreen, spreading shrub. Leaves have 16–24 narrowly oval leaflets. Flower heads consist of many pink-stamened florets from late autumn to spring. Min. 7°C (45°F).

Clerodendrum trichotomum
Deciduous, upright, bushy-headed, tree-like shrub. Clusters of deep pink and greenish-white buds open to fragrant, white flowers above large leaves from late summer to mid-autumn, followed by decorative, blue berries.

Euonymus hαmiltoniαnus subsp. **sieboldiαnus 'Red Elf'**
Deciduous, upright shrub with mid- to dark green foliage. Decorative, deep pink fruits, borne in profusion after tiny, green flowers in early summer, open in autumn to reveal red seeds.

Viburnum fαrreri,
syn. *V. fragrans*
Deciduous, upright shrub. In late autumn and during mild periods in winter and early spring bears fragrant, white or pale pink flowers. Dark green foliage is bronze when young.

Euonymus europαeus 'Red Cascade'
Deciduous, bushy shrub or small tree with narrowly oval, mid-green leaves that redden in autumn as red fruits open to show orange seeds. Has inconspicuous, greenish flowers in early summer.

OTHER RECOMMENDED PLANTS:
Berberis 'Rubrostilla', p.142
Brugmansia arborea
Euonymus cornutus var. *quinquecornutus*

Eupatorium ligustrinum
Symphoricarpos albus var. *laevigatus*
Turraea obtusifolia, p.142
Viburnum opulus 'Compactum', p.142

117

Shrubs/medium AUTUMN INTEREST

■ RED   ■ PURPLE–BLUE   ■ ORANGE

Viburnum betulifolium
Deciduous, upright, arching shrub. Bright green leaves are slightly glossy beneath. Heads of small, white flowers in early summer are succeeded by profuse nodding clusters of decorative, bright red fruits in autumn-winter.

Euonymus alatus (Winged spindle)
Deciduous, bushy, dense shrub with shoots that develop corky wings. Dark green leaves turn brilliant red in autumn. Inconspicuous, greenish flowers in summer are followed by small, purple and red fruits.

Callicarpa bodinieri var. **giraldii**
Deciduous, bushy shrub. Leaves are pale green, often bronze-tinged when young. Tiny, lilac flowers in mid-summer are followed by small, violet berries.

Berberis 'Barbarossa'
Semi-evergreen, arching shrub. Has narrowly oval, dark green leaves and racemes of rounded, yellow flowers in late spring and early summer, followed by globose, orange-scarlet fruits.

Nymania capensis
Evergreen, more or less rounded, rigidly branched shrub or small tree. In spring has flowers with upright, pink to rose-purple petals. Bears papery, inflated, red fruits in autumn. Min. 7–10°C (45–50°F).

Clerodendrum bungei
Evergreen or deciduous, upright, suckering shrub or sub-shrub with heart-shaped, coarsely serrated leaves. Has domed clusters of small, fragrant, red-purple to deep pink flowers in late summer and early autumn.

Zanthoxylum simulans
Deciduous, bushy shrub or small tree with stout spines. Aromatic, glossy, bright green leaves consist of 5 leaflets. Tiny, yellowish-green flowers in late spring and early summer are followed by orange-red fruits.

Euonymus latifolius
Deciduous, open shrub. Mid-green foliage turns brilliant red in late autumn. At the same time large, deep red fruits with prominent wings open to reveal orange seeds.

Cornus alba 'Kesselringii'
Vigorous, deciduous shrub with deep purplish stems. Dark green leaves become flushed reddish-purple in autumn. Creamy-white flowers in late spring and early summer are followed by white fruits.

Ceanothus 'Autumnal Blue'
Fast-growing, evergreen, bushy shrub. Has glossy, bright green foliage and large panicles of pale to mid-blue flowers from late spring to autumn.

Cotoneaster simonsii
Deciduous or semi-evergreen, upright shrub, suitable for hedging. Has oval, glossy, dark green leaves, shallowly cup-shaped, white flowers in early summer and long-lasting, orange-red fruits in autumn.

Berberis x carminea 'Pirate King'
Cotoneaster 'Firebird'
Cotoneaster hupehensis
Cotoneaster 'Hybridus Pendulus'

Disanthus cercidifolius
Grevillea 'Poorinda Constance'
Vaccinium corymbosum 'Pioneer', p.143

Callicarpa bodinieri
Elsholtzia stauntonii, p.143
HEATHERS, pp.148–9
Vitex agnus-castus

Corokia x virgata
Cotoneaster franchetii
Pyracantha angustifolia
Pyracantha rogersiana

Shrubs/medium WINTER INTEREST

ORANGE–YELLOW

☐ WHITE

Colquhounia coccinea
Evergreen or semi-evergreen, open shrub. Has aromatic, sage-green leaves and whorls of scarlet or orange flowers in late summer and autumn.

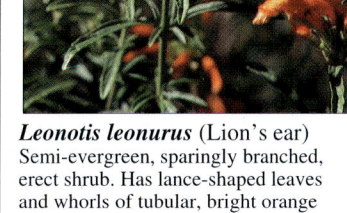

Cotoneaster sternianus
Evergreen or semi-evergreen, arching shrub. Leaves are grey-green, white beneath. Pink-tinged, white flowers in early summer are followed by orange-red fruits.

Rubus biflorus
Deciduous, upright shrub with chalky-white, young shoots in winter. Leaves, consisting of 5–7 oval leaflets, are dark green above, white beneath. White flowers in late spring and early summer are followed by edible, yellow fruits.

Viburnum tinus (Laurustinus)
Evergreen, bushy, dense shrub with oval, dark green leaves. Freely produced flat heads of small, white blooms open from pink buds during late winter and spring.

Leonotis leonurus (Lion's ear)
Semi-evergreen, sparingly branched, erect shrub. Has lance-shaped leaves and whorls of tubular, bright orange flowers in late autumn and early winter.

Rubus thibetanus
Deciduous, arching shrub with white-bloomed, brownish-purple, young shoots in winter and fern-like, glossy, dark green foliage, white beneath. Small, pink flowers from mid- to late summer are followed by black fruits.

Chamelaucium uncinatum [white form] (Geraldton waxflower)
Evergreen, wiry-stemmed, bushy shrub. Each needle-like leaf has a tiny, hooked tip. Flowers ranging from deep rose-purple to pink, lavender or white appear in late winter or spring. Min. 5°C (41°F).

Pyracantha 'Golden Charmer'
Evergreen, bushy, arching, spiny shrub with glossy, bright green leaves. Flattish clusters of white flowers in early summer are succeeded by large, bright orange berries in early autumn.

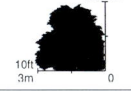

Pyracantha 'Golden Dome'
Evergreen, rounded, very dense, spiny shrub. Dark green foliage sets off white flowers borne in early summer. These are followed by orange-yellow berries in early autumn.

Viburnum foetens
Deciduous, bushy shrub that has aromatic, dark green leaves. Dense clusters of pink buds open to very fragrant, white flowers from mid-winter to early spring.

Calliandra haematocephala [white form]
Evergreen, spreading shrub. Leaves have 16–24 leaflets. Flower heads comprising many white-stamened florets appear from late autumn to spring. Min. 7°C (45°F).

Chaenomeles speciosa
HEATHERS, pp.148–9
HOLLIES, pp.72–3
Nicotiana glauca

Stephanandra incisa

OTHER RECOMMENDED PLANTS:
Buddleja asiatica
Gardenia thunbergia
Lonicera fragrantissima

Lonicera x purpusii, p.143
Lonicera standishii
Rubus cockburnianus
Salix irrorata

Shrubs/medium WINTER INTEREST

WHITE–PINK

Dombeya burgessiae, syn. *D. mastersii*
Evergreen shrub with rounded, 3-lobed, downy leaves and dense clusters of fragrant, white flowers, with pink to red veins, in autumn-winter. Min. 5°C (41°F).

Chamelaucium uncinatum [pink form] (Geraldton waxflower)
Evergreen, wiry-stemmed, bushy shrub. Each needle-like leaf has a tiny, hooked tip. Flowers ranging from deep rose-purple to pink, lavender or white appear in late winter or spring. Min. 5°C (41°F).

Acokanthera oblongifolia, syn. *A. spectabilis*, *Carissa spectabilis* (Wintersweet)
Evergreen, rounded shrub. Has fragrant, white or pinkish flowers in late winter and spring and poisonous, black fruits in autumn. Min. 10°C (50°F).

Viburnum × bodnantense 'Dawn'
Deciduous, upright shrub with oval, bronze, young leaves that mature to dark green. Racemes of deep pink buds open to fragrant, pink flowers during mild periods from late autumn to early spring.

Daphne bholua
Evergreen, occasionally deciduous, upright shrub with leathery, dark green foliage. Terminal clusters of richly fragrant, purplish-pink and white flowers are borne in winter.

Euphorbia pulcherrima (Poinsettia)
Evergreen, sparingly branched shrub. Has small, greenish-red flowers surrounded by bright red, pink, yellow or white bracts from late autumn to spring. Min. 15°C (59°F).

CAMELLIAS, pp.98–9
Daphne odora
Viburnum × bodnantense 'Deben'
Viburnum farreri 'Candidissimum'

Viburnum grandiflorum
Viburnum tinus 'Eve Price'
Viburnum tinus 'Gwenllian'

RED–PURPLE

Cornus alba 'Sibirica'
Deciduous, upright shrub with scarlet, young shoots in winter. Has dark green foliage and heads of creamy-white flowers in late spring and early summer, succeeded by rounded, white fruits.

Ardisia crenata, syn. *A. crenulata* (Coralberry, Spiceberry)
Evergreen, upright, open shrub. Has fragrant, star-shaped, white flowers in early summer, followed by long-lasting, bright red fruits. Min. 10°C (50°F).

Iochroma cyanea, syn. *I. tubulosa*
Evergreen, semi-upright, slender-branched shrub. Tubular, deep purple-blue flowers, with flared mouths, appear in dense clusters from late autumn to early summer. Min. 7–10°C (45–50°F).

Cornus alba
HEATHERS, pp.148–9
Pachystachys coccinea
Skimmia japonica 'Rubella', p.144

YELLOW

Stachyurus praecox
Deciduous, spreading, open shrub with purplish-red shoots. Drooping spikes of pale greenish-yellow flowers open in late winter and early spring, before pointed, deep green leaves appear.

Duranta erecta, syn. *D. plumieri*, *D. repens* (Pigeon berry, Skyflower)
Fast-growing, usually evergreen, bushy shrub. Has spikes of lilac-blue flowers, mainly in summer, followed by yellow fruits. Min. 10°C (50°F).

Chimonanthus praecox
Chimonanthus praecox 'Luteus'
Mahonia bealei
Mahonia lomariifolia

Shrubs/medium ALL YEAR INTEREST

☐ YELLOW

▭ WHITE–PURPLE

Mahonia japonica
Evergreen, upright shrub with deep green leaves consisting of many spiny leaflets. Long, spreading sprays of fragrant, yellow flowers appear from late autumn to spring, succeeded by purple-blue fruits.

Jasminum nudiflorum
(Winter jasmine)
Deciduous, arching shrub with oval, dark green leaves. Bright yellow flowers appear on slender, leafless, green shoots in winter and early spring.

Euonymus japonicus 'Macrophyllus Albus'
Evergreen, upright, bushy and dense shrub with oval, dark green leaves broadly edged with white. Produces clusters of insignificant, greenish-white flowers in late spring.

Fatsia japonica 'Variegata'
Evergreen, rounded, bushy and dense shrub with palmate, glossy, dark green leaves, variegated marginally with creamy-white, and large sprays of small, white flowers in autumn.

Euonymus fortunei 'Silver Queen'
Evergreen, bushy, sometimes scandent shrub with a dense growth of dark green leaves, broadly edged with white. Produces insignificant, greenish-white flowers in spring.

Nandina domestica 'Firepower'
Evergreen or semi-evergreen, elegant, bamboo-like, dwarf shrub. Leaves have dark green leaflets, purplish-red when young and in autumn-winter. Bears small, white flowers in summer followed in warm areas by orange-red fruits.

x Citrofortunella microcarpa, syn. x *C. mitis*, *Citrus mitis* (Calamondin)
Evergreen, bushy shrub with leathery, leaves. Intermittently has tiny, fragrant flowers followed by orange-yellow fruits. Min. 5–10°C (41–50°F).

Dracaena sanderiana
(Ribbon plant)
Evergreen, upright shrub with seldom branching, cane-like stems. Lance-shaped leaves, 15–25cm (6–10in) long, are pale to grey-green, with bold, creamy-white edges. Min. 13°C (55°F).

Dodonaea viscosa 'Purpurea'
Evergreen, bushy shrub or tree. Firm-textured leaves are flushed copper-purple. Has clusters of small, reddish or purplish seed capsules in late summer or autumn. Makes a good hedging plant for windy sites. Min. 5°C (41°F).

Acacia podalyriifolia, p.105
Cornus stolonifera 'Flaviramea'
Cytisus x *spachianus*
Edgeworthia chrysantha

HEATHERS, pp.148–9
Stachyurus chinensis

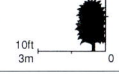

OTHER RECOMMENDED PLANTS:
Acalypha hispida
Cleyera japonica 'Tricolor'
x *Fatshedera lizei* 'Variegata'

Mackaya bella
Pandanus veitchii, p.145
Polyscias filicifolia 'Marginata'
Rhamnus alaternus 'Argenteovariegatus'

121

Shrubs/medium ALL YEAR INTEREST
GREEN

Corokia cotoneaster
(Wire-netting bush)
Evergreen, bushy, open shrub with interlacing shoots. Has small, spoon-shaped, dark green leaves, fragrant, yellow flowers in late spring and red fruits in autumn.

Encephalartos ferox
Slow-growing, evergreen, palm-like plant, almost trunkless for many years. Feather-shaped leaves, 60–180cm (2–6ft) long, have many serrated and spine-tipped, leathery, greyish leaflets. Min. 10–13°C (50–55°F).

Cycas revoluta
(Japanese sago palm)
Slow-growing, evergreen, palm-like plant; may produce several trunks. Leaves have spine-tipped leaflets with rolled margins. Bears tight clusters of reddish fruits in autumn. Min. 13°C (55°F).

Aucuba japonica
Evergreen, dense, bushy shrub with stout, green shoots and glossy, dark green leaves. Small, purplish flowers in mid-spring are followed on female plants by rounded to egg-shaped, bright red berries.

Arctostaphylos patula
Evergreen, rounded shrub with reddish-brown bark and bright grey-green foliage. Urn-shaped, white or pale pink flowers appear from mid- to late spring, followed by brown fruits.

Rhapis excelsa (Bamboo palm, Slender lady palm)
Evergreen fan palm, eventually forming clumps. Leaves are 20–30cm (8–12in) long, composed of 20 or more narrow, glossy, deep green lobes in fan formation. Min. 15°C (59°F).

***Buxus sempervirens* 'Handsworthensis'**
Vigorous, evergreen, bushy, upright shrub or small tree. Has broad, very dark green leaves. A dense habit makes it ideal for hedging or screening.

x ***Fatshedera lizei*** (Tree ivy)
Evergreen, loose-branched shrub that forms a mound of deeply lobed, glossy, deep green leaves. May also be trained as a climber. Sprays of small, white flowers appear in autumn.

Ficus deltoidea (Mistletoe fig)
Slow-growing, evergreen, bushy shrub with bright green leaves, red-brown-tinted beneath. Bears small, greenish-white fruits that mature to dull yellow. Min. 15–18°C (59–64°F).

Philodendron bipinnatifidum, syn. *P. selloum*
Evergreen, unbranched shrub. Glossy leaves, to 60cm (2ft) or more long, are divided into many finger-like lobes. Occasionally produces greenish-white spathes. Min. 15–18°C (59–64°F).

Polyscias filicifolia
(Fern-leaf aralia)
Evergreen, erect, sparingly branched shrub. Leaves are 30cm (12in) long and are divided into many small, serrated, bright green leaflets. Min. 15–18°C (59–64°F).

Chamaedorea elegans, syn. *Neanthe bella* (Dwarf mountain palm, Parlour palm)
Evergreen, slender palm, suckering with age. Feather-shaped leaves of many glossy leaflets are 60–100cm (2–3ft) long. Min. 18°C (64°F).

Aristotelia chilensis
Buxus wallichiana
Chamaerops humilis, p.146
Dioon edule

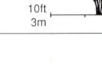

DWARF CONIFERS, pp.84–5
Elaeagnus macrophylla
Eurya emarginata, p.146
Fatsia japonica

HOLLIES, pp.72–3
Illicium floridanum
Ligustrum japonicum 'Rotundifolium'
Lonicera nitida

Lonicera nitida 'Yunnan'
Olearia lacunosa
Phoenix roebelenii

122

Shrubs/small SPRING INTEREST

■ GREEN–YELLOW

☐ WHITE

Portulacaria afra (Elephant bush)
Semi-evergreen, upright shrub with horizontal branches and tiny, fleshy, bright green leaves. Clusters of pale pink flowers appear in late spring and summer. Min. 7–10°C (45–50°F).

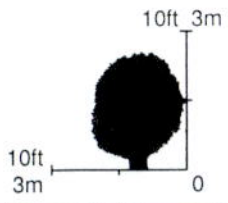

***Elaeagnus* x *ebbingei* 'Limelight'**
Evergreen, bushy, dense shrub with glossy, dark green leaves, silver beneath, centrally marked yellow and pale green. Bears small, fragrant, white flowers in autumn.

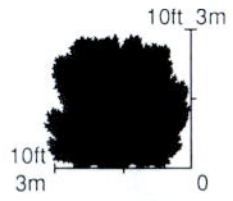

Buxus balearica (Balearic box)
Evergreen, tree-like shrub suitable for hedging in mild areas. Has broadly oval, bright green leaves.

***Ligustrum* 'Vicaryi'**, syn. *L.* x *vicaryi*
Semi-evergreen, bushy, dense shrub with broad, oval, golden-yellow leaves. Dense racemes of small, white flowers appear in mid-summer.

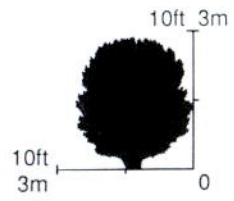

***Salix hastata* 'Wehrhahnii'**
Deciduous, upright-branched shrub with deep purple stems that contrast with silver-grey catkins borne in early spring before foliage appears. Stems later turn yellow. Has oval, bright green leaves.

Yucca aloifolia (Spanish bayonet)
Slow-growing, evergreen shrub or small tree with few branches. Has sword-shaped, deep green leaves, 50–75cm (20–30in) long, and large panicles of purple-tinted, white flowers in summer-autumn. Min. 7°C (45°F).

***Aucuba japonica* 'Crotonifolia'**
Evergreen, bushy, dense shrub with stout, green shoots. Large, glossy, dark green leaves are heavily mottled yellow. Small, purplish flowers in mid-spring are followed by bright red berries.

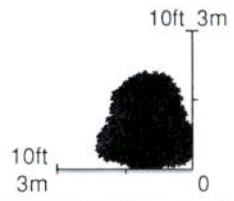

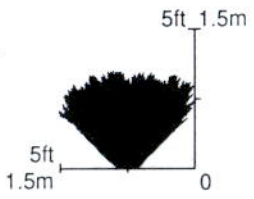

Deutzia gracilis
Deciduous, upright or spreading shrub. Massed, 5-petalled, pure white flowers are borne in upright clusters amid bright green foliage in late spring and early summer.

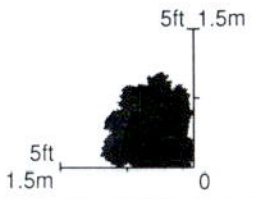

***Prunus glandulosa* 'Alba Plena'**
Deciduous, open shrub, with narrowly oval, mid-green leaves, bearing racemes of double, white flowers in late spring.

Aucuba japonica 'Gold Dust'
Aucuba japonica 'Picturata'
Coprosma repens 'Picturata'
DWARF CONIFERS, pp.84–5

Eurya japonica 'Variegata'
Graptophyllum pictum
HOLLIES, pp.72–3
Sanchezia speciosa, p.147

OTHER RECOMMENDED PLANTS:
Chamaedaphne calyculata
Fothergilla gardenii
HEATHERS, pp.148–9

Loropetalum chinense
Olearia x *mollis*
RHODODENDRONS, pp.102–104
Westringia fruticosa, p.128

123

Shrubs/small SPRING INTEREST

☐ WHITE

▨ WHITE–PINK

Ledum groenlandicum
(Labrador tea)
Evergreen, bushy shrub. Foliage is dark green and aromatic. Rounded heads of small, white flowers are carried from mid-spring to early summer.

Campanula vidalii,
syn. *Azorina vidalii*
Evergreen sub-shrub with erect stems. Has coarsely serrated, glossy, dark green leaves and racemes of bell-shaped, white or pink flowers in spring and summer. Min. 5°C (41°F).

Gaultheria × wisleyensis 'Wisley Pearl', syn. × *Gaulnettya* 'Wisley Pearl'
Evergreen, bushy, dense shrub with oval, deeply veined, dark green leaves. Small, white flowers in late spring and early summer are followed by purplish-red fruits.

Viburnum × juddii
Deciduous, rounded, bushy shrub with dark green foliage. Rounded heads of very fragrant, pink-tinged, white flowers open from pink buds from mid- to late spring.

Spiraea × vanhouttei
(Bridal wreath)
Deciduous, compact shrub with slender, arching shoots. In late spring and early summer abundant, small, dense clusters of white flowers appear amid diamond-shaped, dark green leaves.

Prunus laurocerasus 'Zabeliana'
Evergreen, wide-spreading, open shrub. Leaves are very narrow and glossy, dark green. Spikes of white flowers in late spring are followed by cherry-like, red, then black, fruits.

Deutzia × rosea
Deciduous, bushy, dense shrub. In late spring and early summer produces massed, broad clusters of 5-petalled, pale pink flowers. Leaves are oval and dark green.

Viburnum carlesii
Deciduous, bushy, dense shrub with dark green leaves that redden in autumn. Rounded heads of very fragrant, white and pink flowers, pink in bud, appear from mid- to late spring, followed by decorative, black fruits.

Prunus laurocerasus 'Otto Luyken'
Evergreen, very dense shrub. Has upright, narrow, glossy, dark green leaves, spikes of white flowers in late spring, followed by cherry-like, red, then black, fruits.

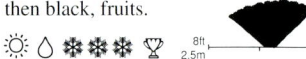

Prunus × cistena
Slow-growing, deciduous, upright shrub with deep reddish-purple leaves, red when young. Small, pinkish-white flowers from mid- to late spring may be followed by purple fruits.

Daphne × burkwoodii 'Somerset'
Semi-evergreen, upright shrub that bears dense clusters of very fragrant, white and pink flowers in late spring and sometimes again in autumn. Leaves are lance-shaped and pale to mid-green.

Bauera rubioides
Centradenia floribunda
Diosma ericoides
Gaultheria × wisleyensis 'Pink Pixie'

HEATHERS, pp.148–9
RHODODENDRONS, pp.102–104
Vinca minor 'Bowles' White'

■■ PINK–RED

Daphne retusa
Evergreen, densely branched, rounded shrub clothed with leathery, glossy leaves notched at the tips. In late spring and early summer, deep purple buds open to very fragrant, pink-flushed, white flowers borne in terminal clusters.

Menziesia ciliicalyx var. *purpurea*
Deciduous, bushy shrub with bright green foliage and racemes of nodding, purplish-pink blooms in late spring and early summer.

Euphorbia milii (Crown of thorns)
Fairly slow-growing, mainly evergreen, spiny, semi-succulent shrub. Clusters of tiny, yellowish flowers, enclosed by 2 bright red bracts, open intermittently during the year.
Min. 8°C (46°F).

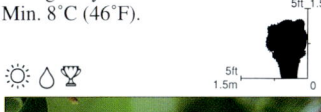

Chaenomeles x *superba* 'Rowallane'
Deciduous, low, spreading shrub. Has glossy, dark green foliage and bears a profusion of large, red flowers during spring.

Prunus tenella
Deciduous, bushy shrub with upright shoots and narrowly oval, glossy leaves. Shallowly cup-shaped, bright pink flowers appear from mid- to late spring.

Ribes sanguineum 'Brocklebankii'
Deciduous, spreading shrub. Has aromatic, pale yellow leaves and pendent clusters of small, pale pink flowers in spring, followed by white-bloomed, black fruits.

Epacris impressa (Australian heath)
Evergreen, usually erect, fairly open, heath-like shrub with short, red-tipped leaves. Tubular, pink or red flowers appear in late winter and spring. Min. 5°C (41°F).

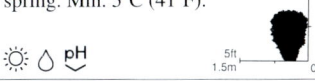

Cantua buxifolia, syn. *C. dependens*
Evergreen, arching, bushy shrub. Has grey-green foliage and drooping clusters of bright red and magenta flowers from mid- to late spring.

Chaenomeles speciosa 'Simonii'
Coleonema pulchrum
Euphorbia fulgens
Gaylussacia baccata

HEATHERS, pp.148–9
Justicia rizzinii
Kalmiopsis leachiana 'M. Le Piniec'
Protea magnifica

RHODODENDRONS, pp.102–104

125

Shrubs/small SPRING INTEREST

■ RED–GREEN

Chaenomeles x superba 'Nicoline'
Deciduous, bushy, dense shrub. Has glossy, dark green leaves and a profusion of large, scarlet flowers in spring, followed by yellow fruits.

Salix lanata (Woolly willow)
Deciduous, bushy, dense shrub with stout, woolly, grey shoots and broad, silver-grey leaves. Large, yellowish-green catkins appear in late spring with foliage.

Boronia megastigma
Evergreen, well branched, wiry-stemmed shrub. Small leaves have 3–5 narrow leaflets. Fragrant, bowl-shaped, brownish-purple and yellow flowers hang from leaf axils in late winter and spring. Min. 7–10°C (45–50°F).

Euphorbia characias subsp. **characias**
Evergreen, upright shrub with clusters of narrow, grey-green leaves. During spring and early summer, bears dense spikes of pale yellowish-green flowers with deep purple centres.

■ GREEN–YELLOW

Euphorbia characias subsp. **wulfenii**
Evergreen, upright shrub. Stems are biennial, producing clustered, grey-green leaves one year and spikes of yellow-green blooms the following spring.

Arctostaphylos 'Emerald Carpet'
Evergreen shrub that, with a low, dense growth of oval, bright green leaves and purple stems, makes excellent ground cover. Bears small, urn-shaped, white flowers in spring.

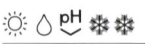

Daphne laureola subsp. **philippi**
Evergreen, dwarf shrub with oval, dark green leaves. Slightly fragrant, tubular, pale green flowers with short, spreading lobes appear in late winter and early spring, followed by black fruits.

Salix repens (Creeping willow)
Deciduous, prostrate or semi-upright and bushy shrub. Silky, grey catkins become yellow from mid- to late spring, before small, narrowly oval leaves, which are grey-green above, silvery beneath, appear.

Cytisus x praecox (Warminster broom)
Deciduous, densely branched shrub. From mid- to late spring, pea-like, creamy-yellow flowers appear in profusion amid tiny, silky, grey-green leaves with 3 leaflets.

Agathosma pulchella
Barleria obtusa
Boronia molloyae
Brunfelsia pauciflora

HEATHERS, pp.148–9
RHODODENDRONS, pp.102–104
Vinca major
Xanthorhiza simplicissima

■ YELLOW

Mahonia aquifolium
(Oregon grape)
Evergreen, open shrub. Leaves, with glossy, bright green leaflets, often turn red or purple in winter. Bunches of small, yellow flowers in spring are followed by blue-black berries.

Caragana arborescens 'Nana'
Deciduous, bushy, dwarf shrub with mid-green leaves consisting of many oval leaflets. Pea-like, yellow flowers are borne in late spring.

Pachystachys lutea
(Lollipop plant)
Evergreen, loose, more or less rounded shrub often grown annually from cuttings. Has tubular, white flowers in tight, golden-bracted spikes in spring-summer. Min. 13–15°C (55–9°F).

Cytisus x praecox 'Allgold'
Deciduous, densely branched shrub with silky, grey-green leaves, divided into 3 leaflets, and a profusion of pea-like, yellow flowers from mid- to late spring.

Ulex europaeus (Gorse)
Leafless or almost leafless, bushy shrub with year-round, dark green shoots and spines that make it appear evergreen. Bears massed, fragrant, pea-like, yellow flowers in spring.

Berberis empetrifolia
Evergreen, arching, prickly shrub with narrow, grey-green leaves, globose, golden-yellow flowers in late spring and black fruits in autumn.

Coronilla valentina subsp. **glauca**
Evergreen, bushy, dense shrub. Has blue-grey leaves with 5 or 7 leaflets. Fragrant, pea-like, yellow flowers are borne from mid-spring to early summer.

Acacia pulchella
(Western prickly Moses)
Semi-evergreen or deciduous shrub of diffuse habit, with spiny twigs and rich green foliage. Tiny, deep yellow flowers appear in dense, globular heads in spring. Min. 5–7°C (41–5°F).

■■ YELLOW–ORANGE

Genista tinctoria
(Dyers' greenweed)
Deciduous, spreading, dwarf shrub that bears dense spires of pea-like, golden-yellow flowers in spring and summer. Leaves are narrow and dark green.

Chorizema ilicifolium
(Holly flame pea)
Evergreen, sprawling or upright shrub, with spiny-toothed, leathery leaves. Has spikes of bicoloured, orange and pinkish-red flowers in spring-summer. Min. 7°C (45°F).

Nematanthus gregarius, syn. *N. radicans*, *Hypocyrta radicans*
Evergreen, prostrate or slightly ascending shrub with fleshy, glossy leaves. Inflated, tubular, orange and yellow flowers appear mainly from spring to autumn. Min. 13–15°C (55–9°F).

Acacia ulicifolia
Berberis calliantha
Caragana frutex 'Globosa'
Genista tinctoria 'Royal Gold'

Hermannia incana
Mahonia 'Heterophylla'
Mahonia repens
RHODODENDRONS, pp.102–104

Daphne giraldii
Halimium lasianthum
Lantana camara
Nematanthus strigillosus

127

Shrubs/small SUMMER INTEREST
☐ WHITE

Deutzia monbeigii
Deciduous, arching, elegant shrub. Clusters of small, 5-petalled, white flowers appear in profusion among small, dark green leaves from early to mid-summer.

Olearia phlogopappa var. **subrepanda**
Evergreen, upright, compact shrub. Heads of daisy-like, white flowers are borne profusely from mid-spring to early summer amid narrow, toothed, grey-green leaves.

Westringia fruticosa, syn. *W. rosmariniformis* (Australian rosemary)
Evergreen, rounded, compact shrub. Crowded leaves, in whorls of 4, are white-felted beneath. White to palest blue flowers open in spring-summer. Min. 5–7°C (41–5°F).

Potentilla 'Abbotswood'
Deciduous, bushy shrub. Large, pure white flowers are borne amid dark blue-green leaves, divided into 5 narrowly oval leaflets, throughout summer-autumn.

Hebe brachysiphon 'White Gem'
Evergreen, rounded shrub that produces a dense mound of small, glossy leaves covered in early summer with tight racemes of small, white flowers.

Gardenia augusta 'Fortuniana'
Fairly slow-growing, evergreen, leafy shrub with oval, glossy leaves up to 10cm (4in) long and fragrant, double, white flowers from summer to winter. Min. 15°C (59°F).

Potentilla 'Manchu', syn. *P. davurica* var. *mandschurica* of gardens
Deciduous, mound-forming shrub with reddish-pink, prostrate shoots. Pure white flowers are borne amid divided, silvery-grey leaves from late spring to early autumn.

Rhodotypos scandens, syn. *R. kerrioides*
Deciduous, upright or slightly arching shrub. In late spring and early summer, amid sharply toothed leaves, bears shallowly cupped, white flowers, followed by small, pea-shaped, black fruits.

Cuphea hyssopifolia (False heather)
Evergreen, rounded, dense shrub with tiny, narrowly lance-shaped, deep green leaves. Rose-purple to lilac or white flowers appear in summer-autumn.

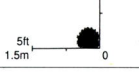

Philadelphus 'Manteau d'Hermine'
Deciduous, bushy, compact shrub. Clusters of fragrant, double, creamy-white flowers appear amid small, pale to mid-green leaves from early to mid-summer.

Convolvulus cneorum
Evergreen, rounded, bushy, dense shrub. Pink-tinged buds opening to white flowers with yellow centres are borne from late spring to late summer among narrow, silky, silvery-green leaves.

OTHER RECOMMENDED PLANTS:
Berzelia lanuginosa
Boenninghausenia albiflora
Cyrilla racemiflora

Cytisus albus
FUCHSIAS, pp.134–5
HEATHERS, pp.148–9
Hebe 'Cranleigh Gem'

Hebe macrantha
Hebe ochracea
Hebe rakaiensis
Hebe vernicosa, p.299

Helichrysum splendidum
RHODODENDRONS, pp.102–104
Rubus tricolor

☐ WHITE

Halimium umbellatum
Evergreen, upright shrub. Narrow, glossy, dark green leaves are white beneath. White flowers, centrally blotched with yellow, are produced in early summer from reddish buds.

Cistus salviifolius
Evergreen, bushy, dense shrub with slightly wrinkled, grey-green foliage. White flowers, with central, yellow blotches, appear in profusion during early summer.

Potentilla 'Farrer's White'
Deciduous, bushy shrub with divided, grey-green leaves. Bears an abundance of white flowers during summer-autumn.

Cistus monspeliensis
Evergreen, bushy shrub with narrow, wrinkled, dark green leaves and small, white flowers freely borne from early to mid-summer.

Cistus × *cyprius*
Evergreen, bushy shrub with sticky shoots and narrow, glossy, dark green leaves. In early summer bears large, white flowers, with a red blotch at each petal base, that appear in succession for some weeks but last only a day.

× *Halimiocistus sahucii*
Evergreen, bushy, dense shrub with narrow, dark green leaves that set off an abundance of pure white flowers in late spring and early summer.

Cistus × *corbariensis*, syn. *C.* × *hybridus*
Evergreen, bushy, dense shrub. Has wrinkled, wavy-edged, dark green leaves and massed white flowers, with central, yellow blotches, carried in late spring and early summer.

Cistus × *aguilarii* 'Maculatus'
Evergreen, bushy shrub with narrow, wavy-edged, slightly sticky, rich green leaves. Large, white flowers, with a central, deep red and yellow pattern, appear from early to mid-summer.

Cistus ladanifer
Evergreen, open, upright shrub. Leaves are narrow, dark green and sticky. Large, white flowers, with red markings around the central tuft of stamens, are borne in profusion during early summer.

Chamaebatiaria millefolium
× *Halimiocistus wintonensis*
Sibiraea altaiensis

129

Shrubs/small SUMMER INTEREST
☐ WHITE

Leptospermum rupestris
Evergreen, semi-prostrate, widely arching shrub with reddish shoots and small, dark green leaves that turn bronze-purple in winter. Small, open cup-shaped, white flowers, red-flushed in bud, appear in early summer.

Vaccinium corymbosum
(Highbush blueberry)
Deciduous, upright, slightly arching shrub. Small, white or pinkish flowers in late spring and early summer are followed by sweet, edible, blue-black berries. Foliage turns red in autumn.

Yucca whipplei, syn. *Hesperoyucca whipplei*
Evergreen, virtually stemless shrub that forms a dense tuft of slender, pointed, blue-green leaves. Very long panicles of fragrant, greenish-white flowers are produced in late spring and early summer.

Yucca flaccida 'Ivory', syn. *Y. filifera* 'Ivory'
Evergreen, very short-stemmed shrub that produces tufts of narrow, dark green leaves and long panicles of bell-shaped, white flowers from mid- to late summer.

Catharanthus roseus, syn. *Vinca rosea* (Rose periwinkle)
Evergreen, spreading shrub, becoming untidy with age. Has white to rose-pink flowers in spring to autumn, also in winter in warm areas.
Min. 5–7°C (41–5°F).

Rhaphiolepis umbellata
Evergreen, bushy shrub with rounded, leathery, dark green leaves and clusters of fragrant, white flowers in early summer.

Weigela florida 'Variegata'
Deciduous, bushy, dense shrub. Carries a profusion of funnel-shaped, pink flowers in late spring and early summer and has mid-green leaves broadly edged with creamy-white.

Ozothamnus ledifolius, syn. *Helichrysum ledifolium*
Evergreen, dense shrub. Yellow shoots are covered with small, aromatic leaves, glossy, dark green above, yellow beneath. Small, white flower heads are borne in early summer.

Bouvardia longiflora
Cistus x lusitanicus
FUCHSIAS, pp.134–5
HEATHERS, pp.148–9
Philadelphus 'Sybille'
RHODODENDRONS, pp.102–104
Spiraea trilobata

☐ WHITE

Lomatia silaifolia
Evergreen, bushy shrub. Spikes of creamy-white flowers, each with 4 narrow, twisted petals, are borne amid deeply divided, dark green leaves from mid- to late summer.

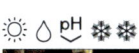

Viburnum acerifolium
Deciduous, upright-branched shrub with bright green leaves that turn orange, red and purple in autumn. Decorative, red fruits, which turn purple-black, follow heads of creamy-white flowers in early summer.

Eriogonum arborescens
Evergreen, sparingly branched shrub. Small leaves have recurved edges and woolly, white undersides. Leafy umbels of small, white or pink flowers appear from spring to autumn.
Min. 5°C (41°F).

Cassinia vauvilliersii, syn. *C. leptophylla* subsp. *vauvilliersii*
Evergreen, upright shrub. Whitish shoots are covered with tiny, dark green leaves and heads of small, white flowers from mid- to late summer.

Hebe recurva
Evergreen, open, spreading shrub. Leaves are narrow, curved and bluegrey. Small spikes of white flowers appear from mid- to late summer.

☐ WHITE–PINK

Hebe albicans
Evergreen shrub that forms a dense mound of blue-grey foliage covered with small, tight clusters of white flowers from early to mid-summer.

Deutzia 'Mont Rose'
Deciduous, bushy shrub that produces clusters of pink or pinkish-purple flowers, in early summer, with yellow anthers and occasionally white markings. Leaves are sharply toothed and dark green.

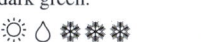

Potentilla 'Daydawn'
Deciduous, bushy, rather arching shrub. Creamy-yellow flowers, flushed with orange-pink, appear among divided, mid-green leaves from early summer to mid-autumn.

Campanula vidalii, p.124
Cistus parviflorus
Cistus 'Peggy Sammons'
Dorycnium hirsutum

Protea cynaroides (King protea)
Evergreen, bushy, rounded shrub. Water lily-shaped flower heads, 13–20cm (5–8in) wide, with silky-haired, petal-like, pink to red bracts, appear in spring-summer. Leaves are oval and mid- to dark green.
Min. 5–7°C (41–5°F).

Abelia 'Edward Goucher'
Deciduous or semi-evergreen, arching shrub. Oval, bright green leaves are bronze when young. Bears a profusion of lilac-pink flowers from mid-summer to autumn.

FUCHSIAS, pp.134–5
HEATHERS, pp.148–9
HYDRANGEAS, p.114
RHODODENDRONS, pp.102–104

131

Shrubs/small SUMMER INTEREST
PINK

Abelia schumannii
Deciduous, arching shrub. Pointed, mid-green leaves are bronze when young. Yellow-blotched, rose-purple and white flowers appear from mid-summer to mid-autumn.

Myoporum parvifolium
Evergreen, spreading to prostrate shrub with semi-succulent leaves. In summer has clusters of small, honey-scented flowers, white or pink with purple spots, and tiny, purple fruits in autumn. Min. 2–5°C (36–41°F).

Cistus × skanbergii
Evergreen, bushy shrub. A profusion of pale pink flowers appears amid narrow, grey-green leaves from early to mid-summer.

Pimelea ferruginea
Evergreen, dense, rounded shrub with tiny, recurved, deep green leaves. Small, tubular, rich pink flowers appear in dense heads in spring or early summer. Min. 7°C (45°F).

Gaultheria shallon (Shallon)
Evergreen, bushy shrub. Red shoots carry broad, sharply pointed, dark green leaves. Racemes of urn-shaped, pink flowers in late spring and early summer are followed by purple berries.

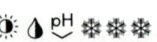

Phlomis italica
Evergreen, upright shrub. In mid-summer, whorls of lilac-pink flowers are borne at the ends of shoots amid narrow, woolly, grey-green leaves.

Indigofera dielsiana
Deciduous, upright, open shrub. Dark green leaves consist of 7–11 oval leaflets. Slender, erect spikes of pale pink flowers are borne from early summer to early autumn.

Deutzia × elegantissima 'Rosealind'
Deciduous, rounded, bushy, dense shrub that produces clusters of 5-petalled, deep pink flowers from late spring to early summer.

Bauera rubioides
Calliandra eriophylla, p.142
FUCHSIAS, pp.134–5
HEATHERS, pp.148–9
HYDRANGEAS, p.114
Indigofera pseudotinctoria
RHODODENDRONS, pp.102–104

PINK

Weigela florida 'Foliis Purpureis'
Deciduous, low, bushy shrub that bears funnel-shaped flowers, deep pink outside, pale pink to white inside, in late spring and early summer. Leaves are dull purple or purplish-green.

Ceanothus 'Perle Rose'
Deciduous, bushy shrub that from mid-summer to early autumn bears dense racemes of bright carmine-pink flowers amid broad, oval, mid-green leaves.

Spiraea japonica 'Little Princess'
Slow-growing, deciduous, mound-forming shrub that produces copious small heads of rose-pink blooms from mid- to late summer. Small, dark green leaves are bronze when young.

Hebe 'Great Orme'
Evergreen, rounded, open shrub. Has deep purplish shoots and glossy, dark green foliage. Slender spikes of deep pink flowers that fade to white are produced from mid-summer to mid-autumn.

Penstemon isophyllus
Slightly untidy, deciduous shrub or sub-shrub that from mid- to late summer carries long sprays of large, white- and red-throated, deep pink flowers above spear-shaped, glossy, mid-green leaves.

Spiraea japonica 'Goldflame'
Deciduous, upright, slightly arching shrub with orange-red, young leaves turning to bright yellow and finally pale green. Bears heads of deep rose-pink flowers from mid- to late summer.

Justicia carnea, syn. *Jacobinia carnea*, *J. pohliana* (King's crown)
Evergreen, sparingly branched shrub with velvety-haired leaves. Has spikes of pink to rose-purple flowers in summer-autumn. Min. 10–15°C (50–59°F).

Pentas lanceolata, syn. *P. carnea* (Egyptian star, Star-cluster)
Mainly evergreen, loosely rounded shrub with hairy, bright green leaves. In summer-autumn produces dense clusters of pink, lilac, red or white flowers. Min. 10–15°C (50–59°F).

Spiraea japonica 'Anthony Waterer'
Deciduous, upright, compact shrub. Red, young foliage matures to dark green. Heads of crimson-pink blooms appear from mid- to late summer.

Cistus 'Silver Pink'
FUCHSIAS, pp.134–5
HEATHERS, pp.148–9
Hebe 'Gauntlettii'
HYDRANGEAS, p.114
Indigofera decora
RHODODENDRONS, pp.102–104

133

Fuchsias

With their vividly coloured blooms and long flowering season (usually throughout summer and well into autumn), fuchsias make outstanding shrubs for the greenhouse and for the garden. Single to double flowers often have flared or elegantly recurved sepals. In mild areas fuchsias may be grown outside all year; in cool climates most are best grown in a greenhouse or as summer bedding, although some are frost hardy.

Fuchsias raised from cuttings are sparingly branched and often become straggly unless pruned from an early stage by pinching out the growing tips. To produce standard plants, the leader shoot is left, supported, but emerging sideshoots are pinched back to one pair of leaves. When the stem has reached the required height and has produced 2 or 3 pairs of leaves above this, it is then pinched out and the plant is left to develop naturally.

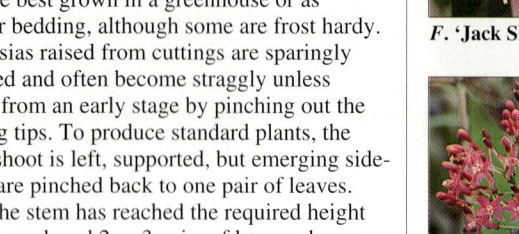

F. 'Jack Shahan'

F. arborescens

F. 'Lady Thumb'

F. 'Peppermint Stick'

F. 'Autumnale'

F. 'Harry Gray'

F. 'Pink Galore'

F. 'Swingtime'

F. 'White Ann'

F. 'Rufus'

F. 'Riccartonii'

F. 'Annabel'

F. 'Leonora'

F. 'Nellie Nuttall'

F. 'Dollar Princess'

F. 'Ann Howard Tripp'

F. 'Golden Dawn'

F. 'Red Spider'

F. 'Gruss aus dem Bodenthal'

F. 'Other Fellow'

F. 'Tom Thumb'

F. 'White Spider'

F. 'Jack Acland'

F. 'Kwintet'

F. magellanica

Shrubs/small SUMMER INTEREST
PINK–RED

F. 'Mrs Popple'

F. 'Cascade'

F. 'La Campanella'

F. 'Golden Marinka'

F. 'Mary Poppins'

F. 'Rose of Castile'

F. 'Celia Smedley'

F. × *bacillaris*

F. 'Estelle Marie'

F. boliviana 'Alba'

F. 'Thalia'

F. procumbens

F. fulgens

F. 'Lye's Unique'

F. 'Koralle'

Cistus creticus, syn. *C. incanus* subsp. *creticus*
Evergreen, bushy shrub. Pink or purplish-pink flowers, each with a central, yellow blotch, appear amid grey-green leaves from early to mid-summer.

Kalmia angustifolia f. *rubra*, syn. *K. a.* 'Rubra' (Sheep laurel)
Evergreen, bushy, mound-forming shrub with oval, dark green leaves and clusters of small, deep red flowers in early summer.

Crossandra nilotica
Evergreen, upright to spreading, leafy shrub with oval, pointed, rich green leaves. Small, tubular, apricot to pale brick-red flowers with spreading petals are carried in short spikes from spring to autumn. Min. 15°C (59°F).

Escallonia rubra 'Woodside'
Evergreen, bushy, dense shrub. Has small, glossy, dark green leaves and short racemes of small, tubular, crimson flowers in summer-autumn.

Sutherlandia frutescens
Evergreen, upright shrub. Has leaves of 13–21 grey-haired, deep green leaflets; bright red flowers in late spring and summer are followed by pale green, later red-flushed, inflated seed pods. Min. 10°C (50°F).

Ixora coccinea
Evergreen, rounded shrub with glossy, dark green leaves to 10cm (4in) long. Small, tubular, red, pink, orange or yellow flowers appear in dense heads in summer. Min. 13–16°C (55–61°F).

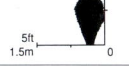

Anisodontea capensis
Cistus × *purpureus*
Coleonema pulchrum
HEATHERS, pp.148–9

Philesia magellanica
Potentilla 'Royal Flush'
RHODODENDRONS, pp.102–104

135

Shrubs/small SUMMER INTEREST
■ RED

Potentilla 'Red Ace'
Deciduous, spreading, bushy, dense shrub. Bright vermilion flowers, pale yellow on the backs of petals, are produced among mid-green leaves from late spring to mid-autumn but fade quickly in full sun.

Salvia microphylla var. **neurepia**
Evergreen, well-branched, upright shrub with pale to mid-green leaves. Has tubular, bright red flowers from purple-tinted, green calyces in late summer and autumn.

Justicia brandegeana, syn. *Beloperone guttata*, *Drejerella guttata* (Shrimp plant)
Evergreen, rounded shrub intermittently, but mainly in summer, producing white flowers surrounded by shrimp-pink bracts. Min. 10–15°C (50–59°F).

Grevillea 'Robyn Gordon'
Evergreen, sprawling shrub with leathery, dark green leaves. At intervals from early spring to late summer, arching stems bear racemes of crimson flowers with protruding, recurved styles. Min. 5–10°C (41–50°F).

Acer palmatum 'Chitoseyama'
Deciduous, arching, mound-forming shrub or small tree with lobed, mid-green foliage that gradually turns brilliant red from late summer to autumn. Produces small, reddish-purple flowers in mid-spring.

Phygelius aequalis
Evergreen or semi-evergreen, upright sub-shrub. Clusters of tubular, pale red flowers with yellow throats appear from mid-summer to early autumn. Leaves are oval and dark green.

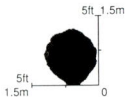

Salvia fulgens
Evergreen, upright sub-shrub. Oval leaves are white and woolly beneath, hairy above. Racemes of tubular, 2-lipped, scarlet flowers appear in late summer.

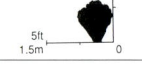

Acer palmatum 'Dissectum Atropurpureum'
Deciduous shrub that forms a mound of deeply divided, bronze-red or purple foliage, which turns brilliant red, orange or yellow in autumn. Has small, reddish-purple flowers in mid-spring.

Bouvardia ternifolia, p.143
FUCHSIAS, pp.134–5
Grevillea alpina
RHODODENDRONS, pp.102–104

PURPLE

Hebe hulkeana 'Lilac Hint'
Evergreen, upright, open-branched shrub with toothed, glossy, pale green leaves. A profusion of small, pale lilac flowers appears in large racemes in late spring and early summer.

Desmodium elegans, syn. *D. tiliifolium*
Deciduous, upright sub-shrub. Mid-green leaves consist of 3 large leaflets. Large racemes of pale lilac to deep pink flowers appear from late summer to mid-autumn.

Heliotropium arborescens, syn. *H. peruvianum*
Evergreen, bushy shrub. Semi-glossy, dark green leaves are finely wrinkled. Purple to lavender flowers are borne in dense, flat clusters from late spring to winter. Min. 7°C (45°F).

Lavandula stoechas (French lavender)
Evergreen, bushy, dense shrub. Heads of tiny, fragrant, deep purple flowers, topped by rose-purple bracts, appear in late spring and summer. Mature leaves are silver-grey and aromatic.

Rosmarinus officinalis (Rosemary)
Evergreen, bushy, dense shrub with aromatic, narrow leaves. Small, purplish-blue to blue flowers appear from mid-spring to early summer and sometimes in autumn. Used as a culinary herb.

Hebe 'E.A. Bowles'
Evergreen, rounded, bushy shrub with narrow, glossy, pale green leaves and slender spikes of lilac flowers produced from mid-summer to late autumn.

Polygala myrtifolia 'Grandiflora'
Evergreen, erect shrub with small, greyish-green leaves. White-veined, rich purple flowers appear from late spring to autumn. Min. 7°C (45°F).

Hebe 'Autumn Glory'
Evergreen shrub that forms a mound of purplish-red shoots and rounded, deep green leaves, over which dense racemes of deep purple-blue flowers appear from mid-summer to early winter.

Lantana montevidensis, syn. *L. delicatissima*, *L. sellowiana*
Evergreen, trailing or mat-forming shrub with serrated leaves. Has heads of rose-purple flowers, each with a yellow eye, intermittently all year but mainly in summer. Min. 10–13°C (50–55°F).

Brunfelsia pauciflora 'Macrantha'
Evergreen, spreading shrub with leathery leaves. Blue-purple flowers, ageing to white in about 3 days, appear from winter to summer. Min. 10–13°C (50–55°F).

Lavandula 'Hidcote', syn. *L. angustifolia* 'Hidcote'
Evergreen, bushy shrub with dense spikes of fragrant, deep purple flowers from mid- to late summer and narrow, aromatic, silver-grey leaves.

Agathosma pulchella
Amorpha canescens
Barleria cristata
Berberis thunbergii 'Atropurpurea Nana'

Berberis thunbergii 'Rose Glow'
Carmichaelia enysii
FUCHSIAS, pp.134–5
HEATHERS, pp.148–9

Hebe 'Alicia Amherst'
Hebe 'La Séduisante'
HYDRANGEAS, p.114
Lavandula dentata

Lavandula 'Grappenhall'
Osbeckia stellata
Perovskia atriplicifolia
RHODODENDRONS, pp.102–104

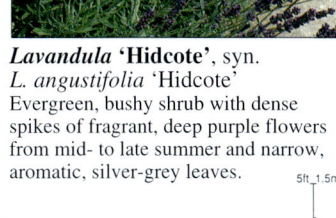

137

Shrubs/small SUMMER INTEREST

PURPLE–BLUE

Hebe 'Purple Queen'
Evergreen, bushy, compact shrub with glossy, deep green leaves that are purple-tinged when young. Dense racemes of deep purple flowers appear from early summer to mid-autumn.

Ceanothus thyrsiflorus var. repens (Creeping blue blossom)
Evergreen, dense shrub that forms a mound of broad, glossy, dark green leaves. Racemes of blue flowers are borne in late spring and early summer.

Felicia amelloides 'Santa Anita'
Evergreen, bushy, spreading shrub. Blue flower heads with bright yellow centres are borne on long stalks from late spring to autumn among round to oval, bright green leaves.

Hyssopus officinalis (Hyssop)
Semi-evergreen or deciduous, bushy shrub with aromatic, narrowly oval, deep green leaves. Small, blue flowers appear from mid-summer to early autumn. Sometimes used as a culinary herb.

Ceanothus 'Gloire de Versailles'
Vigorous, deciduous, bushy shrub. Has broad, oval, mid-green leaves and large racemes of pale blue flowers from mid-summer to early autumn.

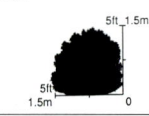

Caryopteris × clandonensis 'Arthur Simmonds'
Deciduous, bushy sub-shrub. Masses of blue to purplish-blue flowers appear amid narrowly oval, irregularly toothed, grey-green leaves from late summer to autumn.

Perovskia atriplicifolia 'Blue Spire'
Deciduous, upright sub-shrub with grey-white stems. Profuse spikes of violet-blue flowers appear from late summer to mid-autumn above aromatic, deeply cut, grey-green leaves.

Brunfelsia pauciflora
Hebe 'Fairfieldii'
Hebe × franciscana 'Blue Gem'
Hebe hulkeana

HYDRANGEAS, p.114
Lavandula 'Munstead'
Perovskia 'Hybrida'
Phlomis cashmeriana

RHODODENDRONS, pp.102–104

GREEN–YELLOW

Symphoricarpos orbiculatus 'Foliis Variegatis', syn. *S.o.* 'Variegatus'
Deciduous, bushy, dense shrub with bright green leaves edged with yellow. Occasionally bears white or pink flowers in summer-autumn.

Justicia brandegeana 'Chartreuse'
Evergreen, arching shrub producing white flowers surrounded by pale yellow-green bracts mainly in summer but also intermittently during the year. Min. 10–15°C (50–59°F).

Weigela middendorffiana
Deciduous, bushy, arching shrub. From mid-spring to early summer funnel-shaped, sulphur-yellow flowers, spotted with orange inside, are borne amid bright green foliage.

Berberis 'Parkjuweel'
Muehlenbeckia complexa
Myrsine africana

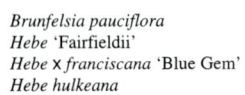

☐ YELLOW

Potentilla 'Vilmoriniana'
Deciduous, upright shrub that bears pale yellow or creamy-white flowers from late spring to mid-autumn. Leaves are silver-grey and divided into narrow leaflets.

Phygelius aequalis 'Yellow Trumpet'
Evergreen or semi-evergreen, upright sub-shrub. Bears clusters of pendent, tubular, pale creamy-yellow flowers from mid-summer to early autumn.

Potentilla 'Friedrichsenii'
Vigorous, deciduous, upright shrub. From late spring to mid-autumn pale yellow flowers are produced amid grey-green leaves.

Potentilla 'Elizabeth'
Deciduous, bushy, dense shrub with small, deeply divided leaves and large, bright yellow flowers that appear from late spring to mid-autumn.

Halimium ocymoides 'Susan'
Evergreen, spreading shrub with narrow, oval, grey-green leaves. Numerous single or semi-double, bright yellow flowers with central, deep purple-red markings are borne in small clusters along branches in summer.

Santolina pinnata subsp. neapolitana 'Sulphurea'
Evergreen, rounded, bushy shrub with aromatic, deeply cut, feathery, grey-green foliage. Produces heads of pale primrose-yellow flowers in mid-summer.

Lupinus arboreus (Tree lupin)
Fast-growing, semi-evergreen, sprawling shrub that in early summer usually bears short spikes of fragrant, clear yellow flowers above hairy, pale green leaves composed of 6–9 leaflets.

Grevillea juniperina f. sulphurea, syn. *G. sulphurea*
Evergreen, rounded, bushy shrub with almost needle-like leaves, recurved and dark green above, silky-haired beneath. Has clusters of small, spidery, pale yellow flowers in spring-summer.

Halimium lasianthum subsp. formosum
Evergreen, spreading, bushy shrub. Has grey-green foliage and golden-yellow flowers with central, deep red blotches, borne in late spring and early summer.

Coronilla valentina subsp. *glauca*, p.127
Cytisus supinus
Genista tinctoria 'Royal Gold'
Helichrysum italicum

Potentilla 'Moonlight'
RHODODENDRONS, pp.102–104
Ruta graveolens 'Jackman's Blue', p.147
Santolina pinnata subsp. *neapolitana*

Santolina rosmarinifolia 'Primrose Gem'
Vinca minor 'Alba Variegata'

139

Shrubs/small SUMMER INTEREST
YELLOW

Phlomis fruticosa
(Jerusalem sage)
Evergreen, spreading shrub with upright shoots. Whorls of deep golden-yellow flowers are produced amid sage-like, grey-green foliage from early to mid-summer.

Hypericum calycinum
(Aaron's beard, Rose of Sharon)
Evergreen or semi-evergreen dwarf shrub that makes good ground cover. Has large, bright yellow flowers from mid-summer to mid-autumn and dark green leaves.

Hypericum kouytchense
Deciduous or semi-evergreen, arching shrub. Golden-yellow flowers with conspicuous stamens are borne among foliage from mid-summer to early autumn and followed by decorative, bronze-red fruit capsules.

Berberis thunbergii 'Aurea'
Deciduous, bushy, spiny shrub with small, golden-yellow leaves. Racemes of small, red-tinged, pale yellow flowers in mid-spring are followed by red berries in autumn.

Genista hispanica (Spanish gorse)
Deciduous, bushy, very spiny shrub with few leaves but dense clusters of golden-yellow flowers borne profusely in late spring and early summer.

Hypericum 'Hidcote'
Evergreen or semi-evergreen, bushy, dense shrub. Bears an abundance of large, golden-yellow flowers from mid-summer to early autumn amid narrowly oval, dark green leaves.

Reinwardtia indica,
syn. *R. trigyna* (Yellow flax)
Evergreen, upright sub-shrub, branching from the base. Has greyish-green leaves and small clusters of yellow flowers mainly in summer but also during the year. Min. 10°C (50°F).

Cytisus nigricans
Deciduous, upright shrub with dark green leaves composed of 3 leaflets. Has a long-lasting display of tall, slender spires of yellow flowers during summer.

Senecio 'Sunshine'
Evergreen, bushy shrub that forms a mound of silvery-grey, young leaves, later becoming dark green. Large clusters of bright yellow flower heads are produced on felted shoots from early to mid-summer.

Hypericum x inodorum 'Elstead'
Deciduous or semi-evergreen, upright shrub. Abundant, small, yellow flowers borne from mid-summer to early autumn are followed by ornamental, orange-red fruits. Dark green leaves are aromatic when crushed.

Euryops pectinatus
Evergreen, upright shrub. Deeply cut, grey-green leaves set off large heads of daisy-like, bright yellow flowers, borne in late spring and early summer and often again in winter.
Min. 5–7°C (41–5°F).

Hermannia incana
Hypericum x moserianum
Lantana camara
Oxalis hedysaroides

Potentilla arbuscula
Potentilla 'Beesii'
Potentilla fruticosa
Potentilla parvifolia 'Gold Drop'

Santolina chamaecyparissus
Santolina rosmarinifolia
Senecio compactus

YELLOW–ORANGE

Senecio monroi
Evergreen, bushy, dense shrub that makes an excellent windbreak in mild, coastal areas. Has small, wavy-edged, dark green leaves with white undersides. Bears heads of bright yellow flowers in mid-summer.

***Abutilon* 'Kentish Belle'**
Semi-evergreen, arching shrub with purple shoots and deeply lobed, purple-veined, dark green leaves. Bears large, pendent, bell-shaped, orange-yellow and red flowers in summer-autumn.

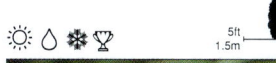

Isoplexis canariensis, syn. *Digitalis canariensis*
Evergreen, rounded, sparingly branched shrub. Bears foxglove-like, yellow to red- or brownish-orange flowers in dense, upright spikes, to 30cm (12in) tall, in summer. Min. 7°C (45°F).

***Potentilla* 'Sunset'**
Deciduous shrub, bushy at first, later arching. Deep orange flowers, fading in hot sun, appear from early summer to mid-autumn. Mid-green leaves are divided into narrowly oval leaflets.

Grindelia chiloensis, syn. *G. speciosa*
Mainly evergreen, bushy shrub with sticky stems. Sticky, lance-shaped, serrated leaves are up to 12cm (5in) long. Has large, daisy-like, yellow flower heads in summer.

Juanulloa mexicana, syn. *J. aurantiaca*
Evergreen, upright, sparingly branched shrub. Leaves are felted beneath. Has orange flowers, each with a ribbed calyx, in short, nodding clusters in summer. Min. 13–15°C (55–9°F).

Cytisus scoparius* f. *andreanus
Deciduous, arching shrub with narrow, dark green leaves that are divided into 3 leaflets. Bears a profusion of bright yellow-and-red flowers along elegant, green branchlets in late spring and early summer.

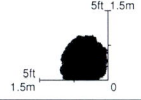

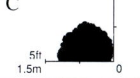

Mimulus aurantiacus, syn. *M. glutinosus*, *Diplacus glutinosus*
Evergreen, domed to rounded shrub with sticky, lance-shaped, glossy, rich green leaves. Has tubular, orange, yellow or red-purple flowers from late spring to autumn.

***Lantana* 'Spreading Sunset'**
Evergreen, rounded to spreading shrub with finely wrinkled, deep green leaves. Has tiny, tubular flowers in a range of colours, carried in dense, rounded heads from spring to autumn. Min. 10–13°C (50–55°F).

Halimium lasianthum
Halimium ocymoides
Hypericum bellum
Hypericum x *cyathiflorum* 'Gold Cup'
Hypericum patulum
Nematanthus gregarius, p.127
Nematanthus strigillosus
Phlomis chrysophylla
Phlomis longifolia var. *bailanica*
Potentilla 'Goldfinger'
Potentilla 'Jackman's Variety'
RHODODENDRONS, pp.102–104
Senecio laxifolius

141

Shrubs/small SUMMER INTEREST

🟧 ORANGE

Cuphea ignea (Cigar flower)
Evergreen, spreading, bushy sub-shrub with bright green leaves. From spring to autumn has tubular, dark orange-red flowers, each with a dark band and white ring at the mouth.
Min. 2°C (36°F).

Cuphea cyanea
Evergreen, rounded sub-shrub with narrowly oval, sticky-haired leaves. Tubular flowers, orange-red, yellow and violet-blue, are carried in summer.

Justicia spicigera,
syn. *J. ghiesbreghtiana* of gardens, *Jacobinia spicigera*
Evergreen, well-branched shrub with spikes of tubular, orange or red flowers in summer and occasionally other seasons. Min. 10–15°C (50–59°F).

Potentilla 'Tangerine'

Shrubs/small AUTUMN INTEREST

WHITE–RED

Turraea obtusifolia
Evergreen, rounded, bushy, arching shrub with oval to lance-shaped leaves. Bears fragrant, white flowers from autumn to spring, followed by orange-yellow fruits like tiny, peeled tangerines.
Min. 13°C (55°F).

Calliandra eriophylla (Fairy duster)
Evergreen, stiff, dense shrub. Leaves have numerous tiny leaflets. From late spring to autumn has pompons of tiny, pink-anthered, white florets, followed by brown seed pods. Min. 13°C (55°F).

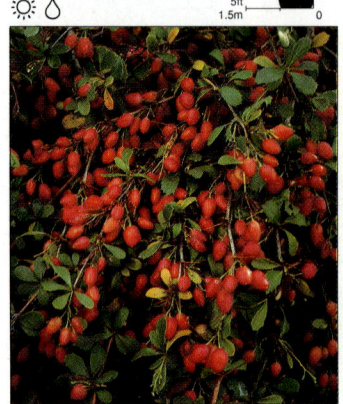

Berberis 'Rubrostilla'
Deciduous, arching shrub. Globose to cup-shaped, pale yellow flowers, appearing in early summer, are followed by a profusion of large, coral-red fruits. Grey-green leaves turn brilliant red in late autumn.

OTHER RECOMMENDED PLANTS:
Anisodontea capensis
Bouvardia longiflora
Cuphea hyssopifolia, p.128

Cotoneaster horizontalis (Wall-spray)
Deciduous, stiff-branched, spreading shrub. Glossy, dark green leaves redden in late autumn. Bears pinkish-white flowers from late spring to early summer, followed by red fruits.

Vaccinium angustifolium var. *laevifolium* (Low-bush blueberry)
Deciduous, bushy shrub with bright green leaves that redden in autumn. Edible, blue fruits follow white, sometimes pinkish, spring flowers.

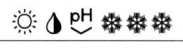

Cyrilla racemiflora
HEATHERS, pp.148–9
Pentas lanceolata, p.133
Salvia leucantha

Viburnum opulus 'Compactum'
Deciduous, dense shrub. Has deep green leaves, red in autumn, and profuse white flowers in spring and early summer followed by bunches of bright red berries.

142

Shrubs/small WINTER INTEREST

■ RED ■■ PURPLE–YELLOW □ WHITE

Bouvardia ternifolia,
syn. *B. triphylla* (Scarlet trompetilla)
Mainly evergreen, bushy, upright shrub with leaves in whorls of 3. Has tubular, bright scarlet flowers from summer to early winter. Min. 7–10°C (45–50°F).

Elsholtzia stauntonii (Mint bush)
Deciduous, open sub-shrub. Sharply toothed, mint-scented, dark green leaves turn red in autumn. Slender spires of pale purplish flowers appear during late summer and autumn.

Vaccinium corymbosum 'Pioneer'
Deciduous, upright, slightly arching shrub. Dark green leaves turn bright red in autumn. Small, white or pinkish flowers in late spring are followed by sweet, edible, blue-black berries.

Ceratostigma willmottianum
Deciduous, open shrub. Has leaves that turn red in late autumn and bright, rich blue flowers from late summer until well into autumn.

Skimmia japonica 'Fructo-alba'
Evergreen, bushy, dense, dwarf shrub. Has aromatic, dark green leaves and dense clusters of small, white flowers from mid- to late spring, succeeded by white berries.

Vaccinium parvifolium
Deciduous, upright shrub. Has small, dark green leaves that become bright red in autumn. Edible, bright red fruits are produced after small, pinkish-white flowers borne in late spring and early summer.

Coriaria terminalis var. **xanthocarpa**
Deciduous, arching sub-shrub. Leaves have oval leaflets and turn red in autumn. Greenish flowers in late spring are followed by decorative, succulent, yellow fruits in late summer and autumn.

Gaultheria mucronata 'Wintertime'
Evergreen, bushy, dense shrub. Has prickly, glossy, dark green leaves and white flowers in late spring and early summer, followed by large, long-lasting, white berries.

Lonicera x purpusii
Semi-evergreen, bushy, dense shrub with oval, dark green leaves. Small clusters of fragrant, short-tubed, white flowers, with spreading petal lobes and yellow anthers, appear in winter and early spring.

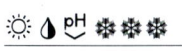

Cotoneaster adpressus
Cotoneaster microphyllus
Crossandra infundibuliformis
Salvia microphylla

Ceratostigma griffithii
HEATHERS, pp.148–9
Heliotropium arborescens, p.137
Rosmarinus officinalis 'Severn Sea'

OTHER RECOMMENDED PLANTS:
Daphne mezereum var. *alba*
HEATHERS, pp.148–9
Sarcococca confusa

Sarcococca hookeriana
Sarcococca ruscifolia
Turraea obtusifolia, p.142

143

Shrubs/small WINTER INTEREST

WHITE–PINK

Sarcococca humilis, syn.
S. hookeriana var. *humilis*
Evergreen, low, clump-forming shrub. Tiny, fragrant, white flowers with pink anthers appear amid glossy, dark green foliage in late winter and are followed by spherical, black fruits.

Sarcococca hookeriana var. ***digyna***
Evergreen, clump-forming, suckering, dense shrub with narrow, bright green leaves. Tiny, fragrant, white flowers, with pink anthers, open in winter and are followed by spherical, black fruits.

***Gaultheria mucronata* 'Mulberry Wine'**
Evergreen, bushy, dense shrub with large, globose, magenta berries that mature to deep purple. These follow white flowers borne in spring-summer. Leaves are glossy, dark green.

RED

***Skimmia japonica* 'Rubella'**
Evergreen, upright, dense shrub with aromatic, red-rimmed, bright green foliage. Deep red flower buds in autumn and winter open to dense clusters of small, white flowers in spring.

***Daphne odora* 'Aureo-marginata'**
Evergreen, bushy shrub with glossy, dark green leaves narrowly edged with yellow. Clusters of very fragrant, deep purplish-pink and white flowers appear from mid-winter to early spring.

Daphne mezereum (Mezereon)
Deciduous, upright shrub. Very fragrant, purple or pink blooms clothe the bare stems in late winter and early spring, followed by red fruits. Mature leaves are narrowly oval and dull grey-green.

Correa pulchella
Evergreen, fairly bushy, slender-stemmed shrub with oval leaves. Small, pendent, tubular, rose-red flowers appear from summer to winter, sometimes at other seasons.

Skimmia japonica subsp. ***reevesiana* 'Robert Fortune'**, syn. *S. reevesiana*
Evergreen, bushy, rather weak-growing shrub with aromatic leaves. Small, white flowers in spring are followed by crimson berries.

Abeliophyllum distichum
Centradenia floribunda
Diosma ericoides
HEATHERS, pp.148–9

Symphoricarpos x chenaultii 'Hancock'

Euphorbia fulgens
Gaultheria mucronata 'Cherry Ripe'
Justicia rizzinii
Solanum pseudocapsicum 'Balloon', p.293

Shrubs/small ALL YEAR INTEREST

🟥🟩 RED–GREEN
⬜🟩 WHITE–GREEN

Skimmia japonica
Evergreen, bushy, dense shrub. Has aromatic, mid- to dark green leaves and dense clusters of small, white flowers from mid- to late spring, followed on female plants by bright red fruits if plants of both sexes are grown.

Viburnum davidii
Evergreen shrub that forms a dome of dark green foliage, over which heads of small, white flowers appear in late spring. If plants of both sexes are grown, female plants bear decorative, metallic-blue fruits.

Vinca major 'Variegata'
Evergreen, prostrate, arching, spreading sub-shrub. Has bright green leaves broadly edged with creamy-white and large, bright blue flowers borne from late spring to early autumn.

Helichrysum petiolare, syn. *H. petiolatum*
Evergreen shrub forming mounds of trailing, silver-green shoots and grey-felted leaves. Has creamy-yellow flowers in summer. Usually grown as an annual for ground cover and edging.

Breynia nivosa, syn. *B. disticha*, *Phyllanthus nivosus* (Snow bush)
Evergreen, well-branched shrub with slender stems. Leaves are green with white marbling; tiny, greenish flowers, borne intermittently, have no petals. Min. 13°C (55°F).

Pandanus veitchii, syn. *P. tectorius* 'Veitchii' (Veitch's screw pine)
Evergreen, upright, arching shrub with rosettes of long, light green leaves that have spiny, white to cream margins. Min. 13–16°C (55–61°F).

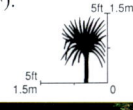

Calocephalus brownii
Evergreen, intricately branched shrub with velvety, grey branches and tiny, scale-like leaves. Clusters of flower heads, silver in bud, yellowish when expanded, appear in summer. Min. 7–10°C (45–50°F).

Ribes laurifolium
Evergreen, spreading shrub. Has leathery, deep green leaves and pendent racemes of greenish-yellow flowers in late winter and early spring. Produces edible, black berries on female plants if plants of both sexes are grown.

Coprosma × *kirkii* 'Variegata'
Evergreen, densely branched shrub, prostrate when young, later semi-erect. White-margined leaves are borne singly or in small clusters. Tiny, translucent, white fruits appear in autumn on female plants if both sexes are grown.

Artemisia arborescens
Evergreen, upright shrub, grown for its finely cut, silvery-white foliage. Heads of small, bright yellow flowers are borne in summer and early autumn.

Brunfelsia pauciflora 'Macrantha', p.137
Eranthemum pulchellum
HEATHERS, pp.148–9
Vinca difformis

OTHER RECOMMENDED PLANTS:
Artemisia arborescens 'Faith Raven'
Breynia nivosa 'Roseo-picta'
DWARF CONIFERS, pp.84–5

Euonymus fortunei 'Coloratus'
Euonymus fortunei 'Emerald Gaiety'
Pseuderanthemum atropurpureum
Pseudowintera colorata

Ruta graveolens 'Jackman's Blue', p.147
Sideritis candicans
Ursinia sericea

145

Shrubs/small ALL YEAR INTEREST
■ GREEN

Ballota acetabulosa
Evergreen sub-shrub that forms a mound of rounded, grey-green leaves, felted beneath. Whorls of small, pink flowers open from mid- to late summer.

Vaccinium glaucoalbum
Evergreen shrub with deep green leaves that when young are pale green above, bluish-white beneath. Pink-tinged, white flowers in late spring and early summer are followed by white-bloomed, blue-black fruits.

Hebe cupressoides
Evergreen, upright, dense shrub with cypress-like, grey-green foliage. On mature plants tiny, pale lilac flowers are borne from early to mid-summer.

Ruscus hypoglossum
Evergreen, clump-forming shrub with arching shoots. Pointed, glossy, bright green 'leaves' are actually flattened shoots that bear tiny, yellow flowers in spring, followed by large, bright red berries.

Vinca minor (Lesser periwinkle)
Evergreen, prostrate, spreading sub-shrub that forms extensive mats of small, glossy, dark green leaves. Bears small, purple, blue or white flowers, mainly from mid-spring to early summer.

Mimosa pudica (Humble plant, Sensitive plant)
Short-lived, evergreen shrub with prickly stems; needs support. Fern-like leaves fold when touched. Has minute, pale mauve-pink flowers in summer-autumn. Min. 13–16°C (55–61°F).

Eurya emarginata
Slow-growing, evergreen, densely branched, rounded shrub with small, leathery, deep green leaves. Small, greenish-white flowers in late spring or summer are followed by tiny, purple-black berries.

Lonicera pileata
Evergreen, low, spreading, dense shrub with narrow, dark green leaves and tiny, short-tubed, creamy-white flowers in late spring, followed by violet-purple berries. Makes good ground cover.

Chamaerops humilis (Dwarf fan palm, European fan palm)
Slow-growing, evergreen palm, suckering with age. Fan-shaped leaves, 60–90cm (2–3ft) across, have green to grey-green lobes. Has tiny, yellow flowers in summer.

Artemisia abrotanum (Lad's love, Old man, Southernwood)
Deciduous or semi-evergreen, moderately bushy shrub. Aromatic, grey-green leaves have many very slender lobes. Has clusters of small, yellowish flower heads in late summer.

Buxus microphylla 'Green Pillow'
Evergreen, compact, dwarf shrub, forming a dense, rounded mass of small, oval, dark green leaves. Bears insignificant flowers in late spring or early summer.

Sabal minor (Dwarf palmetto)
Evergreen, suckering fan palm with stems mainly underground. Has leaves of 20–30 green or grey-green lobes. Erect sprays of small, white flowers are followed by shiny, black berries. Min. 5°C (41°F).

Buxus microphylla
Calothamnus villosus
Coprosma x kirkii
Danäe racemosa

Daphne laureola
DWARF CONIFERS, pp.84–5
Ephedra gerardiana
Euonymus fortunei 'Kewensis'

Euonymus fortunei 'Sarcoxie'
Hymenanthera crassifolia
Macrozamia spiralis
Ruscus aculeatus

Salvia officinalis cvs
Senecio reinoldii
Serissa foetida

146

GREEN–YELLOW

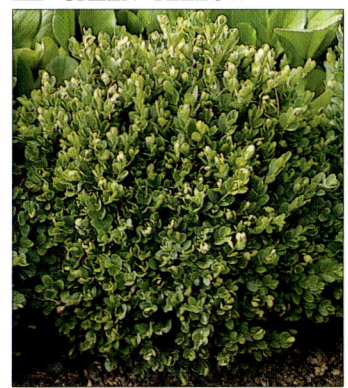

Buxus sempervirens 'Suffruticosa'
Evergreen, dwarf shrub that forms a tight, dense mass of oval, bright green leaves. Bears insignificant flowers in late spring or early summer. Trimmed to about 15cm (6in) is used for edging.

Ruta graveolens 'Jackman's Blue'
Evergreen, bushy, compact sub-shrub. Has aromatic, finely divided, blue foliage. In summer, clusters of small, mustard-yellow flowers are borne.

Leucothöe fontanesiana 'Rainbow'
Evergreen, arching shrub with sharply toothed, leathery, dark green leaves that age from pink- to cream-variegated. Racemes of white flowers open below shoots in spring.

YELLOW–ORANGE

Lonicera nitida 'Baggesen's Gold'
Evergreen, bushy shrub with long, arching shoots covered with tiny, bright yellow leaves. Insignificant, yellowish-green flowers in mid-spring are occasionally followed by mauve fruits.

Salvia officinalis 'Icterina'
Evergreen or semi-evergreen, bushy shrub used as a culinary herb. Has aromatic, grey-green leaves variegated with pale green and yellow. Occasionally bears small spikes of tubular, 2-lipped, purplish flowers.

Codiaeum variegatum (Croton)
Evergreen, erect, sparingly branched shrub. Leathery, glossy leaves vary greatly in size and shape, and are variegated with red, pink, orange or yellow. Min. 10–13°C (50–55°F).

Sanchezia speciosa, syn. *S. nobilis* of gardens
Evergreen, erect, soft-stemmed shrub. Glossy leaves have yellow- or white-banded main veins. Tubular, yellow flowers appear in axils of red bracts, in summer. Min. 15–18°C (59–64°F).

Euonymus fortunei 'Emerald 'n Gold'
Evergreen, bushy shrub with bright green leaves edged with bright yellow and tinged with pink in winter.

Pittosporum tenuifolium 'Tom Thumb'
Evergreen, rounded, dense shrub with pale green, young leaves that contrast with deep reddish-brown, older foliage. Bears cup-shaped, purplish flowers in summer.

DWARF CONIFERS, pp.84–5

Euonymus fortunei 'Gold Tip'
Euonymus fortunei 'Sunspot'

Heathers

As a group of plants, heathers (or heaths) are remarkable in that there are species and cultivars available to provide interest at all times of the year. Several are grown for their golden foliage, which often turns a deep burnt orange in winter, while others flower for a long period during summer, autumn or winter. Flowers are in a variety of colours, and are occasionally bicoloured; those with double flowers may be dried for winter decoration. In habit heathers vary from tree-heaths of up to 6m (20ft) to dwarf, prostrate forms, many of which are excellent for providing ground cover.

There are three genera: *Calluna*, *Daboecia* and *Erica*. All *Calluna* and *Daboecia* cultivars and most *Erica* species must be grown in acid soil but otherwise heathers require little attention. Main seasons of interest are given for each plant.

E. vagans 'Lyonesse' (sum.-aut.) ♀

E. arborea var. *alpina* (win.-spr.) ♀

E. ciliaris 'David McClintock' (sum.) ♀

C. vulgaris 'Silver Queen' (sum.-aut.) ♀

E. x *darleyensis* 'Darley Dale' (win.-spr.)

C. vulgaris 'Elsie Purnell' (sum.-aut.) ♀

E. x *williamsii* 'P.D. Williams' (sum.) ♀

C. vulgaris 'Kinlochruel' (sum.-aut.) ♀

E. cinerea 'Hookstone White' (sum.) ♀

E. canaliculata (win.-spr.)

C. vulgaris 'My Dream' (sum.-aut.)

C. vulgaris 'J.H. Hamilton' (sum.-aut.) ♀

E. mackaiana 'Plena' (sum.)

C. vulgaris 'Spring Cream' (spr.-aut.) ♀

E. mackaiana 'Dr Ronald Gray' (sum.)

E. x *darleyensis* 'White Perfection' (win.-spr.) ♀

D. cantabrica 'Snowdrift' (spr.-aut.)

E. tetralix 'Alba Mollis' (sum.-aut.) ♀

E. x *darleyensis* 'White Glow' (win.-spr.)

C. vulgaris 'Anthony Davis' (sum.-aut.) ♀

E. x *darleyensis* 'Ghost Hills' (win.-spr.) ♀

E. x *veitchii* 'Pink Joy' (win.-spr.)

E. carnea 'Springwood White' (win.-spr.) ♀

E. x *darleyensis* 'Archie Graham' (win.-spr.)

D. cantabrica 'Bicolor' (spr.-aut.) ♀

E. x *veitchii* 'Exeter' (win.-spr.)

E. ciliaris 'White Wings' (sum.)

C. vulgaris 'County Wicklow' (sum.-aut.) ♀

E. x *watsonii* 'Dawn' (sum.) ♀

D. x *scotica* 'William Buchanan' (spr.-aut.) ♀

E. vagans **'Birch Glow'** (sum.-aut.) ♀

E. tetralix **'Pink Star'** (sum.-aut.) ♀

E. tetralix **'Con Underwood'** (sum.-aut.) ♀

E. carnea **'December Red'** (win.)

E. erigena **'Golden Lady'** (all year) ♀

E. cinerea **'Windlebrooke'** (all year) ♀

E. vagans **'Valerie Proudley'** (all year) ♀

E. cinerea **'Rock Pool'** (all year)

E. ciliaris **'Corfe Castle'** (sum.) ♀

E. cinerea **'Eden Valley'** (sum.) ♀

C. vulgaris **'Peter Sparkes'** (sum.-aut.)

E. carnea **'Vivellii'** (win.-spr.) ♀

E. carnea **'Ann Sparkes'** (all year) ♀

C. vulgaris **'Robert Chapman'** (all year) ♀

E. cinerea **'C.D. Eason'** (sum.) ♀

E. erigena **'Brightness'** (spr.)

E. cinerea **'Hookstone Lavender'** (sum.)

C. vulgaris **'Golden Feather'** (all year)

C. vulgaris **'Beoley Gold'** (all year) ♀

E. cinerea **'Romiley'** (sum.)

E. carnea **'C.J. Backhouse'** (win.-spr.)

C. vulgaris **'Tib'** (sum.) ♀

C. vulgaris **'Silver Knight'** (sum.-aut.)

C. vulgaris **'Gold Haze'** (all year) ♀

C. vulgaris **'Multicolor'** (all year)

E. cinerea **'Glencairn'** (spr.-sum.)

C. vulgaris **'Darkness'** (sum.-aut.) ♀

C. vulgaris **'Foxii Nana'** (all year)

C. vulgaris **'Boskoop'** (all year)

E. cinerea **'Purple Beauty'** (sum.)

E. carnea **'Westwood Yellow'** (all year) ♀

ROSES/shrub and old garden
☐ WHITE

R. 'Mme Hardy'
Vigorous, upright Damask rose with plentiful, leathery, matt leaves. Richly fragrant, quartered-rosette, fully double flowers, 10cm (4in) across, white with green eyes, are borne in summer. H 1.5m (5ft), S 1.2m (4ft).

R. 'Boule de Neige'
Upright Bourbon rose with arching stems and very fragrant, cupped to rosette, fully double flowers. White flowers, sometimes tinged with pink, 8cm (3in) across, appear in summer-autumn. Leaves are glossy and dark green. H 1.5m (5ft), S 1.2m (4ft).

R. pimpinellifolia 'Plena'
Dense, spreading, prickly species rose with cupped, double, creamy-white flowers, 4cm (1½in) across, in early summer. Has small, fern-like, dark green leaves and blackish hips. H 1m (3ft), S 1.2m (4ft).

R. 'Penelope'
Dense, bushy shrub rose, with plentiful, dark green foliage, that bears many scented, cupped, double, pink-cream flowers, 8cm (3in) across, in clusters in summer-autumn. H and S 1m (3ft), more if lightly pruned.

Categories of Rose

Grown for the extraordinary beauty of their flowers, roses have been in cultivation for some hundreds of years. They have been widely hybridized, producing a vast number of shrubs suitable for growing as specimen plants, in the border, as hedges and as climbers for training on walls, pergolas and pillars. Roses are classified in three main groups:

Species
Species, or wild, roses and **species hybrids**, which share most of the characteristics of the parent species, bear flowers generally in one flush in summer and hips in autumn.

Old Garden Roses
Alba – large, freely branching roses with clusters of flowers in mid-summer and abundant, greyish-green foliage.
Bourbon – open, remontant shrub roses that may be trained to climb. Flowers are borne, often 3 to a cluster, in summer-autumn.
China – remontant shrubs with flowers borne singly or in clusters in summer-autumn; provide shelter.
Damask – open shrubs bearing loose clusters of usually very fragrant flowers mainly in summer.
Gallica – fairly dense shrubs producing richly coloured flowers, often 3 to a cluster, in the summer months.
Hybrid Perpetual – vigorous, remontant shrubs with flowers borne singly or in 3s in summer-autumn.
Moss – often lax shrubs with a furry, moss-like growth on stems and calyx and flowers in summer.
Noisette – remontant climbing roses that bear large clusters of flowers, with a slight spicy fragrance, in summer-autumn; provide shelter.
Portland – upright, rather dense, remontant shrubs, bearing loose clusters of flowers in summer-autumn.
Provence (Centifolia) – lax, thorny shrubs bearing scented flowers in summer.
Sempervirens – semi-evergreen climbing roses that bear numerous flowers in late summer.
Tea – remontant shrubs and climbers with elegant, pointed buds that open to loose flowers with a spicy fragrance; provide shelter.

Modern Garden roses
Shrub – a diverse group, illustrated here with the Old Garden Roses because of their similar characteristics. Most are remontant and are larger than bush roses, with flowers borne singly or in sprays in summer and/or autumn.

Flower shapes

With the mass hybridization that has occurred in recent years, roses have been developed to produce plants with a wide variety of characteristics, in particular different forms of flower, often with a strong fragrance. These flower types, illustrated below, give a general indication of the shape of the flower at its perfect state (which in some cases may be before it has opened fully). Growing conditions may affect the form of the flower. Flowers may be single (4–7 petals), semi-double (8–14 petals), double (15–30 petals) or fully double (over 30 petals).

Flat – open, usually single or semi-double flowers have petals that are almost flat.

Cupped – open, single to fully double flowers have petals curving outwards gently from the centre.

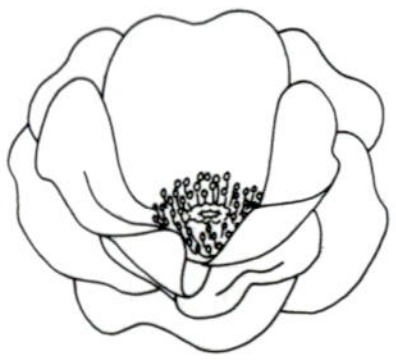

Pointed – elegant, 'Hybrid Tea' shape, semi-double to fully double flowers have high, tight centres.

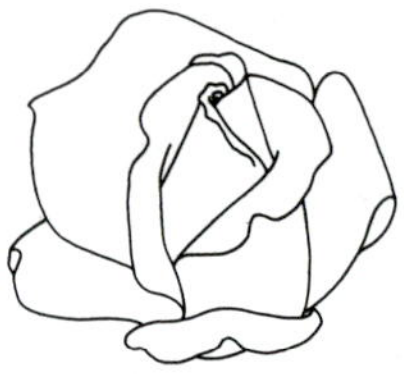

Urn-shaped – classic, curved, flat-topped, semi-double to fully double flowers are of 'Hybrid Tea' type.

Rounded – usually double or fully double flowers have even-sized, overlapping petals that form a bowl-shaped or rounded outline.

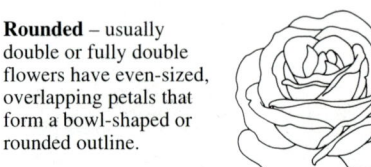

Rosette – usually double or fully double flowers are rather flat with many confused, slightly overlapping petals of uneven size.

Quartered-rosette – rather flat, usually double or fully double flowers have confused petals of uneven size arranged in a quartered pattern.

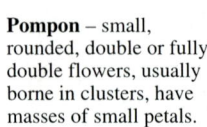

Pompon – small, rounded, double or fully double flowers, usually borne in clusters, have masses of small petals.

Large-flowered bush (Hybrid Tea) – remontant shrubs with large flowers borne in summer-autumn.
Cluster-flowered bush (Floribunda) – remontant shrubs with usually large sprays of flowers in summer-autumn.
Dwarf clustered-flowered bush (Patio) – neat, remontant shrubs with sprays of flowers borne in summer-autumn.
Miniature bush – very small, remontant shrubs with sprays of tiny flowers in summer-autumn.
Polyantha – tough, compact, remontant shrubs with many small flowers in summer-autumn.
Ground cover – trailing and spreading roses, some flowering in summer only, others remontant, flowering in summer-autumn.
Climbing – vigorous climbing roses, diverse in growth and flower, some flowering in summer only, others remontant, flowering in summer-autumn.
Rambler – vigorous climbing roses with flexible stems that bear clusters of flowers mostly in summer.

☐ WHITE–PINK

R. **'Dupontii'** (Snowbush rose)
Upright, bushy shrub rose with abundant, greyish foliage. Bears many clusters of fragrant, flat, single, white flowers, tinged with blush-pink, 6cm (2½in) across, in mid-summer. H and S 2.2m (7ft).

☼ ◊ ❋❋❋

R. **'Nevada'**
Dense, arching shrub rose with abundant, light green leaves. Scented, flat, semi-double, creamy-white flowers, 10cm (4in) across, are borne freely in summer and more sparsely in autumn. H and S 2.2m (7ft).

☼ ◊ ❋❋❋ ♉

R. **'Pearl Drift'**, syn. *R.* 'Leggab'
Bushy, spreading shrub rose that produces clusters of lightly scented, cupped, double, blush-pink flowers, 10cm (4in) across, in summer-autumn. Leaves are plentiful and glossy. H 1m (3ft), S 1.2m (4ft).

☼ ◊ ❋❋❋

R. 'Aimée Vibert'
R. x *alba* 'Semiplena'
R. 'Blanche Moreau'
R. 'Blush Noisette'
R. 'Jacqueline du Pré'
R. 'Mousseline'
R. 'Souvenir de la Malmaison'

☐ PINK

R. **'Fantin-Latour'**
Vigorous, shrubby Provence rose. Flowers appear in summer and are fragrant, cupped to flat, fully double, blush-pink, with neat, green button eyes, and 10cm (4in) across. Has broad, dark green leaves. H 1.5m (5ft), S 1.2m (4ft).

☼ ◊ ❋❋❋ ♉

R. **'Great Maiden's Blush'**, syn. *R.* 'Cuisse de Nymphe', *R.* 'La Séduisante'
Vigorous, upright Alba rose. Very fragrant, rosette, fully double, pinkish-white flowers, 8cm (3in) across, appear in mid-summer. H 2m (6ft), S 1.3m (4½ft).

☼ ◊ ❋❋❋

R. **'Felicia'**
Vigorous shrub rose with abundant, healthy, greyish-green foliage. Scented, cupped, double flowers, 8cm (3in) across, are light pink, tinged with apricot, and are borne in summer-autumn. H 1.5m (5ft), S 2.2m (7ft).

☼ ◊ ❋❋❋ ♉

R. 'Cécile Brunner'
R. 'Félicité Parmentier'
R. 'Kathleen Harrop'
R. 'Omar Khayyám'

R. **'Conrad Ferdinand Meyer'**
Vigorous, arching shrub rose with cupped, fully double, pink flowers, 7cm (3in) across, that are richly fragrant and borne in large numbers in summer, fewer in autumn. Foliage is leathery and prone to rust. H 2.5m (8ft), S 1.2m (4ft).

☼ ◊ ❋❋❋

R. **'Céleste'**, syn. *R.* 'Celestial'
Vigorous, spreading, bushy Alba rose. Fragrant, cupped, double, light pink flowers, 8cm (3in) across, appear in summer. Makes a good hedge. H 1.5m (5ft), S 1.2m (4ft).

☼ ◊ ❋❋❋

R. ***eglanteria***, syn. *R. rubiginosa* (Eglantine, Sweet briar)
Vigorous, arching, thorny species rose that has distinctive, apple-scented foliage. Bears cupped, single, pink flowers, 2.5cm (1in) across, in mid-summer and red hips in autumn. H and S 2.4m (8ft).

☼ ◊ ❋❋❋ ♉

Roses/shrub and old garden
■ PINK

R. 'Rosy Cushion', syn. *R.* 'Interall'
Dense, spreading shrub rose with plentiful, glossy, dark green leaves. Bears clusters of scented, cupped, semi-double flowers, 6cm (2½in) across, that are pink with ivory centres, in summer-autumn. H 1m (3ft), S 1.2m (4ft).

R. glauca, syn. *R. rubrifolia*
Vigorous, arching species rose, grown for its fine, greyish-purple leaves and red stems. Flat, single, cerise-pink flowers, 4cm (1½in) across, with pale centres and gold stamens, appear in early summer, followed by red hips in autumn. H 2m (6ft), S 1.5m (5ft).

R. 'Marguerite Hilling', syn. *R.* 'Pink Nevada'
Dense, arching shrub rose. Many scented, flat, semi-double, rose-pink flowers, 10cm (4in) across, are borne in summer and a few in autumn. Has plentiful, light green foliage. H and S 2.2m (7ft).

R. 'Reine Victoria'
Lax Bourbon rose with slender stems and light green leaves. Sweetly scented, rosette, double flowers, 8cm (3in) across, in shades of pink, are borne in summer-autumn. Grows well on a pillar. H 2m (6ft), S 1.2m (4ft).

R. 'Königin von Dänemark', syn. *R.* 'Belle Courtisanne'
Vigorous, rather open Alba rose. Heavily scented, quartered-rosette, fully double, warm-pink flowers, 8cm (3in) across and with green button eyes, appear in mid-summer. H 1.5m (5ft), S 1.2m (4ft).

R. 'Pink Grootendorst'
Upright, bushy shrub rose with plentiful, small leaves. Rosette, double flowers, 5cm (2in) across, have serrated, clear pink petals. Blooms are carried in sprays in summer-autumn. H 2m (6ft), S 1.5m (5ft).

R. 'Complicata'
Very vigorous Gallica rose, with thorny, arching growth, useful as a large hedge. Slightly fragrant, cupped, single flowers, 11cm (4½in) across, are pink with pale centres and appear in mid-summer. H 2.2m (7ft), S 2.5m (8ft).

R. 'Bonica '82', syn. *R.* 'Meidonomac'
Vigorous, spreading shrub rose bearing large sprays of slightly fragrant, cup-shaped, fully double, rose-pink flowers, 7cm (3in) across, in summer-autumn. Foliage is glossy and plentiful. H 90cm (3ft), S 1.1m (3½ft).

R. 'Constance Spry'
Shrub rose of arching habit that will climb if supported. Cupped, fully double, pink flowers, 12cm (5in) across, with a spicy scent, are borne freely on nodding stems in summer. Leaves are large and plentiful. H 2m (6ft), S 1.5m (5ft).

R. x *centifolia* 'Muscosa'
R. 'Cristata'
R. 'Frühlingsmorgen'
R. 'Gloire des Mousseux'
R. 'Ispahan'
R. 'Louise Odier'
R. 'Mme Ernst Calvat'
R. 'Mme Pierre Oger'
R. 'Mary Rose'
R. 'Perle d'Or'
R. 'St Nicholas'
R. 'Tricolore de Flandres'

■□ PINK–RED

R. 'Mrs John Laing'
Bushy Hybrid Perpetual rose with plentiful, light green foliage. Produces many richly fragrant, rounded, fully double, pink flowers, 12cm (5in) across, in summer and a few in autumn. H 1m (3ft), S 80cm (2½ft).

☼ ◊ ❀❀❀

R. chinensis 'Mutabilis', syn. *R.* x *odorata* 'Mutabilis'
Open species rose with coppery, young foliage. In summer-autumn bears shallowly cup-shaped, single, buff-yellow flowers, 6cm (2½in) across, that age to coppery-pink or -crimson. H and S 1m (3ft), to 2m (6ft) against a wall.

☼ ◊ ❀❀❀ ♕

R. rugosa
(Hedgehog rose, Japanese rose)
Vigorous, dense species rose with wrinkled leaves and large, red hips. Cupped, single, white or purplish-red flowers, 9cm (3½in) across, appear in good succession in summer-autumn. H and S 1–2m (3–6ft).

☼ ◊ ❀❀❀

R. 'Roseraie de l'Haÿ'
Vigorous, dense shrub rose. Bears many strongly scented, cupped to flat, double, reddish-purple flowers, 11cm (4½in) across, in summer-autumn. Leaves are abundant and disease-resistant. H 2.2m (7ft), S 2m (6ft).

☼ ◊ ❀❀❀ ♕

R. 'Old Blush China', syn. *R.* 'Parson's Pink China'
Bushy China rose that may be trained as a climber on a sheltered wall. Cupped, double, pink flowers, 6cm (2½in) across, are produced freely from summer to late autumn. H 1m (3ft), S 80cm (2½ft) or more.

☼ ◊ ❀❀❀

R. 'Geranium', syn. *R. moyesii* 'Geranium'
Vigorous, arching species rose. Flat, single flowers, 5cm (2in) across, are dusky scarlet with yellow stamens and are borne branches in summer. Has small, dark green leaves and in autumn large, red hips. H 3m (10ft), S 2.5m (8ft).

☼ ◊ ❀❀❀ ♕

R. gallica 'Versicolor'
(Rosa mundi)
Neat, bushy Gallica rose. In summer produces striking, slightly scented, flat, semi-double flowers, 5cm (2in) across, very pale blush-pink with crimson stripes. H 75cm (2½ft), S 1m (3ft).

☼ ◊ ❀❀❀ ♕

R. 'Mme Isaac Pereire'
Vigorous, arching Bourbon rose. Fragrant, cupped to quartered-rosette, fully double flowers, 15cm (6in) across, are deep purplish-pink and are produced freely in summer-autumn. H 2.2m (7ft), S 2m (6ft).

☼ ◊ ❀❀❀ ♕

R. 'Henri Martin', syn. *R.* 'Red Moss'
Vigorous, upright Moss rose. Rosette, double, purplish-crimson flowers, 9cm (3½in) across, appear in summer and have a light scent and some furry, green 'mossing' of the calyces underneath. H 1.5m (5ft), S 1m (3ft).

☼ ◊ ❀❀❀

R. 'Armada'
R. 'Commandant Beaurepaire'
R. 'Du Maître d'Ecole'
R. 'Fellenberg'

R. gallica var. officinalis
R. 'Honorine de Brabant'
R. macrophylla
R. moyesii

R. 'Président de Sèze'
R. 'Variegata di Bologna'

153

Roses/shrub and old garden

▨ RED–PURPLE

R. 'Empereur du Maroc'
Compact, shrubby Hybrid Perpetual rose. Fragrant, quartered-rosette, fully double flowers, 8cm (3in) across, are rich purplish-crimson and are borne freely in summer and more sparsely in autumn. H 1.2 m (4ft), S 1m (3ft).

☼ ◊ ✿✿✿

R. 'Belle de Crécy'
Gallica rose of rather lax growth and few thorns. Rosette, fully double flowers, 8cm (3in) across, are pink, tinged greyish-purple with green eyes, have a rich, spicy fragrance and are produced in summer. H 1.2 m (4ft), S 1m (3ft).

☼ ◊ ✿✿✿ ♈

R. 'Cardinal Hume', syn. *R.* 'Harregale'
Bushy, spreading shrub rose. Cupped, fully double, reddish-purple flowers, 7.5cm (3in) across, are borne in dense clusters in summer-autumn and have a musky scent. H 90cm (3ft), S 1m (3ft).

☼ ◊ ✿✿✿

R. 'William Lobb', syn. *R.* 'Duchesse d'Istrie'
Moss rose with strong, arching, prickly stems that will climb if supported. In summer bears rosette, double, deep purplish-crimson flowers, 9cm (3½in) across, that fade to lilac-grey. H and S 2m (6ft).

☼ ◊ ✿✿✿ ♈

R. 'Cardinal de Richelieu'
Vigorous, compact Gallica rose that bears plentiful, dark green foliage and fragrant, rounded, fully double, deep burgundy-purple flowers, 8cm (3in) across, in summer. H 1.2m (4ft), S 1m (3ft).

☼ ◊ ✿✿✿ ♈

R. 'Tour de Malakoff'
Provence rose of open habit. Scented, rosette, double flowers, 12cm (5in) across, are magenta with violet veins, fading to greyish-purple, and appear in summer. H 2m (6ft), S 1.5m (5ft).

☼ ◊ ✿✿✿

▨ YELLOW

R. primula (Incense rose)
Lax, arching species rose that bears scented, cupped, single, primrose-yellow flowers, 4cm (1½in) across, in late spring. Foliage is plentiful, aromatic and fern-like. May die back in hard winters. H and S 2m (6ft).

☼ ◊ ✿✿✿ ♈

R. foetida 'Persiana', syn. *R.* 'Persian Yellow'
Upright, arching, species rose with cupped, double, yellow flowers, 2.5cm (1in) across, in early summer. Glossy leaves are prone to blackspot. Prune spent branches only and shelter from cold winds. H and S 1.5m (5ft).

☼ ◊ ✿✿✿

R. ecae
Erect, wiry species rose. Cupped, single, bright yellow flowers, 2cm (¾in) across, with a musky scent, are borne close to reddish stems in late spring. Foliage is fern-like and graceful. Needs shelter. H 1.5m (5ft), S 1.2m (4ft).

☼ ◊ ✿✿✿

R. 'Assemblage des Beautés'
R. 'Capitaine John Ingram'
R. 'Charles de Mills'
R. 'Nuits de Young'

R. 'Robert le Diable'
R. 'Souvenir d'Alphonse Lavallée'
R. 'Tuscany Superb'
R. 'Zigeunerknabe'

R. 'Buff Beauty'
R. 'Golden Wings'

Roses/modern

YELLOW WHITE–PINK

R. 'Iceberg',
syn. R. 'Schneewittchen'
Cluster-flowered bush rose. Produces many sprays of cupped, fully double, white flowers, 7cm (3in) across, in summer-autumn. Has abundant, glossy leaves. H 75cm (30in), S 65cm (26in), more if not pruned hard.

R. 'Canary Bird'
Vigorous, dense, arching species hybrid with small, fern-like leaves. Cupped, single, yellow flowers, 5cm (2in) across, with a musky scent, appear in late spring and sparsely in autumn. May die back in hard winters. H and S 2.1m (7ft).

R. 'Margaret Merril'
Upright, cluster-flowered bush rose. Very fragrant, double, blush-white or white flowers, are well-formed, urn-shaped and 10cm (4in) across, and are borne singly or in clusters in summer-autumn. H 1m (3ft), S 60cm (2ft).

R. 'Elizabeth Harkness'
Neat, upright, large-flowered bush rose with abundant, dark green foliage. Fragrant, pointed, fully double flowers, 12cm (5in) across, are pale creamy-pink, tinted buff, and are borne in summer-autumn. H 80cm (2½ft), S 60cm (2ft).

R. 'Graham Thomas',
syn. R. 'Ausmas'
Vigorous, arching shrub rose, lax in habit, with glossy, bright green leaves. In summer-autumn bears cupped, fully double, yellow flowers, 11cm (4½in) across, with some scent. H 1.2m (4ft), S 1.5m (5ft).

R. 'Grouse', syn. R. 'Korimro'
Trailing, ground cover rose with very abundant, glossy foliage and flat, single, blush-pink flowers, 4cm (1½in) across, borne close to stems in summer-autumn. Has a pleasant fragrance. H 45cm (18in), S 3m (10ft).

R. 'The Fairy'
Dense, cushion-forming, dwarf cluster-flowered bush rose with abundant, small, glossy leaves. Rosette, double, pink flowers, 2.5cm (1in) across, are borne freely in late summer and autumn. H and S 60cm (2ft).

R. 'Nozomi', syn. R. 'Heideröslein'
Creeping, ground cover rose bearing flat, single, blush-pink and white flowers, 2.5cm (1in) across, close to stems in summer. Has small, dark green leaves. May be used for a container. H 45cm (18in), S 1.2m (4ft).

OTHER RECOMMENDED PLANTS:
R. 'City of London'
R. 'English Miss'
R. 'Hannah Gordon'
R. 'Langford Light'
R. 'Len Turner'
R. 'Ophelia'
R. 'Pascali'
R. 'Peaudouce'
R. 'Sue Lawley'
R. 'Yvonne Rabier'

Roses/modern
■ PINK

R. 'Pink Bells', syn. *R.* 'Poulbells'
Very dense, spreading, ground cover rose with abundant, small, dark green leaves and many pompon, fully double, pink flowers, 2.5cm (1in) across, borne in clusters in summer. H 75cm (2½ft), S 1.2m (4ft).

R. 'Queen Elizabeth'
Upright, cluster-flowered bush rose that bears long-stemmed, rounded, fully double, pink flowers, 10cm (4in) across, singly or in clusters, in summer-autumn. Leaves are large and leathery. H 1.5m (5ft), S 75cm (2½ft), more if not pruned hard.

R. 'Iced Ginger'
Upright, cluster-flowered bush rose with sparse, reddish foliage. Pointed, fully double, buff to copper-pink flowers, 11cm (4½in) across, are borne singly or in clusters in summer-autumn. H 1m (3ft), S 70cm (28in).

R. 'Alpine Sunset'
Compact, large-flowered bush rose with fragrant, rounded, fully double, peach-yellow flowers, 20cm (8in) across, appearing on short stems in summer-autumn. Has large, semi-glossy leaves. May die back in hard winters. H and S 60cm (2ft).

R. 'Sexy Rexy', syn. *R.* 'Macrexy'
Compact, bushy, cluster-flowered bush rose. Bears clusters of slightly fragrant, cupped, camellia-like, fully double, pink flowers, 8cm (3in) across, in summer-autumn. Leaves are dark green. H and S 60cm (2ft).

R. 'Peek-a-boo', syn. *R.* 'Brass Ring', *R.* 'Dicgrow'
Dense, cushion-forming, dwarf cluster-flowered bush rose with many sprays of urn-shaped, double, apricot-pink flowers, 4cm (1½in) across, from summer to early winter. Leaves are narrow and dark green. H and S 45cm (18in).

R. 'Lovely Lady', syn. *R.* 'Dicjubell'
Dense, rounded, large-flowered bush rose. Slightly scented, pointed, fully double, rose-pink flowers, 10cm (4in) across, are produced freely in summer-autumn. H 80cm (30in), S 70cm (28in).

R. 'Rosemary Harkness', syn. *R.* 'Harrowbond'
Vigorous, large-flowered bush rose with abundant, glossy leaves. Bears fragrant, pointed, double flowers, 10 cm (4in) across, in shades of salmon-pink and orange, singly or in clusters in summer-autumn. H 1m (3ft), S 75cm (2½ft).

R. 'Blessings'
Upright, large-flowered bush rose with slightly fragrant, salmon-pink flowers that are urn-shaped and fully double, 10 cm (4in) across, and are borne singly or in clusters in summer-autumn. Has large, dark green leaves. H 1m (3ft), S 75cm (2½ft).

R. 'Gentle Touch'
R. 'Petit Four'
R. 'Pink Parfait'
R. 'Royal Highness'

■ PINK

R. 'Anisley Dickson',
syn. *R.* 'Dickimono', *R.* 'Dicky',
R. 'Münchner Kindl'
Vigorous, cluster-flowered bush rose.
Carries large clusters of slightly
fragrant, pointed, double, salmon-pink
flowers, 8cm (3in) across, in summer-
autumn. H 1m (3ft), S 75cm (2½ft).

☼ ◊ ✻✻✻ ♛

R. 'Silver Jubilee'
Dense, upright, large-flowered bush
rose. Bears slightly scented, pointed,
fully double, soft salmon-pink flowers,
12cm (5in) across, very freely in
summer-autumn. Foliage is abundant
and glossy. H 1.1m (3½ft),
S 75cm (2½ft).

☼ ◊ ✻✻✻ ♛

R. 'Keepsake', syn. *R.* 'Kormalda'
Neat, bushy, large-flowered bush rose
with plentiful, glossy leaves. Slightly
scented, rounded, fully double, pink
flowers, 12cm (5in) across, are freely
produced in summer-autumn. H 75cm
(2½ft), S 60cm (2ft).

☼ ◊ ✻✻✻

R. 'Paul Shirville', syn.
R. 'Harqueterwife', *R.* 'Heartthrob'
Spreading, large-flowered bush rose.
Bears fragrant, pointed, fully double,
rosy salmon-pink flowers, 9cm (3½in)
across, in summer-autumn. Leaves are
glossy, reddish and abundant. H and S
75cm (2½ft).

☼ ◊ ✻✻✻ ♛

R. 'Double Delight'
Large-flowered bush rose of upright,
uneven growth. Fragrant, rounded,
fully double flowers, 12cm (5in)
across, are creamy-white, edged with
red, and are borne in summer-autumn.
H 1m (3ft), S 60cm (2ft).

☼ ◊ ✻✻✻

R. 'Congratulations'
R. 'Fairy Changeling'
R. 'Fragrant Delight'
R. 'Little Woman'

R. 'Mischief'

■ PINK–RED

R. 'Escapade'
Dense, cluster-flowered bush rose.
Fragrant, cupped, semi-double, rose-
violet flowers, 8cm (3in) across, with
white eyes, are borne in sprays in
summer-autumn. Foliage is light green
and glossy. H 75cm (2½ft),
S 60cm (2ft).

☼ ◊ ✻✻✻

R. 'Anna Ford',
syn. *R.* 'Harpiccolo'
Dwarf cluster-flowered bush rose.
Has urn-shaped (opening flat), double,
orange-red flowers, 4cm (1½in)
across, borne in summer-autumn, and
many small, dark green leaves. H 45cm
(18in), S 38cm (15in).

☼ ◊ ✻✻✻ ♛

R. 'Trumpeter', syn. *R.* 'Mactru'
Neat, bushy, cluster-flowered bush rose
with many cupped, fully double, bright
red flowers, 6cm (2½in) across, in
summer-autumn. Leaves are deep
green and semi-glossy. H 60cm (24in),
S 50cm (20in).

☼ ◊ ✻✻✻ ♛

R. 'Buttons'
R. 'Rose Gaujard'
R. 'Yesterday'

Roses/modern

■ RED

R. 'Royal William', syn.
R. 'Duftzauber '84', *R.* 'Korzaun'
Vigorous, large-flowered bush rose with large, dark green leaves. Slightly scented, pointed, fully double, deep crimson flowers, 12cm (5in) across, are carried on long stems in summer-autumn. H 1m (3ft), S 75cm (2½ft).

R. 'Precious Platinum',
syn. *R.* 'Opa Potschke'
Vigorous, large-flowered bush rose with abundant, glossy leaves. Bears slightly scented, rounded, fully double, deep crimson-scarlet flowers, 10cm (4in) across, in summer-autumn. H 1m (3ft), S 60cm (2ft).

R. 'Wee Jock', syn. *R.* 'Cocabest'
Dense, bushy, dwarf cluster-flowered bush rose. Bears rosette, fully double, crimson flowers, 4cm (1½in) across, in summer-autumn. Plentiful leaves are small and dark green. H and S 45cm (18in).

■ YELLOW

R. 'Champagne Cocktail',
syn. *R.* 'Horflash'
Upright, cluster-flowered bush rose. Fragrant, cupped, double, yellow-pink flowers, 9cm (3½in) across, opening wide, are borne in summer-autumn. H 1m (3ft), S 60cm (2ft).

R. 'The Times Rose', syn. *R.* 'Korpeahn'
Spreading, cluster-flowered bush rose. Slightly scented, cupped, double, deep crimson flowers, 8cm (3in) across, are borne in wide clusters in summer-autumn. Foliage is dark green and plentiful. H 60cm (2ft), S 75cm (2½ft).

R. 'Peace', syn. *R.* 'Gioia', *R.* 'Gloria Dei', *R.* 'Mme A. Meilland'
Vigorous, shrubby, large-flowered bush rose with scented, pointed to rounded, fully double flowers, 15cm (6in) across, borne freely in clusters in summer-autumn. Has abundant, large, glossy foliage. H 1.2m (4ft), S 1m (3ft).

R. 'Alexander',
syn. *R.* 'Alexandra'
Vigorous, upright, large-flowered bush rose with abundant, dark green foliage. Slightly scented, pointed, double, bright red flowers, 12cm (5in) across, are borne on long stems in summer-autumn. H 1.5m (5ft), S 75cm (2½ft).

R. 'Alec's Red'
Vigorous, large-flowered bush rose bearing strongly fragrant, deep cherry-red flowers that are pointed and fully double, 15cm (6in) across, in summer-autumn. H 1m (3ft), S 60cm (2ft).

R. 'Rugul', syn. *R.* 'Guletta', *R.* 'Tapis Jaune'
Compact, dense, dwarf cluster-flowered bush rose with cupped to flat, double, yellow flowers, 5cm (2in) across, that are borne in summer-autumn, and rich green leaves. H 30cm (12in), S 40cm (16in).

R. 'Blue Moon'
R. 'Drummer Boy'
R. 'Felicity Kendall'
R. 'Fragrant Cloud'
R. 'Ingrid Bergman'
R. 'Intrigue'
R. 'Invincible'
R. 'Loving Memory'
R. 'Malcolm Sargent'
R. 'National Trust'
R. 'News'
R. 'Shocking Blue'
R. 'Julia's Rose'
R. 'Sheila's Perfume'

☐ YELLOW

R. 'Grandpa Dickson',
syn. R. 'Irish Gold'
Neat, upright, large-flowered bush rose with sparse, pale, glossy foliage. Bears many slightly scented, pointed, fully double, light yellow flowers, 18cm (7in) across, in summer-autumn.
H 80cm (2½ft), S 60cm (2ft).

☼ ◊ ❀❀❀

R. 'Korresia', syn. R. 'Friesia'
Bushy, upright, cluster-flowered bush rose. Bears open sprays of strongly scented, urn-shaped, double flowers, 8cm (3in) across, with waved, yellow petals, in summer-autumn. H 75cm (2½ft), S 60cm (2ft).

☼ ◊ ❀❀❀

R. 'Glenfiddich'
Upright, cluster-flowered bush rose. Slightly fragrant, urn-shaped, double, amber-yellow flowers, 10cm (4in) across, are borne singly or in clusters in summer-autumn. H 75cm (2½ft), S 60cm (2ft).

☼ ◊ ❀❀❀

R. 'Bright Smile',
syn. R. 'Dicdance'
Low, bushy, cluster-flowered bush rose with bright, glossy leaves. Bears clusters of slightly scented, flat, semi-double, yellow flowers, 8cm (3in) across, in summer-autumn. H and S 45cm (18in).

☼ ◊ ❀❀❀

R. 'Simba', syn. R. 'Goldsmith', R. 'Korbelma'
Upright, large-flowered bush rose with lightly fragrant, urn-shaped, fully double, yellow flowers, 9cm (3½in) across, borne freely in summer-autumn. Leaves are large and dark green.
H 75cm (2½ft), S 60cm (2ft).

☼ ◊ ❀❀❀

R. 'Mountbatten',
syn. R. 'Harmantelle'
Shrubby, cluster-flowered bush rose with disease-resistant foliage. Bears scented, rounded, fully double, yellow flowers, 10cm (4in) across, singly or in clusters, in summer-autumn. H 1.2m (4ft), S 75cm (2½ft).

☼ ◊ ❀❀❀ ♛

R. 'Freedom', syn. R. 'Dicjem'
Neat, large-flowered bush rose, with many shoots and abundant, glossy foliage. Bears many lightly scented, rounded, double, bright yellow flowers, 9cm (3½in) across, in summer-autumn. H 75cm (2½ft), S 60cm (2ft).

☼ ◊ ❀❀❀ ♛

R. 'Amber Queen',
syn. R. 'Harroony'
Spreading, cluster-flowered bush rose. Amber flowers are fragrant, rounded and fully double, 8cm (3in) across, and are borne in summer-autumn. Has abundant, reddish foliage. H and S 50cm (20in).

☼ ◊ ❀❀❀ ♛

R. 'Arthur Bell'
R. 'Dutch Gold'
R. 'Goldstar'
R. 'Princess Alice'

R. 'Princess Michael of Kent'

Roses/modern
YELLOW–ORANGE

R. 'Pot o' Gold',
syn. R. 'Dicdivine'
Large-flowered bush rose of neat, even growth. Fragrant, rounded, fully double, golden-yellow flowers, 9cm (3½in) across, are carried singly or in wide sprays in summer-autumn. H 75cm (2½ft), S 60cm (2ft).

R. 'Sweet Magic',
syn. R. 'Dicmagic'
Bushy, dwarf cluster-flowered bush rose. Bears sprays of lightly fragrant, urn-shaped, double, pink-flushed, golden-orange flowers, 4cm (1½in) across, in summer-autumn. H 38cm (15in), S 30cm (12in).

R. 'Troika', syn. R. 'Royal Dane'
Vigorous, dense, large-flowered bush rose with plentiful, semi-glossy leaves. Fragrant, pointed, double flowers, 15cm (6in) across, are orange-red, tinged with pink, and are borne in summer-autumn. H 1m (3ft), S 75cm (2½ft).

R. 'Piccadilly'
Vigorous, bushy, large-flowered bush rose with pointed, double, red and yellow flowers, 12cm (5in) across, produced freely singly or in clusters in summer-autumn. Abundant foliage is reddish and glossy. H 1m (3ft), S 60cm (2ft).

R. 'Southampton',
syn. R. 'Susan Ann'
Upright, cluster-flowered bush rose. Bears fragrant, pointed, double, apricot flowers, 8cm (3in) across, singly or in clusters in summer-autumn. Foliage is glossy and disease-resistant. H 1m (3ft), S 60cm (2ft).

R. 'Doris Tysterman'
Vigorous, upright, large-flowered bush rose with lightly scented, pointed, fully double, orange-red flowers, 10cm (4in) across, borne in summer-autumn. Leaves are large, glossy and dark green. H 1.2m (4ft), S 75cm (2½ft).

R. 'Anne Harkness',
syn. R. 'Harkaramel'
Upright, cluster-flowered bush rose. Urn-shaped, double, amber flowers, 8cm (3in) across, are borne in sprays of many blooms in late summer and autumn. H 1.2m (4ft), S 60cm (2ft).

R. 'Remember Me',
syn. R. 'Cocdestin'
Vigorous, dense, large-flowered bush rose with pointed, fully double, copper-orange flowers, 9cm (3½in) across, freely borne in summer-autumn. Leaves are abundant and glossy. H 1m (3ft), S 75cm (2½ft).

R. 'Just Joey'
Branching, open, large-flowered bush rose with leathery, dark green foliage. Bears rounded, fully double flowers, 12cm (5in) across, with waved, copper-pink petals and some scent, in summer-autumn. H 75cm (2½ft), S 60cm (2ft).

R. 'Clarissa'
R. 'Disco Dancer'
R. 'Matangi'
R. 'Sweet Dream'
R. 'Typhoon'
R. 'Whisky Mac'

Roses/miniature

WHITE–PINK

R. 'Snowball', syn. R. 'Angelita', R. 'Macangel'
Compact, creeping, miniature bush rose with pompon, fully double, white flowers, 2.5cm (1in) across, that are borne in summer-autumn. Leaves are small, glossy and plentiful. H 20cm (8in), S 30cm (12in).

R. 'Baby Masquerade', syn. R. 'Baby Carnival'
Dense, miniature bush rose with plentiful, leathery foliage and clusters of rosette, double, yellow-pink flowers, 2.5cm (1in) across, in summer-autumn. H and S 40cm (16in), more if not pruned.

R. 'Angela Rippon', syn. R. 'Ocarina', R. 'Ocaru'
Miniature bush rose with slightly fragrant, urn-shaped, fully double, salmon-pink flowers, 4cm (1½in) across, in summer-autumn, and many small, dark green leaves. H 45cm (18in), S 30cm (12in).

R. 'Stacey Sue'
Spreading, miniature bush rose with plentiful, dark green foliage and rosette, fully double, pink flowers, 2.5cm (1in) across, that are borne freely in summer-autumn. H and S 38cm (15in).

R. 'Hula Girl'
Wide, bushy, miniature bush rose with glossy, dark green foliage. Slightly scented, urn-shaped, fully double, salmon-orange flowers, 2.5cm (1in) across, are produced freely in summer-autumn. H 45cm (18in), S 40cm (16in).

RED

R. 'Sheri Anne'
Upright, miniature bush rose with glossy, leathery foliage. Slightly scented, rosette, double, light red flowers, 2.5cm (1in) across, are borne in summer-autumn. H 45cm (18in), S 30 cm (12in).

R. 'Fire Princess'
Upright, miniature bush rose with small, glossy leaves. Bears sprays of rosette, fully double, scarlet flowers, 4cm (1½in) across, in summer-autumn. H 45cm (18in), S 30cm (12in).

R. 'Red Ace', syn. R. 'Amruda'
Compact, miniature bush rose with rosette, double, dark red flowers, 4cm (1½in) across, borne in summer-autumn. H 35cm (14in), S 30cm (12in).

OTHER RECOMMENDED PLANTS:
R. 'Easter Morning'
R. 'Pink Sunblaze'
R. 'Snow Carpet'

Roses/miniature

🟥🟧 **RED–ORANGE**

R. 'Orange Sunblaze',
syn. R. 'Meijikitar', R. 'Sunblaze'
Compact, miniature bush rose. Rosette, fully double, bright orange-red flowers, 4cm (1½in) across, are freely produced in summer-autumn. Has plentiful, dark green leaves. H and S 30cm (12in).

☀ ◊ ❊❊❊

R. 'Rise 'n Shine',
syn. R. 'Golden Sunblaze'
Bushy, upright, miniature bush rose with dark green leaves that bears rosette, fully double, yellow flowers, 2.5cm (1in) across, in summer-autumn. H 40cm (16in), S 25cm (10in).

☀ ◊ ❊❊❊

R. 'Colibri '79',
syn. R. 'Meidanover'
Upright, rather open, miniature bush rose. Urn-shaped, double, red-veined, orange flowers, 4cm (1½in) across, are borne in summer-autumn. H 38cm (15in), S 25cm (10in).

☀ ◊ ❊❊

R. 'Baby Gold Star'
R. 'Magic Carousel'

Roses/climbing

☐ **WHITE**

R. 'Albéric Barbier'
Vigorous, semi-evergreen rambler rose. Slightly fragrant, rosette, fully double, creamy-white flowers, 8cm (3in) across, appear in clusters in summer. Leaves are small and bright green. Tolerates a north-facing wall. H to 5m (15ft), S 3m (10ft).

☀ ◊ ❊❊❊ ♧

R. 'Paul's Lemon Pillar'
Stiff, upright climbing rose with large leaves and scented, pointed to rounded, fully double, lemon-white flowers, 15cm (6in) across, that appear in summer. Prefers a sunny, sheltered wall. H 5m (15ft), S 3m (10ft).

☀ ◊ ❊❊❊

R. filipes 'Kiftsgate'
Rampant climbing rose with abundant, glossy, light green foliage. Cupped to flat, single, creamy-white flowers, 2.5cm (1in) across, appear in late summer in spectacular clusters. Use to grow up a tree or in a wild garden. H and S 10m (30ft) or more.

☀ ◊ ❊❊❊ ♧

R. 'Félicité Perpétue'
Sempervirens climbing rose with long, slender stems. Clusters of rosette, fully double, blush-pink to white flowers, 4cm (1½in) across, appear in mid-summer. Small leaves are semi-evergreen. Prune spent wood only. H 5m (15ft), S 4m (12ft).

☀ ◊ ❊❊❊ ♧

R. 'Mme Alfred Carrière'
Noisette climbing rose with slender, smooth stems. Very fragrant, rounded, double flowers are creamy-white, tinged pink, 4cm (1½in) across, and are borne in summer-autumn. H to 5.5m (18ft), S 3m (10ft).

☀ ◊ ❊❊❊ ♧

R. 'Gloire de Dijon'
Stiffly branched Noisette or climbing Tea rose. Fragrant, quartered-rosette, fully double, creamy-buff flowers, 10cm (4in) across, are borne in summer-autumn. H 4m (12ft), S 2.5m (8ft).

☀ ◊ ❊❊❊ ♧

OTHER RECOMMENDED PLANTS:
R. banksiae
R. 'Niphetos'
R. 'Sander's White Rambler'
R. 'Wedding Day'
R. 'White Cockade'

□ PINK

R. 'New Dawn', syn.
R. 'Everblooming Dr W. van Fleet'
Vigorous, very hardy climbing rose. Fragrant, cupped, double, pale pearl-pink flowers, 8cm (3in) across, are borne in clusters in summer-autumn. Tolerates a north-facing wall. H and S 5m (15ft).

R. 'Handel'
Stiff, upright climbing rose. Slightly scented, urn-shaped, double flowers, 8cm (3in) across, are cream, edged with pinkish-red, and produced in clusters in summer-autumn. Has glossy, dark green foliage. H 3m (10ft), S 2.2m (7ft).

R. 'Mme Grégoire Staechelin', syn. *R.* 'Spanish Beauty'
Vigorous, arching climbing rose with large clusters of blooms in summer. Bears rounded to cupped, fully double flowers, 13cm (5in) across, with ruffled, clear pink petals, shaded carmine. H to 6m (20ft), S to 4m (12ft).

R. 'Pink Perpétue'
Stiffly branched climbing rose that may be pruned to grow as a shrub. Bears clusters of cupped to rosette, double, deep pink flowers, 8cm (3in) across, in summer-autumn. Leathery foliage is plentiful. H 2.8m (9ft), S 2.5m (8ft).

R. 'Breath of Life', syn. *R.* 'Harquanne'
Stiff, upright climbing rose with large, lightly scented, rounded, fully double, pinkish-apricot flowers, 10cm (4in) across, borne in summer-autumn. Leaves are semi-glossy. H 2.8m (9ft), S 2.2m (7ft).

R. 'Zéphirine Drouhin'
(Thornless rose)
Lax, arching Bourbon rose that will climb if supported. Bears fragrant, cupped, double, deep pink flowers, 8cm (3in) across, in summer-autumn. Is prone to mildew. May be grown as a hedge. H to 2.5m (8ft), S to 2m (6ft).

R. 'Albertine'
Vigorous rambler rose with arching, thorny, reddish stems. Scented, cup-shaped, fully double, salmon-pink flowers, 8cm (3in) across, are borne in abundant clusters in summer. Is prone to mildew in a dry site. H to 5m (15ft), S 3m (10ft).

R. 'Chaplin's Pink Companion'
Vigorous climbing rose with glossy, dark green foliage and slightly scented, rounded, double, light pink flowers, 5cm (2in) across, borne freely in large clusters in summer. H and S 3m (10ft).

R. 'Rosy Mantle'
Stiff, open-branched climbing rose. Very fragrant, pointed, fully double, rose-pink flowers, 10cm (4in) across, are borne in summer-autumn. Dark green foliage is rather sparse. H 2.5m (8ft), S 2m (6ft).

R. 'Aimée Vibert'
R. 'Aloha'
R. 'Belle Portugaise'
R. 'Blush Noisette'

R. 'Blush Rambler'
R. 'Compassion'
R. 'François Juranville'
R. 'Kathleen Harrop'

R. 'Lady Waterlow'
R. 'Morning Jewel'
R. 'Paul's Himalayan Musk Rambler'
R. 'Paul Transon'

Roses/climbing

■■■ RED–YELLOW

R. 'Dortmund'
Upright climbing rose that may be pruned to make a shrub. Flat, single, red flowers, 10cm (4in) across, with white eyes and a slight scent, are borne freely in clusters in summer-autumn. Has healthy, dark green foliage. H 3m (10ft), S 1.8m (6ft).

☼ ◊ ❊❊❊

R. 'Guinée'
Vigorous, stiffly branched climbing rose. Fragrant, cupped, fully double, blackish-red to maroon flowers, 11cm (4½in) across, are borne in summer. Leaves are large and leathery. H 5m (15ft), S 2.2m (7ft).

☼ ◊ ❊❊❊

□ YELLOW

R. 'Golden Showers'
Stiff, upright climbing rose that may be pruned to grow as a shrub. In summer-autumn produces many fragrant, pointed, double, yellow flowers, 10cm (4in) across, that open flat. H 2m (6ft), S 2.2m (7ft) or more.

☼ ◊ ❊❊❊ ♕

R. banksiae 'Lutea'
(Yellow banksian)
Vigorous climbing rose bearing clusters of many scentless, rosette, fully double, yellow flowers, 2cm (¾in) across, in late spring. Needs a sunny, sheltered wall and pruning of spent wood only. H and S to 10m (30ft).

☼ ◊ ❊❊❊ ♕

R. 'Danse du Feu',
syn. **R. 'Spectacular'**
Vigorous, stiffly branched climbing rose with abundant, glossy foliage. Bears rounded, double, scarlet flowers, 8cm (3in) across, in summer-autumn. H and S 2.5m (8ft).

☼ ◊ ❊❊❊

R. 'Veilchenblau',
syn. **R. 'Blue Rambler'**
Vigorous rambler rose. Rosette, double, violet flowers, streaked white, 2.5cm (1in) across, have a fruity scent and appear in clusters in summer. H 4m (12ft), S 2.2m (7ft).

☼ ◊ ❊❊❊ ♕

R. 'Dublin Bay'
Dense, shrubby climbing rose that may be pruned to grow as a shrub. Bears clusters of cupped, double, bright crimson flowers, 10cm (4in) across, in summer-autumn. Foliage is glossy, dark green and plentiful. H and S 2.2m (7ft).

☼ ◊ ❊❊❊ ♕

R. 'Mermaid'
Slow-growing climbing rose that produces flat, single, primrose-yellow flowers, 12cm (5in) across, in summer-autumn. Has stiff, reddish stems, large, hooked thorns and glossy, dark green leaves. Prefers a sunny, sheltered wall. H and S to 6m (20ft).

☼ ◊ ❊❊❊ ♕

R. 'Maigold'
Vigorous climbing rose with prickly, arching stems that may be pruned to grow as shrub. Fragrant, cupped, semi-double, bronze-yellow flowers, 10cm (4in) across, are borne freely in early summer and sparsely in autumn. H and S 2.5m (8ft).

☼ ◊ ❊❊❊ ♕

R. 'Climbing Ena Harkness'
R. 'Fellenberg'
R. 'John Cabot'
R. 'Ramona'

R. 'Sympathie'

R. 'Alister Stella Gray'
R. 'Climbing Lady Hillingdon'
R. 'Easlea's Golden'
R. 'Emily Gray'

R. 'Fortune's Double Yellow'
R. 'Goldfinch'
R. 'Maréchal Niel'
R. 'Schoolgirl'

CLIMBERS SPRING INTEREST

☐ WHITE

■ PINK–RED

Beaumontia grandiflora
(Herald's trumpet)
Vigorous, evergreen, woody-stemmed, twining climber with rich green leaves that are hairy beneath. Has large, fragrant, white flowers from late spring to summer. H 8m (25ft). Min. 7–10°C (45–50°F).

Mandevilla splendens
Evergreen, woody-stemmed, twining climber. Has lustrous leaves and trumpet-shaped, rose-pink flowers, with yellow centres, appearing in late spring or early summer. H 3m (10ft). Min. 7–10°C (45–50°F).

Clianthus puniceus (Parrot's bill)
Evergreen or semi-evergreen, woody-stemmed, scrambling climber with leaves composed of many leaflets. In spring and early summer bears drooping clusters of unusual, claw-like, brilliant red flowers. H 4m (12ft).

Kennedia rubicunda
(Dusky coral pea)
Fast-growing, evergreen, woody-stemmed, twining climber with leaves divided into 3 leaflets. Coral-red flowers are borne in small trusses in spring-summer. H to 3m (10ft). Min. 5–7°C (41–5°F).

Stephanotis floribunda
(Madagascar jasmine, Wax flower)
Moderately vigorous, evergreen, woody-stemmed, twining climber with leathery, glossy leaves. Scented, waxy, white flowers appear in small clusters from spring to autumn. H 5m (15ft) or more. Min. 13–16°C (55–61°F).

Passiflora coccinea
(Red passion flower)
Vigorous, evergreen, woody-stemmed, tendril climber with rounded, oblong leaves. Has bright deep scarlet flowers, with red, pink and white crowns, from spring to autumn. H 3–4m (10–12ft). Min. 15°C (59°F).

Clianthus puniceus f. *albus*
Evergreen or semi-evergreen, woody-stemmed, scrambling climber, grown for its drooping clusters of claw-like, creamy-white flowers that open in spring and early summer. Mid-green leaves consist of many small leaflets. H 4m (12ft).

Distictis buccinatoria, syn.
Phaedranthus buccinatorius
(Mexican blood flower)
Vigorous, evergreen, woody-stemmed, tendril climber. Has trumpet-shaped, rose-crimson flowers, orange-yellow within, from early spring to summer. H to 5m (15ft) or more. Min. 5°C (41°F).

Tropaeolum tricolorum
Herbaceous climber with delicate stems, small tubers and 5–7-lobed leaves. Small, orange or yellow flowers with black-tipped, reddish-orange calyces are borne from early spring to early summer. H to 1m (3ft). Min. 5°C (41°F).

OTHER RECOMMENDED PLANTS:
CLEMATIS, pp.172–3
Decumaria sinensis

CLEMATIS, pp.172–3
Pandorea jasminoides, p.168
Podranea ricasoliana
Thunbergia coccinea

Climbers SPRING INTEREST

■■■ RED–BLUE

■■ GREEN–YELLOW

Agapetes serpens
Evergreen, arching to pendulous, scandent shrub, best grown with support as a perennial climber. Has small, lance-shaped, lustrous leaves and pendent flowers, rose-red with darker veins, in spring. H 2–3m (6–10ft). Min. 5°C (41°F).

Hardenbergia comptoniana
Evergreen, woody-stemmed, twining climber with leaves of 3 or 5 lance-shaped leaflets. Has racemes of pea-like, deep purple-blue flowers in spring. H to 2.5m (8ft).

Strongylodon macrobotrys (Jade vine)
Fast-growing, evergreen, woody-stemmed, twining climber with claw-like, luminous, blue-green flowers in long, pendent spikes in winter-spring. Leaves have 3 oval, glossy leaflets. H to 20m (70ft). Min. 18°C (64°F).

Mitraria coccinea
Evergreen, woody-stemmed, scrambling climber with oval, toothed leaves. Small, tubular, orange-red flowers are borne singly in leaf axils during late spring to summer. H to 2m (6ft).

Akebia quinata (Chocolate vine)
Woody-stemmed, twining climber, semi-evergreen in mild winters or warm areas, with leaves of 5 leaflets. Vanilla-scented, brownish-purple flowers appear in late spring, followed by sausage-shaped, purplish fruits. H 10m (30ft) or more.

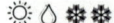

Petrea volubilis
Strong-growing, evergreen, woody-stemmed, twining climber with elliptic, rough-textured leaves and deep violet and lilac-blue flowers carried in simple or branched spikes from late winter to late summer. H 6m (20ft) or more. Min. 13–15°C (55–9°F).

Humulus lupulus 'Aureus'
Herbaceous, twining climber with rough, hairy stems and toothed, yellowish leaves divided into 3 or 5 lobes. Greenish, female flower spikes are borne in pendent clusters in autumn. H to 6m (20ft).

Manettia luteo-rubra, syn. *M. bicolor*, *M. inflata* (Brazilian firecracker)
Fast-growing, evergreen, semi-woody-stemmed, twining climber with glossy leaves. Has small, funnel-shaped, red flowers, with yellow tips, in spring-summer. H 2m (6ft). Min. 5°C (41°F).

Clytostoma callistegioides
Fast-growing, evergreen, woody-stemmed, tendril climber. Each leaf has 2 oval leaflets and a tendril. Small, nodding clusters of purple-veined, lavender flowers, fading to pale pink, are borne in spring-summer. H to 5m (15ft). Min. 10–13°C (50–55°F).

Sollya heterophylla
Evergreen, woody-based, twining climber with narrowly lance-shaped to oval leaves, 2–6cm ($3/4$–$2^1/2$in) long. Nodding clusters of 4–9 broadly bell-shaped, sky-blue flowers are carried from spring to autumn. H to 3m (10ft).

Solandra maxima (Capa de oro, Golden-chalice vine)
Strong-growing, evergreen, woody-stemmed, scrambling climber with glossy leaves. In spring-summer bears fragrant, pale yellow, later golden flowers. H 7–10m (23–30ft) or more. Min. 13–16°C (55–61°F).

Agapetes 'Ludgvan Cross'
Agapetes rugosa
Akebia x *pentaphylla*
Akebia trifoliata

CLEMATIS, pp.172–3
Hardenbergia violacea
Kennedia nigricans
Solanum seaforthianum

Stigmaphyllon ciliatum, p.176

Climbers SUMMER INTEREST

☐ YELLOW

☐ WHITE

Thunbergia mysorensis
Evergreen, woody-stemmed, twining climber. Has narrow leaves and pendent spikes of flowers with yellow tubes and recurved, reddish-brown lobes from spring to autumn. H 6m (20ft). Min. 15°C (59°F).

Bougainvillea glabra **'Snow White'**
Vigorous, evergreen or semi-evergreen, woody-stemmed, scrambling climber with rounded-oval leaves. In summer has clusters of white floral bracts with green veins. H to 5m (15ft). Min. 7–10°C (45–50°F).

Wisteria sinensis **'Alba'**
Vigorous, deciduous, woody-stemmed, twining climber. Leaves are 25–30cm (10–12in) long with 11 leaflets. Has strongly scented, pea-like, white flowers in racemes, 20–30cm (8–12in) long, in early summer. H to 30m (100ft).

Araujia sericofera, syn. *A. sericifera* (Cruel plant)
Evergreen, woody-stemmed, twining climber with leaves that are white-downy beneath. Has scented, white flowers, often striped pale maroon inside, from late summer to autumn. H to 7m (23ft).

Solanum jasminoides **'Album'**
Semi-evergreen, woody-stemmed, scrambling climber. Oval to lance-shaped leaves are sometimes lobed or divided into leaflets. Has star-shaped, white flowers, 2–2.5cm (³/₄–1in) across, in summer–autumn. H to 6m (20ft).

Jasminum mesnyi, syn. *J. primulinum* (Primrose jasmine)
Evergreen or semi-evergreen, woody-stemmed, scrambling climber. Leaves are divided into 3 leaflets; semi-double, pale yellow flowers appear in spring. H to 3m (10ft).

Trachelospermum jasminoides
(Confederate jasmine, Star jasmine)
Evergreen, woody-stemmed, twining climber with oval leaves up to 10cm (4in) long. Very fragrant, white flowers are carried in summer, followed by pairs of pods, up to 15cm (6in) long, containing tufted seeds. H to 9m (28ft).

Wisteria floribunda **'Alba'**
Deciduous, woody-stemmed, twining climber with leaves of 11–19 oval leaflets. Scented, pea-like, white flowers are carried in drooping racemes, up to 60cm (2ft) long, in early summer. H to 9m (28ft).

Canarina canariensis, p.179
Macfadyena unguis-cati, p.176
Pyrostegia venusta, p.178
Tecomaria capensis

OTHER RECOMMENDED PLANTS:
Actinidia chinensis
Actinidia polygama
Anredera cordifolia

Beaumontia grandiflora, p.165
Cionura erecta
CLEMATIS, pp.172–3
Clianthus puniceus f. *albus*, p.165

Cobaea scandens f. *alba*
Ipomoea alba
Jasminum angulare
Jasminum grandiflorum

167

Climbers SUMMER INTEREST

☐ WHITE

Hydrangea petiolaris, syn.
H. anomala subsp. *petiolaris*
(Climbing hydrangea)
Deciduous, woody-stemmed, root climber. Has toothed leaves and lacy heads of small, white flowers in summer, only sparingly borne on young plants. H to 15m (50ft).

Clerodendrum thomsoniae
Vigorous, evergreen, woody-stemmed, scandent shrub with oval, rich green leaves. Flowers with crimson petals and bell-shaped, pure white calyces appear in clusters in summer. H 3m (10ft) or more. Min. 16°C (61°F).

☐ WHITE–PINK

Hoya lanceolata subsp. **bella**,
syn. *H. bella*
Evergreen, woody-stemmed, trailing shrub with narrowly oval, pointed leaves. In summer bears tiny, star-shaped, white flowers, with red centres, in pendulous, flattened clusters. H 45cm (18in). Min. 10–12°C (50–54°F).

Hoya australis
Moderately vigorous, evergreen, woody-stemmed, twining, root climber with fleshy, rich green leaves. Has trusses of 20–50 fragrant, star-shaped flowers, white with red-purple markings, in summer. H to 5m (15ft). Min. 15°C (59°F).

Pileostegia viburnoides,
syn. *Schizophragma viburnoides*
Slow-growing, evergreen, woody-stemmed, root climber. Tiny, white or cream flowers, with many prominent stamens, are borne in heads from late summer to autumn. H to 6m (20ft).

Schizophragma integrifolium
Deciduous, woody-stemmed, root climber with oval or heart-shaped leaves. In summer, white flowers are borne in flat heads up to 30cm (12in) across, marginal sterile flowers each having a large, white bract. H to 12m (40ft).

Pandorea jasminoides, syn.
Bignonia jasminoides (Bower vine)
Evergreen, woody-stemmed, twining climber with leaves of 5–9 leaflets. Has clusters of funnel-shaped, white flowers, with pink-flushed throats, from late winter to summer. H 5m (15ft). Min. 5°C (41°F).

Hoya carnosa (Wax plant)
Fairly vigorous, evergreen, woody-stemmed, twining, root climber. Scented, star-shaped flowers, white, fading to pink, with deep pink centres, are borne in dense trusses in summer-autumn. H to 5m (15ft) or more. Min. 5–7°C (41–5°F).

Jasminum officinale f. **affine**
Semi-evergreen or deciduous, woody-stemmed, twining climber with leaves comprising 7 or 9 leaflets. Clusters of fragrant, 4- or 5-lobed flowers, white inside and pink outside, are borne in summer-autumn. H to 12m (40ft).

Lathyrus odoratus 'Selana'
Vigorous, annual, tendril climber with oval, mid-green leaves and large, fragrant, pink-flushed, white flowers in summer to early autumn. H 2m (6ft).

Mandevilla boliviensis
Mandevilla laxa
Polygonum aubertii
Polygonum baldschuanicum, p.177

Schizophragma hydrangeoides
Stephanotis floribunda, p.165
Wisteria venusta
Wisteria venusta 'Alba Plena'

CLEMATIS, pp.172–3
Dregea sinensis
Hoya coronaria
Pandorea pandorana

Passiflora × allardii

■ PINK

Actinidia kolomikta
Deciduous, woody-stemmed, twining climber with 8–16cm (3–6in) long leaves, the upper sections often creamy-white and pink. Has small, cup-shaped, white flowers in summer, male and female on separate plants. H 4m (12ft).

☼ ◊ ❄❄❄ ♀

Mandevilla x *amoena* 'Alice du Pont', syn. *M.* x *amabilis* 'Alice du Pont'
Vigorous, evergreen, woody-stemmed, twining climber with oval, impressed leaves. Has large clusters of trumpet-shaped, glowing pink flowers in summer. H 3m (10ft). Min. 7–10°C (45–50°F).

☼ ◊ ♀

Lathyrus grandiflorus
(Everlasting pea)
Herbaceous, tendril climber with unwinged stems and neat racemes of pink-purple and red flowers in summer. H to 1.5m (5ft).

☼ ◊ ❄❄❄

Lonicera x *heckrottii*, syn. *L.* 'Gold Flame'
Deciduous, woody-stemmed, twining climber that needs support. Leaves are oblong or oval, bluish beneath, upper ones joined into shallow cups. Scented, orange-throated, pink flowers appear in clusters in summer. H to 5m (15ft).

☼ ◊ ❄❄

Lathyrus odoratus 'Xenia Field'
Moderately fast-growing, slender, annual, tendril climber with oval, mid-green leaves. Large, fragrant, pink-and-cream flowers appear from summer to early autumn. H 2m (6ft).

☼ ◊ ❄❄❄

Antigonon leptopus (Coral vine)
Fast-growing, evergreen, woody-stemmed, tendril climber with crinkly, pale green leaves. Has dense trusses of bright pink, sometimes red or white, flowers mainly in summer but all year in tropical conditions. H 6m (20ft). Min. 15°C (59°F).

☼ ◊

Asarina erubescens
Evergreen, soft-stemmed, scandent, perennial climber, sometimes woody-stemmed, often grown as an annual. Stems and leaves are downy. Rose-pink flowers, 7cm (2¾in) long, are borne in summer-autumn. H to 3m (10ft) or more. Min. 5°C (41°F).

☼ ◊

Ipomoea horsfalliae
Strong-growing, evergreen, woody-stemmed, twining climber. Leaves have 5–7 radiating lobes or leaflets; stalked clusters of deep rose-pink or rose-purple flowers, 6cm (2½in) long, appear from summer to winter. H 2–3m (6–10ft). Min. 7–10°C (45–50°F).

☼ ◊ ♀

Argyreia splendens
CLEMATIS, pp.172–3
Lathyrus sylvestris
Mandevilla splendens, p.165

Mutisia oligodon
Passiflora x *exoniensis*
Passiflora mollissima
x *Philageria veitchii*

Podranea ricasoliana
Sarmienta scandens

Climbers SUMMER INTEREST
PINK–RED

Lonicera sempervirens (Coral honeysuckle)
Evergreen or deciduous, woody-stemmed, twining climber with oval leaves, bluish beneath, upper ones united and saucer-like. Has salmon-red to orange flowers, yellow inside, in whorls on shoot tips in summer. H to 4m (12ft).

Lapageria rosea (Chilean bellflower, Copihue)
Evergreen, woody-stemmed, twining climber with oblong to oval, leathery leaves. Has pendent, fleshy, pink to red flowers, 7–9cm (2¾–3½in) long, with paler flecks, from summer to late autumn. H to 5m (15ft).

Lathyrus odoratus 'Red Ensign'
Vigorous, annual, tendril climber with oval, mid-green leaves and large, sweetly scented, rich scarlet flowers from summer to early autumn. H 2m (6ft).

Lonicera × brownii 'Dropmore Scarlet'
Deciduous, woody-stemmed, twining climber with oval, blue-green leaves. Small, fragrant, red flowers with orange throats are borne throughout summer. H to 4m (12ft).

Bougainvillea 'Miss Manila'
Vigorous, mainly evergreen, woody-stemmed, scrambling climber with rounded-oval leaves. Bears clusters of pink floral bracts in summer. H to 5m (15ft). Min. 7–10°C (45–50°F).

Ipomoea lobata, syn. *I. versicolor*, *Mina lobata*, *Quamoclit lobata*
Deciduous or semi-evergreen, twining climber with 3-lobed leaves, usually grown as an annual. One-sided racemes of small, tubular, dark red flowers fade to orange, then creamy-yellow, in summer. H to 5m (15ft).

Bougainvillea 'Dania'
Vigorous, mainly evergreen, woody-stemmed, scrambling climber. Has rounded-oval, mid-green leaves and bears clusters of deep pink floral bracts in summer. H to 5m (15ft). Min. 7–10°C (45–50°F).

Ipomoea quamoclit, syn. *Quamoclit pinnata* (Cypress vine)
Annual, twining climber with oval, bright green leaves cut into many thread-like segments. Slender, tubular, orange or scarlet flowers are carried in summer-autumn. H 2–4m (6–12ft).

Tropaeolum speciosum (Flame creeper, Flame nasturtium)
Herbaceous, twining climber with a creeping rhizome and lobed, blue-green leaves. Bears scarlet flowers in summer, followed by bright blue fruits surrounded by deep red calyces. Roots should be in shade. H to 3m (10ft).

Asteranthera ovata
Bougainvillea × buttiana 'Scarlet Queen'
Clerodendrum splendens
Clianthus puniceus, p.165
Distictis buccinatoria, p.165
Gloriosa superba 'Rothschildiana', p.345
Ipomoea coccinea
Ipomoea × multifida
Ipomoea nil 'Scarlett O'Hara'
Jasminum beesianum
Kennedia rubicunda, p.165
Manettia cordifolia
Manettia luteo-rubra, p.166
Passiflora antioquiensis
Passiflora coccinea, p.165
Passiflora vitifolia

RED–PURPLE

Quisqualis indica
(Rangoon creeper)
Fairly fast-growing, deciduous or semi-evergreen, scandent shrub, often grown as an annual. From late spring to late summer has fragrant flowers, varying from orange to red, sometimes pink. H 3–5m (10–15ft). Min. 10°C (50°F).

Rhodochiton atrosanguineum, syn. *R. volubile*
Evergreen, leaf-stalk climber, usually grown as an annual, with toothed leaves. Has tubular, blackish-purple flowers, with bell-shaped, red-purple calyces, from late spring to late autumn. H to 3m (10ft). Min. 5°C (41°F).

Aristolochia littoralis, syn. *A. elegans* (Calico flower)
Fast-growing, evergreen, woody-stemmed, twining climber with heart- to kidney-shaped leaves. Heart-shaped, 12cm (5in) wide flowers, maroon with white marbling, are carried in summer. H to 7m (23ft). Min. 13°C (55°F).

Lathyrus latifolius
(Everlasting pea, Perennial pea)
Herbaceous, tendril climber with winged stems. Leaves have broad stipules and a pair of leaflets. Has small racemes of pink-purple flowers in summer and early autumn. H 2m (6ft) or more.

Schisandra rubriflora, syn. *S. grandiflora* var. *rubriflora*
Deciduous, woody-stemmed, twining climber with leathery, toothed leaves, paler beneath. Has small, crimson flowers in spring or early summer and drooping, red fruits in late summer. H to 6m (20ft).

Berberidopsis corallina
(Coral plant)
Evergreen, woody-stemmed, twining climber with oval to heart-shaped, leathery leaves edged with small spines. Pendent clusters of globular, deep red flowers are produced in summer to early autumn. H 4.5m (14ft).

Lablab purpureus, syn. *Dolichos lablab*, *D. lignosus* (Australian pea, Hyacinth bean, Lablab)
Deciduous, woody-stemmed, twining climber, often grown as an annual. Purple, pinkish or white flowers in summer are followed by long pods with edible seeds. H 10m (30ft). Min. 5°C (41°F).

Bougainvillea glabra 'Variegata'
Vigorous, mainly evergreen, woody-stemmed, scrambling climber. Rounded-oval, dark green leaves are edged with creamy-white. Has an abundance of bright purple floral bracts in summer. H to 5m (15ft). Min. 7–10°C (45–50°F).

Aristolochia griffithii
Bougainvillea x *buttiana* 'Mrs Butt'
Bougainvillea spectabilis
CLEMATIS, pp.172–3

Cryptostegia grandiflora
Hoya imperialis
Ipomoea horsfalliae, p.169
Ipomoea horsfalliae 'Briggsii'

Lathyrus rotundifolius
Lonicera henryi
Lonicera periclymenum 'Serotina'
Mucuna pruriens var. *utilis*

Pueraria lobata

171

Clematis

Among the climbers, clematis are unsurpassed in their long period of flowering (with species flowering in almost every month of the year), variety of flower shapes and colours and tolerance of almost any aspect and climate. Some spring-flowering species and cultivars are vigorous and excellent for rapidly covering buildings, old trees and pergolas. Other, less rampant cultivars display often large, exquisite blooms from early summer to autumn in almost every colour. Clematis look attractive when trained on walls or trellises and when grown in association with other climbers, trees or shrubs, treating them as hosts. Less vigorous cultivars may also be left unsupported to scramble at ground level, where their flowers will be clearly visible.

The various types of clematis (see the Plant Dictionary) may be divided into 3 groups, each of which has different pruning requirements. Incorrect pruning may result in cutting out the stems that will produce flowers in the current season, so the following guidelines should be followed closely.

Group 1
Early-flowering species, Alpina, Macropetala and Montana types
Flower stems are produced direct from the previous season's ripened stems. Prune after flowering to allow new growth to be produced and ripened for the next season. Remove dead or damaged stems and cut back other shoots that have outgrown their allotted space.

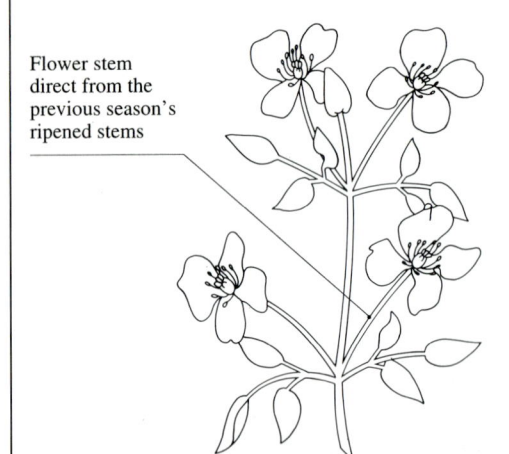

Group 2
Early, large-flowered cultivars
Flowers are produced on short, current season's stems, so prune before new growth starts, in early spring. Remove dead or damaged stems and cut back all others to where strong, leaf-axil buds are visible. (These buds will produce the first crop of flowers.)

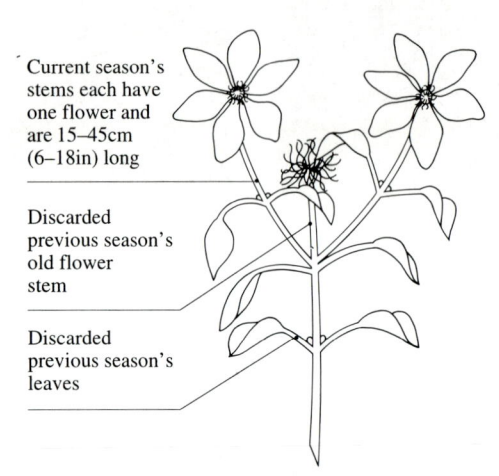

Current season's stems each have one flower and are 15–45cm (6–18in) long

Discarded previous season's old flower stem

Discarded previous season's leaves

Group 3
Late, large-flowered cultivars, Late-flowering species, Small-flowered cultivars and Herbaceous types
Flowers are produced on the current season's growth only, so prune before new growth commences, in early spring. Remove all of the previous season's stems down to a pair of strong, leaf-axil buds, 15–30cm (6–12in) above the soil.

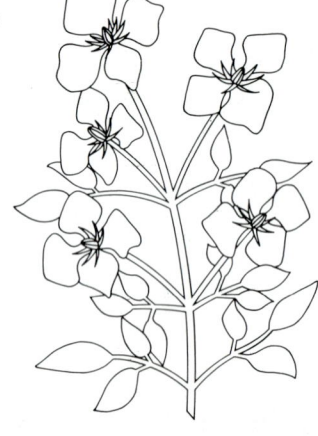

Flowering stems on current season's growth only

Flower stem direct from the previous season's ripened stems

C. florida 'Sieboldii' (3, small-fl.)

C. recta (3, herbaceous)

C. montana var. *rubens* (1, Montana) ♀

C. 'Huldine' (3, late large-fl.)

C. montana 'Tetrarose' (1, Montana) ♀

C. macropetala 'Markham's Pink' (1, Macrop.) ♀

C. 'Hagley Hybrid' (3, late large-fl.)

C. montana (1, Montana)

C. 'Henryi' (2, early large-fl.) ♀

C. 'Mrs George Jackman' (2, early large-fl.) ♀

C. armandii (1, early-fl.)

C. flammula (3, late-fl.)

C. 'Nelly Moser' (2, early large-fl.) ♀

C. **'Lincoln Star'**
(2, early large-fl.)

C. **'Duchess of Albany'**
(3, small-fl.) ♀

C. **'Purpurea Plena Elegans'** (3, late-fl.) ♀

C. **'Proteus'**
(2, early large-fl.)

C. **'Abundance'**
(3, late-fl.)

C. **'Ville de Lyon'**
(3, late large-fl.)

C. **'Mme Julia Correvon'** (3, late-fl.) ♀

C. **'Gravetye Beauty'**
(3, small-fl.)

C. **'Ernest Markham'**
(3, late large-fl.) ♀

C. **'Countess of Lovelace'** (2, early large-fl.)

C. **'Ascotiensis'**
(3, late large-fl.) ♀

C. **'Star of India'**
(3, late large-fl.) ♀

C. **'The President'**
(2, early large-fl.) ♀

C. **'Richard Pennell'**
(2, early large-fl.) ♀

C. **'Elsa Spath'**
(2, early large-fl.) ♀

C. **'H.F. Young'**
(2, early large-fl.) ♀

C. **'Jackmanii'**
(3, late large-fl.) ♀

C. **'Beauty of Worcester'**
(2, early large-fl.)

C. **'Etoile Violette'**
(3, late-fl.) ♀

C. **'Lasurstern'**
(2, early large-fl.) ♀

C. integrifolia
(3, herbaceous)

C. heracleifolia **'Wyevale'** (3, herbaceous) ♀

C. **'William Kennett'**
(2, early large-fl.)

C. **'Vyvyan Pennell'**
(2, early large-fl.) ♀

C. alpina **'Frances Rivis'** (1, Alpina) ♀

C. **'Perle d'Azur'**
(3, late large-fl.) ♀

C. macropetala
(1, Macropetala)

C. rehderiana
(3, late-fl.) ♀

C. cirrhosa (1, early-fl.)

C. tangutica (3, late-fl.)

C. **'Bill MacKenzie'**
(3, late-fl.) ♀

Climbers SUMMER INTEREST
■ PURPLE

Bougainvillea glabra
Vigorous, evergreen or semi-evergreen, woody-stemmed, scrambling climber with rounded-oval leaves. Clusters of floral bracts, in shades of cyclamen-purple, appear in summer. H to 5m (15ft). Min. 7–10°C (45–50°F).

Lathyrus odoratus 'Lady Diana'
Moderately fast-growing, slender, annual, tendril climber with oval, mid-green leaves. Fragrant, pale violet-blue flowers are borne from summer to early autumn. H 2m (6ft).

Solanum crispum 'Glasnevin'
Vigorous, evergreen or semi-evergreen, woody-stemmed, scrambling climber with oval leaves. Has clusters of lilac to purple flowers, 2.5cm (1in) across, in summer. H to 6m (20ft).

Passiflora x caponii 'John Innes'
Strong-growing, evergreen, woody-stemmed, tendril climber with 3-lobed leaves. Has bowl-shaped, nodding, white flowers, flushed claret-purple, with purple-banded, white crowns, in summer-autumn. H 8m (25ft). Min. 7–10°C (45–50°F).

Ipomoea hederacea (Morning glory)
Annual, twining climber with heart-shaped or 3-lobed, mid- to bright green leaves. Has funnel-shaped, red, purple, pink or blue flowers in summer to early autumn. H 3–4m (10–12ft).

Codonopsis convolvulacea, syn. *C. vinciflora*
Herbaceous, twining climber with 5cm (2in) long, oval or lance-shaped leaves. Widely bell- to saucer-shaped, bluish-violet flowers, 2.5–5cm (1–2in) across, are borne in summer. H to 2m (6ft).

Cobaea scandens (Cup-and-saucer vine)
Evergreen or deciduous, woody-stemmed, tendril climber, grown as an annual. From late summer to first frosts has flowers that open yellow-green and age to purple. H 4–5m (12–15ft). Min. 4°C (39°F).

Solanum wendlandii
Robust, mainly evergreen, prickly-stemmed, scrambling climber with oblong, variably lobed leaves. Lavender flowers appear in late summer and autumn. H 3–6m (10–20ft). Min. 10°C (50°F).

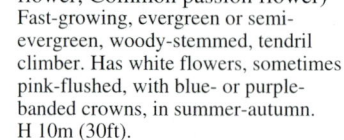

Passiflora caerulea (Blue passion flower, Common passion flower)
Fast-growing, evergreen or semi-evergreen, woody-stemmed, tendril climber. Has white flowers, sometimes pink-flushed, with blue- or purple-banded crowns, in summer-autumn. H 10m (30ft).

Asarina barclaiana
CLEMATIS, pp.172–3
Clytostoma callistegioides, p.166
Ipomoea purpurea

Lathyrus nervosus
Passiflora x caerulea-racemosa
Petrea volubilis, p.166
Solanum seaforthianum

Tropaeolum azureum
Vitis vinifera 'Purpurea', p.178
Wisteria sinensis 'Black Dragon'

PURPLE–BLUE

Passiflora quadrangularis
(Giant granadilla)
Strong-growing, evergreen, woody-stemmed climber with angled, winged stems. White, pink, red or pale violet flowers, the crowns banded white and deep purple, appear mainly in summer. H 5–8m (15–25ft). Min. 10°C (50°F).

Ipomoea tricolor 'Heavenly Blue', syn. *I. rubrocaerulea* 'Heavenly Blue'
Fast-growing, annual, twining climber with heart-shaped leaves and large, funnel-shaped, sky-blue flowers borne from summer to early autumn. H to 3m (10ft).

Wisteria x formosa
Deciduous, woody-stemmed, twining climber with leaves of 9–15 narrowly oval leaflets. Scented, pea-like, mauve and pale lilac flowers are carried in early summer in drooping racemes, 25cm (10in) long, followed by velvety pods. H to 25m (80ft) or more.

Plumbago auriculata,
syn. *P. capensis* (Cape leadwort)
Fast-growing, evergreen, woody-stemmed, scrambling climber. Trusses of sky-blue flowers are carried from summer to early winter. H 3–6m (10–20ft). Min. 7°C (45°F).

Wisteria sinensis, syn.
W. chinensis (Chinese wisteria)
Vigorous, deciduous, woody-stemmed, twining climber. Has leaves of 11 leaflets and fragrant, lilac or pale violet flowers, in racemes 20–30cm (8–12in) long, in early summer, followed by velvety pods. H to 30m (100ft).

Tweedia caerulea,
syn. *Oxypetalum caeruleum*
Herbaceous, twining climber with white-haired stems. Small, fleshy, pale blue flowers, maturing purple, appear in summer and early autumn; has green fruits to 15cm (6in) long. H to 1m (3ft). Min. 5°C (41°F).

CLEMATIS, pp.172–3
Clitoria ternatea
Jacquemontia pentantha
Solanum jasminoides

Sollya heterophylla, p.166
Thunbergia grandiflora
Wisteria floribunda 'Macrobotrys'
Wisteria sinensis 'Prolific'

YELLOW

Lonicera x americana
Very free-flowering, deciduous, woody-stemmed, twining climber. Leaves are oval, upper ones united and saucer-like. Clusters of strongly fragrant, yellow flowers, flushed with red-purple, appear in summer. H to 7m (23ft).

Lonicera japonica 'Halliana'
Evergreen or semi-evergreen, woody-stemmed, twining climber with soft-haired stems and oval, sometimes lobed, bright green leaves. Very fragrant, white flowers, ageing to pale yellow, are borne in summer and autumn. H to 10m (30ft).

Lonicera periclymenum 'Graham Thomas'
Deciduous, woody-stemmed, twining climber. Oval or oblong leaves are bluish beneath; fragrant, white flowers ageing to yellow, are borne in summer. H to 7m (23ft).

Ampelopsis brevipedunculata var. *maximowiczii*
Anemopaegma chamberlaynii
Bomarea andimarcana

Humulus lupulus 'Aureus', p.166
Lonicera japonica 'Aureoreticulata'
Sinofranchetia chinensis
Solandra maxima, p.166

175

Climbers SUMMER INTEREST
YELLOW–ORANGE

Allemanda cathartica 'Hendersonii'
Fast-growing, evergreen, woody-stemmed, scrambling climber. Has lance-shaped leaves in whorls and trumpet-shaped, rich bright yellow flowers in summer-autumn. H to 5m (15ft). Min. 13–15°C (55–9°F).

Thunbergia alata
(Black-eyed Susan)
Moderately fast-growing, annual, twining climber. Has toothed, oval to heart-shaped leaves and rounded, rather flat, small flowers, orange-yellow with very dark brown centres, from early summer to early autumn. H 3m (10ft).

Thladiantha dubia
Fast-growing, herbaceous or deciduous, tendril climber. Oval to heart-shaped, mid-green leaves, 10cm (4in) long, are hairy beneath; bell-shaped, yellow flowers are carried in summer. H 3m (10ft).

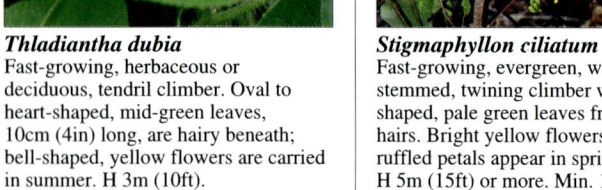

Stigmaphyllon ciliatum
Fast-growing, evergreen, woody-stemmed, twining climber with heart-shaped, pale green leaves fringed with hairs. Bright yellow flowers with ruffled petals appear in spring-summer. H 5m (15ft) or more. Min. 15–18°C (59–64°F).

Macfadyena unguis-cati
(Cat's claw)
Fast-growing, evergreen, woody-stemmed, tendril climber. Leaves have 2 leaflets and a tendril. Has yellow flowers, 10cm (4in) long, in late spring or early summer. H 8–10m (25–30ft). Min. 5°C (41°F).

Tropaeolum tuberosum 'Ken Aslet'
Herbaceous climber with yellowish, red-streaked tubers and blue-green leaves. From mid-summer to autumn has flowers with red sepals and orange petals. In cool areas, lift and store tubers in winter. H to 2.5m (8ft).

Bougainvillea x buttiana 'Golden Glow'
Gelsemium sempervirens
Gloriosa superba
Hibbertia scandens

Lonicera etrusca
Lonicera hildebrandiana
Lonicera tragophylla
Merremia tuberosa

Mussaenda erythrophylla
Thunbergia mysorensis, p.167
Tropaeolum peregrinum
Tropaeolum tuberosum, p.178

Vigna caracalla

◼ ORANGE

Bomarea caldasii,
syn. *B. kalbreyeri* of gardens
Herbaceous, twining climber with rounded clusters of 5–40 tubular to funnel-shaped, orange-red flowers, spotted crimson within, in summer. H 3–4m (10–12ft).

Lonicera x *tellmanniana*
Deciduous, woody-stemmed, twining climber with oval leaves; upper ones are joined and resemble saucers. Bright yellowish-orange flowers are carried in clusters at the ends of shoots in late spring and summer. H to 5m (15ft).

Climbers AUTUMN INTEREST
◼ WHITE–RED

Eccremocarpus scaber
(Chilean glory flower, Glory vine)
Evergreen, sub-shrubby, tendril climber, often grown as an annual. In summer has racemes of small, orange-red flowers, followed by inflated fruit pods containing many winged seeds. H 2–3m (6–10ft).

Thunbergia gregorii
Evergreen, woody-stemmed, twining climber, usually grown as an annual. Triangular-oval leaves have winged stalks. Glowing orange flowers are carried in summer. H to 3m (10ft). Min. 10°C (50°F).

Polygonum baldschuanicum,
syn. *Bilderdykia baldschuanica*,
Fallopia baldschuanica (Mile-a-minute plant, Russian vine)
Vigorous, deciduous, woody-stemmed, twining climber with drooping panicles of pink or white flowers in summer-autumn. H 12m (40ft) or more.

Senecio confusus
(Mexican flame vine)
Evergreen, woody-stemmed, twining climber bearing clusters of daisy-like, orange-yellow flower heads, ageing to orange-red, mainly in summer. H 3m (10ft) or more. Min. 7–10°C (45–50°F).

Mutisia decurrens
Evergreen, tendril climber with narrowly oblong leaves, 7–13cm (2¾–5in) long. Flower heads, 10–13cm (4–5in) across with red or orange ray flowers, are produced in summer. Proves difficult to establish, but is worthwhile. H to 3m (10ft).

Campsis x *tagliabuana* 'Mme Galen'
Deciduous, woody-stemmed, root climber with leaves of 7 or more narrowly oval, toothed leaflets. Trumpet-shaped, orange-pink flowers are borne in pendent clusters from late summer to autumn. H to 10m (30ft).

Passiflora manicata
Fast-growing, evergreen, woody-stemmed, tendril climber with slender, angular stems and 3-lobed leaves. Red flowers, with deep purple and white crowns, appear in summer-autumn. H 3–5m (10–15ft). Min. 7°C (45°F).

Bignonia capreolata
Campsis radicans
Mitraria coccinea, p.166
Streptosolen jamesonii, p.180

Tecomaria capensis

OTHER RECOMMENDED PLANTS:
Asarina erubescens, p.169
Cobaea scandens f. *alba*
Lapageria rosea, p.170

Pileostegia viburnoides, p.168
Stephanotis floribunda, p.165
Tecomanthe speciosa

Climbers AUTUMN INTEREST

RED–PURPLE

Parthenocissus tricuspidata
(Boston ivy, Japanese ivy)
Vigorous, deciduous, woody-stemmed, tendril climber. Has spectacular, crimson, autumn leaf colour and dull blue berries. Will cover large expanses of wall. H to 20m (70ft).

Parthenocissus tricuspidata 'Lowii'
Vigorous, deciduous, woody-stemmed, tendril climber with deeply cut and crinkled, 3–7-lobed leaves that turn crimson in autumn. Has insignificant flowers, followed by dull blue berries. H to 20m (70ft).

Vitis coignetiae
(Crimson glory vine)
Vigorous, deciduous, woody-stemmed, tendril climber. Large leaves, brown-haired beneath, are brightly coloured in autumn. Has tiny, pale green flowers in summer, followed by purplish-bloomed, black berries. H to 15m (50ft).

Parthenocissus thompsonii, syn. *Vitis thompsonii*
Deciduous, woody-stemmed, tendril climber. Has glossy, green leaves with 5 leaflets that turn red-purple in autumn, and black berries. Provide some shade for best autumn colour. H to 10m (30ft).

Vitis vinifera 'Purpurea'
Deciduous, woody-stemmed, tendril climber with toothed, 3- or 5-lobed, purplish leaves, white-haired when young. Has tiny, pale green flowers in summer and tiny, green or purple berries. H to 7m (23ft).

Parthenocissus tricuspidata 'Veitchii', syn. *Ampelopsis veitchii*
Vigorous, deciduous, woody-stemmed, tendril climber. Has spectacular, red-purple, autumn leaf colour and dull blue berries. Greenish flowers are insignificant. H to 20m (70ft).

PURPLE–ORANGE

Billardiera longiflora
Evergreen, woody-stemmed, twining climber with narrow leaves. Small, bell-shaped, sometimes purple-tinged, green-yellow flowers are produced singly in leaf axils in summer, followed by purple-blue fruits in autumn. H to 2m (6ft).

Tropaeolum tuberosum
Herbaceous, tuberous-rooted, leaf-stalk climber. Greyish-green leaves have 3–5 lobes; from mid-summer to late autumn has cup-shaped flowers with orange-yellow petals, orange-red sepals and a long spur. H 2–3m (6–10ft).

Pyrostegia venusta, syn. *P. ignea*
(Flame flower, Flame vine, Golden shower)
Fast-growing, evergreen, woody-stemmed, tendril climber with clusters of tubular, golden-orange flowers from autumn to spring. H 10m (30ft) or more. Min. 13–15°C (55–9°F).

Asarina barclaiana
Cobaea scandens, p.174
Ipomoea horsfalliae 'Briggsii'
Mucuna pruriens var. utilis

Parthenocissus henryana
Passiflora caerulea, p.174
Passiflora x caerulea-racemosa
Passiflora vitifolia

Rhodochiton atrosanguineum, p.171
Solanum seaforthianum
Solanum wendlandii, p.174

Allemanda cathartica 'Hendersonii', p.176
Campsis grandiflora
Senecio mikanioides
Solanum jasminoides

Climbers WINTER INTEREST
◻◻ WHITE–ORANGE

Jasminum polyanthum
Evergreen, woody-stemmed, twining climber. Dark green leaves have 5 or 7 leaflets. Large clusters of fragrant, 5-lobed, white flowers, sometimes reddish on the outside, are carried from late summer to winter. H 3m (10ft) or more.

Agapetes macrantha
Evergreen or semi-evergreen, loose, scandent shrub that may be trained against supports. Has lance-shaped leaves and narrowly urn-shaped, white or pinkish-white flowers, patterned in red, in winter. H 1–2m (3–6ft). Min. 15–18°C (59–64°F).

Canarina canariensis,
syn. *C. campanula*
(Canary Island bellflower)
Herbaceous, tuberous, scrambling climber with triangular, serrated leaves. Has waxy, orange flowers with red veins from late autumn to spring. H 2–3m (6–10ft). Min. 7°C (45°F).

OTHER RECOMMENDED PLANTS:
Calochone redingii
Holmskioldia sanguinea
Pyrostegia venusta, p.178
Senecio macroglossus
Senecio mikanioides
Strongylodon macrobotrys, p.166
Thunbergia coccinea

Climbers ALL YEAR INTEREST
◻ WHITE–GREEN

Asparagus scandens
Evergreen, scrambling climber with lax stems and short, curved, leaf-like shoots in whorls of 3. Tiny, nodding, white flowers appear in clusters of 2–3 in summer, followed by red berries. H 1m (3ft) or more. Min. 10°C (50°F).

Senecio macroglossus 'Variegatus'
Evergreen, woody-stemmed, twining climber with triangular, fleshy leaves, bordered in white to cream, and, mainly in winter, daisy-like, cream flower heads. H 3m (10ft). Min. 7°C (45°F), but best at 10°C (50°F).

Epipremnum aureum 'Marble Queen', syn. *Scindapsus aureus* 'Marble Queen'
Fairly fast-growing, evergreen, woody-stemmed, root climber. Leaves are streaked and marbled with white. Is less robust than the species. H 3–10m (10–30ft). Min. 15–18°C (59–64°F).

Scindapsus pictus 'Argyraeus', syn. *Epipremnum pictum* 'Argyraeus' (Silver vine)
Slow-growing, evergreen, woody-stemmed, root climber. Heart-shaped leaves are dark green with silver markings. H 2–3m (6–10ft) or more. Min. 15–18°C (59–64°F).

Syngonium podophyllum 'Trileaf Wonder'
Evergreen, woody-stemmed, root climber with tufted stems and arrow-head-shaped leaves when young. Mature leaves have 3 glossy leaflets with pale green or silvery-grey veins. H 2m (6ft) or more. Min. 18°C (64°F).

Dioscorea discolor
(Ornamental yam)
Evergreen, woody-stemmed, twining climber. Heart-shaped, olive-green leaves are 12–15cm (5–6in) long, marbled silver, paler green and brown, and are red beneath. H to 2m (6ft). Min. 5°C (41°F).

OTHER RECOMMENDED PLANTS:
Cissus discolor
Cissus striata
Gynura aurantiaca 'Purple Passion'

IVIES, p.181
Lonicera japonica 'Aureoreticulata'
Peperomia scandens
Philodendron 'Burgundy'

Climbers ALL YEAR INTEREST
GREEN–ORANGE

Syngonium podophyllum, syn.
Nephthytis triphylla of gardens
Evergreen, woody-stemmed, root climber with tufted stems and arrow-head-shaped leaves when young. Mature plants have leaves of 7–9 glossy leaflets up to 30cm (12in) long. H 2m (6ft). Min. 16–18°C (61–4°F).

Tetrastigma voinierianum, syn. *Cissus voinieriana* (Chestnut vine)
Strong-growing, evergreen, woody-stemmed, tendril climber. Young stems and leaves are rust-coloured and hairy; mature leaves turn lustrous, deep green above. H 10m (30ft) or more. Min. 15–18°C (59–64°F).

Philodendron melanochrysum
Robust, fairly slow-growing, evergreen, woody-based, root climber. Heart-shaped leaves, to 75cm (30in) long, are lustrous, deep olive green with a coppery sheen and have pale veins. H 3m (10ft) or more. Min. 15–18°C (59–64°F).

Cissus antarctica (Kangaroo vine)
Moderately vigorous, evergreen, woody-stemmed, tendril climber. Oval, pointed, coarsely serrated leaves are lustrous, rich green. H to 5m (15ft). Min. 7°C (45°F).

Philodendron scandens (Heart leaf)
Fairly fast-growing, evergreen, woody-based, root climber. Rich green leaves are 10–15cm (4–6in) long when young, to 30cm (12in) long on mature plants. H 4m (12ft) or more. Min. 15–18°C (59–64°F).

Gynura aurantiaca (Velvet plant)
Evergreen, woody-based, soft-stemmed, semi-scrambling climber or lax shrub with purple-haired stems and leaves. Clusters of daisy-like, orange-yellow flower heads are borne in winter. H 2–3m (6–10ft), less as a shrub. Min. 16°C (61°F).

Monstera deliciosa (Swiss-cheese plant)
Robust, evergreen, woody-stemmed, root climber with large-lobed, holed leaves, 40–90cm (16–36in) long. Mature plants bear cream spathes, followed by scented, edible fruits. H to 6m (20ft). Min. 15–18°C (59–64°F).

Cissus rhombifolia (Grape ivy)
Moderately vigorous, evergreen, woody-stemmed, tendril climber with lustrous leaves divided into 3 coarsely toothed leaflets. H 3m (10ft) or more. Min. 7°C (45°F).

Streptosolen jamesonii (Marmalade bush)
Evergreen or semi-evergreen, loosely scrambling shrub. Has oval, finely corrugated leaves and, mainly in spring-summer, many bright orange flowers. H 2–3m (6–10ft). Min. 7°C (45°F).

Cissus rhombifolia
Ercilla volubilis
Ficus pumila
Holboellia coriacea

Lardizabala biternata
Monstera acuminata
Philodendron cordatum
Philodendron domesticum

Philodendron pedatum
Philodendron pinnatifidum
Philodendron sagittifolium
Rhoicissus capensis

Rubus henryi var. bambusarum
Syngonium auritum
Syngonium erythrophyllum

Ivies

Ivies (*Hedera*) are evergreen, climbing and trailing plants suitable for growing up walls and fences or as ground cover. Plants take a year or so to establish but thereafter growth is rapid. There is a large number of cultivars available, of which the non-variegated forms are shade-tolerant. With height and access to light, the typical, ivy-shaped leaves may become less lobed. Not all ivies are fully hardy.

H. helix 'Gracilis'

H. helix 'Digitata'

H. helix 'Adam'

H. helix 'Merion Beauty'

H. helix 'Nigra'

H. helix 'Heise'

H. helix 'Eva' ♀

H. helix 'Erecta' ♀

H. helix 'Ivalace' ♀

H. helix 'Woerner'

H. helix 'Glacier' ♀

H. helix 'Angularis Aurea' ♀

H. colchica 'Dentata' ♀

H. helix 'Telecurl'

H. helix 'Lobata Major'

H. helix 'Glymii'

H. helix 'Anna Marie'

H. helix 'Goldheart'

H. helix 'Pedata' ♀

H. helix 'Deltoidea'

H. hibernica 'Sulphurea'

H. helix 'Manda's Crested' ♀

H. helix 'Buttercup' ♀

H. helix 'Green Ripple'

H. helix 'Pittsburgh'

H. hibernica

H. canariensis 'Ravensholst'

H. helix 'Parsley Crested'

H. helix 'Atropurpurea' ♀

H. colchica 'Sulphur Heart' ♀

GRASSES, BAMBOOS, RUSHES AND SEDGES
WHITE–GREEN

Cortaderia selloana 'Silver Comet'
Evergreen, clump-forming, perennial grass with very narrow, sharp-edged, recurved leaves, 1m (3ft) long, that have silver margins. Carries plume-like panicles of spikelets from late summer. H 1.2–1.5m (4–5ft), S 1m (3ft).

Pleioblastus variegatus, syn. *Arundinaria fortunei, A. variegata* (Dwarf white-stripe bamboo)
Evergreen, slow-spreading bamboo with narrow, slightly downy, white-striped leaves. Stems are branched near the base. H 80cm (30in), S indefinite.

Lagurus ovatus (Hare's-tail grass)
Tuft-forming, annual grass that in early summer bears dense, egg-shaped, soft panicles of white flower spikes, with golden stamens, that last well into autumn. Leaves are long, narrow and flat. Self seeds readily. H 45cm (18in), S 15cm (6in).

Glyceria maxima 'Variegata', syn. *G. aquatica* 'Variegata'
Herbaceous, spreading, perennial grass with cream-striped leaves, often tinged pink at the base. Bears open panicles of greenish spikelets in summer. H 80cm (30in), S indefinite.

Phalaris arundinacea 'Picta', syn. *P. a.* var. *picta, P. a.* 'Tricolor' (Gardener's garters)
Evergreen, spreading, perennial grass with broad, white-striped leaves. Produces narrow panicles of spikelets in summer. Can be invasive. H 1m (3ft), S indefinite.

Arundo donax 'Versicolor', syn. *A. d.* 'Variegata'
Herbaceous, rhizomatous, perennial grass with strong stems bearing broad, creamy-white-striped leaves. May bear dense, erect panicles of whitish-yellow spikelets in late summer. H 2.5–3m (8–10ft), S 60cm (2ft).

Miscanthus sinensis 'Zebrinus'
Herbaceous, clump-forming, perennial grass. Leaves, hairy beneath, have transverse, yellowish-white ring markings. May carry awned, hairy, white spikelets in fan-shaped panicles in autumn. H 1.2m (4ft), S 45cm (18in).

Holcus mollis 'Albovariegatus', syn. *H. m.* 'Variegatus' (Variegated creeping soft grass)
Evergreen, spreading, perennial grass with white-striped leaves and hairy nodes. In summer carries purplish-white flower spikes. H 30–45cm (12–18in), S indefinite.

Sasa veitchii, syn. *S. albomarginata*
Evergreen, slow-spreading bamboo. Leaves, 25cm (10in) long, soon develop white edges. Stems, often purple, produce a single branch at each node. White powder appears beneath nodes. H to 1.5m (5ft), S indefinite.

Cortaderia selloana 'Sunningdale Silver'
Evergreen, clump-forming, perennial grass with narrow, sharp-edged, recurved leaves, 1.5m (5ft) long. Bears long-lasting, feathery panicles of creamy-white spikelets in late summer. H 2.1m (7ft), S 1.2m (4ft).

Schoenoplectus lacustris subsp. **tabernaemontani 'Zebrinus'**, syn. *S. tabernaemontani* 'Zebrinus'
Evergreen, spreading, perennial sedge with leafless stems, striped horizontally with white, and brown spikelets in summer. Withstands brackish water. H 1.5m (5ft), S indefinite.

OTHER RECOMMENDED PLANTS:
Arrhenatherum elatius 'Variegatum'
Arundo donax
Carex riparia 'Variegata'
Chionochloa conspicua
Cortaderia selloana
Cyperus albostriatus 'Variegatus'
Dactylis glomerata 'Variegata'
Luzula sylvatica 'Marginata'
Scirpoides holoschoenus 'Variegatus'
Stenotaphrum secundatum 'Variegatum'

■ GREEN

Luzula nivea (Snowy woodrush)
Evergreen, slow-spreading, perennial rush with fairly dense clusters of shining, white flower spikes in early summer. Leaves are edged with white hairs. H 60cm (24in), S 45–60cm (18–24in).

☼ ◐ ◊ ❀❀❀

Hordeum jubatum (Foxtail barley, Squirreltail grass)
Tufted, short-lived perennial or annual grass. In summer to early autumn has flat, arching, feathery, plume-like flower spikes with silky awns. H 30–60cm (1–2ft), S 30cm (1ft).

☼ ◊ ❀❀❀

Pennisetum villosum, syn. *P. longistylum* (Feather-top)
Herbaceous, tuft-forming, perennial grass with long-haired stems. In autumn has panicles of creamy-pink spikelets, fading to pale brown, with very long, bearded bristles. H to 1m (3ft), S 50cm (20in).

☼ ◊ ❀❀

Stipa gigantea (Golden oats)
Evergreen, tuft-forming, perennial grass with narrow leaves, 45cm (18in) or more long. In summer carries elegant, open panicles of silvery spikelets, with long awns and dangling, golden anthers, which persist well into winter. H 2.5m (8ft), S 1m (3ft).

☼ ◊ ❀❀❀ ♛

Cyperus papyrus (Paper reed, Papyrus)
Evergreen, clump-forming, perennial sedge with stout, triangular, leafless stems, each carrying huge umbels of spikelets with up to 100 rays in summer. Grows in water. H to 3–5m (10–15ft), S 1m (3ft). Min 7–10°C (45–50°F).

☼ ●

Helictotrichon sempervirens, syn. *Avena candida* of gardens, *A. sempervirens* (Blue oat grass)
Evergreen, tufted, perennial grass with stiff, silvery-blue leaves up to 30cm (12in) or more long. Produces erect panicles of straw-coloured flower spikes in summer. H 1m (3ft), S 60cm (2ft).

☼ ◊ ❀❀❀ ♛

Bambusa multiplex, syn. *B. glaucescens* (Hedge bamboo)
Evergreen, clump-forming bamboo with narrow leaves, 10–15cm (4–6in) long. Useful for hedges and windbreaks. H to 15m (50ft), S indefinite.

☼ ◊ ❀

Bouteloua gracilis, syn. *B. oligostachya* (Blue grama, Mosquito grass)
Semi-evergreen, tuft-forming, narrow-leaved, perennial grass. In summer bears comb-like flower spikes, 4cm (1½in) long, held at right-angles to stems. H 50cm (20in), S 20cm (8in).

☼ ◊ ❀❀❀

Melica altissima 'Atropurpurea'
Evergreen, tuft-forming, perennial grass with broad leaves, short-haired beneath. Purple spikelets in narrow panicles, 10cm (4in) long, hang from the tops of stems during summer. H and S 60cm (2ft).

☼ ◊ ❀❀❀

Briza maxima
Briza media
Bromus ramosus
Carex buchananii

Fargesia nitida
Molinia caerulea subsp. *arundinacea*
Pennisetum alopecuroides
Pennisetum setaceum

Stipa arundinacea

183

Grasses, bamboos, rushes and sedges
■ GREEN

Yushania anceps, syn. *Arundinaria anceps, A. jaunsarensis, Sinarundinaria jaunsarensis* (Anceps bamboo)
Evergreen, spreading bamboo with erect, later arching, stems bearing several branches at each node. H 2–3m (6–10ft), S indefinite.

Semiarundinaria fastuosa, syn. *Arundinaria fastuosa* (Narihira bamboo)
Evergreen, clump-forming bamboo with 15cm (6in) long leaves and short, tufted branches at each node. Culm sheaths open to reveal polished, purplish interiors. H 6m (20ft), S indefinite.

Shibataea kumasasa
Evergreen, clump-forming bamboo with stubby, side branches on greenish-brown stems. Leaves are broad, 5–10cm (2–4in) long. H 1–1.5m (3–5ft), S 30cm (1ft).

Panicum capillare (Old-witch grass)
Tuft-forming, annual grass with broad leaves and hairy stems. Top half of each stem carries a dense panicle of numerous, minute, greenish-brown spikelets on delicate stalks in summer. H 60cm–1m (2–3ft), S 30cm (1ft).

Phyllostachys nigra var. ***henonis,*** syn. *P.* 'Henonis'
Evergreen, clump-forming bamboo with bristled auricles on culm sheaths and a profusion of leaves. H 10m (30ft), S 2–3m (6–10ft).

Juncus effusus f. ***spiralis*** (Corkscrew rush)
Evergreen, tuft-forming, perennial rush with leafless stems that twist and curl and are often prostrate. Fairly dense, greenish-brown flower panicles form in summer. H 1m (3ft), S 60cm (2ft).

Chusquea culeou (Chilean bamboo)
Slow-growing, evergreen, clump-forming bamboo. Bears long-lasting culm sheaths, shining white when young, at the swollen nodes of stout, solid stems. H to 5m (15ft), S 2.5m (8ft) or more.

Cyperus involucratus, syn. *C. alternifolius* of gardens, *C. flabelliformis*
Evergreen, tuft-forming, perennial sedge with leaf-like bracts forming a whorl beneath the clustered flower spikes in summer. H to 1m (3ft), S 30cm (1ft). Min. 4–7°C (39–45°F).

Pseudosasa japonica, syn. *Arundinaria japonica* (Arrow bamboo, Metake)
Evergreen, clump-forming bamboo with long-persistent, roughly pubescent, brown sheaths and broad leaves, 35cm (14in) long. H 5m (15ft), S indefinite.

Phyllostachys bambusoides (Timber bamboo)
Evergreen, clump-forming bamboo with stout, erect, green stems. Bears leaf sheaths with prominent bristles, and large, broad leaves. H 6–8m (20–25ft), S indefinite.

Coix lacryma-jobi (Job's tears)
Tuft-forming, annual grass with broad leaves and insignificant spikelets followed by hard, bead-like, green fruits turning shiny, greyish-mauve in autumn. H 45cm–1m (1½–3ft), S 10–15cm (4–6in).

Phyllostachys flexuosa (Zigzag bamboo)
Evergreen, clump-forming bamboo with slender, markedly zigzag stems that turn black with age. Leaf sheaths have no bristles. Leaves stay fresh green all winter. H 6–8m (20–25ft), S indefinite.

Arundo donax
Carex elata
Cyperus longus
Deschampsia caespitosa
Festuca glauca
Leymus arenarius
Sesleria heufleriana
Stipa calamagrostis

GREEN–YELLOW

***Miscanthus sinensis* 'Gracillimus'**
Herbaceous, clump-forming, perennial grass with very narrow leaves, hairy beneath, often turning bronze. May bear fan-shaped panicles of awned, hairy, white spikelets in early autumn. H 1.2m (4ft), S 45cm (18in).

***Spartina pectinata* 'Aureomarginata'**, syn. *S.p.* 'Aureovariegata'
Herbaceous, spreading, rhizomatous grass with long, arching, yellow-striped leaves, which turn orange-brown in late autumn to winter. H to 2m (6ft), S indefinite.

Pleioblastus auricoma, syn. *P. viridistriatus*, *Arundinaria auricoma*, *A. viridistriata*
Evergreen, slow-spreading bamboo with purple stems and broad, softly downy, bright yellow leaves with green stripes. H 1.5m (5ft), S indefinite.

***Alopecurus pratensis* 'Aureovariegatus'**, syn. *A.p.* 'Aureomarginatus' (Golden foxtail)
Herbaceous, tuft-forming, perennial grass with yellow or yellowish-green-streaked leaves and dense flower spikes in summer. H and S 23–30cm (9–12in).

Carex pendula (Pendulous sedge)
Evergreen, tuft-forming, graceful, perennial sedge with narrow, green leaves, 45cm (18in) long. Solid, triangular stems freely produce pendent, greenish-brown flower spikes in summer. H 1m (3ft), S 30cm (1ft).

***Carex oshimensis* 'Evergold'**
Evergreen, tuft-forming, perennial sedge with narrow, yellow-striped leaves, 20cm (8in) long. Solid, triangular stems may carry insignificant flower spikes in summer. H 20cm (8in), S 15–20cm (6–8in).

Phyllostachys viridiglaucescens
Evergreen, clump-forming bamboo with greenish-brown stems that arch at the base. Has white powder beneath nodes. H 6–8m (20–25ft), S indefinite.

***Carex elata* 'Aurea'**, syn. *C. stricta* 'Aurea' (Bowles' golden sedge)
Evergreen, tuft-forming, perennial sedge with golden-yellow leaves. Solid, triangular stems bear blackish-brown flower spikes in summer. H to 40cm (16in), S 15cm (6in).

***Hakonechloa macra* 'Aureola'**
Slow-growing, herbaceous, shortly rhizomatous grass with purple stems and green-striped, yellow leaves that age to reddish-brown. Open panicles of reddish-brown flower spikes may appear in early autumn and last into winter. H 40cm (16in), S 45–60cm (18–24in).

Eleocharis acicularis
Isolepsis setaceus
Lamarckia aurea
Melica altissima

Milium effusum 'Aureum'
Miscanthus floridulus
Molinia caerulea subsp. *caerulea* 'Variegata'
Phyllostachys aurea

Phyllostachys aureosulcata
Rhynchelytrum repens
Sasa palmata
Stipa arundinacea

FERNS
GREEN

Phlebodium aureum **'Mandaianum'**
Evergreen fern with creeping rhizomes. Has arching, deeply lobed, glaucous fronds with attractive, orange-yellow sporangia on reverses; pinnae are deeply cut and wavy. H 1–1.5m (3–5ft), S 60cm (2ft). Min. 5°C (41°F).

Dryopteris filix-mas (Male fern)
Deciduous or semi-evergreen fern with 'shuttlecocks' of elegantly arching, upright, broadly lance-shaped, mid-green fronds that arise from crowns of large, upright, brown-scaled rhizomes. H 1.2m (4ft), S 1m (3ft).

Dicksonia antarctica (Australian tree fern)
Evergreen, tree-like fern. Stout trunks are covered with brown fibres and crowned by spreading, somewhat arching, broadly lance-shaped, much-divided, palm-like fronds. H 10m (30ft) or more, S 4m (12ft).

Polystichum munitum (Giant holly fern)
Evergreen fern with erect, leathery, lance-shaped, dark green fronds that consist of small, spiny-margined pinnae. H 1.2m (4ft), S 30cm (1ft).

Polypodium glycyrrhiza (Liquorice fern)
Deciduous fern. Has oblong-triangular to narrowly oval, divided, mid-green fronds, with lance-shaped to oblong pinnae, that arise from a liquorice-scented rootstock. H and S 45cm (18in).

Platycerium bifurcatum (Common stag's-horn fern)
Evergreen, epiphytic fern with broad, plate-like sterile fronds and long, arching or pendent, forked, grey-green fertile fronds bearing velvety, brownish spore patches beneath. H and S 1m (3ft). Min. 5°C (41°F).

Polystichum aculeatum **'Pulcherrimum'**
Evergreen or semi-evergreen fern with broadly lance-shaped, daintily cut, sharp-edged fronds that are yellowish-green in spring and mature to a glossy, rich dark green. H 60cm (24in), S 75cm (30in).

Blechnum capense
Evergreen fern with lance-shaped, spreading, dark green fronds arising from a creeping, almost black rhizome. Extreme ends of blade divisions are narrowed and spore-bearing. H 45cm (18in), S 60cm (24in).

Microlepia strigosa
Evergreen fern. Broad, irregularly lance-shaped, deeply cut and divided, pale green fronds arise from a creeping rootstock. H 90cm (3ft), S 60cm (2ft). Min. 5°C (41°F).

OTHER RECOMMENDED PLANTS:
Asplenium bulbiferum
Blechnum penna-marina
Blechnum spicant
Cheilanthes pteridioides
Dryopteris marginalis
Nephrolepis cordifolia
Pellaea atropurpurea
Pellaea pteridioides
Polystichum acrostichoides
Polystichum aculeatum
Polystichum tsus-simense
Pteris ensiformis
Pteris ensiformis 'Arguta'

■ GREEN

Polypodium scouleri
Evergreen, creeping fern with triangular to oval, leathery, divided fronds that arise from a spreading rootstock. H and S 30–40cm (12–16in).

Pteris cretica (Cretan brake)
Evergreen or semi-evergreen fern with triangular to broadly oval, divided, pale green fronds that have finger-like pinnae. H 45cm (18in), S 30cm (12in). Min. 5°C (41°F).

Adiantum pedatum var. ***aleuticum***
Semi-evergreen fern with a short rootstock. Has glossy, dark brown or blackish stems and dainty, divided, finger-like fronds, with blue-green pinnae, that are more crowded than those of *A. pedatum*. Grows well in alkaline soils. H and S to 45cm (18in).

Phlebodium aureum, syn. *Polypodium aureum*
Evergreen fern with creeping, golden-scaled rhizomes. Has arching, deeply lobed, mid-green or glaucous fronds with attractive, orange-yellow sporangia on reverses. H 90cm–1.5m (3–5ft), S 60cm (2ft). Min. 5°C (41°F).

Selaginella martensii
Evergreen, moss-like perennial with dense, much-branched, frond-like sprays of glossy, rich green foliage. H and S 23cm (9in). Min. 5°C (41°F).

Ceterach officinarum
(Rusty-back fern)
Semi-evergreen fern with lance-shaped, leathery, dark green fronds divided into alternate, bluntly rounded lobes. Backs of young fronds are covered with silvery scales that mature to reddish-brown. H and S 15cm (6in).

Cyrtomium falcatum
(Fishtail fern, Holly fern)
Evergreen fern. Fronds are lance-shaped and have holly-like, glossy, dark green pinnae; young fronds are often covered with whitish or brown scales. H 30–60cm (12–24in), S 30–45cm (12–18in).

Asplenium trichomanes
(Maidenhair spleenwort)
Semi-evergreen fern that has long, slender, tapering fronds with glossy, black, later brown midribs bearing many rounded-oblong, bright green pinnae. Is suitable for limestone soils. H 15cm, S 15–30cm (6–12in).

***Polystichum setiferum* 'Divisilobum'**
Evergreen or semi-evergreen fern. Broadly lance-shaped or oval, soft-textured, much-divided, spreading fronds are clothed with white scales as they unfurl. H 60cm (24in), S 45cm (18in).

Adiantum raddianum 'Fritz Luthii'
Adiantum tenerum
Adiantum tenerum 'Glory of Moordrecht'
Asplenium rhizophyllum

Blechnum tabulare
Cheilanthes lanosa
Davallia canariensis
Davallia mariesii

Dennstaedtia punctilobula
Dryopteris carthusiana
Dryopteris dilatata
Dryopteris filix-mas 'Grandiceps'

Lygodium japonicum
Lygodium palmatum
Oreopteris limbosperma
Phegopteris connectilis

Ferns
◼ GREEN

Thelypteris palustris
(Marsh buckler fern, Marsh fern)
Deciduous fern. Has strong, erect, lance-shaped, pale green fronds, with widely separated, deeply cut pinnae, produced from wiry, creeping, blackish rhizomes. Grows well beside a pool or stream. H 75cm (30in), S 30cm (12in).

Nephrolepis exaltata (Sword fern)
Evergreen fern. Has erect, sometimes spreading, lance-shaped, divided, pale green fronds borne on wiry stems.
H and S 90cm (3ft) or more.
Min. 5°C (41°F).

Osmunda regalis (Royal fern)
Deciduous fern with elegant, broadly oval to oblong, divided, bright green fronds, pinkish when young. Mature plants bear tassel-like, rust-brown fertile flower spikes at ends of taller fronds. H 2m (6ft), S 1m (3ft).

Selaginella kraussiana
Evergreen, trailing, more or less prostrate, moss-like perennial with bright green foliage. H 1cm (1/2in).
S indefinite. Min. 5°C (41°F).

Polypodium vulgare 'Cornubiense'
Evergreen fern with narrow, lance-shaped, divided, fresh green fronds; segments are further sub-divided to give an overall lacy effect. H and S 25–30cm (10–12in).

Onoclea sensibilis (Sensitive fern)
Deciduous, creeping fern with handsome, arching, almost triangular, divided, fresh pale green fronds, often suffused pinkish-brown in spring. In autumn, fronds turn an attractive yellowish-brown. H and S 45cm (18in).

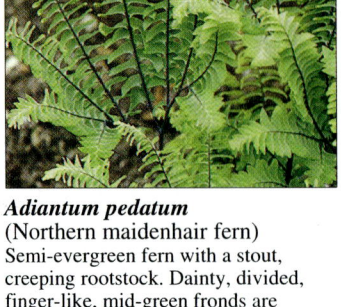

Adiantum pedatum
(Northern maidenhair fern)
Semi-evergreen fern with a stout, creeping rootstock. Dainty, divided, finger-like, mid-green fronds are produced on glossy, dark brown or blackish stems. H and S to 45cm (18in).

Matteuccia struthiopteris
(Ostrich-feather fern, Ostrich fern)
Deciduous, rhizomatous fern. Lance-shaped, erect, divided fronds are arranged like a shuttlecock; outermost, fresh green sterile fronds surround denser, dark brown fertile fronds.
H 1m (3ft), S 45cm (1 1/2ft).

Phegopteris hexagonoptera
Polypodium polypodioides
Polypodium virginianum
Polypodium vulgare 'Cristatum'

Pteris multifida
Selaginella lepidophylla
Woodsia alpina
Woodsia ilvensis

Woodwardia radicans
Woodwardia virginica

■ GREEN

Polystichum setiferum 'Proliferum'
Evergreen or semi-evergreen fern. Broadly lance-shaped or oval, finely divided, soft and moss-like, spreading fronds are clothed with white scales as they unfurl. Largest fronds bear plantlets on upper sides of their midribs. H 60cm (24in), S 45cm (18in).

Asplenium nidus (Bird's-nest fern)
Evergreen fern. Produces broadly lance-shaped, glossy, bright green fronds in a shuttlecock-like arrangement. H 60cm–1.2m (2–4ft), S 30–60cm (1–2ft). Min. 5°C (41°F).

Athyrium nipponicum, syn. *A. goeringianum* (Painted fern)
Deciduous fern. Broad, triangular, divided, purple-tinged, greyish-green fronds arise from a scaly, creeping, brownish or reddish rootstock. H and S 30cm (12in).

Cryptogramma crispa (Parsley fern)
Deciduous fern with broadly oval to triangular, finely divided, bright pale green fronds that resemble parsley. In autumn, fronds turn bright rusty-brown and persist throughout winter. H 15–23cm (6–9in), S 15–30cm (6–12in).

Polypodium vulgare (Common polypody, Polypody)
Evergreen fern with narrow, lance-shaped, divided, herring-bone-like, mid-green fronds, arising from creeping rhizomes covered with copper-brown scales. Suits a rock garden. H and S 25–30cm (10–12in).

Asplenium scolopendrium, syn. *Phyllitis scolopendrium*, *Scolopendrium vulgare* (Hart's-tongue fern)
Evergreen fern with stocky rhizomes and tongue-shaped, leathery, bright green fronds. Is good in alkaline soils. H 45–75cm (18–30in), S to 45cm (18in).

Asplenium scolopendrium 'Marginatum'
Evergreen fern with stocky, upright rhizomes and lobed, slightly frilled, tongue-shaped fronds that are leathery and bright green. Is good in alkaline soils. H and S 30cm (12in) or more.

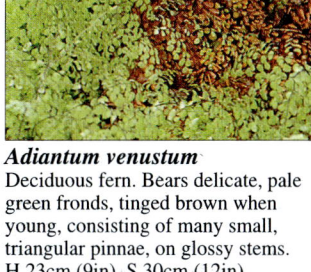

Adiantum venustum
Deciduous fern. Bears delicate, pale green fronds, tinged brown when young, consisting of many small, triangular pinnae, on glossy stems. H 23cm (9in), S 30cm (12in).

Adiantum capillus-veneris
Adiantum raddianum
Adiantum raddianum 'Grandiceps'
Athyrium filix-femina

Cystopteris bulbifera
Cystopteris dickiana
Dryopteris erythrosora
Dryopteris goldiana

Lunathyrium japonicum
Osmunda cinnamomea
Osmunda claytoniana
Pityrogramma chrysophylla

Pityrogramma triangularis
Pteris cretica 'Albolineata'
Pteris cretica 'Mayi'
Selaginella kraussiana 'Variegata'

PERENNIALS/large SUMMER INTEREST
☐ WHITE

Epilobium angustifolium f. *album*
(White rosebay)
Vigorous, upright perennial bearing sprays of pure white flowers along wand-like stems in late summer. Leaves are small and lance-shaped. May spread rapidly. H 1.2–1.5m (4–5ft), S 50cm (20in) or more.

Crambe cordifolia
Robust perennial with clouds of small, fragrant, white flowers borne in branching sprays in summer above mounds of large, crinkled and lobed, dark green leaves. H to 2m (6ft), S 1.2m (4ft).

Eremurus himalaicus
Upright perennial with strap-shaped, basal leaves. In early summer has huge, dense racemes of open cup-shaped, pure white blooms with long stamens. Cover crowns in winter with compost or bracken. Needs staking. H 2–2.5m (6–8ft), S 1m (3ft).

Sanguisorba canadensis
(Canadian burnet)
Clump-forming perennial. In late summer bears slightly pendent spikes of bottlebrush-like, white flowers on stems that arise from toothed, divided, mid-green leaves. H 1.2–2m (4–6ft), S 60cm (2ft).

Nicotiana sylvestris
(Flowering tobacco)
Branching perennial, often grown as an annual, carrying panicles of fragrant, tubular, white flowers at the ends of stems in late summer. Has long, rough, mid-green leaves. H 1.5m (5ft), S 75cm (2½ft).

Eryngium eburneum
Evergreen, arching perennial bearing heads of thistle-like, green flowers with white stamens on branched stems in late summer. Has arching, spiny, grass-like leaves. H 1.5–2m (5–6ft), S 60cm (2ft).

Romneya coulteri
(Californian poppy)
Vigorous, bushy, sub-shrubby perennial, grown for its large, fragrant, white flowers, with prominent centres of golden stamens, that appear in late summer. Has deeply divided, grey leaves. H and S 2m (6ft).

Echinops sphaerocephalus
Massive, bushy perennial with deeply cut, mid-green leaves, pale grey beneath, and grey stems bearing round, greyish-white flower heads in late summer. H 2m (6ft), S 1m (3ft).

Aruncus dioicus, syn. *A. sylvester*, *Spiraea aruncus* (Goat's beard)
Hummock-forming perennial carrying large leaves with lance-shaped leaflets on tall stems and above them, in mid-summer, branching plumes of tiny, creamy-white flowers. H 2m (6ft), S 1.2m (4ft).

OTHER RECOMMENDED PLANTS:
Aconitum 'Ivorine'
Aconitum napellus f. *album*
Campanula pyramidalis
Cimicifuga racemosa
Filipendula kamtschatica
Macleaya cordata
Selinum wallichianum
Veratrum album

WHITE–PINK

Artemisia lactiflora
(White mugwort)
Vigorous, erect perennial. Many sprays of creamy-white buds open to off-white flower heads in summer. Dark green leaves are jagged-toothed. Needs staking and is best as a foil to stronger colours. H 1.2–1.5m (4–5ft), S 50cm (20in).

Lavatera cachemiriana,
syn. *L. cachemirica*
Semi-evergreen, woody-based perennial or sub-shrub with wiry stems bearing panicles of trumpet-shaped, silky, clear pink flowers in summer. Has ivy-shaped, downy, mid-green leaves. H 1.5–2m (5–6ft), S 1m (3ft).

Alpinia zerumbet,
syn. *A. nutans*, *A. speciosa*
(Shell flower, Shell ginger)
Evergreen, clump-forming perennial. Has racemes of white flowers, with yellow lips and pink- or red-marked throats, mainly in summer. H 3m (10ft), S 1m (3ft). Min. 18°C (64°F).

Macleaya microcarpa 'Coral Plume'
Clump-forming perennial that in summer produces branching spikes of rich pink-buff flowers. Large, rounded, lobed leaves are grey-green above, grey-white beneath. H 2–2.5m (6–8ft), S 1–1.2m (3–4ft).

Eremurus robustus
Upright perennial with strap-like leaves that die back during summer as huge racemes of cup-shaped, pink blooms appear. Cover crowns in winter with compost or bracken. Needs staking. H 2.2m (7ft), S 1m (3ft).

PINK–RED

Filipendula rubra
Vigorous, upright perennial with large, jagged leaves and feathery plumes of tiny, soft pink flowers on tall, branching stems in mid-summer. Will rapidly colonize a boggy site. H 2–2.5m, (6–8ft), S 1.2m (4ft).

Meconopsis napaulensis
(Satin poppy)
Clump-forming, short-lived perennial or biennial bearing pale to deep blue, pink or red flowers in late spring and early summer. Deeply cut foliage is covered with bronze hairs. H 2m (6ft), S 1m (3ft).

Rheum palmatum 'Atrosanguineum'
Clump-forming perennial with very large, lobed, deeply cut leaves that are deep red-purple when young. Bears large, fluffy panicles of crimson flowers in early summer.
H and S 2m (6ft).

Anemone x *hybrida* 'Max Vogel'
Eremurus Shelford Hybrids
Kitaibela vitifolia
Lupinus 'The Chatelaine', p.208

Filipendula purpurea, p.209
Kniphofia uvaria var. *nobilis*
Lobelia tupa
PHLOX, p.206

Perennials/large SUMMER INTEREST

■ PURPLE–BLUE

Veratrum nigrum
(Black false hellebore)
Erect, stately perennial that from late summer onwards bears long spikes of chocolate-purple flowers at the ends of stout, upright stems. Stems are clothed with ribbed, oval to narrowly oval leaves. H 2m (6ft), S 60cm (2ft).

Verbena patagonica,
syn. *V. bonariensis*
Perennial with a basal clump of dark green leaves. Upright, wiry stems carry tufts of tiny, purplish-blue flowers in summer-autumn. H 1.5m (5ft), S 60cm (20in).

Cynara cardunculus (Cardoon)
Stately perennial with large clumps of arching, pointed, divided, silver-grey leaves, above which rise large, thistle-like, blue-purple flower heads borne singly on stout, grey stems in summer. Flower heads dry well. H 2m (6ft), S 1m (3ft).

Campanula lactiflora 'Prichard's Variety'
Upright perennial with slender stems carrying branching heads of large, nodding, bell-shaped, violet-blue flowers from early summer to late autumn. May need staking. H 1.2–1.5m (4–5ft), S 60cm (2ft).

Echinops bannaticus
Upright perennial with narrow, deeply cut leaves and globose, pale to mid-blue heads of flowers, borne on branching stems in late summer. Flower heads dry well. H 1.2–1.5m (4–5ft), S 75cm (2½ft).

Galega × hartlandii 'Lady Wilson'
Vigorous, upright perennial with spikes of small, pea-like, blue and pinkish-white flowers in summer above bold leaves divided into oval leaflets. Needs staking. H to 1.5m (5ft), S 1m (3ft).

■ GREEN–YELLOW

Gunnera manicata
Architectural perennial with rounded, prickly-edged leaves, to 1.5m (5ft) across. Has conical, light green flower spikes in early summer, followed by orange-brown seed pods. Needs mulch cover for crowns in winter and a sheltered site. H 2m (6ft), S 2.2m (7ft).

Angelica archangelica (Angelica)
Upright perennial, usually grown as a biennial, with deeply divided, bright green leaves and umbels of white or green flowers in late summer. Stems have culinary usage and when crystallized may be used for confectionery decoration. H 2m (6ft), S 1m (3ft).

Ferula communis (Giant fennel)
Upright perennial. Large, cow parsley-like umbels of yellow flowers are borne from late spring to summer on the tops of stems that arise from a mound of finely cut, mid-green foliage. H 2–2.3m (6–7ft), S 1–1.2m (3ft).

Acanthus spinosus, p.215
Aconitum napellus
Agapanthus inapertus
Campanula pyramidalis

Cicerbita bourgaei
DELPHINIUMS, p.194
Echinops ritro 'Veitch's Blue', p.216
Galega × hartlandii 'His Majesty'

Meconopsis × sheldonii
Thalictrum chelidonii
Thalictrum delavayi 'Hewitt's Double'

Cephalaria gigantea
Eryngium agavifolium
Thalictrum flavum 'Illuminator'

☐YELLOW

Verbascum olympicum
Semi-evergreen, rosette-forming biennial or short-lived perennial. Branching stems, arising from felt-like, grey foliage at the plant base, bear sprays of 5-lobed, bright golden flowers from mid-summer onwards. H 2m (6ft), S 1m (3ft).

☼ ◊ ❄❄❄

Rudbeckia 'Goldquelle'
Erect perennial. In late summer and autumn, daisy-like, double, bright yellow flower heads with green centres are borne singly on stout stems. Has deeply divided, mid-green foliage. H 1.5–2m (5–6ft), S 60–75 cm (2–2½ft).

☼ ◊ ❄❄❄❄ ♈

Achillea filipendulina 'Gold Plate', p.222
CHRYSANTHEMUMS, pp.224–5
Helianthus atrorubens 'Monarch'
Helianthus 'Loddon Gold', p.196

Ligularia przewalskii
Loosely clump-forming perennial with stems clothed in deeply cut, round, dark green leaves. Narrow spires of small, daisy-like, yellow flower heads appear from mid- to late summer. H 1.2–2m (4–6ft), S 1m (3ft).

☼ ♦ ❄❄❄

Inula magnifica
Robust, clump-forming, upright perennial with a mass of lance-shaped to elliptic, rough leaves. Leafy stems bear terminal heads of large, daisy-like, yellow flower heads in late summer. Needs staking. H 1.8m (6ft), S 1m (3ft).

☼ ◊ ❄❄❄

Ligularia 'Gregynog Gold'
Meconopsis paniculata
Rudbeckia 'Herbstsonne', p.196
Rudbeckia laciniata 'Golden Glow'

Ligularia stenocephala
Loosely clump-forming perennial with jagged-edged, round, mid-green leaves. Large heads of daisy-like, yellow-orange flowers open on purplish stems from mid- to late summer. H 1.2m (4ft) or more, S 60cm (2ft).

☼ ♦ ❄❄❄

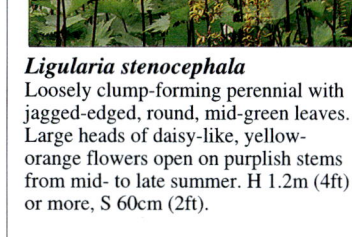

Silphium laciniatum
Telekia speciosa
Thalictrum flavum

☐ORANGE

Heliopsis 'Light of Loddon'
Upright perennial bearing dahlia-like, double, bright orange flower heads on strong stems in late summer. Dark green leaves are coarse and serrated. H 1.2–1.5m (4–5ft), S 60cm (2ft).

☼ ◊ ❄❄❄

Hedychium densiflorum
Clump-forming, rhizomatous perennial bearing a profusion of short-lived, fragrant, orange or yellow flowers in dense spikes during late summer. Broadly lance-shaped leaves are glossy, mid-green. H 1.2–2m (4–6ft), S 60cm (2ft).

☼ ◊ ❄❄

Heliconia psittacorum (Parrot's flower, Parrot's plantain)
Tufted perennial with long-stalked, lance-shaped leaves. In summer, mature plants carry green-tipped, orange flowers with narrow, glossy, orange-red bracts. H to 2m (6ft), S 1m (3ft). Min. 18°C (64°F).

☼ ◊

Heliopsis helianthoides 'Incomparabilis'
Heliopsis helianthoides subsp. scabra
Kniphofia 'Samuel's Sensation'
Wachendorfia thyrsiflora

Delphiniums

Delphiniums are among the most attractive of the tall perennials, with their showy spires of flowers making a spectacular display in the summer herbaceous border. In addition to the classic blues, hybrids are available in a broad range of colours, from white through the pastel shades of dusky-pink and lilac to the richer mauves and violet-purples. Plants should be securely staked to support the heavy flower spikes.

D. 'Butterball'

D. 'Emily Hawkins' ♀

D. 'Bruce' ♀

D. × belladonna 'Blue Bees'

D. 'Loch Leven' ♀

D. 'Fanfare' ♀

D. 'Gillian Dallas' ♀

D. 'Lord Butler' ♀

D. 'Blue Nile' ♀

D. 'Sandpiper' ♀

D. 'Strawberry Fair'

D. 'Mighty Atom' ♀

D. 'Spindrift' ♀

D. grandiflorum 'Blue Butterfly'

D. 'Chelsea Star'

D. 'Olive Poppleton'

D. 'Langdon's Royal Flush'

D. 'Blue Dawn' ♀

D. 'Sungleam' ♀

Perennials/large AUTUMN INTEREST

☐ WHITE–PINK

■ PINK–PURPLE

Cimicifuga simplex
Upright perennial with arching spikes of tiny, slightly fragrant, star-shaped, white flowers in autumn. Leaves are glossy and divided. Needs staking.
H 1.2–1.5m (4–5ft), S 60cm (2ft).

Eupatorium purpureum
(Joe Pye weed)
Stately, upright perennial with terminal heads of tubular, pinkish-purple flowers borne in late summer and early autumn. Coarse, oval leaves are arranged in whorls along purplish stems.
H to 2.2m (7ft), S to 1m (3ft).

Anemone x hybrida 'Honorine Jobert'
Vigorous, branching perennial. Slightly cupped, white flowers with contrasting yellow stamens are carried on wiry stems in late summer and early autumn above deeply divided, dark green leaves. H 1.5m (5ft), S 60cm (2ft).

Anemone hupehensis 'September Charm'
Vigorous, branching perennial. In late summer and early autumn bears slightly cupped, clear pink flowers on wiry stems. Leaves are deeply divided and dark green. H 75cm (30in), S 50cm (20in).

Salvia involucrata 'Bethellii'
Sub-shrubby perennial that produces long racemes of large, cerise-crimson blooms, with pink bracts, in late summer and autumn. Leaves are oval to heart-shaped. H 1.2–1.5m (4–5ft), S 1m (3ft).

Anemone x hybrida 'Bressingham Glow'
Vigorous, branching perennial with slightly cupped, rose-purple flowers borne on wiry stems in late summer and early autumn over clumps of deeply divided, dark green leaves.
H 1.2–1.5m (4–5ft), S 60cm (2ft).

OTHER RECOMMENDED PLANTS:
Anemone hupehensis var. *japonica* 'Prinz Heinrich'
Anemone x *hybrida* 'Max Vogel'

CHRYSANTHEMUMS, pp.224–5
Chrysanthemum uliginosum
Cimicifuga simplex 'Prichard's Giant'
MICHAELMAS DAISIES, p.226

Aconitum carmichaelii 'Arends'
Campanula lactiflora 'Prichard's Variety', p.192
Helenium 'Riverton Gem'

MICHAELMAS DAISIES, p.226
Phytolacca americana
Salvia uliginosa
Verbena patagonica, p.192

195

Perennials/large AUTUMN INTEREST
GREEN–YELLOW

Asclepias physocarpa, syn. *Gomphocarpus physocarpus*
Deciduous, erect, hairy sub-shrub with lance-shaped leaves, 10cm (4in) long. Has umbels of 5-horned, creamy-white flowers in summer, followed by large, inflated, globose seed pods with soft bristles. H to 2m (6ft), S to 60cm (2ft).

Helianthus 'Loddon Gold'
Upright perennial bearing showy, large, vivid deep yellow flower heads with rounded, double centres in late summer and early autumn. Needs staking and may spread quickly. H 1.5m (5ft), S 60cm (2ft).

Hedychium gardnerianum
Upright, rhizomatous perennial. In late summer and early autumn has many spikes of short-lived, fragrant, lemon-yellow and red flowers. Lance-shaped leaves are greyish-green, most markedly when young. H 1.5–2m (5–6ft), S 75cm (2½ft). Min. 5°C (41°F).

Rudbeckia 'Herbstsonne'
Erect perennial bearing daisy-like, yellow flower heads, with conical, green centres, that are carried singly on tall stems in late summer and autumn. Mid-green leaves are shallowly lobed. H 1.5–2.3m (5–7ft), S 60–75cm (2–2½ft).

Helianthus × multiflorus
Upright perennial. Has large, yellow flower heads, with double centres surrounded by larger, rayed segments, that are borne in late summer and early autumn. Needs staking and may spread rapidly. H 1.5m (5ft), S 60cm (2ft).

CHRYSANTHEMUMS, pp.224–5
Helianthus salicifolius
Kniphofia caulescens, p.227
Rudbeckia laciniata 'Golden Glow'
Solidago 'Golden Wings'

Perennials/large WINTER/ALL YEAR
PINK–PURPLE

Strelitzia nicolai
Evergreen, palm-like perennial with a stout trunk. Has leaves, 1.5m (5ft) or more long, on very long stalks and intermittently bears beak-like, white and pale blue flowers in boat-shaped, dark purple bracts. H 8m (25ft), S 5m (15ft). Min. 5–10°C (41–50°F).

Musa ornata (Flowering banana)
Evergreen, palm-like, suckering perennial with oblong, waxy, bluish-green leaves to 2m (6ft) long. In summer has erect, yellow-orange flowers with pinkish bracts and greenish-yellow fruits. H to 3m (10ft), S 2.2m (7ft). Min. 18°C (64°F).

Calathea majestica 'Sanderiana'
Evergreen, clump-forming perennial. Broadly oval, leathery, glossy leaves, to 60cm (2ft) long, are dark green with pink to white lines above, and purple beneath. Intermittently has short spikes of white to mauve flowers. H 1.2–1.5m (4–5ft), S 1m (3ft). Min. 15°C (59°F).

Doryanthes palmeri
Evergreen perennial with a rosette of arching, ribbed leaves, to 2m (6ft) long. Intermittently bears panicles of small, red-bracted, orange-red flowers, white within. Flowers are often replaced by bulbils. H 2–2.5m (6–8ft), S 2.5m (8ft). Min. 10°C (50°F).

Phormium tenax 'Purpureum'
Evergreen, upright perennial with bold, stiff, pointed leaves that are rich reddish-purple to dark copper. In summer, panicles of reddish flowers appear on purplish-blue stems. H 2–2.5m (6–8ft), S 1m (3ft).

OTHER RECOMMENDED PLANTS:
Calathea majestica 'Roseo-lineata'
Costus speciosus
Impatiens sodenii

ORCHIDS, pp.260–63

■ BLUE–GREEN

Perennials/medium SPRING INTEREST
☐ WHITE

Pycnostachys dawei
Strong-growing, bushy perennial with toothed, oblong leaves, 12–30cm (5–12in) long, that are reddish below. Has compact spikes of tubular, 2-lipped, bright blue flowers in winter-spring. H 1.2–1.5m (4–5ft), S 30–90cm (1–3ft). Min. 15°C (59°F).

Musa basjoo, syn. *M. japonica* (Japanese banana)
Evergreen, palm-like, suckering perennial with arching leaves to 1m (3ft) long. Has drooping, pale yellow flowers with brownish bracts in summer followed by green fruits. H 3–5m (10–15ft), S 2–2.5m (6–8ft).

Ranunculus aconitifolius 'Flore Pleno'
Clump-forming perennial with deeply divided, dark green leaves. Double, pure white flowers are borne on strong, branched stems in spring-summer. H 60–75cm (24–30in), S 50cm (20in).

Ranunculus aconitifolius
Vigorous, clump-forming perennial with deeply divided, dark green leaves. Single, white flowers about 3cm (1in) across are borne in spring and early summer. H and S 1m (3ft).

Ensete ventricosum, syn. *Musa arnoldiana*, *M. ensete*
Evergreen, palm-like perennial with small, banana-like fruits. Has 6m (20ft) long leaves with reddish midribs and, intermittently, reddish-green flowers with dark red bracts. H 6m (20ft), S 3m (10ft) or more. Min. 10°C (50°F).

Smilacina racemosa
(False spikenard)
Arching perennial. Has oval, light green leaves terminating in feathery sprays of white flowers that appear from spring to mid-summer and are followed by fleshy, reddish fruits. H 75–90cm (2½–3ft), S 45cm (1½ft).

Aciphylla scott-thomsonii
Artemisia ludoviciana var. albula, p.230
Nepenthes rafflesiana
Phormium cookianum

Phormium cookianum 'Tricolor'
Phormium cookianum 'Variegatum'
Phormium tenax

OTHER RECOMMENDED PLANTS:
Asphodelus aestivus
Asphodelus albus, p.203
Dicentra spectabilis f. alba, p.203

IRISES, pp.198–9
PEONIES, pp. 200–201
Polygonatum hirtum
Polygonatum multiflorum 'Flore Pleno'

197

Irises

These beautiful flowers were originally named after Iris, the Greek goddess of the rainbow, as the shades of their colouring and markings are reminiscent of those of the rainbow. Their distinctive flowers often have 'beards' (short hairs) or crests along the centres of the falls.

The genus is classified into many divisions, some of which are used horticulturally for irises with similar characteristics or cultural requirements. Of these, the easiest to grow are the bearded, crested, Xiphium and dwarf Reticulata groups. Siberian and Japanese types are excellent in a bog garden or by water, but also tolerate drier conditions. Others, such as Juno, Oncocyclus and Regelia irises, may be less easy to cultivate, though their flowers are among the most beautiful. Full details of all groups and guidance on their cultivation are given in the Plant Dictionary.

I. cristata (Evansia)

I. 'Rippling Rose' (bearded)

I. rosenbachiana (Juno)

I. latifolia (Xiphium) ♛

I. 'Wisley White' (Siberian)

I. iberica (Oncocyclus)

I. 'Bold Print' (bearded)

I. 'Krasnia' (bearded)

I. 'Mary Frances' (bearded)

I. 'Dreaming Yellow' (Siberian)

I. 'Geisha Gown' (Japanese)

I. pallida 'Aurea Variegata' (bearded)

I. 'Ruffled Velvet' (Siberian)

I. 'Fulvala' (beardless)

I. douglasiana (Pacific Coast)

I. 'Joette' (bearded)

I. tenax (Pacific Coast)

I. 'Magic Man' (bearded)

I. bucharica (Juno)

I. 'Annabel Jane' (bearded)

I. versicolor (beardless)

I. chrysographes (Siberian)

I. magnifica (Juno)

I. missouriensis (Pacific Coast)

I. 'Paradise Bird' (bearded)

I. tectorum (Evansia)

I. 'Matinata' (bearded)

I. 'Lady of Quality' (Siberian)

I. japonica 'Ledger's Variety' (Evansia)

I. histrioides 'Major' (Reticulata) ♀

I. 'Harmony' (Reticulata)

I. winogradowii (Reticulata)

I. innominata (Pacific Coast)

I. danfordiae (Reticulata)

I. 'Peach Frost' (bearded)

I. 'Carnaby' (bearded)

I. 'Sapphire Star' (Japanese)

I. laevigata (Japanese) ♀

I. 'Joyce' (Reticulata)

I. pseudacorus (beardless)

I. hoogiana (Regelia)

I. xiphium 'Wedgwood' (Xiphium)

I. 'Katharine Hodgkin' (Reticulata)

I. setosa (beardless)

I. 'Early Light' (bearded) ♀

I. 'Sun Miracle' (bearded)

I. fulva (beardless)

I. reticulata 'Cantab' (Reticulata)

I. 'Galathea' (Japanese)

I. 'Bronze Queen' (Xiphium)

I. variegata (bearded)

I. 'Eye Bright' (bearded)

I. 'Blue-eyed Brunette' (bearded) ♀

I. 'Mountain Lake' (Siberian)

I. 'Lavender Royal' (Pacific Coast)

I. 'Butter and Sugar' (Siberian)

I. forrestii (Siberian)

I. 'Shepherd's Delight' (bearded)

I. 'Flamenco' (bearded)

Peonies

Peonies (*Paeonia* species and cultivars) have long been valued for their showy blooms, filling the border with subtle shades of mainly whites, pinks and reds in late spring and early to mid-summer. Peony flowers vary from single to double or anemone form (with broad, outer petals and a mass of petaloids in the centre) and are often heavily scented. The flowers are also good for cutting. Peony foliage is also striking, often tinged bronze when young and assuming rich reddish tints in autumn. Besides the wide variety of border hybrids available, there are many attractive species as well as several tree peonies (cultivars of *P. suffruticosa*), open shrubs often over 2m (6ft) high.

Peonies are long-lived plants that should, if possible, be left undisturbed, as they resent transplanting. If it is necessary to divide the clumps, lift and divide them in autumn or early spring.

P. 'White Wings' (single)

P. 'Shirley Temple' (double)

P. 'Whitleyi Major' (single) ♀

P. 'Mother of Pearl' (single)

P. 'Sarah Bernhardt' (double) ♀

P. 'Baroness Schroeder' (double)

P. 'Kelway's Supreme' (double)

P. 'Ballerina' (double)

P. officinalis 'Alba Plena' (double)

P. suffruticosa 'Hana-kisoi' (double)

P. 'Bowl of Beauty' (anemone) ♀

P. mascula (single)

P. emodi (single)

P. suffruticosa 'Godaishu' (semi-dbl.)

P. suffruticosa 'Rock's Variety' (semi-double)

P. 'Duchesse de Nemours' (dbl.) ♀

P. 'Avant Garde' (single)

P. 'Alice Harding' (double)

P. suffruticosa 'Reine Elizabeth' (double)

P. 'Globe of Light' (anemone)

P. 'Krinkled White' (single)

P. 'Cornelia Shaylor' (double)

P. suffruticosa 'Tamafuyo' (double)

P. suffruticosa 'Kamadu-nishiki' (double)

P. cambessedesii (single) ♀

P. 'Silver Flare' (single)

P. suffruticosa 'Hana-daijin' (double)

P. 'Sir Edward Elgar' (single)

P. wittmanniana (single)

P. lutea 'L'Espérance' (single)

P. veitchii (single)

P. 'Instituteur Doriat' (anemone)

P. 'Laura Dessert' (double) ♀

P. potaninii var. *trollioides* (single)

P. 'Kelway's Gorgeous' (single)

P. suffruticosa 'Kao' (semi-double)

P. suffruticosa 'Kok-kuryu-nishiki' (single)

P. mlokosewitschii (single) ♀

P. 'Magic Orb' (double)

P. × smouthii 'Hibernica'

P. peregrina 'Sunshine' (single) ♀

P. delavayi (single) ♀

P. 'Knighthood' (double)

P. officinalis 'Rubra Plena' (double) ♀

P. 'Argosy' (single)

P. 'Auguste Dessert' (semi-double)

P. tenuifolia (single)

P. 'Souvenir de Maxime Cornu' (double)

P. officinalis 'China Rose' (single)

P. 'Defender' (single) ♀

P. 'Chocolate Soldier' (semi-double) ♀

P. lutea var. *ludlowii* (single)

P. 'Mme Louis Henri' (semi-double)

201

Perennials/medium SPRING INTEREST

WHITE–BLUE

Polygonatum × hybridum
(Solomon's seal)
Arching, leafy perennial with fleshy rhizomes. In late spring, clusters of small, pendent, tubular, greenish-white flowers are produced in axils of neat, oval leaves. H 1.2m (4ft), S 1m (3ft).

Geranium phaeum
(Mourning widow)
Clump-forming perennial with lobed, soft green leaves and maroon-purple flowers, with reflexed petals, borne on rather lax stems in late spring. H 75cm (30in), S 45cm (18in).

Peltiphyllum peltatum
(Umbrella plant)
Spreading perennial with large, rounded leaves. Has clusters of white or pale pink flowers in spring on white-haired stems before foliage appears. H 1–1.2m (3–4ft), S 60cm (2ft).

Symphytum caucasicum
Clump-forming perennial carrying clusters of pendent, azure-blue flowers in spring above rough, hairy, mid-green foliage. Is best suited to a wild garden. H and S 60–90cm (2–3ft).

Tanacetum coccineum 'Brenda', syn. *Pyrethrum* 'Brenda'
Erect perennial with somewhat aromatic, feathery leaves. Daisy-like, single, magenta-pink flower heads are borne in late spring and early summer. H 60cm (24in), S 45cm (18in) or more.

BLUE–YELLOW

Symphytum × uplandicum 'Variegatum'
Perennial with large, hairy, grey-green leaves that have broad, cream margins. In late spring and early summer, pink or blue buds open to tubular, blue or purplish-blue flowers. H 1m (3ft), S 60cm (2ft).

Doronicum pardalianches, syn. *D. cordatum* (Leopard's bane)
Clump-forming perennial with heart-shaped, bright green leaves and, in spring, small, daisy-like, clear yellow flower heads on slender, branching stems. Spreads freely. H 75cm (2½ft), S 60cm (2ft).

Aciphylla aurea (Golden Spaniard)
Evergreen, rosette-forming perennial with long, bayonet-like, yellow-green leaves. Bears spikes of golden flowers up to 2m (6ft) tall from late spring to early summer. H and S in leaf 60–75cm (2–2½ft)

Chelidonium majus 'Flore Pleno'
Upright perennial with divided, bright green leaves and many cup-shaped, double, yellow flowers borne on branching sprays in late spring and early summer. Seeds freely and is best in a wild garden. H 60–90cm (24–30in), S 30cm (12in).

Asphodeline lutea
(Yellow asphodel)
Neat, clump-forming perennial that bears dense spikes of star-shaped, yellow flowers amid narrow, grey-green leaves in late spring. H 1–1.2m (3–4ft), S 60cm–1m (2–3ft).

Anigozanthos manglesii, p.214
Anigozanthos rufa
Columnea gloriosa
Dicentra spectabilis, p.209
Erysimum 'Bowles Mauve'
Geranium maculatum
Meconopsis betonicifolia, p.217
Salvia × superba 'Mainacht', p.216
Anigozanthos flavidus, p.218
DAYLILIES, p.221
Doronicum plantagineum 'Excelsum'
Euphorbia palustris
Trollius 'Alabaster', p.238
Trollius europaeus, p.239
Trollius 'Orange Princess'

Perennials/medium SUMMER INTEREST
☐ WHITE

Gypsophila paniculata 'Bristol Fairy'
Perennial with small, dark green leaves and wiry, branching stems bearing panicles of tiny, double, white flowers in summer. H 60–75cm (2–2½ft), S 1m (3ft).

Libertia grandiflora
(New Zealand satin flower)
Loosely clump-forming, rhizomatous perennial that in early summer produces spikes of white flowers above grass-like, dark green leaves, which turn brown at the tips. Has decorative seed pods in autumn. H 75cm (30in), S 60cm (24in).

Asphodelus albus
(White asphodel)
Upright perennial with clusters of star-shaped, white flowers borne in late spring and early summer. Has narrow, basal tufts of mid-green leaves. H 1m (3ft), S 45cm (18in).

Hesperis matronalis
(Dame's violet, Sweet rocket)
Upright perennial with long spikes of many 4-petalled, white or violet flowers borne in summer. Flowers have a strong fragrance in the evening. Leaves are smooth and narrowly oval. H 75cm (2½ft), S 60cm (2ft).

Anaphalis margaritacea, syn. *A. yedoensis* (Pearl everlasting)
Bushy perennial that has lance-shaped, grey-green or silvery-grey leaves with white margins and many heads of small, white flowers borne on erect stems in late summer. Flower heads dry well. H 60–75cm (2–2½ft), S 60cm (2ft).

Achillea ptarmica 'The Pearl'
Upright perennial with large heads of small, pompon-like, white flowers in summer and tapering, glossy, dark green leaves. May spread rapidly. H and S 75cm (2½ft).

Nicotiana alata, syn. *N. affinis*
Rosette-forming perennial, often grown as an annual, that in late summer bears clusters of tubular, creamy-white flowers, pale brownish-violet externally, which are fragrant at night. Has oval, mid-green leaves. H 75cm (30in), S 30cm (12in).

Dicentra spectabilis f. alba
Leafy perennial forming a hummock of fern-like, deeply cut, light green foliage with arching sprays of pendent, heart-shaped, pure white flowers in late spring and summer. H 60–75cm (2–2½ft), S 60cm (2ft).

OTHER RECOMMENDED PLANTS:
Anemone narcissiflora, p.239
Anemone rivularis, p.239
Caladium x *hortulanum* 'Candidum'
Campanula alliariifolia, p.239
Crambe maritima, p.240
DELPHINIUMS, p.194
Echinacea purpurea 'White Lustre'
Eupatorium rugosum
Filipendula vulgaris 'Flore Pleno'
Heuchera x *brizoides* 'Pearl Drops'
IRISES, pp.198–9
Lysimachia ephemerum
Physostegia virginiana 'Summer Snow'
Ranunculus aconitifolius, p.197
Smilacina racemosa, p.197

Perennials/medium SUMMER INTEREST
☐ WHITE

Salvia argentea
Rosette-forming perennial, grown for its woolly, silver foliage. In summer, branching clusters of sage-like, white flowers are produced on strong, upright stems. H 60cm–1m (2–3ft), S 45cm (18in).

Thalictrum aquilegiifolium 'White Cloud'
Perennial with divided, greyish-green leaves. In summer produces terminal sprays of delicate, fluffy, white flowers. H 1–1.2m (3–4ft), S 30cm (1ft).

Leucanthemum x superbum 'Elizabeth', syn. *Chrysanthemum x superbum* 'Elizabeth'
Robust perennial with large, daisy-like, single, pure white flower heads borne singly in summer. Divide and replant every 2 years. H 1m (3ft), S 60cm (2ft).

Rodgersia podophylla
Clump-forming, rhizomatous perennial with large, many-veined leaves that are bronze when young and later become mid-green, then copper-tinted. Panicles of creamy-white flowers are borne well above foliage in summer. H 1.2m (4ft), S 1m (3ft).

Dictamnus albus (Burning bush)
Upright perennial bearing, in early summer, spikes of fragrant, star-shaped, white flowers with long stamens. Light green leaves are divided into oval leaflets. Dislikes disturbance. H 1m (3ft), S 60cm (2ft).

Rodgersia sambucifolia
Clump-forming, rhizomatous perennial with emerald-green, sometimes bronze-tinged leaves composed of large leaflets. Sprays of creamy-white flowers appear above foliage in summer. H 1–1.2m (3–4ft), S 1m (3ft).

Myrrhis odorata (Sweet Cicely)
Graceful perennial that resembles cow parsley. Has aromatic, fern-like, mid-green foliage and fragrant, bright creamy-white flowers in early summer. H 60cm–1m (2–3ft), S 60cm (2ft).

Chrysanthemum frutescens, syn. *Argyranthemum frutescens* (Marguerite)
Evergreen, woody-based, bushy perennial that bears many daisy-like, white, yellow or pink flower heads throughout summer. Attractive leaves are fresh green. H and S 1m (3ft).

Aruncus dioicus 'Kneiffii'
Hummock-forming perennial that has deeply cut, feathery leaves with lance-shaped leaflets on elegant stems and bears branching plumes of tiny, star-shaped, creamy-white flowers in mid-summer. H 1m (3ft), S 50cm (20in).

☐ WHITE

Lysimachia clethroides
Vigorous, clump-forming, spreading perennial carrying spikes of small, white flowers above mid-green foliage in late summer. H 1m (3ft), S 60cm–1m (2–3ft).

Rodgersia aesculifolia
Clump-forming, rhizomatous perennial that is excellent for a bog garden or pool side. In mid-summer, plumes of fragrant, pinkish-white flowers rise from crinkled, bronze foliage like that of a horse-chestnut tree. H and S 1m (3ft).

Morina longifolia
Evergreen perennial that produces rosettes of large, spiny, thistle-like, rich green leaves. Whorls of hooded, tubular, white flowers, flushed pink within, are borne well above foliage in mid-summer. H 60–75cm (2–2½ft), S 30cm (1ft).

Veronica virginicum f. *album*, syn. *Veronica virginica* f. *alba*
Upright perennial. In late summer, spires of small, white flowers, with pink-flushed bases and pink anthers, crown stems clothed with whorls of narrow, dark green leaves. H 1.2m (4ft), S 45cm (18in).

Papaver orientale 'Perry's White'
Hairy-leaved perennial with deep, fleshy roots. Satiny, white flowers with purple centres appear on strong stems in early summer. May need support. H 80cm (2½ft), S 60cm (2ft).

Digitalis ferruginea
Perennial, best treated as a biennial, with long, slender spikes bearing many funnel-shaped, pale orange-brown and white flowers in mid-summer above basal rosettes of oval, rough leaves. Propagate from seed. H 1–1.2m (3–4ft), S 30cm (1ft).

Gillenia trifoliata
Upright perennial with many wiry, branching stems carrying clusters of dainty, white flowers with reddish-brown calyces in summer. Leaves are dark green and lance-shaped. Needs staking. Thrives in most situations. H 1–1.2m (3–4ft), S 60cm (2ft).

Valeriana officinalis
(Cat's valerian, Common valerian)
Clump-forming, fleshy perennial that bears spikes of white to deep pink flowers in summer. Leaves are deeply toothed and mid-green. Has disadvantage of attracting cats. H 1–1.2m (3–4ft), S 1m (3ft).

Campanula persicifolia 'Fleur de Neige'
Gaura lindheimeri
Leucanthemum x *superbum* 'Aglaia'
Leucanthemum x *superbum* 'Wirral Supreme'
PHLOX, p.206
Ranunculus aconitifolius 'Flore Pleno', p.197
Smilacina racemosa, p.197
Xerophyllum tenax

Phlox

Border phlox (cultivars of *Phlox maculata* and *P. paniculata*) are considered to be some of the most elegant and stately of garden plants and are a mainstay of the herbaceous border in mid- to late summer. Their dome-shaped or conical panicles of flowers, which are often delicately scented, are produced in shades that include white, pink, red and purple. Many of the flowers have attractive, contrasting eyes. Some cultivars have strikingly variegated foliage and make a bold feature before the flowers appear.

Phlox thrive in humus-rich, well-drained soil in sun or partial shade. Taller varieties may need staking. If larger flowers are required, reduce the number of stems in spring, when the plant is about a quarter to a third of its eventual height, by pinching out or cutting back the weakest shoots. To prolong the flowering season, cutting back the central part of the flower head as the blooms fade will encourage the sideshoots to flower. Once the plants have finished flowering in autumn, they should be cut back to the ground.

Border phlox are particularly susceptible to eelworm; to avoid infestation, plant only young, healthy stock propagated from root cuttings taken from clean stock. Any plants that show signs of disease should be dug up and burnt. It is advisable not to replant infested soil with phlox for at least two seasons.

P. paniculata 'Eva Cullum'

P. paniculata 'Brigadier' ♀

P. paniculata 'Eventide' ♀

P. paniculata 'Norah Leigh'

P. paniculata 'Graf Zeppelin'

P. paniculata 'Balmoral'

P. maculata 'Alpha' ♀

P. paniculata 'Harlequin'

P. paniculata 'Fujiyama' ♀

P. paniculata 'Windsor' ♀

P. paniculata 'Sandringham'

P. paniculata 'Le Mahdi' ♀

P. paniculata 'Mia Ruys'

P. paniculata 'White Admiral' ♀

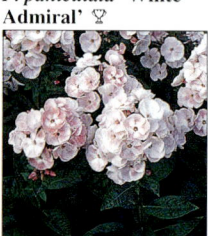

P. maculata 'Omega' ♀

P. paniculata 'Mother of Pearl'

P. paniculata 'Prince of Orange' ♀

P. paniculata 'Amethyst' ♀

P. paniculata 'Hampton Court'

Perennials/medium SUMMER INTEREST
☐ PINK

Linaria purpurea 'Canon Went'
Upright perennial bearing spikes of snapdragon-like, pink blooms with orange-tinged throats from mid- to late summer. Has narrow, grey-green leaves. H 60cm–1m (2–3ft), S 60cm (2ft).

Dictamnus albus var. **purpureus**
Upright perennial. In early summer bears stiff spikes of fragrant, star-shaped, purplish-pink, sometimes paler, flowers with long stamens. Has light green leaves divided into oval leaflets. Dislikes disturbance. H 1m (3ft), S 60cm (2ft).

Monarda didyma 'Croftway Pink'
Clump-forming perennial carrying whorls of hooded, soft pink blooms throughout summer above neat mounds of aromatic foliage. H 1m (3ft), S 45cm (18in).

Chrysanthemum frutescens 'Mary Wootton', syn. *Argyranthemum frutescens* 'Mary Wootton'
Evergreen, woody-based, bushy perennial bearing daisy-like, pink flower heads throughout summer. Has attractive, fern-like, divided, pale green foliage. H and S to 1m (3ft).

Geranium × oxonianum 'Winscombe'
Semi-evergreen, carpeting perennial with dense, dainty, lobed leaves and cup-shaped, deep pink flowers, which fade to pale pink, borne throughout summer. H 60–75cm (24–30in), S 45cm (18in).

Malva moschata
Bushy, branching perennial producing successive spikes of saucer-shaped, rose-pink flowers during early summer. Narrow, lobed, divided leaves are slightly scented. H 60cm–1m (2–3ft), S 60cm (2ft).

Polygonum bistorta 'Superbum'
Vigorous, clump-forming perennial that from early to late summer produces spikes of soft pink flowers above oval leaves. H 60–75cm (2–2½ft), S 60cm (2ft).

Sidalcea 'Jimmy Whittet', syn. *S.* 'Jimmy Whitelet'
Perennial with basal clumps of bright green leaves. In summer bears delicate, single, purplish-pink flowers up erect stems. H 1–1.2m (3–4ft), S 1m (3ft).

Anemone hupehensis, p.223
Astrantia major, p.241
Astrantia maxima, p.241
Campanula lactiflora 'Loddon Anna'

Chelone obliqua, p.223
Lythrum virgatum 'Rose Queen'
PELARGONIUMS, pp.210–11
PEONIES, pp.200–201

Sidalcea 'Loveliness'
Sidalcea 'Oberon'
Tanacetum coccineum 'Brenda', p.202
Tanacetum coccineum 'Eileen May Robinson'

Tanacetum coccineum 'James Kelway'

207

Perennials/medium SUMMER INTEREST
■ PINK

Astilbe 'Venus'
Leafy perennial bearing feathery, tapering plumes of tiny, pale pink flowers in summer. Foliage is broad and divided into leaflets; flowers remain on the plant, dried and brown, well into winter. Prefers humus-rich soil. H and S to 1m (3ft).

Lupinus 'The Chatelaine'
Clump-forming perennial carrying spikes of pink-and-white flowers above divided, mid-green foliage in early summer. H 1.2m (4ft), S 45cm (18in).

Physostegia virginiana 'Variegata'
Erect perennial. In late summer produces spikes of tubular, purplish-pink blooms that can be placed into position. Toothed, mid-green leaves are white-variegated. H 1–1.2m (3–4ft), S 60cm (2ft).

Rehmannia elata
Straggling perennial bearing foxglove-like, yellow-throated, rose-purple flowers in leaf axils of notched, stem-clasping, soft leaves from early to mid-summer. H 1m (3ft), S 45cm (18in). Min. 1°C (34°F).

Astilbe 'Ostrich Plume'
Leafy perennial with handsome, divided foliage and arching, feathery, tapering plumes of tiny, coral-pink flowers in summer. Dry, brown flowers remain on the plant well into winter. Prefers humus-rich soil. H and S to 1m (3ft).

Echinacea purpurea 'Robert Bloom'
Upright perennial. Has lance-shaped, dark green leaves and large, daisy-like, deep crimson-pink flower heads, with conical, brown centres, borne singly on strong stems in summer. Needs humus-rich soil. H 1.2m (4ft), S 50cm (20in).

Centaurea pulcherrima
Upright perennial with deeply cut, silvery leaves. Rose-pink flower heads, with thistle-like centres paler than surrounding star-shaped ray petals, are borne singly on slender stems in summer. H 75cm (2½ft), S 60cm (2ft).

Kohleria digitaliflora
Erect, bushy, rhizomatous perennial with white-haired stems. Has scalloped, hairy leaves and stalked clusters of tubular, very hairy, pink-and-white flowers, with purple-spotted, green lobes, in summer-autumn. H 60cm (24in) or more, S 45cm (18in). Min. 15°C (59°F).

Caladium x hortulanum 'Pink Cloud'
Chrysanthemum rubellum 'Clara Curtis', p.223
DAYLILIES, p.221

Dierama dracomontanum
Erigeron 'Foerster's Liebling'
Heuchera x brizoides 'Coral Cloud'
Heuchera x brizoides 'Scintillation'

Liatris pycnostachya
Lythrum salicaria 'Robert'
Papaver orientale 'Mrs Perry'
PEONIES, pp.200–201

PHLOX, p.206
Rodgersia pinnata 'Superba'
Verbascum 'Pink Domino'

◨ PINK–RED

Dicentra spectabilis (Bleeding heart, Dutchman's trousers)
Leafy perennial forming a hummock of fern-like, deeply cut, mid-green foliage, above which rise arching stems carrying pendent, heart-shaped, pinkish-red and white flowers in late spring and summer. H 75cm (30in), S 50cm (20in).

Lythrum salicaria **'Firecandle'**, syn. *L. s.* **'Feuerkerze'**
Clump-forming perennial for a waterside or bog garden. Bears spikes of intense rose-red blooms from mid- to late summer. Small, lance-shaped leaves are borne on flower stems. H 1m (3ft), S 45cm (18in).

Mirabilis jalapa (Four o'clock flower, Marvel of Peru)
Bushy, tuberous perennial. Fragrant, trumpet-shaped, crimson, pink, white or yellow flowers, opening in evening, cover mid-green foliage in summer. H 60cm–1.2m (2–4ft), S 60–75cm (2–2½ft).

◧ RED

Lupinus **'Inverewe Red'**
Upright perennial with divided, bright green leaves. In early summer bears tall, upright racemes of red flowers. Cut back after flowering. May be short-lived. H 1–1.2m (3–4ft), S 60cm (2ft).

Lythrum virgatum **'The Rocket'**
Clump-forming perennial that carries slender spikes of rose-red flowers above mid-green foliage during summer. Good for a waterside or bog garden. H 1m (3ft), S 45cm (1½ft).

Geranium psilostemon, syn. *G. armenum*
Clump-forming perennial that has broad, deeply cut leaves with good autumn colour and many cup-shaped, single, black-centred, magenta flowers in mid-summer. H and S 1.2m (4ft).

Filipendula purpurea
Upright perennial with deeply divided leaves. Produces large, terminal heads of masses of tiny, rich reddish-purple flowers in summer. Makes a good waterside plant. H 1.2m (4ft), S 60cm (2ft).

Centranthus ruber (Red valerian)
Perennial forming loose clumps of fleshy leaves. Branching heads of small, star-shaped, deep reddish-pink or white flowers are borne above foliage from late spring to autumn. Thrives in poor, exposed conditions. H 60cm–1m (2–3ft), S 45–60cm (1½–2ft) or more.

Asclepias hallii
Centaurea dealbata 'Steenbergii'
Centaurea hypoleuca 'John Coutts', p.243
Francoa sonchifolia

Liatris spicata, p.243
Mimulus lewisii, p.243
PELARGONIUMS, pp.210–11
PENSTEMONS, p.212

PEONIES, pp.200–201
PHLOX, p.206
Sidalcea 'Puck'
Sidalcea 'Sussex Beauty'

Pelargoniums

Pelargoniums are grown throughout the world for their colourful flowers. They grow happily in containers or beds and flower almost continuously in warm climates or under glass. To flower well, they need warmth, sunshine (without too much humidity) and well-drained soil.

Zonal – the common geranium: plants with rounded leaves, distinctively marked with a darker 'zone' and single to double flowers.
Regal – shrubby plants with deeply serrated leaves and exotic, broadly trumpet-shaped flowers that are prone to weather damage in the open.
Ivy-leaved – trailing plants, ideal for hanging baskets, with lobed, somewhat fleshy leaves and single to double flowers.
Scented-leaved and **species** – plants with small, often irregularly star-shaped flowers; scented-leaved forms are grown for their fragrant leaves.
Unique – tall-growing sub-shrubs with regal-like, brightly coloured flowers that are borne continuously through the season. Leaves, which may be scented, vary in shape.

P. 'Dale Queen' (zonal)

P. 'Purple Emperor' (regal)

P. x *fragrans* (scented-leaved)

P. 'Butterfly Lorelei' (zonal)

P. 'Apple Blossom Rosebud' (zonal) ♀

P. 'Timothy Clifford' (zonal)

P. 'Ivalo' (zonal)

P. 'Clorinda' (scented-leaved)

P. 'Mauritania' (zonal)

P. 'Golden Lilac Mist' (zonal)

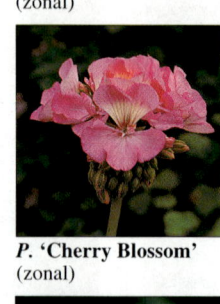

P. 'Cherry Blossom' (zonal)

P. 'Fraicher Beauty' (zonal)

P. 'Fair Ellen' (scented-leaved)

P. 'Autumn Festival' (regal)

P. 'Francis Parrett' (zonal) ♀

P. 'Rica' (zonal)

P. acetosum (species)

P. 'Mr Henry Cox' (zonal) ♀

P. 'Schöne Helena' (zonal)

P. 'Lesley Judd' (regal)

P. 'Mr Everaarts' (zonal)

P. 'Manx Maid' (regal)

P. 'Brookside Primrose' (zonal)

P. 'The Boar' (species) ♀

P. peltatum 'Lachskönigin' (ivy-leaved)

P. 'Lachsball' (zonal)

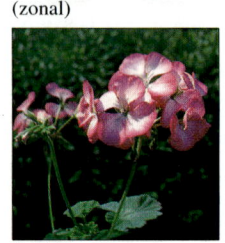

P. 'Alberta' (zonal)

P. 'Amethyst' (ivy-leaved) ♀

P. 'Mini Cascade'
(ivy-leaved)

P. 'Mme Fournier'
(zonal)

P. 'Bredon'
(regal) ♀

P. 'Gustav Emich'
(zonal)

P. L'Elégante'
(ivy-leaved) ♀

P. 'Paton's Unique'
(unique) ♀

P. 'Rouletta'
(ivy-leaved)

P. 'Caligula'
(zonal)

P. 'Dolly Varden'
(zonal) ♀

P. 'Voodoo' (unique)

P. 'Royal Oak'
(scented-leaved)

P. 'Tip Top Duet'
(regal)

P. 'Paul Humphris'
(zonal)

P. 'Flower of Spring'
(zonal) ♀

P. 'Robe' (zonal)

P. capitatum
(scented-leaved)

P. tomentosum
(scented-leaved) ♀

P. 'Friesdorf'
(zonal)

P. 'Irene'
(zonal) ♀

P. 'Mrs Pollock'
(zonal)

P. 'Polka' (unique)

P. 'Purple Unique'
(unique)

P. crispum 'Variegatum' (scented-leaved) ♀

P. peltatum 'Tavira'
(ivy-leaved)

P. 'Purple Wonder'
(zonal)

P. 'Mabel Grey'
(scented-leaved) ♀

P. 'Rollinson's Unique' (unique)

P. 'Capen' (unique)

P. 'Orange Ricard'
(zonal)

P. 'Prince of Orange'
(scented-leaved)

P. 'Mrs Quilter'
(zonal)

Penstemons

Increasingly valued by gardeners for their long racemes of foxglove-like flowers, penstemons are among the most elegant and reliable of border perennials. A large number of excellent cultivars is available, in a subtle range of colours that includes white, pastel and deep pink, warm cherry-red, clear blue and dusky purple. Many penstemon flowers have contrasting white throats or are attractively streaked with other colours.

Penstemons flower most prolifically in summer, but the display can be prolonged into autumn provided the plants are regularly dead-headed throughout the season. Some taller cultivars may benefit from being staked.

Penstemons thrive in well-drained soil preferably with full sun, as some are not fully hardy. Where winters are expected to be severe, the plants should be lifted after flowering and before the first frosts and overwintered in a cold frame. Plants are more likely to survive frost if they are grown in a sheltered spot in free-draining soil.

Penstemons are generally not long-lived plants and tend to become woody at the base after a few seasons; they should be regularly replaced with younger plants propagated from softwood cuttings of non-flowering shoots taken in mid- to late summer.

P. 'Apple Blossom' ♀

P. 'Pennington Gem' ♀

P. 'Andenken an Friedrich Hahn' ♀

P. 'Maurice Gibbs'

P. 'Chester Scarlet' ♀

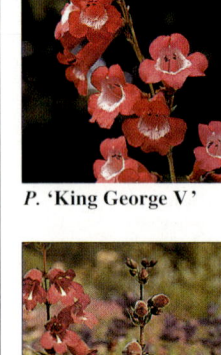

P. 'King George V'

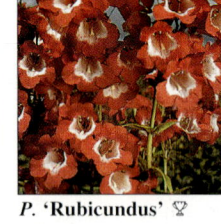

P. 'Rubicundus' ♀

P. 'White Bedder' ♀

P. 'Beech Park' ♀

P. 'Evelyn' ♀

P. 'Burgundy'

P. 'Countess of Dalkeith'

P. 'Sour Grapes'

P. 'Mother of Pearl'

P. 'Pink Endurance'

P. 'Prairie Fire'

P. 'Schoenholzeri' ♀

Perennials/medium SUMMER INTEREST
■ RED

Aquilegia vulgaris 'Nora Barlow'
Leafy perennial that in summer has several short-spurred, funnel-shaped, double, red flowers, pale green at the tips, on long stems. Leaves are grey-green, rounded and deeply divided. H 60–75cm (24–30in), S 50cm (20in).

Polygonum amplexicaule 'Firetail'
Clump-forming perennial that carries slender spikes of bright red flowers above heart-shaped leaves in summer-autumn. H and S 1–1.2m (3–4ft).

Knautia macedonica, syn. *Scabiosa rumelica*
Upright perennial with deeply divided leaves and many rather lax, branching stems bearing double, almost globular, bright crimson flower heads in summer. Needs staking. H 75cm (2½ft), S 60cm (2ft).

Lobelia 'Cherry Ripe'
Clump-forming perennial bearing spikes of cerise-scarlet flowers from mid- to late summer. Leaves, usually fresh green, are often tinged red-bronze. H 1m (3ft), S 23cm (9in).

Astilbe 'Montgomery'
Leafy perennial bearing feathery, tapering plumes of tiny, deep salmon-red flowers in summer. Foliage is broad and divided into leaflets; flowers, brown when dried, remain on the plant well into winter. Prefers humus-rich soil. H 75cm (2½ft), S to 1m (3ft).

Cosmos atrosanguineus, syn. *Bidens atrosanguinea* (Chocolate cosmos)
Upright, tuberous perennial with chocolate-scented, maroon-crimson flower heads in late summer. In warm sites tubers may overwinter if protected. H 60cm (24in) or more, S 45cm (18in).

Lobelia 'Queen Victoria'
Clump-forming perennial. From late summer to mid-autumn spikes of blazing red flowers on branching stems arise from basal, deep red-purple foliage. H 1m (3ft), S 30cm (1ft).

Ruellia graecizans, syn. *R. amoena*
Evergreen, bushy sub-shrub with wide-spreading stems. Oval, pointed leaves are 10cm (4in) long. Intermittently bears clusters of small, tubular, scarlet flowers on stalks to 10cm (4in) long. H and S 60cm (2ft) or more. Min. 15°C (59°F).

DAYLILIES, p.221
Lobelia cardinalis
Lobelia fulgens
Lobelia 'Will Scarlet'

ORCHIDS, pp.260–63
Papaver orientale 'Indian Chief'
PELARGONIUMS, pp.210–11
PENSTEMONS, p.212

PEONIES, pp.200–201
Polygonum amplexicaule
Potentilla 'Etna'
PRIMULAS, pp.236–7

Salvia involucrata
Smithiantha zebrina
Xeronema callistemon

213

Perennials/medium SUMMER INTEREST

■ RED

■ RED–PURPLE

Hedysarum coronarium
(French honeysuckle)
Spreading, shrubby perennial or biennial. Spikes of pea-like, bright red flowers are produced in summer above divided, mid-green leaves. H and S 1m (3ft).

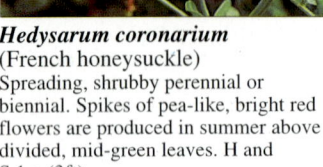

Russelia equisetiformis,
syn. *R. juncea* (Coral plant)
Evergreen, branching, bushy sub-shrub with rush-like stems and tiny leaves. Showy, pendent clusters of tubular, scarlet flowers appear in summer-autumn. H to 1m (3ft) or more, S 60cm (2ft). Min. 15°C (59°F).

Anigozanthos manglesii
(Red-and-green kangaroo paw)
Vigorous, bushy perennial that bears racemes of large, tubular, woolly, red-and-green flowers in spring and early summer. Has long, narrow, grey-green leaves. May suffer from ink disease. H 1m (3ft), S 45cm (18in).

Kohleria eriantha
Robust, bushy, rhizomatous perennial with reddish-haired stems. Oval leaves, to 13cm (5in) long, are edged with red hairs. Has tubular, red flowers, with yellow-spotted lobes, in nodding clusters in summer. H and S 1m (3ft) or more. Min. 15°C (59°F).

Monarda didyma 'Cambridge Scarlet'
Clump-forming perennial that throughout summer bears whorls of hooded, rich red flowers above neat mounds of aromatic, hairy foliage. H 1m (3ft), S 45cm (18in).

Columnea x *banksii*
Evergreen, trailing perennial with oval, fleshy leaves, glossy above, purplish-red below. Tubular, hooded, brilliant red flowers, to 8cm (3in) long, appear from spring to winter. Makes a useful plant for a hanging basket. H 90cm (3ft), S indefinite. Min. 15°C (59°F).

Papaver orientale 'Allegro Viva'
Hairy-leaved perennial with very deep, fleshy roots. Papery, bright scarlet flowers are borne in summer on strong stems. H 60–75cm (24–30in), S 45cm (18in).

Glycyrrhiza glabra (Liquorice)
Upright perennial that has pea-like, purple-blue and white flowers, borne in short spikes on erect stems in late summer, and large leaves divided into oval leaflets. Is grown commercially for production of liquorice. H 1.2m (4ft), S 1m (3ft).

Asclepias curassavica
Calceolaria integrifolia
Gaillardia x *grandiflora* 'Dazzler', p.247
Geum 'Fire Opal'
Geum 'Mrs Bradshaw'
ORCHIDS, pp.260–63
Papaver orientale
Smithiantha zebrina

PELARGONIUMS, pp.210–11

■ PURPLE

Acanthus balcanicus,
syn. *A. hungaricus*, *A. longifolius*
Perennial with long, deeply cut, basal, dark green leaves. Spikes of white or pink-flushed flowers, set in spiny, red-purple bracts, are carried in summer. H 60cm–1m (2–3ft), S 1m (3ft).

Linaria triornithophora
(Three birds toadflax)
Upright perennial that from early to late summer produces spikes of snapdragon-like, purple and yellow flowers above narrow, grey-green leaves. H 1m (3ft), S 60cm (2ft).

Monarda fistulosa
Clump-forming perennial that produces small heads of lilac-purple flowers from mid- to late summer. H 1.2m (4ft), S 45cm (18in).

***Campanula glomerata* 'Superba'**
Vigorous, clump-forming perennial with dense, rounded heads of large, bell-shaped, purple flowers borne in summer. Bears oval leaves in basal rosettes and on flower stems. Must be divided and replanted regularly. H 75cm (2½ft), S 1m (3ft) or more.

***Geranium sylvaticum* 'Mayflower'**
Upright perennial with a basal clump of deeply lobed leaves, above which rise branching stems of cup-shaped, violet-blue flowers in early summer. H 1m (3ft), S 60cm (2ft).

Acanthus spinosus
Stately perennial that has very large, arching, deeply cut and spiny-pointed, glossy, dark green leaves. Spires of funnel-shaped, soft mauve and white flowers are borne freely in summer. H 1.2m (4ft), S 60cm (2ft) or more.

Thalictrum aquilegiifolium
Clump-forming perennial with a mass of finely divided, grey-green leaves, resembling those of maidenhair fern. Bunched heads of fluffy, lilac-purple flowers are borne on strong stems in summer. H 1–1.2m (3–4ft), S 45cm (18in).

***Veronica longifolia* 'Romiley Purple'**
Clump-forming perennial that in summer freely produces large spikes of purple flowers above whorled, mid-green leaves. H 1–1.2m (3–4ft), S 30–60cm (1–2ft).

Campanula 'Burghaltii', p.250
Centaurea dealbata
DELPHINIUMS, p.194
Digitalis x *mertonensis*

Erigeron 'Quakeress'
Erysimum 'Bowles Mauve'
Hesperis matronalis, p.203
Lupinus 'Thundercloud'

Monarda 'Prairie Night'
ORCHIDS, pp.260–63
PHLOX, p.206
Polemonium foliosissimum

Thalictrum diffusiflorum
Tricyrtis hirta

215

Perennials/medium SUMMER INTEREST
■ PURPLE

Campanula trachelium
(Nettle-leaved bellflower)
Upright perennial with rough, serrated, oval, pointed, basal leaves. Wide, bell-shaped, blue or purple-blue flowers are spaced along erect stems in summer. H 60cm–1m (2–3ft), S 30cm (1ft).

Dichorisandra reginae
Evergreen, erect, clump-forming perennial. Glossy, often silver-banded and flecked leaves are purple-red beneath. Has small spikes of densely set, purple-blue flowers in summer-autumn. H 60–75cm (2–2½ft), S to 30cm (1ft). Min. 20°C (68°F).

Echinops ritro 'Veitch's Blue'
Upright perennial with round, thistle-like, purplish-blue heads of flowers carried in late summer on silvery stems. Sharply divided leaves have pale down beneath. H 1.2m (4ft), S 75cm (2½ft).

Galega orientalis
Vigorous, upright but compact perennial that in summer bears spikes of pea-like, blue-tinged, violet flowers above delicate leaves divided into oval leaflets. Needs staking. Spreads freely. H 1.2m (4ft), S 60cm (2ft).

Salvia x superba 'Mainacht', syn. *S. nemorosa* 'Mainacht', *S. n.* 'May Night'
Neat, clump-forming perennial with narrow, wrinkled, mid-green leaves. In late spring and summer bears stiff racemes of violet-blue flowers. H 1m (3ft), S 45cm (18in).

Campanula trachelium 'Bernice'
Upright perennial that has wide, bell-shaped, double, purple-violet flowers carried along erect stems in summer. Leaves are mostly basal and are rough, serrated, oval and pointed. H 75cm (2½ft), S 30cm (1ft).

Aconitum x cammarum 'Bicolor', syn. *A. x bicolor*
Compact, tuberous perennial with violet-blue and white flowers borne in summer along upright stems. Has deeply cut, divided, glossy, dark green leaves and poisonous roots. H 1.2m (4ft), S 50cm (20in).

Baptisia australis (False indigo)
Upright perennial bearing spikes of pea-like, violet-blue flowers in summer. Bright green leaves are divided into oval leaflets. Dark grey seed pods may be used for winter decoration. H 75cm (2½ft), S 60cm (2ft).

Campanula latifolia 'Brantwood'
DELPHINIUMS, p.194
Erigeron 'Darkest of All'
Erigeron 'Serenity', p.248
IRISES, pp.198–9
Lobelia 'Vedrariensis'
ORCHIDS, pp.260–63
PEONIES, pp.200–201
Strobilanthes atropurpureus, p.227
Sutera grandiflora

PURPLE–BLUE

Dianella tasmanica
Upright perennial with nodding, star-shaped, bright blue or purple-blue flowers carried in branching sprays in summer, followed by deep blue berries in autumn. Has untidy, evergreen, strap-shaped leaves. H 1.2m (4ft), S 50cm (20in).

Eryngium alpinum
Upright perennial with basal rosettes of heart-shaped, deeply toothed, glossy foliage, above which rise stout stems bearing, in summer, heads of conical, purplish-blue flower heads, surrounded by blue bracts and soft spines. H 75cm–1m (2½–3ft), S 60cm (2ft).

Eryngium x tripartitum
Perennial with wiry stems above a basal rosette of coarsely toothed, grey-green leaves. Conical, metallic-blue flower heads on blue stems are borne in summer-autumn and may be dried for winter decoration. H 1–1.2m (3–4ft), S 50cm (20in).

Meconopsis grandis 'Branklyn'
Erect perennial that carries large, pendent, poppy-like, blue flowers on stout stems in early summer. Erect, oblong, slightly toothed and hairy leaves are produced at the base of the plant. H 1–1.2m (3–4ft), S 30–45cm (1–1½ft).

Eryngium x oliverianum
Upright perennial that produces large, rounded heads of thistle-like, blue to lavender-blue flowers in late summer. Has heart-shaped, jagged-edged, basal, mid-green leaves. H 60cm–1m (2–3ft), S 45–60cm (1½–2ft).

Campanula persicifolia 'Telham Beauty'
Perennial with basal rosettes of narrow, bright green leaves. In summer, large, nodding, cup-shaped, light blue flowers are borne on slender spikes. H 1m (3ft), S 30cm (1ft).

Meconopsis betonicifolia (Blue poppy)
Clump-forming perennial that bears blue flowers in late spring and early summer. Oblong, mid-green leaves are produced in basal rosettes and in decreasing size up flowering stems. H 1–1.2m (3–4ft), S 45cm (1½ft).

Cichorium intybus (Chicory)
Clump-forming perennial with basal rosettes of light green leaves and daisy-like, bright blue flower heads borne along upper parts of willowy stems in summer. Flowers are at their best before noon. H 1.2m (4ft), S 45cm (18in).

Agapanthus 'Dorothy Palmer'
Clump-forming perennial bearing rounded heads of rich blue flowers, fading to reddish-mauve, on erect stems in late summer. Leaves are narrow and greyish-green. Protect crowns in winter with mulch. H 1m (3ft), S 50cm (20in).

Campanula lactiflora
Campanula latiloba
Campanula latiloba 'Percy Piper'
Campanula persicifolia 'Pride of Exmouth'

DELPHINIUMS, p.194
Dianella caerulea
Eryngium giganteum
Eryngium 'Violetta'

Geranium pratense
Geranium pratense 'Plenum Violaceum'
Linaria purpurea
MICHAELMAS DAISIES, p.226

Salvia haematodes
Trachelium caeruleum, p.282

Perennials/medium SUMMER INTEREST

BLUE

Agapanthus praecox subsp. **orientalis**, syn. *A. orientalis*
Perennial with large, dense umbels of sky-blue flowers borne on strong stems in late summer over clumps of broad, almost evergreen, dark green leaves. Makes a good plant for pots.
H 1m (3ft), S 60cm (2ft).

Cynoglossum nervosum
(Himalayan hound's tongue)
Clump-forming perennial with bright blue flowers, similar to forget-me-nots, carried in small clusters on branching stems in summer over narrow, mid-green leaves. H 75cm (30in), S 50cm (20in).

Anchusa azurea 'Loddon Royalist'
Upright perennial that bears flat, single, deep blue flowers on branching spikes in early summer. Most of the lance-shaped, coarse, hairy leaves are at the base of plant. Needs staking.
H 1.2m (4ft), S 60cm (2ft).

DELPHINIUMS, p.194
Echinops bannaticus 'Taplow Blue'
Gentiana asclepiadea, p.227
IRISES, pp.198–9
Meconopsis grandis
Salvia nemorosa
Veronica exaltata
Veronica longifolia

GREEN–YELLOW

Rheum alexandrae
Clump-forming perennial with panicles of cream flowers, opening in early summer, that are hidden by large, greenish-white to cream bracts, which turn to red. Leaves are glossy and dark green. H 1m (3ft), S 60cm (2ft).

Anigozanthos flavidus
(Yellow kangaroo paw)
Bushy perennial with racemes of large, woolly, tubular, sometimes red-tinged, yellowish-green flowers, with reddish anthers, borne in spring-summer. Narrow leaves, to 60cm (2ft) long, are mid-green. H 1.2m (4ft), S 45cm (18in).

Nepeta govaniana
Upright, well-branched perennial. During summer, sprays of long, tubular, pale yellow flowers are borne above a mass of pointed, grey-green leaves. H 1m (3ft), S 60cm (2ft).

Verbascum 'Gainsborough'
Semi-evergreen, rosette-forming, short-lived perennial bearing branched racemes of 5-lobed, pale sulphur-yellow flowers throughout summer above oval, mid-green leaves borne on flower stems. H 60cm–1.2m (2–4ft), S 30–60cm (1–2ft).

Thalictrum lucidum
Perennial with glossy leaves composed of numerous leaflets. Strong stems bear loose panicles of fluffy, greenish-yellow flowers in summer. H 1–1.2m (3–4ft), S 50cm (20in).

Artemisia pontica, p.254
Euphorbia nicaeensis
HOSTAS, pp.252–3
ORCHIDS, pp.260–63
Paris polyphylla
Polygonum virginianum 'Painter's Palette', p.254

YELLOW

Chrysanthemum frutescens 'Jamaica Primrose'
Evergreen, woody-based, bushy perennial with divided, fern-like, pale green leaves and many daisy-like, single, soft yellow flower heads borne throughout summer. Take stem cuttings in early autumn. H and S to 1m (3ft).

Aconitum lycoctonum subsp. **vulparia**, syn. *A. vulparia* (Wolf's bane)
Upright, fibrous perennial that has hooded, straw-yellow flowers in summer. Leaves are dark green and deeply divided. Needs staking. H 1–1.2m (3–4ft), S 30–60cm (1–2ft).

Gentiana lutea (Great yellow gentian)
Erect, unbranched perennial with oval, stalkless leaves to 30cm (1ft) long. In summer has dense whorls of tubular, yellow flowers in axils of greenish bracts. H 1–1.2m (3–4ft), S 60cm (2ft).

Phlomis russeliana
Evergreen perennial, forming excellent ground cover, with large, rough, heart-shaped leaves. Stout flower stems bear whorls of hooded, butter-yellow flowers in summer. H 1m (3ft), S 60cm (2ft) or more.

Anthemis tinctoria 'E.C. Buxton'
Clump-forming perennial with a mass of daisy-like, lemon-yellow flower heads borne singly in summer on slim stems. Cut back hard after flowering to promote a good rosette of crinkled leaves for winter. H and S 1m (3ft).

Lysimachia punctata (Garden loosestrife)
Clump-forming perennial that in summer produces spikes of bright yellow flowers above mid-green leaves. H 60–75cm (2–2½ft) S 60cm (2ft).

Solidago 'Goldenmosa'
Clump-forming perennial. Sprays of tufted, mimosa-like, yellow flower heads are carried in late summer and autumn above lance-shaped, toothed, hairy, yellowish-green leaves. H 1m (3ft), S 60cm (2ft).

Aciphylla squarrosa, p.230
Chelidonium majus 'Flore Pleno', p.202
CHRYSANTHEMUMS, pp.224–5
DAYLILIES, p.221

Digitalis lutea
Helianthus 'Capenoch Star'
IRISES, pp.198–9
Kirengeshoma palmata, p.227

Kniphofia 'Little Maid', p.254
Linaria genistifolia
Lupinus 'Tom Reeves'
PRIMULAS, pp.236–7

Rudbeckia fulgida var. *deamii*
x *Solidaster luteus*, p.255
Thermopsis villosa
Verbascum chaixii

219

Perennials/medium SUMMER INTEREST
☐ YELLOW

Thermopsis rhombifolia,
syn. *T. montana*
Upright perennial bearing spikes of bright yellow flowers above divided, mid-green leaves in summer.
H 60cm–1m (2–3ft), S 60cm (2ft).

Inula hookeri
Clump-forming perennial with lance-shaped to elliptic, hairy leaves and a mass of slightly scented, daisy-like, greenish-yellow flower heads borne in summer. H 75cm (30in), S 45cm (18in).

Aphelandra squarrosa 'Louisae'
Evergreen, erect perennial. Long, oval, glossy, slightly wrinkled, dark green leaves have white veins and midribs. Bears dense spikes of golden-yellow flowers from axils of yellow bracts in late summer to autumn. H to 1m (3ft), S 60cm (2ft). Min. 13°C (55°F).

Solidago 'Laurin'
Compact perennial bearing spikes of deep yellow flowers in late summer. Foliage is mid-green. H 60–75cm (24–30in), S 45 cm (18in).

Rudbeckia fulgida var. **sullivantii 'Goldsturm',** syn. *R. f.* 'Goldsturm'
Erect perennial. In late summer and autumn, daisy-like, golden flower heads with conical, black centres are borne at the ends of strong stems. Has narrow, rough, mid-green leaves.
H 75cm (30in), S 30cm (12in) or more.

Achillea 'Coronation Gold'
Upright perennial with feathery, silvery leaves. Bears large, flat heads of small, golden flower heads in summer that dry well for winter decoration. Should be divided and replanted every third year.
H 1m (3ft), S 60cm (2ft).

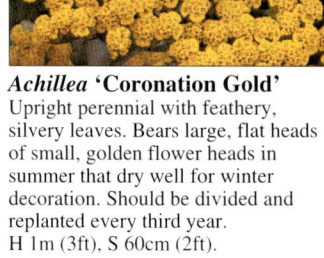

Verbascum nigrum
Semi-evergreen, clump-forming perennial bearing narrow spikes of small, 5-lobed, purple-centred, yellow flowers during summer and autumn. Oblong, mid-green leaves are downy beneath. H 60cm–1m (2–3ft), S 60cm (2ft).

Achillea 'Moonshine', p.255	CHRYSANTHEMUMS, pp.224–5	IRISES, pp.198–9	PEONIES, pp.200–201
Achillea 'Taygetea', p.255	DAYLILIES, p.221	*Isatis tinctoria*	*Potentilla* 'Yellow Queen', p.256
Aciphylla aurea, p.202	*Gaillardia aristata,* p.255	*Mirabilis jalapa,* p.209	PRIMULAS, pp.236–7
Centaurea macrocephala	HOSTAS, pp.252–3	ORCHIDS, pp.260–63	*Scabiosa ochroleuca*

Daylilies

Although they belong to the lily family (Liliaceae), daylilies (*Hemerocallis*) are not true lilies; their common name comes from the lily-like flowers that generally last only a day. Daylilies range in size from compact plants that grow only to 30–38cm (12–15in) tall, to large plants that may reach 1.5m (5ft). Modern cultivars are available in a wide range of colours, from creamy-white, through shades of yellow, orange, red, pink and purple to almost black; some flowers have differently coloured bands on the petals. The flower forms are usually classified as single, double or spider. Some species and a few cultivars are strongly fragrant. Most daylilies flower for four to six weeks; they thrive in almost any soil, except poorly drained clay, in sun or shade, but bloom best if they are in sun for at least half the day. The clumps may be divided in early spring.

H. 'Little Grapette'

H. 'Prairie Blue Eyes'

H. 'Super Purple'

H. 'Golden Chimes' ♕

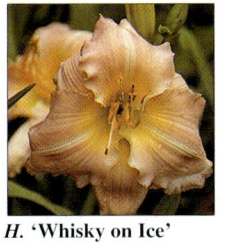

H. 'Gentle Shepherd'

H. 'Whisky on Ice'

H. 'Cat's Cradle'

H. 'Hyperion'

H. 'Siloam Virginia Henson'

H. 'Jolyene Nichole'

H. lilioasphodelus ♕

H. 'Stella d'Oro' ♕

H. 'Betty Woods'

H. 'Scarlet Orbit'

H. 'Stafford'

H. 'Marion Vaughn' ♕

H. 'Cartwheels' ♕

H. 'Ruffled Apricot'

H. 'Lusty Leland'

H. 'Brocaded Gown'

H. 'Eenie Weenie'

H. citrina

H. 'Corky' ♕

H. fulva 'Kwanso Flore Pleno'

221

Perennials/medium SUMMER INTEREST

YELLOW

Berkheya macrocephala
Upright perennial bearing large, daisy-like, yellow flower heads on branched, spiny-leaved stems throughout summer. Prefers rich soil and a warm, sheltered position. H and S 1m (3ft).

Achillea filipendulina 'Gold Plate'
Upright perennial with stout, leafy stems carrying broad, flat, terminal heads of yellow flowers in summer, above filigree foliage. Flower heads retain colour if dried. Divide plants regularly. H 1.2m (4ft) or more, S 60cm (2ft).

Heliopsis 'Ballet Dancer'
Upright perennial flowering freely in late summer and bearing double, yellow flower heads with frilled petals. Dark green leaves are coarse and serrated. H 1–1.2m (3–4ft), S 60cm (2ft).

YELLOW–ORANGE

Kniphofia 'Royal Standard'
Upright perennial with grass-like, basal tufts of leaves and terminal spikes of scarlet buds, opening to lemon-yellow flowers, borne on erect stems in late summer. Protect crowns with winter mulch. H 1–1.2m (3–4ft), S 60cm (2ft).

Asclepias tuberosa
(Butterfly weed)
Erect, tuberous perennial with long, lance-shaped leaves. Small, 5-horned, bright orange-red flowers are borne in summer and followed by narrow, pointed pods, to 15cm (6in) long. H to 75cm (30in), S 45cm (18in).

Sphaeralcea ambigua
Branching, shrubby perennial. Broadly funnel-shaped, orange-coral blooms are produced singly in leaf axils from summer until the onset of cold weather. Leaves are soft, hairy and mid-green. H and S 75cm–1m (2½–3ft).

Euphorbia griffithii 'Fireglow'
Bushy perennial that bears orange-red flowers in terminal umbels in early summer. Leaves are lance-shaped, mid-green and have pale red midribs. H to 1m (3ft), S 50cm (20in).

Kniphofia thomsonii var. snowdenii
Upright perennial with grass-like, basal foliage. In summer bears coral-pink flowers, with yellowish interiors, spaced widely along terminal spikes. Protect crowns with winter mulch. H 1m (3ft), S 50cm (20in).

Lychnis chalcedonica
(Jerusalem cross, Maltese cross)
Neat, clump-forming perennial that bears flat heads of small, vermilion flowers at the tips of stout stems in early summer. Foliage is mid-green. H 1–1.2m (3–4ft), S 30–45cm (12–18in).

Buphthalmum salicifolium, p.256
Chelidonium majus 'Flore Pleno', p.202
CHRYSANTHEMUMS, pp.224–5
Eremurus spectabilis
IRISES, pp.198–9
Linaria genistifolia var. *dalmatica*
ORCHIDS, pp.260–63
PRIMULAS, pp.236–7
DAYLILIES, p.221
Helenium 'Wyndley', p.227
Kniphofia 'Atlanta'
Ligularia dentata 'Desdemona'
Lupinus 'Flaming June'
Papaver orientale 'May Queen'

Perennials/medium AUTUMN INTEREST

□ WHITE–PINK

Actaea alba, syn. *A. pachypoda* (Doll's eyes, White baneberry)
Compact, clump-forming perennial with spikes of small, fluffy, white flowers in summer and clusters of white berries, borne on stiff, fleshy scarlet stalks, in autumn. H 1m (3ft), S 50cm (20in).

Polygonum campanulatum, syn. *Persicaria campanulata*
Compact, mat-forming perennial bearing elegant, branching heads of bell-shaped, pink or white flowers from mid-summer to early autumn. Has oval leaves, brown-felted beneath. H and S 1m (3ft).

Anemone hupehensis
Branching perennial that bears soft pink flowers, with rounded petals, in late summer and early autumn. Leaves are dark green and deeply divided, with toothed leaflets. H 60–75cm (2–2½ft), S 45cm (18in).

Tricyrtis formosana, syn. *T. stolonifera*
Upright, rhizomatous perennial. In early autumn bears spurred flowers, heavily spotted with purplish-pink and with yellow-tinged throats. Glossy, dark green leaves clasp stems. H 60cm–1m (2–3ft), S 45cm (18in).

Chelone obliqua (Turtle-head)
Upright perennial that bears terminal spikes of hooded, lilac-pink flowers in late summer and autumn. Leaves are dark green and lance-shaped. H 1m (3ft), S 50cm (20in).

Chrysanthemum rubellum 'Clara Curtis', syn. *Dendranthema zawadskii* 'Clara Curtis'
Bushy perennial producing many clusters of flat, daisy-like, clear pink flower heads throughout summer and autumn. Divide plants every other spring. H 75cm (30in), S 45cm (18in).

OTHER RECOMMENDED PLANTS:
Centranthus ruber, p.209
CHRYSANTHEMUMS, pp.224–5
Cimicifuga simplex 'Elstead'
Kohleria digitaliflora, p.208
MICHAELMAS DAISIES, p.226
PELARGONIUMS, pp.210–11
Schizostylis coccinea 'Sunrise', p.258

Chrysanthemums

Florists' chrysanthemum hybrids (botanically now classified under the genus *Dendranthema*) are grouped according to their differing flower forms, approximate flowering season (early, mid- or late autumn) and habit (see also the Plant Dictionary under *Chrysanthemum*). The best groups for garden decoration are the sprays, pompons and early reflexed chrysanthemums. The dwarf charms, forming a dense, dome-shaped mass of flowers, look most attractive displayed in pots. Most groups have only one large flower per stem, although the sprays, charms and pompons have several. The various flower forms are described below.

Incurved – fully double, dense, spherical flowers have incurved petals arising from the base of the flower and closing tightly over the crown.

Fully reflexed – fully double flowers have curved, pointed petals reflexing outwards and downwards from the crown, back to touch the stem.

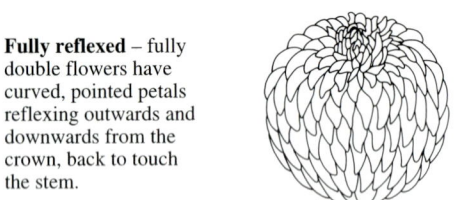

Reflexed – fully double flowers are similar to those of fully reflexed forms except that the petals are less strongly reflexed and form an umbrella-like or spiky outline.

Intermediate – fully double, roughly spherical flowers have loosely incurving petals, which may close at the crown or may reflex for the bottom half of each flower.

Anemone-centred – single flowers each have a central, dome-shaped disc, up to half the diameter of the bloom, and up to 5 rows of flat or occasionally spoon-type ray petals at right angles to the stem.

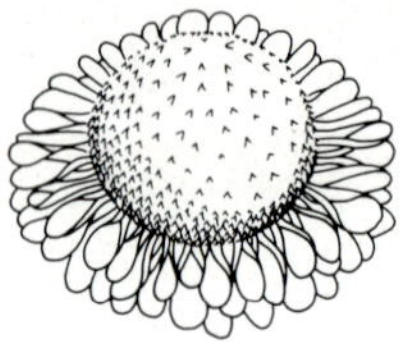

Single – flowers each have about 5 rows of flat petals borne at right angles to the stem, that may incurve or reflex at the tips; the prominent, central disc is golden throughout or has a small, green centre.

Pompon – fully double, dense, spherical, or occasionally hemispherical, flowers, have tubular petals with flat, rounded tips, growing outwards from the crown.

Spoon-type – flowers are similar to those of single forms except that the ray petals are tubular and open out at their tips to form a spoon shape.

C. 'Salmon Fairweather' (incurved)

C. 'Madeleine' (spray, reflexed) ♀

C. 'Roblush' (spray, reflexed)

C. 'Brietner' (reflexed)

C. 'Dorridge Dream' (incurved)

C. 'Pavilion' (intermediate)

C. 'Dawn Mist' (spray, single)

C. 'Alison Kirk' (incurved)

C. 'Duke of Kent' (reflexed)

C. 'Marian Gosling' (reflexed)

C. 'Pennine Flute' (spray, spoon-type) ♀

C. 'Cloudbank' (spray, anemone)

C. 'Michael Fish' (intermediate)

C. 'Pennine Oriel' (spray, anemone)

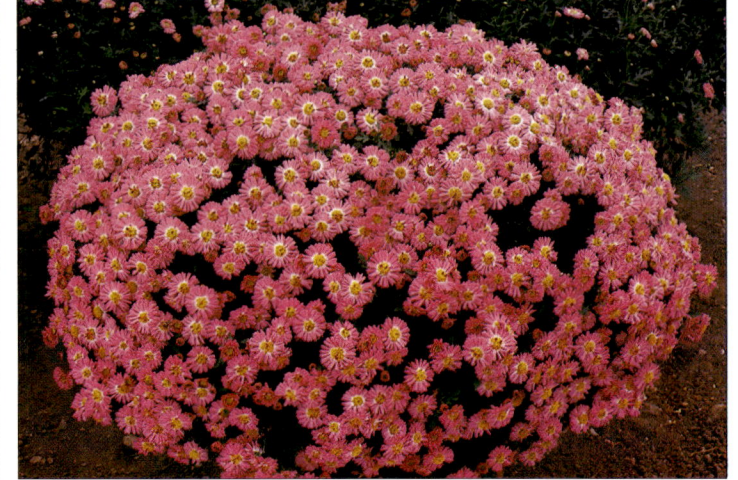

C. 'Ringdove' (charm)

C. 'Talbot Jo'
(spray, single)

C. 'Chippendale'
(reflexed)

C. 'Skater's Waltz'
(intermediate)

C. 'Yvonne Arnaud'
(reflexed)

C. 'Purple Pennine Wine' (spray, reflexed) ⚜

C. 'Maria'
(pompon)

C. 'Cherry Chintz'
(reflexed)

C. 'Rose Yvonne Arnaud' (reflexed)

C. 'Sentry'
(reflexed)

C. 'George Griffiths'
(reflexed)

C. 'Redwing'
(spray, single)

C. 'Green Satin'
(intermediate)

C. 'Marion'
(spray, reflexed)

C. 'Discovery'
(intermediate)

C. 'Marlene Jones'
(intermediate)

C. 'Primrose West Bromwich' (reflexed)

C. 'Pennine Jewel'
(spray, spoon-type)

C. 'Yellow John Hughes' (incurved)

C. 'Edwin Painter'
(single)

C. 'Yellow Brietner'
(reflexed)

C. 'Pennine Alfie'
(spray, spoon-type)

C. 'Golden Woolman's Glory' (single)

C. 'Wendy'
(spray, reflexed) ⚜

C. 'Sally Ball'
(spray, anemone)

C. 'Bronze Fairie'
(pompon)

C. 'Bronze Yvonne Arnaud' (reflexed)

C. 'Salmon Fairie'
(pompon)

C. 'Rytorch'
(spray, single)

C. 'Bronze Hedgerow'
(single)

C. 'Peach Margaret'
(spray, reflexed)

C. 'Peach Brietner'
(reflexed)

C. 'Salmon Margaret'
(spray, reflexed) ⚜

C. 'Autumn Days'
(intermediate)

C. 'Oracle'
(intermediate)

C. 'Buff Margaret'
(spray, reflexed)

Michaelmas daisies

Michaelmas daisies (*Aster* species and cultivars, mostly of *A. novae-angliae* and *A. novi-belgii*) are invaluable border plants as they flower later than most other perennials and continue the display until late autumn. The daisy-like, single or double flowers, usually with yellow centres, range in colour from white, through pink and red to purple and blue and are excellent for cutting. For larger flowers pinch out or cut back weaker shoots in spring; pinching out the top 2.5–5cm (1–2in) of the remaining shoots produces bushier plants that will bear a greater quantity of smaller flowers. Tall cultivars may need staking. Michaelmas daisies thrive in sun or partial shade in well-drained soil; many are susceptible to mildew, so in areas where this is a problem treat the plants regularly with a fungicide or choose from the resistant varieties available.

A. novae-angliae 'Harrington's Pink' ♀

A. novi-belgii 'Fellowship'

A. novi-belgii 'Lassie'

A. novi-belgii 'Patricia Ballard' ♀

A. 'Professor A. Kippenburg'

A. ericoides 'White Heather'

A. novae-angliae 'Barr's Pink'

A. novi-belgii 'Orlando'

A. novi-belgii 'Marie Ballard'

A. novi-belgii 'Peace'

A. novi-belgii 'Carnival'

A. novae-angliae 'Alma Potschke' ♀

A. frikartii 'Wunder von Stäfa' ♀

A. novi-belgii 'Royal Velvet'

A. novae-angliae 'Herbstschnee'

A. lateriflorus 'Delight'

A. novi-belgii 'Royal Ruby'

A. novi-belgii 'Jenny'

A. amellus 'King George' ♀

A. novi-belgii 'Chequers'

A. novi-belgii 'White Swan'

A. cordifolius 'Silver Spray'

A. novi-belgii 'Freda Ballard'

A. thomsonii 'Nanus'

A. lateriflorus var. *horizontalis* ♀

A. novi-belgii 'Apple Blossom'

A. frikartii 'Mönch' ♀

A. turbinellus ♀

A. linosyris

226

Perennials/medium AUTUMN INTEREST

■■■ BLUE–YELLOW

Strobilanthes atropurpureus
Upright, branching perennial with oval, toothed leaves. Spikes of numerous, violet-blue to purple flowers appear in summer-autumn. H to 1.2m (4ft), S to 60cm (2ft).

Gentiana asclepiadea
(Willow gentian)
Arching perennial with narrow, oval leaves to 8cm (3in) long. In late summer to autumn has arching sprays of trumpet-shaped, deep blue flowers, spotted and striped inside. H to 90cm (3ft), S to 60cm (2ft).

Kniphofia 'Percy's Pride'
Upright perennial with large, terminal spikes of creamy flowers, tinged green and yellow, borne on erect stems in autumn. Protect crowns with winter mulch. H 1m (3ft), S 50cm (20in).

■ YELLOW–ORANGE

Kirengeshoma palmata
Upright perennial with rounded, lobed, bright green leaves, above which strong stems bearing clusters of narrowly funnel-shaped, creamy yellow flowers appear in late summer to autumn. H 1m (3ft), S 60cm (2ft).

Helenium 'Wyndley'
Bushy perennial with branching stems bearing sprays of daisy-like, orange-yellow flower heads for a long period in late summer and autumn. Foliage is dark green. Needs regular division in spring or autumn. H 80cm (30in), S 50cm (20in).

Kniphofia caulescens
Stately, evergreen, upright perennial with basal tufts of narrow, blue-green leaves and smooth, stout stems bearing terminal spikes of reddish-salmon flowers in autumn. H 1.2m (4ft), S 60cm (2ft).

Kniphofia triangularis,
syn. *K. galpinii* of gardens
Upright perennial with fine, grass-like, basal leaves and erect, wiry stems bearing small, terminal spikes of flame-red flowers in autumn. Protect crowns with winter mulch. H to 1m (3ft), S 50cm (20in).

Helenium 'Moerheim Beauty'
Upright perennial with strong, branching stems bearing sprays of daisy-like, rich reddish-orange flower heads in early autumn above dark green foliage. Needs regular division in spring or autumn. H 1m (3ft), S 60cm (2ft).

Chirata lavandulacea, p.258
CHRYSANTHEMUMS, pp.224–5
Dichorisandra reginae, p.216
Eryngium x *tripartitum*, p.217

Lobelia siphilitica
ORCHIDS, pp.260–63
Rudbeckia fulgida var. *deamii*

Aphelandra squarrosa 'Dania'
Aphelandra squarrosa 'Louisae', p.220
CHRYSANTHEMUMS, pp.224–5
Helenium 'Bressingham Gold'

Helenium 'Butterpat'
Solidago 'Goldenmosa', p.219
Verbascum nigrum, p.220

Perennials/medium WINTER/ALL YEAR INTEREST

◻ WHITE–PINK   ■ RED

Ctenanthe oppenheimiana 'Tricolor'
Robust, evergreen, bushy perennial. Has leathery, lance-shaped leaves, to over 30cm (12in) long, splashed with large, cream blotches, and, intermittently, spikes of 3-petalled, white flowers. H and S 1m (3ft). Min 15°C (59°F).

Anthurium crystallinum (Crystal anthurium)
Evergreen, erect, tufted perennial. Long, velvety, dark green leaves are distinctively pale green- to white-veined. Has long-lasting, red-tinged, green spathes. H to 75cm (2½ft), S to 60cm (2ft). Min 15°C (59°F).

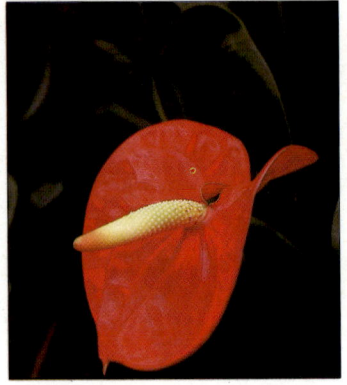

Anthurium andraeanum (Tail flower)
Evergreen, erect perennial. Long-stalked, oval leaves, with a heart-shaped base, are 20cm (8in) long. Has long-lasting, bright red spathes with yellow spadices. H 60–75cm (24–30in), S 50cm (20in). Min. 15°C (59°F).

Nepenthes × hookeriana
Evergreen, epiphytic, insectivorous perennial with oval, leathery leaves to 30cm (12in) long and pendent, pale green pitchers, with reddish-purple markings and a spurred lid, to 13cm (5in) long. H 60–75cm (2–2½ft). Min. 18°C (64°F).

Plectranthus forsteri 'Marginatus', syn. *P. coleoides* 'Variegatus'
Evergreen, bush perennial. Oval leaves, to 6cm (2½in) long, are greyish-green with scalloped, white margins. Irregularly has tubular, white to pale mauve flowers. H and S 60cm (2ft) or more. Min. 10°C (50°F).

Hypoestes phyllostachya, syn. *H. sanguinolenta* of gardens (Freckle face, Polka-dot plant)
Evergreen, bush perennial or sub-shrub. Dark green leaves are covered with irregular, pink spots. Bears small, tubular, lavender flowers intermittently. H and S 75cm (30in). Min. 10°C (50°F).

Dieffenbachia 'Exotica', syn. *D. seguine* 'Exotica', *D. maculata* 'Exotica'
Evergreen, tufted, perennial, sometimes woody at base. Broadly lance-shaped leaves, to 45cm (18in) long, are blotched with creamy-white. H and S 1m (3ft) or more. Min. 15°C (59°F).

Caladium × hortulanum 'Pink Beauty'
Tufted, tuberous perennial. Has long-stalked, triangular, pink-mottled, green leaves, to 45cm (18in) long, with darker pink veins. White spathes appear in summer. H and S 90cm (3ft). Min. 19°C (66°F).

Phormium tenax 'Dazzler'
Evergreen, upright perennial with tufts of bold, stiff, pointed leaves in tones of yellow, salmon-pink, orange-red and bronze. Bluish-purple stems carry panicles of reddish flowers in summer. H 2–2.5m (6–8ft) in flower, S 1m (3ft).

OTHER RECOMMENDED PLANTS:
Alpinia calcarata
Anthurium veitchii
BROMELIADS, p.229

Dieffenbachia seguine 'Memoria Corsii'
Hypoestes aristata
ORCHIDS, pp.260–63

Gerbera jamesonii, p.266
Tellima grandiflora 'Rubra', p.266

Bromeliads

Bromeliads, or plants that belong to the family Bromeliaceae, are distinguished by their bold, usually rosetted foliage and showy flowers. Although many are epiphytes, or air plants (absorbing their food through moisture in the atmosphere and not from the host on which they grow), and are suitable for growing outdoors only in tropical regions, they will grow happily indoors in cooler climates.

Aechmea fasciata ♀

Billbergia nutans

Bromelia balansae

Tillandsia lindenii ♀

Cryptanthus zonatus 'Zebrinus'

Puya alpestris

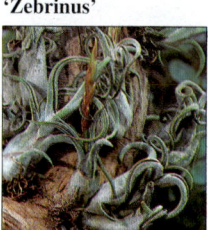

Tillandsia caput-medusae

Ananas bracteatus 'Tricolor' ♀

Aechmea distichantha

Neoregelia carolinae 'Tricolor' ♀

Guzmania monostachia ♀

Dyckia remotiflora

Tillandsia argentea

Cryptanthus 'Pink Starlight' ♀

Aechmea recurvata

Aechmea 'Foster's Favorite' ♀

Guzmania lingulata var. *minor* ♀

Guzmania lingulata ♀

Tillandsia fasciculata

Cryptanthus bivittatus ♀

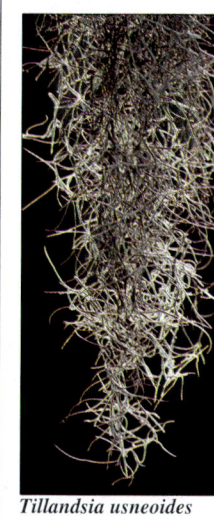

Tillandsia usneoides

Tillandsia stricta

Vriesea splendens ♀

Neoregelia concentrica

Tillandsia cyanea

Puya chilensis

229

Perennials/medium WINTER/ALL YEAR INTEREST
PURPLE–GREEN

Browallia speciosa (Bush violet)
Bushy perennial, usually grown as an annual, propagated by seed each year. Has oval leaves to 10cm (4in) long and showy, violet-blue flowers with white eyes, the season depending when sown. H 60–75cm (24–30in), S 45cm (18in). Min. 10–15°C (50–59°F).

Artemisia ludoviciana var. ***albula***
Bushy perennial, grown for its aromatic, lance-shaped leaves which are silvery-white and woolly on both surfaces and have jagged margins. Bears slender plumes of tiny, greyish-white flower heads in summer. H 1.2m (4ft), S 60cm (2ft).

Aciphylla squarrosa (Bayonet plant)
Evergreen, clump-forming perennial with tufts of pointed, divided leaves. In summer bears spiky, yellow flowers in compound umbels with male and female flowers often mixed. H and S 1–1.2m (3–4ft).

Asparagus densiflorus, syn. *A. sprengeri*
Evergreen, trailing perennial with clusters of narrow, bright green, leaf-like stems. In summer has pink-tinged, white flowers, followed by red berries. Suits a hanging basket. H to 1m (3ft), S 50cm (20in). Min. 10°C (50°F).

Alocasia cuprea
Evergreen, tufted perennial. Oval leaves are 30cm (12in) long, with a metallic sheen and darker, impressed veins above, purple below; leaf stalks arise from the lower surface. Purplish spathes appear intermittently. H and S to 1m (3ft). Min. 15°C (59°F).

Xanthosoma sagittifolium
Wide-spreading, tufted perennial with thick stems. Broadly arrow-shaped leaves, 60cm (2ft) or more on long leaf stalks, are green with a greyish bloom. Has green spathes intermittently during the year. H to 2m (6ft) in flower, S 2m (6ft) or more. Min. 15°C (59°F).

Columnea microphylla **'Variegata'**
Evergreen, trailing perennial. Has rounded leaves narrowly bordered with cream and tubular, hooded, scarlet flowers, with yellow throats, in winter-spring. H 1m (3ft) or more, S indefinite. Min. 15°C (59°F).

Calathea zebrina (Zebra plant)
Robust, evergreen, clump-forming perennial with long-stalked, velvety, dark green leaves, to 60cm (2ft) long (less if pot-grown), with paler veins, margins and midribs. Bears short spikes of white to pale purple flowers. H and S to 90cm (3ft). Min. 15°C (59°F).

Aglaonema pictum, p.268
Alocasia veitchii
Calathea lindeniana
Calathea makoyana, p.268

Helleborus argutifolius, p.268
Musa uranoscopus
Nephthytis afzelii
ORCHIDS, pp.260–63

Peperomia scandens
Xanthorrhoea australis
Xanthosoma violaceum

Perennials/small SPRING INTEREST

GREEN–ORANGE

Asparagus densiflorus 'Myersii', syn. *A. meyeri*, *A. myersii* (Foxtail fern)
Evergreen, erect perennial with spikes of tight, feathery clusters of leaf-like stems and pinkish-white flowers in summer, then red berries. H to 1m (3ft), S 50cm (20in). Min. 10°C (50°F).

Globba winitii
Evergreen, clump-forming perennial with lance-shaped leaves to 20cm (8in) long. Intermittently has pendent racemes of tubular, yellow flowers with large, reddish-purple, reflexed bracts. H 1m (3ft), S 30cm (1ft). Min. 18°C (64°F).

Dieffenbachia seguine 'Rudolph Roehrs', syn. *D. maculata* 'Rudolph Roehrs', *D.s.* 'Roehrsii'
Evergreen, tufted perennial, sometimes woody at the base. Leaves, to 45cm (18in) long, are yellowish-green or white with green midribs and margins. H and S 1m (3ft). Min. 15°C (59°F).

Peristrophe hyssopifolia 'Aureovariegata'
Evergreen, bushy perennial. Small leaves are broadly lance-shaped with long, pointed tips and central, creamy-yellow blotches. Has tubular, rose-pink flowers in winter. H to 60cm (2ft) or more, S 1.2m (4ft). Min. 15°C (59°F).

Sansevieria trifasciata 'Laurentii'
Evergreen, stemless perennial with a rosette of about 5 stiff, erect, lance-shaped and pointed leaves with yellow margins. Occasionally has pale green flowers. Propagate by division to avoid reversion. H 45cm–1.2m (1½–4ft), S 10cm (4in). Min. 10–15°C (50–59°F).

Strelitzia reginae (Bird-of-paradise flower)
Evergreen, clump-forming perennial with long-stalked, bluish-green leaves. Has beak-like, orange-and-blue flowers in boat-shaped, red-edged bracts mainly in spring. H over 1m (3ft), S 75cm (2½ft). Min. 5–10°C (41–50°F).

WHITE

Epimedium × youngianum 'Niveum'
Compact, ground-cover perennial with heart-shaped, serrated, bronze-tinted leaflets that turn green in late spring, when small, cup-shaped, snow-white flowers are borne. H 15–30cm (6–12in), S 30cm (12in).

Pachyphragma macrophyllum, syn. *Thlaspi macrophyllum*
Creeping, mat-forming perennial with rosettes of rounded, long-stalked, glossy, bright green leaves, each to 10cm (4in) long. Bears many racemes of tiny, white flowers in spring. H to 30cm (12in), S indefinite.

Lamium maculatum 'White Nancy'
Semi-evergreen, mat-forming perennial with white-variegated, mid-green foliage and spikes of hooded, white flowers in late spring and summer. H 15cm (6in), S 1m (3ft).

Trillium cernuum f. album
Clump-forming perennial with nodding, maroon-centred, white flowers borne in spring beneath luxuriant, 3-parted, mid-green leaves. H 30–45cm (12–18in), S 30cm (12in).

Pulmonaria 'Sissinghurst White'
Semi-evergreen, clump-forming perennial that bears funnel-shaped, white flowers in spring above long, elliptic, mid-green, paler spotted leaves. H 30cm (12in), S 45–60cm (18–24in).

Lamium maculatum 'Album'
Semi-evergreen, mat-forming perennial that has dark green leaves with central, white stripes. Bears clusters of hooded, white flowers in spring-summer. H 20cm (8in), S 1m (3ft).

Astilbe 'Ostrich Plume', p.208
Astilbe 'Venus', p.208
BROMELIADS, p.229
Ctenanthe lubbersiana

Dasylirion texanum
Dieffenbachia seguine
Sansevieria cylindrica
Sansevieria trifasciata

OTHER RECOMMENDED PLANTS:
Convallaria majalis 'Fortin's Giant'
Helleborus niger, p.265
Helleborus orientalis [white form], p.264

PRIMULAS, pp.236–7
Sanguinaria canadensis 'Plena'
Trillium grandiflorum 'Flore Pleno'

Perennials/small SPRING INTEREST
☐ WHITE

Convallaria majalis
(Lily-of-the-valley)
Low-growing, rhizomatous perennial with narrowly oval, mid- to dark green leaves and sprays of small, very fragrant, pendulous, bell-shaped, white flowers. Likes humus-rich soil. H 15cm (6in), S indefinite.

Trillium ovatum
Clump-forming perennial with white flowers, later turning pink, that are carried singly in spring just above red-stalked, 3-parted, dark green foliage. H 25–38cm (10–15in), S 20cm (8in).

Adonis brevistyla
Clump-forming perennial. Buttercup-like flowers, borne singly at tips of stems in early spring, are white, tinged blue outside. Has finely cut, mid-green leaves. H and S 15–23cm (6–9in).

Trillium chloropetalum
Clump-forming perennial with reddish-green stems carrying 3-parted, grey-marbled, dark green leaves. Flowers vary from purplish-pink to white and appear above foliage in spring. H and S 30–45cm (12–18in).

Podophyllum hexandrum, syn. *P. emodi*
(Himalayan May apple)
Perennial with pairs of 3-lobed, brown-mottled leaves followed by white or pink flowers in spring and fleshy, red fruits in summer. H 30–45cm (12–18in), S 30cm (12in).

Epimedium pubigerum
Evergreen, carpeting perennial, grown for its dense, smooth, heart-shaped, divided foliage and clusters of cup-shaped, creamy-white or pink flowers in spring. H and S 45cm (18in).

Trillium grandiflorum
(Wake-robin)
Clump-forming perennial. Large, pure white flowers that turn pink with age are borne singly in spring just above large, 3-parted, dark green leaves. H 38cm (15in), S 30cm (12in).

Anemone sylvestris
(Snowdrop windflower)
Carpeting perennial that may be invasive. Fragrant, semi-pendent, white flowers with yellow centres are borne in spring and early summer. Has divided, mid-green leaves. H and S 30cm (12in).

Bergenia 'Silberlicht', syn. *B.* 'Silver Light'
Evergreen, clump-forming perennial that has flat, oval, mid-green leaves with toothed margins. Clusters of white flowers, sometimes suffused with pink, are borne on erect stems in spring. H 30cm (12in), S 50cm (20in).

■ PINK

Bergenia ciliata
Evergreen, clump-forming perennial with attractive, large, rounded, hairy leaves. In spring bears clusters of white flowers that age to pink. Leaves are often damaged by frost, although fresh ones will appear in spring. H 30cm (12in), S 50cm (20in).

☼ ◊ ❄❄

Geranium macrorrhizum 'Ingwersen's Variety'
Compact, carpeting perennial, useful as weed-suppressing ground cover. Small, soft rose-pink flowers appear in late spring and early summer. Aromatic leaves turn bronze- and scarlet-tinted in autumn. H 30cm (1ft), S 60cm (2ft).

☼ ◊ ❄❄❄ ♕

Heloniopsis orientalis
Clump-forming perennial with basal rosettes of narrowly lance-shaped leaves, above which rise nodding, rose-pink flowers in spring. H and S 30cm (12in).

☼ ◊ ❄❄❄

Bergenia cordifolia 'Purpurea'
Evergreen, clump-forming perennial, useful for ground cover, with large, rounded, purple-tinged, deep green leaves. Clusters of bell-shaped, rose-pink flowers are carried on red stems from late winter to early spring. H and S 50cm (20in).

☼ ◊ ❄❄❄ ♕

Epimedium grandiflorum 'Rose Queen'
Carpeting perennial with dense, heart-shaped, divided leaves, tinged with copper, and wiry stems bearing clusters of cup-shaped, spurred, deep pink flowers in spring. H and S 30cm (12in).

☼ ◊ ❄❄❄ ♕

Lamium maculatum
Semi-evergreen, mat-forming perennial with mauve-tinged, often pink-flushed leaves that have central, silvery stripes. Clusters of hooded, mauve-pink flowers are borne in mid-spring. H 15cm (6in), S 1m (3ft).

☼ ◊ ❄❄❄

■ RED

Epimedium × rubrum
Carpeting perennial with dense, heart-shaped, divided leaves that are dark brownish-red in spring when clusters of cup-shaped, crimson flowers with yellow spurs appear. H 30cm (12in), S 20cm (8in).

☼ ◊ ❄❄❄ ♕

Trillium erectum (Birthroot, Squawroot)
Clump-forming perennial with 3-lobed, mid-green leaves and bright maroon-purple flowers in spring. H 30–45cm (12–18in), S 30cm (12in).

☼ ◊ ❄❄❄ ♕

Trillium sessile (Toadshade, Wake-robin)
Clump-forming perennial that in spring bears red-brown flowers, nestling in a collar of 3-lobed leaves, marked white, pale green or bronze. H 30–38cm (12–15in), S 30–45cm (12–18in).

☼ ◊ ❄❄❄

Bergenia 'Abendglut'
Bergenia cordifolia
Bergenia 'Morgenröte'
Bergenia × *schmidtii*
Bergenia stracheyi
Bergenia 'Sunningdale'
Dicentra 'Spring Morning', p.242
Dicentra 'Adrian Bloom'
Dicentra 'Stuart Boothman'
Helleborus orientalis [pink form], p.265
PRIMULAS, pp.236–7
Rehmannia glutinosa

Epimedium alpinum
Pulmonaria rubra

233

Perennials/small SPRING INTEREST
■ PURPLE

Anemone nemorosa 'Allenii'
Carpeting perennial with many large, cup-shaped, single, rich lavender-blue flowers appearing in spring over deeply divided, mid-green leaves. H 15cm (6in), S 30cm (12in) or more.

Glaucidium palmatum
Leafy perennial that has large, lobed leaves and, in spring, large, delicate, cup-shaped, lavender flowers. A woodland plant, it requires humus-rich soil and a sheltered position. H and S 50cm (20in).

Geranium nodosum
Clump-forming perennial with lobed, glossy leaves and delicate, cup-shaped, lilac or lilac-pink flowers borne in spring-summer. Thrives in deep shade. H and S 45cm (18in).

Lathyrus vernus
Clump-forming perennial bearing in spring small, pea-like, bright purple and blue flowers veined with red, several on each slender stem. Leaves are soft and fern-like. Proves difficult to transplant successfully. H and S 30cm (12in).

Anemone nemorosa 'Robinsoniana'
Carpeting perennial with flat, star-shaped, lavender-blue flowers, pale creamy-grey beneath, borne singly on maroon stems. Leaves are deeply divided into lance-shaped segments. H 15cm (6in), S 30cm (12in).

Cardamine pentaphyllos, syn. *Dentaria pentaphylla*
Upright perennial spreading by fleshy, horizontal rootstocks. Produces clusters of large, white or pale purple flowers in spring. H 30–60cm (12–24in), S 45–60cm (18–24in).

Cardamine pratensis 'Flore Pleno'
Chirita lavandulacea, p.258
Epimedium grandiflorum f. *violaceum*
Helleborus orientalis [purple form], p.266

Helonias bullata
IRISES, pp.198–9
ORCHIDS, pp.260–63
PEONIES, pp.200–201

PRIMULAS, pp.236–7
Sarracenia purpurea

PURPLE–BLUE

Pulmonaria saccharata
Semi-evergreen, clump-forming perennial. In spring bears funnel-shaped flowers, opening pink and turning to blue. Long, elliptic leaves are variably spotted with creamy-white. H 30cm (12in), S 60cm (24in).

Lathraea clandestina (Toothwort)
Spreading perennial that grows as a parasite on willow or poplar roots. Fleshy, underground stems have colourless scales instead of leaves. Bears bunches of hooded, purple flowers from late winter to early spring. H 10cm (4in), S indefinite.

Lamium orvala
Clump-forming perennial that forms a mound of mid-green leaves, sometimes with central white stripes. Clusters of pink or purple-pink flowers open in late spring to early summer. H and S 30cm (12in).

Pulmonaria angustifolia 'Mawson's Variety'
Clump-forming perennial that in early spring bears clusters of funnel-shaped, blue flowers, tinged red with age, above narrow leaves. H and S 23cm (9in).

Scopolia carniolica
Clump-forming perennial that carries spikes of nodding, purple-brown flowers, yellow inside, in early spring. H and S 60cm (24in).

Mertensia pulmonarioides, syn. **M. virginica**
Elegant perennial with rich blue flowers, hanging in clusters in spring. Leaves are soft blue-green. Dies down in summer. Crowns are prone to slug damage. H 30–60cm (12–24in), S 30–45cm (12–18in).

Ajuga pyramidalis
Brunnera macrophylla
Cyanotis somaliensis, p.267
IRISES, pp.198–9

Meehania urticifolia
Pulmonaria angustifolia
Pulmonaria longifolia
Symphytum 'Hidcote Blue'

BLUE–YELLOW

Brunnera macrophylla 'Dawson's White', syn. *B. m.* 'Variegata'
Ground-cover perennial with heart-shaped leaves, marked creamy-white. In spring bears delicate sprays of small, bright blue flowers. Shelter from wind to prevent leaf damage. H 45cm (18in), S 60cm (24in).

Euphorbia seguieriana
Bushy perennial with large, terminal clusters of yellowish-green flowers in late spring and narrow, lance-shaped, glaucous leaves on slender stems. H and S 45cm (18in).

Meconopsis quintuplinervia (Harebell poppy)
Mat-forming perennial. Lavender-blue flowers, deepening to purple at the bases, are carried singly on hairy stems in late spring and early summer above a dense mat of large, mid-green leaves. H 30–45cm (12–18in), S 30cm (12in).

Euphorbia amygdaloides var. **robbiae**
Evergreen, spreading perennial with rosettes of dark green leaves, useful as ground cover even in poor, dry soil and semi-shade. Bears open, rounded heads of lime-green flowers in spring. H 45–60cm (18–24in), S 60cm (24in).

Euphorbia cyparissias
Rounded, leafy perennial with a mass of slender, grey-green leaves and umbels of small, bright lime-green flowers in late spring. May be invasive. H and S 30cm (12in).

Chirita sinensis
Helleborus argutifolius, p.268
Helleborus cyclophyllus
Helleborus foetidus, p.269

Helleborus viridis, p.269
IRISES, pp.198–9
Linum narbonense, p.251
Sarracenia flava, p.254

Primulas

There are primulas to suit almost every kind of garden situation, ranging from the pool side to the scree, but most have particular needs and care should be taken with their cultivation. Among the various botanical groups, Candelabra and Auricula primulas are the most widely known, and have a distinctive arrangement of their flowers. (For fuller details see the Plant Dictionary.)

P. allionii

P. sonchifolia

P. vulgaris 'Gigha White'

P. sinensis

P. sieboldii 'Wine Lady'

P. polyneura

P. pulverulenta (Candelabra) ♀

P. frondosa ♀

P. 'Craddock White'

P. malacoides [single]

P. warshenewskiana

P. secundiflora

P. petiolaris

P. denticulata f. *alba*

P. farinosa

P. malacoides [double]

P. melanops

P. japonica 'Miller's Crimson' (Candelabra)

P. modesta var. *fauriae*

P. japonica 'Postford White' (Candelabra)

P. vulgaris subsp. *sibthorpii* ♀

P. × *scapeosa*

P. clusiana

P. gracilipes

P. clarkei

P. hirsuta

P. vialii ♀

P. 'Mrs J.H. Wilson' (Auricula)

P. pulverulenta 'Bartley' (Candelabra) ♀

P. rosea ♀

P. sieboldii ♀

P. edgeworthii

P. × *pubescens* 'Janet'

236

P. 'Mark'
(Auricula)

P. 'Linda Pope'

P. aureata

P. florindae ♀

P. bulleyana
(Candelabra) ♀

P. 'Adrian'
(Auricula)

P. reidii var. *williamsii*

P. alpicola var. *luna*

P. denticulata ♀

P. vulgaris

P. sikkimensis

P. forrestii

P. 'Janie Hill'
(Auricula)

P. 'Moonstone'
(Auricula)

P. elatior ♀

P. veris ♀

P. 'Blossom'
(Auricula)

P. marginata
'Prichard's Variety'

P. bhutanica

P. 'Blairside Yellow'
(Auricula)

P. chungensis
(Candelabra)

P. 'Inverewe'
(Candelabra) ♀

P. flaccida ♀

P. 'Margaret Martin'
(Auricula)

P. prolifera
(Candelabra)

P. x *kewensis* ♀

P. marginata ♀

P. 'Chloë'
(Auricula)

P. palinuri

P. verticillata

P. Gold Lace Group

237

Perennials/small SPRING INTEREST
YELLOW

Valeriana phu 'Aurea'
Perennial with rosettes of lemon- to butter-yellow, young foliage that turns mid-green by summer, when heads of insignificant, white flowers appear. H 38cm (15in), S 30–38cm (12–15in).

Petasites japonicus
Spreading, invasive perennial that in early spring produces dense cones of small, daisy-like, yellowish-white flowers before large, light green leaves appear. H 60cm (2ft), S 1.5m (5ft).

Cardamine enneaphyllos, syn. *Dentaria enneaphylla*
Lax perennial spreading by fleshy, horizontal rootstocks. In spring, nodding, pale yellow or white flowers open at the ends of shoots arising from deeply divided leaves. H 30–60cm (12–24in), S 45–60cm (18–24in).

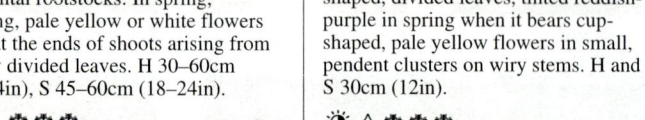

Epimedium x versicolor 'Neo-sulphureum'
Carpeting perennial with dense, heart-shaped, divided leaves, tinted reddish-purple in spring when it bears cup-shaped, pale yellow flowers in small, pendent clusters on wiry stems. H and S 30cm (12in).

Anemone x lipsiensis, syn. *A. x seemannii*
Prostrate, carpeting perennial that in spring has many single, pale yellow flowers with bright yellow stamens. Leaves are deeply cut with long leaflets. H 15cm (6in), S 30cm (12in).

Uvularia grandiflora
(Bellwort, Merry-bells)
Clump-forming perennial. Clusters of long, bell-shaped, yellow flowers hang gracefully from slender stems in spring. H 45–60cm (18–24in), S 30cm (12in).

Trollius 'Alabaster'
Clump-forming perennial producing rounded, yellowish-white flowers in spring. These emerge from a basal mass of rounded, deeply divided, mid-green leaves. H 60cm (24in), S 45cm (18in).

Anemone ranunculoides
Spreading perennial for damp woodland, bearing buttercup-like, single, deep yellow flowers in spring. Divided leaves have short stalks. H and S 20cm (8in).

Asphodeline liburnica
Doronicum austriacum
Doronicum 'Spring Beauty'
Duchesnea indica

Epimedium x *perralchicum*
Epimedium perralderianum
Epimedium pinnatum subsp. *colchicum*
Euphorbia rigida

Hylomecon japonicum, p.298
Meconopsis integrifolia, p.256
PRIMULAS, pp.236–7
Ranunculus acris 'Flore Pleno', p.256

Symphytum grandiflorum
Trollius 'Earliest of All'
Trollius europaeus 'Canary Bird'
Uvularia perfoliata

Perennials/small SUMMER INTEREST

🟨 **YELLOW–ORANGE**

☐ **WHITE**

Trollius europaeus (Globeflower)
Clump-forming perennial that in spring bears rounded, lemon- to mid-yellow flowers above deeply divided, mid-green leaves. H 60cm (24in), S 45cm (18in).

Adonis amurensis
Clump-forming perennial that in late winter and early spring bears buttercup-like, golden blooms singly at the tips of stems. Mid-green foliage is finely cut. H 30cm (12in), S 23–30cm (9–12in).

Anemone rivularis
Perennial with stiff, free-branching stems bearing delicate, cup-shaped, white flowers in summer above deeply divided, dark green leaves. H 60cm (24in), S 30cm (12in).

Euphorbia polychroma, syn. *E. epithymoides*
Rounded, bushy perennial with mid-green leaves and heads of bright yellow flowers carried for several weeks in spring. H and S 50cm (20in).

Meconopsis cambrica (Welsh poppy)
Spreading perennial that in late spring carries lemon-yellow or rich orange blooms. Double forms are available. Has deeply divided, fern-like foliage. H 30–45cm (12–18in), S 30cm (12in).

Campanula alliariifolia
Mound-forming perennial with heart-shaped leaves, above which rise nodding, bell-shaped, creamy-white flowers borne along arching, wiry stems throughout summer. H 60cm (24in), S 50cm (20in).

Anemone narcissiflora
Leafy perennial that in late spring and early summer produces cup-shaped, single, white flowers with a blue or purplish-pink stain on reverse of petals. Leaves are dark green and deeply divided. H to 60cm (24in), S 50cm (20in).

Adonis vernalis
Clump-forming perennial that in early spring produces buttercup-like, greenish-yellow blooms singly at the tips of stems. Mid-green leaves are delicately dissected. H and S 23–30cm (9–12in).

Epimedium x *warleyense*
Carpeting perennial with heart-shaped, divided, light green leaves, tinged purple-red, and cup-shaped, rich orange flowers borne in clusters on wiry stems in spring. H and S 30cm (12in).

Galium odoratum, syn. *Asperula odorata* (Woodruff)
Carpeting perennial that bears whorls of star-shaped, white flowers above neat, whorled leaves in summer. All parts of plant are aromatic. H 15cm (6in), S 30cm (12in) or more.

Leucanthemum x *superbum* 'Esther Read', syn. *Chrysanthemum* x *superbum* 'Esther Read'
Robust perennial with large, daisy-like, double, white flower heads borne singly on strong stems in summer. Divide and replant every two years. H and S 45cm (18in).

Calceolaria 'John Innes', p.257
IRISES, pp.198–9
PRIMULAS, pp.236–37
Trollius 'Goldquelle'

OTHER RECOMMENDED PLANTS:
Lamium maculatum 'White Nancy', p.231
Libertia ixioides
Ophiopogon jaburan 'Variegatus'

Paradisea liliastrum
Polygonatum multiflorum 'Variegatum'
Scabiosa caucasica 'Miss Willmott'
Sinningia 'Mont Blanc'

239

Perennials/small SUMMER INTEREST
☐ WHITE

Anthemis punctata subsp. **cupaniana**
Evergreen, carpeting perennial with dense, finely cut, silvery foliage that turns green in winter. Small, daisy-like, white flower heads with yellow centres are borne singly on short stems in early summer. H and S 30cm (12in).

Anaphalis nepalensis var. **monocephala**, syn. **A. nubigena**
Dwarf, leafy perennial that has woolly, silvery stems and lance-shaped leaves. Carries dense, terminal clusters of white flower heads in late summer. H 20–30cm (8–12in), S 15cm (6in).

Aegopodium podagraria 'Variegatum' (Variegated Bishop's weed, Variegated gout weed)
Vigorous, spreading perennial, excellent for ground cover, with lobed, creamy-white-variegated leaves. Insignificant, white flowers borne in summer are best removed. H 10cm (4in), S indefinite.

Tradescantia 'Osprey'
Clump-forming perennial with narrow, lance-shaped leaves, 15–30cm (6–12in) long. Has clusters of white flowers with purple-blue stamens, surrounded by 2 leaf-like bracts, in summer. H to 60cm (24in), S 45cm (18in).

Anthericum liliago (St Bernard's lily)
Upright perennial that in early summer bears tall racemes of trumpet-shaped, white flowers above clumps of long, narrow, grey-green leaves. H 45–60cm (18–24in), S 30cm (12in).

Mentha suaveolens 'Variegata' (Variegated apple mint)
Spreading perennial with soft, woolly, mid-green leaves, splashed with white and cream, that smell of apples. Seldom produces flowers. H 30–45cm (12–18in), S 60cm (24in).

Astilbe 'Irrlicht'
Leafy perennial bearing tapering, feathery plumes of tiny, white flowers in summer. Foliage is dark green and flowers remain on the plant, dried and brown, well into winter. Prefers humus-rich soil. H 45–60cm (1½–2ft), S to 1m (3ft).

Geranium clarkei 'Kashmir White', syn. **G. pratense 'Kashmir White'**
Carpeting, rhizomatous perennial with divided leaves and loose clusters of cup-shaped flowers, white with pale lilac-pink veins, borne for a long period in summer. H and S 45–60cm (18–24in).

Crambe maritima (Sea kale)
Robust perennial with a mound of wide, curved, lobed, silvery-green leaves. Bears large heads of small, fragrant, white flowers, opening into branching sprays in summer. H and S 60cm (24in).

Heuchera cylindrica 'Greenfinch'
Evergreen, clump-forming perennial with rosettes of lobed, heart-shaped leaves and, in summer, graceful spikes of small, bell-shaped, pale green or greenish-white flowers. H 45–60cm (18–24in), S 50cm (20in).

Streptocarpus caulescens
Erect perennial with small, narrow to oval, fleshy, dark green leaves. Stalked clusters of small, tubular, violet-striped, violet or white flowers are carried in leaf axils intermittently. H and S to 45cm (18in) or more. Min. 10–15°C (50–59°F).

Anemone sylvestris, p.232
PELARGONIUMS, pp.210–11
PRIMULAS, pp.236–7
Tanacetum parthenium, p.271
Tulbaghia natalensis

☐ WHITE

Geranium renardii
Compact, clump-forming perennial with lobed, circular, sage-green leaves and purple-veined, white flowers, borne in early summer. H and S 30cm (12in).

☼ ◊ ❄❄❄ ♛

Diplarrhena moraea
Clump-forming perennial with fans of long, strap-shaped leaves and clusters of iris-like, white flowers, with centres of yellow and purple, borne on wiry stems in early summer. H 45cm (18in), S 23cm (9in).

☼ ◊ ❄

Osteospermum 'Whirligig'
Evergreen, clump-forming, semi-woody perennial of lax habit that bears bluish-white flower heads singly, but in great profusion, during summer. Leaves are grey-green. H 60cm (24in), S 30–45cm (12–18in).

☼ ◊ ❄ ♛

Arctotis stoechadifolia
Arctotis stoechadifolia var. grandis
Osteospermum ecklonis
Osteospermum ecklonis 'Blue Streak'
Sinningia concinna
Tricyrtis hirta var. alba, p.258

☐▨ WHITE–PINK

Ruellia devosiana
Evergreen, bushy sub-shrub with spreading, purplish branches. Leaves are broadly lance-shaped, dark green with paler veins above and purple below. Has mauve-tinged, white flowers in spring-summer. H and S to 45cm (18in) or more. Min. 15°C (59°F).

◐ ◊

Heuchera micrantha 'Palace Purple'
Clump-forming perennial with persistent, heart-shaped, deep purple leaves and sprays of small, white flowers in summer. Cut leaves last well in water. H and S 45cm (18in).

☼ ◊ ❄❄❄ ♛

Melittis melissophyllum
(Bastard balm)
Erect perennial that in early summer bears white flowers with purple lower lips in axils of rough, oval, mid-green leaves. H and S 30cm (12in).

◐ ◊ ❄❄❄

Geranium incanum
Sedum populifolium

Astrantia major subsp. *involucrata*
Clump-forming perennial very similar to *A. major* (below) but with longer bracts surrounding centres of flower heads. H 60cm (24in), S 45cm (18in).

☼ ◊ ❄❄❄

Astrantia major (Masterwort)
Clump-forming perennial producing greenish-white, sometimes pink-tinged flower heads throughout summer-autumn above a dense mass of divided, mid-green leaves. H 60cm (24in), S 45cm (18in).

☼ ◊ ❄❄❄

Astrantia maxima
Clump-forming perennial that bears rose-pink flower heads during summer-autumn. H 60cm (24in), S 30cm (12in).

☼ ◊ ❄❄❄ ♛

241

Perennials/small SUMMER INTEREST
■ PINK

Mimulus 'Andean Nymph'
Spreading perennial, with hairy leaves, that in summer bears snapdragon-like, rose-pink flowers tipped with creamy-yellow and spotted deep pink. H 23cm (9in), S 25cm (10in).

Dicentra 'Spring Morning'
Neat, leafy perennial with small, heart-shaped, pink flowers hanging in arching sprays in late spring and summer. Attractive, fern-like foliage is grey-green and finely cut. H and S 30cm (12in).

x Heucherella tiarelloides
Evergreen, ground-cover perennial that has dense clusters of leaves and feathery sprays of tiny, bell-shaped, pink flowers in early summer. H and S 45cm (18in).

x Heucherella 'Bridget Bloom'
Evergreen, clump-forming perennial with dense, bright green leaves and, in early summer, many feathery sprays of tiny, bell-shaped, rose-pink flowers, which continue intermittently until autumn. H 45cm (18in), S 30cm (12in).

Erigeron 'Charity'
Clump-forming perennial with a mass of daisy-like, light pink flower heads with greenish-yellow centres borne for a long period in summer. May need some support. H and S to 60cm (2ft).

Geranium endressii
Semi-evergreen, compact, carpeting perennial with small, lobed leaves and cup-shaped, rose-pink flowers borne throughout summer. H 45cm (18in), S 60cm (24in).

Achimenes 'Peach Blossom'
Francoa appendiculata
PELARGONIUMS, pp.210–11
PENSTEMONS, p.212
PRIMULAS, pp.236–7
Rehmannia glutinosa
Tradescantia pallida 'Purple Heart', p.267

☐ PINK

Geranium × *oxonianum* **'Wargrave Pink'**
Semi-evergreen, carpeting perennial with dense, dainty, lobed, basal leaves acting as weed-suppressing ground cover. Cup-shaped, bright salmon-pink flowers are borne throughout summer. H 45cm (18in), S 60cm (24in).

Osteospermum jucundum, syn. *O. barberiae*, *Dimorphotheca barberiae*
Evergreen, neat, clump-forming perennial with mid-green leaves. In late summer, soft pink flower heads, mostly dark-eyed, are borne singly but in great abundance. H and S 30cm (12in).

Erodium manescavii
Mound-forming perennial with divided, ferny, blue-green leaves. Produces loose clusters of single, deep pink, darker blotched flowers throughout summer. H 45cm (18in), S 60cm (24in).

Mimulus lewisii
Upright perennial with downy, sticky, grey leaves that provide an excellent foil for snapdragon-like, deep rose-pink flowers borne singly in summer. Tolerates dry soil. H 60cm (24in), S 45cm (18in).

Lychnis flos-jovis
Clump-forming perennial with rounded clusters of deep rose-pink flowers, opening in mid-summer, that are set off by grey foliage. H and S 45cm (18in).

Centaurea hypoleuca **'John Coutts'**
Upright perennial. Deep rose-red flower heads, with thistle-like centres encircled by star-shaped ray petals, are borne singly on slender stems in summer. Deeply divided leaves are white-grey beneath. H 60cm (24in), S 45cm (18in).

Incarvillea mairei
Compact, clump-forming perennial that has short stems bearing several trumpet-shaped, purplish-pink flowers in early summer. Leaves are divided into oval leaflets. Protect crowns with winter mulch. H and S 30cm (12in).

Geranium macrorrhizum
Semi-evergreen, carpeting perennial bearing magenta flowers in early summer. Rounded, divided, aromatic leaves make good, weed-proof ground cover and assume bright tints in autumn. H 30–38cm (12–15in), S 60cm (24in).

Polygonum macrophyllum, syn. *P. sphaerostachyum*, *Persicaria sphaerostachyum*
Compact perennial carrying neat spikes of rich rose-pink blooms above narrow, lance-shaped, glaucous leaves in late summer. H 45–60cm (18–24in), S 30cm (12in).

Liatris spicata, syn. *L. callilepis*
Clump-forming perennial. In late summer bears spikes of crowded, rose-purple flower heads on stiff stems that arise from basal tufts of grassy, mid-green foliage. H 60cm (24in), S 30cm (12in).

Physostegia virginiana **'Vivid'** (Obedient plant)
Erect, compact perennial that in late summer and early autumn bears spikes of tubular, dark lilac-pink flowers that can be placed in postion. Has toothed, mid-green leaves. H and S 30–60cm (12–24in).

BEGONIAS, p.259
CARNATIONS and PINKS, pp.244–5
Dicentra formosa
Drosera spathulata, p.269

Geranium macrorrhizum 'Ingwersen's Variety', p.233
Geranium 'Russell Prichard'
Lamium maculatum 'Beacon Silver'

ORCHIDS, pp.260–63
Osteospermum 'Cannington Roy'
PELARGONIUMS, pp.211–12
PENSTEMONS, p.212

PEONIES, pp.200–201
Sedum spectabile
Sphaeralcea munroana

243

Carnations and Pinks

Although perhaps best known for providing excellent cut flowers, carnations and pinks (*Dianthus* cultivars) are highly ornamental border subjects, valued for their usually fragrant blooms, produced mainly in summer, and their distinctive, silvery- or grey-green foliage. Shorter-growing cultivars – the old-fashioned and modern pinks – are excellent edging plants. The range of colours available is increasingly wide, from the classic whites, pinks and reds to the newer creams, yellows and oranges. Many of the flowers are attractively marked or have fringed petals. Carnations and pinks need an open, sunny position, preferably in alkaline soil. All except the perpetual-flowering carnations are frost hardy. Carnations and pinks are divided into the following groups:

Border carnations – plants are of upright habit and flower prolifically once in mid-summer; each stem bears 5 or more flowers.
Perpetual-flowering carnations – similar in habit to border carnations, they are usually grown for cut flowers and bloom year-round under glass. Plants are normally disbudded, leaving one flower per stem, but spray forms have up to 5 flowers per stem.
Old-fashioned pinks – these have a low, spreading habit and form neat cushions of foliage; masses of fragrant flowers are produced in mid-summer.
Modern pinks – usually more vigorous than old-fashioned pinks, they are repeat-flowering with 2 or 3 main flushes of flowers in summer.

D. 'Eva Humphries' (border carnation)

D. 'London Brocade' (modern pink)

D. 'Fair Folly' (modern pink)

D. 'Alice' (modern pink)

D. 'Prudence' (old-fashioned pink)

D. 'Emile Paré' (old-fashioned pink)

D. 'Mrs Sinkins' (old-fashioned pink)

D. 'Gran's Favourite' (old-fashioned pink)

D. 'Forest Treasure' (border carnation)

D. 'Doris' (modern pink)

D. 'Pierrot' (perpetual-flowering carnation)

D. 'Truly Yours' (perpetual-fl. carnation)

D. 'Haytor' (modern pink)

D. 'White Ladies' (old-fashioned pink)

D. 'Musgrave's Pink' (old-fashioned pink)

D. 'Nives' (perpetual-flowering carnation)

D. 'Dad's Favourite' (old-fashioned pink)

D. 'Becky Robinson' (modern pink)

D. 'Pink Jewel'
(modern pink)

D. 'Bovey Belle'
(modern pink)

D. 'Laced Monarch'
(perpetual-fl. carnation)

D. 'Clara' (perpetual-
flowering carnation)

D. 'Aldridge Yellow'
(border carnation)

D. 'Valencia' (perpetual-
flowering carnation)

D. 'Monica Wyatt'
(modern pink) ♀

D. 'Joy' (modern pink)

D. 'Astor' (perpetual-
flowering carnation)

D. 'Lavender Clove'
(border carnation)

D. 'Crompton Princess'
(perpetual-fl. carnation)

D. 'Raggio di Sole'
(perpetual-fl. carnation)

D. 'Houndspool Ruby'
(modern pink) ♀

D. 'Nina' (perpetual-
flowering carnation)

D. 'Cream Sue'
(perpetual-fl. carnation)

D. 'Christine Hough'
(border carnation)

D. 'Borello' (perpetual-
flowering carnation)

D. 'Albisola' (perpetual-
flowering carnation)

D. 'Christopher'
(modern pink)

D. 'Bookham Perfume'
(border carnation)

D. 'Bookham Fancy'
(border carnation)

D. 'Happiness'
(border carnation)

D. 'Golden Cross'
(border carnation)

245

Perennials/small SUMMER INTEREST

■ PINK

***Achimenes* 'Little Beauty'**
Bushy perennial with oval, toothed leaves. Large, funnel-shaped, deep pink flowers with yellow eyes are carried in summer. H 25cm (10in), S 30cm (12in). Min. 10°C (50°F).

Incarvillea delavayi
Clump-forming perennial with deeply divided leaves and erect stems bearing several trumpet-shaped, pinkish-red flowers in early summer. Has attractive seed pods. H 45–60cm (18–24in), S 30cm (12in).

***Sinningia* 'Red Flicker'**
Short-stemmed, tuberous perennial with rosettes of oval, velvety leaves, to 20cm (8in) long. In summer has fleshy, nodding, funnel-shaped, pinkish-red flowers, pouched on lower sides. H to 30cm (12in), S 45cm (18in). Min. 15°C (59°F).

***Potentilla nepalensis* 'Miss Willmott'**
Clump-forming perennial with palmate, strawberry-like, bright green leaves. Numerous slender, branching stems carry warm cherry-red-centred, pink flowers throughout summer. H 50cm (20in), S 60cm (24in).

***Verbena* 'Sissinghurst'**
Mat-forming perennial that throughout summer bears heads of brilliant pink flowers above mid-green foliage. Is excellent for edging a path or growing in a tub. H 15–20cm (6–8in), S 45cm (18in).

Sedum spectabile 'Brilliant', p.258

■ RED

***Lychnis viscaria* 'Splendens Plena'**
Clump-forming perennial bearing spikes of double, magenta flowers in early summer. Stems and large, oval to lance-shaped, basal leaves are covered in sticky hairs. H 30–45cm (12–18in), S 23cm (9in) or more.

Lychnis coronaria
Clump-forming perennial, often grown as a biennial. From mid- to late summer, brilliant rose-crimson flowers are borne in panicles on branched, grey stems that rise from neat, grey leaves. H 45–60cm (18–24in), S 45cm (18in).

***Sinningia* 'Switzerland'**
Short-stemmed, tuberous perennial with rosettes of oval, velvety leaves, to 20cm (8in) long. In summer has large, fleshy, trumpet-shaped, bright scarlet flowers with ruffled, white borders. H to 30cm (12in), S 45cm (18in). Min. 15°C (59°F).

Alonsoa warscewiczii, p.281
Geum 'Lionel Cox'
Lychnis viscaria
Potentilla 'Monsieur Rouillard'

***Astilbe* 'Fanal'**
Leafy perennial with strong stems. In summer bears neat, tapering, feathery panicles of tiny, crimson-red flowers that turn brown and keep their shape in winter. Broad leaves are divided into leaflets. Prefers humus-rich soil. H 60cm (2ft), S to 1m (3ft).

***Heuchera* x *brizoides* 'Red Spangles'**
Evergreen perennial forming clumps of heart-shaped, purplish-green leaves. Bears spikes of small, bell-shaped, crimson-scarlet flowers in summer. H and S 30cm (12in).

Salvia blepharophylla
Stachys coccinea
Verbena peruviana

246

RED

Lotus berthelotii (Coral gem)
Semi-evergreen, straggling perennial suitable for a hanging basket or large pan in an alpine house. Has hairy, silvery branches and leaves, and clusters of pea-like, scarlet flowers in summer. H 30cm (12in), S indefinite. Min. 5°C (41°F).

Columnea crassifolia
Evergreen, shrubby perennial with fleshy, lance-shaped leaves. Erect, tubular, hairy, scarlet flowers, about 8cm (3in) long, each with a yellow throat, are carried from spring to autumn. H and S to 45cm (18in). Min. 15°C (59°F).

Potentilla atrosanguinea
Clump-forming perennial with hairy, palmate, strawberry-like leaves. Loose clusters of dark red flowers are borne throughout summer. H 50cm (20m), S 60cm (24in).

PURPLE

Polemonium carneum
Clump-forming perennial that carries clusters of cup-shaped, pink or lilac-pink flowers in early summer. Foliage is finely divided. H and S 45cm (18in).

Smithiantha 'Orange King'
Strong-growing, erect, rhizomatous perennial. Large, scalloped, velvety leaves are emerald-green with dark red-marked veins. In summer-autumn has tubular, orange-red flowers, red-spotted within and with yellow lips. H and S to 60cm (24in). Min. 15°C (59°F).

Gaillardia x grandiflora 'Dazzler'
Upright, rather open perennial bearing large, terminal, daisy-like, yellow-tipped, red flower heads for a long period in summer. Leaves are soft and divided. Needs staking and may be short-lived. H 60cm (24in), S 50cm (20in).

Mimulus 'Royal Velvet'
Compact perennial, often grown as an annual, producing in summer many large, snapdragon-like, mahogany-red flowers with mahogany-speckled, gold throats. Leaves are mid-green. H 30cm (12in), S 23cm (9in).

Kaempferia pulchra
Tufted, rhizomatous perennial with horizontal, aromatic, dark green leaves, variegated with paler green above. Short spikes of lilac-pink flowers appear from the centre of tufts in summer. H 15cm (6in), S 30cm (12in). Min. 18°C (64°F).

Tulbaghia violacea
Vigorous, semi-evergreen, clump-forming perennial that in summer-autumn carries umbels of lilac-purple or lilac-pink flowers above a mass of narrow, glaucous, blue-grey leaves. H 45–60cm (18–24in), S 30cm (12in).

Achimenes antirrhina
Achimenes 'Brilliant'
Aeschynanthus pulcher
Impatiens niamniamensis

Kohleria bogotensis
Lychnis x haagena
Potentilla 'Gibson's Scarlet'
Potentilla 'Gloire de Nancy'

Potentilla 'William Rollison'
Smithiantha cinnabarina

Chirita lavandulacea, p.258
Geranium nodosum, p.234
Ophiopogon planiscapus 'Nigrescens', p.267
PELARGONIUMS, pp.210–11

247

Perennials/small SUMMER INTEREST
■ PURPLE

Streptocarpus saxorum
(False African violet)
Evergreen, rounded, woody-based perennial with small, oval, hairy leaves in whorls. Lilac flowers with white tubes arise from leaf axils in summer-autumn. H and S 30cm (12in) or more. Min. 10–15°C (50–59°F).

Verbena rigida, syn. *V. venosa*
Neat, compact perennial bearing heads of pale violet flowers from mid-summer onwards. Has lance-shaped, rough, mid-green leaves borne on flower stems. H 45–60cm (18–24in), S 30cm (12in).

Heterocentron elegans
Evergreen, mat-forming perennial with dense, creeping, mid-green foliage. Massed, bright deep purple flowers open in summer-autumn and, under glass, in winter. H 5cm (2in), S indefinite. Min. 5°C (41°F).

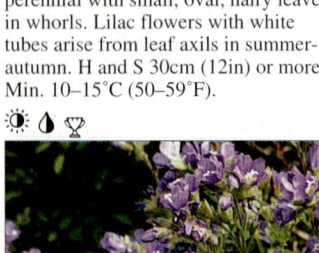

Tradescantia 'Purple Dome'
Clump-forming perennial with narrow, lance-shaped leaves, 15–30cm (6–12in) long. Has clusters of rich purple flowers, surrounded by 2 leaf-like bracts, in summer. H to 60cm (24in), S 45cm (18in).

Centaurea montana
Spreading perennial with many rather lax stems carrying, in early summer, one or more large, purple, blue, white or pink flower heads with thistle-like centres encircled by star-shaped ray petals. H 50cm (20in), S 60cm (24in).

Polemonium pulcherrimum
Vigorous perennial with bright green leaves divided into leaflets. Tubular, purple-blue flowers with throats of yellow or white are borne in summer. H 50cm (20in), S 30cm (12in).

Erigeron 'Serenity'
Clump-forming perennial bearing many daisy-like, violet flower heads, with yellow centres, borne singly for many weeks in summer. Needs some support. H and S to 60cm (2ft).

Platycodon grandiflorus
(Balloon flower)
Neat, clump-forming perennial that in summer bears clusters of large, balloon-like buds opening to bell-shaped, blue or purplish flowers. Stems are clothed with bluish-green leaves. H 45–60cm (18–24in), S 30–45cm (12–18in).

CARNATIONS and PINKS, pp.244–5
Drosera capensis, p.269
Euphorbia amygdaloides 'Purpurea'
Geranium 'Ann Folkard'
Geranium clarkei 'Kashmir Purple'
Geranium palmatum
Geranium procurrens
Lamium orvala, p.235
ORCHIDS, pp.260–63
Osteospermum 'Nairobi Purple'
PENSTEMONS, p.212
PRIMULAS, pp.236–7
Ranzania japonica
Senecio pulcher, p.258
Sinningia speciosa
Stachys macrantha

■ PURPLE

Stachys macrantha 'Superba'
Clump-forming perennial with heart-shaped, soft, wrinkled, mid-green leaves, from which arise stout stems producing whorls of hooded, purple-violet flowers in summer. H 30–45cm (12–18in), S 30–60cm (12–24in).

Geranium × magnificum
Clump-forming perennial with hairy, deeply lobed leaves and cup-shaped, prominently veined, violet-blue flowers borne in small clusters in summer. H 45cm (18in), S 60cm (24in).

Geranium 'Johnson's Blue'
Vigorous, clump-forming perennial with many divided leaves and cup-shaped, deep lavender-blue flowers borne throughout summer. H 30cm (12in), S 60cm (24in).

Geranium himalayense, syn. *G. grandiflorum*
Clump-forming perennial with large, cup-shaped, violet-blue flowers borne on long stalks in summer over dense tufts of neatly cut leaves. H 30cm (12in), S 60cm (24in).

Scabiosa caucasica 'Clive Greaves'
Clump-forming perennial that throughout summer has violet-blue flower heads with pincushion-like centres. Basal, mid-green leaves are lance-shaped and slightly lobed on the stems. H and S 45–60cm (18–24in).

Polemonium caeruleum (Jacob's ladder)
Clump-forming perennial. Clusters of cup-shaped, lavender-blue flowers with orange-yellow stamens open in summer amid finely divided foliage. H and S 45–60cm (18–24in).

Stokesia laevis
Perennial with overwintering, evergreen rosettes. In summer, cornflower-like, lavender- or purple-blue flower heads are borne freely. Leaves are narrow and mid-green. H and S 30–45cm (12–18in).

Catananche caerulea 'Major'
Perennial forming clumps of grassy, grey-green leaves, above which rise wiry, branching stems each carrying a daisy-like, lavender-blue flower head in summer. Propagate regularly by root cuttings. H 45–60cm (18–24in), S 30cm (24in).

Adenophora potaninii
Chirita sinensis
Dracocephalum ruyschianum
Geranium ibericum

IRISES, pp.198–9
Liriope spicata
Meconopsis quintuplinervia, p.235
MICHAELMAS DAISIES, p.226

Nierembergia hippomanica var. violacea 'Purple Robe', p.283
Salvia farinacea 'Victoria' p.284
Salvia jurisicii

Scabiosa caucasica 'Floral Queen'
Stachys officinalis
Tacca leontopetaloides

Perennials/small SUMMER INTEREST

■ PURPLE

■ BLUE

Limonium latifolium 'Blue Cloud'
Clump-forming perennial. In late summer carries diffuse clusters of bluish-mauve flowers that can be dried for indoor decoration. Has large, leathery, dark green leaves.
H 30cm (12in), S 45cm (18in).

Campanula 'Burghaltii'
Mound-forming perennial with long, pendent, funnel-shaped, pale lavender flowers displayed on erect, wiry stems in summer. Leaves are oval, soft and leathery. May need staking.
H 60cm (24in), S 30cm (12in).

Eryngium bourgatii
Clump-forming perennial that from mid- to late summer carries heads of thistle-like, blue-green, then lilac-blue, flowers on branched, wiry stems well above deeply cut, basal, grey-green leaves. H 45–60cm (18–24in), S 30cm (12in).

Amsonia orientalis, syn. *Rhazya orientalis*
Neat, clump-forming perennial. In summer, heads of small, star-shaped, grey-blue flowers open on tops of wiry stems clothed with green, sometimes greyish, leaves. H 45–60cm (18–24in), 30–45cm (12–18in).

Anemonopsis macrophylla (False anemone)
Clump-forming perennial with waxy, nodding, purplish-blue flowers, borne on slender, branching stems in summer above fern-like leaves. H 45–60cm (18–24in), S 50cm (20in).

Nepeta x faassenii (Catmint)
Bushy, clump-forming perennial, useful for edging. Forms mounds of small, greyish-green leaves, from which loose spikes of tubular, soft lavender-blue flowers appear in early summer. H and S 45cm (18in).

Parahebe perfoliata, syn. *Veronica perfoliata* (Digger speedwell)
Evergreen sub-shrub with willowy stems clasped by leathery, glaucous leaves. Elegant, long, branching sprays of blue flowers are borne in summer.
H 45–60cm (18–24in), S 45cm (18in).

Eryngium variifolium
Evergreen, rosette-forming perennial with stiff stems that in late summer bear heads of thistle-like, grey-blue flowers, each with a collar of white bracts. Jagged-edged leaves are mid-green, marbled with white. H 45cm (18in), S 25cm (10in).

Campanula isophylla
Chirita lavandulacea, p.258
IRISES, pp.198–9
Nepeta nervosa

■ BLUE

Veronica gentianoides
Mat-forming perennial with spikes of very pale blue flowers opening in early summer on tops of stems that arise from glossy, basal leaves. H and S 45cm (18in).

☼ ◊ ❋❋❋ ♟

Linum narbonense
Clump-forming, short-lived perennial, best renewed frequently from seed. Has lance-shaped, greyish-green leaves and heads of somewhat cup-shaped, pale to deep blue flowers in spring-summer. H 30–60cm (12–24in), S 30cm (12in).

☼ ◊ ❋❋❋

Amsonia tabernaemontana
syn. *A. salicifolia*
Clump-forming perennial with willowy stems bearing drooping clusters of small, tubular, pale blue flowers in summer. Leaves are small and narrow. H 45–60cm (18–24in), S 30cm (12in).

☼ ◊ ❋❋❋

Myosotidium hortensia
(Chatham Island forget-me-not)
Evergreen, clump-forming perennial bearing large clusters of forget-me-not-like, blue flowers in summer above a basal mound of large, ribbed, glossy leaves. H 45–60cm (18–24in), S 60cm (24in).

☼ ◊ ❋

Geranium wallichianum 'Buxton's Blue', syn. *G.w.* 'Buxton's Variety'
Spreading perennial with luxuriant, white-flecked leaves and large, white-centred, blue or blue-purple flowers from mid-summer to autumn. H 30–45cm (12–18in), S 1m (3ft).

☼ ◊ ❋❋❋ ♟

Salvia patens
Erect, branching perennial that in late summer and autumn produces whorls of pale or deep blue flowers on stems clothed with oval, mid-green leaves. H 45–60cm (18–24in), S 45cm (18in).

☼ ◊ ❋ ♟

Nepeta 'Souvenir d'André Chaudon'
Orthrosanthus chimboracensis
Platycodon grandiflorus var. *mariesii*
Salvia nemorosa 'Lubecca'

Stokesia laevis 'Blue Star'
Veronica spicata

Hostas

The luxuriance of their foliage and attractive habit have made hostas, or plantain lilies, increasingly sought after as plants for every garden, large or small. Native to the East, they also add an exotic touch to any waterside or damp, shady corner.

Hosta species vary in size from plants a few centimetres high to vigorous forms that will make a clump of up to 1.5m (5ft) across in a few years. Their elegant leaves are diverse in shape, texture and coloration, with subtle variegations and shadings. Many hostas also produce decorative spikes of flowers, which rise gracefully above the foliage in mid-summer.

Suitable for a range of situations from containers to borders and pool sides, hostas are essentially shade- and moisture-loving plants, preferring rich, well-drained soils. Leaves must be protected from slugs to avoid damage.

H. sieboldii ♀

H. sieboldiana

H. 'Ginko Craig'

H. 'Halcyon' ♀

H. tokudama 'Aureonebulosa'

H. tardiflora

H. undulata var. *univittata* ♀

H. 'Shade Fanfare' ♀

H. 'Snowden'

H. decorata f. *decorata*

H. 'Thomas Hogg'

H. 'Regal Splendor'

H. 'Krossa Regal' ♀

H. 'Love Pat' ♀

H. sieboldiana var. *elegans* ♀

H. 'Blue Wedgwood'

H. crispula ♀

H. rohdeifolia

H. 'Hadspen Blue'

H. tokudama

H. ventricosa

H. plantaginea

H. 'Royal Standard'

H. sieboldiana 'Frances Williams'

H. 'Golden Tiara'

H. 'Zounds'

H. 'Kabitan'

H. 'Tall Boy'

H. fortunei 'Albopicta'

H. 'Wide Brim'

H. 'Honeybells'

H. 'August Moon'

H. montana 'Aurea Marginata'

H. ventricosa 'Variegata'

H. 'Gold Standard'

H. 'Piedmont Gold'

H. 'Golden Prayers'

H. venusta

H. fortunei 'Aureomarginata'

H. ventricosa 'Aureomaculata'

H. 'Sum and Substance'

253

Perennials/small SUMMER INTEREST

GREEN–YELLOW

Artemisia pontica
(Roman wormwood)
Vigorous, upright perennial with aromatic, feathery, silver-green foliage and tall spikes of small, greyish flower heads in summer. May spread.
H 60cm (24in), S 20cm (8in).

Alchemilla mollis (Lady's mantle)
Clump-forming, ground-cover perennial that has rounded, pale green leaves with crinkled edges. Bears small sprays of tiny, bright greenish-yellow flowers, with conspicuous, outer calyces, in mid-summer that may be dried. H and S 50cm (20in).

Alchemilla conjuncta
Clump-forming perennial that has neat, wavy, star-shaped leaves with pale margins. Loose clusters of tiny, greenish-yellow flowers, with conspicuous, outer calyces, appearing in mid-summer may be dried for winter decoration. H and S 30cm (12in).

Sarracenia flava
(Trumpets, Yellow pitcher plant)
Erect perennial with red-marked, yellow-green pitchers (modified leaves) that have hooded tops. From late spring to early summer bears nodding, yellow or greenish-yellow flowers. H and S 45cm (18in). Min. 5°C (41°F).

Tovara virginiana 'Painter's Palette'
Mounded perennial grown for its attractive leaves, which are green with central, brown zones, ivory-yellow splashes and stripes and an overall deep pink tinge. Seldom flowers in cultivation. H and S 60cm (2ft).

Origanum vulgare 'Aureum'
Woody-based perennial forming a dense mat of aromatic, golden-yellow, young leaves that turn pale yellow-green in mid-summer. Occasionally bears tiny, mauve flowers in summer. H in leaf 8cm (3in), S indefinite.

YELLOW

Sisyrinchium striatum
Semi-evergreen perennial that forms tufts of long, narrow, grey-green leaves. Bears slender spikes of purple-striped, straw-yellow flowers in summer. Self seeds freely. H 45–60cm (18–24in), S 30cm (12in).

Kniphofia 'Little Maid'
Upright perennial with grass-like leaves and short, erect stems bearing terminal spikes of pale creamy-yellow flowers in summer. Protect crowns with winter mulch. H 60cm (24in), S 45cm (18in).

Filipendula ulmaria 'Aurea'
Leafy perennial, grown for its divided, bright golden-yellow foliage in spring which turns pale green in summer. Clusters of creamy-white flowers are carried in branching heads in mid-summer. H and S 30cm (12in).

Osteospermum 'Buttermilk'
Evergreen, upright, semi-woody perennial. Daisy-like, pale yellow flower heads with dark eyes are borne singly amid grey-green foliage from mid-summer to autumn. H 60cm (24in), S 30cm (12in).

Aeschynanthus marmoratus
Clintonia borealis
Mentha x *gracilis* 'Variegata'
Mentha x *piperita* 'Citrata'

Mentha suaveolens 'Variegata', p.240
Nicotiana sanderae 'Lime Green'
Pellionia pulchra

Astrantia major 'Sunningdale Variegated'
IRISES, pp.198–9
Ligularia tussilaginum 'Aureomaculata'
Mimulus moschatus

PRIMULAS, pp.236–7
Scrophularia auriculata 'Variegata'

■ YELLOW

Achillea 'Taygetea'
Perennial with erect stems bearing flat heads of lemon-yellow flowers throughout summer above clumps of feathery, grey leaves. Divide and replant every third year. H 60cm (24in), S 50cm (20in).

Potentilla recta 'Warrenii', syn. *P. r.* 'Macrantha'
Clump-forming perennial with lobed, mid-green leaves. Rich golden-yellow flowers are borne on open, branched stems throughout summer. H 50cm (20in), S 60cm (24in).

Gaillardia aristata (Blanket flower)
Upright, rather open perennial that has large, terminal, daisy-like, single flower heads, rich yellow with red centres, for a long period in summer and soft, aromatic, divided leaves. Needs staking and may be short-lived. H 60cm (24in), S 50cm (20in).

Calceolaria biflora
Evergreen perennial with a basal rosette of soft-haired, oval, toothed leaves about 15cm (6in) long. Clusters of small, pouched, yellow flowers appear in summer. H 30cm (12in), S 15cm (6in).

x Solidaster luteus, syn. **x** *S. hybridus*
Clump-forming perennial. From midsummer onwards, slender stems carry dense heads of bright creamy-yellow flowers above narrow, mid-green leaves. H 60cm (24in), S 75cm (30in).

Helichrysum 'Sulphur Light', syn. *H.* 'Schweffellicht'
Clump-forming perennial that bears silver-grey leaves and a mass of everlasting, fluffy, sulphur-yellow flower heads from mid- to late summer. H 40–60cm (16–24in), S 30cm (12in).

Achillea 'Moonshine'
Upright perennial that bears flat heads of bright yellow flowers throughout summer above a mass of small, feathery, grey-green leaves. Divide plants regularly in spring. H 60cm (24in), S 50cm (20in).

Barbarea vulgaris 'Variegata'
Perennial with rosettes of long, toothed, glossy leaves, blotched with cream, above which rise branching heads of small, silvery-yellow flowers in early summer. H 25–45cm (10–18in), S to 23cm (9in).

Oenothera fruticosa 'Fireworks', syn. *O. tetragona* 'Fireworks'
Clump-forming perennial that from mid- to late summer bears spikes of fragrant, cup-shaped flowers. Has reddish stems and glossy, mid-green foliage. H and S 30–38cm (12–15in).

BEGONIAS, p.259
DAYLILIES, p.221
HOSTAS, pp.252–3
IRISES, pp.198–9

Nautilocalyx bullatus
Oenothera perennis
ORCHIDS, pp.260–63
PELARGONIUMS, pp.210–11

Potentilla argyrophylla
PRIMULAS, pp.236–7
Salvia bulleyana
Sedum aizoon

Sedum rosea
Ursinia chrysanthemoides

255

Perennials/small SUMMER INTEREST
◻ YELLOW

***Ranunculus constantinopolitanus* 'Plenus'**, syn. *R. gouanii* 'Plenus', *R. speciosus* 'Plenus'
Clump-forming perennial with divided, toothed leaves sometimes spotted grey and white. Neat, pompon-like, double, yellow flowers appear in early summer. H 50cm (20in), S 30cm (12in).

***Geum* 'Lady Stratheden'**, syn. *G. chiloense* 'Lady Stratheden'
Clump-forming perennial with lobed leaves and cup-shaped, double, bright yellow flowers with prominent, green stamens borne on slender, branching stems for a long period in summer. H 45–60cm (18–24in), S 45cm (18in).

Gazania rigens* var. *uniflora
Mat-forming perennial, grown as an annual in all except mildest areas. Yellow or orange-yellow flower heads, sometimes with central white spots, are borne singly in early summer above rosettes of narrow, silver-backed leaves. H 23cm (9in), S 20–30cm (8–12in).

***Potentilla* 'Yellow Queen'**
Clump-forming perennial with strawberry-like, dark green leaves and bright yellow flowers in mid-summer. H to 60cm (24in) or more, S 45cm (18in).

Mimulus luteus (Yellow musk)
Spreading perennial. Throughout summer, snapdragon-like, occasionally red-spotted, yellow flowers are freely produced above hairy, mid-green foliage. H and S 30cm (12in).

Potentilla megalantha
Clump-forming perennial with large, palmate, hairy, soft green leaves. Large, rich yellow flowers are produced in summer. H 20cm (8in), S 15cm (6in).

Meconopsis integrifolia (Lampshade poppy)
Rosette-forming biennial or short-lived perennial carrying spikes of large, pale yellow flowers in late spring and early summer. Has large, pale green leaves. H 45–60cm (18–24in), S 60cm (24in).

Buphthalmum salicifolium (Yellow ox-eye)
Spreading perennial that carries daisy-like, deep yellow flower heads singly on willowy stems throughout summer. May need staking. Divide regularly; spreads on rich soil. H 60cm (2ft), S 1m (3ft).

***Ranunculus acris* 'Flore Pleno'** (Double meadow buttercup)
Clump-forming perennial. Wiry stems with lobed and cut leaves act as a foil for rosetted, double, golden-yellow flowers in late spring and early summer. H and S 45–60cm (18–24in).

Arctotheca calendula, p.258
Coreopsis 'Sunray', p.289
Coreopsis tinctoria 'Golden Crown'
DAYLILIES, p.221

Duchesnea indica
HOSTAS, pp.252–3
IRISES, pp.198–9
ORCHIDS, pp.260–63

PRIMULAS, pp.236–7

YELLOW

***Sedum aizoon* 'Aurantiacum'**, syn. *S. a.* 'Euphorbioides'
Erect perennial with red stems carrying fleshy, toothed, dark green leaves. In summer produces gently rounded heads of dark yellow flowers followed by red seed capsules. H and S 45cm (18in).

Hieracium lanatum
Clump-forming perennial that produces mounds of broad, downy, grey leaves, above which dandelion-like, yellow flower heads appear on wiry stems in summer. H 30–45cm (12–18in), S 30cm (12in).

Eriophyllum lanatum, syn. *Bahia lanata*
Perennial forming low cushions of divided, silvery leaves. Daisy-like, yellow flower heads are produced freely in summer, usually singly, on grey stems. H and S 30cm (12in).

Tropaeolum polyphyllum
Prostrate perennial with spurred, short, trumpet-shaped, rich yellow flowers, borne singly in summer above trailing, grey-green leaves and stems. May spread widely once established but is good on a bank. H 5–8cm (2–3in), S 30cm (12in) or more.

Coreopsis verticillata
Bushy perennial with finely divided, dark green foliage and many tiny, star-shaped, golden flower heads borne throughout summer. Divide and replant in spring. H 40–60cm (16–24in), S 30cm (12in).

Inula ensifolia
Clump-forming perennial with small, lance-shaped to elliptic leaves, bearing many daisy-like, yellow flower heads, singly on wiry stalks, in late summer. H and S 30cm (12in).

Coreopsis lanceolata
Bushy perennial that in summer freely produces daisy-like, bright yellow flower heads on branching stems. Lance-shaped leaves are borne on flower stems. Propagate by seed or division. H 45cm (18in), S 30cm (12in).

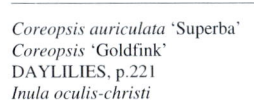

Impatiens repens
Evergreen, creeping perennial with rooting stems. Has small, oval to rounded leaves and, in summer, yellow flowers, each with a large, hairy spur. H to 5cm (2in), S indefinite. Min. 10°C (50°F).

***Calceolaria* 'John Innes'**
Vigorous, evergreen, clump-forming perennial that in spring-summer produces large, pouch-like, reddish-brown-spotted, deep yellow flowers, several to each stem. Has broadly oval, basal, mid-green leaves. H 15–20cm (6–8in), S 25–30cm (10–12in).

Coreopsis auriculata 'Superba'
Coreopsis 'Goldfink'
DAYLILIES, p.221
Inula oculis-christi

IRISES, pp.198–9
Mimulus guttatus
Oenothera fruticosa subsp. *glauca*
PRIMULAS, pp.236–7

Wachendorfia thyrsiflora

ORANGE

Aeschynanthus speciosus, syn. *A. splendens*
Evergreen, trailing perennial with waxy, narrowly oval leaves usually carried in whorls. Erect, tubular, bright orange-red flowers are borne in large clusters in summer. H and S 30–60cm (12–24in). Min. 18°C (64°F).

Geum × borisii
Clump-forming perennial with irregularly lobed leaves, above which in summer rise slender, branching, hairy stems bearing single, orange flowers with prominent, yellow stamens. H and S 30cm (12in).

Sedum rosea* var. *heterodontum, syn. *Rhodiola heterodonta*
Clump-forming perennial with heads of yellow or red, sometimes greenish flowers from spring to early summer. Stems are clothed with toothed, blue-green leaves. H 45cm (18in), S 25cm (10in).

DAYLILIES, p.221
IRISES, pp.198–9
PELARGONIUMS, pp.210–11
PRIMULAS, pp.236–7

Perennials/small AUTUMN INTEREST

▢ WHITE–PINK

Tricyrtis hirta var. **alba**
Upright, rhizomatous perennial that bears clusters of large, bell-shaped, spurred, white flowers, occasionally purple-spotted, in upper leaf axils of hairy, stem-clasping, dark green leaves during late summer and early autumn. H 45–60cm (18–24in), S 45cm (18in).

Sedum spectabile 'Brilliant'
(Ice-plant)
Clump-forming perennial that from late summer to autumn produces flat heads of bright rose-pink flowers. These are borne profusely over a mass of fleshy, grey-green leaves and attract butterflies. H and S 30–45cm (12–18in).

Schizostylis coccinea 'Sunrise'
Clump-forming, rhizomatous perennial that in early autumn produces spikes of large, shallowly cup-shaped, pink flowers above grassy, mid-green foliage. H 60cm (24in), S 23–30cm (9–12in).

▢ RED–YELLOW

Schizostylis coccinea 'Major', syn. *S. c.* 'Grandiflora'
Rhizomatous perennial with long, narrow, grass-like leaves. Gladiolus-like spikes of cup-shaped, bright crimson flowers appear in autumn. H 60cm (2ft) or more, S 30cm (1ft) or more.

Liriope muscari
Evergreen, spreading perennial that in autumn carries spikes of thickly clustered, rounded-bell-shaped, lavender or purple-blue flowers among narrow, glossy, dark green leaves. H 30cm (12in), S 45cm (18in).

Cautleya spicata
Upright perennial that in summer and early autumn bears spikes of light orange or soft yellow flowers in maroon-red bracts. Has handsome, long, mid-green leaves. Needs a sheltered site and rich, deep soil. H 60cm (24in), S 50cm (20in).

Chirita lavandulacea
Evergreen, erect perennial with downy, pale green leaves to 20cm (8in) long. In leaf axils has clusters of lavender-blue flowers with white tubes. May be sown in succession to flower from spring to autumn. H and S 60cm (2ft). Min. 15°C (59°F).

Senecio pulcher
Perennial with leathery, hairy, dark green leaves. In summer-autumn produces handsome, daisy-like, yellow-centred, bright purplish-pink flower heads. H 45–60cm (18–24in), S 50cm (20in).

Arctotheca calendula
(Cape dandelion)
Carpeting perennial. Leaves are woolly below, rough-haired above. Heads of daisy-like, bright yellow flowers, with darker yellow centres, appear from late spring to autumn. H 30cm (12in), S indefinite. Min. 5°C (41°F).

OTHER RECOMMENDED PLANTS:
Arctotis venusta
Astrantia major, p.241
Astrantia major subsp. *involucrata*, p.241
Astrantia maxima, p.241
MICHAELMAS DAISIES, p.226
PELARGONIUMS, pp.210–11
Schizostylis coccinea 'Mrs Hegarty'
Geranium wallichianum 'Buxton's Blue', p.251
Mimulus guttatus
MICHAELMAS DAISIES, p.226
ORCHIDS, pp.260–63
Osteospermum 'Buttermilk', p.254
Salvia patens, p.251
Tricyrtis macrantha

Begonias

The genus *Begonia* is one of the most versatile, providing interest year-round. Semperflorens begonias are excellent for summer bedding, while the Rex Cultorum group has distinctive and handsome foliage. Others, such as the × *tuberhybrida* cultivars with their large and showy blooms, are grown mainly for their flowers. Many also make attractive plants for hanging baskets. (See also the Plant Dictionary.)

B. × *tuberhybrida* 'Billie Langdon'

B. scharffii

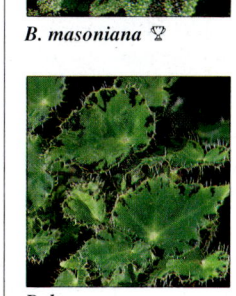

B. 'Merry Christmas' ♀

B. masoniana ♀

B. prismatocarpa

B. albo-picta

B. × *weltoniensis*

B. 'Orpha C. Fox'

B. × *tuberhybrida* 'Flamboyant'

B. bowerae

B. × *tuberhybrida* 'Apricot Cascade'

B. foliosa

B. 'Ingramii'

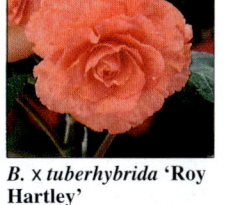

B. × *tuberhybrida* 'Roy Hartley'

B. 'Helen Lewis'

B. × *tuberhybrida* 'Can-Can'

B. olsoniae

B. metallica

B. 'Lucerna'

B. 'Duartei'

B. manicata 'Crispa'

B. 'Orange Rubra'

B. 'Red Ascot'

B. 'Thurstonii'

B. 'Norah Bedson'

B. manicata

B. serratipetala

B. 'Organdy'

B. pustulata 'Argentea'

B. 'Oliver Twist'

B. sutherlandii ♀

Orchids

Flamboyant, exotic, even seductive, the orchid is prized for its unusual flowers. Its aura of mystique and many popular misconceptions may have discouraged gardeners from growing these beautiful plants, but their cultivation is not always difficult and some will thrive happily indoors as house plants.

There are two main groups. Terrestrials grow in a wide range of habitats in the wild; many are at least frost hardy. Epiphytes, the more showy of the two and mostly native to the tropics, cling to tree branches or rocks, obtaining nourishment through their leaves and aerial roots. They need special composts and in cool climates must be grown under glass (see 'Orchids' in the Plant Dictionary).

(Key: x *Brass.* – x *Brassolaeliocattleya*; x *Soph.* – x *Sophrolaeliocattleya*; x *Oda.* – x *Odontioda*; *Odm.* – *Odontoglossum*; e – epiphyte; t – terrestrial)

Coelogyne cristata [e]

Masdevallia infracta [e]

Lemboglossum rossii [e]

Cypripedium acaule [t]

Masdevallia tovarensis [e]

Dendrobium infundibulum [e]

Coelogyne flaccida [e]

Calanthe vestita [t]

Lemboglossum cervantesii [e]

Dendrobium aphyllum [e]

Phalaenopsis Allegria [e]

Odm. Royal Occasion [e]

Paphiopedilum Freckles [t]

Paphiopedilum fairrieanum [t]

Oncidium ornithorrhynchum [e]

Paphiopedilum niveum [t]

Coelogyne nitida [e]

Cymbidium Portlett Bay [e]

Brassavola nodosa [e]

Paphiopedilum callosum [t]

Cypripedium reginae [t]

Spiranthes cernua [t]

Odm. crispum [e]

Miltoniopsis Robert Strauss 'Ardingly' [e]

Cymbidium Strathbraan [e]

Paphiopedilum bellatulum [t]

Angraecum sesquipedale [e]

Paphiopedilum appletonianum [t]

Lemboglossum bictoniense [e]

260

Dendrobium nobile [e]

Ophrys tenthredinifera [t]

Calypso bulbosa [t]

× *Laeliocattleya* Rojo 'Mont Millais' [e]

× *Oda*. Mount Bingham [e]

Cymbidium Strath Kanaid [e]

Lemboglossum cordatum [e]

× *Wilsonara* Hambuhren Stern 'Cheam' [e]

× *Brass*. Hetherington 'Coronation' [e]

Pleione bulbocodioides [t]

Cattleya J.A. Carbone [e]

× *Vuylstekeara* Cambria 'Lensing's Favorite' [e]

Cymbidium Pontac 'Mont Millais' [e]

× *Oda*. Pacific Gold × *Odm. cordatum* [e]

Laelia anceps [e]

Bletilla striata [t]

× *Soph*. Trizac 'Purple Emperor' [e]

× *Brassocattleya* Mount Adams [e]

× *Brass*. St Helier [e]

Miltoniopsis Anjou 'St Patrick' [e]

Epidendrum ibaguense [e]

Rossioglossum grande [e]

Phalaenopsis Lady Jersey × Lippeglut [e]

Cattleya bowringiana [e]

Odm. Le Nez Point [e]

Cymbidium Strathdon 'Cooksbridge Noel' [e] ♀

Paphiopedilum × *maudiae* [t]

× *Odontocidium* Artur Elle 'Colombian' [e]

Masdevallia coccinea [e]

× *Oda*. Petit Port [e]

Phaius tankervilleae [t]

Paphiopedilum Lyric 'Glendora' [t]

× *Odontocidium* Tiger Butter × *Wilsonara* Wigg's 'Kay' [e]

261

Perennials/small WINTER/ALL YEAR INTEREST
☐ WHITE

Spathiphyllum wallisii (Peace lily, White sails)
Evergreen, tufted, rhizomatous perennial. Has clusters of long, lance-shaped leaves. Fleshy, white spadices of fragrant flowers in white spathes are irregularly produced. H and S 30cm (12in) or more. Min. 15°C (59°F).

Pilea cadierei (Aluminium plant)
Evergreen, bushy perennial with broadly oval leaves, each with a sharply pointed tip and raised, silvery patches that appear quilted. Has insignificant, greenish flowers. H and S 30cm (12in). Min. 10°C (50°F).

Tradescantia fluminensis 'Variegata', syn. *T. albiflora* 'Variegata'
Evergreen, trailing perennial with rooting stems and leaves, irregularly striped creamy-white. Intermittently has clusters of white flowers. H 30cm (12in), S indefinite. Min. 15°C (59°F).

Fittonia verschaffeltii var. **argyroneura**, syn. *F. argyroneura* (Silver net-leaf)
Evergreen, creeping perennial with small, oval, white-veined, olive-green leaves. Remove flowers if they form. H to 15cm (6in), S indefinite. Min. 15°C (59°F).

Tradescantia fluminensis 'Albovittata', syn. *T. albiflora* 'Albovittata'
Strong-growing, evergreen perennial with trailing, rooting stems. Bluish-green leaves have broad, white stripes. Bears small, white flowers. H 30cm (12in), S indefinite. Min. 15°C (59°F).

Aglaonema commutatum 'Treubii'
Evergreen, erect, tufted perennial. Lance-shaped leaves, to 30cm (12in) long, are marked with pale green or silver. Occasionally has greenish-white spathes. H and S to 45cm (18in). Min. 15°C (59°F).

Glechoma hederacea 'Variegata' (Variegated ground ivy)
Evergreen, carpeting perennial that has small, heart-shaped leaves, with white marbling, on trailing stems. Bears insignificant flowers in summer. Spreads rapidly but is useful for a container. H 15cm (6in), S indefinite.

Helleborus orientalis [white form]
Evergreen, clump-forming perennial with dense, divided foliage, above which rise nodding, cup-shaped, white, pink or purple flowers, sometimes darker spotted, in winter or early spring. H and S 45cm (18in).

Peperomia caperata (Emerald ripple)
Evergreen, bushy perennial with pinkish leaf stalks. Has oval, fleshy, wrinkled, dark green leaves, to 5cm (2in) long, with sunken veins; spikes of white flowers appear irregularly. H and S to 15cm (6in). Min. 10°C (50°F).

Sonerila margaritacea 'Argentea'
Sonerila margaritacea 'Hendersonii'
Spathiphyllum 'Clevelandii'
Spathiphyllum floribundum

264

WHITE–PINK

Chlorophytum comosum 'Vittatum'
Evergreen, tufted, rosette-forming perennial. Long, narrow, lance-shaped, creamy-white leaves have green stripes and margins. Irregularly has small, star-shaped, white flowers on thin stems. H and S 30cm (12in). Min. 5°C (41°F).

Aspidistra elatior 'Variegata'
Evergreen, rhizomatous perennial with upright, narrow, glossy, dark green leaves which are longitudinally cream-striped. Occasionally has inconspicuous, cream to purple flowers near soil level. H 60cm (24in), S 45cm (18in). Min. 5–10°C (41–50°F).

Oplismenus hirtellus 'Variegatus', syn. *O. africanus* 'Variegatus'
Evergreen, creeping, perennial grass with wiry, rooting stems. White-striped leaves, with wavy margins, are often tinged pink. Bears inconspicuous flowers intermittently. H 20cm (8in) or more, S indefinite. Min. 12°C (54°F).

Helleborus niger (Christmas rose)
Evergreen, clump-forming perennial with divided, deep green leaves and cup-shaped, nodding, white flowers with golden stamens borne in winter or early spring. H and S 30cm (12in).

Helleborus x sternii
Evergreen, clump-forming perennial with divided leaves and cup-shaped, often pink-tinged, pale green flowers borne in terminal clusters in winter and early spring. H and S 45cm (18in).

Helleborus orientalis [pink form]
Evergreen, clump-forming perennial with dense, divided foliage, above which rise nodding, cup-shaped, white, pink or purple flowers, sometimes darker spotted, in winter or early spring. H and S 45cm (18in).

AFRICAN VIOLETS, p.266
Bergenia cordifolia 'Purpurea', p.233
BROMELIADS, p.229
Episcia cupreata 'Metallica'
Episcia cupreata 'Tropical Topaz'
ORCHIDS, pp.260–63
Peristrophe hyssopifolia
Tradescantia zebrina 'Quadricolor'

PINK–RED

Streptocarpus 'Nicola'
Evergreen, stemless perennial with a rosette of strap-shaped, wrinkled leaves. Funnel-shaped, rose-pink flowers are produced intermittently in small clusters. H 25cm (10in), S 50cm (20in). Min. 10–15°C (50–59°F).

Tradescantia zebrina, syn. *Zebrina pendula* (Silver inch plant)
Evergreen, trailing or mat-forming perennial. Bluish-green leaves, purple-tinged beneath, have 2 broad, silver bands. Has pink or violet-blue flowers intermittently during the year. H 15cm (6in), S indefinite. Min. 15°C (59°F).

Episcia cupreata (Flame violet)
Evergreen, creeping perennial. Has small, downy, wrinkled leaves, usually silver-veined or -banded, and, intermittently, scarlet flowers marked yellow within. H 10cm (4in), S indefinite. Min. 15°C (59°F).

Anthurium scherzerianum 'Rothschildianum'
Bergenia 'Ballawley'
Bergenia crassifolia
BROMELIADS, p.229

265

African violets

African violet is the common name for the genus *Saintpaulia*, although frequently used for the many hybrids of *S. ionantha*. These small, rosetted perennials may be grown as summer bedding in warm, humid climates but also make attractive, indoor pot plants, flowering freely if kept in a draught-free, light and humid position. The selection below shows the range of flower colours and forms available.

S. 'Miss Pretty'

S. 'Garden News'

S. 'Fancy Pants'

S. 'Pip Squeak'

S. 'Rococo Pink'

S. 'Colorado'

S. 'Kristi Marie'

S. 'Porcelain'

S. 'Bright Eyes'

S. 'Delft'

Perennials/small WINTER/ALL YEAR
RED–PURPLE

Anthurium scherzerianum
(Flamingo flower)
Evergreen, tufted perennial with erect, leathery, dark green leaves to 20cm (8in) long. Has large, long-lasting, bright red spathes and fleshy, orange to yellow spadices. H and S 30–60cm (12–24in). Min. 15°C (59°F).

Helleborus orientalis [purple form]
Evergreen, clump-forming perennial with dense, divided foliage, above which rise nodding, cup-shaped, white, pink or purple flowers, sometimes darker spotted, in winter or early spring. H and S 45cm (18in).

Gerbera jamesonii
(Barberton daisy)
Evergreen, upright perennial with daisy-like, variably coloured flower heads, borne intermittently on long stems, and basal rosettes of large, jagged leaves. Flowers are excellent for cutting. H 60cm (24in), S 45cm (18in).

Phormium tenax 'Bronze Baby'
Evergreen, upright perennial with tufts of bold, stiff, pointed, wine-red leaves. Panicles of reddish flowers are occasionally produced on purplish stems during summer. H and S 45–60cm (18–24in).

Tellima grandiflora 'Rubra', syn. *T. g.* 'Purpurea'
Semi-evergreen, clump-forming perennial with a mass of hairy, basal, reddish-purple leaves, underlaid dark green. In late spring, erect stems bear spikes of bell-shaped, pinkish-cream flowers. H and S 60cm (24in).

Fittonia verschaffeltii
(Painted net-leaf)
Evergreen, creeping perennial with small, oval, red-veined, olive-green leaves. Flowers are best removed if they form. H to 15cm (6in), S indefinite. Min. 15°C (59°F).

Alternanthera ficoidea var. *amoena*
Alternanthera ficoidea 'Versicolor'
Bergenia purpurascens
Bertolonia marmorata

ORCHIDS, pp.260–63

266

PURPLE

Ophiopogon planiscapus 'Nigrescens'
Evergreen, spreading, clump-forming perennial, grown for its distinctive, grass-like, black leaves. Racemes of lilac flowers in summer are followed by black fruits. H 23cm (9in), S 30cm (12in).

Pellionia repens, syn. *P. daveauana* (Watermelon begonia)
Evergreen, creeping perennial with rooting stems. Broadly oval, olive-green leaves have purplish-brown edges and paler green centres. Flowers are insignificant. H 10cm (4in), S indefinite. Min. 15°C (59°F).

Tetranema roseum, syn. *T. mexicanum* (Mexican foxglove, Mexican violet)
Short-stemmed perennial with crowded, stalkless leaves, bluish-green beneath. Intermittently, has nodding, purple flowers with paler throats. H to 20cm (8in), S 30cm (12in). Min. 13°C (55°F).

PURPLE–GREEN

Cyanotis somaliensis (Pussy ears)
Evergreen, creeping perennial. Small, narrow, glossy, dark green leaves with white hairs surround stems. Has purplish-blue flowers in leaf axils in winter-spring. H 5cm (2in), S indefinite. Min. 10–15°C (50–59°F).

Tradescantia pallida 'Purple Heart'
Evergreen, creeping perennial with dark purple stems and slightly fleshy leaves. Has pink or pink-and-white flowers in summer. H 30–40cm (12–16in), S 30cm (12in) or more. Min. 15°C (59°F).

Tradescantia sillamontana, syn. *T. pexata*, *T. velutina*
Evergreen, erect perennial. Oval, stem-clasping leaves are densely covered with white, woolly hairs. Has clusters of small, bright purplish-pink flowers in summer. H and S to 30cm (12in). Min. 10–15°C (50–59°F).

Ajuga reptans 'Multicolor', syn. *A.r.* 'Rainbow'
Evergreen, mat-forming perennial. Dark green leaves, marked with cream and pink, make good ground cover. Spikes of small, blue flowers appear in spring. H 12cm (5in), S 45cm (18in).

Maranta leuconeura 'Erythroneura', syn. *M.l.* 'Erythrophylla' (Herringbone plant)
Evergreen perennial. Oblong leaves have veins marked red, with paler yellowish-green midribs, and are upright at night, flat by day. H and S to 30cm (12in). Min. 15°C (59°F).

Ajuga reptans 'Atropurpurea'
Evergreen, ground-cover perennial, spreading freely by runners, with small rosettes of glossy, deep bronze-purple leaves. Short spikes of blue flowers appear in spring. H 15cm (6in), S 1m (3ft).

Streptocarpus 'Constant Nymph'
Evergreen, stemless perennial with a rosette of strap-shaped, wrinkled leaves. Funnel-shaped, purplish-blue flowers, darker veined and yellow-throated, are intermittently produced in small clusters. H 25cm (10in), S 50cm (20in). Min. 10–15°C (50–59°F).

Sansevieria trifasciata 'Hahnii'
Evergreen, stemless perennial with a rosette of about 5 stiff, erect, broadly lance-shaped and pointed leaves, banded horizontally with pale green or white. Occasionally has small, pale green flowers. H 15–30cm (6–12in), S 10cm (4in). Min. 15°C (59°F).

Helleborus atrorubens
Heterocentron elegans, p.248
Lathraea clandestina, p.235
ORCHIDS, pp.260–63

Streptocarpus rexii

Aeschynanthus pulcher
BROMELIADS, p.229
Cyanotis kewensis
Tradescantia zebrina 'Purpusii'

Perennials/small WINTER/ALL YEAR INTEREST
□ GREEN

Calathea makoyana
(Peacock plant)
Evergreen, clump-forming perennial. Horizontal leaves, 30cm (12in) long, are dark and light green above, reddish-purple below. Intermittently has short spikes of white flowers. H to 60cm (2ft), S to 1.2m (4ft). Min. 15°C (59°F).

Maranta leuconeura var. **kerchoviana** (Rabbit tracks)
Evergreen perennial that intermittently bears white to mauve flowers. Oblong leaves with dark brown blotches become greener with age and are upright at night, flat by day. H and S to 30cm (12in). Min. 15°C (59°F).

Stachys byzantina, syn. *S. lanata*, *S. olympica* (Bunnies' ears, Lamb's tongue)
Evergreen, mat-forming perennial with woolly, grey foliage excellent for a border front or as ground cover. Bears mauve-pink flowers in summer. H 30–38cm (12–15in), S 60cm (24in).

Aglaonema pictum
Evergreen, erect, tufted perennial. Oval leaves, to 15cm (6in) long, are irregularly marked with greyish-white or grey-green. Has creamy-white spathes in summer. H and S to 60cm (2ft). Min. 15°C (59°F).

Welwitschia mirabilis, syn. *W. bainesii*
Evergreen perennial with a short, woody trunk. Has 2 strap-shaped leaves, to 2.5m (8ft) long, with tips splitting to form many tendril-like strips. Bears small, reddish-brown cones. H to 30cm (12in), S indefinite. Min. 10°C (50°F).

Ophiopogon japonicus
Evergreen, clump or mat-forming perennial with grass-like, glossy, dark green foliage. Spikes of lilac flowers in late summer are followed by blue-black berries. H 30cm (12in), S indefinite.

Soleirolia soleirolii, syn. *Helxine soleirolii* (Baby's tears, Mind-your-own-business)
Usually evergreen, invasive, prostrate perennial with small, round, vivid green leaves that form a carpet. May choke other plants if not controlled. H 5cm (2in), S indefinite.

Aglaonema commutatum 'Silver King'
Evergreen, erect, tufted perennial. Broadly lance-shaped leaves, to 30cm (12in) long, are mid-green, marked with dark and light green. Has greenish-white spathes in summer. H and S to 45cm (18in). Min. 15°C (59°F).

Peperomia marmorata (Silver heart)
Evergreen, bushy perennial with insignificant flowers. Has oval, long-pointed, fleshy, dull green leaves, marked with greyish-white and quilted above, reddish below. H and S to 20cm (8in). Min. 10°C (50°F).

Helleborus argutifolius, syn. *H. corsicus*, *H. lividus* subsp. *corsicus*
Clump-forming perennial with evergreen, divided, spiny, dark green leaves and cup-shaped, pale green flowers borne in large clusters in winter-spring. H 60cm (24in), S 45cm (18in).

Callisia repens
Evergreen, creeping perennial with rooting stems and densely packed leaves, sometimes white-banded and often purplish beneath. Rarely, has inconspicuous, white flowers in winter. H 10cm (4in), S indefinite. Min. 15°C (59°F).

Ajuga pyramidalis 'Metallica Crispa'
Ajuga reptans 'Jungle Beauty'
Aspidistra elatior
Callisia navicularis

Chamaemelum nobile
Chamaemelum nobile 'Treneague'
Chlorophytum comosum
Peperomia clusiifolia

Peperomia griseoargentea
Peperomia magnoliifolia
Peperomia metallica
Peperomia rubella

Plectranthus australis
Plectranthus oertendahlii
Tradescantia fluminensis

◻ GREEN

Drosera spathulata
Evergreen, insectivorous perennial with rosettes of spoon-shaped leaves that have sensitive, red, glandular hairs. Has many small, pink or white flowers on leafless stems in summer. H and S to 8cm (3in). Min. 5–10°C (41–50°F).

☼ ◊

Drosera capensis (Cape sundew)
Evergreen, insectivorous perennial. Rosettes of narrow leaves have sensitive, red, glandular hairs. Many small, purple flowers are borne on leafless stems in summer. H and S to 15cm (6in). Min. 5–10°C (41–50°F).

☼ ◊

Peperomia glabella (Wax privet)
Evergreen perennial with wide-spreading, red stems. Has broadly oval, fleshy, glossy, bright green leaves, to 5cm (2in) long, and insignificant flowers. H to 15cm (6in), S 30cm (12in). Min. 10°C (50°F).

☼ ◊

Peperomia obtusifolia 'Variegata'
Evergreen, bushy perennial with spade-shaped, fleshy leaves, to 20cm (8in) long, that have irregular, yellowish-green to creamy-white margins and usually greyish centres. Flowers are insignificant. H and S to 15cm (6in). Min. 10°C (50°F).

☼ ◊

Helleborus viridis (Green hellebore)
Clump-forming perennial with deciduous, divided, dark green leaves. Bears cup-shaped, green flowers in late winter or early spring. H and S 30cm (12in).

☼ ◊ ❋❋❋

Pilea nummulariifolia (Creeping Charlie)
Evergreen, mat-forming perennial with creeping, rooting, reddish stems. Rounded, pale green leaves, about 2cm (3/4in) wide, have a corrugated surface. Flowers are insignificant. H to 5cm (2in), S 30cm (12in). Min. 10°C (50°F).

☼ ◊

BROMELIADS, p.229
Helleborus lividus
ORCHIDS, pp.260–63
Tolmiea menziesii

◻ GREEN–YELLOW

Helleborus foetidus (Stinking hellebore)
Evergreen, clump-forming perennial with deeply divided, dark green leaves and, in late winter and early spring, panicles of cup-shaped, red-margined, pale green flowers. H and S 45cm (18in).

☼ ◊ ❋❋❋ ♈

Iresine herbstii 'Aureoreticulata'
Evergreen, bushy perennial with red stems and inconspicuous flowers. Rounded, mid-green leaves, 10cm (4in) long, have yellow or red veins and notched tips. H to 60cm (24in), S 45cm (18in). Min. 10–15°C (50–59°F).

☼ ◊

Sansevieria trifasciata 'Golden Hahnii'
Evergreen, stemless perennial with a rosette of about 5 stiff, erect, broadly lance-shaped leaves with wide, yellow borders. Sometimes bears small, pale green flowers. H 15–30cm (6–12in), S 10cm (4in). Min. 15°C (59°F).

☼ ◊ ♈

Dionaea muscipula (Venus flytrap)
Evergreen, insectivorous perennial with rosettes of 6 or more spreading, hinged leaves, pink-flushed inside, edged with stiff bristles. Clusters of tiny, white flowers are carried in summer. H 10cm (4in), S 30cm (12in). Min. 5°C (41°F).

☼ ◊

Nautilocalyx lynchii
Robust, evergreen, erect, bushy perennial. Broadly lance-shaped, slightly wrinkled leaves are glossy, greenish-red above, reddish beneath. In summer has tubular, red-haired, pale yellow flowers with red calyces. H and S to 60cm (24in). Min. 15°C (59°F).

☼ ◊

Adonis amurensis, p.239
BROMELIADS, p.229
ORCHIDS, pp.260–63
Peperomia magnoliifolia 'Green and Gold'

Pilea involucrata
Tradescantia spathacea 'Vittata'

269

ANNUALS AND BIENNIALS
☐ WHITE

Digitalis purpurea f. **alba**, syn.
D. p. 'Alba', *D. p.* var. *albiflora*
Slow-growing, short-lived perennial, grown as a biennial. Has a rosette of large, pointed-oval leaves and erect stems carrying tubular, white flowers in summer. H 1–1.5m (3–5ft), S 30–45cm (1–1½ft).

Omphalodes linifolia
(Venus's navelwort)
Fairly fast-growing, slender, erect annual with lance-shaped, grey-green leaves. Tiny, slightly scented, rounded, white flowers, rarely tinged blue, are carried in summer. H 15–30cm (6–12in), S 15cm (6in).

Gypsophila elegans
Fast-growing, erect, bushy annual. Has lance-shaped, greyish-green leaves and clouds of tiny, white flowers in branching heads from summer to early autumn. H 60cm (24in), S 30cm (12in) or more.

Lavatera trimestris 'Mont Blanc'
Moderately fast-growing, erect, branching annual with oval, lobed leaves. Shallowly trumpet-shaped, brilliant white flowers appear from summer to early autumn. H to 60cm (24in), S 45cm (18in).

Iberis amara
Fast-growing, erect, bushy annual with lance-shaped, mid-green leaves. Has flattish heads of small, scented, 4-petalled, white flowers in summer. H 30cm (12in), S 15cm (6in).

Lobularia maritima 'Little Dorrit', syn. *Alyssum maritimum* 'Little Dorrit'
Fast-growing, compact annual with lance-shaped, greyish-green leaves and heads of scented, 4-petalled, white flowers in summer and early autumn. H 8–15cm (3–6in), S 20–30cm (8–12in).

Matthiola 'Giant Imperial'
Fast-growing, erect, bushy biennial, grown as an annual. Has lance-shaped, greyish-green leaves and long spikes of highly scented, white to creamy-yellow flowers in summer. Produces excellent flowers for cutting. H to 60cm (24in), S 30cm (12in).

Petunia, Recoverer Series [white]
Moderately fast-growing, branching, bushy perennial, grown as an annual. Has oval, mid- to deep green leaves and large, flared, trumpet-shaped, white flowers in summer-autumn. H 15–30cm (6–12in), S 30cm (12in).

OTHER RECOMMENDED PLANTS:
Adlumia fungosa
Ageratum houstonianum 'White Cushion'
Antirrhinum majus and cvs

Bellis perennis 'White Carpet'
Callistephus chinensis and cvs
Campanula medium and cvs
Centaurea cyanus

Consolida ambigua, Hyacinth-flowered Series
Convolvulus tricolor 'Minor'
Dianthus chinensis and cvs
Eustoma grandiflorum, F1 Hybrids [mixed]

Helichrysum bracteatum cvs
Iberis umbellata
Lobelia erinus 'Colour Cascade'
Lobularia maritima

☐ WHITE

Dimorphotheca pluvialis, syn. *D. annua* (Rain daisy)
Branching annual with oval, hairy, deep green leaves. In summer has small, daisy-like flower heads, the rays purple beneath and white above, with brownish-purple centres. H 20–30cm (8–12in), S 15cm (6in).

Tanacetum parthenium, syn. *Chrysanthemum parthenium* (Feverfew)
Moderately fast-growing, short-lived, bushy perennial, grown as an annual. Has aromatic leaves and small, white flower heads in summer and early autumn. H and S 20–45cm (8–18in).

Reseda odorata (Mignonette)
Moderately fast-growing, erect, branching annual with oval leaves. Conical heads of small, very fragrant, somewhat star-shaped, white flowers with orange-brown stamens are carried in summer and early autumn. H 30–60cm (12–24in), S 30cm (12in).

Euphorbia marginata (Snow-in-summer, Snow-on-the-mountain)
Moderately fast-growing, upright, bushy annual. Has pointed-oval, bright green leaves; upper leaves are white-margined. Broad, petal-like, white bracts surround insignificant flowers in summer. H 60cm (24in), S 30cm (12in).

Nemophila maculata (Five-spot baby)
Fast-growing, spreading annual with lobed leaves. Small, bowl-shaped, white flowers with purple-tipped petals are carried in summer. H and S 15cm (6in).

Viola, **Floral Dance Series** [white]
Bushy perennial, grown as an annual or biennial. Has oval, mid-green leaves and rounded, 5-petalled, white flowers in winter. H 15–20cm (6–8in), S 20cm (8in).

Salvia splendens, **Carabiniere Series** [white]
Slow-growing, bushy perennial, grown as an annual. Has oval, serrated, fresh green leaves and dense racemes of tubular, white flowers in summer and early autumn. H to 30cm (12in), S 20–30cm (8–12in).

Nicotiana* x *sanderae, **Domino Series**
Fairly slow-growing, bushy annual with pointed-oval leaves. Somewhat trumpet-shaped flowers, to 8cm (3in) long, appear in a wide range of colours in summer and early autumn. H and S 30cm (12in).

Eustoma grandiflorum, syn. *Lisianthus russellianus*
Slow-growing, upright annual with lance-shaped, deep green leaves. Poppy-like, pink, purple, blue or white flowers, 5cm (2in) wide, are carried in summer. H 60cm (24in), S 30cm (12in). Min. 4–7°C (39–45°F).

Matthiola cvs
Myosotis 'White Ball'
Nemesia strumosa
Nigella damascena

Papaver somniferum 'White Cloud'
PELARGONIUMS, pp.210–11
Petunia cvs
Phlox drummondii cvs

Portulaca grandiflora
Salvia farinacea 'Alba'
Salvia viridis cvs
Senecio x *hybridus* cvs

Verbascum lychnitis
Viola x *wittrockiana* cvs
Zinnia cvs

Annuals and Biennials
WHITE–PINK

Hibiscus trionum
(Flower-of-the-hour)
Fairly fast-growing, upright annual with oval, serrated leaves. Trumpet-shaped, creamy-white or pale yellow flowers, with purplish-brown centres, are borne from late summer to early autumn.
H 60cm (24in), S 30cm (12in).

Martynia annua,
syn. *M. louisiana* (Unicorn plant)
Fairly fast-growing, upright annual with long-stalked leaves. Has foxglove-like, lobed, creamy-white flowers marked red, pink and yellow in summer, followed by horned, green, then brown, fruits. H 60cm (24in), S 30cm (12in).

***Lathyrus odoratus* 'Knee Hi'**
Fast-growing annual with oval, divided, mid-green leaves and large, fragrant flowers, in shades of pink, red, blue or white, that are borne in summer or early autumn. H and S 90cm (3ft).

Crepis rubra
Fairly fast-growing, rosette-forming annual with lance-shaped, serrated leaves. In summer bears dandelion-like, pink, occasionally red or white flower heads. H 30cm (12in), S 15cm (6in).

***Zea mays* 'Gracillima Variegata'**
Fairly fast-growing, erect annual with lance-shaped leaves, striped green and creamy-white. Has tassel-like, silvery flower heads, followed by large, bright yellow seed heads (cobs). H 90cm (3ft), S 30–45cm (1–1½ft).

Impatiens balsamina (Balsam)
Fairly fast-growing, erect, compact, bushy annual with lance-shaped leaves. Small, cup-shaped, spurred, pink or white flowers are borne in summer and early autumn. H to 75cm (30in), S 45cm (18in).

***Chrysanthemum carinatum*
'Monarch Court Jesters'**
Fast-growing, erect, branching annual. Has feathery, grey-green leaves and, in summer, daisy-like, zoned flower heads, to 8cm (3in) wide, in various colour combinations. H 60cm (24in), S 30cm (12in).

Silene coeli-rosa, syn.
Agrostemma coeli-rosa, *Lychnis coeli-rosa*, *Viscaria elegans*
Moderately fast-growing, erect annual with lance-shaped, greyish-green leaves. Has 5-petalled, pinkish-purple flowers with white centres in summer. H 45cm (18in), S 15cm (6in).

Cleome hassleriana
Impatiens balsamina 'Blackberry Ice'

272

■ PINK

Helipterum roseum, syn. *Acroclinium roseum*
Moderately fast-growing, erect annual. Lance-shaped leaves are greyish-green; small, daisy-like, papery, semi-double, pink flower heads appear in summer. Flowers dry well. H 30cm (12in), S 15cm (6in).

Primula, Posy Series
Rosette-forming perennial, normally grown as a biennial, with lance-shaped leaves. Fragrant, flat, almost stemless flowers, in shades of pink, red, yellow, purple or white, with contrasting centres, appear in spring. H 8cm (3in), S 10cm (4in).

Matthiola, Brompton Series
Fast-growing, erect, bushy biennial, grown as an annual. Lance-shaped leaves are greyish-green; long spikes of highly scented flowers in shades of pink, red, purple, yellow or white are borne in summer. H 45cm (18in), S 30cm (12in).

Alcea rosea, syn. *Althaea rosea* (Hollyhock)
Biennial with tall, erect stems and lobed, rough-textured leaves. Spikes of single flowers, in a range of colours including pink, yellow and cream, appear in summer and early autumn. H 1.5–2m (5–6ft), S to 60cm (2ft).

Matthiola 'Giant Excelsior'
Fast-growing, erect, bushy biennial, grown as an annual. Lance-shaped leaves are greyish-green; long spikes of highly scented flowers in shades of pink, red, pale blue or white appear in summer. H to 75cm (30in), S 30cm (12in).

Helipterum manglesii, syn. *Rhodanthe manglesii*
Moderately fast-growing, erect annual. Has pointed-oval, greyish-green leaves and daisy-like, papery, red, pink or white flower heads, in summer and early autumn. Flowers dry well. H 30cm (12in), S 15cm (6in).

Pelargonium, Orbit Series [salmon]
Slow-growing, evergreen, branching, bushy perennial, grown as an annual, with lobed leaves, zoned with bronze or red. Has large, rounded heads of salmon-pink flowers in summer-autumn. H and S 30–60cm (12–24in).

Salvia splendens, Cleopatra Series [salmon]
Slow-growing, bushy perennial, grown as an annual. Oval, serrated leaves are fresh green; dense racemes of tubular, salmon-pink flowers are carried in summer and early autumn. H to 30cm (12in), S 20–30cm (8–12in).

Papaver somniferum, Peony-flowered Series
Fast-growing, erect annual with lobed, pale greyish-green leaves. Has large, rounded, often cup-shaped, double flowers in a mixture of colours – red, pink, purple or white – in summer. H 75cm (30in), S 30cm (12in).

Antirrhinum majus, Madame Butterfly Series
Erect perennial, grown as an annual. Leaves are lance-shaped; spikes of tubular, 2-lipped, double flowers are carried from spring to autumn. Is available in a mixture of colours. H 60–75cm (24–30in), S 45cm (18in).

Impatiens, Novette Series [salmon] (Busy lizzie)
Fast-growing, evergreen, bushy perennial, grown as an annual, with pointed-oval, fresh green leaves. Small, flat, spurred, salmon-pink flowers are carried from spring to autumn. H and S 15cm (6in).

Ageratum houstonianum 'Bengali'
Ageratum houstonianum 'Pinkie'
Silene pendula

Annuals and Biennials
◻ PINK

Zinnia, Thumbelina Series
[Erecta group]
Moderately fast-growing, sturdy, erect annual with oval to lance-shaped leaves. In summer and early autumn has large, daisy-like, double and semi-double flower heads in a range of colours. H 15cm (6in), S 30cm (12in).

☼ ◊ ❋

Papaver rhoeas, Shirley Series
[double]
Fast-growing, slender, erect annual with lobed, light green leaves. Rounded, often cup-shaped, double flowers, in shades of red, pink or white, including bicolours, are borne in summer. H 60cm (24in), S 30cm (12in).

☼ ◊ ❋❋❋ ♉

Cleome hassleriana 'Colour Fountain'
Fast-growing, bushy annual with hairy stems and divided leaves. In summer has heads of narrow-petalled flowers, with long, protruding stamens, in shades of pink, mauve, purple or white. H 1–1.2m (3–4ft), S 45–60cm (1½–2ft).

☼ ◊ ❋

Lobularia maritima 'Wonderland', syn. *Alyssum maritimum* 'Wonderland'
Fast-growing, compact annual with lance-shaped leaves. Bears heads of tiny, scented, deep purplish-pink flowers in summer and early autumn. H to 15cm (6in), S to 30cm (12in).

☼ ◊ ❋❋❋

Eschscholzia californica, Ballerina Series
Fast-growing, slender, erect annual. Has feathery, bluish-green leaves and cup-shaped, 4-petalled, frilled, double flowers, in shades of red, yellow, pink or orange, in summer-autumn. H 30cm (12in), S 15cm (6in).

☼ ◊ ❋❋❋

Schizanthus 'Hit Parade'
Moderately fast-growing, erect, bushy annual with deeply divided leaves. Tubular flowers, in a mixture of rich colours including pink, red, purple or white, often with contrasting markings inside, appear in summer-autumn. H and S 30cm (12in). Min. 5°C (41°F).

☼ ◊

Centaurea cyanus [tall, rose]
(Cornflower)
Fast-growing, erect, branching annual with lance-shaped, grey-green leaves. Branching heads of daisy-like, rose-pink flower heads are carried in summer and early autumn. H to 90cm (3ft), S 30cm (1ft).

☼ ◊ ❋❋❋

Limonium sinuatum
Fairly slow-growing, bushy, upright perennial, grown as an annual. Has lance-shaped, lobed, deep green leaves and, in summer and early autumn, tiny, blue, pink or white flowers borne in clusters on winged stems. H 45cm (18in), S 30cm (12in).

☼ ◊ ❋

Petunia, Victorious Series
Moderately fast-growing, branching, bushy perennial, grown as an annual. Has oval leaves and ruffled, flared, trumpet-shaped, double flowers in a mixture of colours in summer-autumn. H 15–30cm (6–12in), S 30cm (12in).

☼ ◊ ❋

Onopordum acanthium
(Cotton thistle, Scotch thistle)
Slow-growing, erect, branching biennial. Large, lobed, spiny leaves are hairy and bright silvery-grey; winged, branching flower stems bear deep purplish-pink flower heads in summer. H 1.8m (6ft), S 90cm (3ft).

☼ ◊ ❋❋❋

Silybum marianum
(Blessed Mary's thistle)
Biennial with a basal rosette of deeply lobed, very spiny, heavily white-marbled, deep green leaves. Has thistle-like, dark purplish-pink flower heads on erect stems in summer and early autumn. H 1.2m (4ft), S 60cm (2ft).

☼ ◊ ❋❋❋

Iberis umbellata, Fairy Series
Fast-growing, upright, bushy annual with lance-shaped, mid-green leaves. Heads of small, 4-petalled flowers, in shades of pink, red, purple or white, are carried in summer and early autumn. H and S 20cm (8in).

☼ ◊ ❋❋❋

Agrostemma githago
Alcea rosea 'Majorette'
Bellis perennis cvs
Campanula medium and cvs

Centaurea cyanus
Clarkia amoena
Clarkia amoena, Azalea-flowered Series
Consolida ambigua, Hyacinth-flowered Series

Eustoma grandiflorum, F1 Hybrids [mixed]
Impatiens cvs
Matthiola cvs
Salvia viridis cvs

■ PINK

Petunia, Bonanza Series
Moderately fast-growing, branching, bushy perennial, grown as an annual. Has oval leaves and frilled, flared, trumpet-shaped, double flowers in a mixture of colours in summer-autumn. H 15–30cm (6–12in), S 30cm (12in).

Agrostemma githago 'Milas'
Fast-growing, slender, upright, thin-stemmed annual. Has lance-shaped leaves and 5-petalled, purplish-pink flowers, 8cm (3in) wide, in summer. H 60–90cm (2–3ft), S 30cm (1ft).

Malcolmia maritima
(Virginian stock)
Fast-growing, slim, erect annual with oval, greyish-green leaves. Carries tiny, fragrant, 4-petalled, pink, red or white flowers from spring to autumn. Sow seeds in succession for a long flowering season. H 20cm (8in), S 5–8cm (2–3in).

Matthiola, Brompton Series [pink]
Fast-growing, erect, bushy biennial, grown as an annual, with lance-shaped, greyish-green leaves. Long spikes of highly scented, pink flowers are carried in summer. H 45cm (18in), S 30cm (12in).

Lunaria annua 'Variegata'
Fast-growing, erect biennial with pointed-oval, serrated, white-variegated leaves. Heads of small, scented, 4-petalled, deep purplish-pink flowers are borne in spring and early summer followed by rounded, silvery seed pods. H 75cm (30in), S 30cm (12in).

Silene armeria 'Electra'
Moderately fast-growing, erect annual with oval, greyish-green leaves. Heads of 5-petalled, bright rose-pink flowers are carried in summer and early autumn. H 30cm (12in), S 15cm (6in).

Annuals and Biennials
☐ PINK

Lavatera trimestris 'Silver Cup'
Moderately fast-growing, erect, branching annual with oval, lobed leaves. Shallowly trumpet-shaped, rose-pink flowers are carried in summer and early autumn. H 60cm (24in), S 45cm (18in).

Dianthus chinensis, Baby Doll Series
Neat, bushy annual or biennial, grown as an annual. Light or mid-green leaves are lance-shaped; small, single, zoned flowers in various colours are carried in summer and early autumn. H 15cm (6in), S 15–30cm (6–12in).

Clarkia 'Brilliant'
Fast-growing, erect, bushy annual with oval leaves. Large, rosette-like, double, bright reddish-pink flowers are carried in long spikes in summer and early autumn. H to 60cm (24in), S 30cm (12in).

Lathyrus odoratus 'Bijou'
Fast-growing annual with oval, divided, mid-green leaves and large, fragrant flowers, in shades of pink, red or blue, that are carried in summer or early autumn. H and S 45cm (18in).

Callistephus chinensis, Milady Series [rose]
Moderately fast-growing, erect, bushy annual with oval, toothed leaves. Has large, daisy-like, double, rose-pink flower heads in summer and early autumn. H 25–30cm (10–12in), S 30–45cm (12–18in).

Brassica oleracea forms (Ornamental cabbage)
Moderately fast-growing, evergreen, rounded biennial, grown as an annual. Has heads of large, often crinkled leaves, in combinations of red/green, white/pink, pink/green. Do not allow to flower. H and S 30–45cm (12–18in).

Coleus blumei, syn. Solenostemon scutellarioides
Fast-growing, bushy perennial, grown as an annual. Leaves are a mixture of colours, including pink, red, green or yellow. Flower spikes should be removed. H to 45cm (18in), S 30cm (12in) or more. Min. 10°C (50°F).

Malope trifida
Moderately fast-growing, erect, branching annual with rounded, lobed leaves. Flared, trumpet-shaped, reddish-purple flowers, to 8cm (3in) wide and with deep pink veins, are carried in summer and early autumn. H 90cm (3ft), S 30cm (1ft).

Matthiola, Ten-week Series [dwarf]
Fast-growing, erect, bushy biennial, grown as an annual. Lance-shaped leaves are greyish-green; long spikes of highly scented, single to double flowers, in shades of pink, red or white, appear in summer. H and S 30cm (12in).

Antirrhinum majus
Arctotis x hybrida 'China Rose'
Callistephus chinensis and cvs
Celosia argentea var. cristata

Cleome hassleriana 'Rose Queen'
Dianthus chinensis
Digitalis purpurea
Helichrysum bracteatum cvs

Papaver somniferum
PELARGONIUMS, pp.210–11
Petunia cvs
Phlox drummondii cvs

Portulaca grandiflora
Schizanthus cvs
Senecio x hybridus cvs
Zinnia cvs

◻ PINK

Petunia, **Resisto Series**
[rose-pink]
Moderately fast-growing, branching, bushy perennial, grown as an annual. Has oval leaves and rain-resistant, flared, trumpet-shaped, rose-pink flowers in summer-autumn.
H 15–30cm (6–12in), S 30cm (12in).

☼ ◊ ❄

Silene coeli-rosa **'Rose Angel'**
Moderately fast-growing, slim, erect annual. Has lance-shaped, greyish-green leaves and 5-petalled, deep rose-pink flowers in summer. H 30cm (12in), S 15cm (6in).

☼ ◊ ❄❄❄

Clarkia **'Arianna'**
Fast-growing annual with slender, erect, branching stems. Has lance-shaped leaves and spikes of frilled, waved, rosette-like, semi-double, deep rose-pink flowers in summer and early autumn. H 45cm (18in), S 30cm (12in).

☼ ◊ ❄❄❄

Dorotheanthus bellidiformis, **Magic Carpet Series**
Carpeting annual with succulent, lance-shaped, pale green leaves. Daisy-like flower heads, in bright shades of red, pink, yellow or white, open only in summer sunshine.
H 15cm (6in), S 30cm (12in).

☼ ◊ ❄

Antirrhinum majus, **Princess Series** [white, with purple eye]
Erect perennial, grown as an annual, branching from the base. Has lance-shaped leaves and spikes of tubular, 2-lipped, white and pinkish-purple flowers borne from spring to autumn. H and S 45cm (18in).

☼ ◊ ❄

Lunaria annua, syn. *L. biennis* (Honesty)
Fast-growing, erect biennial with pointed-oval, serrated leaves. Heads of scented, 4-petalled, white to deep purple flowers in spring and early summer are followed by rounded, silvery seed pods. H 75cm (30in), S 30cm (12in).

◐ ◊ ❄❄❄

Xeranthemum annuum [double]
Erect annual with lance-shaped, silvery leaves and branching heads of daisy-like, papery, double flower heads in shades of pink, mauve, purple or white, in summer. Produces good dried flowers. H 60cm (24in), S 45cm (18in).

☼ ◊ ❄

Cosmos **'Sensation'**
Moderately fast-growing, bushy, erect annual. Has feathery, mid-green leaves and daisy-like flower heads, to 10cm (4in) wide, in shades of red, pink or white, from early summer to early autumn. H 90cm (3ft), S 60cm (2ft).

☼ ◊ ❄❄❄

Impatiens, **Rosette Series** (Busy lizzie)
Fast-growing, evergreen, bushy perennial, grown as an annual. Has small, flat, spurred, double or semi-double flowers, in shades of red, pink, orange or white, from spring to autumn. H and S 30cm (12in).

◐ ◊ ❄

277

Annuals and Biennials
PINK–RED

Impatiens, Super Elfin Series 'Lipstick' (Busy lizzie)
Fast-growing, evergreen, bushy perennial, grown as an annual. Has pointed-oval, fresh green leaves and small, flat, spurred, rose-red flowers from spring to autumn. H and S 20cm (8in).

Verbena x hybrida 'Showtime'
Fairly slow-growing, bushy perennial, grown as an annual. Has lance-shaped, serrated, mid- to deep green leaves and clusters of small, tubular flowers, in a range of colours, in summer-autumn. H 20cm (8in), S 30cm (12in).

Petunia, Star Series [crimson]
Moderately fast-growing, branching, bushy perennial, grown as an annual. Has oval, mid- to deep green leaves and large, flared, trumpet-shaped, white-striped, deep red flowers in summer-autumn. H 15–30cm (6–12in), S 30cm (12in).

Phlox drummondii, Cecily Series
Moderately fast-growing, slim, erect annual with lance-shaped, pale green leaves. Has heads of large, flat flowers, in red, pink, blue, purple or white, all with contrasting centres, in summer and early autumn. H 15cm (6in), S 10cm (4in).

Portulaca grandiflora, Sundance Series
Slow-growing, semi-trailing annual with lance-shaped, succulent, bright green leaves. Cup-shaped flowers, with conspicuous stamens, appear in a mixture of colours in summer and early autumn. H to 20cm (8in), S 15cm (6in).

Amaranthus caudatus
(Love-lies-bleeding, Tassel flower)
Bushy annual with oval, pale green leaves. Pendulous panicles of tassel-like, red flowers, 45cm (18in) long, are carried in summer-autumn. H to 1.2m (4ft), S 45cm (1½ft).

Phlox drummondii, Twinkle Series
Moderately fast-growing, slim, erect annual. Lance-shaped leaves are pale green; heads of star-shaped flowers in a bright mixture of colours, some with contrasting centres, are carried in summer. H 15cm (6in), S 10cm (4in).

Nicotiana x sanderae, Sensation Series
Fairly slow-growing, erect, branching annual with pointed-oval leaves. Scented, trumpet-shaped flowers, 8cm (3in) long, appear in a wide range of colours in summer and early autumn. H 75–90cm (2½–3ft), S 30cm (1ft).

Petunia 'Mirage Velvet'
Branching, bushy perennial, grown as an annual, with oval, dark green leaves. Large, flared, trumpet-shaped, rich red flowers, with almost black centres, appear in summer-autumn. H 25cm (10in), S 30cm (12in).

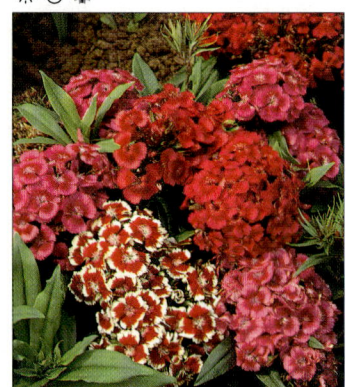

Dianthus barbatus, Roundabout Series [dwarf]
Slow-growing, upright, bushy biennial with lance-shaped leaves. In early summer has flat heads of zoned and eyed flowers in shades of pink, red or white. H 15cm (6in), S 20–30cm (8–12in).

Alcea rosea 'Chater's Double'
Erect biennial with lobed, rough-textured leaves. Spikes of rosette-like, double flowers in several different colours are carried on upright stems in summer and early autumn. H 1.8–2.4m (6–8ft), S to 60cm (2ft).

Bellis perennis cvs
Celosia argentea var. cristata
Myosotis 'Carmine King'
PELARGONIUMS, pp.210–11
Senecio x hybridus cvs
Zinnia cvs

■ RED

Viola 'Roggli Giants'
Fairly fast-growing, bushy perennial, grown as an annual or biennial, with oval, serrated leaves. In summer-autumn has rounded, 5-petalled flowers, 10cm (4in) across, in a wide range of colours. H and S to 20cm (8in).

Impatiens, Novette Series 'Red Star' (Busy lizzie)
Fast-growing, evergreen, bushy perennial, grown as an annual. Has pointed-oval, fresh green leaves and from spring to autumn small, flat, spurred, red flowers with star-shaped, white markings. H and S 15cm (6in).

Impatiens, Confection Series (Busy lizzie)
Fast-growing, evergreen, bushy perennial, grown as an annual. Has fresh green leaves and small, flat, spurred, double or semi-double flowers, in shades of red or pink, from spring to autumn. H and S 20–30cm (8–12in).

Dianthus chinensis 'Fire Carpet'
Slow-growing, bushy annual or biennial, grown as an annual. Lance-shaped leaves are light or mid-green. Small, rounded, single, bright red flowers are carried in summer and early autumn. H 20cm (8in), S 15–30cm (6–12in).

Linum grandiflorum 'Rubrum'
Fairly fast-growing, slim, erect annual. Lance-shaped leaves are grey-green; small, rounded, flattish, deep red flowers are carried in summer. H 45cm (18in), S 15cm (6in).

Dianthus barbatus, Monarch Series [auricula-eyed]
Slow-growing, upright, bushy biennial with lance-shaped leaves. Bicoloured flowers are carried in flat heads, to 12cm (5in) wide, in early summer. H 45cm (18in), S 20–30cm (8–12in).

Bellis perennis 'Pomponette'
Slow-growing, carpeting perennial, grown as a biennial. Has oval leaves and small, daisy-like, double flower heads, in red, pink or white, in spring. H and S 10–15cm (4–6in).

Impatiens, Novette Series [red] (Busy lizzie)
Fast-growing, evergreen, bushy perennial, grown as an annual. Pointed-oval leaves are fresh green; small, flat, spurred, red flowers are carried from spring to autumn. H and S 15cm (6in).

Phlox drummondii, Twinkle Series
Moderately fast-growing, slim, erect annual with lance-shaped, pale green leaves. Heads of star-shaped flowers, some with contrasting centres, appear in summer in a bright mixture of colours. H 15cm (6in), S 10cm (4in).

279

Annuals and Biennials
■ RED

Verbena × hybrida 'Defiance'
Fairly slow-growing, bushy perennial, grown as an annual. Has lance-shaped, serrated, mid- to deep green leaves and clusters of small, fragrant, tubular, white-eyed, deep red flowers in summer-autumn. H 20cm (8in), S 30cm (12in).

Salvia splendens 'Flare Path'
Slow-growing, bushy perennial, grown as an annual. Has oval, serrated, fresh green leaves and dense racemes of tubular, pure scarlet flowers in summer and early autumn. H to 30cm (12in), S 20–30cm (8–12in).

Nemesia strumosa, Carnival Series
Fairly fast-growing, bushy annual with serrated, pale green leaves. In summer has small, somewhat trumpet-shaped flowers in a range of colours, including yellow, red, orange, purple and white. H 20–30cm (8–12in), S 15cm (6in).

Petunia, Picotee Series [red]
Fairly fast-growing, branching, bushy perennial, grown as an annual, with oval leaves. Has flared, somewhat trumpet-shaped, red flowers, edged with white, in summer-autumn. H 15–30cm (6–12in), S 30cm (12in).

Pelargonium, Diamond Series [scarlet]
Slow-growing, evergreen, branching, compact perennial, grown as an annual, with rounded, lobed leaves. Large, rounded heads of weather-resistant, red flowers are carried in summer-autumn. H and S 30–60cm (12–24in).

Petunia, Resisto Series
Fairly fast-growing, bushy perennial, grown as an annual, with oval leaves. Has rain-resistant, flared, somewhat trumpet-shaped flowers in a range of bright colours in summer-autumn. H 15–30cm (6–12in), S 30cm (12in).

Zinnia, Ruffles Series [Erecta group; scarlet]
Moderately fast-growing, sturdy, upright annual with oval to lance-shaped leaves. Ruffled, daisy-like, double, scarlet flower heads are carried in summer and early autumn. H 60cm (24in), S 30cm (12in).

Papaver rhoeas, Shirley Series [single]
Fast-growing, slender, erect annual with lobed, light green leaves. Rounded, often cup-shaped, single flowers, in shades of red, pink, salmon or white, appear in summer. H 60cm (24in), S 30cm (12in).

Primula, Pacific Series [dwarf]
Rosette-forming perennial, normally grown as a biennial, with lance-shaped leaves. Has heads of large, fragrant, flat flowers in shades of blue, yellow, red, pink or white in spring. H 10–15cm (4–6in), S 20cm (8in).

Antirrhinum majus
Antirrhinum majus 'His Excellency'
Calceolaria 'Anytime'
Calceolaria, Monarch Series
Capsicum annuum 'Holiday Cheer'
Chrysanthemum carinatum, Tricolor Series
Coleus blumei, p.276
Cosmos, Bright Lights Series
Cosmos 'Sunny Red'
Erysimum cheiri
Gaillardia pulchella [double, mixed]
Helichrysum bracteatum cvs
Impatiens, Duet Series
Impatiens, Tom Thumb Series
Impatiens walleriana
Ipomoea coccinea

■ RED

Alonsoa warscewiczii
(Mask flower)
Perennial, grown as an annual, with slender, branching, red stems carrying oval, toothed, deep green leaves. Spurred, bright scarlet flowers are produced during summer-autumn.
H 30–60cm (12–24in), S 30cm (12in).

Zinnia, **Ruffles Series**
[Erecta group]
Moderately fast-growing, sturdy, upright annual with pale or mid-green leaves. Ruffled, pompon, double flower heads in a mixture of colours are carried in summer and early autumn.
H 60cm (24in), S 30cm (12in).

Capsicum annuum 'Holiday Time'
Moderately fast-growing, evergreen, bushy perennial, grown as an annual, with oval, mid-green leaves. In autumn-winter, cone-shaped fruits turn from yellow to red. H and S 20–30cm (8–12in). Min. 4°C (39°F).

Viola, **Floral Dance Series**
Fairly fast-growing, bushy perennial, grown as an annual or biennial. Has oval, mid-green leaves and rounded, 5-petalled flowers in a wide range of colours in winter. H 15–20cm (6–8in), S 20cm (8in).

Dahlia, **Coltness Hybrids**
Well-branched, erect, bushy, tuberous perennial, grown as an annual. Has deeply lobed leaves and daisy-like, single flower heads in many colours throughout summer until autumn frosts.
H and S 45cm (18in).

Salpiglossis sinuata 'Splash'
Moderately fast-growing, erect, branching annual. Has lance-shaped, pale green leaves and widely flared, trumpet-shaped, multicoloured flowers in summer and early autumn. H 60cm (24in), S 30cm (12in).

Tagetes patula 'Cinnabar'
Fast-growing, bushy annual with aromatic, very feathery, deep green leaves. Heads of rounded, daisy-like, single, rich rust-red flowers, yellow-red beneath, are carried in summer and early autumn. H and S 30cm (12in).

Chrysanthemum carinatum 'Monarch Court Jesters'
Fast-growing, erect, branching annual. In summer has feathery, grey-green leaves and daisy-like, zoned flower heads, to 8cm (3in) wide, in various colour combinations. H 60cm (24in), S 30cm (12in).

Coleus blumei 'Brightness', syn. *Solenostemon scutellarioides* 'Brightness'
Fast-growing, bushy perennial, grown as an annual. Has rust-red leaves, edged with green. Flower spikes should be removed. H to 45cm (18in), S 30cm (12in) or more. Min 10°C (50°F).

Ipomoea x *multifida*
Ipomoea nil 'Scarlett O'Hara'
Ipomoea quamoclit, p.170
Ipomopsis aggregata

Mimulus, Malibu Series
Papaver commutatum 'Lady Bird'
Petunia cvs
Portulaca grandiflora

Salpiglossis cvs
Salvia splendens
Solanum capsicastrum
Solanum pseudocapsicum 'Fancy'

Tagetes patula
Tropaeolum majus cvs
Verbena x *hybrida* 'Mme du Barry'

281

Annuals and Biennials

■ RED ■ PURPLE

Amaranthus hypochondriacus, syn. *A. hybridus* var. *erythrostachys* (Prince's feather)
Bushy annual with upright, sometimes flattened panicles, 15cm (6in) or more long, of dark red flowers in summer-autumn. Leaves are heavily suffused purple. H to 1.2m (4ft), S 45cm (1½ft).

Trachelium caeruleum (Throatwort)
Moderately fast-growing, erect perennial, grown as an annual. Has oval, serrated leaves and clustered heads of small, tubular, lilac-blue or white flowers in summer. H 60–90cm (2–3ft), S 30cm (1ft).

Salvia sclarea var. ***turkestanica***
Moderately fast-growing, erect biennial, grown as an annual. Has aromatic, oval, hairy leaves and panicles of tubular, white and lavender-purple flowers with prominent, lavender-purple bracts in summer. H 75cm (30in), S 30cm (12in).

***Linaria maroccana* 'Fairy Lights'**
Fast-growing, erect, bushy annual with lance-shaped, pale green leaves. Tiny, snapdragon-like flowers, in shades of red, pink, purple, yellow or white, are borne in summer. H 20cm (8in), S 15cm (6in).

***Ricinus communis* 'Impala'**
Fast-growing, evergreen, erect shrub, usually grown as an annual. Has deeply lobed, bronze leaves to 30cm (12in) wide and clusters of small, red flowers in summer, followed by globular, prickly, red seed heads. H 1.5m (5ft), S 90cm (3ft).

Calomeria amaranthoides, syn. *Humea elegans* (Incense plant)
Erect, branching biennial with a strong fragrance of incense. Has lance-shaped leaves and heads of tiny, pink, brownish-red or crimson flowers in summer-autumn. H to 1.8m (6ft), S 90cm (3ft). Min. 4°C (39°F).

Psylliostachys suworowii, syn. *Statice suworowii* (Statice)
Fairly slow-growing, erect, branching annual with lance-shaped leaves. Bears branching spikes of small, tubular, pink to purple flowers in summer and early autumn. Flowers are good for drying. H 45cm (18in), S 30cm (12in).

Schizanthus pinnatus
Moderately fast-growing, upright, bushy annual with feathery, light green leaves. In summer-autumn has rounded, lobed, multicoloured flowers in shades of pink, purple, white or yellow. H 30cm–1.2m (1–4ft), S 30cm (1ft). Min. 5°C (41°F).

Amaranthus tricolor 'Joseph's Coat'
Atriplex hortensis 'Rubra'
Nicotiana x *sanderae* 'Crimson Rock'
Scabiosa atropurpurea

Antirrhinum majus
Centaurea cyanus
Consolida ambigua, Hyacinth-flowered Series
Iberis umbellata

Impatiens, Tom Thumb Series
Ionopsidium acaule
Lobelia erinus 'Colour Cascade'
Matthiola cvs

Nemesia strumosa
Proboscidea fragrans
Xeranthemum annuum

■ PURPLE

Exacum affine (Persian violet)
Evergreen, bushy biennial, usually grown as an annual. Has oval, glossy leaves and masses of tiny, scented, saucer-shaped, purple flowers, with yellow stamens, in summer and early autumn. H and S 20–30cm (8–12in). Min. 7–10°C (45–50°F).

Campanula medium 'Bells of Holland'
Slow-growing, evergreen, clump-forming, erect biennial with lance-shaped, toothed leaves. In spring and early summer has bell-shaped flowers in a mixture of blue, lilac, pink or white. H to 60cm (24in), S 30cm (12in).

Salvia viridis, syn. *S. horminum*
Moderately fast-growing, upright, branching annual with oval leaves. Tubular, lipped flowers, enclosed by purple, pink or white bracts, are carried in spikes at tops of stems in summer and early autumn. H 45cm (18in), S 20cm (8in).

Gomphrena globosa (Globe amaranth)
Moderately fast-growing, upright, bushy annual with oval, hairy leaves. Has oval, clover-like flower heads in pink, yellow, orange, purple or white in summer and early autumn. H 30cm (12in), S 20cm (8in).

Collinsia grandiflora
Moderately fast-growing, slender-stemmed annual. Upper leaves are lance-shaped; lower are oval. Whorls of pale purple flowers, with purplish-blue lips, are carried in spring-summer. H and S 15–30cm (6–12in).

Echium vulgare [dwarf]
Moderately fast-growing, erect, bushy annual or biennial with lance-shaped, dark green leaves. Spikes of tubular flowers, in shades of white, pink, blue or purple, appear in summer. H 30cm (12in), S 20cm (8in).

Orychophragmus violaceus
Moderately fast-growing, upright annual or biennial with branching flower stems and pointed-oval, pale green leaves. Heads of 4-petalled, purple-blue flowers are carried in spring. H 30–60cm (12–24in), S 30cm (12in).

Petunia 'Blue Frost'
Fairly fast-growing, branching, bushy perennial, grown as an annual. Has oval leaves and in summer-autumn produces flared, somewhat trumpet-shaped, violet-blue flowers, edged with white. H 15–30cm (6–12in), S 30cm (12in).

Scabiosa atropurpurea, Cockade Series
Moderately fast-growing, erect, bushy annual with lobed leaves and large, rounded, double flower heads, in shades of red, pink, purple or blue, on slender stems in summer and early autumn. H 90cm (3ft), S 20–30cm (8–12in).

Nierembergia hippomanica var. **violacea 'Purple Robe'**
Moderately fast-growing, rounded, branching perennial, grown as an annual, with narrow, lance-shaped leaves. Has cup-shaped, dark bluish-purple flowers in summer and early autumn. H and S 15–20cm (6–8in).

Callistephus chinensis, Thousand Wonders Series
Moderately fast-growing, erect, bushy annual with oval, toothed leaves. In summer and early autumn has daisy-like, double flower heads in pink, blue, purple, red or white. H to 20cm (8in), S 30cm (12in).

Salvia splendens, Cleopatra Series [violet]
Slow-growing, bushy perennial, grown as an annual. Oval, serrated leaves are dark green; dense racemes of tubular, deep violet-purple flowers are carried in summer and early autumn. H to 30cm (12in), S 20–30cm (8–12in).

Amaranthus tricolor 'Molten Fire'
Arctotis x *hybrida* 'Bacchus'
Papaver somniferum
Perilla frutescens

Petunia cvs
Phlox drummondii cvs
Schizanthus cvs
Senecio elegans

Viola x *wittrockiana* cvs
Zinnia cvs

Annuals and Biennials
PURPLE–BLUE

Senecio 'Spring Glory' (Cineraria)
Slow-growing, evergreen, mound- or dome-shaped perennial, grown as a biennial, with large, oval, serrated, mid- to deep green leaves. Heads of large, daisy-like flowers in a mixture of colours are carried in spring. H 20cm (8in), S 30cm (12in). Min. 5°C (41°F).

Callistephus chinensis, Milady Series [blue]
Moderately fast-growing, erect, bushy annual with oval, toothed leaves. Has large, daisy-like, double, purplish-blue flower heads in summer and early autumn. H 25–30cm (10–12in), S 30–45cm (12–18in).

Petunia, Resisto Series [blue]
Moderately fast-growing, branching, bushy perennial, grown as an annual. Has oval leaves and rain-resistant, flared, trumpet-shaped, intense blue flowers in summer-autumn.
H 15–30cm (6–12in), S 30cm (12in).

Lobelia erinus 'Sapphire'
Slow-growing, pendulous, spreading annual or occasionally perennial. Oval to lance-shaped leaves are pale green; small, sapphire-blue flowers with white centres are produced continuously in summer and early autumn. H 20cm (8in), S 15cm (6in).

Convolvulus tricolor 'Blue Flash'
Moderately fast-growing, upright, bushy annual with oval to lance-shaped leaves. Has small, saucer-shaped, intense blue flowers with cream and yellow centres in summer. H 20–30cm (8–12in), S 20cm (8in).

Viola, Joker Series
Bushy, spreading perennial, usually grown as an annual or biennial. Large, rounded, 5-petalled, purplish-blue flowers, with black and white 'faces' and yellow eyes, appear in summer. H and S 15cm (6in).

Consolida ambigua, Imperial Series (Giant larkspur)
Fast-growing, upright, branching annual with feathery leaves. Long spikes of rounded, spurred, double flowers, pink, blue or white, are carried in summer. H 1.2 m (4ft), S 30cm (1ft).

Salvia farinacea 'Victoria'
Moderately fast-growing perennial, grown as an annual, with many erect stems. Has oval or lance-shaped leaves and spikes of tubular, violet-blue flowers in summer. H 45cm (18in), S 30cm (12in).

Primula, Super Giants Series [blue]
Rosette-forming perennial, usually grown as a biennial, with lance-shaped leaves. Heads of large, fragrant, flat, blue flowers appear in spring. H and S to 30cm (12in).

Browallia speciosa, p.223
Ipomoea purpurea
Salvia farinacea 'Blue Bedder'

■ BLUE

Torenia fournieri
(Wishbone flower)
Moderately fast-growing, erect, branching annual with serrated, light green leaves. Dark blue-purple flowers, paler and yellow within, are carried in summer and early autumn. H 30cm (12in), S 20cm (8in). Min. 5°C (41°F).

Gilia capitata
Erect, branching annual. Has very feathery, mid-green leaves and tiny, dense, rounded heads of soft lavender-blue flowers in summer and early autumn. Is good for cut flowers. H 45cm (18in), S 20cm (8in).

Ageratum houstonianum 'Blue Mink'
Moderately fast-growing, hummock-forming annual. Has pointed-oval leaves and clusters of feathery, brush-like, pastel blue flower heads in summer-autumn. Makes a useful edging plant. H and S 20–30cm (8–12in).

Nigella damascena 'Miss Jekyll'
Fast-growing, slender, erect annual. Feathery leaves are bright green; small, rounded, many-petalled, semi-double, blue flowers are carried in summer, followed by inflated seed pods which can be cut and dried. H 45cm (18in), S 20cm (8in).

Nigella damascena 'Persian Jewels'
Fast-growing, slender, erect annual with feathery, bright green leaves. Small, semi-double flowers, in shades of blue, pink or white, appear in summer, followed by inflated seed pods which can be cut and dried. H 45cm (18in), S 20cm (8in).

Ageratum houstonianum 'Blue Danube'
Moderately fast-growing, hummock-forming annual with pointed-oval leaves. Has clusters of feathery, brush-like, lavender-blue flower heads in summer-autumn. Makes a useful edging plant. H and S 15cm (6in).

Nemophila menziesii, syn. *N. insignis* (Baby blue-eyes)
Fast-growing, spreading annual with serrated, grey-green leaves. Small, bowl-shaped, blue flowers with white centres are carried in summer. H 20cm (8in), S 15cm (6in).

Sedum caeruleum
Moderately fast-growing annual with branching flower stems. Oval, light green leaves become red-tinged when clusters of small, star-shaped, light blue flowers with white centres are borne in summer. H and S 10–15cm (4–6in).

Viola 'Azure Blue'
Bushy perennial, grown as an annual or biennial. Has oval, mid-green leaves and rounded, 5-petalled, azure-blue flowers in spring. H 15–20cm (6–8in), S 20cm (8in).

Callistephus chinensis
Consolida ambigua, Hyacinth-flowered Series
Lobelia erinus 'Blue Cascade'
Nolana paradoxa
Petunia cvs
Phacelia tanacetifolia
Phlox drummondii cvs
Salvia viridis cvs
Trachymene coerulea
Viola x *wittrockiana* cvs

Annuals and Biennials

■ BLUE　　■ BLUE–GREEN

Centaurea cyanus [tall, blue] (Cornflower)
Fast-growing, erect, branching annual. Has lance-shaped, grey-green leaves and branching heads of daisy-like, blue flowers in summer and early autumn. H to 90cm (3ft), S 30cm (1ft).

Myosotis 'Blue Ball'
Slow-growing, bushy, compact perennial, often grown as a biennial. Has lance-shaped leaves and, in spring and early summer, spikes of tiny, 5-lobed, deep blue flowers. H to 20cm (8in), S 15cm (6in).

Phacelia campanularia (California bluebell)
Moderately fast-growing, branching, bushy annual with oval, serrated, deep green leaves. Bell-shaped, pure blue flowers, 2.5cm (1in) wide, are carried in summer and early autumn. H 20cm (8in), S 15cm (6in).

Lobelia erinus 'Crystal Palace'
Slow-growing, spreading, compact, bushy annual or occasionally perennial. Bronzed leaves are oval to lance-shaped; small, deep blue flowers are produced continuously in summer and early autumn. H 10–20cm (4–8in), S 10–15cm (4–6in).

Brachycome iberidifolia (Swan River daisy)
Moderately fast-growing, thin-stemmed, bushy annual with deeply cut leaves. Has small, fragrant, daisy-like flower heads, usually blue but also pink, mauve, purple or white, in summer and early autumn. H and S to 45cm (18in).

Cynoglossum amabile 'Firmament'
Slow-growing, upright, bushy annual or biennial with lance-shaped, hairy, grey-green leaves. Pendulous, tubular, pure sky-blue flowers are carried in summer. H 45cm (18in), S 30cm (12in).

Anchusa capensis 'Blue Angel'
Bushy biennial, grown as an annual. Has lance-shaped, bristly leaves. Heads of shallowly bowl-shaped, brilliant blue flowers are borne in summer. H and S 20cm (8in).

Senecio cineraria 'Silver Dust', syn. *S. maritima* 'Silver Dust'
Moderately fast-growing, evergreen, bushy sub-shrub, usually grown as an annual, with deeply lobed, silver leaves. Small, daisy-like, yellow flower heads appear in summer but are best removed. H and S 30cm (12in).

Felicia bergeriana (Kingfisher daisy)
Fairly fast-growing, mat-forming annual. Has lance-shaped, hairy, grey-green leaves. Small, daisy-like, blue flower heads with yellow centres open only in sunshine in summer and early autumn. H and S 15cm (6in).

Commelina coelestis (Day flower)
Fairly fast-growing, upright perennial, usually grown as an annual, with lance-shaped, mid-green leaves. Small, 3-petalled, bright pure blue flowers, each lasting only a day, are produced in succession in summer. H to 45cm (18in), S 30cm (12in).

Borago officinalis (Borage)
Spreading, clump-forming, annual herb. Has oval, crinkled, rough-haired leaves and sprays of star-shaped, blue flowers in summer and early autumn. Young leaves are sometimes used as a coolant in drinks. Self seeds prolifically. H 90cm (3ft), S 30cm (1ft).

Nicotiana langsdorfii
Fairly slow-growing, erect, branching perennial, grown as an annual, with oval to lance-shaped leaves. Slightly pendent, bell-shaped, pale green to yellow-green flowers appear in summer. H 1–1.5m (3–5ft), S 30cm (1ft).

Anchusa capensis 'Blue Bird'
Browallia americana
Campanula medium
Centaurea cyanus

Convolvulus tricolor 'Flying Saucers'
Convolvulus tricolor 'Minor'
Eustoma grandiflorum, F1 Hybrids [mixed]
Gilia achilleifolia

Lobelia erinus 'Cambridge Blue'
Lobelia erinus 'Colour Cascade'
Nigella damascena
Verbena x hybrida 'Amethyst'

Senecio cineraria

286

GREEN–YELLOW

Moluccella laevis
(Bells of Ireland, Shell flower)
Fairly fast-growing, erect, branching annual. Rounded leaves are pale green; spikes of small, tubular, white flowers, each surrounded by a conspicuous, pale green calyx, appear in summer.
H 60cm (24in), S 20cm (8in).

☼ ◊ ❋

Ricinus communis
(Castor-oil plant)
Fast-growing, evergreen, erect shrub, usually grown as an annual. Has large, deeply lobed, mid-green leaves and heads of green and red flowers in summer, followed by globular, prickly seed pods. H 1.5m (5ft), S 90cm (3ft).

☼ ◊ ❋

***Zinnia* 'Envy'** [Elegans group]
Moderately fast-growing, sturdy, erect annual. Has oval to lance-shaped, pale or mid-green leaves and large, daisy-like, double, green flower heads in summer and early autumn. H 60cm (24in), S 30cm (12in).

☼ ◊ ❋

Centaurea moschata, syn.
Amberboa moschata (Sweet sultan)
Fast-growing, upright, slender-stemmed annual with lance-shaped, greyish-green leaves. Large, fragrant, cornflower-like flower heads, in a range of colours, are carried in summer and early autumn.
H 45cm (18in), S 20cm (8in).

☼ ◊ ❋❋❋

YELLOW

Smyrnium perfoliatum
Slow-growing, upright biennial. Upper leaves, rounded and yellow-green, encircle stems which bear heads of yellowish-green flowers in summer.
H 60cm–1m (2–3ft), S 60cm (2ft).

☼ ◊ ❋❋❋

Kochia scoparia* f. *trichophylla,
syn. ***Bassia scoparia* f. *trichophylla***
(Burning bush, Summer cypress)
Moderately fast-growing, erect, very bushy annual. Narrow, lance-shaped, light green leaves, 5–8cm (2–3in) long, turn red in autumn. Has insignificant flowers. H 90cm (3ft), S 60cm (2ft).

☼ ◊ ❋ ⚘

Platystemon californicus
(Cream cups)
Moderately fast-growing, upright, compact annual with lance-shaped, greyish-green leaves. Saucer-shaped, cream or pale yellow flowers, about 2.5cm (1in) across, appear in summer.
H 30cm (12in), S 10cm (4in).

☼ ◊ ❋❋❋

Argemone mexicana
(Devil's fig, Prickly poppy)
Spreading perennial, grown as an annual, with leaves divided into white-marked, greyish-green leaflets. In summer has fragrant, poppy-like, yellow or orange flowers, 8cm (3in) wide. H to 60cm (24in), S 30cm (12in).

☼ ◊ ❋

Glaucium flavum (Horned poppy)
Slow-growing, erect biennial with oval, lobed, light greyish-green leaves. Poppy-like, vivid yellow flowers, 8cm (3in) wide, appear in summer and early autumn. H 30–60cm (12–24in), S 45cm (18in).

☼ ◊ ❋❋❋

Angelica archangelica, p.192
Coleus blumei, p.276
Nicandra physalodes
Nicotiana × *sanderae* 'Lime Green'

Tanacetum parthenium 'Aureum'

Calendula officinalis 'Art Shades'
Calendula officinalis, Fiesta Series
Chrysanthemum coronarium
Tagetes 'Lemon Gem'

Zea mays 'Gigantea Quadricolor'

Annuals and Biennials
YELLOW

Calendula officinalis 'Kablouna'
Fast-growing, bushy annual with strongly aromatic, lance-shaped leaves. From spring to autumn has crested, daisy-like, orange, gold or yellow flower heads. H 60cm (24in), S 30–60cm (12–24in).

Mentzelia lindleyi, syn. *Bartonia aurea*
Fairly fast-growing, bushy annual with fleshy stems and lance-shaped, serrated leaves. Has fragrant, cup-shaped, deep yellow flowers, with conspicuous stamens, in summer. H 45cm (18in), S 20cm (8in).

Viola 'Super Chalon Giants'
Fairly fast-growing, bushy perennial, grown as an annual or biennial. Has oval, serrated leaves and, in summer-autumn, 5-petalled, ruffled and waved, bicoloured flowers. H 15–20cm (6–8in), S 20cm (8in).

Limnanthes douglasii (Meadow foam, Poached-egg flower)
Fast-growing, slender, erect annual. Feathery leaves are glossy, light green; slightly fragrant, cup-shaped, white flowers with yellow centres are carried from early to late summer. H 15cm (6in), S 10cm (4in).

Eschscholzia caespitosa
Fast-growing, slender, erect annual with feathery, bluish-green leaves. Cup-shaped, 4-petalled, yellow flowers, 2.5cm (1in) wide, appear in summer and early autumn. H and S 15cm (6in).

Antirrhinum majus, Wedding Bells Series
Erect perennial, branching from the base, grown as an annual. Has lance-shaped leaves and, from late spring to autumn, spikes of small, trumpet-shaped flowers in a range of colours. H 75cm (30in), S 45cm (18in).

Sanvitalia procumbens (Creeping zinnia)
Moderately fast-growing, prostrate annual with pointed-oval leaves. Daisy-like, yellow flower heads, 2.5cm (1in) wide, with black centres, are borne in summer. H 15cm (6in), S 30cm (12in).

Lindheimera texana (Star daisy)
Moderately fast-growing, erect, branching annual with hairy stems and oval, serrated, hairy leaves. Daisy-like, yellow flower heads appear in late summer and early autumn. H 30–60cm (12–24in), S 30cm (12in).

Tagetes 'Gold Coins' [Erecta group]
Fast-growing, erect, bushy annual. Has aromatic, feathery, glossy, deep green leaves and large, daisy-like, double flower heads in shades of yellow and orange in summer and early autumn. H 90cm (3ft), S 30–45cm (1–1½ft).

Helianthus annuus 'Taiyo' [intermediate]
Fast-growing, erect annual with large, rounded-oval, serrated leaves. Large, daisy-like, yellow flower heads with black centres appear in summer. H to 1.2m (4ft), S 30–45cm (1–1½ft).

Coreopsis tinctoria (Tick-seed)
Fast-growing, erect, bushy annual with lance-shaped leaves. Large, daisy-like, bright yellow flower heads with red centres are carried in summer and early autumn. H 60–90cm (2–3ft), S 20cm (8in).

Amaranthus tricolor 'Joseph's Coat'
Antirrhinum majus
Arctotis x *hybrida* 'Sunshine'
Calceolaria 'Anytime'
Calceolaria, Monarch Series
Celosia argentea var. *cristata*
Chrysanthemum carinatum, Tricolor Series
Coleus blumei, p.276
Cosmos, Bright Lights Series
Cosmos 'Sunny Gold'
Erysimum cheiri
Gaillardia pulchella [double, mixed]
Gazania 'Daybreak'
Helianthus annuus 'Russian Giant'
Helianthus annuus 'Teddy Bear'
Helichrysum bracteatum cvs

YELLOW

Viola, Clear Crystals Series [yellow]
Bushy perennial, grown as an annual or biennial. Has oval, mid-green leaves and rounded, 5-petalled, yellow flowers in summer. H 15–20cm (6–8in), S 20cm (8in).

Chrysanthemum segetum
Moderately fast-growing, erect annual with lance-shaped, grey-green leaves. Daisy-like, single flower heads, to 8cm (3in) wide, in shades of yellow, are carried in summer and early autumn. Is excellent for cut flowers. H 45cm (18in), S 30cm (12in).

Coreopsis 'Sunray'
Spreading, clump-forming perennial, grown as an annual by sowing under glass in early spring. Has lance-shaped, serrated leaves and daisy-like, double, bright yellow flower heads in summer. H 45cm (18in), S 30–45cm (12–18in).

Helianthus annuus
Fast-growing, erect annual with oval, serrated, mid-green leaves. Daisy-like, yellow flower heads to 30cm (12in) or more wide, with brown or purplish centres, appear in summer. H 1–3m (3–10ft), S 30–45cm (1–1½ft).

Viola, Icequeen Series [yellow]
Bushy, spreading perennial, grown as an annual or biennial. Has oval, mid-green leaves and large, rounded, 5-petalled, yellow flowers with dark central blotches in winter and early spring. H 10–12cm (4–5in), S 12–15cm (5–6in).

Cladanthus arabicus
Moderately fast-growing, hummock-forming annual with aromatic, feathery, light green leaves. Has fragrant, daisy-like, single, deep yellow flower heads, 5cm (2in) wide, in summer and early autumn. H 60cm (24in), S 30cm (12in).

Viola, Crystal Bowl Series [yellow]
Bushy, spreading perennial, usually grown as an annual or biennial, with oval, mid-green leaves. Large, rounded, 5-petalled, yellow flowers are carried in summer. H and S 15cm (6in).

Antirrhinum majus 'Coronette'
Erect, bushy, compact perennial, grown as an annual, with lance-shaped leaves. Spikes of tubular, 2-lipped flowers in a wide range of colours appear from spring to autumn. H 60cm (24in), S 30cm (12in).

Helianthus, Chrysanthemum-flowered Series
Fast-growing, erect annual with large, oval, serrated, mid-green leaves. Globose, daisy-like, double, deep yellow flower heads, 15cm (6in) wide, are carried in summer. H 1.5m (5ft), S 30–45cm (1–1½ft).

Hunnemannia fumariifolia 'Sunlite'
Layia platyglossa
Mimulus, Malibu Series
Nemesia strumosa

Portulaca grandiflora
Salpiglossis cvs
Schizanthus cvs
Tolpis barbata

Tropaeolum majus cvs
Tuberaria guttata
Viola x *wittrockiana* cvs
Zinnia cvs

Annuals and Biennials
▢ YELLOW

Zinnia 'Belvedere Dwarfs'
[Elegans group]
Moderately fast-growing, sturdy, erect annual with oval to lance-shaped, pale or mid-green leaves. Large, daisy-like, double flower heads in a mixture of colours appear in summer and early autumn. H and S 30cm (12in).

☼ ◊ ✲

Ursinia anthemoides
Moderately fast-growing, bushy annual with feathery, pale green leaves. Small, daisy-like, purple-centred flower heads with orange-yellow rays, purple beneath, appear in summer and early autumn. H 30cm (12in), S 20cm (8in).

☼ ◊ ✲

Gaillardia pulchella 'Lollipops'
Moderately fast-growing, upright annual with lance-shaped, hairy, greyish-green leaves. Daisy-like, double, red-and-yellow flower heads, 5cm (2in) wide, are carried in summer. H and S 30cm (12in).

☼ ◊ ✲✲✲

Calceolaria, Bikini Series
Compact, bushy annual or biennial. Has oval, slightly hairy, mid-green leaves, and heads of small, rounded, pouched flowers in shades of yellow, orange or red in summer. H and S 20cm (8in). Min. 5°C (41°F).

☼ ◊

Tagetes 'Crackerjack'
[Erecta group]
Fast-growing, erect, bushy annual with aromatic, deeply cut, glossy, deep green leaves. Has large, double flower heads, in yellow and orange shades, in summer and early autumn. H 60cm (24in), S 30–45cm (12–18in).

☼ ◊ ✲

Tagetes patula 'Naughty Marietta'
Fast-growing, bushy annual with aromatic, deeply cut, deep green leaves. Heads of daisy-like, bicoloured flowers, deep yellow and maroon, are carried in summer and early autumn. H and S 30cm (12in).

☼ ◊ ✲

Eschscholzia californica
Fast-growing, slender, erect annual with feathery, bluish-green leaves. Cup-shaped, 4-petalled, vivid orange-yellow flowers are borne in summer-autumn. H 30cm (12in), S 15cm (6in).

☼ ◊ ✲✲✲ ♛

Viola 'Redwing'
Bushy, spreading perennial, usually grown as an annual or biennial. Large, rounded, 5-petalled, reddish-brown and yellow flowers appear in summer. H and S 15–20cm (6–8in).

☼ ◊ ✲✲✲

Eschscholzia californica [mixed]
Fast-growing, slender, erect annual. Feathery leaves are bluish-green; cup-shaped, 4-petalled, single flowers, in shades of red, orange, yellow or cream, are borne in summer-autumn. H 30cm (12in), S 15cm (6in).

☼ ◊ ✲✲✲

Rudbeckia hirta
Rudbeckia hirta 'Irish Eyes'
Tagetes 'Galore'
Tagetes, Inca Series

Tagetes, Jubilee Series
Tagetes patula

YELLOW–ORANGE

Calceolaria 'Sunshine'
Evergreen, compact, bushy perennial, grown as an annual. Has oval, mid-green leaves and heads of small, rounded, pouched, bright golden-yellow flowers in late spring and summer. H and S 20cm (8in).

Tropaeolum majus 'Alaska'
Fast-growing, bushy annual with rounded, variegated leaves. Spurred, trumpet-shaped flowers, in shades of red or yellow, appear in summer and early autumn. H and S 30cm (12in).

Primula, Pacific Series (Polyanthus)
Rosette-forming perennial, normally grown as a biennial, with lance-shaped leaves. Has heads of large, fragrant, flat flowers in shades of blue, yellow, red, pink or white in spring. H and S 20–22cm (8–9in).

Nemesia strumosa, Triumph Series
Fairly fast-growing, bushy, compact annual with lance-shaped, serrated, pale green leaves. Small, somewhat trumpet-shaped flowers, in a range of bright colours, are borne in summer. H 20cm (8in), S 15cm (6in).

Viola, Universal Series [apricot]
Bushy, spreading perennial, usually grown as a biennial. Large, rounded, 5-petalled, deep apricot flowers are borne in winter-spring. H and S 15–20cm (6–8in).

ORANGE

Antirrhinum majus 'Trumpet Serenade'
Erect perennial, branching from the base, grown as an annual. Has lance-shaped leaves and, from spring to autumn, spikes of open trumpet-shaped, bicoloured flowers in a mixture of pastel shades. H and S 30cm (12in).

Viola 'Scarlet Clan'
Slow- to fairly fast-growing, bushy perennial, grown as an annual or biennial. Has oval leaves and rounded, 5-petalled, orange-yellow flowers, each with a scarlet blotch, in summer. H 15–20cm (6–8in), S 20cm (8in).

Calendula officinalis (Fiesta Series) 'Gitana'
Fast-growing, bushy annual with strongly aromatic, lance-shaped, pale green leaves. Daisy-like, double flower heads, ranging from cream to orange in colour, are carried from spring to autumn. H and S 30cm (12in).

Calceolaria 'Anytime'
Calceolaria, Monarch Series
Tagetes 'Nell Gwyn'

Calendula officinalis 'Art Shades'
Calendula officinalis, Fiesta Series
Celosia argentea var. *cristata*

Annuals and Biennials
ORANGE

Rudbeckia hirta 'Marmalade'
Moderately fast-growing, erect, branching perennial, grown as an annual, with lance-shaped leaves. In summer-autumn bears daisy-like, deep golden-orange flower heads, 8cm (3in) wide, with black centres. H 45cm (18in), S 30cm (12in).

Rudbeckia hirta 'Goldilocks'
Moderately fast-growing, erect, branching perennial, grown as an annual. Has lance-shaped leaves and daisy-like, double or semi-double, golden-orange flower heads, 8cm (3in) across, in summer-autumn. H 60cm (24in), S 30cm (12in).

Tagetes 'Tangerine Gem'
Fast-growing, bushy annual with aromatic, feathery leaves. Small, single, deep orange flower heads appear in summer and early autumn. H 20cm (8in), S 30cm (12in).

Rudbeckia hirta 'Rustic Dwarfs'
Moderately fast-growing, erect, branching perennial, grown as an annual, with lance-shaped, mid-green leaves. Bears daisy-like, yellow, mahogany or bronze flower heads, 8cm (3in) wide, in summer-autumn. H 60cm (24in), S 30cm (12in).

Tithonia rotundifolia 'Torch'
Slow-growing, erect annual with rounded, lobed leaves. Has daisy-like, bright orange or scarlet flower heads, 5–8 cm (2–3in) wide, in summer and early autumn. H 90cm (3ft), S 30cm (1ft).

Tagetes patula 'Orange Winner'
Fast-growing, bushy annual with aromatic, feathery, deep green leaves. Crested, daisy-like, double, bright orange flower heads are carried in summer and early autumn. H 15cm (6in), S to 30cm (12in).

Tagetes 'Paprika'
Fast-growing, bushy annual with aromatic, feathery leaves. In summer and early autumn has daisy-like, single, red flower heads edged with gold. H 15cm (6in), S 30cm (12in).

Erysimum x allionii 'Orange Bedder'
Slow-growing, short-lived, evergreen, bushy perennial, grown as a biennial. Has lance-shaped, mid-green leaves. Heads of scented, 4-petalled, brilliant orange flowers appear in spring. H and S 30cm (12in).

Helichrysum bracteatum, Monstrosum Series
Moderately fast-growing, erect, branching annual. Has daisy-like, papery, double flower heads, in pink, red, orange, yellow or white, in summer and early autumn. Flowers dry well. H 90cm (3ft), S 30cm (1ft).

Antirrhinum majus
Arctotis x hybrida 'Tangerine'
Arctotis x hybrida 'Torch'
Cosmos, Bright Lights Series

Erysimum cheiri
Gazania 'Daybreak'
Gazania 'Sundance'
Helichrysum bracteatum cvs

Impatiens, Duet Series
Portulaca grandiflora
Salpiglossis cvs
Tagetes, Inca Series

Tagetes, Jubilee Series
Tagetes patula
Tagetes 'Toreador'
Tropaeolum majus cvs

■ ORANGE

Calendula officinalis 'Geisha Girl'
Fast-growing, bushy annual with strongly aromatic, lance-shaped, pale green leaves. Heads of double, orange flowers with incurved petals are borne from late spring to autumn. H 60cm (24in), S 30–60cm (12–24in).

☼ ◊ ❄❄❄

Zinnia, Burpee Hybrids [Elegans group]
Moderately fast-growing, sturdy, erect annual with oval to lance-shaped, pale or mid-green leaves. Has large, cactus dahlia-like, double flower heads in a range of colours in summer and early autumn. H 60cm (24in), S 30cm (12in).

☼ ◊ ❄

Solanum pseudocapsicum 'Balloon'
Evergreen, bushy shrub, grown as an annual. Has lance-shaped leaves and, in summer, small, star-shaped, white flowers. Large, cream fruits turn orange in winter. H 30cm (12in), S 30–45cm (12–18in). Min. 5°C (41°F).

☼ ◊

Sanvitalia procumbens 'Mandarin Orange'
Moderately fast-growing, prostrate annual. Has pointed-oval, mid-green leaves and daisy-like, orange flower heads, 2.5cm (1in) wide, in summer. H 15cm (6in), S 30cm (12in).

☼ ◊ ❄❄❄

Tropaeolum majus, Jewel Series
Fast-growing, bushy annual with rounded leaves. Spurred, trumpet-shaped flowers, in shades of red, yellow or orange, are held well above leaves from early summer to early autumn. H and S 30cm (12in).

☼ ◊ ❄❄❄

Erysimum cheiri 'Fire King', syn. *Cheiranthus cheiri* 'Fire King'
Moderately fast-growing, evergreen, bushy perennial, grown as a biennial. Lance-shaped leaves are mid- to deep green; heads of 4-petalled, reddish-orange flowers are carried in spring. H 38cm (15in), S 30–38cm (12–15in).

☼ ◊ ❄❄❄

Solanum pseudocapsicum 'Red Giant'
Fairly slow-growing, evergreen, bushy shrub, usually grown as an annual. Has lance-shaped, deep green leaves, small, white flowers in summer and large, round, orange-red fruits in winter. H and S 30cm (12in). Min. 5°C (41°F).

☼ ◊

Dahlia 'Dandy'
Well-branched, erect, bushy, tuberous perennial, grown as an annual. Has pointed-oval, serrated leaves and heads of daisy-like flowers, with contrasting central collars of quilled petals, in shades of red, yellow or orange in summer. H and S 60cm (2ft).

☼ ◊ ❄

Emilia coccinea, syn. *E. flammea* of gardens, *E. javanica* of gardens (Tassel flower)
Moderately fast-growing, upright annual with lance-shaped, greyish-green leaves and double, red or yellow flower heads in summer. H 30–60cm (12–24in), S 30cm (12in) or more.

☼ ◊ ❄

Mimulus, Malibu Series [orange]
Fast-growing, branching perennial, grown as an annual, with pointed-oval, pale green leaves. In summer has flared, tubular, vibrant deep orange flowers. H 15cm (6in), S to 30cm (12in).

◐ ◊ ❄

Clarkia amoena, Princess Series [salmon]
Fast-growing annual with slender, upright stems and lance-shaped, mid-green leaves. Spikes of frilled, salmon-pink flowers are carried in summer. H and S 30cm (12in).

☼ ◊ ❄❄❄

Celosia argentea var. *cristata* 'Fairy Fountains'
Moderately fast-growing, erect, bushy perennial, grown as an annual. Has pointed-oval leaves and conical, feathery flower heads, to 15cm (6in) tall, in a wide range of colours in summer-autumn. H and S 30cm (12in).

☼ ◊ ❄

293

ROCK PLANTS/large SPRING INTEREST
☐ WHITE

Leontopodium alpinum (Edelweiss)
Short-lived perennial with tufts of lance-shaped, woolly leaves. Clusters of small, silvery-white flower heads, borne in spring or early summer, are surrounded by petal-like, thick, felted bracts that form a star shape. Dislikes winter and summer wet. H and S 15–20cm (6–8in).

Lithophragma parviflorum
Clump-forming, tuberous perennial that has small, open clusters of campion-like, white or pink flowers in spring above a basal cluster of deeply toothed, kidney-shaped leaves. Lies dormant in summer. H 15–20cm (6–8in), S to 20cm (8in).

Iberis sempervirens
Evergreen, spreading sub-shrub, with narrow, oblong, dark green leaves, bearing dense, rounded heads of white flowers in late spring and early summer. Trim after flowering. H 15–30cm (6–12in), S 45–60cm (18–24in).

Pulsatilla alpina (Alpine anemone)
Tufted perennial with feathery leaves that bears upright, or somewhat nodding, cup-shaped, white, sometimes blue- or pink-flushed flowers singly in spring and early summer, followed by feathery seed heads. H 15–30cm (6–12in), S to 10cm (4in).

Saxifraga granulata (Fair maids of France, Meadow saxifrage)
Clump-forming perennial that loses its kidney-shaped, crumpled, glossy leaves in summer. Sticky stems carry loose panicles of rounded, white flowers in late spring. H 23–38cm (9–15in), S to 15cm (6in) or more.

Chrysanthemum hosmariense, syn. **Pyrethropsis hosmariense**
Evergreen, shrubby perennial with finely cut, bright silvery-green leaves that clothe lax, woody stems. From late spring to early autumn, white flower heads are borne singly above foliage. H 15cm (6in) or more, S 30cm (12in).

Andromeda polifolia 'Alba'
Evergreen, open, twiggy shrub bearing terminal clusters of pitcher-shaped, white flowers in spring and early summer. Glossy, dark green leaves are leathery and lance-shaped. H 45cm (18in), S 60cm (24in).

Cassiope 'Muirhead'
Evergreen, loose, bushy shrub with scale-like, dark green leaves on upright branches. In spring, these bear tiny, virtually stemless, bell-shaped, white flowers along their length. H and S 20cm (8in).

Cassiope 'Edinburgh'
Evergreen, dwarf shrub with tiny, dark green leaves tightly pressed to upright stems. In spring, many small, bell-shaped, white flowers are borne singly in leaf axils. H and S 20cm (8in).

OTHER RECOMMENDED PLANTS:
Achillea clavennae, p.321
Cassiope fastigiata
Cassiope mertensiana, p.312
Cassiope selaginoides
Corydalis ochroleuca, p.299
Cyathodes colensoi, p.308
DAFFODILS, pp.360–62
Daphne alpina
Galax urceolata, p.299
x Halimiocistus wintonensis
Hebe carnosula
Hebe pinguifolia 'Pagei', p.299
Houttuynia cordata 'Flore Pleno'
Iberis sempervirens 'Snowflake'
IRISES, pp.198–9

□ **WHITE**

Saxifraga hirsuta
Evergreen, mound-forming perennial with rosettes of round, hairy leaves and loose panicles of tiny, star-shaped, white flowers, often yellow-spotted at the base of petals, in late spring and early summer. H 15–20cm (6–8in), S 20cm (8in).

Tiarella cordifolia (Foamflower)
Vigorous, evergreen, spreading perennial. Lobed, pale green leaves sometimes have darker marks; veins turn bronze-red in winter. Bears many spikes of profuse white flowers in late spring and early summer. H 15–20cm (6–8in), S to 30cm (12in) or more.

Jeffersonia diphylla
Slow-growing, tufted perennial with distinctive, 2-lobed, light to mid-green leaves. Bears solitary cup-shaped, white flowers with prominent, yellow stamens in late spring. Do not disturb roots. H 15–23cm (6–9in), S to 23cm (9in).

Saxifraga 'Tumbling Waters'
Slow-growing, evergreen, mat-forming perennial with a tight rosette of narrow, lime-encrusted leaves. After several years produces arching sprays of white flowers in conical heads; main rosette then dies but small offsets survive. H to 60cm (24in), S to 20cm (8in).

Daphne blagayana
Evergreen, prostrate shrub with trailing branches each bearing a terminal cluster of oval, leathery leaves and, in early spring, dense clusters of fragrant, tubular, white flowers. Likes humus-rich soil. H 30–40cm (12–16in), S 60–80cm (24–32in) or more.

Vancouveria hexandra
Vigorous, spreading perennial with open sprays of many tiny, white flowers in late spring and early summer. Leathery leaves are divided into almost hexagonal leaflets. Makes good woodland ground cover. H 20cm (8in), S indefinite.

Leiophyllum buxifolium
Ourisia macrocarpa
Parnassia palustris, p.298
PEONIES, pp.200–201

Phyllodoce nipponica
PRIMULAS, pp.236–7
Saxifraga cuneifolia, p.299

□ **WHITE–PINK**

Dodecatheon meadia f. *album*
Clump-forming perennial with basal rosettes of oval, pale green leaves. In spring, strong stems bear several white flowers with dark centres and reflexed petals. Lies dormant in summer. H 20cm (8in), S 15cm (6in).

Andromeda polifolia 'Compacta'
Evergreen, compact, twiggy shrub that bears delicate, terminal clusters of pitcher-shaped, coral-pink flowers, with white undertones, in spring and early summer. Leaves are lance-shaped and glossy, dark green. H 15–23cm (6–9in), S 30cm (12in).

Saxifraga × *geum*
Evergreen, mat-forming perennial with shallow-rooted rosettes of spoon-shaped, hairy leaves. In summer, star-shaped, pink-spotted, white flowers, deep pink in bud, are borne on loose panicles on slender stems. H 15–20cm (6–8in), S 30cm (12in).

Dodecatheon hendersonii
Clump-forming perennial with a flat rosette of kidney-shaped leaves, above which deep pink flowers with reflexed petals appear in late spring. Needs a dry, dormant summer period. H 30cm (12in), S 8cm (3in).

Andromeda polifolia
Evergreen, open, twiggy shrub with narrow, leathery, glossy, mid-green leaves. Bears terminal clusters of pitcher-shaped, pink flowers in spring and early summer. H 30–45cm (12–18in), S 60cm (24in).

Phyllodoce × *intermedia* 'Drummondii'
Evergreen, bushy, dwarf shrub with narrow, heath-like, glossy leaves. From late spring to early summer bears terminal clusters of pitcher-shaped, rich pink flowers on slender, red stalks. H and S 23cm (9in).

Anthyllis montana, p.300
Anthyllis montana 'Rubra'
Arabis caucasica 'Variegata', p.310
DAFFODILS, pp.360–62

Dodecatheon meadia
Saxifraga cotyledon, p.300
Saxifraga 'Southside Seedling', p.300
Shortia galacifolia, p.312

Rock plants/large SPRING INTEREST

■ PINK–PURPLE

Daphne cneorum
Evergreen, low-growing shrub with trailing branches clothed in small, oval, leathery, dark green leaves. Fragrant, deep rose pink flowers are borne in terminal clusters in late spring. Prefers humus-rich soil. H 23cm (9in), S to 2m (6ft).

Phyllodoce empetriformis
Evergreen, mat-forming shrub with fine narrow, heath like leaves and terminal clusters of bell-shaped, purplish-pink flowers in late spring and early summer. H 15–23cm (6–9in), S 20cm (8in).

Dodecatheon 'Red Wings'
Clump-forming perennial with a basal cluster of oblong, soft, pale green leaves. In late spring and early summer bears small, loose clusters of deep magenta flowers, with reflexed petals, on strong stems. Lies dormant in summer. H 20cm (8in), S 10cm (4in).

Phyllodoce caerulea, syn. *P. taxifolia*
Evergreen, dwarf shrub with fine, narrow, heath-like leaves. Bears bell-shaped, purple to purplish-pink flowers, singly or in clusters, in late spring and summer. H and S to 30cm (12in).

Pulsatilla vulgaris (Pasque flower)
Tufted perennial with feathery, light green leaves. In spring bears nodding, cup-shaped flowers, in shades of purple, red, pink or white, with bright yellow centres. Flower stems rapidly elongate as feathery seeds mature. H and S 15–23cm (6–9in).

■ PURPLE–BLUE

Erinacea anthyllis, syn. *E. pungens* (Hedgehog broom)
Slow growing, evergreen sub-shrub with hard, blue-green spines. Pea-like, soft lavender flowers appear in axils of spines in late spring to early summer. H and S 15–25cm (6–10in).

Pulsatilla halleri
Tufted perennial, intensely hairy in all parts, that in spring bears nodding, later erect, cup-shaped flowers in shades of purple. Has feathery leaves and seed heads. H 15–38cm (6–15in), S 15–20cm (6–8in).

Aquilegia alpina (Alpine columbine)
Short-lived, upright perennial with spurred, clear blue or violet-blue flowers on slender stems in spring and early summer. Has basal rosettes of rounded, finely divided leaves. Needs rich soil. H 45cm (18in), S 15cm (6in).

Cortusa matthioli, p.302
Daphne collina
Dicentra eximia 'Spring Morning', p.242
Kalmiopsis leachiana 'M. Le Piniec'

x *Phylliopsis hillieri* 'Pinocchio'
Phyllodoce x *intermedia* 'Fred Stoker'
PRIMULAS, pp.236–7
Rhodothamnus chamaecistus, p.300

IRISES, pp.198–9
Sisyrinchium douglasii
Sisyrinchium graminoides, p.304

296

■ BLUE　　　□■ GREY–YELLOW

Viola cornuta (Horned violet)
Rhizomatous perennial with oval, toothed leaves and flat-faced, rather angular, spurred, pale to deep purplish-blue, occasionally white flowers in spring and much of summer. H 12–20cm (8in), S to 20cm (8in) or more.

Salix helvetica
Deciduous, spreading, much-branched, dwarf shrub that has small, oval, glossy leaves, white-haired beneath. In spring bears short-stalked, silky, grey, then yellow catkins. H 60cm (24in), S 30cm (12in).

Corydalis wilsonii
Evergreen perennial with a fleshy rootstock. Forms rosettes of near-prostrate, divided, bluish-green leaves. Loose racemes of spurred, green-tipped, yellow flowers are produced in spring. H and S 10–25cm (4–10in).

Chiastophyllum oppositifolium, syn. *Cotyledon simplicifolia*
Evergreen, trailing perennial with large, oblong, serrated, succulent leaves. In late spring and early summer bears many tiny, yellow flowers in arching sprays. H 15–20cm (6–8in), S 15cm (6in).

Omphalodes verna
Semi-evergreen, clump-forming perennial that in spring bears long, loose sprays of flat, bright blue flowers with white eyes. Leaves are oval and mid-green. H and S 20cm (8in) or more.

Betula nana (Arctic birch)
Deciduous, bushy, dwarf shrub with small, toothed leaves that turn bright yellow in autumn. Has tiny, yellowish-brown catkins in spring. H 30cm (12in), S 45cm (18in).

Aurinia saxatilis 'Citrina', syn. *Alyssum saxatile* 'Citrinum'
Evergreen, clump-forming perennial with oval, hairy, grey-green leaves. Bears racemes of many, small, pale lemon-yellow flowers in late spring and early summer. H 23cm (9in), S 30cm (12in).

Cytisus × beanii
Deciduous, low-growing shrub with arching sprays of pea-like, golden-yellow flowers that appear in late spring and early summer on previous year's wood. Leaves, divided into 3 leaflets, are small, linear and hairy. H 15–40cm (6–16in), S 30–75cm (12–30in).

Omphalodes cappadocica
Spreading perennial with creeping underground stems and many loose sprays of flat, bright blue flowers in spring-summer above tufts of oval, hairy, basal leaves. H 15–20cm (6–8in), S 25cm (10in) or more.

Corydalis cheilanthifolia
Evergreen perennial with fleshy roots. Produces spreading rosettes of fern-like, near-prostrate, sometimes bronze-tinted, mid-green leaves. Dense spikes of short-spurred, yellow flowers are borne in late spring and early summer. H 20–30cm (8–12in), S 15–20cm (6–8in).

Aurinia saxatilis 'Variegata', syn. *Alyssum saxatile* 'Variegatum'
Evergreen perennial that bears racemes of many small, yellow flowers in spring above a mat of large, oval, soft grey-green leaves with cream margins. H 23cm (9in), S 30cm (12in).

IRISES, pp.198–9
Lithodora zahnii
Pulsatilla halleri subsp. *grandis*

DAFFODILS, pp.360–62
IRISES, pp.198–9
Paeonia mlokosewitschii, p.201
PRIMULAS, pp.236–7

Pulsatilla alpina subsp. *apiifolia*
Salix arbuscula

297

Rock plants/large SPRING INTEREST
☐ YELLOW–ORANGE

Hylomecon japonicum
Vigorous, spreading perennial with large, cup-shaped, bright yellow flowers that are borne singly on slender stems in spring. Soft, dark green leaves are divided into 4 unequal lobes. H to 30cm (12in), S 20cm (8in).

Erysimum 'Bredon', syn. *Cheiranthus* 'Bredon'
Semi-evergreen, rounded, woody perennial clothed in oval, dark green leaves. In late spring bears dense spikes of flat, bright mustard-yellow flowers. H 30–45cm (12–18in), S 45cm (18in).

Stylophorum diphyllum
Perennial with basal rosettes of large, lobed, hairy leaves. Bears open cup-shaped, golden-yellow flowers in spring on upright, branched stems. Prefers rich, woodland conditions. H and S to 30cm (12in) or more.

Erysimum 'Harpur Crewe', syn. *Cheiranthus* 'Harpur Crewe'
Evergreen, shrubby perennial with stiff stems and narrow leaves. Fragrant, double, deep yellow flowers open in succession from late spring to mid-summer. Grows best in poor soil and a sheltered site. H and S 30cm (12in).

Aurinia saxatilis, syn. *Alyssum saxatile* (Gold dust)
Evergreen perennial forming low clumps of oval, hairy, grey-green leaves. Has substantial spikes of small, chrome-yellow flowers in spring. H 23cm (9in), S 30cm (12in).

Berberis × stenophylla 'Corallina Compacta'
Evergreen, neat, dwarf shrub with spiny stems clothed in small, narrowly oval leaves. In late spring bears many tiny, bright orange flowers. Is slow-growing and difficult to propagate. H and S to 25cm (10in).

Alyssoides utriculata
Corydalis lutea, p.307
DAFFODILS, pp.360–62
Euryops acraeus, p.306
Euryops pectinatus, p.140
Genista lydia, p.307
IRISES, pp.198–9

Rock plants/large SUMMER INTEREST
☐ WHITE

Parnassia palustris (Grass of Parnassus)
Perennial with low, basal tufts of heart-shaped, pale to mid-green leaves. Bears saucer-shaped, white flowers, with dark green or purplish-green veins, on erect stems in late spring and early summer. H 20cm (8in), S 6cm (2½in) or more.

Armeria pseudarmeria, syn. *A. latifolia*
Evergreen, clump-forming perennial with large, spherical heads of white flowers occasionally suffused pink; these are borne in summer on stiff stems above long, narrow, glaucous leaves. H and S 30cm (12in).

Celmisia walkeri
Evergreen, loose, spreading perennial with long, oval or lance-shaped leaves, glossy, green above and hairy, white beneath. Has large, daisy-like, white flower heads in summer. H 23cm (9in), S to 2m (6ft).

OTHER RECOMMENDED PLANTS:
Andromeda polifolia 'Alba', p.294
Artemisia frigida
Artemisia pedemontana
Artemisia stelleriana 'Boughton Silver'
Boykinia aconitifolia
CARNATIONS and PINKS, pp.244–5
Chrysanthemum hosmariense, p.294

☐ WHITE

Helianthemum apenninum
Evergreen, spreading, much-branched shrub that bears saucer-shaped, pure white flowers in mid-summer. Stems and small, linear leaves are covered in white down. H and S 45cm (18in).

Saxifraga cuneifolia
Evergreen, carpeting perennial with neat rosettes of rounded leaves. In late spring and early summer bears panicles of tiny, white flowers, frequently with yellow, pink or red spots, on slender stems. H 15–20cm (6–8in), S 30cm (12in) or more.

Cytisus purpureus f. **albus**, syn. **Chamaecytisus purpureus** f. **albus**
Deciduous, low-growing shrub with semi-erect stems clothed in leaves, divided into 3 leaflets. A profusion of pea-like, white flowers appear in early summer on previous year's wood. H 45cm (18in), S 60cm (24in).

Helianthemum 'Wisley White'
Evergreen, spreading shrub, with oblong, grey-green leaves, bearing saucer-shaped, white flowers for a long period in summer. H 23cm (9in), S 30cm (12in) or more.

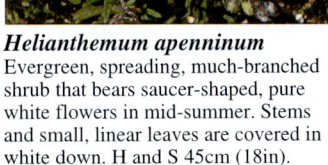

Galax urceolata, syn. **G. aphylla**
Evergreen, clump-forming perennial. Large, round, leathery, mid-green leaves on slender stems turn bronze in autumn-winter. Has dense spikes of small, white flowers in late spring and early summer. H 15–20cm (6–8in), S to 30cm (12in).

Epilobium glabellum
Semi-evergreen, clump-forming perennial. Bears outward-facing, cup-shaped, white flowers singly on slender stems in summer above oval, mid-green leaves. Makes useful ground cover. H to 20cm (8in), S 15cm (6in).

Hebe pinguifolia 'Pagei'
Evergreen, semi-prostrate shrub with small, oblong, slightly cupped, intensely glaucous leaves. Bears short spikes of small, pure white flowers in late spring or early summer. Is excellent for ground or rock cover. H 15–30cm (6–12in), S 60cm–1m (2–3ft).

Corydalis ochroleuca, syn. **Pseudofumaria ochroleuca**
Evergreen, clump-forming perennial with fleshy, fibrous roots and much divided, basal, grey-green leaves. Bears slender, yellow-tipped, creamy-white flowers in late spring and summer. H and S 20–30cm (8–12in).

Hebe vernicosa
Evergreen, bushy, compact shrub with small, oval, glossy, dark green leaves densely packed on stems. In early and mid-summer, spikes of small, 4-lobed, white flowers are freely produced. H 60cm (2ft), S 1.2m (4ft).

Cotoneaster congestus
Deinanthe bifida
Eomecon chionantha
Fabiana imbricata 'Prostrata'

x *Halimiocistus wintonensis*
Hebe buchananii
HOSTAS, pp.252–3
Iberis sempervirens 'Snowflake'

IRISES, pp.198–9
Leontopodium alpinum, p.294
Luetkea pectinata
Papaver burseri

Phyllodoce nipponica
PRIMULAS, pp.236–7
Stephanandra incisa 'Crispa'
Tiarella cordifolia, p.295

Rock plants/large SUMMER INTEREST

WHITE–PINK

Saxifraga cotyledon
Evergreen perennial with large, pale green rosettes of leaves, dying after flowering. In late spring and early summer produces arching, conical panicles of cup-shaped, white flowers sometimes strongly marked red internally. H and S to 30cm (12in).

Onosma alboroseum
Semi-evergreen, clump-forming perennial covered in fine hairs, which may irritate skin. Clusters of long, pendent, tubular flowers, borne for a long period in summer, open white and then turn pink. H 15–30cm (6–12in), S 20cm (8in).

Aethionema grandiflorum, syn. *A. pulchellum* (Persian stone cress)
Short-lived, evergreen or semi-evergreen, lax shrub. Bears tiny, pale to deep rose-pink flowers in loose sprays in spring-summer. Blue-green leaves are narrow and lance-shaped. H 30cm (12in), S 23cm (9in).

***Saxifraga* 'Southside Seedling'**
Evergreen, mat-forming perennial, with large, pale green rosettes of leaves, dying after flowering. In late spring and early summer bears arching panicles of open cup-shaped, white flowers, strongly red-banded within. H and S to 30cm (12in).

Rhodothamnus chamaecistus
Evergreen, low-growing, dwarf shrub with narrow, oval leaves, edged with bristles. In late spring and early summer bears cup-shaped, rose- to lilac-pink flowers, with dark stamens, in leaf axils. H 15–20cm (6–8in), S to 25cm (10in).

PINK

Anthyllis montana
Rounded, bushy or somewhat spreading perennial with loose branches and finely cut foliage. Heads of clover-like, pale pink flowers with red markings are borne in late spring and early summer. H and S 30cm (12in).

Phuopsis stylosa, syn. *Crucianella stylosa*
Low-growing perennial with whorls of pungent, pale green leaves and rounded heads of small, tubular, pink flowers in summer. Is good grown over a bank or large rock. H 30cm (12in), S 30cm (12in) or more.

***Lewisia* 'George Henley'**
Evergreen, clump-forming perennial with rosettes of narrow, fleshy, dark green leaves. Bears dense sprays of open cup-shaped, deep pink flowers, with magenta veins, from late spring to late summer. H 15cm (6in) or more, S 10cm (4in).

Achillea x *kellereri*, p.323
Diascia vigilis
Dorycnium hirsutum
Gaultheria miqueliana

Alyssum spinosum
Saxifraga granulata 'Plena'
Saxifraga longifolia
Saxifraga x *urbium*

Aethionema armenum, p.325
Aethionema 'Warley Rose', p.324
CARNATIONS and PINKS, pp.244–5
Crepis incana

Dianthus superbus
Erigeron glaucus 'Elstead'
Erodium foetidum
Incarvillea mairei, p.243

□ PINK

Origanum 'Kent Beauty'
Prostrate perennial with trailing stems clothed in aromatic, rounded-oval leaves. In summer bears short spikes of tubular, pale pink flowers with darker bracts. Is suitable for a wall or ledge. H 15–20cm (6–8in), S 30cm (12in).

Astilbe 'Perkeo'
Erect, compact perennial bearing small plumes of tiny, salmon-pink flowers from mid- to late summer on fine stems. Has stiff, deeply cut, crinkled leaves. H 15–20cm (6–8in), S 10cm (4in).

Crassula sarcocaulis
Evergreen or, in severe climates, semi-evergreen, bushy sub-shrub with tiny, oval, succulent leaves. Bears terminal clusters of tiny, red buds opening to pale pink flowers in summer. H and S 30cm (12in).

Diascia cordata
Prostrate perennial with stems clothed in heart-shaped, pale green leaves. Bears terminal clusters of spurred, flat-faced, bright pink flowers in summer and early autumn. H 15–20cm (6–8in), S 20cm (8in).

Ononis fruticosa
(Shrubby restharrow)
Deciduous shrub that in summer bears pendent clusters of large, pea-like, purplish-pink blooms with darker streaks. Leaves are divided into 3 serrated leaflets, which are hairy when young. H and S 30–60cm (12–24in).

Helianthemum 'Wisley Pink', syn. *H.* 'Rhodanthe Carneum'
Evergreen, lax shrub with saucer-shaped, soft, pale pink flowers with orange centres borne for a long period in summer. Has oblong, grey-green leaves. H and S 30cm (12in) or more.

Oxalis tetraphylla, syn. *O. deppei*
Tuft-forming, tuberous perennial with brown-marked, basal leaves, usually divided into 4 leaflets. Produces loose sprays of widely funnel-shaped, deep pink flowers in late spring and summer. Needs a sheltered site. H 15–30cm (6–10in), S 10–15cm (4–6in).

Diascia rigescens
Trailing perennial with semi-erect stems covered in heart-shaped, mid-green leaves. Spurred, flat-faced, salmon-pink flowers are borne along stem length in summer and early autumn. H 23cm (9in), S to 30cm (12in).

Geranium orientalitibeticum
Perennial spreading by tuberous, underground runners. Has cup-shaped, pink flowers, with white centres, in summer. Leaves are deeply cut and marbled in shades of green. May be invasive. H in flower 15–25cm (6–10in), S indefinite.

Incarvillea mairei 'Frank Ludlow'
Ononis rotundifolia
Ourisia 'Loch Ewe'
PEONIES, pp.200–201
Potentilla nitida
PRIMULAS, pp.236–7
Teucrium polium, p.327

Rock plants/large SUMMER INTEREST

☐ PINK

☐ RED

Cortusa matthioli
Clump-forming perennial with a basal rosette of rounded, dull green leaves and, in late spring and early summer, one-sided racemes of small, pendent, bell-shaped, reddish- or pinkish-purple flowers. H 15–20cm (6–10in), S 10cm (4in).

Geranium sanguineum
(Bloody cranesbill)
Hummock-forming, spreading perennial with many cup-shaped, deep magenta-pink flowers borne in summer above round, deeply divided, dark green leaves. Makes good ground cover. H to 25cm (10in), S 30cm (12in) or more.

Penstemon newberryi f. **humilior**
Evergreen, mat-forming shrub with arching branches clothed in small, leathery, dark green leaves. Bears short sprays of tubular, lipped, cherry-red to deep pink flowers in early summer. H 15–20cm (6–8in), S 30cm (12in).

Helianthemum 'Ben More'
Evergreen, spreading, twiggy shrub that bears a succession of saucer-shaped, reddish-orange flowers in loose, terminal clusters in late spring and summer. Has small, glossy, dark green leaves. H 23–30cm (9–12in), S 30cm (12in).

Origanum laevigatum
Deciduous, mat-forming sub-shrub with small, aromatic, dark green leaves, branching, red stems and a profusion of tiny, tubular, cerise-pink flowers, surrounded by red-purple bracts, in summer. H 23–30cm (9–12in), S 20cm (8in) or more.

Lewisia, Cotyledon Hybrids
Evergreen, clump-forming perennials with rosettes of large, thick, toothed leaves. In early summer bear clusters of flowers, in various shades of pink to purple, on erect stems. Is good for a rock crevice or an alpine house. H to 30cm (12in), S 15cm (6in) or more.

Helianthemum 'Raspberry Ripple'
Evergreen, spreading shrub with saucer-shaped, red-centred, white flowers that are borne in mid-summer. Has small, linear, grey-green leaves. H 15–23cm (6–9in), S 23–30cm (9–12in).

Penstemon pinifolius
Evergreen, bushy shrub with branched stems clothed in fine, dark green leaves. In summer, very narrow, tubular, orange-red flowers are borne in loose, terminal spikes. H 10–20cm (4–8in), S 15cm (6in).

Dianthus carthusianorum
Evergreen perennial carrying rounded, upward-facing, cherry-red or deep pink flowers on slender stems in summer above small tufts of grass-like leaves. H 20cm (8in), S 6cm (2½in).

Helianthemum 'Fire Dragon'
Evergreen, spreading shrub with saucer-shaped, orange-scarlet flowers in late spring and summer. Leaves are linear and grey-green. H 23–30cm (9–12in), S 45cm (18in).

Anthyllis montana 'Rubra'
Astilbe chinensis 'Pumila'
Dianthus armeria
Micromeria juliana

Origanum rotundifolium
Phyllodoce caerulea, p.296
Phyllodoce empetriformis, p.296
Phyllodoce x intermedia 'Drummondii', p.295

Armeria pseudarmeria 'Bees Ruby'
Helianthemum 'Ben Hope'
PEONIES, pp.200–201
Phyllodoce x intermedia 'Fred Stoker'

PRIMULAS, pp.236–7
Zauschneria californica
Zauschneria cana

302

■ RED

Zauschneria californica 'Glasnevin'
Clump-forming, woody-based perennial with lance-shaped, grey-green leaves. From late summer to early autumn bears terminal clusters of tubular, deep orange-scarlet flowers. H 30cm (12in), S 45cm (18in).

☀ ◊ ❀❀❀ ♛

Punica granatum var. nana, syn. *P. g.* 'Nana' (Dwarf pomegranate)
Slow-growing, deciduous, rounded shrub that, in summer, bears funnel-shaped, red flowers with somewhat crumpled petals, followed by small, rounded, orange-red fruits. H and S 30–90cm (1–3ft).

☀ ◊ ❀

Delphinium nudicaule
Short-lived, upright perennial with erect stems bearing deeply divided, basal leaves and, in summer, spikes of hooded, red or occasionally yellow flowers, with contrasting stamens. H 20cm (8in), S 5–10cm (2–4in).

☀ ◊ ❀❀❀

■ PURPLE

Calceolaria arachnoidea
Evergreen, clump-forming perennial with a basal rosette of wrinkled leaves, covered in white down. Upright stems carry spikes of many pouch-shaped, dull purple flowers in summer. Is best treated as a biennial. H 25cm (10in), S 12cm (5in).

☀ ● ❀

Erodium cheilanthifolium, syn. *E. petraeum* subsp. *crispum*
Compact, mound-forming perennial with pink flowers, veined and marked with purple-red, borne on stiff stems in late spring and summer. Greyish-green leaves are crinkled and deeply cut. H 15–20cm (6–8in), S 20cm (8in) or more.

☀ ◊ ❀❀

Scabiosa lucida
Clump-forming perennial with tufts of oval leaves and rounded heads of pale lilac to deep mauve flowers, borne on erect stems in summer. H 20cm (8in), S 15cm (6in).

☀ ◊ ❀❀❀

Penstemon serrulatus, syn. *P. diffusus*
Semi-evergreen sub-shrub, deciduous in severe climates, that has small, elliptic, dark green leaves and tubular, blue to purple flowers borne in loose spikes in summer. Soil should not be too dry. H 60cm (24in), S 30cm (12in).

☀ ◊ ❀❀❀

Erigeron alpinus (Alpine fleabane)
Clump-forming perennial of variable size that bears daisy-like, lilac-pink flower heads on erect stems in summer. Leaves are long, oval and hairy. Suits a sunny border, bank or large rock garden. H 25cm (10in), S 20cm (8in).

☀ ◊ ❀❀❀

Semiaquilegia ecalcarata
Short-lived, upright perennial with narrow, lobed leaves. In summer each slender stem bears several pendent, open bell-shaped, dusky-pink to -purple flowers, with no spurs. H 20cm (8in), S 6cm (2½in).

☀ ◊ ❀❀❀

Cytisus purpureus
Geranium procurrens
IRISES, pp.198–9
Oxalis hirta

PEONIES, pp.200–201
Scabiosa graminifolia

Rock plants/large SUMMER INTEREST
PURPLE–BLUE

Campanula barbata
(Bearded bellflower)
Evergreen perennial with a basal rosette of oval, hairy, grey-green leaves. In summer bears one-sided racemes of bell-shaped, white to lavender-blue flowers. Is short lived but sets seed freely. H 20cm (8in), S 12cm (5in).

Sisyrinchium graminoides, syn. *S. angustifolium*, *S. bermudiana*
Semi-evergreen, erect perennial with tufts of grass-like leaves. Small, iris-like, pale to dark purplish-blue flowers with yellow bases are borne in terminal clusters in late spring and early summer. H to 30cm (12in), S 8cm (3in).

Parahebe catarractae
Evergreen sub-shrub with oval, toothed, mid-green leaves and, in summer, loose sprays of small, open funnel-shaped, white flowers, heavily zoned and veined pinkish-purple. H and S 30cm (12in).

Phlox divaricata subsp. **laphamii**
Semi-evergreen, creeping perennial with oval leaves and upright stems bearing loose clusters of saucer-shaped, pale to deep violet-blue flowers in summer. H 30cm (12in) or more, S 20cm (8in).

Lithodora oleifolia, syn. *Lithospermum oleifolium*
Evergreen shrub with oval, pointed, silky, mid-green leaves. Curving stems carry loose sprays of several small, funnel-shaped, light blue flowers in early summer. H 15–20cm (6–8in), S to 1m (3ft).

Wulfenia amherstiana
Evergreen perennial with rosettes of narrowly spoon-shaped, toothed leaves. Erect stems bear loose clusters of small, tubular, purple or pinkish-purple flowers in summer. H 15–30cm (6–12in), S to 30cm (12in).

Phlox 'Chatahoochee'
Short-lived, clump-forming perennial that has saucer-shaped, red-eyed, bright lavender flowers throughout summer-autumn. Narrow, pointed leaves are dark reddish-purple when young. H 15–20cm (6–8in), S 30cm (12in).

Convolvulus sabatius, syn. *C. mauritanicus*
Trailing perennial with slender stems clothed in small, oval leaves and open trumpet-shaped, vibrant blue-purple flowers in summer and early autumn. Shelter in a rock crevice in a cold site. H 15–20cm (6–8in), S 30cm (12in).

Linum perenne
Upright perennial with slender stems, clothed in grass-like leaves, that bear terminal clusters of open funnel-shaped, clear blue flowers in succession throughout summer. H 30cm (12in), S to 15cm (6in).

Aster alpinus, p.329
Carmichaelii enysii
CARNATIONS and PINKS, pp.244–5
Deinanthe caerulea

Dracocephalum ruyschianum
Erinacea anthyllis, p.296
Erysimum linifolium
Horminum pyrenaicum

IRISES, pp.198–9
Moltkia petraea
Penstemon hirsutus
Penstemon procerus

Phlox divaricata
Satureja montana
Viola cornuta, p.297
Wulfenia carinthiaca

■ BLUE

Veronica prostrata
(Prostrate speedwell)
Dense, mat-forming perennial that has upright spikes of small, saucer-shaped, brilliant blue flowers in early summer. Foliage is narrow, oval and toothed. H to 30cm (12in), S indefinite.

☼ ◊ ✱✱✱ ♀

Moltkia suffruticosa
Deciduous, upright sub-shrub. In summer bears clusters of funnel-shaped, bright blue flowers, pink in bud, on hairy stems. Leaves are long, pointed and hairy. H 15–40cm (6–16in), S 30cm (12in).

☼ ◊ ✱✱✱

Veronica prostrata 'Trehane'
Dense, mat-forming perennial bearing upright spikes of small, saucer-shaped, deep violet-blue flowers in early summer above narrow, toothed, yellow or yellowish-green leaves. H in flower 15–20cm (6–8in), S indefinite.

☼ ◊ ✱✱✱

Veronica prostrata 'Kapitan'
Dense, mat-forming perennial bearing erect spikes of small, saucer-shaped, bright deep blue flowers in early summer. Foliage is narrow, oval and toothed. H to 30cm (12in), S indefinite.

☼ ◊ ✱✱✱

Mertensia echioides
Clump-forming perennial with basal rosettes of long, oval, hairy, blue-green leaves. Slender stems carry many open funnel-shaped, dark blue flowers in summer. H 15–23cm (6–9in), S 15cm (6in).

☼ ◊ ✱✱✱

Phyteuma scheuchzeri
Tufted perennial with narrow, dark green leaves and terminal heads of spiky, blue flowers that are borne in summer. Seeds freely; dislikes winter wet. H 15–20cm (6–8in), S 10cm (4in).

☼ ◊ ✱✱✱

Symphyandra wanneri
Clump-forming perennial with branching stems and hairy, oval leaves. In summer bears pendent, bell-shaped, blue to violet-blue flowers in loose, terminal spikes. H 15–23cm (6–9in), S 25cm (10in).

☼ ◊ ✱✱✱

Lithodora diffusa 'Heavenly Blue'
Evergreen, prostrate shrub with trailing stems bearing pointed, oblong, hairy leaves and, in summer, a profusion of open funnel-shaped, deep blue flowers in leaf axils. Trim stems hard after flowering. H 15–30cm (6–12in), S to 45cm (18in).

☼ ◊ pH ✱✱✱ ♀

Aphyllanthes monspeliensis
Ceratostigma plumbaginoides, p.308
Delphinium tatsienense
Gentiana septemfida, p.308

IRISES, pp.198–9
Lithodora diffusa 'Grace Ward'
Lithodora zahnii
Moltkia x *intermedia*

Omphalodes cappadocica, p.297
Penstemon heterophyllus 'True Blue'
Rosmarinus officinalis 'Prostratus'
Symphyandra armena

Veronica austriaca subsp. *teucrium* 'Royal Blue'
Veronica pectinata
Veronica prostrata 'Spode Blue'

305

Rock plants/large SUMMER INTEREST
BLUE–YELLOW   YELLOW

Veronica austriaca subsp. **teucrium**, syn. *V. teucrium*
Spreading perennial with narrow spikes of small, flat, outward-facing, bright blue flowers in summer. Leaves are small, divided, hairy and greyish-green. H and S 25–60cm (10–24in).

Verbascum 'Letitia'
Evergreen, stiff-branched shrub with toothed, grey leaves. Bears outward-facing, 5-lobed, bright yellow flowers with reddish-orange centres continuously from late spring to mid-autumn. Hates winter wet; is good in an alpine house. H and S to 25cm (10in).

Helianthemum 'Wisley Primrose'
Fast-growing, evergreen, compact shrub with saucer-shaped, soft pale yellow flowers in summer. Has oblong, grey-green leaves. H 23cm (9in), S 30cm (12in) or more.

Othonna cheirifolia, syn. *Othonnopsis cheirifolia*
Evergreen shrub with narrow, somewhat fleshy, grey leaves. In early summer bears daisy-like, yellow flower heads singly on upright stems. Needs a warm, sheltered site. H 20–30cm (8–12in), S 30cm (12in) or more.

Erodium chrysanthum
Mound-forming perennial, grown for its dense, silvery stems and finely cut, fern-like leaves. Has small sprays of cup-shaped, sulphur- or creamy-yellow flowers in late spring and summer. H and S 23cm (9in).

Euryops acraeus, syn. *E. evansii* of gardens
Evergreen, dome-shaped shrub with stems clothed in toothed, silvery-blue leaves. Bears solitary daisy-like, bright yellow flower heads in late spring and early summer. H and S 30cm (12in).

Hypericum olympicum 'Sulphureum', syn. *H. o.* 'Citrinum'
Deciduous, dense, rounded sub-shrub with tufts of upright stems, clothed in small, oval, grey-green leaves. Bears terminal clusters of lemon-yellow flowers throughout summer. H and S 15–30cm (6–12in).

Linum arboreum
Evergreen, compact shrub with blue-green leaves. In summer has a succession of funnel-shaped, bright yellow flowers opening in sunny weather and borne in terminal clusters. H to 30cm (12in), S 30cm (12in).

Eriogonum umbellatum
Evergreen, prostrate to upright perennial with mats of green leaves, white and woolly beneath. In summer carries heads of tiny, yellow flowers that later turn copper. Dwarf forms are available. H 8–30cm (3–12in), S 15–30cm (6–12in).

Helianthemum 'Jubilee'
HOSTAS, pp.252–3
Lomatium nudicaule
Meum athamanticum

Anthyllis hermanniae
Arnebia pulchra
Arnica montana
Aurinia saxatilis 'Citrina', p.297

Chiastophyllum oppositifolium, p.297
Corydalis cheilanthifolia, p.297
Corydalis nobilis
Craspedia incana

Cremanthodium reniforme
Cytisus × *beanii*, p.297
Erysimum 'Harpur Crewe', p.298
Euryops pectinatus, p.140

306

▢ YELLOW

Corydalis lutea, syn.
Pseudofumaria lutea
Evergreen, clump-forming perennial with fleshy, fibrous roots, semi-erect, basal, grey-green leaves. Bears racemes of slender, yellow flowers, with short spurs, in late spring and summer. H and S 20–30cm (8–12in).

☼ ◊ ❄❄❄

Verbascum dumulosum
Evergreen, mat-forming, shrubby perennial with hairy, grey or grey-green leaves. In late spring and early summer bears a succession of 5-lobed, bright yellow flowers in short racemes. Dislikes winter wet. H 15cm (6in) or more, S 23–30cm (9–12in) or more.

☼ ◊ ❄❄❄ ♀

Chrysogonum virginianum
Mat-forming perennial with daisy-like, yellow flower heads borne on short stems in summer-autumn and oval, toothed, mid-green leaves. Although plant spreads by underground runners, it is not invasive. H 15–20cm (6–8in), S 10–15cm (4–6in) or more.

◐ ◊ ❄❄❄

Sedum reflexum
(Reflexed stonecrop)
Evergreen perennial with loose mats of rooting stems bearing narrow, fleshy leaves. Carries flat, terminal heads of tiny, bright yellow flowers in summer. Makes good ground cover. H 15–20cm (6–8in), S indefinite.

☼ ◊ ❄❄❄

Ranunculus gramineus
Erect, slender perennial with grass-like, blue-green leaves. Bears several cup-shaped, bright yellow flowers in late spring and early summer. Prefers rich soil. Seedlings will vary in height and flower size. H 40–50cm (16–20in), S 8–10cm (3–4in).

☼ ◊ ❄❄❄ ♀

▢ YELLOW–ORANGE

Ononis natrix
(Large yellow restharrow)
Deciduous, compact, erect shrub with pea-like, red-streaked, yellow flowers in pendent clusters in summer. Hairy leaves are divided into 3 leaflets. H and S 30cm (12in) or more.

☼ ◊ ❄❄❄

Genista lydia
Deciduous, domed shrub with slender, arching branches and blue-green leaves. Massed terminal clusters of pea-like, bright yellow flowers appear in late spring and early summer. Will trail over a large rock or wall. H 45–60cm (18–24in), S 60cm (24in) or more.

☼ ◊ ❄❄❄ ♀

Crepis aurea
Clump-forming perennial with a basal cluster of oblong, light green leaves. In summer produces dandelion-like, orange flower heads, singly, on stems covered with black and white hairs. H 10–30cm (4–12in), S 15cm (6in).

☼ ◊ ❄❄❄

Genista pilosa
Helianthemum 'Golden Queen'
HOSTAS, pp.252–3
Hypericum balearicum

Hypericum coris
Hypericum olympicum
IRISES, pp.198–9
Linum flavum

PEONIES, pp.200–201
PRIMULAS, pp.236–7
Tanacetum haradjanii

Helianthemum 'Ben Nevis'
IRISES, pp.198–9
PRIMULAS, pp.236–7

307

Rock plants/large AUTUMN INTEREST
WHITE–BLUE

Gaultheria cuneata
Evergreen, compact shrub with stiff stems clothed in leathery, oval leaves. In summer bears nodding, urn-shaped, white flowers, in leaf axils, followed by white berries in autumn. H and S 30cm (12in).

Ceratostigma plumbaginoides
Bushy perennial that bears small, terminal clusters of single, brilliant blue flowers on reddish, branched stems in late summer and autumn. Oval leaves turn rich red in autumn. H 45cm (18in), S 20cm (8in).

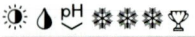

Sorbus reducta
Deciduous shrub forming a low thicket of upright branches. Small, grey-green leaves, divided into leaflets, turn bronze-red in late autumn. In early summer bears loose clusters of flat, white flowers, followed by pink berries. H and S to 30cm (12in) or more.

Gentiana septemfida
Evergreen perennial with many upright, then arching stems clothed with oval leaves. Bears heads of trumpet-shaped, mid-blue flowers in summer-autumn. Likes humus-rich soil but tolerates reasonably drained, heavy clay. H 15–20cm (6–8in), S 30cm (12in).

OTHER RECOMMENDED PLANTS:
Chrysanthemum hosmariense, p.294
Chrysogonum virginianum, p.307
Convolvulus sabatius, p.304
Diascia rigescens, p.301
Gaultheria miqueliana
Polygonum affine 'Darjeeling Red'

Rock plants/large WINTER/ALL YEAR
WHITE–GREY

Ranunculus calandrinioides
Clump-forming perennial that loses its long, oval, blue-green leaves in summer; in a reasonable winter will bear a succession of cup-shaped, pink-flushed, white flowers for many weeks. Needs very sharp drainage. H and S to 20cm (8in).

Celmisia coriacea
Evergreen perennial with sword-like, silver leaves in large clumps and, in summer, daisy-like, white flower heads borne singly on hairy stems. H and S 30cm (12in).

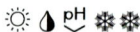

Cyathodes colensoi
Evergreen, low-growing shrub with stiff stems clothed in tiny, grey-green leaves. Bears clusters of small, tubular, white flowers in spring at the ends of new growth. Red or white berries in late summer are rare in cultivation. H and S 30cm (12in).

Tanacetum argenteum, syn. *Achillea argentea*
Mat-forming perennial, usually evergreen, grown for its finely cut, bright silver leaves. Has a profusion of small, daisy-like, white flower heads in summer. H in flower 15–23cm (6–9in), S 20cm (8in).

OTHER RECOMMENDED PLANTS:
Eriogonum crocatum
Eriogonum ovalifolium
Euonymus fortunei 'Emerald Gaiety'
Polygonum affine

GREY–GREEN

Tanacetum densum subsp. **amani**
Clump-forming perennial retaining fern-like, hairy, grey leaves in winter in mild climates. Bears daisy-like, yellow flower heads with woolly bracts in summer. Dislikes winter wet. H and S 20cm (8in).

Hebe cupressoides 'Boughton Dome'
Slow-growing, evergreen, dome-shaped shrub with scale-like, stem-clasping, dark grey-green leaves. Has terminal clusters of small, 4-lobed, blue-tinged, white flowers in summer. H 30cm (12in), S to 60cm (24in).

Helichrysum coralloides, syn. **Ozothamnus coralloides**
Evergreen, upright shrub with grey stems clothed in neat, dark green leaves, marked silver. Occasionally bears fluffy, yellow flower heads. Suits a cold frame or an alpine house. Hates winter wet. H 15–23cm (6–9in), S 15cm (6in).

Helichrysum selago, syn. **Ozothamnus selago**
Evergreen, upright shrub with stiff stems covered in scale-like leaves. Intermittently bears fluffy, creamy-white flower heads. Makes a good foil for spring bulbs. H and S 15–23cm (6–9in).

Salix x boydii
Very slow-growing, deciduous, upright shrub forming a gnarled, branched bush. Has oval, rough-textured leaves; catkins are rarely produced. Will tolerate light shade. H to 15–23cm (6–9in), S to 30cm (12in).

Ballota pseudodictamnus
Evergreen, mound-forming sub-shrub with rounded, grey-green leaves and stems covered with woolly, white hairs. In summer bears whorls of small, pink flowers with conspicuous, enlarged, pale green calyces. H 60cm (2ft), S 90cm (3ft).

Arctostaphylos 'Emerald Carpet', p.126
Buxus microphylla 'Green Pillow', p.146
Euonymus fortunei 'Gold Tip'
Myrsine africana

Pachysandra axillaris
Paxistima canbyi
Rubus tricolor

Rock plants/small SPRING INTEREST
WHITE

Cerastium tomentosum (Snow-in-summer)
Very vigorous, ground-cover perennial, only suitable for a hot, dry bank, with prostrate stems covered by tiny, grey leaves. In late spring and summer bears star-shaped, white flowers above foliage. H 8cm (3in), S indefinite.

Androsace pyrenaica
Evergreen perennial with small rosettes of tiny, hairy leaves, tightly packed to form hard cushions. Minute, stemless, single, white flowers appear in spring. H 4cm (1½in), S to 10cm (4in).

Arenaria tetraquetra
Evergreen perennial that forms a hard, grey-green cushion of small leaves. Stemless, star-shaped, white flowers appear in late spring. Is well-suited for a trough or an alpine house. H 2.5cm (1in), S 15cm (6in) or more.

OTHER RECOMMENDED PLANTS:
Androsace chamaejasme
Androsace cylindrica
Androsace hirtella

Arabis ferdinandi-coburgii 'Variegata', p.336
Callianthemum coriandrifolium
Cassiope wardii
Claytonia megarhiza

309

Rock plants/small SPRING INTEREST
☐ WHITE

Arabis caucasica 'Variegata'
Evergreen, mat-forming perennial with rosettes of oval, cream-splashed, mid-green leaves. Bears bunches of single, sometimes pink-flushed, white flowers from early spring to summer. H and S 15cm (6in).

Cardamine trifolia
Ground-cover perennial with creeping stems clothed in rounded, toothed, 3-parted leaves. In late spring and early summer bears loose heads of open cup-shaped, white flowers on bare stems. H 10–15cm (4–6in), S 30cm (12in).

Saxifraga scardica
Slow-growing, evergreen perennial with hard cushions composed of blue-green rosettes of leaves. In spring bears small clusters of upward-facing, cup-shaped, white flowers. Does best in an alpine house or sheltered scree. H 2.5cm (1in), S 8cm (3in).

Arenaria balearica
Prostrate perennial that is evergreen in all but the most severe winters. Will form a green film over a wet, porous rock face. Minute, white flowers stud mats of foliage in late spring and early summer. H less than 1cm (½in), S indefinite.

Androsace vandellii, syn. *A. imbricata*
Evergreen, dense, cushion-forming perennial with narrow, grey leaves and a profusion of stemless, white flowers in spring. Needs careful cultivation with a deep collar of grit under the cushion. H 2.5cm (1in), S to 10cm (4in).

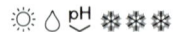

Weldenia candida
Perennial with rosettes of strap-shaped, wavy-margined leaves, growing from tuberous roots. Bears a succession of upright, cup-shaped, pure white flowers in late spring and early summer. H and S 8–15cm (3–6in).

Dicentra cucullaria
(Dutchman's breeches)
Compact perennial with fern-like foliage and arching stems each bearing a few small, yellow-tipped, white flowers, like tiny, inflated trousers, in spring. Lies dormant in summer. H 15cm (6in), S to 30cm (12in).

Celmisia ramulosa, p.322
Clintonia uniflora
Cornus canadensis, p.322
DAFFODILS, pp.360–62
Dodecatheon dentatum
Epigaea asiatica
Haberlea rhodopensis 'Virginalis', p.321
Iberis saxatilis, p.322
IRISES, pp.198–9
Leontopodium stracheyi
Lewisia rediviva [white form], p.321
Ourisia caespitosa, p.322
Pieris nana
Pratia angulata
PRIMULAS, pp.236–7
Saxifraga burseriana 'Crenata'

☐ WHITE

Maianthemum canadense
Vigorous, ground-cover, rhizomatous perennial with large, upright, oval, wavy-edged, glossy leaves. Slender stems bear sprays of small, white flowers in late spring and early summer followed by red berries. H 10cm (4in), S indefinite.

Sanguinaria canadensis
(Bloodroot)
Rhizomatous perennial with fleshy, underground stems that exude red sap when cut. In spring bears white flowers, sometimes pink-flushed or slate-blue on reverses, as blue-grey leaves unfurl. H 10–15cm (4–6in), S 30cm (12in).

Ranunculus alpestris
(Alpine buttercup)
Short-lived, evergreen, clump-forming perennial that bears cup-shaped, white flowers on erect stems from late spring to mid-summer. Glossy, dark green leaves are rounded and serrated. H 2.5–12cm (1–5in), S 10cm (4in).

Cassiope lycopodioides
Evergreen, prostrate, mat-forming shrub with slender stems densely set with minute, scale-like, dark green leaves. In spring, short, reddish stems carry tiny, bell-shaped, white flowers, in red calyces, singly in leaf axils. H 8cm (3in), S 30cm (12in).

Saxifraga burseriana
Slow-growing, evergreen perennial with hard cushions of spiky, grey-green leaves. In spring bears open cup-shaped, white flowers on short stems. H 2.5–5cm (1–2in), S to 10cm (4in).

Pulsatilla vernalis
Tufted perennial with rosettes of feathery leaves. Densely hairy, brown flower buds appear in late winter and open in early spring to somewhat nodding, open cup-shaped, pearl-white flowers. Buds dislike winter wet. H 5–10cm (2–4in), S 10cm (4in).

Androsace villosa
Evergreen, mat-forming perennial with very hairy rosettes of tiny leaves. Bears umbels of small, white flowers, with yellow centres that turn red, in spring. H 2.5cm (1in), S 20cm (8in).

Salix apoda
Slow-growing, deciduous, prostrate shrub. In early spring, male forms bear fat, silky, silver catkins with orange to pale yellow stamens and bracts. Oval, leathery leaves are hairy when young, becoming dark green later. H to 15cm (6in), S 30–60cm (12–24in).

Ranunculus ficaria 'Albus'
Mat-forming perennial bearing in early spring cup-shaped, single, creamy-white flowers with glossy petals. Leaves are heart-shaped and dark green. Can spread rapidly; is good for a wild garden. H 5cm (2in), S 20cm (8in).

Saxifraga burseriana 'Gloria'
Silene alpestris, p.321
Tanakaea radicans
Thlaspi alpinum

311

Rock plants/small SPRING INTEREST
WHITE–PINK

Cassiope mertensiana
Evergreen, dwarf shrub with scale-like, dark green leaves tightly pressed to stems. In early spring carries bell-shaped, creamy-white flowers, with green or red calyces, in leaf axils. H 15cm (6in), S 20cm (8in).

Scoliopus bigelowii, syn. S. bigelovii
Compact perennial with basal, veined leaves, sometimes marked brown. In early spring bears flowers with purple inner petals and greenish-white outer petals with deep purple lines. H 8–10cm (3–4in), S 10–15cm (4–6in).

Gypsophila cerastioides
Prostrate perennial with a profusion of small, saucer-shaped, purple-veined, white flowers borne in late spring and early summer above mats of rounded, velvety, mid-green foliage. H 2cm (³/₄in), S to 10cm (4in) or more.

Paraquilegia anemonoides, syn. P. grandiflora
Tufted perennial with fern-like, blue-green leaves. In spring, pale lavender-blue buds open to pendent, cup-shaped, almost white flowers borne singly on arching stems. May be difficult to establish. H and S 10–15cm (4–6in).

Corydalis popovii
Tuberous perennial with leaves divided into 3–6 bluish-green leaflets. In spring bears loose racemes of deep red-purple and white flowers, each with a long spur. Keep dry when dormant. H and S 10–15cm (4–6in).

Cotula atrata var. **luteola**
Evergreen, mat-forming perennial that in late spring and early summer bears blackish-red flower heads with creamy-yellow stamens. Leaves are small, finely cut and dark green. Needs adequate moisture; best in an alpine house. H 2.5cm (1in), S to 25cm (10in).

Anemonella thalictroides
Perennial with delicate, fern-like leaves growing from a cluster of small tubers. From spring to early summer bears small, cup-shaped, white or pink flowers, singly on finely branched stems. Needs humus-rich soil. H 10cm (4in), S 4cm (1¹/₂in) or more.

Lewisia tweedyi
Evergreen, rosetted perennial with large, fleshy leaves and stout, branched stems that bear open cup-shaped, many-petalled, white to pink flowers in spring. Best grown in an alpine house. H 15cm (6in), S 12–15cm (5–6in).

Shortia galacifolia (Oconee bells)
Evergreen, clump-forming, dwarf perennial with round, toothed, leathery, glossy leaves. In late spring bears cup- to trumpet-shaped, often pink-flushed, white flowers with deeply serrated petals. H to 15cm (6in), S 15–23cm (6–9in).

Daphne jasminea
Evergreen, compact shrub. Bears small, white flowers, pink-flushed externally, in late spring and early summer and again in autumn. Brittle stems are clothed in grey-green leaves. Suits an alpine house or a dry wall. H 8–10cm (3–4in), S to 30cm (12in).

Epigaea gaultherioides, syn. *Orphanidesia gaultherioides*
Evergreen, prostrate sub-shrub with cup-shaped, shell-pink flowers borne in terminal clusters in spring. Hairy stems carry heart-shaped, dark green leaves. Is difficult to grow and propagate. H to 10cm (4in), S to 25cm (10in) or more.

Antennaria dioica
Arctostaphylos uva-ursi 'Point Reyes', p.336
Claytonia virginica
Dryas octopetala, p.323
Epigaea repens
Hypsela reniformis
Kelseya uniflora
Ranunculus glacialis
Shortia uniflora 'Grandiflora'

■ PINK

Trillium rivale
Perennial with oval leaves, divided into 3 leaflets. In spring bears open cup-shaped, white or pale pink flowers with dark-spotted, heart-shaped petals, singly on upright, later arching stems. H to 15cm (6in), S 10cm (4in).

Saxifraga x *irvingii* 'Jenkinsiae', syn. *S.* 'Jenkinsiae'
Slow-growing perennial with very tight, grey-green cushions of foliage. Carries a profusion of open cup-shaped, lilac-pink flowers on slender stems in early spring. H 8–10cm (3–4in), S to 15cm (6in).

Androsace carnea
Evergreen, cushion-forming perennial that has small rosettes of pointed leaves with hairy margins. In spring, 2 or more stems rise above each rosette, bearing tiny, single, pink flowers. Suits a trough. H and S 5cm (2in).

Thlaspi rotundifolium
Clump-forming perennial with dense tufts of round leaves and tight, flat heads of small, open cup-shaped, pale to deep purplish- or lilac-pink flowers in spring. Needs ample moisture and cool conditions. May be short-lived. H 5–8cm (2–3in), S 10cm (4in).

Oxalis acetosella var. **purpurascens**, syn. *O. a.* f. *rosea*
Creeping, rhizomatous perennial forming mats of 3-lobed leaves. Cup-shaped, soft pink flowers, each 1cm (½in) across, with 5 darker-veined petals, are produced in spring. H 5cm (2in), S indefinite.

Armeria juniperifolia, syn. *A. caespitosa*, *A. cespitosa*
Evergreen, cushion-forming perennial composed of loose rosettes of sharp-pointed, mid- to grey-green leaves. Pale pink flowers are borne in spherical umbels in late spring and early summer. H 5–8cm (2–3in), S 15cm (6in).

Arenaria purpurascens
Evergreen, mat-forming perennial with sharp-pointed, glossy leaves, above which rise many small clusters of star-shaped, pale to deep purplish-pink flowers in early spring. H 1cm (½in), S to 15cm (6in).

Silene acaulis (Moss campion)
Evergreen, cushion-forming perennial with minute, bright green leaves studded with tiny, stemless, 5-petalled, pink flowers in spring. May be difficult to bring into flower; prefers a cool climate. H to 2.5cm (1in), S 15cm (6in).

Arabis caucasica 'Rosabella'
Evergreen, mat-forming perennial with a profusion of single, deep pink flowers in spring and early summer and large rosettes of small, oval, soft green leaves. H 15cm (6in), S 30cm (12in).

Aethionema 'Warley Rose', p.324
Androsace hedraeantha
Armeria maritima
Dicentra peregrina

Erodium corsicum, p.324
Ourisia microphylla, p.324
x Phyllothamnus erectus
PRIMULAS, pp.236–7

Rhodohypoxis baurii 'Margaret Rose', p.324

313

Rock plants/small SPRING INTEREST
PINK

Oxalis adenophylla
Mat-forming, fibrous-rooted, tuberous perennial with grey-green leaves divided into narrow, finger-like, wavy lobes. In spring bears rounded, purplish-pink flowers, each 2.5–4cm (1–1½in) across, with darker purple eyes. H to 5cm (2in), S 8–10cm (3–4in).

Shortia soldanelloides
Evergreen, mat-forming perennial with rounded, toothed leaves and small, pendent, bell-shaped and fringed, deep pink flowers in late spring. H 5–10cm (2–4in), S 10–15cm (4–6in).

Erinus alpinus
Semi-evergreen, short-lived perennial with rosettes of soft, mid-green leaves covered, in late spring and summer, with small, purple, pink or white flowers. Self seeds freely. H and S 5–8cm (2–3in).

Antennaria dioica 'Rosea'
Semi-evergreen perennial forming a spreading mat of tiny, oval, woolly leaves. Bears fluffy, rose-pink flower heads in small, terminal clusters in late spring and early summer. Is good as ground cover with small bulbs. H 2.5cm (1in), S to 40cm (16in).

Daphne arbuscula
Evergreen, prostrate shrub. In late spring bears many very fragrant, tubular, deep pink flowers in terminal clusters. Narrow, leathery, dark green leaves are crowded at the ends of the branches. Likes humus-rich soil. H 10–15cm (4–6in), S 50cm (20in).

Saxifraga oppositifolia
(Purple mountain saxifrage)
Evergreen, prostrate perennial with clusters of tiny, white-flecked leaves. Has open cup-shaped, dark purple, purplish-pink or, rarely, white flowers in early spring. Likes an open position. H 2.5–5cm (1–2in), S 15cm (6in).

Daphne petraea 'Grandiflora'
Slow-growing, evergreen, compact shrub that bears terminal clusters of fragrant, rich pink flowers in late spring and tiny, glossy leaves. Suits an alpine house, a sheltered, humus-rich rock garden or a trough. H to 15cm (6in), S to 25cm (10in).

Claytonia megarhiza var. nivalis
Evergreen perennial with a rosette of spoon-shaped, succulent leaves. Bears small heads of tiny, deep pink flowers in spring. Grows best in a deep pot of gritty compost in an alpine house. H 1cm (½in), S 8cm (3in).

Vaccinium vitis-idaea var. minus, syn. *V. v.-i.* 'Minus'
Evergreen, mat-forming sub-shrub with tiny, oval, leathery leaves. In late spring produces small, erect racemes of many tiny, bell-shaped, deep pink or deep pink-and-white flowers. H 5–8cm (2–3in), S 10–15cm (4–6in).

Androsace sarmentosa
Anemonella thalictroides 'Oscar Schoaf'
Antennaria dioica 'Nyewoods'
Arabis blepharophylla

Erinus alpinus 'Dr Hähnle'
Geranium cinereum 'Ballerina', p.327
Lewisia rediviva [pink form], p.327
PRIMULAS, pp.236–7

Salix lindleyana
Saxifraga x angelica
Saxifraga 'Walter Irving'
Shortia soldanelloides var. ilicifolia

Shortia soldanelloides var. *magna*
Viola cazorlensis

■ PINK–RED

Anagallis tenella 'Studland'
Short-lived perennial that forms prostrate mats of tiny, bright green leaves studded in spring with honey-scented, star-shaped, bright pink flowers. H 1cm (½in), S 15cm (6in) or more.

Androsace carnea subsp. **laggeri**
Evergreen, cushion-forming perennial composed of small, tight rosettes of pointed leaves. Cup-shaped, deep pink flowers are borne in small clusters above cushions in spring. H and S 5cm (2in).

Corydalis solida 'George Baker', syn. *C.s.* 'G.P. Baker'
Tuberous perennial with fern-like, divided leaves and dense racemes of spurred, rich deep rose-red flowers in spring. H and S 10–15cm (4–6in).

Saxifraga grisebachii 'Wisley Variety', syn. *S. federici-augusti* subsp. *grisebachii* 'Wisley Variety'
Evergreen perennial with rosettes of lime-encrusted leaves. Crosier-shaped stems with pale pink to bright red hairs, bear dense racemes of dark red flowers in spring. H 10cm (4in), S 15cm (6in).

Saxifraga sempervivum
Evergreen, hummock-forming perennial with tight rosettes of tufted, silvery-green leaves. Crosier-shaped flower stems, covered in silvery hairs and emerging from rosettes, bear racemes of dark red flowers in early spring. H and S 10–15cm (4–6in).

Cotula atrata
Evergreen, mat-forming perennial with small, finely cut, greyish-green leaves and blackish-red flower heads in late spring and early summer. Is uncommon and not easy to grow successfully. H 2.5cm (1in), S to 25cm (10in).

■ PURPLE

Corydalis diphylla
Tuberous perennial with semi-erect, basal leaves, divided into narrow leaflets, and loose racemes of purple-lipped flowers with white spurs in spring. Protect tubers from excess moisture in summer. H 10–15cm (4–6in), S 8–10cm (3–4in).

Polygonatum hookeri
Slow-growing, dense, rhizomatous perennial that bears loose spikes of several small, bell-shaped, lilac-pink flowers in late spring and early summer. Leaves are tiny and lance-shaped. Suits a peat bed. H to 5cm (2in), S to 30cm (12in).

Aubrieta 'Joy'
Vigorous, evergreen, trailing perennial that forms mounds of soft green leaves. In spring bears double, pale mauve flowers on short stems. H 10cm (4in), S 20cm (8in).

Aethionema 'Warley Ruber'
Androsace sempervivoides
Armeria juniperifolia 'Bevan's Variety'
Dodecatheon pulchellum

Penstemon davidsonii
PRIMULAS, pp.236–7
Rhodohypoxis baurii 'Albrighton', p.328
Rhodohypoxis baurii 'Douglas', p.328

PRIMULAS, pp.236–7
Ramonda nathaliae
Ramonda serbica
Viola 'Haslemere', p.329

Rock plants/small SPRING INTEREST
■ PURPLE

Aubrieta deltoidea **'Argenteo-variegata'**
Evergreen, compact perennial, grown for its trailing, green leaves which are heavily splashed with creamy-white. Produces pinkish-lavender flowers in spring. H 5cm (2in), S 15cm (6in).

Mazus reptans
Prostrate perennial that has tubular, purple or purplish-pink flowers, with protruding, white lips, spotted red and yellow, borne singly on short stems in spring. Narrow, toothed leaves are in pairs along stem. H to 5cm (2in), S 30cm (12in) or more.

Soldanella villosa
Evergreen, clump-forming perennial with round, leathery, hairy-stalked leaves and nodding, bell-shaped, fringed, purplish-lavender flowers borne on erect stems in early spring. Dislikes winter wet. H 10cm (4in), S 10–15cm (4–6in).

Aubrieta **'Carnival'**
Vigorous, evergreen, mound-forming perennial that carries many short spikes of large, single, violet-purple flowers in spring above small, soft green leaves. H 10cm (4in), S 30cm (12in).

Soldanella alpina
(Alpine snowbell)
Evergreen, clump-forming perennial with tufts of leaves and short, bell-shaped, fringed, pinkish-lavender or purplish-pink flowers in early spring. Is difficult to flower well. H to 8cm (3in), S 8–10cm (3–4in).

Saxifraga stribrnyi
Evergreen, mound-forming perennial with small, lime-encrusted rosettes of leaves. Crosier-shaped stems, covered in pinkish-buff hairs, bear racemes of deep maroon-red flowers above leaves in late spring and early summer. H 8cm (3in), S 10–12cm (4–5in).

Polygala chamaebuxus var. *grandiflora*, syn. *P.c.* var. *rhodoptera*
Evergreen, woody-based perennial that bears terminal clusters of pea-like, reddish-purple and yellow flowers in late spring and early summer. Leaves are small, oval, leathery and dark green. H to 15cm (6in), S to 30cm (12in).

Aubrieta **'J.S. Baker'**
Evergreen perennial with single, purple flowers in spring borne above mounds of small, soft green leaves. H 10cm (4in), S 20cm (8in).

Viola calcarata
Clump-forming perennial, with oval leaves, that bears flat, outward-facing, single, white, lavender or purple flowers for a long period from late spring to summer. Prefers rich soil. H 10–15cm (4–6in), S to 20cm (8in).

Aubrieta 'Dr Mule's'
IRISES, pp.198–9
Polygala vayredae
Saxifraga x arco-valleyi 'Arco'

Saxifraga oppositifolia 'Ruth Draper'
Soldanella minima
Soldanella montana
Viola obliqua

Viola odorata
Viola palmata

316

PURPLE–BLUE

Aubrieta 'Cobalt Violet'
Evergreen, mound-forming perennial with single, blue-violet flowers carried in short, terminal spikes in spring above a mat of small, soft green leaves. H 10cm (4in), S 20cm (8in).

Viola tricolor
(Heartsease, Wild pansy)
Short-lived perennial or annual with neat, flat-faced flowers in combinations of white, yellow and shades of purple from spring to autumn. Self seeds profusely. H 5–15cm (2–6in), S 5–15cm (2–6in) or more.

Viola labradorica 'Purpurea'
Clump-forming perennial with tiny, flat-faced, purple flowers in spring-summer. Leaves are kidney-shaped and dark purple-green. Is invasive but suits a bank, woodland or wild garden. H 2.5–5cm (1–2in), S indefinite.

Viola pedata (Bird's-foot violet)
Clump-forming perennial with finely divided foliage and yellow-centred, pale violet, rarely white flowers borne singly on slender stems in late spring and early summer. Needs sharp drainage; grow in an alpine house. H 5cm (2in) S 8cm (3in).

Mertensia maritima
Prostrate perennial with oval, fleshy, bright silver-blue or silver-grey leaves. Stout stems carry clusters of pendent, funnel-shaped, sky-blue flowers in spring. Is prone to slug damage. Needs very sharp drainage. H 10–15cm (4–6in), S 12cm (5in).

Hepatica nobilis var. **japonica**
Slow-growing perennial with leathery, lobed leaves, semi-evergreen in all but very cold or arid climates. Bears slightly cupped, lilac-mauve, pink or white flowers in spring. H to 8cm (3in), S to 12cm (5in).

Jeffersonia dubia,
syn. *Plagiorhegma dubia*
Tufted perennial with 2-lobed, blue-green leaves, sometimes flushed pink when unfolding. Bears cup-shaped, pale lilac to purplish blue flowers singly in spring. H 10–15cm (4–6in), S to 23cm (9in).

Jankaea heldreichii
Perennial with rosettes of thick, hairy, silver-green leaves, above which rise slender stems bearing clusters of tiny, lavender-blue flowers in late spring. Is rare and difficult to grow and is best in an alpine house. H and S to 8cm (3in).

Synthyris stellata
Evergreen, mounded, rhizomatous perennial that bears dense spikes of small, violet-blue flowers in spring above rounded, deeply toothed leaves. Tolerates sun if soil remains moist. H 10–15cm (4–6in), S 15cm (6in).

Aubrieta 'Gurgedyke'
Haberlea ferdinandi-coburgii
Haberlea rhodopensis
Hedyotis michauxii, p.331

IRISES, pp.198–9
Omphalogramma vinciflorum
Phlox 'Emerald Cushion', p.329
Thalictrum orientale

Rock plants/small SPRING INTEREST

■ BLUE

■ BLUE–GREEN

Myosotis alpestris
(Alpine forget-me-not)
Short-lived, clump-forming perennial producing dense clusters of tiny, bright blue flowers with creamy-yellow eyes in late spring and early summer, just above tufts of hairy leaves. Prefers gritty soil. H and S 10–15cm (4–6in).

☼ ◊ ✶✶✶

Anchusa caespitosa
Evergreen, mound-forming perennial with rosettes of lance-shaped, dark green leaves. In spring, stemless, white-centred, blue flowers appear in centres of rosettes. Old plants do not flower well; take early summer cuttings.
H 2.5–5cm (1–2in), S to 23cm (9in).

☼ ◊ ✶✶✶ ♀

Gentiana acaulis, syn. *G. excisa*, *G. kochiana* (Stemless gentian)
Evergreen, clump-forming perennial with narrowly oval, glossy leaves. Bears trumpet-shaped, deep blue flowers, with green-spotted throats, on short stems in spring and often in autumn. H in leaf 2cm ($^3/_4$in), S to 10cm (4in) or more.

☼ ◊ ✶✶✶ ♀

Viola tricolor 'Bowles' Black'
Clump-forming perennial with flat-faced, very dark violet, almost black, flowers, borne continuously from spring to autumn. Oval leaves are sometimes lobed and toothed. Is short-lived; treat as biennial. H 5–15cm (2–6in), S 5–8cm (2–3in).

☼ ◊ ✶✶✶

Gentiana verna (Spring gentian)
Evergreen perennial, often short-lived, with small rosettes of oval, dark green leaves. In early spring, tubular, bright blue flowers with white throats are held upright on short stems. H and S to 5cm (2in).

☼ ◊ ✶✶✶

Salix reticulata
(Net-veined willow)
Deciduous, spreading, mat-forming shrub. Carries plump, reddish-brown, then yellow catkins on male plants in spring and rounded, slightly crinkled leaves. Likes cool, peaty soil. H 5–8cm (2–3in), S 20cm (8in) or more.

☼ ◊ ✶✶✶ ♀

Eritrichium nanum, p.331
Hepatica x *media* 'Ballardii'
IRISES, pp.198–9
Omphalodes luciliae

Polygala calcarea, p.332
Polygala calcarea 'Bulley's Variety', p.332
Synthyris reniformis

Hepatica nobilis
Hepatica transsilvanica

■ GREEN–YELLOW   □ YELLOW

Mandragora officinarum
Rosetted, fleshy-rooted perennial with coarse, wavy-edged leaves. Bears funnel-shaped, yellowish- or purplish-white flowers in spring, followed by large, tomato-like, shiny yellow fruits. H 5cm (2in), S 30cm (12in).

Saxifraga 'Hindhead Seedling'
Evergreen perennial that forms a hard dome of small, tufted, spiny, blue-green leaves. In spring bears upward-facing, open, cup-shaped, pale yellow flowers, 2 or 3 to each short stem. H 2.5cm (1in), S 8cm (3in).

Viola aetolica
Clump-forming perennial bearing flat-faced, yellow flowers singly on upright stems in late spring and early summer. Leaves are oval and mid-green. H 5–8cm (2–3in), S 15cm (6in).

Draba longisiliqua
Semi-evergreen, cushion-forming perennial composed of firm rosettes of tiny, silver leaves. Bears sprays of small, yellow flowers on long stalks in spring. Needs plenty of water in growth; is best grown in an alpine house. H 5–8cm (2–3in), S 15cm (6in).

Hacquetia epipactis
Clump-forming perennial spreading by short rhizomes. In late winter and early spring bears yellow or yellow-green flower heads, encircled by apple-green bracts, before rounded, 3-parted leaves appear. H 6cm (2½in), S 15–23cm (6–9in).

Saxifraga x apiculata
Evergreen perennial with a tight cushion of bright green foliage. Bears clusters of open, cup-shaped, pale yellow flowers in early spring. H 10–15cm (4–6in), S 15cm (6in) or more.

Draba rigida
Evergreen perennial with tight hummocks of minute, dark green leaves. Tiny clusters of bright yellow flowers on fine stems cover hummocks in spring. Suits a rough, scree garden or alpine house. Dislikes winter wet. H 4cm (1½in), S 6cm (2½in).

Saxifraga sancta
Evergreen, mat-forming perennial with tufts of bright green leaves. Bears short racemes of upward-facing, open cup-shaped, bright yellow flowers in spring. H 5cm (2in), S 15cm (6in).

Euphorbia myrsinites
Evergreen, prostrate perennial with terminal clusters of bright yellow-green flowers in spring. Woody stems are clothed in small, pointed, fleshy, grey leaves. Is good on a wall or ledge. H 5–8cm (2–3in), S to 20cm (8in) or more.

Saxifraga 'Elizabethae', syn. *S.* 'Carmen'
Evergreen, cushion-forming perennial, composed of densely packed, tiny rosettes of spiny leaves. In spring, tight upward-facing, bright yellow flowers are carried on tops of red-based stems. H 2.5cm (1in), S 10–15cm (4–6in).

Dionysia aretioides
Evergreen perennial forming cushions of soft, hairy, greyish-green leaves that are covered in early spring by scented, stemless, round, bright yellow flowers. H 5–10cm (2–4in), S 15–30cm (6–12in).

Corydalis wilsonii, p.297

Calceolaria darwinii
Cytisus ardoinii, p.334
DAFFODILS, pp.360–62
Dionysia lamingtonii

Draba aizoides
Draba hispanica
Draba polytricha
Draba rigida var. *bryoides*

IRISES, pp.198–9
Onosma stellulatum
Polygala chamaebuxus, p.333
PRIMULAS, pp.236–7

319

Rock plants/small SPRING INTEREST
YELLOW

Draba mollissima
Semi-evergreen, cushion-forming perennial covered in spring with clusters of tiny, yellow flowers on slender stems. Minute leaves form a soft green dome, which should be packed beneath with small stones. Grow in an alpine house. H 4cm (1½in), S 15cm (6in) or more.

Dionysia tapetodes
Evergreen, prostrate perennial producing a tight mat of tiny, grey-green leaves. Bears small, upward-facing, yellow flowers in early spring. H 1cm (½in), S to 15cm (6in).

Morisia monanthos, syn. *M. hypogaea*
Prostrate perennial with flat rosettes of divided, leathery, dark green leaves. Bears stemless, flat, bright yellow flowers in late spring and early summer. Needs very sharp drainage. H 2.5cm (1in), S to 8cm (3in).

Ranunculus ficaria 'Flore Pleno'
Mat-forming perennial with heart-shaped, dark green leaves and, in early spring, double, bright yellow flowers with glossy petals. May spread rapidly. Is good for a wild garden. H 2.5–5cm (1–2in), S 20cm (8in).

Vitaliana primuliflora, syn. *Douglasia vitaliana*
Evergreen, prostrate perennial with a mat of rosetted, mid-green leaves that are covered in spring with many small clusters of stemless, tubular, bright yellow flowers. H 2.5cm (1in), S 20cm (8in).

Viola 'Jackanapes'
Clump-forming perennial with oval, toothed leaves. Produces flat-faced flowers with reddish-brown, upper petals and yellow, lower ones throughout late spring and summer. H 8–12cm (3–5in), S to 20cm (8in) or more.

Ranunculus ficaria 'Brazen Hussy'
Salix herbacea
Sarcocapnos enneaphylla
Saxifraga aizoides

Saxifraga brunonis
Saxifraga burseriana 'Brookside'
Saxifraga ferdinandi-coburgii
Saxifraga 'Valerie Finnis'

Sedum acre 'Aureum', p.333
Townsendia parryi
Viola glabella
Viola lutea

Waldsteinia fragarioides
Waldsteinia ternata, p.333

Rock plants/small SUMMER INTEREST

☐ YELLOW

☐ WHITE

Trollius pumilus
Tufted perennial with leaves divided into 5 segments, each further lobed. Carries solitary cup-shaped, bright yellow flowers in late spring and early summer. H 15cm (6in), S 15cm (6in) or more.

Silene alpestris,
syn. *Heliosperma alpestris*
Perennial with branching stems and narrow leaves. Bears small, rounded, fringed, white, occasionally pink-flushed flowers in late spring and early summer. Self seeds freely. H 10–15cm (4–6in), S 20cm (8in).

Haberlea rhodopensis 'Virginalis'
Evergreen perennial with small, arching sprays of funnel-shaped, pure white flowers borne in late spring and early summer above neat rosettes of oval, toothed, dark green leaves. H and S in flower 10–15cm (4–6in).

Cyananthus lobatus f. *albus*
Prostrate perennial with branched stems clothed in small, wedge-shaped, dull green leaves. Bears funnel-shaped, single, white flowers with spreading lobes in late summer. H 8cm (3in), S 30cm (12in).

Erysimum helveticum,
syn. *E. pumilum*
Semi-evergreen, clump-forming perennial with closely-packed tufts of long, narrow leaves and many fragrant, bright yellow flowers borne in flat heads in late spring and early summer. H 10cm (4in), S 15cm (6in).

Phlox stolonifera 'Ariane'
Evergreen, low-growing perennial with flowering side shoots that bear heads of open, saucer-shaped, white blooms in early summer. Has oval, pale green leaves. Cut back flowered shoots by half after flowering. H to 15cm (6in), S 30cm (12in).

Potentilla alba
Vigorous mat-forming perennial bearing loose sprays of flat, single, white flowers in summer. Leaves are divided into oval leaflets and are silvery beneath. H 5–8cm (2–3in), S 8cm (3in).

Campanula carpatica 'Bressingham White'
Clump-forming perennial bearing open cup-shaped, white flowers, singly on unbranched stems, in summer. Has abundant, rounded, bright green leaves. H 10–15cm (4–6in), S 15cm (6in).

Ranunculus ficaria 'Aurantiacus'
Mat-forming perennial bearing in early spring cup-shaped, single, orange flowers with glossy petals. Leaves are heart-shaped and mid-green. May spread rapidly. Is good for a wild garden. H 5cm (2in), S 20cm (8in).

Achillea clavennae
Semi-evergreen, carpeting perennial that bears loose clusters of white flower heads with gold centres from summer to mid-autumn. Leaves are narrowly oval, many-lobed and covered with fine, white hairs. Dislikes winter wet. H 15cm (6in), S 23cm (9in) or more.

Lewisia rediviva [white form]
(Bitter root)
Tufted, rosetted perennial with clusters of fine, narrow leaves that are summer-deciduous. Bears large, white flowers that open in bright weather in late spring and early summer. H 1–4cm (½–1½in), S to 5cm (2in).

DAFFODILS, pp.360–62
Hypericum cerastoides
IRISES, pp.198–9
Leucogenes grandiceps, p.337

OTHER RECOMMENDED PLANTS:
Aethionema iberideum
Arabis caucasica
Arabis caucasica 'Plena'

Arabis caucasica 'Variegata', p.310
Arctostaphylos nevadensis
Arenaria balearica, p.310
Artemisia schmidtiana 'Nana', p.337

Cardamine trifolia, p.310
Celmisia bellidioides
Celmisia traversii
Cerastium alpinum

Rock plants/small SUMMER INTEREST
☐ WHITE

Arenaria montana
Prostrate perennial that forms loose mats of small, narrowly oval leaves and bears large, round, white flowers in summer. Suits a wall or rock crevice. Must have adequate moisture. H 5cm (2in), S 12cm (5in).

Celmisia ramulosa
Evergreen, shrubby perennial with small, hairy, grey-green leaves. Daisy-like, white flower heads are borne singly on short stems in late spring and early summer. H and S 10cm (4in).

Anacyclus depressus
Short-lived, prostrate perennial that has clusters of white flower heads, with red reverses to ray petals, in summer. Flowers close in dull light. Stems clothed in fine leaves radiate from central root. Dislikes wet. H 2.5–5cm (1–2in) or more, S 10cm (4in).

Epilobium chlorifolium* var. *kaikourense
Clump-forming, woody-based perennial with deciduous but persistent, oval, hairy, bronze and dark green leaves. In summer has short spikes of funnel-shaped, white to pink flowers. H 10cm (4in), S 15cm (6in).

Iberis saxatilis
Evergreen, dwarf sub-shrub that in late spring and early summer produces large heads of numerous small, white flowers, which become tinged violet with age. Glossy, dark leaves are linear and cylindrical. Trim after flowering. H 8–12cm (3–5in), S 30cm (12in).

Ourisia caespitosa
Evergreen, prostrate perennial with creeping rootstocks and stems bearing tiny, oval leaves and many outward-facing, open cup-shaped, white flowers in late spring and early summer. H 2.5cm (1in), S 10cm (4in).

Petrocosmea kerrii
Evergreen perennial with compact rosettes of oval, pointed, hairy, rich green leaves. In summer bears clusters of short, outward-facing, tubular, open-mouthed white flowers. Suits an alpine house. H to 8cm (3in), S 12–15cm (5–6in). Min. 2–5°C (36–41°F).

Gentiana saxosa
Evergreen, hummock-forming perennial clothed in small, spoon-shaped, fleshy, dark green leaves. Produces small, upturned, bell-shaped, white flowers in early summer. Is a short-lived scree plant. H 5cm (2in), S 15cm (6in).

Cornus canadensis
(Creeping dogwood)
Ground-cover perennial with whorls of oval leaves. In late spring and early summer bears green, sometimes purple-tinged flowers, within white bracts, followed by red berries. H 10–15cm (4–6in), S 30cm (12in) or more.

Nierembergia repens
Mat-forming perennial with upright, open bell-shaped, yellow-centred, white flowers, occasionally flushed pink with age, borne for a long period in summer. Leaves are small, oval and light green. Is useful for cracks in paving. H 5cm (2in), S 20cm (8in) or more.

Cerastium tomentosum, p.309
Chrysanthemum alpinum
Cymbalaria muralis
Diapensia lapponica

Dryas drummondii
Dryas x *suendermannii*
Helichrysum bellidioides
Lewisia nevadensis

Maianthemum bifolium
Pentachondra pumila
Petrophytum caespitosum
Petrophytum hendersonii

Phlox douglasii 'May Snow'
Ranunculus alpestris, p.311
Satureja montana 'Prostrate White'
Weldenia candida, p.310

WHITE–PINK

Dryas octopetala
Evergreen, prostrate perennial forming mats of oval, lobed, leathery, dark green leaves on stout stems. In late spring and early summer, cup-shaped, creamy-white flowers are borne just above foliage, followed by attractive, feathery seeds. H 6cm (2½in), S indefinite.

Achillea x kellereri
Semi-evergreen perennial that bears daisy-like, white flower heads in loose clusters in summer. Leaves are feathery and grey-green. Is good for a wall or bank. Dislikes winter wet and must have perfect drainage. H 15cm (6in), S 23cm (9in) or more.

Carlina acaulis (Alpine thistle)
Clump-forming perennial that in summer-autumn bears large, stemless, thistle-like, single, off-white or pale brown flower heads, with papery bracts, on rosettes of long, spiny-margined, deeply-cut leaves. H 8–10cm (3–4in), S 15–23cm (6–9in).

Linnaea borealis (Twin flower)
Evergreen, mat-forming, sub-shrubby perennial with rooting stems bearing small, oval leaves, above which in summer rise thread-like stems bearing pairs of small, fragrant, tubular, pale pink and white flowers. H 2cm (¾in), S 30cm (12in) or more.

Dianthus pavonius, syn. *D. neglectus*
Evergreen, prostrate perennial with comparatively large, rounded, pale to deep pink flowers, buff on reverses, borne on short stems in summer above low mats of spiky leaves. H 5cm (2in), S 8cm (3in).

***Gypsophila repens* 'Dorothy Teacher'**
Semi-evergreen, prostrate perennial. Sprays of small, rounded, white flowers, which age to deep pink, cover mats of narrow, bluish-green leaves in summer. trim stems after flowering. H 2.5–5cm (1–2in), S 30cm (12in) or more.

PINK

Petrorhagia saxifraga, syn. *Tunica saxifraga* (Tunic flower)
Mat-forming perennial with tufts of grass-like leaves. In summer bears a profusion of small, pale pink flowers, veined deeper pink, on slender stems. Grows best on poor soil and self-seeds easily. H 10cm (4in), S 15cm (6in).

Thymus caespititius
Evergreen, mat-forming, aromatic sub-shrub with slender, woody stems covered in minute, hairy, mid-green leaves. Bears tiny, pale lilac or lilac-pink flowers in small clusters in summer. H 2.5cm (1in), S 20cm (8in).

Convolvulus althaeoides
Vigorous perennial with long, trailing stems clothed in heart-shaped, cut, mid-green leaves, overlaid silver. Bears large, open trumpet-shaped pink flowers in summer. May be invasive in a mild climate. H 5cm (2in), S indefinite.

Anemonella thalictroides, p.312
Daphne jasminea, p.312
Erodium reichardii
Geranium cinereum

Gypsophila cerastoides, p.312
Lewisia columbiana
Oenothera acaulis
Ranunculus glacialis

Antennaria dioica
Arctostaphylos uva-ursi, p.336
Arctostaphylos uva-ursi 'Point Reyes', p.336
CARNATIONS and PINKS, pp.244–5

Diascia barberiae 'Ruby Field'
Dicentra peregrina
Gaultheria trichophylla
Geranium traversii var. *elegans*

Rock plants/small SUMMER INTEREST
■ PINK

Geranium sanguineum var. **striatum**, syn. *G. s.* var. *lancastriense*
Hummock-forming, spreading perennial that has cup-shaped, pink flowers, with darker veins, borne singly in summer above round, deeply divided, dark green leaves. H 10–15cm (4–6in), S 30cm (12in) or more.

Erodium corsicum
Compact, clump-forming perennial that has soft, grey-green leaves with wavy margins. Bears flat-faced, pink flowers, with darker veins, on stiff, slender stems in late spring and summer. Grows best in an alpine house as dislikes winter wet. H 8cm (3in), S 15cm (6in).

Saponaria x olivana
Compact perennial with a firm cushion of narrow leaves. Flowering stems, produced around edges of the cushion, bear flat, single, pale pink flowers in summer. Needs very sharp drainage. H 8cm (3in), S 10cm (4in).

Aethionema 'Warley Rose'
Short-lived, evergreen or semi-evergreen, compact sub-shrub with tiny, linear, bluish-green leaves. Bears racemes of small, pink flowers on short stems in profusion in spring-summer. H and S 15cm (6in).

Ourisia microphylla
Semi-evergreen, mat-forming perennial, with neat, scale-like, pale green leaves, bearing a profusion of small, pink flowers in late spring and early summer. Is difficult to grow in an arid climate. H 5–10cm (2–4in), S 15cm (6in).

Phlox adsurgens 'Wagon Wheel'
Evergreen, prostrate perennial forming wide mats of woody stems, clothed in oval leaves. Bears heads of wheel-shaped, pink flowers with narrow petals in summer. Needs humus-rich soil. H 10cm (4in), S 30cm (12in).

Asperula suberosa
Clump-forming perennial with a mound of loose stems bearing tiny, hairy, grey leaves and, in early summer, many tubular, pale pink flowers. Dislikes winter wet but needs moist soil in summer. Is best in an alpine house. H 8cm (3in), S to 30cm (12in).

Rhodohypoxis baurii 'Margaret Rose'
Perennial with a tuber-like rootstock and an erect, basal tuft of narrowly lance-shaped, hairy leaves. Bears a succession of upright, flattish, pale pink flowers on slender stems in spring and early summer. H 5–10cm (2–4in), S 2.5–5cm (1–2in).

Polygonum affine 'Donald Lowndes'
Evergreen, mat-forming perennial that has stout, branching, spreading stems clothed with pointed leaves. In summer bears dense spikes of small, red flowers, which become paler with age. H 8–15cm (3–6in), S to 15cm (6in).

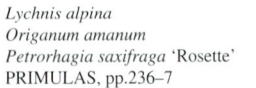

Lychnis alpina
Origanum amanum
Petrorhagia saxifraga 'Rosette'
PRIMULAS, pp.236–7

Sedum spurium
Silene hookeri
Vaccinium myrtillus

□ PINK

Geranium dalmaticum
Prostrate, spreading perennial with outward-facing, almost flat, shell-pink flowers borne in summer above divided, dark green leaves. Will grow taller in partial shade and is evergreen in all but severest winters. H 8–10cm (3–4in) or more, S 12–20cm (5–8in).

Androsace lanuginosa
Evergreen, trailing perennial with loose stems, covered in silky hairs, carrying deep green leaves and, in summer, clusters of small, flat, lilac-pink or pale pink flowers with dark pink or yellow eyes. H 4cm (1½in), S to 18cm (7in).

Loiseleuria procumbens
(Alpine azalea, Trailing azalea)
Evergreen, prostrate shrub with small, oval leaves, hairy and beige beneath. Has terminal clusters of open funnel-shaped, rose-pink to white flowers in early summer. H to 8cm (3in), S 10–15cm (4–6in).

Erigeron karvinskianus, syn. *E. mucronatus*
Spreading perennial with lax stems bearing narrow, lance-shaped, hairy leaves and, in summer-autumn, daisy-like flower heads that open white, turn pink and fade to purple. H 10–15cm (4–6in), S indefinite.

Dianthus 'Little Jock'
Evergreen, compact, clump-forming perennial with spiky, silvery-green foliage. In summer produces strongly fragrant, rounded, semi-double, pink flowers, with darker eyes, above foliage. H and S 10cm (4in).

Aethionema armenum
Short-lived, evergreen or semi-evergreen, dense sub-shrub with narrow, blue-green leaves. Carries loose sprays of tiny, pale to deep pink flowers in summer. H and S 15cm (6in).

Dianthus gratianopolitanus, syn. *D. caesius* (Cheddar pink)
Evergreen perennial with loose mats of narrow, grey-green leaves. In summer, produces very fragrant, flat, pale pink flowers on slender stems. H to 15cm (6in), S to 30cm (12in).

Acantholimon glumaceum
Evergreen, cushion-forming perennial with hard, spiny, dark green leaves and short spikes of small, star-shaped, pink flowers in summer. H 10cm (4in), S 20cm (8in).

Acantholimon venustum
Antennaria dioica 'Rosea', p.314
Arabis caucasica 'Rosabella', p.313
Armeria juniperinifolia, p.313

Armeria maritima
Astilbe simplicifolia 'Gnome'
Bruckenthalia spiculifolia
CARNATIONS and PINKS, pp.244–5

Dianthus deltoides
Epilobium obcordatum
Erodium × variabile 'Ken Aslet'
Geranium farreri

Gypsophila repens
Penstemon rupicola
Phlox adsurgens
Polygonum capitatum

325

Rock plants/small SUMMER INTEREST
■ PINK

Dianthus 'Pike's Pink'
Evergreen, compact, cushion-forming perennial, with spiky, grey-green foliage, that bears fragrant, rounded, double, pink flowers in summer. H and S 10cm (4in).

Saponaria caespitosa
Mat-forming perennial with small, lance-shaped leaves. Tiny, flat, single, pink to purple flowers are borne in small heads in summer. Needs very sharp drainage. H 8cm (3in), S 10cm (4in).

Oxalis depressa, syn. *O. inops*
Tuberous perennial with 3-lobed leaves and short-stemmed, widely funnel-shaped, bright rose-pink flowers, 2cm (³⁄₄in) across, in summer. Needs a sheltered site or cool greenhouse. H 5cm (2in), S 8–10cm (3–4in).

Androsace villosa var. **jacquemontii**
Evergreen, mat-forming perennial with small rosettes of hairy, grey-green leaves. Bears tiny, pinkish-purple flowers on red stems in late spring and early summer. Suits an alpine house. H 1–4cm (¹⁄₂–1¹⁄₂in), S 20cm (8in).

Dianthus microlepis
Evergreen perennial with tiny tufts of minute, fine, grass-like leaves, above which rise numerous small, rounded, pink flowers in early summer. Is best suited to a trough. H 5cm (2in), S 20cm (8in).

Saponaria ocymoides (Tumbling Ted)
Perennial with compact or loose, sprawling mats of hairy, oval leaves, above which a profusion of tiny, flat, pale pink to crimson flowers is carried in summer. Is excellent on a dry bank. H 2.5–8cm (1–3in), S 40cm (16in).

Dianthus myrtinervius
Evergreen, spreading perennial with numerous small, rounded, pink flowers that appear in summer above tiny, grass-like leaves. H 5cm (2in), S 20cm (8in).

Dianthus 'La Bourboule', syn. *D.* 'La Bourbille'
Evergreen perennial with small clumps of tufted, spiky foliage. Bears a profusion of strongly fragrant, small, single, pink flowers in summer. H 5cm (2in), S 8cm (3in).

Dianthus alpinus (Alpine pink)
Evergreen, compact perennial that bears comparatively large, rounded, rose-pink to crimson flowers, singly in summer, above mats of narrow, dark green foliage. Likes humus-rich soil. H 5cm (2in), S 8cm (3in).

Phlox subulata 'Marjorie'
Evergreen, mound-forming perennial with fine leaves and a profusion of flat, star-shaped, bright rose-pink flowers in early summer. Trim after flowering. H 10cm (4in), S 20cm (8in).

PRIMULAS, pp.236–7
Prunella grandiflora 'Pink Loveliness'
Vaccinium nummularia

☐ PINK

Dianthus **'Annabelle'**
Evergreen, compact, clump-forming perennial with spiky, grey-green foliage. In summer bears fragrant, rounded, semi-double, cerise-pink flowers, singly on slender stems. H and S 10cm (4in).

Teucrium polium
Deciduous, dome-shaped sub-shrub that has much-branched, woolly, white or yellowish stems and leaves with scalloped margins. Bears yellowish-white or pinkish-purple flowers in flat heads in summer. Requires very sharp drainage. H and S 15cm (6in).

Silene schafta
Spreading perennial with tufts of narrow, oval leaves. Bears sprays of 5-petalled, rose-magenta flowers from late spring to late autumn. H 10–15cm (4–6in), S 8–10cm (3–4in).

Phlox **'Camla'**
Evergreen mound-forming perennial with wiry, arching stems and fine leaves. Has a profusion of open saucer-shaped, rich pink flowers in early summer. Trim after flowering. Needs humus-rich soil. H 12cm (5in), S 30cm (12in).

Lewisia rediviva [pink form] (Bitter root)
Tufted, rosetted perennial. Clusters of narrow leaves are summer-deciduous. Large, many-petalled, pink flowers open in bright weather in late spring and early summer. Suits an alpine house. H 1–4cm ($^1/_2$–1$^1/_2$in), S to 5cm (2in).

Geranium cinereum **'Ballerina'**
Spreading, rosetted perennial that bears cup-shaped, purplish-pink flowers, with deep purple veins, on lax stems in late spring and summer. Basal leaves are round, deeply divided and soft. H 10cm (4in), S 30cm (12in).

Phlox douglasii **'Crackerjack'**
Evergreen, compact, mound-forming perennial. Has a profusion of saucer-shaped, bright crimson or magenta flowers in early summer. Leaves are lance-shaped and mid-green. Cut back after flowering. H to 8cm (3in), S 20cm (8in).

Geranium cinereum var. **sub-caulescens**, syn. *G. subcaulescens*
Spreading perennial with round, deeply divided, soft leaves. In summer bears brilliant purple-magenta flowers, with striking, black eyes and stamens, on lax stems. H 10cm (4in), S 30cm (12in).

Aethionema 'Warley Ruber'
Anemonella thalictroides 'Oscar Schoaf'
Antennaria dioica 'Nyewoods'
Armeria juniperifolia 'Bevan's Variety'
Erinus alpinus 'Dr Hähnle'
Phlox 'Millstream'
Saponaria 'Bressingham'
Saxifraga 'Primulaize'
Silene elisabethae

327

Rock plants/small SUMMER INTEREST

■ PINK

Rhodohypoxis baurii 'Albrighton'
Perennial with tuber-like rootstock and an erect, basal tuft of narrowly lance-shaped, hairy leaves. Bears a succession of erect, deep pink flowers singly on slender stems in spring and early summer. H 5–10cm (2–4in), S 2.5–5cm (1–2in).

Armeria maritima 'Vindictive'
Evergreen, clump-forming perennial with grass-like, dark blue-green leaves, above which rise stiff stems bearing spherical heads of small, deep rose-pink flowers for a long period in summer. H 10cm (4in), S 15cm (6in).

Dianthus deltoides 'Flashing Light', syn. 'Leuchtfunk'
Evergreen, mat-forming perennial. Many small, flat, upward-facing, brilliant cerise flowers are borne singly above tiny, oblong, pointed leaves. H 10–15cm (4–6in), S 20cm (8in).

■ RED–PURPLE

Rhodohypoxis baurii 'Douglas'
Perennial with a tuber-like rootstock and an erect, basal tuft of narrowly lance-shaped, hairy leaves. Bears a succession of upright, flattish, rich deep red flowers singly on slender stems in spring and early summer. H 5–10cm (2–4in), S 2.5–5cm (1–2in).

Penstemon hirsutus 'Pygmaeus'
Short-lived, evergreen, compact sub-shrub that bears tubular, lipped, hairy, purple- or blue-flushed, white flowers in summer. Has tightly packed, dark green leaves and is suitable for a trough. H and S 8cm (3in).

Pterocephalus perennis subsp. **perennis**, syn. *P.p.* var. *parnassi*
Semi-evergreen, mat-forming perennial with crinkled, hairy leaves. Bears tight, rounded heads of tubular, pinkish-lavender flowers, singly on short stems in summer, followed by feathery seed heads. H 5cm (2in), S 10cm (4in).

Dianthus 'Hidcote'
Dianthus 'Mars'
Erinus alpinus, p.314
Geranium stapfianum
Penstemon menziesii
Phlox douglasii 'Red Admiral'
Sedum sempervivioides
Viola cenisia

■ PURPLE

Phlox douglasii 'Boothman's Variety'
Evergreen, mound-forming perennial with lance-shaped leaves and masses of pale lavender-blue flowers, with violet-blue markings around eyes, in early summer. Cut back after flowering. H to 5cm (2in), S 20cm (8in).

Physoplexis comosa, syn. *Phyteuma comosum*
Tufted perennial with deeply cut leaves and round heads of bottle-shaped, violet-blue, rarely white, flowers in summer. Suits crevices but dislikes winter wet. H 8cm (3in), S 10cm (4in).

Globularia meridionalis, syn. *G. cordifolia* subsp. *bellidifolia*
Evergreen, dome-shaped sub-shrub. In summer, globular, fluffy, lavender to lavender-purple flower heads are borne singly just above glossy leaves. H to 10cm (4in), S to 20cm (8in).

Aster alpinus 'Beechwood'
Cotula atrata, p.315
IRISES, pp.198–9
Linaria alpina

■ PURPLE

Thymus leucotrichus
Evergreen, aromatic, mound-forming sub-shrub with fine, twiggy stems and narrow leaves fringed with white hairs. Bears dense heads of small, pinkish-purple flowers with purple bracts in summer. H 10–12cm (4–5in), S 15cm (6in).

☼ ◊ ❀❀❀

***Phlox* 'Emerald Cushion'**
Evergreen perennial with emerald-green mounds of fine leaves, studded in late spring and early summer with large, saucer-shaped, bright violet-blue flowers. Trim after flowering. H 8cm (3in), S 15cm (6in).

☼ ◊ ❀❀❀

Phlox bifida (Sand phlox)
Evergreen, mound-forming perennial with lance-shaped leaves. Bears a profusion of small heads of star-shaped, lilac or white flowers with deeply cleft petals in summer. Cut back stems by half after flowering. H 10–15cm (4–6in), S 15cm (6in).

☼ ◊ ❀❀❀

Mentha requienii
Origanum dictamnus
Oxalis enneaphylla
Penstemon 'Six Hills'

***Viola* 'Haslemere'**
Clump-forming perennial with small, oval, toothed leaves and flat-faced, lavender-pink flowers borne from late spring to late summer. Soil should not be too dry. H 8–15cm (3–6in), S to 20cm (8in).

☼ ◊ ❀❀❀❀ ♛

Campanula poscharskyana
Rampant, spreading perennial with bell-shaped, violet flowers borne on leafy stems in summer. Leaves are round with serrated edges. Vigorous runners make it suitable for a bank or a wild garden. H 10–15cm (4–6in), S indefinite.

☼ ◊ ❀❀❀

Phlox caespitosa
Polygala vayredae
Polygonatum hookeri, p.315
Saxifraga stribrnyi, p.316

Prunella grandiflora
(Large self-heal)
Semi-evergreen, basal-rosetted, ground-cover perennial with whorls of purple flowers borne in terminal spikes on leafy stems in summer. May be invasive; cut old flower stems before they seed. H 10–15cm (4–6in), S 30cm (12in).

☼ ● ❀❀❀

Sedum anacampseros
Viola calcarata, p.316
Viola hederacea
Viola pedata, p.317

Aster alpinus
Clump-forming, spreading perennial with lance-shaped, dark green leaves. Bears daisy-like, purplish-blue or pinkish-purple flower heads, with yellow centres, from mid- to late summer. H 15cm (6in), S 30–45cm (12–18in).

☼ ◊ ❀❀❀❀ ♛

Thymus herba-barona
(Caraway thyme)
Evergreen sub-shrub with a loose mat of tiny, caraway-scented, dark green leaves. In summer, small, lilac flowers are borne in terminal clusters. H in flower 5–10cm (2–4in), S to 20cm (8in).

☼ ◊ ❀❀

Edraianthus serpyllifolius
Evergreen, prostrate perennial with tight mats of tiny leaves and small, bell-shaped, deep violet flowers, borne on short stems in early summer. Is uncommon and seldom sets seed in gardens. H 1cm (½in), S to 5cm (2in).

☼ ◊ ❀❀❀

Viola tricolor, p.317

329

Rock plants/small SUMMER INTEREST
■ PURPLE

Prunella webbiana
Semi-evergreen, spreading, mat-forming perennial with basal rosettes of leaves. In mid-summer bears short spikes of funnel-shaped, purple flowers in whorls. H 10–15cm (4–6in), S 30cm (12in).

Campanula 'Birch Hybrid'
Vigorous, evergreen perennial with tough, arching, prostrate stems and ivy-shaped, bright green leaves. Bears many open bell-shaped, deep violet flowers in summer. H 10cm (4in), S 30cm (12in) or more.

Campanula portenschlagiana
Vigorous, evergreen, prostrate perennial with dense mats of small, ivy-shaped leaves and large clusters of erect, open bell-shaped, violet flowers in summer. H 15cm (6in), S indefinite.

Ramonda myconi, syn. *R. pyrenaica*
Evergreen, rosette-forming perennial with hairy, crinkled leaves and, in late spring and early summer, flat, blue-mauve, pink or white flowers, borne on branched stems. H 8cm (3in), S to 10cm (4in).

Viola 'Huntercombe Purple'
Perennial forming wide clumps of neat, oval, toothed leaves. Has a profusion of flat-faced, rich violet flowers from spring to late summer. Divide clumps every 3 years. H 10–15cm (4–6in), S 15–30cm (6–12in) or more.

Pinguicula grandiflora
Clump-forming perennial with a basal rosette of sticky, oval, pale green leaves. In summer bears spurred, open funnel-shaped, violet-blue to purple flowers singly on upright, slender stems. H 12–15cm (5–6in), S 5cm (2in).

Campanula 'G.F. Wilson'
Neat, mound-forming perennial with large, upturned, bell-shaped, violet flowers in summer. Has rounded, pale yellow-green leaves. H 8–10cm (3–4in), S 12–15cm (5–6in).

Edraianthus pumilio
Short-lived perennial with low tufts of fine, grass-like leaves. In early summer, upturned, bell-shaped, pale to deep lavender flowers, on very short stems, appear amid foliage. H 2.5cm (1in), S 8cm (3in).

Delphinium brunonianum
IRISES, pp.198–9
Polygala chamaebuxus var. *grandiflora*, p.316
Scutellaria scordiifolia

Thalictrum kiusianum

PURPLE–BLUE

Cyananthus microphyllus
Mat-forming perennial with very fine, red stems clothed in tiny leaves. Bears funnel-shaped, violet-blue flowers at the end of each stem in late summer. Likes humus-rich soil. H 2cm (¾in), S 20cm (8in).

Townsendia grandiflora
Short-lived, evergreen perennial with basal rosettes of small, spoon-shaped leaves. Upright stems carry solitary daisy-like, violet or violet-blue flower heads in late spring and early summer. H to 15cm (6in), S 10cm (4in).

Sisyrinchium bellum
Semi-evergreen, upright, clump-forming perennial that for a long period in summer and early autumn has many flowering stems carrying tiny tufts of iris-like, blue to violet-blue flowers. Foliage is grass-like. Self seeds readily. H to 12cm (5in), S 10cm (4in).

Aquilegia jonesii
Compact perennial that bears short-spurred, violet-blue flowers in summer, a few to each slender stem. Has small rosettes of finely divided, blue-grey or grey-green leaves. Is uncommon, suitable for an alpine house only. H 2.5cm (1in), S to 5cm (2in).

Pratia pedunculata
Vigorous, evergreen, creeping perennial with small leaves and a profusion of star-shaped, pale to mid-blue or occasionally purplish-blue flowers borne in summer. Makes good ground cover in a moist site. H 1cm (½in), S indefinite.

Hedyotis michauxii, syn. *Houstonia serpyllifolia* (Creeping bluets)
Vigorous perennial with rooting stems. Produces mats of mid-green foliage studded with star-shaped, violet-blue flowers in late spring and early summer. H 8cm (3in), S 30cm (12in).

Globularia cordifolia
Evergreen, mat-forming, dwarf shrub with creeping, woody stems clothed in tiny, oval leaves. Bears stemless, round, fluffy, blue to pale lavender-blue flower heads in summer. H 2.5–5cm (1–2in), S to 20cm (8in).

Campanula cochleariifolia, syn. *C. pusilla* (Fairy thimbles)
Spreading perennial. Runners produce mats of rosetted, tiny, round leaves. Bears small clusters of white, lavender or pale blue flowers in summer on many thin stems above foliage. H 8cm (3in), S indefinite.

Trachelium asperuloides, syn. *Diosphaera asperuloides*
Mat-forming perennial with thread-like stems clothed in minute, mid-green leaves, above which rise many tiny, upright, tubular, pale blue flowers in summer. Do not remove old stems in winter. H 8cm (3in), S to 15cm (6in).

Eritrichium nanum
Clump-forming perennial with tufts of hairy, grey-green leaves. Bears small, stemless, flat, pale blue flowers in late spring and early summer. Requires sharp drainage. Is only suitable for an alpine house. H 2cm (¾in), S 2.5cm (1in).

Campanula carpatica 'Jewel'
Campanula carpatica 'Turbinata'
Campanula garganica 'W.H. Paine'
Campanula x *haylodgensis*

Campanula 'Joe Elliott'
Campanula pulla
Campanula raineri
Campanula zoysii

Cyananthus lobatus
Edraianthus serpyllifolius 'Major'
Haberlea ferdinandi-coburgii
Haberlea rhodopensis

Myosotis alpestris, p.318
Omphalodes luciliae
Phlox stolonifera 'Blue Ridge'
Wahlenbergia congesta

331

Rock plants/small SUMMER INTEREST

■ BLUE   ■ GREEN–YELLOW   ■ YELLOW

Parochetus communis
(Shamrock pea)
Evergreen, prostrate perennial with clover-like leaves and pea-like, brilliant blue flowers that are borne almost continuously. Grows best in an alpine house. H 2.5–5cm (1–2in), S indefinite.

Polygala calcarea 'Bulley's Variety'
Evergreen, prostrate perennial with rosettes of small, narrowly oval leaves and loose heads of deep blue flowers in late spring and early summer. Likes humus-rich soil. Suits a trough. H 2.5cm (1in), S 8–10cm (3–4in).

Polygala calcarea
Evergreen, prostrate, occasionally upright, perennial. Has small, narrowly oval leaves and pale to dark blue flowers in loose heads in late spring and early summer. Likes humus-rich soil. Suits a trough. May be difficult to establish. H 2.5cm (1in), S to 15cm (6in).

Gunnera magellanica
Mat-forming perennial, grown for its rounded, toothed leaves, often bronze-tinged when young, on short, creeping stems. Small, green, unisexual flowers, with reddish-bracts, are borne on male and female plants. Likes peaty soil. H 2.5cm (1in), S to 30cm (12in).

Sedum acre (Biting stonecrop, Common stonecrop)
Evergreen, mat-forming perennial with dense, spreading shoots, clothed in tiny, fleshy, pale green leaves, each bearing a flat, terminal head of tiny, yellow flowers in summer. Is invasive but easily controlled. H 2.5–5cm (1–2in), S indefinite.

Mitella breweri
Neat, clump-forming, rhizomatous perennial with slender, hairy stems bearing small, pendent, tubular, greenish-white flowers, with flared mouths, in summer. Has lobed, kidney-shaped, basal leaves. H and S 15cm (6in).

Asarina procumbens, syn. *Antirrhinum asarina*
Semi-evergreen perennial with trailing stems bearing soft, hairy leaves and tubular, pale cream flowers, with yellow palates, throughout summer. Dislikes winter wet. Self seeds freely. H 1–2.5cm (½–1in), S 23–30cm (9–12in).

Papaver miyabeanum, syn. *P. nudicaule* Group
Short-lived, clump-forming perennial with basal rosettes of finely cut, hairy, soft grey leaves. Bears pendent, open cup-shaped, pale yellow flowers in summer. Dislikes winter wet. H and S 5–10cm (2–4in).

Gentiana clusii
Gentiana gracilipes
Wahlenbergia albomarginata

Alchemilla alpina
Arctostaphylos alpinus
Salix lindleyana
Satureja montana 'Prostrate White'

Alyssum montanum
Alyssum wulfenianum
Calceolaria fothergillii
Calceolaria polyrrhiza

■ YELLOW

Polygala chamaebuxus
Evergreen, woody-based perennial with tiny, hard, dark green leaves. In late spring and early summer bears many racemes of small, pea-like, white-and-yellow flowers, sometimes marked brown. Needs humus-rich soil. H 5cm (2in), S 20cm (8in).

Sedum acre 'Aureum'
Evergreen, dense, mat-forming perennial with spreading shoots, yellow-tipped in spring and early summer, and clothed in tiny, fleshy, yellow leaves. Bears flat heads of tiny, bright yellow flowers in summer. Is invasive but easy to control. H 2.5–5cm (1–2in), S indefinite.

Waldsteinia ternata, syn. *W. trifolia*
Semi-evergreen perennial with loose, spreading mats of toothed, 3-parted leaves. Bears saucer-shaped, yellow flowers in late spring and early summer. Is good on a bank. H 10cm (4in), S 20–30cm (8–12in).

Linum flavum 'Compactum'
Shrubby perennial with narrow leaves and terminal clusters of many upward-facing, open funnel-shaped, single, bright yellow flowers in summer. Provide a sunny, sheltered position and protection from winter wet. H and S 15cm (6in).

Scutellaria orientalis
Rhizomatous perennial with hairy, grey, rooting stems. Has terminal spikes of tubular, yellow flowers, with brownish-purple lips, in summer. Leaves are toothed and oval. May be invasive in a small space. H 5–10cm (2–4in), S to 23cm (9in).

Oenothera macrocarpa, syn. *O. missouriensis*
Spreading perennial with stout stems and oval leaves. Throughout summer bears a succession of wide, bell-shaped, yellow flowers, sometimes spotted red, that open at sundown. H to 10cm (4in), S to 40cm (16in) or more.

Potentilla eriocarpa
Clump-forming perennial with tufts of oval, dark green leaves divided into leaflets. Flat, single, pale yellow flowers are borne throughout summer just above leaves. H 5–8cm (2–3in), S 10–15cm (4–6in).

Calceolaria tenella
Vigorous, evergreen, prostrate perennial with creeping, reddish stems and oval, mid-green leaves, above which rise small spikes of pouch-shaped, red-spotted, yellow flowers in summer. H 10cm (4in), S indefinite.

Cytisus demissus
Erigeron aureus
Erysimum helveticum, p.321
Erysimum 'Moonlight'

Genista delphinensis
Helianthemum oelandicum subsp. *alpestre*
Hippocrepis comosa 'E.R. Janes'
Hypericum reptans

IRISES, pp.198–9
Jasminum parkeri
Jovibarba hirta, p.336
Linum 'Gemmell's Hybrid'

Morisia monanthos, p.320
Onosma stellulatum
Oxalis chrysantha
Oxalis lobata, p.335

333

Rock plants/small SUMMER INTEREST

☐ YELLOW

☐ YELLOW–ORANGE

Genista sagittalis, syn. **Chamaespartium sagittale**
Deciduous, semi-prostrate shrub with winged stems bearing a few oval, dark green leaves. Pea-like, yellow flowers appear in dense, terminal clusters in early summer, followed by hairy seed pods. H 8cm (3in), S 30cm (12in) or more.

☼ ◊ ❋❋❋

Lysimachia nummularia 'Aurea'
(Creeping Jenny, Moneywort)
Prostrate perennial. Creeping, rooting stems bear pairs of round, soft yellow leaves, which later turn greenish-yellow or green in dense shade. Has bright yellow flowers in leaf axils in summer. H 2.5–5cm (1–2in), S indefinite.

☼ ◊ ❋❋❋ ♈

Hypericum empetrifolium 'Prostratum'
Evergreen, prostrate shrub with angled branches and bright green leaves that have curled margins. Bears flat heads of small, bright yellow flowers in summer. Needs winter protection. H 2cm (³/₄in), S 30cm (12in).

☼ ◊ ❋❋

Hippocrepis comosa
(Horseshoe vetch)
Vigorous perennial with prostrate, rooting stems bearing small, open spikes of pea-like, yellow flowers in summer and leaves divided into leaflets. Self seeds freely and may spread rapidly. H 5–8cm (2–3in), S indefinite.

☼ ◊ ❋❋❋

Calceolaria 'Walter Shrimpton'
Evergreen, mound-forming perennial with glossy, dark green leaves. In early summer bears short spikes of many pouch-shaped, bronze-yellow flowers, spotted rich brown, with white bands across centres. H 10cm (4in), S 23cm (9in).

☼ ● ❋❋❋

Cytisus ardoinii
Deciduous, hummock-forming, dwarf shrub with arching stems. In late spring and early summer, pea-like, bright yellow flowers are produced in pairs in leaf axils. Leaves are divided into 3 leaflets. H 10cm (4in), S 15cm (6in).

☼ ◊ ❋❋❋ ♈

Potentilla aurea
Rounded perennial, with a woody base, that in late summer bears loose sprays of flat, single, yellow flowers with slightly darker eyes. Leaves are divided into oval, slightly silvered leaflets. H 10cm (4in), S 20cm (8in).

☼ ◊ ❋❋❋

Alstroemeria hookeri
Tuberous perennial with narrow leaves and loose heads of widely flared, orange-suffused, pink flowers in summer; upper petals are spotted and blotched red and yellow. H 10–15cm (4–6in), S 45–60cm (18–24in).

☼ ◊ ❋❋

PRIMULAS, pp.236–7
Ranunculus montanus 'Molten Gold'
Saxifraga aizoides
Trollius pumilus, p.321
Viola aetolica, p.319
Viola biflora
Viola 'Jackanapes', p.320
Viola lutea

Achillea x *lewisii* 'King Edward'
Crepis aurea, p.307
Erigeron aurantiacus
Geum montanum

Helianthemum nummularium 'Amy Baring'
Hypericum cerastoides
Inula acaulis
Potentilla x *tonguei*

334

Rock plants/small AUTUMN INTEREST
PINK–BLUE

Polygonum vacciniifolium, syn.
Bistorta vacciniifolia, *Persicaria vacciniifolium*
Evergreen, perennial with woody, red stems. Leaves are tinged red in autumn. Bears deep pink or rose-red flowers in late summer and autumn. H 10–15cm (4–6in), S to 30cm (12in).

Gaultheria procumbens
Vigorous, evergreen sub-shrub with prostrate stems carrying clusters of oval, leathery leaves that turn red in winter. In summer, solitary bell-shaped, pink-flushed, white flowers appear in leaf axils, followed by scarlet berries. H 5–15cm (2–6in), S indefinite.

Gentiana x macaulayi 'Wellsii', syn. *G. x m.* 'Wells Variety'
Evergreen, prostrate perennial with trumpet-shaped, mid-blue flowers in late summer and autumn. Spreading stems are clothed in narrow, mid-green leaves. Soil should be quite moist. H in flower 5cm (2in), S 20cm (8in).

OTHER RECOMMENDED PLANTS:
Arctostaphylos nevadensis
Cornus canadensis
Sedum cauticola

Sedum ewersii

BLUE–ORANGE

Gentiana sino-ornata
Evergreen, prostrate, spreading perennial that, in autumn, bears trumpet-shaped, rich blue flowers singly at the ends of stems. Leaves are narrow. Lift and divide every 3 years. Needs moist soil. H in flower 5cm (2in), S to 30cm (12in).

Oxalis lobata
Clump-forming perennial with woolly-coated tubers. Mid-green leaves have up to 5 rounded lobes. Produces racemes of widely funnel-shaped, bright yellow flowers, 1–2cm ($1/2$–$3/4$in) across, in late summer and autumn. H 5cm (2in), S 8–10cm (3–4in).

Nertera granadensis, syn.
N. depressa (Bead plant)
Prostrate perennial with dense mats of tiny, bright green leaves. In early summer bears minute, greenish-white flowers, followed by many shiny, orange berries. Needs ample moisture in summer. H to 1cm ($1/2$in), S 10cm (4in).

Carlina acaulis, p.323
Gentiana ornata
Gentiana 'Susan Jane'
Sedum kamtschaticum

Sedum kamtschaticum 'Variegatum', p.339
Solidago virgaurea subsp. *minuta*
Vaccinium myrtillus
Vaccinium nummularia

Rock plants/small WINTER/ALL YEAR INTEREST
WHITE–RED

Arabis ferdinandi-coburgii 'Variegata'
Evergreen, mat-forming perennial with small, oval, green leaves, splashed with cream. Bears small, white flowers in spring and early summer. May revert to type with plain green leaves. H 2cm (¾in), S 30cm (12in).

Arctostaphylos uva-ursi 'Point Reyes'
Evergreen, prostrate shrub with long shoots and glossy leaves. In late spring and early summer bears terminal clusters of urn-shaped, pale pink to white flowers, followed by red berries. H 10cm (4in), S 50cm (20in).

Pachysandra terminalis
Evergreen, creeping perennial that has smooth leaves clustered at the ends of short stems. Bears spikes of tiny, white flowers, sometimes flushed purple, in early summer. Makes excellent ground cover in a moist or dry site. H 10cm (4in), S 20cm (8in).

Sedum lydium
Evergreen, mat-forming perennial with reddish stems and narrow, fleshy, often red-flushed leaves. Bears flat-topped, terminal clusters of tiny, white flowers in summer. H 5cm (2in), S to 15cm (6in).

Jovibarba hirta, syn. *Sempervivum hirtum*
Evergreen, mat-forming perennial with rosettes of hairy, mid-green leaves, often suffused red, and terminal clusters of star-shaped, pale yellow flowers in summer. Dislikes winter wet. H 8–15cm (3–6in), S 10cm (4in).

Arctostaphylos uva-ursi
Evergreen, low-growing shrub with arching, intertwining stems clothed in small, oval, bright green leaves. Bears urn-shaped, pinkish-white flowers in summer followed by scarlet berries. H 10cm (4in), S 50cm (20in).

Sedum obtusatum
Evergreen, prostrate perennial with small, fat, succulent leaves that turn bronze-red in summer. Loose, flat sprays of tiny, bright yellow flowers are borne in summer. Dislikes summer wet. H 5cm (2in), S 10–15cm (4–6in).

Trifolium repens 'Purpurascens'
Vigorous, semi-evergreen, ground-cover perennial, grown for its divided, bronze-green foliage, variably edged bright green. Produces heads of small, pea-like, white blooms throughout summer. Suits a wild bank. H in flower 8–12cm (3–5in), S 20–30cm (8–12in) or more.

OTHER RECOMMENDED PLANTS:
Ajuga reptans 'Multicolor', p.267
Briggsia muscicola
Caltha introloba
Pachysandra terminalis 'Variegata'
Paronychia capitata

RED–GREEN

Acaena microphylla
Compact, mat-forming perennial, usually evergreen, with leaves divided into tiny leaflets, bronze-tinged when young. Heads of small flowers with spiny, dull red bracts are borne in summer and develop into decorative burs. H 5cm (2in), S 15cm (6in).

☼ ◊ ❋❋❋

Raoulia hookeri var. **albo-sericea**
Evergreen, prostrate perennial with tiny rosettes of silver leaves. Flower heads appear briefly in summer as fragrant, yellow fluff. Is best in poor, gritty humus in an alpine house. Dislikes winter wet. H to 1cm (¹/₂in), S 25cm (10in).

☼ ◊ ❋❋

Artemisia schmidtiana 'Nana'
Prostrate perennial with fern-like, silver foliage. Has insignificant sprays of daisy-like, yellow flowers in summer. Is suitable for a wall or bank. H 8cm (3in), S 20cm (8in).

☼ ◊ ❋❋❋ ♀

Sempervivum tectorum (Common houseleek, Roof houseleek)
Vigorous, evergreen perennial with rosettes of purple-tipped leaves, sometimes suffused deep red. In summer has clusters of star-shaped, reddish-purple flowers on stems 30cm (12in) tall. H 10–15cm (4–6in), S to 20cm (8in).

☼ ◊ ❋❋❋ ♀

Sempervivum arachnoideum (Cobweb houseleek)
Evergreen, mat-forming perennial. Rosettes of oval, fleshy leaves with red tips are covered in a web of white hairs. Bears loose clusters of star-shaped, rose-red flowers in summer. H 5–12cm (2–5in), S to 10cm (4in) or more.

☼ ◊ ❋❋❋ ♀

Sedum spathulifolium
Evergreen, mat-forming perennial with rosettes of fleshy, green or silver leaves, usually strongly suffused bronze-red, and small clusters of tiny, yellow flowers borne just above foliage in summer. Tolerates shade. H 5cm (2in), S indefinite.

☼ ◊ ❋❋❋

Leucogenes grandiceps
Evergreen, dense, woody-based perennial with neat rosettes of downy, silver leaves. Yellow flower heads, within woolly, white bracts, are borne singly in spring or early summer. H and S 10–15cm (4–6in).

☼ ◊ ❋❋

Sempervivum ciliosum
Evergreen, mat-forming perennial with rosettes of hairy, grey-green leaves and, in summer, heads of small, star-shaped, yellow flowers. Dislikes winter wet; is best grown in an alpine house. H 8–10cm (3–4in), S 10cm (4in).

☼ ◊ ❋❋❋ ♀

Acaena anserinifolia
Acaena 'Blue Haze'
Acaena buchananii
Acantholimon venustum

Ajuga reptans 'Atropurpurea', p.267
Raoulia eximia
Saxifraga stolonifera 'Tricolor'
Sempervivum 'Commander Hay'

Sempervivum grandiflorum
Thymus pseudolanuginosus
Tiarella wherryi
Viola odorata

337

Rock plants/small WINTER/ALL YEAR INTEREST
GREEN

Raoulia australis
Evergreen, carpeting perennial forming a hard mat of grey-green leaves. Bears tiny, fluffy, sulphur-yellow flower heads in summer. H to 1cm (½in), S 25cm (10in).

Acaena caesiiglauca
Vigorous, ground-cover perennial, usually evergreen. Has hairy, glaucous blue leaves divided into leaflets. Heads of small flowers with spiny, brownish-green bracts, borne in summer, develop into brownish-red burs. H 5cm (2in), S 75cm (30in) or more.

Sempervivum montanum
Evergreen, mat-forming perennial with dark green rosettes of fleshy, hairy leaves. Star-shaped, wine-red flowers are borne in terminal clusters in summer. Is a variable plant that hybridizes freely. H 10–15cm (3–6in), S 10cm (4in).

Azorella trifurcata,
syn. *Bolax glebaria* of gardens
Evergreen perennial forming tight, hard cushions of tiny, leathery, oval leaves in rosettes. Bears many small, stalkless umbels of yellow flowers in summer. H to 10cm (4in), S 15cm (6in).

Asarum europaeum (Asarabacca)
Vigorous, evergreen, prostrate, rhizomatous perennial with large, kidney-shaped, leathery, glossy leaves that hide tiny, brown flowers appearing in spring. H 15cm (6in), S indefinite.

Sagina boydii
Evergreen perennial with hard cushions of minute, stiff, bottle-green leaves in small rosettes. Bears insignificant flowers in summer. Is difficult and slow-growing. H 1cm (½in), S to 20cm (8in).

Acantholimon glumaceum, p.325
Arctostaphylos uva-ursi subsp. hookeri 'Monterey Carpet'
Asarum caudatum
Asarum hartwegii
Asarum shuttleworthii
Euonymus fortunei 'Kewensis'
Homogyne alpina
Jovibarba sobolifera
Mitchella repens
Thymus 'Porlock'
Thymus praecox
Thymus praecox 'Annie Hall'
Thymus praecox 'Elfin'

☐ GREEN–YELLOW ☐ YELLOW

Sempervivum giuseppii, syn.
S. x giuseppii
Vigorous, evergreen, prostrate perennial. Leaves are hairy, especially in spring, and have dark spots at tips. Bears terminal clusters of star-shaped, deep pink or red flowers in summer. H in flower 8–10cm (3–4in), S 10cm (4in).

☼ ◊ ❉❉❉

Bolax gummifera
Very slow-growing, evergreen perennial with neat rosettes of small, blue-green leaves forming extremely hard cushions. Insignificant, yellow flowers are rarely produced. Grows well on tufa. H 2.5cm (1in), S 10cm (4in).

☼ ◊ ❉❉❉

Raoulia haastii
Evergreen perennial forming low, irregular hummocks of minute leaves that are apple-green in spring, dark green in autumn and chocolate-brown in winter. Occasionally has small, fluffy, sulphur-yellow flower heads in summer. H to 1cm (½in), S 25cm (10in).

☼ ◊ ❉❉❉

Sedum kamtschaticum 'Variegatum'
Semi-evergreen, prostrate perennial with fleshy leaves, edged with cream. Has tracery of fleshy stems and leaf buds in winter and loose, terminal clusters of orange-flushed, yellow flowers in early autumn. H 5–8cm (2–3in), S 20cm (8in).

☼ ◊ ❉❉❉❉ ♆

Plantago nivalis
Evergreen perennial with neat rosettes of thick, silver-haired, green leaves. Bears spikes of insignificant, dull grey flowers in summer. Dislikes winter wet. H in leaf 2.5cm (1in), S 5cm (2in).

☼ ◊ ❉❉❉

Paronychia kapela subsp. **serpyllifolia**
Evergreen, very compact, mat-forming perennial with minute, silver leaves. Inconspicuous flowers, borne in summer, are surrounded by papery, silver bracts. Is good for covering tufa. H to 1cm (½in), S 20cm (8in).

☼ ◊ ❉❉❉

Sedum spathulifolium 'Cape Blanco', syn. *Hylotelephium sieboldi* 'Variegatum'
Evergreen perennial with rosettes of fleshy leaves, frequently suffused purple. Tiny, yellow flowers appear above foliage in summer. Tolerates shade. H 5cm (2in), S indefinite.

☼ ◊ ❉❉❉❉ ♆

Saxifraga exarata subsp. **moschata 'Cloth of Gold'**
Evergreen hummock-forming perennial, grown for its small, soft rosettes of bright golden foliage; produces best colour in shade. Has star-shaped, white flowers on slender stems in summer. H 10–15cm (4–6in), S 15cm (6in).

☼ ◊ ❉❉❉

Leucogenes leontopodium
Thymus x citriodorus 'Aureus'

339

BULBS/large SPRING INTEREST
☐ WHITE

Leucojum aestivum
(Summer snowflake)
Spring-flowering bulb with long, strap-shaped, semi-erect, basal leaves. Bears heads of pendent, long-stalked, bell-shaped, green-tipped, white flowers on leafless stems. H 50cm–1m (1½–3ft), S 10–12cm (4–5in).

Fritillaria raddeana
Robust, spring-flowering bulb with lance-shaped leaves in whorls on lower half of stem. Has a head of up to 20 widely conical, pale yellow or greenish-yellow flowers, 3–4cm (1¼–1½in) long, topped by a 'crown' of small leaves. H to 1m (3ft), S 15–23cm (6–9in).

■ PURPLE–ORANGE

Fritillaria persica
Spring-flowering bulb with narrow, lance-shaped, grey-green leaves along stem. Produces a spike of 10–20 or more narrow, bell-shaped, blackish- or brownish-purple flowers, 1.5–2cm (⅝–¾in) long. H to 1.5m (5ft), S 10cm (4in).

Fritillaria recurva
(Scarlet fritillary)
Spring-flowering bulb with whorls of narrow, lance-shaped, grey-green leaves. Bears a spike of up to 10 narrow, yellow-chequered, orange or red flowers with flared tips. H to 1m (3ft), S 8–10cm (3–4in).

Fritillaria verticillata
Spring-flowering bulb with slender leaves in whorls up stem, which bears a loose spike of 1–15 bell-shaped, white flowers, 2–4cm (¾–1½in) long and chequered green or brown. H to 1m (3ft), S 8–10cm (3–4in).

Fritillaria imperialis
(Crown imperial)
Spring-flowering bulb with glossy, pale green leaves carried in whorls on leafy stems. Has up to 5 widely bell-shaped, orange flowers crowned by small, leaf-like bracts. H to 1.5m (5ft), S 23–30cm (9–12in).

OTHER RECOMMENDED PLANTS:
Albuca major
Canna iridiflora, p.347
Chasmanthe aethiopica
Fritillaria imperialis 'Lutea'
Fritillaria imperialis 'Rubra Maxima'
Hymenocallis x *festalis*
Phaedranassa carmioli, p.335

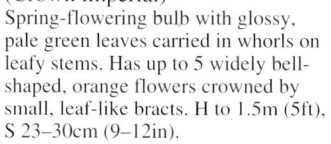

Bulbs/large SUMMER INTEREST
☐ WHITE

Crinum × powellii 'Album'
Late summer- or autumn-flowering bulb, with a long neck, producing a group of semi-erect, strap-shaped leaves. Leafless flower stems carry heads of fragrant, widely funnel-shaped, white flowers. H to 1m (3ft), S 60cm (2ft).

Hymenocallis × macrostephana
Evergreen, spring- or summer-flowering bulb with strap-shaped, semi-erect, basal leaves. Bears fragrant, white or cream- to greenish-yellow flowers, 15–20cm (6–8in) wide. H 80cm (32in), S 30–45cm (12–18in). Min. 15°C (59°F).

Eucomis pallidiflora (Giant pineapple flower, Giant pineapple lily)
Summer-flowering bulb with sword-shaped, crinkly edged, semi-erect, basal leaves. Bears a dense spike of star-shaped, greenish-white flowers, crowned with a cluster of leaf-like bracts. H to 75cm (30in), S 30–60cm (12–24in).

Zantedeschia aethiopica 'Green Goddess'
Very robust, summer-flowering tuber with arrow-shaped, semi-erect, basal, deep green leaves. Bears a succession of green spathes each with a large, central, green-splashed, white area. H 45cm–1m (1½–3ft), S 45–60cm (1½–2ft).

Galtonia candicans (Summer hyacinth)
Late summer- or autumn-flowering bulb with widely strap-shaped, fleshy, semi-erect, basal, grey-green leaves. Leafless stem has a spike of up to 30 pendent, short-tubed, white flowers. H 1–1.2m (3–4ft), S 18–23cm (7–9in).

Zantedeschia aethiopica 'Crowborough'
Early to mid-summer-flowering tuber with arrow-shaped, semi-erect, basal, deep green leaves. Produces a succession of arum-like, white spathes, each with a yellow spadix. H 45cm–1m (1½–3ft), S 35–45cm (14–18in).

Camassia leichtlinii
Tuft-forming bulb with long, narrow, erect, basal leaves. Each leafless stem bears a dense spike of 6-petalled, star-shaped, bluish-violet or white flowers, 4–8in (1½–3in) across, in summer. H 1–1.5m (3–5ft), S 20–30cm (8–12in).

Camassia leichtlinii 'Semiplena'
Tuft-forming bulb with long, narrow, erect, basal leaves. Each leafless stem carries a dense spike of narrow-petalled, double, creamy-white flowers, 4–8cm (1½–3in) across, in summer. H 1–1.5m (3–5ft), S 20–30cm (8–12in).

OTHER RECOMMENDED PLANTS:
Camassia quamash
Cardiocrinum giganteum, p.342
Chlorogalum pomeridianum
DAHLIAS, pp.350–51
GLADIOLI, p.343
LILIES, p.338
Polianthes tuberosa
Nomocharis saluenensis

Bulbs/large SUMMER INTEREST
WHITE–PINK

Crinum moorei
Summer-flowering bulb with a long neck, up to 1m (3ft) tall, and strap-shaped, semi-erect, grey-green leaves grouped at neck top. Leafless flower stems bear heads of long-tubed, funnel-shaped, white to deep pink flowers. H 50–70cm (20–28in), S 60cm (24in).

Nectaroscordum siculum subsp. **bulgaricum**, syn. *N. dioscoridis*
Late spring- to early summer-flowering bulb with pendent, bell-shaped, white flowers, flushed purple-red and green. In seed, stalks bend upwards, holding dry seed pods erect. H to 1.2m (4ft), S 30–45cm (1–1½ft).

Cardiocrinum giganteum
(Giant lily)
Stout, leafy-stemmed bulb. In summer has long spikes of fragrant, slightly pendent, cream flowers, 15cm (6in) long, with purple-red streaks inside, then brown seed pods. H to 3m (10ft), S 75cm–1.1m (2½–3½ft).

Eucomis comosa
Clump-forming bulb with strap-shaped, wavy-margined leaves, spotted purple beneath. Purple-spotted stem bears a spike of white or greenish-white, sometimes pink-tinted flowers, with purple ovaries, crowned by a tuft of bracts. H to 70cm (28in), S 30–60cm (12–24in).

Nomocharis pardanthina, syn. *N. mairei*
Summer-flowering bulb with stems bearing whorls of lance-shaped leaves and up to 15 outward-facing, white or pale pink flowers, each with purple blotches and a dark purple eye. H to 1m (3ft), S 12–15cm (5–6in).

Crinum × powellii
Late summer- or autumn-flowering bulb with a long neck producing a group of strap-shaped, semi-erect leaves. Leafless flower stems bear heads of fragrant, widely funnel-shaped, pink flowers. H to 1m (3ft), S 60cm (2ft).

Crinum bulbispermum
DAHLIAS, pp.350–51
GLADIOLI, p.343
LILIES, p.348–9

Gladioli

Gladiolus hybrids produce excellent flowers for garden decoration, flower arrangements and exhibition. They are divided into the Grandiflorus group, with long, densely packed spikes of flowers, categorized as small, medium-sized, large or giant according to the width of their lowest flowers, and the Primulinus group, with fairly loose spikes of small flowers. (See also the Plant Dictionary.)

G. 'Ice Cap' (large)

G. 'Inca Queen' (large)

G. 'Robin' (Primulinus group)

G. 'Moon Mirage' (giant)

G. 'Peter Pears' (large)

G. 'Dancing Queen' (large)

G. 'Miss America' (medium)

G. 'Pink Lady' (large)

G. 'Black Lash' (small)

G. 'Tesoro' (medium)

G. 'Café au Lait' (Primulinus group)

G. × *colvillei* 'The Bride' (small) ♀

G. 'Gigi' (small)

G. 'Mexicali Rose' (large)

G. 'Rutherford' (Primulinus group)

G. 'Green Woodpecker' (medium) ♀

G. 'Rose Supreme' (giant)

G. 'Deliverance' (giant)

G. 'Drama' (large)

G. 'Renegade' (large)

G. 'Victor Borge' (large)

G. 'Melodie' (small)

G. 'Carioca' (medium)

Bulbs/large SUMMER INTEREST
PINK–RED

Notholirion campanulatum
Early summer-flowering bulb with long, narrow leaves in a basal tuft. Leafy stem bears a spike of 10–40 pendent, funnel-shaped flowers, each 4–5cm (1½–2in) long, with green-tipped, deep rose-purple petals. H to 1m (3ft), S 8–10cm (3–4in).

Watsonia pyramidata
Very robust, summer-flowering corm with narrowly sword-shaped leaves both at base and on stem. Produces a loose, branched spike of rich pink flowers, with 6 spreading, pointed, rose-red lobes. H 1–1.5m (3–5ft), S 45–60cm (1½–2ft).

Dierama pulcherrimum
Upright, summer-flowering corm with long, narrow, strap-like, evergreen leaves, above which rise elegant, arching, wiry stems bearing funnel-shaped, deep pink flowers. Prefers deep, rich soil. H 1.5m (5ft), S 30cm (1ft).

Watsonia beatricis
Summer-flowering corm with long, sword-shaped, erect leaves, some basal and some on stem. Stem carries a dense, branched spike of tubular, orange-red flowers, each 6–8cm (2½–3in) long, with 6 short lobes. H to 1m (3ft), S 30–45cm (1–1½ft).

Gladiolus italicus, syn. *G. segetum*
Early summer-flowering corm with a fan of erect, sword-shaped leaves from the basal part of stem. Carries a loose spike of up to 20 pinkish-purple flowers, 4–5cm (1½–2in) long. H to 1m (3ft), S 10–15cm (4–6in).

Gladiolus communis subsp. **byzantinus**
Early summer-flowering corm with a dense spike of up to 20 deep purplish-red or purplish-pink flowers, 4–6cm (1½–2½in) long. Produces a fan of sword-shaped, erect, basal leaves. H to 70cm (28in), S 10–15cm (4–6in).

Phaedranassa carmioli
Spring- and summer-flowering bulb with upright, elliptic or lance-shaped, basal leaves. Bears a head of 6–10 pendent, pinkish-red flowers, with green bases and yellow-edged, green lobes at each apex. H 50–70cm (20–28in), S 30–45cm (12–18in).

Alstroemeria 'Margaret'
Mid- to late summer-flowering tuber with narrow, lance-shaped, twisted, bright green leaves. Stout, leafy stems bear widely flared, funnel-shaped, deep red flowers. H 1m (3ft), S 60cm–1m (2–3ft).

x *Amarcrinum memoria-corsii*, p.352
DAHLIAS, pp.350–51
LILIES, p.348–9
Watsonia meriana

RED

Dracunculus vulgaris, syn. *Arum dracunculus* (Dragon arum)
Spring- and summer-flowering tuber with deeply divided leaves at apex of thick, blotched stem. A blackish-maroon spadix protrudes from a deep maroon-purple spathe, 35cm (14in) long. H to 1m (3ft), S 45–60cm (1½–2ft).

☼ ◊ ❈ ❈

Gloriosa superba 'Rothschildiana' (Glory lily)
Deciduous, summer-flowering, tuberous, tendril climber. Upper leaf axils each bear a large flower that has 6 reflexed, red petals with scalloped, yellow edges. H to 2m (6ft), S 30–45cm (1–1½ft). Min. 8°C (46°F).

☼ ◊ ♀

Crocosmia 'Bressingham Blaze'
Clump-forming, late summer-flowering corm with sword-shaped, pleated, basal, erect leaves. Branched stem bears widely funnel-shaped, fiery-red flowers. H 75cm (30in), S 15–20cm (6–8in).

☼ ◊ ❈ ❈

RED–PURPLE

Scadoxus multiflorus subsp. **katherinae**, syn. *Haemanthus katherinae* (Blood flower)
Very robust, clump-forming bulb with lance-shaped, wavy-edged leaves. Bears an umbel of up to 200 red flowers in summer. H to 1.2m (4ft), S 30–45cm (1–1½ft). Min. 10°C (50°F).

☼ ◊

Crocosmia 'Lucifer'
Robust, clump-forming corm with sword-shaped, erect, basal, bright green leaves. Bears funnel-shaped, deep rich red flowers in dense, branching spikes in mid-summer. H to 1m (3ft), S 20–25cm (8–10in).

☼ ◊ ❈ ❈ ❈ ♀

Canna x generalis 'Assault'
Summer-flowering, rhizomatous perennial with stout, leafy stems bearing wide, purple-green leaves. Has a spike of scarlet flowers surrounded by purple bracts. H to 1.2m (4ft), S 45–60cm (1½–2ft). Min. 15°C (59°F).

☼ ◊

Crocosmia masonorum
Robust, clump-forming corm with erect, basal, deep green leaves, pleated lengthways. Erect, branched stem has a horizontal, upper part which carries upright, reddish-orange flowers in summer-autumn. H to 1.5m (5ft), S 30–45cm (1–1½ft).

☼ ◊ ❈ ❈ ❈ ♀

Allium stipitatum
Summer-flowering bulb with stout stems and strap-like, semi-erect, basal leaves. Carries 50 or more star-shaped, purplish-pink flowers in a spherical umbel, 8–12cm (3–5in) across. H to 1–1.5m (3–4ft), S 15–20cm (6–8in).

☼ ◊ ❈ ❈

Dierama pendulum (Angel's fishing rod)
Clump-forming, late summer-flowering corm with arching, basal leaves. Bears pendulous, loose racemes of bell-shaped, pinkish-purple flowers, 2.5cm (1in) long. H to 1.5m (5ft), S 15–20cm (6–8in).

☼ ◊ ❈ ❈

Canna x *generalis* 'Dazzler'
Canna x *generalis* 'Lucifer'
Chasmanthe aethiopica
Chasmanthe floribunda

DAHLIAS, pp.350–51
Dichelostemma ida-maia
GLADIOLI, p.343
LILIES, p.348–9

Scadoxus multiflorus
Watsonia fourcadei

DAHLIAS, pp.350–51

Bulbs/large SUMMER INTEREST

■ PURPLE–BLUE

■ GREEN–YELLOW

Allium giganteum
Robust, summer-flowering bulb with long, wide, semi-erect, basal leaves. Produces a stout stem with a dense, spherical umbel, 12cm (5in) across, of 50 or more star-shaped, purple flowers. H to 2m (6ft), S 30–35cm (12–14in).

Allium aflatunense
Summer-flowering bulb with semi-erect, basal leaves dying away by flowering time. Carries 50 or more star-shaped, purple flowers in a large, tight, spherical umbel, 10cm (4in) across. H to 75cm (30in), S 15–20cm (6–8in).

Aristea major, syn. *A. thyrsiflora*
Robust, evergreen, clump-forming rhizome with sword-shaped, erect leaves, to 2.5cm (1in) across, and dense spikes of purple-blue flowers on short stalks in summer. H to 1m (3ft), S 45–60cm (1½–2ft).

Dietes bicolor
Evergreen, tuft-forming, summer-flowering rhizome with tough, long and narrow, erect, basal leaves. Branching stems each bear a succession of flattish, iris-like, pale to mid-yellow flowers; each large petal has a brown patch. H to 1m (3ft), S 30–60cm (1–2ft).

Dichelostemma congestum, syn. *Brodiaea congesta*
Early summer-flowering bulb with semi-erect, basal leaves dying away when a dense head of funnel-shaped, purple flowers, each 1.5–2cm (⅝–¾in) long, appears. H to 1m (3ft), S 8–10cm (3–4ft).

Neomarica caerulea
Summer-flowering rhizome with sword-shaped, semi-erect leaves in basal fans. Stems each bear a leaf-like bract and a succession of iris-like, blue flowers, with white, yellow and brown central marks. H to 1m (3ft), S 1–1.5m (3–5ft). Min. 10°C (50°F).

Galtonia viridiflora
Clump-forming, summer-flowering bulb with widely strap-shaped, fleshy, semi-erect, basal, grey-green leaves. Leafless stem bears a spike of up to 30 pendent, short-tubed, funnel-shaped, pale green flowers. H 1–1.2m (3–4ft), S 18–23cm (7–9in).

Arisaema consanguineum
Summer-flowering tuber with robust, spotted stems and erect, umbrella-like leaves with narrow leaflets. Produces purplish-white- or white-striped, green spathes, 15–20cm (6–8in) long, and bright red berries. H to 1m (3ft), S 30–45cm (1–1½ft).

Allium caeruleum, p.366
Arisaema tortuosum
Camassia quamash
DAHLIAS, pp.350–51

GLADIOLI, p.343
IRISES, pp.198–9

Bowiae volubilis, p.396
Stenanthium gramineum

☐ YELLOW

Crocosmia 'Citronella'
Clump-forming, late summer-flowering corm with sword-shaped, erect, basal, grey-green leaves. Flowers are funnel-shaped and clear golden-yellow. H 60–75cm (24–30in), S 15–20cm (6–8in).

☼ ◊ ❊❊

Moraea huttonii
Summer-flowering corm with long, narrow, semi-erect, basal leaves. Tough stem bears a succession of iris-like, yellow flowers, 5–7cm (2–3in) across, with brown marks near the centre. H 75cm–1m (2½–3ft), S 15–25cm (6–10in).

☼ ◊ ❊❊

Zantedeschia elliottiana
(Golden arum lily)
Summer-flowering tuber with heart-shaped, semi-erect, basal leaves with transparent marks. Bears a 15cm (6in) long, yellow spathe surrounding a yellow spadix. H 60cm–1m (2–3ft), S 45–60cm (1½–2ft). Min. 10°C (50°F).

☼ ◊ ♈

Alstroemeria 'Walter Fleming'
DAHLIAS, pp.350–51
GLADIOLI, p.343
Gladiolus papilio, p.352

Gloriosa superba
Moraea spathulata

☐ ORANGE

Alstroemeria aurea,
syn. *A. aurantiaca*
Summer-flowering, tuberous perennial with narrow, lance-shaped, twisted leaves and loose heads of orange flowers, tipped with green and streaked dark red. H to 1m (3ft), S 60cm–1m (2–3ft).

☼ ◊ ❊❊

Littonia modesta
Deciduous, summer-flowering, tuberous, scandent climber. Has slender stems bearing lance-shaped leaves with tendrils at apex. Leaf axils bear pendent, bell-shaped, orange flowers, 4–5cm (1½–2in) across. H 1–2m (3–6ft), S 10–15cm (4–6in). Min. 16°C (61°F).

☼ ◊

Canna iridiflora
Very robust, spring- or summer-flowering, rhizomatous perennial with broad, oblong leaves and spikes of pendent, long-tubed, reddish-pink or orange flowers, each 10–15cm (4–6in) long, with reflexed petals. H 3m (10ft), S 45–60cm (1½–2ft). Min. 15°C (59°F).

☼ ◊

Belamcanda chinensis
Canna x *generalis* 'Orange Perfection'
Crocosmia aurea
Crocosmia paniculata

DAHLIAS, pp.350–51
GLADIOLI, p.343

Lilies

Lilies (*Lilium* species and cultivars) are graceful plants and bring elegance to the summer border. Their attractive, flamboyant flowers are in various shapes, some nodding, others upright or trumpet-shaped, others in the distinctive turkscap form (with recurving petals), and are borne usually several per stem. Many have a distinctive, powerful fragrance, though a few species are unpleasantly scented. Most widely grown are the numerous hybrids, available in a dazzling array of colours – white, pink, red and rich shades of yellow and orange – but among the species are several that have been undeservedly neglected. Lily flowers are often attractively spotted with a darker or contrasting colour, or have conspicuous stamens or pollen. Lilies thrive in sun and well-drained soil. Once established, they are best left undisturbed, as the bulbs are easily damaged.

L. 'Mont Blanc'

L. 'Bronwen North'

L. 'Sterling Star'

L. martagon

L. 'Star Gazer'

L. 'Casa Blanca' ♀

L. martagon var. *album* ♀

L. 'Bright Star'

L. rubellum

L. 'Côte d'Azur'

L. candidum ♀

L. Olympic Hybrids

L. Imperial Gold Group

L. mackliniae ♀

L. 'Black Magic'

L. 'Corsage'

L. 'Montreux'

L. 'Journey's End'

L. longiflorum ♀

L. duchartrei

L. 'Magic Pink'

L. speciosum var. *rubrum*

L. regale ♀

L. auratum var. *platyphyllum* ♀

L. 'Angela North'

L. × *dalhansonii*

L. 'Black Beauty'

L. nepalense

L. 'Rosemary North' ♛

L. 'Amber Gold'

L. lancifolium var. *splendens*

L. 'Lady Bowes Lyon'

L. 'Karen North' ♛

L. 'Golden Splendor'

L. hansonii ♛

L. 'Brushmarks'

L. tsingtauense

L. canadense

L. 'Roma'

L. monadelphum ♛

L. 'Apollo'

L. bulbiferum var. *croceum* ♛

L. pyrenaicum var. *rubrum*

L. Golden Clarion Hybrids

L. 'Destiny'

L. pyrenaicum ♛

L. 'Enchantment' ♛

L. 'Harmony'

L. chalcedonicum ♛

L. pardalinum

L. 'Connecticut King'

L. superbum ♛

L. medeoloides

L. davidii var. *willmottiae*

349

Dahlias

The variety of border hybrids now available offers a dazzling display of colour and form to every gardener, and no special skills are required to cultivate or propagate them. In colour the flowers range from vibrant pinks and crimsons, through rich hues of mauves and purples, to the pastel shades of lilacs, pinks and creams, in size from the tiny pompons to huge exhibition blooms up to 30cm (1ft) or more across.

Dahlias are excellent for providing cut flowers and will bloom vigorously throughout summer until the first frosts – in the right conditions a single plant may produce up to 100 blooms. Their various flower types (shown below) form the basis of the recognized groups.

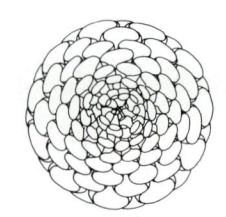

Single – each flower usually has 8–10 broad petals surrounding an open, central disc.

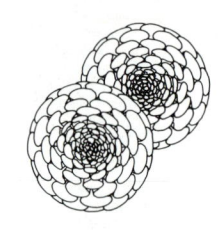

Anemone – fully double flowers each have one or more rings of flattened ray petals surrounding a dense group of shorter, tubular petals, usually longer than disc petals found in single dahlias.

Collerette – single flowers each have broad, outer petals, usually 8–10, and an inner 'collar' of smaller petals surrounding an open, central disc.

Water-lily – fully double flowers have large, generally sparse ray petals, which are flat or with slightly incurved or recurved margins, giving the flower a flat appearance.

Decorative – fully double flowers have broad, flat petals that incurve slightly at their margins and usually reflex to the stem.

Ball – flattened to spherical, fully double flowers have densely packed, almost tubular petals.

Pompon – flattened to spherical, fully double flowers are no more than 5cm (2in) across – a miniature form of ball flowers.

Cactus – fully double flowers have narrow, pointed petals that can be straight or curl inwards and have recurved edges for more than two-thirds of their length.

Semi-cactus – fully double flowers are similar to cactus flowers but have broader-based petals, the edges of which are generally recurved towards their tips.

Miscellaneous – flowers are in a wide range of unclassified types, including orchid-like (shown right), single and double forms.

D. **'Majestic Kerkrade'** (cactus)

D. **'Pink Symbol'** (semi-cactus)

D. **'Rhonda'** (pompon)

D. **'Candy Keene'** (semi-cactus)

D. **'Athalie'** (cactus)

D. **'Vicky Crutchfield'** (water-lily)

D. **'Gilt Edge'** (decorative)

D. **'Angora'** (decorative)

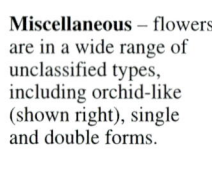

D. **'Easter Sunday'** (collerette)

D. **'White Klankstad'** (cactus)

D. **'Monk Marc'** (cactus)

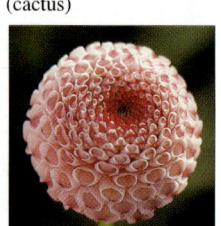

D. **'Gay Princess'** (decorative)

D. **'Nina Chester'** (decorative)

D. **'Small World'** (pompon)

D. **'Noreen'** (pompon)

D. **'By the Cringe'** (cactus)

D. 'Wootton Cupid' (ball) ♛

D. 'Pontiac' (cactus)

D. 'Fascination' (miscellaneous)

D. 'Jocondo' (decorative)

D. 'Betty Bowen' (decorative)

D. 'Whale's Rhonda' (pompon)

D. 'Comet' (anemone)

D. 'Brunton' (decorative)

D. 'Bishop of Llandaff' (miscellaneous)

D. 'Corona' (semi-cactus)

D. 'Scarlet Beauty' (water-lily)

D. 'Flutterby' (decorative)

D. 'Chimborazo' (collerette)

D. 'Bassingbourne Beauty' (decorative)

D. 'Hamari Katrina' (semi-cactus)

D. 'Yellow Hammer' (single)

D. 'Clair de Lune' (collerette)

D. 'Butterball' (decorative) ♛

D. 'Early Bird' (decorative)

D. 'Cortez Sovereign' (semi-cactus)

D. 'Davenport Sunlight' (semi-cactus)

D. 'East Anglian' (decorative)

D. 'Gay Mini' (decorative)

D. 'So Dainty' (semi-cactus) ♛

D. 'Shandy' (cactus)

D. 'Corton Olympic' (decorative)

D. 'Frank Hornsey' (decorative)

D. 'Paul Chester' (cactus)

D. 'Highgate Torch' (semi-cactus)

D. 'Chinese Lantern' (decorative)

D. 'Biddenham Sunset' (decorative)

D. 'Quel Diable' (semi-cactus)

351

Bulbs/large AUTUMN INTEREST
WHITE–PURPLE

Amaryllis belladonna 'Hathor'
Autumn-flowering bulb with a stout, purple stem bearing fragrant, pure white flowers, 10cm (4in) long, with yellow throats. Strap-shaped, semi-erect, basal leaves appear in late winter or spring. H 50–80cm (20–32in), S 30–45cm (12–18in).

x Amarcrinum memoria-corsii, syn. x *A. howardii*, x *Crinodonna corsii*
Evergreen, clump-forming bulb with wide, semi-erect, basal leaves. Stout stems carry fragrant, rose-pink flowers in loose heads in late summer and autumn. H and S to 1m (3ft).

Amaryllis belladonna (Belladonna lily)
Autumn-flowering bulb with a stout, purple stem bearing fragrant, funnel-shaped, pink flowers, 10cm (4in) long. Forms strap-shaped, semi-erect, basal leaves after flowering. H 50–80cm (20–32in), S 30–45cm (12–18in).

x Amarygia parkeri, syn. x *Brunsdonna parkeri*
Early autumn-flowering bulb. Stout stem carries a large head of funnel-shaped, deep rose flowers with yellow and white throats. Produces strap-shaped, semi-erect, basal leaves after flowering. H to 1m (3ft), S 60cm–1m (2–3ft).

Gladiolus papilio, syn. *G. purpureoauratus*
Clump-forming, summer- or autumn-flowering corm with stolons. Bears up to 10 yellow or white flowers, suffused violet, with hooded, upper petals and darker yellow patches on lower petals. H to 1m (3ft), S 15cm (6in).

OTHER RECOMMENDED PLANTS:
Crinum macowanii
Crinum x *powellii*, p.342
Crinum x *powellii* 'Album', p.341
Crocosmia masonorum, p.345
Galtonia candicans, p.341
Urginea maritima
Worsleya rayneri

Bulbs/medium SPRING INTEREST
WHITE

Erythronium oregonum
Clump-forming, spring-flowering tuber with 2 semi-erect, mottled, basal leaves. Has up to 3 pendent, white flowers, with yellow eyes and often brown rings near centre; petals reflex as flowers open. Increases rapidly by offsets. H to 35cm (14in), S 12cm (5in).

Allium neapolitanum, syn. *A. cowanii*
Spring-flowering bulb with narrow, semi-erect leaves on the lower quarter of flower stems. Stems each develop an umbel, 5–10cm (2–4in) across, of up to 40 white flowers. H 20–50cm (8–20in), S 10–12cm (4–5in).

Calochortus venustus
Late spring-flowering bulb with 1 or 2 narrow, erect leaves near the base of the branched stem. Bears 1–4 white, yellow, purple or red flowers, with a dark red, yellow-margined blotch on each large petal. H 20–60cm (8–24in), S 5–10cm (2–4in).

OTHER RECOMMENDED PLANTS:
Crinum asiaticum, p.363
DAFFODILS, pp.360–62
Freesia 'White Swan'
Hippeastrum 'White Dazzler'
Hymenocallis narcissiflora, p.364
Leucojum aestivum, p.340
Ornithogalum umbellatum

☐ WHITE

**Erythronium californicum
'White Beauty'**
Vigorous, clump-forming tuber with basal, mottled leaves. In spring has a loose spike of 1–10 reflexed, white flowers, each with a brown ring near the centre. Increases rapidly. H 20–30cm (8–12in), S 10–12cm (4–5in).

☼ ◊ ✶✶✶ ♛

Calochortus albus
(Fairy lantern, Globe lily)
Spring-flowering bulb with long, narrow, erect, grey-green leaves near the base of the loosely branched stem. Each branch carries a pendent, globose, white or pink flower. H 20–50cm (8–20in), S 5–10cm (2–4in).

☼ ◊ ✶✶

Pamianthe peruviana
Evergreen, spring-flowering bulb with a stem-like neck and strap-shaped, semi-erect leaves with drooping tips. Stem has a head of 2–4 fragrant, white flowers, each with a bell-shaped cup and 6 spreading petals. H 50cm (20in), S 45–60cm (18–24in). Min. 12°C (54°F).

☼ ◊

☐ PINK

Erythronium hendersonii
Spring-flowering tuber with 2 semi-erect, basal, brown- and green-mottled leaves. Flower stem carries up to 10 lavender or lavender-pink flowers, with reflexed petals and deep purple, central eyes. H 20–30cm (8–12in), S 10–12cm (4–5in).

☼ ◊ ✶✶

Allium unifolium
Late spring-flowering bulb with one semi-erect, basal, grey-green leaf. Each flower stem carries a domed umbel, 5cm (2in) across, of up to 30 purplish-pink flowers. H to 30cm (12in), S 8–10cm (3–4in).

☼ ◊ ✶✶

Erythronium revolutum
Freesia carymbosa
Hippeastrum 'Apple Blossom', p.369
Hippeastrum 'Bouquet'
Rhodophiala advenum, p.365

Tulips

Tulips suit a wide range of planting schemes and are excellent in the rock garden, in formal bedding, as elegant cut flowers and for containers. Their bold flowers are generally simple in outline and held upright, with colours that are often bright and strong. Many of the species deserve to be more widely grown, alongside the large variety of hybrids now available. *Tulipa* is classified in 15 divisions, described below.

Div.1 Single early – cup-shaped, single flowers, often opening wide in the sun, are borne from early to mid-spring.

Div.2 Double early – long-lasting, double flowers open wide in early and mid-spring.

Div.3 Triumph – sturdy stems bear rather conical, single flowers, becoming more rounded, in mid- and late spring.

Div.4 Darwin hybrids – large, single flowers are borne on strong stems from mid- to late spring.

Div.5 Single late – single flowers, variable but usually with pointed petals, are borne in late spring and very early summer.

Div.6 Lily-flowered – strong stems bear narrow-waisted, single flowers, with long, pointed, often reflexed petals, in late spring.

Div.7 Fringed – flowers are similar to those in Div.6, but have fringed petals.

Div.8 Viridiflora – variable, single flowers, with partly greenish petals, are borne in late spring.

Div.9 Rembrandt – flowers are similar to those in Div.6, but have striped or feathered patterns caused by virus, and appear in late spring.

Div.10 Parrot – has large, variable, single flowers, with frilled or fringed and usually twisted petals, in late spring.

Div.11 Double late (peony-flowered) – usually bowl-shaped, double flowers appear in late spring.

Div.12 Kaufmanniana hybrids – single flowers are usually bicoloured, open flat in the sun and appear in early spring. Leaves are usually mottled or striped.

Div.13 Fosteriana hybrids – large, single flowers open wide in the sun from early to mid-spring. Leaves are often mottled or striped.

Div.14 Greigii hybrids – large, single flowers appear in mid- and late spring. Mottled or striped leaves are often wavy-edged.

Div.15 Miscellaneous – a diverse category of other species and their cultivars and hybrids. Flowers appear in spring and early summer.

T. turkestanica (Div.15) ♀

T. 'Diana' (Div.1)

T. 'Angélique' (Div.11)

T. 'Don Quichotte' (Div.3)

T. 'Purissima' (Div.13)

T. 'Groenland' (Div.8)

T. saxatilis (Div.15)

T. 'Page Polka' (Div.3)

T. 'Spring Green' (Div.8) ♀

T. 'Ballade' (Div.6) ♀

T. 'White Triumphator' (Div.6)

T. 'White Parrot' (Div.10)

T. 'Carnaval de Nice' (Div.11)

T. 'Menton' (Div.5)

T. 'White Dream' (Div.3)

T. biflora (Div.15)

T. 'Fancy Frills' (Div.10)

T. 'New Design' (Div.3)

T. 'China Pink' (Div.6)

T. 'Gordon Cooper' (Div.4)

T. 'Union Jack' (Div.5)

T. 'Ad Rem' (Div.4)

T. 'Red Riding Hood' (Div.14) ♀

T. 'Juan' (Div.13)

T. 'Peer Gynt' (Div.3)

T. 'Garden Party' (Div.3)

T. 'Bing Crosby' (Div.3)

T. 'Dreamland' (Div.5)

T. 'Red Parrot' (Div.10)

T. 'Uncle Tom' (Div.11)

T. 'Oranje Nassau' (Div.2) ♀

T. undulatifolia (Div.15)

T. 'Keizerskroon' (Div.1) ♀

T. 'Attila' (Div.3)

T. 'Mme Lefèbre' (Div.13)

T. 'Plaisir' (Div.14) ♀

T. acuminata (Div.15)

T. 'Bird of Paradise' (Div.10)

T. sprengeri (Div.15) ♀

T. 'Kingsblood' (Div.5) ♀

T. 'Margot Fonteyn' (Div.3)

T. 'Apeldoorn's Elite' (Div.4) ♀

T. clusiana (Div.15)

T. linifolia (Div.15) ♀

T. 'Merry Widow' (Div.3)

T. 'Balalaika' (Div.5)

T. 'Glück' (Div.12)

T. 'Estella Rijnveld' (Div.10)

T. hageri (Div.15)

T. praestans 'Unicum' (Div.15)

T. 'Fringed Beauty' (Div.7)

T. marjolettii (Div.15)

355

T. 'Dreaming Maid' (Div.3)

T. batalinii (Div.15) ♀

T. 'West Point' (Div.6)

T. sylvestris (Div.15)

T. 'Candela' (Div.13)

T. 'Dreamboat' (Div.13)

T. urumiensis (Div.15) ♀

T. clusiana var. *chrysantha* (Div.15) ♀

T. 'Artist' (Div.8)

T. 'Shakespeare' (Div.12)

T. 'Greuze' (Div.5)

T. violacea (Div.15)

T. kaufmanniana (Div.15)

T. 'Dillenburg' (Div.5)

T. orphanidea (Div.15)

T. humilis (Div.15)

T. tarda (Div.15) ♀

T. 'Golden Apeldoorn' (Div.4)

T. 'Prinses Irene' (Div.1) ♀

T. praestans 'Van Tubergen's Variety' (Div.15)

T. 'Blue Parrot' (Div.10)

T. 'Maja' (Div.7)

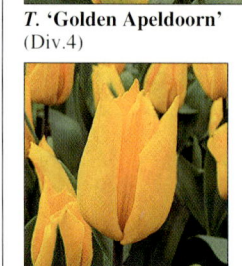

T. 'Yokohama' (Div.1)

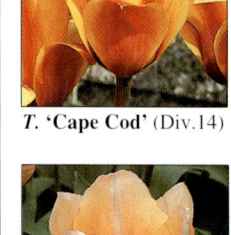

T. 'Cape Cod' (Div.14)

T. 'Queen of Night' (Div.5)

T. 'Bellona' (Div.1)

T. 'Golden Artist' (Div.8)

T. 'Apricot Beauty' (Div.1)

T. whittallii (Div.15)

356

Bulbs/medium SPRING INTEREST

■■ RED–PURPLE

Anemone pavonina
Leafy tuber with cup-shaped, single, dark-centred, scarlet, purple or blue flowers rising above divided, frilly leaves in early spring. H 40cm (16in), S 20cm (8in).

Sprekelia formosissima
(Aztec lily, Jacobean lily)
Clump-forming, spring-flowering bulb with strap-shaped, semi-erect, basal leaves. Stem bears a deep red flower, 12cm (5in) wide, that has 6 narrow petals with green-striped bases. H 15–35cm (6–14in), S 12–15cm (5–6in).

Sauromatum venosum, syn. *S. guttatum*
(Monarch-of-the-East, Voodoo lily)
Early spring-flowering tuber. Bears a large, acrid, purple-spotted spathe, then a lobed leaf on a long, spotted stalk. H 30–45cm (12–18in), S 30–35cm (12–14in). Min. 5–7°C (41–5°F).

Fritillaria meleagris
(Snake's-head fritillary)
Spring-flowering bulb with slender stems producing scattered, narrow, grey-green leaves. Has solitary bell-shaped, prominently chequered flowers, in shades of pinkish-purple or white. H to 30cm (12in), S 5–8cm (2–3in).

Fritillaria pyrenaica
Spring-flowering bulb with scattered, lance-shaped leaves, often rather narrow. Develops 1, or rarely 2, broadly bell-shaped flowers with flared-tipped, chequered, deep brownish- or blackish-purple petals. H 15–30cm (6–12in), S 5–8cm (2–3in).

■ PURPLE

Muscari latifolium
Spring-flowering bulb with one strap-shaped, semi-erect, basal, grey-green leaf. Has a dense spike of tiny, bell-shaped, blackish-violet to -blue flowers with constricted mouths; upper ones are paler and smaller. H to 25cm (10in), S 5–8cm (2–3in).

Fritillaria camschatcensis
(Black sarana)
Spring-flowering bulb. Stout stems carry lance-shaped, glossy leaves, mostly in whorls. Bears up to 8 bell-shaped, deep blackish-purple or brown flowers. Needs humus-rich soil. H 15–60cm (6–24in), S 8–10cm (3–4in).

Anemone x *fulgens*, p.374
Arum dioscoridis
Freesia 'Romany'
Fritillaria tubiformis
Hippeastrum 'Red Lion', p.369
Rhodophiala pratense
Scadoxus puniceus
TULIPS, pp.354–6

357

Bulbs/medium SPRING INTEREST

■ BLUE

■■ BLUE–GREEN

Hyacinthoides hispanica,
syn. *Endymion hispanicus*, *Scilla campanulata*, *S. hispanica* (Spanish bluebell)
Spring-flowering bulb with strap-shaped, glossy leaves and pendent, bell-shaped, blue, white or pink flowers. H to 30cm (12in), S 10–15cm (4–6in).

Hyacinthoides non-scripta,
syn. *Scilla non-scripta* (English bluebell)
Tuft-forming, spring-flowering bulb with strap-shaped leaves. An erect stem, arching at the apex, bears fragrant, blue, pink or white flowers. H 20–40cm (8–16in), S 8–10cm (3–4in).

Fritillaria pontica
Spring-flowering bulb with stems carrying lance-shaped, grey-green leaves, the topmost in a whorl of 3. Has solitary broadly bell-shaped, green flowers, 3–4.5cm (1¼–1¾in) long, often suffused brown. H 15–45cm (6–18in), S 5–8cm (2–3in).

Ixia viridiflora
Spring- to early summer-flowering corm with very narrow, erect leaves mostly at stem base. Carries a spike of flattish, jade-green flowers, 2.5–5cm (1–2in) across, with purple-black eyes. H 30–60cm (12–24in), S 2.5–5cm (1–2in).

Leucocoryne ixioides
(Glory-of-the-sun)
Spring-flowering bulb with long, narrow, semi-erect, basal leaves that are withered by flowering time. Wiry, slender flower stem has a loose head of up to 10 soft blue flowers. H 30–40cm (12–16in), S 8–10cm (3–4in).

Ixiolirion tataricum,
syn. *I. montanum*
Spring- to early summer-flowering bulb with long, narrow, semi-erect leaves on the lower part of stem. Has a loose cluster of blue flowers with a darker, central line along each petal. H to 40cm (16in), S 8–10cm (3–4in).

Hermodactylus tuberosus,
syn. *Iris tuberosa* (Widow iris)
Spring-flowering perennial with finger-like tubers. Very long, narrow, grey-green leaves are square in cross-section. Has a fragrant, yellowish-green flower with large, blackish-brown-tipped petals. H 20–40cm (8–16in), S 5–8cm (2–3in).

Fritillaria cirrhosa
Spring-flowering bulb with slender stems and narrow, whorled leaves; upper leaves have tendril-like tips. Produces up to 4 widely bell-shaped flowers, purple or yellowish-green with dark purple chequered patterns. H to 60cm (24in), S 5–8cm (2–3in).

Aristea ecklonii
IRISES, pp.198–9
Moraea polystachya

Bellevalia pycnantha

GREEN–YELLOW

Fritillaria pallidiflora
Robust, spring-flowering bulb with broadly lance-shaped, grey-green leaves, scattered or in pairs on stem. Has 1–5 widely bell-shaped, yellow to greenish-yellow flowers, usually faintly chequered brownish-red within. H 15–70cm (6–28in), S 8–10cm (3–4in).

Arum creticum
Spring-flowering tuber that bears white or yellow spathes, each bottle-shaped at the base, slightly reflexed at the apex and with a protruding, yellow spadix. Has arrow-shaped, semi-erect, deep green leaves in autumn. H 30–50cm (12–20in), S 20–30cm (8–12in).

Ferraria crispa, syn. *F. undulata*
Spring-flowering corm with leafy stem bearing a succession of upward-facing, brown or yellowish-brown flowers, 4–5cm (1½–2in) across, with 6 wavy-edged, spreading petals that are conspicuously lined and blotched. H 20–40cm (8–16in), S 8–10cm (3–4in).

Gladiolus 'Christabel'
Spring-flowering corm with a wiry stem producing a loose spike of fragrant, widely funnel-shaped, primrose-yellow flowers, 6–8cm (2½–3in) across, with purple-brown-veined, upper petals. H to 45cm (18in), S 8–10cm (3–4in).

Erythronium 'Pagoda'
Robust, spring-flowering tuber with 2 semi-erect, basal, faintly mottled, glossy leaves. Flower stem produces up to 10 pendent, pale yellow flowers with reflexed petals. H 25–35cm (10–14in), S 15–20cm (6–8in).

Calochortus luteus
(Yellow mariposa)
Late spring-flowering bulb with long, narrow, erect leaves near the base of the loosely branched stem. Each branch bears a 3-petalled, yellow flower with central, brown blotches. H 20–45cm (8–18in), S 5–10cm (2–4in).

Bloomeria crocea
Calochortus amabilis, p.378
DAFFODILS, pp.360–62
Erythronium grandiflorum
Erythronium tuolumnense
Freesia 'Rijnveld's Yellow'
Freesia 'Yellow River', p.369
Fritillaria acmopetala
Homeria ochroleuca
Hymenocallis 'Sulphur Queen'
IRISES, pp.198–9
TULIPS, pp.354–6

Daffodils

The charm and beauty of the daffodil (botanically known as *Narcissus*) grace the garden early in the year. The diversity of its flowers provides infinite variation, from the tiny Cyclamineus daffodil, with its swept back petals, to the stately trumpet daffodil. Many may be naturalized, forming a golden carpet in grass or a wild garden, but dwarf forms are best in rock gardens or troughs, brightening the landscape before most other plants emerge.

The genus is classified in 12 divisions. Their flower forms are illustrated below, with the exception of Div.10, the wild species, and Div.12, the miscellaneous category. Both have varying flowers, including hoop-petticoat forms, produced between autumn and early summer.

Div.1 Trumpet – usually solitary flowers each have a trumpet that is as long as, or longer, than the petals. Early to late spring-flowering.

Div.2 Large-cupped – solitary flowers each have a cup at least one-third the length of, but shorter than, the petals. Spring-flowering.

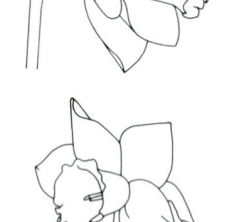

Div.3 Small-cupped – flowers are often borne singly; each has a cup not more than one-third the length of the petals. Spring- or early summer-flowering.

Div.4 Double – most have solitary large, fully or semi-double flowers with the cup and petals, or just the cup, replaced by petaloid structures. Some have smaller flowers in clusters of 4 or more. Spring- or early summer-flowering.

Div.5 Triandrus – nodding flowers, with short, sometimes straight-sided cups and narrow, reflexed petals, are borne 2–6 per stem. Spring-flowering.

Div.6 Cyclamineus – flowers are borne usually 1 or 2 per stem with cups that are sometimes flanged and often longer than those of Div.5. Petals are narrow, pointed and reflexed. Early to mid-spring flowering.

Div.7 Jonquil – sweetly scented flowers are borne usually 2 or more per stem. Cups are short, sometimes flanged; petals are often flat, fairly broad and rounded. Spring-flowering.

Div.8 Tazetta – flowers are borne in clusters of either 12 or more small, fragrant flowers per stem or 3 or 4 large ones. Cups are small and often straight-sided, petals broad and mostly pointed. Late autumn- to mid-spring-flowering.

Div.9 Poeticus – flowers each have a small, coloured cup and glistening white petals. They are borne usually 1 but sometimes 2 per stem and may be sweetly fragrant. Late spring- or early summer-flowering.

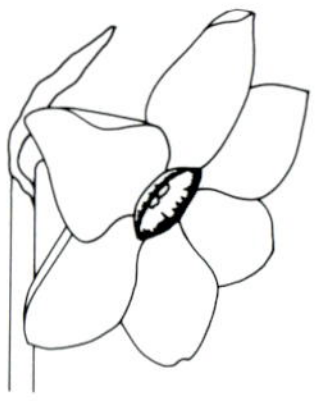

Div.11 Split-cupped – usually solitary flowers have cups that are typically split for more than half of their length. Cup segments lie back on the petals and may be ruffled. Spring-flowering.

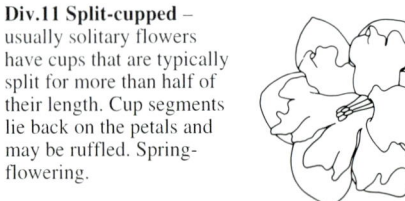

N. 'Cool Crystal' (Div.3)

N. 'Aircastle' (Div.3)

N. 'Mount Hood' (Div.1)

N. 'Ice Follies' (Div.3)

N. 'Silver Chimes' (Div.8)

N. 'Irene Copeland' (Div.4)

N. 'Cheerfulness' (Div.4)

N. 'Dover Cliffs' (Div.2)

N. 'Trousseau' (Div.1)

N. 'Passionale' (Div.2)

N. watieri (Div.10)

N. 'Portrush' (Div.3)

N. 'Thalia' (Div.5)

N. 'Panache' (Div.1)

N. 'Empress of Ireland' (Div.1)

N. 'Satin Pink' (Div.2)

N. 'Actaea' (Div.9)

N. 'Merlin' (Div.3)

N. 'Pride of Cornwall' (Div.8)

N. 'Kilworth' (Div.2)

N. 'Bridal Crown' (Div.4)

N. 'Salome' (Div.2)

N. 'Rainbow' (Div.2)

N. 'Rockall' (Div.3)

N. 'Acropolis' (Div.4)

N. 'Broadway Star' (Div.11)

N. poeticus var. *recurvus* (Div.10)

N. 'Woodland Star' (Div.3)

N. poeticus 'Praecox' (Div.10)

N. 'Dove Wings' (Div.6)

N. 'February Silver' (Div.6)

N. 'Jack Snipe' (Div.6)

N. canaliculatus (Div.8)

N. 'Avalanche' (Div.8)

N. romieuxii (Div.10)

N. 'Little Beauty' (Div.1)

N. pseudonarcissus (Div.10)

N. triandrus (Div.10)

N. 'Lemon Glow' (Div.1)

N. 'Cassata' (Div.11)

361

N. 'Pipit' (Div.7)

N. 'Rip van Winkle' (Div.4)

N. 'Sweetness' (Div.7) ♀

N. 'Golden Ducat' (Div.4)

N. 'Chanterelle' (Div.11)

N. rupicola (Div.10)

N. 'Bob Minor' (Div.1)

N. 'Bartley' (Div.6)

N. 'Ambergate' (Div.2)

N. 'Daydream' (Div.2)

N. 'Hawera' (Div.5)

N. cyclamineus (Div.10) ♀

N. 'February Gold' (Div.6) ♀

N. 'Suzy' (Div.7) ♀

N. 'Charity May' (Div.6) ♀

N. minor (Div.10)

N. 'Stratosphere' (Div.7)

N. 'Sealing Wax' (Div.2)

N. 'Yellow Cheerfulness' (Div.4)

N. jonquilla (Div.10)

N. 'Pencrebar' (Div.4)

N. 'St Keverne' (Div.2) ♀

N. 'Kingscourt' (Div.1) ♀

N. 'Home Fires' (Div.2)

N. 'Binkie' (Div.2)

N. 'Fortune' (Div.2)

N. 'Tête-à-Tête' (Div.6) ♀

N. 'Scarlet Gem' (Div.8)

N. 'Liberty Bells' (Div.5)

N. x *odorus* 'Rugulosus' (Div.10) ♀

N. 'Jumblie' (Div.6)

N. 'Tahiti' (Div.4)

N. 'Altruist' (Div.3)

Bulbs/medium SPRING INTEREST
◻ ORANGE

Stenomesson variegatum, syn. *S. incarnatum*
Clump-forming bulb. Bears reddish-yellow, pink or white flowers, with 6 green lobes at the apex, in winter or spring. H 30–60cm (12–24in), S 30cm (12in). Min. 10°C (50°F).

Stenomesson miniata, syn. *Urceolina peruviana*
Late spring-flowering bulb with strap-shaped, semi-erect, basal leaves. Bears a head of red or orange flowers, 2–4cm (³⁄₄–1¹⁄₂in) long, with yellow anthers. H 20–30cm (8–12in), S 10–15cm (4–6in). Min. 5°C (41°F).

Clivia miniata
Evergreen, tuft-forming rhizome with strap-shaped, semi-erect, basal, dark green leaves, 40–60cm (16–24in) long. Stems each produce a head of 10–20 orange or red flowers in spring or summer. H 40cm (16in), S 30–60cm (12–24in). Min. 10°C (50°F).

DAFFODILS, pp.360–62
Dipcadi serotinum, p.378
Hippeastrum 'Orange Sovereign', p.369
Ixia maculata
Tritonia crocata
TULIPS, pp.354–6

Bulbs/medium SUMMER INTEREST
◻ WHITE

Crinum asiaticum
Clump-forming bulb with strap-shaped, semi-erect, basal, dark green leaves, 1m (3ft) long. Leafless flower stems produce heads of long-tubed, white flowers, with narrow petals, in spring or summer. H 45–60cm (1¹⁄₂–2ft), S 60cm–1m (2–3ft). Min. 16°C (61°F).

Pancratium illyricum
Summer-flowering bulb with strap-shaped, semi-erect, basal, greyish-green leaves. Leafless stem has a head of 5–12 fragrant, 6-petalled, white flowers, 8cm (3in) across. H to 45cm (18in), S 25–30cm (10–12in).

Triteleia hyacinthina, syn. *Brodiaea hyacinthina*
Late spring- to early summer-flowering corm with long, narrow, semi-erect or spreading, basal leaves. Heads of white, sometimes purple-tinged flowers are borne on wiry stems. H 30–50cm (12–20in), S 8–10cm (3–4in).

Ornithogalum narbonense
Clump-forming, late spring- to summer-flowering bulb with long, narrow, semi-erect, basal, grey-green leaves. Leafless stem produces a spike of star-shaped, white flowers, 2cm (³⁄₄in) wide. H 30–40cm (12–16in), S 10–15cm (4–6in).

Arisaema sikokianum
Early summer-flowering tuber with erect leaves divided into 3–5 leaflets. Produces deep brownish-purple and white spathes, 15cm (6in) long, with club-like, white spadices protruding from the mouths. H 30–50cm (12–20in), S 30–45cm (12–18in).

OTHER RECOMMENDED PLANTS:
BEGONIAS, p.259
Cyrtanthus mackenii
Dietes iridioides
Haemanthus albiflos
Hesperocallis undulata
LILIES, p.348–9
Pancratium maritimum

Bulbs/medium SUMMER INTEREST

☐ WHITE

Ornithogalum thyrsoides
(Chincherinchee)
Summer-flowering bulb with strap-shaped, semi-erect, basal leaves. Bears a dense, conical spike of cup-shaped, white flowers, 2–3cm (¾–1¼in) across. H 30–45cm (12–18in), S 10–15cm (4–6in).
☼ ◊ ❄

Zigadenus fremontii
Clump-forming, early summer-flowering bulb with long, strap-shaped, semi-erect, basal leaves. Stem produces a spike of star-shaped, pale creamy-green flowers with darker green nectaries on petal bases. H 30–50cm (12–20in), S 10–15cm (4–6in).
☼ ◊ ❄❄

Hymenocallis narcissiflora,
syn. *Ismene calathina*
(Peruvian daffodil)
Spring- or summer-flowering bulb with semi-erect, basal leaves, dying down in winter. Bears a loose head of 2–5 fragrant, white flowers. H to 60cm (24in), S 30–45cm (12–18in).
◐ ◊ ❄

☐☐ WHITE–RED

Ornithogalum arabicum
Early summer-flowering bulb with strap-shaped, semi-erect leaves in a basal cluster. Has a flattish head of up to 15 scented, white or creamy-white flowers, 4–5cm (1½–2in) across, with black ovaries in centres. H 30–45cm (12–18in), S 10–15cm (4–6in).
☼ ◊ ❄

Allium cernuum
Clump-forming, summer-flowering bulb with narrow, semi-erect, basal leaves. Each stem produces up to 30 cup-shaped, pink or white flowers in a loose, nodding umbel, 2–4cm (¾–1½in) across. H 30–70cm (12–28in), S 8–12cm (3–5in).
☼ ◊ ❄❄❄ ♀

Alstroemeria pelegrina,
syn. *A. gayana*
Summer-flowering tuber with narrow, lance-shaped leaves. Each leafy stem has 1–3 white flowers, stained pinkish-mauve and spotted yellow and brownish-purple. H 30–60cm (1–2ft), S 60cm–1m (2–3ft).
☼ ◊ ❄❄

Tritonia disticha subsp.
rubrolucens, syn. *T. rosea*, *T. rubrolucens*, *Crocosmia rosea*
Late summer-flowering corm with narrowly sword-shaped, erect leaves in a flattish, basal fan. Has pink flowers in a loose, one-sided spike. H 30–50cm (12–20in), S 8–10cm (3–4in).
☼ ◊ ❄❄

Lycoris radiata (Red spider lily)
Late summer-flowering bulb with a head of 5 or 6 bright rose-red flowers with narrow, wavy-margined, reflexed petals and conspicuous anthers. Has strap-shaped, semi-erect, basal leaves after flowering time. H 30–40cm (12–16in), S 10–15cm (4–6in).
☼ ◊ ❄❄

Allium campanulatum
LILIES, p.348–9

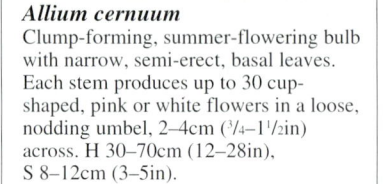

BEGONIAS, p.259
Bessera elegans
GLADIOLI, p.343
Haemanthus coccineus, p.379

Haemanthus sanguineus
Zantedeschia rehmannii
Zephyranthes grandiflora, p.368

■ RED–PURPLE

Rhodophiala advenum,
syn. *Hippeastrum advenum*
Clump-forming, spring- to summer-flowering bulb with basal, grey-green leaves. Leafless stem carries a head of 2–8 narrowly funnel-shaped, red flowers, 5cm (2in) long. H to 40cm (16in), S 15–20cm (6–8in).

Ranunculus asiaticus [red form]
(Persian buttercup)
Early summer-flowering perennial with claw-like tubers and long-stalked leaves both at base and on stem. Has single or double flowers in red, white, pink, yellow or orange. H 45–55cm (18–22in), S 10cm (4in).

Allium schubertii
Early summer-flowering bulb with widely strap-shaped, semi-erect, basal leaves. Bears large umbels of 40 or more star-shaped, pink or purple flowers on very unequal stalks, followed by brown seed capsules. H 30–60cm (12–24in), S 15–20cm (6–8in).

Allium beesianum
Allium cyaneum
Allium macranthum
Allium narcissiflorum, p.380
Cyrtanthus obliquus
Cyrtanthus sanguineus
DAHLIAS, pp.350–51
Dichelostemma pulchellum

■ PURPLE

Allium christophii,
syn. *A. albopilosum*
Summer-flowering bulb with semi-erect, hairy, grey leaves that droop at tips. Has a large, spherical umbel of 50 or more star-shaped, purplish-violet flowers, which dry well. H 15–40cm (6–16in), S 15–20cm (6–8in).

Patersonia umbrosa
Evergreen, clump-forming, spring- and early summer-flowering rhizome with erect, basal leaves. Tough flower stems each carry a succession of iris-like, purple-blue flowers, 3–4cm (1¼–1½in) across. H 30–45cm (12–18in), S 30–60cm (12–24in).

Gloxinia perennis
Late summer- to autumn-flowering rhizome with heart-shaped, toothed, hairy leaves on spotted stems. Has bell-shaped, lavender-blue flowers, with rounded lobes and purple-blotched throats. H to 60cm (24in), S 30–35cm (12–14in). Min. 10°C (50°F).

Oxalis bowiei
Roscoea purpurea

365

Bulbs/medium SUMMER INTEREST
■ BLUE

□ GREEN–YELLOW

Allium caeruleum, syn. *A. azureum*
Clump-forming, summer-flowering bulb with narrow, erect leaves on the lower third of slender flower stems, which bear 30–50 star-shaped, blue flowers in a dense, spherical umbel, 3–4cm (1¼–1½in) across. H 20–80cm (8–32in), S 10–15cm (4–6in).

Triteleia laxa, syn. *Brodiaea laxa*
Early summer-flowering corm with narrow, semi-erect, basal leaves. Stem carries a large, loose umbel of funnel-shaped, deep to pale purple-blue flowers, 2–5cm (¾–2in) long, mostly held upright. H 10–50cm (4–20in), S 8–10cm (3–4in).

Arisaema triphyllum, syn. *A. atrorubens* (Jack-in-the-pulpit)
Summer-flowering tuber with 3-lobed, erect leaves. Produces green or purple spathes, hooded at tips, followed by bright red berries. H 40–50cm (16–20in), S 30–45cm (12–18in).

Arisaema griffithii
Summer-flowering tuber with large, erect leaves above a green or purple spathe, 20–25cm (8–10in) long, strongly netted with paler veins and expanded like a cobra's hood. Protect in winter or lift for frost-free storage. H to 60cm (24in), S 45–60cm (18–24in).

Arisaema jacquemontii
Summer-flowering tuber with 1 or 2 erect leaves, divided into wavy-edged leaflets. Produces slender, white-lined, green spathes that are hooded at tips and drawn out into long points. H 30–50cm (12–20in), S 30–38cm (12–15in).

Eucomis bicolor
Summer-flowering bulb with wavy-margined, semi-erect, basal leaves. Stem, often spotted purple, bears a spike of green or greenish-white flowers, with 6 purple-edged petals, topped by a cluster of leaf-like bracts. H 30–50cm (12–20in), S 30–60cm (12–24in).

Cyrtanthus mackenii var. *cooperi*
Clump-forming, summer-flowering bulb with long, narrow, semi-erect, basal leaves. Leafless stems each carry a head of up to 10 fragrant, tubular, cream or yellow flowers, 5cm (2in) long and slightly curved. H 30–40cm (12–16in), S 8–10cm (3–4in).

Aristea ecklonii
Ixiolirion tataricum, p.358

Allium flavum
Cyrtanthus breviflorus
Hymenocallis 'Sulphur Queen'
Ixia viridiflora, p.355

Lycoris aurea
Triteleia ixioides

366

◻ YELLOW ◻ ORANGE

***Calochortus barbatus**,*
syn. *Cyclobothra lutea*
Summer-flowering bulb with narrow, erect leaves near the base of the loosely branched stem. Each branch bears a pendent, yellow or greenish-yellow flower that is hairy inside. H 30–60cm (12–24in), S 5–10cm (2–4in).

Ranunculus asiaticus
[yellow form] (Persian buttercup)
Early summer-flowering perennial with claw-like tubers and long-stalked, palmate leaves at base and on stem. Has single or double flowers in yellow, white, pink, red or orange. H 45–55cm (18–22in), S 8–10cm (3–4in).

***Polianthes geminiflora**,*
syn. *Bravoa geminiflora*
Summer-flowering tuber with narrowly strap-shaped, semi-erect leaves in a basal tuft. Stems each carry long spikes of downward-curving, tubular, red or orange flowers in pairs. H 20–40cm (8–16in), S 10–15cm (4–6in).

***Alstroemeria**,* **Ligtu Hybrids**
Summer-flowering tuber with narrow, twisted leaves and heads of widely flared flowers in shades of pink, yellow or orange, often spotted or streaked with contrasting colours. H 45–60cm (1½–2ft), S 60cm–1m (2–3ft).

Cypella herbertii
Summer-flowering bulb with a fan of narrow, sword-shaped, erect, basal leaves. Branched flower stem carries a succession of short-lived, iris-like, orange-yellow flowers, each spotted purple in the centre. H 30–50cm (12–20in), S 8–10cm (3–4in).

Sandersonia aurantiaca
(Chinese-lantern lily)
Deciduous, summer-flowering, tuberous climber with a slender stem bearing scattered, lance-shaped leaves, some tendril-tipped. Orange flowers are produced in axils of upper leaves. H 60cm (24in), S 25–30cm (10–12in).

***Crocosmia* 'Jackanapes'**
Clump-forming, late summer-flowering corm with sword-shaped, erect, basal leaves. Produces striking bicoloured, yellow and orange-red flowers. H 40–60cm (16–24in), S 15–20cm (6–8in).

Tigridia pavonia
(Peacock flower, Tiger flower)
Summer-flowering bulb with sword-shaped, pleated, erect leaves near stem base. A succession of short-lived flowers vary from white to orange, red or yellow, often with contrasting spots. H to 45cm (18in), S 12–15cm (4–6in).

Clivia miniata, p.363
Crocosmia 'Emily McKenzie'
LILIES, p.348–9

Bulbs/medium AUTUMN INTEREST
WHITE–PINK
PINK–ORANGE

Nerine bowdenii f. **alba**
Autumn-flowering bulb with a stout stem and strap-shaped, semi-erect, basal leaves. Produces a head of 5–10 white, often pink-flushed flowers; petals widen slightly towards wavy-margined, recurved tips. H 45–60cm (18–24in), S 12–15cm (5–6in).

Nerine undulata
Autumn-flowering bulb with narrowly strap-shaped, semi-erect, basal leaves. Flower stem carries a head of pink flowers with very narrow petals crinkled for their whole length. H 30–45cm (12–18in), S 10–12cm (4–5in).

Nerine 'Orion'
Autumn-flowering bulb with strap-shaped, semi-erect, basal leaves. Stout, leafless stem bears a head of pale pink flowers with very wavy-margined petals that have recurved tips. H 30–50cm (12–20in), S 20–25cm (8–10in).

Nerine bowdenii
Autumn-flowering bulb with a stout stem and strap-shaped, semi-erect, basal leaves. Carries a head of 5–10 glistening, pink flowers with petals that widen slightly towards wavy-margined, recurved tips. H 45–60cm (18–24in), S 12–15cm (5–6in).

Zephyranthes grandiflora, syn. *Z. carinata*, *Z. rosea*
Late summer- to early autumn-flowering bulb with narrowly strap-shaped, semi-erect, basal leaves. Each stem bears a funnel-shaped, pink flower, held almost erect. H 20–30cm (8–12in), S 8–10cm (3–4in).

Scilla scilloides, syn. *S. chinensis*
Late summer- and autumn-flowering bulb with 2–4 narrowly strap-shaped, semi-erect, basal leaves. Stem bears a slender, dense spike of up to 30 flattish, pink flowers, 0.5–1cm (¼–½in) across. H to 30cm (12in), S 5cm (2in).

Nerine 'Brian Doe'
Autumn-flowering bulb with strap-shaped, semi-erect, basal leaves. Stout, leafless stem bears a head of salmon flowers with 6 reflexed, wavy-margined petals. H 30–50cm (12–20in), S 20–25cm (8–10in).

Nerine sarniensis (Guernsey lily)
Autumn-flowering bulb with strap-shaped, semi-erect, basal leaves. Leafless stem carries a spherical head of up to 20 deep orange-pink flowers, 6–8cm (2½–3in) across, with wavy-margined petals. H 45–60cm (18–24in), S 12–15cm (5–6in).

OTHER RECOMMENDED PLANTS:
Amaryllis belladonna, p.352
Crinum macowanii
Habranthus robustus, p.381
Nerine 'Blanchefleur'
Nerine filifolia
Nerine flexuosa
Nerine 'Margaret Rose'
Brunsvigia josephinae
Eucomis autumnalis
Gloxinia perennis, p.365
Nerine 'Bagdad'
Nerine 'Corusca Major'
Nerine 'Fothergillii Major'

Bulbs/medium WINTER INTEREST
WHITE–RED

Eucharis × grandiflora, syn. *E. amazonica*
Evergreen, clump-forming bulb with strap-shaped, semi-erect, basal leaves. Bears a head of up to 6 fragrant, slightly pendent, white flowers at almost any season. H 40–60cm (16–24in), S 60cm–1m (2–3ft). Min. 15°C (59°F).

Veltheimia bracteata, syn. *V. undulata*, *V. viridifolia*
Clump-forming, winter-flowering bulb with semi-erect, strap-shaped, basal, glossy leaves and dense spikes of pendent, tubular, pink, red or yellowish-red flowers. H 30–45cm (12–18in), S 25–38cm (10–15in). Min. 10°C (50°F).

Hippeastrum 'Red Lion'
Tuft-forming, winter- and spring-flowering bulb with a stout stem bearing a head of 2–6 dark red flowers with yellow anthers. Strap-shaped leaves appear with or just after flowers. H 30–50cm (12–20in), S 30cm (12in). Min. 13°C (55°F).

RED–YELLOW

Hippeastrum 'Orange Sovereign'
Winter- to spring-flowering bulb with strap-shaped, semi-erect, basal, grey-green leaves produced as, or just after, flowers form. Stout stem carries a head of 2–6 rich orange-red flowers. H 30–50cm (12–20in), S 30cm (12in). Min. 13°C (55°F).

Hippeastrum 'Apple Blossom'
Winter- to spring-flowering bulb with strap-shaped, semi-erect, basal leaves produced as, or just after, flowers form. Stout stem has a head of 2–6 white flowers, becoming pink at petal tips. H 30–50cm (12–20in), S 30cm (12in). Min. 13°C (55°F).

Hippeastrum 'Striped'
Winter- to spring-flowering bulb with strap-shaped, semi-erect, basal leaves produced with or just after flowers. Stout stem has a head of 2–6 widely funnel-shaped flowers, striped white and red. H 50cm (20in), S 30cm (12in). Min. 13°C (55°F).

Freesia 'Everett'
Winter- and spring-flowering corm with narrow, erect, bright green leaves in basal fans and a spike of large, fragrant, pinkish-red flowers. H to 30cm (12in), S 1.5–2.5cm (⅝–1in).

Lachenalia archioides var. **glaucina**, syn. *L. glaucina*
Late winter- and early spring-flowering bulb with 2 strap-shaped, semi-erect, basal leaves, usually spotted purple. Has a spike of fragrant, whitish-blue or pale lilac flowers. H to 30cm (12in), S 5–8cm (2–3in).

Freesia 'Yellow River'
Winter- and spring-flowering corm with narrow, erect, bright green leaves in basal fans and a spike of large, fragrant, bright yellow flowers. H to 30cm (12in), S 1.5–2.5cm (⅝–1in).

OTHER RECOMMENDED PLANTS:
Freesia alba
Freesia carymbosa
Hippeastrum aulicum

Hippeastrum 'Belinda'
Hippeastrum 'Bouquet'
Hippeastrum 'White Dazzler'
Hymenocallis speciosa

Lachenalia rubida
Stenomesson variegatum, p.363
Veltheimia capensis

Freesia 'Rijnveld's Yellow'
Freesia 'Romany'
Lachenalia mutabilis
Moraea polystachya

Bulbs/small SPRING INTEREST
☐ WHITE

Anemone blanda 'White Splendour'
Knobbly tuber with semi-erect leaves that have 3 deeply toothed lobes. Bears upright, flattish, white flowers, 4–5cm (1½–2in) across, with 9–14 narrow petals, in early spring. H 5–10cm (2–4in), S 10–15cm (4–6in).

Puschkinia scilloides 'Alba'
Spring-flowering bulb with usually 2 strap-shaped, semi-erect, basal leaves. Produces a dense spike of star-shaped, white flowers, 1.5–2cm (⅝–¾in) across. H 15cm (6in), S 2.5–5cm (1–2in).

Ornithogalum balansae
Spring-flowering bulb with 2 narrow, green or slightly greyish-green leaves, becoming wider towards tips. Has a broad head of 2–5 white flowers, green outside, that open wide. H 5–15cm (2–6in), S 5–8cm (2–3in).

Leucojum vernum (Spring snowflake)
Spring-flowering bulb with strap-shaped, semi-erect, basal leaves. Leafless stem carries 1 or 2 pendent, bell-shaped flowers, 1.5–2cm (⅝–¾in) long, with 6 green-tipped, white petals. H 10–15cm (4–6in), S 8–10cm (3–4in).

Sternbergia candida
Spring-flowering bulb. Strap-shaped, semi-erect, basal, greyish-green leaves appear together with a fragrant, funnel-shaped, white flower, 4–5cm (1½–2in) long, borne on a leafless stem. H 10–20cm (4–8in), S 8–10cm (3–4in).

Ornithogalum montanum
Clump-forming, spring-flowering bulb with strap-shaped, semi-erect, basal, grey-green leaves. Leafless stem produces a head of star-shaped, white flowers, 3–4cm (1¼–1½in) across, striped green outside. H and S 10–15cm (4–6in).

Ornithogalum lanceolatum
Spring-flowering, dwarf bulb with a flattish rosette of prostrate, lance-shaped, basal leaves. Carries a head of flattish, star-shaped, white flowers, 3–4cm (1¼–1½in) across, broadly striped green outside. H 5–10cm (2–4in), S 10–15cm (4–6in).

OTHER RECOMMENDED PLANTS:
CROCUSES, pp.372–3
Cyclamen creticum
DAFFODILS, pp.360–62
Fritillaria bucharica
Gagea graeca
Galanthus 'Atkinsii', p.383
Galanthus elwesii, p.383
Galanthus gracilis, p.383
Galanthus ikariae, p.384
Galanthus nivalis 'Flore Pleno', p.383
Galanthus nivalis 'Lutescens', p.384
Galanthus nivalis 'Pusey Green Tips', p.383
Galanthus nivalis 'Scharlockii', p.384
Galanthus plicatus subsp. byzantinus, p.384
Galanthus rizehensis, p.383

☐ WHITE–PINK

Erythronium californicum
Clump-forming, spring-flowering tuber. Bears 2 semi-erect, basal, mottled leaves. Up to 3 white or creamy-white flowers, sometimes red-brown externally, have reflexed petals, yellow eyes and often brown rings near centres. H 15–35cm (6–14in), S 10–12cm (4–5in).

☐ PINK

Chionodoxa siehei **'Pink Giant'**
Early spring-flowering bulb with 2 narrow, semi-erect, basal leaves. Leafless stem produces a spike of 5–10 flattish, white-eyed, pink flowers, 2–2.5cm (³⁄₄–1in) across, that face outwards or slightly downwards. H 10–25cm (4–10in), S 2.5–5cm (1–2in).

Allium akaka
Spring-flowering bulb with 1–3 broad, prostrate and basal, grey-green leaves and an almost stemless, spherical umbel, 5–7cm (2–3in) across of 30–40 star-shaped, white to pinkish-white flowers with red centres. H 15–20cm (6–8in), S 12–15cm (5–6in).

Allium karataviense
Late spring-flowering bulb with narrowly elliptic to elliptic, prostrate, basal, greyish-purple leaves. Stem bears 50 or more star-shaped, pale purplish-pink flowers in a spherical umbel, 15cm (6in) or more across. H to 20cm (8in), S 25–30cm (10–12in).

Cyclamen libanoticum
Spring-flowering tuber with ivy-shaped, dull green leaves with lighter patterns and purplish-green undersides. Has musty-scented, clear pink flowers, each with deep carmine marks at the mouth. Grows best in an alpine house. H to 10cm (4in), S 10–15cm (4–6in).

Anemone biflora
Spring-flowering tuber with deeply lobed, toothed, semi-erect leaves. Bears 5-petalled, saucer-shaped, bright red, coppery-pink or yellowish-red flowers, 3–4cm (1¹⁄₄–1¹⁄₂in) across. Needs warm, dry, summer dormancy. H 5–10cm (2–4in), S 5–8cm (2–3in).

HYACINTHS, p.375
Hypoxis capensis
IRISES, pp.198–9
Ornithogalum nutans

Ornithogalum umbellatum
TULIPS, pp.354–6

Cyclamen persicum, p.384
Cyclamen persicum 'Pearl Wave', p.384
Merendera robusta
TULIPS, pp.354–6

Crocuses

Crocuses (*Crocus* species and cultivars) are possibly the most versatile and reliable of dwarf bulbous plants, most flowering in late winter or early spring, though some species flower in autumn. A wide colour range is available, from white through pinkish-lilac to rich purples, creams and yellows. Many are also attractively striped or feathered with other colours. Their usually goblet-shaped flowers, sometimes produced before the leaves, open wide in full sun, in some cases to reveal contrasting centres or conspicuous stamens. Most crocuses are also fragrant. To appreciate their scent to the full, plant in bowls or shallow pans for indoor display; start them into growth in a bulb frame outside, then bring them inside a week or two before they would normally bloom in the garden, so that the warmth stimulates them into flower.

Outside, crocuses may be used in rock gardens in association with other early-flowering dwarf bulbs or perennials, or planted *en masse* in drifts or beneath deciduous trees and shrubs, where they will rapidly colonize large areas and provide pleasure for many years. To promote vigour, feed the plants once the flowers have faded and while the leaves are in full growth. If naturalized in grass, delay mowing until the leaves have died down. Congested clumps may be dug up after flowering and divided.

C. 'Eyecatcher'

C. 'Blue Bird'

C. hadriaticus

C. tommasinianus 'Albus'

C. kotschyanus ♀

C. dalmaticus

C. minimus

C. vernus subsp. *albiflorus*

C. sieberi 'Bowles' White' ♀

C. laevigatus

C. pulchellus ♀

C. sieberi subsp. *sublimis* f. *tricolor* ♀

C. etruscus 'Zwanenburg'

C. malyi ♀

C. boryi ♀

C. goulimyi ♀

C. medius ♀

C. 'Snow Bunting' ♀

C. imperati 'De Jager'

C. 'Blue Pearl' ♀

C. nudiflorus

C. banaticus ♧

C. etruscus ♧

C. longiflorus

C. baytopiorum

C. speciosus 'Conqueror'

C. vernus 'Queen of the Blues'

C. biflorus

C. 'Advance'

C. vernus

C. tommasinianus 'Ruby Giant'

C. vernus 'Prinses Juliana'

C. vernus 'Pickwick'

C. 'Dorothy'

C. 'Ladykiller' ♧

C. cvijicii

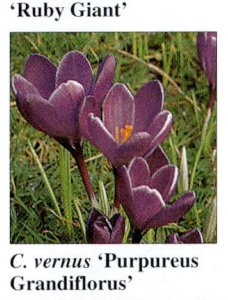

C. tommasinianus 'Whitewell Purple'

C. speciosus ♧

C. vernus 'Purpureus Grandiflorus'

C. cartwrightianus ♧

C. 'Cream Beauty' ♧

C. vernus 'Remembrance'

C. speciosus 'Oxonian'

C. 'E. A. Bowles' ♧

C. gargaricus

373

Bulbs/small SPRING INTEREST

◻ PINK

◼◼ RED–PURPLE

Erythronium dens-canis
(Dog's-tooth violet)
Spring-flowering tuber with 2 basal, mottled leaves. Stem has a pendent, pink, purple or white flower, with bands of brown, purple and yellow near the centre and reflexed petals. H 15–25cm (6–10in), S 8–10cm (3–4in).

Anemone x fulgens
Spring- or early summer-flowering tuber with deeply divided, semi-erect, basal leaves. Stout stems each carry an upright, bright red flower, 5–7cm (2–3in) across, with 10–15 petals. H 10–30cm (4–12in), S 8–10cm (3–4in).

Sparaxis tricolor
Spring-flowering corm with erect, lance-shaped leaves in a basal fan. Stem produces a loose spike of up to 5 flattish, orange, red, purple, pink or white flowers, 5–6cm (2–2½in) across, with black or red centres. H 10–30cm (4–12in), S 8–12cm (3–5in).

Allium acuminatum,
syn. *A. murrayanum*
Spring-flowering bulb with 2–4 long, narrow, semi-erect, basal leaves. Stem bears an umbel, 5cm (2in) across, of up to 30 small, purplish-pink flowers. H 10–30cm (4–12in), S 5–8cm (2–3in).

Allium oreophilum,
syn. *A. ostrowskianum*
Spring- and summer-flowering, dwarf bulb with 2 narrow, semi-erect, basal leaves. Has loose, domed umbels of up to 10 widely bell-shaped, deep rose-pink flowers, 1.5–2cm (⅝–¾in) across. H 5–10cm (2–4in), S 8–10cm (3–4in).

***Anemone blanda* 'Radar'**
Knobbly tuber with semi-erect, deep green leaves with 3 deeply toothed lobes. In early spring, stems each bear an upright, flattish, white-centred, deep reddish-carmine flower with 9–14 narrow petals. H 5–10cm (2–4in), S 10–15cm (4–6in).

Bulbocodium vernum
Spring-flowering corm with stemless, widely funnel-shaped, reddish-purple flowers. Narrow, semi-erect, basal leaves appear with flowers but do not elongate until later. Dies down in summer. H 3–4cm (1¼–1½in), S 3–5cm (1¼–2in).

Cyclamen persicum 'Cardinal'
Cyclamen persicum, Dwarf Scented Series
Cyclamen persicum 'Renown', p.385
Cyclamen trochopteranthum

Erythronium hendersonii, p.353
HYACINTHS, p.375
TULIPS, pp.354–6

Anemone coronaria
Arisarum proboscideum
Corydalis bulbosa
CROCUSES, pp.372–3

■ PURPLE

Babiana rubro-cyanea (Winecups)
Spring-flowering corm with lance-shaped, erect, folded leaves in a basal fan. Carries short spikes of 5–10 flowers, each with 6 petals, purple-blue at the top and red at the base.
H 15–20cm (6–8in), S 5–8cm (2–3in). Min. 10°C (50°F).

Ipheion uniflorum **'Froyle Mill'**
Spring-flowering bulb with narrow, semi-erect, basal, pale green leaves that smell of onions if crushed. Each leafless stem carries a star-shaped, violet-blue flower, 3–4cm (1¼–1½in) across. H 10–15cm (4–6in), S 5–8cm (2–3in).

Romulea bulbocodium
Spring-flowering corm with long, semi-erect, thread-like leaves in a basal tuft. Slender flower stems each carry 1–6 upward-facing flowers, usually pale lilac-purple with yellow or white centres. H 5–10cm (2–4in), S 2.5–5cm (1–2in).

Hyacinths

Grown for their sweet, penetrating scent, hyacinths (*Hyacinthus*) are deservedly popular spring bulbs. The most widely offered for sale are showy cultivars derived from *H. orientalis* and are available in an increasingly wide colour range: white, many shades of pink, lilac, blue, purple, yellow and warm salmon-orange. In the garden they are most effective if planted *en masse* in blocks of a single colour, or in containers positioned near the house where their fragrance can be appreciated to the full. They are also excellent for growing in bowls indoors; the bulbs may later be planted out in the garden. Specially treated bulbs for indoor culture are available that flower in mid-winter. (For details see the Plant Dictionary.) These forced bulbs may afterwards be planted outdoors in a sheltered spot, where they should flower again in later years. To keep the bulbs healthy and productive, feed them well after the flower spikes have faded.

H. orientalis **'Delft Blue'** ♀

H. orientalis **'White Pearl'**

H. orientalis **'Lady Derby'**

H. orientalis **'Ostara'** ♀

H. orientalis **'Queen of Pinks'**

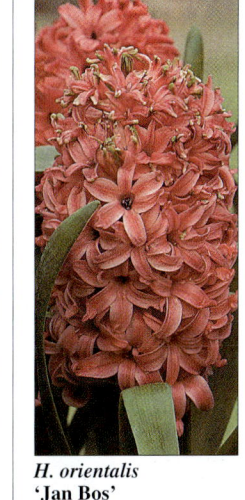

H. orientalis **'Jan Bos'**

H. orientalis **'Blue Jacket'** ♀

H. orientalis **'Distinction'**

H. orientalis **'Violet Pearl'**

H. orientalis **'City of Haarlem'** ♀

H. orientalis **'Princess Maria Christina'**

Bulbs/small SPRING INTEREST
■ PURPLE–BLUE

Gynandriris sisyrinchium
Spring-flowering corm with 1 or 2 semi-erect, narrow, basal leaves. Wiry stems each carry a succession of lavender- to violet-blue flowers, 3–4cm (1¼–1½in) across, with white or orange patches on the 3 larger petals. H 10–20cm (4–8in), S 8–10cm (3–4in).

Hyacinthella leucophaea
Spring-flowering bulb with 2 narrowly strap-shaped, semi-erect, basal leaves and a thin, wiry, leafless flower stem. Carries a short spike of tiny, bell-shaped, very pale blue, almost white flowers. H 10cm (4in), S 2.5–5cm (1–2in).

Bellevalia hyacinthoides, syn. *Strangweia spicata*
Spring-flowering bulb with prostrate, narrow leaves in a basal cluster. Bears a dense spike of up to 20 bell-shaped, pale lavender-blue, almost white flowers with darker, central veins. H 5–15cm (2–6in), S 5cm (2in).

Scilla mischtschenkoana, syn. *S. tubergeniana*
Early spring-flowering bulb with 2 or 3 strap-shaped, semi-erect, basal, mid-green leaves. Stems elongate as cup-shaped or flattish, pale blue flowers, with darker blue veins, open. H 5–10cm (2–4in), S 5cm (2in).

Muscari comosum 'Plumosum', syn. *M.c.* 'Monstrosum' (Feather grape hyacinth)
Spring-flowering bulb with up to 5 strap-shaped, semi-erect, basal, grey-green leaves. Sterile flowers are replaced by a fluffy mass of purple threads. H to 25cm (10in), S 10–12cm (4–5in).

Tecophilaea cyanocrocus var. **leichtlinii**, syn. *T.c.* 'Leichtlinii'
Spring-flowering corm with 1 or 2 narrowly lance-shaped, semi-erect, basal leaves and solitary upward-facing, widely funnel-shaped, pale blue flowers with large, white centres. H 8–10cm (3–4in), S 5–8cm (2–3in).

Puschkinia scilloides, syn. *P. libanotica* (Striped squill)
Spring-flowering bulb with usually 2 strap-shaped, semi-erect, basal leaves. Carries a dense spike of star-shaped, pale blue flowers with a darker blue stripe down each petal centre. H 15cm (6in), S 2.5–5cm (1–2in).

Brimeura amethystina, syn. *Hyacinthus amethystinus*
Late spring-flowering bulb with very narrow, semi-erect, basal leaves. Each leafless stem bears a spike of up to 15 pendent, tubular, blue flowers. H 10–25cm (4–10in), S 2.5–5cm (1–2in).

Scilla siberica 'Atrocoerulea'
Early spring-flowering bulb with 2–4 strap-shaped, semi-erect, basal, glossy leaves, widening towards tips. Bell-shaped, deep rich blue flowers, 1–1.5cm (½–⅝in) long, are borne in a short spike. H 10–15cm (4–6in), S 5cm (2in).

Anemone apennina
Anemone blanda
Babiana plicata
Babiana stricta

Brodiaea coronaria
Chionodoxa sardensis
Corydalis ambigua
CROCUSES, pp.272–3

Hyacinthoides italica
Ipheion uniflorum 'Wisley Blue'
IRISES, pp.198–9
Muscari armeniacum 'Blue Spike'

Muscari azureum
Muscari botryoides
Muscari comosum

■ BLUE

Muscari aucheri,
syn. *M. tubergenianum*
Spring-flowering bulb with 2 strap-shaped, greyish-green leaves. Bears small, almost spherical, bright blue flowers with white-rimmed mouths; upper flowers are often paler.
H 5–15cm (2–6in), S 5–8cm (2–3in).

Chionodoxa siehei, syn.
C. luciliae of gardens, *C. tmolusii*
Early spring-flowering bulb with 2 semi-erect, narrow, basal leaves. Bears a spike of 5–10 outward-facing, rich blue-lilac flowers with white eyes. H 10–25cm (4–10in), S 2.5–5cm (1–2in).

Muscari neglectum,
syn. *M. racemosum*
Spring-flowering bulb. Bears 4–6 often prostrate leaves from autumn to early summer. Has small, ovoid, deep blue or blackish-blue flowers with white-rimmed mouths. Increases rapidly.
H 10–20cm (4–8in), S 8–10cm (3–4in).

Tecophilaea cyanocrocus
(Chilean blue crocus)
Spring-flowering corm with 1 or 2 lance-shaped, semi-erect, basal leaves. Carries upward-facing, funnel-shaped, deep gentian-blue flowers, 4–5cm (1½–2in) across, with white throats.
H 8–10cm (3–4in), S 5–8cm (2–3in).

Anemone blanda 'Atrocaerulea'
Knobbly tuber with semi-erect, dark green leaves that have 3 deeply toothed lobes. In early spring, stems each bear an upright, flattish, deep blue flower, 4–5cm (1½–2in) across, with 9–14 narrow petals. H 5–10cm (2–4in), S 10–15cm (4–6in).

Chionodoxa luciliae,
syn. *C. gigantea*
Early spring-flowering bulb with 2 somewhat curved, semi-erect, basal leaves. Leafless stem bears 1–3 upward-facing, blue flowers with white eyes.
H 5–10cm (2–4in), S 2.5–5cm (1–2in).

Muscari armeniacum
Spring-flowering bulb with 3–6 long, narrow, semi-erect, basal leaves. Carries a dense spike of small, fragrant, bell-shaped, deep blue flowers with constricted mouths that have a rim of small, paler blue or white 'teeth'.
H 15–20cm (6–8in), S 8–10cm (3–4in).

× *Chionoscilla allenii*
Early spring-flowering bulb with 2 narrow, semi-erect, basal, dark green leaves and flattish, star-shaped, deep blue flowers, 1–2cm (½–¾in) across, in a loose spike. H 10–15cm (4–6in), S 2.5–5cm (1–2in).

HYACINTHS, p.375

Bulbs/small SPRING INTEREST

☐ GREEN–YELLOW

☐ YELLOW–ORANGE

Ledebouria socialis, syn. *Scilla violacea*
Evergreen, spring-flowering bulb with lance-shaped, semi-erect, basal, dark-spotted, grey or green leaves. Produces a short spike of bell-shaped, purplish-green flowers. H 5–10cm (2–4in), S 8–10cm (3–4in).

Muscari macrocarpum
Spring-flowering bulb with 3–5 semi-erect, basal, greyish-green leaves. Carries a dense spike of fragrant, brown-rimmed, bright yellow flowers. Upper flowers may initially be brownish-purple. H 10–20cm (4–8in), S 10–15cm (4–6in).

Calochortus amabilis (Golden fairy lantern, Golden globe tulip)
Spring-flowering bulb with one long, narrow, erect leaf, near the base of the loosely branched stem. A fringed, deep yellow, sometimes green-tinged flower hangs from each branch. H 10–30cm (4–12in), S 5–10cm (2–4in).

Colchicum luteum
Spring-flowering corm with wineglass-shaped, yellow flowers – the only known yellow *Colchicum*. Semi-erect, basal leaves are short at flowering time but later expand. H 5–10cm (2–4in), S 5–8cm (2–3in).

Erythronium americanum
Spring-flowering tuber with 2 semi-erect, basal leaves, mottled green and brown, and a pendent, yellow flower, often bronze outside, with petals reflexing in sunlight. Forms clumps by stolons. H 5–25cm (2–10in), S 5–8cm (2–3in).

Fritillaria pudica (Yellow fritillary)
Spring-flowering bulb with stems bearing scattered, narrowly lance-shaped, grey-green leaves. Has 1 or 2 deep yellow, sometimes red-tinged flowers, 1–2.5cm (½–1in) long. H 5–20cm (2–8in), S 5cm (2in).

Arum italicum 'Pictum'
Late spring-flowering tuber. Produces semi-erect leaves, with cream or white veins, in autumn, followed by pale green or creamy-white spathes, then red berries in autumn. Is good for flower arrangements. H 15–25cm (6–10in), S 20–30cm (8–12in).

Dipcadi serotinum
Spring-flowering bulb with 2–5 very narrow, semi-erect, basal leaves. Leafless stem has a loose spike of nodding, tubular, brown or dull orange flowers, 1–1.5cm (½–⅝in) long. H 10–30cm (4–12in), S 5–8cm (2–3in).

Albuca canadensis
CROCUSES, pp.372–3
Eranthis hyemalis, p.385
Erythronium grandiflorum
Fritillaria pallidiflora, p.359
Gagea peduncularis
HYACINTHS, p.375
TULIPS, pp.354–6

CROCUSES, pp.372–3
Sparaxis elegans
TULIPS, pp.354–6

Bulbs/small SUMMER INTEREST

WHITE–PINK

Lloydia serotina
Early summer-flowering bulb with wiry stems bearing scattered, thread-like, semi-erect leaves near stem base. Carries 1 or 2 bell-shaped, white flowers, 1–1.5cm (1/2–5/8in) long, with purple or purple-red veins. H 5–15cm (2–6in), S 2.5–5cm (1–2in).

Albuca humilis
Summer-flowering, dwarf bulb with very narrow, basal, dark green leaves. Carries a loose head of 1–3 cup-shaped, white flowers, 1cm (1/2in) long, striped green, later reddish, outside. H 5–10cm (2–4in), S 5–8cm (2–3in).

Calochortus subalpinus
Summer-flowering bulb with one long, narrow, semi-erect, basal leaf. Has 1–3 erect, saucer-shaped, creamy-white flowers with yellow hairs and usually a small, purple mark at the base of the 3 smaller petals. H 15–20cm (6–8in), S 5–7cm (2–3in).

Arisaema candidissimum
Early summer-flowering tuber with large, cowl-like, pink-striped, white spathes, enclosing tiny, fragrant flowers on spadices, followed by broad, 3-parted, semi-erect leaves, 30cm (12in) long. H 10–15cm (4–6in), S 30–45cm (12–18in).

Cyclamen purpurascens, syn. *C. europaeum*, *C. fatrense*
Summer- and autumn-flowering tuber with rounded, silver-patterned leaves. Bears very fragrant, lilac-pink to reddish-purple flowers. H to 10cm (4in), S 10–15cm (4–6in).

RED

Hippeastrum striatum, syn. *H. rutilum*
Spring- and summer-flowering bulb with strap-shaped, semi-erect, basal, bright green leaves. Funnel-shaped flowers have pointed, scarlet petals with central, green stripes. H 30cm (12in), S 20–25cm (8–10in). Min. 15°C (59°F).

Cyrtanthus brachyscyphus, syn. *C. parviflorus*
Clump-forming, summer-flowering bulb with strap-shaped, semi-erect, basal, bright green leaves. Leafless stem bears a head of 6–12 tubular, orange- or brilliant red flowers with 6 lobes. H 20–30cm (8–12in), S 10–15cm (4–6in).

Haemanthus coccineus (Blood lily)
Summer-flowering bulb with 2 elliptic leaves, hairy beneath, that lie flat on the ground. Spotted stem, forming before leaves, bears a cluster of tiny, red flowers with prominent stamens, within fleshy, red or pink bracts. H to 30cm (12in), S 20–30cm (8–12in). Min. 10°C (50°F).

OTHER RECOMMENDED PLANTS:
Allium campanulatum
Allium oreophilum, p.374
Caloscordum neriniflorum
Habranthus robustus, p.381
Triteleia peduncularis
Zephyranthes atamasco

Anomatheca laxa
BEGONIAS, p.259
Habranthus tubispathus

379

Bulbs/small SUMMER INTEREST

■ PURPLE   ■■ PURPLE–YELLOW

Allium schoenoprasum (Chives)
Clump-forming, summer-flowering bulb with narrow, hollow, erect, dark green leaves at base. Stems each carry up to 20 tiny, bell-shaped, pale purple or pink flowers in a dense umbel up to 5cm (2in) across. H 12–25cm (5–10in), S 5–10cm (2–4in).

Allium narcissiflorum, syn. *A. pedemontanum*
Clump-forming, summer-flowering bulb with very narrow, erect, grey-green leaves on the lower part of the flower stem. Has an umbel of up to 15 bell-shaped, pinkish-purple flowers. H 15–30cm (6–12in), S 8–10cm (3–4in).

Allium cyathophorum var. *farreri*
Clump-forming, summer-flowering bulb with tufts of narrow, erect, basal leaves. Each stem bears a small, loose umbel, 1.5–4cm (⅝–1½in) wide, of up to 30 bell-shaped, dark reddish-purple flowers with sharply pointed petals. H 15–30cm (6–12in), S 10–15cm (4–6in).

Chlidanthus fragrans
Summer-flowering bulb with narrow, semi-erect leaves in a basal tuft. Leafless stem carries a head of 3–5 fragrant, funnel-shaped, yellow flowers, 4–7cm (1½–2¾in) long. H 10–30cm (4–12in), S 8–10cm (3–4in).

Scilla peruviana
Early summer-flowering bulb with a basal cluster of up to 10 lance-shaped, semi-erect leaves. Stem bears a broadly conical head of up to 50 flattish, violet-blue flowers, 1.5–3cm (⅝–1¼in) across. H 10–25cm (4–10in), S 15–20cm (6–8in).

Allium moly
Clump-forming, summer-flowering bulb with 1–3 broad, semi-erect, basal, grey-green leaves. Stems each bear up to 40 star-shaped, yellow flowers in a fairly dense umbel, 4–8cm (1½–3in) across. H 10–35cm (4–14in), S 10–12cm (4–5in).

Roscoea humeana
Summer-flowering tuber. Erect, broadly lance-shaped, rich green leaves form a stem-like sheath at base. Carries up to 10 long-tubed, purple flowers, each with a hooded, upper petal, a wide, pendent lip and 2 narrower petals. H 15–25cm (6–10in), S 15–20cm (6–8in).

Roscoea cautleoides
Summer-flowering tuber. Erect, lance-shaped leaves, semi-erect, basal leaves. Has up to 5 long-tubed, yellow flowers, each with a hooded, upper petal, a broad, 2-lobed, lower lip and 2 narrower petals. H 15–25cm (6–10in), S 10–15cm (4–6in).

Hypoxis angustifolia
Summer-flowering corm with slender, hairy, semi-erect, basal leaves. Stems each carry 3–7 star-shaped, yellow flowers, 1.5–2cm (⅝–¾in) across. H 10–20cm (4–8in), S 5–8cm (2–3in).

Allium christophii, p.365
Biarum tenuifolium, p.383
Ledebouria cooperi
Scilla litardieri

Triteleia laxa, p.366

Allium cyaneum
Allium flavum
Allium sikkimense
BEGONIAS, p.259

Pinellia ternata
Romulea macowanii

Bulbs/small AUTUMN INTEREST
☐ WHITE

☐ WHITE–PINK

Cyclamen hederifolium var. **album**
Autumn-flowering tuber. Pure white flowers, with reflexed petals, appear before or with leaves, which vary but are often ivy-shaped with silvery-green patterns. H to 10cm (4in), S 10–15cm (4–6in).

Colchicum speciosum 'Album'
Vigorous, autumn-flowering corm with large, semi-erect, basal leaves in late winter or spring. Cup-shaped, white flowers successfully withstand bad weather. H and S 15–20cm (6–8in).

Leucojum autumnale
(Autumn snowflake)
Autumn-flowering bulb with thread-like, erect, basal leaves appearing with, or just after, flowers. Slender stems each produce a head of 1–4 bell-shaped, white flowers, tinged pink at bases. H 10–15cm (4–6in), S 2.5–5cm (1–2in).

Zephyranthes candida
Autumn-flowering bulb with narrow, erect, basal leaves forming rush-like tufts. Each leafless stem carries crocus-like, white flowers, to 6cm (2½in) across. H 15–25cm (6–10in), S 5–8cm (2–3in).

Cyclamen africanum
Autumn-flowering tuber with ivy-shaped, deep green leaves with lighter patterns. Bears pendent, white or pink flowers, with reflexed petals and darker stains around mouths, as or just before leaves appear. H to 10cm (4in), S 10–15cm (4–6in).

Cyclamen mirabile
Autumn-flowering tuber with pale pink flowers with toothed petals and dark purple-stained mouths. Heart-shaped, patterned leaves, purplish-green beneath, are minutely toothed on margins. H to 10cm (4in), S 5–8cm (2–3in).

Habranthus robustus
Late summer- to early autumn-flowering bulb with narrowly strap-shaped, semi-erect, basal leaves. Leafless flower stems each bear a funnel-shaped, pink flower inclined at an angle. H 20–30cm (8–12in), S 8–10cm (3–4in).

Colchicum autumnale
(Autumn crocus, Meadow saffron)
Autumn-flowering corm with up to 8 long-tubed, wineglass-shaped, purple, pink or white flowers, followed by 3–5 large, strap-shaped, semi-erect, basal, glossy leaves in spring. H and S 10–15cm (4–6in).

Colchicum byzantinum
Robust, autumn-flowering corm with up to 20 large, funnel-shaped, pale purplish-pink flowers, 10–15cm (4–6in) long. In spring produces very broad, semi-erect, basal leaves, ribbed lengthways. H and S 15–20cm (6–8in).

OTHER RECOMMENDED PLANTS:
Colchicum autumnale 'Alboplenum'
Cyclamen cyprium
Galanthus reginae-olgae

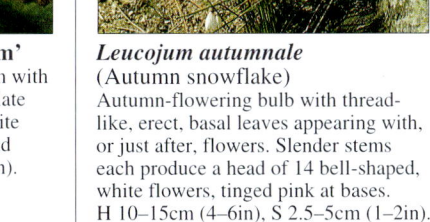

Allium mairei
Leucojum roseum

381

Bulbs/small AUTUMN INTEREST
◾ PINK

Cyclamen graecum
Autumn-flowering tuber with heart-shaped, toothed, velvety, dark green leaves, patterned silver or light green. Flowers are pink or white, with purple stains around mouths. Grows best in an alpine house. H to 10cm (4in), S 10–15cm (4–6in).

Cyclamen rohlfsianum
Autumn-flowering tuber with coarsely toothed leaves, zoned with light and dark green patterns, and pale pink-lilac flowers, stained darker at mouths. H to 10cm (4in), S 10–15cm (4–6in).

Cyclamen cilicium
Autumn-flowering tuber with broadly heart-shaped leaves that have light and dark green zones. Has white or pink flowers, each with a dark purple stain at the mouth, just before or with leaves. H to 10cm (4in), S 5–10cm (2–4in).

Colchicum cilicicum
Autumn-flowering corm with large, cup-shaped, pale pink to deep rose-purple flowers, sometimes slightly chequered. Very broad, semi-erect, basal leaves, ribbed lengthways, appear soon after flowers have faded. H and S 15–20cm (6–8in).

Cyclamen hederifolium, syn. *C. neapolitanum*
Autumn-flowering tuber. Pale to deep pink flowers, stained darker at mouths, appear before or with foliage. Leaves vary but are often ivy-shaped with silvery-green patterns. H to 10cm (4in), S 10–15cm (4–6in).

Colchicum bivonae, syn. *C. bowlesianum, C. sibthorpii*
Autumn-flowering corm with large, funnel-shaped, pinkish-purple flowers, strongly chequered darker purple and with purple anthers. Produces 8–10 erect leaves in spring. H 10–15cm (4–6in), S 15–20cm (6–8in).

Bulbs/small WINTER INTEREST

■■ PINK–PURPLE   ■■■ PURPLE–YELLOW   □ WHITE

Colchicum agrippinum
Early autumn-flowering corm. Narrow, slightly waved, semi-erect, basal leaves develop in spring. Bears erect, funnel-shaped, bright purplish-pink flowers with a darker chequered pattern and pointed petals. H 10–15cm (4–6in), S 8–10cm (3–4in).

Colchicum 'Waterlily'
Autumn-flowering corm with rather broad, semi-erect, basal leaves in winter or spring. Tightly double flowers have 20–40 pinkish-lilac petals. H 10–15cm (4–6in), S 15–20cm (6–8in).

Merendera montana, syn. *M. bulbocodium*
Autumn-flowering corm with narrowly strap-shaped, semi-erect, basal leaves, produced just after upright, broad-petalled, funnel-shaped, rose- or purple-lilac flowers appear. H to 5cm (2in), S 5–8cm (2–3in).

Biarum eximium
Colchicum 'The Giant'
CROCUSES, p.372–3
Cyclamen purpurascens, p.379

Biarum tenuifolium
Late summer- or autumn-flowering tuber producing clusters of acrid, narrow, erect, basal leaves after stemless, upright and often twisted, blackish-purple spathes appear. H to 20cm (8in), S 8–10cm (3–4in).

Arum pictum
Autumn-flowering tuber. Arrow-shaped, semi-erect, glossy leaves, with cream veins, appear at the same time as a cowl-like, deep purple-brown spathe and dark purple spadix. H 15–25cm (6–10in), S 15–20cm (6–8in).

Sternbergia lutea
Autumn-flowering bulb with strap-shaped, semi-erect, basal, deep green leaves appearing together with a funnel-shaped, bright yellow flower, 2.5–6cm (1–2½in) long, on a leafless stem. H 2.5–15cm (1–6in), S 8–10cm (3–4in).

CROCUSES, pp.372–3
Sternbergia clusiana
Sternbergia sicula
Zephyranthes citrina

Galanthus nivalis 'Flore Pleno' (Double common snowdrop)
Late winter- and early spring-flowering bulb with semi-erect, basal, grey-green leaves. Bears rosetted, many-petalled, double, white flowers, some inner petals having a green mark at the apex. H 10–15cm (4–6in), S 5–8cm (2–3in).

Galanthus nivalis 'Pusey Green Tips'
Late winter- and early spring-flowering bulb with narrowly strap-shaped, semi-erect, basal, grey-green leaves. Each stem bears a white flower with many mostly green-tipped petals. H 10–15cm (4–6in), S 5–8cm (2–3in).

Galanthus gracilis, syn. *G. graecus*
Late winter- and early spring-flowering bulb with slightly twisted, strap-shaped, semi-erect, basal, grey-green leaves. Bears white flowers with 3 inner petals, each marked with a green blotch at the apex and base. H 10–15cm (4–6in), S 5–8cm (2–3in).

OTHER RECOMMENDED PLANTS:
Galanthus nivalis
Galanthus plicatus

Galanthus rizehensis
Late winter- and early spring-flowering bulb with very narrow, strap-shaped, semi-erect, basal, dark green leaves. Produces white flowers, 1.5–2cm (⅝–¾in) long, with a green patch at the apex of each inner petal. H 10–20cm (4–8in), S 5cm (2in).

Galanthus elwesii
Late winter- and early spring-flowering bulb with semi-erect, basal, grey-green leaves that widen gradually towards tips. Each inner petal of the white flowers bears green marks at the apex and base, which may merge. H 10–30cm (4–12in), S 5–8cm (2–3in).

Galanthus 'Atkinsii'
Vigorous, late winter- and early spring-flowering bulb with strap-shaped, semi-erect, basal, grey-green leaves. Each stem carries a slender, white flower with a green mark at the apex of each inner petal. H 10–25cm (4–10in), S 5–9cm (2–3½in).

383

Bulbs/small WINTER INTEREST
☐ WHITE

Galanthus nivalis 'Lutescens'
Late winter- and early spring-flowering bulb with narrowly strap-shaped, semi-erect, basal, grey-green leaves. Flowers are 1.5–2cm (⅝–¾in) long and white with yellow patches at the apex of each inner petal. H 10cm (4in), S 2.5–5cm (1–2in).

Galanthus ikariae, syn. *G. latifolius*
Late winter- and early spring-flowering bulb with strap-shaped, semi-erect, basal, glossy, bright green leaves. Produces one white flower, 1.5–2.5cm (⅝–1in) long, marked with a green patch at the apex of each inner petal. H 10–25cm (4–10in), S 5–8cm (2–3in).

Galanthus nivalis 'Scharlockii'
Vigorous, late winter- and early spring-flowering bulb with semi-erect, basal, grey-green leaves. Has white flowers, with green marks at the apex of inner petals, overtopped by 2 narrow spathes that resemble donkeys' ears. H 10–15cm (4–6in), S 5–8cm (2–3in).

Galanthus plicatus subsp. **byzantinus**
Late winter- and early spring-flowering bulb. Semi-erect, basal, deep green leaves have a grey bloom and reflexed margins. White flowers have green marks at bases and tips of inner petals. H 10–20cm (4–8in), S 5–8cm (2–3in).

Cyclamen coum subsp. **coum** 'Album'
Winter-flowering tuber with rounded, deep green leaves, sometimes silver-patterned. Carries white flowers, each with a maroon mark at the mouth. H to 10cm (4in), S 5–10cm (2–4in).

■ PINK

Cyclamen persicum
Winter- or spring-flowering tuber with heart-shaped leaves, marked light and dark green and silver. Bears fragrant, slender, white or pink flowers, 3–4cm (1¼–1½in) long and stained carmine at mouths. H 10–20cm (4–8in), S 10–15cm (4–6in). Min. 5–7°C (41–5°F).

Cyclamen persicum 'Pearl Wave'
Winter- and spring-flowering tuber with heart-shaped leaves, marked light and dark green and silver. Produces fragrant, slender, deep pink flowers, with 5–6cm (2–2½in) long, frilly-edged petals. H 10–20cm (4–8in), S 10–15cm (4–6in). Min. 5–7°C (41–5°F).

Cyclamen coum subsp. **coum**
Winter-flowering tuber with rounded leaves, plain deep green or silver-patterned. Produces bright carmine flowers with dark stains at mouths. H to 10cm (4in), S 5–10cm (2–4in).

Cyclamen coum subsp. *caucasicum*
Cyclamen persicum 'Cardinal'
Cyclamen persicum, Dwarf Scented Series
HYACINTHS, p.375

■ RED

Cyclamen persicum 'Esmeralda'
Winter-flowering tuber with heart-shaped, silver-patterned leaves and broad-petalled, carmine-red flowers. H 10–20cm (4–8in), S 15–20cm (6–8in). Min. 5–7°C (41–5°F).

Cyclamen persicum 'Renown'
Winter- and spring-flowering tuber with heart-shaped, bright silver-green leaves, each with a central, dark green mark. Carries fragrant, slender, scarlet flowers, 5–6cm (2–2½in) long. H 10–20cm (4–8in), S 10–15cm (4–6in). Min. 5–7°C (41–5°F).

☼ ◊

■ YELLOW

Lachenalia aloides var. quadricolor, syn. *L. a.* 'Quadricolor'
Winter- to spring-flowering bulb with 2 strap-shaped, semi-erect, basal leaves. Has a spike of 10–20 purplish-red buds opening to greenish-yellow or -orange flowers. H 15–25cm (6–10in), S 5–8cm (2–3in).

Lachenalia 'Nelsonii'
Winter- to spring-flowering bulb with 2 strap-shaped, purple-spotted, semi-erect, basal leaves. Has a spike of 10–20 pendent, tubular, green-tinged, bright yellow flowers, 3cm (1¼in) long. H 15–25cm (6–10in), S 5–8cm (2–3in).

☼ ◊ ❄

Cyclamen persicum, Kaori Series
Winter-flowering tuber with neat, heart-shaped, silver-marbled leaves. Produces fragrant flowers, 4cm (1½in) long, in a wide range of colours. H 10–20cm (4–8in), S 10–15cm (4–6in). Min. 5–7°C (41–5°F).

☼ ◊

HYACINTHS, p.375
Lachenalia aloides
Lachenalia rubida

Eranthis hyemalis, syn. *E. cilicicus* (Winter aconite)
Clump-forming tuber. Bears stalkless, cup-shaped, yellow flowers, 2–2.5cm (¾–1in) across, from late winter to early spring. A dissected, leaf-like bract forms a ruff beneath each bloom. H 5–10cm (2–4in), S 8–10cm (3–4in).

◐ ◊ ❄❄❄ ▽

Eranthis x *tubergenii* 'Guinea Gold'
HYACINTHS, p.375

385

WATER PLANTS
☐ WHITE

Menyanthes trifoliata
(Bog bean, Buckbean)
Deciduous, perennial, marginal water plant that has 3-parted, mid-green leaves and fringed, white flowers borne in spring. H 23cm (9in), S 30cm (12in).

Alisma plantago-aquatica
(Water plantain)
Deciduous, perennial, marginal water plant with upright, oval, bright green leaves that emerge well above water. Loose, conical panicles of small, pinkish to white flowers appear in summer. H 75cm (30in), S 45cm (18in).

Lysichiton camtschatcensis
Vigorous, deciduous, perennial, marginal water or bog plant. Pure white spathes, surrounding spikes of small, insignificant flowers, are borne in spring, before oblong to oval, bright green leaves emerge. H 75cm (30in), S 60cm (24in).

Hydrocharis morsus-ranae
(Frogbit)
Deciduous, perennial, floating water plant with rosettes of kidney-shaped, olive-green leaves and small, white flowers during summer. S 10cm (4in), but young plantlets remain attached to form a mass up to 1m (3ft) across.

Calla palustris (Bog arum)
Deciduous or semi-evergreen, perennial, spreading, marginal water plant with heart-shaped, glossy, mid- to dark green leaves. In spring produces large, white spathes usually followed by red or orange fruits. H 25cm (10in), S 30cm (12in).

Sagittaria latifolia
(American arrowhead, Duck potato)
Deciduous, perennial, marginal water plant with curved, soft green leaves and sprays of white flowers in summer. H 1.5m (5ft), S 60cm (2ft).

OTHER RECOMMENDED PLANTS:
Luronium natans
Nelumbo nucifera 'Alba Grandiflora'
Sagittaria sagittifolia
Sagittaria sagittifolia 'Flore Pleno'
WATER LILIES, p.390

☐ WHITE

Saururus cernuus (Lizard's tail, Swamp lily, Water dragon)
Deciduous, perennial, marginal water or bog plant. Has clumps of heart-shaped, mid-green leaves and racemes of creamy flowers in summer. H 23cm (9in), S 30cm (12in).

☼ ◊ ❄❄❄

Hottonia palustris (Water violet)
Deciduous, perennial, submerged water plant. Dense whorls of much-divided, light green leaves form a spreading mass of foliage. Lilac or whitish flowers appear above water surface in summer. S indefinite.

☼ ◊ ❄❄❄

Acorus gramineus 'Variegatus'
Semi-evergreen, perennial, marginal or submerged water plant. Narrow, stiff, grass-like leaves are dark green with cream variegation. H 25cm (10in), S 15cm (6in).

☼ ◊ ❄❄

☐ WHITE–PINK

Stratiotes aloides (Water soldier)
Semi-evergreen, perennial, submerged, free-floating water plant. Spiny, olive-green leaves are arranged in rosettes. Produces cup-shaped, white, sometimes pink-tinged flowers in summer. Increases by producing small water buds. S 30cm (12in).

☼ ◊ ❄❄❄

Aponogeton distachyos (Cape pondweed, Water hawthorn)
Deciduous, perennial, deep-water plant with floating, oblong, mid- to dark green leaves, often splashed with purple. Very fragrant, 'forked', white flowers with black stamens are produced throughout summer. S 1.2m (4ft).

☼ ◊ ❄❄

Caltha leptosepala
Deciduous, perennial, marginal water plant with heart-shaped, dark green leaves and buttercup-like, white flowers produced in spring. H and S 30cm (12in).

☼ ◊ ❄❄❄

Acorus calamus 'Variegatus' (Myrtle flag, Sweet flag)
Semi-evergreen, perennial, marginal water plant. Sword-like, tangerine-scented, mid-green leaves have cream variegation and are flushed rose-pink in spring. H 75cm (30in), S 60cm (24in).

☼ ◊ ❄❄❄

Houttuynia cordata 'Chamaeleon', syn. *H. c.* 'Variegata'
Vigorous, deciduous, perennial, ground-cover, marginal water plant. Aromatic, leathery leaves are splashed yellow and red. Has small sprays of white flowers in summer. Needs some sun to enhance variegation. H 10cm (4in), S indefinite.

☼ ◊ ❄❄❄

Nelumbo nucifera (Sacred lotus)
Vigorous, deciduous, perennial, marginal water plant. Sturdy stems carry very large, plate-like, blue-green leaves and, in summer, large, vivid rose-pink flowers, maturing to flesh-pink. H 1–1.5m (3–5ft) above water surface, S 1.2m (4ft). Min. 7°C (45°F).

☼ ◊

Nelumbo nucifera 'Alba Striata'
Nelumbo nucifera 'Rosea Plena'
WATER LILIES, p.390

Water plants
■■ PINK–BLUE

■■ BLUE–GREEN

Butomus umbellatus
(Flowering rush)
Deciduous, perennial, rush-like, marginal water plant with narrow, twisted, mid-green leaves and umbels of pink to rose-red flowers in summer. H 1m (3ft), S 45cm (1½ft).

Azolla caroliniana
(Fairy moss, Water fern)
Deciduous, perennial, floating water fern with divided fronds that vary from red to purple in full sun and from pale green to blue-green in shade. Helps reduce algae by lessening light in water. S indefinite.

Pontederia cordata
(Pickerel weed)
Deciduous, perennial, marginal water plant. In late summer, dense spikes of blue flowers emerge between lance-shaped, glossy, dark green leaves. H 75cm (30in), S 45cm (18in).

Myosotis scorpioides 'Mermaid'
Deciduous, perennial, marginal water plant for mud or very shallow water. Narrow, mid-green leaves form sprawling mounds. Bears small, blue, forget-me-not flowers throughout summer. H 15cm (6in), S 30cm (12in).

Eichhornia crassipes
(Water hyacinth)
Evergreen or semi-evergreen, perennial water plant, with glossy leaves, floating on air-filled leaf stalks. Bears spikes of blue-and-lilac flowers in summer. May be invasive in warm conditions. S 23cm (9in). Min. 1°C (34°F).

Colocasia esculenta (Taro)
Deciduous, perennial, marginal water plant with large, bold, mid- to dark green leaves, often with prominent, white veins. Bears insignificant spathes in summer. May be grown in wet soil in a pot. H 1.1m (3½ft), S 60cm (2ft). Min. 1°C (34°F).

Typha latifolia
Deciduous, perennial, marginal water plant with large clumps of mid-green foliage. Produces spikes of beige flowers in late summer, followed by decorative, cylindrical, dark brown seed heads. Is invasive. H to 2.5m (8ft), S 60cm (2ft).

Myriophyllum aquaticum,
syn. *M. proserpinacoides*
(Parrot's feather)
Deciduous, perennial, partially or completely submerged water plant. Spreading, finely divided, blue-green foliage turns reddish in autumn if it surfaces. S indefinite.

Potamogeton crispus
(Curled pondweed)
Deciduous, perennial, submerged water plant that produces spreading colonies of seaweed-like, bronze- or mid-green foliage. Insignificant, reddish flowers are borne in summer. Prefers cool water. S indefinite.

Colocasia esculenta 'Illustris'
Cryptocoryne spiralis
Thalia dealbata
Thalia geniculata

WATER LILIES, p.390

WATER LILIES, p.390

GREEN

Pistia stratiotes (Water lettuce)
Deciduous, perennial, floating water plant for a pool or aquarium, evergreen in tropical conditions. Hairy, soft green foliage is lettuce-like in arrangement. Produces tiny, greenish flowers at varying times. H and S 10cm (4in). Min. 10–15°C (50–59°F).

☼ ●

Lagarosiphon major,
syn. ***Elodea crispa*** of gardens
Semi-evergreen, perennial, spreading, submerged water plant that forms dense, underwater swards of foliage. Ascending stems are covered in narrow, reflexed, dark green leaves. Bears insignificant flowers in summer. S indefinite.

☼ ● ❄❄❄

Euryale ferox
Annual, deep-water plant. Has floating, rounded, spiny, olive-green leaves with rich purple undersides and bears small, violet-purple flowers in summer. Is suitable only for a tropical pool. S 1.5m (5ft). Min. 5°C (41°F).

☼ ●

GREEN–YELLOW

Sparganium erectum, syn. ***S. ramosum*** (Branched bur reed)
Vigorous, deciduous or semi-evergreen, perennial, marginal water plant with narrow, mid-green leaves. Bears small, greenish-brown burs in summer. H 1m (3ft), S 60cm (2ft).

☼ ● ❄❄❄

Salvinia auriculata
Deciduous, perennial, floating water plant, evergreen in tropical conditions, that forms spreading colonies. Has rounded, pale to mid-green leaves, sometimes suffused purplish-brown, in pairs on branching stems. S indefinite. Min. 10–15°C (50–59°F).

☼ ●

Trapa natans
(Jesuit's nut, Water chestnut)
Annual, floating water plant with diamond-shaped, mid-green leaves, often marked purple, arranged in neat rosettes. Bears white flowers in summer. S 23cm (9in).

☼ ● ❄❄

Typha minima
Deciduous, perennial, marginal water plant with grass-like leaves. Spikes of rust-brown flowers in late summer are succeeded by decorative, cylindrical seed heads. H 45–60cm (18–24in), S 30cm (12in).

☼ ● ❄❄❄

Myriophyllum verticillatum
(Whorled water milfoil)
Deciduous, perennial, spreading, submerged water plant, overwintering by club-shaped winter buds. Slender stems are covered with whorls of finely divided, olive-green leaves. S indefinite.

☼ ● ❄❄❄

Hydrocleys nymphoides
(Water poppy)
Deciduous, perennial, deep-water plant, evergreen in tropical conditions, with floating, oval, mid-green leaves. Poppy-like, yellow flowers are held above foliage during summer. S to 60cm (2ft). Min. 1°C (34°F).

☼ ●

Baldellia ranunculoides
Cabomba caroliniana
Ceratophyllum demersum
Colocasia esculenta 'Fontanesii'
Cryptocoryne ciliata
Egeria densa
Fontinalis antipyretica
Myriophyllum hippuroides
Potamogeton lucens
Salvinia natans
Sparganium minimus
Vallisneria spiralis

389

CACTI AND OTHER SUCCULENTS/large
☐ WHITE

Selenicereus grandiflorus
(Queen-of-the-night)
Climbing, perennial cactus. Has 7-ribbed, 1–2cm (½–¾in) wide, green stems with yellow spines. White flowers, 18–30cm (7–12in) across, open at night in summer. H 3m (10ft), S indefinite. Min. 5°C (41°F).

Echinopsis bridgesii,
syn. *Trichocerus bridgesii*
Columnar, perennial cactus with 4–8-ribbed stems branching at base. Areoles each produce up to 6 spines. Scented, funnel-shaped, white flowers open at night in summer. H to 5m (15ft), S 1m (3ft). Min. 10°C (50°F).

Cereus validus, syn. *C. forbesii*
Columnar, perennial cactus with a branching, blue-green stem bearing dark spines on 4–7 prominent ribs. Has 25cm (10in) long, cup-shaped, white flowers at night in summer, followed by red fruits. H 7m (22ft), S 3m (10ft). Min. 7°C (45°F).

Myrtillocactus geometrizans
(Blue candle)
Columnar, perennial cactus with a much-branched, 5- or 6-ribbed, blue-green stem. Bears short, black spines, on plants over 30cm (1ft) tall, and white flowers at night in summer. H to 4m (12ft), S 2m (6ft). Min. 12°C (54°F).

Cereus peruvianus
Columnar, perennial cactus. Has a branching, silvery-blue stem with golden spines on 4–8 sharply indented ribs. Carries cup-shaped, white flowers, 10cm (4in) across, at night in summer, followed by pear-shaped, red fruits. H 5m (15ft), S 4m (12ft). Min. 7°C (45°F).

Browningia hertlingiana,
syn. *Azureocereus hertlingianus*
Slow-growing, columnar, perennial cactus with a silvery-blue stem, golden spines and tufted areoles. Nocturnal, white flowers appear in summer, only on plants over 1m (3ft) high. H 8m (25ft), S 4m (12ft). Min. 7°C (45°F).

Pachypodium lamerei
Tree-like, perennial succulent with a spiny, pale green stem crowned by linear leaves. Has fragrant, trumpet-shaped, creamy-white flowers in summer, on plants over 1.5m (5ft) tall. Stems branch after each flowering. H 6m (20ft), S 2m (6ft). Min. 11°C (52°F).

Carnegiea gigantea (Saguaro)
Very slow-growing, perennial cactus with a thick, 12–24-ribbed, spiny, green stem. Tends to branch and bears short, funnel-shaped, fleshy, white flowers at stem tips in summer, only when over 4m (12ft) high. H to 12m (40ft), S 3m (10ft). Min. 7°C (45°F).

OTHER RECOMMENDED PLANTS:
Agave americana
Echinopsis candicans, p.397
Epiphyllum anguliger, p.397
Rhipsalis tucumanensis, p.397
Rhipsalis warmingiana, p.397

☐ WHITE

Lemaireocereus marginatus, syn. *Marginatocereus marginatus* (Organ-pipe cactus)
Columnar, perennial cactus with a 5- or 6-ribbed, branching, shiny stem. Areoles bear minute spines. Produces funnel-shaped, white flowers in summer. H 7m (22ft), S 3m (10ft). Min. 11°C (52°F).
☼ ◊

Agave parviflora
Basal-rosetted, perennial succulent. Has narrow, white-marked, dark green leaves with white fibres peeling from edges. Produces white flowers in summer. H 1.5m (5ft), S 50cm (20in). Min. 5°C (41°F).
☼ ◊

Stetsonia coryne
Tree-like, perennial cactus with a short, swollen trunk bearing 8- or 9-ribbed, blue-green stems. Black spines fade with age to white with black tips. Funnel-shaped, white flowers appear at night in summer. H 8m (25ft), S 4m (12ft). Min. 10°C (50°F).
☼ ◊

Crassula ovata, syn. *C. argentea*, *C. portulacea* (Friendship tree, Jade tree, Money tree)
Perennial succulent with a swollen stem crowned by glossy, green leaves, at times red-edged. Bears 5-petalled, white flowers in autumn-winter. H 4m (12ft), S 2m (6ft). Min. 5°C (41°F).
☼ ◊ ♛

Pereskia aculeata (Barbados gooseberry, Lemon vine)
Fast-growing, deciduous, climbing cactus with broad, glossy leaves. Orange-centred, creamy-white flowers appear in autumn, only on plants over 1m (3ft) high. H to 10m (30ft), S 5m (15ft). Min. 5°C (41°F).
☼ ◊

Trichocereus spachianus (Torch cactus)
Clump-forming, perennial cactus with glossy, green stems bearing 10–15 ribs and pale golden spines. Fragrant, funnel-shaped, white flowers open at night in summer. H and S 2m (6ft). Min. 8°C (46°F).
☼ ◊

Espostoa lanata (Cotton ball, Peruvian old-man cactus)
Very slow-growing, columnar, perennial cactus with a branching, woolly, green stem. Foul-smelling, white flowers appear in summer, only on plants over 1m (3ft) high. H to 4m (12ft), S 2m (6ft). Min. 10°C (50°F).
☼ ◊

Agave americana var. ***stricta***, syn. *A. a.* 'Variegata'
Basal-rosetted, perennial succulent. Has sharply pointed, sword-shaped, blue-green leaves with yellow edges. Stem carries white flowers, each 9cm (3½in) long, in spring-summer. Offsets freely. H and S 2m (6ft).
☼ ◊ ❄

Agave americana var. *mediopicta*
Ceropegia haygarthii
Epiphyllum crenatum
Epiphyllum oxypetalum
Lemaireocereus thurberi
Pereskia aculeata var. *godseffiana*

393

Cacti and other Succulents/large
WHITE–PINK

Haageocereus versicolor
Columnar, perennial cactus. Dense, radial spines, golden, red or brown, at times form coloured bands around a longer, central spine up the green stem. Long-tubed, white flowers appear near crown of plant in summer. H to 2m (6ft), S 1m (3ft). Min. 11°C (52°F).

Crassula arborescens
(Silver jade plant)
Perennial succulent with a thick, robust stem crowned by branches bearing rounded, silvery-blue leaves, often with red edges. Has 5-petalled, pink flowers in autumn-winter. H 4m (12ft), S 2m (6ft). Min. 7°C (45°F).

Pereskia grandifolia,
syn. *Rhodocactus grandifolius*
(Rose cactus)
Deciduous, bushy, perennial cactus with black spines. Single rose-like, pink flowers form in summer-autumn only on plants over 30cm (1ft) high. H 5m (15ft), S 3m (10ft). Min. 10°C (50°F).

Adenium obesum
Tree-like, perennial succulent with a fleshy, tapering, green trunk and stems crowned by oval, glossy, green leaves, dull green beneath. Carries funnel-shaped, pink to pinkish-red flowers, white inside, in summer. H 2m (6ft), S 50cm (20in). Min. 15°C (59°F).

Pachycereus schottii,
syn. *Lophocereus schottii*
Columnar, perennial cactus, branching with age. Olive- to dark green stem, covered with small, white spines, bears 4–15 ribs. Funnel-shaped, pink flowers are produced at night in summer. H 7m (22ft), S 2m (6ft). Min. 10°C (50°F).

Pilosocereus leucocephalus,
syn. *P. palmeri*
Columnar, perennial cactus with a 10–12-ribbed stem and white-haired crown. Bears tubular, pink flowers, with cream anthers, at night in summer, on plants over 1.5m (5ft) tall. H to 6m (20ft), S 1m (3ft). Min. 11°C (52°F).

Ceropegia linearis subsp. *woodii*, p.398
Cleistocactus celsianus, p.416
Haageocereus chosicensis
Pachycereus schottii 'Monstrosus'

Senecio rowleyanus, p.398

■ RED

Aloe arborescens var. *variegatum*
Evergreen, bushy, succulent-leaved shrub. Each stem is crowned by rosettes of long, slender, blue-green leaves with toothed edges and cream stripes. Produces numerous spikes of red flowers in late winter and spring. H and S 2m (6ft). Min. 7°C (45°F).

Aloe ferox
Evergreen, succulent tree with a woody stem crowned by a dense rosette of sword-shaped, blue-green leaves that have spined margins. Carries an erect spike of bell-shaped, orange-scarlet flowers in spring. H to 3m (10ft), S 1.5–2m (5–6ft). Min. 7°C (45°F).

Cyphostemma juttae, syn. *Cissus juttae*
Perennial succulent. Swollen stem has peeling bark and deciduous, scandent branches with broad leaves. Bears inconspicuous, yellow-green flowers in summer. Green fruits turn yellow or red. H and S 2m (6ft). Min. 10°C (50°F).

Pedilanthus tithymaloides 'Variegata' (Redbird flower)
Bushy, perennial succulent with stems angled at each node. Leaves have white or pink marks. Stem tips carry small, greenish flowers in red to yellowish-green bracts in summer. H to 3m (10ft), S 30cm (1ft). Min. 10°C (50°F).

Aloe ciliaris
Climbing, perennial succulent with a slender stem crowned by a rosette of narrow, green leaves. Has white teeth where leaf base joins stem. Bears bell-shaped, scarlet flowers, with yellow and green mouths, in spring. H 5m (15ft), S 30cm (1ft). Min. 7°C (45°F).

Lemaireocereus euphorbioides, syn. *Rooksbya euphorbioides*
Columnar, perennial cactus. Has grey-green to dark green stems, 10cm (4in) across, with 8–10 ribs and 1 or 2 black spines per areole. Funnel-shaped, wine-red flowers appear in summer. H to 3m (10ft), S 1m (3ft). Min. 11°C (52°F).

Cleistocactus strausii (Silver torch)
Fast-growing, columnar, perennial cactus with 8cm (3in) wide stems carrying short, dense, white spines. Masses of long, tubular, red flowers appear in spring, only on plants over 60cm (2ft) high. H 3m (10ft), S 1–2m (3–6ft). Min. 5°C (41°F).

Aloe arborescens
Cereus peruvianus 'Monstrosus'
Crassula perfoliata var. *falcata*, p.400
Opuntia cylindrica
Pedilanthus tithymaloides
Tylecodon paniculatus

Cacti and other Succulents/large
GREEN–YELLOW

Cephalocereus senilis
(Old-man cactus)
Very slow-growing, columnar, perennial cactus with a green stem covered in long, white hairs, masking short, white spines. Is unlikely to flower in cultivation. H 15m (50ft), S 15cm (6in). Min. 5°C (41°F).

Furcraea foetida 'Mediopicta', syn. *F. gigantea* 'Mediopicta'
Basal-rosetted, perennial succulent with broad, sword-shaped, green leaves, striped with creamy-white, to 2.5m (8ft) long. Has bell-shaped, green flowers, with white interiors, in summer. H 3m (10ft), S 5m (15ft). Min. 6°C (43°F).

Bowiae volubilis
Bulbous succulent with climbing, much-branched, slender stems and no proper leaves. Produces small, star-shaped, green flowers at tips of stems in summer. Provide support. H 1–2m (3–6ft), S 45–60cm (1½–2ft). Min. 10°C (50°F).

Kalanchoe beharensis
Bushy, perennial succulent with triangular to lance-shaped, olive-green leaves, covered with fine, brown hairs. Bell-shaped, yellow flowers appear in late winter, only on plants over 2m (6ft) high. H and S to 4m (12ft). Min. 10°C (50°F).

Pachycereus pringlei
Slow-growing, columnar, perennial cactus with a branched, bluish-green stem that has 10–15 ribs. Large areoles each have 15–25 black-tipped, white spines. Is unlikely to flower in cultivation. H 11m (35ft), S 3m (10ft). Min. 10°C (50°F).

Echinocactus grusonii (Golden barrel cactus, Mother-in-law's seat)
Slow-growing, hemispherical, perennial cactus. Spined, green stem has 30 ribs. Woolly crown bears a ring of straw-coloured flowers in summer, only on stems over 38cm (15in) wide. H and S to 2m (6ft). Min. 11°C (52°F).

Ferocactus acanthodes
Slow-growing, columnar, perennial cactus, spherical when young. Green, 10–20-ribbed stem is covered with large, hooked, red or yellow spines. Funnel-shaped, yellow flowers form in summer on plants over 25cm (10in) across. H 3m (10ft), S 80cm (32in). Min. 5°C (41°F).

Opuntia robusta
Bushy, perennial cactus. Silvery-blue stem has flattened, oval segments with either no spines or 8–12 white ones, to 5cm (2in) long, per areole. Saucer-shaped, yellow flowers, 7cm (3in) across, appear in spring-summer. H and S 5m (15ft). Min. 5°C (41°F).

Agave attenuata, p.403
Agave filifera, p.401
Ceropegia sandersoniae
Cleistocactus smaragdiflorus
Echinocactus ingens
Euphorbia candelabrum
Ferocactus wislizenii
Furcraea foetida
Opuntia brasiliensis
Opuntia ficus-indica
Pachycereus pecten-aboriginum
Parodia leninghausii, p.404

Cacti and other Succulents/**medium**
☐ WHITE

Gymnocalycium gibbosum
Spherical to columnar, perennial cactus that has a dark green stem with 12–19 rounded ribs, pale yellow spines, darkening with age, and white flowers, to 7cm (3in) long, in summer. H 30cm (12in), S 20cm (8in). Min. 5°C (41°F).

Echinopsis candicans, syn. ***Trichocereus candicans***
Clump-forming, branching, perennial cactus with up to 11 ribs. Areoles each have 10–15 radial spines and 4 central ones. Fragrant, funnel-shaped, white flowers open at night in summer. H 1m (3ft), S indefinite. Min. 8°C (46°F).

Echinopsis oxygona, syn. ***E. multiplex***
Spherical to columnar, perennial cactus with a 13–15-ribbed, green stem and long spines. Has 10cm (4in) wide, tubular, white to lavender flowers, to 20cm (8in) long, in spring-summer. H and S 30cm (12in). Min. 5°C (41°F).

Agave victoriae-reginae
(Royal agave)
Very slow-growing, domed, perennial succulent with a basal rosette of spineless, white-striped and -edged leaves. Has cream flowers on a 4m (12ft) tall stem in spring-summer after 20–30 years. H and S 60cm (2ft). Min. 5°C (41°F).

Rhipsalis cereuscula (Coral cactus)
Pendent, perennial cactus with 4- or 5-angled or cylindrical, green stems and branches, to 3cm (1¼in) long in whorls. Stem tips each bear bell-shaped, white flowers in winter-spring. H 60cm (24in), S 50cm (20in). Min. 10°C (50°F).

Epiphyllum lauii
Bushy, perennial cactus, usually with strap-shaped, red-tinged, glossy stems which may also be spiny, cylindrical or 4-angled. Has fragrant, white flowers, with brown sepals, in spring-summer. H 30cm (12in), S 50cm (20in). Min. 10°C (50°F).

Epiphyllum anguliger
(Fishbone cactus)
Erect, then pendent, perennial cactus. Has strap-shaped, flattened, green stems with indented margins. Produces tubular, 10cm (4in) wide, white flowers in summer. H 1m (3ft), S 40cm (16in). Min. 11°C (52°F).

Rhipsalis warmingiana
Erect, then pendent, perennial cactus with slender, notched, cylindrical, green branches, sometimes tinged red or brown, with 2–4 angles. Has green-white flowers in winter-spring, followed by violet berries. H 1m (3ft), S 50cm (20in). Min. 11°C (52°F).

Rhipsalis tucumanensis
Pendent, perennial cactus with cylindrical, green stems, to 1cm (½in) across, branching less than many other *Rhipsalis* species. Has masses of very pale pink flowers in early summer, then pinkish-white berries. H 1m (3ft), S 50cm (20in). Min. 10°C (50°F).

OTHER RECOMMENDED PLANTS:
Echinopsis eyriesii
Haageocereus decumbens
Hatiora clavata
Hylocereus undatus
Neoporteria chilensis
Rhipsalis capilliformis
Rhipsalis crispata
Rhipsalis paradoxa
Ruschia crassa

Cacti and other Succulents/medium
◻◼ WHITE–PINK   ◼ PINK

Senecio rowleyanus
(String-of-beads)
Pendent, perennial succulent. Very slender, green stems bear cylindrical, green leaves. Has heads of fragrant, tubular, white flowers from spring to autumn. Suits a hanging pot. H 1m (3ft), S indefinite. Min. 5°C (41°F).

Ceropegia linearis subsp. **woodii**, syn. *C. woodii* (Heart vine, Rosary vine, String-of-hearts)
Semi-evergreen, trailing, succulent sub-shrub with tuberous roots. Leaves redden in sun. Has hairy, pinkish-green flowers from spring to autumn. H 1m (3ft), S indefinite. Min. 7°C (45°F).

Senecio articulatus var. **variegata**, syn. *S. a.* 'Variegatus'
Deciduous, spreading, perennial succulent with grey-marked stems. Has cream- and pink-marked, blue-green leaves in summer and yellow flower heads from spring to autumn. H 60cm (2ft), S indefinite. Min. 10°C (50°F).

Hesperaloe parviflora, syn. *Yucca parviflora*
Basal-rosetted, perennial succulent, often with peeling, white fibres at leaf edges. Flower stems each bear a raceme of bell-shaped, pink to red flowers in summer-autumn. H 1m (3ft) or more, S 2m (6ft). Min. 3°C (37°F).

Nopalxochia phyllanthoides 'Deutsche Kaiserin'
Pendent, epiphytic, perennial cactus with flattened, toothed, glossy, green stems, each 5cm (2in) across. Stem margins each bear pink flowers, to 10cm (4in) across, in spring. H 60cm (2ft), S 1m (3ft). Min. 10°C (50°F).

Cleistocactus celsianus, syn. *Borzicactus celsianus*, *Oreocereus celsianus* (Old man of the Andes)
Slow-growing, perennial cactus. Has heavy and wispy, spines. Mature plants bear pink flowers in summer. H 1m (3ft), S 30cm (1ft). Min. 10°C (50°F).

Kalanchoe fedtschenkoi f. **variegata**
Bushy, perennial succulent. Blue-green and cream leaves also colour red. Bears a new plantlet in each leaf notch. Has brownish-pink flowers in late winter. H and S to 1m (3ft). Min. 10°C (50°F).

Echinocereus schmollii, syn. *Wilcoxia schmollii* (Lamb's-tail cactus)
Erect to prostrate, tuberous cactus with 8–10-ribbed, purplish-green stems and mostly white spines. Has pinkish-purple flowers in spring-summer. H and S 30cm (12in). Min. 8°C (46°F).

Kalanchoe daigremontiana (Mexican hat plant)
Erect, perennial succulent with a stem bearing fleshy, boat-shaped, toothed leaves. Produces a plantlet in each leaf notch. Umbels of pink flowers appear at stem tops in winter. H to 1m (3ft), S 30cm (1ft). Min. 7°C (45°F).

Echinocereus reichenbachii var. **baileyi**, syn. *E. baileyi*
Columnar, perennial cactus with a slightly branched stem bearing 12–23 ribs and yellowish-white, 3cm (1¼in) long spines. Produces pink flowers with darker bases in spring. H 30cm (12in), S 20cm (8in). Min. 7°C (45°F).

Echinopsis rhodotricha

398

☐ PINK

Kalanchoe 'Wendy'
Semi-erect, perennial succulent with narrowly oval, glossy, green leaves, 7cm (3in) long. In late winter bears bell-shaped, pinkish-red flowers, 2cm (³/4in) long, with yellow tips. Is ideal for a hanging basket. H and S 30cm (12in). Min. 10°C (50°F).

Aporocactus flagelliformis
(Rat's-tail cactus)
Pendent, perennial cactus with pencil-thick, green stems bearing short, golden spines. Has double, cerise flowers along stems in spring. Keep dry in winter. Is good for a hanging basket. H 1m (3ft), S indefinite.

Epicactus 'Gloria',
syn. *Epiphyllum* 'Gloria'
Erect, then pendent, perennial cactus. Strap-shaped, flattened, green stems have toothed edges. Produces pinkish-red flowers, 10cm (4in) across, in spring. H 30cm (1ft), S 1m (3ft). Min. 5°C (41°F).

Epicactus 'M.A. Jeans',
syn. *Epiphyllum* 'M.A. Jeans'
Erect, then pendent, perennial cactus. Strap-shaped, flattened, green stems have shallowly toothed edges. In spring has deep pink flowers, 8cm (3in) across, with white anthers. H 30cm (12in), S 50cm (20in). Min. 5°C (41°F).

Echinocereus pentalophus
Clump-forming, perennial cactus with spined, green stems, 3–4cm (1¹/2–1¹/2in) wide, that have 4–8 ribs, later rounded. Has trumpet-shaped, bright pink flowers, paler at base, to 12cm (5in) across, in spring. H 60cm (2ft), S 1m (3ft). Min. 5°C (41°F).

Lampranthus spectabilis
Spreading, perennial succulent with erect stems and narrow, cylindrical, grey-green leaves. In summer produces daisy-like flowers, cerise with yellow centres or golden-yellow throughout. H 30cm (1ft), S indefinite. Min. 5°C (41°F).

Cleistocactus trollii
Echinocereus cinerascens
Echinocereus pectinatus
Echinocereus reichenbachii
Kalanchoe fedtschenkoi
Lampranthus haworthii
Neoporteria chilensis

Cacti and other Succulents/medium
PINK–RED

Kalanchoe blossfeldiana Hybrids
Bushy, perennial succulent with oval to oblong, toothed, glossy, green leaves. Produces clusters of yellow, orange, pink, red or purple flowers, year-round. Makes an excellent house plant. H and S 30cm (1ft). Min. 10°C (50°F).

Gymnocalycium mihanovichii 'Red Head', syn. *G.m.* var. *hibotan*
Perennial cactus with a red stem, 8 angular ribs and curved spines. Must be grafted on to any fast-growing stock as it contains no chlorophyll. Has pink flowers in spring-summer. H and S as per graft stock. Min. 10°C (50°F).

Aloe variegata (Partridge-breasted aloe)
Humped, perennial succulent. Has triangular, white-marked, dark green leaves with pronounced keels beneath. Bears a spike of pinkish-red flowers in spring. Is a good house plant. H 30cm (12in), S 10cm (4in). Min. 7°C (45°F).

Echeveria pulvinata (Plush plant)
Bushy, perennial succulent with brown-haired stems each crowned by a rosette of thick, rounded, green leaves that become red-edged in autumn. Leaves have short, white hairs. Bears red flowers in spring. H 30cm (12in), S 50cm (20in). Min. 5°C (41°F).

Oroya peruviana, syn. *O. neoperuviana*
Spherical, perennial cactus with a much-ribbed stem covered in yellow spines, 1.5cm (5/8in) long, with darker bases. Pink flowers, with yellow bases, open in spring-summer. H 25cm (10in), S 20cm (8in). Min. 10°C (50°F).

Dudleya pulverulenta
Basal-rosetted, perennial succulent with strap-shaped, pointed, silvery-grey leaves. Bears masses of star-shaped, red flowers in spring-summer. H 60cm (2ft), S 30cm (1ft). Min. 7°C (45°F).

Kalanchoe 'Tessa'
Prostrate to pendent, perennial succulent with narrowly oval, green leaves, 3cm (1 1/4in) long. Bears tubular, orange-red flowers, 2cm (3/4in) long, in late winter. H 30cm (1ft), S 60cm (2ft). Min. 10°C (50°F).

Ferocactus haematacanthus, syn. *Hamatocactus hamatacanthus*
Slow-growing, spherical to columnar, perennial cactus with a 13-ribbed stem that bears hooked, red spines, to 12cm (5in) long. Has yellow blooms in summer, then spherical, red fruits. H and S 60cm (2ft). Min. 5°C (41°F).

Crassula perfoliata var. *falcata*, syn. *C. falcata* (Aeroplane propeller)
Bushy, perennial succulent that branches freely. Long leaves each twist like a propeller. Has large clusters of fragrant, red flowers in late summer. H and S 1m (3ft). Min. 7°C (45°F).

Cleistocactus haynei
Echeveria gibbiflora
Mammillaria magnimamma
Neoporteria subgibbosa

Pachypodium succulentum

RED–GREEN

Mammillaria hahniana (Old-lady cactus, Old-woman cactus)
Spherical to columnar, perennial cactus with a green stem bearing long, woolly, white hairs. Carries cerise flowers in spring and spherical, red fruits in autumn. H 40cm (16in), S 15cm (6in). Min. 5°C (41°F).

☼ ◊ ♀

Opuntia erinacea
Bushy, perennial cactus with a green stem consisting of 15cm (6in) long, flattened segments. Areoles bear 6–15 flattened, 20cm (8in) long, hair-like spines. Has masses of saucer-shaped, red or yellow flowers in summer. H 50cm (20in), S 2m (6ft). Min. 5°C (41°F).

☼ ◊

Beschorneria yuccoides
Clump-forming, perennial succulent with a basal rosette of up to 20 rough, greyish-green leaves, to 1m (3ft) long and 5cm (2in) across. Produces pendent, tubular, bright red flowers in summer in spikes over 2m (6ft) tall. H 1m (3ft), S 3m (10ft).

☼ ◊ ❄ ♀

Tylecodon reticulatus, syn. ***Cotyledon reticulata*** (Barbed-wire plant)
Deciduous, bushy, succulent shrub. Swollen branches bear cylindrical leaves in winter. Has tubular, green-yellow flowers on a woody stem in autumn. H and S 30cm (1ft). Min. 7°C (45°F).

☼ ◊

Mammillaria geminispina
Clump-forming, perennial cactus. Has a spherical, green stem densely covered with short, white, radial spines and very long, white, central spines. Has red flowers, 1–2cm (½–¾in) across, in spring. H 25cm (10in), S 50cm (20in). Min. 5°C (41°F).

☼ ◊ ♀

Agave filifera (Thread agave)
Basal-rosetted, perennial succulent with narrow, green leaves, each spined at the tip. White leaf margins gradually break away, leaving long, white fibres. Carries yellow-green flowers on a 2.5m (8ft) tall stem in summer. Offsets freely. H 1m (3ft), S 2m (6ft).

☼ ◊ ❄

Drosanthemum hispidum
Echeveria gibbiflora
Echinocereus triglochidiatus
Hoodia gordonii

Kalanchoe blossfeldiana
Mammillaria elegans
Mammillaria rhodantha
Nopalxochia ackermannii

Rochea coccinea
Tylecodon wallichii

Cacti and other Succulents/medium

☐ YELLOW

Aloe vera, syn. *A. barbadensis*
Clump-forming, perennial succulent with basal rosettes of tapering, thick leaves, mottled green, later grey-green. Flower stems carry bell-shaped, yellow flowers in summer. Propagate by offsets as plant is sterile. H 60cm (2ft), S indefinite. Min. 10°C (50°F).

x *Pachyveria glauca*
Clump-forming, perennial succulent with a dense, basal rosette of fleshy, incurved, oval, silvery-blue leaves, to 6cm (2½in) long, with darker marks. Bears star-shaped, yellow flowers, each with a red tip, in spring. H and S 30cm (12in).

Agave parryi
Basal-rosetted, perennial succulent with stiff, broad, grey-green leaves, each to 30cm (12in) long with a solitary dark spine at its pointed tip. Flower stem, to 4m (12ft) long, bears creamy-yellow flowers in summer. H 50cm (20in), S 1m (3ft). Min. 5°C (41°F).

Leuchtenbergia principis
Basal-rosetted, perennial cactus with narrow, angular, dull grey-green tubercles, each 10cm (4in) long and crowned by papery spines to 10cm (4in) long. Crown bears yellow flowers, to 7cm (3in) across, in summer. H and S 30cm (12in). Min. 6°C (43°F).

Astrophytum myriostigma
(Bishop's cap, Bishop's mitre)
Slow-growing, spherical to slightly elongated, perennial cactus. A fleshy stem has 4–6 ribs and is flecked with tiny tufts of white spines. Bears yellow flowers in summer. H 30cm (12in), S 20cm (8in). Min. 5°C (41°F).

Aeonium haworthii (Pinwheel)
Bushy, perennial succulent. Freely branching stems bear rosettes, 12cm (5in) across, of blue-green leaves, often with red margins. Has a terminal spike of star-shaped, pink-tinged, pale yellow flowers in spring. H 60cm (2ft), S 1m (3ft). Min. 5°C (41°F).

Copiapoa cinerea
Very slow-growing, clump-forming, perennial cactus. Blue-green stem bears up to 25 ribs and black spines. Has a woolly, white-grey crown and, on plants over 10cm (4in) across, yellow flowers in spring-summer. H 50cm (20in), S 2m (6ft). Min. 10°C (50°F).

Copiapoa marginata
Graptopetalum amethystinum
Pelargonium carnosum
Senecio articulatus

◻ YELLOW

Opuntia microdasys var. *alpispina*
Bushy, perennial cactus with green, flattened, oval segments. Spineless areoles, with slender, barbed, white hairs, are set in diagonal rows. Funnel-shaped, yellow flowers appear in summer. H 60cm (2ft), S 30cm (1ft). Min. 10°C (50°F).

Kalanchoe tomentosa
(Panda plant, Pussy ears)
Bushy, perennial succulent with thick, oval, grey leaves, covered with velvety bristles and often edged with brown at tips. Has yellowish-purple flowers in winter. H 50cm (20in), S 30cm (12in). Min. 10°C (50°F).

Agave attenuata
Perennial succulent with a thick stem crowned by a rosette of sword-shaped, spineless, pale green leaves. Arching flower stem, to 1.5m (5ft) long, is densely covered with yellow flowers in spring-summer. H 1m (3ft), S 2m (6ft). Min. 5°C (41°F).

Ferocactus setispinus,
syn. *Hamatocactus setispinus*
Slow-growing, perennial cactus with a 13-ribbed stem and yellow or white spines. Fragrant, yellow flowers with red throats appear in summer, only on plants over 5cm (2in) across. H and S 30cm (12in). Min. 5°C (41°F).

Dioscorea elephantipes,
syn. *Testudinaria elephantipes*
(Elephant's foot)
Very slow-growing, deciduous, perennial succulent with a domed, woody trunk, annual, climbing stems and yellow flowers in autumn. H 50cm (20in), S 1m (3ft). Min. 10°C (50°F).

Lobivia haageana
Columnar, perennial cactus with a 20–25-ribbed, bluish- to dark green stem that has yellow, radial spines with longer, darker, central ones. In summer produces yellow flowers, 7cm (3in) across, with red throats. H 30cm (12in), S 15cm (6in). Min. 5°C (41°F).

Epicactus 'Jennifer Ann',
syn. *Epiphyllum* 'Jennifer Ann'
Erect, then pendent, perennial cactus. Has strap-shaped, flattened, green stems with toothed margins. Bears yellow flowers, 15cm (6in) across, in spring. H 30cm (12in), S 50cm (20in). Min. 5°C (41°F).

Aeonium arboreum
Copiapoa coquimbana
Cyphostemma bainesii
Ferocactus chrysacanthus
Hoodia bainii
Notocactus scopa
Opuntia microdasys

403

Cacti and other Succulents/medium
YELLOW–ORANGE

Parodia chrysacanthion
Spherical, perennial cactus with a much-ribbed, green stem densely covered with bristle-like, golden spines, each 1–2cm (½–¾in) long. Crown bears yellow flowers in spring and, often, pale yellow wool. H and S 30cm (1ft). Min. 10°C (50°F).

Aeonium arboreum 'Schwarzkopf'
Bushy, perennial succulent with stems each crowned by a rosette, to 15cm (6in) across, of narrow, purple leaves. Bears golden pyramids of flowers in spring on 2–3-year-old stems, which then die. H to 60cm (2ft), S 1m (3ft). Min. 5°C (41°F).

Aloe striata
Basal-rosetted, perennial succulent. Has broad, blue-green leaves, with white margins and marks, that become suffused red in full sun. Has a panicle of reddish-orange flowers in spring. Makes a good house plant. H and S 1m (3ft). Min. 7°C (45°F).

Hatiora salicornioides
(Bottle plant, Drunkard's dream) Bushy, perennial, epiphytic cactus with freely branching, 3cm (1¼in) long stems. Has joints with expanded tips and terminal, bell-shaped, golden-yellow flowers in spring. H and S 30cm (1ft). Min. 11°C (52°F).

Cotyledon undulata
Evergreen, upright, succulent sub-shrub with a swollen stem bearing oval, green leaves, densely coated in white wax, with flat, wavy tips. Each 70cm (28in) long flower stem bears a bell-shaped, orange flower in autumn. H and S 50cm (20in). Min. 7°C (45°F).

Parodia leninghausii, syn. *Notocactus leninghausii* (Golden ball cactus)
Clump-forming, perennial cactus. Woolly crown always slopes towards sun. In summer, on plants over 10cm (4in) tall, yellow blooms open flat. H 1m (3ft), S 30cm (1ft). Min. 10°C (50°F).

Opuntia tunicata
Mounded, perennial cactus. Has cylindrical, green stem segments densely covered with 5cm (2in) long, golden spines that are enclosed in a papery sheath. Bears shallowly saucer-shaped, yellow flowers in spring-summer. H 60cm (2ft), S 1m (3ft). Min. 10°C (50°F).

Kalanchoe tubiflora
Erect, perennial succulent with long, almost cylindrical, grey-green leaves with reddish-brown mottling and flattened, notched tips that form plantlets. Bears an umbel of orange-yellow flowers in late winter. H to 1m (3ft), S 30cm (1ft). Min. 8°C (46°F).

Lampranthus aurantiacus
Erect, then prostrate, sparse-branching perennial succulent with short, cylindrical, tapering, grey-green leaves. Masses of daisy-like, bright orange flowers, 5cm (2in) wide, open in summer sun. H 50cm (20in), S 70cm (28in). Min. 5°C (41°F).

Cleistocactus baumannii
Cotyledon orbiculata
Drosanthemum speciosum

Cacti and other Succulents/small
☐ WHITE

Mammillaria plumosa
Clump-forming, perennial cactus. Has a spherical, green stem completely covered with feathery, white spines. Carries cream flowers in mid-winter. Is difficult to grow. Add calcium to soil. H 12cm (5in), S 40cm (16in). Min. 10°C (50°F).

Gibbaeum velutinum
Clump-forming, perennial succulent with paired, finger-like, velvety, bluish grey-green leaves, to 6cm (2½in) long. Produces daisy-like, pink, lilac or white flowers, 5cm (2in) across, in spring. H 8cm (3in), S 30cm (12in). Min. 5°C (41°F).

Lithops marmorata
Egg-shaped, perennial succulent, divided into 2 unequal-sized, swollen, pale grey leaves with dark grey marks on convex, upper surfaces. Bears a white flower in late summer or early autumn. H 2–3cm (¾–1¼in), S 5cm (2in). Min. 5°C (41°F).

Echinocereus leucanthus, syn. *Wilcoxia albiflora*
Clump-forming, tuberous cactus with spined, 6- or 7-ribbed, prostrate stems. In spring bears often terminal, dark-throated, white flowers, softly streaked purple, with green stigmas. H 20cm (8in), S 30cm (12in). Min. 8°C (46°F).

Trichodiadema mirabile
Bushy to prostrate, perennial succulent with cylindrical, dark green leaves tipped with dark brown bristles and covered in papillae. Stem tip bears white flowers, 4cm (1½in) across, from spring to autumn. H 15cm (6in), S 30cm (12in). Min. 5°C (41°F).

Haworthia truncata
Clump-forming, perennial succulent with a basal fan of broad, erect, rough, blue-grey leaves with pale grey lines and flat ends. Produces small, tubular, white flowers, with spreading petals, from spring to autumn. H 2cm (¾in), S 10cm (4in). Min. 10°C (50°F).

Lithops karasmontana
Egg-shaped, perennial succulent, divided into 2 unequal-sized, grey leaves, that have pink, upper surfaces with sunken, darker pink marks. Bears a white flower in late summer or early autumn. H to 4cm (1½in), S 5cm (2in). Min. 5°C (41°F).

Adromischus maculatus
Clump-forming, perennial succulent with rounded, glossy, green leaves with purple marks. Leaf tips are often wavy. Carries tubular, purplish-white flowers, on a 30cm (12in) tall stem, in summer. H 6cm (2½in), S 10–15cm (4–6in). Min. 7°C (45°F).

Haworthia attenuata var. *clariperla*
Clump-forming, perennial succulent with a basal rosette of triangular, 3cm (1¼in) long, dark green leaves, that have pronounced, white dots. Has tubular, white flowers, with spreading petals, from spring to autumn. H 7cm (3in), S 25cm (10in). Min. 5°C (41°F).

OTHER RECOMMENDED PLANTS:
Cheiridopsis denticulata
Gymnocalycium quehlianum
Lithops fulleri
Lithops julii
Orostachys chanetii

Cacti and other Succulents/small
☐ WHITE

Crassula socialis
Spreading, perennial succulent with short, dense rosettes, to 1cm (½in) across, of fleshy, triangular, green leaves. Produces clusters of star-shaped, white flowers on 3cm (1¼in) tall stems in spring. H 5cm (2in), S indefinite. Min. 5°C (41°F).

Strombocactus disciformis
Very slow-growing, hemispherical, perennial cactus with a grey-green to brown stem set with a spiral of blunt tubercles. Woolly crown has bristle-like spines, which soon fall off, and cream flowers in summer. H 3cm (1¼in), S 10cm (4in). Min. 5°C (41°F).

Mammillaria schiedeana
Clump-forming, perennial cactus. Green stem is covered with short, feathery, yellow spines that turn white. Produces cream flowers and narrow, red seed pods in late summer. H 10cm (4in), S 30cm (12in). Min. 10°C (50°F).

Neoporteria villosa
Clump-forming, perennial cactus with a branched, green to dark grey-green stem. Has dense, sometimes curved, grey spines, 3cm (1¼in) long. Produces tubular, pink or white flowers in spring or autumn. H 15cm (6in), S 10cm (4in). Min. 8°C (46°F).

Oophytum nanum
Clump-forming, perennial succulent with 2 united, very fleshy, bright green leaves. Has daisy-like, white flowers, 1cm (½in) across, in autumn. Is covered in a dry, paper-like sheath, except in spring. H 2cm (¾in), S 1cm (½in). Min. 5°C (41°F).

Lithops lesliei var. **albinica**
Egg-shaped, perennial succulent, divided into 2 unequal-sized leaves; convex, pale green, upper surfaces have dark green and yellow marks. Bears a white flower in late summer or early autumn. H 2–3cm (¾–1¼in), S 5cm (2in). Min. 5°C (41°F).

Haworthia arachnoidea, syn. *H. setata*
Slow-growing, clump-forming, perennial succulent with a basal rosette of triangular leaves. Bears soft, white teeth along leaf margins. Has white flowers from spring to autumn. H 5cm (2in), S 10cm (4in). Min. 6°C (43°F).

Conophytum truncatum
Slow-growing, clump-forming, perennial succulent with pea-shaped, dark spotted, blue-green leaves, each with a sunken fissure at the tip. Produces cream flowers, 1.5cm (⅝in) across, in autumn. H 1.5cm (⅝in), S 15cm (6in). Min. 4°C (39°F).

Mammillaria elongata
(Lace cactus)
Clump-forming, perennial cactus. Has a columnar, green stem, 3cm (1¼in) across, densely covered with yellow, golden or brown spines. Bears cream flowers in summer. Offsets freely. H 15cm (6in), S 30cm (12in). Min. 5°C (41°F).

Aptenia cordifolia 'Variegata'
Crassula lactea
Delosperma tradescantioides
Haworthia attenuata
Haworthia cuspidata
Haworthia fasciata
Lithops karasmontana subsp. bella
Mammillaria gracilis
Mammillaria gracilis var. fragilis
Mammillaria prolifera
Ophthalmophyllum maughanii

☐ WHITE–PINK   ☐ PINK

Mammillaria bocasana
(Powder-puff cactus)
Clump-forming, perennial cactus. Long, white hairs cover a hemispherical stem. Has cream or rose-pink flowers in summer and red seed pods the following spring-summer. H 10cm (4in), S 30cm (12in). Min. 5°C (41°F).

Crassula multicava
Bushy, perennial succulent with oval, grey-green leaves, 8cm (3in) across. Carries numerous clusters of small, star-shaped, pink flowers on elongated stems in spring, followed by small plantlets. H 15cm (6in), S 1m (3ft). Min. 7°C (45°F).

Neoporteria napina var. *mitis*
Flattened spherical, perennial cactus with very short, grey spines pressed flat against a greenish-brown stem. Produces white, pink, carmine or brown flowers, 5cm (2in) across, from crown in summer. H 2cm (¾in), S 3.5cm (1½in). Min. 8°C (46°F).

Oscularia deltoides
Spreading, perennial succulent. Has chunky, triangular, blue-green leaves, to 1cm (½in) long, with small-toothed, often reddened leaf margins. Fragrant, pink flowers, 1–2cm (½–¾in) wide, appear in early summer. H 15cm (6in), S 1m (3ft). Min. 3°C (37°F).

Parodia ructilans, syn.
Notocactus rutilans
Columnar, perennial cactus. Areoles each have about 15 radial spines and 2 upward- or downward-pointing, central spines. Has cream-centred, pink flowers in summer. H 10cm (4in), S 5cm (2in). Min. 10°C (50°F).

Lophophora williamsii
(Dumpling cactus, Mescal button)
Very slow-growing, clump-forming, perennial cactus with an 8-ribbed, blue-green stem. Masses of pink flowers appear in summer on plants over 3cm (1¼in) high. H 5cm (2in), S 8cm (3in). Min. 10°C (50°F).

Epithelantha micromeris
Slow-growing, spherical, perennial cactus with a green stem completely obscured by close-set areoles bearing tiny, white spines. Bears funnel-shaped, pale pinkish-red flowers, 0.5cm (¼in) across, on a woolly crown in summer. H and S 4cm (1½in). Min. 10°C (50°F).

Echeveria elegans
Clump-forming, perennial succulent with a basal rosette of broad, fleshy, pale silvery-blue leaves, edged with red, and yellow-tipped, pink flowers in summer. Keep dry in winter. Makes a good bedding plant. H 5cm (2in), S 50cm (20in). Min. 5°C (41°F).

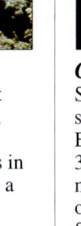

Coryphantha vivipara
Spherical, perennial cactus with a green stem densely covered with grey spines. Bears funnel-shaped, pink flowers, 3.5cm (1½in) across, in summer. Is much more difficult to grow than many other species in this genus. H and S 5cm (2in). Min. 5°C (41°F).

Crassula cooperi
Gymnocalycium schickendantzii
Mammillaria candida
Ophthalmophyllum longum

Schlumbergera 'Wintermärchen'

Ruschia acuminata
Stenocactus lamellosus

Cacti and other Succulents/small
■ PINK

Schlumbergera 'Gold Charm'
Erect, then pendent, perennial cactus. Has flattened, oblong, green stem segments with toothed margins. Yellow flowers in early autumn turn pinkish-orange in winter. H 15cm (6in), S 30cm (12in). Min. 10°C (50°F).

Mammillaria sempervivi
Slow-growing, spherical, perennial cactus. Has a dark green stem with short, white spines. Has white wool between short, angular tubercles on plants over 4cm (1½in) high. Bears cerise flowers in spring. H and S 7cm (3in). Min. 5°C (41°F).

Melocactus intortus,
syn. *M. communis* (Melon cactus)
Flattened spherical, perennial cactus. Has an 18–20-ribbed stem with yellow-brown spines. Crown matures to a white column with fine, brown spines. Bears pink flowers in summer. H 20cm (8in), S 25cm (10in). Min. 15°C (59°F).

Ariocarpus fissuratus
Very slow-growing, flattened spherical, perennial cactus. Grey stem is covered with rough, triangular tubercles each producing a tuft of wool. Has 4cm (1½in) wide, pink-red flowers in autumn. H 10cm (4in), S 15cm (6in). Min. 5°C (41°F).

Stenocactus violaciflorus,
syn. *Echinofossulocactus violaciflorus*
Spherical, perennial cactus. Areoles each have a flat, upper, radial spine and rounded, lower ones. Bears pink to violet flowers in spring. H and S 10cm (4in). Min. 5°C (41°F).

Aptenia cordifolia
Fast-growing, prostrate, perennial succulent with oval, glossy, green leaves and, in summer, daisy-like, bright pink flowers. Is ideal for ground cover. H 5cm (2in), S indefinite. Min. 7°C (45°F).

Rhipsalidopsis rosea
Bushy, perennial cactus with slender, 3- or 4-angled, bristly, green stem segments, usually tinged purple, to 5cm (2in) long. Has masses of bell-shaped, pink flowers, to 4cm (1½in) across, in spring. H and S 10cm (4in). Min. 10°C (50°F).

Ophthalmophyllum villetii
Clump-forming, perennial succulent with 2 fleshy, grey-green leaves that are broad, erect and united for most of their length but have distinctly divided, upper lobes. Pale pink flowers appear in late summer. H 2.5cm (1in), S 1cm (½in). Min. 5°C (41°F).

Thelocactus bicolor
Spherical to columnar, perennial cactus with an 8–13-ribbed stem. Areoles each have 4 usually flattened, yellow, central spines, or bicoloured yellow and red, and numerous shorter, radial spines. Flowers are purple-pink. H and S 20cm (8in). Min. 7°C (45°F).

Rebutia violaciflora
Clump-forming, perennial cactus with a tuberculate, dark green stem. Areoles each produce 15–20 brown spines, to 0.5cm (¼in) long. Has trumpet-shaped, deep pink to violet flowers, to 2cm (¾in) across, in spring. H 5cm (2in), S 15cm (6in). Min. 5°C (41°F).

Aptenia cordifolia 'Variegata'
Carpobrotus edulis
Kalanchoe pumila
Melocactus bahiensis
Melocactus matanzanus
Rebutia pygmaea
Rebutia senilis
Ruschia macowanii

☐ PINK

Mammillaria zeilmanniana
(Rose pincushion)
Clump-forming, perennial cactus with a spherical, green stem that has hooked spines and bears a ring of deep pink to purple flowers in spring. H 15cm (6in), S 30cm (12in). Min. 10°C (50°F).

☼ ◊ ♉

Echinopsis backebergii,
syn. *Lobivia backebergii*
Clump-forming, almost spherical, perennial cactus with a 10–15-ribbed, spined, dark green stem. Has funnel-shaped, pink, red or purple flowers, with paler throats, in summer. H 10cm (4in), S 15cm (6in). Min. 5°C (41°F).

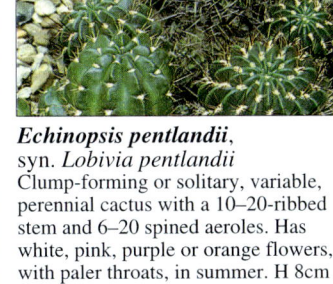

☼ ◊

Echinopsis pentlandii,
syn. *Lobivia pentlandii*
Clump-forming or solitary, variable, perennial cactus with a 10–20-ribbed stem and 6–20 spined aeroles. Has white, pink, purple or orange flowers, with paler throats, in summer. H 8cm (3in), S 10cm (4in). Min. 5°C (41°F).

☼ ◊

Melocactus oaxacensis

☐ ■ PINK–RED

Frithia pulchra
Basal-rosetted, perennial succulent with erect, rough, grey leaves, cylindrical with flattened tips. Produces masses of stemless, daisy-like, bright pink flowers, with paler centres, in summer. H 3cm (1¼in), S 6cm (2½in). Min. 10°C (50°F).

☼ ◊

Crassula schmidtii
Carpeting, perennial succulent with dense rosettes of linear, dark green leaves, pitted and marked, each 3–4cm (1¼–1½in) long. Bears masses of star-shaped, bright pinkish-red flowers in clusters in winter. H 10cm (4in), S 30cm (12in). Min. 7°C (45°F).

◐ ◊

Echeveria secunda
Clump-forming, perennial succulent with short stems each crowned by a rosette of broad, fleshy, light green to grey leaves, reddened towards tips. Bears cup-shaped, red-and-yellow flowers in spring-summer. H 4cm (1½in), S 30cm (12in). Min. 5°C (41°F).

☼ ◊

Cheiridopsis purpurea
Echinopsis cinnabarina
Gibbaeum petrense
Huernia pillansii

Graptopetalum bellum,
syn. *Tacitus bellus*
Basal-rosetted, perennial succulent with triangular to oval, grey leaves, 5cm (2in) long. Has clusters of deep pink to red flowers, 2cm (¾in) across, in spring-summer. H 3cm (1¼in), S 15cm (6in). Min. 10°C (50°F).

☼ ◊

Argyroderma pearsonii,
syn. *A. schlechteri*
Prostrate, egg-shaped, perennial succulent. A united pair of very fleshy, silvery-grey leaves has a deep fissure in which a red flower, 3cm (1¼in) across, appears in summer. H 3cm (1¼in), S 5cm (2in). Min. 5°C (41°F).

☼ ◊

Schlumbergera truncata,
syn. *Zygocactus truncatus*
(Crab cactus, Lobster cactus)
Erect, then pendent, perennial cactus. Oblong stem segments have toothed margins. Bears purple-red flowers in early autumn and winter. H 15cm (6in), S 30cm (12in). Min. 10°C (50°F).

◐ ◊

Lampranthus multiradiatus
Neoporteria nidus
Trichodiadema densum

409

Cacti and other Succulents/small
RED

Cephalophyllum alstonii
(Red spike)
Prostrate, perennial succulent with cylindrical, grey-green leaves, to 7cm (3in) long. Carries daisy-like, dark red flowers, 8cm (3in) across, in summer. H 10cm (4in), S 1m (3ft). Min. 5°C (41°F).

Parodia sanguiniflora
Clump-forming, perennial cactus. Has a much-ribbed, green stem densely covered with brown, radial spines and red, central spines, some of which are hooked. Bears blood-red, occasionally yellow flowers in spring. H 8cm (3in), S 30cm (12in). Min. 10°C (50°F).

Rebutia krainziana
Clump-forming, perennial cactus with a tuberculate, dark green stem. Bears prominent, white areoles with very short, white spines. Trumpet-shaped, bright red flowers, to 5cm (2in) across, appear at stem base in spring. H 5cm (2in), S 20cm (8in). Min. 5°C (41°C).

Parodia nivosa
Ovoid, perennial cactus that has a much-ribbed, green stem with stiff, white spines, each 1–2cm ($^1/_2$–$^3/_4$in) long. Has a white, woolly crown and bright red flowers, to 5cm (2in) across, in summer. H to 15cm (6in), S 10cm (4in). Min. 10°C (50°F).

Rebutia tiraquensis,
syn. *Sulcorebutia tiraquensis*
Variable, perennial cactus with a green stem. Elongated areoles bear spines of gold or bicoloured red and white. Has dark pink- or orange-red flowers in spring. H 15cm (6in), S 10cm (4in). Min. 10°C (50°F).

Argyroderma fissum,
syn. *A. brevipes*
Clump-forming, perennial succulent with finger-shaped, fleshy leaves, 5–10cm (2–4in) long and often reddish at the tip. Has light red flowers between leaves in summer. H 15cm (6in), S 10cm (4in). Min. 5°C (41°F).

Cotyledon ladismithensis
Evergreen, freely branching, later prostrate, succulent sub-shrub with fleshy, green leaves, swollen and blunt at tips and covered with short, golden-brown hairs. Clusters of tubular, brownish-red flowers appear in autumn. H and S 20cm (8in). Min. 5°C (41°F).

Opuntia verschaffeltii
Clump-forming, perennial cactus with cylindrical, usually spineless stems, to 25cm (10in) long. Stem tips each bear short-lived, cylindrical leaves from spring to autumn. Has orange-red flowers in spring. H 15cm (6in), S 1–2m (3–6ft). Min. 5°C (41°F).

Echeveria agavoides
Basal-rosetted, perennial succulent with tapering, light green leaves, often red-margined. Carries cup-shaped, red flowers, 1cm ($^1/_2$in) long, in summer. H 15cm (6in), S 30cm (12in). Min. 5°C (41°F).

Aloe brevifolia
Ferocactus latispinus
Rebutia minuscula
Rebutia senilis

■ RED

Rebutia spegazziniana
Clump-forming, perennial cactus with a spherical, spined, green stem, to 4cm (1½in) across, becoming columnar with age. Bears masses of slender-tubed, orange-red flowers at base in late spring. H 10cm (4in), S 20cm (8in). Min. 5°C (41°F).

Parodia haselbergii,
syn. *Noctocactus haselbergii*
(Scarlet ball cactus)
Slow-growing, perennial cactus with a stem covered in white spines. Slightly sunken crown bears red flowers, with yellow stigmas, in spring. H 10cm (4in), S 25cm (10in). Min. 10°C (50°F).

Echinocereus triglochidiatus
var. *paucispina*
Clump-forming, perennial cactus with a 10cm (4in) wide, dark green stem that has 6 or 7 ribs, and 4–6 spines, 3–4cm (1¼–1½in) long, per areole. Has orange-red flowers in spring. H 20cm (8in), S 50cm (20in). Min. 5°C (41°C).

■ PURPLE

Schlumbergera **'Bristol Beauty'**
Erect, then pendent, perennial cactus with flattened, green stem segments with toothed margins. Bears reddish-purple flowers, with silvery-white tubes, in early autumn and winter. H 15cm (6in), S 30cm (12in). Min. 10°C (50°F).

Lobivia silvestrii,
syn. *Chamaecereus silvestri*
(Peanut cactus)
Clump-forming, perennial cactus with spined stems, initially erect, then prostrate. Has funnel-shaped, oranged-red flowers in late spring. H 10cm (4in), S indefinite. Min. 3°C (37°F).

Pachyphytum oviferum
(Moonstones, Sugared-almond plum)
Clump-forming, perennial succulent with a basal rosette of oval, pinkish-blue leaves. Stem bears 10–15 bell-shaped flowers, with powder-blue calyces and orange-red petals, in spring. H 10cm (4in), S 30cm (12in). Min. 10°C (50°F).

Neolloydia conoidea
Clump-forming, perennial cactus. Has a columnar, blue-green stem densely covered with white, radial spines and longer, black, central spines. Bears funnel-shaped, purple-violet flowers in summer. H 10cm (4in), S 15cm (6in). Min. 10°C (50°F).

Rhipsalidopsis gaertneri
(Easter cactus)
Bushy, perennial cactus with flat, oblong, glossy, green stem segments, each to 5cm (2in) long, often tinged red at edges. Segment ends each bear orange-red flowers in spring. H 15cm (6in), S 20cm (8in). Min. 6°C (43°F).

Caralluma joannis
Clump-forming, perennial succulent with blue-grey stems and rudimentary leaves on stem angles. Bears clusters of star-shaped, purple flowers, with short, fine hairs on petal tips, in late summer near stem tips. H 20cm (8in), S 1m (3ft). Min. 11°C (52°F).

Echeveria harmsii
Echeveria setosa
Schlumbergera 'Zara'

Schlumbergera 'Buckleyi'
Thelocactus mcdowellii

Cacti and other Succulents/small

■ PURPLE

■ GREEN–YELLOW

Argyroderma delaetii, syn. *A. blandum*
Prostrate, egg-shaped, perennial succulent with 2 very fleshy, silvery-green leaves between which daisy-like, pink-purple flowers, 5cm (2in) across, appear in late summer. H 3cm (1½in), S 5cm (2in). Min. 5°C (41°F).

Duvalia corderoyi
Clump-forming, perennial succulent. Has a prostrate, leafless stem with 6 often purple, indistinct ribs. Bears star-shaped, dull green flowers, 1cm (½in) across and covered in purple hairs, in summer-autumn. H 5cm (2in), S 60cm (24in). Min. 10°C (50°F).

Crassula deceptor, syn. *C. deceptrix*
Slow-growing, clump-forming, perennial succulent with branching stems surrounded by fleshy, grey leaves set in 4 rows. Each leaf has minute lines around raised dots. Bears insignificant flowers in spring. H and S 10cm (4in). Min. 5°C (41°F).

Stenocactus pentacanthus, syn. *Echinofossulocactus pentacanthus*
Spherical, perennial cactus. The stem bears 25–50 ribs. Areoles each have an upper, radial spine and lower ones that form a cross. Bears a funnel-shaped, white-edged, violet flower in spring. H and S 8cm (3in). Min. 5°C (41°F).

Orbea variegata, syn. *Stapelia variegata* (Star flower)
Clump-forming, branching, perennial succulent with 4-angled, indented stems. Flowers, variable in colour and blotched yellow, purple- or red-brown, appear in summer-autumn. H to 10cm (4in), S indefinite. Min. 11°C (52°F).

Huernia macrocarpa var. **arabica**
Clump-forming, perennial succulent with finger-shaped, 4- or 5-sided, green stems. Produces short-lived, deciduous leaves and, in autumn, bell-shaped, white-haired, dark purple flowers with recurved petal tips. H and S 10cm (4in). Min. 8°C (46°F).

Stapelia flavirostris
Clump-forming, perennial succulent with 4-angled, hairy, toothed, green stems. In summer-autumn carries star-shaped, purple-brown flowers, to 10cm (4in) across, ridged with white or purple hairs. H to 20cm (8in), S indefinite. Min. 11°C (52°F).

Stapelia gigantea, syn. *Orbea variegata*
Clump-forming, perennial succulent. In summer-autumn bears star-shaped, red-marked, yellow-brown flowers, 30cm (12in) across, with white-haired, recurved edges. H to 20cm (8in), S indefinite. Min. 11°C (52°F).

Rebutia rauschii

Ancistrocactus uncinatus
Huernia zebrina
Parodia graessneri

◻ YELLOW

Ancistrocactus scheeri,
syn. *A. megarhizus*
Globose to columnar, perennial cactus. Stem bears spines and, in spring, funnel-shaped, straw-coloured flowers. Lowest and longest spines are darker and hooked. H 10cm (4in), S 6cm (2½in). Min. 2°C (36°F).

Frailea pulcherrima
Columnar, perennial cactus with a much-ribbed, dark green stem bearing white to light brown spines. Buds, which rarely open to flattish, yellow flowers in summer, become tufts of spherical, spiny seed pods. H to 5cm (2in), S 2cm (¾in). Min. 5°C (41°F).

Maihuenia poeppigii
Slow-growing, clump-forming, perennial cactus. Has a cylindrical, branched, spiny, green-brown stem. Most branches produce a spike of cylindrical, green leaves at the tip, crowned by a funnel-shaped, yellow flower in summer. H 6cm (2½in), S 30cm (12in).

Aloinopsis schooneesii
Dwarf, mounded, perennial succulent with tuberous roots and fleshy, almost spherical, blue-green leaves arranged tightly in tufts. Produces flattish, yellow flowers in winter-spring. H 3cm (1½in), S to 7cm (3in). Min. 7°C (45°F).

Astrophytum ornatum
Elongated, spherical, perennial cactus with a very fleshy, 8-ribbed stem. Crown of each rib bears 5–11cm (2–4½in) long spines on each raised areole. Has yellow flowers, 8cm (3in) across, in summer. H 15cm (6in), S 12cm (5in). Min. 5°C (41°F).

Aichryson x *domesticum*
'Variegata' (Cloud grass)
Prostrate, perennial succulent with stems crowned by rosettes of hairy, cream-marked, green leaves, sometimes pure cream. Has star-shaped, yellow flowers in spring. H 15cm (6in), S 40cm (16in). Min. 5°C (41°F).

Parodia mammulosus,
syn. *Noctocactus mammulosus*
Spherical, perennial cactus. Green stem has about 20 ribs and straight, stiff, yellow-brown to white spines, to 1cm (½in) long. Woolly crown produces masses of golden flowers in summer. H and S 10cm (4in). Min. 5°C (41°F).

Gymnocalycium andreae
Clump-forming, spherical, perennial cactus with a glossy, dark green stem bearing 8 rounded ribs and up to 8 pale yellow-white spines per areole. Has 5cm (2in) wide, yellow flowers in spring-summer. H 6cm (2½in), S 10cm (4in). Min. 10°C (50°F).

Euphorbia obesa
(Gingham golf ball)
Spherical, perennial succulent. Spineless, dark green stem, often chequered light green, has 8 low ribs. Crown bears rounded heads of cupped, yellow flowers in summer. H 12cm (5in), S 15cm (6in). Min. 10°C (50°F).

Carpobrotus edulis
Conophytum meyeri
Copiapoa echinoides
Dudleya brittonii

Echinopsis aurea
Ferocactus latispinus
Huernia macrocarpa
Huernia thurettii var. *primulina*

Mammillaria densispina
Neoporteria napina
Stomatium agninum
Stomatium patulum

Titanopsis schwantesii

413

Cacti and other Succulents/small
◻ YELLOW

Rhombophyllum rhomboideum
Clump-forming, perennial succulent. Linear, glossy, grey-green leaves have expanded middles and white margins. Stems, 2–5cm (¾–2in) long, bear 3–7 yellow flowers, to 4cm (1½in) across, in summer. H 5cm (2in), S 15cm (6in). Min. 5°C (41°F).

Schwantesia ruedebuschii
Mat-forming, perennial succulent with cylindrical, bluish-green leaves, 3–5cm (1–2in) long, with expanded tips. Leaf edges each produce 3–7 minute, blue teeth with brown tips. Has yellow flowers in summer. H 5cm (2in), S 20cm (8in). Min. 5°C (41°F).

Pleiospilos compactus, syn *P. simulans*
Clump-forming, perennial succulent with 1 or 2 pairs of thick, grey leaves, to 8cm (3in) long. Bears coconut-scented, yellow flowers in early autumn. H 10cm (4in), S 30cm (12in). Min. 5°C (41°F).

Lithops pseudotruncatella var. *pulmonuncula*
Egg-shaped, perennial succulent, divided into 2 unequal-sized, grey leaves with dark green and red marks on upper surfaces. Bears a yellow flower in summer or autumn. H 2–3cm (¾–1¼in), S 4cm (1½in). Min. 5°C (41°F).

Aeonium tabuliforme
Prostrate, almost stemless, short-lived, perennial succulent with a basal rosette, to 30cm (12in) across, like a flat, bright green plate. Has star-shaped, yellow flowers in spring, then dies. Propagate from seed. H 5cm (2in), S 30cm (12in). Min. 5°C (41°F).

Lithops schwantesii var. *kuibisensis*
Egg-shaped, perennial succulent, divided into 2 unequal-sized leaves with blue or red marks on upper surface. Has a yellow flower in late summer or autumn. H 2–3cm (¾–1¼in), S 3cm (1¼in). Min. 5°C (41°F).

Lithops dorotheae
Egg-shaped, perennial succulent, divided into 2 unequal-sized leaves, pale pink-yellow to green with darker marks and red dots and lines on upper surfaces. Produces a daisy-like, yellow flower in summer or autumn. H 2–3cm (¾–1¼in), S 5cm (2in). Min. 5°C (41°F).

Mammillaria microhelia
Columnar, perennial cactus with a 5cm (2in) wide, green stem bearing cream or brown spines, discolouring with age. Has 1.5cm (⅝in) wide, yellow or pink flowers in spring. Offsets slowly with age. H 20cm (8in), S 40cm (16in). Min. 5°C (41°F).

Opuntia humifusa
Prostrate, perennial cactus. Each areole bears up to 3 spines, 3cm (1¼in) long. Has flat, rounded to oval, purple-tinged, dark green stem segments, 7–18cm (3–7in) long. Bears 8cm (3in) wide, yellow flowers in spring-summer. Keep dry in winter. H 15cm (6in), S 1m (3ft).

Agave utahensis
Basal-rosetted, perennial succulent with rigid, blue-grey leaves, each with spines up margins and a long, dark spine at tip. Flower stem, to 1.5m (5ft) long, carries yellow flowers in summer. H 23cm (9in) or more, S 2m (6ft).

Graptopetalum paraguayense
(Mother-of-pearl plant)
Clump-forming, perennial succulent with a basal rosette, 15cm (6in) across, of grey-green leaves, often tinged pink. Bears star-shaped, yellow-and-red flowers in summer. H 10cm (4in), S 1m (3ft). Min. 5°C (41°F).

Astrophytum asterias
Caralluma europaea
Cephalophyllum pillansii
Crassula lycopodioides

Euphorbia gorgonis
Lithops aucampiae
Lithops hookeri
Lithops insularis

Lithops lesliei
Lithops olivacea
Lithops otzeniana
Lithops pseudotruncatella

Lithops schwantesii

☐ YELLOW

Coryphantha cornifera, syn. *C. radians*
Spherical to columnar, perennial cactus with angular tubercles, each bearing a curved, dark, central spine and shorter, radial spines. Has funnel-shaped, yellow flowers in summer. H 15cm (6in), S 10cm (4in). Min. 5°C (41°F).
☼ ◊

Conophytum bilobum
Slow-growing, clump-forming, perennial succulent with 2-lobed, fleshy, green leaves, each 4cm (1½in) long and 2cm (¾in) wide. Has a flared, yellow flower, 3cm (1¼in) across, in autumn. H 4cm (1½in), S 15cm (6in). Min. 4°C (39°F).
☼ ◊

Faucaria tigrina (Tiger-jaws)
Clump-forming, stemless, perennial succulent. Fleshy, green leaves, 5cm (2in) long, have 9 or 10 teeth along each margin. Bears daisy-like, yellow flowers, 5cm (2in) across, in autumn. H 10cm (4in), S 50cm (20in). Min. 6°C (43°F).
☼ ◊

Parodia vorwerkiana
(Colombian ball cactus)
Slow-growing, flattened spherical, perennial cactus with a glossy stem, up to 20 wart-like ribs, bearing yellow-white spines, and yellow blooms in summer. H 8cm (3in), S 9cm (3½in). Min. 5°C (41°F).
☼ ◊

Glottiphyllum nelii
Clump-forming, perennial succulent with semi-cylindrical, fleshy, green leaves, to 5cm (2in) long. Carries daisy-like, golden-yellow flowers, 4cm (1½in) across, in spring-summer. H 5cm (2in), S 30cm (12in). Min. 5°C (41°F).
☼ ◊

Lobivia shaferi
Columnar, perennial cactus. Has narrow, much-ribbed, green stems covered with pale, radial spines often surrounded by 1–3 very stout, central spines, to 2.5cm (1in) long. Produces yellow flowers in summer. H and S 10cm (4in). Min. 5°C (41°F).
☼ ◊

Fenestraria aurantiaca
var. *aurantiaca*, syn. *F. a.* f. *rhopalophylla* (Baby's toes)
Clump-forming, perennial succulent with a basal rosette of glossy leaves. Has yellow flowers in late summer and autumn. H 5cm (2in), S 30cm (12in). Min. 6°C (43°F).
☼ ◊

Titanopsis calcarea
Clump-forming, perennial succulent with a basal rosette of very fleshy, triangular, blue-grey leaves covered in wart-like, grey-white and beige tubercles. Has yellow flowers from autumn to spring. H 3cm (1¼in), S 10cm (4in). Min. 8°C (46°F).
☼ ◊

Pleiospilos bolusii
(Living rock, Mimicry plant)
Clump-forming, perennial succulent with 1 or 2 pairs of grey leaves, often wider than long and narrowing at incurved tips. Has golden-yellow flowers in early autumn. H 10cm (4in), S 20cm (8in). Min. 5°C (41°F).
☼ ◊ ♆

Glottiphyllum semicylindricum
Parodia apricus
Parodia ottonis
Thelocactus leucacanthus

Cacti and other Succulents/small
YELLOW–ORANGE

Rebutia arenacea,
syn. *Sulcorebutia arenacea*
Spherical, perennial cactus. Has a brown-green stem with white spines on spirally arranged tubercles. Has golden-yellow blooms, to 3cm (1¼in) across, in spring. H 5cm (2in), S 6cm (2½in). Min. 10°C (50°F).

Gasteria bicolor var. **liliputana**,
syn. *G. liliputana*
Perennial succulent that forms rosettes of dark green leaves blotched with white. Flower stems, to 15cm (6in) long, bear spikes of bell-shaped, orange-green flowers in spring. H 7cm (3in), S 10cm (4in). Min. 5°C (41°F).

Pachyphytum compactum
Clump-forming, perennial succulent with a basal rosette of green leaves, each narrowing to a blunt point, with angular, paler edges. Stems each bear 3–10 flowers with green to pink calyces and orange petals in spring. H 15cm (6in), S indefinite. Min. 5°C (41°F).

Conophytum notabile
Slow-growing, spherical, perennial succulent forming clumps of 2-lobed, very fleshy, grey-green leaves, often with a red spot on edge of fissure between the lobes. Carries copper-orange flowers in autumn. H 3cm (1¼in). S indefinite. Min. 4°C (39°F).

Malephora crocea
Erect or spreading, perennial succulent with semi-cylindrical, blue-green leaves on short shoots. Carries solitary daisy-like, orange-yellow flowers, reddened on outsides, in spring-summer. H 20cm (8in), S 1m (3ft). Min. 5°C (41°F).

Cleistocactus aurantiacus,
syn. *Borzicactus aurantiacus*,
Submatucana aurantiaca
Spherical, perennial cactus with a 15–17-ribbed stem. Elongated areoles each bear up to 30 spines. Has orange-yellow flowers in summer. H 12cm (5in), S 40cm (16in). Min. 10°C (50°F).

Aloe aristata
(Lace aloe, Torch plant)
Clump-forming, perennial succulent that has a basal rosette of pointed, dark green leaves with white spots and soft-toothed edges. Has orange flowers in spring. Offsets freely. H 10cm (4in), S 30cm (12in). Min. 7°C (45°F).

Rebutia aureiflora
Clump-forming, perennial cactus with a dark green stem, often tinged violet-red, that has stiff, radial spines and longer, soft, central spines. Has masses of yellow, violet or red flowers in late spring. H 10cm (4in), S 20cm (8in). Min. 5°C (41°F).

Gasteria carinata var. **verrucosa**,
syn. *G. verrucosa*
Clump-forming, perennial succulent with stiff, dark green leaves, with raised, white dots and incurved edges. Has spikes of bell-shaped, orange-green flowers in spring. H 10cm (4in), S 30cm (12in). Min. 5°C (41°F).

Rebutia muscula
Clump-forming, perennial cactus. Has a dark green stem densely covered with soft, white spines, to 0.5cm (¼in) long. Bears bright orange flowers, 2–3cm (¾–1¼in) across, in late spring. H 10cm (4in), S 15cm (6in). Min. 5°C (41°F).

Aloe humilis
Echeveria derenbergii
Gasteria caespitosa
Rebutia neocumingii

Rebutia senilis

The Award of Garden Merit

The Royal Horticultural Society's Award of Garden Merit (AGM) recognizes plants of outstanding excellence for garden decoration or use, whether grown in the open or under glass. Besides being the highest accolade the Society can give to a plant, the AGM is of practical value for ordinary gardeners, helping them in making a choice from the many thousands of plants currently available. The Award's remit is broad, encompassing fruit and vegetables as well as ornamental plants. All plant types can be considered, be they the most minuscule alpines or majestic trees, whether they are hardy throughout the British Isles or need year-round protection under heated glass. In each case the purpose is the same: to highlight the best plants available.

An award is made either by trial, usually at the Society's garden at Wisley, or on the recommendation of a panel of experts. The AGM is not graded, so there is no attempt to differentiate the good from the better or best, but in those groups in which there are many cultivars standards need to be set especially high if the resulting selection is to offer helpful guidance to the gardener. The following lists are of the plants described in this book, some 2000 in all, that have received the Award.

In order to receive the Award, a plant must fulfil the following criteria:

- It should be excellent for ordinary garden use (either in the open or under glass).

- It should be of good constitution.

- It should be available in the horticultural trade, or be available for propagation.

- It should not be susceptible to any pest or disease.

- It should not require highly specialized care other than the provision of appropriate growing conditions (e.g. lime-free soil when required).

- It should not be subject to an unreasonable degree of reversion in its vegetative or floral characteristics.

Trees

Acacia baileyana, p.71
Acacia dealbata, p.57
Acer capillipes, p.56
Acer cappadocicum 'Aureum'
Acer circinatum
Acer davidii var. *grosseri*
Acer griseum, p.74
Acer japonicum 'Aconitifolium', p.69
Acer japonicum 'Vitifolium', p.69
Acer negundo var. *violaceum*

Acer negundo var. *violaceum*

Acer palmatum var. *coreanum*, p.68
Acer pensylvanicum, p.58
Acer pensylvanicum 'Erythrocladum'
Acer platanoides
Acer platanoides 'Crimson King', p.39
Acer platanoides 'Drummondii'
Acer platanoides 'Schwedleri'
Acer pseudoplatanus 'Brilliantissimum', p.63
Acer rubrum 'October Glory'
Acer rubrum 'Scanlon', p.44
Acer rufinerve, p.56
Acer saccharinum
Acer shirasawanum 'Aureum', p.66
Aesculus x *carnea* 'Briotii', p.38
Aesculus flava, p.56
Aesculus hippocastanum, p.38
Aesculus hippocastanum 'Baumannii'
Aesculus indica

Aesculus x *neglecta* 'Erythroblastos', p.63
Aesculus pavia
Ailanthus altissima
Albizia lophantha
Alnus cordata, p.40
Alnus glutinosa 'Imperialis'
Aralia elata
Aralia elata 'Variegata'
Arbutus x *andrachnoides*, p.57
Arbutus menziesii
Arbutus unedo, p.67
Archontophoenix cunninghamiana
Beaucarnea recurvata, p.74
Betula albosinensis, p.48
Betula 'Jermyns'
Betula pendula
Betula pendula 'Tristis'
Betula pendula 'Youngii', p.66
Carpinus betulus
Carpinus betulus 'Fastigiata', p.74
Carya cordiformis
Carya ovata, p.43
Castanea sativa
Catalpa bignonioides, p.52
Catalpa bignonioides 'Aurea'
Catalpa x *erubescens* 'Purpurea'
Cercidiphyllum japonicum, p.45
Cercis canadensis 'Forest Pansy', p.65
Cercis siliquastrum, p.62
Chrysalidocarpus lutescens, p.74
Cladrastis lutea, p.56
Cordyline australis
Cornus controversa 'Variegata', p.63
Cornus 'Eddie's White Wonder', p.70
Cornus kousa var. *chinensis*
Cornus 'Norman Hadden'
Cornus 'Porlock', p.60
Crataegus laevigata 'Paul's Scarlet', p.65
Crataegus x *lavallei* 'Carrierei'
Crataegus x *prunifolia*
Cydonia oblonga 'Vranja', p.66
Davidia involucrata, p.51
Eucalyptus coccifera, p.46
Eucalyptus dalrympleana, p.46
Eucalyptus globulus
Eucalyptus gunnii, p.46
Eucalyptus niphophila, p.58
Eucryphia glutinosa, p.64
Fagus sylvatica, p.42
Fagus sylvatica 'Dawyck'
Fagus sylvatica 'Dawyck Purple'
Fagus sylvatica f. *pendula*, p.40
Fagus sylvatica 'Riversii'

Ficus benjamina
Ficus elastica 'Decora'
Ficus elastica 'Doescheri', p.46
Ficus lyrata
Ficus rubiginosa
Fraxinus excelsior
Fraxinus excelsior 'Jaspidea'
Fraxinus excelsior 'Pendula'
Fraxinus ornus, p.50
Genista aetnensis, p.67
Gleditsia triacanthos 'Sunburst', p.51
Grevillea robusta
Hoheria 'Glory of Amlwch'
Hoheria lyallii, p.64
Hoheria sexstylosa
Howea forsteriana
Ilex x *altaclerensis* 'Belgica Aurea'
Ilex x *altaclerensis* 'Camelliifolia', p.72
Ilex x *altaclerensis* 'Hodginsii'
Ilex x *altaclerensis* 'Wilsonii'
Ilex aquifolium, p.72
Ilex aquifolium 'Argentea Marginata', p.72
Ilex aquifolium 'Mme Briot', p.73
Ilex aquifolium 'Pyramidalis', p.72
Juglans nigra, p.41
Juglans regia, p.40
Koelreuteria paniculata, p.66
Laburnum x *watereri* 'Vossii', p.66
Lagerstroemia indica, p.65
Laurus nobilis
Liquidambar styraciflua 'Lane Roberts'
Liriodendron tulipifera, p.39
Liriodendron tulipifera 'Aureomarginatum', p.43
Livistona chinensis, p.59
Magnolia campbellii 'Charles Raffill'

Magnolia campbellii 'Charles Raffill'

Magnolia cylindrica, p.49
Magnolia denudata
Magnolia grandiflora 'Exmouth', p.49
Magnolia hypoleuca, p.49
Magnolia salicifolia, p.49
Magnolia 'Wada's Memory', p.49
Magnolia wilsonii, p.49
Malus floribunda, p.62
Malus 'Golden Hornet', p.71
Malus hupehensis, p.48
Malus 'John Downie', p.69
Malus 'Katherine'
Malus 'Neville Copeman'
Malus x *robusta* 'Yellow Siberian'
Malus transitoria
Malus tschonoskii
Morus nigra
Nyssa sinensis, p.56
Nyssa sylvatica, p.45
Olea europaea
Parrotia persica, p.56
Paulownia tomentosa, p.51
Phoenix canariensis
Phoenix roebelenii
Pittosporum eugenioides 'Variegatum', p.73
Pittosporum tobira
Platanus x *hispanica*, p.41
Platanus orientalis
Populus lasiocarpa
Populus nigra 'Italica', p.40
Populus tremula
Prunus 'Accolade', p.62
Prunus 'Amanogawa'
Prunus avium, p.45
Prunus avium 'Plena', p.50
Prunus cerasifera 'Nigra', p.65
Prunus 'Kanzan', p.51
Prunus 'Kiku-shidare', p.62
Prunus 'Kursar'
Prunus 'Okame'
Prunus padus 'Watereri'
Prunus 'Pandora', p.61
Prunus 'Pink Perfection', p.61
Prunus sargentii, p.61
Prunus serrula
Prunus 'Shirofugen', p.61
Prunus 'Shirotae', p.60
Prunus 'Shogetsu', p.60
Prunus 'Spire', p.61
Prunus subhirtella 'Autumnalis'
Prunus subhirtella 'Pendula Rubra', p.62
Prunus 'Taihaku', p.60

Prunus triloba 'Multiplex'
Prunus 'Ukon', p.60
Prunus x *yedoensis*, p.61
Pterocarya fraxinifolia
Pterostyrax hispida
Pyrus calleryana 'Chanticleer', p.50
Pyrus salicifolia 'Pendula', p.65
Quercus canariensis, p.40
Quercus cerris
Quercus coccinea 'Splendens'
Quercus x *hispanica* 'Lucombeana', p.47
Quercus ilex
Quercus palustris, p.43
Quercus petraea
Quercus phellos, p.45
Quercus robur
Quercus rubra, p.43
Robinia pseudoacacia
Robinia pseudoacacia 'Frisia', p.55
Salix alba 'Britzensis'
Salix alba var. *sericea*
Salix alba var. *vitellina*, p.48
Salix 'Chrysocoma', p.48
Salix matsudana 'Tortuosa', p.59
Salix x *rubens* 'Basfordiana'
Schefflera actinophylla, p.58
Sophora japonica
Sophora tetraptera, p.63
Sorbus aria 'Lutescens', p.52
Sorbus aria 'Majestica'
Sorbus aucuparia 'Fructu Luteo'
Sorbus aucuparia 'Sheerwater Seedling'
Sorbus cashmiriana, p.67
Sorbus commixta 'Embley'
Sorbus hupehensis
Sorbus 'Joseph Rock', p.56
Sorbus sargentiana
Sorbus scalaris
Sorbus thibetica 'John Mitchell', p.53
Sorbus vilmorinii, p.68
Stuartia pseudocamellia, p.52
Stuartia sinensis
Styrax japonica, p.51
Styrax obassia
Tilia cordata
Tilia cordata 'Greenspire'
Tilia 'Euchlora'
Tilia mongolica
Tilia 'Petiolaris', p.42
Trachycarpus fortunei, p.58
Washingtonia filifera
Zelkova serrata, p.45

Conifers

Abies balsamea f. *hudsonia*
Abies concolor
Abies concolor 'Compacta', p.84
Abies grandis, p.78
Abies lasiocarpa var. *arizonica* 'Compacta', p.84
Abies nordmanniana
Abies nordmanniana 'Golden Spreader', p.85
Abies procera
Abies veitchii, p.76
Araucaria heterophylla
Calocedrus decurrens, p.80
Cedrus deodara
Cedrus deodara 'Aurea', p.83
Chamaecyparis lawsoniana 'Ellwoodii'
Chamaecyparis lawsoniana 'Fletcheri'
Chamaecyparis lawsoniana 'Intertexta', p.76
Chamaecyparis lawsoniana 'Kilmacurragh'
Chamaecyparis lawsoniana 'Lanei', p.81
Chamaecyparis lawsoniana 'Pembury Blue', p.79
Chamaecyparis lawsoniana 'Wisselii'
Chamaecyparis nootkatensis 'Pendula'
Chamaecyparis obtusa 'Crippsii', p.83
Chamaecyparis obtusa 'Kosteri'
Chamaecyparis obtusa 'Nana'
Chamaecyparis obtusa 'Nana Aurea', p.85
Chamaecyparis obtusa 'Nana Gracilis'
Chamaecyparis obtusa 'Tetragona Aurea'
Chamaecyparis pisifera 'Boulevard'
Chamaecyparis pisifera 'Filifera Aurea', p.85
Chamaecyparis thyoides 'Andelyensis'
Cryptomeria japonica
Cryptomeria japonica 'Bandai-sugi'
Cryptomeria japonica 'Elegans Compacta', p.85
Cryptomeria japonica 'Vilmoriniana'
x *Cupressocyparis leylandii* 'Robinson's Gold'
Cupressus cashmeriana
Cupressus macrocarpa 'Goldcrest', p.83
Ginkgo biloba, p.77
Juniperus chinensis 'Aurea'
Juniperus chinensis 'Kaizuka'
Juniperus chinensis 'Obelisk', p.82
Juniperus chinensis 'Pyramidalis'
Juniperus communis 'Compressa'
Juniperus communis 'Hibernica', p.85
Juniperus communis 'Hornibrookii'
Juniperus horizontalis 'Plumosa'
Juniperus horizontalis 'Wiltonii'
Juniperus x *media* 'Blaauw'
Juniperus x *media* 'Pfitzeriana', p.85
Juniperus x *media* 'Plumosa Aurea', p.85
Juniperus procumbens 'Nana', p.84
Juniperus squamata 'Blue Carpet'
Juniperus squamata 'Blue Star', p.84
Juniperus squamata 'Holger', p.84
Juniperus virginiana 'Grey Owl', p.84
Larix decidua
Larix kaempferi
Metasequoia glyptostroboides, p.76
Microbiota decussata, p.84
Picea abies 'Little Gem'
Picea abies 'Nidiformis'
Picea breweriana, p.79
Picea glauca var. *albertiana* 'Conica', p.85
Picea glauca 'Echiniformis'
Picea mariana 'Nana', p.84
Picea omorika, p.77
Picea orientalis
Picea orientalis 'Aurea'
Picea pungens 'Hoopsii'
Picea pungens 'Koster', p.79
Picea smithiana
Pinus cembra, p.80
Pinus coulteri, p.76
Pinus heldreichii var. *leucodermis*, p.78
Pinus heldreichii var. *leucodermis* 'Schmidtii', p.85
Pinus jeffreyi, p.77
Pinus mugo 'Mops'
Pinus muricata, p.77
Pinus nigra

Pinus nigra

Pinus nigra var. *maritima*
Pinus parviflora, p.79
Pinus parviflora 'Adcock's Dwarf'
Pinus pinaster, p.77
Pinus pinea, p.83
Pinus ponderosa, p.77
Pinus radiata, p.78
Pinus strobus 'Radiata'
Pinus sylvestris
Pinus sylvestris 'Aurea', p.85
Pinus sylvestris 'Beuvronensis'
Pinus wallichiana, p.77
Podocarpus salignus, p.80
Pseudolarix amabilis, p.81
Pseudotsuga menziesii
Sciadopitys verticillata, p.80
Sequoia sempervirens
Sequoiadendron giganteum, p.76
Taxodium distichum, p.78
Taxus baccata
Taxus baccata 'Dovastoniana'
Taxus baccata 'Dovastonii Aurea', p.85
Taxus baccata 'Fastigiata'
Taxus baccata 'Repandens'
Taxus baccata 'Semperaurea'
Taxus x *media* 'Hicksii'
Thuja occidentalis 'Holmstrup'
Thuja occidentalis 'Lutea Nana'
Thuja occidentalis 'Rheingold'
Thuja occidentalis 'Smaragd'
Thuja orientalis 'Aurea Nana', p.85
Thuja plicata 'Atrovirens'
Thuja plicata 'Aurea'
Thuja plicata 'Stoneham Gold', p.85
Thujopsis dolabrata
Tsuga canadensis f. *pendula*
Tsuga heterophylla

Shrubs

Abelia 'Edward Goucher', p.131
Abelia floribunda
Abelia x *grandiflora*, p.91
Abelia x *grandiflora* 'Francis Mason'
Abutilon x *hybridum* 'Ashford Red'
Abutilon 'Kentish Belle', p.141
Abutilon megapotamicum
Acalypha hispida
Acer palmatum 'Bloodgood', p.113
Acer palmatum 'Butterfly', p.110
Acer palmatum 'Chitoseyama', p.136
Acer palmatum 'Osakazuki'
Acer palmatum 'Senkaki', p.94
Aesculus parviflora, p.89
Amelanchier lamarckii, p.86
Artemisia abrotanum, p.146
Aucuba japonica 'Crotonifolia', p.123
Azara microphylla, p.95
Berberis calliantha
Berberis darwinii, p.87
Berberis x *ottawensis* 'Superba'
Berberis 'Parkjuweel'
Berberis 'Rubrostilla', p.142
Berberis x *stenophylla*, p.105
Berberis thunbergii
Berberis thunbergii 'Atropurpurea Nana'
Berberis thunbergii 'Rose Glow'
Berberis verruculosa, p.104
Berberis wilsoniae
Brugmansia 'Grand Marnier', p.93
Brugmansia sanguinea, p.93
Brunfelsia pauciflora
Buddleja alternifolia, p.92
Buddleja asiatica
Buddleja davidii 'Black Knight'
Buddleja davidii 'Empire Blue'
Buddleja davidii 'Royal Red', p.91

Buddleja davidii 'Royal Red'

Buddleja fallowiana var. *alba*
Buddleja globosa, p.93
Buddleja 'Lochinch'
Buxus sempervirens
Buxus sempervirens 'Suffruticosa', p.147
Callistemon citrinus 'Splendens', p.113
Callistemon sieberi
Calluna vulgaris 'Allegro'
Calluna vulgaris 'Anthony Davis', p.148
Calluna vulgaris 'Beoley Gold', p.149
Calluna vulgaris 'County Wicklow', p.148
Calluna vulgaris 'Darkness', p.149
Calluna vulgaris 'Elsie Purnell', p.148
Calluna vulgaris 'Finale'
Calluna vulgaris 'Gold Haze', p.149
Calluna vulgaris 'J.H. Hamilton', p.148
Calluna vulgaris 'Joy Vanstone'
Calluna vulgaris 'Kinlochruel', p.148
Calluna vulgaris 'Mair's Variety'
Calluna vulgaris 'Mullion'
Calluna vulgaris 'Orange Queen'
Calluna vulgaris 'Robert Chapman', p.149
Calluna vulgaris 'Silver Queen', p.148
Calluna vulgaris 'Sir John Charrington'
Calluna vulgaris 'Sister Anne'
Calluna vulgaris 'Spring Cream', p.148
Calluna vulgaris 'Sunset'
Calluna vulgaris 'Tib', p.149
Calluna vulgaris 'White Lawn'
Calluna vulgaris 'Wickwar Flame'
Camellia 'Anticipation', p.99
Camellia 'Captain Rawes', p.99
Camellia 'Cornish Snow', p.98
Camellia 'Dr Clifford Parks', p.99
Camellia 'Inspiration', p.99
Camellia japonica 'Adolphe Audusson', p.99
Camellia japonica 'Alexander Hunter', p.99
Camellia japonica 'Berenice Boddy', p.98
Camellia japonica 'Donckelaeri'
Camellia japonica 'Elegans', p.99
Camellia japonica 'Gloire de Nantes', p.99
Camellia japonica 'Guilio Nuccio', p.99
Camellia japonica 'Jupiter', p.99
Camellia japonica 'Lavinia Maggi', p.98
Camellia japonica 'Magnoliiflora'
Camellia japonica 'Mrs D.W. Davis', p.98
Camellia japonica 'R.L. Wheeler', p.99
Camellia japonica 'Rubescens Major', p.99
Camellia 'Leonard Messel', p.99
Camellia sasanqua 'Narumigata', p.98
Camellia tsaii, p.98
Camellia x *williamsii* 'Brigadoon', p.99

Camellia x *williamsii* 'Brigadoon'

Camellia x *williamsii* 'Donation', p.98
Camellia x *williamsii* 'George Blandford', p.99
Camellia x *williamsii* 'J.C. Williams', p.98
Camellia x *williamsii* 'Mary Christian', p.99
Camellia x *williamsii* 'St Ewe', p.99
Camellia x *williamsii* 'Water Lily', p.99
Caragana arborescens 'Lorbergii'
Carpenteria californica, p.108
Caryopteris x *clandonensis* 'Heavenly Blue'
Catharanthus roseus, p.130
Ceanothus 'Autumnal Blue', p.118
Ceanothus 'Blue Mound'
Ceanothus 'Burkwoodii'
Ceanothus 'Cascade'
Ceanothus 'Delight'
Ceanothus 'Gloire de Versailles', p.138
Ceanothus 'Italian Skies'
Ceanothus 'Southmead'
Ceanothus thyrsiflorus var. *repens*, p.138
Ceanothus 'Trewithen Blue'
Ceratostigma willmottianum, p.143
Cestrum elegans, p.112

Cestrum 'Newellii'
Cestrum parqui
Chaenomeles speciosa 'Moerloosii', p.100
Chaenomeles x *superba* 'Crimson and Gold'
Chaenomeles x *superba* 'Knap Hill Scarlet'
Chaenomeles x *superba* 'Nicoline', p.126
Chaenomeles x *superba* 'Rowallane', p.125
Chamaedorea elegans, p.122
Chamaerops humilis, p.146
Chimonanthus praecox 'Luteus'
Choisya ternata, p.97
Choisya ternata 'Sundance'
Cistus x *aguilarii* 'Maculatus', p.129
Cistus x *corbariensis*, p.129
Cistus x *cyprius*, p.129
Cistus ladanifer, p.129
Cistus laurifolius
Cistus 'Peggy Sammons'
Cistus x *purpureus*
Cistus x *skanbergii*, p.132

Cistus x *skanbergii*

x *Citrofortunella microcarpa*, p.121
Clerodendrum bungei, p.118
Clethra barbinervis, p.108
Clethra delavayi, p.91
Colquhounia coccinea, p.119
Convolvulus cneorum, p.128
Cornus alba 'Elegantissima', p.108
Cornus alba 'Sibirica', p.120
Cornus alba 'Spaethii', p.116
Cornus alternifolia 'Argentea', p.63
Cornus mas
Cornus mas 'Variegata', p.89
Cornus stolonifera 'Flaviramea'
Coronilla valentina subsp. *glauca*, p.127
Correa backhousiana
Correa 'Mannii'
Corylopsis pauciflora, p.101
Corylopsis sinensis
Corylus avellana 'Contorta', p.95
Corylus colurna
Corylus maxima 'Purpurea', p.92
Cotinus coggygria
Cotinus coggygria 'Flame', p.94
Cotinus coggygria 'Royal Purple'
Cotinus obovatus
Cotoneaster adpressus
Cotoneaster 'Cornubia', p.94
Cotoneaster horizontalis, p.142
Cotoneaster lacteus, p.95
Cotoneaster microphyllus
Cotoneaster 'Rothschildianus'
Cotoneaster simonsii, p.118
Cotoneaster x *watereri* 'John Waterer'
Crinodendron hookerianum, p.112
Cuphea hyssopifolia, p.128
Cuphea ignea, p.142
Cycas revoluta, p.122
Cytisus battandieri, p.93
Cytisus demissus
Cytisus x *kewensis*
Cytisus x *praecox* 'Allgold', p.127
Cytisus purpureus
Cytisus 'Zeelandia'
Daboecia cantabrica 'Bicolor', p.148
Daboecia x *scotica* 'Jack Drake'
Daboecia x *scotica* 'William Buchanan', p.148
Danäe racemosa
Daphne cneorum 'Eximia'
Daphne collina
Daphne retusa, p.125
Daphne tangutica
Desfontainia spinosa, p.113
Deutzia x *elegantissima* 'Rosealind', p.132
Deutzia longifolia 'Veitchii', p.110

Disanthus cercidifolius
Dracaena deremensis 'Warneckii', p.96
Dracaena fragrans 'Massangeana'
Dracaena marginata
Dracaena sanderiana, p.121
Elaeagnus x *ebbingei* 'Gilt Edge'
Elaeagnus pungens 'Maculata', p.97
Enkianthus campanulatus, p.87
Enkianthus cernuus f. *rubens*, p.101
Enkianthus perulatus, p.97
Erica arborea 'Albert's Gold'
Erica arborea var. *alpina*, p.148
Erica australis
Erica australis 'Mr Robert'
Erica australis 'Riverslea'
Erica carnea 'Ann Sparkes', p.149
Erica carnea 'Foxhollow'
Erica carnea 'Loughrigg'
Erica carnea 'Myretoun Ruby'
Erica carnea 'Pink Spangles'
Erica carnea 'R.B. Cooke'
Erica carnea 'Springwood White', p.148
Erica carnea 'Vivellii', p.149
Erica carnea 'Westwood Yellow', p.149
Erica ciliaris 'Corfe Castle', p.149
Erica ciliaris 'David McClintock', p.148
Erica ciliaris 'Mrs C.H. Gill'
Erica cinerea 'C.D. Eason', p.149
Erica cinerea 'C.G. Best'
Erica cinerea 'Cevennes'
Erica cinerea 'Eden Valley', p.149
Erica cinerea 'Hookstone White', p.148
Erica cinerea 'P. S. Patrick'
Erica cinerea 'Pentreath'
Erica cinerea 'Pink Ice'
Erica cinerea 'Stephen Davis'
Erica cinerea 'Velvet Night'
Erica cinerea 'Windlebrooke', p.149
Erica x *darleyensis* 'Arthur Johnson'
Erica x *darleyensis* 'Ghost Hills', p.148
Erica x *darleyensis* 'J.W. Porter'
Erica x *darleyensis* 'White Perfection', p.148
Erica erigena 'Golden Lady', p.149
Erica erigena 'Irish Dusk'
Erica erigena 'W.T. Rackliff'
Erica lusitanica
Erica x *stuartii* 'Irish Lemon'
Erica terminalis
Erica tetralix 'Alba Mollis', p.148
Erica tetralix 'Con Underwood', p.149
Erica tetralix 'Pink Star', p.149
Erica vagans 'Birch Glow', p.149
Erica vagans 'Lyonesse', p.148
Erica vagans 'Mrs D.F. Maxwell'
Erica vagans 'Valerie Proudley', p.149
Erica x *veitchii* 'Gold Tips'
Erica watsonii 'Dawn', p.148
Erica x *williamsii* 'P.D. Williams', p.148
Eriobotrya japonica
Erythrina crista-galli, p.113
Escallonia 'Apple Blossom', p.111
Escallonia 'Edinensis'
Escallonia 'Iveyi', p.88
Escallonia 'Langleyensis', p.112
Escallonia rubra var. *macrantha* 'Crimson Spire'
Euonymus alatus, p.118
Euonymus alatus 'Compactus'
Euonymus europaeus 'Red Cascade', p.117
Euonymus fortunei 'Emerald 'n Gold', p.147
Euonymus fortunei 'Emerald Gaiety'
Euonymus fortunei 'Silver Queen', p.121
Euonymus japonicus 'Ovatus Aureus'
Euonymus planipes
Euphorbia characias subsp. *characias*, p.126
Euphorbia characias subsp. *wulfenii*, p.126
Euphorbia milii, p.125
Euryops pectinatus, p.140
Exochorda x *macrantha* 'The Bride', p.105
Fabiana imbricata f. *violacea*, p.115
x *Fatshedera lizei*, p.122
x *Fatshedera lizei* 'Variegata'
Fatsia japonica
Fatsia japonica 'Variegata', p.121
Felicia amelloides 'Santa Anita', p.138
Ficus deltoidea, p.122

Forsythia x *intermedia* 'Lynwood'
Forsythia suspensa, p.101
Fothergilla major, p.97
Fremontodendron 'California Glory', p.93
Fuchsia 'Annabel', p.134
Fuchsia boliviana 'Alba', p.135
Fuchsia 'Brutus'
Fuchsia 'Celia Smedley', p.135
Fuchsia 'Cloverdale Pearl'
Fuchsia 'Display'
Fuchsia 'Dollar Princess', p.134
Fuchsia 'Flash'
Fuchsia fulgens, p.135
Fuchsia 'Genii'
Fuchsia 'Golden Marinka', p.135
Fuchsia 'Jack Shahan', p.134
Fuchsia 'Joy Patmore'
Fuchsia 'La Campanella', p.135
Fuchsia 'Lady Thumb', p.134
Fuchsia 'Lena'
Fuchsia 'Leonora', p.134
Fuchsia 'Marin Glow'
Fuchsia 'Marinka'
Fuchsia 'Mrs Popple', p.135
Fuchsia 'Nellie Nuttall', p.134
Fuchsia 'Pacquesa'
Fuchsia 'Phyllis'
Fuchsia 'Riccartonii', p.134
Fuchsia 'Rose of Castile', p.135
Fuchsia 'Royal Velvet'
Fuchsia splendens
Fuchsia 'Swingtime', p.134
Fuchsia 'Thalia', p.135
Fuchsia 'Tom Thumb', p.134
Fuchsia 'Westminster Chimes'
Fuchsia 'White Ann', p.134
Garrya elliptica 'James Roof'
Gaultheria mucronata 'Mulberry Wine', p.144
Gaultheria mucronata 'Wintertime', p.143
Genista x *spachiana*
Genista tenera 'Golden Showers'
Genista tinctoria 'Royal Gold'
Grevillea juniperina f. *sulphurea*, p.139
Grevillea rosmarinifolia
Griselinia littoralis
x *Halimiocistus sahucii*, p.129
x *Halimiocistus wintonensis*
Halimium lasianthum
Halimium ocymoides
Halimium ocymoides 'Susan'
Hamamelis x *intermedia* 'Arnold Promise', p.96
Hamamelis x *intermedia* 'Diane', p.95
Hamamelis x *intermedia* 'Jelena'
Hamamelis mollis
Hamamelis mollis 'Pallida'
Hamamelis vernalis 'Sandra', p.94
Hebe albicans, p.131
Hebe 'Alicia Amherst'
Hebe x *franciscana* 'Blue Gem'
Hebe 'Great Orme', p.133
Hebe hulkeana
Hebe 'La Séduisante'
Hebe macrantha
Hebe 'Midsummer Beauty'
Hebe rakaiensis
Hedera helix 'Erecta', p.181
Helianthemum 'Jubilee'
Helichrysum italicum
Helichrysum milfordiae
Helichrysum petiolare, p.145
Helichrysum splendidum
Hibiscus schizopetalus
Hibiscus syriacus 'Blue Bird', p.115
Hibiscus syriacus 'Diana'
Hibiscus syriacus 'Red Heart', p.109
Hibiscus syriacus 'Woodbridge', p.111

Hibiscus syriacus 'Woodbridge'

Hippophäe rhamnoides, p.94
Hydrangea arborescens 'Annabelle', p.114
Hydrangea arborescens 'Grandiflora', p.114
Hydrangea aspera subsp. *aspera*, p.114
Hydrangea heteromalla 'Bretschneideri', p.114
Hydrangea involucrata 'Hortensis', p.114
Hydrangea macrophylla 'Altona', p.114
Hydrangea macrophylla 'Ami Pasquier'
Hydrangea macrophylla 'Blue Wave', p.114
Hydrangea macrophylla 'Générale Vicomtesse de Vibraye', p.114
Hydrangea macrophylla 'Lanarth White', p.114
Hydrangea macrophylla 'Mme E. Mouillère'
Hydrangea macrophylla 'Veitchii', p.114
Hydrangea paniculata 'Floribunda', p.114
Hydrangea paniculata 'Grandiflora'
Hydrangea paniculata 'Praecox', p.114
Hydrangea paniculata 'Unique', p.114
Hydrangea 'Preziosa'
Hydrangea quercifolia, p.114
Hydrangea sargentiana
Hydrangea serrata 'Bluebird', p.114
Hypericum 'Hidcote', p.140
Hypericum kouytchense, p.140
Hypericum x *moserianum*
Hypericum olympicum
Hypericum 'Rowallane'
Ilex x *altaclerensis* 'Golden King'
Ilex x *altaclerensis* 'Lawsoniana', p.72
Ilex aquifolium 'Ferox Argentea'
Ilex aquifolium 'Golden Milkboy'
Ilex aquifolium 'Golden Queen'
Ilex aquifolium 'Handsworth New Silver'
Ilex aquifolium 'J.C. Van Tol'
Ilex aquifolium 'Pyramidalis Fructu Luteo'
Ilex aquifolium 'Silver Milkmaid', p.72
Ilex aquifolium 'Silver Queen', p.73
Ilex crenata 'Convexa', p.72
Ilex x *meserveae* 'Blue Princess', p.72
Indigofera heterantha, p.111
Itea ilicifolia, p.116
Jasminum humile 'Revolutum'
Jasminum nudiflorum, p.121
Justicia brandegeana, p.136
Justicia rizzinii
Kalmia angustifolia f. *rubra*, p.135
Kalmia latifolia, p.111
Kalmia latifolia 'Ostbo Red'
Kerria japonica var. *japonica*, p.101
Kolkwitzia amabilis 'Pink Cloud', p.91
Lavandula 'Hidcote', p.137
Lavandula stoechas, p.137
Lavatera olbia 'Rosea', p.111
Leiophyllum buxifolium
Leptospermum scoparium 'Keatleyi'
Leptospermum scoparium 'Red Damask', p.101
Leucothöe fontanesiana
Ligustrum lucidum
Ligustrum lucidum 'Excelsum Superbum', p.97
Ligustrum ovalifolium 'Aureum'
Lindera obtusiloba
Lonicera nitida 'Baggesen's Gold', p.147
Lupinus arboreus, p.139
Mackaya bella
Magnolia 'Heaven Scent', p.49
Magnolia liliiflora 'Nigra', p.49
Magnolia x *loebneri* 'Leonard Messel', p.49
Magnolia x *loebneri* 'Merrill'
Magnolia x *soulangeana* 'Brozzonii'
Magnolia x *soulangeana* 'Etienne Soulange-Bodin', p.49
Magnolia x *soulangeana* 'Lennei'
Magnolia x *soulangeana* 'Rustica Rubra', p.49
Magnolia stellata, p.49
Magnolia stellata 'Waterlily', p.49
Mahonia japonica, p.121
Mahonia lomariifolia
Mahonia x *media* 'Buckland', p.95
Mahonia x *media* 'Charity', p.95
Melianthus major

419

Mimulus aurantiacus, p.141
Moltkia x *intermedia*
Myrtus communis, p.100
Myrtus communis var. *tarentina*
Myrtus luma, p.89
Nandina domestica
Olearia macrodonta
Olearia x *mollis* 'Zennorensis'
Osmanthus x *burkwoodii*, p.86
Osmanthus delavayi, p.86
Osmanthus heterophyllus 'Gulftide'
Ozothamnus ledifolius, p.130
Pachystachys lutea, p.127
Paeonia delavayi
Paeonia lutea var. *ludlowii*, p.201
Pandanus veitchii, p.145
Penstemon isophyllus, p.133
Penstemon newberryi
Perovskia atriplicifolia 'Blue Spire', p.138
Philadelphus 'Beauclerk', p.105
Philadelphus 'Belle Etoile', p.106
Philadelphus coronarius 'Aureus'
Philadelphus coronarius 'Variegatus', p.109
Philadelphus 'Manteau d'Hermine', p.128
Philadelphus 'Sybille'
Philadelphus 'Virginal'
Phlomis chrysophylla
Phlomis fruticosa, p.140
Photinia x *fraseri* 'Red Robin'
Photinia villosa
Physocarpus opulifolius 'Dart's Gold'
Pieris 'Forest Flame'
Pieris formosa var. *forrestii* 'Wakehurst', p.112
Pieris japonica 'Mountain Fire'
Pittosporum 'Garnettii', p.96
Pittosporum tenuifolium, p.97
Pittosporum tenuifolium 'Tom Thumb', p.147
Potentilla 'Abbotswood', p.128
Potentilla 'Beesii'
Potentilla 'Daydawn', p.131
Potentilla 'Elizabeth', p.139
Potentilla 'Goldfinger'
Potentilla 'Moonlight'
Potentilla 'Tangerine'
Prostanthera rotundifolia, p.115
Prunus x *blireana*
Prunus x *cistena*, p.124
Prunus glandulosa 'Alba Plena', p.123
Prunus glandulosa 'Sinensis'
Prunus laurocerasus
Prunus laurocerasus 'Otto Luyken', p.124
Prunus lusitanica

Prunus lusitanica

Prunus lusitanica subsp. *azorica*, p.96
Prunus tenella 'Fire Hill'
Ptelea trifoliata
Ptelea trifoliata 'Aurea', p.115
Pyracantha 'Orange Glow'
Pyracantha rogersiana
Pyracantha x *watereri*, p.106
Rhamnus alaternus 'Argenteovariegatus'
Rhaphiolepis umbellata, p.130
Rhapis excelsa, p.122
Rhododendron albrechtii
Rhododendron 'Azuma-kagami', p.102
Rhododendron 'Beauty of Littleworth', p.102
Rhododendron 'Blue Peter', p.103
Rhododendron calophytum, p.102
Rhododendron x *cilpinense*
Rhododendron cinnabarinum var. *xanthocodon*
Rhododendron 'Coccinea Speciosa'
Rhododendron 'Corneille', p.102

Rhododendron 'Curlew', p.104
Rhododendron 'Cynthia', p.103
Rhododendron davidsonianum, p.103
Rhododendron 'Dora Amateis'
Rhododendron 'Fabia', p.104
Rhododendron 'Fastuosum Flore Pleno'
Rhododendron fortunei subsp. *discolor*
Rhododendron fulvum, p.102
Rhododendron 'Gomer Waterer'
Rhododendron 'Hatsugiri', p.103
Rhododendron 'Hinodegiri', p.103
Rhododendron 'Hino-mayo', p.103
Rhododendron 'Homebush', p.103
Rhododendron 'Hotei'
Rhododendron 'Hydon Hunter'
Rhododendron impeditum
Rhododendron 'Irouha-yama', p.103
Rhododendron 'John Cairns', p.103
Rhododendron kaempferi, p.103
Rhododendron 'Kilimanjaro'
Rhododendron 'Kirin', p.102
Rhododendron kiusianum
Rhododendron 'Lady Alice Fitzwilliam'
Rhododendron 'Lady Clementine Mitford'
Rhododendron 'Lem's Cameo'
Rhododendron luteum, p.104
Rhododendron macabeanum, p.104
Rhododendron 'May Day', p.103
Rhododendron moupinense
Rhododendron 'Nancy Waterer'
Rhododendron 'Narcissiflorum', p.104
Rhododendron 'Norma'
Rhododendron occidentale, p.102
Rhododendron 'Palestrina', p.102
Rhododendron 'Percy Wiseman', p.102
Rhododendron 'Ptarmigan'
Rhododendron 'Purple Splendour'
Rhododendron 'Queen Elizabeth II', p.103
Rhododendron 'Queen of Hearts'
Rhododendron 'Romany Chai'
Rhododendron 'Roza Stevenson'
Rhododendron schlippenbachii, p.102
Rhododendron 'Seven Stars', p.102
Rhododendron 'Spek's Orange'
Rhododendron 'Strawberry Ice', p.102
Rhododendron 'Susan', p.103
Rhododendron 'Temple Belle'
Rhododendron 'Vuyk's Scarlet', p.103
Rhododendron williamsianum, p.102
Rhododendron 'Yellow Hammer', p.104
Rhus hirta
Rhus hirta 'Laciniata', p.94
Ribes sanguineum 'Brocklebankii', p.125
Ribes sanguineum 'Pulborough Scarlet', p.100
Ribes sanguineum 'Tydeman's White'
Ribes speciosum
Robinia hispida, p.111
Rosmarinus officinalis 'Miss Jessop's Upright'
Rosmarinus officinalis 'Prostratus'
Rosmarinus officinalis 'Severn Sea'
Rubus 'Benenden', p.106
Rubus cockburnianus
Rubus thibetanus, p.119
Ruta graveolens 'Jackman's Blue', p.147
Salix caprea 'Kilmarnock'
Salix elaeagnos
Salix hastata 'Wehrhahnii', p.123
Salix lanata, p.126
Salix magnifica
Salix 'Melanostachys'
Salvia fulgens, p.136
Salvia leucantha
Salvia officinalis 'Icterina', p.147

Salvia officinalis 'Icterina'

Salvia officinalis 'Purpurascens'
Sambucus nigra 'Aurea'
Sambucus nigra 'Guincho Purple'
Santolina chamaecyparissus
Santolina pinnata subsp. *neapolitana* 'Sulphurea', p.139
Santolina rosmarinifolia 'Primrose Gem'
Sarcococca confusa
Sarcococca hookeriana
Sarcococca hookeriana var. *digyna*, p.144
Sarcococca ruscifolia var. *chinensis*
Schefflera arboricola
Senecio monroi, p.141
Senecio 'Sunshine', p.140
Skimmia japonica 'Rubella', p.144
Skimmia japonica subsp. *reevesiana* 'Robert Fortune', p.144
Sorbaria aitchisonii
Sparmannia africana, p.88
Spartium junceum, p.117
Spiraea japonica 'Anthony Waterer', p.133
Spiraea japonica 'Goldflame', p.133
Spiraea nipponica 'Snowmound', p.109
Spiraea thunbergii
Spiraea x *vanhouttei*, p.124
Stachyurus praecox, p.120
Syringa 'Bellicent'
Syringa 'Charles Joly', p.90
Syringa 'Esther Staley', p.90
Syringa 'Katherine Havemeyer'
Syringa 'Mme Antoine Buchner', p.90
Syringa 'Mme Lemoine', p.90
Syringa meyeri 'Palibin', p.90
Syringa microphylla 'Superba', p.90
Syringa 'Mrs Edward Harding', p.90
Syringa x *persica*, p.90
Syringa 'Souvenir de Louis Späth'
Tetrapanax papyrifer, p.96
Teucrium fruticans 'Azureum'
Tibouchina urvilleana, p.92
Ulex europaeus 'Flore Pleno'
Vaccinium corymbosum, p.130
Vaccinium glaucoalbum, p.146
Vestia foetida
Viburnum x *bodnantense* 'Dawn', p.120
Viburnum x *bodnantense* 'Deben'
Viburnum x *burkwoodii* 'Anne Russell'
Viburnum x *burkwoodii* 'Park Farm Hybrid'
Viburnum x *carlcephalum*, p.87
Viburnum cinnamomifolium
Viburnum davidii, p.145
Viburnum farreri, p.117
Viburnum x *juddii*, p.124
Viburnum opulus 'Compactum', p.142
Viburnum opulus 'Xanthocarpum'
Viburnum plicatum 'Mariesii', p.86
Viburnum plicatum 'Pink Beauty', p.100
Viburnum 'Prägense', p.109
Viburnum sargentii 'Onondaga'
Viburnum tinus 'Eve Price'
Viburnum tinus 'Gwenllian'
Vinca major 'Variegata', p.145
Vinca minor 'Gertrude Jekyll'
Vinca minor 'La Grave'
Weigela florida 'Foliis Purpureis', p.133
Weigela florida 'Variegata', p.130
Weigela praecox 'Variegata'
Westringia fruticosa, p.128
Xanthoceras sorbifolium, p.88
Yucca filamentosa
Yucca flaccida 'Ivory', p.130
Yucca gloriosa, p.107

Roses

Rosa x *alba* 'Semiplena'
Rosa 'Albéric Barbier', p.162
Rosa 'Albertine', p.163
Rosa 'Alexander', p.158
Rosa 'Amber Queen', p.159
Rosa 'Anisley Dickson', p.157
Rosa 'Anna Ford', p.157
Rosa 'Arthur Bell'
Rosa banksiae 'Lutea', p.164
Rosa 'Belle de Crécy', p.154
Rosa 'Blessings', p.156

Rosa 'Bonica '82', p.152
Rosa 'Buff Beauty'
Rosa californica 'Plena'
Rosa 'Canary Bird', p.155
Rosa 'Capitaine John Ingram'
Rosa 'Cardinal de Richelieu', p.154
Rosa 'Céleste', p.151
Rosa x *centifolia* 'Muscosa'
Rosa 'Charles de Mills'
Rosa chinensis 'Mutabilis', p.153
Rosa 'City of London'
Rosa 'Climbing Lady Hillingdon'
Rosa 'Compassion'
Rosa 'Complicata', p.152
Rosa 'Comte de Chambord'
Rosa 'Constance Spry', p.152
Rosa 'Cristata'
Rosa 'Du Maître d'Ecole'
Rosa 'Dublin Bay', p.164
Rosa eglanteria, p.151
Rosa 'Fantin-Latour', p.151
Rosa 'Felicia', p.151
Rosa 'Félicité Parmentier'
Rosa 'Félicité Perpétue', p.162
Rosa filipes 'Kiftsgate', p.162
Rosa 'François Juranville'
Rosa 'Freedom', p.159
Rosa gallica var. *officinalis*
Rosa gallica 'Versicolor', p.153
Rosa 'Gentle Touch'
Rosa 'Geranium', p.153
Rosa glauca, p.152
Rosa 'Gloire de Dijon', p.162
Rosa 'Golden Showers', p.164
Rosa 'Golden Wings'
Rosa 'Graham Thomas', p.155
Rosa 'Handel', p.163
Rosa 'Henri Martin', p.153
Rosa 'Iceberg', p.155
Rosa 'Ingrid Bergman'
Rosa 'Ispahan'
Rosa 'Just Joey', p.160
Rosa 'Königin von Dänemark', p.152
Rosa 'Lovely Lady', p.156
Rosa 'Mme Alfred Carrière', p.162
Rosa 'Mme Grégoire Staechelin', p.163
Rosa 'Mme Hardy', p.150
Rosa 'Mme Isaac Pereire', p.153
Rosa 'Maigold', p.164
Rosa 'Margaret Merril', p.155
Rosa 'Marguerite Hilling', p.152
Rosa 'Mermaid', p.164
Rosa 'Morning Jewel'
Rosa 'Mountbatten', p.159
Rosa 'Nevada', p.151
Rosa 'New Dawn', p.163
Rosa 'Nozomi', p.155
Rosa 'Paul Shirville', p.157
Rosa 'Paul Transon'
Rosa 'Paul's Himalayan Musk Rambler'
Rosa 'Peace', p.158

Rosa 'Peace'

Rosa 'Peaudouce'
Rosa 'Penelope', p.150
Rosa 'Perle d'Or'
Rosa 'Pink Grootendorst', p.152
Rosa 'Président de Sèze'
Rosa primula, p.154
Rosa 'Queen Elizabeth', p.156
Rosa 'Remember Me', p.160
Rosa 'Roseraie de l'Haÿ', p.153
Rosa 'Rosy Cushion', p.152
Rosa 'Royal William', p.158
Rosa rugosa 'Alba'
Rosa 'Sander's White Rambler'
Rosa 'Sexy Rexy', p.156
Rosa 'Silver Jubilee', p.157
Rosa 'Snow Carpet'
Rosa 'Southampton', p.160
Rosa 'Stacey Sue', p.161

Rosa 'Sweet Dream'
Rosa 'Sweet Magic', p.160
Rosa 'The Fairy', p.155
Rosa 'The Times Rose', p.158
Rosa 'Troika', p.160

Rosa 'Troika'

Rosa 'Trumpeter', p.157
Rosa 'Tuscany Superb'
Rosa 'Veilchenblau', p.169
Rosa 'White Cockade'
Rosa 'William Lobb', p.154
Rosa 'Yesterday'
Rosa 'Yvonne Rabier'
Rosa 'Zéphirine Drouhin', p.163
Rosa 'Zigeunerknabe'

Climbers

Actinidia kolomikta, p.169
Agapetes serpens, p.166
Allemanda cathartica 'Hendersonii', p.176
Aristolochia littoralis, p.171
Asarina erubescens, p.169
Bougainvillea x *buttiana* 'Mrs Butt'
Bougainvillea glabra, p.174
Campsis x *tagliabuana* 'Mme Galen', p.177
Canarina canariensis, p.179
Celastrus orbiculatus
Cissus antarctica, p.180
Cissus rhombifolia, p.180
Clematis alpina
Clematis alpina 'Frances Rivis', p.173
Clematis 'Ascotiensis', p.173
Clematis 'Bees Jubilee'
Clematis 'Bill Mackenzie', p.173
Clematis 'Comtesse de Bouchaud'
Clematis 'Daniel Deronda'
Clematis 'Duchess of Albany', p.173
Clematis 'Edith'
Clematis 'Elsa Spath', p.173
Clematis 'Ernest Markham', p.173
Clematis 'Etoile Violette', p.173
Clematis 'General Sikorski'
Clematis 'Gipsy Queen'
Clematis 'H.F. Young', p.173
Clematis 'Henryi', p.172
Clematis 'Horn of Plenty'
Clematis 'Jackmanii', p.173
Clematis x *jouiniana* 'Praecox'
Clematis 'Lasurstern', p.173
Clematis macropetala 'Markham's Pink', p.172
Clematis 'Mme Julia Correvon', p.173
Clematis 'Marie Boisselot'
Clematis 'Miss Bateman'
Clematis montana 'Elizabeth'
Clematis montana var. *rubens*, p.172
Clematis montana 'Tetrarose', p.172
Clematis 'Mrs Cholmondeley'
Clematis 'Mrs George Jackman', p.172
Clematis 'Nelly Moser', p.172

Clematis 'Nelly Moser'

Clematis 'Niobe'
Clematis 'Perle d'Azur', p.173
Clematis 'Purpurea Plena Elegans', p.173
Clematis rehderiana, p.173
Clematis 'Richard Pennell', p.173

Clematis 'Star of India', p.173
Clematis 'The President', p.173
Clematis 'Vyvyan Pennell', p.173
Clerodendrum splendens
Clerodendrum thomsoniae, p.168
Clianthus puniceus, p.165
Clianthus puniceus f. *albus*, p.165
Cobaea scandens, p.174
Cobaea scandens f. *alba*
Codonopsis convolvulacea, p.174
Eccremocarpus scaber, p.177
Ficus pumila
Gelsemium sempervirens
Gynura aurantiaca 'Purple Passion'
Hardenbergia comptoniana
Hardenbergia violacea
Hedera canariensis 'Gloire de Marengo'
Hedera canariensis 'Ravensholst', p.181
Hedera colchica
Hedera colchica 'Dentata', p.181
Hedera colchica 'Dentata Variegata'
Hedera colchica 'Sulphur Heart', p.181
Hedera helix 'Angularis Aurea', p.181
Hedera helix 'Atropurpurea', p.181
Hedera helix 'Buttercup', p.181
Hedera helix 'Eva', p.181
Hedera helix 'Glacier', p.181
Hedera helix 'Ivalace', p.181
Hedera helix 'Manda's Crested', p.181
Hedera helix 'Pedata', p.181
Hibbertia scandens
Hoya carnosa, p.168
Hoya lanceolata subsp. *bella*, p.168
Humulus lupulus 'Aureus', p.166
Hydrangea petiolaris, p.168
Ipomoea horsfalliae, p.169
Ipomoea tricolor 'Heavenly Blue', p.175
Jasminum mesnyi, p.167
Jasminum officinale
Jasminum polyanthum, p.179
Lapageria rosea, p.170
Lathyrus latifolius, p.171
Lathyrus odoratus
Lonicera x *americana*, p.175
Lonicera japonica 'Halliana', p.175
Lonicera periclymenum 'Graham Thomas', p.175
Lonicera periclymenum 'Serotina'
Lonicera sempervirens, p.170
Lonicera x *tellmanniana*, p.177
Lonicera tragophylla
Mandevilla x *amoena* 'Alice du Pont', p.169
Monstera deliciosa, p.180
Parthenocissus henryana
Parthenocissus quinquefolia
Parthenocissus tricuspidata, p.178
Passiflora antioquiensis
Passiflora caerulea, p.174
Passiflora x *caerulea-racemosa*
Passiflora x *exoniensis*
Passiflora mollissima
Passiflora quadrangularis, p.175
Passiflora racemosa
Philodendron 'Burgundy'
Philodendron erubescens
Philodendron scandens, p.180
Pileostegia viburnoides, p.168
Plumbago auriculata, p.175
Polygonum baldschuanicum, p.177
Rhodochiton atrosanguineum, p.171
Rhoicissus capensis
Schizophragma integrifolium, p.168
Scindapsus pictus 'Argyraeus'
Senecio macroglossus 'Variegatus', p.179
Solanum crispum 'Glasnevin', p.174
Solanum jasminoides 'Album', p.167
Sollya heterophylla, p.166
Stephanotis floribunda, p.165
Streptosolen jamesonii, p.180
Syngonium podophyllum, p.180
Tecomaria capensis
Tetrastigma voinierianum, p.180
Thunbergia grandiflora
Thunbergia mysorensis, p.167
Trachelospermum asiaticum
Trachelospermum jasminoides, p.167
Tropaeolum speciosum, p.170
Tropaeolum tricolorum, p.165
Tropaeolum tuberosum 'Ken Aslet', p.176
Tweedia caerulea, p.175

Vitis 'Brant'
Vitis coignetiae, p.178
Vitis vinifera 'Purpurea'
Wisteria floribunda 'Alba', p.167
Wisteria floribunda 'Macrobotrys'
Wisteria sinensis, p.175
Wisteria sinensis 'Alba', p.167

Grasses, Bamboos, Rushes and Sedges

Carex elata 'Aurea', p.185
Chusquea culeou, p.184
Cortaderia selloana 'Sunningdale Silver'
Cyperus involucratus, p.184
Fargesia nitida
Fargesia spathacea
Hakonechloa macra 'Aureola', p.185
Helictotrichon sempervirens, p.183
Miscanthus sacchariflorus
Molinia caerulea subsp. *caerulea* 'Variegata'
Oplismenus hirtellus 'Variegatus', p.265
Phalaris arundinacea 'Picta', p.182
Phyllostachys aurea
Phyllostachys nigra
Phyllostachys nigra var. *henonis*, p.184
Phyllostachys viridiglaucescens, p.185
Pleioblastus variegatus, p.182
Pleioblastus viridistriatus, p.185
Pseudosasa japonica, p.184
Sasa palmata
Semiarundinaria fastuosa, p.184
Stenotaphrum secundatum 'Variegatum'
Stipa gigantea, p.183

Ferns

Adiantum pedatum, p.188
Adiantum venustum, p.189
Asplenium bulbiferum
Asplenium nidus, p.189
Asplenium scolopendrium, p.189
Athyrium filix-femina
Blechnum spicant
Blechnum tabulare
Cyrtomium falcatum, p.187
Davallia canariensis
Davallia mariesii
Dicksonia antarctica, p.186
Dryopteris erythrosora
Dryopteris filix-mas, p.186
Matteuccia struthiopteris, p.188
Onoclea sensibilis, p.188
Osmunda regalis, p.188
Pellaea rotundifolia
Phlebodium aureum, p.187
Platycerium bifurcatum, p.186
Polystichum aculeatum
Pteris cretica, p.187
Pteris cretica 'Albolineata'
Selaginella kraussiana
Selaginella kraussiana 'Variegata'
Selaginella martensii, p.187
Woodwardia radicans

Perennials

Acanthus spinosus, p.215
Achillea 'Coronation Gold', p.220
Achillea filipendulina 'Gold Plate', p.220
Achillea x *lewisii* 'King Edward'
Achillea 'Moonshine', p.255
Aconitum 'Bressingham Spire'
Aconitum x *cammarum* 'Bicolor'
Aconitum henryi 'Sparks' Variety'
Actaea alba, p.223
Actaea rubra
Aechmea fasciata, p.229

Aechmea 'Foster's Favorite', p.229
Aechmea nudicaulis
Aeschynanthus marmoratus
Aeschynanthus pulcher
Aeschynanthus speciosus, p.257
Agapanthus africanus
Agapanthus 'Loch Hope'
Ajuga reptans 'Atropurpurea', p.267
Alchemilla mollis, p.254
Amsonia orientalis, p.250
Ananas bracteatus 'Tricolor', p.229
Anchusa azurea 'Loddon Royalist', p.218
Anemone apennina
Anemone hupehensis var. *japonica* 'Prinz Heinrich'
Anemone hupehensis 'September Charm', p.195
Anemone x *hybrida* 'Honorine Jobert', p.195
Anemone nemorosa
Anemone nemorosa 'Allenii', p.234
Anemone nemorosa 'Robinsoniana', p.234
Anemone nemorosa 'Vestal'
Anemone ranunculoides, p.238
Anigozanthos manglesii, p.214
Anthemis punctata subsp. *cupaniana*, p.240
Anthericum liliago, p.240
Aphelandra squarrosa 'Louisae', p.220
Aquilegia canadensis
Aquilegia flabellata
Aquilegia vulgaris 'Nora Barlow', p.213
Artemisia absinthium 'Lambrook Silver'
Artemisia alba 'Canescens'
Artemisia frigida
Artemisia lactiflora, p.191
Artemisia pontica, p.254
Artemisia 'Powis Castle'
Aruncus dioicus, p.190
Asparagus densiflorus, p.230
Asparagus densiflorus 'Myersii', p.231
Aspidistra elatior
Aspidistra elatior 'Variegata', p.265
Aster amellus 'King George', p.226
Aster amellus 'Violet Queen'
Aster frikartii 'Mönch', p.226
Aster frikartii 'Wunder von Stäfa'
Aster lateriflorus var. *horizontalis*, p.226
Aster novae-angliae 'Alma Potschke', p.226

Aster novae-angliae 'Harrington's Pink'

Aster novae-angliae 'Harrington's Pink', p.226
Aster turbinellus, p.226
Astilbe chinensis 'Pumila'
Astilbe 'Fanal', p.246
Astilbe 'Ostrich Plume'
Astrantia major 'Sunningdale Variegated'
Astrantia maxima, p.241
Baptisia australis
Begonia 'Cocktail'
Begonia fuchsioides
Begonia masoniana, p.259
Begonia 'Merry Christmas', p.259
Bergenia cordifolia 'Purpurea', p.233
Bergenia 'Morgenröte'
Bergenia purpurascens
Bergenia x *schmidtii*
Bergenia 'Silberlicht', p.232
Billbergia x *windii*
Brunnera macrophylla
Calathea makoyana, p.268
Calathea zebrina, p.230
Campanula betulifolia
Campanula 'Burghaltii', p.250
Campanula carpatica
Campanula garganica
Campanula garganica 'W.H. Paine'

421

Campanula glomerata 'Superba', p.215
Campanula isophylla
Campanula lactiflora 'Loddon Anna'
Campanula lactiflora 'Prichard's Variety'
Campanula latiloba 'Percy Piper'
Campanula persicifolia 'Fleur de Neige'
Cardamine pratensis 'Flore Pleno'
Catananche caerulea 'Major', p.249
Chirita sinensis
Chlorophytum comosum 'Vittatum', p.265
Chrysanthemum frutescens 'Jamaica Primrose'
Chrysanthemum 'Madeleine', p.224
Chrysanthemum 'Pennine Flute', p.224
Chrysanthemum 'Purple Pennine Wine'
Chrysanthemum 'Salmon Margaret', p.225
Chrysanthemum uliginosum
Chrysanthemum 'Wendy', p.225
Cimicifuga racemosa
Cimicifuga simplex 'Elstead'
Clematis x durandii
Clematis heracleifolia 'Wyevale', p.173
Convallaria majalis, p.232
Corydalis bulbosa
Corydalis solida
Crambe cordifolia, p.190
Crepis incana
Cryptanthus bivittatus, p.229
Cryptanthus bromelioides 'Tricolor'
Cryptanthus 'Pink Starlight'
Cryptanthus zonatus
Ctenanthe lubbersiana
Ctenanthe oppenheimiana 'Tricolor', p.228
Cyananthus lobatus
Cyananthus microphyllus, p.331
Cyanotis kewensis
Cyanotis somaliensis, p.267
Cymbidium Strathdon 'Cooksbridge Noel', p.261
Cynara cardunculus, p.192
Dactylorhiza elata
Dactylorhiza foliosa
Delphinium 'Blue Dawn', p.194
Delphinium 'Blue Nile', p.194
Delphinium 'Bruce', p.194
Delphinium 'Emily Hawkins', p.194
Delphinium 'Fanfare', p.194
Delphinium 'Fenella'
Delphinium 'Gillian Dallas', p.194
Delphinium 'Loch Leven', p.194
Delphinium 'Lord Butler', p.194
Delphinium 'Mighty Atom', p.194
Delphinium 'Sandpiper', p.194
Delphinium 'Spindrift', p.194
Delphinium 'Sungleam', p.194
Dianthus 'Becky Robinson', p.244
Dianthus 'Doris', p.244
Dianthus 'Gran's Favourite', p.244
Dianthus 'Houndspool Ruby', p.245
Dianthus 'Monica Wyatt', p.245
Dianthus 'Valda Wyatt'
Diascia barberiae 'Ruby Field'
Diascia vigilis
Dicentra spectabilis, p.209
Dicentra spectabilis f. alba, p.203
Dictamnus albus, p.204
Dictamnus albus var. purpureus, p.207
Digitalis grandiflora
Digitalis x mertonensis
Echinops bannaticus 'Taplow Blue'
Edraianthus pumilio
Epimedium grandiflorum 'Rose Queen', p.233
Epimedium x perralchicum
Epimedium pinnatum subsp. colchicum
Epimedium x rubrum, p.233
Epimedium x youngianum 'Niveum', p.231
Erigeron 'Darkest of All'
Erigeron 'Foerster's Liebling'
Eryngium alpinum, p.217
Eryngium giganteum
Eryngium x oliverianum, p.217
Eryngium x tripartitum, p.217
Erysimum 'Bowles' Mauve'
Erysimum 'Bredon', p.298
Erysimum 'Harpur Crewe', p.298
Euphorbia amygdaloides var. robbiae, p.235
Euphorbia palustris
Euphorbia polychroma, p.239
Farfugium japonicum 'Aureomaculata'

Filipendula purpurea, p.209
Fittonia verschaffeltii, p.266
Fittonia verschaffeltii var. argyroneura, p.264
Gaillardia x grandiflora 'Dazzler'
Gaura lindheimeri
Gentiana asclepiadea, p.227
Geranium endressii
Geranium farreri
Geranium 'Johnson's Blue', p.249
Geranium macrorrhizum 'Ingwersen's Variety', p.233
Geranium maderense
Geranium x magnificum, p.249

Geranium x magnificum

Geranium x oxonianum 'Wargrave Pink', p.242
Geranium palmatum
Geranium pratense 'Plenum Violaceum'
Geranium psilostemon, p.209
Geranium renardii, p.241
Geranium x riversleaianum 'Russell Prichard'
Geranium sylvaticum 'Mayflower', p.215
Geranium x wallichianum 'Buxton's Blue', p.251
Geum 'Lady Stratheden', p.256
Geum 'Mrs Bradshaw'
Gillenia trifoliata, p.205
Glaucidium palmatum, p.234
Gunnera manicata, p.192
Guzmania lingulata, p.229
Guzmania lingulata var. minor, p.229
Guzmania monostachia, p.229
Guzmania sanguinea
Gypsophila paniculata 'Bristol Fairy', p.203
Haberlea rhodopensis
Hedychium gardnerianum, p.196
Helianthus atrorubens 'Monarch'
Helianthus 'Capenoch Star'
Helianthus 'Loddon Gold', p.196
Helleborus argutifolius, p.268
Helleborus foetidus, p.269
Helleborus lividus
Helleborus niger, p.265
Hemerocallis 'Cartwheels', p.221
Hemerocallis 'Corky', p.221
Hemerocallis 'Golden Chimes', p.221
Hemerocallis lilioasphodelus, p.221
Hemerocallis 'Marion Vaughn', p.221
Hemerocallis 'Stella d'Oro', p.221
Hepatica x media 'Ballardii'
Hepatica nobilis
Hepatica transsilvanica
Heuchera x brizoides 'Red Spangles', p.246
x Heucherella tiarelloides, p.242
Hosta crispula, p.252
Hosta fortunei 'Albopicta', p.253
Hosta fortunei 'Aureomarginata', p.253
Hosta 'Golden Tiara', p.253
Hosta 'Halcyon', p.252

Hosta 'Halcyon'

Hosta 'Honeybells', p.253
Hosta 'Krossa Regal', p.252
Hosta lancifolia
Hosta 'Love Pat', p.252
Hosta 'Royal Standard', p.253
Hosta 'Shade Fanfare', p.252
Hosta sieboldiana var. elegans, p.252
Hosta sieboldiana 'Frances Williams', p.253
Hosta sieboldii, p.252
Hosta 'Sum and Substance', p.253
Hosta undulata var. undulata
Hosta undulata var. univittata, p.252
Hosta undulata var. erromena
Hosta ventricosa, p.253
Hosta ventricosa 'Variegata', p.253
Hosta venusta, p.253
Hosta 'Wide Brim', p.253
Impatiens repens, p.257
Iresine lindenii
Iris 'Blue-eyed Brunette', p.199
Iris 'Early Light', p.199
Iris laevigata, p.199
Iris 'Stepping Out'
Iris unguicularis
Kirengeshoma palmata, p.227
Kniphofia caulescens, p.227
Kniphofia 'Little Maid', p.254
Kniphofia 'Royal Standard', p.222
Kniphofia 'Samuel's Sensation'
Kniphofia triangularis, p.227
Kohleria digitaliflora, p.208
Kohleria eriantha, p.214
Lamium maculatum 'White Nancy', p.231
Lathyrus vernus, p.234
Leucanthemum x superbum 'Aglaia'
Leucanthemum x superbum 'Wirral Supreme'
Ligularia dentata 'Desdemona'
Ligularia 'Gregynog Gold'
Liriope muscari, p.258
Lobelia cardinalis
Lobelia 'Queen Victoria', p.213
Lotus berthelotii, p.247
Lychnis chalcedonica, p.222
Lychnis viscaria 'Splendens Plena', p.246
Lysimachia clethroides, p.205
Lythrum salicaria 'Firecandle', p.209
Macleaya cordata
Macleaya microcarpa 'Coral Plume', p.191
Maranta leuconeura var. kerchoviana, p.268
Meconopsis betonicifolia, p.217
Meconopsis grandis
Meconopsis quintuplinervia, p.235
Meconopsis x sheldonii
Meconopsis x sheldonii 'Slieve Donard'
Mertensia pulmonarioides, p.235
Mimulus 'Andean Nymph', p.242
Mimulus lewisii
Mimulus 'Whitecroft Scarlet'
Monarda didyma 'Cambridge Scarlet', p.214
Monarda didyma 'Croftway Pink', p.207
Musa ornata, p.196
Musa uranoscopus
Neoregelia carolinae 'Tricolor', p.229
Nicotiana sylvestris, p.190
Oenothera fruticosa 'Fireworks', p.255
Oenothera fruticosa subsp. glauca
Ophiopogon planiscapus 'Nigrescens'
Origanum vulgare 'Aureum', p.254
Osteospermum 'Buttermilk', p.254
Osteospermum jucundum, p.243
Osteospermum 'Whirligig', p.241
Paeonia 'Bowl of Beauty', p.200
Paeonia cambessedesii, p.201
Paeonia 'Defender', p.201
Paeonia 'Duchesse de Nemours', p.200
Paeonia 'Félix Crousse'
Paeonia 'Festiva Maxima'
Paeonia 'Laura Dessert', p.201
Paeonia mlokosewitschii, p.201
Paeonia officinalis 'Rubra Plena', p.201
Paeonia peregrina 'Sunshine', p.201
Paeonia 'Sarah Bernhardt', p.200
Paeonia 'Whitleyi Major', p.200
Papaver orientale 'Mrs Perry'
Paradisea liliastrum
Parahebe perfoliata, p.250
Pelargonium 'Amethyst', p.210
Pelargonium 'Apple Blossom Rosebud', p.210
Pelargonium 'Bredon', p.211
Pelargonium crispum 'Variegatum', p.211

Pelargonium 'Dolly Varden', p.211
Pelargonium 'Flower of Spring', p.211
Pelargonium 'Francis Parrett', p.210
Pelargonium 'Irene', p.211
Pelargonium 'L'Elégante', p.211
Pelargonium 'Mabel Grey', p.211
Pelargonium 'Mr Henry Cox', p.210
Pelargonium 'Paton's Unique', p.210
Pelargonium 'The Boar', p.210
Pelargonium tomentosum, p.211
Pelargonium, Video Series
Pellaea rotundifolia
Pellionia repens
Peltiphyllum peltatum
Penstemon 'Andenken an Friedrich Hahn', p.212
Penstemon 'Apple Blossom', p.212
Penstemon 'Beech Park', p.212

Penstemon 'Beech Park'

Penstemon 'Chester Scarlet', p.212
Penstemon 'Evelyn', p.212
Penstemon hartwegii
Penstemon 'Pennington Gem', p.212
Penstemon 'Rubicundus', p.212
Penstemon 'Schoenholzeri', p.212
Penstemon 'White Bedder', p.212
Peperomia argyreia
Peperomia scandens
Phlomis russeliana, p.219
Phlox maculata 'Alpha'
Phlox maculata 'Omega', p.206
Phlox paniculata 'Amethyst', p.206
Phlox paniculata 'Brigadier', p.206
Phlox paniculata 'Bright Eyes'
Phlox paniculata 'Eventide', p.206
Phlox paniculata 'Fujiyama', p.206
Phlox paniculata 'Le Mahdi', p.206
Phlox paniculata 'Mother of Pearl', p.206
Phlox paniculata 'Prince of Orange', p.206
Phlox paniculata 'White Admiral', p.206
Phlox paniculata 'Windsor', p.206
Phormium cookianum
Phormium cookianum 'Tricolor'
Phormium tenax
Physalis alkekengi
Physostegia virginiana 'Summer Snow'
Physostegia virginiana 'Vivid', p.243
Pilea cadierei, p.264
Platycodon grandiflorus, p.248
Platycodon grandiflorus var. mariesii
Plectranthus oertendahlii
Polianthes tuberosa
Polygonujm amplexicaule 'Firetail'
Polygonum bistorta 'Superbum'
Potentilla 'Gibson's Scarlet'
Potentilla megalantha, p.256
Potentilla nepalensis 'Miss Willmott', p.246
Potentilla 'William Rollison'
Primula auricula
Primula bulleyana, p.237
Primula denticulata, p.237
Primula elatior, p.237
Primula flaccida, p.237
Primula florindae, p.237
Primula frondosa, p.236
Primula 'Garryarde Guinevere'
Primula 'Inverewe', p.237
Primula japonica
Primula x kewensis, p.237
Primula 'Linda Pope'
Primula marginata, p.237
Primula prolifera
Primula pulverulenta, p.236
Primula pulverulenta 'Bartley', p.236
Primula rosea, p.236

422

Primula sieboldii
Primula veris, p.237
Primula vialii, p.236
Primula vulgaris subsp. *sibthorpii*, p.236
Primula 'Wanda'
Pulmonaria angustifolia
Pulmonaria rubra
Pulmonaria 'Sissinghurst White', p.231
Ranunculus aconitifolius 'Flore Pleno', p.197
Rehmannia glutinosa
Rheum palmatum
Rodgersia aesculifolia, p.205
Rodgersia pinnata 'Superba'
Rodgersia podophylla, p.204
Romneya coulteri, p.190
Roscoea humeana
Rudbeckia fulgida var. *deamii*
Rudbeckia fulgida var. *sullivantii* 'Goldsturm', p.220
Rudbeckia 'Goldquelle', p.193
Russelia equisetiformis, p.214
Saintpaulia ionantha 'Bright Eyes', p.266
Saintpaulia ionantha 'Colorado', p.266
Saintpaulia ionantha 'Delft', p.266
Saintpaulia ionantha 'Garden News', p.266
Saintpaulia ionantha 'Rococo Pink', p.266
Salvia argentea, p.204
Salvia haematodes
Salvia involucrata
Salvia patens, p.251
Salvia uliginosa
Sansevieria trifasciata 'Golden Hahnii', p.269
Sansevieria trifasciata 'Hahnii'
Sansevieria trifasciata 'Laurentii', p.231
Sarmienta scandens
Sarracenia flava, p.254
Scabiosa caucasica 'Clive Greaves', p.249
Scabiosa caucasica 'Miss Willmott'
Schizostylis coccinea 'Major', p.258
Schizostylis coccinea 'Sunrise', p.258
Sedum morganianum
Sedum spectabile
Sedum spectabile 'Brilliant', p.258
Sisyrinchium douglasii
Smilacina racemosa, p.197
Solidago 'Goldenmosa', p.219
Sonerila margaritacea
Spathiphyllum 'Mauna Loa', p.263
Strelitzia reginae, p.231
Symphytum x *uplandicum* 'Variegatum', p.202
Tanacetum coccineum 'Brenda', p.202
Tanacetum coccineum 'Eileen May Robinson'
Tanacetum coccineum 'James Kelway'
Tetranema roseum, p.267
Thalictrum delavayi
Thalictrum delavayi 'Hewitt's Double'
Tillandsia lindenii, p.229

Tillandsia lindenii

Tradescantia cerinthoides 'Variegata'
Tradescantia fluminensis 'Variegata', p.264
Tradescantia 'J.C. Weguelin'
Tradescantia 'Osprey', p.240
Tradescantia pallida 'Purple Heart', p.267
Tradescantia sillamontana, p.267
Tradescantia spathacea 'Vittata'
Tradescantia zebrina, p.265
Tradescantia zebrina 'Purpusii'

Tradescantia zebrina 'Quadricolor'
Tricyrtis formosana, p.223
Trillium chloropetalum, p.232
Trillium erectum, p.233
Trillium grandiflorum, p.232
Trillium grandiflorum 'Flore Pleno'
Trollius 'Goldquelle'
Trollius 'Orange Princess'
Uvularia grandiflora, p.238
Veratrum nigrum, p.192
Verbascum 'Gainsborough', p.218
Verbascum 'Pink Domino'
Verbena rigida
Verbena 'Sissinghurst', p.246
Veronica gentianoides, p.251
Vriesea fosteriana
Vriesea psittacina
Vriesea splendens, p.229

Annuals and Biennials

Alonsoa warscewiczii, p.281
Antirrhinum majus 'Coronette', p.289
Bellis perennis 'Pomponette', p.279
Calceolaria integrifolia
Calceolaria 'Sunshine', p.291
Calendula officinalis (Fiesta Series) 'Gitana', p.291

Calendula officinalis (Fiesta Series) 'Gitana'

Callistephus, Milady Series
Digitalis purpurea f. *alba*, p.270
Eschscholzia caespitosa, p.288
Eschscholzia californica, p.290
Helichrysum bracteatum 'Bright Bikini'
Impatiens, Super Elfin Series
Ionopsidium acaule
Kochia scoparia f. *trichophylla*, p.287
Limnanthes douglasii, p.288
Lobelia erinus 'Cambridge Blue'
Lobelia erinus 'Colour Cascade'
Lobelia erinus 'Crystal Palace'
Myosotis 'Blue Ball', p.286
Nicotiana langsdorffii, p.286
Nicotiana x *sanderae*, Domino Series, p.271
Nicotiniana x *sanderae* 'Lime Green'
Nigella damascena 'Miss Jekyll', p.285
Omphalodes linifolia, p.270
Papaver rhoeas, Shirley Series (single), p.280
Papaver somniferum
Salvia farinacea 'Victoria'
Senecio cineraria 'Silver Dust', p.286
Tagetes patula, Boy-o-boy Series
Tagetes patula 'Honeycomb'
Trachelium caeruleum, p.282
Tropaeolum majus, Whirlybird Series
Verbascum bombyciferum
Viola, Joker Series, p.284
Viola x *wittrockiana*, Universal Series, p.291

Rock Plants

Acaena microphylla, p.337
Aethionema grandiflorum, p.300
Aethionema 'Warley Rose', p.324
Anagallis tenella 'Studland', p.315
Anchusa caespitosa, p.318
Andromeda polifolia 'Compacta', p.295
Androsace carnea subsp. *laggeri*, p.315
Androsace lanuginosa
Androsace sarmentosa
Androsace sempervivoides
Antennaria dioica 'Rosea', p.314

Anthyllis montana 'Rubra'
Arabis caucasica 'Plena'
Arabis ferdinandi-coburgii 'Variegata', p.336
Arenaria montana, p.322
Armeria juniperifolia, p.313
Armeria juniperifolia 'Bevans Variety'
Armeria maritima 'Vindictive'
Armeria pseudarmeria 'Bees Ruby'
Artemisia schmidtiana
Artemisia schmidtiana 'Nana', p.337
Asperula suberosa, p.324
Aster alpinus, p.329
Astilbe 'Perkeo', p.301
Aubrieta 'Dr Mule's'
Aurinia saxatilis, p.298
Aurinia saxatilis 'Citrina'
Ballota pseudodictamnus, p.309
Berberis x *stenophylla* 'Corallina Compacta', p.298
Campanula 'Birch Hybrid', p.330
Campanula cochleariifolia, p.331
Campanula 'G.F. Wilson', p.330
Campanula 'Joe Elliott'
Campanula portenschlagiana, p.330
Campanula raineri
Cassiope 'Edinburgh', p.294
Cassiope lycopodioides, p.311
Cassiope 'Muirhead', p.294
Ceratostigma plumbaginoides, p.308
Chiastophyllum oppositifolium, p.297
Convolvulus sabatius, p.304
Cornus canadensis, p.322
Corydalis solida 'George Baker', p.315
Cytisus ardoinii, p.334
Cytisus x *beanii*, p.297
Daphne arbuscula
Daphne petraea 'Grandiflora'
Dianthus alpinus, p.326
Dianthus armeria
Dianthus deltoides
Dianthus gratianopolitanus, p.325
Dianthus 'La Bourboule', p.326
Dianthus pavonius, p.323
Dianthus 'Pike's Pink', p.326
Diascia rigescens, p.301
Dionysia aretioides, p.319
Dodecatheon hendersonii, p.295
Dodecatheon meadia
Dodecatheon meadia f. *album*, p.295
Dodecatheon pulchellum
Draba longisiliqua, p.319
Dryas octopetala, p.323
Dryas x *suendermannii*
Erigeron karvinskianus, p.325
Erinacea anthyllis, p.296
Erinus alpinus, p.314
Erysimum 'Harpur Crewe', p.298
Euphorbia myrsinites, p.319
Euryops acraeus, p.306
Gaultheria cuneata, p.308
Gaultheria procumbens, p.335
Gentiana acaulis, p.318
Gentiana septemfida, p.308
Gentiana sino-ornata, p.335
Geranium cinereum 'Ballerina', p.327
Geranium cinereum var. *subcaulescens*, p.327
Geranium dalmaticum, p.325
Geranium sanguineum var. *striatum*, p.324
Geum montanum
Globularia cordifolia, p.331
Gypsophila repens
Gypsophila repens 'Dorothy Teacher', p.323
Hacquetia epipactis, p.319
Hebe cupressoides 'Boughton Dome', p.309
Hebe pinguifolia 'Pagei', p.299
Hedera helix 'Congesta'
Helianthemum 'Fire Dragon', p.302
Helianthemum 'Wisley Pink', p.301
Helianthemum 'Wisley Primrose', p.306
Helichrysum coralloides, p.309
Hypericum olympicum 'Sulphureum', p.306
Iberis sempervirens, p.294
Iberis sempervirens 'Snowflake'
Lewisia tweedyi, p.312
Linum arboreum
Linum 'Gemmell's Hybrid'

Lithodora diffusa 'Heavenly Blue', p.305
Lithodora oleifolia, p.304
Lysimachia nummularia 'Aurea', p.334
Oenothera macrocarpa
Omphalodes cappadocica, p.297
Origanum amanum
Origanum laevigatum, p.302
Origanum rotundifolium
Oxalis adenophylla, p.314
Oxalis enneaphylla
Oxalis 'Ione Hecker'
Pachysandra terminalis, p.336
Pachysandra terminalis 'Variegata'
Parahebe catarractae, p.304
Parochetus communis, p.332
Penstemon menziesii
Penstemon pinifolius, p.302
Penstemon rupicola
Penstemon scouleri
Petrorhagia saxifraga, p.323
Phlox adsurgens
Phlox 'Chatahoochee', p.304
Phlox divaricata
Phlox douglasii 'Boothman's Variety', p.328
Phlox douglasii 'Crackerjack', p.327
Phlox douglasii 'Red Admiral'
Phlox x *procumbens* 'Millstream'
Phlox stolonifera 'Blue Ridge'
Phyllodoce caerulea, p.296
Phyllodoce nipponica
Physoplexis comosa, p.328
Polygala chamaebuxus, p.333
Polygala chamaebuxus var. *grandiflora*, p.316
Polygonum affine 'Darjeeling Red'
Polygonum affine 'Donald Lowndes', p.324
Polygonum vacciniifolium, p.335
Potentilla x *tonguei*
Pulsatilla alpina subsp. *apiifolia*
Pulsatilla halleri, p.296
Pulsatilla vernalis, p.311
Pulsatilla vulgaris, p.296
Ramonda myconi, p.330
Ramonda nathaliae
Ranunculus calandrinioides, p.308
Ranunculus gramineus, p.307
Ranunculus montanus 'Molten Gold'
Rhodohypoxis baurii
Salix x *boydii*, p.309
Salix helvetica, p.297
Salix reticulata, p.318
Sanguinaria canadensis 'Plena'
Saponaria 'Bressingham'
Saponaria ocymoides, p.326
Saxifraga x *angelica* 'Cranbourne'
Saxifraga x *apiculata* 'Gregor Mendel', p.319
Saxifraga burseriana 'Gloria'
Saxifraga callosa
Saxifraga cochlearis 'Minor'
Saxifraga ferdinandi-coburgii
Saxifraga fortunei
Saxifraga grisebachii 'Wisley Variety'
Saxifraga x *irvingii* 'Jenkinsiae', p.313
Saxifraga 'Southside Seedling', p.300
Saxifraga stolonifera
Saxifraga stolonifera 'Tricolor'
Saxifraga 'Tumbling Waters', p.295
Saxifraga x *urbium*
Scadoxus multiflorus subsp. *katherinae*
Sedum cauticola
Sedum kamtschaticum
Sedum kamtschaticum 'Variegatum', p.339
Sedum morganianum
Sedum spathulifolium 'Cape Blanco', p.339
Sempervivum arachnoideum, p.337
Sempervivum ciliosum, p.337

Sempervivum ciliosum

Sempervivum 'Commander Hay'
Sempervivum tectorum, p.337
Silene schafta, p.327
Sorbus reducta, p.308
Thymus x *citriodorus* 'Aureus'
Thymus x *citriodorus* 'Silver Queen'
Tiarella cordifolia, p.295
Tiarella wherryi
Trillium rivale, p.313
Verbascum dumulosum, p.307
Verbascum 'Letitia', p.306
Veronica austriaca subsp. *teucrium* 'Royal Blue'
Veronica cinerea
Veronica prostrata, p.305
Viola cornuta, p.297
Viola 'Haslemere', p.329
Viola 'Huntercombe Purple', p.330
Viola 'Jackanapes', p.320
Zauschneria californica 'Glasnevin', p.303

Bulbs

Allium beesianum
Allium caeruleum, p.366
Allium carinatum subsp. *pulchellum*
Allium cernuum, p.364
Allium christophii, p.365
Allium cyaneum
Allium flavum
Allium karataviense, p.371
Allium moly
Allium oreophilum, p.374
Alstroemeria, Ligtu Hybrids, p.367
Anemone blanda
Anemone blanda 'Atrocaerulea', p.377
Anemone blanda 'Radar', p.374
Anemone blanda 'White Splendour', p.370
Arisaema candidissimum, p.379
Arum italicum 'Pictum', p.378
Cardiocrinum giganteum, p.342
Chionodoxa luciliae, p.377
Chionodoxa sardensis
Chionodoxa siehei, p.377
Clivia miniata, p.363
Colchicum agrippinum, p.383
Colchicum byzantinum, p.381
Colchicum speciosum
Colchicum speciosum 'Album', p.381

Arisaema candidissimum

Crinum x *powellii*, p.342
Crocosmia 'Lucifer', p.345
Crocosmia masonorum, p.345
Crocus angustifolius
Crocus banaticus, p.373
Crocus 'Blue Pearl', p.372
Crocus boryi, p.372
Crocus cartwrightianus, p.373
Crocus chrysanthus
Crocus 'Cream Beauty', p.373
Crocus 'E.A. Bowles', p.373
Crocus etruscus, p.373
Crocus flavus
Crocus 'Golden Yellow'
Crocus goulimyi, p.372
Crocus imperati
Crocus kotschyanus, p.372
Crocus 'Ladykiller', p.373

Crocus malyi, p.372
Crocus medius, p.372
Crocus pulchellus, p.372
Crocus sieberi 'Bowles' White', p.372
Crocus sieberi subsp. *sublimis* f. *tricolor*, p.372
Crocus 'Snow Bunting', p.372
Crocus speciosus, p.373
Crocus tommasinianus
Crocus tournefortii
Crocus 'Zwanenburg Bronze'
Cyclamen cilicium, p.382
Cyclamen hederifolium, p.382
Cyclamen libanoticum, p.371
Cyclamen mirabile, p.381
Cyclamen purpurascens, p.379
Cyrtanthus elatus
Dahlia 'Butterball', p.351
Dahlia 'Evening Mail'
Dahlia 'Hamari Bride'
Dahlia 'Hamari Gold'
Dahlia 'So Dainty', p.351
Dahlia 'Wootton Cupid', p.351
Eranthis hyemalis, p.385
Eranthis x *tubergenii* 'Guinea Gold'
Erythronium californicum 'White Beauty'
Erythronium dens-canis, p.374
Erythronium 'Pagoda', p.359
Erythronium revolutum
Erythronium tuolumnense
Fritillaria acmopetala
Fritillaria meleagris, p.357
Fritillaria michailovskyi
Fritillaria pallidiflora, p.359
Fritillaria pyrenaica, p.357
Galanthus 'Atkinsii', p.383
Galanthus elwesii, p.383
Galanthus ikariae, p.384
Galanthus nivalis
Galanthus nivalis 'Flore Pleno'
Galanthus plicatus subsp. *byzantinus*, p.384
Galanthus plicatus subsp. *plicatus*
Galtonia viridiflora, p.346
Gladiolus callianthus
Gladiolus x *colvillei* 'The Bride', p.343
Gladiolus communis subsp. *byzantinus*, p.344
Gladiolus 'Green Woodpecker', p.343
Gloriosa superba 'Rothschildiana', p.345
Hippeastrum 'Belinda'
Hippeastrum 'Orange Sovereign', p.369
Hyacinthus orientalis 'Blue Jacket', p.362
Hyacinthus orientalis 'City of Haarlem', p.362
Hyacinthus orientalis 'Delft Blue'
Hyacinthus orientalis 'L'Innocence'
Hyacinthus orientalis 'Ostara', p.362
Hyacinthus orientalis 'Pink Pearl', p.362
Hymenocallis x *macrostephana*, p.341
Ipheion uniflorum 'Froyle Mill', p.375
Ipheion uniflorum 'Wisley Blue'
Iris histrioides 'Major', p.199
Iris latifolia, p.198
Iris reticulata
Leucojum autumnale, p.381
Leucojum vernum, p.370
Lilium auratum var. *platyphyllum*, p.348
Lilium 'Black Dragon'
Lilium bulbiferum var. *croceum*, p.349
Lilium candidum, p.348
Lilium 'Casa Blanca', p.348
Lilium chalcedonicum, p.349
Lilium davidii
Lilium 'Enchantment', p.349
Lilium hansonii, p.349
Lilium henryi
Lilium 'Karen North', p.349
Lilium longiflorum, p.348
Lilium mackliniae, p.348
Lilium 'Marhan'
Lilium martagon var. *album*, p.348
Lilium monadelphum, p.349
Lilium pumilum
Lilium pyrenaicum, p.349
Lilium regale, p.348
Lilium 'Rosemary North', p.349
Lilium speciosum
Lilium superbum, p.349
Lilium x *testaceum*
Muscari armeniacum, p.377
Muscari aucheri, p.377

Muscari azureum
Narcissus 'Actaea', p.361
Narcissus 'Arctic Gold'
Narcissus 'Charity May', p.362
Narcissus cyclamineus, p.362
Narcissus 'Dove Wings', p.361
Narcissus 'Empress of Ireland', p.360
Narcissus 'February Gold', p.362
Narcissus 'Ice Follies', p.360
Narcissus 'Jenny'
Narcissus 'Kingscourt', p.362
Narcissus 'Merlin', p.361
Narcissus x *odorus* 'Rugulosus'
Narcissus 'Passionale', p.360
Narcissus poeticus var. *recurvus*, p.361
Narcissus 'St Keverne'
Narcissus 'Spellbinder'
Narcissus 'Suzy', p.362
Narcissus 'Sweetness', p.362
Narcissus 'Tête-à-Tête', p.362
Nerine bowdenii, p.368
Ornithogalum nutans
Roscoea cautleoides, p.380
Roscoea humeana, p.380
Scilla bifolia
Scilla mischtschenkoana
Sprekelia formosissima, p.357
Tecophilaea cyanocrocus, p.377
Tecophilaea cyanocrocus var. *leichtlinii*, p.376
Tulipa 'Ancilla'
Tulipa 'Apeldoorn's Elite', p.355
Tulipa aucheriana
Tulipa 'Ballade', p.354
Tulipa batalinii, p.356
Tulipa clusiana var. *chrysantha*, p.356
Tulipa 'Keizerskroon', p.355
Tulipa 'Kingsblood', p.355
Tulipa linifolia, p.355
Tulipa 'Monte Carlo'
Tulipa 'Oranje Nassau', p.355
Tulipa 'Plaisir', p.355
Tulipa praestans 'Fusilier'
Tulipa 'Prinses Irene', p.356
Tulipa 'Red Riding Hood', p.355
Tulipa sprengeri, p.355
Tulipa 'Spring Green', p.354
Tulipa tarda, p.356
Tulipa 'Toronto'
Tulipa turkestanica, p.354
Tulipa urumiensis, p.356
Veltheimia bracteata, p.369
Veltheimia capensis
Zantedeschia aethiopica 'Green Goddess', p.341
Zantedeschia elliottiana, p.347
Zantedeschia rehmannii

Water Plants

Butomus umbellatus, p.388
Caltha palustris, p.391
Caltha palustris 'Flore Pleno', p.391
Colocasia esculenta, p.390
Lysichiton americanum, p.391
Lysichiton camtschatcensis, p.386
Nymphaea 'Escarboucle', p.390

Nymphaea 'Escarboucle'

Nymphaea 'Gladstoniana', p.390
Nymphaea 'Gonnère'
Nymphaea 'James Brydon', p.390
Nymphaea Marliacea Group 'Chromatella', p.390
Pontederia cordata, p.388

Cacti and Succulents

Aeonium arboreum
Aeonium arboreum 'Schwarzkopf', p.404
Aeonium haworthii, p.402
Aeonium tabuliforme, p.414
Aichryson x *domesticum* 'Variegata', p.413
Aloe barbadensis, p.402
Aloe variegata, p.400
Astrophytum myriostigma, p.402

Astrophytum myriostigma

Beschorneria yuccoides, p.401
Ceropegia linearis subsp. *woodii*, p.398
Cleistocactus strausii, p.395
Crassula ovata, p.393
Crassula perfoliata var. *falcata*, p.400
Echeveria agavoides, p.410
Echeveria derenbergii
Echeveria elegans, p.407
Echeveria harmsii
Echeveria pulvinata, p.400
Echeveria setosa
Echinocereus reichenbachii
Echinopsis aurea
Echinopsis cinnabarina
Ferocactus setispinus, p.403
Gymnocalycium andreae, p.413
Gymnocalycium quehlianum
Hatiora salicornioides, p.404
Kalanchoe pumila
Kalanchoe 'Tessa', p.400
Kalanchoe tomentosa, p.403
Kalanchoe 'Wendy', p.399
Lobivia aurea
Lobivia cinnabarina
Mammillaria bocasana, p.407
Mammillaria candida
Mammillaria elongata, p.406
Mammillaria geminispina, p.401
Mammillaria hahniana, p.401
Mammillaria plumosa, p.405
Mammillaria prolifera
Mammillaria zeilmanniana, p.409
Nopalxochia phyllanthoides 'Deutsche Kaiserin', p.398
Orbea variegata, p.412
Parodia apricus
Parodia mammulosus, p.413
Parodia scopa
Pleiospilos bolusii, p.415
Rebutia arenacea, p.416
Rebutia aureiflora, p.416
Rebutia muscula, p.416
Rebutia violaciflora, p.408
Rhipsalidopsis gaertneri, p.411
Rhipsalidopsis rosea, p.408
Stapelia gigantea, p.412
Thelocactus bicolor, p.408

Plants that have been given the Award are identified by the symbol ♀ in the Plant Catalogue and ♀ in the Planter's Guide and Plant Dictionary, enabling the reader to locate them at a glance. The list of AGM plants is periodically reviewed by the Royal Horticultural Society to ensure that it remains as relevant and up-to-date as possible. New plants are added if appropriate and any plant that no longer measures up to the required standards or has dropped out of circulation may be removed.

The PLANT DICTIONARY

A complete guide to the characteristics and
cultivation of over 8000 plants, suitable for
growing in temperate gardens worldwide.

A

ABELIA (Caprifoliaceae)
Genus of deciduous, semi-evergreen or evergreen shrubs, grown for their foliage and freely borne flowers. Fully to half hardy, but in cold areas does best against a south- or west-facing wall. Requires a sheltered, sunny position and fertile, well-drained soil. Remove dead wood in late spring and prune out older branches after flowering to restrict growth, if required. Propagate by softwood cuttings in summer.
♥ *A.* **'Edward Goucher'** illus. p.131.
♥ *A. floribunda.* Evergreen, arching shrub. H 3m (10ft), S 4m (12ft). Half hardy. Has oval, glossy, dark green leaves and, in early summer, drooping, tubular, bright red flowers.
♥ *A.* x *grandiflora* illus. p.91.
♥ **'Francis Mason'** is a vigorous, semi-evergreen, arching shrub. H 2m (6ft), S 3m (10ft). Frost hardy. Has coppery-yellow, young shoots and oval, yellowish-green leaves, darker in centres. Bears a profusion of fragrant, bell-shaped, white flowers, tinged with pink, from mid-summer to mid-autumn.
A. schumannii illus. p.132.
A. triflora illus. p.89.

ABELIOPHYLLUM (Oleaceae)
Genus of one species of deciduous shrub, grown for its winter flowers. Fully hardy, but in cold areas grow against a south- or west-facing wall. Requires plenty of sun and fertile, well-drained soil. Thin out excess older shoots after flowering each year to encourage vigorous, young growth. Propagate by softwood cuttings in summer.
A. distichum. Deciduous, open shrub. H and S 1.2m (4ft). In late winter has fragrant, star-shaped, white flowers, tinged with pink, on bare stems; flowers may be damaged by hard frosts. Leaves are oval and dark green.

ABIES (Pinaceae)
Silver fir
Genus of tall conifers with whorled branches. Spirally arranged leaves are needle-like, flattened, usually soft and often have silvery bands beneath. Bears erect cones that ripen in their first autumn to release seeds and scales. See also CONIFERS.
A. alba. Fast-growing, conical conifer. H 15–25m (50–80ft), S 5–8m (15–25ft). Fully hardy. Has silvery-grey bark and dull green leaves, silvery beneath. Cylindrical cones, 10–15cm (4–6in) long, ripen to red-brown.
A. amabilis (Pacific fir). Conical conifer. H 15m (50ft), S 4–5m (12–15ft). Fully hardy. Dense, notched, square-tipped, glossy, dark green leaves, banded with white beneath, are borne on hairy, grey shoots. Has oblong, violet-blue cones, 9–15cm (3½–6in) long. **'Spreading Star'**, H 50cm (20in), S 4–5m (12–15ft), is a procumbent form suitable for ground cover.
A. balsamea (Balsam fir).
♥ f. *hudsonia* is a dense, dwarf conifer of flattened to globose habit. H and S 60cm–1m (2–3ft). Fully hardy. Has smooth, grey bark and grey-green leaves that are semi-spirally arranged. **'Nana'** (illus. p.84) is another dwarf form that makes a dense, globose mound with spirally arranged leaves.
A. cephalonica (Greek fir). Upright conifer with a conical crown; old trees have massive, spreading, erect branches. H 20–30m (70–100ft), S 5–10m (15–30ft). Fully hardy. Sharp, stiff, glossy, deep green leaves are whitish-green beneath. Cylindrical, tapered cones, 10–15cm (4–6in) long, are brown when ripe. **'Meyer's Dwarf'** (syn. *A.c.* 'Nana'; illus. p.84), H 50cm (20in), S 1.5m (5ft), has short leaves and forms a spreading, flat-topped mound.
♥ *A. concolor* (White fir). Upright conifer. H 15–30m (50–100ft), S 5–8m (15–25ft). Fully hardy. Has widely spreading, blue-green or grey leaves and cylindrical, green or pale blue cones, 8–12cm (3–5in) long. **'Argentea'** (syn. *A.c.* 'Candicans'; illus. p.75). ♥ **'Compacta'** (syn. *A.c.* 'Glauca Compacta'; illus. p.84), H to 2m (6ft), S 2–3m (6–10ft), is a cultivar with steel-blue foliage.
A. delavayi (Delavay fir). Upright conifer with tiered, spreading branches. H 10–15m (30–50ft), S 4–6m (12–20ft). Fully hardy. Has maroon shoots and curved, bright deep green leaves, spirally arranged, with vivid silver bands beneath and rolled margins. Narrowly cylindrical cones, 6–15cm (2½–6in) long, are violet-blue.
♥ *A. grandis* illus. p.78.
A. homolepis (Nikko fir). Conifer that is conical when young, later columnar. H 15m (50ft), S 6m (20ft). Fully hardy. Pink-grey bark peels in fine flakes. Has pale green leaves, silver beneath, and cylindrical, violet-blue cones, 8–12cm (3–5in) long. Tolerates urban conditions.
A. koreana illus. p.83.
A. lasiocarpa (Subalpine fir). Narrowly conical conifer. H 10–15m (30–50ft), S 3–4m (10–12ft). Fully hardy. Has grey or blue-green leaves and cylindrical, violet-blue cones, 6–10cm (2½–4in) long.
var. *arizonica* **'Compacta'** (illus. p.84), H 4–5m (12–15ft), S 1.5–2m (5–6ft), is a slow-growing, ovoid to conical tree with corky bark and blue foliage. **'Roger Watson'** (illus. p.84), H and S 75cm (2½ft), is dwarf and conical, with silvery-grey leaves.
♥ *A. nordmanniana* (Caucasian fir). Columnar, dense conifer. H 15–25m (50–80ft), S 5m (15ft). Fully hardy. Luxuriant foliage is rich green. Cylindrical cones, 10–15cm (4–6in) long, are green-brown, ripening to brown. ♥ **'Golden Spreader'** (illus. p.85), H and S 1m (3ft), is a dwarf form with a spreading habit and bright golden-yellow leaves.
♥ *A. procera* (Noble fir). Narrowly conical conifer. H 15m (50ft), S 5m (15ft). Fully hardy. Has attractive, smooth, silvery-grey bark and grey-green or bright blue-grey leaves. Produces stoutly cylindrical, green cones, 15–25cm (6–10in) long, that ripen to brown. **'Glauca'** illus. p.79.
♥ *A. veitchii* illus. p.76.

ABUTILON (Malvaceae)
Genus of evergreen, semi-evergreen or deciduous shrubs, perennials and annuals, grown for their flowers and foliage. Frost hardy to frost tender, min. 5–7°C (41–5°F). Needs full sun or partial shade and fertile, well-drained soil. Water containerized specimens freely when in full growth, less at other times. In the growing season, young plants may need tip pruning to promote bushy growth. Mature specimens may have previous season's stems cut back hard annually in early spring. Tie lax-growing species to a support if necessary. Propagate by seed in spring or by softwood, greenwood or semi-ripe cuttings in summer. Whitefly and red spider mite may be troublesome.
A. globosum of gardens. See *A.* x *hybridum*.
A. x *hybridum*, syn. *A. globosum* of gardens. ♥ **'Ashford Red'** is a strong-growing, evergreen, rounded shrub. H and S 2–3m (6–10ft). Half hardy. Has maple- to heart-shaped, serrated, rich green leaves. Pendent, bell-shaped, crimson flowers are carried from spring to autumn. **'Golden Fleece'** has yellow flowers.
♥ *A.* **'Kentish Belle'** illus. p.141.
♥ *A. megapotamicum*. Evergreen shrub with long, slender branches normally trained against a wall. H and S to 3m (10ft). Half hardy. Pendent, bell-shaped, yellow-and-red flowers appear from late spring to autumn. Leaves are oval, with heart-shaped bases, and dark green.
A. pictum, syn. *A. striatum* of gardens. **'Thompsonii'** illus. p.117.
A. striatum of gardens. See *A. pictum*.
A. x *suntense.* Fast-growing, deciduous, upright, arching shrub. H 5m (15ft), S 3m (10ft). Frost hardy. Has oval, lobed, toothed, dark green leaves. Produces an abundance of large, bowl-shaped, pale to deep purple, occasionally white, flowers from late spring to early summer. **'Violetta'** illus. p.115.
A. vitifolium. Fast-growing, deciduous, upright shrub. H 4m (12ft), S 2.5m (8ft). Frost hardy. Masses of large, bowl-shaped, purplish-blue flowers are produced in late spring and early summer. Has oval, lobed, sharply toothed, grey-green leaves. **'Album'** illus. p.89.

ACACIA (Leguminosae)
Mimosa
Genus of evergreen, semi-evergreen or deciduous trees and shrubs, grown for their tiny flowers, composed of massed stamens, and for their foliage. Many species have phyllodes instead of true leaves. Frost hardy to frost tender, min. 5–7°C (41–5°F). Requires full sun and well-drained soil. Propagate by seed in spring. Red spider mite and mealy bug may be problematic.
♥ *A. baileyana* illus. p.71.
♥ *A. dealbata* illus. p.57.
A. juniperina. See *A. ulicifolia.*
A. longifolia (Sydney golden wattle). Evergreen, spreading tree. H and S 6m (20ft). Frost hardy. Has narrowly oblong, dark green phyllodes. Bears cylindrical clusters of golden-yellow flowers in early spring.
A. neriifolia. Evergreen, bushy shrub or tree. H and S 3m (10ft). Half hardy. Has narrowly lance-shaped, usually grey-green phyllodes. Produces dense, globular heads of bright yellow flowers in spring.
A. podalyriifolia illus. p.105.
A. pravissima illus. p.71.
A. pulchella illus. p.127.
A. ulicifolia, syn. *A. juniperina.* Evergreen, bushy shrub. H 1m (3ft), S 1.5m (5ft). Frost hardy. Has very narrow, cylindrical, spine-like, rich green phyllodes and, in mid-spring, globular clusters of pale yellow flowers.
A. verticillata (Prickly Moses). Evergreen, spreading tree or bushy shrub. H and S 9m (28ft). Half hardy. Has needle-like, dark green phyllodes and, in spring, dense, bottlebrush-like spikes of bright yellow flowers.

ACAENA (Rosaceae)
Genus of mainly summer-flowering sub-shrubs and perennials, evergreen in all but the severest winters, grown for their leaves and coloured burs and as ground cover. Has tight, rounded heads of small flowers. Is good for a rock garden but some species may be invasive. Fully to frost hardy. Needs sun or partial shade and well-drained soil. Propagate by division in early spring or by seed in autumn.
A. anserinifolia. Vigorous, evergreen, prostrate sub-shrub. H 10cm (4in), S 75cm (30in) or more. Fully hardy. Has brown-green leaves, divided into 9–13 oval, toothed leaflets. In summer, red-spined, brownish burs develop from spherical heads of greenish-brown flowers.
A. **'Blue Haze'**. Vigorous, evergreen, prostrate perennial. H 10cm (4in), S 75cm (30in) or more. Fully hardy. Leaves are divided into 9–15 oval, toothed, steel-blue leaflets. Produces spherical, brownish-red flower heads that develop in autumn to dark

red burs with pinkish-red spines.
A. buchananii. Vigorous, evergreen, prostrate perennial. H 2cm (³/₄in), S 75cm (30in) or more. Fully hardy. Bears glaucous leaves, composed of 11–17 oval, toothed leaflets. Globose, green flower heads are borne in summer and develop into spiny, yellow-green burs.
A. caesiiglauca illus. p.338.
♀ *A. microphylla* illus. p.337.

ACALYPHA (Euphorbiaceae)
Genus of evergreen shrubs and perennials, grown for their flowers and foliage. Frost tender, min. 10–13°C (50–55°F), but best at min. 16°C (61°F). Needs partial shade and humus-rich, well-drained soil. Water containerized plants freely when in full growth, much less at other times and in low temperatures. Stem tips may be removed in growing season to promote branching of young plants. Propagate by softwood, greenwood or semi-ripe cuttings in summer. Red spider mite, whitefly and mealy bug may be troublesome.
♀ *A. hispida* (Red-hot cat's tail). Evergreen, upright, soft-stemmed shrub. H 2m (6ft) or more, S 1–2m (3–6ft). Has oval, toothed, lustrous, deep green leaves. Tiny, crimson flowers hang in long, dense, catkin-like spikes, intermittently year-round. May be grown as a short-lived cordon.
A. wilkesiana illus. p.113.

ACANTHOLIMON (Plumbaginaceae)
Genus of evergreen perennials, grown for their flowers and tight cushions of spiny leaves. Is suitable for rock gardens and walls. Fully hardy. Prefers sun and well-drained soil. Dislikes damp winters. Seed is rarely set in cultivation. Propagate by softwood cuttings in late spring.
A. glumaceum illus. p.325.
A. venustum. Evergreen, cushion-forming perennial. H and S 10cm (4in). Small spikes of star-shaped, pink flowers, on 3cm (1¹/₄in) stems, appear from late spring to early summer amid rosetted, spear-shaped, spiny, blue-green leaves that are edged with silver. Needs a very hot, well-drained site. Makes an excellent alpine house plant.

ACANTHOPANAX. See ELEUTHEROCOCCUS.
A. ricinifolium. See *Kalopanax septemlobus.*

ACANTHUS (Acanthaceae)
Bear's breeches
Genus of perennials, some of which are semi-evergreen, grown for their large, deeply cut leaves and their spikes of flowers. Fully hardy. Prefers full sun, warm conditions and well-drained soil, but will tolerate shade. Protect crowns in first winter after planting. Long, thong-like roots make plants difficult to eradicate if wrongly placed. Propagate by seed or division in early autumn or spring or by root cuttings in winter.
A. balcanicus, syn. *A. hungaricus*, *A. longifolius*, illus. p.215.
A. dioscoridis. Upright, architectural perennial. H to 1m (3ft), S 45cm (18in). Has oval, deeply cut, rigid, basal leaves and hairy stems. Bears dense spikes of small, funnel-shaped, purple-and-white flowers in summer.
A. hungaricus. See *A. balcanicus.*
A. longifolius. See *A. balcanicus.*
A. mollis. Semi-evergreen, stately, upright perennial. H 1.2m (4ft), S 45cm (18in). Has long, oval, deeply cut, bright green leaves and, in summer, produces many spikes of funnel-shaped, mauve-and-white flowers.
♀ *A. spinosus* illus. p.215.

Acca sellowiana. See *Feijoa sellowiana.*

ACER (Aceraceae)
Maple
Genus of deciduous or evergreen trees and shrubs, grown for their foliage, which often colours brilliantly in autumn and, in some cases, for their ornamental bark or stems. Small, but often attractive flowers are followed by 2-winged fruits. Fully to frost hardy. Requires sun or semi-shade and fertile, well-drained soil. Many acers produce their best autumn colour on neutral to acid soil. Propagate species by seed as soon as ripe or in autumn; cultivars by various grafting methods in late winter or early spring, or by budding in summer. Leaf-eating caterpillars or aphids sometimes infest plants, and maple tar spot may affect *A. platanoides* and *A. pseudoplatanus.*
A. buergerianum (Trident maple). Deciduous, spreading tree. H 10m (30ft) or more, S 8m (25ft). Fully hardy. Has 3-lobed, glossy, dark green leaves, usually providing a long-lasting display of red, orange and purple in autumn.
♀ *A. capillipes* illus. p.56.
A. cappadocicum (Cappadocian maple). Deciduous, spreading tree. H 20m (70ft), S 15m (50ft). Fully hardy. Has 5-lobed, bright green leaves that turn yellow in autumn.
♀ '**Aureum**' has bright yellow, young leaves that turn light green in summer and later assume yellow autumn tints.
A. carpinifolium illus. p.66.
♀ *A. circinatum* (Vine maple). Deciduous, spreading, bushy tree or shrub. H 5m (15ft) or more, S 6m (20ft). Fully hardy. Rounded, 7–9-lobed, mid-green leaves turn brilliant orange and red in autumn. Bears clusters of small, purple-and-white flowers in spring.
A. cissifolium. Deciduous, spreading tree. H 8m (25ft), S 12m (40ft). Fully hardy. Leaves consist of 3 oval, toothed leaflets, bronze-tinged when young, dark green in summer, turning red and yellow in autumn. Does best in semi-shade and on neutral to acid soil.
A. cissifolium subsp. *henryi*, syn. *A. henryi*, illus. p.56.
A. crataegifolium (Hawthorn maple). Deciduous, arching tree. H and S 10m (30ft). Fully hardy. Branches are streaked green and white. Small, oval, mid-green leaves turn orange in autumn. '**Veitchii**' illus. p.63.
A. davidii (Père David's maple, Snake-bark maple). Deciduous tree with upright branches. Fully hardy. H and S 15m (50ft). Branches are striped green and white. Oval, glossy, dark green leaves often turn yellow or orange in autumn. '**Madeline Spitta**' illus. p.55.
♀ var. *grosseri*, syn. *A. grosseri*, *A. hersii* (Hers's maple, Snake-bark maple) is a deciduous, upright and spreading tree. H and S 10m (30ft). Fully hardy. Has white-striped trunk and branches. Broadly oval, deeply lobed, bright green leaves turn red in autumn.
A. ginnala illus. p.69.
A. giraldii. Deciduous, spreading tree. H and S 10m (30ft). Frost hardy. Shoots have a blue-grey bloom. Large, sycamore-like, shallowly lobed leaves, with long, pink stalks, are dark green above, blue-white beneath.
A. grandidentatum. Deciduous, spreading tree. H and S 8m (25ft) or more. Fully hardy. Has broad 3- or 5-lobed, bright green leaves that turn bright orange-red in early autumn.
♀ *A. griseum* illus. p.74.
A. grosseri. See *A. davidii* var. *grosseri.*
A. henryi. See *A. cissifolium* subsp. *henryi.*
A. hersii. See *A. davidii* var. *grosseri.*
A. japonicum (Full-moon maple, Japanese maple). Deciduous, bushy tree or shrub. H and S 10m (30ft). Fully hardy. Rounded, lobed leaves are mid-green, turning red in autumn. Clusters of small, reddish-purple flowers open in mid-spring. Shelter from strong winds. ♀ '**Aconitifolium**' and ♀ '**Vitifolium**' illus. p.69.
'**Aureum**'. See *A. shirasawanum* '**Aureum**'.
A. laxiflorum. See *A. pectinatum* subsp. *laxiflorum.*
A. lobelii illus. p.40.
A. macrophyllum illus. p.38.
A. maximowiczianum, syn. *A. nikoense* (Nikko maple). Slow-growing, deciduous, round-headed tree. H and S 12m (40ft). Fully hardy. Leaves have 3 oval, bluish-green leaflets that turn brilliant red and yellow in autumn.
A. monspessulanum (Montpelier maple). Deciduous, usually compact, round-headed tree or shrub. H and S 12m (40ft). Fully hardy. Small, 3-lobed, glossy, dark green leaves remain on tree until late autumn.
A. negundo (Ash-leaved maple, Box elder). Fast-growing, deciduous, spreading tree. H 15m (50ft), S 8m (25ft). Fully hardy. Bright green leaves have 3–5 oval leaflets. Clusters of inconspicuous, greenish-yellow flowers are borne in late spring. '**Variegatum**' illus. p.53. ♀ var. *violaceum* has purplish branchlets covered in a glaucous bloom and bears prominent clusters of tassel-like, purplish-pink flowers.
A. nikoense. See *A. maximowiczianum.*
A. opalus (Italian maple). Deciduous, round-headed tree. H and S 15m (50ft). Fully hardy. Clusters of small, yellow flowers emerge from early to mid-spring before foliage. Leaves are broad, 5-lobed and dark green, turning yellow in autumn.
A. palmatum (Japanese maple). Deciduous, bushy-headed shrub or tree. H and S 6m (20ft) or more. Fully hardy. Palmate, deeply lobed, mid-green leaves turn brilliant orange, red or yellow in autumn. Clusters of small, reddish-purple flowers are borne in mid-spring. f. *atropurpureum* illus. p.91. ♀ '**Bloodgood**' illus. p.113.
♀ '**Butterfly**' illus. p.110.
♀ '**Chitoseyama**' illus. p.136.
'**Corallinum**' illus. p.100.
♀ var. *coreanum* illus. p.68.
'**Dissectum Atropurpureum**' illus. p.136.
var. *heptalobum* illus. p.94. var. *heptalobum* '**Lutescens**' illus. p.92.
var. *heptalobum* '**Rubrum**' illus. p.92.
♀ '**Osakazuki**' has larger leaves, with 7 lobes, that turn brilliant scarlet in autumn. ♀ '**Senkaki**' (syn. *A.p.* '**Sango Kaku**') illus. p.94.
A. pectinatum subsp. *laxiflorum*, syn. *A. laxiflorum*, illus. p.73.
♀ *A. pensylvanicum* (Snake-bark maple) illus. p.58.
♀ '**Erythrocladum**' is a deciduous, upright tree. H 10m (30ft), S 6m (20ft). Fully hardy. Has brilliant candy-pink, young shoots in winter and large, boldly lobed, mid-green leaves that turn bright yellow in autumn.
♀ *A. platanoides* (Norway maple). Vigorous, deciduous, spreading tree. H 25m (80ft), S 15m (50ft). Fully hardy. Has large, broad, sharply lobed, bright green leaves that turn yellow or orange in autumn and clusters of yellow flowers borne in mid-spring before the leaves appear. '**Columnare**', H 12m (40ft), S 8m (25ft), is dense and columnar. ♀ '**Crimson King**' illus. p.39. ♀ '**Drummondii**' has leaves broadly edged with creamy-white. '**Emerald Queen**' is upright when young. '**Globosum**', H 8m (25ft), S 10m (30ft), has a dense, round crown. '**Lorbergii**' illus. p.45. '**Royal Red**' has deep reddish-purple leaves. Those of ♀ '**Schwedleri**' are bright red when young, mature to purplish-green in summer and turn orange-red in autumn. '**Summershade**' has dark green leaves.
A. pseudoplatanus (Sycamore). Fast-growing, deciduous, spreading tree. H 30m (100ft), S 15m (50ft). Fully hardy. Has broadly 5-lobed, dark green leaves. Makes a fine specimen tree and is good for an exposed position.
♀ '**Brilliantissimum**' illus. p.63.
f. *erythrocarpum* illus. p.44. '**Simon Louis Frères**' illus. p.53.
A. rubrum (Red maple) illus. p.44.
'**Columnare**' illus. p.55. ♀ '**October Glory**' is a deciduous, spreading tree. H 20m (70ft), S 12m (40ft). Fully hardy. Has 3- or 5-lobed, glossy, dark green leaves that become intense red in autumn, particularly on neutral to acid soil. In spring, bare branches are covered with clusters of tiny, red flowers. '**Red Sunset**' has dense growth that also turns brilliant red in autumn. ♀ '**Scanlon**' and '**Schlesingeri**' illus. p.44.
♀ *A. rufinerve* (Snake-bark maple) illus. p.56. f. *albolimbatum* is a deciduous, arching tree. H 10m (30ft), S 8m (25ft). Fully hardy. Branches are striped green and white. Has 3-lobed, mid-green leaves, mottled and edged with white, that turn brilliant orange and red in autumn.
♀ *A. saccharinum* (Silver maple).

Fast-growing, deciduous, spreading tree. H 25m (80ft), S 15m (50ft). Fully hardy. Deeply lobed, mid-green leaves, with silver undersides, turn yellow in autumn. 'Wieri' has pendent, lower branches and very deeply lobed leaves.
A. saccharum (Sugar maple). '**Green Mountain**' is a deciduous, spreading tree with an oval crown. H 20m (70ft), S 12m (40ft). Fully hardy. Large, 5-lobed, bright green leaves turn brilliant scarlet in autumn. '**Temple's Upright**' illus. p.56.
♈ *A. shirasawanum* '**Aureum**', syn. *A. japonicum* 'Aureum', illus. p.66.
A. triflorum illus. p.70.
A. velutinum. Deciduous, spreading tree. H 20m (70ft), S 15m (50ft). Fully hardy. Produces large, sycamore-like, lobed, dark green leaves, with undersides covered with pale brown down. var. *vanvolxemii* (Van Volxem's maple) has even larger leaves, slightly glaucous and smooth beneath.

ACHILLEA (Compositae)
Genus of mainly upright perennials, some of which are semi-evergreen, suitable for borders and rock gardens. Has fern-like foliage and large, usually plate-like, flower heads mainly in summer. Flower heads may be dried for winter decoration. Fully hardy. Tolerates most soils but does best in a sunny, well-drained site. Tall species and cultivars need staking. Propagate by division in early spring or autumn or by softwood cuttings in early summer.
A. argentea. See *Tanacetum argenteum.*
A. clavennae illus. p.321.
A. clypeolata. Semi-evergreen, upright perennial. H 45cm (18in), S 30cm (12in). Has divided, hairy, silver leaves and dense, flat heads of small, yellow flowers in summer. Divide plants regularly in spring.
♈ *A.* '**Coronation Gold**' illus. p.220.
♈ *A. filipendulina* '**Gold Plate**' illus. p.220.
A. x *kellereri* illus. p.323.
♈ *A.* x *lewisii* '**King Edward**'. Semi-evergreen, rounded, compact, woody-based perennial. H 10cm (4in), S 23cm (9in) or more. Has feathery, soft, grey-green leaves. Bears compact heads of minute, buff-yellow flower heads in summer. Is suitable for a rock garden, wall or bank.
A. millefolium (Yarrow). '**Fire King**' is a vigorous, upright perennial. H and S 60cm (24in). Has a mass of feathery, dark green leaves and flat heads of rich red flowers in summer.
♈ *A.* '**Moonshine**' illus. p.255.
A. ptarmica '**The Pearl**' illus. p.203.
A. '**Taygetea**' illus. p.255.

ACHIMENES (Gesneriaceae)
Hot-water plant
Genus of erect or trailing perennials with small rhizomes and showy flowers. Frost tender, min. 10°C (50°F). Prefers bright light, but not direct sun, and well-drained soil. Use tepid water for watering pot-grown plants. Allow to dry out after flowering and store rhizomes in a frost-free place over winter. Propagate by division of rhizomes or by seed, if available, in spring or by stem cuttings in summer.
A. antirrhina. Erect perennial. H and S 35cm (14in) or more. Has oval, toothed leaves, to 5cm (2in) or more long and of unequal size in each opposite pair. In summer bears funnel-shaped, red-orange flowers, to 4cm (1 1/2in) long, with yellow throats.
A. '**Brilliant**'. Erect, compact perennial. H and S 30cm (1ft). Has oval, toothed leaves and, in summer, large, funnel-shaped, scarlet flowers.
A. coccinea. See *A. erecta.*
A. erecta, syn. *A. coccinea*, *A. pulchella*. Erect, bushy, branching perennial. H and S 45cm (18in). Has narrowly oval, toothed leaves, often in whorls of 3. Tubular, scarlet flowers with yellow eyes appear in summer.
A. grandiflora. Erect perennial. H and S to 60cm (2ft). Oval, toothed leaves are often reddish below. In summer has tubular, dark pink to purple flowers with white eyes.
A. '**Little Beauty**' illus. p.246.
A. '**Paul Arnold**'. Erect, compact, free-flowering perennial. H and S 30cm (1ft). Has oval, toothed leaves. Bears large, funnel-shaped, purple flowers in summer.
A. '**Peach Blossom**'. Trailing perennial. H and S to 25cm (10in). Has oval, toothed leaves. Large, funnel-shaped, peach-coloured flowers appear in summer.
A. pulchella. See *A. erecta.*

Achnatherum calamagrostis. See *Stipa calamagrostis.*

Acidanthera bicolor. See *Gladiolus callianthus.*

ACIPHYLLA (Umbelliferae)
Genus of evergreen perennials, grown mainly for the architectural value of their spiky foliage but also for their flowers, which are produced more freely on male plants. Frost hardy. Requires sun and well-drained soil. Protect neck of plant from winter wet with a deep layer of stone chippings. Propagate by seed when fresh, in late summer, or in early spring.
A. aurea illus. p.202.
A. scott-thomsonii (Giant Spaniard). Evergreen, rosette-forming perennial. H to 4.5m (14ft), S 60cm–1m (2–3ft). Much-dissected, spiny foliage is bronze when young, maturing to silver-grey. Prickly spikes of tiny, creamy-yellow flowers are rarely produced. Prefers a moist but well-drained site.
A. squarrosa illus. p.230.

ACOKANTHERA (Apocynaceae)
Genus of evergreen shrubs and trees, grown for their flowers and overall appearance. Frost tender, min. 10°C (50°F). Requires full light and good drainage. Water containerized plants moderately, less when not in full growth. Propagate by seed in spring or autumn or by semi-ripe cuttings in summer.
A. oblongifolia, syn. *A. spectabilis*, *Carissa spectabilis*, illus. p.120.
A. spectabilis. See *A. oblongifolia.*

ACONITUM (Ranunculaceae)
Monkshood, Wolf's bane
Genus of perennials with poisonous, tuberous or fibrous roots and upright, sometimes scandent, stems, bearing hooded flowers in summer. Leaves are mostly rounded in outline. Is good for rock gardens and borders. Fully hardy. Prefers sun, but tolerates some shade and this may enhance flower colour. Requires fertile, well-drained soil. Propagate by division in autumn, every 2–3 years, or by seed in autumn.
A. anthora. Compact, tuberous perennial. H 60cm (24in), S 50cm (20in). Has erect, leafy stems that bear several hooded, yellow flowers in summer. Leaves are divided and dark green.
♈ *A.* '**Bressingham Spire**'. Compact, upright, tuberous perennial. H 1m (3ft), S 50cm (20in). Hooded, violet-blue flowers are borne in very erect spikes in summer. Leaves are deeply divided, glossy and dark green.
♈ *A.* x *cammarum* '**Bicolor**', syn. *A.* x *bicolor*, illus. p.216.
A. carmichaelii '**Arends**'. Erect, tuberous perennial. H 1.5m (5ft), S 30cm (1ft). Has divided, rich green leaves and, in autumn, spikes of hooded, rich deep blue flowers. Upright stems may need staking, particularly if in a shady site.
A. hemsleyanum, syn. *A. volubile.* Wiry, scandent, fibrous perennial. H 2–2.5m (6–8ft), S 1–1.2m (3–4ft). Hooded, lilac flowers are borne in drooping clusters in late summer. Leaves are divided and mid-green. Is best grown where it can scramble through a shrub or be supported.
♈ *A.* x *henryi* '**Sparks' Variety**'. Upright, tuberous perennial. H 1.2m (4ft), S 50cm (20in). Bears violet-blue flowers on branching stems in summer and has deeply divided, glossy, dark green leaves.
A. '**Ivorine**'. Upright, tuberous perennial. H 1.5m (5ft), S 50cm (20in). Bears hooded, creamy-white flowers in erect spikes in early summer. Strong stems bear deeply divided, glossy, green leaves.
A. lycoctorum subsp. *vulparia*, syn. *A. vulparia*, illus. p.219.
A. napellus (Helmet flower, Monkshood). Upright, tuberous perennial. H 1.5m (5ft), S 30cm (1ft). Has tall, slender spires of hooded, light indigo-blue flowers in late summer and deeply cut, mid-green leaves. f. *album* has white flowers.
A. '**Newry Blue**'. Upright, tuberous perennial. H 1.2m (4ft), S 50cm (20in). Bears hooded, dark blue flowers on erect stems in summer and has deeply divided, glossy, dark green leaves.
A. volubile. See *A. hemsleyanum.*
A. vulparia. See *A. lycoctorum* subsp. *vulparia.*

ACORUS (Araceae)
Genus of semi-evergreen, perennial, marginal and submerged water plants, grown for their often aromatic foliage. Fully to frost hardy. Needs an open, sunny position. *A. calamus* requires up to 25cm (10in) depth of water. Tidy up fading foliage in autumn and lift and divide every 3 or 4 years, in spring, as clumps become congested.
A. calamus '**Variegatus**' illus. p.389.
A. gramineus '**Pusillus**'. Semi-evergreen, perennial, marginal water plant or submerged aquarium plant. Frost hardy. H and S 10cm (4in). Has narrow, grass-like, stiff leaves. Rarely, insignificant, greenish flower spikes are produced in summer. '**Variegatus**' illus. p.389.

ACRADENIA (Rutaceae)
Genus of evergreen shrubs, grown for their foliage and flowers. Half hardy. Requires a sheltered position in sun or semi-shade and fertile, well-drained soil. Does best against a south- or west-facing wall. Propagate by semi-ripe cuttings in summer.
A. frankliniae. Evergreen, upright, stiffly branched shrub. H and S 2m (6ft). Has aromatic, dark green leaves with 3 narrowly lance-shaped leaflets and, from late spring to early summer, bears small clusters of star-shaped, white flowers.

Acroclinium roseum. See *Helipterum roseum.*

ACTAEA (Ranunculaceae)
Baneberry
Genus of clump-forming perennials, grown for their colourful, poisonous berries. Fully hardy. Likes woodland conditions – moist, peaty soil and shade. Propagate by division in spring or by seed in autumn.
♈ *A. alba*, syn. *A. pachypoda*, illus. p.223.
A. pachypoda. See *A. alba.*
♈ *A. rubra* (Red baneberry). Clump-forming perennial. H 50cm (20in), S 30cm (12in). Small, fluffy, white flowers are followed in autumn by clusters of poisonous, rounded, scarlet berries, borne above oval, divided, bright green leaves.

ACTINIDIA (Actinidiaceae)
Genus of mainly deciduous, woody-stemmed, twining climbers. Fully hardy to frost tender, min. 10°C (50°F). Prefers partial shade (*A. kolomikta* will withstand full sun). Grow in any well-drained soil that does not dry out. Prune in winter if necessary. Propagate by seed in spring or autumn, by semi-ripe cuttings in mid-summer or by layering in winter.
A. chinensis, syn. *A. deliciosa* (Chinese gooseberry, Kiwi fruit). Vigorous, mainly deciduous, woody-stemmed, twining climber. H 9–10m (28–30ft). Frost hardy. Heart-shaped leaves are 13–20cm (5–8in) long. In summer bears clusters of cup-shaped, white flowers that later turn yellowish, followed by edible, hairy, brown fruits. To obtain fruits, male and female plants must be grown.
A. deliciosa. See *A. chinensis.*
♈ *A. kolomikta* illus. p.169.
A. polygama (Silver vine). Mainly deciduous, woody-stemmed, twining climber. H 4–6m (12–20ft). Frost hardy. Heart-shaped leaves, 7–13cm (3–5in) long, are bronze when young and sometimes have creamy, upper sections. In summer has scented, cup-shaped, white flowers, usually in groups of 3 male, female or bisexual, followed by edible but not very tasty, egg-shaped, bright yellow fruits.

ADA. See ORCHIDS.
A. aurantiaca (illus. p.263). Evergreen, epiphytic orchid for a cool greenhouse. H 23cm (9in). Bears sprays of tubular, orange flowers, 2.5cm (1in) long, in early spring. Has narrowly oval leaves, 10cm (4in) long. Needs shade in summer.

ADANSONIA (Bombacaceae)
Baobab
Genus of deciduous or semi-evergreen, mainly spring-flowering trees, grown for their characteristically swollen trunks, their foliage and for shade. Has flowers only on large, mature specimens. Frost tender, min. 13–16°C (55–61°F). Requires full light and sharply drained soil. Allow soil of containerized specimens almost to dry out between waterings. Propagate by seed in spring. Pot specimens under glass are prone to red spider mite.
A. digitata. Slow-growing, semi-evergreen, rounded tree. H and S 15m (50ft) or more. Has palmate leaves of 5–7 lustrous, green leaflets. Produces fragrant, pendent, long-stalked, white flowers, with 5 reflexed petals, in spring, followed by edible, sausage-shaped, brown fruits.

ADENIUM (Apocynaceae)
Desert rose
Genus of perennial succulents with fleshy, swollen trunks. Frost tender, min. 15°C (59°F). Needs sun or partial shade and well-drained soil; plants are very prone to rotting. Propagate by seed in spring or summer.
A. obesum illus. p.394.

ADENOCARPUS (Leguminosae)
Genus of deciduous or semi-evergreen shrubs, grown for their profuse, broom-like, yellow flowers in spring or early summer. Frost to half hardy. Requires full sun and well-drained soil. Does best grown against a south- or west-facing wall. Propagate by seed in autumn.
A. viscosus. Semi-evergreen, arching shrub. H and S 1m (3ft). Frost hardy. Grey-green leaves with 3 narrowly lance-shaped leaflets densely cover shoots. Produces dense, terminal racemes of orange-yellow flowers in late spring.

ADENOPHORA (Campanulaceae)
Gland bellflower
Genus of summer-flowering, fleshy-rooted perennials. Fully hardy. Requires full sun and rich, well-drained but not over-dry soil. May become invasive but resents disturbance. Propagate by basal cuttings in early spring or by seed in autumn.
A. potaninii. Rosette-forming perennial. H 45cm (18in) or more, S 60cm (24in). Bears arching sprays of bell-shaped, pale bluish-lavender flowers in late summer. Has oval to lance-shaped, basal, mid-green leaves.

Adhatoda duvernoia. See *Duvernoia adhatodoides.*
Adhatoda vasica. See *Justicia adhatoda.*

ADIANTUM (Polypodiaceae)
Genus of deciduous, semi-evergreen or evergreen ferns. Fully hardy to frost tender, min. 7–13°C (45–55°F). Prefers semi-shade and moist, neutral to acid soil (*A. pedatum* var. *aleuticum* prefers alkaline soil). Remove fading fronds regularly. Propagate by spores in summer.
A. capillus-veneris (Maidenhair fern). Semi-evergreen or evergreen fern. H and S 30cm (12in). Half hardy. Has dainty, triangular to oval, segmented, arching, light green fronds borne on black stems.
♟ *A. pedatum* illus. p.188.
var. *aleuticum* illus. p.187.
A. raddianum (Delta maidenhair). Semi-evergreen or evergreen fern. H and S 30cm (12in). Frost tender, min. 7°C (45°F). Triangular, divided, pale green segments are borne on finely dissected fronds with purplish-black stems. **'Fritz-Luthii'** has bright green fronds. **'Grandiceps'** (Tassel maidenhair) has elegant, tasselled fronds.
A. tenerum. Semi-evergreen or evergreen fern. H 30cm–1m (1–3ft), S 60cm–1m (2–3ft). Frost tender, min. 13°C (55°F). Broadly lance-shaped, much-divided, spreading, mid-green fronds consist of rounded or diamond-shaped pinnae. **'Glory of Moordrecht'** has more erect fronds and is hardy down to 10°C (50°F).
♟ *A. venustum* illus. p.189.

ADLUMIA (Fumariaceae)
Genus of one species of herbaceous, biennial, leaf-stalk climber, grown for its leaves and flowers. Frost hardy. Grow in any soil, in semi-shade. Propagate by seed in spring.
A. fungosa (Allegheny vine, Climbing fumitory). Herbaceous, biennial, leaf-stalk climber. H 3–4m (10–12ft). Delicate leaves have numerous leaflets. Tiny, tubular, spurred, white or purplish flowers are carried in drooping panicles in summer.

ADONIS (Ranunculaceae)
Genus of spring-flowering perennials, grown for their foliage and flowers. Fully hardy. Does best in semi-shade and in moist but well-drained soil. Propagate by seed when fresh, in late summer, or by division after flowering.
A. amurensis illus. p.239.
A. brevistyla illus. p.232.
A. vernalis illus. p.239.

ADROMISCHUS (Crassulaceae)
Genus of perennial succulents and evergreen sub-shrubs with rounded, thin or fat leaves. Frost tender, min. 7°C (45°F). Needs partial shade and very well-drained soil. Propagate by leaf or stem cuttings in spring or summer.
A. festivus (Plovers' eggs). Slow-growing, clump-forming, perennial succulent. H 10cm (4in), S 15cm (6in). Has egg-shaped, purple-blotched, grey-green leaves, each with a wavy-edged, compressed tip. Flower stem, to 30cm (12in) long, bears small, tubular, pink flowers in summer.
A. maculatus illus. p.405.

AECHMEA (Bromeliaceae)
Genus of evergreen, rosette-forming, epiphytic perennials, grown for their foliage, flowers and fruits. Frost tender, min. 10–15°C (50–59°F). May be grown in full light or semi-shade. Provide a rooting medium of equal parts humus-rich soil and either sphagnum moss or bark or plastic chips used for orchid culture. Using soft water, water moderately in summer, sparingly at other times, and keep cup-like, rosette centres filled with water from spring to autumn. Propagate by offsets in late spring.
A. distichantha (illus. p.229). Evergreen, basal-rosetted, epiphytic perennial. H and S to 1m (3ft). Forms dense rosettes of narrowly oblong, round-tipped, arching leaves, dull green above, grey and scaly beneath. Bears panicles of small, tubular, purple or blue flowers among white-felted, pink bracts, usually in summer.
♟ *A. fasciata*, syn. *Billbergia rhodocyanea* (Silver vase plant, Urn plant; illus. p.229). Evergreen, tubular-rosetted, epiphytic perennial. H 40–60cm (16–24in), S 30–50cm (12–20in). Has loose rosettes of broadly oblong, round-tipped, incurved, arching leaves with dense, grey scales and silver cross-banding. From spring to autumn bears dense, pyramidal panicles of tubular, blue-purple flowers among pink bracts, just above foliage.
♟ *A.* **'Foster's Favorite'** (Lacquered wine-cup; illus. p.229). Evergreen, basal-rosetted, epiphytic perennial. H and S 30–60cm (12–24in). Has loose rosettes of strap-shaped, arching, lustrous, wine-red leaves. Drooping spikes of small, tubular, deep purple-blue flowers, in summer, are followed by pear-shaped, red fruits.
A. fulgens (Coral berry). Evergreen, basal-rosetted, epiphytic perennial. H and S 40–75cm (16–30in). Forms loose rosettes of broadly oblong, arching, glossy, mid-green leaves with grey scales beneath and rounded or pointed tips. In summer produces, above foliage, erect panicles of small, tubular, violet-purple flowers that turn red with age. These are succeeded by small, rounded to ovoid, red fruits on red stalks.
♟ *A. nudicaulis.* Evergreen, basal-rosetted, epiphytic perennial. H and S 40–75cm (16–30in). Produces loose rosettes of a few broadly strap-shaped, arching, olive-green leaves with spiny edges and usually banded with grey scales beneath. Spikes of small, tubular, yellow flowers open above large, red bracts in summer.
A. recurvata (illus. p.229). Evergreen, basal-rosetted, epiphytic perennial. H and S 15–20cm (6–8in). Narrowly triangular, tapered, spiny-edged, arching, red-flushed, mid-green leaves are produced in dense rosettes. In summer bears a short, dense spike of tubular, red-and-white flowers, with red bracts, just above leaves.

AEGOPODIUM (Umbelliferae)
Bishop's weed, Gout weed, Ground elder
Genus of invasive, rhizomatous perennials, most of which are weeds although *A. podagraria* 'Variegata' provides excellent ground cover. Fully hardy. Tolerates sun or shade and any well-drained soil. Propagate by division of rhizomes in spring or autumn.
A. podagraria **'Variegatum'** illus. p.240.

AEONIUM (Crassulaceae)
Genus of perennial succulents, some of which are short-lived, and evergreen, succulent shrubs, grown for their rosettes of bright green or blue-green, occasionally purple, leaves. Frost tender, min. 5°C (41°F). Prefers partial shade and very well-drained soil. Most species grow from autumn to spring and are semi-dormant in mid-summer. Propagate by seed in summer or, for branching species, by stem cuttings in spring or summer.
♟ *A. arboreum.* Bushy, perennial succulent. H to 60cm (2ft), S 1m (3ft). Branched stems are each crowned by a rosette, up to 15cm (6in) across, of broadly lance-shaped, glossy, bright green leaves. In spring produces cones of small, star-shaped, golden flowers on 2–3-year-old stems, which then die back. ♟ **'Schwarzkopf'** illus. p.404.
♟ *A. haworthii* illus. p.402.
♟ *A. tabuliforme* illus. p.414.

AESCHYNANTHUS (Gesneriaceae)
Genus of evergreen, climbing, trailing or creeping perennials, useful for growing in hanging baskets. Frost tender, min. 18°C (64°F). Needs a fairly humid atmosphere and a position out of direct sun. Water sparingly in low temperatures. Propagate by tip cuttings in spring or summer.
♟ *A. marmoratus*, syn. *A. zebrinus.* Evergreen, trailing perennial. H and S to 60cm (2ft). Oval, waxy leaves are dark green, veined yellowish-green above, purplish below. Produces tubular, greenish flowers, marked dark brown, in terminal clusters in summer.
♟ *A. pulcher* (Lipstick plant, Royal red bugler). Evergreen, climbing or trailing perennial. H and S indefinite. Has thick, oval leaves and small, tubular, hooded, bright red flowers, with yellow throats, borne in terminal clusters from summer to winter.
♟ *A. speciosus*, syn. *A. splendens*, illus. p.257.
A. splendens. See *A. speciosus.*
A. x *splendidus.* Vigorous, evergreen, trailing perennial. H and S 1m (3ft) or more. Leaves are thick, narrowly oval, to 13cm (5in) long. Has clusters of erect, tubular, bright orange-red flowers, marked with brown-red, in summer-autumn.
A. zebrinus. See *A. marmoratus.*

AESCULUS (Hippocastanaceae)
Buckeye, Horse-chestnut
Genus of deciduous trees and shrubs, grown for their bold, divided leaves and conspicuous, upright panicles or clusters of flowers, followed by fruits (horse-chestnuts) sometimes with spiny outer casings. Fully to frost hardy. Requires sun or semi-shade and fertile,

well-drained soil. Propagate species by seed in autumn, cultivars by budding in late summer or by grafting in late winter. Leaf spot may affect young foliage, and coral spot fungus may attack damaged wood.
A. californica illus. p.59.
A. x *carnea* (Red horse-chestnut).
♥ '**Briotii**' illus. p.38.
A. chinensis illus. p.38.
♥ *A. flava*, syn. *A. octandra*, illus. p.56.
A. glabra (Ohio buckeye). Deciduous, round-headed, sometimes shrubby tree. H and S 10m (30ft). Fully hardy. Leaves, usually composed of 5 narrowly oval leaflets, are dark green. Bears 4-petalled, greenish-yellow flowers in upright clusters in late spring and early summer.
♥ *A. hippocastanum* (Horse-chestnut) illus. p.38. ♥ '**Baumannii**' is a vigorous, deciduous, spreading tree. H 30m (100ft), S 15m (50ft). Fully hardy. Large, dark green leaves, consisting of 5 or 7 narrowly oval leaflets, turn yellow in autumn. Has large panicles of long-lasting, double, yellow- or red-marked, white flowers from mid- to late spring.
♥ *A. indica* (Indian horse-chestnut). Deciduous, spreading, elegant tree. H 20m (70ft), S 12m (40ft). Frost hardy. Glossy, dark green leaves with usually 7 narrowly oval leaflets are bronze when young, orange or yellow in autumn. Upright panicles of 4-petalled, pink-tinged, white flowers, marked with red or yellow, appear in mid-summer. '**Sydney Pearce**' illus. p.53.
A. x *neglecta* (Sunrise horse-chestnut).
♥ '**Erythroblastos**' illus. p.63.
A. octandra. See *A. flava*.
♥ *A. parviflora* illus. p.89.
♥ *A. pavia* (Red buckeye). Deciduous, round-headed, sometimes shrubby tree. H 5m (15ft), S 3m (10ft). Fully hardy. Glossy, dark green leaves consist of 5 narrowly oval leaflets. Has panicles of 4-petalled, red flowers in early summer. '**Atrosanguinea**' illus. p.65.
A. turbinata (Japanese horse-chestnut). Deciduous, spreading, stout-branched tree. H 20m (70ft), S 12m (40ft). Fully hardy. Large, dark green leaves consist of 5 or 7 narrowly oval leaflets. Panicles of creamy-white flowers appear in late spring and early summer.

AETHIONEMA (Cruciferae)
Genus of short-lived, evergreen or semi-evergreen shrubs, sub-shrubs and perennials, grown for their prolific flowers. Fully hardy. Needs sun and well-drained soil. Propagate by softwood cuttings in spring or by seed in autumn. Most species self seed readily.
A. armenum illus. p.325.
♥ *A. grandiflorum*, syn. *A. pulchellum*, illus. p.300.
A. iberideum. Evergreen or semi-evergreen, rounded, compact shrub. H and S 15cm (6in). Bears small, lance-shaped, grey-green leaves and, in summer, 2cm (3/4in) stems each bear a raceme of small, saucer-shaped, white flowers.
A. pulchellum. See *A. grandiflorum*.

♥ *A.* '**Warley Rose**' illus. p.324.
A. '**Warley Ruber**'. Evergreen or semi-evergreen, rounded, compact sub-shrub. H and S 15cm (6in). Has tiny, linear, bluish-green leaves. Racemes of small, deep rose-pink flowers appear on 2–3cm (3/4–1 1/4in) stems in spring-summer.

AGAPANTHUS (Liliaceae)
Genus of clump-forming perennials, some of which are evergreen, with erect stems that carry large umbels of bell- to tubular-bell-shaped or trumpet-shaped flowers, usually blue and often fading to purple with age. Leaves are strap-shaped. Narrow-leaved forms are frost hardy, broad-leaved ones half hardy. Grow in full sun and in moist but well-drained soil. Protect crowns in winter with ash or mulch. Plants increase slowly but may be propagated by division in spring; may also be raised from seed in autumn or spring. Named cultivars will not come true from seed.
♥ *A. africanus* (African lily). Evergreen, clump-forming perennial. H 1m (3ft), S 50cm (20in). Half hardy. In late summer has rounded umbels of deep blue flowers on upright stems, above broad, dark green leaves.
A. '**Alice Gloucester**'. Clump-forming perennial. H 1m (3ft), S 50cm (20in). Frost hardy. Bears large, dense, rounded umbels of white flowers in summer, above narrow, mid-green leaves.
A. '**Ben Hope**'. Clump-forming perennial. H 1–1.2m (3–4ft), S 50cm (20in). Frost hardy. Erect stems support dense, rounded umbels of deep blue flowers in late summer and early autumn, over narrow, greyish-green leaves.
A. campanulatus. Clump-forming perennial. H 60cm–1.2m (2–4ft), S 50cm (20in). Frost hardy. Rounded umbels of blue flowers are borne on strong stems in summer, above narrow, greyish-green leaves.
A. '**Cherry Holley**'. Clump-forming perennial. H 1m (3ft), S 50cm (20in). Frost hardy. Rounded umbels of dark blue flowers, carried in summer above narrow leaves, do not fade to purple with age.
A. '**Dorothy Palmer**' illus. p.217.
A. inapertus. Clump-forming perennial. H 1.5m (5ft), S 60cm (2ft). Frost hardy. Pendent, narrowly tubular, blue flowers are borne on very erect stems, above narrow, bluish-green leaves, in late summer and autumn.
A. '**Lilliput**'. Compact, clump-forming perennial. H 80cm (32in), S 50cm (20in). Frost hardy. Produces small, rounded umbels of dark blue flowers in summer. Leaves are narrow and mid-green.
♥ *A.* '**Loch Hope**'. Clump-forming perennial. H 1–1.2m (3–4ft), S 50cm (20in). Frost hardy. Bears large, rounded umbels of deep blue flowers in late summer and early autumn, above narrow, greyish-green leaves.
A. orientalis. See *A. praecox* subsp. *orientalis*.
A. praecox subsp. *orientalis*, syn. *A. orientalis*, illus. p.218.

AGAPETES, syn. **PENTAPTERYGIUM** (Ericaceae)
Genus of evergreen or deciduous, scandent shrubs and semi-scrambling climbers, grown for their flowers. Frost tender, min. 5–18°C (41–64°F). Provide a humus-rich, well-drained but not dry, neutral to acid soil and full light or partial shade. Water potted specimens freely in full growth, moderately at other times. Overlong stems may be cut back to promote branching, but are best tied to supports. Propagate by seed in spring or by semi-ripe cuttings in late summer.
A. '**Ludgvan Cross**', syn. *A. rugosa* x *serpens*. Evergreen, scandent shrub with arching or pendulous stems. H and S 2–3m (6–10ft). Min. 5°C (41°F). Lance-shaped leaves are dark green. Urn-shaped, red flowers with darker patterns are produced in spring.
A. macrantha illus. p.179.
A. rugosa. Evergreen, loose shrub with arching or spreading stems. H and S to 3m (10ft). Min. 5°C (41°F). Leaves are lance-shaped, wrinkled and bright green. In spring, clusters of pendent, urn-shaped, white flowers, patterned with purple-red, are produced from leaf axils.
A. rugosa x *serpens*. See *A.* '**Ludgvan Cross**'.
♥ *A. serpens* illus. p.166.

AGASTACHE (Labiatae)
Mexican giant hyssop
Genus of summer-flowering perennials with aromatic leaves. Half hardy. Needs fertile, well-drained soil and full sun. Plants are short-lived and should be propagated each year by softwood or semi-ripe cuttings in late summer.
A. mexicana, syn. *Brittonastrum mexicanum*, *Cedronella mexicana*. Upright perennial with aromatic leaves. H to 1m (3ft), S to 30cm (1ft). In summer bears whorls of small, tubular flowers in shades of pink to crimson. Leaves are oval, pointed, toothed and mid-green.

Agathaea coelestis. See *Felicia amelloides*.

AGATHOSMA (Rutaceae)
Genus of evergreen shrubs, grown for their flowers and overall appearance. Frost tender, min. 5–7°C (41–5°F). Needs full light and well-drained, acid soil. Water containerized specimens moderately, less when not in full growth. Propagate by semi-ripe cuttings in late summer.
A. pulchella, syn. *Barosma pulchella*. Evergreen, rounded, wiry, aromatic shrub. H and S to 1m (3ft). Has a dense mass of small, oval, leathery leaves. Small, 5-petalled, purple flowers are freely produced in terminal clusters in spring-summer.

AGAVE (Agavaceae)
Genus of rosetted, perennial succulents with sword-shaped, sharp-toothed leaves. Small species, to 30cm (1ft) high, flower only after 5–10 years; tall species, to 5m (15ft) high, may take 20–40 years to flower. Most species with hard, blue-grey leaves are half

hardy; grey-green- or green-leaved species are usually frost tender, min. 5°C (41°F). Needs full sun and well-drained soil. Propagate by seed or offsets in spring or summer.
A. americana (Century plant). Basal-rosetted, perennial succulent. H 1–2m (3–6ft), S 2–3m (6–10ft) or more. Half hardy. Has sharply pointed, toothed leaves, to 1.5–2m (5–6ft) long. Branched flower stem, to 8m (25ft) long, bears dense, tapering spikes of bell-shaped, white to pale creamy-yellow flowers, each 9cm (3 1/2in) long, in spring-summer. Offsets freely. var. *mediopicta*, H and S 2m (6ft), has central, yellow stripes along leaves. var. *striata* (syn. *A.a.* '**Variegata**') illus. p.393.
A. attenuata illus. p.403.
A. filifera illus. p.401.
A. parryi illus. p.402.
A. parviflora illus. p.393.
A. utahensis illus. p.414.
A. victoriae-reginae illus. p.397.

Ageratina ligustrina. See *Eupatorium rugosum*.

AGERATUM (Compositae)
Floss flower
Genus of annuals and biennials. Half hardy. Grow in sun and in fertile, well-drained soil, which should not be allowed to dry out otherwise growth and flowering will be poor. Dead-head plants regularly to ensure continuous flowering. Propagate by seed sown outdoors in late spring.
A. houstonianum. Moderately fast-growing, hummock-forming annual. Tall cultivars, H and S 30cm (12in); medium, H and S 20cm (8in); dwarf, H and S 15cm (6in). All have oval, mid-green leaves and clusters of feathery, brush-like flower heads throughout summer and into autumn. Is useful for edging. '**Bengali**' (medium) has light pink flowers, deepening with age. '**Blue Danube**' (dwarf) and '**Blue Mink**' (tall) illus. p.285. '**Pinkie**' (tall) is a warm shade of pink; and '**White Cushion**' (medium) has white flowers.

AGLAONEMA (Araceae)
Chinese evergreen
Genus of evergreen, erect, tufted perennials, grown mainly for their foliage. Frost tender, most species requiring min. 15°C (59°F). Tolerates shade, although the variegated forms need more light, and prefers moist but well-drained soil. Water moderately when in full growth, less in winter. Propagate by division or stem cuttings in summer. Mealy bug may be a problem.
A. commutatum. Evergreen, erect, tufted perennial. H and S to 45cm (18in) or more. Broadly lance-shaped leaves are 30cm (1ft) long and dark green with irregular, greyish-white patches along lateral veins. Has greenish-white spathes in summer. '**Silver King**' illus. p.268; '**Treubii**' illus. p.264.
A. '**Malay Beauty**', syn. *A.* '**Pewter**'. Evergreen, erect, tufted perennial. H and S 30cm (1ft) or more. Oval leaves, to 30cm (1ft) long, are very dark green, mottled greenish-white and cream.

Has greenish-white spathes in summer.
A. **'Pewter'**. See *A.* **'Malay Beauty'**.
A. pictum illus. p.268.

AGONIS (Myrtaceae)
Willow myrtle
Genus of evergreen, mainly spring-flowering shrubs and trees, grown for their foliage, flowers and graceful appearance. Frost tender, min. 10°C (50°F). Needs full light and well-drained but moisture-retentive soil. Water containerized specimens moderately, scarcely at all in winter. Pruning is tolerated when necessary. Propagate by seed in spring or by semi-ripe cuttings in summer.
A. flexuosa illus. p.64.

AGROSTEMMA (Caryophyllaceae)
Corn cockle
Genus of summer-flowering annuals. Fully hardy. Grow in sun; flowers best in very well-drained soil that is not very fertile. Support with sticks and dead-head to prolong flowering. Propagate by seed sown *in situ* in spring or early autumn.
A. coeli-rosa. See *Silene coeli-rosa.*
A. githago. Fast-growing, erect annual with thin stems. H 60cm–1m (2–3ft), S 30cm (1ft). Has lance-shaped, mid-green leaves and, in summer, 5-petalled, open trumpet-shaped, pink flowers, 8cm (3in) wide. Seeds are tiny, rounded, dark brown and poisonous. **'Milas'** illus. p.275.

AICHRYSON (Crassulaceae)
Genus of annual and perennial succulents, often shrub-like, grown for their fleshy, spoon-shaped to rounded, hairy leaves. Most species are short-lived, dying after flowering. Frost tender, min. 5°C (41°F). Needs full sun or partial shade and very well-drained soil. Propagate by seed or stem cuttings in spring or summer.
⚑ *A. x domesticum* **'Variegata'** illus. p.413.

AILANTHUS (Simaroubaceae)
Genus of deciduous trees, grown for their foliage and 3–5-winged fruits; is extremely tolerant of urban pollution. Fully hardy. Needs sun or semi-shade and deep, fertile, well-drained soil. To grow as shrubs, cut back hard in spring, after which vigorous shoots bearing very large leaves are produced. Propagate by seed in autumn or by suckers or root cuttings in winter.
⚑ *A. altissima* (Tree of heaven). Fast-growing, deciduous, spreading tree. H 25m (80ft), S 15m (50ft). Has large, dark green leaves consisting of 15–30 paired, oval leaflets. Large clusters of small, green flowers in mid-summer are followed by attractive, winged, green, then reddish-brown fruits.

AJUGA (Labiatae)
Genus of annuals and perennials, some of which are semi-evergreen or evergreen and excellent as ground cover. Fully hardy. Tolerates sun or shade and any soil, but grows more vigorously in moist conditions. Propagate by division in spring.
A. pyramidalis (Pyramidal bugle). Semi-evergreen perennial. H 15cm (6in), S 45cm (18in). Forms a creeping carpet of oblong to spoon-shaped, deep green leaves, above which appear spikes of whorled, 2-lipped, blue flowers in spring. **'Metallica Crispa'** has crisp, curled leaves, with a metallic-bronze lustre, and dark blue flowers.
⚑ *A. reptans* **'Atropurpurea'** illus. p.267. **'Jungle Beauty'** is a semi-evergreen, mat-forming perennial. H 38cm (15in), S 60cm (24in). Has large, oval, toothed or slightly lobed, dark green leaves, sometimes suffused purple, and, in spring, spikes of whorled, 2-lipped, blue flowers. **'Multicolor'** (syn. *A.r.* **'Rainbow'**) illus. p.267.

AKEBIA (Lardizabalaceae)
Genus of deciduous or semi-evergreen, woody-stemmed, twining climbers, grown for their leaves and flowers. Individual plants seldom produce fruits; cross-pollination between 2 individuals is required for fruit formation. Frost hardy. Prefers full sun and any good, well-drained soil. Tolerates an east- or north-facing position. Dislikes disturbance. Propagate by seed in autumn or spring, by semi-ripe cuttings in summer or by layering in winter.
A. x pentaphylla. Mainly deciduous, woody-stemmed, twining climber. H to 10m (30ft). Mid-green leaves, bronze-tinted when young, have 3 or 5 oval leaflets. Drooping racemes of small, 3-petalled, purple flowers (larger, female at base of plant, smaller, male at apex) are borne in spring.
A. quinata illus. p.166.
A. trifoliata. Deciduous, woody-stemmed, twining climber. H to 10m (30ft) or more. Mid-green leaves, bronze-tinted when young, have 3 oval leaflets; drooping racemes of 3-petalled, purple flowers appear in spring, followed by sausage-shaped, purplish fruits.

ALANGIUM (Alangiaceae)
Genus of deciduous or evergreen trees and shrubs, grown for their foliage and flowers. Frost hardy. Needs full sun and any fertile, well-drained soil. Propagate by seed in spring or by softwood cuttings in summer.
A. platanifolium. Deciduous, upright, tree-like shrub. H 3m (10ft), S 2m (6ft). Has maple-like, 3-lobed, mid-green leaves. Fragrant, tubular, white flowers are borne from early to mid-summer.

ALBIZIA (Leguminosae)
Genus of deciduous or semi-evergreen trees, grown for their feathery foliage and unusual flower heads, composed of numerous stamens and resembling bottlebrushes. Half hardy: grow against a south- or west-facing wall; in cold areas do not plant out until late spring. Requires full sun and well-drained soil. *A. julibrissin* may be grown as a summer bedding plant for its foliage. Propagate by seed in autumn.
A. distachya. See *A. lophantha.*
A. julibrissin illus. p.64.
⚑ *A. lophantha,* syn. *A. distachya, Paraserianthes distachya,* illus. p.67.

ALBUCA (Liliaceae)
Genus of spring- or summer-flowering bulbs. Half hardy to frost tender, min. 10°C (50°F). Needs an open, sunny position and well-drained soil. Dies down in spring or late summer after flowering. Propagate by seed in spring or by offsets when dormant.
A. canadensis. Spring-flowering bulb. H 15cm (6in), S 8–10cm (3–4in). Half hardy. Has 3–6 narrowly lance-shaped, erect, basal leaves. Produces a loose spike of tubular, yellow flowers, 1.5–2cm (⅝–¾in) long, with a green stripe on each petal.
A. humilis illus. p.379.
A. major. Early spring-flowering bulb. H 50cm–1m (20in–3ft), S 12–15cm (5–6in). Frost tender. Has long, lance-shaped, erect, basal leaves and robust stems, each with a loose spike of up to 12 tubular, yellow flowers, 1.5–2cm (⅝–¾in) long, with green or brown stripes outside.

ALCEA (Malvaceae)
Hollyhock
Genus of biennials and short-lived perennials, grown for their tall spikes of flowers. Fully hardy. Needs full sun and well-drained soil. Propagate by seed in late summer or spring. Rust may be a problem.
A. rosea, syn. *Althaea rosea* (biennial), illus. p.273. **'Chater's Double'** (biennial) illus. p.278. **'Majorette'** is an erect biennial, grown as an annual. H 60cm (2ft), S to 30cm (1ft). Rounded, lobed, pale green leaves are rough-textured. Spikes of rosette-like, double flowers in several different colours are produced in summer and early autumn. **'Summer Carnival'** (annual or biennial), H 1.8–2.4m (6–8ft), S to 60cm (2ft), has double flowers in mixed colours.

ALCHEMILLA (Rosaceae)
Lady's mantle
Genus of perennials that bear sprays of tiny, greenish-yellow flowers, with conspicuous, outer calyces, in summer. Some are good for ground cover. Fully hardy. Grows in all but boggy soils, in sun or partial shade. Propagate by seed or division in spring or autumn.
A. alpina (Alpine lady's mantle). Mound-forming perennial. H 15cm (6in), S 60cm (24in) or more. Rounded, lobed, pale green leaves are covered in silky hairs. Bears upright spikes of tiny, greenish-yellow flowers, with conspicuous, green, outer calyces, in summer. Is suitable for ground cover and a dry bank.
A. conjuncta illus. p.254.
⚑ *A. mollis* illus. p.254.

x ALICEARA. See ORCHIDS.
x *A.* **Dark Warrior** (illus. p.254). Evergreen, epiphytic orchid for a cool greenhouse. H 25cm (10in). Produces sprays of wispy, mauve-brown, cream-yellow or green flowers, 4cm (1½in) across; flowering season varies. Leaves, 10cm (4in) long, are narrowly oval. Grow in semi-shade in summer.

ALISMA (Alismataceae)
Genus of deciduous, perennial, marginal water plants, grown for their foliage and flowers. Fully to frost hardy. Requires an open, sunny position in mud or up to 25cm (10in) depth of water. Tidy up fading foliage in autumn and remove dying flower spikes before ripening seeds are dispersed. May be propagated by division in spring or by seed in late summer.
A. natans. See *Luronium natans.*
A. plantago-aquatica illus. p.388.
A. ranunculoides. See *Baldellia ranunculoides.*

ALLEMANDA (Apocynaceae)
Genus of evergreen, woody-stemmed, scrambling climbers, grown for their trumpet-shaped flowers. Frost tender, min. 13–15°C (55–9°F). Prefers partial shade in summer and humus-rich, well-drained, neutral to acid soil. Water regularly, less when not in full growth. Stems must be tied to supports. Prune previous season's growth back to 1 or 2 nodes in spring. Propagate by softwood cuttings in spring or summer. Whitefly and red spider mite may be troublesome.
A. cathartica (Golden trumpet). ⚑ **'Hendersonii'** illus. p.176.

ALLIUM (Liliaceae)
Onion
Genus of perennials, some of which are edible, with bulbs, rhizomes or fibrous rootstocks. Nearly all have narrow, basal leaves smelling of onions when crushed, and most have small flowers packed together in a dense, spherical or shuttlecock-shaped umbel. Dried umbels of tall border species are good for winter decoration. Fully to frost hardy. Requires an open, sunny situation and well-drained soil; is best left undisturbed to form clumps. Plant in autumn. Propagate by seed in autumn or by division of clumps – spring-flowering varieties in late summer and summer-flowering ones in spring.
A. acuminatum, syn. *A. murrayanum,* illus. p.374.
A. aflatunense illus. p.346.
A. akaka illus. p.371.
A. albopilosum. See *A. christophii.*
A. azureum. See *A. caeruleum.*
⚑ *A. beesianum.* Clump-forming, late summer-flowering bulb. H 20–30cm (8–12in), S 5–10cm (2–4in). Fully hardy. Has linear, grey-green leaves and, in late summer, pendent heads of bell-shaped, blue flowers.
⚑ *A. caeruleum,* syn. *A. azureum,* illus. p.366.
A. campanulatum. Clump-forming, summer-flowering bulb. H 10–30cm (4–12in), S 5–10cm (2–4in). Frost hardy. Linear, semi-erect, basal leaves die away before flowering time. Bears a domed umbel, 2.5–7cm (1–3in) wide, of up to 30 small, star-shaped, pale pink or white flowers.
⚑ *A. carinatum* subsp. *pulchellum,* syn. *A. pulchellum.* Clump-forming, summer-flowering bulb. H 30–60cm (1–2ft), S 8–10cm (3–4in). Fully hardy. Has linear semi-erect leaves sheathing stem in lower two-thirds. Bears an umbel of pendent, cup-shaped, purple flowers.
⚑ *A. cernuum* illus. p.364.

♀ *A. christophii*, syn. *A. albopilosum*, illus. p.365.
A. cowanii. See *A. neapolitanum*.
♀ *A. cyaneum*. Tuft-forming, summer-flowering bulb. H 10–30cm (4–12in), S 5–8cm (2–3in). Fully hardy. Leaves are thread-like and erect. Stems each bear a small, dense umbel of 5 or more pendent, cup-shaped, blue or violet-blue flowers, 0.5cm (¼in) long.
A. cyathophorum var. *farreri* illus. p.380.
♀ *A. flavum*. Clump-forming, summer-flowering bulb. H 10–30cm (4–12in), S 8–10cm (3–4in). Fully hardy. Leaves are linear and semi-erect on lower half of slender flower stem. Produces a loose umbel of up to 30 bell-shaped, yellow flowers, each 0.5cm (¼in) long, on thin, arching stalks.
A. giganteum illus. p.346.
A. kansuense. See *A. sikkimense*.
♀ *A. karataviense* illus. p.371.
A. macranthum. Tuft-forming, summer-flowering bulb. H 20–30cm (8–12in), S 10–12cm (4–5in). Fully hardy. Has linear leaves on lower part of flower stem, which bears a loose umbel of up to 20 bell-shaped, deep purple flowers, each 1cm (½in) long, on slender stalks.
A. mairei. Clump-forming, late summer- to autumn-flowering bulb. H 10–20cm (4–8in), S 10–12cm (4–5in). Fully hardy. Leaves are erect, thread-like and basal. Wiry stems each carry a small, shuttlecock-shaped umbel of up to 20 upright, bell-shaped, pink flowers, each 1cm (½in) long.
♀ *A. moly* illus. p.380.
A. murrayanum. See *A. acuminatum*.
A. narcissiflorum, syn. *A. pedemontanum*, illus. p.380.
A. neapolitanum, syn. *A. cowanii*, illus. p.352.
♀ *A. oreophilum*, syn. *A. ostrowskianum*, illus. p.374.
A. ostrowskianum. See *A. oreophilum*.
A. pedemontanum. See *A. narcissiflorum*.
A. pulchellum. See *A. carinatum* subsp. *pulchellum*.
A. schoenoprasum illus. p.380.
A. schubertii illus. p.365.
A. sikkimense, syn. *A. kansuense*. Tuft-forming, summer-flowering bulb. H 10–25cm (4–10in), S 5–10cm (2–4in). Fully hardy. Leaves are linear, erect and basal. Up to 15 bell-shaped, blue flowers, 0.5–1cm (¼–½in) long, are borne in a small, pendent umbel.
A. sphaerocephalon. Clump-forming, summer-flowering bulb. H to 60cm (24in), S 8–10cm (3–4in). Fully hardy. Has linear, semi-erect leaves on basal third of slender, wiry stems and a very dense umbel, 2–4cm (¾–1½in) across, of up to 40 small, bell-shaped, pinkish-purple flowers.
A. stipitatum illus. p.345.
A. unifolium illus. p.353.

ALNUS (Betulaceae)
Alder
Genus of deciduous trees and shrubs, grown mainly for their ability to thrive in wet situations. Flowers are borne in catkins in late winter or early spring, males conspicuous and attractive, females forming persistent, woody, cone-like fruits. Fully hardy. Most do best in sun and any moist or even waterlogged soil, but *A. cordata* will also grow well on poor, dry soils. Propagate species by seed in autumn, cultivars by budding in late summer or by hardwood cuttings in early winter.
♀ *A. cordata* illus. p.40.
A. glutinosa (Black alder, Common alder). 'Aurea' is a slow-growing, deciduous, conical tree. H to 25m (80ft), S 10m (30ft). Has rounded leaves, bright yellow until mid-summer, later becoming pale green. Bears yellow-brown catkins in early spring. Is useful in a boggy area.
♀ 'Imperialis', H 10m (30ft), S 4m (12ft), is slow-growing and has deeply cut, lobed leaves.
A. incana (Grey alder) illus. p.40. 'Aurea' is a deciduous, conical tree. H 20m (70ft), S 8m (25ft). Has reddish-yellow or orange shoots in winter and broadly oval, yellow leaves. Reddish-yellow or orange catkins are borne in late winter and early spring. Is useful for cold, wet areas and poor soils. 'Ramulis Coccineis' has red, winter shoots and buds, and orange catkins.

ALOCASIA (Araceae)
Genus of evergreen perennials with underground rhizomes, grown for their attractive foliage. Produces tiny flowers on a spadix enclosed in a leaf-like spathe. Frost tender, min. 15°C (59°F). Needs high humidity, partial shade and well-drained soil. Propagate by seed, stem cuttings or division of rhizomes in spring.
A. cuprea illus. p.230.
A. lowii var. *veitchii*. See *A. veitchii*.
A. macrorrhiza (Giant elephant's ear, Taro). Evergreen, tufted perennial with a thick, trunk-like stem. H to 3m (10ft) or more, S 2m (6ft). Broad, arrow-shaped, glossy, green leaves, to 1m (3ft) long, are carried on stalks 1m (3ft) long. Has yellowish-green spathes to 20cm (8in) high.
A. picta. See *A. veitchii*.
A. veitchii, syn. *A. lowii* var. *veitchii*, *A. picta*. Evergreen, tufted perennial. H 1m (3ft) or more, S 75cm (30in). Narrow leaves, triangular with arrow-shaped bases, are 45cm (18in) long and green with greyish midribs, veins and margins, purple beneath. Spathes are greenish.

ALOE (Liliaceae)
Genus of evergreen, rosetted trees, shrubs, perennials and scandent climbers with succulent foliage and tubular to bell-shaped flowers. Frost tender, min. 7–10°C (45–50°F). Tree aloes and shrubs with a spread over 30cm (1ft) prefer full sun; most smaller species prefer partial shade. Needs very well-drained soil. Propagate by seed, stem cuttings or offsets in spring or summer.
A. arborescens. Evergreen, bushy, succulent-leaved shrub. H and S 2m (6ft). Stems are each crowned by rosettes of widely spreading, long, slender, curved, dull blue-green leaves with toothed margins. Long flower stems produce masses of tubular to bell-shaped, red flowers in late winter and spring. var. *variegatum* illus. p.395.
A. aristata illus. p.416.
♀ *A. barbadensis*. See *A. vera*.
A. brevifolia. Basal-rosetted, perennial succulent, producing many offsets. H 15cm (6in), S 30cm (12in). Has broadly sword-shaped, fleshy, blue-green leaves with a few teeth along edges. In spring, flower stems, 50cm (20in) long, carry narrowly bell-shaped, bright red flowers.
A. ciliaris illus. p.395.
A. ferox illus. p.395.
A. humilis. Rosetted, perennial succulent. H 10cm (4in), S 30cm (12in). Has a dense, basal rosette of narrowly sword-shaped, spine-edged, fleshy, blue-green leaves, often erect, with incurving tips. Flower stems are 30cm (12in) long and each bears a spike of narrowly bell-shaped, orange flowers in spring. Offsets freely.
A. striata illus. p.404.
♀ *A. variegata* illus. p.400.
A. vera, syn. *A. barbadensis*, illus. p.402.

ALOINOPSIS (Aizoaceae)
Genus of dwarf, tuberous, perennial succulents with daisy-like flowers from late summer to early spring. Frost tender, min. 7°C (45°F). Requires a sunny site and very well-drained soil. Is very susceptible to overwatering. Propagate by seed in summer.
A. schooneesii illus. p.413.

ALONSOA (Scrophulariaceae)
Genus of perennials, grown as annuals. May be used for cut flowers. Half hardy. Grow in sun and in rich, well-drained soil. Flowering may be poor outdoors in a wet summer. Young plants should have growing shoots pinched out to encourage bushy growth. Propagate by seed sown outdoors in late spring. Aphids may be troublesome, particularly under glass.
♀ *A. warscewiczii* illus. p.281.

ALOPECURUS (Gramineae). See GRASSES, BAMBOOS, RUSHES and SEDGES.
A. pratensis 'Aureovariegatus', syn. *A.p.* 'Aureomarginatus', illus. p.185.

ALOYSIA (Verbenaceae)
Genus of deciduous or evergreen, summer-flowering shrubs, grown for their aromatic foliage and sprays of tiny flowers. Frost to half hardy; in cold areas plant against a south- or west-facing wall or raise afresh each year. Needs full sun and well-drained soil. Cut out any dead wood in early summer. Propagate by softwood cuttings in summer.
A. triphylla, syn. *Lippia citriodora*, illus. p.113.

ALPINIA (Zingiberaceae)
Genus of mainly evergreen perennials with fleshy rhizomes, grown for their flowers. Frost tender, min. 18°C (64°F). Needs well-drained soil with plenty of humus, partial shade and a moist atmosphere. Is not easy to grow successfully in pots. Propagate by division in late spring or early summer. Red spider mite may be a problem.
A. calcarata (Indian ginger). Evergreen, upright, clump-forming perennial. H and S to 1m (3ft). Has stalkless, aromatic, lance-shaped leaves, to 30cm (1ft) long. At any time of year may bear horizontal spikes of whitish flowers, with 2.5cm (1in) long, yellow lips marked reddish-purple.
A. nutans. See *A. zerumbet*.
A. speciosa. See *A. zerumbet*.
A. zerumbet, syn. *A. nutans*, *A. speciosa*, illus. p.191.

Alsobia dianthiflora. See *Episcia dianthiflora*.

ALSOPHILA. See CYATHEA.

ALSTROEMERIA (Alstroemeriaceae)
Genus of mostly summer-flowering, tuberous perennials with showy, multicoloured flowers. Flowers are good for cutting as they last well. Frost hardy, but in very cold winters protect by covering dormant tubers with dry bracken or loose peat. Needs well-drained soil and a sunny, sheltered site. Propagate by seed or division in early spring.
A. aurantiaca. See *A. aurea*.
A. aurea, syn. *A. aurantiaca*, illus. p.347.
A. gayana. See *A. pelegrina*.
A. hookeri illus. p.334.
♀ *A.*, Ligtu Hybrids illus. p.367.
A. 'Margaret' illus. p.344.
A. pelegrina, syn. *A. gayana*, illus. p.364.
A. 'Walter Fleming'. Summer-flowering, tuberous perennial. H to 1m (3ft), S 60cm–1m (2–3ft). Each leafy stem produces narrowly lance-shaped, twisted leaves and widely funnel-shaped, deep yellow flowers, 5–6cm (2–2½in) across, flushed purple with reddish-purple spots.

ALTERNANTHERA (Amaranthaceae)
Genus of bushy perennials, grown for their attractive, coloured foliage. Is useful for carpeting or bedding. Frost tender, min. 15–18°C (59–64°F). Needs sun or partial shade and moist but well-drained soil. Propagate by tip cuttings or division in spring.
A. amoena. See *A. ficoidea* var. *amoena*.
A. ficoidea (Parrot leaf). var. *amoena* (syn. *A. amoena*) is a mat-forming perennial. H 5cm (2in), S indefinite. Has narrowly oval, green leaves, marked red, yellow and orange, with wavy margins. 'Versicolor' (syn. *A. versicolor*) is an erect form, H and S to 30cm (1ft), with rounded to spoon-shaped leaves shaded brown, red and yellow.
A. versicolor. See *A. ficoidea* 'Versicolor'.

Althaea rosea. See *Alcea rosea*.

ALYSSOIDES (Cruciferae)
Genus of one species of short-lived, evergreen sub-shrub, grown for its flowers and swollen fruits. Is suitable for dry banks and rock gardens. Frost hardy. Needs sun and well-drained soil. Propagate by seed in autumn.

A. utriculata. Evergreen, rounded sub-shrub. H and S 30cm (12in). Has oval, glossy, dark green leaves. Bears loose sprays of small, bright yellow flowers in spring, then balloon-like, buff seed pods.

ALYSSUM (Cruciferae)
Genus of perennials, some of which are evergreen, and annuals, grown for their flowers. Fully hardy. Needs sun and well-drained soil. Cut back lightly after flowering. Propagate by softwood cuttings in late spring or by seed in autumn.
A. maritimum. See *Lobularia maritima*.
A. montanum. Evergreen, prostrate perennial. H and S 15cm (6in). Leaves are small, oval, hairy and grey. Flower stems, 15cm (6in) long, each bear an open, spherical raceme of small, very fragrant, soft yellow flowers in summer. Is good in a rock garden.
A. saxatile. See *Aurinia saxatilis*.
A. spinosum, syn. *Ptilotrichum spinosum*. Semi-evergreen, rounded, compact shrub. H 20cm (8in) or more, S 30cm (12in). Intricate branches bear spines and narrowly oval to linear, silver leaves. Spherical heads of tiny, 4-petalled, white to purple-pink flowers appear in early summer.
A. wulfenianum. Prostrate perennial. H 2cm (³/4in), S 20cm (8in). Loose heads of small, bright yellow flowers appear in summer above small, oval, grey leaves.

AMARANTHUS (Amaranthaceae)
Genus of annuals, grown for their dense panicles of tiny flowers or for their colourful foliage. Half hardy. Grow in sun and in rich or fertile, well-drained soil. Propagate from seed sown outdoors in late spring. Aphids may sometimes be troublesome.
A. caudatus illus. p.278.
A. hybridus var. *erythrostachys*. See *A. hypochondriacus*.
A. hypochondriacus, syn. *A. hybridus* var. *erythrostachys*, illus. p.282.
A. tricolor **'Joseph's Coat'**. Bushy annual. H to 1m (3ft), S 45cm (1¹/2ft) or more. Has oval, scarlet, green and yellow leaves, to 20cm (8in) long, and small panicles of tiny, red flowers in summer. Leaves of **'Molten Fire'** are crimson, bronze and purple.

× AMARCRINUM (Amaryllidaceae)
Hybrid genus (*Amaryllis* × *Crinum*) of one robust, evergreen bulb, grown for its large, funnel-shaped flowers. Frost hardy. Needs sun and well-drained soil. Plant with neck just covered by soil. Propagate by division in spring.
× *A. howardii*. See × *A. memoria-corsii*.
× *A. memoria-corsii*, syn. × *A. howardii*, × *Crinodonna corsii*, illus. p.342.

× AMARYGIA (Amaryllidaceae)
Hybrid genus (*Amaryllis* × *Brunsvigia*) of stout, autumn-flowering bulbs, which are cultivated for their large, showy flowers. Frost hardy. Needs full sun and, preferably, the shelter of a wall. Plant bulbs just beneath surface of well-drained soil.

Propagate by division in spring.
× *A. parkeri*, syn. × *Brunsdonna parkeri*, illus. p.342.

AMARYLLIS (Amaryllidaceae)
Genus of autumn-flowering bulbs, grown for their funnel-shaped flowers. Frost hardy, but in cool areas should be grown against a south-facing wall for protection. Needs well-drained soil and a sheltered, sunny situation. Plant bulbs under at least 8cm (3in) of soil. Propagate by division in late spring, as leaves die down, or in late summer, before growth recommences.
A. belladonna and **'Hathor'** illus. p.352.

Amberboa moschata. See *Centaurea moschata*.

AMELANCHIER (Rosaceae)
Juneberry, Serviceberry, Shadbush
Genus of deciduous, spring-flowering trees and shrubs, grown for their profuse flowers and their foliage, which is frequently brightly coloured in autumn. Fully hardy. Requires sun or semi-shade and well-drained but not too dry, preferably neutral to acid soil. Propagate in autumn by seed, in late autumn to early spring by layering or, in the case of suckering species, by division. Fireblight is sometimes a problem.
A. alnifolia. Deciduous, upright, suckering shrub. H 4m (12ft) or more, S 3m (10ft) or more. Leaves are oval to rounded and dark green. Erect spikes of star-shaped, creamy-white flowers are borne in late spring, followed by small, edible, juicy, rounded, purple-black fruits.
A. arborea. Deciduous, spreading, sometimes shrubby, tree. H 10m (30ft), S 12m (40ft). Clusters of star-shaped, white flowers appear in mid-spring as oval, white-haired, young leaves unfold. Foliage matures to dark green, turning to red or yellow in autumn. Rounded fruits are small, dry and reddish-purple.
A. asiatica. Deciduous, spreading tree or shrub of elegant habit. H 8m (25ft), S 10m (30ft). Leaves are oval and dark green, usually woolly when young and turning yellow or red in autumn. Star-shaped, white flowers are borne profusely in late spring, followed by edible, juicy, rounded, blackcurrant-like fruits.
A. canadensis. Deciduous, upright, dense shrub. H 6m (20ft), S 3m (10ft). Star-shaped, white flowers are borne from mid- to late spring amid unfolding, oval, white-haired leaves that mature to dark green and turn orange-red in autumn. Fruits are edible, rounded, blackish-purple, sweet and juicy.
A. laevis illus. p.59.
❦ *A. lamarckii* illus. p.86.

Amomyrtus luma. See *Myrtus luma*.
Amomyrtus luma **'Glanleam Gold'**. See *Myrtus luma* 'Glanleam Gold'.

AMORPHA (Leguminosae)
Genus of deciduous shrubs and sub-shrubs, grown for their flowers and foliage. Is useful for cold, dry, exposed

positions. Fully hardy. Requires full sun and well-drained soil. Propagate by softwood cuttings in summer or by seed in autumn.
A. canescens (Lead plant). Deciduous, open sub-shrub. H 1m (3ft), S 1.5m (5ft). Dense spikes of tiny, pea-like, purple flowers, with orange anthers, are produced in late summer and early autumn, amid oval, grey-haired leaves divided into 21–41 narrowly oval leaflets.

AMORPHOPHALLUS (Araceae)
Genus of tuberous perennials, cultivated for their huge and dramatic, but foul-smelling, spathes, which surround tiny flowers on stout spadices. Frost tender, min. 10°C (50°F). Requires partial shade and humus-rich soil kept continuously moist during growing season. Keep tubers dry in winter. Propagate by seed in spring or by offsets in spring or summer.
A. rivieri. Summer-flowering, tuberous perennial. H to 40cm (16in), S 60cm–1m (2–3ft). Has a flattish, wavy-edged, dark reddish-brown spathe, to 40cm (16in) long, from which protrudes an erect, dark brown spadix. Brownish-green-mottled, pale green stem, 1m (3ft) long, bears one large, deeply lobed leaf after flowering.

AMPELOPSIS (Vitaceae)
Genus of deciduous, woody-stemmed, tendril climbers, some of which are twining, grown for their leaves. Frost hardy. Grow in any soil in a sheltered position, in sun or partial shade. Needs plenty of room as grows quickly and can cover a large area. Propagate by greenwood or semi-ripe cuttings in mid-summer.
A. aconitifolia, syn. *Vitis aconitifolia*. Fast-growing, deciduous, woody-stemmed, twining, tendril climber. H to 12m (40ft). Rounded leaves have 3 or 5 toothed, lobed, dark green leaflets; inconspicuous, greenish flowers appear in late summer, followed by orange berries.
A. brevipedunculata var. *maximowiczii*, syn. *A. heterophylla*, *Vitis heterophylla*. Vigorous, deciduous, woody-stemmed, twining, tendril climber with hairy, young stems. H to 5m (15ft) or more. Has dark green leaves that vary in size and shape and are almost hairless beneath. Inconspicuous, greenish flowers are produced in summer, followed by bright blue berries.
A. heterophylla. See *A. brevipedunculata* var. *maximowiczii*.
A. sempervirens. See *Cissus striata*.
A. tricuspidata. See *Parthenocissus tricuspidata*.
A. veitchii. See *Parthenocissus tricuspidata* 'Veitchii'.

AMSONIA (Apocynaceae)
Blue star
Genus of slow-growing, clump-forming, summer-flowering perennials. Fully hardy. Grow in well-drained soil and in semi-shade. Is best left undisturbed for some years. May be propagated by division in spring, by softwood cuttings in

summer or by seed in autumn.
❦ *A. orientalis*, syn. *Rhazya orientalis*, illus. p.250.
A. salicifolia. See *A. tabernaemontana*.
A. tabernaemontana, syn. *A. salicifolia*, illus. p.251.

Anacharis densa. See *Egeria densa*.

ANACYCLUS (Compositae)
Genus of summer-flowering, prostrate perennials with stems radiating from a central rootstock. Frost hardy. Needs full sun and well-drained soil. Propagate by softwood cuttings in spring or by seed in autumn.
A. depressus illus. p.322.

ANAGALLIS (Primulaceae)
Genus of annuals and creeping perennials, grown for their flowers. Fully to frost hardy. Plant in an open, sunny site in fertile, moist soil. Propagate by seed or division in spring. Raise *A. tenella* by soft tip cuttings in spring or early summer.
A. tenella (Bog pimpernel).
❦ **'Studland'** illus. p.315.

ANANAS (Bromeliaceae)
Genus of evergreen, rosette-forming perennials, grown for their foliage and edible fruits (pineapples). Frost tender, min. 13–15°C (55–9°F). Prefers full light, but tolerates some shade. Needs fertile, well-drained soil. Water moderately during growing season, sparingly at other times. Propagate by suckers or cuttings of 'leafy' fruit tops in spring or summer.
A. bracteatus (Red pineapple, Wild pineapple). ❦ **'Tricolor'** (syn. *A.b.* 'Striatus', *A.b.* var. *tricolor*; illus. p.229) is an evergreen, basal-rosetted perennial. H and S 1m (3ft). Forms dense rosettes of strap-shaped, spiny-edged, arching, deep green leaves, longitudinally yellow-striped and often with marginal, red spines. Dense spikes of small, tubular, lavender-violet flowers, with prominent, reddish-pink bracts, appear usually in summer. These are followed by brownish-orange-red fruits that are 15cm (6in) or more long.
A. comosus **'Variegatus'**. Evergreen, basal-rosetted perennial. H and S 60cm (24in) or more. Produces very narrowly strap-shaped, channelled, rigid, grey-green leaves that are suffused pink, have cream margins, are grey-scaled beneath and sometimes have spiny edges. Produces tubular, purple-blue flowers with inconspicuous, green bracts; fruits are the edible pineapples grown commercially, but are much smaller on pot-grown plants.

ANAPHALIS (Compositae)
Pearl everlasting
Genus of perennials with heads of small, papery flowers, used dried for winter decoration. Fully hardy. Prefers sun but will grow in semi-shade. Soil should be well-drained but not too dry. Propagate by seed in autumn or by division in winter or spring.
❦ *A. margaritacea*, syn. *A. yedoensis*, illus. p.203.

ANCHUSA

A. nepalensis var. *monocephala*, syn. *A. nubigena*, illus. p.240.
A. nubigena. See *A. nepalensis* var. *monocephala*.
A. yedoensis. See *A. margaritacea*.

ANCHUSA (Boraginaceae)
Genus of annuals, biennials and perennials, some of which are evergreen, with usually blue flowers. Fully to frost hardy. Needs sun and well-drained soil; resents too much winter wet. Tall, perennial species may need to be staked and allowed room to spread. Propagate perennials by root cuttings in winter, annuals and biennials by seed in autumn or spring.
A. azurea, syn. *A. italica*. **'Little John'** is a coarse-growing, basal-rosetted perennial. H 50cm (20in), S 60cm (24in). Fully hardy. Produces branching racemes of large, open cup-shaped, dark blue flowers in early summer. Leaves are narrowly oval and hairy. ♦ **'Loddon Royalist'** illus. p.218. **'Opal'**, H 1.2m (4ft), has paler blue flowers.
♦ *A. caespitosa* illus. p.318.
A. capensis **'Blue Angel'** illus. p.286.
'Blue Bird' is a bushy biennial, grown as an annual. H to 45cm (18in), S 20cm (8in). Frost hardy. Has lance-shaped, bristly, mid-green leaves and, in summer, heads of shallowly bowl-shaped, sky-blue flowers.
A. italica. See *A. azurea*.

ANCISTROCACTUS (Cactaceae)
Genus of perennial cacti, grown for their spherical to short, globose to columnar stems, each with 10–20 tuberculate ribs. Flowers are sometimes prevented from opening by density of spines. Frost tender, min. 2°C (36°F) if completely dry. Needs sun and very well-drained soil. May rot if overwatered. Propagate by seed in spring.
A. megarhizus. See *A. scheeri*.
A. scheerii, syn. *A. megarhizus*, illus. p.413.
A. uncinatus. Globose to columnar, perennial cactus. H 20cm (8in), S 10cm (4in). Stem is blue-green. Areoles each produce a very long, hooked, reddish spine and 15–18 straight ones. Has cup-shaped, brown-green or reddish flowers, 2cm (^{3}/4in) across, in spring.

ANDROMEDA (Ericaceae)
Genus of evergreen shrubs with an open, twiggy habit. Fully hardy. Needs full light and humus-rich, moist, acid soil. Propagate by semi-ripe cuttings in late summer or by seed in spring.
A. polifolia illus. p.295. **'Alba'** illus. p.294. ♦ **'Compacta'** illus. p.295.

ANDROSACE (Primulaceae)
Genus of annuals and evergreen perennials, usually compact cushion-forming and often with soft, hairy leaves. Many species are suitable for cold greenhouses and troughs with winter cover. Fully to frost hardy. Needs sun and very well-drained soil; some species prefer acid soil. Resents wet foliage in winter. Propagate by tip cuttings in summer or by seed in autumn. Is prone to botrytis and attack by aphids.
A. carnea illus. p.313. ♦ subsp. *laggeri* illus. p.315.
A. chamaejasme. Evergreen, basal-rosetted, variable perennial with easily rooted stolons. H 3–6cm (1^1/4–2^1/2in), S to 15cm (6in). Fully hardy. Has open, hairy rosettes of oval leaves. In spring bears clusters of 2–8 flattish, white flowers, each with a yellow eye that sometimes turns red with age.
A. cylindrica. Evergreen, basal-rosetted perennial. H 1–2cm (^{1}/2–^{3}/4in), S 10cm (4in). Fully hardy. Leaves are linear and glossy. Flower stems each carry up to 10 small, flattish, white flowers, each with a yellow-green eye, in early spring. Is suitable for a cold greenhouse.
A. hedraeantha. Evergreen, tight cushion-forming perennial. H 1–2cm (^{1}/2–^{3}/4in), S to 10cm (4in). Fully hardy. Bears loose rosettes of narrowly oval, glossy leaves. Umbels of 5–10 flattish, yellow-throated, pink flowers are borne in spring. Is best in a cold greenhouse.
A. hirtella. Evergreen, tight cushion-forming perennial. H 1cm (^{1}/2in), S to 10cm (4in). Fully hardy. Produces rosettes of small, thick, linear to oblong, hairy leaves. Almond-scented, flattish, white flowers appear in spring on very short stems, 1 or 2 per rosette.
A. imbricata. See *A. vandellii*.
♦ *A. lanuginosa* illus. p.325.
A. pyrenaica illus. p.309.
♦ *A. sarmentosa.* Evergreen, mat-forming perennial, spreading by runners. H 4–10cm (1^1/2–4in), S 30cm (12in). Fully hardy. Has open rosettes of small, narrowly elliptic, hairy leaves. Large clusters of flattish, yellow-eyed, bright pink flowers open in spring. Is a good rock plant in all but very wet areas.
♦ *A. sempervivoides.* Evergreen, mat-forming, rosetted perennial with stolons. H 1–7cm (^{1}/2–3in), S 30cm (12in). Fully hardy. Has leathery, oblong or spoon-shaped leaves. In spring produces small heads of 4–10 flattish, pink flowers, with yellow, then red, eyes. Is a good rock plant.
A. vandellii, syn. *A. imbricata*, illus. p.310.
A. villosa illus. p.311. var. *jacquemontii* illus. p.326.

ANEMONE (Ranunculaceae)
Windflower
Genus of spring-, summer- and autumn-flowering perennials, sometimes tuberous or rhizomatous, with mainly rounded, shallowly cup-shaped flowers. Leaves are rounded to oval, often divided into 3–15 leaflets. Fully to frost hardy. Most species thrive in humus-rich, well-drained soil in full light or semi-shade. Propagate by division in spring, by seed sown in late summer, when fresh, or by root cuttings in winter.
♦ *A. apennina* (Apennine anemone). Spreading, spring-flowering, rhizomatous perennial. H and S 15–20cm (6–8in). Fully hardy. Fern-like leaves have 3 deeply toothed lobes. Each stem carries a large, upright, flattish, blue, white or pink flower, with 10–20 narrow petals.
A. biflora illus. p.371.
♦ *A. blanda.* Spreading, early spring-flowering perennial with a knobbly tuber. H 5–10cm (2–4in), S 10–15cm (4–6in). Fully hardy. Leaves are broadly oval and semi-erect, with 3 deeply toothed lobes. Each stem bears an upright, flattish, blue, white or pink flower, 4–5cm (1^1/2–2in) across, with 9–14 narrow petals.
♦ **'Atrocaerulea'** illus. p.377.
♦ **'Radar'** illus. p.374.
♦ **'White Splendour'** illus. p.370.
A. coronaria. Spring-flowering perennial with a misshapen tuber. H 5–25cm (2–10in), S 10–15cm (4–6in). Frost hardy. Produces parsley-like, divided, semi-erect leaves. Each stiff stem carries a large, 5–8-petalled, shallowly cup-shaped flower in shades of red, pink, blue or purple. Garden groups include **De Caen Series** and **St Brigid Series**, which have larger flowers varying in colour from white through red to blue.
A. × *fulgens* illus. p.374.
A. hepatica. See *Hepatica nobilis*.
A. hupehensis illus. p.223.
♦ var. *japonica* **'Prinz Heinrich'** (syn. *A.* 'Prince Henry') has single, deep pink flowers on slender stems.
♦ **'September Charm'** (single) illus. p.195.
A. × *hybrida*, syn. *A. japonica* of gardens. Group of vigorous, branching, perennials. H 1.5m (5ft), S 60cm (2ft). Fully hardy. Bears shallowly cup-shaped, single, semi-double or double flowers in late summer and early autumn. Leaves are deeply divided and dark green. **'Bressingham Glow'** (semi-double) and ♦ **'Honorine Jobert'** (single) illus. p.195. **'Max Vogel'** has semi-double, pinkish-mauve flowers on wiry stems.
A. japonica of gardens. See *A.* × *hybrida*.
A. × *lipsiensis*, syn. *A.* × *seemannii*, illus. p.238.
A. narcissiflora illus. p.239.
♦ *A. nemorosa* (Wood anemone). Vigorous, carpeting, rhizomatous perennial. H 15cm (6in), S 30cm (12in). Fully hardy. Produces masses of star-shaped, single, white flowers, with prominent, yellow stamens, in spring and early summer, above deeply cut, mid-green leaves. Likes woodland conditions. ♦ **'Allenii'** and ♦ **'Robinsoniana'** illus. p.234.
♦ **'Vestal'** has anemone-centred, double, white flowers. **'Wilk's Giant'** has larger, single, white flowers.
A. pavonina illus. p.357.
A. **'Prince Henry'**. See *A. hupehensis* var. *japonica* 'Prinz Heinrich'.
♦ *A. ranunculoides* illus. p.238.
'Flore Pleno' is a spreading, rhizomatous perennial. H and S 20cm (8in). Fully hardy. Bears buttercup-like, double, yellow flowers in spring. Leaves are divided. Likes damp, woodland conditions.
A. rivularis illus. p.239.
A. × *seemannii.* See *A.* × *lipsiensis*.
A. sylvestris illus. p.232. **'Macrantha'**, (syn. *A.s.* 'Grandiflora'), is a carpeting perennial that can be invasive. H and S 30cm (12in). Fully hardy. Has large, fragrant, semi-pendent, shallowly cup-shaped, white flowers in spring and early summer. Leaves are divided and mid-green.
A. vitifolia. Branching, clump-forming perennial. H 1.2m (4ft), S 50cm (20in). Fully hardy. In summer bears open cup-shaped, occasionally pink-flushed, white flowers with yellow stamens. Vine-like leaves are woolly beneath.

ANEMONELLA (Ranunculaceae)
Genus of one species of tuberous perennial, grown for its flowers. Fully hardy. Needs shade and humus-rich, moist soil. Propagate by seed when fresh or by division every 3–5 years in autumn.
A. thalictroides illus. p.312. **'Oscar Schoaf'** (syn. *A.t.* 'Schoaf's Double') is a slow-growing, tuberous perennial. H 10cm (4in), S 4cm (1^1/2in) or more. Has delicate, fern-like leaves. From spring to early summer bears small, cup-shaped, double, strawberry-pink flowers, singly on finely branched, slender stems.

ANEMONOPSIS (Ranunculaceae)
False anemone
Genus of one species of perennial, related to *Anemone*. Fully hardy. Likes a sheltered, semi-shaded position and humus-rich, moist but well-drained soil. Propagate by division in spring or by seed sown in late summer, when fresh.
A. macrophylla illus. p.250.

ANEMOPAEGMA (Bignoniaceae)
Genus of evergreen, tendril climbers, grown for their flowers. Frost tender, min. 13–15°C (55–9°F). Needs humus-rich, well-drained soil and partial shade in summer. Water regularly and freely when in full growth, less at other times. Provide support and in summer thin out stems at intervals; shorten all growths by half in spring. Propagate by softwood or semi-ripe cuttings in spring or summer.
A. chamberlaynii. Fast-growing, evergreen, tendril climber. H to 6m (20ft). Leaves have 2 pointed, oval leaflets and a 3-hooked tendril. Foxglove-like, primrose-yellow flowers are carried in pairs from upper leaf axils in summer.

ANGELICA (Umbelliferae)
Genus of summer-flowering, often short-lived perennials, some of which have culinary and medicinal uses. Fully hardy. Grows in sun or shade and any well-drained soil. Remove seed heads when produced, otherwise plants may die. Propagate by seed when ripe.
A. archangelica illus. p.192.

ANGRAECUM. See ORCHIDS.
A. sesquipedale, syn. *Macroplectrum sesquipedale* (Star-of-Bethlehem orchid; illus. p.260). Evergreen, epiphytic orchid for an intermediate greenhouse. H 30cm (12in) or more. Waxy, apple-white flowers, 8cm (3in) across, each with a 30cm (12in) long spur, are borne, usually 2 to a stem, in winter. Has narrowly oval, semi-rigid, horizontal leaves, 15cm (6in) long. Needs shade in summer.

ANGULOA

ANGULOA. See ORCHIDS.
A. clowesii (Cradle orchid).
Deciduous, epiphytic orchid for a cool greenhouse. H 60cm (24in). Fragrant, erect, cup-shaped, lemon-yellow flowers, 10cm (4in) long, each with a loosely hinged, yellow lip, are borne singly in early summer. Broadly oval, ribbed leaves are 45cm (18in) long. Grow in semi-shade in summer.

ANIGOZANTHOS (Haemodoraceae)
Kangaroo paw
Genus of perennials, with thick rootstocks and fans of sword-shaped leaves, grown for their curious flowers. Half hardy. Needs an open, sunny position and does best in well-drained, peaty or leafy, acid soil. Propagate by division in spring or by seed when fresh, in late summer.
A. flavidus illus. p.218.
❧ *A. manglesii* illus. p.214.
A. rufa. Tufted perennial. H 1m (3ft), S 60cm (2ft). Panicles of 2-lipped, rich burgundy flowers, covered with purple hairs, appear in spring. Has long, sword-shaped, stiff, mid-green leaves.

ANISODONTEA (Malvaceae)
Genus of evergreen shrubs and perennials, grown for their flowers. Frost tender, min. 3–5°C (37–41°F). Needs full light and well-drained soil. Water containerized plants freely when in full growth, very little at other times. In growing season, young plants may need tip pruning to promote a bushy habit. Propagate by seed in spring or by greenwood or semi-ripe cuttings in late summer.
A. capensis, syn. *A. × hypomadarum* of gardens, *Malvastrum capensis*, *M. hypomadarum* of gardens. Evergreen, erect, bushy shrub. H to 1m (3ft), S 60cm (2ft) or more. Each oval leaf has 3–5 deep lobes. Bowl-shaped, 5-petalled, rose-magenta flowers, with darker veins, appear from spring to autumn.
A. × hypomadarum of gardens. See *A. capensis*.

ANNONA (Annonaceae)
Cherimoya, Custard apple, Sweet sop
Genus of deciduous or evergreen shrubs and trees, grown for their edible fruits and ornamental appearance. Frost tender, min. 15°C (59°F), preferably higher. Needs full light or partial shade and fertile, moisture-retentive but well-drained soil. Water containerized specimens moderately when in full growth, sparingly in winter. Propagate by seed in spring or by semi-ripe cuttings in late summer. Red spider mite may be a nuisance.
A. reticulata (Bullock's heart, Custard apple). Mainly deciduous, rounded tree. H 6m (20ft) or more, S 3–5m (10–15ft). Has oblong to lance-shaped, 13–25cm (5–10in) long leaves. Cup-shaped, olive-green flowers, often flushed purple, appear in summer, followed by edible, heart-shaped, red-flushed, greenish-brown fruits, each 13cm (5in) long.

Anoiganthus breviflorus. See *Cyrtanthus breviflorus*.

ANOMATHECA (Iridaceae)
Genus of upright, summer-flowering corms, grown for their red flowers, followed by egg-shaped seed pods that split to reveal red seeds. Frost hardy. Plant 5cm (2in) deep in an open, sunny situation and in well-drained soil. In cold areas, lift corms and store dry for winter. Propagate by seed in spring.
A. laxa, syn. *Lapeirousia cruenta*, *L. laxa*. Upright, summer-flowering corm. H 10–30cm (4–12in), S 5–8cm (2–3in). Narrowly sword-shaped, erect, basal leaves form a flat fan. Each stem bears a loose spike of long-tubed, red flowers, 2.5cm (1in) across. Lower petals have basal, darker red spots.

ANOPTERUS (Grossulariaceae, Saxifragaceae)
Genus of evergreen shrubs or small trees, grown for their foliage and flowers. Half hardy. Needs shade or semi-shade and moist but well-drained soil. Propagate by semi-ripe cuttings in summer.
A. glandulosus illus. p.86.

ANREDERA (Basellaceae)
Madeira vine, Mignonette vine
Genus of evergreen, tuberous, twining climbers, grown for their luxuriant foliage and small, scented flowers. Frost tender, min. 7°C (45°F). If grown in cool areas will die down in winter. Provide well-drained soil and full light. Water moderately in growing season, sparingly at other times. Provide support. Cut back previous season's growth by half or to just above ground level in spring. Propagate by tubers, produced at stem bases, in spring or by softwood cuttings in summer.
A. cordifolia, syn. *Boussingaultia basilloides* of gardens. Fast-growing, evergreen, tuberous, twining climber. H to 6m (20ft). Has oval to lance-shaped, fleshy leaves and tiny, fragrant, white flowers borne in clusters from upper leaf axils in summer.

ANTENNARIA (Compositae)
Cat's ears
Genus of evergreen or semi-evergreen perennials, grown for their almost stemless flower heads and mats of often woolly leaves. Makes good ground cover. Fully hardy. Needs sun and well-drained soil. Propagate by seed or division in spring.
A. dioica. Semi-evergreen, mat-forming, dense perennial. H 2.5cm (1in), S 25cm (10in). Leaves are tiny, oval, usually woolly and greenish-white. Short stems carry fluffy, white or pale pink flower heads in late spring and early summer. Is good for a rock garden. Compact **'Nyewoods'** has very deep rose-pink flowers. ❧ **'Rosea'** illus. p.314.

ANTHEMIS (Compositae)
Chamomile, Dog fennel
Genus of carpeting and clump-forming perennials, some of which are evergreen, grown for their daisy-like flower heads and fern-like foliage. Fully to frost hardy. Prefers sun and well-drained soil. May need staking. Cut to ground level after flowering for good leaf rosettes in winter. Propagate by division in spring or, for some species, by basal cuttings in late summer, autumn or spring.
A. nobile **'Treneague'.** See *Chamaemelum nobile* 'Treneague'.
❧ *A. punctata* subsp. *cupaniana* illus. p.240.
A. sancti-johannis. Evergreen, spreading, bushy perennial. H and S 60cm (2ft). Frost hardy. In summer bears many daisy-like, bright orange flower heads among fern-like, shaggy, mid-green leaves.
A. tinctoria. Evergreen, clump-forming perennial. H and S 1m (3ft). Fully hardy. Has a mass of daisy-like, yellow flower heads in mid-summer, borne singly above a basal clump of fern-like, crinkled, mid-green leaves. Propagate by basal cuttings in spring or late summer. **'E.C. Buxton'** illus. p.219.

ANTHERICUM (Liliaceae)
Spider plant
Genus of upright perennials with saucer- or trumpet-shaped flowers rising in spike-like racemes from clumps of leaves. Fully hardy. Likes a sunny site and fertile, well-drained soil that does not dry out in summer. Propagate by division in spring or by seed in autumn.
❧ *A. liliago* illus. p.240.
A. ramosum. Upright perennial. H 1m (3ft), S 30cm (1ft). Erect racemes of small, saucer-shaped, white flowers are borne in summer above a clump of grass-like, greyish-green leaves.

Antholyza paniculata. See *Crocosmia paniculata*.

ANTHURIUM (Araceae)
Genus of evergreen, erect, climbing or trailing perennials, some grown for their foliage and others for their bright flower spathes. Frost tender, min. 15°C (59°F). Prefers bright light in winter and indirect sun in summer; needs a fairly moist atmosphere and moist, but not waterlogged, peaty soil. Propagate by division in spring.
A. andraeanum illus. p.228.
A. crystallinum illus. p.228.
A. **'Rothschildianum'**, syn. *A. × rothschildianum*. Evergreen, erect, short-stemmed perennial. H and S 30cm (12in). Produces upright, oblong leaves, to 20cm (8in) long. Intermittently bears a long-lasting, red spathe, spotted with white, that surrounds a yellow spadix.
A. × *rothschildianum.* See *A.* 'Rothschildianum'.
A. scherzerianum illus. p.266.
A. veitchii (King anthurium). Evergreen, erect, short-stemmed perennial. H 1m (3ft) or more, S to 1m (3ft). Glossy, corrugated leaves, to 1m (3ft) long, are oval, with heart-shaped bases on 60cm–1m (2–3ft) long leaf stalks. Intermittently bears a long-lasting, leathery, green to white spathe that surrounds a cream spadix.

ANTHYLLIS (Leguminosae)
Genus of rounded, bushy perennials, grown for their flowers and finely divided leaves. Frost hardy. Needs sun and well-drained soil. Propagate by softwood cuttings in summer or by seed in autumn.
A. hermanniae. Rounded, bushy perennial. H and S to 60cm (2ft). Spiny, tangled stems bear simple or 3-parted, bright green leaves. Has small, pea-like, yellow flowers in summer. Is good for a rock garden.
A. montana illus. p.300. ❧ **'Rubra'** is a rounded or spreading, woody-based perennial. H and S 30cm (12in). Divided leaves consist of 17–41 narrowly oval leaflets. Heads of clover-like, bright pink flowers are borne in late spring and early summer. Is good for a rock garden.

ANTIGONON (Polygonaceae)
Genus of evergreen, woody-stemmed, tendril climbers, grown for their foliage and profuse clusters of small flowers. Frost tender, min. 15°C (59°F). Grow in any fertile, well-drained soil with full light. Water freely in growing season, sparingly at other times. Needs tropical conditions to flower well. Provide support. Thin out congested growth in early spring. Propagate by seed in spring or by softwood cuttings in summer.
A. leptopus illus. p.169.

ANTIRRHINUM (Scrophulariaceae)
Snapdragon
Genus of perennials and semi-evergreen sub-shrubs, usually grown as annuals, flowering from spring to autumn. Fully to half hardy. Needs sun and rich, well-drained soil. Dead-head to prolong flowering season. Propagate by seed sown outdoors in late spring or by stem cuttings in early autumn or spring. Rust disease may be a problem with *A. majus*, but rust-resistant cultivars are available.
A. asarina. See *Asarina procumbens*.
A. majus. Erect perennial that branches from the base. Cultivars are grown as annuals and are grouped according to size and flower type: tall, H 60cm–1m (2–3ft), S 30–45cm (12–18in); intermediate, H and S 45cm (18in); dwarf, H 20–30cm (8–12in), S 30cm (12in); peloric, regular tubular-shaped flowers; penstemon, trumpet-shaped flowers; double; and irregular tubular-shaped flowers. Half hardy. All have lance-shaped leaves and, from spring to autumn, carry spikes of usually 2-lipped, sometimes double, flowers in a variety of colours, including white, pink, red, purple, yellow and orange. ❧ **'Coronette'** (tall, peloric) illus. p.289. **'His Excellency'** (intermediate, irregular tubular-shaped) has scarlet flowers. **Hyacinth-flowered Series** (intermediate, penstemon) has dense spikes of large, wide flowers. **'Little Darling'** (dwarf, penstemon) has trumpet-shaped flowers. **Madame Butterfly Series** (tall, peloric) illus. p.273. **Princess Series** (intermediate, peloric) has flowers in a mixture of colours (white with purple eye, illus. p.277). **Royal Carpet Series** (dwarf, irregular tubular-shaped) is a compact form. **Supreme Series** (tall, peloric) has double, ruffled flowers. **'Trumpet Serenade'** (dwarf, penstemon) illus.

435

APHELANDRA

p.291. **Wedding Bells Series** (tall, penstemon) illus. p.288.

APHELANDRA (Acanthaceae)
Genus of evergreen shrubs and perennials with showy flowers. Frost tender, min. 13°C (55°F). Grows best in bright light but out of direct sun in summer. Use soft water and keep soil moist but not waterlogged. Benefits from feeding when flower spikes are forming. Propagate by seed or tip cuttings from young stems in spring.
A. squarrosa (Zebra plant). **'Dania'** is an evergreen, compact perennial. H 1m (3ft), S slightly less. Oval, glossy, dark green leaves, with white veins and midribs, are nearly 30cm (1ft) long. Has dense, 4-sided spikes, to 15cm (6in) long, of 2-lipped, bright yellow flowers in axils of yellow bracts in autumn. ❦ **'Louisae'** illus. p.220.

APHYLLANTHES (Liliaceae)
Genus of one species of summer-flowering perennial. Frost hardy, but shelter from cold wind. Grow in a sunny, warm, sheltered corner, preferably in an alpine house, and in well-drained, sandy, peaty soil. Resents disturbance. Propagate by seed in autumn or spring.
A. monspeliensis. Tuft-forming perennial. H 15–20cm (6–8in), S 5cm (2in). Star-shaped, pale to deep blue flowers are borne singly or in small groups at tops of wiry, glaucous green stems from early to mid-summer. Leaves are reduced to red-brown sheaths surrounding stems.

APONOGETON (Aponogetonaceae)
Genus of deciduous, perennial, deep-water plants, grown for their floating foliage and often heavily scented flowers. Frost hardy to frost tender, min. 16°C (61°F). Requires an open, sunny position. Fading foliage needs tidying up in autumn. Propagate by division in spring or by seed while still fresh.
A. distachyos illus. p.389.

APOROCACTUS (Cactaceae)
Genus of perennial cacti, grown for their pendent, slender, fleshy stems and bright flowers. Is suitable for hanging baskets. Half hardy to frost tender, min. 5°C (41°F). Needs partial shade and very well-drained soil. Propagate by stem cuttings in spring or summer.
A. flagelliformis illus. p.399.

APTENIA (Aizoaceae)
Genus of fast-growing, perennial succulents, with trailing, freely branching stems, that make good ground cover. Frost tender, min. 7°C (45°F). Requires full sun and very well-drained soil. Keep dry in winter. Propagate by seed or stem cuttings in spring or summer.
A. cordifolia illus. p.408. **'Variegata'** is a fast-growing, prostrate, perennial succulent. H 5cm (2in), S indefinite. Has oval, glossy, bright green leaves, with creamy-white margins, and small, daisy-like, bright pink flowers in summer.

AQUILEGIA (Ranunculaceae)
Columbine
Genus of graceful, clump-forming, short-lived perennials, grown for their mainly bell-shaped, spurred flowers in spring and summer. Is suitable for rock gardens. Fully to frost hardy. Prefers well-drained soil in an open, sunny site. Propagate species by seed in autumn or spring. Selected forms only occasionally come true from seed (e.g. *A. vulgaris* 'Nora Barlow') as they cross freely; they should be widely segregated. Is prone to aphid attack.
A. alpina illus. p.296.
❦ *A. canadensis* (Canadian columbine). Clump-forming, leafy perennial. H 60cm (24in), S 30cm (12in). Fully hardy. In early summer bears semi-pendent, bell-shaped flowers, with yellow sepals and red spurs, several per slender stem, above fern-like, dark green foliage.
A. chrysantha. Vigorous, clump-forming perennial. H 1.2m (4ft), S 60cm (2ft). Fully hardy. Bears semi-pendent, bell-shaped, soft yellow flowers, with long spurs, several per stem, in early summer. Has fern-like, divided, mid-green leaves.
❦ *A. flabellata*. Clump-forming perennial. H 25cm (10in), S 10cm (4in). Fully hardy. Bell-shaped, soft blue flowers, each with fluted petals and a short spur, are produced in summer. Rounded, finely divided leaves form an open, basal rosette. Needs semi-shade and moist soil. **'Alba'** (syn. *A. f.* 'Nana Alba') has white flowers.
A. jonesii illus. p.331.
A. longissima. Clump-forming, leafy perennial. H 60cm (24in), S 50cm (20in). Fully hardy. Bell-shaped, pale yellow flowers, with very long, bright yellow spurs, are borne, several per stem, in early summer, above fern-like, divided, mid-green leaves.
A. **'Mrs Scott Elliott'**. Clump-forming, leafy perennial. H 1m (3ft), S 50cm (20in). Fully hardy. Bell-shaped flowers of various colours, often bicoloured, have long spurs and appear in early summer on branching, wiry stems. Has fern-like, divided, bluish-green leaves.
A. scopulorum. Clump-forming perennial. H 6cm (2½in), S 9cm (3½in). Frost hardy. In summer produces bell-shaped, fluted, pale blue, or rarely pink, flowers, each with a cream centre and very long spurs. Leaves are divided into 9 oval, glaucous leaflets.
A. vulgaris (Granny's bonnets). Clump-forming, leafy perennial. H 1m (3ft), S 50cm (20in). Fully hardy. Many funnel-shaped, short-spurred flowers, in shades of pink, crimson, purple and white, are borne, several per long stem, in early summer. Leaves are grey-green, rounded and divided into leaflets. ❦ **'Nora Barlow'** illus. p.213.

ARABIS (Cruciferae)
Genus of robust, evergreen perennials. Makes excellent ground cover in a rock garden. Fully to frost hardy. Needs sun and well-drained soil. Propagate by softwood cuttings in summer or by seed in autumn.

A. albida. See *A. caucasica*.
A. blepharophylla. Short-lived, evergreen, mat-forming perennial. H 12cm (5in), S 15cm (6in). Fully hardy, but dislikes winter wet. Has oval, toothed, dark green leaves, with hairy, grey edges, in loose rosettes. Fragrant, 4-petalled, bright pink to white flowers appear in spring.
A. caucasica, syn. *A. albida*. Evergreen, mat-forming perennial. H 15cm (6in), S 25cm (10in). Fully hardy. Bears loose rosettes of oval, toothed, mid-green leaves and, in late spring and summer, fragrant, 4-petalled, white, occasionally pink, flowers. Is excellent on a dry bank. Trim back after flowering. ❦ **'Plena'** has double, white flowers. **'Rosabella'** illus. p.313. **'Variegata'** illus. p.310.
❦ *A. ferdinandi-coburgii* **'Variegata'** illus. p.336.

ARALIA (Araliaceae)
Genus of deciduous trees, shrubs and perennials, grown for their bold leaves and small, but profusely borne flowers. Fully hardy. Requires sun or semi-shade, some shelter and fertile, well-drained soil. Propagate those listed below by seed in autumn or by suckers or root cuttings in late winter.
❦ *A. elata* (Japanese angelica tree). Deciduous tree or suckering shrub with sparse, stout, prickly stems. H and S 10m (30ft). Has large, dark green leaves with numerous oval, paired leaflets. Billowing heads of tiny, white flowers, forming a large panicle, 30–60cm (1–2ft) long, are borne in late summer and autumn. **'Aureovariegata'** has leaflets broadly edged with yellow. Leaflets of
❦ **'Variegata'** have creamy-white margins.
A. elegantissima. See *Schefflera elegantissima*.
A. japonica. See *Fatsia japonica*.
A. sieboldii. See *Fatsia japonica*.

ARAUCARIA (Araucariaceae). See CONIFERS.
A. araucana illus. p.77.
❦ *A. heterophylla* (Norfolk Island pine). Upright conifer. H 30m (100ft), S 5–8m (15–25ft). Half hardy. Has spirally set, needle-like, incurved, fresh green leaves. Cones are seldom produced in cultivation. Is often grown as a shade-tolerant house plant.

ARAUJIA (Asclepiadaceae)
Genus of evergreen, twining climbers with woody stems that exude milky juice when cut. Half hardy. Grow in sun and fertile, well-drained soil. Propagate by seed in spring or by stem cuttings in late summer or early autumn.
A. sericifera. See *A. sericofera*.
A. sericofera, syn. *A. sericifera*, illus. p.167.

ARBUTUS (Ericaceae)
Genus of evergreen trees and shrubs, grown for their leaves, clusters of small, urn-shaped flowers, ornamental bark and strawberry-like fruits, which are edible but insipid. Frost hardy, but protect from strong, cold winds when young. Prefers full sun and needs fertile, well-drained soil; *A. menziesii*

ARCTOSTAPHYLOS

requires acid soil. Propagate by semi-ripe cuttings in late summer or by seed in autumn.
A. andrachne (Grecian strawberry tree). Evergreen, spreading tree or shrub. H and S 6m (20ft). Has oval, glossy, dark green leaves and peeling, reddish-brown bark. Panicles of urn-shaped, white flowers in late spring are followed by orange-red fruits. Prefers a sheltered position.
❦ *A.* × *andrachnoides* illus. p.57.
❦ *A. menziesii* (Madroña, Madroñe). Evergreen, spreading tree. H and S 15m (50ft). Has smooth, peeling, reddish bark and oval, dark green leaves. Large, upright, terminal panicles of urn-shaped, white flowers in early summer are followed by orange or red fruits.
❦ *A. unedo* illus. p.67.

ARCHONTOPHOENIX (Palmae)
King palm
Genus of evergreen palms, grown for their majestic appearance. Frost tender, min. 15°C (59°F). Needs full light or partial shade and humus-rich, well-drained soil. Water containerized specimens moderately, much less when temperatures are low. Propagate by seed in spring at not less than 24°C (75°F). Red spider mite may be troublesome.
A. alexandrae illus. p.46.
❦ *A. cunninghamiana* (Illawarra palm, Piccabeen palm). Evergreen palm. H 15–20m (50–70ft), S 2–5m (6–15ft). Has long, arching, feather-shaped leaves. Mature trees produce large clusters of small, lavender or lilac flowers in summer, followed by large, egg-shaped, red fruits.

Arcterica nana. See *Pieris nana*.

ARCTOSTAPHYLOS (Ericaceae)
Manzanita
Genus of evergreen trees and shrubs, grown for their foliage, flowers and fruits. Some species are also grown for their bark, others for ground cover. Fully hardy to frost tender, min. 7°C (45°F). Provide shelter from strong winds. Does best in full sun and well-drained, acid soil. Propagate by semi-ripe cuttings in summer or by seed in autumn.
A. alpinus, syn. *Arctous alpinus*. Deciduous, creeping shrub. H 5cm (2in), S to 12cm (5in). Has drooping, terminal clusters of tiny, urn-shaped, pink-flushed, white flowers in late spring, followed by rounded, purple-black berries. Leaves are oval, toothed, glossy and bright green.
A. diversifolia, syn. *Comarostaphylis diversifolia* (Summer holly). Evergreen, upright shrub or tree. H 5m (15ft), S 3m (10ft). Half hardy. Leaves are oblong, glossy and dark green. Terminal racemes of fragrant, urn-shaped, white flowers appear from early to mid-spring, followed by spherical, red fruits.
A. **'Emerald Carpet'** illus. p.126.
A. manzanita. Evergreen, upright shrub. H and S 2m (6ft) or more. Frost hardy. Has peeling, reddish-brown bark and oval, leathery, grey-green leaves. From early to mid-spring produces

small, urn-shaped, deep pink flowers.
A. nevadensis (Pine-mat manzanita). Evergreen, prostrate shrub. H 10cm (4in), S 1m (3ft). Frost hardy. Has small, oval leaves. In summer, pendent, urn-shaped, white flowers are borne in clusters in leaf axils, followed by globose, brownish-red fruits. Is useful as ground cover.
A. nummularia. Evergreen, erect to prostrate shrub. H 30cm (1ft) or more, S 1m (3ft). Frost hardy. Leaves are small, rounded, leathery and toothed. Pendent, urn-shaped, white flowers are borne in clusters from leaf axils in summer, followed by globose, green fruits. Makes good ground cover.
A. patula illus. p.122.
A. stanfordiana (Stanford manzanita). Evergreen, erect shrub. H and S 1.5m (5ft). Half hardy. Bark is smooth and reddish-brown. Has narrowly oval, glossy, bright green leaves. Drooping clusters of urn-shaped, pink flowers are borne from early to mid-spring.
A. uva-ursi illus. p.336. **'Point Reyes'** illus. p.336. **'Vancouver Jade'** is an evergreen, trailing, sometimes arching shrub. H 10cm (4in), S 50cm (20in). Fully hardy. Has small, oval, bright green leaves and bears urn-shaped, white flowers in summer. subsp. **hookeri 'Monterey Carpet'** is an open, half hardy shrub, H 10–15cm (4–6in), S 40cm (16in) or more. Has hairy branchlets bearing glossy, pale green leaves and, in early summer, urn-shaped, white flowers, sometimes flushed pink, that are followed by globose, red fruits.

ARCTOTHECA (Compositae)
Genus of creeping perennials. Frost tender, min. 5°C (41°F). Needs bright light and fertile, well-drained soil; dislikes humid conditions. Propagate by seed or division in spring.
A. calendula illus. p.258.

ARCTOTIS (Compositae)
Genus of annuals and perennials, grown for their flower heads and foliage. Half hardy to frost tender, min. 1–5°C (34–41°F). Requires full sun and leafy loam with sharp sand added. Propagate by seed in autumn or spring or by stem cuttings year-round.
A. x hybrida, syn. x *Venidio-arctotis*. Fairly slow-growing, upright, branching perennial, usually grown as an annual. H and S 45cm (18in). Half hardy. Lance-shaped, lobed leaves are greyish-green above, white below. In summer has large, daisy-like flower heads in many shades, including yellow, orange, bronze, purple, pink, cream and red. **'Bacchus'** has purple flower heads; **'China Rose'** deep pink; **'Sunshine'** yellow; **'Tangerine'** orange-yellow; and **'Torch'** bronze.
A. stoechadifolia (African daisy). Compact perennial, often grown as an annual. H 50cm (20in) or more, S 40cm (16in). Half hardy. Daisy-like, creamy-white flower heads with blue centres are borne singly throughout summer and into autumn. Chrysanthemum-like leaves are dark green above, grey beneath. var. **grandis** is larger and has longer leaves.

Arctous alpinus. See *Arctostaphylos alpinus.*

ARDISIA (Myrsinaceae)
Genus of evergreen shrubs and trees, grown for their fruits and foliage. Half hardy to frost tender, min. 10°C (50°F). Needs partial shade and humus-rich, well-drained but not dry soil. Water potted plants freely when in full growth, moderately at other times. Cut back old plants in early spring if required. Propagate by seed in spring or by semi-ripe cuttings in summer.
A. crenata, syn. *A. crenulata*, illus. p.120.
A. crenulata. See *A. crenata.*

Areca lutescens. See *Chrysalidocarpus lutescens.*

Arecastrum romanzoffianum. See *Syagrus romanzoffianum.*

Aregelia carolinae. See *Neoregelia carolinae.*

ARENARIA (Carophyllaceae)
Sandwort
Genus of spring- and summer-flowering annuals and perennials, some of which are evergreen. Fully to frost hardy. Most need sun and well-drained, sandy soil. Propagate by division or softwood cuttings in early summer or by seed in autumn or spring.
A. balearica illus. p.310.
♁ **A. montana** illus. p.322.
A. purpurascens illus. p.313.
A. tetraquetra illus. p.309.

ARGEMONE (Papaveraceae)
Genus of robust perennials, most of which are best treated as annuals. Fully to half hardy. Grow in sun and in very well-drained soil. Supports should not be provided as they may damage the rather fleshy stems. Dead-head plants to prolong the flowering season. Propagate by seed sown outdoors in late spring.
A. mexicana illus. p.287.

Argyranthemum frutescens. See *Chrysanthemum frutescens.*

ARGYREIA (Convolvulaceae)
Genus of evergreen, twining climbers, closely allied to *Ipomoea* and grown for their showy flowers. Frost tender, min. 13°C (55°F). Plant in any fertile, well-drained soil, in full light. Water freely when in full growth, sparingly at other times. Support is needed. Thin out previous season's growth in spring. Propagate by seed in spring or by softwood or greenwood cuttings in summer. Red spider mite and whitefly may be troublesome.
A. splendens (Silver morning glory). Vigorous, evergreen, twining climber. H 4m (12ft) or more. Oval leaves are 18cm (7in) long with silky, white hairs beneath. Large clusters of funnel-shaped, rose-pink flowers appear in summer-autumn.

Argyrocytisus battandieri. See *Cytisus* x *beanii.*

ARGYRODERMA (Aizoaceae)
Genus of perennial succulents, grown for their very fleshy, grey-green leaves united in a prostrate, egg shape. In summer, daisy-like flowers appear in central split between leaves. Frost tender, min. 5°C (41°F). Needs full sun and very well-drained soil. Over-watering will cause leaves to split or plant to rot. Propagate by seed in summer.
A. blandum. See *A. delaetii.*
A. brevipes. See *A. fissum.*
A. delaetii, syn. *A. blandum*, illus. p.412.
A. fissum, syn. *A. brevipes*, illus. p.410.
A. pearsonii, syn. *A. schlechteri*, illus. p.409.
A. schlechteri. See *A. pearsonii.*

ARIOCARPUS (Cactaceae)
Living rock
Genus of extremely slow-growing, perennial cacti with large, swollen roots. Produces flattened, spherical, green stems with angular tubercles and tufts of wool. Frost tender, min. 5°C (41°F). Prefers full sun and extremely well-drained, lime-rich soil. Is very prone to rotting. Propagate by seed in spring or summer.
A. fissuratus illus. p.408.

ARISAEMA (Araceae)
Genus of tuberous perennials, grown for their large, curious, hooded spathes, each enclosing a pencil-shaped spadix. Forms spikes of fleshy, red fruits in autumn, before plant dies down. Fully to half hardy. Needs sun or partial shade and humus-rich soil. Plant tubers 15cm (6in) deep in spring. Propagate by seed in autumn or spring or by offsets in spring.
A. atrorubens. See *A. triphyllum.*
♁ **A. candidissimum** illus. p.379.
A. consanguineum illus. p.346.
A. griffithii illus. p.366.
A. jacquemontii illus. p.366.
A. ringens. Early spring-flowering, tuberous perennial. H 25–30cm (10–12in), S 30–45cm (12–18in). Half hardy. Bears 2 erect leaves, each with 3 long-pointed lobes, and a widely hooded, green spathe, enclosing the spadix, that has paler green stripes and is edged with dark brown-purple.
A. sikokianum illus. p.363.
A. tortuosum. Summer-flowering, tuberous perennial. H 30cm–1m (1–3ft), S 30–45cm (1–1½ft). Half hardy. Each dark green-mottled, pale green stem bears 2–3 erect leaves, divided into several oval leaflets. A hooded, green or purple spathe, with a protruding, S-shaped spadix, overtops leaves. Produces spikes of attractive, fleshy, red fruits in autumn.
A. triphyllum, syn. *A. atrorubens*, illus. p.366.

ARISARUM (Araceae)
Genus of tuberous perennials, grown mainly for their curious, hooded spathes enclosing spadices with minute flowers. Frost hardy. Needs partial shade and humus-rich, well-drained soil. Propagate in autumn by division of an established clump of tubers, which produce offsets freely.

A. proboscideum (Mouse plant). Clump-forming, spring-flowering, tuberous perennial. H to 10cm (4in), S 20–30cm (8–12in). Leaves are arrow-shaped and prostrate. Produces a spadix of minute flowers concealed in a hooded, dark brown spathe that is drawn out into a tail up to 15cm (6in) long – a mouse-like effect.

ARISTEA (Iridaceae)
Genus of evergreen, clump-forming, rhizomatous perennials, grown for their spikes of blue flowers in spring or summer. Half hardy. Prefers sun and well-drained soil. Established plants cannot be moved satisfactorily. Propagate by seed in autumn or spring.
A. ecklonii. Evergreen, clump-forming, rhizomatous perennial. H 30–60cm (12–24in), S 20–40cm (8–16in). Has long, sword-shaped, tough leaves, overtopped in summer by loosely branched spikes of saucer-shaped, blue flowers, produced in long succession.
A. major, syn. *A. thyrsiflora*, illus. p.346.
A. thyrsiflora. See *A. major.*

ARISTOLOCHIA (Aristolochiaceae)
Birthwort
Genus of evergreen or deciduous, woody-stemmed, twining and scrambling climbers, grown for their foliage and flowers. Frost hardy to frost tender, min. 10–13°C (50–55°F). Needs well-drained soil and partial shade in summer. Water regularly, less when not in full growth. Provide support. Cut back previous season's growth to 2 or 3 nodes in spring. Propagate by seed in spring or by semi-ripe cuttings in summer. Red spider mite and whitefly may be a nuisance.
♁ **A. elegans.** See *A. littoralis.*
A. gigas. See *A. grandiflora.*
A. grandiflora, syn. *A. gigas* (Pelican flower, Swan flower). Fast-growing, evergreen, woody-stemmed, twining climber. H 7m (22ft) or more. Frost tender. Leaves are broadly oval, 15–25cm (6–10in) long. In summer bears large, unpleasant-smelling, tubular, purple-veined, white flowers, each with a long tail and expanding at the mouth into a heart-shaped lip.
A. griffithii, syn. *Isotrema griffithii*. Moderately vigorous, evergreen, woody-stemmed, twining climber. H 5–6m (15–20ft). Half hardy; deciduous in cold winters. Has heart-shaped leaves and tubular, dark red flowers, each with an expanded, spreading lip, in summer.
♁ **A. littoralis**, syn. *A. elegans*, illus. p.171.

ARISTOTELIA (Elaeocarpaceae)
Genus of evergreen shrubs and deciduous trees, grown for their foliage. Needs separate male and female plants in order to obtain fruits. Frost hardy, but in most areas protect by growing against other shrubs or a south- or west-facing wall. Needs sun or semi-shade and fertile, well-drained soil. Propagate by semi-ripe cuttings in summer.

A. chilensis. Evergreen, spreading shrub. H 3m (10ft), S 5m (15ft). Leaves are oval, glossy and deep green. Tiny, star-shaped, green flowers are borne in summer, followed by small, spherical, black fruits.

ARMERIA (Plumbaginaceae)
Genus of evergreen perennials and, occasionally, sub-shrubs, grown for their tuft-like clumps or rosettes of leaves and their flower heads. Fully to frost hardy. Requires sun and well-drained soil. Propagate by semi-ripe cuttings in summer or by seed in autumn.
A. caespitosa. See *A. juniperifolia.*
A. cespitosa. See *A. juniperifolia.*
♛ *A. juniperifolia,* syn. *A. caespitosa, A. cespitosa,* illus. p.313. ♛ **'Bevans Variety'** is an evergreen, densely cushioned sub-shrub. H 5–8cm (2–3in), S 15cm (6in). Fully hardy. Has narrow, pointed, mid- to grey-green leaves in loose rosettes. Round heads of small, deep pink flowers are borne in late spring and early summer.
A. latifolia. See *A. pseudarmeria.*
A. maritima (Sea pink, Thrift). Evergreen, clump-forming perennial or dwarf sub-shrub. H 10cm (4in), S 15cm (6in). Fully hardy. Leaves are narrow, grass-like and dark green. Stiff stems carry round heads of many small, white to pink flowers in summer. Makes a good edging plant.
♛ **'Vindictive'** illus. p.328.
A. pseudarmeria, syn. *A. latifolia,* illus. p.298. ♛ **'Bees Ruby'** is an evergreen, clump-forming, dwarf sub-shrub. H and S 30cm (12in). Fully hardy. Round heads of many small, ruby-red flowers are produced in summer on stiff stems above narrow, grass-like, dark green leaves.

ARNEBIA (Boraginaceae)
Genus of perennials with hairy leaves, suitable for rock gardens and banks. Fully hardy. Needs sun and gritty, well-drained soil. Propagate by seed in autumn, by root cuttings in winter or by division in spring.
A. echioides. See *A. pulchra.*
A. pulchra, syn. *A. echioides, Echioides longiflorum, Macrotomia echioides* (Prophet flower). Clump-forming perennial. H 23–30cm (9–12in), S 25cm (10in). Leaves are lance-shaped to narrowly oval, hairy and light green. In summer bears loose racemes of tubular, bright yellow flowers, each with 5 spreading lobes and fading, dark spots at petal bases.

ARNICA (Compositae)
Genus of rhizomatous perennials, grown for their large, daisy-like flower heads. Is suitable for large rock gardens. Fully hardy. Prefers sun and humus-rich, well-drained soil. Propagate by division or seed in spring.
A. montana. Tufted, rhizomatous perennial. H 30cm (12in), S 15cm (6in). Bears narrowly oval to oval, hairy, grey-green leaves and, in summer, solitary daisy-like, golden flower heads, 5cm (2in) wide. Prefers acid soil.

ARONIA (Rosaceae)
Chokeberry
Genus of deciduous shrubs, grown for their flowers, fruits and colourful autumn foliage. Fully hardy. Needs sun (for best autumn colour) or semi-shade and fertile, well-drained soil. Propagate by softwood or semi-ripe cuttings in summer, by seed in autumn or by division from early autumn to spring.
A. arbutifolia illus. p.100.
A. melanocarpa illus. p.106.
A. x *prunifolia.* Deciduous, upright shrub. H 3m (10ft), S 2.5m (8ft). Oval, glossy, dark green leaves redden in autumn. Has star-shaped, white flowers in late spring and early summer, followed by spherical, purplish-black fruits.

ARRHENATHERUM (Gramineae). See GRASSES, BAMBOOS, RUSHES and SEDGES.
A. elatius (False oat grass). **'Variegatum'** is a loosely tuft-forming, herbaceous, perennial grass. H 50cm (20in), S 20cm (8in). Fully hardy. Has a basal stem swelling, hairless, grey-green leaves, with white margins, and open panicles of brownish spikelets in summer.

ARTEMISIA (Compositae)
Wormwood
Genus of perennials and spreading, dwarf sub-shrubs and shrubs, some of which are evergreen or semi-evergreen, grown mainly for their fern-like, silvery foliage that is sometimes aromatic. Fully to half hardy. Prefers an open, sunny, well-drained site; dwarf types benefit from a winter protection of sharp grit or gravel. Trim lightly in spring. Propagate by division in spring or autumn or by softwood or semi-ripe cuttings in summer.
♛ *A. abrotanum* illus. p.146.
♛ *A. absinthium* **'Lambrook Silver'**. Evergreen, bushy perennial, woody at base. H 80cm (32in), S 50cm (20in). Frost hardy. Has a mass of finely divided, aromatic, silvery-grey leaves. Tiny, insignificant, grey flower heads are borne in long panicles in summer. Needs protection in an exposed site.
♛ *A. alba* **'Canescens'**. Semi-evergreen, bushy perennial. H 50cm (20in), S 30cm (12in). Fully hardy. Has delicate, finely cut, curling, silvery-grey leaves. In summer, insignificant, yellow flower heads are borne on erect, silver stems. Makes good ground cover.
A. arborescens illus. p.145. **'Faith Raven'** is an evergreen, upright shrub. H 1.2m (4ft), S 1m (3ft). Differs from the species only in that it is frost hardy. Has finely cut, aromatic, silvery-white foliage and, in summer and early autumn, rounded heads of small, bright yellow flowers.
A. assoana. See *A. pedemontana.*
♛ *A. frigida.* Semi-evergreen, mat-forming perennial with a woody base. H 30cm (12in) in flower, S 30cm (12in) or more. Fully hardy. Has small, fern-like, aromatic, silky, grey-white leaves, divided into many linear lobes. In summer produces narrow panicles of small, rounded, yellow flower heads.
♛ *A. lactiflora* illus. p.191.
A. lanata. See *A. pedemontana.*
A. ludoviciana var. *albula* illus. p.230.
A. pedemontana, syn. *A. assoana, A. lanata.* Evergreen or semi-evergreen, prostrate perennial. H and S 30cm (12in). Fully hardy. Fern-like foliage is densely covered with silvery-white hairs. Small clusters of small, rounded, yellow flower heads are borne in summer. Suits a rock garden or wall.
♛ *A. pontica* illus. p.254.
♛ *A.* **'Powis Castle'**. Vigorous, evergreen sub-shrub. H 1m (3ft), S 1.2m (4ft). Frost hardy. Has abundant, finely cut, aromatic, silvery-grey foliage and sprays of insignificant, yellowish-grey flower heads in summer.
♛ *A. schmidtiana.* Semi-evergreen, hummock-forming perennial with creeping stems. H 8–30cm (3–12in), S 60cm (24in). Frost hardy. Has fern-like, very finely and deeply cut, silver leaves and, in summer, short racemes of small, rounded, pale yellow flower heads. Is good for a large rock garden, wall or bank. Needs sandy, peaty soil.
♛ **'Nana'** illus. p.337.
A. stelleriana. Evergreen, rounded, rhizomatous perennial with a woody base. H 30–60cm (1–2ft), S 60cm–1m (2–3ft). Fully hardy. White-haired, silver leaves are deeply lobed or toothed. Bears slender sprays of small, yellow flower heads in summer. Needs light soil. **'Boughton Silver'**, S 1m (3ft), is vigorous and arching in habit.

ARTHROPODIUM (Liliaceae)
Genus of tufted perennials, grown for their flowers. Half hardy. Grow in fertile soil against a sunny, sheltered wall. Propagate by division in spring or by seed in spring or autumn.
A. cirrhatum (Rienga lily, Rock lily). Branching perennial. H 1m (3ft), S 30cm (1ft). Bears sprays of nodding, shallowly cup-shaped, white flowers on wiry stems in early summer. Has a basal tuft of narrowly sword-shaped leaves and fleshy roots.

ARUM (Araceae)
Cuckoo pint, Lords and ladies
Genus of tuberous perennials, grown for their ornamental leaves and attractive spathes, each enclosing a pencil-shaped spadix of tiny flowers. Fully to half hardy. Prefers sun or partial shade and moist but well-drained soil. Propagate by seed in autumn or by division in early autumn.
A. creticum illus. p.359.
A. dioscoridis. Spring-flowering, tuberous perennial. H 20–35cm (8–14in), S 30–45cm (12–18in). Frost hardy. Has a sail-like, green or purple spathe, blotched dark purple, surrounding a blackish-purple spadix. Arrow-shaped, semi-erect leaves appear in autumn. Needs a sheltered, sunny site.
A. dracunculus. See *Dracunculus vulgaris.*
♛ *A. italicum* **'Pictum'**, syn. *A.i.* **'Marmoratum'**, illus. p.378.
A. pictum illus. p.383.

ARUNCUS (Rosaceae)
Genus of perennials, grown for their hummocks of broad, fern-like leaves and their plumes of white flowers in summer. Fully hardy. Thrives in any well-drained soil and in full light. Propagate by seed in autumn or by division in spring or autumn.
♛ *A. dioicus,* syn. *A. sylvester, Spiraea aruncus,* illus. p.190. **'Kneiffii'** illus. p.204.
A. sylvester. See *A. dioicus.*

ARUNDINARIA (Bambusoideae). See GRASSES, BAMBOOS, RUSHES and SEDGES.
A. anceps. See *Yushania anceps.*
A. auricoma. See *Pleioblastus auricoma.*
A. falconeri. See *Drepanostachyum falconeri.*
A. fastuosa. See *Semiarundinaria fastuosa.*
A. fortunei. See *Pleioblastus variegatus.*
A. japonica. See *Pseudosasa japonica.*
A. jaunsarensis. See *Yushania anceps.*
A. murielae. See *Thamnocalamus spathaceus.*
A. nitida. See *Fargesia nitida.*
A. variegata. See *Pleioblastus variegatus.*
A. viridistriata. See *Pleioblastus auricoma.*

ARUNDO (Gramineae). See GRASSES, BAMBOOS, RUSHES and SEDGES.
A. donax (Giant reed). Herbaceous, rhizomatous, perennial grass. H to 6m (20ft), S 1m (3ft). Half hardy. Thick stems bear broad, floppy, blue-green leaves. Produces dense, erect panicles of whitish-yellow spikelets in summer. May also be grown in moist soil. **'Versicolor'** (syn. *A.d.* **'Variegata'**) illus. p.182.

ASARINA, syn. MAURANDIA, MAURANDYA (Scrophulariaceae)
Genus of evergreen climbers and perennials, often with scandent stems, grown for their flowers. Is herbaceous in cold climates. Frost hardy to frost tender, min. 5°C (41°F). Grow in full light and in any well-drained soil. Propagate by seed in spring.
A. barclaiana. Evergreen, soft-stemmed, scandent climber, herbaceous in cold climates. H to 2m (6ft). Frost tender. Has angular, heart-shaped, hairless leaves. Trumpet-shaped, purple flowers, each with a whitish throat, 6–7cm (2½–3in) long, are produced in summer-autumn.
♛ *A. erubescens* illus. p.169.
A. procumbens, syn. *Antirrhinum asarina,* illus. p.332.

ASARUM, syn. HEXASTYLIS (Aristolochiaceae)
Wild ginger
Genus of rhizomatous perennials, some of which are evergreen, with pitcher-shaped flowers carried under kidney- or heart-shaped leaves. Makes good ground cover, although leaves may become damaged in severe weather. Fully hardy. Prefers shade and humus-rich, moist but well-drained soil. Propagate by division in spring. Self seeds readily.
A. caudatum. Evergreen, prostrate, rhizomatous perennial. H 8cm (3in), S 25cm (10in) or more. Heart-shaped,

leathery, glossy, dark green leaves, 5–10cm (2–4in) across, conceal small, pitcher-shaped, reddish-brown or brownish-purple flowers, with tail-like lobes, in early summer.
A. europaeum illus. p.338.
A. hartwegii. Evergreen, prostrate, rhizomatous perennial. H 8cm (3in), S 25cm (10in) or more. Pitcher-shaped, very dark brown, almost black, flowers, with tail-like lobes, appear in early summer beneath heart-shaped, silver-marked, mid-green leaves, 5–10cm (2–4in) wide.
A. shuttleworthii. Evergreen, prostrate, rhizomatous perennial. H 8cm (3in), S 25cm (10in) or more. Has broadly heart-shaped, usually silver-marked, mid-green leaves, 8cm (3in) across. Bears pitcher-shaped, dark brown flowers, mottled violet inside, in early summer.

ASCLEPIAS (Asclepiadaceae)
Silk weed
Genus of tuberous perennials or sub-shrubs, some of which are evergreen, grown for their flowers. Stems exude milky, white latex when cut. Fully hardy to frost tender, min. 5–10°C (41–50°F). Fully to half hardy species prefer sun and a humus-rich, well-drained soil. Propagate by division or seed in spring. Frost tender species need sun and a moist atmosphere; cut back during periods of growth to keep bushy. Water very sparingly in low temperatures. Propagate by tip cuttings or seed in spring.
A. curassavica (Blood flower). Evergreen, bushy, tuberous sub-shrub. H and S 1m (3ft). Frost tender. Has narrowly oval leaves to 15cm (6in) long. Umbels of small, but showy, 5-horned, orange-red flowers with yellow centres appear in summer-autumn and are followed by narrowly ovoid, pointed fruits, 8cm (3in) long, with silky seeds.
A. hallii. Upright, tuberous perennial. H to 1m (3ft), S 60cm (2ft). Fully hardy. Has oblong leaves, to 13cm (5in) long. Umbels of small, 5-horned, dark pink flowers are carried in summer; tightly packed silky seeds are enclosed in narrowly ovoid fruits, to 15cm (6in) long.
A. physocarpa, syn. *Gomphocarpus physocarpus*, illus. p.196.
A. syriaca. Upright, tuberous perennial. H and S 1m (3ft) or more. Fully hardy. Bears oval leaves to 20cm (8in) long. Umbels of small, 5-horned, purplish-pink flowers are carried on drooping flower stalks in summer, followed by narrowly ovoid fruits, to 15cm (6in) long and filled with silky seeds.
A. tuberosa illus. p.222.

ASIMINA (Annonaceae)
Genus of deciduous or evergreen shrubs and trees, grown for their foliage and flowers. Fully hardy. Prefers full sun and fertile, deep, moist but well-drained soil. Propagate by seed in autumn or by layering or root cuttings in winter.
A. triloba (Pawpaw). Deciduous, open shrub. H and S 4m (12ft). Large, oval, mid-green leaves emerge in late spring or early summer, just after, or at the same time as, 6-petalled, purplish-brown flowers. Later it produces small, globular, brownish fruits.

ASPARAGUS (Liliaceae)
Genus of perennials and scrambling climbers and shrubs, some of which are evergreen, grown for their foliage. Fully hardy to frost tender, min. 10°C (50°F). Grow in partial shade or bright light, but not direct sun, in any fertile, well-drained soil. Propagate by seed or division in spring.
♛ *A. densiflorus*, syn. *A. sprengeri*, illus. p.230. ♛ 'Myersii' (syn. *A. meyeri*, *A. myersii*) illus. p.231.
A. meyeri. See *A. densiflorus* 'Myersii'.
A. myersii. See *A. densiflorus* 'Myersii'.
A. scandens illus. p.179.
A. sprengeri. See *A. densiflorus*.

ASPERULA (Rubiaceae)
Genus of annuals and perennials; some species make good alpine house plants. Fully to frost hardy. Most species need sun and well-drained soil with moisture at roots. Dislikes winter wet on crown. Propagate by softwood cuttings or seed in early summer.
A. odorata. See *Galium odoratum*.
♛ *A. suberosa* illus. p.324.

ASPHODELINE (Liliaceae)
Genus of perennials with thick, fleshy roots. Frost to half hardy. Requires sun and not over-rich soil. Propagate by division in early spring, taking care not to damage roots, or by seed in autumn or spring.
A. liburnica. Neat, clump-forming perennial. H 25–60cm (10–24in), S 30cm (12in). Frost hardy. In spring produces racemes of shallowly cup-shaped, yellow flowers on slender stems above linear, grey-green leaves.
A. lutea illus. p.202.

ASPHODELUS (Liliaceae)
Genus of spring- or summer-flowering annuals and perennials. Frost to half hardy. Requires sun; most prefer fertile, well-drained soil. *A. albus* prefers light, well-drained soil. Propagate by division in spring or by seed in autumn.
A. acaulis. Prostrate perennial. H 5cm (2in), S 23cm (9in). Half hardy. In spring or early summer, stemless, funnel-shaped, flesh-pink flowers appear in the centre of each cluster of grass-like, mid-green leaves. Is suitable for an alpine house.
A. aestivus, syn. *A. microcarpus* (Asphodel). Upright perennial. H 1m (3ft), S 30cm (1ft). Frost hardy. Dense panicles of star-shaped, white flowers are borne in late spring. Has basal rosettes of upright, then spreading, grass-like, channelled, leathery, mid-green leaves.
A. albus illus. p.203.
A. microcarpus. See *A. aestivus*.

ASPIDISTRA (Liliaceae)
Genus of evergreen, rhizomatous perennials that spread slowly, grown mainly for their glossy foliage. Frost tender, min. 5–10°C (41–50°F). Very tolerant, but is best grown in a cool, shady position away from direct sunlight and in well-drained soil. Water frequently when in full growth, less at other times. Propagate by division of rhizomes in spring.
♛ *A. elatior* (Cast-iron plant). Evergreen, rhizomatous perennial. H 60cm (2ft), S 45cm (18in). Has upright, narrow, pointed-oval leaves, to 60cm (2ft) long; inconspicuous, cream to purple flowers are occasionally produced on short stalks near soil level.
♛ 'Variegata' illus. p.265.

ASPLENIUM (Polypodiaceae)
Genus of evergreen or semi-evergreen ferns. Fully hardy to frost tender, min. 5°C (41°F). Plants described prefer partial shade, but *A. trichomanes* tolerates full sun. Grow in any moist soil, although containerized plants should be grown in a compost including chopped sphagnum moss or coarse peat. Remove fading fronds regularly. Propagate by spores or bulbils, if produced, in late summer.
♛ *A. bulbiferum* (Hen-and-chicken fern, Mother spleenwort). Semi-evergreen or evergreen fern. H 15–30cm (6–12in), S 30cm (12in). Frost tender. Lance-shaped, finely divided, dark green fronds produce bulbils from which young plants develop.
♛ *A. nidus* illus. p.189.
A. rhizophyllum, syn. *Camptosorus rhizophyllus* (Walking fern). Semi-evergreen or deciduous fern. H 15–23cm (6–9in), S 15–30cm (6–12in). Heart-shaped, mid-green fronds are borne at ends of naked, reddish-green stems. Is suitable for a rock garden.
♛ *A. scolopendrium*, syn. *Phyllitis scolopendrium*, *Scolopendrium vulgare*, illus. p.189. 'Marginatum', illus. p.189.
A. trichomanes illus. p.187.

ASTELIA (Liliaceae)
Genus of clump-forming perennials, grown mainly for their foliage. Frost to half hardy. Prefers full sun or semi-shade and fertile soil that does not dry out readily. Propagate by division in spring.
A. nervosa. Clump-forming perennial. H 60cm (2ft), S 1.5m (5ft). Frost hardy. Has long, sword-shaped, arching, silvery-grey leaves, above which, in summer, rise graceful, branching panicles of small, star-shaped, pale brown flowers.

ASTER (Compositae)
Michaelmas daisy
Genus of perennials and deciduous or evergreen sub-shrubs with daisy-like flower heads borne in summer-autumn. Fully to half hardy. Prefers sun or partial shade and fertile, well-drained soil, with adequate moisture throughout summer. Tall species and cultivars require staking. Propagate by softwood cuttings in spring or by division in spring or autumn. Modern forms of *A. novi-belgii* require rigorous spraying against mildew and insect attack. Other species may also suffer.
A. acris. See *A. sedifolius*.
A. albescens, syn. *Microglossa albescens.* Deciduous, upright, slender-stemmed sub-shrub. H 1m (3ft), S 1.5m (5ft). Frost hardy. Has narrowly lance-shaped, grey-green leaves. Produces flattish sprays of lavender-blue flower heads, with yellow centres, in mid-summer.
♛ *A. alpinus* illus. p.329.
'Beechwood' is a clump-forming perennial. H 15cm (6in), S 30–45cm (12–18in). Fully hardy. Leaves are lance-shaped and dark green. Purple flower heads with yellow centres are borne from mid- to late summer. Is suitable for a rock garden.
♛ *A. amellus* 'King George' (illus. p.226). Bushy perennial. H and S 50cm (20in). Fully hardy. In autumn, carries many large, terminal, daisy-like, deep blue-violet flower heads with yellow centres. Leaves are oval and rough. 'Mauve Beauty' bears clusters of large, violet flower heads, with yellow centres, in autumn. Leaves are lance-shaped, coarse and mid-green. 'Rudolph Goethe' with large, violet-blue flower heads, 'Sonia' with pink flower heads and ♛ 'Violet Queen' with deep violet flower heads are other good cultivars.
A. capensis. See *Felicia amelloides*.
A. cordifolius 'Silver Spray' (illus. p.226). Bushy perennial. H 1.2m (4ft), S 1m (3ft). Fully hardy. Dense, arching stems carry sprays of small, pink-tinged, white flower heads in autumn. Mid-green leaves are lance-shaped. Needs staking.
A. ericoides 'White Heather' (illus. p.226). Bushy perennial. H 75cm (30in), S 50cm (20in). Fully hardy. Produces long-lasting sprays of neat, daisy-like, white flower heads in late autumn. Has small, lance-shaped, fresh green leaves and wiry, branching stems, which may need support.
♛ *A. frikartii* 'Mönch' (illus. p.226). Bushy perennial. H 75cm (30in), S 45cm (18in). Fully hardy. Bears daisy-like, single, soft lavender-blue flower heads with yellowish-green centres continuously from mid-summer to late autumn. Leaves are oval and rough. May need staking.
♛ 'Wunder von Stäfa' (illus. p.226) is similar but has lavender flowers.
A. lateriflorus. Branching perennial. H 60cm (24in), S 50cm (20in). Fully hardy. Bears sprays of tiny, mauve flower heads, with pinkish-brown centres, in autumn. Lance-shaped leaves are small and dark green.
'Delight' (syn. *A. vimineus* 'Delight'; illus. p.226) is a spreading perennial. H 1.2m (4ft), S 1m (3ft) or more. Fully hardy. In autumn, has small, daisy-like, white flower heads. Leaves are narrow and lance-shaped. Needs staking.
♛ var. *horizontalis* (illus. p.226) has flowers heads that are sometimes tinged pink, with darker pink centres.
A. linosyris (Goldilocks; illus. p.226). Upright, unbranched perennial. H 60cm (24in), S 30cm (12in). Fully hardy. Bears numerous small, dense, single, golden-yellow flower heads in late summer-autumn. Leaves are narrowly lance-shaped.
♛ *A. novae-angliae* 'Alma Potschke' (illus. p.226). Vigorous, upright perennial. H 75cm (30in), S to 60cm (24in). Fully hardy. In autumn

produces clusters of single, pink flower heads on stiff stems. Has lance-shaped, rough leaves. May need staking. **'Barr's Pink'** (illus. p.226) bears semi-double, bright rose-pink flower heads in summer-autumn. ☸ **'Harrington's Pink'** (illus. p.226), H 1.2–1.5m (4–5ft), has single, clear pink flower heads with yellow centres. Those of **'Herbstschnee'** (illus. p.226), H 75cm–1.1m (2½–3½ft), are white with yellow centres.
A. novi-belgii. **'Apple Blossom'** (illus. p.226) is a vigorous, spreading perennial. H 90cm (3ft), S 60–75cm (24–30in). Fully hardy. Panicles of single, pale soft pink flowers are borne in autumn amid lance-shaped, mid-green leaves. **'Carnival'** (illus. p.226), H 75cm (30in), S to 45cm (18in), bears double, cerise-red flower heads with yellow centres. Leaves are dark green. Is prone to mildew. **'Chequers'** (illus. p.226), H 90cm (3ft), S 60–75cm (2–2½ft), has single, purple flowers. **'Climax'**, H 1.5m (5ft), S 60cm (2ft), bears single, light blue flowers. Is mildew-resistant. The flower heads of **'Fellowship'** (illus. p.226), H 1.2m (4ft), S 50cm (20in), are large, double and clear, deep pink; those of **'Freda Ballard'** (illus. p.226) are semi-double and rich rose-red. **'Jenny'** (illus. p.226), H 45cm (18in), S 30cm (12in), has double, red flowers. **'Lassie'** (illus. p.226), H 1.2m (4ft), S 75cm (30in), produces large, single, clear pink flowers. **'Little Pink Beauty'**, H 45cm (18in), S 50cm (20in), is a good dwarf semi-double, pink cultivar. **'Marie Ballard'** (illus. p.226), H to 1m (3ft), S to 45cm (18in), has double, mid-blue flowers. Is prone to mildew. **'Orlando'** (illus. p.226), H 1m (3ft), S to 45cm (18in), has large, single, bright pink flower heads with golden centres. Leaves are dark green. Mildew may be a problem. **'Patricia Ballard'** (illus. p.226), H 1.2m (4ft), S 75cm (30in), produces semi-double, pink flowers. Large, single flowers of **'Peace'** (illus. p.226) are mauve; those of **'Raspberry Ripple'**, H 75cm (30in), S 60cm (24in), are smaller and reddish-violet. **'Royal Ruby'** (illus. p.226), H and S to 45cm (18in), bears semi-double, rich red flower heads with yellow centres. Is prone to mildew. **'Royal Velvet'** (illus. p.226), H 1.2m (4ft), S 75cm (30in), has single deep violet flowers. **'White Swan'** (illus. p.226), H 1.2m (4ft), S 60cm (2ft), is a useful, white cultivar.
A. **'Professor A. Kippenburg'** (illus. p.226) is a compact, bushy perennial. H 30cm (12in), S to 45cm (18in). Fully hardy. Carries large clusters of daisy-like, yellow-centred, clear blue flower heads in autumn.
A. sedifolius, syn. *A. acris.* Bushy perennial. H 1m (3ft), S 60cm (2ft). Fully hardy. Produces clusters of almost star-shaped, lavender-blue flower heads, with yellow centres, in autumn. Has small, narrowly oval, bright green leaves. **'Nanus'** (illus. p.226), H and S 50cm (20in), makes a compact dome of blooms.
A. thomsonii. Upright perennial. H 1m (3ft), S 50cm (20in). Fully hardy. Produces long-petalled, pale lilac flower heads, freely in autumn. Leaves are slightly heart-shaped. **'Nanus'** (illus. p.226) is more compact, H 45cm (18in), S 23cm (9in).
A. tongolensis. Mat-forming perennial. H 50cm (20in), S 30cm (12in). Fully hardy. Large, lavender-blue flower heads, with orange centres, are borne singly in early summer. Has lance-shaped, hairy, dark green leaves.
A. tradescantii. Erect perennial. H 1.2m (4ft), S 50cm (20in). Fully hardy. Has lance-shaped, mid-green leaves. In autumn, clusters of small, white flower heads appear on wiry, leafy stems and provide a good foil to bright, autumn leaf colours.
☸ *A. turbinellus* (illus. p.226). Upright perennial. H 1.2 (4ft), S 60cm (2ft). Fully hardy. In autumn has small, daisy-like, violet flower heads on small stems. Leaves are narrow and lance-shaped. Is easy to grow and disease-resistant.
A. vimineus **'Delight'**. See *A. lateriflorus* 'Delight'.

ASTERANTHERA (Gesneriaceae)
Genus of one species of evergreen, root climber. May be grown up mossy tree-trunks, trained against walls or used as ground cover. Frost hardy. Needs a dampish, semi-shaded position and neutral to slightly acid soil. Propagate by tip cuttings in summer or by stem cuttings in late summer or early autumn.
A. ovata. Evergreen, root climber with stems covered in white hairs. H to 4m (12ft). Has small, oblong, toothed leaves. Tubular, reddish-pink flowers, 5–6cm (2–2½in) long, often with yellow-striped, lower lips, appear singly or in pairs in leaf axils in summer.

ASTILBE (Saxifragaceae)
Genus of summer-flowering perennials, grown for their panicles of flowers that remain handsome even when dried brown in winter. Is suitable for borders and rock gardens. Fully hardy. Needs rich, moist soil and, in most species, partial shade. Leave undisturbed if possible, and give an annual spring mulch of well-rotted compost. Propagate species by seed in autumn, others by division in spring or autumn.
A. **'Bressingham Beauty'**. Leafy, clump-forming perennial. H and S to 1m (3ft). In summer bears feathery, tapering panicles of small, star-shaped, rich pink flowers on strong stems. Broad leaves are divided into oblong to oval, toothed leaflets.
☸ *A. chinensis* **'Pumila'**. Clump-forming perennial. H 30cm (12in), S 20cm (8in). Lower two-thirds of flower stem bears deeply dissected, coarse, toothed, hairy, dark green leaves. Dense, fluffy spikes of tiny, star-shaped, deep raspberry-red flowers appear in summer. Is good for a shaded, moist rock garden.
☸ *A.* **'Fanal'** illus. p.246.
A. **'Granat'**. Clump-forming, leafy perennial. H 60cm (2ft), S to 1m (3ft). Produces pyramidal trusses of tiny, star-shaped, deep red flowers in summer above broad, bronze-flushed, rich green leaves, which are divided into oblong to oval, toothed leaflets.
A. **'Irrlicht'** illus. p.240.
A. **'Montgomery'** illus. p.213.
☸ *A.* **'Ostrich Plume'** illus. p.208.
☸ *A.* **'Perkeo'** illus. p.301.
A. simplicifolia **'Gnome'**. Arching, clump-forming, slender-stemmed perennial. H 15cm (6in), S 10cm (4in). Has oval, deeply lobed or cut, crimped, reddish-green leaves in a basal rosette. Produces dense racemes of tiny, star-shaped, pink flowers in summer. Is good for a shaded, moist rock garden or peat bed. Self seeds in damp places but will not come true.
A. **'Sprite'**. Clump-forming, dwarf, leafy perennial. H 50cm (20in), S to 1m (3ft). Has feathery, tapering panicles of tiny, star-shaped, shell-pink flowers in summer, borne above broad leaves divided into narrowly oval, toothed leaflets.
A. **'Venus'** illus. p.208.

ASTRANTIA (Umbelliferae)
Masterwort
Genus of perennials, widely used in flower arrangements. Fully hardy. Needs sun or semi-shade and well-drained soil. Propagate by division in spring or by seed when fresh, in late summer.
A. major and subsp. *involucrata* illus. p.241.
☸ *A.* **'Sunningdale Variegated'** is a clump-forming perennial. H 60cm (24in), S 45cm (18in). Produces rounded, sometimes pink-tinged, greenish-white flower heads on wiry, branched stems throughout summer-autumn. Palmate, deeply 3–5-lobed, mid-green leaves are variegated with yellow and cream.
☸ *A. maxima* illus. p.241.

ASTROPHYTUM (Cactaceae)
Genus of slow-growing, perennial cacti, grown for their freely produced, flattish, yellow flowers, some with red centres. Frost tender, min. 5°C (41°F). Prefers sun and very well-drained, lime-rich soil. Dry completely in winter. Is prone to rot if wet. Propagate by seed in spring or summer.
A. asterias (Sea urchin, Silver dollar cactus). Slow-growing, slightly domed, perennial cactus. H 8–10cm (3–4in), S 10cm (4in). Spineless stem has about 8 low ribs bearing small, tufted areoles. Produces bright yellow flowers, to 6cm (2½in) across, in summer.
☸ *A. myriostigma* illus. p.402.
A. ornatum illus. p.413.

Asystasia bella. See *Mackaya bella.*

ATHEROSPERMA (Monimiaceae, syn. Atherospermataceae)
Genus of evergreen trees, grown for their foliage and flowers in summer. Frost tender, min. 3–5°C (37–41°F). Needs full light or partial shade and well-drained soil. Water containerized specimens moderately, less in winter. Pruning is tolerated if necessary. Propagate by seed in spring or by semi-ripe cuttings in summer.
A. moschatum (Australian sassafras, Tasmanian sassafras). Evergreen, spreading tree, conical when young. H 15–25m (50–80ft), S 5–10m (15–30ft). Has lance-shaped, nutmeg-scented, glossy leaves, slightly toothed and covered with white down beneath. Produces small, saucer-shaped, creamy-white flowers in summer.

ATHROTAXIS (Taxodiaceae)
Genus of conifers with awl-shaped leaves that clasp stems. See also CONIFERS.
A. selaginoides (King William pine). Irregularly conical conifer. H 15m (50ft) or more, S 5m (15ft). Half hardy. Has tiny, thick-textured, loosely overlapping, dark green leaves and insignificant, globular cones.

ATHYRIUM (Polypodiaceae)
Genus of deciduous or, occasionally, semi-evergreen ferns. Fully hardy to frost tender, min. 5°C (41°F). Needs shade and humus-rich, moist soil. Remove fading fronds regularly. Propagate by spores in late summer or by division in autumn or winter.
☸ *A. filix-femina* (Lady fern). Deciduous fern. H 60cm–1.2m (2–4ft), S 30cm–1m (1–3ft). Fully hardy. Dainty, lance-shaped, much-divided, arching fronds are pale green. Has very variable frond dissection.
A. goeringianum. See *A. nipponicum.*
A. nipponicum, syn. *A. goeringianum,* illus. p.189.

ATRIPLEX (Chenopodiaceae)
Genus of annuals, perennials and evergreen or semi-evergreen shrubs, grown for their foliage. Grows well by the coast. Fully to half hardy. Needs full sun and well-drained soil. Propagate by softwood cuttings in summer or by seed in autumn.
A. halimus (Tree purslane). Semi-evergreen, bushy shrub. H 2m (6ft), S 3m (10ft). Frost hardy. Oval leaves are silvery-grey. Produces flowers very rarely.
A. hortensis **'Rubra'** (Red mountain spinach, Red orach). Fast-growing, erect annual. H 1.2m (4ft), S 30cm (1ft). Half hardy. Triangular, deep red leaves, to 15cm (6in) long, are edible. Produces insignificant flowers in summer.

AUBRIETA (Cruciferae)
Genus of evergreen, trailing and mound-forming perennials. Is useful on dry banks, walls and in rock gardens. Fully hardy. Thrives in sun and in any well-drained soil. To maintain a compact shape, cut back hard after flowering. Propagate by greenwood cuttings in summer or by semi-ripe cuttings in late summer or autumn.
A. **'Carnival'** illus. p.316.
A. **'Church Knowle'**. Evergreen, trailing, very compact perennial. H 8cm (3in), S 15–23cm (6–9in). Has small, stemless, 4-petalled, lavender-blue flowers in spring. Leaves are rounded and toothed.
A. **'Cobalt Violet'** illus. p.317.
A. deltoidea **'Argenteovariegata'** illus. p.316.
☸ *A.* **'Dr Mule's'**. Vigorous, evergreen, mound-forming perennial. H 5–8cm (2–3in), S 30cm (12in). Has rounded, toothed, soft green leaves and, in spring, large, double, rich purple flowers on short spikes.

A. **'Gurgedyke'**. Evergreen, mound-forming perennial. H 10cm (4in), S 20cm (8in). Bears rounded, toothed, soft green leaves. Produces 4-petalled, deep purple flowers in spring.
A. **'Joy'** illus. p.315.
A. **'J.S. Baker'** illus. p.316.

AUCUBA (Cornaceae)
Genus of evergreen shrubs, grown for their foliage and fruits. In order to obtain fruits, grow both male and female plants. Makes good house plants when kept in a cool, shaded position. Fully to frost hardy. Tolerates full sun through to dense shade. Grow in any but waterlogged soil. To restrict growth, cut old shoots back hard in spring. Propagate by semi-ripe cuttings in summer.
A. japonica illus. p.122.
♈ **'Crotonifolia'** (male) illus. p.123. **'Gold Dust'** is an evergreen, bushy, dense, female shrub. H and S 2.5m (8ft). Frost hardy. Has stout, green shoots and oval, glossy, gold-speckled, dark green leaves. Small, star-shaped, purple flowers in mid-spring are followed by egg-shaped, bright red fruits. Bright green leaves of **'Picturata'** (male) each have a central, golden blotch. Some plants of 'Crotonifolia' and 'Picturata' are known to be female and have produced fruits.

AURINIA (Cruciferae)
Genus of evergreen perennials, grown for their grey-green foliage and showy flower sprays. Is suitable for rock gardens, walls and banks. Fully hardy. Needs sun and well-drained soil. Propagate by softwood or greenwood cuttings in early summer or by seed in autumn.
♈ *A. saxatilis*, syn. *Alyssum saxatile*, illus. p.298. ♈ **'Citrina'** and **'Variegata'** illus. p.297. **'Dudley Neville'** is an evergreen, clump-forming perennial. H 23cm (9in), S 30cm (12in). Has oval, hairy, grey-green leaves and, in late spring and early summer, bears racemes of many small, 4-petalled, buff-yellow flowers.

AUSTROCEDRUS (Cupressaceae)
Genus of conifers with flattish sprays of scale-like leaves. See also CONIFERS.

A. chilensis, syn. *Libocedrus chilensis*, illus. p.80.

Avena candida of gardens. See *Helictotrichon sempervirens*.
Avena sempervirens. See *Helictotrichon sempervirens*.

AZARA (Flacourtiaceae)
Genus of evergreen shrubs and trees, grown for their foliage and yellow flowers that are composed of a mass of stamens. Frost to half hardy; in cold climates, plants are best situated against a south- or west-facing wall for added protection. Grows in sun or shade and in fertile, well-drained soil. Propagate by semi-ripe cuttings in summer.
A. lanceolata. Evergreen, bushy shrub or spreading tree. H and S 6m (20ft). Frost hardy. Has narrowly oval, sharply toothed, bright green leaves. Small, rounded clusters of pale yellow flowers are carried in late spring or early summer.
♈ *A. microphylla* illus. p.95.
A. serrata illus. p.105.

AZOLLA (Salviniaceae)
Genus of deciduous, perennial, floating water ferns, grown for their decorative foliage and to control algae by reducing light beneath water. Frost to half hardy. Grows in sun or shade. If not kept in check, may be invasive; reduce spread by removing portions with a net. Propagate by redistributing clusters of plantlets when they appear.
A. caroliniana illus. p.390.

AZORELLA (Hydrocotylaceae)
Genus of evergreen, tufted or spreading perennials, grown for their neat, rosetted foliage and their flowers. Is useful as alpine house plants. Fully hardy. Thrives in full light and well-drained soil. Propagate by division in spring.
A. trifurcata, syn. *Bolax glebaria* of gardens, illus. p.338.

Azorina vidalii. See *Campanula vidalii*.

Azureocereus hertlingianus. See *Browningia hertlingiana*.

B

BABIANA (Iridaceae)
Genus of spring- and early summer-flowering corms, valued for their brightly coloured flowers, somewhat like freesias. Frost tender, min. 10°C (50°F). Needs sun and well-drained soil. Propagate in autumn by seed or natural division of corms.
B. plicata. Spring-flowering corm. H 10–20cm (4–8in), S 5–8cm (2–3in). Has a fan of lance-shaped, erect, basal leaves and short spikes of funnel-shaped, violet-blue flowers, 4–5cm (1½–2in) long, with yellow-patched petals.
B. rubrocyanea illus. p.375.
B. stricta. Spring-flowering corm. H 10–20cm (4–8in), S 5–8cm (2–3in). Produces a fan of narrowly lance-shaped, erect, basal leaves and short spikes of up to 10 funnel-shaped, purple, blue, cream or pale yellow flowers, 2.5–4cm (1–1½in) long and sometimes red-centred.

BACCHARIS (Compositae)
Genus of evergreen or deciduous, mainly autumn-flowering shrubs, grown for their foliage and fruits. Is useful for exposed, coastal gardens and dry soil. Fully hardy. Needs full sun and well-drained soil. Propagate by softwood cuttings in summer.
B. halimifolia (Bush groundsel). Vigorous, deciduous, bushy shrub. H and S 4m (12ft). Oval leaves are grey-green and sharply toothed. Large clusters of minute, white flower heads in mid-autumn are succeeded by fluffy, white heads of tiny fruits.

Bahia lanata. See *Eriophyllum lanatum.*

BALDELLIA (Alismataceae)
Genus of deciduous or evergreen, perennial, bog plants and submerged water plants, grown for their foliage. Frost hardy. Prefers sun, but tolerates shade. Remove fading foliage regularly and excess growth as required. Propagate by division in spring or summer.
B. ranunculoides, syn. *Alisma ranunculoides*, *Echinodorus ranunculoides.* Deciduous, perennial, bog plant or submerged water plant. H 23cm (9in), S 15cm (6in). Has lance-shaped, mid-green leaves and, in summer, umbels of small, 3-parted, pink or white flowers with basal, yellow marks.

BALLOTA (Labiatae)
Genus of perennials and evergreen or deciduous sub-shrubs, grown for their foliage and flowers. Frost hardy. Requires very well-drained soil and full sun. Cut back in spring before growth commences. Propagate by semi-ripe cuttings in summer.
B. acetabulosa illus. p.146.
♈ ***B. pseudodictamnus*** illus. p.309.

BAMBUSA (Bambusoideae). See GRASSES, BAMBOOS, RUSHES and SEDGES.
B. glaucescens. See *B. multiplex.*
B. multiplex, syn. *B. glaucescens*, illus. p.183.

BANKSIA (Proteaceae)
Genus of evergreen shrubs and trees, grown for their flowers, foliage and overall appearance. Frost tender, min. 7–10°C (45–50°F). Requires full light and sharply drained, sandy soil that contains little phosphates or nitrates. Water containerized plants moderately when in full growth, sparingly at other times. Freely ventilate plants grown under glass. Propagate by seed in spring.
B. baxteri. Evergreen, spreading, open shrub. H and S 2–3m (6–10ft). Leathery, mid-green leaves are strap-shaped, cut from the midrib into triangular, sharply pointed lobes. Produces dense, spherical heads of small, tubular, yellow flowers in summer.
B. coccinea illus. p.100.
B. ericifolia (Heath banksia). Evergreen, irregularly rounded, wiry, freely branching shrub. H and S to 3m (10ft). Has small, needle-like leaves and dense, upright, bottlebrush-like spikes, each 10–15cm (4–6in) long, of small, tubular, bronze-red or occasionally yellow flowers in late winter and spring.
B. serrata. Evergreen, bushy, upright shrub or tree. H 3–10m (10–30ft), S 1.5–3m (5–10ft). Oblong to lance-shaped, saw-toothed, leathery leaves are mid- to deep green. Small, tubular, reddish-budded, cream flowers appear in dense, upright, bottlebrush-like spikes, each 10–15cm (4–6in) long, from spring to late summer.

BAPTISIA (Leguminosae)
Genus of summer-flowering perennials, grown for their flowers. Fully hardy. Requires full sun and deep, well-drained, preferably neutral to acid soil. Is best not disturbed once planted. May be propagated by division in early spring or by seed in autumn.
♈ ***B. australis*** illus. p.216.

Barbacenia elegans. See *Talbotia elegans.*

BARBAREA (Cruciferae)
Genus of summer-flowering perennials, biennials and annuals. Most species are weeds or winter salad plants, but the variegated form of *B. vulgaris* is grown for decorative purposes. Fully hardy. Grows in sun or shade and in any well-drained but not very dry soil. Propagate by seed or division in spring.
B. vulgaris (Winter cress, Yellow rocket). **'Variegata'** illus. p.255.

BARLERIA (Acanthaceae)
Genus of evergreen shrubs and perennials, grown for their flowers. Frost tender, min. 7–18°C (45–64°F). Needs full light or partial shade and fertile soil. Water potted plants well when in full growth, moderately at other times. In the growing season, prune tips of young plants to encourage branching. For a more compact habit, shorten long stems after flowering. Propagate by seed in spring or by greenwood or semi-ripe cuttings in summer.
B. cristata (Philippine violet). Evergreen, semi-erect shrub. H and S 60cm–1.2m (2–4ft). Min. 15–18°C (59–64°F). Has elliptic, coarsely haired leaves. Tubular, light violet flowers, sometimes pale pink or white, are produced from upper leaf axils in summer.
B. obtusa. Evergreen, erect, spreading shrub. H and S to 1m (3ft). Min. 7–10°C (45–50°F). Leaves are elliptic. Tubular, mauve flowers are borne from upper leaf axils during winter-spring.

Barosma pulchella. See *Agathosma pulchella.*

Bartlettina sordida. See *Eupatorium sordidum.*

Bartonia aurea. See *Mentzelia lindleyi.*

BASSIA. See KOCHIA.

BAUERA (Cunoniaceae, syn. Baueraceae)
Genus of evergreen shrubs, grown mainly for their flowers. Frost tender, min. 3–5°C (37–41°F). Needs full sun and humus-rich, well-drained, neutral to acid soil. Water potted plants moderately, less when not in full growth. Remove straggly stems after flowering. Propagate by seed in spring or by semi-ripe cuttings in late summer.
B. rubioides. Evergreen, bushy, wiry-stemmed shrub, usually of spreading habit. H and S 30–60cm (1–2ft). Leaves each have 3 oval to lance-shaped, glossy leaflets. Bowl-shaped, pink or white flowers appear in early spring and summer.

BAUHINIA (Leguminosae)
Genus of evergreen, semi-evergreen or deciduous trees, shrubs and scandent climbers, grown for their flowers. Frost tender, min. 5–18°C (41–64°F). Needs full light and fertile, well-drained soil. Water containerized specimens freely when in full growth, less in winter. Congested growth may be thinned out after flowering. Propagate by seed in spring.
B. galpinii, syn. *B. punctata*, illus. p.112.
B. punctata. See *B. galpinii.*
B. purpurea of gardens. See *B. variegata.*
B. variegata, syn. *B. purpurea* of gardens, and **'Candida'** illus. p.71.

BEAUCARNEA (Agavaceae)
Genus of evergreen shrubs and trees, grown mainly for their intriguing, overall appearance. Frost tender, min. 7°C (45°F). Needs full light and sharply drained, fertile soil; drought conditions are tolerated. Water potted specimens moderately; allow compost almost to dry out between waterings. Propagate by seed or suckers in spring or by stem-tip cuttings in summer.
♈ ***B. recurvata***, syn. *Nolina recurvata, N. tuberculata*, illus. p.74.

BEAUMONTIA (Apocynaceae)
Genus of evergreen, woody-stemmed, twining climbers, grown for their large, fragrant flowers and handsome leaves. Frost tender, min. 7–10°C (45–50°F). Requires fertile, well-drained soil and full light. Water freely in growing season, sparingly otherwise. Provide support. Thin out previous season's growth after flowering. Propagate by semi-ripe cuttings in late summer.
B. grandiflora illus. p.165.

BEGONIA (Begoniaceae)
Genus of evergreen or deciduous shrubs and small, tree-like plants, perennials and annuals, grown for their colourful flowers and/or ornamental leaves. Prefers slightly acidic soil. Is susceptible to powdery mildew and botrytis from late spring to early autumn. Commonly cultivated begonias are divided into the following groups, each with varying cultivation requirements.

Cane-stemmed group
Evergreen, woody perennials, many known as 'Angelwings', with usually erect, cane-like stems bearing regularly spaced, swollen nodes and flowers in large, pendulous panicles. Encourage branching by pinching out growing tips. New growth develops from base of plant. Frost tender, min. 10°C (50°F). Grow under glass in good light but not direct sun (poor light reduces quantity of flowers) and in free-draining, loam-based compost. Stake tall plants. Propagate in spring by seed or tip cuttings.

Rex Cultorum and rhizomatous group
Mostly evergreen, rhizomatous begonias, though some are tuberous. Rex Cultorum begonias are grown for their foliage; within the rhizomatous group flowers are also important. From horizontal (creeping) or erect rhizomes arise plain or crested, green or brown leaves, 2.5–30cm (1–12in) long, that are sometimes spirally twisted. Creeping rhizomatous types are more freely branched than erect ones and are useful for hanging baskets. Unless otherwise stated, frost tender, min. 13–15°C (55–9°F), but preferably 18°C (64°F), with 40–75% relative humidity. Grow under glass in cool climates, in partial shade and in well-drained soil;

water only sparingly. Do not allow water to remain on the leaves, otherwise they become susceptible to botrytis. Propagate in spring by seed, leaf cuttings or division of rhizomes.

Semperflorens group
Evergreen, bushy perennials, often grown as half-hardy bedding annuals. Stems are soft, succulent and branch freely, bearing generally rounded, green, bronze or variegated leaves, 5cm (2in) long. Flowers are single or double. Pinch out growing tips to produce bushy plants. Frost tender, min. 13°C (55°F). Requires sun or partial shade and well-drained soil. Propagate in spring by seed or stem cuttings.

Shrub-like group
Evergreen, multi-stemmed, bushy perennials, usually freely branched with flexible, erect or pendent stems, often hairy. Leaves may be hairy or glabrous and up to 15cm (6in) across, 10–30cm (4–12in) long. Single flowers are pink, cream or white. Frost tender, min. 7°C (45°F) with 55% relative humidity. Grow under glass in good light and moist but well-drained soil. Propagate in spring by seed or stem cuttings.

Tuberous group
Tuberous and semi-tuberous perennials, grown as annuals for their single flowers produced mainly in summer. Large-flowered, double types are known as *B.* x *tuberhybrida*. Most form bushy plants with leaves to 25cm (10in) long. Frost tender, min. 13°C (55°F). Outdoors, grow in dappled shade and moist conditions; under glass, plant in cool shade with 65–70% relative humidity. Tubers are dormant in winter. Start into growth in spring for mid-summer to early autumn flowering. Remove all buds until stems show at least 3 pairs of leaves; with large-flowered types allow only central male bud to flower, so remove flanking buds. Plants require staking. Propagate in spring by seed, stem or basal cuttings or division of tubers.

Winter-flowering group
Evergreen, low-growing, very compact perennials, with succulent, thin stems, that are often included in the tuberous group. Hiemalis types have single, semi-double or double flowers in a wide colour range; Cheimantha types differ in that flowers are single and white or pink; Rieger cultivars are improved Hiemalis types. Leaves are green or bronze, 5cm (2in) long. Flowers are borne mainly from late autumn to mid-spring. Frost tender, min. 18°C (64°F) with 40% relative humidity. Indirect sun and moist soil are preferred. Cut back old stems to 10cm (4in) after flowering. Propagate in spring by seed or stem cuttings.

B. albo-picta (illus. p.259). Fast-growing, evergreen, cane-stemmed begonia. H to 1m (3ft), S 30cm (1ft). Freely branched, green stems turn brown-green when mature. Narrowly oval to lance-shaped, wavy-edged, green leaves are silver-spotted. Has clusters of single, green-white flowers in summer-autumn.

B. angularis. See *B. stipulacea*.
B. 'Apricot Cascade'. See *B.* x *tuberhybrida*.
B. 'Beatrice Hadrell'. Evergreen, creeping, rhizomatous begonia. H 20–30cm (8–12in), S 25–30cm (10–12in). Oval leaves are deeply cleft, 8–15cm (3–6in) long and dark green with paler veins. Produces single, pale pink flowers, above foliage, in early spring.
B. 'Bethlehem Star'. Evergreen, creeping, rhizomatous begonia. H 20–30cm (8–12in), S 25–30cm (10–12in). Oval, slightly indented, almost black leaves, less than 8cm (3in) long, each have a central, creamy-green star. Bears masses of single, pale pink flowers, with darker pink spots, from late winter to early spring.
B. 'Billie Langdon'. See *B.* x *tuberhybrida*.
B. 'Bokit'. See *B.* 'Bowkit'.
B. bowerae (Eyelash begonia; illus. p.259). Evergreen, creeping, rhizomatous begonia. H 25–30cm (10–12in), S 20–25cm (8–10in). Produces 2.5cm (1in) long, oval, bright green leaves with chocolate marks and bristles around edges. Flowers, produced freely in winter, are single and white, tinted pink. 'Major' has larger leaves.
B. 'Bowkit', syn. *B.* 'Bokit'. Evergreen, erect, rhizomatous begonia. H 20–30cm (8–12in), S 25–35cm (10–14in). Has oval, spirally twisted, yellow-green leaves with brown tiger stripes. Bears masses of single, white flowers, flecked with pink, in winter.
B. 'Bridal Cascade'. See *B.* x *tuberhybrida*.
B. 'Can-Can'. See *B.* x *tuberhybrida*.
B. x *cheimantha* 'Gloire de Lorraine' (Christmas begonia, Lorraine begonia). Evergreen, Cheimantha-type, winter-flowering begonia. H 30cm (12in), S 30–35cm (12–14in). Is well-branched with rounded, bright green leaves and single, white to pale pink flowers. Male flowers are sterile, female, highly infertile.
B. 'City of Ballarat'. See *B.* x *tuberhybrida*.
B. coccinea (Angelwing begonia). Evergreen, cane-stemmed begonia. H 1.2m (4ft), S 30cm (1ft). Produces narrowly oval, glossy, green leaves, buff-coloured beneath, and, in spring, profuse, single, pink or coral-red flowers.
♀ *B.* 'Cocktail', syn. *B. semperflorens* 'Cocktail'. Slow-growing evergreen begonia. H and S 20–30cm (8–12in), has rounded, wavy, green-bronze leaves and pink, red or white flowers from summer until autumn frosts.
B. 'Corallina de Lucerna'. See *B.* 'Lucerna'.
B. 'Crimson Cascade'. See *B.* x *tuberhybrida*.
B. 'Curly Merry Christmas'. See *B.* Rex Cultorum group.
B. dichroa. Evergreen, cane-stemmed begonia. H 35cm (14in), S 25cm (10in). Oval leaves are mid-green, 12cm (5in) long; occasionally new leaves bear silver spots. Produces small, single, orange flowers, each with a white ovary, in summer.

B. dregei (Mapleleaf begonia). Semi-tuberous begonia. H 75cm (30in), S 35cm (14in). Has small, maple-like, lobed, purple-veined, bronze leaves, with red beneath and occasionally silver-speckled when young. Profuse, pendent, single, white flowers are borne in summer. Needs winter rest.
B. 'Duartei'. See *B.* Rex Cultorum group.
B. x *erythrophylla*, syn. *B.* 'Feastii'. Evergreen, creeping, rhizomatous begonia. H 20cm (8in), S 23–30cm (9–12in). Thick, mid-green leaves, 8–15cm (3–6in) long, are almost rounded, with leaf stalks attached to centre of red undersides; slightly wavy margins have white hairs. Produces single, light pink flowers well above foliage in early spring.
B. 'Feastii'. See *B.* 'Erythrophylla'.
B. 'Flamboyant'. See *B.* x *tuberhybrida*.
B. foliosa (illus. p.259). Evergreen, shrub-like begonia. H 30–50cm (12–20in), S 30–35cm (12–14in). Bears erect, then arching stems and oval, toothed, dark green leaves, 1cm (½in) long. Has very small, single, white flowers in spring and autumn. Is susceptible to whitefly. var. *miniata* see *B. fuchsioides*.
♀ *B. fuchsioides*, syn. *B. foliosa* var. *miniata* (Fuchsia begonia). Evergreen, shrub-like begonia. H to 1.2m (4ft), S 30cm (1ft). Oval, toothed leaves are numerous and dark green, 4cm (1½in) long. Pendent, single, bright red flowers are borne in winter.
B. 'Gold Cascade'. See *B.* x *tuberhybrida*.
B. gracilis var. *martiana*, syn. *B. martiana*. Tuberous begonia. H 60–75cm (24–30in), S 40cm (16in). Small, oval to lance-shaped, lobed, pale green or brown-green leaves have tapering tips. Produces large, fragrant, single, pink flowers, 2.5cm (1in) across, in summer.
B. haageana. See *B. scharffii*.
B. 'Helen Lewis'. See *B.* Rex Cultorum group.
B. 'Helene Harms'. See *B.* x *tuberhybrida*.
B. hiemalis 'Krefeld'. Evergreen, Rieger-type, winter-flowering begonia. H 25cm (10in) S 30cm (12in). Is semi-tuberous with succulent stems, oval, mid-green leaves and masses of single, vivid orange or bright crimson flowers. Is very susceptible to botrytis and mildew at base of stems, so water by pot immersion.
B. 'Ingramii' (illus. p.259). Evergreen, shrub-like begonia. H 70cm (28in), S 45cm (18in). Produces elliptic, toothed, bright green leaves, 8cm (3in) long, and, intermittently from spring to autumn, masses of single, pink flowers on spreading branches.
B. 'Iron Cross'. See *B. masoniana*.
B. 'Lucerna', syn. *B.* 'Corallina de Lucerna' (illus. p.259). Vigorous, evergreen, cane-stemmed begonia. H 2–2.2m (6–7ft), S 45–60cm (1½–2ft). Has oval, silver-spotted, bronze-green leaves, 25–35cm (10–14in) long, with tapered tips and, year-round, large panicles of single, deep pink flowers; male flowers remain almost closed.

B. 'Mac's Gold'. Evergreen, creeping, rhizomatous begonia. H and S 20–25cm (8–10in). Star-shaped, lobed, yellow leaves, 8–15cm (3–6in) long, have chocolate-brown marks. Has single, pink flowers intermittently during spring-summer but in moderate quantity.
B. 'Mme Richard Galle'. See *B.* x *tuberhybrida*.
B. manicata (illus. p.259). Evergreen, erect, rhizomatous begonia. H 60cm (24in), S 30–40cm (12–16in). Bears large, oval, brown-mottled, green leaves and, below each leaf base, a collar of stiff, red hairs around leaf stalk. Produces single, pale pink flowers in very early spring. Propagate by plantlets during growing season. 'Crispa' (illus. p.259) has deeper pink flowers and light green leaves with crested margins.
B. martiana. See *B. gracilis* var. *martiana*.
♀ *B. masoniana*, syn. *B.* 'Iron Cross' (Iron cross begonia; illus. p.259). Evergreen, creeping, rhizomatous begonia. H 45–60cm (18–24in), S 30–45cm (12–18in). Bears oval, toothed, rough, bright green leaves, 15cm (6in) long, with tapering tips and cross-shaped, black or dark brown centres. Has single, pink-flushed, white flowers during summer.
B. 'Masquerade'. See *B.* x *tuberhybrida*.
B. mazae. Evergreen, trailing, rhizomatous begonia. H to 23cm (9in), S indefinite. Rounded leaves are bronze-green with red-brown veins. Has fragrant, single, pink flowers, with red spots, in early spring. Is good for a hanging basket.
♀ *B.* 'Merry Christmas'. See *B.* Rex Cultorum group.
B. metallica (Metal-leaf begonia; illus. p.259). Evergreen, shrub-like begonia. H 50cm–1.2m (20in–4ft), S 45cm (18in). Bears white-haired stems and oval, toothed, silver-haired, bronze-green leaves, 18cm (7in) long, with dark green veins, red beneath. Has single, pink flowers, with red bristles, in summer-autumn.
B. 'Midas'. See *B.* x *tuberhybrida*.
B. 'Norah Bedson' (illus. p.259). Evergreen, creeping, rhizomatous begonia. H to 23cm (9in), S 25–30cm (10–12in). Produces rounded, bright green leaves, to 15cm (6in) long, with dark brown splashes, and single, pink flowers in early spring.
B. 'Oliver Twist' (illus. p.259). Evergreen, creeping, rhizomatous begonia. H 45–60cm (18–24in), S 25–45cm (10–18in). Oval leaves are pale to mid-green, to 30cm (12in) long, with heavily crested edges. Has single, pink flowers in early spring.
B. olsoniae (illus. p.259). Evergreen, compact, shrub-like begonia. H 23–30cm (9–12in), S 30cm (12in). Rounded, satiny, bronze-green leaves have cream veins. Bears single, very pale pink flowers, year-round, on arching, 30cm (12in) long stems.
B. 'Orange Cascade'. See *B.* x *tuberhybrida*.
B. 'Orange Rubra' (illus. p.259). Slow-growing, evergreen, cane-stemmed begonia. H 50cm (20in),

443

S 45cm (18in). Oval leaves are light green. Produces abundant clusters of single, orange flowers throughout the year.

B. 'Organdy'. See *B. semperflorens*.

B. 'Orpha C. Fox' (illus. p.259). Evergreen, cane-stemmed begonia. H 1m (3ft), S 30cm (1ft). Oval, silver-spotted, olive-green leaves, 15cm (6in) long, are maroon beneath. Produces large clusters of single, bright pink flowers year-round.

B. paulensis. Evergreen, creeping, rhizomatous begonia. H and S 25–30cm (10–12in). Erect stems bear rounded, mid-green leaves, 15cm (6in) long, with 'seersucker' surfaces crisscrossed with a spider's web of veins. Produces single, cream-white flowers, with wine-coloured hairs, in late spring.

B. 'Président Carnot'. Vigorous, evergreen, cane-stemmed begonia. H to 2.2m (7ft), S 45cm (1½ft). Erect stems bear 28cm (11in) long, 'angelwing', green leaves, with lighter spots. Produces large panicles of single, pink flowers, each 4cm (1½in) across, year-round.

B. 'Princess of Hanover'. See *B.* Rex Cultorum group.

B. prismatocarpa (illus. p.259). Evergreen, creeping, rhizomatous begonia. H 15–20cm (6–8in), S 20–25cm (8–10in). Leaves are oval, lobed, light green and less than 8cm (3in) long. Produces single, bright yellow flowers year-round. Needs 60–65% relative humidity.

B. pustulata. Evergreen, creeping, rhizomatous begonia. H 15–20cm (6–8in), S 20–25cm (8–10in). Bears oval, fine-haired, dark green leaves, with small blisters or pustules, and single, rose-pink flowers in summer. Prefers min. 22–4°C (72–5°F) and 70–75% relative humidity. **'Argentea'** (syn. *B.* 'Silver'; illus. p.259) has silver-splashed leaves and creamy-white flowers.

B. 'Red Ascot'. See *B. semperflorens*.

B. Rex Cultorum group. Group of evergreen, creeping, rhizomatous begonias. H 30–40cm (12–16in), S 35cm (14in). Has 20–25cm (8–10in) long, heart-shaped, deep green leaves, with a metallic sheen and purple-tinged edges, arising vertically from rhizome. Produces sparse, pale rose flowers in late winter and early spring. Includes the following hybrids.

'Curly Merry Christmas' is a sport of *B.* 'Merry Christmas', but has spirally twisted leaves.

'Duartei' (illus. p.259), H and S 45–60cm (18–24in), has spirally twisted, red-haired, very dark green leaves, over 15cm (6in) long, with silver-grey streaks and almost black edges. Is difficult to grow to maturity.

'Helen Lewis' (illus. p.259), H and S 45–60cm (18–24in), has an erect rhizome and silky, deep royal purple leaves, 15–20cm (6–8in) long, with silver bands. Slightly hairy, single, cream flowers are produced in early summer.

☙ 'Merry Christmas' (syn. *B.* 'Ruhrtal'; illus. p.259), H and S 25–30cm (10–12in), has satiny, red leaves, 15–20cm (6–8in) long, each with an outer, broad band of emerald-green and a deep velvet-red centre, sometimes edged with grey.

'Princess of Hanover', H and S 25–30cm (10–12in), has spirally twisted, emerald leaves, 20cm (8in) long, with bands of silver edged with ruby-red; entire leaf surfaces are covered with fine, pink hairs.

'Silver Helen Teupel', H and S 30–35cm (12–14in), has long, deeply cut, silver leaves, each with a glowing pink centre, giving a feathered effect.

B. 'Roy Hartley'. See *B.* x *tuberhybrida*.

B. 'Ruhrtal'. See *B.* 'Merry Christmas' (under *B. rex*).

B. scharffii, syn. *B. haageana* (illus. p.259). Evergreen, shrub-like begonia. H 60cm–1.2m (2–4ft), S 60cm (2ft). Stems are often covered with white hairs. Has oval, fine-haired, dark metallic-green leaves, 28cm (11in) long, with very tapered tips and reddish-green undersides. Produces single, pinkish-white flowers, each with a pink beard, from autumn to summer.

B. 'Scherzo'. Evergreen, creeping, rhizomatous begonia. H 25–30cm (10–12in), S 30–35cm (12–14in). Oval leaves are small, highly serrated and yellow with black marks. Bears single, white flowers in early spring.

B. Semperflorens Cultorum Group. Slow-growing, evergreen begonia. H 30cm (12in), S 25cm (10in). Has fleshy, freely branched, green stems and rounded, fleshy, dark green leaves, 8cm (3in) long, with light green undersides. Leaf axils each bear a small, single, pink flower, sometimes with a white centre, year-round, but especially in summer. Is the parent of the following hybrids.

'Cocktail'. See *B.* 'Cocktail'.

'Organdy' (illus. p.259), H and S 15cm (6in), has rounded, waxy, green-bronze leaves and pink, red or white flowers throughout summer until autumn frosts.

'Red Ascot' (illus. p.259), H and S 15cm (6in), has rounded, emerald-green leaves and masses of crimson-red flowers in summer.

B. serratipetala (illus. p.259). Evergreen, trailing, shrub-like begonia. H and S 45cm (18in). Obliquely oval leaves are highly serrated and bronze-green, with raised, deep pink spots. Produces mostly female, single, deep pink flowers intermittently throughout the year. Prefers 60% relative humidity, but with fairly dry roots.

B. 'Silver'. See *B. pustulata*.

B. 'Silver Helen Teupel'. See *B.* Rex Cultorum group.

B. stipulacea, syn. *B. angularis*, *B. zebrina*. Evergreen, cane-stemmed begonia. H 60cm–1.2m (2–4ft), S 30cm (1ft). Bears well-branched, angular stems and oval, wavy-edged, 20cm (8in) long, grey-green leaves, with silver-grey veins, pale green beneath. Carries single, white flowers in winter-spring.

B. sutherlandii (illus. p.259). Trailing, tuberous begonia. H 1m (3ft), S indefinite. Slender stems carry small, lance-shaped, lobed, bright green leaves, with red veins, and, in summer, loose clusters of single, orange flowers in profusion. Makes an excellent hanging-basket plant. Is particularly susceptible to mildew.

B. 'Thurstonii' (illus. p.259). Evergreen, shrub-like begonia. H to 1.2m (4ft), S 45cm (1½ft). Has rounded to oval, smooth, glossy, bronze-green leaves, with dark red veins, and, in summer, single, pink flowers.

B. x **tuberhybrida.** Group of large-flowered, double, tuberous begonias. H and S 75cm (30in). Bears male flowers, each of more than 10cm (4in) diameter, singly on stiff, succulent flower stems, from early summer to mid-autumn. Each flower resembles a double camellia flower or has a rose-bud centre. Oval leaves are green, 20cm (8in) long. Multiflora cultivars, H and S 30cm (12in), are more bushy and have 8cm (3in) long leaves and single, semi-double or double, male flowers, each 4–5cm (1½–2in) across, in summer; tolerates full sun. Pendula cultivars, H to 1m (3ft), have long, thin, trailing stems; leaves are 6–8cm (2½–3in) long. Masses of single or double flowers are borne in summer.

'Apricot Cascade' (Pendula group; illus. p.259) bears emerald-green leaves and double, orange-apricot flowers. Other cascades are **'Bridal Cascade'** (pink-edged petals), **'Crimson Cascade'**, **'Gold Cascade'** and **'Orange Cascade'**.

'Billie Langdon' (illus. p.259) has masses of heavily veined, double, white flowers, each 18cm (7in) across, with a perfect rose-bud centre.

'Can-Can' (illus. p.259), H 1m (3ft), has double, yellow flowers, each 20cm (8in) wide, with rough-edged, red petals. Produces few side shoots.

'City of Ballarat' carries double, glowing orange flowers, each 18cm (7in) across, with broad petals and a formal centre. Leaves are rich dark green.

'Flamboyant' (Multiflora group; illus. p.251) produces single, scarlet flowers in profusion. Leaves are slender and bright green.

'Helene Harms' (Multiflora group) has numerous semi-double, canary-yellow flowers. Is excellent in a hanging basket.

'Mme Richard Galle' (Multiflora group), H 30–45cm (12–18in), produces masses of small, double, soft apricot flowers.

'Masquerade', S to 1m (3ft), has double, white flowers, each 15cm (6in) across, with crinkled, red edges to broad petals. Makes an excellent pot plant, being prolific with side shoots.

'Midas' bears double, pale golden-yellow flowers, each 15cm (6in) across, with broad petals. Is particularly susceptible to mildew. Early flowers sometimes develop leaf-like petals.

'Roy Hartley' (illus. p.259) has double, salmon-coloured flowers, tinged with soft pink. Depth of colour is very dependent on light intensity. Has few side shoots.

B. versicolor. Evergreen, creeping, rhizomatous begonia. H to 30cm (12in), S 15cm (6in). Produces broadly oval or oblong, velvety leaves, 8cm (3in) long, in shades of mahogany, apple-green and maroon, and, in spring-summer, single, salmon-pink flowers. Provide min. 16–19°C (61–6°F) with 65–70% relative humidity.

B. x **weltoniensis** (Mapleleaf begonia; illus. p.251). Semi-tuberous begonia with a shrub-like habit. H 30–50cm (12–20in), S 30cm (12in). Has small, oval, long-pointed, toothed, dark green leaves. Heads of 5–8 single, pink or white flowers appear from leaf axils in summer.

B. xanthina. Evergreen, bushy, creeping, rhizomatous begonia. H 25–30cm (10–12in), S 30–35cm (12–14in). Bears oval, dark green leaves, 15–23cm (6–9in) long, with yellow veins, purple and hairy beneath. Pendent, single, orange-yellow flowers are borne in summer. Provide min. 21–4°C (70–75°F) with 75% relative humidity. var. *pictifolia* has pale yellow flowers and silver-zoned leaves.

B. zebrina. See *B. stipulacea*.

BELAMCANDA (Iridaceae)
Genus of summer-flowering bulbs, grown for their iris-like flowers. Frost hardy, but protect in cold winters. Needs sun and well-drained soil with added humus. Propagate by seed in spring.

B. chinensis. Summer-flowering bulb. H 45cm–1m (1½–3ft), S 15–25cm (6–10in). Carries a fan of sword-shaped, semi-erect leaves. A loosely branched stem bears a succession of flattish, orange-red flowers, 4–5cm (1½–2in) across, with darker blotches. Seeds are shiny and black.

BELLEVALIA (Liliaceae)
Genus of spring-flowering bulbs, similar to *Muscari*, but with longer, more tubular flowers. Some species have ornamental value, but most are uninteresting horticulturally. Frost hardy. Needs an open, sunny position and well-drained soil that dries out in summer. Propagate by seed, preferably in autumn.

B. hyacinthoides, syn. *Strangweia spicata*, illus. p.376.

B. pycnantha, syn. *Muscari paradoxum*, *M. pycnantha*. Spring-flowering bulb. H to 40cm (16in), S 5–8cm (2–3in). Has strap-shaped, semi-erect, basal, greyish-green leaves. Tubular, deep dusky-blue flowers, 0.5cm (¼in) long and with yellow tips, appear in a dense, conical spike.

BELLIS (Compositae)
Daisy
Genus of perennials, some grown as biennials. Fully hardy. Grow in sun or semi-shade and in fertile, very well-drained soil. Dead-head regularly. Propagate by seed in early summer or by division after flowering.

B. perennis (Double daisy). Slow-growing, carpeting perennial. Cultivars are grown as biennials. H and S 15–20cm (6–8in). All have oval, mid-green leaves and fully double flower heads in spring. Large-flowered (flower heads to 8cm (3in) wide) and miniature-flowered (flower heads to

2.5cm (1in) wide) cultivars are available. **Carpet Series** (large-flowered), H and S 15cm (6in), which is compact, and **'Goliath'** (large-flowered), H 20cm (8in), are both available in a mixture of red, pink and white. ♚ **'Pomponette'** (miniature-flowered) illus. p.279. **'White Carpet'** (large-flowered), H and S 15cm (6in), is a compact, white-flowered cultivar.

Beloperone guttata. See *Justicia brandegeana.*

BERBERIDOPSIS (Flacourtiaceae)
Genus of one species of evergreen, woody-stemmed, twining climber. Frost hardy. Dislikes strong winds and strong sun and is best grown in a north or west aspect. Soil, preferably lime-free, should be well-drained. Cut out dead growth in spring; train to required shape. Propagate by seed in spring or by stem cuttings or layering in late summer or autumn.
B. corallina illus. p.171.

BERBERIS (Berberidaceae)
Barberry
Genus of deciduous, semi-evergreen or evergreen, spiny shrubs, grown mainly for their rounded to cup-shaped flowers, with usually yellow sepals and petals, and for their fruits. The evergreens are also cultivated for their leaves, the deciduous shrubs for their colourful autumn foliage. Fully to frost hardy. Requires sun or semi-shade and any but waterlogged soil. Propagate species by seed in autumn, deciduous hybrids and cultivars by softwood or semi-ripe cuttings in summer, evergreen hybrids and cultivars by semi-ripe cuttings in summer.
B. aggregata. Deciduous, bushy shrub. H and S 1.5m (5ft). Fully hardy. Oblong to oval, mid-green leaves redden in autumn. Dense clusters of pale yellow flowers appear in late spring or early summer and are followed by egg-shaped, white-bloomed, red fruits.
B. **'Barbarossa'** illus. p.118.
B. buxifolia. Semi-evergreen or deciduous, arching shrub. H 2.5m (8ft), S 3m (10ft). Fully hardy. Has oblong to oval, spine-tipped, leathery, dark green leaves. Deep orange-yellow flowers appear from early to mid-spring and are followed by spherical, black fruits with a white bloom.
♚ *B. calliantha.* Evergreen, bushy shrub. H and S 1–1.5m (3–5ft). Fully hardy. Has oblong, sharply spiny, glossy, green leaves, white beneath, and large, pale yellow flowers in late spring, followed by egg-shaped, black fruits with a white bloom.
B. candidula. Evergreen, bushy, compact shrub. H and S 1m (3ft). Fully hardy. Leaves are narrowly oblong, glossy, dark green, white beneath. Has bright yellow flowers in late spring, then egg-shaped, blue-purple fruits.
B. x *carminea* **'Pirate King'**. Deciduous, arching shrub. H 2m (6ft), S 3m (10ft). Fully hardy. Produces oblong, dark green leaves and, in late spring and early summer, clusters of yellow flowers. Bears a profusion of spherical, pale red fruits.

B. **'Chenaultii'**. Evergreen, bushy shrub. H 1.5m (5ft), S 2m (6ft). Fully hardy. Narrowly oblong, wavy-edged, glossy, dark green leaves set off golden-yellow flowers in late spring and early summer. Bears egg-shaped, blue-black fruits.
B. coxii. Evergreen, bushy, dense shrub. H 2m (6ft), S 3m (10ft). Fully hardy. Produces narrowly oval, glossy, dark green leaves with white undersides and, in late spring, yellow flowers. Egg-shaped, blue-black fruits have a grey-blue bloom.
♚ *B. darwinii* illus. p.87.
B. empetrifolia illus. p.127.
B. gagnepainii illus. p.101.
B. jamesiana. Vigorous, deciduous, arching shrub. H and S 4m (12ft). Fully hardy. Yellow flowers in late spring are followed by pendent racemes of spherical, red berries. Oval, dark green leaves redden in autumn.
B. julianae illus. p.104.
B. linearifolia **'Orange King'** illus. p.105.
B. x *lologensis.* Vigorous, evergreen, arching shrub. H 3m (10ft), S 5m (15ft). Fully hardy. Has broadly oblong, glossy, dark green leaves. Profuse clusters of orange flowers are borne from mid- to late spring. **'Stapehill'** illus. p.105.
♚ *B.* x *ottawensis* **'Superba'**, syn. *B.* x *o.* **'Purpurea'**. Deciduous, arching shrub. H and S 2.5m (8ft). Fully hardy. Produces rounded to oval, deep reddish-purple leaves. Bears small, red-tinged, yellow flowers in late spring, then egg-shaped, red fruits in autumn.
♚ *B.* **'Parkjuweel'**. Semi-evergreen, bushy, rounded shrub. H and S 1m (3ft). Fully hardy. Leaves are oval, glossy and bright green; some become red in autumn. Flowers are of little value.
B. prattii. Deciduous, bushy shrub. H and S 3m (10ft). Fully hardy. Has oblong, glossy, dark green leaves. Large clusters of small, yellow flowers in late summer are followed by a profusion of long-lasting, egg-shaped, coral-pink fruits.
♚ *B.* **'Rubrostilla'** illus. p.142.
B. sargentiana. Evergreen, bushy shrub. H and S 2m (6ft). Fully hardy. Leaves are oblong, glossy, bright green. Yellow flowers in late spring and early summer are succeeded by egg-shaped, blue-black fruits.
♚ *B.* x *stenophylla* illus. p.105.
♚ **'Corallina Compacta'** illus. p.298.
♚ *B. thunbergii.* Deciduous, arching, dense shrub. H 2m (6ft), S 3m (10ft). Fully hardy. Broadly oval, pale to mid-green leaves turn brilliant orange-red in autumn. Small, red-tinged, pale yellow flowers appear in mid-spring, followed by egg-shaped, bright red fruits. f. *atropurpurea* illus. p.101.
♚ **'Atropurpurea Nana'** (syn. *B.t.* 'Crimson Pygmy'), H and S 60cm (24in), bears reddish-purple foliage. **'Aurea'** illus. p.140. Upright branches of **'Erecta'** spread with age.
♚ **'Rose Glow'** has reddish-purple leaves marbled with pink and white.
♚ *B. verruculosa* illus. p.104.
♚ *B. wilsoniae.* Deciduous or semi-evergreen, bushy shrub. H 1m (3ft),

S 1.5m (5ft). Fully hardy. Narrowly oblong, grey-green leaves become bright orange-red in autumn. Produces yellow flowers in late spring and early summer, then masses of showy, spherical, coral-red fruits.

BERCHEMIA (Rhamnaceae)
Genus of deciduous, twining climbers, grown for their leaves and fruit. Is useful for covering walls, fences and tree stumps. Fully hardy. Grow in sun or shade, in any well-drained soil. Propagate by seed in autumn or spring, by semi-ripe cuttings in summer or by layering or root cuttings in winter.
B. racemosa **'Variegata'**. Deciduous, twining climber. H 5m (15ft) or more. Has heart-shaped, green leaves, 3–8cm (1¼–3in) long and paler beneath, that are variegated creamy-white, especially towards ends of shoots. Small, bell-shaped, greenish-white flowers in summer, are followed in autumn by rounded, green fruits that turn red, then black.

BERGENIA, syn. MEGASEA (Saxifragaceae)
Genus of evergreen perennials with thick, usually large, rounded to oval or spoon-shaped, leathery leaves, with indented veins, that make ideal ground cover. Fully to frost hardy. Tolerates sun or shade and any well-drained soil, but leaf colour is best on poor soil and in full sun. Propagate by division in spring after flowering.
B. **'Abendglut'**, syn. *B.* 'Evening Glow'. Evergreen, clump-forming perennial. H 23cm (9in), S 30cm (12in). Fully hardy. Produces rosettes of oval, crinkled, short-stemmed, maroon leaves, from which arise racemes of open cup-shaped, semi-double, deep magenta flowers in spring.
B. **'Ballawley'**. Evergreen, clump-forming perennial. H and S 60cm (24in). Fully hardy. Large, rounded to oval, flat, deep green leaves turn deep red in winter. Racemes of cup-shaped, bright crimson flowers are produced on red stems in spring. Needs shelter from cold winds.
B. ciliata illus. p.233.
B. cordifolia. Evergreen, clump-forming perennial. H 45cm (18in), S 60cm (24in). Fully hardy. Leaves are rounded, puckered and crinkle-edged. Produces racemes of open cup-shaped, light pink flowers in spring.
♚ **'Purpurea'** illus. p.233.
B. crassifolia. Evergreen, clump-forming perennial. H 30cm (12in), S 45cm (18in). Fully hardy. Has oval- or spoon-shaped, fleshy, flat leaves that turn mahogany in winter. Bears spikes of open cup-shaped, lavender-pink flowers in spring.
B. **'Evening Glow'**. See *B.* 'Abendglut'.
♚ *B.* **'Morgenröte'**. Evergreen, clump-forming perennial. H 45cm (18in), S 30cm (12in). Fully hardy. Leaves are rounded, crinkled and deep green. Spikes of open cup-shaped, deep carmine flowers in spring are often followed by a second crop in summer.
♚ *B. purpurascens.* Evergreen, clump-forming perennial. H 45cm

(18in), S 30cm (12in). Fully hardy. Oval to spoon-shaped, flat, dark green leaves turn beetroot-red in late autumn. In spring bears racemes of open cup-shaped, rich red flowers.
♚ *B.* x *schmidtii.* Evergreen, clump-forming perennial. H 30cm (12in), S 60cm (24in). Fully hardy. Oval, flat leaves have toothed margins. Sprays of open cup-shaped, soft pink flowers are produced in early spring on short stems.
♚ *B.* **'Silberlicht'**, syn. *B.* 'Silver Light', illus. p.232.
B. **'Silver Light'**. See *B.* 'Silberlicht'.
B. stracheyi. Evergreen, clump-forming perennial. H 23cm (9in), S 30cm (12in). Fully hardy. Small, rounded, flat leaves form neat rosettes, among which nestle branched heads of open cup-shaped, white or pink flowers in spring.
B. **'Sunningdale'**. Evergreen, clump-forming perennial. H 60cm (24in), S 30cm (12in). Fully hardy. Rounded, slightly crinkled, deep green leaves are mahogany beneath. Bears racemes of open cup-shaped, lilac-carmine flowers on red stalks in spring.

BERKHEYA (Compositae)
Genus of summer-flowering perennials. Frost to half hardy, but, except in mild areas, grow most species against a south- or west-facing wall. Requires full sun and fertile, well-drained soil. Propagate by division in spring or by seed in autumn.
B. macrocephala illus. p.222.

BERTOLONIA (Melastomataceae)
Genus of evergreen perennials, grown mainly for their foliage. Frost tender, min. 15°C (59°F) but preferably warmer. Requires a fairly shaded position and high humidity, although soil should not be waterlogged. Propagate by tip or leaf cuttings in spring or summer.
B. marmorata. Evergreen, rosette-forming perennial. H 15cm (6in) or more in flower, S 45cm (18in). Broadly oval, slightly fleshy leaves have heart-shaped bases, silvery midribs and puckered surfaces, and are reddish-purple below, velvety green above. Intermittently produces spikes of saucer-shaped, pinkish-purple flowers.

BERZELIA (Bruniaceae)
Genus of evergreen, heather-like, summer-flowering shrubs, grown for their flowers. Frost tender, min. 7°C (45°F). Requires full sun and well-drained, neutral to acid soil. Water containerized plants moderately, less when not in full growth. Plants may be cut back lightly after flowering. Propagate by seed in spring or by semi-ripe cuttings in late summer.
B. lanuginosa. Evergreen, erect shrub with soft-haired, young shoots. H and S to 1m (3ft). Has small, heather-like leaves. Compact, spherical heads of tiny, creamy-white flowers are carried in dense, terminal clusters in summer.

BESCHORNERIA (Agavaceae)
Genus of perennial succulents with narrowly lance-shaped leaves forming

445

erect, almost stemless, basal rosettes. Half hardy. Needs full sun and very well-drained soil. Propagate by seed or division in spring or summer.
♞ *B. yuccoides* illus. p.401.

BESSERA (Liliaceae)
Genus of summer-flowering bulbs, grown for their striking, brightly coloured flowers. Half hardy. Needs an open, sunny situation and well-drained soil. Propagate by seed in spring.
B. elegans (Coral drops). Summer-flowering bulb. H to 60cm (24in), S 8–10cm (3–4in). Bears long, narrow, erect, basal leaves and slender, leafless stems, each producing a loose head of pendent, bell-shaped, bright red flowers on long, slender stalks.

Betonica officinalis. See *Stachys officinalis.*

BETULA (Betulaceae)
Birch
Genus of deciduous trees and shrubs, grown for their bark and autumn colour. Fully hardy. Plant in a sunny position and in any moist but well-drained soil; some species prefer acid soil. Young trees should be transplanted in autumn. Propagate by grafting in late winter or by softwood cuttings in early summer.
♞ *B. albosinensis* illus. p.48.
B. alleghaniensis, syn. *B. lutea* (Yellow birch). Deciduous, upright, open tree, often multi-stemmed. H 12m (40ft) or more, S 3m (10ft). Smooth, glossy, amber or golden-brown bark peels in thin shreds. Oval, mid- to pale green leaves rapidly turn gold in autumn. Bears yellow-green catkins in spring.
B. ermanii illus. p.46.
♞ *B.* 'Jermyns'. Deciduous, upright, open tree. H 15m (50ft), S 10m (30ft). Has peeling, bright white bark, which is particularly striking in winter, and long, elegant, yellow, male catkins in spring. Oval, mid-green leaves have serrated edges.
B. lutea. See *B. alleghaniensis.*
B. maximowicziana (Monarch birch). Fast-growing, deciduous, broad-headed tree. H 18m (60ft), S 3m (10ft). Has orange-brown or pink bark that does not peel and racemes of pendulous, yellowish catkins in spring. Large, oval, mid-green leaves turn bright butter-yellow in autumn.
B. nana illus. p.297.
B. papyrifera illus. p.46.
♞ *B. pendula* (Silver birch). Deciduous, broadly columnar or conical, graceful tree. H 20m (70ft) or more, S 10m (30ft). Has slender, drooping shoots and silver-white bark that becomes black and rugged at base of trunk with age. Yellow-green catkins appear in spring. Oval, bright green leaves turn yellow in autumn.
♞ 'Laciniata' has a narrow crown.
♞ 'Tristis' illus. p.47.
♞ 'Youngii' illus. p.66.
B. szechuanica (Szechuan birch). Vigorous, deciduous, open tree with stiff branches. H 14m (46ft), S 2.5m (8ft). Bark is strikingly chalky-white when mature. Has triangular to oval, serrated, leathery, deep green leaves that turn brilliant gold in autumn. Bears yellow-green catkins in spring.
B. utilis (Himalayan birch). Deciduous, upright, open tree. H 18m (60ft), S 2.5m (8ft). Paper-thin, peeling bark is creamy-white or dark copper-brown. Yellow-brown catkins are borne in spring. Oval, mid-green leaves, hairy beneath when young, turn golden-yellow in autumn.
var. *jacquemontii* illus. p.57.

BIARUM (Araceae)
Genus of mainly autumn-flowering, tuberous perennials with tiny flowers carried on a pencil-shaped spadix, enclosed within a tubular spathe. Upper part of spathe is hooded or flattened out and showy. Frost hardy, but during cold, wet winters protect in a cold frame or greenhouse. Needs a sunny position and well-drained soil. Dry out tubers when dormant in summer. Propagate in autumn by seed or offsets.
B. eximium. Early autumn-flowering, tuberous perennial. H and S 8–10cm (3–4in). Lance-shaped, semi-erect, basal leaves follow stemless, tubular, velvety, blackish-maroon spathe, up to 15cm (6in) long and often lying flat on ground. Upper part is flattened out. Spadix is upright and black.
B. tenuifolium illus. p.383.

Bidens atrosanguinea. See *Cosmos atrosanguineus.*

BIFRENARIA. See ORCHIDS.
B. harrisoniae. Evergreen, epiphytic orchid for a cool greenhouse. H 10cm (4in). Fragrant, rounded, creamy-white flowers, 8cm (3in) across, each with a hairy, reddish-purple lip, are borne in spring-summer, 1 or 2 to a stem. Leaves are broadly oval, semi-rigid and 15cm (6in) long. Provide semi-shade in summer.

BIGNONIA (Bignoniaceae)
Genus of one species of evergreen, tendril climber. Frost hardy; in cool areas may lose its leaves in winter. Needs sun and fertile soil to flower well. If necessary, prune in spring. Propagate by stem cuttings in summer or autumn or by layering in winter.
B. capensis. See *Tecomaria capensis.*
B. capreolata, syn. *Doxantha capreolata* (Cross vine, Trumpet flower). Evergreen, tendril climber. H 10m (30ft) or more. Each leaf has 2 narrowly oblong leaflets and a branched tendril. In summer, funnel-shaped, reddish-orange flowers appear in clusters in leaf axils. Pea-pod-shaped fruits, to 15cm (6in) long, are produced in autumn.
B. grandiflora. See *Campsis grandiflora.*
B. jasminoides. See *Pandorea jasminoides.*
B. pandorana. See *Pandorea pandorana.*
B. radicans. See *Campsis radicans.*
B. stans. See *Tecoma stans.*

Bilderdykia aubertii. See *Polygonum aubertii.*
Bilderdykia baldschuanica. See *Polygonum baldschuanicum.*

BILLARDIERA (Pittosporaceae)
Genus of evergreen, woody-stemmed, twining climbers, grown mainly for their fruits. Half hardy. Grow in any well-drained soil, in a sheltered position and partial shade. Propagate by seed in spring or stem cuttings in summer or autumn.
B. longiflora illus. p.178.

BILLBERGIA (Bromeliaceae)
Genus of evergreen, rosette-forming perennials, grown for their flowers and foliage. Frost tender, min. 5–7°C (41–5°F). Requires semi-shade and well-drained soil, ideally with the addition of sphagnum moss or plastic chips used for orchid culture. Water moderately when in full growth, sparingly at other times. Propagate by division or offsets after flowering or in late spring.
B. nutans (Queen's tears; illus. p.229). Evergreen, clump-forming, tubular-rosetted perennial. H and S to 40cm (16in). Strap-shaped leaves are usually dark green. In spring bears pendent clusters of tubular, purple-blue-edged, lime-green flowers that emerge from pink bracts.
B. rhodocyanea. See *Aechmea fasciata.*
♞ *B.* x *windii* (Angel's tears). Evergreen, clump-forming, tubular-rosetted perennial. H and S to 40cm (16in). Is similar to *B. nutans*, but has broader, spreading, grey-green leaves and larger bracts. Flowers intermittently from spring to autumn.

Biota orientalis. See *Thuja orientalis.*

Bistorta affine. See *Polygonum affine.*
Bistorta amplexicaule. See *Polygonum amplexicaule.*
Bistorta macrophylla. See *Polygonum macrophyllum.*
Bistorta major. See *Polygonum bistorta.*
Bistorta milletii. See *Polygonum milletii.*
Bistorta vacciniifolia. See *Polygonum vacciniifolium.*

BLECHNUM (Polypodiaceae)
Genus of evergreen or semi-evergreen ferns. Fully hardy to frost tender, min. 5°C (41°F). Most species prefer semi-shade. Requires moist, neutral to acid soil. Remove fading fronds regularly. Propagate *B. penna-marina* by division in spring, other species by spores in late summer.
B. capense illus. p.186.
B. magellanicum. See *B. tabulare.*
B. penna-marina. Fast-growing, evergreen or semi-evergreen, carpeting fern. H 15–30cm (6–12in), S 30–45cm (12–18in). Frost hardy. Has linear, dark green fronds, the outermost sterile narrow, indented and spreading, central ones fertile, short and upright. Is suitable for a rock or peat garden.
♞ *B. spicant* (Hard fern). Evergreen fern. H 30–75cm (12–30in), S 30–45cm (12–18in). Fully hardy. Has narrowly lance-shaped, indented, leathery, spreading, dark green fronds. Prefers shade and peaty or leafy soil.
♞ *B. tabulare*, syn. *B. magellanicum.* Evergreen or semi-evergreen fern. H 30cm–1m (1–3ft), S 30–60cm (1–2ft). Half hardy. Outer, mid-green sterile fronds are broadly lance-shaped, heavily indented and arranged symmetrically. Inner fertile fronds are brown and fringed.

BLETILLA. See ORCHIDS.
B. striata (illus. p.261). Deciduous, terrestrial orchid. H to 60cm (24in). Half hardy. In late spring or early summer produces loose spikes of magenta or white flowers, 3cm (1¼in) long. Has broadly lance-shaped leaves, 50cm (20in) long. Needs shade in summer.

BLOOMERIA (Liliaceae)
Genus of onion-like, spring-flowering bulbs, with spherical flower heads on leafless stems, which die down in summer. Frost hardy. Needs a sheltered, sunny situation and well-drained soil. Propagate by seed in autumn or by division in late summer or autumn.
B. crocea. Late spring-flowering bulb. H to 30cm (12in), S to 10cm (4in). Long, narrow, semi-erect, basal leaves die away at flowering time. Each leafless stem carries a loose, spherical head, 10–15cm (4–6in) across, of star-shaped, yellow flowers with dark stripes.

Bocconia cordata. See *Macleaya cordata.*

BOENNINGHAUSENIA (Rutaceae)
Genus of one species of deciduous sub-shrub, usually with soft, herbaceous stems and grown for its foliage and flowers. Frost hardy, although cut to ground level in winter. Requires full sun and fertile, well-drained but not too dry soil. Propagate by softwood cuttings in summer or by seed in autumn.
B. albiflora. Deciduous, bushy sub-shrub. H and S 1m (3ft). Has pungent, mid-green leaves, divided into oval leaflets. Loose panicles of small, cup-shaped, white flowers appear from mid-summer to early autumn.

BOLAX (Hydrocotylaceae)
Genus of evergreen, hummock- and cushion-forming perennials, often included in *Azorella.* Is grown for its small, thick, tough leaves in symmetrical rosettes. Flowers only rarely in cultivation. Is suitable for gritty screes, troughs and alpine houses. Fully hardy. Needs sun and humus-rich, well-drained soil. Propagate by rooting rosettes in summer.
B. glebaria of gardens. See *Azorella trifurcata.*
B. gummifera illus. p.339.

BOMAREA (Alstroemeriaceae)
Genus of herbaceous or evergreen, tuberous-rooted, scrambling and twining climbers, grown for their tubular or bell-shaped flowers. Half hardy to frost tender, min. 5°C (41°F). Grow in any well-drained soil and in full light. Water regularly in growing season, sparingly when dormant. Provide support. Cut out old flowering

stems at ground level when leaves turn yellow. Propagate by seed or division in early spring.
B. andimarcana, syn. *B. pubigera*. Evergreen, scrambling climber with straight, slender stems. H 2–3m (6–10ft). Frost tender. Has lance-shaped leaves, white and hairy beneath. Produces nodding, tubular, green-tipped, pale yellow flowers, suffused pink, from early summer to autumn.
B. caldasii, syn. *B. kalbreyeri* of gardens, illus. p.177.
B. kalbreyeri of gardens. See *B. caldasii*.
B. pubigera. See *B. andimarcana*.

BORAGO (Boraginaceae)
Borage
Genus of annuals and perennials, grown for culinary use as well as for their flowers. Fully hardy. Requires sun and fertile, well-drained soil. For culinary use gather only young leaves. Propagate by seed sown outdoors in spring. Some species will self seed prolifically and may become a nuisance.
B. officinalis illus. p.286.

BORONIA (Rutaceae)
Genus of evergreen shrubs, grown primarily for their flowers. Frost tender, min. 7–10°C (45–50°F). Requires full light and sandy, neutral to acid soil. Water potted specimens moderately, less when not in full growth. To maintain a compact habit, long stems may be shortened after flowering. Propagate by seed in spring or by semi-ripe cuttings in late summer. Red spider mite may be a problem.
B. elatior. See *B. molloyae*.
B. megastigma illus. p.126.
B. molloyae, syn. *B. elatior*. Evergreen, bushy, slender-stemmed shrub. H and S 1–1.5m (3–5ft). Leaves are divided into 5–13 linear leaflets. In spring, upper leaf axils produce almost spherical, 4-petalled, carmine-red or yellow flowers.

BORZICACTUS. See CLEISTOCACTUS.

BOUGAINVILLEA (Nyctaginaceae)
Genus of deciduous or evergreen, woody-stemmed, scrambling climbers, grown for their showy floral bracts. Frost tender, min. 7–10°C (45–50°F). Grow in fertile, well-drained soil and in full light. Water moderately in the growing season; keep containerized plants almost dry when dormant. Needs tying to a support. Cut back previous season's lateral growths in spring, leaving 2–3cm (³⁄₄–1¹⁄₄in) long spurs. Propagate by semi-ripe cuttings in summer or by hardwood cuttings when dormant. Whitefly and mealy bug may cause problems.
❀ **B. x buttiana 'Mrs Butt'**. Vigorous, evergreen, woody-stemmed, scrambling climber. H to 5m (15ft). Has elliptic leaves, 4–8cm (1¹⁄₂–3in) long, and clusters of crimson-magenta floral bracts in summer. Floral bracts of **'Golden Glow'** are orange-yellow; those of **'Scarlet Queen'** are scarlet.

B. 'Dania' illus. p.170.
❀ **B. glabra** illus. p.174. **'Snow White'** illus. p.167. **'Variegata'** illus. p.171.
B. 'Miss Manila' illus. p.170.
B. spectabilis. Strong-growing, mainly evergreen, woody-stemmed, scrambling climber; stems usually have a few spines. H to 7m (22ft). Has elliptic to oval leaves and, in summer, large trusses of red-purple floral bracts.

Boussingaultia basselloides. See *Anredera cordifolia*.

BOUTELOUA (Gramineae). See GRASSES, BAMBOOS, RUSHES and SEDGES.
B. gracilis, syn. *B. oligostachya*, illus. p.183.
B. oligostachya. See *B. gracilis*.

BOUVARDIA (Rubiaceae)
Genus of deciduous, semi-evergreen or evergreen shrubs and perennials, grown for their flowers. Frost tender, min. 7–10°C (45–50°F), but 13–15°C (55–9°F) for winter-flowering species. Prefers full light and fertile, well-drained soil. Water freely when in full growth, moderately at other times. Cut back stems by half to three-quarters after flowering to maintain a neat habit. Propagate by softwood cuttings in spring or by greenwood or semi-ripe cuttings in summer. Whitefly and mealy bug may be troublesome.
B. humboldtii. See *B. longiflora*.
B. longiflora, syn. *B. humboldtii*. Semi-evergreen, spreading shrub. H and S 1m (3ft) or more. Min. 13–15°C (55–9°F) until flowering ceases, then 7°C (45°F). Has small, lance-shaped leaves. Fragrant, white flowers, with slender tubes and 4 petal lobes, are borne in terminal clusters from summer to early winter.
B. ternifolia, syn. *B. triphylla*, illus. p.143.
B. triphylla. See *B. ternifolia*.

BOWIAE (Liliaceae)
Genus of summer-flowering, bulbous succulents with scrambling, branched, green stems that produce no proper leaves; is grown mainly for botanical interest. Frost tender, min. 10°C (50°F). Needs sun and well-drained soil; plant with half of bulb above soil level. Support with sticks or canes. Propagate by seed, sown under glass in winter or spring. May produce offsets.
B. volubilis illus. p.396.

BOYKINIA (Saxifragaceae)
Genus of mound-forming perennials. Fully hardy. Most species require shade and humus-rich, moist but well-drained, acid soil. Propagate by division in spring or by seed in autumn.
B. aconitifolia. Mound-forming perennial. H 1m (3ft), S 15cm (6in). Has rounded to kidney-shaped, lobed leaves. In summer, flower stems carry very small, bell-shaped, white flowers.
B. jamesii, syn. *Telesonix jamesii*. Mound-forming, rhizomatous perennial. H and S 15cm (6in). Almost woody stems each bear a rosette of kidney-shaped leaves with lacerated edges. In early summer produces open

bell-shaped, frilled, pink flowers with green centres.

BRACHYCHILUM (Zingiberaceae)
Genus of aromatic perennials, grown for their flowers and ornamental fruits. Frost tender, min. 18°C (64°F). Prefers a humid atmosphere, humus-rich, moist but well-drained soil and partial shade. May be grown in pots if roots are not too restricted. Propagate by division in spring or summer.
B. horsfieldii. Clump-forming, tufted perennial. H and S to 1m (3ft). Has short-stalked, lance-shaped, leathery leaves, to 30cm (1ft) long. Showy, tubular, yellow-and-white flowers, to 8cm (3in) across, appear in summer, followed by orange fruits that open to reveal red seeds.

BRACHYCHITON (Sterculiaceae)
Genus of evergreen or deciduous, mainly spring- and summer-flowering trees, grown for their flowers and overall appearance. Frost tender, min. 7–10°C (45–50°F). Needs full light and humus-rich, well-drained, preferably acid soil. Water containerized specimens moderately, much less in winter. Pruning is tolerated if needed. Propagate by seed in spring. Red spider mite may be a nuisance.
B. acerifolius, syn. *Sterculia acerifolia*, illus. p.39.
B. populneus, syn. *Sterculia diversifolia* (Kurrajong). Evergreen, conical tree, pyramidal when young. H and S 15–20m (50–70ft). Pointed or 3–5-lobed, glossy, deep green leaves are chartreuse when young. In spring-summer has panicles of saucer-shaped, cream or greenish-white flowers with red, purple or yellow throats.

BRACHYCOME (Compositae)
Genus of annuals and perennials with daisy-like flower heads. Fully hardy. Requires sun, a sheltered position and rich, well-drained soil. Pinch out growing shoots of young plants to encourage a bushy habit. Propagate by seed sown under glass in spring or outdoors in late spring.
B. iberidifolia illus. p.286.

BRACHYGLOTTIS (Compositae)
Genus of evergreen shrubs and trees, grown for their bold foliage and overall effect. Some species are currently included in *Senecio*. Frost tender, min. 3°C (37°F). Needs full light or partial shade and well-drained soil. Water containerized plants freely in summer, moderately at other times. Propagate by semi-ripe cuttings in late summer.
B. compacta. See *Senecio compactus*.
B. laxifolia. See *Senecio laxifolius*.
B. monroi. See *Senecio monroi*.
B. repanda illus. p.97.
B. rotundifolia. See *Senecio reinoldii*.

Bracteantha bracteata. See *Helichrysum bracteatum*.

BRASSAIA. See SCHEFFLERA.

BRASSAVOLA. See ORCHIDS.
B. nodosa (Lady-of-the-night; illus. p.260). Evergreen, epiphytic orchid for

an intermediate greenhouse. H 23cm (9in). Narrow-petalled, pale green flowers, 5cm (2in) across and each with a white lip, are borne, 1–3 to a stem, in spring; they are fragrant at night. Leaves, 8–10cm (3–4in) long, are thick and cylindrical. Is best grown on a bark slab. Provide good light in summer.

BRASSICA (Cruciferae)
Genus of annuals and evergreen biennials and perennials. Most are edible vegetables, e.g. cabbages and kales, but forms of *B. oleracea* are grown for their ornamental foliage. Fully hardy. Grow in sun and in fertile, well-drained soil. Lime-rich soil is recommended, though not essential. Propagate by seed sown outdoors in spring or under glass in early spring. Is susceptible to club root.
B. oleracea forms illus. p.276.

x BRASSOCATTLEYA. See ORCHIDS.
x **B. Mount Adams** (illus. p.253). Evergreen, epiphytic orchid for an intermediate greenhouse. H 45cm (18in). Intermittently produces lavender-pink flowers, to 15cm (6in) across and each with a darker lip marked yellow and red, up to 4 per stem. Has oval, stiff leaves, 10–15cm (4–6in) long. Needs good light in summer.

x BRASSOLAELIOCATTLEYA. See ORCHIDS.
x **B. St Helier** (illus. p.253). Evergreen, epiphytic orchid for an intermediate greenhouse. H 45cm (18in). Pinkish-purple flowers, to 10cm (4in) across, each with a yellow-marked, rich red lip, are produced 1–4 to a stem, mainly in spring. Bears oval, stiff leaves, 10–15cm (4–6in) long. Grow in good light in summer.
x **B. Hetherington 'Coronation'** (illus. p.253). Evergreen, epiphytic orchid for an intermediate greenhouse. H 45cm (18in). Bears fragrant, light pink flowers, 10cm (4in) across and each with a deep pink and yellow lip, up to 4 to a stem, mainly in spring. Leaves, 10–15cm (4–6in) long, are stiff and oval. Provide good light in summer.

Bravoa geminiflora. See *Polianthes geminiflora*.

BREYNIA (Euphorbiaceae)
Genus of evergreen shrubs and trees, grown for their foliage. Frost tender, min. 13°C (55°F). Needs full light or partial shade and fertile, well-drained soil. Water containerized plants freely when in full growth, moderately at other times. Large bushes should be cut back hard after flowering. Propagate by greenwood or semi-ripe cuttings in summer. Whitefly, red spider mite and mealy bug may be troublesome.
B. disticha. See *B. nivosa*.
B. nivosa, syn. *B. disticha*, *Phyllanthus nivosus*, illus. p.145. **'Roseo-picta'** is an evergreen, rounded, well-branched shrub with slender stems. H and S to 1m (3ft). Has broadly oval, green

447

leaves variably bordered and splashed with white and flushed pink. Insignificant, petalless flowers are borne in spring-summer.

BRIGGSIA (Gesneriaceae)
Genus of evergreen perennials, grown for their rosettes of hairy leaves. Frost tender, min. 2–5°C (36–41°F). Needs shade and peaty soil with plenty of moisture in summer and good air circulation in winter. Protect against damp in winter. Propagate by seed in spring.
B. muscicola. Evergreen, basal-rosetted perennial. H 8–10cm (3–4in), S 23cm (9in). Leaves are oval, silver-haired and pale green. Arching flower stems bear loose clusters of tubular, pale yellow flowers, with protruding tips, in early summer. Is best grown in an alpine house.

BRIMEURA (Liliaceae)
Genus of spring-flowering bulbs, similar to miniature bluebells, grown for their flowers. Is good in rock gardens and shrub borders. Frost hardy. Requires partial shade. Prefers humus-rich, well-drained soil. Propagate by seed in autumn or by division in late summer.
B. amethystina, syn. *Hyacinthus amethystinus*, illus. p.376.

Brittonastrum mexicanum. See *Agastache mexicana.*

BRIZA (Gramineae), Quaking grass. See GRASSES, BAMBOOS, RUSHES and SEDGES.
B. maxima (Greater quaking grass). Robust, tuft-forming, annual grass. H to 50cm (20in), S 8–10cm (3–4in). Fully hardy. Mid-green leaves are mainly basal. Has loose panicles of up to 10 pendent, purplish-green spikelets, in early summer, that dry well for winter decoration. Self seeds readily.
B. media (Common quaking grass). Evergreen, tuft-forming, rhizomatous, perennial grass. H 30–60cm (12–24in), S 8–10cm (3–4in). Fully hardy. Mid-green leaves are mainly basal. In summer bears open panicles of up to 30 pendent, purplish-brown spikelets that dry well for winter decoration.

BRODIAEA (Liliaceae)
Genus of mainly spring-flowering bulbs with colourful flowers produced in loose heads on leafless stems. Frost hardy. Needs a sheltered, sunny situation and light, well-drained soil. Dies down in summer. Propagate in autumn by seed or in late summer and autumn by offsets, which are produced freely.
B. congesta. See *Dichelostemma congestum.*
B. coronaria. Late spring- to early summer-flowering bulb. H 10–25cm (4–10in), S 8–10cm (3–4in). Long, narrow, semi-erect, basal leaves die down by flowering time. Leafless stems each carry a loose head of erect, funnel-shaped, violet-blue flowers on long, slender stalks.
B. hyacinthina. See *Triteleia hyacinthina.*

B. ida-maia. See *Dichelostemma ida-maia.*
B. ixioides. See *Triteleia ixioides.*
B. laxa. See *Triteleia laxa.*
B. peduncularis. See *Triteleia peduncularis.*
B. pulchella. See *Dichelostemma pulchellum.*

BROMELIA (Bromeliaceae)
Genus of evergreen, rosette-forming perennials, grown for their overall appearance. Frost tender, min. 5–7°C (41–5°F). Needs full light and well-drained soil. Water moderately in summer, sparingly at other times. Propagate by suckers in spring.
B. balansae (Heart of flame; illus. p.229). Evergreen, clump-forming, basal-rosetted perennial. H 1m (3ft), S 1.5m (5ft). Has narrowly strap-shaped, arching, mid- to grey-green leaves with large, hooked spines. Club-shaped panicles of tubular, red- or violet-purple flowers, with long, bright red bracts, are borne in spring-summer or sometimes later.

BROMUS (Gramineae). See GRASSES, BAMBOOS, RUSHES and SEDGES.
B. ramosus (Hairy brome grass). Evergreen, tuft-forming, perennial grass. H to 2m (6ft), S 30cm (1ft). Fully hardy. Mid-green leaves are lax and hairy. Produces long, arching panicles of nodding, grey-green spikelets in summer. Prefers shade.

BROUSSONETIA (Moraceae)
Genus of deciduous trees and shrubs, grown for their foliage and unusual flowers. Male and female flowers are produced on different plants. Frost hardy. Needs full sun and well-drained soil. Propagate by softwood cuttings in summer or by seed in autumn.
B. papyrifera illus. p.53.

BROWALLIA (Solanaceae)
Genus of shrubby perennials, usually grown as annuals, with showy flowers. Frost tender, min. 4–15°C (39–59°F). Grows best in sun or partial shade and in fertile, well-drained soil that should not dry out completely. Feed when flowering if pot-grown and pinch out young shoots to encourage bushiness. Propagate by seed in spring; for winter flowers, sow in late summer.
B. elata. Moderately fast-growing, bushy perennial, usually grown as an annual. H 30cm (12in), S 15cm (6in). Min. 4°C (39°F). Has oval, mid-green leaves and, in summer, trumpet-shaped, blue flowers, 4cm (1½in) wide.
B. speciosa illus. p.230.

BROWNINGIA (Cactaceae)
Genus of slow-growing, eventually tree-like, perennial cacti. Spiny, silvery- or green-blue stems, with up to 20 or more ribs, are crowned by stiff, erect, green-blue branches. Frost tender, min. 7°C (45°F). Requires sun and very well-drained soil. Propagate by seed in spring or summer.
B. hertlingiana, syn. *Azureocereus hertlingianus*, illus. p.392.

BRUCKENTHALIA (Ericaceae)
Genus of one species of evergreen shrub, grown for its bell-shaped flowers. Is related to *Calluna* and *Erica* and is suitable for rock gardens and peat beds. Fully hardy. Needs sun and well-drained, peaty, acid soil. Propagate in late summer by semi-ripe cuttings or in spring by seed.
B. spiculifolia (Spike heath). Evergreen, heath-like shrub. H and S 15cm (6in). Tiny, needle-like, glossy, dark green leaves clothe stiff stems. Terminal clusters of tiny, pink flowers appear in summer.

BRUGMANSIA, syn. DATURA (Solanaceae)
Angels' trumpets
Genus of evergreen or semi-evergreen shrubs, trees and annuals, grown for their flowers borne mainly in summer-autumn. Frost hardy to frost tender, min. 7–10°C (45–50°F). Prefers full light and fertile, well-drained soil. Water containerized specimens freely in full growth, moderately at other times. May be pruned hard in early spring. Propagate by seed in spring or by greenwood or semi-ripe cuttings in early summer or later. Whitefly and red spider mite may be troublesome.
B. arborea, syn. *B. cornigera, Datura arborea.* Evergreen or semi-evergreen, rounded, robust shrub. H and S to 3m (10ft). Frost tender. Bears narrowly oval leaves, each 20cm (8in) or more long. Has strongly fragrant, pendent, trumpet-shaped, white flowers, each 16–20cm (6–8in) long with a spathe-like calyx in summer-autumn.
B. aurea, syn. *Datura aurea.* Evergreen, rounded shrub or tree. H and S 6–11m (20–35ft). Frost tender. Has oval leaves, 15cm (6in) long, and, in summer-autumn, pendent, trumpet-shaped, white or yellow flowers, 15–25cm (6–10in) long.
B. bicolor. See *B. sanguinea.*
B. x *candida*, syn. *Datura* x *candida*, illus. p.89.
B. cornigera. See *B. arborea.*
☆ *B. 'Grand Marnier'* illus. p.93.
B. rosei, syn. *B. sanguinea* of gardens, *Datura bicolor* of gardens. Evergreen, rounded shrub or tree. H and S to 6m (20ft). Frost tender. Leaves are 18cm (7in) long and oval, sometimes with small, rounded lobes. From late summer to following spring bears pendent, trumpet-shaped, red, saffron-yellow or orange flowers, 15–18cm (6–7in long).
☆ *B. sanguinea*, syn. *B. bicolor, Datura sanguinea*, illus. p.93.
B. sanguinea of gardens. See *B. rosei.*

BRUNFELSIA (Solanaceae)
Genus of evergreen shrubs, grown for their flowers. Frost tender, min. 10–13°C (50–55°F), but 15–18°C (59–64°F) for good winter flowering. Needs semi-shade and humus-rich, well-drained soil. Water containerized plants moderately, much less in low temperatures. Remove stem tips to promote branching in growing season. Propagate by semi-ripe cuttings in summer. Mealy bug and whitefly may be a problem.
B. calycina. See *B. pauciflora.*

☆ *B. pauciflora*, syn. *B. calycina* (Yesterday-today-and-tomorrow). Evergreen, spreading shrub. H and S 60cm (2ft) or more. Bears oblong to lance-shaped, leathery, glossy leaves. Blue-purple flowers, each with a tubular base and 5 overlapping, wavy-edged petals, are carried from winter to summer. 'Macrantha' illus. p.137.

BRUNNERA (Boraginaceae)
Genus of spring-flowering perennials. Fully hardy. Prefers light shade and moist soil. Propagate by division either in spring or autumn or by seed in autumn.
☆ *B. macrophylla* (Siberian bugloss). Clump-forming perennial. H 45cm (18in), S 60cm (24in). Delicate sprays of small, star-shaped, forget-me-not-like, bright blue flowers in early spring are followed by heart-shaped, rough, long-stalked leaves. Makes good ground cover. 'Dawson's White' (syn. *B.m.* 'Variegata') illus. p.235.

x *Brunsdonna parkeri.* See x *Amarygia parkeri.*

BRUNSVIGIA (Amaryllidaceae)
Genus of autumn-flowering bulbs with heads of showy flowers. Half hardy. Needs sun and well-drained soil. Water in autumn to bring bulbs into growth and continue watering until summer, when leaves will die away and dormant bulbs should be kept fairly dry and warm. Propagate by seed in autumn or by offsets in late summer.
B. josephinae (Josephine's lily). Autumn-flowering bulb. H to 45cm (18in), S 45–60cm (18–24in). Bears a stout, leafless stem with a spherical head of 20–30 funnel-shaped, red flowers, 7–9cm (3–3½in) long, with recurved petal tips. Semi-erect, oblong leaves appear after flowering.

BRYOPHYLLUM. See KALANCHOE.

BUDDLEJA (Loganiaceae)
Genus of deciduous, semi-evergreen or evergreen shrubs and trees, grown for their clusters of small, often fragrant flowers. Fully to half hardy. Requires full sun and fertile, well-drained soil. *B. crispa, B. davidii, B. fallowiana, B.* 'Lochinch' and *B.* x *weyeriana* should be cut back hard in spring. Prune *B. alternifolia* by removing shoots that have flowered. Other species may be cut back lightly after flowering. Propagate by semi-ripe cuttings in summer.
☆ *B. alternifolia* illus. p.92.
☆ *B. asiatica.* Evergreen, arching shrub. H and S 3m (10ft). Half hardy. Long plumes of very fragrant, tubular, white flowers appear amid long, narrow, dark green leaves in late winter and early spring. Grow against a south- or west-facing wall.
B. colvilei illus. p.91. 'Kewensis' is a deciduous, arching shrub, often tree-like with age. H and S 5m (15ft). Frost hardy. Has lance-shaped, dark green leaves. Large, tubular, white-throated, deep red flowers hang in drooping clusters in early summer.
B. crispa illus. p.113.

B. davidii (Butterfly bush). ♛ **'Black Knight'** is a vigorous, deciduous, arching shrub. H and S 5m (15ft). Fully hardy. Leaves are long, lance-shaped and dark green with white-felted undersides. Dense clusters of fragrant, tubular, dark violet-purple flowers are borne from mid-summer to autumn. Flowers of ♛ **'Empire Blue'** are rich violet-blue. **'Harlequin'** illus. p.91. **'Peace'** illus. p.88. **'Pink Pearl'** produces pale lilac-pink flowers. ♛ **'Royal Red'** illus. p.91.
B. fallowiana. Deciduous, arching shrub. H 2m (6ft), S 3m (10ft). Frost hardy. Shoots and lance-shaped leaves, when young, are covered with white hairs; foliage then becomes dark grey-green. Fragrant, tubular, lavender-purple flowers appear in late summer and early autumn. Is often damaged in very severe winters; grow against a wall in cold areas. ♛ var. *alba* has white flowers.
♛ *B. globosa* illus. p.93.
♛ *B.* **'Lochinch'**. Deciduous, arching shrub. H and S 3m (10ft). Frost hardy. Long plumes of fragrant, tubular, lilac-blue flowers are borne above lance-shaped, grey-green leaves in late summer and autumn.
B. madagascariensis. Evergreen, arching shrub. H and S 4m (12ft) or more. Half hardy. Has narrowly lance-shaped, dark green leaves, white beneath, and, in late winter and spring, long clusters of tubular, orange-yellow flowers. Grow against a south- or west-facing wall.
B. x *weyeriana.* Deciduous, arching shrub. H and S 4m (12ft). Fully hardy. Leaves are lance-shaped and dark green. Bears loose, rounded clusters of tubular, orange-yellow flowers, often tinged purple, from mid-summer to autumn.

BULBOCODIUM (Liliaceae)
Genus of spring-flowering corms, related to *Colchicum* and suitable for rock gardens and cool greenhouses. Fully hardy. Needs an open, sunny situation and well-drained soil. Propagate by seed in autumn or by division in late summer and early autumn.
B. vernum illus. p.374.

BULBOPHYLLUM. See ORCHIDS.
B. careyanum (illus. p.262). Evergreen, epiphytic orchid for an intermediate greenhouse. H 8cm (3in). In spring produces tight sprays of many slightly fragrant, brown flowers, 0.5cm (¼in) across. Oval leaves are 8–10cm (3–4in) long. Is best grown in a hanging basket. Requires semi-shade in summer.

BUPHTHALMUM (Compositae)
Genus of summer-flowering perennials. Fully hardy. Requires full sun; grows well in any but rich soil. Propagate by seed in spring or autumn or by division in autumn. Needs frequent division to curb invasiveness.
B. salicifolium illus. p.256.
B. speciosum. See *Telekia speciosa*.

BUPLEURUM (Umbelliferae)
Genus of perennials and evergreen shrubs, grown for their foliage and flowers. Grows well in coastal gardens. Frost hardy. Needs full sun and well-drained soil. Propagate by semi-ripe cuttings in summer.
B. fruticosum illus. p.116.

BUTIA (Palmae)
Yatay palm
Genus of evergreen palms, grown for their overall appearance. Frost hardy to frost tender, min. 5°C (41°F). Grow in any fertile, well-drained soil and in full light or partial shade. Water regularly, less in winter. Propagate by seed in spring at not less than 24°C (75°F). Red spider mite may be a problem.
B. capitata, syn. *Cocos capitata,* illus. p.74.

BUTOMUS (Butomaceae)
Genus of one species of deciduous, perennial, rush-like, marginal water plant, grown for its flowers. Fully hardy. Requires an open, sunny situation in up to 25cm (10in) depth of water. Propagate by division in spring or by seed in spring or late summer.
♛ *B. umbellatus* illus. p.390.

BUXUS (Buxaceae)
Box
Genus of evergreen shrubs and trees, grown for their foliage and habit. Is excellent for edging, hedging and topiary work. Flowers are insignificant. Fully to frost hardy. Requires sun or semi-shade and any but waterlogged soil. Trim hedges during summer. Promote new growth by cutting back stems to 30cm (12in) or less in late spring. Propagate by semi-ripe cuttings in summer.
B. balearica illus. p.123.
B. microphylla (Small-leaved box). Evergreen, bushy shrub. H 1m (3ft), S 1.5m (5ft). Fully hardy. Forms a dense, rounded mass of small, oblong, dark green leaves. **'Green Pillow'** illus. p.146.
♛ *B. sempervirens* (Common box). Evergreen, bushy shrub or tree. H and S 5m (15ft). Fully hardy. Leaves are oblong, glossy, dark green. Is useful for hedging and screening. **'Handsworthensis'** illus. p.122.
♛ **'Suffruticosa'** illus. p.147.
B. wallichiana (Himalayan box). Slow-growing, evergreen, bushy, open shrub. H and S 2m (6ft). Frost hardy. Produces long, narrow, glossy, bright green leaves.

C

CABOMBA (Cabombaceae)
Genus of deciduous or semi-evergreen, perennial, submerged water plants with finely divided foliage. Is suitable for aquariums. Frost tender, min. 5°C (41°F). Prefers partial shade. Propagate by stem cuttings in spring or summer.
C. caroliniana (Fanwort, Fish grass, Washington grass). Deciduous or semi-evergreen, perennial, submerged water plant. S indefinite. Forms dense, spreading hummocks of fan-shaped, coarsely cut, bright green leaves. Is used as an oxygenating plant.

CAESALPINIA (Leguminosae)
Genus of deciduous or evergreen shrubs, trees and scrambling climbers, grown for their foliage and flowers. Frost hardy to frost tender, min. 5–10°C (41–50°F). Requires full sun and fertile, well-drained soil. Propagate by softwood cuttings in summer or by seed in autumn or spring.
C. gilliesii illus. p.93.
C. pulcherrima, syn. *Poinciana pulcherrima* (Barbados pride). Evergreen shrub or tree of erect to spreading habit. H and S 3–6m (10–20ft). Frost tender, min. 5°C (41°F). Has fern-like leaves composed of many small, mid-green leaflets. In summer bears cup-shaped, yellow flowers, 3cm (1¼in) wide, with very long, red anthers, in short, dense, erect racemes.

CALADIUM (Araceae)
Genus of perennials with tubers from which arise long-stalked, ornamental leaves. Frost tender, min. 18–19°C (64–6°F). Requires partial shade and moist, humus-rich soil. After leaves die down, store tubers in a frost-free, dark place. Propagate by separating small tubers when replanting in spring.
C. x *hortulanum* (Angels' wings). 'Candidum' is a tufted perennial. H and S to 90cm (3ft). Triangular, green-veined, white leaves, to 45cm (18in) long, have arrow-shaped bases and long leaf stalks. Intermittently bears white spathes; small flowers clustered on spadix sometimes produce whitish berries. 'John Peed' has purple stems and waxy, green leaves with metallic orange-red centres and scarlet veins. 'Pink Beauty' illus. p.228. 'Pink Cloud' has large, dark green leaves with the central areas mottled pink, and pink to white areas along the veins.

CALANTHE. See ORCHIDS.
C. vestita (illus. p.260). Deciduous, terrestrial orchid. H 60cm (24in). Frost tender, min. 18°C (64°F). In winter bears sprays of many white flowers, 4cm (1½in) across, each with a large, red-marked lip. Has broadly oval, ribbed, soft leaves, 30cm (12in) long. In summer requires semi-shade and regular feeding.

CALATHEA (Marantaceae)
Genus of evergreen perennials with brightly coloured and patterned leaves. Frost tender, min. 15°C (59°F). Prefers a shaded, humid position, without fluctuations of temperature, in humus-rich, well-drained soil. Water with soft water, sparingly in low temperatures, but do not allow to dry out completely. Propagate by division in spring.
C. lindeniana. Evergreen, clump-forming perennial. H 1m (3ft), S 60cm (2ft). Lance-shaped, long-stalked, more or less upright leaves, over 30cm (1ft) long, are dark green, with paler green, feathered midribs above and marked with reddish-purple below. Intermittently bears short, erect spikes of 3-petalled, pale yellow flowers.
C. majestica 'Roseo-lineata', syn. *C. ornata* 'Roseo-lineata'. Evergreen, clump-forming, stemless perennial. H to 2m (6ft), S to 1.5m (5ft). Narrowly oval, leathery leaves, to 60cm (2ft) long, are dark green, with close-set, fine, pink stripes along the lateral veins and reddish-purple below. Intermittently bears short, erect spikes of 3-petalled, white to mauve flowers. 'Sanderiana' illus. p.196.
❦ *C. makoyana* illus. p.268.
C. ornata 'Rosea-lineata'. See *C. majestica* 'Rosea-lineata'.
❦ *C. zebrina* illus. p.230.

CALCEOLARIA (Scrophulariaceae)
Genus of annuals, biennials and evergreen perennials, sub-shrubs and scandent climbers, some of which are grown as annuals. Fully hardy to frost tender, min. 5–7°C (41–5°F). Most prefer full sun but some like a shady, cool site and moist but well-drained soil, incorporating sharp sand and compost, and dislike wet conditions in winter. Propagate by softwood cuttings in late spring or summer or by seed in autumn.
C. 'Anytime'. Compact, bushy annual or biennial. H 20cm (8in), S 15cm (6in). Half hardy. Has oval, slightly hairy, mid-green leaves. In spring-summer bears heads of 5cm (2in) long, rounded, pouched flowers in shades of red and yellow, including bicolours.
C. arachnoidea illus. p.303.
C. biflora illus. p.255.
C., Bikini Series illus. p.290.
C. darwinii. Evergreen, clump-forming, short-lived perennial. H 8cm (3in), S 10cm (4in). Fully hardy. Bears rounded, wrinkled, glossy, dark green leaves. In late spring, flower stems carry pendent, pouch-shaped, yellow flowers with dark brown spots on lower lips and central, white bands. Is difficult to grow. Needs a sheltered, sunny site in moist, gritty, peaty soil. Requires frequent spraying against aphids.
C. fothergillii. Evergreen, clump-forming, short-lived perennial. H and S 12cm (5in). Frost hardy. Has a rosette of rounded, light green leaves with hairy edges and, in summer, solitary pouch-shaped, sulphur-yellow flowers with crimson spots. Is good for a sheltered rock ledge or trough or in an alpine house. Needs gritty, peaty soil. Is prone to aphid attack.
❦ *C. integrifolia*. Evergreen, upright sub-shrub, sometimes grown as an annual. H to 1.2m (4ft), S 60cm (2ft). Half hardy. In summer bears crowded clusters of pouch-shaped, yellow to red-brown flowers above oblong to elliptic, mid-green leaves, sometimes rust-coloured beneath.
C. 'John Innes' illus. p.257.
C., Monarch Series. Group of bushy annuals or biennials. H and S 30cm (12in). Half hardy. Has oval, lightly hairy, mid-green leaves and, in spring-summer, bears heads of large, rounded, pouched flowers, 5cm (2in) long, in a wide range of colours.
C. pavonii. Robust, evergreen, scandent climber. H 2m (6ft) or more. Frost tender, min. 7°C (45°F). Has oval, serrated, soft-haired leaves with winged stalks. Pouched, yellow flowers with brown marks appear in large trusses from late summer to winter.
C. polyrrhiza. Evergreen, prostrate perennial. H 2.5cm (1in), S 15cm (6in). Frost hardy. Has rounded, hairy, mid-green leaves along flower stem, which bears pouch-shaped, purple-spotted, yellow flowers in summer. Is good for a shady rock garden. May also be propagated by division in autumn or spring.
❦ *C.* 'Sunshine' illus. p.291.
C. tenella illus. p.333.
C. 'Walter Shrimpton' illus. p.334.

CALENDULA (Compositae)
Marigold
Genus of annuals and evergreen shrubs. Annuals are fully hardy; shrubs are frost tender, min. 4°C (39°F). Grow in sun and in any well-drained soil. Dead-head to prolong flowering. Propagate annuals by seed sown outdoors in spring or autumn, shrubs by stem cuttings in summer. Annuals may self seed. Cucumber mosaic virus and powdery mildew may cause problems.
C. officinalis (Pot marigold). Fast-growing, bushy annual. Tall cultivars, H and S 60cm (2ft); dwarf forms, H and S 30cm (1ft). All have lance-shaped, strongly aromatic, pale green leaves. Daisy-like, single or double flower heads are produced from spring to autumn. 'Art Shades' (tall) has double, apricot, orange or cream flowers. Fiesta Series (dwarf) has double flowers in a mixture of colours ranging from cream to orange (see ❦ 'Gitana' illus. p.291). 'Geisha Girl' (tall) illus. p.293. 'Kablouna' (tall) illus. p.288.

CALLA (Araceae)
Genus of one species of deciduous or semi-evergreen, perennial, spreading, marginal water plant, grown for its foliage and showy spathes that surround insignificant flower clusters. Fully hardy. Requires a sunny position, in mud or up to 25cm (10in) depth of water. Propagate by division in spring or by seed in late summer.
C. palustris illus. p.388.

CALLIANDRA (Leguminosae)
Genus of evergreen trees, shrubs and scandent semi-climbers, grown for their flowers and overall appearance. Frost tender, min. 7–18°C (45–64°F). Needs full light or partial shade and well-drained soil. Water containerized plants freely when in full growth, much less when temperatures are low. To restrict growth, cut back stems by one-half to two-thirds after flowering. Propagate by seed in spring. Whitefly and mealy bug may be troublesome.
C. eriophylla illus. p.142.
C. haematocephala [pink form] illus. p.117, [white form] illus. p.119.

CALLIANTHEMUM (Ranunculaceae)
Genus of perennials, grown for their daisy-like flowers and thick, dissected leaves. Is excellent for rock gardens and alpine houses. Fully hardy. Needs sun and moist but well-drained soil. Propagate by seed when fresh.
C. coriandrifolium. Prostrate perennial with upright flower stems. H 8cm (3in), S 20cm (8in). Leaves, forming open rosettes, are long-stalked, very dissected and blue-green. In spring carries short-stemmed, many-petalled, white flowers with yellow centres. Is susceptible to slugs.

CALLICARPA (Verbenaceae)
Genus of deciduous, summer-flowering shrubs, grown for their small but striking, clustered fruits. Fully hardy. Does best in full sun and fertile, well-drained soil. Propagate by softwood cuttings in summer.
C. bodinieri. Deciduous, bushy shrub. H 3m (10ft), S 2.5m (8ft). Has oval, dark green leaves. Tiny, star-shaped, lilac flowers in mid-summer are followed by dense clusters of spherical, violet fruits. var. *giraldii* illus. p.118.

CALLISIA (Commelinaceae)
Genus of evergreen, prostrate perennials, grown for their ornamental foliage and trailing habit. Frost tender, min. 10–15°C (50–59°F). Grow in full light, but out of direct sunlight, in fertile, well-drained soil. Propagate by tip cuttings in spring, either annually or when plants become straggly.
C. navicularis, syn. *Tradescantia navicularis*. Evergreen, low-growing perennial with creeping, rooting shoots, 50cm (20in) or more long. H 5–8cm (2–3in), S indefinite. Has 2 rows of oval, keeled leaves, 2.5cm (1in) long, sheathing the stem, and stalkless clusters of small, 3-petalled,

pinkish-purple flowers in leaf axils in summer-autumn.
C. repens illus. p.268.

CALLISTEMON (Myrtaceae)
Bottlebrush
Genus of evergreen shrubs, usually with narrow, pointed leaves, grown for their clustered flowers, which, with their profusion of long stamens, resemble bottlebrushes. Frost to half hardy, but in cool areas all except *C. sieberi* need protection of a south- or west-facing wall. Requires full sun and fertile, well-drained soil. Propagate by semi-ripe cuttings in summer or by seed in autumn or spring.
♈ *C. citrinus* '**Splendens**' illus. p.113.
C. pallidus illus. p.116.
C. pityoides. See *C. sieberi*.
C. rigidus illus. p.115.
♈ *C. sieberi*, syn. *C. pityoides*. Evergreen, bushy, dense shrub. H 1.5m (5ft), S 1m (3ft). Frost hardy. Has short, narrowly lance-shaped, rigid, mid-green leaves and, from mid- to late summer, small clusters of pale yellow flowers.
C. speciosus (Albany bottlebrush). Evergreen, bushy shrub. H and S 3m (10ft). Half hardy. Produces long, narrow, grey-green leaves. Cylindrical clusters of bright red flowers appear in late spring and early summer.
C. subulatus. Evergreen, arching shrub. H and S 1.5m (5ft). Frost hardy. Leaves are narrowly oblong and bright green. Dense spikes of crimson flowers are produced in summer.
C. viminalis. Evergreen, arching shrub. H and S 5m (15ft). Half hardy. Narrowly oblong, bronze, young leaves mature to dark green. Bears clusters of bright red flowers in summer.

CALLISTEPHUS (Compositae)
China aster
Genus of one species of annual. Half hardy. Requires sun, a sheltered position and fertile, well-drained soil. Tall cultivars need support; all should be dead-headed. Propagate by seed sown under glass in spring; seed may also be sown outdoors in mid-spring. Wilt disease, virus diseases, foot rot, root rot and aphids may be a problem.
C. chinensis. Moderately fast-growing, erect, bushy annual. Tall cultivars, H 60cm (24in), S 45cm (18in); intermediate, H 45cm (18in), S 30cm (12in); dwarf, H 25–30cm (10–12in), S 30–45cm (12–18in); very dwarf, H 20cm (8in), S 30cm (12in). All have oval, toothed, mid-green leaves and flower in summer and early autumn. Different forms are available in a wide colour range, including pink, red, blue and white. **Andrella Series** (tall) has single, daisy-like flower heads; those of **Duchess Series** (tall) are incurved and chrysanthemum-like.
♈ **Milady Series** (dwarf) has incurved, fully double flower heads available either in mixed or single colours (blue, illus. p.284; rose, illus. p.276). **Ostrich Plume Series** (tall) bears feathery, reflexed, fully double flower heads. **Pompon Series** (tall) has small, double flower heads; those of '**Powderpuffs**' (tall) are large and double. **Princess Series** (tall) has double flower heads with quilled petals. **Thousand Wonders Series** (very dwarf) illus. p.283.

CALLUNA (Ericaceae). See HEATHERS.
C. vulgaris (Ling, Scotch heather). Evergreen, bushy shrub. H to 60cm (24in), S 45cm (18in). Fully hardy. Slightly fleshy, linear leaves, in opposite and overlapping pairs, may range in colour from bright green to many shades of grey, yellow, orange and red. Spikes of bell- to urn-shaped, single or double flowers are produced from mid-summer to late autumn. Unlike *Erica*, most of the flower colour derives from the sepals. The following cultivars are H 45cm (18in), have mid-green leaves and bear single flowers in late summer and early autumn, unless otherwise stated.
'**Alba Plena**', H 30–45cm (12–18in), bears double, white flowers.
♈ '**Allegro**', H 60cm (24in), is compact in habit and produces purple-red flowers. '**Alportii**', H 60–90cm (24–36in), has purple-red flowers.
♈ '**Anthony Davis**' (illus. p.148) has grey leaves and white flowers. '**August Beauty**' bears long, pendulous stems and white flowers. '**Beechwood Crimson**' is bushy with crimson-purple flowers. '**Beoley Gold**' (illus. p.149), S 50cm (20in), has golden foliage and white flowers. '**Beoley Silver**', H 40cm (16in), has silver foliage and white flowers. '**Bonfire Brilliance**', H 30cm (12in), has bright, flame-coloured foliage and mauve-pink flowers. '**Boskoop**' (illus. p.149), H 30cm (12in), is compact with golden foliage that turns deep orange in winter and lilac-pink flowers. ♈ '**County Wicklow**' (illus. p.148), H 30cm (12in), S 35cm (14in), is compact with double, shell-pink flowers. ♈ '**Darkness**' (illus. p.149), H 40cm (16in), S 35cm (14in), is compact with crimson flowers. '**Drum-Ra**' is a typical form of the Scotch white heather. ♈ '**Elsie Purnell**' (illus. p.148) is a spreading cultivar with greyish-green leaves and double, pale pink flowers. ♈ '**Finale**' bears dark pink flowers from late autumn to early winter. '**Foxii Nana**' (illus. p.149), H 15cm (6in), forms low mounds of bright green foliage and produces a few mauve-pink flowers. '**Fred J. Chapple**' has bright pink- and coral-tipped foliage in spring; mauve-pink flowers are borne on long stems.
♈ '**Gold Haze**' (illus. p.149) has bright golden foliage and white flowers. '**Golden Feather**' (illus. p.149) has bright yellow foliage, turning orange in winter, and mauve-pink flowers. '**H.E. Beale**', H 50cm (20in), is one of the best double-flowered heathers, with pale pink flowers on long stems. '**Highland Rose**' has golden-bronze foliage and bears deep pink flowers. '**Humpty-Dumpty**', H 15cm (6in), is a hummock-forming cultivar that has emerald-green foliage and a few white flowers. ♈ '**J.H. Hamilton**' (illus. p.148), H 20cm (8in), S 40cm (16in), is compact with double, salmon-pink flowers. ♈ '**Joy Vanstone**' has golden foliage, turning to orange and bronze, and mauve-pink flowers.
♈ '**Kinlochruel**' (illus. p.148), H 30cm (12in), S 35cm (14in), bears an abundance of large, double, white flowers. '**Loch Turret**', H 30cm (12in), has emerald-green foliage and produces white flowers in early summer. ♈ '**Mair's Variety**', an old cultivar, has white flowers on long spikes. '**Marleen**' is unusual in that its long-lasting, dark mauve flower buds, borne from early to late autumn, do not open fully.
♈ '**Mullion**', H 25cm (10in), S 50cm (20in), is a spreading cultivar with rich mauve-pink flowers. '**Multicolor**' (illus. p.149), H 20cm (8in), is compact with foliage in shades of yellow, orange, red and green year-round; flowers are mauve-pink. '**My Dream**' (illus. p.148), H 50cm (20in), produces double, white flowers that are borne on long, tapering stems. ♈ '**Orange Queen**' has yellow and golden foliage and mauve flowers. '**Peter Sparkes**' (illus. p.149), H 50cm (20in), S 55cm (22in), bears double, deep pink flowers. ♈ '**Robert Chapman**' (illus. p.149) is a spreading cultivar and grown mainly for its foliage, which is golden-yellow in summer, turning orange and brilliant red in winter; flowers are mauve-pink. '**Ruth Sparkes**', H 25cm (10in), has golden foliage and white flowers. '**Silver Knight**' (illus. p.149), H 30cm (12in), is of upright habit with grey leaves and mauve-pink flowers. ♈ '**Silver Queen**' (illus. p.148), H 40cm (16in), S 55cm (22in), is a spreading cultivar with dark mauve-pink flowers. ♈ '**Sir John Charrington**' has bright-coloured foliage, varying from golden-yellow in summer to orange and red in winter, and dark mauve-pink flowers.
♈ '**Sister Anne**', H 15cm (6in), has grey leaves and pale mauve-pink flowers. '**Soay**', H 10cm (4in), has tawny foliage and mauve-pink flowers.
♈ '**Spring Cream**' (illus. p.148) has bright green leaves, which have cream tips in spring, and white flowers.
♈ '**Sunset**', H 25cm (10in), has brightly coloured foliage, changing from golden-yellow in spring to orange in summer and fiery red in winter; flowers are mauve-pink. ♈ '**Tib**' (illus. p.149), H 30cm (12in), S 40cm (16in), is the earliest-flowering double cultivar, producing small, double, deep pink flowers in early summer.
♈ '**White Lawn**', H 10cm (4in), is a creeping cultivar with bright green foliage and white flowers on long stems; is suitable for a rock garden.
♈ '**Wickwar Flame**' is primarily a foliage plant with leaves in shades of yellow, orange and flame that are particularly effective in winter; flowers are mauve-pink.

CALOCEDRUS (Cupressaceae). See CONIFERS.
♈ *C. decurrens*, syn. *Libocedrus decurrens*, illus. p.80.

CALOCEPHALUS (Compositae)
Genus of annuals and evergreen shrubs and perennials, often grown annually from cuttings and used as summer bedding. Frost tender, min. 7–10°C (45–50°F). Requires full light and well-drained soil. Water containerized plants moderately when in full growth, sparingly at other times. Remove stem tips of young plants to promote a bushy habit. Propagate by semi-ripe cuttings in late summer. Botrytis may be troublesome if plants are kept too cool and damp in winter.
C. brownii illus. p.145.

CALOCHONE (Rubiaceae)
Genus of evergreen, scrambling climbers, grown for their showy flowers. Frost tender, min. 18°C (64°F). Requires humus-rich, well-drained soil and full light. Water regularly, less in cold weather. Needs tying to a support. Thin out crowded stems after flowering. Propagate by semi-ripe cuttings in summer.
C. redingii. Moderately vigorous, evergreen, scrambling climber. H 3–5m (10–15ft). Has oval, pointed, hairy leaves, 7–12cm (3–5in) long. Trusses of primrose-shaped, red to orange-pink flowers appear in winter.

CALOCHORTUS (Liliaceae)
Cat's ears, Fairy lantern, Mariposa tulip
Genus of bulbs, grown for their spring and summer flowers. Frost hardy. Needs a sheltered, sunny position and well-drained soil. In cold, damp climates, cover or lift spring-flowering species when dormant (in summer), or grow in cold frames or cold houses. After flowering, remove bulbils formed in leaf axils for propagation. Propagate by seed or bulbils – spring-flowering species in autumn, summer-flowering species in spring.
C. albus illus. p.353.
C. amabilis illus. p.378.
C. barbatus, syn. *Cyclobothra lutea*, illus. p.367.
C. luteus illus. p.359.
C. splendens. Late spring-flowering bulb. H 20–60cm (8–24in), S 5–10cm (2–4in). Bears 1 or 2 linear, erect leaves near base of branched stem and 1–4 upward-facing, saucer-shaped, pale purple flowers, 5–7cm (2–3in) across, with a darker blotch at the base of each of the 3 large petals.
C. subalpinus illus. p.379.
C. superbus. Late spring-flowering bulb. H 20–60cm (8–24in), S 5–10cm (2–4in). Is similar to *C. splendens*, but flowers are white, cream or pale lilac, with a brown mark near the base of each of the 3 large petals.
C. venustus illus. p.352.
C. vestae. Late spring-flowering bulb. H 20–60cm (8–24in), S 5–10cm (2–4in). Is similar to *C. splendens*, but flowers are white or purple, with a rust-brown mark near the base of each of the 3 large petals.
C. weedii. Summer-flowering bulb. H 30–60cm (1–2ft), S 5–10cm (2–4in). Has a linear, erect leaf near base of stem. Carries usually 2 upright, saucer-shaped flowers, 4–5cm (1 1/2–2in) across, that are orange-yellow with brown lines and flecks and are hairy inside.

CALODENDRUM (Rutaceae)
Genus of evergreen trees, grown for their flowers that are produced mainly in spring-summer. Frost tender, min. 7–10°C (45–50°F). Needs full light and fertile, well-drained but moisture-retentive soil. Water containerized specimens freely when in full growth, less at other times. Tolerates some pruning. Propagate by seed in spring or by semi-ripe cuttings in summer.
C. capense (Cape chestnut). Fairly fast-growing, evergreen, rounded tree. H and S to 15m (50ft) or more. Has oval leaves patterned with translucent dots. Terminal panicles of 5-petalled, light pink to deep mauve flowers appear from spring to early summer.

CALOMERIA, syn. HUMEA (Compositae)
Genus of perennials and evergreen shrubs. Only one species, *C. elegans*, is cultivated, usually as a biennial. Frost tender, min. 4°C (39°F). Grow in sun and in fertile, well-drained soil. Propagate by seed sown under glass in mid-summer.
C. elegans illus. p.282.

Calonyction aculeatum. See *Ipomoea alba*.

CALOSCORDUM (Liliaceae)
Genus of one species of summer-flowering bulb, related and similar to *Allium*. Is suitable for a rock garden. Frost hardy. Needs an open, sunny situation and well-drained soil. Lies dormant in winter. Propagate in early spring by seed or division before growth starts.
C. neriniflorum. Clump-forming bulb. H 10–25cm (4–10in), S 8–10cm (3–4in). Thread-like, semi-erect, basal leaves die down at flowering time. Each leafless stem produces a loose head of 10–20 small, funnel-shaped, pinkish-red flowers in late summer.

CALOTHAMNUS (Myrtaceae)
Genus of evergreen, summer-flowering shrubs, grown for their flowers and overall appearance. Thrives in a dryish, airy environment. Frost tender, min. 10–15°C (50–59°F). Requires full sun and well-drained, sandy soil. Water containerized plants moderately when in full growth, less at other times. Propagate by seed or semi-ripe cuttings in summer.
C. villosus (Woolly netbush). Evergreen, bushy shrub. H and S 60cm–1.2m (2–4ft). Has small, dense, needle-like leaves covered with thick, grey down. Flowers appear in summer in dense, lateral clusters, each composed of bundles of branched, deep red stamens.

CALTHA (Ranunculaceae)
Genus of deciduous, perennial, marginal water plants, bog plants and rock garden plants, grown for their attractive flowers. Fully hardy. Smaller growing species are suitable for rock gardens, troughs and alpine houses and require moist but well-drained soil; larger species are best in marginal conditions. Most prefer an open, sunny position. Propagate species by seed in autumn or by division in autumn or early spring, selected forms by division in autumn or early spring.
C. introloba. Tuft-forming perennial. H and S 5cm (2in). Bears oval, lobed, glossy, dark green leaves. Erect, goblet-shaped, white flowers, flushed purple outside, open in late winter. Is good for a trough or alpine house. Prefers shade and humus-rich soil. Is difficult to grow in hot, dry climates.
C. leptosepala illus. p.389.
♀ *C. palustris* and ♀ '**Flore Pleno**' illus. p.391.

CALYCANTHUS (Calycanthaceae)
Genus of deciduous, summer-flowering shrubs, grown for their purplish- or brownish-red flowers with strap-shaped petals. Fully hardy. Requires sun or light shade and fertile, deep, moist but well-drained soil. Propagate by softwood cuttings in summer or by seed in autumn.
C. floridus (Carolina allspice). Deciduous, bushy shrub. H and S 2m (6ft). Has oval, aromatic, dark green leaves and, from early to mid-summer, fragrant, brownish-red flowers with masses of spreading petals.
C. occidentalis illus. p.112.

CALYPSO. See ORCHIDS.
C. bulbosa (illus. p.261). Deciduous, terrestrial orchid. H 5–20cm (2–8in). Fully hardy. Corm-like stem produces a single, oval, pleated leaf, 3–10cm (1¼–4in) long. Purplish-pink flowers, 1.5–2cm (⅝–¾in) long, with hairy, purple-blotched, white or pale pink lips, are borne singly in late spring or early summer. Requires a damp, semi-shaded position with a mulch of leaf mould.

CAMASSIA (Liliaceae)
Genus of summer-flowering bulbs, suitable for borders and pond margins. Frost hardy. Requires sun or partial shade and deep, moist soil. Plant bulbs in autumn, 10cm (4in) deep. Lies dormant in autumn-winter. Propagate by seed in autumn or by division in late summer. If seed is not required, cut off stems after flowering.
C. leichtlinii illus. p.341. '**Semiplena**' illus. p.341.
C. quamash (Common camassia, Quamash). Clump-forming, summer-flowering bulb. H 20–80cm (8–32in), S 20–30cm (8–12in). Produces long, narrow, erect, basal leaves. Leafless stems each bear a dense spike of star-shaped, blue, violet or white flowers, to 7cm (3in) across.

CAMELLIA (Theaceae)
Genus of evergreen shrubs and trees, grown for their flowers and foliage. Flowers are classified according to the following types: single, semi-double, anemone-form, peony-form, rose-form, formal double and irregular double (see p.98 for illustrations and descriptions). Grows well against walls and in containers. Fully hardy to frost tender, min. 7–10°C (45–50°F). Most forms prefer a sheltered position and semi-shade. Well-drained, neutral to acid soil is essential. Prune to shape after flowering. Propagate by semi-ripe or hardwood cuttings from mid-summer to early winter or by grafting in late winter or early spring. Aphids, thrips and scale insects may cause problems under glass.
♀ *C.* '**Anticipation**' (illus. p.99). Robust, evergreen, upright shrub. H 3m (10ft), S 1.5m (5ft). Has lance-shaped, dark green leaves. Large, peony-form, deep rose-pink blooms are freely produced in spring.
C. chrysantha. Fast-growing, evergreen, open shrub or tree. H 6m (20ft) or more, S 3m (10ft). Half hardy. Has large, oval, leathery, veined leaves. Small, stalked, cup-shaped, single, clear yellow flowers are produced from leaf axils in spring.
♀ *C.* '**Cornish Snow**' (illus. p.98). Fast-growing, evergreen, upright, bushy shrub. H 3m (10ft), S 1.5m (5ft). Frost hardy. Has lance-shaped leaves, bronze when young, maturing to dark green. In early spring bears a profusion of small, cup-shaped, single, white flowers.
C. cuspidata. Evergreen, upright shrub becoming bushy with age. H 3m (10ft), S 1.5m (5ft). Frost hardy. Has small, lance-shaped leaves, bronze when young, maturing to purplish-green. Small, cup-shaped, single, pure white flowers are freely produced from leaf axils in early spring.
♀ *C.* '**Dr Clifford Parks**' (illus. p.99). Evergreen, spreading shrub. H 4m (12ft), S 2.5m (8ft). Frost hardy. In mid-spring has large, flame-red flowers, often semi-double, peony- and anemone-form on the same plant. Leaves are large, oval and dark green.
C. granthamiana. Evergreen, open shrub. H to 3m (10ft), S 2m (6ft). Half hardy. Oval, leathery leaves are crinkly and glossy, deep green. In late autumn bears large, saucer-shaped, single, white flowers, to 18cm (7in) across, with up to 8 broad petals.
C. x *hiemalis.* Evergreen, upright, bushy shrub. H 2–3m (6–10ft), S 1.5m (5ft). Frost hardy. Has small, lance-shaped leaves and fragrant, single, cup-shaped, semi- or irregular double, white, pink or red flowers borne in late autumn and winter. Is good for hedging.
C. hongkongensis. Evergreen, bushy shrub or tree. H to 3m (10ft), S 2m (6ft). Half hardy. Lance-shaped leaves, 10cm (4in) long, are dark red when young, maturing to dark green. Cup-shaped, single, deep crimson flowers, velvety on reverse of petals, are borne in late spring.
C. '**Innovation**' (illus. p.99). Evergreen, open, spreading shrub. H 5m (15ft), S 3m (10ft). Frost hardy. Has large, oval, leathery leaves and, in spring, large, peony-form, lavender-shaded, claret-red flowers with twisted petals.
♀ *C.* '**Inspiration**' (illus. p.99). Evergreen, upright shrub. H 4m (12ft), S 2m (6ft). Frost hardy. Leaves are oval and leathery. Saucer-shaped, semi-double, phlox-pink flowers are freely produced in spring.
C. japonica (Common camellia). Evergreen shrub that is very variable in habit, foliage and floral form. H 10m (30ft), S 8m (25ft). Frost hardy. Numerous cultivars are available; they are spring-flowering unless otherwise stated. ♀ '**Adolphe Audusson**' (illus. p.99) is a very reliable, old cultivar that is suitable for all areas and will withstand lower temperatures than most other variants. Produces large, saucer-shaped, semi-double, dark red flowers with prominent, yellow stamens. Leaves are broadly lance-shaped and dark green. '**Alba Simplex**' (illus. p.98) is bushy in habit with broadly lance-shaped, mid- to yellow-green leaves and cup-shaped, single, white flowers in early spring.
♀ '**Alexander Hunter**' (illus. p.99), an upright, compact shrub, has flattish, single, deep crimson flowers, with some petaloids, and lance-shaped, dark green leaves. '**Althaeiflora**' (illus. p.99) has a vigorous, bushy habit, large, peony-form, dark red flowers and broadly oval, very dark green leaves. '**Apollo**' (syn. *C.j.* '**Paul's Apollo**'; illus. p.99) is a vigorous, branching shrub that produces semi-double, red flowers sometimes blotched with white. Leaves are glossy, dark green. '**Ave Maria**' (illus. p.98) has peony-form, pale pink flowers and light green leaves. ♀ '**Berenice Boddy**' (illus. p.98) is a vigorous shrub that bears semi-double, light pink flowers amid lance-shaped, dark green leaves. '**Betty Sheffield Supreme**' (illus. p.98) is upright in habit with lance-shaped, mid-green leaves and has irregular double flowers, basically white, but with each petal bordered with several shades of rose-pink.
♀ '**Donckelaeri**' is slow-growing, bushy and pendulous with saucer-shaped, semi-double, red flowers, often white-marbled. Has lance-shaped, dark green leaves. ♀ '**Elegans**' (illus. p.99) has a spreading habit and anemone-form, deep rose-pink flowers with central petaloids often variegated white. Leaves are broadly lance-shaped and dark green. '**Elegans Supreme**' (illus. p.99) is a mutant of *C.j.* '**Elegans**' and has wine-red flowers. '**Fleur de Pêche**' (syn. *C.j.* '**Peachblossom**') is bushy, with narrowly lance-shaped, dark green leaves and cup-shaped, semi-double, delicate pink flowers with deeper pink centres. ♀ '**Gloire de Nantes**' (illus. p.99) is an upright shrub, becoming bushy with age, that bears flattish to cup-shaped, semi-double, bright rose-pink flowers over a long period. Has oval to lance-shaped, glossy, dark green leaves. '**Grand Sultan**' see *C.j.* '**Mathotiana**'.
♀ '**Guilio Nuccio**' (illus. p.99) is an upright, free-flowering cultivar that spreads with age. Produces large, cup-shaped, semi-double, rose-red flowers that each have wavy petals and often a confused centre of petaloids and golden stamens. Dark green leaves are lance-shaped and occasionally have 'fish-tail' tips. '**Hagoromo**' see *C.j.* '**Magnoliiflora**'. '**Janet Waterhouse**' (illus. p.98) is strong-growing and has semi-double, white flowers with golden anthers borne amid dark green foliage. '**Jessie Burgess**' (illus. p.99) has a very strong, upright growth and lance-shaped, mid-green

leaves. Bears large, cup-shaped, semi-double, rose-red flowers with a silver tint. ❀ **'Julia Drayton'** (illus. p.99) has an upright habit and large, crimson flowers varying from formal double to rose-form. Dark green leaves are oval to lance-shaped and slightly twisted. ❀ **'Jupiter'** (illus. p.99) is an upright shrub that bears lance-shaped, dark green leaves and large, saucer-shaped, single, pinkish-red flowers with golden stamens. **'Kumasaka'** has an upright habit with narrowly lance-shaped, mid-green leaves and produces formal double, or occasionally peony-form, deep rose-pink flowers. **'Lady Vansittart'** (illus. p.98) is upright, with unusual, holly-like, twisted, mid-green foliage. Saucer-shaped, semi-double, white flowers are flushed rose-pink; flower colour is variable and often self-coloured mutations appear. ❀ **'Lavinia Maggi'** (illus. p.98) has an upright habit and formal double, white flowers striped with pink and carmine. Sometimes sports red flowers. **'Magnoliiflora'** (syn. *C.j.* 'Hagoromo') has a bushy habit and flattish to cup-shaped, semi-double, blush-pink flowers. Twisted, light green leaves point downwards. **'Margaret Davis'** (illus. p.98) is a spreading cultivar, with oval to lance-shaped, dark green leaves. Has irregular double blooms with ruffled, creamy-white petals, often lined with pink. Edges of each petal are bright rose-red. **'Mathotiana'** (syn. *C.j.* 'Grand Sultan', *C.j.* 'Te Deum'; illus. p.99) is of spreading habit; very large, formal double, velvety, dark crimson flowers become purplish with age and in warm climates often have rose-form centres. Leaves are lance-shaped to oval, slightly twisted and dark green. ❀ **'Mrs D.W. Davis'** (illus. p.98) is a dense, spreading cultivar that bears very large, pendulous, cup-shaped, semi-double, delicate pink flowers that are backed by oval to lance-shaped, dark green leaves. **'Paul's Apollo'** see *C.j.* 'Apollo'. **'Peachblossom'** see *C.j.* 'Fleur de Pêche'. **'Pink Dawn'** has an upright habit, lance-shaped, mid-green leaves and formal double, deep pink flowers. ❀ **'R.L. Wheeler'** (illus. p.99) has a robust, upright growth, large, broadly oval, leathery, very dark green leaves and very large, flattish, anemone-form to semi-double, rose-pink flowers, with distinctive rings of golden stamens, often including some petaloids. ❀ **'Rubescens Major'** (illus. p.99) is an upright cultivar, becoming bushy with age, with oval to lance-shaped, dark green leaves and bears formal double, crimson-veined, rose-red flowers. **'Te Deum'** see *C.j.* 'Mathotiana'. **'Tomorrow Park Hill'**, one of the best of many mutations of 'Tomorrow', is of vigorous, upright habit. Has lance-shaped, mid-green leaves and bears irregular double flowers with deep pink, outer petals gradually fading to soft pink centres that are often variegated with white. **'Tomorrow's Dawn'** (illus. p.98) is similar, but with pale pink flowers, each with a white border and often red-streaked.

'Yamato-nishiki' has a low, spreading habit and has thick petals forming flattish, single flowers with large, central bosses of many golden stamens; petals are white, streaked and blotched pink and red. Small, lance-shaped leaves are dark green. ❀ *C.* **'Leonard Messel'** (illus. p.99). Evergreen, open shrub. H 4m (12ft), S 2.5m (8ft). Frost hardy. Has large, oval, leathery, dark green leaves. In spring bears a profusion of large, flattish to cup-shaped, semi-double, rose-pink flowers.
C. maliflora. Evergreen, upright, bushy shrub. H 2m (6ft), S 1m (3ft). Frost hardy. Has small, lance-shaped, thin-textured, light green leaves and, in spring, produces flattish to cup-shaped, semi-double, pale pink- or white-centred flowers with rose-pink margins.
C. oleifera. Evergreen, bushy shrub. H 2m (6ft), S 1.5m (5ft). Frost hardy. Leaves are oval and dull green. Has cup-shaped, single, sometimes pinkish, white flowers in early spring.
C. reticulata. Evergreen, open, tree-like shrub. H 10m (30ft) or more, S 5m (15ft). Half hardy. Has large, oval, leathery leaves; large, saucer-shaped, single, rose-pink and salmon-red flowers are borne in spring. Needs shelter. **'Butterfly Wings'** see *C.r.* 'Houye Diechi'. **'Captain Rawes'** (illus. p.99) has a profusion of large, semi-double, carmine-rose blooms. **'Early Crimson'** see *C.r.* 'Zaotaohong'. **'Early Peony'** see *C.r.* 'Zaomudan'. **'Houye Diechi'** (syn. *C.r.* 'Butterfly Wings'; illus. p.99) produces very large, flattish to cup-shaped, semi-double, rose-pink flowers with wavy, central petals. **'Mudan Cha'** (syn. *C.r.* 'Moutancha', *C.r.* 'Peony Camellia'; illus. p.98) has very large, cup-shaped, semi- to irregular double, deep pink flowers that have curled petals. **'Robert Fortune'** (syn. *C.r.* 'Songzilin') is upright in habit and has large, formal double, deep red flowers. **'Zaomudan'** (syn. *C.r.* 'Early Peony'; illus. p.99) has large, irregular double, rose-pink flowers, each with inner petals forming a peony centre. Flowers of **'Zaotaohong'** (syn. *C.r.* 'Early Crimson'; illus. p.99) are large, cup-shaped, semi- to formal double and crimson.
C. rosiflora. Evergreen, spreading shrub. H and S 1m (3ft). Frost hardy. Leaves are oval and dark green. In spring bears small, saucer-shaped, single, rose-pink flowers.
C. saluenensis (illus. p.98). Fast-growing, evergreen, bushy shrub. H to 4m (12ft), S to 2.5m (8ft). Half hardy. Has lance-shaped, stiff, dull green leaves. Cup-shaped, single, white to rose-red flowers are freely produced in early spring. Some forms may withstand lower temperatures.
C. sasanqua. Fast-growing, evergreen, dense, upright shrub. H 3m (10ft), S 1.5m (5ft). Frost hardy. Has lance-shaped, glossy, bright green leaves. In autumn bears a profusion of fragrant, flattish to cup-shaped, single, rarely semi-double, white flowers; they may occasionally be pink or red. Does best in a hot, sunny site. ❀ **'Narumigata'**

(illus. p.98) has large, cup-shaped, single, white flowers, sometimes pink-flushed. **'Onigoromo'** has cup-shaped, single, pink-edged, white flowers and is useful for hedging.
C. **'Shiro-wabisuke'** (illus. p.98). Slow-growing, compact shrub. H 2.5m (8ft), S 1.5m (5ft). Fully hardy. Has narrow, mid-green leaves. From mid-winter to early spring produces small, single, bell-shaped, white flowers.
❀ *C. tsaii* (illus. p.98). Evergreen, bushy shrub. H 4m (12ft), S 3m (10ft). Half hardy. Small, lance-shaped, light green leaves turn bronze with age. Small, cup-shaped, single, white flowers are freely produced in spring.
C. x *vernalis.* Fast-growing, evergreen, upright shrub. H to 3m (10ft), S 1.5m (5ft). Frost hardy. Has lance-shaped, bright green leaves and, in late winter, fragrant, flattish to cup-shaped, single, white, pink or red flowers. Some forms produce irregular double flowers.
C. **'William Hertrich'** (illus. p.99). Strong-growing, evergreen, open shrub. H 5m (15ft), S 3m (10ft). Half hardy. Is free-flowering with large, flattish to cup-shaped, semi-double blooms of a bright cherry-red in spring. Petal formation is very irregular, and petals often form a confused centre with only a few golden stamens. Leaves are large, oval and deep green.
C. x *williamsii* **'Bow Bells'** (illus. p.99). Evergreen, upright, spreading shrub. H 4m (12ft), S 2.5m (8ft). Fully hardy. Has small, lance-shaped, mid-green leaves and, in early spring, masses of cup-shaped, single, rose-pink flowers with deeper pink centres and veins. ❀ **'Brigadoon'** (illus. p.99) is a bushy shrub, bearing semi-double, rose-pink flowers with broad, down-curving petals. **'Caerhays'** (illus. p.99) is arching in habit with large, glossy leaves and large, anemone- to peony-form, crimson-pink flowers that become purplish with age. **'Clarrie Fawcett'** (illus. p.98) has an upright habit and cup-shaped, semi-double, rose-pink flowers; foliage is glossy. ❀ **'Donation'** (illus. p.98) is a compact, upright plant that is very floriferous, with large, cup-shaped, semi-double, pink flowers. **'E.G. Waterhouse'** (illus. p.98) is an upright, free-flowering cultivar bearing formal double, pink flowers among pale green foliage. **'Elizabeth de Rothschild'** is vigorous and upright; cup-shaped, semi-double, rose-pink flowers appear among glossy foliage. **'Francis Hangar'** (illus. p.98) has an upright habit and carries single, white flowers with gold stamens. ❀ **'George Blandford'** (illus. p.99) is spreading and bears semi-double, bright crimson-pink flowers in early spring. **'Golden Spangles'** (illus. p.99) is a cup-shaped, single, deep pink cultivar with unusual, variegated foliage, yellowish in centres of leaves with dark green margins. ❀ **'J.C. Williams'** (illus. p.98) is of pendulous habit when mature and bears cup-shaped, single, pink flowers from early winter to late spring. ❀ **'Mary Christian'** (illus. p.99) is vigorous and free-flowering; has dark green foliage and cup-shaped, single, phlox-pink flowers with deeper pink veins.

Vigorous, upright **'Mary Larcom'** (illus. p.98) produces cup-shaped, single, cerise-pink flowers from mid-winter to late spring. Free-flowering ❀ **'St Ewe'** (illus. p.99) has glossy, light green foliage and funnel-shaped, single, deep pink flowers. ❀ **'Water Lily'** (illus. p.99) is an upright, compact cultivar with dark green leaves that bears formal double mid-pink flowers with incurving petals in mid- to late spring.

CAMPANULA (Campanulaceae)
Bellflower
Genus of spring- and summer-flowering annuals, biennials and perennials, some of which are evergreen. Fully to half hardy. Grows in sun or shade, but delicate flower colours are preserved best in shade. Most forms prefer moist but well-drained soil. Propagate by softwood or basal cuttings in summer or by seed or division in autumn or spring. Is prone to slug attack, and rust may be a problem in autumn.
C. alliariifolia illus. p.239.
C. barbata illus. p.304.
❀ *C. betulifolia.* Prostrate, slender-stemmed perennial. H 2cm ($^3/_4$in), S 30cm (12in). Fully hardy. In summer, long, branching flower stems each carry a cluster of open bell-shaped, single, white to pink flowers, deep pink outside. Leaves are wedge-shaped.
❀ *C.* **'Birch Hybrid'** illus. p.330.
❀ *C.* **'Burghaltii'**, syn. *C.* x *burghaltii*, illus. p.250.
❀ *C. carpatica.* Clump-forming perennial. H 8–10cm (3–4in), S to 30cm (12in). Fully hardy. Leafy, branching stems bear rounded to oval, toothed leaves and, in summer, broadly bell-shaped, blue or white flowers. **'Bressingham White'** illus. p.321. **'Jewel'** has deep violet flowers. Flowers of **'Turbinata'** are pale lavender.
❀ *C. cochleariifolia*, syn. *C. pusilla*, illus. p.331.
❀ *C. garganica.* Spreading perennial. H 5cm (2in), S 30cm (12in). Fully hardy. Has small, ivy-shaped leaves along stems. Clusters of star-shaped, single, pale lavender flowers are produced from leaf axils in summer. Makes an excellent wall or bank plant. **'W.H. Paine'** has bright lavender-blue flowers, each with a white eye.
❀ *C.* **'G.F. Wilson'** illus. p.330.
❀ *C. glomerata* **'Superba'** illus. p.215.
C. x *haylodgensis.* Spreading perennial. H 5cm (2in), S 20cm (8in). Fully hardy. Has small, heart-shaped leaves and, in summer, bears pompon-like, double, deep lavender-blue flowers. Is suitable for a rock garden or wall.
❀ *C. isophylla.* Evergreen, dwarf, trailing perennial. H 10cm (4in), S 30cm (12in). Half hardy. In summer, star-shaped, blue or white flowers are borne above small, heart-shaped, toothed leaves. Is ideal for a hanging basket.
❀ *C.* **'Joe Elliott'**. Mound-forming perennial. H 8cm (3in), S 12cm (5in). Fully hardy. In summer, large, funnel-shaped, mid-lavender-blue flowers

453

almost obscure small, heart-shaped, downy, grey-green leaves. Is good for an alpine house, trough or rock garden. Needs well-drained, alkaline soil. Protect from winter wet. Is prone to slug attack.
C. lactiflora. Upright, branching perennial. H 1.2m (4ft), S 60cm (2ft). Fully hardy. In summer, slender stems bear racemes of large, nodding, bell-shaped, blue, occasionally pink or white flowers. Leaves are narrowly oval. Needs staking on a windy site. ♛ **'Loddon Anna'** has soft dusty-pink flowers. ♛ **'Prichard's Variety'** illus. p.192.
C. latifolia **'Brantwood'**. Clump-forming, spreading perennial. H 1.2m (4ft), S 60cm (2ft). Fully hardy. Strong stems are clothed with bell-shaped, rich violet-purple flowers in summer. Oval leaves are rough-textured.
C. latiloba. Rosette-forming perennial. H 1m (3ft), S 45cm (1½ft). Fully hardy. Leaves are oval. Widely cup-shaped flowers, in shades of blue, occasionally white, are borne in summer. ♛ **'Percy Piper'** has lavender flowers.
C. medium (Canterbury bell). Slow-growing, evergreen, erect, clump-forming biennial. Tall cultivars, H 1m (3ft), S 30cm (1ft); dwarf, H 60cm (2ft), S 30cm (1ft). Fully hardy. All have lance-shaped, toothed, fresh green leaves. Bell-shaped, single or double flowers, white or in shades of blue and pink, are produced in spring and early summer. **'Bells of Holland'** illus. p.283.
C. morettiana. Tuft-forming perennial. H 2.5cm (1in), S 7cm (3in). Frost hardy. Leaves are ivy-shaped with fine hairs. Arching flower stems each carry a solitary erect, bell-shaped, violet-blue flower in late spring and early summer. Needs gritty, alkaline soil and a dry but not arid winter climate. Red spider mite may be troublesome.
C. persicifolia. Rosette-forming, spreading perennial. H 1m (3ft), S 30cm (1ft). Fully hardy. In summer, nodding, bell-shaped, papery, white or blue flowers are borne above narrowly lance-shaped, bright green leaves. ♛ **'Fleur de Neige'** has double, white flowers. **'Pride of Exmouth'** bears double, powder-blue flowers. **'Telham Beauty'** illus. p.217.
♛ *C. portenschlagiana* illus. p.330.
C. poscharskyana illus. p.329.
C. pulla. Often short-lived perennial that spreads by underground runners. H 2.5cm (1in), S 10cm (4in). Fully hardy. Tiny, rounded leaves form 1cm (½in) wide rosettes, each bearing a flower stem with a solitary pendent, bell-shaped, deep violet flower from late spring to early summer. Is good for a scree or rock garden. Needs gritty, alkaline soil that is not too dry. Slugs may prove troublesome.
C. pusilla. See *C. cochleariifolia.*
C. pyramidalis (Chimney bellflower). Erect, branching biennial. H 2m (6ft), S 60cm (2ft). Half hardy. Produces long racemes of star-shaped, blue or white flowers in summer. Leaves are heart-shaped. Needs staking.
♛ *C. raineri.* Perennial that spreads by underground runners. H 4cm (1½in),
S 8cm (3in). Frost hardy. Leaves are oval, toothed and grey-green. Flower stems each carry a large, upturned, bell-shaped, pale lavender flower in summer. Is suitable for an alpine house or trough that is protected from winter wet. Requires semi-shade.
C. trachelium and **'Bernice'** illus. p.216.
C. vidalii, syn. *Azorina vidalii*, illus. p.124.
C. zoysii. Tuft-forming perennial. H 5cm (2in), S 10cm (4in). Frost hardy. Has tiny, rounded, glossy green leaves. In summer, flower stems each bear a bottle-shaped, lavender flower held horizontally. Needs gritty, alkaline soil. Is difficult to grow and flower well, dislikes winter wet and is prone to slug attack.

CAMPSIS (Bignoniaceae)
Genus of deciduous, woody-stemmed, root climbers, grown for their flowers. Frost hardy; in cooler areas needs protection of a sunny wall. Grow in sun and fertile, well-drained soil and water regularly in summer. Prune in spring. Propagate by semi-ripe cuttings in summer or by layering in winter.
C. chinensis. See *C. grandiflora.*
C. grandiflora, syn. *Bignonia grandiflora*, *Campsis chinensis*, *Tecoma grandiflora* (Chinese trumpet creeper, Chinese trumpet vine). Deciduous, woody-stemmed, root climber. H 7–10m (22–30ft). Leaves have 7 or 9 oval, toothed leaflets, hairless beneath. Drooping clusters of trumpet-shaped, deep orange or red flowers, 5–8cm (2–3in) long, are produced in late summer and autumn, abundantly in warm areas.
C. radicans, syn. *Bignonia radicans*, *Tecoma radicans* (Trumpet creeper, Trumpet honeysuckle, Trumpet vine). Deciduous, woody-stemmed, root climber. H to 12m (40ft). Leaves of 7–11 oval, toothed leaflets are downy beneath. Small clusters of trumpet-shaped, orange, scarlet or yellow flowers, 6–8cm (2½–3in) long, open in late summer and early autumn.
♛ *C. x tagliabuana* **'Mme Galen'** illus. p.177.

Camptosorus rhizophyllus. See *Asplenium rhizophyllus.*

CANARINA (Campanulaceae)
Genus of herbaceous, tuberous, scrambling climbers, grown for their flowers. Frost tender, min. 7°C (45°F). Grow in any fertile, well-drained soil and in full light. Water moderately from early autumn to late spring, then keep dry. Needs tying to a support. Remove dead stems when dormant. Propagate by basal cuttings or seed sown in spring or autumn.
C. campanula. See *C. canariensis.*
♛ *C. canariensis*, syn. *C. campanula*, illus. p.179.

Candollea cuneiformis. See *Hibbertia cuneiformis.*

CANNA (Cannaceae)
Genus of robust, showy, rhizomatous perennials, grown for their striking flowers and ornamental foliage. Is generally used for summer-bedding displays and container growing. Frost tender, min. 10–15°C (50–59°F). Requires a warm, sunny position and humus-rich, moist soil. If grown under glass or for summer bedding, encourage into growth in spring at 16°C (61°F) and store rhizomes in slightly damp soil or peat in winter. Propagate in spring by division or in winter by seed sown at 20°C (68°F) or more.
C. x generalis **'Assault'** illus. p.345. **'Dazzler'** is a rhizomatous perennial. H to 1.2m (4ft), S 45–60cm (1½–2ft). Min. 15°C (59°F). Stout stems bear bold, broadly lance-shaped, bronze-green leaves, to 30cm (12in) wide. In summer produces a spike of large, orchid-like, brilliant cardinal-red flowers, each with 3 large petals and a lower lip. **'Lucifer'**, H 1m (3ft), has purple leaves and yellow-edged, red petals. **'Orange Perfection'**, H 60–80cm (24–32in), has orange flowers.
C. iridiflora illus. p.347.

CANTUA (Polemoniaceae)
Genus of evergreen shrubs, grown for their showy flowers in spring. Only one species, however, is in general cultivation. Half hardy; is best grown against a south- or west-facing wall. Requires full sun and fertile, well-drained soil. Propagate by semi-ripe cuttings in summer.
C. buxifolia, syn. *C. dependens*, illus. p.125.
C. dependens. See *C. buxifolia.*

CAPSICUM (Solanaceae)
Genus of evergreen shrubs, sub-shrubs and short-lived perennials, usually grown as annuals. Some species produce edible fruits (e.g. sweet peppers), others small, ornamental ones. Frost tender, min. 4°C (39°F). Grow in sun and in fertile, well-drained soil. Spray flowers with water to encourage fruit to set. Propagate by seed sown under glass in spring. Red spider mite may sometimes cause problems.
C. annuum (Ornamental pepper). **'Holiday Cheer'** is a moderately fast-growing, evergreen, bushy perennial, grown as an annual. H and S 20–30cm (8–12in). Has oval, mid-green leaves. Produces small, star-shaped, white flowers in summer and, in autumn-winter, spherical fruits that change from green to red. **'Holiday Time'** illus. p.281.

CARAGANA (Leguminosae)
Genus of deciduous shrubs, grown for their foliage and flowers. Fully hardy. Needs full sun and fertile but not over-rich, well-drained soil. Propagate species by softwood cuttings in summer or by seed in autumn, cultivars by softwood or semi-ripe cuttings or budding in summer or by grafting in winter.
C. arborescens. Fast-growing, deciduous, upright shrub. H 6m (20ft), S 4m (12ft). Has spine-tipped, dark green leaves, each composed of 8–12 oblong leaflets. Produces clusters of pea-like, yellow flowers in late spring.
Arching ♛ **'Lorbergii'**, H 3m (10ft), S 2.5m (8ft), has very narrow leaflets and smaller flowers and is often grown as a tree by top-grafting; **'Nana'** illus. p.127. **'Walker'**, H 30cm (1ft), S 2–3m (6–10ft), is prostrate but is usually top-grafted to form a weeping tree 2m (6ft) high and 75cm (2½ft) across.
C. frutex **'Globosa'**. Slow-growing, deciduous, upright shrub. H and S 30cm (1ft). Mid-green leaves each have 4 oblong leaflets. Pea-like, bright yellow flowers are borne only rarely in late spring.

CARALLUMA (Asclepiadaceae)
Genus of perennial succulents with 4–6-ribbed, finger-like, blue-grey or -green to purple stems. Frost tender, min. 11°C (52°F). Needs sun and extremely well-drained soil. Water sparingly, only in the growing season. May be difficult to grow. Propagate by seed or stem cuttings in summer.
C. europaea. Clump-forming, perennial succulent. H 20cm (8in), S 1m (3ft). Rough, 4-angled, erect to procumbent, grey stems often arch over and root. Produces clusters of small, star-shaped, yellow and brownish-purple flowers near stem crown from mid- to late summer, followed by twin-horned, grey seed pods. Flowers smell faintly of rotten meat. Is one of the easier species to grow.
C. joannis illus. p.411.

CARDAMINE (Cruciferae)
Bitter cress
Genus of spring-flowering annuals and perennials. Some are weeds, but others are suitable for informal and woodland gardens. Fully hardy. Requires sun or semi-shade and moist soil. Propagate by seed or division in autumn.
C. enneaphyllos, syn. *Dentaria enneaphylla*, illus. p.238.
C. pentaphyllos, syn. *Dentaria pentaphylla*, illus. p.234.
C. pratensis (Cuckoo flower, Lady's smock). ♛ **'Flore Pleno'** is a neat, clump-forming perennial. H 45cm (18in), S 30cm (12in). Bears dense sheaves of double, lilac flowers in spring. Mid-green leaves are divided into rounded leaflets. May also be propagated by leaf-tip cuttings in mid-summer. Prefers moist or wet conditions.
C. trifolia illus. p.310.

CARDIOCRINUM (Liliaceae)
Genus of summer-flowering, lily-like bulbs, grown for their spectacular flowers. Frost hardy. Needs partial shade and deep, humus-rich, moist soil. Plant bulbs just below soil surface, in autumn. Water well in summer and mulch with humus. After flowering, main bulb dies, but produces offsets. To produce flowers in up to 5 years, propagate by offsets in autumn; may also be propagated by seed in autumn or winter and will then flower in 7 years.
♛ *C. giganteum* (Giant lily) illus. p.342. var. *yunnanense* is a stout, leafy-stemmed bulb. H 1.5–2m (5–6ft), S 75cm–1m (2½–3ft). Has bold, heart-shaped, bronze-green leaves. Fragrant,

pendent, trumpet-shaped, cream flowers, 15cm (6in) long, with purple-red streaks inside, are borne in long spikes in summer and are followed by decorative seed heads.

CARDIOSPERMUM (Sapindaceae)
Genus of herbaceous or deciduous, shrubby climbers, grown mainly for their attractive fruits. Is useful for covering bushes or trellises. Frost tender, min. 5°C (41°F). Grow in full light and any soil. Propagate by seed in spring.
C. halicacabum (Balloon vine, Heart pea, Heart seed, Winter cherry). Deciduous, shrubby, scandent, perennial climber, usually grown as an annual or biennial. H to 3m (10ft). Has toothed leaves of 2 oblong leaflets; inconspicuous, whitish flowers appear in summer, followed by downy, spherical, inflated, 3-angled, straw-coloured fruits containing black seeds, each with a heart-shaped, white spot.

CAREX (Cyperaceae). See GRASSES, BAMBOOS, RUSHES and SEDGES.
C. buchananii (Leatherleaf sedge). Evergreen, tuft-forming, perennial sedge. H to 60cm (24in), S 20cm (8in). Fully hardy. Very narrow, copper-coloured leaves turn red towards base. Solid, triangular stems bear insignificant, brown spikelets in summer.
C. elata, syn. *C. stricta* (Tufted sedge). Evergreen, tuft-forming, perennial sedge. H to 1m (3ft), S 15cm (6in). Fully hardy. Leaves are somewhat glaucous. Solid, triangular stems bear blackish-brown spikelets in summer.
❦ **'Aurea'** illus. p.185.
C. grayi (Mace sedge). Evergreen, tuft-forming, perennial sedge. H to 60cm (24in), S 20cm (8in). Fully hardy. Has bright green leaves. Large, female spikelets, borne in summer, mature to pointed, knobbly, greenish-brown fruits.
C. morrowii of gardens. See *C. oshimensis*.
C. oshimensis, syn. *C. morrowii* of gardens. Evergreen, tuft-forming, perennial sedge. H 20–50cm (8–20in), S 20–25cm (8–10in). Fully hardy. Has narrow, mid-green leaves. Solid, triangular stems bear insignificant spikelets in summer. **'Evergold'** illus. p.185.
C. pendula illus. p.185.
C. riparia (Greater pond sedge). **'Variegata'** is a vigorous, evergreen, perennial sedge. H 60cm–1m (2–3ft), S indefinite. Fully hardy. Has broad, white-striped, mid-green leaves and solid, triangular stems that bear narrow, bristle-tipped, dark brown spikelets in summer.
C. stricta. See *C. elata*.

CARISSA (Apocynaceae)
Genus of evergreen, spring- to summer-flowering shrubs, grown for their flowers and overall appearance. Frost tender, min. 10–13°C (50–55°F). Needs partial shade and well-drained soil. Water containerized specimens moderately, less when temperatures are low. Propagate by seed when ripe or in spring or by semi-ripe cuttings in summer.
C. grandiflora. See *C. macrocarpa*.
C. macrocarpa, syn. *C. grandiflora* (Natal plum). **'Tuttlei'** illus. p.105.
C. spectabilis. See *Acokanthera oblongifolia*.

CARLINA (Compositae)
Thistle
Genus of annuals, biennials and perennials, grown for their ornamental flower heads. Fully hardy. Needs sun and well-drained soil. Propagate by seed: annuals in spring, perennials in autumn.
C. acaulis illus. p.323.

CARMICHAELIA (Leguminosae)
Genus of deciduous, usually leafless shrubs, grown for their profusion of tiny flowers in summer. Flattened, green shoots assume function of leaves. Frost to half hardy. Needs full sun and well-drained soil. Cut out dead wood in spring. Propagate by semi-ripe cuttings in summer or by seed in autumn or spring.
C. australis. Deciduous, upright shrub. H 2m (6ft), S 1.5m (5ft). Frost hardy. Small clusters of pea-like, pale lilac flowers appear from early to mid-summer. May need staking when mature.
C. enysii. Deciduous, mound-forming, dense shrub. H and S 30cm (1ft). Frost hardy. Shoots are rigid. Pea-like, violet flowers are borne in mid-summer. Is best grown in a rock garden.

CARNEGIEA (Cactaceae)
Genus of one species of very slow-growing, perennial cactus with thick, 12–24-ribbed, spiny stems. Is unlikely to flower or branch at less than 4m (12ft) high. Frost tender, min. 7°C (45°F). Requires full sun and very well-drained soil. Propagate by seed in spring or summer.
C. gigantea illus. p.392.

CARPENTERIA (Hydrangeaceae)
Genus of one species of evergreen, summer-flowering shrub, cultivated for its flowers and foliage. Frost hardy. Grows well against a south- or west-facing wall. Prefers full sun and a fairly moist but well-drained soil. Propagate by greenwood cuttings in summer or by seed in autumn.
❦ *C. californica* illus. p.108.

CARPINUS (Carpinaceae)
Hornbeam
Genus of deciduous trees, grown for their foliage, autumn colour and clusters of small, winged nuts. Fully hardy. Needs sun or semi-shade and fertile, well-drained soil. Propagate species by seed in autumn, cultivars by budding in late summer.
❦ *C. betulus* (Common hornbeam). Deciduous, round-headed tree. H 25m (80ft), S 20m (70ft). Has a fluted trunk and oval, prominently veined, dark green leaves that turn yellow and orange in autumn. Bears green catkins from late spring to autumn, when clusters of winged nuts appear.
❦ **'Fastigiata'** illus. p.74.
C. caroliniana (American hornbeam). Deciduous, spreading tree with branches that droop at tips. H and S 10m (30ft). Has a fluted, grey trunk, green catkins in spring and oval, bright green leaves that turn orange and red in autumn, when clusters of winged nuts appear.
C. tschonoskii. Deciduous, rounded tree of elegant habit, with branches drooping at tips. H and S 12m (40ft). Has oval, sharply toothed, glossy, dark green leaves. Green catkins are carried in spring and clusters of small, winged nuts appear in autumn.
C. turczaninowii. Deciduous, spreading tree of graceful habit. H 12m (40ft), S 10m (30ft). Green catkins are borne in spring. Produces clusters of small, winged nuts in autumn, when small, oval, glossy, deep green leaves turn orange.

CARPOBROTUS (Aizoaceae)
Genus of mat-forming, perennial succulents with triangular, fleshy, dark green leaves and daisy-like flowers. Is excellent for binding sandy soils. Half hardy to frost tender, min. 5°C (41°F). Needs full sun and well-drained soil. Propagate by seed or stem cuttings in spring or summer.
C. edulis (Hottentot fig, Kaffir fig). Carpeting, perennial succulent. H 15cm (6in), S indefinite. Frost tender. Prostrate, rooting branches bear leaves 1.5cm (5/8in) thick and 12cm (5in) long. Yellow, purple or pink flowers, 12cm (5in) across, open in spring-summer from about noon in sun. Bears edible, fig-like, brownish fruits in late summer and autumn.

CARRIEREA (Flacourtiaceae)
Genus of deciduous trees. Only *C. calycina*, grown for its flowers, is in general cultivation. Frost hardy. Needs full sun and fertile, well-drained soil. Propagate by softwood cuttings in summer.
C. calycina. Deciduous, spreading tree. H 8m (25ft), S 10m (30ft). Oval, glossy, mid-green leaves set off upright clusters of cup-shaped, creamy-white or greenish-white flowers in early summer.

CARYA (Juglandaceae)
Hickory
Genus of deciduous trees, grown for their stately habit, divided leaves, autumn colour and, in some cases, edible nuts. Has insignificant flowers in spring. Fully hardy. Requires sun or semi-shade and deep, fertile soil. Plant young seedlings in a permanent position during their first year as older plants resent transplanting. Propagate by seed in autumn.
❦ *C. cordiformis* (Bitternut, Bitternut hickory). Vigorous, deciduous, spreading tree. H 25m (80ft), S 15m (50ft). Bark is smooth at first, later fissured. Bright yellow, winter leaf buds develop into large, dark green leaves, with usually 7 oval to oblong leaflets; these turn yellow in autumn. Nuts are pear-shaped or rounded, 2–4cm ($^3/_4$–$1^1/_2$in) long, each with a bitter kernel.
C. glabra (Pignut, Pignut hickory). Deciduous, spreading tree. H 25m (80ft), S 20m (70ft). Dark green leaves, with usually 5 narrowly oval leaflets, turn bright yellow and orange in autumn. Pear-shaped or rounded nuts, 2–4cm ($^3/_4$–$1^1/_2$in) long, each have a bitter kernel.
❦ *C. ovata* illus. p.43.

CARYOPTERIS (Verbenaceae)
Genus of deciduous sub-shrubs, grown for their foliage and small, but freely produced, blue flowers. Frost hardy. Prefers full sun and light, well-drained soil. Cut back hard in spring. Propagate species by greenwood or semi-ripe cuttings in summer or by seed in autumn, cultivars by cuttings only in summer.
C. x *clandonensis* **'Arthur Simmonds'** illus. p.138. ❦ **'Heavenly Blue'** is a deciduous, bushy sub-shrub. H and S 1m (3ft). Forms an upright, compact mass of lance-shaped, grey-green leaves. Dense clusters of tubular, blue to purplish-blue flowers, with prominent stamens, are borne from late summer to autumn.
C. incana. Deciduous, bushy sub-shrub. H and S 1.2m (4ft). Tubular, violet-blue flowers, with prominent stamens, are produced amid lance-shaped, grey-green leaves from late summer to early autumn.

CASSIA (Leguminosae)
Genus of annuals, perennials and evergreen or deciduous trees and shrubs, grown for their flowers mainly produced from winter to summer. Fully hardy to frost tender, min. 7–18°C (45–64°F). Needs full light and fertile, well-drained soil. Water containerized specimens freely when in full growth, moderately to sparingly in winter. Pruning is tolerated, severe if need be, but trees are best left to grow naturally. Propagate by seed in spring.
C. artemisioides, syn. *Senna artemisioides* (Silver cassia, Wormwood cassia). Evergreen, erect to spreading, wiry shrub. H and S 1–2m (3–6ft). Frost tender, min. 10–13°C (50–55°F). Leaves each have 6–14 linear leaflets covered with silky, white down. Axillary spikes of cup-shaped, yellow flowers appear from winter to early summer.
C. corymbosa, syn. *Senna corymbosa*, illus. p.117. var. *plurijuga* of gardens. See *C.* x *floribunda*.
C. didymobotrya, syn. *Senna didymobotrya*, illus. p.117.
C. fistula (Golden shower, Indian laburnum, Pudding pipe-tree). Fast-growing, almost deciduous, ovoid tree. H 8–10m (25–30ft), S 4–6m (12–20ft). Frost tender, min. 16°C (61°F). Has 30–45cm (12–18in) long leaves, each with 4–8 pairs of oval leaflets, coppery when young. In spring produces racemes of small, fragrant, 5-petalled, cup-shaped, bright yellow flowers. Cylindrical, dark brown pods, to 60cm (2ft) long, yield cassia pulp.
C. x *floribunda*, syn. *C. corymbosa* var. *plurijuga* of gardens. Vigorous, evergreen or deciduous, rounded shrub with robust stems. H and S 1.5–2m (5–6ft). Frost tender, min. 7°C (45°F). Bright green leaves consist of 4–6 oval leaflets. Carries very large clusters of

bowl-shaped, rich yellow flowers in late summer.
C. siamea, syn. *Senna siamea*. Fast-growing, evergreen, rounded tree. H and S 8–10m (25–30ft) or more. Frost tender, min. 16–18°C (61–4°F). Leaves, 15–30cm (6–12in) long, have 7–12 pairs of elliptic leaflets. Large, terminal panicles of small, cup-shaped, bright yellow flowers are borne in spring, followed by flat, dark brown pods, to 23cm (9in) long.

CASSINIA (Compositae)
Genus of evergreen shrubs, grown for their foliage and flowers. Frost hardy, but avoid cold, exposed positions. Needs full sun and fertile, well-drained soil. Propagate by softwood cuttings in summer.
C. fulvida, syn. *C. leptophylla* subsp. *fulvida*. Evergreen, bushy shrub. H and S 2m (6ft). Has yellow shoots, small, oblong, dark green leaves and, in mid-summer, clustered heads of minute, white flowers. Is useful as a coastal hedging plant.
C. vauvilliersii, syn. *C. leptophylla* subsp. *vauvilliersii*, illus. p.131.

CASSIOPE (Ericaceae)
Genus of evergreen, spring-flowering shrubs, suitable for peat beds and walls and for rock gardens. Fully hardy. Needs a sheltered, shaded or semi-shaded site and moist, peaty, acid soil. Propagate by semi-ripe or greenwood cuttings in summer or by seed in autumn or spring.
♣ *C. 'Edinburgh'* illus. p.294.
C. fastigiata. Evergreen, upright, loose shrub. H 30cm (12in), S 15–20cm (6–8in). In spring, bell-shaped, creamy-white flowers, resting in green or red calyces, are borne on short stalks in leaf axils. Leaves are tiny and scale-like. Needs semi-shade.
♣ *C. lycopodioides* illus. p.311.
C. mertensiana illus. p.312.
♣ *C. 'Muirhead'* illus. p.294.
C. selaginoides. Evergreen, spreading shrub. H 25cm (10in), S 15cm (6in). Stem is hidden by dense, scale-like, mid-green leaves. Bears solitary relatively large, pendent, bell-shaped, white flowers in spring. Needs a shaded site.
C. tetragona. Evergreen, upright shrub. H 10–25cm (4–10in), S 10–15cm (4–6in). Dense, scale-like, dark green leaves conceal branched stems. In spring, leaf axils bear solitary pendent, bell-shaped, white flowers in red calyces. Needs a semi-shaded site.
C. wardii. Evergreen, upright to spreading, loose shrub. H 15cm (6in), S 20cm (8in). Semi-upright stems are densely clothed with scale-like, dark green leaves that give them a squared appearance. Bell-shaped, white flowers, set close to stems, open in spring. Needs shade in all but cool areas. May also be propagated by division of runners in spring.

CASTANEA (Fagaceae)
Chestnut
Genus of deciduous, summer-flowering trees and shrubs, grown for their foliage, stately habit, flowers and edible fruits (chestnuts). Fully hardy.

Requires sun or semi-shade; does particularly well in hot, dry areas. Needs fertile, well-drained soil; grows poorly on shallow, chalky soil. Propagate species by seed in autumn, cultivars by budding in summer or by grafting in late winter.
C. dentata (American chestnut). Deciduous, spreading tree with rough bark. H 30m (100ft), S 15m (50ft). Oblong, toothed, dull green leaves turn orange-yellow in autumn. Has catkins of greenish-white flowers in summer, followed by typical spiny 'chestnut' fruits.
♣ *C. sativa* (Spanish chestnut, Sweet chestnut). Deciduous, spreading tree. H 30m (100ft), S 15m (50ft). Bark becomes spirally ridged with age. Oblong, glossy, dark green leaves turn yellow in autumn. Spikes of small, creamy-yellow flowers in summer are followed by edible fruits in rounded, spiny husks. **'Albomarginata'** illus. p.39.

CASTANOPSIS (Fagaceae)
Genus of evergreen shrubs and trees, grown for their habit and foliage. Flowers are insignificant. Frost hardy. Needs a sheltered position in sun or semi-shade and fertile, well-drained but not too dry, acid soil. Propagate by seed when ripe, in autumn.
C. cuspidata. Evergreen, bushy, spreading shrub or tree with drooping shoots. H and S 8m (25ft) or more. Bears long, oval, slender-tipped, leathery leaves, glossy, dark green above, bronze beneath.

CASTANOSPERMUM (Leguminosae)
Black bean tree, Moreton Bay chestnut
Genus of one species of evergreen tree, grown for its overall ornamental appearance and for shade. Frost tender, min. 10–15°C (50–59°F). Requires full light and fertile, moisture-retentive but well-drained soil. Water containerized specimens freely when in full growth, moderately at other times. Propagate by seed in spring.
C. australe. Strong-growing, evergreen, rounded tree. H 15m (50ft) or more, S 8m (25ft) or more. Has 45cm (18in) long leaves of 8–17 oval leaflets. Racemes of large, pea-like, yellow flowers, borne in autumn on mature trees, age to orange and red. Cylindrical, reddish-brown pods, each 25cm (10in) long, contain large, chestnut-like seeds.

CATALPA (Bignoniaceae)
Genus of deciduous, summer-flowering trees and shrubs, extremely resistant to urban pollution, grown for their foliage and bell- or trumpet-shaped flowers with frilly lobes. Trees are best grown as isolated specimens. Fully hardy. Prefers full sun and does best in hot summers. Needs deep, fertile, well-drained but not too dry soil. Propagate species by seed in autumn, cultivars by softwood cuttings in summer or by budding in late summer.
♣ *C. bignonioides* (Indian bean tree) illus. p.52. ♣ **'Aurea'** is a deciduous, spreading tree. H and S 10m (30ft). Has broadly oval, bright yellow leaves,

bronze when young. Bell-shaped, white flowers, marked with yellow and purple, appear in summer, followed by long, pendent, cylindrical pods, often persisting after leaf fall.
♣ *C.* x *erubescens* **'Purpurea'**. Deciduous, spreading tree. H and S 15m (50ft). Broadly oval or 3-lobed, very dark purple, young leaves mature to dark green. Fragrant, bell-shaped, white flowers, marked with yellow and purple, are borne from mid- to late summer.
C. ovata. Deciduous, spreading tree. H and S 10m (30ft). Has 3-lobed, purplish leaves when young, maturing to pale green. Large clusters of bell-shaped, white flowers, spotted with red and yellow, are borne from mid- to late summer.
C. speciosa illus. p.52.

CATANANCHE (Compositae)
Blue cupidone
Genus of perennials with daisy-like flower heads that may be successfully dried for winter flower arrangements. Fully hardy. Needs sun and light, well-drained soil. Propagate by seed in spring or by root cuttings in winter.
♣ *C. caerulea* **'Major'** illus. p.249.

CATHARANTHUS (Apocynaceae)
Genus of evergreen shrubs, grown for their flowers. *C. roseus* is often grown annually from seed or cuttings and used as a summer bedding plant in cool climates. Frost tender, min. 5–7°C (41–5°F). Needs full light and well-drained soil. Water potted specimens moderately, less when temperatures are low. Prune long or straggly stems in early spring to promote a more bushy habit. Propagate by seed in spring or by greenwood or semi-ripe cuttings in summer.
♣ *C. roseus*, syn. *Vinca rosea*, illus. p.130.

CATTLEYA. See ORCHIDS.
C. bowringiana (illus. p.261). Evergreen, epiphytic orchid for a cool greenhouse. H 45cm (18in). In autumn bears large heads of rose-purple-lipped, magenta flowers, 8cm (3in) across. Has oval, stiff leaves, 8–10cm (3–4in) long. Grow in semi-shade during summer and do not spray from overhead.
C. J.A. Carbone (illus. p.261). Evergreen, epiphytic orchid for an intermediate greenhouse. H 45cm (18in). Large heads of fragrant, pinkish-mauve flowers, 10cm (4in) across and each with a yellow-marked, deep pink lip, open in early summer. Has oval, stiff leaves, 10–15cm (4–6in) long. Avoid spraying from overhead.

CAUTLEYA (Zingiberaceae)
Genus of summer- and autumn-flowering perennials. Frost hardy. Grow in a sunny, wind-free position and in deep, rich, moist but well-drained soil. Propagate by seed or division in spring.
C. spicata illus. p.258.

CEANOTHUS (Rhamnaceae)
Genus of evergreen or deciduous shrubs and small trees, grown for their small but densely clustered, mainly

blue flowers. Frost to half hardy; in cold areas plant against a south- or west-facing wall. Needs a sheltered site in full sun and light, well-drained soil. Cut dead wood from evergreens in spring and trim their side-shoots after flowering. Cut back shoots of deciduous species to basal framework in early spring. Propagate by semi-ripe cuttings in summer.
♣ *C. 'Autumnal Blue'* illus. p.118.
♣ *C. 'Blue Mound'*. Evergreen, bushy, dense shrub. H 1.5m (5ft), S 2m (6ft). Frost hardy. Forms a mound of oblong, glossy, dark green leaves, covered, in late spring, with rounded clusters of deep blue flowers.
♣ *C. 'Burkwoodii'*. Evergreen, bushy, dense shrub. H 1.5m (5ft), S 2m (6ft). Frost hardy. Has oval, glossy, dark green leaves, downy and grey beneath. Produces dense panicles of bright blue flowers from mid-summer to mid-autumn.
C. 'Burtonensis'. Evergreen, bushy, spreading shrub. H 2m (6ft) or more, S 4m (12ft). Frost hardy. Has small, rounded, almost spherical, crinkled leaves that are lustrous and dark green. Small, deep blue flowers appear in clusters, 2cm (3/4in) wide, from mid-spring to early summer.
♣ *C. 'Cascade'*. Vigorous, evergreen, arching shrub. H and S 4m (12ft). Frost hardy. Leaves are narrowly oblong, glossy and dark green. Large panicles of powder-blue flowers open in late spring and early summer.
♣ *C. 'Delight'*. Fast-growing, evergreen, bushy shrub. H 3m (10ft), S 5m (15ft). Frost hardy. Bears oblong, glossy, deep green leaves. Long clusters of rich blue flowers appear in late spring.
C. dentatus. Evergreen, bushy, dense shrub. H 1.5m (5ft), S 2m (6ft). Frost hardy. Produces small, oblong, glossy, dark green leaves and is covered, in late spring, with rounded clusters of bright blue flowers.
♣ *C. 'Gloire de Versailles'* illus. p.138.
C. gloriosus. Evergreen, prostrate shrub. H 30cm (1ft), S 2m (6ft). Frost hardy. Leaves are oval and dark green. Rounded clusters of deep blue or purplish-blue flowers appear from mid- to late spring. May suffer from chlorosis on chalky soils.
C. impressus illus. p.115.
C. incanus illus. p.108.
♣ *C. 'Italian Skies'*. Evergreen, bushy, spreading shrub. H 1.5m (5ft), S 3m (10ft). Frost hardy. Has small, oval, glossy, dark green leaves. Produces dense, conical clusters of bright blue flowers during late spring.
C. x *lobbianus*. Evergreen, bushy, dense shrub. H and S 2m (6ft). Frost hardy. Rounded clusters of bright deep blue flowers are borne in late spring and early summer amid oval, dark green leaves.
C. 'Marie Simon'. Deciduous, bushy shrub. H and S 1.5m (5ft). Frost hardy. Has broadly oval, mid-green leaves. Conical clusters of soft pink flowers are carried in profusion from mid-summer to early autumn.
C. papillosus. Evergreen, arching shrub. H 3m (10ft), S 5m (15ft). Frost

hardy. Leaves are narrowly oblong, glossy, dark green and sticky. Produces dense racemes of blue or purplish-blue flowers during late spring.
C. **'Perle Rose'** illus. p.133.
C. rigidus (Monterey ceanothus). Evergreen, bushy shrub of dense, spreading habit. H 1.2m (4ft), S 2.5m (8ft). Frost hardy. Bears oblong to rounded, glossy, dark green leaves and, from mid-spring to early summer, rounded clusters of deep purplish-blue flowers.
♀ *C.* **'Southmead'**. Evergreen, bushy, dense shrub. H and S 1.5m (5ft). Frost hardy. Has small, oblong, glossy, dark green leaves. Deep blue flowers are produced in rounded clusters in late spring and early summer.
C. thyrsiflorus. Evergreen, bushy shrub or spreading tree. H and S 6m (20ft). Frost hardy. Has broadly oval, glossy, mid-green leaves and, in late spring and early summer, bears rounded clusters of pale blue flowers.
♀ var. *repens* illus. p.138.
♀ *C.* **'Trewithen Blue'**, syn. *C. arboreus* 'Trewithen Blue'. Vigorous, evergreen, bushy, spreading shrub. H 6m (20ft), S 8m (25ft) or more. Frost hardy. In spring and early summer, large, pyramidal clusters of rich blue flowers are borne amid broadly oval to rounded, dark green leaves.
C. x *veitchianus*. Vigorous, evergreen, bushy shrub. H and S 3m (10ft). Frost hardy. Dense, oblong clusters of deep blue flowers are borne in late spring and early summer amid oblong, glossy, dark green leaves.

Cedrela sinensis. See *Toona sinensis*.

Cedronella mexicana. See *Agastache mexicana*.

CEDRUS (Pinaceae), Cedar. See CONIFERS.
C. atlantica, syn. *C. libani* subsp. *atlantica* (Atlas cedar). Conifer that is conical when young, broadening with age. H 15–25m (50–80ft), S 5–10m (15–30ft). Fully hardy. Leaves are spirally arranged, needle-like, dull green or bright blue-grey. Has ovoid cones, males pale brown, females pale green, ripening to brown. f. *fastigiata*, S 4–5m (12–15ft), has a narrower, more upright habit. f. *glauca* illus. p.75.
♀ *C. deodara* (Deodar). Fast-growing conifer, densely conical with weeping tips when young, broader when mature. H 15–25m (50–80ft), S 5–10m (15–30ft). Fully hardy. Has spirally arranged, needle-like, grey-green leaves and barrel-shaped, glaucous cones, 8–12cm (3–5in) long, ripening to brown. ♀ **'Aurea'** illus. p.83.
C. libani (Cedar of Lebanon; illus. p.77). Open conifer with tiered, arching branches. H 25m (80ft), S 15m (50ft). Fully hardy. Spirally arranged, needle-like, grey-green foliage is produced in dense, flat layers. Has greyish-pink cones. subsp. *atlantica* see *C. atlantica*. **'Comte de Dijon'**, H 1–2m (3–6ft), S 60cm–1.2m (2–4ft), is a dwarf form that grows only 5cm (2in) a year. **'Sargentii'** (illus. p.85) H and S 1–1.5m (3–5ft), has horizontal, then weeping branches and makes a bush that is rounded in shape.

CEIBA (Bombacaceae)
Genus of evergreen, semi-evergreen or deciduous trees, grown for their overall appearance and for shade. Frost tender, min. 15°C (59°F). Requires full light or light shade and fertile, moisture-retentive but well-drained soil. Water potted specimens freely while in full growth, less at other times. Pruning is tolerated if necessary. Propagate by seed in spring or by semi-ripe cuttings in summer.
C. pentandra (Kapok, Silk cotton tree). Fast-growing, semi-evergreen tree with a spine-covered trunk. H and S 25m (80ft) or more. Hand-shaped leaves have 5–9 elliptic leaflets, red when young, becoming mid-green. Bears clusters of 5-petalled, white, yellow or pink flowers in summer, followed by woody, brownish seed pods containing silky kapok fibre.

CELASTRUS (Celastraceae)
Genus of deciduous shrubs and twining climbers, grown for their attractive fruits. Most species bear male and female flowers on separate plants, so both sexes must be grown to obtain fruits; hermaphrodite forms of *C. orbiculatus* are available. Fully to frost hardy. Grow in any soil and in full or partial shade. Likes regular feeding. Prune in spring to cut out old wood and maintain shape. Propagate by seed in autumn or spring or by semi-ripe cuttings in summer.
C. articulatus. See *C. orbiculatus*.
♀ *C. orbiculatus*, syn. *C. articulatus* (Oriental bittersweet, Staff vine). Vigorous, deciduous, twining climber. H to 14m (46ft). Frost hardy. Has small, rounded, toothed leaves. Clusters of 2–4 small, green flowers are produced in summer; tiny, long-lasting, spherical fruits begin green, turn black in autumn, then split and show yellow insides and red seeds.
C. scandens (American bittersweet, Staff tree). Deciduous, twining climber. H to 10m (30ft). Frost hardy. Oval leaves are 5–10cm (2–4in) long; tiny, greenish flowers are borne in small clusters in leaf axils in summer; long-lasting, spherical fruits are produced in bunches, 5–8cm (2–3in) long; each splits to show an orange interior and scarlet seeds.

CELMISIA (Compositae)
Genus of evergreen, late spring- and summer-flowering perennials, grown for their foliage and daisy-like flower heads. Is suitable for rock gardens and peat beds, but may be difficult to grow in hot, dry climates. Frost hardy. Needs a sheltered, sunny site and humus-rich, moist but well-drained, sandy, acid soil. Propagate by division in early summer or by seed when fresh.
C. bellidioides. Evergreen, mat-forming perennial. H 2cm (³⁄₄in), S to 15cm (6in). Dark green leaves are rounded and leathery. Bears almost stemless, 1cm (¹⁄₂in) wide, daisy-like, white flower heads in early summer.
C. coriacea illus. p.308.
C. ramulosa illus. p.322.

C. traversii. Slow-growing, evergreen, clump-forming perennial. H 15cm (6in), S 20cm (8in). Sword-shaped, dark green leaves have reddish-brown margins and cream undersides. In summer carries 6–7cm (2¹⁄₂–3in) wide, daisy-like, white flower heads. Is difficult to establish.
C. walkeri illus. p.298.

CELOSIA (Amaranthaceae)
Genus of erect perennials, grown as annuals. Half hardy. Grows best in a sunny, sheltered position and in fertile, well-drained soil. Propagate by seed sown under glass in spring.
C. argentea var. *cristata*. Moderately fast-growing, erect, bushy perennial, grown as an annual. H 30–60cm (1–2ft), S 30cm (1ft). Has oval, mid-green leaves and, in summer-autumn, pyramid-shaped, feathery flower heads, to 15cm (6in) tall, in red, yellow, pink or apricot. Dwarf cultivars, H 30cm (1ft), include **'Fairy Fountains'**, illus. p.293, and **Geisha Series**, which is available in a wide colour range.

CELTIS (Ulmaceae)
Hackberry, Nettle tree
Genus of deciduous trees, with inconspicuous flowers in spring, grown for their foliage and small fruits. Fully hardy. Needs full sun (doing best in hot summers) and fertile, well-drained soil. Propagate by seed in autumn.
C. australis illus. p.41.
C. occidentalis (Common hackberry). Deciduous, spreading tree. H and S 20m (70ft). Oval, sharply toothed, glossy, bright green leaves turn yellow in autumn, when they are accompanied by globose, yellowish-red, then red-purple fruits.
C. sinensis. Deciduous, rounded tree. H and S 10m (30ft). Has oval, glossy, dark green leaves, with fine teeth, and small, globose, orange fruits.

CENTAUREA (Compositae)
Knapweed
Genus of annuals and perennials, grown for their flower heads that each have a thistle-like centre surrounded by a ring of slender ray petals. Fully hardy. Requires sun; grows in any well-drained soil, even poor soil. Propagate by seed or division in autumn or spring.
C. cyanus (Bluebottle, Cornflower). Fast-growing, upright, branching annual. H 30cm–1m (1–3ft), S 30cm (1ft). Has lance-shaped, grey-green leaves and, in summer and early autumn, branching stems with usually double, daisy-like flower heads in shades of blue, pink, red, purple or white. Flowers are excellent for cutting. Tall (blue, illus. p.286; rose, illus. p.274) and dwarf cultivars are available. **'Jubilee Gem'** (dwarf) has large, double, deep blue flower heads.
C. dealbata. Erect perennial. H 1m (3ft), S 60cm (2ft). Lilac-purple flower heads are borne freely in summer, one or more to each stem. Has narrowly oval, finely cut, light green leaves. **'Steenbergii'**, H 60cm (2ft), has carmine-lilac flowers.
C. hypoleuca **'John Coutts'** illus. p.243.

C. macrocephala. Robust, clump-forming perennial. H 1m (3ft), S 60cm (2ft). In summer, stout stems bear large, yellow flower heads, enclosed in papery, silvery-brown bracts. Mid-green leaves are narrowly oval and deeply cut.
C. montana illus. p.248.
C. moschata, syn. *Amberboa moschata*, illus. p.287.
C. pulcherrima illus. p.208.

CENTRADENIA (Melastomataceae)
Genus of evergreen perennials and shrubs, grown for their flowers and foliage. Frost tender, min. 13°C (55°F). Needs light shade and fertile, well-drained soil. Water containerized plants freely when in full growth, moderately at other times. Tip prune young plants to promote a bushy habit; old plants become straggly unless they are trimmed each spring. Propagate from early spring to early summer by seed or softwood or greenwood cuttings. If grown as pot plants, propagate annually.
C. floribunda. Evergreen, loosely rounded, soft-stemmed shrub. H and S to 60cm (24in). Lance-shaped leaves are prominently veined, glossy, green above, bluish-green beneath. Large, terminal clusters of 4-petalled, pink or white flowers develop from pink buds in late winter and spring.

CENTRANTHUS (Valerianaceae)
Genus of late spring- to autumn-flowering perennials. Fully hardy. Requires sun. Thrives in an exposed position and in poor, especially alkaline, soil. Propagate by seed in autumn or spring.
C. ruber illus. p.209.

CEPHALARIA (Dipsacaceae)
Genus of coarse, summer-flowering perennials, best suited to large borders and wild gardens. Fully hardy. Prefers sun and well-drained soil. Propagate by division in spring or by seed in autumn.
C. gigantea, syn. *C. tatarica* (Giant scabious, Yellow scabious). Robust, branching perennial. H 2m (6ft), S 1.2m (4ft). In early summer, wiry stems bear pincushion-like heads of primrose-yellow flowers well above lance-shaped, deeply cut, dark green leaves.
C. tatarica. See *C. gigantea*.

CEPHALOCEREUS (Cactaceae)
Genus of slow-growing, columnar, perennial cacti with 20–30-ribbed, green stems. Frost tender, min. 5°C (41°F). Prefers full sun and extremely well-drained, lime-rich soil. Is prone to rot if overwatered. Propagate by seed in spring or summer.
C. senilis illus. p.396.

CEPHALOPHYLLUM (Aizoaceae)
Genus of clump-forming, bushy, perennial succulents with semi-cylindrical to cylindrical, green leaves. Flowers are borne after 1 or 2 years. Frost tender, min. 5°C (41°F). Requires sun and well-drained soil. Propagate by seed in spring or summer.
C. alstonii illus. p.410.
C. pillansii. Clump-forming, perennial

succulent. H 8cm (3in), S 60cm (24in). Leaves are cylindrical, 6cm (2½in) long, dark green and covered in darker dots. Short flower stems produce daisy-like, red-centred, yellow flowers, 6cm (2½in) across, from spring to autumn.

CEPHALOTAXUS
(Cephalotaxaceae). See CONIFERS.
C. harringtonia (Cow's-tail pine, Plum yew). Bushy, spreading conifer. H 5m (15ft), S 3m (10ft). Frost hardy. Needle-like, flattened leaves are glossy, dark green, greyish beneath, radiating around erect shoots. Bears ovoid, fleshy, green fruits that ripen to brown.

CERASTIUM (Carophyllaceae)
Genus of annuals and perennials with star-shaped flowers. Some species are useful as ground cover. Fully hardy. Needs sun and well-drained soil. Propagate by division in spring.
C. alpinum (Alpine mouse-ear). Prostrate perennial. H 8cm (3in), S 40cm (16in). Tiny, oval, grey leaves cover stems. Flower stems carry solitary 1cm (½in) wide, star-shaped, white flowers throughout summer.
C. tomentosum illus. p.309.

CERATOPHYLLUM
(Ceratophyllaceae)
Genus of deciduous, perennial, submerged water plants, grown for their foliage. Is suitable for pools and cold-water aquariums. Fully to half hardy. Prefers an open, sunny position, but tolerates shade better than most submerged plants. Propagation occurs naturally when scaly, young shoots or winter buds separate from main plants. Alternatively, take stem cuttings in the growing season.
C. demersum (Hornwort). Deciduous, perennial, spreading, submerged water plant that occasionally floats. S indefinite. Fully hardy. Has small, dark green leaves with 3 linear lobes. Is best suited to a cool-water pool.

CERATOPTERIS (Parkeriaceae)
Genus of deciduous or semi-evergreen, perennial, floating water ferns, grown for their attractive foliage. Is suitable for aquariums. Frost tender, min. 10°C (50°F). Prefers a sunny position. Remove fading fronds regularly. Propagate in summer by division or by buds that develop on the leaves.
C. thalictroides (Water fern). Semi-evergreen, perennial, spreading, floating water fern that sometimes roots and becomes submerged. S indefinite. Lance- or heart-shaped, soft green fronds are wavy-edged.

CERATOSTIGMA (Plumbaginaceae)
Genus of deciduous, semi-evergreen or evergreen shrubs and perennials, grown for their blue flowers and autumn colour. Fully hardy to frost tender, min. 10°C (50°F). Requires sun and well-drained soil. Cut out dead wood from shrubs in spring. Propagate shrubs by softwood cuttings in summer, perennials by division in spring.
C. griffithii. Evergreen or semi-evergreen, bushy, dense shrub. H 1m (3ft), S 1.5m (5ft). Half hardy. Spoon-shaped, bristly, purple-edged, dull green leaves redden in autumn. Clusters of tubular, bright blue flowers, with spreading petal lobes, appear in late summer and autumn.
℗ *C. plumbaginoides* illus. p.308.
℗ *C. willmottianum* illus. p.143.

CERCIDIPHYLLUM
(Cercidiphyllaceae)
Genus of deciduous trees, grown for their foliage and often spectacular autumn colour. Fully hardy. Late frosts may damage young foliage, but do not usually cause lasting harm. Needs sun or semi-shade and fertile, moist but well-drained soil. Propagate by seed in autumn.
℗ *C. japonicum* illus. p.45.

CERCIS (Leguminosae)
Judas tree, Redbud
Genus of deciduous shrubs and trees with sometimes shrubby growth, cultivated for their foliage and profuse, small, pea-like flowers. Fully hardy. Requires full sun and deep, fertile, well-drained soil. Plant out as young specimens. Resents transplanting. Propagate species by seed in autumn, cultivars by budding in late summer.
C. canadensis (Eastern redbud). Deciduous, spreading tree or shrub. H and S 10m (30ft). Heart-shaped, dark green leaves turn yellow in autumn. Pea-like flowers are magenta in bud, opening to pale pink in mid-spring before leaves emerge. ℗ **'Forest Pansy'** illus. p.65.
℗ *C. siliquastrum* illus. p.62.

CEREUS (Cactaceae)
Genus of columnar, perennial cacti with spiny stems, most having 4–10 pronounced ribs. Cup-shaped flowers usually open at night. Frost tender, min. 7°C (45°F). Needs full sun and very well-drained soil. Propagate in spring by seed or, for branching species, by stem cuttings.
C. forbesii. See *C. validus*.
C. peruvianus illus. p.392.
'Monstrosus' is a columnar, perennial cactus. H 5m (15ft), S 4m (12ft). Swollen, occasionally fan-shaped, silvery-blue stems bear golden spines on 4–8 (or more) uneven ribs. Is unlikely to flower in cultivation.
C. validus, syn. *C. forbesii*, illus. p.392.

CEROPEGIA (Asclepiadaceae)
Genus of semi-evergreen, succulent shrubs and sub-shrubs, most with slender, climbing or pendent stems, grown for their unusual flowers. Frost tender, min. 7–11°C (45–52°F). Needs partial shade and very well-drained soil. Propagate by seed or stem cuttings in spring or summer. *C. woodii* is often used as grafting stock for difficult asclepiads.
C. haygarthii. Semi-evergreen, climbing, succulent sub-shrub. H 2m (6ft) or more, S indefinite. Min. 11°C (52°F). Bears oval or rounded, dark green leaves, 1–2cm (½–¾in) long, and, in summer, masses of small, white or pinkish-white flowers, each with a pitcher-shaped tube, widening towards the top and then united at the tip by purplish-spotted petals that form a short stem ending in 5 'knobs' edged with fine hairs. The whole resembles an insect hovering over a flower.
℗ *C. linearis* subsp. *woodii*, syn. *C. woodii*, illus. p.398.
C. sandersonae (Fountain flower, Parachute plant). Semi-evergreen, scrambling, succulent sub-shrub. H 2m (6ft), S indefinite. Min. 11°C (52°F). Leaves are triangular to oval, fleshy and 2cm (¾in) long. In summer-autumn has tubular, green flowers, 5cm (2in) long, with paler green to white marks; the petals are flared widely at tips to form 'parachutes'.
C. woodii. See *C. linearis* subsp. *woodii*.

CESTRUM (Solanaceae)
Genus of deciduous or evergreen shrubs and semi-scrambling climbers, grown for their showy flowers. Foliage has an unpleasant scent. Frost hardy to frost tender, min. 7–10°C (45–50°F); in cold areas grow frost hardy species against a south- or west-facing wall or in a greenhouse. Requires a sheltered, sunny position and fertile, well-drained soil. Water containerized specimens freely when in full growth, moderately at other times. Support is needed for scrambling species. Propagate frost hardy species by softwood cuttings in summer, tender species by seed in spring or by semi-ripe cuttings in summer.
C. aurantiacum. Mainly evergreen semi-scrambler that remains a rounded shrub if cut back annually. H and S to 2m (6ft). Frost tender, min. 7–10°C (45–50°F); is deciduous at low temperatures. Bears oval, bright green leaves. Tubular, bright orange flowers are carried in large, terminal trusses in summer and may be followed by spherical, white fruits. Prune annually, cutting out old stems to near base after flowering.
℗ *C. elegans* illus. p.112.
℗ *C.* **'Newellii'**, syn. *C. fasciculatum* 'Newellii'. Evergreen, arching shrub. H and S 3m (10ft). Frost hardy. Bears clusters of tubular, crimson flowers in late spring and summer. Leaves are large, broadly lance-shaped and dark green.
℗ *C. parqui*. Deciduous, open shrub. H and S 2m (6ft). Frost hardy. Large clusters of tubular, yellowish-green flowers, fragrant at night, are borne in profusion in summer amid narrowly lance-shaped, mid-green leaves.

CETERACH (Polypodiaceae)
Genus of evergreen or semi-evergreen ferns, useful for planting in crevices. Fully hardy. Needs dappled shade and moist but well-drained, chalky soil. Remove fading fronds regularly. Propagate by division in spring or by spores in late summer.
C. officinarum illus. p.187.

CHAENOMELES (Rosaceae)
Flowering quince, Japonica
Genus of deciduous, usually thorny, spring-flowering shrubs, grown for their showy flowers and fragrant fruits, used for preserves. Fully hardy. Prefers sun and well-drained soil. On wall-trained shrubs cut back side-shoots after flowering to 2 or 3 buds and shorten shoots growing away from wall during growing season. Propagate species by softwood or greenwood cuttings in summer or by seed in autumn, cultivars by cuttings only in summer. Fireblight and, on chalk soils, chlorosis are common problems.
C. cathayensis. Deciduous, spreading, open shrub with thorns. H and S 3m (10ft) or more. Produces long, narrow, pointed, mid-green leaves. Small, 5-petalled, pink-flushed, white flowers appear from early to mid-spring, followed by large, egg-shaped, yellow-green fruits.
C. japonica (Japanese quince, Japonica). Deciduous, bushy, spreading shrub with thorns. H 1m (3ft), S 2m (6ft). Has oval, mid-green leaves and, in spring, a profusion of 5-petalled, red or orange-red flowers, then spherical, yellow fruits.
C. speciosa. Vigorous, deciduous, bushy shrub with thorns. H 2.5m (8ft), S 5m (15ft). Leaves are oval, glossy and dark green. Clustered, 5-petalled, red flowers are borne from early to mid-spring, followed by spherical, greenish-yellow fruits. ℗ **'Moerloosii'** illus. p.100. Flowers of **'Nivalis'** are pure white. **'Simonii'**, H 1m (3ft), S 2m (6ft), bears masses of semi-double, deep red flowers.
℗ *C.* x *superba* **'Crimson and Gold'**. Deciduous, bushy, dense shrub with thorns. H 1m (3ft), S 2m (6ft). Has oval, glossy, dark green leaves. Bears masses of 5-petalled, deep red flowers, with conspicuous, golden-yellow anthers, in spring, followed by round, yellow fruits. Flowers of **'Etna'**, H 1.5m (5ft), S 3m (10ft), are scarlet. ℗ **'Knap Hill Scarlet'**, H 1.5m (5ft), S 3m (10ft), produces large, brilliant red flowers. ℗ **'Nicoline'** illus. p.126. ℗ **'Rowallane'** illus. p.125.

CHAMAEBATIARIA (Rosaceae)
Genus of one species of deciduous shrub, grown for its foliage and summer flowers. Frost hardy. Needs a sheltered, sunny position and well-drained soil. Propagate by semi-ripe cuttings in summer.
C. millefolium. Deciduous, upright, open shrub. H and S 1m (3ft). Has finely divided, aromatic, grey-green leaves. Shallowly cup-shaped, white flowers, with yellow stamens, are borne in terminal, branching panicles from mid- to late summer.

Chamaecereus silvestri. See *Lobivia silvestrii*.

CHAMAECYPARIS (Cupressaceae),
False cypress. See CONIFERS.
C. lawsoniana (Lawson cypress). Upright, columnar conifer with branches drooping at tips. H 15–25m (50–80ft), S 3–4m (10–12ft). Fully hardy. Bears flattened sprays of scale-like, aromatic, dark green leaves and globular cones, the males brick-red, the females insignificant and green. **'Columnaris'** illus. p.82.
℗ **'Ellwoodii'**, H 3m (10ft), S 1.5m (5ft), is erect with incurved, blue-grey leaves. ℗ **'Fletcheri'**, H 5–12m (15–40ft), S 2–3m (6–10ft), has grey leaves that are incurved. **'Gnome'**

(illus. p.85), H and S 50cm (20in), is a dwarf, bun-shaped form with blue foliage. **'Green Pillar'** illus. p.81. ♀ **'Intertexta'** illus. p.76. ♀ **'Kilmacurragh'**, H 10–15m (30–50ft), S 1m (3ft), has very bright green foliage. ♀ **'Lanei'** illus. p.81. **'Minima'** (illus. p.85), H and S 1m (3ft), is dwarf and globular, and has light green foliage. ♀ **'Pembury Blue'** illus. p.79. **'Tamariscifolia'**, H 3m (10ft), S 4m (12ft), is a dwarf, spreading form. **'Triomf van Boskoop'**, H 20m (70ft), is broadly columnar, with grey-blue foliage. ♀ **'Wisselii'**, H 15m (50ft), S 2–3m (6–10ft), is fast-growing, with erect branches and blue-green leaves.
C. nootkatensis (Nootka cypress). Almost geometrically conical conifer. H 15m (50ft), S 6m (20ft). Fully hardy. Bears long, pendent sprays of scale-like, aromatic, grey-green leaves and globular, hooked, dark blue and green cones that ripen to brown. ♀ **'Pendula'** has a gaunt crown of arching, weeping foliage.
C. obtusa (Hinoki cypress). Conical conifer. H 15–20m (50–70ft), S 5m (15ft). Fully hardy. Has stringy, red-brown bark and scale-like, aromatic, dark green leaves with bright silver lines at sides and incurving tips. Small, rounded cones ripen to yellow-brown. **'Coralliformis'**, H to 50cm (20in), S 1m (3ft), is dwarf, with thread-like shoots. ♀ **'Crippsii'** illus. p.83. **'Intermedia'** (illus. p.85), H to 30cm (12in), S 40cm (16in), is a globular, open, dwarf shrub with downward-spreading, light green foliage. ♀ **'Kosteri'**, H 1–2m (3–6ft), S 2–3m (6–10ft), forms a sprawling bush with twisted, lustrous foliage. Extremely slow-growing ♀ **'Nana'**, eventual H 1m (3ft), S 1.5–2m (5–6ft), makes a flat-topped bush. ♀ **'Nana Aurea'** (illus. p.85), H and S 2m (6ft), is a form with golden-yellow leaves. ♀ **'Nana Gracilis'**, H 2m (6ft), S 1.5–2m (5–6ft), is a form with glossy foliage. **'Nana Pyramidalis'** (illus. p.85), H and S to 60cm (24in), is a slow-growing, dense, conical, dwarf cultivar with horizontal, cup-shaped leaves. ♀ **'Tetragona Aurea'**, H 10m (30ft), S 2–3m (6–10ft), has golden- or bronze-yellow leaves.
C. pisifera (Sawara cypress). Conical conifer with horizontal branches. H 15m (50ft), S 5m (15ft). Fully hardy. Has ridged, peeling, red-brown bark, scale-like, aromatic, fresh green leaves, white at sides and beneath, and angular, yellow-brown cones. ♀ **'Boulevard'** has silver-blue foliage. **'Filifera'** has whip-like, hanging shoots and dark green foliage. ♀ **'Filifera Aurea'** (illus. p.85), H 12m (40ft), S 3–5m (10–15ft), also has whip-like shoots, but with golden-yellow leaves. **'Filifera Nana'**, H 60cm (2ft), S 1m (3ft), is a dwarf form with whip-like branches. **'Nana'**, H and S 50cm (20in), is also dwarf, with dark bluish-green foliage. **'Plumosa'** is broadly conical to columnar, with yellowish-grey-green leaves. **'Plumosa Rogersii'**, H 2m (6ft), S 1m (3ft), has yellow foliage. Slow-growing **'Squarrosa'**, H to 20m (70ft), has a broad crown and soft, blue-grey foliage.
C. thyoides illus. p.81. ♀ **'Andelyensis'** is a slow-growing, conical, dwarf conifer. H 3m (10ft), S 1m (3ft). Fully hardy. Has wedge-shaped tufts of scale-like, aromatic, blue-green leaves. Globular cones are glaucous blue-grey.

Chamaecytisus hirsutus. See *Cytisus demissus.*
Chamaecytisus purpureus. See *Cytisus purpureus.*
Chamaecytisus purpureus f. *albus.* See *Cytisus purpureus* f. *albus.*

CHAMAEDAPHNE (Ericaceae)
Genus of one species of evergreen shrub, grown for its white flowers. Fully hardy. Needs sun or semi-shade and moist, peaty, acid soil. Propagate by semi-ripe cuttings in summer.
C. calyculata (Leatherleaf). Evergreen, arching, open shrub. H 75cm (2½ft), S 1m (3ft). Leaves are small, oblong, leathery and dark green. Leafy racemes of small, urn-shaped flowers are borne on slender branches from mid- to late spring.

CHAMAEDOREA (Palmae)
Genus of evergreen palms, grown for their overall appearance. Frost tender, min. 18°C (64°F). Needs shade or semi-shade and humus-rich, well-drained soil. Water containerized plants moderately, less when temperatures are low. Propagate by seed in spring at not less than 25°C (77°F). Red spider mite may be troublesome.
♀ *C. elegans*, syn. *Neanthe bella*, illus. p.122.

CHAMAEMELUM (Compositae)
Genus of evergreen perennials, suitable as ground cover or to make a lawn. Flowers may be used to make tea. Fully hardy. Needs sun and well-drained soil. Propagate by division in spring or by seed in autumn.
C. nobile (Chamomile). Evergreen, mat-forming, invasive perennial. H 10cm (4in), S 45cm (18in). Has finely divided, aromatic leaves and daisy-like heads of white flowers, with yellow centres, borne in late spring or summer. **'Treneague'** (syn. *Anthemis nobile* 'Treneague') is a non-flowering, less invasive cultivar that, requiring less mowing and lacking flower heads, is better for a lawn.

CHAMAENERION. See EPILOBIUM.

CHAMAEROPS (Palmae)
Genus of evergreen palms, cultivated for their overall appearance. Half hardy to frost tender, min. 7°C (45°F). Needs full light and fertile, well-drained soil. Water containerized plants moderately, less when not in full growth. Propagate by seed in spring at not less than 22°C (72°F) or by suckers in late spring. Red spider mite may be a nuisance.
♀ *C. humilis* illus. p.146.

Chamaespartium sagittale. See *Genista sagittalis.*

CHAMELAUCIUM (Myrtaceae)
Genus of evergreen shrubs, grown for their flowers and overall appearance. Frost tender, min. 5°C (41°F), but best at 7–10°C (45–50°F). Requires full sun and well-drained, sandy, neutral to acid soil. Water containerized specimens moderately, sparingly when not in full growth. To maintain a more compact habit, cut back flowered stems by half when the last bloom falls. Propagate by seed in spring or by semi-ripe cuttings in summer.
C. uncinatum [pink form] illus. p.120, [white form] illus. p.119.

CHASMANTHE (Iridaceae)
Genus of corms, grown for their showy flowers. Frost to half hardy. Needs full sun or partial shade and well-drained soil with plenty of water in growing season (late winter and early spring). Reduce watering in summer-autumn. Propagate by division in autumn.
C. aethiopica. Spring- and early summer-flowering corm. H to 80cm (32in), S 12–18cm (5–7in). Frost hardy. Has narrowly sword-shaped, erect, basal leaves in a flat fan. Produces a spike of scarlet flowers, all facing one way, with yellow tubes, 5–6cm (2–2½in) long, and hooded, upper lips.
C. floribunda. Summer-flowering corm. H to 80cm (32in), S 12–18cm (5–7in). Half hardy. Is similar to *C. aethiopica*, but the leaves are much wider, and the longer, orange or scarlet flowers do not all face in the same direction.

CHEILANTHES (Polypodiaceae)
Genus of evergreen ferns. Half hardy. Needs full light and humus-rich, well-drained soil. Do not overwater containerized plants or splash water on fronds. Remove fading foliage regularly. Propagate by spores in summer.
C. lanosa. Evergreen fern. H and S 15–23cm (6–9in). Leaves are triangular or lance-shaped and have much divided, soft green fronds on hairy, black stems.
C. pteriodes, syn. *C. fragrans.* Evergreen fern. H 8–23cm (3–9in), S 15–23cm (6–9in). Lance-shaped, dark green fronds, with many rounded pinnae, are slightly hairy beneath and have the fragrance of fresh violets.

CHEIRANTHUS. See ERYSIMUM.

CHEIRIDOPSIS (Aizoaceae)
Genus of clump-forming, perennial succulents with semi-cylindrical, green, glaucous-green or blue-grey leaves in pairs. Frost tender, min. 5°C (41°F). Needs sun and well-drained soil. Water in autumn to encourage flowers. Propagate by seed or stem cuttings in spring or summer.
C. candidissima. See *C. denticulata.*
C. denticulata, syn. *C. candidissima.* Clump-forming, perennial succulent. H 10cm (4in), S 20cm (8in). Has semi-cylindrical, slender, fleshy, blue-grey leaves, each 10cm (4in) long with a flat top, joined in pairs for almost half their length. Daisy-like, shiny, white flowers, to 6cm (2½in) across, appear in spring.
C. purpurata. See *C. purpurea.*
C. purpurea, syn. *C. purpurata.* Carpeting, perennial succulent. H 10cm (4in), S 30cm (12in). Has semi-cylindrical, thick, short, glaucous green leaves, each with a flat top. In early spring produces daisy-like, purple-pink flowers, 4cm (1½in) across.

CHELIDONIUM (Papaveraceae)
Celandine, Greater celandine
Genus of one species of perennial that rapidly forms ground cover. Fully hardy. Grows in sun or shade and in any but very wet soil. Propagate by seed or division in autumn.
C. majus **'Flore Pleno'** illus. p.202.

CHELONE (Scrophulariaceae)
Turtle-head
Genus of summer- and autumn-flowering perennials. Fully hardy. Requires semi-shade and moist soil. Propagate by soft-tip cuttings in summer or by division or seed in autumn or spring.
C. barbata. See *Penstemon barbatus.*
C. obliqua illus. p.223.

CHIASTOPHYLLUM (Crassulaceae)
Genus of one species of evergreen perennial, grown for its succulent leaves and attractive sprays of small, yellow flowers. Thrives in rock crevices. Fully hardy. Needs shade and well-drained soil that is not too dry. Propagate by side-shoot cuttings in early summer or by seed in autumn.
♀ *C. oppositifolium*, syn. *Cotyledon simplicifolia*, illus. p.297.

CHIMONANTHUS (Calycanthaceae)
Genus of deciduous or evergreen, winter-flowering shrubs, grown for their flowers. Frost hardy, but in cold areas reduce susceptibility of flowers to frost by training plants against a south- or west-facing wall. Needs full sun and fertile, well-drained soil. Propagate species by seed when ripe, in late spring and early summer, cultivars by softwood cuttings in summer.
C. praecox (Wintersweet). Deciduous, bushy shrub. H 2.5m (8ft) or more, S 3m (10ft). Has oval, rough, glossy, dark green leaves. Bears very fragrant, many-petalled, cup-shaped, yellow flowers, with purple centres, on bare branches in mild periods during winter. ♀ **'Luteus'** has pure yellow flowers.

CHIONANTHUS (Oleaceae)
Genus of deciduous shrubs, grown for their profuse, white flowers. Flowers more freely in areas with hot summers. Fully hardy. Prefers full sun and fertile, well-drained but not too dry soil. Propagate by seed in autumn.
C. retusus (Chinese fringe tree). Deciduous, often tree-like, arching shrub. H and S 3m (10ft). From early to mid-summer, star-shaped, pure white flowers appear in large clusters amid oval, bright green leaves.
C. virginicus illus. p.89.

CHIONOCHLOA (Gramineae). See GRASSES, BAMBOOS, RUSHES and SEDGES.
C. conspicua (Hunangemoho grass). Evergreen, tussock-forming, perennial

grass. H 1.2–1.5m (4–5ft), S 1m (3ft). Fully hardy. Very long, mid-green leaves are tinged reddish-brown. Has stout, arching stems with long, loose, open panicles of cream spikelets in summer.

CHIONODOXA (Liliaceae)
Glory-of-the-snow
Genus of spring-flowering bulbs, related to *Scilla*. Is suitable for rock gardens and for naturalizing under shrubs, in sun or partial shade. Fully hardy. Requires well-drained soil, top dressed with leaf mould or mature garden compost in autumn. Propagate by seed in autumn or by division in late summer or autumn.
C. gigantea. See *C. luciliae.*
❦ *C. luciliae*, syn. *C. gigantea*, illus. p.377.
C. luciliae of gardens. See *C. siehei.*
❦ *C. sardensis.* Early spring-flowering bulb. H 10–20cm (4–8in), S 2.5–5cm (1–2in). Has 2 narrowly lance-shaped, semi-erect, basal leaves. Leafless stem produces 5–10 flattish, slightly pendent or outward-facing, deep rich blue flowers, 1.5–2cm ($5/8$–$3/4$in) across and without white eyes.
❦ *C. siehei*, syn. *C. luciliae* of gardens, *C. tmolusii*, illus. p.377. **'Pink Giant'** illus. p.371.
C. tmolusii. See *C. siehei.*

x CHIONOSCILLA (Liliaceae)
Hybrid genus (*Chionodoxa* x *Scilla*) of spring-flowering bulbs, suitable for rock gardens. Fully hardy. Needs full sun or partial shade and humus-rich, well-drained soil. Propagate by division in late summer or autumn.
x *C. allenii* illus. p.377.

CHIRITA (Gesneriaceae)
Genus of evergreen perennials or sub-shrubs, grown for their flowers. Frost tender, min. 15°C (59°F). Requires well-drained soil, a fairly humid atmosphere and a light position out of direct sunlight. Propagate by tip cuttings in summer or, if available, seed in late winter or spring.
C. lavandulacea illus. p.258.
❦ *C. sinensis.* Evergreen, stemless, rosetted perennial. H to 15cm (6in), S 25cm (10in) or more. Has oval, almost fleshy leaves, the corrugated, hairy surfaces usually patterned with silver marks. In spring-summer, clusters of tubular, lavender flowers are held above leaves.

CHLIDANTHUS (Amaryllidaceae)
Genus of one species of summer-flowering bulb, grown for its showy, funnel-shaped flowers. Half hardy. Needs a sunny site and well-drained soil. Plant in the open in spring and after flowering, if necessary, lift and dry off for winter. Propagate by offsets in spring.
C. fragrans illus. p.380.

CHLOROGALUM (Liliaceae)
Genus of summer-flowering bulbs, grown more for botanical interest than for floral display. Frost hardy, but in cold areas plant in a sheltered site. Requires sun and well-drained soil. Propagate by seed in autumn or spring.

C. pomeridianum. Summer-flowering bulb. H to 2.5m (8ft), S 15–20cm (6–8in). Semi-erect, basal leaves are long, narrow and grey-green, with wavy margins. Carries a large, loosely branched head of small, saucer-shaped, white flowers, with a central, green or purple stripe on each petal, that open after midday.

CHLOROPHYTUM (Liliaceae)
Genus of evergreen, stemless perennials with short rhizomes, grown for their foliage. Frost tender, min. 5°C (41°F). Grow in a light position, away from direct sun, in fertile, well-drained soil. Water freely in growing season but sparingly at other times if pot-grown. Propagate by seed, division or plantlets (produced on flower stems of some species) at any time except winter.
C. capense of gardens. See *C. comosum.*
C. comosum, syn. *C. capense* of gardens (Spider plant). Evergreen, tufted perennial. H 30cm (12in), S indefinite. Very narrow leaves, to 45cm (18in) long, spread from a rosette. Racemes of many small, star-shaped, white flowers are carried on thin stems, to 60cm (2ft) or more long, at any time. Small rosettes of leaves may appear on flower stems, forming plantlets. ❦ **'Vittatum'** illus. p.265.

CHOISYA (Rutaceae)
Genus of evergreen shrubs, grown for their foliage and flowers. Frost to half hardy; in most areas needs some shelter. Requires full sun and fertile, well-drained soil. Propagate by semi-ripe cuttings in late summer.
❦ *C. ternata* (Mexican orange blossom) illus. p.97. ❦ **'Sundance'** is an evergreen, rounded, dense shrub. H and S 2.5m (8ft). Frost hardy. Aromatic, glossy, bright yellow leaves each consist of 3 oblong leaflets. Fragrant, star-shaped, white flowers are produced in clusters in late spring and often again in autumn.

CHORDOSPARTIUM (Leguminosae)
Genus of one species of deciduous, almost leafless shrub, grown for its habit and flowers. Slender, green shoots assume function of leaves. Frost hardy. Requires a sheltered, sunny position and fertile, well-drained soil. Propagate by seed in autumn.
C. stevensonii. Deciduous, almost leafless, arching shrub. H 3m (10ft), S 2m (6ft). Produces small, pea-like, purplish-pink flowers in cylindrical racemes in mid-summer.

CHORISIA (Bombacaceae)
Genus of deciduous trees, usually with spine-covered trunks, grown mainly for their flowers in autumn and winter and their overall appearance. Frost tender, min. 15°C (59°F). Needs full light and well-drained soil. Water containerized specimens freely when in full growth, very little when leafless. Pruning is tolerated if necessary. Propagate by seed in spring. Red spider mite may be troublesome.
C. speciosa illus. p.44.

CHORIZEMA (Leguminosae)
Genus of evergreen sub-shrubs, shrubs and scandent climbers, grown mainly for their flowers. Frost tender, min. 7°C (45°F). Requires full light and humus-rich, well-drained, sandy soil, preferably neutral to acid. Water potted plants moderately, less when not in full growth. Tie climbers to supports, or grow in hanging baskets. Propagate by seed in spring or by semi-ripe cuttings in summer.
C. ilicifolium illus. p.127.

CHRYSALIDOCARPUS (Palmae)
Genus of evergreen palms, grown for their elegant appearance. Frost tender, min. 16°C (61°F). Needs full light or partial shade and fertile, well-drained soil. Water potted specimens moderately, much less when temperatures are low. Propagate by seed in spring at not less than 26°C (79°F). Red spider mite may sometimes be a nuisance.
❦ *C. lutescens*, syn. *Areca lutescens*, illus. p.74.

CHRYSANTHEMUM (Compositae)
Genus of annuals, perennials, some of which are evergreen, and evergreen sub-shrubs, grown for their flowers. Each flower head is referred to horticulturally as a flower, even though it does in fact comprise a large number of individual flowers or florets; this horticultural usage has been followed in the descriptions below. Leaves are usually deeply lobed or cut, often feathery, oval to lance-shaped. Florists' chrysanthemums (nowadays considered to belong to the genus *Dendranthema*) comprise the vast majority of chrysanthemums now cultivated and are perennials grown for garden decoration, cutting and exhibition. Annuals are fully to half hardy. Florists' chrysanthemums are half hardy and should be lifted and stored in a frost-free place over winter. Other perennial chrysanthemums are fully to half hardy. Provide a sunny site and reasonably fertile, well-drained soil. If grown for exhibition will require regular feeding. Pinch out growing tips to encourage lateral growths on which flowers will be borne, and stake tall plants with canes. Propagate annuals by seed sown in position in spring; thin out, but do not transplant. Propagate hardy perennials by division in autumn, after flowering, or in early spring. Florists' chrysanthemums should be propagated from basal softwood cuttings in spring. Spray regularly to control aphids, capsids, froghoppers, earwigs, mildew and white rust.

Florists' chrysanthemums
Florists' chrysanthemums are grouped according to their widely varying flower forms, approximate flowering season (early, mid- or late autumn) and habit. They are divided into disbudded and non-disbudded types.

Disbudded types – single, anemone-centred, incurved, intermediate and reflexed – are so called as all buds, except the one that is to flower, are removed from each stem. To produce exhibition flowers, incurved, intermediate and reflexed chrysanthemums may be restricted to only 2 blooms per plant by removing all except the 2 most vigorous laterals. In gardens, allow 4 or 5 blooms per plant to develop. Single and anemone-centred flowers should be reduced to 4–8 blooms per plant for exhibition, according to their vigour, and 10 or more for garden decoration or cutting. For descriptions and illustrations of flower forms see p.224.

Non-disbudded types – charm, pompon and spray chrysanthemums – have several flowers per stem.
Charm chrysanthemums are dwarf plants that produce hundreds of star-shaped, single flowers, 2.5cm (1in) across, densely covering each plant to form a hemispherical to almost spherical head. For exhibition, finish growing in at least 30cm (12in) pots. Plants for indoor decoration are grown in smaller pots and have correspondingly smaller, though equally dense, heads of blooms.
Pompon chrysanthemums are also dwarf. Each plant has 50 or more dense, spherical or occasionally hemispherical, fully double flowers that have tubular petals (for illustration see p.224). They are excellent for growing in borders.
Spray chrysanthemums have a variety of flower forms: single, anemone-centred, intermediate, reflexed, pompon and spoon-type (in which each straight, tubular floret opens out like a spoon at its tip). Each plant should be allowed to develop 4 or 5 stems with at least 5 flowers per stem. Grow late-flowering sprays on up to 3 stems per plant. With controlled day length, to regulate flowering dates for exhibition purposes, late sprays should be allowed to develop at least 12 flowers per stem; without day length control, 6 or 7 flowers per stem.

Those most suitable for garden decoration are sprays, pompons and early reflexed chrysanthemums. All are suitable for cutting, except for charms. Late-flowering chrysanthemums are only suitable for growing under glass as flowers need protection from poor weather; they should be grown in pots and placed in a greenhouse in early autumn, when the flower buds have developed. Intermediate cultivars are also less suitable for garden decoration as florets may collect and retain rain and thus become damaged. Those cultivars suitable for exhibition are noted below. Measurements of flowers given are the greatest normally achieved and may vary considerably depending on growing conditions.

C. **'Alison Kirk'** (illus. p.224). Incurved florists' chrysanthemum. H 1.2m (4ft), S 30–60cm (1–2ft). Produces white flowers, to 12–15cm (5–6in) across, in early autumn. Is more suitable for exhibition than for garden use.
C. alpinum, syn. *Leucanthemopsis*

CHRYSANTHEMUM

alpinum. Tuft-forming, short-lived perennial. H 7cm (3in), S 12cm (5in). Fully hardy. Small tufts of deeply cut leaves are produced from short, rhizomatous stems. Has large, daisy-like, white flower heads, with yellow centres, in summer. Is good for a rock garden or scree. Propagate by division of underground stems in autumn or early spring.

C. 'Amber Yvonne Arnaud'. Reflexed florists' chrysanthemum. H 1.2m (4ft), S 60–75cm (2–2½ft). Is a sport of C. 'Yvonne Arnaud' with fully reflexed, amber flowers in early autumn.

C. 'Autumn Days' (illus. p.225). Intermediate florists' chrysanthemum. H 1.1–1.2m (3½–4ft), S to 75cm (2½ft). Bears loosely incurving, bronze flowers, 12cm (5in) across, in early autumn.

C. 'Bill Wade'. Intermediate florists' chrysanthemum. H 1.35m (4½ft), S 60cm (2ft). Has loosely incurving, white flowers, 18–20cm (7–8in) across, in early autumn. Is more suitable for exhibition than for garden use.

C. 'Brietner' (illus. p.224). Reflexed florists' chrysanthemum. H 1.1–1.2m (3½–4ft), S 75cm (2½ft). Fully reflexed, pink flowers, to 12cm (5in) wide, appear in early autumn.

C. 'Bronze Fairie' (illus. p.225). Pompon florists' chrysanthemum. H 30–60cm (1–2ft), S 60cm (2ft). Has bronze flowers, 4cm (1 in) across, in early autumn.

C. 'Bronze Hedgerow' (illus. p.225). Single florists' chrysanthemum. H 1.5m (5ft), S 75cm–1m (2½–3ft). Produces bronze flowers, 12cm (5in) across, in late autumn.

C. 'Bronze Yvonne Arnaud' (illus. p.225). Reflexed florists' chrysanthemum. H 1.2m (4ft), S 60–75cm (2–2½ft). Is a sport of C. 'Yvonne Arnaud' with fully reflexed, bronze flowers in early autumn.

C. 'Buff Margaret' (illus. p.225). Spray florists' chrysanthemum. H 1.2m (4ft), S to 75cm (2½ft). Is similar to C. 'Peach Margaret', but has pale bronze flowers.

C. carinatum, syn. C. tricolor. 'Monarch Court Jesters' (red with yellow centres) illus. p.281, (white with red centres) illus. p.272. Tricolor Series is a group of fast-growing, upright, branching annuals. H 30–60cm (12–24in), S 30cm (12in). Half hardy. Has feathery, light green leaves and, in summer, daisy-like, single or double flower heads, to 8cm (3in) wide, in many colour combinations. Tall cultivars, H 60cm (24in), S 30cm (12in), and dwarf, H and S 30cm (12in), are available.

C. 'Cherry Chintz' (illus. p.225). Reflexed florists' chrysanthemum. H 1.2m (4ft), S 45cm (1½ft). Has fully reflexed, red flowers, to 15cm (6in) across, in early autumn. Is excellent for exhibition.

C. 'Chessington'. Intermediate florists' chrysanthemum. H 2–2.2m (6–7ft), S 75cm (2½ft). Produces fairly tightly incurving, white flowers, 18–20cm (7–8in) across, in early autumn. Is more suitable for exhibition than for garden use.

C. 'Chippendale' (illus. p.225). Reflexed florists' chrysanthemum. H 1.2m (4ft), S 45cm (1½ft). Fully reflexed, pink flowers, 18cm (7in) or more wide, appear in mid-autumn. Is more suitable for exhibition than for garden use.

C. 'Christina'. Intermediate florists' chrysanthemum. H 1.35–1.5m (4½–5ft), S 60–75cm (2–2½ft). Bears loosely incurving, white flowers, to 14cm (5½in) wide, in early autumn. Is suitable for exhibition.

C. 'Claire Louise'. Reflexed florists' chrysanthemum. H 1.2–1.35m (4–4½ft), S 75cm (2½ft). Produces fully reflexed, bronze flowers, to 15cm (6in) across, in early autumn. Is ideal for exhibition.

C. 'Cloudbank' (illus. p.224). Spray florists' chrysanthemum. H 1.2m (4ft), S 75cm (2½ft). Has anemone-centred, white flowers, 8–9cm (3–3½in) wide, in late autumn. Is good for exhibition.

C. coccineum. See Tanacetum coccineum.

C. coronarium. Fast-growing, upright, branching annual. H 30cm–1m (1–3ft), S 38cm (15in). Fully hardy. Has feathery, divided, light green leaves. In summer bears single or semi-double, daisy-like, yellow or yellow-and-white flower heads, to 5cm (2in) across.

C. 'Dawn Mist' (illus. p.224). Spray florists' chrysanthemum. H 1.2m (4ft), S 75cm–1m (2½–3ft). Single, very pale pink flowers, to 8cm (3in) across, appear in early autumn.

C. 'Discovery' (illus. p.225). Intermediate florists' chrysanthemum. H 1.2m (4ft), S 75cm (2½ft). Produces loosely incurving, light yellow flowers, 10–12cm (4–5in) across, in early autumn.

C. 'Dorridge Dream' (illus. p.224). Incurved florists' chrysanthemum. H 1.2m (4ft), S 60cm (2ft). Has rose-pink flowers, 12cm (5in) wide, in early autumn. Is excellent for exhibition.

C. 'Duke of Kent' (illus. p.224). Reflexed florists' chrysanthemum. H 1.5m (5ft), S 30cm (1ft). Bears almost fully reflexed, white flowers, 25cm (10in) or more across, in late autumn. Use only for exhibition.

C. 'Edwin Painter' (illus. p.225). Single florists' chrysanthemum. H 1.35–1.5m (4½–5ft), S 75cm–1m (2½–3ft). Produces yellow flowers, 14cm (5½in) across, in late autumn. Is good for exhibition.

C. 'Fiona Lynn'. Reflexed florists' chrysanthemum. H 1.5m (5ft), S 75cm (2½ft). Fully reflexed, pink flowers, to 18–20cm (7–8in) across, appear in early autumn. Is ideal for exhibition.

C. frutescens, syn. Argyranthemum frutescens (Marguerite), illus. p.204.

♥ 'Jamaica Primrose' illus. p.219.

'Mary Wootton' illus. p.207.

C. 'George Griffiths' (illus. p.225). Reflexed florists' chrysanthemum. H 1.2–1.35m (4–4½ft), S 75cm (2½ft). Produces fully reflexed, deep red flowers, to 14cm (5½in) wide, in early autumn. Is excellent for exhibition.

C. 'Ginger Nut'. Intermediate florists' chrysanthemum. H 1.2m (4ft), S 60–75cm (2–2½ft). Bears tightly incurving, light bronze flowers, to 14cm (5½in) across, occasionally closing at top to form a true incurved flower, in early autumn. Is good for exhibition.

C. 'Golden Woolman's Glory' (illus. p.225). Single florists' chrysanthemum. H 1.5m (5ft), S 1m (3ft). Golden flowers, to 18cm (7in) across, appear in late autumn. Is excellent for exhibition.

C. 'Green Satin' (illus. p.225). Intermediate florists' chrysanthemum. H 1.2m (4ft), S 60cm (2ft). Produces loosely incurving, green flowers, to 12cm (5in) wide, in early autumn.

C. haradjanii. See Tanacetum haradjanii.

C. hosmariense, syn. Pyrethropsis hosmariense, illus. p.294.

♥ C. 'Madeleine' (illus. p.224). Spray florists' chrysanthemum. H 1.2m (4ft), S 75cm (2½ft). Has reflexed, pink flowers, to 8cm (3in) across, in early autumn. Is good for exhibition.

C. 'Maria' (illus. p.225). Pompon florists' chrysanthemum. H 45cm (1½ft), S 30–60cm (1–2ft). Bears masses of pink flowers, to 4cm (1½in) across, in early autumn.

C. 'Marian Gosling' (illus. p.224). Reflexed florists' chrysanthemum. H 1.2–1.35m (4–4½ft), S 60cm (2ft). Fully reflexed, pale pink flowers, to 14cm (5½in) wide, appear in early autumn. Is good for exhibition.

C. 'Marion' (illus. p.225). Spray florists' chrysanthemum. H 1.2m (4ft), S 75cm (2½ft). Produces reflexed, pale yellow flowers, to 8cm (3in) wide, from late summer.

C. 'Marlene Jones' (illus. p.225). Intermediate florists' chrysanthemum. H 1m (3ft), S to 60cm (2ft). Has loosely incurving, pale yellow flowers, 12–14cm (5–5½in) wide, in early autumn. Is good for exhibition.

C. 'Mason's Bronze'. Single florists' chrysanthemum. H 1.35–1.5m (4½–5ft), S to 1m (3ft). Has bronze flowers, to 12cm (5in) wide, in late autumn. Is excellent for exhibition.

C. maximum of gardens. See Leucanthemum × superbum.

C. 'Michael Fish' (illus. p.224). Intermediate florists' chrysanthemum. H 1.2m (4ft), S 60–75cm (2–2½ft). Bears fairly tightly incurving, white flowers, to 15cm (6in) wide, in early autumn. Is suitable for exhibition.

C. 'Oracle' (illus. p.225). Intermediate florists' chrysanthemum. H 1.2m (4ft), S 60–75cm (2–2½ft). Produces loosely incurving, pale bronze flowers, to 12cm (5in) wide, in early autumn. Is useful for exhibition.

C. parthenium. See Tanacetum parthenium.

C. 'Pavilion' (illus. p.224). Intermediate florists' chrysanthemum. H 1.35m (4½ft), S 60–75cm (2–2½ft). Loosely incurving, white flowers, 18cm (7in) wide, appear in early autumn. Is more suitable for exhibition than for garden use.

C. 'Peach Brietner' (illus. p.225). Reflexed florists' chrysanthemum. H 1.1–1.2m (3½–4ft), S 75cm (2½ft). Is a sport of C. 'Brietner' with fully reflexed, peach-coloured flowers.

C. 'Peach Margaret' (illus. p.225). Spray florists' chrysanthemum. H 1.2m (4ft), S to 75cm (2½ft). Bears reflexed, pale salmon flowers, 8–9cm (3–3½in) wide, in early autumn. Is excellent for exhibition.

C. 'Pennine Alfie' (illus. p.225). Spray florists' chrysanthemum. H 1.2m (4ft), S 60–75cm (2–2½ft). Spoon-type, pale bronze flowers, to 6–8cm (2½–3in) wide, appear in early autumn. Is suitable for exhibition.

♥ C. 'Pennine Flute' (illus. p.224). Spray florists' chrysanthemum. H 1.2m (4ft), S 60–75cm (2–2½ft). Is similar to C. 'Pennine Alfie', but has pink flowers.

C. 'Pennine Jewel' (illus. p.225). Spray florists' chrysanthemum. H 1.2m (4ft), S 60–75cm (2–2½ft). Spoon-type, pale bronze flowers, 6–8cm (2½–3in) wide, appear in early autumn. Is suitable for exhibition.

C. 'Pennine Oriel' (illus. p.224). Spray florists' chrysanthemum. H 1.2m (4ft), S 60–75cm (2–2½ft). Has anemone-centred, white flowers, to 9cm (3½in) across, in early autumn. Is very good for exhibition.

C. 'Peter Rowe'. Incurved florists' chrysanthemum. H 1.35m (4½ft), S 60–75cm (2–2½ft). Produces yellow flowers, to 14cm (5½in) across, in early autumn. Is ideal for exhibition.

C. 'Primrose Fairweather'. Incurved florists' chrysanthemum. H 1–1.1m (3–3½ft), S to 75cm (2½ft). Produces pale yellow flowers, to 14–15cm (5½–6in) wide, in late autumn. Is good for exhibition.

C. 'Primrose West Bromwich' (illus. p.225). Reflexed florists' chrysanthemum. H 2.2m (7ft), S 45–60cm (1½–2ft). Fully reflexed, pale yellow flowers, to 18cm (7in) or more wide, appear in mid-autumn. Use only for exhibition.

♥ C. 'Purple Pennine Wine' (illus. p.225). Spray florists' chrysanthemum. H 1.2m (4ft), S 60–75cm (2–2½ft). Bears reflexed, purplish-red flowers, to 8cm (3in) wide, in early autumn. Is very good for exhibition.

C. 'Redwing' (illus. p.225). Spray florists' chrysanthemum. H 1.2–1.35m (4–4½ft), S 60cm (2ft). Bears single, red flowers, to 8cm (3in) wide, in early autumn. Is suitable for exhibition.

C. 'Ringdove' (illus. p.224). Charm florists' chrysanthemum. H and S 1m (3ft). Has masses of pink flowers, 2.5cm (1in) across, in late autumn. Is excellent for exhibition.

C. 'Roblush' (illus. p.224). Spray florists' chrysanthemum. H 1.5m (5ft), S 75cm–1m (2½–3ft). Has reflexed, pale pink flowers, to 8cm (3in) wide, in late autumn. Is good for exhibition.

C. 'Rose Yvonne Arnaud' (illus. p.225). Reflexed florists' chrysanthemum. H 1.2m (4ft), S 60–75cm (2–2½ft). Is a sport of C. 'Yvonne Arnaud' with fully reflexed, red flowers in early autumn.

C. rubellum, syn. C. zawadskii var. latilobum, Dendranthema zawadskii. 'Clara Curtis' illus. p.223.

C. 'Rytorch' (illus. p.225). Spray florists' chrysanthemum. H 1.5m (5ft), S 75cm–1m (2½–3ft). In late autumn produces single, pale bronze flowers, to 8cm (3in) wide with a yellow ring surrounding the central disc. Is good for exhibition.

C. **'Sally Ball'** (illus. p.225). Spray florists' chrysanthemum. H 1.2m (4ft), S 75cm (2½ft). Anemone-centred, bronze flowers, to 8cm (3in) wide with a yellow-bronze, central disc, appear in early autumn. Is suitable for exhibition.
C. **'Salmon Fairie'** (illus. p.225). Pompon florists' chrysanthemum. H 30–60cm (1–2ft), S 60cm (2ft). Is similar to *C.* 'Bronze Fairie', but has salmon flowers.
C. **'Salmon Fairweather'** (illus. p.224). Incurved florists' chrysanthemum. H 1–1.1m (3–3½ft), S to 75cm (2½ft). Is similar to *C.* 'Primrose Fairweather', but has pale salmon flowers.
♥ *C.* **'Salmon Margaret'** (illus. p.225). Spray florists' chrysanthemum. H 1.2m (4ft), S to 75cm (2½ft). Is similar to *C.* 'Peach Margaret', but has salmon flowers.
C. segetum illus. p.289.
C. **'Sentry'** (illus. p.225). Reflexed florists' chrysanthemum. H 1.2–1.35m (4–4½ft), S 60–75cm (2–2½ft). Bears fully reflexed, deep red flowers, to 12cm (5in) wide, in early autumn. Is excellent for exhibition.
C. **'Skater's Waltz'** (illus. p.225). Intermediate florists' chrysanthemum. H 1.5m (5ft), S 60–75cm (2–2½ft). Produces loosely incurving, pink flowers, to 15–18cm (6–7in) wide, in late autumn.
C. × *superbum*. See *Leucanthemum* × *superbum*.
C. **'Talbot Jo'** (illus. p.225). Spray florists' chrysanthemum. H 1.35m (4½ft), S 60–75cm (2–2½ft). Bears single, pink flowers, to 8cm (3in) wide, in early autumn. Is suitable for exhibition.
C. tricolor. See *C. carinatum*.
♥ *C. uliginosum* (sometimes included in *Leucanthemella* as *L. serotina*). Erect perennial. H 2.2m (7ft), S 60cm (2ft). Fully hardy. Lance-shaped leaves are lobed and dark green. Leafy stems carry sprays of large, daisy-like, green-centred, white flower heads in late autumn.
C. **'Venice'**. Reflexed florists' chrysanthemum. H 1.2m (4ft), S 60–75cm (2–2½ft). Has reflexed, pink flowers, to 15cm (6in) wide, in early autumn. Is good for exhibition.
♥ *C.* **'Wendy'** (illus. p.225). Spray florists' chrysanthemum. H 1.2m (4ft), S 60–75cm (2–2½ft). Produces reflexed, pale bronze flowers, to 8cm (3in) wide, in early autumn. Is excellent for exhibition.
C. **'Yellow Brietner'** (illus. p.225). Reflexed florists' chrysanthemum. H 1.1–1.2m (3½–4ft), S 75cm (2½ft). Is a sport of *C.* 'Brietner' with fully reflexed, yellow flowers in early autumn.
C. **'Yellow John Hughes'** (illus. p.225). Incurved florists' chrysanthemum. H 1.2m (4ft), S 60–75cm (2–2½ft). Yellow flowers, to 12–14cm (5–5½in) wide, appear in late autumn. Is excellent for exhibition.
C. **'Yvonne Arnaud'** (illus. p.225). Reflexed florists' chrysanthemum. H 1.2m (4ft), S 60–75cm (2–2½ft). Bears fully reflexed, purple flowers, to 12cm (5in) wide, in early autumn.

C. zawadskii var. *latilobum*. See *C. rubellum*.

CHRYSOGONUM (Compositae)
Genus of one species of summer- to autumn-flowering perennial. Suits a rock garden. Fully hardy. Needs partial shade and moist but well-drained, peaty, sandy soil. Propagate by division in spring or by seed when fresh.
C. virginianum illus. p.307.

CHUSQUEA (Bambusoideae). See GRASSES, BAMBOOS, RUSHES and SEDGES.
♥ *C. culeou* illus. p.184.

CICERBITA, syn. **MULGEDIUM** (Compositae)
Genus of perennials, grown for their attractive flower heads. Fully hardy. Requires shade and damp but well-drained soil. Propagate by division in spring or by seed in autumn. Some species may be invasive.
C. alpina, syn. *Lactuca alpina* (Mountain sow thistle). Branching, upright perennial. H to 2m (6ft), S 60cm (2ft). Mid-green leaves are lobed with a large, terminal lobe. Elongated panicles of thistle-like, pale blue flower heads appear in summer.
C. bourgaei, syn. *Lactuca bourgaei*. Rampant, erect perennial. H to 2m (6ft), S 60cm (2ft). Leaves are oblong to lance-shaped, toothed and light green. Bears many-branched panicles of thistle-like, mauve-blue or purplish-blue flower heads in summer.

CICHORIUM (Compositae)
Chicory
Genus of annuals, biennials and perennials, grown mainly as ornamental plants (*C. intybus* has edible leaves). Fully hardy. Needs full sun and well-drained soil. Propagate by seed in autumn or spring.
C. intybus illus. p.217.

CIMICIFUGA (Ranunculaceae)
Bugbane
Genus of perennials, grown for their flowers, which have an unusual, slightly unpleasant smell. Fully hardy. Grow in light shade and moist soil. Needs staking. Propagate by seed when fresh or by division in spring.
♥ *C. racemosa*. Clump-forming perennial. H 1.5m (5ft), S 60cm (2ft). Spikes of bottlebrush-like, pure white flowers are borne in mid-summer above broadly oval to lance-shaped, divided, fresh green leaves. var. *cordifolia* has feathery plumes of star-shaped, creamy-white flowers and dissected, light green leaves.
C. simplex illus. p.195. ♥ **'Elstead'** is an upright perennial. H 1.2m (4ft), S 60cm (2ft). Purple stems have arching racemes of fragrant, bottlebrush-like, white flowers in autumn. Has broadly oval to lance-shaped, divided, glossy leaves. **'Pritchard's Giant'** (syn. *C.s.* var. *ramosa*), H 2.2m (7ft), has large, much-divided leaves and white flowers on arching panicles.

Cineraria × *hybrida*. See *Senecio* × *hybridus*.

CINNAMOMUM (Lauraceae)
Genus of evergreen trees, grown for their foliage and to provide shade. Frost tender, min. 10°C (50°F). Requires full light or partial shade and fertile, moisture-retentive but well-drained soil. Water containerized specimens freely when in full growth but less at other times. May be pruned if necessary. Propagate by seed in spring or by semi-ripe cuttings in summer.
C. camphora. Moderately fast-growing, evergreen, rounded tree. H and S 12m (40ft) or more. Oval, lustrous, rich green leaves, blue-grey-tinted beneath, reddish or coppery when young, are camphor-scented when bruised. Has insignificant flowers in spring.

CIONURA (Asclepiadaceae)
Genus of one species of deciduous, twining climber, grown for its flowers. Half hardy. Grow in any soil and in full sun. Cut stems exude milky sap that may cause blisters. Prune after flowering. Propagate by seed in spring or by stem cuttings in late summer or early autumn.
C. erecta, syn. *Marsdenia erecta*. Deciduous, twining climber. H 3m (10ft) or more. Heart-shaped, greyish-green leaves are 3–6cm (1¼–2½in) long. In summer, clusters of fragrant, white flowers, with 5 spreading petals, are borne in leaf axils, followed by 7cm (3in) long fruits, containing many silky seeds, in autumn.

CIRSIUM (Compositae)
Genus of annuals, biennials and perennials. Most species are not cultivated – indeed some are pernicious weeds – but *C. rivulare* has decorative flower heads. Fully hardy. Tolerates sun or shade and any but wet soil. Propagate by division in spring or by seed in autumn.
C. rivulare **'Atropurpureum'**. Erect perennial. H 1.2m (4ft), S 60cm (2ft). Heads of pincushion-like, deep crimson flowers are borne on erect stems in summer. Leaves are narrowly oval to oblong or lance-shaped and deeply cut, with weakly spiny margins.

CISSUS (Vitaceae)
Genus of evergreen, woody-stemmed, mainly tendril climbers, grown for their attractive foliage. Bears insignificant, greenish flowers, mainly in summer. Half hardy to frost tender, min. 7–18°C (45–64°F). Provide fertile, well-drained soil, with semi-shade in summer. Water regularly, less in cold weather. Needs tying to supports. Thin out crowded stems in spring. Propagate by semi-ripe cuttings in summer.
♥ *C. antarctica* illus. p.180.
C. bainesii. See *Cyphostemma bainesii*.
C. discolor (Rex begonia vine). Moderately vigorous, evergreen, tendril climber with slender, woody stems. H to 3m (10ft). Frost tender, min. 18°C (64°F). Has oval, pointed leaves, 10–15cm (4–6in) long, that are deep green with silver bands above, maroon beneath.
C. hypoglauca. Evergreen, woody-stemmed, scrambling climber. H 2–3m (6–10ft). Frost tender, min. 7°C (45°F). Leaves are divided into 4 or 5 oval leaflets, pale green above and blue-grey beneath.
C. juttae. See *Cyphostemma juttae*.
♥ *C. rhombifolia*, syn. *Rhoicissus rhombifolia*, illus. p.180.
C. striata, syn. *Ampelopsis sempervirens*, *Vitis striata* (Ivy of Uruguay, Miniature grape ivy). Fast-growing, evergreen, woody-stemmed, tendril climber. H 10m (30ft) or more. Half hardy. Has leaves of 3–5 oval, serrated, lustrous, green leaflets. Mature plants may produce pea-shaped, glossy, black berries in autumn.
C. voinieriana. See *Tetrastigma voinierianum*.

CISTUS (Cistaceae)
Rock rose
Genus of evergreen shrubs, grown for their succession of freely borne, short-lived, showy flowers. Is good in coastal areas, withstanding sea winds well. Frost to half hardy; in cold areas needs shelter. Does best in full sun and light, well-drained soil. Resents being transplanted. Cut out any dead wood in spring, but do not prune hard. Propagate species by softwood or greenwood cuttings in summer or by seed in autumn, hybrids and cultivars by cuttings only in summer.
♥ *C.* × *aguilarii* **'Maculatus'** illus. p.129.
C. albidus. Evergreen, bushy shrub. H and S 1m (3ft). Half hardy. Leaves are oblong and white-felted. Saucer-shaped, pale rose-pink flowers, each with a central, yellow blotch, open in early summer.
♥ *C.* × *corbariensis*, syn. *C.* × *hybridus*, illus. p.129.
C. creticus, syn. *C. incanus* subsp. *creticus*, illus. p.135.
♥ *C.* × *cyprius* illus. p.129.
C. × *dansereaui*. See *C.* × *lusitanicus*.
C. × *hybridus*. See *C.* × *corbariensis*.
♥ *C. ladanifer* illus. p.129.
♥ *C. laurifolius*. Evergreen, bushy, dense shrub. H and S 2m (6ft). Frost hardy. Has oval, aromatic, dark green leaves and, in summer, saucer-shaped, white flowers, each with a central, yellow blotch.
C. × *lusitanicus*, syn. *C.* × *dansereaui*. Evergreen, bushy, compact shrub. H and S 1m (3ft). Frost hardy. Leaves are narrowly oblong and dark green. Saucer-shaped, white flowers, each with a central, deep red blotch, appear from early to mid-summer.
C. monspeliensis illus. p.129.
C. parviflorus. Evergreen, bushy, dense shrub. H and S 1m (3ft). Frost hardy. Small, saucer-shaped, pale pink flowers appear among oval, grey-green leaves in early summer.
♥ *C.* **'Peggy Sammons'**. Evergreen, bushy shrub. H and S 1m (3ft). Frost hardy. Has oval, grey-green leaves. Saucer-shaped, pale pink flowers are produced freely during early summer.
♥ *C.* × *purpureus*. Evergreen, bushy, rounded shrub. H and S 1m (3ft). Frost hardy. Produces saucer-shaped, deep purplish-pink flowers, each blotched with deep red, from early to mid-

summer. Leaves are narrowly lance-shaped and grey-green.
C. salviifolius illus. p.129.
C. 'Silver Pink'. Evergreen, bushy shrub. H 60cm (2ft), S 1m (3ft). Frost hardy. Oval, dark green leaves set off large, saucer-shaped, clear pink flowers, each with conspicuous, yellow stamens, from early to mid-summer.
♥ *C.* x *skanbergii* illus. p.132.

x **CITROFORTUNELLA** (Rutaceae)
Hybrid genus (*Citrus* x *Fortunella*) of evergreen shrubs and trees, grown for their flowers, fruits and overall appearance. Frost tender, min. 5–10°C (41–50°F). Requires full light and fertile, well-drained but not dry soil. Water containerized specimens freely when in full growth, moderately at other times. Propagate by seed when ripe or by greenwood or semi-ripe cuttings in summer. Whitefly, red spider mite, mealy bug, lime-induced and magnesium-deficiency chlorosis may be troublesome.
♥ x *C. microcarpa*, syn. x *C. mitis*, *Citrus mitis*, illus. p.121.
x *C. mitis*. See x *C. microcarpa*.

Citrus mitis. See x *Citrofortunella mitis*.

CLADANTHUS (Compositae)
Genus of one species of annual, grown for its fragrant foliage and daisy-like flower heads. Fully hardy. Grow in sun and in reasonably fertile, very well-drained soil. Dead-head to prolong flowering. Propagate by seed sown outdoors in mid-spring.
C. arabicus illus. p.289.

CLADRASTIS (Leguminosae)
Genus of deciduous, summer-flowering trees, grown for their pendent, wisteria-like flower clusters and autumn foliage. Fully hardy. Requires full sun and fertile, well-drained soil. Propagate by seed in autumn or by root cuttings in late winter. The wood is brittle: old trees are prone to damage by strong winds.
♥ *C. lutea* illus. p.56.

CLARKIA, syn. GODETIA (Onagraceae)
Genus of annuals, grown for their flowers, which are good for cutting. Fully hardy. Grow in sun and in reasonably fertile, well-drained soil. Avoid rich soil as this encourages vegetative growth at the expense of flowers. Propagate by seed sown outdoors in spring, or in early autumn in mild areas. Botrytis may be troublesome.
C. amoena. Fast-growing annual with upright, thin stems. H to 60cm (2ft), S 30cm (1ft). Has lance-shaped, mid-green leaves. Spikes of 5-petalled, single or double flowers, in shades of pink, are produced in summer. Tall forms, H 60cm (2ft), have double flowers in shades of pink or red.
Azalea-flowered Series, intermediate, H 38cm (15in), has semi-double flowers in shades of pink. **Princess Series**, dwarf, H 30cm (12in), has frilled flowers in shades of pink, including salmon (illus. p.293).

C. 'Arianna' illus. p.277.
C. 'Brilliant' illus. p.276.

CLAYTONIA (Portulacaceae)
Genus of mainly evergreen perennials with succulent leaves; is related to *Lewisia*. Grows best in alpine houses. Fully hardy. Prefers well-drained soil and tolerates sun or shade. Propagate by seed or division in autumn. May be difficult to grow.
C. megarhiza. Evergreen, basal-rosetted perennial with a long tap root. H 1cm (½in), S 8cm (3in). Leaves are spoon-shaped and fleshy. Bears small heads of tiny, bowl-shaped, white flowers in spring. Prefers sun and gritty soil. Is prone to aphid attack.
var. *nivalis* illus. p.314.
C. virginica (Spring beauty). Clump-forming perennial with flat, black tubers. H 10cm (4in), S 20cm (8in) or more. Narrowly spoon-shaped leaves, reddish when young, later turn green and glossy. Branched stems bear cup-shaped, white or pink flowers, striped deep pink, in early spring. Needs shade.

CLEISTOCACTUS, syn. BORZICACTUS (Cactaceae)
Genus of columnar, perennial cacti with branched, cylindrical, much-ribbed stems with spines. Is one of the faster-growing cacti, some reaching 2m (6ft) in 5 years or less. Tubular flowers contain plenty of nectar and are pollinated by hummingbirds. Frost tender, min. 5°C (41°F). Needs full sun and very well-drained soil. Propagate by seed or stem cuttings in spring or summer.
C. aurantiacus, syn. *Borzicactus aurantiacus*, *Submatucana aurantiaca*, illus. p.416.
C. baumannii. Erect, then prostrate, perennial cactus. H 1m (3ft) or more, S 5m (15ft). Thick stems produce long, uneven, variable-coloured spines. Has S-shaped, tubular, bright orange-red flowers in spring-summer.
C. celsianus, syn. *Borzicactus celsianus*, *Oreocereus celsianus*, illus. p.398.
C. haynei, syn. *Borzicactus haynei*, *Matucana haynei*. Slow-growing, spherical to columnar, perennial cactus. H 60cm (24in), S 10cm (4in). Has a cylindrical, much-ribbed, green stem densely covered with short, white spines. Tubular, pinkish- or orange-red flowers appear in summer from stem crown, on plants over 15cm (6in) high.
C. smaragdiflorus. Erect, then prostrate, perennial cactus. H 1.5m (5ft), S 6m (20ft). Is similar to *C. baumannii*, but has straight, tubular flowers with green-tipped petals.
♥ *C. strausii* illus. p.395.
C. trollii, syn. *Borzicactus trollii*, *Oreocereus trollii* (Old man of the Andes). Slow-growing, columnar, perennial cactus. H 70cm (28in), S 10cm (4in). Cylindrical, green stem, 7–10cm (3–4in), with thick, golden spines is almost hidden by long, wispy, hair-like, white spines. Tubular, pink flowers, recurved at tips and 10cm (4in) long, appear in summer on fully mature plants.

CLEMATIS (Ranunculaceae)
Old man's beard, Travellers' joy
Genus of evergreen or deciduous, mainly twining climbers and herbaceous perennials, cultivated for their mass of flowers, often followed by decorative seed heads, and grown on walls and trellises and together with trees, shrubs and other host plants. Only early-flowering species are evergreen, although some later-flowering species are semi-evergreen. Most species have nodding, bell-shaped flowers, with 4 petals (botanically known as perianth segments), or flattish flowers, each usually with 4–6 generally pointed petals. Large-flowered cultivars also bear flattish flowers, but with 4–10 petals. Flower colour may vary according to climatic conditions: in general, the warmer the climate, the darker the flower colour. Fully to half hardy. May be grown in shade or full sun, but prefers rich, well-drained soil with roots shaded. Propagate cultivars in early summer by softwood or semi-ripe cuttings or layering, species from seed sown in autumn. Aphids, mildew and clematis wilt may cause problems.
  Clematis may be divided into groups according to their approximate flowering seasons, habit and pruning needs.

Group 1
Early-flowering species prefer a south- or south-west-facing wall. Small, single flowers, either bell-shaped, 3cm (1¼in) long, or flattish, 4–5cm (1½–2in) wide, are borne on the previous season's ripened shoots in spring or, occasionally, in late winter. Leaves are evergreen, glossy and divided into 3 lance-shaped, 12cm (5in) long leaflets or into 3 fern-like, 5cm (2in) long leaflets. Frost hardy.
Alpina and **macropetala** types tolerate exposed, north- or north-east-facing positions. Small, bell-shaped, single, semi-double or double flowers, 8cm (3in) or less across, are borne on the previous season's ripened shoots in spring, occasionally also on the current season's shoots in summer. Pale to mid-green leaves are divided into 3–5 lance-shaped to broadly oval leaflets, 3cm (1¼in) long with serrated margins. Fully hardy.
Montana types are vigorous climbers, suitable for growing over large buildings and trees. Small, flattish, mainly single flowers, 4–7cm (1½–3in) across, are borne on the previous season's ripened shoots in late spring. Leaves are mid- to purplish-green and divided into 3 lance-shaped to broadly oval, serrated leaflets, 8cm (3in) long with pointed tips. Frost hardy.
  Prune after flowering to allow new growth to be produced and ripened for the following season. Remove dead or damaged stems and cut back other shoots that have outgrown their allotted space.

Group 2
Early, large-flowered cultivars bear flattish, single, semi-double or double flowers, 8–20cm (3–8in) across, that are borne on the previous season's ripened shoots and on new shoots, from early to late summer. Generally the second flush of flowers on semi-double and double forms produces single flowers. Usually pale to mid-green leaves are simple and oval, to 10cm (4in) long, or are divided into 3 oval or lance-shaped leaflets, 15–18cm (6–7in) long. Fully to frost hardy.
  Prune before new growth starts, in early spring. Remove dead or damaged stems and cut back all remaining shoots to where strong, leaf-axil buds are visible. (These buds will produce first crop of flowers.)

Group 3
Late, large-flowered cultivars produce outward-facing, flattish, single flowers, 6–15cm (2½–6in) across, borne on new shoots in summer or early autumn. Leaves are similar to those of early cultivars (group 2), described above. Frost hardy.
Late-flowering species and **small-flowered cultivars** have small, single or double flowers borne on the current season's shoots in summer-autumn. Flowers vary in shape and may be star-shaped, tubular, bell-shaped, flattish or resembling nodding lanterns; they vary in size from 1cm (½in) to 10cm (4in) across. Have pale to dark green or grey-green leaves divided into 3 lance-shaped to broadly oval leaflets, each 1cm (½in) long, or hairy and/or toothed leaves divided into 5 or more lance-shaped to broadly oval leaflets, each 1–10cm (½–4in) long. Fully to half hardy.
Herbaceous types produce single flowers that are either flattish, 1–2cm (½–¾in) wide, or bell-shaped or tubular, 1–4cm (½–1½in) long, and are produced on the current season's shoots in summer. Mid- to dark green or grey-green leaves are simple and lance-shaped to elliptic, 2.5–15cm (1–6in) long, or are divided into 3–5 rounded, serrated leaflets, each 10–15cm (4–6in) long with a pointed tip. Fully to frost hardy.
  Prune before new growth begins, in early spring. Remove all of the previous season's stems down to a pair of strong, leaf-axil buds, 15–30cm (6–12in) above soil level.

C. 'Abundance'. See under *C. viticella*.
♥ *C. alpina*. Alpina clematis (group 1). H 2–3m (6–10ft), S 1.5m (5ft). Fully hardy. Has lantern-shaped, single, blue flowers, 4–7cm (1½–3in) long, in spring and, occasionally, summer. Forms attractive, fluffy, silvery seed heads in summer. Is ideal for a north-facing or very exposed site. 'Columbine' has 5cm (2in) long, pale blue petals. ♥ 'Frances Rivis' (illus. p.173) is free-flowering, with slightly twisted, 7cm (3in) long, mid-blue petals. 'Ruby' has purplish-pink flowers.
C. armandii (illus. p.172). Strong-growing, evergreen, early-flowering clematis (group 1). H 3–5m (10–15ft), S 2–3m (6–10ft). Frost hardy. Bears scented, flattish, single, white flowers, 4cm (1½in) across, in early spring. Needs a sheltered, south- or south-west-facing site.

♚ *C.* **'Ascotiensis'** (illus. p.173). Vigorous, late, large-flowered clematis (group 3). H 3–4m (10–12ft), S 1m (3ft). Frost hardy. In summer produces single, bright violet-blue flowers, 9–12cm (3½–5in) across, with pointed petals and brownish-green anthers.
C. **'Barbara Jackman'**. Compact, early, large-flowered clematis (group 2). H 2.5–3m (8–10ft), S 1m (3ft). Frost hardy. In summer bears single, blue flowers, 10cm (4in) across, with magenta stripes and creamy-yellow anthers. Prefers partial shade.
C. **'Beauty of Worcester'** (illus. p.173). Early, large-flowered clematis (group 2). H 2.5–3m (8–10ft), S 1m (3ft). Frost hardy. Has double, rich deep violet-blue flowers, 10cm (4in) across with cream anthers, in early summer and then single flowers in late summer.
♚ *C.* **'Bees Jubilee'**. Compact, early, large-flowered clematis (group 2). H 2.5m (8ft), S 1m (3ft). Frost hardy. In early summer bears a profusion of single, deep pink flowers, 10–12cm (4–5in) across with brown anthers and a central, rose-madder stripe on each petal. Prefers partial shade.
♚ *C.* **'Bill Mackenzie'** (illus. p.173). Vigorous, late-flowering clematis (group 3). H 7m (22ft), S 3–4m (10–12ft). Fully hardy. Has dark green leaves. Flowers are yellow and 6–7cm (2½–3in) wide. Is best pruned with shears.
C. **'Carnaby'**. Compact, early, large-flowered clematis (group 2). H 2.5m (8ft), S 1m (3ft). Frost hardy. In early summer has a profusion of single, deep pink flowers, 8–10cm (3–4in) across, with a darker stripe on each petal and red anthers. Prefers partial shade.
C. cirrhosa (illus. p.173). Evergreen, early-flowering clematis (group 1). H 2–3m (6–10ft), S 1–2m (3–6ft). Frost hardy. Produces bell-shaped, cream flowers, 3cm (1¼in) across and spotted red inside, in late winter and early spring during frost-free weather.
♚ *C.* **'Comtesse de Bouchaud'**. Strong-growing, late, large-flowered clematis (group 3). H 2–3m (6–10ft), S 1m (3ft). Frost hardy. In summer has masses of single, bright mauve-pink flowers, 8–10cm (3–4in) across, with yellow anthers.
C. **'Countess of Lovelace'** (illus. p.173). Early, large-flowered clematis (group 2). H 2.5m (8ft), S 1m (3ft). Frost hardy. Has double, bluish-lilac flowers, 10cm (4in) across, with pointed petals and cream anthers, in early summer, then single flowers in late summer.
C. **'Crimson King'**. Late, large-flowered clematis (group 3). H 2.5–3m (8–10ft), S 1m (3ft). Frost hardy. Is shy-flowering, bearing single, crimson flowers, 10cm (4in) across with red anthers, in summer.
♚ *C.* **'Daniel Deronda'**. Vigorous, early, large-flowered clematis (group 2). H 3m (10ft), S 1m (3ft). Frost hardy. Has double and semi-double, deep purple-blue flowers, 10–14cm (4–5½in) across, with cream anthers, then single flowers in late summer.
♚ *C.* **'Duchess of Albany'** (illus. p.173). Vigorous, small-flowered clematis (group 3). H 2.5m (8ft), S 1m (3ft). Frost hardy. In summer and early autumn has masses of small, tulip-like, single, soft pink flowers, 6cm (2½in) long, with brown anthers and a deeper pink stripe inside each petal.
C. **'Duchess of Edinburgh'**. Early, large-flowered clematis (group 2). H 2–3m (6–10ft), S 1m (3ft). Frost hardy. In summer produces double, white flowers, 8–10cm (3–4in) across, with yellow anthers and green, outer petals. May sometimes be weak-growing.
♚ *C.* × *durandii*. Semi-herbaceous, late-flowering clematis (group 3). H 1–2m (3–6ft), S 45cm–1.5m (1½–5ft). Frost hardy. In summer produces flattish, single, deep blue flowers, 6–8cm (2½–3in) across, with 4 petals and yellow anthers. Leaves are elliptic.
♚ *C.* **'Edith'**. Compact, early, large-flowered clematis (group 2). H 2.5m (8ft), S 1m (3ft). Frost hardy. Single, 10cm (4in) wide flowers, with white petals and red anthers, are borne in summer. Is good for a patio garden.
♚ *C.* **'Elsa Spath'** (illus. p.173). Early, large-flowered clematis (group 2). H 2–3m (6–10ft), S 1m (3ft). Frost hardy. Bears masses of single, 12cm (5in) wide flowers, with overlapping, rich mauve-blue petals and red anthers, throughout summer.
♚ *C.* **'Ernest Markham'** (illus. p.173). Vigorous, late, large-flowered clematis (group 3). H 3–4m (10–12ft), S 1m (3ft). Frost hardy. In summer bears 10cm (4in) wide, single flowers with blunt-tipped, vivid magenta petals and chocolate anthers. Thrives in full sun.
C. **'Etoile Violette'**. See under *C. viticella*.
C. flammula (illus. p.172). Vigorous, late-flowering clematis; may be semi-evergreen (group 3). H 3–5m (10–15ft), S 2m (6ft). Frost hardy. Produces masses of almond-scented, flattish, single, white flowers, 2cm (¾in) across, in summer and early autumn.
C. florida **'Sieboldii'** (illus. p.172). Weak-growing, small-flowered clematis (group 2). H 2–3m (6–10ft), S 1m (3ft). Frost hardy. In summer has passion-flower-like, single blooms, each 8cm (3in) wide with creamy-white petals and a domed boss of petal-like, rich purple stamens. Needs a sheltered aspect.
♚ *C.* **'General Sikorski'**. Early, large-flowered clematis (group 2). H 3m (10ft), S 1m (3ft). Frost hardy. Has numerous 10cm (4in) wide, single flowers, with large, overlapping, blue petals and cream anthers, in summer.
♚ *C.* **'Gipsy Queen'**. Vigorous, late, large-flowered clematis (group 3). H 3m (10ft), S 1m (3ft). Frost hardy. Bears single, 10cm (4in) wide flowers, with velvety, violet-purple petals and red anthers, in summer.
C. **'Gravetye Beauty'** (illus. p.173). Vigorous, small-flowered clematis (group 3). H 2.5m (8ft), S 1m (3ft). Frost hardy. In summer and early autumn has masses of small, tulip-like, single, bright red flowers, 6cm (2½in) long, with brown anthers. Is similar to *C.* 'Duchess of Albany', but flowers are more open.
C. **'Hagley Hybrid'** (illus. p.172). Vigorous, late, large-flowered clematis (group 3). H 2.5m (8ft), S 1m (3ft). Frost hardy. Produces 8–10cm (3–4in) wide, single flowers with boat-shaped, rose-mauve petals and red anthers, in summer. Prefers partial shade.
♚ *C.* **'Henryi'** (illus. p.172). Vigorous, early, large-flowered clematis (group 2). H 3m (10ft), S 1m (3ft). Frost hardy. Has 12cm (5in) wide, single flowers, with white petals and dark chocolate anthers, in summer.
C. heracleifolia var. *davidiana*. Herbaceous clematis (group 3). H 1m (3ft), S 75cm (2½ft). Fully hardy. In summer, thick stems each bear axillary clusters of scented, tubular, single, pale blue flowers, 2–3cm (¾–1¼in) long, with reflexed petal tips. ♚ **'Wyevale'** (illus. p.173) has strongly scented, dark blue flowers.
♚ *C.* **'H.F. Young'** (illus. p.173). Compact, early, large-flowered clematis (group 2). H 2.5m (8ft), S 1m (3ft). Frost hardy. Bears 10cm (4in) wide, single flowers, with violet-tinged, blue petals and cream anthers, in summer. Is ideal for a container or patio garden.
♚ *C.* **'Horn of Plenty'**. Vigorous, compact, early, large-flowered clematis (group 2). H 2.5m (8ft), S 1m (3ft). Frost hardy. In early summer produces 12cm (5in) wide, single flowers, with rose-mauve petals, fading to mauve-blue, and dark red anthers. Is good for a patio garden.
C. **'Huldine'** (illus. p.172). Very vigorous, late, large-flowered clematis (group 3). H 3–4m (10–12ft), S 2m (6ft). Frost hardy. Has 6cm (2½in) wide, single, white flowers, mauve beneath and with cream anthers, in summer. Is ideal for an archway or pergola.
C. integrifolia (illus. p.173). Herbaceous clematis (group 3). H and S 75cm (30in). Fully hardy. Leaves are narrowly lance-shaped. In summer bears bell-shaped, single, deep blue flowers, 3cm (1¼in) long with cream anthers, followed by attractive, grey-brown seed heads.
♚ *C.* **'Jackmanii'** (illus. p.173). Vigorous, late, large-flowered clematis (group 3). H 3m (10ft), S 1m (3ft). Frost hardy. Masses of velvety, single, dark purple flowers, fading to violet, 8–10cm (3–4in) across, with light brown anthers, are borne in mid-summer.
C. **'Jackmanii Superba'**. Vigorous, late, large-flowered clematis (group 3). Is similar to *C.* 'Jackmanii', but has more rounded, darker flowers.
C. **'John Huxtable'**. Late, large-flowered clematis (group 3). H 2–3m (6–10ft), S 1m (3ft). Frost hardy. Bears masses of 8cm (3in) wide, single, white flowers, with cream anthers, in mid-summer.
C. **'John Warren'**. Early, large-flowered clematis (group 2). H 2.5–3m (8–10ft), S 1m (3ft). Frost hardy. Produces 12–14cm (5–5½in) wide, single, French grey flowers, with overlapping, pointed, carmine-edged petals and red anthers, in early summer.
C. × *jouiniana*. Sprawling, sub-shrubby, late-flowering clematis (group 3). H 1m (3ft), S 3m (10ft). Frost hardy. Has coarse foliage and, in summer, masses of tubular, single, soft lavender or off-white flowers, 2cm (¾in) wide with reflexed petal tips. Is non-clinging. ♚ **'Praecox'** has slightly darker flowers, 2 weeks earlier.
C. **'Kathleen Wheeler'**. Early, flowered clematis (group 2). H 2.5–3m (8–10ft), S 1m (3ft). Frost hardy. Has single, plum-mauve flowers, 12–14cm (5–5½in) across with yellow anthers, in early summer.
C. **'Lady Betty Balfour'**. Vigorous, late, large-flowered clematis (group 3). H 3m (10ft), S 2m (6ft). Frost hardy. Has 12cm (5in) wide, single flowers with rich purple petals, fading to purple-blue, and yellow anthers, in late summer and early autumn. Prefers a south- or south-west-facing aspect.
♚ *C.* **'Lasurstern'** (illus. p.173). Vigorous, early, large-flowered clematis (group 2). H 2–3m (6–10ft), S 1m (3ft). Frost hardy. In summer bears single, blue flowers, 10–12cm (4–5in) across, with overlapping, wavy-edged petals and cream anthers.
C. **'Lincoln Star'** (illus. p.173). Early, large-flowered clematis (group 2). H 2–3m (6–10ft), S 1m (3ft). Frost hardy. Has 10–12cm (4–5in) wide, single, raspberry-pink flowers, with red anthers, in early summer. Early flowers are darker than late ones, which have very pale pink petal edges. Prefers partial shade.
C. macropetala (illus. p.173). Macropetala clematis (group 1). H 3m (10ft), S 1.5m (5ft). Fully hardy. During late spring and summer has masses of semi-double, mauve-blue flowers, 5cm (2in) long and lightening in colour towards the centre, followed by fluffy, silvery seed heads. ♚ **'Markham's Pink'** (illus. p.172) has pink flowers.
C. **'Mme Julia Correvon'**. See under *C. viticella*.
C. **'Mme Le Coultre'**. See *C.* 'Marie Boisselot'.
♚ *C.* **'Marie Boisselot'**, syn. *C.* 'Mme Le Coultre'. Vigorous, early, large-flowered clematis (group 2). H 3m (10ft), S 1m (3ft). Frost hardy. Bears single flowers, 12cm (5in) across, with overlapping, white petals and cream anthers, in summer.
♚ *C.* **'Miss Bateman'**. Compact, early, large-flowered clematis (group 2). H 2.5m (8ft), S 1m (3ft). Frost hardy. Masses of single, white flowers, 8–10cm (3–4in) across with red anthers, are produced in summer. Is good for a container or patio garden.
C. montana (illus. p.172). Vigorous, Montana clematis (group 1). H 7–12m (22–40ft), S 2–3m (6–10ft). Fully hardy. In late spring bears masses of single, white flowers, 4–5cm (1½–2in) across, with yellow anthers. ♚ **'Elizabeth'**, H 10–12m (30–40ft), has scented, soft pink flowers with widely spaced petals. Flowers of ♚ var. *rubens* (illus. p.172) are pale pink. ♚ **'Tetrarose'** (illus. p.172), H 7–8m (22–25ft), has coarse, 8cm (3in) long leaflets and 6–7cm (2½–3in) wide, deep satin-pink flowers.

♧ *C.* **'Mrs Cholmondeley'**. Early, large-flowered clematis (group 2). H 2–3m (6–10ft), S 1m (3ft). Frost hardy. In summer has single, light bluish-lavender flowers, 10–12cm (4–5in) across with widely spaced petals and light chocolate anthers.
♧ *C.* **'Mrs George Jackman'** (illus. p.172). Early, large-flowered clematis (group 2). H 2–3m (6–10ft), S 1m (3ft). Frost hardy. Bears 10cm (4in) wide, semi-double flowers, with creamy-white petals and light brown anthers, in early summer.
C. **'Mrs N. Thompson'**. Compact, early, large-flowered clematis (group 2). H 2.5m (8ft), S 1m (3ft). Frost hardy. Produces masses of 8–10cm (3–4in) wide, single, magenta flowers, with a central, slightly darker stripe on each bluish-purple-edged petal and red anthers, in summer. Is good for a container or patio garden.
♧ *C.* **'Nelly Moser'** (illus. p.172). Early, large-flowered clematis (group 2). H 3.5m (11ft), S 1m (3ft). Frost hardy. In early summer has 12–16cm (5–6½in) wide, single, rose-mauve flowers, with reddish-purple anthers and, on each petal, a carmine stripe that fades in strong sun. Prefers a shaded, east-, west- or north-facing situation.
♧ *C.* **'Niobe'**. Early, large-flowered clematis (group 2). H 2–3m (6–10ft), S 1m (3ft). Frost hardy. Throughout summer produces masses of single, rich deep red flowers, 10–14cm (4–5½in) across with yellow anthers.
C. orientalis. Late-flowering clematis (group 3). H 3–4m (10–12ft), S 1.5m (5ft). Fully hardy. Leaves are grey- to dark green. In summer has lantern-shaped, single, greenish-yellow flowers, 3cm (1¼in) wide with recurved petal tips, followed by feathery seed heads.
♧ *C.* **'Perle d'Azur'** (illus. p.173). Late, large-flowered clematis (group 3). H 3m (10ft), S 1m (3ft). Frost hardy. Single, azure-blue flowers, 8cm (3in) across with recurved petal tips and creamy-green anthers, open in summer.
C. **'Proteus'** (illus. p.173). Early, large-flowered clematis (group 2). H 2.5–3m (8–10ft), S 1m (3ft). Frost hardy. In early summer bears 10–12cm (4–5in) wide, double, mauve-pink flowers, becoming pinker towards centres and with cream anthers, then single flowers in late summer. Double flowers sometimes have green, outer petals.
C. **'Purpurea Plena Elegans'**. See under *C. viticella*.
C. **'Ramona'**. Early, large-flowered clematis (group 2). H 3m (10ft), S 1m (3ft). Frost hardy. Has coarse, dark green leaves offset, in summer, by single, pale blue flowers, 10–12cm (4–5in) across with red anthers. Prefers a south- or south-west-facing position.
C. recta (illus. p.172). Clump-forming, herbaceous clematis (group 3). H 1–2m (3–6ft), S 50cm (20in). Fully hardy. Leaves are dark or grey-green. Bears masses of sweetly scented, flattish, single, white flowers, 2cm (¾in) across, in mid-summer.
♧ *C. rehderiana* (illus. p.173). Vigorous, late-flowering clematis (group 3). H 6–7m (20–22ft), S 2–3m (6–10ft). Frost hardy. Bears loose clusters of fragrant, tubular, single, yellow flowers, 1–2cm (½–¾in) long, in late summer and early autumn. Leaves are coarse-textured.
♧ *C.* **'Richard Pennell'** (illus. p.173). Early, large-flowered clematis (group 2). H 2–3m (6–10ft), S 1m (3ft). Frost hardy. Produces 10–12cm (4–5in) wide, single flowers, with rich purple-blue petals and golden-yellow anthers, in summer.
C. **'Rouge Cardinal'**. Early, large-flowered clematis (group 2). H 2–3m (6–10ft), S 1m (3ft). Frost hardy. In summer has single, velvety, crimson flowers 8–10cm (3–4in) across with red anthers.
C. **'Serenata'**. Early, large-flowered clematis (group 2). H 2–3m (6–10ft), S 1m (3ft). Frost hardy. Produces 10cm (4in) wide, single flowers, with yellow anthers, in summer. Dusky-purple petals each have a central stripe, slightly darker.
C. **'Souvenir de Capitaine Thuilleaux'**. Compact, early, large-flowered clematis (group 2). H 2.5m (8ft), S 1m (3ft). Frost hardy. In early summer bears 8–10cm (3–4in) wide, single flowers, with red anthers and deep pink-striped, cream-pink petals. Is ideal for a container or patio garden.
♧ *C.* **'Star of India'** (illus. p.173). Vigorous, late, large-flowered clematis (group 3). H 3m (10ft), S 1m (3ft). Frost hardy. Bears masses of 8–10cm (3–4in) wide, single, deep purple-blue flowers, with light brown anthers, in mid-summer; each petal has a deep carmine-red stripe.
C. tangutica (illus. p.173). Vigorous, late-flowering clematis (group 3). H 5–6m (15–20ft), S 2–3m (6–10ft). Fully hardy. Has lantern-shaped, single, yellow flowers, 4cm (1½in) long, throughout summer and early autumn, followed by fluffy, silvery seed heads.
♧ *C.* **'The President'** (illus. p.173). Early, large-flowered clematis (group 2). H 2–3m (6–10ft), S 1m (3ft). Frost hardy. In early summer bears masses of single, rich purple flowers, silver beneath, 10cm (4in) wide with red anthers.
C. **'Ville de Lyon'** (illus. p.173). Late, large-flowered clematis (group 3). H 2–3m (6–10ft), S 1m (3ft). Frost hardy. In mid-summer has single, bright carmine-red flowers, 8–10cm (3–4in) across, with darker petal edges and yellow anthers. Lower foliage tends to become scorched by late summer.
C. viticella. Late-flowering clematis (group 3). H 2–3m (6–10ft), S 1m (3ft). Fully hardy. Produces nodding, open bell-shaped, single, purple-mauve flowers, 3.5cm (1½in) long, in summer. The following are hybrids of *C. viticella* with other species, but are placed here as they are of similar growth and flower character to this species. **'Abundance'** (illus. p.173) has flattish, rose-pink flowers, 5cm (2in) across with yellow anthers.
♧ **'Etoile Violette'** (illus. p.173), H 3–4m (10–12ft), S 1.5m (5ft), is vigorous, with masses of flattish, 4–6cm (1½–2½in) wide, violet-purple flowers. ♧ **'Mme Julia Correvon'** (illus. p.173), H 2.5–3.5m (8–11ft), S 1m (3ft), has flattish, 5–7cm (2–3in) wide, wine-red flowers with twisted petals. ♧ **'Purpurea Plena Elegans'** (illus. p.173), H 3–4m (10–12ft), S 1.5m (5ft), has double, 4–6cm (1½–2½in) wide, rose-purple flowers with petals formed in tight rosettes and, occasionally, green, outer petals.
♧ *C.* **'Vyvyan Pennell'** (illus. p.173). Early, large-flowered clematis (group 2). H 2–3m (6–10ft), S 1m (3ft). Frost hardy. Has double, lilac flowers, each 10–12cm (4–5in) wide with a central, lavender-blue rosette of petals and golden-yellow anthers, in early summer, then single, blue-mauve flowers.
C. **'W.E. Gladstone'**. Vigorous, early, large-flowered clematis (group 2). H 3–4m (10–12ft), S 1.5m (5ft). Frost hardy. Produces single, lavender flowers, 15cm (6in) wide with red anthers, in summer.
C. **'William Kennett'** (illus. p.173). Early, large-flowered clematis (group 2). H 2–3m (6–10ft), S 1m (3ft). Frost hardy. In summer has masses of single flowers, 10–12cm (4–5in) across, with red anthers and tough, lavender-blue petals each bearing a central, darker stripe that fades as the flower matures.

CLEOME (Capparidaceae)
Spider flower
Genus of annuals and a few evergreen shrubs, grown for their unusual, spidery flowers. Half hardy to frost tender, min. 4°C (39°F). Grow in sun and in fertile, well-drained soil. Remove dead flowers. Propagate by seed sown outdoors in late spring. Aphids may be a problem.
C. hassleriana, syn. *C. spinosa*. Fast-growing, bushy annual. H to 1.2m (4ft), S 45cm (18in). Half hardy. Has hairy, spiny stems and mid-green leaves divided into lance-shaped leaflets. Large, rounded heads of narrow-petalled, pink-flushed, white flowers, with long, protruding stamens, appear in summer. **'Colour Fountain'** illus. p.274. **'Rose Queen'** has rose-pink flowers.
C. spinosa. See *C. hassleriana*.

CLERODENDRUM (Verbenaceae)
Genus of evergreen or deciduous, small trees, shrubs, sub-shrubs and woody-stemmed, twining climbers, grown for their showy flowers. Fully hardy to frost tender, min. 10–16°C (50–61°F). Needs humus-rich, well-drained soil and full sun, with partial shade in summer. Water freely in growing season, less at other times. Stems require support. Thin out crowded growth in spring. Propagate by seed in spring, by softwood cuttings in late spring or by semi-ripe cuttings in summer. Whitefly, red spider mite and mealy bug may be a problem.
♧ *C. bungei* illus. p.118.
C. fallax. See *C. speciosissimum*.
C. fragrans. See *C. philippinum*.
C. philippinum, syn. *C. fragrans*, *C.f.* var. *pleniflorum*. Evergreen or deciduous, bushy shrub. H and S to 2.5m (8ft). Frost tender, min. 10°C (50°F). Leaves are broadly oval, coarsely and shallowly toothed and downy. Fragrant, double, pink or white flowers are borne in domed, terminal clusters in summer.
C. speciosissimum, syn. *C. fallax*. Evergreen, erect to spreading, sparingly branched shrub. H and S to 3m (10ft). Frost tender, min. 15°C (59°F). Bears broadly heart-shaped, wavy-edged leaves, each to 30cm (1ft) across, on long stalks and, from late spring to autumn, tubular, scarlet flowers, with spreading petal lobes, in 30cm (1ft) long, terminal clusters. Makes a good pot plant.
♧ *C. splendens.* Vigorous, evergreen, woody-stemmed, twining climber. H 3m (10ft) or more. Frost tender, min. 15°C (59°F). Has oval to elliptic, rich green leaves. Clusters of 5-petalled, tubular, scarlet flowers, 2.5cm (1in) wide, are produced in summer.
♧ *C. thomsoniae* illus. p.168.
C. trichotomum illus. p.117.

CLETHRA (Clethraceae)
Genus of deciduous or evergreen shrubs and trees, grown for their fragrant, white flowers. Fully to half hardy. Needs semi-shade and moist, peaty, acid soil. Propagate by softwood cuttings in summer or by seed in autumn.
C. alnifolia (Sweet pepper-bush). Deciduous, bushy shrub. H and S 2.5m (8ft). Fully hardy. Has oval, toothed, mid-green leaves and, in late summer and early autumn, slender spires of small, bell-shaped flowers.
C. arborea (Lily-of-the-valley tree). Evergreen, bushy, dense shrub or tree. H 8m (25ft), S 6m (20ft). Half hardy. Long, nodding clusters of small, strongly fragrant, bell-shaped flowers are produced among oval, toothed, rich green leaves from late summer to mid-autumn.
♧ *C. barbinervis* illus. p.108.
♧ *C. delavayi* illus. p.91.

CLEYERA (Theaceae)
Genus of evergreen, summer-flowering shrubs and trees, grown for their foliage and flowers. Frost to half hardy. Requires a sheltered position in sun or semi-shade and moist, acid soil. Propagate by semi-ripe cuttings in summer.
C. fortunei. See *C. japonica* 'Tricolor'.
C. japonica. Evergreen, bushy shrub. H and S 3m (10ft). Frost hardy. Small, fragrant, saucer-shaped, white flowers are borne from early to mid-summer amid narrowly oval, glossy, dark green leaves. Has small, spherical, black fruits. **'Tricolor'** (syn. *C. fortunei*), H and S 2m (6ft), is half hardy and produces pink-flushed, young leaves, later dark green edged with creamy-white. Is sometimes confused in cultivation with *Eurya japonica* 'Variegata'.

CLIANTHUS (Leguminosae)
Genus of evergreen or semi-evergreen, woody-stemmed, scrambling climbers, grown for their attractive flowers. Half hardy to frost tender, min. 7°C (45°F).

465

Grow outdoors in warm areas in well-drained soil and full sun. In cooler areas needs to be under glass. In spring prune out growing tips to give a bushier habit and cut out any dead wood. Propagate by seed in spring or stem cuttings in late summer.
♥ *C. puniceus* and ♥ f. *albus* illus. p.165.

CLINTONIA (Liliaceae)
Genus of late spring- or summer-flowering, rhizomatous perennials. Fully hardy. Prefers shade and moist but well-drained, peaty, neutral to acid soil. Propagate by division in spring or by seed in autumn.
C. andrewsiana. Clump-forming, rhizomatous perennial. H 60cm (24in), S 30cm (12in). In early summer produces clusters of small, bell-shaped, pinkish-purple flowers at tops of stems, above sparse, broadly oval, glossy, rich green leaves. Bears globose, blue fruits in autumn.
C. borealis. Clump-forming, rhizomatous perennial. H and S 30cm (12in). Is similar to *C. andrewsiana*, but has nodding, yellowish-green flowers and small, globose, blackish fruits.
C. uniflora (Queencup). Spreading, rhizomatous perennial. H 15cm (6in), S 30cm (12in). Has oval, glossy, green leaves. Slender stems bear solitary star-shaped, white flowers in late spring, then large, globose, blue-black fruits.

CLITORIA (Leguminosae)
Genus of perennials and evergreen shrubs and twining climbers, grown for their large, pea-like flowers. Frost tender, min. 15°C (59°F). Grow in any fertile, well-drained soil and in full light. Water moderately, less when not in full growth. Provide support for stems. Thin out crowded stems in spring. Propagate by seed in spring or by softwood cuttings in summer. Whitefly and red spider mite may be a problem.
C. ternatea. Evergreen, twining climber with slender stems. H 3–5m (10–15ft). Leaves are divided into 3 or 5 oval leaflets. Clear bright blue flowers, 7–12cm (3–5in) wide, are carried in summer.

CLIVIA (Amaryllidaceae)
Genus of robust, evergreen, rhizomatous perennials, cultivated for their funnel-shaped flowers. Suits borders and large containers. Frost tender, min. 10°C (50°F). Needs partial shade and well-drained soil. Water well in summer, less in winter. Propagate by seed in winter or spring or by division in spring or summer after flowering. Mealy bugs may cause problems.
♥ *C. miniata* illus. p.363.
C. nobilis. Evergreen, spring- or summer-flowering, rhizomatous perennial. H 30–40cm (12–16in), S 30–60cm (12–24in). Has strap-shaped, semi-erect, basal leaves, 40–60cm (16–24in) long. Each leafless stem bears a dense, semi-pendent head of over 20 narrowly funnel-shaped, red flowers, with green tips and yellow margins to petals.

CLUSIA (Guttiferae)
Genus of evergreen, mainly summer-flowering climbers, shrubs and trees, grown for their foliage and flowers. Frost tender, min. 16–18°C (61–4°F). Needs partial shade and well-drained soil. Water potted specimens moderately, very little when temperatures are low. Pruning is tolerated if necessary. Propagate by layering in spring or by semi-ripe cuttings in summer. Whitefly and red spider mite may be a problem.
C. major, syn. *C. rosea* (Autograph tree, Copey, Fat pork tree, Pitch apple). Slow-growing, evergreen, rounded tree or shrub. H and S to 15m (50ft). Bears oval, lustrous, deep green leaves. Has cup-shaped, pink flowers, 5cm (2in) wide, in summer, followed by globose, greenish fruits that yield a sticky resin.

CLYTOSTOMA (Bignoniaceae)
Genus of evergreen, woody-stemmed, tendril climbers, grown for their flowers. Frost tender, min. 10–13°C (50–55°F). Grow in well-drained soil, with partial shade in summer. Water freely in summer, less at other times. Provide support for stems. Thin out congested growth after flowering or in spring. Propagate by semi-ripe cuttings in summer.
C. callistegioides illus. p.166.

COBAEA (Cobaeaceae)
Genus of evergreen or deciduous, woody-stemmed, tendril climbers. Only one species is generally cultivated. Frost tender, min. 4°C (39°F). Grow outdoors in warm areas in any well-drained soil and in full light. In cool regions may be grown under glass or treated as an annual. Propagate by seed in spring.
♥ *C. scandens* illus. p.174. ♥ f. *alba* is an evergreen, woody-stemmed, tendril climber. H 4–5m (12–15ft). Has long-stalked, bell-shaped, green, then white flowers from late summer until first frosts. Leaves have 4 or 6 oval leaflets.

Cocos capitata. See *Butia capitata*.

CODIAEUM (Euphorbiaceae)
Genus of evergreen shrubs, grown for their foliage. Frost tender, min. 10–13°C (50–55°F). Prefers partial shade and fertile, moist but well-drained soil. Remove tips from young plants to promote a branched habit. Propagate by greenwood cuttings from firm stem tips in spring or summer. Mealy bug and soft scale may be a nuisance.
C. variegatum illus. p.147.

CODONOPSIS (Campanulaceae)
Genus of perennials and mostly herbaceous, twining climbers, grown for their bell- or saucer-shaped flowers. Fully to frost hardy. Grow in semi-shade and in light, well-drained soil. Train over supports or leave to scramble through other, larger plants. Propagate by seed in autumn or spring.
C. clematidea. Herbaceous, twining climber. H to 1.5m (5ft). Frost hardy. Has small, oval, mid-green leaves. In summer produces nodding, bell-shaped flowers, 2.5cm (1in) long; they are white, tinged with blue, and marked inside with darker veining and 2 purple rings.
♥ *C. convolvulacea*, syn. *C. vinciflora*, illus. p.174.
C. ovata. Upright perennial with scarcely twining stems. H to 30cm (12in). Frost hardy. Has small, oval leaves and, in summer, small, bell-shaped, pale blue flowers, often with darker veins.
C. vinciflora. See *C. convolvulacea*.

COELOGYNE. See ORCHIDS.
C. cristata (illus. p.260). Evergreen, epiphytic orchid for a cool greenhouse. H 15cm (6in). In winter produces sprays of crisp, white flowers, 5cm (2in) across and marked orange on each lip. Narrowly oval leaves are 8–10cm (3–4in) long. Needs good light in summer.
C. flaccida (illus. p.260). Evergreen, epiphytic orchid for a cool greenhouse. H 15cm (6in). During spring bears drooping spikes of fragrant, star-shaped, light buff flowers, 4cm (1½in) across, with yellow and brown marks on each lip. Has narrowly oval, semi-rigid leaves, 8–10cm (3–4in) long. Grow in semi-shade in summer.
C. nitida, syn. *C. ochracea* (illus. p.260). Evergreen, epiphytic orchid for a cool greenhouse. H 12cm (5in). In spring produces sprays of very fragrant, white flowers, 2.5cm (1in) across and with a yellow mark on each lip. Narrowly oval, semi-rigid leaves are 8–10cm (3–4in) long. Requires semi-shade in summer.
C. ochracea. See *C. nitida*.
C. speciosa (illus. p.262). Vigorous, evergreen, epiphytic orchid for an intermediate greenhouse. H 25cm (10in). In summer, pendent, light green flowers, 6cm (2½in) across, with brown- and white-marked lips, open in succession along stems. Has broadly oval leaves, 23–5cm (9–10in) long. Grow in good light in summer.

COIX (Gramineae). See GRASSES, BAMBOOS, RUSHES and SEDGES.
C. lacryma-jobi illus. p.184.

COLCHICUM (Liliaceae)
Genus of spring- and autumn-flowering corms, grown for their mainly goblet-shaped blooms, up to 20cm (8in) long, most of which emerge before leaves. Each corm bears 2–7 narrowly strap-shaped to broadly elliptic, basal leaves. Fully to frost hardy. Needs an open, sunny situation and well-drained soil. Propagate by seed or division in autumn.
♥ *C. agrippinum* illus. p.383.
C. autumnale illus. p.381. 'Albo-plenum' is an autumn-flowering corm. H and S 10–15cm (4–6in). Fully hardy. In spring has 3–5 large, semi-erect, basal, glossy, green leaves. Produces a bunch of up to 8 long-tubed, rounded, double, white flowers with 15–30 narrow petals.
C. 'Beaconsfield'. Robust, autumn-flowering corm. H and S 15–20cm (6–8in). Fully hardy. Bears large, goblet-shaped, rich pinkish-purple flowers, faintly chequered and white in centres. Large, semi-erect, basal leaves appear in spring.
C. bivonae, syn. *C. bowlesianum*, *C. sibthorpii*, illus. p.382.
C. bowlesianum. See *C. bivonae*.
♥ *C. byzantinum* illus. p.381.
C. cilicicum illus. p.382.
C. 'Lilac Wonder'. Vigorous, autumn-flowering corm. H and S 15–20cm (6–8in). Fully hardy. Produces goblet-shaped, deep lilac-pink flowers, 15–20cm (6–8in) long. Broad, semi-erect, basal leaves appear in spring.
C. luteum illus. p.378.
C. sibthorpii. See *C. bivonae*.
♥ *C. speciosum.* Vigorous, autumn-flowering corm. H and S 15–20cm (6–8in). Fully hardy. Bears goblet-shaped, pale to deep pinkish-purple flowers, 15–20cm (6–8in) long, often with white throats. Large, semi-erect, basal leaves develop in winter or spring. ♥ 'Album' illus. p.381.
C. 'The Giant'. Robust, autumn-flowering corm. H and S 15–20cm (6–8in). Fully hardy. Produces up to 5 funnel-shaped, deep mauve-pink flowers, each 15–20cm (6–8in) long and fading to white in the centre. Broad, semi-erect, basal leaves form in winter or spring.
C. variegatum. Autumn-flowering corm. H 10–15cm (4–6in), S 8–10cm (3–4in). Frost hardy. Bears widely funnel-shaped, reddish-purple flowers with strong chequered patterns. More or less horizontal, basal leaves with wavy margins appear in spring. Needs a hot, sunny site.
C. 'Waterlily' illus. p.383.

COLEONEMA (Rutaceae)
Genus of evergreen, heath-like shrubs, grown for their flowers and overall appearance. Frost tender, min. 3–5°C (37–41°F). Needs full sun and well-drained, neutral to acid soil. Water potted plants moderately when in full growth, sparingly at other times. For a more compact habit, clip after flowering. Propagate by seed in spring or by semi-ripe cuttings in late summer.
C. pulchrum. Evergreen, spreading to domed shrub with wiry stems. H 60cm–1.2m (2–4ft), S 1–1.5m (3–5ft). Has soft, needle-like, bright green leaves. Carries 5-petalled, pale pink to red flowers in spring-summer.

COLEUS (Labiatae)
Genus of perennials, annuals and evergreen sub-shrubs, grown for their colourful leaves and flowers. Makes excellent pot plants. Frost tender, min. 4–10°C (39–50°F). Grow in sun or partial shade and in fertile, well-drained soil, choosing a sheltered position. Water freely in summer, much less at other times. Pinch out growing shoots of young plants to encourage a bushy habit. Propagate by seed sown under glass in late winter or by softwood cuttings in spring or summer. Mealy bug and whitefly may cause problems.
C. blumei, syn. *Solenostemon scutellariordes*, illus. p.276. 'Brightness' illus. p.281. 'Fashion Parade' is a fast-growing, bushy perennial, grown as an annual. H to

45cm (18in), S 30cm (12in) or more. Min. 10°C (50°F). Has multicoloured, serrated leaves of various shapes, from oval and unlobed to deeply lobed. Spikes of tiny, blue flowers are produced in summer and should be removed. **'Scarlet Poncho'**, with a pendulous habit and oval, serrated, bright red leaves, and **Wizard Series**, also with oval, serrated leaves, but in a very wide range of leaf colours, are both dwarf forms, H 30cm (12in).
C. thyrsoideus, syn. *Plectranthus thyrsoideus*. Fast-growing, bushy perennial, grown as an annual. H to 1m (3ft), S 60cm (2ft). Min. 4°C (39°F). Has oval, serrated, mid-green leaves. Bears panicles of tubular, bright blue flowers in winter.

COLLETIA (Rhamnaceae)
Genus of deciduous, usually leafless shrubs, grown for their curious, spiny shoots and profuse, small flowers. Shoots assume function of leaves. Frost hardy. Requires a sheltered, sunny site and well-drained soil. Propagate by semi-ripe cuttings in late summer.
C. armata. See *C. hystrix*.
C. cruciata. See *C. paradoxa*.
C. hystrix, syn. *C. armata*, illus. p.117. **'Rosea'** is a deciduous, stoutly branched shrub. H 2.5m (8ft), S 5m (15ft). Shoots have rigid, grey-green spines. Bears fragrant, tubular, pink flowers in late summer and early autumn.
C. paradoxa, syn. *C. cruciata*. Deciduous, arching shrub with stiff branches. H 3m (10ft), S 5m (15ft). Bears stout, flattened, blue-green spines. Fragrant, tubular, white flowers are produced in late summer and early autumn.

COLLINSIA (Scrophulariaceae)
Genus of spring- to summer-flowering annuals. Fully hardy. Grow in partial shade and in fertile, well-drained soil. Support with thin sticks. Propagate by seed sown outdoors in spring or early autumn.
C. grandiflora illus. p.283.

COLOCASIA (Araceae)
Genus of deciduous or evergreen, perennial, marginal water plants, grown for their foliage. Has edible tubers, known as 'taros', for which it is widely cultivated. Is suitable for the edges of frost-free pools; may also be grown in wet soil in pots. Frost tender, min. 1°C (34°F); with min. 25°C (77°F) is evergreen. Grows in sun or light shade and in mud or shallow water. Propagate by division in spring.
♀ *C. esculenta* illus. p.388.
'Fontanesii' is a deciduous, perennial, marginal water plant. H 1.1m (3½ft), S 60cm (2ft). Has large, bold, oval, mid-green leaves with dark green veins and margins and blackish-violet leaf stalks and spathe tubes. **'Illustris'** has brownish-purple leaf stalks and dark green leaf blades with purple spots.

COLQUHOUNIA (Labiatae)
Genus of evergreen or semi-evergreen shrubs, grown for their flowers in late summer and autumn. Frost hardy, but is cut to ground level in cold winters.

Needs a sheltered, sunny position and well-drained soil. Propagate by softwood cuttings in summer.
♀ *C. coccinea* illus. p.119.

COLUMNEA (Gesneriaceae)
Genus of evergreen, creeping or trailing perennials or sub-shrubs, grown for their showy flowers. Trailing species are useful for hanging baskets. Frost tender, min. 15°C (59°F). Needs bright but indirect light, a fairly humid atmosphere and moist soil, except in winter. Propagate by tip cuttings after flowering.
C. x *banksii* illus. p.214.
C. crassifolia illus. p.247.
C. gloriosa (Goldfish plant). Evergreen, trailing perennial with more or less unbranched stems. H or S to 90cm (3ft). Oval leaves have reddish hairs. Has tubular, hooded, scarlet flowers, to 8cm (3in) long with yellow throats, in winter-spring.
C. microphylla. Evergreen perennial, sparsely branched on each trailing stem. H or S 1m (3ft) or more. Has small, rounded leaves with brown hairs; hooded, tubular, scarlet flowers, to 8cm (3in) long with yellow throats, are produced in winter-spring.
'Variegata' illus. p.230.

COLUTEA (Leguminosae)
Genus of deciduous, summer-flowering shrubs, grown for their foliage, pea-like flowers and bladder-shaped seed pods. Fully hardy. Needs full sun and any but waterlogged soil. Propagate by softwood cuttings in summer or by seed in autumn.
C. arborescens illus. p.116.
C. x *media* illus. p.117.
C. orientalis. Deciduous, bushy shrub. H and S 2m (6ft). Has blue-grey leaves consisting of 7 or 9 oval leaflets. Clusters of yellow-marked, coppery-red flowers in summer are followed by inflated, green, then pale brown seed pods.

Comarostaphylis diversifolia. See *Arctostaphylos diversifolia*.

COMBRETUM (Combretaceae)
Genus of evergreen trees, shrubs and scandent to twining climbers, grown for their small, showy flowers. Frost tender, min. 16°C (61°F). Provide humus-rich, well-drained soil, with partial shade in summer. Water freely in summer, less at other times. Support for stems is necessary. Thin out and spur back congested growth after flowering. Propagate by semi-ripe cuttings in summer. Red spider mite may be a problem.
C. grandiflorum. Moderately vigorous, evergreen, scandent to twining climber. H to 6m (20ft). Has oblong to elliptic, pointed leaves, 10–20cm (4–8in) long. Tubular, bright red flowers with long stamens are produced in summer in one-sided spikes, 10–13cm (4–5in) long.

COMMELINA (Commelinaceae)
Genus of perennials, usually grown as annuals. Half hardy. Grow in a sunny, sheltered position and in fertile, well-drained soil. Crowns should be lifted

before the frosts and overwintered in slightly moist, frost-free conditions. Propagate by seed sown under glass or by division in spring.
C. coelestis illus. p.286.

CONANDRON (Gesneriaceae)
Genus of one species of tuberous perennial, grown for its fleshy leaves and drooping flower clusters. Grow in alpine houses. Frost hardy. Needs shade and humus-rich, well-drained soil. Containerized plants need moist soil in summer, dry when dormant in winter. Propagate by division or seed in spring.
C. ramondioides. Hummock-forming, tuberous perennial. H 30cm (12in), S 20cm (8in). Bears broadly oval, fleshy, wrinkled, mid-green leaves with toothed edges. In mid-summer, each flower stem carries 5–25 tubular flowers, usually lilac, but white, purple or pink forms also occur.

CONIFERS
Group of trees and shrubs distinguished botanically from others by producing seeds exposed or uncovered on the scales of fruits. Most conifers are evergreen, have needle-like leaves and bear woody fruits (cones). All genera in the Cupressaceae family, however, have needle-like juvenile leaves and, excepting many junipers and some other selected forms, scale-like adult leaves. Conifers described in this book are evergreen unless otherwise stated.
 Conifers make excellent garden plants. Most provide year-round foliage, which may be green, blue, grey, bronze, gold or silver. They range in height from trees 30m (100ft) or more to dwarf shrubs that grow less than 5cm (2in) every 10 years. Tall conifers may be planted as specimen trees or to provide shelter, screening or hedging. Dwarf conifers make good features in their own right as well as in groups; they also associate well with heathers, add variety to rock gardens and provide excellent ground cover. They may also be grown in containers.

Hardiness
Nearly all conifers described in this book are fully hardy. Some, such as *Picea sitchensis*, *P. omorika* and *Pinus contorta*, are extremely tough and thrive in the coldest, most windswept positions. *Araucaria*, *Cupressus* and *Pinus* are good for coastal conditions. *Athrotaxis*, *Austrocedrus*, *Cephalotaxus*, *Podocarpus* and certain species noted in other genera are frost to half hardy and flourish only in mild localities. Elsewhere they need a very sheltered position or may be grown indoors as dwarf plants.
 Frost may damage new growth of several genera, especially *Abies*, *Larix*, *Picea* and *Pseudotsuga*, and severe cold, dry spells in winter may temporarily harm mature foliage.

Position and soil
x *Cupressocyparis*, *Cupressus*, *Larix* and *Pinus* need full sun. *Cedrus*, *Juniperus* and *Pseudolarix* do not tolerate shade. All other conifers will thrive in sun or shade, and most *Abies*

and all *Cephalotaxus*, *Podocarpus*, *Taxus*, *Thuja*, *Torreya* and *Tsuga* will grow in deep shade once established.
 Conifers grow well on most soils, but certain genera and species will not do well on soils over chalk or limestone. In this book such conifers are: *Abies*, *Pseudolarix*, *Pseudotsuga* and *Tsuga*; also *Picea*, except *P. likiangesis*, *P. omorika* and *P. pungens*; and *Pinus*, except *P. aristata*, *P. armandii*, *P. cembroides*, *P. halepensis*, *P. heldreichii* var. *leucodermis*, *P. nigra*, *P. peuce* and *P. wallichiana*.
 Certain conifers tolerate extreme conditions. *Abies alba*, *A. homolepis*, *A. nordmanniana*, *Cryptomeria*, *Cunninghamia*, *Metasequoia*, *Pinus coulteri*, *P. peuce*, *P. ponderosa*, *Sciadopitys*, *Sequoia*, *Sequoiadendron* and *Taxodium* will grow on heavy clay soils. *Picea omorika*, *P. sitchensis*, *Pinus contorta*, *Sciadopitys verticillata* and *Thuja plicata* are all happy on wet soil, and *Metasequoia* and *Taxodium* thrive in waterlogged conditions. *Cupressus*, *Juniperus* and *Pinus* grow well on dry, sandy soil.

Pruning
If a conifer produces more than one leader, remove all but one. Bear in mind when trimming hedges that most conifers will not make new growth when cut back into old wood or from branches that have turned brown. This does not, however, apply to *Cephalotaxus*, *Cryptomeria*, *Cunninghamia*, *Sequoia*, *Taxus* and *Torreya*, and these conifers may be kept to a reasonable size in the garden by cutting back the main stem, which will later coppice (make new growth). Young specimens of *Araucaria*, *Ginkgo*, *Metasequoia* and *Taxodium* will sometimes do the same.

Propagation
Seed is the easiest method of propagation, but forms selected for leaf colour (other than blue in some species) do not come true. Sow in autumn or spring. All genera apart from *Abies*, *Cedrus*, *Picea* (except young plants or dwarf forms), *Pinus*, *Pseudolarix*, *Pseudotsuga* and *Tsuga* (except young plants or dwarf forms) may be raised fairly easily from cuttings: current growth from autumn to spring for evergreens, softwood cuttings in summer for deciduous conifers. Tall-growing forms of Pinaceae (*Abies*, *Cedrus*, *Picea*, *Pinus*, *Pseudolarix*, *Pseudotsuga* and *Tsuga*) are usually propagated by grafting in late summer, winter or early spring. Layering may be possible for some dwarf conifers.

Pests and diseases
Honey fungus attacks many conifers, especially young plants. Most resistant to the disease are *Abies*, *Calocedrus*, *Larix*, *Pseudotsuga* and *Taxus*. Green spruce aphid may be a problem on *Picea*, and conifer spinning mite may defoliate *Abies*, *Picea* and some *Pinus*.

Tall-growing conifers are illustrated on pp.75–83, dwarf forms on pp.84–5. See also *Abies*, *Araucaria*, *Athrotaxis*, *Austrocedrus*, *Calocedrus*, *Cedrus*, *Cephalotaxus*, *Chamaecyparis*,

Cryptomeria, Cunninghamia,
x *Cupressocyparis, Cupressus, Fitzroya, Ginkgo, Juniperus, Larix, Metasequoia, Microbiota, Phyllocladus, Picea, Pinus, Podocarpus, Pseudolarix, Pseudotsuga, Saxegothaea, Sciadopitys, Sequoia, Sequoiadendron, Taxodium, Taxus, Thuja, Thujopsis, Torreya* and *Tsuga.*

CONOPHYTUM (Aizoaceae)
Genus of slow-growing, clump-forming, perennial succulents with spherical or 2-eared leaves that grow for only 2 months each year, after flowering. In early summer, old leaves gradually shrivel to papery sheaths from which new leaves and flowers emerge in late summer. Frost tender, min. 4°C (39°F) if dry. Requires full sun and well-drained soil. Keep dry in winter. Propagate by seed from spring to autumn or by division in late summer.
C. bilobum illus. p.415.
C. meyeri. Slow-growing, clump-forming, perennial succulent. H 2–3cm ($^3/_4$–1$^1/_4$in), S 10cm (4in). Egg-shaped, grey-green leaves each have a 2-eared, fissured tip. Carries 1.5–2cm ($^5/_8$–$^3/_4$in) wide, daisy-like, yellow flowers in autumn.
C. notabile illus. p.416.
C. truncatum illus. p.406.

CONSOLIDA (Ranunculaceae)
Larkspur
Genus of annuals, providing excellent cut flowers. Fully hardy. Grow in sun and in fertile, well-drained soil. Support stems of tall-growing plants with sticks. Propagate by seed sown outdoors in spring, or in early autumn in mild areas. Protect young plants from slugs and snails.
C. ajacis. See *C. ambigua.*
C. ambigua, syn. *C. ajacis, Delphinium consolida.* Fast-growing, upright, branching annual. Giant forms, H to 1.2m (4ft), S 30cm (1ft); dwarf, H and S 30cm (1ft). All have feathery, mid-green leaves and, throughout summer, spikes of rounded, spurred flowers. **Hyacinth-flowered Series** (dwarf) has spikes of tubular flowers in shades of pink, mauve, blue or white. **Imperial Series** illus. p.284.

CONVALLARIA (Liliaceae)
Lily-of-the-valley
Genus of spring-flowering, rhizomatous perennials. Fully hardy. Prefers partial shade and will grow in any soil but does best in humus-rich, moist soil. Propagate by division after flowering or in autumn.
♀ *C. majalis* illus. p.232. **'Flore Pleno'** is a low-growing, rhizomatous perennial. H 23–30cm (9–12in), S indefinite. Sprays of small, very fragrant, pendent, bell-shaped flowers that are double and white open in spring. Narrowly oval leaves are mid- to dark green. **'Fortin's Giant'**, H 45cm (18in), has larger flowers and leaves that appear a little earlier.

CONVOLVULUS (Convolvulaceae)
Genus of dwarf, bushy and climbing annuals, perennials and evergreen shrubs and sub-shrubs. Fully hardy to frost tender, min 2°C (36°F). Grow in sun and in poor to fertile, well-drained soil. Dead-head to prolong flowering. Propagate by seed sown outdoors in mid-spring for hardy plants or under glass in spring for tender plants, perennials and sub-shrubs by softwood cuttings in late spring or summer.
C. althaeoides illus. p.323.
♀ *C. cneorum* illus. p.128.
C. mauritanicus. See *C. sabatius.*
C. minor. See *C. tricolor.*
C. purpureus. See *Ipomoea purpurea.*
♀ *C. sabatius,* syn. *C. mauritanicus,* illus. p.304.
C. tricolor, syn. *C. minor.* Moderately fast-growing, upright, bushy or climbing annual. H 20–30cm (8–12in), S 20cm (8in). Fully hardy. Has oval to lance-shaped, mid-green leaves. In summer bears saucer-shaped, blue or white flowers. 2.5cm (1in) wide, with yellowish-white throats. Tall, climbing forms, H to 3m (10ft), are half hardy and have flowers to 10cm (4in) wide. **'Blue Flash'** (bushy) illus. p.284. **'Flying Saucers'** (climber) has blue-and-white-striped flowers. **'Major'** (climber) and **'Minor'** (bushy) have flowers in shades of blue, red or white.

COPIAPOA (Cactaceae)
Genus of slow-growing, perennial cacti with funnel-shaped, yellow flowers. Many species have large tap roots. Frost tender, min. 8–10°C (46–50°F). Needs partial shade and very well-drained soil. Propagate by seed or grafting in spring or summer.
C. cinerea illus. p.402.
C. coquimbana. Clump-forming, spherical, then columnar, perennial cactus. H to 30cm (1ft), S 1m (3ft). Min. 8°C (46°F). Dark grey-green stem has 10–17 ribs. Areoles each bear 8–10 dark brown radial spines and 1 or 2 stouter central spines. Yellow flowers, 3cm (1$^1/_4$in) across, appear in summer. Is slow to form clumps.
C. echinoides. Flattened spherical, perennial cactus, ribbed like a sea-urchin. H 15cm (6in), S 10cm (4in). Min. 10°C (50°F). Solitary grey-green stem bears dark brown spines, 3cm (1$^1/_4$in) long, which soon fade to grey. In summer produces pale yellow flowers, 4cm (1$^1/_2$in) across.
C. marginata. Clump-forming, perennial cactus. H 60cm (2ft), S 30cm (1ft). Min. 10°C (50°F). Grey-green stem bears very close-set areoles with dark-tipped, pale brown spines, to 3cm (1$^1/_4$in) long. Has yellow flowers, 2–5cm ($^3/_4$–2in) across, in spring-summer.

COPROSMA (Rubiaceae)
Genus of evergreen shrubs and trees, grown for their foliage and fruits. Separate male and female plants are needed to obtain fruits. Half hardy to frost tender, min. 2–5°C (36–41°F). Prefers full light and well-drained soil. Water containerized specimens freely in summer, moderately at other times. Propagate by seed in spring or by semi-ripe cuttings in late summer.
C. x *kirkii.* Evergreen, prostrate, then semi-erect, densely branched shrub. H to 1m (3ft), S 1.2–2m (4–6ft). Half hardy. Narrowly oblong to lance-shaped, leathery, glossy leaves are borne singly or in small clusters. In late spring has insignificant flowers, followed on female plants by tiny, egg-shaped, translucent, white fruits with red speckles. **'Variegata'** illus. p.145.
C. repens. Evergreen, spreading, then erect shrub. H and S to 2m (6ft). Frost tender, min. 2°C (36°F). Has broadly oval, leathery, lustrous, rich green leaves. Carries insignificant flowers in late spring, followed on female plants by egg-shaped, orange-red fruits from late summer to autumn. Leaves of **'Picturata'** each have a central, cream blotch.

CORDYLINE (Agavaceae)
Genus of evergreen shrubs and trees, grown primarily for their foliage, although some also have decorative flowers. Half hardy to frost tender, min. 5–16°C (41–61°F). Provide fertile, well-drained soil and full light or partial shade. Water potted plants moderately, less in winter. Propagate by seed or suckers in spring or by stem cuttings in summer. Red spider mite may be a nuisance.
♀ *C. australis* (New Zealand cabbage palm). Slow-growing, evergreen, sparsely branched tree. H 15m (50ft) or more, S 5m (15ft) or more. Half hardy. Each stem is crowned by a rosette of strap-shaped, 30cm–1m (1–3ft) long leaves. Has small, scented, white flowers in large, open panicles in summer and, in autumn, globose, white fruits. **'Atropurpurea'** illus. p.73. Long, sword-shaped leaves of **'Veitchii'** have red bases and midribs.
C. fruticosa, syn. *C. terminalis* (Good-luck plant, Ti tree). Slow-growing, evergreen, upright shrub, sparingly branched and suckering. H 2–4m (6–12ft), S 1–2m (3–6ft). Frost tender, min. 13°C (55°F). Broadly lance-shaped, glossy, deep green leaves are 30–60cm (1–2ft) long. Produces branched panicles of small, white, purplish or reddish flowers in summer. Foliage of **'Baptisii'** is deep green with pink and yellow stripes and spots. **'Imperialis'** has red- or pink-marked, deep green leaves.
C. indivisa, syn. *Dracaena indivisa.* Slow-growing, evergreen, erect tree or shrub. H 3m (10ft) or more, S 2m (6ft). Half hardy. Bears lance-shaped, 60cm–2m (2–6ft) long, green leaves, orange-brown veined above, blue-grey tinted beneath. In summer has tiny, star-shaped, white flowers in dense clusters, 60cm (2ft) or more long, followed by tiny, spherical, blue-purple fruits.
C. terminalis. See *C. fruticosa.*

COREOPSIS (Compositae)
Tickseed
Genus of annuals and perennials, grown for their daisy-like flower heads. Fully to frost hardy. Needs full sun and fertile, well-drained soil. Propagate annuals by seed in spring; *C. lanceolata* by seed or division in spring; *C. auriculata* 'Superba', *C.* 'Goldfink' and *C. grandiflora* 'Badengold' by softwood cuttings or division in spring or summer; and *C. verticillata* by division in spring.
C. auriculata **'Superba'**. Bushy perennial. H and S 45cm (18in). Fully hardy. Daisy-like, rich yellow flower heads, with central, purple blotches, are produced throughout summer. Oval to lance-shaped leaves are lobed and light green. Some plants grown as *C. auriculata* are the closely related annual *C. basalis.*
C. **'Goldfink'**. Short-lived, dwarf, bushy perennial. H and S 30cm (12in). Fully hardy. Sprays of daisy-like, deep yellow flower heads appear in summer above narrowly oval, deep green leaves.
C. grandiflora **'Badengold'**. Short-lived, erect perennial with lax stems. H 75cm (30in), S 60cm (24in). Fully hardy. Bears large, daisy-like, rich buttercup-yellow flower heads in summer and broadly lance-shaped, divided, bright green leaves.
C. lanceolata illus. p.257.
C. **'Sunray'** illus. p.289.
C. tinctoria illus. p.288. **'Golden Crown'** is a fast-growing, upright, bushy annual. H 60cm (24in), S 20cm (8in). Fully hardy. Has lance-shaped, deep green leaves and, in summer and early autumn, large, daisy-like, deep yellow flower heads with brown centres.
C. verticillata illus. p.257.

CORIARIA (Coriariaceae)
Genus of deciduous, spring- or summer-flowering shrubs and sub-shrubs, grown for their habit, foliage and fruits. Frost to half hardy. Needs full sun and fertile, well-drained soil. Propagate by softwood cuttings in summer or by seed in autumn.
C. terminalis. Deciduous, arching sub-shrub. H 1m (3ft), S 2m (6ft). Frost hardy. Broadly lance-shaped, fern-like, mid-green leaves turn red in autumn. Minute, green flowers in late spring are succeeded by small, spherical, black fruits. var. **xanthocarpa** illus. p.143.

CORNUS (Cornaceae)
Dogwood
Genus of deciduous shrubs and deciduous or evergreen trees, grown for their flowers, foliage or brightly coloured winter stems. Fully to half hardy. Needs sun or semi-shade and fertile, well-drained soil. Those grown for winter stem colour do best in full sun. *C. florida, C. kousa* and *C. nuttallii* dislike shallow, chalky soil. *C. canadensis* prefers acid soil. Plants grown for their stems should be cut back almost to ground level each year in early spring. Propagate *C. alba* and *C. stolonifera* 'Flaviramea' by softwood cuttings in summer or by hardwood cuttings in autumn or winter; variegated forms of *C. alternifolia* and *C. controversa* by grafting in winter; *C. canadensis* by division in spring or autumn; *C. capitata, C. florida* and *C. kousa* by seed in autumn or by softwood cuttings in summer; *C. nuttallii* by seed in autumn; all others described here by softwood cuttings in summer.
C. alba (Red-barked dogwood). Vigorous, deciduous, upright, then spreading shrub. H and S 3m (10ft).

Fully hardy. Young shoots are bright red in winter. Has oval, dark green leaves, often becoming red or orange in autumn. Flattened heads of star-shaped, creamy-white flowers in late spring and early summer are succeeded by spherical, sometimes blue-tinted, white fruits. ♛ **'Elegantissima'** illus. p.108. **'Gouchaultii'** has pink-flushed leaves broadly edged with yellow. **'Kesselringii'** illus. p.118.
♛ **'Sibirica'** illus. p.120. ♛ **'Spaethii'** illus. p.116.
C. alternifolia. Deciduous, spreading tree or bushy shrub, with tiered branches. H and S 6m (20ft). Fully hardy. Oval, bright green leaves, which each taper to a point, often turn red in autumn. Clusters of tiny, star-shaped, creamy-white flowers in early summer are followed by small, rounded, blue-black fruits. ♛ **'Argentea'** illus. p.63.
♛ *C. canadensis* illus. p.322.
C. capitata, syn. *Dendrobenthamia capitata* (Bentham's cornel). Evergreen or semi-evergreen, spreading tree. H and S up to 12m (40ft). Frost to half hardy. Pale yellow bracts, surrounding insignificant flowers, appear in early summer, followed by large, strawberry-like, red fruits. Has oval, grey-green leaves. Is good for mild coastal areas.
C. controversa. Deciduous tree with layered branches. H and S 15m (50ft). Fully hardy. Clusters of small, star-shaped, white flowers appear in summer. Leaves are oval, pointed and bright green, turning purple in autumn.
♛ **'Variegata'** illus. p.63.
♛ *C.* **'Eddie's White Wonder'** illus. p.70.
C. florida (Flowering dogwood). Deciduous, spreading tree. H 6m (20ft), S 8m (25ft). Fully hardy. Flower heads are borne in late spring, consisting of white or pinkish-white bracts around tiny, insignificant flowers. Oval, pointed, dark green leaves turn red and purple in autumn. **'Apple Blossom'** has pale pink bracts. f. *rubra* bears pink or red bracts.
'Spring Song' illus. p.64. **'Welchii'** illus. p.67. **'White Cloud'** illus. p.59.
C. kousa. Deciduous, vase-shaped tree or shrub. H 7m (22ft), S 5m (15ft). Fully hardy. Flower heads of large, white bracts, surrounding insignificant flowers, appear in early summer, followed, after a hot summer, by strawberry-like fruits. Oval, glossy, dark green leaves turn bright red-purple in autumn. ♛ var. *chinensis* has larger flower heads and more narrowly pointed bracts.
C. macrophylla illus. p.52.
♛ *C. mas* (Cornelian cherry). Deciduous, spreading, open shrub or tree. H and S 5m (15ft). Fully hardy. Oval, dark green leaves change to reddish-purple in autumn. Produces small, star-shaped, yellow flowers on bare shoots in late winter and early spring, then edible, oblong, bright red fruits. **'Aureoelegantissima'**, syn. *C. m.* 'Elegantissima', H 2m (6ft), S 3m (10ft), has pink-tinged leaves edged with yellow. ♛ **'Variegata'** illus. p.89.
♛ *C.* **'Norman Hadden'.** Deciduous, spreading tree. H and S 8m (25ft). Fully hardy. Creamy-white bracts around tiny flowers turn to deep pink in summer. These are often followed by heavy crops of strawberry-like fruits in autumn.
C. nuttallii illus. p.50.
♛ *C.* **'Porlock'** illus. p.60.
♛ *C. stolonifera* **'Flaviramea'.** Vigorous, deciduous, spreading shrub with creeping, underground stems. H 2m (6ft), S 4m (12ft). Fully hardy. Has bright greenish-yellow, young shoots in winter and oval, dark green leaves. Small, star-shaped, white flowers appear in late spring and early summer, followed by spherical, white fruits.

COROKIA (Cornaceae)
Genus of evergreen shrubs, grown for their habit, foliage, flowers and fruits. Is good in mild, coastal areas, where it is very wind-tolerant. Frost to half hardy; in cold areas protect from strong winds. Needs full sun and fertile, well-drained soil. Propagate by softwood cuttings in summer.
C. buddleioides. Evergreen, upright shrub. H 3m (10ft), S 2m (6ft). Half hardy. Has slender, grey shoots and narrowly oblong, glossy, dark green leaves. Produces panicles of star-shaped, yellow flowers in late spring, followed by spherical, blackish-red fruits.
C. cotoneaster illus. p.122.
C. x *virgata.* Evergreen, upright, dense shrub. H and S 3m (10ft). Frost hardy. Leaves are oblong and glossy, dark green above, white beneath. Produces star-shaped, yellow flowers in mid-spring, then egg-shaped, bright orange fruits. Makes a good hedge, especially in coastal areas.

CORONILLA (Leguminosae)
Genus of deciduous or evergreen shrubs and perennials, grown for their foliage and flowers. Fully to half hardy; in cold areas grow half hardy species against a south- or west-facing wall. Requires full sun and light, well-drained soil. Propagate by softwood cuttings in summer.
♛ *C. valentina* subsp. *glauca* illus. p.127.

CORREA (Rutaceae)
Genus of evergreen shrubs, grown for their flowers. Half hardy to frost tender, min. 3–5°C (37–41°F). Prefers full light or partial shade and fertile, well-drained, neutral to acid soil. Water potted specimens moderately, less when not in flower. Propagate by seed in spring or by semi-ripe cuttings in late summer.
♛ *C. backhousiana.* Evergreen, rounded, well-branched shrub. H and S 2m (6ft). Half hardy. Leaves are oval to elliptic and dark green, with dense, pale buff down beneath. Tubular, pale yellow-green to white flowers appear in spring and intermittently until autumn.
C. **'Harrisii'.** See *C.* 'Mannii'.
C. x *harrisii.* See *C.* 'Mannii'.
♛ *C.* **'Mannii'**, syn. *C.* x *harrisii*, *C.* 'Harrisii'. Evergreen, bushy, slender-stemmed shrub. H and S 2m (6ft). Frost tender. Has narrowly oval leaves with short hairs beneath. Tubular, scarlet flowers are carried in summer-autumn, sometimes in other seasons.
C. pulchella illus. p.144.
C. reflexa, syn. *C. speciosa.* Evergreen, bushy, slender-stemmed shrub. H and S to 2m (6ft). Frost tender. Oval leaves have thick down beneath. Bears tubular, greenish-yellow to crimson or rose flowers, with greenish-white petal tips, in summer-autumn, sometimes in other seasons.
C. speciosa. See *C. reflexa.*

CORTADERIA (Gramineae). See GRASSES, BAMBOOS, RUSHES and SEDGES.
C. selloana (Pampas grass). Evergreen, clump-forming, stately, perennial grass. H to 2.5m (8ft), S 1.2m (4ft). Frost hardy. Has narrow, very sharp-edged, outward-curving leaves, 1.5m (5ft) long. In late summer, erect, plume-like, silvery panicles, up to 60cm (2ft) long, are borne above mid-green leaves. Sexes are produced on separate plants; female flowers, with long, silky hairs, are more decorative. **'Silver Comet'** and ♛ **'Sunningdale Silver'** illus. p.182.

CORTUSA (Primulaceae)
Genus of clump-forming, spring- and summer-flowering perennials, related to *Primula,* with one-sided racemes of bell-shaped flowers. Fully hardy. Is not suited to hot, dry climates as needs shade and humus-rich, moist soil. Propagate by seed when fresh or by division in autumn.
C. matthioli illus. p.302.

CORYDALIS (Papaveraceae)
Genus of spring- and summer-flowering annuals and tuberous or fibrous-rooted perennials, some of which are evergreen, grown for their tubular, spurred, 2-lipped flowers or for their fern-like leaves. Fully to frost hardy. Needs full sun or partial shade and well-drained soil; some require humus-rich soil and cool growing conditions. Propagate by seed in autumn or by division when dormant: autumn for spring-flowering species, spring for summer-flowering species.
C. ambigua. Spring-flowering, tuberous perennial. H 10–15cm (4–6in), S 5–8cm (2–3in). Frost hardy. Stem bears divided, semi-erect leaves and a short spike of 2-lipped, short-spurred, bright blue or purplish-blue flowers. Dies down in summer.
♛ *C. bulbosa,* syn. *C. cava.* Spring-flowering, tuberous perennial. H 10–20cm (4–8in), S 8–10cm (3–4in). Fully hardy. Leaves are semi-erect, basal and much divided. Carries dense spikes of tubular, dull purple flowers. Dies down in summer.
C. cashmeriana. Tuft-forming, fibrous-rooted perennial. H 10–25cm (4–10in), S 8–10cm (3–4in). Fully hardy. Has divided, semi-erect, basal leaves and, in summer, dense spikes of 2-lipped, brilliant blue flowers. Needs cool, partially shaded, humus-rich, neutral to acid soil. Is good for a rock garden. Dies down in winter.
C. cava. See *C. bulbosa.*
C. cheilanthifolia illus. p.297.
C. diphylla illus. p.315.
C. lutea, syn. *Pseudofumaria lutea,* illus. p.307.
C. nobilis. Perennial with long, fleshy, fibrous roots. H and S 20–35cm (8–14in). Fully hardy. Bears much divided leaves on lower part of flower stems, each of which carries a dense spike of long-spurred, pale yellow flowers, with lips tipped green or brown, in early summer.
C. ochroleuca, syn. *Pseudofumaria ochroleuca,* illus. p.299.
C. popovii illus. p.312.
♛ *C. solida.* Tuft-forming, tuberous perennial. H 10–20cm (4–8in), S 8–12cm (3–5in). Fully hardy. Leaves alternate on flower stem, which carries a dense spike of dull purplish-red flowers in spring. Dies down in summer. ♛ **'George Baker'** (syn. *C.s.* 'G.P. Baker') illus. p.315.
C. wilsonii illus. p.297.

CORYLOPSIS (Hamamelidaceae)
Genus of deciduous shrubs and trees, grown for their fragrant, yellow flowers, which are produced before hazel-like leaves emerge. Fully hardy, but late frosts may damage flowers. Prefers semi-shade and fertile, moist but well-drained, acid soil. Propagate by softwood cuttings in summer or by seed in autumn.
C. glabrescens illus. p.87.
♛ *C. pauciflora* illus. p.101.
♛ *C. sinensis,* syn. *C. willmottiae.* Vigorous, deciduous, spreading, open shrub. H and S 4m (12ft). Leaves are bright green above, blue-green beneath. Clusters of bell-shaped, pale yellow flowers open from early to mid-spring. **'Spring Purple'** has deep plum-purple, young leaves.
C. spicata. Deciduous, spreading, open shrub. H 2m (6ft), S 3m (10ft). Bristle-toothed leaves are dull, pale green above, blue-green beneath. Drooping clusters of bell-shaped, pale yellow flowers are borne in mid-spring.
C. willmottiae. See *C. sinensis.*

CORYLUS (Corylaceae)
Hazel
Genus of deciduous trees and shrubs, grown for their habit, catkins and often edible fruits (nuts). Fully hardy. Prefers sun or semi-shade and fertile, well-drained soil. Cut out suckers as they arise. Propagate species by seed in autumn, cultivars by grafting in late summer or by suckers or layering in late autumn to early spring. Mildew may cause defoliation; other fungi and insects may spoil nuts.
C. avellana (Cobnut). ♛ **'Contorta'** illus. p.95.
♛ *C. colurna* (Turkish hazel). Deciduous, conical tree. H 20m (70ft), S 7m (22ft). Has broadly oval, strongly toothed, almost lobed, dark green leaves. Long, yellow catkins are borne in late winter. Clusters of nuts are set in fringed husks.
C. maxima (Filbert). Vigorous, deciduous, bushy, open shrub or tree. H 6m (20ft), S 5m (15ft). Bears oval, toothed, mid-green leaves, long, yellow catkins in late winter and edible, egg-shaped, brown nuts. ♛ **'Purpurea'** illus. p.92.

469

CORYNOCARPUS
(Corynocarpaceae)
Genus of evergreen trees, grown for their foliage and overall appearance. Frost tender, min. 7–10°C (45–50°F). Needs full light or partial shade and fertile, moisture-retentive but well-drained soil. Water containerized specimens moderately, less when temperatures are low. Pruning is tolerated if necessary. Propagate by seed when ripe or by semi-ripe cuttings in summer.
C. laevigata illus. p.59.

CORYPHANTHA (Cactaceae)
Genus of perennial cacti with roughly spherical, spiny, green stems. Stems have elongated areoles in grooves running along upper sides of tubercles; many species only show this groove on very old plants. Has funnel-shaped flowers in summer, then cylindrical, green seed pods. Frost tender, min. 5°C (41°F). Needs full sun and very well-drained soil. Propagate by seed in spring or summer.
C. cornifera, syn. *C. radians*, illus. p.415.
C. radians. See *C. cornifera*.
C. vivipara illus. p.407.

COSMOS (Compositae)
Genus of summer- and early autumn-flowering annuals and tuberous perennials. Fully to half hardy. Needs sun and does best in moist but well-drained soil. In mild areas, tubers of half hardy *C. atrosanguineus* may be overwintered in ground if protected with a deep mulch. Propagate half-hardy species by basal cuttings in spring, annuals by seed in autumn or spring.
C. atrosanguineus, syn. *Bidens atrosanguinea*, illus. p.213.
C., **Bright Lights Series** (Yellow cosmos). Group of moderately fast-growing, upright, bushy annuals. H 60cm (2ft), S 45cm (1½ft). Half hardy. Has feathery, deep green leaves and, in summer and early autumn, large, daisy-like, single or semi-double, yellow, orange or red flower heads.
C. 'Sensation' illus. p.277.
C. 'Sunny Gold'. Moderately fast-growing, upright, bushy annual. H and S 45cm (18in). Half hardy. Has feathery, deep green leaves; large, daisy-like, golden-yellow flower heads appear in summer and early autumn.
C. 'Sunny Red'. Moderately fast-growing, upright, bushy annual. H 30cm (12in), S 45cm (18in). Half hardy. Feathery leaves are deep green; large, daisy-like, orange-vermilion flower heads appear in summer and early autumn.

COSTUS (Zingiberaceae)
Spiral flag, Spiral ginger
Genus of mostly clump-forming, rhizomatous perennials with showy flowers. Frost tender, min. 18°C (64°F). Grow in a humid atmosphere, out of direct sunlight, in humus-rich soil. Propagate by division in spring. Pot-grown plants may be attacked by red spider mite.
C. speciosus (Malay ginger). Clump-forming, rhizomatous perennial. H 2m (6ft) or more, S 1m (3ft). Has narrowly oval, downy leaves, to 25cm (10in) long. Reddish bracts are spine-tipped, each surrounding one white or pink-flushed flower, to 10cm (4in) wide with a broad, yellow-centred lip; flowers are produced intermittently throughout the year.

COTINUS (Anacardiaceae)
Genus of deciduous shrubs and trees, grown for their foliage, flower heads and autumn colour. Individual flowers are inconspicuous. Fully hardy. Requires full sun or semi-shade and fertile but not over-rich soil. Purple-leaved forms need full sun to bring out their best colours. Propagate species by softwood or greenwood cuttings in summer or by seed in autumn, cultivars by cuttings only in summer.
♈ *C. coggygria*, syn. *Rhus cotinus* (Smoke tree, Venetian sumach). Deciduous, bushy shrub. H and S 5m (15ft). Leaves are rounded or oval and light green, becoming yellow or red in autumn. From late summer, as insignificant fruits develop, masses of tiny flower stalks form showy, pale fawn, later grey, plume-like clusters.
♈ 'Flame' illus. p.94. 'Notcutt's Variety' illus. p.91. ♈ 'Royal Purple' has deep pink plumes and deep purplish-red leaves.
♈ *C. obovatus*, syn. *Rhus cotinoides*. Vigorous, deciduous, bushy shrub or tree. H 10m (30ft), S 8m (25ft). Has large, oval leaves that are bronze-pink when young, maturing to mid-green and turning orange, red and purple in autumn.

COTONEASTER (Rosaceae)
Genus of deciduous, semi-evergreen or evergreen shrubs and trees, grown for their foliage, flowers and fruits. Some species make fine specimen plants; others may be used for hedging or ground cover. Fully to frost hardy. Deciduous species and cultivars prefer full sun, but evergreens do well in either sun or semi-shade. All resent waterlogged soil and are particularly useful for dry sites. Propagate species by cuttings in summer or by seed in autumn, hybrids and cultivars by cuttings only, in summer. Take semi-ripe cuttings for evergreens and semi-evergreens, softwood cuttings for deciduous plants. Fireblight is a common problem.
♈ *C. adpressus*. Deciduous, arching shrub. H 30cm (1ft), S 2m (6ft). Fully hardy. Rounded, wavy-edged, dark green leaves redden in autumn. Produces small, 5-petalled, pink flowers in early summer, then spherical, red fruits.
C. 'Autumn Fire', syn. *C.* 'Herbstfeuer'. Evergreen, prostrate or arching shrub. H 30cm (1ft), S 2m (6ft). Fully hardy. Has lance-shaped, bright green leaves. Small, 5-petalled white flowers in early summer are followed by spherical, bright red fruits. May be grown as ground cover or a weeping standard.
C. bullatus var. *macrophyllus*. See *C. rehderi*.
C. cochleatus, syn. *C. microphyllus* var. *cochleatus*. Evergreen, prostrate shrub. H to 45cm (18in), S 2m (6ft). Has small, oval, notched, dark green leaves. Small, white flowers are produced in late spring, followed by spherical, red fruits.
C. congestus. Evergreen, prostrate shrub. H 20cm (8in), S 2m (6ft). Fully hardy. Forms dense mounds of oval, dull green leaves. Produces small, 5-petalled, pinkish-white flowers in early summer and spherical, bright red fruits. Is good for a rock garden.
C. conspicuus, syn. *C.c.* var. *decorus*. Evergreen, prostrate, arching shrub. H 30cm (1ft), S 2–3m (6–10ft). Fully hardy. Leaves are oblong, glossy, very dark green. Small, 5-petalled, white flowers in late spring are succeeded by large, spherical, scarlet or orange-red fruits.
C. 'Coral Beauty'. Evergreen, arching, dense shrub. H 1m (3ft), S 2m (6ft). Fully hardy. Has small, oval, glossy, dark green leaves and, in early summer, small, 5-petalled, white flowers. Fruits are spherical and bright orange-red.
♈ *C.* 'Cornubia' illus. p.94.
C. dielsianus. Deciduous, arching shrub. H and S 2.5m (8ft). Fully hardy. Slender shoots are clothed in oval, dark green leaves. Produces small, 5-petalled, pink flowers in early summer, then spherical, glossy, red fruits.
C. divaricatus illus. p.100.
C. 'Exburiensis'. Evergreen or semi-evergreen, arching shrub. H and S 5m (15ft). Frost hardy. Has narrowly lance-shaped, bright green leaves, small, 5-petalled, white flowers, in early summer, and spherical, yellow fruits, sometimes tinged pink later.
C. 'Firebird'. Deciduous, bushy, open shrub. H and S 3m (10ft). Fully hardy. Large, oval, deeply veined, dark green leaves redden in autumn. Small, 5-petalled, white flowers in early summer are followed by masses of spherical, bright red fruits.
C. franchetii. Evergreen or semi-evergreen, arching shrub. H and S 3m (10ft). Fully hardy. Oval, grey-green leaves are white beneath. Bears small, 5-petalled, pink-tinged, white flowers in early summer, then a profusion of oblong, bright orange-red fruits.
var. *sternianus*. See *C. sternianus*.
C. frigidus (Tree cotoneaster). Vigorous, deciduous tree, upright when young, arching when mature. H and S 10m (30ft). Fully hardy. Has large, broadly oval, wavy-edged, dull green leaves and broad heads of small, 5-petalled, white flowers borne in early summer, followed by large clusters of long-lasting, small, spherical, bright red fruits.
C. glaucophyllus. Evergreen, arching, open shrub. H and S 3m (10ft). Fully hardy. Leaves are oval, dark green, bluish-white beneath. Produces small, 5-petalled, white flowers in mid-summer, followed by small, spherical, deep red fruits in autumn. var. *serotinus*. See *C. serotinus*.
C. 'Gnom', syn. *C.* 'Gnome'. Evergreen, prostrate shrub. H 20cm (8in), S 2m (6ft). Fully hardy. Bears narrowly lance-shaped, dark green leaves, small, 5-petalled, white flowers, in early summer, and clusters of small, spherical, red fruits. Makes good ground cover.
C. 'Gnome'. See *C.* 'Gnom'.
C. 'Herbstfeuer'. See *C.* 'Autumn Fire'.
♈ *C. horizontalis* illus. p.142.
C. hupehensis. Deciduous, arching shrub. H 2m (6ft), S 3m (10ft). Fully hardy. Oval, bright green leaves become yellow in autumn. Masses of small, 5-petalled, white flowers in late spring are succeeded by large, spherical, bright red fruits.
C. 'Hybridus Pendulus'. Evergreen, prostrate shrub, almost always grown as a weeping standard. H 2m (6ft), S 1.5m (5ft). Frost hardy. Has oblong, dark green leaves. Small, 5-petalled, white flowers in early summer are followed by spherical, deep red fruits.
♈ *C. lacteus* illus. p.95.
C. linearifolius, syn. *C. microphyllus* var. *thymifolius*. Evergreen, prostrate shrub. H 60cm (2ft), S 2m (6ft). Fully hardy. Rigid branches bear tiny, narrow, blunt-ended, glossy leaves. Bears small, white flowers in late spring, followed by spherical red fruits.
♈ *C. microphyllus*. Evergreen, spreading, dense shrub. H 1m (3ft), S 2m (6ft). Fully hardy. Rigid shoots are clothed in small, oval, dark green leaves. Small, 5-petalled, white flowers in late spring are followed by spherical, red fruits. var. *cochleatus* see *C. cochleatus*. var. *thymifolius* see *C. linearifolius*.
C. prostratus, syn. *C. rotundifolius*. Evergreen, arching shrub. H 1.5m (5ft), S 2.5m (8ft). Fully hardy. Has small, oval, glossy, dark green leaves. Produces small, 5-petalled, white flowers in early summer, followed by spherical, deep red fruits.
C. rehderi, syn. *C. bullatus* var. *macrophyllus*. Deciduous, bushy, open shrub. H 5m (15ft), S 3m (10ft). Fully hardy. Very large, oval, deeply veined, dark green leaves change to red in autumn. Clusters of small, 5-petalled, pink flowers appear in late spring and early summer, then spherical, bright red fruits.
♈ *C.* 'Rothschildianus'. Evergreen or semi-evergreen, arching shrub. H and S 5m (15ft). Frost hardy. Has narrowly oval, bright green leaves, small, 5-petalled, white flowers, in early summer, and large clusters of spherical, golden-yellow fruits.
C. rotundifolius. See *C. prostratus*.
C. salicifolius. Vigorous, evergreen, arching shrub. H and S 5m (15ft). Fully hardy. Has narrowly lance-shaped, dark green leaves. Small, 5-petalled, white flowers, in early summer, are followed by clusters of small, spherical, red fruits.
C. serotinus, syn. *C. glaucophyllus* var. *serotinus*. Evergreen, arching, open shrub. H and S 6m (20ft). Fully hardy. Has oval, dark green leaves. Small white flowers are borne from mid- to late summer and the fruits last until spring.
♈ *C. simonsii* illus. p.118.
C. 'Skogholm', syn. *C.* 'Skogsholmen'. Evergreen, arching, wide-spreading shrub. H 60cm (2ft), S 3m (10ft). Fully hardy. Leaves are small, oval and glossy, dark green. Bears

small, 5-petalled, white flowers during early summer, then rather sparse, spherical, red fruits. Makes good ground cover.
C. **'Skogsholmen'**. See *C.* 'Skogholm'.
C. sternianus, syn. *C. franchetii* var. *sternianus*, illus. p.119.
❦ *C.* x *watereri* **'John Waterer'**. Vigorous, evergreen or semi-evergreen, arching shrub. H and S 5m (15ft). Frost hardy. Has lance-shaped, dark green leaves, small, 5-petalled, white flowers, in early summer, and a profusion of spherical, red fruits in large clusters.

COTULA (Compositae)
Genus of perennials and a few marginal water plants, most of which are evergreen, grown for their neat foliage and button-like flower heads. Many species are useful for cracks in paving stones, but may be invasive. Fully to frost hardy. Most need sun and well-drained soil that is not too dry. Propagate by division in spring.
C. atrata illus. p.315. var. *luteola* illus. p.312.
C. coronopifolia (Brass buttons). Short-lived, deciduous, perennial, marginal water plant. H 15cm (6in), S 30cm (12in). Frost hardy. Has fleshy stems, small, lance-shaped, mid-green leaves and, in summer, button-like, yellow flower heads.

COTYLEDON (Crassulaceae)
Genus of evergreen, succulent shrubs and sub-shrubs, grown for their diverse foliage that ranges from large, oval, grey leaves to small, cylindrical, mid-green leaves. Frost tender, min. 5–7°C (41–5°F). Likes a sunny or partially shaded site and very well-drained soil. Propagate by seed or stem cuttings in spring or summer.
C. ladismithensis illus. p.410.
C. orbiculata. Evergreen, upright, succulent shrub. H and S 50cm (20in) or more. Min. 7°C (45°F). Swollen stem bears thin, oval, mid-green leaves, densely coated in white wax and sometimes red-edged. Flower stems, to 70cm (28in) long, have pendent, tubular, orange flowers in autumn.
C. paniculata. See *Tylecodon paniculata*.
C. reticulata. See *Tylecodon reticulata*.
C. simplicifolia. See *Chiastophyllum oppositifolium*.
C. undulata illus. p.404.
C. wallichii. See *Tylecodon wallichii*.

CRAMBE (Cruciferae)
Genus of annuals and perennials, grown for their bold leaves and large sprays of white flowers in summer. Leaf shoots of *C. maritima* (Sea kale) are eaten as a spring vegetable. Fully hardy. Will grow in any well-drained soil; prefers an open position in full sun but tolerates some shade. Propagate by division in spring or by seed in autumn or spring.
❦ *C. cordifolia* illus. p.190.
C. maritima illus. p.240.

CRASPEDIA (Compositae)
Genus of basal-rosetted, summer-flowering perennials, some of which are best treated as annuals. Half hardy to frost tender, min. 5°C (41°F). Needs sun and well-drained soil. Propagate by seed when very fresh in summer.
C. incana. Basal-rosetted perennial. H 20–30cm (8–12in), S 10cm (4in). Frost tender. Has narrowly oval, basal leaves, with dense, woolly, white hairs beneath, and smaller leaves on flower stem. In summer bears many domed heads of 3–10 tiny, tubular, yellow flowers in large, terminal clusters.

CRASSULA (Crassulaceae)
Genus of perennial succulents and evergreen, succulent shrubs and sub-shrubs, ranging from 2cm (3/4in) high, very succulent-leaved species to 5m (15ft), shrubby ones. Most are easy to grow. Frost hardy to frost tender, min. 5–7°C (41–5°F). Most prefer full sun; others like partial shade. Needs very well-drained soil and a little water in winter. Propagate by seed or stem cuttings in spring or autumn.
C. arborescens illus. p.394.
C. argentea. See *C. ovata*.
C. coccinea. See *Rochea coccinea*.
C. cooperi. Carpeting, perennial succulent. H 2cm (3/4in), S 30cm (12in). Frost tender, min. 7°C (45°F). Has small, spoon- to lance-shaped, light green leaves, pitted with darker green or blackish-green marks. Produces clusters of minute, 5-petalled, white to pale pink flowers in winter. Prefers full sun.
C. deceptor, syn. *C. deceptrix*, illus. p.412.
C. deceptrix. See *C. deceptor*.
C. falcata. See *C. perfoliata* var. *falcata*.
C. lactea. Prostrate to semi-erect, perennial succulent. H 20cm (8in), S 1m (3ft). Frost tender, min. 5°C (41°F). Leaves are triangular-oval, glossy and dark green. In winter produces masses of small, 5-petalled, white flowers in terminal clusters. Likes partial shade.
C. lycopodioides. Dense, bushy, woody-based, perennial succulent. H 15cm (6in), S 30cm (12in). Frost tender, min. 5°C (41°F). Bears small, scale-like, neatly overlapping, mid-green leaves arranged in 4 rows around erect stems and, in spring, tiny, 5-petalled, greenish-yellow flowers. Likes partial shade.
C. multicava illus. p.407.
❦ *C. ovata*, syn. *C. argentea*, *C. portulacea*, illus. p.393.
❦ *C. perfoliata* var. *falcata*, syn. *C. falcata*, illus. p.400.
C. portulacea. See *C. ovata*.
C. sarcocaulis illus. p.301.
C. schmidtii illus. p.409.
C. socialis illus. p.406.

+ CRATAEGOMESPILUS
(Rosaceae)
Group of grafted, hybrid, deciduous trees (*Crataegus* and *Mespilus*), grown for their flowers, foliage and fruits. Fully hardy. Requires sun or semi-shade and fertile, well-drained soil. Propagate by grafting in late summer.
+ *C. dardarii* (Bronvaux medlar). **'Jules d'Asnières'** is a deciduous, spreading tree. H and S 6m (20ft). Has drooping branches and spiny shoots. Variable, oval or deeply lobed, dark green leaves, grey when young, turn orange and yellow in autumn. Clusters of saucer-shaped, white, sometimes rose-tinted, flowers in late spring or early summer are followed by small, rounded, red-brown fruits.

CRATAEGUS (Rosaceae)
Hawthorn, Thorn
Genus of deciduous, or more rarely semi-evergreen, spiny, often spreading trees and shrubs, grown for their clustered, 5-petalled, occasionally double flowers in spring-summer, ornamental fruits and, in some cases, autumn colour. Fully hardy. Prefers full sun but suits most sites and may be grown in any but very wet soil. Is useful for growing in polluted urban areas, exposed sites and coastal gardens. Propagate species by seed in autumn, cultivars by budding in late summer. Fireblight is sometimes a problem.
C. crus-galli (Cockspur thorn). Deciduous, flat-topped tree. H 8m (25ft), S 10m (30ft). Has shoots armed with long, curved thorns and oval, glossy, dark green leaves that turn bright crimson in autumn. Clusters of white flowers, with pink anthers, in late spring are followed by long-lasting, rounded, bright red fruits.
C. ellwangeriana. Deciduous, spreading tree. H and S 6m (20ft). Broadly oval, dark green leaves are shallowly toothed and lobed. Bears clusters of white flowers, with pink anthers, in late spring, followed by rounded, glossy, crimson fruits.
C. flava illus. p.64.
C. laciniata, syn. *C. orientalis*, illus. p.59.
C. laevigata, syn. *C. oxyacantha* (Hawthorn, May). ❦ **'Paul's Scarlet'** illus. p.65. **'Punicea'** is a deciduous, spreading tree. H and S 6m (20ft). In late spring and early summer, oval, lobed, toothed, glossy, dark green leaves set off clusters of crimson flowers, which are followed by rounded, red fruits.
❦ *C.* x *lavallei* **'Carrierei'**. Vigorous, deciduous, spreading tree. H 7m (22ft), S 10m (30ft). Oval, glossy, dark green leaves turn red in late autumn. Has clusters of white flowers in late spring, followed by long-lasting, rounded, orange-red fruits.
C. macrosperma var. *acutiloba* illus. p.68.
C. mollis. Deciduous, spreading tree. H 10m (30ft), S 12m (40ft). Large, broadly oval, lobed, dark green leaves have white-haired undersides when young. Bears heads of large, white flowers in late spring, followed by short-lived, rounded, red fruits.
C. monogyna (Common hawthorn). Deciduous, round-headed tree. H 10m (30ft), S 8m (25ft). Has broadly oval, deeply lobed, glossy, dark green leaves. Clusters of fragrant, white flowers are borne from late spring to early summer, followed by rounded, red fruits. Makes a dense hedge. **'Biflora'** (Glastonbury thorn) has flowers and leaves in mild winters as well as in spring.
C. orientalis. See *C. laciniata*.
C. oxyacantha. See *C. laevigata*.
C. pedicellata illus. p.69.
C. persimilis **'Prunifolia'**. See *C.* x *prunifolia*.

C. phaenopyrum (Washington thorn). Deciduous, round-headed tree. H and S 10m (30ft). Broadly oval leaves are sharply lobed, glossy and dark green. Clusters of white flowers, with pink anthers, are produced from early to mid-summer, followed by rounded, glossy, red fruits that last through winter.
C. x *prunifolia*, syn. *C. persimilis* **'Prunifolia'**. Deciduous, spreading, thorny tree. H 8m (25ft), S 10m (30ft). Oval, glossy, dark green leaves turn red or orange in autumn. Clusters of white flowers, with pink anthers, in early summer are followed by rounded, dark red fruits.
C. tanacetifolia (Tansy-leaved thorn). Deciduous, upright, usually thornless tree. H 10m (30ft), S 8m (25ft). Has oval to diamond-shaped, deeply cut, grey-green leaves, clusters of fragrant, white flowers, with red anthers, in mid-summer and small, apple-shaped, yellow fruits.

CREMANTHODIUM (Compositae)
Genus of basal-rosetted perennials, grown for their pendent, half-closed, daisy-like flower heads. Is often very difficult to grow in all but very cool areas with snow cover. Dislikes winter wet. Fully to frost hardy. Needs shade and humus-rich, moist but well-drained soil. Propagate by seed when fresh.
C. reniforme. Basal-rosetted perennial. H and S 20cm (8in). Fully hardy. Leaves are large and kidney-shaped. Stout stems each carry a large, daisy-like, yellow flower head in summer.

CREPIS (Compositae)
Dandelion
Genus of summer-flowering annuals, biennials and perennials, some of which are evergreen, with long tap roots and leaves in flat rosettes. Many species are persistent weeds, but some are grown for their many-petalled, dandelion-like flower heads. Fully hardy. Tolerates sun or shade and prefers well-drained soil. Propagate annuals and biennials by seed in autumn, perennials by root cuttings (not from tap root) in late winter, although most species self seed freely.
C. aurea illus. p.307.
❦ *C. incana* (Pink dandelion). Basal-rosetted perennial. H 20cm (8in), S 10cm (4in). Bears oblong, divided, hairy, greyish-green leaves and, in summer, uneven discs of ragged, pink flower heads on stiff stems. Is good for a sunny rock garden or border.
C. rubra illus. p.272.

CRINODENDRON (Elaeocarpaceae)
Genus of evergreen shrubs and trees, grown for their flowers and foliage. Frost to half hardy. Requires shade or semi-shade, with plant base in cool shade. Soil should be fertile, moist but well-drained, and acid. Propagate by softwood cuttings in summer or by seed in autumn.
❦ *C. hookerianum*, syn. *Tricuspidaria lanceolata*, illus. p.112.

x *Crinodonna corsii*. See x *Amarcrinum memoria-corsii*.

CRINUM (Amaryllidaceae)
Genus of robust bulbs, grown for their often fragrant, funnel-shaped flowers. Frost hardy to frost tender, min. 16°C (61°F). Needs full sun, shelter and rich, well-drained soil. Propagate by offsets in spring or by seed when fresh or in spring.

C. americanum. Tuft-forming, spring- and summer-flowering bulb. H 40–75cm (16–30in), S 60cm (24in). Half hardy. Has 6–10 strap-shaped, semi-erect, basal leaves. Leafless stem bears a head of up to 6 fragrant, long-tubed, white flowers with narrow petals.
C. asiaticum illus. p.363.
C. bulbispermum, syn. *C. longifolium*. Summer-flowering bulb. H to 1m (3ft), S 60cm (2ft). Half hardy. Leafless flower stem has a head of fragrant, long-tubed, white or pinkish-red flowers with darker red stripes. Bears long, strap-shaped, semi-erect leaves grouped in a tuft on a short stalk.
C. longifolium. See *C. bulbispermum*.
C. macowanii. Autumn-flowering bulb. H and S 60cm (2ft) or more. Half hardy. Is similar to *C. bulbispermum*, but leaves are wavy-edged.
C. moorei illus. p.342.
♀ *C. x powellii* illus. p.342. **'Album'** illus. p.341.

CROCOSMIA (Iridaceae)
Montbretia
Genus of corms, grown for their brightly coloured flowers produced mainly in summer. Forms dense clumps of sword-shaped, erect leaves. Frost hardy. Requires well-drained soil and an open, sunny site. In very cold areas, plant in a sheltered position or lift and store for winter. Propagate by division as growth commences in spring.
C. aurea. Tuft-forming, summer-flowering corm. H 50–75cm (20–30in), S 15–20cm (6–8in). Erect, basal leaves are long, narrow and sword-shaped. Carries a loosely branched spike of tubular, orange or yellow flowers, each 3–5cm (1–2in) long and with 6 spreading petals.
C. 'Bressingham Blaze' illus. p.345.
C. 'Citronella' illus. p.347.
C. 'Emily McKenzie'. Compact, late summer-flowering corm. H to 60cm (24in), S 15–20cm (6–8in). Leaves are erect, basal and sword-shaped. Bears a dense spike of widely funnel-shaped, deep orange flowers, each with a dark mahogany throat.
C. 'Jackanapes' illus. p.367.
♀ *C. 'Lucifer'* illus. p.345.
♀ *C. masonorum* illus. p.345.
C. paniculata, syn. *Antholyza paniculata*. Summer-flowering corm. H to 1.5m (5ft), S 30–45cm (12–18in). Has sword-shaped, erect, basal leaves, pleated lengthways. Carries long-tubed, orange flowers on branched stems, which are strongly zig-zag in shape.
C. rosea. See *Tritonia rubrolucens*.

CROCUS (Iridaceae)
Genus of mainly spring- or autumn-flowering corms with funnel-shaped to rounded, long-tubed flowers. Has long, very narrow, semi-erect, basal leaves, each with a white line along centre, usually 1–5 per corm. Some autumn-flowering species have no leaves at flowering time, these appearing in winter or spring. Most species are less than 10cm (4in) tall when in flower and have a spread of 2.5–8cm (1–3in). Is ideal for rock gardens and for slight forcing in bowls for an early indoor display. Fully to frost hardy. Most require well-drained soil and a sunny situation; *C. banaticus* prefers moist soil and semi-shade. Plant 5–6cm (2–2½in) deep, in late summer or early autumn. Propagate in early autumn by seed or division if clumps of corms have formed.

C. 'Advance' (illus. p.373). Late winter- to mid-spring flowering corm. Fully hardy. Funnel-shaped flowers are butter-cup yellow inside and paler yellow outside, suffused with violet-bronze.
C. aerius of gardens. See *C. biflorus* subsp. *pulchricolor*.
C. ancyrensis. Spring-flowering corm. Frost hardy. Produces up to 7 fragrant, bright orange-yellow flowers, 5–6cm (2–2½in) long.
♀ *C. angustifolius*, syn. *C. susianus* (Cloth-of-gold crocus). Spring-flowering corm. Fully hardy. Fragrant flowers are bright golden-yellow, striped or stained bronze outside.
C. aureus. See *C. flavus*.
♀ *C. banaticus* (illus. p.373). Autumn-flowering corm. Fully hardy. Usually has one long-tubed, pale violet flower; outer 3 petals are much larger than inner 3. Very narrow, semi-erect, basal leaves, each with a paler line along the centre, appear in spring.
C. baytopiorum (illus. p.373). Spring-flowering corm. Frost hardy. Each corm bears 1 or 2 rounded, clear torquoise-blue, slightly darker-veined flowers.
C. biflorus (illus. p.373). Early spring-flowering corm. Fully hardy. Has narrow, semi-erect, basal leaves, each with a white line along the centre. Bears fragrant, white or purplish-white flowers, with yellow throats, vertically striped purple outside. subsp. *alexandri* carries fragrant, deep violet flowers, with white insides. subsp. *pulchricolor* (syn. *C. aerius* of gardens) has rich deep blue flowers with golden-yellow centres.
C. 'Blue Bird' (illus. p.373). Late winter- to mid-spring-flowering corm. Fully hardy. Funnel-shaped flowers are white inside with deep yellow throats and violet margined with white outside.
♀ *C. 'Blue Pearl'* (illus. p.372). Early spring-flowering corm. Fully hardy. Produces narrow, semi-erect, basal leaves, with white lines along the centres. Fragrant, long-tubed, funnel-shaped, soft lavender-blue flowers, bluish-white within, have golden-yellow throats.
♀ *C. boryi* (illus. p.372). Autumn-flowering corm. Frost hardy. Flowers are ivory-white, sometimes veined or flushed with mauve outside.
C. cancellatus. Autumn-flowering corm. Frost hardy. Slender flowers are pale blue, slightly striped outside. Leaves form after flowering, in spring.
♀ *C. cartwrightianus.* Autumn-flowering corm. Frost hardy. Produces leaves at same time as strongly veined, violet or white flowers, each 4–6cm (1½–2½in) across and with 3 long, bright red stigmas, similar to those of *C. sativus*.
♀ *C. chrysanthus.* Spring-flowering corm. Fully hardy. Scented flowers are orange-yellow throughout with deeper orange-red stigmas.
♀ *C. 'Cream Beauty'* (illus. p.373). Spring-flowering corm. Fully hardy. Scented rich cream flowers, with deep yellow throats, are stained purplish-brown outside at base. Bears very narrow, semi-erect, basal, dark green leaves, each with a white line along the centre.
C. cvijicii (illus. p.373). Spring-flowering corm. Fully hardy. Usually has one funnel-shaped, yellow flower. Produces very narrow, semi-erect, basal leaves, each with a white line along the centre, which scarcely show at flowering time.
C. dalmaticus (illus. p.372). Spring-flowering corm. Fully hardy. Very narrow, semi-erect leaves have central, white lines. Bears 1-3 purple-veined, pale violet flowers, with yellow centres, overlaid with silver or yellow wash outside.
C. 'Dorothy' (illus. p.373). Spring-flowering corm. Fully hardy. Scented flowers are pale lemon-yellow.
♀ *C. 'E.A. Bowles'* (illus. p.373). Early spring-flowering corm. Fully hardy. Has scented, funnel-shaped, deep yellow flowers, stained bronze near base on outside. Leaves are narrow, semi-erect and basal, each with a central white line. Increases well by offsets.
♀ *C. etruscus* (illus. p.373). Spring-flowering corm. Frost hardy. Has very narrow, semi-erect, basal, dark green leaves that have central, white lines. Bears long-tubed, funnel-shaped, pale purple-blue flowers, washed silver outside, with violet veining. **'Zwanenburg'** is a spring-flowering corm. Frost hardy. Bears pale purple-blue flowers, washed with biscuit-brown and flecked violet outside.
C. 'Eyecatcher' (illus. p.372). Late winter- to mid-spring-flowering corm. Fully hardy. Produces funnel-shaped, grey-white, yellow-throated flowers with white-edged, deep purple outer segments.
♀ *C. flavus*, syn. *C. aureus.* Spring-flowering corm. Fully hardy. Fragrant flowers are bright yellow or orange-yellow throughout; often several flowers are produced together or in quick succession.
C. gargaricus. Spring-flowering corm. Frost hardy. Bears yellow flowers, 4–5cm (1½–2in) long. Increases by stolons. Tolerates slightly damper conditions than most crocuses.
♀ *C. 'Golden Yellow'.* Very vigorous, clump-forming, spring-flowering corm. Fully hardy. Bears yellow flowers, 8–10cm (3–4in) long and faintly striped outside at bases. Naturalizes well in grass.
♀ *C. goulimyi* (illus. p.372). Autumn-flowering corm. Frost hardy. Usually has one long-tubed, pale lilac to pinkish-lilac flower, with a white throat and 3 inner petals usually paler than the 3 outer ones. Leaves and flowers appear together. Needs a warm site.

C. hadriaticus (illus. p.372). Autumn-flowering corm. Frost hardy. Leaves appear at the same time as the white flowers, which usually have yellow throats and may be lilac-feathered at the base.
♀ *C. imperati.* Strikingly bicoloured, spring-flowering corm. Frost hardy. Develops 1 or 2 scented, purple flowers, 6–8cm (2½–3in) long, fawn with purple striping outside and with yellow throats. In **'De Jager'** (illus. p.372) flowers are rich violet-purple inside and biscuit-coloured with striking, violet feathering outside.
C. korolkowii (Celandine crocus). Spring-flowering corm. Fully hardy. Produces up to 20 narrow leaves. Carries fragrant, yellow flowers, speckled or stained brown or purple outside. When open in sun, petals have glossy surfaces.
♀ *C. kotschyanus* (syn. *C. zonatus*; illus. p.372). Autumn-flowering corm. Fully hardy. Pinkish-lilac or purplish-blue flowers have yellow centres and white anthers. Narrow, semi-erect, basal leaves, with white lines along centres, appear in winter-spring. var. *leucopharynx* has pale lilac-blue flowers with white centres and white anthers. Leaves appear in winter-spring.
♀ *C. 'Ladykiller'* (illus. p.373). Late winter- to mid-spring-flowering corm. Fully hardy. Funnel-shaped flowers, white or pale lilac within and deep violet-purple with white margins externally.
C. laevigatus (illus. p.372). Very variable corm, flowering intermittently for a month or more in autumn or winter depending on the form. Frost hardy. Fragrant flowers appear with leaves and are usually lilac-purple with bold stripes on outside; inside each has a yellow eye and cream-white anthers.
C. longiflorus. Autumn-flowering corm. Frost hardy. Produces fragrant, slender, purple flowers, which are striped darker purple outside, at the same time as leaves. Flowers have yellow centres and anthers and red stigmas.
♀ *C. malyi* (illus. p.372). Spring-flowering corm. Fully hardy. Has 1 or 2 funnel-shaped, white flowers that have yellow throats, brown or purple tubes and showy, bright orange stigmas. Leaves are very narrow, semi-erect and basal with central, white lines.
♀ *C. medius* (illus. p.372). Autumn-flowering corm. Frost hardy. Has 1 or 2 funnel-shaped, uniform rich purple flowers, with contrasting, yellow anthers and red stigmas cut into many thread-like branches. Linear, basal leaves appear in winter-spring, after flowering.
C. minimus (illus. p.372). Late spring-flowering corm. Frost hardy. Has very narrow, semi-erect, basal dark green leaves that have central, white lines. Bears 1 or 2 flowers, purple inside and stained darker violet or sometimes darker striped on outside.
C. niveus. Autumn-flowering corm. Frost hardy. Produces 1 or 2 white or pale lavender flowers, 10–15cm (4–6in) long, with conspicuous, yellow

throats. Leaves appear with flowers or just afterwards. Needs a warm, sunny site.
C. nudiflorus (Autumn crocus; illus. p.372). Autumn-flowering corm. Fully hardy. Has linear, basal leaves in winter-spring. Usually bears one slender long-tubed, rich purple flower, with a frilly, bright orange or yellow stigma. Naturalizes in grass.
C. olivieri. Spring-flowering corm. Frost hardy. Bears rounded, bright orange flowers. Flowers of subsp. *balansae* are stained or striped bronze-brown outside.
♚ *C. pulchellus* (illus. p.372). Autumn-flowering corm. Fully hardy. Bears long-tubed, pale lilac-blue flowers with darker veins, conspicuous, yellow throats and white anthers. Leaves are very narrow, semi-erect and basal, with white lines along centres.
C. salzmannii. See *C. serotinus* subsp. *salzmannii*.
C. sativus, syn. *C.s.* var. *cashmirianus* (Saffron crocus). Autumn-flowering corm. Frost hardy. Leaves appear with saucer-shaped, purple flowers, 5–7cm (2–3in) across, that each have darker veining and 3 long, bright red stigmas that yield saffron.
C. serotinus subsp. *salzmannii*, syn. *C. salzmannii*. Autumn-flowering corm. Frost hardy. Carries lilac-blue flowers, to 10cm (4in) long, sometimes with yellow throats. Leaves appear with flowers.
♚ *C. sieberi* subsp. *sieberi*. Spring-flowering corm. Fully hardy. Has white flowers with yellow throats and purple staining outside, either in horizontal bands or vertical stripes. subsp. *atticus* has pale lilac to violet-blue flowers with frilly, orange stigmas. ♚ subsp. *sublimis* f. *tricolor* (illus. p.372) has unusual flowers, divided into 3 distinct bands of lilac, white and golden yellow. ♚ 'Bowles' White' (illus. p.372) produces pure white flowers with large, deep yellow areas in throats. 'Hubert Edelsten' has yellow-throated, pale lilac flowers, the outer segments of which are white, tipped, centrally marked and feathered with rich purple.
♚ *C.* 'Snow Bunting' (illus. p.372). Spring-flowering corm. Fully hardy. Fragrant long-tubed, funnel-shaped, white flowers have mustard-yellow centres and orange stigmas. Very narrow, semi-erect, basal leaves are dark green with white lines across centres.
♚ *C. speciosus* (illus. p.373). Autumn-flowering corm. Fully hardy. Produces lilac-blue to deep purple-blue flowers, usually with a network of darker veins and a much-divided, orange stigma. Leaves appear in winter-spring. 'Conqueror' (illus. p.373) has large, deep sky-blue flowers. 'Oxonian' (illus. p.373) produces dark violet-blue flowers with prominent darker veining externally.
C. susianus. See *C. angustifolius*.
♚ *C. tommasinianus*. Spring-flowering corm. H to 10cm (4in), S 2.5-8cm (1-3in). Fully hardy. Bears slender, long-tubed, funnel-shaped flowers, very variable in colour from lilac or purple to violet,

sometimes with darker tips to petals and occasionally silver outside. Naturalizes well. 'Albus' (illus. p.372) has white flowers. 'Ruby Giant' (illus. p.373) bears clusters of large rich reddish-purple flowers. 'Whitewell Purple' (illus. p. 373) has slender, reddish-purple flowers.
♚ *C. tournefortii*. Autumn-flowering corm. Frost hardy. Leaves appear at same time as 1 or 2 pale lilac-blue flowers that open flattish to reveal a much-divided, orange stigma and white anthers. Requires a warm, sunny situation.
C. vernus (Dutch crocus, Spring crocus; illus. p.373). Spring-flowering corm. H to 10cm (4in), S 2.5-8cm (1-3in). Fully hardy. Variable in colour from white to purple or violet and often striped and feathered. Stigmas are large, frilly and orange or yellow. Is suitable for naturalizing. subsp. *albiflorus* (illus. p.372) has small, white flowers sometimes slightly marked or striped purple. 'Jeanne d'Arc' has white flowers with a deep purple base. 'Pickwick' (illus. p.373) has pale, greyish-white flowers, with dark violet stripes and purplish bases. 'Prinses Juliana' (illus. p.373) produces mid-purple flowers with darker veins. 'Purpureus Grandiflorus' (illus. p.373) has shiny, violet-purple flowers. 'Queen of the Blues' (illus. p.373) has rich blue flowers that have higher margins and a darker base. 'Remembrance' (illus. p.373) has shiny, violet flowers. 'Vanguard', a very early cultivar, has bluish-lilac flowers, paler and silvered outside.
C. 'Zephyr'. Autumn-flowering corm. Fully hardy. Bears very pale silver-blue flowers, veined darker, each with a conspicuous, yellow throat and white anthers.
C. zonatus. See *C. kotschyanus*.
♚ *C.* 'Zwanenburg Bronze'. Spring-flowering corm. H to 10cm (4in), S 2.5–8cm (1–3in). Fully hardy. Has bicoloured flowers, rich yellow inside, stained bronze outside.

CROSSANDRA (Acanthaceae)
Genus of evergreen perennials, sub-shrubs and shrubs, grown mainly for their flowers. Frost tender, min. 15°C (59°F). Needs partial shade or full light and humus-rich, well-drained soil. Water potted plants freely when in full growth, moderately at other times. For a strong branch system, cut back flowered growth by at least half in late winter. Propagate by seed in spring or by greenwood cuttings in late spring or summer. Whitefly may be troublesome.
C. infundibuliformis, syn. *C. undulifolia*. Evergreen, erect to spreading, soft-stemmed shrub or sub-shrub. H to 1m (3ft), S 60cm (2ft). Has oval to lance-shaped, glossy, deep green leaves and, in summer-autumn or earlier, fan-shaped, salmon-red flowers in conical spikes, each 10cm (4in) long.
C. nilotica illus. p.135.
C. undulifolia. See *C. infundibuliformis*.

CROTALARIA (Leguminosae)
Genus of evergreen shrubs, perennials and annuals, grown mainly for their flowers. Frost tender, min. 10–15°C (50–59°F). Requires full light and well-drained soil. Water containerized specimens freely when in full growth, moderately at other times. For a more compact habit, cut back old stems by half after flowering. Propagate by seed in spring or by semi-ripe cuttings in summer. Red spider mite may be troublesome.
C. agatiflora illus. p.93.

Crucianella stylosa. See *Phuopsis stylosa*.

CRYPTANTHUS (Bromeliaceae)
Genus of evergreen, rosette-forming perennials, grown for their attractive foliage. Frost tender, min. 10–13°C (50–55°F). Needs semi-shade and well-drained soil, preferably mixed with sphagnum moss. Water moderately during the growing season, sparingly at other times. Propagate by offsets or suckers in late spring.
C. acaulis (Green earth star). Evergreen, clump-forming, basal-rosetted perennial. H to 10cm (4in), S 15–30cm (6–12in). Loose, flat rosettes of lance-shaped to narrowly triangular, wavy, mid-green leaves have serrated edges. A cluster of fragrant, tubular, white flowers appears from the heart of each rosette, usually in summer. var. *ruber* has red-flushed foliage.
♚ *C. bivittatus* (illus. p.229). Evergreen, clump-forming, basal-rosetted perennial. H to 15cm (6in), S 25–38cm (10–15in). Loose, flat rosettes of broadly lance-shaped, wavy, mid- to yellowish-green leaves have finely toothed margins and are striped lengthways with 2 coppery-fawn to buff bands. Small clusters of tubular, white flowers appear from centre of each rosette, usually in summer.
C. bromelioides (Rainbow star). Evergreen, spreading, basal-rosetted perennial. H 20cm (8in) or more, S 35cm (14in) or more. Strap-shaped, wavy, finely toothed, arching, mid- to bright green leaves are produced in dense rosettes. Occasionally bears clusters of tubular, white flowers in centre of each rosette, usually in summer. ♚ 'Tricolor' has carmine-suffused, white-striped foliage.
♚ *C.* 'Pink Starlight' (illus. p.229). Vigorous, evergreen, spreading, basal-rosetted perennial. H 20cm (8in) or more, S 35cm (14in) or more. Strap-shaped, wavy, finely toothed, arching, green leaves are striped yellowish-green, and heavily suffused deep pink. Clusters of tubular, white flowers occasionally appear from each rosette centre in summer.
♚ *C. zonatus*. Evergreen, basal-rosetted perennial. H 10–15cm (4–6in), S 30–40cm (12–16in). Forms loose, flat rosettes of strap-shaped, wavy, finely toothed, sepia-green leaves, cross-banded with grey-buff and with greyish-white scales beneath. A cluster of tubular, white flowers opens at centre of each rosette, usually in

summer. 'Zebrinus' (illus. p.229) has silver-banded foliage.

× CRYPTBERGIA (Bromeliaceae)
Hybrid genus (*Cryptanthus* × *Billbergia*) of evergreen, rosette-forming perennials, grown for their foliage. Frost tender, min. 8–10°C (46–50°F). Needs semi-shade and fertile, well-drained soil. Water moderately during growing season, sparingly in winter. Propagate by suckers or offsets in spring.
× *C. rubra*. Evergreen, clump-forming, basal-rosetted perennial. H and S 15–30cm (6–12in). Loose rosettes comprise strap-shaped, pointed, bronze-red leaves. Rarely, small, tubular, white flowers are produced in rosette centres in summer.

CRYPTOCORYNE (Araceae)
Genus of semi-evergreen, perennial, submerged water plants and marsh plants, grown for their foliage. Is suitable for tropical aquariums. Frost tender, min. 10°C (50°F). Needs a sunny position and rich soil. Remove fading foliage regularly and divide plants periodically. Propagate by division in spring or summer.
C. ciliata, syn. *Cryptocoryne beckettii* var. *ciliata*. Semi-evergreen, perennial, submerged water plant. S 15cm (6in). Lance-shaped, deep green leaves have paler midribs. Small, hooded, fringed, purplish spathes appear intermittently at base of plant.
C. spiralis. Semi-evergreen, perennial, submerged water plant. S 15cm (6in). Small, hooded, purplish spathes are borne intermittently among lance-shaped, purplish-green leaves.

CRYPTOGRAMMA (Polypodiaceae)
Genus of deciduous or semi-evergreen ferns. Fully hardy. Needs partial shade and moist but well-drained, neutral or acid soil. Remove fading fronds. Propagate by spores in late summer.
C. crispa illus. p.189.

CRYPTOMERIA (Taxodiaceae). See CONIFERS.
♚ *C. japonica* (Japanese cedar). Fast-growing, columnar to conical, open conifer. H 15–20m (50–70ft), S 5–8m (15–25ft). Fully hardy. Has soft, fibrous, red-brown bark, needle-like, incurved, mid- to dark green leaves, spirally arranged, and globular, brown cones. ♚ 'Bandai-sugi', H and S 2m (6ft), makes an irregularly rounded shrub with foliage that turns bronze in winter. 'Cristata' illus. p.83.
♚ 'Elegans Compacta' (syn. *C.j.* 'Elegans Nana'; illus. p.85), H 2–5m (6–15ft), S 2m (6ft), is a dwarf form. 'Pyramidata' illus. p.83. 'Sekkan-sugi' (illus. p.85), H 10m (30ft), S 3–4m (10–12ft), has semi-pendulous branches and light golden-cream foliage. 'Spiralis' (illus. p.85), H and S 2–3m (6–10ft), forms a tree or dense shrub with spirally twisted foliage and is very slow-growing.
♚ 'Vilmoriniana', H and S 1m (3ft), forms a globular mound of yellow-green foliage that becomes bronze in winter.

CRYPTOSTEGIA

CRYPTOSTEGIA (Asclepiadaceae)
Genus of evergreen, twining climbers, grown for their flowers. Frost tender, min. 15°C (59°F). Provide fertile, well-drained soil and full light. Water regularly, less when not in full growth. Stems require support. Spur back previous season's old flowering stems in spring. Propagate by seed in spring or by softwood cuttings in summer.
C. grandiflora (Rubber vine). Strong-growing, evergreen, twining climber. H 10m (30ft) or more. Has thick-textured, oval, glossy leaves. Funnel-shaped, reddish to lilac-purple flowers appear in summer. Stems yield a poisonous latex.

CTENANTHE (Marantaceae)
Genus of evergreen, bushy perennials, grown for their ornamental foliage. Frost tender, min. 15°C (59°F). Requires a humid atmosphere, even temperature and partial shade. Prefers moist but well-drained soil and soft water; do not allow to dry completely. Propagate by division in spring.
♀ *C. lubbersiana*. Evergreen, clump-forming, bushy perennial. H and S to 75cm (30in) or more. Long-stalked, lance-shaped, sharply pointed leaves are 25cm (10in) long, green above, irregularly marked and striped with pale yellowish-green, and pale greenish-yellow below. Intermittently bears dense, one-sided spikes of many small, 3-petalled, white flowers.
C. oppenheimiana. Robust, evergreen, bushy perennial. H and S 1m (3ft) or more. Lance-shaped, leathery leaves are over 30cm (1ft) long, red below, dark green above with pale green or white bands along veins on either side of midribs. Dense, one-sided spikes of many small, 3-petalled, white flowers are produced intermittently.
♀ 'Tricolor' illus. p.228.

CUDRANIA (Moraceae)
Genus of deciduous or evergreen trees and shrubs, grown for their foliage. Fully hardy. Needs full sun, doing best in hot summers. Prefers any fertile, well-drained soil. Propagate by softwood cuttings in summer or by seed in autumn.
C. tricuspidata. Deciduous, spreading tree. H 7m (22ft), S 6m (20ft). Oval, dark green leaves are sometimes 3-lobed. Bears small, rounded clusters of tiny, green flowers in mid-summer.

CUNNINGHAMIA (Taxodiaceae).
See CONIFERS.
C. lanceolata illus. p.80.

CUNONIA (Cunoniaceae)
Genus of evergreen, summer-flowering trees, grown for their foliage, flowers and overall appearance. Frost tender, min. 10°C (50°F). Requires full light and well-drained soil. Water potted plants moderately, less in winter. Pruning is tolerated. Propagate by seed in spring or by semi-ripe cuttings in summer.
C. capensis (African red alder). Moderately fast-growing, evergreen, rounded tree. H and S 10–15m (30–50ft), more in rich soil. Has lustrous, dark green leaves, divided into pairs of lance-shaped, serrated leaflets. Tiny, long-stamened, white flowers appear in dense, bottlebrush-like spikes, each 10–13cm (4–5in) long, in late summer.

CUPHEA (Lythraceae)
Genus of annuals, perennials and evergreen shrubs and sub-shrubs, grown for their flowers. Half hardy to frost tender, min. 2–7°C (36–45°F). Prefers full sun and fertile, well-drained soil. Water freely when in full growth, moderately at other times. Remove flowered shoots after flowering to maintain a bushy habit. Propagate by seed in spring or by greenwood cuttings in spring or summer. Red spider mite may be troublesome.
C. cyanea illus. p.142.
♀ *C. hyssopifolia* illus. p.128.
♀ *C. ignea* illus. p.142.

× CUPRESSOCYPARIS
(Cupressaceae). See CONIFERS.
× *C. leylandii* illus. p.75. 'Castlewellan' and 'Harlequin' illus. p.78. 'Leighton Green' is a very fast-growing, columnar conifer with a conical tip. H 25–35m (80–120ft), S 4–5m (12–15ft). Fully hardy. Bears flattened sprays of paired, scale-like, rich green leaves and globular, glossy, dark brown cones. ♀ 'Robinson's Gold', H 15–20m (50–70ft), has bright golden leaves.

CUPRESSUS (Cupressaceae),
Cypress. See CONIFERS.
C. arizonica var. *glabra*. See *C. glabra*.
♀ *C. cashmeriana*, syn. *C. torulosa* 'Cashmeriana', illus. p.75.
C. glabra, syn. *C. arizonica* var. *glabra* (Arizona cypress, Smooth cypress). Conical conifer. H 10–15m (30–50ft), S 3–5m (10–15ft). Fully hardy. Has smooth, flaking, reddish-purple bark and upright, spirally arranged sprays of scale-like, aromatic, glaucous blue-grey leaves flecked with white resin. Globular cones are chocolate-brown.
C. lusitanica (Cedar of Goa, Mexican cypress). Conical conifer. H 20m (70ft), S 5–8m (15–25ft). Fully hardy. Has fissured bark, spreading, spirally arranged sprays of scale-like, aromatic, grey-green leaves and small, globular cones, glaucous blue when young, ripening to glossy brown.
C. macrocarpa (Monterey cypress). Fast-growing, evergreen conifer, columnar when young, often wide-spreading with age. H 20m (70ft), S 6–25m (20–80ft). Fully hardy. Bark is shallowly fissured. Scale-like, aromatic, bright to dark green leaves are borne in plume-like sprays. Globular cones are glossy and brown.
♀ 'Goldcrest' illus. p.83.
C. sempervirens illus. p.79.
C. torulosa 'Cashmeriana'. See *C. cashmeriana*.

CYANANTHUS (Campanulaceae)
Genus of late summer-flowering perennials, suitable for rock gardens, walls and troughs. Fully hardy. Needs partial shade and humus-rich, moist but well-drained soil. Propagate by softwood cuttings in spring or by seed in autumn.
♀ *C. lobatus*. Prostrate perennial. H 2cm (³⁄₄in), S 20cm (8in). Branched stems are clothed in small, wedge-shaped, dull green leaves. In late summer, each stem carries a funnel-shaped, blue flower. f. *albus* illus. p.321.
♀ *C. microphyllus* illus. p.331.

CYANOTIS (Commelinaceae)
Genus of evergreen, creeping perennials, grown for their foliage. Frost tender, min. 10–15°C (50–59°F). Prefers humus-rich, well-drained soil and sun or partial shade. Propagate by tip cuttings from spring to autumn.
♀ *C. kewensis* (Teddy-bear vine). Evergreen perennial forming rosettes with trailing stems. H 5cm (2in), S 30cm (12in). Clasping the stem are 2 rows of overlapping, oval leaves, to 5cm (2in) long, dark green above, purple with velvety, brown hairs below. Has stalkless clusters of 3-petalled, purplish-pink flowers in axils of leaf-like bracts almost all year round.
♀ *C. somaliensis* illus. p.267.

CYATHEA, syn. ALSOPHILA, SPHAEROPTERIS (Cyatheaceae)
Genus of evergreen tree ferns, grown for their foliage and overall appearance. Frost tender, min. 10–13°C (50–55°F). Needs a humid atmosphere, sun or partial shade and humus-rich, moisture-retentive but well-drained soil. Water potted plants freely in summer, moderately at other times. Propagate by spores in spring.
C. arborea (West Indian tree fern). Evergreen, upright tree fern with a slender trunk. H 7–10m (22–30ft), S 2–3m (6–10ft). Produces 2–3m (6–10ft) long, arching, bright yellow-green fronds, delicately divided into small, oblong, serrated leaflets.
C. australis illus. p.74.
C. medullaris (Black tree fern, Mamaku). Evergreen, upright tree fern with a slender, black trunk. H 7–16m (22–52ft), S 6–12m (20–40ft). Has arching fronds, each to 7m (22ft) long, divided into small, oblong, glossy, dark green leaflets, paler beneath.

CYATHODES (Epacridaceae)
Genus of evergreen, heath-like shrubs, suitable for rock gardens and peat beds. Frost hardy to frost tender, min. 7°C (45°F). Needs a sheltered, shaded site and gritty, moist, peaty soil. Propagate in summer by seed or semi-ripe cuttings.
C. colensoi illus. p.308.

CYBISTAX (Bignoniaceae)
Genus of deciduous trees, grown for their spring flowers and for shade. Frost tender, min. 16–18°C (61–4°F). Needs full light and fertile, moisture-retentive but well-drained soil. Will not bloom when confined to a container. Young plants may be pruned to shape when leafless; otherwise pruning is not required. Propagate by seed or air-layering in spring or by semi-ripe cuttings in summer.
C. donnell-smithii, syn. *Tabebuia donnell-smithii*. Fairly fast-growing, deciduous, rounded tree. H and S 10m (30ft) or more. Leaves have 5–7 oval, 5–20cm (2–8in) long leaflets. Bell-shaped, 5-lobed, bright yellow flowers appear in spring before the leaves, often in great profusion.

CYCLAMEN

CYCAS (Cycadaceae)
Genus of slow-growing, evergreen, woody-stemmed perennials, grown for their palm-like appearance. Frost tender, min. 10–13°C (50–55°F). Prefers full light and humus-rich, well-drained soil. Water potted specimens moderately, less when not in full growth. Propagate in spring by seed or suckers taken from mature plants.
♀ *C. revoluta* illus. p.122.

CYCLAMEN (Primulaceae)
Genus of tuberous perennials, some of which are occasionally evergreen, grown for their pendent flowers, each with 5 reflexed petals and a mouth often stained with a darker colour. Fully hardy to frost tender, min. 5–7°C (41–5°F). Requires humus-rich, well-drained soil and sun or partial shade. If pot-grown, in summer dry off tubers of all except *C. purpurascens* (which is evergreen and flowers in summer); repot in autumn and water to restart growth. Propagate by seed in late summer or autumn. *C. persicum* and its cultivars are susceptible to black root rot.
C. africanum illus. p.381.
C. alpinum. See *C. trochopteranthum*.
♀ *C. cilicium* illus. p.382.
C. coum subsp. *coum* and 'Album' illus. p.370. subsp. *caucasicum* is a winter-flowering, tuberous perennial. H to 10cm (4in), S 5–10cm (2–4in). Frost hardy. Has heart-shaped, silver-patterned leaves and produces a succession of bright carmine flowers, each with a dark stain at the mouth.
C. creticum. Spring-flowering, tuberous perennial. H to 10cm (4in), S 5–10cm (2–4in). Frost hardy. Produces heart-shaped, dark green leaves, sometimes silver-patterned, and fragrant, white flowers.
C. cyprium. Autumn-flowering, tuberous perennial. H to 10cm (4in), S 5–10cm (2–4in). Frost hardy. Heart-shaped, toothed, dark green leaves, with lighter patterns, appear with or just after fragrant, white flowers, each with carmine marks around the mouth.
C. europaeum. See *C. purpurascens*.
C. fatrense. See *C. purpurascens*.
C. graecum illus. p.382.
♀ *C. hederifolium*, syn. *C. neapolitanum*, illus. p.382. var. *album* illus. p.381.
♀ *C. libanoticum* illus. p.371.
♀ *C. mirabile* illus. p.384.
C. neapolitanum. See *C. hederifolium*.
C. persicum illus. p.370. 'Cardinal' is a winter- or spring-flowering, tuberous perennial. H 10–20cm (4–8in), S 15–20cm (6–8in). Frost tender. Leaves are heart-shaped, marked light and dark green and silver. Produces usually fragrant, slender, bright pinkish-scarlet flowers, each 5–6cm (2–2¹⁄₂in) long and stained carmine at the mouth. Needs plenty of light in winter. **Dwarf Scented Series**

CYCLOBOTHRA

(syn. *C.p.*, Dwarf Fragrant Series), S 10–15cm (4–6in), has white, pink or red flowers, 4cm (1½in) long. 'Esmeralda', **Kaori Series** and 'Renown' illus. p.371. 'Pearl Wave' illus. p.384.
C. pseudibericum. Spring-flowering, tuberous perennial. H to 10cm (4in), S 10–15cm (4–6in). Frost hardy. Has heart-shaped, toothed leaves patterned with silvery- and dark green zones. Flowers are deep carmine-purple with darker, basal stains and white-rimmed mouths.
♛ *C. purpurascens*, syn. *C. europaeum*, *C. fatrense*, illus. p.379.
C. repandum. Spring-flowering, tuberous perennial. H to 10cm (4in), S 10–15cm (4–6in). Frost hardy. Has heart-shaped, jagged-toothed, dark green leaves with lighter patterns. Bears fragrant, slender, reddish-purple flowers.
C. rohlfsianum illus. p.382.
C. trochopteranthum, syn. *C. alpinum*. Spring-flowering, tuberous perennial. H 10cm (4in), S 5–10cm (2–4in). Fully hardy. Bears rounded or heart-shaped leaves, zoned with silver. Produces musty-scented, pale carmine or white flowers, stained dark carmine at mouths; petals are twisted and propeller-shaped.

Cyclobothra lutea. See *Calochortus barbatus*.

CYDONIA (Rosaceae)
Genus of one species of deciduous, spring-flowering tree, grown for its flowers and fruits, which are used as a flavouring and for preserves. Fully hardy, but grow against a south- or west-facing wall in cold areas. Requires sun and fertile, well-drained soil. Propagate species by seed in autumn, cultivars by softwood cuttings in summer. Mildew, brown rot and fireblight are sometimes a problem.
♛ *C. oblonga* (Quince). 'Lusitanica' is a deciduous, spreading tree. H and S 5m (15ft). Broadly oval, dark green leaves are grey-felted beneath. Has a profusion of large, 5-petalled, pale pink flowers in late spring, followed by fragrant, pear-shaped, deep yellow fruits. ♛ 'Vranja' illus. p.66.
C. sinensis. See *Pseudocydonia sinensis*.

CYMBALARIA (Scrophulariaceae)
Genus of annuals, biennials and short-lived perennials, related to *Linaria*, grown for their tiny flowers on slender stems. Is good for rock gardens, walls and banks, but may be invasive. Fully hardy. Needs shade and moist soil. Propagate by seed in autumn. Self seeds readily.
C. muralis (Ivy-leaved toadflax, Kenilworth ivy). Spreading perennial. H 5cm (2in), S 12cm (5in). Bears small, ivy-shaped, pale green leaves and, in summer, masses of tiny, tubular, spurred, sometimes purple-tinted, white flowers.

CYMBIDIUM. See ORCHIDS.
C. **Caithness Ice 'Trinity'** (illus. p.262). Evergreen, epiphytic orchid for a cool greenhouse. H 75cm (30in). Sprays of green flowers, 10cm (4in) across, each with a red-marked, white lip, are borne in early spring. Has narrowly oval leaves, to 60cm (24in) long. Grow in semi-shade in summer.
C. **Christmas Angel 'Cooksbridge Sunburst'** (illus. p.263). Evergreen, epiphytic orchid for a cool greenhouse. H 75cm (30in). In winter produces sprays of yellow flowers, 10cm (4in) across and with red-spotted lips. Narrowly oval leaves are up to 60cm (24in) long. Grow in semi-shade in summer.
C. devonianum (illus. p.262). Evergreen, epiphytic orchid for a cool greenhouse. H 60cm (24in). In early summer bears pendent spikes of 2.5cm (1in) wide, olive-green flowers overlaid with purple and with purple lips. Has broadly oval, semi-rigid leaves, to 30cm (12in) long. Needs semi-shade in summer.
C. elegans, syn. *Cyperorchis elegans* (illus. p.262). Evergreen, epiphytic orchid for a cool greenhouse. H 75cm (30in). Dense, pendent sprays of fragrant, tubular, yellow flowers, 4cm (1½in) across, appear in early summer. Has narrowly oval leaves, to 60cm (24in) long. Requires semi-shade in summer.
C. grandiflorum (illus. p.262). Evergreen, epiphytic orchid for a cool greenhouse. H 75cm (30in). In winter produces sprays of deep green flowers, 8cm (3in) across, each with a hairy, brown-spotted, creamy-white lip. Narrowly oval leaves are up to 60cm (24in) long. Grow in semi-shade in summer.
C. **King's Lock 'Cooksbridge'** (illus. p.262). Evergreen, epiphytic orchid for a cool greenhouse. H 60cm (24in). Sprays of green flowers, 5cm (2in) across and each with a purple-marked, white lip, open in spring. Leaves are narrowly oval and up to 60cm (24in) long. Provide semi-shade in summer.
C. **Pontac 'Mont Millais'** (illus. p.261). Evergreen, epiphytic orchid for a cool greenhouse. H 75cm (30in). Bears sprays of 8cm (3in) wide, rich deep red flowers, edged and marked with white, in spring. Has narrowly oval leaves, to 60cm (24in) long. Grow in semi-shade in summer.
C. **Portlett Bay** (illus. p.260). Evergreen, epiphytic orchid for a cool greenhouse. H 75cm (30in). Red-lipped, white flowers, 10cm (4in) across, are borne in sprays in spring. Has narrowly oval leaves, to 60cm (24in) long. Provide semi-shade in summer.
C. **Strath Kanaid** (illus. p.261). Evergreen, epiphytic orchid for a cool greenhouse. H 60cm (24in). In spring bears arching spikes of deep red flowers, 5cm (2in) across. Lips are white, marked deep red. Narrowly oval leaves are up to 60cm (24in) long. Requires semi-shade in summer.
C. **Strathbraan** (illus. p.260). Evergreen, epiphytic orchid for a cool greenhouse. H 60cm (24in). In spring produces slightly arching spikes of off-white flowers, 5cm (2in) across, with red marks on each lip. Leaves are narrowly oval, to 60cm (24in) long. Requires semi-shade in summer.
♛ *C.* **Strathdon 'Cooksbridge Noel'** (illus. p.261). Evergreen, epiphytic orchid for a cool greenhouse. H 1m (3ft). Sprays of rich pink flowers, 5cm (2in) across, with red-spotted, yellow-tinged lips, appear in winter. Has narrowly oval leaves, up to 60cm (24in) long. Needs semi-shade in summer.
C. tracyanum (illus. p.262). Evergreen, epiphytic orchid for a cool greenhouse. H 75cm (30in). In autumn produces long spikes of fragrant, olive-green flowers, 8cm (3in) across, overlaid with reddish dots and dashes. Has narrowly oval leaves, to 60cm (24in) long. Grow in semi-shade in summer.

CYNARA (Compositae)
Genus of architectural perennials, grown for their large heads of flowers. The plant described is grown both as a vegetable and as a decorative border plant. Frost hardy. Requires sun and fertile, well-drained soil. Propagate by seed or division in spring.
♛ *C. cardunculus* illus. p.192.

CYNOGLOSSUM (Boraginaceae)
Hound's tongue
Genus of annuals, biennials and perennials, grown for their long flowering period from late spring to early autumn. Fully hardy. Needs sun and fertile but not over-rich soil. Propagate by division in spring or by seed in autumn or spring.
C. amabile 'Firmament' illus. p.286.
C. nervosum illus. p.218.

CYPELLA (Iridaceae)
Genus of summer-flowering bulbs, grown for their short-lived, iris-like flowers that have 3 large, spreading outer petals and 3 small, incurved inner ones. Half hardy; may survive outdoors in cool areas if planted near a sunny wall. Needs well-drained soil and full sun. Lift bulbs when dormant; partially dry off in winter. Propagate by seed in spring.
C. herbertii illus. p.367.

Cyperorchis elegans. See *Cymbidium elegans*.

CYPERUS (Cyperaceae). See GRASSES, BAMBOOS, RUSHES and SEDGES.
C. albostriatus, syn. *C. diffusus* of gardens, *C. elegans* of gardens. Evergreen, perennial sedge. H 60cm (2ft), S indefinite. Frost tender, min. 7°C (45°F). Stem has prominently veined, mid-green leaves and up to 8 leaf-like, green bracts surrounding a well-branched umbel of brown spikelets in summer. 'Variegatus' has white-striped leaves and bracts.
C. alternifolius of gardens. See *C. involucratus*.
C. diffusus of gardens. See *C. albostriatus*.
C. elegans of gardens. See *C. albostriatus*.
C. flabelliformis. See *C. involucratus*.
♛ *C. involucratus*, syn. *C. alternifolius* of gardens, *C. flabelliformis*, illus. p.184.
C. isocladus of gardens. See *C. papyrus* 'Nanus'.
C. longus (Galingale). Evergreen, spreading, perennial sedge. H 1.5m (5ft), S indefinite. Fully hardy. Bears rough-edged, glossy, dark green leaves and, in summer, attractive umbels of narrow, flattened, milk-chocolate-coloured spikelets that keep their colour well. Tolerates its roots in water.
C. papyrus illus. p.183. 'Nanus' (syn. *C. isocladus* of gardens) is an evergreen, spreading, perennial sedge with a red rhizome; it is a dwarf variant of the species, sometimes considered distinct, and is often grown under misapplied names. H 80cm (32in), S indefinite. Frost tender, min. 7–10°C (45–50°F). Triangular, leafless stems bear umbels of brown spikelets on 8–10cm (3–4in) stalks in summer.

CYPHOMANDRA (Solanaceae)
Genus of evergreen shrubs and trees, grown for their fruits and foliage. Frost tender, min. 10°C (50°F). Grow in full light or partial shade and in well-drained soil. Water containerized specimens freely when in growth, sparingly at other times, when some leaves may fall. Tip prune at intervals while young to promote branching. Propagate by seed in spring. Whitefly and red spider mite may cause problems.
C. betacea. See *C. crassicaulis*.
C. crassicaulis, syn. *C. betacea*, illus. p.95.

CYPHOSTEMMA (Vitaceae)
Genus of deciduous, perennial succulents with very thick, fleshy, almost woody caudices and branches. Leaf undersides often exude small droplets of resin. Frost tender, min. 10°C (50°F). Needs full sun and very well-drained soil. Keep dry in winter. Is difficult to grow. Propagate by seed in spring.
C. bainesii, syn. *Cissus bainesii*. Deciduous, perennial succulent. H and S 60cm (2ft). Has a thick, swollen, bottle-shaped trunk, often unbranched, covered in peeling, papery, yellow bark. Fleshy, silvery-green leaves, with deeply serrated edges, are divided into 3 oval leaflets and are silver-haired when young. Bears tiny, cup-shaped, yellow-green flowers in summer, then grape-like, red fruits.
C. juttae, syn. *Cissus juttae*, illus. p.395.

CYPRIPEDIUM (Slipper orchid). See ORCHIDS.
C. acaule (Moccasin flower; illus. p.260). Deciduous, terrestrial orchid. H to 40cm (16in). Fully hardy. Yellowish-green or purple flowers, 4–6cm (1½–2½in) long, each with a pouched, pink or white lip, are borne singly in spring-summer. Leaves are broadly lance-shaped, pleated and 10–30cm (4–12in) long. Does best in partial shade.
C. calceolus (Lady's slipper orchid, Yellow lady's slipper orchid; illus. p.262). Deciduous, terrestrial orchid. H 75cm (30in). Fully hardy. In spring-summer bears paired or solitary yellow-pouched, purple flowers, 3–7cm (1¼–3in) long. Broadly

475

lance-shaped leaves, 5–20cm (2–8in) long, are arranged in a spiral up stem. Stems and leaves are slightly hairy. Prefers partial shade. var. *pubescens* (illus. p.263) has larger, purple-marked, greenish-yellow flowers, larger leaves and is more hairy.

C. macranthon (illus. p.262). Deciduous, terrestrial orchid. H 50cm (20in). Fully hardy. Pouched, violet or purplish-red flowers, 4–6cm (1½–2½in) long, usually borne singly, open in spring-summer. Stems and oval leaves, 4–7cm (1½–3in) long, are slightly hairy. Prefers partial shade.

C. reginae (Showy lady's slipper orchid; illus. p.260). Deciduous, terrestrial orchid. H to 1m (3ft). Fully hardy. In spring-summer, white flowers, 2–5cm (¾–2in) long, each with a pouched, white-streaked, pink lip, are borne singly or in groups of 2 or 3. Stem and oval leaves, 10–25cm (4–10in) long, are hairy. Does best in partial shade.

CYRILLA (Cyrillaceae)
Genus of one very variable species of deciduous or evergreen shrub, grown for its flowers in late summer and autumn. Fully to half hardy. Prefers full sun and needs peaty, acid soil. Propagate by semi-ripe cuttings in summer.

C. racemiflora (Leatherwood). Deciduous or evergreen, bushy shrub. H and S 1.2m (4ft). Oblong, glossy, dark green leaves redden in autumn. Slender spires of small, 5-petalled, white flowers are borne in late summer and autumn.

CYRTANTHUS (Amaryllidaceae)
Genus of bulbs with brightly coloured flowers, usually in summer. Frost hardy to frost tender, min. 15°C (59°F). Requires free-draining, light soil and full sun. In frost-free areas may flower for much of the year. Plant in spring. Water freely in the growing season. Propagate by seed or offsets in spring.

C. brachyscyphus, syn. *C. parviflorus*, illus. p.379.

C. breviflorus, syn. *Anoiganthus breviflorus*. Clump-forming summer-flowering bulb. H 20–30cm (8–12in), S 8–10cm (3–4in). Frost hardy. Has narrowly strap-shaped, semi-erect, basal leaves. Leafless flower stem bears up to 6 funnel-shaped, yellow flowers, 2–3cm (¾–1¼in) long. Prefers a warm, sheltered situation.

C. mackenii. Clump-forming, summer-flowering bulb. H 30–40cm (12–16in), S 8–10cm (3–4in). Half hardy. Bears strap-shaped, semi-erect, basal leaves. Leafless stems each carry an umbel of up to 10 fragrant, tubular, white flowers, 5cm (2in) long and slightly curved. var. *cooperi* illus. p.366.

C. obliquus. Clump-forming, summer-flowering bulb. H 20–60cm (8–24in), S 12–15cm (5–6in). Half hardy. Bears widely strap-shaped, semi-erect, basal, greyish-green leaves, twisted lengthways. Carries a head of up to 12 pendent, tubular, red-and-yellow flowers, each 7cm (3in) long.

C. parviflorus. See *C. brachyscyphus*.

♀ *C. elatus*, syn. *C. purpureus*, *Vallota speciosa*. Clump-forming, summer-flowering bulb. H 30–50cm (12–20in), S 12–15cm (5–6in). Half hardy. Bears widely strap-shaped, semi-erect, basal, bright green leaves. Stout stem produces a head of up to 5 widely funnel-shaped, scarlet flowers, 8–10cm (3–4in) long. Makes an excellent house plant.

C. sanguineus. Clump-forming, summer-flowering bulb. H 30–50cm (12–20in), S 12–15cm (5–6in). Half hardy. Has strap-shaped, semi-erect, basal, bright green leaves. Stout stem bears 1 or 2 long-tubed, scarlet flowers, 8–10cm (3–4in) long.

CYRTOMIUM (Polypodiaceae)
Genus of evergreen ferns. Half hardy. Does best in semi-shade and humus-rich, moist soil. Remove fading fronds regularly. Propagate by division in spring or summer or by spores in summer.

♀ *C. falcatum* illus. p.187.

CYSTOPTERIS (Polypodiaceae)
Genus of deciduous ferns, suitable for rock gardens. Fully hardy. Does best in semi-shade and in soil that is never allowed to dry out. Remove fronds as they fade. Propagate by division in spring, by spores in summer or by bulbils when available.

C. bulbifera (Berry bladder fern). Deciduous fern. H 15cm (6in), S 23cm (9in). Broadly lance-shaped, much divided, dainty, pale green fronds produce tiny bulbils along their length. Propagate by bulbils as soon as mature.

C. dickieana. Deciduous fern. H 15cm (6in), S 23cm (9in). Has broadly lance-shaped, divided, delicate, pale green fronds, with oblong, blunt, indented pinnae, that arch downwards.

C. fragilis (Brittle bladder fern). Deciduous fern. H 15cm (6in), S 23cm (9in). Broadly lance-shaped, pale green fronds are delicate and much-divided into oblong, pointed, indented pinnae.

CYTISUS (Leguminosae)
Broom
Genus of deciduous or evergreen shrubs, grown for their abundant, pea-like flowers. Fully to half hardy. Prefers full sun and fertile, but not over-rich, well-drained soil. Resents being transplanted. Propagate species by semi-ripe cuttings in summer or by seed in autumn, hybrids and cultivars by semi-ripe cuttings only in late summer.

C. albus, syn. *C. leucanthus*. Deciduous, spreading shrub. H 30cm (1ft), S 1m (3ft). Fully hardy. Has oval leaves, each with 3 tiny leaflets and, from early to mid-summer, creamy-white flowers borne in dense clusters.

♀ *C. ardoinii* illus. p.334.

♀ *C. battandieri*, syn. *Argyrocytisus battandieri*, illus. p.93.

♀ *C.* x *beanii* illus. p.297.

C. canariensis of gardens. See *Genista* x *spachiana*.

♀ *C. demissus*, syn. *Chamaecytisus hirsutus*. Slow-growing, deciduous, prostrate shrub. H 7cm (3in), S 20–30cm (8–12in). Fully hardy. Densely haired stems bear tiny, bright green leaves with 3 oval leaflets. Leaf axils each bear a terminal cluster of 3 bright yellow flowers, each with a brown pouch, in early summer. Is good for a rock garden or bank.

C. 'Firefly'. Deciduous, bushy shrub with slender, arching shoots. H and S 1.5–2m (5–6ft). Fully hardy. Small, mid-green leaves are oblong and have 3 tiny leaflets. Produces masses of yellow flowers, marked with red, from late spring to early summer.

♀ *C.* x *kewensis*. Deciduous, arching shrub. H 30cm (1ft), S to 2m (6ft). Fully hardy. Has leaves, with 3 tiny leaflets, along downy stems. In late spring bears creamy-white flowers. Is good for a bank or large rock garden.

C. leucanthus. See *C. albus*.

C. nigricans illus. p.140.

C. x *praecox* illus. p.126.

♀ 'Allgold' illus. p.127.

♀ *C. purpureus*, syn. *Chamaecytisus purpureus* (Purple broom). Deciduous, arching shrub. H 45cm (18in), S 60cm (24in). Fully hardy. Semi-erect stems are clothed with leaves of 3 tiny leaflets. Masses of purple flowers open in early summer on previous year's wood. Is good for a bank or sunny border. f. *albus* (syn. *Chamaecytisus purpureus* f. *albus*) illus. p.299.

C. racemosus. See *Genista* x *spachiana*.

C. scoparius (Common broom). f. *andreanus* illus. p.141. var. *prostratus* (syn. *C.s.* subsp. *maritimus*) is a deciduous, prostrate shrub forming dense mounds of interlocking shoots. H 20cm (8in), S 1.2–2m (4–6ft). Fully hardy. Small, grey-green leaves usually have 3 oblong leaflets, but may be reduced to a single leaflet. Has masses of golden-yellow flowers in late spring and early summer.

C. x *spachianus*. See *Genista* x *spachiana*.

C. supinus. Deciduous, bushy, rounded shrub. H and S 1m (3ft). Fully hardy. Dense heads of large, yellow flowers are borne from mid-summer to autumn amid grey-green leaves with 3 oval leaflets.

C. 'Windlesham Ruby'. Deciduous, bushy shrub with slender, arching shoots. H and S 1.5–2m (5–6ft). Fully hardy. Small, mid-green leaves have 3 oblong leaflets. Large, rich red flowers are borne in profusion in late spring and early summer.

♀ *C.* 'Zeelandia'. Deciduous, bushy shrub with slender, arching shoots. H and S 1.5–2m (5–6ft). Fully hardy. Small, mid-green leaves have 3 oblong leaflets. Has masses of bicoloured, creamy-white and lilac-pink flowers from late spring to early summer.

D

DABOECIA (Ericaceae). See HEATHERS.
D. azorica. Evergreen, compact shrub. H to 15cm (6in), S to 60cm (24in). Half hardy. Lance-shaped leaves are dark green above, silver-grey beneath. Urn- to bell-shaped flowers are vivid red and open in late spring or early summer.
D. cantabrica (St Dabeoc's heath). Evergreen, straggling shrub. H to 45cm (18in), S 60cm (24in). Frost hardy; top growth may be damaged by frost and cold winds, but plants respond well to hard pruning and produce new growth from base. Leaves are lance-shaped to oval, dark green above, silver-grey beneath. Bears bell- to urn-shaped, single or double, white, purple or mauve flowers from late spring to mid-autumn. ♀ **'Bicolor'** (illus. p.148) bears white, purple and striped flowers on the same plant. **'Praegerae'**, H 35cm (14in), has glowing deep pink flowers. **'Snowdrift'** (illus. p.148) has bright green foliage and long racemes of large, white flowers.
D. × scotica. Evergreen, compact shrub. H to 15cm (6in), S to 60cm (2ft). Frost hardy. Lance-shaped to oval leaves are dark green above, silver-grey beneath. Bears bell- to urn-shaped, white, purple or mauve flowers from late spring to mid-autumn. ♀ **'Jack Drake'**, H 20cm (8in), has small, dark green leaves and ruby-coloured flowers. ♀ **'William Buchanan'** (illus. p.148), H 45cm (18in), is a vigorous cultivar with dark green leaves and deep purple flowers.

DACTYLIS (Gramineae). See GRASSES, BAMBOOS, RUSHES and SEDGES.
D. glomerata (Cock's-foot, Orchard grass). **'Variegata'** is an evergreen, tuft-forming, perennial grass. H 1m (3ft), S 20–25cm (8–10in). Fully hardy. Silver-striped, grey-green leaves arise from tufted rootstock. In summer bears panicles of densely clustered, awned, purplish-green spikelets.

DACTYLORHIZA. See ORCHIDS.
♀ ***D. elata.*** Deciduous, terrestrial orchid. H 1.1m (3½ft). Frost hardy. Spikes of pink or purple flowers, 1–2cm (½–¾in) long, open in spring-summer. Lance-shaped leaves, 15–25cm (6–10in) long, are spotted with brownish-purple and arranged spirally on stem. Requires shade outdoors; keep pot plants semi-shaded in summer.
♀ ***D. foliosa***, syn. *D. maderensis*, *Orchis maderensis*. Deciduous, terrestrial orchid. H 70cm (28in). Frost hardy. Spikes of bright purple or pink flowers, 1–2cm (½–¾in) long, are borne in spring-summer. Has lance-shaped or triangular leaves, 10–20cm (4–8in) long, arranged spirally on stem. Cultivate as for *D. elata*.
D. maderensis. See *D. foliosa*.

DAHLIA (Compositae)
Genus of bushy, summer- and autumn-flowering, tuberous perennials, grown as bedding plants or for their flower heads, which are good for cutting or exhibition. Dwarf forms are used for mass-planting and are also suitable for containers. Half hardy. Needs a sunny position and well-drained soil. All apart from dwarf forms require staking. After flowering, lift tubers and store in a frost-free place; replant once all frost danger has passed. In frost-free areas, plants may be left in ground as normal herbaceous perennials, but they benefit from regular propagation to maintain vigour. Propagate dwarf forms by seed sown under glass in late winter, others in spring by seed, basal shoot cuttings or division of tubers. Dahlias may be subject to attack by aphids, red spider mite and thrips; they also succumb quickly to virus infection.

Border dahlias
Prolific and long-flowering, various species of *Dahlia* have been hybridized and, with constant breeding and selection, have developed into many forms and have a wide colour range (although there is no blue). Shoots may be stopped, or pinched out, to promote vigorous growth and a bushy shape. Spread measurements depend on the amount of stopping carried out and the time at which it is done: early stopping encourages a broader shape, stopping later in the growing season results in a taller plant with much less spread, even in the same cultivar. Leaves are generally mid-green and divided into oval leaflets, some with rounded tips and some with toothed margins.
Dahlias are divided into groups, according to the size and type of their flower heads, although the latter may vary in colour and shape depending on soil and weather conditions. The groups are single, anemone, collerette, water-lily, decorative, ball, pompon, cactus, semi-cactus and miscellaneous (including orchid); for illustrations and descriptions see p.350. Each flower head is referred to horticulturally as a flower, even though it does in fact comprise a large number of individual flowers. This horticultural usage has been followed in the descriptions below. All forms with flower heads to 15cm (6in) across are suitable for cutting; those suitable for exhibition are so noted.

D. **'Abridge Taffy'**. Decorative-flowered dahlia. H and S 1.2m (4ft). Produces white flowers, 8–10cm (3–4in) across, from mid-summer to late autumn. Is suitable for exhibition.
D. **'Alvas Supreme'**. Decorative-flowered dahlia. H and S 1.1m (3½ft). In summer-autumn bears 25–30cm (10–12in) wide, pale yellow flowers that are good for exhibition.
D. **'Andries Orange'**. Semi-cactus-flowered dahlia. H and S 1m (3ft). In summer-autumn produces orange flowers, 8–10cm (3–4in) across.
D. **'Angora'** (illus. p.350). Decorative-flowered dahlia. H and S 1.1m (3½ft). Has 10cm (4in) wide, white flowers, resembling carnations with split petal tips, in mid-summer.
D. **'Athalie'** (illus. p.350). Cactus-flowered dahlia. H and S 1.2m (4ft). Has glossy, dark green leaves and dark pinkish-bronze flowers, 12–15cm (5–6in) across, in mid-summer. Is good for exhibition.
D. **'Banker'**. Cactus-flowered dahlia. H and S 1.2–1.5m (4–5ft). In summer-autumn produces red flowers, 15–20cm (6–8in) across, that are good for exhibition.
D. **'Bassingbourne Beauty'** (illus. p.351). Decorative-flowered dahlia. H and S 1.2m (4ft). Produces pale pinkish-yellow flowers, 10–15cm (4–6in) across, reflexing to stem, in summer-autumn. Is good for exhibition.
D. **'B.B.C.'**. Semi-cactus-flowered dahlia. H and S 1.5–2m (5–6ft). In late summer bears flame-orange flowers, 25cm (10in) across, that are good for exhibition.
D. **'Betty Bowen'** (illus. p.351). Decorative-flowered dahlia. H and S 1.5m (5ft). Leaves are glossy, dark green. Has rich purple flowers, 10–15cm (4–6in) across, with neat, symmetrical petalling, reflexing to stem, in summer-autumn. Is suitable for exhibition.
D. **'Biddenham Sunset'** (illus. p.351). Decorative-flowered dahlia. H and S 1–1.2m (3–4ft). Orange-red flowers, 10–12cm (4–5in) across, are borne in mid-summer.
D. **'Bishop of Llandaff'** (illus. p.351). Miscellaneous-group dahlia. H and S 1m (3ft). Has bronze-green leaves and single, open-centred, dark red flowers, 10–12cm (4–5in) across, in summer-autumn. It is excellent as a bedding plant.
D. **'Brunton'** (illus. p.351). Decorative-flowered dahlia. H and S 1.1m (3½ft). In summer-autumn bears red flowers, 8–10cm (3–4in) across, that are good for exhibition.
♀ *D.* **'Butterball'** (illus. p.351). Decorative-flowered dahlia. H and S 1m (3ft). Produces 8–10cm (3–4in) wide, bright yellow flowers in early summer.
D. **'By the Cringe'** (illus. p.350). Cactus-flowered dahlia. H and S 1.1m (3½ft). Bears lilac-lavender flowers, 12cm (5in) across, in summer-autumn.
D. **'Candy Keene'** (illus. p.350). Semi-cactus-flowered dahlia. H and S 1.2–1.5m (4–5ft). Is a sport of *D.* 'Reginald Keene' with pink flowers.
D. **'Catherine Ireland'**. Decorative-flowered dahlia. H and S 1.1m (3½ft). In summer-autumn has 8–10cm (3–4in) wide flowers, each white at base with a lilac-lavender flush to petal tips.
D. **'Cherida'**. Ball-flowered dahlia. H and S 1m (3ft). Has bronze-lilac flowers, less than 8cm (3in) across, in summer-autumn.
D. **'Chimborazo'** (illus. p.351). Collerette-flowered dahlia. H and S 1m (3ft). Leaves are glossy, dark green. Has 10cm (4in) wide flowers, with red, outer petals and yellow, inner petals, in summer-autumn. Flowers are good for exhibition.
D. **'Chinese Lantern'** (illus. p.351). Decorative-flowered dahlia. H and S 1.1m (3½ft). In mid-summer produces flame-yellow flowers, 10–15cm (4–6in) across.
D. **'Clair de Lune'** (illus. p.351). Collerette-flowered dahlia. H and S 1m (3ft). Has 8–10cm (3–4in) wide flowers, with lemon-yellow, outer petals and paler yellow, inner petals, in summer-autumn. Is a good exhibition cultivar.
D. **Coltness Hybrids** illus. p.281.
D. **'Comet'** (illus. p.351). Anemone-flowered dahlia. H and S 1.1m (3½ft). Leaves are glossy, dark green. Dark red flowers, 10–15cm (4–6in) across, are produced in summer-autumn.
D. **'Corona'** (illus. p.351). Semi-cactus-flowered dahlia. H and S 45–60cm (18–24in). Produces flame-red flowers, 8–10cm (3–4in) across, in summer-autumn. Does not require staking.
D. **'Cortez Sovereign'** (illus. p.351). Semi-cactus-flowered dahlia. H and S 1–1.2m (3–4ft). Bears 10–15cm (4–6in) wide, pale yellow flowers in summer-autumn.
D. **'Corton Olympic'** (illus. p.351). Decorative-flowered dahlia. H and S 1m (3ft). Has bronze-coloured flowers, 25–30cm (10–12in) across, in late summer.
D. **'Cryfield Bryn'**. Cactus-flowered dahlia. H and S 1.1m (3½ft). In summer-autumn has butter-yellow flowers, 10–15cm (4–6in) across, that are excellent for exhibition.
D. **'Cryfield Rosie'**. Ball-flowered dahlia. H and S 1.1m (3½ft). Produces 10–15cm (4–6in) wide, yellow flowers, tinged with red, in summer-autumn. Is good for exhibition.
D. **'Daleko Jupiter'**. Semi-cactus-flowered dahlia. H and S 1.2m (4ft). Produces reddish-yellow flowers, 25–30cm (10–12in) across, in summer-autumn. Is a leading exhibition cultivar.
D. **'Dandy'** illus. p.293.
D. **'Davenport Sunlight'** (illus. p.351). Semi-cactus-flowered dahlia. H and S 1.2m (4ft). Has 15–20cm (6–8in) wide, bright yellow flowers in summer-autumn. Flowers are good for exhibition.
D. **'David Howard'**. Decorative-flowered dahlia. H and S 75cm (30in). Dark, bronze-coloured foliage offsets orange-bronze flowers, 8–10cm (3–4in) across, produced in summer-autumn.

D., Disco Series. Group of well-branched, erect, perennial dahlias, grown as annuals. H and S 45cm (18in). Leaves are divided into oval leaflets. Daisy-like, semi-double and double flower heads, with quilled florets, in a wide colour range, appear throughout summer and until autumn frosts.

D. 'Downham Royal'. Ball-flowered dahlia. H and S 1m (3ft). Bears deep purple flowers, 8–10cm (3–4in) across, in summer-autumn. Is good for exhibition.

D. 'Early Bird' (illus. p.351). Decorative-flowered dahlia. H and S 1.2m (4ft). Produces 10–15cm (4–6in) wide, mid-yellow flowers in summer-autumn.

D. 'East Anglian' (illus. p.351). Decorative-flowered dahlia. H and S 1m (3ft). Has orange-yellow flowers, 10–15cm (4–6in) across, in summer-autumn.

D. 'Easter Sunday' (illus. p.350). Collerette-flowered dahlia. H and S 1m (3ft). Leaves are glossy, dark green. Produces 12cm (5in) wide flowers, with white, inner and outer petals and dark yellow centres, in summer-autumn. Is good for exhibition.

D. 'Eastwood Moonlight'. Semi-cactus-flowered dahlia. H and S 1–1.2m (3–4ft). Bears bright yellow flowers, 15–20cm (6–8in) across, in summer-autumn. Is a very good exhibition cultivar.

D. 'Edinburgh'. Decorative-flowered dahlia. H and S 1m (3ft). Produces compact, white-tipped, rich purple flowers, 10–15cm (4–6in) across, in summer-autumn.

⚘ ***D. 'Evening Mail'.*** Semi-cactus-flowered dahlia. H and S 1.2–1.5m (4–5ft). Has yellow flowers, 25–30cm (10–12in) across, in summer-autumn. Is good for exhibition.

D. 'Evita'. Semi-cactus-flowered dahlia. H and S 1.2m (4ft). In summer-autumn has 15–20cm (6–8in) wide, soft yellow flowers that are good for exhibition.

D. 'Fascination' (illus. p.351). Dwarf, miscellaneous-group dahlia. H and S 30–45cm (12–18in). Has single, light purple flowers, 8cm (3in) across, in summer-autumn. Is useful for bedding.

D. 'Flutterby' (illus. p.351). Decorative-flowered dahlia. H and S 1m (3ft). Bears reddish-yellow flowers, 10cm (4in) across, in summer-autumn.

D. 'Frank Hornsey' (illus. p.351). Decorative-flowered dahlia. H and S 1.1m (3½ft). Leaves are glossy, dark green. Produces 10–15cm (4–6in) wide, orange-yellow flowers, with reflexed petals, in mid-summer. Is ideal for exhibition.

D. 'Gay Mini' (illus. p.351). Decorative-flowered dahlia. H and S 1m (3ft). Has bronze-yellow flowers, 8–10cm (3–4in) across, in summer-autumn.

D. 'Gay Princess' (illus. p.350). Decorative-flowered dahlia. H and S 1m (3ft). Produces lilac-lavender flowers, 10–15cm (4–6in) across, in summer-autumn.

D. 'Gerrie Hoek'. Water-lily-flowered dahlia. H and S 1.1m (3½ft). Produces 12cm (5in) wide flowers, with pale pink, outer petals and dark pink, inner petals, in summer-autumn.

D. 'Gilt Edge' (illus. p.350). Decorative-flowered dahlia. H and S 1m (3ft). Bears 15–20cm (6–8in) wide flowers, with reflexed, gold-margined, dark pink petals, in summer-autumn. Is suitable for exhibition.

D. 'Giraffe'. Miscellaneous-group dahlia. H and S 75cm (30in). Has sparse, mid-green foliage and double, orchid-like, bronze-yellow flowers, 8cm (3in) across, in summer-autumn.

D. 'Hallmark'. Pompon-flowered dahlia. H and S 75cm (30in). Has spherical, 5cm (2in) wide, pink flowers in summer-autumn. Is good for exhibition.

⚘ ***D. 'Hamari Bride'.*** Semi-cactus-flowered dahlia. H and S 1–1.2m (3–4ft). Bears clear white flowers, 15–20cm (6–8in) across, in summer-autumn. Is a good exhibition cultivar.

D. 'Hamari Fiesta'. Decorative-flowered dahlia. H and S 1.2m (4ft). Has scarlet-yellow flowers, 10–15cm (4–6in) across, in summer-autumn.

⚘ ***D. 'Hamari Gold'.*** Decorative-flowered dahlia. H and S 75cm–1m (2½–3ft). Has 25–30cm (10–12in) wide, golden orange-bronze flowers in summer-autumn. Is suitable for exhibition.

D. 'Hamari Katrina' (illus. p.351). Semi-cactus-flowered dahlia. H and S 1.2–1.5m (4–5ft). Bears 20–25cm (8–10in) wide, deep butter-yellow flowers in summer-autumn. Flowers are good for exhibition.

D. 'Hayley Jane'. Semi-cactus-flowered dahlia. H and S 1m (3ft). Produces 10–15cm (4–6in) wide, white flowers, evenly tipped purple-red, in summer-autumn.

D. 'Highgate Gold'. Semi-cactus-flowered dahlia. H and S 1.2–1.5m (4–5ft). Has deep orange-yellow flowers, 15–20cm (6–8in) across, in summer-autumn. Is suitable for exhibition.

D. 'Highgate Torch' (illus. p.351). Semi-cactus-flowered dahlia. H and S 1–1.2m (3–4ft). Produces bright flame flowers, 15–20cm (6–8in) across, in summer-autumn.

D. 'Holland Festival'. Decorative-flowered dahlia. H and S 1.2m (4ft). Bears 30cm (12in) or more wide flowers, with bright orange petals evenly tipped with white, in summer-autumn. Is suitable for exhibition.

D. 'Jescot Julie'. Miscellaneous-group dahlia. H and S 60cm–1m (2–3ft). Has sparse, mid-green foliage and orchid-like, double, orange-purple flowers, 8–10cm (3–4in) across, with purple-backed petals, in summer-autumn.

D. 'Jocondo' (illus. p.351). Decorative-flowered dahlia. H and S 1–1.2m (3–4ft). In summer-autumn produces 30cm (12in) or more wide, purple-red flowers that are suitable for exhibition.

D. 'Klankstad Kerkrade'. Cactus-flowered dahlia. H and S 1.2m (4ft). Has glossy, dark green leaves and soft sulphur-yellow flowers, each 10–15cm (4–6in) across, in summer-autumn. Is a good exhibition cultivar

D. 'La Cierva'. Collerette-flowered dahlia. H and S 1–1.2m (3–4ft). Leaves are glossy, dark green. Bears 10–15cm (4–6in) wide flowers, with white-tipped, purple, outer petals, and pure white, inner petals, in summer-autumn. Is suitable for exhibition.

D. 'L'Ancresse'. Ball-flowered dahlia. H and S 1m (3ft). Bears pure white flowers, 8–10cm (3–4in) across, in summer-autumn. Is a good exhibition cultivar.

D. 'Lavender Athalie'. Cactus-flowered dahlia. H and S 1.2m (4ft). Is a sport of *D.* 'Athalie' with soft lilac-lavender flowers, 8–10cm (3–4in) wide.

D. 'Majestic Kerkrade' (illus. p.350). Cactus-flowered dahlia. H and S 1.2m (4ft). Is a sport of *D.* 'Klankstad Kerkrade' with dark pinkish-yellow flowers.

D. 'Mariposa'. Collerette-flowered dahlia. H and S 1m (3ft). In summer-autumn has 10–15cm (4–6in) wide, white flowers; outer petals are tinged with lilac. Is good for exhibition.

D. 'Mark Damp'. Semi-cactus-flowered dahlia. H and S 1–1.2m (3–4ft). Bears 20–25cm (8–10in) wide, bronze-orange flowers, tinged with peach, in summer-autumn. Is suitable for exhibition.

D. 'Monk Marc' (illus. p.350). Cactus-flowered dahlia. H and S 60cm–1m (2–3ft). Produces dark pink flowers, 10–15cm (4–6in) across, in summer-autumn. Is suitable for exhibition.

D. 'Moor Place'. Pompon-flowered dahlia. H and S 1m (3ft). Leaves are glossy, dark green. Has 5cm (2in) wide, royal purple flowers in summer-autumn. Is a good exhibition cultivar.

D. 'Nationwide'. Decorative-flowered dahlia. H and S 1–1.2m (3–4ft). Bears 10–15cm (4–6in) wide, yellowish-orange flowers in summer-autumn.

D. 'Nina Chester' (illus. p.350). Decorative-flowered dahlia. H and S 1m (3ft). Has 10–15cm (4–6in) wide, white flowers, occasionally lilac-tinged, in summer-autumn. Is good for exhibition.

D. 'Noreen' (illus. p.350). Pompon-flowered dahlia. H and S 1m (3ft). In summer-autumn produces dark pinkish-purple flowers, 5cm (2in) across, that are good for exhibition.

D. 'Paul Chester' (illus. p.351). Cactus-flowered dahlia. H and S 1m (3ft). In summer-autumn has yellowish-orange flowers, 10–15cm (4–6in) across, that are suitable for exhibition.

D. 'Paul Damp'. Semi-cactus-flowered dahlia. H and S 1.2–1.5m (4–5ft). Bears 15–20cm (6–8in) wide, dark pink flowers in summer-autumn. Is a good exhibition cultivar.

D. 'Pink Giraffe'. Miscellaneous-group dahlia. H and S 60cm–1m (2–3ft). Is a sport of *D.* 'Giraffe' with pink flowers. Flowers are suitable for exhibition.

D. 'Pink Symbol' (illus. p.350). Semi-cactus-flowered dahlia. H and S 1–1.2m (3–4ft). Is a sport of *D.* 'Symbol' with pink flowers that are good for exhibition.

D. 'Pontiac' (illus. p.351). Cactus-flowered dahlia. H and S 1m (3ft). Leaves are glossy, dark green. Bears dark pinkish-purple flowers, 10–15cm (4–6in) across, in summer-autumn.

D. 'Porcelain'. Water-lily-flowered dahlia. H and S 1m (3ft). Produces whitish-lilac flowers, 10–15cm (4–6in) across, in summer-autumn. Is useful for exhibition.

D. 'Quel Diable' (illus. p.351). Semi-cactus-flowered dahlia. H and S 1–1.2m (3–4ft). Bears flame-coloured flowers, 15–20cm (6–8in) across, in summer-autumn. Is a good exhibition cultivar.

D. 'Redskin'. Well-branched, perennial dahlia, grown as an annual. H and S 30cm (12in). Has bronze-green to maroon leaves, divided into oval leaflets, and, throughout summer and until autumn frosts, daisy-like, double flower heads, in many colours.

D. 'Reginald Keene'. Semi-cactus-flowered dahlia. H and S 1–1.2m (3–4ft). In summer-autumn has 20–25cm (8–10in) wide, orange-flame flowers that are good for exhibition.

D. 'Rhonda' (illus. p.350). Pompon-flowered dahlia. H and S 1m (3ft). In summer-autumn produces 5cm (2in) wide, whitish-lilac flowers that are suitable for exhibition.

D. 'Ruby Wedding'. Decorative-flowered dahlia. H and S 1m (3ft). Bears deep purple flowers, 8–10cm (3–4in) across, in summer-autumn. Flowers are suitable for exhibition.

D. 'Scarlet Beauty' (illus. p.351). Water-lily-flowered dahlia. H and S 1m (3ft). Is a sport of *D.* 'Gerrie Hoek' with deep red flowers, 10–15cm (4–6in) across, that are good for exhibition.

D. 'Shandy' (illus. p.351). Cactus-flowered dahlia. H and S 1m (3ft). Produces 10–15cm (4–6in) wide, pale brown flowers in summer-autumn.

D. 'Sherwood Standard'. Decorative-flowered dahlia. H and S 1.2m (4ft). Has orange flowers, 15–20cm (6–8in) across, in summer-autumn. Is suitable for exhibition.

D. 'Shy Princess'. Cactus-flowered dahlia. H and S 1.2–1.5m (4–5ft). In summer-autumn bears 15–20cm (6–8in) wide, white flowers that are good for exhibition.

D. 'Small World' (illus. p.350). Pompon-flowered dahlia. H and S 1m (3ft). Leaves are glossy, dark green. Has 5cm (2in) wide, white flowers in summer-autumn. Is suitable for exhibition.

⚘ ***D. 'So Dainty'*** (illus. p.351). Semi-cactus-flowered dahlia. H and S 60cm–1m (2–3ft). Produces bronze-coloured flowers, 8–10cm (3–4in) across, in summer-autumn. Flowers are suitable for exhibition.

D. 'Spencer'. Decorative-flowered dahlia. H and S 1m (3ft). Bears purplish-pink flowers, 10–15cm (4–6in) across, in summer-autumn. Is good for exhibition.

D. 'Super'. Semi-cactus-flowered dahlia. H and S 1.2–1.5m (4–5ft). Produces orange-flame flowers, 25cm (10in) or more across, in summer-autumn. Is suitable for exhibition.

D. 'Symbol'. Semi-cactus-flowered dahlia. H and S 1–1.2m (3–4ft). Has

15–20cm (6–8in) wide, bronze-orange flowers in summer-autumn. Is a good exhibition cultivar.
D. 'Temptress'. Cactus-flowered dahlia. H and S 1m (3ft). Produces whitish-lilac flowers, 10–15cm (4–6in) across, in summer-autumn. Is a good exhibition cultivar.
D. 'Vicky Crutchfield' (illus. p.350). Water-lily-flowered dahlia. H and S 1m (3ft). Bears pink flowers, 10–15cm (4–6in) across, in summer-autumn. Is suitable for exhibition.
D. 'Whale's Rhonda' (illus. p.351). Pompon-flowered dahlia. H and S 1m (3ft). Leaves are glossy, very dark green. In summer-autumn has 5cm (2in) wide, purple flowers that are good for exhibition.
D. 'White Klankstad' (illus. p.350). Cactus-flowered dahlia. H and S 1–1.2m (3–4ft). Is a sport of *D.* 'Klankstad Kerkrade' with white flowers.
D. 'Willo's Violet'. Pompon-flowered dahlia. H and S 1m (3ft). Produces glossy, dark green leaves and 5cm (2in) wide, violet-purple flowers, in summer-autumn. Is suitable for exhibition.
♥ **D. 'Wootton Cupid'** (illus. p.351). Ball-flowered dahlia. H and S 1–1.2m (3–4ft). In summer-autumn bears pink flowers, 8–10cm (3–4in) across, that are good for exhibition.
D. 'Yellow Hammer' (illus. p.351). Dwarf, single-flowered dahlia. H and S 30–45cm (12–18in). Bears rich yellow flowers, 8cm (3in) across, in summer-autumn.

DAIS (Thymelaeaceae)
Genus of deciduous, summer-flowering shrubs, grown for their flowers and overall appearance. Frost tender, min. 5°C (41°F). Requires full sun and well-drained soil. Water containerized plants freely when in full growth, sparingly when leafless. Propagate by seed in spring or by semi-ripe cuttings in summer.
D. cotinifolia. Deciduous, bushy, neat shrub. H and S 2–3m (6–10ft). Has small, oval to oblong, lustrous leaves. In summer bears scented, star-shaped, rose-lilac flowers in flattened clusters, 8cm (3in) across. Bark yields fibres strong enough to be used as thread.

Daiswa polyphylla. See *Paris polyphylla.*

DANÄE (Liliaceae)
Genus of one species of evergreen shrub, with inconspicuous flowers, grown for its attractive, flattened, leaf-like shoots. Frost hardy. Grows in sun or shade and in moist soil. Propagate by seed in autumn or by division from autumn to spring.
♥ **D. racemosa** (Alexandrian laurel). Evergreen, arching, dense shrub. H and S 1m (3ft). Has slender stems, lance-shaped, leaf-like, glossy, green shoots and pointed, glossy, bright green 'leaves'. Occasionally bears spherical, red berries.

DAPHNE (Thymelaeaceae)
Genus of evergreen, semi-evergreen or deciduous shrubs, grown for their usually fragrant, tubular flowers, each with 4 spreading lobes, and, in some species, for their foliage or fruits (seeds are poisonous). Dwarf species and cultivars are good for rock gardens. Fully to frost hardy. Most need full sun (although *D. alpina*, *D. arbuscula* and *D. blagayana* may be grown in semi-shade and *D. laureola* tolerates deep shade) and fertile, well-drained but not over-dry soil. Resents being transplanted. Propagate species by seed when fresh or by semi-ripe cuttings in summer, cultivars by cuttings only. Is susceptible to viruses that cause leaf mottling.
D. alpina. Deciduous, erect shrub. H 50cm (20in), S 40cm (16in). Fully hardy. Leaves are oval, downy and grey-green. Carries terminal clusters of fragrant, white flowers in late spring. Is suitable for a rock garden.
♥ **D. arbuscula** illus. p.314.
D. bholua illus. p.120.
D. blagayana illus. p.295.
D. × burkwoodii 'Somerset' illus. p.124. **'Variegata'** is a semi-evergreen, upright shrub. H 1.5m (5ft), S 1m (3ft). Fully hardy. Bears dense clusters of very fragrant, white-throated, pink flowers in late spring, sometimes again in autumn. Narrowly oblong, grey-green leaves are edged with creamy-white or pale yellow.
D. cneorum illus. p.296. ♥ **'Eximia'** is an evergreen, prostrate shrub. H 10cm (4in), S to 50cm (20in) or more. Fully hardy. Has small, oval, leathery, dark green leaves and, in late spring, terminal clusters of fragrant, white flowers, crimson outside and often pink-flushed within.
♥ **D. collina.** Evergreen, domed, compact shrub. H and S 50cm (20in). Frost hardy. Oval, dark green leaves densely cover upright branches. Has terminal clusters of small, fragrant, purple-rose flowers in late spring. Is good for a rock garden or shrubbery.
D. genkwa. Deciduous, upright, open shrub. H and S 1.5m (5ft). Fully hardy after a hot summer, otherwise frost hardy. Oval, dark green leaves are bronze when young. Large, faintly scented, lilac flowers are borne from mid- to late spring.
D. giraldii. Deciduous, upright shrub. H and S 60cm (2ft). Fully hardy. Clusters of fragrant, golden-yellow flowers are produced amid oblong, pale blue-green leaves in late spring and early summer and are followed by egg-shaped, red fruits.
D. jasminea illus. p.312.
D. laureola (Spurge laurel). Evergreen, bushy shrub. H 1m (3ft), S 1.5m (5ft). Fully hardy. Has oblong, dark green leaves. Slightly fragrant, pale green flowers are borne from late winter to early spring, followed by spherical, black fruits. subsp. **philippi** illus. p.126.
D. mezereum illus. p.144. var. **alba** is a deciduous, upright to spreading shrub. H and S 1.2m (4ft). Fully hardy. Bears very fragrant, white or creamy-white flowers that clothe bare stems in late winter and early spring. Has spherical, yellow fruits. Leaves are oblong and dull grey-green.
D. odora. Evergreen, bushy shrub. H and S 1.5m (5ft). Frost hardy. Has oval, glossy, dark green leaves and, from mid-winter to early spring, very fragrant, deep purplish-pink-and-white flowers. **'Aureo-marginata'** illus. p.144.
♥ **D. petraea 'Grandiflora'** illus. p.314.
♥ **D. retusa** illus. p.125.
♥ **D. tangutica.** Evergreen, bushy shrub with stout shoots. H and S 1m (3ft). Fully hardy. Narrowly oval, leathery leaves are dark green. Bears clusters of fragrant, white-flushed, purple-pink flowers in mid- to late spring.

DAPHNIPHYLLUM (Daphniphyllaceae)
Genus of evergreen trees and shrubs, grown for their habit and foliage. Male and female flowers are borne on separate plants. Frost hardy. Needs a sheltered position in sun or semi-shade and deep, fertile, well-drained but not too dry soil. Propagate by semi-ripe cuttings in summer.
D. macropodum illus. p.87.

DARMERA. See **PELTIPHYLLUM**.

DARWINIA (Myrtaceae)
Genus of evergreen, spring-flowering shrubs, grown for their flowers and overall appearance. Frost tender, min. 7–10°C (45–50°F). Needs full light and moist, neutral to acid soil, not rich in nitrogen. Water moderately when in full growth, sparingly at other times. Propagate by seed in spring or by semi-ripe cuttings in late summer. Is difficult to root and to grow under glass.
D. citriodora. Evergreen, rounded, well-branched shrub. H and S 60cm–1.2m (2–4ft). Oblong to broadly lance-shaped, blue-green leaves are lemon-scented when bruised. In spring produces pendent, terminal heads of usually 4 small, tubular, yellow or red flowers, each surrounded by 2 red or yellowish bracts.

DASYLIRION (Liliaceae)
Bear grass
Genus of evergreen, palm-like perennials, grown for their foliage and flowers. Male and female flowers are produced on separate plants. Frost tender, min. 10°C (50°F). Requires well-drained soil and a sunny position. Water well when in full growth, less at other times. Propagate by seed in spring.
D. texanum. Evergreen, palm-like perennial with a 75cm (30in) high trunk. H over 1m (3ft), S 3m (10ft). Has a rosette of narrow, drooping, green leaves, each 60–90cm (2–3ft) long, with yellowish prickles along margins. Stems, 5m (15ft) long, emerge from centre of plant carrying dense, narrow panicles of small, bell-shaped, whitish flowers in summer. Dry, 3-winged fruits appear in autumn.

DATURA. See **BRUGMANSIA**.

DAVALLIA (Polypodiaceae)
Genus of evergreen or semi-evergreen, often epiphytic ferns, best suited to growing in pots and hanging baskets. Half hardy to frost tender, min. 5°C (41°F). Requires semi-shade and very fibrous, moist, peaty soil. Remove fading fronds regularly. Propagate by division in spring or summer or by spores in summer.
♥ **D. canariensis** (Hare's-foot fern). Semi-evergreen fern. H and S 30cm (12in). Frost tender. Broadly lance-shaped, leathery, mid-green fronds, with numerous triangular pinnae, arise from a scaly, brown rootstock.
♥ **D. mariesii** (Squirrel's-foot fern). Evergreen fern. H 15cm (6in), S 23cm (9in). Half hardy. Broadly triangular, delicately divided, leathery, mid-green fronds arise from a creeping, scaly, brown rootstock.

DAVIDIA (Davidiaceae)
Genus of one species of deciduous, spring- and summer-flowering tree, grown for its habit and showy, white bracts surrounding insignificant flower heads. Fully hardy, but needs shelter from strong winds. Requires sun or semi-shade and fertile, well-drained but preferably moist soil. Propagate by semi-ripe cuttings in spring or by seed when ripe in autumn.
♥ **D. involucrata** illus. p.51.

DECAISNEA (Lardizabalaceae)
Genus of deciduous, summer-flowering shrubs, grown for their foliage, flowers and sausage-shaped fruits. Frost hardy. Requires a sheltered, sunny situation and fertile soil that is not too dry. Propagate by seed in autumn.
D. fargesii illus. p.92.

DECUMARIA (Hydrangeaceae)
Genus of evergreen or deciduous, woody-stemmed, root climbers. Half hardy. Prefers sun and loamy, well-drained soil that does not dry out. Prune, if necessary, after flowering. Propagate by stem cuttings in late summer or early autumn.
D. sinensis. Evergreen, woody-stemmed, root climber. H to 2m (6ft) or more. Oval leaves, 2.5–8cm (1–3in) long, are often toothed. Conical clusters of small, honey-scented, cream flowers are produced in late spring and early summer.

DEINANTHE (Hydrangeaceae)
Genus of slow-growing perennials with creeping, underground rootstocks. Is useful for rock gardens and peat beds. Fully hardy. Needs shaded, moist soil. Propagate by division in spring or by seed when fresh.
D. bifida. Slow-growing, mound-forming perennial. H 20cm (8in), S 10–20cm (4–8in). Nodding, star-shaped, white flowers are borne in summer amid rounded, crinkled leaves with 2-lobed, lacerated tips.
D. caerulea. Slow-growing, mound-forming perennial. H 20cm (8in), S to 15cm (6in). Stems, each carrying a cluster of nodding, bowl-shaped, pale violet-blue flowers, rise above 3–4 oval, toothed leaves in summer.

Delairea odorata. See *Senecio mikanoides.*

DELOSPERMA (Aizoaceae)
Genus of densely branched, trailing, perennial, sometimes shrubby

479

succulents, some with tuberous roots. Frost tender, min. 5°C (41°F). Requires full sun and very well-drained soil. Propagate by seed or stem cuttings in spring or summer.

D. tradescantioides. Spreading, perennial succulent. H 10cm (4in), S indefinite. Has long, trailing branches with broad, almost cylindrical, 3-angled, fleshy, light green leaves, 3cm (1¼in) long. Bears daisy-like, white flowers in summer-autumn. Is easy to grow.

DELPHINIUM (Ranunculaceae) Genus of perennials and annuals, grown for their spikes of irregularly cup-shaped, sometimes hooded, spurred flowers. Fully to half hardy. Needs an open, sunny position and fertile or rich, well-drained soil. Tall cultivars need staking and ample feeding and watering in spring and early summer. In spring remove thin growths from well-established plants, leaving 5–7 strong shoots. If flower spikes are removed after they fade, a second flush may be produced in late summer, provided plants are fed and watered well. Propagate species and some *D. elatum* selections by seed in autumn or spring; *D.* x *belladonna* cultivars by division or basal cuttings of young shoots in spring; *D. elatum* hybrids by cuttings only.

D. x *belladonna* **'Blue Bees'** (illus. p.194). Upright, branching perennial. H 1.1–1.5m (3½–5ft), S 60cm (2ft). Fully hardy. Has palmate, divided leaves. In summer bears loose spikes, 30cm (12in) long, of widely spaced, single, occasionally semi-double, sky-blue flowers, 2cm (¾in) or more wide, on freely branching spurs. **'Lamartine'** has purple-blue flowers; those of **'Wendy'** are gentian-blue.

D. brunonianum. Upright perennial. H and S to 15cm (6in). Fully hardy. Hairy stems bear rounded, 3- or 5-lobed leaves. In early summer, flower stems each produce a spike, to 15cm (6in) long, of hooded, single, pale blue-purple flowers, 4cm (1½in) wide, with short, black spurs. Is good for a rock garden.

D. cardinale. Short-lived, upright perennial. H 1–2m (3–6ft), S 60cm (2ft). Half hardy. In summer produces single, scarlet flowers, 4cm (1½in) wide, with yellow eyes, on spikes, 30–45cm (12–18in) long, above palmate, finely divided leaves.

D. chinense. See *D. grandiflorum.*
D. consolida. See *Consolida ambigua.*

D. elatum hybrids. Erect perennials. H 1.35–2.2m (4½–7ft), S 75cm–1m (2½–3ft). Fully hardy. All have large, palmate leaves and, in summer, produce closely packed spikes, 40cm–1.2m (16in–4ft) long, of regularly spaced, semi-double, rarely fully double flowers, 8–10cm (3–4in) wide, in a range of colours from white to blue and purple, usually with contrasting eyes.

♛ **'Blue Dawn'** (illus. p.194), H 2–2.2m (6–7ft), bears pinkish-purple-flushed, pale blue flowers, with dark brown eyes, on spikes to 1.2m (4ft) long.

♛ **'Blue Nile'** (illus. p.194), H 1.5–1.7m (5–5½ft), has rich blue flowers, with lightly blue-streaked, white eyes, on spikes to 68cm (27in) long.

♛ **'Bruce'** (illus. p.194), H 1.7–2.2m (5½–7ft), has deep violet-purple flowers, silver-flushed towards centres and with grey-brown eyes, on spikes to 1.2m (4ft) long.

'Butterball' (illus. p.194), H 1.5–1.7m (5–5½ft), produces cream-eyed, white flowers, overlaid with very pale greenish-yellow, on spikes to 50cm (20in) long.

'Chelsea Star' (illus. p.194), H 2m (6ft), produces rich deep violet flowers, with white eyes, on spikes to 1.1m (3½ft) long.

♛ **'Emily Hawkins'** (illus. p.194), H 2–2.2m (6–7ft), has lilac-flushed, pale blue flowers, with fawn eyes, on spikes to 80cm (32in) long.

♛ **'Fanfare'** (illus. p.194), H 2–2.2m (6–7ft), has pale blue to silvery-mauve flowers, with white-and-violet eyes, on spikes 60–75cm (2–2½ft) long.

♛ **'Fenella'**, H 1.1–1.5m (3½–5ft), bears purple-flushed, gentian-blue flowers, with black eyes, on spikes to 1m (3ft) long.

♛ **'Gillian Dallas'** (illus. p.194), H 2m (6ft), produces spikes, to 80cm (32in) long, of white-eyed, pale lilac flowers, shading to even paler lilac on outer edges.

'Langdon's Royal Flush' (illus. p.194), H 2m (6ft), has cream-eyed, pinkish-purple flowers on spikes to 85cm (34in) long; upper petals are a darker shade than lower ones.

♛ **'Loch Leven'** (illus. p.194), H 1.7m (5½ft), produces pale blue flowers, with white eyes, on spikes to 1m (3ft) long.

'Loch Nevis', H 2m (6ft), has mid-blue flowers, with white eyes, on spikes to 1.1m (3½ft) long.

♛ **'Lord Butler'** (illus. p.194), H 1.5–1.7m (5–5½ft), produces pale blue flowers, lightly flushed with pale lilac and with blue-marked, white eyes, on spikes to 75cm (30in) long.

♛ **'Mighty Atom'** (illus. p.194), H 1.5–2m (5–6ft), has mid-violet flowers, with violet-marked, yellowish brown eyes, on spikes to 75cm (2½ft) long.

'Olive Poppleton' (illus. p.194), H 2–2.5m (6–8ft), produces white flowers, with fawn eyes, on spikes to 1m (3ft) long.

Pacific Series (Pacific Hybrids) is a group of short-lived perennials, H variable, often grown as annuals and available in single colours and as a mixture. Seedlings may not come true. Semi-double flowers are borne in spikes, 45cm–1m (1½–3ft) long.

♛ **'Sandpiper'** (illus. p.194), H 1–1.35m (3–4½ft), has white flowers with creamy-brown eyes, on spikes to 75cm (2½ft) long.

♛ **'Spindrift'** (illus. p.194), H 1.7–2m (5½–6ft), has spikes, to 1m (3ft) long, of pinkish-purple flowers, overlaid with pale blue and with creamy-white eyes; towards centres, the pinkish-purple becomes paler and the blue darker. Colour varies from soil to soil; on acid soil, flowers are greenish.

'Strawberry Fair' (illus. p.194), H 1.7m (5½ft), has white-eyed, mulberry-pink flowers on spikes to 78cm (31in) long.

♛ **'Sungleam'** (illus. p.194), H 1.7–2m (5½–6ft), produces spikes, 40–75cm (16–30in) long, of white flowers, overlaid with pale yellow and with yellow eyes.

D. grandiflorum, syn. *D. chinense*. **'Blue Butterfly'** (illus. p.194) is a short-lived, erect perennial, usually grown as an annual. H 45cm (1½ft), S 30cm (1ft). Fully hardy. Has palmate, divided leaves. In summer produces loose, branching spikes, to 15cm (6in) long, of single, deep blue flowers, 3.5cm (1½in) wide. Is useful as a bedding plant.

D. nudicaule illus. p.303.
D. tatsienense. Short-lived, upright perennial. H 30cm (12in), S 5–10cm (2–4in). Fully hardy. Loose spikes, to 15cm (6in) long, of small-spurred, single, bright blue flowers, 2.5cm (1in) long, are borne in summer. Leaves are rounded to oval and deeply cut. Suits a rock garden. Needs gritty soil.

Dendranthema zawadskii. See *Chrysanthemum rubellum.*

DENDROBIUM. See ORCHIDS.
D. aphyllum, syn. *D. pierardii* (illus. p.260). Deciduous, epiphytic orchid for an intermediate greenhouse. H to 60cm (24in). In early spring produces pairs of soft pink flowers, 4cm (1½in) across and each with a large, cream lip. Has oval leaves, 5–8cm (2–3in) long. Requires semi-shade in summer. Is best grown hanging from a bark slab.

D. chrysotoxum (illus. p.263). Deciduous, epiphytic orchid for an intermediate greenhouse. H 60cm (24in). Trusses of cup-shaped, deep yellow flowers, 2cm (¾in) across and with hairy, red-marked lips, are produced in spring. Oval leaves are 5–8cm (2–3in) long. Provide good light in summer.

D. infundibulum (illus. p.260). Evergreen, epiphytic orchid for a cool greenhouse. H 30cm (12in). In spring, stems each produce up to 6 pure white flowers, 8cm (3in) wide and each with a yellow-marked lip. Has oval leaves, 5–8cm (2–3in) long. Grow in semi-shade in summer.

D. nobile (illus. p.261). Deciduous, epiphytic orchid (often evergreen in cultivation) for a cool greenhouse. H 30cm (12in). Trusses of delicate, rose-pink flowers, 5cm (2in) across and each with a prominent, maroon lip, are borne along stems in spring. Oval leaves are 5–8cm (2–3in) long. Requires semi-shade in summer.

D. pierardii. See *D. aphyllum*.

DENDROCHILUM. See ORCHIDS.
D. glumaceum (Silver chain). Evergreen, epiphytic orchid for a cool greenhouse. H 10cm (4in). Pendent sprays of fragrant, pointed, orange-lipped, creamy-white flowers, 1cm (½in) long, are produced in autumn. Narrowly oval leaves are 15cm (6in) long. Grow in semi-shade in summer.

DENDROMECON (Papaveraceae) Genus of evergreen shrubs, grown for their foliage and showy flowers. Frost to half hardy. Plant against a south- or west-facing wall in cold areas. Requires full sun and very well-drained soil. Propagate by softwood cuttings in summer, by seed in autumn or spring or by root cuttings in winter.

D. rigida illus. p.117.

DENNSTAEDTIA (Polypodiaceae) Genus of deciduous or semi-evergreen ferns. Fully hardy. Grows best in shade and in humus-rich, moist soil. Remove fading fronds regularly. Propagate by division in spring.

D. punctilobula. Deciduous, creeping fern. H and S 30cm (12in). Has oval to triangular, much-divided, delicate, lace-like, mid-green fronds that die back in winter.

Dendrobenthamia capitata. See *Cornus capitata.*

Dentaria enneaphylla. See *Cardamine enneaphyllos.*
Dentaria pentaphylla. See *Cardamine pentaphyllos.*

Deparia petersenii. See *Lunathyrium japonicum.*

DESCHAMPSIA (Gramineae). See GRASSES, BAMBOOS, RUSHES and SEDGES.
D. caespitosa (Tufted hair grass). Evergreen, tuft-forming, perennial grass. H to 1m (3ft), S 25–30cm (10–12in). Fully hardy. Has dense, narrow, rough-edged, dark green leaves. In summer produces dainty, open panicles of minute, pale brown spikelets that last well into winter. Tolerates sun or shade.

DESFONTAINIA (Loganiaceae) Genus of evergreen shrubs, grown for their foliage and showy, tubular flowers. Frost to half hardy. Provide shelter in cold areas. Needs some shade, particularly in dry regions, and moist, peaty, preferably acid soil. Propagate by semi-ripe cuttings in summer.

♛ *D. spinosa* illus. p.113.

DESMODIUM (Leguminosae) Genus of perennials and deciduous shrubs and sub-shrubs, grown for their flowers. Fully to frost hardy. Needs full sun and well-drained soil. Propagate by softwood cuttings in late spring or by seed in autumn. May also be divided in spring.

D. elegans, syn. *D. tiliifolium*, illus. p.137.

DEUTZIA (Hydrangeaceae) Genus of deciduous shrubs, grown for their profuse, 5-petalled flowers. Fully to frost hardy. Needs full sun and fertile, well-drained soil. Plants benefit from regular thinning out of old shoots after flowering. Propagate by softwood cuttings in summer.

D. x *elegantissima* **'Fasciculata'**. Deciduous, upright shrub. H 2m (6ft), S 1.5m (5ft). Fully hardy. From late spring to early summer produces large clusters of 5-petalled, pale pink

flowers. Leaves are oval, toothed and mid-green. ♥ 'Rosealind' illus. p.132.
D. gracilis illus. p.123.
D. 'Joconde'. Deciduous, upright shrub. H and S 1.5m (5ft). Fully hardy. Bears 5-petalled, white flowers, striped purple outside, in early summer. Oval, mid-green leaves are long-pointed.
D. longifolia. Deciduous, arching shrub. H 2m (6ft), S 3m (10ft). Fully hardy. Large clusters of 5-petalled, deep pink flowers are produced from early to mid-summer. Narrowly lance-shaped leaves are grey-green.
♥ 'Veitchii' illus. p.110.
D. 'Magicien'. Deciduous, upright shrub. H and S 1.5m (5ft). Fully hardy. In early summer produces large, 5-petalled flowers with wavy petals, pink inside, white with deep pink stripes outside. Has oval, mid-green leaves.
D. × *magnifica*. Vigorous, deciduous, upright shrub. H 2.5m (8ft), S 2m (6ft). Fully hardy. Produces dense clusters of 5-petalled, pure white flowers in early summer. Leaves are narrowly oval and bright green. 'Staphyleoides' illus. p.106.
D. monbeigii illus. p.128.
D. 'Mont Rose' illus. p.131.
D. pulchra. Vigorous, deciduous, upright shrub. H 2.5m (8ft), S 2m (6ft). Frost hardy. Has peeling, orange-brown bark and lance-shaped, dark green leaves. Slender, pendulous panicles of 5-petalled, pink-tinged, white flowers appear in late spring and early summer.
D. × *rosea* illus. p.124.
D. scabra illus. p.106. 'Plena' (syn. *D.s.* 'Flore Pleno') is a deciduous, upright shrub. H 3m (10ft), S 2m (6ft). Fully hardy. Narrowly oval, dark green leaves set off dense, upright clusters of double, white flowers, purplish-pink outside, from early to mid-summer.
♥ *D. setchuenensis*. Deciduous, upright shrub. H 2m (6ft), S 1.5m (5ft). Frost hardy. Small, 5-petalled, white flowers are borne in broad clusters in early and mid-summer. Produces narrowly oval, rough, dark grey-green leaves.

DIANELLA (Liliaceae)
Flax lily
Genus of evergreen, summer-flowering perennials. Frost to half hardy; is suitable outdoors only in mild areas and elsewhere requires a cold greenhouse or frame. Needs sun and well-drained, neutral to acid soil. Propagate by division or seed in spring.
D. caerulea. Evergreen, tuft-forming perennial. H 75cm (30in), S 30cm (12in). Half hardy. In summer produces panicles of small, star-shaped, blue flowers, above grass-like leaves, followed by blue berries.
D. tasmanica illus. p.217.

DIANTHUS (Caryophyllaceae)
Carnation, Pink
Genus of evergreen or semi-evergreen, mainly summer-flowering perennials, annuals and biennials, grown for their mass of flowers, often scented, some of which are excellent for cutting. Carnations and pinks (see below) are excellent for cut flowers and border decoration, the biennial *D. barbatus* (Sweet William) is suitable for bedding and smaller, tuft-forming species and cultivars are good for rock gardens. Fully to half hardy. Needs an open, sunny position and well-drained, slightly alkaline soil, except for *D. pavonius*, which prefers acid soil. Dead-heading of repeat-flowering types is beneficial. Tall forms of carnations and pinks have a loose habit and need staking. Propagate border carnations by layering in late summer, other named forms by softwood cuttings in late spring and species by seed at any time. Is susceptible to rust, red spider mite and virus infection through aphids, but many cultivars are available from virus-free stock.
Carnations and pinks have narrowly lance-shaped, silvery- or grey-green leaves, scattered up flower stems, which may coil outwards on carnations. They are divided into the following groups, all with self-coloured and bicoloured cultivars.

Carnations
Border carnations are annuals or evergreen perennials that flower prolifically once in mid-summer and are good for border decoration and cutting. Each stem bears 5 or more often scented, semi-double or double flowers, to 8cm (3in) across; picotee forms (with petals outlined in a darker colour) are available. H 75cm–1.1m (2½–3½ft), S to 30cm (1ft). Frost hardy.
Perpetual-flowering carnations are evergreen perennials that flower year-round if grown in a greenhouse, but more prolifically in summer. They are normally grown for cut flowers: flower stems should be disbudded, leaving one terminal bud per stem. Fully double flowers, to 10cm (4in) across, are usually unscented and are often flecked or streaked. H 1–1.5m (3–5ft), S 30cm (1ft) or more. Half hardy.
Spray forms are not disbudded so have 5 or more flowers per stem, each 5–6cm (2–2½in) across. H 60cm–1m (2–3ft), S to 30cm (1ft).

Pinks
Evergreen, clump-forming perennials, grown for border decoration and cutting, that in summer produce a succession of basal shoots, each bearing 4–6 fragrant, single to fully double flowers, 3.5–6cm (1½–2½in) across. H 30–45cm (12–18in), S 23–30cm (9–12in) or more. Frost hardy.
Old-fashioned pinks have a low, spreading habit and produce masses of flowers in one flowering period in mid-summer. Mule types (a border carnation crossed with a Sweet William) and laced types (in which the central colour extends as a loop around each petal) are available.
Modern pinks, obtained by crossing an old-fashioned pink with a perpetual-flowering carnation, are more vigorous than old-fashioned pinks, and are repeat-flowering with two or three main flushes of flowers in summer.

D. 'A.J. Macself'. See *D.* 'Dad's Favourite'.
D. 'Albisola' (illus. p.245). Perpetual-flowering carnation. Fully double flowers are clear tangerine-orange.
D. 'Aldridge Yellow' (illus. p.245). Border carnation. Semi-double flowers are clear yellow.
D. 'Alice' (illus. p.244). Modern pink. Has clove-scented, semi-double, ivory-white flowers, each with a bold, crimson eye.
♥ *D. alpinus* illus. p.326.
D. 'Annabelle' illus. p.327.
♥ *D. armeria* (Deptford pink). Evergreen, tuft-forming perennial, sometimes grown as an annual. H 30cm (12in), S 45cm (18in). Fully hardy. Has narrowly lance-shaped, dark green leaves. In summer, tall stems each carry small, 5-petalled, cerise-pink flowers in small bunches. Is good for a rock garden or bank.
D. 'Arctic'. Perpetual-flowering, spray carnation. Fully double flowers are white with pink flecks.
D. 'Astor' (illus. p.245). Perpetual-flowering carnation. Has fully double, scarlet flowers; is one of the few scented, perpetual-flowering cultivars.
D. barbatus (Sweet William). **Monarch Series** (auricula-eyed) illus. p.279. **Roundabout Series** (dwarf) illus. p.278.
♥ *D.* 'Becky Robinson' (illus. p.244). Modern pink. Bears clove-scented, fully double, rose-pink flowers with ruby-red lacing. Is good for exhibition.
D. 'Bombardier'. Evergreen, tuft-forming perennial. H and S 10cm (4in). Frost hardy. Has a basal tuft of linear, grey-green leaves and, in summer, small, double, scarlet flowers. Is good for a rock garden.
D. 'Bookham Fancy' (illus. p.245). Border carnation. Produces bright yellow flowers, edged and flecked carmine-purple, on short, stiff stems.
D. 'Bookham Perfume' (illus. p.245). Perennial border carnation. Has scented, semi-double, crimson flowers.
D. 'Borello' (illus. p.245). Perpetual-flowering carnation. Fully double flowers are yellow.
D. 'Bovey Belle' (illus. p.245). Modern pink. Has clove-scented, fully double, bright purple flowers that are excellent for cutting.
D. 'Brympton Red'. Old-fashioned pink. Flowers are single, bright crimson with deeper shading.
D. caesius. See *D. gratianopolitanus*.
D. carthusianorum illus. p.302.
D. 'Charles Musgrave'. See *D.* 'Musgrave's Pink'.
D. chinensis (Indian pink). Slow-growing, bushy annual. H and S 15–30cm (6–12in). Fully hardy. Lance-shaped leaves are pale or mid-green. Tubular, single or double flowers, 2.5cm (1in) or more wide and with open, spreading petals, in shades of pink, red or white, are produced in summer and early autumn. **Baby Doll Series** illus. p.276; 'Fire Carpet' illus. p.279; 'Heddewigii', H 30cm (12in), has flowers in mixed colours.
D. 'Christine Hough' (illus. p.245). Perennial border carnation. Semi-double flowers are apricot, overlaid and streaked with rose-pink.
D. 'Christopher' (illus. p.245). Modern pink. Produces lightly scented, fully double, bright salmon-red flowers.
D. 'Clara' (illus. p.245). Perpetual-flowering carnation. Fully double flowers are yellow with salmon flecks.
D. 'Constance Finnis'. See *D.* 'Fair Folly'.
D. 'Cream Sue' (illus. p.245). Perpetual-flowering carnation. Flowers are cream. Is a sport of *D.* 'Apricot Sue'.
D. 'Crompton Princess' (illus. p.245). Perpetual-flowering carnation. Flowers are pure white.
D. 'Dad's Favourite', syn. *D.* 'A.J. Macself' (illus. p.244). Old-fashioned pink. Bears scented, semi-double, white flowers with chocolate-brown lacing.
♥ *D. deltoides* (Maiden pink). Evergreen, mat-forming, basal-tufted perennial. H 15cm (6in), S 30cm (12in). Fully hardy. In summer, small, 5-petalled, white, pink or cerise flowers are borne singly above tiny, lance-shaped leaves. Is good for a rock garden or bank. Trim back after flowering. 'Flashing Light' (syn. *D.d.* 'Leuchtfunk') illus. p.328.
D. 'Denis'. Modern pink. Strongly clove-scented, fully double, magenta flowers are freely produced.
♥ *D.* 'Doris' (illus. p.244). Modern pink. Has compact growth and an abundance of fragrant, semi-double, pale pink flowers, each with a salmon-red ring towards base of flower. Is one of the most widely grown cultivars and provides excellent flowers for cutting.
D. 'Edenside'. Perennial border carnation. Semi-double flowers are clear white.
D. 'Emile Paré' (illus. p.244). Old-fashioned, mule pink. Has clusters of semi-double, salmon-pink flowers and, unusually for a pink, mid-green foliage.
D. 'Eva Humphries' (illus. p.244). Perennial border carnation. Has fragrant, semi-double flowers with white petals outlined in purple.
D. 'Fair Folly', syn. *D.* 'Constance Finnis' (illus. p.244). Modern pink. Flowers are single and of variable colour, usually dusky-pink to dusky-purple with 2 white splashes on each petal.
D. 'Forest Treasure' (illus. p.244). Perennial border carnation. Has double, white flowers with reddish-purple splashes on each petal.
D. 'Freckles'. Modern pink. A compact cultivar, it has fully double flowers that are red-speckled and dusky-pink.
D. 'Golden Cross' (illus. p.245). Border carnation. Produces bright yellow flowers on short, stiff stems.
♥ *D.* 'Gran's Favourite' (illus. p.244). Old-fashioned pink. Bears fragrant, semi-double, white flowers with deep raspberry lacing.
♥ *D. gratianopolitanus*, syn. *D. caesius*, illus. p.325.
D. 'Green Eyes'. See *D.* 'Musgrave's Pink'.
D. haematocalyx. Evergreen, tuft-forming perennial. H 12cm (5in), S 10cm (4in). Fully hardy. Leaves are lance-shaped and usually glaucous. Bears 5-petalled, toothed, beige-backed, deep pink flowers on slender

481

stems in summer. Suits a rock garden or scree.
D. 'Happiness' (illus. p.245). Perennial border carnation. Semi-double flowers are yellow, striped scarlet-orange.
D. 'Haytor' (illus. p.244). Modern pink. Fully double, white flowers, borne on strong stems, have a good scent. Is widely grown, especially to provide cut flowers.
D. 'Hidcote'. Evergreen, tufted, compact perennial. H and S 10cm (4in). Fully hardy. Bears a basal tuft of linear, spiky, grey-green leaves and, in summer, double, red flowers. Suits a rock garden.
♥ D. 'Houndspool Ruby', syn. D. 'Ruby', D. 'Ruby Doris' (illus. p.245). Modern pink. Is a sport of D. 'Doris' with strongly scented, ruby-pink flowers that each have a deeper eye.
D. 'Ibiza'. Perpetual-flowering, spray carnation. Fully double flowers are shell-pink.
D. 'Iceberg'. Modern pink. Fragrant flowers are semi-double and pure white. Has a somewhat looser habit than D. 'Haytor'.
D. 'Joy' (illus. p.245). Modern pink. Bears semi-double, pink flowers that are strongly scented and good for cutting.
D. 'La Bourbille'. See D. 'La Bourboule'.
♥ D. 'La Bourboule', syn. D. 'La Bourbille', illus. p.326.
D. 'Laced Monarch' (illus. p.245). Perpetual flowering carnation. Double flowers are pink, laced with maroon-red.
D. 'Lavender Clove' (illus. p.245). Vigorous border carnation. Bears lavender-grey flowers on long stems.
D. 'Little Jock' illus. p.325.
D. 'London Brocade' (illus. p.244). Modern pink. Has clove-scented, crimson-laced, pink flowers.
D. 'London Delight'. Old-fashioned pink. Fragrant flowers are semi-double and lavender, laced with purple.
D. 'Manon'. Perpetual-flowering carnation. Is one of the best deep pink cultivars with frilly double flowers.
D. 'Mars'. Evergreen, tuft-forming perennial. H and S 10cm (4in). Frost hardy. Has small, double, cherry-red flowers in summer. Bears a basal tuft of linear, grey-green leaves. Is good in a rock garden.
D. 'Master Stuart'. Perennial border carnation. Has striking, semi-double flowers that are white with scarlet stripes.
D. microlepis illus. p.326.
♥ D. 'Monica Wyatt' illus. p.245. Modern pink. Very fragrant, fully double flowers are cyclamen-pink, each with a magenta eye. Is very free-flowering and provides excellent cut flowers.
D. monspessulanus. Evergreen, mat-forming perennial. H 30cm (12in), S 10–15cm (4–6in). Fully hardy. In summer, masses of strongly fragrant, 5-petalled, deeply fringed, pale lavender flowers rise on slender stems above short tufts of fine, grass-like leaves. Is good for a rock garden. Needs gritty soil.

D. 'Mrs Sinkins' (illus. p.244). Old-fashioned pink. Flowers are heavily scented, fringed, fully double and white.
D. 'Murcia'. Perpetual-flowering carnation. Has fully double, deep golden-yellow flowers.
D. 'Musgrave's Pink', syn. D. 'Charles Musgrave', D. 'Green Eyes' (illus. p.244). Old-fashioned pink. An old cultivar, it bears single, white flowers with green eyes.
D. myrtinervius illus. p.326.
D. neglectus. See D. pavonius.
D. 'Nina' (illus. p.245). Perpetual-flowering carnation. Is one of the best crimson cultivars. Fully double flowers have smooth-edged petals.
D. 'Nives' (illus. p.244). Perpetual-flowering carnation. Fully double flowers are clear white and are borne on short stems.
♥ D. pavonius, syn. D. neglectus, illus. p.323.
D. 'Pierrot' (illus. p.244). Perpetual-flowering carnation. Attractive, fully double flowers are light rose-lavender with purple-edged petals.
♥ D. 'Pike's Pink' illus. p.326.
D. 'Pink Calypso'. See D. 'Truly Yours'.
D. 'Pink Jewel' (illus. p.245). Modern pink. Has strongly scented, semi-double, pink flowers.
D. 'Prudence' (illus. p.244). Old-fashioned pink. Fragrant flowers are semi-double and pinkish-white with purple lacing. Has a spreading habit.
D. 'Raggio di Sole' (illus. p.245). Perpetual-flowering carnation. Fully double flowers are bright orange.
D. 'Red Barrow'. Perpetual-flowering, spray carnation. Fully double, bright scarlet flowers are borne in abundance.
D. 'Ruby'. See D. 'Houndspool Ruby'.
D. 'Ruby Doris'. See D. 'Houndspool Ruby'.
D. 'Sam Barlow'. Old-fashioned pink. Bears very fragrant, frilly, fully double, white flowers with chocolate-brown centres.
D. 'Show Ideal'. Modern pink. Flat-petalled, semi-double flowers are white with red eyes and are strongly scented. Is excellent for exhibition.
D. 'Sops-in-Wine'. Old-fashioned pink. Bears fragrant, single, maroon flowers with white markings.
D. superbus. Evergreen, mat-forming perennial. H to 20cm (8in), S 15cm (6in). Fully hardy. Has narrowly lance-shaped, pale green leaves. In summer, slender stems bear very fragrant, 5-petalled, deeply fringed, pink flowers with darker centres. Suits a rock garden.
D. 'Tigré'. Perpetual-flowering carnation. Has fully double, yellow flowers with a uniform, pinkish-purple stripe and edging to each petal.
D. 'Tony'. Perpetual-flowering, spray carnation. Fully double flowers are yellow with red stripes.
D. 'Truly Yours', syn. D. 'Pink Calypso' (illus. p.244). Perpetual-flowering carnation. Fully double flowers are a good pink.
♥ D. 'Valda Wyatt'. Modern pink. Flowers are very fragrant, fully double and rose-lavender.

D. 'Valencia' (illus. p.245). Perpetual-flowering carnation. Has fully double, golden-orange flowers.
D. 'White Ladies' (illus. p.244). Old-fashioned pink. Bears very fragrant, fully double, white flowers with greenish centres.
D. 'Widecombe Fair'. Modern pink. Semi-double flowers, borne on strong stems, are of unusual colouring – peach-apricot, opening to blush-pink.

DIAPENSIA (Diapensiaceae)
Genus of evergreen, spreading sub-shrubs, suitable for rock gardens and troughs. Fully hardy. Needs partial shade and peaty, sandy, acid soil. Is very difficult to grow in hot, dry climates at low altitudes. Propagate by seed in spring or by semi-ripe cuttings in summer.
D. lapponica. Evergreen, spreading sub-shrub. H and S 7cm (3in). Has tufts of small, rounded, leathery leaves. Carries solitary tiny, bowl-shaped, white flowers in early summer.

DIASCIA (Scrophulariaceae)
Genus of summer- and autumn-flowering annuals and perennials, some of which are semi-evergreen, grown for their tubular, pink flowers. Is suitable for banks and borders. Frost hardy. Needs sun and humus-rich, well-drained soil that is not too dry. Cut back old stems in spring. Propagate by softwood cuttings in late spring, by semi-ripe cuttings in summer or by seed in autumn.
D. cordata illus. p.301.
♥ D. barberiae 'Ruby Field'. Mat-forming perennial. H 8cm (3in), S 15cm (6in). Heart-shaped, pale green leaves clothe short, wiry stems. Produces tubular, wide-lipped, salmon-pink flowers throughout summer.
♥ D. rigescens illus. p.301.
♥ D. vigilis. Prostrate perennial. H 30–40cm (12–16in), S 60cm (24in). Leaves are small, rounded, toothed and pale green. Upright branchlets carry loose spikes of flattish, outward-facing, pale pink flowers in summer.

DICENTRA (Papaveraceae)
Genus of perennials, grown for their elegant sprays of pendent flowers. Fully hardy. Does best in semi-shade and humus-rich, moist but well-drained soil. Propagate by division when dormant in late winter, species also by seed in autumn.
D. 'Adrian Bloom'. Spreading, tuft-forming perennial. H 45cm (18in), S 30cm (12in). Produces sprays of pendent, heart-shaped, rich carmine-pink flowers above oval, grey-green leaves.
D. cucullaria illus. p.310.
D. formosa. Spreading, tufted perennial. H 45cm (18in), S 30cm (12in). In spring-summer produces slender, arching sprays of pendent, heart-shaped, pink or dull red flowers above oval, finely cut, grey-green leaves.
D. peregrina. Tuft-forming perennial. H 8cm (3in), S to 5cm (2in). Locket-shaped, pink flowers appear in spring-summer above fern-like, blue-green leaves. Needs gritty soil. Is suitable

for an alpine house.
♥ D. spectabilis illus. p.209.
♥ f. alba illus. p.203.
D. 'Spring Morning' illus. p.242.
D. 'Stuart Boothman'. Tufted perennial. H 45cm (18in), S 30cm (12in). Arching sprays of heart-shaped, carmine flowers are borne in spring-summer above oval, finely cut, deep grey-green leaves.

DICHELOSTEMMA (Liliaceae)
Genus of summer-flowering bulbs, grown for their dense flower heads on leafless stems. Is related to Brodiaea and is similar to Allium in appearance. Frost hardy, but in cold areas grow in a sheltered position. Needs a sunny site and well-drained soil. Water freely in spring, but dry out after flowering. Propagate by seed in autumn or spring or by offsets in autumn before growth commences.
D. congestum, syn. Brodiaea congesta, illus. p.346.
D. ida-maia, syn. Brodiaea ida-maia. Early summer-flowering bulb. H to 1m (3ft), S 8–10cm (3–4in). Long, narrow leaves are semi-erect and basal. Leafless stem carries a dense head of 2–2.5cm ($^3/_4$–1in) long flowers, each with a red tube and 6 green petals.
D. pulchellum, syn. Brodiaea pulchella. Early summer-flowering bulb. H 30–60cm (12–24in), S 8–10cm (3–4in). Long, narrow leaves are semi-erect and basal. Leafless stem produces a dense head of narrowly funnel-shaped, pale to deep violet flowers, 1–2cm ($^1/_2$–$^3/_4$in) long, with violet bracts.

DICHORISANDRA (Commelinaceae)
Genus of erect, clump-forming, evergreen perennials, grown for their ornamental foliage. Frost tender, min. 15–20°C (59–68°F). Prefers fertile, moist but well-drained soil, humid conditions and partial shade. Propagate by division in spring or by stem cuttings in summer.
D. reginae illus. p.216.

DICKSONIA (Cyatheaceae)
Genus of evergreen or semi-evergreen, tree-like ferns that resemble palms and that are sometimes used to provide height in fern plantings. Half hardy to frost tender, min. 5°C (41°F). Needs semi-shade and humus-rich, moist soil. Remove faded fronds regularly. Propagate by spores in summer.
♥ D. antarctica illus. p.186.

DICTAMNUS (Rutaceae)
Genus of summer-flowering perennials. Fully hardy. Requires full sun and fertile, well-drained soil. Resents disturbance. Propagate by seed sown in late summer when fresh.
♥ D. albus illus. p.204.
♥ var. purpureus illus. p.207.

Didiscus coeruleus. See Trachymene coerulea.

DIEFFENBACHIA (Araceae)
Dumb cane, Leopard lily
Genus of evergreen, tufted perennials, grown for their foliage. Frost tender,

min. 15°C (59°F). Grow in fertile, well-drained soil and in partial shade. Propagate in spring or summer by stem cuttings or pieces of leafless stem placed horizontally in compost. Contains poisonous sap that should be kept away from mouth, eyes and skin. Scale insect or red spider mite may be troublesome.

D. 'Amoena', syn. *D. seguine* 'Amoena'. Evergreen, robust, tufted perennial. H to 2m (6ft), S 1m (3ft). Broadly lance-shaped, glossy leaves, to 45cm (18in) long, are dark green with creamy-white bars along lateral veins. Has insignificant, greenish-white flowers clustered on the spadix, surrounded by a narrow, leaf-like spathe, produced intermittently.
D. 'Exotica', syn. *D. maculata* 'Exotica', *D. seguine* 'Exotica', illus. p.228.
D. maculata 'Exotica' see *D.* 'Exotica'. **'Rudolph Roehrs'** see *D. seguine* 'Rudolph Roehrs'.
D. 'Memoria'. See *D. seguine* 'Memoria Corsii'.
D. seguine. Evergreen, tufted perennial. H and S 1m (3ft) or more. Broadly lance-shaped leaves, to 45cm (18in) long, are glossy and dark green. Insignificant, tiny, greenish-white flowers, clustered on the spadix, are surrounded by a narrow, leaf-like spathe that appears intermittently. **'Amoena'**, see *D.* 'Amoena'. **'Exotica'**, see *D.* 'Exotica'. **'Memoria Corsii'** (syn. *D.* 'Memoria') has grey-green leaves, marked dark green and spotted white. **'Rudolph Roehrs'** (syn. *D. maculata* 'Rudolph Roehrs', *D.s.* 'Roehrsii') illus. p.231.

DIERAMA (Iridaceae)
Angel's fishing rod, Wandflower
Genus of evergreen, clump-forming, summer-flowering corms with pendent, funnel- or bell-shaped flowers on long, arching, wiry stems. Flourishes near pools. Frost to half hardy. Prefers a warm, sheltered, sunny site and well-drained soil that should be kept moist in summer when in growth. Dies down partially in winter. Propagate by division of corms in spring or by seed in autumn or spring. Resents disturbance, and divisions take a year or more to settle and start flowering again.
D. 'Blackbird'. Evergreen, upright perennial. H 1.5m (5ft), S 30cm (1ft). Frost hardy. Produces cascades of nodding, funnel-shaped, violet-mauve flowers on wiry, pendulous stems in summer above grass-like leaves.
D. dracomontanum, syn. *D. pumilum*. Vigorous, evergreen, upright perennial. H 75cm (30in), S 30cm (12in). Frost hardy. In summer freely produces nodding, funnel-shaped flowers, in shades of pink and violet, on wiry stems. Leaves are grass-like.
D. pendulum illus. p.345.
D. pulcherrimum illus. p.344.
D. pumilum. See *D. dracomontanum*.

DIERVILLA (Caprifoliaceae)
Genus of deciduous, summer-flowering shrubs, grown for their overall appearance. Is similar to *Weigela*. Frost hardy. Tolerates full light or partial shade and moderately fertile, well-drained soil. For a more shapely shrub remove 2- and 3-year-old stems in winter or after flowering. Propagate by semi-ripe cuttings in late summer or by hardwood cuttings in autumn.
D. sessilifolia. Deciduous, spreading shrub. H and S 1–1.5m (3–5ft). Narrowly oval, pointed, serrated, green leaves are often copper-tinted when young. Has terminal and lateral clusters of tubular, pale yellow flowers in summer. To treat as an herbaceous perennial, cut back to ground level each spring and apply a mulch and a fertilizer.

DIETES (Iridaceae)
Genus of evergreen, iris-like, rhizomatous perennials, grown for their attractive flowers in spring or summer. Half hardy. Needs sun or partial shade and humus-rich, well-drained soil that does not dry out excessively. Propagate by seed in autumn or spring or by division in spring (although divisions do not become re-established very readily).
D. bicolor illus. p.346.
D. iridioides, syn. *D. vegeta* of gardens. Evergreen, spring- and summer-flowering, rhizomatous perennial. H to 60cm (2ft), S 30–60cm (1–2ft). Bears sword-shaped, semi-erect, basal leaves in a spreading fan. Branching, wiry stems bear iris-like, white flowers, 6–8cm (2½–3in) across. Each of the 3 large petals has a central, yellow mark.
D. vegeta of gardens. See *D. iridioides*.

DIGITALIS (Scrophulariaceae)
Foxglove
Genus of biennials and perennials, some of which are evergreen, grown for their flower spikes in summer. Fully to frost hardy. Species mentioned grow in most conditions, even dry, exposed sites, but do best in semi-shade and moist but well-drained soil. Propagate by seed in autumn.
D. ambigua. See *D. grandiflora*.
D. canariensis. See *Isoplexis canariensis*.
D. ferruginea illus. p.205.
♥ **D. grandiflora**, syn. *D. ambigua* (Yellow foxglove). Evergreen, clump-forming perennial. H 75cm (30in), S 30cm (12in). Fully hardy. Racemes of downward-pointing, tubular, creamy-yellow flowers appear in summer above a rosette of oval to oblong, smooth, strongly veined leaves.
D. lutea. Upright perennial. H 75cm (30in), S 30cm (12in). Fully hardy. In summer, delicate spires of downward-pointing, narrowly tubular, creamy-yellow flowers are borne above a rosette of oval, smooth, mid-green leaves.
♥ **D. × mertonensis.** Clump-forming perennial. H 75cm (30in), S 30cm (12in). Fully hardy. Bears spikes of downward-pointing, tubular, rose-mauve to coppery flowers in summer, above a rosette of oval, hairy, soft leaves. Divide after flowering.
D. purpurea. Upright, short-lived perennial, grown as a biennial. H 1–1.5m (3–5ft), S 60cm (2ft). Fully hardy. Has a rosette of oval, rough, deep green leaves and, in summer, tall spikes of tubular flowers in shades of pink, red, purple or white. ♥ **f. alba** (syn. *D.p.* var. *albiflora*, *D.p.* 'Alba'), illus. p.270.

DILLENIA (Dilleniaceae)
Genus of evergreen or briefly deciduous, spring-flowering trees, grown for their flowers and foliage and for shade. Frost tender, min. 16°C (61°F). Needs moisture-retentive, fertile soil and full light. Water potted plants freely while in full growth, less in winter. Propagate by seed in spring.
D. indica (Elephant apple). Briefly deciduous, spreading tree. H and S 8–12m (25–40ft). Has oval, serrated, boldly parallel-veined, glossy leaves, each 30cm (1ft) long. Nodding, cup-shaped, white flowers, each 15–20cm (6–8in) wide, are produced in spring, followed by edible, globular, greenish fruits.

DIMORPHOTHECA (Compositae)
African daisy, Cape marigold
Genus of annuals, perennials and evergreen sub-shrubs. Half hardy. Grow in sun and in fertile, very well-drained soil. Dead-head to prolong flowering. Propagate annuals by seed sown under glass in mid-spring, perennials by semi-ripe cuttings in summer. Is susceptible to botrytis in wet summers.
D. annua. See *D. pluvialis*.
D. barberiae. See *Osteospermum jucundum*.
D. pluvialis, syn. *D. annua*, illus. p.271.

DIONAEA (Droseraceae)
Genus of evergreen, insectivorous, rosette-forming perennials. Frost tender, min. 5°C (41°F). Needs partial shade and a humid atmosphere; grow in a mixture of peat and moss, kept constantly moist. Propagate by seed or division in spring.
D. muscipula illus. p.269.

DIONYSIA (Primulaceae)
Genus of evergreen, cushion-forming perennials. Fully hardy. Grow in an alpine house in sun and very gritty, well-drained soil. Position deep collar of grit under cushion and ensure good ventilation at all times. Dislikes winter wet. Propagate by softwood cuttings in summer. Plants are susceptible to botrytis.
♥ **D. aretioides** illus. p.319.
D. lamingtonii. Evergreen, prostrate perennial. H 2.5cm (1in), S 15cm (6in). Oval, grey-green leaves in rosettes form tight cushions. Small, stemless, 5-petalled, bright yellow flowers appear in early spring. Is diffficult to grow.
D. tapetodes illus. p.320.

DIOON (Zamiaceae)
Genus of evergreen shrubs, grown for their palm-like appearance. Frost tender, min. 13–18°C (55–64°F). Needs full sun and fertile, well-drained soil. Water containerized specimens moderately, less when not in full growth. Propagate by seed in spring.
D. edule (Virgin's palm). Very slow-growing, evergreen, palm-like shrub, eventually with a thick, upright trunk. H 2–4m (6–12ft), S 1.5–3m (5–10ft). Leaves are feather-like, 60cm–1.2m (2–4ft) long, with spine-tipped, deep blue-green leaflets.

DIOSCOREA (Dioscoreaceae)
Genus of tuberous perennials, some of which are succulent, and herbaceous or evergreen, twining climbers, grown mainly for their decorative leaves. Insignificant flowers are generally yellow. Frost tender, min. 5–13°C (41–55°F). Prefers full sun or partial shade and fertile, well-drained soil. Propagate by division or by cutting off sections of tuber in spring or autumn, or by seed in spring.
D. discolor illus. p.179.
D. elephantipes, syn. *Testudinaria elephantipes*, illus. p.403.

DIOSMA (Rutaceae)
Genus of evergreen, wiry-stemmed shrubs, grown for their flowers and overall appearance. Frost tender, min. 7°C (45°F). Needs full light and well-drained, neutral to acid soil. Water potted specimens moderately, less when not in full growth. To create a compact habit shorten flowered stems after flowering. Propagate by seed in spring or by semi-ripe cuttings in late summer.
D. ericoides (Breath of heaven). Fast-growing, evergreen, loosely rounded shrub. H and S 30–60cm (1–2ft). Aromatic, needle-like leaves are crowded on stems. In winter-spring carries a profusion of small, fragrant, 5-petalled, white flowers, sometimes tinted red.

DIOSPHAERA. See TRACHELIUM.

DIOSPYROS (Ebenaceae)
Genus of deciduous or evergreen trees and shrubs, grown for their foliage and fruits. Fully to frost hardy. Needs full sun and does best in hot summers. Requires fertile, well-drained soil. To obtain fruits, plants of both sexes should be grown. Propagate by seed in autumn.
D. kaki (Chinese persimmon, Kaki, Persimmon). Deciduous, spreading tree. H 10m (30ft), S 7m (22ft). Frost hardy. Oval, glossy, dark green leaves turn orange, red and purple in autumn. Tiny, yellowish-white flowers in summer are followed on female trees by large, edible, rounded, yellow or orange fruits.
D. lotus (Date plum). Deciduous, spreading tree. H 10m (30ft), S 6m (20ft). Fully hardy. Has oval, glossy, dark green leaves, tiny, red-tinged, green flowers from mid- to late summer and, on female trees, unpalatable, rounded, purple or yellow fruits.

DIPCADI (Liliaceae)
Genus of spring-flowering bulbs, grown mainly for botanical interest. Frost hardy, but will not tolerate cold, wet winters, so is best grown in a cold frame or alpine house. Needs a warm, sunny situation and light, well-drained

soil. Is dormant in summer. Propagate by seed in autumn.
D. serotinum illus. p.378.

DIPELTA (Caprifoliaceae)
Genus of deciduous shrubs, with bold, long-pointed leaves, grown for their showy, tubular flowers and peeling bark. After flowering, bracts beneath flowers enlarge and become papery and brown, surrounding the fruits. Fully hardy. Requires sun or semi-shade and fertile, well-drained soil. Benefits from the occasional removal of old shoots after flowering. Propagate by softwood cuttings in summer.
D. floribunda illus. p.87.
D. yunnanensis illus. p.86.

DIPHYLLEIA (Berberidaceae)
Genus of perennials with creeping rootstocks and umbrella-like leaves. Is best suited to woodland gardens. Fully hardy. Needs semi-shade and moist soil. Propagate by division in spring or by seed in autumn.
D. cymosa (Umbrella leaf). Rounded perennial. H 60cm (24in), S 30cm (12in). Has large, rounded, 2-lobed leaves. In spring bears loose heads of inconspicuous, white flowers followed by indigo-blue berries on red stalks.

DIPIDAX. See ONIXOTIS.

Diplacus glutinosus. See *Mimulus aurantiacus.*

DIPLADENIA. See MANDEVILLA.

DIPLARRHENA (Iridaceae)
Genus of one species of summer-flowering perennial. Half hardy. Needs sun and well-drained soil. Propagate by seed or division in spring.
D. moraea illus. p.241.

Diplazium japonicum. See *Lunathyrium japonicum.*

DIPTERONIA (Aceraceae)
Genus of deciduous trees, grown for their foliage and fruits. Fully hardy. Needs full sun and fertile, well-drained soil. Propagate by softwood cuttings in summer or by seed in autumn.
D. sinensis. Deciduous, spreading, sometimes shrubby tree. H 10m (30ft), S 6m (20ft). Large, mid-green leaves have 7–11 oval to lance-shaped leaflets. Inconspicuous, greenish-white flowers in summer are followed by large clusters of winged, red fruits.

DISA. See ORCHIDS.
D. uniflora. Deciduous, terrestrial orchid. H 45–60cm (1½–2ft). Frost tender, min. 7–10°C (45–50°F). Has narrowly lance-shaped, glossy, dark green leaves, to 22cm (9in) long, that clasp stems. In early summer each stem bears up to 7 hooded, scarlet flowers, 8–10 cm (3–4in) long, that have darker veins and are suffused yellow. Is the parent of many hybrids grown commercially for cut flowers. Needs partial shade and moist soil that should not be allowed to dry out. May be raised from seed or propagated by division of offsets when dormant.

DISANTHUS (Hamamelidaceae)
Genus of one species of deciduous, autumn-flowering shrub, grown for its overall appearance and autumn colour. Frost hardy. Needs partial shade and humus-rich, moist but not wet, neutral to acid soil. Propagate by layering in spring or by seed when ripe or in spring.
♀ *D. cercidifolius.* Deciduous, rounded shrub. H and S to 3m (10ft). Bears broadly oval to almost circular, bluish-green leaves that turn red, purple, orange or yellow in autumn. Has small, dark red flowers, like pairs of back-to-back spiders, in autumn as the leaves fall, or later.

DISCARIA (Rhamnaceae)
Genus of deciduous or almost leafless shrubs and trees, grown for their habit and flowers. Spiny, green shoots assume function of leaves. Frost hardy. Needs a sheltered, sunny site and fertile, well-drained soil. Propagate by softwood cuttings in summer.
D. toumatou (Wild Irishman). Deciduous or almost leafless, bushy shrub. H and S 2m (6ft). Shoots have sharp, rigid spines. Tiny, star-shaped, greenish-white flowers are borne in dense clusters in late spring.

DISPORUM (Liliaceae)
Fairy bells
Genus of spring- or early summer-flowering perennials. Is best suited to woodland gardens. Fully hardy. Requires a cool, semi-shaded position and humus-rich soil. Propagate by division in spring or by seed in autumn.
D. hookeri. Clump-forming perennial. H 75cm (30in), S 30cm (12in). Leaves are narrowly oval and mid-green. Clusters of drooping, open bell-shaped, greenish-white flowers in spring are followed by orange-red berries in autumn.
D. sessile 'Variegatum'. Rapidly spreading, clump-forming perennial. H 45cm (18in), S 30cm (12in). Solitary tubular-bell-shaped to bell-shaped, creamy-white flowers are borne in spring. Narrowly oval leaves are pleated and irregularly striped with green and white.

DISTICTIS (Bignoniaceae)
Genus of evergreen, woody-stemmed, tendril climbers, grown for their colourful, trumpet-shaped flowers. Frost tender, min. 5–7°C (41–5°F). Well-drained soil is suitable with full light. Water freely in summer, less at other times. Support for stems is necessary. Thin out congested growth in spring. Propagate by softwood cuttings in early summer or by semi-ripe cuttings in late summer.
D. buccinatoria, syn. *Phaedranthus buccinatorius*, illus. p.165.

DISTYLIUM (Hamamelidaceae)
Genus of evergreen shrubs and trees, grown for their foliage and flowers. Frost hardy. Prefers a sheltered, partially shaded position and moist, peaty soil. Propagate by semi-ripe cuttings in summer.
D. racemosum. Evergreen, arching shrub. H 2m (6ft), S 3m (10ft). Leaves are oblong, leathery, glossy and dark green. Small flowers, with red calyces and purple anthers, are borne in late spring and early summer.

Dizygotheca elegantissima. See *Schefflera elegantissima.*

DOCYNIA (Rosaceae)
Genus of evergreen or semi-evergreen, spring-flowering trees, grown for their flowers and foliage; is related to *Cydonia*. Half hardy. Requires full light and well-drained soil. Other than shaping while young, pruning is not necessary. Propagate by budding in summer, by seed in autumn or spring or by grafting in winter. Is usually trouble-free, though caterpillars may be troublesome.
D. delavayi. Evergreen or semi-evergreen, spreading tree. H and S 8m (25ft) or more. Oval to lance-shaped leaves are white-felted beneath. In spring has fragrant, 5-petalled, white flowers, pink in bud, followed by ovoid, downy, yellow fruits in autumn.

DODECATHEON (Primulaceae)
Shooting stars
Genus of spring- and summer-flowering perennials, grown for their distinctive flowers, with reflexed petals and prominent stamens. Once fertilized, flowers turn skywards – hence their common name. Is dormant after flowering. Fully to frost hardy. Prefers sun or partial shade and moist but well-drained soil. Propagate by seed in autumn or by division in winter.
D. dentatum. Clump-forming perennial. H 7cm (3in), S 25cm (10in). Fully hardy. Leaves are long, oval and toothed. In late spring, frail stems produce white flowers with prominent, dark stamens and reflexed petals. Prefers a partially shaded position.
♀ *D. hendersonii* illus. p.295.
♀ *D. meadia.* Clump-forming perennial. H 20cm (8in), S 15cm (6in). Fully hardy. Leaves are oval and pale green. In spring, strong stems carry pale pink flowers, with reflexed petals, above foliage. Prefers a partially shaded position. ♀ f. *album* illus. p.295.
♀ *D. pulchellum.* Clump-forming perennial. H 15cm (6in), S 10cm (4in). Fully hardy. Is similar to *D. meadia*, but flowers are usually deep cerise.
D. 'Red Wings' illus. p.296.

DODONAEA (Sapindaceae)
Genus of evergreen trees and shrubs, grown mainly for their foliage and overall appearance. Half hardy to frost tender, min. 3–5°C (37–41°F). Prefers full sun and well-drained soil. Water potted plants freely when in full growth, less at other times. Cut back in late summer, and again in spring, if needed, to maintain a balanced shape. Propagate by seed in spring or by semi-ripe cuttings in late summer.
D. viscosa 'Purpurea' illus. p.121.

Dolichos lablab. See *Lablab purpureus.*
Dolichos lignosus. See *Lablab purpureus.*

DOMBEYA (Sterculiaceae)
Genus of evergreen shrubs and trees, grown for their flowers. Frost tender, min. 5–13°C (41–55°F). Needs full light or partial shade and fertile, well-drained soil. Water potted specimens freely when in full growth, less when temperatures are low. May be cut back after flowering. Propagate by seed in spring or by semi-ripe cuttings in summer. Whitefly and red spider mite may be a nuisance.
D. burgessiae, syn. *D. mastersii*, illus. p.120.
D. x *cayeuxii* illus. p.62.
D. mastersii. See *D. burgessiae.*

DONDIA. See HACQUETIA.

DORONICUM (Compositae)
Leopard's bane
Genus of perennials, grown for their daisy-like flower heads that are good for cutting. Fully hardy. Grows in full light or shade and any well-drained soil. Propagate by division in autumn.
D. austriacum. Clump-forming perennial. H 45cm (18in), S 30cm (12in). Daisy-like, pure yellow flower heads are borne singly on slender stems in spring. Heart-shaped, bright green leaves are hairy and wavy-edged.
D. cordatum. See *D. pardalianches.*
D. 'Frühlingspracht'. See *D.* 'Spring Beauty'.
D. pardalianches, syn. *D. cordatum*, illus. p.202.
D. plantagineum 'Excelsum'. Elegant, clump-forming perennial. H 1m (3ft), S 60cm (2ft). Large, daisy-like, buttercup-yellow flower heads are borne, 3 or 4 to a stem, in spring. Leaves are heart-shaped and bright green. Is good for a dry, shaded site.
D. 'Spring Beauty', syn. *D.* 'Frühlingspracht'. Clump-forming perennial. H 45cm (18in), S 30cm (12in). Produces daisy-like, double, bright yellow flower heads in spring. Bears heart-shaped, bright green leaves.

DOROTHEANTHUS (Aizoaceae)
Genus of succulent annuals, suitable for hot, dry places such as rock gardens, banks and gaps in paving. Half hardy. Needs sun and grows well in poor, very well-drained soil. Deadhead to prolong flowering. Propagate by seed sown under glass in early spring, or outdoors in mid-spring. Protect from slugs and snails.
D. bellidiformis, syn. *Mesembryanthemum criniflorum* (Iceplant, Livingstone daisy). **Magic Carpet Series** illus. p.277.

DORYANTHES (Liliaceae)
Genus of evergreen, rosette-forming perennials, grown for their flowers. Frost tender, min. 10°C (50°F). Needs a sunny position and humus-rich, well-drained soil. Propagate by seed in spring, by mature bulbils or by suckers after flowering.
D. palmeri illus. p.196.

DORYCNIUM (Leguminosae)
Genus of perennials and deciduous or semi-evergreen sub-shrubs, grown for their foliage, flowers and fruits. Frost

hardy. Needs full sun and dry, very well-drained soil. Propagate by softwood cuttings in summer or by seed in autumn.
D. hirsutum. Deciduous, upright sub-shrub. H and S 60cm (24in). Bears silver-grey leaves with 3 oval leaflets. Dense clusters of pea-like, pink-tinged, white flowers in summer and early autumn are followed by oblong to ovoid, reddish-brown seed pods.

Douglasia vitaliana. See *Vitaliana primuliflora*.

DOXANTHA. See **MACFADYENA**.
D. capreolata. See *Bignonia capreolata*.

DRABA (Cruciferae)
Genus of spring-flowering annuals and evergreen or semi-evergreen, cushion- or mat-forming perennials with extensive root systems. Some species form soft, green cushions that in winter turn brown except at the tips, thus appearing almost dead. Is suitable for alpine houses. Fully to frost hardy. Needs sun and gritty, well-drained soil. Dislikes winter wet. Propagate by softwood cuttings of the rosettes in late spring or by seed in autumn.
D. aizoides (Yellow whitlow grass). Semi-evergreen, mat-forming perennial. H 2.5cm (1in), S 15cm (6in). Fully hardy. Has lance-shaped, stiff-bristled leaves in rosettes and, in spring, 4-petalled, bright yellow flowers. Suits a scree.
D. hispanica. Semi-evergreen, cushion-forming perennial. H 5cm (2in), S 10cm (4in). Frost hardy. Leaves are oval, soft, fragile and pale green. Clusters of flat, 4-petalled, pale yellow flowers are borne in spring.
♥ ***D. longisiliqua*** illus. p.319.
D. mollissima illus. p.320.
D. polytricha. Semi-evergreen, cushion-forming perennial. H 6cm (2½in), S 15cm (6in). Fully hardy. Has minute, rounded leaves in neat, symmetrical rosettes. Frail rosettes carry flat, 4-petalled, golden-yellow flowers in spring. Is difficult to grow. Keep stones under cushion at all times. Remove dead rosettes at once.
D. rigida illus. p.319. var. ***bryoides*** is an evergreen, tight hummock-forming perennial. H 4cm (1½in), S 6cm (2½in). Fully hardy. Leaves are tiny, rounded, hard and dark green. Produces small clusters of almost stemless, 4-petalled, bright yellow flowers that cover hummocks in spring. Also suited to a trough or scree.

DRACAENA (Agavaceae)
Genus of evergreen trees and shrubs, grown for their foliage and overall appearance. Frost tender, min. 13–18°C (55–64°F). Needs full light or partial shade and well-drained soil. Water containerized plants moderately, much less in low temperatures. Rejuvenate leggy plants by cutting back to near soil level in spring. Propagate by seed or air-layering in spring or by tip or stem cuttings in summer. Mealy bug may be a nuisance.

***D. cincta* 'Tricolor'.** See *D. marginata* 'Tricolor'.
D. deremensis. Slow-growing, evergreen, erect, sparsely branched shrub. H 2m (6ft) or more, S 1m (3ft) or more. Has lance-shaped, erect to arching, glossy, deep green leaves, to 45cm (18in) long. Mature plants may occasionally bear large panicles of small, red-and-white flowers in summer. ♥ **'Warneckii'** (syn. *D.d.* 'Souvenir de Schriever', *D. fragrans* 'Warneckii') illus. p.96.
D. draco illus. p.74.
D. fragrans (Corn plant). **'Warneckii'** see *D. deremensis* 'Warneckii'.
♥ **'Massangeana'** is an evergreen, erect, sparsely branched shrub. H 3–6m (10–20ft), S 1–3m (3–10ft). Has strap-shaped, arching leaves, to 60cm (2ft) long, with longitudinal bands of yellow and pale green. In early summer, fragrant, star-shaped, yellow flowers, rarely produced, are followed by rounded-oblong, orange-red fruits.
D. indivisa. See *Cordyline indivisa*.
♥ ***D. marginata*** (Madagascar dragon tree). Slow-growing, evergreen, erect shrub or tree. H 3m (10ft) or more, S 1–2m (3–6ft) or more. Leaves are narrowly strap-shaped and rich green with red margins. Flowers are rarely produced. **'Tricolor'** (syn. *D. cincta* 'Tricolor') illus. p.73.
♥ ***D. sanderiana*** illus. p.121.

DRACOCEPHALUM (Labiatae)
Dragon's head
Genus of summer-flowering annuals and perennials, suitable for rock gardens and borders. Fully hardy. Prefers sun and fertile, well-drained soil. Propagate by basal cuttings of young growth in spring or by seed or division in spring or autumn.
D. ruyschianum. Erect perennial. H 45–60cm (18–24in), S 30cm (12in). Freely produces whorled spikes of 2-lipped, violet-blue flowers from early to mid-summer. Mid-green leaves are linear to lance-shaped.

DRACUNCULUS (Araceae)
Genus of robust, tuberous perennials that produce roughly triangular, foul-smelling spathes. Frost hardy, but in severe winters protect dormant tubers with a cloche or dead bracken. Needs sun and well-drained soil that dries out in summer. Propagate by freely produced offsets in late summer or by seed in autumn.
D. vulgaris, syn. *Arum dracunculus*, illus. p.345.

DREGEA (Asclepiadaceae)
Genus of evergreen, woody-stemmed, twining climbers, grown for botanical interest. Frost hardy. Grow in sun and in any well-drained soil. Propagate by seed in spring or by stem cuttings in summer or autumn.
D. corrugata. See *D. sinensis*.
D. sinensis, syn. *D. corrugata*, *Wattakaka sinensis*. Evergreen, woody-stemmed, twining climber. H to 3m (10ft). Oval, mid-green leaves, heart-shaped at base, 3–10cm (1¼–4in) long, are greyish beneath. In summer has clusters of 10–25 small, fragrant, star-shaped flowers, white or cream with red dots and streaks, followed by pairs of slender pods, 5–7cm (2–3in) long.

Drejerella guttata. See *Justicia brandegeana*.

DREPANOSTACHYUM
(Bambusoideae). See GRASSES, BAMBOOS, RUSHES and SEDGES.
D. falconeri, syn. *Arundinaria falconeri*, *Thamnocalamus falconeri*. Evergreen, clump-forming bamboo. H 5–10m (15–30ft), S 1m (3ft). Half hardy. Greenish-brown stems have a dark purple ring beneath each node. Has 10–15cm (4–6in) long, yellowish-green leaves, without visible tessellation, and unimportant flower spikes.

DRIMYS (Winteraceae)
Genus of evergreen trees and shrubs, grown for their foliage and star-shaped flowers. Frost hardy, but in cold areas grow against a south- or west-facing wall. Needs sun or semi-shade and fertile, moist but well-drained soil. Propagate by semi-ripe cuttings in summer or by seed in autumn.
D. axillaris. See *Pseudowintera axillaris*.
D. colorata. See *Pseudowintera colorata*.
D. lanceolata (Mountain pepper). Evergreen, upright, dense shrub or tree. H 4m (12ft), S 2.5m (8ft). Has deep red shoots and oblong, dark green leaves. Clusters of star-shaped, white flowers are borne from mid- to late spring.
D. winteri illus. p.52.

DROSANTHEMUM (Aizoaceae)
Genus of erect or prostrate, succulent shrubs with slender stems and masses of flowers in summer. Leaves are finely covered in papillae. Frost tender, min. 5°C (41°F). Needs full sun and very well-drained soil. Propagate by seed or stem cuttings in spring or summer.
D. hispidum. Succulent shrub, with arching or spreading branches that root down. H 60cm (2ft), S 1m (3ft). Has cylindrical, light green leaves, 1.5–2.5cm (⅝–1in) long. In summer, masses of shiny, daisy-like, purple flowers, to 3cm (1¼in) across, are borne.
D. speciosum. Erect, shrubby succulent. H 60cm (2ft), S 1m (3ft). Produces semi-cylindrical leaves, 1–2cm (½–¾in) long. Masses of daisy-like, green-centred, orange-red flowers, to 5cm (2in) across, appear in summer.

DROSERA (Droseraceae)
Sundew
Genus of evergreen, insectivorous perennials. Fully hardy to frost tender, min. 5–10°C (41–50°F). Grow in sun, in a mixture of peat and moss that is not allowed to dry out. Propagate by seed or division in spring.
D. capensis illus. p.269.
D. spathulata illus. p.269.

DRYANDRA (Proteaceae)
Genus of evergreen, spring- to summer-flowering shrubs and trees, grown for their flowers, foliage and overall appearance. Frost tender, min. 7°C (45°F). Needs full light and well-drained, sandy soil that contains few nitrates or phosphates. Is difficult to grow. Water containerized specimens moderately, much less in low temperatures. Plants under glass must be freely ventilated. Propagate by seed in spring.
D. formosa. Evergreen, bushy shrub. H 2–5m (6–15ft), S 1.5–3m (5–10ft). Strap-shaped leaves are divided into triangular, closely set lobes, creating a saw-blade effect. In spring carries small, scented, tubular, orange-yellow flowers in domed, terminal heads.

DRYAS (Rosaceae)
Mountain avens
Genus of evergreen, prostrate, woody-based perennials with oak-like leaves and cup-shaped flowers. Is useful on banks and walls, in rock gardens and as ground cover. Fully hardy. Prefers sun and gritty, well-drained, peaty soil. Propagate by seed when fresh or by semi-ripe cuttings in summer.
D. drummondii. Evergreen, prostrate, woody-based perennial. H 5cm (2in), S indefinite. Stout stems are clothed in small, oval, lobed, leathery, dark green leaves. Nodding, creamy-white flowers are borne in early summer but never fully open.
♥ ***D. octopetala*** illus. p.323.
♥ ***D.* x *suendermannii.*** Evergreen, prostrate, woody-based perennial. H 5cm (2in), S indefinite. Is similar to *D. drummondii*, but has slightly nodding, pale cream flowers that open horizontally.

DRYOPTERIS (Polypodiaceae)
Genus of deciduous or semi-evergreen ferns, many of which form regular, shuttlecock-like crowns. Fully to half hardy. Requires shade and moist soil. Regularly remove fading fronds. Propagate by spores in summer or by division in autumn or winter.
D. austriaca. See *D. dilatata*.
D. carthusiana (Narrow buckler fern). Deciduous or semi-evergreen, creeping, rhizomatous fern. H 1m (3ft), S 45cm (18in). Fully hardy. Produces lance-shaped, much-divided, mid-green fronds with triangular to oval pinnae.
D. dilatata, syn. *D. austriaca* (Broad buckler fern). Deciduous or semi-evergreen fern. H 1m (3ft), S 45cm (18in). Fully hardy. Has much-divided, arching, mid-green fronds, with triangular to oval pinnae, on stout, dark brown stems.
♥ ***D. erythrosora*** (Japanese shield fern). Deciduous fern. H 45cm (18in), S 30cm (12in). Frost to half hardy. Broadly triangular, divided, coppery-pink fronds, divided into triangular to oval pinnae, persist until mid-winter.
♥ ***D. filix-mas*** illus. p.186. **'Grandiceps'** is a deciduous or semi-evergreen fern. H 1.2m (4ft), S 1m (3ft). Fully hardy. Has 'shuttlecocks' of broadly lance-shaped, tasselled, elegantly arching, mid-green fronds, arising from crowns of large, upright, brown-scaled rhizomes.
D. goldiana (Giant wood fern). Deciduous fern. H 1m (3ft), S 60cm (2ft). Fully hardy. Has broadly oval,

light green fronds divided into numerous oblong, indented pinnae.
D. hexagonoptera. See *Phegopteris hexagonoptera.*
D. marginalis. Deciduous fern. H 60cm (24in), S 30cm (12in). Fully hardy. Fronds are lance-shaped, dark green and divided into numerous oblong, slightly indented pinnae.

DUCHESNEA (Rosaceae)
Genus of perennials, some of which are semi-evergreen, grown as ground cover as well as for their flowers. May be used in hanging baskets. Fully hardy. Grow in well-drained soil and in sun or partial shade. Propagate by division in spring, by rooting plantlets formed at ends of runners in summer or by seed in autumn.
D. indica, syn. *Fragaria indica.* Semi-evergreen, trailing perennial. H to 10cm (4in), S indefinite. Dark green leaves have 3 toothed leaflets like those of strawberries. Solitary 5-petalled, bright yellow flowers, to 2.5cm (1in) wide and with leafy, green frills of sepals, appear from spring to early summer. Strawberry-like, tasteless, red fruits appear in late summer.

DUDLEYA (Crassulaceae)
Genus of basal-rosetted, perennial succulents, closely related to *Echeveria.* Frost tender, min. 7°C (45°F). Requires full sun and very well-drained soil. Water sparingly when plants are semi-dormant in mid-summer. Propagate by seed or division in spring or summer.
D. brittonii. Basal-rosetted, perennial succulent. H 20cm (8in)–60cm (24in) or more when in flower, S 50cm (20in). Has narrowly lance-shaped, tapering, fleshy, silvery-white leaves. Masses of star-shaped, pale yellow flowers are produced in spring-summer.
D. pulverulenta illus. p.400.

DURANTA (Verbenaceae)
Genus of fast-growing, evergreen or partially deciduous trees and shrubs, grown for their flowers and overall appearance. Frost tender, min. 10–13°C (50–55°F). Needs full light and fertile, well-drained soil. Water potted plants freely when in full growth, moderately at other times. Prune as necessary to curb vigour. Propagate by seed in spring or by semi-ripe cuttings in summer. Whitefly may be troublesome.

D. erecta, syn. *D. plumieri, D. repens,* illus. p.120.
D. plumieri. See *D. erecta.*
D. repens. See *D. erecta.*

DUVALIA (Asclepiadaceae)
Genus of clump-forming or carpeting, perennial succulents with short, thick, leafless stems; is closely related to *Stapelia.* Star-shaped flowers have thick, fleshy petals recurved at tips. Frost tender, min. 10°C (50°F), but best at 20°C (68°F). Requires partial shade and very well-drained soil. Propagate by seed or stem cuttings in spring or summer.
D. corderoyi illus. p.412.

DUVERNOIA (Acanthaceae)
Genus of evergreen shrubs, grown primarily for their flowers. Frost tender, min. 10–15°C (50–59°F). Prefers full light and humus-rich, well-drained soil. Water potted specimens freely when in full growth, less at other times. In spring cut back flowered stems by half to create a well-branched plant. Propagate by seed in spring or by softwood or greenwood cuttings in late spring or summer. Whitefly may cause problems.

D. adhatodoides, syn. *Adhatoda duvernoia* (Snake bush). Evergreen, erect shrub. H 2–3m (6–10ft), S 1–2m (3–6ft). Bears elliptic, dark green leaves. Fragrant, tubular, white or mauve flowers, with pink, red or purple marks, appear in summer-autumn. Is sometimes confused in cultivation with *Justicia adhatoda.*

DYCKIA (Bromeliaceae)
Genus of evergreen, rosette-forming perennials, grown for their overall appearance. Frost tender, min. 7–10°C (45–50°F). Requires full light and well-drained soil containing sharp sand or grit. Water moderately in summer, scarcely or not at all in winter, sparingly at other times. Propagate by offsets or division in spring.
D. remotiflora (illus. p.229). Evergreen, basal-rosetted perennial. H and S 30–50cm (12–20in). Has dense rosettes of very narrowly triangular, pointed, thick-textured, arching, dull green leaves with hooked spines and grey scales beneath. Woolly spikes of tubular, orange-yellow flowers appear above foliage in summer-autumn.

E

ECCREMOCARPUS (Bignoniaceae)
Genus of evergreen, sub-shrubby, tendril climbers, grown for their attractive flowers that are produced over a long season. One species only is commonly grown. Half hardy; in cold areas treat as an annual. Grow in full light and in any well-drained soil. Propagate by seed in early spring.
♟ *E. scaber* illus. p.177.

ECHEVERIA (Crassulaceae)
Genus of rosetted, perennial succulents with long-lasting flowers. Leaves assume their brightest colours from autumn to spring. Frost tender, min. 5–7°C (41–5°F). Needs sun, good ventilation and very well-drained soil. Propagate by seed, stem or leaf cuttings, division or offsets in spring or summer.
♟ *E. agavoides* illus. p.410.
♟ *E. derenbergii*. Clump-forming, perennial succulent. H 4cm (1½in), S 30cm (12in). Min. 5°C (41°F). Produces a short-stemmed rosette of rounded, grey-green leaves. Flower stem, 8cm (3in) long, produces cup-shaped, yellow-and-red or orange flowers in spring. Offsets freely. Is often used as a parent in breeding.
♟ *E. elegans* illus. p.391.
E. gibbiflora. Rosetted, perennial succulent. H 1m (3ft), S 15cm (6in). Min. 7°C (45°F). Stems are crowned by rounded, grey-green leaves, often tinged red or brown. Flower stems, 60cm (2ft) long, bear cup-shaped, red flowers, yellow within, in autumn-winter.
♟ *E. harmsii*, syn. *Oliveranthus elegans*. Bushy, perennial succulent. H 20cm (8in), S 30cm (12in). Min. 7°C (45°F). Erect stems are each crowned by a 6cm (2½in) wide rosette of short, narrowly lance-shaped, pale green leaves, covered in short hairs. In spring bears cup-shaped, orange-tipped, red flowers, yellow within.
♟ *E. pulvinata* illus. p.400.
E. secunda illus. p.409.
♟ *E. setosa* (Mexican firecracker). Basal-rosetted, perennial succulent. H 4cm (1½in), S 30cm (12in). Min. 7°C (45°F). Has long, narrow, mid-green leaves covered in short, thick, white hairs. Bears cup-shaped, red-and-yellow flowers in spring. Is prone to rotting: do not water foliage.

ECHINACEA (Compositae)
Coneflower
Genus of summer-flowering perennials. Fully hardy. Prefers sun and humus-rich, moist but well-drained soil. Propagate by division or root cuttings in spring.
E. purpurea, syn. *Rudbeckia purpurea*. 'Robert Bloom' illus. p.208. 'White Lustre' is a vigorous, upright perennial. H 1.2m (4ft), S 45cm (1½ft). Large, daisy-like, white flower heads, each with a prominent, central, orange-brown cone, are borne singly on strong stems in summer. Has lance-shaped, dark green leaves.

ECHINOCACTUS (Cactaceae)
Genus of slow-growing, hemi-spherical, perennial cacti. Frost tender, min. 11°C (52°F); lower temperatures cause yellow patches on *E. grusonii*. Requires full sun and very well-drained soil. Yellow-flowered species are easy to grow. Propagate by seed in spring.
E. grusonii illus. p.396.
E. ingens. Slow-growing, hemi-spherical, perennial cactus. H 3m (10ft), S 2m (6ft). Grey-blue stem has a woolly crown and up to 50 ribs. Funnel-shaped, yellow flowers, 3cm (1¼in) across, appear in summer only on plants over 40cm (16in) in diameter.

ECHINOCEREUS (Cactaceae)
Genus of spherical to columnar, perennial cacti, freely branching with age, some with tuberous rootstocks. Buds, formed inside spiny stem, burst through skin, producing long-lasting flowers, with reflexed petal tips and prominent, green stigmas, followed by pear-shaped, spiny seed pods. Frost tender, min. 5–8°C (41–6°F); some species tolerate light frost if dry. Needs full sun and very well-drained soil. Propagate by seed or stem cuttings in spring or summer.
E. baileyi. See *E. reichenbachii* var. *baileyi*.
E. cinerascens. Clump-forming, perennial cactus. H 30cm (1ft), S 1m (3ft). Min. 5°C (41°F). Has 7cm (3in) wide stems, each with 5–12 ribs. Areoles each bear 8–15 yellowish-white spines. Masses of trumpet-shaped, bright pink or purple flowers, 12cm (5in) across and with paler petal bases, appear in spring only on fully mature plants.
E. leucanthus, syn. *Wilcoxia albiflora*, illus. p.405.
E. pectinatus. Columnar, perennial cactus. H 35cm (14in), S 20cm (8in). Min. 7°C (45°F). Has sparsely branched, green stems with 12–23 ribs and short, comb-like spines, often variably coloured. In spring produces trumpet-shaped, purple, pink or yellow flowers, 12cm (5in) across, with paler petal bases.
E. pentalophus illus. p.399.
♟ *E. reichenbachii*. Columnar, perennial cactus. H 35cm (14in), S 20cm (8in). Min. 7°C (45°F). Has a slightly branched, multicoloured stem with 12–23 ribs and comb-like spines, 1.5cm (⅝in) long. Carries trumpet-shaped, pink or purple flowers, 12cm (5in) across, with darker petal bases, in spring. var. *baileyi* (syn. *E. baileyi*) illus. p.398.
E. schmollii, syn. *Wilcoxia schmollii*, illus. p.398.
E. triglochidiatus. Clump-forming, perennial cactus. H 30cm (12in), S 15cm (6in). Min. 5°C (41°F). Has a short, thick, dark green stem with 3–5 spines, each to 2.5cm (1in) long, per areole. In spring bears funnel-shaped, bright red flowers, 7cm (3in) across, with prominent, red stamens and green stigmas. var. *paucispinas* illus. p.411.

Echinodorus ranunculoides. See *Baldellia ranunculoides*.

ECHINOFOSSULOCACTUS. See STENOCACTUS.

Echinomastus mcdowellii. See *Thelocactus mcdowellii*.

ECHINOPS (Compositae)
Globe thistle
Genus of summer-flowering perennials, grown for their globe-like, spiky flower heads. Fully hardy. Does best in full sun and in poor soil. Propagate by division or seed in autumn or by root cuttings in winter.
E. bannaticus illus. p.192. ♟ 'Taplow Blue' is an erect perennial, H 1.2m (4ft), S 1m (3ft). Wiry stems produce thistle-like, rounded heads of powder-blue flowers in summer. Narrowly oval leaves are divided and greyish-green.
E. ritro 'Veitch's Blue' illus. p.216.
E. sphaerocephalus illus. p.190.

ECHINOPSIS (Cactaceae)
Genus of spherical to columnar, perennial cacti, mostly freely branching; it is sometimes held to include *Trichocereus*. Frost tender, min. 5°C (41°F). Requires full sun and well-drained soil. Is easy to grow and well adapted for long periods of neglect. Propagate by seed or offsets in spring or summer.
♟ *E. aurea*, syn. *Lobivia aurea*. Clump-forming, perennial cactus. H 12cm (5in), S 20cm (8in). Green stem, with 14 or 15 ribs, is densely covered with white, radial spines and a longer, darker, central spine. Produces funnel-shaped to flattish, yellow flowers, 8cm (3in) across, in summer.
E. backebergii, syn. *Lobivia backebergii*, illus. p.409.
E. bridgesii, syn. *Trichocereus bridgesii*, illus. p.392.
E. candicans, syn. *Trichocereus candicans*, illus. p.397.
♟ *E. cinnabarina*, syn. *Lobivia cinnabarina*. Spherical, perennial cactus. H and S 15cm (6in). Glossy, dark green stem has about 20 warty ribs and mostly curved, dark spines. In summer bears funnel-shaped to flattish, carmine-red flowers, 8cm (3in) across.
E. eyriesii. Flattened spherical, perennial cactus. H 30cm (12in), S 50cm (20in). Has slowly branching, mid-green stems with 11–18 ribs and very short spines. Tubular, white flowers appear in spring-summer.
E. multiplex. See *E. oxygona*.
E. oxygona, syn. *E. multiplex*, illus. p.397.
E. pentlandii, syn. *Lobivia pentlandii*, illus. p.409.
E. rhodotricha. Spherical to columnar, perennial cactus. H 60cm (2ft), S 20cm (8in). Produces branching, 8–13-ribbed, dark green stems, 9cm (3½in) across. Curved, dark spines, 2cm (¾in) long, later turn pale. Has tubular, white to pink flowers in spring-summer.
E. spachianus, syn. *Trichocereus spachianus*, illus. p.393.

Echioides longiflorum. See *Arnebia pulchra*.

ECHIUM (Boraginaceae)
Genus of annuals and evergreen shrubs, biennials and perennials, grown for their flowers. Fully hardy to frost tender, min. 3°C (37°F). Needs full sun and fertile, well-drained soil. Water containerized specimens freely in summer, moderately at other times. Propagate by seed in spring or by greenwood or semi-ripe cuttings in summer. Whitefly may sometimes be troublesome.
E. bourgaeanum. See *E. wildpretii*.
E. vulgare [dwarf] illus. p.283.
E. wildpretii, syn. *E. bourgaeanum*. Evergreen, erect, unbranched biennial that dies after fruiting. H 2.5m (8ft) or more, S 60cm (2ft). Half hardy. Narrowly lance-shaped, silver-haired leaves, 30cm (1ft) long, form a dense rosette. Has compact spires, 1–1.5m (3–5ft) long, of small, funnel-shaped, red flowers in late spring and early summer.

EDGEWORTHIA (Thymelaeaceae)
Genus of deciduous shrubs, grown for their flowers in late winter and early spring. Frost hardy, but flowers are susceptible to frost damage. Is best grown against a south- or west-facing wall in most areas. Requires full sun and well-drained soil. Dislikes being transplanted. Propagate by semi-ripe cuttings in summer or by seed in autumn.
E. chrysantha, syn. *E. papyrifera*. Deciduous, rounded, open shrub. H and S 1.5m (5ft). Very supple shoots produce terminal, rounded heads of fragrant, tubular, yellow flowers in late winter and early spring. Has oval, dark green leaves.
E. papyrifera. See *E. chrysantha*.

EDRAIANTHUS (Campanulaceae)
Genus of short-lived perennials, some of which are evergreen, usually growing from central rootstocks. In winter, a small, resting bud is just visible from each rootstock. In spring, prostrate stems radiate to carry leaves and flowers. Is suitable for rock gardens, screes and troughs. Fully hardy. Needs sun and well-drained soil. Propagate by softwood cuttings from side shoots in early summer or by seed in autumn.

E. dalmaticus. Upright, then arching perennial. H 10cm (4in), S 15cm (6in). Bears narrowly lance-shaped, pale green leaves and, in early summer, terminal clusters of bell-shaped, violet-blue flowers, 2.5cm (1in) across.
♟ *E. pumilio* illus. p.330.
E. serpyllifolius illus. p.329. **'Major'** is an evergreen, prostrate perennial. H 1cm ($^1/_2$in), S to 5cm (2in). Has tight mats of tiny, oval, dark green leaves. In early summer, bell-shaped, deep violet flowers, 1.5cm ($^5/_8$in) wide, are borne on very short stems. Needs a sheltered site. Seldom sets seed.

Edwardsia microphylla. See *Sophora microphylla*.

EGERIA (Hydrocharitaceae)
Genus of semi-evergreen or evergreen, perennial, floating or submerged water plants, grown for their foliage. Is similar to *Elodea*, but has more conspicuous flowers, held above water surface. In an aquarium, plants are useful for oxygenating water and provide a suitable depository for fish spawn. Frost tender, min. 1°C (34°F). Needs a sunny position. Thin regularly to keep under control. Propagate by stem cuttings in spring or summer.
E. densa, syn. *Anacharis densa*, *Elodea densa*. Semi-evergreen, perennial, spreading, submerged water plant. S indefinite. Forms a dense mass of whorled, small, lance-shaped, dark green leaves borne on long, wiry stems. Small, 3-parted, white flowers appear in summer.

EHRETIA (Ehretiaceae)
Genus of deciduous, summer-flowering trees, grown for their foliage and star-shaped flowers. Frost hardy, but is susceptible to frost damage when young. Requires sun or semi-shade and fertile, well-drained soil. Propagate by softwood cuttings in summer.
E. dicksonii illus. p.65.

EICHHORNIA (Pontederiaceae)
Genus of evergreen or semi-evergreen, perennial, floating and marginal water plants. Frost tender, min. 1°C (34°F). Needs an open, sunny position in warm water. Grows prolifically and requires regular thinning year-round. Propagate by detaching young plants as required.
E. crassipes illus. p.388.

ELAEAGNUS (Elaeagnaceae)
Genus of deciduous or evergreen shrubs and trees, grown for their foliage and small, usually very fragrant flowers, often followed by ornamental fruits. Evergreen species are good for providing shelter or for hedging, particularly in coastal areas. Fully to frost hardy. Most evergreen species thrive in sun or shade, but those with silver leaves and deciduous species prefer full sun. Needs fertile, well-drained soil. Trim hedges in late summer. Propagate species by seed in autumn, evergreen forms also by semi-ripe cuttings in summer, deciduous forms by softwood or semi-ripe cuttings in summer.
E. angustifolia illus. p.92.
E. x *ebbingei*. Evergreen, bushy, dense shrub. H and S 5m (15ft). Fully hardy. Has oblong to oval, glossy, dark green leaves, silvery beneath. Fragrant, bell-shaped, silvery-white flowers are borne from mid- to late autumn. Leaves of
♟ **'Gilt Edge'** have golden-yellow margins. **'Limelight'** illus. p.123.
E. macrophylla. Evergreen, bushy, dense shrub. H and S 3m (10ft). Frost hardy. Broadly oval leaves are silvery-grey when young, becoming glossy, dark green above, but remaining silvery-grey beneath, when mature. Fragrant, bell-shaped, creamy-yellow flowers, silvery outside, appear from mid- to late autumn, followed by egg-shaped, red fruits.
♟ *E. pungens* **'Maculata'** illus. p.97.
E. umbellata. Vigorous, deciduous, bushy shrub. H and S 5m (15ft). Fully hardy. Oblong, wavy-edged, bright green leaves are silvery when young. Produces fragrant, bell-shaped, creamy-yellow flowers in late spring and early summer, then egg-shaped, red fruits.

ELAEOCARPUS (Elaeocarpaceae)
Genus of evergreen, spring- and summer-flowering shrubs and trees, grown for their flowers and foliage. Half hardy to frost tender, min. 5°C (41°F). Requires full light or partial shade and fertile, well-drained but not dry soil. Water containerized specimens freely when in full growth, less in winter. Current season's growth may be cut back in winter. Propagate by seed in spring or by semi-ripe cuttings in summer. Red spider mite and whitefly may cause problems.
E. cyaneus, syn. *E. reticulatus* (Blueberry ash). Evergreen, rounded shrub or tree. H and S 3m (10ft), sometimes to 12m (40ft) or more. Frost tender. Bears elliptic to lance-shaped, toothed, lustrous leaves and, in summer, axillary racemes of bell-shaped, fringed, white flowers. Has globular, deep blue fruits in autumn.
E. reticulatus. See *E. cyaneus*.

ELATOSTEMA. See **PELLIONIA**.

ELEOCHARIS (Cyperaceae). See **GRASSES, BAMBOOS, RUSHES** and **SEDGES**.
E. acicularis (Needle spike-rush). Evergreen, spreading, rhizomatous, perennial sedge. H to 10cm (4in), S indefinite. Fully hardy. Basal leaves are very narrow and mid-green. Hairless, unbranched, square stems bear solitary minute, brown spikelets in summer.

ELEUTHEROCOCCUS, syn. **ACANTHOPANAX** (Araliaceae)
Genus of deciduous shrubs and trees, grown for their foliage and fruits. Produces tiny, usually greenish-white flowers. Fully hardy. Prefers full sun and needs well-drained soil. Propagate by seed in spring or by root cuttings in late winter.
E. sieboldianus illus. p.115.

Elodea crispa of gardens. See *Lagarosiphon major*.
Elodea densa. See *Egeria densa*.

ELSHOLTZIA (Labiatae)
Genus of perennials and deciduous shrubs and sub-shrubs, grown for their flowers, generally in autumn. Frost hardy. Needs full sun and fertile, well-drained soil. Cut back old shoots hard in early spring. Propagate by softwood cuttings in summer.
E. stauntonii illus. p.143.

Elymus arenarius. See *Leymus arenarius*.

EMBOTHRIUM (Proteaceae)
Genus of evergreen or semi-evergreen trees, grown for their flowers. Frost hardy, but shelter from cold winds. Needs semi-shade and moist but well-drained, lime-free soil. Propagate by suckers in spring or autumn or by seed in autumn.
E. coccineum illus. p.67.

EMILIA (Compositae)
Genus of annuals and perennials with flower heads that are good for cutting. Is ideal for hot, dry areas and coastal soils. Half hardy. Requires sun and very well-drained soil. Propagate by seed sown under glass in spring, or outdoors in late spring.
E. coccinea, syn. *E. flammea* of gardens, *E. javanica* of gardens, illus. p.293.
E. flammea. See *E. coccinea*.
E. javanica of gardens. See *E. coccinea*.

EMMENOPTERYS (Rubiaceae)
Genus of deciduous trees, grown mainly for their foliage; flowers appear only rarely, during hot summers. Frost hardy, but young growths may be damaged by late frosts. Needs full sun and deep, fertile, well-drained soil. Propagate by softwood cuttings in summer.
E. henryi illus. p.54.

ENCEPHALARTOS (Zamiaceae)
Genus of evergreen shrubs and trees, grown for their palm-like appearance. Frost tender, min. 10–13°C (50–55°F). Needs full light and well-drained soil. Water containerized specimens moderately when in full growth, less at other times. Propagate by seed in spring.
E. ferox illus. p.122.
E. longifolius. Slow-growing, evergreen, palm-like tree, sometimes branched with age. H 3m (10ft) or more, S 1.5–2.5m (5–8ft). Has feather-shaped leaves, each 60cm–1.5m (2–5ft) long, divided into narrowly lance-shaped to oval, blue-green leaflets, usually with hook-tipped teeth. Cone-like, brownish flower heads appear intermittently.

ENCYCLIA. See **ORCHIDS.**
E. cochleata. Evergreen, epiphytic orchid for a cool greenhouse. H 30cm (12in). Upright spikes of green flowers, 5cm (2in) long, with dark purple lips at the top and ribbon-like sepals and petals, are produced in summer and, on mature plants, intermittently throughout the year. Leaves are narrowly oval and 15cm (6in) long. Requires semi-shade in summer.
E. radiata. Evergreen, epiphytic orchid for a cool greenhouse. H 25cm (10in). Bears upright spikes of very fragrant, well-rounded, creamy-white flowers, 1cm ($^1/_2$in) across, in summer. Flower lips are white, lined with red. Narrowly oval leaves are 10–15cm (4–6in) long. Grow in semi-shade in summer.

ENDYMION. See **HYACINTHOIDES**.

ENKIANTHUS (Ericaceae)
Genus of deciduous or semi-evergreen, spring-flowering shrubs and trees, grown for their mass of small, bell- or urn-shaped flowers and their autumn colour. Fully to frost hardy. Needs sun or semi-shade and moist, peaty, acid soil. Propagate by semi-ripe cuttings in summer or by seed in autumn.
♟ *E. campanulatus* illus. p.87.
♟ *E. cernuus* f. *rubens* illus. p.101.
♟ *E. perulatus* illus. p.97.

ENSETE (Musaceae)
Genus of evergreen perennials, grown for their foliage, which resembles that of bananas, and fruits. Has false stems made of overlapping leaf sheaths that die after flowering. Frost tender, min. 10°C (50°F). Grow in sun or partial shade and humus-rich soil. Propagate by seed in spring or by division year-round.
E. ventricosum, syn. *Musa arnoldiana*, *M. ensete*, illus. p.197.

EOMECON (Papaveraceae)
Genus of one species of perennial that spreads rapidly and deeply underground. Is suitable for large rock gardens. Fully hardy. Needs sun and well-drained soil. Propagate by seed or runners in spring.
E. chionantha (Snow poppy). Vigorous, spreading perennial. H to 40cm (16in), S indefinite. Leaves are large, palmate and grey. Erect stems each carry a long panicle of small, poppy-like, white flowers in summer.

EPACRIS (Epacridaceae)
Genus of evergreen, heath-like shrubs, grown for their flowers. Frost tender, min. 5°C (41°F). Needs full sun and humus-rich, well-drained, neutral to acid soil. Water potted plants moderately when in full growth, less at other times. Flowered stems may be shortened after flowering to maintain a neat habit. Propagate by seed in spring or semi-ripe cuttings in late summer.
E. impressa illus. p.125.

EPHEDRA (Ephedraceae)
Genus of evergreen shrubs, grown for their habit and green shoots. Makes good ground cover in dry soil. Grow male and female plants together in order to obtain fruits. Fully hardy. Requires full sun and well-drained soil. Propagate by seed in autumn or by division in autumn or spring.
E. gerardiana. Evergreen, spreading shrub with slender, erect, rush-like, green shoots. H 60cm (2ft), S 2m (6ft). Leaves and flowers are inconspicuous. Bears small, spherical, red fruits.

EPICACTUS (Cactaceae)
Genus of perennial cacti with strap-shaped, flattened, green stems that have notched edges. Flowers are produced at notches. Frost tender, min. 5–11°C (41–52°F). Grow in sun or partial shade and in rich, well-drained soil. Propagate by stem cuttings in spring or summer.
E. 'Gloria', syn. *Epiphyllum* 'Gloria', illus. p.399.
E. 'Jennifer Ann', syn. *Epiphyllum* 'Jennifer Ann', illus. p.403.
E. 'M.A. Jeans', syn. *Epiphyllum* 'M.A. Jeans', illus. p.399.

EPIDENDRUM. See ORCHIDS.
E. difforme (illus. p.262). Evergreen, epiphytic orchid for an intermediate greenhouse. H 23cm (9in). Large heads of semi-translucent, green flowers, 0.5cm (¼in) across, open in autumn. Has oval, rigid leaves, 2.5–5cm (1–2in) long. Requires shade in summer. Avoid spraying, which can cause spotting of leaves. Propagate by division in spring.
E. ibaguense, syn. *E. radicans* (illus. p.261). Evergreen, epiphytic orchid for a cool greenhouse. H 2m (6ft) or more. Flowers more or less constantly, bearing a succession of feathery-lipped, deep red blooms, 0.5cm (¼in) across. Leaves, 2.5–5cm (1–2in) long, are oval and rigid. Grow in semi-shade during summer. Propagate by tip cuttings in spring.
E. radicans. See *E. ibaguense*.

EPIGAEA (Ericaceae)
Genus of evergreen, prostrate, spring-flowering sub-shrubs. Fully to frost hardy. Needs shade and humus-rich, moist, acid soil. Most are difficult to cultivate. Propagate by seed in spring or by softwood cuttings in early summer.
E. asiatica. Evergreen, creeping sub-shrub. H to 10cm (4in), S to 20cm (8in). Fully hardy. Stems and heart-shaped, deep green leaves are covered with brown hairs. Bears terminal clusters of 3–6 tiny, slightly fragrant, urn-shaped, white or pink flowers in spring.
E. gaultherioides, syn. *Orphanidesia gaultherioides*, illus. p.312.
E. repens (Mayflower, Trailing arbutus). Evergreen, creeping sub-shrub. H 10cm (4in), S 30cm (12in). Fully hardy. Hairy stems, bearing heart-shaped, leathery leaves, root at intervals. In spring produces terminal clusters of 4–6 cup-shaped, white flowers, sometimes flushed pink. Is relatively easy to grow.

EPIGENEIUM. See ORCHIDS.
E. amplum. Evergreen, epiphytic orchid for a cool greenhouse. H 60cm (24in). In spring bears a star-shaped, single, olive-green flower, lined with reddish-brown and with a blackish-brown lip, 5cm (2in) long. Leaves are oval, to 8cm (3in) long. Grow on a bark slab. Requires semi-shade in summer.

EPILOBIUM, syn. **CHAMAENERION** (Onagraceae)
Willow herb
Genus of annuals, biennials, perennials and deciduous sub-shrubs, grown for their deep pink to white flowers in summer. Is useful on dry banks; many species are invasive. Fully to frost hardy. Tolerates sun or shade and prefers moist but well-drained soil. Propagate species by seed in autumn, selected forms by softwood cuttings from side-shoots in spring.
E. angustifolium f. *album* illus. p.190.
E. canum. See *Zauschneria californica* and *Z. cana*.
E. chlorifolium var. *kaikourense* illus. p.322.
E. glabellum illus. p.299.
E. obcordatum. Clump-forming perennial. H 15cm (6in), S 10cm (4in). Frost hardy. Oval leaves are glossy green. Spikes of open cup-shaped, deep rose-pink flowers are borne in summer. Is good for a rock garden or alpine house. Needs a sheltered site and full sun. In cultivation may not retain character, especially in mild climates.

EPIMEDIUM (Berberidaceae)
Genus of spring-flowering perennials, some of which are evergreen. Flowers are cup-shaped with long or short spurs. Makes good ground cover. Fully hardy. Does best in partial shade and humus-rich, moisture-retentive but well-drained soil. Cut back just before new growth appears in spring. Propagate by division in spring or autumn.
E. alpinum (Barrenwort). Evergreen, carpeting perennial. H 23cm (9in), S to 30cm (12in). Racemes of pendent, short-spurred flowers, with crimson sepals and yellow petals, appear in spring. Has finely toothed, glossy leaves divided into oval, angled, mid-green leaflets, bronze when young.
⚘ *E. grandiflorum* 'Rose Queen' illus. p.233. f. *violaceum* is a carpeting perennial. H and S 30cm (12in). Produces racemes of pendent, long-spurred, deep lilac flowers in spring at same time as heart-shaped, dark brownish-red, young leaves, divided into oval leaflets, that mature to mid-green.
⚘ *E.* x *perralchicum*. Evergreen, carpeting perennial. H 45cm (18in), S 30cm (12in). Short spires of pendent, yellow flowers, with short spurs, are borne on slender stems in spring. Leaves, divided into rounded to oval leaflets, are dark green.
E. perralderianum. Semi-evergreen, carpeting perennial. H 30cm (12in), S 45cm (18in). Clusters of small, pendent, short-spurred, bright yellow flowers are borne in spring. Has large, toothed, glossy, deep green leaves, divided into rounded to oval leaflets.
⚘ *E. pinnatum* subsp. *colchicum*. Evergreen, carpeting perennial. H and S 30cm (12in). In spring, clusters of small, pendent, bright yellow flowers, with short spurs, are produced above dark green leaves, divided into rounded to oval leaflets, that are hairy when young.
E. pubigerum illus. p.232.
⚘ *E.* x *rubrum* illus. p.233.
E. x *versicolor*. Carpeting perennial. H and S 30cm (12in). Small, pendent clusters of yellow flowers, with long, red-tinged spurs, appear in spring. Heart-shaped, fresh green leaves are divided into oval leaflets that are tinted reddish-purple. 'Neo-sulphureum' illus. p.238.
E. x *warleyense* illus. p.239.
⚘ *E.* x *youngianum* 'Niveum' illus. p.231.

EPIPHYLLUM (Cactaceae)
Orchid cactus, Strap cactus
Genus of perennial cacti with strap-shaped, flattened, green stems that have notched edges. Flowers are produced at notches. Frost tender, min. 5–11°C (41–52°F). Grow in sun or partial shade and in rich, well-drained soil. Propagate by stem cuttings in spring or summer.
E. anguliger illus. p.397.
E. crenatum. Erect, then pendent, perennial cactus. H and S 3m (10ft). Min. 11°C (52°F). Has a flattened stem. Bears lightly perfumed, funnel-shaped, broad-petalled, white flowers, 20cm (8in) across, in spring-summer. Is often used as a parent for breeding.
E. 'Gloria'. See *Epicactus* 'Gloria'.
E. 'Jennifer Ann'. See *Epicactus* 'Jennifer Ann'.
E. lauii illus. p.397.
E. 'M.A. Jeans'. See *Epicactus* 'M.A. Jeans'.
E. oxypetalum. Erect, then pendent, perennial cactus. H 3m (10ft), S 1m (3ft). Min. 11°C (52°F). Produces freely branching, flattened stems, 12cm (5in) across. In spring-summer bears nocturnal, tubular, white flowers, 25cm (8in) long. Makes a good house plant.

EPIPREMNUM (Araceae)
Genus of evergreen, woody-stemmed, root climbers, including *Pothos*, grown for their handsome leaves. Frost tender, min. 15–18°C (59–64°F). Grow in light shade away from direct sun; any well-drained, moisture-retentive soil is suitable. Water regularly, less in cold weather. Stems need good supports. Remove shoot tips to induce branching at any time. Propagate by leaf-bud or stem-tip cuttings in late spring or by layering in summer.
E. aureum 'Marble Queen', syn. *Scindapsus aureus* 'Marble Queen', illus. p.179.
E. pictum 'Argyraeus'. See *Scindapsus pictus* 'Argyraeus'.

EPISCIA (Gesneriaceae)
Genus of evergreen, low-growing and creeping perennials, grown for their ornamental leaves and colourful flowers. Is useful as ground cover or in hanging baskets. Frost tender, min. 15°C (59°F). Requires high humidity and a fairly shaded position in humus-rich, well-drained soil. Keep well watered, but avoid waterlogging. Propagate in summer by stem cuttings, division or removing rooted runners.
E. cupreata (Flame violet) illus. p.265. 'Metallica' is an evergreen, creeping perennial. H 10cm (4in), S indefinite. Has oval, downy, wrinkled leaves, tinged pink to copper and with broad, silvery bands along midribs. Funnel-shaped, orange-red flowers, marked yellow within, are borne intermittently. 'Tropical Topaz' has yellow flowers.
E. dianthiflora, syn. *Alsobia dianthiflora*, illus. p.263.
E. lilacina. Evergreen, low-growing perennial, with runners bearing plantlets. H 10cm (4in), S indefinite. Has oval, hairy, pale green leaves, to 8cm (3in) long. Funnel-shaped, white flowers, tinged mauve and with yellow eyes, are produced in small clusters from autumn to spring. Leaves of 'Cuprea' are bronze-tinged.

EPITHELANTHA (Cactaceae)
Genus of very slow-growing, spherical, perennial cacti densely covered with very short spines. Frost tender, min. 10°C (50°F). Needs full sun and well-drained soil, and is prone to rot if overwatered. Is easier to cultivate if grafted. Propagate by seed or stem cuttings in spring or summer.
E. micromeris illus. p.407.

ERANTHEMUM (Acanthaceae)
Genus of perennials and evergreen shrubs, grown for their flowers. Frost tender, min. 15–18°C (59–64°F). Requires full light or partial shade and fertile, well-drained soil. Water containerized plants freely when in full growth, moderately at other times. In spring or after flowering remove at least half of each spent flowering stem to encourage a bushier habit. Propagate by softwood cuttings in late spring. Whitefly may be a nuisance.
E. atropurpureum. See *Pseuderanthemum atropurpureum*.
E. nervosum. See *E. pulchellum*.
E. pulchellum, syn. *E. nervosum*. Evergreen, erect shrub. H 1–1.2m (3–4ft), S 60cm (2ft) or more. Produces elliptic to oval, prominently veined, deep green leaves. Blue flowers, each with a 3cm (1¼in) long tube and rounded petal lobes, are produced in winter-spring.

ERANTHIS (Ranunculaceae)
Genus of clump-forming perennials, with knobbly tubers, grown for their cup-shaped flowers surrounded by leaf-like ruffs of bracts. Fully to frost hardy. Prefers partial shade and humus-rich soil, well-drained but not drying out excessively. Dies down in summer. Propagate by seed in autumn or by division of clumps immediately after flowering while still in leaf.
⚘ *E. hyemalis* illus. p.385.
⚘ *E.* x *tubergenii* 'Guinea Gold'. Late winter- or early spring-flowering, tuberous perennial. H 8–10cm (3–4in), S 4–6cm (1½–2½in). Frost hardy. Stems each bear a stalkless, deep golden-yellow flower, 3–4cm (1¼–1½in) across, surrounded by a bronze-green bract, cut into narrow lobes. Rounded leaves are divided into finger-shaped lobes.

ERCILLA (Phytolaccaceae)
Genus of one species of evergreen, root climber, grown for its neat, green leaves that densely clothe stems. Frost hardy. Grow in sun or partial shade and in any well-drained soil. Prune after flowering, if required. Propagate by stem cuttings in late summer or autumn.
E. spicata. See *E. volubilis*.

489

E. volubilis, syn. *E. spicata*. Evergreen, root climber. H to 15m (50ft) or more. Oval to heart-shaped, mid-green leaves are 2.5–5cm (1–2in) long. Spikes of petalless flowers, each consisting of 5 greenish sepals and 6–8 white stamens, are borne in spring. Very occasionally, may be followed by red berries.

EREMURUS (Liliaceae)
Foxtail lily, King's spear
Genus of perennials, with fleshy, finger-like roots, grown for their stately spires of shallowly cup-shaped flowers in summer. Fully to frost hardy. Requires a sunny, warm position and well-drained soil. Tends to come into growth very early, and young shoots may be frosted. Provide a covering of dry bracken in late winter to protect the crowns when shoots are first developing. Stake tall species and hybrids. Propagate by division in spring or early autumn or by seed in autumn.
E. himalaicus illus. p.190.
E. robustus illus. p.191.
E. **Shelford Hybrids.** Group of perennials of varying habit and flower colour. H 1.5m (5ft), S 60cm (2ft). Frost hardy. Long racemes of orange, buff, pink or white flowers are borne freely in mid-summer. Leaves are strap-shaped, in basal rosettes.
E. spectabilis. Erect perennial. H 1.2m (4ft), S 60cm (2ft). Frost hardy. Bears long racemes of pale yellow flowers, with brick-red anthers, in early summer. Leaves are strap-shaped, in basal rosettes.

ERIA. See ORCHIDS.
E. coronaria. Evergreen, epiphytic orchid for a cool greenhouse. H 23cm (9in). Sprays of fragrant, rounded, creamy-white flowers, 1cm (½in) across, each with a red- and yellow-marked lip, open in autumn. Has broadly oval, glossy leaves, 10cm (4in) long. Needs semi-shade in summer and moist compost year-round.

ERICA (Ericaceae). See HEATHERS.
E. arborea (Tree heath). Evergreen, upright, shrub-like tree heath. H 6m (20ft), S 1.5m (5ft). Frost hardy, but liable to damage from frost and cold winds. Has needle-like, bright green leaves in whorls of 3 or 4 and bears scented, bell-shaped, white flowers from late winter to late spring. Is lime-tolerant. ♛ **'Albert's Gold'**, H 2m (6ft), retains its golden foliage year-round. ♛ var. *alpina* (illus. p.148) has vivid green foliage that contrasts well with compact racemes of white flowers. May be pruned hard to keep its shape and to encourage new growth. ♛ *E. australis* (Spanish heath, Spanish tree heath). Evergreen, shrub-like tree heath. H to 2.2m (7ft), S 1m (3ft). Frost hardy, but stems may be damaged by snow and frost. Has needle-like leaves in whorls of 4 and tubular to bell-shaped, white or purplish-pink flowers in spring. ♛ **'Mr Robert'** has white flowers. ♛ **'Riverslea'** has bright purple-pink flowers, mostly in clusters of 4.
E. canaliculata (Channelled heath; illus. p.148). Evergreen, erect shrub. H to 3m (10ft), S 1m (3ft). Half hardy. Dark green leaves are narrow and needle-like in whorls of 3. Cup-shaped, pearl-white flowers, sometimes rose-tinted, with dark brown, almost black anthers, are borne in winter (under glass) or early spring (in the open). Requires acid soil.
E. carnea, syn. *E. herbacea* (Alpine heath, Winter heath). Evergreen, spreading shrub. H to 30cm (12in), S to 45cm (18in) or more. Fully hardy. Produces whorls of needle-like, mid- to dark green leaves and bears tubular to bell-shaped flowers that are in shades of pink and red, occasionally white, from early winter to late spring. Tolerates lime and some shade. Makes good ground cover. **'Altadena'** has golden foliage and pale pink flowers. ♛ **'Ann Sparkes'** (illus. p.149), H 15cm (6in), has golden foliage, turning to bronze in winter, and rose-pink flowers. **'Cecilia M. Beale'**, H 15cm (6in), bears an abundance of white flowers from mid-winter to early spring. **'C.J. Backhouse'** (illus. p.149) produces pale pink flowers from mid-winter to early spring. **'December Red'** (illus. p.149) has a spreading habit and vigorous growth. Deep rose-pink flowers are borne in winter. **'Eileen Porter'** is slow-growing and bears deep red flowers with paler sepals; is often in flower in late autumn. ♛ **'Foxhollow'**, a vigorous, spreading cultivar, has foliage that is golden-yellow in summer, with orange tips in spring, and a few pale pink flowers. **'King George'**, H 20cm (8in), has dark green foliage and deep rose-pink flowers from early winter to mid-spring. ♛ **'Loughrigg'**, H 15cm (6in), produces dark purplish-red flowers from late winter to spring. **'March Seedling'** has a spreading habit, dark green foliage and rich, rose-purple flowers. ♛ **'Myretoun Ruby'**, H 20cm (8in), is vigorous but compact with brilliant deep purple-red flowers in late winter and early spring. ♛ **'Pink Spangles'**, H 15cm (6in), is vigorous with flowers that have shell-pink sepals and deeper pink corollas. **'Pirbright Rose'** is very floriferous with bright pink flowers from early winter to early spring. ♛ **'R.B. Cooke'**, H 20cm (8in), bears clear pink flowers from early winter to early spring. ♛ **'Springwood White'** (illus. p.148), H 15cm (6in), the most vigorous white cultivar, makes excellent ground cover and bears large, white flowers, with brown anthers, from late winter to spring. ♛ **'Vivellii'** (illus. p.149), H 15cm (6in), has dark bronze-green foliage and bears deep purple-pink flowers from late winter to spring. ♛ **'Westwood Yellow'** (illus. p.149) is compact with golden-yellow foliage and deep pink flowers.
E. ciliaris (Dorset heath). Evergreen, loose shrub. H to 30cm (12in), S 40cm (16in). Fully hardy, but may be damaged in severe weather. Has needle-like, dark green leaves in whorls of 3. Bears long racemes of bell-shaped, bright pink flowers in tiers of 3 or 4 in summer. Requires acid soil and prefers warm, moist conditions. **'Aurea'** has somewhat straggly growth with golden foliage and clear pink flowers. ♛ **'Corfe Castle'** (illus. p.149) produces salmon-pink flowers from summer to early autumn. ♛ **'David McClintock'** (illus. p.148) has light grey-green foliage and bears white flowers, with deep pink tips, from summer to early autumn. ♛ **'Mrs C.H. Gill'** has dark grey-green foliage and clear red flowers. **'White Wings'** (illus. p.148), a sport of 'Mrs C.H. Gill', has dark grey-green foliage and white flowers.
E. cinerea (Bell heather). Evergreen, compact shrub. H 30cm (12in), S 45–60cm (18–24in). Fully hardy. Has needle-like, mid- to deep green leaves and bears bell-shaped flowers that are in shades of pink and dark red, occasionally white, from early summer to early autumn. Prefers a warm, dry position. Requires acid soil. **'Alba Major'** has mid-green foliage and clear white flowers. **'Atropurpurea'** has deep purple flowers in long racemes. **'Atrorubens'**, a free-flowering cultivar, produces ruby-red flowers. ♛ **'C.D. Eason'** (illus. p.149) has distinctive, dark green foliage and bright red flowers. ♛ **'Cevennes'** is upright in habit and bears a profusion of mauve flowers. ♛ **'C.G. Best'** has mid-green foliage and rose-pink flowers. **'Contrast'** bears very dark purple flowers. **'Domino'** produces white flowers that contrast with dark brown stems and sepals and almost black stigmas. ♛ **'Eden Valley'** (illus. p.149), H 20cm (8in), bears white flowers with lavender-mauve tips. **'Foxhollow Mahogany'** has very dark green foliage and deep ruby-red flowers. **'Frances'** produces rose-pink flowers. **'Glencairn'** (illus. p.149), H 20cm (8in), has foliage tipped with pink and red, particularly in spring, and magenta flowers. **'Hookstone Lavender'** (illus. p.149), H 38cm (15in), is somewhat straggling in habit and bears an abundance of pale lavender flowers. ♛ **'Hookstone White'** (illus. p.148), H 35cm (14in), has bright green foliage and bears long racemes of large, white flowers. **'Janet'** is compact with pale shell-pink flowers from mid-summer. ♛ **'Pentreath'** has rich purple flowers. ♛ **'Pink Ice'**, H 20cm (8in), is compact with soft pink flowers. ♛ **'P.S. Patrick'** is a vigorous cultivar with purple flowers and dark green foliage. **'Purple Beauty'** (illus. p.149) has purple flowers and dark foliage. **'Purple Robe'**, H 20cm (8in), bears dusky-purple flowers in long racemes. **'Rock Pool'** (illus. p.149), H 15cm (6in), is compact and has golden foliage, tinted orange and red in winter, and mauve flowers. **'Romiley'** (illus. p.149), H 25cm (10in), bears masses of ruby-red flowers from early to late summer. ♛ **'Stephen Davis'**, H 25cm (10in), has brilliant, almost fluorescent, red flowers. ♛ **'Velvet Night'**, H 25cm (10in), produces very dark purple, almost black flowers. ♛ **'Windlebrooke'** (illus. p.149), H 25cm (10in), is vigorous, with golden foliage, turning bright orange-red in winter, and mauve flowers.

E. x darleyensis. Evergreen, bushy shrub. H 45cm (18in), S 1m (3ft) or more. Fully hardy. Has needle-like, mid-green foliage, with cream, pink or red, young growth in late spring. Bell-shaped, white, pink or purple flowers are borne in racemes from early winter to late spring. Tolerates lime. **'Archie Graham'** (illus. p.148), H 50cm (20in), is vigorous with mauve-pink flowers. ♛ **'Arthur Johnson'**, H 1m (3ft), has young foliage with cream and pink tips in spring and long racemes of mauve-pink flowers from mid-winter to spring. **'Darley Dale'** (illus. p.148) bears pale mauve flowers from mid-winter to spring. **'George Rendall'** carries deep pink flowers from early winter to early spring. ♛ **'Ghost Hills'** (illus. p.148) has cream-tipped foliage in spring and a profusion of pink flowers from mid-winter to spring. **'Jack H. Brummage'**, H 30cm (12in), has golden foliage, with yellow and orange tints, and mauve flowers. ♛ **'J.W. Porter'**, H 30cm (12in), has reddish, young shoots in spring and mauve-pink flowers from mid-winter to late spring. **'Silberschmelze'** (syn. *E. x d.* 'Molten Silver') is vigorous and produces young shoots with creamy-pink tips in spring and white flowers. **'White Glow'** (illus. p.148), H 30cm (12in), bears white flowers. ♛ **'White Perfection'** (illus. p.148) has bright green foliage and white flowers.
E. erigena, syn. *E. mediterranea.* Evergreen, upright shrub. H to 2.5m (8ft), S to 1m (3ft). Frost hardy; top growth may be damaged in severe weather, but plant recovers well from the base. Has needle-like, mid-green leaves and, usually, bell-shaped, mauve-pink flowers from early winter to late spring. Tolerates lime. Flowers of some cultivars have a pronounced scent of honey. **'Brightness'** (illus. p.149), H 45cm (18in), has bronze-green foliage and mauve-pink flowers in spring. ♛ **'Golden Lady'** (illus. p.149), H 30cm (12in), has a neat, compact habit with year-round, golden foliage and white flowers in late spring. ♛ **'Irish Dusk'**, H 45cm (18in), has dark green foliage and salmon-pink flowers from mid-winter to early spring. **'Superba'**, H 2m (6ft), bears strongly scented, rose-pink flowers during spring. ♛ **'W.T. Rackliff'**, H 60cm (2ft), has dark green foliage and produces thick clusters of white flowers from late winter to late spring.
E. gracilis. Evergreen, compact shrub. H and S to 30cm (12in). Frost tender, min. 5°C (41°F). Has needle-like, mid-green leaves and clusters of small, bell-shaped, cerise flowers from early autumn to early spring. Is usually grown as a pot plant; may be planted outdoors in summer in a sheltered position.
E. herbacea. See *E. carnea*.
E. x hiemalis. Evergreen, bushy shrub. H and S 30cm (12in). Half hardy. Has needle-like, mid-green foliage and racemes of tubular to bell-shaped, pink-tinged, white flowers from late autumn to mid-winter.
♛ *E. lusitanica* (Portuguese heath).

ERIGERON

Evergreen, upright, bushy, tree heath. H to 3m (10ft), S 1m (3ft). Frost hardy. Has feathery, bright green leaves and, from late autumn to late spring, bears tubular to bell-shaped flowers that are pink in bud but pure white when fully open. 'George Hunt' has golden foliage; is frost hardy but needs a sheltered position.

E. mackaiana (Mackay's heath). Evergreen, spreading shrub. H to 25cm (10in), S 40cm (16in). Fully hardy. Has needle-like, mid-green leaves and bears umbels of rounded, pink, mauve-pink or white flowers from mid-summer to early autumn. Likes damp, acid soil. 'Dr Ronald Gray' (illus. p.148), H 15cm (6in), has dark green foliage and pure white flowers. 'Plena' (illus. p.148), H 15cm (6in), has double, deep-pink flowers, shading to white in centres.

E. mediterranea. See *E. erigena*.
E. pageana. Evergreen, bushy shrub. H to 60cm (2ft), S 30cm (1ft). Half hardy. Has needle-like, mid-green leaves and, from late spring to early summer, bell-shaped, rich yellow flowers.
E. × *praegeri.* See *E.* × *stuartii*.
E. scoparia (Besom heath). Evergreen, bushy shrub. H to 3m (10ft), S 1m (3ft). Frost hardy. Has needle-like, dark green leaves. Clusters of rounded, bell-shaped, greenish-brown flowers appear in late spring and early summer. Requires acid soil. 'Minima', H 30cm (12in), has bright green foliage.
E. × *stuartii*, syn. *E.* × *praegeri*. Evergreen, compact shrub. H 15cm (6in), S 30cm (12in). Fully hardy. Has needle-like, dark green leaves. Numerous umbels of bell-shaped, pink flowers are produced in late spring and summer. Prefers moist, acid soil.
♆ 'Irish Lemon' produces young foliage with lemon-yellow tips in spring and bright pink flowers. 'Irish Orange' has orange-tipped young foliage and dark pink flowers.
♆ *E. terminalis* (Corsican heath). Evergreen, shrub-like tree heath with stiff, upright growth. H and S to 2.5m (8ft). Frost hardy. Has needle-like, mid-green foliage. Bell-shaped, mauve-pink flowers, borne from early summer to early autumn, turn russet as they fade in winter. Tolerates lime.
E. tetralix (Cross-leaved heath). Evergreen, spreading shrub. H to 30cm (12in), S 45cm (18in). Fully hardy. Has needle-like, grey-green leaves in whorls of 4. Large umbels of bell-shaped, pink flowers appear from summer to early autumn. Requires acid, preferably moist soil. ♆ 'Alba Mollis' (illus. p.148) has silver-grey foliage and bears white flowers from early summer to late autumn. ♆ 'Con Underwood' (illus. p.149) has dark red flowers. 'Hookstone Pink' has silver-grey foliage and bears rose-pink flowers from late spring to early autumn. ♆ 'Pink Star' (illus. p.149) produces pink flowers held upright in a star-like pattern.
E. umbellata. Evergreen, bushy shrub. H and S 60cm (2ft). Frost hardy. Has needle-like, mid-green foliage and bell-shaped, mauve flowers, with chocolate-brown anthers, in late spring.
E. vagans (Cornish heath). Vigorous, evergreen, bushy shrub. H and S 75cm (30in). Fully hardy. Leaves are needle-like and mid-green. Rounded, bell-shaped, pink, mauve or white flowers appear from mid-summer to late autumn. Tolerates some lime. Responds well to hard pruning. ♆ 'Birch Glow' (illus. p.149), H 45cm (18in), has bright green foliage and glowing rose-pink flowers. ♆ 'Lyonesse' (illus. p.148), H 45cm (18in), has dark green foliage and long, tapering spikes of white flowers with brown anthers. ♆ 'Mrs D.F. Maxwell', H 45cm (18in), has dark green foliage and glowing deep pink flowers. 'St Keverne', H 45cm (18in), is a neat, bushy shrub with rose-pink flowers; may be used for a low hedge. ♆ 'Valerie Proudley' (illus. p.149), H 45cm (18in), has golden foliage year-round when grown in full light and sparse, white flowers in late summer and autumn. 'Viridiflora' has feathery, green bracts in place of flowers and is useful for flower arrangements.
E. × *veitchii.* Evergreen, bushy, shrub-like tree heath. H to 2m (6ft), S 1m (3ft). Frost hardy. Has needle-like, mid-green leaves. Scented, tubular to bell-shaped, white flowers are produced in dense clusters from mid-winter to spring. 'Exeter' (illus. p.148) has a profusion of white flowers, almost obscuring the foliage. ♆ 'Gold Tips' is similar to 'Exeter', but young foliage has golden tips in spring. 'Pink Joy' (illus. p.148) has pink flower buds that open to clear white.
E. × *watsonii.* Evergreen, compact shrub. H 30cm (12in), S 38cm (15in). Fully hardy. Needle-like, mid-green leaves often have bright-coloured tips in spring. Bears rounded, bell-shaped, pink flowers from mid- to late summer. 'Cherry Turpin' has long racemes of pale pink flowers from mid-summer to mid-autumn. ♆ 'Dawn' (illus. p.148) produces young foliage with orange-yellow tips and bears deep mauve-pink flowers in compact clusters all summer.
E. × *williamsii.* Evergreen, spreading shrub. H 30cm (12in), S 60cm (24in). Fully hardy. Has needle-like, dark green leaves, with bright yellow tips when young in spring. Bears bell-shaped, mauve or pink flowers in mid-summer. Prefers acid soil. 'Gwavas' has pale pink flowers on a neat, compact plant from mid-summer to autumn. ♆ 'P.D. Williams' (illus. p.148), H 45cm (18in), has dark mauve-pink flowers; sometimes keeps its golden foliage tips throughout summer.

ERIGERON (Compositae)
Fleabane
Genus of mainly spring- and summer-flowering annuals, biennials and perennials, grown for their daisy-like flower heads. Is good for rock gardens or herbaceous borders. Fully to frost hardy. Prefers sun and well-drained soil. Resents winter damp, but should not be allowed to dry out during growing season. Propagate by division in spring or early autumn or by seed in autumn, selected forms by softwood cuttings in early summer.
E. alpinus illus. p.303.
E. aurantiacus. Clump-forming perennial. H 15cm (6in), S 30cm (12in). Fully hardy. Has long, oval, grey-green leaves and, in summer, daisy-like, brilliant orange flower heads. Propagate by seed or division in spring.
E. aureus. Clump-forming perennial. H 5cm (2in), S 10cm (4in). Fully hardy. Bears small, spoon-shaped, hairy leaves. Fine stems each carry a relatively large, daisy-like, golden-yellow flower head in summer. Dislikes winter wet with no snow cover. Is excellent for a scree, trough or alpine house; is prone to aphid attack.
E. 'Charity' illus. p.242.
♆ *E.* 'Darkest of All', syn. *E.* 'Dunkelste Aller'. Clump-forming perennial. H 80cm (32in), S 60cm (24in) or more. Fully hardy. Produces a mass of daisy-like, deep purple flower heads, with yellow centres, in summer. Has narrowly oval, greyish-green leaves.
E. 'Dunkelste Aller'. See *E.* 'Darkest of All'.
♆ *E.* 'Foerster's Liebling'. Clump-forming perennial. H 80cm (32in), S 60cm (24in). Fully hardy. In summer, daisy-like, semi-double, pink flower heads, with yellow centres, are borne above narrowly oval, greyish-green leaves.
E. glaucus 'Elstead'. Tufted perennial. H 30cm (12in), S 15cm (6in). Fully hardy. Daisy-like, dark lilac-pink flower heads appear throughout summer above oval, grey-green leaves.
♆ *E. karvinskianus*, syn. *E. mucronatus*, illus. p.325.
E. mucronatus. See *E. karvinskianus*.
E. 'Quakeress'. Clump-forming perennial. H 80cm (32in), S 60cm (24in). Fully hardy. Daisy-like, delicate lilac-pink flower heads, with yellow centres, are borne in abundance during summer. Narrowly oval leaves are greyish-green.
E. 'Serenity' illus. p.248.

ERINACEA (Leguminosae)
Genus of one species of slow-growing, evergreen sub-shrub with hard, sharp, blue-green spines and pea-like flowers. In spring produces short-lived, soft leaves. Frost hardy. Requires a sheltered position with full sun and deep, gritty, well-drained soil. Propagate by seed when available or by softwood cuttings in late spring or summer.
♆ *E. anthyllis*, syn. *E. pungens*, illus. p.296.
E. pungens. See *E. anthyllis*.

ERINUS (Scrophulariaceae)
Fairy foxglove
Genus of short-lived, semi-evergreen perennials, suitable for rock gardens, walls and troughs. Fully hardy. Needs sun and well-drained soil. Propagate species by seed in autumn (but seedlings will vary considerably), selected forms by softwood cuttings in early summer. Self seeds freely.
♆ *E. alpinus* illus. p.314. 'Dr Hähnle' (syn. *E. a.* 'Dr Haenele') is a semi-evergreen, basal-rosetted perennial. H and S 5–8cm (2–3in). Small, flat, 2-lipped, deep pink flowers open in late spring and summer. Leaves are small, oval and mid-green.

ERITRICHIUM

ERIOBOTRYA (Rosaceae)
Genus of evergreen, autumn-flowering trees and shrubs, grown for their foliage, flowers and edible fruits. Frost hardy, but in cold areas grow against a south- or west-facing wall. Fruits, which ripen in spring, may be damaged by hard, winter frosts. Requires sunny, fertile, well-drained soil. Propagate by seed in autumn or spring.
♆ *E. japonica* (Loquat). Evergreen, bushy shrub or spreading tree. H and S 8m (25ft). Has stout shoots and large, oblong, prominently veined, glossy, dark green leaves. Fragrant, 5-petalled, white flowers are borne in large clusters in early autumn, followed by pear-shaped, orange-yellow fruits.

ERIOGONUM (Polygonaceae)
Wild buckwheat
Genus of annuals, biennials and evergreen perennials, sub-shrubs and shrubs, grown for their rosetted, hairy, often silver or white leaves. Fully hardy to frost tender, min. 5°C (41°F). Needs full sun and well-drained, even poor soil. In cool, wet-winter areas protect shrubby species and hairy-leaved perennials. Water potted specimens moderately in summer, less in spring and autumn, very little in winter. Remove spent flower heads after flowering unless seed is required. Propagate by seed in spring or autumn or by semi-ripe cuttings in summer. Divide perennial root clumps in spring.
E. arborescens illus. p.131.
E. crocatum. Evergreen, sub-shrubby perennial. H to 20cm (8in), S 15cm (6in). Frost hardy. Oval, hairy leaves have woolly, white undersides. Heads of minute, sulphur-yellow flowers are borne in summer. Is a good alpine house plant.
E. giganteum illus. p.109.
E. ovalifolium. Evergreen, domed perennial. H 30cm (12in), S 10cm (4in). Fully hardy. In summer carries tiny, bright yellow flowers in umbels above branched stems bearing tiny, spoon-shaped, hairy, grey leaves. Is excellent for an alpine house.
E. umbellatum illus. p.306.

ERIOPHYLLUM (Compositae)
Genus of summer-flowering perennials and evergreen sub-shrubs, with, usually, silvery foliage and attractive, daisy-like flower heads, suitable for rock gardens as well as front of borders. Frost hardy. Requires sun and well-drained soil. Propagate by division in spring or by seed in autumn.
E. lanatum, syn. *Bahia lanata*, illus. p.257.

ERITRICHIUM (Boraginaceae)
Genus of short-lived perennials with soft, grey-green leaves and forget-me-not-like flowers. Is suitable for rock gardens and alpine houses. Fully hardy. Needs sun and well-drained, peaty, sandy soil with a deep collar of grit; dislikes damp conditions. Is extremely difficult to grow. Propagate by seed when available or by softwood cuttings in summer.

491

ERODIUM

E. elongatum. Tuft-forming perennial. H 2cm (³/₄in), S 3cm (1¹/₄in). Leaves are oval, hairy and grey-green. Short flower stems each carry small, rounded, flat, blue flowers in early summer.
E. nanum illus. p.331.

ERODIUM (Geraniaceae)
Genus of mound-forming perennials, suitable for rock gardens. Fully to half hardy. Needs sun and well-drained soil. Propagate by semi-ripe cuttings in summer or by seed when available.
E. chamaedryoides. See *E. reichardii*.
E. cheilanthifolium, syn. *E. petraeum* subsp. *crispum,* illus. p.303.
E. chrysanthum illus. p.306.
E. corsicum illus. p.324.
E. foetidum, syn. *E. petraeum*. Compact, mound-forming perennial. H 15–20cm (6–8in), S 20cm (8in). Frost hardy. Produces saucer-shaped, single, red-veined, pink flowers in summer. Oval, grey leaves have deeply cut edges.
E. manescavii illus. p.243.
E. petraeum. See *E. foetidum*. subsp. *crispum* see *E. cheilanthifolium*.
E. reichardii, syn. *E. chamaedryoides.* Mound-forming perennial. H 2.5cm (1in), S 6–8cm (2¹/₂–3in). Half hardy. In summer, saucer-shaped, single flowers, either white or pink with darker veins, are borne above tiny, oak-like leaves. Is good for a rock garden or trough.
E. × variabile **'Flore Pleno'.** Variable, cushion-forming or spreading perennial. H 10cm (4in), S 30cm (12in). Fully hardy. Has oval to narrowly oval, dark to grey-green leaves with scalloped edges and long stalks. From spring to autumn, flower stems each bear 1 or 2 rounded, double, pink flowers with darker veins; outer petals are rounded, inner petals narrower. **'Ken Aslet'** has prostrate stems, mid-green leaves and single, deep pink flowers.

Erpetion reniforme. See *Viola hederacea*.

ERYNGIUM (Umbelliferae)
Sea holly
Genus of biennals and perennials, some of which are evergreen, grown for their flowers, foliage and habit. Fully to half hardy. Does best in sun and in fertile, well-drained soil. Propagate species by seed in autumn, selected forms by division in spring or by root cuttings in winter.
E. agavifolium. Evergreen, clump-forming perennial. H 1.5m (5ft), S 60cm (2ft). Half hardy. Forms rosettes of sword-shaped, sharply toothed, rich green leaves. Thistle-like, greenish-white flower heads are produced on branched stems in summer.
♆ *E. alpinum* illus. p.217.
E. amethystinum. Rosette-forming perennial. H and S 60cm (24in). Fully hardy. Produces much-branched stems that, in summer, bear heads of small, thistle-like, blue flowers surrounded by spiky, darker blue bracts. Oval leaves are divided, spiny and mid-green.

E. bourgatii illus. p.250.
E. eburneum illus. p.190.
♆ *E. giganteum.* Clump-forming biennial or short-lived perennial that dies after flowering. H 1–1.2m (3–4ft), S 75cm (2¹/₂ft). Fully hardy. Bears large, rounded heads of thistle-like, blue flowers, surrounded by broad, spiny, silvery bracts, in late summer. Heart-shaped, basal leaves are mid-green.
♆ *E. × oliverianum* illus. p.217.
♆ *E. × tripartitum* illus. p.217.
E. variifolium illus. p.250.
E. **'Violetta'.** Upright perennial. H 75cm (30in), S 60cm (24in). Fully hardy. Loose heads of thistle-like, deep violet flowers, surrounded by narrow, spiny, silvery-blue bracts, appear in late summer. Has rounded, mid-green leaves divided into 3–5 segments.

ERYSIMUM (Cruciferae)
Genus of annuals, biennials, evergreen or semi-evergreen, short-lived perennials and sub-shrubs, grown for their flowers. Is closely related to *Cheiranthus* and is suitable for borders, banks and rock gardens. Fully to frost hardy. Needs sun and well-drained soil. Propagate by seed in spring or autumn or by softwood cuttings in summer.
E. × allionii, syn. *Cheiranthus × allionii* (Siberian wallflower). **'Orange Bedder'** illus. p.292.
♆ *E.* **'Bowles Mauve',** syn. *Cheiranthus* 'Bowles Mauve'. Bushy perennial. H to 75cm (30in), S 45cm (18in). Half hardy. Narrowly lance-shaped, dark green leaves are 5cm (2in) long. Many clusters of small, rich mauve flowers, each with 4 spreading petals, are borne in spring and early summer.
♆ *E.* **'Bredon',** syn. *Cheiranthus* 'Bredon' illus. p.298.
E. cheiri, syn. *Cheiranthus cheiri.* Moderately fast-growing, evergreen, bushy perennial, grown as a biennial. Tall cultivars, H to 60cm (24in), S 38cm (15in); intermediate, H 30–45cm (12–18in), S 30cm (12in); dwarf, H and S 20cm (8in). Fully hardy. All have lance-shaped, mid- to deep green leaves. Heads of fragrant, 4-petalled flowers in many colours, including red, yellow, bronze, white and orange, are produced in spring. **Fair Lady Series** (intermediate) has flowers in a mixture of pale colours; **'Fire King'** (intermediate) illus. p.293. **'Rose Queen'** (tall) bears rose and pink flowers; flowers of **Tom Thumb Series** (dwarf) are in a mixture of colours.
♆ *E.* **'Harpur Crewe',** syn. *Cheiranthus* 'Harpur Crewe', illus. p.298.
E. helveticum, syn. *E. pumilum,* illus. p.321.
E. linifolium. Short-lived, semi-evergreen, open, dome-shaped sub-shrub. H to 30cm (12in), S 20cm (8in) or more. Frost hardy. Leaves are narrowly lance-shaped and blue-grey. Tight heads of small, 4-petalled, pale violet flowers are produced in early summer.
E. **'Moonlight',** syn. *Cheiranthus* 'Moonlight'. Evergreen, mat-forming perennial. H 5cm (2in), S 20cm (8in) or more. Fully hardy. Leaves are small and narrowly oval. Leafy stems each carry clusters of 4-petalled, pale yellow flowers, opening in succession during summer. Makes a good rock garden plant. Needs a sheltered, sunny site and poor, gritty soil.
E. pumilum. See *E. helveticum*.

ERYTHRINA (Leguminosae)
Genus of deciduous or semi-evergreen trees, shrubs and perennials, grown for their flowers from spring to autumn. Half hardy to frost tender, min. 5°C (41°F). Requires full light and well-drained soil. Water containerized plants moderately, very little in winter or when leafless. Propagate by seed in spring or by semi-ripe cuttings in summer. Red spider mite may be troublesome.
E. americana, syn. *E. coralloides* (Flame coral tree, Naked coral tree). Deciduous, untidily rounded shrub or tree with somewhat prickly stems. H and S 3–6m (10–20ft). Frost tender. Has leaves of 3 triangular leaflets, the largest central one 11cm (4¹/₂in) long. Racemes of pea-like, red flowers are borne on leafless stems in early spring and summer.
E. × bidwillii illus. p.113.
E. coralloides. See *E. americana*.
♆ *E. crista-galli* illus. p.113.

ERYTHRONIUM (Liliaceae)
Genus of spring-flowering, tuberous perennials, grown for their pendent flowers and in some cases attractively mottled leaves. Fully to frost hardy. Requires partial shade and humus-rich, well-drained soil, where tubers will not become too hot and dry in summer while dormant. Prefers cool climates. Propagate by seed in autumn. Some species increase by offsets, which should be divided in late summer. Do not allow tubers to dry out before replanting, 15cm (6in) deep.
E. americanum illus. p.378.
E. californicum illus. p.371.
♆ **'White Beauty'** illus. p.353.
♆ *E. dens-canis* illus. p.374.
E. grandiflorum. Spring-flowering, tuberous perennial. H 10–30cm (4–12in), S 5–8cm (2–3in). Fully hardy. Has 2 lance-shaped, semi-erect, basal, plain bright green leaves. Stem carries 1–3 pendent, bright yellow flowers with reflexed petals.
E. hendersonii illus. p.353.
E. oregonum illus. p.352.
♆ *E.* **'Pagoda'** illus. p.359.
♆ *E. revolutum.* Spring-flowering, tuberous perennial. H 20–30cm (8–12in), S 15cm (6in). Frost hardy. Produces 2 lance-shaped, semi-erect, basal, brown-mottled, green leaves and a loose spike of 1–4 pendent, pale to deep pink flowers with reflexed petals.
♆ *E. tuolumnense.* Spring-flowering, tuberous perennial. H to 30cm (12in), S 12–15cm (5–6in). Frost hardy. Has 2 lance-shaped, semi-erect, basal, glossy, plain green leaves. Carries a spike of up to 10 pendent, bright yellow flowers with reflexed petals. Increases rapidly by offsets.

ESCALLONIA (Crassulariaceae)
Genus of evergreen, semi-evergreen or deciduous shrubs and trees, grown for their profuse, 5-petalled flowers and glossy foliage. Thrives in mild areas, where *Escallonia* is wind-resistant and ideal for hedging in coastal gardens. Frost hardy, but in cold areas protect from strong winds and grow against a south- or west-facing wall. Requires full sun and fertile, well-drained soil. Trim hedges and wall-trained plants after flowering. Propagate by softwood cuttings in summer.
♆ *E.* **'Apple Blossom'** illus. p.111.
E. **'Donard Beauty'.** Evergreen, arching shrub with slender shoots. H and S 1.5m (5ft). Bears deep pink flowers from early to mid-summer among small, oval, dark green leaves.
E. **'Donard Seedling'** illus. p.110.
♆ *E.* **'Edinensis'.** Vigorous, evergreen, arching shrub. H 2m (6ft), S 3m (10ft). Bears small, oblong, bright green leaves. Small, pink flowers are produced from early to mid-summer. Is one of the more hardy escallonias.
♆ *E.* **'Iveyi'** illus. p.88.
♆ *E.* **'Langleyensis'** illus. p.112.
E. leucantha illus. p.88.
♆ *E. rubra* var. *macrantha* **'Crimson Spire'.** Very vigorous, evergreen, upright shrub. H and S 3m (10ft). Has oval, rich green leaves and, throughout summer, tubular, deep red flowers. **'Woodside'** illus. p.135.
E. virgata illus. p.107.

ESCHSCHOLZIA (Papaveraceae)
Genus of annuals, grown for their bright, poppy-like flowers. Is suitable for rock gardens and gaps in paving. Fully hardy. Requires sun and grows well in poor, very well-drained soil. Dead-head regularly to ensure a long flowering period. Propagate by seed sown outdoors in spring or early autumn.
♆ *E. caespitosa* illus. p.288.
♆ *E. californica* illus. p.290, [mixed] illus. p.290. **Ballerina Series** illus. p.290.

ESPOSTOA (Cactaceae)
Genus of columnar, perennial cacti, each with a 10–30-ribbed stem, eventually becoming bushy or tree-like with age. Most species are densely covered in woolly, white hairs masking short, sharp spines. Bears cup-shaped flowers, as well as extra wool down the side of stems facing the sun, only after about 30 years. Frost tender, min. 10°C (50°F). Needs full sun and very well-drained soil. Propagate by seed in spring or summer.
E. lanata illus. p.393.

EUCALYPTUS (Myrtaceae)
Gum tree
Genus of evergreen trees and shrubs, grown for their bark, flowers and aromatic foliage. Frost hardy to frost tender, min. 1–10°C (34–50°F). Requires full sun, shelter from strong, cold winds and fertile, well-drained soil. Plant smallest obtainable trees. Water containerized plants moderately, less in winter. Attractive, young foliage of some species, normally lost with

age, may be retained by cutting growth back hard in spring. Propagate by seed in spring or autumn.
E. camaldulensis (Murray red gum, River red gum). Fast-growing, evergreen, irregularly rounded tree. H 30m (100ft) or more, S 20m (70ft) or more. Frost tender, min. 3–5°C (37–41°F). Young bark is grey, brown and cream; leaves are lance-shaped, slender, green or blue-green. Has umbels of small, cream flowers in summer. Is a drought-resistant tree.
⚘ *E. coccifera* illus. p.46.
⚘ *E. dalrympleana* illus. p.46.
E. ficifolia (Flowering gum). Moderately fast-growing, evergreen, rounded tree. H and S to 8m (25ft). Frost tender, min. 1–3°C (34–7°F). Has broadly lance-shaped, glossy, deep green leaves. Produces large panicles of many-stamened, pale to deep red flowers in spring-summer. Is best in acid soil.
E. glaucescens (Tingiringi gum). Evergreen, spreading tree. H 12m (40ft), S 8m (25ft). Frost hardy. Young bark is white. Leaves are silvery-blue and rounded when young, long, narrow and blue-grey when mature. In autumn bears clusters of many-stamened, white flowers.
⚘ *E. globulus* (Blue gum, Tasmanian blue gum). Very fast-growing, evergreen, spreading tree. H 30m (100ft), S 12m (40ft). Half hardy. Bark peels in ribbons. Large, oval to oblong, silvery-blue leaves are long, narrow and glossy, mid-green when mature. White flowers, consisting of tufts of stamens, appear in summer-autumn, often year-round.
⚘ *E. gunnii* illus. p.46.
⚘ *E. niphophila*, syn. *E. pauciflora* subsp. *niphophila*, illus. p.58.
E. pauciflora illus. p.58.
subsp. *niphophila.* See *E. niphophila.*
E. perriniana illus. p.74.
E. viminalis (Manna gum, Ribbon gum). Vigorous, evergreen, spreading tree. H 30m (100ft), S 15m (50ft). Frost tender, min. 5°C (41°F). Bark peels on upper trunk. Lance-shaped, dark green leaves become very long, narrow and pale green when mature. Bears clusters of many-stamened, white flowers in summer.

EUCHARIS (Amaryllidaceae)
Genus of evergreen bulbs, grown for their fragrant, white flowers that resemble large, white daffodils, with a cup and 6 spreading petals. Frost tender, min. 15°C (59°F). Prefers at least 50% relative humidity. Needs partial shade and humus-rich soil. Water freely in summer. Propagate by seed when ripe or by offsets in spring.
E. amazonica of gardens. See *E. grandiflora.*
E. x *grandiflora,* syn. *E. amazonica* of gardens, illus. p.369.

EUCOMIS (Liliaceae)
Pineapple flower
Genus of summer- and autumn-flowering bulbs, grown for their dense spikes of flowers, which are overtopped by a tuft of small, leaf-like bracts, as in a pineapple. Frost hardy, but in severe winters protect with dead bracken or loose, rough peat. Needs full sun and well-drained soil. Plant in spring and water freely in summer growing period. Propagate by seed or division of clumps in spring.
E. autumnalis, syn. *E. undulata.* Late summer- to autumn-flowering bulb. H 20–30cm (8–12in), S 60–75cm (24–30in). Has strap-shaped, wavy-edged leaves in a semi-erect, basal tuft. Leafless stem bears small, star-shaped, pale green or white flowers in a dense spike, with a cluster of leaf-like bracts at apex.
E. bicolor illus. p.366.
E. comosa illus. p.342.
E. pallidiflora illus. p.341.
E. undulata. See *E. autumnalis.*

EUCOMMIA (Eucommiaceae)
Genus of one species of deciduous tree, grown for its unusual foliage. Fully hardy. Needs full sun and fertile, well-drained soil. Propagate by softwood cuttings in summer.
E. ulmoides (Gutta-percha tree). Deciduous, spreading tree. H 12m (40ft), S 8m (25ft). Drooping, oval leaves are pointed and glossy, dark green; when pulled apart, leaf pieces remain joined by rubbery threads. Inconspicuous flowers appear in late spring, before leaves emerge.

EUCRYPHIA (Eucryphiaceae)
Genus of evergreen, semi-evergreen or deciduous trees and shrubs, grown for their foliage and often fragrant, white flowers. Frost hardy. Needs a sheltered, semi-shaded position in all but mild, wet areas, where it will withstand more exposure. Does best with roots in a cool, moist, shaded position and crown in sun. Requires fertile, well-drained soil that is lime-free, except in the case of *E. cordifolia* and *E.* x *nymansensis.* Propagate by semi-ripe cuttings in late summer.
E. cordifolia (Ulmo). Evergreen, columnar tree. H 15m (50ft), S 8m (25ft). Has oblong, wavy-edged, dull green leaves, with grey down beneath. Large, rose-like, white flowers are borne in late summer and autumn.
⚘ *E. glutinosa* illus. p.64.
E. lucida illus. p.64.
E. milliganii illus. p.107.
E. x *nymansensis* illus. p.55.

Eugenia australis. See *Syzygium paniculatum.*
Eugenia paniculata. See *Syzygium paniculatum.*
Eugenia ugni. See *Myrtus ugni.*

EUODIA. See **TETRADIUM.**

EUONYMUS (Celastraceae)
Genus of evergreen or deciduous shrubs and trees, sometimes climbing, grown for their foliage, autumn colour and fruits. Fully to frost hardy. Needs sun or semi-shade and any well-drained soil, although, if grown in full sun, soil for evergreen species should not be very dry. Propagate by semi-ripe cuttings in summer or by seed in autumn. *E. europaeus* and *E. japonicus* may be attacked by caterpillars; *E. japonicus* is susceptible to mildew.
⚘ *E. alatus* illus. p.118.
⚘ **'Compactus'** is a deciduous, bushy, dense shrub. H 1m (3ft), S 3m (10ft). Fully hardy. Shoots have corky wings. Oval, dark green leaves turn brilliant red in autumn. Inconspicuous, greenish-white flowers in summer are followed by small, 4-lobed, purple or red fruits.
E. cornutus var. *quinquecornutus.* Deciduous, spreading, open shrub. H 2m (6ft), S 3m (10ft). Frost hardy. Has narrowly lance-shaped, dark green leaves. Small, purplish-green flowers in summer are followed by showy, 5-horned, pink fruits that open to reveal orange-red seeds.
E. europaeus (Spindle tree). ⚘ **'Red Cascade'** illus. p.117.
E. fortunei. Evergreen shrub, grown only as var. *radicans* and its cultivars, which are climbing or creeping and prostrate. H 5m (15ft) if supported, S indefinite. Fully hardy. Bears oval, dark green leaves and inconspicuous, greenish-white flowers from early to mid-summer. Makes good ground cover. Foliage of **'Coloratus'** turns reddish-purple in autumn-winter.
⚘ **'Emerald 'n Gold'** illus. p.147.
⚘ **'Emerald Gaiety'**, H 1m (3ft), S 1.5m (5ft), is bushy, with rounded, white-edged, deep green leaves. Young, rounded leaves of **'Gold Tip'** are edged bright yellow, ageing to creamy-white. **'Kewensis'**, H 10cm (4in) or more, has slender stems and tiny leaves, and forms dense mats of growth. **'Sarcoxie'**, H and S 1.2m (4ft), is vigorous, upright and bushy, with glossy, dark green leaves.
⚘ **'Silver Queen'** illus. p.121.
'Sunspot' bears deep green leaves, each marked in centre with golden-yellow.
E. hamiltonianus subsp. *sieboldianus,* syn. *E. yedoensis.* Deciduous, tree-like shrub. H and S 6m (20ft) or more. Fully hardy. Oval, mid-green leaves often turn pink and red in autumn. Tiny, green flowers in late spring and early summer are followed by 4-lobed, pink fruits. subsp. *sieboldianus* **'Red Elf'** illus. p.117.
E. japonicus (Japanese spindle). **'Macrophyllus'** is an evergreen, upright, dense shrub. H 4m (12ft), S 2m (6ft). Frost hardy. Has very large, oval, glossy, dark green leaves and, in summer, small, star-shaped, green flowers, sometimes succeeded by spherical, pink fruits with orange seeds. Is good for hedging, particularly in coastal areas. **'Macrophyllus Albus'** illus. p.121. Leaves of ⚘ **'Ovatus Aureus'** are broadly edged with golden-yellow.
E. latifolius illus. p.118.
E. myrianthus illus. p.94.
E. oxyphyllus. Deciduous, upright shrub or tree. H and S 2.5m (8ft). Fully hardy. Oval, dull green leaves become purplish-red in autumn. Produces tiny, greenish-white flowers in late spring, followed by globose, 4- or 5-lobed, deep red fruits with orange-scarlet seeds.
⚘ *E. planipes,* syn. *E. sachalinensis* of gardens. Deciduous, upright, open shrub. H and S 3m (10ft). Fully hardy. Bears oval, mid-green leaves that turn to brilliant red in autumn, as large, 4- or 5-lobed, red fruits open to reveal bright orange seeds. Star-shaped, green flowers open in late spring.
E. sachalinensis of gardens. See *E. planipes.*
E. yedoensis. See *E. hamiltonianus* var. *sieboldianus.*

EUPATORIUM (Compositae)
Genus of perennials, sub-shrubs and shrubs, many of which are evergreen, grown mainly for their flowers, some also for their architectural foliage. Fully hardy to frost tender, min. 5–13°C (41–55°F). Requires full light or partial shade. Will grow in any conditions, although most species prefer moist but well-drained soil. Water containerized plants freely when in full growth, moderately at other times. Prune shrubs lightly after flowering or in spring. Propagate by seed in spring; shrubs and sub-shrubs may also be propagated by softwood or greenwood cuttings in summer, perennials by division in early spring or autumn. Red spider mite and whitefly may be troublesome.
E. ageratoides. See *E. rugosum.*
E. ianthinum. See *E. sordidum.*
E. ligustrinum, syn. *E. micranthum, E. weinmannianum.* Evergreen, rounded shrub. H and S 2–4m (6–12ft). Half hardy. Has elliptic to lance-shaped, bright green leaves and, in autumn, fragrant, groundsel-like, white or pink flowers in flattened clusters, 10–20cm (4–8in) wide.
E. micranthum. See *E. ligustrinum.*
E. purpureum illus. p.195.
E. rugosum, syn. *Ageratina ligustrina, E. ageratoides, E. urticifolium* (Hardy age, Mist flower, White snakeroot). Erect perennial. H 1.2m (4ft), S 45cm (1½ft). Fully hardy. In late summer bears dense, flat, thistle-like, pure white flower heads. Has nettle-like, light green leaves.
E. sordidum, syn. *Bartlettina sordida, E. ianthinum.* Evergreen, rounded, robust-stemmed shrub. H and S 1–2m (3–6ft). Frost tender, min. 10–13°C (50–55°F). Oval, serrated, deep green leaves are red haired. Produces fragrant, pompon-like, violet-purple flower heads in flattened clusters, 10cm (4in) wide, mainly in winter.
E. urticifolium. See *E. rugosum.*
E. weinmannianum. See *E. ligustrinum.*

EUPHORBIA (Euphorbiaceae)
Milkweed, Spurge
Genus of shrubs, succulents and perennials, some of which are semi-evergreen or evergreen, and annuals. Flower heads consist of cup-shaped bracts, in various colours and usually each containing several flowers lacking typical sepals and petals. Fully hardy to frost tender, min. 5–15°C (41–59°F). Does best in sun or partial shade and in moist but well-drained soil. Propagate by basal cuttings in spring or summer, by division in spring or early autumn or by seed in autumn or spring. Milky sap may irritate skin.
E. amygdaloides (Wood spurge). **'Purpurea'** is a semi-evergreen, erect perennial. H and S 30cm (1ft). Fully hardy. Stems and narrowly oval leaves are green, heavily suffused purple-red.

493

Has flower heads of cup-shaped, yellow bracts in spring. Is susceptible to mildew. ♀ var. *robbiae* illus. p.235.
E. biglandulosa. See *E. rigida*.
E. candelabrum. Deciduous, tree-like, perennial succulent. H 10m (30ft), S 5m (15ft). Frost tender, min. 10°C (50°F). Erect, 3–5-angled, deeply indented, glossy, dark green stems, often marbled white, branch and rebranch candelabra-like. Has short-lived, spear-shaped leaves and rounded heads of small flowers with cup-shaped, yellow bracts in spring.
♀ *E. characias* subsp. *characias* and ♀ subsp. *wulfenii* illus. p.126.
E. cyparissias illus. p.235.
E. epithymoides. See *E. polychroma*.
E. fulgens (Scarlet plume). Evergreen shrub of erect and arching habit. H 1–1.5m (3–5ft), S 60cm–1m (2–3ft). Frost tender, min. 5–7°C (41–5°F). Has elliptic to lance-shaped, mid- to deep green leaves, to 10cm (4in) long. Bears leafy, wand-like sprays of small flowers, each cluster surrounded by 5 petal-like, bright scarlet bracts, 2–3cm (³/₄–1¹/₄in) across, in winter-spring.
E. gorgonis (Gorgon's head). Deciduous, hemispherical, perennial succulent. H 8cm (3in), S 10cm (4in). Frost tender, min. 10°C (50°F). Has a much-ribbed, green main stem crowned by 3–5 rows of prostrate, 1cm (¹/₂in) wide stems that are gradually shed. In spring, crown also bears rounded heads of small, fragrant flowers with cup-shaped, yellow bracts.
E. griffithii 'Fireglow' illus. p.222.
E. marginata illus. p.271.
♀ *E. milii* illus. p.125. var. *splendens* (syn. *E. splendens*) is a slow-growing, mainly evergreen, spreading, spiny, semi-succulent shrub. H to 2m (6ft), S to 1m (3ft). Frost tender, min. 5–7°C (41–5°F). Has oblong to oval leaves and, intermittently year-round but especially in spring, clusters of tiny flowers enclosed in large, petal-like, red bracts.
♀ *E. myrsinites* illus. p.319.

E. nicaeensis. Clump-forming perennial with a woody base. H 75cm (30in), S 45cm (18in). Fully hardy. Umbels of greenish-yellow flower heads with cup-shaped bracts are borne throughout summer. Has narrowly oval, fleshy, grey-green leaves.
E. obesa illus. p.413.
♀ *E. palustris*. Bushy perennial. H and S 1m (3ft). Fully hardy. Clusters of yellow-green flower heads with cup-shaped bracts appear in spring above oblong to lance-shaped, yellowish-green leaves.
♀ *E. polychroma*, syn. *E. epithymoides*, illus. p.239.
E. pulcherrima illus. p.120. 'Paul Mikkelson' is a mainly evergreen, erect, freely branching shrub. H and S 3–4m (10–12ft). Frost tender, min. 5–7°C (41–5°F). Bears oval to lance-shaped, shallowly lobed leaves. From late autumn to spring, flattened heads of small, greenish-white flowers are surrounded by large, leaf-like, bright red bracts.
E. rigida, syn. *E. biglandulosa*. Evergreen, erect perennial. H and S 45cm (18in). Frost hardy. In early spring produces terminal heads of yellow flowers with cup-shaped bracts above oblong to oval, pointed, grey-green leaves.
E. seguieriana illus. p.235.
E. splendens. See *E. milii* var. *splendens*.

EUPTELEA (Eupteleaceae)
Genus of deciduous trees, grown for their foliage. Fully to half hardy. Needs full sun and fertile, well-drained soil. Propagate by seed in autumn.
E. polyandra. Deciduous, bushy-headed tree. H 8m (25ft), S 6m (20ft). Fully hardy. Long-stalked, narrowly oval, pointed, sharply toothed leaves are glossy and bright green, turning red and yellow in autumn. Has inconspicuous flowers in spring before leaves emerge.

EURYA (Theaceae)
Genus of evergreen shrubs and trees, grown for their foliage. Insignificant flowers are produced in summer. Half hardy to frost tender, min. 7°C (45°F). Tolerates partial shade or full light and needs fertile, well-drained soil. Water potted specimens freely when in full growth, less at other times. Propagate by seed when ripe or in spring or by semi-ripe cuttings in late summer.
E. emarginata illus. p.146.
E. japonica 'Variegata'. Evergreen, bushy shrub or tree. H and S 1.5–3m (5–10ft). Half hardy. Has elliptic to lance-shaped, bluntly toothed, lustrous leaves, deep green edged with creamy-white above, paler green or yellowish-green beneath. Inconspicuous, green flowers, borne in summer, are followed by tiny, spherical, purple-black fruits. Is sometimes confused in cultivation with *Cleyera japonica* 'Tricolor'.

EURYALE (Nymphaeaceae)
Genus of one species of annual, deep-water plant, grown for its floating foliage; is suitable only for a tropical pool. Frost tender, min. 5°C (41°F). Needs full light, constant warmth and heavy feeding. Propagate by seed in spring.
E. ferox illus. p.389.

EURYOPS (Compositae)
Genus of evergreen shrubs and sub-shrubs, grown for their attractive leaves and showy, daisy-like flower heads. Is suitable for borders and rock gardens. Fully hardy to frost tender, min. 5–7°C (41–5°F). Needs sun and moist but well-drained soil. May not tolerate root disturbance. Propagate by softwood cuttings in summer.
♀ *E. acraeus*, syn. *E. evansii* of gardens, illus. p.306.
E. evansii of gardens. See *E. acraeus*.
♀ *E. pectinatus* illus. p.140.

EUSTOMA (Gentianaceae)
Genus of annuals and perennials with poppy-like flowers that are good for cutting. Makes good pot plants. Frost tender, min. 4–7°C (39–45°F). Needs sun and well-drained soil. Propagate by seed sown under glass in late winter.
E. grandiflorum, syn. *Lisianthus russellianus*, illus. p.271. **F1 Hybrids** (mixed) is a group of slow-growing, upright annuals. H 45cm (18in), S 30cm (12in). Has lance-shaped, green leaves. Poppy-like, pink, blue or white flowers are produced in summer.

EVODIA. See **TETRADIUM**.

EXACUM (Gentianaceae)
Genus of annuals, biennials and perennials, grown for their profusion of flowers, that are excellent as pot plants. Frost tender, min. 7–10°C (45–50°F). Grow in sun and in well-drained soil. Propagate by seed sown in early spring for flowering the same year or in late summer for the following year.
E. affine illus. p.283.

EXOCHORDA (Rosaceae)
Genus of deciduous shrubs, grown for their abundant, showy, white flowers. Fully hardy. Does best in full sun and fertile, well-drained soil. Improve vigour and flowering by thinning out old shoots after flowering. Propagate by softwood cuttings in summer or by seed in autumn. Chlorosis may be a problem on shallow, chalky soil.
E. giraldii. Deciduous, widely arching shrub. H and S 3m (10ft). Has pinkish-green, young growths, oblong leaves and, in late spring, upright racemes of large, 5-petalled, white flowers.
♀ *E.* x *macrantha* 'The Bride' illus. p.105.
E. racemosa. Deciduous, arching shrub. H and S 4m (12ft). Has upright clusters of 5-petalled, white flowers in late spring. Leaves are oblong and deep blue-green. Prefers acid soil.

F

FABIANA (Solanaceae)
Genus of evergreen shrubs, grown for their foliage and flowers. Frost hardy, but in cold areas plant in a sheltered position. Requires full sun and fertile, well-drained soil. Propagate by softwood cuttings in summer.
F. imbricata **'Prostrata'**. Evergreen, mound-forming, very dense shrub. H 1m (3ft), S 2m (6ft). Shoots are densely covered with tiny, heath-like, deep green leaves. Bears a profusion of tubular, white flowers in early summer. ❦ f. *violacea*, syn. *F. i.* 'Violacea', illus. p.115.

FAGUS (Fagaceae)
Beech
Genus of deciduous trees, grown for their habit, foliage and autumn colour. Insignificant flowers appear in late spring and hairy fruits ripen in autumn to release edible, triangular nuts. Fully hardy. Requires sun or semi-shade; purple-leaved forms prefer full sun, yellow-leaved forms a little shade. Grows well in any but waterlogged soil. *F. sylvatica*, when used as hedging, should be trimmed in summer. Propagate species by seed in autumn, selected forms by budding in late summer. Problems may be caused by bracket fungi, canker-causing fungi, aphids and beech coccus.
F. americana. See *F. grandifolia*.
F. grandifolia, syn. *F. americana* (American beech). Deciduous, spreading tree. H and S 10m (30ft). Oval leaves are pale green and silky when young; they mature to dark green in summer and turn golden-brown in autumn.
F. orientalis (Oriental beech). Deciduous, spreading tree. H 20m (70ft), S 15m (50ft). Has large, oval, wavy-edged, dark green leaves that turn yellow in autumn.
❦ *F. sylvatica* (Common beech) illus. p.42. **'Aurea Pendula'** is a deciduous, slender tree with pendulous branches. H 30m (100ft), S 25m (80ft). Oval, wavy-edged leaves are bright yellow and become rich yellow and orange-brown in autumn. ❦ **'Dawyck'**, S 7m (22ft), is columnar, with erect branches; pale green leaves mature to mid- to dark green, and turn rich yellow and orange-brown in autumn. ❦ **'Dawyck Purple'** is similar, but has deep purple foliage. Leaves of f. *laciniata* are deeply cut.
❦ f. *pendula* illus. p.40. f. *purpurea* (syn. *F. s.* f. *atropunicea*) illus. p.39. **'Purpurea Pendula'**, H and S 3m (10ft), has stiff, weeping branches and blackish-purple foliage. Leaves of ❦ **'Riversii'** are very dark purple. Those of **'Rohanii'** are deeply cut and reddish-purple. **'Zlatia'** produces yellow, young foliage that later becomes mid- to dark green.

Fallopia aubertii. See *Polygonum aubertii*.

Fallopia baldschuanica. See *Polygonum baldschuanicum*.

FALLUGIA (Rosaceae)
Genus of one species of deciduous shrub, grown for its flowers and showy fruit clusters. Frost hardy, but in cold areas protect in winter. Needs a hot, sunny position and well-drained soil. Propagate by softwood cuttings in summer or by seed in autumn.
F. paradoxa illus. p.106.

FARFUGIUM (Compositae)
Genus of perennials, grown for their foliage and daisy-like flower heads. Frost hardy. Grow in sun or semi-shade and in moist but well-drained soil. Propagate by division in spring or by seed in autumn or spring.
F. japonicum, syn. *Ligularia tussilagineum*. Loosely clump-forming perennial. H and S 60cm (24in). Frost hardy. Has large, rounded, toothed, basal, mid-green leaves, above which rise woolly, branched stems bearing clusters of daisy-like, pale yellow flower heads in late summer.
❦ **'Aureomaculata'** (Leopard plant) has variegated, gold-and-white leaves and is half hardy.

FARGESIA (Bambusoideae). See GRASSES, BAMBOOS, RUSHES and SEDGES.
❦ *F. nitida*, syn. *Arundinaria nitida*, *Sinarundinaria nitida*. Evergreen, clump-forming bamboo. H 5m (15ft), S indefinite. Frost hardy. Has small, pointed, mid-green leaves on dark purple stalks and several branches at each node. Stems are often purple with close sheaths.
❦ *F. spathacea*, syn. *Arundinaria murielae*, *Sinarundinaria murielae*, *Thamnocalamus murielae*, *T. spathaceus* (Muriel bamboo). Evergreen, clump-forming bamboo. H 4m (12ft), S indefinite. Frost hardy. Has attractive, grey young culms with loose, light brown sheaths and broad, apple-green leaves, each very long, drawn-out at its tip. Flower spikes are unimportant.

FASCICULARIA (Bromeliaceae)
Genus of evergreen, rosette-forming perennials, grown for their overall appearance. Half hardy. Prefers full light; any well-drained soil is suitable. Water moderately from spring to autumn, sparingly in winter. Propagate by offsets or division in spring.
F. bicolor. Evergreen, rosetted perennial forming congested hummocks. H to 45cm (18in), S to 60cm (24in). Has dense rosettes of linear, tapered, arching, mid- to deep green leaves. In summer produces a cluster of tubular, pale blue flowers, surrounded by bright red bracts, at the heart of each mature rosette. Is best grown at not less than 2°C (36°F).

x **FATSHEDERA** (Araliaceae)
Hybrid genus (*Fatsia japonica* 'Moseri' x *Hedera helix* 'Hibernica') of one evergreen, autumn-flowering shrub, grown for its foliage. Is good trained against a wall or pillar or, if supported by canes, grown as a house plant. Frost hardy. Thrives in sun or shade and in fertile, well-drained soil. Propagate by semi-ripe cuttings in summer.
❦ x *F. lizei* illus. p.122.
❦ **'Variegata'** is an evergreen, mound-forming, loose-branched shrub. H 1.5m (5ft), or more if trained as a climber, S 3m (10ft). Has rounded, deeply lobed, glossy, deep green leaves, narrowly edged with creamy-white. From mid- to late autumn, sprays of small, white flowers are produced.

FATSIA (Araliaceae)
Genus of one species of evergreen, autumn-flowering shrub, grown for its foliage, flowers and fruits. Is excellent for conservatories. Frost hardy, but in cold areas shelter from strong winds. Tolerates sun or shade and requires fertile, well-drained soil. Propagate by semi-ripe cuttings in summer or by seed in autumn or spring.
❦ *F. japonica*, syn. *Aralia japonica*, *A. sieboldii* (Japanese aralia). Evergreen, rounded, dense shrub. H and S 3m (10ft). Has stout shoots and very large, rounded, deeply lobed, glossy, dark green leaves. Dense clusters of tiny, white flowers in mid-autumn are followed by rounded, black fruits. ❦ **'Variegata'** illus. p.121.
F. papyrifera. See *Tetrapanax papyrifera*.

FAUCARIA (Aizoaceae)
Genus of clump-forming, stemless, perennial succulents with semi-cylindrical or 3-angled, fleshy, bright green leaves and yellow flowers that open in late afternoons in autumn. Buds and dead flowers may appear orange or red. Frost tender, min. 6°C (43°F). Needs full sun and well-drained soil. Keep dry in winter and water sparingly in spring. Propagate by seed or stem cuttings in spring or summer.
F. tigrina illus. p.415.

FEIJOA (Myrtaceae)
Genus of evergreen, summer-flowering shrubs and trees, grown for their showy flowers and edible fruits; the latter are usually produced only after a hot summer and often only if more than one plant is grown. Frost hardy. Needs a sheltered, sunny site and light, well-drained soil. Propagate by softwood cuttings in summer.
F. sellowiana, syn. *Acca sellowiana*, illus. p.113.

FELICIA (Compositae)
Genus of annuals, evergreen sub-shrubs and (rarely) shrubs, grown for their daisy-like, mainly blue flower heads. Fully hardy to frost tender, min. 5–7°C (41–5°F). Needs full sun and well-drained soil. Water potted plants moderately, less when not in full growth; will not tolerate wet conditions, particularly when temperatures are low. Remove dead flowering stems and cut back straggly shoots regularly. Propagate by seed in spring or by greenwood cuttings in summer or early autumn.
F. amelloides, syn. *Agathaea coelestis*, *Aster capensis* (Blue marguerite).
❦ **'Santa Anita'** illus. p.138.
F. bergeriana illus. p.286.

FENESTRARIA (Aizoaceae)
Genus of clump-forming, perennial succulents with basal rosettes of fleshy leaves that have grey 'windows' in their flattened tips. Frost tender, min. 6°C (43°F). Needs full sun and very well-drained soil. Keep bone dry in winter. Propagate by seed in spring or summer.
F. aurantiaca. Clump-forming, perennial succulent. H 5cm (2in), S 30cm (12in). Produces an erect, basal rosette of cylindrical, glossy, glaucous to mid-green leaves, each with a flattened tip. Has daisy-like, white flowers on long flower stems in late summer and autumn. var. *aurantiaca* (syn. *F. a.* f. *rhopalophylla*) illus. p.415.

FEROCACTUS (Cactaceae)
Barrel cactus
Genus of slow-growing, spherical, perennial cacti, becoming columnar after many years. Frost tender, min. 5°C (41°F). Needs full sun and very well-drained soil. Propagate by seed in spring or summer. Treat blackened areoles with systemic fungicide and ensure plants have good ventilation.
F. acanthodes illus. p.396.
F. chrysacanthus. Slow-growing, spherical, perennial cactus. H 1m (3ft), S 60cm (2ft). Green stem, with 15–20 ribs, is fairly densely covered with curved, yellow-white spines. In summer bears funnel-shaped, yellow, rarely red, flowers, 5cm (2in) across, only on plants 25cm (10in) or more in diameter.
F. haematacanthus, syn. *Hamatocactus hamatacanthus*, illus. p.400.
F. latispinus. Slow-growing, flattened spherical, perennial cactus. H 20cm (8in), S 40cm (16in). Green stem, with 15–20 ribs, bears very broad, hooked, yellow or red spines. Funnel-shaped, pale yellow or red flowers appear in summer only on plants over 10cm (4in) wide.
❦ *F. setispinus*, syn. *Hamatocactus setispinus*, illus. p.403.
F. wislizenii. Slow-growing, spherical, perennial cactus. H 2m (6ft), S 1m (3ft). Green stem, with up to 25 ribs, is covered in flattened, fish-hook,

usually reddish-brown spines, to 5cm (2in) long. Funnel-shaped, orange or yellow flowers, 6cm (2½in) across, appear in late summer only on plants over 25cm (10in) wide, which should attain this size 10–15 years after raising from seed.

FERRARIA (Iridaceae)
Genus of spring-flowering corms, grown for their curious flowers with 3 large, outer petals and 3 small, inner ones, with very wavy edges. Is unpleasant-smelling to attract flies, which pollinate flowers. Half hardy. Requires full sun and well-drained soil. Plant in autumn, water during winter and dry off after flowering. Dies down in summer. Propagate by division in late summer or by seed in autumn.
F. crispa, syn. *F. undulata*, illus. p.359.
F. undulata. See *F. crispa*.

FERULA (Umbelliferae)
Giant fennel
Genus of mainly summer-flowering perennials, grown for their bold, architectural form. Should not be confused with culinary fennel, *Foeniculum*. Frost hardy. Requires sun and well-drained soil. Propagate by seed when fresh, in late summer.
F. communis illus. p.192.

FESTUCA (Gramineae). See GRASSES, BAMBOOS, RUSHES and SEDGES.
F. glauca, syn. *F. ovina* var. *glauca* (Blue fescue). Group of evergreen, tuft-forming, perennial grasses. H and S 10cm (4in). Fully hardy. Has narrow leaves in various shades of blue-green to silvery-white. Bears unimportant panicles of spikelets in summer. Is good for bed edging. Divide every 2–3 years in spring.
F. ovina var. *glauca.* See *F. glauca*.

FICUS (Moraceae)
Genus of evergreen or deciduous trees, shrubs and scrambling or root climbers, grown for their foliage and for shade; a few species also for fruit. All bear insignificant clusters of flowers in spring or summer. Frost hardy to frost tender, min. 5–18°C (41–64°F). Prefers full light or partial shade and fertile, well-drained soil. Water potted specimens moderately, very little when temperatures are low. Propagate by seed in spring or by leaf-bud or stem-tip cuttings or air-layering in summer. Red spider mite may be a nuisance.
F. benghalensis illus. p.46.
⚘ *F. benjamina* (Weeping fig). Evergreen, weeping tree, often with aerial roots. H and S 18–20m (60–70ft). Frost tender, min. 10°C (50°F). Has slender, oval leaves, 7–13cm (3–5in) long, in lustrous, rich green. 'Variegata' illus. p.58.
⚘ *F. deltoidea* illus. p.122.
F. elastica (India rubber tree, Rubber plant). ⚘ 'Decora' is a strong-growing, evergreen, irregularly ovoid tree. H to 30m (100ft), S 15–20m (50–70ft). Frost tender, min. 10°C (50°F). Has broadly oval, leathery, lustrous, deep green leaves, pinkish-bronze when young. ⚘ 'Doescheri' illus. p.46. Leaves of 'Variegata' are cream-edged, mottled with grey-green.
⚘ *F. lyrata* (Fiddle-leaf fig). Evergreen, ovoid, robust-stemmed tree. H 15m (50ft) or more, S to 10m (30ft). Frost tender, min. 15–18°C (59–64°F). Fiddle-shaped leaves, 30cm (1ft) or more long, are lustrous, deep green.
F. macrophylla (Australian banyan, Moreton Bay fig). Evergreen, wide-spreading, dense tree with a buttressed trunk when mature. H 20–30m (70–100ft), S 30–40m (100–130ft). Frost tender, min. 15–18°C (59–64°F). Oval leaves, to 20cm (8in) long, are leathery, glossy, deep green.
⚘ *F. pumila* (Creeping fig). Evergreen, root climber. H 8m (25ft); 1.5m (5ft) as a pot-grown plant. Frost tender, min. 5°C (41°F). Bright green leaves are heart-shaped and 2–3cm (¾–1¼in) long when young, 3–8cm (1¼–3in) long, leathery and oval when mature. Unpalatable fruits are 6cm (2½in) long, orange at first, then flushed red-purple. Only reaches adult stage in very warm regions or under glass. Pinch out branch tips to encourage branching. Young leaves of 'Minima' are shorter and narrower.
F. religiosa (Bo, Peepul, Sacred fig tree). Mainly evergreen, rounded to wide-spreading tree with prop roots from branches. H and S 20–30m (70–100ft). Frost tender, min. 15–18°C (59–64°F). Leaves, 10–15cm (4–6in) long, are broadly oval to almost triangular with long, thread-like tips, pink-flushed when expanding.
⚘ *F. rubiginosa* (Port Jackson fig, Rusty-leaved fig). Evergreen, dense-headed tree with a buttressed trunk. H and S 20–30m (70–100ft) or more. Frost tender, min. 15–18°C (59–64°F). Elliptic, blunt-pointed leaves, to 10cm (4in) long, are glossy, dark green above, usually with rust-coloured down beneath.

FILIPENDULA (Rosaceae)
Meadowsweet
Genus of spring- and summer-flowering perennials. Fully hardy. Does well in full sun when soil is moisture-retentive, elsewhere is best grown in semi-shade and in moist conditions; some species, e.g. *F. rubra*, will grow in boggy sites. Propagate by seed in autumn or by division in autumn or winter.
F. camtschatica. Clump-forming perennial. H to 1.5m (5ft), S 1m (3ft). Frothy, flat heads of scented, star-shaped, white or pale pink flowers are produced in mid-summer above large, lance-shaped, divided and cut leaves.
F. hexapetala. See *F. vulgaris*.
⚘ *F. purpurea* illus. p.209.
F. rubra illus. p.191.
F. ulmaria, syn. *Spiraea ulmaria*. 'Aurea' illus. p.254.
F. vulgaris, syn. *F. hexapetala* (Dropwort). 'Flore Pleno' is an upright perennial with fleshy, swollen roots. H 1m (3ft), S 45cm (1½ft). In summer bears flat panicles of rounded, double, white flowers, sometimes flushed pink, above fern-like foliage.

FIRMIANA (Sterculiaceae)
Genus of mainly deciduous trees and shrubs, grown for their foliage and to provide shade. Half hardy, but to reach tree proportions needs min. 2–5°C (36–41°F). Requires well-drained but moisture-retentive, fertile soil and full light and partial shade. Water containerized specimens freely when in full growth, less in winter. Pruning is tolerated if necessary. Propagate by seed when ripe or in spring.
F. platanifolia. See *F. simplex*.
F. simplex, syn. *F. platanifolia*, *Sterculia platanifolia*, illus. p.42.

FITTONIA (Acanthaceae)
Genus of evergreen, creeping perennials, grown mainly for their foliage. Is useful as ground cover. Frost tender, min. 15°C (59°F). Needs a fairly humid atmosphere. Grow in a shaded position and in well-drained soil; keep well watered but avoid waterlogging, especially in winter. If plant becomes too straggly, cut back in spring. Propagate in spring or summer, with additional heat, by division or stem cuttings.
F. argyroneura. See *F. verschaffeltii* var. *argyroneura*.
⚘ *F. verschaffeltii* illus. p.266.
⚘ var. *argyroneura* (syn. *F. argyroneura*) illus. p.264.

FITZROYA (Cupressaceae). See CONIFERS.
F. cupressoides, syn. *F. patagonica*, illus. p.80.
F. patagonica. See *F. cupressoides*.

FOENICULUM (Umbelliferae)
Genus of summer-flowering biennials and perennials, some of which are grown for their umbels of yellow flowers. Is also grown for its leaves, which are both decorative in borders and used for culinary flavouring. Fully to frost hardy. Grow in an open, sunny position and in fertile, well-drained soil. Remove flower heads after fading to prevent self seeding. Propagate by seed in autumn.
F. vulgare (Fennel). 'Purpureum' is an erect, branching perennial. H 2m (6ft), S 45cm (1½ft). Fully hardy. Has very finely divided, hair-like, bronze leaves and, in summer, large, flat umbels of small, yellow flowers.

FONTINALIS (Fontinalaceae)
Genus of evergreen, perennial, submerged water plants, grown for their foliage, which provides dense cover for fish and a good site for the deposit of spawn. Fully hardy. Grows in sun or semi-shade in streams and other running water; tolerates still water if cool, but then does not grow to full size. Propagate by division in spring.
F. antipyretica (Water moss, Willow moss). Evergreen, perennial, submerged water plant. H 2.5cm (1in), S indefinite. Forms spreading colonies of moss-like, dark olive-green leaves.

FORSYTHIA (Oleaceae)
Genus of deciduous, spring-flowering shrubs, grown for their usually profuse, yellow flowers, which open before the leaves emerge. *F.* x *intermedia* 'Beatrix Farrand' and *F.* x *i*. 'Lynwood' make attractive, flowering hedges. Fully hardy. Prefers full sun and fertile, well-drained soil. Thin out old shoots and trim hedges immediately after flowering. Propagate by softwood cuttings in summer or by hardwood cuttings in autumn or winter.
F. x *intermedia* 'Arnold Giant'. Deciduous, bushy shrub. H 1.5m (5ft), S 2.5m (8ft). Has stout shoots and oblong, sharply toothed, mid-green leaves. Large, 4-lobed, deep yellow flowers are borne sparsely from early to mid-spring. 'Beatrix Farrand' illus. p.105. 'Karl Sax', H 2.5m (8ft), is dense-growing, with an abundance of flowers, and leaves that turn red or purple in autumn. ⚘ 'Lynwood', H 3m (10ft), is very free-flowering, vigorous and upright. 'Minigold', H and S 2m (6ft), has oblong, mid-green leaves with masses of small, 4-lobed, yellow flowers from early to mid-spring. 'Spectabilis' illus. p.104. 'Spring Glory', H 2m (6ft), S 1.5m (5ft), has clusters of large, 4-lobed, pale yellow flowers borne in mid-spring and oblong, toothed, bright green leaves.
F. ovata. Deciduous, bushy shrub. H and S 1.5m (5ft). Leaves are broadly oval, toothed and dark green. Produces small, 4-lobed, bright yellow flowers in early spring. 'Ottawa' has an upright habit and flowers profusely.
⚘ *F. suspensa* illus. p.101.

FOTHERGILLA (Hamamelidaceae)
Genus of deciduous, spring-flowering shrubs, grown for their autumn colour and fragrant flowers, each with a dense, bottlebrush-like cluster of stamens, which open before or as leaves emerge. Fully hardy. Grows in sun or semi-shade, but colours best in full sun. Needs moist, peaty, acid soil. Propagate by softwood cuttings in summer.
F. gardenii (Witch alder). Deciduous, bushy, dense shrub. H and S 1m (3ft). Dense clusters of tiny, fragrant, white flowers appear from mid- to late spring, usually before broadly oval, dark blue-green leaves emerge. Leaves turn brilliant red in autumn.
⚘ *F. major*, syn. *F. monticola*, illus. p.97.
F. monticola. See *F. major*.

Fragaria indica. See *Duchesnea indica*.

FRAILEA (Cactaceae)
Genus of spherical to columnar, perennial cacti with tuberculate ribs. Bears short spines, mostly bristle-like. In summer produces masses of buds, most of which develop into small, spherical, shiny pods without opening. Frost tender, min. 5°C (41°F). Needs partial shade and very well-drained soil. Is not well-adapted to long periods of drought. Propagate by seed in spring or summer.
F. pulcherrima illus. p.413.

FRANCOA (Saxifragaceae)
Genus of summer- and early autumn-flowering perennials. Frost hardy. Needs

FRANKLINIA

full sun and fertile, well-drained soil. Propagate by seed or division in spring.
F. appendiculata (Bridal wreath). Clump-forming perennial. H 60cm (24in), S 45cm (18in). Racemes of small, bell-shaped, pale pink flowers, spotted with deep pink at base, appear on graceful, erect stems from summer to early autumn, above oblong to oval, lobed, hairy, crinkled, dark green leaves.
F. sonchifolia. Clump-forming perennial. H 75cm (30in), S 45cm (18in). Bears racemes of cup-shaped, red-marked, pink flowers from summer to early autumn. Lobed leaves each have a large, terminal lobe.

FRANKLINIA (Theaceae)
Genus of one species of deciduous tree or shrub, grown for its flowers and autumn colour. Fully hardy, but thrives only during hot summers. Needs full sun and moist but well-drained, neutral to acid soil. Propagate by softwood cuttings in summer, by seed in autumn or by hardwood cuttings in early winter.
F. alatamaha. Deciduous, upright tree or shrub. H and S 5m (15ft) or more. Large, shallowly cup-shaped, white flowers with yellow stamens open in late summer and early autumn. Oblong, glossy, bright green leaves turn red in autumn.

FRAXINUS (Oleaceae)
Ash
Genus of deciduous trees and shrubs, grown mainly for their foliage of paired leaflets; flowers are usually insignificant. Fully hardy. Needs sun and fertile, well-drained but not too dry soil. Propagate species by seed in autumn, selected forms by budding in summer.
F. americana (White ash). Fast-growing, deciduous, spreading tree. H 25m (80ft), S 15m (50ft). Leaves are dark green, with 5–9 oval to lance-shaped leaflets, sometimes turning yellow or purple in autumn.
F. angustifolia (Narrow-leaved ash). Deciduous, spreading, elegant tree. H 25m (80ft), S 12m (40ft). Leaves usually consist of 9–11 slender, lance-shaped, glossy, dark green leaflets.
♣ **F. excelsior** (Common ash). Vigorous, deciduous, spreading tree. H 30m (100ft), S 20m (70ft). Dark green leaves, with usually 9–11 oval leaflets, sometimes become yellow in autumn. Black leaf buds are conspicuous in winter. f. *diversifolia* has leaves that are simple or with only 3 leaflets. Leaves of ♣ **'Jaspidea'** are yellow in spring and turn golden in autumn; shoots are yellow in winter.
♣ **'Pendula'**, H 15m (50ft), S 8–10m (25–30ft), has branches weeping to the ground.
F. mariesii. Slow-growing, deciduous, compact-headed tree. H 6m (20ft), S 5m (15ft). Leaves consist of 3–5 oval, dark green leaflets each on a purple stalk. Has clusters of small, fragrant, star-shaped, creamy-white flowers in early summer, followed by narrowly oblong, purple fruits.
♣ **F. ornus** illus. p.50.

F. oxycarpa 'Raywood' (Claret ash). Vigorous, deciduous, spreading tree of elegant habit. H 20m (70ft), S 15m (50ft). Leaves have 5–7 narrowly oval, glossy, dark green leaflets that turn bright reddish-purple in autumn.
F. pennsylvanica (Green ash, Red ash). Fast-growing, deciduous, spreading tree. H and S 20m (70ft). Leaves of usually 7 or 9 narrowly oval, dull green leaflets are often velvety beneath, like the shoots. **'Patmore'** is disease-resistant, with long-lasting, glossy leaves.
F. velutina illus. p.54.

FREESIA (Iridaceae)
Genus of winter- and spring-flowering corms, grown for their usually fragrant, funnel-shaped flowers, which are popular for cutting. Half hardy. Requires full sun and well-drained soil. Plant in autumn and water throughout winter. Support with twigs or small canes. Dry off corms after flowering. Plant specially prepared corms outdoors in spring for flowering in summer. Propagate by offsets in autumn or by seed in spring.
F. alba, syn. *F. lactea*, *F. refracta* var. *alba*. Late winter- and spring-flowering corm. H 20–30cm (8–12in), S 4–6cm (1½–2½in). Has narrowly sword-shaped, erect leaves in a basal fan. Leafless stems bear loose spikes of very fragrant, white flowers, each 5–8cm (2–3in) long.
F. armstrongii. See *F. corymbosa*.
F. corymbosa, syn. *F. armstrongii*. Late winter- and spring-flowering corm. H to 30cm (12in), S 4–6cm (1½–2½in). Has narrowly sword-shaped, erect, basal leaves. Flower stem bends horizontally near the top and bears a spike of unscented, upright, pink flowers, 3–3.5cm (1¼–1½in) long, with yellow bases.
F. 'Everett' illus. p.369.
F. lactea. See *F. alba*.
F. refracta var. *alba*. See *F. alba*.
F. 'Rijnveld's Yellow'. Winter- and spring-flowering corm. H to 30cm (12in), S 4–6cm (1½–2½in). Is similar to *F. armstrongii*, but has larger, fragrant flowers, yellow throughout.
F. 'Romany'. Winter- and spring-flowering corm. H to 30cm (12in), S 4–6in (1½–2½in). Is similar to *F. armstrongii*, but has fragrant, double, pale mauve flowers.
F. 'White Swan'. Winter- and spring-flowering corm. H to 30cm (12in), S 4–6cm (1½–2½in). Is similar to *F. armstrongii*, but has very fragrant, white flowers with cream throats.
F. 'Yellow River' illus. p.369.

FREMONTIA. See FREMONTODENDRON.

FREMONTODENDRON,
syn. FREMONTIA (Sterculiaceae)
Flannel flower
Genus of vigorous, evergreen or semi-evergreen shrubs, grown for their large, very showy flowers. Frost hardy, but in cold areas plant against a south- or west-facing wall. Needs full sun and light, not too rich, well-drained soil. In mild areas may be grown as a spreading shrub, but needs firm staking when young. Resents being transplanted. Propagate by semi-ripe cuttings in summer or by seed in autumn or spring.
♣ **F. 'California Glory'** illus. p.93.
F. californicum. Vigorous, evergreen or semi-evergreen, upright shrub. H 6m (20ft), S 4m (12ft), when grown against a wall. Large, saucer-shaped, bright yellow flowers are borne amid dark green leaves, each with 3 rounded lobes, from late spring to mid-autumn.
F. mexicanum. Vigorous, evergreen or semi-evergreen, upright shrub. H 6m (20ft), S 4m (12ft) when grown against a wall. Dark green leaves have 5 deep, rounded lobes. Bears a profusion of large, saucer-shaped, deep golden-yellow flowers from late spring to mid-autumn.

FRITHIA (Aizoaceae)
Genus of one species of rosette-forming, perennial succulent. Frost tender, min. 10°C (50°F). Needs sun and well-drained soil. Propagate by seed in spring or summer.
F. pulchra illus. p.409.

FRITILLARIA (Liliaceae)
Genus of spring-flowering bulbs, grown for their pendent, mainly bell-shaped flowers on leafy stems. Fully to frost hardy; protect smaller, 5–15cm (2–6in) high species in cold frames or cold greenhouses. Needs full sun or partial shade and well-drained soil that dries out slightly in summer when bulbs are dormant but that does not become sunbaked. Grow *F. meleagris*, which is good for naturalizing in grass, in moisture-retentive soil. Propagate by offsets in summer or by seed in autumn or winter.
♣ **F. acmopetala.** Spring-flowering bulb. H 15–35cm (6–14in), S 5–8cm (2–3in). Frost hardy. Slender stems bear narrowly lance-shaped, scattered leaves and 1 or 2 broadly bell-shaped, green flowers, 2.5–4cm (1–1½in) long, with brown-stained petals flaring outwards at tips.
F. bucharica. Spring-flowering bulb. H 10–35cm (4–14in), S 5cm (2in). Frost hardy. Stems each bear scattered, lance-shaped, grey-green leaves and a raceme of up to 10 cup-shaped, green-tinged, white flowers, 1.5–2cm (⅝–¾in) long.
F. camschatcensis illus. p.357.
F. cirrhosa illus. p.358.
F. crassifolia. Spring-flowering bulb. H 10–20cm (4–8in), S 5cm (2in). Frost hardy. Has scattered, lance-shaped, grey leaves. Stems each produce 1–3 bell-shaped, green flowers, 2–2.5cm (¾–1in) long and chequered with brown.
F. delphinensis. See *F. tubiformis*.
F. imperialis illus. p.340. **'Lutea'** is a spring-flowering bulb. H to 1.5m (5ft), S 23–30cm (9–12in). Fully hardy. Leafy stems each bear lance-shaped, glossy, pale green leaves in whorls and a head of up to 5 widely bell-shaped, yellow flowers, 5cm (2in) long, crowned by small, leaf-like bracts. **'Rubra Maxima'** is very robust with red flowers.
♣ **F. meleagris** illus. p.357.
♣ **F. michailovskyi.** Spring-flowering bulb. H 10–20cm (4–8in), S 5cm (2in). Frost hardy. Has lance-shaped, grey leaves scattered on stem. Bears 1–4 bell-shaped, 2–3cm (¾–1¼in) long flowers, coloured purplish-brown with upper third of petals bright yellow.
♣ **F. pallidiflora** illus. p.359.
F. persica illus. p.340. **'Adiyaman'** is a spring-flowering bulb. H to 1.5m (5ft), S 10cm (4in). Frost hardy. Has narrowly lance-shaped, grey leaves along stem. Produces a spike of 10–20 or more narrowly bell-shaped, deep blackish-purple flowers, 1.5–2cm (⅝–¾in) long.
F. pontica illus. p.358.
F. pudica illus. p.378.
♣ **F. pyrenaica** illus. p.357.
F. raddeana illus. p.340.
F. recurva illus. p.340.
F. sewerzowii, syn. *Korolkowia sewerzowii*. Spring-flowering bulb. H 15–25cm (6–10in), S 8–10cm (3–4in). Frost hardy. Stems bear scattered, broadly lance-shaped leaves. Produces a spike of up to 10 narrowly bell-shaped, green or metallic purplish-blue flowers, 2.5–3.5cm (1–1½in) long, with flared mouths.
F. tubiformis, syn. *F. delphinensis*. Spring-flowering bulb. H 15–35cm (6–14in), S 5–8cm (2–3in). Frost hardy. Stems carry scattered, narrowly lance-shaped, grey leaves and a solitary broadly bell-shaped, purplish-pink flower, 3.5–5cm (1½–2in) long, conspicuously chequered and suffused grey outside.
F. verticillata illus. p.340.

FUCHSIA (Onagraceae)
Genus of deciduous or evergreen shrubs and trees, grown for their flowers, usually borne from early summer to early autumn. Frost hardy to frost tender, min. 5°C (41°F). If temperature remains above 4°C (39°F), deciduous plants are evergreen, but temperatures above 32°C (90°F) should be avoided. Prolonged low temperatures cause loss of top growth. If top growth dies in winter, cut back to ground level in spring. Needs a sheltered, partially shaded position, except where stated otherwise, and fertile, moist but well-drained soil. When grown as pot plants in a greenhouse, fuchsias also need high-nitrogen feeds and, when flowering, plenty of potash. Propagate by softwood cuttings in any season.

Tubular flowers are almost always pendulous and often bicoloured, with petals of one hue, and a tube and 4 sepals of another. Leaves are oval and mid-green unless otherwise stated. Spherical to cylindrical, usually blackish-purple fruits are edible, but mostly poor-flavoured. Upright types may be trained as compact bushes or standards or, with more difficulty, as pyramids. Lax or trailing plants are good for hanging baskets, but may be trained on trellises; if they are used for summer bedding they require staking.

Heights given in descriptions below are of plants grown in frost-free conditions.

F. 'Alice Hoffman'. Deciduous, compact shrub. H and S 75cm (2½ft).

Frost hardy. Has bronze foliage and small, semi-double flowers with rose-red tubes and sepals and rose-veined, white petals.
F. **'Ann Howard Tripp'** (illus. p.134). Vigorous, deciduous, upright shrub. H and S 75cm (2½ft). Half hardy. Bears single or semi-double, white flowers. Foliage is pale green.
♚ *F.* **'Annabel'** (illus. p.134). Deciduous, upright shrub. H 1m (3ft), S 75cm (2½ft). Half hardy. Produces large, double, pink-tinged, creamy-white flowers amid pale green leaves. Makes an excellent standard.
F. arborescens (Tree fuchsia; illus. p.134). Evergreen, upright tree. H 8m (25ft), S 2.5m (8ft). Frost tender. Erect heads of tiny, pale mauve to pink flowers, borne year-round, are followed by black fruits with grey-blue bloom. Foliage is mid- to dark green. May also be grown as a pot plant.
F. **'Autumnale'** (illus. p.134). Deciduous, lax shrub, grown mainly for its foliage. H 2m (6ft), S 50cm (20in). Half hardy. Bears variegated red, gold and bronze leaves. Flowers have red tubes and sepals with reddish-purple petals. Is suitable for a hanging basket or for training as a weeping standard.
F. × *bacillaris* (illus. p.135). Group of deciduous, lax shrubs. H and S 75cm (2½ft). Frost hardy. Bears minute, white, pink or crimson flowers (colour varying according to sun), sometimes followed by glossy, black fruits. Small leaves are mid- to dark green. Is suitable for a rock garden or hanging basket.
F. boliviana. Fast-growing, deciduous, upright shrub. H 3m (10ft), S 1m (3ft). Frost tender. Has large, soft, grey-green leaves with reddish midribs. Long-tubed, scarlet flowers, bunched at ends of branches, are followed by pleasantly flavoured, black fruits. Needs a large pot and plenty of space to grow well. Resents being pinched back. Is very susceptible to whitefly. ♚ **'Alba'** (syn. *F.b.* var. *luxurians* 'Alba', *F. corymbiflora* 'Alba'; illus. p.135) has flowers with white tubes and sepals and scarlet petals, followed by green fruits.
F. **'Bon Accord'**, syn. *F.* 'Bon Accorde'. Vigorous, deciduous, upright shrub. H 1.5m (5ft), S 50cm (20in). Half hardy. Small, erect flowers have white tubes and sepals and pale purple petals.
♚ *F.* **'Brutus'**. Vigorous, deciduous, upright shrub. H 1.5m (5ft), S 1m (3ft). Frost hardy. Single or semi-double flowers have crimson-red tubes and sepals and deep purple petals.
F. **'Cambridge Louie'**. Vigorous, deciduous, upright shrub. H 1m (3ft), S 75cm (2½ft). Half hardy. Produces flowers in various shades of pink.
F. **'Cascade'** (illus. p.135). Deciduous, trailing shrub. H 2m (6ft), S indefinite. Half hardy. Bears red-tinged, white tubes and sepals and deep carmine petals. Is excellent grown in a hanging basket.
♚ *F.* **'Celia Smedley'** (illus. p.135). Vigorous, deciduous, upright shrub. H 1.5m (5ft), S 1m (3ft). Half hardy. Large, single or semi-double flowers have greenish-white tubes, pale pinkish-white sepals and currant-red petals. Is best when trained as a standard.
♚ *F.* **'Cloverdale Pearl'**. Deciduous, upright shrub. H 1m (3ft), S 75cm (2½ft). Half hardy. Flowers have pinkish-white tubes, pink-veined, white petals and green-tipped, pink sepals. Foliage is mid-green with crimson midribs. Is readily trained as a standard.
F. **'Coquet Bell'**. Vigorous, deciduous, upright shrub. H 1.5m (5ft), S 1m (3ft). Half hardy. Has a profusion of single or semi-double flowers with pinkish-red tubes and sepals and red-veined, pale mauve petals.
F. corymbiflora 'Alba'. See *F. boliviana* 'Alba'.
F. denticulata. Deciduous, straggling shrub. H 4m (12ft), S indefinite. Frost tender. Flowers have long, crimson tubes, green-tipped, pale pink sepals and vermilion petals. Leaves are glossy, dark green above and reddish-green beneath. With good cultivation under glass, flowers appear throughout autumn and winter.
♚ *F.* **'Display'**. Deciduous, upright shrub. H 1m (3ft), S 75cm (2½ft). Frost hardy. Bears saucer-shaped flowers in shades of pink.
♚ *F.* **'Dollar Princess'** (illus. p.134). Deciduous, upright shrub. H 1m (3ft), S 75cm (2½ft). Frost hardy. Small, double flowers have cerise-red tubes and sepals and purple petals.
F. **'Eden Lady'**. Deciduous, upright shrub. H 1m (3ft), S 75cm (2½ft). Half hardy. Flowers have reddish-pink tubes and sepals and pale violet-blue petals with red veins. Makes a good standard.
F. **'Estelle Marie'** (illus. p.135). Deciduous, upright shrub. H 1m (3ft), S 50cm (20in). Half hardy. Flowers with white tubes, green-tipped, white sepals and mauve petals are borne above foliage. Is excellent for summer bedding.
♚ *F.* **'Flash'**. Fast-growing, deciduous, stiffly erect shrub. H 2.5m (8ft), S 50cm (20in). Frost hardy. Produces small, red flowers amid small leaves.
F. **'Flirtation Waltz'**. Vigorous, deciduous, upright shrub. H 1m (3ft), S 75cm (2½ft). Half hardy. Has large, double flowers with petals in shades of pink, and white tubes and sepals.
♚ *F. fulgens* (illus. p.135). Deciduous, upright shrub with tubers. H 2m (6ft), S 1m (3ft). Frost tender. Long-tubed, orange flowers hang in short clusters amid large, pale green leaves and are followed by edible but acidic, green fruits. Tubers may be stored dry for winter. May also be propagated by division of tubers in spring. Is very susceptible to whitefly.
♚ *F.* **'Genii'**. Deciduous, erect shrub. H 1.5m (5ft), S 75cm (2½ft). Frost hardy. Produces small flowers with cerise-red tubes and sepals and reddish-purple petals. Has golden-green foliage. Makes a good standard.
F. **'Golden Dawn'** (illus. p.134). Deciduous, upright shrub. H 1.5m (5ft), S 75cm (2½ft). Half hardy. Flowers are salmon-pink. Is good for training as a standard.
♚ *F.* **'Golden Marinka'** (illus. p.135). Deciduous, trailing shrub. H 2m (6ft), S indefinite. Half hardy. Has red flowers and variegated golden-yellow leaves with red veins. Is excellent for a hanging basket.
F. **'Gruss aus dem Bodenthal'** (illus. p.134). Deciduous, upright shrub. H 1m (3ft), S 75cm (2½ft). Half hardy. Small, single or semi-double, crimson flowers open almost black, becoming larger and paler with age.
F. **'Harry Gray'** (illus. p.134). Deciduous, lax shrub. H 2m (6ft), S indefinite. Half hardy. Bears a profusion of double flowers with pale pink tubes, green-tipped, white sepals and white to pale pink petals. Is excellent in a hanging basket.
F. **'Heidi Ann'**. See *F.* 'White Ann'.
F. **'Heidi Weiss'**. See *F.* 'White Ann'.
F. **'Hula Girl'**. Deciduous, trailing shrub. H 2m (6ft), S indefinite. Half hardy. Bears large, double flowers with deep rose-pink tubes and sepals and pink-flushed, white petals. Does best in a large hanging basket or when trained against a trellis.
F. **'Jack Acland'** (illus. p.134). Deciduous, upright shrub. H 1.5m (5ft), S 1m (3ft). Half hardy. Has large, pale to deep pink flowers.
♚ *F.* **'Jack Shahan'** (illus. p.134). Vigorous, deciduous, trailing shrub. H 2m (6ft), S indefinite. Half hardy. Has large, pale to deep pink flowers. Is excellent for a hanging basket or for training into a weeping standard or upright against a trellis.
F. **'Joe Kusber'**. Vigorous, deciduous, lax shrub. H and S 1m (3ft). Half hardy. Bears large, double flowers with white tubes, long, pink-tipped, white sepals and bluish-purple petals.
♚ *F.* **'Joy Patmore'**. Vigorous, deciduous, upright shrub. H 1.5m (5ft), S 1m (3ft). Half hardy. Flowers have white tubes, green-tipped, white sepals and cerise petals with white bases. Makes a good standard.
F. **'Koralle'** (illus. p.135). Deciduous, upright shrub. H and S 1m (3ft). Frost tender. Salmon-orange flowers, with long, narrow tubes and small sepals and petals, are bunched at ends of branches. Foliage is velvety and deep green. Is useful as summer bedding and makes an excellent specimen plant. Prefers sun.
F. **'Kwintet'** (illus. p.134). Vigorous, deciduous, upright shrub. H 1.5m (5ft), S 1m (3ft). Half hardy. Flowers are in shades of deep pink.
♚ *F.* **'La Campanella'** (illus. p.134). Deciduous, trailing shrub. H 1.5m (5ft), S indefinite. Half hardy. Has small, semi-double flowers with white tubes, pink-flushed, white sepals and cerise-purple petals. Does best in a hanging basket or when trained against a trellis.
♚ *F.* **'Lady Thumb'** (illus. p.134). Deciduous, upright, dwarf shrub. H and S 50cm (20in). Frost hardy. Has small, semi-double flowers with reddish-pink tubes and pink-veined, white petals. May be trained as a miniature standard.
♚ *F.* **'Lena'**. Deciduous, lax shrub. H and S 1m (3ft). Frost hardy. Bears double flowers with pale pink sepals and tubes and pink-flushed, purple petals. Makes a good standard.
♚ *F.* **'Leonora'** (illus. p.134). Vigorous, deciduous, upright shrub. H 1.5m (5ft), S 1m (3ft). Half hardy. Flowers are pink with green-tipped sepals. Is good for training as a standard.
F. **'Lindisfarne'**. Vigorous, deciduous, upright shrub. H 1.5m (5ft), S 1m (3ft). Half hardy. Produces single or semi-double flowers with pale pink tubes and sepals and violet petals.
F. **'Lye's Unique'** (illus. p.135). Vigorous, deciduous, upright shrub. H 1.5m (5ft), S 1m (3ft). Half hardy. Has small flowers with long, white tubes and sepals and orange-red petals. Is excellent for training as a large pyramid.
F. magellanica. (Lady's eardrops; illus. p.134). Deciduous, upright shrub. H 3m (10ft), S 2m (6ft). Frost hardy. Small flowers with red tubes, long, red sepals and purple petals are followed by black fruits. var. *molinae* (syn. *F.m.* 'Alba') has very pale pink flowers.
F. **'Margaret Roe'**. Deciduous, upright shrub. H 1m (3ft), S 75cm (2½ft). Half hardy. Erect flowers with red tubes and sepals and pale violet petals are borne above foliage. Is particularly good for summer bedding.
♚ *F.* **'Marin Glow'**. Vigorous, deciduous, upright shrub. H 1.5m (5ft), S 1m (3ft). Half hardy. Flowers have white tubes, green-tipped, white sepals and purplish-blue petals. Makes an excellent standard.
♚ *F.* **'Marinka'**. Deciduous, trailing shrub. H 2m (6ft), S indefinite. Half hardy. Red flowers with darker petals that are folded at outer edges are produced amid dark green leaves with crimson midribs. Foliage becomes discoloured in full sun or cold winds. Is excellent in a hanging basket.
F. **'Mary Poppins'** (illus. p.135). Deciduous, upright shrub. H 1.5m (5ft), S 75cm (2½ft). Half hardy. Flowers have apricot-pink tubes and sepals and vermilion petals.
F. **'Micky Gault'**. Vigorous, deciduous, upright shrub. H 1m (3ft), S 75cm (2½ft). Half hardy. Small flowers, with white tubes, pink-tinged, white sepals and pale purple petals, are produced amid pale green foliage.
F. **'Mieke Meursing'**. Deciduous, upright shrub. H 1m (3ft), S 75cm (2½ft). Half hardy. Single to semi-double flowers have red tubes and sepals and pale pink petals with cerise veins.
♚ *F.* **'Mrs Popple'** (illus. p.135). Vigorous, deciduous, upright shrub. H 1.5m (5ft), S 75cm (2½ft). Frost hardy. Has flowers with red tubes, overhanging, red sepals and purple petals. In a sheltered area may be grown as a hedge.
F. **'Mrs Rundle'**. Vigorous, deciduous, lax shrub. H and S 75cm (2½ft). Frost tender. Produces large flowers with long, pink tubes, green-tipped, pink sepals and vermilion petals. Is good for training as a standard or growing in a large hanging basket.
F. **'Nancy Lou'**. Vigorous, deciduous, upright shrub. H and S 1m (3ft). Half

hardy. Large, double flowers have pink tubes, upright, green-tipped, pink sepals and bright white petals.

♛ *F.* 'Nellie Nuttall' (illus. p.134). Vigorous, deciduous, upright shrub. H 1m (3ft), S 75cm (2½ft). Half hardy. Flowers, with rose-red tubes and sepals and white petals, are borne well above foliage. Is especially suitable for summer bedding; is also good as a standard.

F. 'Other Fellow' (illus. p.134). Deciduous, upright shrub. H 1.5m (5ft), S 75cm (2½ft). Half hardy. Has small flowers with white tubes and sepals and pink petals.

♛ *F.* 'Pacquesa'. Vigorous, deciduous, upright shrub. H 1m (3ft), S 75cm (2½ft). Half hardy. Has flowers with deep red tubes and sepals and red-veined, white petals. Is good for training as a standard.

F. 'Peppermint Stick' (illus. p.134). Deciduous, upright shrub. H 1.5m (5ft), S 1m (3ft). Half hardy. Large, double flowers have carmine-red tubes and sepals and pink-splashed, purple petals. Makes a good standard.

♛ *F.* 'Phyllis'. Deciduous, upright shrub. H 2m (6ft), S 1m (3ft). Frost hardy. Single to semi-double flowers, with rose-red tubes and sepals and crimson petals, are followed by masses of black fruits. In a sheltered area may be grown as a hedge.

F. 'Pink Galore' (illus. p.134). Deciduous, trailing shrub. H 1.5m (5ft), S indefinite. Half hardy. Has large, double, pale pink flowers. Grows best in a large hanging basket or when trained against a trellis.

F. procumbens (illus. p.135). Deciduous, prostrate shrub. H 10cm (4in), S indefinite. Half hardy. Produces tiny, erect, petalless, yellow-tubed flowers with purple sepals and bright blue pollen. Has small, dark green leaves and large, red fruits. Suits a rock garden as well as a hanging basket. Encourage flowering by root restriction or growing in poor, sandy soil.

F. 'Red Spider' (illus. p.134). Deciduous, trailing shrub. H 1.5m (5ft), S indefinite. Half hardy. Has long, red flowers with long, narrow, spreading sepals and darker petals. Is best in a large hanging basket or when trained against a trellis.

♛ *F.* 'Riccartonii' (illus. p.134). Deciduous, stiff, upright shrub. H 2m (6ft), S 1.5m (5ft). Frost hardy, but, with good drainage and wind protection, is sometimes fully hardy. Has small flowers with red tubes, broad, overhanging, red sepals and purple petals. In a sheltered area may be grown as a hedge. Many plants sold under name of *F.* 'Riccartonii' are lax hybrids of *F. magellanica*.

♛ *F.* 'Rose of Castile' (illus. p.135). Vigorous, deciduous, upright shrub. H 1.5m (5ft), S 1m (3ft). Frost hardy. Produces small flowers with white tubes, green-tipped, white sepals and purple-flushed, pink petals. Makes a good standard.

F. 'Rough Silk'. Vigorous, deciduous, trailing shrub. H 2m (6ft), S indefinite. Half hardy. Bears large flowers with pink tubes, long, spreading, pink sepals and wine-red petals. Grows best in a large hanging basket or when trained against a trellis.

♛ *F.* 'Royal Velvet'. Vigorous, deciduous, upright shrub. H 1.5m (5ft), S 75cm (2½ft). Half hardy. Has large, double flowers with red tubes and sepals and deep purple petals, splashed with deep pink. Makes an excellent standard.

F. 'Rufus' (illus. p.134). Vigorous, deciduous, upright shrub. H 1.5m (5ft), S 75cm (2½ft). Half hardy. Has a profusion of small, bright red flowers. Is easily trained as a standard.

♛ *F. splendens*. Deciduous, upright shrub. H 2m (6ft), S 1m (3ft). Half hardy. Small flowers, with broad, orange tubes, pinched in their middles, and short, green sepals and petals, appear in spring amid pale green foliage. Is very susceptible to whitefly.

F. 'Strawberry Delight'. Deciduous, lax shrub. H and S 1m (3ft). Half hardy. Large, double flowers have red tubes and sepals and pink-flushed, white petals. Leaves are yellowish-green and slightly bronzed. Makes an excellent standard or hanging basket plant.

♛ *F.* 'Swingtime' (illus. p.134). Vigorous, deciduous, lax shrub. H and S 1m (3ft). Half hardy. Has large, double flowers with red tubes and sepals and red-veined, creamy-white petals. Is good for training as a standard or growing in a hanging basket.

F. 'Temple Bells'. Vigorous, deciduous, upright shrub. H 1m (3ft), S 75cm (2½ft). Frost hardy. Flowers have red tubes and sepals and mauve petals.

F. 'Texas Longhorn'. Deciduous, lax shrub. H and S 75cm (2½ft). Half hardy. Very large, double flowers have red tubes, long, spreading, red sepals and white petals with cerise veins. May be grown as a standard or in a hanging basket.

♛ *F.* 'Thalia' (illus. p.135). Deciduous, upright shrub. H and S 1m (3ft). Frost tender. Long, slender flowers, with long, red tubes, small, red sepals and small, orange-red petals, are bunched at ends of branches. Foliage is dark maroon and velvety. Makes an excellent specimen plant in summer bedding schemes. Prefers full sun.

F. thymifolia. Deciduous, lax shrub. H and S 1m (3ft). Half hardy. Has pale green foliage and a few minute, greenish-white flowers that age to purplish-pink. Bears black fruits on female plants if pollen-bearing plants of this species or of *F.* x *bacillaris* are also grown.

♛ *F.* 'Tom Thumb' (illus. p.134). Deciduous, upright shrub. H and S 50cm (20in). Frost hardy. Bears small flowers with red tubes and sepals and mauve-purple petals. May be trained as a miniature standard.

F. triphylla. Deciduous, upright shrub, sometimes confused with *F.* 'Thalia'. H and S 50cm (20in). Frost tender. Spikes of narrow, long-tubed, bright reddish-orange flowers, with small petals and sepals, are borne above dark bronze-green leaves that are purple beneath. Is very difficult to grow.

F. 'Tristesse'. Deciduous, upright shrub. H 1m (3ft), S 75cm (2½ft). Half hardy. Has double flowers with pink tubes, reflexed, green-tipped, pink sepals and pale mauve petals. Makes a good standard.

♛ *F.* 'Westminster Chimes'. Deciduous, trailing shrub. H 2m (6ft), S indefinite. Half hardy. Small, semi-double flowers have deep pink tubes, green-tipped, paler pink sepals and purple petals. Is best in a hanging basket.

♛ *F.* 'White Ann', syn. *F.* 'Heidi Ann', *F.* 'Heidi Weiss' (illus. p.134). Deciduous, upright shrub. H 1m (3ft), S 75cm (2½ft). Half hardy. Has double flowers with red tubes and sepals and cerise-veined, white petals. Is good for training as a standard.

F. 'White Spider' (illus. p.134). Vigorous, deciduous, trailing shrub. H 2m (6ft), S indefinite. Half hardy. Bears large, very pale pink flowers with long, narrow, spreading, green-tipped sepals. Makes an excellent hanging basket plant or may be trained as a weeping standard or against a trellis.

FURCRAEA (Agavaceae)
Genus of perennial succulents with basal rosettes of sword-shaped, toothed, fleshy leaves; rosettes die after flowering. Resembles *Agave*, but differs in its short-tubed flowers. Bears bulbils on lower stems. Frost tender, min. 6°C (43°F). Needs sun and very well-drained soil. Propagate by bulbils when developed.

F. foetida, syn. *F. gigantea*. Basal-rosetted, perennial succulent. H 3m (10ft), S 5m (15ft). Has broadly sword-shaped, fleshy, mid-green leaves, to 2.5m (8ft) long, with edges toothed only at the base. Flower stems, to 8m (25ft), produce strongly scented, bell-shaped, green flowers, white within, in summer. 'Mediopicta' illus. p.396.

F. gigantea. See *F. foetida*.

G

GAGEA (Liliaceae)
Genus of spring-flowering bulbs, grown for their clusters of funnel- or star-shaped, white or yellow flowers. Is suitable for rock gardens. Frost to half hardy. Prefers full light and well-drained soil that does not become too hot and dry. Dies down in summer. Propagate by division in spring or autumn or by seed in autumn.
G. graeca, syn. *Lloydia graeca*. Spring-flowering bulb. H 5–10cm (2–4in), S 3–5cm (1¼–2in). Half hardy. Thread-like, semi-erect leaves form at ground level and on wiry stems. Bears up to 5 widely funnel-shaped, purple-veined, white flowers, 1–1.5cm (½–⅝in) long.
G. peduncularis. Spring-flowering bulb. H 5–15cm (2–6in), S 2.5–5cm (1–2in). Frost hardy. Has thread-like, semi-erect leaves at base and on stem. Produces a loose head of flat, star-shaped, yellow flowers, each 1.5–3cm (⅝–1¼in) across, with green stripes outside.

GAILLARDIA (Compositae)
Blanket flower
Genus of summer-flowering annuals and perennials that tend to be short-lived. Fully to frost hardy. Requires sun and prefers well-drained soil. May need staking. Propagate species by seed in autumn or spring, selected forms by root cuttings in winter.
G. aristata illus. p.255.
❀ *G.* × *grandiflora* 'Dazzler' illus. p.247. 'Wirral Flame' is a clump-forming, short-lived perennial. H 60cm (24in), S 50cm (20in). Fully hardy. Produces large, terminal, daisy-like, deep cardinal-red flower heads during summer. Leaves are lance-shaped, lobed and soft green.
G. pulchella [double, mixed]. Moderately fast-growing, upright annual or short-lived perennial. H 45cm (18in), S 30cm (12in). Fully hardy. Has lance-shaped, hairy, greyish-green leaves and, in summer, daisy-like, double, crimson-zoned, yellow, pink or red flower heads. 'Lollipops' illus. p.290.

GALANTHUS (Amaryllidaceae)
Snowdrop
Genus of bulbs, grown for their pendent, white flowers, one on each slender stem between 2 basal leaves. Is easily recognized by its 3 large, outer petals and 3 small, inner ones forming a cup, which is green-marked. Fully to frost hardy. Needs a cool, partially shaded position and humus-rich, moist soil. Do not allow bulbs to dry out excessively. Propagate by division in spring after flowering or in late summer or autumn when bulbs are dormant.
❀ *G.* 'Atkinsii' illus. p.383.
❀ *G. elwesii* illus. p.383.
G. gracilis, syn. *G. graecus*, illus. p.383.
G. graecus. See *G. gracilis*.

❀ *G. ikariae*, syn. *G. latifolius*, illus. p.384.
G. latifolius. See *G. ikariae*.
❀ *G. nivalis* (Common snowdrop). Late winter- and early spring-flowering bulb. H 10–15cm (4–6in), S 5–8cm (2–3in). Fully hardy. Produces narrowly strap-shaped, semi-erect, basal, grey-green leaves. Flowers are 2–2.5cm (¾–1in) long with a green mark at the tip of each inner petal.
❀ 'Flore Pleno' and 'Pusey Green Tips' illus. p.383. 'Lutescens' and 'Scharlockii' illus. p.384.
❀ *G. plicatus* subsp. *plicatus*. Late winter- and early spring-flowering bulb. H 10–20cm (4–8in), S 5–8cm (2–3in). Fully hardy. Bears broadly strap-shaped, semi-erect, basal, deep green leaves with a grey bloom and reflexed margins. White flowers, 2–3cm (¾–1¼in) long, have a green patch at the tip of each inner petal.
❀ subsp. *byzantinus* illus. p.384.
G. reginae-olgae (Autumn snowdrop). Autumn-flowering bulb. H 10–20cm (4–8in), S 5cm (2in). Frost hardy. Has 1.5–2.5cm (⅝–1in) long flowers, with a green patch at the apex of each inner petal, before or just as narrowly strap-shaped, deep green leaves, each with a central, grey stripe, appear.
G. rizehensis illus. p.383.

GALAX (Diapensiaceae)
Genus of one species of evergreen perennial, grown for its foliage and flowers. Is useful for underplanting shrubs. Fully hardy. Needs shade and moist, peaty, acid soil. Propagate by division of rooted runners in spring.
G. aphylla. See *G. urceolata*.
G. urceolata, syn. *G. aphylla*, illus. p.299.

GALEGA (Leguminosae)
Goat's rue
Genus of summer-flowering perennials. Fully hardy. Grow in an open, sunny position and in any well-drained soil. Requires staking. Propagate by seed in autumn or by division in winter.
G. × *hartlandii* 'His Majesty'. Vigorous, upright perennial. H to 1.5m (5ft), S 1m (3ft). In summer produces spikes of small, pea-like, clear lilac-mauve and white flowers. Bold, oblong to lance-shaped leaves consist of oval leaflets. 'Lady Wilson' illus. p.192.
G. orientalis illus. p.216.

Galeobdolon argentatum. See *Lamium galeobdolon* 'Variegatum'.

GALIUM (Rubiaceae)
Bedstraw
Genus of spring- and summer-flowering perennials, many of which are weeds; *G. odoratum* is cultivated as ground cover. Fully hardy. Grows well in partial shade, but tolerates sun and thrives in any well-drained soil. Propagate by division in early spring or autumn.

G. odoratum, syn. *Asperula odorata*, illus. p.239.

GALTONIA (Liliaceae)
Genus of summer- and autumn-flowering bulbs, grown for their elegant spikes of pendent, funnel-shaped, white or green flowers. Frost hardy. Needs a sheltered, sunny site and fertile, well-drained soil that does not dry out in summer. Dies down in winter. May be lifted for replanting in spring. Propagate by seed in spring or by offsets in autumn or spring.
G. candicans illus. p.341.
❀ *G. viridiflora* illus. p.346.

GARDENIA (Rubiaceae)
Genus of evergreen shrubs and trees, grown for their flowers and foliage. Frost tender, min. 15°C (59°F). Prefers partial shade and humus-rich, well-drained, neutral to acid soil. Water containerized specimens freely when in full growth, moderately at other times. After flowering, shorten strong shoots to maintain a shapely habit. Propagate by greenwood cuttings in spring or by semi-ripe cuttings in summer. Whitefly and mealy bug may cause problems.
G. augusta, syn. *G. florida*, *G. grandiflora*, *G. jasminoides* (Cape jasmine, Common gardenia). 'Fortuniana' illus. p.128.
G. capensis. See *Rothmannia capensis*.
G. florida. See *G. augusta*.
G. grandiflora. See *G. augusta*.
G. jasminoides. See *G. augusta*.
G. rothmannia. See *Rothmannia capensis*.
G. thunbergia. Evergreen, bushy shrub with white stems. H and S to 2m (6ft) or more. Has elliptic, glossy, deep green leaves. Fragrant, 7–9-petalled, white flowers, 6–10cm (2½–4in) wide, are borne in winter-spring.

GARRYA (Garryaceae)
Genus of evergreen shrubs and trees, grown for their catkins in winter and spring, which are longer and more attractive on male plants. Frost hardy. Requires a sheltered, sunny site and tolerates any poor soil. Is suitable for growing against a south- or west-facing wall. Dislikes being transplanted. Propagate by semi-ripe cuttings in summer. Hard frosts may damage catkins.
G. elliptica illus. p.95. ❀ 'James Roof' is an evergreen, bushy, dense shrub. H and S 4m (12ft). Has oval, wavy-edged, leathery, dark green leaves. Very long, grey-green catkins, with yellow anthers, are borne from mid- or late winter to early spring.

GASTERIA (Liliaceae)
Genus of perennial succulents with thick, fleshy leaves, usually arranged in a fan, later becoming a tight rosette. Frost tender, min. 5°C (41°F). Is easy to grow, needing sun or partial shade and very well-drained soil. Propagate by

seed, leaf cuttings or division in spring or summer.
G. bicolor var. *liliputana*, syn. *G. liliputana*, illus p.416.
G. caespitosa. Fan-shaped, perennial succulent. H 15cm (6in), S 30cm (12in). Produces triangular, thick, dark green leaves, 15cm (6in) long, with horny borders. Upper leaf surfaces have numerous white or pale green dots, usually in diagonal rows. Bears spikes of bell-shaped, orange-green flowers in spring.
G. carinata var. *verrucosa*, syn. *G. verrucosa*, illus p.416.
G. liliputana. See *G. bicolor* var. *liliputana*.
G. verrucosa. See *G. carinata* var. *verrucosa*.

× *Gaulnettya* 'Pink Pixie'. See *Gaultheria* × *wisleyensis* 'Pink Pixie'.
× *Gaulnettya* 'Wisley Pearl'. See *Gaultheria* × *wisleyensis* 'Wisley Pearl'.

GAULTHERIA (Ericaceae)
Genus of evergreen shrubs and sub-shrubs, grown for their foliage, flowers and fruits. Fully to half hardy. Requires shade or semi-shade and moist, peaty, acid soil. Provided soil is permanently moist, will tolerate sun. Propagate by semi-ripe cuttings in summer, by seed in autumn or, for *G. shallon* and *G. trichophylla*, by division in autumn or spring.
❀ *G. cuneata* illus. p.308.
G. forrestii. Evergreen, rounded shrub. H and S 1.5m (5ft). Half hardy. Has oblong, glossy, dark green leaves and racemes of small, fragrant, rounded, white flowers, in spring, followed by rounded, blue fruits.
G. miqueliana. Evergreen, compact shrub. H and S 25cm (10in). Frost hardy. Has oval, leathery leaves clothing stiff stems. In late spring produces bell-shaped, pink-tinged, white flowers, up to 6 per stem, followed by rounded, white or pink fruits.
G. mucronata, syn. *Pernettya mucronata*. Evergreen, bushy, dense shrub, spreading by underground stems. H and S 1.2m (4ft). Fully hardy. Oval, prickly, glossy, dark green leaves set off tiny, urn-shaped, white flowers in late spring and early summer. Has spherical, fleshy fruit varying in colour between cultivars. Sprays of fruit are good for indoor display. Fruits of 'Cherry Ripe' (female) are large and bright cherry-red. 'Edward Balls' (male) bears stout, upright, red shoots and sharply spined, bright green leaves.
❀ 'Mulberry Wine' (female) illus. p.144. ❀ 'Wintertime' (female) illus. p.143.
G. myrsinoides, syn. *Pernettya prostrata*. Evergreen, spreading shrub. H 15–30cm (6–12in), S 30cm (12in) or more. Fully hardy. Bears oval, leathery, dark green leaves. Urn-shaped, white

flowers are produced in early summer and are followed by large, rounded, blue-purple fruits. Is good for a rock garden or peat bed.
G. nummularioides. Evergreen, compact shrub. H 10–15cm (4–6in), S 20cm (8in). Frost hardy. Leaves are oval to heart-shaped and leathery. Egg-shaped, pink-flushed, white flowers are produced from the upper leaf axils in late spring or summer. Produces rounded, blue-black fruits only rarely.
♛ *G. procumbens* illus. p.335.
G. pumila, syn. *Pernettya pumila.* Evergreen, mat-forming, creeping shrub. H 5cm (2in), S 30–60cm (12–24in). Fully hardy. Prostrate branches bear tiny, bell-shaped, white flowers in early summer among tiny, rounded, leathery leaves. Rounded fruits are pink or white. Is good for a rock garden or peat bed.
G. shallon illus. p.132.
G. tasmanica, syn. *Pernettya tasmanica.* Evergreen, mat-forming shrub. H 5–8cm (2–3in), S 20cm (8in). Frost hardy. Has oval, toothed, leathery leaves with wavy edges. Bell-shaped, white flowers in early summer are followed by rounded, red fruits. Is good for a rock garden or peat bed.
G. trichophylla. Evergreen, compact shrub with creeping, underground stems. H 7–15cm (3–6in), S 20cm (8in). Frost hardy. Bell-shaped, pink flowers in early summer are followed by egg-shaped, blue fruits produced from leaf axils. Leaves are small and oval.
G. x *wisleyensis* '**Pink Pixie**', syn. x *Gaulnettya* 'Pink Pixie'. Evergreen, dense, bushy shrub. H and S 1m (3ft). Bears broadly oval, deeply veined, dark green leaves. Small, urn-shaped, pale pink flowers that are produced in late spring and early summer are followed by spherical, purplish-red fruits. '**Wisley Pearl**', syn. x *Gaulnettya* 'Wisley Pearl', illus. p.124.

GAURA (Onagraceae)
Genus of summer-flowering annuals and perennials that are sometimes short-lived. Fully hardy. Prefers full sun and light, well-drained soil. Propagate by softwood or semi-ripe cuttings in summer or by seed in autumn or spring.
♛ *G. lindheimeri.* Bushy perennial. H 1.2m (4ft), S 1m (3ft). In summer produces racemes of tubular, pink-suffused, white flowers. Leaves are lance-shaped and mid-green.

GAYLUSSACIA (Ericaceae)
Huckleberry
Genus of deciduous, occasionally evergreen, shrubs, grown for their flowers, fruits and autumn colour. Fully hardy. Needs sun or semi-shade and moist, peaty, acid soil. Propagate by softwood cuttings in summer or by seed in autumn.
G. baccata (Black huckleberry). Deciduous, bushy shrub. H and S 1m (3ft). Oval, sticky, dark green leaves redden in autumn. Has clusters of small, urn-shaped, dull red flowers in late spring, then edible, spherical, black fruits.

GAZANIA (Compositae)
Genus of perennials, often grown as annuals and useful for summer bedding, pots and tubs. Half hardy. Requires sun and sandy soil. Propagate by seed in spring or by heel cuttings in spring or summer.
G. '**Daybreak**'. Carpeting perennial, grown as an annual. H and S 20cm (8in). Has lance-shaped leaves and, in summer, large, daisy-like flower heads in a mixture of orange, yellow, pink, bronze and white. Flowers remain open in dull weather.
G. pinnata. Mat-forming perennial. H 15cm (6in), S 30cm (12in). Daisy-like, orange-red flower heads, with central, black rings, appear singly in early summer above oval, finely cut, hairy, bluish-grey leaves.
G. rigens var. *uniflora*, syn. *G. uniflora*, illus. p.256.
G. '**Sundance**'. Carpeting perennial, grown as an annual. H and S 30cm (12in). Has lance-shaped leaves. In summer bears very large, daisy-like flower heads, over 8cm (3in) wide, in brilliant flame shades, some with red-and-yellow stripes.
G. uniflora. See *G. rigens* var. *uniflora*.

GELSEMIUM (Loganiaceae)
Genus of evergreen, twining climbers, grown for their fragrant, jasmine-like flowers. Half hardy. In cool climates best grown under glass. Provide fertile, well-drained soil and full light. Water regularly, less in cold weather. Stems require support and should be thinned out after flowering or in spring. Propagate by seed in spring or by semi-ripe cuttings in summer.
♛ *G. sempervirens* (Carolina jasmine, False jasmine). Moderately vigorous, evergreen, twining climber. H 3m (10ft) or more. Has oval to lance-shaped, pointed, lustrous leaves. Clusters of fragrant, funnel-shaped, pale to deep yellow flowers, each with 5 petal lobes, are borne from late spring to late summer.

GENISTA (Leguminosae)
Broom
Genus of deciduous, sometimes almost leafless, shrubs and trees, grown for their mass of small, pea-like flowers. Fully to half hardy. Does best in full sun and not over-rich, well-drained soil. Resents being transplanted. Propagate species by softwood or semi-ripe cuttings in summer or by seed in autumn, selected forms by softwood cuttings only in summer.
♛ *G. aetnensis* illus. p.67.
G. cinerea illus. p.93.
G. delphinensis, syn. *G. sagittalis* subsp. *delphinensis.* Deciduous, prostrate shrub. H 1cm (1/2in), S 20cm (8in). Frost hardy. Has tangled, winged branches, covered with minute, oval, dark green leaves. Masses of golden-yellow flowers are borne along stems in early summer. Suits a rock garden or wall.
G. fragrans. See *G.* x *spachiana*.
G. hispanica illus. p.140.
G. lydia illus. p.307.
G. monosperma, syn. *Retama monosperma.* Deciduous, almost leafless, graceful, arching shrub. H 1m (3ft), S 1.5m (5ft). Half hardy. Slender, silky-grey shoots bear clusters of very fragrant, white flowers in early spring. Has a few inconspicuous leaves. Grow against a south- or west-facing wall.
G. pilosa. Deciduous, domed shrub. H and S 30cm (12in). Fully hardy. Narrowly oval leaves are silky-haired beneath. Bright yellow flowers on short stalks are borne in leaf axils in summer. Is useful on a bank or as ground cover. Propagate by semi-ripe cuttings in summer.
G. sagittalis, syn. *Chamaespartium sagittale,* illus. p.334. subsp. *delphinensis.* See *G. delphinensis*.
♛ *G.* x *spachiana,* syn. *G. fragrans, Cytisus canariensis* of gardens, *Cytisus racemosus, Cytisus* x *spachianus.* Vigorous, evergreen, arching shrub. H and S 3m (10ft). Half hardy. Has dark green leaves with 3 oval leaflets. Produces long, slender clusters of fragrant, golden-yellow flowers in winter and early spring. Is often grown as a houseplant.
♛ *G. tenera* '**Golden Showers**'. Vigorous, deciduous, arching shrub. H 3m (10ft), S 5m (15ft). Frost hardy. Narrowly oblong leaves are grey-green. Bears racemes of fragrant, golden-yellow flowers in early to mid-summer.
G. tinctoria illus. p.127. ♛ '**Royal Gold**'. H and S 1m (3ft). Fully hardy. Produces long, conical panicles of golden-yellow flowers in spring-summer and leaves that are narrowly lance-shaped and dark green.

GENTIANA (Gentianaceae)
Gentian
Genus of annuals, biennials and perennials, some of which are semi-evergreen or evergreen, grown for their usually blue flowers. Is excellent for rock gardens and peat beds. Fully hardy. Prefers sun or semi-shade and humus-rich, well-drained, moist, neutral to acid soil. Some species grow naturally on limestone soils. Propagate by division or offshoots in spring or by seed in autumn. Divide autumn-flowering species and *G. clusii* every 3 years in early spring and replant in fresh soil.
♛ *G. acaulis,* syn. *G. excisa, G. kochiana,* illus. p.318.
G. angustifolia. Evergreen, clump-forming perennial. H 10cm (4in), S 20cm (8in). Has rosettes of oblong, dull green leaves and, in summer, solitary tubular, sky-blue flowers on 7cm (3in) stems. Tolerates alkaline soils.
♛ *G. asclepiadea* illus. p.227.
G. clusii (Trumpet gentian). Evergreen, clump-forming perennial. H 5cm (2in), S 15–23cm (6–9in). Has rosettes of oval, glossy, dark green leaves. Trumpet-shaped, azure-blue flowers, with green-spotted, paler throats, are borne on 2.5–10cm (1–4in) stems in early summer. Tolerates alkaline soils.
G. excisa. See *G. acaulis*.
G. gracilipes. Semi-evergreen, tufted perennial with arching stems. H 15cm (6in), S 20cm (8in). Forms a central rosette of long, strap-shaped, dark green leaves from which lax flower stems bearing tubular, dark purplish-blue flowers, greenish within, are produced in summer. Tolerates some shade.
G. kochiana. See *G. acaulis*.
G. lutea illus. p.219.
G. x *macaulayi* '**Wellsii**', syn. *G.* x *m.* 'Wells Variety' illus. p.335.
G. ornata. Semi-evergreen, clump-forming perennial with small, over-wintering rosettes. H 5cm (2in), S 10cm (4in). Forms a central rosette of grass-like leaves. In autumn, each stem tip carries an upright, bell-shaped, mid-blue flower, with a white throat and deep blue stripes shading to creamy-white outside. Requires acid soil and a moist atmosphere.
G. saxosa illus. p.322.
♛ *G. septemfida* illus. p.308.
♛ *G. sino-ornata* illus. p.335.
G. '**Susan Jane**'. Vigorous, semi-evergreen, spreading perennial with small, overwintering rosettes. H 5cm (2in), S 30cm (12in). Prostrate stems bear grass-like leaves. Large, trumpet-shaped, white-throated, deep blue flowers, greenish within, appear in autumn. Requires acid soil.
G. verna illus. p.318.

GERANIUM (Geraniaceae)
Cranesbill
Genus of perennials, some of which are semi-evergreen, grown for their attractive flowers and often as ground cover. Compact species are suitable for rock gardens. Fully to half hardy. Most species prefer sun, but some do better in shade. Will grow in any but waterlogged soil. Propagate by semi-ripe cuttings in summer or by seed or division in autumn or spring. Cultivars should be propagated by division or cuttings only.
G. '**Ann Folkard**'. Spreading perennial. H 50cm (20in), S 1m (36in). Fully hardy. Has rounded, deeply cut, yellowish-green leaves and, in summer-autumn, masses of shallowly cup-shaped, rich magenta flowers with black veins.
G. armenum. See *G. psilostemon*.
G. cinereum. Semi-evergreen, rosetted perennial with spreading flowering stems. H 15cm (6in), S 30cm (12in). Fully hardy. Has cup-shaped flowers, either white to pale pink, strongly veined with purple, or pure white, on lax stems in late spring and summer. Basal leaves are rounded, deeply divided, soft and grey-green. Is good for a large rock garden. ♛ '**Ballerina**' and ♛ var. *subcaulescens* (syn. *G. subcaulescens*) illus. p.327.
G. clarkei '**Kashmir Purple**', syn. *G. pratense* 'Kashmir Purple'. Carpeting, rhizomatous perennial. H and S 45–60cm (18–24in). Fully hardy. Bears loose clusters of cup-shaped, deep purple flowers in summer. Rounded leaves are deeply divided and finely veined. '**Kashmir White**' (syn. *G. pratense* 'Kashmir White') illus. p.240.
♛ *G. dalmaticum* illus. p.325.
♛ *G. endressii* illus. p.242.
'**Wargrave Pink**'. See *G.* x *oxonianum* 'Wargrave Pink'.

GERBERA

♣ *G. farreri.* Rosetted perennial with a tap root. H 10cm (4in), S 10–15cm (4–6in) or more. Fully hardy. Outward-facing, flattish, very pale mauve-pink flowers set off blue-black anthers in early summer. Has kidney-shaped, matt green leaves. Flower and leaf stems are red.
G. grandiflorum. See *G. himalayense*.
G. himalayense, syn. *G. grandiflorum,* illus. p.249.
G. ibericum. Clump-forming perennial. H and S 60cm (24in). Fully hardy. In summer produces sprays of 5-petalled, saucer-shaped, violet-blue flowers. Has heart-shaped, lobed or cut, hairy leaves.
G. incanum. Semi-evergreen, spreading, mounded perennial. H 30–38cm (12–15in), S 60cm–1m (2–3ft). Frost hardy. Shallowly cup-shaped flowers are variable, but usually deep pink, and are borne singly in summer above aromatic, deeply divided, grey-green leaves with linear segments.
♣ *G. 'Johnson's Blue'* illus. p.249.
G. macrorrhizum illus. p.243.
♣ *'Ingwersen's Variety'* illus. p.233.
G. maculatum. Clump-forming perennial. H 75cm (30in), S 45cm (18in). Fully hardy. In spring bears heads of flattish, pinkish-lilac flowers above rounded, lobed or scalloped, mid-green leaves that turn fawn and red in autumn.
♣ *G. maderense.* Vigorous, semi-evergreen, bushy perennial with a woody base. H and S 1m (3ft). Half hardy. Produces large sprays of shallowly cup-shaped, deep magenta flowers in summer above palmate, finely cut, dark green leaves.
♣ *G. x magnificum* illus. p.249.
G. nodosum illus. p.234.
G. orientalitibeticum illus. p.301.
G. x oxonianum 'Claridge Druce'. Vigorous, semi-evergreen, carpeting perennial. H and S 60–75cm (24–30in). Fully hardy. Bears clusters of cup-shaped, darker-veined, mauve-pink flowers throughout summer. Has dainty, rounded, lobed leaves.
♣ *'Wargrave Pink'* (syn. *G. endressii* 'Wargrave Pink') illus. p.243.
'Winscombe' illus. p.207.
♣ *G. palmatum.* Vigorous, semi-evergreen, bushy perennial with a woody base. H 45cm (18in), S 60cm (24in). Half hardy. Has palmate, deeply lobed, dark green leaves and, in late summer, large sprays of shallowly cup-shaped, purplish-red flowers.
G. phaeum illus. p.202.
G. pratense (Meadow cranesbill). Clump-forming perennial. H 75cm (30in), S 60cm (24in). Fully hardy. Produces 5-petalled, saucer-shaped, violet-blue flowers on branching stems in summer. Rounded, lobed to deeply divided, mid-green leaves become bronze in autumn. **'Kashmir Purple'** see *G. clarkei* 'Kashmir Purple'. **'Kashmir White'** see *G. clarkei* 'Kashmir White'. ♣ **'Plenum Violaceum'** is more compact than the species with double, deep violet flowers.
G. procurrens. Carpeting perennial. H 30cm (12in), S 60cm (24in). Fully hardy. Has rounded, lobed, glossy leaves and, in summer, clusters of saucer-shaped, deep rose-purple flowers.
♣ *G. psilostemon,* syn. *G. armenum,* illus. p.209.
G. pylzowianum. Spreading perennial with underground runners and tiny tubers. H 12–25cm (5–10in), S 25cm (10in) or more. Fully hardy. Bears semi-circular, deeply cut, dark green leaves and, in late spring and summer, trumpet-shaped, green-centred, deep rose-pink flowers. May be invasive.
♣ *G. renardii* illus. p.241.
♣ *G. x riversleianum 'Russell Prichard'.* Semi-evergreen, clump-forming perennial. H 30cm (1ft), S 1m (3ft). Frost hardy. Saucer-shaped, clear pink flowers are borne singly or in small clusters from early summer to autumn. Rounded leaves are lobed and grey-green.
G. sanguineum illus. p.302.
♣ var. *striatum* (syn. *G.s.* var. *lancastriense*) illus. p.324.
G. stapfianum. Semi-prostrate perennial with underground stolons. H 12cm (5in), S 15cm (6in) or more. Frost hardy. Leaves are kidney-shaped to rounded, deeply cut to scalloped and green marbled white with red leaf stalks and margins. In summer has solitary upward-facing, saucer-shaped, reddish-purple flowers with red centres and veins. Is best grown in an alpine house or in scree conditions as it dislikes winter wet.
G. subcaulescens. See *G. cinereum* var. *subcaulescens*.
♣ *G. sylvaticum 'Mayflower'* illus. p.215.
G. traversii var. *elegans.* Semi-evergreen, rosetted perennial with spreading stems. H 10cm (4in), S 25cm (10in). Frost hardy. Large, upward-facing, saucer-shaped, pale pink flowers, with darker veins, rise above rounded, lobed, grey-green leaves in summer. Is suitable for a sheltered ledge or rock garden. Protect from winter wet. Needs gritty soil.
♣ *G. wallichianum 'Buxton's Blue',* syn. *G.w.* 'Buxton's Variety', illus. p.251.
G. wlassovianum. Clump-forming perennial. H and S 60cm (24in). Fully hardy. Has velvety stems and rounded, lobed, dark green leaves. Saucer-shaped, deep purple flowers are borne singly or in small clusters in summer.

GERBERA (Compositae)
Genus of perennials, flowering from summer to winter depending on growing conditions. Half hardy. Grow in full sun and in light, sandy soil. Propagate by heel cuttings from side shoots in summer or by seed in autumn or early spring.
G. jamesonii illus. p.266.

GEUM (Rosaceae)
Avens
Genus of summer-flowering perennials. Fully hardy. Does best in sun and prefers moist but well-drained soil. Propagate by division or seed in autumn.
G. x borisii illus. p.257.
G. chiloense 'Lady Stratheden'. See *G.* 'Lady Stratheden'.
G. chiloense 'Mrs Bradshaw'. See *G.* 'Mrs Bradshaw'.
G. 'Fire Opal'. Clump-forming perennial. H 80cm (32in), S 45cm (18in). Rounded, double, bronze-scarlet flowers are borne in small clusters in summer above oblong to lance-shaped, lobed, fresh green leaves.
♣ *G. 'Lady Stratheden',* syn. *G. chiloense* 'Lady Stratheden', illus. p.256.
G. 'Lionel Cox'. Clump-forming perennial. H and S 30cm (12in). In early summer produces small clusters of 5-petalled, cup-shaped, shrimp-red flowers above oblong to lance-shaped, lobed, fresh green leaves.
♣ *G. montanum* (Alpine avens). Dense, clump-forming, rhizomatous perennial that spreads slowly. H 10cm (4in), S 23cm (9in). Shallowly cup-shaped, golden-yellow flowers in early summer are followed by fluffy, buff-coloured seed heads. Leaves are pinnate, each with a large, rounded, terminal lobe. Suits a rock garden.
♣ *G. 'Mrs Bradshaw',* syn. *G. chiloense* 'Mrs Bradshaw'. Clump-forming perennial. H 80cm (32in), S 45cm (18in). Rounded, double, crimson flowers are borne in small sprays in summer. Fresh green leaves are oblong to lance-shaped and lobed.

GEVUINA (Proteaceae)
Genus of evergreen trees, grown for their foliage and flowers in summer. Frost hardy. Needs semi-shade and fertile, moist but well-drained soil. Propagate by semi-ripe cuttings in late summer or by seed in autumn.
G. avellana (Chilean hazel). Evergreen, conical tree. H and S 10m (30ft). Has large, glossy, dark green leaves divided into numerous oval, toothed leaflets. Slender spires of spidery, white flowers in late summer are followed by cherry-like, red, then black fruits.

GIBBAEUM (Aizoaceae)
Genus of clump-forming, perennial succulents with pairs of small, swollen leaves, often of unequal size. Frost tender, min. 5°C (41°F). Requires full sun and very well-drained soil. Water very lightly and occasionally in early winter. Propagate by seed or stem cuttings in spring or summer.
G. petrense. Carpeting, perennial succulent. H 3cm (1¼in), S 30cm (12in) or more. Each branch carries 1 or 2 pairs of thick, triangular, pale grey-green leaves, 1cm (⅜in) long. Bears daisy-like, pink-red flowers, 1.5cm (⅝in) across, in spring.
G. velutinum illus. p.405.

GILIA (Polemoniaceae)
Genus of summer- and autumn-flowering annuals. Fully hardy. Grows best in sun and in fertile, very well-drained soil. Stems may need support, especially on windy sites. Propagate by seed sown outdoors in spring, or in early autumn for early flowering the following year.
G. achilleifolia. Fast-growing, upright, bushy annual. H 60cm (24in), S 20cm (8in). Finely divided, mid-green leaves are hairy and sticky. Heads of funnel-shaped, blue flowers, 2.5cm (1in) wide, are produced in summer.
G. capitata illus. p.285.

GILLENIA (Rosaceae)
Genus of summer-flowering perennials. Fully hardy. Tolerates sun or shade and any well-drained soil. Needs staking. Propagate by seed in autumn or spring.
♣ *G. trifoliata* illus. p.205.

GINKGO (Ginkgoaceae). See CONIFERS.
♣ *G. biloba* illus. p.77.

GLADIOLUS (Iridaceae)
Genus of corms, each producing a spike of funnel-shaped flowers and a fan of erect, sword-shaped leaves on basal part of flower stem. Is suitable for cutting or for planting in mixed borders; most hybrids are also good for exhibition. Frost to half hardy. Needs a sunny and fertile, well-drained site. Plant 10–15cm (4–6in) deep and the same distance apart in spring. Water well in summer and support tall cultivars with canes. In autumn lift half-hardy types, cut off stems and dry corms in a frost-free but cool place. Pot up spring-flowering species and cultivars in autumn and place in a cool greenhouse; after flowering, dry off for summer months and repot in autumn.

Propagate by seed or by removal of young cormlets from parent. Seed sown in early spring in a cool greenhouse will take 2–3 years to flower and may not breed true to type. Cormlets, removed after lifting, should be stored in frost-free conditions and then be planted out 5cm (2in) deep in spring; lift in winter as for mature corms. They will flower in 1–2 years.

While in store, corms may be attacked by various rots. Protect sound, healthy corms by dusting with a fungicide or soaking in a fungicide solution before drying; store in an airy, cool, frost-free place. Gladiolus scab causes blotches on leaves; gladiolus yellows shows as yellowing stripes, then death of leaves; in both cases destroy affected corms and plant healthy ones in a new site each year.

Gladiolus hybrids
Most hybrids are derived from *G. x hortulanus*. Each spike has flowers arranged on opposite sides of stem, either vertically, touching each other (formally placed), or alternating up the stem with a slight space between each flower (ladder placed). All have stiff leaves, 20–50cm (8–20in) long, ranging from pale willow-green or steely blue-green to almost bottle-green. Half hardy. They are divided into Grandiflorus and Primulinus groups.

Grandiflorus group produces long, densely packed spikes of funnel-shaped flowers, with ruffled, thick-textured petals or plain-edged, thin-textured ones. Giant-flowered hybrids have a bottom flower of over 14cm (5½in) across (flower head is 65–80cm (26–32in) long); large-flowered 11–14cm (4½–5½in) across (flower head 60cm–1m (24–36in) long); medium-flowered 9–11cm

GLADIOLUS

(3½–4½in) across (flower head 60–80cm (24–32in) long); and small-flowered 6–9cm (2½–3½in) across (flower head 50–70cm (20–28in) long). **Primulinus group** has fairly loose spikes of plain-edged, funnel-shaped flowers, 6–8cm (2½–3in) across, each with a strongly hooded, upper petal over the stigma and anthers. Flower heads are 30cm (12in) long.

G. **'Amsterdam'**. Grandiflorus group, giant-flowered gladiolus. H 1.7m (5½ft), S 30cm (1ft). Has up to 10 formally placed, slightly upward-facing, finely ruffled, white flowers in late summer. Is good for exhibition.

G. **'Amy Beth'**. Grandiflorus group, small-flowered gladiolus. H 1.2m (4ft), S 20–25cm (8–10in). Bears 7 formally placed, heavily ruffled, lavender flowers, with thick, waxy, cream-lipped petals, in late summer.

G. **'Atlantis'**. Grandiflorus group, medium-flowered gladiolus. H 1.5m (5ft), S 20–25cm (8–10in). Carries 8 lightly ruffled, deep violet-blue flowers, with small, white throats, in ladder placement up stem during late summer. Good in flower arrangements.

G. **'Black Lash'** (illus. p.343). Grandiflorus group, small-flowered gladiolus. H 1.35m (4½ft), S 15–20cm (6–8in). Bears 8 or 9 formally placed, lightly ruffled, deep black-rose flowers, with pointed, slightly reflexed petals, from late summer to early autumn.

G. blandus. See *G. carneus*.

G. **'Café au Lait'** (illus. p.343). Primulinus group gladiolus. H 1.1m (3½ft), S 15–20cm (6–8in). Bears 4 or 5 pale coffee-coloured flowers, each with a beige throat and a hooded, central, upper petal, in ladder placement up stem in late summer. Is excellent in flower arrangements.

♛ *G. callianthus*, syn. *Acidanthera bicolor*. Late summer-flowering corm. H to 1m (3ft), S 10–15cm (4–6in). Half hardy. Has a loose spike of up to 10 fragrant flowers, each with a curved, 10cm (4in) long tube and 6 white petals, each with a deep purple blotch at the base.

G. cardinalis. Summer-flowering corm. H to 1.2m (4ft), S 10–15cm (4–6in). Half hardy. Arching stem bears a spike of up to 6 widely funnel-shaped flowers, each 8cm (3in) long and bright red with spear-shaped, white marks on lower 3 petals.

G. **'Carioca'** (illus. p.343). Grandiflorus group, medium-flowered gladiolus. H 1.35m (4½ft), S 30cm (1ft). Produces a spike of 8 or 9 formally placed, heavily ruffled, orange flowers, with bright red throat marks, in mid-summer.

G. carneus, syn. *G. blandus*. Spring-flowering corm. H 20–40cm (8–16in), S 8–10cm (3–4in). Half hardy. Stem bears a loose spike of up to 8 widely funnel-shaped, white or pink flowers, 4–6cm (1½–2½in) long, marked on lower petals with darker red or yellow blotches.

G. **'Christabel'** illus. p.359.

G. × *colvillei*, syn. *G. nanus*. **'Amanda Mahy'** is an early summer-flowering corm. H 30–50cm (12–20in), S 8–10cm (3–4in). Half hardy. Produces a loose spike of 5–8 salmon-pink flowers, 7cm (3in) long, marked violet on lower petals. Plant in autumn. ♛ **'The Bride'** (illus. p.343) has white flowers with green-marked throats.

♛ *G. communis* subsp. *byzantinus* illus. p.344.

G. dalenii, syn. *G. natalensis*, *G. psittacinus*. Vigorous, summer-flowering corm. H to 1.5m (5ft), S 10–15cm (4–6in). Half hardy. Produces up to 25 red or yellow-orange flowers, 8–12cm (3–5in) long, each with a hooded, upper petal and often flecked or streaked red.

G. **'Dancing Queen'** (illus. p.343). Grandiflorus group, large-flowered gladiolus. H 1.35m (4½ft), S 20–25cm (8–10in). Has 9 formally placed, ivory-white flowers, with 2 wine-red blotches in each throat, in late summer; petals are thin and silky.

G. **'Deliverance'** (illus. p.343). Grandiflorus group, giant-flowered gladiolus. H 1.7m (5½ft), S 25–30cm (10–12in). In late summer produces 9 formally placed flowers that have heavily ruffled, salmon-pink petals overlaid with orange.

G. **'Drama'** (illus. p.343). Grandiflorus group, large-flowered gladiolus. H 1.7m (5½ft), S 25–30cm (10–12in). In late summer bears 8–10 formally placed, lightly ruffled, deep watermelon-pink flowers with red-marked, yellow throats. Is superb for exhibition.

G. **'Esta Bonita'**. Grandiflorus group, giant-flowered gladiolus. H 1.7m (5½ft), S 30cm (1ft). Produces 7 formally placed, apricot-orange flowers, slightly darker towards petal edges, in late summer. Is good for exhibition.

G. **'Frank's Perfection'**. Primulinus group gladiolus. H 1.1m (3½ft), S 15–20cm (6–8in). Has 6 or 7 ladder-placed, vibrant red flowers, each with a hooded, upper petal and with a central, golden line down ribs, in late summer.

G. **'Gigi'** (illus. p.343). Grandiflorus group, small-flowered gladiolus. H 1m (3ft), S 15–20cm (6–8in). Has 7 formally placed, lightly ruffled, dark pink flowers, each with a small, white throat, in late summer.

G. **'Green Isle'**. Grandiflorus group, medium-flowered gladiolus. H 1.35m (4½ft), S 20–25cm (8–10in). Bears 7 or 8 slightly informal flowers, lime-green throughout with chiselled ruffling, in late summer.

♛ *G.* **'Green Woodpecker'** (illus. p.343). Grandiflorus group, medium-flowered gladiolus. H 1.5m (5ft), S 30cm (1ft). Has 10 formally placed, uranium-green flowers, with wine-red throats, in late summer. Is very good for exhibition.

G. **'Hastings'**. Primulinus group gladiolus. H 1.35m (4½ft), S 15–20cm (6–8in). Bears 6 ladder-placed, pale coffee-coloured flowers, with reddish-brown marks, in late summer. Is superb in flower arrangements.

G. **'Ice Cap'** (illus. p.343). Grandiflorus group, large-flowered gladiolus. H 1.7m (5½ft), S 25–30cm (10–12in). Carries 10–12 formally placed, heavily ruffled, ice-white flowers from late summer to early autumn.

G. **'Inca Queen'** (illus. p.343). Grandiflorus group, large-flowered gladiolus. H 1.5m (5ft), S 20–25cm (8–10in). Produces 9 formally placed, heavily ruffled, waxy, deep salmon-pink flowers, with lemon-yellow lip petals and throats, in late summer.

G. italicus, syn. *G. segetum*, illus. p.344.

G. **'Krystal'**. Grandiflorus group, small-flowered gladiolus. H 1.2m (4ft), S 20–25cm (8–10in). Produces 8 formally placed, very lightly ruffled, cerise-red flowers, with prominent, white stamens and white-edged lip petals, in late summer.

G. **'Leonore'**. Primulinus group gladiolus. H 1.1m (3½ft), S 15–20cm (6–8in). Bears 5–7 ladder-placed, buttercup-yellow flowers, each with a hooded, central, upper petal, in late summer.

G. **'Melodie'** (illus. p.343). Grandiflorus group, small-flowered gladiolus. H 1.2m (4ft), S 15–20cm (6–8in). Produces 6 ladder-placed, salmon-rose flowers, with longitudinal, spear-like, red-orange marks in throats, in late summer. Is excellent in flower arrangements.

G. **'Mexicali Rose'** (illus. p.343). Grandiflorus group, large-flowered gladiolus. H 1.5m (5ft), S 30cm (1ft). Has 8 formally placed, ruffled, deep rose flowers, with very narrow, silvery-white petal margins, in late summer.

G. **'Miss America'** (illus. p.343). Grandiflorus group, medium-flowered gladiolus. H 1.5m (5ft), S 30cm (1ft). In late summer produces 10 formally placed, deep pink flowers that are heavily ruffled. Is excellent for exhibition.

G. **'Moon Mirage'** (illus. p.343). Grandiflorus group, giant-flowered gladiolus. H 1.7m (5½ft), S 25–35cm (10–14in). Produces 9–11 formally placed, moderately ruffled, pale canary-yellow flowers, with slightly darker lip petals, from late summer to early autumn.

G. nanus. See *G.* × *colvillei*.

G. natalensis. See *G. dalenii*.

G. papilio, syn. *G. purpureoauratus*, illus. p.352.

G. **'Parade'**. Grandiflorus group, giant-flowered gladiolus. H 1.7m (5½ft), S 25–35cm (10–14in). Produces 10 formally placed, finely ruffled, salmon-pink flowers, with small, cream throats, in early autumn. Is superb for exhibition.

G. **'Peter Pears'** (illus. p.343). Grandiflorus group, large-flowered gladiolus. H 1.7m (5½ft), S 35cm (14in). In late summer bears 10 formally placed, apricot-salmon flowers, which have red throat marks. Is excellent for exhibition and in flower arrangements.

G. **'Pink Lady'** (illus. p.343). Grandiflorus group, large-flowered gladiolus. H 1.5m (5ft), S 25–30cm (10–12in). Has 9 formally placed, lightly ruffled, deep rose-pink flowers, with large, white throats, in late summer and early autumn.

G. **'Pink Slipper'**. Grandiflorus group, medium-flowered gladiolus. H 1.35m (4½ft), S 20–25cm (8–10in). Produces 9 or 10 formally placed, very ruffled, pink flowers, with white throats, in late summer. Is a fine exhibition cultivar.

G. primulinus. Summer-flowering corm, closely related to *G. natalensis* and sometimes included in it. H to 1m (3ft), S 15–20cm (6–8in). Half hardy. Bears up to 20 soft yellow flowers, each 8–12cm (3–5in) long with a hooded, upper petal.

G. psittacinus. See *G. dalenii*.

G. purpureoauratus. See *G. papilio*.

G. **'Rajah'**. Grandiflorus group, medium-flowered gladiolus. H 1.5m (5ft), S 20–25cm (8–10in). Produces 10 formally placed, rich plum-purple flowers, each with slightly darker lips and white midrib lines, from late summer to early autumn. Petals are pointed and reflexed. Is excellent for exhibition.

G. **'Renegade'** (illus. p.343). Grandiflorus group, large-flowered gladiolus. H 1.5m (5ft), S 20–25cm (8–10in). Bears 7 or 8 formally placed, crisply ruffled, dark red flowers in late summer.

G. **'Robin'** (illus. p.343). Primulinus group gladiolus. H 1.2m (4ft), S 20cm (8in). Produces 5 or 6 ladder-placed, rose-purple flowers, each with a hooded, central, upper petal, in late summer. Is good in flower arrangements.

G. **'Rose Supreme'** (illus. p.343). Grandiflorus group, giant-flowered gladiolus. H 1.7m (5½ft), S 25–30cm (10–12in). Has 9 formally placed, rose-pink flowers, flecked and streaked darker pink towards petal tips, and with cream throats, in late summer.

G. **'Royal Dutch'**. Grandiflorus group, large-flowered gladiolus. H 1.7m (5½ft), S 25–30cm (10–12in). Produces 10 formally placed flowers, each pale lavender blending into a white throat, from late summer to early autumn. Is very good for exhibition.

G. **'Rutherford'** (illus. p.343). Primulinus group gladiolus. H 1m (3ft), S 15–20cm (6–8in). Bears 6 ladder-placed, dark red flowers, splashed with cream, each with a hooded, central, upper petal, in late summer.

G. segetum. See *G. italicus*.

G. **'Shari'**. Grandiflorus group, small-flowered gladiolus. H 1.2m (4ft), S 15–20cm (6–8in). Produces 7 or 8 formally placed, ruffled, dark rose-pink flowers, with very large, cream throats, in late summer.

G. **'Tendresse'**. Grandiflorus group, medium-flowered gladiolus. H 1.5m (5ft), S 20–25cm (8–10in). In late summer has 10 formally placed, slightly ruffled, dark pink flowers, with small, cream throats marked with longitudinal, faint rose-pink 'spears'.

G. **'Tesoro'** (illus. p.343). Grandiflorus group, medium-flowered gladiolus. H 1.5m (5ft), S 20–25cm (8–10in). Bears 10 formally placed, silky flowers, slightly ruffled and glistening yellow, in early autumn. Is among the top exhibition gladioli.

G. **'Victor Borge'** (illus. p.343). Grandiflorus group, large-flowered gladiolus. H 1.7m (5½ft), S 35cm (14in). Produces 8 formally placed, vermilion-orange flowers, with pale cream throat marks, in late summer.

GLAUCIDIUM

G. 'Video'. Grandiflorus group, large-flowered gladiolus. H 1.7m (5½ft), S 20–30cm (8–12in). Has 9 or 10 formally placed, purple flowers, each with a small, ivory-white throat, in late summer. Is good for exhibition.
G. 'White City'. Primulinus group gladiolus. H 1.1m (3½ft), S 15–20cm (6–8in). Bears 6 or 7 ladder-placed, paper-white flowers, each with a hooded, central, upper petal, in late summer.

GLAUCIDIUM (Paeoniaceae)
Genus of one species of spring-flowering perennial. Is excellent in woodland gardens. Fully hardy. Needs a partially shaded, sheltered position and moist, peaty soil. Propagate by seed in autumn.
♀ *G. palmatum* illus. p.234.

GLAUCIUM (Papaveraceae)
Horned poppy
Genus of annuals, biennials and perennials, grown for their bright, poppy-like flowers. Fully hardy. Grow in sun and in fertile, well-drained soil. Propagate annuals by seed sown outdoors in spring; perennials by seed sown outdoors in spring or autumn; biennials by seed sown under glass in late spring or early summer.
G. flavum illus. p.287.

GLECHOMA (Labiatae)
Genus of evergreen, summer-flowering perennials. Makes good ground cover, but may be invasive. Fully hardy. Tolerates sun or shade. Prefers moist but well-drained soil. Propagate by division in spring or autumn or by softwood cuttings in spring.
G. hederacea (Ground ivy). **'Variegata'** illus. p.264.

GLEDITSIA (Leguminosae)
Genus of deciduous, usually spiny trees, grown for their foliage. Has inconspicuous flowers, often followed by large seed pods after hot summers. Fully hardy, but young plants may suffer frost damage. Requires plenty of sun and fertile, well-drained soil. Propagate species by seed in autumn, selected forms by budding in late summer.
G. caspica (Caspian locust). Deciduous, spreading tree. H 12m (40ft), S 10m (30ft). Trunk is armed with long, branched spines. Has fern-like, glossy, mid-green leaves.
G. japonica illus. p.54.
G. triacanthos (Honey locust). Deciduous, spreading tree. H 20m (70ft), S 15m (50ft). Trunk is very thorny. Fern-like, glossy, dark green leaves turn yellow in autumn. f. *inermis* is thornless. **'Shademaster'** is vigorous, with long-lasting leaves. **'Skyline'** is thornless, broadly conical and has golden-yellow foliage in autumn. ♀ **'Sunburst'** illus. p.51.

GLOBBA (Zingiberaceae)
Genus of evergreen, aromatic, clump-forming perennials, grown for their flowers. Frost tender, min. 18°C (64°F). Requires high humidity, partial shade and humus-rich, well-drained soil. Plants should remain dry when dormant in winter. Propagate by division or seed in spring or by mature bulbils that fall off plants.
G. winitii illus. p.231.

GLOBULARIA (Globulariaceae)
Genus of mainly evergreen, summer-flowering shrubs and sub-shrubs, grown for their dome-shaped hummocks and usually blue or purple flower heads. Fully to frost hardy. Needs full sun and well-drained soil. Propagate by division in spring, by softwood or semi-ripe cuttings in summer or by seed in autumn.
♀ *G. cordifolia* illus. p.331.
subsp. *bellidifolia* see *G. meridionalis*.
G. meridionalis, syn. *G. cordifolia* subsp. *bellidifolia*, illus. p.328.

GLORIOSA (Liliaceae)
Genus of deciduous, summer-flowering, tendril climbers with finger-like tubers. Frost tender, min. 8–10°C (46–50°F). Needs full sun and rich, well-drained soil. Water freely in summer and liquid feed every 2 weeks. Provide support. Dry off tubers in winter and keep cool but frost-free. Propagate by seed or division in spring.
G. superba (Glory lily). Deciduous, tendril climber with tubers. H to 2m (6ft), S 30–45cm (1–1½ft). Min. 8°C (46°F). Slender stems produce scattered, broadly lance-shaped leaves. In summer, upper leaf axils carry large, yellow or red flowers, each with 6 sharply reflexed, wavy-edged petals, changing to dark orange or deep red. Stamens protrude prominently.
♀ **'Rothschildiana'** illus. p.345.

GLOTTIPHYLLUM (Aizoaceae)
Genus of clump-forming, perennial succulents with semi-cylindrical leaves often broadened at tips. Frost tender, min. 5°C (41°F). Grow in very poor, well-drained soil and in full sun to make tight, compact plants. Propagate by seed or stem cuttings in spring or summer.
G. nelii illus. p.415.
G. semicylindricum. Clump-forming, perennial succulent. H 8cm (3in), S 30cm (12in) or more. Has semi-cylindrical, bright green leaves, 6cm (2½in) long, with a tooth half-way along each margin. Has daisy-like, golden-yellow flowers, 4cm (1½in) across, on short stems in spring-summer.

GLOXINIA (Gesneriaceae)
Genus of late summer- to autumn-flowering, rhizomatous perennials. Frost tender, min. 10°C (50°F). Needs partial shade and humus-rich, well-drained soil. Dies down in late autumn or winter; then keep rhizomes nearly dry. Propagate by division or seed in spring or by stem or leaf cuttings in summer.
G. perennis illus. p.365.
G. speciosa. See *Sinningia speciosa*.

GLYCERIA (Gramineae). See GRASSES, BAMBOOS, RUSHES and SEDGES.
G. aquatica **'Variegata'**. See *G. maxima* 'Variegata'.
G. maxima **'Variegata'**, syn. *G. aquatica* 'Variegata', illus. p.182.

GLYCYRRHIZA (Leguminosae)
Liquorice
Genus of summer-flowering perennials. Fully hardy. Needs sun and deep, rich, well-drained soil. Propagate by division in spring or by seed in autumn or spring.
G. glabra illus. p.214.

GODETIA. See CLARKIA.

GOMESA. See ORCHIDS.
G. planifolia (illus. p.262). Evergreen, epiphytic orchid for a cool greenhouse. H 23cm (9in). Sprays of star-shaped, pea-green flowers, 0.5cm (¼in) across, are produced in autumn. Narrowly oval leaves are 15cm (6in) long. Grow in semi-shade during summer.

Gomphocarpus physocarpus. See *Asclepias physocarpa*.

GOMPHRENA (Amaranthaceae)
Genus of annuals, biennials and perennials. Only one species, *G. globosa*, is usually cultivated; its flower heads are good for cutting and drying. Half hardy. Grows best in sun and in fertile, well-drained soil. Propagate by seed sown under glass in spring.
G. globosa illus. p.283.

GONGORA. See ORCHIDS.
G. quinquenervis (illus. p.262). Evergreen, epiphytic orchid for an intermediate greenhouse. H 25cm (10in). In summer, fragrant, brown, orange and yellow flowers, 1cm (½in) across, which resemble birds in flight, are borne in long, pendent spikes. Has oval, ribbed leaves, 12–15cm (5–6in) long. Is best grown in a hanging basket. Requires semi-shade in summer.

GORDONIA (Theaceae)
Genus of evergreen shrubs and trees, grown for their flowers and overall appearance. Half hardy, but best at min. 3°C (37°F). Prefers humus-rich, acid soil and sun or partial shade. Water potted plants moderately, less in winter. Propagate by semi-ripe cuttings in late summer or by seed when ripe, in autumn, or in spring.
G. axillaris. Evergreen, bushy shrub or tree. H and S 3–5m (10–15ft), sometimes much more. Bears lance-shaped, leathery, glossy leaves, each with a blunt tip, and, from autumn to spring, saucer-shaped, white flowers.
G. lasianthus (Loblolly bay). Evergreen, upright tree. H to 20m (70ft), S to 10m (30ft). Lance-shaped to elliptic leaves are shallowly serrated. Has fragrant, saucer- to bowl-shaped, white flowers in summer. Needs sub-tropical summer warmth to grow and flower well.

GRAPTOPETALUM (Crassulaceae)
Genus of rosetted, perennial succulents very similar to *Echeveria*, with which it hybridizes. Frost tender, min. 5–10°C (41–50°F). Is easy to grow and flower, needing sun or partial shade and very well-drained soil. Propagate by seed or by stem or leaf cuttings in spring or summer.
G. amethystinum. Clump-forming, prostrate, perennial succulent. H 40cm (16in), S 1m (3ft). Min. 10°C (50°F). Produces thick, rounded, blue-grey to red leaves, 7cm (3in) long, in terminal rosettes and star-shaped, yellow-and-red flowers, 1–2cm (½–¾in) across, in spring-summer.
G. bellum, syn. *Tacitus bellus*, illus. p.409.
G. paraguayense illus. p.414.

GRAPTOPHYLLUM (Acanthaceae)
Genus of evergreen shrubs, grown mainly for their foliage. Frost tender, min. 16–18°C (61–4°F). Needs partial shade and fertile, well-drained soil. Water potted plants freely when in full growth, much less when temperatures are low. Young plants need tip pruning after flowering to promote branching; leggy specimens may be cut back hard after flowering or in spring. Propagate by greenwood or semi-ripe cuttings in spring or summer.
G. pictum (Caricature plant). Evergreen, erect, loose shrub. H to 2m (6ft), S 60cm (2ft) or more. Oval, pointed, glossy, green leaves have central, yellow blotches. Short, terminal spikes of tubular, red to purple flowers appear mainly in spring and early summer.

GRASSES, BAMBOOS, RUSHES and SEDGES
Group of evergreen or herbaceous, perennial and annual grasses or grass-like plants belonging to the Gramineae (including Bambusoideae), Juncaceae and Cyperaceae families. They are grown mainly as foliage plants, adding grace and contrast to borders and rock gardens, although several grasses have attractive flower heads in summer that may be dried for winter decoration. Dead foliage may be cut back on herbaceous perennials, when dormant. Propagate species by seed in spring or autumn or by division in spring, selected forms by division only. Pests and diseases generally give little trouble.

Grasses (Gramineae)
Family of evergreen, semi-evergreen or herbaceous, sometimes creeping perennials, annuals and marginal water plants, usually with rhizomes or stolons, that form tufts, clumps or carpets. All have basal leaves and rounded flower stems that bear alternate, long, narrow leaves. Flowers are bisexual (males and females in same spikelet) and are arranged in panicles, racemes or spikes. Each flower head comprises spikelets, with one or more florets, that are covered with glumes (scales) from which awns (long, slender bristles) may grow. Fully hardy to frost tender, min. 5–12°C (41–54°F). Unless otherwise stated, grasses will tolerate a range of light conditions and flourish in any well-drained soil. Many genera, such as *Briza*, self seed readily.
See also *Alopecurus, Arrhenatherum, Arundo, Bouteloua, Briza, Bromus, Chionochloa, Coix, Cortaderia, Dactylis, Deschampsia,*

Festuca, Glyceria, Hakonechloa, Helictotrichon, Holcus, Hordeum, Lagurus, Lamarckia, Leymus, Melica, Milium, Miscanthus, Molinia, Oplismenus, Panicum, Pennisetum, Phalaris, Rhynchelytrum, Sesleria, Setaria, Spartina, Stenotaphrum, Stipa, Zea and *Zizania*.

Bamboos (Bambusoideae)
Sub-family of Gramineae, comprising evergreen, rhizomatous perennials, sometimes grown as hedging as well as for ornamentation. Most bamboos differ from other perennial grasses in that they have woody stems (culms). These are hollow (except in *Chusquea*), mostly greenish-brown and, due to their silica content, very strong with a circumference of up to 15cm (6in) in some tropical species. Leaves are lance-shaped with cross veins that give a tessellated appearance, which may be obscured in the more tender species. Flowers are produced at varying intervals but are not decorative. After flowering, stems die down but few plants die completely. Fully to half hardy. Bamboos thrive in a sheltered, not too dry situation in sun or shade, unless stated otherwise.

See also *Bambusa, Chusquea, Drepanostachyum, Fargesia, Phyllostachys, Pleioblastus, Pseudosasa, Sasa, Semiarundinaria, Shibataea* and *Yushania*.

Rushes (Juncaceae)
Family of evergreen, tuft-forming or creeping, mostly rhizomatous annuals and perennials. All have either rounded, leafless stems or stems bearing long, narrow, basal leaves that are flat and hairless except *Luzula* (the woodrushes) which have flat leaves, edged with white hairs. Rounded flower heads are generally unimportant. Fully to half hardy. Most rushes prefer sun or partial shade and a moist or wet situation, but *Luzula* prefers drier conditions.

See also *Juncus* and *Luzula*.

Sedges (Cyperaceae)
Family of evergreen, rhizomatous perennials that form dense tufts. Typically, stems are triangular and bear long, narrow leaves, sometimes reduced to scales. Spikes or panicles of florets covered with glumes are produced and contain both male and female flowers, although some species of *Carex* have separate male and female flower heads on the same stem. Fully hardy to frost tender, min. 4–7°C (39–45°F). Grow in sun or partial shade. Some sedges grow naturally in water, but many may be grown in any well-drained soil.

See also *Carex, Cyperus, Eleocharis, Schoenoplectus* and *Scirpoides*.

GREVILLEA (Proteaceae)
Genus of evergreen shrubs and trees, grown for their flowers and foliage. Half hardy to frost tender, min. 5–10°C (41–50°F). Grow in well-drained, preferably acid soil and in full sun. Water containerized specimens moderately, very little in winter. Pruning is tolerated if necessary. Propagate by seed in spring or by semi-ripe cuttings in summer.
G. alpestris. See *G. alpina*.
G. alpina, syn. *G. alpestris*. Evergreen, rounded, wiry-stemmed shrub. H and S 30–60cm (1–2ft). Half hardy. Has narrowly oblong or oval leaves, dark green above, silky-haired beneath. Bears tubular, red flowers in small clusters in spring-summer.
G. banksii illus. p.73.
❀ *G. juniperina* f. *sulphurea*, syn. *G. sulphurea*, illus. p.139.
G. 'Poorinda Constance'. Evergreen, bushy, rounded shrub. H and S to 2m (6ft). Half hardy. Has small, lance-shaped, mid- to deep green leaves with prickly toothed margins. Tubular, bright red flowers in conspicuous clusters are borne from spring to autumn, sometimes longer.
❀ *G. robusta* (Silky oak). Fast-growing, evergreen, upright to conical tree. H 30m (100ft), S to 15m (50ft). Frost tender, min. 5°C (41°F). Fern-like leaves are 15–25cm (6–10in) long. Mature specimens bear upturned bell-shaped, bright yellow or orange flowers in dense, one-sided spikes, 10cm (4in) or more long, in spring-summer.
G. 'Robyn Gordon' illus. p.136.
❀ *G. rosmarinifolia*. Evergreen, rounded, well-branched shrub. H and S to 2m (6ft). Half hardy. Dark green leaves are needle-shaped with reflexed margins, silky-haired beneath. Has short, dense clusters of tubular, red, occasionally pink or white flowers in summer.
G. sulphurea. See *G. juniperina* f. *sulphurea*.

GREYIA (Greyiaceae)
Genus of evergreen, semi-evergreen or deciduous, spring-flowering shrubs and trees, grown for their flowers and overall appearance. Frost tender, min. 7–10°C (45–50°F). Needs full light and well-drained soil. Water containerized specimens moderately, less when not in full growth. Remove or shorten flowered stems after flowering. Propagate by seed in spring or by semi-ripe cuttings in summer. Plants grown under glass need plenty of ventilation.
G. sutherlandii illus. p.101.

GRINDELIA (Compositae)
Genus of annuals, biennials, evergreen perennials and sub-shrubs, grown for their flower heads. Frost to half hardy, but in cold areas grow in a warm, sheltered site. Requires sun and well-drained soil. Water potted specimens moderately, less when not in full growth. Remove spent flowering stems either as they die or in following spring. Propagate by seed in spring or by semi-ripe cuttings in late summer.
G. chiloensis, syn. *G. speciosa*, illus. p.141.
G. speciosa. See *G. chiloensis*.

GRISELINIA (Cornaceae)
Genus of evergreen shrubs and trees, with inconspicuous flowers, grown for their foliage. Thrives in mild, coastal areas where it is effective as a hedge or windbreak as it is very wind- and salt-resistant. Frost to half hardy; in cold areas provide shelter. Requires sun and fertile, well-drained soil. Restrict growth and trim hedges in early summer. Propagate by semi-ripe cuttings in summer.
❀ *G. littoralis* (Broadleaf). Fast-growing, evergreen, upright shrub of dense habit. H 6m (20ft), S 5m (15ft). Frost hardy. Bears oval, leathery leaves that are bright apple-green. Tiny, inconspicuous, yellow-green flowers are borne in late spring. 'Dixon's Cream', H 3m (10ft), S 2m (6ft), is slower-growing and has central, creamy-white leaf variegation. 'Variegata' illus. p.96.
G. lucida. Fast-growing, evergreen, upright shrub. H 6m (20ft), S 5m (15ft). Half hardy. Is similar to *G. littoralis*, but has larger, glossy, dark green leaves.

GUNNERA (Gunneraceae)
Genus of summer-flowering perennials, grown mainly for their foliage. Some are clump-forming with very large leaves; others are mat-forming with smaller leaves. Frost hardy, but shelter from wind in summer and cover with bracken or compost in winter. Requires sun and moist soil. Propagate by seed in autumn or spring; small species by division in spring.
G. chilensis. See *G. tinctoria*.
G. magellanica illus. p.322.
❀ *G. manicata* illus. p.192.
G. tinctoria, syn. *G. chilensis*. Robust, rounded, clump-forming perennial. H and S 1.5m (5ft) or more. Has very large, rounded, puckered and lobed leaves, 45–60cm (1½–2ft) across. In early summer produces dense, conical clusters of tiny, dull reddish-green flowers.

GUZMANIA (Bromeliaceae)
Genus of evergreen, rosette-forming, epiphytic perennials, grown for their overall appearance. Frost tender, min. 10–15°C (50–59°F). Needs semi-shade and a rooting medium of equal parts humus-rich soil and either sphagnum moss, or bark or plastic chips used for orchid culture. Using soft water, water moderately during growing season, sparingly at other times, and keep rosette centres filled with water from spring to autumn. Propagate by offsets in spring or summer.
❀ *G. lingulata* (illus. p.229). Evergreen, basal-rosetted, epiphytic perennial. H and S 30–45cm (12–18in). Forms loose rosettes of broadly strap-shaped, arching, mid-green leaves. Bears a cluster of tubular, white to yellow flowers, surrounded by a rosette of bright red bracts, usually in summer.
❀ var. *minor* (illus. p.229). H and S 15cm (6in), has yellow-green leaves and red or yellow bracts.
❀ *G. monostachia*, syn. *G. monostachya, G. tricolor* (Striped torch; illus. p.229). Evergreen, basal-rosetted, epiphytic perennial. H and S 30–40cm (12–16in). Has dense rosettes of strap-shaped, erect to arching, pale to yellowish-green leaves. Usually in summer, elongated spikes of tubular, white flowers emerge from axils of oval bracts, the upper ones red, the lower green with purple-brown stripes.
G. monostachya. See *G. monostachia*.
❀ *G. sanguinea*. Evergreen, basal-rosetted, epiphytic perennial. H 20cm (8in), S 30–35cm (12–14in). Has dense, slightly flat rosettes of broadly strap-shaped, arching, mid- to deep green leaves. In summer, a compact cluster of tubular, yellow flowers, surrounded by red bracts, appears at the heart of each mature rosette.
G. tricolor. See *G. monostachia*.
G. vittata. Evergreen, basal-rosetted, epiphytic perennial. H and S 35–60cm (14–24in). Produces fairly loose rosettes of strap-shaped, erect, dark green leaves with pale green cross-bands and recurved tips. Stem bears a compact, egg-shaped head of small, tubular, white flowers in summer.

GYMNOCALYCIUM (Cactaceae)
Genus of perennial cacti with masses of funnel-shaped flowers in spring-summer. Crowns generally bear smooth, scaly buds. Frost tender, min. 5–10°C (41–50°F). Needs full sun or partial shade and very well-drained soil. Propagate by seed or offsets in spring or summer.
❀ *G. andreae* illus. p.413.
G. gibbosum illus. p.397.
G. mihanovichii 'Red Head', syn. *G. m.* var. *hibotan*, illus. p.400.
❀ *G. quehlianum*. Flattened spherical, perennial cactus. H 5cm (2in), S 7cm (3in). Min. 5°C (41°F). Grey-blue to brown stem has 11 or so rounded ribs. Areoles each produce 5 curved spines. Has white flowers, 5cm (2in) across, with red throats, in spring-summer. Is easy to flower.
G. schickendantzii. Spherical, perennial cactus. H and S 10cm (4in). Min. 5°C (41°F). Dark green stem has 7–14 deeply indented ribs and long, red-tipped, grey-brown spines. Bears greenish-white to pale pink flowers, 5cm (2in) across, in summer.

GYMNOCLADUS (Leguminosae)
Genus of deciduous trees, grown for their foliage. Fully hardy. Needs full sun and deep, fertile, well-drained soil. Propagate by seed in autumn.
G. dioica (Kentucky coffee tree). Slow-growing, deciduous, spreading tree. H 20m (70ft), S 15m (50ft). Very large leaves, with 4–7 pairs of oval leaflets, are pinkish when young, dark green in summer, then yellow in autumn. Small, star-shaped, white flowers are borne in early summer.

GYNANDRIRIS (Iridaceae)
Genus of iris-like, spring-flowering corms, grown mainly for botanical interest, with very short-lived blooms. Frost hardy. Dormant corms require warmth and dryness, so plant in a sunny site that dries out in summer. Needs well-drained soil, but with plenty of moisture in winter-spring. Propagate by seed or cormlets in autumn.
G. sisyrinchium illus. p.376.

GYNURA (Compositae)
Genus of evergreen perennials, shrubs and semi-scrambling climbers, grown

505

for their ornamental foliage or flower heads. Frost tender, min. 16°C (61°F). Any fertile, well-drained soil is suitable, with light shade in summer. Water moderately throughout the year, less in cool conditions; do not overwater. Provide support for stems. Remove stem tips to encourage branching. Propagate by softwood or semi-ripe cuttings in spring or summer.

G. aurantiaca illus. p.180. ♛ **'Purple Passion'**, syn. *G. sarmentosa* of gardens. Evergreen, erect, woody-based, soft-stemmed shrub or semi-scrambling climber. H 60cm (2ft) or more. Stems and lance-shaped, lobed, serrated leaves are covered with velvety, purple hairs. Leaves are purple-green above, deep red-purple beneath. In winter bears clusters of daisy-like, orange-yellow flower heads that become purplish with age.
G. sarmentosa of gardens. See *G. aurantiaca* 'Purple Passion'.

GYPSOPHILA (Caryophyllaceae) Genus of spring- to autumn-flowering annuals and perennials, some of which are semi-evergreen. Fully hardy. Needs sun. Will grow in dry, sandy and stony soils, but does best in deep, well-drained soil. Resents being disturbed. Cut back after flowering for a second flush of flowers. Propagate *G. paniculata* cultivars by grafting in winter; others by softwood cuttings in summer or by seed in autumn or spring.

G. cerastioides illus. p.312.
G. elegans illus. p.270.
♛ *G. paniculata* **'Bristol Fairy'** illus. p.203. **'Flamingo'** is a spreading, short-lived perennial. H 60–75cm (2–2½ft), S 1m (3ft). In summer bears panicles of numerous, small, rounded, double, pale pink flowers on wiry, branching stems. Has small, linear, mid-green leaves.
♛ *G. repens*. Semi-evergreen, prostrate perennial with much-branched rhizomes. H 2.5–5cm (1–2in) or more, S 30cm (12in) or more. In summer has sprays of small, rounded, white, lilac or pink flowers on slender stems that bear narrowly oval, bluish-green leaves. Is excellent for a rock garden, wall or dry bank. May also be propagated by division in spring.
♛ **'Dorothy Teacher'** illus. p.323.

H

HAAGEOCEREUS (Cactaceae)
Genus of perennial cacti with ribbed, densely spiny, columnar, green stems branching from the base. Frost tender, min. 11°C (52°F). Needs full sun and very well-drained soil. Propagate by seed or stem cuttings in spring or summer.
H. chosicensis. Upright, perennial cactus. H 1.5m (5ft), S 1m (3ft). Green stem, 10cm (4in) across, with 19 or so ribs, bears white, golden or red, central spines and shorter, dense, bristle-like, white, radial ones. Has tubular, white, lilac-white or pinkish-red flowers, 7cm (3in) long, near crown in summer.
H. decumbens. Prostrate, perennial cactus. H 30cm (1ft), S 1m (3ft). Stems, 6cm (2½in) across, with 20 or so ribs, have dark brown, central spines, 5cm (2in) long, and shorter, dense, golden, radial spines. Tubular, white flowers, 8cm (3in) across, form in summer near crowns, only on mature plants.
H. versicolor illus. p.394.

HABERLEA (Gesneraceae)
Genus of evergreen, rosetted perennials, grown for their elegant sprays of flowers. Is useful on walls. Fully hardy. Needs partially shaded, moist soil. Resents disturbance to roots. Propagate by seed in spring or by leaf cuttings or offsets in early summer.
H. ferdinandi-coburgii. Evergreen, dense, basal-rosetted perennial. H 10–15cm (4–6in), S 30cm (12in). Has oblong, toothed, dark green leaves, hairy below, almost glabrous above. Sprays of funnel-shaped, blue-violet flowers, each with a white throat, appear on long stems in late spring and early summer.
♀ *H. rhodopensis.* Evergreen, dense, basal-rosetted perennial. H 10cm (4in), S 15cm (6in) or more. Is similar to *H. ferdinandi-coburgii*, but leaves are soft-haired on both surfaces. **'Virginalis'** illus. p.321.

HABRANTHUS (Amaryllidaceae)
Genus of summer- and autumn-flowering bulbs with funnel-shaped flowers. Frost to half hardy. Needs a sheltered, sunny site and fertile soil that is well supplied with moisture in growing season. Propagate by seed or offsets in spring.
H. andersonii. See *H. tubispathus.*
H. brachyandrus. Summer-flowering bulb. H to 30cm (12in), S 5–8cm (2–3in). Half hardy. Long, linear, semi-erect, narrow leaves form a basal cluster. Each flower stem bears a semi-erect, widely funnel-shaped, pinkish-red flower, 7–10cm (3–4in) long.
H. robustus illus. p.381.
H. tubispathus, syn. *H. andersonii.* Summer-flowering bulb. H to 15cm (6in), S 5cm (2in). Frost hardy. Has linear, semi-erect, basal leaves and a succession of flower stems each bearing solitary 2.5–3.5cm (1–1½in) long, funnel-shaped flowers, yellow inside, copper-red outside.

HACQUETIA, syn. DONDIA (Umbelliferae)
Genus of one species of clump-forming, rhizomatous perennial that creeps slowly, grown for its yellow or yellow-green flower heads borne on leafless plants in late winter and early spring. Is good in rock gardens. Fully hardy. Prefers shade and humus-rich, moist soil. Resents disturbance to roots. Propagate by division in spring, by seed when fresh in autumn or by root cuttings in winter.
♀ *H. epipactis* illus. p.319.

HAEMANTHUS (Amaryllidaceae)
Genus of summer-flowering bulbs with dense heads of small, star-shaped flowers, often brightly coloured. Frost tender, min. 10°C (50°F). Prefers full sun or partial shade and well-drained soil or sandy compost. Liquid feed in the growing season. Leave undisturbed as long as possible before replanting. Propagate by offsets or seed before growth commences in early spring.
H. albiflos (Paintbrush). Summer-flowering bulb. H 5–30cm (2–12in), S 20–30cm (8–12in). Has 2–6 almost prostrate, broadly elliptic leaves with hairy edges. Flower stem, appearing between leaves, bears a brush-like head of up to 50 white flowers with very narrow petals and protruding stamens.
H. coccineus illus. p.379.
H. katherinae. See *Scadoxus multiflorus* subsp. *katherinae.*
H. magnificus. See *Scadoxus puniceus.*
H. multiflorus. See *Scadoxus multiflorus.*
H. natalensis. See *Scadoxus puniceus.*
H. puniceus. See *Scadoxus puniceus.*
H. sanguineus. Summer-flowering bulb. H to 30cm (12in), S 20–30cm (8–12in). Bears 2 prostrate, elliptic, rough, dark green leaves, hairy beneath. Brownish-purple-spotted, green flower stem, forming before leaves, bears a dense head of small, narrow-petalled, red flowers, surrounded by whorls of narrow, leaf-like, red or pink bracts.

HAKEA (Proteaceae)
Genus of evergreen shrubs and trees, grown for their often needle-like leaves and their flowers. Is very wind-resistant, except in cold areas. Frost hardy to frost tender, min. 5–7°C (41–5°F). Requires full sun and fertile, well-drained soil. Water containerized specimens moderately when in full growth, sparingly in winter. Propagate by semi-ripe cuttings in summer or by seed in autumn.
H. lissosperma, syn. *H. sericea* of gardens. Evergreen, upright, densely branched shrub of pine-like appearance. H 5m (15ft), S 3m (10ft). Frost hardy. Has long, slender, sharply pointed, grey-green leaves and, in late spring and early summer, clusters of small, spidery, white flowers.
H. sericea of gardens. See *H. lissosperma.*
H. suaveolens. Evergreen, rounded shrub. H and S 2m (6ft) or more. Frost tender. Leaves are divided into cylindrical, needle-like leaflets or occasionally are undivided and lance-shaped. Small, fragrant, tubular, white flowers are carried in short, dense clusters from summer to winter.

HAKONECHLOA (Gramineae). See GRASSES, BAMBOOS, RUSHES and SEDGES.
♀ *H. macra* **'Aureola'** illus. p.185.

HALESIA (Styracaceae)
Silver bell, Snowdrop tree
Genus of deciduous, spring-flowering trees and shrubs, grown for their showy, pendent, bell-shaped flowers and curious, winged fruits. Fully hardy, but needs a sunny, sheltered position. Prefers moist but well-drained, neutral to acid soil. Propagate by softwood cuttings in summer or by seed in autumn.
H. carolina. See *H. tetraptera.*
♀ *H. monticola* illus. p.50.
H. tetraptera, syn. *H. carolina.* Deciduous, spreading tree or shrub. H 8m (25ft), S 10m (30ft). Masses of bell-shaped, white flowers hang from bare shoots in late spring and are followed by 4-winged, green fruits. Oval leaves are mid-green.

x HALIMIOCISTUS (Cistaceae)
Hybrid genus (*Cistus* x *Halimium*) of evergreen shrubs, grown for their flowers. Frost hardy, but in cold areas needs shelter. Requires full sun and well-drained soil. Propagate by semi-ripe cuttings in summer.
♀ x *H. sahucii* illus. p.129.
♀ x *H. wintonensis.* Evergreen, bushy shrub. H 60cm (2ft), S 1m (3ft). Saucer-shaped, white flowers, each with deep red bands and a yellow centre, open amid lance-shaped, grey-green leaves in late spring and early summer.

HALIMIUM (Cistaceae)
Genus of evergreen shrubs, grown for their showy flowers. Is good for coastal gardens. Frost hardy, but in cold areas needs shelter. Does best in full sun and light, well-drained soil. Propagate by semi-ripe cuttings in summer.
♀ *H. lasianthum.* Evergreen, bushy, spreading shrub. H 1m (3ft), S 1.5m (5ft). Leaves are oval and grey-green. In late spring and early summer bears saucer-shaped, golden-yellow flowers, sometimes with small, central, red blotches. subsp. *formosum* illus. p.139.
♀ *H. ocymoides.* Evergreen, bushy shrub. H 60cm (2ft), S 1m (3ft). Narrowly oval leaves, covered in white hairs when young, mature to dark green. In early summer produces upright clusters of saucer-shaped, golden-yellow flowers, conspicuously blotched with black or purple.
♀ **'Susan'** illus. p.139.
H. umbellatum illus. p.129.

HAMAMELIS (Hamamelidaceae)
Witch hazel
Genus of deciduous, autumn- to early spring-flowering shrubs, grown for their autumn colour and fragrant, frost-resistant flowers, each with 4 narrowly strap-shaped petals. Fully hardy. Flourishes in sun or semi-shade and fertile, well-drained, peaty, acid soil, although tolerates good, deep soil over chalk. Propagate species by seed in autumn, selected forms by softwood cuttings in summer, by budding in late summer or by grafting in winter.
♀ *H.* x *intermedia* **'Arnold Promise'** illus. p.96. ♀ **'Diane'** illus. p.95.
♀ **'Jelena'** is a deciduous, upright shrub. H and S 4m (12ft) or more. Broadly oval, glossy, bright green leaves turn bright orange or red in autumn. Bears masses of large, fragrant, orange flowers, along bare branches, from early to mid-winter.
H. japonica (Japanese witch hazel). Deciduous, upright, open shrub. H and S 4m (12ft). Broadly oval, glossy, mid-green leaves turn yellow in autumn. Fragrant, yellow flowers, with crinkled petals, are borne on bare branches from mid- to late winter. **'Sulphurea'** illus. p.96. **'Zuccariniana'** bears paler, lemon-yellow flowers in early spring and has orange-yellow leaves in autumn.
♀ *H. mollis* (Chinese witch hazel). Deciduous, upright, open shrub. H and S 4m (12ft) or more. Broadly oval, mid-green leaves turn yellow in autumn. Produces extremely fragrant, yellow flowers, along bare branches, in mid- and late winter. **'Coombe Wood'** illus. p.96. ♀ **'Pallida'**, S 3m (10ft), is more upright than the species and bears dense clusters of large, sulphur-yellow flowers.
♀ *H. vernalis* **'Sandra'** illus. p.94.
H. virginiana illus. p.95.

Hamatocactus hamatacanthus. See *Ferocactus hamatacanthus.*
Hamatocactus setispinus. See *Ferocactus setispinus.*

HARDENBERGIA (Leguminosae)
Genus of evergreen, woody-stemmed, twining climbers or sub-shrubs, grown for their curtains of leaves and racemes of pea-like flowers. Half hardy to frost tender, min. 7°C (45°F). Grows best in sun and in well-drained soil that does not dry out. Propagate by stem cuttings in late summer or autumn or by seed (soaked before sowing) in spring.
♀ *H. comptoniana* illus. p.166.
H. monophylla. See *H. violacea.*

HATIORA

❦ *H. violacea*, syn. *H. monophylla* (Australian sarsparilla, Coral pea, Vine lilac). Evergreen, woody-stemmed, twining climber. H to 3m (10ft). Frost tender. Narrowly oval leaves are 2.5–12cm (1–5in) long. Violet, occasionally pink or white, flowers, with yellow blotches on upper petals, are borne in spring. Brownish pods, 3–4cm (1¼–1½in) long, are produced in autumn.

HATIORA (Cactaceae)
Genus of perennial, epiphytic cacti with short, jointed, cylindrical stems each swollen at one end like a bottle. Frost tender, min. 10–11°C (50–52°F). Requires partial shade and very well-drained soil. Keep damp in summer; water a little in winter. Propagate by stem cuttings in spring or summer.

H. clavata, syn. *Rhipsalis clavata*. Pendent, perennial, epiphytic cactus. H 60cm (2ft), S 1m (3ft). Min. 11°C (52°F). Multi-branched, cylindrical, dark green stems each widen towards tips. Masses of terminal, bell-shaped, white flowers, 1.5cm (⅝in) wide, appear in late winter and early spring on plants over 30cm (1ft) high.

❦ *H. salicornioides* illus. p.404.

HAWORTHIA (Liliaceae)
Genus of basal-rosetted, clump-forming, perennial succulents with triangular to rounded, green leaves. Roots tend to wither in winter or during long periods of drought. Frost tender, min. 5–10°C (41–50°F). Needs partial shade to stay green and grow quickly; if planted in full sun turns red or orange and grows slowly. Requires very well-drained soil. Keep dry in winter. Propagate by seed or division from spring to autumn.

H. arachnoidea, syn. *H. setata*, illus. p.406.

H. attenuata. Clump-forming, perennial succulent. H 7cm (3in), S 25cm (10in). Min. 5°C (41°F). Bears a basal rosette of triangular, dark green leaves, 3cm (1¼in) long, covered in raised, white dots. Has tubular to bell-shaped, white flowers, on long, slender stems, from spring to autumn. var. *clariperla* illus. p.405.

H. cuspidata. Clump-forming, perennial succulent. H 5cm (2in), S 25cm (10in). Min. 5°C (41°F). Produces a basal rosette of smooth, rounded, fleshy, light green leaves with translucent marks. Tubular to bell-shaped, white flowers appear from spring to autumn on long, slender stems.

H. fasciata. Slow-growing, clump-forming, perennial succulent. H 15cm (6in), S 30cm (12in). Min. 5°C (41°F). Has raised, white dots, mostly in bands, on undersides of triangular, slightly incurved leaves, to 8cm (3in) long, which are arranged in a basal rosette. Bears tubular to bell-shaped, white flowers, on long, slender stems, from spring to autumn.

H. setata. See *H. arachnoidea*.
H. truncata illus. p.405.

HEATHERS (Ericaceae)
Heathers (otherwise known as heaths) are evergreen, woody-stemmed shrubs, grown for their flowers and foliage, both of which may provide colour in the garden all year round. There are 3 genera: *Calluna*, *Daboecia* and *Erica*. *Calluna* has only one species, *C. vulgaris*, but it contains a large number of cultivars that flower mainly from mid-summer to late autumn. *Daboecia* has 2 species, both of which are summer-flowering. The largest genus is *Erica*, which, although broadly divided into 2 groups – winter- and summer-flowering species – has some species also flowering in spring and autumn. They vary in height from tree heaths, which may grow to 6m (20ft), to dwarf, prostrate plants that, if planted 30–45cm (12–18in) apart, soon spread to form a thick mat of ground cover.

Heathers are fully hardy to frost tender, min. 5–7°C (41–5°F). They prefer an open, sunny position and require humus-rich, well-drained soil. *Calluna* and *Daboecia* dislike lime and must be grown in acid soil; some species of *Erica* are lime-tolerant but all are better grown in acid soils. Prune lightly after flowering each year to keep plants bushy and compact. Propagate species by seed in spring or by softwood cuttings, division or layering in summer. Seed cannot be relied on to come true. All cultivars should be vegetatively propagated. Heathers are illustrated on pp.148–9.

HEBE (Scrophulariaceae)
Genus of evergreen shrubs, grown for their often dense spikes, panicles or racemes of flowers and their foliage. Grows well in coastal areas. Smaller species and cultivars are good for rock gardens. Fully to half hardy. Requires full sun and well-drained soil. Growth may be restricted, or leggy plants tidied, by cutting back in spring. Propagate by semi-ripe cuttings in summer.

❦ *H. albicans* illus. p.131.
❦ *H. 'Alicia Amherst'*. Fast-growing, evergreen, upright shrub. H and S 1.2m (4ft). Frost hardy. Has large, oblong, glossy, dark green leaves and, in late summer and autumn, large spikes of small, 4-lobed, deep violet-purple flowers.

H. 'Andersonii Variegata'. Evergreen, bushy shrub. H and S 2m (6ft). Half hardy. Leaves are oblong and dark green, each with a grey-green centre and creamy-white margins. Has dense spikes of small, 4-lobed, lilac flowers from mid-summer to autumn.

H. 'Autumn Glory' illus. p.137.
H. brachysiphon. Evergreen, bushy, dense shrub. H and S 2m (6ft). Fully hardy. Has oblong, dark green leaves. Produces dense spikes of small, 4-lobed, white flowers in mid-summer. 'White Gem' illus. p.128.

H. buchananii. Evergreen, dome-shaped shrub. H and S 15cm (6in) or more. Frost hardy. Very dark stems bear oval, bluish-green leaves. In summer produces clusters of small, 4-lobed white flowers at stem tips. 'Minor', H 5–10cm (2–4in), has smaller leaves.

H. cantaburiensis. Evergreen, low growing, spreading shrub. H and S 30–90cm (1–3ft). Frost hardy. Small, oval, glossy, dark green leaves are densely packed on stems. In early summer, short racemes of small, white flowers are freely produced in leaf axils.

H. 'Carl Teschner'. Evergreen, prostrate then dome-shaped shrub. H 15cm (6in), S 30cm (12in) or more. Frost hardy. Blackish-brown stems are covered in small, oval, glossy, dark green leaves. Short racemes of tiny, 4-lobed, white-throated, purple flowers are produced in summer. Is excellent as a border plant.

H. carnosula. Evergreen, prostrate shrub. H 15–30cm (6–12in), S 30cm (12in) or more. Frost hardy. Has small, oblong to oval, slightly convex, fleshy, glaucous leaves. Terminal clusters of many small, white flowers, with 4 pointed lobes, are borne in late spring or early summer.

H. 'Cranleigh Gem'. Evergreen, rounded shrub. H 60cm (2ft), S 1m (3ft). Frost hardy. Produces dense spikes of small, 4-lobed, white flowers, with conspicuous, black anthers, amid narrowly oval, grey-green leaves in early summer.

H. cupressoides illus. p.146.
❦ 'Boughton Dome' illus. p.309.
H. 'E.A. Bowles' illus. p.137.
H. 'Fairfieldii'. Evergreen, upright shrub. H and S 60cm (2ft). Frost hardy. Oval, toothed, glossy, bright green leaves are red-margined. Large, open panicles of small, 4-lobed, pale lilac flowers are produced in late spring and early summer.

❦ *H. × franciscana 'Blue Gem'*. Evergreen, spreading shrub. H 60cm (2ft), S 1.2m (4ft). Frost hardy. Has oblong, densely arranged, mid-green leaves. Bears dense spikes of small, 4-lobed, violet-blue flowers from mid-summer until early winter.

H. 'Gauntlettii'. Evergreen, upright shrub. H and S 1m (3ft). Frost hardy. Long spikes of small, 4-lobed, pink flowers, with purplish tubes, are produced amid oblong, rich green leaves from late summer to late autumn.

❦ *H. 'Great Orme'* illus. p.133.
❦ *H. hulkeana*. Evergreen, upright, open shrub. H and S 1m (3ft). Frost hardy. Oval, toothed, glossy, dark green leaves have red margins. Has masses of small, 4-lobed, pale lilac flowers in large, open panicles in late spring and early summer. 'Lilac Hint' illus. p.137.

❦ *H. 'La Séduisante'*. Evergreen, upright shrub. H and S 1m (3ft). Frost hardy. Oval, glossy, deep green leaves are purple beneath. Produces small, 4-lobed, deep purplish-red flowers in large spikes from late summer to late autumn.

❦ *H. macrantha*. Evergreen, bushy shrub. H 60cm (2ft), S 1m (3ft). Frost hardy. Has oval, toothed, fleshy, bright green leaves and racemes of large, 4-lobed, pure white flowers in early summer. May become bare at base.

❦ *H. 'Midsummer Beauty'*. Evergreen, rounded, open shrub. H 2m (6ft), S 1.5m (5ft). Frost hardy. Long, narrow, glossy, bright green leaves are reddish-purple beneath. Long spikes of small, 4-lobed, lilac flowers that fade to white are borne from mid-summer to late autumn.

H. ochracea. Evergreen, bushy, dense shrub. H and S 1m (3ft). Fully hardy. Slender shoots are densely covered with tiny, scale-like, ochre-tinged, olive-green leaves. Clusters of small, 4-lobed, white flowers appear in late spring and early summer.

H. pinguifolia (Disk-leaved hebe).
❦ 'Pagei' illus. p.299.
❦ *H. 'Purple Queen'* illus. p.138.
❦ *H. rakaiensis*. Evergreen, rounded, compact shrub. H 1m (3ft), S 1.2m (4ft). Fully hardy. Small, dense spikes of small, 4-lobed, white flowers are produced amid small, oblong, mid-green leaves from early to mid-summer.

H. recurva illus. p.131.
H. salicifolia. Evergreen, upright shrub. H and S 2.5m (8ft). Frost hardy. Has long, narrow, pointed, pale green leaves and, in summer, slender spikes of small, 4-lobed, white or pale lilac flowers.

H. vernicosa illus. p.299.

HEDERA (Araliaceae)
Ivy
Genus of evergreen, woody-stemmed, trailing perennials and self-clinging climbers with adventitious rootlets, used for covering walls and fences and as ground cover. Takes a year or so to become established, but thereafter growth is rapid. On the ground and while climbing, mostly bears roughly triangular, usually lobed leaves. Given extra height and access to light, leaves become less lobed and, in autumn, umbels of small, yellowish-green flowers are produced, followed by globose, black, occasionally yellow, fruits. Fully to half hardy. Ivies with green leaves are very shade tolerant and do well against a north-facing wall. Those with variegated or yellow leaves prefer more light, are usually less hardy and may sustain frost and wind damage in severe winters. All prefer well-drained, alkaline soil. Prune in spring to control height and spread, and to remove any damaged growth. Propagate in late summer by softwood cuttings or rooted layers. Red spider mite may be a problem when plants are grown against a south-facing wall or in dry conditions.

H. algeriensis. See *H. canariensis*.
H. canariensis, syn. *H. algeriensis* (Canary Island ivy). Fast-growing, evergreen, self-clinging climber. H to 6m (20ft), S 5m (15ft). Half hardy; may be damaged in severe winters but soon recovers. Has oval to triangular, unlobed, glossy, mid-green leaves and reddish-purple stems. Is suitable for growing against a wall in a sheltered area. ❦ 'Gloire de Marengo' has silver-variegated leaves.
❦ 'Ravensholst' (illus. p.181) is vigorous with large leaves; makes good ground cover.

❦ *H. colchica* (Persian ivy). Evergreen, self-clinging climber or trailing perennial. H 10m (30ft), S 5m (15ft). Fully hardy. Has large, oval, unlobed, dark green leaves. Is suitable for growing against a wall.
❦ 'Dentata' (Elephant's ears; illus.

p.181) is more vigorous and has large, light green leaves that droop, hence its common name. Is good when grown against a wall or for ground cover.
☙ **'Dentata Variegata'**, H 5m (15ft), has variegated, cream-yellow leaves; is useful to brighten a shady corner.
☙ **'Sulphur Heart'** (Paddy's pride; illus. p.181), H 5m (15ft), S 3m (10ft), has variegated, yellow and light green leaves.
H. helix (Common English ivy). Vigorous, evergreen, self-clinging climber or trailing perennial. H 10m (30ft), S 5m (15ft). Fully hardy. Has 5-lobed, dark green leaves. Makes good ground and wall cover, but may be invasive; for a small garden, the more decorative cultivars are preferable. **'Adam'** (illus. p.181), H 1.2m (4ft), S 1m (3ft), is half hardy and has small, light green leaves variegated cream-yellow; may suffer leaf damage in winter, but will recover.
☙ **'Angularis Aurea'** (illus. p.181), H 4m (12ft), S 2.5m (8ft), has glossy, light green leaves, with bright yellow variegation; is not suitable as ground cover. **'Anna Marie'** (illus. p.181), H 1.2m (4ft), S 1m (3ft), is frost hardy and has light green leaves with cream variegation, mostly at margins; may suffer leaf damage in winter.
☙ **'Atropurpurea'** (syn. *H.h.* 'Purpurea'; Purple-leaved ivy; illus. p.181), H 4m (12ft), S 2.5m (8ft), has dark green leaves that turn deep purple in winter. **'Baltica'**, an exceptionally hardy cultivar, has small leaves and makes good ground cover in an exposed area. ☙ **'Buttercup'** (illus. p.181), H 2m (6ft), S 2.5m (8ft), is frost hardy and has light green leaves that turn rich butter-yellow in full sun.
'Caenwoodiana' see *H.h.* 'Pedata'.
☙ **'Congesta'**, H 45cm (1½ft), S 60cm (2ft), is a non-climbing, erect cultivar with spire-like shoots and small leaves; is suitable for a rock garden.
'Conglomerata' (Clustered ivy), H and S 1m (3ft), will clamber over a low wall or grow in a rock garden; has small, curly, unlobed leaves. **'Cristata'** see *H.h.* 'Parsley Crested'. **'Curly Locks'** see *H.h.* 'Manda's Crested'. **'Deltoidea'** (Shield ivy, Sweetheart ivy; illus. p.181), H 5m (15ft), S 3m (10ft), has heart-shaped leaves; is suitable only for growing against a wall. **'Digitata'** (Finger-leaved ivy; illus. p.181), H 6m (20ft), has large leaves; is not suitable for ground cover. ☙ **'Erecta'** (illus. p.181), H 1m (3ft), S 1.2m (4ft), is a non-climbing, erect cultivar similar to *H.h.* 'Congesta'.
☙ **'Eva'** (illus. p.181), H 1.2m (4ft), S 1m (3ft), is frost hardy and has small, grey-green leaves with cream variegation; may suffer leaf damage in winter. ☙ **'Glacier'** (illus. p.181), H 3m (10ft), S 2m (6ft), is frost hardy and has silvery-grey-green leaves. **'Glymii'** (illus. p.181), H 2.5m (8ft), S 2m (6ft), has glossy, dark green leaves that turn deep purple in winter; is not suitable for ground cover. **'Goldheart'** (illus. p.181), H 6m (20ft), has dark green leaves with bright yellow centres; is slow to establish, then grows rapidly; is not suitable for ground cover.
'Gracilis' (illus. p.181), H 5m (15ft), has sharply lobed, dark green leaves that turn bronze-purple in winter; is not suitable for ground cover. **'Green Feather'** see *H.h.* 'Triton'. **'Green Ripple'** (illus. p.181), H and S 1.2m (4ft), is frost hardy and has mid-green leaves with prominent, light green veins; is good for ground cover or for growing against a low wall. **'Hahn's Self-branching'** see *H.h.* 'Pittsburgh'. **'Heise'** (illus. p.181), H 30cm (1ft), S 60cm (2ft), is frost hardy and has small, grey-green leaves with cream variegation; is suitable as ground cover for a small, sheltered area. var. *hibernica*. See *H. hibernica*.
☙ **'Ivalace'** (syn. *H.h.* 'Mini Green'; illus. p.181), H 1m (3ft), S 1.2m (4ft), is frost hardy and has curled and crimped, glossy leaves; is good for ground cover and for growing against a low wall. **'Königers Auslese'** (syn. *H.h.* 'Sagittifolia'), H 1.2m (4ft), S 1m (3ft), is frost hardy and has finger-like, deeply cut leaves; is not suitable for ground cover. **'Lobata Major'** (illus. p.181), H 5m (15ft), is vigorous with large, 3-lobed leaves.
☙ **'Manda's Crested'** (syn. *H.h.* 'Curly Locks'; illus. p.181), H and S 2m (6ft), is frost hardy and has elegant, wavy-edged, mid-green leaves that turn a coppery shade in winter. **'Merion Beauty'** (illus. p.181), H 1.2m (4ft), S 1m (3ft), is frost hardy with delicately lobed leaves; is not suitable for ground cover. **'Mini Green'** see *H.h.* 'Ivalace'. **'Nigra'** (illus. p.181), H and S 1.2m (4ft), has small, very dark green leaves that turn purple-black in winter. **'Parsley Crested'** (syn. *H.h.* 'Cristata'; illus. p.181), H 2m (6ft), S 1.2m (4ft), is frost hardy and has light green leaves, crested at margins; is not suitable for ground cover. ☙ **'Pedata'** (syn. *H.h.* 'Caenwoodiana'; Bird's-foot ivy; illus. p.181), H 4m (12ft), S 3m (10ft), has grey-green leaves shaped like a bird's foot; is not suitable for ground cover.
'Pittsburgh' (syn. *H.h.* 'Hahn's Self-branching'; illus. p.181), H 1m (3ft), S 1.2m (4ft), is frost hardy and has mid-green leaves; is suitable for growing against a low wall and for ground cover. **'Purpurea'** see *H.h.* 'Atropurpurea'. **'Sagittifolia'** see *H.h.* 'Königers Auslese'. **'Telecurl'** (illus. p.181), H and S 1m (3ft), is frost hardy and has elegantly twisted, light green leaves. **'Triton'** (syn. *H.h.* 'Green Feather'), H 45cm (1½ft), S 1m (3ft), is a frost hardy, non-climbing cultivar that has leaves with deeply incised lobes that resemble whips; makes good ground cover. **'Woerneri'** (illus. p.181), H 4m (12ft), S 3m (10ft), is a vigorous cultivar that has bluntly lobed, grey-green leaves, with lighter coloured veins, that turn purple in winter.
H. hibernica, syn. *H. helix* var. *hibernica* (Irish ivy; illus. p.181). Vigorous, evergreen climber. H 5m (15ft), S 6m (20ft). Fully hardy. Has large, mid-green leaves. Is good for covering a large area, either on the ground or against a wall. **'Sulphurea'** (illus. p.181), H and S 3m (10ft), has medium-sized leaves with sulphur-yellow variegation; is good when grown against a wall or for ground cover, and as foil for brightly coloured plants.
H. nepalensis (Nepalese ivy). Evergreen, self-clinging climber. H 4m (12ft), S 2.5m (8ft). Half hardy; young growth may be damaged by late spring frosts. Produces oval to triangular, toothed, olive-green leaves. Is suitable only for growing against a sheltered wall.
H. pastuchovii. Moderately vigorous, evergreen, self-clinging climber. H 2.5m (8ft), S 2m (6ft). Fully hardy. Has shield-shaped, glossy, dark green leaves. Is suitable only for growing against a wall.
H. rhombea (Japanese ivy). Evergreen, self-clinging climber. H and S 1.2m (4ft). Frost hardy. Has small, fairly thick, diamond-shaped, unlobed, mid-green leaves. Is suitable only for growing against a low wall. **'Variegata'** has leaves with narrow, white margins.

HEDYCHIUM (Zingiberaceae)
Garland flower, Ginger lily
Genus of perennials with stout, fleshy rhizomes. Fragrant, showy flowers are individually short-lived, but borne profusely. Is ideal for sheltered borders and conservatories. Frost hardy to frost tender, min. 5°C (41°F). Needs sun and rich, moist soil. Propagate by division in spring; rhizomes should not be divided when dormant.
H. coronarium (White ginger lily). Upright, rhizomatous perennial. H 1.5m (5ft), S 60cm–1m (2–3ft). Frost tender. Dense spikes of very fragrant, butterfly-like, white flowers with basal, yellow blotches are borne in mid-summer. Lance-shaped, mid-green leaves are downy beneath.
H. densiflorum illus. p.193.
☙ *H. gardnerianum* illus. p.196.

HEDYOTIS, syn. **HOUSTONIA** (Rubiaceae)
Genus of mat-forming, summer-flowering perennials. Fully hardy. Thrives in shady sites on moist, sandy leaf mould. Propagate by division in spring or by seed in autumn.
H. michauxii, syn. *Houstonia serpyllifolia*, illus. p.331.

HEDYSARUM (Leguminosae)
Genus of perennials, biennials and deciduous sub-shrubs. Fully hardy. Prefers sun and well-drained soil. Resents being disturbed. Propagate by seed in autumn or spring.
H. coronarium illus. p.214.

HEERIA. See **HETEROCENTRON**.

Heimerliodendron brunonianum. See *Pisonia umbellifera.*

HELENIUM (Compositae)
Sneezeweed
Genus of late summer- and autumn-flowering perennials, grown for their sprays of daisy-like flower heads, each with a prominent, central disc. Fully hardy. Requires full sun and any well-drained soil. Propagate by division in spring or autumn.
H. **'Bressingham Gold'**. Erect, bushy perennial with stout stems clothed in lance-shaped, mid-green leaves. H 1m (3ft), S 60cm (2ft). Sprays of bright yellow flower heads are produced in late summer and autumn.
H. **'Bruno'**. Erect, bushy perennial. H 1.2m (4ft), S 75cm (2½ft). Bears sprays of deep bronze-red flower heads in late summer and autumn. Stout stems are clothed in lance-shaped leaves.
H. **'Butterpat'**. Compact perennial. H 1m (3ft), S 60cm (2ft). Has stout stems clothed in lance-shaped leaves. In late summer and autumn produces sprays of rich deep yellow flower heads.
H. **'Moerheim Beauty'** illus. p.227.
H. **'Riverton Gem'**. Erect, bushy perennial. H 1.4m (4½ft), S 1m (3ft). Has sprays of red-and-gold flower heads in late summer and autumn. Stout stems are clothed in lance-shaped leaves.
H. **'Wyndley'** illus. p.227.

HELIANTHEMUM (Cistaceae)
Rock rose
Genus of evergreen, spring- to autumn-flowering shrubs and sub-shrubs, grown for their flowers. Is useful for rock gardens and dry banks. Fully to frost hardy. Needs full sun and well-drained soil. Cut back lightly after flowering. Propagate by semi-ripe cuttings in early summer.
H. apenninum illus. p.299.
H. **'Ben Hope'**. Evergreen, domed shrub. H 23–30cm (9–12in), S 45cm (18in). Fully hardy. Leaves are small, linear and grey-green. Produces saucer-shaped, carmine-red flowers in mid-summer.
H. **'Ben More'** illus. p.302.
H. **'Ben Nevis'**. Evergreen, hummock-forming, compact shrub. H and S 15–23cm (6–9in). Fully hardy. Has small, linear, dark green leaves and, in mid-summer, saucer-shaped, orange flowers with bronze centres.
☙ *H.* **'Fire Dragon'** illus. p.302.
H. **'Golden Queen'**. Evergreen, domed, compact shrub. H 23cm (9in), S 30cm (12in). Fully hardy. Saucer-shaped, golden-yellow flowers appear amid small, linear, dark green leaves in mid-summer.
H. guttatum. See *Tuberaria guttata.*
☙ *H.* **'Jubilee'**. Evergreen, domed, compact shrub. H 15–23cm (6–9in), S 23–30cm (9–12in). Fully hardy. Bears small, linear, dark green leaves. Saucer-shaped, double, pale yellow flowers appear from spring to late summer.
H. nummularium **'Amy Baring'**. Evergreen, spreading shrub. H 10–15cm (4–6in), S 60cm (24in). Fully hardy. Small, oblong, light grey leaves are hairy beneath. In summer bears a succession of saucer-shaped, orange-centred, deep yellow flowers in loose, terminal clusters.
H. oelandicum subsp. *alpestre*. Evergreen, open, twiggy shrub. H 7–12cm (3–5in), S 15cm (6in) or more. Fully hardy. Produces terminal clusters of 3–6 saucer-shaped, bright yellow flowers from early to mid-summer. Leaves are tiny, oblong and mid-green. Is suitable for a trough.
H. **'Raspberry Ripple'** illus. p.302.

H. **'Rhodanthe Carneum'**. See *H.* 'Wisley Pink'.
☙ *H.* **'Wisley Pink'**, syn. *H.* 'Rhodanthe Carneum' illus. p.301.
☙ *H.* **'Wisley Primrose'** illus. p.306.
☙ *H.* **'Wisley White'** illus. p.299.

HELIANTHUS (Compositae)
Sunflower
Genus of summer- and autumn-flowering annuals and perennials, grown for their large, daisy-like, usually yellow flower heads. May be invasive. Fully hardy. Prefers sun and well-drained soil. Needs staking. Propagate by seed or division in autumn or spring.
H. annuus (illus. p.289). Fast-growing, upright annual. H 1–3m (3–10ft) or more; S 30–45cm (12–18in). Has large, oval, serrated, mid-green leaves. Very large, daisy-like, brown- or purplish-centred, yellow flower heads, 30cm (12in) or more wide, are produced in summer. Tall, intermediate and dwarf cultivars are available. **'Italian White'** (intermediate), H 1.2m (4ft), has black-centred, creamy-white flower heads. **'Russian Giant'** (tall), H 3m (10ft) or more, has yellow flower heads with green-brown centres. **'Taiyo'** (intermediate) illus. p.288. **'Teddy Bear'** (dwarf), H 60cm (2ft), has 15cm (6in) wide, fully double, deep yellow flower heads.
☙ *H. atrorubens* **'Monarch'**. Erect perennial. H 2.2m (7ft), S 1m (3ft). Bears terminal, daisy-like, semi-double, golden-yellow flower heads on branching stems in late summer. Has lance-shaped, coarse, mid-green leaves. Replant each spring to keep in check.
☙ *H.* **'Capenoch Star'**. Erect perennial. H 1.2m (4ft), S 60cm (2ft). In summer, daisy-like, lemon-yellow flower heads are borne terminally on branching stems. Leaves are lance-shaped, coarse and mid-green.
H., **Chrysanthemum-flowered Series** illus. p.289.
☙ *H.* **'Loddon Gold'** illus. p.196.
H. x *multiflorus* illus. p.196.
H. salicifolius (Willow-leaved sunflower). Upright perennial. H 2.2m (7ft), S 60cm (2ft). Bears small, daisy-like, yellow flower heads at ends of stout, branching stems in autumn. Has narrow, willow-like, drooping, deep green leaves.

HELICHRYSUM (Compositae)
Genus of summer- and autumn-flowering perennials, annuals and evergreen sub-shrubs and shrubs. When dried, flower heads are 'ever-lasting'. Fully hardy to frost tender, min. 5°C (41°F). Needs sun and well-drained soil. Propagate shrubs and sub-shrubs by heel or semi-ripe cuttings in summer; perennials by division or seed in spring; annuals by seed in spring.
H. angustifolium. See *H. italicum*.
H. bellidioides. Evergreen, prostrate shrub. H 5cm (2in), S 23cm (9in). Fully hardy. Has small, rounded, fleshy, dark green leaves and, in early summer, terminal clusters of daisy-like, white flower heads.
H. bracteatum, syn. *Bracteantha bracteata* (Everlasting flower, Immortelle, Strawflower). ☙ **'Bright Bikini'** is a moderately fast-growing, upright, branching annual. H and S 30cm (12in). Half hardy. Has lance-shaped, mid-green leaves and, from summer to early autumn, papery, daisy-like flower heads in many colours, including red, pink, orange, yellow and white. **Monstrosum Series** illus. p.292.
☙ *H. coralloides*, syn. *Ozothamnus coralloides* illus. p.309.
☙ *H. italicum*, syn. *H. angustifolium* (Curry plant). Evergreen, bushy sub-shrub. H 60cm (2ft), S 1m (3ft). Frost hardy. Has linear, aromatic, silvery-grey leaves. Broad clusters of small, oblong, bright yellow flower heads are produced on long, upright, white shoots during summer. subsp. *serotinum* (syn. *H. serotinum*), H and S 15cm (6in), is dome-shaped; stems and oval leaves are densely felted with white hairs. Dislikes winter wet and cold climates.
Helichrysum ledifolium. See *Ozothamnus ledifolius*.
H. marginatum. See *H. milfordiae*.
☙ *H. milfordiae*, syn. *H. marginatum*. Evergreen, mat-forming, dense sub-shrub. H 5cm (2in), S 23cm (9in). Frost hardy. On sunny days in early summer, large, conical, red buds open into daisy-like, white flower heads with red-backed petals; they close in dull or wet weather. Has basal rosettes of oval, hairy, silver leaves. Prefers very gritty soil. Dislikes winter wet. Propagate in spring by rooting single rosettes.
☙ *H. petiolare*, syn. *H. petiolatum*, illus. p.145.
H. petiolatum. See *H. petiolare*.
H. rosmarinifolium. See *Ozothamnus rosmarinifolius*.
H. **'Schweffellicht'**. See *H.* 'Sulphur Light'.
H. selago, syn. *Ozothamnus selago*, illus. p.309.
H. serotinum. See *H. italicum* subsp. *serotinum*.
☙ *H. splendidum*. Evergreen, bushy, dense shrub. H and S 1.2m (4ft). Frost hardy. Woolly, white shoots are clothed in small, oblong, silvery-grey leaves. Small, oblong, bright yellow flower heads appear in clusters from mid-summer to autumn or into winter.
H. **'Sulphur Light'**, syn. *H.* 'Schweffellicht', illus. p.255.

HELICONIA (Heliconiaceae)
Lobster claws
Genus of tufted perennials, evergreen in warm climates, grown for their spikes of colourful flowers and for the attractive foliage on younger plants. Frost tender, min. 18°C (64°F). Needs partial shade and humus-rich, well-drained soil. Water generously in growing season, very sparingly when plants die down in winter. Propagate by seed or division of rootstock in spring.
H. metallica. Tufted perennial. H to 3m (10ft), S 1m (3ft). Oblong, long-stalked leaves, to 60cm (2ft) long, are velvety-green above with paler veins, sometimes purple below. In summer, mature plants bear erect stems with tubular, glossy, greenish-white-tipped, red flowers enclosed in narrow, boat-shaped, green bracts.
H. psittacorum illus. p.193.

HELICTOTRICHON (Gramineae).
See GRASSES, BAMBOOS, RUSHES and SEDGES.
☙ *H. sempervirens*, syn. *Avena candida* of gardens, *A. sempervirens*, illus. p.183.

HELIOPSIS (Compositae)
Genus of summer-flowering perennials. Fully hardy. Requires sun and any well-drained soil. Propagate by seed or division in autumn or spring.
H. **'Ballet Dancer'** illus. p.222.
H. helianthoides **'Incomparabilis'**. Upright perennial. H to 1.5m (5ft), S 60cm (2ft). Bears daisy-like, single, orange flower heads in late summer. Leaves are narrowly oval, coarsely toothed and mid-green. **'Patula'** bears flattish, semi-double, orange-yellow flower heads. subsp. *scabra* (syn. *H. scabra*) has very rough stems and leaves and double, orange-yellow flower heads.
H. **'Light of Loddon'** illus. p.193.
H. scabra. See *H. helianthoides* subsp. *scabra*.

Heliosperma alpestris. See *Silene alpestris*.

HELIOTROPIUM (Boraginaceae)
Genus of annuals, evergreen sub-shrubs and shrubs, grown for their fragrant flowers. Frost hardy to frost tender, min. 5–7°C (41–5°F). Requires full sun and fertile, well-drained soil. Water potted plants freely when in full growth, moderately at other times. In spring, tip prune young plants to promote a bushy habit and cut leggy, older plants back hard. Propagate by seed in spring, by greenwood cuttings in summer or by semi-ripe cuttings in early autumn.
H. arborescens, syn. *H. peruvianum*, illus. p.137.
H. peruvianum. See *H. arborescens*.

HELIPTERUM (Compositae)
Genus of annuals and perennials; flower heads of annuals are suitable for cutting and drying. Half hardy. Grow in sun and in poor, very well-drained soil. Propagate by seed sown outdoors in mid-spring. Aphids may cause problems.
H. manglesii, syn. *Rhodanthe manglesii*, illus. p.273.
H. roseum, syn. *Acroclinium roseum*, illus. p.273.

HELLEBORUS (Ranunculaceae)
Christmas rose, Lenten rose
Genus of perennials, some of which are evergreen, grown for their winter and spring flowers. Most deciduous species retain their old leaves over winter. These should be cut off in early spring as flower buds develop. Is excellent in woodlands. Fully to half hardy. Prefers semi-shade and moisture-retentive, well-drained soil. Propagate by fresh seed or division in autumn or very early spring. Is prone to aphid attack in early summer.
☙ *H. argutifolius*, syn. *H. corsicus*, *H. lividus* subsp. *corsicus*, illus p.268.
H. atrorubens. Clump-forming perennial. H and S 30cm (1ft). Fully hardy. Shallowly cup-shaped, deep purple flowers are borne in late winter. Has palmate, deeply divided, toothed, glossy, dark green leaves.
H. corsicus. See *H. argutifolius*.
H. cyclophyllus. Clump-forming perennial. H to 60cm (24in), S 45cm (18in). Fully hardy. In early spring produces shallowly cup-shaped, yellow-green flowers with prominent, yellowish-white stamens. Has palmate, deeply divided, bright green leaves.
☙ *H. foetidus* illus. p. 269.
☙ *H. lividus*. Evergreen, clump-forming perennial. H and S 45cm (18in). Half hardy. Has 3-parted, mid-green leaves, marbled pale green, purplish-green below, with obliquely oval, slightly toothed or entire leaflets. Large clusters of cup-shaped, purple-suffused, yellow-green flowers open in late winter. subsp. *corsicus* see *H. argutifolius*.
☙ *H. niger* illus. p.265.
H. orientalis [white form] illus. p.264, [pink form] illus. p.265, [purple form] illus. p.266.
H. purpurascens. Neat, clump-forming perennial. H and S 30cm (1ft). Fully hardy. Small, nodding, cup-shaped, pure deep purple or green flowers, splashed with deep purple on outside, appear in early spring. Leaves are dark green and palmate, deeply divided into narrowly lance-shaped, toothed segments.
H. x *sternii* illus. p.265.
H. viridis illus. p.269.

HELONIAS (Liliaceae)
Genus of one species of spring-flowering perennial. Fully hardy. Is excellent in bog gardens. Needs an open, sunny position and moist to wet soil. Propagate by division in spring or by seed in autumn.
H. bullata (Swamp pink). Rosetted, clump-forming perennial. H 38–45cm (15–18in), S 30cm (12in). Has basal rosettes of strap-shaped, fresh green leaves, above which dense racemes of small, fragrant, star-shaped, pinkish-purple flowers are borne in spring.

HELONIOPSIS (Liliaceae)
Genus of spring-flowering, rosette-forming perennials. Fully hardy. Grow in semi-shade and in moist soil. Propagate by division in autumn or by seed in autumn or spring.
H. orientalis illus. p.233.

HELWINGIA (Cornaceae)
Genus of deciduous shrubs, bearing flowers and showy fruits directly on leaf surfaces, grown mainly for botanical interest. Needs separate male and female plants in order to produce fruits. Fully hardy. Requires sun or semi-shade and moist soil. Propagate by softwood cuttings in summer.
H. japonica. Deciduous, bushy, open shrub. H and S 1.5m (5ft). Oval, bright green leaves have bristle-like teeth. In early summer, tiny, star-shaped, green flowers appear on each leaf centre and are followed by spherical, black fruits.

Helxine soleirolii. See *Soleirolia soleirolii*.

HEMEROCALLIS

HEMEROCALLIS (Liliaceae)
Daylily
Genus of perennials, some of which are semi-evergreen. Flowers, borne in succession, each last for only a day. Fully hardy. Does best in full sun and fertile, moist soil. Propagate by division in autumn or spring. Cultivars raised from seed will not come true to type; species may come true if grown in isolation from other daylilies. Slug and snail control is essential in early spring when young foliage first appears.

H. **'Betty Woods'** (illus. p.221). Robust, spreading evergreen, clump-forming perennial. H 65cm (26in), S 60cm (2ft). Large, peony-like, yellow flowers are borne in mid- and late summer. Leaves are arching, strap-shaped and mid-green.

H. **'Bold Courtier'**. Clump-forming perennial. H 1m (3ft), S 75cm (2½ft). Trumpet-shaped flowers, with rose-pink, inner petals and rose-flushed, rich cream, outer petals, are borne from early to mid-summer above strap-shaped, mid-green leaves.

H. **'Brocaded Gown'** (illus. p.221). Semi-evergreen, clump-forming perennial. H and S to 60cm (2ft). In summer has rounded, ruffled creamy-yellow flowers amid broad, strap-shaped, mid-green leaves.

♀ *H.* **'Cartwheels'** (illus. p.221). Clump-forming perennial. H and S 75cm (2½ft). In mid-summer bears large, light golden-orange flowers that open wide amid strap-shaped leaves.

H. **'Catherine Woodbury'**. Robust, clump-forming perennial. H 70cm (28in), S 75cm (30in). Fragrant, trumpet-shaped, delicate, pale lavender flowers, with soft lime-green throats, are produced above strap-shaped, mid-green leaves in late summer.

H. **'Cat's Cradle'** (illus. p.221). Semi-evergreen, clump-forming perennial. H 1m (3ft), S 75cm (2½ft). In summer has large, spider-like, bright yellow flowers amid slender grassy leaves.

H. citrina (illus. p.221). Vigorous, coarse-growing, clump-forming perennial. H and S 75cm (2½ft). Many large, very fragrant, trumpet-shaped, rich lemon-yellow flowers, lasting only a day, open at night in mid-summer. Strap-shaped leaves are dark green.

♀ *H.* **'Corky'** (illus. p.221). Clump-forming perennial. H and S 45cm (18in). Trumpet-shaped, lemon-yellow flowers, brown on outsides, open in late spring and early summer above slender, strap-shaped, mid-green leaves. Flowers, borne prolifically, last only a day.

H. dumortieri. Compact, clump-forming perennial. H 45cm (1½ft), S 60cm (2ft). In early summer produces fragrant, trumpet-shaped, brown-backed, golden-yellow flowers. Mid-green leaves are strap-shaped, stiff and coarse.

H. **'Eenie Weenie'** (illus. p.221). Clump-forming perennial. H and S 30cm (1ft). Produces an abundance of clear yellow flowers in early summer. Leaves are narrow and arching.

H. flava. See *H. lilioasphodelus*.

H. **'Frank Gladney'**. Robust, semi-evergreen, clump-forming perennial. In summer produces very large, rounded, vivid coral-pink flowers with golden throats; anthers are black. Broad, strap-shaped leaves are mid-green.

H. fulva (Fulvous daylily, Tawny daylily). Vigorous, clump-forming perennial. H 1m (3ft), S 75cm (2½ft). Trumpet-shaped, tawny-orange flowers appear from mid- to late summer above a mound of strap-shaped, light green leaves. **'Kwanso Flore Pleno'** (illus. p.221) has double flowers. **'Kwanso Variegata'** has leaves variably marked with white.

H. **'Gentle Shepherd'** (illus. p.221). Semi-evergreen, clump-forming perennial. H 70cm (28in), S 60cm (24in). From early to mid-summer carries ruffled, white flowers with green throats above light green leaves.

♀ *H.* **'Golden Chimes'** (illus. p.221). Clump-forming perennial of graceful habit. H 75cm (2½ft), S 60cm (2ft). Bears small, delicate, trumpet-shaped, golden-yellow flowers with a brown reverse, lasting only a day, from early to mid-summer. Has narrow strap-shaped mid-green leaves.

H. **'Hyperion'** (illus. p.221). Clump-forming perennial. H and S 90cm (3ft). In mid-summer, very fragrant, lily-like, pale lemon-yellow flowers are borne above strap-shaped leaves.

H. **'Joan Senior'**. Vigorous, semi-evergreen, clump-forming perennial. H 63cm (25in), S 1m (3ft). Open trumpet-shaped, almost pure white flowers are borne on well-branched stems from mid- to late summer. Leaves are strap-shaped and mid-green.

H. **'Jolyene Nichole'** (illus. p.221). Semi-evergreen, clump-forming perennial. H and S 50cm (20in). Rounded, ruffled, rose-pink flowers are borne amid lush, blue-green leaves.

♀ *H. lilioasphodelus* (syn. *H. flava*; illus. p.221). Robust, clump-forming, spreading perennial. H and S 60cm (2ft) or more. Very fragrant, delicate, lemon- to chrome-yellow flowers last only 1 or 2 days in late spring and early summer. Strap-shaped leaves are mid-green.

H. **'Little Grapette'** (illus. p.221). Vigorous, semi-evergreen, clump-forming perennial. H 50cm (20in), S 60cm (24in). From early spring to mid-summer bears grape-purple flowers with green throats and darker purple eyes. Leaves are slender and mid-green.

H. **'Lullaby Baby'**. Vigorous, semi-evergreen, clump-forming perennial. H and S 50cm (20in). Bears very pale pink flowers with green throats in mid-summer amid strap-shaped, mid-green leaves.

H. **'Lusty Leland'** (illus. p.221). Vigorous, clump-forming perennial. H 1m (3ft), S 75cm (2½ft). In mid-summer bears many large, trumpet-shaped, velvety, deep red flowers with greenish-yellow throats. Has thick, strap-shaped, mid-green leaves.

H. **'Luxury Lace'**. Compact, clump-forming perennial. H 75cm (2½ft), S 60cm (2ft). In late summer produces small, trumpet-shaped, creamy-lavender-pink flowers with lime-green throats and prominent anthers. Each petal is curled and ruffled and has a near-white midrib. Leaves are strap-shaped and mid-green.

♀ *H.* **'Marion Vaughn'** (illus. p.221). Clump-forming perennial. H 1m (3ft), S 60cm (2ft). In mid-summer bears fragrant, trumpet-shaped, green-throated, pale lemon-yellow flowers, lasting only a day. Each petal has a raised, near-white midrib. Strap-shaped leaves are mid-green.

H. minor (Grass-leaved daylily). Compact, clump-forming perennial. H 40cm (16in), S 45cm (18in). In early summer, fragrant, trumpet-shaped, lemon-yellow flowers, with tawny-backed, outer petals, overtop narrowly strap-shaped, mid-green leaves that die back in early autumn.

H. **'Norton Orange'**. Clump-forming perennial. H 1m (3ft), S 75cm (2½ft). Large, trumpet-shaped, rich clear orange flowers, with ruffled-edged petals and black anthers, appear from mid- to late summer. Has strap-shaped, dark green leaves.

H. **'Prairie Blue Eyes'** (illus. p.221). Semi-evergreen, clump-forming perennial. H 80cm (32in), S 90cm (36in). In mid-summer produces lavender flowers, banded with blue-purple, that have green throats. Slender, strap-shaped leaves are mid-green.

H. **'Ruffled Apricot'** (illus. p.221). Slow-growing, clump-forming perennial. Large, deep apricot flowers, with lavender-pink midribs, are ruffled at edges. Leaves are stiff, arching and mid-green.

H. **'Scarlet Orbit'** (illus. p.221). Low-growing, semi-evergreen, clump-forming perennial. H 50cm (20in), S 65cm (26in). In mid-summer, scarlet flowers with green throats open flat amid lush, strap-shaped, mid-green leaves.

H. **'Siloam Virginia Henson'** (illus. p.221). Clump-forming perennial. H 45cm (18in), S 65cm (26in). In early summer bears rounded, ruffled, creamy-pink flowers banded with rose-pink and with green throats. Broad, strap-shaped leaves are mid-green.

H. **'Stafford'** (illus. p.221). Vigorous clump-forming perennial. H 75cm (2½ft), S 60cm (2ft). Bears trumpet-shaped, bright red flowers, with maroon and yellow throats and a narrow, yellow midrib on each petal, that last only a day, from mid- to late summer. Leaves are strap-shaped.

♀ *H.* **'Stella d'Oro'** (illus. p.221). Vigorous, compact, clump-forming perennial. H 40cm (16in), S 45cm (18in). Abundant, small, widely bell-shaped, pale orange-yellow flowers, with green-tipped, outer petals, are borne from mid-summer to first frosts. Narrowly strap-shaped leaves are mid-green.

H. **'Super Purple'** (illus. p.221). Clump-forming perennial. H 68cm (27in), S 65cm (26in). Rounded, ruffled, velvety, red-purple flowers with lime-green throats are produced in mid-summer amid strap-shaped, light-green leaves.

H. **'Whisky on Ice'** (illus. p.221). Semi-evergreen, clump-forming perennial. Flowers, produced from early to mid-summer, are an unusual combination of amber, lemon and paler yellow and have rich green throats. Leaves are lush and bluish-green.

HEMIGRAPHIS (Acanthaceae)
Genus of annuals and evergreen perennials, usually grown as foliage plants. Frost tender, min. 15°C (59°F). Grow in bright but not direct sunlight and in moist but well-drained soil. Water frequently in growing season, less in winter. Cut back straggly stems. Propagate by stem cuttings in spring or summer.

H. repanda illus. p.263.

HEPATICA (Ranunculaceae)
Genus of very variable perennials, some of which are semi-evergreen, flowering in early spring before new leaves are properly formed. Fully hardy. Requires partial shade and deep, humus-rich, moist soil. Stout, much-branched rootstock resents disturbance. Propagate by seed when fresh or by division or removing side shoots in spring.

H. angulosa. See *H. transsilvanica*.

♀ *H.* x *media* **'Ballardii'**. Slow-growing, dome-shaped perennial. H 10cm (4in), S 30cm (12in). Has rounded, 3-lobed, stalked, soft green leaves and, in early spring, shallowly cup-shaped, many-petalled, intense blue flowers. Fully double, coloured forms are also known. Propagate by division only.

♀ *H. nobilis*, syn. *Anemone hepatica*. Slow-growing, semi-evergreen, dome-shaped perennial. H 8cm (3in), S 10–12cm (4–5in). Bears rounded, 3-lobed, fleshy, mid-green leaves. Shallowly cup-shaped, many-petalled flowers – white through pink to carmine, pale to deep blue or purple – appear in early spring. Fully double, coloured forms are also known. Is excellent for woodland or a rock garden. var. *japonica* illus. p.317.

♀ *H. transsilvanica*, syn. *H. angulosa*. Semi-evergreen, spreading perennial. H 8cm (3in), S 20cm (8in). Shallowly cup-shaped, many-petalled flowers, varying from blue to white or pink, are borne in early spring amid rounded, 3-lobed, hairy, green leaves. Fully double, coloured forms are also known.

HEPTAPLEURUM. See SCHEFFLERA.

HERBERTIA (Iridaceae)
Genus of spring-flowering bulbs, grown for their iris-like flowers. Half hardy. Requires full sun and well-drained soil. Reduce watering when bulb dies down after flowering. Propagate by seed in autumn.

H. pulchella. Spring-flowering bulb. H 10–15cm (4–6in), S 3–5cm (1¼–2in). Leaves are narrowly lance-shaped, pleated, erect and basal. Produces a succession of upward-facing, violet-blue flowers, 5–6cm (2–2½in) wide and usually with dark-spotted centres.

HERMANNIA (Sterculiaceae)
Genus of evergreen sub-shrubs and shrubs, grown mainly for their flowers. Frost tender, min. 7°C (45°F). Prefers full light and fertile, well-drained soil.

HERMODACTYLUS

Water potted plants freely when in full growth, moderately at other times. Young plants may need tip pruning to produce well-branched specimens. Propagate by softwood or greenwood cuttings in late spring or summer.
H. candicans. See *H. incana.*
H. incana, syn. *H. candicans.* Evergreen, bushy sub-shrub. H and S 60cm (24in) or more. Oval to oblong leaves have white down beneath. Small, nodding, bell-shaped, bright yellow flowers are carried in terminal clusters, to 15cm (6in) long, in spring-summer.

HERMODACTYLUS (Iridaceae)
Genus of one species of spring-flowering tuber, with an elongated, finger-like rootstock, grown for its iris-like flowers. Fully hardy. Requires a hot, sunny situation, where tubers will ripen well in summer, and well-drained soil. Is particularly successful on hot, chalky soils. Propagate by division in late summer.
H. tuberosus, syn. *Iris tuberosa*, illus. p.358.

HESPERALOE (Agavaceae)
Genus of basal-rosetted, perennial succulents with very narrow, strap-shaped, grooved, dark green leaves, often with peeling, white fibres at margins. Is closely related to *Agave* and *Yucca*. Offsets freely at base. Frost tender, min. 3°C (37°F). Needs a sunny position and very well-drained soil. Propagate by seed or division in spring or summer.
H. parviflora, syn. *Yucca parviflora*, illus. p.398.

HESPERANTHA (Iridaceae)
Genus of spring-flowering corms with spikes of small, funnel- or cup-shaped flowers. Half hardy. Needs full sun and well-drained soil. Plant in autumn, water through winter and dry off corms after flowering. Propagate by seed in autumn or spring.
H. buhrii. See *H. cucullata.*
H. cucullata, syn. *H. buhrii.* Spring-flowering corm. H 20–30cm (8–12in), S 3–5cm (1¼–2in). Has linear, erect leaves on lower part of branched stems, each of which produces a spike of up to 7 cup-shaped, white flowers, flushed pink or purple outside, that open in evenings.

HESPERIS (Cruciferae)
Genus of late spring- or summer-flowering annuals and perennials. Fully hardy. Requires sun and well-drained soil. *H. matronalis* tolerates poor soil. Tends to become woody at base, so raise new stock from seed every few years. Propagate by basal cuttings in spring or by seed in autumn or spring.
H. matronalis illus. p.203.

HESPEROCALLIS (Liliaceae)
Genus of spring- to summer-flowering bulbs. Half hardy. Needs a sunny, well-drained site. Is difficult to cultivate in all but warm, dry areas; in cool, damp climates, protect in a cool greenhouse. Requires ample water in spring, followed by a hot, dry period during its summer dormancy. Propagate by seed in autumn.
H. undulata. Spring- to summer-flowering bulb. H 20–50cm (8–20in), S 10–15cm (4–6in). Has a cluster of long, narrow, wavy-margined leaves, semi-erect or prostrate, at base. Stout stems each bear a spike of upward-facing, funnel-shaped, white flowers, with a central, green stripe along each of the 6 petals.

HESPEROYUCCA. See **YUCCA.**

HETEROCENTRON, syn. HEERIA, SCHIZOCENTRON (Melastomataceae)
Genus of evergreen, summer- and autumn-flowering perennials and shrubs. Frost tender, min. 5°C (41°F). Requires sun and well-drained soil. Propagate by softwood or stem-tip cuttings in late winter or early spring.
H. elegans illus. p.248.

HETEROMELES (Rosaceae)
Genus of one species of evergreen, summer-flowering shrub or tree, grown for its foliage, flowers and fruits. Frost hardy, but in cold areas needs protection. Requires sun or semi-shade and fertile, well-drained soil. Propagate by semi-ripe cuttings in summer or by seed in autumn.
H. arbutifolia, syn. *Photinia arbutifolia* (Christmas berry, Toyon). Evergreen, bushy, spreading shrub or tree. H 6m (20ft), S 8m (25ft). Has oblong, sharply toothed, leathery, glossy, dark green leaves. Broad, flat heads of small, 5-petalled, white flowers in late summer are succeeded by large clusters of rounded, red fruits.

HEUCHERA (Saxifragaceae)
Alum root
Genus of evergreen, summer-flowering perennials forming large clumps of leaves, often tinted bronze or purple. Makes good ground cover. Fully to frost hardy. Prefers semi-shade and moisture-retentive but well-drained soil. Propagate species by seed in autumn or by division in autumn or spring, cultivars by division only, using young, outer portions of crown.
H. x *brizoides* **'Coral Cloud'**. Evergreen, clump-forming perennial. H 45–75cm (18–30in), S 30–45cm (12–18in). Fully hardy. In summer bears long, feathery sprays of small, pendent, bell-shaped, pink flowers. Leaves are rounded, lobed, toothed and dark green. **'Firebird'** has crimson-scarlet flowers; those of **'Pearl Drops'** are white; ❀ **'Red Spangles'** illus. p.246; those of **'Scintillation'** are bright pink.
H. cylindrica **'Greenfinch'** illus. p.240.
H. micrantha **'Palace Purple'** illus. p.241.

x **HEUCHERELLA** (Saxifragaceae)
Hybrid genus (*Heuchera* x *Tiarella*) of evergreen, mainly late spring- and summer-flowering perennials. Fully hardy. Prefers semi-shade and needs fertile, well-drained soil. Propagate by basal cuttings in spring or by division in spring or autumn.
x *H.* **'Bridget Bloom'** illus. p.242.
❀ x *H. tiarelloides* illus. p.242.

HEXASTYLIS. See **ASARUM.**

HIBBERTIA (Dilleniaceae)
Genus of evergreen shrubs and twining climbers, grown for their flowers. Frost tender, min. 5–10°C (41–50°F). Grow in well-drained soil, in full light or semi-shade. Water freely in summer, less at other times. Provide stems with support. Thin out congested growth in spring. Propagate by semi-ripe cuttings in summer.
H. cuneiformis, syn. *Candollea cuneiformis*, illus. p.116.
❀ *H. scandens.* Vigorous, evergreen, twining climber. H 6m (20ft). Has 4–9cm (1½–3½in) long, oblong to lance-shaped, glossy, deep green leaves. Saucer-shaped, bright yellow flowers, 4cm (1½in) across, appear mainly in summer.

HIBISCUS (Malvaceae)
Genus of evergreen or deciduous shrubs, trees, perennials and annuals, grown for their flowers. Fully hardy to frost tender, min. 5–15°C (41–59°F). Needs full sun and humus-rich, well-drained soil. Water containerized specimens freely when in full growth, moderately at other times. Tip prune young plants to promote bushiness; cut old plants back hard in spring. Propagate by seed in spring; shrubs and trees by greenwood cuttings in late spring or by semi-ripe cuttings in summer; perennials by division in autumn or spring. Whitefly may cause problems.
H. mutabilis (Confederate rose, Cotton rose). Evergreen, erect to spreading shrub or tree. H and S 3–5m (10–15ft). Frost tender, min 5°C (41°F). Rounded leaves have 5–7 shallow lobes. In summer-autumn bears funnel-shaped, sometimes double, white or pink flowers, 7–10cm (3–4in) wide, that age from pink to deep red. In light frost dies back to ground level.
H. rosa-sinensis. Evergreen, rounded, leafy shrub. H and S 1.5–3m (5–10ft) or more. Frost tender, min. 10–13°C (50–55°F). Oval, glossy leaves are coarsely serrated. Produces funnel-shaped, bright crimson flowers, 10cm (4in) wide, mainly in summer but also in spring and autumn. Many colour selections are grown including **'The President'** illus. p.111.
❀ *H. schizopetalus.* Evergreen, upright, spreading, loose shrub. H to 3m (10ft), S 2m (6ft) or more. Frost tender, min. 10–13°C (50–55°F). Has oval, serrated leaves and, in summer, pendent, long-stalked flowers, 6cm (2½in) wide, with deeply fringed, reflexed, pink or red petals. May be trained as a climber.
H. sinosyriacus **'Lilac Queen'** illus. p.113.
❀ *H. syriacus* **'Blue Bird'**, syn. *H.s.* 'Oiseau Bleu', illus. p.115. ❀ **'Diana'** is a deciduous, upright shrub. H 3m (10ft), S 2m (6ft). Fully hardy. Very large, trumpet-shaped, pure white flowers, with wavy-edged petals, are borne from late summer to mid-autumn. Oval, lobed leaves are deep green.

HIPPEASTRUM

❀ **'Red Heart'** illus. p.109.
❀ **'Woodbridge'** illus. p.111.
H. trionum illus. p.272.

HIDALGOA (Compositae)
Climbing dahlia
Genus of evergreen, leaf-stalk climbers, grown for their single, dahlia-like flower heads. Frost tender, min. 10°C (50°F). Requires humus-rich, well-drained soil and full light. Water freely when in full growth, less at other times. Provide support and thin out crowded stems or cut back all growth to ground level in spring. Propagate by softwood cuttings in spring. Aphids, red spider mite and whitefly may prove troublesome.
H. wercklei. Moderately vigorous, evergreen, leaf-stalk climber. H 5m (15ft) or more. Oval leaves are divided into 3, 5 or more coarsely serrated leaflets. In summer bears dahlia-like, scarlet flower heads, yellowish in bud.

HIERACIUM (Compositae)
Hawkweed
Genus of perennials; most are weeds, but the species described is grown for its foliage. Fully hardy. Does best in sun and in poor, well-drained soil. Propagate by seed or division in autumn or spring.
H. lanatum illus. p.257.

HIPPEASTRUM (Amaryllidaceae)
Genus of bulbs, grown for their huge, funnel-shaped flowers. Is often incorrectly cultivated as *Amaryllis*. Frost hardy to frost tender, min. 13–15°C (55–9°F). Needs full sun or partial shade and well-drained soil. Plant large-flowered hybrids in autumn, half burying bulb; after leaves die away, dry off bulb until following autumn. Smaller, summer-flowering species should be kept dry while dormant in winter. Propagate by seed in spring or by offsets in spring (summer-flowering species) or autumn (large-flowered hybrids).
H. advenum. See *Rhodophiala advenum.*
H. **'Apple Blossom'** illus. p.369.
H. aulicum. Winter- and spring-flowering bulb with a basal leaf cluster. H 30–50cm (12–20in), S 30cm (12in). Frost tender, min. 13°C (55°F). Produces strap-shaped, semi-erect leaves as, or just after, flowers form. Stout stem has a head of 2–6 red flowers, each to 25cm (10in) across with a green blotch in the throat near the base of each petal.
❀ *H.* **'Belinda'**. Winter- and spring-flowering bulb with a basal leaf cluster. H 30–50cm (12–20in), S 30cm (12in). Frost tender, min. 13°C (55°F). Is similar to *H. aulicum*, but flowers are deep velvety-red throughout, stained darker towards centres.
H. **'Bouquet'**. Winter- and spring-flowering bulb with a basal leaf cluster. H 30–50cm (12–20in), S 30cm (12in). Frost tender, min. 13°C (55°F). Is similar to *H. aulicum*, but has very wide, salmon-pink flowers with deep red veins and red centres.
❀ *H.* **'Orange Sovereign'** illus. p.369.
H. pratense. See *Rhodophialum pratense.*

H. procerum. See *Worsleya rayneri.*
H. **'Red Lion'** illus. p.369.
H. reginae. Summer-flowering bulb. H to 50cm (20in), S 20–25cm (8–10in). Frost tender, min. 13°C (55°F). Flower stem produces a head of 2–4 scarlet flowers, each 10–15cm (4–6in) across and with a star-shaped, green mark in the throat. Long, strap-shaped, semi-erect leaves develop at base after flowering.
H. rutilum. See *H. striatum.*
H. striatum, syn. *H. rutilum,* illus. p.379.
H. **'Striped'** illus. p.369.
H. vittatum. Vigorous, spring-flowering bulb. H 1m (3ft), S 30cm (1ft). Frost tender, min. 13°C (55°F). Leaves are broadly strap-shaped, semi-erect and basal. Stout, leafless stem precedes leaves and terminates in a head of 2–6 red-striped, white flowers, each 12–20cm (5–8in) across.
H. **'White Dazzler'.** Winter- and spring-flowering bulb with a basal leaf cluster. H 30–50cm (12–20in), S 30cm (12in). Frost tender, min. 13°C (55°F). Is similar to *H. aulicum,* but has pure white flowers.

HIPPOCREPIS (Leguminosae)
Vetch
Genus of annuals and perennials, grown for their pea-like flowers. Fully hardy. Needs full sun and well-drained soil. Propagate by seed in spring or autumn. Self seeds readily. May be invasive.
H. comosa (Horseshoe vetch) illus. p.334. **'E.R. Janes'** is a vigorous, prostrate, woody-based perennial. H 5–8cm (2–3in), S 15cm (6in) or more. Rooting stems bear small, loose spikes of pea-like, yellow flowers from late spring to late summer. Has divided leaves with 3–8 pairs of narrowly oval leaflets.

HIPPOPHAË (Elaeagnaceae)
Genus of deciduous shrubs and trees, with inconspicuous flowers, grown for their foliage and showy fruits. Needs separate male and female plants in order to obtain fruits. Is suitable for coastal areas, where it is wind-resistant and excellent as hedging. Fully hardy. Needs sun and is particularly useful for poor, dry or very sandy soil. Propagate by softwood cuttings in summer or by seed in autumn.
♀ *H. rhamnoides* illus. p.94.

HOHERIA (Malvaceae)
Genus of deciduous, semi-evergreen or evergreen trees and shrubs, grown for their flowers mainly in summer. Frost hardy, but in cold areas grow against a south- or west-facing wall. Needs sun or semi-shade and fertile, well-drained soil. Propagate by semi-ripe cuttings in summer or by seed in autumn.
H. angustifolia illus. p.64.
♀ *H.* **'Glory of Amlwch'.** Semi-evergreen, spreading tree. H 7m (22ft), S 6m (20ft). Has long, narrowly oval, glossy, bright green leaves and a profusion of large, 5-petalled, white flowers from mid- to late summer.
♀ *H. lyallii* illus. p.64.
H. populnea (Lace-bark). Evergreen, spreading tree. H 12m (40ft), S 10m (30ft). Bears narrowly oval, glossy, dark green leaves and dense clusters of 5-petalled, white flowers in late summer and early autumn. Bark on mature trees is pale brown and white and often flaky.
♀ *H. sexstylosa* (Ribbonwood). Fast-growing, evergreen, upright tree or shrub. H 8m (25ft), S 6m (20ft). Glossy, pale green leaves are narrowly oval and sharply toothed. Star-shaped, 5-petalled, white flowers are borne in clusters from mid- to late summer.

HOLBOELLIA (Lardizabalaceae)
Genus of evergreen, twining climbers, grown mainly for their fine foliage. Half hardy. Male and female flowers are borne on the same plant. Grow in any well-drained soil and in shade or full light. Propagate by stem cuttings in late summer or autumn.
H. coriacea. Evergreen, twining climber. H to 7m (22ft) or more. Glossy, green leaves have 3 leaflets. In spring produces clusters of tiny, mauve, male flowers and, lower down stems, larger, purple-tinged, green, female flowers. Has sausage-shaped, purple fruits, 4–6cm (1½–2½in) long.

HOLCUS (Gramineae). See GRASSES, BAMBOOS, RUSHES and SEDGES.
H. mollis (Creeping soft grass). **'Albovariegatus'** (syn. *H.m.* 'Variegatus') illus. p.182.

HOLMSKIOLDIA (Verbenaceae)
Genus of evergreen shrubs or scrambling climbers. Frost tender, min. 16°C (61°F). Any fertile, well-drained soil is suitable with full light. Water freely when in full growth, less at other times. Needs tying to supports. Thin out crowded growth in spring or after flowering. Propagate by seed in spring or by softwood or semi-ripe cuttings in summer. Whitefly and red spider mite may be troublesome.
H. sanguinea (Chinese hat plant, Mandarin's hat plant). Evergreen, straggly shrub. H to 5m (15ft), S 2m (6ft). Oval, serrated leaves are 5–10cm (2–4in) long. Showy, red or orange flowers, with saucer-shaped calyces and central, 5-lobed tubes, appear in autumn-winter.

HOLODISCUS (Rosaceae)
Genus of deciduous shrubs, grown for their flowers in summer. Fully hardy. Needs sun or semi-shade and any but very dry soil. Propagate by softwood cuttings in summer.
H. discolor illus. p.89.

HOMERIA (Iridaceae)
Genus of spring- or summer-flowering corms with widely funnel-shaped, cup-shaped or flattish flowers. Half hardy. Needs a sunny site and well-drained soil. For flowers in spring, pot in autumn in a cool greenhouse, water until after flowering, then dry off for summer. For flowers in summer, plant in the open in spring. Propagate by seed, division or offsets in autumn.
H. ochroleuca. Spring- or summer-flowering corm. H to 55cm (22in), S 5–8cm (2–3in). Slender, wiry stems each bear 1 or 2 long, narrow, semi-erect leaves on lower part of stem. Bears a succession of upright, cup-shaped to flattish, yellow flowers, each sometimes with a central, orange stain.

HOMOGYNE (Compositae)
Genus of evergreen perennials, useful for ground cover, especially in rock gardens and woodland. Fully hardy. Needs shade and moist soil. Propagate by division in spring or by seed when fresh.
H. alpina (Alpine coltsfoot). Evergreen, mat-forming, rhizomatous perennial. H 8–15cm (3–6in), S 15cm (6in) or more. Has kidney-shaped, toothed, glossy leaves and, in summer, stems, 8–15cm (3–6in) or more long, each carry a daisy-like, rose-purple flower head.

HOODIA (Asclepiadaceae)
Genus of branching, perennial succulents with firm, erect, green stems, generally branching from the base. Frost tender, min. 10–15°C (50–59°F). Needs full sun and very well-drained soil. Is difficult to cultivate. Water sparingly at all times. Propagate by seed or grafting in spring or summer.
H. bainii. Clump-forming, perennial succulent. H 30cm (12in), S 15cm (6in). Min. 15°C (59°F). Dull green stem has spiral rows of tubercles, each terminating in a sharp thorn. Produces 5-lobed, dull yellow flowers, 7cm (3in) across, in summer-autumn. Keep dry in winter.
H. gordonii. Variable, erect, clump-forming, perennial succulent. H 80cm (32in), S 30cm (12in). Min 10°C (50°F). Green stem is covered with short, spine-tipped tubercles in distorted rows. Often branches into clumps. In late summer bears 5-lobed, flesh-coloured to brownish flowers.

HORDEUM (Gramineae). See GRASSES, BAMBOOS, RUSHES and SEDGES.
H. jubatum illus. p.183.

HORMINUM (Labiatae)
Genus of one species of basal-rosetted perennial, suitable for rock gardens. Fully hardy. Needs sun and well-drained soil. Propagate by division in spring or by seed in autumn.
H. pyrenaicum (Dragon's mouth). Basal-rosetted perennial. H and S 20cm (8in). In summer carries whorls of nodding, short-stalked, funnel-shaped, blue-purple or white flowers above oval, leathery, dark green leaves, 8–10cm (3–4in) long.

HOSTA (Funkiaceae)
Funkia, Plantain lily
Genus of perennials, grown mainly for their decorative foliage. Forms large clumps that are excellent for ground cover (heights given are those of foliage). Fully hardy. Most species prefer shade and rich, moist but well-drained, neutral soil. Propagate by division in early spring. Seed-raised plants (with the exception of *H. ventricosa*) very rarely come true to type. Slug and snail control is essential.

H. albomarginata. See *H. sieboldii.*
H. **'August Moon'** (illus. p.253). Fast-growing, robust, clump-forming perennial. H 60cm (2ft), S 1m (3ft). Oval, soft golden-yellow leaves each have a faint glaucous blue bloom. Racemes of trumpet-shaped, pale greyish-mauve flowers crown foliage in mid-summer. Requires semi-shade.
H. **'Blue Moon'.** Slow-growing, compact, clump-forming perennial. H 12cm (5in), S 30cm (12in). Oval to rounded, greyish-blue leaves taper to a point. In mid-summer, dense clusters of trumpet-shaped, mauve flowers are borne just above leaves. Is suitable for a rock garden. Prefers partial shade.
H. **'Blue Wedgwood'** (illus. p.252). Slow-growing, clump-forming perennial. H 30cm (1ft), S 45cm (1½ft). Has wedge-shaped, deeply quilted, blue leaves and, in summer, lavender flowers.
♀ *H. crispula* (illus. p.252). Slow-growing, clump-forming perennial. H 75cm (2½ft), S 1m (3ft). Large, oval to heart-shaped, wavy-edged leaves are dark green with irregular, white margins. Racemes of trumpet-shaped, pale mauve flowers appear well above foliage in mid-summer. Is excellent for waterside planting. Leaves are sometimes damaged by late frost. Needs protection from strong wind. Is prone to virus.
H. decorata f. *decorata* (illus p.252). Slow-growing, clump-forming perennial. H 45cm (1½ft), S 1m (3ft). Oval to rounded leaves, tapering to a point, are dark green with narrow, regular, white margins. Dense racemes of trumpet-shaped, violet flowers in mid-summer are followed by large, ovoid, glossy, dark green, then brown seed heads that are much used in flower arrangements. Prefers sandy soil and semi-shade and is best grown in a woodland garden. f. *normalis* has plain green leaves.
H. fortunei. Group of vigorous, clump-forming, hybrid perennials. H 75cm–1m (2½–3ft), S 1m (3ft) or more. Leaves are oval to heart-shaped.
♀ **'Albopicta'** (syn. *H.f.* var. *albopicta;* illus. p.253) has pale green leaves, with creamy-yellow centres, fading to dull green from mid-summer. Racemes of trumpet-shaped, pale violet flowers open above foliage in early summer. ♀ **'Aureomarginata'** (syn. *H.f.* 'Yellow Edge'; illus. p.253) has mid-green leaves with irregular, creamy-yellow edges. In mid-summer, trumpet-shaped, violet flowers are carried in racemes above foliage. Mass planting produces the most effective results. Tolerates full sun. **'Marginato-alba'** has thin-textured, mid- to dark green leaves that have irregular, white margins. Racemes of trumpet-shaped, violet flowers are borne on tall stems in mid-summer. Is ideal for a waterside. **'Yellow Edge'** see *H.f.* 'Aureomarginata'.
H. **'Ginko Craig'** (illus. p.252). Low-growing, clump-forming perenial. H and S 30cm (1ft). Has small, narrow, dark green leaves irregularly edged with white. In summer produces spikes of bell-shaped, deep mauve flowers. Is a good edging plant.

H. 'Gold Standard' (illus. p.253). Vigorous, clump-forming perennial. H 75cm (2½ft), S 1m (3ft). Oval to heart-shaped leaves are pale green, turning to gold from mid-summer, with narrow, regular, dark green margins. Racemes of trumpet-shaped, violet flowers are produced above leaves in mid-summer. Prefers partial shade.
H. 'Golden Prayers' (illus. p.253). Upright, clump-forming perennial. H 15cm (6in), S 30cm (12in). Cupped leaves are puckered and bright golden green. Flowers are white suffused with pale lavender. Suits a rock garden.
▼ *H.* 'Golden Tiara' (illus. p.253). Clump-forming perennial. H 15cm (6in), S 30cm (12in). Neat, broadly heart-shaped, dark green leaves have well-defined, chartreuse-yellow edges. In summer produces tall spikes of lavender-purple flowers.
H. 'Hadspen Blue' (illus. p.252). Slow-growing, clump-forming perennial. H and S 30cm (12in). Smooth leaves are heart-shaped and deep glaucous blue. Produces short spikes of lavender flowers in summer.
▼ *H.* 'Halcyon' (illus. p.252). Robust, clump-forming perennial. H 30cm (1ft), S 1m (3ft). Has heart-shaped, tapering, greyish-blue leaves that fade to muddy-green in full sun; texture may be spoiled by heavy rain. Heavy clusters of trumpet-shaped, violet-mauve flowers open just above foliage in mid-summer.
▼ *H.* 'Honeybells' (illus. p.253). Clump-forming perennial. H 1m (3ft), S 60cm (2ft). Light green leaves are blunt at the tips and have wavy edges. In late summer bears fragrant, pale lilac flowers.
H. hypoleuca (White-backed hosta). Clump-forming perennial. H 45cm (1½ft), S 1m (3ft). Broadly oval leaves have widely spaced veins and are pale green above, striking white beneath. In late summer bears drooping racemes of trumpet-shaped, milky-violet flowers with mauve-flecked, pale green bracts. Tolerates full sun.
H. 'Kabitan' (illus. p.253). Clump-forming perennial, spreading by stolons. H to 30cm (1ft), S 60cm (2ft). Lance-shaped, thin-textured, glossy leaves are yellow-centred and have narrow, undulating, dark green margins. In early summer produces a short raceme of small, trumpet-shaped, pale violet flowers. Is suitable for a shaded rock garden. Needs establishing in a pot for first few years.
▼ *H.* 'Krossa Regal' (illus. p.252). Vase-shaped, clump-forming perennial. H and S 1m (3ft). Arching, deeply ribbed leaves are greyish blue. Produces tall spikes of pale lilac flowers in summer. Tolerates sun.
▼ *H. lancifolia*. Arching, clump-forming perennial. H 45cm (1½ft), S 75cm (2½ft). Has narrowly lance-shaped, thin-textured, glossy, mid-green leaves. Racemes of trumpet-shaped, deep violet flowers are borne above foliage in late summer and autumn.
▼ *H.* 'Love Pat' (illus p.252). Vigorous, clump-forming perennial. H and S to 60cm (2ft). Produces rounded, deeply puckered, deep glaucous blue leaves. Racemes of pale lilac flowers are produced during summer.
H. montana. Vigorous, clump-forming perennial. H 1.1m (3½ft), S 1m (3ft). Has oval, prominently veined, glossy, dark green leaves. Racemes of trumpet-shaped, pale violet flowers open well above foliage in mid-summer. Slower-growing 'Aurea Marginata' (illus. p.253) has leaves irregularly edged with golden-yellow. Is always the first hosta to appear in spring.
H. 'Piedmont Gold' (illus. p.253). Slow-growing, clump-forming perennial. H 60cm (2ft), S 75cm (2½ft). Smooth leaves are bright yellowish-green with fluted margins. Racemes of white flowers are produced in summer. Is best in light shade.
H. plantaginea (August lily; illus. p.253). Lax, clump-forming perennial. H 60cm (2ft), S 1.2m (4ft). Leaves are oval and glossy, pale green. Rising well above these are flower stems crowned in late summer and early autumn with fragrant, trumpet-shaped, white flowers that open in the evening. Prefers sunny conditions.
H. rectifolia. Upright, clump-forming perennial. H 1m (3ft), S 75cm (2½ft). Has oval to lance-shaped, dark green leaves and racemes of large, trumpet-shaped, violet flowers from mid- to late summer.
H. 'Regal Splendor' (illus. p.252). Clump-forming perennial. H and S 1m (3ft). Arching, greyish-blue leaves are suffused white or yellow at the margins. Talls spikes of lilac flowers are produced in summer.
H. rohdeifolia (illus. p. 252). Clump-forming perennial. H and S to 1m (3ft). Has shining, dark olive-green leaves that are usually edged with pale yellow. In summer, erect stems carry nodding, pale mauve flowers.
▼ *H.* 'Royal Standard' (illus. p.253). Upright, clump-forming perennial. H 60cm (2ft), S 1.2m (4ft). Leaves are broadly oval and glossy, pale green. Slightly fragrant, trumpet-shaped, pure white flowers are borne well above foliage and open in the evening. Prefers sun.
▼ *H.* 'Shade Fanfare' (illus. p.252). Vigorous, clump-forming perennial. H 45cm (1½ft), S 75cm (2½ft). Heart-shaped leaves are pale green with cream margins. In summer has an abundance of lavender flowers.
H. sieboldiana (illus. p.252). Robust, clump-forming perennial. H 1m (3ft) or more, S 1.5m (5ft). Large, heart-shaped, deeply ribbed, puckered leaves are bluish-grey. Racemes of trumpet-shaped, very pale lilac flowers open in early summer, just above foliage. Makes good ground cover. Tolerates sun, but leaves may then turn dull green. ▼ var. *elegans* (illus. p252) has larger, bluer leaves. ▼ 'Frances Williams' (illus. p.253) has yellow-edged leaves, is slower-growing and should not be grown in full sun.
▼ *H. sieboldii* (syn. *H. albomarginata*; illus. p.252). Vigorous, clump-forming perennial. H 45cm (1½ft), S 60cm (2ft). Lance-shaped, round-tipped leaves are mid- to dark green with narrow, irregular, white edges. Racemes of trumpet-shaped, violet flowers appear at tops of stems in late summer and are followed by ovoid, glossy, dark green, then brown seed heads that are useful for flower arrangements.
H. 'Snowden' (illus. p.252). Clump-forming perennial. H and S 1m (3ft) or more. Large, pointed glaucous blue leaves age to sage green. Tall stems bear white tinged with green flowers in summer.
▼ *H.* 'Sum and Substance' (illus. p.253). Vigorous, clump-forming perennial. H and S to 1m (3ft). Produces large, greenish-gold leaves that are thick in texture and, in mid-summer, tall spikes of pale lavender flowers. Tolerates full sun.
H. 'Tall Boy' (illus. p.253). Clump-forming perennial. H and S 60cm (2ft). Large, bright green leaves end in long points. In summer produces an abundance of rich lilac flowers on stems 1.2m (4ft) tall or more.
H. tardiflora (illus. p.252). Slow-growing, clump-forming perennial. H 30cm (1ft), S 75cm (2½ft). Has narrowly lance-shaped, thick-textured, dark green leaves. Dense racemes of trumpet-shaped, lilac-purple flowers open just above foliage from late summer to early autumn.
H. 'Thomas Hogg' (syn. *H. undulata* 'Albomarginata'; illus. p.252). Clump-forming perennial. H 75cm (2½ft), S 1m (3ft). Has oval to lance-shaped, round-tipped, mid-green leaves, each with an irregular, pale cream to white margin that continues as a narrow line down leaf stalk. Racemes of trumpet-shaped, pale violet flowers rise well above foliage in mid-summer. Grows well under trees.
H. tokudama. Very slow-growing, clump-forming perennial. H 45cm (1½ft), S 75cm (2½ft). Produces cup-shaped, puckered, blue leaves. Racemes of trumpet-shaped, pale lilac-grey flowers appear just above foliage in mid-summer. 'Aureonebulosa' (syn. *H.t.* 'Variegata'; illus. p252) has irregular, cloudy-yellow centres to leaves. 'Flavocircinalis', often mistaken for a juvenile *H.* 'Frances Williams', has heart-shaped leaves with wide, irregular, creamy-yellow margins. 'Variegata' see *H.t.* 'Aureonebulosa'.
H. undulata 'Albomarginata' see *H.* 'Thomas Hogg'. ▼ var. *erromena* is a robust, clump-forming perennial. H 45cm (1½ft), S 60cm (2ft). Bears trumpet-shaped, pale mauve flowers high above oblong, wavy, mid-green leaves. Leaves of ▼ var. *undulata* have irregular, white centres on first crop, white streaks on second. Leaf stalks and flower bracts are white with narrow, green margins.
▼ var. *univittata* (illus. p252) is vigorous with oval, mid-green leaves that have narrow, white centres.
▼ *H. ventricosa* (illus. p.253). Clump-forming perennial. H 70cm (28in), S 1m (3ft) or more. Has heart-shaped to oval, slightly wavy-edged, glossy, dark green leaves. Racemes of bell-shaped, deep purple flowers appear above foliage in late summer. Usually comes true from seed. 'Aureomaculata' (illus. p.253) has leaves with irregular, central, creamy-yellow variegation.
▼ 'Variegata' (syn. *H.v.* 'Aureomarginata', illus. p.253) has leaves with irregular, cream margins.
▼ *H. venusta* (illus. p.253). Vigorous, mat-forming perennial. H 2.5cm (1in), S to 30cm (12in). Has oval to lance-shaped, mid- to dark green leaves and abundant racemes of trumpet-shaped, purple flowers, borne well above foliage in mid-summer. Suits a rock garden.
▼ *H.* 'Wide Brim' (illus. p.253). Vigorous, clump-forming perennial. H and S to 75cm (2½ft). Leaves are heavily puckered and dark blue-green with wide irregular, creamy-white margins. Produces white or very pale lavender flowers in summer.
H. 'Zounds' (illus. p.253). Slow-growing, clump-forming perennial. H and S to 1m (3ft). Large, bright gold leaves are heavily puckered and have metallic sheen. White or pale lavender flowers are produced in early summer.

HOTTONIA (Primulaceae)
Genus of deciduous, perennial, submerged water plants, grown for their handsome foliage and flowers. Fully hardy. Needs sun and clear, cool water, still or running. Periodically thin overcrowded growth. Propagate by stem cuttings in spring or summer.
H. palustris illus. p.387.

HOUSTONIA. See HEDYOTIS.

HOUTTUYNIA (Saururaceae)
Genus of one species of perennial or deciduous marginal water plant, with far-spreading rhizomes. Is suitable for ground cover, although invasive. Fully hardy. Prefers semi-shade and moist soil or shallow water, beside streams and ponds. Propagate by runners in spring.
H. cordata 'Chamaeleon', syn. *H.c.* 'Variegata', illus. p.387. 'Flore Pleno' (syn. *H.c.* 'Plena') is a spreading perennial. H 15–60cm (6–24in), S indefinite. In spring, spikes of insignificant flowers, surrounded by 8 or more oval, white bracts, appear above heart-shaped, pointed, aromatic, fleshy, leathery leaves.

HOVENIA (Rhamnaceae)
Genus of one species of deciduous, summer-flowering tree, grown for its foliage. Fully hardy, but young, unripened growth is susceptible to frost damage. Does best in full sun and requires fertile, well-drained soil. Propagate by softwood cuttings in summer or by seed in autumn.
H. dulcis illus. p.53.

HOWEA, syn. HOWEIA, KENTIA (Palmae)
Genus of evergreen palms, grown for their ornamental appearance. Frost tender, min. 16–18°C (61–4°F). Needs partial shade and humus-rich, well-drained soil. Water potted specimens freely in summer, minimally in winter, moderately at other times. Propagate by seed in spring at not less than 26°C (79°F). Is prone to red spider mite.
▼ *H. forsteriana* (Paradise palm, Sentry palm, Thatch-leaf palm). Evergreen, upright palm with a slender

stem. H 10m (30ft), S 3–4m (10–12ft). Has spreading, feather-shaped leaves, 1.5–2.5m (5–8ft) long, of strap-shaped leaflets. Produces branching clusters of several spikes of small, greenish-brown flowers in winter.

HOWEIA. See **HOWEA.**

HOYA (Asclepiadaceae)
Genus of evergreen, woody-stemmed, twining and/or root climbers and loose shrubs, grown for their flowers and foliage. Frost tender, min. 5–18°C (41–64°F). Grow in humus-rich, well-drained soil with semi-shade in summer. Water moderately when in full growth, sparingly at other times. Stems require support. Cut back and thin out crowded stems after flowering or in spring. Propagate by semi-ripe cuttings in summer.
H. australis illus. p.168.
H. bella. See *H. lanceolata* subsp. *bella*.
❦ *H. carnosa* illus. p.168.
H. coronaria. Slow-growing, evergreen, woody-stemmed, twining and root climber. H 2–3m (6–10ft). Min. 16–18°C (61–4°F). Has thick, leathery, oblong to oval leaves. Bell-shaped, yellow to white flowers, spotted with red, are produced in summer.
H. imperialis. Vigorous, evergreen, woody-stemmed, twining and root climber. H to 6m (20ft). Min. 16–18°C (61–4°F). Oval, leathery, downy leaves are 10–23cm (4–9in) long. In summer has large, star-shaped, brown-purple to deep magenta flowers with cream centres.
❦ *H. lanceolata* subsp. *bella*, syn. *H. bella*, illus. p.168.

HUERNIA (Asclepiadaceae)
Genus of clump-forming, perennial succulents with finger-like, usually 4-angled stems. Produces minute, short-lived, deciduous leaves on new growth. Frost tender, min. 8–11°C (46–52°F). Requires sun or partial shade and extremely well-drained soil. Is one of easiest stapeliads to grow. Propagate by seed or stem cuttings in spring or summer.
H. macrocarpa. Deciduous, clump-forming, perennial succulent. H and S 10cm (4in). Min. 8°C (46°F). Has finger-like, 4- or 5-sided, grey-green stems. In summer-autumn produces bell-shaped, yellow flowers, with narrow, purple bands and recurved tips, at base of new growth. var. *arabica* illus. p.412.
H. pillansii. Deciduous, clump-forming, perennial succulent. H 5cm (2in), S 10cm (4in). Min. 11°C (52°F). Has a finger-like, light green stem densely covered with short tubercles that have hair-like tips. In summer-autumn bears bell-shaped, creamy-red flowers, with red spots, at base of new growth.
H. primulina. See *H. thuretii* var. *primulina*.
H. thuretii var. *primulina*, syn. *H. primulina.* Deciduous, clump-forming, perennial succulent. H 10cm (4in), S 15cm (6in). Min. 11°C (52°F). Stems are short, thick and grey-green.

In summer-autumn carries bell-shaped, dull yellow flowers, 2cm (³⁄₄in) across, with reflexed, blackish tips, at base of new growth.
H. zebrina (Owl-eyes). Deciduous, clump-forming, perennial succulent. H 10cm (4in), S 15cm (6in). Min. 11°C (52°F). Is similar to *H. thuretii* var. *primulina*, but has pale yellow-green flowers with bands of red-brown.

HUMEA. See **CALOMERIA.**

HUMULUS (Cannabidaceae)
Hop
Genus of herbaceous, twining climbers. Is useful for concealing unsightly sheds or tree-stumps. Male and female flowers are produced on separate plants; female flower spikes become drooping clusters of 'hops'. Fully hardy. Grow in sun or semi-shade and in any well-drained soil. Propagate by tip cuttings in spring.
H. lupulus (Common hop).
❦ **'Aureus'** illus. p.166.

HUNNEMANNIA (Papaveraceae)
Genus of poppy-like perennials, usually grown as annuals. Half hardy. Grow in sun and in poor to fertile, very well-drained soil. Dead-head plants regularly. Provide support, especially in windy areas. Propagate by seed sown under glass in early spring, or outdoors in mid-spring.
H. fumariifolia (Mexican tulip poppy). **'Sunlite'** is a fast-growing, upright perennial, grown as an annual. H 60cm (24in), S 20cm (8in). Has oblong, very divided, bluish-green leaves and, in summer and early autumn, poppy-like, semi-double, bright yellow flowers, to 8cm (3in) wide.

HYACINTHELLA (Liliaceae)
Genus of spring-flowering bulbs with short spikes of small, bell-shaped flowers, suitable for rock gardens and cold greenhouses. Frost hardy. Requires an open, sunny situation and well-drained soil, which partially dries out while bulbs are dormant in summer. Propagate by seed in autumn.
H. leucophaea illus. p.376.

HYACINTHOIDES, syn. **ENDYMION** (Liliaceae)
Bluebell
Genus of spring-flowering bulbs, grown for their bluebell flowers. Is suitable for growing in borders and for naturalizing in grass beneath trees and shrubs. Fully hardy. Requires partial shade and plenty of moisture. Prefers heavy soil. Plant bulbs in autumn 10–15cm (4–6in) deep. Propagate by division in late summer or by seed in autumn.
H. hispanica, syn. *Scilla campanulata, S. hispanica,* illus. p.358.
H. italica. Spring-flowering bulb. H 15–20cm (6–8in), S 5–8cm (2–3in). Produces a basal cluster of narrowly strap-shaped, semi-erect leaves. Leafless stem produces a conical spike of many flattish, star-shaped, blue flowers, 1cm (¹⁄₂in) across.
H. non-scripta, syn. *Scilla non-scripta,* illus. p.358.

HYACINTHUS (Liliaceae)
Hyacinth
Genus of bulbs, grown for their dense spikes of fragrant, tubular flowers; is ideal for spring bedding displays and for pot cultivation indoors. Frost hardy. Needs an open, sunny situation or partial shade and well-drained soil. Plant in autumn. For winter flowers, force large-size, specially 'treated' bulbs of *H. orientalis* cultivars by potting in early autumn, then keep cool and damp for several weeks to ensure adequate root systems develop. When shoot tips are visible, move into max. 10°C (50°F) at first, raising temperature as more shoot appears and giving as much light as possible. After forcing, keep in a cool place to finish growth, then plant out to recover. Propagate by offsets in late summer or early autumn.
H. amethystinus. See *Brimeura amethystina.*
H. azureus. See *Muscari azureum.*
H. orientalis **'Amsterdam'** (illus. p.375). Winter- or spring-flowering bulb. H 10–20cm (4–8in), S 6–10cm (2¹⁄₂–4in). Frost hardy. Has strap-shaped, channelled, semi-erect, glossy, basal leaves that develop fully only after flowering. Flower stem carries a dense, cylindrical spike of fragrant, tubular, bright rose-red flowers, each with 6 recurving petals.
❦ **'Blue Jacket'** (illus. p.375) has very large spikes of navy-blue flowers with purple veining. ❦ **'City of Haarlem'** (illus. p.375) bears pale yellow flowers in a dense spike. ❦ **'Delft Blue'** (illus. p.375) has violet-flushed, soft blue flowers. **'Distinction'** (illus. p.375) produces slender, open spikes of reddish-purple flowers; those of **'Jan Bos'** (illus. p.375) are crimson. **'Lady Derby'** (illus. p.375) bears rose-pink flowers. ❦ **'L'Innocence'** has ivory-white flowers. ❦ **'Ostara'** (illus. p.375) has a large spike of blue flowers, with a dark stripe along each petal centre. ❦ **'Pink Pearl'** has a dense spike of carmine-pink flowers. Flowers of **'Princess Maria Christina'** (illus. p.375) are salmon-pink; those of **'Queen of Pinks'** (illus. p.375) are soft pink. **'Violet Pearl'** (illus. p.375) produces spikes of violet flowers. **'White Pearl'** (illus. p.375) is pure white.

HYDRANGEA (Hydrangeaceae)
Genus of deciduous shrubs and deciduous or evergreen, root climbers, grown for their mainly domed or flattened flower heads. Each head usually consists of masses of small, inconspicuous, fertile flowers, surrounded by or mixed with much larger, sterile flowers bearing showy, petal-like sepals. However, in some forms, all or most of the flowers are sterile. Fully to frost hardy. Prefers sun or semi-shade and fertile, moist but well-drained soil. Requires more shade in dry areas. Propagate by softwood cuttings in summer.
H. anomala subsp. *petiolaris.* See *H. petiolaris.*
❦ *H. arborescens* **'Annabelle'** (illus. p.114). Deciduous, open shrub. H and S 2.5m (8ft). Fully hardy. Long-stalked,

broadly oval leaves are glossy, dark green above, paler beneath. Very large rounded heads of mainly sterile, white flowers are borne in summer.
❦ **'Grandiflora'** (illus. p.114) has smaller flower heads but larger sterile flowers.
❦ *H. aspera* subsp. *aspera* (syn. *H.a.* Villosa group, *H. villosa*; illus. p.114). Deciduous, upright shrub. H and S 3m (10ft). Fully hardy. Has peeling bark and from late summer to mid-autumn heads of small, blue or purple, central flowers and larger, white, sometimes flushed purplish-pink, outer ones. subsp. *sargentiana* see *H. sargentiana.*
H. heteromalla. Deciduous, arching shrub. H 5m (15ft), S 3m (10ft). Fully hardy. Narrowly oval, dark green leaves turn yellow in autumn. Broad, flat, open heads of white flowers are borne in mid- and late summer, the outer ones ageing to deep pink.
❦ **'Bretschneideri'** illus. p.114.
H. involucrata. Deciduous, spreading, open shrub. H 1m (3ft), S 2m (6ft). Frost hardy. Has broadly heart-shaped, bristly, mid-green leaves. During late summer and autumn bears heads of small, blue, inner flowers surrounded by large, pale blue to white, outer ones.
❦ **'Hortensis'** (illus. p.114) is smaller and has clusters of cream, pink and green flowers.
H. macrophylla. Deciduous, bushy shrub. H 1.5–2m (5–6ft), S 2–2.5m (6–8ft). Frost hardy. Has oval, toothed, glossy, light green leaves. In mid- to late summer, blue or purple flowers are produced in acid soils with a pH of up to about 5.5. In neutral or alkaline soils above this level, flowers are pink or red. White flowers are not affected by pH. Prune older shoots back to base in spring. Trim back winter-damaged growths to new growth and remove spent flower heads in spring. Is divided into 2 groups: Hortensias, which have domed, dense heads of mainly sterile flowers; and Lacecaps, which have flat, open heads, each with fertile flowers in the centre and larger, sterile flowers on the outside that are green in bud. ❦ **'Altona'** (Hortensia; illus. p.114), H 1m (3ft), S 1.5m (5ft), has large heads of rich pink to deep purple-blue flowers. ❦ **'Ami Pasquier'** (Hortensia), H 60cm (2ft), S 1m (3ft), is compact, with deep crimson or blue-purple flowers. **'Blue Bonnet'** (Hortensia; illus. p.114) and ❦ **'Blue Wave'** (syn. *H.m.* 'Mariesii Perfecta', Lacecap; illus. p.114), H 2m (6ft), S to 2.5m (8ft), produce heads of rich blue or lilac to pink flowers. Flower heads of ❦ **'Générale Vicomtesse de Vibraye'** (Hortensia; illus. p.114), H and S 1.5m (5ft), are rounded and pale blue or pink. Foliage is light green. **'Hamburg'** (Hortensia; illus. p.114), H 1m (3ft), S 1.5m (5ft), is vigorous and has large, deep pink to deep blue flowers with serrated sepals. ❦ **'Lanarth White'** (Lacecap; illus. p.114), H and S 1.5m (5ft), has pink or blue fertile flowers edged with pure white sterile flowers. **'Lilacina'** (Lacecap; illus. p.114), H and S 2m (6ft), has deep, lilac, central flowers and pinkish-purple outer flowers. ❦ **'Mme E. Mouillère'**

515

(Hortensia) has white flowers, becoming pale pink, and prefers partial shade. subsp. *serrata* see *H. serrata*. subsp. *serrata* 'Preziosa' see *H.* 'Preziosa'. ♀ **'Veitchii'** (Lacecap; illus. p.114) has lilac-blue flowers.
H. paniculata **'Brussels Lace'** (illus. p.114). Deciduous, upright, open shrub. H and S to 3m (10ft). Fully hardy. Has large, pointed, dark green leaves. Bears delicate, open panicles of white flowers in late summer and early autumn. ♀ **'Floribunda'** (illus. p.114) has dense conical heads of small, fertile, central flowers surrounded by large, white ray flowers.
♀ **'Grandiflora'** has large, oval and dark green leaves. Large, conical panicles of mostly sterile, white flowers turn pink or red from late summer. Prune back hard in spring to obtain largest panicles. Flower heads of **'Pink Diamond'** (illus. p.114) turn pink with age. ♀ **'Praecox'** (illus. p.114) flowers from mid-summer. ♀ **'Unique'** is similar to 'Grandiflora' but is more vigorous and has larger flowers. **'Tardiva'** has both fertile and sterile flowers from early to mid-autumn.
♀ *H. petiolaris*, syn. *H. anomala* subsp. *petiolaris*, illus. p.168.
♀ *H.* **'Preziosa'**, syn. *H. macrophylla* subsp. *serrata* 'Preziosa', *H. serrata* 'Preziosa'. Deciduous, bushy shrub. H 1.5–2m (5–6ft), S 2–2.5m (6–8ft). Frost hardy. Has oval, toothed, light green leaves. Bears pink flowers, becoming deep crimson.
♀ *H. quercifolia* (Oak-leaved hydrangea; illus. p.114). Deciduous, bushy, mound-forming shrub. H and S 2m (6ft). Frost hardy. Deeply lobed, dark green leaves turn red and purple in autumn. Has white flower heads from mid-summer to mid-autumn.
♀ *H. sargentiana*, syn. *H. aspera* subsp. *sargentiana*. Deciduous, upright, gaunt shrub. H 2.5m (8ft), S 2m (6ft). Frost hardy. Has peeling bark, stout shoots and very large, narrowly oval, dull green leaves that are bristly and have grey down beneath. From late summer to mid-autumn bears broad heads of flowers, the inner ones small and blue or deep purple, the outer ones larger and white, sometimes flushed purplish-pink.
H. serrata, syn. *H. macrophylla* subsp. *serrata* (illus. p.114). Deciduous, bushy, dense shrub. H and S 1.2m (4ft). Frost hardy. Has slender stems and light green leaves. From mid- to late summer bears flat heads of pink, lilac or white inner and pink or blue outer flowers. ♀ **'Bluebird'** (illus. p.114) has pale pink, pale purple or blue flowers. **'Preziosa'** see *H.* 'Preziosa'.
H. villosa. See *H. aspera* subsp. *aspera*.

HYDROCHARIS (Hydrocharitaceae) Genus of one species of deciduous, perennial, floating water plant, grown for its foliage and flowers. Fully hardy. Requires an open, sunny position in still water. Propagate by detaching young plantlets as required.
H. morsus-ranae illus. p.386.

HYDROCLEIS. See HYDROCLEYS.

HYDROCLEYS, syn. HYDROCLEIS (Butomaceae or Limnocharitaceae) Genus of deciduous or evergreen, annual or perennial, water plants, grown for their floating foliage and attractive flowers. Frost tender, min. 5°C (41°F). Is best grown in large aquariums and tropical pools with plenty of light. Propagate by seed when ripe or by tip cuttings year-round.
H. nymphoides illus. p.389.

HYGROPHILA (Acanthaceae) Genus of deciduous or evergreen, perennial, submerged water plants and marsh plants, grown for their foliage. Frost tender, min. 13°C (55°F). Regularly remove fading leaves. Propagate by stem cuttings in spring or summer.
H. polysperma. Deciduous, perennial, submerged water plant. S indefinite. Forms spreading colonies of lance-shaped, pale green leaves on woody stems. Given water above 16°C (61°F), is evergreen. Is suitable for a tropical aquarium.

HYLOCEREUS (Cactaceae) Genus of fast-growing, perennial cacti with erect, slender, climbing stems, jointed into sections, and many aerial roots. Makes successful grafting stock except in northern Europe. Frost tender, min. 11°C (52°F). Requires sun or partial shade and very well-drained soil. Propagate by stem cuttings in spring or summer.
H. undatus (Night-blooming cereus, Queen-of-the-night). Fast-growing, climbing, perennial cactus. H 1m (3ft), S indefinite. Bears freely branching, 3-angled, weakly spined, dark green stems, 7cm (3in) wide and jointed into sections. In summer produces flattish, white flowers, 30cm (12in) across, that last only one night.

HYLOMECON (Papaveraceae) Genus of one species of vigorous perennial, grown for its large, cup-shaped flowers. Is good for rock gardens, borders and woodlands but may be invasive. Fully hardy. Prefers partial shade and humus-rich, moist soil. Propagate by division in spring or by seed in autumn.
H. japonicum illus. p.298.

Hylotelephium anacampseros. See *Sedum anacampseros*.
Hylotelephium cauticola. See *Sedum cauticola*.
Hylotelephium ewersii. See *Sedum ewersii*.
Hylotelephium populifolium. See *Sedum populifolium*.
Hylotelephium sieboldii **'Variegatum'**. See *Sedum spathulifolium* 'Cape Blanco'.
Hylotelephium spectabile. See *Sedum spectabile*.
Hylotelephium tartarinowii. See *Sedum tartarinowii*.

HYMENANTHERA (Violaceae) Genus of evergreen shrubs, grown for their overall appearance and botanical interest. Frost to half hardy. Prefers full light and well-drained soil. Containerized specimens should be watered freely when in full growth, moderately at other times. Propagate by seed when ripe or in spring or by semi-ripe cuttings in late summer.
H. crassifolia. Evergreen, densely twiggy shrub of irregular outline. H and S to 1.2m (4ft). Frost hardy. Bears narrowly oval to oblong, leathery, mid-green leaves. Carries tiny, bell-shaped, 5-petalled, yellow flowers in spring-summer, followed by egg-shaped, purple fruits.

HYMENOCALLIS (Amaryllidaceae) Genus of bulbs, some of which are evergreen, grown for their fragrant flowers, somewhat like those of large daffodils. Half hardy to frost tender, min. 15°C (59°F). Needs a sheltered site, full sun or partial shade and well-drained soil. Plant in early summer, lifting for winter in cold districts. Alternatively, grow in a heated greenhouse; reduce water in winter, without drying out completely, then repot in spring. Propagate by offsets in spring or early summer.
H. x *festalis*. Spring- or summer-flowering bulb with a basal leaf cluster. H to 80cm (32in), S 30–45cm (12–18in). Frost tender. Bears strap-shaped, semi-erect leaves. Produces a head of 2–5 scented, white flowers, each 20cm (8in) across with a deep, central cup and 6 narrow, reflexed petals.
♀ *H.* x *macrostephana* illus. p.341.
H. narcissiflora, syn. *Ismene calathina*, illus. p.364.
H. speciosa. Evergreen, winter-flowering bulb. H and S 30–45cm (12–18in). Frost tender. Has broadly elliptic, semi-erect, basal leaves. Produces a head of 5–10 fragrant, white or green-white flowers, each 20–30cm (8–12in) wide with a funnel-shaped cup and 6 long, narrow petals.
H. **'Sulphur Queen'**. Spring- or summer-flowering bulb. H 60cm (24in), S 30–45cm (12–18in). Frost tender. Produces widely strap- or lance-shaped, semi-erect, basal leaves. Has a loose head of 2–5 fragrant, yellow-green flowers, each 16–20cm (6–8in) wide with a frilly-edged cup and 6 spreading petals.

HYMENOSPORUM (Pittosporaceae) Genus of one species of evergreen shrub or tree, grown for its flowers and overall appearance. Frost tender, min. 5–7°C (41–5°F). Prefers full sun, though some shade is tolerated. Requires humus-rich, well-drained soil, ideally neutral to acid. Water containerized specimens freely when in full growth, less at other times. Propagate by seed when ripe, in autumn, or in spring or by semi-ripe cuttings in late summer.
H. flavum (Native Australian frangipani). Evergreen, erect shrub or tree, gradually spreading with age. H 10m (30ft) or more, S 5m (15ft) or more. Has oval to oblong, lustrous, rich green leaves. In spring-summer bears terminal panicles of very fragrant, tubular, 5-petalled, cream flowers that age to deep sulphur-yellow.

HYPERICUM (Hypericaceae) Genus of perennials and deciduous, semi-evergreen or evergreen sub-shrubs and shrubs, grown for their conspicuous, yellow flowers with prominent stamens. Fully to half hardy. Large species and cultivars need sun or semi-shade and fertile, not too dry soil. Smaller types, which are good in rock gardens, do best in full sun and well-drained soil. Propagate species sub-shrubs and shrubs by softwood cuttings in summer or by seed in autumn, cultivars by softwood cuttings only in summer; perennials by seed or division in autumn or spring. Is generally trouble-free but *H.* x *inodorum* 'Elstead' is susceptible to rust, which produces orange spots on leaves, *H.* 'Hidcote' to a virus that makes leaves narrow and variegated.
H. balearicum. Evergreen, compact shrub. H and S to 60cm (2ft). Frost hardy. Small, oval, green leaves have wavy edges and rounded tips. Solitary large, fragrant, shallowly cup-shaped, yellow flowers are borne at stem tips above foliage from early summer to autumn.
H. beanii **'Gold Cup'**. See *H.* x *cyathiflorum* 'Gold Cup'.
H. bellum. Semi-evergreen, arching, graceful shrub. H 1m (3ft), S 1.5m (5ft). Fully hardy. Cup-shaped, golden-yellow flowers are borne from mid-summer to early autumn. Shoots are red. Oval, wavy-edged, mid-green leaves redden in autumn.
H. calycinum illus. p.140.
H. cerastoides, syn. *H. rhodoppeum*. Vigorous, evergreen sub-shrub with upright and arching branches. H 15cm (6in) or more, S 40–50cm (16–20in). Fully hardy. Leaves are oval, hairy and soft greyish-green. In late spring and early summer produces masses of saucer-shaped, bright yellow flowers in terminal clusters. Cut back hard after flowering. Suits a large rock garden.
H. coris. Evergreen, open, dome-shaped, occasionally prostrate, sub-shrub. H 15–30cm (6–12in), S 20cm (8in) or more. Frost hardy. Bears long-stemmed whorls of 3 or 4 pointed-oval leaves. Produces panicles of shallowly cup-shaped, bright yellow flowers, streaked red, in summer. Suits a sheltered rock garden.
H. x *cyathiflorum* **'Gold Cup'**, syn. *H. beanii* 'Gold Cup'. Semi-evergreen, arching shrub. H and S 1m (3ft). Frost hardy. Bears pinkish-brown shoots, oval, dark green leaves and, from mid-summer to early autumn, large, cup-shaped, golden-yellow flowers.
H. empetrifolium **'Prostratum'** illus. p.334.
♀ *H.* **'Hidcote'** illus. p.140.
H. x *inodorum* **'Elstead'** illus. p.140.
♀ *H. kouytchense* illus. p.140.
♀ *H.* x *moserianum*. Deciduous, arching shrub. H 30cm (12in), S 60cm (24in). Frost hardy. Small, bowl-shaped, yellow flowers are produced above oval, dark green leaves from mid-summer to mid-autumn.
'Tricolor' has leaves edged with white and pink. Prefers a sheltered position.
♀ *H. olympicum*. Deciduous, upright, slightly spreading, dense sub-shrub. H 15–30cm (6–12in), S to 15cm (6in).

Fully hardy. Tufts of upright stems are covered in small, oval, grey-green leaves. Produces terminal clusters of up to 5 cup-shaped, bright yellow flowers in summer. ♥ **'Sulphureum'** (syn. *H.o.* 'Citrinum') illus. p.306.
H. patulum. Evergreen or semi-evergreen, upright shrub. H and S 1m (3ft). Frost hardy. Large, cup-shaped, golden-yellow flowers open above oval, dark green leaves from mid-summer to mid-autumn.
H. reptans. Deciduous, mat-forming shrub. H 5cm (2in), S 20cm (8in). Frost hardy. Oval, green leaves turn yellow or bright red in autumn. In summer has flattish, golden-yellow flowers, crimson-flushed outside. Suits a rock garden.
H. rhodoppeum. See *H. cerastoides*.
♥ **H. 'Rowallane'.** Semi-evergreen, arching shrub. H and S 1.5m (5ft). Frost hardy, but is cut to ground level in severe winters. Large, bowl-shaped, deep golden-yellow flowers are borne from mid-summer to mid- or late autumn. Leaves are oval and rich green.

Hypocyrta radicans. See *Nematanthus gregarius*.
Hypocyrta strigillosa. See *Nematanthus strigillosus*.

HYPOESTES (Acanthaceae)
Genus of mainly evergreen perennials, shrubs and sub-shrubs, grown for their flowers and foliage. Frost tender, min. 10°C (50°F). Grow in bright light and in well-drained soil. Water frequently in growing season, less in winter. Cut back straggly stems. Propagate by stem cuttings in spring or summer. *H. phyllostachya* may be treated as an annual and propagated by seed in spring.
H. aristata. Evergreen, bushy perennial or sub-shrub. H to 1m (3ft), S 60cm (2ft). Has oval, mid-green leaves to 8cm (3in) long. Small, tubular, deep pink to purple flowers are produced in terminal spikes in late winter.
H. phyllostachya, syn. *H. sanguinolenta* of gardens, illus. p.228.
H. sanguinolenta of gardens. See *H. phyllostachya*.

HYPOXIS (Hypoxidaceae)
Genus of spring- or summer-flowering corms, grown for their flat, star-shaped flowers. Suits rock gardens. Frost to half hardy. Requires full sun and light, well-drained soil. Propagate by seed in autumn or spring.
H. angustifolia illus. p.380.
H. capensis, syn. *H. stellata*, *Spiloxene capensis*. Spring-flowering corm with a basal leaf cluster. H 10–20cm (4–8in), S 5–8cm (2–3in). Half hardy. Has very slender, narrowly lance-shaped, erect leaves. Stems each produce an upward-facing flower with pointed, white or yellow petals and a purple eye.
H. stellata. See *H. capensis*.

HYPSELA (Campanulaceae)
Genus of vigorous, creeping perennials, grown for their flowers and heart-shaped leaves. Good as ground cover, especially in rock gardens. Frost hardy. Needs shade and moist soil. Propagate by division in spring.
H. reniformis. Vigorous, creeping, stemless perennial. H 2cm (³/₄in), S indefinite. Has tiny, heart-shaped, fleshy leaves and, in spring-summer, small, star-shaped, pink-and-white flowers.

HYSSOPUS (Labiatae)
Genus of perennials and semi-evergreen or deciduous shrubs, grown for their flowers, which attract bees and butterflies, and for their aromatic foliage, which has culinary and medicinal uses. May be grown as a low hedge. Fully hardy. Requires full sun and fertile, well-drained soil. Cut back hard or, if grown as a hedge, trim lightly, in spring. Propagate by softwood cuttings in summer or by seed in autumn.
H. officinalis (Hyssop) illus. p.138. subsp. ***aristatus*** is a semi-evergreen or deciduous, upright, dense shrub. H 60cm (2ft), S 1m (3ft). Aromatic, narrowly lance-shaped leaves are bright green. From mid-summer to early autumn has densely clustered, small, 2-lipped, dark blue flowers.

I

IBERIS (Cruciferae)
Genus of annuals, perennials, evergreen sub-shrubs and shrubs, grown for their flowers and excellent for rock gardens. Some species are short-lived, flowering themselves to death. Fully to half hardy. Requires sun and well-drained soil. Propagate by seed in spring, sub-shrubs and shrubs by semi-ripe cuttings in summer.
I. amara illus. p.270. **Hyacinth-flowered Series** is a group of fast-growing, upright, bushy annuals. H 30cm (12in), S 15cm (6in). Fully hardy. Has lance-shaped, mid-green leaves and, in summer, flattish heads of large, scented, 4-petalled flowers in a variety of colours.
I. saxatilis illus. p.322.
♀ *I. sempervirens* illus. p.294.
♀ **'Snowflake'** (syn. *I. s.* 'Schneeflocke') is an evergreen, spreading sub-shrub. H 15–30cm (6–12in), S 45–60cm (18–24in). Fully hardy. Leaves are narrowly oblong, glossy and dark green. Dense, semi-spherical heads of 4-petalled, white flowers are borne in late spring and early summer. Trim after flowering.
I. umbellata. Fast-growing, upright, bushy annual. H 15–30cm (6–12in), S 20cm (8in). Fully hardy. Has lance-shaped, mid-green leaves. Heads of small, 4-petalled, white or pale purple flowers, sometimes bicoloured, are carried in summer and early autumn.
Fairy Series illus. p.274.

IDESIA (Flacourtiaceae)
Genus of one species of deciduous, summer-flowering tree, grown for its foliage and fruits. Both male and female plants are required to obtain fruits. Fully hardy. Needs sun or semi-shade and fertile, moist but well-drained soil, preferably neutral to acid. Propagate by softwood cuttings in summer or by seed in autumn.
I. polycarpa illus. p.54.

ILEX (Aquifoliaceae)
Holly
Genus of evergreen or deciduous trees and shrubs, grown for their foliage and fruits (berries). Mainly spherical berries, ranging in colour from red through yellow to black, are produced in autumn, following insignificant, usually white, flowers borne in spring. Almost all plants are unisexual, and to obtain fruits on a female plant a male also needs to be grown. Fully to half hardy. All prefer well-drained soil. Grow in sun or shade, but deciduous plants and those with variegated foliage do best in sun or semi-shade. Hollies resent being transplanted, but respond well to hard pruning and pollarding, which should be carried out in late spring. Propagate by seed in spring or by semi-ripe cuttings from late summer to early winter. Holly leaf miner and holly aphid may cause problems.
I. x *altaclerensis*. Group of vigorous, evergreen shrubs and trees. Frost hardy. Is resistant to pollution and coastal exposure.
'Balearica' (illus. p.72) is an erect, female tree. H 12m (40ft), S 5m (15ft). Has green to olive-green, young branches. Large, broadly oval leaves are spiny- or smooth-edged and glossy, dark green. Freely produces large, bright red berries.
'Belgica' (illus. p.72) is an erect, dense, female tree. H 12m (40ft), S 5m (15ft). Young branches are green to yellowish-green. Has large, lance-shaped to oblong, spiny- or smooth-edged, glossy, mid-green leaves. Large, orange-red fruits are freely produced.
♀ **'Belgica Aurea'** (syn. *I.* x *a.* 'Silver Sentinel', *I. perado* 'Aurea'; illus. p.73) is an upright, female tree. H 8m (25ft), S 3m (10ft). Young branches are green with yellow streaks. Has large, lance-shaped, mainly spineless, dark green leaves, mottled with grey-green and irregularly edged with yellow. Red berries are produced only rarely.
♀ **'Camelliifolia'** (illus. p.72) is a narrow, pyramidal, female tree. H 14m (46ft), S 3m (10ft). Has purple, young branches and large, oblong, mainly smooth-edged, glossy, dark green leaves. Reliably produces large, scarlet fruits; makes an excellent specimen tree.
'Camelliifolia Variegata' (illus. p.73). H 8m (25ft), S 3m (10ft). Is similar to *I.* x *a.* 'Camelliifolia', but leaves have broad, yellow margins.
♀ **'Golden King'** is a bushy, female shrub. H 6m (20ft), S 5m (15ft). Young branches are green with a purplish flush. Has large, oblong to oval, sometimes slightly spiny, dark green leaves, each splashed with grey-green in the centre and with a bright yellow margin that turns to cream on older leaves. Is not a good fruiter, bearing only a few reddish-brown berries, but is excellent as a hedge or a specimen plant.
♀ **'Hodginsii'** is a vigorous, dense, male tree. H 14m (46ft), S 10m (30ft). Shoots are purple; leaves are broadly oval, sparsely spiny and glossy, blackish-green.
'Lawsoniana' (illus. p.72) is a bushy, female shrub. H 6m (20ft), S 5m (15ft). Is similar to *I.* x *a.* 'Golden King', but has leaves splashed irregularly in the centre with gold and lighter green. Foliage tends to revert to plain green.
'N.F. Barnes' (illus. p.72) is a dense, female shrub. H 5.5m (18ft), S 4m (12ft). Has purple shoots and oval, mainly entire but spine-tipped, glossy, dark green leaves and red berries.
'Silver Sentinel' see *I.* x *a.* 'Belgica Aurea'.
♀ **'Wilsonii'** is a vigorous, female tree. H 8m (25ft), S 5m (15ft). Has purplish-green, young branches and large, oblong to oval, glossy, mid-green leaves with prominent veins and large spines. Freely produces large, scarlet fruits and makes a good hedging or specimen plant.
♀ *I. aquifolium* (Common holly; illus. p.72). Evergreen, much-branched, erect shrub or tree. H 20m (70ft), S 6m (20ft). Frost hardy. Has variably shaped, wavy, sharply spined, glossy, dark green leaves and bright red berries.
♀ **'Argentea Longifolia'** is a spreading, male tree. H 10m (30ft), S 6m (20ft). Shoots are purplish-green. Narrowly oval, spiny, dark green leaves, narrowly edged with creamy-white, are pink-tinged when young.
'Argentea Marginata' (Silver-margined holly; illus. p.72) is a columnar, female tree. H 14m (46ft), S 5m (15ft). Young branches are green, streaked with cream. Broadly oval, spiny, dark green leaves, with wide, cream margins, are shrimp-pink when young. Bears an abundance of bright red berries. Is good for hedging.
'Argentea Marginata Pendula' (Perry's weeping silver; illus. p.72) is a slow-growing, weeping, female tree. H 6m (20ft), S 5m (15ft). Has purple, young branches and broadly oval, spiny, dark green leaves, mottled with grey-green and broadly edged with cream. Bears red fruits. Makes a good specimen plant for a small garden.
'Atlas' is an erect, male shrub. H 5m (15ft), S 3m (10ft). Has green, young branches and oval, spiny, glossy, dark green leaves. Is useful for landscaping and hedging.
'Aurifodina' (illus. p.73) is an erect, dense, female shrub. H 6m (20ft), S 3m (10ft). Young branches are purplish. Oval, spiny leaves are olive-green with golden-yellow margins that turn tawny-yellow in winter. Produces a good crop of deep scarlet berries.
f. *bacciflava* is a much-branched, usually erect shrub or tree. H 20m (70ft), S 6m (20ft). Has variably shaped, wavy, sharply spined, glossy, dark green leaves and yellow fruits.
'Crispa Aurea Picta' (illus. p.73) is a male tree of open habit. H 10m (30ft), S 6m (20ft). Narrowly oval, twisted, sparsely spiny, blackish-green leaves are centrally blotched with golden-yellow. Foliage tends to revert to plain green.
'Elegantissima' (illus. p.73) is a bushy, dense, compact, male shrub. H 6m (20ft), S 5m (15ft). Young branches are green, streaked with yellow. Small, oval, spiny, bright green leaves, with cream margins, are bright pink when young.
'Ferox' (Hedgehog holly) is an open, male shrub. H 6m (20ft), S 4m (12ft). Has purple, young branches and oval, dark green leaves with spines over entire leaf surface.
♀ **'Ferox Argentea'** (Silver hedgehog holly) is similar to *I. a.* 'Ferox', but has leaves with cream margins.
'Flavescens' (Moonlight holly) is a columnar, female shrub. H 6m (20ft), S 5m (15ft). Young branches are purplish-red. Variably shaped leaves are dark green with a yellowish flush, when young, that will last year-round when grown in good light. Produces plentiful, red berries.
♀ **'Golden Milkboy'** is a dense, male shrub. H 6m (20ft), S 4m (12ft). Has purplish-green, young branches and oval, very spiny, bright green leaves with heavily blotched, bright yellow centres. Leaves tend to revert to plain green.
♀ **'Golden Queen'** is a dense tree that, despite its name, is male. H 10m (30ft), S 6m (20ft). Broadly oval, very spiny, mid-green leaves are edged with golden-yellow.
'Golden Van Tol', a sport of *I. a.* 'J.C. Van Tol', is an upright, female shrub. H 4m (12ft), S 3m (10ft). Young branches are purple. Oval, puckered, slightly spiny, dark green leaves have irregular, clear yellow margins. Produces a sparse crop of red fruits. Is good for hedging or as a specimen plant.
♀ **'Handsworth New Silver'** is a dense, columnar, female shrub. H 8m (25ft), S 5m (15ft). Branches are purple. Oblong to oval, spiny, dark green leaves have broad, cream margins. Bears a profusion of bright red fruits. Is excellent as a hedge or specimen plant and is good for a small garden.
'Hascombensis' is a slow-growing, dense shrub of unknown sex. H 1.5m (5ft), S 1–1.2m (3–4ft). Has purplish-green, young branches and small, oval, spiny, dark green leaves. Does not produce berries. Suits a rock garden.
♀ **'J.C. Van Tol'** is an open, female shrub that does not require cross-fertilization to produce fruits. H 6m (20ft), S 4m (12ft). Branches are dark purple when young. Oval, puckered, slightly spiny leaves are dark green. Produces a good crop of red berries. Is useful as a hedge or for a tub.
♀ **'Mme Briot'** (illus. p.73) is a vigorous, bushy, female tree. H 10m (30ft), S 5m (15ft). Young branches are purplish-green. Leaves are large, broadly oval, spiny and dark green with bright golden borders. Bears scarlet berries.
'Ovata Aurea' (illus. p.73) is a dense, male shrub. H 5m (15ft), S 4m (12ft). Has reddish-brown, young branches and oval, regularly spiny, dark green leaves with bright golden margins.
♀ **'Pyramidalis'** (illus. p.72) is a dense, female tree that does not require cross-fertilization to produce fruits. H 6m (20ft), S 5m (15ft). Has green, young branches and narrowly elliptic, slightly spiny, mid-green leaves. Produces masses of scarlet fruits. Is suitable for a small garden.

'Pyramidalis Aurea Marginata' (illus. p.73) is an upright, female shrub. H 6m (20ft), S 5m (15ft). Young branches are green. Has narrowly elliptic, mid-green leaves with prominent, golden margins and spines on upper half. Bears a large crop of red berries.

♀ 'Pyramidalis Fructu Luteo' is a conical, female shrub that broadens with age. H 6m (20ft), S 4m (12ft). Branches are green when young. Has oval, often spineless, dark green leaves and bears yellow berries. Is excellent for a small garden.

'Scotica' (illus. p.72) is a large, stiff, compact, female shrub. H 6m (20ft), S 4m (12ft). Oval, usually spineless, glossy, very dark green leaves are slightly twisted. Bears red fruits.

'Silver Milkboy' see *I. a.* 'Silver Milkmaid'.

♀ 'Silver Milkmaid' (syn. *I. a.* 'Silver Milkboy'; illus. p.72) is a dense, female shrub. H 5.5m (18ft), S 4m (12ft). Oval, wavy-edged, very spiny leaves are bronze when young, maturing to bright green, each with a central, creamy-white blotch, but tend to revert to plain green. Produces an abundance of scarlet berries. Makes a very attractive specimen plant.

♀ 'Silver Queen' (illus. p.73) is a dense shrub that, despite its name, is male. H 5m (15ft), S 4m (12ft). Has purple, young branches. Oval, spiny leaves, pink when young, mature to very dark green, almost black, with broad, cream edging.

'Watereriana' (Waterer's gold holly; illus. p.73) is a dense, male bush. H and S 5m (15ft). Young branches are green, streaked with yellow. Oval, spiny- or smooth-edged leaves are greyish-green, with broad, golden margins. Is best grown as a specimen plant.

I. x *aquipernyi* (illus. p.72). Evergreen, upright shrub. H 5m (15ft), S 3m (10ft). Frost hardy. Has small, oval, spiny, glossy, dark green leaves with long tips. Berries are large and red.

I. chinensis (illus. p.73). Evergreen, upright tree. H 12m (40ft), S 6m (20ft). Half hardy. Oval, thin-textured, glossy, dark green leaves have rounded teeth. Lavender flowers are followed by egg-shaped, glossy, scarlet fruits.

I. ciliospinosa (illus. p.72). Evergreen, upright shrub or tree. H 6m (20ft), S 4m (12ft). Frost hardy. Has small, oval, weak-spined, dull green leaves and red berries.

I. cornuta (Horned holly). Evergreen, dense, rounded shrub. H 4m (12ft), S 5m (15ft). Frost hardy. Rectangular, dull green leaves are spiny except on older bushes. Produces large, red berries. 'Burfordii' (illus. p.72) is female, S 2.5m (8ft), has glossy leaves with only a terminal spine and bears a profusion of fruits. 'Rotunda', H 2m (6ft), S 1.2m (4ft), is also female and produces a small crop of fruits; is useful for a tub or small garden.

I. crenata (Box-leaved holly, Japanese holly). Evergreen, spreading shrub or tree. H 5m (15ft), S 3m (10ft). Fully hardy. Has very small, oval, dark green leaves with rounded teeth. Bears glossy, black fruits. Is useful for landscaping or as hedging.

'Bullata' see *I. c.* 'Convexa'.

♀ 'Convexa' (syn. *I. c.* 'Bullata'; illus. p.72) is a dense, female shrub. H 2.5m (8ft), S 1.2–1.5m (4–5ft). Has purplish-green, young branches and oval, puckered, glossy leaves. Bears glossy, black fruits.

'Helleri' (illus. p.72) is a spreading, female shrub. H 1.2m (4ft), S 1–1.2m (3–4ft). Has green, young branches and oval leaves with few spines. Has glossy, black fruits. Is much used for landscaping.

'Latifolia' (illus. p.72) is a spreading to erect, female shrub or tree. H 6m (20ft), S 3m (10ft). Young branches are green and broadly oval leaves have tiny teeth. Produces glossy, black berries.

var. *paludosa* (illus. p.72) is a prostrate shrub or tree. H 15–30cm (6–12in), S indefinite. Has very small, oval, dark green leaves with rounded teeth. Bears glossy, black fruits.

'Variegata' (illus. p.73) is an open, male shrub. H 4m (12ft), S 2.5m (8ft). Oval leaves are spotted or blotched with yellow, but tend to revert to plain green.

I. dipyrena (Himalayan holly). Evergreen, dense, upright tree. H 12m (40ft), S 8m (25ft). Frost hardy. Elliptic, dull green leaves are spiny when young, later smooth-edged. Bears large, red fruits.

I. fargesii (illus. p.72). Evergreen, broadly conical tree or shrub. H 6m (20ft), S 5m (15ft). Frost hardy. Has green or purple shoots and oval, small-toothed, mid- to dark green leaves. Produces red berries. var. *brevifolia* (illus. p.72), H 4m (12ft), is dense and rounded.

I. georgei. Evergreen, compact shrub. H 5m (15ft), S 4m (12ft). Half hardy. Has small, lance-shaped or oval, weak-spined, glossy, dark green leaves with long tips. Berries are red.

I. glabra (Inkberry). Evergreen, dense, upright shrub. H 2.5m (8ft), S 2m (6ft). Fully hardy. Small, oblong to oval, dark green leaves are smooth-edged or may have slight teeth near tips. Produces black fruits.

I. insignis. See *I. kingiana*.

I. integra. Evergreen, dense, bushy shrub or tree. H 6m (20ft), S 5m (15ft). Frost hardy. Has oval, blunt-tipped, bright green leaves with smooth edges. Bears large, deep red berries.

I. 'Jermyns Dwarf'. See *I. pernyi* 'Jermyns Dwarf'.

I. kingiana, syn. *I. insignis*. Evergreen, upright tree. H 6m (20ft), S 4m (12ft). Half hardy. Very large, oblong, leathery, dark green leaves have small spines. Berries are bright red.

I. x *koehneana* (illus. p.72). Evergreen, conical shrub. H 6m (20ft), S 5m (15ft). Fully hardy. Young branches are green. Has very large, oblong, spiny, mid-green leaves and red fruits.

I. laevigata (Smooth winterberry). Deciduous, spreading shrub. H 2.5m (8ft), S 2m (6ft). Fully hardy. Oval, finely toothed leaves are pale green and berries are orange-red.

I. latifolia (Tarajo holly). Evergreen, upright shrub. H 6m (20ft), S 5m (15ft). Half hardy. Has stout, olive-green, young branches, very large, oblong, dark green leaves with short spines and plentiful, red fruits.

I. macrocarpa (illus. p.72). Deciduous, upright tree. H 10m (30ft), S 6m (20ft). Frost hardy. Has large, oval, saw-toothed, mid-green leaves and very large, black berries.

I. x *meserveae* (Blue holly). Group of vigorous, evergreen, dense shrubs. Fully hardy, but does not thrive in a maritime climate. Has oval, glossy, greenish-blue leaves. ♀ 'Blue Princess' (illus. p.72), H 3m (10ft), S 1.2m (4ft), is female and has purplish-green, young branches, small, oval, wavy, spiny leaves and an abundance of red fruits.

I. opaca (American holly; illus. p.72). Evergreen, erect tree. H 14m (46ft), S 1.2m (4ft). Fully hardy, but does not thrive in a maritime climate. Oval leaves are dull green above, yellow-green beneath and spiny- or smooth-edged. Has red fruits.

I. pedunculosa (illus. p.73). Evergreen, upright shrub or tree. H 10m (30ft), S 6m (20ft). Fully hardy. Oval, dark green leaves are smooth-edged. Bright red berries are borne on very long stalks.

I. perado 'Aurea'. See *I.* x *altaclerensis* 'Belgica Aurea'.

I. pernyi (illus. p.72). Slow-growing, evergreen, stiff shrub. H 8m (25ft), S 4m (12ft). Fully hardy. Has pale green, young branches and small, oblong, spiny, dark green leaves. Produces red berries. 'Jermyns Dwarf' (syn. *I.* 'Jermyns Dwarf'), H 60cm (2ft), S 1.2m (4ft), is low-growing and female, with glossy, very spiny leaves.

I. serrata. Deciduous, bushy shrub. H 4m (12ft), S 2.5m (8ft). Fully hardy. Small, oval, finely toothed, bright green leaves are downy when young. Pink flowers are followed by small, red fruits. f. *leucocarpa* (illus. p.73) bears white berries.

I. verticillata (Winterberry; illus. p.72). Deciduous, dense, suckering shrub. H 2m (6ft), S 1.2–1.5m (4–5ft). Fully hardy. Young branches are purplish-green. Produces oval or lance-shaped, saw-toothed, bright green leaves. Bears masses of long-lasting, red berries that remain on bare branches during winter.

I. yunnanensis. Evergreen, spreading to erect shrub. H 4m (12ft), S 2.5m (8ft). Frost hardy. Branches are downy. Small, oval leaves, with rounded teeth, are brownish-green when young, glossy, dark green in maturity. Produces red berries.

ILLICIUM (Illiciaceae)
Genus of evergreen, spring- to early summer-flowering trees and shrubs, grown for their foliage and unusual flowers. Frost to half hardy. Does best in semi-shade or shade and moist, neutral to acid soil. Propagate by semi-ripe cuttings in summer.

I. anisatum (Chinese anisatum, Chinese anise). Slow-growing, evergreen, conical tree or shrub. H and S 6m (20ft). Frost hardy. Produces oval, aromatic, glossy, dark green leaves. Star-shaped, greenish-yellow flowers, with numerous narrow petals, are carried in mid-spring.

I. floridanum (Purple anise). Evergreen, bushy shrub. H and S 2m (6ft). Half hardy. Lance-shaped, leathery, deep green leaves are very aromatic. Star-shaped, red or purplish-red flowers, with numerous, narrow petals, are produced in late spring and early summer.

IMPATIENS (Balsaminaceae)
Genus of annuals and mainly evergreen perennials and sub-shrubs, often with succulent but brittle stems. In cold climates some may be herbaceous. Frost hardy to frost tender, min. 10°C (50°F). Prefers sun or semi-shade and moist but not waterlogged soil. Propagate by seed or by stem cuttings in spring or summer. Red spider mite, aphids and whitefly may cause problems under glass.

I. balsamina illus. p.272. 'Blackberry Ice' is a fast-growing, upright, bushy annual. H 45–60cm (18–24in), S 45cm (18in). Half hardy. Has lance-shaped, pale green leaves and, in summer and early autumn, large, double, white flowers, splashed with purple.

I., Confection Series illus. p.279.

I., Duet Series (Busy lizzie). Group of fast-growing, evergreen, bushy perennials, grown as annuals. H and S 30cm (12in). Half hardy. Has oval, fresh green leaves. From spring to autumn produces flattish, spurred, semi-double and double flowers, in shades of red and orange, suffused with white.

I. niamniamensis. Evergreen, bushy perennial, uncommon in cultivation. H to 60cm (2ft), S 30cm (1ft). Frost tender. Has reddish-green stems and oval, toothed leaves to 20cm (8in) long. Showy, 5-petalled, yellowish-green flowers, 2.5cm (1in) long and each with a long, orange, red, crimson or purple spur, appear in summer-autumn. 'Congo Cockatoo' has red, green and yellow flowers.

I., Novette Series (Busy lizzie). Group of fast-growing, evergreen, bushy perennials, grown as annuals. H and S 15cm (6in). Half hardy. Oval leaves are fresh green. Flattish, 5-petalled, spurred flowers, in a mixture of colours, are borne from spring to autumn (red, illus. p.279; salmon, illus. p.273). 'Red Star' illus. p.279.

I. oliveri. See *I. sodenii*.

♀ *I. repens* illus. p.257.

I., Rosette Series illus. p.277.

I. sodenii, syn. *I. oliveri*. Evergreen, strong-growing, bushy perennial. H 1.2m (4ft) or more, S 60cm (2ft). Frost tender. Narrowly oval, toothed leaves, in whorls of 4–10, are 15cm (6in) or more long. Almost flat, white or pale pink to mauve flowers, 5cm (2in) or more wide, are produced in autumn-winter.

I. sultanii. See *I. walleriana*.

♀ *I.*, Super Elfin Series (Busy lizzie). Group of fast-growing, evergreen, bushy perennials, grown as annuals. H and S 20cm (8in). Half hardy. Has oval, fresh green leaves and, from spring to autumn, flattish, 5-petalled, spurred flowers in mixed colours. 'Lipstick' illus. p.278.

INCARVILLEA

I., **Tom Thumb Series** (Busy lizzie). Group of fast-growing, evergreen, bushy perennials, grown as annuals. H and S 20–30cm (8–12in). Half hardy. Has oval, fresh green leaves and, from spring to autumn, bears 8cm (3in) wide, rounded, spurred, double flowers in shades of red, purple, pink or white.
I. walleriana, syn. *I. sultanii* (Busy lizzie). Fast-growing, evergreen, bushy perennial, usually grown as an annual. H and S to 60cm (2ft). Half hardy. Has oval, fresh green leaves. Flattish, 5-petalled, spurred, bright red flowers appear from spring to autumn.

INCARVILLEA (Bignoniaceae)
Genus of late spring- or summer-flowering perennials, suitable for rock gardens and borders. Fully to frost hardy, but protect crowns with bracken or compost in winter. Requires sun and fertile, well-drained soil. Propagate by seed in autumn or spring.
I. delavayi illus. p.246.
I. mairei illus. p.243. **'Frank Ludlow'** is a compact, clump-forming perennial. H and S 30cm (12in). Fully hardy. Short stems bear trumpet-shaped, rich deep pink flowers in early summer. Divided leaves consist of up to 4 pairs of narrowly oval, dark green leaflets.

INDIGOFERA (Leguminosae)
Genus of perennials and deciduous shrubs, grown for their foliage and small, pea-like flowers. Fully to frost hardy; in cold areas, hard frosts may cut plants to ground, but they usually regrow from base in spring. Needs full sun and fertile, well-drained soil. Cut out dead wood in spring. Propagate by softwood cuttings in summer or by seed in autumn.
I. decora. Deciduous, bushy shrub. H 45cm (18in), S 1m (3ft). Frost hardy. Glossy, dark green leaves each have 7–13 oval leaflets. Long spikes of pink or white flowers appear from mid- to late summer.
I. dielsiana illus. p.132.
I. gerardiana. See *I. heterantha*.
♥ *I. heterantha*, syn. *I. gerardiana*, illus. p.111.
I. pseudotinctoria. Deciduous, arching shrub. H 1m (3ft) or more, S 2m (6ft). Fully hardy. Dark green leaves consist of oval leaflets, usually 7–9 to each leaf. Small, pale pink flowers are produced in long, dense racemes from mid-summer to early autumn.

INULA (Compositae)
Genus of summer-flowering, clump-forming, sometimes rhizomatous perennials. Fully hardy. Needs sun and any well-drained soil. Propagate by seed or division in spring or autumn.
I. acaulis. Tuft-forming, rhizomatous perennial. H 5–10cm (2–4in), S 15cm (6in). Has lance-shaped to elliptic, hairy leaves. Solitary, almost stemless, daisy-like, golden-yellow flower heads are produced in summer. Is good for a rock garden.
I. ensifolia illus. p.257.
I. hookeri illus. p.220.
I. magnifica illus. p.193.
I. oculis-christi. Spreading, rhizomatous perennial. H 45cm (18in), S 60cm (24in). Stems each bear 2 or 3 daisy-like, yellow flower heads, which appear in summer. Has lance-shaped to elliptic, hairy, mid-green leaves.

IOCHROMA (Solanaceae)
Genus of evergreen shrubs, grown for their flowers. Frost tender, min. 7–10°C (45–50°F). Needs full light or partial shade and fertile, well-drained soil. Water potted plants freely when in full growth, moderately at other times. Tip prune young plants to stimulate a bushy habit. Cut back flowered stems by half in late winter. Propagate by greenwood or semi-ripe cuttings in summer. Whitefly and red spider mite are sometimes troublesome.
I. cyanea, syn. *I. tubulosa*, illus. p.120.
I. tubulosa. See *I. cyanea*.

IONOPSIDIUM (Cruciferae)
Genus of annuals. Only one species is usually cultivated: this is suitable for rock gardens and as edging. Frost hardy. Grow in semi-shade and in fertile, well-drained soil. Propagate by seed sown outdoors in spring, early summer or early autumn.
♥ *I. acaule* (Violet cress). Fast-growing, upright annual. H 5–8cm (2–3in), S 2.5cm (1in). Rounded leaves are mid-green. Tiny, 4-petalled, lilac or white flowers, flushed with deep blue, are produced in summer and early autumn.

IPHEION (Liliaceae)
Genus of bulbs that freely produce many star-shaped, blue, white or yellow flowers in spring and make excellent pot plants in cold greenhouses. Frost hardy. Prefers a sheltered situation in dappled sunlight and well-drained soil. Plant in autumn; after flowering, dies down for summer. Propagate by offsets in late summer or early autumn.
♥ *I. uniflorum* **'Froyle Mill'** illus. p.375. ♥ **'Wisley Blue'** is a spring-flowering bulb. H 10–15cm (4–6in), S 5–8cm (2–3in). Bears linear, semi-erect, basal, pale green leaves, which smell of onions if damaged. Leafless stems each produce an upward-facing, pale blue flower, 3–4cm (1¼–1½in) across.

IPOMOEA, syn. PHARBITIS (Convolvulaceae)
Genus of mainly evergreen shrubs, perennials, annuals and soft- or woody-stemmed, twining climbers. Half hardy to frost tender, min. 7–10°C (45–50°F). Provide humus-rich, well-drained soil and full light. Water freely when in full growth, less at other times. Support is needed. Thin out or cut back congested growth in spring. Propagate by seed in spring or by softwood or semi-ripe cuttings in summer. Whitefly and red spider mite may cause problems.
I. alba, syn. *I. bona-nox*, *Calonyction aculeatum* (Moon flower). Evergreen, soft-stemmed, twining climber with prickly stems that exude milky juice when cut. H 7m (22ft) or more. Frost tender, min. 10°C (50°F). Oval or sometimes 3-lobed leaves are 20cm (8in) long. Fragrant, tubular, white flowers, to 15cm (6in) long and expanded at the mouths to 15cm (6in) across, open at night in summer.
I. bona-nox. See *I. alba*.
I. coccinea, syn. *Quamoclit coccinea* (Red morning glory, Star ipomoea). Annual, twining climber. H to 3m (10ft). Frost tender, min. 10°C (50°F). Arrow- or heart-shaped leaves are long-pointed and often toothed. Fragrant, tubular, scarlet flowers, with yellow throats and expanded mouths, are produced in late summer and autumn.
I. hederacea illus. p.174.
♥ *I. horsfalliae* illus. p.169. **'Briggsii'** is a strong-growing, evergreen, woody-stemmed, twining climber. H 2–3m (6–10ft). Frost tender, min. 7–10°C (45–50°F). Has leaves with 5–7 radiating lobes or leaflets. Stalked clusters of funnel-shaped, deep rose-pink or rose-purple flowers are produced from summer to winter; flowers are larger and more richly coloured than those of the species.
I. imperialis. See *I. nil*.
I. lobata, syn. *I. versicolor*, *Mina lobata*, *Quamoclit lobata*, illus. p.170.
I. × *multifida*, syn. *I.* × *sloteri* (Cardinal climber, Hearts-and-honey vine). Annual, twining climber. H 3m (10ft). Frost tender, min. 10°C (50°F). Triangular-oval leaves are divided into 7–15 segments. Tubular, wide-mouthed, crimson flowers with white eyes appear in summer.
I. nil, syn. *I. imperialis*. **'Early Call'** is a short-lived, soft-stemmed, perennial, twining climber with hairy stems, best grown as an annual. H to 4m (12ft). Half hardy. Leaves are heart-shaped or 3-lobed. From summer to early autumn bears large, funnel-shaped flowers in a range of colours, with white tubes. **'Scarlett O'Hara'** has deep red flowers.
I. purpurea, syn. *Convolvulus purpureus* (Common morning glory). Short-lived, soft-stemmed, perennial, twining climber, best grown as an annual, with hairy stems. H to 5m (15ft). Half hardy. Leaves are heart-shaped or 3-lobed. From summer to early autumn has funnel-shaped, deep purple to bluish-purple or reddish flowers with white throats and bristly sepals.
I. quamoclit, syn. *Quamoclit pinnata*, illus. p.170.
I. rubrocaerulea **'Heavenly Blue'**. See *I. tricolor* 'Heavenly Blue'.
I. × *sloteri*. See *I.* × *multifida*.
♥ *I. tricolor* **'Heavenly Blue'**, syn. *I. rubrocaerulea* 'Heavenly Blue', illus. p.175.
I. tuberosa. See *Merremia tuberosa*.
I. versicolor. See *I. lobata*.

IPOMOPSIS (Polemoniaceae)
Genus of perennials and biennials, often grown as pot plants for greenhouses and conservatories. Half hardy. Grow in cool, airy conditions with bright light and in fertile, well-drained soil. Propagate by seed sown under glass in early spring or early summer.
I. aggregata. Slow-growing biennial with upright, slender, hairy stems. H to 1m (3ft), S 30cm (1ft). Mid-green leaves are divided into linear leaflets. Fragrant, trumpet-shaped flowers borne in summer are usually brilliant red, sometimes spotted yellow, but may be rose, yellow or white.

IRESINE (Amaranthaceae)
Genus of perennials, grown for their colourful leaves. Frost tender, min. 10–15°C (50–59°F). Needs a good, loamy, well-drained soil and bright light to retain leaf colour. Pinch out tips in growing season to obtain bushy plants. Propagate by stem cuttings in spring.
I. herbstii (Beefsteak plant). Bushy perennial. H to 60cm (2ft), S 45cm (18in). Has red stems and rounded, purplish-red leaves, notched at their tips and 10cm (4in) long, with paler or yellowish-red veins. Flowers are insignificant. **'Aureoreticulata'** illus. p.269.
♥ *I. lindenii* (Blood leaf). Bushy perennial. H 60cm (2ft), S 45cm (18in). Has lance-shaped, dark red leaves, 5–10cm (2–4in) long. Flowers are insignificant.

IRIS (Iridaceae)
Genus of upright, rhizomatous or bulbous (occasionally fleshy-rooted) perennials, some of which are evergreen, grown for their distinctive and colourful flowers. Each flower has 3 usually large 'falls' (pendent or semi-pendent petals), which in a number of species have conspicuous beards or crests; 3 generally smaller 'standards' (erect, horizontal or, occasionally, pendent petals); and a 3-branched style. In many irises the style branches are petal-like. Unless otherwise stated below, flower stems are unbranched. Green, then brown seed pods are ellipsoid to cylindrical and often ribbed. Irises are suitable for borders, rock gardens, woodlands, watersides, bog gardens, alpine houses, cold frames and containers. Species and cultivars described are fully hardy unless otherwise stated, but some groups may thrive only in the specific growing conditions mentioned below. Propagate species by division of rhizomes or offsets in late summer or by seed in autumn, named cultivars by division only. Botanically, irises are divided into a number of Subgenera and Sections, and it is convenient, for horticultural purposes, to use some of these botanical names for groups of irises with similar characteristics and requiring comparable cultural treatment.

Rhizomatous
These irises have rhizomes as rootstocks; leaves are sword-shaped and usually in a basal fan.

Bearded irises are rhizomatous and have 'beards', consisting of numerous often coloured hairs, along the centre of each fall. The group covers the vast majority of irises, including many named cultivars, grown in gardens; all are derived from *I. pallida* and related species. Bearded irises thrive in fairly rich, well-drained, preferably slightly alkaline soil in full sun. Some are very tolerant and will grow and flower reasonably in poorer soil in partial shade. For horticultural purposes, various groupings of hybrid, bearded irises are recognized, based mainly on the height of the plants in flower. These include **Miniature Dwarf**, H to 20cm (8in); **Standard Dwarf**, H 20–40cm (8–16in); **Intermediate**, H 40–70cm (16–28in); and **Standard Tall**, H 70cm (28in) or more (this last category may be further subdivided). In general, the shorter the iris, the earlier the flowering season (from early spring to early summer). **Oncocyclus** irises are rhizomatous, with very large and often bizarrely coloured flowers, one to each stem, that have bearded falls. They require sharply drained but fairly rich soil, full sun and, after flowering, a dry period of dormancy in summer and early autumn. Difficult to cultivate successfully, they are best grown in an alpine house or covered frame in climates subject to summer rains. **Regelia** irises are closely related to Oncocyclus irises, differing in having bearded standards as well as falls and in having 2 flowers to each stem. They require similar conditions of cultivation, although a few species, such as *I. hoogiana*, have proved easier to grow than Oncocyclus irises. Hybrids between the 2 groups have been raised and are known as **Regeliocyclus** irises.

Beardless irises, also rhizomatous, lack hairs on the falls; most have very similar cultural requirements to bearded irises but some prefer heavier soil. Various groupings are recognized, of which the following are the most widely known. **Pacific Coast** irises, a group of Californian species and their hybrids, prefer acid to neutral soil and grow well in sun or partial shade, appreciating some humus in the soil; they are best grown from seed as they resent being moved. **Spuria** irises (*I. spuria* and its relatives) grow in sun or semi-shade and well-drained but moist soil. A number of species and hybrids prefers moist, waterside conditions; these include the well-known **Siberian** irises (*I. sibirica* and its relatives) and the **Japanese** water irises such as *I. ensata* and *I. laevigata*, which may also be grown as border plants, but succeed best in humus-rich, moist, open, sunny sites.

Crested irises, also rhizomatous, have ridges, or cockscomb-like crests, instead of beards. They include the **Evansia** irises, with often widely spreading, creeping stolons. Most have very similar cultivation requirements to bearded irises but some prefer damp, humus-rich conditions; a few are half hardy to frost tender, min. 5°C (41°F).

Bulbous
These irises are distinguished by having bulbs as storage organs, sometimes with thickened, fleshy roots, and leaves that are either, lance-shaped and channelled, 4-sided (more or less square in cross section), or almost cylindrical, rather than flat and usually sword-shaped like the leaves of the rhizomatous irises.

Xiphium irises include the commonly grown Spanish, English and Dutch irises, which are excellent both for garden decoration and as cut flowers. All are easy to cultivate in sunny, well-drained sites, preferring slightly alkaline conditions, but also growing well on acid soil. **Spanish** irises are derived from *I. xiphium*, which is variable in flower colour, from blue and violet to yellow and white, and produces its channelled leaves in autumn. **English** irises have been produced from *I. latifolia*, which varies from blue to violet (occasionally white) and produces its channelled leaves in spring. **Dutch** irises are hybrids of *I. xiphium*, the related pale to deep blue *I. tingitana* and *I. latifolia*. They are extremely variable in flower colour.

Juno irises have bulbs with thickened, fleshy roots, channelled leaves and very small standards that are sometimes only bristle-like and usually horizontally placed. Although very beautiful in flower, they are mostly difficult to grow successfully, requiring the same cultivation conditions as Oncocyclus irises to thrive. Care must be taken not to damage the fleshy roots when transplanting or dividing clumps.

Reticulata irises include the dwarf, bulbous irises valuable for flowering early in the year. Unlike other bulbous irises, they have net-like bulb tunics and leaves that are 4-sided, or occasionally cylindrical. With few exceptions (not described here), reticulata irises grow well in open, sunny, well-drained sites.

I. acutiloba. Rhizomatous Oncocyclus iris. H 8–25cm (3–10in), S 30–38cm (12–15in). Has narrowly sickle-shaped, mid-green leaves. In late spring produces solitary strongly purple-violet- or brownish-purple-veined, white flowers, 5–7cm (2–3in) across, with a dark brown blaze around the beard of each fall.
I. 'Annabel Jane' (illus. p.198). Vigorous, rhizomatous, bearded iris (Standard Tall). H 1.2m (4ft), S indefinite. Well-branched stem bears 8–12 flowers, 15–25cm (6–10in) across, with pale lilac falls and paler standards. Flowers in early summer.
I. 'Anniversary'. Rhizomatous, beardless Siberian iris. H 60cm (2ft), S indefinite. From late spring to early summer, branched stem bears 1–4 white flowers, 5–10cm (2–4in) across, with a creamy-white stripe in the throat of each fall. Grows well in moist soil or a bog garden.
I. aphylla. Rhizomatous, bearded iris. H 15–30cm (6–12in), S indefinite. Branched stem produces up to 5 pale to dark purple or blue-violet flowers, 6–7cm (2½–3in) across, in late spring and sometimes again in autumn if conditions suit.
I. aucheri. Bulbous Juno iris. H 15–25cm (6–10in), S 15cm (6in). Has channelled, mid-green leaves packed closely together on stem, looking somewhat leek-like. In late spring bears up to 6 blue to white flowers, 6–7cm (2½–3in) across with yellow-ridged falls, in leaf axils.
I. aurea. See *I. crocea*.
I. bakeriana. Bulbous Reticulata iris. H 10cm (4in), S 5–6cm (2–2½in). In early spring bears a solitary long-tubed, pale blue flower, 5–6cm (2–2½in) across, with each fall having a dark blue blotch at the tip and a spotted, deep blue centre. Has narrow, almost cylindrical leaves that are very short at flowering time but elongate later.
I. 'Banbury Beauty'. Rhizomatous, beardless Pacific Coast iris. H 45cm (18in), S indefinite. In late spring and early summer, branched stem produces 2–10 light lavender flowers, 10–15cm (4–6in) across, with a purple zone on each fall.
I. 'Bibury'. Rhizomatous, bearded iris (Standard Dwarf). H 30cm (12in), S indefinite. Has 2–4 cream flowers, 10cm (4in) wide, on a branched stem in late spring.
♀ *I.* 'Blue-eyed Brunette' (illus. p.199). Rhizomatous, bearded iris (Standard Tall). H 1m (3ft), S indefinite. Well-branched stem produces 7–10 brown flowers, 10–15cm (4–6in) wide, with a blue blaze and a golden beard on each fall, in early summer.
I. 'Bold Print' (illus. p.198). Rhizomatous, bearded iris (Intermediate). H 55cm (22in), S indefinite. In late spring or early summer, branched stem bears up to 6 flowers, 13cm (5in) wide, with purple-edged, white standards and white falls that are each purple-stitched at the edge and have a bronze-tipped, white beard.
I. 'Bristol Gem'. Rhizomatous, bearded iris (Standard Tall). H 1m (3ft), S indefinite. Well-branched stem produces 6–10 deep blue flowers, 15–18cm (6–7in) wide, in early summer.
I. 'Bronze Queen' (illus. p.199). Bulbous Xiphium iris (Dutch). H to 80cm (32in), S 15cm (6in). In spring and early summer produces 1 or 2 golden-brown flowers, 8–10cm (3–4in) wide, flushed bronze and purple. Lance-shaped, channelled, mid-green leaves are scattered up flower stem.
I. 'Brown Lasso'. Rhizomatous, bearded iris (Intermediate). H 55cm (22in), S indefinite. In early summer, sturdy, well-branched stem bears 6–10 flowers, 10–13cm (4–5in) across, with deep butterscotch standards and brown-edged, light violet falls.
I. bucharica (illus. p.198). Vigorous, bulbous Juno iris. H 20–40cm (8–16in), S 12cm (5in). In late spring produces 2–6 flowers, 6cm (2½in) across, golden-yellow to white with yellow falls, from leaf axils. Has narrowly lance-shaped, channelled, glossy, mid-green leaves scattered up flower stem. Is easier to grow than most Juno irises.
I. 'Butter and Sugar' (illus. p.199). Rhizomatous beardless Siberian iris. H to 1m (3ft), S indefinite. From late spring to early summer produces large, yellow and white flowers.
I. 'Carnaby' (illus. p.199). Rhizomatous, bearded iris (Standard Tall). H to 1m (3ft), S indefinite. Well-branched stem bears 6–8 flowers, 15–18cm (6–7in) wide, with pale pink standards and deep rose-pink falls with orange beards, in early summer.
I. chamaeiris. See *I. lutescens*.
I. chrysographes (illus. p.198). Rhizomatous, beardless Siberian iris. H 40cm (16in), S indefinite. From late spring to early summer, branched stem bears 1–4 deep red-purple or purple-black flowers, 5–10cm (2–4in) across, with gold etching down falls. Prefers moist conditions.
I. clarkei. Rhizomatous, beardless Siberian iris. H 60cm (2ft), S indefinite. From late spring to early summer, solid stem produces 2–3 branches each with 2 blue to red-purple flowers, 5–10cm (2–4in) across, with a violet-veined, white blaze on each fall. Prefers moist conditions.
I. colchica. See *I. graminea*.
I. confusa. Evergreen or semi-evergreen, rhizomatous Evansia iris with short stolons. H 30cm–1m (1–3ft), S indefinite. Half hardy. Bamboo-like, erect stem is crowned by a fan of lax leaves. In mid-spring, widely branched flower stem produces a long succession of up to 30 white flowers, 4–5cm (1½–2in) across, with yellow and purple spots around a yellow crest on each fall. Prefers well-drained soil and the protection of a south-facing wall.
I. cristata (illus. p.198). Evansia iris with much-branched rhizomes. H 10cm (4in), S indefinite. Has neat fans of lance-shaped leaves. In early summer produces 1 or 2 virtually stemless, long-tubed, lilac, blue, lavender or white flowers, 3–4cm (1¼–1½in) across, with a white patch and orange crest on each fall. Prefers semi-shade and moist soil; is ideal for peat banks.
I. crocea, syn. *I. aurea.* Rhizomatous, beardless Spuria iris. H 1–1.2m (3–4ft), S indefinite. Has long leaves. Strong, erect, sparsely branched stem produces terminal clusters of 2–10 golden-yellow flowers, 12–18cm (5–7in) across, with wavy-edged falls, in early summer. Resents being disturbed.
I. cuprea. See *I. fulva*.
I. 'Custom Design'. Rhizomatous, beardless Spuria iris. H 1m (3ft), S indefinite. Strong, erect-branched stem produces 2–10 deep maroon-brown flowers, each 5–12cm (2–5in) wide, with a heavily veined, bright yellow blaze on each fall, from early to mid-summer.
I. danfordiae (illus. p.199). Bulbous Reticulata iris. H 5–10cm (2–4in), S 5cm (2in). In early spring bears usually one yellow flower, 3–5cm (1¼–2in) across, with green spots on each fall. Standards are reduced to short bristles. Narrow, squared leaves are very short at flowering time but elongate later. Tends to produce masses of small bulblets and requires deeper planting than other Reticulata irises to maintain bulbs at flowering size.
I. douglasiana (illus. p.198). Evergreen, rhizomatous, beardless Pacific Coast iris. H 25–70cm (10–28in), S indefinite. Leathery, dark

green leaves are stained red-purple at base. Branched stem produces 1–3 lavender to purple, occasionally white, flowers, 7–12cm (3–5in) wide, with variable, central, yellowish zones on the falls, in late spring and early summer.

I. **'Dreaming Spires'**. Rhizomatous, beardless Siberian iris. H 1m (3ft), S indefinite. From late spring to early summer, branched stem produces 1–4 flowers, 5–10cm (2–4in) wide, with lavender standards and royal-blue falls. Prefers moist soil.

I. **'Dreaming Yellow'** (illus. p.198). Rhizomatous, beardless Siberian iris. H 1m (3ft), S indefinite. From late spring to early summer, branched stem produces 1–4 flowers, 5–10cm (2–4in) across. Standards are white, falls creamy-yellow fading to white with age. Prefers moist soil.

♀ *I.* **'Early Light'** (illus. p.199). Rhizomatous, bearded iris (Standard Tall). H 1m (3ft), S indefinite. In early summer, well-branched stem bears 8–10 flowers, 15–18cm (6–7in) wide, with lemon-flushed, cream standards and slightly darker falls with a yellow beard.

I. **'Elmohr'**. Rhizomatous, bearded iris. H 1m (3ft), S indefinite. In early summer, well-branched stem produces 2–5 strongly veined, red-purple flowers, 15–20cm (6–8in) across.

I. ensata, syn. *I. kaempferi* (Japanese flag). Rhizomatous, beardless Japanese iris. H 60cm–1m (2–3ft), S indefinite. Branched stem produces 3–15 purple or red-purple flowers, 8–15cm (3–6in) across, with a yellow blaze on each fall, from early to mid-summer. May be distinguished from the related, smooth-leaved *I. laevigata* by the prominent midrib on the leaves. Has produced many hundreds of garden forms, some with double flowers, in shades of purple, pink, lavender and white, sometimes bicoloured. Prefers partial shade and thrives in a water or bog garden.

I. **'Eye Bright'** (illus. p.199). Rhizomatous, bearded iris (Standard Dwarf). H 30cm (12in), S indefinite. In late spring produces 2–4 bright yellow flowers, 7–10cm (3–4in) wide, each with a brown zone on the falls surrounding the beard, on usually an unbranched stem.

I. **'Flamenco'** (illus. p.199). Rhizomatous, bearded iris (Standard Tall). H 1m (3ft), S indefinite. In early summer, well-branched stem produces 6–9 flowers, 15cm (6in) wide, with gold standards, infused red, and white to yellow falls with red borders.

I. foetidissima (Gladwin, Roast-beef plant, Stinking iris). Evergreen, rhizomatous, beardless iris. H 30cm–1m (1–3ft), S indefinite. Branched stem bears up to 9 yellow-tinged, dull purple or occasionally pure yellow flowers, 5–10cm (2–4in) wide, from early to mid-summer. Cylindrical seed pods open to reveal rounded, bright scarlet fruits throughout winter. Thrives in a bog or water garden, although tolerates drier conditions.

I. forrestii (illus. p.199). Rhizomatous, beardless Siberian iris. H 15–40cm (6–16in), S indefinite. From late spring to early summer, unbranched stem produces 1 or 2 fragrant, yellow flowers, 5–6cm (2–2½in) across, with black lines on each fall and occasionally brownish-flushing on standards. Has linear, glossy, mid-green leaves, grey-green below. Prefers moist, lime-free soil.

I. fosteriana. Bulbous Juno iris. H 10–15cm (4–6in), S 6cm (2½in). In spring produces 1 or 2 long-tubed flowers, 4–5cm (1½–2in) wide, with downward-turned, rich purple standards, which are larger than those of most Juno irises, and creamy-yellow falls. Has narrowly lance-shaped, channelled, silver-edged, mid-green leaves scattered on flower stem. Is difficult to grow and is best in an alpine house or cold frame.

I. **'Frank Elder'**. Bulbous Reticulata iris. H 6–10cm (2½–4in), S 5–7cm (2–3in). Has a solitary very pale blue flower, 6–7cm (2½–3in) wide, suffused pale yellow and veined and spotted darker blue, in early spring. Narrow, squared leaves are very short at flowering time but elongate later.

I. fulva, syn. *I. cuprea* (illus. p.199). Rhizomatous, beardless iris. H 45–80cm (18–32in), S indefinite. Frost hardy. In late spring or summer produces a slender, slightly branched stem with 4–6 (occasionally more) copper- or orange-red flowers, 5–7cm (2–3in) across, with 2 flowers per leaf axil. Thrives in a bog or water garden.

I. **'Fulvala'** (illus. p.198). Rhizomatous, beardless iris. H 45cm (18in), S indefinite. Frost hardy. In summer, zigzag stem produces 4–6 (occasionally more) velvety, deep red-purple flowers, 5–12cm (2–5in) across, with 2 flowers per leaf axil. Has a yellow blaze on each fall. Thrives in a bog or water garden.

I. **'Galathea'** (illus. p.199). Rhizomatous, beardless Japanese iris. H 80cm (32in), S indefinite. In summer has blue-purple flowers with a yellow blaze on each fall.

I. **'Geisha Gown'** (illus. p.198). Rhizomatous, beardless Japanese iris. H 80cm (32in), S indefinite. In summer, branched stem produces 3–5 double, rose-purple-veined, white flowers, 15–30cm (6–12in) across, with purple styles and a gold blaze on each fall. Leaves are ridged. Prefers sun or semi-shade. Thrives in a bog or water garden.

I. germanica (Common German flag). Rhizomatous, bearded iris. H to 60cm–1.2m (2–4ft), S indefinite. Sparsely branched stem produces up to 6 yellow-bearded, blue-purple or blue-violet flowers, 10–15cm (4–6in) wide, in late spring and early summer.

I. **'Golden Harvest'**. Bulbous Xiphium iris (Dutch). H to 80cm (32in), S 15cm (6in). Bears 1 or 2 deep rich yellow flowers, 6–8cm (2½–3in) wide, in spring and early summer. Has scattered, narrowly lance-shaped, channelled, mid-green leaves.

I. gracilipes. Clump-forming, rhizomatous Evansia iris with short stolons. H 15–20cm (6–8in), S indefinite. In late spring and early summer, slender, branched stem produces a succession of 4 or 5 lilac-blue flowers, each 3–4cm (1¼–1½in) across, with a violet-veined, white zone surrounding a yellow-and-white crest. Has narrow, grass-like leaves. Prefers semi-shade and peaty soil.

I. graeberiana. Bulbous Juno iris. H 15–35cm (6–14in), S 6–8cm (2½–3in). In late spring produces 4–6 bluish-lavender flowers, 6–8cm (2½–3in) across, with a white crest on each fall, from leaf axils. Lance-shaped, channelled leaves are white-margined, glossy, mid-green above, greyish-green below and scattered up flower stem. Is easier to grow than most Juno irises.

I. graminea, syn. *I. colchica*. Rhizomatous, beardless Spuria iris. H 20–40cm (8–16in), S indefinite. In late spring, narrowly lance-shaped leaves partially hide up to 10 plum-scented flowers, 5–12cm (2–5in) wide, with wine-purple standards and heavily veined, violet-blue falls borne on flattened, angled stem. Resents being disturbed.

I. **'Harmony'** (illus. p.199). Bulbous Reticulata iris. H 6–10cm (2½–4in), S 6–7cm (2½–3in). In early spring bears a solitary fragrant, long-tubed, clear pale blue flower, 5–6cm (2–2½in) across, with white marks and a yellow ridge down each fall centre. Narrow, squared leaves are very short at flowering time but elongate later.

I. histrioides. Bulbous Reticulata iris. H 6–10cm (2½–4in), S 6–7cm (2½–3in). In early spring produces solitary flowers, 6–7cm (2½–3in) across, which vary from light to deep violet-blue. Each fall is lightly to strongly spotted with dark blue and has white marks and a yellow ridge down centre. Narrow, squared leaves are very short at flowering time but elongate later. **'Lady Beatrix Stanley'** has light blue flowers and heavily spotted falls. ♀ **'Major'** (illus. p.199) has darker blue-violet flowers.

I. **'Holden Clough'**. Rhizomatous, beardless iris. H 50–70cm (20–28in), S indefinite. In early summer, branched stem bears 6–12 yellow flowers, each 5cm (2in) wide, with very heavy, burnt-sienna veining. Is excellent in a bog or water garden, but also grows well in any rich, well-drained soil.

I. hoogiana (illus. p.199). Regelia iris with stout rhizomes. H 40–60cm (16–24in), S indefinite. Produces 2 or 3 scented, delicately veined, lilac-blue flowers, 7–10cm (3–4in) across, in late spring and early summer. Is relatively easy to cultivate.

I. iberica (illus. p.198). Rhizomatous Oncocyclus iris. H 15–20cm (6–8in), S indefinite. Has narrow, strongly curved, grey-green leaves. Bears solitary bicoloured flowers, 10–12cm (4–5in) across, in late spring. Standards are white, pale yellow, or pale blue with slight brownish-purple veining; spoon-shaped falls are white or pale lilac, spotted and strongly veined brownish-purple. Grows best in a frame or alpine house.

I. innominata (illus. p.199). Evergreen or semi-evergreen, rhizomatous, beardless Pacific Coast iris. H 16–25cm (6–10in), S indefinite. Stem bears 1 or 2 flowers, 6.5–7.5cm (2½–3in) across, from late spring to early summer. Varies greatly in colour from cream to yellow or orange and from lilac-pink to blue or purple; falls are often veined with maroon or brown.

I. japonica. Vigorous, rhizomatous Evansia iris with slender stolons. H 45–80cm (18–32in), S indefinite. Frost hardy. Has fans of broadly lance-shaped, glossy leaves. In late spring produces branched flower stem with a long succession of flattish, frilled or ruffled, pale lavender or white flowers, 1–8cm (½–3in) across, marked violet around an orange crest on each fall. Prefers the protection of a sheltered, sunny wall. **'Ledger's Variety'** (illus. p.199) has white flowers marked with purple and is hardier than the species.

I. **'Joette'** (illus. p.198). Rhizomatous, bearded iris (Intermediate). H 45cm (18in), S indefinite. In late spring or early summer, branched stems carry uniformly lavender-blue flowers with yellow beards. Is excellent in flower arrangements.

I. **'Joyce'** (illus. p.199). Bulbous Reticulata iris. H 6–10cm (2½–4in), S 6–7cm (2½–3in). In early spring bears a solitary fragrant, long-tubed, clear blue flower, 5–6cm (2–2½in) across, with white marks and a yellow ridge down each fall centre. Narrow, squared leaves are very short at flowering time but elongate later.

I. **'June Prom'**. Vigorous, rhizomatous, bearded iris (Intermediate). H 50cm (20in), S indefinite. In late spring or early summer, branched stem bears up to 6 pale blue flowers, 8–10cm (3–4in) wide, with a green tinge on each fall.

I. kaempferi. See *I. ensata*.

I. **'Katharine Hodgkin'** (illus. p.199). Bulbous Reticulata iris. H 6–10cm (2½–4in), S 5–7cm (2–3in). Is similar to *I.* 'Frank Elder', but has yellower flowers, 6–7cm (2½–3in) wide, suffused pale blue, lined and dotted dark blue. Flowers in early spring.

I. kerneriana. Rhizomatous, beardless Spuria iris. H 25cm (10in), S indefinite. Has very narrow, grass-like leaves. Strong, erect-branched stem bears 2–4 soft lemon- or creamy-yellow flowers, 5–12cm (2–5in) across, from each pair of bracts, in early summer. Resents being disturbed.

I. korolkowii. Regelia iris with stout rhizomes. H 40–60cm (16–24in), S indefinite. From late spring to early summer, each spathe encloses 2 or 3 delicately blackish-maroon- or olive-green-veined, creamy-white or light purple flowers, 6–8cm (2½–3in) across. Is best grown in a bulb frame.

I. **'Krasnia'** (illus. p.198). Rhizomatous, bearded iris (Standard Tall). H 1m (3ft), S indefinite. In early summer, well-branched stem produces 8–12 flowers, 13–18cm (5–7in) wide, with purple standards and purple-edged, white falls.

I. **'Lady of Quality'** (illus. p.198). Rhizomatous, beardless Siberian iris. H to 1m (3ft), S indefinite. Flowers, produced in mid- or late spring, have light blue-violet standards and lighter blue falls.

♀ *I. laevigata* (illus. p.199). Rhizomatous, beardless Japanese iris. H 60cm–1m (2–3ft) or more, S indefinite. Sparsely branched stem produces 2–4 blue, blue-purple or white flowers, 5–12cm (2–5in) across, from early to mid-summer. Is related to *I. ensata* but has smooth, not ridged leaves. Grows well in sun or semi-shade in moist conditions or in shallow water. **'Regal'** bears single, cyclamen-red flowers. Flowers of **'Snowdrift'** are double and white. **'Variegata'**, H 25cm (10in), has white-and-green-striped leaves and often flowers a second time in early autumn.

I. latifolia, syn. *I. xiphioides* (English iris; illus. p.198). Bulbous Xiphium iris (English). H 80cm (32in), S 15cm (6in). In late spring and summer, 1 or 2 blue to deep violet flowers, 8–10cm (3–4in) wide, with a yellow stripe down centre of each very broad fall, are produced from the bracts. Lance-shaped, channelled, mid-green leaves are scattered up flower stem. **'Blue Giant'** has darker-flecked, bluish-purple standards and dark blue falls. **'La Nuit'** bears deep purple-red flowers. Flowers of **'Mont Blanc'** are pure white.

I. **'Lavender Royal'** (illus. p.199). Rhizomatous, beardless, Pacific Coast iris. H 45cm (18in), S indefinite. In late spring to early summer, branched stems carry lavender flowers with darker flushes.

I. lutescens, syn. *I. chamaeiris*. Fast-growing, very variable, rhizomatous, bearded iris. H 5–30cm (2–12in), S indefinite. Branched stem produces 1 or 2 yellow-bearded, violet, purple, yellow, white or bicoloured flowers, 6–8 cm (2½–3in) across, in early summer. **'Nancy Lindsay'** has scented, yellow flowers.

I. **'Magic Man'** (illus. p.198). Rhizomatous, bearded iris (Standard Tall). H 1m (3ft), S indefinite. In early summer, branched stems carry flowers that have light blue standards and velvety purple falls with light blue edges; beards are orange.

I. magnifica (illus. p.198). Bulbous Juno iris. H 30–60cm (1–2ft), S 15cm (6in). In late spring produces 3–7 very pale lilac flowers, 6–8cm (2½–3in) across, with a central, yellow area on each fall, from leaf axils. Bears scattered, lance-shaped, channelled, glossy, mid-green leaves.

I. **'Margot Holmes'**. Rhizomatous, beardless Siberian iris. H 25cm (10in), S indefinite. Frost hardy. In early summer produces 2 or 3 purple-red flowers, 10–15cm (4–6in) across, with yellow veining on each fall.

I. **'Marhaba'**. Rhizomatous, bearded iris (Miniature Dwarf). H 15cm (6in), S indefinite. Bears 1, rarely 2 deep blue flowers, 5–8cm (2–3in) wide, in mid-spring.

I. **'Mary Frances'** (illus. p.198). Rhizomatous, bearded iris (Standard Tall). H 1m (3ft), S indefinite. In early summer, well-branched stem bears 6–9, occasionally to 12 pink-lavender flowers, 15cm (6in) wide.

I. **'Matinata'** (illus. p.198). Rhizomatous, bearded iris (Standard Tall). H 1m (3ft), S indefinite. In early summer, well-branched stem produces 6–9, occasionally to 12 flowers, 15cm (6in) wide, that are dark purple-blue throughout.

I. missouriensis, syn. *I. tolmeiana* (Missouri flag; illus. p.198). Very variable, rhizomatous, beardless Pacific Coast iris. H to 75cm (2½ft), S indefinite. Branched stem produces 2 or 3 pale blue, lavender, lilac, blue or white flowers, 5–8cm (2–3in) wide, in each spathe, in late spring or early summer. Falls are veined and usually have a yellow blaze.

I. **'Mountain Lake'** (illus. p.199). Rhizomatous, beardless Siberian iris. H 1m (3ft), S indefinite. From late spring to early summer, branched stem produces 1–4 mid-blue flowers, 5–10cm (2–4in) across, with darker veining on falls. Prefers moist soil.

I. orientalis. See *I. sanguinea*.

I. pallida (Dalmatian iris). Rhizomatous, bearded iris. H 70cm–1m (28–36in) or more, S indefinite. In late spring and early summer produces 2–6 scented, lilac-blue flowers, 8–12cm (3–5in) across and with yellow beards, from silvery spathes on strong, branched stems. Leaves of **'Aurea Variegata'** (illus. p.198) are striped green and yellow.

I. **'Paradise Bird'** (illus. p.198). Rhizomatous, bearded iris (Standard Tall). H 85cm (34in), S indefinite. In early summer, well-branched stem produces 8–10 flowers, 14–15cm (5½–6in) wide, with magenta falls and paler standards.

I. **'Peach Frost'** (illus. p.199). Rhizomatous, bearded iris (Standard Tall). H 1m (3ft), S indefinite. Well-branched stem bears 6–10 flowers, 15cm (6in) wide, in early summer. Standards are peach-pink, falls white with peach-pink borders and tangerine beards.

I. **'Piona'**. Rhizomatous, bearded iris (Intermediate). H 45cm (18in), S indefinite. In late spring and early summer, branched stem bears up to 6 deep violet flowers, 8–10cm (3–4in) wide, with golden beards. Mid-green leaves have purple bases.

I. **'Professor Blaauw'**. Bulbous Xiphium iris (Dutch). H 80cm (32in), S 15cm (6in). From spring to early summer produces 1 or 2 rich violet-blue flowers, 6–8cm (2½–3in) across. Narrowly lance-shaped, channelled, mid-green leaves are scattered up flower stem.

I. **'Promise'**. Rhizomatous, bearded iris (Miniature Dwarf). H 18cm (7in), S indefinite. In mid-spring, well-branched stem bears 2–5 veined, red flowers, 5–8cm (2–3in) wide, with a yellow stripe in the throat of each fall.

I. pseudocorus (Yellow flag; illus. p.199). Robust, rhizomatous, beardless iris. H to 2m (6ft), S indefinite. Branched stem produces 4–12 golden-yellow flowers, 5–12cm (2–5in) wide, usually with brown or violet veining and a darker yellow patch on the falls, from early to mid-summer. Leaves are broad, ridged and greyish-green. Prefers semi-shade and thrives in a water garden. **'Variegata'** has yellow-and-green-striped foliage in spring, often turning green before flowering.

I. pumila (Dwarf bearded iris). Rhizomatous, bearded iris. H 10–15cm (4–6in), S indefinite. In mid-spring has a 1cm (½in) long flower stem bearing 2 or 3 long-tubed flowers, 2.5–5cm (1–2in) wide, varying from violet-purple to white, yellow or blue, with yellow or blue beards on the falls. Prefers very well-drained, slightly alkaline soil.

I. reticulata. Bulbous Reticulata iris. H 10–15cm (4–6in), S 4–6cm (1½–2½in). In early spring bears a solitary fragrant, long-tubed, deep violet-purple flower, 4–6cm (1½–2½in) wide, with a yellow ridge down each fall centre. Narrow, squared leaves elongate after flowering time. **'Cantab'** (illus. p.199) has clear pale blue flowers with a deep yellow ridge on each fall. **'Clairette'** bears pale blue flowers with each fall having a dark blue blotch at the tip and a dark blue-spotted centre. Flowers of **'J.S. Dijt'** are reddish-purple with an orange ridge on each fall. **'Violet Beauty'** has deep violet-purple flowers with an orange ridge down the centre of each fall.

I. **'Rippling Rose'** (illus. p.198). Rhizomatous, bearded iris (Standard Tall). H 1m (3ft), S indefinite. In early summer, well-branched stem has 6–10 white flowers, 15cm (6in) wide, with purple marks and lemon-yellow beards.

I. rosenbachiana (illus. p.198). Bulbous Juno iris. H 10–15cm (4–6in), S 6cm (2½in). In spring produces 1 or 2 long-tubed flowers, 4–5cm (1½–2in) wide, with small, downward-turned, rich purple standards and reddish-purple falls, each with a yellow ridge in the centre. Has lance-shaped, channelled, mid-green leaves in a basal tuft. Is difficult to grow and is best in an alpine house or cold frame.

I. **'Ruffled Velvet'** (illus. p.198). Rhizomatous, beardless Siberian iris. H to 1m (3ft), S indefinite. In early summer produces 2 or 3 red-purple flowers marked with yellow.

I. **'Saffron Jewel'**. Rhizomatous, bearded iris (Intermediate). H 75cm (30in), S indefinite. In early summer, branched stem produces 2–5 flowers, 5–10cm (2–4in) across, with oyster falls, veined chartreuse, and paler standards. Falls each have a blue blaze and beard.

I. sanguinea, syn. *I. orientalis*. Rhizomatous, beardless Siberian iris. H to 1m (3ft), S indefinite. From late spring to early summer, branched stem produces 2 or 3 deep purple or red-purple flowers, 5–10cm (2–4in) wide, from each set of bracts. Falls are red-purple with white throats finely veined purple.

I. **'Sapphire Star'** (illus. p.199). Rhizomatous, beardless Japanese iris. H 1.2m (4ft), S indefinite. In summer, branched stem bears 3–5 white-veined, lavender flowers, 15–30cm (6–12in) wide, pencilled with a white halo around a yellow blaze on each fall. Prefers moist soil.

I. setosa (Bristle-pointed iris; illus. p.197). Rhizomatous, beardless iris, very variable in stature. H 10cm–1m (4in–3ft), S indefinite. Bears 2–13 deep blue or purple-blue flowers, 5–8cm (2–3in) across, from each spathe in late spring and early summer. Falls have paler blue or white marks; each standard is reduced to a bristle.

I. **'Shepherd's Delight'** (illus. p.199). Rhizomatous, bearded iris (Standard Tall). H 1m (3ft), S indefinite. In early summer, well-branched stem produces 6–10 clear pink flowers, 15–18cm (6–7in) wide, with a yellow cast.

I. sibirica (Siberian flag). Rhizomatous, beardless Siberian iris. H 50cm–1.2m (20in–4ft), S indefinite. From late spring to early summer, branched stem bears 2 or 3 dark-veined, blue or blue-purple flowers, 5–10cm (2–4in) across, from each spathe. Prefers moist or boggy soil.

I. **'Splash Down'**. Rhizomatous, beardless Siberian iris. H 1m (3ft), S indefinite. From late spring to early summer, branched stem produces 1–4 flowers, 5–10cm (2–4in) across. Standards are pale blue and falls speckled blue on a pale ground. Prefers moist soil.

I. spuria. Very variable, rhizomatous, beardless Spuria iris. H 50cm–1m (20in–3ft), S indefinite. Strong, erect-branched stem produces 2–5 pale blue-purple, sky-blue, violet-blue, white or yellow flowers, 5–12cm (2–5in) across, in early summer. Prefers moist soil.

I. **'Stepping Out'**. Rhizomatous, bearded iris (Standard Tall). H 1m (3ft), S indefinite. Well-branched stem produces 8–11 white flowers, 14–15cm (5½–6in) wide, with deep blue-purple marks in early summer.

I. **'Sun Miracle'** (illus. p.199). Rhizomatous, bearded iris (Standard Tall). H 1m (3ft), S indefinite. Well-branched stem produces 7–10 pure yellow flowers, 15–18cm (6–7in) wide, in early summer.

I. susiana (Mourning iris). Rhizomatous Oncocyclus iris. H 35–40cm (14–16in), S indefinite. In late spring produces a solitary greyish-white flower, 8–15cm (3–6in) wide, heavily veined deep purple. Standards appear larger than incurved falls, which each carry a black blaze and a deep purple beard. Grows best in a frame or alpine house.

I. tectorum (Japanese roof iris, Wall flag; illus. p.198). Evansia iris with stout rhizomes. H 25–35cm (10–14in), S indefinite. Frost hardy. Has fans of broadly lance-shaped, ribbed leaves. In early summer, sparsely branched stem produces 2–3 darker-veined, bright lilac flowers, 1–8cm (½–3in) across with a white crest on each fall, from each spathe. Prefers a sheltered, sunny site near a south- or west-facing wall.

I. tenax (illus. p.198). Rhizomatous, beardless Pacific Coast iris. H 15–30cm (6–12in), S indefinite. From late spring to early summer produces 1 or 2 deep purple to lavender-blue flowers, 8–12cm (3–5in) across, often with yellow-and-white marking on falls. White, cream and yellow variants also occur. Narrow, dark green leaves are stained pink at base.

I. **'Theseus'**. Rhizomatous Regeliocyclus iris. H 45cm (18in), S indefinite. From late spring to early summer produces usually 2 flowers, 10–15cm (4–6in) across, with violet standards and violet-veined, cream

falls. Grows best in a frame or alpine house.
I. tolmeiana. See *I. missouriensis.*
I. tuberosa. See *Hermodactylus tuberosus.*
♛ *I. unguicularis*, syn. *I. stylosa* (Algerian iris, Algerian winter iris, Winter iris). Evergreen, rhizomatous, beardless iris. H to 20cm (8in), S indefinite. Has narrow, tough leaves. Almost stemless, primrose-scented, lilac flowers, 5–8cm (2–3in) across with yellow centres to the falls and with very long tubes, appear from late autumn to early spring. Buds are prone to slug attack. Is excellent for cutting. Prefers a sheltered site against a south- or west-facing wall. **'Mary Barnard'** has deep violet-blue flowers. Flowers of **'Walter Butt'** are pale silvery-lavender.
I. variegata (Variegated iris; illus. p.199). Rhizomatous, bearded iris. H 30–50cm (12–20in), S indefinite. In early summer, branched stem produces 3–6 flowers, 5–8cm (2–3in) across, with bright yellow standards and white or pale yellow falls, heavily veined red-brown and appearing striped.
I. verna. Rhizomatous, beardless iris. H 5cm (2in), S indefinite. In mid-spring bears 1, occasionally 2 lilac-blue flowers, 2.5–5cm (1–2in) across, with a narrow, orange stripe in the centre of each fall. Prefers semi-shade and moist but well-drained soil.
I. versicolor (Blue flag, Wild iris; illus. p.198). Robust, rhizomatous, beardless iris. H 60cm (2ft), S indefinite. Branched stem produces 3–5 or more purple-blue, reddish-purple, lavender or slate-purple flowers, 5–10cm (2–4in) across, from early to mid-summer. Falls usually have a central, white area veined purple. Prefers partial shade and thrives in moist soil or in shallow water. **'Kermesina'** has red-purple flowers.
I. warleyensis. Bulbous Juno iris. H 20–45cm (8–18in), S 7–8cm (3in). In spring produces up to 5 pale lilac or violet-blue flowers, 5–7cm (2–3in) across, in leaf axils. Each fall has a darker blue apex and a yellow stain in the centre. Bears scattered, lance-shaped, channelled, mid-green leaves. Is best in an unheated greenhouse.

I. **'White Excelsior'**. Bulbous Xiphium iris (Dutch). H to 80cm (32in), S 15cm (6in). From spring to early summer bears 1 or 2 white flowers, 6–8cm (2½–3in) wide, with a yellow stripe down each fall centre. Narrowly lance-shaped, channelled, mid-green leaves are scattered on flower stem.
I. winogradowii (illus. p.199). Bulbous Reticulata iris. H 6–10cm (2½–4in), S 6–7cm (2½–3in). Solitary pale primrose-yellow flower, 6–7cm (2½–3in) wide, spotted green on falls, appears in early spring. Narrow, squared leaves are very short at flowering time but elongate later.
I. **'Wisley White'** (illus. p.198). Rhizomatous, beardless Siberian iris. H to 1m (3ft), S indefinite. Each stem carries 2 or 3 white flowers, held well above the foliage, in early summer.
I. xiphioides. See *I. latifolia.*
I. xiphium. Bulbous Xiphium iris (Spanish). H to 80cm (32in), S 15cm (6in). Has 1 or 2 blue or violet, occasionally yellow or white, flowers, 6–8cm (2½–3in) across, with central, orange or yellow marks on the falls, in spring and early summer. Narrowly lance-shaped, channelled, mid-green leaves are scattered on flower stem. **'Blue Angel'** has bright mid-blue flowers with a yellow mark in the centre of each fall. Flowers of **'Lusitanica'** are pure yellow. **'Queen Wilhelmina'** produces white flowers in spring. **'Wedgwood'** (illus. p.199) has bright blue flowers.

ISATIS (Cruciferae)
Genus of summer-flowering annuals, biennials and perennials. Fully hardy. Needs sun and fertile, well-drained soil. Propagate by seed in autumn or spring.
I. tinctoria (Woad). Vigorous, upright biennial. H to 1.2m (4ft), S 45cm (1½ft). Has oblong to lance-shaped, glaucous leaves and, in summer, large, terminal panicles of 4-petalled, yellow flowers.

Ismene calathina. See *Hymenocallis narcissiflora.*

ISOLEPIS (Cyperaceae). See GRASSES, BAMBOOS, RUSHES and SEDGES.
I. setaceus, syn. *Scirpus setaceus* (Bristle club-rush). Tuft-forming, annual or short-lived, perennial rush. H 10–15cm (4–6in), S 8cm (3in). Fully hardy. Has very slender, lax, basal leaves. Very slender, unbranched stems each bear 1–3 minute, egg-shaped, green spikelets in summer.

ISOPLEXIS (Scrophulariaceae)
Genus of evergreen, mainly summer-flowering shrubs, grown for their flowers. Is closely related to *Digitalis*. Frost tender, min. 7°C (45°F). Tolerates full light or partial shade and well-drained soil. Water potted specimens freely when in full growth, moderately at other times. Remove spent flower spikes. Propagate by seed in spring or by semi-ripe cuttings in late summer.
I. canariensis, syn. *Digitalis canariensis*, illus. p.141.

ISOPYRUM (Ranunculaceae)
Genus of spring-flowering perennials, grown for their small flowers and delicate foliage. Is suitable for woodlands, peat beds and rock gardens. Fully hardy. Requires shade and humus-rich, moist soil. Propagate by seed when fresh or by division in autumn. Self seeds readily.
I. thalictroides. Dainty, clump-forming perennial. H and S 25cm (10in). Central stalk bears fern-like, 3-parted leaves, each leaflet being cut into 3. Has small, nodding, cup-shaped, white flowers in spring.

Isotrema griffithii. See *Aristolochia griffithii.*

ITEA (Grossulariaceae)
Genus of deciduous or evergreen trees and shrubs, grown for their foliage and flowers. Frost hardy, but in most areas protect by growing against a south- or west-facing wall. Needs sun or semi-shade and fertile, well-drained but not too dry soil. Propagate by softwood cuttings in summer.
♛ *I. ilicifolia* illus. p.116.

IXIA (Iridaceae)
Genus of spring- and summer-flowering corms with wiry stems and spikes of flattish flowers. Half hardy. Grow in an open, sunny situation and in well-drained soil. Plant in autumn for spring and early summer flowers; plant in spring for later summer display. Dry off after flowering. Propagate in autumn by seed or by offsets at replanting time.
I. maculata. Spring- to early summer-flowering corm. H 40cm (16in), S 2.5–5cm (1–2in). Leaves are linear, erect and mostly basal. Wiry stem bears a spike of flattish, orange or yellow flowers, 2.5–5cm (1–2in) across, with brown or black centres.
I. monadelpha. Spring- to early summer-flowering corm. H 30cm (12in), S 2.5–5cm (1–2in). Linear, erect leaves are mostly basal. Stem produces a dense spike of 5–10 flattish, white, pink, purple or blue flowers, 3–4cm (1¼–1½in) across, often with differently coloured eyes.
I. viridiflora illus. p.358.

IXIOLIRION (Amaryllidaceae)
Genus of bulbs, grown for their funnel-shaped flowers mainly in spring. Fully hardy. Needs a sheltered, sunny position and well-drained soil that becomes hot and dry in summer to ripen the bulb. Propagate by seed or offsets in autumn.
I. montanum. See *I. tataricum.*
I. tataricum, syn. *I. montanum*, illus. p.358.

IXORA (Rubiaceae)
Genus of evergreen, summer-flowering shrubs, grown primarily for their flowers, some also for their foliage. Frost tender, min. 13–16°C (55–61°F). Prefers full sun and humus-rich, well-drained soil. Water containerized specimens freely when in full growth, moderately at other times. Propagate by seed in spring or by semi-ripe cuttings in summer.
I. coccinea illus. p.135.

J

JACARANDA (Bignoniaceae)
Genus of deciduous or evergreen trees, grown for their flowers in spring-summer and their foliage. Frost tender, min. 7–10°C (45–50°F). Grows in any fertile, well-drained soil and in full light. Water potted specimens freely when in full growth, sparingly at other times. Potted plants grown for their foliage only may be cut back hard in late winter. Propagate by seed in spring or by semi-ripe cuttings in summer.
J. acutifolia of gardens. See *J. mimosifolia*.
J. mimosifolia, syn. *J. acutifolia* of gardens, *J. ovalifolia*, illus. p.53.
J. ovalifolia. See *J. mimosifolia*.

Jacobinia carnea. See *Justicia carnea*.
Jacobinia coccinea. See *Pachystachys coccinea*.
Jacobinia pohliana. See *Justicia carnea*.
Jacobinia spicigera. See *Justicia spicigera*.

JACQUEMONTIA (Convolvulaceae)
Genus of evergreen, twining climbers, grown for their flowers. Frost tender, min. 16–18°C (61–4°F). Any well-drained soil is suitable with full light. Water freely except in cold weather. Provide support and thin out by cutting old stems to ground level in spring. Propagate by seed in spring or by semi-ripe cuttings in summer. Red spider mite and whitefly may cause problems.
J. pentantha. Fast-growing, evergreen, twining climber. H 2–3m (6–10ft). Has heart-shaped, pointed leaves and 2.5cm (1in) wide, funnel-shaped, rich violet-blue or pure blue flowers in long-stalked clusters in summer-autumn.

JAMESIA (Hydrangeaceae)
Genus of one species of deciduous shrub, grown for its flowers. Fully hardy. Needs full sun and fertile, well-drained soil. Propagate by softwood cuttings in summer.
J. americana. Deciduous, bushy shrub. H 1.5m (5ft), S 2.5m (8ft). Rounded, grey-green leaves are grey-white beneath. Clusters of small, slightly fragrant, star-shaped, white flowers are produced during late spring.

JANCAEA. See **JANKAEA**.

JANKAEA, syn. **JANCAEA** (Gesneriaceae)
Genus of one species of evergreen, rosetted perennial, grown for its flowers and silver-green leaves. Makes a good alpine house plant. Frost hardy. Is difficult to grow, as needs shade from mid-day sun in high summer, a humus-rich, gritty, moist, alkaline soil and a gritty collar. Dislikes winter wet. Propagate by seed in spring or by leaf cuttings in mid-summer.
J. heldreichii illus. p.317.

JASIONE (Campanulaceae)
Genus of summer-flowering annuals, biennials and perennials, grown for their attractive flower heads. Fully hardy. Needs sun and sandy soil. Remove old stems in autumn. Propagate by seed in autumn or by division in spring.
J. laevis, syn. *J. perennis* (Sheep's bit). Tufted perennial. H 5–30cm (2–12in), S 10–20cm (4–8in). Has narrowly oblong, very hairy or glabrous, grey-green leaves and, in summer, spiky, spherical, blue flower heads borne on erect stems. Is good for a rock garden.
J. perennis. See *J. laevis*.

JASMINUM (Oleaceae)
Jasmine
Genus of deciduous or evergreen shrubs and woody-stemmed, scrambling or twining climbers, grown for their often fragrant flowers and their foliage. Fully hardy to frost tender, min. 7–18°C (45–64°F). Requires full sun and fertile, well-drained soil. *J. nudiflorum*, which is not self-supporting, benefits from having old shoots thinned out after flowering; prune others as required after flowering. Propagate by semi-ripe cuttings in summer.
J. angulare. Evergreen, woody-stemmed, scrambling climber. H 2m (6ft) or more. Frost tender, min. 7–10°C (45–50°F). Dark green leaves have 3 oval leaflets. Small clusters of fragrant, tubular, 5-lobed, white flowers are carried in late summer.
J. beesianum. Evergreen, woody-stemmed, scrambling climber, deciduous in cool areas. H to 5m (15ft). Frost hardy. Has lance-shaped leaves. Fragrant, tubular, usually 6-lobed, pinkish-red flowers, 1–3 together, are borne in early summer, followed by shiny, black berries.
J. grandiflorum (Royal jasmine, Spanish jasmine). Evergreen, woody-stemmed, scrambling climber. H 2m (6ft). Frost tender, min. 7–10°C (45–50°F). Leaves have 7 or 9 leaflets. Clusters of up to 50 fragrant, tubular, 5- or 6-lobed, sometimes red-tinged, white flowers are carried in summer-autumn.
J. humile (Yellow jasmine) illus. p.116. ❦ '**Revolutum**' is an evergreen, bushy shrub. H 2.5m (8ft), S 3m (10ft). Fully hardy. Bears large, fragrant, tubular, upright, bright yellow flowers, with 5 spreading lobes, on long, slender, green shoots from early spring to late autumn. Glossy, bright green leaves each consist of 3–7 oval leaflets. f. *wallichianum* has semi-pendent flowers and 7–13 leaflets.
❦ *J. mesnyi*, syn. *J. primulinum*, illus. p.167.
❦ *J. nudiflorum* illus. p.121.
❦ *J. officinale* (Common jasmine, Jessamine). Semi-evergreen or deciduous, woody-stemmed, twining climber. H to 12m (40ft). Leaves comprise 7 or 9 leaflets. Has clusters of fragrant, 4- or 5-lobed, white flowers in summer-autumn. f. *affine* illus. p.168.
J. parkeri. Evergreen, domed shrub. H 15cm (6in), S 38cm (18in) or more. Frost hardy. Produces a tangled mass of fine stems and twigs bearing minute, oval leaves. Masses of tiny, tubular, 5-lobed, yellow flowers appear from leaf axils in early summer.
❦ *J. polyanthum* illus. p.179.
J. primulinum. See *J. mesnyi*.
J. rex. Evergreen, woody-stemmed, twining climber. H 3m (10ft). Frost tender, min. 18°C (64°F). Has broadly oval, leathery, deep green leaves, 10–20cm (4–8in) long. Scentless, tubular, 5-lobed, pure white flowers are pink-tinged in bud and appear intermittently all year if warm enough.

JEFFERSONIA (Berberidaceae)
Genus of spring-flowering perennials. Fully hardy. Needs shade or partial shade and humus-rich, moist soil. Extensive root systems resent disturbance. Top-dress crown in late autumn. Propagate by seed as soon as ripe.
J. diphylla illus. p.295.
J. dubia, syn. *Plagiorhegma dubia*, illus. p.317.

JOVIBARBA (Crassulaceae)
Genus of evergreen perennials that spread by short stolons and are grown for their symmetrical rosettes of oval to strap-shaped, pointed, fleshy leaves. Makes ground-hugging mats, suitable for rock gardens, screes, walls, banks and alpine houses. Fully hardy. Needs sun and gritty soil. Takes several years to reach flowering size. Rosettes die after plants have flowered, but leave numerous offsets. Propagate by offsets in summer.
J. hirta, syn. *Sempervivum hirtum*, illus. p.336.
J. sobolifera, syn. *Sempervivum soboliferum*. Vigorous, evergreen, mat-forming perennial. H 10cm (4in), S 20cm (8in). Rounded, greyish-green or olive-green rosettes are often red-tinged. Flower stems bear terminal clusters of small, cup-shaped, 6-petalled (rarely 5 or 7), pale yellow flowers in summer.

JUANULLOA (Solanaceae)
Genus of evergreen, summer-flowering shrubs, grown for their flowers. Frost tender, min. 13–15°C (55–9°F). Low temperatures cause leaf drop. Prefers full light and fertile, freely draining soil. Water potted specimens moderately, less when not in full growth. To encourage a branching habit, tip prune young plants. Propagate by semi-ripe cuttings in summer. Whitefly, red spider mite and mealy bug may be troublesome.
J. aurantiaca. See *J. mexicana*.
J. mexicana, syn. *J. aurantiaca*, illus. p.141.

JUBAEA (Palmae)
Genus of one species of evergreen palm, grown for its overall appearance. Frost tender, min. 10°C (50°F). Needs full light and fertile, well-drained soil. Water potted specimens moderately, less frequently in winter. Propagate by seed in spring at not less than 25°C (77°F). Red spider mite may be a nuisance.
J. chilensis, syn. *J. spectabilis*, illus. p.59.
J. spectabilis. See *J. chilensis*.

JUGLANS (Juglandaceae)
Walnut
Genus of deciduous trees, with aromatic leaves, grown for their foliage, stately habit and, in some species, edible nuts (walnuts). Produces greenish-yellow catkins in spring and early summer. Fully hardy, but young plants are prone to frost damage. Requires full sun and deep, fertile, well-drained soil. Propagate by seed, when ripe, in autumn.
J. ailantifolia (Japanese walnut). Deciduous, spreading tree with stout shoots. H and S 15m (50ft). Very large leaves consist of 11–17 oblong, glossy, bright green leaflets. Bears edible walnuts in autumn. var. *cordiformis* illus. p.42.
J. cathayensis (Chinese walnut). Deciduous, spreading tree. H and S 20m (70ft). Has very large leaves, consisting of 11–17 oval to oblong, dark green leaflets. Bears edible walnuts in autumn.
J. cinerea (Butternut). Fast-growing, deciduous, spreading tree. H 25m (80ft), S 20m (70ft). Leaves are large and very aromatic, with 7–19 oval to oblong, pointed, bright green leaflets. Bears dense clusters of large, rounded nuts in autumn.
J. microcarpa, syn. *J. rupestris*, illus. p.66.
❦ *J. nigra* illus. p.41.
❦ *J. regia* illus. p.40.
J. rupestris. See *J. microcarpa*.

JUNCUS (Juncaceae). See GRASSES, BAMBOOS, RUSHES and SEDGES.
J. effusus f. *spiralis* illus. p.184.

JUNIPERUS (Cupressaceae), Juniper. See CONIFERS.
J. chinensis (Chinese juniper). Conical conifer, making a tree, H 15m (50ft), S 2–3m (6–10ft), or a spreading shrub, H 1–5m (3–15ft), S 3–5m (10–15ft). Fully hardy. Has peeling bark. Both scale- and needle-like, aromatic, dark green leaves, paired or in 3s, are borne on same shoot. Globose, fleshy, berry-like fruits are glaucous white. Many cultivars commonly listed under *J. chinensis* are forms of *J.* x *media*.
❦ '**Aurea**', H 10–15m (30–50ft), S 3–4m (10–12ft), is a slow-growing, oval or conical form with gold foliage and abundant yellow, male cones.

♈ 'Kaizuka', H 5m (15ft), S 3–5m (10–15ft), forms a sprawling, irregular bush and produces a profusion of cones. 'Keteleeri' illus. p.79.
♈ 'Obelisk' illus. p.82.
♈ 'Pyramidalis', H 10m (30ft), S 1–2m (3–6ft), is a columnar, dense form with ascending branches bearing needle-like, blue-green leaves.
'Robusta Green' see *J. virginiana* 'Robusta Green'. 'Stricta' (illus. p.84), H to 5m (15ft), S to 1m (3ft), is conical, with soft, blue-green, young foliage.
J. communis (Common juniper). Conifer, ranging from a spreading shrub to a narrow, upright tree. H 30cm–8m (1–25ft), S 1–4m (3–12ft). Fully hardy. Has needle-like, aromatic, glossy, mid- or yellow-green leaves in 3s and bears globular to ovoid, fleshy, greenish berries that become glaucous blue, then ripen to black in their third year. ♈ 'Compressa', H 75cm (30in), S 15cm (6in), is a dwarf, erect form.
♈ 'Hibernica' (illus. p.85), H 3–5m (10–15ft), S 30cm (12in), is columnar.
♈ 'Hornibrookii', H 50cm (20in), S 2m (6ft), and 'Prostrata', H 20–30cm (8–12in), S 1–2m (3–6ft), are carpeting plants.
J. conferta (Shore juniper). Prostrate, shrubby conifer. H 15cm (6in), S 1–2m (3–6ft). Fully hardy. Spreading branches bear dense, needle-like, aromatic, glossy, bright green leaves, glaucous beneath. Produces glaucous black berries. Tolerates salty, coastal air.
J. davurica (Dahurian juniper). 'Expansa Variegata' (illus. p.85) is a conifer with trailing or ascending branchlets. H 75cm (30in), S 1.5–2m (5–6ft). Fully hardy. Bears scale- and needle-like, aromatic, yellow-variegated, bluish-green leaves.
J. drupacea (Syrian juniper). Columnar conifer. H 10–15m (30–50ft), S 1–2m (3–6ft). Fully hardy. Has needle-like, aromatic, light green leaves, in 3s, and ovoid or almost globose, fleshy, brown berries.
J. horizontalis (Creeping juniper). Prostrate, wide-spreading, shrubby conifer, eventually forming mats up to 50cm (20in) thick. Fully hardy. Has scale- or needle-like, aromatic, blue-green or -grey leaves and pale blue berries. 'Douglasii' (illus. p.84) has glaucous blue foliage that turns plum-purple in winter. ♈ 'Plumosa' has grey-green leaves, becoming purple during winter. 'Plumosa Compacta' is denser than 'Plumosa' and turns bronze-purple in winter. 'Prince of Wales' has bright green foliage, tinged blue when young and turning purple-brown in winter. 'Turquoise Spreader' (illus. p.84) has turquoise-green foliage.
♈ 'Wiltonii' has bluish-grey leaves that retain their colour over winter.
J. × *media*. Group of conical conifers. H 15m (50ft), S 2–3m (6–10ft). Fully hardy. Has peeling bark. Mainly scale-like, dark green leaves exude a fetid smell when crushed. Fruits are globose to rounded, white or blue-black. Cultivars are suitable as ground cover or as specimen plants in a small garden. Some forms are commonly listed under *J. chinensis*. ♈ 'Blaauw', H and S 2m (6ft), is a spreading shrub with blue-green foliage. 'Blue Gold' (illus. p.85), H to 1m (3ft), S 1m (3ft), is a spreading form with leaves variegated sky-blue and gold. 'Hetzii' (syn. *J. virginiana* 'Hetzii'), H 3–4m (10–12ft), S 4m (12ft), has tiers of grey-green foliage.
♈ 'Pfitzeriana' (illus. p.85), H 3m (10ft), S 3–5m (10–15ft), is a spreading, flat-topped shrub with grey-green leaves. 'Pfitzeriana Aurea' (illus. p.85) has golden foliage. 'Pfitzeriana Glauca' (illus. p.84) has grey-blue leaves. 'Plumosa', H 1m (3ft), S 2–3m (6–10ft), is a spreading shrub with drooping sprays of mid-green foliage. ♈ 'Plumosa Aurea' (illus. p.85) is more erect, with green-gold foliage, turning bronze in winter.
J. procumbens (Bonin Isles juniper; illus. p.84). Spreading, prostrate, shrubby conifer. H 75cm (30in), S 2m (6ft). Fully hardy. Has red-brown bark. Thick branches carry needle-like, aromatic, light green or yellow-green leaves and globose, fleshy, brown or black berries. ♈ 'Nana' (illus. p.84), H 15–20cm (6–8in), S 75cm (30in), is less vigorous and is mat-forming.
J. recurva (Drooping juniper, Himalayan weeping juniper). Slow-growing, conical conifer. H to 15m (50ft), S to 7m (22ft). Fully hardy. Smooth bark flakes in thin sheets. Weeping sprays of needle-like, aromatic, incurved leaves are grey-green or blue-green. Globose or ovoid, fleshy berries are black. var. *coxii* (Coffin juniper) has longer, bright green leaves. 'Densa' (illus. p.84), H 30cm (1ft), S 1m (3ft), is a spreading shrub with sprays of green leaves that are erect at tips.
J. rigida (Temple juniper). Sprawling, shrubby conifer. H and S 8m (25ft). Fully hardy. Grey or brown bark peels in strips. Very sharp, needle-like, aromatic, bright green leaves, in 3s, are borne in nodding sprays. Globose, fleshy fruits are purplish-black.
J. sabina (Savin). Spreading, shrubby conifer. H to 4m (12ft), S 3–5m (10–15ft). Fully hardy. Has flaking, red-brown bark. Slender shoots bear mainly scale-like, aromatic, dark green leaves that give off a fetid smell when crushed. Bears rounded, blue-black berries. 'Blue Danube', H 2m (6ft), S 2–4m (6–12ft), is a spreading form with branch tips curved upwards and grey-blue foliage. 'Cupressifolia' (illus. p.84), H 2m (6ft), S 4m (12ft), is a free-fruiting, female form with horizontal or ascending branches and blue-green leaves. 'Mas' (illus. p.84), has ascending branches. Leaves are blue above, green below, purplish in winter. var. *tamariscifolia* (illus. p.84), H 1m (3ft), S 2m (6ft), has tiered layers of mainly needle-like, bright green or blue-green leaves.
J. scopulorum (Rocky Mountain juniper). Slow-growing, round-crowned conifer. H 10m (30ft), S 4m (12ft). Fully hardy. Reddish-brown bark is furrowed into strips or squares and peels on branches. Scale-like, aromatic leaves are grey-green to dark green. Bears globose, fleshy, blue berries. 'Skyrocket' (syn. *J. virginiana* 'Skyrocket'; illus. p.84), H 8m (25ft), S 75cm (2½ft), is very narrow in habit with glaucous blue foliage. 'Springbank' (illus. p.84) is narrowly conical with drooping branch tips and intense silvery-blue foliage. 'Tabletop', H 2m (6ft), S 5m (15ft), has a flat-topped habit and silvery-blue leaves.
J. squamata (Flaky juniper). Prostrate to sprawling, shrubby conifer. H 30cm–4m (1–12ft), S 1–5m (3–15ft). Fully hardy. Bark is red-brown and flaking. Needle-like, aromatic, fresh green or bluish-green leaves spread at tips of shoots. Produces ovoid, fleshy, black berries. ♈ 'Blue Carpet', H 30cm (1ft), S 2–3m (6–10ft), is vigorous and prostrate, with glaucous blue foliage. ♈ 'Blue Star' (illus. p.84), H 50cm (20in), S 60cm (24in), forms a dense, rounded bush and has blue foliage. 'Chinese Silver' (illus. p.84), H and S 3–4m (10–12ft), has branches with nodding tips and bluish leaves with bright silver undersides.
♈ 'Holger' (illus. p.84), H and S 2m (6ft), has sulphur-yellow young leaves that contrast with steel-blue old foliage. 'Meyeri', H and S 5m (15ft), sprawls and has steel-blue foliage.
J. virginiana (Pencil cedar). Slow-growing, conical or broadly columnar conifer. H 15–20m (50–70ft), S 6–8m (20–25ft). Fully hardy. Both scale- and needle-like, aromatic, grey-green leaves are borne on same shoot. Ovoid, fleshy berries are brownish-violet and very glaucous. 'Burkii' illus. p.82.
♈ 'Grey Owl' (illus. p.84), H 3m (10ft), S 3–5m (10–15ft), is a low, spreading cultivar with ascending branches and silvery-grey foliage. 'Hetzii' see *J.* × *media* 'Hetzii'. 'Robusta Green' (syn. *J. chinensis* 'Robusta Green') illus. p.82. 'Skyrocket' see *J. scopulorum* 'Skyrocket'.

JUSTICIA (Acanthaceae)
Genus of evergreen perennials, sub-shrubs and shrubs, grown mainly for their flowers. Frost tender, min. 10–15°C (50–59°F). Requires full light or partial shade and fertile, well-drained soil. Water containerized specimens freely when in full growth, moderately at other times. Some species need regular pruning. Propagate by softwood or greenwood cuttings in spring or early summer. Whitefly may cause problems.
J. adhatoda, syn. *Adhatoda vasica*. Evergreen, erect, sparingly branched shrub. H 2–3m (6–10ft), S 1–2m (3–6ft). Min. 10–15°C (50–59°F). Has oval, pointed, prominently veined, mid-green leaves and, in summer, dense, terminal spikes of tubular, white flowers, veined pink or purple on lower lip. May be cut back hard in early spring to reduce height. Is sometimes confused in cultivation with *Duvernoia adhatodoides*.
♈ *J. brandegeana*, syn. *Beloperone guttata*, *Drejerella guttata*, illus. p.136. 'Chartreuse' illus. p.138.
J. carnea, syn. *Jacobinia carnea*, *J. pohliana*, illus. p.133.
J. coccinea. See *Pachystachys coccinea*.
J. floribunda. See *J. rizzinii*.
J. ghiesbreghtiana of gardens. See *J. spicigera*.
J. pauciflora. See *J. rizzinii*.
♈ *J. rizzinii*, syn. *J. floribunda*, *J. pauciflora*, *Libonia floribunda*. Evergreen, rounded, freely branching shrub. H and S 30–60cm (1–2ft). Min. 15°C (59°F) to flower well in winter. Leaves are oval and mid-green. Has nodding clusters of tubular, yellow-tipped, scarlet flowers mainly from autumn to spring. Is best repropagated every few years.
J. spicigera, syn. *J. ghiesbreghtiana* of gardens, *Jacobinia spicigera*, illus. p.142.

K

KADSURA (Schisandraceae)
Genus of evergreen, twining climbers, grown for their foliage and fruits. Male and female flowers are borne on separate plants, so plants of both sexes must be grown to obtain fruits. Frost hardy. Grow in semi-shade and in any soil. Propagate by stem cuttings in late summer.
K. japonica. Evergreen, twining climber. H 3–4m (10–12ft). Has oval or lance-shaped, mid-green leaves. Solitary small, fragrant, cream flowers are produced in leaf axils in summer, followed by bright red berries.

KAEMPFERIA (Zingiberaceae)
Genus of tufted, rhizomatous perennials, grown for their aromatic leaves and their flowers. Frost tender, min. 18°C (64°F). Needs a moist atmosphere, partial shade and moist, humus-rich soil. Allow to dry out when plants become dormant. Propagate by division in late spring.
K. pulchra illus. p.247.
K. roscoeana. Rhizomatous perennial without an obvious stem. H 5–10cm (2–4in), S 20–25cm (8–10in). Usually has only 2 almost round, aromatic leaves, to 10cm (4in) long, dark green with pale green variegation above, reddish-green beneath, that are held horizontally. A short spike of pure white flowers, each with a deeply lobed lip, appears from the centre of leaf tuft in autumn.

KALANCHOE, syn. BRYOPHYLLUM (Crassulaceae)
Genus of perennial succulents or shrubs with very fleshy, mainly cylindrical, oval or linear leaves and bell-shaped to tubular flowers. Many species produce new plantlets from indented leaf margins. Frost tender, min. 7–15°C (45–59°F). Needs full sun or partial shade and well-drained soil. Keep moist from spring to autumn. Water lightly and only occasionally in winter. Propagate by seed, offsets or stem cuttings in spring or summer.
K. beharensis illus. p.396.
K. blossfeldiana (Flaming Katy). Bushy, perennial succulent. H and S 30cm (12in). Min. 10°C (50°F). Bears oval to oblong, glossy, dark green leaves with toothed edges and clusters of tubular, scarlet flowers, 0.5cm (^{1}/$_4$in) across, in spring. Prefers partial shade. Many hybrids are available in a range of colours (salmon pink, illus. p.400).
K. daigremontiana illus. p.398.
K. fedtschenkoi (South American air plant). Bushy, perennial succulent. H and S 1m (3ft). Min. 10°C (50°F). Produces oval, indented, blue-grey leaves with new plantlets in each notch. Bell-shaped, brownish-pink flowers, 2cm (^{3}/$_4$in) long, appear in late winter. Prefers a sunny position.
f. *variegata* illus. p.398.
♀ *K. pumila.* Creeping, perennial succulent. H 10cm (4in), S indefinite. Min. 10°C (50°F). Has oval, powdery grey-white leaves with indented margins. Tubular, pink flowers, 1cm (^{1}/$_2$in) long, appear in spring. Suits a hanging basket in a sunny position.
♀ *K.* 'Tessa' illus. p.400.
♀ *K. tomentosa* illus. p.403.
K. tubiflora illus. p.404.
K. uniflora, syn. *Kitchingia uniflora.* Creeping, perennial succulent. H 6cm (2^1/$_2$in), S indefinite. Min. 15°C (59°F). Produces rounded, mid-green leaves, 0.5–3cm (^{1}/$_4$–1^1/$_4$in) long, and bell-shaped, yellow-flushed, reddish-purple flowers, 1cm (^{1}/$_2$in) long, in late winter. Prefers partial shade.
♀ *K.* 'Wendy' illus. p.399.

KALMIA (Ericaceae)
Genus of evergreen, summer-flowering shrubs, grown for their clusters of distinctive, usually cup-shaped flowers. Fully hardy. Needs sun or semi-shade and moist, peaty, acid soil. Propagate species by softwood cuttings in summer or by seed in autumn, selected forms by softwood cuttings in summer.
K. angustifolia (Sheep laurel).
♀ f. *rubra* (syn. *K. angustifolia* 'Rubra') illus. p.135.
♀ *K. latifolia* (Calico bush) illus. p.111. ♀ 'Ostbo Red' is an evergreen, bushy, dense shrub. H and S 3m (10ft). Has oval, glossy, rich green leaves. Large, showy clusters of deep pink flowers open in early summer from distinctively crimped, deep red buds. Prefers full sun.

KALMIOPSIS (Ericaceae)
Genus of one species of evergreen, spring-flowering shrub, grown for its flowers. Fully hardy. Requires semi-shade and moist, peaty, acid soil. Propagate by softwood or semi-ripe cuttings in summer.
K. leachiana 'M. Le Piniec', syn. *K.l.* 'La Piniec'. Evergreen, bushy shrub. H and S 30cm (12in). Terminal clusters of small, widely bell-shaped, purplish-pink flowers are produced from early to late spring. Has small, oval, glossy, dark green leaves.

KALOPANAX (Araliaceae)
Genus of one species of deciduous, autumn-flowering tree, grown for its foliage and fruits. Fully hardy, but unripened wood on young plants is susceptible to frost damage. Does best in sun or semi-shade and in fertile, moist but well-drained soil. Propagate by softwood cuttings in summer.
K. pictus. See *K. septemlobus.*
K. ricinifolius. See *K. septemlobus.*
K. septemlobus, syn. *K. pictus, K. ricinifolius, Acanthopanax ricinifolium,* illus. p.54.

KELSEYA (Rosaceae)
Genus of one species of extremely small, evergreen sub-shrub. Is difficult to grow and is best in an alpine house as foliage deeply resents both summer and winter wet. Fully hardy. Needs full sun and moist, alkaline soil. Propagate by soft-tip cuttings in late spring or by seed in autumn. Is susceptible to moulds, so remove any dead rosettes at once.
K. uniflora. Slow-growing, evergreen, rosetted sub-shrub. H 1cm (^{1}/$_2$in), S to 20cm (8in). Forms a hard mat of closely packed, small rosettes of tiny, oval, dark green leaves. In early spring carries stemless, star-shaped, occasionally pink-flushed, white flowers.

KENNEDIA, syn. KENNEDYA (Leguminosae)
Genus of evergreen, woody-stemmed, trailing and twining climbers, grown for their pea-like flowers. Frost tender, min. 5–7°C (41–5°F). Provide full light and moderately fertile, sandy soil. Water regularly when in full growth, sparingly in cold weather. Requires support. Thin out congested growth after flowering or in spring. Propagate by seed in spring or by semi-ripe cuttings in summer.
K. nigricans (Black bean). Vigorous, evergreen, woody-stemmed, twining climber. H to 2m (6ft). Leaves are divided into 3 leaflets with notched tips. Has small trusses of pea-like, velvety, black-purple flowers, with yellow blazes, in spring-summer.
K. rubicunda illus. p.165.

KENNEDYA. See KENNEDIA.

KENTIA. See HOWEA.

KERRIA (Rosaceae)
Genus of one species of deciduous shrub, grown for its showy, yellow flowers. Fully hardy. Needs sun or semi-shade and fertile, well-drained soil. Thin out old shoots after flowering. Propagate by softwood cuttings in summer or by division in autumn.
♀ *K. japonica* var. *japonica,* syn. *K.j.* 'Pleniflora', illus. p.101. var. *simplex* is a deciduous, arching, graceful shrub. H and S 2m (6ft). Has bright green foliage. Single, buttercup-like, golden yellow flowers are borne from mid- to late spring.

KIGELIA (Bignoniaceae)
Genus of one species of evergreen tree, grown for its flowers, curious, sausage-like fruits and for shade. Frost tender, min. 16°C (61°F). Requires full light and humus-rich, well-drained soil. Water potted specimens moderately, very little when temperatures are low. Propagate by seed in spring at not less than 23°C (73°F).
K. africana, syn. *K. pinnata* (Sausage tree). Evergreen, spreading, fairly bushy tree. H and S 8m (25ft) or more. Leaves have 7–11 oblong to oval leaflets. Scented, bell-shaped, purplish-red flowers open at night from autumn to spring. Inedible, cylindrical, hard-shelled, brown fruits, 30–45cm (12–18in) long, are long-lasting.
K. pinnata. See *K. africana.*

KIRENGESHOMA (Hydrangeaceae)
Genus of late summer- and autumn-flowering perennials. Fully hardy. Grow in light shade and in deep, moist, lime-free soil. Propagate by seed or division in autumn or spring.
♀ *K. palmata* illus. p.227.

KITAIBELA (Malvaceae)
Genus of one species of summer-flowering perennial. Fully hardy. Needs full sun and fertile, preferably dry soil. Propagate by seed in autumn or spring.
K. vitifolia. Bushy, upright perennial. H to 1.5m (5ft), S 60cm (2ft). In summer bears small clusters of open cup-shaped, white or rose-pink flowers. Has palmately lobed, coarsely toothed leaves.

Kitchingia uniflora. See *Kalanchoe uniflora.*

Kleinia articulata. See *Senecio articulatus.*

KNAUTIA (Dispacaceae)
Genus of summer-flowering annuals and perennials. Fully hardy. Needs sun and well-drained soil. Requires staking. Propagate by basal cuttings in spring or by seed in autumn.
K. arvensis (Scabious). Erect perennial. H 1.2m (4ft), S 45cm (1^1/$_2$ft). Heads of pincushion-like, bluish-lilac flowers appear in summer. Stems are clothed in narrowly oval to lyre-shaped, deeply divided leaves.
K. macedonica, syn. *Scabiosa rumelica,* illus. p.213.

KNIGHTIA (Proteaceae)
Genus of evergreen, summer-flowering trees, grown for their flowers, foliage and overall appearance. Half hardy, but is best at min. 3–5°C (37–41°F). Grows in any reasonably fertile, well-drained soil and in sun or partial shade. Water potted specimens moderately, less in winter. Propagate by seed in spring.
K. excelsa (New Zealand honeysuckle, Rewa rewa). Evergreen, upright tree. H 20m (70ft) or more, S 2–4m (6–12ft). Has oblong to lance-shaped, coarsely serrated, leathery leaves, glossy and deep green. Dense racemes of slender, tubular, deep red flowers are produced in summer.

KNIPHOFIA (Liliaceae or Aloeaceae)
Red-hot poker, Torch lily
Genus of perennials, some of which are evergreen. Fully to half hardy. Needs full sun and well-drained conditions, with constantly moist soil in summer.

Propagate species by seed or division in spring, cultivars by division only in spring.
K. **'Atlanta'**. Evergreen, upright perennial. H to 1m (3ft), S 45cm (1½ft). Fully hardy. In summer, stout stems bear dense, terminal racemes of tubular, bright orange-yellow flowers. Has thick, grass-like, channelled leaves. Does well in a coastal area.
K. **'Bee's Lemon'**. Upright perennial. H 1m (3ft), S 45cm (1½ft). Fully hardy. Has dense, terminal racemes of tubular, green-tinged, citron-yellow flowers on stout stems in late summer and autumn. Grass-like, deep green leaves have serrated edges.
♛ *K. caulescens* illus. p.227.
K. galpinii of gardens. See *K. triangularis*.
♛ *K.* **'Little Maid'** illus. p.254.
K. **'Maid of Orleans'**. Upright perennial. H 1.2m (4ft), S 45cm (1½ft). Frost hardy. In summer, slender stems are each crowned with a dense raceme of yellow buds that open to tubular, creamy-white flowers. Leaves are fresh green, basal and strap-shaped.
K. **'Percy's Pride'** illus. p.227.
♛ *K.* **'Royal Standard'** illus. p.222.
♛ *K.* **'Samuel's Sensation'**. Upright perennial. H 1.5m (5ft), S 60cm (2ft). Fully hardy. In late summer, dense, terminal racemes of tubular, deep orange flowers are produced on stout stems. Has strap-shaped, basal, dark green leaves.
K. thomsonii var. *snowdenii* illus. p.222.
♛ *K. triangularis*, syn. *K. galpinii* of gardens, illus. p.227.
K. uvaria (Red-hot poker). var. *nobilis* is an upright perennial with erect, then spreading leaves. H 2m (6ft), S 1m (3ft). Fully hardy. In late summer and autumn, stout stems each bear a dense, terminal raceme of tubular, bright red flowers. Has strap-shaped, channelled, dark green leaves.

KOCHIA, syn. BASSIA (Chenopodiaceae)
Genus of annuals and perennials, sometimes included in *Bassia*. Only *K. scoparia* f. *trichophylla* is usually cultivated. Half hardy. Does best in sun and in fertile, well-drained soil. May require support in very windy areas. Propagate by seed sown under glass in early spring, or outdoors in mid-spring.
♛ *K. scoparia* f. *trichophylla* illus. p.287.

KOELREUTERIA (Sapindaceae)
Genus of deciduous, summer-flowering trees, grown for their foliage, flowers and fruits. Fully hardy to frost tender, min. 10°C (50°F). Requires full sun, doing best in hot summers, and fertile, well-drained soil. Propagate by seed in autumn or by root cuttings in late winter.
♛ *K. paniculata* illus. p.66.

KOHLERIA (Gesneriaceae)
Genus of erect perennials with scaly rhizomes, grown for their showy, tubular flowers borne mainly in summer. Frost tender, min. 15°C (59°F). Grow in moist but well-drained soil and in full sun or semi-shade. Water sparingly in winter; over-watering will cause rhizomes to rot. Propagate in spring by division of rhizomes or by seed if available.
K. amabilis. Rhizomatous perennial. H 8–16cm (3–6in), S 60cm (2ft). Oval, hairy leaves, to 8cm (3in) long, are often marked with silver and brown above. Small, nodding, tubular, deep pink flowers, with red-marked lobes, appear in summer. Is useful for a hanging basket.
K. bogotensis. Erect, rhizomatous perennial. H and S 45cm (18in) or more. Oval, velvety, green leaves, to 8cm (3in) long, are sometimes marked with paler green above. In summer has small, tubular flowers, red with a yellow base outside, red-dotted, yellow within.
♛ *K. digitaliflora* illus. p.208.
♛ *K. eriantha* illus. p.214.

KOLKWITZIA (Caprifoliaceae)
Genus of one species of deciduous shrub, grown for its abundant flowers. Fully hardy. Prefers full sun and fertile, well-drained soil. Cut out old shoots after flowering. Propagate by softwood cuttings in summer.
K. amabilis (Beauty bush). Deciduous, arching shrub. H and S 3m (10ft). Has peeling bark and oval, dark green leaves. Bell-shaped, yellow-throated, white or pink flowers are borne in late spring and early summer. ♛ **'Pink Cloud'** illus. p.91.

Korolkowia sewerzowii. See *Fritillaria sewerzowii*.

KUNZEA (Myrtaceae)
Genus of evergreen shrubs and trees, grown for their flowers and overall appearance. Frost tender, min. 5–7°C (41–5°F). Prefers full light and sandy, well-drained, neutral to acid soil. Water potted specimens moderately, less when not in full growth. Propagate by semi-ripe cuttings in late summer or by seed in spring.
K. baxteri. Evergreen, rounded, wiry-stemmed shrub. H and S to 2m (6ft). Has narrow, cylindrical, pointed leaves and, in early summer, deep red flowers, each with a brush of stamens, in 5cm (2in) long spikes.

L

LABLAB (Leguminosae)
Genus of one species of deciduous, woody-stemmed, twining climber, grown for its attractive, pea-like flowers (in tropics is grown for green manure and animal feed, and for its edible pods and seeds). Is often raised as an annual. Frost tender, min. 5–10°C (41–50°F). Grow in sun and in any well-drained soil. Propagate by seed in spring.
L. purpureus, syn. *Dolichos lablab*, *D. lignosus*, illus. p.171.

+ LABURNOCYTISUS (Leguminosae)
Deciduous tree, grown for its flowers. Is a graft hybrid between *Laburnum anagyroides* and *Cytisus purpureus*. Fully hardy. Needs full sun; grows in any but waterlogged soil. Propagate by grafting on laburnum in late summer.
+ *L. adamii*. Deciduous, spreading tree. H 8m (25ft), S 6m (20ft). In late spring and early summer bears 3 types of blooms: yellow, laburnum flowers; purple, cytisus flowers; and laburnum-like, yellow and pinkish-purple flowers. Leaves, with 3 oval leaflets, are dark green.

LABURNUM (Leguminosae)
Genus of deciduous trees, grown for their profuse, pendent flower clusters in spring and summer. Fully hardy. Does best in full sun; grows in any but waterlogged soil. Seeds are very poisonous. Propagate species by seed in autumn, hybrids by budding in summer.
L. alpinum illus. p.67.
L. anagyroides (Common laburnum, Golden chain). Deciduous, spreading tree. H and S 7m (22ft). Leaves have 3 oval leaflets and are grey-green. Short, pendent, dense clusters of large, pea-like, yellow flowers appear in late spring and early summer.
L. x *watereri* (Voss's laburnum). ♣ **'Vossii'** illus. p.66.

LACHENALIA (Liliaceae)
Genus of winter- and spring-flowering bulbs with tubular or bell-shaped flowers; some have attractively mottled leaves. Is useful as pot plants and in open borders. Half hardy. Requires light, well-drained soil and a sunny site. Plant in early autumn; dry off in summer when foliage has died down. Propagate in autumn by seed or freely produced offsets.
L. aloides, syn. *L. tricolor*, *L.* 'Tricolor'. Winter- and spring-flowering bulb. H 15–25cm (6–10in), S 5–8cm (2–3in). Produces 2 strap-shaped, semi-erect, basal, purple-spotted, green leaves. Has a spike of 10–20 pendent flowers, each 3cm (1¼in) long with a yellow tube shading to red at the apex and with flared, green tips. var. *quadricolor* (syn. *L.a.* 'Quadricolor') illus. p.385.
L. archoides var. *glaucina*, syn. *L. glaucina*, illus. p.369.
L. contaminata. Winter- and spring-flowering bulb. H to 20cm (8in), S 5–8cm (2–3in). Has narrowly strap-shaped, semi-erect leaves in a basal cluster. Bears a spike of bell-shaped, white flowers, 0.5cm (¼in) long, suffused and tipped with red and green.
L. glaucina. See *L. archoides* var. *glaucina*.
L. mutabilis. Winter- and spring-flowering bulb. H to 30cm (12in), S 5–8cm (2–3in). Has 2 strap-shaped, semi-erect, basal leaves. Stem bears a loose spike of up to 25 tubular, 1cm (½in) long flowers that are purple or lilac in bud and open to reddish-brown-tipped petals with a green tube base.
L. 'Nelsonii' illus. p.385.
L. orchioides. Winter- and spring-flowering bulb. H 15–30cm (6–12in), S 5–8cm (2–3in). Has 2 strap-shaped, semi-erect, basal, green leaves, sometimes spotted blackish- or purple-brown. Stem produces a dense spike of fragrant, semi-erect, tubular, white flowers, 1cm (½in) long, blue-tinged and tipped with green.
L. rubida. Winter-flowering bulb. H to 25cm (10in), S 5–8cm (2–3in). Bears 2 strap-shaped, purple-spotted, green leaves, semi-erect and basal, and a loose spike of pendent, tubular, red flowers, 2–3cm (¾–1¼in) long, shading to yellow at tips.
L. tricolor. See *L. aloides*.
L. 'Tricolor'. See *L. aloides*.

Lactuca alpina. See *Cicerbita alpina*.
Lactuca bourgaei. See *Cicerbita bourgaei*.

LAELIA. See ORCHIDS.
L. anceps (illus. p. 261). Evergreen, epiphytic orchid for a cool greenhouse. H 25cm (10in). Lilac-pink flowers, 6cm (2½in) wide, each with a deep mauve lip, are carried in tall spikes in autumn. Has oval, rigid leaves, 10–15cm (4–6in) long. Needs semi-shade in summer.
L. cinnabarina (illus. p.263). Evergreen, epiphytic orchid for an intermediate greenhouse. H 15cm (6in). Produces sprays of slender, orange flowers, 5cm (2in) or more across, usually in winter. Has narrowly oval, rigid leaves, 8–10cm (3–4in) long. Needs good light in summer.

x LAELIOCATTLEYA. See ORCHIDS.
x *L.* Rojo 'Mont Millais' (illus. p.253). Evergreen, epiphytic orchid for an intermediate greenhouse. H 30cm (12in). In winter-spring bears arching heads of slender, reddish-orange flowers, 2cm (¾in) across. Oval leaves are up to 15cm (6in) long. Provide good light in summer.

LAGAROSIPHON (Hydrocharitaceae)
Genus of semi-evergreen, perennial, spreading, submerged water plants grown for their decorative foliage. Oxygenates water. Fully hardy. Needs full sun. Thin regularly to keep under control. Propagate by stem cuttings in spring or summer.
L. major, syn. *Elodea crispa* of gardens, illus. p.389.

LAGERSTROEMIA (Lythraceae)
Genus of deciduous or evergreen, summer-flowering shrubs and trees, grown for their flowers. Frost hardy to frost tender, min. 3–5°C (37–41°F). Prefers fertile, well-drained soil and full light. Water containerized specimens freely when in full growth, less at other times. To maintain as shrubs, cut back hard the previous season's stems each spring. Propagate by seed in spring or by semi-ripe cuttings in summer.
♣ *L. indica* illus. p.65.
L. speciosa (Pride of India, Queen's crape myrtle). Deciduous, rounded tree. H 15–20m (50–70ft), S 10–15m (30–50ft). Frost tender. Mid- to deep green leaves are narrowly oval, 8–18cm (3–7in) long. Has panicles of funnel-shaped, rose-pink to rose-purple flowers in summer-autumn, often when leafless.

LAGUNARIA (Malvaceae)
Genus of one species of evergreen tree, grown for its flowers in summer-autumn and its overall appearance. Frost tender, min. 3–5°C (37–41°F). Prefers fertile, well-drained soil and full light. Water containerized plants freely when in full summer growth, moderately at other times. Pruning is tolerated if required. Propagate by seed in spring or by semi-ripe cuttings in summer. Under cover, red spider mite may be troublesome.
L. patersonii (Norfolk Island hibiscus, Queensland pyramidal tree). Fast-growing, evergreen, upright tree, pyramidal when young. H 10–14m (30–46ft), S 5–7m (15–22ft). Oval, rough-textured leaves are matt-green above, whitish-green beneath. Has hibiscus-like, rose-pink flowers, 5cm (2in) wide, in summer.

LAGURUS (Gramineae). See GRASSES, BAMBOOS, RUSHES and SEDGES.
L. ovatus illus. p.182.

LAMARCKIA (Gramineae). See GRASSES, BAMBOOS, RUSHES and SEDGES.
L. aurea (Golden top). Tuft-forming, annual grass. H and S 20cm (8in). Fully hardy. Wiry stems bear scattered, pale green leaves and, in summer, erect, dense, one-sided, golden panicles. Needs sun.

Lamiastrum galeobdolon 'Variegatum'. See *Lamium galeobdolon* 'Variegatum'.

LAMIUM (Labiatae)
Deadnettle
Genus of spring- or summer-flowering perennials, most of which are semi-evergreen, including a number of weeds; some species make useful ground cover. Fully hardy. Prefers full or partial shade and moist but well-drained soil. Resents excessive winter wet. Propagate by stem-tip cuttings of non-flowering shoots in mid-summer or by division in autumn or early spring.
L. galeobdolon 'Variegatum', syn. *Galeobdolon argentatum*, *Lamiastrum galeobdolon* 'Variegatum'. Semi-evergreen, carpeting perennial. H to 30cm (12in), S indefinite. Oval, mid-green leaves are marked with silver. Racemes of tubular, 2-lipped, lemon-yellow flowers appear in summer.
L. maculatum illus. p.233. **'Album'** illus. p.231. **'Beacon Silver'** is a semi-evergreen, mat-forming perennial. H 20cm (8in), S 1m (3ft). Has oval, mauve-tinged, silver leaves, sometimes with narrow, green margins. Whorls of hooded, mauve-pink flowers appear on short stems in summer. ♣ **'White Nancy'** illus. p.231.
L. orvala illus. p.235.

LAMPRANTHUS (Aizoaceae)
Genus of creeping, bushy, perennial succulents and sub-shrubs with daisy-like flowers. Becomes woody after several years, when is best replenished. Plants are good for summer bedding, particularly in arid conditions. Leaves redden in strong sun. Frost tender, min. 5°C (41°F) if dry. Requires full sun and very well-drained soil. Propagate by seed or stem cuttings in spring or autumn.
L. aurantiacus illus. p.404.
L. haworthii. Erect to creeping, perennial succulent. H 50cm (20in), S indefinite. Blue-grey leaves are cylindrical and 5cm (2in) long. In spring bears masses of daisy-like, cerise flowers, 7cm (3in) across, that only open in sun.
L. multiradiatus, syn. *L. roseus*. Creeping, perennial succulent. H 15cm (6in), S indefinite. Produces solid, 3-angled, mid- to glaucous green leaves, 5cm (2in) long. Daisy-like, dark rose-red flowers, 4cm (1½in) across, open only in sun from spring to autumn.
L. roseus. See *L. multiradiatus*.
L. spectabilis illus. p.399.

LANTANA (Verbenaceae)
Genus of evergreen perennials and shrubs, grown for their flowers. Frost tender, min. 10–13°C (50–55°F). Needs full light and fertile, well-drained soil. Water containerized specimens freely when in full growth, moderately at other times. Tip prune young plants to promote a bushy habit. Propagate by seed in spring or by semi-ripe cuttings in summer. Red spider mite and whitefly may be troublesome.

LAPAGERIA

L. camara. Evergreen, rounded to spreading shrub. H and S 1–2m (3–6ft). Bears oval, finely wrinkled, deep green leaves. From spring to autumn, tiny, tubular, 5-lobed flowers, in dense, domed heads, open yellow, then turn red. Many colour forms have been selected.
L. delicatissima. See *L. montevidensis.*
L. montevidensis, syn. *L. delicatissima, L. sellowiana,* illus. p.137.
L. sellowiana. See *L. montevidensis.*
L. 'Spreading Sunset' illus. p.141.

LAPAGERIA (Liliaceae)
Genus of one species of evergreen, woody-stemmed, twining climber, grown for its large, waxy blooms. Half hardy. Requires humus-rich, well-drained soil and partial shade. Water moderately, scarcely at all when not in full growth. Provide support. Thin out congested growth in spring. Propagate in spring by seed, soaked for 2 days before sowing, or in spring or autumn by layering.
❦ *L. rosea* illus. p.170.

Lapeirousia cruenta. See *Anomatheca laxa.*
Lapeirousia laxa. See *Anomatheca laxa.*

LARDIZABALA (Lardizabalaceae)
Genus of evergreen, woody-stemmed, twining climbers, grown for their foliage. Male and female flowers are borne on the same plant in late autumn to winter. Is useful for growing on trellises or pergolas. Frost to half hardy. Grow in sun or partial shade and in any well-drained soil. Propagate by seed in spring or by stem cuttings in late summer or autumn.
L. biternata. Evergreen, woody-stemmed, twining climber. H 3–4m (10–12ft). Half hardy. Rounded leaves have broadly oval, leathery, dark green leaflets. In winter has brown flowers with tiny, whitish petals, the males in drooping spikes, the females solitary. In winter-spring bears many-seeded, berry-like, purple fruits, 5–8cm (2–3in) long.

LARIX (Pinaceae). See CONIFERS.
❦ *L. decidua* (European larch). Fast-growing, deciduous conifer with a conical crown when young, broadening on maturity, and spaced branches. H 25–30m (80–100ft), S 5–15m (16–50ft). Fully hardy. Shoots are yellow-brown in winter. Has light green leaves and small, erect, conical cones.
❦ *L. kaempferi,* syn. *L. leptolepis* (Japanese larch). Fast-growing, deciduous, columnar conifer with a conical tip. H 25–30m (80–100ft), S 5–8m (15–25ft). Fully hardy. Shoots are purplish-red and leaves are needle-like, flattened, greyish-green or bluish. Small cones have reflexed scales.
L. leptolepis. See *L. kaempferi.*

LATHRAEA (Scrophulariaceae)
Genus of spreading perennials that grow as parasites on the roots of trees, in the case of *L. clandestina* on willow or poplar. True leaves are not produced. Fully hardy. Grows in dappled shade cast by host tree and prefers moist conditions. Roots resent being disturbed. Propagate by seed when fresh, in late summer.
L. clandestina illus. p.235.

LATHYRUS (Leguminosae)
Genus of annuals and perennials, many of them tendril climbers, grown for their racemes of attractive flowers. Flowers are followed by long, thin seed pods. Fully to frost hardy. Grow in humus-rich, fertile, well-drained soil and in full light. Provide support and remove dead flowers regularly. Cut down perennials in late autumn. Propagate annuals by seed (soaked before sowing) in early spring or early autumn, perennials by seed in autumn or by division in spring. Botrytis and mildew may cause problems.
L. grandiflorus illus. p.169.
❦ *L. latifolius* illus. p.171.
L. nervosus (Lord Anson's blue pea). Herbaceous, tendril climber. H to 5m (15ft). Frost hardy. Grey-green leaves each have a pair of leaflets, a 3-branched tendril and large stipules. Fragrant, purplish-blue flowers are borne in long-stalked racemes in summer.
❦ *L. odoratus* (Sweet pea). Moderately fast-growing, annual, tendril climber. H to 3m (10ft). Fully hardy. Has oval, mid-green leaves with tendrils. Scented flowers are produced in shades of pink, blue, purple or white, from summer to early autumn. Dwarf, non-climbing cultivars are available. **'Bijou'** illus. p.276; **'Knee Hi'** illus. p.272; **'Lady Diana'** illus. p.174; **'Red Ensign'** illus. p.170; **'Selana'** illus. p.168; **'Xenia Field'** illus. p.169.
L. rotundifolius (Persian everlasting pea). Herbaceous, tendril climber with winged stems. H to 1m (3ft). Fully hardy. Leaves each have narrow stipules, a pair of leaflets and a 3-branched tendril. Has small racemes of 3–8 pink to purplish flowers in summer.
L. sylvestris (Everlasting pea, Perennial pea). Herbaceous, tendril climber with winged stems. H to 2m (6ft). Fully hardy. Leaves each have narrow stipules, a pair of leaflets and a terminal, branched tendril. In summer and early autumn bears racemes of 4–10 rose-pink flowers, marked with green and purple.
❦ *L. vernus* illus. p.234. **'Alboroseus'** is a clump-forming perennial. H and S 30cm (12in). Fully hardy. In spring, slender stems each bear 3–5 white-and-deep-pink flowers. Has fern-like, much-divided, soft leaves.

LAURELIA (Atherospermataceae or Monimiaceae)
Laurel
Genus of evergreen trees and shrubs, grown for their aromatic foliage. Frost hardy, but needs shelter from cold winds. Requires sun or semi-shade; grows in any but very dry soil. Propagate by semi-ripe cuttings in summer.
L. sempervirens, syn. *L. serrata* (Chilean laurel). Evergreen, broadly conical tree or shrub. H and S to 15m (50ft). Oval, leathery leaves are glossy, dark green and very aromatic. In summer bears inconspicuous flowers.
L. serrata. See *L. sempervirens.*

LAURUS (Lauraceae)
Bay tree, Laurel
Genus of evergreen trees, grown for their foliage. Frost hardy, but foliage may be scorched by extremely cold weather or strong, cold winds. Needs a sheltered position in sun or semi-shade and fertile, well-drained soil. In tubs may be grown well as standards, which should be trimmed during summer. Propagate by semi-ripe cuttings in summer or by seed in autumn.
❦ *L. nobilis* (Bay laurel, Sweet bay). Evergreen, broadly conical tree. H 12m (40ft), S 10m (30ft). Narrowly oval, leathery, glossy, dark green leaves are very aromatic and used in cooking. Has small, star-shaped, pale yellow flowers in spring, followed by globose to ovoid, green, then black fruits.

LAVANDULA (Labiatae)
Lavender
Genus of evergreen, mainly summer-flowering shrubs, with entire or divided, often grey-green leaves, grown for their aromatic foliage and flowers. Makes an effective, low hedge. Fully to half hardy. Needs full sun and fertile, well-drained soil. Trim hedges lightly in spring to maintain a compact habit. Propagate by semi-ripe cuttings in summer.
❦ *L. angustifolia* 'Hidcote' see *L.* 'Hidcote'. **'Munstead'** see *L.* 'Munstead'.
L. dentata (French lavender). Evergreen, bushy shrub. H and S 1m (3ft). Frost hardy. Aromatic leaves are fern-like, toothed and grey-green. Dense spikes of small, slightly fragrant, tubular, lavender-blue flowers and purple bracts are borne from mid- to late summer.
L. 'Grappenhall'. Evergreen, bushy shrub. H 1m (3ft), S 1.5m (5ft). Frost hardy. Bears narrowly oblong, aromatic, grey-green leaves. Produces long-stalked spikes of tiny, slightly fragrant, tubular, blue-purple flowers in mid- and late summer.
❦ *L.* 'Hidcote', syn. *L. angustifolia* 'Hidcote', illus. p.137.
L. 'Munstead', syn. *L. angustifolia* 'Munstead'. Evergreen, bushy, compact shrub. H and S 60cm (2ft). Fully hardy. Narrowly oblong, aromatic leaves are grey-green. Produces dense spikes of tiny, fragrant, tubular, blue flowers from mid- to late summer.
❦ *L. stoechas* illus. p.137.

LAVATERA (Malvaceae)
Tree mallow
Genus of mainly summer-flowering annuals, biennials, perennials and semi-evergreen sub-shrubs and shrubs. Fully to frost hardy. Needs sun and well-drained soil. Propagate perennials, sub-shrubs and shrubs by softwood cuttings in early spring or summer, annuals and biennials by seed in spring or early autumn.
L. assurgentiflora illus. p.111.
L. cachemiriana, syn. *L. cachemirica,* illus. p.191.

LEMAIREOCEREUS

L. cachemirica. See *L. cachemiriana.*
❦ *L. olbia* 'Rosea', syn. *L.* 'Rosea', illus. p.111.
L. 'Rosea'. See *L. olbia* 'Rosea'.
L. trimestris 'Mont Blanc' illus. p.270. **'Silver Cup'** illus. p.276.

LAYIA (Compositae)
Genus of annuals, useful for hot, dry places. Fully hardy. Grow in sun and in poor to fertile, very well-drained soil. Propagate by seed sown outdoors in spring or early autumn.
L. elegans. See *L. platyglossa.*
L. platyglossa, syn. *L. elegans* (Tidy tips). Fast-growing, upright, bushy annual. H 45cm (18in), S 30cm (12in). Has lance-shaped, greyish-green leaves. Daisy-like flower heads, 5cm (2in) wide, with white-tipped, yellow ray petals and yellow centres, are produced from early summer to early autumn. Is suitable for cutting.

LECHENAULTIA. See LESCHENAULTIA.

LEDEBOURIA (Liliaceae)
Genus of bulbs, some of which are evergreen, with ornamental, narrowly lance-shaped leaves. Flowers are very small with reflexed tips. Makes good pot plants in cool greenhouses. Half hardy. Needs full light, to allow leaf marks to develop well, and loose, open soil. Propagate by offsets in spring.
L. cooperi, syn. *Scilla adlamii.* Summer-flowering bulb. H 5–10cm (2–4in), S 2.5–5cm (1–2in). Semi-erect, basal, green leaves, with brownish-purple stripes, die away in winter. Stem carries a short spike of small, bell-shaped, greenish-purple flowers.
L. socialis, syn. *Scilla violacea,* illus. p.378.

LEDUM (Ericaceae)
Genus of evergreen shrubs, grown for their aromatic foliage and small, white flowers. Fully hardy. Needs shade or semi-shade and moist, peaty, acid soil. Benefits from dead-heading. Propagate by semi-ripe cuttings in summer or by seed in autumn.
L. groenlandicum illus. p.124.

LEIOPHYLLUM (Ericaceae)
Genus of one species of evergreen shrub with an extensive, spreading root system. Fully hardy. Prefers semi-shade and well-drained, peaty, acid soil. Top-dress regularly with peaty soil. Propagate by seed in spring or by semi-ripe cuttings in summer.
❦ *L. buxifolium.* Evergreen, dome-shaped shrub. H 25cm (10in), S 45cm (18in). Stems are covered with tiny, oval, leathery, dark green leaves. In late spring, terminal clusters of deep pink buds develop into small, star-shaped, white flowers, with prominent stamens.

LEMAIREOCEREUS (Cactaceae)
Genus of columnar, perennial cacti with ribbed, spiny, dark green stems. Reaches 1m (3ft) in 5–10 years. Flowers appear in summer, only on plants over 2m (6ft) high. Frost tender, min. 11°C (52°F), otherwise plants may

become badly marked. Needs full sun and very well-drained soil. Propagate by seed or stem cuttings in spring or summer.
L. euphorbioides, syn. *Rooksbya euphorbioides*, illus. p.395.
L. marginatus, syn. *Marginatocereus marginatus*, illus. p.393.
L. thurberi. Columnar, perennial cactus, branching from low down. H to 7m (22ft), S 3m (10ft). Has 5- or 6-ribbed, glossy, dark green stems with very short-spined areoles set in close rows down each rib. Produces funnel-shaped, white flowers in summer.

LEMBOGLOSSUM. See ORCHIDS.
L. bictoniense, syn. *Odontoglossum bictoniense* (illus. p.260). Evergreen, epiphytic orchid for a cool greenhouse. H 23cm (9in). Olive-green flowers, 4cm (1½in) across, barred with dark brown and each with a sometimes pink-flushed, white lip, are produced in spikes in late summer. Leaves are narrowly oval and 10–15cm (4–6in) long. Requires shade in summer.
L. cervantesii, syn. *Odontoglossum cervantesii* (illus. p.260). Evergreen, epiphytic orchid for a cool greenhouse. H 8cm (3in). In winter bears sprays of papery, white flowers, 2.5cm (1in) across, with cobweb-like, light brown marks. Has narrowly oval leaves, 10–15cm (4–6in) long. Grow in shade in summer.
L. cordatum, syn. *Odontoglossum cordatum* (illus. p.261). Evergreen, epiphytic orchid for a cool greenhouse. H 12cm (5in). Sprays of brown-marked, corn-yellow flowers, 2.5cm (1in) across, open in spring. Leaves are narrowly oval and 10–15cm (4–6in) long. Provide shade in summer and keep very dry in winter.
L. rossii, syn. *Odontoglossum rossii* (illus. p.260). Evergreen, epiphytic orchid for a cool greenhouse. H 8cm (3in). In autumn-winter, white to mushroom-pink flowers, 2.5cm (1in) across and speckled with beige-brown, are borne in spikes. Narrowly oval leaves are 10–15cm (4–6in) long. Needs shade in summer.

LEONOTIS (Labiatae)
Genus of annuals, evergreen and semi-evergreen perennials, sub-shrubs and shrubs, grown for their flowers and overall appearance. Half hardy to frost tender, min. 5–7°C (41–5°F). Needs full sun and rich, well-drained soil. Water containerized specimens freely when in full growth, much less at other times. Cut back perennials, sub-shrubs and shrubs to within 15cm (6in) of ground in early spring. Propagate by seed in spring or by greenwood cuttings in early summer.
L. leonurus illus. p.119.

LEONTOPODIUM (Compositae)
Edelweiss
Genus of short-lived, spring-flowering, woolly perennials, grown for their flower heads. Is suitable for rock gardens. Fully hardy. Needs sun, gritty, well-drained soil and a deep collar of grit. Shelter from prevailing, rain-bearing winds, as crowns are very intolerant of winter wet. Propagate by division in spring or by seed when fresh. Many seeds are not viable.
L. alpinum illus. p.294.
L. stracheyi. Mound-forming, spreading, woolly perennial. H and S 10cm (4in). Star-shaped, glistening, white flower heads are produced among thick, oval, silver leaves in spring. Makes a good alpine house plant.

LEPTOSPERMUM (Myrtaceae)
Genus of evergreen trees and shrubs, grown for their foliage and small, often profuse flowers. Grows well in coastal areas if not too exposed. Frost to half hardy, but in cold areas plant against a south- or west-facing wall. Needs full sun and fertile, well-drained soil. Propagate by semi-ripe cuttings in summer.
L. flavescens. See *L. polygalifolium*.
L. humifusum. See *L. rupestris*.
L. polygalifolium, syn. *L. flavescens*, illus. p.109.
L. rupestris, syn. *L. humifusum*, illus. p.130.
L. scoparium (Manuka, New Zealand tea-tree). ♀ '**Keatleyi**' is an evergreen, rounded shrub. H and S 3m (10ft). Half hardy. Narrowly lance-shaped, aromatic, grey-green leaves set off a profusion of large, star-shaped, pale pink flowers during late spring and summer. '**Nicholsii**' produces bronze-purple leaves and smaller, crimson flowers. ♀ '**Red Damask**' illus. p.101.

LESCHENAULTIA, syn. LECHENAULTIA (Goodeniaceae)
Genus of evergreen shrubs, grown for their flowers. Frost tender, min. 7–10°C (45–50°F). Needs full light and peaty, well-drained soil with few phosphates and nitrates. Water containerized plants moderately during growing season, sparingly at other times. Shorten over-long stems after flowering. Propagate by seed in spring or by semi-ripe cuttings in summer. Most species are not easy to grow under glass; good ventilation is essential.
L. floribunda. Evergreen, domed, wiry-stemmed shrub. H and S 30–60cm (12–24in). Has narrow, cylindrical, pointed leaves and, in spring-summer, short, tubular, pale blue flowers, each with 5 angular petals, in terminal clusters.

LEUCADENDRON (Proteaceae)
Genus of evergreen shrubs and trees, grown for their flower heads from autumn to spring and for their foliage. Frost tender, min. 5–7°C (41–5°F). Needs full light and sharply drained soil, mainly of sand and peat, ideally with very little nitrogen and phosphates. Water potted specimens moderately while in growth, sparingly at other times. Propagate by seed in spring.
L. argenteum illus. p.74.

Leucanthemella serotina. See *Chrysanthemum uliginosum*.
Leucanthemopsis alpina. See *Chrysanthemum alpina*.

LEUCANTHEMUM (Compositae)
Genus of annuals and perennials, grown for their attractive flowers. Fully hardy. Cultivars of L. x *superbum* are valued for their profusion of large daisy-like summer flowers. Some species are suitable for rock gardens. Needs full sun and well-drained soil. Propagate species by seed or division, cultivars by division only.
L. x superbum, syn. *Chrysanthemum maximum* of gardens (Shasta daisy). ♀ '**Aglaia**' is a robust perennial. H 1m (3ft), S 60cm (2ft). Fully hardy. Large, daisy-like, semi-double, white flower heads are borne singly in summer. Has spoon-shaped, coarse, lobed, toothed leaves. Divide every 2 years. '**Elizabeth**' illus. p.204. '**Esther Read**' illus. p.239 ♀ '**Wirral Supreme**' is double with short, central florets.

LEUCHTENBERGIA (Cactaceae)
Genus of one species of perennial cactus. Looks like *Agave* in foliage, but its flowers, seed pods and seeds are similar to *Ferocactus*. Tubercles eventually form on short, rough, woody stems. Frost tender, min. 6°C (43°F). Needs full sun and very well-drained soil. Keep completely dry in winter; water sparingly from spring to autumn. Propagate by seed in spring or summer.
L. principis illus. p.402.

LEUCOCORYNE (Liliaceae)
Genus of spring-flowering bulbs with loose heads of flattish flowers. Half hardy. Needs sun and well-drained soil. Plant in autumn and water until after flowering. Lies dormant in summer. Propagate by seed or offsets in autumn.
L. ixioides illus. p.358.

LEUCOGENES (Compositae)
New Zealand edelweiss
Genus of evergreen, woody-based perennials, grown mainly for their foliage. Is excellent for alpine houses in areas where summers are cool. Frost to half hardy. Needs sun and gritty, well-drained, peaty soil. Resents winter wet and may be difficult to grow. Propagate by seed when fresh or by softwood cuttings in late spring or early summer.
L. grandiceps illus. p.337.
L. leontopodium (North Island edelweiss). Evergreen, rosetted perennial. H and S 12cm (5in). Half hardy. Has oblong to oval, overlapping, silvery-white to yellowish leaves. In early summer bears up to 15 small, star-shaped, woolly, silvery-white flower heads surrounded by thick, felted, white bracts.

LEUCOJUM (Amaryllidaceae)
Snowflake
Genus of bulbs, grown for their pendent, bell-shaped, white or pink flowers in autumn or spring. Fully to frost hardy. Some species prefer a moist, partially shaded site, others do best in a sunny position and well-drained soil. Propagate by division in spring or early autumn or by seed in autumn.
L. aestivum illus. p.340.
♀ *L. autumnale* illus. p.381.

L. roseum. Early autumn-flowering bulb. H to 10cm (4in), S 2.5–5cm (1–2in). Frost hardy. Slender stems produce usually solitary pale pink flowers, 1cm (½in) long. Thread-like, erect, basal leaves appear with, or just after, flowers. Prefers sun and well-drained soil.
♀ *L. vernum* illus. p.376.

LEUCOSPERMUM (Proteaceae)
Genus of evergreen shrubs, grown for their flower heads. Frost tender, min. 7–10°C (45–50°F). Requires full light and sandy, well-drained soil with few phosphates and nitrates. Water containerized specimens moderately when in growth, sparingly at other times. Propagate by seed in spring. Is not easy to cultivate long term under glass; good ventilation is essential.
L. cordifolium, syn. *L. nutans*. Evergreen, rounded to spreading, well-branched shrub. H and S 1.2m (4ft). Elongated, heart-shaped, blue-grey leaves each have a 3-toothed tip. In summer, very slender, tubular, brick-red to orange flowers, each with a long style, are borne in tight heads that resemble single blooms.
L. nutans. See *L. cordifolium*.
L. reflexum illus. p.101.

LEUCOTHOË (Ericaceae)
Genus of evergreen, semi-evergreen or deciduous shrubs, grown for their white flowers and their foliage. Fully to frost hardy. Needs shade or semi-shade and moist, peaty, acid soil. Propagate by semi-ripe cuttings in summer.
♀ *L. fontanesiana*. Evergreen, arching shrub. H 1.5m (5ft), S 3m (10ft). Fully hardy. Lance-shaped, leathery, glossy, dark green leaves have long points and sharp teeth. Short racemes of small, urn-shaped, white flowers are borne beneath shoots from mid- to late spring. '**Rainbow**' illus. p.147.
L. keiskei. Evergreen shrub with erect or semi-procumbent stems. H 15–60cm (6–24in), S 30–60cm (12–24in). Frost hardy. Oval, thin-textured, glossy, dark green leaves have a red flush when young and a leathery appearance. Bears pendent, urn-shaped, white flowers from leaf axils in summer. Is good for a rock garden, peat bed or alpine house. Prefers mild, damp climates.

LEWISIA (Portulacaceae)
Genus of perennials, some of which are evergreen, with rosettes of succulent leaves and deep tap roots. Most species are good in alpine houses, troughs and rock gardens. Fully to frost hardy. Evergreen species need semi-shaded, humus-rich, moist or well-drained, neutral to acid soil and resent water in their rosettes at all times. Herbaceous species shed their leaves in summer and require sun and well-drained, neutral to acid soil; dry off after flowering. Propagate herbaceous species by seed in spring or autumn, evergreen species by seed in spring or by offsets in summer. Seed of *L. Cotyledon Hybrids* may not come true.

L. columbiana. Evergreen, basal-rosetted perennial. H 15cm (6in) or more, S 10–15cm (4–6in). Fully hardy. Bears thick, narrowly oblong, flat, glossy, green leaves and, in early summer, terminal sprays of small, cup-shaped, deeply veined, white to deep pink flowers. Prefers moist soil.
L. Cotyledon Hybrids illus. p.302.
L. 'George Henley' illus. p.300.
L. nevadensis. Loose, basal-rosetted perennial. H 4–6cm (1½–2½in), S 8cm (3in). Fully hardy. In summer, large, almost stemless, cup-shaped, white flowers appear above small clusters of strap-shaped, dark green leaves.
L. rediviva [pink form] illus. p.327; [white form] illus. p.321.
♆ *L. tweedyi* illus. p.312.

LEYCESTERIA (Caprifoliaceae)
Genus of deciduous shrubs, grown for their showy flower clusters. Frost to half hardy. Needs full sun and fertile, well-drained soil. Propagate by softwood cuttings in summer or by seed or division in autumn.
L. formosa (Himalayan honeysuckle). Deciduous, upright shrub. H and S 2m (6ft). Frost hardy. Has blue-green shoots and slender, oval, dark green leaves. In summer and early autumn, small, funnel-shaped, white flowers are borne at tip of each pendent cluster of purplish-red bracts and are followed by spherical, reddish-purple fruits. Cut weak shoots to ground level in early spring.

LEYMUS (Gramineae). See GRASSES, BAMBOOS, RUSHES and SEDGES.
L. arenarius, syn. *Elymus arenarius* (Lyme grass). Vigorous, spreading, herbaceous, rhizomatous, perennial grass. H to 1.5m (5ft), S indefinite. Fully hardy. Has broad, glaucous leaves. Produces stout, terminal spikes of greyish-green flowers on erect stems in late summer. Is useful for binding coastal dunes.

LIATRIS (Compositae)
Gay feathers
Genus of summer-flowering perennials with thickened, corm-like rootstocks. Fully hardy. Prefers sun and well-drained soil. Propagate by division in spring.
L. callilepis. See *L. spicata.*
L. pycnostachya (Kansas gay feather). Clump-forming perennial. H 1.2m (4ft), S 30cm (1ft). In summer bears tall spikes of clustered, feathery, mauve-pink flower heads. Grass-like, dark green leaves form basal tufts.
L. spicata, syn. *L. callilepis*, illus. p.243.

LIBERTIA (Iridaceae)
Genus of rhizomatous perennials, grown for their foliage, decorative seed pods and flowers. Frost to half hardy. Needs a sheltered, sunny or partially shaded site and well-drained soil. Propagate by division in spring or by seed in autumn or spring.
L. grandiflora illus. p.203.
L. ixioides. Clump-forming, rhizomatous perennial. H and S 60cm (24in). Frost hardy. Produces panicles of saucer-shaped, white flowers in summer. Grass-like, dark green leaves turn orange-brown during winter.

Libocedrus chilensis. See *Austrocedrus chilensis.*
Libocedrus decurrens. See *Calocedrus decurrens.*

Libonia floribunda. See *Justicia rizzinii.*

LIGULARIA (Compositae)
Genus of perennials, grown for their foliage and large, daisy-like flower heads. Fully to half hardy. Grow in sun or semi-shade and in moist but well-drained soil. Propagate by division in spring or by seed in autumn or spring. Is prone to damage by slugs and snails.
♆ *L. dentata* 'Desdemona', syn. *Senecio clivorum* 'Desdemona'. Compact, clump-forming perennial. H 1.2m (4ft), S 60cm (2ft). Fully hardy. Has heart-shaped, long-stalked, leathery, basal, dark brownish-green leaves, almost mahogany beneath, and bears terminal clusters of large, daisy-like, vivid orange flower heads on branching stems from mid- to late summer.
♆ *L.* 'Gregynog Gold'. Clump-forming perennial. H 2m (6ft), S 60cm (2ft). Fully hardy. Leaves are large, heart-shaped and deep green. Conical panicles of daisy-like, orange-yellow flower heads are borne from mid- to late summer.
L. przewalskii illus. p.193.
L. stenocephala illus. p.193.
L. tussilagineum. See *Farfugium japonicum.*

LIGUSTRUM (Oleaceae)
Privet
Genus of deciduous, semi-evergreen or evergreen shrubs and trees, grown for their foliage and, in some species, flowers. Fully to frost hardy. Requires sun or semi-shade, the variegated forms doing best in full sun. Thrives on any well-drained soil, including chalky soil. All except *L. lucidum* occasionally need cutting back in mid-spring to restrict growth. Propagate by semi-ripe cuttings in summer.
L. japonicum (Japanese privet). Evergreen, bushy, dense shrub. H 3m (10ft), S 2.5m (8ft). Frost hardy. Has oval, glossy, very dark green leaves and, from mid-summer to early autumn, large, conical panicles of small, tubular, white flowers with 4 lobes. 'Rotundifolium' is slow-growing, with a dense mass of rounded, leathery leaves.
♆ *L. lucidum* (Chinese privet). Evergreen, upright shrub or tree. H 10m (30ft), S 8m (25ft). Frost hardy. Bears large, oval, glossy, dark green leaves. Produces large panicles of small, tubular, white flowers, with 4 lobes, in late summer and early autumn. ♆ 'Excelsum Superbum' illus. p.97.
L. ovalifolium illus. p.96.
♆ 'Aureum' is a vigorous, evergreen or semi-evergreen, upright, dense shrub. H 4m (12ft), S 3m (10ft). Fully hardy. Leaves are oval, glossy and mid-green, broadly edged with bright yellow. Dense panicles of small, rather unpleasantly scented, tubular, white flowers, with 4 lobes, appear in mid-summer and are succeeded by spherical, black fruits. Cut back hedges to 30cm (1ft) after planting and prune hard for first 2 years; then trim as necessary during the growing season.
L. sinense illus. p.88.
L. 'Vicaryi', syn. *L.* x *vicaryi*, illus. p.123.
L. x *vicaryi.* See *L.* 'Vicaryi'.
L. vulgare. Deciduous or semi-evergreen, bushy shrub. H and S 3m (10ft). Fully hardy. Leaves are narrowly lance-shaped and dark green. Produces panicles of small, strongly scented, tubular, white flowers, with 4 lobes, from early to mid-summer, then spherical, black fruits. Cut back hedges to 30cm (1ft) after planting and prune hard for first 2 years; then trim as necessary during the growing season. 'Aureum', H and S 2m (6ft), has golden-yellow foliage.

LILIUM (Liliaceae)
Lily
Genus of mainly summer-flowering bulbs, grown for their often fragrant, brightly coloured flowers. Each fleshy-scaled bulb produces one unbranched, leafy stem, in some cases with annual roots in lower part. Mostly lance-shaped or linear leaves, to 22cm (9in) long, are scattered or in whorls, sometimes with bulbils in axils. Flowers, usually several per stem, are mainly trumpet- to bowl-shaped or with the 6 petals strongly reflexed to form a turkscap shape. (Petals of *Lilium* are known botanically as perianth segments.) Each plant has a spread of up to 30cm (12in). Frost hardy, unless otherwise stated. Needs sun and any well-drained soil, unless otherwise stated. Propagate by seed in autumn or spring, by bulb scales in summer or by stem bulbils (where present) in autumn. Virus and fungal diseases (such as botrytis) and lily beetle may cause problems.
L. amabile. Summer-flowering bulb with stem roots. H 30cm–1m (1–3ft). Scattered leaves are lance-shaped. Bears up to 10 unpleasant-smelling, nodding, turkscap, red flowers; each petal is 5–5.5cm (2–2¼in) long, with black spots.
L. 'Amber Gold' (illus. p.349). Summer-flowering bulb. H 1.2–1.5m (4–5ft). Has nodding, turkscap, deep yellow flowers, each with maroon spots in throat.
L. 'Angela North' (illus. p.348). Mid-summer flowering bulb. H to 1m (3ft). Has slightly fragrant, dark red flowers, spotted darker red, that have stronger, recurved petals.
L. 'Apollo' (illus. p.349). Summer-flowering bulb. H 1.2m (4ft). Produces downward-facing, turkscap, pale orange flowers with strongly reflexed petals.
L. auratum (Golden-rayed lily of Japan). Summer- and autumn-flowering bulb with stem roots. H 60cm–1.5m (2–5ft). Has long, scattered, lance-shaped leaves. Produces up to 10, sometimes more, fragrant, outward-facing, widely bowl-shaped, white flowers; each petal is 12–18cm (5–7in) long with a central, red or yellow band and often red or yellow spots. Requires semi-shade and neutral to acid soil. ♆ var. *platyphyllum* (illus. p.348) has broader leaves; petals have a central, yellow band and fewer spots.
L. 'Black Beauty' (illus. p.349). Summer-flowering bulb. H 1.5–2m (5–6ft). Produces outward-facing, flattish, green-centred, very deep red flowers with recurved, white-margined petals.
♆ *L.* 'Black Dragon'. Summer-flowering bulb. H 1.5m (5ft). Has outward-facing, trumpet-shaped flowers with dark purplish-red outsides and white insides.
L. 'Black Magic' (illus. p.348). Summer-flowering bulb. H 1.2–2m (4–6ft). Scented, outward-facing, trumpet-shaped flowers are purplish-brown outside and white inside.
L. 'Bonfire'. Late summer-flowering bulb. H 1.2–1.5m (4–5ft). Bears outward-facing, bowl-shaped flowers with broad petals, white outside flushed with pink, and dark crimson inside spotted paler crimson.
L. 'Bright Star' (illus. p.348). Summer-flowering bulb. H 1–1.5m (3–5ft). Produces flattish, white flowers; petals have recurved tips and a central, orange streak inside.
L. 'Bronwen North' (illus. p.348). Mid-summer-flowering bulb. H to 1m (3ft). In mid-summer, each stem carries 7 or more slightly fragrant flowers with strongly recurving, pale mauve-pink petals, paler at the tips, and pale pink throats with dark spots and lines; nectaries are reddish-black.
L. 'Brushmarks' (illus. p.349). Early summer-flowering bulb. H 1.35m (4½ft). Upward-facing, cup-shaped, orange flowers are green-throated. Petals each have deep red blotches and sometimes spots.
L. bulbiferum (Fire lily, Orange lily). Summer-flowering bulb with stem roots. H 40cm–1.5m (16in–5ft). Stem bears scattered, lance-shaped leaves and, usually, bulbils in leaf axils. Bears 1–5 or more upward-facing, shallowly cup-shaped, orange-red flowers. Each petal is 6–8.5cm (2½–3¼in) long and spotted black or deep red. ♆ var. *croceum* (illus. p.349) has orange flowers and does not normally produce bulbils.
L. canadense (Canada lily, Meadow lily, Wild yellow lily; illus. p.349). Summer-flowering bulb with stem roots. H to 1.5m (5ft). Narrowly to broadly lance-shaped leaves are mainly in whorls. Bears about 10 nodding, bell-shaped, yellow or red flowers; each petal is 5–8cm (2–3in) long, with dark red or purple spots in lower part.
♆ *L. candidum* (Madonna lily; illus. p.348). Summer-flowering bulb. H 1–2m (3–6ft). Flower stem bears scattered, lance-shaped leaves and 5–20 fragrant, outward-facing, broadly funnel-shaped, white flowers. Each petal is 5–8cm (2–3in) long with a yellow base and a slightly recurved tip. In autumn produces basal leaves, which remain throughout winter but die

off as flowering stems mature. Prefers lime-rich soil.

L. carniolicum. See *L. pyrenaicum* var. *carniolicum.*

☙ *L.* 'Casa Blanca' (illus. p.348). Late summer-flowering bulb. H 90cm (3ft). Waxy, white flowers, have yellowish-white midribs and violet-red nectaries.

L. cernuum. Summer-flowering bulb with stem roots. H to 60cm (2ft). Long, linear leaves are scattered. Has 7–15 fragrant, nodding, turkscap flowers, usually pinkish-purple with purple spots. Each petal is 3.5–5cm (1^{1}/$_{2}$–2in) long.

☙ *L. chalcedonicum,* syn. *L. heldreichii* (Scarlet turkscap lily; illus. p.349). Summer-flowering bulb with stem roots. H 50cm–1.5m (20in–5ft). Leaves are scattered and mostly lance-shaped, lower ones spreading, upper ones smaller and closer to stem. Bears up to 12 slightly scented, nodding, turkscap flowers with red or reddish-orange petals, each 5–7cm (2–3in) long.

L. 'Connecticut King' (illus. p.349). Early to mid-summer-flowering bulb. H 1m (3ft). Flowers are upward-facing, cup-shaped and bright yellow.

L. 'Corsage' (illus. p.348). Summer-flowering bulb. H 1.2m (4ft). Produces outward-facing, bowl-shaped flowers with recurved petals, pink-flushed outside and pink inside with white centres and maroon spots.

L. 'Côte d'Azur' (illus. p.348). Summer-flowering bulb. H 40cm (16in). Strong stems bear deep rose-pink flowers with darker-spotted throats.

L. x *dalhansonii* (illus. p.348). Variable, summer-flowering bulb. H 1.5–2m (5–6ft). Produces unpleasant-smelling, turkscap flowers that are chestnut brown or dark maroon with gold spots.

☙ *L. davidii.* Summer-flowering bulb with stem roots. H 1–1.4m (3–4^{1}/$_{2}$ft). Linear leaves are scattered. Has 5–20 nodding, turkscap, red or reddish-orange flowers; each petal is 5–8cm (2–3in) long with dark purple spots. var. *willmottiae* (illus. p.349) differs in its slender, arching stems to 2m (6ft) and pendant flower stalks.

L. 'Destiny' (illus. p.349). Early summer-flowering bulb. H 1–1.2m (3–4ft). Flowers are upward-facing, cup-shaped and yellow with brown spots.

L. duchartrei (illus. p.348). Summer-flowering bulb. H 60cm–1m (2–3ft). Lance-shaped leaves are scattered up stems. Has up to 12 fragrant, nodding, turkscap, white flowers that are flushed purple outside and spotted deep purple inside.

☙ *L.* 'Enchantment' (illus. p.349). Early summer-flowering bulb. H 1m (3ft). Produces upward-facing, cup-shaped, orange-red flowers with black-spotted throats.

L. 'First Love'. Summer-flowering bulb. H 1.5m (5ft). Bears slightly fragrant, outward-facing, bowl-shaped flowers. Each petal has pink edges merging into a central, yellow stripe; base is pale green.

L. Golden Clarion Hybrids (illus. p.349). Late spring to early summer-flowering bulb. H 1–2m (3–6ft). Has outward-facing, trumpet-shaped, pale to deep yellow flowers that may be flushed with reddish-purple outside.

L. 'Golden Splendor' (illus. p.349). Vigorous, mid- to late summer-flowering bulb with stem roots. H 1.5–2m (5–6ft). Large, funnel-shaped flowers are deep golden yellow with maroon stripes on the back of each petal.

☙ *L. hansonii* (illus. p.349). Summer-flowering bulb with stem roots. H 1–1.5m (3–5ft). Long leaves in whorls are lance-shaped to oval. Bears 3–12 scented, nodding, turkscap, orange-yellow flowers. Each thick petal is 3–4cm (1^{1}/$_{4}$–1^{1}/$_{2}$in) long with brown-purple spots towards base.

L. 'Harmony' (illus. p.349). Summer-flowering bulb. H 50cm–1m (1^{1}/$_{2}$–3ft). Orange flowers are upward-facing, cup-shaped and spotted with maroon.

L. heldreichii. See *L. chalcedonicum.*

☙ *L. henryi.* Late summer-flowering bulb with stem roots. H 1–3m (3–10ft). Has scattered, lance-shaped leaves. Produces 5–20, sometimes up to 70, nodding, turkscap, orange flowers; petals are 6–8cm (2^{1}/$_{2}$–3in) long with dark spots and prominent warts towards bases. Prefers lime-rich soil.

L. Imperial Crimson Group. Late summer-flowering bulb. H 1.5m (5ft). Fragrant, flattish, deep crimson flowers have white throats and white-edged petals.

L. Imperial Gold Group (illus. p.348). Summer-flowering bulb. H 2m (6ft). Bears fragrant, flattish, white flowers, spotted maroon, and with a yellow stripe up each petal centre.

L. 'Journey's End' (illus. p.348). Late summer-flowering bulb. H 2m (6ft). Outward-facing, bowl-shaped, maroon-spotted, deep pink flowers have recurved petals, white at tips and edges.

☙ *L.* 'Karen North' (illus. p.349). Summer-flowering bulb. H to 1.4m (4^{1}/$_{2}$ft). Turkscap flowers are downward-facing, with orange-pink petals sparsely spotted with deep pink.

L. 'Lady Bowes Lyon' (illus. p.349). Summer-flowering bulb. H 1–1.2m (3–4ft). Downward-facing, black-spotted, rich red flowers have reflexed petals.

L. lancifolium, syn. *L. tigrinum* (Tiger lily). Summer- to early autumn-flowering bulb with stem roots. H 60cm–1.5m (2–5ft). Has long, scattered, narrowly lance-shaped leaves. Produces 5–10, sometimes up to 40, nodding, turkscap, pink- to red-orange flowers; each petal is 7–10cm (3–4in) long and spotted with purple. var. *flaviflorum* has yellow flowers. Vigorous var. *splendens* (illus. p.349) bears larger, brighter red-orange flowers.

L. lankongense. Summer-flowering bulb with stem roots. H to 1.2m (4ft). Leaves are scattered and lance-shaped. Has up to 15 scented, nodding, turkscap, pink flowers. Petals, each 4–6.5cm (1^{1}/$_{2}$–2^{1}/$_{2}$in) long with a central, green stripe and red-purple spots mainly on edges, are often mauve-flushed. Needs partial shade in warm areas.

L. leichtlinii. Summer-flowering bulb with stem roots. H to 1.2m (4ft). Scattered leaves are linear to narrowly lance-shaped. Produces 1–6 nodding, turkscap, yellow flowers; each petal is 6–8.5cm (2^{1}/$_{2}$–3^{1}/$_{4}$in) long with dark reddish-purple spots. Needs semi-shade.

L. 'Limelight'. Summer-flowering bulb. H 1–1.5m (3–5ft). Flowers are outward-facing or slightly nodding, trumpet-shaped and yellow-green.

☙ *L. longiflorum* (Bermuda lily, Easter lily, White trumpet lily; illus. p.348). Summer-flowering bulb with stem roots. H 30cm–1m (1–3ft). Leaves are scattered and lance-shaped. Produces 1–6 fragrant, outward-facing, funnel-shaped, white flowers. Each petal is 13–20cm (5–8in) long with slightly recurved tips.

☙ *L. mackliniae* (Manipur lily; illus. p.348). Late spring- to summer-flowering bulb with stem roots. H to 40cm (16in). Small, narrowly lance-shaped to narrowly oval leaves are scattered or whorled near top of stem. Has 1–6 usually nodding, broadly bell-shaped, purplish-pink flowers; each petal is 4.5–5cm (1^{3}/$_{4}$–2in) long. Needs semi-shade.

L. maculatum. Summer-flowering bulb with stem roots. H to 60cm (2ft). Fully hardy. Scattered leaves are lance-shaped or oval. Produces 1–6 upward-facing, cup-shaped, yellow, orange or red flowers with darker spots; each petal is 8–10cm (3–4in) long.

L. 'Magic Pink' (illus. p.348). Early summer-flowering bulb. H 1–1.2m (3–4ft). Flowers are satin pink with darker pink spots.

☙ *L.* 'Marhan'. Early summer-flowering bulb. H 1.2–2m (4–6ft). Nodding, turkscap, deep orange flowers are spotted red-brown.

L. martagon (Martagon lily; illus. p.348). Summer-flowering bulb with stem roots. H 1–2m (3–6ft). Fully hardy. Has lance-shaped to oval leaves in whorls and up to 50 scented, nodding, turkscap flowers. Petals are 3–4.5cm (1^{1}/$_{4}$–1^{3}/$_{4}$in) long and pink or purple, often with darker spots. ☙ var. *album* (illus. p. 348) has pure white flowers.

L. medeoloides (illus. p.349). Summer-flowering bulb. H to 75cm (2^{1}/$_{2}$ft). Has lance-shaped leaves and up to 10 turkscap, apricot to orange-red flowers, usually with darker spots.

☙ *L. monadelphum,* syn. *L. szovitsianum* (illus. p.349). Summer-flowering bulb with stem roots. H 50cm–2m (1^{1}/$_{2}$–6ft). Has scattered, lance-shaped to oval leaves. Produces usually 1–5, sometimes up to 30, scented, nodding, turkscap, yellow flowers, usually with deep red or purple spots inside. Each petal is 6–10cm (2^{1}/$_{2}$–4in) long.

L. 'Mont Blanc' (illus. p.348). Summer-flowering bulb. H 90cm (3ft). Has upward-facing, creamy-white flowers, spotted with brown.

L. 'Montreux' (illus. p.348). Mid-summer-flowering bulb. H 1m (3ft). Bears about 8 pink flowers with darker pink midribs; orange-pink throats are spotted with brown.

L. nanum. Late spring- or summer-flowering bulb. H 6–45cm (2^{1}/$_{2}$–18in). Scattered leaves are linear. Bears a usually nodding, broadly bell-shaped, purplish-pink flower with 4.5–5cm (1^{3}/$_{4}$–2in) long petals. Needs partial shade. var. *flavidum* has pale yellow flowers.

L. nepalense (illus. p.349). Summer-flowering bulb with stem roots. H 70cm–1m (28–36in). Has scattered, lance-shaped leaves. Produces often unpleasant-smelling, nodding, funnel-shaped, greenish-white or greenish-yellow flowers, each with a dark reddish-purple base inside and petals to 15cm (6in) long.

L. Olympic Hybrids (illus. p.348). Summer-flowering bulb. H 1.2–2m (4–6ft). Scented, outward-facing, trumpet-shaped flowers are usually pink or purple outside with yellow throats; white, cream, yellow or pink variants also occur.

L. pardalinum (Leopard lily, Panther lily; illus. p.349). Summer-flowering bulb. H 2–3m (6–10ft). Long, narrowly elliptic leaves are mainly in whorls. Produces up to 10 often scented, nodding, turkscap flowers. Each petal is 5–9cm (2–3^{1}/$_{2}$in) long with red, upper parts. Orange, lower parts have maroon spots, some of which are encircled with yellow.

L. ponticum. See *L. pyrenaicum* var. *ponticum.*

☙ *L. pumilum,* syn. *L. tenuifolium.* Summer-flowering bulb with stem roots. H 15cm–1m (6–36in). Small, scattered leaves are linear. Bears usually up to 7 but occasionally up to 30 slightly scented, nodding, turkscap flowers; each petal is 3–3.5cm (1^{1}/$_{4}$–1^{1}/$_{2}$in) long and scarlet with or without basal, black spots.

☙ *L. pyrenaicum* (Yellow turkscap lily; illus. p.349). Late spring to early summer-flowering bulb often with stem roots. H 30cm–1.35m (1–4^{1}/$_{2}$ft). Has scattered, linear to narrowly elliptic, hairless leaves. Produces up to 12 unpleasant-smelling, nodding, turkscap flowers. Each petal is 4–6.5cm (1^{1}/$_{2}$–2^{1}/$_{2}$in) long and yellow or green-yellow with deep purple spots and lines. var. *carniolicum* (syn. *L. carniolicum*) has red- or orange-spotted flowers. Leaves may be hairless or downy. var. *ponticum* (syn. *L. ponticum*) bears deep yellow flowers, densely lined and spotted with red-brown or purple; leaves are downy beneath. var. *rubrum* (illus. p. 349) has orange-red or dark red flowers.

☙ *L. regale* (Regal lily; illus. p.348). Summer-flowering bulb with stem roots. H 50cm–2m (20in–6ft). Linear leaves are scattered. Produces up to 25 fragrant, outward-facing, funnel-shaped flowers. Petals are each 12–15cm (5–6in) long, white inside with a yellow base and pinkish-purple outside.

L. 'Roma' (illus. p.349). Early summer-flowering bulb. H 1.5m (5ft). Green buds open to cream flowers that sometimes age to pale greenish-yellow.

☙ *L.* 'Rosemary North' (illus. p.349). Mid- to late summer-flowering bulb. H to 1m (3ft). Produces 12 or more

slightly fragrant, rich orange flowers that sometimes have darker spots.
L. rubellum (illus. p.348). Early summer-flowering bulb with stem roots. H 30–80cm (12–32in). Has scattered, narrowly oval leaves and up to 9 scented, outward-facing, broadly funnel-shaped, pink flowers with dark red spots at bases; each petal is 6–8cm (2½–3in) long.
L. 'Shuksan'. Summer-flowering bulb. H 1.2–2m (4–6ft). Nodding, turkscap, yellowish-orange flowers are flushed red at petal tips and sparsely spotted with black.
♥ *L. speciosum*. Late summer-flowering bulb with stem roots. H 1–1.7m (3–5½ft). Has long, scattered, broadly lance-shaped leaves. Produces up to 12 scented, nodding, turkscap, white or pink flowers; each petal is up to 10cm (4in) long, with pink or crimson spots. Requires neutral to acid soil. var. *album* has white flowers and purple stems. Flowers of var. *rubrum* (illus. p.348) are carmine, stems are purple.
L. 'Star Gazer' (illus. p.348). Late summer-flowering bulb. H 90cm (3ft). Fragrant, rich crimson flowers are spotted maroon.
L. 'Sterling Star' (illus. p.348). Summer-flowering bulb. H 1–1.2m (3–4ft). Has upward-facing, cup-shaped, white flowers with tiny, brown spots.
♥ *L. superbum* (Swamp lily, Turkscap lily; illus. p.349). Late summer- to early autumn-flowering bulb with stem roots. H 1.5–3m (5–10ft). Lance-shaped to elliptic leaves are mainly in whorls. Bears up to 40 nodding, turkscap, orange flowers. Each petal is 6–10cm (2½–4in) long, with a green base inside and usually flushed red and spotted maroon. Requires neutral to acid soil.
L. szovitsianum. See *L. monadelphum*.
L. tenuifolium. See *L. pumilum*.
♥ *L.* × *testaceum* (Nankeen lily). Summer-flowering bulb. H 1–1.5m (3–5ft). Has scattered, linear, often twisted leaves. Produces 6–12 fragrant, nodding, turkscap, light orange to brownish-yellow flowers; each petal is 8cm (3in) long, usually with reddish spots inside.
L. tigrinum. See *L. lancifolium*.
L. tsingtauense (illus. p.349). Summer-flowering bulb with stem roots. H 1m (3ft). Lance-shaped leaves are mainly in whorls. Produces 1–5 upward-facing, cup-shaped, orange to orange-red flowers; petals are up to 5cm (2in) long and spotted with maroon.
L. wallichianum. Late summer- to autumn-flowering bulb with stem roots. H to 2m (6ft). Half hardy. Long, scattered leaves are linear or lance-shaped. Bears 1–4 fragrant, outward-facing, funnel-shaped, white or cream flowers that are green or yellow towards bases. Each petal is 15–30cm (6–12in) long.

LIMNANTHES (Limnanthaceae)
Genus of annuals, useful for rock gardens and for edging. Fully hardy. Prefers a sunny situation and fertile, well-drained soil. Propagate by seed sown outdoors in spring or early autumn. Self seeds freely.
♥ *L. douglasii* illus. p.288.

LIMONIUM (Plumbaginaceae)
Sea lavender
Genus of summer- and autumn-flowering perennials, sometimes grown as annuals, and sub-shrubs, some of which are evergreen. Fully hardy to frost tender, min. 7–10°C (45–50°F). Grows in full sun and in well-drained soil. Propagate by division in spring, by seed in autumn or early spring or by root cuttings in winter.
L. bellidifolium. Evergreen, dome-shaped perennial with a woody base. H 15–20cm (6–8in), S 10cm (4in). Frost hardy. Has basal rosettes of rounded, dark green leaves. Much-branched flower stems bear masses of small, 'everlasting', trumpet-shaped blue flowers in summer-autumn. Is excellent for a rock garden.
L. latifolium 'Blue Cloud' illus. p.250.
L. perezii. Evergreen, rounded sub-shrub. H and S 1m (3ft) or more. Frost tender, min. 7–10°C (45–50°F). Has long-stalked, oval to diamond-shaped, deep green leaves. Dense clusters, 20cm (8in) wide, of tiny, tubular, deep mauve-blue flowers are carried well above the leaves in autumn. Needs good ventilation if grown under glass.
L. sinuatum illus. p.274. 'Fortress' is a slow-growing, upright, bushy perennial, grown as an annual. H 45cm (18in), S 30cm (12in). Half hardy. Has lance-shaped, lobed and often wavy-margined, deep green leaves and, in summer and early autumn, clusters of small, tubular flowers in a mixture of shades such as pink, yellow or blue.

LINARIA (Scrophulariaceae)
Toadflax
Genus of spring-, summer- or autumn-flowering annuals, biennials and perennials, useful for rock gardens and borders. Fully to frost hardy. Prefers sun or light shade; thrives in any well-drained soil. Propagate by seed in autumn or spring. Self seeds freely.
L. alpina (Alpine toadflax). Tuft-forming, compact, annual, biennial or short-lived perennial with a sparse root system. H 15cm (6in), S 10–15cm (4–6in). Fully hardy. Has whorls of linear to lance-shaped, fleshy, grey-green leaves. A succession of snapdragon-like, yellow-centred, purple-violet flowers is borne in loose racemes in summer.
L. dalmatica. See *L. genistifolia* var. *dalmatica*.
L. genistifolia. Upright perennial. H 60cm–1.2m (2–4ft), S 23cm (9in). Fully hardy. From mid-summer to autumn produces racemes of small, snapdragon-like, orange-marked, yellow flowers. Lance-shaped, glossy, mid-green leaves clasp the stems. var. *dalmatica* (syn. *L. dalmatica*; Dalmatian toadflax), H 1–1.2m (3–4ft), S 60cm (2ft), bears much larger, golden-yellow flowers, from mid- to late summer, and has broader, more glaucous leaves.
L. maroccana 'Fairy Lights' illus. p.282.

L. purpurea (Purple toadflax). Upright perennial. H 60cm–1m (2–3ft), S 60cm (2ft). Fully hardy. From mid- to late summer, racemes of snapdragon-like, purplish-blue flowers, touched with white at throats, are borne above narrowly oval, grey-green leaves. 'Canon Went' illus. p.207.
L. triornithophora illus. p.215.

LINDERA (Lauraceae)
Genus of deciduous or evergreen shrubs and trees, grown for their foliage, which is often aromatic, and their autumn colour. Fruits are produced on female plants if male plants are also grown. Fully to frost hardy. Needs semi-shade and moist, acid soil. Propagate by softwood cuttings in summer or by seed in autumn.
L. benzoin illus. p.101.
♥ *L. obtusiloba*. Deciduous, bushy shrub. H and S 6m (20ft). Fully hardy. Bears 3-lobed, aromatic, glossy, dark green leaves, becoming butter-yellow in autumn. Clusters of small, star-shaped, deep yellow flowers, produced on bare shoots from early to mid-spring, are followed by small, spherical, black fruits.

LINDHEIMERA (Compositae)
Genus of late summer- and early autumn-flowering annuals. Fully hardy. Grow in sun and in fertile, well-drained soil. Propagate by seed sown under glass in early spring or outdoors in late spring.
L. texana illus. p.288.

LINNAEA (Caprifoliaceae)
Twin flower
Genus of one species of evergreen, creeping, summer-flowering, sub-shrubby perennial that makes an extensive, twiggy mat. Is useful as ground cover on peat beds and rock gardens. Fully hardy. Requires partial shade and moist, peaty, acid soil. Propagate by rooted runners in spring, by softwood cuttings in summer or by seed in autumn.
L. borealis illus. p.323.

LINUM (Linaceae)
Genus of annuals, biennials, perennials, sub-shrubs and shrubs, some of which are evergreen or semi-evergreen, grown for their flowers. Is suitable for rock gardens. Fully to half hardy, but in cold areas some species need a sheltered position. Prefers sun and humus-rich, well-drained, peaty soil. Propagate sub-shrubs and shrubs by semi-ripe cuttings in summer or by seed in autumn, annuals, biennials and perennials by seed in autumn.
♥ *L. arboreum* illus. p.306.
L. flavum (Golden flax, Yellow flax). Bushy perennial with a woody rootstock. H 30cm (12in), S 15cm (6in). Fully hardy. Has narrowly oval, green leaves and, in summer, upward-facing, funnel-shaped, yellow flowers in terminal clusters. 'Compactum' illus. p.333.
♥ *L.* 'Gemmell's Hybrid'. Semi-evergreen, domed perennial with a woody rootstock. H 15cm (6in), S 20cm (8in). Frost hardy. Leaves are

oval and grey-green. In summer, short-stalked, broadly funnel-shaped, bright chrome-yellow flowers are produced in terminal clusters. Prefers alkaline soil.
L. grandiflorum 'Rubrum' illus. p.279.
L. narbonense illus. p.251.
L. perenne illus. p.304.
L. salsoloides. See *L. suffruticosum* subsp. *salsoloides*.
L. suffruticosum subsp. *salsoloides*, syn. *L. salsoloides*. Perennial with spreading, sometimes woody-based, stems. H 5–20cm (2–8in), S 8cm (3in). Frost hardy. Slender stems bear fine, heath-like, grey-green leaves and, in summer, a succession of short-lived, saucer-shaped, pearl-white flowers, flushed blue or pink, in terminal clusters.

Lippia citriodora. See *Aloysia triphylla*.

LIQUIDAMBAR (Hamamelidaceae)
Genus of deciduous trees, with inconspicuous flowers, grown for their maple-like foliage and autumn colour. Fully to frost hardy. Requires sun or semi-shade and fertile, moist but well-drained soil; grows poorly on shallow, chalky soil. Propagate by softwood cuttings in summer or by seed in autumn.
L. formosana, syn. *L.f.* var. *monticola*. Deciduous, broadly conical tree. H 12m (40ft), S 10m (30ft). Frost hardy. Has large, 3-lobed, toothed leaves, purple when young, dark green in summer and turning orange, red and purple in autumn.
L. orientalis (Oriental sweet gum). Slow-growing, deciduous, bushy tree. H 6m (20ft), S 4m (12ft). Frost hardy. Small, 5-lobed, mid-green leaves turn orange in autumn.
L. styraciflua (Sweet gum) illus. p.44.
♥ 'Lane Roberts' is a deciduous, broadly conical to spreading tree. H 25m (80ft), S 12m (40ft). Fully hardy. Shoots usually have corky ridges. Glossy, green leaves, each with 5 lobes, turn deep reddish-purple in autumn.

LIRIODENDRON (Magnoliaceae)
Genus of deciduous trees, grown for their foliage and flowers in summer. Flowers are almost hidden by unusual leaves and are not produced on young trees. Fully hardy. Requires sun or semi-shade and deep, fertile, well-drained, preferably slightly acid, soil. Propagate species by seed in autumn, selected forms by budding in late summer.
L. chinense (Chinese tulip tree). Fast-growing, deciduous, spreading tree. H 25m (80ft), S 12m (40ft). Bears large, deep green leaves, cut off at the tips and with a deep lobe on each side; leaves become yellow in autumn. Cup-shaped, orange-based, greenish-white flowers appear in mid-summer.
♥ *L. tulipifera* illus. p.39.
♥ 'Aureomarginatum' illus. p.43.

LIRIOPE (Liliaceae)
Lilyturf
Genus of evergreen perennials with swollen, fleshy rhizomes. Some are

grown as ground cover. Fully to frost hardy. Requires sun and well-drained soil. Propagate by division in spring or by seed in autumn.

⚇ *L. muscari* illus. p.258. **'Majestic'** is an evergreen, spreading, rhizomatous perennial. H 30cm (12in), S 45cm (18in). Frost hardy. In late autumn produces spikes of thickly clustered, rounded-bell-shaped, violet flowers among linear, glossy, bright green leaves.

L. spicata. Evergreen, spreading, rhizomatous perennial. H 30cm (12in), S 30–40cm (12–16in). Fully hardy. Grass-like, glossy, dark green leaves make good ground cover. Bears spikes of rounded-bell-shaped, pale lavender flowers in late summer.

Lisianthus russellianus. See *Eustoma grandiflorum.*

LITHOCARPUS (Fagaceae)
Genus of evergreen trees, grown for their foliage. Frost hardy. Needs sun or semi-shade. Prefers well-drained, neutral to acid soil. Shelter from strong winds. Propagate by seed, when ripe, in autumn.
L. densiflorus (Tanbark oak). Evergreen, spreading tree. H and S 10m (30ft). Has sweet chestnut-like, leathery, glossy, dark green leaves and upright, pale yellow flower spikes borne in spring and often again in autumn.
L. henryi illus. p.74.

LITHODORA (Boraginaceae)
Genus of evergreen sub-shrubs and shrubs, grown for their flowers. Is excellent in rock gardens. Fully to frost hardy. Needs full sun and moist, well-drained soil; some species are lime haters and need acid conditions. Resents root disturbance. Propagate by semi-ripe cuttings in mid-summer or by seed in autumn.
L. diffusa, syn. *Lithospermum diffusum.* **'Grace Ward'** is an evergreen, compact, semi-prostrate shrub. H 15–30cm (6–12in), S to 30cm (12in). Frost hardy. Trailing stems bear lance-shaped, hairy, dull green leaves. In early summer bears masses of funnel-shaped, deep blue flowers in terminal clusters. Needs acid soil. Plants should be trimmed back after flowering. ⚇ **'Heavenly Blue'** illus. p.305.
⚇ *L. oleifolia,* syn. *Lithospermum oleifolium,* illus. p.304.
L. zahnii, syn. *Lithospermum zahnii.* Evergreen, much-branched, upright shrub. H and S 30cm (12in) or more. Frost hardy. Stems are covered in oval, hairy, dark green or greyish-green leaves. Funnel-shaped, azure-blue flowers, with spreading lobes, open in succession from early spring to mid-summer. Sets buds and flowers intermittently until mid-autumn. Prefers alkaline soil.

LITHOPHRAGMA (Saxifragaceae)
Genus of tuberous perennials, grown for their campion-like flowers. Is dormant in summer. Fully hardy. Tolerates all but deepest shade and prefers humus-rich, moist soil.

Propagate by seed or division in spring or autumn.
L. parviflorum illus. p.294.

LITHOPS (Aizoaceae)
Living stones, Stone plant
Genus of prostrate, egg-shaped, perennial succulents, with almost united pairs of swollen, erect leaves that are separated on upper surface by a fissure from which a daisy-like flower emerges. Each pair of old leaves splits and dries away to papery skin in spring to reveal a pair of new leaves growing at right angles to old ones. Slowly forms clumps after 3–5 years. Frost tender, min. 5°C (41°F). Needs full sun and extremely well-drained soil or gritty compost. Water regularly in growing season (mid-summer to early autumn), not at all in winter. Propagate by seed in spring or summer.
L. aucampiae. Egg-shaped, perennial succulent. H 1cm (^{1}/$_2$in), S 3cm (1^1/$_4$in). Pairs of brown leaves have flat, upper surfaces bearing darker marks. Produces a yellow flower in late summer or early autumn.
L. dorotheae illus. p.414.
L. fulleri. Egg-shaped, perennial succulent. H and S 2cm (^{3}/$_4$in). Pairs of leaves are dove-grey to brown-yellow. Convex, upper surfaces have sunken, darker marks. In late summer or early autumn bears a white flower.
L. hookeri, syn. *L. turbiniformis.* Egg-shaped, perennial succulent. H 4cm (1^1/$_2$in), S 2cm (^{3}/$_4$in). Has a flattish, upper surface with, usually, sunken, dark brown marks on paired, brown leaves. Yellow flower appears in late summer or early autumn.
L. insularis. Egg-shaped, perennial succulent. H 2–3cm (^{3}/$_4$–1^1/$_4$in), S 2cm (^{3}/$_4$in). Slightly convex, upper surfaces of paired, brown leaves have dark green windows and red dots and lines. Produces a yellow flower in late summer or early autumn.
L. julii. Egg-shaped, perennial succulent. H 2–3cm (^{3}/$_4$–1^1/$_4$in), S 5cm (2in). Has paired, pearl- to pink-grey leaves, each with a slightly convex, darker-marked, upper surface. In late summer or autumn produces a white flower.
L. karasmontana illus. p.405. subsp. *bella* is an egg-shaped, perennial succulent. H 2–3cm (^{3}/$_4$–1^1/$_4$in), S 1.5cm (^{5}/$_8$in). Has pairs of brown to brown-yellow leaves with darker marks on convex, upper surfaces. Produces a white flower in late summer or early autumn.
L. lesliei. Egg-shaped, perennial succulent. H 1cm (^{1}/$_2$in), S 2cm (^{3}/$_4$in). Is similar to *L. aucampiae,* but upper leaf surfaces are convex. var. *albinica* illus. p.406.
L. marmorata illus. p.405.
L. olivacea. Egg-shaped, perennial succulent. H and S 2cm (^{3}/$_4$in). Paired, dark olive-green leaves have darker windows on convex, upper surfaces. Yellow flower appears in late summer or early autumn.
L. otzeniana. Egg-shaped, perennial succulent. H 3cm (1^1/$_4$in), S 2cm (^{3}/$_4$in). Paired, grey-violet leaves each have a convex, upper surface with a distinctive, light border and large,

semi-translucent windows. In late summer or early autumn bears a yellow flower.
L. pseudotruncatella. Egg-shaped, perennial succulent. H 3cm (1^1/$_4$in), S 4cm (1^1/$_2$in). Bears pairs of pale grey or blue to lilac leaves with darker marks on convex, upper surfaces. Fissure reaches from side to side only on mature plants. Has a yellow flower in late summer or early autumn. var. *pulmonuncula* illus.p.414.
L. schwantesii. Very variable, egg-shaped, perennial succulent. H and S 3cm (1^1/$_4$in). Has paired, usually rough leaves often with sunken, dark red or blue lines or dots on their flat or slightly convex, upper surfaces. Bears a yellow flower in late summer or early autumn. var. *kuibisensis* illus. p.414.
L. turbiniformis. See *L. hookeri.*

Lithospermum diffusum. See *Lithodora diffusa.*
Lithospermum oleifolium. See *Lithodora oleifolia.*
Lithospermum zahnii. See *Lithodora zahnii.*

LITTONIA (Liliaceae)
Genus of deciduous, perennial, scandent, tuberous climbers, grown for their pendent, bell-shaped flowers in summer. Frost tender, min. 8–16°C (46–61°F). Requires full sun and rich, well-drained soil. Provide support. Dies down in winter; lift and dry off tubers and store in a frost-free place. Propagate by seed in spring; tubers sometimes will divide naturally.
L. modesta illus. p.347.

LIVISTONA (Palmae)
Genus of evergreen palms, grown for their overall appearance. Has clusters of insignificant flowers in summer. Frost tender, min. 7°C (45°F). Needs full light and partial shade and fertile, well-drained soil, ideally neutral to acid. Water potted specimens moderately, less in winter. Propagate by seed in spring at not less than 23°C (73°F). Red spider mite may be a nuisance on containerized plants.
L. australis (Australian cabbage palm, Gippsland fountain palm). Slow-growing, evergreen palm with a fairly slender trunk. H 15–20m (50–70ft), S 3–6m (10–20ft). Has fan-shaped leaves, 1.2–2.5m (4–8ft) wide, divided into narrow, slender-pointed, glossy, green leaflets. Leaf stalks are spiny.
⚇ *L. chinensis* illus. p.59.

LLOYDIA (Liliaceae)
Genus of summer-flowering bulbs, grown for their small, graceful, bell-shaped flowers. Fully to half hardy. Is not easy to grow. Requires partial shade and well-drained, peaty soil; provide plenty of moisture in summer but, preferably, keep fairly dry in winter. Propagate by seed in spring.
L. graeca. See *Gagea graeca.*
L. serotina illus. p.379.

LOBELIA (Campanulaceae)
Genus of annuals, perennials and deciduous or evergreen shrubs, grown for their flowers. Some are suitable for wild gardens or by the waterside. Fully

hardy to frost tender, min. 5°C (41°F). Prefers sun and moist but well-drained soil. Resents wet conditions in winter; in cold areas some perennials and shrubs are therefore best lifted in autumn and placed in well-drained compost in frames. Propagate annuals by seed in spring, perennial species by seed or division in spring, perennial cultivars by division only and shrubs by semi-ripe cuttings in summer.
⚇ *L. cardinalis* (Cardinal flower). Clump-forming perennial. H 1m (3ft), S 23cm (9in). Frost hardy. Bears racemes of 2-lipped, brilliant scarlet flowers from mid- to late summer. Lance-shaped leaves may be fresh green or red-bronze.
L. 'Cherry Ripe' illus. p.213.
L. 'Dark Crusader'. Clump-forming perennial. H 1m (3ft), S 23cm (9in). Half hardy. From mid- to late summer bears racemes of 2-lipped, dark red flowers above lance-shaped, fresh green or red-bronze leaves.
L. erinus 'Blue Cascade'. Slow-growing, pendulous, spreading annual, occasionally perennial. H 10–20cm (4–8in), S 10–15cm (4–6in). Half hardy. Oval to lance-shaped leaves are pale green. Small, 2-lipped, pale blue flowers are produced continuously in summer and early autumn.
⚇ 'Cambridge Blue' is compact and has blue flowers. ⚇ 'Colour Cascade' has flowers in a mixture of colours, such as blue, red, pink, mauve or white. ⚇ 'Crystal Palace' illus. p.286. 'Red Cascade' produces white-eyed, purple-red flowers. 'Sapphire' illus. p.284.
L. fulgens. Clump-forming perennial. H 1m (3ft), S 23cm (9in). Half hardy. Racemes of 2-lipped, brilliant scarlet flowers are produced from mid- to late summer. Lance-shaped leaves are reddish-green.
⚇ *L.* 'Queen Victoria' illus. p.213.
L. siphilitica. Clump-forming perennial. H 1m (3ft), S 23cm (9in). Fully hardy. Racemes of 2-lipped, blue flowers appear in late summer and autumn above narrowly oval, green leaves. Thrives in damp, heavy soil.
L. tupa. Clump-forming perennial. H 1.5–2m (5–6ft), S 1m (3ft). Half hardy. Bears large spikes of 2-lipped, brick-red flowers in late summer, above narrowly oval, hairy, light green leaves. Does best in a sheltered, sunny site.
L. 'Vedrariensis'. Clump-forming perennial. H 1m (3ft), S 30cm (1ft). Frost hardy. In late summer produces racemes of 2-lipped, purple flowers. Has lance-shaped, dark green leaves.
L. 'Will Scarlet'. Clump-forming perennial. H 1m (3ft), S 30cm (1ft). Frost hardy. Racemes of 2-lipped, bright red flowers are borne in summer. Lance-shaped leaves are coppery-green.

LOBIVIA (Cactaceae)
Genus of spherical to columnar, perennial cacti, forming clumps with age, with flowers that each last only 1–2 days. Frost tender, min. 5°C (41°F). Is easy to grow, requiring full sun and well-drained soil. Propagate by seed or stem cuttings from spring to autumn.

LOBULARIA

L. aurea. See *Echinopsis aurea*.
L. backebergii. See *Echinopsis bachebergii*.
L. cinnabarina. See *Echinopsis cinnabarina*.
L. haageana illus. p.403.
L. pentlandii. See *Echinopsis pentlandii*.
L. pygmaea. See *Rebutia pygmaea*.
L. shaferi illus. p.415.
L. silvestrii, syn. *Chamaecereus silvestrii,* illus. p.411.

LOBULARIA (Cruciferae)
Genus of summer- and early autumn-flowering annuals. Fully hardy. Grow in sun and in fertile, well-drained soil. Dead-head to encourage continuous flowering. Propagate by seed sown under glass in spring, or outdoors in late spring.
L. maritima, syn. *Alyssum maritimum* (Sweet alyssum). Fast-growing, spreading annual. H 8–15cm (3–6in), S 20–30cm (8–12in). Has lance-shaped, greyish-green leaves. Rounded heads of tiny, scented, 4-petalled, white flowers are produced in summer and early autumn. **'Little Dorrit'** illus. p.270. **'Wonderland'** illus. p.274.

LOISELEURIA (Ericaceae)
Genus of one species of evergreen, creeping, prostrate shrub, grown for its flowers. Fully hardy. Requires full light and humus-rich, well-drained, acid soil. Is difficult to grow. Propagate by seed in spring or by softwood or semi-ripe cuttings in summer.
L. procumbens illus. p.325.

LOMATIA (Proteaceae)
Genus of evergreen shrubs and trees, grown for their foliage and flowers, which have 4 narrow, twisted petals. Frost hardy, but in cold areas needs shelter from strong winds. Requires sun or semi-shade and moist but well-drained, acid soil. Propagate by softwood or semi-ripe cuttings in summer.
L. ferruginea. Evergreen, upright shrub or tree. H 10m (30ft), S 5m (15ft). Stout, brown-felted shoots bear oblong to oval, dark green leaves, deeply cut into 6–15 oblong lobes. Racemes of yellow-and-red flowers are borne in mid-summer. Thrives outside only in mild, moist areas.
L. silaifolia illus. p.131.

LOMATIUM (Umbelliferae)
Genus of perennials with long, thick roots. Is useful on banks or in wild gardens. Fully to frost hardy. Needs sun and fertile, well-drained soil. Propagate by seed when fresh, in spring.
L. nudicaule (Pestle parsnip). Tufted perennial with a tap root. H and S 60cm (24in). Frost hardy. Has fern-like, deeply dissected leaves, 20cm (8in) long, and, in summer, flat heads of inconspicuous, yellow flowers.

LONICERA (Caprifoliaceae)
Honeysuckle
Genus of deciduous, semi-evergreen or evergreen shrubs and woody-stemmed, twining climbers, grown mainly for their flowers, which are often fragrant. Flowers are tubular, with spreading, 2-lipped petal lobes. Climbers may be trained into large shrubs. Fully hardy to frost tender, min. 5°C (41°F). Grows in any fertile, well-drained soil in sun or semi-shade. Prune out flowered wood of climbers after flowering. Prune shrubs only to remove dead shoots or restrain growth. Propagate by seed in autumn or spring, by semi-ripe cuttings in summer or by hardwood cuttings in late autumn. Aphids may be a problem.
L. x americana illus. p.175.
L. x brownii (Scarlet trumpet honeysuckle). **'Dropmore Scarlet'** illus. p.170.
L. etrusca (Etruscan honeysuckle). Deciduous or semi-evergreen, woody-stemmed, twining climber. H to 4m (12ft). Half hardy. Oval, mid-green leaves are blue-green beneath, the upper ones united into cups. Fragrant, long-tubed, pale yellow flowers, borne in summer-autumn, turn deeper yellow and become red-flushed with age. Grow in sun.
L. fragrantissima. Deciduous or semi-evergreen, bushy, spreading shrub. H 2m (6ft), S 4m (12ft). Fully hardy. Bears oval, dark green leaves. Fragrant, short-tubed, creamy-white flowers open in winter and early spring.
L. **'Gold Flame'**. See *L. x heckrottii*.
L. x heckrottii, syn. *L.* **'Gold Flame'**, illus. p.169.
L. henryi. Evergreen or semi-evergreen, woody-stemmed, twining climber. H to 10m (30ft). Frost hardy. Narrowly oval, dark green leaves are paler beneath. Terminal clusters of long-tubed, red-purple flowers appear in summer-autumn, followed by black berries.
L. hildebrandiana (Giant Burmese honeysuckle). Evergreen or semi-evergreen, woody-stemmed, twining climber. H to 20m (70ft). Frost tender. Oval or rounded, mid-green leaves are paler beneath. Long-tubed, white or cream flowers, ageing to creamy-orange or brownish-yellow, appear in pairs in leaf axils or at shoot tips in summer. Grow in sun.
L. japonica (Japanese honeysuckle). **'Aureoreticulata'** is an evergreen or semi-evergreen, twining climber with soft-haired, woody stems. H to 10m (30ft). Frost hardy. Oval, sometimes lobed, leaves are bright green with bright yellow veins. Fragrant, long-tubed, white flowers, becoming yellowish, are produced in summer-autumn. Is useful for hiding a tree stump or an unsightly wall or fence. **'Halliana'** illus. p.175.
L. ledebourii illus. p.112.
L. maackii. Vigorous, deciduous, bushy shrub. H and S 5m (15ft). Fully hardy. Leaves are oval and dark green. Fragrant, short-tubed, white, later yellow flowers, in early summer, are followed by spherical, bright red fruits.
L. morrowii. Deciduous, spreading shrub with arching branches. H 2m (6ft), S 3m (10ft). Fully hardy. Has oval, dark green leaves and, in late spring and early summer, small, short-tubed, creamy-white flowers that age to yellow.
L. nitida. Evergreen, bushy, dense shrub. H 2m (6ft), S 3m (10ft). Fully hardy. Leaves are small, oval, glossy and dark green. Tiny, fragrant, short-tubed, creamy-white flowers appear in late spring and are followed by small, spherical, purple fruits. Is good for hedging. **'Baggesen's Gold'** illus. p.147. **'Yunnan'** is more upright, has stouter shoots and larger leaves and flowers more freely.
L. periclymenum (Common honeysuckle, Woodbine). **'Graham Thomas'** illus. p.175. **'Serotina'** (Late Dutch honeysuckle) is a deciduous, woody-stemmed, twining climber. H to 7m (22ft). Fully hardy. Oval or oblong, mid-green leaves are grey-green beneath. Very fragrant, long-tubed, dark purple flowers, pinkish within, are borne in mid- and late summer. Grow in sun or shade.
L. pileata illus. p.146.
L. x purpusii illus. p.143.
L. sempervirens illus. p.170.
L. standishii. Evergreen, bushy shrub. H and S 2m (6ft). Fully hardy. Has peeling bark, oblong, bristly, dark green leaves and, in winter, fragrant, short-tubed, creamy-white flowers.
L. tatarica illus. p.110. **'Hack's Red'** is a deciduous, bushy shrub. H and S 2.5m (8ft). Fully hardy. Produces short-tubed, deep pink flowers in late spring and early summer, followed by spherical, red fruits. Leaves are oval and dark green.
L. x tellmanniana illus. p.177.
L. tragophylla. Deciduous, woody-stemmed, twining climber. H 5–6m (15–20ft). Frost hardy. Oval leaves are bluish-green, the uppermost pair united into a cup. Produces clusters of up to 20 long-tubed, bright yellow flowers in early summer.
L. x xylosteoides **'Clavey's Dwarf'**. Deciduous, upright, dense shrub. H 2m (6ft), S 1m (3ft). Fully hardy. Leaves are oval and grey-green. Bears short-tubed, pink flowers in late spring, then spherical, red fruits.
L. xylosteum illus. p.110.

Lophocereus schottii. See *Pachycereus schottii*.

Lophomyrtus bullata. See *Myrtus bullata*.

LOPHOPHORA (Cactaceae)
Peyote
Genus of very slow-growing, perennial cacti that resemble small, blue dumplings, with up to 10 ribs, each separated by an indented line. Has long tap roots. Flowering areoles each produce tufts of short, white hairs. Frost tender, min. 5–10°C (41–50°F). Needs sun and well-drained soil. Is very prone to rotting, so water lightly from spring to autumn. Propagate by seed in spring or summer.
L. williamsii illus. p.407.

Lophostemon confertus. See *Tristania conferta*. **'Variegatus'** see *Tristania conferta* **'Variegata'**.

LOROPETALUM (Hamamelidaceae)
Genus of evergreen shrubs, grown for their flowers. Half hardy, but needs min. 5°C (41°F) to flower well. Requires full light or semi-shade and rich, well-drained, neutral to acid soil. Water containerized plants freely when in full growth, moderately at other times. Propagate by layering or seed in spring or by semi-ripe cuttings in late summer.
L. chinense. Evergreen, rounded, well-branched shrub. H and S 1.2m (4ft). Asymmetrically oval leaves are deep green. White flowers, each with 4 strap-shaped petals, are borne in tufted, terminal clusters, mainly in winter-spring.

LOTUS (Leguminosae)
Genus of summer-flowering perennials, some of which are semi-evergreen, and evergreen sub-shrubs, grown for their foliage and flowers. Fully hardy to frost tender, min. 5°C (41°F). Prefers sun and well-drained soil. Propagate by softwood cuttings from early to mid-summer or by seed in autumn or spring.
L. berthelotii illus. p.247.

LUCULIA (Rubiaceae)
Genus of evergreen shrubs, grown for their flowers and foliage. Frost tender, min. 5–10°C (41–50°F). Needs full light and partial shade and fertile, well-drained soil. Water potted specimens freely when in full growth, moderately at other times. Cut back flowered stems hard in spring, if container-grown. Propagate by seed in spring or by semi-ripe cuttings in summer.
L. grandifolia. Evergreen, rounded to upright, robust shrub. H and S 3–6m (10–20ft). Min. 5°C (41°F). Oval, green leaves have red veins and stalks. Fragrant, tubular, white flowers, each 6cm (2½in) long, with 5 rounded, petal lobes, appear in terminal clusters in summer.

LUETKEA (Rosaceae)
Genus of one species of deciduous sub-shrub, grown for its fluffy flower heads. Is suitable for banks and rock gardens. Fully hardy. Requires shade and well-drained but not too dry soil. Propagate by division or seed in spring.
L. pectinata. Deciduous, spreading, decumbent sub-shrub. H to 30cm (12in), S 20cm (8in). Stems are clothed in finely dissected, very dark green leaves. In summer has terminal racemes of small, fluffy, off-white flower heads.

LUNARIA (Cruciferae)
Honesty
Genus of biennials and perennials, grown for their flowers and silvery seed pods. Fully hardy. Will grow in sun or shade, but prefers partial shade and well-drained soil. Propagate perennials by seed in autumn or spring or by division in spring, biennials by seed only. Self seeds prolifically.
L. annua, syn. *L. biennis,* illus. p.277. **'Variegata'** illus. p.275.
L. biennis. See *L. annua*.
L. rediviva. Rosette-forming perennial. H 60–75cm (24–30in), S 30cm (12in). Produces racemes of 4-petalled, lilac or white flowers in spring, followed by elliptical, silvery seed pods that are useful for indoor decoration. Has oval, coarse, sometimes maroon-tinted, mid-green leaves.

LUNATHYRIUM (Polypodiaceae)
Genus of deciduous or semi-evergreen ferns. Frost tender, min. 5°C (41°F). Needs a partially shaded position and humus-rich, moist soil. Remove fading fronds regularly. Propagate by division in spring or by spores in summer.
L. japonicum, syn. *Deparia petersenii*, *Diplazium japonicum*. Semi-evergreen, creeping fern. H 30cm (12in), S 23cm (9in). Has lance-shaped, papery, light green fronds divided into oblong, indented, blunt pinnae.

LUPINUS (Leguminosae)
Lupin
Genus of annuals, perennials and semi-evergreen shrubs, grown for their large, imposing racemes of pea-like flowers. Fully to frost hardy. Prefers sun and well-drained soil. Remove seed heads of most varieties to prevent self seeding. Propagate species by seed when fresh, in autumn, selected forms by cuttings from non-flowering side-shoots in spring or early summer.
♥ *L. arboreus* illus. p.139.
L. 'Flaming June'. Clump-forming perennial. H 1.2m (4ft), S 45cm (1½ft). Fully hardy. In early summer, spikes of orange-red flowers arise from palmate, deeply divided, mid-green leaves.
L. 'Inverewe Red' illus. p.209.
L. 'The Chatelaine' illus. p.208.
L. 'Thundercloud'. Clump-forming perennial. H 1.2m (4ft), S 45cm (1½ft). Fully hardy. Bears spikes of deep violet-blue flowers above palmate, deeply divided, mid-green leaves in early summer.
L. 'Tom Reeves'. Clump-forming perennial. H 75cm–1m (2½–3ft), S 75cm (2½ft). Fully hardy. During early summer, racemes of pure yellow flowers arise from palmate, deeply divided, mid-green leaves.

LURONIUM (Alismataceae)
Genus of deciduous, perennial, marginal water plants and marsh plants, grown for their foliage and flowers. Fully hardy. Requires shallow water and full sun. Thin plants when overcrowded. Propagate in spring by seed or division.
L. natans, syn. *Alisma natans* (Floating water plantain). Deciduous, perennial, marginal water plant. H 2.5–5cm (1–2in), S 30cm (12in). Has small, elliptic to lance-shaped, mid-green leaves and, in summer, small, 3-lobed, yellow-spotted, white flowers.

LUZULA (Juncaceae), Woodrush. See GRASSES, BAMBOOS, RUSHES and SEDGES.
L. maxima. See *L. sylvatica*.
L. nivea illus. p.183.
L. sylvatica, syn. *L. maxima* (Greater woodrush). 'Marginata' (syn. *L.s.* 'Aureo-marginata') is a slow-growing, evergreen, spreading, rhizomatous, perennial grass. H to 30cm (12in), S indefinite. Fully hardy. Produces thick tufts of broad, hairy-edged, mid-green leaves, with white margins. Leafy stems bear terminal, open, brown flower spikes in summer. Tolerates shade.

LYCASTE. See ORCHIDS.
L. cruenta (illus. p.263). Vigorous, deciduous, epiphytic orchid for a cool greenhouse. H 30cm (12in). Fragrant, triangular, green-and-yellow flowers, 5cm (2in) across, are produced singly in spring. Has broadly oval, ribbed, soft leaves, to 30cm (12in) long. Grow in semi-shade during summer and avoid spraying, which can mark leaves.

LYCHNIS (Caryophyllaceae)
Genus of summer-flowering annuals, biennials and perennials. Fully hardy. Requires sun and well-drained soil. Propagate by division or seed in autumn or spring.
L. 'Abbotswood Rose', syn. *L.* x *walkeri* 'Abbotswood Rose'. Neat, clump-forming perennial. H 30–38cm (12–15in), S 23cm (9in). Has oval, grey leaves and grey, branching stems that, from mid- to late summer, bear sprays of rounded, 5-petalled, bright rose-pink flowers.
L. alpina, syn. *Viscaria alpina* (Alpine catchfly). Tuft-forming perennial. H 5–15cm (2–6in), S 10–15cm (4–6in). Has dense tufts of thick, linear, deep green leaves. In summer, sticky stems each bear a rounded head of pale to deep pink or, rarely, white flowers with spreading, frilled petals. Suits a rock garden.
♥ *L. chalcedonica* illus. p.222.
L. coeli-rosa. See *Silene coeli-rosa*.
L. coronaria illus. p.246.
L. flos-jovis illus. p.243.
L. x *haagena*, syn. *L.* x *haageana*. Short-lived, clump-forming perennial. H 45cm (18in), S 30cm (12in). Produces clusters of large, 5-petalled, white, orange or red flowers in summer. Oval leaves are mid-green. Is best raised regularly from seed.
L. x *walkeri* 'Abbotswood Rose'. See *L.* 'Abbotswood Rose'.
L. viscaria. Clump-forming perennial. H 30cm (12in), S 30–45cm (12–18in). From early to mid-summer, rather sticky, star-shaped, reddish-purple flowers are borne in dense clusters above narrowly oval to oblong, dark green leaves. Is suitable for the front of a border or a rock garden.
♥ 'Splendens Plena' illus. p.246.

Lycianthes rantonetii. See *Solanum rantonetii*.

LYCIUM (Solanaceae)
Genus of deciduous shrubs, sometimes with long, scandent branches, grown for their habit, flowers and fruits. Is useful for poor, dry soil and coastal gardens. May be grown as a hedge. Fully hardy. Prefers full sun and not too rich, well-drained soil. Remove dead wood in winter and cut back to restrict growth if necessary. Cut back hedges hard in spring. Propagate by softwood cuttings in summer, by seed in autumn or by hardwood cuttings in winter.
L. barbarum, syn. *L. halimifolium* (Chinese box thorn, Duke of Argyll's tea-tree). Deciduous, arching, often spiny shrub. H 2.5m (8ft), S 5m (15ft). Funnel-shaped, purple or pink flowers in late spring and summer are followed by spherical, orange-red berries. Leaves are lance-shaped, bright green or grey-green.
L. halimifolium. See *L. barbarum*.

LYCORIS (Amaryllidaceae)
Genus of late summer- and early autumn-flowering bulbs with showy flower heads on leafless stems. Frost hardy; in cool areas is best grown in pots or planted in greenhouse borders. Needs sun, well-drained soil and a warm period in summer to ripen bulbs so they flower. Provide regular liquid feed while in growth. After summer dormancy, water from early autumn until following summer, when foliage dies away. Propagate by seed when ripe or in spring or summer or by offsets in late summer.
L. aurea (Golden spider lily). Late summer- and early autumn-flowering bulb. H 30–40cm (12–16in), S 10–15cm (4–6in). Bears a head of 5 or 6 bright yellow flowers that have narrow, reflexed petals, with very wavy margins, and conspicuous stamens. Strap-shaped, semi-erect, basal leaves appear after flowering.
L. radiata illus. p.364.
L. squamigera. Late summer- or early autumn-flowering bulb. H 45–60cm (18–24in), S 10–15cm (4–6in). Carries a head of 6–8 fragrant, funnel-shaped, rose-pink flowers, 10cm (4in) long, with reflexed petal tips. Strap-shaped, semi-erect, basal leaves form after flowers.

LYGODIUM (Schizaeaceae)
Genus of deciduous or semi-evergreen, climbing ferns, usually with 2 kinds of fronds: vegetative and fertile. Half hardy to frost tender, min. 5°C (41°F). Needs shade or semi-shade and humus-rich, moist, peaty soil. Is best grown among shrubby plants that can provide support. Plants grown under glass in pots need strong, twiggy supports. Remove faded fronds regularly. Propagate by division in spring or by spores in summer.
L. japonicum (Japanese climbing fern). Deciduous, climbing fern. H 2m (6ft), S indefinite. Frost tender. Mid-green, vegetative fronds consist of delicate, finger-shaped pinnae; fertile fronds are broader and 3–5 lobed, with a longer, terminal lobe.
L. palmatum. Deciduous, climbing fern. H 2m (6ft), S indefinite. Half hardy. Has hand-shaped, 5–7 lobed, mid-green vegetative fronds and linear, more or less palmate, fertile pinnae.

LYONIA (Ericaceae)
Genus of deciduous, semi-evergreen or evergreen shrubs and trees, grown for their racemes of small, urn-shaped flowers. Fully hardy. Needs shade or semi-shade and moist, peaty, acid soil. Propagate by semi-ripe cuttings in summer.
L. ligustrina. Deciduous, bushy shrub. H and S 2m (6ft). Oval, dark green leaves set off dense racemes of globular urn-shaped, white flowers from mid- to late summer.
L. ovalifolia. Deciduous or semi-evergreen, bushy shrub. H and S 2m (6ft). Has red shoots and oval, dark green leaves. Racemes of urn-shaped, white flowers appear in late spring and early summer.

LYONOTHAMNUS (Rosaceae)
Genus of one species of evergreen tree, grown for its foliage and flowers. Frost hardy. Needs sun or semi-shade, a warm, sheltered position and fertile, well-drained soil. Propagate by softwood cuttings in summer or by seed in autumn.
L. floribundus (Catalina ironwood). Evergreen tree grown only in the form var. *aspleniifolius*. This slender tree, H 12m (40ft), S 6m (20ft), has rather stringy, reddish-brown bark and fern-like, dark green leaves. Large, flattened heads of 5-petalled, white flowers are produced in early summer.

LYSICHITON (Araceae)
Genus of deciduous, perennial, marginal water plants and bog plants, grown for their handsome spathes and foliage. Fully hardy. Prefers full sun, but tolerates semi-shade. Tolerates both still and running water. Propagate by seed when fresh, in late summer.
♥ *L. americanum* illus. p.391.
♥ *L. camtschatcensis* illus. p.386.

LYSIMACHIA (Primulaceae)
Loosestrife
Genus of summer-flowering annuals and perennials, suitable for the border or rock garden. Fully to half hardy. Prefers sun or semi-shade and moist but well-drained soil. Propagate by division in spring or by seed in autumn.
♥ *L. clethroides* illus. p.205.
L. ephemerum. Neat, clump-forming perennial. H 1m (3ft), S 30cm (1ft). Fully hardy. Erect, terminal racemes of star-shaped, greyish-white flowers are produced on slender stems in summer, followed by light green seed heads. Lance-shaped leaves are leathery and glaucous.
♥ *L. nummularia* 'Aurea' illus. p.334.
L. punctata illus. p.219.

LYTHRUM (Lythraceae)
Purple loosestrife
Genus of summer-flowering perennials that thrive by the waterside and in bog gardens. Fully hardy. Grows in full sun or semi-shade and in moist or wet soil. Propagate cultivars by division in spring, species by seed or division in spring or autumn.
♥ *L. salicaria* 'Firecandle', syn. *L.s.* 'Feuerkerze', illus. p.209. 'Robert' is a clump-forming perennial. H 75cm (30in), S 45cm (18in). Has racemes of 4-petalled, clear pink flowers from mid- to late summer. Leaves are mid-green and lance-shaped.
L. virgatum 'Rose Queen'. Clump-forming perennial. H 1m (3ft), S 60cm (2ft). Racemes of 4-petalled, light pink flowers are borne from mid- to late summer above lance-shaped, mid-green leaves. 'The Rocket' illus. p.209.

M

MAACKIA (Leguminosae)
Genus of deciduous, summer-flowering trees, grown for their foliage and flowers. Fully hardy. Requires full sun and fertile, well-drained soil. Propagate by seed in autumn.
M. amurensis illus. p.64.

MACADAMIA (Proteaceae)
Genus of evergreen trees, grown for their foliage and fruits. Frost tender, min. 10–13°C (50–55°F). Prefers full light, though some shade is tolerated. Provide humus-rich, moisture-retentive but well-drained soil. Water freely while in full growth, moderately at other times. Pruning is not usually necessary, but is tolerated in autumn. Propagate by seed when ripe, in autumn, or in spring.
M. integrifolia illus. p.47.

MACFADYENA, syn. DOXANTHA (Bignoniaceae)
Genus of evergreen, woody-stemmed, tendril climbers, grown for their foxglove-like flowers. Frost tender, min. 5°C (41°F). Any fertile, well-drained soil is suitable with full light. Water regularly, less when not in full growth. Provide support for stems. Thin out crowded shoots after flowering or in spring. Propagate by semi-ripe cuttings in summer.
M. unguis-cati illus. p.176.

MACKAYA (Acanthaceae)
Genus of one species of evergreen shrub, grown for its flowers and overall appearance. Frost tender, min. 7–10°C (45–50°F). Requires full light or partial shade and fertile, well-drained soil. Water potted plants freely when in full growth, moderately at other times. Pruning is tolerated in winter if necessary. Propagate by greenwood cuttings in spring or by semi-ripe cuttings in summer.
♀ *M. bella*, syn. *Asystasia bella*. Evergreen, erect, then spreading, well-branched shrub. H to 1.5m (5ft), S 1.2–1.5m (4–5ft). Leaves are oval, pointed, glossy and mid- to deep green. Has spikes of tubular, dark-veined, lavender flowers, each with 5 large, flared petal lobes, from spring to autumn. In warm conditions, above 13°C (55°F), will flower into winter.

MACLEANIA (Ericaceae)
Genus of evergreen, spring- to summer-flowering shrubs and scrambling climbers, grown primarily for their flowers. Frost tender, min. 10°C (50°F). Needs partial shade and humus-rich, freely draining, neutral to acid soil. Water potted specimens moderately, less when not in full growth. Long shoots may be shortened in winter or after flowering. Propagate by seed in spring, by semi-ripe cuttings in summer or by layering in autumn.
M. insignis. Evergreen, scrambling climber with erect, sparingly branched, wand-like stems. H 3m (10ft), S 1–3m (3–10ft). Has oval, leathery, deep green leaves, red-flushed when young. Tubular, waxy, scarlet flowers, with white tips, hang in clusters in summer. Needs support.

MACLEAYA (Papaveraceae)
Plume poppy
Genus of summer-flowering perennials, grown for their overall appearance. Fully hardy. Grows in sun and in well-drained soil. May spread rapidly. Propagate by division in early spring or by root cuttings in winter.
♀ *M. cordata*, syn. *Bocconia cordata*. Spreading, clump-forming perennial. H 1.5m (5ft) or more, S 60cm (2ft) or more. Large, rounded, lobed, grey-green leaves, grey-white beneath, are produced at base of plant and up lower parts of stems. Large, feathery panicles of dainty, creamy-white flowers are produced in summer.
♀ *M. microcarpa* '**Coral Plume**', syn. *M.m.* 'Kelway's Coral Plume', illus. p.191.

MACLURA (Moraceae)
Genus of one species of deciduous tree, grown for its foliage and unusual fruits. Both male and female trees need to be planted to obtain fruits. Fully hardy, but young plants are susceptible to frost damage. Requires full sun and needs hot summers to thrive in cold areas. Grows in any but waterlogged soil. Propagate by softwood cuttings in summer, by seed in autumn or by root cuttings in late winter.
M. pomifera (Osage orange). Deciduous, spreading tree. H 15m (50ft), S 12m (40ft). Has spiny shoots and oval, dark green leaves that turn yellow in autumn. Tiny, yellow flowers in summer are followed by large, rounded, wrinkled, pale green fruits.

Macroplectrum sesquipedale. See *Angraecum sesquipedale*.

Macrotomia echioides. See *Arnebia pulchra*.

MACROZAMIA (Zamiaceae)
Genus of slow-growing, evergreen shrubs and small trees, with or without trunks, grown for their palm-like appearance. Mature plants may produce conical, green flower spikes. Frost tender, min. 13–16°C (55–61°F). Needs full light or partial shade and well-drained soil. Water containerized plants moderately when in full growth, less at other times. Propagate by seed in spring.
M. spiralis. Evergreen, palm-like shrub with a very short, mainly underground trunk. H and S 60cm–1m (2–3ft). Has a rosette of deep green leaves, each with a spirally twisted mid-rib and very narrow, leathery leaflets.

MAGNOLIA (Magnoliaceae)
Genus of deciduous, semi-evergreen or evergreen trees and shrubs, grown for their showy, usually fragrant flowers. Leaves are mainly oval. Fully to frost hardy. Flowers and buds of early-flowering magnolias may be damaged by late frosts. Needs sun or semi-shade and shelter from strong winds. Does best in fertile, well-drained soil. *M. delavayi*, *M. kobus*, *M. sieboldii* and *M. wilsonii* grow on chalky soil. Other species ideally prefer neutral to acid soil, but will grow in alkaline soil if deep and humus-rich. Dry, sandy soils should be generously enriched with manure and leaf mould before planting. Propagate species by semi-ripe cuttings in summer or by seed, when ripe, in autumn, selected forms by semi-ripe cuttings in summer or by grafting in winter.
M. acuminata (Cucumber tree). Vigorous, deciduous tree, conical when young, later spreading. H 20m (70ft), S 10m (30ft). Fully hardy. Fragrant, cup-shaped, bluish-green flowers appear from early to mid-summer amid large, oval, pale green leaves, followed by small, egg-shaped, green, later red fruits.
M. campbellii illus. p.49. Deciduous tree, upright when young, later spreading. H 15m (50ft), S 10m (30ft). Frost hardy. Large, slightly fragrant, pale to deep pink flowers are borne on leafless branches from late winter to mid-spring on trees 15–20 years old or more. ♀ '**Charles Raffill**' bears large, fragrant, cup-shaped, purplish-pink flowers are borne from late winter to mid-spring on trees at least 15 years old. Leaves are large, oval and mid-green. '**Darjeeling**' (illus. p.49) is similar to the species but has large, very deep pink flowers. '**Kew's Surprise**' bears deep purplish-pink flowers. var. *mollicomata* (illus. p.49), has lilac-pink flowers slightly earlier in the year.
M. '**Charles Coates**' (illus. p.49). Deciduous, rounded, open, spreading tree. H 19m (70ft), S 8m (25ft). Fully hardy. Extremely fragrant, creamy-white flowers with conspicuous, red stamens are produced in late spring and early summer amid large, light green leaves.
♀ *M. cylindrica* (illus. p.49). Deciduous, spreading tree or large shrub. H and S 5m (15ft). Fully hardy. Fragrant, upright, creamy-white flowers are produced in mid-spring, after which the young leaves turn dark green.
M. dawsoniana. Deciduous tree or shrub, with a broadly oval head. H 15m (50ft), S 10m (30ft). Frost hardy. In early spring, large, fragrant, pendent, open cup-shaped, pale lilac-pink flowers are carried profusely on older plants (20 years from seed, 10 years from grafting). Leaves are oval, leathery and deep green.
M. delavayi. Evergreen, rounded, dense shrub or tree. H and S 10m (30ft). Frost hardy. Large, slightly fragrant, bowl-shaped, parchment-white flowers are short-lived and open intermittently from mid-summer to early autumn. Large, oval leaves are deep blue-green above and bluish-white beneath.
♀ *M. denudata*, syn. *M. heptapeta* (Lily tree, Yulan; illus. p.49). Deciduous, rounded, bushy shrub or spreading tree. H and S 10m (30ft). Fully hardy. Produces masses of fragrant, cup-shaped, white flowers from mid- to late spring before oval, mid-green leaves appear.
M. fraseri (illus. p.49). Deciduous, spreading, open tree. H 10m (30ft), S 8m (25). Fully hardy. Fragrant white or pale yellow flowers open in late spring and early summer amid large, pale green leaves.
M. globosa. Deciduous, bushy shrub. H and S 5m (15ft). Frost hardy. In early summer, large, oval, glossy, dark green leaves set off fragrant, cup-shaped, creamy-white flowers with red anthers.
M. grandiflora (Bull bay). Evergreen, broadly conical or rounded, dense tree. H and S 10m (30ft). Frost hardy. Bears large, very fragrant, bowl-shaped, white flowers intermittently from mid-summer to early autumn. Has oblong, glossy, mid- to dark green leaves. ♀ '**Exmouth**' (illus. p.49) has creamy-white flowers and narrow, leathery leaves. '**Ferruginea**' has dark green leaves, rust-brown beneath.
♀ *M.* '**Heaven Scent**' (illus. p.49). Vigorous, deciduous shrub or tree. H and S 10m (30ft). Fully hardy. Fragrant, vase-shaped flowers, each with usually 9 petals, that are pink outside, white within, are borne from mid-spring to early summer. Leaves are broadly elliptic and glossy green.
M. heptapeta. See *M. denudata*.
♀ *M. hypoleuca*, syn. *M. obovata* (Japanese big-leaf magnolia; illus. p.49). Vigorous, deciduous, upright tree. H 15m (50ft), S 10m (30ft). Fully hardy. Large, fragrant, pink-flushed, white or pale cream flowers with crimson stamens appear in early summer.
M. insignis. See *Manglietia insignis*.
M. x *kewensis* '**Wada's Memory**'. See *M.* 'Wada's Memory'.
M. kobus (illus. p.49). Deciduous, broadly conical tree. H 10m (30ft), S 8m (25ft). Fully hardy. Bears a profusion of fragrant, pure white flowers in mid-spring before small, slightly aromatic, dark green leaves appear.
M. liliiflora. Deciduous, bushy shrub. H 3m (10ft), S 4m (12ft). Fully hardy. Has fragrant, upright, vase-shaped, purplish-pink flowers that are borne amid oval, very dark green leaves from mid-spring to mid-summer. ♀ '**Nigra**' (illus. p.49) has large, deep purple flowers.

♣ *M.* x *loebneri* 'Leonard Messel' (illus. p.49). Deciduous, upright shrub or small tree. H 8m (25ft), S 6m (20ft). Fully hardy. In mid-spring, fragrant flowers with many pale lilac-pink petals appear before and after oval, deep green leaves emerge. 'Merrill' has funnel-shaped, white flowers.
M. macrophylla. Deciduous, broadly upright tree, becoming rounded with age. H and S 10m (30ft). Frost hardy. Produces stout, blue-grey shoots and very large, oval, bright green leaves. Large, fragrant, bowl-shaped, parchment-white flowers are borne in early summer.
M. 'Manchu Fan' (illus. p.49). Vigorous, deciduous shrub or tree. H 6m (20ft), S 5m (15ft). Fully hardy. In late spring has large, goblet-shaped, creamy-white flowers with usually 9 petals, the inner ones flushed purple-pink at the base. Leaves are ovate.
M. 'Norman Gould' (illus. p.49). Deciduous, spreading tree or bushy shrub. H and S 5m (15ft). Fully hardy. Silky buds open into fragrant, star-shaped, white flowers in mid-spring. Leaves are oblong and dark green.
M. obovata. See *M. hypoleuca.*
♣ *M. salicifolia* (Willow-leaved magnolia; illus. p.49). Deciduous, conical tree. H 10m (30ft), S 5m (15ft). Fully hardy. Has aromatic, oval leaves, mid-green above, grey-white beneath. Fragrant, pure white flowers open in mid-spring before foliage appears.
M. sargentiana. Deciduous, broadly conical tree. H 15m (50ft), S 10m (30ft). Fully hardy. Large, fragrant, narrowly bowl-shaped, many-petalled flowers, white inside, purplish-pink outside, open from mid- to late spring, before oval, dark green leaves emerge.
M. sieboldii. Deciduous, arching shrub or wide-spreading tree. H 8m (25ft), S 12m (40ft). Frost hardy. Fragrant, cup-shaped, white flowers, with crimson anthers, are carried above oval, dark green leaves from late spring to late summer. subsp. *sinensis* see *M. sinensis.*
M. sinensis, syn. *M. sieboldii* subsp. *sinensis.* Deciduous, spreading shrub. H 6m (20ft), S 8m (25ft). Fully hardy. In late spring and early summer bears fragrant, nodding, cup-shaped, white flowers with crimson anthers. Oval, glossy, bright green leaves have velvety undersides.
♣ *M.* x *soulangeana* 'Etienne Soulange-Bodin' (syn. *M.* x *soulangeana* illus. p.49). Deciduous, rounded, spreading shrub or tree. H and S 6m (20ft). Fully hardy. Bears large, fragrant, tulip-like, purple-flushed, white blooms from mid-spring to early summer, the first before mid- to dark green leaves emerge. 'Alba Superba' has white flowers faintly flushed with pink. ♣ 'Brozzonii', H 8m (25ft), S 6m (25ft), is tree-like, with large, purple-flushed, white flowers. Flowers of ♣ 'Lennei' are large, goblet-shaped and deep rose-purple. 'Picture', H 8m (25ft), S 6m(20ft), is vigorous, compact and upright, with large, erect, deep reddish-purple flowers.
♣ 'Rustica Rubra' (syn. *M.* x *s.* 'Rubra'; illus. p.49) has purplish-red blooms suffused pink.

M. sprengeri (illus. p.49). Deciduous, spreading tree. H 15m (50ft), S 10m (30ft). Frost hardy. In mid-spring has fragrant, bowl-shaped, white flowers sometimes fringed with red or pale pink, before oval, dark green leaves appear. 'Wakehurst' (illus. p.49) has deep purplish-pink flowers.
♣ *M. stellata* (Star magnolia; illus. p.49). Deciduous, bushy, dense shrub. H 3m (10ft), S 4m (12ft). Fully hardy. Fragrant, star-shaped flowers with many narrow petals open from silky buds during early to mid-spring. Leaves are narrow and deep green.
♣ 'Water Lily' (illus. p.49) has large, white flowers with many petals.
M. tripetala (illus. p.49). Deciduous, spreading, open tree, conical when young. H 10m (30ft), S 8m (25ft). Fully hardy. Has large, dark green leaves, clustered about shoot tips, and rather unpleasantly scented, creamy-white flowers with narrow petals in late spring and early summer.
M. x *veitchii* 'Peter Veitch' (illus. p.49). Fast-growing, deciduous, spreading tree. H 20m (80ft), S 15m (52ft). Frost hardy. Bears large, fragrant, pale pink and white flowers in mid-spring, before dark green leaves emerge. Usually flowers within 10 years of planting
M. virginiana (Sweet bay). Deciduous or semi-evergreen, conical shrub or tree. H 9m (28ft), S 6m (20ft). Fully hardy. Has very fragrant, cup-shaped, creamy-white flowers from early summer to early autumn. Oblong, glossy, mid- to dark green leaves are bluish-white beneath.
♣ *M.* 'Wada's Memory' (syn. *M.* x *kewensis* 'Wada's Memory'; illus. p.49). Deciduous, conical tree. H 10m (30ft), S 5m (15ft). Fully hardy. Has aromatic, dark green foliage and a profusion of large, fragrant white flowers borne from mid- to late spring before oval leaves appear.
M. x *watsonii.* See *M.* x *wieseneri.*
M. x *wieseneri* (syn. *M.* x *watsonii*; illus. p.49). Deciduous, spreading, open tree or shrub. H 8m (25ft), S 5m (15ft). Fully hardy. Rounded, white buds open in late spring to early summer to fragrant, creamy-white flowers, flushed pink outside and with crimson stamens.
♣ *M. wilsonii* (illus. p.49). Deciduous, spreading tree or shrub. H 8m (25ft), S 7m (22ft). Frost hardy. In late spring and early summer, fragrant, cup-shaped, white flowers with crimson stamens hang from arching branches amid narrow, dark green leaves.

x MAHOBERBERIS (Berberidaceae)
Hybrid genus (*Berberis* x *Mahonia*) of evergreen shrubs, grown for their foliage, flowers and botanical interest. Fully hardy. Needs sun or semi-shade and fertile, well-drained soil. Propagate by semi-ripe cuttings in summer.
x *M. aquisargentii.* Evergreen, upright, densely leaved shrub. H and S 2m (6ft). Leaves are bright green, often with 3 leaflets, some oblong and finely toothed, others holly-shaped. Terminal clusters of berberis-like, yellow flowers are sparsely produced in late spring.

MAHONIA (Berberidaceae)
Genus of evergreen shrubs, grown for their foliage, their usually short racemes of often fragrant, rounded, bell-shaped, yellow flowers and, on tall species and cultivars, for their deeply fissured bark. Large mahonias make good specimen plants; low-growing ones are excellent for ground cover. Fully to half hardy. Prefers shade or semi-shade and fertile, well-drained but not too dry soil. Propagate species by leaf-bud or semi-ripe cuttings in summer or by seed in autumn, selected forms by leaf-bud or semi-ripe cuttings only.
M. acanthifolia. Evergreen, upright shrub. H 4m (12ft), S 2m (6ft). Frost hardy. Has large, dark green leaves, composed of 17–27 holly-like, spiny leaflets. Long, dense racemes of rich yellow flowers are borne in late autumn and early winter.
M. aquifolium illus. p.127.
M. bealei. Evergreen, upright shrub. H and S 2m (6ft). Fully hardy. Blue-green leaves consist of 13–17 broad, holly-like, spiny leaflets. Stout, upright racemes of fragrant, yellow flowers are produced from late winter to early spring.
M. 'Heterophylla'. Evergreen, upright shrub. H 1m (3ft), S 1.5m (5ft). Frost hardy. Has reddish-purple shoots and glossy, bright green leaves, each composed of 5 or 7 narrowly lance-shaped, wavy-edged or curled leaflets that turn reddish-purple in winter. Small clusters of yellow flowers appear in spring.
♣ *M. japonica* illus. p.121.
♣ *M. lomariifolia.* Evergreen, very upright shrub. H 3m (10ft), S 2m (6ft). Frost hardy. Large, dark green leaves each have 19–37 narrow, holly-like, spiny leaflets. Fragrant, bright yellow flowers are produced in dense, upright racemes during late autumn and winter.
M. x *media.* ♣ 'Buckland' and ♣ 'Charity' illus. p.95.
M. napaulensis. Evergreen, upright, open shrub. H 2.5m (8ft), S 3m (10ft). Frost hardy. Leaves are composed of up to 15 holly-like, spiny, dark green leaflets. Produces long, slender racemes of yellow flowers in early and mid-spring.
M. repens. Evergreen, upright shrub that spreads by underground stems. H 30cm (1ft), S 2m (6ft). Fully hardy. Blue-green leaves each consist of 3–7 oval leaflets, with bristle-like teeth. Dense clusters of deep yellow flowers are borne from mid- to late spring.
M. 'Undulata'. Evergreen, upright shrub. H and S 2m (6ft). Fully hardy. Glossy, dark green leaves each have 5–9 holly-like, wavy-edged leaflets that become bronzed in winter. Bears dense clusters of deep yellow flowers in mid- and late spring.

MAIANTHEMUM (Liliaceae)
May lily
Genus of perennials with extensive, spreading rhizomes. Is useful as ground cover in woodlands and wild areas. Fully hardy. Prefers shade and humus-rich, moist, sandy, neutral to acid soil. Propagate by seed in autumn or by division in any season.

M. bifolium. Spreading, rhizomatous perennial. H 10cm (4in), S indefinite. Pairs of large, oval, glossy, dark green leaves, with wavy edges, arise direct from rhizomes. Slender stems each produce a raceme of 4-petalled, white flowers in early summer, then small, spherical, red berries. May be invasive.
M. canadense illus. p.311.

MAIHUENIA (Cactaceae)
Genus of slow-growing, summer-flowering, alpine cacti, clump-forming with age, with cylindrical stems. Fully to frost hardy. Requires sun and well-drained soil. Protect from winter rain. Propagate by seed or stem cuttings in spring or summer.
M. poeppigii illus. p.413.

MALCOLMIA (Cruciferae)
Genus of spring- to autumn-flowering annuals. Fully hardy. Grow in sun and in fertile, well-drained soil. Propagate by seed sown outdoors in spring, summer or early autumn. Self seeds freely.
M. maritima illus. p.275.

MALEPHORA (Aizoaceae)
Genus of erect or spreading, perennial succulents with semi-cylindrical leaves. Frost tender, min. 5°C (41°F). Needs sun and very well-drained soil. Propagate by seed or stem cuttings in spring or summer.
M. crocea illus. p.416.

MALOPE (Malvaceae)
Genus of annuals, grown for their showy flowers that are ideal for cutting. Fully hardy. Grow in sun and in fertile, well-drained soil. Propagate by seed sown outdoors in spring. Self seeds freely.
M. trifida illus. p.276.

MALUS (Rosaceae)
Crab apple
Genus of deciduous, mainly spring-flowering trees and shrubs, grown for their shallowly cup-shaped flowers, fruits, foliage or autumn colour. Crab apples may be used to make preserves. Fully hardy. Prefers full sun, but tolerates semi-shade; grows in any but waterlogged soil. In winter, cut out dead or diseased wood and prune to maintain a balanced branch system. Propagate by budding in late summer or by grafting in mid-winter. Trees are sometimes attacked by aphids, caterpillars and red spider mite, and are susceptible to fireblight and apple scab.
M. 'Almey'. Deciduous, rounded tree. H and S 8m (25ft). Oval leaves are reddish-purple when young, maturing to dark green. Single, deep pink flowers, with paler pink, almost white centres, in late spring are followed by long-lasting, rounded, orange-red crab apples, which are subject to apple scab.
M. x *arnoldiana* illus. p.61.
M. x *atrosanguinea.* Deciduous, spreading tree. H and S 6m (20ft). Produces oval, glossy, dark green leaves. Red flower buds open to single, rich pink blooms in late spring. Bears small, rounded, red-flushed, yellow crab apples.

M. baccata (Siberian crab). Deciduous, spreading tree. H and S 15m (50ft). Has oval, dark green leaves, a profusion of single, white flowers from mid- to late spring and tiny, rounded, red or yellow crab apples. var. *mandschurica* illus. p.48.
M. 'Chilko'. Deciduous, spreading tree. H and S 8m (25ft). Oval, dark green leaves are reddish-purple when young. Has single, rose-pink flowers in mid-spring, followed by large, rounded, bright crimson crab apples.
M. coronaria 'Charlottae'. Deciduous, spreading tree. H and S 9m (28ft). Broadly oval, lobed or deeply toothed leaves are dark green, turning red in autumn. Semi-double, pale pink flowers are borne in late spring and early summer.
M. 'Cowichan' illus. p.68.
M. 'Dorothea'. Deciduous, spreading tree. H and S 8m (25ft). Semi-double, silvery-pink flowers, red in bud, are borne in late spring, followed by rounded, yellow crab apples. Oval leaves are mid-green. Is subject to apple scab.
M. 'Eleyi'. Deciduous, spreading tree. H and S 8m (25ft). Oval leaves are dark reddish-purple when young, dark green when mature. Bears single, deep purplish-red flowers from mid- to late spring and rounded, purplish-red crab apples.
♛ *M. floribunda* illus. p.62.
M. 'Frettingham's Victoria'. Deciduous, upright tree. H 8m (25ft), S 4m (12ft). Single, white flowers, borne amid oval, dark green leaves in late spring, are followed by rounded, red-flushed, yellow crab apples.
♛ *M.* 'Golden Hornet', syn. *M.* × *zumi* 'Golden Hornet', illus. p.71.
M. × *hartwigii* 'Katherine'. See *M.* 'Katherine'.
M. 'Hopa'. Deciduous, spreading tree. H and S 10m (30ft). Oval, dark green leaves are reddish-purple when young. Single, deep pink flowers in mid-spring are succeeded by rounded, orange-and-red crab apples.
♛ *M. hupehensis* illus. p.48.
♛ *M.* 'John Downie' illus. p.69.
♛ *M.* 'Katherine', syn. *M.* × *hartwigii* 'Katherine'. Deciduous, round-headed tree. H and S 6m (20ft). Has oval, mid-green leaves, large, double, pale pink flowers, fading to white, from mid- to late spring and tiny, rounded, yellow-flushed, red crab apples.
M. 'Lemoinei' illus. p.63.
M. 'Magdeburgensis' illus. p.62.
M. 'Marshall Oyama' illus. p.70.
♛ *M.* 'Neville Copeman', syn. *M.* × *purpurea* 'Neville Copeman'. Deciduous, spreading tree. H and S 9m (28ft). Oval, dark green leaves are purplish-red when young. Single, dark purplish-pink flowers, borne from mid- to late spring, are followed by rounded, orange-red to carmine crab apples.
M. niedzwetskyana, syn. *M. pumila* var. *niedzwetskyana*, *M.* 'Niedzwetskyana'. Deciduous, spreading tree. H 6m (20ft), S 8m (25ft). Oval leaves are red when young, later purple. Produces clusters of single, deep reddish-purple flowers in late spring, then very large, conical, reddish-purple crab apples.

M. 'Niedzwetskyana'. See *M. niedzwetskyana*.
M. prattii. Deciduous tree, upright when young, later spreading. H and S 10m (30ft). Oval, red-stalked, glossy, mid-green leaves become orange and red in autumn. Single, white flowers in late spring are followed by small, rounded or egg-shaped, white-flecked, red crab apples.
M. 'Professor Sprenger' illus. p.70.
M. 'Profusion' illus. p.51.
M. prunifolia illus. p.69.
M. pumila var. *niedzwetskyana*. See *M. niedzwetskyana*.
M. × *purpurea* (Purple crab). Deciduous, spreading tree. H 8m (25ft), S 10m (30ft). Oval, young leaves are reddish, maturing to green. Single, deep ruby-red flowers, which become paler with age, are produced in late spring and are followed by rounded, reddish-purple crab apples. 'Neville Copeman' see *M.* 'Neville Copeman'.
M. 'Red Jade'. Deciduous, weeping tree. H 4m (12ft), S 6m (20ft). In late spring has single, white flowers, sometimes pale pink-flushed, then long-lasting, rounded to egg-shaped, red crab apples. Leaves are dark green and oval.
M. × *robusta*. Vigorous, deciduous, spreading tree. H 12m (40ft), S 10m (30ft). Bears masses of single, white or pink flowers above oval, dark green leaves in late spring. These are followed by long-lasting, rounded, yellow or red crab apples. ♛ 'Yellow Siberian' produces white flowers, which are sometimes pink-tinged, and yellow crab apples.
M. 'Royalty' illus. p.63.
M. sargentii illus. p.86.
M. sieboldii illus. p.100.
M. spectabilis. Deciduous, round-headed tree. H and S 10m (30ft). Has oval, dark green leaves, large, single, blush-pink flowers, rose-red in bud, from mid- to late spring and large, rounded, yellow crab apples.
M. toringoides. Deciduous, spreading tree. H 8m (25ft), S 10m (30ft). Oval, deeply lobed, glossy, bright green leaves turn yellow in autumn. Bears single, white flowers in late spring and rounded or egg-shaped, red-flushed, yellow crab apples in autumn.
♛ *M. transitoria*. Deciduous, spreading, elegant tree. H 8m (25ft), S 10m (30ft). Oval, deeply lobed, mid-green leaves turn yellow in autumn. Has masses of single, white flowers in late spring, followed by small, rounded, pale yellow crab apples.
M. trilobata. Deciduous, conical tree. H 15m (50ft), S 7m (22ft). Has maple-like, lobed, glossy, bright green leaves that often become brightly coloured in autumn. Bears single, white flowers in early summer, followed by small, rounded or pear-shaped, red or yellow crab apples.
♛ *M. tschonoskii*. Deciduous, conical tree. H 12m (40ft), S 7m (22ft). Broadly oval, glossy, mid-green leaves turn brilliant shades of orange, red and purple in autumn. Single, pink-tinged, white flowers, borne in late spring, are succeeded by rounded, red-flushed, yellowish-green crab apples.

M. 'Van Eseltine'. Deciduous, upright tree. H 6m (20ft), S 4m (12ft). Bears double, pink flowers in late spring and rounded, yellow crab apples in autumn. Has oval, dark green leaves.
M. 'Veitch's Scarlet' illus. p.68.
M. yunnanensis var. *veitchii* illus. p.65.
M. × *zumi* 'Calocarpa' illus. p.70.
'Golden Hornet'. See *M.* 'Golden Hornet'.

MALVA (Malvaceae)
Mallow
Genus of annuals, biennials and free-flowering, short-lived perennials. Fully hardy. Requires sun and fertile, well-drained soil. Propagate species by seed in autumn, selected forms by cuttings from firm, basal shoots in late spring or summer. These shoots may be encouraged by cutting plant back after first flowers have faded.
M. moschata illus. p.207.

Malvastrum capensis. See *Anisodontea capensis*.
Malvastrum hypomadarum of gardens. See *Anisodontea capensis*.

MALVAVISCUS (Malvaceae)
Genus of evergreen shrubs and trees, grown for their flowers. Frost tender, min. 13–16°C (55–61°F). Grow in full light and in fertile, well-drained soil. Water containerized plants freely during growing season, moderately at other times. To maintain shape, flowered stems may be cut back hard in late winter. Propagate by seed in spring or by semi-ripe cuttings in summer. Whitefly and red spider mite may be troublesome.
M. arboreus illus. p.91.

MAMMILLARIA (Cactaceae)
Pincushion cactus
Genus of hemispherical, spherical or columnar cacti, grown for their rings of funnel-shaped flowers that develop near crowns. Flowers, offsets and long, slender to spherical seed pods grow between tubercles on a spiny, green stem with extended areoles. Frost tender, min. 5–10°C (41–50°F). Requires full sun and very well-drained soil. Keep completely dry in winter, otherwise plants rot easily. Propagate by seed in spring or summer.
♛ *M. bocasana* illus. p.407.
♛ *M. candida* (Snowball pincushion). Slow-growing, clump-forming, perennial cactus. H and S 15cm (6in). Min. 5°C (41°F). Columnar, green stem is densely covered with short, stiff, white spines. Produces cream to rose flowers, 1–2cm ($^1/_2$–$^3/_4$in) across, in spring. Water sparingly in summer.
M. densispina. Slow-growing, spherical, perennial cactus. H 10cm (4in), S 20cm (8in). Min. 5°C (41°F). Has a green stem densely covered with stout, golden spines and, in spring, yellow flowers, 1–2cm ($^1/_2$–$^3/_4$in) wide.
M. elegans. Spherical to columnar, perennial cactus. H 30cm (12in), S 20cm (8in). Min. 5°C (41°F). Bears a green stem densely covered with short, bristly spines and bright red flowers, 1cm ($^1/_2$in) across, in spring. Offsets occasionally.
♛ *M. elongata* illus. p.406.

♛ *M. geminispina* illus. p.401.
M. gracilis. Clump-forming, perennial cactus. H 5cm (2in), S 20cm (8in). Min. 5°C (41°F). Produces a columnar, green stem densely covered with pure white spines. In early summer carries pale cream flowers, 1–2cm ($^1/_2$–$^3/_4$in) across. Stem is shallow-rooted and reroots readily. var. *fragilis*, H 4cm ($1^1/_2$in), is more fragile and has off-white spines.
♛ *M. hahniana* illus. p.401.
M. magnimamma. Clump-forming, perennial cactus. H 30cm (1ft), S 60cm (2ft). Min. 5°C (41°F). Green stem has very pronounced, angular, dark green tubercles with white spines of variable length. Bears cream, pink or red flowers, 1–2cm ($^1/_2$–$^3/_4$in) wide, in spring and possibly again in late summer.
M. microhelia illus. p.414.
♛ *M. plumosa* illus. p.405.
♛ *M. prolifera* (Strawberry cactus). Clump-forming, perennial cactus. H 10cm (4in), S 30cm (12in). Min 5°C (41°F). Green stem bears dense, golden to white spines. Produces masses of cream or yellow flowers, 1–2cm ($^1/_2$–$^3/_4$in) wide, in summer, followed by red berries that taste like strawberries.
M. rhodantha. Spherical to columnar, perennial cactus. H and S 60cm (2ft). Min. 5°C (41°F). Green stem, branching from crown with age, is densely covered with brown to yellow spines. In late summer produces bright red flowers, 1–2cm ($^1/_2$–$^3/_4$in) across.
M. schiedeana illus. p.406.
M. sempervivi illus. p.408.
♛ *M. zeilmanniana* illus. p.409.

MANDEVILLA, syn. DIPLADENIA (Apocynaceae)
Genus of evergreen, semi-evergreen or deciduous, woody-stemmed, twining climbers, grown for their large, trumpet-shaped flowers. Half hardy to frost tender, min. 7–10°C (45–50°F). Grow in any well-drained soil, with light shade in summer. Water freely when in full growth, sparingly at other times. Provide support and thin out and spur back congested growth in early spring. Propagate by seed in spring or by semi-ripe cuttings in summer. Whitefly and red spider mite may cause problems.
M. × *amabilis* 'Alice du Pont'. See *M.* × *amoena* 'Alice du Pont'.
♛ *M.* × *amoena* 'Alice du Pont', syn. *M.* × *amabilis* 'Alice du Pont', illus. p.169.
M. boliviensis. Vigorous, evergreen, woody-stemmed, twining climber. Frost tender. H to 4m (12ft). Oblong, pointed leaves are lustrous green. Large, trumpet-shaped, white flowers with gold eyes are produced in small clusters in summer.
M. laxa, syn. *M. suaveolens* (Chilean jasmine). Fast-growing, deciduous or semi-evergreen, woody-stemmed, twining climber. H 5m (15ft) or more. Half hardy. Oval leaves have heart-shaped bases. Clusters of fragrant, white flowers are borne in summer.
M. splendens illus. p.165.
M. suaveolens. See *M. laxa*.

MANDRAGORA

MANDRAGORA (Solanaceae)
Mandrake
Genus of rosetted perennials with large, deep, fleshy roots. Fully to frost hardy. Needs sun or partial shade and deep, humus-rich, well-drained soil. Resents being transplanted. Propagate by seed in autumn.
M. officinarum illus. p.319.

MANETTIA (Rubiaceae)
Genus of evergreen, soft- or semi-woody-stemmed, twining climbers, grown for their small but showy flowers. Frost tender, min. 5°C (41°F), but 7–10°C (45–50°F) is preferred. Grow in any humus-rich, well-drained soil, with partial shade in summer. Water regularly, sparingly when temperatures are low. Stems need support. Cut back if required in spring. Propagate by softwood or semi-ripe cuttings in summer. Whitefly is sometimes a problem.
M. bicolor. See *M. luteo-rubra.*
M. cordifolia (Firecracker vine). Fast-growing, evergreen, soft-stemmed, twining climber. H 2m (6ft) or more. Has narrowly heart-shaped, glossy leaves. Funnel-shaped, red flowers appear in small clusters in summer.
M. inflata. See *M. luteo-rubra.*
M. luteo-rubra, syn. *M. bicolor, M. inflata,* illus. p.166.

MANGLIETIA (Magnoliaceae)
Genus of evergreen trees, grown for their foliage and flowers. Half hardy, but is best at min. 3–5°C (37–41°F). Provide humus-rich, moisture-retentive but well-drained, acid soil and full light or partial shade. Water potted plants freely when in full growth, less at other times. Pruning is tolerated if necessary. Propagate by seed in spring.
M. insignis, syn. *Magnolia insignis.* Evergreen, broadly conical tree. H 8–12m (25–40ft) or more, S 3–5m (10–15ft) or more. Leaves are narrowly oval, lustrous, dark green above, bluish-green beneath. In early summer has solitary magnolia-like, pink to carmine flowers that are cream-flushed.

MARANTA (Marantaceae)
Genus of evergreen perennials, grown for their distinctively patterned, coloured foliage. Frost tender, min. 10–15°C (50–59°F). Needs constant, high humidity and a shaded position away from draughts or wind. Grow in humus-rich, well-drained soil. Propagate by division in spring or summer or by stem cuttings in summer.
M. leuconeura (Prayer plant). 'Erythroneura' (syn. *M.l.* 'Erythrophylla') illus. p.267. ❦ var. *kerchoviana* illus. p.268. var. *massangeana* is an evergreen, short-stemmed perennial, branching at the base. H and S 30cm (1ft). Oblong, velvety, dark green leaves, each 15cm (6in) long with a wide, irregular, pale midrib, white, lateral veins and often purplish-green below, stand upright at night but lie flat during the day. Small, 3-petalled, white to mauve flowers appear in slender, upright spikes year-round.

Marginatocereus marginatus. See *Lemaireocereus marginatus.*

MARGYRICARPUS (Rosaceae)
Genus of evergreen shrubs, grown for their fruits. Is good for rock gardens. Frost hardy. Needs a sheltered, sunny position and well-drained soil. Propagate by softwood cuttings in early summer or by seed in autumn.
M. pinnatus, syn. *M. setosus* (Pearl berry). Evergreen, prostrate shrub. H 23–30cm (9–12in), S 1m (3ft). Has dark green leaves divided into linear, silky leaflets. Bears tiny, inconspicuous, green flowers in early summer, then small, globose, glossy, white fruits.
M. setosus. See *M. pinnatus.*

Marsdenia erecta. See *Cionura erecta.*

MARTYNIA (Pedaliaceae, syn. Martyniaceae)
Genus of annuals, grown for their flowers and curious, horned fruits. Half hardy. Grow in a sunny, sheltered position and in fertile, well-drained soil. Propagate by seed sown under glass in early spring.
M. annua, syn. *M. louisiana,* illus. p.272.
M. louisiana. See *M. annua.*

MASDEVALLIA. See ORCHIDS.
M. coccinea (illus. p.261). Evergreen, epiphytic orchid for a cool greenhouse. H 15cm (6in). Rich cerise flowers, 8cm (3in) long, are borne singly in summer. Has narrowly oval leaves, 10cm (4in) long. Needs shade in summer.
M. infracta (illus. p.260). Evergreen, epiphytic orchid for a cool greenhouse. H 15cm (6in). In summer bears rounded, red-and-white flowers, 5cm (2in) long, with tail-like, greenish sepals. Narrowly oval leaves are 10cm (4in) long. Requires shade in summer.
M. tovarensis (illus. p.260). Evergreen, epiphytic orchid for a cool greenhouse. H 15cm (6in). In autumn, 4cm (1½in) long, milky-white flowers, with short-tailed sepals, appear singly or up to 3 to a stem. Has narrowly oval leaves, 10cm (4in) long. Grow in shade in summer.
M. wagneriana (illus. p.262). Evergreen, epiphytic orchid for a cool greenhouse. H 8cm (3in). Pale yellow flowers, 4cm (1½in) long, with long, tail-like sepals, are produced singly or in pairs in summer. Leaves are narrowly oval and 10cm (4in) long. Needs shade in summer.

MATTEUCCIA (Polypodiaceae)
Genus of deciduous, rhizomatous ferns. Fully hardy. Prefers semi-shade and wet soil. Remove faded fronds regularly and divide plants when crowded. Propagate by division in autumn or winter.
❦ *M. struthiopteris* illus. p.188.

MATTHIOLA (Cruciferae)
Stock
Genus of annuals, biennials, perennials and evergreen sub-shrubs. Flowers of most annual or biennial stocks are highly scented and excellent for cutting. Fully hardy to frost tender, min. 4°C (39°F). Grow in sun or semi-shade and in fertile, well-drained, ideally lime-rich soil. Tall cultivars may need support. If grown as biennials outdoors, provide cloche protection during winter. To produce flowers outdoors the same summer, sow seed of annuals under glass in early spring, or outdoors in mid-spring. Sow seed of perennials under glass in spring. Propagate sub-shrubs by semi-ripe cuttings in summer. Is prone to aphids, flea beetle, club root, downy mildew and botrytis.
M., Brompton Series (mixed) illus. p.273, (pink) illus. p.275.
M., East Lothian Series. Group of fast-growing, upright, bushy biennials and short-lived perennials, grown as annuals. H and S 30cm (1ft). Fully hardy. Has lance-shaped, greyish-green leaves and, in summer, spikes, 15cm (6in) or more long, of scented, 4-petalled, single or double flowers, in shades of pink, red, purple, yellow or white.
M. 'Giant Excelsior' illus. p.273.
M. 'Giant Imperial' illus. p.270.
M. incana (Brompton stock). Fast-growing, upright, bushy biennial or short-lived perennial, grown as an annual. H 30–60cm (1–2ft), S 30cm (1ft). Fully hardy. Has lance-shaped, greyish-green leaves and, in summer, scented, 4-petalled, light purple flowers borne in spikes, 7–15cm (3–6in) long.
M. 'Mammoth Column'. Fast-growing, upright, bushy biennial or short-lived perennial, grown as an annual. H to 75cm (2½ft), S 30cm (1ft). Fully hardy. Has lance-shaped, greyish-green leaves and, in summer, 30–38cm (12–15in) long spikes, of scented, 4-petalled flowers, available in mixed or single colours. Flowers are excellent for cutting.
M., Park Series. Group of fast-growing, upright, bushy biennials and short-lived perennials, grown as annuals. H and S to 30cm (1ft). Fully hardy. Lance-shaped leaves are greyish-green. In summer, spikes, at least 15cm (6in) long, of scented, 4-petalled flowers are borne in a wide range of colours.
M., Ten-week Series. Group of fast-growing, upright, bushy biennials and short-lived perennials, grown as annuals. H and S to 30cm (1ft). Fully hardy. Has lance-shaped, greyish-green leaves. Scented, 4-petalled flowers, in spikes at least 15cm (6in) long, are produced in a wide range of colours in summer. Dwarf (illus. p.276) and 'selectable' cultivars have double flowers.
M. 'Trysomic'. Fast-growing, upright, bushy biennial or short-lived perennial, grown as an annual. H and S to 30cm (1ft). Fully hardy. Lance-shaped leaves are greyish-green. Spikes, at least 15cm (6in) long, of scented, mostly double flowers are produced in a wide range of colours in summer.

Matucana haynei. See *Cleistocactus haynei.*

MAURANDIA. See ASARINA.

MAXILLARIA. See ORCHIDS.
M. picta. Evergreen, epiphytic orchid for a cool greenhouse. H 23cm (9in).

MECONOPSIS

Fragrant, yellow flowers, 2.5cm (1in) across and marked reddish-brown outside, are produced singly beneath foliage in winter. Has narrowly oval leaves, 15–23cm (6–9in) long. Requires semi-shade in summer.
M. porphyrostele (illus. p.263). Evergreen, epiphytic orchid for a cool greenhouse. H 8cm (3in). White- and red-lipped, yellow flowers, 1cm (½in) across, are borne singly in summer-autumn. Narrowly oval leaves are 8cm (3in) long. Grow in good light during summer.
M. tenuifolia. Evergreen, epiphytic orchid for a cool greenhouse. H 15cm (6in). Fragrant, yellow flowers, 2.5cm (1in) across, heavily overlaid with red and with white lips, are borne singly throughout summer. Has narrowly oval leaves, 15cm (6in) long. Needs good light in summer.

MAYTENUS (Celastraceae)
Genus of evergreen trees, grown for their neat foliage. Frost hardy, but needs shelter from strong, cold winds when young. Requires sun or semi-shade and fertile, well-drained soil. Propagate by semi-ripe cuttings in summer or by suckers in autumn or spring.
M. boaria (Maiten). Evergreen, bushy-headed, elegant tree. H 10m (30ft), S 8m (25ft). Narrowly oval, glossy, dark green leaves are produced on slender shoots. Bears tiny, star-shaped, green flowers in late spring.

MAZUS (Scrophulariaceae)
Genus of creeping, spring-flowering perennials. Is useful for rock gardens and in paving. Frost hardy. Needs a sheltered, sunny site and moist soil. Propagate by division in spring or by seed in autumn.
M. reptans illus. p.316.

MECONOPSIS (Papaveraceae)
Blue poppy
Genus of perennials, some short-lived, others monocarpic (die after flowering), grown for their flowers. Fully hardy. Needs shade and, in warm areas, a cool position. Most prefer humus-rich, moist, neutral to acid soil. Propagate all but *M.* x *sheldonii* by seed when fresh, in late summer; *M. cambrica, M. grandis, M. quintuplinervia, M.* x *sheldonii* and their cultivars may also be propagated by division after flowering.
❦ *M. betonicifolia* illus. p.217.
M. cambrica illus. p.239.
❦ *M. grandis.* Erect perennial. H 1–1.5m (3–5ft), S 30cm (1ft). Stout stems bear slightly nodding, cup-shaped, deep blue flowers in early summer. Oblong, slightly toothed, hairy, erect, mid-green leaves are produced in rosettes at base. Divide every 2–3 years to maintain vigour. 'Branklyn' illus. p.217.
M. integrifolia illus. p.256.
M. napaulensis illus. p.191.
M. paniculata. Short-lived, clump-forming perennial that dies after flowering. H 1.5m (5ft), S 60cm (2ft). Bears racemes of nodding, cup-shaped, yellow flowers in late spring or early summer. Has large rosettes of oblong

to lance-shaped, deeply lobed and cut, hairy, yellowish-green leaves.
♥ *M. quintuplinervia* illus. p.235.
♥ *M.* x *sheldonii*. Clump-forming perennial. H 1.2–1.5m (4–5ft), S 45–60cm (1½–2ft). Clusters of cup-shaped, clear deep blue flowers are produced in early summer. Has rosettes of oblong to oval, toothed, hairy, mid-green leaves. Divide every three years to maintain vigour. ♥ **'Slieve Donard'** has slightly smaller flowers and leaves with fewer teeth.

MEDICAGO (Leguminosae)
Genus of annuals, perennials and evergreen shrubs, grown for their flowers. Is good in mild, coastal areas as is very wind-resistant. Frost hardy, but in cold areas plant against a south- or west-facing wall. Requires sun and well-drained soil. Cut out dead wood in spring. Propagate shrubs by semi-ripe or softwood cuttings in summer or by seed in autumn or spring, annuals and perennials by seed in autumn or spring.
M. arborea (Moon trefoil, Tree medick). Evergreen, bushy, dense shrub. H and S 2m (6ft). Clusters of small, pea-like, yellow flowers appear from mid-spring to late autumn or winter. These are followed by curious, flattened, snail-like, green, then brown seed pods. Has dark green leaves, each composed of 3 narrowly triangular leaflets, which are silky-haired when young.

MEDINILLA (Melastomataceae)
Genus of evergreen shrubs and scrambling climbers, grown for their flowers and foliage. Frost tender, min. 16–18°C (61–4°F). Needs partial shade and humus-rich, well-drained soil. Water potted plants freely when in full growth, moderately at other times. Propagate by greenwood cuttings in spring or summer.
M. magnifica illus. p.111.

MEEHANIA (Labiatae)
Genus of perennials often with creeping stems, grown mainly as ground cover. Frost hardy. Prefers shade and well-drained but not dry, humus-rich soil. Propagate by seed, division or stem cuttings in spring.
M. urticifolia. Trailing, hairy perennial with long, creeping, leafy stems and erect flowering stems. H to 30cm (1ft), S indefinite. Oval to triangular, toothed leaves are 10cm (4in) or more long on the creeping stems – smaller on flowering stems. Whorls of fragrant, 2-lipped, purplish-blue flowers, to 5cm (2in) long, are carried in erect spikes in late spring.

MEGASEA. See **BERGENIA.**

MELALEUCA (Myrtaceae)
Genus of evergreen, spring- and summer-flowering trees and shrubs, grown for their flowers and overall appearance. Half hardy to frost tender, min. 4–7°C (39–45°F). Needs full light and well-drained soil, preferably without much nitrogen. Some species tolerate waterlogged soils. Water containerized specimens moderately, less in low temperatures. Propagate by seed in spring or by semi-ripe cuttings in summer.
M. armillaris (Bracelet honey myrtle). Evergreen, rounded, wiry-stemmed shrub or tree. H 3–6m (10–20ft), S 1.2–3m (4–10ft). Frost tender. Has needle-like, deep green leaves and, in summer, dense, bottlebrush-like clusters, 3–6cm (1¼–2½in) long, each flower consisting of a small brush of white stamens.
M. elliptica illus. p.112.
M. hypericifolia. Evergreen, rounded shrub. H and S 2–5m (6–15ft). Frost tender. Leaves are oblong to elliptic and mid- to pale green. Flowers, each composed of a 2–2.5cm (¾–1in) long brush of crimson stamens, are borne in summer, mainly in bottlebrush-like spikes, 4–8cm (1½–3in) long.
M. nesophyia illus. p.115.
M. quinquenervia. See *M. viridiflora* var. *rubriflora.*
M. squarrosa (Scented paper-bark). Evergreen, erect, wiry-stemmed shrub or tree. H 3–6m (10–20ft), S 2–4m (6–12ft). Frost tender. Has tiny, oval, deep green leaves. Bears 4cm (1½in) long spikes of scented flowers, each comprising a tiny brush of cream stamens, in late spring and summer.
M. viridiflora var. *rubriflora,* syn. *M. quinquenervia* (Paper-bark tree). Strong-growing, evergreen, rounded tree. H 6–12m (20–40ft), S 3–6m (10–20ft). Frost tender. Leaves are elliptic and lustrous. Has peeling, papery, tan-coloured bark and, in spring, small, white or creamy-pink flowers in bottlebrush-like clusters. Tolerates waterlogged soil.

MELASPHAERULA (Iridaceae)
Genus of one species of spring-flowering corm, grown mainly for botanical interest. Half hardy. Needs sun and well-drained soil. Plant in autumn and keep watered until after flowering, then dry off. Propagate by seed or offsets in autumn.
M. graminea. See *M. ramosa.*
M. ramosa, syn. *M. graminea.* Spring-flowering corm. H to 60cm (24in), S 10–15cm (4–6in). Has narrowly sword-shaped, semi-erect leaves in a basal fan. Wiry, branched stem bears loose sprays of small, pendent, funnel-shaped, yellowish-green flowers with pointed petals.

MELASTOMA (Melastomataceae)
Genus of evergreen, mainly summer-flowering shrubs and trees, grown for their flowers and foliage. Frost tender, min. 10–13°C (50–55°F). Requires full light or partial shade and fertile, well-drained soil. Water containerized specimens freely when in full growth, moderately at other times. Pruning is tolerated in late winter if necessary. Propagate by softwood or greenwood cuttings in spring or summer. Red spider mite and whitefly may cause problems.
M. candidum. Evergreen, rounded, bristly-stemmed shrub. H and S 1–2m (3–6ft). Has oval, leathery, bristly leaves. Small, terminal clusters of fragrant, 5–7-petalled, white or pink flowers are borne profusely in summer.

MELIA (Meliaceae)
Genus of deciduous, spring-flowering trees, grown for their foliage, flowers and fruits. Is useful for very dry soil and does well in coastal gardens in mild areas. Frost hardy. Needs full sun; grows in any well-drained soil. Propagate by seed in autumn.
M. azedarach illus. p.50.

MELIANTHUS (Melianthaceae)
Genus of evergreen perennials and shrubs, grown primarily for their foliage. Half hardy to frost tender, min. 5°C (41°F). Requires sun and fertile, well-drained soil. Water potted specimens freely in summer, moderately at other times. Long stems may be shortened in early spring. Propagate by seed in spring or by greenwood cuttings in summer. Red spider mite may be troublesome.
♥ *M. major* (Honeybush). Evergreen, sprawling, sparingly branched shrub. H and S 2–3m (6–10ft). Half hardy, but best at min. 5°C (41°F). Leaves are 25–45cm (10–18in) long, with 7–13 oval, toothed, blue-grey leaflets. Has tubular, rich brownish-red flowers in terminal spikes, 30cm (12in) long, in spring-summer.

MELICA (Gramineae). See **GRASSES, BAMBOOS, RUSHES** and **SEDGES.**
M. altissima (Siberian melic, Tall melic). Evergreen, tuft-forming, perennial grass. H 60cm (24in), S 20cm (8in). Fully hardy. Has slender stems and broad, mid-green leaves, rough beneath. In summer bears pendent, tawny spikelets in narrow panicles. **'Atropurpurea'** illus. p.183.

MELICYTUS (Violaceae)
Genus of evergreen shrubs and trees, grown for their overall appearance and ornamental fruits. Half hardy to frost tender, min. 3–5°C (37–41°F). Grows in well-drained soil and in full light or partial shade. Water pot plants moderately, less in winter. Pruning is tolerated if required. Propagate by seed when ripe, in autumn, or in spring.
M. ramiflorus (Mahoe, Whiteywood). Evergreen, spreading shrub or tree. H and S 6–10m (20–30ft). Frost tender. Bark is grey-white. Bears lance-shaped, bluntly serrated, bright green leaves. Has small, rounded, greenish flowers in axillary clusters in summer, followed by tiny, violet to purple-blue berries.

MELIOSMA (Sabiaceae)
Genus of deciduous trees and shrubs, grown for their habit, foliage and flowers, which, however, do not appear reliably. Frost hardy. Prefers full sun and deep, fertile, well-drained soil. Propagate by seed in autumn.
M. oldhamii. See *M. pinnata.*
M. pinnata, syn. *M. oldhamii.* Deciduous, stout-branched tree, upright when young, spreading when mature. H 10m (30ft), S 6m (20ft). Has very large, dark green leaves divided into 5–13 oval leaflets. Bears large clusters of small, fragrant, star-shaped, white flowers in early summer.
M. veitchiorum illus. p.54.

MELITTIS (Labiatae)
Bastard balm
Genus of one species of summer-flowering perennial. Fully hardy. Does best in light shade and requires fertile, well-drained soil. Propagate by seed in autumn or by division in spring or autumn.
M. melissophyllum illus. p.241.

MELOCACTUS (Cactaceae)
Turk's cap
Genus of spherical, ribbed, perennial cacti. On reaching flowering size, usually 15cm (6in) high, stems produce woolly crowns; then stems appear to stop growing while woolly crowns develop into columns. Has funnel-shaped flowers in summer, followed by elongated or rounded, red, pink or white seed pods. Frost tender, min. 11–15°C (52–9°F). Requires full sun and extremely well-drained soil. Propagate by seed in spring or summer.
M. bahiensis. Spherical, perennial cactus. H and S 15cm (6in). Min. 15°C (59°F). Dull green stem bears 10–15 ribs. Produces stout, slightly curved, dark brown spines that become paler with age. Crown bears brown bristles and pink flowers, 1–2cm (½–¾in) across, in summer.
M. communis. See *M. intortus.*
M. cunispinus, syn. *M. oaxacensis.* Spherical to columnar, perennial cactus. H 20cm (8in), S 15cm (6in). Min. 15°C (59°F). Green stem has 15 rounded ribs. Areoles each bear a straight central spine and curved radial spines. Flat, woolly crown bears deep pink flowers, 1cm (½in) across, in summer.
M. intortus, syn. *M. communis,* illus. p.408.
M. matanzanus. Spherical, perennial cactus. H and S 10cm (4in). Min. 15°C (59°F). Dark green stem has neat, short spines and develops a woolly crown about 5 years from seed. In summer bears pink flowers, 1cm (½in) across.
M. oaxacensis. See *M. cunispinus.*

MENISPERMUM (Menispermaceae)
Moonseed
Genus of deciduous, woody or semi-woody, twining climbers, grown for their attractive fruits that each contain a crescent-shaped seed – hence the common name. Male and female flowers are carried on separate plants; to produce fruits, plants of both sexes must be grown. Frost hardy. Grow in sun and in any well-drained soil. Propagate by seed or suckers in spring.
M. canadense (Canada moonseed, Yellow parilla). Vigorous, deciduous, woody-stemmed, twining climber, producing a dense tangle of stems and spreading by underground suckers. H to 5m (15ft). Oval to heart-shaped leaves are usually 3–7-lobed. Small, cup-shaped, greenish-yellow flowers are carried in summer, followed by clusters of poisonous, spherical, blackish fruits.

MENTHA (Labiatae)
Mint
Genus of perennials, some of which are semi-evergreen, grown for their

aromatic foliage, which is both decorative and used as a culinary herb. Plants are invasive, however, and should be used with caution. Fully to frost hardy. Grow in sun or shade and in well-drained soil. Propagate by division in spring or autumn.

M. x *gentilis* 'Variegata'. See *M.* x *gracilis* 'Variegata'.

M. x *gracilis* 'Variegata', syn. *M.* x *gentilis* 'Variegata'. Spreading perennial. H 45cm (18in), S 60cm (24in). Fully hardy. Forms a mat of oval, dark green leaves that are speckled and striped with yellow, most conspicuously in full sun. Produces stems that carry whorls of small, 2-lipped, pale mauve flowers in summer.

M. x *piperita* (Peppermint). 'Citrata' (Eau-de-Cologne mint) is a spreading perennial. H 30–60cm (12–24in), S 60cm (24in). Fully hardy. Reddish-green stems, bearing terminal spikes of small, 2-lipped, purple flowers in summer, arise from a carpet of oval, slightly toothed, mid-green leaves that have a scent which is similar to Eau de Cologne.

M. requienii (Corsican mint). Semi-evergreen, creeping perennial. H to 1cm (1/2in), S indefinite. Frost hardy. When crushed, rounded, bright apple-green leaves exude a strong peppermint fragrance. Carries tiny, stemless, lavender-purple flowers in summer. Suits a rock garden or paved path. Needs shade and moist soil.

M. rotundifolia. See *M. suaveolens*.

M. suaveolens, syn. *M. rotundifolia* (Apple mint). 'Variegata' illus. p.240.

MENTZELIA (Loasaceae)
Genus of annuals, perennials and evergreen shrubs. Fully hardy to frost tender, min. 4°C (39°F). Grow in sun and in fertile, very well-drained soil; tender species are best grown in pots under glass. Propagate by seed in spring; shrubs may also be propagated by semi-ripe cuttings in summer.
M. lindleyi, syn. *Bartonia aurea*, illus. p.288.

MENYANTHES (Gentianaceae)
Genus of deciduous, perennial, marginal water plants, grown for their foliage and flowers. Fully hardy. Prefers an open, sunny position. Remove fading flower heads and foliage, and divide overcrowded clumps in spring. Propagate by stem cuttings in spring.
M. trifoliata illus. p.386.

MENZIESIA (Ericaceae)
Genus of deciduous shrubs, grown for their small, urn-shaped flowers. Fully hardy. Needs semi-shade and fertile, moist, peaty, acid soil. Propagate by softwood cuttings in summer or by seed in autumn.
M. ciliicalyx var. *purpurea* illus. p.125.

MERENDERA (Liliaceae)
Genus of corms similar to *Colchicum* but with less showy flowers. Fully to frost hardy. Needs a sunny position and well-drained soil. In cool, damp areas grow in an unheated greenhouse or frame where corms can dry out in summer. Plant in autumn and keep watered through winter and spring. Propagate by seed or offsets in autumn.
M. bulbocodium. See *M. montana*.
M. montana, syn. *M. bulbocodium*, illus. p.383.
M. robusta. Spring-flowering corm. H 8cm (3in), S 5–8cm (2–3in). Frost hardy. Narrowly lance-shaped, semi-erect, basal leaves appear at same time as upright, funnel-shaped flowers, 5–6cm (2–2 1/2in) wide, which have narrow, pale purplish-pink or white petals.

MERREMIA (Convolvulaceae)
Genus of evergreen, twining climbers, grown for their flowers and fruits. Frost tender, min. 7–10°C (45–50°F). Prefers fertile, well-drained soil and full light. Water moderately, much less when not in full growth. Provide support. Thin out congested stems during spring. Propagate by seed in spring. Red spider mite may be a problem.
M. tuberosa, syn. *Ipomoea tuberosa* (Wood rose, Yellow morning glory). Fast-growing, evergreen, twining climber. H 6m (20ft) or more. Leaves have 7 radiating lobes. Funnel-shaped, yellow flowers are borne mainly in summer, followed by semi-woody, globose, ivory-brown fruits.

MERTENSIA (Boraginaceae)
Genus of perennials, grown for their funnel-shaped flowers. Fully hardy. Requires sun or shade and deep, well-drained soil. Propagate by division in spring or by seed in autumn.
M. echioides illus. p.305.
M. maritima illus. p.317.
❦ *M. pulmonarioides*, syn. *M. virginica*, illus. p.235.
M. virginica. See *M. pulmonarioides*.

MERYTA (Araliaceae)
Genus of evergreen trees, grown for their handsome foliage. Frost tender, min. 5°C (41°F). Requires full light or partial shade and humus-rich, moisture-retentive but moderately drained soil. Water containerized plants freely when in full growth, less at other times. Propagate by semi-ripe cuttings in summer or by seed when ripe (dried seed is short-lived) in late summer.
M. sinclairii illus. p.74.

Mesembryanthemum criniflorum. See *Dorotheanthus bellidiformis*.

MESPILUS (Rosaceae)
Medlar
Genus of one species of deciduous tree or shrub, grown for its habit, flowers, foliage and edible fruits. Fully hardy. Needs sun or semi-shade and fertile, well-drained soil. Propagate species by seed in autumn and named forms (selected for their fruit) by budding in late summer.
M. germanica illus. p.59.

METASEQUOIA (Taxodiaceae). See CONIFERS.
❦ *M. glyptostroboides* illus. p.76.

METROSIDEROS (Myrtaceae)
Genus of evergreen, winter-flowering shrubs, trees and scrambling climbers, grown for their flowers, the trees also for their overall appearance and for shade. Frost tender, min. 5°C (41°F). Grows in fertile, well-drained soil and in full light. Water containerized specimens freely when in full growth, moderately at other times. Pruning is tolerated if necessary. Propagate by seed in spring or by semi-ripe cuttings in summer.
M. excelsa illus. p.57.
M. robusta (Rata). Robust, evergreen, rounded tree. H 20–25m (70–80ft) or more, S 10–15m (30–50ft). Oblong to elliptic, leathery leaves are dark green and lustrous. Produces large clusters of flowers, which are mostly composed of long, dark red stamens, during winter.

MEUM (Umbelliferae)
Genus of summer-flowering perennials, grown for their aromatic leaves. Is useful on banks and in wild gardens. Fully hardy. Needs sun and well-drained soil. Propagate by seed when fresh, in autumn.
M. athamanticum (Baldmoney, Spignel). Upright, clump-forming perennial. H 15–45cm (6–18in), S 10–15cm (4–6in). Mainly basal and deeply dissected leaves have narrowly linear leaflets. In summer produces flattish flower heads consisting of clusters of tiny, white or purplish-white flowers.

MICHELIA (Magnoliaceae)
Genus of evergreen, winter- to summer-flowering shrubs and trees, grown for their flowers and foliage. Half hardy to frost tender, min. 5°C (41°F). Provide humus-rich, well-drained, neutral to acid soil and full light or partial shade. Water potted specimens freely when in full growth, less in winter. Pruning is seldom necessary. Propagate by semi-ripe cuttings in summer or by seed when ripe, in autumn, or in spring.
M. doltsopa illus. p.57.
M. figo illus. p.63.

MICROBIOTA (Cupressaceae). See CONIFERS.
❦ *M. decussata* (illus. p.84). Spreading, shrubby conifer. H 50cm (20in), S 2–3m (6–10ft). Fully hardy. Bears flat sprays of scale-like, yellow-green leaves that turn bronze in winter. Globose, yellow-brown cones each contain only one seed.

Microglossa albescens. See *Aster albescens*.

MICROLEPIA (Polypodiaceae)
Genus of deciduous, semi-evergreen or evergreen ferns, best grown in pans and hanging baskets. Frost tender, min. 5°C (41°F). Requires shade or semi-shade and moist soil. Remove faded fronds regularly. Propagate by division in spring or by spores in summer.
M. strigosa illus. p.186.

MICROMERIA (Labiatae)
Genus of evergreen or semi-evergreen shrubs, sub-shrubs and perennials, suitable for rock gardens and banks. Frost hardy. Needs sun and well-drained soil. Propagate by seed in spring or by softwood cuttings in early summer.
M. juliana. Evergreen or semi-evergreen, bushy shrub or sub-shrub. H and S 30cm (12in). Has small, oval, aromatic, green leaves pressed close to stems. In summer, minute, tubular, bright deep pink flowers are carried in whorls on upper parts of stems.

MIKANIA (Compositae)
Genus of evergreen or herbaceous, scrambling or twining climbers, shrubs and erect perennials, grown for their foliage and flower heads. Half hardy to frost tender, min. 7°C (45°F). Any fertile, well-drained soil is suitable, with partial shade in summer. Water regularly, less when not in full growth. Support for stems is needed and ties may be necessary. Thin out congested growth in spring. Propagate by semi-ripe or softwood cuttings in summer. Aphids may be a problem.
M. scandens. Herbaceous, twining climber. H 3–5m (10–15ft). Half hardy. Has oval to triangular, mid-green leaves with 2 basal lobes. Compact clusters of tiny, groundsel-like, pink to purple flower heads are produced in summer-autumn.

MILIUM (Gramineae). See GRASSES, BAMBOOS, RUSHES and SEDGES.
M. effusum (Wood millet). 'Aureum' is an evergreen, tuft-forming, perennial grass. H 1m (3ft), S 30cm (1ft). Fully hardy. Has flat, golden-yellow leaves. Produces open, tiered panicles of greenish-yellow spikelets in summer. Self seeds readily in shade.

MILLA (Liliaceae)
Genus of summer-flowering bulbs, grown for their fragrant flowers, each comprising a slender tube with 6 spreading, star-shaped petals at the tip. Half hardy. Needs a sheltered, sunny position and well-drained soil. Plant in spring. Lift bulbs after flowering and partially dry off for winter. Propagate by seed or offsets in spring.
M. biflora. Summer-flowering bulb. H 30–45cm (12–18in), S 8–10cm (3–4in). Has long, narrow, semi-erect, basal leaves. Stem bears a loose head of 2–6 erect, white flowers, 3–6cm (1 1/4–2 1/2in) across, each on a slender stalk to 20cm (8in) long.

MILTONIA. See ORCHIDS.
M. candida var. *grandiflora* (illus. p.262). Evergreen, epiphytic orchid for a cool or intermediate greenhouse. H 20cm (8in). Cream-lipped, green-patterned, brown flowers, 5cm (2in) across, are borne in spikes in autumn. Has narrowly oval leaves, 10–12cm (4–5in) long. Grow in semi-shade in summer.
M. clowesii (illus. p.262). Evergreen, epiphytic orchid for an intermediate greenhouse. H 20cm (8in). In early summer produces large spikes of 4cm (1 1/2in) wide, yellow flowers, barred with reddish-brown and each with a white-and-mauve lip. Has broadly oval

543

leaves, 30cm (12in) long. Grow in semi-shade in summer.

MILTONIOPSIS (Pansy orchid). See ORCHIDS.
M. **Anjou 'St Patrick'** (illus. p.261). Evergreen, epiphytic orchid for a cool greenhouse. H 15cm (6in). Carries sprays of deep crimson flowers, 10cm (4in) across, with red and yellow patterns on each lip, mainly in summer. Has narrowly oval, soft leaves, 10–12cm (4–5in) long. Needs shade in summer.
M. **Robert Strauss 'Ardingly'** (illus. p.260). Evergreen, epiphytic orchid for a cool greenhouse. H 15cm (6in). Bears sprays of white flowers, 10cm (4in) across, marked reddish-brown and purple; flowering season varies. Narrowly oval, soft leaves are 10–12cm (4–5in) long. Requires shade in summer.

MIMOSA (Leguminosae)
Genus of annuals, evergreen perennials, shrubs, trees and scrambling climbers, cultivated for their flowers and foliage. *M. pudica* is usually grown as an annual. Frost tender, min. 13–16°C (55–61°F). Needs partial shade and fertile, well-drained soil. Water potted specimens freely when in full growth, moderately at other times. Propagate by seed in spring, shrubs also by semi-ripe cuttings in summer. Red spider mite may be a nuisance.
M. pudica illus. p.146.

MIMULUS (Scrophulariaceae)
Monkey musk
Genus of annuals, perennials and evergreen shrubs. Small species suit damp pockets in rock gardens. Fully to half hardy. Most prefer full sun and wet or moist soil. Propagate perennials by division in spring, sub-shrubs by softwood cuttings in late summer; annuals and all species by seed in autumn or early spring.
♀ *M.* **'Andean Nymph'** illus. p.242.
♀ *M. aurantiacus*, syn. *M. glutinosus*, *Diplacus glutinosus*, illus. p.141.
M. glutinosus. See *M. aurantiacus*.
M. guttatus. Spreading, mat-forming perennial. H and S 60cm (24in). Frost hardy. Snapdragon-like, bright yellow flowers, spotted with reddish-brown on lower lobes, are borne in succession in summer and early autumn. Oval leaves are toothed and mid-green.
♀ *M. lewisii* illus. p.243.
M. luteus illus. p.256.
M., **Malibu Series.** Group of fast-growing, branching perennials, grown as annuals. H 15cm (6in), S 30cm (12in). Half hardy. Has oval, mid-green leaves. Flared, tubular, red, yellow or orange flowers are borne in summer (orange, illus. p.293).
M. moschatus (Musk). Spreading, mat-forming perennial. H and S 15–30cm (6–12in). Fully hardy. Bears snapdragon-like, pale yellow flowers, lightly speckled with brown, in summer-autumn. Leaves are oval, hairy and pale green.
M. **'Royal Velvet'** illus. p.247.
♀ *M.* **'Whitecroft Scarlet'**. Short-lived, spreading perennial. H 20–30cm (8–12in), S 30cm (12in). Half hardy. Snapdragon-like, scarlet flowers are produced freely from early to late summer. Leaves are oval, toothed and mid-green.

Mina lobata. See *Ipomoea lobata*.

MIRABILIS (Nyctaginaceae)
Four o'clock flower, Marvel of Peru
Genus of summer-flowering annuals and tuberous perennials. Half hardy. Is best grown in a sheltered position in fertile, well-drained soil and in full sun. Tubers are best lifted and stored over winter in frost-free conditions. Propagate by seed or division of tubers in early spring.
M. jalapa illus. p.209.

MISCANTHUS (Gramineae). See GRASSES, BAMBOOS, RUSHES and SEDGES.
♀ *M. sacchariflorus* (Amur silver grass). Vigorous, herbaceous, rhizomatous, perennial grass that spreads slowly. H 3m (10ft), S indefinite. Frost hardy. Hairless, mid-green leaves last well into winter, often turning bronze. Rarely, bears open, branched panicles of hairy, purplish-brown spikelets in summer. **'Robustus'** is more likely to flower.
M. sinensis **'Gracillimus'** illus. p.185.
'Zebrinus' illus. p.182.

MITCHELLA (Rubiaceae)
Genus of evergreen, trailing sub-shrubs, grown for their foliage and fruits. Makes excellent ground cover, especially in woodlands, although is sometimes difficult to establish. Fully hardy. Prefers shade and humus-rich, neutral to acid soil. Propagate by division of rooted runners in spring or by seed in autumn.
M. repens (Partridge berry). Evergreen, trailing, mat-forming sub-shrub. H 5cm (2in), S indefinite. Bears small, oval, white-striped, green leaves with heart-shaped bases. In early summer has pairs of tiny, fragrant, tubular, white flowers, sometimes purple-tinged, followed by spherical, bright red fruits. Suits a rock garden or peat bed.

MITELLA (Saxifragaceae)
Genus of clump-forming, summer-flowering, slender-stemmed, rhizomatous perennials. Fully hardy. Requires shade and humus-rich, moist soil. Propagate by division in spring or by seed in autumn.
M. breweri illus. p.332.

MITRARIA (Gesneriaceae)
Genus of one species of evergreen, woody-stemmed, scrambling climber. Half hardy. Grow in semi-shade and in peaty, acid soil. Propagate by seed in spring or by stem cuttings in summer.
M. coccinea illus. p.166.

MOLINIA (Gramineae). See GRASSES, BAMBOOS, RUSHES and SEDGES.
M. caerulea subsp. *arundinacea*. Tuft-forming, herbaceous, perennial grass. H 2.5m (8ft), S 60cm (2ft) Fully hardy. Has broad, flat, grey-green leaves. Produces spreading panicles of purple spikelets on stiff, erect stems in summer. Needs a dry, sunny position and acid soil. ♀ subsp. *caerulea* **'Variegata'** (Variegated purple moor grass), H 60cm (2ft), has yellow-striped, mid-green leaves and, in late summer, panicles of purplish spikelets.

MOLTKIA (Boraginaceae)
Genus of deciduous, semi-evergreen or evergreen sub-shrubs and perennials, grown for their funnel-shaped flowers in summer. Fully to frost hardy. Prefers sun and well-drained, neutral to acid soil. Propagate by semi-ripe cuttings in summer or by seed in autumn.
♀ *M.* × *intermedia*. Evergreen, open, dome-shaped sub-shrub. H 30cm (12in), S 50cm (20in). Fully hardy. Stems are clothed in narrowly linear, dark green leaves. Masses of loose spikes of small, open funnel-shaped, bright blue flowers appear in summer.
M. petraea. Semi-evergreen, bushy shrub. H 30cm (12in), S 60cm (24in). Fully hardy. Hairy leaves are long and narrow. Clusters of pinkish-purple buds open into funnel-shaped, violet-blue flowers in summer.
M. suffruticosa illus. p.305.

MOLUCCELLA (Labiatae)
Genus of annuals and perennials, grown for their flowers that may be dried successfully. Half hardy. Grow in sun and in rich, very well-drained soil. Propagate by seed sown under glass in spring, or outdoors in late spring.
M. laevis illus. p.287.

MONARDA (Labiatae)
Bergamot
Genus of annuals and perennials, grown for their aromatic foliage as well as their flowers. Fully hardy. Requires sun and moist soil. Propagate species and cultivars by division in spring, species only by seed in spring.
M. **'Adam'**. Clump-forming perennial. H 75cm (30in), S 45cm (18in). Bears dense whorls of 2-lipped, cerise flowers throughout summer. Oval, usually toothed, mid-green leaves are aromatic and hairy.
M. didyma (Bee balm, Bergamot).
♀ **'Cambridge Scarlet'** illus. p.214.
♀ **'Croftway Pink'** illus. p.207.
M. fistulosa illus. p.215.
M. **'Prairie Night'**. Clump-forming perennial. H 1.2m (4ft), S 45cm (1½ft). Produces dense whorls of 2-lipped, rich violet-purple flowers from mid- to late summer. Oval, toothed leaves are mid-green.

MONSTERA (Araceae)
Genus of evergreen, woody-stemmed, root climbers, grown for their large, handsome leaves. Bears insignificant, creamy-white flowers with hooded spathes intermittently. Frost tender, min. 15–18°C (59–64°F). Provide humus-rich, well-drained soil and light shade in summer. Water moderately, less when temperatures are low. Provide support. If necessary, shorten long stems in spring. Propagate by leaf-bud or stem-tip cuttings in summer.
M. acuminata (Shingle plant). Evergreen, woody-stemmed, root climber with robust stems. H 3m (10ft) or more. Has lopsided-oval, pointed, rich green leaves with a heart-shaped base, sometimes cleft into a few large lobes, to 25cm (10in) long.
♀ *M. deliciosa* illus. p.180.

MORAEA (Iridaceae)
Genus of corms with short-lived, iris-like flowers. Divides into 2 groups: winter- and summer-growing species. Winter-growing species are half hardy, need full sun and well-drained soil; keep dry in summer during dormancy and start into growth by watering in autumn. Summer-growing species are frost hardy and dormant in winter; grow in a sheltered, sunny site and well-drained soil. Propagate by seed in autumn (winter growers) or spring (summer growers).
M. huttonii illus. p.347.
M. polystachya. Winter-growing corm. H to 30cm (12in), S 5–8cm (2–3in). Bears long, narrow, semi-erect, basal leaves. Stem produces a succession of erect, flattish, blue or lilac flowers, 8cm (3in) wide, in winter-spring. Outer petals each have a central, yellow mark.
M. spathacea. See *M. spathulata*.
M. spathulata, syn. *M. spathacea*. Summer-growing corm. H to 1m (3ft), S 10–15cm (4–6in). Has one long, narrow, semi-erect, basal leaf. Tough flower stem carries a succession of up to 5 upward-facing, yellow flowers, 5–7cm (2–3in) wide, with reflexed, outer petals, in summer.

MORINA (Morinaceae)
Whorl flower
Genus of evergreen perennials, only one species of which is in general cultivation: this is grown for its thistle-like foliage and its flowers. Frost hardy, but needs protection from drying spring winds. Requires full sun and well-drained, preferably sandy soil. Propagate by division directly after flowering or, preferably, by seed when fresh, in late summer.
M. longifolia illus. p.205.

MORISIA (Cruciferae)
Genus of one species of rosetted perennial with a long tap root. Is good for screes, rock gardens and alpine houses. Fully hardy. Needs sun and gritty, well-drained soil. Propagate by seed in autumn or by root cuttings in winter.
M. hypogaea. See *M. monanthos*.
M. monanthos, syn. *M. hypogaea*, illus. p.320.

MORUS (Moraceae)
Mulberry
Genus of deciduous trees, grown for their foliage and edible fruits. Inconspicuous flowers appear in spring. Fully hardy. Requires full sun and fertile, well-drained soil. Propagate by softwood cuttings in summer or by seed in autumn.
M. alba (White mulberry). **'Laciniata'** illus. p.66. **'Pendula'** is a deciduous, weeping tree. H 3m (10ft), S 5m (15ft). Rounded, sometimes lobed, glossy, deep green leaves turn yellow in autumn. Edible, oval, fleshy, pink, red or purple fruits ripen in summer.

MUCUNA

⚘ *M. nigra* (Black mulberry). Deciduous, round-headed tree. H 12m (40ft), S 15m (50ft). Heart-shaped, dark green leaves turn yellow in autumn. Bears edible, oval, succulent, dark purplish-red fruits in late summer or early autumn.

MUCUNA (Leguminosae)
Genus of vigorous, evergreen, twining climbers, grown for their large, pea-like flowers. Frost tender, min. 18°C (64°F). Humus-rich, moist but well-drained soil is essential, with partial shade in summer. Water freely when in full growth, less at other times. Needs plenty of space to climb; provide support. Thin out crowded stems in spring. Propagate by seed in spring or by layering in late summer. Whitefly and red spider mite may cause problems.
M. bennettii. Strong- and fast-growing, evergreen, twining climber. H 15–25m (50–80ft). Leaves are divided into 3 oval leaflets. Pea-like, orange-scarlet flowers appear in large, pendent clusters in summer.
M. deeringiana. See *M. pruriens* var. *utilis.*
M. pruriens var. *utilis*, syn. *M. deeringiana.* Vigorous, fast-growing, evergreen, twining climber. H 15m (50ft) or more. Has pea-like, both green- and red-purple flowers in long, pendent clusters in summer-autumn. Leaves, of 3 oval leaflets, are used for fodder and green manure. May be short-lived.

MUEHLENBECKIA (Polygonaceae)
Genus of deciduous or evergreen, slender-stemmed, summer-flowering shrubs and woody-stemmed, scrambling climbers, grown for their foliage. Frost hardy. Grow in sun or shade and in well-drained soil. Propagate by semi-ripe cuttings in summer.
M. complexa. Deciduous, mound-forming shrub or twining climber. H 60cm–1m (2–3ft), S 1m (3ft). Slender, wiry stems bear variably shaped (oval to fiddle-shaped), dark green leaves. Tiny, star-shaped, greenish-white flowers in mid-summer are followed by small, spherical, waxy, white fruits.

MULGEDIUM. See **CICERBITA.**

MURRAYA (Rutaceae)
Genus of evergreen trees and shrubs, grown for their overall appearance. Frost tender, min. 13–15°C (55–9°F). Needs full light or partial shade and humus-rich, well-drained soil. Water containerized plants freely when in full growth, moderately at other times. Pruning is tolerated in late winter if necessary. Propagate by seed in spring or by semi-ripe cuttings in summer. Whitefly may be troublesome.
M. exotica. See *M. paniculata.*
M. paniculata, syn. *M. exotica* (Orange jasmine). Evergreen, rounded shrub or tree. H and S 2–4m (6–12ft). Pungently aromatic, edible (used for flavouring curries), glossy, rich green leaves each have 9 or more oval leaflets. Fragrant, 5-petalled, white flowers are carried in terminal clusters year-round. Tiny, egg-shaped fruits are red.

MUSA (Musaceae)
Banana
Genus of evergreen, palm-like, suckering perennials, with false stems formed from overlapping leaf sheaths, grown for their foliage, flowers and fruits (bananas), not all of which are edible. Frost hardy to frost tender, min. 18°C (64°F). Grow in sun or partial shade and in humus-rich, well-drained soil. Propagate by division year-round, by offsets in summer or by suckers after flowering.
M. arnoldiana. See *Ensete ventricosum.*
M. basjoo, syn. *M. japonica*, illus. p.197.
M. coccinea. See *M. uranoscopus.*
M. ensete. See *Ensete ventricosum.*
M. japonica. See *M. basjoo.*
⚘ *M. ornata* illus. p.196.
⚘ *M. uranoscopus*, syn. *M. coccinea* (Scarlet banana). Evergreen, palm-like perennial. H to 1m (3ft), S 1.5m (5ft). Frost tender. Oblong to oval, dark green leaves, to 1m (3ft) long, are paler below. In summer produces erect spirals of tubular, yellow flowers, enclosed in red bracts; banana-like, orange-yellow fruits are 5cm (2in) long.

MUSCARI (Liliaceae)
Grape hyacinth
Genus of spring-flowering bulbs, each with a cluster of narrowly strap-shaped, basal leaves, usually appearing in spring just before flowers. Leafless flower stems bear dense spikes of small flowers, most of which have constricted mouths. Fully to half hardy. Needs a sunny position and fairly well-drained soil. Plant in autumn. Propagate by division in late summer or by seed in autumn.
⚘ *M. armeniacum* illus. p.377. 'Blue Spike' is a spring-flowering bulb. H 15–20cm (6–8in), S 8–10cm (3–4in). Frost hardy. Produces 3–6 long, narrow, semi-erect, basal leaves. Bears dense spikes of fragrant, bell-shaped, deep blue flowers; constricted mouths have rims of paler blue or white 'teeth'.
⚘ *M. aucheri*, syn. *M. tubergenianum*, illus. p.377.
⚘ *M. azureum*, syn. *Hyacinthus azureus*. Spring-flowering bulb. H 10–15cm (4–6in), S 5–8cm (2–3in). Frost hardy. Has 2 or 3 narrow, semi-erect, basal, greyish-green leaves, slightly wider towards tips. Bears a very dense spike of bell-shaped, pale clear blue flowers; mouths have small 'teeth' with central, dark blue stripes.
M. botryoides. Spring-flowering bulb. H 15–20cm (6–8in), S 5–8cm (2–3in). Frost hardy. Bears 2–4 narrow, semi-erect, basal leaves that widen slightly towards tips, and minute, nearly spherical, bright blue flowers, each with a constricted mouth and white-toothed rim.
M. comosum (Tassel grape hyacinth). Late spring-flowering bulb. H 20–30cm (8–12in), S 10–12cm (4–5in). Frost hardy. Has up to 5 strap-shaped, semi-erect, basal, grey-green leaves. Produces a loose spike of bell-shaped, fertile, brownish-yellow flowers with a tuft of thread-like, sterile, purplish-blue flowers at the tip. 'Plumosum' (syn. *M.c.* 'Monstrosum') illus. p.376.
M. latifolium illus. p.357.
M. macrocarpum illus. p.378.
M. neglectum, syn. *M. racemosum*, illus. p.377.
M. paradoxum. See *Bellevalia pycnantha.*
M. pycnantha. See *Bellevalia pycnantha.*
M. racemosum. See *M. neglectum.*
M. tubergenianum. See *M. aucheri.*

MUSSAENDA (Rubiaceae)
Genus of evergreen shrubs and scrambling climbers, grown for their flowers. Frost tender, min. 16–18°C (61–4°F). Needs fertile, well-drained soil and full light. Water freely when in full growth, less at other times. Provide support and thin out crowded stems in spring. Propagate by seed in spring or by air-layering in summer. Whitefly and red spider mite cause problems.
M. erythrophylla. Moderately vigorous, evergreen, scrambling climber. H 6–10m (20–30ft). Has broadly oval, bright green leaves and flowers in summer-autumn. Each flower has one greatly enlarged, oval, bract-like, red sepal, a red tube and yellow petal lobes.

MUTISIA (Compositae)
Genus of evergreen, tendril climbers, grown for their long-lasting flower heads. Frost to half hardy. Grow in well-drained soil; is best planted with roots in shade and leafy parts in sun. Propagate by seed in spring, by stem cuttings in summer or by layering in autumn.
M. decurrens illus. p.177.
M. oligodon. Evergreen, tendril climber. H to 1.5m (5ft). Frost hardy. Oblong leaves with toothed margins are 2.5–3.5cm (1–1½in) long. In summer-autumn has long-stalked, daisy-like, pink flower heads. Is effective when grown against a low wall or through a shrub.

MYOPORUM (Myoporaceae)
Genus of evergreen shrubs and trees, grown for their overall appearance and as hedges and windbreaks. Frost tender, min. 2–5°C (36–41°F). Prefers full light and well-drained soil; will tolerate poor soil. Water potted specimens moderately. Propagate by seed when ripe or in spring or by semi-ripe cuttings in late summer.
M. laetum. Evergreen, rounded to upright shrub or tree. H 3–10m (10–30ft), S 2–5m (6–15ft). Bears lance-shaped to oval, lustrous, bright green leaves. Axillary clusters of small, bell-shaped, white flowers, with purple dots, appear in spring-summer. Tiny, oblong fruits are red-purple.
M. parvifolium illus. p.132.

MYOSOTIDIUM (Boraginaceae)
Chatham Island forget-me-not
Genus of one species of evergreen perennial that is suitable for mild, coastal areas. Half hardy. Prefers semi-shade and moist soil. Seaweed is often recommended as a mulch. Is not easy to cultivate, and once established should not be disturbed. Propagate by division in spring or by seed when ripe, in summer or autumn.
M. hortensia, syn. *M. nobile*, illus. p.251.
M. nobile. See *M. hortensia.*

MYOSOTIS (Boraginaceae)
Forget-me-not
Genus of annuals, biennials and perennials, grown for their flowers. Most species are good for rock gardens, banks and screes; *M. scorpioides* is best grown as a marginal water plant. Fully hardy. Most prefer sun or semi-shade and fertile, well-drained soil. Propagate by seed in autumn.
M. alpestris illus. p.318.
M. australis. Short-lived, tuft-forming perennial. H 12cm (5in), S 8cm (3in). Has oval, rough-textured leaves and, in summer, tight sprays of open funnel-shaped, yellow or white flowers. Is good for a scree.
⚘ *M.* 'Blue Ball' illus. p.286.
M. caespitosa. See *M. laxa* subsp. *caespitosa.*
M. 'Carmine King'. Slow-growing, short-lived, bushy perennial, grown as a biennial. H to 20cm (8in), S 15cm (6in). Has lance-shaped leaves and, in spring and early summer, produces sprays of tiny, 5-lobed, dark pink-red flowers.
M. laxa subsp. *caespitosa*, syn. *M. caespitosa.* Clump-forming annual or short-lived perennial. H 12cm (5in), S 15cm (6in). Leaves are lance-shaped, leathery and dark green. Produces sprays of rounded, bright blue flowers in summer.
M. palustris. See *M. scorpioides.*
M. scorpioides, syn. *M. palustris* (Water forget-me-not). 'Mermaid' illus. p.388.
M. 'White Ball'. Slow-growing, short-lived, bushy, compact perennial, grown as a biennial. H to 20cm (8in), S 15cm (6in). Leaves are lance-shaped. Sprays of tiny, 5-lobed, pure white flowers are produced in early summer.

Myrceugenia apiculata. See *Myrtus luma.*

MYRIOPHYLLUM (Haloragidaceae)
Genus of deciduous, perennial, submerged water plants, grown for their foliage. Most species are ideal as depositories for fish spawn. Fully hardy to frost tender, min. 5°C (41°F). Requires full sun. Spreads widely: keep in check by removing excess growth as required. Propagate by stem cuttings in spring or summer.
M. aquaticum, syn. *M. proserpinacoides*, illus. p.388.
M. hippuroides. Deciduous, perennial, spreading, submerged water plant. S indefinite. Half hardy. Produces a dense mass of small, feathery, pale green leaves. Insignificant, greenish-cream flowers are borne in summer. Is suitable for a cold-water aquarium.
M. proserpinacoides. See *M. aquaticum.*
M. verticillatum illus. p.389.

MYRRHIS (Umbelliferae)
Sweet Cicely
Genus of one species of summer-flowering perennial. Fully hardy. Grow in sun or shade and in well-drained soil. Propagate by seed in autumn or spring.
M. odorata illus. p.204.

MYRSINE (Myrsinaceae)
Genus of evergreen shrubs and trees, with inconspicuous flowers, grown mainly for their foliage. Also bears decorative fruits, to obtain which plants of both sexes must be grown. Is suitable for rock and peat gardens. Frost hardy, but in cold areas requires shelter. Grow in sun or shade and in any fertile, well-drained soil other than a shallow, chalky one. Propagate by semi-ripe cuttings in summer.
M. africana (Cape myrtle). Very slow-growing, evergreen, bushy, dense shrub. H and S 75cm (30in). Small, glossy, dark green leaves are aromatic and rounded. Tiny, yellowish-brown flowers in late spring are succeeded by spherical, pale blue fruits.

MYRTILLOCACTUS (Cactaceae)
Genus of branching, perennial cacti with ribbed, spiny, blue-green stems. Bears star-shaped flowers that open at night. Frost tender, min. 11–12°C (52–4°F). Needs a sunny, well-drained site. Propagate by seed or stem cuttings in spring or summer.
M. geometrizans illus. p.392.

MYRTUS (Myrtaceae)
Myrtle
Genus of evergreen shrubs, sometimes tree-like, grown for their flowers, fruits and aromatic foliage. Frost to half hardy; in cold areas plant against a south- or west-facing wall. Needs full sun and fertile, well-drained soil. May be pruned in spring. Propagate by semi-ripe cuttings in late summer.
M. apiculata. See *M. luma*.
M. bullata, syn. *Lophomyrtus bullata*. Evergreen, upright shrub. H 5m (15ft), S 3m (10ft). Half hardy. Rounded, puckered leaves are purple when young, maturing to reddish-brown. Produces saucer-shaped, white flowers in late spring and early summer, then egg-shaped, dark red fruits.
♛ *M. communis* (Common myrtle) illus. p.97. ♛ var. *tarentina* is an evergreen, bushy shrub. H and S 2m (6ft). Frost hardy. Small leaves are narrowly oval, glossy and dark green. Fragrant, saucer-shaped, white flowers, each with a dense cluster of stamens, are borne from mid-spring to early summer, followed by spherical, white fruits. Is very wind-resistant and good for hedging in mild areas.
♛ *M. luma*, syn. *M. apiculata*, *Amomyrtus luma*, *Myrceugenia apiculata*, illus. p.89. **'Glanleam Gold'** (syn. *Amomyrtus luma* 'Glanleam Gold') is a strong-growing, evergreen, upright shrub. H and S 10m (30ft). Half hardy. Has stout stems, peeling, brown-and-white bark and oval, bright green leaves edged with creamy-yellow. Slightly fragrant, saucer-shaped, white flowers from mid-summer to mid-autumn are followed by spherical, red fruits, ripening to purple.
M. ugni, syn. *Eugenia ugni*, *Ugni molinae*. Evergreen, upright, densely branched shrub. H 1.5m (5ft), S 1m (3ft). Half hardy. Glossy, dark green leaves are oval. Has fragrant, slightly nodding, cup-shaped, flesh-pink flowers in late spring, then aromatic, edible, spherical, dark red fruits. Is good for hedging in mild areas.

N

NANDINA (Berberidaceae)
Genus of one species of evergreen or semi-evergreen, summer-flowering shrub, grown for its foliage and flowers. Frost hardy. Prefers a sheltered, sunny site and fertile, well-drained but not too dry soil. On established plants prune untidy, old stems to base in spring. Propagate by semi-ripe cuttings in summer.
♚ *N. domestica* (Heavenly bamboo, Sacred bamboo). Evergreen or semi-evergreen, upright, elegant shrub. H and S 2m (6ft). Leaves have narrowly lance-shaped, dark green leaflets, purplish-red when young and in autumn-winter. Large panicles of small, star-shaped, white flowers in mid-summer are followed in warm climates by spherical, red fruits. 'Firepower' illus. p.121.

NARCISSUS (Amaryllidaceae)
Daffodil
Genus of bulbs, grown for their ornamental flowers. Daffodils have usually linear leaves and a spread of up to 20cm (8in). Each flower has a trumpet or cup (the corona) and petals (botanically known as perianth segments). Fully hardy, except where otherwise stated. Prefers sun or light shade and well-drained soil, but Div.8 cultivars (see below) prefer a sunny site and tolerate lighter soils. Dead-head flowers and remove foliage during mid- or late summer. Most cultivars increase naturally by offsets; dense clumps should be divided no sooner than 6 weeks after flowering every 3–5 years. Species may be propagated by fresh seed in late summer or autumn. Narcissus yellow stripe virus, basal rot, slugs, large narcissus fly and bulb and stem eelworm may cause problems.

Horticulturally, *Narcissus* is split into the following divisions:

Div.1 Trumpet – usually solitary flowers each have a trumpet that is as long as, or longer than, the petals. Early to late spring-flowering.
Div.2 Large-cupped – solitary flowers each have a cup at least one-third the length of, but shorter than, the petals. Spring-flowering.
Div.3 Small-cupped – flowers are often borne singly; each has a cup not more than one-third the length of the petals. Spring- or early summer-flowering.
Div.4 Double – most have solitary large, fully or semi-double flowers, rarely scented, with both cup and petals or cup alone replaced by petaloid structures. Some have smaller flowers, produced in clusters of 4 or more, which are often sweetly fragrant. Spring- or early summer-flowering.
Div.5 Triandrus – nodding flowers, with short, sometimes straight-sided cups and narrow, reflexed petals, are borne 2–6 per stem. Spring-flowering.
Div.6 Cyclamineus – flowers are borne usually 1 or 2 per stem, each with a cup sometimes flanged and often longer than those of Div.5. Petals are narrow, pointed and reflexed. Early to mid-spring-flowering.
Div.7 Jonquil – sweetly scented flowers are borne usually 2 or more per stem. Cup is short, sometimes flanged; petals are often flat, fairly broad and rounded. Spring-flowering.
Div.8 Tazetta – sweetly fragrant flowers of small-flowered cultivars are borne in clusters of 12 or more per stem; large-flowered cultivars have 3 or 4 flowers per stem. All have a small, often straight-sided cup and broad, mostly pointed petals. Late autumn- to mid-spring-flowering. Most are frost to half hardy. Autumn-flowering hybrids provide valuable cut flowers; 'prepared' bulbs may be grown in pots for mid-winter flowering.
Div.9 Poeticus – flowers, sometimes borne 2 per stem, may be sweetly fragrant. Each has a small, coloured cup and glistening white petals. Many Poeticus hybrids are categorized as Div.3 or 8. Late spring- or early summer-flowering.
Div.10 Wild species – usually solitary flowers vary from diminutive, hoop-petticoat daffodils, with funnel-shaped, flanged cups, to stately trumpet daffodils. Early autumn- to early summer-flowering.
Div.11 Split-cupped – usually solitary flowers each have a cup that is typically split for more than half of its length; the number of splits varies. Cup segment edges lie back on petals and may be ruffled. Spring-flowering.
Div.12 Miscellaneous – a miscellaneous category containing hybrids with varying, intermediate flower shapes. Hybrids of hoop-petticoat species are also placed here. Autumn- to spring-flowering.

N. 'Acropolis' (illus. p.361), Div.4. Mid- to late spring-flowering bulb. H 42cm (17in). Large, double flowers have white, outer petals and petaloids; white, inner petals are interspersed with shorter, orange-red petaloids. Is suitable for exhibition.
♚ *N.* 'Actaea' (illus. p.361), Div.9. Late spring-flowering bulb. H 40cm (16in). Fragrant flowers have glistening white petals and shallow, flanged, rich lemon cups with narrow, orange-red rims.
N. 'Aircastle' (illus. p.360), Div.3. Mid-spring-flowering bulb. H 40cm (16in). Flowers have white petals that age greenish; small, flat, lemon-yellow coronas deepen in colour at the rim.
N. albus var. *plenus-odoratus*. See *N. poeticus* 'Flore Pleno'.
N. 'Altruist' (illus. p.362), Div.3. Mid-spring flowering bulb. H 45cm (18in). Medium-sized flowers have pointed, pale apricot petals and small, fluted, bright red cups. Fades rapidly in sunlight.
N. 'Ambergate' (illus. p.362), Div.2. Mid-spring-flowering bulb. H 45cm (18in). Flowers each have a shallow, widely expanded, fiery scarlet cup and soft tangerine petals.
♚ *N.* 'Arctic Gold', Div.1. Mid-spring-flowering bulb. H 40cm (16in). Rich golden-yellow flowers have broad petals and well-proportioned, flanged trumpets with neatly serrated rims. Is suitable for exhibition.
N. assoanus, syn. *N. juncifolius*, *N. requienii*, Div.10. Mid-spring-flowering bulb. H 15cm (6in). Is similar to *N. jonquilla*, but has thin, cylindrical leaves and rounded, bright clear yellow flowers with a sweet, slightly lemony fragrance. Thrives in sunny, gritty soil.
N. asturiensis, syn. *N. minimus*, Div.10. Late winter- or early spring-flowering bulb. H 8cm (3in). Small, lemon flowers have waisted trumpets and slender petals. Prefers full sun.
N. 'Avalanche' (illus. p.361), Div.8. Mid-spring-flowering bulb. H 35cm (14in). Has up to 10 small, sweetly scented flowers with white petals and small, lemon-yellow cups. Leaves are strap-shaped.
N. 'Bartley' (illus. p.362), Div.6. Early spring-flowering bulb. H 35cm (14in). Long, slender, golden flowers have reflexed petals and narrow, angled trumpets. Flowers are long-lasting.
N. 'Binkie' (illus. p.362), Div.2. Early spring-flowering bulb. H 30cm (12in). Flowers open clear pale lemon but cups turn sulphur-white with ruffled, lemon rims.
N. 'Birma', Div.3. Mid-spring-flowering bulb. H 45cm (18in). Flowers have soft yellow petals and fiery orange cups with heavily ruffled rims.
N. 'Bob Minor' (illus. p.362), Div.1. Robust, early spring-flowering bulb. H 20cm (8in). Small, rich golden flowers have twisted petals and sturdy trumpets.
N. 'Bridal Crown' (illus. p.361), Div.4. Late spring-flowering bulb. H 40cm (16in). Long-lasting, small, sweetly scented flowers are semi-double, with rounded, milk-white petals and white petaloids interspersed with shorter, saffron-orange ones towards centre.
N. 'Broadway Star' (illus. p.361), Div.11. Mid-spring-flowering bulb. H 38cm (15in). Double, sweetly-scented flowers are white with short, orange-cup segments clustered in the centre. Is good for forcing.
N. 'Brunswick', Div.2. Early spring-flowering bulb. H 40cm (16in). Long-lasting flowers have white petals and long, flared, primrose cups, which fade to lemon, with darker rims. Foliage is a striking bluish-green. Is suitable for cutting.
N. bulbocodium subsp. *bulbocodium* (Hoop-petticoat daffodil), Div.10. Vigorous, spring-flowering bulb. H 8–15cm (3–6in). Flowers are golden-yellow with conical cups and narrow, pointed petals. Thrives in moist turf in full sun. var. *citrinus*, H 15cm (6in), has slender, dark green leaves and clear pale lemon flowers.
N. campernellii. See *N.* x *odorus*.
N. canaliculatus, syn. *N. tazetta* 'Canaliculatus' (illus. p.361), Div.8. Mid-spring-flowering bulb. H 23cm (9in). Frost hardy. Produces a cluster of 4 or more fragrant flowers per stem, each with reflexed, white petals and a shallow, straight-sided, dark yellow cup.
N. cantabricus, Div.10. Spring- and sometimes late autumn-flowering bulb. H 10cm (4in). Is similar in form to *N. bulbocodium* subsp. *bulbocodium*, but is less robust. Flowers are milk- or ice-white. Thrives in an alpine house or greenhouse.
N. 'Cantatrice', Div.1. Mid-spring-flowering bulb. H 40cm (16in). Flowers have pure white petals and slender, milk-white trumpets.
N. 'Capax Plenus'. See *N.* 'Eystettensis'.
N. 'Cassata' (illus. p.361), Div.11. Mid-spring-flowering bulb. H 40cm (16in). Cups are soft primrose and distinctly split into segments with ruffled margins, while petals are broad and milk-white.
N. 'Chanterelle' (illus. p.362), Div.11. Mid-spring-flowering bulb. H 40cm (16in). Flowers have creamy-white petals and bright yellow split cups.
♚ *N.* 'Charity May' (illus. p.362), Div.6. Early to mid-spring-flowering bulb. H 30cm (12in). Small, pale lemon flowers each have broad, reflexed petals and slightly darker cups.
N. 'Cheerfulness' (illus. p.360), Div.4. Late spring-flowering bulb. H 40cm (16in). Is similar to *N.* 'Bridal Crown', but has fully double flowers with milk-white petals and petaloids interspersed with shorter, orange-yellow ones at the centre. Is excellent for cutting.
N. 'Cool Crystal' (illus. p.360), Div.3. Mid-spring-flowering bulb. H 50cm (20in). Has large, white flowers with bowl-shaped cups, each with a green eye.
♚ *N. cyclamineus* (illus. p.362), Div.10. Late winter- to early spring-flowering bulb. H 15cm (6in). Slender, nodding, clear gold flowers have narrow, reflexed petals and long, slender, flanged, waisted trumpets.
N. 'Daydream' (illus. p.362), Div.2. Mid-spring-flowering bulb. H 35cm (14in). Long-lasting, rounded flowers have broad, rich sulphur-yellow petals, each with a halo of near white, and flared, lemon-yellow cups that turn sulphur-white.
♚ *N.* 'Dove Wings' (illus. p.361), Div.6. Mid-spring-flowering bulb. H 30cm (12in). Has small flowers with milk-white petals and fairly long, soft primrose cups.

N. **'Dover Cliffs'** (illus. p.360), Div.2. Mid-spring-flowering bulb. H 40cm (16in). Large, well-proportioned flowers are white; petals tend to lean forwards over the corona.

♈ *N.* **'Empress of Ireland'** (illus. p.360), Div.1. Mid-spring-flowering bulb. H 40cm (16in). Large, white flowers have very broad, overlapping, triangular petals and narrow trumpets.

N. **'Eystettensis'**, syn. *N.* **'Capax Plenus'** (Queen Anne's double daffodil), Div.4. Mid-spring-flowering bulb. H 20cm (8in). Dainty, double flowers are composed of pointed, soft pale primrose petaloids neatly arranged in whorls.

♈ *N.* **'February Gold'** (illus. p.362), Div.6. Early spring-flowering bulb. H 32cm (13in). Solitary long-lasting flowers have clear golden petals and long, flanged, slightly darker trumpets. Is useful for borders and naturalizing.

N. **'February Silver'** (illus. p.361), Div.6. Robust, early spring-flowering bulb. H 32cm (13in). Has long-lasting flowers with milk-white petals and long, sturdy, nodding trumpets that open rich lemon and age to creamy-yellow.

N. **'Fortune'** (illus. p.362), Div.2. Early to mid-spring-flowering bulb. H 40cm (16in). Flowers have ribbed, dark lemon petals and bold, flared, copper-orange cups; they are good for cutting.

N. **'Foxfire'**, Div.2. Mid-spring-flowering bulb. H 35cm (14in). Has very rounded flowers with conspicuously white petals. Small, greenish-cream cups each have a small, green eye zone and a coral-orange rim.

N. **'Golden Ducat'** (illus. p.362), Div.4. Mid-spring-flowering bulb. H 38cm (15in). Produces variable, sometimes poorly formed, double, rich golden flowers. Is suitable for cutting.

N. **'Grand Primo Citronière'**, Div.8. Late autumn- to early spring-flowering bulb. H 32cm (13in). half hardy. Bears 8 or more fragrant flowers, each with milk-white petals and a clear lemon cup, which fades to cream. 'Treated' bulbs may be forced for mid-winter flowering. Is good for cutting.

N. **'Hawera'** (illus. p.362), Div.5. Mid-spring-flowering bulb. H 20cm (8in). Nodding flowers are a delicate lemon-yellow. Requires a sunny position. Makes a good pot plant.

N. **'Home Fires'** (illus. p.362), Div.2. Early spring-flowering bulb. H 45cm (18in). Flowers each have pointed, rich lemon petals and an orange-scarlet cup with a lobed and frilled rim.

♈ *N.* **'Ice Follies'** (illus. p.360), Div.3. Early spring-flowering bulb. H 40cm (16in). Rather coarse flowers have milk-white petals and very wide, almost flat, primrose-yellow cups, fading to cream.

N. **'Irene Copeland'** (illus. p.360), Div.4. Mid-spring-flowering bulb. H 35cm (14in). Bears large, fully double flowers of neatly arranged, milk-white petaloids interspersed with shorter, pale creamy-yellow ones. Is excellent for cutting.

N. **'Jack Snipe'** (illus. p.361), Div.6. Sturdy, early to mid-spring-flowering bulb. H 23cm (9in). Long-lasting, milk-white flowers are similar to those of *N.* 'Dove Wings', but have narrower petals with incurved margins and medium-length cups of rich dark lemon-yellow.

♈ *N.* **'Jenny'**, Div.6. Early to mid-spring-flowering bulb. H 30cm (12in). Bears long-lasting flowers, each with milk-white petals and a medium-length, flanged, soft lemon trumpet that turns creamy-white.

N. jonquilla (illus. p.362), Div.10. Mid-spring-flowering bulb. H 30cm (12in). Richly fragrant flowers are borne in a cluster of 6 or more; each has tapering, yellow petals and a shallow, dark gold cup. Distinctive foliage is dark, shining and grooved. **'Flore Pleno'** (Queen Anne's jonquil) has loosely double flowers; broad, incurved, yellow petals are interspersed with short, darker ones.

N. **'Jumblie'** (Jonquil; illus. p.362), Div.6. Early spring-flowering bulb. H 20cm (8in). Bears 2 or 3 long-lasting flowers, each with broad, golden petals and a sturdy, flanged, orange-yellow cup. Is ideal as a pot plant.

N. juncifolius. See *N. assoanus.*

N. **'Kilworth'** (illus. p.361), Div.2. Vigorous, late spring-flowering bulb. H 38cm (15in). Flowers have pointed, milk-white petals and dark reddish-orange cups with green eyes. Is effective in large groups.

♈ *N.* **'Kingscourt'** (illus. p.362), Div.1. Sturdy, mid-spring-flowering bulb. H 42cm (17in). Flowers have flanged, flared, rich gold trumpets with broad, rounded, paler gold petals.

N. **'Lemon Glow'** (illus. p.361), Div.1. Mid-spring-flowering bulb. H 45cm (18in). Has pale primrose-yellow flowers with broad trumpets that have conspicuously recurved, lobed, clear lemon-yellow rims.

N. **'Liberty Bells'** (illus. p.362), Div.5. Sturdy, mid-spring-flowering bulb. H 32cm (13in). Flowers are slightly fragrant and clear lemon.

N. **'Little Beauty'** (illus. p.361), Div.1. Early spring-flowering bulb. H 10cm (4in). Flowers have milk-white petals and bright lemon trumpets. Is suitable for naturalizing.

♈ *N.* **'Merlin'** (illus. p.361), Div.3. Mid-spring-flowering bulb. H 35cm (14in). Flowers have broad, rounded, glistening white petals and relatively large, almost flat, rich gold cups, each with a small, green eye and a broad, lightly ruffled, orange-red rim. Is excellent for exhibition.

N. minimus. See *N. asturiensis.*

N. **'Minnow'**, Div.8. Robust, early to mid-spring-flowering bulb. H 18cm (7in). Frost hardy. Has a cluster of 4 or more fragrant flowers per stem, each with rounded, creamy-yellow petals and a lemon cup. Increases freely. Is suitable for a container or rock garden.

N. minor, syn. *N. nanus* of gardens (illus. p.362), Div.10. Early spring-flowering bulb. H 20cm (8in). Flowers have slightly overlapping, soft yellow petals and almost straight, darker yellow trumpets with frilled rims.

N. minor of gardens see *N. pumilus.* var. *pumilus* see *N. pumilus.*

N. **'Mount Hood'** (illus. p.360), Div.1. Vigorous, mid-spring-flowering bulb. H 42cm (17in). Long-lasting flowers have milk-white petals and bold, flanged, creamy-yellow, then milk-white trumpets with reflexed, serrated rims.

N. nanus, Div.10. Early spring-flowering bulb. H 12cm (5in). Flowers each have twisted, cream petals and a stout, straight, dull yellow trumpet with a frilled rim. Leaves are broad. Is suitable for naturalizing.

N. nanus of gardens. See *N. minor.*

N. obvallaris (Tenby daffodil), Div.10. Sturdy, early spring-flowering bulb. H 30cm (12in). Gold flowers have short petals and broad trumpets and are borne on stiff stems.

N. x *odorus*, syn. *N. campernellii* (Campernelle jonquil), Div.10. Robust, mid-spring-flowering bulb. H 20–30cm (8–12in). Has usually 2 richly fragrant, dark gold flowers. ♈ **'Rugulosus'** (illus. p.362), H 28cm (11in), is more vigorous and produces up to 4 small-cupped, rich gold flowers.

N. **'Panache'** (illus. p.360), Div.1. Mid-spring-flowering bulb. H 45cm (18in). White flowers are tinged with green where the broad, overlapping petals unite with the trumpets.

N. **'Paper White'**, syn. *N. papyraceus* 'Grandiflorus', *N.p.* 'Paper White Snowflake', Div.8. Winter- to mid-spring-flowering bulb. H 35cm (14in). Half hardy. Produces 10 or more long-lived, heavily fragrant, star-shaped, glistening white flowers, each with long, spreading petals and a small, flanged cup containing conspicuous, saffron-yellow stamens. Produces flowers continuously through winter indoors.

N. papyraceus **'Grandiflorus'**. See *N.* 'Paper White'. **'Paper White Snowflake'**. See *N.* 'Paper White'.

♈ *N.* **'Passionale'** (illus. p.360), Div.2. Mid-spring-flowering bulb. H 40cm (16in). Each flower has milk-white petals and a long, flanged, apricot-tinged, pink cup.

N. **'Pencrebar'** (illus. p.362), Div.4. Mid-spring-flowering bulb. H 18cm (7in). Fragrant flowers are small, rounded and fully double, often in pairs. Outer petaloids and large, inner ones are pale gold and are evenly interspersed with darker ones.

N. **'Pipit'** (illus. p.362), Div.7. Mid-spring-flowering bulb. H 10cm (4in). Fragrant, greenish-yellow flowers have cups that age to white.

N. poeticus (Poet's daffodil, Poet's narcissus), Div.10. Variable, late spring-flowering bulb. H 22–42cm (9–17in). Each fragrant flower comprises glistening white petals and a small, shallow, yellow or orange cup with a red rim. Is ideal for naturalizing in moist turf although slow to establish. **'Flore Pleno'**, syn. *N. albus* var. *plenus-odoratus*, H 40cm (16in), has loosely double, pure white flowers, with inconspicuous, greenish-yellow or orange centres, in late spring or early summer. Is good for cutting.

♈ var. *recurvus* (Pheasant's eye; illus. p.361), H 42cm (17in), bears larger, long-lasting flowers with strongly swept-back petals and very shallow, greenish-yellow cups, with crimson rims, in early summer. **'Praecox'** (illus. p.361), similar, but flowers about 4 weeks earlier.

N. **'Portrush'** (illus. p.360), Div.3. Late spring- to early summer-flowering bulb. H 35cm (14in). Produces small flowers, each with green-tinged, glistening milk-white petals and a small, shallow, flanged, creamy-white cup with a bright green eye.

N. **'Pride of Cornwall'** (illus. p.361), Div.8. Mid-spring-flowering bulb. H 38cm (15in). Bears large, fragrant flowers, each with milk-white petals and a rich yellow cup shading to an orange-red rim outside. Is excellent for cutting.

N. pseudonarcissus (Lent lily, Wild daffodil; illus. p.361), Div.10. Extremely variable, early spring-flowering bulb. H 15–30cm (6–12in). Nodding flowers have overlapping, straw-yellow petals and large, darker yellow trumpets. Is ideal for naturalizing.

N. pumilus, syn. *N. minor* of gardens, *N. minor* var. *pumilus*, Div.10. Early spring-flowering bulb. H 15–22cm (6–9in). Bears bright gold flowers with separated, slightly paler petals and large trumpets with lobed and frilled rims. **'Plenus'** see *N.* 'Rip van Winkle'.

N. **'Rainbow'** (illus. p.362), Div.2. Vigorous, mid-spring-flowering bulb. H 45cm (18in). White flowers have coronas that are broadly bonded with coppery-pink at the mouths.

N. requienii. See *N. assoanus.*

N. **'Rip van Winkle'**, syn. *N. pumilus* 'Plenus' (illus. p.362), Div.4. Early spring-flowering bulb. H 15cm (6in). Shaggy, double flowers have densely arranged, flat, tapering, greenish-lemon petals with incurving tips.

N. **'Rockall'** (illus. p.361), Div.3. Mid-spring-flowering bulb. H 50cm (20in). Produces large flowers that have overlapping, pointed, clear white petals and finely fluted, rich orange-red cups.

N. romieuxii (illus. p.361), Div.10. Early spring-flowering bulb. H 10cm (4in). Frost hardy, but is best grown in a frame or an alpine house. Is similar to *N. bulbocodium* subsp. *bulbocodium*, but each fragrant flower has a large, almost flat, flanged cup of glistening pale primrose.

N. rupicola (illus. p.362), Div.10. Mid-spring-flowering bulb. H 8cm (3in). Is similar to *N. assoanus*, but has more angled, bluish-green foliage and solitary less scented, lemon flowers, each with a 6-lobed cup.

♈ *N.* **'St Keverne'** (illus. p.362), Div.2. Sturdy, early to mid-spring-flowering bulb. H 42cm (17in). Solitary flowers have clear rich golden petals and slightly darker cups of almost trumpet proportions.

N. **'Salome'** (illus. p.361), Div.2. Sturdy, late spring-flowering bulb. H 40cm (16in). Produces long-lasting flowers with broad, rounded, milk-white petals and long, flared, warm apricot-pink cups that have lightly ruffled, gold-tinged rims.

N. **'Satin Pink'** (illus. p.360), Div.2. Mid-spring-flowering bulb. H 42cm (17in). Each flower has broad, ribbed,

milk-white petals and a long, barely flared, flanged, soft buff-pink cup of almost trumpet proportions.
N. 'Scarlet Gem' (illus. 362), Div.8. Mid-spring-flowering bulb. H 35cm (14in). Frost hardy. Produces 4–6 small, scented flowers with golden petals and scarlet or deep orange-red cups.
N. 'Sealing Wax' (illus. p.362), Div.2. Mid- to late spring-flowering bulb. H 45cm (18in). Petals are golden-yellow and cups are bold, fiery red; flowers are borne singly.
N. 'Silver Chimes' (illus. p. 360), Div.8. Sturdy, mid- to late spring-flowering bulb. H 32cm (13in). Frost hardy. Has up to 10 fragrant flowers, each with broad, milk-white petals and a straight, shallow, creamy-primrose cup. Foliage is dark green. Thrives in a warm site.
N. 'Soleil d'Or', Div.8. Late autumn- to early spring-flowering bulb. H 35cm (14in). Half hardy. Flowers are sweetly scented with a dash of lemon. Each has rich golden petals and a clear tangerine cup. May be forced for mid-winter flowering, but staking is needed. Is excellent for cutting.
⚘ *N.* 'Spellbinder', Div.1. Early spring-flowering bulb. H 42cm (17in). Long-lasting, bright sulphur-yellow flowers each have a slender, flanged trumpet, reversing to palest sulphur-white inside, except for the lobed, rolled-back rim, which is tinged with lemon.
N. 'Stratosphere' (illus. p.362), Div.7. Mid-spring-flowering bulb. H 40cm (16in). Bears usually 3 fragrant flowers, each with rich golden petals and a darker gold cup. Is excellent for exhibition.
⚘ *N.* 'Suzy' (illus. p.362), Div.7. Robust, mid-spring-flowering bulb. H 38cm (15in). Produces 3–4 long-lasting, large, fragrant flowers, each with clear golden petals and a large, flanged, rich tangerine cup.
⚘ *N.* 'Sweetness' (illus. p.362), Div.7. Early spring-flowering bulb. H 38cm (15in). Each solitary fragrant, rich yellow flower has a slender, straight, flanged cup.
N. 'Tahiti' (illus. p.362), Div.4. Robust, mid-spring-flowering bulb. H 38cm (15in). Solitary loosely double flowers have golden petals and petaloids, interspersed with short, fiery orange, inner petaloids.
N. tazetta (Bunch-flowered daffodil, Polyanthus daffodil), Div.10. Extremely variable, late autumn- to mid-spring-flowering bulb. H 30–40cm (12–16in). Bears usually 12 or more fragrant flowers, generally with slender, white or yellow petals and shallow, white or yellow cups.
'Canaliculatus'. See *N. canaliculatus*.
⚘ *N.* 'Tête-à-Tête' (illus. p.362), Div.6. Early spring-flowering bulb. H 15–30cm (6–12in). Long-lasting flowers each have reflexed, rich golden petals and a square, flanged, warm yellowish-orange cup. Is very susceptible to viruses.
N. 'Thalia' (illus. p.360), Div.5. Vigorous, mid-spring-flowering bulb. H 38cm (15in). Has 3 or more long-lived, milk-white flowers per stem,

each with irregularly formed, often propeller-shaped petals and a flanged, bold cup.
N. 'Tresamble', Div.5. Sturdy, early spring-flowering bulb. H 40cm (16in). Bears up to 6 flowers per stem, each with milk-white petals and a flanged, paler-rimmed, creamy-white cup.
N. 'Trevithian', Div.7. Vigorous, early to mid-spring-flowering bulb. H 45cm (18in). Produces 2 or 3 large, fragrant flowers, rounded and soft primrose, each with broad petals and a short cup.
N. triandrus (Angel's tears; illus. p.361), Div.10. Early spring-flowering bulb. H 12cm (5in). Bears nodding, milk-white flowers, each with narrow, reflexed petals and a fairly long, straight-sided cup. Makes a good container plant.
N. 'Trousseau', Div.1. Early spring-flowering bulb. H 42cm (17in). Flowers each have milk-white petals and a straight, flanged, soft lemon trumpet, with a flared, lobed rim turning rich creamy-buff tinged with pale pink.
N. 'Tudor Minstrel', Div.2. Mid-spring-flowering bulb. H 42cm (17in). Petals are white and pointed. Chrome-yellow cups are slender and flanged.
N. 'Waterperry', Div.1. Mid-spring-flowering bulb. H 25cm (10in). Flowers have dull creamy-white petals; lightly flanged, spreading, primrose cups turn rich buff-yellow, shading to pinkish-apricot rims.
N. watieri (illus. p.360), Div.10. Mid-spring-flowering bulb. H 10cm (4in). Half hardy. Produces relatively large, fragrant, crystalline-textured, white flowers with shallow, lobed cups. Foliage is greyish-blue and angled.
N. 'W.P. Milner', Div.1. Early spring-flowering bulb. H 23cm (9in). Nodding flowers each have slender, twisted, light creamy-yellow petals and a flared, pale lemon trumpet, which fades to palest sulphur.
N. 'Woodland Star' (illus. p.361), Div.3. Mid-spring-flowering bulb. H 50cm (20in). Large flowers have white petals and small, bowl-shaped, deep red cups.
N. 'Yellow Cheerfulness' (illus. p.362), Div.4. Mid-spring-flowering bulb. H 40cm (16in). Produces 4 or more fragrant, loosely double flowers, each with soft primrose-yellow petals and petaloids, interspersed with shorter, yellowish-orange petaloids at the centre.

NAUTILOCALYX (Gesneriaceae)
Genus of evergreen, erect, bushy perennials, grown for their flowers and foliage. Frost tender, min. 15°C (59°F). Requires high humidity, partial shade and well-drained soil; avoid waterlogging, especially in winter. Propagate by stem cuttings in summer or by seed, if available, in spring.
N. bullatus, syn. *N. tessellatus*. Evergreen, erect, bushy perennial. H and S 60cm (2ft). Narrowly oval, wrinkled leaves, to 23cm (9in) long, are dark green with a bronze sheen above, reddish-green beneath. Clusters of small, tubular, white-haired, pale yellow flowers are produced in the leaf axils mainly in summer.

N. lynchii illus. p.269.
N. tessellatus. See *N. bullatus*.

Neanthe bella. See *Chamaedorea elegans*.

NECTAROSCORDUM (Liliaceae)
Genus of flowering bulbs, related to *Allium*, with long, linear, erect leaves. Exudes a very strong onion smell when bruised. Stems with erect, shuttlecock-like seed heads may be dried for winter decoration. Frost hardy. Needs partial shade. Grow in rough grass or borders in any soil that is neither too dry nor waterlogged. Propagate by freely produced offsets in late summer or by seed in autumn.
N. dioscoridis. See *N. siculum* subsp. *bulgaricum*.
N. siculum subsp. *bulgaricum*, syn. *N. dioscoridis*, illus. p.342.

NEILLIA (Rosaceae)
Genus of deciduous shrubs, grown for their graceful habit and profuse clusters of small flowers. Fully hardy. Requires sun or semi-shade and fertile, well-drained soil. Established plants benefit from having some older shoots cut to base after flowering. Propagate by softwood cuttings in summer or by suckers in autumn.
N. sinensis. Deciduous, arching shrub. H and S 2m (6ft). Has peeling bark and oval, sharply toothed, mid-green leaves. Nodding racemes of small, tubular, pinkish-white flowers are borne in late spring and early summer.
N. thibetica illus. p.110.

NELUMBO (Nelumbonaceae)
Genus of deciduous, perennial, marginal water plants, grown for their foliage and flowers. Half hardy to frost tender, min. 1–7°C (34–45°F). Needs an open, sunny position and 60cm (24in) depth of water. Remove fading foliage; flowers may be left to develop into decorative seed pods. Divide overgrown plants in spring. Propagate species by seed in spring, selected forms by division in spring.
N. lutea (American lotus). Vigorous, deciduous, perennial, marginal water plant. H and S 1m (3ft). Half hardy. Rounded, blue-green leaves develop on stout, 30–60cm (1–2ft) long stems. Large, chalice-shaped, yellow flowers open in summer.
N. nucifera illus. p.387. 'Alba Striata' is a vigorous, deciduous, perennial, marginal water plant. H 2.5m (8ft), S 1.2m (4ft). Frost tender, min. 1°C (34°F). Sturdy stems carry very large, rounded, blue-green leaves. Large, fragrant, chalice-shaped, white flowers, 15cm (6in) across and edged with crimson, are borne in summer. 'Alba Grandiflora' bears pure white flowers and 'Rosa Plena' produces double, soft pink flowers to 30cm (12in) across.

NEMATANTHUS (Gesneriaceae)
Genus of perennials and soft-stemmed, evergreen shrubs, grown for their flowers and foliage. Frost tender, min. 13–15°C (55–9°F). Requires partial shade and humus-rich, moist but well-drained soil. Water potted specimens moderately, allowing soil almost to dry

out between applications. Tip prune young plants to stimulate branching. Propagate by softwood or greenwood cuttings in summer.
N. gregarius, syn. *N. radicans*, *Hypocyrta radicans*, illus. p.127.
N. radicans. See *N. gregarius*.
N. strigillosus, syn. *Hypocyrta strigillosa*. Evergreen, prostrate shrub. H 15–30cm (6–12in), S 60cm–1m (2–3ft). Elliptic, slightly cupped leaves are clothed in dense down. Small, tubular, orange or orange-red flowers appear in leaf axils mainly from spring to autumn.

NEMESIA (Scrophulariaceae)
Genus of annuals, perennials and evergreen sub-shrubs, grown for summer bedding and as greenhouse plants. Half hardy. Prefers sun and fertile, well-drained soil. Cut back stems after flowering. Pinch out growing shoots of young plants to ensure a bushy habit. Propagate by seed sown under glass in early spring, or outdoors in late spring.
N. strumosa. Fast-growing, bushy annual. H 20–45cm (8–18in), S 15cm (6in). Has lance-shaped, serrated, pale green leaves and, in summer, trumpet-shaped, yellow, white or purple flowers, 2.5cm (1in) across, that are suitable for cutting. **Carnival Series** illus. p.280. **Triumph Series** illus. p.291.

NEMOPHILA (Hydrophyllaceae)
Genus of annuals, useful for rock gardens and for edging. Fully hardy. Grow in sun or semi-shade and in fertile, well-drained soil. Propagate by seed sown outdoors in spring or early autumn. Is prone to aphids.
N. insignis. See *N. menziesii*.
N. maculata illus. p.271.
N. menziesii, syn. *N. insignis*, illus. p.285.

NEOLITSEA (Lauraceae)
Genus of evergreen trees and shrubs, grown for their foliage. Frost to half hardy. In cold areas needs shelter from strong winds; does best against a south- or west-facing wall. Requires sun or semi-shade and fertile, well-drained soil. Propagate by semi-ripe cuttings in late summer.
N. sericea. Evergreen, broadly conical, dense tree or shrub. H and S 6m (20ft). Frost hardy. Narrowly oval, pointed leaves are glossy, mid-green above, white beneath and, when young, are densely covered with silky, brown hairs. Small, star-shaped, yellow flowers are borne in autumn.

NEOLLOYDIA (Cactaceae)
Genus of spherical to columnar, perennial cacti with dense spines and short tubercles in spirals. Most species are exceptionally difficult to cultivate unless grafted. Frost tender, min. 10°C (50°F). Needs full sun and well-drained soil. Water sparingly from spring to autumn; keep dry in winter. Propagate by seed in spring or summer.
N. conoidea illus. p.411.
N. mcdowellii. See *Thelocactus mcdowellii*.

549

NEOMARICA (Iridaceae)
Genus of evergreen, summer-flowering, iris-like, rhizomatous perennials with clusters of short-lived flowers. Frost tender, min. 10°C (50°F). Needs partial shade and fertile, moist, preferably humus-rich soil. Water freely in summer; reduce water in winter but do not dry out. Propagate by seed in spring or by division in spring or summer.
N. caerulea illus. p.346.

NEOPANAX. See **PSEUDOPANAX**.

NEOPORTERIA (Cactaceae)
Genus of spherical to columnar, perennial cacti. Egg-shaped, red, brown or green seed pods are similar to those of *Wigginsia*. Frost tender, min. 8°C (46°F). Requires full sun and very well-drained soil. Propagate by seed in spring or summer.
N. chilensis. Spherical, then columnar, perennial cactus. H 30cm (12in), S 10cm (4in). Pale green stem has a dense covering of stout, golden spines of varying lengths. Crown bears flattish, pink-orange or white flowers, to 5cm (2in) across, in summer.
N. napina. Flattened spherical, perennial cactus. H 3cm (1¼in), S 5cm (2in). Has very short, black spines pressed flat against a chocolate-brown stem. In summer, crown produces flattish, yellow flowers, 5cm (2in) across. var. *mitis* illus. p.407.
N. nidus. Spherical to columnar, perennial cactus. H 10cm (4in), S 8cm (3in). Long, soft, grey spines completely encircle a dark green-brown stem. Crown produces tubular, pink to cerise flowers that are 3–5cm (1¼–2in) long, with paler bases, and only open at tips, in spring or autumn.
N. subgibbosa. Spherical to columnar, perennial cactus. H 30cm (12in), S 10cm (4in). Light green to dark grey-green stem bears large, woolly areoles and stout, amber spines. In late summer, crown bears flattish, carmine-pink flowers, 4cm (1½in) across.
N. villosa illus. p.406.

NEOREGELIA (Bromeliaceae)
Genus of evergreen, rosette-forming, epiphytic perennials, grown for their overall appearance. Frost tender, min. 10°C (50°F). Requires semi-shade and a rooting medium of equal parts humus-rich soil and either sphagnum moss or bark or plastic chips used for orchid culture. Using soft water, water moderately during growing season, sparingly at other times, and keep rosette centres filled with water from spring to autumn. Propagate by offsets in spring or summer.
N. carolinae, syn. *Aregelia carolinae, Nidularium carolinae* (Blushing bromeliad). Evergreen, spreading, basal-rosetted, epiphytic perennial. H 20–30cm (8–12in), S 40–60cm (16–24in). Strap-shaped, finely spine-toothed, lustrous, bright green leaves are produced in dense rosettes. A compact cluster of tubular, blue-purple flowers, surrounded by red bracts, is borne at the heart of each mature rosette, usually in summer.
♥ **'Tricolor'** (syn. *N.c.* f. *tricolor*; illus. p.229) has leaves, striped with ivory-white, that flush pink with age.
N. concentrica (illus. p.229). Evergreen, spreading, basal-rosetted, epiphytic perennial. H 20–30cm (8–12in), S to 70cm (28in). Very broadly strap-shaped to oval, glossy, dark green leaves, with spiny, black teeth and usually with dark blotches, are produced in dense rosettes. In summer, a compact cluster of tubular, pale blue flowers, surrounded by pinkish-lilac bracts, is produced at the heart of each mature rosette. **'Plutonis'** has bracts flushed with red.

NEPENTHES (Nepenthaceae)
Pitcher plant
Genus of evergreen, insectivorous, mostly epiphytic perennials, with leaves adapted to form pendulous, lidded, coloured pitchers that trap and digest insects. Is suitable for hanging baskets. Frost tender, min. 18°C (64°F). Requires a humid atmosphere, partial shade and moist, fertile soil with added peat and moss. Propagate by seed in spring or by stem cuttings in spring or summer.
N. x *hookeriana* illus. p.228.
N. rafflesiana. Evergreen, epiphytic, insectivorous perennial. H 3m (10ft), S 1–1.2m (3–4ft). Has lance-shaped, dark green leaves. Greenish-yellow pitchers, to 25cm (10in) long, are mottled purple and brown and have spurred lids. Inconspicuous, green flowers are borne in racemes and produced intermittently.

NEPETA (Labiatae)
Catmint
Genus of summer-flowering perennials, useful for edging, particularly where they can tumble over paving. Fully hardy. Prefers sun and moist but well-drained soil. Propagate by division in spring or by stem-tip or softwood cuttings in spring or summer, species only by seed in autumn.
N. **'Blue Beauty'.** See *N.* 'Souvenir d'André Chaudon'.
N. x *faassenii* illus. p.250.
N. govaniana illus. p.218.
N. grandiflora. Neat, erect perennial. H 40–80cm (16–32in), S 45–60cm (18–24in). Has slightly hairy stems, oval, round-toothed, light green leaves, with heart-shaped bases, and, in summer, racemes of small, hooded, blue flowers.
N. nervosa. Clump-forming perennial. H 35cm (14in), S 30cm (12in). Forms a mound of narrowly oblong to lance-shaped, pointed, prominently veined, mid-green leaves. Produces dense racemes of small, tubular, pale blue flowers from early to mid-summer.
N. **'Souvenir d'André Chaudon'**, syn. *N.* 'Blue Beauty'. Spreading, clump-forming perennial. H and S 45cm (18in). Tubular, blue flowers are borne throughout summer above oval to lance-shaped, toothed, grey leaves.

NEPHROLEPIS (Polypodiaceae)
Genus of evergreen or semi-evergreen ferns. Frost tender, min. 5°C (41°F). Needs a shady position. Prefers moist soil, but is extremely tolerant of both drought and waterlogging. Remove fading fronds and divide regularly. Propagate by division in summer or early autumn.
N. cordifolia (Ladder fern, Sword fern). Semi-evergreen fern. H 45cm (18in), S 30cm (12in). Has narrowly lance-shaped, arching, dark green fronds with rounded, finely serrated pinnae.
N. exaltata illus. p.188.

NEPHTHYTIS (Araceae)
Genus of evergreen, tufted perennials, with horizontal, creeping rhizomes, grown for their foliage. Frost tender, min. 18°C (64°F). Requires a humid atmosphere, moist, humus-rich soil and partial shade. Propagate by division in spring or summer.
N. afzelii. Evergreen, creeping, rhizomatous perennial. H to 75cm (30in), S indefinite. Has tufts of arrow-shaped, lobed, dark green leaves, to 25cm (10in) long. Intermittently bears a hooded, greenish spathe, enclosing a green spadix, followed by spherical, orange fruits.
N. triphylla of gardens. See *Syngonium podophyllum*.

NERINE (Amaryllidaceae)
Genus of bulbs, some of which are semi-evergreen, grown for their spherical heads of wavy-petalled, pink to red, occasionally white, flowers. Most flower in autumn before leaves appear. Frost to half hardy. Needs full sun and light, sandy soil. Plant in early autumn. Dislikes being disturbed. Water until leaves die down, then dry off. Propagate by seed when fresh or divide offsets in autumn or when leaves have died down.
N. **'Bagdad'.** Autumn-flowering bulb. H 60cm (24in), S 15–20cm (6–8in). Half hardy. Leaves are strap-shaped, semi-erect and basal. Has crimson flowers, paler towards centres; long, narrow petals have recurved tips and crisped margins.
N. **'Blanchefleur'.** Autumn-flowering bulb. H 30–50cm (12–20in), S 15–20cm (6–8in). Half hardy. Produces strap-shaped, semi-erect, basal leaves and a tight head of 5–10 pure white flowers. Upper parts of petals are twisted.
♥ *N. bowdenii* illus. p.368. f. *alba* illus. p.368.
N. **'Brian Doe'** illus. p.368.
N. **'Corusca Major'.** Autumn-flowering bulb. H 60cm (24in), S 12–15cm (5–6in). Half hardy. Forms strap-shaped, semi-erect, basal leaves. Stout stem bears 10–15 scarlet-red flowers with narrow petals. Is useful for cutting.
N. filifolia. Autumn-flowering bulb. H to 25cm (10in), S 8–10cm (3–4in). Half hardy. Has thread-like, semi-erect leaves in a basal tuft. Slender stem produces pale pink flowers with narrow petals.
N. flexuosa. Semi-evergreen, autumn-flowering bulb. H 40–50cm (16–20in), S 12–15cm (5–6in). Half hardy. Bears strap-shaped, semi-erect, basal leaves and 10–15 pink flowers; each petal has a deeper pink mid-vein and a recurved, wavy upper half.
N. **'Fothergillii Major'.** Late summer- to early autumn-flowering bulb. H 45–60cm (18–24in), S 12–15cm (5–6in). Half hardy. Leaves are strap-shaped, semi-erect and basal. Very strong stem has about 10 bright scarlet-salmon flowers with recurved petals.
N. **'Margaret Rose'.** Autumn-flowering bulb. H 30–50cm (12–20in), S 15–20cm (6–8in). Half hardy. Bears strap-shaped, semi-erect, basal leaves. Stem has a tight head of bright rose-pink flowers with recurved petals.
N. masonorum. Autumn-flowering bulb. H 15–20cm (6–8in), S 8–10cm (3–4in). Half hardy. Produces thread-like, semi-erect leaves in a basal tuft. Stem bears pink flowers with very crisped petal margins.
N. **'Orion'** illus. p.368.
N. sarniensis illus. p.368.
N. undulata illus. p.368.

NERIUM (Apocynaceae)
Genus of evergreen shrubs, grown for their flowers. Frost tender, min. 10°C (50°F). Requires full sun and well-drained soil. Water containerized plants freely when in full growth, sparingly at other times. Tip prune young plants to promote branching. Propagate by seed in spring or by semi-ripe cuttings in summer.
N. oleander illus. p.91.

NERTERA (Rubiaceae)
Genus of creeping perennials, grown for their mass of spherical, bead-like fruits in autumn. Makes excellent alpine house plants. Half hardy. Needs a sheltered, semi-shaded site in gritty, moist but well-drained, sandy soil. Resents winter wet. Propagate in spring by seed, division or tip cuttings.
N. depressa. See *N. granadensis*.
N. granadensis, syn. *N. depressa,* illus. p.335.

NICANDRA (Solanaceae)
Genus of one species of annual with short-lived flowers. Fully hardy. Grow in sun and in rich, well-drained soil. Propagate by seed sown in spring.
N. physalodes (Apple of Peru, Shoo-fly). Fast-growing, upright, branching annual. H 1m (3ft), S 30cm (1ft) or more. Has oval, serrated, mid-green leaves. In summer and early autumn bears bell-shaped, white-throated, light violet-blue flowers, over 2.5cm (1in) wide, each lasting only one day. Spherical, green fruits, 5cm (2in) wide, are surrounded by purple and green calyces. Is thought to repel flies, hence its name.

NICOTIANA (Solanaceae)
Genus of annuals, perennials, that are usually grown as annuals, and semi-evergreen shrubs. Frost hardy to frost tender, min. 1°C (34°F). Needs sun and fertile, well-drained soil. Propagate annuals and perennials by seed in early spring, shrubs by seed in spring or by semi-ripe cuttings in summer.
N. affinis. See *N. alata*.
N. alata, syn. *N. affinis,* illus. p.203.
N. glauca. Semi-evergreen, upright shrub. H and S 2.5–3m (8–10ft). Half hardy. Stout, blue-grey shoots bear narrowly oval, fleshy, blue-grey

leaves. Showy, tubular, bright yellow flowers are produced in summer and early autumn.
❀ *N. langsdorfii* illus. p.286.
N. × *sanderae*. **'Crimson Rock'** is a fairly slow-growing, bushy annual. H 60cm (2ft), S 30cm (1ft). Half hardy. Oval leaves are mid-green. Evening-scented, trumpet-shaped, bright crimson flowers, to 8cm (3in) long, open throughout summer and early autumn. ❀ **Domino Series** illus. p.271. ❀ **'Lime Green'** bears racemes of open trumpet-shaped, greenish-yellow flowers that are fragrant at night in late summer and autumn. **Nicki Series**, H 38cm (15in), produces fragrant flowers in a good colour range that includes white, pink, red and purple. **Sensation Series** illus. p.278.
❀ *N. sylvestris* illus. p.190.

NIDULARIUM (Bromeliaceae)
Genus of evergreen, rosette-forming, epiphytic perennials, grown for their overall appearance. Frost tender, min. 10–15°C (50–59°F). Requires a postion in semi-shade and a rooting medium of equal parts humus-rich soil and either sphagnum moss or bark or plastic chips generally used for orchid culture. Using soft water, water moderately during the growing season, sparingly at other times, and keep centres of rosettes filled with water from spring to autumn. Propagate by offsets in spring or summer.
N. carolinae. See *Neoregelia carolinae.*
N. fulgens (Blushing bromeliad). Evergreen, spreading, basal-rosetted, epiphytic perennial. H 20cm (8in) or more, S 40–50cm (16–20in). Has dense rosettes of strap-shaped, spiny-toothed, arching, glossy, rich green leaves. Tubular, white-and-purple flowers, almost hidden in a rosette of bright scarlet bracts, are mainly produced in summer.
N. innocentii (Bird's-nest bromeliad). Evergreen, spreading, basal-rosetted, epiphytic perennial. H 20–30cm (8–12in), S 60cm (24in). Has dense rosettes of strap-shaped, prickle-toothed, arching, dark green, sometimes reddish-green leaves with reddish-purple undersides. Tubular, white flowers, partially hidden in a rosette of bright red bracts, appear mainly in summer.
N. procerum. Evergreen, spreading, basal-rosetted, epiphytic perennial. H 20–30cm (8–12in), S 50–75cm (20–30in). Strap-shaped, spiny-toothed, bright green leaves are produced in dense rosettes. Clusters of small, tubular, blue flowers are produced in summer.

NIEREMBERGIA (Solanaceae)
Genus of summer-flowering perennials, sometimes grown as annuals, and deciduous or semi-evergreen sub-shrubs. Frost to half hardy. Prefers sun and moist but well drained soil. Propagate by division in spring, by semi-ripe cuttings in summer or by seed in autumn.
N. caerulea. See *N. hippomanica* var. *violacea.*

N. hippomanica var. *violacea,* syn. *N. caerulea.* **'Purple Robe'** illus. p.283.
N. repens illus. p.322.

NIGELLA (Ranunculaceae)
Genus of annuals, grown for their attractive flowers, which are suitable for cutting, and their ornamental seed pods. Fully hardy. Grows best in sun and in fertile, well-drained soil. Dead-head plants to prolong flowering if seed heads are not required. Propagate by seed sown outdoors in spring or early autumn.
N. damascena (Love-in-a-mist). Fast-growing, upright annual. H 60cm (24in), S 20cm (8in). Has feathery, bright green leaves. Spurred, many-petalled, blue or white flowers appear in summer, followed by inflated, rounded, green, then brown seed pods that may be cut and dried. ❀ **'Miss Jekyll'** and **'Persian Jewels'** illus. p.285.

NOLANA (Nolanaceae)
Genus of annuals, useful for growing in hot, dry sites and rock gardens and as edging. Frost hardy. Grow in sun and in fertile, well-drained soil. Propagate by seed sown outdoors in spring.
N. atriplicifolia. See *N. paradoxa.*
N. grandiflora. See *N. paradoxa.*
N. paradoxa, syn. *N. atriplicifolia, N. grandiflora.* Moderately fast-growing, prostrate annual. H 8cm (3in), S 15cm (6in). Has oval, mid-green leaves and, in summer, funnel-shaped, purplish-blue flowers, to 5cm (2in) wide, that have white-zoned, yellow throats.

Nolina recurvata. See *Beaucarnea recurvata.*
Nolina tuberculata. See *Beaucarnea recurvata.*

NOMOCHARIS (Liliaceae)
Genus of bulbs with a lily-like habit and, in summer, loose spikes of flattish flowers that are often conspicuously spotted. Fully hardy. Requires partial shade and rich, well-drained soil with a high humus content. When in growth in summer needs moist but not water-logged soil. Is dormant throughout winter. Propagate by seed in winter or spring.
N. mairei. See *N. pardanthina.*
N. pardanthina, syn. *N. mairei,* illus. p.342.
N. saluenensis. Summer-flowering bulb. H 85cm (34in), S 12–15cm (5–6in). Leafy stems bear lance-shaped, scattered leaves. Produces a loose spike of 2–6 saucer-shaped, white or pink flowers, with dark purple eyes and purple spots.

NOPALXOCHIA (Cactaceae)
Genus of epiphytic, perennial cacti with flattened, strap-shaped stems. Is closely related to *Epiphyllum,* with which it hybridizes. Spines are insignificant. Stems may die back after flowering well. Frost tender, min. 10°C (50°F). Needs partial shade and rich, well-drained soil. Is easy to grow. Propagate by stem cuttings in spring or summer.

N. ackermannii (Red orchid cactus). Erect, then pendent, epiphytic, perennial cactus. H 30cm (1ft), S 60cm (2ft). Has fleshy, toothed, green stems, to 7cm (3in) across and 40cm (16in) long. Bears 15cm (6in) wide, funnel-shaped, red flowers in spring-summer along indented edges of stems.
❀ *N. phyllanthoides* **'Deutsche Kaiserin'**, illus. p.398.

NOTHOFAGUS (Fagaceae)
Southern beech
Genus of deciduous or evergreen trees, grown for their habit, foliage and, in the case of deciduous species, autumn colour. Has inconspicuous flowers in late spring. Fully to frost hardy. Needs sun or semi-shade and, as it is not very resistant to strong winds, should have the shelter of other trees. Prefers deep, fertile, moist but well-drained soil; is not suitable for shallow, chalky soil. Propagate by seed in autumn.
N. antarctica (Antarctic beech, Nirre). Deciduous, broadly conical tree, sometimes with several main stems. H 15m (50ft), S 10m (30ft). Fully hardy. Small, oval, crinkly-edged, glossy, dark green leaves turn yellow in autumn.
N. betuloides illus. p.48.
N. dombeyi illus. p.46.
N. menziesii (Silver beech). Evergreen, conical tree. H 20m (70ft), S 12m (40ft). Frost hardy. Has tiny, rounded, sharply toothed, glossy, dark green leaves.
N. obliqua illus. p.42.
N. procera illus. p.41.

NOTHOLIRION (Liliaceae)
Genus of summer-flowering bulbs, related to *Lilium,* grown for their funnel-shaped flowers. Frost hardy. Often produces early leaves, which may be damaged by spring frosts, so grow in a cool greenhouse in areas subject to alternating mild and cold periods in spring. Prefers partial shade or full sun and humus-rich, well-drained soil. Bulb dies after flowering. Propagate in spring or autumn by offsets, which take 2–3 years to reach flowering size. Alternatively propagate by seed in winter or spring.
N. campanulatum illus. p.344.

NOTHOPANAX. See **PSEUDOPANAX.**

Notocactus apricus. See *Parodia apricus.*
Notocactus graessneri. See *Parodia graessneri.*
Notocactus haselbergii. See *Parodia haselbergii.*
Notocactus leninghausii. See *Parodia leninghausii.*
Notocactus mammulosus. See *Parodia mammulosus.*
Notocactus ottonis. See *Parodia ottonis.*
Notocactus rutilans. See *Parodia rutilans.*
Notocactus scopa. See *Parodia scopa.*

NOTOSPARTIUM (Leguminosae)
Genus of leafless, summer-flowering shrubs, grown for their habit, green shoots and flowers. Frost hardy, but in

cold areas does best against a south- or west-facing wall. Requires a sheltered, sunny position and well-drained soil. Older plants may need staking. Propagate by semi-ripe cuttings in summer or by seed in autumn.
N. carmichaeliae (Pink broom). Leafless, arching shrub. H 2m (6ft), S 1.5m (5ft). Short, dense spikes of pea-like, purple-blotched, pink flowers are produced in mid-summer on slender, drooping, green shoots.

NUPHAR (Nymphaeaceae)
Genus of deciduous, perennial, deep-water plants, grown for their floating foliage and spherical flowers. Fully to frost hardy. Grows in shade or sun and in running or still water; is often grown for a water-lily effect in conditions where true water lilies would not thrive. Remove fading foliage and flowers, and periodically divide crowded plants. Propagate by division in spring.
N. advena (American spatterdock, Yellow pond lily). Deciduous, perennial, deep-water plant. S 1.2m (4ft). Fully hardy. Has broadly oval, floating, mid-green leaves; central ones are occasionally erect. Small, purple-tinged, yellow flowers in summer are followed by decorative seed heads.
N. lutea illus. p.391.

NUTTALLIA. See **OEMLERIA.**

NYMANIA (Aitoniaceae)
Genus of one species of evergreen, spring-flowering shrub, grown for its flowers and fruits. Frost tender, min. 7–10°C (45–50°F). Needs full light and fertile, well-drained soil. Water potted specimens moderately, less when not in full growth. Propagate by seed in spring or by semi-ripe cuttings in summer.
N. capensis illus. p.118.

NYMPHAEA (Nymphaeaceae)
Water lily
Genus of deciduous, summer-flowering, perennial water plants, grown for their floating, usually rounded leaves and brightly coloured flowers. Fully hardy to frost tender, min. 10°C (50°F). Needs an open, sunny position and still water. Remove fading foliage to prevent it polluting water. Plants have tuber-like rhizomes and require dividing and replanting in spring or early summer every 3 or 4 years. Most frost tender plants may be treated as annuals. May also be propagated by seed or by separating plantlets in spring or early summer.
N. alba (White water lily; illus. p.376). Deciduous, perennial water plant with floating leaves. S to 3m (10ft). Fully hardy. Has dark green leaves and, in summer, cup-shaped, semi-double, pure white flowers, 10cm (4in) across, with golden centres.
N. **'American Star'** (illus. p.390). Deciduous, perennial water plant with floating leaves. S to 1.2m (4ft). Frost hardy. Young leaves are purplish-green or bronze, maturing to bright green. Star-shaped, semi-double flowers, 10cm (4in) across, are deep pink and are held above water throughout summer.

551

N. **'Attraction'** (illus. p.390). Deciduous, perennial water plant with floating leaves. S to 2m (6ft). Fully hardy. Has dark green leaves. In summer bears cup-shaped, semi-double, garnet-red flowers, 15cm (6in) across and flecked with white.
N. **'Aurora'**. Deciduous, perennial water plant with floating leaves. S to 75cm (30in). Frost hardy. Olive-green leaves are mottled with purple. In summer has star-shaped, semi-double flowers, 5cm (2in) across, cream in bud, opening to yellow, then passing through orange to blood-red. Suits a small- to medium-sized pool.
N. **'Aviator Pring'**. Deciduous, perennial water plant with floating leaves. S to 1.5m (5ft). Frost tender. Mid-green leaves have toothed, wavy edges. Rounded, semi-double, bright yellow flowers, to 30cm (12in) across, are held above water throughout summer.
N. **'Blue Beauty'** (illus. p.390). Deciduous, perennial water plant with floating leaves. S to 2.5m (8ft). Frost tender. Leaves are brown-freckled, dark green above, purplish-green beneath. Fragrant, rounded, semi-double, deep blue flowers, to 30cm (12in) across, are produced in summer.
N. capensis (Cape blue water lily). Deciduous, perennial water plant with floating leaves. S to 2m (6ft). Frost tender. Large, mid-green leaves are often splashed with purple beneath. Star-shaped, semi-double, bright blue flowers, 15–20cm (6–8in) across, appear in summer.
N. caroliniana **'Nivea'** (illus. p.390). Deciduous, perennial water plant with floating leaves. S to 1.2m (4ft). Fully hardy. Has pale green leaves and, in summer, fragrant, cup-shaped, semi-double, pure white flowers, 10–15cm (4–6in) across.
N. **'Emily Grant Hutchings'**. Deciduous, perennial water plant with floating leaves. S to 1.2m (4ft). Frost tender. Has small, green leaves overlaid with bronze-crimson. Cup-shaped, semi-double, pinkish-red flowers, 15–20cm (6–8in) across, open at night in summer.
♛ *N.* **'Escarboucle'** (illus. p.390). Deciduous, perennial water plant with floating leaves. S to 3m (10ft). Fully hardy. Leaves are dark green. In summer has cup-shaped, semi-double, deep crimson flowers, 10–15cm (4–6in) across, with golden centres.

N. **'Fire Crest'** (illus. p.390). Deciduous, perennial water plant with floating leaves. S to 1.2m (4ft). Fully hardy. Dark green leaves are suffused with purple. In summer bears star-shaped, semi-double, deep pink flowers, 15–20cm (6–8in) across, with red-tipped stamens.
♛ *N.* **'Gladstoniana'** (illus. p.390). Deciduous, perennial water plant with floating leaves. S to 3m (10ft). Frost hardy. Leaves are mid-green. Star-shaped, semi-double, white flowers, 15–30cm (6–12in) across, appear in summer.
♛ *N.* **'Gonnère'**. Deciduous, perennial water plant with floating leaves. S to 1.5m (5ft). Fully hardy. Has bright pea-green leaves and, in summer, rounded, fully double, white flowers, 15–20cm (6–8in) across.
N. **'Green Smoke'**. Deciduous, perennial water plant with floating leaves. S to 2m (6ft). Frost tender. Bronze-green leaves have bronze speckling. Star-shaped, single flowers, 10–20cm (4–8in) across, are chartreuse, shading to blue.
♛ *N.* **'James Brydon'** (illus. p.390). Deciduous, perennial water plant with floating leaves. S to 2.5m (8ft). Frost hardy. Fragrant, peony-shaped, double, orange-suffused, crimson flowers, 15–20cm (6–8in) across, are borne in summer above dark green leaves.
N. **Laydekeri Group, 'Fulgens'** (illus. p.390). Deciduous, perennial water plant with floating leaves. S to 1m (3ft). Fully hardy. Dark green leaves have purplish-green undersides. Star-shaped, semi-double, bright crimson flowers, 5–10cm (2–4in) across, appear in summer.
N. **'Mme Auguste Tézier'**. Deciduous, perennial water plant with floating leaves. S to 1.5m (5ft). Frost tender. Purplish-green leaves are splashed and spotted with brown. In summer produces star-shaped, semi-double, lavender-blue flowers, 10–15cm (4–6in) across and each with a brown centre, that open at night.
N. **'Mme Wilfron Gonnère'**. Deciduous, perennial water plant with floating leaves. S to 1.5m (5ft). Frost hardy. Has mid-green leaves and, in summer, cup-shaped, almost fully double, white flowers, 15cm (6in) across, spotted with deep rose-pink.
N. **'Margaret Mary'**. Deciduous, perennial water plant with floating leaves. S to 45cm (18in). Half hardy. Leaves are dark green above, light brown beneath. Star-shaped, single, blue flowers, 5–8cm (2–3in) across, are borne year-round in frost-free conditions.
N. **Marliacea Group, 'Albida'** (illus. p.390). Deciduous, perennial water plant with floating leaves. S to 2m (6ft). Fully hardy. Deep green leaves have red or purplish-green undersides. Bears fragrant, cup-shaped, semi-double, pure white flowers, 15–20cm (6–8in) across, in summer. **'Carnea'** (illus. p.376) has dark green leaves and star-shaped, semi-double, soft pink flowers, 15–25cm (6–10in) across, with golden centres.
♛ **'Chromatella'** (illus. p.390) has olive-green leaves, heavily mottled with maroon and bronze, and cup-shaped, semi-double, canary-yellow flowers, 15–20cm (6–8in) across.
N. **'Midnight'**. Deciduous, perennial water plant with floating leaves. S to 1.2m (4ft). Frost tender. Has small leaves, dark green flecked with brown above, purple beneath. Bears star-shaped, semi-double, rich purple flowers, 10–15cm (4–6in) across, in summer.
N. **'Mrs George H. Pring'**. Deciduous, perennial water plant with floating leaves. S to 1.5m (5ft). Frost tender. Has large leaves, mid-green with reddish-brown splashes above, purplish-green beneath. Produces fragrant, star-shaped, semi-double, creamy-white flowers, 15–25cm (6–10in) across, throughout summer.
N. odorata **'Sulphurea Grandiflora'**. Deciduous, perennial water plant with floating leaves. S to 1m (3ft). Fully hardy. Dark green leaves are heavily mottled with maroon. Bears fragrant, star-shaped, semi-double, yellow flowers, 10–15cm (4–6in) across, throughout summer.
N. pygmaea. See *N. tetragona*.
N. **'Rose Arey'** (illus. p.390). Deciduous, perennial water plant with floating leaves. S to 1.5m (5ft). Frost hardy. Leaves are reddish-green. In summer produces star-shaped, semi-double, deep rose-pink flowers, 10–15cm (4–6in) across, that pale with age and have a strong aniseed fragrance.
N. **'St Louis'**. Deciduous, perennial water plant with floating leaves. S to 2m (6ft). Frost tender. Bright green leaves are spotted with brown when young. Produces star-shaped, semi-double, bright yellow flowers, 15–25cm (6–10in) across, in summer.
N. **'Sunrise'** (illus. p.390). Deciduous, perennial water plant with floating leaves. S to 2m (6ft). Frost hardy. Mid-green leaves have downy stalks and undersides. Bears star-shaped, semi-double, yellow flowers, 10–15cm (4–6in) across, in summer.
N. tetragona **'Alba'**, syn. *N. pygmaea* (illus. p.390). Deciduous, perennial water plant with floating leaves. S to 30cm (12in). Fully hardy. Has small, dark green leaves, purplish-green beneath, and, in summer, star-shaped, single, white flowers, 2–3cm ($^3/_4$–$1^1/_4$in) across. **'Helvola'** (illus. p.390), S to 45cm (18in), is frost hardy and has small, olive-green leaves with heavy purple or brown mottling. Produces star-shaped, semi-double, yellow flowers, 2–4cm ($^3/_4$–$1^1/_2$in) across, in summer.
N. **'Virginia'** (illus. p.390). Deciduous, perennial water plant with floating leaves. S to 1.5m (5ft). Fully hardy. Has purplish-green leaves and, in summer, star-shaped, semi-double, white flowers, 10–15cm (4–6in) across.
N. **'Wood's White Knight'**. Deciduous, perennial water plant with floating leaves. S to 2m (6ft). Frost tender. Leaves are mid-green, dappled with darker green beneath. In summer, star-shaped, semi-double, creamy-white flowers, 10–20cm (4–8in) across and with prominent, gold stamens, open at night.

NYMPHOIDES (Menyanthaceae)
Genus of deciduous, perennial, deep-water plants, with floating foliage, grown for their flowers. Fully hardy to frost tender, min. 5°C (41°F). Requires an open, sunny position. Propagate by division in spring or summer.
N. peltata illus. p.391.

NYSSA (Nyssaceae)
Tupelo
Genus of deciduous trees grown for their foliage and brilliant autumn colour. Fully hardy. Needs sun or semi-shade; does best in hot summers. Requires moist, neutral to acid soil. Resents being transplanted. Propagate by softwood cuttings in summer or by seed in autumn.
♛ *N. sinensis* illus. p.56.
♛ *N. sylvatica* illus. p.45.

O

OCHNA (Ochnaceae)
Genus of mainly evergreen trees and shrubs, grown mostly for their flowers and fruits. Frost tender, min. 10°C (50°F). Prefers full light and well-drained soil. Water containerized specimens moderately, less when not in full growth. Prune, if necessary, in early spring. Propagate by seed in spring or by semi-ripe cuttings in summer.
O. serrulata (Mickey-mouse plant). Evergreen, irregularly rounded, twiggy shrub that is semi-evergreen in low temperatures. H to 2m (6ft), S 1–2m (3–6ft) or more. Leaves are narrowly elliptic, toothed and glossy. Has 5-petalled, bright yellow flowers in spring-summer, then shuttlecock-shaped, red fruits, each with 1–5 berry-like seeds clustered on top.

x **ODONTIODA**. See ORCHIDS.
x *O.* (*O. Chantos* x *O. Marzorka*)
x *Odontoglossum* **Buttercrisp** (illus. p.263). Evergreen, epiphytic orchid for a cool greenhouse. H 23cm (9in). Produces arching spikes of intricately patterned, red, tan, orange and yellow flowers, 8cm (3in) across; flowering season varies. Has narrowly oval leaves, 10–15cm (4–6in) long. Needs shade in summer.
x *O.* **Mount Bingham** (illus. p.261). Evergreen, epiphytic orchid for a cool greenhouse. H 23cm (9in). Pink-edged, red flowers, 9cm (3½in) across, are borne in spikes; flowering season varies. Has narrowly oval leaves, 10–15cm (4–6in) long. Needs shade in summer.
x *O.* **Pacific Gold** x *Odontoglossum cordatum* (illus. p.261). Evergreen, epiphytic orchid for a cool greenhouse. H 23cm (9in). Bears long spikes of yellow-striped and -marked, rich chocolate-brown flowers, 7cm (3in) across; flowering season varies. Leaves are narrowly oval and 10–15cm (4–6in) long. Grow in shade in summer.
x *O.* **Petit Port** (illus. p.261). Evergreen, epiphytic orchid for a cool greenhouse. H 23cm (9in). Bears spikes of rich red flowers, 8cm (3in) across, each with a pink-and-yellow-marked lip; flowering season varies. Narrowly oval leaves are 10–15cm (4–6in) long. Shade in summer.

x **ODONTOCIDIUM**. See ORCHIDS.
x *O.* **Artur Elle 'Colombian'** (illus. p.261). Evergreen, epiphytic orchid for a cool greenhouse. H 23cm (9in). Produces tall spikes of pale yellow flowers, 6cm (2½in) across and intricately patterned with brown; flowering season varies. Has narrowly oval leaves, 10–15cm (4–6in) long. Requires shade in summer.
x *O.* **Tiger Butter** x *Wilsonara* **Wigg's 'Kay'** (illus. p.261). Evergreen, epiphytic orchid for a cool greenhouse. H 23cm (9in). Bears spikes of mottled, deep reddish-brown flowers, 5cm (2in) across, each with a rich golden-yellow lip; flowering season varies. Narrowly oval leaves are 10–15cm (4–6in) long. Grow in shade in summer.
x *O.* **Tiger Hambuhren** (illus. p.263). Evergreen, epiphytic orchid for a cool greenhouse. H 23cm (9in). Deep yellow flowers, 8cm (3in) across and heavily patterned with chestnut-brown, are borne in tall spikes; flowering season varies. Has narrowly oval leaves, 10–15cm (4–6in) long. Needs shade in summer.
x *O.* **Tigersun 'Orbec'** (illus. p.262). Evergreen, epiphytic orchid for a cool greenhouse. Is very similar to x *O.* Tiger Hambuhren, but flowers are slightly smaller and have lighter patterning.

ODONTOGLOSSUM. See ORCHIDS.
O. bictoniense. See *Lemboglossum bictoniense*.
O. cervantesii. See *Lemboglossum cervantesii*.
O. cordatum. See *Lemboglossum cordatum*.
O. crispum (illus. p.260). Evergreen, epiphytic orchid for a cool greenhouse. H 15cm (6in). Bears long sprays of rounded flowers, 8cm (3in) across, white or spotted or flushed with pink, each with a red-and-yellow-marked lip; flowering season varies. Has narrowly oval leaves, 10–15cm (4–6in) long. Requires shade in summer.
O. **Eric Young** (illus. p.262). Evergreen, epiphytic orchid for a cool greenhouse. H 15cm (6in). White-lipped, pale yellow flowers, 8cm (3in) across and spotted with rich yellow, are carried in spikes; flowering season varies. Has narrowly oval leaves, 10–15cm (4–6in) long. Grow in shade in summer.
O. grande. See *Rossioglossum grande*.
O. **Le Nez Point** (illus. p.261). Evergreen, epiphytic orchid for a cool greenhouse. H 15cm (6in). Bears spikes of crimson flowers, 6cm (2½in) across; flowering season varies. Has narrowly oval leaves, 10–15cm (4–6in) long. Needs shade in summer.
O. rossii. See *Lemboglossum rossii*.
O. **Royal Occasion** (illus. p.260). Evergreen, epiphytic orchid for a cool greenhouse. H 15cm (6in). Produces spikes of white flowers, 8cm (3in) across, with yellow-blotched lips, in autumn-winter. Leaves are narrowly oval and 10–15cm (4–6in) long. Shade in summer.

x **ODONTONIA**. See ORCHIDS.
x *O.* **Olga**. Evergreen, epiphytic orchid for a cool greenhouse. H 15cm (6in). Pure white flowers, 10cm (4in) across, with large, reddish-brown-blotched lips, are borne in spikes, mainly in autumn. Has narrowly oval leaves, 12cm (5in) long. Is best grown in shade during the summer.

OEMLERIA, syn. NUTTALLIA, OSMARONIA (Rosaceae)
Genus of one species of deciduous, early spring-flowering shrub, grown for its fragrant flowers and decorative fruits. Separate male and female plants are needed in order to obtain fruits. Fully hardy. Prefers sun or semi-shade and moist soil. To restrict growth remove suckers and cut old shoots back or down to base in late winter. Propagate by suckers in autumn.
O. cerasiformis (Indian plum, Oso berry). Deciduous, upright, then arching shrub that forms dense thickets. H 2.5m (8ft), S 4m (12ft). Leaves are narrowly oval and dark blue-green. Has nodding clusters of small, fragrant, bell-shaped, white flowers in early spring, then small, plum-shaped, purple fruits.

OENOTHERA (Onagraceae)
Evening primrose
Genus of annuals, biennials and perennials, grown for their profuse but short-lived flowers in summer. Fully to frost hardy. Needs full sun and well-drained, sandy soil. Propagate by seed or division in autumn or spring or by softwood cuttings in late spring.
O. acaulis. Tuft-forming perennial. H 15cm (6in), S 20cm (8in). Fully hardy. Has oblong to oval, deeply toothed or lobed leaves. Cup-shaped, white flowers, turning pink, open at sunset in summer. Suits a rock garden.
O. caespitosa. Clump-forming, stemless perennial. H 12cm (5in), S 20cm (8in). Fully hardy. Has narrowly oval, entire or toothed, mid-green leaves. Flowers, opening at sunset in summer, are fragrant, cup-shaped and white, becoming pink with age. Suits a rock garden.
♈ *O. fruticosa* **'Fireworks'**, syn. *O. tetragona* 'Fireworks', illus. p.255.
♈ subsp. *glauca* (syn. *O. tetragona*) is a clump-forming perennial. H 45–60cm (18–24in), S 45cm (18in). Fully hardy. Dense spikes of fragrant, cup-shaped, bright yellow flowers appear from mid- to late summer. Leaves, borne on reddish-green stems, are narrowly oval to lance-shaped and glossy, mid-green.
♈ *O. macrocarpa*, syn. *O. missouriensis*, illus. p.333.
O. missouriensis. See *O. macrocarpa*.
O. perennis, syn. *O. pumila*. Clump-forming perennial. H 15–60cm (6–24in), S 30cm (12in). Fully hardy. In summer, loose spikes of nodding buds open to fragrant, funnel-shaped, yellow flowers above a basal mass of narrowly spoon-shaped, mid-green leaves.
O. pumila. See *O. perennis*.
O. speciosa (White evening primrose). Often short-lived, clump-forming perennial with running rhizomes. H 45cm (18in), S 30cm (12in) or more. Frost hardy. In summer bears spikes of fragrant, saucer-shaped, green-centred, pure white flowers that age to pink and open flat. Leaves are narrowly spoon-shaped, deeply cut and mid-green.
O. tetragona. See *O. fruticosa* subsp. *glauca*. **'Fireworks'** see *O. fruticosa* 'Fireworks'.

OLEA (Oleaceae)
Genus of evergreen trees, grown for their foliage and edible fruits. Frost to half hardy; in cold areas requires the protection of a sheltered, south- or west-facing wall. Needs full sun and deep, fertile, very well-drained soil. Propagate by semi-ripe cuttings in summer or by seed in autumn.
♈ *O. europaea* (Olive). Slow-growing, evergreen, spreading tree. H and S 10m (30ft). Frost hardy. Is very long-lived. Narrowly oblong leaves are grey-green above, silvery beneath. Tiny, fragrant, white flowers, borne in short racemes in late summer, are followed by edible, oval, green, later purple fruits.

OLEARIA (Compositae)
Daisy bush
Genus of evergreen shrubs and trees, grown for their foliage and daisy-like flower heads. In mild, coastal areas provides good, very wind-resistant shelter. Frost to half hardy. Needs full sun and well-drained soil. Cut out dead wood in spring. Propagate by semi-ripe cuttings in summer.
O. albida of gardens. See *O.* 'Talbot de Malahide'.
O. avicenniifolia. Evergreen, rounded, dense shrub. H 3m (10ft), S 5m (15ft). Frost hardy. Oval to lance-shaped, dark grey-green leaves are white beneath. Bears wide heads of fragrant, white flowers in late summer and early autumn.
O. x *haastii* illus. p.109.
O. **'Henry Travers'**, syn. *O. semidentata* of gardens. Evergreen, rounded, compact shrub. H and S 3m (10ft). Half hardy. Has white shoots and narrowly lance-shaped, leathery, grey-green leaves. Large heads of purple-centred, lilac flowers appear from early to mid-summer.
O. ilicifolia. Evergreen, bushy, dense shrub. H and S 3m (10ft). Frost hardy. Narrowly oblong, rigid leaves are sharply toothed, grey-green and musk-scented. Fragrant, white flower heads are borne in clusters in early summer.
O. lacunosa. Evergreen, upright, dense shrub. H and S 3m (10ft). Frost hardy. Narrowly oblong, pointed, rigid leaves have rust-brown hairs when young and mature to glossy, dark green with central, white veins. Produces white flowerheads only rarely.
♈ *O. macrodonta*. Vigorous, evergreen, upright shrub, often tree-like. H 6m (20ft), S 5m (15ft). Frost hardy. Has holly-shaped, sharply toothed, grey-green leaves, silvery-white beneath. Large heads of fragrant, white flowers appear in early summer.
O. x *mollis*. Evergreen, rounded, dense shrub. H 1m (3ft), S 1.5m (5ft). Frost

hardy. Has oval, wavy-edged, silvery-grey leaves. Large heads of small, white flowers are borne profusely in late spring. ❦ **'Zennorensis'**. Evergreen, rounded, dense shrub. H and S 2m (6ft). Frost hardy. Has narrowly oblong, sharply toothed, grey-green leaves. Large heads of small, white flowers are produced in abundance in late spring.
O. nummulariifolia illus. p.106.
O. phlogopappa. Evergreen, upright, compact shrub. H and S 2m (6ft). Half hardy. Leaves are grey-green and oblong, with wavy edges. Massed, white flower heads are carried in late spring. var. *subrepanda* illus. p.128.
O. x scilloniensis. Evergreen, upright, then rounded, dense shrub. H and S 2m (6ft). Frost hardy. Narrowly oblong, wavy-edged, dark grey-green leaves set off masses of white flower heads in late spring.
O. semidentata of gardens. See *O. 'Henry Travers'*.
O. 'Talbot de Malahide', syn. *O. albida* of gardens. Evergreen, bushy, dense shrub. H 3m (10ft), S 5m (15ft). Frost hardy. Oval leaves are dark green above, silvery beneath. Broad heads of fragrant, white flowers are borne in late summer. Is excellent for exposed, coastal gardens.
O. virgata illus. p.88.

Oliveranthus elegans. See *Echeveria harmsii*.

Olsynium douglasii. See *Sisyrinchium douglasii*.

OMPHALODES (Boraginaceae)
Genus of annuals and perennials, some of which are evergreen or semi-evergreen. Makes good ground cover, especially in rock gardens. Fully to half hardy. Needs shade or semi-shade and moist but well-drained soil, except for *O. linifolia* and *O. luciliae*, which prefer sun. Propagate by seed or division in spring.
❦ *O. cappadocica* illus. p.297.
❦ *O. linifolia* illus. p.270.
O. luciliae. Semi-evergreen, mound-forming perennial. H 7cm (3in), S 15cm (6in). Half hardy. Has oval, blue-grey leaves. In spring-summer, loose sprays of pink buds develop into flattish, sky-blue flowers. Resents winter wet, so plant in a sheltered site or alpine house. Prefers sun and very gritty soil.
O. verna illus. p.297.

OMPHALOGRAMMA (Primulaceae)
Genus of perennials, closely related to *Primula*, grown for their flowers. Makes good rock garden plants, but is difficult to grow, especially in hot, dry areas. Frost hardy. Needs shade and gritty, moist but well-drained, peaty soil. Propagate by seed in spring.
O. vinciflorum. Basal-rosetted perennial. H 15cm (6in), S 10cm (4in). Has oval, hairy, leaves that are mid-green in colour. In spring produces nodding, funnel-shaped, violet flowers, each with a deeper violet throat and a flat, flared mouth.

ONCIDIUM. See ORCHIDS.
O. flexuosum (Dancing-doll orchid; illus. p.263). Evergreen, epiphytic orchid for a cool or intermediate greenhouse. H 23cm (9in). In autumn produces terminal sprays of many small, large-lipped, bright yellow flowers, 0.5cm ($1/4$in) across. Has narrowly oval leaves, 10cm (4in) long. Is best grown on a bark slab. Keep in semi-shade in summer.
O. ornithorrhynchum (illus. p.260). Evergreen, epiphytic orchid for a cool greenhouse. H 15cm (6in). Dense, arching sprays of very fragrant, rose-lilac flowers, 0.5cm ($1/4$in) across, with a yellow highlight, are borne freely in autumn. Has narrowly oval leaves, 10cm (4in) long. Requires semi-shade in summer.
O. papilio. See *Psychopsis papilio*.
O. tigrinum (illus. p.262). Evergreen, epiphytic orchid for a cool or intermediate greenhouse. H 23cm (9in). Branching spikes of fragrant, yellow-marked, brown flowers, 5cm (2in) across, each with a large, yellow lip, open in autumn. Has oval leaves, 15cm (6in) long. Requires semi-shade in summer.

ONIXOTIS, syn. DIPIDAX. (Liliaceae)
Genus of spring-flowering corms, cultivated mainly for botanical interest. Half hardy. Requires sun and well-drained soil. Plant corms in early autumn and keep them watered until after flowering. Dry off in summer. Propagate by seed in autumn.
O. triquetra. Spring-flowering corm. H 20–30cm (8–12in), S 5–8cm (2–3in). Long, narrow leaves are semi-erect and basal. Carries a spike of flattish, star-shaped, white flowers, each narrow petal having a basal, red mark.

ONOCLEA (Polypodiaceae)
Genus of one species of deciduous fern that rapidly colonizes wet areas. Fully hardy. Grows in sun or shade and in wet soil. Remove fronds as they fade. Propagate by division in autumn or winter.
❦ *O. sensibilis* illus. p.188.

ONONIS (Leguminosae)
Genus of summer-flowering annuals, perennials and deciduous or semi-evergreen shrubs and sub-shrubs, grown for their pea-like flowers. Is good for rock gardens, walls and banks. Fully hardy. Needs a sunny position in well-drained soil. Propagate by seed in autumn or spring, shrubs also by softwood cuttings in summer.
O. fruticosa illus. p.301.
O. natrix illus. p.307.
O. rotundifolia. Deciduous or semi-evergreen, glandular, upright sub-shrub. H 20–60cm (8–24in), S 20–30cm (8–12in) or more. Bears small, rounded, 3-parted, toothed, hairy, green leaves. Flowers that are relatively large, red-streaked, and rose-pink appear in small clusters in summer.

ONOPORDON. See ONOPORDUM.

ONOPORDUM, syn. ONOPORDON (Compositae)
Genus of annuals, biennials and perennials, ranging from stemless to tall, branching plants. Fully hardy. Grow in sun or semi-shade and in rich, well-drained soil. To prevent self seeding remove dead flower heads. Propagate by seed sown outdoors in autumn or spring. Leaves are prone to slug and snail damage.
O. acanthium illus. p.274.

ONOSMA (Boraginaceae)
Genus of summer-flowering annuals, semi-evergreen biennials, perennials and sub-shrubs, grown for their long, pendent, tubular flowers. Is suitable for rock gardens and banks. Fully to frost hardy. Requires full sun and well-drained soil. Dislikes wet summers. Propagate by softwood cuttings in summer or by seed in autumn.
O. alboroseum illus. p.300.
O. stellulatum. Semi-evergreen, upright sub-shrub. H and S 15cm (6in). Fully hardy. Leaves are oblong and covered in hairs which may irritate the skin. Clusters of yellow flowers open in late spring and summer.

OOPHYTUM (Aizoaceae)
Genus of clump-forming, egg-shaped, perennial succulents with 2 united, very fleshy leaves. These are covered in dry, papery sheaths, except in spring when sheaths split open, revealing a new pair of leaves. Flowers are produced from a slight central fissure on upper surface. Is difficult to grow. Frost tender, min. 5°C (41°F). Requires sun and well-drained soil. Propagate by seed or stem cuttings in spring or summer.
O. nanum illus. p.406.

OPHIOPOGON (Liliaceae)
Genus of evergreen perennials, grown mainly for their grass-like foliage. Fully to half hardy. Grows in sun or partial shade and in fertile, well-drained soil. Propagate by division in spring or by seed in autumn.
O. jaburan. Evergreen, clump-forming perennial. H 15cm (6in), S 30cm (12in). Frost hardy. Has dark green foliage. Racemes of bell-shaped, white flowers in early summer are followed by deep blue berries. **'Variegatus'** is half hardy, has white- or yellow-striped foliage and is much less robust.
O. japonicus illus. p.268.
❦ *O. planiscapus* **'Nigrescens'** illus. p.267.

OPHRYS. See ORCHIDS.
O. aranifera. See *O. sphegodes*.
O. fuciflora. See *O. holoserica*.
O. fusca (illus. p.262). Deciduous, terrestrial orchid. H 10–40cm (4–16in). Frost hardy. Spikes of greenish, yellow or brown flowers, 0.5cm ($1/4$in) long, each with a yellow-edged, bluish, brown or purple lip, are produced in spring. Has oval or lance-shaped leaves, 8–12cm (3–5in) long. Grow in shade outdoors. Containerized plants require semi-shade in summer.
O. holoserica, syn. *O. fuciflora*. Deciduous, terrestrial orchid. H 15–55cm (6–22in). Frost hardy. Spikes of flowers, 1cm ($1/2$in) long, varying from white through pink to blue and green, appear in spring-summer. Leaves are oval to oblong, 5–10cm (2–4in) long. Cultivate as for *O. fusca*.
O. lutea var. *lutea* (illus. p.262). Deciduous, terrestrial orchid. H 8–30cm (3–12in). Frost hardy. In spring bears short spikes of flowers, 1cm ($1/2$in) long, with greenish sepals, yellow petals and brown-centred, bright yellow lips. Has oval, basal leaves, 5–10cm (2–4in) long. Cultivate as for *O. fusca*.
O. speculum. See *O. vernixia*.
O. sphegodes, syn. *O. aranifera* (Spider orchid). Deciduous, terrestrial orchid. H 10–45cm (4–18in). Frost hardy. In spring-summer carries spikes of flowers, 1cm ($1/2$in) long, that vary from green to yellow and have spider-like, blackish-brown marks on lips. Leaves are oval to lance-shaped and 4–8cm ($1 1/2$–3in) long. Cultivate as for *O. fusca*.
O. tenthredinifera (Sawfly orchid; illus. p.261). Deciduous, terrestrial orchid. H 15–55cm (6–22in). Frost hardy. In spring produces spikes of flowers, 1cm ($1/2$in) long, in a variety of colours, including white to pink, or blue and green, each with a violet or bluish lip edged with pale green. Has a basal rosette of oval to oblong leaves, 5–9cm (2–$3 1/2$in) long. Cultivate as for *O. fusca*.
O. vernixia, syn. *O. speculum*. Deciduous, terrestrial orchid. H 8–30cm (3–12in). Frost hardy. In spring produces dense spikes of flowers, 1cm ($1/2$in) long, with greenish or yellow sepals, purple petals and 3-centred, brown lips. Has oblong to lance-shaped leaves, 4–7cm ($1 1/2$–3in) long. Cultivate as for *O. fusca*.

OPHTHALMOPHYLLUM (Aizoaceae)
Genus of clump-forming, perennial succulents related to *Lithops*. Each plant has 2 cylindrical, very fleshy leaves, erect and united for most of their length, but with a fissure between them at the top. Leaves each have a rounded upper surface with a translucent window. Frost tender, min. 5°C (41°F). Requires sun and well-drained soil. Water in late summer and early autumn; thereafter keep almost dry. Propagate by seed or stem cuttings in spring or summer.
O. herrei. See *O. longum*.
O. longum, syn. *O. herrei*. Clump-forming, perennial succulent. H 3cm ($1 1/4$in), S 1.5cm ($5/8$in). Has 2 almost united, cylindrical, very fleshy, erect, grey-green to brown leaves. Bears daisy-like, white to pale pink flowers, 2cm ($3/4$in) across, in late summer.
O. maughanii. Clump-forming, perennial succulent. H 4cm ($1 1/2$in), S 10cm (4in). Produces 2 almost united, cylindrical, very fleshy, erect, yellowish-green leaves with upper surface bearing 0.5–1cm ($1/4$–$1/2$in) deep fissure. In late summer bears daisy-like, white flowers, 1.5cm ($5/8$in) across.
O. villetii illus. p.408.

OPLISMENUS

OPLISMENUS (Gramineae). See GRASSES, BAMBOOS, RUSHES and SEDGES.
O. hirtellus (Basket grass).
♀ **'Variegatus'** (syn. *O. africanus* 'Variegatus') illus. p.265.

OPLOPANAX (Araliaceae)
Genus of deciduous, summer-flowering shrubs, grown for their habit, fruits and spiny foliage. Fully hardy, but young growths may be damaged by late frosts. Does best in a cool, partially shaded position and in moist soil. Propagate by seed in autumn or by root cuttings in late winter.
O. horridus (Devil's club). Deciduous, spreading, open, sparsely branched shrub. H and S 2m (6ft). Prickly stems bear large, oval, 7–9-lobed, toothed, mid-green leaves. Dense umbels of small, star-shaped, greenish-white flowers appear from mid- to late summer; these are followed by spherical, red fruits.

OPUNTIA (Cactaceae)
Prickly pear
Genus of perennial cacti, ranging from small, alpine, ground-cover plants to large, evergreen, tropical trees, with at times insignificant glochids – short, soft, barbed spines produced on areoles. Mature plants carry masses of short-spined, pear-shaped, green, yellow, red or purple fruits (prickly pears), edible in some species. Fully hardy to frost tender, min. 5–10°C (41–50°F). Needs sun and well-drained soil. Water containerized specimens when in full growth. Propagate by seed or stem cuttings in spring or summer.
O. brasiliensis. Tree-like, perennial cactus. H 5.5m (18ft), S 3m (10ft). Frost tender, min. 10°C (50°F). Has a cylindrical, green stem bearing bright green branches of flattened, oval, spiny segments. Sheds 2–3-year-old side branches. Masses of shallowly saucer-shaped, yellow flowers, 4cm (1½in) across, appear in spring-summer, only on plants over 60cm (2ft) tall, and are followed by small, yellow fruits.
O. cylindrica. Bushy, perennial cactus. H 4–6m (12–20ft). S 1m (3ft). Frost tender, min. 10°C (50°F). Cylindrical stems, 4–5cm (1½–2in) across, bear short-lived, cylindrical, dark green leaves, to 2cm (¾in) long, on new growth. Areoles may lack spines or each produce 2 or 3 barbed ones. Masses of shallowly saucer-shaped, pink-red flowers appear in spring-summer, only on plants over 2m (6ft) tall, and are followed by greenish-yellow fruits.
O. engelmannii. See *O. ficus-indica.*
O. erinacea illus. p.401.
O. ficus-indica, syn. *O. engelmannii*, *O. megacantha* (Edible prickly pear, Indian fig). Bushy to tree-like, perennial cactus. H and S 5m (15ft). Frost tender, min. 10°C (50°F). Bears flattened, oblong, blue-green stem segments, 50cm (20in) long and spineless. In spring-summer has masses of shallowly saucer-shaped, yellow flowers, 10cm (4in) across, followed by edible, purple fruits.
O. humifusa illus. p.414.
O. megacantha. See *O. ficus-indica.*

O. microdasys (Bunny ears). Bushy, perennial cactus. H and S 60cm (2ft). Frost tender, min. 10°C (50°F). Has flattened, oval, green stem segments, 8–18cm (3–7in) long, that develop brown marks in low temperatures. Bears spineless areoles, with white, yellow, brown or red glochids, closely set in diagonal rows. Masses of funnel-shaped, yellow flowers, 5cm (2in) across, appear in summer on plants over 15cm (6in) tall, and are followed by small, dark red fruits. var. **albispina** illus. p.403.
O. robusta illus. p.396.
O. tunicata illus. p.404.
O. verschaffeltii illus. p.410.

ORBEA (Asclepiadaceae)
Genus of clump-forming, perennial succulents with erect, 4-angled stems. Stem edges are often indented and may produce small leaves that drop after only a few weeks. Frost tender, min. 11°C (52°F). Needs sun or partial shade and well-drained soil. Propagate by seed or stem cuttings in spring or summer.
♀ *O. variegata*, syn. *Stapelia variegata*, illus. p.412.

ORCHIDS (Orchidaceae)
Family of perennials, some of which are evergreen or semi-evergreen, grown for their beautiful, unusual flowers. These consist of 3 outer sepals and 3 inner petals, the lowest of which, known as the lip, is usually enlarged and different from the others in shape, markings and colour. There are about 750 genera and 17,500 species, together with an even greater number of hybrids, bred partly for their vigour and ease of care. They are divided into epiphytic and terrestrial plants. (The spread of an orchid is indefinite.)

Epiphytic orchids
Epiphytes have more flamboyant flowers than terrestrial orchids and are more commonly grown. In the wild, they grow on tree branches or rocks (lithophytes), obtaining nourishment through clinging roots and moisture through aerial roots. Most consist of a horizontal rhizome, from which arise vertical, water-storing, often swollen stems known as pseudobulbs. Flowers and foliage are produced from the newest pseudobulbs. Other epiphytes consist of a continuously growing upright rhizome; on these, flower spikes appear in the axils of leaves growing from the rhizome. In temperate climates, epiphytes need to be grown under glass.

Cultivation of epiphytes
For cultivation purposes, epiphytes, which are all frost tender, may be divided into 3 groups: cool-greenhouse types, which require min. 10°C (50°F) and max. 24°C (75°F); intermediate-greenhouse types, needing a range of 13–27°C (55–80°F); and warm-greenhouse types, requiring 18–27°C (65–80°F). In summer, temperatures need to be controlled by shading the glass and by ventilation. Cool-greenhouse orchids may be placed outdoors in summer; this improves flowering. Other types may also be grown outdoors if the air temperature remains within these ranges.
The amount of light required in summer is given in individual plant entries. All epiphytic orchids, however, need to be kept out of direct sun in summer, to avoid scorching, and require full light in winter.
Epiphytic orchids, whether grown indoors or outside, require a special soil-free compost obtained from an orchid nursery or made by mixing 2 parts fibrous material (such as bark chippings and/or peat) with 1 part porous material (such as sphagnum moss and/or expanded clay pellets). Most epiphytes may be grown in pots, although some may be successfully cultivated in a hanging basket or on a slab of bark (with moss around their roots) suspended in the greenhouse.
In summer, water plants freely and spray regularly. Those suspended on bark slabs need a constantly moist atmosphere. In winter, water moderately and, if plants are in growth, spray occasionally. Some orchids rest in winter and require scarcely any water or none at all. Orchids benefit from weak foliar feeds; apply as for watering. Repot plants every other year, in spring; if they are about to flower, repot after flowering.

Terrestrial orchids
Terrestrial orchids, some of which also produce pseudobulbs, grow in soil or leaf mould, sustaining themselves in the normal way through roots or tubers. Some may be grown in borders, but many in temperate climates need to be cultivated in pots and protected under glass during winter.

Cultivation of terrestrial orchids
Terrestrial orchids are fully hardy to frost tender, min. 18°C (65°F). *Cypripedium* species may be grown outdoors in any area, preferably in neutral to acid soil, but cannot withstand severe frost, if frozen solid in pots or without snow cover, or tolerate very wet soil in winter. Other terrestrial orchids, except in very mild areas, are best grown in pots; use the same compost as for epiphytes but add 1 part grit to 2 parts compost. Place pots outdoors in a peat bed or in a glasshouse in the growing season. Keep dry when dormant. Under glass, light requirements, watering, feeding and repotting are as for epiphytes.

Orchid propagation
Orchids with pseudobulbs may be increased by removing and replanting old, leafless pseudobulbs when repotting in spring. Take care to retain at least 4 pseudobulbs on the parent plant. Some genera that may be propagated in this way are: *Ada*, x *Aliceara*, *Anguloa*, *Bifrenaria*, *Bletilla*, *Brassavola* (large plants only and retaining at least 6 pseudobulbs on the parent), x *Brassocattleya*, x *Brassolaeliocattleya*, *Bulbophyllum*, *Calanthe*, *Cattleya*, *Coelogyne*, *Cymbidium*, *Dendrobium*, *Dendrochilum*, *Encyclia*, *Epigeneium*, *Gomesa*, *Gongora*, *Laelia*, x *Laeliocattleya*, *Lycaste*, *Maxillaria*, *Miltonia*, *Miltoniopsis*, x *Odontioda*, x *Odontocidium*, *Odontoglossum*, x *Odontonia*, *Oncidium*, *Peristeria*, *Phaius*, *Pleione*, x *Potinara*, x *Sophrolaeliocattleya*, *Stanhopea*, x *Vuylstekeara*, x *Wilsonara* and *Zygopetalum*.
Some orchids without pseudobulbs produce new growth from the base. When a plant has 6 new growths, divide it in spring into 2 and repot both portions. Propagate *Disa*, *Paphiopedilum* and *Phragmipedium* in this way. Large specimens of *Eria*, *Masdevallia* and *Pleurothallis* may be divided in spring, leaving 4–6 stems on each portion.
Propagation of *Phalaenopsis* is by stem cuttings taken soon after flowering. *Vanda* may be increased by removing the top half of the stem once it has produced aerial roots and leaves; new growths will develop from the leafless base. With both these methods achieving success is difficult and not recommended for the beginner.
Terrestrial orchids with tubers may be propagated by division of the tubers. Genera that may be increased in this way are: *Cypripedium* (in spring), *Dactylorrhiza* (spring), *Ophrys* (autumn), *Orchis* (spring), *Serapias* (autumn) and *Spiranthes* (spring). *Calypso* is rarely propagated successfully in cultivation. *Angraecum* should not be propagated in cultivation, as the parent plant is easily endangered.
The most easily increased orchids are *Cymbidium*. Propagation of *Epidendrum* may be extremely difficult; see genus for specific details.

Orchids are illustrated on pp.260–63.
See also *Ada*, x *Aliceara*, *Angraecum*, *Anguloa*, *Bifrenaria*, *Bletilla*, *Brassavola*, x *Brassocattleya*, x *Brassolaeliocattleya*, *Bulbophyllum*, *Calanthe*, *Calypso*, *Cattleya*, *Coelogyne*, *Cymbidium*, *Cypripedium*, *Dactylorhiza*, *Dendrobium*, *Dendrochilum*, *Encyclia*, *Epidendrum*, *Epigeneium*, *Eria*, *Gomesa*, *Gongora*, *Laelia*, x *Laeliocattleya*, *Lemboglossum*, *Lycaste*, *Masdevallia*, *Maxillaria*, *Miltonia*, *Miltoniopsis*, x *Odontioda*, x *Odontocidium*, *Odontoglossum*, x *Odontonia*, *Oncidium*, *Ophrys*, *Orchis*, *Paphiopedilum*, *Peristeria*, *Phaius*, *Phalaenopsis*, *Phragmipedium*, *Pleione*, *Pleurothallis*, x *Potinara*, *Psychopsis*, *Rossioglossum*, *Serapias*, x *Sophrolaeliocattleya*, *Spiranthes*, *Stanhopea*, *Vanda*, x *Vuylstekeara*, x *Wilsonara* and *Zygopetalum*.

ORCHIS. See ORCHIDS.
O. morio (Gandergoose, Green-veined orchid; illus. p.262). Deciduous, terrestrial orchid. H 40cm (16in). Half hardy. Green-veined, reddish-purple, mauve or rarely white flowers, 1cm (½in) long, open along stems in spring. Has a basal cluster of lance-shaped or broadly oblong leaves, 10–16cm (4–6in) long. Requires sun or semi-shade.

Oreocereus celsianus. See *Cleistocactus celsianus.*
Oreocereus trollii. See *Cleistocactus trollii.*

OREOPTERIS

OREOPTERIS (Polypodiaceae)
Genus of deciduous ferns. Fully hardy. Tolerates sun or semi-shade. Grow in moist or very moist soil. Remove fading fronds regularly. Propagate by division in spring.
O. limbosperma, syn. *Thelypteris oreopteris* (Mountain buckler fern, Mountain fern, Mountain wood fern). Deciduous fern. H 60cm–1m (2–3ft), S 30cm (1ft). Produces broadly lance-shaped, much-divided fronds, with oblong to lance-shaped, mid-green pinnae.

ORIGANUM (Labiatae)
Dittany
Genus of deciduous sub-shrubs and perennials, sometimes with overwintering leaf rosettes. Some species are grown as culinary herbs, others for their clusters of tubular, usually pink flowers. Most species have arching, prostrate stems and are useful for trailing over rocks, banks and walls. Fully to frost hardy. Prefers sun and well-drained, alkaline soil. Propagate by division in spring, by cuttings of non-flowering shoots in early summer or by seed in autumn or spring.
♥ *O. amanum.* Deciduous, rounded, compact sub-shrub. H and S 15–20cm (6–8in). Frost hardy. Open funnel-shaped, pale pink or white flowers are borne all summer above small, heart-shaped, pale green leaves. Makes a good alpine house plant; dislikes a damp atmosphere.
O. dictamnus (Cretan dittany). Prostrate perennial. H 12–15cm (5–6in), S 40cm (16in). Frost hardy. Arching stems are clothed in rounded, aromatic, hairy, grey-white leaves. Has pendent heads of open funnel-shaped, purplish-pink flowers in summer.
O. 'Kent Beauty' illus. p.301.
♥ *O. laevigatum* illus. p.302.
♥ *O. rotundifolium.* Deciduous, prostrate sub-shrub. H 23–30cm (9–12in), S 30cm (12in). Fully hardy. Throughout summer bears whorls of pendent, funnel-shaped, pale pink flowers, surrounded by yellow-green bracts. Has small, rounded, mid-green leaves.
O. vulgare (Wild marjoram). Mat-forming, woody-based perennial. H and S 45cm (18in). Fully hardy. Has oval, aromatic, dark green leaves, above which branched, wiry stems bear clusters of tiny, tubular, 2-lipped, mauve flowers in summer.
♥ *'Aureum'* illus. p.254.

ORNITHOGALUM (Liliaceae)
Star-of-Bethlehem
Genus of bulbs, grown for their mostly star-shaped, white flowers, usually backed with green. Fully hardy to frost tender, min. 7°C (45°F). Needs sun or partial shade and well-drained soil. Lift and dry tender species for winter, if grown outside in summer, and replant in spring. Propagate by seed or offsets, in autumn for spring-flowering bulbs, in spring for summer-flowering bulbs.
O. arabicum illus. p.364.
O. balansae illus. p.370.
O. lanceolatum illus. p.370.
O. montanum illus. p.370.

O. narbonense illus. p.363.
♥ *O. nutans* (Drooping star-of-Bethlehem). Spring-flowering bulb. H 15–35cm (6–14in), S 8–10cm (3–4in). Frost hardy. Has a cluster of linear, channelled, semi-erect, basal leaves. Stem bears a spike of pendent, bell-shaped, translucent, white flowers, 2–3cm (³⁄₄–1¹⁄₄in) long with pale green outsides. Prefers partial shade.
O. saundersiae. Summer-flowering bulb. H to 1m (3ft), S 15–20cm (6–8in). Half hardy. Produces a basal cluster of strap- or lance-shaped, semi-erect leaves. Stem bears a flat-topped head of erect, flattish, white or cream flowers, each with a blackish-green ovary forming a dark eye.
O. thyrsoides illus. p.364.
O. umbellatum. Spring-flowering bulb. H 10–30cm (4–12in), S 10–15cm (4–6in). Frost hardy. Has linear, channelled, semi-erect, green leaves each have a white line on upper surface. Bears a loose, flat-topped head of star-shaped, white flowers, backed with green.

ORONTIUM (Araceae)
Genus of one species of deciduous, perennial, deep-water plant, grown for its floating foliage and flower spikes. Fully hardy. Needs full sun. Remove faded flower spikes. Propagate by seed when fresh, in mid-summer.
O. aquaticum illus. p.391.

OROSTACHYS (Crassulaceae)
Genus of short-lived, basal-rosetted, perennial succulents with very fleshy, sword-shaped leaves. Produces flowers 3 years from sowing seed, then dies. Frost tender, min. 8°C (46°F). Requires sun and well-drained soil. Propagate by seed or division in spring or summer.
O. chanetii. Basal-rosetted, perennial succulent. H 4cm (1¹⁄₂in), S 8cm (3in). Grey-green leaves are shorter in rosette centre. Flower stem bears a dense, tapering spike of star-shaped, white or pink flowers, 1–2cm (¹⁄₂–³⁄₄in) across, in spring-summer.

OROYA (Cactaceae)
Genus of spherical, perennial cacti. Inner flower petals form a tube and outer ones open fully. Frost tender, min. 10°C (50°F). Needs a sunny, well-drained site. Propagate by seed in spring or summer.
O. peruviana, syn. *O. neoperuviana*, illus. p.400.

Orphanidesia gaultherioides. See *Epigaea gaultherioides.*

ORTHROSANTHUS (Iridaceae)
Genus of perennials with short, woody rhizomes, grown for their flowers. Frost tender, min. 5°C (41°F). Prefers sun and well-drained soil. Propagate by division or seed in spring.
O. chimboracensis. Tufted, rhizomatous perennial. H 60cm (2ft) in flower, S 15cm (6in). Has very narrow, grass-like, ribbed, stiff leaves, to 45cm (18in) long, with finely toothed margins. In summer, flower stems bear clusters of short-lived, long-stalked, lavender-blue flowers, each enclosed in 2 leaf-like bracts.

ORYCHOPHRAGMUS (Cruciferae)
Genus of late spring- to summer-flowering annuals. Half hardy. Grow in sun and in fertile, well-drained soil. Propagate by seed in spring.
O. violaceus illus. p.283.

OSBECKIA (Melastomataceae)
Genus of evergreen, summer-flowering perennials, sub-shrubs and shrubs, grown for their flowers and foliage. Frost tender, min. 16°C (61°F). Needs full light or partial shade and humus-rich, well-drained soil. Water potted specimens freely when in full growth, moderately at other times. Cut back flowered stems by at least half in early spring to maintain vigour and to produce large flower trusses. Propagate by seed in spring or by greenwood cuttings in summer.
O. stellata. Evergreen, rounded, stiff-stemmed shrub. H and S 1–2m (3–6ft). Produces narrowly oval, hairy, prominently veined leaves. Has terminal clusters of 4-petalled, rose-purple flowers in late summer.

OSCULARIA (Aizoaceae)
Genus of bushy, spreading, perennial succulents with chunky, triangular leaves and masses of fragrant, daisy-like flowers. Frost tender, min. 3°C (37°F). Needs sun and well-drained soil. Propagate by seed or stem cuttings in spring or summer.
O. deltoides illus. p.407.

OSMANTHUS (Oleaceae)
Genus of evergreen shrubs and trees, grown for their foliage and small, fragrant flowers. *O.* x *burkwoodii* and *O. heterophyllus* may be used for hedging. Fully to half hardy. Tolerates sun or shade and fertile, well-drained soil. Restrict growth by cutting back after flowering; trim hedges in mid-summer. Propagate by semi-ripe cuttings in summer.
O. armatus. Evergreen, bushy, dense shrub. H and S 4m (12ft). Frost hardy. Large, oblong, dark green leaves are rigid and sharply toothed. Has tubular, 4-lobed, white flowers in autumn, followed by egg-shaped, dark violet fruits.
♥ *O.* x *burkwoodii*, syn. x *Osmarea burkwoodii*, illus. p.86.
O. decorus, syn. *Phillyrea decora.* Evergreen, upright, rounded, dense shrub. H 3m (10ft), S 5m (15ft). Fully hardy. Has large, oblong, glossy, dark green leaves and, in mid-spring, tubular, 4-lobed, white flowers, followed by egg-shaped, blackish-purple fruits.
♥ *O. delavayi*, syn. *Siphonosmanthus delavayi*, illus. p.86.
O. forrestii. See *O. yunnanensis.*
O. fragrans (Fragrant olive). Evergreen, upright shrub or tree. H and S 6m (20ft). Half hardy. Extremely fragrant, tubular, 4-lobed, white flowers are borne amid oblong, glossy, dark green leaves from early to late summer. Is suitable only for very mild areas. f. *aurantiacus* has orange flowers.
O. heterophyllus **'Aureomarginatus'** illus. p.97. ♥ **'Gulftide'** is an evergreen, bushy, dense shrub. H 2.5m

OSTEOSPERMUM

(8ft), S 3m (10ft). Frost hardy. Holly-shaped, sharply toothed, glossy, dark green leaves set off tubular, 4-lobed, white flowers in autumn.
O. yunnanensis, syn. *O. forrestii.* Evergreen, tree-like, upright, then spreading shrub. H and S 10m (30ft). Frost hardy. Has large, oblong, glossy, bright green leaves, bronze when young. Produces tubular, 4-lobed, creamy-white flowers in clusters in late winter or early spring.

x *Osmarea burkwoodii.* See *Osmanthus* x *burkwoodii.*

OSMARONIA. See **OEMLERIA.**

OSMUNDA (Osmundaceae)
Genus of deciduous ferns. Fully hardy. Requires shade, except for *O. regalis*, which also tolerates sun. *O. cinnamomea* and *O. claytoniana* need moist soil; *O. regalis* does best in very wet conditions. Remove fading fronds regularly. Propagate by division in autumn or winter or by spores as soon as ripe.
O. cinnamomea (Cinnamon fern). Deciduous fern. H 1m (3ft), S 45cm (18in). Outer, lance-shaped, divided, pale green sterile fronds, with deeply cut pinnae, surround brown fertile fronds, all arising from a fibrous rootstock.
O. claytoniana (Interrupted fern). Deciduous fern. H 60cm (2ft), S 30cm (1ft). Has lance-shaped, pale green fronds, divided into oblong, blunt pinnae; outer sterile fronds are larger than fertile ones at centre of plant.
♥ *O. regalis* illus. p.188.

OSTEOMELES (Rosaceae)
Genus of evergreen, summer-flowering shrubs, grown for their habit, foliage and flowers. Frost to half hardy. In most areas plant against a south- or west-facing wall. Requires sun and fertile, well-drained soil. Propagate by semi-ripe cuttings in summer.
O. schweriniae illus. p.108.

OSTEOSPERMUM (Compositae)
Genus of evergreen, semi-woody perennials. Frost to half hardy; does best in warm areas. Requires sun and well-drained soil. Propagate by cuttings of non-flowering shoots in mid-summer.
O. barberiae. See *O. jucundum.*
♥ *O.* **'Buttermilk'** illus. p.254.
O. **'Cannington Roy'**. Evergreen, clump-forming, prostrate perennial. H 30cm (12in), S 45cm (18in). Half hardy. Large, daisy-like, pink flower heads, with darker eyes, are borne profusely in summer-autumn. Leaves are linear and grey.
O. ecklonis. Evergreen, upright or somewhat straggling perennial. H and S 45cm (18in). Half hardy. In summer-autumn, daisy-like, white flower heads, with dark blue centres, are borne singly above lance-shaped, grey-green leaves. **'Blue Streak'** has petals that are suffused with blue outside.
♥ *O. jucundum*, syn. *O. barberiae, Dimorphotheca barberiae*, illus. p.243.
O. **'Nairobi Purple'**. Evergreen, semi-prostrate perennial. H 30cm (12in),

556

OSTROWSKIA

S 30–45cm (12–18in). Half hardy. Daisy-like, velvety, deep purple-red flower heads, with darker streaks on outside of ray petals, are borne in summer. Leaves are fresh green and lance-shaped. Will not flower freely in rich soils.
♈ *O.* '**Whirligig**' illus. p.241.

OSTROWSKIA (Campanulaceae)
Genus of one species of summer-flowering perennial. Fully hardy. Prefers a warm, sunny situation and rich, moist but well-drained soil. May be difficult to grow as needs a resting period after flowering, so cover with a frame until late autumn to keep dry. Propagate by seed in autumn or spring.
O. magnifica. Erect perennial. H 1.5m (5ft), S 45cm (1½ft). From early to mid-summer produces very large, bell-shaped blooms of delicate light blue-purple, veined with darker purple. Has whorls of oval, blue-grey leaves.

OSTRYA (Betulaceae)
Genus of deciduous trees, grown for their foliage, catkins and fruits. Fully hardy. Needs sun or semi-shade and fertile, well-drained soil. Propagate by seed in autumn.
O. carpinifolia (Hop hornbeam). Deciduous, rounded tree. H and S 15m (50ft). Has grey bark and oval, glossy, dark green leaves that turn yellow in autumn. Yellow catkins in mid-spring are followed by hop-like, greenish-white fruit clusters that become brown in autumn.
O. virginiana illus. p.51.

OTHONNA, syn. OTHONNOPSIS (Compositae)
Genus of evergreen shrubs, grown for their daisy-like flower heads in summer. Half hardy. Needs sun and well-drained soil. Propagate by softwood cuttings in early summer.
O. cheirifolia illus. p.306.

OTHONNOPSIS. See OTHONNA.

OURISIA (Scrophulariaceae)
Genus of evergreen perennials with creeping rootstocks. Excellent for peat beds and walls. Fully to frost hardy. Needs shade and moist, peaty soil. Propagate by division or seed in spring.
O. caespitosa illus. p.322.
O. '**Loch Ewe**'. Vigorous, evergreen, rosetted perennial. H and S 30cm (12in). Frost hardy. Prostrate stems have heart-shaped, leathery, green leaves. Bears dense spikes of outward-facing, tubular, salmon-pink flowers in late spring and early summer.
O. macrocarpa. Vigorous, evergreen, prostrate perennial. H 60cm (24in), S 20cm (8in). Frost hardy. Has rosettes of heart-shaped, leathery, dark green leaves. Produces spikes of open cup-shaped, yellow-centred, white flowers in late spring.
O. magellanica. Evergreen, straggling perennial. H 4cm (1½in), S to 15cm (6in). Frost hardy. In summer bears tubular, scarlet flowers above broadly heart-shaped leaves.
O. microphylla illus. p.324.

OXALIS (Oxalidaceae)
Genus of tuberous, rhizomatous or fibrous-rooted perennials and semi-evergreen sub-shrubs, grown for their colourful flowers, which in bud are rolled like an umbrella, and their often attractive leaves. Leaves are mostly less than 2cm (¾in) across and are divided into 3 or more leaflets. Some species may be invasive; smaller species and cultivars suit a rock garden. Fully hardy to frost tender, min. 5°C (41°F). Needs full sun or semi-shade and well-drained soil. Propagate by division in autumn or early spring.
O. acetosella (Wood sorrel). Creeping, spring-flowering, rhizomatous perennial. H 5cm (2in), S 30–45cm (12–18in). Fully hardy. Forms mats of clover-like, 3-lobed leaves. Delicate stems bear cup-shaped, white flowers, each 1cm (½in) across with 5 purple-veined petals. Prefers semi-shade. var. *purpurascens* (syn. *O.a.* f. *rosea*) illus. p.313.
♈ *O. adenophylla* illus. p.314.
O. bowiei, syn. *O. purpurata* var. *bowiei.* Spring- to summer-flowering, tuberous perennial. H to 30cm (12in), S 15cm (6in). Half hardy. Has long-stalked, clover-like, 3-lobed leaves. Stems each produce a loose head of 3–10 widely funnel-shaped, pinkish-purple flowers, 3–4cm (1¼–1½in) across. Needs a sheltered, sunny site.
O. chrysantha. Creeping, fibrous-rooted perennial. H 4–5cm (1½–2in), S 15–30cm (6–12in). Half hardy. Forms mats of clover-like, 3-lobed leaves. Stems each produce a funnel-shaped, bright yellow flower, 2–3cm (¾–1¼in) across, in summer. Needs a sheltered site.
O. deppei. See *O. tetraphylla.*
O. depressa, syn. *O. inops,* illus. p.326.
♈ *O. enneaphylla* (Scurvy grass). Tuft-forming, rhizomatous perennial. H 5–7cm (2–3in), S 8–10cm (3–4in). Frost hardy. Grey-green leaves are divided into narrowly oblong to oval leaflets. In summer, stems bear widely funnel-shaped, 3–4cm (1¼–1½in) wide, lilac-pink or white flowers.
O. hedysaroides. Semi-evergreen, bushy sub-shrub. H 1m (3ft), S 30–45cm (1–1½ft). Half hardy. Leafy stems bear clover-like, 3-lobed, green leaves. Leaf axils produce clusters of widely funnel-shaped, yellow flowers, 2–3cm (¾–1¼in) across, in spring-summer.
O. hirta. Late summer-flowering, tuberous perennial. H 30cm (12in), S 10–15cm (4–6in). Half hardy. Stem produces scattered leaves, with 3 narrowly lance-shaped leaflets. Leaf axils each produce a widely funnel-shaped, rose-purple flower, 2–3cm (¾–1¼in) wide, with a yellow centre.
O. inops. See *O. depressa.*

♈ *O.* '**Ione Hecker**'. Tuft-forming, rhizomatous perennial. H 5cm (2in), S 5–8cm (2–3in). Frost hardy. Grey leaves are composed of narrowly oblong, wavy leaflets. In summer bears funnel-shaped, pale purple-blue flowers, 4cm (1½in) across, with darker veins.
O. laciniata. Tuft-forming, rhizomatous perennial. H 5cm (2in), S 5–8cm (2–3in). Frost hardy. Has blue-grey leaves with narrowly oblong, crinkly-edged leaflets. Widely funnel-shaped, steel-blue flowers, 4cm (1½in) wide, with darker veins, are borne in summer.
O. lobata illus. p.335.
O. purpurata var. *bowiei.* See *O. bowiei.*
O. tetraphylla, syn. *O. deppei,* illus. p.301.

OXYDENDRUM (Ericaceae)
Genus of one species of deciduous tree, grown for its flowers and spectacular autumn colour. Fully hardy. For good autumn colouring plant in an open position in sun or semi-shade. Needs moist, acid soil. Propagate by softwood cuttings in summer or by seed in autumn.
O. arboreum illus. p.52.

Oxypetalum caeruleum. See *Tweedia caerulea.*

OZOTHAMNUS (Compositae)
Genus of evergreen, summer-flowering shrubs, grown for their foliage and small, densely clustered flower heads. Frost to half hardy. Requires full sun and well-drained soil. Propagate by semi-ripe cuttings in summer.
O. coralloides. See *Helichrysum coralloides.*
♈ *O. ledifolius,* syn. *Helichrysum ledifolium,* illus. p.130.
O. rosmarinifolius, syn. *Helichrysum rosmarinifolium,* illus. p.108.
O. selago. See *Helichrysum selago.*

P

PACHISTIMA. See PAXISTIMA.

PACHYCEREUS (Cactaceae)
Genus of slow-growing, columnar, perennial cacti, branching with age. Flowers, which are funnel-shaped, are unlikely to appear in cultivation as they are produced only on plants over 3m (10ft) high. Frost tender, min. 10°C (50°F). Requires sun and well-drained soil. Propagate by seed in spring or summer.
P. pecten-aboriginum. Columnar, perennial cactus. H 11m (35ft), S 3m (10ft). Dark green stems bear 9–11 deep ribs. Areoles each produce 8 radial spines, 1cm (^{1}/$_2$in) long, and longer central spines. All spines are dark brown with red bases and fade to grey.
P. pringlei illus. p.396.
P. schottii, syn. *Lophocereus schottii,* illus. p.394. 'Monstrosus' is a columnar, perennial cactus. H 7m (22ft), S 2m (6ft). Irregular, olive- to dark green stems have 4–15 ribs and no spines. Bears funnel-shaped, pink flowers, 3cm (1^1/$_4$in) wide, at night in summer.

PACHYPHRAGMA (Cruciferae)
Genus of perennials with rosettes of basal leaves, often grown as ground cover under shrubs. Fully hardy. Needs moist soil and sun or partial shade. Propagate by division in spring or by stem cuttings in late spring or by seed in autumn.
P. macrophyllum, syn. *Thlaspi macrophyllum,* illus. p.231.

PACHYPHYTUM (Crassulaceae)
Genus of rosetted, perennial succulents, closely related to *Echeveria,* with which it hybridizes. Frost tender, min. 5–10°C (41–50°F). Needs sun and well-drained soil. Propagate by seed, leaf or stem cuttings in spring or summer.
P. compactum illus. p.416.
P. oviferum illus. p.411.

PACHYPODIUM (Apocynaceae)
Genus of bushy or tree-like, perennial succulents, mostly with swollen stems, closely related to *Adenium,* except that most species have spines. Frost tender, min. 10–15°C (50–59°F). Requires full sun and very well-drained soil. May be very difficult to grow. Propagate by seed in spring or summer.
P. lamerei illus. p.392.
P. succulentum. Tree-like, perennial succulent. H 60cm (2ft), S 30cm (1ft). Min. 10°C (50°F). Swollen trunk, 15cm (6in) across, produces narrower, vertical, green to grey-brown stems. Bears trumpet-shaped, pink-crimson flowers, 2cm (^{3}/$_4$in) across, near stem tips in summer.

PACHYSANDRA (Buxaceae)
Genus of evergreen, creeping perennials and sub-shrubs, grown for their tufted foliage. Is useful for ground cover. Fully hardy. Tolerates dense shade and well-drained soil. Propagate by division in spring.
P. axillaris. Evergreen, mat-forming sub-shrub. H 20cm (8in), S 25cm (10in). Stems are each crowned by 3–6 oval, toothed, leathery leaves. Carries small, white flowers in erect spikes in late spring.
♥ *P. terminalis* illus. p.336.
♥ 'Variegata' is an evergreen, creeping perennial. H 10cm (4in), S 20cm (8in). Diamond-shaped, cream-variegated leaves are clustered at stem tips. In early summer bears spikes of tiny, white flowers, sometimes flushed purple.

PACHYSTACHYS (Acanthaceae)
Genus of evergreen perennials and shrubs, grown for their flowers. Frost tender, min. 13–18°C (55–64°F). Needs partial shade and fertile, well-drained soil. Water potted plants freely when in full growth, moderately at other times. Cut back flowered stems in late winter to maintain a bushy habit. Propagate by greenwood cuttings in early summer. Whitefly and red spider mite may cause problems.
P. cardinalis. See *P. coccinea.*
P. coccinea, syn. *P. cardinalis, Jacobinia coccinea, Justicia coccinea* (Cardinal's guard). Evergreen, erect, robust shrub. H 1.2–2m (4–6ft), S 60cm–1m (2–3ft). Min. 15–18°C (59–64°F) to flower well. Leaves are oval and deep green. Has tubular, bright red flowers in tight, green-bracted spikes, 15cm (6in) long, in winter.
♥ *P. lutea* illus. p.127.

x PACHYVERIA (Crassulaceae)
Hybrid genus (*Echeveria* x *Pachyphytum*) of clump-forming, rosetted, perennial succulents, sometimes almost stemless. Frost tender, min. 5–7°C (41–5°F). Needs full sun or partial shade and very well-drained soil. Propagate by leaf or stem cuttings in spring or summer.
x *P. glauca* illus. p.402.

PAEONIA (Paeoniaceae)
Peony
Genus of late spring-flowering perennials and deciduous shrubs ('tree peonies'), valued for their bold foliage, showy blooms and, in some species, colourful seed pods. Fully hardy, unless otherwise stated, although young growth (especially on tree peonies) may be damaged by late spring frosts. Prefers sun (but tolerates light shade) and rich, well-drained soil. Tall and very large-flowered cultivars need support. Propagate all species by seed in autumn (may take up to 3 years to germinate), tuberous species by root cuttings in winter, tree peonies by semi-ripe cuttings in late summer or by grafting in winter. Perennials may also be propagated by division in autumn or early spring. Is prone to peony wilt.

Flower forms
Unless stated otherwise, peonies described below flower between late spring and early to mid-summer and have large, alternate leaves divided into oval to lance-shaped or linear leaflets. Flowers are either single, semi-double, double or anemone-form.
Single – flowers are mostly cup-shaped, each with 1 or 2 rows of large, often lightly ruffled, incurving petals and a conspicuous central boss of stamens.
Semi-double – flowers are similar to single ones, but have 2 or 3 rows of petals.
Double – flowers are rounded, usually composed of 1 or 2 outer rows of large, often lightly ruffled, incurving petals, the remaining petals being smaller, usually becoming more densely arranged and diminishing in size towards the centre. Stamens are few, inconspicuous, or absent.
Anemone-form (Imperial or Japanese) – flowers usually have 1 or 2 rows of broad, incurving, outer petals; the centre of the flower is often filled entirely with numerous densely arranged, sometimes deeply cut, narrow petaloids derived from stamens.

P. 'Alice Harding' (illus. p.200). Clump-forming perennial. H and S to 1m (3ft). Bears very large, fragrant, double, creamy-white flowers.
P. 'America'. Clump-forming perennial. H and S to 1m (3ft). Has large, single flowers with very broad, deep crimson petals, lightly ruffled at margins.
P. 'Argosy' (illus. p.201). Deciduous, upright shrub (tree peony). H and S to 1.5m (5ft). Magnificent, large, single flowers are lemon-yellow, each with a crimson-purple blotch at base. Is hard to propagate.
P. arietina. See *P. mascula* subsp. *arietina.*
P. 'Auguste Dessert' (illus. p.201). Clump-forming perennial. H and S to 75cm (2^1/$_2$ft). Foliage provides rich autumn colour. Has masses of fragrant, semi-double flowers; carmine petals are tinged salmon-pink and have slightly ruffled, striking silvery-white margins.
P. 'Avant Garde' (illus. p.200). Clump-forming perennial. H and S to 1m (3ft). Has luxuriant foliage. Medium-sized to large, fragrant, single flowers are pale rose-pink with darker veins and bright golden anthers that have yellow-red filaments. Flowers are borne on stiff, straight stems in mid-spring and are ideal for cutting.
P. 'Ballerina' (illus. p.200). Clump-forming perennial. H and S 1m (3ft). Foliage provides autumn colour. Fragrant, double flowers are soft blush-pink, tinged lilac at first, later fading to white. Outer rows of petals are loosely arranged, very broad and incurving; inner petals are also incurving, but more densely arranged, narrower, more uneven in size and often have slightly ruffled margins.
P. 'Baroness Schroeder' (illus. p.200). Vigorous, clump-forming perennial. H and S to 1m (3ft). Is very free-flowering with large, fragrant, globe-shaped, double flowers, tinged with pale flesh-pink on opening but fading to almost pure white. Has several rows of almost flat, outer petals; inner petals are incurving, ruffled and very tightly arranged. Is one of the best peonies for cutting.
P. 'Barrymore'. Clump-forming perennial. H and S to 85cm (34in). Has very large, anemone-form flowers with broad, outer petals that are palest blush-pink on opening, later white. Clear pale golden-yellow petaloids are very narrow, relatively short and are neatly and densely arranged.
♥ *P.* 'Bowl of Beauty' (illus. p.200). Clump-forming perennial. H and S to 1m (3ft). Has very large, striking, anemone-form flowers with pale carmine-pink, outer petals and numerous narrow, densely arranged, ivory-white petaloids.
♥ *P. cambessedesii* (Majorcan peony; illus. p.201). Clump-forming perennial. H and S 45cm (1^1/$_2$ft). Half hardy. Has especially attractive foliage, dark green above with veins, stalks and under-surfaces suffused purple-red. Single, deep rose-pink flowers are borne in mid-spring.
P. 'Cheddar Cheese'. Clump-forming perennial. H and S to 1m (3ft). Has well-formed, large, double flowers in mid-summer. Neatly and densely arranged, slightly ruffled, ivory-white petals, the inner ones incurving, are interspersed with shorter, yellow petals.
P. 'Chocolate Soldier' (illus. p.201). Clump-forming perennial. H and S to 1m (3ft). Has mid- to dark green leaves that are often tinged bronze-red when young. Semi-double, purple-red flowers, borne in early summer, have yellow-mottled centres.
P. 'Colonel Heneage'. Clump-forming perennial of upright habit. H and S to 85cm (34in). Has masses of anemone-form flowers with both outer petals and inner petaloids of clear dark rose-crimson.
P. corallina. See *P. mascula.*
P. 'Cornelia Shaylor' (illus. p.200). Erect, clump-forming perennial. H and S to 85cm (34in). Fragrant, double flowers, flushed rose-pink on opening and gradually fading to blush-white, are borne freely from early to mid-summer. Ruffled petals are neatly and densely arranged.
P. 'Dayspring'. Clump-forming perennial. H and S to 70cm (28in). Has an abundance of fragrant, single, clear pink flowers borne in trusses.
♥ *P.* 'Defender' (illus. p.201). Clump-forming, vigorous perennial. H and S to 1m (3ft). Single, satiny crimson flowers, to 15cm (6in) across, with a central boss of golden

anthers, are carried on strong stems.

❦ *P. delavayi* (illus. p.201). Deciduous, upright, open shrub (tree peony). H to 2m (6ft), S to 1.2m (4ft). Leaves are divided into pointed-oval leaflets, often with reddish stalks. Single flowers are small, bowl-shaped and rich dark red, with conspicuous, leafy bracts below flowers.

P. **'Dresden'**. Robust, clump-forming perennial. H and S to 85cm (34in). Foliage provides autumn colour. Single flowers are ivory-white, tinged with soft blush-rose-pink.

❦ *P.* **'Duchesse de Nemours'**, syn. *P.* 'Mrs Gwyn Lewis' (illus. p.200). Vigorous, clump-forming perennial. H and S to 70cm (28in). Produces masses of richly fragrant, double flowers with very large, incurving, outer petals, tinged palest green at first, soon fading to pure white; inner petals with irregular margins are densely arranged towards the centre and are creamy-yellow at their base.

P. emodi (illus. p.200). Clump-forming perennial. H to 1.2m (4ft), S to 1m (3ft). Glossy, green foliage is topped by tall stems bearing several large, fragrant, single, pure white flowers with golden-yellow anthers.

P. **'Evening World'**. Clump-forming perennial. H and S to 1m (3ft). Has abundant, large, anemone-form flowers with soft blush-pink, outer petals and very tightly arranged, pale flesh-pink petaloids.

❦ *P.* **'Félix Crousse'**, syn. *P.* 'Victor Hugo'. Vigorous, clump-forming perennial. H and S to 75cm (30in). Bears a profusion of fragrant, double, rich carmine-pink flowers with darker red centres. Petals are ruffled, very numerous and tightly arranged, with margins sometimes tipped silvery-white.

❦ *P.* **'Festiva Maxima'**. Clump-forming perennial. H and S to 1m (3ft). Has dense, spreading foliage and huge, fragrant, double flowers borne on strong stems. Rather loosely arranged petals are large with irregular margins; outer petals are pure white, inner ones each have a basal, crimson blotch.

P. **'Flamingo'**. Clump-forming perennial. H and S to 85cm (34in). Foliage provides autumn colour. Double flowers are large and clear pale salmon-pink.

P. **'Globe of Light'** (illus. p.200). Clump-forming perennial. H and S to 1m (3ft). Has large, fragrant, anemone-form flowers. Outer petals are pure rose-pink, petaloids clear golden-yellow.

P. **'Heirloom'**. Compact, clump-forming perennial. H and S to 70cm (28in). Bears masses of large, fragrant, double, pale lilac-pink flowers.

P. **'Instituteur Doriat'** (illus. p.201). Clump-forming perennial. H and S to 1m (3ft). Foliage provides autumn colour. Has abundant, large, anemone-form flowers with reddish-carmine, outer petals and densely arranged, relatively broad petaloids, paler and more pink than outer petals, with ruffled, silvery-white margins.

P. **'Kelway's Gorgeous'** (illus. p.201). Clump-forming perennial. H and S to 85cm (34in). Single, intense clear carmine flowers, with a hint of salmon-pink, are borne very freely.

P. **'Kelway's Majestic'**. Clump-forming perennial. H and S to 1m (3ft). Freely borne, large, fragrant, anemone-form flowers have bright cherry rose-pink, outer petals and lilac-pink petaloids flecked with silver or pale gold.

P. **'Kelway's Supreme'** (illus. p.200). Clump-forming perennial. H and S to 1m (3ft). Foliage provides autumn colour. Has large, strongly fragrant, double flowers, produced over a long period, sometimes borne in clusters on well-established plants. Petals are broad, incurving, soft blush-pink, fading to milk-white. Single or semi-double axillary flowers are often produced.

P. **'Knighthood'** (illus. p.201). Clump-forming perennial. H and S to 75cm (30in). Double flowers have densely arranged, rather narrow, ruffled petals of unusually rich burgundy-red.

P. **'Krinkled White'** (illus. p.200). Robust, clump-forming perennial. H and S to 80cm (32in). Large, bowl-shaped, single, milk-white flowers are sometimes flushed palest pink. Petals are large with ruffled margins.

❦ *P.* **'Laura Dessert'** (illus. p.201). Clump-forming perennial. H and S to 75cm (30in). Produces fragrant, double flowers with creamy blush-white, outer petals. Densely arranged, incurving, inner petals are flushed rich lemon-yellow, and their margins are sometimes deeply cut.

P. lobata. See *P. peregrina*.

P. lutea. Deciduous, upright shrub (tree peony). H and S to 1.5m (5ft). Single flowers are usually vivid yellow; brownish- or purplish-yellow forms also occur. **'L'Espérance'** (illus. p.201) has very large, semi-double, primrose-yellow flowers with a carmine blotch at the base of each petal. ❦ var. *ludlowii* (illus. p.201), H and S to 2.5m (8ft), is grown as much for its large, bright green leaves, divided into deeply cut, pointed leaflets, as for its bowl-shaped, golden-yellow flowers.

P. **'Mme Louis Henri'** (illus. p.201). Deciduous, upright shrub (tree peony). H and S to 1.5m (5ft). Has loosely semi-double, whitish-yellow flowers with large, incurving, outer petals very heavily suffused with rusty-red. Smaller, often darker, inner petals each have a basal, dull red blotch.

P. **'Magic Orb'** (illus. p.201). Clump-forming perennial. H and S to 1m (3ft). Foliage provides autumn colour. Bears masses of large, strongly fragrant, double flowers, each with several outer whorls of fairly broad, ruffled, intense cherry-pink petals and a centre of densely arranged, smaller, incurving petals. Outermost rows of central petals are blush-white, heavily shaded with mid-rose-carmine; the innermost petals are mostly creamy-white.

P. mascula, syn. *P. corallina* (illus. p.200). Clump-forming perennial. H and S to 1m (3ft). Foliage is shiny, hairless and dark green; stems are dark red. Produces single, purple- or carmine-red, occasionally white, flowers with bosses of golden-yellow anthers borne on purple filaments. Seed capsules with 2–5 boat-shaped sections split to reveal purplish-black seeds. subsp. *arietina* (syn. *P. arietina*) is a tuberous perennial. H to 75cm (2^1/2ft), S to 60cm (2ft). Has single, reddish-pink flowers. Leaflets are hairy underneath.

❦ *P. mlokosewitschii* (illus. p.201). Clump-forming perennial. H and S to 75cm (30in). Soft bluish-green foliage, sometimes edged reddish-purple, is topped by large, single, lemon-yellow flowers.

P. **'Mother of Pearl'** (illus. p.200). Clump-forming perennial. H to 75cm (2^1/2ft), S to 60cm (2ft). Greyish-green leaves provide an attractive foil for the single, dog rose-pink flowers.

P. **'Mrs Gwyn Lewis'**. See *P.* 'Duchesse de Nemours'.

P. officinalis. Clump-forming, tuberous perennial. H and S to 60cm (2ft). This single, red apothecaries' peony has long been in cultivation, but is seldom seen today, having been superseded by larger, often double-flowered hybrids, such as the following. **'Alba Plena'** (illus. p.200), H and S to 75cm (30in), has double white flowers that are sometimes tinged with pink. **'China Rose'** (illus. p.201), H and S to 45cm (18in), has handsome, dark green foliage and single flowers with incurving, clear dark salmon-rose petals contrasting with central bosses of orange-yellow anthers. ❦ **'Rubra Plena'** (illus. p.201), H and S to 75cm (30in), is long-lived and has distinctive foliage, divided into broadly oval leaflets, and double, vivid pinkish-crimson flowers with ruffled petals.

P. peregrina, syn. *P. lobata*. Clump-forming, tuberous perennial. H and S to 1m (3ft). Produces bowl-shaped, single, ruby-red flowers. ❦ **'Sunshine'** (syn. *P.p.* 'Otto Froebel'; illus. p.201) has glossy, bright green leaves and bears large, single, vermilion flowers, tinged with salmon-rose.

P. potaninii. Deciduous, rhizomatous shrub (tree peony). H to 60cm (2ft), S indefinite. Has deeply divided leaves with narrowly oval, irregularly cut or lobed leaflets and small, single, deep maroon-red, or occasionally white flowers. Is similar to *P. delavayi*, but is more invasive and has less conspicuous, leafy bracts below the flowers. var. *trollioides* (illus. p.201) has finely divided leaves and yellow flowers.

P. **'Président Poincaré'**. Clump-forming perennial. H and S to 1m (3ft). Foliage provides autumn colour. Fragrant, double, clear rich ruby-crimson flowers are borne very freely.

❦ *P.* **'Sarah Bernhardt'** (illus. p.200). Vigorous, erect, clump-forming perennial. H and S to 1m (3ft). Produces an abundance of huge, fragrant, fully double flowers with large, ruffled, slightly dull rose-pink petals, fading to silvery blush-white at margins.

P. **'Shirley Temple'** (illus. p.200). Clump-forming perennial. H and S to 85cm (34in). Profuse, soft rose-pink flowers, fading to palest buff-white, are fully double, with broad petals arranged in whorls; innermost petals are smaller and more loosely packed.

P. **'Silver Flare'** (illus. p.201). Clump-forming perennial. H and S to 1m (3ft). Foliage provides autumn colour. Stems are flushed dull reddish-brown. Produces masses of fragrant, single flowers with rather long, slender, rich carmine-pink petals, each feathering to a striking silvery-white margin.

P. **'Sir Edward Elgar'** (illus. p.201). Clump-forming perennial. H and S to 75cm (30in). Foliage provides autumn colour. Has an abundance of single, chocolate-brown-tinged, rich crimson flowers with bosses of loosely arranged, clear lemon-yellow anthers.

P. × *smouthii* (illus. p.201). Clump-forming perennial. H and S to 60cm (2ft). Produces an abundance of fragrant, single, glistening, dark crimson flowers, to 10cm (4in) across, with conspicuous, yellow stamens, although both flowers and foliage may vary in colour.

P. **'Souvenir de Maxime Cornu'** (illus. p.201). Deciduous, upright shrub (tree peony). H and S to 1.5m (5ft). Large, richly fragrant flowers are fully double with warm golden-yellow petals densely arranged towards centres; ruffled margins are dull reddish-orange.

P. suffruticosa (Moutan). Deciduous, upright shrub (tree peony). H and S to 2.2m (7ft). Bears variable, large, cup-shaped flowers, single or semi-double, with incurving, rose-pink or white petals, each sometimes with a basal, usually chocolate-maroon blotch. Has given rise to many cultivars with semi-double and double flowers. **'Godaishu'** ('large globe'; illus. p.200) bears semi- or fully double, white flowers with yellow centres amid light green leaves that are fringed and twisted. **'Hana-daijin'** ('magnificent flower'; illus. p.201), H and S 2m (6ft) or more, is a vigorous cultivar that bears masses of double, purple, flowers. **'Hana-kisoi'** ('floral rivalry'; illus. p.200), has double, pale cerise-pink flowers. **'Joseph Rock'** see *P.s.* 'Rock's Variety'. **'Kamadu-nishiki'** ('kamada brocade'; illus. p.200), H and S to 1.2m (4ft), produces large, double flowers, to 20cm (8in) across, that are lilac-pink striped white at the edge of each petal. **'Kao'** ('king of flowers'; illus. p.201) freely produces large, semi-double or double, bright scarlet flowers. Flowers of **'Kokkuryu-nishiki'** ('black dragon brocade'; illus. p.201) are single and deep purple-red with lighter stripes. **'Reine Elizabeth'** (illus. p.200), H and S to 2m (6ft), has large, fully double flowers with broad, salmon-pink petals, flushed with bright copper-red and lightly ruffled at margins. **'Renkaku'** ('flight of cranes'), H and S to 1m (3ft), bears double flowers, each with broad, incurving, slightly ruffled, ivory-white petals, loosely arranged in 3 or more whorls, that surround a large boss of long, golden-yellow anthers. **'Rock's Variety'** (syn. *P.s.* 'Joseph Rock', *P.s.* subsp. *rockii*; illus. p.200) has large, spreading, semi-double, white flowers; inner petals each have a basal, dark maroon blotch. Is difficult to

PALIURUS

propagate. **'Tamafuyo'** ('jewelled lotus'; illus. p.200) is vigorous and freely produces double, pink flowers earlier than most other cultivars.
P. tenuifolia (illus. p.201). Clump-forming perennial. H and S to 45cm (18in). Elegant leaves are finely divided into many linear segments. Has single, dark crimson flowers, with golden-yellow anthers.
P. veitchii (illus. p.201). Clump-forming perennial. H and S to 75cm (30in). Shiny, bright green leaves are divided into oblong to elliptic leaflets. In early summer has nodding, cup-shaped, single, purple-pink flowers.
P. **'Victor Hugo'**. See *P.* 'Félix Crousse'.
P. **'White Wings'** (illus. p.200). Clump-forming perennial. H and S to 85cm (34in). Glossy, dark green foliage also provides autumn colour. In mid-summer bears masses of large, fragrant, single flowers with broad, white petals, sometimes tinged sulphur-yellow, that are each slightly ruffled at the apex.
♀ *P.* **'Whitleyi Major'** (illus. p.200). Clump-forming perennial. H to 1m (3ft), S to 60cm (2ft). Foliage and stems are flushed rich reddish-brown. Large, single, ivory-white flowers have a satin sheen and central bosses of clear yellow anthers.
P. wittmanniana (illus. p.201). Clump-forming perennial. H and S to 1m (3ft). Has large, single, pale primrose-yellow flowers each with a large, central boss of yellow anthers on purple-red filaments. Leaves are divided into broadly oval leaflets, shiny dark green above, paler beneath.

PALIURUS (Rhamnaceae)
Genus of deciduous, spiny, summer-flowering shrubs and trees, grown for their foliage and flowers. *P. spina-christi* is also grown for its religious association, reputedly being the plant from which Christ's crown of thorns was made. Frost hardy. Requires full sun and well-drained soil. Propagate by softwood cuttings in summer or by seed in autumn.
P. spina-christi illus. p.92.

PAMIANTHE (Amaryllidaceae)
Genus of one species of evergreen, spring-flowering bulb, grown for its large, strongly fragrant, showy flowers. Frost tender, min. 12°C (54°F). Needs partial shade and rich, well-drained soil. Feed with high-potash liquid fertilizer in summer. Reduce watering in winter but do not dry out. Propagate by seed in spring or by offsets in late winter.
P. peruviana illus. p.353.

PANCRATIUM (Amaryllidaceae)
Genus of bulbs with large, fragrant, daffodil-like flowers in summer. Frost to half hardy. Needs sun and well-drained soil that is warm and dry in summer when bulbs are dormant. Plant at least 15cm (6in) deep. Feed with a high-potash liquid fertilizer every 2 weeks from autumn to spring. Propagate by seed in autumn or by offsets detached in early autumn.
P. illyricum illus. p.363.

P. maritimum (Sea daffodil, Sea lily). Late summer-flowering bulb. H 45cm (18in), S 25–30cm (10–12in). Half hardy. Has strap-shaped, erect, basal, greyish-green leaves. Produces a head of 5–12 white flowers, each with a large, deep cup in the centre and 6 spreading petals. Is shy-flowering in cultivation.

PANDANUS (Pandanaceae)
Screw pine
Genus of evergreen trees, shrubs and scramblers, grown for their foliage and overall appearance. Flowers and fruits only appear on large, mature specimens. Frost tender, min. 13–16°C (55–61°F). Requires full light or partial shade and fertile, well-drained soil. Water containerized plants freely when in full growth, moderately at other times. Propagate by seed or suckers in spring or by cuttings of lateral shoots in summer. Red spider mite may be troublesome.
P. odoratissimus. See *P. tectorius*.
P. tectorius, syn. *P. odoratissimus*. Evergreen, rounded tree. H to 6m (20ft), S 3m (10ft) or more. Has rosettes of strap-shaped, deep green leaves, each 1–1.5m (3–5ft) long, with spiny margins and a spiny midrib beneath. Small flowers, the males in clusters, each with a lance-shaped, white bract, appear mainly in summer. Fruits are like round pineapples.
'Veitchii' see *P. veitchii*.
♀ *P. veitchii*, syn. *P. tectorius* 'Veitchii', illus. p.145.

PANDOREA (Bignoniaceae)
Genus of evergreen, woody-stemmed, twining climbers, grown for their handsome flowers and attractive leaves. Frost tender, min. 5°C (41°F). Grow in sun and in any well-drained soil. Prune after flowering to restrain growth. Propagate by seed in spring or by stem cuttings or layering in summer.
P. jasminoides, syn. *Bignonia jasminoides*, illus. p.168.
P. pandorana, syn. *Bignonia pandorana*, *Tecoma australis* (Wonga-wonga vine). Fast-growing, evergreen, woody-stemmed, twining climber. H 6m (20ft) or more. Leaves have 3–9 scalloped leaflets. Small, funnel-shaped, cream flowers, which are streaked and often spotted with red, brown or purple, are borne in clusters in summer.
P. ricasoliana. See *Podranea ricasoliana*.

PANICUM (Gramineae). See GRASSES, BAMBOOS, RUSHES and SEDGES.
P. capillare illus. p.184.

PAPAVER (Papaveraceae)
Poppy
Genus of annuals, biennials and perennials, some of which are semi-evergreen, grown for their cup-shaped flowers. Fully hardy. Needs sun or semi-shade and prefers moist but well-drained soil. Propagate by seed in autumn or spring. *P. orientale* and its cultivars are best propagated by root cuttings in winter. Self seeds readily.

P. alpinum group. See *P. burseri*.
P. atlanticum. Clump-forming, short-lived perennial. H and S 10cm (4in). Has oval, toothed, hairy leaves and, in summer, single, dull orange flowers. Is good for a rock garden.
P. burseri (*P. alpinum* group; Alpine poppy). Semi-evergreen, tuft-forming, short-lived perennial, best treated as an annual or biennial. H 15–20cm (6–8in), S 10cm (4in). Leaves are finely cut and grey. Carries single, white flowers throughout summer. Suits a rock garden, wall or bank.
P. commutatum **'Lady Bird'**. Fast-growing, erect, branching annual. H and S 45cm (18in). Has elliptic, deeply lobed, mid-green leaves and, in summer, single, red flowers, each with a black blotch in centre.
P. miyabeanum, syn. *P. nudicaule* Group, illus. p.332.
P. nudicaule (Iceland poppy). Tuft-forming perennial. H to 30cm (12in), S 10cm (4in). Hairy stems each produce a fragrant, single, white-and-yellow flower, sometimes marked green outside, in summer. Many colour forms have been selected. Leaves are oval, toothed and soft green. Needs partial shade. Is good for a rock garden.
P. orientale (Oriental poppy). Rosetted perennial. H 1m (3ft), S 30cm–1m (1–3ft). Single, brilliant vermilion flowers, with dark blotches at bases of petals, are borne in early summer. Has broadly lance-shaped, toothed or cut, rough, mid-green leaves. Flowering stems need support. **'Allegro Viva'** illus. p.214. **'Indian Chief'** has deep mahogany-red flowers. **'May Queen'** bears double, orange flowers. ♀ **'Mrs Perry'** has large, salmon-pink flowers. **'Perry's White'** illus. p.205.
P. rhoeas (Corn poppy, Field poppy).
♀ **Shirley Series** (double) illus. p.274; ♀ (single) illus. p.280.
♀ *P. somniferum* (Opium poppy). Fast-growing, upright annual. H 75cm (2½ft), S 30cm (1ft). Has oblong, lobed, light greyish-green leaves. Large, single flowers, to 10cm (4in) wide, in shades of red, pink, purple or white, are produced in summer. Several double-flowered forms are available, including **Carnation-flowered Series**, with fringed flowers in mixed colours; **Peony-flowered Series** illus. p.273; **'Pink Beauty'**, which has salmon-pink flowers; and **'White Cloud'**, with large, white flowers.

PAPHIOPEDILUM. See ORCHIDS.
P. appletonianum (illus. p.260). Evergreen, terrestrial orchid. H 8cm (3in). Frost tender, min. 13°C (55°F). In spring, green flowers, 6cm (2½in) across and each with a pouched, brownish lip and pink-flushed petals, are borne singly on tall, slender stems. Has oval, mottled leaves, 10cm (4in) long. Needs shade in summer.
P. bellatulum (illus. p.260). Evergreen, terrestrial orchid. H 5cm (2in). Frost tender, min. 18°C (64°F). Almost stemless, rounded, pouch-lipped, white flowers, 8cm (3in) across, spotted with dark maroon, are borne singly in spring. Oval, marbled leaves are 8cm (3in) long. Grow in shade in summer.

P. Buckhurst **'Mont Millais'** (illus. p.262). Evergreen, terrestrial orchid. H 10cm (4in). Frost tender, min. 13°C (55°F). Rounded, yellow-and-white flowers, to 12cm (5in) across and lined and spotted with red, are produced singly in winter. Has oval leaves, 10cm (4in) long. Requires shade in summer.
P. callosum (illus. p.260). Evergreen, terrestrial orchid. H 8cm (3in). Frost tender, min. 13°C (55°F). Purple- and green-veined, white flowers, 8cm (3in) across, are borne on tall stems in spring-summer. Has oval, mottled leaves, 10cm (4in) long. Needs shade in summer.
P. fairrieanum (illus. p.260). Evergreen, terrestrial orchid. H 8cm (3in). Frost tender, min. 10°C (50°F). Rich purple- and green-veined flowers, 5cm (2in) across, with curved petals and orange-brown pouches, are borne singly in autumn. Oval leaves are 8cm (3in) long. Grow in shade in summer.
P. Freckles (illus. p.260). Evergreen, terrestrial orchid. H 10cm (4in). Frost tender, min. 13°C (55°F). Rounded, reddish-brown-spotted and pouched, white flowers, 10cm (4in) across, are produced singly in winter. Has oval leaves, 10cm (4in) long. Grow in shade in summer.
P. haynaldianum (illus. p.262). Evergreen, terrestrial orchid. H 12cm (5in). Frost tender, min. 13°C (55°F). In summer, long-petalled, brown-marked, green-, pink-and-white flowers, to 15cm (6in) across, are produced singly. Has oval leaves, 20–23cm (8–9in) long. Requires shade in summer.
P. Lyric **'Glendora'** (illus. p.261). Evergreen, terrestrial orchid. H 10cm (4in). Frost tender, min. 13°C (55°F). Rounded, glossy, white-, red-and-green flowers, 10cm (4in) across, appear singly in winter. Has oval leaves, 15cm (6in) long. Needs shade in summer.
P. Maudiae grex, syn. *P.* x *maudiae* (illus. p.261). Evergreen, terrestrial orchid. H 10cm (4in). Frost tender, min. 13°C (55°F). Clear apple-green or deep reddish-purple flowers, 10cm (4in) across, appear singly on long stems in spring or early summer. Has oval, mottled leaves, 10cm (4in) long. Requires shade in summer.
P. niveum (illus. p.260). Evergreen, terrestrial orchid. H 5cm (2in). Frost tender, min. 13–18°C (55–64°F). White flowers, 4cm (1½in) across, are produced singly, mainly in spring. Oval, marbled leaves are 8cm (3in) long. Needs shade in summer.
P. sukhakulii (illus. p.262). Evergreen, terrestrial orchid. H 8cm (3in). Frost tender, min. 13°C (55°F). In spring-summer, purple-pouched, black-spotted, green flowers, 8cm (3in) across, appear singly on tall stems. Has oval, mottled leaves, 10cm (4in) long. Grow in shade in summer.
P. venustum (illus. p.263). Evergreen, terrestrial orchid. H 10cm (4in). Frost tender, min. 13°C (55°F). Variably coloured flowers, ranging from pink to orange with green veins and darker spots, are 6cm (2½in) across and borne singly in autumn. Has oval, mottled leaves, 10cm (4in) long. Needs shade in summer.

PARADISEA

PARADISEA (Liliaceae)
Genus of perennials, grown for their flowers and foliage. Fully hardy. Requires a sunny site and fertile, well-drained soil. Propagate by division in spring or by seed in autumn. After division may not flower for a season.
⚘ *P. liliastrum* (St Bruno's lily). Clump-forming, fleshy-rooted perennial. H 30–60cm (12–24in), S 30cm (12in). Slender stems, bearing racemes of saucer-shaped, white flowers in early summer, arise above broad, grass-like, greyish-green leaves.

PARAHEBE (Scrophulariaceae)
Genus of evergreen or semi-evergreen, summer-flowering perennials, sub-shrubs and shrubs, similar to *Hebe* and *Veronica*. Is suitable for rock gardens. Frost hardy. Needs sun and well-drained, peaty, sandy soil. Propagate by semi-ripe cuttings in early summer.
⚘ *P. catarractae* illus. p.304.
P. lyallii. Semi-evergreen, prostrate shrub. H 15cm (6in), S 20–25cm (8–10in). Has oval, toothed, leathery leaves and, in early summer, erect stems bearing loose sprays of flattish, pink-veined, white flowers.
⚘ *P. perfoliata*, syn. *Veronica perfoliata*, illus. p.250.

PARAQUILEGIA (Ranunculaceae)
Genus of tufted perennials, grown for their cup-shaped flowers and fern-like foliage. Is difficult to cultivate and flower successfully. Prefers dry winters and cool climates. Is good in alpine houses and troughs. Fully hardy. Needs sun and gritty, well-drained, alkaline soil. Propagate by seed in autumn.
P. anemonoides, syn. *P. grandiflora*, illus. p.312.
P. grandiflora. See *P. anemonoides*.

Paraserianthes distachya. See *Albizia distachya*.

PARIS (Liliaceae, Trilliaceae)
Herb Paris
Genus of summer-flowering, rhizomatous perennials. Fully hardy. Requires shade or semi-shade and humus-rich soil. Propagate by division in spring or by seed in autumn.
P. polyphylla, syn. *Daiswa polyphylla*. Erect, rhizomatous perennial. H 60cm–1m (2–3ft), S to 30cm (1ft). In early summer, at tips of slender stems, bears unusual flowers consisting of a ruff of green sepals, with another ruff of greenish-yellow petals, marked with crimson above, crowned by a violet-purple stigma. Leaves, borne in whorls at stem tips, are lance-shaped to oval and mid-green.

PARKINSONIA (Leguminosae)
Genus of evergreen, spring-flowering shrubs and trees, grown for their flowers and overall appearance. Frost tender, min. 15°C (59°F). Needs fertile, free-draining soil, a dry atmosphere and as much sunlight as possible to thrive. Water potted specimens moderately when in full growth, sparingly at other times. Pruning is tolerated, but spoils the natural habit. Propagate by seed in spring.

P. aculeata (Jerusalem thorn, Mexican palo verde). Evergreen, feathery shrub or tree with a spiny, green stem. H and S 3–6m (10–20ft) or more. Long, linear leaves have winged midribs bearing tiny, elliptic, short-lived leaflets. Has fragrant, 5-petalled, yellow flowers in arching racemes in spring.

PARNASSIA (Saxifragaceae)
Genus of rosetted, mainly summer-flowering perennials, grown for their saucer-shaped flowers. Is suitable for rock gardens. Fully hardy. Needs sun and wet soil. Propagate by seed in autumn.
P. palustris illus. p.298.

PAROCHETUS (Leguminosae)
Genus of one species of evergreen perennial. Grows best in alpine houses. Half hardy. Needs semi-shade and gritty, moist soil. Propagate by division of rooted runners in any season.
⚘ *P. communis* illus. p.332.

PARODIA (Cactaceae)
Genus of rounded, perennial cacti with tubercles arranged in ribs that often spiral around green stems. Crown forms woolly buds, then funnel-shaped flowers. Frost tender, min. 10°C (50°F). Requires full sun or partial shade and very well-drained soil. Water occasionally and very lightly in winter; tends to lose roots during a long period of drought. Propagate by seed in spring or summer.
⚘ *P. apricus*, syn. *Notocactus apricus*. Flattened spherical, perennial cactus. H 7cm (3in), S 10cm (4in). Min. 10°C (50°F). Much-ribbed, pale green stem is densely covered with short, soft, golden-brown spines. In summer, crown produces flattish, glossy, bright yellow flowers, 8cm (3in) across, with purple stigmas. Prefers partial shade.
P. chrysacanthion illus. p.404.
P. graessneri, syn. *Notocactus graessneri*. Slow-growing, flattened spherical, perennial cactus. H 10cm (4in), S 25cm (10in). Min 10°C (50°F). Bristle-like, golden spines completely cover much-ribbed, green stem. Slightly sunken crown bears funnel-shaped, glossy, greenish-yellow flowers, with yellow stigmas, in early spring. Prefers partial shade.
P. haselbergii, syn. *Notocactus haselbergii*, illus. p.411.
P. leninghausii, syn. *Notocactus leninghausii*, illus. p.404.
⚘ *P. mammulosus*, syn. *Notocactus mammulosus*, illus. p.413.
P. nivosa illus. p.410.
P. ottonis, syn. *Notocactus ottonis*. Variable, spherical, perennial cactus. H and S 10cm (4in). Min. 5°C (41°F). Has pale to dark green stem with 8–12 rounded ribs bearing stiff, golden radial spines and longer, soft, red central spines. In summer, crown carries flattish, glossy, golden flowers, 8cm (3in) across, with purple stigmas. Offsets freely from stolons. Prefers a sunny site.
P. rutilans, syn. *Notocactus rutilans*, illus. p.407.
P. sanguiniflora illus. p.410.
⚘ *P. scopa*, syn. *Notocactus scopa* (Silver ball cactus). Spherical to columnar, perennial cactus. H 25cm (10in), S 15cm (6in). Min. 10°C (50°F). Stem, with 30–35 ribs, is densely covered with white radial spines and longer, red central spines, 3 or 4 per areole. Crown bears funnel-shaped, glossy, yellow flowers, 4cm (1½in) across, with purple stigmas, in summer. Prefers a sunny position.
P. vorwerkiana, syn. *Wigginsia vorwerkiana*, illus. p.415.

PARONYCHIA (Caryophyllaceae)
Genus of evergreen perennials making extensive, loose mats of prostrate stems. Is useful for rock gardens and walls. Fully to frost hardy. Needs sun and well-drained soil. Propagate by division in spring.
P. capitata. Vigorous, evergreen, mat-forming perennial. H 1cm (½in), S 40cm (16in). Fully hardy. Silvery leaves are small and oval. In summer produces inconspicuous flowers surrounded by papery bracts. Makes good ground cover.
P. kapela subsp. *serpyllifolia* illus. p.339.

PARROTIA (Hamamelidaceae)
Genus of one species of deciduous tree, grown for its flowers and autumn colour. Fully hardy, but flower buds may be killed by hard frosts. Requires full sun and grows best in fertile, moist but well-drained soil. Is lime-tolerant, but usually colours best in acid soil. Propagate by softwood cuttings in summer or by seed in autumn.
⚘ *P. persica* illus. p.56.

PARROTIOPSIS (Hamamelidaceae)
Genus of one species of deciduous tree or shrub, grown for its ornamental, dense flower heads surrounded by conspicuous bracts. Fully hardy. Needs sun or semi-shade. Grows in any fertile, well-drained soil except very shallow soil over chalk. Propagate by softwood cuttings in summer or by seed in autumn.
P. jacquemontiana. Deciduous, shrubby or upright tree. H 6m (20ft), S 4m (12ft). Has witch-hazel-like, dark green leaves that turn yellow in autumn. From mid- to late spring and in summer bears clusters of minute flowers, with tufts of yellow stamens, surrounded by white bracts.

PARTHENOCISSUS (Vitaceae)
Genus of deciduous, woody-stemmed, tendril climbers, grown for their leaves, which often turn beautiful colours in autumn. Broad tips of tendrils have sucker-like pads that cling to supports. Has insignificant, greenish flowers in summer. Will quickly cover north- or east-facing walls or fences and may be grown up large trees. Fully to half hardy. Grow in semi-shade or shade and in well-drained soil. Propagate by softwood or greenwood cuttings in summer or by hardwood cuttings in early spring.
⚘ *P. henryana*, syn. *Vitis henryana*. Deciduous, tendril climber with 4-angled, woody stems. H to 10m (30ft). Frost hardy. Leaves have 3–5 toothed, oval leaflets, each 4–13cm (1½–5in) long, and are velvety, deep green or bronze with white or pinkish veins. Small, dark blue berries are produced in autumn. Leaf colour is best with a north or east aspect.
⚘ *P. quinquefolia*, syn. *Vitis quinquefolia* (Five-leaved ivy, Virginia creeper). Deciduous, woody-stemmed, tendril climber. H 15m (50ft) or more. Frost hardy. Leaves have 5 oval, toothed, dull green leaflets, paler beneath, that turn a beautiful crimson in autumn. Blue-black berries are produced in autumn. Is ideal for covering a high wall or building.
P. thompsonii, syn. *Vitis thompsonii*, illus. p.178.
⚘ *P. tricuspidata*, illus. p.178. 'Lowii' and 'Veitchii' (syn. *Ampelopsis veitchii*) illus. p.178.

PASSIFLORA

PASSIFLORA (Passifloraceae)
Passion flower
Genus of evergreen or semi-evergreen, woody-stemmed, tendril climbers, grown for their unique flowers, each one with a central corona of filaments. Many species have egg-shaped to rounded, fleshy, edible fruits that mature to orange or yellow in autumn. Half hardy to frost tender, min. 5–16°C (41–61°F). Grow in any fertile, well-drained soil and in full sun or partial shade. Water freely in full growth, less at other times. Stems need support. Thin out and spur back crowded growth in spring. Propagate by seed in spring or by semi-ripe cuttings in summer.
P. x *allardii*. Strong-growing, evergreen, woody-stemmed, tendril climber. H 7–10m (22–30ft). Frost tender, min. 7°C (45°F). Has 3-lobed leaves. Flowers, 7–10cm (3–4in) wide, are white, tinted pink, with purple-banded crowns, and are carried in summer-autumn.
⚘ *P. antioquiensis*, syn. *Tacsonia van-volxemii* (Banana passion fruit). Fast-growing, evergreen, woody-stemmed, tendril climber. H 5m (15ft) or more. Frost tender, min. 7°C (45°F). Leaves, with 3 deep lobes, are softly downy. Long-tubed, rose-red flowers, 10–12cm (4–5in) across, with purplish-blue centres, appear in summer-autumn.
⚘ *P. caerulea* illus. p.174.
⚘ *P.* x *caerulea-racemosa*. Vigorous, evergreen, woody-stemmed, tendril climber. H 10m (30ft). Frost tender, min. 7–10°C (45–50°F). Has 3-lobed leaves. Purple flowers, 8cm (3in) across, appear in summer-autumn.
P. x *caponii* 'John Innes' illus. p.174.
P. coccinea illus. p.165.
⚘ *P.* x *exoniensis*. Fast-growing, evergreen, woody-stemmed, tendril climber. H 8m (25ft) or more. Frost tender, min. 7°C (45°F). Leaves have 3 deep lobes and are softly downy. Rose-pink flowers, 8cm (3in) across, with purplish-blue crowns, are produced in summer-autumn.
P. manicata illus. p.177.
⚘ *P. mollissima*, syn. *Tacsonia mollissima*. Fast-growing, evergreen, woody-stemmed, tendril climber. H 5m (15ft) or more. Frost tender, min. 7°C (45°F). Softly downy leaves have 3 deep lobes. Long-tubed, pink flowers, to 8cm (3in) wide, each with

561

PATERSONIA

a purplish-blue crown, appear in summer-autumn.
P. quadrangularis illus. p.175.
P. racemosa (Red passion flower). Fast-growing, evergreen, tendril climber with slender, woody stems. H 5m (15ft). Frost tender, min. 15°C (59°F). Has wavy, leathery leaves with 3 deep lobes. In summer-autumn bears terminal racemes of pendent, crimson flowers, 8–10cm (3–4in) across, with short, white- and purple-banded crowns.
P. sanguinea. See *P. vitifolia.*
P. vitifolia, syn. *P. sanguinea.* Evergreen, woody-stemmed, tendril climber; slender stems have fine, brown hairs. H to 5m (15ft). Frost tender, min. 16°C (61°F). Has 3-lobed, lustrous leaves. In summer-autumn bears bright scarlet flowers, 13cm (5in) wide, each with a short crown, banded red, yellow and white.

PATERSONIA (Iridaceae)
Genus of evergreen, clump-forming, spring- and early summer-flowering, rhizomatous perennials. Half hardy. Needs full sun and light, well-drained soil. Leave undisturbed once planted. Propagate by seed in autumn.
P. umbrosa illus. p.365.

PATRINIA (Valerianaceae)
Genus of perennials, forming neat clumps, grown for their flowers. Is suitable for rock gardens and peat beds. Fully hardy. Needs partial shade and moist soil. Propagate by division in spring or by seed in autumn. Self seeds freely.
P. triloba. Clump-forming perennial. H 20–50cm (8–20in), S 15–30cm (6–12in). Flower stems produce flat heads of small, golden-yellow flowers throughout summer. Rounded, 3- to 5-lobed, green leaves turn gold in autumn.

PAULOWNIA (Scrophulariaceae)
Genus of deciduous trees, grown for their large leaves and foxglove-like flowers, borne before the foliage emerges. Fully to frost hardy, but flower buds and young growth of small plants may be damaged by very hard frosts. Requires full sun and fertile, moist but well-drained soil. In cold areas may be grown for foliage only by cutting back young shoots hard in spring and cutting out all but one of the subsequent shoots; this results in very large leaves. Propagate by seed in autumn or spring or by root cuttings in winter.
P. fortunei. Deciduous, spreading tree. H and S 8m (25ft). Fully hardy. Has large, oval, mid-green leaves. In late spring bears large, fragrant flowers, purple-spotted and white inside, pale purple outside.
P. imperialis. See *P. tomentosa.*
P. tomentosa, syn. *P. imperialis,* illus. p.51.

PAVONIA (Malvaceae)
Genus of evergreen, mainly summer-flowering perennials and shrubs, grown usually for their flowers. Frost tender, min. 16–18°C (61–4°F). Needs full light or partial shade and humus-rich, well-drained soil. Water freely when in full growth, moderately at other times. Leggy stems may be cut back hard in spring. Propagate by seed in spring or by greenwood cuttings in summer. Whitefly and red spider mite may be troublesome.
P. hastata. Evergreen, erect shrub. H 2–3m (6–10ft), S 1–2m (3–6ft). Has lance-shaped to oval, mid-green leaves, each with 2 basal lobes. Funnel-shaped, pale red to white flowers, with darker basal spotting, appear in summer.

PAXISTIMA, syn. **PACHISTIMA** (Celastraceae)
Genus of evergreen, spreading shrubs and sub-shrubs, grown for their foliage. Is suitable for ground cover. Fully hardy. Prefers shade and humus-rich, moist soil. Propagate by division in spring or by semi-ripe cuttings in summer.
P. canbyi. Evergreen, spreading sub-shrub. H 15–30cm (6–12in), S 20cm (8in). Has linear or oblong leaves and, in summer, short, pendent spikes of tiny, greenish-white flowers.

PEDILANTHUS (Euphorbiaceae)
Genus of bushy, summer-flowering, perennial succulents containing poisonous, milky sap. Produces small, yellowish-green, pink, red or brown bracts that are each shaped like a bird's head. Frost tender, min. 10–11°C (50–52°F). Needs sun or partial shade and well-drained soil. Propagate by seed or stem cuttings in spring or summer.
P. tithymaloides. Bushy, perennial succulent. H 3m (10ft), S 30cm (1ft). Min 10°C (50°F). Has thin, erect stems zigzagging at each node. Leaves are mid-green and boat-shaped, with prominent ribs beneath. Stem tips each bear red to yellowish-green bracts in summer. Prefers partial shade.
'Variegata' illus. p.395.

PELARGONIUM (Geraniaceae)
Geranium
Genus of mainly summer-flowering perennials, most of which are evergreen, often cultivated as annuals. Is grown for its colourful flowers and is useful in pots or as bedding plants; in warm conditions flowers are borne almost continuously. Frost tender, min. 1°C (34°F), unless otherwise stated. Dislikes very hot, humid conditions. Well-drained, neutral to alkaline soil is preferred, and a sunny site with 12 hours of daylight is required for good flowering. Dead-head frequently and fertilize regularly if grown in pots; do not overwater. Plants may be kept through winter in the greenhouse by cutting back in autumn-winter to 12cm (5in) and repotting. Propagate by softwood cuttings from spring to autumn.

Pelargoniums may be divided into 5 groups; all flower in summer-autumn unless stated otherwise:
Zonal – plants with rounded leaves, distinctively marked with a darker 'zone', and single (5-petalled), semi-double or fully double flowers;
Regal – shrubby plants with rounded to oval, deeply serrated leaves and broadly trumpet-shaped, exotic-coloured flowers that are prone to weather-damage in the open;
Ivy-leaved – trailing plants, ideal for hanging baskets, with rounded, lobed leaves and flowers similar to those of zonal pelargoniums;
Scented-leaved and **species** – plants with small, often irregularly star-shaped flowers; scented-leaved forms are grown for their fragrant leaves.
Unique – Tall-growing sub-shrubs with regal-like, single, brightly coloured flowers that are borne continuously through the season. Leaves, which may be scented, vary in shape.

P. acetosum (illus. p.210). Species pelargonium. H 50–60cm (20–24in), S 20–25cm (8–10in). Stems are succulent with fleshy, grey-green leaves that are often margined red. Bears single, salmon-pink flowers. Is good as a pot plant in a greenhouse.
P. 'Alberta' (illus. p.210). Evergreen zonal pelargonium. H 45cm (18in), S 30cm (12in). Bears clusters of small, single, crimson-and-white flowers. Is best grown as a bedding plant.
P. 'Amethyst' (illus. p.210). Evergreen, trailing ivy-leaved pelargonium. H and S to 1.5m (5ft). Leaves are fleshy with pointed lobes. Bears fully double, light mauve-purple flowers.
P. 'Apple Blossom Rosebud' (illus. p.210). Evergreen zonal pelargonium. H 30cm (12in), S 23cm (9in). Fully double, pinkish-white flowers, edged with red, look like miniature rosebuds.
P. 'Autumn Festival' (illus. p.210). Evergreen, bushy regal pelargonium. H and S 30cm (12in). Salmon-pink flowers have pronounced, white throats.
P. 'Bredon' (illus. p.211). Strong-growing, evergreen regal pelargonium. H 45cm (18in), S to 30cm (12in). Has large, maroon flowers.
P. 'Brookside Primrose' (illus. p.210). Dwarf zonal pelargonium. H 10–12cm (4–5in), S 7–10cm (3–4in). Bears double, pale pink flowers. Leaves have a butterfly mark in the centre of each leaf. Is good as a pot plant or for bedding.
P. 'Butterfly Lorelei' (illus. p.210). Fancy-leaved, zonal pelargonium. H 25–30cm (10–12in), S 15–20cm (6–8in). Has butterfly-shaped leaves and double, pale salmon-pink flowers. Is suitable as a pot plant in a greenhouse.
P. 'Caligula' (illus. p.211). Evergreen, miniature zonal pelargonium. H 15–20cm (6–8in), S 10cm (4in). Has small, double, crimson flowers and tiny, dark green leaves. Suits a windowsill.
P. 'Capen' (illus. p.211). Bushy zonal pelargonium. H 38–45cm (15–18in), S 15–20cm (6–8in). Bears coral-pink, semi-double flowers. Is good as a pot plant.
P. capitatum (illus. p.210). Evergreen scented-leaved pelargonium. H 30–60cm (12–24in), S 30cm (12in). Has mauve flowers and irregularly 3-lobed leaves that smell faintly of roses. Is mainly used to produce geranium oil for the perfume industry, but may be grown as a pot plant.
P. carnosum. Deciduous, shrubby pelargonium (unclassified), with thick, succulent stems and a woody, swollen, tuber-like rootstock. H and S 30cm (1ft). Min. 10°C (50°F). Has long, grey-green leaves with triangular, deeply lobed leaflets. Produces branched, umbel-like flower heads with white or greenish-yellow flowers, the upper petals streaked red and shorter than the green sepals.
P. 'Cherry Blossom' (illus. p.210). Vigorous, evergreen zonal pelargonium. H and S to 45cm (18in). Semi-double flowers are mauve-pink with white centres.
P. 'Chew Magna'. Evergreen regal pelargonium. H 30–45cm (12–18in), S to 30cm (12in). Each petal of the pale pink flowers has a wine-red blaze.
P. 'Clorinda' (illus. p.210). Vigorous, scented-leaved pelargonium. H 45–50cm (18–20in), S 20–25cm (8–10in). Leaves smell of cedar and are 3-lobed. Bears large, single, rose-pink flowers and is excellent in a large pot. Is suitable for a greenhouse or patio.
P. crispum 'Variegatum' (illus. p.211). Evergreen, upright scented-leaved pelargonium. H to 1m (3ft), S 30–45cm (1–1½ft). Has gold-variegated leaves and small, pale lilac flowers. Foliage tends to become creamy-white in winter.
P. 'Dale Queen' (illus. p.210). Evergreen, bushy zonal pelargonium. H 23–30cm (9–12in), S 23cm (9in). Single flowers are delicate salmon-pink. Is particularly suitable for a pot.
P., Diamond Series. Group of slow-growing, evergreen, compact, bushy zonal pelargoniums, grown as annuals. H and S 30–60cm (1–2ft). Half hardy. Has rounded, lobed, light to mid-green leaves, often zoned with bronze or red. Large, domed heads of weather-resistant, single flowers are available in a range of colours, including shades of red and rose-pink (scarlet, illus. p.280).
P. 'Dolly Varden' (illus. p.211). Evergreen zonal pelargonium. H 30cm (12in), S 23cm (9in). Green leaves have handsome, purple-brown, white and crimson markings. Single, scarlet flowers are insignificant.
P. 'Emma Hossler'. Evergreen, dwarf zonal pelargonium. H 20–25cm (8–10in), S 15cm (6in). Bears large, fully double, mauve-pink flowers. Is useful for a window box.
P. 'Fair Ellen' (illus. p.210). Compact scented-leaved pelargonium. H and S 30cm (12in). Has dark green leaves and pale pink flowers marked with red.
P. 'Flower of Spring' (Silver-leaved geranium; illus. p.211). Vigorous, evergreen zonal pelargonium. H 60cm (2ft), S 30cm (1ft). Has green-and-white leaves and single, red flowers.
P. x fragrans (illus. p.210). Evergreen, very bushy scented-leaved pelargonium. H and S 30cm (12in). Rounded, shallowly lobed, grey-green leaves smell strongly of pine. Bears small, white flowers.
P. 'Fraicher Beauty' (illus. p.210). Evergreen zonal pelargonium.

H 30cm (12in), S 23cm (9in). Flowers are fully double and perfectly formed with delicate colouring: white with a thin, red edge to each petal. Is excellent as a pot plant.

♀ *P.* 'Francis Parrett' (illus. p.210). Evergreen, short-jointed zonal pelargonium. H 15–20cm (6–8in), S 10cm (4in). Bears fully double, purplish-mauve flowers and small, green leaves. Is good for a windowsill.

P. 'Friesdorf' (illus. p.211). Evergreen zonal pelargonium. H 25cm (10in), S 15cm (6in). Has dark green foliage and narrow-petalled, single, orange-scarlet flowers. Is good for a window box or planted in a large group.

P. 'Golden Lilac Mist' (illus. p.210). Bushy zonal pelargonium. H 25–30cm (10–12in), S 15–20cm (6–8in). Leaves are gold marked with bronze. Bears double, lavender-pink flowers. Is a good window-box plant.

P. 'Gustav Emich' (illus. p.211). Vigorous, evergreen, zonal pelargonium. H and S to 60cm (24in). Semi-double flowers are vivid scarlet.

♀ *P.* 'Irene' (illus. p.211). Evergreen zonal pelargonium. H 45cm (18in), S 23–30cm (9–12in). Bears large, semi-double, light crimson blooms.

P. 'Ivalo' (illus. p.210). Evergreen, bushy, short-jointed zonal pelargonium. H 23–30cm (9–12in), S 23cm (9in). Large, semi-double flowers are pale pink with crimson-dotted, white centres.

P. 'Lachsball' (illus. p.210). Vigorous zonal pelargonium. H 45–50cm (18–20in), S 15–20cm (6–8in). Bears semi-double, salmon-pink flowers each with a scarlet eye. Is good for summer bedding.

♀ *P.* 'L'Elégante' (illus. p.211). Evergreen, trailing ivy-leaved pelargonium. H and S to 60m (24in). Foliage is variegated with creamy-white margins, sometimes turning pink at the edges; semi-double flowers are pale mauve. Is best grown in a hanging basket.

P. 'Lesley Judd' (illus. p.210). Vigorous, evergreen, bushy regal pelargonium. H 30–45cm (12–18in), S to 30cm (12in). Flowers are soft salmon-pink with a central, red blotch. Pinch out growing tips before flowering to control shape.

♀ *P.* 'Mabel Grey' (illus. p.211). Evergreen scented-leaved pelargonium. H 45–60cm (18–24in), S 30–45cm (12–18in). Has diamond-shaped, rough-textured, toothed, strongly lemon-scented leaves, with 5–7 pointed lobes, and mauve flowers.

P. 'Mme Fournier' (illus. p.211). Evergreen, short-jointed zonal pelargonium. H 15–20cm (6–8in), S 10cm (4in). Small, single, scarlet flowers contrast well with almost black leaves. Is useful for a pot or as a summer bedding plant.

P. 'Manx Maid' (illus. p.210). Evergreen regal pelargonium. H 30–38cm (12–15in), S 25cm (10in). Flowers and leaves are small for regal type. Pink flowers are veined and blotched with burgundy.

P. 'Mauritania' (illus. p.210). Evergreen zonal pelargonium. H 30cm (12in), S 23cm (9in). Single, white flowers are ringed towards centres with pale salmon-pink.

P. 'Mini Cascade' (illus. p.211). Evergreen, trailing, short-jointed ivy-leaved pelargonium. H and S 30–45cm (12–18in). Bears many single, red flowers. Regular dead-heading is essential for continuous display.

P. 'Mr Everaarts' (illus. p.210). Bushy dwarf zonal pelargonium. H 15–20cm (6–8in), S 10–12cm (4–5in). Bears double, bright pink flowers. Is good in a window box.

♀ *P.* 'Mr Henry Cox' (illus. p.210). Evergreen zonal pelargonium. H 30cm (12in), S 15cm (6in). Mid-green leaves are marked with red, yellow and purple-brown. Flowers are single and pink.

P. 'Mrs Pollock' (illus. p.211). Evergreen zonal pelargonium. H 30cm (12in), S 15cm (6in). Each golden leaf has a grey-green butterfly mark in centre, with a bronze zone running through it. Bears small, single, orange-red flowers.

P. 'Mrs Quilter' (illus. p.211). Evergreen zonal pelargonium. H 30cm (12in), S 23cm (9in). Has yellow leaves with wide, chestnut-brown zones and single, pink flowers.

P. 'Orange Ricard' (illus. p.211). Vigorous, robust, evergreen zonal pelargonium. H 45–60cm (18–24in), S 30cm (12in). Produces masses of large, semi-double, orange blooms.

P., Orbit Series. Group of slow-growing, evergreen, bushy zonal pelargoniums, grown as annuals. H and S 30–60cm (1–2ft). Half hardy. Has rounded, lobed, bronze- or red-zoned, mid-green leaves and large, domed, single flower heads in mixed or separate colours, including shades of white, pink, red and orange (salmon, illus. p.273).

♀ *P.* 'Paton's Unique' (illus. p.211). Vigorous unique pelargonium with pungent-smelling leaves. H 38–45cm (15–18in), S 15–20cm (6–8in). Flowers are single, red or pale pink, each with a small, white eye.

P. 'Paul Humphris' (illus. p.211). Evergreen, bushy, compact zonal pelargonium. H 30cm (12in), S 23cm (9in). Has fully double, deep wine-red flowers. Is good as a pot plant.

P. peltatum. Evergreen, trailing, brittle-jointed pelargonium from which ivy-leaved cultivars have been derived. H and S to 1.5m (5ft). Has fleshy leaves, with pointed lobes, and single, mauve or white flowers. Cultivars suit hanging baskets and window boxes. Flowers of 'Lachskönigin' (illus. p.210) are semi-double and mauve-purple. 'Tavira' (illus. p.211) bears single, soft crimson flowers.

P. 'Polka' (illus. p.211). Vigorous unique pelargonium. H 45–50cm (18–20in), S 20–25cm (8–10in). Flowers are semi-double. Upper petals are orange-red, blotched and feathered deep purple; lower ones are salmon-orange.

P. 'Prince of Orange' (illus. p.211). Scented-leaved pelargonium. H 25–30cm (10–12in), S 15–20cm (6–8in). Small, rounded leaves smell of orange. Bears single, mauve flowers. Is good as a pot plant indoors.

P. 'Purple Emperor' (illus. p.210). Evergreen regal pelargonium. H 45cm (18in), S 30cm (12in). Pink-mauve flowers have a deeper, central coloration. Flowers well into autumn.

P. 'Purple Unique' (illus. p.211). Vigorous, evergreen, upright, shrubby unique pelargonium. H and S 1m (3ft) or more. Rounded, large-lobed leaves are very aromatic. Has single, open trumpet-shaped, light purple flowers. Does well when trained against a sunny wall.

P. 'Purple Wonder' (illus. p.211). Vigorous zonal pelargonium. H 38–45cm (15–18in), S 38–45cm (8–10in). Bears semi-double, cerise-purple flowers. Is good as a pot plant in the greenhouse or in the garden.

P. 'Rica' (illus. p.210). Bushy zonal pelargonium. H 30–38cm (12–15in), S 15–20cm (6–8in). Flowers are semi-double and deep rose-pink, each with a large, white eye. Is good in a window box or as a greenhouse pot plant.

P. 'Robe' (illus. p.211). Vigorous zonal pelargonium. H 38–45cm (15–18in), S 15–20cm (6–8in). Bears semi-double, cerise-crimson flowers. Is suitable as a pot plant in a greenhouse, or as a bedding plant, and is good for exhibition.

P. 'Rollinson's Unique' (illus. p.211). Evergreen, shrubby pelargonium (unique). H 60cm (2ft) or more, S 30cm (1ft). Has oval, notched, pungent leaves and small, single, open trumpet-shaped, wine-red flowers with purple veins.

P. 'Rouletta' (illus. p.211). Vigorous, evergreen, trailing ivy-leaved pelargonium. H and S 60cm–1m (2–3ft). Bears semi-double, red-and-white flowers. To control shape, growing tips should be pinched out regularly.

P. 'Royal Oak' (illus. p.211). Evergreen, bushy, compact scented-leaved pelargonium. H 38cm (15in), S 30cm (12in). Oak-like, slightly sticky leaves have a spicy fragrance and are dark green with central, brown markings. Flowers are small and mauve-pink.

P. 'Schöne Helena' (illus. p.210). Evergreen zonal pelargonium. H 30–45cm (12–18in), S 23cm (9in). Produces masses of large, semi-double, salmon-pink blooms.

P., Sprinter Series. Group of slow-growing, evergreen, branching, bushy zonal pelargoniums, grown as annuals. H and S 30–60cm (1–2ft). Half hardy. Has rounded, lobed, light to mid-green leaves. Bears large, domed, single flower heads in shades of red. Is very free-flowering.

♀ *P.* 'The Boar' (illus. p.210). Evergreen, trailing pelargonium (species). H and S to 60cm (24in). Has unusual, 5-lobed, notched leaves, each with a central, dark brown blotch, and long-stemmed, single, salmon-pink flowers. Is useful for a hanging basket.

P. 'Timothy Clifford' (illus. p.210). Evergreen, short-jointed zonal pelargonium. H 15–20cm (6–8in), S 10cm (4in). Has dark green leaves and fully double, salmon-pink flowers. Suits a windowsill.

P. 'Tip Top Duet' (illus. p.211). Evergreen, bushy, free-branching regal pelargonium. H 30–38cm (12–15in), S 25cm (10in). Leaves and blooms are small for regal type. Bears pink-veined, white flowers; uppermost petals have dark burgundy blotches.

♀ *P. tomentosum* (Peppermint geranium; illus. p.211). Evergreen, bushy scented-leaved pelargonium. H 30–60cm (1–2ft), S 1m (3ft). Large, rounded, shallowly lobed, velvety, grey-green leaves have a strong peppermint aroma. Bears clusters of small, white flowers. Growing tips should be pinched out to control spread. Dislikes full sun.

♀ *P.*, Video Series. Group of slow-growing, evergreen, branching, bushy zonal pelargoniums, grown as annuals. H and S 30–60cm (1–2ft). Half hardy. Has rounded, lobed, bronze-zoned, deep green leaves and large, domed, single flower heads in white and shades of pink or red.

P. 'Voodoo' (illus. p. 211). Unique pelargonium. H 50–60cm (20–24in), S 20–25cm (8–10in). Flowers are single and pale burgundy with a purple-black blaze on each petal. Is suitable as a greenhouse pot plant.

PELLAEA (Polypodiaceae)
Genus of deciduous, semi-evergreen or evergreen ferns. Half hardy to frost tender, min. 5°C (41°F). Grow in semi-shade and gritty, moist but well-drained soil. Remove fading fronds regularly. Propagate by spores in summer.

P. atropurpurea (Purple rock brake, Purple-stemmed cliff brake). Semi-evergreen or evergreen fern. H and S 30cm (12in). Frost tender. Small, narrowly lance-shaped, divided fronds have oblong, blunt pinnae and are dark green with a purplish tinge.

♀ *P. rotundifolia* (Button-fern). Evergreen fern. H and S 15cm (6in). Frost tender. Small, narrowly lance-shaped, divided fronds are dark green and have rounded pinnae.

PELLIONIA, syn. ELATOSTEMA (Urticaceae)
Genus of evergreen, creeping perennials with attractive foliage that tends to lie flat, making useful ground cover. Frost tender, min. 15°C (59°F). Requires a humid atmosphere, away from draughts, indirect light and moist soil. Propagate from rooting stems or stem cuttings in spring or summer.

P. daveauana. See *P. repens*.

P. pulchra. Evergreen, slightly fleshy perennial with rooting, creeping stems. H 8–10cm (3–4in), S 60cm (2ft) or more. Broadly oval leaves, 5cm (2in) long, are green with dark brown veins above, purple below. Flowers are insignificant.

♀ *P. repens*, syn. *P. daveauana*, illus. p.267.

PELTIPHYLLUM, syn. DARMERA (Saxifragaceae)
Umbrella plant
Genus of one species of perennial, grown for its unusual foliage. Is sometimes now included in *Darmera*. Makes fine marginal water plants. Fully hardy. Grows in sun or shade and

requires moist soil. Propagate by division in spring or by seed in autumn or spring.
♚ *P. peltatum* illus. p.202.

PENNISETUM (Gramineae). See GRASSES, BAMBOOS, RUSHES and SEDGES.
P. alopecuroides, syn. *P. compressum* (Chinese fountain grass). Tuft-forming, herbaceous, perennial grass. H 1m (3ft), S 45cm (1½ft). Frost hardy. Has narrow, mid-green leaves; leaf sheaths each have a hairy tip. In late summer bears arching, cylindrical panicles with decorative, purple bristles that last well into winter.
P. compressum. See *P. alopecuroides*.
P. longistylum. See *P. villosum*.
P. ruppelii. See *P. setaceum*.
P. setaceum, syn. *P. ruppellii* (African fountain grass). Tuft-forming, herbaceous, perennial grass. H 1m (3ft), S 45cm (1½ft). Frost hardy. Has very rough, mid-green leaves and stems. In summer bears dense, cylindrical panicles of copper-red spikelets, with decorative, bearded bristles, that last well into winter.
P. villosum, syn. *P. longistylum*, illus. p.183.

PENSTEMON (Scrophulariaceae) Genus of annuals, perennials, sub-shrubs and shrubs, most of which are semi-evergreen or evergreen. Fully to half hardy. Prefers full sun and fertile, well-drained soil. Propagate species by seed in autumn or spring or by softwood or semi-ripe cuttings of non-flowering shoots in mid-summer, cultivars by cuttings only.
♚ *P. 'Andenken an Friedrich Hahn'*, syn. *P. 'Garnet'* (illus. p.212). H 60–75cm (2–2½ft), S 60cm (2ft). Vigorous, semi-evergreeen, bushy perennial. Frost hardy. Bears sprays of tubular, deep wine-red flowers from mid-summer to autumn. Has narrow, fresh green leaves.
♚ *P. 'Apple Blossom'* (illus. p.212). Semi-evergreen, bushy perennial. H and S 60cm (2ft). Frost hardy. Carries sprays of small, tubular, pale pink flowers from mid-summer onwards above narrow, fresh green foliage.
P. barbatus, syn. *Chelone barbata*. Semi-evergreen, rosette-forming perennial. H 1m (3ft), S 30cm (1ft). Frost hardy. From mid-summer to early autumn bears racemes of slightly nodding, tubular, 2-lipped, rose-red flowers. Flower stems rise from rosettes of oblong to oval, mid-green leaves.
P. 'Barbara Barker'. See *P. 'Beech Park'*.
♚ *P. 'Beech Park'*, syn. *P. 'Barbara Barker'* (illus. p.212). Semi-evergreen perennial. H and S 60cm (2ft). Frost hardy. Bears bright pink and white flowers. Leaves are linear and light green.
P. 'Burford Seedling'. See *P. 'Burgundy'*.
P. 'Burford White'. See *P. 'White Bedder'*.
P. 'Burgundy', syn. *P. 'Burford Seedling'* (illus. p.212). Robust, semi-evergreen perennial. H 1.2m (4ft), S 60cm (2ft). Frost hardy. Produces purplish-red flowers with white throats streaked dark red. Leaves are linear and light green.
P. campanulatus. Semi-evergreen, upright perennial. H 30–60cm (12–24in), S 30cm (12in). Frost hardy. Long racemes of bell-shaped, dark purple, violet, or, occasionally, white flowers appear in early summer above lance-shaped, toothed, mid-green leaves.
♚ *P. 'Chester Scarlet'* (illus. p.212). Semi-evergreen perennial. H and S 90cm (3ft). Frost hardy. Large, bright red flowers are borne above narrowly lance-shaped, light green leaves.
P. confertus. Semi-evergreen, neat, clump-forming perennial. H 45cm (18in), S 30cm (12in). Frost hardy. Bears spikes of tubular, creamy-yellow flowers above long, lance-shaped, mid-green leaves in early summer.
♚ *P. 'Countess of Dalkeith'* (illus. p.212). Erect, semi-evergreen perennial. H 1m (3ft), S 60cm (2ft). Frost hardy. Bears large, deep purple flowers each with a pure white throat. Leaves are linear and light green.
P. davidsonii. Evergreen, prostrate shrub. H 8cm (3in), occasionally more, S 15cm (6in) or more. Frost hardy. In late spring and early summer, funnel-shaped, violet to ruby-red flowers, with protruding lips, develop from leaf axils. Leaves are small, oval to rounded and leathery. Trim after flowering. subsp. *menziesii*. See *P. menziesii*.
P. diffusus. See *P. serrulatus*.
♚ *P. 'Evelyn'* (illus. p.212). Semi-evergreen, bushy perennial. H and S 45cm (18in). Frost hardy. Racemes of small, tubular, pink flowers open from mid-summer onwards. Has broadly lance-shaped, mid-green leaves.
P. fruticosus. Evergreen, upright, woody-based sub-shrub. H and S 15–30cm (6–12in). Frost hardy. Has lance-shaped to oval, toothed leaves and, in early summer, funnel-shaped, lipped, lavender-blue flowers. Is suitable for a rock garden. Trim back after flowering. subsp. *scouleri*. See *P. scouleri*.
P. 'Garnet'. See *P. 'Andenken an Friedrich Hahn'*.
♚ *P. hartwegii*. Semi-evergreen, erect perennial. H 60cm (24in) or more, S 30cm (12in). Frost hardy. Bears sprays of slightly pendent, tubular to bell-shaped, scarlet flowers from mid- to late summer. Lance-shaped leaves are mid-green.
P. heterophyllus 'True Blue'. Semi-evergreen, upright shrub. H and S 25cm (10in) or more. Frost hardy. Bears linear to lance-shaped, pale green leaves. Funnel-shaped, pure blue flowers are borne on short side-shoots in summer. Trim back after flowering. Suits a rock garden.
P. hirsutus. Short-lived, evergreen, open sub-shrub. H 60cm–1m (2–3ft), S 30–60cm (1–2ft). Frost hardy. In summer produces hairy, tubular, lipped, purple- or blue-flushed, white flowers. Leaves are oval and dark green. Is suitable for a rock garden.
'Pygmaeus' illus. p.328.
♚ *P. isophyllus* illus. p.133.
P. 'King George V' (illus. p.212). Semi-evergreen, upright perennial. H 75cm (30in), S 45–60cm (18–24in). Frost hardy. Trumpet-shaped, white-throated, bright crimson flowers are produced in racemes from mid-summer until first frosts. Narrowly oval leaves are mid-green.
P. 'Maurice Gibbs' (illus. p.212). Semi-evergreen perennial. H 90cm (3ft), S 60cm (2ft). Bears claret-red flowers with white throats. Leaves are narrowly lance-shaped and light green.
♚ *P. menziesii*, syn. *P. davidsonii* subsp. *menziesii*. Evergreen, prostrate shrub. H 5cm (2in), S 20cm (8in). Frost hardy. In late spring and early summer bears funnel-shaped, lavender-blue flowers. Leaves are small, oval to rounded and toothed.
P. 'Mother of Pearl' (illus. p.212). Semi-evergreen perennial. H to 1m (3ft), S 60cm (2ft). Frost hardy. Produces numerous small, pearly-mauve flowers tinted pink and white. Leaves are linear and light green.
♚ *P. newberryi* (Mountain pride). Evergreen, mat-forming shrub. H 15–20cm (6–8in), S 30cm (12in). Frost hardy. Branches are covered in small, oval, leathery, dark green leaves. Bears short sprays of tubular, lipped, deep rose-pink flowers in early summer. Trim back after flowering. Is good for a rock garden. f. *humilior* illus. p.302.
♚ *P. 'Pennington Gem'* (illus. p.212). Vigorous, semi-evergreen perennial. H 1m (3ft), S 45cm (18in). Frost hardy. Produces sprays of tubular, pink flowers from mid-summer to autumn. Leaves are narrow and fresh green.
♚ *P. pinifolius* illus. p.302.
P. 'Pink Endurance' (illus. p.212). Semi-evergreen dwarf perennial. H and S 30–38cm (12–15in). Frost hardy. Produces short sprays of of rose-pink flowers above narrow, mid-green leaves in summer.
P. 'Prairie Fire' (illus. p.212). Semi-evergreen, rosette-forming perennial. H 1.5m (5ft), S 60cm (2ft). Bears clusters of pinkish-red flowers. Leaves are narrowly lance-shaped and mid-green.
P. procerus. Upright, semi-evergreen perennial. H 50cm (20in), S 20cm (8in). Frost hardy. Leaves are oblong to lance-shaped. Produces slim spikes of funnel-shaped, blue-purple flowers in summer. Is suitable for a rock garden.
P. 'Royal White'. See *P. 'White Bedder'*.
♚ *P. 'Rubicundus'* (illus. p.212). Erect, semi-evergreen perennial. H 1.2m (4ft), S 60cm (2ft). Frost hardy. Bears very large, bright red flowers each with a white throat. Leaves are linear and light green.
♚ *P. rupicola*. Evergreen, prostrate shrub. H 5cm (2in), S 15cm (6in). Frost hardy. Has rounded to oval, fleshy, blue-grey leaves and, in summer, variable, funnel-shaped, pale to deep pink flowers. Is best grown in a rock garden. Seed may not come true.
♚ *P. 'Schoenholzeri'* (illus. p.212). Vigorous, semi-evergreen, upright perennial. H 1m (3ft), S 30–45cm (1–1½ft). Frost hardy. Produces racemes of trumpet-shaped, brilliant scarlet flowers from mid-summer to autumn. Lance-shaped to narrowly oval leaves are mid-green.
♚ *P. scouleri*, syn. *P. fruticosus* subsp. *scouleri*. Evergreen, upright sub-shrub. H and S 15–30cm (6–12in). Frost hardy. Has narrow, lance-shaped, toothed leaves. In summer produces long, funnel-shaped, lavender-blue flowers. Is suitable for a rock garden.
P. serrulatus, syn. *P. diffusus*, illus. p.303.
P. 'Six Hills'. Evergreen, prostrate shrub. H 5cm (2in), S 15cm (6in). Frost hardy. Has rounded, fleshy, grey-green leaves. In summer carries funnel-shaped, cool lilac flowers at stem tips. Is suitable for a rock garden.
P. 'Snow Storm'. See *P. 'White Bedder'*.
P. 'Sour Grapes' (illus. p.212). Semi-evergreen perennial. H 90cm (3ft), S 60cm (2ft). Bears deep purple-blue flowers suffused violet. Light green leaves are narrowly lance-shaped.
♚ *P. 'White Bedder'*, syn. *P. 'Burford White'*, *P. 'Royal White'*, *P. 'Snow Storm'* (illus p.212). Semi-evergreen, free-flowering perennial. H 70cm (28in), S 60cm (24in). Frost hardy. Produces white flowers with dark anthers. Leaves are linear and fresh green.

PENTACHONDRA (Epacridaceae) Genus of evergreen, spreading shrubs with heath-like leaves. Frost hardy. Needs full light and gritty, moist, peaty soil. Is difficult to grow, especially in hot, dry areas. Propagate by rooted offsets in spring, by semi-ripe cuttings in summer or by seed in autumn.
P. pumila. Evergreen, mat-forming, dense shrub. H 3–10cm (1¼–4in), S 20cm (8in) or more. Leaves are oblong to narrowly oval and purplish-green. Small, tubular, white flowers, with reflexed lobes, open in early summer, followed, though rarely in cultivation, by small, spherical, orange fruits.

PENTAPTERYGIUM. See AGAPETES.

PENTAS (Rubiaceae) Genus of mainly evergreen perennials and shrubs, grown for their flowers. Frost tender, min. 10–15°C (50–59°F). Needs full light or partial shade and fertile, well-drained soil. Water freely when in full growth, moderately at other times. May be hard pruned in winter. Propagate by softwood cuttings in summer or by seed in spring. Is prone to whitefly.
P. carnea. See *P. lanceolata*.
P. lanceolata, syn. *P. carnea*, illus. p.133.

PEPEROMIA (Piperaceae) Genus of annuals and evergreen perennials, grown for their ornamental foliage. Frost tender, min. 10°C (50°F). Grow in bright light or partial shade, ideally in a peat-based compost. Do not overwater. Propagate by division, seed or leaf or stem cuttings in spring or summer.
♚ *P. argyreia*, syn. *P. sandersii* (Watermelon plant). Evergreen, bushy,

compact perennial. H and S 20cm (8in). Has red-stalked, oval, fleshy, dark green leaves, to 10cm (4in) or more long, striped with broad bands of silver. Flowers are insignificant.
P. caperata illus. p.264.
P. clusiifolia (Baby rubber plant). Evergreen perennial with branching, sometimes prostrate, reddish-green stems. H to 20cm (8in), S 25cm (10in). Narrowly oval, fleshy leaves, 8–15cm (3–6in) long, are dark green, edged with red. Flowers are insignificant. Leaves of **'Variegata'** have cream-and-red margins.
P. glabella illus. p.269.
P. griseoargentea, syn. *P. hederifolia* (Ivy peperomia, Silver-leaf peperomia). Evergreen, bushy perennial. H to 15cm (6in), S 20cm (8in). Oval, fleshy leaves, 5cm (2in) or more long, each have a heart-shaped base, a quilted green surface and a silvery sheen. Flowers are insignificant.
P. hederifolia. See *P. griseoargentea.*
P. magnoliifolia (Desert privet). Evergreen, bushy perennial with erect, branching stems. H to 25cm (10in), S 20cm (8in) or more. Has almost rounded, thick, fleshy, glossy, green leaves, 10cm (4in) or more long. Flowers are insignificant. **'Green and Gold'** has green leaves blotched with yellow.
P. marmorata illus. p.268.
P. metallica. Evergreen perennial with erect, branching, reddish-green stems. H and S to 15cm (6in). Narrowly oval, dark green leaves, to 2.5cm (1in) long, have a metallic sheen, and wide, pale midribs above, reddish-green veins below. Flowers are insignificant.
P. nummulariifolia. See *P. rotundifolia.*
P. obtusifolia **'Variegata'** illus. p.269.
P. rotundifolia, syn. *P. nummulariifolia*. Evergreen, creeping perennial. H 5–8cm (2–3in), S 30cm (1ft) or more. Very slender stems produce tiny, rounded, fleshy, bright green leaves, 1cm (½in) wide. Flowers are insignificant. Is useful for a hanging basket.
P. rubella. Evergreen perennial with erect, branching, red stems. H and S 15cm (6in). Leaves, in whorls of 4, are 1cm (½in) long, narrowly oval, fleshy and dark green above, crimson below. Flowers are insignificant.
P. sandersii. See *P. argyreia.*
☙ *P. scandens* (Cupid peperomia). Evergreen, climbing or trailing perennial with pinkish-green stems. H and S to 1m (3ft). Oval, pointed, fleshy leaves, to over 5cm (2in) long, are waxy and bright green. Flowers are insignificant.

PERESKIA (Cactaceae)
Genus of deciduous cacti, some of which are climbing, with fleshy leaves and woody, green, then brown stems. Is considered to be the most primitive genus of the *Cactaceae*, producing true leaves unlike the majority of members of the family. Frost tender, min. 5–10°C (41–50°F). Needs sun and well-drained soil. Water moderately in summer. Propagate by stem cuttings in spring or summer.
P. aculeata illus. p.393. var. *godseffiana* (syn. *P.a.* 'Godseffiana') is a fast-growing, deciduous, erect, then climbing cactus. H to 10m (30ft), S 5m (15ft). Min. 5°C (41°F). Broadly oval, slightly fleshy, orange-brown leaves, usually purplish beneath and 9cm (3½in) long, mature to glossy, green. Short flower stems, carrying rose-like, single, orange-centred, cream flowers, 5cm (2in) across, appear in autumn only on plants over 1m (3ft) high. Cut back hard to main stems in autumn.
P. grandifolia, syn. *Rhodocactus grandifolius*, illus. p.394.

Pericallis x *hybrida.* See *Senecio* x *hybrida.*

PERILLA (Labiatae)
Genus of annuals, grown for their foliage. Half hardy. Grow in sun and in fertile, well-drained soil. Pinch out growing tips of young plants to encourage a bushy habit. Propagate by seed sown under glass in early spring.
P. frutescens. Moderately fast-growing, upright, bushy annual. H 60cm (2ft), S 30cm (1ft). Has oval, serrated, aromatic, reddish-purple leaves. In summer produces spikes of very small, tubular, white flowers.

PERIPLOCA (Asclepiadaceae)
Genus of deciduous or evergreen, twining climbers, grown for their leaves. Stems exude milky juice if cut. Frost hardy. Grow in sun and in any well-drained soil. Propagate by seed in spring or by semi-ripe cuttings in summer.
P. graeca (Silk vine). Deciduous, twining climber. H to 9m (28ft). Oval, glossy leaves are 2.5–5cm (1–2in) long. In summer produces clusters of 8–12 greenish-yellow flowers, purplish-brown inside, each with 5 lobes. Pairs of narrowly cylindrical seed pods, 12cm (5in) long, contain winged, tufted seeds. Scent of the flowers is thought by some to be unpleasant.

PERISTERIA. See ORCHIDS.
P. elata. Semi-evergreen, epiphytic orchid for a cool or intermediate greenhouse. H 1m (3ft). Bears tall spikes of cup-shaped, waxy, creamy-white flowers, 8cm (3in) across, in summer. Lips are small and have faint, purple marks. Has broadly oval, ribbed leaves, 45cm (18in) long. Needs shade in summer.

PERISTROPHE (Acanthaceae)
Genus of mainly evergreen perennials and sub-shrubs, grown usually for their flowers. Frost tender, min. 15°C (59°F). Grow in sun or partial shade and in well-drained soil; do not overwater in winter. Propagate by stem cuttings in spring or summer.
P. angustifolia. See *P. hyssopifolia.*
P. hyssopifolia, syn. *P. angustifolia.* Evergreen, bushy perennial. H to 60cm (2ft), S 1–1.2m (3–4ft). Broadly lance-shaped leaves, with long-pointed tips, are 8cm (3in) long. Small clusters of tubular, deep rose-pink flowers are borne in winter. **'Aureovariegata'** illus. p.231.

Pernettya mucronata. See *Gaultheria mucronata.*
Pernettya prostrata. See *Gaultheria myrsinoides.*
Pernettya pumila. See *Gaultheria pumila.*
Pernettya tasmanica. See *Gaultheria tasmanica.*

PEROVSKIA (Labiatae)
Genus of deciduous sub-shrubs, grown for their aromatic, grey-green foliage and blue flowers. Fully hardy. Needs full sun and very well-drained soil. Cut plants back hard, almost to base, in spring, as new growth starts. Propagate by softwood cuttings in late spring.
P. atriplicifolia. Deciduous, upright sub-shrub. H 1.2m (4ft), S 1m (3ft). Grey-white stems bear narrowly oval, coarsely toothed leaves. Bears 2-lipped, violet-blue flowers in long, slender spikes from late summer to mid-autumn. ☙ **'Blue Spire'** illus. p.138.
P. **'Hybrida'.** Deciduous, upright sub-shrub. H 1m (3ft), S 75cm (2½ft). Has oval, deeply lobed and toothed leaves and, from late summer to mid-autumn, tall spires of 2-lipped, deep lavender-blue flowers.

Persicaria affinis. See *Polygonum affine.*
Persicaria amplexicaulis. See *Polygonum amplexicaule.*
Persicaria bistorta. See *Polygonum bistorta.*
Persicaria campanulata. See *Polygonum campanulatum.*
Persicaria capitata. See *Polygonum capitatum.*
Persicaria sphaerostachya. See *Polygonum macrophyllum.*
Persicaria vacciniifolium. See *Polygonum vacciniifolium.*

PETASITES (Compositae)
Genus of invasive perennials, grown for their usually large leaves and value as ground cover. Fully hardy. Tolerates sun and prefers moist but well-drained soil. Propagate by division in spring or autumn.
P. fragrans (Winter heliotrope). Spreading, invasive perennial. H 23–30cm (9–12in), S 1.2m (4ft). Has rounded to heart-shaped, dark green leaves. Small, vanilla-scented, daisy-like, pinkish-white flower heads are produced in late winter before foliage.
P. japonicus illus. p.238.

PETREA (Verbenaceae)
Genus of evergreen shrubs and woody-stemmed, twining climbers, grown for their flowers. Frost tender, min. 13–15°C (55–9°F). Needs full light and fertile, well-drained soil. Water regularly, less when not in full growth. Provide support. Thin out and spur back crowded growth in spring. Propagate by semi-ripe cuttings in summer. Mealy bug and whitefly may cause problems.
P. volubilis illus. p.166.

PETROCOSMEA (Gesneriaceae)
Genus of evergreen, rhizomatous perennials. Frost tender, min. 2–5°C (36–41°F). Requires shade and well-drained, peaty soil. Propagate by seed in early spring or by leaf cuttings in early summer.
P. kerrii illus. p.322.

PETROPHYTON. See PETROPHYTUM.

PETROPHYTUM, syn. PETROPHYTON (Rosaceae)
Genus of evergreen, summer-flowering shrubs, grown for their spikes of small, fluffy flowers. Is good for growing on tufa or in alpine houses. Fully hardy. Needs sun and gritty, well-drained, alkaline soil.
May be difficult to grow. Propagate by softwood or semi-ripe cuttings in summer or by seed in autumn. Aphids and red spider mite may be troublesome in hot weather.
P. caespitosum. Evergreen, mat-forming shrub. H 5–8cm (2–3in), S 10–15cm (4–6in). Has clusters of small, oval leaves. Flower stems, 2cm (¾in) long, each carry a conical spike of small, fluffy, white flowers in summer.
P. hendersonii. Evergreen, mound-forming shrub. H 5–10cm (2–4in), S 10–15cm (4–6in). Has branched stems covered in rounded, blue-green leaves. Conical spikes of small, fluffy, white flowers are produced on 2.5cm (1in) stems in summer.

PETRORHAGIA (Caryophyllaceae)
Genus of annuals and perennials, grown for their flowers. Is suitable for rock gardens and banks. Fully hardy. Prefers sun and well-drained, sandy soil. Propagate by seed in autumn. Self seeds readily.
☙ *P. saxifraga*, syn. *Tunica saxifraga* (Tunic flower), illus. p.323. **'Rosette'** is a mat-forming perennial. H 10cm (4in), S 15cm (6in). Has tufts of grass-like leaves. In summer, slender stems carry a profusion of cup-shaped, double, white to pale pink flowers, sometimes veined deeper pink.

PETTERIA (Leguminosae)
Genus of one species of deciduous shrub, grown for its flowers. Is related to *Laburnum*, differing in its erect racemes. Fully hardy. Requires full sun and fertile, well-drained soil. Propagate by softwood cuttings in summer or by seed in autumn.
P. ramentacea (Dalmatian laburnum). Deciduous, upright shrub. H 2m (6ft), S 1m (3ft). Dense, upright spikes of fragrant, laburnum-like, yellow flowers appear in late spring and early summer. Mid-green leaves are each composed of 3 oval leaflets.

PETUNIA (Solanaceae)
Genus of perennials, grown as annuals, with showy, colourful flowers. Half hardy. Grow in a sunny position that is sheltered from wind and in fertile, well-drained soil. Dead-head regularly. Propagate by seed sown under glass in early spring. May suffer from viruses, including cucumber mosaic and tomato spotted wilt.
P. x *hybrida.* Moderately fast-growing, branching, bushy perennial, grown

PHACELIA

as an annual. H 15–30cm (6–12in), S 30cm (12in). Has oval, mid- to deep green leaves and, in summer-autumn, flared, trumpet-shaped, single or double flowers in a wide range of colours (available in mixtures or singly), including blue, violet, purple, red, pink and white. Flowers vary in form and size: those of Grandiflora hybrids are large (8–10cm (3–4in) wide), but they are often marked by rain and so should be protected; those of Multiflora hybrids are smaller (5cm (2in) wide) and are more rain-resistant.

Grandiflora types include the following:
    'Blue Frost' illus. p.283.
    Cascade Series has trailing stems and single flowers in a wide range of colours.
    'Colour Parade' has a wide colour range of single flowers that are ruffled.
    Flash Series has single flowers in a range of bright colours.
    'Magic Cherry' is a compact cultivar with single, cherry-red flowers.
    Picotee Series bears a mixed colour range of single, brightly coloured flowers that have contrasting margins (red, illus. p.280).
    'Razzle Dazzle' has single flowers, striped with white, in a wide range of colours.
    Recoverer Series bears single flowers in a range of single or mixed colours (white, illus. p.270).
    Star Series has a wide colour range of single, white-striped flowers (crimson, illus. p.278).
    Victorious Series illus. p.274.

Multiflora types include the following:
    Bonanza Series illus. p.275.
    'Cherry Tart' bears double, deep pink-and-white flowers.
    'Gypsy' is a cultivar with single, salmon-red flowers.
    Jamboree Series has pendulous stems bearing single flowers in a range of colours.
    'Mirage Velvet' illus. p.278.
    Pearl Series is a dwarf variety with small, single flowers in a wide range of colours.
    Picotee Ruffled Series has ruffled, single flowers, edged with white, in a range of colours.
    'Plum Crazy' bears single flowers that have contrasting veins and throats. Colours available include white, with yellow throat and veins, and shades of violet, pink and magenta, all with darker throats and veins.
    'Red Satin' has single, brilliant scarlet flowers.
    Resisto Series bears single flowers that are particularly rain-resistant and are available in a range of colours (blue, illus. p.280; red, illus. p.284; rose-pink, illus. p.277).

PHACELIA (Hydrophyllaceae)
Genus of annuals, biennials and perennials. Fully hardy. Grow in sun and in fertile, well-drained soil. Tall species may need support. Propagate by seed sown outdoors in spring or early autumn.
P. campanularia illus. p.286.
P. tanacetifolia. Moderately fast-growing, upright annual. H 60cm (2ft) or more, S 30cm (1ft). Has feathery, deep green leaves and, in summer, spikes of bell-shaped, lavender-blue flowers.

PHAEDRANASSA (Amaryllidaceae)
Genus of bulbs with tubular, often pendent flowers. Half hardy. Needs full sun or partial shade and fairly rich, well-drained soil. Feed with high-potash fertilizer in summer. Reduce watering in winter. Propagate by seed or offsets in spring.
P. carmioli illus. p.344.

Phaedranthus buccinatorius. See *Distictis buccinatoria.*

PHAIOPHLEPS (Iridaceae)
Genus of rush-like, rhizomatous perennials, related to *Sisyrinchium*, grown for their fragrant, trumpet-shaped flowers in spring. Frost hardy. Requires gritty, peaty soil and cool growing conditions, such as a peat garden would provide, and full sun or partial shade. Do not allow soil to dry out. Propagate by seed in winter or spring.
P. biflora, syn. *Sisyrinchium odoratissimum.* Clump-forming, spring- to summer-flowering, rhizomatous perennial. H 25–35cm (10–14in), S 5–8cm (2–3in). Has cylindrical, rush-like, erect, basal leaves. Bears a small head of pendent, white flowers that are striped and veined red.

PHAIUS. See ORCHIDS.
P. tankervilleae (illus. p.261). Semi-evergreen, terrestrial orchid. H 75cm (30in). Frost tender, min. 10°C (55°F). Tall spikes of flowers, 9cm (3½in) across, brown within, silvery-grey outside and each with a long, red-marked, pink lip, open in early summer. Leaves are broadly oval, ribbed and 60cm (24in) long. Provide semi-shade in summer.

PHALAENOPSIS. See ORCHIDS.
P. Allegria (illus. p.260). Evergreen, epiphytic orchid for a warm greenhouse. H 15cm (6in). Carries sprays of white flowers, to 12cm (5in) across; flowering season varies. Broadly oval, fleshy leaves are 15cm (6in) long. Needs shade in summer.
P. cornu-cervi (illus. p.262). Evergreen, epiphytic orchid for a warm greenhouse. H 15cm (6in). Yellowish-green flowers, 5cm (2in) across, with brown marks, are produced successively, either singly or in pairs, in summer. Has broadly oval leaves, 10cm (4in) long. Needs shade in summer.
P. Lady Jersey x Lippeglut (illus. p.261). Evergreen, epiphytic orchid for a warm greenhouse. H 15cm (6in). Bears tall, pendent spikes of pink flowers, 9cm (3½in) across; flowering season varies. Broadly oval leaves are 10cm (4in) long. Requires shade in summer.
P. Lundy (illus. p.262). Evergreen, epiphytic orchid for a warm greenhouse. H 15cm (6in). Bears sprays of red-striped, yellow flowers, 8cm (3in) across; flowering season varies. Has broadly oval leaves, 23cm (9in) long. Grow in shade in summer.

PHALARIS (Gramineae). See GRASSES, BAMBOOS, RUSHES and SEDGES.
♥ *P. arundinacea* 'Picta', syn. *P.a.* var. *picta*, *P.a.* 'Tricolor', illus. p.182.

PHARBITIS. See IPOMOEA.

Phaseolus caracalla. See *Vigna caracalla.*

PHEGOPTERIS (Polypodiaceae)
Genus of deciduous ferns. Fully hardy. Grow in semi-shade and in humus-rich, moist but well-drained soil. Propagate by division in spring or by spores in summer.
P. connectilis, syn. *Thelypteris phegopteris* (Beech fern). Deciduous fern. H 23cm (9in), S 30cm (12in). Broadly lance-shaped, mid-green fronds, each consisting of tiny, triangular pinnae on wiry stalks, arise from a creeping rootstock. Is useful for ground cover.
P. hexagonoptera, syn. *Dryopteris hexagonoptera*, *Thelypteris hexagonoptera* (Broad beech fern). Deciduous fern. H 45cm (18in), S 30cm (12in). Broadly lance-shaped, much-divided, mid-green fronds, with oblong to triangular, indented pinnae, arise from a creeping rootstock. Needs a shaded position.

PHELLODENDRON (Rutaceae)
Genus of deciduous trees, grown for their foliage, which colours well in autumn. Male and female flowers are produced on different plants. Fully hardy, but young growth is susceptible to damage by late frosts. Needs full sun and fertile, well-drained soil. Does best during hot summers. Propagate by softwood cuttings in summer, by seed in autumn or by root cuttings in late winter.
P. amurense (Amur cork tree). Deciduous, spreading tree. H 12m (40ft), S 15m (50ft). Has corky, dark bark when old. Aromatic leaves, with 5 to 11 oblong leaflets, are glossy, dark green, becoming yellow in autumn. Tiny, green flowers in early summer are followed by small, rounded, black fruits.
P. chinense illus. p.55.

PHILADELPHUS (Hydrangeaceae)
Genus of deciduous, mainly summer-flowering shrubs, grown for their usually fragrant flowers. Fully to frost hardy. Needs sun and fertile, well-drained soil. After flowering, cut some older shoots back to young growths, leaving young shoots to flower the following year. Propagate by softwood cuttings in summer. May become infested with aphids.
♥ *P.* 'Beauclerk' illus. p.105.
♥ *P.* 'Belle Etoile' illus. p.106.
P. 'Boule d'Argent' illus. p.107.
P. coronarius (Mock orange).
♥ 'Aureus' is a deciduous, upright shrub. H 2.5m (8ft), S 1.5m (5ft). Fully hardy. Clusters of very fragrant, 4-petalled, creamy-white flowers are produced in late spring and early summer. Oval, golden-yellow, young leaves turn yellow-green in summer. Protect from full sun. ♥ 'Variegatus' illus. p.109.
P. 'Dame Blanche' illus. p.107.
P. delavayi. Deciduous, upright shrub. H 3m (10ft), S 2.5m (8ft). Frost hardy. Dense clusters of very fragrant, 4-petalled, white flowers, with sometimes purple-flushed, green sepals, open from early to mid-summer. Leaves are dark green, oval and toothed. f. *melanocalyx* illus. p.109.
P. 'Lemoinei' illus. p.108.
P. magdalenae. Deciduous, bushy shrub. H and S 4m (12ft). Fully hardy. Bark peels on older shoots. Narrowly oval, dark green leaves set off fragrant, 4-petalled, white flowers in late spring and early summer.
♥ *P.* 'Manteau d'Hermine' illus. p.128.
♥ *P.* 'Sybille'. Deciduous, arching shrub. H 1.2m (4ft), S 2m (6ft). Fully hardy. Very fragrant, 4-petalled, white flowers, each with a central, pink stain, are borne profusely in early and mid-summer. Leaves are mid-green and oval.
♥ *P.* 'Virginal'. Vigorous, deciduous, upright shrub. H 3m (10ft), S 2.5m (8ft). Fully hardy. Has oval, dark green leaves. Produces masses of large, very fragrant, double or semi-double, white flowers from early to mid-summer.

x PHILAGERIA (Liliaceae)
Hybrid genus (*Philesia* x *Lapageria*) of one evergreen, scrambling or twining shrub. Frost tender, min. 5°C (41°F). Grow in semi-shade and in well-drained, preferably acid soil. Propagate by layering in late summer or autumn.
x *P. veitchii.* Evergreen, scrambling or twining shrub. H 3–4m (10–12ft). Has oblong, slightly toothed leaves. Nodding, tubular, rose-pink flowers are produced in leaf axils in summer.

PHILESIA (Liliaceae)
Genus of one species of evergreen shrub, grown for its showy flowers. Frost hardy, but thrives only in mild, moist areas. Needs semi-shade and humus-rich, moist, acid soil. Benefits from an annual dressing of leaf mould. Propagate by semi-ripe cuttings in summer or by suckers in autumn.
P. magellanica. Evergreen, erect shrub. H 1m (3ft), S 2m (6ft). Bears trumpet-shaped, waxy, crimson-pink flowers, in leaf axils, from mid-summer to late autumn. Narrowly oblong, dark green leaves are bluish-white beneath.

PHILLYREA (Oleaceae)
Genus of evergreen shrubs and trees, with inconspicuous flowers, grown for their foliage. Frost hardy, but in cold areas requires shelter. Does best in full sun and in fertile, well-drained soil. To restrict growth, cut back in spring. Propagate by semi-ripe cuttings in summer.
P. angustifolia. Evergreen, bushy, dense shrub. H and S 3m (10ft). Leaves are narrowly oblong and dark green. Small, fragrant, 4-lobed, greenish-white flowers in late spring and early summer are followed by spherical, blue-black fruits.

P. decora. See *Osmanthus decorus*.
P. latifolia. Evergreen, rounded shrub or tree. H and S 8m (25ft). Has oval, glossy, dark green leaves. Bears tiny, fragrant, 4-lobed, greenish-white flowers from late spring to early summer, then spherical, blue-black fruits.

PHILODENDRON (Araceae)
Genus of evergreen shrubs and woody-based, root climbers, grown for their handsome leaves. Intermittently bears insignificant flowers. Frost tender, min. 15–18°C (59–64°F). Needs partial shade and humus-rich, well-drained soil. Water moderately, sparingly in cold weather. Provide support. Young stem tips may be removed to promote branching. Propagate by leaf-bud or stem-tip cuttings in summer.
P. auritum of gardens. See *Syngonium auritum*.
P. bipinnatifidum, syn. *P. selloum*, illus. p.122.
♀ *P.* '**Burgundy**'. Slow-growing, evergreen, woody-based, root climber. H 2m (6ft) or more. Leaves are narrowly oblong, red-flushed, deep green above, wine-red beneath, and up to 30cm (12in) long.
P. cordatum (Heart leaf). Moderately vigorous, evergreen, woody-based, root climber. H 3m (10ft) or more. Has heart-shaped, lustrous, rich green leaves, to 45cm (18in) long.
P. domesticum, syn. *P. hastatum* of gardens (Elephant's ear, Spade leaf). Fairly slow-growing, evergreen, woody-based, root climber. H 2–3m (6–10ft). Lustrous, bright green leaves, 30–40cm (12–16in) long, are arrow-shaped on young plants and later have prominent, basal lobes.
♀ *P. erubescens* (Blushing philodendron). Evergreen, erect, woody-based, root climber. H to 3m (10ft). Oval to triangular leaves, 15–25cm (6–10in) long, have long, red stalks and are dark green with a lustrous, coppery flush.
P. hastatum of gardens. See *P. domesticum*.
P. laciniatum. See *P. pedatum*.
P. melanochrysum illus. p.180.
P. pedatum, syn. *P. laciniatum*. Slow-growing, evergreen, woody-based, root climber. H 2–3m (6–10ft). Has oval, lustrous, deep green leaves, 30–80cm (12–32in) long, cut into 5 or 7 prominent lobes.
P. pinnatifidum. Evergreen, erect, robust, unbranched shrub. H to 3m (10ft), S 1–2m (3–6ft). Glossy, deep green leaves are broadly oval in outline, 40–60cm (16–24in) long and divided into 15 or more finger-like lobes.
P. sagittatum. See *P. sagittifolium*.
P. sagittifolium, syn. *P. sagittatum*. Slow-growing, evergreen, woody-based, root climber. H 2–3m (6–10ft). Oval leaves with basal lobes are up to 40–60cm (16–24in) long and glossy, bright green.
♀ *P. scandens* illus. p.180.
P. selloum. See *P. bipinnatifidum*.
P. trifoliatum. See *Syngonium auritum*.

PHLEBODIUM (Polypodiaceae)
Genus of evergreen or semi-evergreen ferns. Frost tender, min. 5°C (41°F). Needs full light or semi-shade and humus-rich, moist but well-drained soil. Remove fading fronds regularly. Propagate by division in spring or by spores in summer.
♀ *P. aureum*, syn. *Polypodium aureum*, illus. p.187. '**Mandaianum**' illus. p.186.

PHLOMIS (Labiatae)
Genus of evergreen, summer-flowering shrubs and perennials, grown for their conspicuous, hooded flowers, which are borne in dense whorls, and for their foliage. Fully to frost hardy. Prefers full sun and well-drained soil. Propagate by seed in autumn; shrubs may also be increased from softwood cuttings in summer, perennials by division in spring.
P. cashmeriana. Evergreen, upright shrub. H 60cm (24in), S 45cm (18in). Frost hardy. Produces masses of 2-lipped, pale lilac flowers in summer. Narrowly oval, mid-green leaves have woolly, white undersides.
♀ *P. chrysophylla.* Evergreen, rounded, stiffly branched shrub. H and S 1m (3ft). Frost hardy. Bears 2-lipped, golden-yellow flowers in early summer. Oval leaves are grey-green when young, becoming golden-green.
♀ *P. fruticosa* illus. p.140.
P. italica illus. p.132.
P. longifolia var. *bailanica.* Evergreen, bushy shrub. H 1.2m (4ft), S 1m (3ft). Frost hardy. Leaves are oblong to heart-shaped, deeply veined and bright green. Has 2-lipped, deep yellow flowers from early to mid-summer.
♀ *P. russeliana* illus. p.219.

PHLOX (Polemoniaceae)
Genus of mainly late spring- or summer-flowering annuals and perennials, some of which are semi-evergreen or evergreen, grown for their terminal panicles or profusion of brightly coloured flowers. Fully to half hardy. Does best in sun or semi-shade and in fertile, moist but well-drained soil; some species prefer acid soil; in light, dry soils is better grown in partial shade. Trim back rock garden species after flowering. Propagate rock garden species and hybrids by cuttings from non-flowering shoots in spring or summer; species by seed in autumn or spring; *P. maculata*, *P. paniculata* and their cultivars also by division in early spring or by root cuttings in winter; and annuals by seed in spring. *P. maculata*, *P. paniculata* and their cultivars are susceptible to eelworm.
♀ *P. adsurgens.* Evergreen, mat-forming, prostrate perennial. H 10cm (4in), S 30cm (12in). Fully hardy. Woody-based stems are clothed in oval, light to mid-green leaves. In summer produces terminal clusters of short-stemmed, saucer-shaped, purple, pink or white flowers with overlapping petals. Is good for a rock garden or peat bed. Prefers partial shade and gritty, peaty, acid soil. '**Wagon Wheel**' illus. p.324.
P. amoena '**Variegata**'. See *P.* x *procumbens* '**Folio-variegata**'.
P. bifida illus. p.329.
P. caespitosa. Evergreen, mound-forming, compact perennial. H 8cm (3in), S 12cm (5in). Fully hardy. Leaves are narrow and needle-like. Solitary almost stemless, saucer-shaped, lilac or white flowers are borne in summer. Suits a rock garden or trough. Needs sun and very well-drained soil.
P. '**Camla**' illus. p.327.
♀ *P.* '**Chatahoochee**' illus. p.304.
♀ *P. divaricata.* Semi-evergreen, creeping perennial. H 30cm (12in) or more, S 20cm (8in). Fully hardy. In early summer, upright stems carry saucer-shaped, lavender-blue flowers in loose clusters. Leaves are oval. Suits a rock garden or peat bed. Prefers semi-shade and moist but well-drained, peaty soil. subsp. *laphamii* illus. p.304.
♀ *P. douglasii* '**Boothman's Variety**' illus. p.328. ♀ '**Crackerjack**' illus. p.327. '**May Snow**' is an evergreen, mound-forming perennial. H 8cm (3in), S 20cm (8in). Fully hardy. Masses of saucer-shaped, white flowers are carried in early summer. Leaves are lance-shaped and mid-green. Is suitable for a rock garden, wall or bank. Vigorous, compact ♀ '**Red Admiral**', H 15cm (6in), has crimson flowers.
P. drummondii (Annual phlox). **Beauty Series** is a group of moderately fast-growing, compact, upright annuals. H 15cm (6in), S 10cm (4in). Half hardy. Has lance-shaped, pale green leaves and, from summer to early autumn, heads of star-shaped flowers in many colours, including red, pink, blue, purple and white. '**Carnival**' has larger flowers with contrasting centres. **Cecily Series** illus. p.278. '**Petticoat**' has bicoloured flowers. **Twinkle Series** purplish-red, illus. p.278; vivid red, illus. p.279.
P. '**Emerald Cushion**' illus. p.329.
P. hoodii. Evergreen, compact, prostrate perennial. H 5cm (2in), S 10cm (4in). Fully hardy. Solitary flat, white flowers open in early summer above fine, needle-like, hairy leaves. Suits a rock garden. Needs sun and very well-drained soil.
P. maculata. Erect perennial. H 1m (3ft), S 45cm (1½ft). Fully hardy. In summer produces cylindrical panicles of tubular, 5-lobed, mauve-pink flowers above oval, mid-green leaves. ♀ '**Alpha**' (illus. p.206) has rose-pink flowers. ♀ '**Omega**' (illus. p.206) has white flowers, each with a lilac eye.
P. paniculata. Upright perennial, seldom grown, as is replaced in gardens by its more colourful cultivars. H 1.2m (4ft), S 60cm (2ft). Fully hardy. Tubular, 5-lobed flowers are borne in conical heads above oval, mid-green leaves in late summer. '**Aida**' (illus. p.206) is purple-red, each flower with a purple eye. Flowers of ♀ '**Amethyst**' (illus. p.206) are pale lilac with paler-edged petals. '**Balmoral**' (illus. p.206) has large, rosy-mauve flowers. ♀ '**Brigadier**' (illus. p.206) has deep orange-red flowers. ♀ '**Bright Eyes**' has pale pink flowers, each with a red eye. '**Eva Cullum**' (illus. p.206) has clear pink flowers with magenta eyes. ♀ '**Eventide**' (illus. p.206) produces lavender-blue flowers.
♀ '**Fujiyama**' (illus. p.206) has star-shaped, white flowers. Flowers of '**Graf Zeppelin**' (illus. p.206) are white with red centres. '**Hampton Court**' (illus. p.206) is a mauve-blue cultivar, with dark green foliage. '**Harlequin**' (illus. p.206) has reddish-purple flowers. Leaves are variegated ivory-white. ♀ '**Le Mahdi**' (illus. p.206) has deep purple flowers. '**Mia Ruys**' (illus. p.206), H 45cm (18in), has large, white flowers, and is shorter than most other cultivars. ♀ '**Mother of Pearl**' (illus. p.206) has white flowers tinted pink. '**Norah Leigh**' (illus. p.206) has pale lilac flowers and ivory-variegated leaves. ♀ '**Prince of Orange**' (illus. p.206) is orange-red. '**Russian Violet**' (illus. p.206) is of open habit and has pale lilac-blue flowers. Flowers of '**Sandringham**' (illus. p.206) have widely spaced petals and are pink with darker centres. '**Sir John Falstaff**' has large, deep salmon flowers, each with a cherry-red eye. ♀ '**White Admiral**' (illus. p.206) bears pure white flowers. Those of ♀ '**Windsor**' (illus. p.206) are carmine-rose with red eyes.
P. x *procumbens* '**Folio-variegata**', syn. *P. amoena* '**Variegata**'. Evergreen, prostrate perennial. H 2.5cm (1in), S 25cm (10in). Fully hardy. Has oval, glossy, green leaves with white margins. In early summer produces small, saucer-shaped, bright cerise-pink flowers. Is suitable for a rock garden.
♀ '**Millstream**', H to 15cm (6in), S 30cm (12in), has deep pink flowers with white eyes.
P. stolonifera (Creeping phlox). Evergreen, prostrate, spreading perennial. H 10–15cm (4–6in), S 30cm (12in) or more. Fully hardy. Produces small, cup-shaped, pale blue flowers in early summer. Leaves are oblong to oval. Prefers moist, peaty, acid soil; is good for a peat bed or rock garden. '**Ariane**' illus. p.321. ♀ '**Blue Ridge**' has masses of lavender-blue flowers.
P. subulata. Evergreen, mound-forming perennial. H 10cm (4in), S 20cm (8in). Fully hardy. Bears fine, needle-like leaves. Masses of star-shaped, white, pink or mauve flowers appear in early summer. Is good for a sunny rock garden. '**Marjorie**' illus. p.326.

PHOENIX (Palmae)
Genus of evergreen palms, grown for their overall appearance and their edible fruits. Frost tender, min. 10–15°C (50–59°F). Grows in any fertile, well-drained soil and in full light, though tolerates partial shade. Water potted specimens moderately, less during winter. Propagate by seed in spring at not less than 24°C (75°F). Red spider mite may be a nuisance.
♀ *P. canariensis* (Canary Island date palm). Evergreen, upright palm with a robust trunk. H 18m (60ft) or more, S 10m (30ft) or more. Min. 10°C (50°F). Feather-shaped, arching leaves, each to 5m (15ft) long, are divided into narrowly lance-shaped, leathery, bright green leaflets. Bears large, pendent clusters of tiny, yellowish-brown flowers that on mature specimens arefollowed by shortly oblong, yellow to red fruits in autumn-winter.

♟ *P. roebelenii* (Miniature date palm, Pygmy date palm). Evergreen palm with a slender trunk. H 2–4m (6–12ft), S 1–2m (3–6ft). Min. 15°C (59°F). Has feather-shaped, arching, glossy, dark green leaves, 1–1.2m (3–4ft) long, and, in summer, large panicles of tiny, yellow flowers. Egg-shaped, black fruits are borne in pendent clusters, 45cm (18in) long, in autumn.

PHORMIUM (Agavaceae)
New Zealand flax
Genus of evergreen perennials, grown for their bold, sword-shaped leaves. Frost hardy. Requires sun and moist but well-drained soil. Propagate by division or seed in spring.
P. colensoi. See *P. cookianum.*
♟ *P. cookianum*, syn. *P. colensoi* (Mountain flax). Evergreen, upright perennial. H 1–2m (3–6ft), S 30cm (1ft). Has tufts of sword-shaped, upright, dark green leaves. Panicles of tubular, pale yellowish-green flowers are borne in summer. ♟ **'Tricolor'** has leaves striped vertically with red, yellow and green. **'Variegatum'** has cream-striped leaves.
♟ *P. tenax.* Evergreen, upright perennial. H 3m (10ft), S 1–2m (3–6ft). Has tufts of sword-shaped, stiff, dark green leaves. Panicles of tubular, dull red flowers are produced on short, slightly glaucous green stems in summer. Thrives by the sea. **'Aurora'** has leaves vertically striped with red, bronze, salmon-pink and yellow. **'Bronze Baby'** illus. p.266. **'Dazzler'** illus. p.228. **'Purpureum'** illus. p.196. **'Veitchii'** has broad, creamy-white-striped leaves.

PHOTINIA (Rosaceae)
Genus of evergreen or deciduous shrubs and trees, with small, white flowers, grown for their foliage and, in the case of deciduous species, for their autumn colour and fruits. Fully to frost hardy, but protect evergreen species from strong, cold winds. Requires sun or semi-shade and fertile, well-drained soil; some species prefer acid soil. Propagate evergreen and deciduous species by semi-ripe cuttings in summer, deciduous species also by seed in autumn.
P. arbutifolia. See *Heteromeles arbutifolia.*
P. davidiana, syn. *Stranvaesia davidiana*, illus. p.68.
P. x fraseri. Group of evergreen, hybrid shrubs. Frost hardy. Has bold, oblong leaves and good resistance to damage by late frosts. Young growths are attractive over a long period. **'Birmingham'** illus. p.87. ♟ **'Red Robin'** is upright and dense. H 6m (20ft), H 4m (12ft). Glossy, dark green leaves are brilliant red when young. Broad heads of 5-petalled flowers are borne in late spring.
P. nussia, syn. *Stranvaesia nussia.* Evergreen, spreading tree. H and S 6m (20ft). Half hardy. Has oblong, leathery, glossy, dark green leaves and saucer-shaped, 5-petalled, white flowers in mid-summer, followed by rounded, orange-red fruits.
P. serratifolia, syn. *P. serrulata.* Evergreen, upright shrub or bushy-headed tree. H 10m (30ft), S 8m (25ft). Frost hardy. Oblong, often sharply toothed leaves are red when young, maturing to glossy, dark green. Small, 5-petalled flowers from mid- to late spring are sometimes followed by spherical, red fruits. Young growth may be damaged by late frosts.
P. serrulata. See *P. serratifolia.*
♟ *P. villosa.* Deciduous, upright shrub or spreading tree. H and S 5m (15ft). Fully hardy. Oval, dark green leaves, bronze-margined when young, are brilliant orange-red in autumn. Has clusters of 5-petalled flowers in late spring, then spherical, red fruits. Prefers acid soil.

PHRAGMIPEDIUM. See **ORCHIDS**.
P. caudatum. Evergreen, epiphytic orchid for an intermediate greenhouse. H 23cm (9in). In summer bears sprays of flowers with light green and tan sepals and pouches and drooping, ribbon-like, yellow and brownish-crimson petals to 30cm (12in) long. Has narrowly oval leaves, 30cm (12in) long. Needs shade in summer.

PHUOPSIS (Rubiaceae)
Genus of one species of mat-forming, summer-flowering perennial, grown for its small, pungent, tubular flowers. Is good for ground cover, especially on banks and in rock gardens. Fully hardy. Needs sun and well-drained soil. Propagate by division in spring, by semi-ripe cuttings in summer or by seed in autumn.
P. stylosa, syn. *Crucianella stylosa*, illus. p.300.

PHYGELIUS (Scrophulariaceae)
Genus of evergreen or semi-evergreen shrubs and sub-shrubs, grown for their showy, tubular flowers. Frost hardy, but in most areas plant in a sheltered position; will attain a considerably greater height when grown against a south- or west-facing wall. Needs sun and fertile, well-drained but not too dry soil. Usually loses leaves or has shoots cut to ground by frosts. Cut back to just above ground level in spring, or, if plants have woody bases, prune to live wood. Propagate by softwood cuttings in summer.
P. aequalis illus. p.136. **'Yellow Trumpet'** illus. p.139.
P. capensis **'Coccineus'.** Evergreen or semi-evergreen, upright sub-shrub. H 1.5m (5ft), S 2m (6ft). Tubular, curved, bright orange-red flowers, each with a red mouth and a yellow throat, are produced from mid-summer to early autumn in tall, slender spires amid triangular, dark green leaves.
P. x rectus **'Winchester Fanfare'.** Evergreen or semi-evergreen, upright sub-shrub. H 1.5m (5ft), S 2m (6ft). Has pendulous, tubular, dusky, reddish-pink flowers, each with scarlet lobes and a yellow throat, from mid-summer to early autumn, and triangular, dark green leaves.

Phyllanthus nivosus. See *Breynia nivosa.*

x PHYLLIOPSIS (Ericaceae)
Hybrid genus (*Phyllodoce* x *Kalmiopsis*) of one species of evergreen shrub, grown for its flowers. Is suitable for peat beds and rock gardens. Fully hardy. Needs partial shade and peaty, acid soil. Trim back after flowering to maintain a compact habit. Propagate by semi-ripe cuttings in late summer.
x *P. hillieri* **'Pinocchio'.** Evergreen, upright shrub. H 20cm (8in), S 25cm (10in). Branched stems bear thin, oval leaves. Long, open clusters of bell-shaped, very deep pink flowers appear in spring and intermittently thereafter.

Phylitis scolopendrium. See *Asplenium scolopendrium.*
Phylitis scolopendrium **'Marginatum'.** See *Asplenium scolopendrium* **'Marginatum'.**

PHYLLOCLADUS (Podocarpaceae).
See **CONIFERS**.
P. aspleniifolius (Tasman celery pine). Slow-growing, upright conifer. H 5–10m (15–30ft), S 3–5m (10–15ft). Half hardy. Instead of true leaves has flattened, leaf-like structures known as phylloclades; these are dull dark green and resemble celery leaves in outline. Produces inedible, white-coated nuts with fleshy, red bases.
P. trichomanoides illus. p.80.

PHYLLODOCE (Ericaceae)
Genus of evergreen shrubs, grown for their fine, heath-like leaves and attractive flowers. Fully to frost hardy. Needs semi-shade and moist, peaty, acid soil. Propagate by semi-ripe cuttings in late summer or by seed in spring.
♟ *P. caerulea*, syn. *P. taxifolia*, illus. p.296.
P. empetriformis illus. p.296.
P. x intermedia **'Drummondii'** illus. p.295. **'Fred Stoker'** is an evergreen, upright shrub. H and S 23cm (9in). Fully hardy. Has narrow, glossy, green leaves. From late spring to early summer carries terminal clusters of pitcher-shaped, bright reddish-purple flowers on slender, red stalks.
♟ *P. nipponica.* Evergreen, upright shrub. H 10–20cm (4–8in), S 10–15cm (4–6in). Frost hardy. Freely branched stems bear fine, linear leaves and, in late spring and summer, stalked, bell-shaped, white flowers from their tips.
P. taxifolia. See *P. caerulea.*

PHYLLOSTACHYS
(Bambusoideae). See **GRASSES, BAMBOOS, RUSHES** and **SEDGES**.
♟ *P. aurea* (Fishpole bamboo, Golden bamboo). Evergreen, clump-forming bamboo. H 6–8m (20–25ft), S indefinite. Frost hardy. Erect, grooved stems have cup-shaped swellings beneath most nodes, which, towards the base, are often close together and distorted. Bears mid-green leaves and unimportant flower spikes.
P. aureosulcata (Golden-groove bamboo). Evergreen, clump-forming bamboo. H 6–8m (20–25ft), S indefinite. Frost hardy. Bears striped sheaths and yellow grooves on rough, brownish-green stems. Mid-green leaves are up to 15cm (6in) long; flower spikes are unimportant.
P. bambusoides illus. p.184.
P. flexuosa illus. p.184.
P. **'Henonis'.** See *P. nigra* var. *henonis.*
♟ *P. nigra* (Black bamboo). Evergreen, clump-forming bamboo. H 6–8m (20–25ft), S indefinite. Frost hardy. Grooved, greenish-brown stems turn black in second season. Almost unmarked culm sheaths bear bristled auricles, mid-green leaves and unimportant flower spikes.
♟ var. *henonis* (syn. *P.* 'Henonis') illus. p.184.
♟ *P. viridiglaucescens* illus. p.185.

x PHYLLOTHAMNUS (Ericaceae)
Hybrid genus (*Phyllodoce* x *Rhodothamnus*) of one species of evergreen shrub, grown for its foliage and flowers. Is good for peat beds and rock gardens. Fully hardy. Needs a sheltered, semi-shaded site and moist, acid soil. Propagate by semi-ripe cuttings in late summer.
x *P. erectus.* Evergreen, upright shrub. H and S 15cm (6in). Bears small, linear, deep green leaves. Clusters of slender-stalked, bell-shaped, soft rose-pink flowers are produced in late spring.

PHYSALIS (Solanaceae)
Chinese lantern
Genus of summer-flowering perennials and annuals, grown mainly for their decorative, lantern-like calyces and fruits, produced in autumn. Fully to half hardy. Grows in sun or shade and in well-drained soil. Propagate by division or softwood cuttings in spring, annuals by seed in spring or autumn.
♟ *P. alkekengi* (Bladder cherry, Winter cherry). Spreading perennial, sometimes grown as an annual. H 45cm (18in), S 60cm (24in). Fully hardy. Inconspicuous, nodding, star-shaped, white flowers in summer are followed, in autumn, by rounded, bright orange-red fruits, surrounded by inflated, orange calyces. Leaves are mid-green and oval.
P. peruviana (Cape gooseberry, Strawberry tomato). Spreading perennial. H and S to 1.2m (4ft). Half hardy. Inconspicuous, star-shaped, purple-marked, yellow flowers in summer are followed by rounded, yellow fruits enclosed in inflated, pale cream calyces. Leaves are mid-green and oval to triangular.

PHYSOCARPUS (Rosaceae)
Genus of deciduous, mainly summer-flowering shrubs, grown for their foliage and flowers. Fully hardy. Requires sun and fertile, not too dry soil. Prefers acid soil and does not grow well on shallow, chalky soil. Thin established plants occasionally by cutting some older shoots back to ground level after flowering. Propagate by softwood cuttings in summer.
P. opulifolius (Ninebark). Deciduous, arching, dense shrub. H 3m (10ft), S 5m (15ft). Has peeling bark and broadly oval, toothed and lobed, mid-green leaves. Clusters of tiny, sometimes pink-tinged, white flowers

are borne in early summer. ♥ **'Dart's Gold'** illus. p.116.

PHYSOPLEXIS (Campanulaceae)
Genus of one species of tufted perennial, grown for its flowers. Is good grown on tufa, in rock gardens, troughs and alpine houses. Fully hardy. Needs sun and very well-drained, alkaline soil, but should face away from midday sun. Keep fairly dry in winter. Propagate by seed in autumn or by softwood cuttings in early summer. Is susceptible to slug damage.
♥ **P. comosa**, syn. *Phyteuma comosum*, illus. p.328.

PHYSOSTEGIA (Labiatae)
Obedient plant
Genus of summer- to early autumn-flowering perennials. Fully hardy. Needs sun and fertile, moist but well-drained soil. Propagate by division in spring.
P. virginiana. Erect perennial. H 1m (3ft), S 60cm (2ft). In late summer produces spikes of hooded, 2-lipped, rose-purple flowers with hinged stalks that allow flowers to remain in position once moved. Has lance-shaped, toothed, mid-green leaves.
♥ **'Summer Snow'** has pure white flowers. **'Variegata'** illus. p.208.
♥ **'Vivid'** illus. p.243.

PHYTEUMA (Campanulaceae)
Genus of early- to mid-summer-flowering perennials that are useful for rock gardens. Fully hardy. Needs sun and well-drained soil. Propagate by seed in autumn.
P. comosum. See *Physoplexis comosa*.
P. scheuchzeri illus. p.305.

PHYTOLACCA (Phytolaccaceae)
Genus of perennials and evergreen shrubs and trees, grown for their overall appearance and decorative but poisonous fruits. Fully hardy to frost tender, min. 5°C (41°F). Tolerates sun or shade and requires fertile, moist soil. Propagate by seed in autumn or spring.
P. americana (Red-ink plant, Virginian pokeweed). Upright, spreading perennial. H and S 1.2–1.5m (4–5ft). Fully hardy. Oval to lance-shaped, mid-green leaves are tinged purple in autumn. Shallowly cup-shaped, sometimes pink-flushed, white-and-green flowers, borne in terminal racemes in summer, are followed by poisonous, rounded, fleshy, blackish-purple berries.
P. clavigera. Stout, upright perennial. H and S 1.2m (4ft). Fully hardy. Has brilliant crimson stems, oval to lance-shaped, mid-green leaves that turn yellow in autumn and, in summer, clusters of shallowly cup-shaped, pink flowers, followed by poisonous, rounded, blackish-purple berries.

PICEA (Pinaceae)
Spruce
Genus of conifers with needle-like leaves set on a pronounced peg on the shoots and arranged spirally. Cones are pendulous and ripen in their first autumn; scales are woody and flexible. See also CONIFERS.
P. abies (Common spruce, Norway spruce; illus. p.78). Fast-growing conifer, narrowly conical when young, broader with age. H 20–30m (70–100ft), S 5–7m (15–22ft). Fully hardy. Has needle-like, dark green leaves and bears pendulous cones. **'Clanbrassiliana'**, H 5m (15ft), S 3–5m (10–15ft), is slow-growing, rounded and spreading. **'Gregoryana'** (illus. p.85), H and S 60cm (2ft), is slow-growing, with a dense, globose form. **'Inversa'**, H 5–10m (15–30ft), S 2m (6ft), has an erect leader, but pendent side branches. ♥ **'Little Gem'**, H and S 30–50cm (12–20in), has a nest-shaped, central depression caused by spreading branches.
♥ **'Nidiformis'**, H 1m (3ft), S 1–2m (3–6ft), is similar, but larger and faster-growing. **'Ohlendorffii'** (illus. p.84), H and S 1m (3ft), is slow-growing, initially rounded, becoming conical with age. **'Reflexa'** (illus. p.84), H 30cm (1ft), S 5m (15ft), is prostrate and ground-hugging, but may be trained up a stake, to form a mound of weeping foliage.
♥ **P. breweriana** illus. p.79.
♥ **P. engelmannii** illus. p.79.
P. glauca (White spruce). Narrowly conical conifer. H 10–15m (30–50ft), S 4–5m (12–15ft). Fully hardy. Glaucous shoots produce blue-green leaves. Ovoid, light brown cones fall after ripening. ♥ var. **albertiana 'Conica'** (syn. *P.g.* 'Albertiana Conica'; illus. p.85), H 2–5m (6–15ft), S 1–2m (3–6ft), is of neat, pyramidal habit and slow-growing, with longer leaves and smaller cones. **'Coerulea'** illus. p.79. ♥ **'Echiniformis'**, H 50cm (20in), S 1m (36in), is a dwarf, flat-topped, rounded form.
P. likiangensis (Lijiang spruce). Upright conifer. H 15m (50ft), S 5–10m (15–30ft). Fully hardy. Bluish-white leaves are well-spaced. Cones, 8–15cm (3–6in) long, are cylindrical, females bright red when young, ripening to purple, males pink.
P. mariana (Black spruce). Conical conifer, whose lowest branches often layer naturally, forming a ring of stems around the parent plant. H 10–15m (30–50ft), S 3m (10ft). Fully hardy. Leaves are bluish-green or bluish-white. Oval cones are dark grey-brown. **'Doumetii'** illus. p.82. ♥ **'Nana'** (illus. p.84), H 50cm (20in), S 50–80cm (20–32in), is a neat shrub with blue-grey foliage.
P. morrisonicola illus. p.81.
♥ **P. omorika** illus. p.77. **'Gnom'** (illus. p.84) is a shrub-like conifer with pendent branches that arch out at tips. H to 1.5m (5ft), S 1–2m (3–6ft). Fully hardy. Leaves are dark green above, white beneath. **'Nana'**, H and S 1m (3ft), is a slow-growing, rounded or oval cultivar.
♥ **P. orientalis** (Caucasian spruce, Oriental spruce). Columnar, dense conifer. H 20m (70ft), S 5m (15ft). Fully hardy. Glossy, deep green leaves are very short. Has ovoid to conical cones, 6–10cm (2½–4in) long, dark purple, ripening to brown, the males brick-red in spring. ♥ **'Aurea'** has golden, young foliage in spring, later turning green. **'Skylands'** illus. p.78.
P. pungens (Colorado spruce). Columnar conifer. H 15m (50ft), S 5m (15ft). Fully hardy. Has scaly, grey bark and very sharp, stout, greyish-green or bright blue leaves. Cylindrical, light brown cones have papery scales. **'Glauca Globosa'** see *P.p.* 'Montgomery'. ♥ **'Hoopsii'**, H 10–15m (30–50ft), has silvery-blue foliage. ♥ **'Koster'** illus. p.79. **'Montgomery'** (syn. *P.p.* 'Glauca Globosa'; illus. p.84), H and S 1m (3ft), is dwarf, compact, spreading or conical, with grey-blue leaves.
P. sitchensis (Sitka spruce). Very vigorous, broadly conical conifer. H 30–50m (100–160ft) in damp locations, 15–20m (50–70ft) in dry situations, S 6–10m (20–30ft). Fully hardy. Bark scales on old trees. Has prickly, bright deep green leaves and cylindrical, papery, pale brown or whitish cones, 5–10cm (2–4in) long. Is good on an exposed or poor site.
♥ **P. smithiana** (Morinda spruce, West Himalayan spruce). Slow-growing conifer, conical when young, columnar with horizontal branches and weeping shoots when mature. H 25–30m (80–100ft), S 6m (20ft). Fully hardy. Has dark green leaves and produces cylindrical, bright brown cones, 10–20cm (4–8in) long.

PICRASMA (Simaroubaceae)
Genus of deciduous trees, grown for their brilliant autumn colour. Produces insignificant flowers in late spring. Fully hardy. Requires sun or semi-shade and fertile, well-drained soil. Propagate by seed in autumn.
P. ailanthoides, syn. *P. quassioides*, illus. p.71.
P. quassioides. See *P. ailanthoides*.

PIERIS (Ericaceae)
Genus of evergreen shrubs, grown for their foliage and small, profuse, urn-shaped flowers. Fully to frost hardy. Needs a sheltered site in semi-shade or shade and in moist, peaty, acid soil. *P. floribunda*, however, grows well in any acid soil. Young shoots are sometimes frost-killed in spring and should be cut back as soon as possible. Dead-heading after flowering improves growth. Propagate by soft tip or semi-ripe cuttings in summer.
P. 'Bert Chandler'. Evergreen, bushy shrub. H 2m (6ft), S 1.5m (5ft). Frost hardy. Lance-shaped leaves are bright pink when young, becoming creamy-yellow, then white and finally dark green. Produces white flowers only very rarely. Likes an open position.
P. floribunda illus. p.97.
♥ **P. 'Forest Flame'.** Evergreen, upright shrub. H 4m (12ft), S 2m (6ft). Frost hardy. Narrowly oval, glossy leaves are brilliant red when young, become pink, then cream and finally turn dark green. White flowers are borne with the young leaves from mid- to late spring.
P. formosa. Evergreen, bushy, dense shrub. H and S 4m (12ft). Frost hardy. Large, oblong, glossy, dark green leaves are bronze when young. Bears large clusters of white flowers from mid- to late spring. ♥ var. **forrestii 'Wakehurst'** illus. p.112. **'Henry Price'** has deep-veined leaves, which are bronze-red when young.
P. japonica illus. p.86. **'Daisen'** is an evergreen, rounded, dense shrub. H and S 3m (10ft). Fully hardy. Narrowly oval, bronze leaves mature to glossy, dark green. Drooping clusters of red-budded, deep pink flowers appear in spring. **'Dorothy Wyckoff'** has deep crimson buds, opening to pink blooms; foliage is bronze in winter. Young foliage of ♥ **'Mountain Fire'** is brilliant red. **'Scarlett O'Hara'** illus. p.97. **Taiwanensis Group** see *P. taiwanensis*. **'Variegata'** is slow-growing, with small leaves, edged with white.
P. nana, syn. *Arcterica nana*. Evergreen, prostrate, dwarf shrub. H 2.5–5cm (1–2in), S 10–15cm (4–6in). Fully hardy. Has tiny, oval, leathery, dark green leaves, usually in whorls of 3, on fine stems that root readily. In early spring bears small, terminal clusters of white flowers with green or red calyces. Is excellent for binding a peat wall or in a rock garden.
P. taiwanensis, syn. *P. japonica* Taiwanensis Group. Evergreen, bushy, dense shrub. H 3m (10ft), S 5m (15ft). Fully hardy. Narrowly oval, dark green leaves are bronze-red when young. Produces clusters of white flowers in early and mid-spring.

PILEA (Urticaceae)
Genus of bushy or trailing annuals and evergreen perennials, grown for their ornamental foliage. Frost tender, min. 10°C (50°F). Grow in any well-drained soil out of direct sunlight and draughts; do not overwater in winter. Pinch out tips in growing season to avoid straggly plants. Propagate perennials by stem cuttings in spring or summer, annuals by seed in spring or autumn. Red spider mite may be a problem.
♥ **P. cadierei** illus. p.264.
P. involucrata (Friendship plant). Evergreen, bushy perennial. H 15cm (6in), S 30cm (1ft). Oval to rounded leaves, to 5cm (2in) long, have corrugated surfaces and are bronze above, reddish-green below; leaves are green above when grown in shade. Has insignificant flowers.
P. nummulariifolia illus. p.269.

PILEOSTEGIA (Hydrangeaceae)
Genus of evergreen, woody-stemmed, root climbers. Frost hardy. Grows in sun or shade and in any well-drained soil; is therefore useful for planting against a north wall. Prune in spring, if required. Propagate by semi-ripe cuttings in summer.
♥ **P. viburnoides**, syn. *Schizophragma viburnoides*, illus. p.168.

PILOSOCEREUS (Cactaceae)
Genus of columnar, summer-flowering, perennial cacti with wool-like spines in flowering zones at crowns. Some species are included in *Cephalocereus*. Frost tender, min. 11°C (52°F). Needs full sun and very well-drained soil. Propagate by seed or stem cuttings in spring or summer.
P. palmeri. See *P. leucocephalus*.
P. leucocephalus, syn. *P. palmeri*, illus. p.394.

PIMELEA

PIMELEA (Thymelaeaceae)
Genus of evergreen shrubs, grown for their flowers and overall appearance. Frost hardy to frost tender, min. 5–7°C (41–5°F). Needs full sun and well-drained, neutral to acid soil. Water potted plants moderately, less when temperatures are low. Needs good winter light and ventilation in northern temperate greenhouses. Propagate by seed in spring or by semi-ripe cuttings in late summer.
P. ferruginea illus. p.132.

PINELLIA (Araceae)
Genus of summer-flowering, tuberous perennials that produce slender, hood-like, green spathes, each enclosing and concealing a pencil-shaped spadix. Frost hardy. Needs partial shade or sun and humus-rich soil. Water well in spring-summer. Is dormant in winter. Propagate in early spring by offsets or in late summer by bulbils borne in leaf axils.
P. ternata. Summer-flowering, tuberous perennial. H 15–25cm (6–10in), S 10–15cm (4–6in). Has erect stems crowned by oval, flat, 3-parted leaves. Leafless stem bears a tubular, green spathe, 5–6cm (2–2½in) long, with a hood at the tip.

PINGUICULA (Lentibulariaceae)
Genus of summer-flowering perennials with sticky leaves that trap insects and digest them for food. Is useful in pots under glass among plants subject to aphid damage. Fully hardy to frost tender, min. 7°C (45°F). Needs sun and wet soil. Propagate by division in early spring or by seed in autumn.
P. caudata. See *P. moranensis*.
P. moranensis, syn. *P. caudata*. Basal-rosetted perennial. H 12–15cm (5–6in), S 5cm (2in). Frost tender. Leaves are narrowly oval and dull green with inrolled, purplish margins. In summer, long stems carry 5-petalled, deep carmine flowers.
P. grandiflora illus. p.330.

PINUS (Pinaceae)
Pine
Genus of small to large conifers with spirally arranged leaves in bundles, usually of 2, 3 or 5 needles. Cones ripen over 2 years and are small in the first year. See also CONIFERS.
P. aristata illus. p.82.
P. armandii (Armand pine, David pine). Conical, open conifer. H 10–15m (30–50ft), S 5–8m (15–25ft). Fully hardy. Has pendent, glaucous blue leaves and conical, green cones, 8–25cm (3–10in) long, ripening to brown.
P. banksiana illus. p.81.
P. bungeana illus. p.82.
♀ *P. cembra* illus. p.80.
P. cembroides illus. p.83.
P. chylla. See *P. wallichiana*.
P. contorta illus. p.81. var. *latifolia* illus. p.80. 'Spaan's Dwarf' is a conical, open, dwarf conifer with short, stiffly erect shoots. H and S 75cm (30in). Fully hardy. Has erect branches with bright green leaves in 2s.
♀ *P. coulteri* illus. p.76.
P. densiflora (Japanese red pine). Flat-topped conifer. H 15m (50ft), S 5–7m (15–22ft). Fully hardy. Has scaling, reddish-brown bark, bright green leaves and conical, yellow or pale brown cones. 'Alice Verkade', H and S 75cm (30in), is a diminutive, rounded form with fresh green leaves. 'Umbraculifera' (syn. *P.d.* 'Tagyosho'), H 4m (12ft), S 6m (20ft), is a slow-growing, rounded or umbrella-shaped form.
P. excelsa. See *P. wallichiana*.
P. griffithii. See *P. wallichiana*.
P. halepensis illus. p.81.
♀ *P. heldreichii* var. *leucodermis*, syn. *P. leucodermis*, illus. p.78. 'Compact Gem' (illus. p.85) is a broadly conical, dense, dwarf conifer. H and S 25–30cm (10–12in). Fully hardy. Has very dark green leaves in 2s. Grows only 2.5cm (1in) a year.
♀ 'Schmidtii' (illus. p.85) is a dwarf form with an ovoid habit and sharp, dark green leaves.
P. x *holfordiana* illus. p.75.
P. insignis. See *P. radiata*.
♀ *P. jeffreyi* illus. p.77.
P. leucodermis. See *P. heldreichii* var. *leucodermis*.
P. montezumae illus. p.76.
P. mugo (Dwarf pine, Mountain pine, Swiss mountain pine). Spreading, shrubby conifer. H 3–5m (10–15ft), S 5–8m (15–25ft). Fully hardy. Has bright to dark green leaves in 2s and ovoid, brown cones. 'Gnom', H and S to 2m (6ft), and ♀ 'Mops', H 1m (3ft), S 2m (6ft), are rounded cultivars.
♀ *P. muricata* illus. p.77.
♀ *P. nigra* (Black pine). Upright, later spreading conifer, generally grown in one of the following forms. 'Hornibrookiana', H 1.5–2m (5–6ft), S 2m (6ft), is a fully hardy, shrubby cultivar with stout, spreading or erect branches and dark green leaves in 2s.
♀ var. *maritima* (syn. subsp. *laricio*; Corsican pine), H 25–30m (80–100ft), S 8m (25ft), is fast-growing and narrowly conical with an open crown; bears grey-green leaves, in 2s, and ovoid to conical, yellow- or pale grey-brown cones. var. *nigra* illus. p.78.
♀ *P. parviflora* illus. p.79.
♀ 'Adcock's Dwarf' is a slow-growing, rounded, dense, dwarf conifer. H 2–3m (6–10ft), S 1.5–2m (5–6ft). Fully hardy. Bears grey-green leaves in 5s.
P. peuce illus. p.75.
♀ *P. pinaster* illus. p.77.
♀ *P. pinea* illus. p.83.
♀ *P. ponderosa* illus. p.77.
P. pumila (Dwarf Siberian pine). Spreading, shrubby conifer. H 2–3m (6–10ft), S 3–5m (10–15ft). Fully hardy. Has bright blue-green leaves in 5s. Ovoid cones are violet-purple, ripening to red-brown or yellow-brown, the males bright red-purple in spring. 'Globe', H and S 50cm–1m (20–36in), is a rounded cultivar with blue foliage.
♀ *P. radiata*, syn. *P. insignis*, illus. p.78.
P. rigida illus. p.80.
P. strobus illus. p.76. ♀ 'Radiata' (syn. *P.s.* f. *nana*, *P.s.* 'Nana') is a rounded, dwarf conifer with an open, sparse, whorled crown. H 1–2m (3–6ft), S 2–3m (6–10ft). Fully hardy. Grey bark is smooth at first, later fissured. Bears grey-green leaves in 5s.
♀ *P. sylvestris* (Scots pine). Conifer, upright and with whorled branches when young, that develops a spreading, rounded crown with age. H 15–25m (50–80ft), S 8–10m (25–30ft). Fully hardy. Bark is flaking and red-brown on upper trunk, fissured and purple-grey at base. Has blue-green leaves in 2s and conical, green cones that ripen to pale grey- or red-brown. ♀ 'Aurea' (illus. p.85), H 10m (30ft), S 4m (12ft), has golden-yellow leaves in winter-spring, otherwise blue-green.
♀ 'Beuvronensis', H and S 1m (3ft), is a rounded shrub. 'Doone Valley' (illus. p.84), H and S 1m (3ft), is an upright, irregularly shaped shrub.
f. *fastigiata* illus. p.82. 'Gold Coin' (illus. p.85), H and S 2m (6ft), is a dwarf version of *P.s.* 'Aurea'. 'Nana' (illus. p.84), H and S 50cm (20in), is a very dense cultivar with widely spaced leaves.
P. thunbergii illus. p.80.
P. virginiana illus. p.81.
♀ *P. wallichiana*, syn. *P. chylla*, *P. excelsa*, *P. griffithii*, illus. p.77.

PIPTANTHUS (Leguminosae)
Genus of deciduous or semi-evergreen shrubs, grown for their foliage and flowers. Frost hardy. In cold areas needs the protection of a south- or west-facing wall. Requires sun and fertile, well-drained soil. In spring cut some older shoots back to ground level and prune any frost-damaged growths back to healthy wood. Propagate by seed in autumn.
P. laburnifolius. See *P. nepalensis*.
P. nepalensis, syn. *P. laburnifolius*, illus. p.116.

PISONIA (Nyctaginaceae)
Genus of evergreen shrubs and trees, grown for their foliage and overall appearance. Frost tender, min. 10–15°C (50–59°F). Needs full light or partial shade and humus-rich, well-drained soil. Water containerized specimens freely when in full growth, moderately at other times. Pruning is tolerated if required. Propagate by seed in spring or by semi-ripe cuttings in summer.
P. umbellifera, syn. *Heimerliodendron brunonianum* (Bird-catcher tree, Para para). Evergreen, rounded shrub or tree. H and S 3–6m (10–20ft). Has oval, leathery, lustrous leaves and, in spring, clusters of tiny, green or pink flowers, followed by 5-winged, sticky, brownish fruits.

PISTACIA (Anacardiaceae)
Genus of evergreen or deciduous trees, grown for their foliage and overall appearance. Frost tender, min. 10°C (50°F). Requires full light and free-draining, even dry soil. Water containerized plants moderately when in full growth, sparingly at other times. Pruning is tolerated if necessary. Propagate by seed in spring or by semi-ripe cuttings in summer.
P. lentiscus (Mastic tree). Evergreen, irregularly rounded shrub or tree. H 5m (15ft), S to 3m (10ft). Leaves are divided into 2–5 pairs of oval, leathery, glossy leaflets. Axillary clusters of insignificant flowers from spring to early summer develop into globose, red, then black fruits in autumn.
P. terebinthus (Cyprus turpentine, Terebinth tree). Deciduous, rounded to ovoid tree. H 6–9m (20–28ft), S 3–6m (10–20ft). Leaves have 5–9 oval leaflets, usually lustrous, rich green. In spring and early summer bears axillary clusters of insignificant flowers that develop into tiny, globular to ovoid, red, then purple-brown fruits in autumn.

PISTIA (Araceae)
Genus of one species of deciduous, perennial, floating water plant, grown for its foliage. In water above 19–21°C (66–70°F) is evergreen. Is suitable for tropical aquariums and frost-free pools. Frost tender, min 10–15°C (50–59°F). Grows in sun or semi-shade. Remove fading foliage and thin plants out as necessary. Propagate by separating plantlets during growing period in summer.
P. stratiotes illus. p.389.

PITCAIRNIA (Bromeliaceae)
Genus of evergreen, rosette-forming perennials, grown for their overall appearance. Frost tender, min. 10°C (50°F). Needs semi-shade and well-drained soil. Water moderately during growing season, sparingly at other times. Propagate by offsets or division in spring.
P. andreana. Evergreen, clump-forming, basal-rosetted perennial. H 20cm (8in), S 30cm (12in) or more. Loose rosettes comprise narrowly lance-shaped, strongly arching, green leaves, grey-scaled beneath. Racemes of tubular, orange-and-red flowers are borne in summer.
P. heterophylla. Evergreen, basal-rosetted perennial with swollen, much-branched rhizomes. H 10cm (4in) or more, S to 30cm (12in). Forms loose rosettes; outer leaves resemble barbed spines, inner leaves are strap-shaped, low-arching and green, with downy, white undersides. Almost stemless spikes of tubular, bright red, or rarely white, flowers appear in summer.

PITTOSPORUM (Pittosporaceae)
Genus of evergreen trees and shrubs, grown for their ornamental foliage and fragrant flowers. Frost hardy to frost tender, min. 7°C (45°F). Does best in mild areas; in cold regions grow against a south- or west-facing wall. *P. crassifolium* and *P. ralphii* make good, wind-resistant hedges in mild, coastal areas and, like forms with variegated or purple leaves, prefer sun. Others will grow in sun or semi-shade. All need well-drained soil. Propagate *P. dallii* by budding in summer, other species by seed in autumn or spring or by semi-ripe cuttings in summer and selected forms by semi-ripe cuttings only in summer.
P. crassifolium (Karo). Evergreen, bushy-headed, dense tree or shrub. H 5m (15ft), S 3m (10ft). Frost hardy. Has oblong, dark green leaves, grey-felted beneath. Clusters of small,

fragrant, star-shaped, dark reddish-purple flowers are borne in spring.
'Variegatum' illus. p.73.
P. dallii illus. p.74.
P. eugenioides. Evergreen, columnar tree. H 10m (30ft), S 5m (15ft). Frost hardy. Narrowly oval, wavy-edged leaves are glossy, dark green. Honey-scented, star-shaped, pale yellow flowers are produced in spring.
♛ 'Variegatum' illus. p.73.
♛ *P.* 'Garnettii' illus. p.96.
P. ralphii. Evergreen, bushy-headed tree or shrub. H 4m (12ft), S 3m (10ft). Frost hardy. Large leaves are oblong, leathery and grey-green, very hairy beneath. Small, fragrant, star-shaped, dark red flowers are borne in spring.
♛ *P. tenuifolium* illus. p.97. **'James Stirling'** is an evergreen, columnar, then rounded, shrub or tree. H 12m (40ft), S 5m (15ft). Frost hardy. Has deep purple shoots and small, rounded, wavy-edged, silvery-green leaves. Honey-scented, tubular, purple flowers are borne in late spring. ♛ **'Tom Thumb'** illus. p.147.
♛ *P. tobira* (Japanese pittosporum, Mock orange). Evergreen, bushy-headed, dense tree or shrub. H 6m (20ft), S 4m (12ft). Frost hardy. Has oblong to oval, glossy, dark green leaves. Very fragrant, star-shaped, white flowers, opening in late spring, later become creamy-yellow.
P. undulatum (Victorian box). Evergreen, broadly conical tree. H 12m (40ft), S 8m (25ft). Half hardy. Has long, narrowly oval, pointed, wavy-edged, dark green leaves. Fragrant, star-shaped, white flowers are borne in late spring and early summer, followed by rounded, orange fruits.

PITYROGRAMMA (Polypodiaceae)
Genus of semi-evergreen or evergreen ferns, suitable for hanging baskets. Frost tender, min. 10°C (50°F). Needs semi-shade and humus-rich, moist but well-drained soil. Remove fading fronds regularly. Water carefully to avoid spoiling farina (meal-like powder) on fronds. Propagate by spores in late summer.
P. chrysophylla. Semi-evergreen or evergreen fern. H and S 45cm (18in). Lance-shaped, delicately filigreed, spreading, mid-green fronds, with yellow farina on undersides, are produced on brown stems.
P. triangularis. Semi-evergreen or evergreen fern. H and S 45cm (18in). Has broadly triangular, delicately divided, mid-green fronds with orange or creamy-white farina.

PLAGIANTHUS (Malvaceae)
Genus of evergreen or deciduous trees and shrubs, grown for their habit and inconspicuous, fragrant flowers. Frost to half hardy. In cold areas plant against a warm wall. Needs sun and fertile, well-drained soil. Propagate by semi-ripe cuttings in summer.
P. divaricatus. Deciduous, bushy shrub. H 2.5m (8ft), S 4m (12ft). Half hardy. Slender, interlacing, dark brown branches produce linear, dark green leaves. Has very fragrant, tiny, yellowish-white flowers in late spring.

Plagiorhegma dubia. See *Jeffersonia dubia.*

PLANTAGO (Plantaginaceae)
Genus of summer-flowering annuals, biennials and evergreen perennials and shrubs. Many species are weeds, but a few are grown for their foliage and architectural value. Fully hardy to frost tender, min. 7–10°C (45–50°F). Needs full sun and well-drained soil. Water potted plants moderately, sparingly in winter. Propagate by seed or division in spring.
P. nivalis illus. p.339.

PLATANUS (Platanaceae)
Plane
Genus of deciduous trees, grown for their habit, foliage and flaking bark. Flowers are inconspicuous. Spherical fruit clusters hang from shoots in autumn. Fully to half hardy. Needs full sun and deep, fertile, well-drained soil. Propagate species by seed in autumn, *P.* x *hispanica* by hardwood cuttings in early winter. All except *P. orientalis* are susceptible to the fungal disease plane anthracnose.
P. x *acerifolia.* See *P.* x *hispanica.*
♛ *P.* x *hispanica,* syn. *P.* x *acerifolia,* illus. p.41. **'Suttneri'** is a vigorous, deciduous, spreading tree. H 20m (70ft), S 15m (50ft). Fully hardy. Has flaking bark and large, palmate, 5-lobed, sharply toothed, bright green leaves that are blotched with creamy-white.
♛ *P. orientalis* (Oriental plane). Deciduous, spreading tree. H and S 25m (80ft) or more. Fully hardy. Produces large, palmate, glossy, pale green leaves with 5 deep lobes.

PLATYCARYA (Juglandaceae)
Genus of one species of deciduous tree, grown for its foliage and catkins. Fully hardy. Requires full sun and fertile, well-drained soil. Propagate by seed in autumn.
P. strobilacea. Deciduous, spreading tree. H and S 10m (30ft). Has ash-like, bright green leaves with 5–15 leaflets. Upright, green catkins are borne from mid- to late summer; males are slender and cylindrical, often drooping at tips, females are cone-like, become brown and persist through winter.

PLATYCERIUM (Polypodiaceae)
Stag's-horn fern
Genus of evergreen, epiphytic ferns, best grown in hanging baskets and fastened to or suspended from pieces of wood. Produces 2 kinds of fronds: permanent, broad, sterile 'nest leaves' forming the main part of the plant; and strap-shaped, usually partly bifurcated, arching fertile fronds. Frost tender, min. 5°C (41°F). Needs warm, humid conditions in semi-shade and fibrous, peaty compost with hardly any soil. Propagate by detaching buds in spring or summer and planting in compost or by spores in summer or early autumn.
♛ *P. bifurcatum* illus. p.186.

Platycladus orientalis. See *Thuja orientalis.*

PLATYCODON (Campanulaceae)
Balloon flower
Genus of one species of perennial, grown for its flowers in summer. Fully hardy. Needs sun and light, sandy soil. Propagate by basal cuttings of non-flowering shoots in summer, preferably with a piece of root attached, or by seed in autumn.
♛ *P. grandiflorus* illus. p.248.
♛ var. *mariesii* is a neat, clump-forming perennial. H and S 30–45cm (12–18in). In mid-summer produces solitary terminal, large, balloon-like flower buds opening to bell-shaped, blue or purplish-blue flowers. Has oval, sharply toothed, bluish-green leaves.

PLATYSTEMON (Papaveraceae)
Genus of one species of summer-flowering annual. Fully hardy. Grow in sun and in fertile, well-drained soil. Propagate by seed sown outdoors in spring or early autumn.
P. californicus illus. p.287.

PLECTRANTHUS (Labiatae)
Genus of evergreen, trailing or bushy perennials, grown for their foliage. Frost tender, min. 10°C (50°F). Is easy to grow if kept moist in partial shade or bright light. Cut back tips in growing season if plants become too straggly. Propagate by stem cuttings or division in spring or summer.
P. australis (Swedish ivy). Evergreen, trailing perennial with square stems. H to 15cm (6in), S indefinite. Has rounded, waxy, glossy, green leaves with scalloped edges. Intermittently produces racemes of tubular, white or pale mauve flowers.
P. coleoides 'Variegatus'. See *P. forsteri* 'Marginatus'.
P. forsteri 'Marginatus', syn. *P. coleoides* 'Variegatus', illus. p.228.
♛ *P. oertendahlii* (Prostrate coleus, Swedish ivy). Evergreen, prostrate perennial. H to 15cm (6in), S indefinite. Rounded, scalloped, dark green leaves are reddish-green below, with white veins above. Racemes of tubular, white or pale mauve flowers are borne at irregular intervals throughout the year.

Plectranthus thyrsoideus. See *Coleus thyrsoideus.*

PLEIOBLASTUS (Bambusoideae). See GRASSES, BAMBOOS, RUSHES and SEDGES.
♛ *P. auricoma,* syn. *P. viridistriatus, Arundinaria auricoma, A. viridistriata,* illus. p.185.
♛ *P. variegatus,* syn. *Arundinaria fortunei, A. variegata,* illus. p.182.
P. viridistriatus. See *P. auricoma.*

PLEIONE. See ORCHIDS.
P. bulbocodioides (illus. p.261). Deciduous, terrestrial orchid. H 20cm (8in). Frost hardy. In spring, usually before solitary leaf appears, bears pink, rose or magenta flowers, 5–12cm (2–5in) across, with darker purple marks on lips. Leaf is narrowly lance-shaped, 14cm (5½in) long. Is often difficult to flower: regular feeding helps to increase pseudobulbs to flowering size.
P. x *confusa.* Deciduous, terrestrial orchid. H 15cm (6in). Frost hardy. Canary-yellow flowers, 5–8cm (2–3in) across, with brown or purple blotches on lips, appear singly in spring, before foliage. Has lance-shaped leaves, 10–18cm (4–7in) long. Does best in an alpine house. Needs semi-shade.
P. hookeriana. Deciduous, terrestrial orchid. H 8–15cm (3–6in). Frost hardy. Lilac-pink, rose or white flowers, 5–7cm (2–3in) across, each with a brown- or purplish-spotted lip, are borne singly in spring with lance-shaped, 5–20cm (2–8in) long leaves. Cultivate as for *P.* x *confusa.*
P. humilis. Deciduous, terrestrial orchid. H 5–8cm (2–3in). Frost hardy. In winter, before foliage appears, white flowers, 7–9cm (3–3½in) across, each with a crimson-spotted lip, are borne singly or in pairs. Lance-shaped leaves are 18–25cm (7–10in) long. Cultivate as for *P.* x *confusa.*
P. praecox. Deciduous, terrestrial orchid. H 8–13cm (3–5in). Frost hardy. Flowers, to 8cm (3in) across, in pairs, appear in autumn, after foliage. They are white to pinkish-purple or lilac-purple, with violet marks. Leaves are oblong to lance-shaped and 15–25cm (6–10in) long. Cultivate as for *P.* x *confusa.*

PLEIOSPILOS (Aizoaceae)
Genus of clump-forming, perennial succulents with almost stemless rosettes bearing up to 4 pairs of fleshy, erect leaves, like pieces of granite, each with a flat upper surface and each pair united at the base. Flowers are daisy-like. Individual species are very similar, and many are difficult to identify. Frost tender, min. 5°C (41°F). Needs sun and well-drained soil. Propagate by seed or division in spring or summer.
♛ *P. bolusii* illus. p.415.
P. compactus illus. p.414.

PLEUROTHALLIS. See ORCHIDS.
P. grobyi. Evergreen, epiphytic orchid for a cool greenhouse. H 2.5cm (1in). In summer produces sprays of minute, white flowers, 0.25cm (⅛in) long. Leaves are oval, fleshy and 0.5cm (¼in) long. Shade in summer.

PLUMBAGO (Plumbaginaceae)
Genus of annuals, evergreen or semi-evergreen shrubs, perennials and woody-stemmed, scrambling climbers, grown for their primrose-shaped flowers. Frost hardy to frost tender, min. 7°C (45°F). Grow in full light or semi-shade and in fertile, well-drained soil. Water regularly, less when not in full growth. Tie stems to supports. Thin out or spur back all previous year's growth in early spring. Propagate by semi-ripe cuttings in summer. Whitefly may be a problem.
♛ *P. auriculata,* syn. *P. capensis,* illus. p.175.
P. capensis. See *P. auriculata.*
P. indica, syn. *P. rosea.* Evergreen or semi-evergreen, spreading shrub or semi-climber. H 2m (6ft), S 1–2m (3–6ft). Frost tender. Leaves are oval to elliptic and mid-green. Has terminal racemes of primrose-shaped, red or

pink flowers, 2.5cm (1in) long, in summer, if hard pruned annually in spring, or from late winter onwards, if left unpruned and trained as a climber.
P. rosea. See *P. indica.*

PLUMERIA (Apocynaceae)
Frangipani
Genus of mainly deciduous, fleshy-branched shrubs and trees, grown for their flowers in summer-autumn. Has poisonous sap. Frost tender, min. 13°C (55°F). Requires full sun and freely draining soil. Water potted specimens moderately while in growth, keep dry in winter when leafless. Stem tips may be cut out to induce branching. Propagate by seed or leafless stem-tip cuttings in late spring. Red spider mite may be a nuisance.
P. acuminata. See *P. rubra* var. *acutifolia.*
P. acutifolia. See *P. rubra* var. *acutifolia.*
P. alba (West Indian jasmine). Deciduous, rounded, sparingly branched tree. H to 6m (20ft), S to 4m (12ft). Leaves are lance-shaped and slender-pointed, to 30cm (1ft) long. Terminal clusters of fragrant, yellow-eyed, white flowers, each with 5 spreading petals and a tubular base, appear in summer.
P. rubra illus. p.70. f. **acutifolia** (syn. *P. acuminata, P. acutifolia*) is a deciduous, spreading, sparsely branched tree or shrub. H and S 4m (12ft) or more. Produces fragrant, yellow-centred, white flowers, with 5 spreading petals, in summer-autumn. Leaves are lance-shaped to oval and 20–30cm (8–12in) long.

PODALYRIA (Leguminosae)
Genus of evergreen, mainly summer-flowering shrubs, grown for their flowers and overall appearance. Frost tender, min. 7–10°C (45–50°F). Requires full light and fertile, well-drained soil. Water containerized plants moderately, less when not in full growth. Prune, if necessary, after flowering. Propagate by seed in spring or by semi-ripe cuttings in summer.
P. calyptrata. Vigorous, evergreen, rounded shrub. H and S 1.2–3m (4–10ft). Has oval, downy, mid-green leaves and sweet pea-like, pink flowers, 3–4cm (1¼–1½in) wide, in summer.

PODOCARPUS (Podocarpaceae).
See CONIFERS.
P. alpinus (Tasmanian podocarp). Rounded, spreading, shrubby conifer. H 2m (6ft), S 3–5m (10–15ft). Frost hardy. Has linear, dull green leaves and rounded, egg-shaped, fleshy, bright red fruits.
P. andinus. See *Prumnopitys andinus.*
P. macrophyllus (Kusamaki). Erect conifer. H 10m (30ft), S 3–5m (10–15ft). Half hardy. Long, linear leaves are bright green above, glaucous beneath. May be grown as a shrub, H and S 1–2m (3–6ft), and planted in a tub in hot climates.
P. nivalis (Alpine totara; illus. p.84). Rounded, spreading, shrubby conifer. H 2m (6ft), S 3–5m (10–15ft). Frost hardy. Is very similar to *P. alpinus*, but has longer, broader, more rigid leaves.
☙ *P. salignus* illus. p.80.

PODOPHYLLUM (Berberidaceae)
Genus of spring-flowering, rhizomatous perennials. Fully hardy, but young leaves may be damaged by frost. Does best in semi-shade and moist, peaty soil. Propagate by division in spring or by seed in autumn.
P. emodi. See *P. hexandrum.*
P. hexandrum, syn. *P. emodi*, illus. p.232.
P. peltatum (May apple). Vigorous, spreading, rhizomatous perennial. H 30–45cm (12–18in), S 30cm (12in). Palmate, sometimes brown-mottled, light green leaves, with 3–5 deep lobes, push up through soil, looking like closed umbrellas, and are followed, in spring, by nodding, cup-shaped, white flowers. Produces large, fleshy, plum-like, glossy, deep rose-pink fruits in autumn.

PODRANEA (Bignoniaceae)
Genus of evergreen, twining climbers, grown for their foxglove-like flowers. Frost tender, min. 5–10°C (41–50°F). Grow in any fertile, well-drained soil, with full light. Water regularly, less in cold weather. Provide support. Thin out crowded growth in winter or early spring. Propagate by seed in spring or by semi-ripe cuttings in summer.
P. ricasoliana, syn. *Pandorea ricasoliana, Tecoma ricasoliana.* Fast-growing, evergreen, twining climber. H 4m (12ft) or more. Has leaves of 7 or 9 lance-shaped to oval, wavy, deep green leaflets. Loose clusters of fragrant, pink flowers with darker veins appear from spring to autumn.

Poinciana pulcherrima. See *Caesalpinia pulcherrima.*

POLEMONIUM (Polemoniaceae)
Jacob's ladder
Genus of late spring- or summer-flowering annuals and perennials, some perennials tending to be short-lived. Fully hardy. Prefers sun and fertile, well-drained soil. Propagate by division in spring or by seed in autumn.
P. caeruleum illus. p.249.
P. carneum illus. p.247.
P. foliosissimum. Vigorous, clump-forming perennial. H 75cm (30in), S 60cm (24in). Terminal clusters of cup-shaped, lilac flowers, with yellow stamens, are borne in summer above oblong to lance-shaped, mid-green leaves with numerous, small leaflets.
P. pulcherrimum illus. p.248.

POLIANTHES (Amaryllidaceae)
Genus of tuberous perennials, grown for their fragrant flowers in summer. Half hardy to frost tender, min. 15–20°C (59–68°F). Needs a sheltered site in full sun and well-drained soil. Water well in spring-summer; feed liquid fertilizer every 2 weeks when in growth. Dry off after leaves die down in winter. Propagate by seed or offsets in spring.
P. geminiflora, syn. *Bravoa geminiflora*, illus. p.367.
☙ *P. tuberosa* (Tuberose). Summer-flowering, tuberous perennial. H 60cm–1m (2–3ft), S 10–15cm (4–6in). Half hardy. Has a basal cluster of strap-shaped, erect leaves; flower stem also bears leaves on lower part. Produces a spike of funnel-shaped, single, white flowers with 6 spreading petals. A double form is also available.

POLIOTHYRSIS (Flacourtiaceae)
Genus of one species of deciduous tree, grown for its foliage and flowers. Fully hardy. Needs sun or semi-shade and fertile, well-drained soil. Propagate by softwood cuttings in summer.
P. sinensis. Deciduous, spreading tree. H 10m (30ft), S 6m (20ft). Bears long, oval, sharply toothed leaves, glossy and dark green, with wine-red stalks. Fragrant, star-shaped, white, later yellow flowers are produced in late summer and early autumn.

POLYGALA (Polygalaceae)
Genus of annuals, evergreen perennials, shrubs and trees, grown mainly for their pea-like flowers. Fully hardy to frost tender, min. 7°C (45°F). Needs full light or partial shade and moist, well-drained soil. Water potted specimens freely when in full growth, moderately at other times. Lanky stems may be cut back hard in late winter. Propagate by seed in spring or by semi-ripe cuttings in late summer. Is susceptible to whitefly.
P. calcarea illus. p.332. **'Bulley's Variety'** illus. p.332.
☙ *P. chamaebuxus* illus. p.333.
☙ var. **grandiflora** (syn. var. *rhodoptera*) illus. p.316.
P. dalmaisiana, syn. *P. myrtifolia* 'Grandiflora' of gardens, illus. p.137.
P. myrtifolia 'Grandiflora' of gardens. See *P. dalmaisiana.*
P. vayredae. Evergreen, mat-forming shrub. H 5–10cm (2–4in), S 20–30cm (8–12in). Frost hardy. Slender, prostrate stems bear small, linear leaves. Pea-like, reddish-purple flowers, each with a yellow lip, appear in late spring and early summer. Suits a rock garden or alpine house.

POLYGONATUM (Liliaceae)
Solomon's seal
Genus of spring- or early summer-flowering, rhizomatous perennials. Fully hardy to frost tender, min. 5°C (41°F). Requires a cool, shady situation and fertile, well-drained soil. Propagate by division in early spring or by seed in autumn. Sawfly caterpillar is a common pest.
P. biflorum, syn. *P. canaliculatum, P. commutatum, P. giganteum* (Great Solomon's seal). Arching, rhizomatous perennial. H 1.5m (5ft) or more, S 60cm (2ft). Fully hardy. Bears pendent clusters of bell-shaped, white flowers in leaf axils during late spring. Oval to oblong leaves are mid-green.
P. canaliculatum. See *P. biflorum.*
P. commutatum. See *P. biflorum.*
P. giganteum. See *P. biflorum.*
P. hirtum, syn. *P. latifolium.* Upright, then arching, rhizomatous perennial. H 1m (3ft), S 30cm (1ft). Fully hardy. Clusters of 2–5 drooping, tubular, green-tipped, white flowers open in late spring. Undersides of stems, leaf stalks and oval to lance-shaped, mid-green leaves are hairy.
P. hookeri illus. p.315.
P. x hybridum illus. p.202.
P. latifolium. See *P. hirtum.*
P. multiflorum. Arching, leafy perennial with fleshy rhizomes. H 1m (3ft), S 30cm (1ft). Fully hardy. In late spring, clusters of 2–6 pendent, tubular, green-tipped, white flowers appear in upper leaf axils. Has oval to lance-shaped, mid-green leaves. **'Flore Pleno'** has double flowers that resemble ballet dancers' skirts. **'Variegatum'**, H 60cm (2ft), has leaves with creamy-white stripes.
P. odoratum (Angled Solomon's seal). Arching, rhizomatous perennial. H 60cm (2ft), S 30cm (1ft). Fully hardy. Produces pairs of fragrant, tubular to bell-shaped, green-tipped, white flowers in upper leaf axils in late spring. Oval to lance-shaped leaves are mid-green.
P. verticillatum (Whorled Solomon's seal). Upright, rhizomatous perennial. H 1.2m (4ft), S 45cm (1½ft). Fully hardy. Produces narrowly bell-shaped, greenish-white flowers in upper leaf axils in early summer. Has whorls of lance-shaped, mid-green leaves.

POLYGONUM (Polygonaceae)
Knotweed
Genus of annuals, perennials, some of which are evergreen, and deciduous, woody-stemmed, twining climbers. Some species are invasive. Fully to half hardy. Tolerates sun or shade and well-drained soil, but does particularly well in a damp position. Propagate by seed or division in autumn or spring, climbers only by semi-ripe cuttings in summer.
P. affine, syn. *Bistorta affine, Persicaria affinis.* Evergreen, mat-forming perennial. H 15–30cm (6–12in), S 30cm (12in) or more. Fully hardy. Stout stems bear small, lance-shaped, glossy, green leaves that turn bronze in winter. In autumn carries dense spikes of small, funnel-shaped, rose-red flowers. Is good on a bank or in a rock garden. ☙ **'Darjeeling Red'**, H 20–25cm (8–10in), has long spikes of deep red flowers. ☙ **'Donald Lowndes'** illus. p.324.
P. amplexicaule, syn. *Bistorta amplexicaule, Persicaria amplexicaulis.* Clump-forming, leafy perennial. H and S 1.2m (4ft). Fully hardy. Bears profuse spikes of small, rich red flowers in summer-autumn. Has oval to heart-shaped, mid-green leaves. ☙ **'Firetail'** illus. p.213.
P. aubertii, syn. *Bilderdykia aubertii, Fallopia aubertii* (Mile-a-minute plant, Russian vine). Vigorous, deciduous, woody-stemmed, twining climber. H to 12m (40ft) or more. Fully hardy. Leaves are broadly heart-shaped. Panicles of small, white or greenish flowers, ageing to pink, are carried in summer-autumn; they are followed by angled, pinkish-white fruits.
☙ *P. baldschuanicum*, syn. *Bilderdykia baldschuanica, Fallopia baldschuanica*, illus. p.177.
P. bistorta, syn. *Bistorta major*,

POLYPODIUM

Persicaria bistorta (Bistort).
♚ **'Superbum'** illus. p.207.
P. campanulatum, syn. *Persicaria campanulata*, illus. p.223.
P. capitatum, syn. *Persicaria capitata*. Compact, spreading perennial. H 5cm (2in), S 15–20cm (6–8in). Half hardy. Small, oval leaves are green with darker marks. Small, spherical heads of pink flowers are borne in summer. Is suitable for a rock garden or bank.
P. macrophyllum, syn. *Bistorta macrophylla*, *Persicaria sphaerostachya*, *Polygonum sphaerostachyum*, illus. p.243.
P. milletii, syn. *Bistorta milletii*. Compact perennial. H and S 60cm (2ft). Fully hardy. Produces slender spikes of rich crimson flowers from mid-summer to early autumn. Narrow, lance-shaped leaves are mid-green.
P. sphaerostachyum. See *P. macrophyllum*.
♚ *P. vacciniifolium*, syn. *Bistorta vacciniifolia*, *Persicaria vacciniifolium*, illus. p.335.
P. virginianum **'Painter's Palette'**, syn. *Tovara virginiana* 'Painter's Palette', illus. p.254.

POLYPODIUM (Polypodiaceae)
Genus of deciduous, semi-evergreen or evergreen ferns, grown for their sculptural fronds. Fully hardy to frost tender, min. 10°C (50°F). Grow in semi-shade and fibrous, moist but well-drained soil. Propagate by division in spring or by spores in late summer.
P. aureum. See *Phlebodium aureum*.
P. glycyrrhiza illus. p.186.
P. polypodioides. Semi-evergreen, creeping fern. H 10cm (4in), S 15cm (6in). Frost tender. Lance-shaped, mid-green fronds have widely spaced pinnae and emerge from a scaly base and rhizome.
P. scouleri illus. p.187.
P. virginianum (American wall fern). Semi-evergreen, creeping fern. H 30cm (12in), S 23cm (9in). Frost hardy. Has narrowly lance-shaped, divided, mid-green fronds.
P. vulgare illus. p.189. **'Cornubiense'** illus. p.188. **'Cristatum'** is an evergreen, creeping fern. H and S 25–30cm (10–12in). Fully hardy. Narrowly lance-shaped, divided, mid-green fronds, with semi-pendulous, terminal crests, grow from creeping rhizomes covered with copper-brown scales.

POLYSCIAS (Araliaceae)
Genus of evergreen trees and shrubs, grown for their foliage. Sometimes has insignificant flowers in summer, but only on large, mature specimens. Frost tender, 15–18°C (59–64°F). Needs partial shade and humus-rich, well-drained soil. Water containerized plants freely when in full growth, moderately at other times. Straggly stems may be cut out in spring. Propagate by seed in spring or by stem-tip or leafless stem-section cuttings in summer. Red spider mite may be troublesome.
P. filicifolia illus. p.122. **'Marginata'** is an evergreen, erect, sparsely branched shrub. H 2m (6ft) or more, S 1m (3ft) or more. Has 30cm (1ft) long leaves with many small, oval to lance-shaped, serrated, bright green leaflets with white edges.
P. guilfoylei (Wild coffee). Slow-growing, evergreen, rounded tree. H 3–8m (10–25ft), S to 2m (6ft) or more. Leaves are 25–40cm (10–16in) long and divided into oval to rounded, serrated, deep green leaflets.
'Victoriae' illus. p.96.

POLYSTICHUM (Polypodiaceae)
Genus of evergreen, semi-evergreen or deciduous ferns. Fully to frost hardy. Does best in semi-shade and moist but well-drained soil enriched with fibrous organic matter. Remove faded fronds regularly. Propagate species by division in spring or by spores in summer, selected forms by division in spring.
P. acrostichoides (Christmas fern). Evergreen fern. H 60cm (24in), S 45cm (18in). Fully hardy. Slender, lance-shaped, deep green fronds have small, holly-like pinnae. Is excellent for cutting.
♚ *P. aculeatum* (Hard shield fern, Prickly shield fern). Semi-evergreen fern. H 60cm (24in), S 75cm (30in). Fully hardy. Broadly lance-shaped, yellowish-green, then deep green fronds, with oblong to oval, spiny-edged, glossy pinnae, are produced on stems often covered in brown scales.
'Pulcherrimum' illus. p.186.
P. munitum illus. p.186.
P. setiferum (Soft shield fern). **'Divisilobum'** illus. p.187.
'Proliferum' illus. p.189.
P. tsus-simense. Semi-evergreen fern. H 30cm (12in), S 23cm (9in). Frost hardy. Has broadly lance-shaped, dull green fronds divided into oblong, finely spiny-edged pinnae. Is suitable for a peat garden or alpine house.

PONCIRUS (Rutaceae)
Genus of one species of very spiny, deciduous shrub or small tree, grown for its foliage, showy flowers and orange-like fruits. Is very effective as a protective hedge. Fully hardy. Needs sun and fertile, well-drained soil. Cut out dead wood in spring, and trim hedges in early summer. Propagate by semi-ripe cuttings in summer or by seed when ripe, in autumn.
P. trifoliata (Japanese bitter orange). Deciduous, bushy shrub or tree. H and S 5m (15ft). Stout, spiny, green shoots bear dark green leaves each composed of 3 oval leaflets. Fragrant, white flowers, with 4 or 5 large petals, borne in late spring and often again in autumn, are followed by rounded, 2–3cm (³/₄–1¹/₄in) wide fruits.

PONTEDERIA (Pontederiaceae)
Genus of deciduous, perennial, marginal water plants, grown for their foliage and flower spikes. Fully to frost hardy. Needs full sun and up to 23cm (9in) depth of water. Remove fading flowers regularly. Propagate in spring by division or seed.
♚ *P. cordata* illus. p.388.

POPULUS (Salicaceae)
Poplar
Genus of deciduous trees, grown for their habit, foliage and very quick growth. Has catkins in late winter or spring. Female trees produce copious amounts of fluffy, white seeds. Fully hardy. Prefers full sun and needs deep, fertile, moist but well-drained soil; resents dry soil, apart from *P. alba*, which thrives in coastal gardens. Extensive root systems can undermine foundations and so make poplars unsuitable for planting close to buildings, particularly on clay soil. Propagate by hardwood cuttings in winter. Is susceptible to bacterial canker and fungal diseases.
P. alba illus. p.38. Is much confused with the commoner *P. canescens*. **'Pyramidalis'** is a vigorous, deciduous, upright tree. H 20m (70ft), S 5m (15ft). Broadly oval, wavy-margined or lobed, dark green leaves, white beneath, turn yellow in autumn. **'Raket'** (syn. *P.a.* 'Rocket') illus. p.43. **'Richardii'**, H 15m (50ft), S 12m (40ft), has leaves that are golden-yellow above.
P. balsamifera (Balsam poplar, Tacamahac). Fast-growing, deciduous, upright tree. H 30m (100ft), S 8m (25ft). Oval, glossy, dark green leaves, whitish beneath, have a strong fragrance of balsam when young.
P. x *berolinensis* (Berlin poplar). Deciduous, columnar tree. H 25m (80ft), S 8m (25ft). Has broadly oval, bright green leaves with white undersides.
P. x *canadensis* (Canadian poplar). **'Eugenei'** is a deciduous, columnar tree. H 30m (100ft), S 12m (40ft). Has broadly oval, bronze, young leaves, maturing to dark green, and red catkins in spring. **'Robusta'** illus. p.39. **'Serotina de Selys'** (syn. *P.* x *c.* 'Serotina Erecta') illus. p.40.
P. x *candicans*, syn. *P. gileadensis*, *P.* x *jackii* 'Gileadensis' (Balm of Gilead, Ontario poplar). Very fast-growing, deciduous, conical tree. H 25m (80ft), S 10m (30ft). Oval leaves are dark green and, when young, balsam-scented. Is very susceptible to canker. **'Aurora'**, H 15m (50ft) or more, S 6m (20ft), has leaves heavily but irregularly blotched with creamy-white.
P. x *canescens* illus. p.39.
P. deltoides (Cottonwood, Eastern cottonwood, Necklace poplar). Very fast-growing, deciduous, spreading tree. H 30m (100ft), S 20m (70ft). Has lush growth of broadly oval, glossy, bright green leaves.
P. gileadensis. See *P. candicans*.
P. x *jackii* **'Gileadensis'**. See *P.* x *candicans*.
♚ *P. lasiocarpa* (Chinese necklace poplar). Very fast-growing, deciduous, spreading tree. H 15m (50ft), S 12m (40ft). Has stout shoots and very large, heart-shaped, mid-green leaves with red veins, on long, red stalks. Stout, drooping, yellow catkins are borne in spring.
P. maximowiczii illus. p.38.
P. nigra (Black poplar). Very fast-growing, deciduous, spreading tree. H 25m (80ft), S 20m (70ft). Has dark bark. Diamond-shaped leaves are bronze when young, bright green when mature and yellow in autumn. Male trees bear red catkins in mid-spring.

PORTULACARIA

♚ **'Italica'** illus. p.40.
P. szechuanica. Very fast-growing, deciduous, conical tree. H 25m (80ft), S 10m (30ft). Has flaking, pinkish-grey bark and large, heart-shaped, dark green leaves.
♚ *P. tremula* (Aspen). Vigorous, deciduous, spreading tree. H 15m (50ft), S 10m (30ft). Rounded leaves are bronze-red when young, grey-green when mature and yellow in autumn. Flattened stalks make foliage tremble and rattle in wind. **'Erecta'**, S 5m (15ft), has an upright habit. **'Pendula'** illus. p.53.
P. tremuloides (American aspen, Quaking aspen). Very fast-growing, deciduous, spreading tree. H 15m (50ft) or more, S 10m (30ft). Has rounded, finely toothed, glossy, dark green leaves that flutter in the wind and turn yellow in autumn.
P. trichocarpa (Black cottonwood, Western balsam poplar). Very fast-growing, deciduous, conical tree. H 30m (100ft) or more, S 10m (30ft). Bears dense growth of oval, glossy, dark green leaves with green-veined, white undersides, strongly balsam-scented when young. Foliage turns yellow in autumn.

PORANA (Convolvulaceae)
Genus of evergreen or deciduous, twining climbers, grown for their flowers. Frost tender, min. 5–7°C (41–5°F), 10–13°C (50–55°F) for good winter blooms. Provide fertile, moisture-retentive, well-drained soil and full light. Water freely in full growth, sparingly in cold weather. Stems require support. Thin out evergreen species and cut back deciduous ones to just above ground level in late winter or early spring. Propagate by basal, softwood cuttings in late spring or early summer or by seed in spring.
P. paniculata (Bridal bouquet, Snow creeper). Vigorous, evergreen, twining climber. H 6–10m (20–30ft). Large, loose panicles of small, elder-scented, trumpet-shaped, white flowers are produced from late summer to mid-winter. Leaves are heart-shaped.

PORTULACA (Portulacaceae)
Genus of fleshy annuals and perennials with flowers that open in sun and close in shade. Half hardy. Needs full light and any well-drained soil. Propagate by seed sown under glass in early spring, or outdoors in late spring. Is prone to attack by aphids.
P. grandiflora (Sun plant). Slow-growing, partially prostrate annual. H 15–20cm (6–8in), S 15cm (6in). Has lance-shaped, succulent, bright green leaves. In summer and early autumn bears shallowly bowl-shaped flowers, 2.5cm (1in) wide and with conspicuous stamens, in shades of yellow, red, orange, pink or white. **'Cloudbeater'** is a double-flowered cultivar. **Sundance Series** illus. p.278. **Sunnyside Series** has rose-like, double flowers.

PORTULACARIA (Portulacaceae)
Genus of one species of evergreen or semi-evergreen, succulent-leaved

573

POTAMOGETON

shrub, grown for its foliage and overall appearance. Frost tender, min. 7–10°C (45–50°F). Needs full sun and well-drained soil. Water potted plants moderately when in full growth, sparingly at other times. Propagate by semi-ripe cuttings in summer.
P. afra illus. p.123. **'Variegata'** is an evergreen or semi-evergreen, erect shrub with more or less horizontal branches. H and S 2–3m (6–10ft). Has oval to rounded, fleshy, cream-edged, bright green leaves and, from late spring to summer, tiny, star-shaped, pale pink flowers that are borne in small clusters.

POTAMOGETON
(Potamogetonaceae)
Genus of deciduous, perennial, submerged water plants, grown for their foliage. Is suitable for cold-water pools and aquariums. Fully hardy. Prefers sun. Remove fading foliage and thin colonies of plants as necessary. Propagate by stem cuttings in spring or summer.
P. crispus illus. p.388.
P. lucens (Shining pondweed). Deciduous, perennial, submerged water plant. S indefinite. Oblong, deep olive-green leaves form spreading colonies. Inconspicuous, greenish flowers are borne in summer. Is suitable for a medium- to large-sized pool.

POTENTILLA (Rosaceae)
Genus of perennials and deciduous shrubs, grown for their clusters of small, flattish to saucer-shaped flowers and for their foliage. Tall species – particularly the shrubs – are useful in borders. Dwarf potentillas are good for rock gardens. Fully hardy. Does best in full sun, but flower colour is better on orange-, red- and pink-flowered cultivars if they are shaded from hottest sun. Needs well-drained soil. Propagate perennial species by seed in autumn or by division in spring or autumn; selected forms by division only in spring or autumn. Shrubby species may be raised by seed in autumn or by softwood or greenwood cuttings in summer, selected forms by softwood or greenwood cuttings during summer.
⚑ *P.* **'Abbotswood'** illus. p.128.
P. alba illus. p.321.
P. arbuscula, syn. *P. fruticosa* var. *arbuscula*. Deciduous, bushy, dense shrub. H 1m (3ft), S 1.2m (4ft). Saucer-shaped, golden-yellow flowers are produced from mid-summer to autumn amid grey-green to silver-grey leaves divided into 3 or 5 narrowly oblong leaflets.
P. argyrophylla, syn. *P. atrosanguinea* var. *argyrophylla*. Clump-forming perennial. H 45cm (18in), S 60cm (24in). Saucer-shaped, clear yellow flowers are produced in profusion from early to late summer above strawberry-like, silvery leaves.
P. atrosanguinea illus. p.247.
var. *argyrophylla*. See *P. argyrophylla*.
P. aurea illus. p.334.
⚑ *P.* **'Beesii'**, syn. *P.* 'Nana Argentea'. Slow-growing, deciduous, compact shrub. H 75cm (30in), S 1m (3ft). Has saucer-shaped, golden-yellow flowers in summer-autumn. Has silver leaves comprising 3 or 5 narrowly oblong leaflets.
P. **'Coronation Triumph'**. Deciduous, mound-forming, dense shrub. H 1m (3ft), S 1.5m (5ft). Has mid-green leaves divided into 5 narrowly oblong leaflets. Saucer-shaped, bright yellow flowers are borne profusely over a long period in summer-autumn.
P. crantzii (Alpine cinquefoil). Upright perennial with a thick, woody rootstock. H and S 10–20cm (4–8in). Produces wedge-shaped, 5-lobed leaves and, in spring, flattish, yellow flowers with orange centres. Is good in a rock garden.
P. davurica var. *mandschurica* of gardens. See *P.* 'Manchu'.
⚑ *P.* **'Daydawn'** illus. p.131.
⚑ *P.* **'Elizabeth'** illus. p.139.
P. eriocarpa illus. p.333.
P. **'Etna'**. Clump-forming perennial. H 75cm (30in), S 45cm (18in). In mid-summer produces saucer-shaped, maroon flowers above strawberry-like, dark green leaves.
P. **'Farrer's White'** illus. p.129.
P. **'Friedrichsenii'** illus. p.139.
P. fruticosa. Deciduous, bushy, dense shrub. H 1m (3ft), S 1.5m (5ft). From late spring to late summer produces saucer-shaped, bright yellow flowers. Dark green leaves have 5 narrowly oblong leaflets. var. *arbuscula*. See *P. arbuscula* **'Gold Drop'**. See *P. parvifolia* 'Gold Drop'.
⚑ *P.* **'Gibson's Scarlet'**. Clump-forming perennial. H and S 45cm (18in). Bears saucer-shaped, brilliant scarlet flowers from mid- to late summer. Dark green leaves are strawberry-like.
P. **'Gloire de Nancy'**, syn. *P.* 'Glory of Nancy'. Clump-forming perennial. H and S 45cm (18in). Very large, saucer-shaped, semi-double, orange and coppery-red flowers appear throughout summer. Has strawberry-like, dark green leaves.
P. **'Glory of Nancy'**. See *P.* 'Gloire de Nancy'.
⚑ *P.* **'Goldfinger'**. Deciduous, bushy, dense shrub. H and S 1.5m (5ft). Bears large, saucer-shaped, rich yellow flowers in profusion from late spring to autumn. Leaves are deep green and divided into 5 narrowly oblong leaflets.
P. **'Jackman's Variety'**. Deciduous, upright, dense shrub. H 1.2m (4ft), S 1.5m (5ft). Large, saucer-shaped, bright yellow flowers are produced from late spring to mid-autumn amid dark green leaves comprising 5 narrowly oblong leaflets.
P. **'Maanelys'**. See *P.* 'Moonlight'.
P. **'Manchu'**, syn. *P. davurica* var. *mandschurica* of gardens, illus. p.128.
P. **'Manelys'**. See *P.* 'Moonlight'.
⚑ *P. megalantha* illus. p.256.
P. **'Monsieur Rouillard'**. Clump-forming perennial. H and S 45cm (18in). Saucer-shaped, double, deep blood-red flowers are borne in summer above strawberry-like, dark green leaves.
⚑ *P.* **'Moonlight'**, syn. *P.* 'Maanelys', *P.* 'Manelys'. Deciduous, upright, dense shrub. H 1.2m (4ft), S 2m (6ft). Produces saucer-shaped, soft yellow flowers from late spring to early autumn. Bears grey-green leaves divided into 5 narrowly oblong leaflets.
P. **'Nana Argentea'**. See *P.* 'Beesii'.
⚑ *P. nepalensis* **'Miss Willmott'** illus. p.246.
P. nitida. Dense, mat-forming perennial. H 2.5–5cm (1–2in), S 20cm (8in). Has rounded, 3-lobed, silver leaves. Flower stems each carry 1–2 flattish, rose-pink flowers with dark centres in early summer. Is often shy-flowering. Suits a rock garden or trough.
P. parvifolia **'Gold Drop'**, syn. *P. fruticosa* 'Gold Drop'. Deciduous, upright, dense shrub. H and S 1.2m (4ft). Bears a mass of saucer-shaped, golden-yellow flowers from late spring to early autumn amid bright green leaves composed of 5 narrowly oblong leaflets.
P. recta **'Warrenii'**, syn. *P.r.* 'Macrantha', illus. p.255.
P. **'Red Ace'** illus. p.136.
P. **'Royal Flush'**. Deciduous, bushy shrub. H 45cm (18in), S 75cm (30in). Leaves are mid-green and divided into 5 narrowly oblong leaflets. From late spring to autumn bears saucer-shaped, sometimes semi-double, yellow-stamened, rich pink flowers that fade to white in full sun.
P. **'Sunset'** illus. p.141.
⚑ *P.* **'Tangerine'**. Deciduous, arching, dense shrub. H 1.2m (4ft), S 1.5m (5ft). Saucer-shaped, yellow flowers, flushed with pale orange-red, are produced from early summer to autumn. Bears mid-green leaves consisting of 5 or 7 narrowly oblong leaflets.
⚑ *P.* x *tonguei*. Mat-forming perennial. H 5cm (2in), S 25cm (10in). Leaves are rounded, 3–5 lobed and green. Prostrate branches bear flattish, orange-yellow flowers with red centres throughout summer. Is good for a rock garden.
⚑ *P.* **'Vilmoriniana'** illus. p.139.
⚑ *P.* **'William Rollison'**. Clump-forming perennial. H and S 45cm (18in). From mid- to late summer bears saucer-shaped, semi-double, scarlet-suffused, deep orange flowers with yellow centres. Has strawberry-like, dark green leaves.
P. **'Yellow Queen'** illus. p.256.

POTHOS. See EPIPREMNUM.

x **POTINARA.** See ORCHIDS.
x *P.* **Cherub 'Spring Daffodil'** (illus. p.255). Evergreen, epiphytic orchid for an intermediate greenhouse. H 15cm (6in). Sprays of yellow flowers, 5cm (2in) across, open in spring. Broadly oval, rigid leaves are 10cm (4in) long. Provide good light in summer.

PRATIA (Campanulaceae)
Genus of evergreen, mat-forming perennials with small leaves, grown for their mass of star-shaped flowers; is suitable for rock gardens. Is sometimes included in *Lobelia*. Some species may be invasive. Frost to half hardy. Prefers shade and moist soil. Propagate by division or seed in autumn.
P. angulata. Evergreen, creeping perennial. H 1cm (½in), S indefinite. Frost hardy. Bears small, broadly oval, dark green leaves. Star-shaped, white flowers, with 5 unevenly spaced petals, are carried in leaf axils in late spring and are followed by globose, purplish-red fruits in autumn.
P. pedunculata illus. p.331. **'County Park'** is a vigorous, evergreen, creeping perennial. H 1cm (½in), S indefinite. Frost hardy. Has small, rounded to oval leaves and, in summer, a profusion of star-shaped, rich violet-blue flowers. Makes good ground cover.

PRIMULA (Primulaceae)
Genus of annuals, biennials and perennials, some of which are evergreen. All species have rosettes of basal leaves and tubular, bell- or primrose-shaped (flat) flowers. Flower stems, leaves, sepals and, occasionally, sections of the petals are, in some species and hybrids, covered with farina – a waxy powder. There are primulas suitable for almost every type of site: the border, scree garden, rock garden, peat garden, bog garden, pool margin, greenhouse and alpine house. Some, however, may be difficult to grow as they dislike winter damp or summer heat. Fully hardy to frost tender, min. 7–10°C (45–50°F). Repot pot-grown plants annually. Tidy up fading foliage and dead-head as flowering ceases. Propagate species by seed when fresh or in spring; increase selected forms during dormancy, either by division or by root cuttings. Auriculas should be propagated by offsets in early spring or early autumn. Border cultivars may be prone to slug damage in damp situations and to attack by root aphids when grown in very dry conditions or in pots.

Primula groups
Primulas are classified according to various botanical sections, of which the following are in common horticultural usage.
Candelabra is used as a common descriptive name for those with tubular, flat-faced flowers borne in tiered whorls up the stem.
Polyanthus is a descriptive name for primulas derived from *P. vulgaris*, crossed with *P. veris*, *P. juliae* and other species, that are usually grown as biennials; their flowers are produced in large umbels on stout stems.
Auricula primulas may be considered as 3 sub-groups: alpine, border and show. Flowers of all Auriculas are carried in an umbel on a stem above the foliage and are individually flat and smooth. Show Auriculas need to be grown as single-stem specimens under glass to protect the flowers from rain. Their flowers have white centres (known as 'paste') and, for edged cultivars, an outer ring in a contrasting colour, often green, grey or white, with a black body colour. Self-coloured cultivars may be red, yellow, blue or violet. Show Auriculas have white farina on their foliage (except those with green-edged flowers), on their flower eyes and, sometimes, on their petal margins. Farina is totally absent on cultivars of the alpine sub-group.

Some border Auriculas have farina on flower stems and leaves, but many have none at all.

Cultivation
Primulas have varying cultivation requirements. For ease of reference, these have been grouped as follows:
1 – Full sun and soil that does not dry out.
2 – Full sun or partial shade and moist but well-drained soil.
3 – Partial shade and moist but well-drained soil.
4 – Partial shade and well-drained, gritty loam.
5 – Full sun or partial shade and gritty, alkaline soil.
6 – Partial shade and gritty, alkaline soil.
7 – Full or partial shade and moist, peaty soil.
8 – Full sun or partial shade and gritty, peaty soil.
9 – Partial shade and gritty, peaty soil.

P. **'Adrian'** (illus. p.237). Basal-rosetted perennial (alpine Auricula). H 22–25cm (9–10in), S 15–20cm (6–8in). Fully hardy. Has flat, light to dark blue flowers, with light centres, in mid- to late spring. Leaves are oval and mid-green. Is useful for exhibition. Cultivation group 2.
P. allionii (illus. p.236). Clump-forming perennial. H 8cm (3in), S 8–15cm (3–6in). Frost hardy. Tubular, rose, mauve or white flowers cover a tight cushion of oval, mid-green leaves in spring. Cultivation group 5.
P. alpicola. Compact perennial. H 15–60cm (6–24in), S 15–30cm (6–12in). Fully hardy. Produces terminal clusters of pendent, bell-shaped, yellow to white or purple flowers on slender stems in early summer. Mid-green leaves are oval to lance-shaped. Cultivation group 7.
var. *luna* (illus. p.237) has soft sulphur-yellow flowers.
P. aurantiaca. Neat, upright perennial (Candelabra primula). H 60cm (2ft), S 30cm (1ft). Fully hardy. Tubular, reddish-orange flowers are borne in early summer. Has long, broadly oval to lance-shaped, coarse, mid-green leaves. Cultivation group 1 or 7.
P. aureata (illus. p.237). Clump-forming perennial. H and S 8–15cm (3–6in). Frost hardy. Produces small umbels of flat, cream to yellow flowers in spring. In summer, oval, toothed, mid-green leaves have striking purple-red midribs; in winter, leaves form tight buds covered with whitish farina. Cultivation group 4.
♉ *P. auricula*. Clump-forming perennial (alpine Auricula). H 15–23cm (6–9in), S 15cm (6in). Fully hardy. Fragrant, flat, yellow flowers are borne in large umbels in spring. Oval, soft, pale green to grey-green leaves are densely covered with white farina. Cultivation group 5 or 6.
P. bhutanica (illus. p.237). Neat, clump-forming perennial. H 15cm (6in), S 15–23cm (6–9in). Fully hardy, but often short-lived. In spring produces neat umbels of tubular, pale purplish-blue flowers, each with a white or creamy-white eye, close to

oval to lance-shaped, crinkled, mid-green leaves. Cultivation group 8.
P. **'Blairside Yellow'** (illus. p.237). Basal-rosetted perennial (alpine Auricula). H 2.5cm (1in), S 15cm (6in). Fully hardy. In early spring, bell-shaped, golden-yellow flowers nestle in a rosette of tiny, rounded to oval, pale green leaves. Cultivation group 2.
P. **'Blossom'** (illus. p.237). Basal-rosetted perennial (alpine Auricula). H 23–25cm (9–10in), S 15–25cm (6–10in). Fully hardy. Flat, deep crimson to bright red flowers with golden centres are borne profusely in spring. Has oval, dark green leaves. Is suitable for exhibition. Cultivation group 2.
♉ *P. bulleyana* (illus. p.237). Neat, upright perennial (Candelabra primula). H 60cm (2ft), S 30cm (1ft). Fully hardy. Tubular, deep orange flowers appear in early summer. Leaves are oval to lance-shaped, toothed and dark green. Cultivation group 1 or 7.
P. **'Chloë'** (illus. p.237). Basal-rosetted perennial (show Auricula). H 20–24cm (8–9½in), S 15–20cm (6–8in). Fully hardy. In late spring produces flat, dark-green-edged flowers with a black body colour and brilliant white paste centres. Oval leaves are dark green and have no farina. Is good for exhibition. Cultivation group 2.
P. chungensis (illus. p.237). Neat, upright perennial (Candelabra primula). H 60cm (2ft), S 30cm (1ft). Fully hardy. In summer bears tiered whorls of tubular, orange flowers among oval to lance-shaped, mid-green leaves. Cultivation group 1 or 7.
P. clarkei (illus. p.236). Clump-forming perennial. H and S 5–10cm (2–4in). Fully hardy. In spring bears flat, rose-pink flowers, with yellow eyes, just above a clump of rounded to oval, pale green leaves. Cultivation group 8. Divide in late winter.
P. clusiana (illus. p.236). Clump-forming perennial. H 15–23cm (6–9in), S 15cm (6in). Fully hardy. In spring bears umbels of tubular, rose-pink flowers with white eyes. Leaves are oval, glossy and mid-green. Cultivation group 8.
P. **'Craddock White'** (illus. p.236). Clump-forming perennial. H to 8cm (3in), S 13cm (5in). Fully hardy. Fragrant, upward-facing, flat, white flowers, with yellow eyes, are borne in spring just above long, oval, red-veined, dark green leaves. Cultivation group 1 or 3.
P. **'David Green'**. Clump-forming perennial. H 10–15cm (4–6in), S 15–20cm (6–8in). Fully hardy. In spring produces flat, bright crimson-purple flowers amid oval, coarse, mid-green leaves. Cultivation group 1 or 3.
♉ *P. denticulata* (Drumstick primula; illus. p.231). Vigorous, neat, upright, clump-forming perennial. H 30–60cm (12–24in), S 30–45cm (12–18in). Fully hardy. From early to mid-spring, dense, rounded heads of flat, lilac, purple or pink flowers are borne on tops of stout stems. Mid-green leaves are broadly lance-shaped and toothed. Cultivation group 1 or 3. f. *alba* (illus. p.236) has white flowers.

P. edgeworthii (illus. p.236). Rosetted perennial. H 5–10cm (2–4in), S 10–15cm (4–6in). Fully hardy. Flat, pale mauve flowers with white eyes appear singly among oval, toothed, pale green leaves in spring. Cultivation group 8.
♉ *P. elatior* (Oxlip; illus. p.237). Clump-forming perennial. H 15–30cm (6–12in), S 15cm (6in). Fully hardy. Produces umbels of small, fragrant, tubular, yellow flowers in spring, above neat, oval to lance-shaped, toothed, mid-green leaves. Cultivation group 1, 7 or 9.
P. **'E.R. Janes'**. Clump-forming perennial. H 10–15cm (4–6in), S 15–20cm (6–8in). Fully hardy. Flat, pale rose-pink flowers, flushed with orange, are borne singly in spring amid broadly oval, toothed, mid-green leaves. Cultivation group 1 or 3.
P. farinosa (Bird's-eye primrose; illus. p.230). Clump-forming perennial. H 15–30cm (6–12in), S 15cm (6in). Fully hardy. In spring, umbels of tubular, lilac-pink, occasionally white, flowers are borne on short, stout stems. Oval, toothed, mid-green leaves are densely covered with white farina. Cultivation group 8.
♉ *P. flaccida*, syn. *P. nutans* (illus. p.237). Lax, clump-forming, short-lived perennial. H 15–30cm (6–12in), S 15–23in (6–9in). Fully hardy. In early summer, each stout stem produces a conical head of pendent, bell-shaped, lavender or violet flowers above narrowly oval, pale to mid-green leaves. Cultivation group 7 or 8.
♉ *P. florindae* (Giant cowslip; illus. p.237). Bold, upright, clump-forming perennial. H 60cm–1m (2–3ft), S 30–60cm (1–2ft). Fully hardy. In summer, large heads of pendent, bell-shaped, sulphur-yellow flowers appear above broadly lance-shaped, toothed, mid-green leaves. Cultivation group 1 or 7.
P. forrestii (illus. p.237). Clump-forming perennial. H 15–23cm (6–9in), S 15–30cm (6–12in). Frost hardy. Dense umbels of flat, yellow flowers with orange eyes are borne in late spring or early summer. Has oval, toothed, dark green leaves. Cultivation group 8.
♉ *P. frondosa* (illus. p.236). Compact, clump-forming perennial. H 5–15cm (2–6in), S 10–15cm (4–6in). Fully hardy. In spring bears umbels of flat, yellow-eyed, lilac-rose to reddish-purple flowers on short stems above neat, oval, mid-green leaves, densely covered with white farina. Cultivation group 8.
♉ *P.* **'Garryarde Guinevere'**. Clump-forming perennial. H 10–15cm (4–6in), S 15–20cm (6–8in). Fully hardy. Flat, purplish-pink flowers with yellow eyes are produced in spring among oval, toothed, bronze-green leaves. Cultivation group 1 or 3.
P. **Gold Lace Group** (illus. p.237). Group of basal-rosetted perennials. H 20cm (8in), S 25cm (10in). Frost hardy. Produces flat flowers, in a variety of colours, with gold-laced margins, from mid- to late spring. Leaves are oval and mid-green, sometimes tinged red. Raise annually by seed. Cultivation group 2 or 3.

P. gracilipes (illus. p.236). Neat, clump-forming perennial. H 10–15cm (4–6in), S 15–23cm (6–9in). Fully hardy. Tubular, purplish-pink flowers with greenish-yellow eyes are borne singly in spring or early summer among oval, wavy, toothed, mid-green leaves. Cultivation group 8.
P. **'Harlow Car'**. Neat, clump-forming perennial. H 10–15cm (4–6in), S 15–23cm (6–9in). Fully hardy. In spring produces flat, white flowers on short stems above oval, soft, mid-green leaves. Cultivation group 1, 3, 5 or 8.
♉ *P. helodoxa*. See *P. prolifera*.
P. hirsuta, syn. *P. rubra* (illus. p.236). Clump-forming perennial. H 5–15cm (2–6in), S 10–15cm (4–6in). Fully hardy. Produces small umbels of flat, rose or lilac flowers in spring. Has small, rounded to oval, sticky, mid-green leaves. Cultivation group 8.
♉ *P.* **'Inverewe'** (illus. p.237). Upright, clump-forming perennial (Candelabra primula). H 45–75cm (18–30in), S 30–45cm (12–18in). Fully hardy. Tubular, bright orange-red flowers are produced in summer on stems coated with white farina. Has oval to lance-shaped, toothed, coarse, mid-green leaves. Cultivation group 1 or 7.
P. ioessa. Clump-forming perennial. H 10–30cm (4–12in), S 15–30cm (6–12in). Fully hardy. Clustered heads of funnel-shaped, pink or pinkish-mauve, or sometimes white, flowers are borne in spring or early summer above oval to lance-shaped, toothed, mid-green leaves. Cultivation group 7.
P. **'Janie Hill'** (illus. p.237). Basal-rosetted perennial (alpine Auricula). H 23–25cm (9–10in), S 15–20cm (6–8in). Fully hardy. Flat, dark to golden-brown flowers, with golden centres, open in mid- to late spring. Has oval, mid-green leaves. Is useful for exhibition. Cultivation group 2.
♉ *P. japonica*. Bold, upright, clump-forming perennial (Candelabra primula). H 30–45cm (12–24in), S 30–45cm (12–18in). Fully hardy. In early summer produces tubular, deep red flowers on stout stems above oval to lance-shaped, toothed, coarse, pale green leaves. Cultivation group 1 or 7. **'Miller's Crimson'** (illus. p.236) has intense crimson flowers. **'Postford White'** (illus. p.236) bears white flowers.
♉ *P.* x *kewensis* (illus. p.237). Upright, clump-forming perennial. H 30cm (12in), S 15–30cm (6–12in). Frost tender. Produces whorls of fragrant, tubular, bright yellow flowers in winter and early spring. Oval to lance-shaped, toothed, pale green leaves are covered with white farina. Cultivation group 4.
♉ *P.* **'Linda Pope'** (illus. p.237). Neat, clump-forming perennial. H 10–15cm (4–6in), S 15–23cm (6–9in). Fully hardy. In spring bears flat, mauve-blue flowers on short stems above oval, toothed, mid-green leaves covered with white farina. Cultivation group 5.
P. malacoides (single, double, illus. p.236). Neat, clump-forming perennial.

H and S 20–30cm (8–12in). Frost tender. Dense whorls of small, flat, single or double, pink, purplish-pink or white flowers open in winter-spring. Leaves are oval, hairy, soft and pale green. Cultivation group 4.
P. 'Margaret Martin' (illus. p.237). Basal-rosetted perennial (show Auricula). H 20cm (8in), S 15cm (6in). Fully hardy. Bears flat, grey-edged flowers, with a black body colour and white centres, in mid- to late spring. Has spoon-shaped, grey-green leaves covered with white farina. Is excellent for exhibition. Cultivation group 2.
♛ **P. marginata** (illus. p.237). Neat, clump-forming perennial. H 10–15cm (4–6in), S 10–20cm (4–8in). Fully hardy. In spring, clusters of funnel-shaped, blue-lilac flowers appear above oval, toothed, mid-green leaves densely covered with white farina. Cultivation group 5. **'Prichard's Variety'** (illus. p.237) has lilac-purple flowers with white eyes.
P. 'Mark' (illus. p.237). Basal-rosetted perennial (alpine Auricula). H 23–28cm (9–11in), S 15–23cm (6–9in). Fully hardy. Produces flat, pink flowers, with light yellow centres, in spring. Leaves are oval and vibrant green. Is a leading exhibition cultivar. Cultivation group 2.
P. melanops (illus. p.236). Clump-forming, short-lived perennial. H 20–35cm (8–14in), S 15cm (6in). Fully hardy. In summer has umbels of pendent, narrowly funnel-shaped, deep violet-purple flowers, with black eyes, above long, strap-shaped, mid-green leaves. Cultivation group 1 or 7.
P. modesta. Clump-forming perennial. H 5–10cm (2–4in), S 10cm (4in). Fully hardy. Dense heads of small, tubular, pinkish-purple flowers appear on short stems in spring. Rounded to oval, mid-green leaves are covered with yellow farina. Cultivation group 8. var. **fauriae** (illus. p.236), H and S 5cm (2in), has yellow-eyed, pinkish-purple flowers and leaves covered with white farina.
P. 'Moonstone' (illus. p.237). Basal-rosetted perennial (border Auricula). H 23–25cm (9–10in), S 15–20cm (6–8in). Fully hardy. Rounded, double, whitish- or greenish-yellow flowers are produced in profusion in spring. Leaves are oval and mid-green. Preferably, grow under glass. Cultivation group 2.
P. 'Mrs J.H. Wilson' (illus. p.236). Neat, clump-forming perennial (alpine Auricula). H 10–15cm (4–6in), S 15–20cm (6–8in). Fully hardy. Bears small umbels of flat, white-centred, purple flowers in spring. Oval leaves are greyish-green. Cultivation group 2 or 3.
P. nutans. See **P. flaccida**.
P. obconica. Neat, clump-forming perennial. H and S 15–30cm (6–12in). Frost tender. Flat, purple, lilac or white flowers, with yellow eyes, are borne in dense umbels during winter-spring. Leaves are oval, toothed, hairy and pale green. Cultivation group 4.
P. 'Orb'. Basal-rosetted perennial (show Auricula). H 20cm (8in), S 15cm (6in). Fully hardy. Flat, dark-green-edged flowers, each with a black body colour and a central zone of white paste, are produced from mid- to late spring. Has spoon-shaped, dark green leaves without farina. Is good for exhibition. Cultivation group 2.
P., Pacific Series (Polyanthus) illus. p.291 (dwarf) illus. p 280; cultivation group 2.
P. palinuri (illus. p.237). Clump-forming perennial. H 10–20cm (4–8in), S 15cm (6in). Frost hardy. One-sided clusters of semi-pendent, narrowly funnel-shaped, yellow flowers appear on thick stems in early summer. Has rounded to oval, lightly toothed, thick-textured, powdered, green leaves. Cultivation group 5.
P. petiolaris (illus. p.236). Rosette-forming perennial. H 5–10cm (2–4in), S 10–15cm (4–6in). Fully hardy. Tubular, purplish-pink flowers, with toothed petals, are borne singly in spring. Has small, oval, toothed, mid-green leaves. Cultivation group 8.
P. polyneura (illus. p.236). Neat, clump-forming perennial. H 20–30cm (8–12in), S 15–20cm (6–8in). Frost hardy. Dense heads of tubular, pale rose, rich rose or purple-rose flowers are produced in late spring or early summer. Rounded to oval, shallowly lobed, downy, soft leaves are mid-green. Cultivation group 7.
P., Posy Series (Polyanthus) illus. p.273; cultivation group 2.
♛ **P. prolifera**, syn **P. helodoxa** (illus. p.237). Upright, clump-forming perennial (Candelabra primula). H 60cm–1m (2–3ft), S 30–45cm (1–1½ft). Fully hardy. Bell-shaped, yellow flowers are borne in summer. Leaves are oval, toothed and pale green. Cultivation group 1 or 7.
P. x pubescens 'Janet' (illus. p.236). Evergreen, clump-forming perennial. H 20–35cm (8–14in), S 15cm (6in). Fully hardy. In spring, clusters of outward-facing, flat, purplish-pink flowers are produced above a rosette of oval to rounded, soft, mid-green leaves. Cultivation group 2 or 8. Propagate by offsets after flowering.
♛ **P. pulverulenta** (illus. p.236). Bold, upright, clump-forming perennial (Candelabra primula). H 60cm–1m (2–3ft), S 30–45cm (1–1½ft). Fully hardy. In early summer bears tubular, deep red flowers with purple-red eyes on stems covered with white farina. Has broadly lance-shaped, toothed, coarse, mid-green leaves. Cultivation group 1 or 7. ♛ **'Bartley'** (illus. p.236) has pink flowers.
P. reidii. Clump-forming perennial. H 5–10cm (2–4in), S 10–15cm (4–6in). Frost hardy. Produces dense clusters of bell-shaped, pure white flowers on slender stems in early summer. Has oval, hairy, pale green leaves. Cultivation group 8. var. **williamsii** (illus. p.237) is more robust and has purplish-blue to pale blue flowers.
♛ **P. rosea** (illus. p.236). Clump-forming perennial. H 10–15cm (4–6in), S 15–20cm (6–8in). Fully hardy. In early spring bears small clusters of flat, glowing rose-pink flowers on short stems, among oval to lance-shaped, mid-green leaves, often bronze-flushed when young. Cultivation group 1 or 7.
P. 'Royal Velvet'. Vigorous, basal-rosetted perennial (border Auricula). H 15–35cm (6–14in), S 20–35cm (8–14in). Fully hardy. Flat, velvety, blue-tinged, maroon flowers, with frilled petals and large, creamy-yellow centres, are produced in spring. Has large, spoon-shaped, pale green leaves. Cultivation group 2.
P. rubra. See **P. hirsuta**.
P. x scapeosa (illus. p.236). Clump-forming perennial. H and S 15cm (6in). Fully hardy. Clusters of outward-facing, flat, mauve-pink flowers, in early spring, are initially hidden by broadly oval, sharply toothed, mid-green leaves covered at first with slight farina; later, flower stem elongates above leaves. Cultivation group 8.
P. secundiflora (illus. p.236). Neat, upright, clump-forming perennial. H 30–45cm (12–18in), S 30cm (12in). Fully hardy. Has clusters of pendent, funnel-shaped, reddish-purple flowers in summer above lance-shaped, toothed leaves. Cultivation group 1 or 7.
♛ **P. sieboldii** (illus. p.236). Spreading, clump-forming perennial. H and S 15–20cm (6–8in). Fully hardy. Umbels of flat, white, pink or purple flowers, with white eyes, open above oval, round-toothed, downy, soft, pale green leaves in early summer. Cultivation group 7. **'Dancing Ladies'** is a variable, seed-raised selection that has upward-facing flowers with deeply cleft petals. Petals are either white with reverses suffused pale pink or blue, or pink with reverses suffused blue; leaves are slightly toothed. **'Wine Lady'** (illus. p.236) has white flowers, strongly suffused with purplish-red.
P. sikkimensis (illus. p.237). Bold, upright, clump-forming perennial. H 45–75cm (18–30in), S 30–45cm (12–18in). Fully hardy. Pendent clusters of funnel-shaped, yellow flowers are borne in summer. Has rounded to oval, toothed, pale green leaves. Cultivation group 1 or 7.
P. sinensis (illus. p.236). Neat, compact perennial. H and S 15–20cm (6–8in). Frost tender. Flat, purple, purple-rose, pink or white flowers, with yellow eyes, are produced in neat whorls in winter-spring. Leaves are oval, toothed, hairy and mid-green. Cultivation group 4.
P. sonchifolia (illus. p.236). Rosette-forming perennial. H and S 20–30cm (8–12in). Fully hardy. Dense umbels of tubular, blue-purple flowers with white eyes and yellow edges open in spring. Leaves are oval to lance-shaped, toothed and mid-green. Cultivation group 8.
P., Super Giants Series. Rosette-forming perennial (Polyanthus), usually grown as a biennial. H and S to 30cm (12in). Fully hardy. Produces large, fragrant, flat flowers in a wide range of colours in spring (blue, illus. p.284). Cultivation group 2.
P. 'Tawny Port'. Clump-forming perennial. H 10–15cm (4–6in), S 15–20cm (6–8in). Fully hardy. Bears flat, port-wine-coloured flowers on short stems in spring. Rounded to oval, toothed leaves are reddish-green. Cultivation group 1 or 3.
♛ **P. veris** (Cowslip; illus. p.237). Neat, clump-forming perennial. H and S 15–20cm (6–8in). Fully hardy. Tight clusters of fragrant, tubular, yellow flowers are produced on stout stems in spring. Leaves are oval to lance-shaped, toothed and mid-green. Cultivation group 1 or 3.
P. verticillata (illus. p.237). Clump-forming perennial. H 20–25cm (8–10in), S 15–20cm (6–8in). Half hardy. Fragrant, bell-shaped, yellow flowers are borne in whorls in spring. Has oval, toothed, mid-green leaves. Cultivation group 4.
♛ **P. vialii** (illus. p.236). Clump-forming, short-lived perennial. H 30–45cm (12–18in), S 20–30cm (8–12in). Frost hardy. Dense, conical spikes of tubular, bluish-purple-and-red flowers open in late spring. Has lance-shaped, toothed, soft, mid-green leaves. Cultivation group 1 or 3.
P. vulgaris (Primrose; illus. p.237). Neat, clump-forming perennial. H and S 15–20cm (6–8in). Fully hardy. Flat, soft yellow flowers, with darker eyes, are borne singly among oval to lance-shaped, toothed, bright green leaves in spring. Cultivation group 1 or 3. **'Alba Plena'** has double, white flowers. **'Gigha White'** (illus. p.236) is very floriferous and has yellow-eyed, white flowers. ♛ subsp. **sibthorpii** (illus. p.236) has pink or purplish-pink flowers.
♛ **P. 'Wanda'.** Neat, clump-forming perennial. H 10–15cm (4–6in), S 15–20cm (6–8in). Fully hardy. In spring, flat, crimson-purple flowers are produced singly amid oval, toothed, purplish-green foliage. Cultivation group 1 or 3.
P. warshenewskiana (illus. p.236). Clump-forming perennial. H and S 2.5cm (1in). Fully hardy. Tiny, flat, white-eyed, bright pink flowers sit just above spoon-shaped, dark green leaves in early spring. Cultivation group 8. Divide clumps regularly in late winter before flowering.

PRINSEPIA (Rosaceae)
Genus of deciduous, usually spiny, spring- and early summer-flowering shrubs, grown for their habit, flowers and fruits. Fully hardy. Needs sun and any not too dry soil. Does well against a south- or west-facing wall. Propagate by softwood cuttings in summer or by seed in autumn.
P. uniflora illus. p.107.

PROBOSCIDEA (Pedaliaceae, syn. Martyniaceae)
Genus of annuals and perennials. Half hardy. Grow in a sunny, sheltered position and in fertile, well-drained soil. Propagate by seed sown under glass in early spring.
P. fragrans. Moderately fast-growing, upright annual. H 60cm (2ft), S 30cm (1ft). Has rounded, serrated or lobed leaves. Fragrant, bell-shaped, crimson-purple flowers, to 5cm (2in) long, appear in summer-autumn, followed by rounded, horned, brown fruits, 8–10cm (3–4in) long, which, if gathered young, may be pickled and eaten.

PROSTANTHERA (Labiatae)
Mint bush
Genus of evergreen shrubs, grown for their flowers and mint-scented foliage.

Half hardy to frost tender, min. 5°C (41°F). Requires full light or partial shade and fertile, well-drained soil. Water containerized specimens freely when in full growth, moderately at other times. Leggy stems may be cut back after flowering. Propagate by seed in spring or by semi-ripe cuttings in late summer.
P. ovalifolia illus. p.115.
♥ *P. rotundifolia* illus. p.115.

PROTEA (Proteaceae)
Genus of evergreen shrubs and trees, grown mainly for their colourfully bracted flower heads. Is difficult to grow. Frost tender, min. 5–7°C (41–5°F). Requires full light and well-drained, neutral to acid soil, low in phosphates and nitrates. Water containerized specimens moderately, less when not in full growth. Plants under glass must have plenty of ventilation throughout the year. Prune, if necessary, in early spring. Propagate by seed in spring or by semi-ripe cuttings in summer.
P. barbigera. See *P. magnifica.*
P. cynaroides illus. p.131.
P. magnifica, syn. *P. barbigera.* Evergreen, rounded to spreading shrub. H and S 1m (3ft). Has oblong to elliptic, leathery, mid- to greyish-green leaves. Spherical flower heads, 15–20cm (6–8in) wide, with petal-like, pink, red, yellow or white bracts, appear in spring-summer.
P. mellifera. See *P. repens.*
P. neriifolia illus. p.110.
P. repens, syn. *P. mellifera* (Sugar bush). Evergreen, ovoid to rounded shrub. H and S 2–3m (6–10ft). Mid-green leaves are narrowly oblong to elliptic and tinted blue-grey. In spring-summer produces cup-shaped, 13cm (5in) long flower heads, with petal-like, pink, red or white bracts.

PRUMNOPITYS (Podocarpaceae)
See CONIFERS.
P. andinus, syn. *Podocarpus andinus* (Plum-fruited yew, Plum yew). Conifer with a domed crown on several stems. H 15m (50ft), S 8m (25ft). Frost hardy. Has smooth, grey-brown bark, needle-like, flattened, bluish-green leaves and edible, yellowish-white fruits like small plums.

PRUNELLA (Labiatae)
Self-heal
Genus of semi-evergreen perennials with spreading mats of leaves from which arise short, stubby flower spikes in mid-summer. Suits rock gardens. Fully hardy. Grows well in sun or shade and in moist but well-drained soil. Propagate by division in spring.
P. grandiflora illus. p.329. '**Pink Loveliness**' is a semi-evergreen, rosetted perennial that spreads by short runners. H 10–15cm (4–6in), S 30cm (12in). Forms a dense mat of narrowly oval leaves. Whorls of funnel-shaped, soft pink flowers are borne in terminal spikes in summer. Makes good ground cover, but may be invasive. Cut off old flower stems before they produce seed. '**White Loveliness**' has white flowers.
P. webbiana illus. p.330.

PRUNUS (Rosaceae)
Cherry
Genus of deciduous or evergreen shrubs and trees. The trees are grown mainly for their single (5-petalled) to double flowers and autumn colour; the shrubs for their autumn colour, bark, flowers or fruits. All have oval to oblong leaves. Plants described here are fully hardy, unless otherwise stated. Evergreen species tolerate sun or shade; deciduous species prefer full sun. All may be grown in any but waterlogged soil. Trim deciduous hedges after flowering, evergreen ones in early or mid-spring. Propagate deciduous species by seed in autumn, deciduous hybrids and selected forms by softwood cuttings in summer. Increase evergreens by semi-ripe cuttings in summer. Bullfinches may eat flower buds and foliage may be attacked by aphids, caterpillars and the fungal disease silver leaf. Flowering cherries are prone to a fungus that causes 'witches' brooms' (abnormal, crowded shoots).
♥ *P.* '**Accolade**' illus. p.62.
♥ *P.* '**Amanogawa**'. Deciduous, upright tree. H 10m (30ft), S 4m (12ft). Bears fragrant, semi-double, pale pink flowers in late spring. Oblong to oval, taper-pointed, dark green leaves turn orange and red in autumn.
P. x *amygdalo-persica* '**Pollardii**'. Deciduous, spreading tree. H and S 7m (22ft). Large, saucer-shaped, 5-petalled, bright pink flowers open from early to mid-spring, before oval, glossy, mid-green leaves emerge. Green, then brown fruits are similar to large almonds in shape and taste.
♥ *P. avium* illus. p.45. ♥ '**Plena**' illus. p.50.
♥ *P.* x *blireana.* Deciduous, spreading shrub or small tree. H and S 4m (12ft). Bears double, pink flowers in mid-spring and has oval, purple leaves.
P. campanulata (Bell-flowered cherry, Taiwan cherry). Deciduous, spreading tree. H and S 8m (25ft). Frost hardy. Shallowly bell-shaped, deep rose-red flowers are produced from early to mid-spring, before or with oval, taper-pointed, dark green leaves. Fruits are small, rounded and reddish.
P. cerasifera (Cherry plum, Myrobalan). ♥ '**Nigra**' illus. p.65. '**Pissardii**' is a deciduous, round-headed tree. H and S 10m (30ft). Small, 5-petalled, pale pink flowers open from early to mid-spring and are often followed by edible, plum-like, red fruits. Has oval, red, young leaves turning deeper red, then purple. May be used for hedging.
P. '**Cheal's Weeping**', syn. *P.* '**Kiku-shidare**' of gardens, illus. p.62.
♥ *P.* x *cistena* illus. p.124.
P. davidiana (David's peach). Deciduous, spreading tree. H and S 8m (25ft). Saucer-shaped, 5-petalled, white or pale pink flowers are carried on slender shoots in late winter and early spring, but are susceptible to late frosts. Leaves are narrowly oval and dark green. Fruits are small, rounded and reddish.
P. dulcis (Almond). '**Roseoplena**' is a deciduous, spreading tree. H and S 8m (25ft). Bears double, pink flowers in late winter and early spring, before oblong, pointed, toothed, dark green leaves.
♥ *P. glandulosa* '**Alba Plena**' illus. p.123. ♥ '**Sinensis**' (syn. *P.g.* 'Rosea Plena') is a deciduous, rounded, open shrub. H and S 1.5m (5ft). Produces double, bright rose-pink flowers in late spring and oval, mid-green leaves. Flowers best when grown against a south- or west-facing wall. Cut back young shoots to within a few buds of old wood after flowering.
P. '**Hally Jolivette**'. Deciduous, rounded, compact tree. H and S 5m (15ft). Double, white flowers open from pink buds in late spring. Leaves are oval and dark green.
P. x *hillieri* '**Spire**'. See *P.* 'Spire'.
P. '**Hokusai**', syn. *P.* 'Uzu-zakura', illus. p.61.
P. incisa (Fuji cherry) illus. p.60. '**February Pink**' is a deciduous, spreading tree. H and S 8m (25ft). Oval, sharply toothed, dark green leaves are reddish when young, orange-red in autumn. During mild, winter periods bears 5-petalled, pale pink flowers. Has tiny, rounded, reddish fruits.
♥ *P.* '**Kanzan**' illus. p.51.
♥ *P.* '**Kiku-shidare**', syn. *P.* 'Cheal's Weeping', illus. p.62.
♥ *P.* '**Kursar**'. Deciduous, spreading tree. H and S 8m (25ft). Bears masses of small, 5-petalled, deep pink flowers in early spring. Has oval, dark green leaves that turn brilliant orange in autumn.
P. laurocerasus (Cherry laurel, Laurel). Evergreen, dense, bushy shrub becoming spreading and open. H 6m (20ft), S 10m (30ft). Frost hardy. Has long spikes of small, single, white flowers from mid- to late spring, large, oblong, glossy, bright green leaves and cherry-shaped, red, then black fruits. Restrict growth by cutting back hard in spring. ♥ '**Otto Luyken**' illus. p.124. '**Schipkaensis**', H 2m (6ft), S 3m (10ft), is fully hardy and of elegant, spreading habit, with narrow leaves and freely borne flowers in upright spikes. '**Zabeliana**' illus. p.124.
♥ *P. lusitanica* (Laurel, Portugal laurel). Evergreen, bushy, dense shrub or spreading tree. H and S 6–10m (20–30ft). Frost hardy. Reddish-purple shoots bear oval, glossy, dark green leaves. Slender spikes of small, fragrant, 5-petalled, white flowers appear in early summer, followed by egg-shaped, fleshy, deep purple fruits. Restrict growth by pruning hard in spring. ♥ subsp. *azorica* illus. p.96. '**Variegata**' illus. p.96.
P. maackii illus. p.58.
P. mahaleb illus. p.50.
P. '**Mount Fuji**'. See *P.* 'Shirotae'.
P. mume (Japanese apricot). '**Beni-shidori**' (syn. *P.m.* 'Beni-shidon') illus. p.100. '**Omoi-no-wae**' (syn. *P.* 'Omoi-no-mama') illus. p.100. '**Pendula**' is a deciduous, weeping tree with slender, arching branches. H and S 6m (20ft). Fragrant, 5-petalled, pink flowers appear in late winter or early spring, before broadly oval, bright green leaves, and are sometimes succeeded by edible, apricot-like, yellow fruits.

♥ *P.* '**Okame**'. Deciduous, bushy-headed tree. H 10m (30ft), S 8m (25ft). Bears masses of 5-petalled, carmine-pink flowers in early spring. Oval, sharply toothed, dark green leaves turn orange-red in autumn.
P. padus (Bird cherry) illus. p.50. var. *commutata* is a deciduous, spreading tree, conical when young. H 15m (50ft), S 10m (30ft). Produces pendent racemes of fragrant, star-shaped, 5-petalled, white flowers in mid-spring, followed by small, pea-shaped, black fruits. Oval, dark green leaves, often fully open by early spring, turn yellow in autumn. '**Plena**' has long-lasting, double flowers and no fruits; ♥ '**Watereri**' (syn. *P. padus* 'Grandiflora') bears long racemes of flowers from mid- to late spring.
♥ *P.* '**Pandora**' illus. p.61.
P. pensylvanica (Pin cherry). Deciduous, spreading tree. H 15m (50ft), S 10m (30ft). Has peeling, red-banded bark and oval, taper-pointed, bright green leaves. Produces clusters of small, star-shaped, 5-petalled, white flowers from mid- to late spring, followed by small, pea-shaped, red fruits.
P. persica (Peach). '**Klara Meyer**' is a deciduous, spreading tree. H 5m (15ft), S 6m (20ft). Bears double, bright pink flowers in mid-spring. Has slender, lance-shaped, bright green leaves. Is susceptible to the fungal disease peach leaf curl. '**Prince Charming**' illus. p.62.
♥ *P.* '**Pink Perfection**' illus. p.61.
P. '**Pink Star**'. See *P. subhirtella* 'Stellata'.
♥ *P. sargentii* illus. p.61.
P. serotina illus. p.39.
♥ *P. serrula.* Deciduous, round-headed tree. H and S 10m (30ft). Has gleaming, coppery-red bark that peels. In late spring bears small, 5-petalled, white flowers amid oval, tapering, toothed, dark green leaves that turn yellow in autumn. Fruits are tiny, rounded and reddish-brown.
P. serrulata var. *spontanea* illus. p.50.
P. '**Shimidsu**'. See *P.* 'Shogetsu'.
♥ *P.* '**Shirofugen**' illus. p.61.
♥ *P.* '**Shirotae**', syn. *P.* 'Mount Fuji', illus. p.60.
♥ *P.* '**Shogetsu**', syn. *P.* 'Shimidsu', illus. p.60.
P. spinosa (Blackthorn, Sloe). '**Purpurea**' illus. p.89.
♥ *P.* '**Spire**', syn. *P.* x *hillieri* 'Spire', illus. p.61.
P. subhirtella (Higan cherry, Rosebud cherry). Deciduous, spreading tree. H and S 8m (25ft). From early to mid-spring, a profusion of small, 5-petalled, pale pink flowers appear before oval, taper-pointed, dark green leaves, which turn yellow in autumn. Has small, round, reddish-brown fruits. ♥ '**Autumnalis**' produces semi-double, white flowers, pink in bud, in mild periods in winter.
♥ '**Pendula Rubra**' illus. p.62. '**Stellata**' (syn. *P.* 'Pink Star') illus. p.61.
♥ *P.* '**Taihaku**', syn. *P.* 'Tai Haku', illus. p.60.
P. tenella illus. p.125. ♥ '**Fire Hill**' is a deciduous, bushy shrub with upright, then spreading branches. H and S 2m

(6ft). Narrowly oval, glossy, dark green leaves are a foil for small, almond-like, single, very deep pink flowers borne profusely from mid- to late spring, followed by small, almond-like fruits.
P. tomentosa (Downy cherry). Deciduous, bushy, dense shrub. H 1.5m (5ft), S 2m (6ft). Bears small, 5-petalled, pale pink flowers from early to mid-spring before oval, slightly downy, dark green leaves appear. Fruits are spherical and bright red. Thrives in hot summers.
P. 'Trailblazer'. Deciduous, spreading tree. H and S 5m (15ft). Bears 5-petalled, white flowers from early to mid-spring, sometimes followed by edible, plum-like, red fruits. Oval, light green, young leaves mature to deep red-purple.
❦ *P. triloba* 'Multiplex'. Deciduous, bushy, spreading tree or shrub. H and S 4m (12ft). Double, pink flowers are borne in mid-spring. Has oval, dark green leaves, often 3-lobed, that turn yellow in autumn. Does best against a sunny wall. Cut back young shoots to within a few buds of old wood after flowering.
❦ *P.* 'Ukon' illus. p.60.
P. 'Uzu-zakura'. See *P.* 'Hokusai'.
P. virginiana (Virginian bird cherry). 'Shubert' is a deciduous, conical tree. H 10m (30ft), S 8m (25ft). Produces dense spikes of small, star-shaped, white flowers from mid- to late spring, followed by globose, dark purple-red fruits. Has oval, pale green, young leaves, becoming deep reddish-purple in summer.
P. 'Yae-murasaki', syn.
P. 'Yae-marasatizatura', illus. p.62.
❦ *P.* x *yedoensis* illus. p.61.

PSEUDERANTHEMUM
(Acanthaceae)
Genus of evergreen perennials and shrubs, grown mainly for their foliage. Frost tender, min. 16°C (61°F). Requires partial shade and fertile, well-drained soil. Water potted plants freely when in full growth, moderately at other times. Tip prune young plants to promote a bushy habit. Cut leggy plants back hard in spring. Propagate annually or biennially as a pot plant by greenwood cuttings in spring or summer. Whitefly may sometimes be troublesome.
P. atropurpureum, syn. *Eranthemum atropurpureum*. Evergreen, erect shrub. H 1–1.2m (3–4ft), S 30–60cm (1–2ft). Has oval, strongly purple-flushed leaves and, mainly in summer, short spikes of tubular, purple-marked, white flowers.

PSEUDOCYDONIA (Rosaceae)
Genus of one species of deciduous or semi-evergreen, spring-flowering tree, grown for its bark, flowers and fruits. Frost hardy, but in cool areas grow against a south- or west-facing wall. Requires full sun and does well only in hot summers. Needs well-drained soil. Propagate by seed in autumn.
P. sinensis, syn. *Cydonia sinensis*. Deciduous or semi-evergreen, spreading tree. H and S 6m (20ft). Has decorative, flaking bark. Shallowly cup-shaped, pink flowers, borne from mid- to late spring, are followed after hot summers by large, egg-shaped, yellow fruits. Oval, finely toothed leaves are dark green.

Pseudofumaria lutea. See *Corydalis lutea*.
Pseudofumaria ochroleuca. See *Corydalis ochroleuca*.

Pseudogynoxys chenopodioides. See *Senecio confusus*.

PSEUDOLARIX (Pinaceae). See CONIFERS.
❦ *P. amabilis*, syn. *P. kaempferi*, illus. p.81.
P. kaempferi. See *P. amabilis*.

PSEUDOPANAX, syn. NEOPANAX, NOTHOPANAX (Araliaceae)
Genus of evergreen trees and shrubs, grown for their unusual foliage and fruits. Is excellent for landscaping and may also be grown in large containers. Insignificant flowers are produced in summer. Frost to half hardy. Grows in sun or semi-shade and in fertile, well-drained soil. Propagate by semi-ripe cuttings in summer or by seed in autumn or spring.
P. arboreus (Five fingers). Evergreen, round-headed, stout-branched tree. H 6m (20ft), S 4m (12ft). Frost hardy. Large, glossy, dark green leaves are divided into 5 or 7 oblong leaflets. Tiny, honey-scented, green flowers in summer are followed by rounded, purplish-black fruits on female plants.
P. crassifolius (Lancewood). Evergreen tree, unbranched for many years, then becoming round-headed. H 6m (20ft), S 2m (6ft). Frost hardy. Dark green leaves are extremely variable in shape on young trees, but eventually become long, narrow, rigid and downward-pointing on older specimens. Female plants bear small, rounded, black fruits.
P. ferox illus. p.66.
P. laetus. Evergreen, round-headed, stout-branched tree or shrub. H and S 3m (10ft). Half hardy. Has large, long-stalked, leathery leaves composed of 5 or 7 oblong, dark green leaflets. Produces tiny, greenish-purple flowers in summer, followed by rounded, purplish-black fruits in autumn on female plants.

PSEUDOSASA (Bambusoideae). See GRASSES, BAMBOOS, RUSHES and SEDGES.
❦ *P. japonica*, syn. *Arundinaria japonica*, illus. p.184.

PSEUDOTSUGA (Pinaceae). See CONIFERS.
P. douglasii. See *P. menziesii*.
❦ *P. menziesii*, syn. *P. douglasii*, *P. taxifolia* (Douglas fir). Fast-growing, conical conifer. H 25m (80ft), S 8–12m (25–40ft). Fully hardy. Has thick, corky, fissured, grey-brown bark. Spirally arranged, aromatic, needle-like, slightly flattened leaves, which develop from sharply pointed buds, are dark green with white bands beneath. Elliptic cones, 8–10cm (3–4in) long, with projecting bracts, are dull brown. 'Fletcheri', H 3m (10ft), S 2–3m (6–10ft), makes a flat-topped shrub. 'Fretsii' (illus. p.84), H 6m (20ft) or more, S 3–4m (10–12ft), is slow-growing, with very short, dull green leaves. subsp. *glauca* illus. p.76. 'Oudemansii' (illus. p.85) is very slow-growing, with ascending branches and short, glossy leaves, dark green all over.
P. taxifolia. See *P. menziesii*.

PSEUDOWINTERA (Winteraceae)
Genus of evergreen shrubs and trees, grown for their foliage. Frost to half hardy. Needs full light or partial shade and humus-rich, well-drained but moisture-retentive soil, ideally neutral to acid. Water containerized plants freely when in full growth, only moderately at other times. Pruning is tolerated if needed. Propagate by semi-ripe cuttings taken in summer or by seed when ripe, in autumn, or in spring.
P. axillaris, syn. *Drimys axillaris* (Heropito, Pepper-tree). Evergreen, rounded shrub or tree. H and S 3–8m (10–25ft). Half hardy. Has oval, lustrous, mid-green leaves, blue-grey beneath. Axillary clusters of tiny, star-shaped, greenish-yellow flowers appear in spring-summer, followed by tiny, globular, bright red fruits.
P. colorata, syn. *Drimys colorata*. Evergreen, bushy, spreading shrub. H 1m (3ft), S 1.5m (5ft). Half hardy. Has oval, pale yellow-green leaves, blotched with pink and narrowly edged with deep red-purple; undersides are bluish-white. Small, star-shaped, greenish-yellow flowers appear in mid-spring. Provide shelter in all but the mildest areas.

PSYCHOPSIS. See ORCHIDS.
P. papilio, syn. *Oncidium papilio* (Butterfly orchid; illus. p.263). Evergreen, epiphytic orchid for a warm greenhouse. H 15cm (6in). In summer, rich yellow-marked, orange-brown flowers, 8cm (3in) long, are borne singly and in succession on tops of stems. Has oval, semi-rigid, mottled leaves, 10–15cm (4–6in) long. Grow in good light in summer.

PSYLLIOSTACHYS
(Plumbaginaceae)
Statice
Genus of annuals, perennials and evergreen sub-shrubs, grown for cut flowers and for drying. Is suitable for coastal areas. Fully to half hardy. Grow in sun and fertile, well-drained soil. If required for drying, cut flowers before they are fully open. Cut down dead stems of perennials in autumn. Propagate by seed sown under glass in early spring; perennials and sub-shrubs may also be increased by softwood cuttings in spring. Botrytis and powdery mildew may be troublesome.
P. suworowii, syn. *Statice suworowii*, illus. p.282.

PTELEA (Rutaceae)
Genus of deciduous trees and shrubs, grown for their foliage and fruits. Fully hardy. Requires sun and fertile soil. Propagate species by softwood cuttings in summer or by seed in autumn, selected forms by softwood cuttings only in summer.
❦ *P. trifoliata* (Hop tree). Deciduous, bushy, spreading tree or shrub. H and S 7m (22ft). Has aromatic, dark green leaves with 3 narrowly oval leaflets. Clusters of small, star-shaped, green flowers from early to mid-summer are succeeded by clusters of winged, pale green fruits. ❦ 'Aurea' illus. p.115.

PTERIS (Polypodiaceae)
Genus of deciduous, semi-evergreen or evergreen ferns. Frost tender, min. 5°C (41°F). Tolerates sun or shade. Grow in moist, peaty soil. Remove faded fronds regularly. Propagate by division in spring or by spores in summer.
❦ *P. cretica* illus. p.187.
❦ 'Albolineata' is an evergreen or semi-evergreen fern. H 45cm (18in), S 30cm (12in). Wiry stems produce triangular to broadly oval, divided, pale green fronds, centrally variegated with creamy-white, that have finger-shaped pinnae. Variegated 'Mayi', H 30cm (12in), has crested frond tips.
P. ensiformis (Snow brake). Deciduous or semi-evergreen fern. H 30cm (12in), S 23cm (9in). Dark green fronds, often greyish-white around the midribs, are coarsely divided into finger-shaped pinnae. 'Arguta', H 45cm (18in), has deeper green fronds with central, silver-white marks.
P. multifida (Spider fern). Deciduous or semi-evergreen fern. H 30cm (12in), S 23cm (9in). Narrow, much-divided, wispy fronds are mid-green.

PTEROCARYA (Juglandaceae)
Wing nut
Genus of deciduous trees, grown for their foliage and catkins. Fully hardy. Needs full sun and any deep, moist but well-drained soil. Suckers should be removed regularly. Propagate by softwood cuttings in summer or by suckers or seed, when ripe, in autumn.
❦ *P. fraxinifolia* (Caucasian wing nut). Deciduous, spreading tree. H 25m (80ft), S 20m (70ft). Large, ash-like leaves are glossy, dark green and turn yellow in autumn. Long, green catkins are produced in summer, the female developing winged, green, then brown fruits.
P. x *rehderiana* illus. p.43.
P. stenoptera (Chinese wing nut). Deciduous, spreading tree. H 20m (70ft), S 15m (50ft). Ash-like, bright green leaves, each with a winged stalk, turn yellow in autumn. Produces long, green catkins in summer, the female developing winged, pink-tinged, green fruits.

PTEROCELTIS (Ulmaceae)
Genus of one species of deciduous tree, with inconspicuous flowers in summer, grown for its foliage and fruits. Fully hardy. Needs full sun and does best in hot summers. Requires well-drained soil. Propagate by seed in autumn.
P. tatarinowii. Deciduous, spreading tree. H 12m (40ft), S 10m (30ft). Has peeling, grey bark, oval, dark green leaves and, in autumn, small, spherical, green fruits, each with a broad, circular wing.

PTEROCEPHALUS

PTEROCEPHALUS (Dipsaceae)
Genus of compact, summer-flowering annuals, perennials and semi-evergreen sub-shrubs, grown for their scabious-like flower heads and feathery seed heads. Is useful for rock gardens. Fully hardy. Requires sun and well-drained soil. Propagate by softwood or semi-ripe cuttings in summer or by seed in autumn. Self seeds moderately.
P. perennis subsp. *perennis*, syn. *P.p.* var. *parnassi*, illus. p.328.

PTEROSTYRAX (Styracaceae)
Genus of deciduous trees and shrubs, grown for their foliage and fragrant flowers. Fully hardy. Requires sun or semi-shade and deep, well-drained, neutral to acid soil. Propagate by softwood or semi-ripe cuttings in summer or by seed in autumn.
♚ *P. hispida* (Epaulette tree). Deciduous, spreading tree or shrub. H 15m (50ft), S 12m (40ft). Bears large, drooping panicles of small, bell-shaped, white flowers from early to mid-summer. Oblong to oval leaves are mid-green.

Ptilotrichum spinosum. See *Alyssum spinosum*.

PUERARIA (Leguminosae)
Genus of deciduous, woody-stemmed or herbaceous, twining climbers. Half hardy. Grow in sun and in any well-drained soil. Propagate by seed in spring.
P. hirsuta. See *P. lobata*.
P. lobata, syn. *P. hirsuta*, *P. thunbergiana* (Kudzu vine). Deciduous, woody-stemmed, twining climber with hairy stems. H to 5m (15ft) or to 30m (100ft) in the wild. Leaves have 3 broadly oval leaflets. In summer produces racemes, to 30cm (12in) long, of small, scented, sweet pea-like, reddish-purple flowers, followed by long, slender, hairy pods, 6–8cm (2½–3in) long. In cold areas is best grown as an annual.
P. thunbergiana. See *P. lobata*.

PULMONARIA (Boraginaceae)
Lungwort
Genus of mainly spring-flowering perennials, some of which are semi-evergreen with small, overwintering rosettes of leaves. Fully hardy. Prefers shade; grows in any moist but well-drained soil. Propagate by division in spring or autumn.
♚ *P. angustifolia.* Clump-forming perennial. H 23cm (9in), S 20–30cm (8–12in) or more. Has narrowly lance-shaped, mid-green leaves. In early spring produces heads of tubular, 5-lobed, borage-like, sometimes pink-tinged, deep blue flowers. 'Mawson's Variety' illus. p.235.
P. longifolia. Clump-forming perennial. H 30cm (12in), S 45cm (18in). Very narrowly lance-shaped, dark green leaves are spotted with white. Heads of tubular, 5-lobed, borage-like, vivid blue flowers appear in late spring.
P. montana. See *P. rubra*.
♚ *P. rubra*, syn. *P. montata*. Semi-evergreen, clump-forming perennial. H 30cm (12in), S 60cm (24in). Has broadly oval, velvety, mid-green leaves. Heads of tubular, 5-lobed, borage-like, brick-red flowers open from late winter to early spring.
P. saccharata illus. p.235.
♚ *P.* 'Sissinghurst White' illus. p.231.

PULSATILLA (Ranunculaceae)
Genus of perennials, some of which are evergreen, grown for their large, feathery leaves, upright or pendent, bell- or cup-shaped flowers, covered in fine hairs, and feathery seed heads. Has fibrous, woody rootstocks. Leaves increase in size after flowering time. Is suitable for large rock gardens. Fully hardy. Needs full sun and humus-rich, well-drained soil. Resents disturbance to roots. Propagate by root cuttings in winter or by seed when fresh.
P. alpina illus. p.294.
♚ subsp. *apiifolia* (syn. *P.a.* subsp. *sulphurea*) is a clump-forming perennial. H 15–30cm (6–12in), S to 10cm (4in). Has feathery, soft green leaves. Bears upright, bell-shaped, soft pale yellow flowers in spring, followed by feathery, silvery seed heads.
♚ *P. halleri* illus. p.296.
subsp. *grandis* (syn. *P. vulgaris* subsp. *grandis*) is a clump-forming perennial. H and S 15–23cm (6–9in). In spring, mainly before feathery, light green leaves appear, produces large, upright, shallowly bell-shaped, lavender-blue flowers, 5cm (2in) wide, with bright yellow centres. Flower stems rapidly elongate as feathery, silvery seed heads mature.
P. occidentalis. Clump-forming perennial. H 20cm (8in), S 15cm (6in). In late spring to early summer, solitary nodding buds develop into erect, goblet-shaped, white flowers, stained blue-violet at base outside and sometimes flushed pink, followed by feathery, silvery seed heads. Bears feathery leaves. Is extremely difficult to grow and flower well at low altitudes.
♚ *P. vernalis* illus. p.311.
♚ *P. vulgaris* illus. p.296.
subsp. *grandis.* See *P. halleri* subsp. *grandis*.

PUNICA (Punicaceae)
Pomegranate
Genus of deciduous, summer-flowering shrubs and trees, grown for their bright red flowers and yellow to orange-red fruits, which ripen and become edible only in warm climates. Frost to half hardy. Needs a sheltered, sunny position and well-drained soil. Propagate by seed in spring or by semi-ripe cuttings in summer.
P. granatum. Deciduous, rounded shrub or tree. H and S 2–8m (6–25ft). Half hardy. Has narrowly oblong leaves and, in summer, funnel-shaped, bright red flowers, with crumpled petals. Fruits are spherical and deep yellow to orange. May be grown in a southern or eastern aspect, either free-standing or, in frost-prone climates, against a wall. var. *nana* (syn. *P.g.* 'Nana') illus. p.303.

PUSCHKINIA (Liliaceae)
Genus of dwarf, *Scilla*-like bulbs, grown for their early spring flowers. Fully hardy. Needs sun or partial shade and humus-rich soil that has grit or sand added to ensure good drainage. Plant in autumn. Dies down in summer. Propagate by offsets in late summer or by seed in autumn.
P. libanotica. See *P. scilloides*.
P. scilloides, syn. *P. libanotica*, illus. p.376. 'Alba' illus. p.370.

PUYA (Bromeliaceae)
Genus of evergreen, rosette-forming perennials and shrubs, grown for their overall appearance. Half hardy to frost tender, min. 5–7°C (41–5°F). Requires full light and well-drained soil. Water moderately during the growing season, sparingly at other times. Propagate by seed or offsets in spring.
P. alpestris (illus. p.229). Evergreen perennial with stout, branched, prostrate stems. H to 2m (6ft), S 3m (10ft). Half hardy. Each stem is topped by a dense rosette of linear, tapering, arching, bright green leaves that are fleshy and have hooked, spiny teeth along the edges and dense, white scales beneath. Tubular, deep metallic-blue flowers, ageing to purple-red, are borne in stiff, erect panicles and are produced in early summer.
P. chilensis (illus. p.229). Evergreen, upright perennial with a short, woody stem. H and S to 2m (6ft). Half hardy. Stem is crowned by a dense rosette of linear, tapering, arching, grey-green leaves that are fleshy and have margins of hooked, spiny teeth. Bears tubular, metallic-yellow or greenish-yellow flowers in erect, much-branched panicles during summer.

PYCNOSTACHYS (Labiatae)
Genus of bushy perennials, grown for their whorled clusters of flowers. Frost tender, min. 15°C (59°F). Grow in bright light and in fertile, well-drained soil. Propagate by stem cuttings in early summer.
P. dawei illus. p.197.
P. urticifolia. Strong-growing, erect perennial with square stems. H 1–2m (3–6ft), S 20–60cm (8–24in). Has oval, sharply toothed, hairy, mid-green leaves, 10cm (4in) long. Whorls of small, tubular, bright blue flowers are carried in racemes, 10cm (4in) long, in winter.

PYRACANTHA (Rosaceae)
Firethorn
Genus of evergreen, spiny, summer-flowering shrubs, grown for their foliage, flowers and fruits. Fully to frost hardy. Requires a sheltered site in sun or semi-shade and fertile soil. To produce a compact habit on a plant grown against a wall, train and cut back long shoots after flowering. Propagate by semi-ripe cuttings in summer. Is susceptible to scab and fireblight.
P. angustifolia. Evergreen, bushy, dense shrub. H and S 3m (10ft). Frost hardy. Has narrowly oblong leaves, dark green above, grey and felt-like beneath. Produces clusters of small, 5-petalled, white flowers in early summer, followed by spherical, orange-yellow fruits in autumn.
P. atalantioides. Vigorous, evergreen shrub, part upright, part arching. H 5m (15ft), S 4m (12ft). Frost hardy. Oblong leaves are glossy and dark green. Large clusters of small, 5-petalled, white flowers in early summer are followed by spherical, red fruits in early autumn. 'Aurea' illus. p.94.
P. coccinea. Evergreen, dense, bushy shrub. H and S 4m (12ft). Fully hardy. Dense clusters of small, 5-petalled, white flowers open amid oval, dark green leaves in early summer and are succeeded by spherical, bright red fruits. 'Lalandei' has larger leaves and larger, orange-red fruits.
P. 'Golden Charmer' illus. p.119.
P. 'Golden Dome' illus. p.119.
P. 'Mohave'. Vigorous, evergreen, bushy shrub. H 4m (12ft), S 5m (15ft). Frost hardy. Produces clusters of small, 5-petalled, white flowers in early summer, then spherical, orange-red fruits. Leaves are broadly oval and dark green. Is disease-resistant.
♚ *P.* 'Orange Glow'. Evergreen, upright, dense shrub. H 5m (15ft), S 3m (10ft). Frost hardy. Has oblong, glossy, dark green leaves, clusters of small, 5-petalled, white flowers, in early summer, and spherical, orange fruits.
♚ *P. rogersiana.* Evergreen, upright, then arching shrub. H and S 3m (10ft). Frost hardy. Leaves are narrowly oblong, glossy and bright green. Clusters of small, 5-petalled, white flowers in early summer are followed by round, orange-red or yellow fruits.
♚ *P.* x *watereri*, syn. *P.* 'Watereri', *P.* 'Waterer's Orange', illus. p.106.
P. 'Watereri'. See *P.* x *watereri*.
P. 'Waterer's Orange'. See *P.* x *watereri*.

Pyrethropsis hosmariense. See *Chrysanthemum hosmariense*.

Pyrethrum 'Brenda'. See *Tanacetum coccineum* 'Brenda'.
Pyrethrum roseum. See *Tanacetum coccineum*.

PYROLA (Pyrolaceae)
Wintergreen
Genus of evergreen, spreading, spring- and summer-flowering perennials. Fully hardy. Needs partial shade, cool conditions and well-drained, peaty, acid soil; is best suited to light woodlands. Resents disturbance. Propagate by seed in autumn or spring or by division in spring.
P. asarifolia. Evergreen, rosette-forming perennial. H 15–25cm (6–10in), S 15cm (6in) or more. Has kidney-shaped, leathery, glossy, light green leaves. Each flower stem bears open tubular, pale to deep pink flowers in spring.
P. rotundifolia (Round-leaved wintergreen, Wild lily-of-the-valley). Evergreen, rosette-forming perennial. H 23cm (9in), S 30cm (12in). Produces rounded, leathery, glossy, mid-green leaves and, in late spring and early summer, sprays of fragrant, white flowers that resemble lily-of-the-valley.

PYROSTEGIA (Bignoniaceae)
Genus of evergreen, woody-stemmed, tendril climbers, grown for their flowers. Frost tender, min. 13–15°C (55–9°F). Needs full light and fertile, well-drained soil. Water regularly, less in winter. Provide support. Thin stems after flowering. Propagate by semi-ripe cuttings or layering in summer.
P. ignea. See *P. venusta.*
P. venusta, syn. *P. ignea*, illus. p.178.

PYRUS (Rosaceae)
Pear
Genus of deciduous, spring-flowering trees, grown for their habit, foliage, flowers and edible fruits (pears). Fully hardy. Does best in full sun and needs well-drained soil. Propagate species by seed in autumn, cultivars by budding in summer or by grafting in winter. Many species are susceptible to fireblight and scab and, in North America, pear decline.
P. amygdaliformis. Deciduous, spreading tree. H 10m (30ft), S 8m (25ft). Lance-shaped leaves are grey when young, maturing to glossy, dark green. Clusters of 5-petalled, white flowers are produced in mid-spring, and are followed by small, brownish fruits.
P. calleryana (Callery pear). Deciduous, broadly conical tree. H and S to 15m (50ft). Oval, glossy, dark green leaves often turn red in autumn. Produces 5-petalled, white flowers from mid- to late spring. Fruits are small and brownish. **'Bradford'**, S 10m (30ft), is resistant to fireblight. ♛ **'Chanticleer'** (syn. *P.c.* 'Cleveland Select') illus. p.50.
♛ *P. communis* (Common pear). **'Beech Hill'** is a deciduous, narrowly conical tree. H 10m (30ft), S 7m (22ft). Oval, glossy, dark green leaves often turn orange and red in autumn. From mid- to late spring bears 5-petalled, white flowers as the leaves emerge, followed by small, brownish fruits.
P. elaeagrifolia. Deciduous, spreading, thorny tree. H and S 8m (25ft). Has lance-shaped, grey-green leaves and loose clusters of 5-petalled, white flowers in mid-spring. Has small, brownish fruits.
P. salicifolia. Deciduous, mound-shaped tree with slightly drooping branches. H 5–8m (15–25ft), S 4m (12ft). White flowers, with 5 petals, open as lance-shaped, grey leaves emerge in mid-spring. Fruits are small and brownish. ♛ **'Pendula'** illus. p.65.

Q R

Quamoclit coccinea. See *Ipomoea coccinea.*
Quamoclit lobata. See *Ipomoea lobata.*
Quamoclit pinnata. See *Ipomoea quamoclit.*

QUERCUS (Fagaceae)
Oak
Genus of deciduous or evergreen trees and shrubs, grown for their habit, foliage and, in some deciduous species, autumn colour. Produces insignificant flowers from late spring to early summer, followed by egg-shaped to rounded, brownish fruits (acorns). Fully to frost hardy. Does best in sun or semi-shade and in deep, well-drained soil. Except where stated otherwise, will tolerate lime. Propagate species by seed in autumn, selected forms and hybrids by grafting in late winter. May be affected, though not usually seriously, by mildew and various galls, and, in North America, by oak wilt.
Q. acutissima (Sawtooth oak). Deciduous, round-headed tree. H and S 15m (50ft). Fully hardy. Has sweet-chestnut-like, glossy, dark green leaves, edged with bristle-tipped teeth, that last until late in the year.
Q. agrifolia illus. p.59.
Q. alba illus. p.45.
Q. aliena (Oriental white oak). Deciduous, spreading tree. H 15m (50ft), S 12m (40ft). Fully hardy. Has large, oblong, prominently toothed, glossy, dark green leaves.
Q. alnifolia (Golden oak of Cyprus). Evergreen, spreading tree. H 6m (20ft), S 5m (15ft). Frost hardy. Rounded, leathery leaves are glossy, dark green above, with mustard-yellow or greenish-yellow felt beneath.
♈ *Q. canariensis* illus. p.40.
Q. castaneifolia illus. p.42.
♈ *Q. cerris* (Turkey oak). Fast-growing, deciduous, spreading tree of stately habit. H 30m (100ft), S 25m (80ft). Fully hardy. Oblong, glossy, dark green leaves are deeply lobed. Thrives on shallow, chalky soil. **'Argenteovariegata'** (syn. *Q.c.* 'Variegata') illus. p.52.
Q. coccifera (Kermes oak). Evergreen, bushy, compact tree or shrub. H and S 5m (15ft). Frost hardy. Holly-like leaves are glossy, dark green and rigid with spiny margins.
Q. coccinea illus. p.44. ♈ **'Splendens'** is a deciduous, round-headed tree. H 20m (70ft), S 15m (50ft). Fully hardy. Oblong, glossy, mid-green leaves, with deep, tooth-like lobes, turn deep scarlet in autumn. Prefers acid soil.
Q. dentata (Daimio oak). Deciduous, spreading, stout-branched tree of rugged habit. H 15m (50ft), S 10m (30ft). Fully hardy. Has oval, lobed, dark green leaves, 30cm (12in) or more long. Prefers acid soil.
Q. ellipsoidalis illus. p.44.
Q. frainetto illus. p.43.
Q. garryana illus. p.54.
Q. x *heterophylla* illus. p.56.

♈ *Q.* x *hispanica* **'Lucombeana'** illus. p.47.
♈ *Q. ilex* (Holm oak). Evergreen, round-headed tree. H 25m (80ft), S 20m (70ft). Frost hardy. Glossy, dark green leaves are silvery-grey when young and very variably shaped, but are most often oval. Thrives on shallow chalk and is excellent for an exposed, coastal position.
Q. imbricaria (Shingle oak). Deciduous, spreading tree. H 20m (70ft), S 15m (50ft). Fully hardy. Bears long, narrow leaves that are yellowish when young, dark green in summer and yellowish-brown in autumn.
Q. ithaburensis subsp. *macrolepis.* See *Q. macrolepis.*
Q. laurifolia illus. p.43.
Q. macranthera illus. p.40.
Q. macrocarpa illus. p.54.
Q. macrolepis, syn. *Q. ithaburensis* subsp. *macrolepis,* illus. p.54.
Q. marilandica illus. p.54.
Q. mongolica var. *grosseserrata.* Deciduous, spreading tree. H 20m (70ft), S 15m (50ft). Fully hardy. Has large, oblong, lobed, dark green leaves with prominent, triangular teeth.
Q. muehlenbergii illus. p.41.
Q. myrsinifolia illus. p.58.
Q. nigra illus. p.42.
♈ *Q. palustris* illus. p.43.
♈ *Q. petraea* (Durmast oak, Sessile oak). Deciduous, spreading tree. H 30m (100ft), S 25m (80ft). Fully hardy. Has oblong, lobed, leathery, dark green leaves with yellow stalks. **'Columna'** illus. p.42.
♈ *Q. phellos* illus. p.45.
Q. pontica (Armenian oak, Pontine oak). Deciduous, sometimes shrubby tree with upright, stout branches and broadly oval head. H 6m (20ft), S 5m (15ft). Fully hardy. Large, oval, toothed, glossy, bright green leaves turn yellow in autumn.
♈ *Q. robur* (Common oak, Pedunculate oak). Deciduous, spreading, rugged tree. H and S 25m (80ft). Fully hardy. Bears oblong, wavy, lobed, dark green leaves. **'Concordia',** H 10m (30ft), is slow-growing and has golden-yellow, young foliage that becomes yellowish-green in mid-summer. f. *fastigiata* illus. p.40.
♈ *Q. rubra* illus. p.43. **'Aurea'** illus. p.54.
Q. suber illus. p.46.
Q. x *turneri* illus. p.46.
Q. velutina (Black oak). Fast-growing, deciduous, spreading tree. H 30m (100ft), S 25m (80ft). Fully hardy. Large, oblong, lobed, glossy, dark green leaves turn reddish-brown in autumn.

QUISQUALIS (Combretaceae)
Genus of evergreen or deciduous, scandent shrubs and twining climbers, grown for their flowers. Frost tender, min. 10–18°C (50–64°F). Provide humus-rich, moist but well-drained soil and full light or semi-shade. Water

freely when in full growth, less in cold weather. Stems need support. Thin out crowded growth in spring. Propagate by seed in spring or by semi-ripe cuttings in summer.
Q. indica illus. p.171.

RAMONDA (Gesneriaceae)
Genus of evergreen perennials, grown for their rosettes of rounded, crinkled, hairy leaves and for their flowers. Is useful for rock gardens and peat walls. Fully hardy. Prefers shade and moist soil. Water plants well if they curl up during a dry spell. Propagate by rooting offsets in early summer or by leaf cuttings or seed in early autumn.
♈ *R. myconi,* syn. *R. pyrenaica,* illus. p.330.
♈ *R. nathaliae.* Evergreen, basal-rosetted perennial. H and S 10cm (4in). Bears small, pale green leaves and, in late spring and early summer, umbels of small, outward-facing, flattish, white or lavender flowers, with yellow anthers.
R. pyrenaica. See *R. myconi.*
R. serbica. Evergreen, basal-rosetted perennial. H and S 10cm (4in). Is similar to *R. nathaliae,* but has cup-shaped, lilac-blue flowers and dark violet-blue anthers. May be difficult to grow.

RANUNCULUS (Ranunculaceae)
Buttercup
Genus of annuals, aquatics and perennials, some of which are evergreen or semi-evergreen, grown mainly for their flowers. Many species grow from a thickened rootstock or a cluster of tubers. Some are invasive. Aquatic species are seldom cultivated. Fully to half hardy. Grows in sun or shade and in moist but well-drained soil. Propagate by seed when fresh or by division in spring or autumn.
R. aconitifolius and ♈ **'Flore Pleno'** illus. p.197.
R. acris (Meadow buttercup). **'Flore Pleno'** illus. p.256.
R. alpestris illus. p.311.
R. amplexicaulis. Upright perennial. H 25cm (10in), S 10cm (4in). Fully hardy. Has narrowly oval, blue-grey leaves. In early summer produces clusters of shallowly cup-shaped, white flowers with yellow anthers. Needs humus-rich soil.
R. asiaticus [red form] illus. p.365, [yellow form] illus. p.367.
R. bullatus. Clump-forming perennial with thick, fibrous roots. H 5–8cm (2–3in), S 8–10cm (3–4in). Half hardy. Produces fragrant, shallowly cup-shaped, bright yellow flowers in autumn above neat mounds of foliage. Oblong to oval, green leaves have sharply toothed tips and are puckered. Suits an alpine house or rock garden.
♈ *R. calandrinioides* illus. p.308.
R. constantinopolitanus **'Plenus',** syn. *R. gouanii* 'Plenus', *R. speciosus* 'Plenus', illus. p.256.

R. crenatus. Semi-evergreen, rosetted perennial with thick, fibrous roots. H and S 10cm (4in). Fully hardy. Produces rounded, toothed, green leaves and, in summer, short stems bearing 1 or 2 shallowly cup-shaped, white flowers just above foliage. May also be propagated by removing a flower stem at its first joint in summer; rosettes will form and may then be rooted. Rarely sets seed in cultivation. Suits an alpine house or an acid, scree rock garden.
R. ficaria (Lesser celandine). **'Albus'** illus. p.311. **'Aurantiacus'** illus. p.321. **'Brazen Hussy'** is a mat-forming, tuberous perennial. H 5cm (2in), S to 20cm (8in). Fully hardy. Is grown for its heart-shaped, purple-bronze leaves produced in spring. Shallowly cup-shaped, glossy, sulphur-yellow flowers, with bronze reverses, appear in early spring. All *R. ficaria* forms die down in late spring. May spread rapidly; is good for a wild garden. **'Flore Pleno'** illus. p.320.
R. glacialis. Hummock-forming perennial with fibrous roots. H 5–25cm (2–10in), S 5cm (2in) or more. Fully hardy. Bears rounded, deeply lobed, glossy, dark green leaves and, in late spring and early summer, clusters of shallowly cup-shaped, white or pink flowers. Is very difficult to grow at low altitudes. Suits a scree or alpine house. Prefers humus-rich, moist, acid soil that is drier in winter. Slugs may be troublesome.
R. gouanii **'Plenus'.** See *R. constantinopolitanus* 'Plenus'.
♈ *R. gramineus* illus. p.307.
R. lingua illus. p.391. **'Grandiflorus'** is a deciduous, perennial, marginal water plant. H 1m (3ft), S 30cm (1ft). Fully hardy. Has stout, pinkish-green stems, lance-shaped, glaucous leaves and, in late spring, racemes of large, saucer-shaped, yellow flowers.
R. lyallii (Giant buttercup). Evergreen, stout, upright, tufted perennial. H and S 30cm (12in) or more. Frost hardy. Has rounded, leathery, dark green leaves, each 15cm (6in) or more across, and, in summer, bears panicles of large, shallowly cup-shaped, white flowers. Is very difficult to flower in hot, dry climates. Is suitable for an alpine house. Rarely sets seed in cultivation.
♈ *R. montanus* **'Molten Gold'.** Clump-forming, compact perennial. H 15cm (6in), S 10cm (4in). Fully hardy. Leaves are rounded and 3-lobed. Flower stems each produce a shallowly cup-shaped, shiny, bright golden-yellow flower in early summer. Is useful for a sunny rock garden.
R. speciosus **'Plenus'.** See *R. constantinopolitanus* 'Plenus'.

RANZANIA (Berberidaceae)
Genus of one species of perennial, grown for its unusual appearance as well as its flowers. Is ideal for woodland gardens. Fully hardy. Prefers

shade or semi-shade and humus-rich, moist soil. Propagate by division in spring or by seed in autumn.
R. japonica. Upright perennial. H 45cm (18in), S 30cm (12in). Produces 3-parted, fresh green leaves and, in early summer, small clusters of nodding, shallowly cup-shaped, pale mauve flowers.

RAOULIA (Compositae)
Genus of evergreen, mat-forming perennials, grown for their foliage. Some species are suitable for alpine houses, others for rock gardens. Fully to frost hardy. Needs sun or semi-shade and gritty, moist but well-drained, peaty soil. Propagate by seed when fresh or by division in spring.
R. australis illus. p.338.
R. eximia. Evergreen, cushion-forming perennial. H 2.5cm (1in), S 5cm (2in). Fully hardy. Has oblong to oval, overlapping, woolly, grey leaves and, in late spring or summer, small, rounded heads of yellowish-white flowers. Suits an alpine house. Prefers some shade.
R. haastii illus. p.339.
R. hookeri var. *albo-sericea* illus. p.337.

RAVENALA (Strelitziaceae)
Genus of one species of evergreen, palm-like tree, grown for its foliage and overall appearance. Is related to *Strelitzia*. Frost tender, min. 16°C (61°F). Requires full light and humus-rich, well-drained soil. Water potted specimens freely in summer, less in winter or when temperatures are low. Propagate by seed in spring. Red spider mite may be troublesome.
R. madagascariensis (Traveller's tree). Evergreen, upright, fan-shaped tree. H and S to 10m (30ft). Has banana-like, long-stalked leaves, each 3–6m (10–20ft) long, with expanded stalk bases. Groups of boat-shaped spathes with 6-parted, white flowers emerge from leaf axils in summer.

REBUTIA (Cactaceae)
Genus of mostly clump-forming, spherical to columnar, perennial cacti with flowers produced in profusion from plant bases, usually 2–3 years after raising from seed. Much-ribbed, tuberculate, green stems have short spines. A few species are sometimes included in *Aylostera*. Frost tender, min. 5°C (41°F). Requires sun or partial shade and well-drained soil. Is easy to grow. Propagate by seed in spring or summer.
❦ *R. arenacea*, syn. *Sulcorebutia arenacea*, illus. p.416.
❦ *R. aureiflora* illus. p.416.
R. krainziana illus. p.410.
R. minuscula. Clump-forming, perennial cactus. H 5cm (2in), S 15cm (6in). Pale brown spines, 1cm (½in) long, uniformly cover 5cm (2in) wide, light green stem. Carries masses of trumpet-shaped, red flowers, to 4cm (1½in) wide, in early spring.
❦ *R. muscula* illus. p.416.
R. neocumingii, syn. *Weingartia neocumingii.* Spherical, perennial cactus. H and S 10cm (4in). Has a tuberculate, green stem. Areoles each bear dense clusters of yellow spines, to 1.5cm (⅝in) long, some thicker than others, and several cup-shaped, dark yellow flowers, 3cm (¼in) long, in spring.
R. pygmaea, syn. *Lobivia pygmaea.* Clump-forming, columnar, perennial cactus. H 5cm (2in), S 10cm (4in). Very short, comb-like spines are pressed against grey- to purple-green stem. Trumpet-shaped, pink to salmon or rose-purple flowers, to 2cm (¾in) across, appear in spring. Prefers a sunny position.
R. rauschii, syn. *Sulcorebutia rauschii.* Flattened spherical, perennial cactus. H 5cm (2in), S 10cm (4in). Grey-green stem bears very short, comb-like, golden or black spines. Carries 3cm (1¼) wide, flattish, deep purple flowers in spring. Grows better when grafted.
R. senilis. Clump-forming, perennial cactus. H 5cm (2in), S 15cm (6in). Has soft, white spines, 3cm (1¼in) long, matted around a 5cm (2in) wide stem. In spring bears trumpet-shaped, red, yellow, pink or orange flowers, 5cm (2in) across. Prefers a sunny site.
R. spegazziniana illus. p.411.
R. tiraquensis, syn. *Sulcorebutia tiraquensis*, illus. p.410.
❦ *R. violaciflora* illus. p.408.

REHDERODENDRON (Styracaceae)
Genus of deciduous, spring-flowering trees, grown for their flowers and fruits. Frost hardy. Needs sun or semi-shade, some shelter and fertile, moist but well-drained, acid soil. Propagate by semi-ripe cuttings in summer or by seed in autumn.
R. macrocarpum. Deciduous, spreading tree. H 10m (30ft), S 7m (22ft). Young shoots are red. Pendent clusters of lemon-scented, cup-shaped, pink-tinged, white flowers are borne amid oblong, taper-pointed, red-stalked, glossy, dark green leaves in late spring. Bears cylindrical, woody, red, then brown fruits in autumn.

REHMANNIA (Gesneriaceae)
Genus of spring- and summer-flowering perennials. Half hardy to frost tender, min. 1–5°C (34–41°F). Needs a warm, sunny position and light soil. Propagate by seed in autumn or spring or by root cuttings in winter.
R. angulata. Upright perennial. H 75cm (30in), S 45cm (18in). Frost tender, min. 2–5°C (36–41°F). In spring-summer produces racemes of tubular to funnel-shaped, orange-marked, purplish-red or purplish-brown flowers. Oblong, hairy, mid-green leaves have paired leaflets that are irregularly toothed.
R. elata illus. p.208.
❦ *R. glutinosa.* Rosette-forming perennial. H 30cm (12in), S 25cm (10in). Frost tender, min. 1°C (34°F). Tubular, pink, red-brown or yellow flowers, with purple veins, are borne on leafy shoots in late spring and early summer. Leaves are oval to lance-shaped, toothed, hairy and light green.

REINWARDTIA (Linaceae)
Genus of evergreen sub-shrubs, grown for their flowers. Frost tender, min. 7–10°C (45–50°F). Requires full light or partial shade and fertile, well-drained soil. Water freely when in full growth, moderately at other times. Tip prune young plants to promote branching; cut back hard after flowering. Raise annually by softwood cuttings in late spring. Red spider mite may cause problems.
R. indica, syn. *R. trigyna*, illus. p.140.
R. trigyna. See *R. indica*.

RESEDA (Resedaceae)
Mignonette
Genus of annuals and biennials with flowers that attract bees and that are also suitable for cutting. Fully hardy. Grow in sun and in any fertile, well-drained soil. Dead-head to ensure a long flowering period. Propagate by seed sown outdoors in spring or early autumn.
R. odorata illus. p.271.

Retama monosperma. See *Genista monosperma*.

RHAMNUS (Rhamnaceae)
Buckthorn
Genus of deciduous or evergreen shrubs and trees, with inconspicuous flowers, grown for their foliage and fruits. Fully to frost hardy. Grows in sun or semi-shade and in fertile soil. Propagate deciduous species by seed in autumn, evergreens by semi-ripe cuttings in summer.
R. alaternus (Italian buckthorn).
❦ 'Argenteovariegatus' is an evergreen, bushy shrub. H and S 3m (10ft). Frost hardy. Has oval, leathery, glossy, grey-green leaves edged with creamy-white. Tiny, yellowish-green flowers from early to mid-summer are followed by spherical, red, then black fruits.
R. imeretinus. Deciduous, spreading, open shrub. H 3m (10ft), S 5m (15ft). Fully hardy. Stout shoots bear large, broadly oblong, prominently veined, dark green leaves that turn bronze-purple in autumn. Small, green flowers are borne in summer.

RHAPHIOLEPIS (Rosaceae)
Genus of evergreen shrubs, grown for their flowers and foliage. Frost to half hardy. In most areas does best against a sheltered wall; *R. umbellata* is the most hardy. Needs sun and fertile, well-drained soil. Propagate by semi-ripe cuttings in late summer.
R. x *delacourii* 'Coates' Crimson'. Evergreen, rounded shrub. H 2m (6ft), S 2.5m (8ft). Frost hardy. Clusters of fragrant, star-shaped, deep pink flowers in spring or summer are set off by oval, leathery, dark green leaves.
R. indica (Indian hawthorn). Evergreen, bushy shrub. H 1.5m (5ft), S 2m (6ft). Half hardy. Clusters of fragrant, star-shaped, white flowers, flushed with pink, are borne during spring or early summer amid narrowly lance-shaped, glossy, dark green leaves.
❦ *R. umbellata* illus. p.130.

RHAPIS (Palmae)
Genus of evergreen fan palms, grown for their foliage and overall appearance. May have tiny, yellow flowers in summer. Frost tender, min. 15°C (59°F). Needs partial shade and humus-rich, well-drained soil. Water potted plants freely when in full growth, moderately at other times. Propagate by seed, suckers or division in spring. Is susceptible to red spider mite.
❦ *R. excelsa* illus. p.122.

Rhazya orientalis. See *Amsonia orientalis*.

RHEUM (Polygonaceae)
Rhubarb
Genus of perennials, grown for their foliage and striking overall appearance. Includes the edible rhubarb and various ornamental plants. Some species are extremely large and require plenty of space in which to grow. Fully hardy. Prefers sun or semi-shade and deep, rich, well-drained soil. Propagate by division in spring or by seed in autumn.
R. alexandrae illus. p.218.
R. nobile. Clump-forming perennial. H 1.5m (5ft), S 1m (3ft). Oblong to oval, leathery, basal, mid-green leaves are 60cm (2ft) long. In late summer produces long stems bearing conical spikes of large, overlapping, pale cream bracts that hide insignificant flowers.
❦ *R. palmatum.* Clump-forming perennial. H and S 2m (6ft). Has 60–75cm (2–2½ft) long, rounded, 5-lobed, mid-green leaves. In early summer produces broad panicles of small, creamy-white flowers. 'Atrosanguineum' illus. p.191.

RHIPSALIDOPSIS (Cactaceae)
Genus of perennial cacti with trumpet- or bell-shaped flowers in spring. Is easy to grow, but may shed stem segments for no apparent reason. Frost tender, min. 6–10°C (43–50°F). Requires partial shade and rich, well-drained soil. Water well in summer, but allow to become almost dry between waterings. Water only occasionally in winter. Propagate by seed or stem cuttings in spring or summer.
❦ *R. gaertneri* illus. p.411.
❦ *R. rosea* illus. p.408.

RHIPSALIS (Cactaceae)
Mistletoe cactus
Genus of epiphytic, perennial cacti with usually pendent, variously formed stems. Flowers are followed by spherical, translucent berries. Frost tender, min. 10–11°C (50–52°F). Needs partial shade and rich, well-drained soil. Prefers 80% relative humidity – higher than for most cacti. Give only occasional, very light watering in winter. Propagate by seed or stem cuttings in spring or summer.
R. capilliformis. Pendent, perennial cactus. H 1m (3ft), S 50cm (20in). Min. 10°C (50°F). Produces freely branching, cylindrical, green stems and, in winter-spring, short, funnel-shaped, white flowers, to 1cm (½in) wide, with recurved tips, followed by white berries.
R. cereuscula illus. p.397.
R. clavata. See *Hatiora clavata*.
R. crispata. Bushy, then pendent,

perennial cactus. H 1m (3ft), S indefinite. Min. 11°C (52°F). Has leaf-like, elliptic to oblong, pale green stem segments, to 12cm (5in) long, with undulating edges that produce short, funnel-shaped, cream or pale yellow flowers, to 1cm (½in) across, with recurved tips, in winter-spring, then white berries.
R. paradoxa (Chain cactus). Bushy, then pendent, perennial cactus. H 1m (3ft), S indefinite. Min. 11°C (52°F). Triangular, green stems have segments alternately set at different angles. Short, funnel-shaped, white flowers, 2cm (¾in) across, with recurved tips, appear from stem edges in winter-spring and are followed by red berries.
R. tucumanensis illus. p.397.
R. warmingiana illus. p.397.

Rhodanthe manglesii. See *Helipterum manglesii.*

Rhodiola heterodonta. See *Sedum rosea* var. *heterodontum.*
Rhodiola rosea. See *Sedum rosea.*

Rhodocactus grandifolius. See *Pereskia grandifolia.*

RHODOCHITON (Scrophulariaceae)
Genus of one species of evergreen, leaf-stalk climber, grown for its unusual flowers. Does best when grown as an annual. May be planted against fences and trellises or used as ground cover. Frost tender, min. 5°C (41°F). Grow in sun and in any well-drained soil. Propagate by seed in early spring.
♛ *R. atrosanguineum,* syn. *R. volubile,* illus. p.171.
R. volubile. See *R. atrosanguineum.*

RHODODENDRON (Ericaceae)
Genus of evergreen, semi-evergreen or deciduous shrubs, ranging from a dwarf habit to a tree-like stature, grown mainly for beauty of flower. Fully hardy to frost tender, min. 4–7°C (39–45°F). Most prefer dappled shade, but a considerable number tolerates full sun, especially in cool climates. Needs neutral to acid soil – ideally, humus-rich and well-drained. Shallow planting is essential, as plants are surface-rooting. Dead-head spent flowers, wherever practical, to encourage energy into growth rather than seed production. Propagate by layering or semi-ripe cuttings in late summer. Yellowing leaves are usually caused by poor drainage, excessively deep planting or lime in soil. Weevils and powdery mildew may also cause problems.

Rhododendrons and **azaleas**
The genus *Rhododendron* includes not only evergreen, large-leaved and frequently large-flowered species and hybrids but also dwarf, smaller-leaved shrubs, both evergreen and deciduous, with few-flowered clusters of usually small blooms. 'Azalea' is the common name given to the deciduous species and hybrids as well as to a group of compact, evergreen shrubs derived mainly from Japanese species. They are valued for their mass of small colourful blooms in late spring. Many of the evergreen azaleas (sometimes known as Belgian azaleas) may also be grown as house plants. Botanically, however, all are classified as *Rhododendron.* The flowers are usually single, but may be semi-double or double, including hose-in-hose (one flower tube inside the other). Unless otherwise stated below, flowers are single and leaves are mid- to dark green and oval.

R. aberconwayi. Evergreen rhododendron with distinct, erect habit. H to 2.5m (8ft), S 1.2m (4ft). Frost hardy. Broadly lance-shaped leaves are small, rigid and deep green. Saucer-shaped, white flowers appear in late spring.
♛ *R. albrechtii.* Deciduous, upright, bushy azalea. H to 3m (10ft), S 2m (6ft). Fully hardy. Has spoon-shaped leaves clustered at branch tips and, in spring, bell-shaped, green-spotted, purple or pink flowers in loose clusters of 3–5.
R. 'Alison Johnstone'. Evergreen, bushy, compact rhododendron. H and S 2m (6ft). Frost hardy. Produces an abundance of exquisite, bell-shaped, peach-pink flowers in spring and bears attractive, waxy, grey-green leaves.
R. 'Angelo'. Evergreen, bushy rhododendron. H and S to 4m (12ft). Frost hardy. Has bold foliage and large, fragrant, bell-shaped, white flowers that are valuable in mid-summer in light woodland.
♛ *R. arboreum* (illus. p.103). Evergreen, tree-like rhododendron. H to 12m (40ft), S 3m (10ft). Frost hardy. Undersides of broadly lance-shaped leaves are silver, fawn or cinnamon. In spring has dense clusters of bell-shaped flowers in colours ranging from red (most tender form) through pink to white.
R. argyrophyllum (illus. p.102). Evergreen, spreading rhododendron. H and S to 5m (15ft). Fully hardy. Oblong leaves are silvery-white on undersides. Loose bunches of bell-shaped, rich pink flowers, sometimes with deeper coloured spots, are borne in spring. Is ideal for a light woodland.
R. 'Ascot Brilliant'. Evergreen, bushy rhododendron. H and S 3m (10ft). Frost hardy. Leaves are broadly oval. In spring produces loose bunches of funnel-shaped, waxy, rose-red blooms with darker margins.
R. augustinii (illus. p.103). Evergreen, bushy rhododendron. H and S to 4m (12ft). Fully hardy. Has lance-shaped to oblong, light green leaves and, in spring, bears an abundance of multi-stemmed, widely funnel-shaped, pale to deep blue or lavender flowers.
R. auriculatum (illus. p.102). Evergreen, bushy, widely branching rhododendron. H and S to 6m (20ft). Fully hardy. Has large, oblong, hairy leaves with distinct, ear-like lobes at their base. In late summer bears loose bunches of 7–15 large, heavily scented, tubular to funnel-shaped, white flowers. Is best in light woodland.
R. 'Azor'. Evergreen, upright rhododendron. H and S to 4m (12ft). Frost hardy. Leaves are broadly oval. Produces large, fragrant, funnel-shaped, salmon-pink flowers. Is especially valuable as it flowers in mid-summer.
♛ *R.* 'Azuma-kagami' (illus. p.102). Evergreen, compact azalea. H and S 1.2m (4ft). Frost hardy. Many small, hose-in-hose, deep pink flowers are borne in mid-spring. Is best in semi-shade.
R. barbatum. Evergreen, upright rhododendron. H and S to 10m (30ft). Fully hardy. Has lance-shaped, dark green leaves with bristles on stems; bark is plum-coloured and peeling. Produces tight bunches of tubular to bell-shaped, bright scarlet flowers in early spring.
R. 'Beauty of Littleworth' (illus. p.102). Evergreen, open, shrubby rhododendron. H and S 4m (12ft). Frost hardy. Bears huge, conical bunches of scented, funnel-shaped, crimson-spotted, white flowers in late spring.
R. 'Beefeater'. Evergreen, bushy rhododendron. H and S to 2.5m (8ft). Frost hardy. Leaves are broadly lance-shaped. Produces striking, flat-topped bunches of bell-shaped, scarlet flowers in late spring and early summer.
R. 'Blue Diamond'. Evergreen, upright rhododendron. H and S to 1.5m (5ft). Fully hardy. Small, neat leaves contrast with funnel-shaped, bright blue flowers borne in mid- to late spring. Likes full sun.
♛ *R.* 'Blue Peter' (illus. p.103). Evergreen, bushy rhododendron. H and S to 4m (12ft). Fully hardy. In early summer produces bold, open funnel-shaped, 2-tone lavender-purple flowers with frilled petal edges.
R. calendulaceum (Flame azalea). Deciduous, bushy azalea. H and S 2–3m (6–10ft). Fully hardy. Has funnel-shaped, scarlet or orange flowers in bunches of 5–7 in early summer.
♛ *R. calophytum* (illus. p.102). Evergreen, widely-branched rhododendron. H and S to 6m (20ft). Frost hardy. Produces large, lance-shaped leaves and, in early spring, huge bunches of bell-shaped, white or pale pink flowers with crimson spots.
R. calostrotum (illus. p.103). Evergreen, compact rhododendron. H to 1m (3ft). Fully hardy. Has attractive, blue-green leaves and, in late spring, saucer-shaped, purple or scarlet flowers in clusters of 2–5.
R. 'Catawbiense Album'. Evergreen, rounded rhododendron. H and S to 3m (10ft). Fully hardy. Bears glossy leaves and, in early summer, dense, rounded bunches of bell-shaped, white flowers.
R. 'Catawbiense Boursault'. Evergreen, rounded rhododendron. H and S to 3m (10ft). Fully hardy. Has glossy leaves. Dense, rounded bunches of bell-shaped, lilac-purple blooms are borne in early summer.
♛ *R.* x *cilpinense.* Semi-evergreen, compact rhododendron. H and S to 1.5m (5ft). Frost hardy. Leaves are dark green and glossy. Produces masses of large, bell-shaped, blush-pink flowers, flushed deeper in bud, in early spring that are vulnerable to frost damage.
R. cinnabarinum (illus. p.103). Evergreen, upright rhododendron. H and S 1.5–4m (5–12ft). Frost hardy. Has blue-green leaves with small scales. Narrowly tubular, waxy, orange to red flowers are borne in loose, drooping bunches in late spring.
♛ subsp. *xanthocodon* (syn. *R. xanthocodon;* illus. p.103) is of open, upright habit. H and S 1.5–4m (5–12ft). Aromatic leaves are blue-green when young, maturing to mid-green. Produces loose clusters of bell-shaped, waxy, yellow flowers during late spring.
♛ *R.* 'Coccinea Speciosa'. Deciduous, bushy azalea. H and S 1.5–2.5m (5–8ft). Fully hardy. Produces open funnel-shaped, brilliant rich orange-red blooms in early summer. Broadly lance-shaped leaves provide good autumn colour.
♛ *R.* 'Corneille' (illus. p.102). Deciduous, bushy azalea. H and S 1.5–2.5m (5–8ft). Fully hardy. In early summer bears fragrant, honeysuckle-like, cream flowers, flushed pink outside. Has attractive autumn foliage.
♛ *R.* 'Curlew' (illus. p.104). Evergreen rhododendron of compact, spreading habit. H and S 30cm (1ft). Fully hardy. Produces dull green leaves and, in late spring, relatively large, open funnel-shaped, yellow flowers.
♛ *R.* 'Cynthia' (illus. p.103). Vigorous, evergreen, dome-shaped rhododendron. H and S to 6m (20ft). Fully hardy. Bears conical bunches of bell-shaped, magenta-purple flowers, marked blackish-red within, in late spring. Is excellent for sun or shade.
R. dauricum. Evergreen, upright rhododendron. H and S to 1.5m (5ft). Fully hardy. Produces funnel-shaped, vivid purple flowers in loose clusters throughout winter. Green leaves turn purple-brown in frosty conditions.
♛ *R. davidsonianum* (illus. p.103). Deciduous, upright rhododendron. H 1.5–4m (5–12ft). Fully hardy. Aromatic leaves are lance-shaped to oblong. In late spring has clusters of funnel-shaped flowers, ranging from pale pink to mid-lilac-mauve.
R. decorum. Evergreen, bushy rhododendron. H and S 4m (12ft). Frost hardy. Oblong to lance-shaped leaves are mid-green above, paler beneath. Has large, fragrant, funnel-shaped, white or shell-pink flowers, green- or pink-spotted within, in early summer.
R. degronianum, syn. *R. metternichii.* Evergreen, upright rhododendron. H and S 1.5–4m (5–12ft). Fully hardy. Has attractive, oblong leaves, glossy and green above, reddish-brown-felted beneath. Bell-shaped, rose-red flowers, borne in spring, are in rounded bunches of 10–15, often subtly spotted within.
R. 'Doncaster'. Evergreen, compact rhododendron. H and S 2–2.5m (6–8ft). Frost hardy. Has leathery, glossy leaves and, in late spring, funnel-shaped, dark red flowers in dense bunches.
♛ *R.* 'Dora Amateis'. Evergreen, compact rhododendron. H and S 60cm (2ft). Fully hardy. Leaves are slender, glossy and pointed. Masses of broadly funnel-shaped, white flowers, tinged with pink and marked with green, appear in late spring. Is sun tolerant.

R. **'Elizabeth'** (illus. p.103). Evergreen, dome-shaped rhododendron. H and S to 1.5m (5ft). Frost hardy. Leaves are oblong. Has large, trumpet-shaped, brilliant red flowers in late spring. Is good in sun or partial shade.

R. **'Elizabeth Lockhart'**. Evergreen, dome-shaped rhododendron. H and S 60cm (2ft). Frost hardy. Produces shiny, purple-green leaves that become darker in winter. Bell-shaped, deep pink flowers are carried in spring.

R. **'English Roseum'**. Vigorous, evergreen, bushy rhododendron. H and S to 2.5m (8ft). Fully hardy. Bears handsome, dark green leaves, paler on undersides, and, in late spring, compact bunches of funnel-shaped, lilac-rose flowers.

R. **'Fabia'** (illus. p.104). Evergreen, dome-shaped rhododendron. H and S 2m (6ft). Frost hardy. Leaves are lance-shaped. Loose, flat trusses of funnel-shaped, orange-tinted, scarlet flowers are produced in early summer.

R. **'Fastuosum Flore Pleno'**. Evergreen, dome-shaped rhododendron. H and S 1.5–4m (5–12ft). Fully hardy. In early summer produces loose bunches of funnel-shaped, double, rich mauve flowers with red-brown marks and wavy edges.

R. **'Firefly'**. See *R.* 'Hexe'.

R. fortunei subsp. *discolor*. Evergreen, tree-like rhododendron. H and S to 8m (25ft). Frost hardy. Leaves are oblong to oval. Bears fragrant, funnel-shaped, pink flowers that give a magnificent display in mid-summer. Is ideal in a light woodland.

R. **'Freya'** (illus. p.104). Deciduous azalea of compact, shrubby habit. H and S 1.5m (5ft). Fully hardy. Fragrant, funnel-shaped, pink-flushed, orange-salmon flowers appear from late spring to early summer.

R. **'Frome'** (illus. p.104). Deciduous azalea of shrubby habit. H and S 1.5m (5ft). Fully hardy. In spring bears trumpet-shaped, saffron-yellow flowers, overlaid red in throats; petals are frilled and wavy-margined.

R. fulvum (illus. p.102). Evergreen, bushy rhododendron. H and S 1.5–4m (5–12ft). Frost hardy. Oblong to oval, polished, deep green leaves are brown-felted beneath. In early spring has loose bunches of bell-shaped, pink flowers, fading to white, each with a basal, deep red blotch.

R. **'George Reynolds'** (illus. p.104). Deciduous, bushy azalea. H and S to 2m (6ft). Fully hardy. Large, funnel-shaped, yellow flowers, flushed pink in bud, are borne with or before the leaves in spring.

R. **'Gloria Mundi'** (illus. p.104). Deciduous azalea of twiggy habit. H and S to 2m (6ft). Fully hardy. Has fragrant, honeysuckle-like, yellow-flared, orange flowers, with frilled margins, in early summer.

R. **'Glory of Littleworth'** (illus. p.104). Evergreen or semi-evergreen, bushy hybrid between a rhododendron and an azalea. H and S 1.5m (5ft). Frost hardy. Compact bunches of fragrant, bell-shaped, orange-marked, creamy-white flowers are borne abundantly in late spring and early summer. Is not easy to cultivate.

R. **'Gomer Waterer'**. Evergreen, compact rhododendron. H and S 1.5–4m (5–12ft). Fully hardy. Leaves are curved back at margins. Bell-shaped flowers, borne in dense bunches in early summer, are white, flushed mauve, each with a basal, mustard blotch. Likes sun or partial shade.

R. **'Hatsugiri'** (illus. p.103). Evergreen, compact azalea. H and S 60cm (2ft). Frost hardy. Has small, but very numerous, funnel-shaped, bright crimson-purple flowers in spring. Flowers very reliably.

R. **'Hawk Crest'** (illus. p.104). Evergreen rhododendron of open habit. H and S 1.5–4m (5–12ft). Frost hardy. Has broadly lance-shaped leaves. Bell-shaped flowers are borne in loose, flat-topped bunches, and are apricot in bud, opening to clear sulphur-yellow in late spring.

R. **'Hexe'**, syn. *R.* 'Firefly'. Evergreen azalea of neat habit. H and S 60cm (2ft). Frost hardy. Has numerous relatively large, hose-in-hose, glowing, crimson flowers in spring.

R. **'Hinodegiri'** (illus. p.103). Evergreen, compact azalea. H and S 1.5m (5ft). Frost hardy. Funnel-shaped, bright crimson flowers are small, but produced in abundance in late spring. Likes sun or light shade.

R. **'Hino-mayo'** (illus. p.103). Evergreen, compact azalea. H and S 1.5m (5ft). Frost hardy. Small, funnel-shaped, clear pink flowers are produced in great abundance in spring. Likes sun or light shade.

R. hippophaeoides (illus. p.103). Evergreen, erect rhododendron. H and S 1.5m (5ft). Fully hardy. Narrowly lance-shaped, aromatic leaves are grey-green. Small, funnel-shaped, lavender or lilac flowers are borne in spring. Is tolerant of wet, but not stagnant, soil conditions.

R. **'Homebush'** (illus. p.103). Deciduous, compact azalea. H and S 1.5m (5ft). Frost hardy. In late spring bears tight, rounded heads of trumpet-shaped, semi-double, rose-purple flowers with paler shading.

R. **'Hotei'**. Evergreen rhododendron of neat, compact habit. H and S 1.5–2.5m (5–8ft). Fully hardy. Has excellent foliage. Large, funnel-shaped, deep yellow flowers are freely produced in late spring.

R. **'Humming Bird'**. Evergreen, dome-shaped rhododendron of neat, compact habit. H and S to 1.5m (5ft). Fully hardy. From mid- to late spring produces bell-shaped, rose-pink flowers in loose, nodding bunches, above rounded, glossy leaves.

R. **'Hydon Hunter'**. Evergreen rhododendron of neat habit. H and S to 1.5m (5ft). Fully hardy. In late spring or early summer has masses of large, narrowly bell-shaped, red-rimmed flowers, paler towards the centre and orange-spotted within.

R. impeditum. Slow-growing, evergreen rhododendron. H and S to 60cm (2ft). Fully hardy. Aromatic leaves are grey-green. Funnel-shaped, purplish-blue flowers appear in spring. Is ideal for a rock garden.

R. **'Irouha-yama'** (illus. p.103). Evergreen, compact azalea. H and S to 1.5m (5ft). Frost hardy. Has abundant, small, funnel-shaped, white flowers, with pale lavender margins and faint brown eyes, in spring. Does well in light shade.

R. **'Jalisco'**. Deciduous, open, bushy rhododendron. H and S 1.5–4m (5–12ft). Frost hardy. Bears bunches of narrowly bell-shaped, straw-coloured flowers, tinted orange-rose at tips, in early summer.

R. **'Jeanette'**. Semi-evergreen azalea of upright habit. H and S 1.5–2m (5–6ft). Frost hardy. Funnel-shaped, vivid phlox-pink flowers, with darker blotches, appear in spring. Is good in light shade or full sun.

R. **'John Cairns'** (illus. p.103). Evergreen, upright, compact azalea. H and S 1.5–2m (5–6ft). Fully hardy. Funnel-shaped, orange-red flowers are produced abundantly in spring. Grows consistently and reliably in sun or semi-shade.

R. kaempferi (illus. p.103). Semi-evergreen, erect, loosely branched azalea. H and S 1.5–2.5m (5–8ft). Fully hardy. Leaves are lance-shaped. Has an abundance of funnel-shaped flowers in various shades of orange or red in late spring and early summer.

R. **'Kilimanjaro'**. Evergreen, bushy rhododendron. H and S 1.5–4m (5–12ft). Frost hardy. Bears broadly lance-shaped leaves. Produces large, rounded bunches of funnel- to bell-shaped, wavy-edged, maroon-red flowers, spotted chocolate within, in late spring and early summer.

R. **'Kirin'** (illus. p.102). Evergreen, compact azalea. H and S to 1.5m (5ft). Frost hardy. In spring has numerous hose-in-hose flowers that are deep rose, shaded a delicate silvery-rose. Looks best in light shade.

R. kiusianum. Semi-evergreen azalea of compact habit. H and S to 60cm (2ft). Fully hardy. Leaves are narrowly oval. Produces clusters of 2–5 funnel-shaped flowers, usually lilac-rose or mauve-purple, in late spring. Prefers full sun.

R. **'Lady Alice Fitzwilliam'**. Evergreen, bushy rhododendron. H and S 1.5–4m (5–12ft). Half hardy. Leaves are glossy, dark green. Loose bunches of heavily scented, broadly funnel-shaped, white flowers, flushed pale pink, are borne in mid- and late spring. Grows best against a south- or west-facing wall.

R. **'Lady Clementine Mitford'**. Evergreen, rounded, dense rhododendron. H and S 4m (12ft). Fully hardy. Has broadly oval, glossy, dark green leaves that are silvery when young and, in late spring and early summer, bold bunches of tubular- to bell-shaped flowers, peach-pink fading to white in the centre, with V-shaped, pink, green and brown marks within.

R. **'Lady Rosebery'**. Evergreen, stiffly branched rhododendron. H and S 1.5–4m (5–12ft). Frost hardy. In late spring produces clusters of drooping, narrowly bell-shaped, waxy flowers that are deep pink, paler towards petal edges. Is ideal for a light woodland margin.

R. **'Lem's Cameo'**. Evergreen, rounded, bushy rhododendron. H and S 1.5–2.5m (5–8ft). Frost hardy. Leaves are rounded. In spring has large-domed bunches of open funnel-shaped, pale peach flowers, deep pink in bud, shaded to pink at edges and with basal, deep rose-coloured blotches.

R. **'Loderi'**. Evergreen rhododendron of open habit. H and S 4m (12ft). Frost hardy. In spring bears huge bunches of richly fragrant, trumpet-shaped flowers that are usually soft pink, but at times paler, even white.

R. lutescens (illus. p.103). Semi-evergreen, upright rhododendron. H and S 1.5–3m (5–10ft). Fully hardy. Has oval to lance-shaped leaves that are bronze-red when young. Produces funnel-shaped, primrose-yellow flowers in early spring. Elegant and delicate, it is effective in a light woodland.

R. luteum (illus. p.104). Deciduous azalea of open habit. H and S 1.5–2.5m (5–8ft). Fully hardy. Leaves are oblong to lance-shaped. Bold, heavily fragrant, funnel-shaped, yellow blooms appear in spring. Autumn foliage is rich and colourful.

R. macabeanum (illus. p.104). Evergreen, tree-like rhododendron. H and S up to 13.5m (45ft). Frost hardy. Has bold, broadly oval leaves, dark green above, grey-felted beneath, and, in early spring, large bunches of bell-shaped, yellow flowers, blotched purple within.

R. mallotum. Evergreen, upright, open rhododendron, occasionally tree-like. H and S to 4m (12ft). Frost hardy. Oblong to oval leaves are deep green above, red-brown-felted beneath. Showy, tubular, crimson flowers in loose bunches are borne in early spring.

R. **'May Day'** (illus. p.103). Evergreen, spreading rhododendron. H and S to 1.5m (5ft). Frost hardy. Produces masses of loose bunches of long-lasting, funnel-shaped, scarlet flowers in late spring; calyces are petal-like and match the flower colour. Leaves are fresh green above, whitish-felted beneath.

R. **'Medway'** (illus. p.104). Deciduous, bushy, open azalea. H and S 1.5–2.5m (5–8ft). Fully hardy. In late spring has large, trumpet-shaped, pale pink flowers with darker edges and orange-flashed throats; petals have frilled margins.

R. metternichii. See *R. degronianum*.

R. **'Moonshine Crescent'** (illus. p.104). Evergreen, rounded to upright rhododendron. H 2–2.5m (6–8ft), S 2m (6ft). Frost hardy. In late spring produces compact trusses of bell-shaped, yellow flowers. Leaves are oblong to oval and dark green.

R. moupinense. Evergreen, rounded, compact rhododendron. H and S to 1.5m (5ft). Frost hardy. Produces funnel-shaped, pink blooms in loose bunches in late winter and early spring. Leaves are glossy, dark green above, paler beneath. Is best grown in a sheltered situation to reduce risk of frosted flowers.

R. **'Mrs G.W. Leak'** (illus. p.102). Evergreen, upright, compact rhododendron. H and S 4m (12ft). Fully hardy. Has compact, conical bunches of funnel-shaped, pink flowers, with black-brown and crimson marks within, in late spring.

R. nakaharae. Evergreen, mound-forming azalea. H and S 60cm (2ft). Frost hardy. Shoots and oblong to oval leaves are densely hairy. Funnel-shaped, dark brick-red flowers are borne in small clusters. Is valuable for mid-summer flowering and is ideal for a rock garden.

♛ *R.* 'Nancy Waterer'. Deciduous, twiggy azalea. H and S 1.5–2.5m (5–8ft). Fully hardy. Has large, long-tubed and honeysuckle-like, brilliant golden-yellow flowers in early summer. Is ideal in a light woodland or full sun.

♛ *R.* 'Narcissiflorum' (illus. p.104). Vigorous, deciduous, compact azalea. H and S 1.5–2.5m (5–8ft). Fully hardy. Sweetly scented, hose-in-hose, pale yellow flowers, darker outside and in centre, are produced in late spring or early summer. Has bronze autumn foliage.

R. 'Nobleanum' (illus. p.102). Evergreen, upright shrub or tree-like rhododendron. H and S to 5m (15ft). Frost hardy. Bears large, compact bunches of broadly funnel-shaped, rose-red, pink or white flowers in winter or early spring. Will flower for long periods in mild weather; is best in a sheltered position.

♛ *R.* 'Norma'. Vigorous, deciduous, compact azalea. H and S to 1.5m (5ft). Fully hardy. Bears masses of hose-in-hose, rose-red flowers with a salmon glow in spring. Grows well in sun or light shade.

R. 'Nova Zembla'. Vigorous, evergreen, upright rhododendron. H and S 1.5–4m (5–12ft). Fully hardy. Has funnel-shaped, dark red flowers in closely set bunches from late spring to early summer.

♛ *R. occidentale* (illus. p.102). Deciduous, bushy azalea. H and S 1.5–2.5m (5–8ft). Fully hardy. Leaves are glossy and turn yellow or orange in autumn. Flowers are fragrant, funnel-shaped, white or pale pink, each with a basal, yellow-orange blotch, and are borne from early to mid-summer.

R. orbiculare (illus. p.102). Evergreen rhododendron of compact habit. H and S to 3m (10ft). Fully hardy. Produces rounded, bright green leaves. Bell-shaped, rose-pink flowers are borne in loose bunches in late spring.

R. oreotrephes (illus. p.103). Deciduous, upright shrub or tree-like rhododendron. H and S to 5m (15ft). Fully hardy. Has attractive, scaly, grey-green foliage. In spring bears loose bunches of 3–10 broadly funnel-shaped flowers, usually mauve or purple, but variable, often with crimson spots.

♛ *R.* 'Palestrina' (illus. p.102). Evergreen or semi-evergreen, compact, free-flowering azalea. H and S to 1.2m (4ft). Frost hardy. Has large, open funnel-shaped, white flowers, with faint, green marks, in late spring. Grows well in light shade.

♛ *R.* 'Percy Wiseman' (illus. p.102). Evergreen rhododendron with a domed, compact habit. H and S to 2m (6ft). Fully hardy. In late spring produces open funnel-shaped, peach-yellow flowers that fade to white.

R. 'Pink Pearl' (illus. p.103). Vigorous, evergreen, upright, open rhododendron. H and S 4m (12ft) or more. Frost hardy. Tall bunches of open funnel-shaped, pink flowers give a spectacular display in late spring.

R. 'P. J. Mezitt'. Evergreen, compact rhododendron. H and S up to 1.5m (5ft). Fully hardy. Aromatic leaves are small, dark green in summer, bronze-purple in winter. Frost-resistant, funnel-shaped, lavender-pink flowers are borne in early spring. Is good in full sun.

R. 'President Roosevelt' (illus. p.103). Slow-growing, evergreen, open, weakly branched rhododendron. H and S to 2m (6ft). Frost hardy. Leaves are yellow-variegated. Open bell-shaped, pale pink flowers, lighter towards centres and frilled at margins, appear from mid- to late spring. Leaves have a tendency to revert to plain green.

♛ *R.* 'Ptarmigan'. Evergreen, spreading rhododendron that forms a compact mound. H to 30cm (1ft), S 75cm (2½ft) or more. Fully hardy. Funnel-shaped, pure white flowers are borne in early spring. Prefers full sun.

♛ *R.* 'Purple Splendour'. Evergreen, bushy rhododendron. H and S to 3m (10ft). Fully hardy. Has well-formed bunches of open funnel-shaped, rich royal-purple flowers, with prominent, black marks in throats, in late spring or early summer.

♛ *R.* 'Queen Elizabeth II' (illus. p.103). Evergreen, bushy rhododendron. H and S 1.5–4m (5–12ft). Frost hardy. Funnel-shaped, greenish-yellow flowers are borne in loose bunches in late spring. Produces narrowly oval or lance-shaped, glossy leaves that are mid-green above, paler beneath.

♛ *R.* 'Queen of Hearts'. Evergreen, open rhododendron. H and S 1.5–4m (5–12ft). Frost hardy. Has masses of domed bunches of funnel-shaped, deep crimson flowers, black-speckled within, in mid-spring.

R. racemosum (illus. p.102). Evergreen, upright, stiffly branched rhododendron. H and S to 2.5m (8ft). Fully hardy. Has clusters of widely funnel-shaped, bright pink flowers carried along the stems in spring. Small, broadly oval, aromatic leaves are dull green above, grey-green below.

R. rex. Vigorous, evergreen, upright shrub or tree-like rhododendron. H and S 4m (12ft) or more. Frost hardy. Leaves are pale buff-felted beneath and pink or white flowers each have a crimson blotch and a spotted throat. subsp. *fictolacteum* (illus. p.102), H to 13.5m (45ft), has large leaves that are green above, brown-felted beneath. Bold bunches of bell-shaped, white flowers, each with a maroon blotch and often with a spotted throat, are produced during spring.

♛ *R.* 'Romany Chai'. Vigorous, evergreen rhododendron, open when young, becoming denser with age. H and S 1.5–4m (5–12ft). Frost hardy. Has dark green foliage, tinged bronze, and, in early summer, large, compact bunches of broadly funnel-shaped, rich brown-red flowers, each with a basal, maroon blotch. Prefers a light woodland.

R. 'Rosalind' (illus. p.103). Vigorous, evergreen rhododendron of open-branched habit. H and S 4m (12ft). Frost hardy. Leaves are dull green. Bears broadly funnel-shaped, pink flowers in loose bunches in spring.

R. 'Roseum Elegans'. Vigorous, evergreen, rounded rhododendron. H and S 2.5m (8ft) or more. Fully hardy. In early summer has rounded bunches of broadly funnel-shaped, reddish-purple flowers, marked yellow-brown. Foliage is bold and glossy, deep green.

♛ *R.* 'Roza Stevenson'. Vigorous, evergreen, upright rhododendron of open habit. H and S 1.5–4m (5–12ft). Frost hardy. Produces masses of fine, loose bunches of saucer-shaped, lemon flowers in mid- to late spring. Is excellent in light shade.

R. rubiginosum. Vigorous, evergreen, upright, well-branched rhododendron. H 6m (20ft), S 2.5m (8ft). Frost hardy. Aromatic leaves are lance-shaped, dull green above, reddish-brown beneath. Has funnel-shaped, lilac-purple flowers in loose bunches in mid-spring.

♛ *R. schlippenbachii* (illus. p.102). Deciduous, rounded, open azalea. H and S 2.5m (8ft). Fully hardy. Spoon-shaped leaves are in whorls at branch ends. Saucer-shaped, pink flowers in loose bunches of 3–6 appear in mid-spring. Is ideal in a light woodland.

R. 'Seta' (illus. p.102). Evergreen, erect rhododendron. H 1.5m (5ft), S 1–1.5m (3–5ft). Frost hardy. In early spring bears loose bunches of tubular, shiny, vivid pink-and-white-striped flowers, fading to white at bases.

♛ *R.* 'Seven Stars' (illus. p.102). Vigorous, evergreen, upright, dense rhododendron. H and S 2–3m (6–10ft). Fully hardy. Has yellowish-green foliage and, in spring, masses of bunches of large, bell-shaped, wavy-margined, white flowers, flushed with apple-blossom pink, opening from pink buds.

R. 'Silver Moon' (illus. p.102). Evergreen azalea of broad, spreading habit. H and S 1.5–2.5m (5–8ft). Frost hardy. Funnel-shaped, white flowers, with pale green-blotched throats and frilled petal edges, appear in spring. Grows best in partial shade.

R. 'Snowdrift'. Deciduous, bushy azalea. H and S to 2.5m (8ft). Fully hardy. Bears bunches of large, slender-tubed flowers in spring before the leaves appear. Flowers are white with yellow marks that deepen to orange.

R. souliei (illus. p.102). Evergreen, open rhododendron. H and S 1.5–4m (5–12ft). Fully hardy. Has rounded leaves and, in late spring, saucer-shaped, soft pink flowers. Grows best in areas of low rainfall.

♛ *R.* 'Spek's Orange'. Deciduous, bushy azalea. H and S to 2.5m (8ft). Fully hardy. In late spring carries bold bunches of large, slender-tubed blooms that are bright reddish-orange with greenish marks within.

♛ *R.* 'Strawberry Ice' (illus. p.102). Deciduous, bushy azalea. H and S 1.5–2.5m (5–8ft). Fully hardy. Flowers are trumpet-shaped, deep pink in bud, opening flesh-pink, and mottled deeper pink at petal edges with deep yellow-marked throats; they appear in late spring.

R. 'Surprise'. Evergreen, dense azalea. H and S to 1.5m (5ft). Frost hardy. Has abundant, small, funnel-shaped, light orange-red flowers in mid-spring. Looks effective when mass planted and is ideal in light shade or full sun.

♛ *R.* 'Susan' (illus. p.103). Evergreen, close-growing rhododendron. H and S 1.5–4m (5–12ft). Fully hardy. In spring bears large bunches of open funnel-shaped flowers in 2 shades of blue-mauve and spotted purple within. Has handsome, glossy, dark green foliage.

R. sutchuenense (illus. p.102). Evergreen, spreading shrub or tree-like rhododendron. H and S to 5m (16ft). Frost hardy. Has large leaves and, in early spring, bears large bunches of broadly funnel-shaped, pink flowers, spotted deeper within. Prefers a light woodland.

♛ *R.* 'Temple Belle'. Evergreen rhododendron of neat, compact habit. H and S 1.5–2.5m (5–8ft). Fully hardy. Loose bunches of bell-shaped, clear pink flowers are produced in spring. Rounded leaves are dark green above, grey-green beneath.

R. thomsonii (illus. p.103). Evergreen, rounded rhododendron of open habit. H and S to 5.5m (18ft). Frost hardy. Leaves are waxy, dark green above, whiter beneath. Peeling, fawn-coloured bark contrasts well with bell-shaped, waxy, red flowers in spring.

♛ *R.* 'Vuyk's Scarlet' (illus. p.103). Evergreen, compact azalea. H and S to 60cm (2ft). Frost hardy. In spring bears an abundance of relatively large, open funnel-shaped, brilliant red flowers, with wavy petals, that completely cover the glossy foliage.

R. wardii (illus. p.103). Evergreen, compact rhododendron. H and S 1.5–4m (5–12ft). Fully hardy. Has rounded leaves and, in late spring, loose bunches of saucer-shaped, clear yellow flowers, each with a basal, crimson blotch.

♛ *R. williamsianum* (illus. p.102). Evergreen rhododendron of compact, spreading habit. H and S 1.5m (5ft). Fully hardy. Leaves are bronze when young, maturing to mid-green. Has loosely clustered, bell-shaped, pink flowers in spring. Is ideal for a small garden.

R. 'Woodcock'. Evergreen rhododendron of compact, spreading habit. H and S 1.5–2.5m (5–8ft). Fully hardy. Has semi-glossy, dark green leaves and, in spring, masses of loose bunches of funnel-shaped, rose-red flowers.

R. xanthocodon. See *R. cinnabarinum* subsp. *xanthocodon*.

R. yakushimanum (illus. p.102). Evergreen, dome-shaped rhododendron of neat, compact habit. H 1m (3ft), S 1.5m (5ft). Fully hardy. Broadly oval leaves are silvery at first, maturing to deepest green, and are brown-felted beneath. In late spring has open funnel-shaped, pink flowers that fade to near white and are flecked green within.

♛ *R.* 'Yellow Hammer' (illus. p.104). Evergreen, erect, bushy rhododendron. H and S to 2m (6ft). Fully hardy.

Abundant clusters of tubular, bright yellow flowers are borne in spring; frequently flowers again in autumn. *R. yunnanense* (illus. p.102). Semi-evergreen, open rhododendron. H and S 1.5–4m (5–12ft). Fully hardy. In spring produces masses of butterfly-like, pale pink or white flowers with spotted or blotched throats. Aromatic leaves are grey-green.

RHODOHYPOXIS (Hypoxidaceae)
Genus of dwarf, spring- to summer-flowering, tuberous perennials, grown for their pink, red or white flowers, each comprising 6 petals that meet at the centre, so the flower has no eye. Frost hardy, if kept fairly dry when dormant in winter. Requires full sun and sandy, peaty soil with plenty of moisture in summer. Propagate by seed or offsets in spring.
♀ *R. baurii*. Spring- and early summer-flowering, tuberous perennial. H 5–10cm (2–4in), S 2.5–5cm (1–2in). Has an erect, basal tuft of narrowly lance-shaped, hairy leaves. Slender stems produce a succession of erect, flattish, white, pale pink or red flowers, 2cm (3/4in) across. **'Albrighton'** and **'Douglas'** illus. p.328. **'Margaret Rose'** illus. p.324. var. *platypetala* has 2.5cm (1in) wide, white or very pale pink flowers.

RHODOLEIA (Hamamelidaceae)
Genus of evergreen, mainly spring-flowering trees, grown for their foliage and flowers. Frost tender, min. 7–10°C (45–50°F). Needs full light or partial shade and humus-rich, well-drained, neutral to acid soil. Water potted specimens freely when in full growth, moderately at other times. Pruning is tolerated if necessary. Propagate by semi-ripe cuttings in summer or by seed when ripe, in autumn, or in spring.
R. championii. Evergreen, bushy tree. H and S 4–8m (12–25ft). Elliptic to oval, bright green leaves, each to 9cm (3 1/2in) long, are borne towards the shoot tips. Clusters of insignificant flowers, surrounded by petal-like, pink bracts, appear in spring.

RHODOPHIALA (Amaryllidaceae)
Genus of bulbs, grown for their large, funnel-shaped flowers. Frost hardy to frost tender, min. 13–15°C (55–9°F). Needs full sun or partial shade and well-drained soil. Keep bulbs dry while dormant in winter. Propagate by seed in spring or by offsets in spring (summer-flowering species).
R. advenum, syn. *Hippeastrum advenum*, illus. p.365. Spring- and summer-flowering bulb. H to 60cm (24in), S 15–20cm (6–8in). Half hardy. Leaves are narrowly strap-shaped, semi-erect and basal. Flower stem produces a head of 2–5 red flowers, each 8–12cm (3–5in) across, with yellow-veined centres.

RHODOTHAMNUS (Ericaceae)
Genus of one species of evergreen, semi-prostrate, open shrub, grown for its flowers. Is suitable for rock gardens and peat beds. Fully hardy. Needs sun and humus-rich, well-drained, acid soil. Propagate by seed in spring or by semi-ripe cuttings in summer.
R. chamaecistus illus. p.300.

RHODOTYPOS (Rosaceae)
Genus of one species of deciduous shrub, grown for its flowers. Fully hardy. Needs sun or semi-shade and fertile soil. On established plants cut some older shoots back or to ground level after flowering. Propagate by softwood cuttings in summer or by seed in autumn.
R. kerrioides. See *R. scandens*.
R. scandens, syn. *R. kerrioides*, illus. p.128.

Rhoeo discolor. See *Tradescantia spathacea*.
Rhoeo spathacea. See *Tradescantia spathacea*.

RHOICISSUS (Vitaceae)
Genus of evergreen, tendril climbers, grown for their handsome foliage. Bears inconspicuous flowers intermittently during the year. Frost tender, min. 7–10°C (45–50°F). Grow in any fertile, well-drained soil with light shade in summer. Water regularly, less in cold weather. Provide support. Remove crowded stems when necessary or in early spring. Propagate by seed in spring or by semi-ripe cuttings in summer.
♀ *R. capensis* (Cape grape). Vigorous, evergreen, tendril climber. H to 5m (15ft). Rounded, toothed, lustrous, mid- to deep green leaves, to 20cm (8in) wide, have a deeply rounded, heart-shaped base.
R. rhomboidea. See *Cissus rhombifolia*.

RHOMBOPHYLLUM (Aizoaceae)
Genus of mat-forming, perennial succulents with dense, basal rosettes of linear or semi-cylindrical leaves, each expanded towards middle or tip; leaf tip is also reflexed or incurved. Frost tender, min. 5°C (41°F). Needs sun and very well-drained soil. Propagate by seed or stem cuttings in spring or summer.
R. rhomboideum illus. p.414.

RHUS (Anacardiaceae)
Sumach
Genus of deciduous trees, shrubs and scrambling climbers, grown for their divided, ash-like foliage, autumn colour and, in some species, showy fruit clusters. Fully to frost hardy. Requires sun and well-drained soil. Propagate by semi-ripe cuttings in summer, by seed in autumn or by root cuttings in winter. May be attacked by coral spot fungus.
R. aromatica. Deciduous, bushy shrub. H 1m (3ft), S 1.5m (5ft). Fully hardy. Deep green leaves, each composed of 3 oval leaflets, turn orange or reddish-purple in autumn. Tiny, yellow flowers are borne in mid-spring, before foliage, and are followed by spherical, red fruits.
R. copallina (Dwarf sumach). Deciduous, upright shrub. H and S 1–1.5m (3–5ft), sometimes more. Fully hardy. Has glossy, dark green leaves, with numerous lance-shaped leaflets, that turn red-purple in autumn. Minute, greenish-yellow flowers, borne in dense clusters from mid- to late summer, develop into narrowly egg-shaped, bright red fruits.
R. cotinoides. See *Cotinus obovatus*.
R. cotinus. See *Cotinus coggygria*.
R. glabra illus. p.113.
♀ *R. hirta*, syn. *R. typhina* (Stag's horn sumach). Deciduous, spreading, suckering, open shrub or tree. H 5m (15ft), S 6m (20ft). Fully hardy. Velvety shoots are clothed in dark green leaves with oblong leaflets. Produces minute, greenish-white flowers from mid- to late summer. Leaves become brilliant orange-red in autumn, accompanying clusters of spherical, deep red fruits on female plants. ♀ **'Laciniata'** (syn. *R.h.* 'Dissecta', *R. typhina* 'Laciniata') illus. p.94.
R. potaninii. Deciduous, round-headed tree. H 12m (40ft), S 8m (25ft). Fully hardy. Has large, dark green leaves, with usually 7–11 oval leaflets that turn red in autumn. In summer bears dense clusters of tiny, yellow-green flowers. Female flower clusters develop into tiny, spherical, black or brownish fruits.
R. succedanea (Wax tree). Deciduous, spreading tree. H and S 10m (30ft). Frost hardy. Bears dense clusters of tiny, yellow-green flowers in summer. Large, glossy, dark green leaves, consisting of 9–15 oval leaflets, turn red in autumn. Female flower clusters develop into tiny, spherical, black or brownish fruits.
R. trichocarpa, syn. *Toxicodendron succedaneum* illus. p.69.
R. typhina. See *R. hirta*. **'Laciniata'** see *R. hirta* 'Laciniata'.
R. verniciflua, syn. *Toxicodendron verniciflua* (Varnish tree). Deciduous, spreading tree. H 15m (50ft), S 10m (30ft). Fully hardy. Large, glossy, bright green leaves, with 7–13 oval leaflets, redden in autumn. Bears dense clusters of tiny, yellow-green flowers in summer, followed by berry-like, brownish-yellow fruits. Sap may severely irritate the skin.

RHYNCHELYTRUM (Gramineae).
See GRASSES, BAMBOOS, RUSHES and SEDGES.
R. repens, syn. *R. roseum* (Natal grass, Ruby grass). Tuft-forming, annual or short-lived, perennial grass. H 1.2–2m (4–6ft), S 60cm–1m (2–3ft). Frost tender, min. 5°C (41°F). Leaves are mid-green, flat and finely pointed. Produces loose panicles of awned, pink spikelets in summer.
R. roseum. See *R. repens*.

RIBES (Grossulariaceae)
Currant
Genus of deciduous or evergreen, mainly spring-flowering shrubs, grown for their edible fruits (currants and gooseberries) or their flowers. Fully to frost hardy. Needs full sun and fertile, well-drained soil, but *R. laurifolium* also tolerates shade. Cut out some older shoots after flowering and prune straggly, old plants hard in winter or early spring. Propagate deciduous species by hardwood cuttings in winter, evergreens by semi-ripe cuttings in summer. Aphids may attack young foliage.
R. laurifolium illus. p.145.
R. odoratum (Buffalo currant). Deciduous, upright shrub. H and S 2m (6ft). Fully hardy. Clusters of fragrant, tubular, golden-yellow flowers are borne from mid- to late spring, followed by rounded, purple fruits. Rounded, 3-lobed, bright green leaves turn red and purple in autumn.
R. sanguineum (Flowering currant).
♀ **'Brocklebankii'** illus. p.125. **'King Edward VII'** is a deciduous, upright, compact shrub. H and S 2m (6ft). Fully hardy. Small, tubular, deep reddish-pink flowers are freely borne amid rounded, 3–5-lobed, aromatic, dark green leaves, from mid- to late spring, and are sometimes succeeded by spherical, black fruits with a white bloom. Is useful for hedging.
♀ **'Pulborough Scarlet'** illus. p.100.
♀ **'Tydeman's White'**, H and S 2.5m (8ft), is less compact and has pure white flowers.
♀ *R. speciosum* (Fuchsia-flowered currant). Deciduous, bushy, spiny shrub. H and S 2m (6ft). Frost hardy. Slender, drooping, tubular, red flowers, with long, red stamens, open from mid- to late spring. Fruits are spherical and red. Has red, young shoots and oval, 3–5-lobed, glossy, bright green leaves. Flourishes when trained against a south- or west-facing wall.

RICHEA (Epacridaceae)
Genus of evergreen, summer-flowering shrubs, grown for their foliage and densely clustered flowers. Frost to half hardy. Needs sun or semi-shade and moist, peaty, neutral to acid soil. Propagate by semi-ripe cuttings in summer or by seed in autumn.
R. scoparia. Evergreen, upright shrub. H and S 2m (6ft). Frost hardy. Shoots are densely covered with narrowly lance-shaped, sharp-pointed, dark green leaves. Dense, upright spikes of small, egg-shaped, pink, white, orange or maroon flowers appear in early summer.

RICINUS (Euphorbiaceae)
Genus of one species of fast-growing, evergreen, tree-like shrub, grown for its foliage. Is often grown in gardens as an annual. In cool climates is grown as an annual. Half hardy. Needs sun and fertile to rich, well-drained soil. May require support in exposed areas. Propagate by seed sown under glass in early spring.
R. communis illus. p.287. **'Impala'** illus. p.282.

ROBINIA (Leguminosae)
Genus of deciduous, mainly summer-flowering trees and shrubs, grown for their foliage and clusters of pea-like flowers. Is useful for poor, dry soil. Fully hardy. Needs a sunny position. Grows in any but waterlogged soil. Branches are brittle and may be damaged by strong winds. Propagate by seed or suckers in autumn or by root cuttings in winter.
R. x *ambigua* **'Decaisneana'**. Deciduous, spreading tree. H 15m (50ft), S 10m (30ft). Dark green leaves

have numerous oval leaflets. Long, hanging clusters of pea-like, pink flowers are borne in early summer.
❀ *R. hispida* illus. p.111.
R. kelseyi. Deciduous, spreading, open shrub. H 2.5m (8ft), S 4m (12ft). Clusters of pea-like, rose-pink flowers open in late spring or early summer and are followed by pendent, red seed pods. Dark green leaves each consist of 9 or 11 oval leaflets.
❀ *R. pseudoacacia* (False acacia, Locust). Fast-growing, deciduous, spreading tree. H 25m (80ft), S 15m (50ft). Dark green leaves consist of 11–23 oval leaflets. Dense, drooping clusters of fragrant, pea-like, white flowers are borne in late spring and early summer. ❀ 'Frisia' illus. p.55. 'Umbraculifera' (Mop-head acacia), H and S 6m (20ft), has a rounded, dense head but flowers are rarely produced.

ROCHEA (Crassulaceae)
Genus of evergreen, succulent sub-shrubs and shrubs. Frost tender, min. 10°C (50°F). Requires sun and well-drained soil. Propagate by seed or stem cuttings in spring or summer.
R. coccinea, syn. *Crassula coccinea*. Evergreen, erect, succulent shrub. H to 60cm (2ft), S 30cm (1ft) or more. Alternate pairs of fleshy, oval to oblong-oval, hairy-margined, dull green leaves, each united at the base, are arranged at right angles in 4 rows up the woody, green stems. Has umbels of tubular, bright red flowers in summer or autumn.

RODGERSIA (Saxifragaceae)
Genus of summer-flowering, rhizomatous perennials. Is ideal for pond sides. Fully to frost hardy. Grows in sun or semi-shade and in moist soil in a position sheltered from strong winds, which may damage foliage. Propagate by division in spring or by seed in autumn.
❀ *R. aesculifolia* illus. p.205.
❀ *R. pinnata* 'Superba'. Clump-forming, rhizomatous perennial. H 1–1.2m (3–4ft), S 75cm (2½ft). Frost hardy. Has divided, bronze-tinged, emerald-green leaves with 5–9 narrowly oval leaflets. Long, much-branched, dense panicles of star-shaped, bright pink flowers appear in mid-summer.
❀ *R. podophylla* illus. p.204.
R. sambucifolia illus. p.204.

ROMNEYA (Papaveraceae)
Californian poppy
Genus of summer-flowering, woody-based perennials and deciduous sub-shrubs. Frost hardy. Requires a warm, sunny position and deep, well-drained soil. Is difficult to establish, resents being moved and in very cold areas roots may need protection in winter. Once established, may spread rapidly. Propagate by softwood cuttings of basal shoots in early spring, by seed in autumn (transplanting seedlings without disturbing rootballs) or by root cuttings in winter.
❀ *R. coulteri* illus. p.190.
R. 'White Cloud'. Vigorous, bushy, woody-based perennial. H and S 1m (3ft). Throughout summer produces large, slightly fragrant, shallowly cup-shaped, white flowers with prominent golden stamens. Leaves are oval, deeply lobed and grey.

ROMULEA (Iridaceae)
Genus of crocus-like corms, grown for their funnel-shaped flowers. Frost to half hardy. Needs full light and well-drained, sandy soil. Water freely during the growing period. Most species die down in summer and then need warmth and dryness. *R. macowanii*, however, is dormant in winter. Propagate by seed in autumn, or in spring for *R. macowanii*.
R. bulbocodioides of gardens. See *R. flava*.
R. bulbocodium illus. p.375.
R. flava, syn. *R. bulbocodioides* of gardens. Early spring-flowering corm. H to 10cm (4in), S 2.5–5cm (1–2in). Half hardy. Has a thread-like, erect, basal leaf and 1–5 upright, widely funnel-shaped, usually yellow flowers, 2–4cm (³/₄–1½in) across, with deeper yellow centres.
R. longituba. See *R. macowanii*.
R. macowanii, syn. *R. longituba*. Summer-flowering corm. H and S 1–2cm (½–¾in). Half hardy. Leaves are thread-like, erect and basal. Bears 1–3 upright, yellow flowers, each 3cm (1¼in) across with a long tube expanding into a wide funnel shape.
R. sabulosa. Early spring-flowering corm. H 5–15cm (2–6in), S 2.5–5cm (1–2in). Half hardy. Forms thread-like, erect, basal leaves. Stems bear 1–4 upward-facing, funnel-shaped, black-centred, bright red flowers that open flattish in sun to 4–5cm (1½–2in) across.

RONDELETIA (Rubiaceae)
Genus of evergreen, mainly summer-flowering trees and shrubs, grown primarily for their flowers. Frost tender, min. 13–16°C (55–61°F). Requires full light or partial shade and fertile, well-drained soil. Water containerized specimens freely when in full growth, moderately at other times. Stems may be shortened in early spring if necessary. Propagate by seed in spring or by semi-ripe cuttings in summer.
R. amoena. Evergreen, rounded shrub. H and S 2–4m (6–12ft). Oval, dark green leaves have dense, brown down on undersides. Has dense clusters of tubular, 4- or 5-lobed, pink flowers in summer.

Rooksbya euphorbioides. See *Lemaireocereus euphorbioides*.

ROSA (Rosaceae)
Rose
Genus of deciduous or semi-evergreen, open shrubs and scrambling climbers, grown for their profusion of flowers, often fragrant, and sometimes for their fruits (rose hips). Leaves are divided into usually 5 or 7 oval leaflets, with rounded or pointed tips, that are sometimes toothed. Stems usually bear thorns, or prickles. Fully hardy, unless otherwise stated below. Prefers an open, sunny site and needs fertile, moist but well-drained soil. Avoid planting in an area where roses have been grown in recent years, as problems due to harmful organisms may occur: either exchange the soil, which may be used satisfactorily elsewhere, or choose another site for the new rose. To obtain blooms of high quality, feed in late winter or early spring with a balanced fertilizer and apply a mulch. In spring-summer feed at 3-weekly intervals. Remove spent flower heads from plants that are 'remontant' ('rising up again'; other terms used are repeat- or perpetual-flowering). May be trimmed for tidiness in early winter. To improve health, flower quality and shape of bush, prune in the dormant season or, preferably, in spring, before young shoots develop from dormant growth buds: remove dead, damaged and dying wood; lightly trim Old Garden and Ground cover roses (see below); remove two-thirds of previous summer's growth of Modern bush, including miniature, roses. Correct treatment of Modern shrub and climbing roses, ramblers and species roses depends on the variety but in general prune only lightly. Propagate by budding in summer or by hardwood cuttings in autumn. All roses are prone to attack by various pests and diseases, including aphids, blackspot, powdery mildew, rust and sawfly.

Roses are officially classified in three groups, each comprising different types, based, it is claimed, on the functional qualities of each plant, such as whether it is remontant, rather than on any historical, botanical or genetical relationships. Flowers occur in a variety of forms (illustrated and described on p.150) and are single (4–7 petals), semi-double (8–14 petals), double (15–30 petals) or fully double (over 30 petals). Roses are illustrated on pp.150–164.

Species roses
Species, or wild, roses and **species hybrids** that share most of the characteristics of the parent species. Includes shrubs and climbing roses. Produces usually single flowers mainly in summer, borne generally in one flush, followed by red or black hips in autumn.

Old Garden roses
Alba – large, freely branching shrubs with clusters of, usually, 5–7 semi-double to double flowers in mid-summer. Has abundant, greyish-green leaves. Is very hardy and is good for borders and as specimen plants.
Bourbon – open, remontant shrubs that may be trained to climb. Produces usually fully double flowers, borne commonly in 3s, in summer-autumn. Is suitable for borders and for training over fences, walls and pillars.
China – spindly, remontant shrubs that produce single to double flowers, borne singly or in clusters of 2–13, in summer-autumn. Has pointed, shiny leaflets. Needs a sheltered position. Is suitable for borders and walls.
Damask – open shrubs with usually very fragrant, semi- to fully double flowers, borne singly or in loose clusters of 5–7 mainly in summer. Is suitable for borders.
Gallica – shrubs of fairly dense, free-branching growth. Leaves are dull green. Produces single to fully double, richly coloured flowers, often in clusters of 3, in summer. Is suitable for borders and as hedging.
Hybrid Perpetual – vigorous, free-branching, remontant shrubs that bear fully double flowers, held singly or in 3s, in summer-autumn. Leaves are usually olive-green. Is suitable for beds and borders.
Moss – often lax shrubs with a furry, moss-like growth on stems and calyces. Leaves are usually dark green. Has double to fully double flowers in summer.
Noisette – remontant climbing roses that bear clusters of up to 9 usually double flowers, with a slight spicy fragrance, in summer-autumn. Has generally smooth stems and glossy leaves. Needs a sheltered site. Is suitable for a south- or west-facing wall.
Portland – upright, remontant shrubs with semi-double to double flowers, held singly or in 3s, in summer-autumn. Suits beds and borders.
Provence (Centifolia) – lax, thorny shrubs that produce scented, usually double to fully double flowers, borne singly or in 3s, in summer. Leaves are often dark green. Is suitable for borders.
Sempervirens – semi-evergreen climbing roses with shiny, light green leaves and, in late summer, numerous semi- to fully double flowers. Is ideal for naturalizing or for growing on fences and pergolas.
Tea – remontant shrubs and climbing roses that produce spicy-scented, slender-stemmed, pointed, semi- to fully double flowers, borne singly or in 3s, in summer-autumn. Leaves are shiny and pale green. Frost hardy. Needs a sheltered position. Is suitable for beds and borders.

Modern roses
Shrub – a diverse group of modern roses, most of which are remontant, that grow larger (mostly H 1–2m (3–6ft)) than most bush roses. Has single to fully double flowers, held singly or in sprays, in summer and/or autumn. Is suitable for beds and borders and for growing as specimen plants.
Large-flowered bush (Hybrid Tea) – remontant shrubs with mostly pointed, double flowers, 8cm (3in) or more across, borne singly or in 3s, in summer-autumn. Is excellent for beds, borders, hedges and cutting.
Cluster-flowered bush (Floribunda) – remontant shrubs that produce sprays of usually 3–25 single to fully double flowers in summer-autumn. Is excellent for beds, borders and hedges.
Dwarf cluster-flowered bush (Patio) – neat, remontant shrubs, H 38–60cm (15–24in), S 30–60cm (12–24in), that bear sprays of generally 3–11 single to double flowers in summer-autumn. Is ideal for beds, borders and hedges and for growing in containers.
Miniature bush – remontant shrubs,

H to 45cm (18in), S to 40cm (16in), with sprays of usually 3–11 tiny, single to fully double flowers in summer-autumn. Has tiny leaves. Is suitable for rock gardens, small spaces and for growing in containers.
Polyantha – tough, compact, remontant shrubs with sprays of usually 7–15 small, 5-petalled, single to double flowers in summer-autumn. Is suitable for beds.
Ground cover – trailing and spreading roses, many of which are remontant, with single to fully double flowers, borne mostly in clusters of 3–11, in summer and/or autumn. Is suitable for beds, banks and walls.
Climbing – vigorous climbing roses, some of which are remontant, with stiff stems and single to fully double flowers, borne singly or in clusters, from late spring to autumn. Is suitable for training over walls, fences and pergolas.
Rambler – vigorous climbing roses with lax stems. Has clusters of 3–21 single to fully double flowers, mainly in summer. Is suitable for training over walls, fences, pergolas and trees.

R. 'Aimée Vibert', syn. *R.* 'Bouquet de la Mariée'. Noisette climbing rose with long, smooth stems. H 5m (15ft), S 3m (10ft). Bears clusters of lightly scented, cupped, fully double, blush-pink to white flowers, 8cm (3in) across, in summer-autumn. Leaves are glossy, dark green. May be grown as a shrub.
R. 'Alba Semiplena'. See *R.* x *alba* 'Semiplena'.
♛ *R.* x *alba* 'Semiplena', syn. *R.* 'Alba Semiplena'. Vigorous, bushy Alba rose. H 2m (6ft), S 1.5m (5ft). Bears sweetly scented, flat, semi-double, white flowers, 8cm (3in) across, in mid-summer. Has greyish-green leaves. May be grown as a hedge.
♛ *R.* 'Albéric Barbier' illus. p.162.
♛ *R.* 'Albertine' illus. p.163.
R. 'Alec's Red' illus. p.158.
R. 'Alexander', syn. *R.* 'Alexandra', illus. p.158.
R. 'Alexandra'. See *R.* 'Alexander'.
R. 'Alfred de Dalmas'. See *R.* 'Mousseline'.
R. 'Alister Stella Gray', syn. *R.* 'Golden Rambler'. Noisette climbing rose with long, vigorous, upright stems. H 5m (15ft), S 3m (10ft). Bears clusters of musk-scented, quartered, fully double, yolk-yellow flowers, 6cm (2½in) across, in summer-autumn. Has glossy, mid-green leaves.
R. 'Aloha'. Stiff, bushy climbing rose. H and S 2.5m (8ft). Fragrant, cupped, fully double, rose- and salmon-pink flowers, 9cm (3½in) across, appear in summer-autumn. Leaves are leathery, dark green. May be grown as a shrub.
R. 'Alpine Sunset' illus. p.156.
♛ *R.* 'Amber Queen', syn. *R.* 'Harroony', illus. p.159.
R. 'Amruda'. See *R.* 'Red Ace'.
R. 'Angela Rippon', syn. *R.* 'Ocarina', *R.* 'Ocaru', illus. p.161.
R. 'Angelita'. See *R.* 'Snowball'.
♛ *R.* 'Anisley Dickson', syn. *R.* 'Dicki-mono', *R.* 'Dicky', *R.* 'Münchner Kindl', illus. p.157.

♛ *R.* 'Anna Ford', syn. *R.* 'Harpiccolo', illus. p.157.
R. 'Anne Harkness', syn. *R.* 'Harkaramel', illus. p.160.
R. 'Apothecary's Rose'. See *R. gallica* var. *officinalis*.
R. 'Armada', syn. *R.* 'Haruseful'. Vigorous, free-branching shrub rose. H 1.5m (5ft), S 1.2m (4ft). Sprays of spice-scented, cupped, double, deep pink flowers, 8cm (3in) across, are borne in summer-autumn. Has plentiful, glossy, deep green leaves.
♛ *R.* 'Arthur Bell'. Upright, cluster-flowered bush rose. H 1m (3ft), S 60cm (2ft). Clusters of fragrant, cupped, double, yellow flowers, 8cm (3in) across, are borne in summer-autumn. Foliage is bright green.
R. 'Assemblage des Beautés', syn. *R.* 'Rouge Eblouissante'. Upright, dense Gallica rose. H 1.2m (4ft), S 1m (3ft). Faintly scented, rounded, fully double, green-eyed, cerise to crimson-purple flowers, 8cm (3in) across, are borne in summer. Has rich green leaves.
R. 'Ausmary'. See *R.* 'Mary Rose'.
R. 'Ausmas'. See *R.* 'Graham Thomas'.
R. 'Baby Carnival'. See *R.* 'Baby Masquerade'.
R. 'Baby Gold Star', syn. *R.* 'Estrellita de Oro'. Miniature bush rose of uneven habit. H 45cm (18in), S 40cm (16in). Bears slightly scented, cupped, double, yellow flowers, 5cm (2in) across, in summer-autumn. Leaves are small, glossy and dark green.
R. 'Baby Masquerade', syn. *R.* 'Baby Carnival', illus. p.161.
R. banksiae, syn. *R.b.* var. *normalis* (Banksian rose). Climbing species rose. H and S 10m (30ft). Half hardy. Dense clusters of fragrant, flat, single, white flowers, 2.5cm (1in) across, are borne on slender, thornless, light green stems in late spring. Leaves are small and pale green. Is uncommon in cultivation.
♛ *'Lutea'* illus. p.164.
R. 'Beauty of Glazenwood'. See *R.* 'Fortune's Double Yellow'.
R. 'Belle Courtisanne'. See *R.* 'Königin von Dänemark'.
♛ *R.* 'Belle de Crécy' illus. p.154.
R. 'Belle de Londres'. See *R.* 'Compassion'.
R. 'Belle of Portugal'. See *R.* 'Belle Portugaise'.
R. 'Belle Portugaise', syn. *R.* 'Belle of Portugal'. Very vigorous, climbing Tea rose. H 6m (20ft), S 3m (10ft). Bears fragrant, pointed, double, light salmon-pink flowers, 12cm (5in) across, in summer. Leaves are large and glossy.
R. 'Bizarre Triomphant'. See *R.* 'Charles de Mills'.
R. 'Blanche Moreau'. Moss rose of rather lax growth. H 1.5m (5ft), S 1.2m (4ft). Fragrant, cupped, fully double, white flowers, 10cm (4in) across, with brownish 'mossing', appear in summer. Has dull green leaves.
♛ *R.* 'Blessings' illus. p.156.
R. 'Blue Moon', syn. *R.* 'Mainzer Fastnacht', *R.* 'Sissi'. Large-flowered bush rose of open habit. H 1m (3ft), S 60cm (2ft). Sweetly scented, pointed, fully double, lilac flowers, 10cm (4in) across, are borne in summer-autumn. Leaves are large and dark green.

R. 'Bluenette'. See *R.* 'Blue Peter'.
R. 'Blue Peter', syn. *R.* 'Bluenette', *R.* 'Ruiblun'. Miniature bush rose of neat habit. H 35cm (14in), S 30cm (12in). Slightly scented, cupped, double, purple flowers, 5cm (2in) across, are produced in summer-autumn. Leaves are small and plentiful.
R. 'Blue Rambler'. See *R.* 'Veilchenblau'.
R. 'Blush Noisette'. Noisette climbing rose of branching habit and lax growth. H 2–4m (6–12ft), S 2–2.5m (6–8ft). In summer-autumn, smooth stems bear clusters of spice-scented, cupped, double, blush-pink flowers, 4cm (1½in) across. Has matt foliage. May be grown as a shrub.
R. 'Blush Rambler'. Vigorous rambler rose. H 3m (10ft), S 4m (12ft). Clusters of delicately fragrant, cupped, semi-double, light pink flowers, 4cm (1½in) across, are borne in summer. Has an abundance of glossy leaves. Is a particularly good scrambler for an arch, pergola or tree.
♛ *R.* 'Bonica '82', syn. *R.* 'Meidonomac', illus. p.152.
R. 'Boule de Neige' illus. p.150.
R. 'Bouquet de la Mariée'. See *R.* 'Aimée Vibert'.
R. 'Brass Ring'. See *R.* 'Peek-a-boo'.
R. 'Breath of Life', syn. *R.* 'Harquanne', illus. p.163.
R. 'Bright Smile', syn. *R.* 'Dicdance', illus. p.159.
♛ *R.* 'Buff Beauty'. Rounded shrub rose. H and S 1.2m (4ft). Slightly fragrant, cupped, fully double, apricot-buff flowers, 9cm (3½in) across, are borne freely in summer, sparsely in autumn. Has plentiful, glossy, dark green leaves.
R. 'Buttons', syn. *R.* 'Dicmickey'. Upright, dwarf cluster-flowered bush rose. H 45cm (18in), S 35cm (14in). Bears well-spaced sprays of urn-shaped, double, light reddish-salmon flowers, 4cm (1½in) across, in summer-autumn. Has small, dark green leaves. Is good as a low hedge.
R. californica. Shrubby species rose. H 2.2m (7ft), S 2m (6ft). Fragrant, flat, single, lilac-pink flowers, 4cm (1½in) across, are borne freely in mid-summer, sparsely in autumn. Has small, dull green leaves. ♛ 'Plena' (syn. *R.c.* var. *plena*) is semi-double and more pink-toned.
♛ *R.* 'Canary Bird' illus. p.155.
♛ *R.* 'Capitaine John Ingram'. Vigorous, bushy Moss rose. H and S 1.2m (4ft). In summer bears fragrant, cupped, fully double, rich maroon-crimson flowers, 8cm (3in) across; petals are paler on reverses. Foliage is dark green.
♛ *R.* 'Cardinal de Richelieu' illus. p.154.
R. 'Cardinal Hume', syn. *R.* 'Harregale', illus. p.154.
R. 'Cécile Brunner', syn. *R.* 'Mignon'. Upright, spindly China rose with fairly smooth stems. H 75cm (30in), S 60cm (24in). Produces slightly scented, urn-shaped, fully double, light pink flowers, 4cm (1½in) across, in summer-autumn. Small, dark green leaves are sparse.
♛ *R.* 'Céleste', syn. *R.* 'Celestial', illus. p.151.

R. 'Celestial'. See *R.* 'Céleste'.
R. centifolia var. *cristata*. See *R.* 'Cristata'.
♛ *R.* x *centifolia* 'Muscosa', syn. *R.* 'Common Moss', *R.* 'Old Pink Moss'. Vigorous, lax Moss rose. H 1.5m (5ft), S 1.2m (4ft). Fragrant, rounded to cupped, fully double, mossed, pink flowers, 8cm (3in) across, are produced in summer. Leaves are matt, dull green. Is best grown on a support.
R. 'Champagne Cocktail', syn. *R.* 'Horflash', illus. p.158.
R. 'Chapeau de Napoléon'. See *R.* 'Cristata'.
R. 'Chaplin's Pink Companion' illus. p.163.
♛ *R.* 'Charles de Mills', syn. *R.* 'Bizarre Triomphant'. Upright, arching Gallica rose with fairly smooth stems. H 1.2m (4ft), S 1m (3ft). Very fragrant, quartered-rosette, fully double, crimson-purple flowers, 10cm (4in) across, appear in summer. Leaves are plentiful and mid-green. May be grown on a support.
♛ *R. chinensis* 'Mutabilis', syn. *R.* x *odorata* 'Mutabilis', illus. p.153.
♛ *R.* 'City of London', syn. *R.* 'Harukfore'. Rounded, cluster-flowered bush rose. H 1m (3ft), S 75cm (2½ft). In summer-autumn, sweet-smelling, urn-shaped, double, blush-pink flowers, 8cm (3in) across, are produced in dainty sprays amid bright green foliage.
R. 'Clarissa', syn. *R.* 'Harprocrustes'. Upright, cluster-flowered bush rose. H 75cm (30in), S 45cm (18in). Dense sprays of slightly scented, urn-shaped, fully double, apricot flowers, 5cm (2in) across, appear in summer-autumn. Has many small, glossy leaves. Makes a good, narrow hedge.
R. 'Climbing Ena Harkness'. Stiff, branching climbing rose. H and S 2.5m (8ft). Fragrant, large, pointed, fully double, scarlet-crimson flowers, 10cm (4in) across, are borne on nodding stems in summer-autumn. Leaves are mid-green.
♛ *R.* 'Climbing Lady Hillingdon'. Stiff climbing rose. H 4m (12ft), S 2m (6ft). Dark green leaves are produced on reddish-green stems. Bears spice-scented, pointed, double, apricot-yellow flowers, 10cm (4in) across, in summer-autumn. Is best grown in a sheltered site.
R. 'Climbing Mrs Sam McGredy'. Vigorous, stiff, branching climbing rose. H and S 3m (10ft). Leaves are glossy, rich reddish-green. Bears faintly fragrant, large, urn-shaped, fully double, coppery salmon-pink flowers, 11cm (4½in) across, in summer and again, sparsely, in autumn.
R. 'Cocabest'. See *R.* 'Wee Jock'.
R. 'Cocdestin'. See *R.* 'Remember Me'.
R. 'Colibri '79', syn. *R.* 'Meidanover', illus. p.162.
R. 'Commandant Beaurepaire', syn. *R.* 'Panachée d'Angers'. Vigorous, spreading Bourbon rose. H and S 1.2m (4ft). Fragrant, cupped, double flowers, 10cm (4in) across, are borne in summer-autumn. They are a rich mixture of blush-pink, splashed with mauve, purple, crimson and scarlet.

Light green leaflets have wavy margins.
R. 'Common Moss'. See
R. x centifolia 'Muscosa'.
☙ R. 'Compassion', syn. R. 'Belle de Londres'. Upright, freely-branching climbing rose. H 3m (10ft), S 2.5m (8ft). Glossy, dark green leaves are produced on reddish stems. Sweetly fragrant, rounded, double, pink-tinted, salmon-apricot flowers, 10cm (4in) across, are borne freely in summer-autumn. May be pruned to grow as a shrub.
☙ R. 'Complicata' illus. p.152.
☙ R. 'Comte de Chambord', syn. R. 'Mme Knorr'. Vigorous, erect Portland rose. H 1.2m (4ft), S 1m (3ft). In summer-autumn, fragrant, quartered-rosette, fully double, lilac-tinted, pink flowers, 10cm (4in) across, appear amid plentiful, light green foliage. Is suitable for a hedge.
R. 'Congratulations', syn. R. 'Korlift', R. 'Sylvia'. Upright, vigorous, large-flowered bush rose. H 1.2m (4ft), S 1m (3ft). Produces neat, urn-shaped, fully double, deep rose-pink flowers, 11cm (4½in) across, on long stems in summer-autumn. Leaves are large and dark green. Makes a tall hedge.
R. 'Conrad Ferdinand Meyer' illus. p.151.
☙ R. 'Constance Spry' illus. p.152.
R. 'Crested Moss'. See R. 'Cristata'.
☙ R. 'Cristata', syn. R. centifolia var. cristata, R. 'Chapeau de Napoléon', R. 'Crested Moss'. Bushy, lanky Provence rose. H 1.5m (5ft), S 1.2m (4ft). In summer, very fragrant, cupped, fully double, pink flowers, 9cm (3½in) across and with tufted sepals, are borne on nodding stems amid dull green foliage. May be grown on a support.
R. 'Cuisse de Nymphe'. See R. 'Great Maiden's Blush'.
R. 'Cuthbert Grant'. Vigorous, bushy shrub rose. H and S 1m (3ft). Bears slightly scented, cupped, semi-double, deep purplish-red flowers, 12cm (5in) across, in summer-autumn. Foliage is glossy. May be grown as a hedge.
R. 'Danse du Feu', syn. R. 'Spectacular', illus. p.164.
R. 'Dicdance'. See R. 'Bright Smile'.
R. 'Dicdivine'. See R. 'Pot o' Gold'.
R. 'Dicgrow'. See R. 'Peek-a-boo'.
R. 'Dicinfra'. See R. 'Disco Dancer'.
R. 'Dicjana'. See R. 'Peaudouce'.
R. 'Dicjeep'. See R. 'Len Turner'.
R. 'Dicjem'. See R. 'Freedom'.
R. 'Dicjubell'. See R. 'Lovely Lady'.
R. 'Dickimono'. See R. 'Anisley Dickson'.
R. 'Dicky'. See R. 'Anisley Dickson'.
R. 'Diclittle'. See R. 'Little Woman'.
R. 'Diclulu'. See R. 'Gentle Touch'.
R. 'Dicmagic'. See R. 'Sweet Magic'.
R. 'Dicmickey'. See R. 'Buttons'.
R. 'Disco Dancer', syn. R. 'Dicinfra'. Dense, rounded, cluster-flowered bush rose. H 75cm (30in), S 60cm (24in). Dense sprays of slightly fragrant, cupped, double, bright orange-red flowers, 6cm (2½in) across, are carried in summer-autumn. Produces a mass of glossy foliage.
R. 'Doris Tysterman' illus. p.160.
R. 'Dortmund' illus. p.164.
R. 'Double Delight' illus. p.157.
R. 'Double Velvet'. See R. 'Tuscany Superb'.

R. 'Drummer Boy', syn. R. 'Harvacity'. Dwarf cluster-flowered bush rose of bushy, spreading habit. H and S 50cm (20in). Faintly scented, cupped, double, bright crimson flowers, 5cm (2in) across, appear in dense sprays amid a mass of small, dark green leaves in summer-autumn. Makes a good, low hedge.
☙ R. 'Dublin Bay' illus. p.164.
R. 'Duchesse d'Istrie'. See R. 'William Lobb'.
R. 'Duftzauber '84'. See R. 'Royal William'.
☙ R. 'Du Maître d'Ecole'. Bushy, spreading Gallica rose. H 1.2m (4ft), S 1m (3ft). Bears fragrant, quartered-rosette, fully double, carmine to light pink flowers, 10cm (4in) across, in summer. Foliage is dull green.
R. 'Dupontii' illus. p.151.
R. 'Dutch Gold'. Vigorous, upright, large-flowered bush rose. H 1.1m (3½ft), S 60cm (2ft). Produces fragrant, rounded, fully double, yellow flowers, 15cm (6in) across, in summer-autumn. Has large, dark green leaves.
R. 'Easlea's Golden', syn. R. 'Easlea's Golden Rambler'. Vigorous, arching climbing rose. H 5m (15ft), S 3m (10ft). Pleasantly scented, cupped, fully double, yellow flowers, 10cm (4in) across and flecked with red, appear, usually in clusters, during summer. Has plentiful, leathery foliage.
R. 'Easlea's Golden Rambler'. See R. 'Easlea's Golden'.
R. 'Easter Morn'. See R. 'Easter Morning'.
R. 'Easter Morning', syn. R. 'Easter Morn'. Upright, miniature bush rose. H 40cm (16in), S 25cm (10in). During summer-autumn, faintly fragrant, urn-shaped, fully double, ivory-white flowers, 3cm (1¼in) across, are borne freely amid glossy, dark green leaves.
R. ecae illus. p.154.
☙ R. eglanteria, syn. R. rubiginosa, illus. p.151.
R. 'Elina'. See R. 'Peaudouce'.
R. 'Elizabeth Harkness' illus. p.155.
R. 'Emily Gray'. Semi-evergreen rambler rose with long, lax stems. H 5m (15ft), S 3m (10ft). Small trusses of slightly fragrant, cupped, fully double, butter-yellow flowers, 5cm (2in) across, are borne in summer. Leaves are lustrous, dark green. May die back in a hard winter, and is prone to mildew in dry conditions.
R. 'Empereur du Maroc' illus. p.154.
R. 'English Miss'. Cluster-flowered bush rose. H 75cm (30in), S 60cm (24in). In summer-autumn bears fragrant, cupped, camellia-shaped, fully double, blush-pink flowers, 8cm (3in) across. Has leathery, dark green foliage.
R. 'Escapade' illus. p.157.
R. 'Estrellita de Oro'. See R. 'Baby Gold Star'.
R. 'Everblooming Dr W. van Fleet'. See R. 'New Dawn'.
R. 'Fairy Changeling', syn. R. 'Harnumerous'. Compact, spreading Polyantha bush rose. H 45cm (18in), S 50cm (20in). Produces dense sprays of slightly scented, pompon, fully double, deep pink flowers, 4cm (1½in) across, in summer-autumn. Has an abundance of glossy, dark green leaves.

☙ R. 'Fantin-Latour' illus. p.151.
☙ R. 'Felicia' illus. p.151.
☙ R. 'Félicité Parmentier'. Vigorous, compact, upright Alba rose. H 1.2m (4ft), S 1m (3ft). Fragrant, cupped to flat, fully double, pale flesh-pink flowers, 6cm (2½in) across, are borne in mid-summer. Has abundant, greyish-green leaves. Makes a good hedge.
☙ R. 'Félicité Perpétue' illus. p.162.
R. 'Felicity Kendall', syn. R. 'Lanken'. Sturdy, well-branched, large-flowered bush rose. H 1.1m (3½ft), S 75cm (2½ft). Lightly fragrant, rounded, fully double, bright red flowers, 11cm (4½in) across, appear among a mass of dark green foliage in summer-autumn.
R. 'Fellemberg'. See R. 'Fellenberg'.
R. 'Fellenberg', syn. R. 'Fellemberg'. Vigorous, shrubby Noisette or China rose. H 2.5m (8ft), S 1.2m (4ft). Leaves are purplish-green. Clusters of faintly scented, rounded to cupped, fully double flowers, 5cm (2in) across, in shades of light crimson, appear in summer-autumn. Prune to grow as a bedding rose or support as a climber.
☙ R. filipes 'Kiftsgate' illus. p.162.
R. 'Fire Princess' illus. p.161.
R. foetida 'Persiana', syn. R. 'Persian Yellow', illus. p.154.
R. 'Fortune's Double Yellow', syn. R. 'Beauty of Glazenwood', R. 'Gold of Ophir', R. 'San Rafael Rose'. Lax climbing rose of restrained growth. H 2.5m (8ft), S 1.5m (5ft). Half hardy. In summer bears small clusters of pleasantly scented, pointed to cupped, semi-double, copper-suffused, yellow flowers, 5cm (2in) across. Leaves are glossy, light green. Prune very lightly.
R. 'Fragrant Cloud'. Bushy, dense, large-flowered bush rose. H 75cm (30in), S 60cm (24in). Very fragrant, rounded, double, dusky-scarlet flowers, 12cm (5in) across, are borne freely in summer-autumn. Has plentiful, dark green foliage.
R. 'Fragrant Delight'. Bushy, cluster-flowered bush rose of uneven habit. H 1m (3ft), S 75cm (2½ft). Produces an abundance of reddish-green foliage, amid which clusters of fragrant, urn-shaped, double, salmon-pink flowers, 8cm (3in) across, are borne freely in summer-autumn.
☙ R. 'François Juranville'. Vigorous, arching rambler rose. H 6m (20ft), S 5m (15ft). Bears clusters of apple-scented, rosette, fully double, rosy-salmon-pink flowers, 8cm (3in) across, during summer. Produces a mass of glossy leaves. Is prone to mildew in a dry site.
☙ R. 'Freedom', syn. R. 'Dicjem', illus. p.159.
R. 'Friesia'. See R. 'Korresia'.
R. 'Frühlingsmorgen', syn. R. 'Spring Morning'. Open, free-branching shrub rose. H 2m (6ft), S 1.5m (5ft). Foliage is greyish-green. In late spring produces hay-scented, cupped, single, pink flowers, 12cm (5in) across, with a primrose centre and reddish stamens.
R. 'Fryminicot'. See R. 'Sweet Dream'.
☙ R. gallica var. officinalis, syn. R. 'Apothecary's Rose', R. officinalis, R. 'Red Rose of Lancaster'. Bushy

species rose of neat habit. H to 80cm (32in), S 1m (36in). In summer bears flat, semi-double, pinkish-red flowers, 8cm (3in) across, with a moderate scent. ☙ 'Versicolor' illus. p.153.
☙ R. 'Gentle Touch', syn. R. 'Diclulu'. Upright, dwarf cluster-flowered bush rose. H 50cm (20in), S 30cm (12in). Sprays of faintly scented, urn-shaped, semi-double, pale salmon-pink flowers, 5cm (2in) across, appear in summer-autumn. Leaves are small and dark green. Makes a good, low hedge.
☙ R. 'Geranium', syn. R. moyesii 'Geranium', illus. p.153.
R. 'Gioia'. See R. 'Peace'.
R. 'Gipsy Boy'. See R. 'Zigeunerknabe'.
☙ R. glauca, syn. R. rubrifolia, illus. p.152.
R. 'Glenfiddich' illus. p.159.
R. 'Gloire de Dijon' illus. p.162.
R. 'Gloire des Mousseux'. Vigorous, bushy Moss rose. H 1.2m (4ft), S 1m (3ft). Has plentiful, light green foliage. In summer bears fragrant, cupped, fully double flowers, 15cm (6in) across. These are bright pink, paling to blush-pink, with light green 'mossing'.
R. 'Gloria Dei'. See R. 'Peace'.
R. 'Gold of Ophir'. See R. 'Fortune's Double Yellow'.
R. 'Golden Rambler'. See R. 'Alister Stella Gray'.
☙ R. 'Golden Showers' illus. p.164.
R. 'Golden Sunblaze'. See R. 'Rise 'n Shine'.
☙ R. 'Golden Wings'. Bushy, spreading shrub rose. H 1.1m (3½ft), S 1.35m (4½ft). Fragrant, cupped, single, pale yellow flowers, 12cm (5in) across, appear amid light green foliage in summer-autumn. Is suitable for a hedge.
R. 'Goldfinch'. Vigorous, arching rambler rose. H 2.7m (9ft), S 2m (6ft). In summer produces lightly scented, rosette, double, yolk-yellow flowers, 4cm (1½in) across, that fade to white. Has plentiful, bright light green leaves.
R. 'Goldsmith'. See R. 'Simba'.
R. 'Goldstar'. Neat, upright, large-flowered bush rose. H 1m (3ft), S 60cm (2ft). Amid glossy, dark green leaves, lightly scented, urn-shaped, fully double, yellow flowers, 8cm (3in) across, are produced in summer-autumn.
☙ R. 'Graham Thomas', syn. R. 'Ausmas', illus. p.155.
R. 'Grandpa Dickson', syn. R. 'Irish Gold', illus. p.159.
R. 'Great Maiden's Blush', syn. R. 'Cuisse de Nymphe', R. 'La Séduisante', illus. p.151.
R. 'Grouse', syn. R. 'Korimro', illus. p.155.
R. 'Guinée' illus. p.164.
R. 'Guletta'. See R. 'Rugul'.
☙ R. 'Handel' illus. p.163.
R. 'Hannah Gordon', syn. R. 'Korweiso'. Bushy, open, cluster-flowered bush rose. H 75cm (30in), S 60cm (2ft). Sprays of slightly fragrant, cupped, double, blush-pink flowers, 8cm (3in) across, margined with reddish-pink, appear in summer-autumn. Leaves are dark green.
R. 'Harkaramel'. See R. 'Anne Harkness'.

589

R. 'Harlightly'. See R. 'Princess Michael of Kent'.
R. 'Harmantelle'. See R. 'Mountbatten'.
R. 'Harnumerous'. See R. 'Fairy Changeling'.
R. 'Harpiccolo'. See R. 'Anna Ford'.
R. 'Harprocrustes'. See R. 'Clarissa'.
R. 'Harquanne'. See R. 'Breath of Life'.
R. 'Harqueterwife'. See R. 'Paul Shirville'.
R. 'Harregale'. See R. 'Cardinal Hume'.
R. 'Harroony'. See R. 'Amber Queen'.
R. 'Harrowbond'. See R. 'Rosemary Harkness'.
R. 'Harsherry'. See R. 'Sheila's Perfume'.
R. 'Hartanna'. See R. 'Princess Alice'.
R. 'Harukfore'. See R. 'City of London'.
R. 'Haruseful'. See R. 'Armada'.
R. 'Harvacity'. See R. 'Drummer Boy'.
R. 'Harwanna'. See R. 'Jacqueline du Pré'.
R. 'Harwanted'. See R. 'Prima'.
R. 'Harwharry'. See R. 'Malcolm Sargent'.
R. 'Harwotnext'. See R. 'Sheila Macqueen'.
R. 'Heartthrob'. See R. 'Paul Shirville'.
R. 'Heideröslein'. See R. 'Nozomi'.
✥ R. 'Henri Martin', syn. R. 'Red Moss', illus. p.153.
R. 'Honorine de Brabant'. Vigorous, bushy, sprawling Bourbon rose. H and S 2m (6ft). Fragrant, quartered, double flowers, 10cm (4in) across, lilac-pink, marked with light purple and crimson, are produced in summer-autumn. Has plentiful, light green foliage.
R. 'Horflash'. See R. 'Champagne Cocktail'.
R. 'Hula Girl' illus. p.161.
✥ R. 'Iceberg', syn. R. 'Schneewittchen', illus. p.155.
R. 'Iced Ginger' illus. p.156.
✥ R. 'Ingrid Bergman'. Upright, branching, large-flowered bush rose. H 75cm (30in), S 60cm (24in). Bears slightly scented, urn-shaped, double, dark red flowers, 11cm (4^1/$_2$in) across, in summer-autumn. Has leathery, semi-glossy, dark green foliage.
R. 'Interall'. See R. 'Rosy Cushion'.
R. 'Interfour'. See R. 'Petit Four'.
R. 'Intrigue', syn. R. 'Korlech'. Bushy, dense, cluster-flowered bush rose. H 75cm (30in), S 60cm (24in). In summer-autumn bears dense sprays of unscented, cupped, camellia-shaped, double, blackish-crimson flowers, 6cm (2^1/$_2$in) across. Has plentiful, dark green foliage.
R. 'Invincible'. Upright, cluster-flowered bush rose. H 1m (3ft), S 60cm (2ft). Faintly scented, cupped, fully double, bright crimson flowers, 9cm (3^1/$_2$in) across, appear in open clusters in summer-autumn. Leaves are semi-glossy.
R. 'Irish Gold'. See R. 'Grandpa Dickson'.
✥ R. 'Ispahan', syn. R. 'Pompon des Princes', R. 'Rose d'Isfahan'. Vigorous, bushy, dense Damask rose. H 1.5m (5ft), S 1.2m (4ft). Produces fragrant, cupped, double, clear pink flowers, 8cm (3in) across, amid greyish-green foliage throughout summer and autumn.
R. 'Jacqueline du Pré', syn. R. 'Harwanna'. Vigorous, arching shrub rose. H 2m (6ft), S 1.5m (5ft). In summer-autumn bears musk-scented, cupped, double, ivory-white flowers, 10cm (4in) across, with scalloped petals and red stamens. Produces an abundance of glossy leaves. Makes a large hedge.
R. 'John Cabot'. Vigorous shrub rose. H 1.5m (5ft), S 1.2m (4ft). Leaves are yellow-green. Clusters of fragrant, cupped, double, magenta flowers, 6cm (2^1/$_2$in) across, are borne in summer-autumn. May be grown as a climber or hedge.
R. 'Julia's Rose'. Spindly, branching, large-flowered bush rose. H 75cm (30in), S 45cm (18in). In summer-autumn produces faintly scented, urn-shaped, double, brownish-pink to buff flowers, 10cm (4in) across. Foliage is reddish-green. Is good for flower arrangements.
✥ R. 'Just Joey' illus. p.160.
R. 'Kathleen Harrop'. Arching, lax Bourbon rose. H 2.5m (8ft), S 2m (6ft). Fragrant, double, cupped, pale pink flowers, 8cm (3in) across, are borne in summer-autumn. Plentiful, dark green foliage is susceptible to mildew. May be grown as a climber or hedge.
R. 'Keepsake', syn. R. 'Kormalda', illus. p.157.
✥ R. 'Königin von Dänemark', syn. R. 'Belle Courtisanne', illus. p.152.
R. 'Königliche Hoheit'. See R. 'Royal Highness'.
R. 'Korbelma'. See R. 'Simba'.
R. 'Korblue'. See R. 'Shocking Blue'.
R. 'Korgund'. See R. 'Loving Memory'.
R. 'Korimro'. See R. 'Grouse'.
R. 'Korlech'. See R. 'Intrigue'.
R. 'Korlift'. See R. 'Congratulations'.
R. 'Kormalda'. See R. 'Keepsake'.
R. 'Korpeahn'. See R. 'The Times'.
R. 'Korresia', syn. R. 'Friesia', illus. p.159.
R. 'Korweiso'. See R. 'Hannah Gordon'.
R. 'Korzaun'. See R. 'Royal William'.
R. 'La Séduisante'. See R. 'Great Maiden's Blush'.
R. 'La Sevillana', syn. R. 'Meigekanu'. Dense, bushy shrub rose. H 75cm (2^1/$_2$ft), S 1m (3ft). Clusters of faintly scented, cupped, double, bright red flowers, 8cm (3in) across, are borne freely in summer-autumn. Produces an abundance of dark green leaves. Is suitable for growing as a hedge or ground cover.
R. 'Lady Waterlow'. Stiff climbing rose. H 4m (12ft), S 2m (6ft). Pleasantly scented, pointed to cupped, double, light pink shaded, salmon flowers, 12cm (5in) across, are borne mainly in summer, but some may also appear in autumn. Leaves are mid-green.
R. 'Langford Light', syn. R. 'Lannie'. Spreading, cluster-flowered bush rose. H and S 60cm (2ft). Sprays of fragrant, cupped, wide-opening, double, white flowers, 5cm (2in) across, appear in summer-autumn. Has dark green leaves.
R. 'Lanken'. See R. 'Felicity Kendall'.
R. 'Lannie'. See R. 'Langford Light'.
R. 'Leggab'. See R. 'Pearl Drift'.
R. 'Len Turner', syn. R. 'Dicjeep'. Compact, cluster-flowered bush rose. H and S 50cm (20in). Slightly scented, pompon, fully double, red-rimmed, ivory-white flowers, 6cm (2^1/$_2$in) across, are borne in clusters amid plentiful, dark green foliage during summer-autumn. Makes a good, low hedge.
R. 'Little Woman', syn. R. 'Diclittle'. Upright, dwarf cluster-flowered bush rose. H 50cm (20in), S 40cm (16in). In summer-autumn bears sprays of faintly fragrant, urn-shaped, double, salmon-pink flowers, 5cm (2in) across. Has small, dark green leaves. Is suitable for a narrow hedge.
R. 'Louise Odier', syn. R. 'Mme de Stella'. Elegant, upright Bourbon rose. H 2m (6ft), S 1.2m (4ft). Has light greyish-green foliage and fragrant, cupped, fully double, warm rose-pink flowers, 12cm (5in) across, borne in summer-autumn.
✥ R. 'Lovely Lady', syn. R. 'Dicjubell', illus. p.156.
R. 'Loving Memory', syn. R. 'Korgund'. Upright, robust, large-flowered bush rose. H 1.1m (3^1/$_2$ft), S 75cm (2^1/$_2$ft). Produces slightly fragrant, pointed, fully double, deep red flowers, 12cm (5in) across, on stiff stems in summer-autumn. Foliage is dull green.
R. 'Macangel'. See R. 'Snowball'.
R. 'Maccarpe'. See R. 'Snow Carpet'.
R. 'Macman'. See R. 'Matangi'.
R. 'Macrexy'. See R. 'Sexy Rexy'.
R. macrophylla. Vigorous species rose. H 4m (12ft), S 3m (10ft). Bears moderately fragrant, flat, single, red flowers, 5cm (2in) across, in summer, followed by flask-shaped, red hips. Has red stems and large, mid-green leaves.
R. 'Macspash'. See R. 'Sue Lawley'.
R. 'Mactru'. See R. 'Trumpeter'.
✥ R. 'Mme Alfred Carrière' illus. p.162.
R. 'Mme A. Meilland'. See R. 'Peace'.
R. 'Mme de Stella'. See R. 'Louise Odier'.
R. 'Mme Ernst Calvat'. Vigorous, arching Bourbon rose. H 2–3m (6–10ft), S 2m (6ft). In summer-autumn, fragrant, cupped to quartered-rosette, fully double, rose-pink flowers, 15cm (6in) across, are borne freely. Has plentiful, large leaves.
✥ R. 'Mme Grégoire Staechelin', syn. R. 'Spanish Beauty', illus. p.163.
✥ R. 'Mme Hardy' illus. p.150.
R. 'Mme Hébert'. See R. 'Président de Sèze'.
✥ R. 'Mme Isaac Pereire' illus. p.153.
R. 'Mme Knorr'. See R. 'Comte de Chambord'.
R. 'Mme Pierre Oger'. Lax Bourbon rose. H 2m (6ft), S 1.2m (4ft). In summer-autumn, slender stems carry sweetly scented, cupped or bowl-shaped, double, pink flowers, 8cm (3in) across, with rose-lilac tints. Has light green leaves. Grows well on a pillar.
R. 'Magic Carousel'. Bushy, miniature bush rose. H 40cm (16in), S 30cm (12in). Slightly scented, rosette, fully double, yellow-and-red flowers, 4cm (1^1/$_2$in) across, with petals arranged in diminishing circles, appear in summer-autumn. Has small, glossy leaves.
✥ R. 'Maigold' illus. p.164.
R. 'Mainzer Fastnacht'. See R. 'Blue Moon'.
R. 'Malcolm Sargent', syn. R. 'Harwharry'. Shrubby, large-flowered bush rose. H 1.1m (3^1/$_2$ft), S 1m (3ft). In summer-autumn, slightly scented, rounded, double, bright scarlet-crimson flowers, 9cm (3^1/$_2$in) across, are borne singly or in open sprays. Produces an abundance of glossy, dark green foliage.
R. 'Maréchal Davoust'. Vigorous, bushy Moss rose. H 1.5m (5ft), S 1.2m (4ft). In summer bears moderately fragrant, cupped, fully double, deep reddish-pink to purple flowers, 10cm (4in) across, with a green eye and brownish 'mossing'. Leaves are dull green and lance-shaped.
R. 'Maréchal Niel'. Vigorous, spreading Noisette or Tea climbing rose. H 3m (10ft), S 2m (6ft). Drooping stems carry rich green foliage and moderately scented, pointed, fully double, clear yellow flowers, 10cm (4in) across, in summer-autumn.
✥ R. 'Margaret Merril' illus. p.155.
✥ R. 'Marguerite Hilling', syn. R. 'Pink Nevada', illus. p.152.
R. 'Mary Rose', syn. R. 'Ausmary'. Bushy, spreading shrub rose. H and S 1.2m (4ft). Produces moderately fragrant, cupped, fully double, rose-pink flowers, 9cm (3^1/$_2$in) across, in summer-autumn. Has plentiful leaves.
R. 'Matangi', syn. R. 'Macman'. Dense, bushy, cluster-flowered bush rose. H 1m (3ft), S 60cm (2ft). Bears sprays of slightly scented, open cupped, fully double flowers, 9cm (3^1/$_2$in) across, in summer-autumn; they are orange-red with yellowish-white at the base and on reverse of petals. Has plentiful, dark green leaves.
R. 'Meidanover'. See R. 'Colibri '79'.
R. 'Meidonomac'. See R. 'Bonica '82'.
R. 'Meigekanu'. See R. 'La Sevillana'.
R. 'Meijidiro'. See R. 'Pink Sunblaze'.
R. 'Meijikitar'. See R. 'Orange Sunblaze'.
✥ R. 'Mermaid' illus. p.164.
R. 'Mignon'. See R. 'Cécile Brunner'.
R. 'Mischief'. Upright, large-flowered bush rose. H 1m (3ft), S 60cm (2ft). Moderately fragrant, urn-shaped, double, salmon-pink flowers, 10cm (4in) across, are borne freely in summer-autumn. Leaves are plentiful but prone to rust.
✥ R. 'Morning Jewel'. Free-branching climbing rose. H 2.5m (8ft), S 2.2m (7ft). Cupped, double, bright pink flowers, 9cm (3^1/$_2$in) across, are freely borne, usually in clusters, in summer-autumn. Has plentiful, glossy foliage. May be pruned to grow as a shrub.
✥ R. 'Mountbatten', syn. R. 'Harmantelle', illus. p.159.
R. 'Mousseline', syn. R. 'Alfred de Dalmas'. Bushy Moss rose with twiggy growth. H and S 1m (3ft). Mainly in summer bears pleasantly scented, cupped, fully double, blush-pink

flowers, 8cm (3in) across, with little 'mossing'. Has matt green leaves.

❦ *R. moyesii.* Vigorous, arching species rose. H 4m (12ft), S 3m (10ft). In summer, faintly scented, flat, single, dusky-scarlet flowers, 5cm (2in) across, with yellow stamens, are borne close to branches. Produces long, red hips in autumn. Sparse, small, dark green leaves are composed of 7–13 leaflets. **'Geranium'** see *R.* 'Geranium'.

R. **'Mrs John Laing'** illus. p.153.

R. **'Münchner Kindl'.** See *R.* 'Anisley Dickson'.

R. **'National Trust'.** Compact, large-flowered bush rose. H 75cm (30in), S 60cm (24in). Slightly scented, urn-shaped, fully double, scarlet-crimson flowers, 10cm (4in) across, are borne freely in summer-autumn. Produces plentiful, dark green foliage. Makes a good, low hedge.

❦ *R.* **'Nevada'** illus. p.151.

❦ *R.* **'New Dawn'**, syn. *R.* 'Everblooming Dr W. van Fleet', illus. p.163.

R. **'News'.** Upright, cluster-flowered bush rose. H 60cm (24in), S 50cm (20in). Slightly fragrant, cupped, wide-opening, double, bright reddish-purple flowers, 8cm (3in) across, are borne in clusters in summer-autumn. Has dark green leaves.

R. **'Niphetos'.** Branching, climbing Tea rose. H 3m (10ft), S 2m (6ft). Long, pointed buds on nodding stems open to rounded, double, white flowers, 12cm (5in) across, mainly in summer, a few later. Has pointed, pale green leaves.

❦ *R.* **'Nozomi'**, syn. *R.* 'Heideröslein', illus. p.155.

R. **'Nuits de Young'**, syn. *R.* 'Old Black'. Erect Moss rose with wiry stems. H 1.2m (4ft), S 1m (3ft). In summer bears slightly scented, flat, double, dark maroon-purple flowers, 5cm (2in) across, with brownish 'mossing'. Leaves are small and dark green.

R. **'Ocarina'.** See *R.* 'Angela Rippon'.
R. **'Ocaru'.** See *R.* 'Angela Rippon'.
R. x *odorata* **'Mutabilis'.** See *R. chinensis* 'Mutabilis'.
R. officinalis. See *R. gallica* var. *officinalis*.
R. **'Old Black'.** See *R.* 'Nuits de Young'.
R. **'Old Blush China'**, syn. *R.* 'Parson's Pink China', illus. p.153.
R. **'Old Pink Moss'.** See *R.* x *centifolia* 'Muscosa'.

R. **'Omar Khayyám'.** Dense, prickly Damask rose. H and S 1m (3ft). Fragrant, quartered-rosette, fully double, light pink flowers, 8cm (3in) across, are borne amid downy, greyish foliage in summer.

R. **'Opa Potschke'.** See *R.* 'Precious Platinum'.

R. **'Ophelia'.** Upright, open, large-flowered bush rose. H 1m (3ft), S 60cm (2ft). In summer-autumn produces sweetly fragrant, urn-shaped, double, creamy-blush-pink flowers, 10cm (4in) across, singly or in clusters. Dark green foliage is sparse.

R. **'Orange Sunblaze'**, syn. *R.* 'Meijikitar', *R.* 'Sunblaze', illus. p.162.

R. **'Panachée d'Angers'.** See *R.* 'Commandant Beaurepaire'.

R. **'Parson's Pink China'.** See *R.* 'Old Blush China'.

R. **'Pascali'.** Upright, large-flowered bush rose. H 1m (3ft), S 60cm (2ft). Faintly scented, neat, urn-shaped, fully double, white flowers, 9cm (3½in) across, are borne in summer-autumn. Has deep green leaves.

❦ *R.* **'Paul Shirville'**, syn. *R.* 'Harqueterwife', *R.* 'Heartthrob', illus. p.157.

❦ *R.* **'Paul Transon'.** Vigorous, rather lax rambler rose. H 4m (12ft), S 1.5m (5ft). In summer bears slightly fragrant, flat, double, faintly coppery, salmon-pink flowers, 8cm (3in) across, with pleated petals. Plentiful foliage is glossy, dark green.

R. **'Paul's Himalayan Musk'.** See *R.* 'Paul's Himalayan Musk Rambler'.

❦ *R.* **'Paul's Himalayan Musk Rambler'**, syn. *R.* 'Paul's Himalayan Musk', *R.* 'Paul's Himalayan Rambler'. Very vigorous climbing rose. H and S 10m (30ft). Slightly fragrant, rosette, double, blush-pink flowers, 4cm (1½in) across, are borne in late summer, freely in large clusters. Has thorny, trailing shoots and drooping leaves. Is suitable for growing up a tree or in a wild garden.

R. **'Paul's Himalayan Rambler'.** See *R.* 'Paul's Himalayan Musk Rambler'.

R. **'Paul's Lemon Pillar'** illus. p.162.

❦ *R.* **'Peace'**, syn. *R.* 'Gioia', *R.* 'Gloria Dei', *R.* 'Mme A. Meilland', illus. p.158.

R. **'Pearl Drift'**, syn. *R.* 'Leggab', illus. p.151.

❦ *R.* **'Peaudouce'**, syn. *R.* 'Dicjana', *R.* 'Elina'. Vigorous, shrubby, large-flowered bush rose. H 1.1m (3½ft), S 75cm (2½ft). Lightly scented, rounded, fully double, ivory-white flowers, 15cm (6in) across, with lemon-yellow centres, are borne freely in summer-autumn. Has abundant, reddish foliage.

R. **'Peek-a-boo'**, syn. *R.* 'Brass Ring', *R.* 'Dicgrow', illus. p.156.

❦ *R.* **'Penelope'** illus. p.150.

❦ *R.* **'Perle d'Or'.** China rose that forms a twiggy, leafy, small shrub. H 75cm (2½ft), S 60cm (2ft). Small, slightly scented, urn-shaped, fully double, honey-pink flowers, 4cm (1½in) across, are borne in summer-autumn. Leaves have pointed, glossy leaflets.

R. **'Persian Yellow'.** See *R. foetida* 'Persiana'.

R. **'Petit Four'**, syn. *R.* 'Interfour'. Compact, dwarf cluster-flowered bush rose. H and S 40cm (16in). In summer-autumn produces dense clusters of slightly fragrant, flat, wide-opening, semi-double, pink-and-white flowers, 4cm (1½in) across. Has plentiful, mid-green foliage.

R. **'Piccadilly'** illus. p.160.

R. pimpinellifolia, syn. *R. spinosissima* (Burnet rose, Scotch rose). **'Plena'** illus. p.150.

R. **'Pink Bells'**, syn. *R.* 'Poulbells', illus. p.156.

❦ *R.* **'Pink Grootendorst'** illus. p.152.

R. **'Pink Nevada'.** See *R.* 'Marguerite Hilling'.

R. **'Pink Parfait'.** Bushy, cluster-flowered bush rose. H 75cm (2½ft), S 60cm (2ft). In summer-autumn, slightly fragrant, urn-shaped, double flowers, 9cm (3½in) across, in shades of light pink, are produced freely. Has plentiful foliage.

R. **'Pink Perpétue'** illus. p.163.

R. **'Pink Sunblaze'**, syn. *R.* 'Meijidiro'. Neat, compact, miniature bush rose. H and S 30cm (12in). Amid plentiful, bronze-green foliage, rosette, fully double, salmon-pink flowers, 4cm (1½in) across, are borne freely in summer-autumn.

R. **'Pompon des Princes'.** See *R.* 'Ispahan'.

R. **'Pot o' Gold'**, syn. *R.* 'Dicdivine', illus. p.160.

R. **'Poulbells'.** See *R.* 'Pink Bells'.

R. **'Precious Platinum'**, syn. *R.* 'Opa Potschke', illus. p.158.

❦ *R.* **'Président de Sèze'**, syn. *R.* 'Mme Hébert'. Vigorous, rather open Gallica rose. H and S 1.2m (4ft). Fragrant, quartered-rosette, fully double, magenta-pink to pale lilac-pink flowers, 10cm (4in) across, appear in summer.

R. **'Prima'**, syn. *R.* 'Harwanted'. Bushy, spreading shrub rose. H 1m (3ft), S 1.2m (4ft). In summer-autumn, lightly fragrant, cupped, double, blush-pink flowers, 10cm (4in) across, are produced freely among abundant, glossy, dark green leaves.

❦ *R. primula* illus. p.154.

R. **'Princess Alice'**, syn. *R.* 'Hartanna', *R.* 'Zonta Rose'. Upright, cluster-flowered bush rose. H 1.1m (3½ft), S 60cm (2ft). Slightly scented, rounded, double, yellow flowers, 6cm (2½in) across, appear in many-flowered sprays during late summer and autumn. Has healthy foliage.

R. **'Princess Michael of Kent'**, syn. *R.* 'Harlightly'. Neat, compact, cluster-flowered bush rose. H 60cm (24in), S 50cm (20in). In summer-autumn produces pleasantly scented, rounded, fully double, yellow flowers, 9cm (3½in) across, singly or in clusters. Leaves are glossy, bright green and very healthy. Makes a good, low hedge.

❦ *R.* **'Queen Elizabeth'** illus. p.156.

R. **'Queen of the Violets'.** See *R.* 'Reine des Violettes'.

R. **'Ramona'**, syn. *R.* 'Red Cherokee'. Rather stiff, open climbing rose. H 2.7m (9ft), S 3m (10ft). Fragrant, flat, single, carmine-red flowers, 10cm (4in) across, with a greyish-red reverse and gold stamens, appear mainly in summer, a few later. Has sparse foliage. Does best against a warm wall.

R. **'Red Ace'**, syn. *R.* 'Amruda', illus. p.161.

R. **'Red Cherokee'.** See *R.* 'Ramona'.
R. **'Red Moss'.** See *R.* 'Henri Martin'.
R. **'Red Rose of Lancaster'.** See *R. gallica* var. *officinalis*.

R. **'Reine des Violettes'**, syn. *R.* 'Queen of the Violets'. Spreading, vigorous Hybrid Perpetual rose. H and S 2m (6ft). Has greyish-toned leaves and, in summer-autumn, bears fragrant, quartered-rosette, fully double, violet to purple flowers, 8cm (3in) across. May be grown on a support.

R. **'Reine Victoria'** illus. p.152.

R. **'Remember Me'**, syn. *R.* 'Cocdestin', illus. p.160.

R. **'Rise 'n Shine'**, syn. *R.* 'Golden Sunblaze', illus. p.162.

R. **'Robert le Diable'.** Lax, bushy Provence rose. H and S 1m (3ft). In summer bears slightly scented, pompon, double flowers, 8cm (3in) across, in mixed bright and dull purple. Has narrowly oval, dark green leaves. Does best trailing over a low support.

R. **'Rose d'Isfahan'.** See *R.* 'Ispahan'.

R. **'Rose Gaujard'.** Upright, strong, large-flowered bush rose. H 1.1m (3½ft), S 75cm (2½ft). Slightly scented, urn-shaped, double, cherry-red and blush-pink flowers, 10cm (4in) across, are borne freely in summer-autumn. Foliage is glossy and plentiful.

R. **'Rosemary Harkness'**, syn. *R.* 'Harrowbond', illus. p.156.

❦ *R.* **'Roseraie de l'Haÿ'** illus. p.153.

❦ *R.* **'Rosy Cushion'**, syn. *R.* 'Interall', illus. p.152.

R. **'Rosy Mantle'** illus. p.163.

R. **'Rouge Eblouissante'.** See *R.* 'Assemblage des Beautés'.

R. **'Royal Dane'.** See *R.* 'Troika'.

R. **'Royal Highness'**, syn. *R.* 'Königliche Hoheit'. Upright, large-flowered bush rose. H 1.1m (3½ft), S 60cm (2ft). In summer-autumn, leathery, dark green foliage sets off large, fragrant, pointed, fully double, pearl-pink flowers, 12cm (5in) across, borne on firm stems.

❦ *R.* **'Royal William'**, syn. *R.* 'Duftzauber '84', *R.* 'Korzaun', illus. p.158.

R. rubiginosa. See *R. eglanteria*.
R. rubrifolia. See *R. glauca*.

R. rugosa (illus. p.153). ❦ **'Alba'** is a dense, vigorous, species hybrid rose. H and S 1–2m (3–6ft). Bears a succession of fragrant, cupped, single, white flowers, 9cm (3½in) across, in summer-autumn. They are followed by large, tomato-shaped hips. Abundant foliage is leathery, wrinkled and glossy.

R. **'Rugul'**, syn. *R.* 'Guletta', *R.* 'Tapis Jaune', illus. p.158.

R. **'Ruiblun'.** See *R.* 'Blue Peter'.

R. **'St Nicholas'.** Vigorous, erect Damask rose. H and S 1.2m (4ft). In summer produces lightly scented, cupped, semi-double, rose-pink flowers, 12cm (5in) across, with golden stamens. These are followed by red hips in autumn. Has plentiful, dark green foliage.

❦ *R.* **'Sander's White Rambler'.** Vigorous rambler rose of lax growth. H 3m (10ft), S 2.5m (8ft). Fragrant, rosette, fully double, white flowers, 5cm (2in) across, appear in clusters during late summer. Small, glossy leaves are plentiful.

R. **'San Rafael Rose'.** See *R.* 'Fortune's Double Yellow'.

R. **'Schneewittchen'.** See *R.* 'Iceberg'.

R. **'Schoolgirl'.** Stiff, rather lanky, large-flowered climbing rose. H 2.7m (9ft), S 2.2m (7ft). Large, deep green leaves set off moderately fragrant, rounded, fully double, apricot-orange flowers, 10cm (4in) across, borne in summer-autumn.

❦ *R.* **'Sexy Rexy'**, syn. *R.* 'Macrexy', illus. p.156.

R. **'Sheila Macqueen'**, syn. *R.* 'Harwotnext'. Narrow, upright,

cluster-flowered bush rose. H 75cm (30in), S 45cm (18in). In summer-autumn produces dense clusters of peppery-scented, cupped, double, green flowers, 6cm (2½in) across, tinged apricot-pink. Has leathery leaves.
R. 'Sheila's Perfume', syn.
R. 'Harsherry'. Upright, cluster-flowered bush rose. H 75cm (2½ft), S 60cm (2ft). Has glossy, reddish foliage. Fragrant, urn-shaped, double, red-and-yellow flowers, 9cm (3½in) across, appear singly or in clusters in summer-autumn.
R. 'Sheri Anne' illus. p.161.
R. 'Shocking Blue', syn. R. 'Korblue'. Bushy, cluster-flowered bush rose. H 75cm (2½ft), S 60cm (2ft). In summer-autumn bears fragrant, pointed, well-formed, fully double, purple flowers, 10cm (4in) across, singly or in clusters. Foliage is dark green.
♚ R. 'Silver Jubilee' illus. p.157.
R. 'Simba', syn. R. 'Goldsmith', R. 'Korbelma', illus. p.159.
R. 'Sissi'. See R. 'Blue Moon'.
R. 'Snowball', syn. R. 'Angelita', R. 'Macangel', illus. p.161.
♚ R. 'Snow Carpet', syn.
R. 'Maccarpe'. Prostrate, creeping, miniature bush rose. H 15cm (6in), S 50cm (20in). Pompon, fully double, white flowers, 3cm (1¼in) across, appear mainly in summer, a few in autumn. Plentiful leaves are small and glossy. Makes good, compact ground cover.
♚ R. 'Southampton', syn. R. 'Susan Ann', illus. p.160.
R. 'Souvenir d'Alphonse Lavallée'. Vigorous, sprawling Hybrid Perpetual rose. H 2.2m (7ft), S 2m (6ft). Fragrant, cupped, double, burgundy-red to maroon-purple flowers, 10cm (4in) across, are borne in summer-autumn. Leaves are small and mid-green. Is best grown on a light support.
R. 'Souvenir de la Malmaison'. Dense, spreading Bourbon rose. H and S 1.5m (5ft). Bears spice-scented, quartered-rosette, fully double, blush-pink to white flowers, 12cm (5in) across, in summer-autumn. Rain spoils flowers. Has large, dark green leaves.
R. 'Spanish Beauty'. See
R. 'Mme Grégoire Staechelin'.
R. 'Spectacular'. See R. 'Danse du Feu'.
R. spinosissima. See
R. pimpinellifolia.
R. 'Spring Morning'. See
R. 'Frühlingsmorgen'.
♚ R. 'Stacey Sue' illus. p.161.
R. 'Sue Lawley', syn. R. 'Macspash'. Open, cluster-flowered bush rose. H 75cm (2½ft), S 60cm (2ft). Bears sprays of faintly fragrant, cupped, wide-opening, double, pink-and-white flowers, 9cm (3½in) across, in summer-autumn. Leaves are dark green.
R. 'Sunblaze'. See R. 'Orange Sunblaze'.
R. 'Susan Ann'. See R. 'Southampton'.
♚ R. 'Sweet Dream', syn.
R. 'Fryminicot'. Compact, dwarf cluster-flowered bush rose. H 40cm (16in), S 35cm (14in). Clusters of slightly scented, pompon, fully double, peach-apricot flowers, 6cm (2½in) across, appear in summer-autumn. Leaves are small.

♚ R. 'Sweet Magic', syn.
R. 'Dicmagic', illus. p.160.
R. 'Sylvia'. See R. 'Congratulations'.
R. 'Sympathie'. Vigorous, freely branching climbing rose. H 3m (10ft), S 2.5m (8ft). Slightly scented, cupped, fully double, bright deep red flowers, 8cm (3in) across, are borne in summer-autumn, usually in clusters. Has plentiful, glossy, dark green foliage.
R. 'Taifun'. See R. 'Typhoon'.
R. 'Tanky'. See R. 'Whisky Mac'.
R. 'Tapis d'Orient'. See
R. 'Yesterday'.
R. 'Tapis Jaune'. See R. 'Rugul'.
♚ R. 'The Fairy' illus. p.155.
♚ R. 'The Times Rose', syn.
R. 'Korpeahn', illus. p.158.
R. 'Tour de Malakoff' illus. p.154.
R. 'Tricolore de Flandres'. Vigorous, upright Gallica rose. H and S 1m (3ft). Fragrant, pompon, fully double, blush-pink flowers, 6cm (2½in) across, striped with pink and purple, open in summer. Has dull green leaves.
♚ R. 'Troika', syn. R. 'Royal Dane', illus. p.160.
♚ R. 'Trumpeter', syn. R. 'Mactru', illus. p.157.
♚ R. 'Tuscany Superb', syn.
R. 'Double Velvet'. Vigorous, upright Gallica rose. H 1.1m (3½ft), S 1m (3ft). In summer, slightly scented, cupped to flat, double flowers, 5cm (2in) across, are deep crimson-maroon, ageing to purple, with gold stamens. Leaves are dark green.
R. 'Typhoon', syn. R. 'Taifun'. Spreading, large-flowered bush rose. H and S 75cm (30in). Has plentiful, burnished, dark green foliage. Fragrant, rounded, fully double, salmon- to orange-pink flowers, 10cm (4in) across, are borne freely in summer-autumn.
R. 'Variegata di Bologna'. Upright, arching Bourbon rose. H 2m (6ft), S 1.4m (4½ft). In summer-autumn bears fragrant, quartered-rosette, fully double flowers, 8cm (3in) across, blush-pink, striped with rose-purple. Leaves are small. Needs fertile soil and is prone to blackspot.
♚ R. 'Veilchenblau', syn. R. 'Blue Rambler', illus. p.164.
R. 'Wedding Day'. Rampant climbing rose. H 8m (25ft), S 4m (12ft). Large clusters of fruity-scented, flat, single, creamy-white flowers, 2.5cm (1in) across, ageing to blush-pink, are borne in late summer. Is ideal for growing up a tree or in a wild garden.
R. 'Wee Jock', syn. R. 'Cocabest', illus. p.158.
R. 'Whisky Mac', syn. R. 'Tanky'. Neat, upright, large-flowered bush rose. H 75cm (2½ft), S 60cm (2ft). Fragrant, rounded, fully double, amber flowers, 9cm (3½in) across, appear freely in summer-autumn. Reddish foliage is prone to mildew. May die back during a hard winter.
♚ R. 'White Cockade'. Slow-growing, bushy, upright, large-flowered climber. H 2–3m (6–10ft), S 1.5m (5ft). Bears slightly fragrant, rounded, well-formed, fully double, white flowers, 9cm (3½in) across, in summer-autumn. May be pruned and grown as a shrub.
♚ R. 'William Lobb', syn.

R. 'Duchesse d'Istrie', illus. p.154.
♚ R. 'Yesterday', syn. R. 'Tapis d'Orient'. Bushy, arching Polyantha bush rose. H and S 75cm (30in), or more if lightly pruned. Fragrant, rosette, semi-double, lilac-pink flowers, 2.5cm (1in) across, are borne, mainly in clusters, from summer to early winter. Has small, dark green leaves. Makes a good hedge.
♚ R. 'Yvonne Rabier'. Dense, bushy Polyantha bush rose. H 45cm (18in), S 40cm (16in). In summer-autumn, moderately scented, rounded, double, creamy-white flowers, 5cm (2in) across, are borne amid plentiful, bright green foliage.
♚ R. 'Zéphirine Drouhin' illus. p.163.
♚ R. 'Zigeunerknabe', syn. R. 'Gipsy Boy'. Vigorous, thorny Bourbon rose of lanky habit. H and S 2m (6ft). Faintly scented, cupped to flat, double, purplish-crimson flowers, 8cm (3in) across, are borne in summer. Leaves are dark green.
R. 'Zonta Rose'. See R. 'Princess Alice'.

ROSCOEA (Zingiberaceae)
Genus of late summer- and early autumn-flowering, tuberous perennials, related to ginger, grown for their orchid-like flowers. Is suited to open borders, rock gardens and woodland gardens. Frost hardy. Grows in sun or partial shade and in cool, humus-rich soil that should always be kept moist in summer. Dies down in winter, when a top dressing of leaf mould or well-rotted compost is beneficial. Propagate by division in spring or by seed in autumn or winter; expose seed to frost for best germination.
♚ R. cautleoides illus. p.380.
♚ R. humeana illus. p.380.
R. procera. See R. purpurea.
R. purpurea, syn. R. procera. Summer-flowering, tuberous perennial. H 20–30cm (8–12in), S 15–20cm (6–8in). Lance-shaped, erect leaves are long-pointed and wrap around each other at base to form a false stem. Has long-tubed, purple flowers, each with a hooded, upper petal, wide-lobed, lower lip and 2 narrower petals.

ROSMARINUS (Labiatae)
Genus of evergreen shrubs, grown for their flowers and aromatic foliage, which can be used as a culinary herb. Frost hardy, but in cold areas grow against a south- or west-facing wall. Requires sun and well-drained soil. Cut back frost-damaged plants to healthy wood in spring; straggly, old plants may be cut back hard at same time. Trim hedges after flowering. Propagate by semi-ripe cuttings in summer.
R. lavandulaceus of gardens. See R. officinalis 'Prostratus'.
R. officinalis (Rosemary) illus. p.137.
♚ 'Miss Jessop's Upright' is an evergreen, compact, upright shrub. H and S 2m (6ft). From mid- to late spring and sometimes again in autumn bears small, 2-lipped, blue flowers amid narrowly oblong, aromatic, dark green leaves. Is good for hedging.
♚ 'Prostratus' (syn. R. lavandulaceus of gardens), H 15cm (6in), is prostrate and the least hardy form. ♚ 'Severn

Sea', H 1m (3ft), is arching, with bright blue flowers.

ROSSIOGLOSSUM. See ORCHIDS.
R. grande, syn. Odontoglossum grande (illus. p.261). Evergreen, epiphytic orchid for a cool greenhouse. H 15cm (6in). Spikes of rich yellow flowers, to 15cm (6in) across and heavily marked chestnut-brown, are produced in autumn. Has broadly oval, stiff leaves, 15cm (6in) long. Provide shade in summer and keep very dry in winter.

ROTHMANNIA (Rubiaceae)
Genus of evergreen, summer-flowering shrubs and trees, grown for their flowers. Is related to Gardenia. Frost tender, min. 16°C (61°F). Needs full light or partial shade and humus-rich, well-drained, neutral to acid soil. Water potted plants freely when in full growth, moderately at other times. Propagate by seed in spring or by semi-ripe cuttings in summer.
R. capensis, syn. Gardenia capensis, G. rothmannia. Evergreen, ovoid shrub or tree. H 6m (20ft) or more, S 3m (10ft) or more. Leaves are oval, lustrous and rich green. Has fragrant, tubular flowers, each with 5 arching, white to creamy-yellow petal lobes and a purple-dotted throat, in summer.

ROYSTONEA (Palmae)
Royal palm
Genus of evergreen palms, grown for their majestic appearance. Has racemes of insignificant flowers in summer. Frost tender, min. 16–18°C (61–64°F). Needs full light or partial shade and fertile, well-drained but moisture-retentive soil. Water containerized plants freely when in full growth, less at other times, especially when temperatures are low. Propagate by seed in spring at not less than 27°C (81°F). Red spider mite may be troublesome.
R. regia (Cuban royal palm). Evergreen palm with an upright stem, sometimes thickened about the middle. H 20m (70ft) or more, S to 6m (20ft). Feather-shaped leaves, 3m (10ft) long, upright at first, then arching and pendent, are divided into narrowly oblong, leathery, bright green leaflets.

RUBUS (Rosaceae)
Blackberry, Bramble
Genus of deciduous, semi-evergreen or evergreen shrubs and woody-stemmed, scrambling climbers. Some species are cultivated solely for their edible fruits, which include raspberries and blackberries. Those described here are grown mainly for their foliage, flowers or ornamental, often prickly stems, though some may also bear edible fruits. Fully to frost hardy. Deciduous species grown for their winter stems prefer full sun; other deciduous species need sun or semi-shade; evergreens and semi-evergreens tolerate sun or shade. All Rubus require fertile, well-drained soil. Cut old stems of R. biflorus, R. cockburnianus and R. thibetanus to ground after fruiting. Propagate by seed or cuttings (semi-ripe for evergreens, softwood or hardwood for deciduous species) in summer or winter.

Alternatively, *R. odoratus* may be increased by division and *R.* 'Benenden' and *R. ulmifolius* 'Bellidiflorus' by layering in spring.
♚ *R.* **'Benenden'**, syn. *R.* 'Tridel', illus. p.106.
R. biflorus illus. p.119.
♚ *R. cockburnianus*. Deciduous, arching shrub. H and S 2.5m (8ft). Fully hardy. Prickly shoots are brilliant blue-white in winter. Dark green leaves, white beneath, each have usually 9 oval leaflets. Bears panicles of 5-petalled, purple flowers in early summer, then unpalatable, spherical, black fruits.
R. henryi var. ***bambusarum***. Fast-growing, vigorous, evergreen, woody-stemmed, scrambling climber, grown mainly for its attractive foliage. H to 6m (20ft). Fully hardy. Leaves have 3 broadly oval leaflets, white-felted beneath. Tiny, pink flowers are borne in small clusters in summer.
R. odoratus (Flowering raspberry, Thimbleberry). Vigorous, deciduous, upright, thicket-forming shrub. H and S 2.5m (8ft). Fully hardy. Thornless, peeling shoots bear large, velvety, dark green leaves, each with 5 broadly triangular lobes. Large, fragrant, 5-petalled, rose-pink flowers appear from early summer to early autumn, and are sometimes followed by unpalatable, flattened, red fruits.
♚ *R. thibetanus* illus. p.119.
R. tricolor. Evergreen shrub with both prostrate and arching shoots covered in red bristles. H 60cm (2ft), S 2m (6ft). Fully hardy. Oval, toothed, glossy, dark green leaves set off cup-shaped, 5-petalled, white flowers borne in mid-summer. Has edible, raspberry-like, red fruits. Makes good ground cover.
R. **'Tridel'**. See *R.* 'Benenden'.
R. ulmifolius **'Bellidiflorus'**. Vigorous, deciduous or semi-evergreen, arching shrub. H 2.5m (8ft),
S 4m (12ft). Fully hardy. Has prickly stems and dark green leaves, each with 3 or 5 oval leaflets. Produces large panicles of daisy-like, double, pink flowers from mid- to late summer.

RUDBECKIA (Compositae)
Coneflower
Genus of annuals, biennials and perennials grown for their flowers. Fully hardy. Thrives in sun or shade and well-drained or moist soil. Propagate by division in spring or by seed in autumn or spring.
R. fulgida (Black-eyed Susan).
♚ var. ***deamii*** is an erect perennial. H 1m (3ft), S 60cm (2ft) or more. In late summer and autumn produces daisy-like, yellow flower heads with central, black cones. Has narrowly lance-shaped, mid-green leaves. Prefers moist soil. ♚ var. ***sullivantii*** **'Goldsturm'** illus. p.220.
♚ *R.* **'Goldquelle'** illus. p.193.
R. **'Herbstsonne'** illus. p.196.
R. hirta. Moderately fast-growing, upright, branching, short-lived perennial, grown as an annual. H 30cm–1m (1–3ft), S 30–45cm (1–1½ft). Has lance-shaped, mid-green leaves. Large, daisy-like, deep yellow flower heads, with conical, purple centres, appear in summer-autumn. Prefers sun and well-drained soil. **'Goldilocks'** illus. p.292. **'Irish Eyes'**, H to 75cm (2½ft), has yellow flower heads with conical, olive-green centres. **'Marmalade'** and **'Rustic Dwarfs'** illus. p.292.
R. laciniata **'Golden Glow'**. Erect perennial. H 2–2.2m (6–7ft), S 60cm–1m (2–3ft). Bears daisy-like, double, golden-yellow flower heads, with green centres, in late summer and autumn. Mid-green leaves are divided into lance-shaped leaflets, themselves further cut. Prefers well-drained soil.
R. purpurea. See *Echinacea purpurea*.

RUELLIA (Acanthaceae)
Genus of perennials and evergreen sub-shrubs and shrubs with showy flowers. Frost tender, min. 15°C (59°F). Grow in a humid atmosphere, partial shade and in moist but well-drained soil. Propagate by stem cuttings or seed, if available, in spring.
R. amoena. See *R. graecizans*.
R. devosiana illus. p.241.
R. graecizans, syn. *R. amoena*, illus. p.213.

RUSCHIA (Aizoaceae)
Genus of mostly small, tufted, perennial succulents and evergreen shrubs with leaves united up to one-third of their lengths around stems or with very short sheaths. Frost tender, min. 5°C (41°F). Needs sun and well-drained soil. Propagate by seed or stem cuttings in spring or summer.
R. acuminata. Evergreen, erect, succulent shrub. H 20cm (8in), S 50cm (20in). Has woody stems as well as non-woody, bluish-green stems with darker dots. Produces solid, 3-angled, 3cm (1¼in) long leaves, each with a blunt keel and a short sheath. Daisy-like, white to pale pink flowers, 3cm (1¼in) across, appear in summer.
R. crassa. Evergreen, erect, succulent shrub. H and S 50cm (20in). Bears solid, 3-angled, bluish-green leaves, 2cm (¾in) long, with short, white hairs; the undersides are keeled, each with a single tooth. Has 2.5cm (1in) wide, daisy-like, white flowers in summer.
R. macowanii. Erect, then spreading, perennial succulent. H 15cm (6in), S 1m (3ft). Has solid, slightly keeled, 3-angled, bluish-green leaves, to 3cm (1¼in) long. In summer carries masses of daisy-like, bright pink flowers, 3cm (1¼in) across, with darker stripes.

RUSCUS (Liliaceae)
Genus of evergreen, clump-forming, spring-flowering shrubs, grown for their foliage and fruits. The apparent leaves are actually flattened shoots, on which flowers and fruits are borne. Usually, separate male and female plants are required for fruiting. Is particularly useful for dry, shady sites. Fully to frost hardy. Tolerates sun or shade and any other than waterlogged soil. Cut dead shoots to base in spring. Propagate by division in spring.
R. aculeatus (Butcher's broom). Evergreen, erect, thicket-forming shrub. H 75cm (2½ft), S 1m (3ft). Fully hardy. Spine-tipped 'leaves' are glossy and dark green. Tiny, star-shaped, green flowers in spring are succeeded by large, spherical, bright red fruits.
R. hypoglossum illus. p.146.

RUSSELIA (Scrophulariaceae)
Genus of evergreen shrubs and sub-shrubs with showy flowers. Frost tender, min. 10–15°C (50–59°F). Grow in sun or partial shade. Needs humus-rich, light, well-drained soil. Propagate by stem cuttings or division in spring.
♚ *R. equisetiformis*, syn. *R. juncea*, illus. p.214.
R. juncea. See *R. equisetiformis*.

RUTA (Rutaceae)
Rue
Genus of evergreen, summer-flowering sub-shrubs, with deeply divided, aromatic leaves, grown for their foliage and flowers and sometimes used as a medicinal herb. Fully hardy. Requires sun and well-drained soil. Cut back to old wood in spring to keep compact. Propagate by semi-ripe cuttings in summer.
R. graveolens (Common rue).
♚ **'Jackman's Blue'** illus. p.147.

S

SABAL (Palmae)
Genus of evergreen, tree-sized and dwarf fan palms, grown for their foliage and overall appearance. Half hardy to frost tender, min. 5–7°C (41–5°F). Prefers full sun and fertile, well-drained soil. Water moderately, less when not in full growth. Propagate by seed in spring. Red spider mite may be troublesome.
S. minor illus. p.146.

SAGINA (Caryophyllaceae)
Genus of mat-forming annuals and evergreen perennials, grown for their foliage. Is suitable for banks and in paving. Some species may be very invasive. Fully hardy. Prefers sun and gritty, moist soil; dislikes hot, dry conditions. Propagate by division in spring, by seed in autumn or, for *S. boydii*, by tip cuttings in summer. Aphids and red spider mite may cause problems.
S. boydii illus. p.338.

SAGITTARIA (Alismataceae)
Arrowhead
Genus of deciduous, perennial, submerged and marginal water plants, grown for their foliage and flowers. Fully hardy to frost tender, min. 5°C (41°F). Some species are suitable for pools, others for aquariums. All require full sun. Remove fading foliage as necessary. Propagate by division in spring or summer or by breaking off turions (scaly, young shoots) in spring.
S. japonica. See *S. sagittifolia* 'Flore Pleno'.
S. latifolia illus. p.386.
S. sagittifolia (Common arrowhead). Deciduous, perennial, marginal water plant. H 45cm (18in), S 30cm (12in). Fully hardy. Upright, mid-green leaves are acutely arrow-shaped. In summer produces 3-petalled, white flowers with dark purple centres. May be grown in up to 23cm (9in) depth of water. **'Flore Pleno'** (syn. *S. japonica*; Japanese arrowhead) has double flowers.

SAINTPAULIA (Gesneriaceae)
African violet
Genus of evergreen, rosette-forming perennials, grown for their showy flowers. Frost tender, min. 15°C (59°F). Needs a constant temperature, a humid atmosphere, partial shade and fertile soil. Propagate by leaf cuttings in summer. Whitefly and mealy bug may cause problems with plants grown indoors.
S. ionantha. Evergreen, stemless, rosette-forming perennial, often forming clumps. H to 10cm (4in), S 25cm (10in). Almost rounded, scalloped, long-stalked, fleshy, usually hairy leaves, to 8cm (3in) long, are mid-green above and often reddish-green below. Loose clusters of 2–8 tubular, 5-lobed, violet-blue flowers, to 2.5cm (1in) across, are produced on stems held above leaves and appear year-round. Numerous cultivars are available, some with variegated or quilted leaves. Flowers are in a range of colours from white, pink or red to purple or blue, plain or bicoloured and in various forms (single, semi-double or double), some with frilled or ruffled petals. ♛ **'Bright Eyes'** (illus. p.266) has dark green leaves and single, deep violet-blue flowers with yellow centres. ♛ **'Colorado'** (illus. p.266) bears dark green leaves and frilled, single, magenta flowers. Leaves of ♛ **'Delft'** (illus. p.266) are dark green, flowers are semi-double and violet-blue. **'Fancy Pants'** (illus. p.266) produces single, white flowers with frilled, red edges, held well above mid-green leaves. ♛ **'Garden News'** (illus. p.266) has bright green leaves and double, pure white flowers. **'Kristi Marie'** (illus. p.266) produces dark green leaves and semi-double, dusky-red flowers, edged with white. **'Miss Pretty'** (illus. p.266) bears pale green leaves and large, single, pink-flushed, white flowers with frilled petals. **'Pip Squeak'** (illus. p.266), H to 8cm (3in), S 10cm (4in), has oval, unscalloped, deep green leaves, 1–2cm ($^{1}/_{2}$–$^{3}/_{4}$in) long, and bell-shaped, pale pink flowers, 1cm ($^{1}/_{2}$in) wide. **'Porcelain'** (illus. p.266) has semi-double, purple-blue-edged, white flowers. Flowers of ♛ **'Rococo Pink'** (illus. p.266) are double and iridescent pink.

SALIX (Salicaceae)
Willow
Genus of deciduous trees and shrubs, grown for their habit, foliage, catkins and, in some cases, colourful winter shoots. Male catkins are more striking than female; each plant usually bears catkins of only one sex. Fully to frost hardy. Most prefer full sun. Most species grow well in any but very dry soil; *S. caprea*, *S. purpurea* and their variants also thrive in dry soil. Plants grown for their colourful winter shoots should be cut back hard in early spring, every 1–3 years. Propagate by semi-ripe cuttings in summer or by hardwood cuttings in winter. Fungal diseases may cause canker, particularly in *S. babylonica* and *S.* 'Chrysocoma'. Willows may become infested with such pests as caterpillars, aphids and gall mites.
S. aegyptiaca (Musk willow). Vigorous, deciduous, bushy shrub or tree. H 4m (12ft), S 5m (15ft). Fully hardy. Grey catkins that turn to yellow appear on bare, stout shoots in late winter or early spring, before large, narrowly oval, deep green leaves emerge.
S. alba (White willow). ♛ var. *sericea* (syn. f. *argentea*, *S.a.* 'Sericea'; Silver willow) is a fast-growing, deciduous, spreading tree, conical when young. H 15m (50ft), S 8m (25ft). Fully hardy. Has narrowly lance-shaped, bright silver-grey leaves and, in early spring, insignificant, pendent, yellowish-green catkins. ♛ **'Britzensis'** (syn. *S.a.* 'Chermesina'), H 25m (80ft), S 10m (30ft), which has green leaves and bright orange-red, young shoots, is usually cut back to near ground level to provide winter colour. **'Caerulea'** (Cricket-bat willow), H 25m (80ft), S 10m (30ft), is very fast-growing and conical, with long, narrow, deep bluish-green leaves. **'Chermesina'** see *S.a.* 'Britzensis'. f. *argentea* and **'Sericea'** see *S.a.* var. *sericea*. **'Tristis'** see *S.* 'Chrysocoma'. ♛ var. *vitellina* illus. p.48.
S. apoda illus. p.311.
S. arbuscula (Mountain willow). Deciduous, spreading shrub. H and S 60cm (2ft) or more. Fully hardy. In spring, dark brown stems produce narrowly oval, toothed leaves and white-haired, sometimes red-tinged, yellow catkins. Suits a rock garden.
S. babylonica (Weeping willow). Deciduous, weeping tree with slender, pendent shoots that reach almost to the ground. H and S 12m (40ft). Fully hardy. Bears narrowly lance-shaped, long-pointed leaves. Has yellowish-green catkins in early spring. Is susceptible to canker and has been largely replaced in cultivation by *S.* 'Chrysocoma'. var. *pekinenses* **'Tortuosa'** see *S. matsudana* 'Tortuosa'.
♛ *S.* x *boydii* illus. p.309.
S. caprea (Goat willow, Pussy willow). Deciduous, bushy shrub or tree. H 10m (30ft), S 8m (25ft). Fully hardy. Oval leaves are dark green above, grey beneath. Catkins are borne in spring before foliage emerges: females are silky and grey, males are grey with yellow anthers.
♛ **'Kilmarnock'** (Kilmarnock willow), H 1.5–2m (5–6ft), S 2m (6ft), is dense-headed and weeping. From early to mid-spring has grey, later yellow catkins.
♛ *S.* **'Chrysocoma'**, syn. *S. alba* 'Tristis', *S.* x *sepulcralis* 'Chrysocoma', illus. p.48.
S. daphnoides illus. p.48.
♛ *S. elaeagnos* (Hoary willow). Deciduous, upright, dense shrub. H 3m (10ft), S 5m (15ft). Fully hardy. In spring, long shoots bear slender, yellow catkins as leaves appear. These are narrowly oblong and dark green, with white undersides, and turn yellow in autumn.
S. fargesii, syn. *S. moupinensis* of gardens. Deciduous, upright, open shrub. H and S 3m (10ft). Fully hardy. Has purplish-red winter shoots and buds. Slender, erect, green catkins are carried in spring, at same time as bold, oblong, glossy, dark green leaves.
S. fragilis (Crack willow). Deciduous tree with a broad, bushy head. H 15m (50ft), S 12–15m (40–45ft). Fully hardy. Has long, narrow, pointed, glossy, bright green leaves. Catkins, borne in early spring, are yellow on male plants, green on females.
S. gracilistyla. Deciduous, bushy shrub. H 3m (10ft), S 4m (12ft). Fully hardy. Large, silky, grey catkins with red, then bright yellow anthers appear from early to mid-spring, and are followed by lance-shaped, silky, grey, young leaves that mature to bright, glossy green. **'Melanostachys'** see *S.* 'Melanostachys'.
♛ *S. hastata* **'Wehrhahnii'** illus. p.123.
♛ *S. helvetica* illus. p.297.
S. herbacea (Dwarf willow, Least willow). Deciduous, creeping shrub. H 2.5cm (1in), S 20cm (8in) or more. Fully hardy. Has small, rounded to oval leaves and, in spring, small, yellow or yellowish-green catkins. Is good for a rock garden. Needs moist soil.
S. hylematica of gardens. See *S. lindleyana*.
S. irrorata. Deciduous, upright shrub. H 3m (10ft), S 5m (15ft). Fully hardy. Purple, young shoots are white-bloomed in winter. Catkins with red, then yellow anthers appear from early to mid-spring before narrowly oblong, glossy, bright green leaves emerge.
♛ *S. lanata* illus. p.126. **'Stuartii'** see *S.* 'Stuartii'.
S. lindleyana, syn. *S. hylematica* of gardens. Deciduous, prostrate shrub with long, creeping stems. H 2–3cm ($^{3}/_{4}$–$1^{1}/_{4}$in), S 40cm (16in) or more. Frost hardy. Small, narrowly oval to linear, pale green leaves are densely set on short branchlets that produce brownish-pink catkins, 1cm ($^{1}/_{2}$in) long, in spring. Suits a rock garden or bank. Needs partial shade and damp soil.
♛ *S. magnifica.* Deciduous, upright shrub. H 5m (15ft), S 3m (10ft). Fully hardy. Produces very long, slender, green catkins on stout, red shoots in spring, as large, magnolia-like, blue-green leaves emerge.
♛ *S. matsudana* **'Tortuosa'**, syn. *S. babylonica* var. *pekinenses* 'Tortuosa', illus. p.59.
♛ *S.* **'Melanostachys'**, syn. *S. gracilistyla* 'Melanostachys' (Black willow). Deciduous, bushy, spreading shrub. H 3m (10ft), S 4m (12ft). Fully hardy. Bears almost black catkins, with red anthers, in early spring, before lance-shaped, bright green leaves emerge.
S. moupinensis of gardens. See *S. fargesii*.
S. pentandra (Bay willow). Deciduous, large shrub, then small tree with broad, bushy head. H and S 10m (30ft). Fully hardy. Oval, glossy, green leaves are blue-white beneath. Catkins – males bright yellow, females grey-green – open in early summer when the tree is in full leaf.
S. purpurea (Purple osier). Deciduous, bushy, spreading shrub. H and S 5m (15ft). Fully hardy. Grey, male catkins, with yellow anthers, and insignificant, female catkins are both borne on

slender, often purple shoots in spring, before narrowly oblong, deep green leaves emerge. **'Nana'** (syn. *S.p.* f. *gracilis*, *S.p.* 'Gracilis'), H and S 1.5m (5ft), is dense, with silver-grey leaves; is good as a hedge.
S. repens illus. p.126.
♣ **S. reticulata** illus. p.318.
♣ **S. × rubens 'Basfordiana'.** Deciduous, spreading tree. H 15m (50ft), S 10m (30ft). Fully hardy. Has bright orange-yellow, young shoots in winter and long, narrow leaves, grey-green when young, becoming glossy, bright green in summer. Yellowish-green catkins appear in early spring.
S. sachalinensis 'Sekka'. See *S. udensis* 'Sekka'.
S. × sepulcralis 'Chrysocoma'. See *S.* 'Chrysocoma'.
S. 'Stuartii', syn. *S. lanata* 'Stuartii'. Slow-growing, deciduous, spreading shrub. H 1m (3ft), S 2m (6ft). Fully hardy. Has yellow winter shoots. Stout, grey-green catkins open from orange buds in spring, as oval, woolly, grey leaves emerge.
S. udensis 'Sekka', syn. *S. sachalinensis* 'Sekka'. Deciduous, spreading shrub. H 5m (15ft), S 10m (30ft). Fully hardy. Has flattened shoots that are red in winter and lance-shaped, glossy, bright green leaves. Silver catkins are produced in early spring. Shoots are useful for flower arranging.

SALPIGLOSSIS (Solanaceae)
Genus of annuals and biennials. Usually only annuals are cultivated, either for colour in borders or as greenhouse plants. Half hardy. Grow in sun and in rich, well-drained soil. Stems need support. Dead-head regularly. Propagate by seed sown under glass in early spring, or in early autumn for winter flowering indoors. Aphids may be troublesome.
S. sinuata. Bolero Series is a group of moderately fast-growing, branching, upright annuals. H 60cm (2ft), S 30cm (1ft). Has lance-shaped, pale green leaves. Outward-facing, widely flared, trumpet-shaped, conspicuously veined flowers, 5cm (2in) across, appear in summer and early autumn. Is available in a mixture of rich colours, such as red, yellow, orange and blue. **'Friendship'** has upward-facing flowers in a range of colours. **'Splash'** illus. p.281.

SALVIA (Labiatae)
Sage
Genus of annuals, biennials, perennials and evergreen or semi-evergreen shrubs and sub-shrubs, grown for their tubular, 2-lipped, often brightly coloured flowers and aromatic foliage. Leaves of some species may be used for flavouring foods. Fully hardy to frost tender, min. 5°C (41°F). Needs sun and fertile, well-drained soil. Propagate perennials by division in spring, perennials, shrubs and sub-shrubs by softwood cuttings in mid-summer. Sow seed of half-hardy annuals under glass in early spring and of fully-hardy species outdoors in mid-spring.
♣ **S. argentea** illus. p.204.
S. blepharophylla. Spreading, rhizomatous perennial. H and S 45cm (18in). Frost tender. Has oval, glossy, dark green leaves and slender racemes of bright red flowers, with maroon calyces, in summer-autumn.
S. bulleyana. Rosette-forming perennial. H and S 60cm (24in). Fully hardy. Racemes of nettle-like, yellow flowers, with maroon lips, are borne in summer above a basal mass of broadly oval, coarse, prominently veined, dark green leaves.
S. farinacea. 'Alba' is a moderately fast-growing, upright perennial, grown as an annual. H 1m (3ft), S 30cm (1ft). Half hardy. Has lance-shaped, mid-green leaves. Spikes of white flowers are produced in summer. **'Blue Bedder'**, H 45cm (1½ft), has dark violet-blue flowers. ♣ **'Victoria'** illus. p.284. Dwarf forms are also available.
♣ **S. fulgens** illus. p.136.
S. grahamii. See *S. microphylla*.
S. greggii. Evergreen, erect sub-shrub. H to 1m (3ft), S to 60cm (2ft). Frost tender. Leaves are narrowly oblong and matt, deep green. Produces terminal racemes of bright red-purple flowers in autumn.
♣ **S. haematodes**, syn. *S. pratensis* Haematodes Group. Short-lived, rosette-forming perennial. H 1m (3ft), S 45cm (1½ft). Fully hardy. In early summer produces panicles massed with lavender-blue flowers above large, broadly oval, wavy-edged, toothed, rough, dark green leaves.
S. horminum. See *S. viridis*.
♣ **S. involucrata.** Bushy, woody-based perennial. H 60–75cm (2–2½ft) or more, S 1m (3ft). Half hardy. Has oval, mid-green leaves and, in late summer and autumn, racemes of large, rose-crimson flowers. **'Bethellii'** illus. p.195.
S. jurisicii. Rosette-forming perennial. H 45cm (18in), S 30cm (12in). Fully hardy. Stems are clothed with mid-green leaves, divided into 4–6 pairs of linear leaflets. In early summer produces racemes of inverted, violet-blue flowers.
♣ **S. leucantha** (Mexican bush sage). Evergreen, erect, well-branched shrub. H and S to 60cm (2ft) or more. Frost tender. Narrowly lance-shaped, finely wrinkled leaves are deep green above, white-downy beneath. In summer-autumn produces terminal spikes of hairy, white flowers, each from a woolly, violet calyx.
S. microphylla, syn. *S. grahamii*. Evergreen, erect, well-branched shrub. H and S 1–1.2m (3–4ft). Half hardy, but best at 5°C (41°F). Has oval to elliptic, mid- to deep green leaves. Racemes of dark crimson flowers, ageing to purple and with purple-tinted calyces, appear in late summer and autumn. var. *neurepia* illus. p.136.
S. nemorosa, syn. *S. virgata* var. *nemorosa*. Neat, clump-forming perennial. H 1m (3ft), S 45cm (1½ft). Fully hardy. Has narrowly oval, rough, mid-green leaves and, in summer, branching racemes densely set with violet-blue flowers. **'East Friesland'** (syn. *S.n.* 'Ostfriesland', H 75cm (2½ft), is smaller. **'Lubecca'**, H 45cm (1½ft), is a dwarf form. **'Mainacht'** see *S. × superba* 'Mainacht'. **'May Night'** see *S. × superba* 'Mainacht'.
S. officinalis (Sage). ♣ **'Icterina'** illus. p.147. ♣ **'Purpurascens'** is an evergreen or semi-evergreen, bushy shrub. H 60cm (2ft), S 1m (3ft). Frost hardy. Oblong, grey-green leaves are used as a culinary herb and are purple-flushed when young. Racemes of blue-purple flowers are produced during summer.
♣ **S. patens** illus. p.251.
S. pratensis Haematodes Group. See *S. haematodes*.
S. sclarea var. *turkestanica* illus. p.282.
S. splendens. Slow-growing, bushy perennial or evergreen sub-shrub, grown as an annual. H to 30cm (12in), S 20–30cm (8–12in). Half hardy. Has oval, serrated, fresh green leaves. Dense racemes of scarlet flowers are borne in summer and early autumn. **'Blaze of Fire'** has brilliant scarlet flowers. **Carabiniere Series** (white, illus. p.271) and **Cleopatra Series** (salmon, illus. p.273; violet, illus. p.283) are available in mixed or single colours. **'Fireworks'** has red-and-white striped flowers. **'Flare Path'** illus. p.280. **'Rodeo'**, H 20cm (8in), has brilliant scarlet flowers.
S. × superba 'Mainacht', syn. *S. nemorosa* 'Mainacht', *S.* 'May Night', illus. p.216.
♣ **S. uliginosa** (Bog sage). Graceful, upright, branching perennial. H 2m (6ft), S 45cm (1½ft). Half hardy. Has oblong to lance-shaped, saw-edged, mid-green leaves and, in autumn, long racemes with whorls of bright blue flowers. Prefers moist soil.
S. virgata var. *nemorosa*. See *S. nemorosa*.
S. viridis, syn. *S. horminum*, illus. p.283. **Art Shades Series** is a group of moderately fast-growing, upright, branching annuals. H 45cm (18in), S 20cm (8in). Half hardy. Has oval, mid-green leaves. Tiny flowers, hidden inside large, blue, pink or white bracts, are produced in dense spikes in summer and early autumn. Bracts of **Claryssa Series** are in a wide range of brilliant colours, including white, pink, purple and blue.

SALVINIA (Salviniaceae)
Genus of deciduous, perennial, floating water ferns, evergreen in tropical conditions. Is popular for tropical aquariums. Frost tender, min. 10–15°C (50–59°F). Does best in warm water, with plenty of light. Remove fading foliage, and thin plants when overcrowded. Propagate by separating young plants in summer.
S. auriculata illus. p.389.
S. natans. Deciduous, perennial, floating water plant. S indefinite. Oval, elongated, mid-green leaves are borne on branching stems. Tolerates colder conditions than other species and is often used in a cold-water aquarium.

SAMBUCUS (Caprifoliaceae)
Elder
Genus of perennials, deciduous shrubs and trees, grown for their foliage, flowers and fruits. Fully hardy. Needs sun and fertile, moist soil. For best foliage effect, either cut all shoots to ground in winter or prune out old shoots and reduce length of young shoots by half. Propagate species by softwood cuttings in summer, by seed in autumn or by hardwood cuttings in winter, selected forms by cuttings only.
S. canadensis (American elder). **'Aurea'** is a deciduous, upright shrub. H and S 4m (12ft). Has large, golden-yellow leaves, each with usually 7 oblong leaflets. In mid-summer produces large, domed heads of small, star-shaped, creamy-white flowers, then spherical, red fruits. **'Maxima'** bears very large, mid-green leaves and huge flower heads.
S. nigra (Common elder). ♣ **'Aurea'** (Golden elder) is a deciduous, bushy shrub. H and S 6m (20ft). Has stout, corky shoots and golden-yellow leaves, each composed of usually 5 oval leaflets. Flattened heads of fragrant, star-shaped, creamy-white flowers in early summer are followed by spherical, black fruits. Dark green foliage of ♣ **'Guincho Purple'** matures to deep blackish-purple. Produces purple-stalked flowers, pink in bud and opening to white within, pink outside.
S. racemosa (Red-berried elder). Deciduous, bushy shrub. H and S 3m (10ft). Mid-green leaves each have usually 5 oval leaflets. Star-shaped, creamy-yellow flowers, borne in dense, conical clusters in mid-spring, are succeeded by spherical, red fruits. **'Plumosa'** has leaves with finely cut leaflets as does **'Plumosa Aurea'**, but those of the latter are bronze when young, maturing to golden-yellow.

SANCHEZIA (Acanthaceae)
Genus of evergreen, mainly summer-flowering perennials, shrubs and scrambling climbers, grown for their flowers and foliage. Frost tender, min. 15–18°C (59–64°F). Needs full light or partial shade and fertile, well-drained soil. Water potted specimens freely when in full growth, moderately at other times. Tip prune young plants to promote a branching habit. Propagate by greenwood cuttings in spring or summer. Is prone to whitefly and soft scale.
S. nobilis of gardens. See *S. speciosa*.
S. speciosa, syn. *S. nobilis* of gardens, illus. p.147.

SANDERSONIA (Liliaceae)
Genus of one species of deciduous, tuberous climber with urn-shaped flowers in summer. Half hardy. Needs a sheltered, sunny site and well-drained soil. Support with pea sticks or canes. Lift tubers for winter. Propagate in spring by seed or by naturally divided tubers.
S. aurantiaca illus. p.367.

SANGUINARIA (Papaveraceae)
Genus of one species of spring-flowering, rhizomatous perennial. Fully hardy. Grow in sun or semi-shade and in humus-rich, moist but well-drained soil. Propagate by clean division of rhizomes in summer or by seed in autumn.
S. canadensis illus. p.311. ♣ **'Plena'**

SANGUISORBA

(syn. *S.c.* 'Flore Pleno', *S.c.* 'Multiplex') is a clump-forming, rhizomatous perennial with fleshy, underground stems that exude red sap when cut. H 15cm (6in), S 30–45cm (12–18in). Short-lived, rounded, double, white flowers emerge in spring before large, rounded to heart-shaped, scalloped, grey-green leaves, with glaucous undersides, appear.

SANGUISORBA (Rosaceae)
Burnet
Genus of perennials, grown for their bottlebrush-like flower spikes. Fully hardy. Requires sun and moist soil. Propagate by division in spring or by seed in autumn.
S. canadensis illus. p.190.
S. obtusa. Clump-forming perennial. H 1–1.2m (3–4ft), S 60cm (2ft). Arching stems bear spikes of rose-crimson flowers in mid-summer. Leaves, composed of pairs of oval leaflets, are pale green above, blue-green beneath.
S. officinalis (Great burnet). **'Rubra'** is a clump-forming perennial. H 1.2m (4ft), S 60cm (2ft). Produces small spikes of red-brown flowers in late summer above a mass of mid-green leaves, divided into oval leaflets.

SANSEVIERIA (Agavaceae)
Genus of evergreen, rhizomatous perennials, grown for their rosettes of stiff, fleshy leaves. Frost tender, min. 10–15°C (50–59°F). Tolerates sun and shade and is easy to grow in most soil conditions if not overwatered. Propagate by leaf cuttings or division in summer.
S. cylindrica. Evergreen, stemless, rhizomatous perennial. H 45cm–1.2m (1½–4ft), S 10cm (4in). Has a rosette of 3–4 cylindrical, stiff, fleshy, erect leaves, to 1.2m (4ft) long, in dark green with paler horizontal bands. Racemes of small, tubular, 6-lobed, pink or white flowers are occasionally produced.
S. trifasciata (Mother-in-law's tongue). Evergreen, stemless, rhizomatous perennial. H 45cm–1.2m (1½–4ft), S 10cm (4in). Has a rosette of about 5 lance-shaped, pointed, stiff, fleshy, erect leaves, to 1.2m (4ft) long, banded horizontally with pale green and yellow. Occasionally carries racemes of tubular, 6-lobed, green flowers.
♀ **'Golden Hahnii'** illus. p.269.
♀ **'Hahnii'** illus. p.267.
♀ **'Laurentii'** illus. p.231.

SANTOLINA (Compositae)
Genus of evergreen, summer-flowering shrubs, grown for their aromatic foliage and their button-like flower heads, each on a long stem. Frost hardy. Requires sun and not too rich, well-drained soil. Cut off old flower heads and reduce long shoots in autumn. Cut straggly, old plants back hard each spring. Propagate by semi-ripe cuttings in summer.
♀ *S. chamaecyparissus* (Cotton lavender, Lavender cotton). Evergreen, rounded, dense shrub. H 75cm (2½ft), S 1m (3ft). Shoots are covered with woolly, white growth, and narrowly oblong, finely toothed leaves are also white. Bright yellow flower heads are borne in mid- and late summer.
S. neapolitana. See *S. pinnata* subsp. *neapolitana*.
S. pinnata. Evergreen shrub, mainly grown as subsp. *neapolitana* (syn. *S. neapolitana*), which is of rounded and bushy habit. H 75cm (2½ft), S 1m (3ft). Slender flower stems each carry a head of lemon-yellow flowers in mid-summer, among feathery, deeply cut, grey-green foliage.
♀ subsp. *neapolitana* **'Sulphurea'** illus. p.139.
S. rosmarinifolia, syn. *S. virens* (Holy flax). Evergreen, bushy, dense shrub. H 60cm (2ft), S 1m (3ft). Has finely cut, bright green leaves. Each slender stem produces a head of bright yellow flowers in mid-summer. ♀ **'Primrose Gem'** has pale yellow flower heads.
S. virens. See *S. rosmarinifolia*.

SANVITALIA (Compositae)
Genus of perennials and annuals. Fully hardy. Grow in sun and in fertile, well-drained soil. Propagate by seed sown outdoors in spring or early autumn.
S. procumbens illus. p.288.
'Mandarin Orange' illus. p.293.

SAPIUM (Euphorbiaceae)
Genus of evergreen trees, grown for their ornamental appearance. Has poisonous, milky sap. Frost tender, min. 5°C (41°F). Prefers fertile, well-drained soil and full light. Water containerized plants freely when in full growth, less at other times. Pruning is tolerated if necessary. Propagate by seed in spring or by semi-ripe cuttings in summer.
S. sebiferum (Chinese tallow tree). Fast-growing, evergreen, erect to spreading tree. H to 8m (25ft), S 4m (12ft) or more. Rhombic to oval, mid-green leaves turn red with age. Clusters of tiny, greenish-yellow flowers develop into rounded, black fruits covered by a layer of white wax.

SAPONARIA (Caryophyllaceae)
Soapwort
Genus of summer-flowering annuals and perennials, grown for their flowers. Is good for rock gardens, screes and banks. Fully hardy. Needs sun and well-drained soil. Propagate by seed in spring or autumn or by softwood cuttings in early summer.
♀ *S.* **'Bressingham'**, syn. *S.* 'Bressingham Hybrid'. Loose, mat-forming perennial. H 8cm (3in), S 10cm (4in). Has small, narrowly oval leaves. Flattish, deep vibrant pink flowers are produced in clustered heads in summer. Is good for a trough.
S. caespitosa illus. p.326.
♀ *S. ocymoides* illus. p.326.
S. officinalis **'Rubra Plena'** (Double soapwort). Upright perennial. H to 1m (3ft), S 30cm (1ft). Has oval, rough, mid-green leaves on erect stems. Clusters of ragged, double, red flowers are produced from leaf axils on upper part of flower stems in summer.
S. x olivana illus. p.324.

SARCOCAPNOS (Fumariaceae)
Genus of spring-flowering perennials. Is useful for rock gardens. Frost hardy. Prefers sun and well-drained, alkaline soil. Propagate by seed in spring.
S. enneaphylla. Loose, upright perennial. H and S 15cm (6in). Slender, much-branched stems bear small, much-divided, glaucous green leaves with oval to rounded segments. In spring, small, spurred, yellowish-white flowers, tipped with purple, are produced in short racemes. Protect from winter wet.

SARCOCOCCA (Buxaceae)
Christmas box, Sweet box
Genus of evergreen shrubs, grown for their foliage, fragrant, winter flowers and spherical fruits. Flowers are tiny – the only conspicuous part being the anthers. Is useful for cutting in winter. Fully to frost hardy. Grows in sun or shade and in fertile, not too dry soil. Propagate by semi-ripe cuttings in summer or by seed in autumn.
♀ *S. confusa.* Evergreen, bushy, dense shrub. H and S 1m (3ft). Fully hardy. Leaves are small, oval, taper-pointed, glossy and dark green. Tiny, white flowers in winter are followed by black fruits.
♀ *S. hookeriana.* Evergreen, upright, dense, suckering shrub. H 1.5m (5ft), S 2m (6ft). Fully hardy. Forms clumps of narrowly oblong, pointed, dark green leaves and has tiny, white flowers in the leaf axils during winter. Fruits are black. ♀ var. *digyna* illus. p.144. var. *humilis.* See *S. humilis*.
S. humilis, syn. *S. hookeriana* var. *humilis*, illus. p.144.
S. ruscifolia. Evergreen, upright, arching shrub. H and S 1m (3ft). Frost hardy. Has oval, glossy, dark green leaves and, in winter, creamy-white flowers, then red fruits.
♀ var. *chinensis* has narrower leaves.

SARMIENTA (Gesneriaceae)
Genus of one species of evergreen, woody-stemmed, scrambling or trailing perennial. Suits hanging baskets. Half hardy. Likes semi-shade and humus-rich soil that does not dry out. Propagate by seed in spring or by stem cuttings in summer or autumn.
S. repens. See *S. scandens*.
♀ *S. scandens*, syn. *S. repens*. Evergreen, slender-stemmed, scrambling perennial. H and S 60cm (2ft) or more. Tips of oval leaves each have 3–5 teeth. In summer produces small, tubular, coral-pink flowers, each narrowed at the base and towards the mouth, which has 5 deeper pink lobes.

SARRACENIA (Sarraceniaceae)
Pitcher plant
Genus of insectivorous perennials, some of which are evergreen, with pitchers formed from modified leaves with hooded tops. Frost tender, min. 5°C (41°F). Grow in sun or partial shade and in peat and moss. Keep very wet, except in winter, when cool and slightly drier conditions are needed. Propagate by seed in spring.
♀ *S. flava* illus. p.254.
S. purpurea (Common pitcher plant, Huntsman's cup). Evergreen, erect to semi-prostrate, rosette-forming perennial. H 30cm (12in), S 30–38cm (12–15in). Inflated, green pitchers, to 15cm (6in) long, are tinged and veined purplish-red. In spring, 5-petalled, purple flowers, 5cm (2in) or wide, are carried well above pitchers.

SAXEGOTHAEA

SASA (Bambusoideae). See GRASSES, BAMBOOS, RUSHES and SEDGES.
S. albomarginata. See *S. veitchii*.
♀ *S. palmata.* Evergreen, spreading bamboo. H 2m (6ft), S indefinite. Frost hardy. A fine foliage plant, it produces very broad, rich green leaves, to 40cm (16in) long. Hollow, purple-streaked stems have one branch at each node. Flower spikes are unimportant.
S. veitchii, syn. *S. albomarginata*, illus. p.182.

SASSAFRAS (Lauraceae)
Genus of deciduous trees, with inconspicuous flowers, grown for their aromatic foliage. Fully hardy. Needs sun or light shade and deep, fertile, well-drained, preferably acid soil. Propagate by seed or suckers in autumn or by root cuttings in winter.
S. albidum illus. p.42.

SATUREJA (Labiatae)
Genus of summer-flowering annuals, semi-evergreen perennials and sub-shrubs, grown for their highly aromatic leaves and attractive flowers. Is useful for rock gardens and dry banks. Fully hardy. Needs sun and well-drained soil. Propagate by seed in winter or spring or by softwood cuttings in summer.
S. montana (Winter savory). Semi-evergreen, upright perennial or sub-shrub. H 30cm (12in), S 20cm (8in) or more. Leaves are linear to oval, aromatic and green or greyish-green. Carries loose whorls of tubular, 2-lipped, lavender flowers in summer. **'Prostrate White'**, H 7–15cm (3–6in), has a prostrate habit and white flowers.

SAUROMATUM (Araceae)
Genus of spring-flowering, tuberous perennials with tubular spathes that expand into waved, twisted blades. Tubers will flower without soil or moisture, and before leaves appear. Frost tender, min. 5–7°C (41–5°F). Needs a sheltered, semi-shaded position and humus-rich, well-drained soil. Water well in summer. Dry off or lift when dormant in winter. Propagate by offsets in spring.
S. guttatum. See *S. venosum*.
S. venosum, syn. *S. guttatum*, illus. p.357.

SAURURUS (Saururaceae)
Genus of deciduous, perennial, bog and marginal water plants, grown for their foliage. Fully hardy. Prefers full sun, but tolerates some shade. Remove faded leaves and divide plants as required to maintain vigour. Propagate by division in spring.
S. cernuus illus. p.387.

SAXEGOTHAEA (Podocarpaceae).
See CONIFERS.
S. conspicua (Prince Albert's yew). Conifer that is conical in mild areas, more bushy in cold districts. H 5–15m (15–50ft), S 4–5m (12–15ft). Fully hardy. Needle-like, flattened, dark

SAXIFRAGA

green leaves are produced in whorls at ends of shoots. Bears globose, fleshy, glaucous green cones.

SAXIFRAGA (Saxifragaceae)
Saxifrage
Genus of often rosetted perennials, most of which are evergreen or semi-evergreen, grown for their flowers and attractive foliage. Is excellent in rock gardens, raised beds and alpine houses. Fully to half hardy. Propagate by seed in autumn or by rooted offsets in winter. For cultivation, saxifrages may be grouped as follows:

1 – Needs protection from midday sun and moist soil.
2 – Needs semi-shaded, well-drained soil. Is good among rocks and screes.
3 – Thrives in well-drained rock pockets, troughs, alpine-house pans etc, shaded from midday sun. Must never be dry at roots. Most form tight cushions and flower in early spring, flower stems being barely visible above leaves.
4 – Needs full sun and well-drained, alkaline soil. Suits rock pockets. Most have hard leaves encrusted in lime.

S. aizoides. Evergreen perennial forming a loose mat. H 15cm (6in), S 30cm (12in) or more. Fully hardy. Has small, narrowly oval, fleshy, shiny, green leaves and, in spring-summer, terminal racemes of star-shaped, bright yellow or orange flowers, often spotted red, on 8cm (3in) stems. Cultivation group 1.
S. aizoon. See *S. paniculata.*
♀ *S.* x *angelica* 'Cranbourne', syn. *S.* 'Cranbourne'. Evergreen, cushion-forming perennial. H and S 12cm (5in). Fully hardy. In early spring produces solitary cup-shaped, bright purplish-lilac flowers on short stems just above tight rosettes of linear, green leaves. Flower stems are longer if plant is grown in an alpine house. Cultivation group 3.
♀ *S.* x *apiculata* 'Gregor Mendel', illus. p.319; cultivation group 2.
S. x *arco-valleyi* 'Arco'. Evergreen perennial forming a tight cushion. H and S 10cm (4in). Fully hardy. In early spring has upturned, cup-shaped to flattish, pale lilac flowers almost resting on tight rosettes of oblong to linear leaves. Cultivation group 3.
S 'Aretiastrum'. See *S.* 'Valerie Finnis'.
S. 'Bob Hawkins'. Evergreen perennial with a loose rosette of leaves. H 2.5–5cm (1–2in), S 15cm (6in). Fully hardy. Carries small, upturned, rounded, greenish-white flowers in summer on 5cm (2in) stems. Oval, green leaves are white-splashed. Cultivation group 1.
S. brunoniana. See *S. brunonis.*
S. brunonis, syn. *S. brunoniana.* Semi-evergreen, rosetted perennial. H 10cm (4in), S 20cm (8in). Frost hardy. Small, soft green rosettes of lance-shaped, rigid leaves produce masses of long, thread-like, red runners. Many of the rosettes die down to large terminal buds in winter. Short racemes of 5-petalled, spreading, pale yellow flowers are borne in late spring and summer on 5–8cm (2–3in) stems. Is difficult to grow; cultivation group 1.
S. burseriana illus. p.311. 'Brookside' is a slow-growing, evergreen perennial forming a hard cushion. H 2.5–5cm (1–2in), S to 10cm (4in). Fully hardy. Has broadly linear, spiky, grey-green leaves. In early spring bears upturned, rounded, shallowly cup-shaped, bright yellow flowers on short, red stems. Flowers of 'Crenata' have fringed, white petals and red sepals.
♀ 'Gloria' has dark reddish-brown stems, each bearing 1 or 2 flowers, with red sepals and white petals, in late spring. Cultivation group 3.
♀ *S. callosa*, syn. *S. lingulata.* Evergreen, tightly rosetted perennial. H 25cm (10in), S to 20cm (8in). Fully hardy. Bears long, linear, stiff, lime-encrusted leaves and, in early summer, upright, then arching panicles of star-shaped, white flowers with red-spotted petals. Rosettes die after flowering; new ones are produced annually from short stolons. Cultivation group 4.
S. 'Carmen'. See *S.* 'Elizabethae'.
S. cochlearis. Evergreen, rosetted perennial. H 20cm (8in), S 25cm (10in). Fully hardy. Has spoon-shaped, green leaves with white-encrusted edges. Produces loose panicles of rounded, white flowers, often with red-spotted petals, in early summer.
♀ 'Minor', H and S 12cm (5in), has smaller leaf rosettes and loose panicles of red-spotted, white flowers on red stems. Is ideal for a trough. Cultivation group 4.
S. cortusifolia var. *fortunei.* See *S. fortunei.*
S. cotyledon illus. p.300; cultivation group 2.
S. 'Cranbourne'. See *S.* x *angelica* 'Cranbourne'.
S. cuneifolia illus. p.299; cultivation group 1.
S. 'Elizabethae', syn. *S.* 'Carmen', illus. p.319; cultivation group 2.
S. exarata subsp. *moschata*, syn. *S. moschata.* Evergreen perennial forming a loose to tight hummock. H and S 10cm (4in). Fully hardy. Rosettes comprise small, lance-shaped, sometimes 3-toothed, green leaves. Bears 2–5 star-shaped, creamy-white or dull yellow flowers on slender stems in summer. 'Cloth of Gold' illus. p.339. Cultivation group 1.
S. federici-augustii subsp. *grisebachii* 'Wisley Variety'. See *S. grisebachii* 'Wisley Variety'.
♀ *S. ferdinandi-coburgii.* Evergreen, cushion-forming perennial. H and S 15cm (6in). Fully hardy. Forms rosettes of linear, spiny, glaucous green leaves and, in early spring, bears racemes of open cup-shaped, rich yellow flowers on stems 3–10cm (1–4in) long. Cultivation group 3.
♀ *S. fortunei,* syn. *S. cortusifolia* var. *fortunei.* Semi-evergreen or herbaceous, clump-forming perennial. H and S 30cm (12in). Frost hardy. Has rounded, 5- or 7-lobed, fleshy, green or brownish-green leaves, red beneath. In autumn produces panicles of tiny, moth-like, white flowers, with 4 equal-sized petals and one elongated petal, on upright stems. Propagate by division in spring. 'Rubrifolia' has dark red flower stems and dark reddish-green leaves with beetroot-red undersides. Cultivation group 1.
S. x *geum* illus. p.295; cultivation group 1.
S. granulata (Fair maids of France, Meadow saxifrage) illus. p.294. 'Plena' is a clump-forming perennial. H 23–38cm (9–15in), S to 15cm (6in) or more. Fully hardy. Loses its kidney-shaped, glossy, pale to mid-green leaves soon after flowering. Produces a loose panicle of large, rounded, double, white flowers in late spring or early summer. Has bulbils or resting buds forming at base of foliage. Cultivation group 1.
♀ *S. grisebachii* 'Wisley Variety', syn. *S. federici-augusti* subsp. *grisebachii* 'Wisley Variety', illus. p.315; cultivation group 4.
S. 'Hindhead Seedling' illus. p.319; cultivation group 2.
S. hirsuta illus. p.295; cultivation group 1.
S. 'Irvingii'. See *S.* 'Walter Irving'.
♀ *S.* x *irvingii* 'Jenkinsiae', syn. *S.* 'Jenkinsiae', illus. p. 313; cultivation group 2.
S. 'Jenkinsiae'. See *S.* x *irvingii* 'Jenkinsiae'.
S. lingulata. See *S. callosa.*
S. longifolia. Rosetted perennial. H 60cm (24in), S 20–25cm (8–10in). Fully hardy. Has long, narrow, lime-encrusted leaves forming attractive rosettes that, after 3–4 years, develop long, arching, conical to cylindrical panicles bearing numerous rounded, 5-petalled, white flowers in late spring and summer. Rosettes die after flowering, and no daughter rosettes are formed, so propagate by seed in spring or autumn. In cultivation, hybridizes readily with other related species. Cultivation group 4.
S. moschata. See *S. exarata* subsp. *moschata.*
S. oppositifolia (Purple mountain saxifrage) illus. p.314. 'Ruth Draper' is an evergreen, loose mat-forming perennial. H 2.5–5cm (1–2in), S 15cm (6in). Fully hardy. Has small, opposite, oblong to oval, white-flecked, dark green leaves closely set along prostrate stems. Large, cup-shaped, deep purple-pink flowers appear in early spring just above foliage. Prefers peaty soil. Cultivation group 1.
S. paniculata, syn. *S. aizoon.* Evergreen, tightly rosetted perennial. H 15–30cm (6–12in), S 20cm (8in). Fully hardy. In summer produces loose panicles of rounded, usually white flowers, with or without purplish-red spots, on upright stems above rosettes of oblong to oval, lime-encrusted leaves. Is very variable in size. Pale yellow or pale pink forms also occur. Cultivation group 4.
S. 'Primulaize', syn. *S.* x *primulaize.* Evergreen, loosely rosetted perennial. H 15cm (6in). Fully hardy. In summer, branched flower stems, 5–8cm (2–3in) long, produce star-shaped, salmon-pink flowers. Leaves are tiny, narrowly oval, slightly indented and fleshy. Cultivation group 1.
S. sancta illus. p.319; cultivation group 2.
S. sarmentosa. See *S. stolonifera.*
S. scardica illus. p.310; cultivation group 3.
S. sempervivum illus. p.315; cultivation group 3.
♀ *S.* 'Southside Seedling' illus. p.300; cultivation group 4.
♀ *S. stolonifera,* syn. *S. sarmentosa* (Mother of thousands). Evergreen, prostrate perennial with runners. H 15cm (6in) or more, S 30cm (12in) or more. Frost hardy. Has large, rounded, shallowly lobed, hairy, silver-veined, olive-green leaves that are reddish-purple beneath. Loose panicles of tiny, moth-like, white flowers, each with 4 equal-sized petals and one elongated petal, appear in summer on slender, upright stems. Makes good ground cover. ♀ 'Tricolor' (Strawberry geranium) has green-and-red leaves with silver marks and is half hardy. Cultivation group 1.
S. stribrnyi illus. p.316; cultivation group 3.
♀ *S.* 'Tumbling Waters' illus. p.295; cultivation group 4.
♀ *S.* x *urbium* (London pride). Evergreen, rosetted, spreading perennial. H 30cm (12in), S indefinite. Fully hardy. Has spoon-shaped, toothed, leathery, green leaves. Flower stems bear tiny, star-shaped, at times pink-flushed, white flowers, with red spots, in summer. Is useful as ground cover. Cultivation group 1.
S. 'Valerie Finnis', syn. *S.* 'Aretiastrum'. Evergreen, hard cushion-forming perennial. H and S 10cm (4in). Fully hardy. Short, red stems carry upturned, cup-shaped, sulphur-yellow flowers above tight rosettes of oval, green leaves in spring. Cultivation group 3.
S. 'Walter Irving', syn. *S.* 'Irvingii'. Very slow-growing, evergreen, hard-domed perennial. H 2cm (³/₄in), S 8cm (3in). Fully hardy. Bears minute leaves in rosettes. Stemless, cup-shaped, lilac-pink flowers open in early spring. Cultivation group 3.

SCABIOSA (Dipsacaceae)
Scabious
Genus of annuals and perennials, some of which are evergreen, with flower heads that are good for cutting. Fully to frost hardy. Prefers sun and fertile, well-drained, alkaline soil. Propagate annuals by seed in spring and perennials by cuttings of young, basal growths in summer, by seed in autumn or by division in early spring.
S. atropurpurea (Sweet scabious). Moderately fast-growing, upright, bushy annual. H to 1m (3ft), S 20–30cm (8–12in). Fully hardy. Has lance-shaped, lobed, mid-green leaves. Domed heads of scented, pincushion-like, deep crimson flower heads, 5cm (2in) wide, are produced on wiry stems in summer and early autumn. Tall forms, H 1m (3ft), and dwarf, H 45cm (18in), are available with flower heads in shades of blue, purple, red, pink or white. Cockade Series illus. p.283.
♀ *S. caucasica.* 'Clive Greaves' illus. p.249. 'Floral Queen' is a clump-forming perennial. H and S 60cm (24in). Fully hardy. Large, frilled, violet-blue flower heads, with pincushion-like centres, are borne

throughout summer. Light green leaves are lance-shaped at base of plant and segmented on stems. ❦ **'Miss Willmott'** has creamy-white flowers.
S. columbaria var. *ochroleuca*. See *S. ochroleuca*.
S. graminifolia. Evergreen, clump-forming perennial, often with a woody base. H and S 15–25cm (6–10in). Frost hardy. Has tufts of narrow, grass-like, pointed, silver-haired leaves. In summer produces stiff stems with spherical, bluish-violet to lilac flower heads like pincushions. Resents disturbance. Suits a rock garden.
S. lucida illus. p.303.
S. ochroleuca, syn. *S. columbaria* var. *ochroleuca*. Clump-forming perennial. H and S 1m (3ft). Fully hardy. In late summer, branching stems carry many heads of frilled, sulphur-yellow flower heads with pincushion-like centres. Has narrowly oval, toothed, grey-green leaves.
S. rumelica. See *Knautia macedonica*.

SCADOXUS (Amaryllidaceae)
Genus of bulbs with dense, mainly spherical, umbels of red flowers. Frost tender, min. 10–15°C (50–59°F). Requires partial shade and humus-rich, well-drained soil. Reduce watering in winter, when not in active growth. Propagate by seed or offsets in spring.
S. multiflorus, syn. *Haemanthus multiflorus*. Summer-flowering bulb. H to 70cm (28in), S 30–45cm (12–18in). Has broadly lance-shaped, semi-erect, basal leaves. Produces a spherical umbel, 10–15cm (4–6in) wide, of up to 200 narrow-petalled flowers. ❦ subsp. *katherinae* (syn. *Haemanthus katherinae*) illus. p.345.
S. puniceus, syn. *Haemanthus magnificus*, *H. natalensis*, *H. puniceus* (Royal paintbrush). Spring- and summer-flowering bulb. H 30–40cm (12–16in), S 30–45cm (12–18in). Has elliptic, semi-erect leaves in a basal cluster. Leaf bases are joined, forming a false stem. Flower stem bears up to 100 tubular, orange-red flowers in a conical umbel surrounded by a whorl of red bracts.

SCHEFFLERA, syn. **BRASSAIA**, **HEPTAPLEURUM** (Araliaceae)
Genus of evergreen shrubs and trees, grown mainly for their handsome foliage. Half hardy to frost tender, min. 3–16°C (37–61°F). Grows in any fertile, well-drained but moisture-retentive soil and in full light or partial shade. Water potted specimens freely when in full growth, moderately at other times. Pruning is tolerated if needed. Propagate by air-layering in spring, by semi-ripe cuttings in summer or by seed as soon as ripe, in late summer.
❦ *S. actinophylla* illus. p.58.
❦ *S. arboricola*. Evergreen, erect, well-branched shrub or tree. H 2–5m (6–15ft), S 1–3m (3–10ft). Frost tender, min. 15°C (59°F). Leaves each have 7–16 oval, stalked, glossy, deep green leaflets. Mature plants carry small, spherical heads of tiny, green flowers in spring-summer.
S. digitata. Evergreen, rounded to ovoid shrub or bushy tree. H and S 3–8m (10–25ft). Frost tender, min. 5°C (41°F). Leaves are hand-shaped, with 5–10 oval, glossy, rich green leaflets. Has tiny, greenish flowers in large, terminal panicles in spring and tiny, globular, dark violet fruits in autumn.
S. elegantissima, syn. *Aralia elegantissima*, *Dizygotheca elegantissima*, illus. p.97.

SCHIMA (Theaceae)
Genus of one species of very variable, evergreen tree or shrub, grown for its foliage and flowers. Is related to *Camellia*. Frost tender, min. 3–5°C (37–41°F). Prefers humus-rich, well-drained, neutral to acid soil and sun or partial shade. Water potted plants freely in full growth, moderately at other times. Pruning is tolerated if necessary. Propagate by seed as soon as ripe or by semi-ripe cuttings in summer.
S. wallichii. Robust, evergreen, ovoid tree or shrub. H 25–30m (80–100ft), S 12m (40ft) or more. Elliptic to oblong, red-veined, dark green leaves are 10–18cm (4–7in) long, red-flushed beneath. In late summer has solitary fragrant, cup-shaped, white flowers, 4cm (1½in) wide, that are red-flushed in bud.

SCHINUS (Anacardiaceae)
Genus of evergreen shrubs and trees, grown mainly for their foliage and for shade. Frost tender, min. 5°C (41°F). Grows in any freely draining soil and in full light. Water potted specimens moderately, hardly at all in winter. Propagate by seed in spring or by semi-ripe cuttings in summer.
S. molle (Californian pepper-tree, Peruvian mastic tree, Peruvian pepper-tree). Fast-growing, evergreen, weeping tree. H and S to 8m (25ft). Fern-like leaves are divided into many narrowly lance-shaped, glossy, rich green leaflets. Has open clusters of tiny, yellow flowers from late winter to summer, followed by pea-sized, pink-red fruits.
S. terebinthifolius. Evergreen shrub or tree, usually of bushy, spreading habit. H 3m (10ft) or more, S 2–3m (6–10ft) or more. Leaves have 3–13 oval, mid- to deep green leaflets. Tiny, white flowers are borne in clusters in summer-autumn, followed by pea-sized, red fruits, but only if plants of both sexes are grown close together.

SCHISANDRA (Schisandraceae)
Genus of deciduous, woody-stemmed, twining climbers. Male and female flowers are produced on separate plants, so grow plants of both sexes if fruits are required. Is useful for growing against shady walls and training up pillars and fences. Frost hardy. Grow in sun or partial shade and rich, well-drained soil. Propagate by greenwood or semi-ripe cuttings in summer.
S. grandiflora var. *rubriflora*. See *S. rubriflora*.
S. henryi. Deciduous, woody-stemmed, twining climber, with stems that are angled and winged when young. H 3–4m (10–12ft). Glossy, green leaves are oval or heart-shaped. Small, cup-shaped, white flowers appear in spring. Pendent spikes, 5–7cm (2–3in) long, of spherical, fleshy, red fruits are produced in late summer on female plants.
S. rubriflora, syn. *S. grandiflora* var. *rubriflora* illus. p.171.

SCHIZANTHUS (Solanaceae)
Butterfly flower, Poor man's orchid
Genus of annuals, grown for their showy flowers. Makes excellent pot plants. Half hardy to frost tender, min. 5°C (41°F). Grow in a sunny, sheltered position and in fertile, well-drained soil. Pinch out growing tips of young plants to ensure a bushy habit. Propagate by seed sown under glass in early spring for summer-autumn flowers and in late summer to provide plants to flower in pots in late winter or spring. Is prone to damage by aphids.
S., **Bouquet Series**. Group of moderately fast-growing, upright, bushy annuals. H and S 30cm (12in). Half hardy. Has feathery, light green leaves and, in summer-autumn, orchid-like, rounded, lobed flowers in a mixture of colours, including pink, purple and yellow.
S., **Giant Hybrids**. Group of moderately fast-growing, upright, bushy annuals. H 60cm–1.2m (2–4ft), S 30cm (1ft). Half hardy. Feathery leaves are light green. Orchid-like, rounded, lobed flowers in a mixture of colours, including pink, purple and yellow, are produced in summer-autumn.
S. **'Hit Parade'** illus. p.274.
S., **Pansy-flowered Series**. Group of moderately fast-growing, upright, bushy annuals. H 60cm–1m (2–3ft), S 30cm (1ft). Half hardy. Has feathery, light green leaves and, in summer-autumn, large, pansy-like, bicoloured flowers in a mixture of colours, including pink, purple and yellow.
S. pinnatus illus. p.282.

SCHIZOCENTRON. See **HETEROCENTRON**.

SCHIZOPETALON (Cruciferae)
Genus of annuals. Half hardy. Grow in sun and in well-drained, fertile soil. Propagate by seed sown under glass in spring.
S. walkeri. Moderately fast-growing, upright, slightly branching annual. H 45cm (18in), S 20cm (8in). Has deeply divided, mid-green leaves and, in summer, almond-scented, white flowers with deeply cut and fringed petals.

SCHIZOPHRAGMA (Hydrangeaceae)
Genus of deciduous, woody-stemmed, root climbers, useful for training up large trees. Frost hardy. Flowers best in sun, but will grow against a north-facing wall. Needs well-drained soil. Tie young plants to supports. Propagate by seed in spring or by greenwood or semi-ripe cuttings in summer.
S. hydrangeoides (Japanese hydrangea vine). Deciduous, woody-stemmed, root climber. H to 12m (40ft). Broadly oval leaves are 10–15cm (4–6in) long. Small, white or creamy-white flowers, in flat heads 20–25cm (8–10in) across, are produced on pendent side-branches in summer; these are surrounded by marginal, sterile flowers, which each have an oval or heart-shaped, pale yellow sepal, 2–4cm (¾–1½in) long.
❦ *S. integrifolium* illus. p.168.
S. viburnoides. See *Pileostegia viburnoides*.

SCHIZOSTYLIS (Iridaceae)
Kaffir lily
Genus of rhizomatous perennials with flowers that are excellent for cutting. Frost hardy. Requires sun and fertile, moist soil. Plants rapidly become congested, so should be divided in spring every few years.
❦ *S. coccinea*. **'Major'** (syn. *S. c.* 'Grandiflora') illus. p.258. **'Mrs Hegarty'** is a vigorous, clump-forming, rhizomatous perennial. H 60cm (24in), S 23–30cm (9–12in). In mid-autumn produces spikes of shallowly cup-shaped, pale pink flowers above tufts of grass-like, mid-green leaves. ❦ **'Sunrise'** illus. p.258. **'Viscountess Byng'** has pink flowers that last until late autumn.

SCHLUMBERGERA (Cactaceae)
Genus of bushy, perennial cacti with erect, then pendent stems and flattened, oblong stem segments with indented notches at margins – like teeth in some species. Stem tips produce flowers with prominent stigmas and stamens and with petals of different lengths set in 2 rows. In the wild, often grows over mossy rocks, rooting at ends of stem segments. Frost tender, min. 10°C (50°F). Needs partial shade and rich, well-drained soil. Propagate by stem cuttings in spring or early summer.
S. bridgesii. See *S.* 'Buckleyi'.
S. **'Bristol Beauty'** illus. p.411.
S. **'Buckleyi'**, syn. *S. bridgesii* (Christmas cactus). Erect, then pendent, perennial cactus. H 15cm (6in), S 1m (3ft). Has glossy, green stem segments and produces red-violet flowers in mid-winter.
S. **'Gold Charm'** illus. p.408.
S. truncata, syn. *Zygocactus truncatus*, illus. p.409.
S. **'Wintermärchen'**. Erect, then pendent, perennial cactus. H 15cm (6in), S 30cm (12in). Has glossy, green stem segments. In early autumn bears white flowers that become pink-and-white in winter.
S. **'Zara'**. Erect, then pendent, perennial cactus. H 15cm (6in), S 30cm (12in). Has glossy, green stem segments. Bears deep orange-red flowers in early autumn and winter.

SCHOENOPLECTUS (Cyperaceae)
See **GRASSES, BAMBOOS, RUSHES** and **SEDGES**.
S. lacustris subsp. *tabernaemontani* **'Zebrinus'**, syn. *Scirpus lacustris* var. *tabernaemontani* 'Zebrinus', *Scirpus tabernaemontani* 'Zebrinus', illus. p.182.

SCHWANTESIA (Aizoaceae)
Genus of cushion-forming, perennial succulents with stemless rosettes of

unequal-sized pairs of keeled leaves and daisy-like, yellow flowers. Frost tender, min. 5°C (41°F). Needs full sun and well-drained soil. Propagate by seed or stem cuttings in spring or summer.
S. ruedebuschii illus. p.414.

SCIADOPITYS (Taxodiaceae). See CONIFERS.
♚ S. verticillata illus. p.80.

SCILLA (Liliaceae)
Genus of mainly spring- and summer-flowering bulbs with leaves in basal clusters and spikes of small, often blue flowers. Fully to half hardy. Needs an open site, sun or partial shade and well-drained soil. Propagate by division in late summer or by seed in autumn.
S. adlamii. See Ledebouria cooperi.
♚ S. bifolia. Early spring-flowering bulb. H 5–15cm (2–6in), S 2.5–5cm (1–2in). Fully hardy. Has 2 narrowly strap-shaped, semi-erect, basal leaves that widen towards tips. Stem produces a one-sided spike of up to 20 star-shaped, purple-blue, pink or white flowers.
S. campanulata. See Hyacinthoides hispanica.
S. chinensis. See S. scilloides.
S. hispanica. See Hyacinthoides hispanica.
S. litardieri, syn. S. pratensis. Clump-forming, early summer-flowering bulb. H 10–25cm (4–10in), S 5–8cm (2–3in). Fully hardy. Bears up to 5 narrowly strap-shaped, semi-erect, basal leaves. Stem has a dense spike of flat, star-shaped, violet flowers, 1–1.5cm (1/2–5/8in) across.
♚ S. mischtschenkoana, syn. S. tubergeniana, illus. p.376.
S. natalensis. Clump-forming, summer-flowering bulb. H 30–45cm (12–18in), S 15–20cm (6–8in). Half hardy. Lance-shaped, semi-erect, basal leaves lengthen after flowering. Has a long spike of up to 100 flattish, blue flowers, each 1.5–2cm (5/8–3/4in) across.
S. non-scripta. See Hyacinthoides non-scriptus.
S. peruviana illus. p.380.
S. pratensis. See S. litardieri.
S. scilloides, syn. S. chinensis, illus. p.368.
S. siberica (Siberian squill). **'Atrocoerulea'** illus. p.376.
S. tubergeniana. See S. mischtschenkoana.
S. violacea. See Ledebouria socialis.

SCINDAPSUS (Araceae)
Genus of evergreen climbers. Frost tender, min. 15–18°C (59–64°F). Needs light shade and well-drained soil. Propagate by leaf-bud or stem-tip cuttings in late spring or by layering in summer.
S. aureus 'Marble Queen'. See Epipremnum aureum 'Marble Queen'.
♚ S. pictus 'Argyraeus', syn. Epipremnum pictum 'Argyraeus', illus. p.179.

Scirpus holoschoenus. See Scirpoides holoschoenus.
Scirpus lacustris subsp. tabernaemontani 'Zebrinus'. See Schoenoplectus lacustris subsp. tabernaemontani 'Zebrinus'.
Scirpus setaceus. See Isolepis setaceus.
Scirpus tabernaemontani 'Zebrinus'. See Schoenoplectus lacustris subsp. tabernaemontani 'Zebrinus'.

SCIRPOIDES (Cyperceae). See GRASSES, BAMBOOS, RUSHES and SEDGES.
S. holoschoenus, syn. Scirpus holoschoenus (Round-headed club-rush). **'Variegatus'** is an evergreen, tuft-forming, perennial rush. H 1–1.2m (3–4ft), S 30cm (1ft). Fully hardy. Rounded, leafless, green stems are striped horizontally with cream and bear egg-shaped, awned, brown spikelets on stalked, spherical heads in summer.

SCOLIOPUS (Liliaceae)
Genus of one species of spring-flowering perennial. Is usually grown in alpine houses, where its neat habit and curious flowers, which arise directly from buds on the rootstock early in the season, may be better appreciated. Is also suitable for rock gardens and peat beds. Frost hardy. Requires sun or partial shade and moist but well-drained soil. Propagate by seed when fresh, in summer or autumn.
S. bigelowii, syn. S. bigelorii, illus. p.312.

Scolopendrium vulgare. See Phyllitis scolopendrium.

SCOPOLIA (Solanaceae)
Genus of spring-flowering perennials. Fully hardy. Prefers shade and fertile, very well-drained soil. Propagate by division in spring or by seed in autumn.
S. carniolica illus. p.235.

SCROPHULARIA (Scrophulariaceae)
Figwort
Genus of perennials and sub-shrubs, some of which are semi-evergreen or evergreen. Most species are weeds, but some are grown for their variegated foliage. Fully hardy. Does best in semi-shade and moist soil. Propagate by division in spring or by softwood cuttings in summer.
S. aquatica 'Variegata'. See S. auriculata 'Variegata'.
S. auriculata 'Variegata', syn. S. aquatica 'Variegata' (Water figwort). Evergreen, clump-forming perennial. H 60cm (24in), S 30cm (12in) or more. Has attractive, oval, toothed, dark green leaves with cream marks. Remove spikes of insignificant, maroon flowers, borne in summer.

SCUTELLARIA (Labiatae)
Skullcap
Genus of variable, summer-flowering, rhizomatous perennials, grown for their tubular flowers. Fully hardy to frost tender, min. 7–10°C (45–50°F). Needs sun and well-drained soil. Propagate by softwood cuttings in summer or by seed in autumn.
S. indica. Upright, rhizomatous perennial. H 15–30cm (6–12in), S 10cm (4in) or more. Frost hardy. Leaves are oval, toothed and hairy. Has dense racemes of long-tubed, 2-lipped, slate-blue, occasionally white flowers in summer. Suits a rock garden.
S. orientalis illus. p.333.
S. scordiifolia. Mat-forming, rhizomatous perennial. H and S 15cm (6in) or more. Fully hardy. Bears narrowly oval, wrinkled leaves. In summer-autumn produces racemes of tubular, hooded, purple flowers, each with a white-streaked lip. Propagate by division in spring.

SEDUM (Crassulaceae)
Stonecrop
Genus of often fleshy or succulent annuals, evergreen biennials, mostly evergreen or semi-evergreen perennials and evergreen shrubs and sub-shrubs, suitable for rock gardens and borders. Fully hardy to frost tender, min. 5°C (41°F). Needs sun. Does best in fertile, well-drained soil. Propagate perennials, sub-shrubs and shrubs either by softwood cuttings of non-flowering shoots or by division from spring to mid-summer or by seed in autumn or spring. Propagate annuals and biennials by seed, sown under glass in early spring or outdoors in mid-spring.
S. acre illus. p.332. **'Aureum'** illus. p.333.
S. aizoon. Evergreen, erect perennial. H and S 45cm (18in). Fully hardy. Mid-green leaves are oblong to lance-shaped, fleshy and toothed. In summer bears flat heads of star-shaped, yellow flowers. **'Aurantiacum'** (syn. S. a. 'Euphorbioides') illus. p.257.
S. anacampseros, syn. Hylotelephium anacampseros. Semi-evergreen, trailing perennial with overwintering foliage rosettes. H 10cm (4in), S 25cm (10in) or more. Frost hardy. Prostrate, loosely rosetted, brown stems bear oblong to oval, fleshy, glaucous green leaves. Dense, sub-globose, terminal heads of small, cup-shaped, purplish-pink flowers appear in summer.
S. caeruleum illus. p.277.
♚ S. cauticola, syn. Hylotelephium cauticola. Trailing, shallow-rooted perennial with stolons. H 5cm (2in), S 20cm (8in). Fully hardy. Has oval to oblong, stalked, fleshy, blue-green leaves on procumbent, purplish-red stems. Produces leafy, branched, flattish heads of star-shaped, pale purplish-pink flowers in early autumn. Cut back old stems in winter.
S. ewersii, syn. Hylotelephium ewersii. Trailing perennial. H 5cm (2in), S 15cm (6in). Fully hardy. Is similar to S. cauticola, but has more rounded, stem-clasping leaves that are often tinted red and dense, rounded flower heads.
♚ S. kamtschaticum. Semi-evergreen, prostrate perennial with overwintering foliage rosettes. H 5–8cm (2–3in), S 20cm (8in). Fully hardy. Bears narrowly oval, toothed, fleshy, mid-green leaves. Spreading, terminal clusters of star-shaped, orange-flushed, yellow flowers appear in summer-autumn. **'Variegatum'** illus. p.339.
S. lydium illus. p.336.
♚ S. morganianum (Burro's tail, Donkey-tail). Evergreen, prostrate, succulent perennial. H 30cm (12in) or more, S indefinite. Frost tender. Stems are clothed in masses of oblong to lance-shaped, almost cylindrical, fleshy, waxy, white leaves. In summer bears terminal clusters of star-shaped, rose-pink flowers.
S. obtusatum illus. p.336.
S. palmeri. Evergreen, clump-forming perennial. H 20cm (8in), S 30cm (12in). Half hardy. Sprays of star-shaped, yellow or orange flowers open in early summer above oblong-oval to spoon-shaped, fleshy, grey-green leaves.
S. populifolium, syn. Hylotelephium populifolium. Semi-evergreen, bushy perennial. H 30–45cm (12–18in), S 30cm (12in). Fully hardy. Terminal clusters of hawthorn-scented, star-shaped, pale pink or white flowers are borne in late summer. Has broadly oval, irregularly toothed, fleshy, mid-green leaves.
S. reflexum illus. p.307.
S. rosea, syn. Rhodiola rosea (Roseroot). Clump-forming perennial. H and S 30cm (12in). Fully hardy. Stems are clothed with oval, toothed, fleshy, glaucous leaves and, in late spring or early summer, bear dense, terminal heads of pink buds that open to small, star-shaped, greenish-, yellowish- or purplish-white flowers. var. **heterodontum** (syn. Rhodiola heterodonta) illus. p.257.
S. sempervivoides. Evergreen, basal-rosetted biennial. H 8–10cm (3–4in), S 5cm (2in). Half hardy. Has rosettes that are similar to those of Sempervivum; oval to strap-shaped, leathery, glaucous green leaves are strongly marked red-purple. Produces domed heads of star-shaped, scarlet flowers in summer. Dislikes winter wet. Is good for an alpine house.
S. sieboldii 'Variegatum', syn. S.s. 'Folius Medio-variegatis'. Evergreen, spreading, tuberous perennial with long, tapering tap roots. H 10cm (4in), S 20cm (8in) or more. Frost tender. Rounded, fleshy, blue-green leaves, splashed cream and occasionally red-edged, appear in whorls of 3. Bears open, terminal heads of star-shaped, pink flowers in late summer. Is good for an alpine house.
S. spathulifolium illus. p.337.
♚ **'Cape Blanco'** (syn. S. s. 'Cappa Blanca', Hylotelephium sieboldii 'Variegatum'), illus. p.339.
♚ S. spectabile, syn. Hylotelephium spectabile (Ice-plant). Clump-forming perennial. H and S 45cm (18in). Fully hardy. Has oval, indented, fleshy, grey-green leaves, above which flat heads of small, star-shaped, pink flowers that attract butterflies are borne in late summer. ♚ **'Brilliant'** illus. p.250.
S. spurium. Semi-evergreen, mat-forming, creeping perennial. H 10cm (4in) or more, S indefinite. Frost hardy. Oblong to oval, toothed leaves are borne along hairy stems. Large, slightly rounded heads of small, star-shaped, usually purplish-pink flowers are produced in summer. Flower colour varies from deep purple to white.
S. tartarinowii, syn. Hylotelephium tartarinowii. Arching, spreading, tuberous perennial. H 10cm (4in), S 20cm (8in). Fully hardy. Rounded, terminal heads of star-shaped,

pink-flushed, white flowers appear in late summer above small, oval, toothed, green leaves borne along purplish stems. Suits an alpine house.

SELAGINELLA (Selaginellaceae)
Genus of evergreen, moss-like perennials, grown for their foliage. Frost tender, min. 5°C (41°F). Prefers semi-shade and needs moist but well-drained, peaty soil. Remove faded foliage regularly. Propagate from pieces with roots attached that have been broken off plant in any season.
♥ *S. kraussiana* illus. p.188. **'Aurea'** is an evergreen, moss-like perennial. H 1cm (1/2in), S indefinite. Spreading, filigreed, bright yellowish-green fronds are much-branched, denser towards the growing tips and easily root on soil surface. ♥ **'Variegata'** has foliage splashed with creamy-yellow.
S. lepidophylla (Resurrection plant, Rose of Jericho). Evergreen, moss-like perennial. H and S 10cm (4in). Bluntly rounded, emerald-green fronds, ageing red-brown or grey-green, are produced in dense tufts. On drying, fronds curl inwards into a tight ball; they unfold when placed in water.
♥ *S. martensii* illus. p.187.

SELENICEREUS (Cactaceae)
Genus of summer-flowering, perennial cacti with climbing, 4–10-ribbed, green stems, to 2cm (3/4in) across. Nocturnal, funnel-shaped flowers eventually open flat. Frost tender, min. 5°C (41°F). Needs sun or partial shade and rich, well-drained soil. Propagate by seed or stem cuttings in spring or summer.
S. grandiflorus illus. p.392.

SELINUM (Umbelliferae)
Genus of summer-flowering perennials, ideal for informal gardens and backs of borders. Fully hardy. Prefers sun, but will grow in semi-shade, and any well-drained soil. Once established, roots resent disturbance. Propagate by seed when fresh, in summer or autumn.
S. tenuifolium. See *S. wallichianum.*
S. wallichianum, syn *S. tenuifolium.* Upright, architectural perennial. H 1.5m (5ft), S 60cm (2ft). In summer produces small, star-shaped, white flowers, borne in large, flat heads, one above another. Has very finely divided, mid-green leaves.

SEMELE (Liliaceae)
Genus of one species of evergreen, twining climber. Male and female flowers are produced on the same plant. Frost tender, min. 5°C (41°F). Needs partial shade and prefers rich, well-drained soil. Propagate by division or seed in spring.
S. androgyna (Climbing butcher's broom). Evergreen climber, twining in upper part, branched and bearing oval cladodes, 5–10cm (2–4in) long. H to 7m (22ft). Star-shaped, cream flowers appear in early summer, in notches on cladode margins, followed by orange-red berries.

SEMIAQUILEGIA (Ranunculaceae)
Genus of perennials, grown for their flowers that differ from those of *Aquilegia*, with which it is sometimes included, by having no spurs. Is good for rock gardens. Fully hardy. Requires sun and moist but well-drained soil. Propagate by seed in autumn.
S. ecalcarata illus. p.303.

SEMIARUNDINARIA
(Bambusoideae). See GRASSES, BAMBOOS, RUSHES and SEDGES.
♥ *S. fastuosa*, syn. *Arundinaria fastuosa*, illus. p.184.

SEMPERVIVUM (Crassulaceae)
Houseleek
Genus of evergreen perennials that spread by short stolons and are grown for their symmetrical rosettes of oval to strap-shaped, pointed, fleshy leaves. Makes ground-hugging mats, suitable for rock gardens, screes, walls, banks and alpine houses. Flowers are star-shaped with 8–16 spreading petals. Fully hardy. Needs sun and gritty soil. Takes several years to reach flowering size. Rosettes die after flowering but leave numerous offsets. Propagate by offsets in summer.
♥ *S. arachnoideum* illus. p.337.
♥ *S. ciliosum* illus. p.337.
♥ *S.* **'Commander Hay'**. Evergreen, basal-rosetted perennial. H 15cm (6in), S to 30cm (12in). Is mainly grown for its very large, dark red rosettes to 10cm (4in) across. Terminal clusters of dull greenish-red flowers are produced in summer.
S. giuseppii, syn. *G.* x *giuseppii*, illus. p.339.
S. grandiflorum. Evergreen, basal-rosetted perennial. H 10cm (4in), S to 20cm (8in). Variable, densely haired, red-tinted, dark green rosettes exude a goat-like smell when crushed. Produces loose, terminal clusters of yellow-green flowers, stained purple in centres, on long flower stems in summer. Prefers humus-rich, acid soil.
S. hirtum. See *Jovibarba hirta.*
S. montanum illus. p.338.
S. soboliferum. See *Jovibarba sobolifera.*
♥ *S. tectorum* illus. p.337.

SENECIO (Compositae)
Genus of annuals, succulent and non-succulent perennials and evergreen shrubs, sub-shrubs and twining climbers, grown for their foliage and usually daisy-like flower heads. Some shrubby species are now referred to the genus *Brachyglottis*. Shrubs are excellent for coastal gardens. Fully hardy to frost tender, min. 5–10°C (41–50°F). Most prefer full sun and well-drained soil (*S. articulatus* and *S. rowleyanus* tolerate partial shade and need very well-drained soil). Propagate shrubs and climbers by semi-ripe cuttings in summer, annuals by seed in spring, perennials by division in spring (*S. articulatus* and *S. rowleyanus* by seed or stem cuttings in spring or summer).
S. articulatus, syn. *Kleinia articulata* (Candle plant). Deciduous, spreading, perennial succulent. H 60cm (2ft), S indefinite. Frost tender, min. 10°C (50°F). Branching, grey-marked, blue stems have weak joints. Bears rounded to oval, 3–5-lobed, grey leaves and flattish heads of small, cup-shaped, yellow flowers from spring to autumn. Offsets freely from stolons. var. *variegata* (syn. *S. a.* 'Variegatus') illus. p.398.
S. cineraria, syn. *S. maritima*. Moderately fast-growing, evergreen, bushy sub-shrub, often grown as an annual. H and S 30cm (1ft). Half hardy. Has long, oval, very deeply lobed, hairy, silver-grey leaves. Rounded, yellow flower heads appear in summer, but are best removed.
♥ **'Silver Dust'** illus. p.286.
S. clivorum **'Desdemona'**. See *Ligularia dentata* 'Desdemona'.
S. compactus, syn. *Brachyglottis compacta*. Evergreen, bushy, dense shrub. H 1m (3ft), S 2m (6ft). Frost hardy. Has shoots covered with felt-like, white growth and oval, white-edged, dark green leaves, white beneath. Clustered heads of daisy-like, bright yellow flowers are borne from mid- to late summer.
S. confusus, syn. *Pseudogynoxys chenopodioides* illus. p.177.
S. elegans. Moderately fast-growing, upright annual. H 45cm (18in), S 15cm (6in). Half hardy. Has oval, deeply lobed, deep green leaves. Daisy-like, purple flower heads appear on branching stems in summer.
S. grandifolius, syn. *Telanthophora grandiflora*. Evergreen, erect, robust-stemmed shrub. H 3–5m (10–15ft), S 2–3m (6–10ft). Frost tender, min. 10°C (50°F) to flower well. Has oval, toothed, boldly veined leaves, 20–45cm (8–18in) long, glossy, rich green above, red-brown-haired beneath. Carries terminal clusters, 30cm (12in) wide, of small, daisy-like, yellow flower heads in winter-spring.
S. x *hybridus*, syn. *Cineraria* x *hybrida*, *Pericallis* x *hybrida* (Cineraria). Slow-growing, evergreen, mound- or dome-shaped perennial. Cultivars are grown as biennials. Half hardy. All have oval, serrated, mid- to deep green leaves. Large, daisy-like, single, semi-double or double flower heads, in shades of blue, red, pink or white, are produced in winter or spring.
Mini-starlet Series, H and S 15cm (6in), has small, single flower heads.
Saucer Series, H 40cm (16in), S 30cm (12in), has extra-large, single flower heads. Flower heads of **Superb Series**, H 40cm (16in), S 30cm (12in), are large and single.
S. laxifolius, syn. *Brachyglottis laxifolia*. Evergreen, bushy, spreading shrub. H 1m (3ft), S 2m (6ft). Oval, grey-white leaves become dark green. Bears daisy-like, golden-yellow flower heads in large clusters in summer.
S. macroglossus (Natal ivy, Wax vine). Evergreen, woody-stemmed, twining climber. H 3m (10ft). Frost tender, min. 7°C (45°F), best at 10°C (50°F). Leaves are sharply triangular, fleshy-textured and glossy. Loose clusters of daisy-like flower heads, each with a few white ray petals and a central, yellow disc, are borne mainly in winter. ♥ **'Variegatus'** illus. p.179.
S. maritima. See *S. cineraria.*
S. mikanioides, syn. *Delairea odorata* (German ivy). Evergreen, semi-woody, twining climber. H 2–3m (6–10ft). Frost tender, min. 5°C (41°F), best at 7–10°C (45–50°F). Has fleshy leaves with 5–7 broad, pointed, radiating lobes. Mature plants carry large clusters of small, yellow flower heads in autumn-winter.
♥ *S. monroi*, syn. *Brachyglottis monroi* illus. p.141.
S. pulcher illus. p.258.
S. reinoldii, syn. *Brachyglottis rotundifolia*. Evergreen, rounded, dense shrub. H and S 1m (3ft). Frost hardy. Leaves are rounded, leathery, glossy and dark green. Bears inconspicuous, yellow flower heads from early to mid-summer. Withstands full exposure to salt winds in mild, coastal areas.
S. rowleyanus illus. p.398.
S. smithii. Bushy perennial. H 1–1.2m (3–4ft), S 75cm–1m (2 1/2–3ft). Fully hardy. Woolly stems are clothed with long, oval, toothed, leathery, dark green leaves. Daisy-like, white flower heads, with yellow centres, are borne in terminal clusters, up to 15cm (6in) across, in early summer. Likes boggy conditions.
S. **'Spring Glory'** illus. p.284.
♥ *S.* **'Sunshine'** illus. p.140.
S. tamoides. Evergreen, woody-stemmed, twining climber. H 5m (15ft) or more. Frost tender, min. 5–10°C (41–50°F). Has ivy-shaped, light green leaves. In autumn-winter bears yellow flower heads that are daisy-like, but with only a few ray petals.

SENNA. See CASSIA.

SEQUOIA (Taxodiaceae). See CONIFERS.
♥ *S. sempervirens* (Coast redwood, Coastal redwood, Redwood). Very vigorous, columnar to conical conifer with horizontal branches. H 20–30m (70–100ft), S 5–8m (15–25ft), although one specimen – the tallest tree in the world – has reached 112m (375ft). Fully hardy. Has thick, soft, fibrous, red-brown bark and needle-like, flattened, pale green leaves, spirally arranged. Rounded to cylindrical cones are green, ripening to dark brown. Will regrow if cut back. Very cold winters kill foliage, but without affecting tree.

SEQUOIADENDRON (Taxodiaceae). See CONIFERS.
♥ *S. giganteum* illus. p.76.
'Pendulum' is a weeping conifer. H 10m (30ft), S 2m (6ft) or more. Fully hardy. Bark is thick, soft, fibrous and red-brown. Has spiralled, needle-like, incurved, grey-green leaves that darken and become glossy.

SERAPIAS. See ORCHIDS.
S. cordigera. Deciduous, terrestrial orchid. H 40cm (16in). Half hardy. Spikes of reddish or dark purple flowers, 4cm (1 1/2in) long, are borne in spring. Has lance-shaped, red-spotted leaves, 15cm (6in) long. Grow in semi-shade.

SERENOA (Palmae)
Genus of one species of evergreen fan palm, grown for its foliage. Frost

tender, min. 10–13°C (50–55°F). Needs full light or partial shade and well-drained soil. Water potted plants moderately during growing season, less at other times. Propagate by seed or suckers in spring. Red spider mite may be troublesome.
S. repens (Saw palmetto, Scrub palmetto). Evergreen, rhizomatous fan palm, usually stemless. H 60cm–1m (2–3ft), S 2m (6ft) or more. Palmate leaves, 45–75cm (18–30in) wide, are divided into 6–20 strap-shaped, grey to blue-green lobes. Clusters of tiny, fragrant, cream flowers are hidden among leaves in summer, and are followed by egg-shaped, purple-black fruits.

SERISSA (Rubiaceae)
Genus of one species of evergreen shrub, grown for its overall appearance. Frost tender, min. 7–10°C (45–50°F). Needs sun or partial shade and fertile, well-drained soil. Water containerized specimens moderately, less when not in growth. May be trimmed after flowering. Propagate by semi-ripe cuttings in summer.
S. foetida, syn. *S. japonica*. Evergreen, spreading to rounded, freely branching shrub. H to 60cm (2ft), S 60cm–1m (2–3ft). Tiny, oval leaves are lustrous and deep green. Small, funnel-shaped, 4- or 5-lobed, white flowers are produced from spring to autumn.
S. japonica. See *S. foetida*.

SERRATULA (Compositae)
Genus of perennials, grown for their thistle-like flower heads. Fully hardy. Needs sun and well-drained soil. Propagate by seed or division in spring.
S. seoanii, syn. *S. shawii*. Upright, compact perennial. H 23cm (9in), S 12–15cm (5–6in). Stems bear feathery, finely cut leaves and, in autumn, terminal panicles of small, thistle-like, purple flower heads. Is useful for a rock garden.
S. shawii. See *S. seoanii*.

SESLERIA (Gramineae). See GRASSES, BAMBOOS, RUSHES and SEDGES.
S. heufleriana (Balkan blue grass). Evergreen, tuft-forming, perennial grass. H 50cm (20in), S 30–45cm (12–18in). Fully hardy. Bears rich green leaves, glaucous beneath, and, in spring, compact panicles of purple spikelets.

SETARIA (Gramineae). See GRASSES, BAMBOOS, RUSHES and SEDGES.
S. italica (Foxtail millet, Italian millet). Moderately fast-growing, annual grass with stout stems. H 1.5m (5ft), S to 1m (3ft). Half hardy. Has lance-shaped, mid-green leaves, to 45cm (1½ft) long, and loose panicles of white, cream, yellow, red, brown or black flowers in summer-autumn.

Setcreasea purpurea. See *Tradescantia pallida*.

SHEPHERDIA (Elaeagnaceae)
Genus of deciduous or evergreen shrubs, grown for their foliage and fruits. Separate male and female plants are needed in order to obtain fruits. Fully hardy. Requires sun and well-drained soil. Propagate by softwood cuttings in summer or by seed in autumn.
S. argentea (Buffalo berry). Deciduous, bushy, often tree-like shrub. H and S 4m (12ft). Bears tiny, inconspicuous, yellow flowers amid oblong, silvery leaves in spring, followed by small, egg-shaped, bright red fruits.

SHIBATAEA (Bambusoideae). See GRASSES, BAMBOOS, RUSHES and SEDGES.
S. kumasasa illus. p.184.

SHORTIA (Diapensiaceae)
Genus of evergreen, spring-flowering perennials with leaves that often turn red in autumn-winter. Fully hardy, but buds may be frosted in areas without snow cover. Is difficult to grow in hot, dry climates. Needs shade or semi-shade and well-drained, peaty, sandy, acid soil. Propagate by runners in summer or by seed when available.
S. galacifolia illus. p.312.
S. soldanelloides illus. p.314. var. *ilicifolia* is an evergreen, mat-forming perennial. H 5–10cm (2–4in), S 10–15cm (4–6in). Has rounded, toothed leaves. In late spring each flower stem carries 4–6 small, pendent, bell-shaped flowers with fringed edges and rose-pink centres shading to white. Flowers of var. *magna* are rose-pink throughout.
S. uniflora 'Grandiflora'. Vigorous, evergreen, mat-forming perennial with a few rooted runners. H 8cm (3in), S 20cm (8in). Leaves are rounded, toothed, leathery and glossy. Flower stems bear cup-shaped, 5cm (2in) wide, white-pink flowers, with serrated petals, in spring.

SIBIRAEA (Rosaceae)
Genus of deciduous shrubs, grown for their foliage and flowers. Fully hardy. Needs sunny, well-drained soil. Established plants benefit from having old or weak shoots cut to base after flowering. Propagate by softwood cuttings in summer.
S. altaiensis, syn. *S. laevigata*. Deciduous, spreading, open shrub. H 1m (3ft), S 1.5m (5ft). Has narrowly oblong, blue-green leaves and, in late spring and early summer, dense, terminal clusters of tiny, star-shaped, white flowers.
S. laevigata. See *S. altaiensis*.

SIDALCEA (Malvaceae)
Genus of summer-flowering perennials, grown for their hollyhock-like flowers. Fully hardy. Requires sun and any well-drained soil. Propagate by division in spring.
S. 'Jimmy Whitelet'. See *S.* 'Jimmy Whittet'.
S. 'Jimmy Whittet', syn. *S.* 'Jimmy Whitelet', illus. p.207.
S. 'Loveliness'. Upright perennial. H 1m (3ft), S 45cm (1½ft). Has buttercup-like, divided leaves with narrowly oblong segments. In summer bears racemes of shallowly cup-shaped, shell-pink flowers.
S. 'Oberon'. Upright perennial. H 60cm (2ft), S 45cm (1½ft). Has buttercup-like, divided leaves, with narrowly oblong segments, and, in summer, racemes of shallowly cup-shaped, clear pink flowers.
S. 'Puck'. Upright perennial. H 60cm (2ft), S 45cm (1½ft). Has buttercup-like, divided leaves with narrowly oblong segments. In summer produces racemes of shallowly cup-shaped, deep pink flowers.
S. 'Sussex Beauty'. Upright perennial. H 1.2m (4ft), S 45cm (1½ft). Has buttercup-like, divided leaves, with narrowly oblong segments, and, in summer, shallowly cup-shaped, deep rose-pink flowers.

SIDERITIS (Labiatae)
Genus of evergreen perennials, sub-shrubs and shrubs, grown mainly for their foliage. Half hardy to frost tender, min. 7–10°C (45–50°F). Needs full light and well-drained soil. Water containerized plants moderately, less when temperatures are low. Remove spent flower spikes after flowering. Propagate by seed in spring or by semi-ripe cuttings in summer.
S. candicans. Evergreen, erect, well-branched shrub. H to 75cm (2½ft), S to 60cm (2ft). Frost tender. Lance-shaped to narrowly oval or triangular leaves bear dense, white wool. Produces leafy, terminal spikes of tubular, pale yellow-and-light-brown or orange-red flowers in summer.

SILENE (Caryophyllaceae)
Campion, Catchfly
Genus of annuals and perennials, some of which are evergreen, grown for their mass of 5-petalled flowers. Fully to half hardy. Needs sun and fertile, well-drained soil. Propagate by softwood cuttings in spring or by seed in spring or early autumn.
S. acaulis illus. p.313.
S. alpestris, syn. *Heliosperma alpestris*, illus. p.321.
S. armeria 'Electra' illus. p.275.
S. coeli-rosa, syn. *Agrostemma coeli-rosa*, *Lychnis coeli-rosa*, *Viscaria elegans*, illus. p.272. 'Rose Angel' illus. p.277.
S. elisabetha. Basal-rosetted perennial. H 10cm (4in), S 20cm (8in). Fully hardy. Has rosettes of strap-shaped, mid-green leaves. In summer, stems bear large, often solitary deep rose-red flowers with green centres and long-clawed petals. Is suitable for a rock garden.
S. hookeri. Short-lived, trailing, prostrate, late summer-deciduous perennial with a long, slender tap root. H 5cm (2in), S 20cm (8in). Fully hardy. Slender stems bear oval, grey leaves and, in late summer, soft pink, salmon or orange flowers, deeply cleft to base.
S. pendula (Nodding catchfly). Moderately fast-growing, bushy annual. H and S 15–20cm (6–8in). Half hardy. Has oval, hairy, mid-green leaves and, in summer and early autumn, clusters of light pink flowers.
✿ *S. schafta* illus. p.327.
S. uniflora 'Flore Pleno', syn. *S. vulgaris* subsp. *maritima* 'Flore Pleno' (Double sea campion). Lax perennial with deep, wandering roots. H and S 20cm (8in). Fully hardy. Leaves are lance-shaped and grey-green. Produces pompon-like, double, white flowers on branched stems in summer.
S. vulgaris subsp. *maritima* 'Flore Pleno'. See *S. uniflora* 'Flore Pleno'.

SILPHIUM (Compositae)
Genus of fairly coarse, summer-flowering perennials. Fully hardy. Does best in sun or semi-shade and in moist but well-drained soil. Propagate by division in spring or by seed when fresh, in autumn.
S. laciniatum (Compass plant). Clump-forming perennial. H 2m (6ft), S 60cm (2ft). Mid-green leaves, composed of opposite pairs of oblong to lance-shaped leaflets, face north and south wherever the plant is grown, hence the common name. Large clusters of slightly pendent, daisy-like, yellow flower heads are borne in late summer.

SILYBUM (Compositae)
Genus of thistle-like biennials, grown for their spectacular foliage. Fully hardy. Grow in sun and in any well-drained soil. Propagate by seed in late spring or early summer. Is prone to slug and snail damage.
S. marianum illus. p.274.

Sinarundinaria jaunsarensis. See *Yushania anceps*.
Sinarundinaria murielae. See *Fargesia spathacea*.
Sinarundinaria nitida. See *Fargesia nitida*.

SINNINGIA (Gesneriaceae)
Genus of usually summer-flowering, tuberous perennials and deciduous sub-shrubs with showy flowers. Frost tender, min. 15°C (59°F). Grow in bright light but not direct sun. Prefers a humid atmosphere and moist but not waterlogged, peaty soil. When leaves die down after flowering, allow tubers to dry out; then store in a frost-free area. Propagate in spring by seed or in late spring or summer by stem cuttings or dividing tubers into sections, each with a young shoot.
S. barbata. Bushy, tuberous perennial with square, red stems. H and S 60cm (2ft) or more. Broadly lance-shaped leaves, to 15cm (6in) long, are glossy, mid-green above, reddish-green beneath. In summer has 5-lobed, pouched, white flowers, 4cm (1½in) long.
S. concinna. Rosetted perennial with very small tubers. H and S to 15cm (6in). Oval to almost round, scalloped, velvety, red-veined, mid-green leaves, 2cm (¾in) long, are red below. Trumpet-shaped, bicoloured, purple and white or yellowish-white flowers, to 2cm (¾in) long, are produced in summer.
S. 'Etoile du Feu'. Short-stemmed, rosetted, tuberous perennial. H 30cm (12in), S 40cm (16in) or more. Has oval, velvety leaves, 20–24cm (8–9½in) long. Upright, trumpet-

shaped, carmine-red flowers appear in summer.
S. 'Mont Blanc'. Short-stemmed, rosetted, tuberous perennial. H 30cm (12in), S 40cm (16in) or more. Oval, velvety, mid-green leaves are 20–24cm (8–9¹/₂in) long. In summer bears upright, trumpet-shaped, pure white flowers.
S. 'Red Flicker' illus. p.246.
S. regina. See *S. speciosa*.
S. speciosa, syn. *S. regina*, *Gloxinia speciosa* (Gloxinia). Short-stemmed, rosetted, tuberous perennial. H and S to 30cm (1ft). Oval, velvety, green leaves are 20cm (8in) long. Nodding, funnel-shaped, fleshy, violet, red or white flowers, to 5cm (2in) long and pouched on lower sides, are produced in summer. Is a parent of many named hybrids, of which a selection is included above and below.
S. 'Switzerland' illus. p.246.
S. 'Waterloo'. Short-stemmed, rosetted, tuberous perennial. H 30cm (12in), S 40cm (16in) or more. Oval leaves, 20–24cm (8–9¹/₂in) long, are velvety and green. Upright, trumpet-shaped, bright scarlet flowers open in summer.

SINOFRANCHETIA
(Lardizabalaceae)
Genus of one species of deciduous, twining climber, grown mainly for its handsome leaves. Is suitable for covering buildings and growing up large trees. Male and female flowers are produced on separate plants. Frost hardy. Grow in semi-shade and in any well-drained soil. Propagate by semi-ripe cuttings in summer.
S. chinensis. Deciduous, twining climber. H to 15m (50ft). Mid- to dark green leaves have 3 oblong to oval leaflets, each 5–15cm (2–6in) long. In late spring has small, dull white flowers in pendent racemes, to 10cm (4in) long. Pale purple berries, containing many seeds, follow in summer.

SINOJACKIA (Styracaceae)
Genus of deciduous shrubs and trees, grown for their flowers. Fully hardy. Requires a sheltered position in sun or partial shade and fertile, humus-rich, moist, acid soil. Propagate by softwood cuttings in summer.
S. rehderiana. Deciduous, bushy shrub or spreading tree. H and S 6m (20ft). Nodding, saucer-shaped, white flowers, each with a central cluster of yellow anthers, are produced in late spring and early summer. Leaves are dark green and oval.

SINOWILSONIA (Hamamelidaceae)
Genus of one species of deciduous tree, grown for its foliage and catkins. Fully hardy. Requires sun or semi-shade and fertile, moist but well-drained soil. Propagate by seed in autumn.
S. henryi. Deciduous, spreading, sometimes shrubby tree. H and S 8m (25ft). Has oval, toothed, glossy, bright green leaves, and long, pendent, green catkins in late spring.

Siphonosmanthus delavayi. See *Osmanthus delavayi*.

SISYRINCHIUM (Iridaceae)
Genus of annuals and perennials, some of which are semi-evergreen. Fully to half hardy. Prefers sun, although tolerates partial shade, and well-drained or moist soil. Propagate by division in early spring or by seed in spring or autumn.
S. angustifolium. See *S. graminoides*.
S. bellum illus. p.331.
S. bermudiana. See *S. graminoides*.
S. brachypus. See *S. californicum*.
S. californicum (Golden-eyed grass). Semi-evergreen, upright perennial. H 30–60cm (12–24in), S 30cm (12in). Frost hardy. Has grass-like tufts of basal, light green leaves. For a long period in spring-summer produces flattish, bright yellow flowers, with slightly darker veins, on winged stems. Outer leaves may die off and turn black in autumn. Dwarf forms are known as *S. brachypus*. Prefers moist soil.
♀ **S. douglasii**, syn. *Olsynium douglasii*, *Sisyrinchium grandiflorum* (Grass widow, Spring bell). Stiff, upright, summer-deciduous perennial. H 25cm (10in), S 15cm (6in). Fully hardy. Has grass-like leaves sheathing very short, thread-like flowering stems and, in early spring, a succession of pendent, bell-shaped, violet to red-purple, or sometimes white, flowers. Suits a rock garden or alpine house.
S. graminoides, syn. *S. angustifolium*, *S. bermudiana*, illus. p.304.
S. grandiflorum. See *S. douglasii*.
S. odoratissimum. See *Phaiophleps biflora*.
S. striatum illus. p.254. **'Aunt May'** (syn. *S.s.* 'Variegatum') is a semi-evergreen, upright perennial. H 45–60cm (18–24in), S 30cm (12in). Fully hardy. Produces tufts of long, narrow, cream-striped, greyish-green leaves. Slender spikes of trumpet-shaped, purple-striped, straw-yellow flowers are borne in summer.

SKIMMIA (Rutaceae)
Genus of evergreen, spring-flowering shrubs and trees, grown for their flowers, aromatic foliage and their fruits. Except with *S. japonica* subsp. *reevesiana*, separate male and female plants are needed in order to obtain fruits. Fully to frost hardy. Needs shade or semi-shade and fertile, moist soil. Too much sun or poor soil may cause chlorosis. Propagate by semi-ripe cuttings in late summer or by seed in autumn.
S. anquetilia. Evergreen, bushy, open shrub. H 1.2m (4ft), S 2m (6ft). Fully hardy. Produces small clusters of tiny, yellow flowers from mid- to late spring, then spherical, scarlet fruits. Leaves are oblong to oval, pointed, strongly aromatic and dark green.
S. x foremanii of gardens. See *S. japonica* 'Veitchii'.
S. japonica illus. p.145. **'Fructo-alba'** (female) illus. p.143.
♀ subsp. *reevesiana* **'Robert Fortune'** (syn. *S. reevesiana*; hermaphrodite) and ♀ **'Rubella'** (male) illus. p.144. **'Veitchii'** (syn. *S.* x *foremanii* of gardens) is a vigorous, evergreen, upright, dense, female shrub. H and S 1.5m (5ft). Fully hardy. In mid- and late spring produces dense clusters of small, star-shaped, white flowers, followed by large, spherical, bright red fruits. Broadly oval leaves are rich green.
S. reevesiana. See *S. japonica* subsp. *reevesiana*.

SMILACINA (Liliaceae)
Genus of perennials, grown for their graceful appearance. Fully hardy. Prefers semi-shade and humus-rich, moist, neutral to acid soil. Propagate by division in spring or by seed in autumn.
♀ **S. racemosa** illus. p.197.

SMILAX (Liliaceae)
Genus of deciduous or evergreen, woody-stemmed or herbaceous, scrambling climbers with tubers or rhizomes. Male and female flowers are borne on separate plants. Frost hardy to frost tender, min. 5°C (41°F). Grow in any well-drained soil and in sun or semi-shade. Propagate by division or seed in spring or by semi-ripe cuttings in summer.
S. china. Deciduous, woody-based, scrambling climber with straggling, sometimes spiny stems. H to 5m (15ft). Frost hardy. Leaves are broadly oval to rounded. Umbels of yellow-green flowers are produced in spring; tiny, red berries appear in autumn.

SMITHIANTHA (Gesneriaceae)
Genus of bushy, erect perennials with tuber-like rhizomes, grown for their flowers. Frost tender, min. 15°C (59°F). Grow in humus-rich, well-drained soil and in bright light but out of direct sun. Reduce watering after flowering and water sparingly in winter. Propagate by division of rhizomes in early spring.
S. cinnabarina (Temple bells). Robust, erect, rhizomatous perennial. H and S to 60cm (2ft). Broadly oval to almost rounded, toothed leaves, to 15cm (6in) long, are dark green with dark red hairs. Bell-shaped, orange-red flowers, lined with pale yellow, are produced in summer-autumn.
S. 'Orange King' illus. p.247.
S. zebrina. Bushy, rhizomatous perennial with velvety-haired stems. H and S to 1m (3ft). Oval, toothed, hairy leaves, to 18cm (7in) long, are deep green marked with reddish-brown. In summer produces tubular flowers, scarlet above, yellow below, spotted red inside and with orange-yellow lobes.

SMYRNIUM (Umbelliferae)
Genus of biennials. Fully hardy. Grow in sun and in fertile, well-drained soil. Propagate by seed sown outdoors in autumn or spring.
S. perfoliatum illus. p.287.

SOLANDRA (Solanaceae)
Genus of evergreen, woody-stemmed, scrambling climbers, grown for their large, trumpet-shaped flowers. Frost tender, min. 10°C (50°F), but prefers 13–16°C (55–61°F). Needs full light and fertile, well-drained soil. Water freely when in full growth, sparingly in cold weather. Tie to supports. Thin out crowded stems after flowering. Propagate by semi-ripe cuttings in summer.
S. maxima illus. p.166.

SOLANUM (Solanaceae)
Genus of annuals, perennials (some of which are evergreen) and evergreen, semi-evergreen or deciduous sub-shrubs, shrubs (occasionally scandent) and woody-stemmed, scrambling or leaf-stalk climbers, grown for their flowers and ornamental fruits. Frost hardy to frost tender, min. 5–10°C (41–50°F). Requires full sun and fertile, well-drained soil. Water regularly but sparingly in winter. Support scrambling climbers. Thin out and spur back crowded growth of climbers in spring. Propagate by seed in spring or by semi-ripe cuttings in summer. Red spider mite, whitefly and aphids may cause problems.
S. capsicastrum (Winter cherry). Fairly slow-growing, evergreen, bushy sub-shrub, grown as an annual. H and S 30–45cm (1–1¹/₂ft). Half hardy. Has lance-shaped, deep green leaves. In summer bears small, star-shaped, white flowers, followed by egg-shaped, pointed, orange-red or scarlet fruits, at least 1cm (¹/₂in) in diameter, which are at their best in winter.
♀ **S. crispum 'Glasnevin'** illus. p.174.
S. jasminoides (Potato vine). Semi-evergreen, woody-stemmed, scrambling climber. H to 6m (20ft). Half hardy. Oval to lance-shaped leaves may be lobed or have leaflets at base. Small, 5-petalled, pale grey-blue flowers are produced in summer-autumn; tiny, purple berries appear in autumn. ♀ **'Album'** illus. p.167.
S. pseudocapsicum (Jerusalem cherry). Fairly slow-growing, evergreen, bushy shrub, usually grown as an annual. H and S to 1.2m (4ft). Half hardy. Has oval or lance-shaped, bright green leaves. Small, star-shaped, white flowers appear in summer and are followed by spherical, scarlet fruits. Has several smaller selections: **'Balloon'** illus. p.293; **'Dwarf Red'**, H 38cm (15in), with bright orange fruits; **'Fancy'**, H 30cm (1ft), with scarlet fruits; **'Red Giant'** illus. p.293; and **'Snowfire'**, H 30cm (1ft), with white fruits that later turn red.
S. rantonnetii, syn. *Lycianthes rantonetii* (Blue potato bush). **'Royal Robe'** illus. p.115.
S. seaforthianum (Potato creeper). Evergreen, slender-stemmed, scrambling climber. H 2–3m (6–10ft). Frost tender, min. 7°C (45°F). Leaves are divided into several pairs of oval to lance-shaped leaflets. Nodding clusters of star-shaped, mauve-violet flowers, each with a cone of yellow stamens, appear from spring to autumn, followed by small, scarlet berries.
S. wendlandii illus. p.174.

SOLDANELLA (Primulaceae)
Snowbell
Genus of evergreen, early spring-flowering perennials, grown for their flowers. Is good for rock gardens, troughs and alpine houses. Fully hardy, but flower buds are set in autumn and may be destroyed by frost if there is no snow cover. Requires

partial shade and humus-rich, well-drained, peaty soil. Propagate by seed in spring or by division in late summer. Slugs may attack flower buds.
S. alpina illus. p.316.
S. minima (Least snowbell). Evergreen, prostrate perennial. H 2.5cm (1in), S 10cm (4in). Forms a mat of minute, rounded leaves on soil surface. In early spring produces solitary almost stemless, bell-shaped, pale lavender-blue or white flowers with fringed mouths.
S. montana (Mountain tassel). Evergreen, mound-forming perennial. H 10cm (4in), S 15cm (6in). In early spring has tall flower stems each carrying long, pendent, bell-shaped, lavender-blue flowers with fringed mouths. Leaves are rounded and leathery.
S. villosa illus. p.316.

SOLEIROLIA (Urticaceae)
Baby's tears, Mind-your-own-business, Mother of thousands
Genus of one species of, usually evergreen, prostrate perennial that forms a dense carpet of foliage. Frost hardy, but leaves are killed by winter frost. Recovers to grow vigorously again in spring. Tolerates sun or shade. Prefers moist soil. Propagate by division from spring to mid-summer.
S. soleirolii, syn. *Helxine soleirolii*, illus. p.268.

Solenostemon scutellarioides. See *Coleus blumei.*

SOLIDAGO (Compositae)
Golden rod
Genus of summer- and autumn-flowering perennials, some species of which are vigorous, coarse plants that tend to crowd out others in borders. Fully hardy. Most tolerate sun or shade and any well-drained soil. Propagate by division in spring. Occasionally self seeds.
S. 'Golden Wings'. Upright perennial. H 1.5m (5ft), S 1m (3ft). Bears large, feathery panicles of small, bright yellow flower heads in early autumn. Has lance-shaped, toothed, slightly hairy, mid-green leaves.
♥ *S.* 'Goldenmosa' illus. p.219.
S. 'Laurin' illus. p.220.
S. virgaurea subsp. *minuta*, syn. *S.v.* subsp. *alpestris*. Mound-forming perennial. H and S 10cm (4in). Has small, lance-shaped, green leaves and, in autumn, neat spikes of small, yellow flower heads. Is suitable for a rock garden, trough or alpine house. Needs shade and moist soil.

× SOLIDASTER (Compositae)
Hybrid genus (*Solidago* × *Aster*) of one summer-flowering perennial. Fully hardy. Grows in sun or shade and any fertile soil. Propagate by division in spring.
× *S. hybridus.* See × *S. luteus.*
× *S. luteus*, syn. × *S. hybridus*, illus. p.246.

SOLLYA (Pittosporaceae)
Bluebell creeper
Genus of evergreen, woody-based, twining climbers, grown for their attractive, blue flowers. Half hardy. Grow in sun and well-drained soil. Propagate by seed in spring or by softwood or greenwood cuttings in summer.
♥ *S. heterophylla* illus. p.166.

SONERILA (Melastomataceae)
Genus of evergreen, bushy perennials and shrubs, grown for their foliage and flowers. Frost tender, min. 15°C (59°F). Prefers a humid atmosphere in semi-shade and peaty soil. Propagate by tip cuttings in spring.
♥ *S. margaritacea.* Evergreen, bushy, semi-prostrate perennial. H and S 20–25cm (8–10in). Red stems bear oval, dark green leaves, 5–8cm (2–3in) long, reddish below, silver-patterned above. Has racemes of 3-petalled, rose-pink flowers in summer. 'Argentea' has more silvery leaves with green veins; 'Hendersonii' is more compact with white-spotted leaves.

SOPHORA (Leguminosae)
Kowhai
Genus of deciduous or semi-evergreen trees and shrubs, grown for their habit, foliage and clusters of flowers. Fully to frost hardy. Requires full sun (*S. microphylla* and *S. tetraptera* usually need to be grown against a south- or west-facing wall) and fertile, well-drained soil. Propagate by seed in autumn; semi-evergreens may also be raised from softwood cuttings in summer.
S. davidii, syn. *S. viciifolia*, illus. p.115.
♥ *S. japonica* (Pagoda tree). Deciduous, spreading tree. H and S 20m (70ft). Fully hardy. Dark green leaves consist of 9–15 oval leaflets. On mature trees, long clusters of pea-like, creamy-white flowers appear in late summer and early autumn. Does best in hot summers. 'Pendula', H and S 3m (10ft), has long, hanging shoots clothed with dark green foliage. 'Violacea' illus. p.45.
S. microphylla, syn. *Edwardsia microphylla.* Semi-evergreen, spreading tree. H and S 8m (25ft). Frost hardy. Dark green leaves are composed of numerous tiny, oblong leaflets. Produces clusters of pea-like, deep yellow flowers in late spring.
♥ *S. tetraptera* illus. p.63.
S. viciifolia. See *S. davidii.*

× SOPHROLAELIOCATTLEYA.
See ORCHIDS.
× *S.* Hazel Boyd 'Apricot Glow' (illus. p.255). Evergreen, epiphytic orchid for an intermediate greenhouse. H 10cm (4in). In spring and early summer produces small heads of apricot-orange flowers, 9cm (3½in) across, with crimson marks on lips. Has oval, rigid leaves, 10cm (4in) long. Grow in good light in summer.
× *S.* Trizac 'Purple Emperor' (illus. p.253). Evergreen, epiphytic orchid for an intermediate greenhouse. H 10cm (4in). In spring, has crimson-lipped, pinkish-purple flowers, 6cm (2½in) across, in small heads. Has oval, rigid leaves, 10cm (4in) long. Provide good light in summer.

SORBARIA (Rosaceae)
Genus of deciduous, summer-flowering shrubs, grown for their foliage and large panicles of small, white flowers. Fully hardy. Prefers sun and deep, fertile, moist soil. In winter cut out some older stems on mature plants and prune back remaining shoots to growing points. Remove suckers at base to prevent *Sorbaria* spreading too widely. Propagate by softwood cuttings in summer, by division in autumn or by root cuttings in late winter.
S. aitchsonii. See *S. tomentosa* var. *angustifolia.*
S. arborea. See *S. kirilowii.*
S. kirilowii, syn. *S. arborea*, *Spiraea arborea.* Vigorous, deciduous, arching shrub. H and S 6m (20ft). Leaves are composed of 13–17 lance-shaped, taper-pointed, deep green leaflets. Nodding panicles of star-shaped, white flowers are produced in mid- and late summer.
S. sorbifolia, syn. *Spiraea sorbifolia*, illus. p.107.
S. tomentosa var. *angustifolia*, syn. *S. aitchsonii*, *Spiraea aitchisonii.* Deciduous, arching shrub. H and S 3m (10ft). Shoots are red when young. Leaves have 11–23 narrowly lance-shaped, taper-pointed, dark green leaflets. Upright panicles of star-shaped, white flowers are produced from mid- to late summer.

SORBUS (Rosaceae)
Genus of deciduous trees and shrubs, grown for their foliage, small, 5-petalled flowers, attractive fruits and, in some species, autumn colour. Leaves may be whole or divided into leaflets. Fully to frost hardy. Needs sun or semi-shade and fertile, well-drained but moist soil. Species with leaves composed of leaflets do not grow well in very dry soil. Propagate by softwood cuttings or budding in summer, by seed in autumn or by grafting in winter. Is susceptible to fireblight.
S. alnifolia (Korean mountain ash). Deciduous, conical, then spreading tree. H 15m (50ft), S 8m (25ft). Fully hardy. Oval, toothed, bright green leaves turn orange and red in autumn. Small, white flowers in late spring are followed by egg-shaped, orange-red fruits.
S. americana (American mountain ash). Deciduous, round-headed tree. H 10m (30ft), S 7m (22ft). Fully hardy. Light green leaves, divided into 11–17 narrowly oval leaflets, usually colour well in autumn. Bears small, white flowers in early summer, then rounded, bright red fruits, ripening in early autumn.
S. aria (Whitebeam). Deciduous, spreading tree. H 15m (50ft), S 10m (30ft). Fully hardy. Oval, toothed leaves are silver-grey when young, maturing to dark green above, white-felted beneath. Clusters of small, white flowers in late spring are followed by rounded, brown-speckled, deep red fruits. 'Chrysophylla', H 10m (30ft), S 7m (22ft), bears golden-yellow, young leaves. 'Decaisneana' see *S.a.* 'Majestica'. ♥ 'Lutescens' illus. p.52. ♥ 'Majestica' (syn. *S.a.* 'Decaisneana') has larger leaves, white-haired when young, and larger fruits.
S. aucuparia (Mountain ash, Rowan) illus. p.55. ♥ 'Fructu Luteo' is a deciduous, spreading tree. H 15m (50ft), S 8m (25ft). Fully hardy. Leaves consist of 13–15 narrowly oval, mid-green leaflets that turn yellow or red in autumn. Bears small, white flowers in late spring, followed by rounded, orange-yellow fruits in autumn. Fruits of 'Rossica Major' (syn. *S.a.* 'Rossica') are large and deep red. ♥ 'Sheerwater Seedling', S 4m (12ft), has a narrow, upright habit.
♥ *S. cashmiriana* illus. p.67.
S. commixta, syn. *S. discolor* of gardens, illus. p.55. ♥ 'Embley' is a vigorous, deciduous, elegant tree with steeply ascending branches. H 12m (40ft), S 9m (28ft). Fully hardy. Glossy, deep green leaves, each with 13–17 slender, lance-shaped leaflets, turn orange and red in late autumn. Bears small, white flowers in late spring, followed by rounded, bright red fruits in autumn.
S. cuspidata. See *S. vestita.*
S. decora. Deciduous, spreading, sometimes shrubby tree. H 10m (30ft), S 8m (25ft). Fully hardy. Leaves are composed of oblong, blue-green leaflets. Small, white flowers in late spring are succeeded by rounded, orange-red fruits.
S. discolor of gardens. See *S. commixta.*
S. esserteauiana. Deciduous, spreading tree. H and S 10m (30ft). Fully hardy. Dark green leaves, with broadly oblong leaflets, redden in autumn. Has small, white flowers in late spring, followed by large clusters of rounded, bright red, sometimes orange-yellow fruits.
♥ *S. hupehensis* (Hupeh rowan). Deciduous, spreading tree. H 12m (40ft), S 8m (25ft). Fully hardy. Leaves have 9–17 oblong, blue-green leaflets that turn orange-red in late autumn. Small, white flowers in late spring are followed by clusters of rounded, pink-tinged, white fruits. 'Rosea' (syn. *S.h.* var. *obtusa*) illus. p.55.
S. insignis. Deciduous, spreading tree. H 8m (25ft), S 6m (20ft). Frost hardy. Leaves consist of usually 9–21 large, oblong, glossy, dark green leaflets. Large clusters of small, creamy-white flowers in late spring are followed by rounded, pink fruits that become white in winter.
S. intermedia (Swedish whitebeam). Deciduous, broad-headed, dense tree. H and S 12m (40ft). Fully hardy. Has broadly oval, deeply lobed, dark green leaves. Carries clusters of small, white flowers in late spring, then rounded, red fruits.
♥ *S.* 'Joseph Rock' illus. p.56.
S. latifolia (Service tree of Fontainebleau). Deciduous, spreading tree. H 12m (40ft), S 10m (30ft). Fully hardy. Has peeling bark and broadly oval, sharply lobed, glossy, dark green leaves. Small, white flowers in late spring are succeeded by rounded, brownish-red fruits.
S. 'Mitchellii'. See *S. thibetica* 'John Mitchell'.
S. pohuashanensis. Deciduous, spreading tree. H 10m (30ft),

603

S 8m (25ft). Fully hardy. Dark green leaves are divided into 11–15 oblong leaflets. Has small, white flowers in late spring, followed by dense clusters of rounded, red fruits.

S. prattii. Deciduous, spreading tree. H and S 6m (20ft). Fully hardy. Dark green leaves are divided into 21–9 oblong, sharply toothed leaflets. Produces small, white flowers in late spring, followed by rounded, white fruits.

♣ *S. reducta* illus. p.308.
♣ *S. sargentiana* (Sargent's rowan). Deciduous, sparsely branched, spreading tree. H and S 6m (20ft). Fully hardy. Has stout shoots and large, mid-green leaves, consisting of 7–11 oblong leaflets, that turn brilliant red in autumn. Small, white flowers in late spring are succeeded by rounded, red fruits.

♣ *S. scalaris.* Deciduous, spreading, graceful tree. H and S 10m (30ft). Fully hardy. Produces leaves with 21–33 narrowly oblong, glossy, deep green leaflets that become deep red and purple in autumn. Has small, white flowers in late spring, followed by rounded, red fruits in large, dense clusters.

S. scopulina. Deciduous, upright shrub. H and S 4m (12ft). Fully hardy. Bears dark green leaves divided into 11–15 oblong leaflets, clusters of small, white flowers in late spring or early summer and rounded, bright red fruits in autumn.

S. thibetica. Deciduous, conical tree. H 20m (70ft), S 15m (50ft). Fully hardy. Large, broadly oval, dark green leaves are silvery-white when young and remain so on undersides. Heads of small, white flowers in late spring are followed by rounded, brown fruits.
♣ **'John Mitchell'** (syn. *S.* 'Mitchellii') illus. p.53.

S. × thuringiaca. Deciduous, broadly conical, compact tree. H 12m (40ft), S 8m (25ft). Fully hardy. Oval, dark green leaves are deeply lobed and have basal leaflets. Small, white flowers appear in late spring, followed by rounded, bright red fruits. **'Fastigiata'** has upright branches and a broad, oval, dense crown.

S. vestita, syn. *S. cuspidata,* illus. p.52.
♣ *S. vilmorinii* illus. p.68.
S. **'Wilfred Fox'.** Deciduous tree, upright when young, later with a dense, oval head. H 15m (50ft), S 10m (30ft). Fully hardy. Has broadly oval, glossy, dark green leaves and small, white flowers in late spring, followed by rounded, orange-brown fruits.

SPARAXIS (Iridaceae)
Harlequin flower
Genus of spring- and early summer-flowering corms, grown for their very gaudy flowers. Half hardy. Needs a sunny, well-drained site. Plant in autumn. Dry off corms after flowering. Propagate by offsets in late summer or by seed in autumn.

S. elegans, syn. *Streptanthera cuprea, S. elegans.* Spring-flowering corm. H 10–25cm (4–10in), S 8–12cm (3–5in). Has lance-shaped leaves in an erect, basal fan. Stem produces a loose spike of 1–5 flattish, orange or white blooms, each 3–4cm (1¼–1½in) wide and with a yellow centre surrounded by a purple-black band. subsp. *grandiflora* (syn. *S. grandiflora*), H 15–40cm (6–16in), S 8–12cm (3–5in) has sword-shaped leaves in an erect, basal fan. Stem bears a loose spike of up to 5 flattish, deep purple flowers, each 4–5cm (1½–2in) across.
S. grandiflora. See *S. elegans* subsp. *grandiflora.*
S. tricolor illus. p.374.

SPARGANIUM (Sparganiaceae)
Bur reed
Genus of deciduous or semi-evergreen, perennial, marginal water plants, grown for their foliage. Fully hardy. Tolerates deep shade and cold water. Remove faded foliage and cut plants back regularly to control growth. Propagate by seed or division in spring.
S. erectum, syn. *S. ramosum,* illus. p.389.
S. minimus, syn. *S. minimum* (Least bur reed). Vigorous, deciduous or semi-evergreen, perennial, marginal water plant. H 30cm–1m (1–3ft), S 30cm (1ft). Mid-green leaves are grass-like, some erect, some floating. Insignificant, brownish-green flowers, in the form of burs, appear in summer.
S. ramosum. See *S. erectum.*

SPARMANNIA (Tiliaceae)
Genus of evergreen trees and shrubs, grown for their flowers and foliage. Frost tender, min. 7°C (45°F). Prefers full light and fertile, well-drained soil. Water freely when in full growth, moderately at other times. Flowered stems may be cut back after flowering to promote a more compact habit. Propagate by greenwood cuttings in late spring. Is prone to whitefly.
♣ *S. africana* illus. p.88.

SPARTINA (Gramineae). See GRASSES, BAMBOOS, RUSHES and SEDGES.
S. pectinata **'Aureomarginata'**, syn. *S.p.* 'Aureovariegata', illus. p.185.

SPARTIUM (Leguminosae)
Genus of one species of deciduous, almost leafless shrub, grown for its green shoots and showy flowers. Frost hardy. Needs sun and not too rich, well-drained soil. To maintain a compact habit, trim in early spring. Propagate by seed in autumn.
♣ *S. junceum* illus. p.117.

SPATHIPHYLLUM (Araceae)
Genus of evergreen perennials, with rhizomes, grown for their foliage and flowers. Frost tender, min. 15°C (59°F). Prefers a humid atmosphere, humus-rich, moist soil and partial shade. Propagate by division in spring or summer.
S. **'Clevelandii'**, syn. *S. wallisii* 'Clevelandii'. Evergreen, tufted perennial. H and S to 60cm (2ft). Has broadly lance-shaped, semi-erect, glossy, mid-green leaves, 30cm (1ft) or more long. Intermittently bears oval, white spathes, each 15cm (6in) long with a central, green line, that surround fragrant, white spadices.
S. floribundum. Evergreen, tufted, short-stemmed perennial. H and S to 30cm (1ft). Has clusters of lance-shaped, long-pointed, long-stalked, glossy, dark green leaves, to 15cm (6in) long. Intermittently, bears narrowly oval, white spathes, to 8cm (3in) long, each enclosing a green-and-white spadix.
♣ *S.* **'Mauna Loa'** illus. p.263.
S. wallisii illus. p.264. **'Clevelandii'**. See *S.* 'Clevelandii'.

SPATHODEA (Bignoniaceae)
Genus of evergreen trees, grown for their flowers, mainly from autumn to spring, and their overall appearance. Frost tender, min. 16–18°C (61–5°F). Needs full light and fertile, well-drained but moisture-retentive soil. Container-grown and immature plants seldom produce flowers. Propagate by seed in spring or by semi-ripe cuttings in summer.
S. campanulata illus. p.45.

SPHAERALCEA (Malvaceae)
Genus of perennials and deciduous sub-shrubs that are evergreen in warm climates. Half hardy. Requires a warm, sunny situation and fertile, well-drained soil. Propagate by seed or division in spring or by softwood cuttings in mid-summer.
S. ambigua illus. p.222.
S. munroana. Branching, woody-based perennial. H and S 45cm (18in). Broadly funnel-shaped, brilliant coral-pink flowers are borne singly in leaf axils from summer until first frosts. Has oval, round-toothed, hairy, mid-green leaves.

SPHAEROPTERIS. See CYATHEA.

Spiloxene capensis. See *Hypoxis capensis.*

SPIRAEA (Rosaceae)
Genus of deciduous or semi-evergreen shrubs, grown for their mass of small flowers and, in some species, their foliage. Fully hardy. Requires sun and fertile, well-drained but not over-dry soil. On species and cultivars that flower on the current year's growth – *S. × billiardii, S. douglasii* and *S. japonica* and its cultivars – cut young stems back and remove some very old ones in early spring. On species that flower on old wood, cut out older shoots in early spring, leaving young shoots to flower that year. Propagate *S. douglasii* by division between late autumn and early spring, other species and cultivars by softwood cuttings in summer.
S. aitchisonii. See *Sorbaria aitchisonii.*
S. arborea. See *Sorbaria arborea.*
S. **'Arguta'** (Bridal wreath, Foam of May). Deciduous, arching, dense shrub. H and S 2.5m (8ft). Produces clusters of 5-petalled, white flowers from mid- to late spring. Leaves are narrowly oblong and bright green.
S. aruncus. See *Aruncus dioicus.*
S. × billiardii. Deciduous, upright, dense shrub. H and S 2.5m (8ft). Bears dense panicles of 5-petalled, pink flowers in summer and oval, finely toothed, dark green leaves. **'Triumphans'** has large, broadly conical panicles of bright purplish-pink flowers.
S. × bumalda. See *S. japonica.*
S. canescens illus. p.106.
S. douglasii. Vigorous, deciduous, upright shrub. H and S 2m (6ft). Dense, narrow panicles of 5-petalled, purplish-pink flowers are borne from early to mid-summer among oblong, mid-green leaves with grey-white undersides. Leaves of var. *menziesii*, H 1m (3ft), are green on both sides.
S. japonica, syn. *S. × bumalda.*
♣ **'Anthony Waterer'**, ♣ **'Goldflame'** and **'Little Princess'** illus. p.133.
S. nipponica. Deciduous, arching shrub. H and S 2.5m (8ft). Bears dense clusters of 5-petalled, white flowers in early summer. Stout, red shoots carry small, rounded, dark green leaves. **'Halward's Silver'**, H and S 1m (3ft), is slow-growing, very dense and flowers profusely. ♣ **'Snowmound'** (syn. *S.n.* var. *tosaensis* of gardens) illus. p.109.
S. prunifolia. Deciduous, arching, graceful shrub. H and S 2m (6ft). In mid- and late spring has clusters of rosette-like, double, white flowers amid rounded to oblong, bright green leaves, colouring to bronze-yellow in autumn.
S. **'Snow White'**, syn. *S. trichocarpa* 'Snow White'. Deciduous, arching shrub. H and S 2m (6ft). Leaves are oblong and mid-green. Dense clusters of 5-petalled, white flowers are borne along shoots in late spring and early summer.
S. sorbifolia. See *Sorbaria sorbifolia.*
♣ *S. thunbergii.* Deciduous or semi-evergreen, arching, dense shrub. H 1.5m (5ft), S 2m (6ft). Small clusters of 5-petalled, white flowers are borne along slender stems from early to mid-spring. Has narrowly oblong, pale green leaves.
S. trichocarpa **'Snow White'**. See *S.* 'Snow White'.
S. trilobata. Deciduous, arching, graceful shrub. H 1m (3ft), S 1.5m (5ft). In early summer produces 5-petalled, white flowers in clusters along slender shoots. Has rounded, shallowly lobed, toothed, blue-green leaves.
S. ulmaria. See *Filipendula ulmaria.*
♣ *S. × vanhouttei* illus. p.124.
S. veitchii. Vigorous, deciduous, upright shrub. H and S 3m (10ft). Has arching, red branches and oblong, dark green leaves. Produces heads of 5-petalled, white flowers from early to mid-summer.

SPIRANTHES. See ORCHIDS.
S. cernua (illus. p.260). Deciduous, terrestrial orchid. H 50cm (20in). Frost hardy. Spikes of delicate, white flowers, 1cm (½in) long, with pale yellow centres, appear in autumn. Has narrowly lance-shaped leaves, 5–12cm (2–5in) long. Requires semi-shade in summer.

SPREKELIA (Amaryllidaceae)
Genus of one species of bulb, grown for its showy, red flowers in spring. Half hardy. Needs an open, sunny site and well-drained soil. Keep dry in

winter; start into growth by watering in spring. Propagate by offsets in early autumn.
⚑ *S. formosissima* illus. p.357.

STACHYS (Labiatae)
Genus of late spring- or summer-flowering perennials, shrubs and sub-shrubs, some of which are evergreen. Fully hardy to frost tender, min. 5°C (41°F). Grows in any well-drained soil and is particularly tolerant of poor soil. Species mentioned below prefer an open, sunny position; others are woodland plants and grow better in semi-shade. Propagate by division in spring.
S. byzantina, syn. *S. lanata*, *S. olympica*, illus. p.268. **'Silver Carpet'** is an evergreen, mat-forming perennial. H 15cm (6in), S 60cm (24in). Fully hardy. Has oval, woolly, grey leaves. Rarely produces flowers. Makes an excellent front-of-border or ground-cover plant.
S. coccinea. Clump-forming perennial. H 60cm (24in), S 45cm (18in). Frost tender. Has oval, mid-green leaves with a pronounced network of veins. From early to late summer, spikes of small, hooded, bright scarlet flowers, protruding from purple calyces, arise from leaf axils.
S. lanata. See *S. byzantina*.
S. macrantha. Clump-forming perennial. H and S 30cm (12in). Fully hardy. Has heart-shaped, crinkled, round-toothed, soft green leaves. Whorls of large, hooded, rose-purple flowers are produced in summer. **'Superba'** illus. p.249.
S. officinalis, syn. *Betonica officinalis* (Betony). Mat-forming perennial. H 45–60cm (18–24in), S 30–45cm (12–18in). Fully hardy. Whorls of hooded, tubular, purple, pink or white flowers are borne on sturdy stems, arising, in summer, from mats of oval to oblong, round-toothed, mid-green leaves. **'Rosea'** has flowers of clearer pink.
S. olympica. See *S. byzantina*.

STACHYURUS (Stachyuraceae)
Genus of deciduous shrubs, grown for their flowers borne before the leaves. Fully to half hardy; flower spikes, formed in autumn, are usually unharmed by hard frosts. Needs sun or semi-shade and fertile, not too heavy soil, preferably peaty and acid. Does well when trained against a south- or west-facing wall. Propagate by softwood cuttings in summer.
S. chinensis. Deciduous, spreading, open shrub. H 2m (6ft), S 4m (12ft). Fully hardy. Pendent spikes of small, bell-shaped, pale yellow flowers open in late winter and early spring. Leaves are oval and deep green. **'Magpie'** has grey-green leaves, broadly edged with creamy-white.
⚑ *S. praecox* illus. p.120.

STANHOPEA. See ORCHIDS.
S. tigrina. Evergreen, epiphytic orchid for a cool greenhouse. H 23cm (9in). Pendent spikes of fragrant, waxy, rich yellow and maroon flowers, 15cm (6in) across, with red-spotted, white lips, appear in summer. Has broadly oval, ribbed leaves, 30cm (12in) long. Is best grown in a hanging, slatted basket. Provide semi-shade in summer.

STAPELIA (Asclepiadaceae)
Genus of clump-forming, perennial succulents with erect, 4-angled stems. Stem edges are often indented and may produce small leaves that drop after only a few weeks. Flowers are often foul-smelling. Frost tender, min. 11°C (52°F). Needs sun or partial shade and well-drained soil. Propagate by seed or stem cuttings in spring or summer.
S. flavirostris illus. p.412.
⚑ *S. gigantea* illus. p.412.
S. variegata. See *Orbea variegata*.

STAPHYLEA (Staphyleaceae)
Bladder nut
Genus of deciduous, spring-flowering shrubs and trees, grown for their flowers and bladder-like fruits. Fully hardy. Requires sun or semi-shade and fertile, moist soil. Propagate species by softwood or greenwood cuttings in summer or by seed in autumn, selected forms by softwood or greenwood cuttings in summer.
S. colchica. Deciduous, upright shrub. H and S 3.5m (11ft). Erect panicles of bell-shaped, white flowers are borne in late spring and are followed by inflated, greenish-white fruits. Bright green leaves each consist of 3–5 oval leaflets.
S. holocarpa var. *rosea*, syn. *S.h.* 'Rosea', illus. p.87.
S. pinnata illus. p.86.

Statice suworowii. See *Psylliostachys suworowii*.

STAUNTONIA (Lardizabalaceae)
Genus of evergreen, woody-stemmed, twining climbers. Male and female flowers are produced on separate plants. Frost hardy. Grow in any well-drained soil and in sun or semi-shade. To keep under control, prune in early spring. Propagate by seed in spring or by stem cuttings in summer or autumn.
S. hexaphylla. Evergreen, woody-stemmed, twining climber. H to 10m (30ft) or more. Leaves have 3–7 oval leaflets, each 5–13cm (2–5in) long. In spring has racemes of small, fragrant, cup-shaped, pale violet flowers, followed by egg-shaped, edible, fleshy, purple fruits, 2.5–5cm (1–2in) long, if plants of both sexes are grown together.

STENANTHIUM (Liliaceae)
Genus of summer-flowering bulbs, attractive but seldom cultivated. Frost hardy. Needs an open, sunny position in any well-drained soil. In cool areas, plant in a warm, sheltered site in light soil that does not dry out excessively. Propagate by seed in autumn or by division in spring.
S. gramineum. Summer-flowering bulb. H to 1.5m (5ft), S 45–60cm (1½–2ft). Has long, narrowly strap-shaped, semi-erect, basal leaves. Stem produces a dense, branched, often arching spike of fragrant, star-shaped, white or green flowers, each 1–1.5cm (½–⅝in) across.

STENOCACTUS, syn. ECHINOFOSSULOCACTUS (Cactaceae)
Genus of spherical, perennial cacti with spiny, green stems that have very narrow, wavy ribs. Frost tender, min. 5°C (41°F). Needs a sunny position and well-drained soil. Propagate by seed in spring or summer.
S. lamellosus. Spherical, perennial cactus. H and S 8cm (3in). Green stem has 30–35 ribs. Funnel-shaped, flesh-coloured or red flowers, 1–3cm (½–1¼in) across, are produced from crown in spring. Has flattened upper radial spines, shorter, more rounded lower ones and longer, rounded central spines with darker tips.
S. pentacanthus illus. p.412.
S. violaciflorus illus. p.408.

STENOCARPUS (Proteaceae)
Genus of evergreen, summer- and autumn-flowering trees, grown for their flowers and foliage. Frost tender, min. 5–7°C (41–5°F). Needs full light and fertile, well-drained soil. Water containerized plants moderately, less in winter. Pruning is rarely necessary. Propagate by seed in spring or by semi-ripe cuttings in summer.
S. sinuatus (Australian firewheel tree). Slow-growing, evergreen, upright tree. H 12m (40ft) or more, S 5m (15ft). Has lustrous, deep green leaves, each 12–25cm (5–10in) long, lance-shaped and entire or with pairs of oblong lobes. Bears bottle-shaped, bright scarlet flowers, clustered like the spokes of a wheel, from late summer to autumn.

Stenolobium stans. See *Tecoma stans*.

STENOMESSON (Amaryllidaceae)
Genus of bulbs, grown for their long, often pendent, tubular flowers. Frost tender, min. 10°C (50°F). Needs an open, sunny situation and well-drained soil. Propagate by offsets in autumn.
S. incarnatum. See *S. variegatum*.
S. miniata, syn. *Urceolina peruviana*, illus. p.363.
S. variegatum, syn. *S. incarnatum*, illus. p.363.

STENOTAPHRUM (Gramineae). See GRASSES, BAMBOOS, RUSHES and SEDGES.
S. secundatum (St Augustine grass).
⚑ **'Variegatum'** is an evergreen, spreading, rhizomatous, perennial grass. H 15cm (6in), S indefinite. Frost tender, min. 5°C (41°F). Leaves are mid-green with cream stripes and last well into winter. Has erect racemes of brownish-green spikelets in summer. In warm climates is used for lawns.

STEPHANANDRA (Rosaceae)
Genus of deciduous, summer-flowering shrubs, grown for their habit, foliage, autumn colour and winter shoots. Fully hardy. Needs sun or semi-shade and fertile, not too dry soil. On established plants cut out some older shoots after flowering. Propagate by softwood cuttings in summer or by division in autumn.
S. incisa. Deciduous, arching shrub. H 1.5m (5ft), S 3m (10ft). Oval, deeply lobed and toothed, bright green leaves turn orange-yellow in autumn and stems become rich brown in winter. Produces crowded panicles of tiny, star-shaped, greenish-white flowers in early summer. **'Crispa'**, H 60cm (2ft), has wavy-edged and more deeply lobed leaves.
S. tanakae illus. p.110.

STEPHANOTIS (Asclepiadaceae)
Genus of evergreen, woody-stemmed, twining climbers, grown for their scented, waxy flowers. Frost tender, min. 13–16°C (55–61°F). Provide a humus-rich, well-drained soil and partial shade in summer. Water moderately, less in cold weather. Provide stems with support. Shorten over-long or crowded stems in spring. Propagate by seed in spring or by semi-ripe cuttings in summer.
⚑ *S. floribunda* illus. p.165.

Sterculia acerifolia. See *Brachychiton acerifolius*.
Sterculia diversifolia. See *Brachychiton populneus*.
Sterculia platanifolia. See *Firmiana simplex*.

STERNBERGIA (Amaryllidaceae)
Genus of spring- or autumn-flowering bulbs, grown for their large, crocus-like flowers. Frost hardy, but in cool areas grow against a sunny wall. Needs a hot, sunny site and any well-drained, heavy or light soil that dries out in summer, when bulbs die down and need warmth and dryness. Leave undisturbed to form clumps. Propagate by division in spring or autumn.
S. candida illus. p.370.
S. clusiana. Autumn-flowering bulb. H to 2cm (¾in), S 8–10cm (3–4in). Strap-shaped, semi-erect, basal, greyish-green leaves, often twisted lengthways, appear after flowering. Stems carry erect, goblet-shaped, yellow or greenish-yellow flowers, 4–8cm (1½–3in) long.
S. lutea illus. p.383.
S. sicula. Autumn-flowering bulb. H 2.5–7cm (1–3in), S 5–8cm (2–3in). Narrowly strap-shaped, semi-erect, basal, deep green leaves, each with a central, paler green stripe, appear with flowers. Stem bears a funnel-shaped, bright yellow flower, 2–4cm (¾–1½in) long.

STETSONIA (Cactaceae)
Genus of one species of tree-like, perennial cactus with a stout trunk. Nocturnal, funnel-shaped flowers are 15cm (6in) long. Frost tender, min. 10°C (50°F). Needs a sunny, well-drained position. Propagate by seed in spring or summer.
S. coryne illus. p.393.

STEWARTIA. See STUARTIA.

STIGMAPHYLLON (Malpighiaceae)
Genus of evergreen, woody-stemmed, twining climbers, grown for their flowers. Frost tender, min. 15–18°C (59–64°F). Fertile, well-drained soil is needed with partial shade in summer. Water freely when in full growth, less in low temperatures. Provide stems with support. Thin out crowded stems

605

in spring. Propagate by semi-ripe cuttings in summer.
S. ciliatum illus. p.176.

STIPA (Gramineae). See GRASSES, BAMBOOS, RUSHES and SEDGES.
S. arundinacea (Pheasant grass). Evergreen, tuft-forming, perennial grass. H 1.5m (5ft), S 1.2m (4ft). Frost hardy. Brownish-green leaves, 30cm (12in) long, turn soft orange in late summer. Has decorative, pendent, open panicles of purplish-green flower spikes in autumn.
S. calamagrostis, syn. *Achnatherum calamagrostis*. Evergreen, tuft-forming, perennial grass. H 1m (3ft), S 45cm (1½ft). Fully hardy. Leaves are bluish-green and inrolled. In summer bears decorative, large, loose panicles of sand-brown spikelets that dry and last well into winter.
☘ *S. gigantea* illus. p.183.

STOKESIA (Compositae)
Genus of one species of evergreen, summer-flowering perennial. Fully hardy. Requires sun or semi-shade and fertile, well-drained soil. Propagate by division in spring or by seed in autumn.
S. laevis illus. p.249. **'Blue Star'** is an evergreen, basal-rosetted perennial. H and S 30–45cm (12–18in). Bears cornflower-like, deep blue flower heads singly at stem tips in summer. Has rosettes of narrowly lance-shaped, dark green leaves.

STOMATIUM (Aizoaceae)
Genus of mat-forming, perennial succulents with short stems, each bearing 4–6 pairs of solid, 3-angled or semi-cylindrical leaves often with toothed edges and incurved tips. Frost tender, min. 5°C (41°F). Needs sun and well-drained soil. Propagate by seed or stem cuttings in spring or summer.
S. agninum. Mat-forming, perennial succulent. H 5cm (2in), S 1m (3ft) or more. Has solid, 3-angled or semi-cylindrical, soft grey-green leaves, 4–5cm (1½–2cm) long, often without teeth. In summer, fragrant, daisy-like, yellow flowers, 2–5cm (¾–2in) across, open in evening.
S. patulum. Mat-forming, perennial succulent. H 3cm (1¼in), S 1m (3ft). Has semi-cylindrical, grey-green leaves, each 2cm (¾in) long, with rough dots and 2–9 teeth-like tubercles on upper surface. Bears 2cm (¾in) wide, fragrant, daisy-like, pale yellow flowers in evening in summer.

Strangweia spicata. See *Bellevalia hyacinthoides*.

STRANVAESIA. See **PHOTINIA**.

STRATIOTES (Hydrocharitaceae)
Genus of semi-evergreen, perennial, submerged, free-floating water plants, grown for their foliage. Fully hardy. Requires sun. Grows in any depth of cool water. Thin plants as required to prevent overcrowding. Propagate by separating young plants from runners in summer.
S. aloides illus. p.387.

STRELITZIA (Strelitziaceae)
Bird-of-paradise flower
Genus of large, evergreen, tufted, clump-forming, palm-like perennials, grown for their showy flowers. Frost tender, min. 5–10°C (41–50°F). Grow in fertile, well-drained soil and in bright light shaded from direct sun in summer. Reduce watering in low temperatures. Propagate by seed or division of suckers in spring.
S. nicolai illus. p.196.
☘ *S. reginae* illus. p.231.

Streptanthera cuprea. See *Sparaxis elegans*.
Streptanthera elegans. See *Sparaxis elegans*.

STREPTOCARPUS (Gesneriaceae)
Genus of perennials, some of which are evergreen, with showy flowers. Frost tender, min. 10–15°C (50–59°F). Grow in a humid atmosphere in humus-rich, moist soil and in bright light away from direct sunlight. Avoid wetting leaves when watering; water less during cold periods. Propagate by seed, if available, in spring, by division after flowering or by tip cuttings from bushy species or leaf cuttings from stemless species in spring or summer.
S. caulescens illus. p.240.
S. **'Constant Nymph'** illus. p.267.
S. **'Nicola'** illus. p.265.
S. rexii (Cape primrose). Stemless perennial. H to 25cm (10in), S to 50cm (20in). Has a rosette of strap-shaped, wrinkled, green leaves. Stems, each 15cm (6in) or more long, bear loose clusters of funnel-shaped, pale blue or mauve flowers, 5cm (2in) long and with darker lines. They appear intermittently at any time of year.
S. saxorum illus. p.248.

STREPTOSOLEN (Solanaceae)
Genus of one species of evergreen or semi-evergreen, loosely scrambling shrub, grown for its flowers. Frost tender, min. 7–10°C (45–50°F). Requires full sun and humus-rich, well-drained soil. Water freely when in full growth, less at other times. After flowering or in spring, remove flowered shoots and tie in new growths. Propagate by softwood or semi-ripe cuttings in summer.
☘ *S. jamesonii* illus. p.180.

STROBILANTHES (Acanthaceae)
Genus of perennials and evergreen sub-shrubs, grown for their flowers. Frost hardy to frost tender, min. 15°C (59°F). Prefers semi-shade. Grow in fertile, well-drained soil. Propagate by seed, basal stem cuttings or division in spring.
S. atropurpureus illus. p.227.

STROMANTHE (Marantaceae)
Genus of evergreen, creeping perennials, grown mainly for their foliage. Frost tender, min. 15°C (59°F). Prefers high humidity and partial shade. Grow in open soil or compost, use soft water if possible and do not allow to dry out completely. Propagate by division in spring.
S. sanguinea. Strong-growing, evergreen, creeping perennial. H and

S to 1.5m (5ft). Lance-shaped leaves, to 45cm (18in) long, are glossy, green above with paler midribs, reddish below. Bears panicles of small, 3-petalled, white flowers in axils of showy, bright red bracts, usually in spring but also in summer-autumn.

STROMBOCACTUS (Cactaceae)
Genus of extremely slow-growing, hemispherical to cylindrical, perennial cacti. Takes 5 years from seed to reach 1cm (½in) high. Funnel-shaped flowers are 4cm (1½in) across. Frost tender, min. 5°C (41°F). Needs sun and very well-drained soil. Is difficult to grow and very susceptible to overwatering. Propagate by seed or grafting in spring or summer.
S. disciformis illus. p.406.

STRONGYLODON (Leguminosae)
Genus of evergreen, woody-stemmed, twining climbers, grown for their large, claw-like flowers. Frost tender, min. 18°C (64°F). Needs humus-rich, moist but well-drained soil and partial shade in summer. Water freely when in full growth, less at other times. Provide support. If necessary, thin crowded stems in spring. Propagate by seed or stem cuttings in summer or by layering in spring.
S. macrobotrys illus. p.166.

STUARTIA, syn. **STEWARTIA** (Theaceae)
Genus of deciduous trees and shrubs, grown for their flowers, autumn colour and usually peeling bark. Fully to frost hardy. Needs a sunny position, but preferably with roots in shade, and shelter from strong winds. Requires fertile, moist but well-drained, neutral to acid soil. Resents being transplanted. Propagate by softwood cuttings in summer or by seed in autumn.
S. malacodendron. Deciduous, spreading tree or shrub. H 4m (12ft), S 3m (10ft). Frost hardy. Rose-like, purple-stamened, white flowers, sometimes purple-streaked, are borne in mid-summer amid oval, dark green leaves.
S. monadelpha illus. p.56.
☘ *S. pseudocamellia* illus. p.52.
☘ *S. sinensis*. Deciduous, spreading tree. H 12m (40ft), S 7m (22ft). Fully hardy. Has peeling bark and oval, bright green leaves that turn brilliant red in autumn. Produces fragrant, rose-like, white flowers in mid-summer.

STYLIDIUM (Stylidiaceae)
Genus of perennials with grass-like leaves, grown for their unusual flowers that have fused, 'triggered' stamens adapted for pollination by insects. Frost tender, min. 10°C (50°F). Grow in fertile soil and in bright light. Propagate by division or seed in spring.
S. graminifolium (Trigger plant). Rosetted perennial. H and S to 15cm (6in) or more. Grass-like, stiff, dark green leaves, with toothed margins, rise from ground level. Tiny, pale pinkish-mauve flowers are carried in narrow spikes, to over 30cm (12in) long, in summer.

STYLOPHORUM (Papaveraceae)
Genus of spring-flowering perennials with large, deeply lobed leaves, nearly all as basal rosettes. Fully hardy. Needs semi-shade and humus-rich, moist, peaty soil. Propagate by division in spring or by seed in autumn.
S. diphyllum illus. p.298.

STYRAX (Styracaceae)
Genus of deciduous, summer-flowering trees and shrubs, grown for their foliage and flowers. Fully hardy to frost tender, min. 7–10°C (45–50°F). Prefers a sheltered position in sun or semi-shade and moist, neutral to acid soil. Propagate by softwood cuttings in summer or by seed in autumn.
☘ *S. japonica* illus. p.51.
☘ *S. obassia* (Fragrant snowbell). Deciduous, spreading tree. H 12m (40ft), S 7m (22ft). Fully hardy. Bears long, spreading clusters of fragrant, bell- to funnel-shaped, white flowers in early summer. Has broad, rounded, deep green leaves.
S. officinales illus. p.88.
S. wilsonii illus. p.108.

Submatucana aurantiaca. See *Cleistocactus aurantiacus*.

Sulcorebutia arenacea. See *Rebutia arenacea*.
Sulcorebutia rauschii. See *Rebutia rauschii*.
Sulcorebutia tiraquensis. See *Rebutia tiraquensis*.

SUTERA (Scrophulariaceae)
Genus of annuals, perennials and evergreen shrubs, suitable as summer bedding plants. Frost tender, min. 5°C (41°F). Needs a warm, sunny situation and any well-drained soil. Propagate by seed or division in spring or by softwood cuttings taken during summer.
S. grandiflora. Much-branched, sub-shrubby perennial. H 1m (3ft), S 30–45cm (12–18in). Produces terminal racemes of tubular, 5-lobed, frilled, deep purple flowers from mid-summer to autumn. Has oval to oblong, round-toothed leaves.

SUTHERLANDIA (Leguminosae)
Genus of evergreen shrubs, grown for their flowers and fruits. Frost tender, min. 7–10°C (45–50°F). Requires full light and fertile, well-drained soil. Water containerized specimens freely when in full growth, moderately at other times. Remove old, twiggy stems at ground level in late winter. Propagate by seed in spring. Red spider mite may be troublesome.
S. frutescens illus. p.135.

SWAINSONA (Leguminosae)
Genus of annuals, evergreen perennials, sub-shrubs and shrubs, grown for their flowers. Frost tender, min. 5–7°C (41–5°F). Needs full light or partial shade and humus-rich, well-drained soil. Water freely when in active growth, moderately at other times. Propagate by seed in spring or by semi-ripe cuttings in summer.
S. galegifolia (Darling pea). Evergreen, sprawling, loose sub-shrub.

H 60cm–1.2m (2–4ft), S 30–60cm (1–2ft). Leaves have 11–25 narrowly oval, mid- to deep green leaflets. Bears pea-like, red, pink, purple, blue or yellow flowers in late spring and summer. Remove old, flowered shoots in late winter.

SYAGRUS (Palmae)
Queen palm
Genus of one species of evergreen palm, grown for its majestic appearance. Frost tender, min. 18°C (64°F). Requires full light or partial shade and humus-rich, well-drained soil. Water containerized specimens moderately, less when temperatures are low. Propagate by seed in spring at not less than 24°C (75°F). Red spider mite may be a nuisance.
S. romanzoffianum illus. p.47.

SYCOPSIS (Hamamelidaceae)
Genus of evergreen trees and shrubs, grown for their foliage and flowers. Frost hardy. Needs a sheltered position in sun or semi-shade and fertile, not too dry, peaty soil. Propagate by semi-ripe cuttings in summer.
S. sinensis. Evergreen, upright shrub. H 5m (15ft), S 4m (12ft). Leaves are oval, glossy and dark green. Flowers lack petals but have showy, dense clusters of red-tinged, yellow anthers in late winter or early spring.

SYMPHORICARPOS (Caprifoliaceae)
Genus of deciduous shrubs, with inconspicuous, small, bell-shaped flowers, grown mainly for their clusters of showy, long-persistent fruits. Fully hardy. Requires sun or semi-shade and fertile soil. Propagate by softwood cuttings in summer or by division in autumn.
S. albus (Snowberry). var. *laevigatus* is a vigorous, deciduous, dense shrub, part upright, part arching. H and S 2m (6ft). Large, marble-like, white fruits follow pink flowers borne in summer. Rounded leaves are dark green.
S. x chenaultii 'Hancock'. Deciduous, procumbent, dense shrub. H 1m (3ft), S 3m (10ft). Has oval, bronze leaves maturing to bright green. White flowers appear from early to mid-summer. Small, spherical, deep lilac-pink fruits are sparsely borne. Makes excellent ground cover.
S. orbiculatus (Coralberry, Indian currant). Deciduous, bushy, dense shrub. H and S 2m (6ft). Has white or pink flowers in late summer and early autumn, then spherical, deep purplish-red fruits. Oval leaves are dark green. Does best after a hot summer. 'Foliis Variegatis' (syn. *S.o.* 'Variegatus'), illus. p.138.

SYMPHYANDRA (Campanulaceae)
Genus of short-lived, summer-flowering perennials, best grown as biennials. Suits large rock gardens and bases of banks. Fully hardy. Needs sun and well-drained soil. Propagate by seed in autumn. Self seeds readily.
S. armena. Upright or spreading perennial. H 30–60cm (1–2ft), S 30cm (1ft). Produces panicles of upright, bell-shaped, blue or white flowers in summer. Leaves are oval, irregularly toothed, hairy and mid-green.
S. pendula. Arching perennial. H 30–60cm (1–2ft), S 30cm (1ft). Produces panicles of pendent, bell-shaped, cream flowers in summer. Has oval, hairy, pale green leaves. Becomes woody at base with age.
S. wanneri illus. p.305.

SYMPHYTUM (Boraginaceae)
Comfrey
Genus of vigorous, coarse perennials, best suited to wild gardens. Fully hardy. Prefers sun or semi-shade and moist soil. Propagate by division in spring or by seed in autumn; usually self seeds naturally. Propagate named cultivars by division only.
S. caucasicum illus. p.202.
S. grandiflorum. Clump-forming perennial. H 25cm (10in), S 60cm (24in). Has lance-shaped, hairy, rich green leaves. Bears one-sided racemes of tubular, creamy flowers in spring. Makes good ground cover.
S. 'Hidcote Blue'. Clump-forming perennial. H 50cm (20in), S 60cm (24in). Is similar to *S. grandiflorum*, but has pale blue flowers.
S. x uplandicum (Russian comfrey). ♥ 'Variegatum' illus. p.202.

SYMPLOCOS (Symplocaceae)
Genus of evergreen or deciduous trees and shrubs, of which only the species described is in general cultivation. This is grown for its flowers and fruits. Fruits are most prolific when several plants are grown together. Fully hardy. Needs full sun and fertile soil. Propagate by seed in autumn.
S. paniculata illus. p.108.

SYNADENIUM (Euphorbiaceae)
Genus of evergreen, semi-succulent shrubs, grown for their foliage. Frost tender, min. 7–10°C (45–50°F). Requires full light and fertile, freely draining soil. Water containerized plants moderately, less in winter. Prune in late winter if necessary. Propagate by seed in spring or by softwood cuttings in summer.
S. grantii (African milkbush). Evergreen, erect, robust-stemmed shrub. H 3–4m (10–12ft), S 2m (6ft) or more. Has very small, red flowers in autumn, largely concealed by lance-shaped to oval, glossy, rich green leaves. Glowing leaves of 'Rubrum' are red-purple beneath, purplish-green above.

SYNGONIUM (Araceae)
Genus of evergreen, woody-stemmed, root climbers, grown for their ornamental foliage. Flowers are seldom produced in cultivation. Frost tender, min. 16–18°C (61–4°F). Needs partial shade and humus-rich, well-drained soil. Water moderately, less in low temperatures. Provide support, ideally moss poles. Remove young stem tips to promote branching. Propagate by leaf-bud or stem-tip cuttings in summer.
S. auritum, syn. *Philodendron auritum* of gardens, *P. trifoliatum* (Five fingers). Fairly slow-growing, evergreen, woody-stemmed, root climber. H 1–2m (3–6ft). Has glossy, rich green leaves divided into 3, sometimes 5, oval leaflets, the central one the largest.
S. erythrophyllum. Slow-growing, evergreen, root climber with slender, woody stems. H 1m (3ft) or more. Young plants have arrowhead-shaped leaves, flushed purple beneath. Leaves on mature plants have 3 lobes or leaflets and thicker, longer stems.
S. hoffmannii. Moderately vigorous, evergreen, woody-stemmed, root climber. H 2–3m (6–10ft). Young plants have arrowhead-shaped leaves; mature ones have leaves divided into 3 grey-green leaflets with silvery-white veins.
♥ *S. podophyllum*, syn. *Nephthytis triphylla* of gardens, illus. p.180. 'Trileaf Wonder' illus. p.179.

SYNNOTIA (Iridaceae)
Genus of spring-flowering corms, with fans of lance-shaped leaves, grown for their loose spikes of flowers, each with 6 unequal petals, hooded like a small gladiolus. Half hardy. Needs sun and well-drained soil. Plant in autumn. Dry off after flowering. Propagate by seed or offsets in autumn.
S. variegata. Spring-flowering corm. H 10–35cm (4–14in), S 8–10cm (3–4in). Produces erect leaves in a basal fan. Flowers are long-tubed with upright, purple, upper petals and narrower, pale yellowish-purple, lower ones curving downwards. var. *metelerkampiae* has smaller flowers.

SYNTHYRIS (Scrophulariaceae)
Genus of evergreen or deciduous, spring-flowering perennials with gently spreading, rhizomatous rootstocks. Is useful for rock gardens and peat beds. Fully hardy. Prefers partial shade and moist soil. Propagate in late spring by seed or division.
S. reniformis. Evergreen, clump-forming perennial. H 8–10cm (3–4in), S 15cm (6in). Has kidney-shaped to rounded, toothed, dark green leaves and, in spring, short, dense racemes of small, bell-shaped, blue flowers.
S. stellata illus. p.317.

SYRINGA (Oleaceae)
Lilac
Genus of deciduous shrubs and trees, grown for their dense panicles of small, tubular flowers, which are usually extremely fragrant. Fully hardy. Needs sun and deep, fertile, well-drained, preferably alkaline soil. Obtain plants on their own roots, as grafted plants usually sucker freely. Remove flower heads from newly planted lilacs, and dead-head for first few years. Cut out weak shoots in winter and, to maintain shape, prune after flowering. Straggly, old plants may be cut back hard in winter, but the following season's flowers will then be lost. Propagate by softwood cuttings in summer. Leaf miners, leaf spot and lilac blight may be troublesome.
S. 'Belle de Nancy'. Deciduous, upright, then spreading shrub. H and S 5m (15ft). Large, dense panicles of fragrant, tubular, double, mauve-pink flowers open in late spring from purple-red buds. Leaves are heart-shaped and dark green.
♥ *S.* 'Bellicent'. Deciduous, upright, then arching shrub. H 4m (12ft), S 5m (15ft). Large panicles of fragrant, tubular, single, clear pink flowers are borne above oval, dark green leaves in late spring and early summer.
S. 'Blue Hyacinth' (illus. p.90). Deciduous, bushy shrub, upright when young, later spreading. H and S 3m (10ft). Bears large, loose panicles of fragrant, pale lilac-blue flowers from mid-spring to early summer and has broadly heart-shaped, mid-green leaves.
♥ *S.* 'Charles Joly' (illus. p.90). Deciduous, bushy shrub, upright when young, later spreadng. H and S 3m (10ft). From mid-spring to early summer carries dense panicles of large, fragrant, double, deep purple-red flowers above heart-shaped, dark green leaves.
S. x chinensis (Rouen lilac). Deciduous, arching shrub. H and S 4m (12ft). Large, arching panicles of fragrant, tubular, single, lilac-purple flowers appear in late spring. Oval leaves are dark green. 'Alba' (illus. p.90) has white flowers.
S. 'Clarke's Giant' (illus p.90). Vigorous, deciduous, upright, then spreading shrub. H and S 5m (15ft). Large panicles of fragrant, tubular, single, lavender flowers, mauve-pink within, open from mauve-pink buds from mid- to late spring. Has heart-shaped, dark green leaves.
S. 'Congo' (illus. p.90). Deciduous, upright, then spreading shrub. H and S 5m (15ft). Fragrant, tubular, single, deep lilac-purple flowers, purplish-red in bud, are borne in large panicles above heart-shaped, dark green leaves in the spring.
S. 'Cora Brandt' (illus. p.90). Vigorous, deciduous shrub. H and S 5m (15ft). In late spring produces large panicles of scented, double, white flowers amid heart-shaped, mid-green leaves.
S. 'Decaisne' (illus. p.90). Deciduous, compact shrub. H and S 5m (15ft). Fully hardy. Masses of single, fragrant, dark blue flowers that are shaded purple are produced in late spring and early summer amid heart-shaped, mid-green leaves.
S. emodi (Himalayan lilac). Vigorous, deciduous, upright shrub. H 5m (15ft), S 4m (12ft). Produces unpleasantly scented, tubular, single, very pale lilac flowers in large, upright panicles in early summer. Has large, oval, dark green leaves.
♥ *S.* 'Esther Staley' (illus. p.90). Deciduous, bushy shrub, upright when young, later spreading. H 4m (12ft), S 3m (10ft). Broadly conical panicles of red buds open to fragrant, lilac-pink flowers from mid-spring to early summer. Leaves are broad, heart-shaped and mid-green.
S. 'Fountain'. Vigorous, deciduous, arching, open shrub. H 4m (12ft), S 5m (15ft). Large, nodding panicles of fragrant, tubular, single, deep pink flowers open above large, oval, dark green leaves in early summer.
S. 'Isabella'. Vigorous, deciduous,

upright shrub. H and S 4m (12ft). Has large, nodding panicles of fragrant, tubular, single, lilac-purple flowers, almost white within, in early summer, and large, oval, dark green leaves.
S. 'Jan van Tol' (illus. p.90). Deciduous, upright, then spreading shrub. H and S 5m (15ft). Fully hardy. Long, semi-pendent panicles of very fragrant, single, narrow-petalled pure white flowers are borne in late spring and early summer. Heart-shaped leaves are mid-green.
♚ *S.* 'Katherine Havemeyer'. Deciduous, upright, then spreading shrub. H and S 5m (15ft). Large, fragrant, tubular, double, lavender-purple, then lavender-pink flowers are borne in dense, conical panicles in late spring. Leaves are heart-shaped and mid-green.
♚ *S.* 'Mme Antoine Buchner' (illus. p.90). Deciduous, bushy shrub, upright when young, later spreading. H 4m (12ft), S 3m (10ft). In late spring and eary summer, long, narrow panicles of deep purple-red buds open to fragrant, double, pinkish-mauve flowers, fading with age. Has heart-shaped leaves.
S. 'Mme F. Morel' (illus. p.90). Deciduous, upright shrub that broadens with age. Large panicles of fragrant, single, light violet-purple flowers that are purple in bud are borne in very large clusters from late spring to early summer. Leaves are heart-shaped and mid-green.
S. 'Mme Florent Stepman' (illus. p.90). Deciduous, upright, then spreading, shrub. H and S 5m (15ft). Fully hardy. From late spring to early summer bears large panicles of scented, single, white flowers. Leaves are heart-shaped and mid-green.

♚ *S.* 'Mme Lemoine' (illus. p.90). Deciduous, bushy shrub, upright when young, later spreading. H 4m (12ft), S 3m (10ft). In late spring and early summer bears compact panicles of large, fragrant, double, white flowers above heart-shaped, mid-green leaves.
S. 'Maréchal Foch' (illus. p.90). Deciduous, bushy shrub, upright when young, later spreading. H 4m (12ft), S 3m (10ft). Broad, open panicles of large fragrant, single, carmine-pink flowers appear in late spring to early summer. Leaves are heart-shaped and mid-green.
S. 'Masséna' (illus. p.90). Deciduous, spreading shrub. H and S 5m (15ft). Fully hardy. Loose panicles of large, fragrant, deep red-purple flowers are borne from late spring to early summer amid heart-shaped, mid-green leaves.
S. 'Maud Notcutt'. Deciduous, upright, then spreading shrub. H and S 5m (15ft). Produces large panicles of fragrant, tubular, single, pure white flowers above heart-shaped, dark green leaves in late spring.
♚ *S. meyeri* 'Palibin' (syn. *S. palibianina* of gardens, *S. palibiniana* of gardens; illus. p.90). Slow-growing, deciduous, bushy, dense shrub. H and S 1.5m (5ft). Bears dense panicles of fragrant, tubular, single, lilac-pink flowers in late spring and early summer. Has small, oval deep green leaves.
S. 'Michel Buchner' (illus. p.90). Deciduous, open shrub. H and S 5m (15ft). Fully hardy. In late spring and early summer has large panicles of fragrant, double, pink-lilac flowers, each with a white eye. Leaves are heart shaped and mid-green.
S. microphylla. Deciduous, bushy shrub. H and S 2m (6ft). Small panicles of very fragrant, tubular, single, pink flowers appear in early summer, and often again in autumn, amid oval, mid-green leaves. ♚ 'Superba' illus. p.90.
S. 'Monge' (illus. p.90). Deciduous, upright, then spreading, shrub. H and S 5m (15ft). Fully hardy. Produces masses of very large, fragrant, single, deep purple-red flowers from late spring to early summer. Leaves are heart-shaped and mid-green.
♚ *S.* 'Mrs Edward Harding' (illus. p.90). Deciduous, open shrub. H and S 5m (15ft). Fully hardy. In late spring and early summer has large panicles of scented, double or semi-double, purple-red flowers that fade to pink amid heart-shaped, mid-green leaves.
S. palibiniana of gardens. See *S. meyeri* 'Palibin'.
S. 'Paul Thirion' (illus. p.90). Vigorous, compact, deciduous shrub. H and S 5m (15ft). Fully hardy. Strongly scented, double lilac-pink flowers open from deep purple-red buds in late spring and early summer. Heart-shaped leaves are mid-green.
♚ *S.* x *persica* (Persian Lilac; illus. p.90). Deciduous, bushy, dense shrub. H and S 2m (6ft). Produces small, dense panicles of fragrant, purple flowers in late spring. Leaves are narrow, pointed and dark green.
S. 'Président Grévy' (illus. p.90). Deciduous, upright, then spreading, shrub. H and S 5m (15ft). Fully hardy. In late spring and early summer, large panicles of double, scented lilac-blue flowers open from red-violet buds. Leaves are heart-shaped and mid-green.
♚ *S.* 'Primrose' (illus. p.90). Deciduous, bushy shrub, upright when young, later spreading. H and S 5m (15ft). Small, dense panicles of faintly fragrant, pale yellow flowers are produced in late spring and early summer. Has heart-shaped, mid-green leaves.
S. reticulata. Deciduous, broadly conical tree or shrub. H 10m (30ft), S 6m (20ft). Large panicles of fragrant, tubular, single, creamy-white flowers open above oval, taper-pointed, bright green leaves from early to mid-summer.
♚ *S.* 'Souvenir de Louis Späth'. Deciduous, upright, then spreading shrub. H and S 5m (15ft). Long, slender panicles of fragrant, tubular, single, deep purplish-red flowers are borne profusely above heart-shaped, dark green leaves in late spring.
S. velutina of gardens. See *S. meyeri* 'Palibin'.
S. yunnanensis (illus. p.90). Deciduous, upright shrub. H 3m (10ft), S to 3m (10ft). In early summer, large, oval, pointed, dark green leaves set off slender panicles of 4-petalled, pale pink or white flowers.

SYZYGIUM (Myrtaceae)
Genus of evergreen shrubs and trees, grown for their overall appearance. Frost tender, min. 10–13°C (50–56°F). Prefers full light (but tolerates some shade) and fertile, well-drained soil. Water containerized plants freely when in full growth, moderately at other times. Is very tolerant of pruning, but is best grown naturally. Propagate by seed in spring or by semi-ripe cuttings in summer.
S. paniculatum, syn. *Eugenia australis*, *E. paniculata*, illus. p.55.

T

TABEBUIA (Bignoniaceae)
Genus of deciduous or evergreen, mainly spring-flowering trees, grown for their flowers and for shade. Frost tender, min. 16–18°C (61–4°F). Requires full light and fertile, well-drained but not dry soil. Pot-grown plants are unlikely to flower. Pruning, other than shaping while young in autumn, is not needed. Propagate by seed or air-layering in spring or by semi-ripe cuttings in summer.
T. chrysotricha illus. p.71.
T. donnell-smithii. See *Cybistax donnell-smithii.*
T. rosea (Pink trumpet tree). Fast-growing, evergreen, rounded tree, deciduous in cool climates. H and S 15m (50ft) or more. Leaves have 5 oval leaflets. Bears trumpet-shaped, rose- to lavender-pink or white flowers, with yellow throats, in terminal clusters in spring.

TACCA (Taccaceae)
Genus of perennials with rhizomes, grown for their curious flowers. Frost tender, min. 18°C (64°F). Needs a fairly humid atmosphere, partial shade and peaty soil. Water sparingly during resting period in winter. Propagate by division or seed, if available, in spring.
T. chantrieri (Bat flower, Cat's whiskers). Clump-forming, rhizomatous perennial. H and S 30cm (1ft). Narrowly oblong, stalked, arching leaves are 45cm (1½ft) or more long. In summer produces flower umbels with green or purplish bracts on stems up to 60cm (2ft) long. Individual flowers are nodding, bell-shaped, 6-petalled and green, turning purple with long, pendent, maroon to purple threads.
T. leontopetaloides (East Indian arrowroot, South Sea arrowroot). Clump-forming, rhizomatous perennial. H and S 45cm (1½ft). Green leaves, to 1m (3ft) long, are deeply 3-lobed, each lobe also divided, on stalks to over 1m (3ft). In summer, on stems up to 1m (3ft) long, flower umbels are produced with 4–12 purple or brown bracts and 20–40 small, 6-petalled, yellow or purplish-green flowers, with long, purple to brown threads. Rhizomes yield edible starch.

Tacitus bellus. See *Graptopetalum bellum.*

Tacsonia mollissima. See *Passiflora mollissima.*
Tacsonia van-volxemii. See *Passiflora antioquiensis.*

TAGETES (Compositae)
Genus of annuals that flower continuously throughout summer and until the autumn frosts. Is useful as bedding plants and for edging. Half hardy. Grow in sun and in fertile, well-drained soil. Dead-head to ensure a long flowering period. Propagate by seed sown under glass in mid-spring. Is prone to slugs, snails and botrytis.
T. erecta (African marigold, Aztec marigold). Fast-growing, upright, bushy annual. Tall, H 60cm–1m (2–3ft), S 30–45cm (1–1½ft); intermediate, H and S 45cm (1½ft); and dwarf, H 30cm (1ft), S 30–45cm (1–1½ft), Erecta group hybrids are available. All have very deeply divided, aromatic, glossy, deep green leaves. Daisy-like, double flower heads, 5cm (2in) wide, are carried in summer and early autumn.
 '**Crackerjack**' (tall) illus. p.290.
 '**Galore**' (intermediate) has very large, double, golden-yellow flower heads.
 '**Gold Coins**' (tall) illus. p.288.
 Inca Series (dwarf) has double flower heads in single or mixed colours, including yellow, orange and gold.
 Jubilee Series (tall) has large, double flower heads in white, yellow, orange or cream.
 '**Toreador**' (tall) has very large, double, deep orange flower heads.
T. '**Lemon Gem**'. Fast-growing, bushy annual. H and S 15cm (6in). Very finely divided, pale green leaves are pleasantly aromatic. Daisy-like, single, lemon-yellow flower heads, 2.5cm (1in) wide, appear in summer and early autumn.
T. '**Nell Gwyn**'. Fast-growing, bushy annual. H and S to 30cm (1ft). Has very feathery, aromatic, deep green leaves. Large, daisy-like, single, deep yellow flower heads, with dark red centres, are produced in summer and early autumn.
T. '**Paprika**' illus. p.292.
T. patula (French marigold). Fast-growing, bushy annual. H and S to 30cm (1ft). Has deeply divided, aromatic, deep green leaves. Single or carnation-like, double flower heads, in shades of yellow, orange, red or mahogany, are borne in summer and early autumn. **Bonita Series**, H 25cm (10in), has double flower heads in a wide range of colours. ♥ **Boy-o-boy Series**, H and S 15cm (6in), has double flower heads in shades of yellow, mahogany or orange. '**Cinnabar**' illus. p.281. Flower heads of ♥ '**Honeycomb**', H 25cm (10in), S 20cm (8in), are crested, double and yellow- and reddish-orange. '**Naughty Marietta**' illus. p.290. '**Orange Winner**' illus. p.292. '**Spanish Brocade**', H and S 20cm (8in), has large, double, red-and-gold flower heads.
T. '**Tangerine Gem**' illus. p.292.

TALBOTIA (Velloziaceae)
Genus of evergreen perennials and shrubs, grown for their showy flowers. Frost tender, min. 10°C (50°F). Grow in sun or partial shade and in well-drained soil. Propagate by seed or division in spring.
T. elegans, syn. *Barbacenia elegans, Vellozin elegans.* Evergreen, mat-forming perennial with slightly woody stems. H to 15cm (6in), S 15–30cm (6–12in). Lance-shaped, leathery, dark green leaves, to 20cm (8in) long, each have a V-shaped keel. Solitary small, star-shaped, white flowers are held on slender stems above leaves in late spring.

TALINUM (Portulacaceae)
Genus of summer-flowering perennials, some of which are evergreen, grown for their flowers and succulent foliage. Is useful for rock gardens, troughs and alpine houses and as pot plants. Fully hardy to frost tender, min. 7°C (45°F). Needs sun and gritty, not too dry, well-drained soil. Propagate by seed in autumn.
T. okanoganense. Cushion- or mat-forming, prostrate perennial. H to 4cm (1½in), S to 10cm (4in). Fully hardy. Succulent stems bear tufts of cylindrical, succulent, greyish-green leaves and, in summer, produce tiny, cup-shaped, white flowers. Is excellent for cultivating in a trough or alpine house.

TAMARINDUS (Leguminosae)
Genus of one species of evergreen tree, grown for its edible fruits and overall appearance as well as for shade. Frost tender, min. 15–18°C (59–64°F). Needs full light and well-drained soil. Propagate by seed or air-layering in spring.
T. indica (Tamarind). Slow-growing, evergreen, rounded tree. H and S to 25m (80ft). Leaves have 10–15 pairs of oblong to elliptic, bright green leaflets. Produces profuse racemes of asymmetric, 5-petalled, pale yellow flowers, veined red, in summer, then long, brownish pods containing edible but acidic pulp.

TAMARIX (Tamaricaceae)
Tamarisk
Genus of deciduous or evergreen shrubs and trees, grown for their foliage, habit and abundant racemes of small flowers. In mild areas is very wind-resistant and thrives in exposed, coastal positions, making excellent hedges. Fully to frost hardy. Requires sun and fertile, well-drained soil. Restrict growth by cutting back in spring; trim hedges at same time. Propagate by semi-ripe cuttings in summer or by hardwood cuttings in winter.
T. gallica. Deciduous, spreading shrub or tree. H 4m (12ft), S 6m (20ft). Frost hardy. Purple, young shoots are clothed with tiny, scale-like, blue-grey leaves. Star-shaped, pink flowers are borne in slender racemes in summer.
T. pentandra. See *T. ramosissima.*
T. ramosissima, syn. *T. pentandra,* illus. p.91.

TANACETUM (Compositae)
Genus of perennials, some of which are evergreen, often with aromatic foliage, grown for their daisy-like flower heads. Fully to frost hardy. Grow in sun and in fertile, well-drained soil. Propagate by division in spring.
T. argenteum, syn. *Achillea argentea,* illus. p.308.
T. coccineum, syn. *Chrysanthemum coccineum, Pyrethrum roseum* (Pyrethrum). ♥ '**Brenda**' (syn. *Pyrethrum* 'Brenda') illus. p.202.
♥ '**Eileen May Robinson**' is an upright perennial. H to 75cm (30in), S 45cm (18in). Fully hardy. Has slightly aromatic, feathery, recurved leaves, 5cm (2in) long. In summer bears strong-stemmed, pink flower heads, 5cm (2in) wide, with yellow centres. Is useful for cut flowers.
♥ '**James Kelway**' has beige flower heads ageing to pink.
T. densum subsp. *amani* illus. p.309.
T. haradjanii, syn. *Chrysanthemum haradjanii.* Evergreen, mat-forming, woody-based perennial with a tap root. H and S 23–38cm (9–15in). Frost hardy. Has broadly lance-shaped, much-divided, silvery-grey leaves. Has terminal clusters of bright yellow flower heads in summer. Is useful for a rock garden or alpine house.
T. parthenium, syn. *Chrysanthemum parthenium,* illus. p.271. '**Aureum**' is a moderately fast-growing, short-lived, bushy perennial, grown as an annual. H and S 20–45cm (8–18in). Half hardy. Has oval, lobed, aromatic, green-gold leaves and, in summer and early autumn, daisy-like, white flower heads.

TANAKAEA (Saxifragaceae)
Genus of one species of evergreen, spreading perennial, grown for its foliage and flowers. Is suitable for rock gardens and peat beds. Fully hardy. Needs partial shade and well-drained, peaty, sandy soil. Propagate by runners in spring.
T. radicans. Evergreen, dense, basal-rosetted perennial. H 6–8cm (2½–3in), S 20cm (8in). Leaves are narrowly oval to heart-shaped, leathery and mid- to dark green. Bears small panicles of tiny, outward-facing, star-shaped, white flowers in late spring.

TAPEINOCHILUS (Zingiberaceae)
Genus of mostly evergreen perennials, grown for their colourful, leaf-like bracts. Frost tender, min. 18°C (64°F). Needs high humidity, partial shade and humus-rich soil. Is not easy to grow successfully in pots. Propagate by division in spring. Red spider mite may be a problem with pot-grown plants.
T. ananassae. Evergreen, tufted perennial. H to 2m (6ft), S 75cm (2½ft). Non-flowering stems are erect and unbranched, with narrowly oval, long-pointed leaves, to 15cm (6in) long. Flowering stems are leafless, to over 1m (3ft) long, and, in summer,

bear ovoid, dense spikes, 15cm (6in) or more long, of small, tubular, yellow flowers. Showy, recurved, hard, scarlet bracts enclose and almost hide flowers.

TAXODIUM (Taxodiaceae). See CONIFERS.
❦ *T. distichum* illus. p.78.

TAXUS (Taxaceae). See CONIFERS.
❦ *T. baccata* (Yew). Slow-growing conifer with a broadly conical, later domed crown. H 10–15m (30–50ft), S 5–10m (15–30ft). Fully hardy. Needle-like, flattened leaves are dark green. Female plants bear cup-shaped, fleshy, bright red fruits; only the red part, not the seed, is edible. Will regrow if cut back. The following forms are H 6–10m (20–30ft), S 5–8m (15–25ft) unless otherwise stated. **'Adpressa'** is a shrubby, female form with short, broad leaves. **'Aurea'** (illus. p.85) has golden-yellow foliage. ❦ **'Dovastoniana'** is spreading, with weeping branchlets. ❦ **'Dovastonii Aurea'** (illus. p.85) is similar to *T. b.* 'Dovastoniana', but has golden shoots and yellow-margined leaves. ❦ **'Fastigiata'**, H 10–15m (30–50ft), S 4–5m (12–15ft), has erect branches and dark green foliage that stands out all around shoots; **'Fastigiata Aurea'** is similar to *T. b.* 'Fastigiata', but has gold-variegated leaves. ❦ **'Repandens'**, H 60cm (2ft), S 5m (15ft), is a spreading form. ❦ **'Semperaurea'**, H 3m (10ft), S 5m (15ft), has ascending branches with dense, golden foliage.
T. cuspidata illus. p.83. **'Aurescens'** is a spreading, bushy, dwarf conifer. H 30cm (1ft), S 1m (3ft). Fully hardy. Is hardier than *T. baccata* forms. Needle-like, flattened leaves are deep golden-yellow in their first year and mature to dark green. **'Capitata'**, H 10m (30ft), S 2m (6ft), is upright in habit. **'Densa'**, H 1.2m (4ft), S 6m (20ft), is a female form with short, erect shoots.
T. × media. Dense conifer that is very variably shaped. H and S 3–6m (10–20ft). Fully hardy. Has needle-like, flattened leaves, spreading either side of olive-green shoots. Leaves are stiff, broad and widen abruptly at the base. Fruits are similar to those of *T. baccata*. **'Brownii'**, H 2.5m (8ft), S 3.5m (11ft), is a dense, globose form with dark green foliage. **'Densiformis'**, H 2–3m (6–10ft), is dense and rounded, with masses of shoots that have bright green leaves. ❦ **'Hicksii'**, H to 6m (20ft), is columnar and has ascending branches. Male and female forms exist. **'Hillii'**, H and S 3m (10ft), is a broadly conical to rounded, dense bush with glossy, green leaves. **'Wardii'**, H 2m (6ft), S 6m (20ft), is a flat, globose, female form.

TECOMA (Bignoniaceae)
Genus of mainly evergreen shrubs and trees, grown for their flowers from spring to autumn. Frost tender, min. 10–13°C (50–55°F). Prefers well-drained soil and full light. Water potted specimens moderately, hardly at all in winter. May be pruned annually after flowering to maintain as a shrub. Propagate by seed in spring or by semi-ripe cuttings in summer. Red spider mite may be troublesome.
T. australis. See *Pandorea pandorana*.
T. capensis. See *Tecomaria capensis*.
T. grandiflora. See *Campsis grandiflora*.
T. radicans. See *Campsis radicans*.
T. ricasoliana. See *Podranea ricasoliana*.
T. stans, syn. *Bignonia stans*, *Stenolobium stans*, illus. p.70.

TECOMANTHE (Bignoniaceae)
Genus of evergreen, twining climbers, grown for their flowers. Frost tender, min. 16–18°C (61–4°F). Provide humus-rich, well-drained soil and light shade in summer. Water freely when in full growth, less at other times. Provide stems with support. If necessary, thin out crowded stems in spring. Propagate by seed in spring or by semi-ripe cuttings in summer.
T. speciosa. Strong-growing, evergreen, twining climber. H to 10m (30ft) or more. Has leaves of 3 or 5 oval leaflets. Dense clusters of foxglove-like, fleshy-textured, cream flowers, tinged with green, appear in autumn.

TECOMARIA (Bignoniaceae)
Genus of evergreen, shrubby, scrambling climbers, grown for their showy flowers. Frost tender, min. 10°C (50°F). Provide fertile, well-drained soil and full light. Water regularly, less when temperatures are low. Provide support. Thin out crowded stems in spring. Propagate by seed in spring or by semi-ripe cuttings in summer.
❦ *T. capensis*, syn. *Bignonia capensis*, *Tecoma capensis* (Cape honeysuckle). Evergreen, scrambling climber, shrub-like when young. H 2–3m (6–10ft). Leaves have 5–9 rounded, serrated, glossy, dark green leaflets. Tubular, orange-red flowers are carried in short spikes mainly in spring-summer.

TECOPHILAEA (Tecophilaeaceae)
Genus of spring-flowering corms, rare in cultivation and extinct in the wild, grown for their beautiful flowers. Fully hardy, but because of rarity usually grown in a cold greenhouse or cold frame. Requires sun and well-drained soil. Water in winter and spring. Keep corms dry, but not sunbaked, from early summer to autumn, then replant. Propagate in autumn by seed or offsets.
❦ *T. cyanocrocus* illus. p.377.
❦ var. *leichtlinii* (syn. *T. c.* 'Leichtlinii') illus. p.376.

TELEKIA (Compositae)
Genus of summer-flowering perennials, grown for their bold foliage and large flower heads. Fully hardy. Grows in sun or shade and in moist soil. Propagate by division in spring or by seed in autumn.
T. speciosa, syn. *Buphthalmum speciosum*. Upright, spreading perennial. H 1.2–1.5m (4–5ft), S 1–1.2m (3–4ft). Mid-green leaves are heart-shaped at base of plant, oval on stems. In late summer, branched stems bear large, daisy-like, rich gold flower heads. Is ideal for a pool side or woodland.

Telesonix jamesii. See *Boykinia jamesii*.

TELLIMA (Saxifragaceae)
Genus of one species of semi-evergreen, late spring-flowering perennial. Makes good ground cover and is ideal for cool, semi-shaded woodland gardens and beneath shrubs in sunny borders. Fully hardy. Grows in any well-drained soil. Propagate by division in spring or by seed in autumn.
T. grandiflora (Fringecups). Semi-evergreen, clump-forming perennial. H and S 60cm (24in). Has heart-shaped, toothed, hairy, purple-tinted, bright green leaves. Bears racemes of small, bell-shaped, fringed, cream flowers, well above foliage, in late spring. **'Rubra'** (syn. *T.g.* 'Purpurea') illus. p.266.

TELOPEA (Proteaceae)
Genus of evergreen trees and shrubs, grown mainly for their flower heads. Half hardy to frost tender, min. 5°C (41°F). Requires full sun or semi-shade and humus-rich, moist but well-drained, neutral to acid soil. Water containerized plants freely when in full growth, moderately at other times. Propagate by seed in spring or by layering in winter.
T. speciosissima illus. p.112.
T. truncata illus. p.101.

TEMPLETONIA (Leguminosae)
Genus of evergreen shrubs, grown for their flowers. Frost tender, min. 7°C (45°F). Prefers full light and freely draining, alkaline soil. Water potted specimens moderately, less in winter. Propagate by seed in spring or by semi-ripe cuttings in summer.
T. retusa (Coral bush). Evergreen, erect, irregularly branched shrub. H 2m (6ft), S 1–1.5m (3–5ft). Has oval to elliptic, leathery, bluish-green leaves. Pea-like, red flowers, sometimes pink or cream, appear in spring-summer.

TERMINALIA (Combretaceae)
Genus of evergreen trees and shrubs, grown for their overall appearance, edible seeds (nuts) and for shade. Frost tender, min. 16–18°C (61–4°F). Requires full light and well-drained soil. Water potted specimens moderately, scarcely at all when temperatures are low. Pruning is seldom necessary. Propagate by seed in spring.
T. catappa (Indian almond, Tropical almond). Evergreen, rounded tree. H and S 15m (50ft) or more. Has broadly oval, lustrous, green leaves at stem tips. Small, greenish-white flowers appear in spring, followed by flattened ovoid, keeled, green to red fruits, each with an edible seed.

TERNSTROEMIA (Theaceae)
Genus of evergreen trees and shrubs, grown for their overall appearance. Half hardy. Requires full sun or semi-shade and humus-rich, well-drained, neutral to acid soil. Water containerized specimens copiously when in full growth, moderately at other times. Prune in spring if necessary. Propagate by seed when ripe or in spring or by semi-ripe cuttings in late summer.
T. gymnanthera. Evergreen, rounded, dense shrub. H and S 2m (6ft). Oval leaves are lustrous, mid- to deep green. In summer, pendent, 5-petalled, white flowers are borne singly from leaf axils. Pea-sized, berry-like, bright red fruits appear in autumn. Leaves of **'Variegata'** are white-bordered with a pink tinge.

Testudinaria elephantipes. See *Dioscorea elephantipes*.

TETRACENTRON (Tetracentraceae)
Genus of one species of deciduous tree, grown for its foliage and catkins. Fully hardy. Needs sun or partial shade and fertile, well-drained soil. Propagate by seed in autumn.
T. sinense. Deciduous, spreading tree of graceful habit. H and S 10m (30ft) or more. Has rounded to oval, finely toothed, dark green leaves and long, slender, yellow catkins in early summer.

TETRADIUM, syn. EUODIA, EVODIA (Rutaceae)
Genus of deciduous trees, grown for their foliage, late flowers and fruits. Fully hardy. Needs full sun and fertile, well-drained soil. Propagate by softwood cuttings in summer, by seed in autumn or by root cuttings in late winter.
T. daniellii. Deciduous, spreading tree. H and S 15m (50ft). Ash-like, dark green leaves consisting of 5–10 oval to oblong leaflets turn yellow in autumn. Has large clusters of small, fragrant, 5-petalled, white flowers in early autumn, then beaked, red fruits.

TETRANEMA (Scrophulariaceae)
Genus of perennials, grown for their flowers. Frost tender, min. 13°C (55°F). Grow in a light position, shaded from direct sunlight, and in well-drained soil; avoid waterlogging and a humid atmosphere. Propagate by division, or seed if available, in spring.
T. mexicanum. See *T. roseum*.
❦ *T. roseum*, syn. *T. mexicanum*, illus. p.267.

TETRAPANAX (Araliaceae)
Genus of one species of evergreen, summer- to autumn-flowering shrub, grown for its foliage. Half hardy. Requires full sun or partial shade and humus-rich, moist but well-drained soil. Water containerized specimens freely, less in winter. Leggy stems may be cut back to near ground level in winter. Propagate by suckers or seed in early spring.
❦ *T. papyrifer*, syn. *Fatsia papyrifera*, illus. p.96.

TETRASTIGMA (Vitaceae)
Genus of evergreen, woody-stemmed, tendril climbers, grown for their handsome leaves. Frost tender, min.

15–18°C (59–64°F). Grow in any fertile, well-drained soil, with shade in summer. Water freely while in active growth, less in low temperatures. Provide stems with support; cut out crowded stems in spring. Propagate by layering in spring or by semi-ripe cuttings in summer.
♥ *T. voinierianum*, syn. *Cissus voinieriana*, illus. p.180.

TEUCRIUM (Labiatae)
Genus of evergreen or deciduous shrubs, sub-shrubs and perennials, grown for their flowers, foliage (sometimes aromatic) or habit. Fully to half hardy. Needs full sun and well-drained soil. Propagate shrubs and sub-shrubs by softwood or semi-ripe cuttings in summer, perennials by seed or division in spring.
T. aroanum. Evergreen, procumbent, much-branched sub-shrub. H 2.5cm (1in), S 10–15cm (4–6in). Frost hardy. Has white-haired twigs and oblong to oval, slightly hairy leaves, which are densely hairy below, and, in summer, whorls of small, tubular, 2-lipped, purple flowers. Is good for a trough.
T. fruticans (Shrubby germander, Tree germander). ♥ **'Azureum'** is an evergreen, arching shrub. H 2m (6ft), S 4m (12ft). Half hardy. Has oval, aromatic, grey-green leaves with white undersides and, in summer, tubular, 2-lipped, deep blue flowers with prominent stamens. Cut out dead wood in spring.
T. polium illus. p.327.

THALIA (Marantaceae)
Genus of deciduous, perennial, marginal water plants, grown for their foliage and flowers. Frost tender, min. 7°C (45°C). Needs an open, sunny position in up to 45cm (18in) depth of water. Some species tolerate cool water. Remove fading foliage regularly. Propagate in spring by division or seed.
T. dealbata. Deciduous, perennial, marginal water plant. H 1.5m (5ft), S 60cm (2ft). Oval, long-stalked, blue-green leaves have a mealy, white covering. Spikes of narrowly tubular, violet flowers in summer are followed by decorative seed heads. Tolerates cool water.
T. geniculata. Deciduous, perennial, marginal water plant. H 2m (6ft), S 60cm (2ft). Has oval, long-stalked, blue-green leaves and, in summer, spikes of narrowly tubular, violet flowers. Needs a warm pool.

THALICTRUM (Ranunculaceae)
Meadow rue
Genus of perennials, grown for their divided foliage and fluffy flower heads. Flowers lack petals, but each has prominent tufts of stamens and 4 or 5 sepals, which rapidly fall. Does well at edges of woodland gardens. Tall species and cultivars make excellent foils in borders for perennials with bolder leaves and flowers. Fully hardy. Requires sun or light shade. Grows in any well-drained soil, although some species prefer cool, moist conditions. Propagate by seed when fresh, in autumn, or by division in spring.

T. aquilegiifolium illus. p.215. **'White Cloud'** illus. p.204.
T. chelidonii. Clump-forming perennial. H 1–1.5m (3–5ft), S 60cm (2ft). Has finely divided, mid-green leaves and, in summer, produces panicles of fluffy, 4- or 5-sepalled, mauve flowers. Prefers cool soil that does not dry out.
♥ *T. delavayi*, syn. *T. dipterocarpum* of gardens. Elegant, clump-forming perennial. H 1.5–2m (5–6ft), S 60cm (2ft). Has much-divided, mid-green leaves, above which large panicles of nodding, lilac flowers, with 4 or 5 sepals and prominent, yellow stamens, appear from mid- to late summer. ♥ **'Hewitt's Double'** has double flowers.
T. diffusiflorum. Clump-forming perennial. H 1m (3ft), S 30–60cm (1–2ft). Has much-divided, basal, mid-green leaves. Slender stems produce large sprays of delicate, drooping, mauve flowers in summer. Prefers cool, moist soil.
T. dipterocarpum of gardens. See *T. delavayi*.
T. flavum. Clump-forming perennial. H 1.2m–1.5m (4–5ft), S 60cm (2ft). Has much-divided, glaucous blue-green leaves and, from mid- to late summer, clusters of fluffy, pale yellow flowers on slender stems. **'Illuminator'** is a pale yellow cultivar with bright green foliage.
T. kiusianum. Mat-forming perennial with short runners. H 8cm (3in), S 15cm (6in). Has small, fern-like, 3-lobed leaves and, throughout summer, loose clusters of tiny, purple flowers. Is excellent in a peat bed, rock garden, trough or alpine house. Is difficult to grow in hot, dry areas. Prefers shade and moist, sandy, peaty soil.
T. lucidum illus. p.218.
T. orientale. Spreading perennial with short runners. H 15cm (6in), S 20cm (8in). Leaves are fern-like with oval to rounded, lobed leaflets. Bears small, saucer-shaped, blue-mauve to violet flowers, with yellow stamens and large sepals, in late spring.

Thamnocalamus falconeri. See *Drepanostachyum falconeri*.
Thamnocalamus spathaceus. See *Fargesia spathacea*.

THELOCACTUS (Cactaceae)
Genus of spherical to columnar, perennial cacti with tuberculate or ribbed stems. Elongated areoles in crowns produce funnel-shaped flowers. Frost tender, min. 7°C (45°F). Requires sun and well-drained soil. Propagate by seed in spring or summer.
♥ *T. bicolor* illus. p.408.
T. leucacanthus. Clump-forming, perennial cactus. H 10cm (4in), S 30cm (12in). Spherical to columnar, dark green stem has 8–13 tuberculate ribs. Areoles each bear up to 20 short, golden spines and yellow flowers, 5cm (2in) across, in summer.
T. mcdowellii, syn. *Echinomastus mcdowellii*. Also sometimes included in *Neolloydia*. Spherical, perennial cactus. H and S 15cm (6in). Has a tuberculate, dark green stem densely covered with white spines, to 3cm (1¼in) long. Violet-red flowers, 4cm (1½in) across, appear in spring-summer.

THELYPTERIS (Polypodiaceae)
Genus of deciduous ferns. Fully hardy. Tolerates sun or semi-shade. Grow in moist or very moist soil. Remove fading fronds regularly. Propagate by division in spring.
T. hexagonoptera. See *Phegopteris hexagonoptera*.
T. oreopteris. See *Oreopteris limbosperma*.
T. palustris illus. p.188.
T. phegopteris. See *Phegopteris connectilis*.

THERMOPSIS (Leguminosae)
Genus of summer-flowering perennials. Fully hardy. Prefers sun and rich, light soil. Propagate by division in spring or by seed in autumn.
T. caroliniana. See *T. villosa*.
T. montana. See *T. rhombifolia*.
T. rhombifolia, syn. *T. montana*, illus. p.220.
T. villosa, syn. *T. caroliniana*. Straggling perennial. H 1m (3ft) or more, S 60cm (2ft). Produces racemes of pea-like, yellow flowers in late summer. Glaucous leaves are divided into 3 oval leaflets.

THESPESIA (Malvaceae)
Genus of evergreen perennials, shrubs and trees, grown for their flowers. Frost tender, min. 16–18°C (61–4°F). Needs full light and well-drained soil. Water containerized plants freely when in full growth, less when temperatures are low. Prune annually in early spring to maintain as a shrub. Propagate by seed in spring or by semi-ripe cuttings in summer. Whitefly and red spider mite may be a nuisance.
T. populnea (Mahoe, Portia oil nut). Evergreen tree, bushy when young, thinning with age. H 12m (40ft) or more, S 3–6m (10–20ft). Leaves are heart-shaped. Intermittently, or all year round if warm enough, produces cup-shaped, yellow flowers, each with a maroon eye, that age to purple. Grows well by the sea.

THEVETIA (Apocynaceae)
Genus of evergreen shrubs and trees, grown for their flowers from winter to summer. Is related to *Frangipani*. Has poisonous, milky sap. Frost tender, min. 16–18°C (61–4°F). Needs full light and well-drained soil. Water containerized specimens moderately, less in winter. Young stems may be tip pruned in winter to promote branching. Propagate by seed in spring or by semi-ripe cuttings in summer.
T. neriifolia. See *T. peruviana*.
T. peruviana, syn. *T. neriifolia*, illus. p.67.

THLADIANTHA (Cucurbitaceae)
Genus of herbaceous or deciduous, tendril climbers, grown for their bell-shaped, yellow flowers and oval to heart-shaped, mid-green leaves. Frost hardy to frost tender, min. 4°C (39°F). Requires a sheltered position in full sun and fertile, well-drained soil. Propagate by seed sown under glass in spring or by division in early spring.
T. dubia illus. p.176.

THLASPI (Cruciferae)
Genus of annuals and perennials, some of which are evergreen, grown for their flowers. Small plants may flower themselves to death, so remove buds for 2 years, to encourage a larger plant. Is difficult to grow at low altitudes and may require frequent renewal from seed. Is good for screes and troughs. Fully hardy. Needs sun and moist but well-drained soil. Propagate by seed in autumn.
T. alpestre. See *T. alpinum*.
T. alpinum, syn. *T. alpestre* (Alpine penny-cress). Evergreen, mat-forming perennial. H 5cm (2in), S 10cm (4in). Has small, oval, mid-green leaves. Produces racemes of small, 4-petalled, white flowers in spring.
T. bulbosum. Clump-forming, tuberous perennial. H 8cm (3in), S 15–20cm (6–8in). Bears broadly oval, glaucous leaves and, in summer, racemes of 4-petalled, dark violet flowers. Suits a rock garden.
T. macrophyllum. See *Pachyphragma macrophyllum*.
T. rotundifolium illus. p.313.

THUJA (Cupressaceae). See CONIFERS.
T. koraiensis (Korean thuja). Upright conifer, sometimes sprawling and shrubby. H 3–10m (10–30ft), S 3–5m (10–15ft). Fully hardy. Scale-like foliage is bright green or yellow-green above, glaucous silver beneath, and smells of almonds when crushed.
T. occidentalis (American arbor-vitae, Eastern white cedar, White cedar). Slow-growing conifer with a narrow crown. H 15m (50ft), S 3–5m (10–15ft). Fully hardy. Has orange-brown bark and flat sprays of scale-like, yellowish-green leaves, pale or greyish-green beneath, smelling of apples when crushed. Ovoid cones are yellow-green, ripening to brown. **'Caespitosa'** (illus. p.84), H 30cm (12in), S 40cm (16in), is a cushion-shaped, dwarf cultivar. **'Fastigiata'**, H to 15m (50ft), S to 5m (15ft), is broadly columnar, with erect, spreading branches and light green leaves. **'Filiformis'** (illus. p.85), H 1.5m (5ft), S 1.5–2m (5–6ft), forms a mound with pendent, whip-like shoots. **'Hetz Midget'**, H and S 50cm (20in), growing only 2.5cm (1in) each year, is a globose, dwarf form with blue-green foliage. ♥ **'Holmstrup'**, H 3–4m (10–12ft), S 1m (3ft), is slow-growing, dense and conical, with rich green foliage. **'Little Champion'**, H and S 50cm (20in) or more, is a globose, dwarf form, conical when young, with foliage turning brown in winter. ♥ **'Lutea Nana'**, H 2m (6ft), S 1–2m (3–6ft), is a dwarf form with golden-yellow foliage.
♥ **'Rheingold'**, H 3–4m (10–12ft), S 2–4m (6–12ft), is slow-growing, with golden-yellow foliage that becomes bronze in winter. ♥ **'Smaragd'**, H 2–2.5m (6–8ft), S 60–75cm (2–2½ft), is slow-growing and conical, with erect sprays of bright green

leaves. **'Spiralis'**, H 10–15m (30–50ft), S 2–3m (6–10ft), produces foliage in twisted, fern-like sprays. **'Woodwardii'**, H 2.5m (8ft), S to 5m (15ft), is very slow-growing and globose, with mid-green foliage.
T. orientalis, syn. *Biota orientalis*, *Platycladus orientalis* (Biota, Chinese arbor-vitae, Chinese thuja). Conifer with an irregularly rounded crown. H 10–15m (30–50ft), S 5m (15ft). Fully hardy. Has fibrous bark and flattened, vertical sprays of scale-like, scentless, dark green leaves. Egg-shaped cones are glaucous grey. ♛ **'Aurea Nana'** (illus. p.85), H and S 60cm (24in), is a dwarf cultivar with yellow-green foliage that turns bronze in winter. **'Semperaurea'** (illus. p.85), H 3m (10ft), S 2m (6ft), is compact, with golden leaves.
T. plicata (Western red cedar). Fast-growing, conical conifer that has great, curving branches low down. H 20–30m (70–100ft), S 5–8m (15–25ft), greater if lower branches self-layer. Fully hardy. Has red-brown bark, scale-like, glossy, dark green leaves, which have a pineapple aroma when crushed, and erect, ovoid, green cones, ripening to brown. ♛ **'Atrovirens'** has darker green foliage. ♛ **'Aurea'** has golden-yellow foliage. **'Collyer's Gold'** (illus. p.85), H to 2m (6ft), S 1m (3ft), is a dwarf form with yellow, young foliage turning light green. **'Cuprea'**, H 1m (3ft), S 75cm–1m (2½–3ft), is a conical shrub with copper- to bronze-yellow leaves. **'Hillieri'** (illus. p.85), H and S to 1m (3ft), is a slow-growing, dense, rounded, dwarf shrub with moss-like, rich green foliage. ♛ **'Stoneham Gold'** (illus. p.85), H 1–2m (3–6ft), S 1m (3ft), is a conical, dwarf form with bright gold foliage. **'Zebrina'**, H 15m (50ft), has leaves banded with yellowish-white.

THUJOPSIS (Cupressaceae). See CONIFERS.
♛ ***T. dolabrata*** (Hiba). Conical or bushy conifer with a mass of stems. H 10–20m (30–70ft), S 8–10m (25–30ft). Fully hardy. Produces heavy, flat sprays of scale-like leaves, glossy, bright green above, silvery-white beneath. Small, rounded cones are blue-grey. **'Variegata'** illus. p.83.

THUNBERGIA (Acanthaceae)
Genus of annual or mainly evergreen, perennial, twining climbers, perennials and shrubs, grown for their flowers. Half hardy to frost tender, min. 10–15°C (50–59°F). Any fertile, well-drained soil is suitable, with full sun or light shade in summer. Water freely when in full growth, less at other times. Requires support. Thin out crowded stems in early spring. Propagate by seed in spring or by softwood or semi-ripe cuttings in summer.
T. alata illus. p.176.
T. coccinea. Evergreen, woody-stemmed, perennial, twining climber with narrowly oval leaves. H 6m (20ft) or more. Frost tender, min. 15°C (59°F). Pendent racemes of tubular, scarlet flowers are produced in winter-spring.
♛ ***T. grandiflora*** (Blue trumpet vine).

Evergreen, woody-stemmed, perennial, twining climber. H 6–10m (20–30ft). Frost tender, min. 10°C (50°F). Oval leaves, 10–20cm (4–8in) long, usually have a few tooth-like lobes. Trumpet-shaped, pale to deep violet-blue flowers appear in summer.
T. gregorii illus. p.177.
♛ ***T. mysorensis*** illus. p.167.

THYMUS (Labiatae)
Thyme
Genus of evergreen, mat-forming and dome-shaped shrubs, sub-shrubs and woody-based perennials with aromatic leaves. Is useful for growing on banks and in rock gardens, troughs and paving. Fully to half hardy. Requires sun and moist but well-drained soil. Propagate by softwood or semi-ripe cuttings in summer.
T. caespititius illus. p.323.
T. carnosus, syn. *T. nitidus* of gardens. Evergreen, spreading shrub. H and S 20cm (8in). Frost hardy. Has tiny, narrowly oval, aromatic leaves. Erect flowering stems bear whorls of small, 2-lipped, white flowers in summer. Needs a sheltered position.
♛ ***T. x citriodorus*** **'Aureus'**. Evergreen, spreading shrub. H 10cm (4in), S 10–25cm (4–10in). Frost hardy. Golden-yellow leaves are tiny, rounded to oval and very fragrant when crushed. Produces terminal clusters of small, 2-lipped, lilac flowers in summer. Cut back in spring. ♛ **'Silver Queen'** bears silvery-white foliage.
T. herba-barona illus. p.329.
T. leucotrichus illus. p.329.
T. nitidus of gardens. See *T. carnosus*.
T. 'Porlock'. Evergreen, dome-shaped perennial. H 8cm (3in), S 20cm (8in). Fully hardy. Thin stems are covered in small, rounded to oval, very aromatic, glossy, green leaves. In summer bears clusters of small, 2-lipped, pink flowers.
T. praecox. Evergreen, creeping perennial. H 1cm (½in), S indefinite. Fully hardy. Prostrate, woody stems are clothed in minute, oval to oblong, aromatic, green leaves. Small, 2-lipped, purple, mauve or white flowers in small clusters are borne in summer. **'Annie Hall'**, H 5cm (2in), S 20cm (8in), has pale pink flowers and light green leaves. **'Elfin'**, H 5cm (2in), S 10cm (4in), produces emerald-green leaves in dense hummocks; occasionally bears purple flowers.
T. pseudolanuginosus. Evergreen, prostrate shrub. H 2.5–5cm (1–2in), S 20cm (8in) or more. Fully hardy. Has dense mats of very hairy stems bearing tiny, aromatic, grey leaves. Produces 2-lipped, pinkish-lilac flowers in leaf axils in summer.

TIARELLA (Saxifragaceae)
Foamflower
Genus of perennials, some of which are evergreen, that spread by runners. Is excellent as ground cover. Fully hardy. Tolerates deep shade and prefers moist but well-drained soil. Propagate by division in spring.
♛ ***T. cordifolia*** illus. p.295.
♛ ***T. wherryi***. Slow-growing, clump-forming perennial. H 10cm (4in), S 15cm (6in). Forms mounds of

triangular, lobed, hairy, basal, green leaves, stained dark red and with heart-shaped bases. Produces racemes of tiny, star-shaped, soft pink or white flowers from late spring to early summer.

TIBOUCHINA (Melastomataceae)
Genus of evergreen perennials, sub-shrubs, shrubs and scandent climbers, grown for their flowers and leaves. Frost tender, min. 5–7°C (41–5°F). Prefers full sun and fertile, well-drained, neutral to acid soil. Water potted specimens freely when in full growth, moderately at other times. Cut back flowered stems, each to 2 pairs of buds, in spring. Tip prune young plants to promote branching. Propagate by greenwood or semi-ripe cuttings in late spring or summer.
T. semidecandra. See *T. urvilleana*.
♛ ***T. urvilleana***, syn. *T. semidecandra*, illus. p.92.

TIGRIDIA (Iridaceae)
Genus of summer-flowering bulbs, grown for their highly colourful but short-lived flowers, rather iris-like in shape, with 3 large, outer petals. Half hardy. Needs sun and well-drained soil, with ample water in summer. Plant in spring. Lift in autumn; then partially dry bulbs and store in peat or sand at 8–12°C (46–54°F). Propagate by seed in spring.
T. pavonia illus. p.367.

TILIA (Tiliaceae)
Lime, Linden
Genus of deciduous trees, grown for their small, fragrant, cup-shaped flowers and stately habit. Flowers attract bees, but are toxic to them in some cases. Fully hardy. Requires sun or semi-shade and fertile, well-drained soil. Propagate species by seed in autumn, selected forms and hybrids by grafting in late summer. Except for *T. x euchlora*, trees are usually attacked by aphids, which cover growth and ground beneath with sticky honeydew.
T. americana (American lime, Basswood). Deciduous, spreading tree. H 25m (80ft), S 12m (40ft). Has large, rounded, sharply toothed, glossy, dark green leaves. Small, yellowish-white flowers appear in summer.
♛ ***T. cordata*** (Small-leaved lime). Deciduous, spreading tree. H 30m (100ft), S 12m (40ft). Has small, rounded, glossy, dark green leaves and small, yellowish-white flowers in mid-summer. **'Greenspire'**, S 8m (25ft), is very vigorous and pyramidal in habit, even when young. **'Rancho'** illus. p.54.
♛ ***T. 'Euchlora'***, syn. *T. x euchlora* (Caucasian lime, Crimean lime). Deciduous, spreading tree with lower branches that droop with age. H 20m (70ft), S 10m (30ft). Rounded, very glossy, deep green leaves turn yellow in autumn. Bears small, yellowish-white flowers, toxic to bees, in summer. Is relatively pest-free.
T. x europaea. See *T. x vulgaris*.
T. henryana. Deciduous, spreading tree. H 20m S 10m (30ft). Broadly heart-shaped, glossy, bright green

leaves, fringed with long teeth, are often tinged red when young. Produces masses of small, creamy-white flowers in autumn.
♛ ***T. mongolica*** (Mongolian lime). Deciduous, spreading, graceful tree. H 15m (50ft), S 12m (40ft). Young shoots are red. Broadly heart-shaped, coarsely toothed, glossy, dark green leaves turn yellow in autumn. Small, yellowish-white flowers appear in summer.
T. oliveri illus. p.41.
♛ ***T. 'Petiolaris'*** illus. p.42.
T. platyphyllos (Broad-leaved lime, Large-leaved lime). Deciduous, spreading tree. H 30m (100ft), S 20m (70ft). Has rounded, dark green leaves and small, dull yellowish-white flowers in mid-summer. **'Princes Street'** is upright, with bright red shoots in winter.
T. tomentosa (European white lime, Silver lime). Deciduous, spreading tree. H 25m (80ft), S 20m (70ft). Leaves are large, rounded, sharply toothed, dark green above and white beneath. Very fragrant, small, dull white flowers, toxic to bees, are borne in late summer.
T. x vulgaris, syn. *T. x europaea* (Common lime). Vigorous, deciduous, spreading tree. H 35m (120ft), S 15m (50ft). Trunk develops many burs. Has rounded, dark green leaves. Small, yellowish-white flowers, toxic to bees, appear in summer. Periodically remove shoots from burs at base.

TILLANDSIA (Bromeliaceae)
Genus of evergreen, epiphytic perennials, often rosette-forming, some with branching stems and spirally arranged leaves, all grown for their flowers or overall appearance. Frost tender, min. 7–10°C (45–50°F). Requires semi-shade. Provide a rooting medium of equal parts humus-rich soil and either sphagnum moss or bark or plastic chips used for orchid culture. May also be grown on slabs of bark or sections of trees. Using soft water, water moderately in summer, sparingly at other times; spray plants grown on bark or tree sections with water several times a week from mid-spring to mid-autumn. Propagate by offsets or division in spring.
T. argentea (illus. p.229). Evergreen, basal-rosetted, epiphytic perennial. H and S 10–15cm (4–6in). Very narrow, almost thread-like leaves, covered with white scales, are produced in dense, near-spherical rosettes, each with a fleshy, bulb-like base. In summer, small, loose racemes of tubular, red flowers are produced.
T. caput-medusae (illus. p.229). Evergreen, basal-rosetted, epiphytic perennial. H and S 15cm (6in) or more. Linear, channelled, twisted and rolled, incurved leaves, covered in grey scales, develop in loose rosettes that have hollow, bulb-like bases. In summer, spikes of tubular, violet-blue flowers appear above foliage.
T. cyanea (illus. p.229). Evergreen, basal-rosetted, epiphytic perennial. H and S 25cm (10in). Forms dense rosettes of linear, pointed, channelled, arching, usually deep green leaves. In summer, broadly oval, blade-like

TIPUANA

spikes of pansy-shaped, deep purple-blue flowers, emerging from pink or red bracts, are produced among foliage.
T. fasciculata (illus. p.229). Evergreen, basal-rosetted, epiphytic perennial. H and S 30cm (12in) or more. Has dense rosettes of narrowly triangular, tapering, arching, mid-green leaves. In summer, flat spikes of tubular, purple-blue flowers emerge from red or reddish-yellow bracts, just above leaf tips. Bracts require strong light to develop reddish tones.
T. ionantha (Sky plant). Evergreen, clump-forming, basal-rosetted, epiphytic perennial. H and S 12cm (5in). Linear, incurved, arching leaves, covered in grey scales, are produced in dense rosettes; the inner leaves turn red at flowering time. Spikes of tubular, violet-blue flowers, emerging in summer from narrow, white bracts, are borne just above foliage.
⚘ **T. lindenii** (Blue-flowered torch; illus. p.229). Evergreen, basal-rosetted, epiphytic perennial. H and S 40cm (16in). Linear, pointed, channelled, arching, mid-green leaves, with red-brown lines, form dense rosettes. In summer, blade-like spikes of widely pansy-shaped, deep blue flowers, emerging from sometimes pink-tinted, green bracts, are borne just above leaves.
T. recurvata. Evergreen, basal-rosetted, epiphytic perennial. H and S 10–20cm (4–8in). Produces long, loose, stem-like, sometimes branched, rosettes of linear, arching to recurved leaves, densely covered in silvery-grey scales. In summer, short, dense spikes of small, tubular, pale blue or pale green flowers appear above leaves.
T. stricta (illus. p.229). Evergreen, clump-forming, basal-rosetted, epiphytic perennial. H and S 20–30cm (8–12in). Narrowly triangular, tapering, arching, mid-green leaves, usually with grey scales, are produced in dense rosettes. Large, tubular, blue flowers emerge from drooping, cone-like spikes of bright red bracts, usually in summer.
T. usneoides (Spanish moss; illus. p.229). Evergreen, pendent, epiphytic perennial. H 1m (3ft) or more, S 10–20cm (4–8in). Slender, branched, drooping stems bear linear, incurved leaves densely covered in silvery-white scales. Inconspicuous, tubular, greenish-yellow or pale blue flowers, hidden among foliage, are produced in summer.

TIPUANA (Leguminosae)
Genus of one species of evergreen, spring-flowering tree, grown for its flowers and overall appearance when mature and for shade. In certain conditions, may be deciduous. Frost tender, min. 10–13°C (50–55°F). Requires full light and fertile, well-drained soil. Container-grown plants will not produce flowers. Young specimens may be pruned in winter. Propagate by seed in spring.
T. speciosa. See *T. tipu*.
T. tipu, syn. *T. speciosa* (Pride of Bolivia, Tipa tree, Tipu tree). Fast-growing, mainly evergreen, bushy tree. H 10m (30ft), S 8–10m (25–30ft). Leaves, 25cm (10in) long, have 11–25 oval leaflets. Bears profuse racemes of pea-like, bright orange-yellow flowers, 3cm (1¼in) wide, in spring and short, woody, winged, brownish pods in autumn-winter.

TITANOPSIS (Aizoaceae)
Genus of basal-rosetted, perennial succulents eventually forming small, dense clumps. Produces 6–8 opposite pairs of fleshy, triangular leaves, 2–3cm (¾–1¼in) long, narrow at stems and expanding to straight tips. Frost tender, min. 8°C (46°F). Needs sun and well-drained soil. Propagate by seed in spring or summer.
T. calcarea illus. p.415.
T. schwantesii. Clump-forming, perennial succulent. H 3cm (1¼in), S 10cm (4in). Has a basal rosette of triangular, grey-blue leaves, with rounded corners and covered with small, wart-like, yellow-brown tubercles. Carries daisy-like, light yellow flowers, 2cm (¾in) wide, in summer-autumn.

TITHONIA (Compositae)
Genus of annuals. Half hardy. Grow in sun and in fertile, well-drained soil. Provide support and dead-head regularly. Propagate by seed sown under glass in late winter or early spring.
T. rotundifolia (Mexican sunflower). 'Torch' illus. p.292.

TOLMIEA (Saxifragaceae)
Genus of one species of perennial that is sometimes semi-evergreen and is grown as ground cover. Is suitable for cool woodland gardens. Fully hardy. Prefers shade and requires well-drained, neutral to acid soil. Propagate by division in spring or by seed in autumn.
T. menziesii (Pick-a-back plant, Youth-on-age). Mat-forming perennial, sometimes semi-evergreen. H 45–60cm (18–24in), S 30cm (12in) or more. Young plantlets develop where ivy-shaped, mid-green leaves join stem. Spikes of tiny, nodding, tubular to bell-shaped, green and chocolate-brown flowers appear in spring.

TOLPIS (Compositae)
Genus of summer-flowering annuals and perennials. Fully hardy. Grow in sun and in fertile, well-drained soil. Propagate by seed sown outdoors in spring.
T. barbata. Moderately fast-growing, upright, branching annual. H 45–60cm (1½–2ft), S 30cm (1ft). Has lance-shaped, serrated, mid-green leaves. Daisy-like, bright yellow flower heads, 2.5cm (1in) or more wide, with maroon centres, are produced in summer.

TOONA (Meliaceae)
Genus of deciduous trees, grown for their foliage, autumn colour and flowers. Fully hardy. Prefers full sun and requires fertile, well-drained soil. Propagate by seed in autumn or by root cuttings in winter.
T. sinensis, syn. *Cedrela sinensis*, illus. p.53.

TORENIA (Scrophulariaceae)
Genus of annuals and perennials. Half hardy to frost tender, min. 5°C (41°F). Grow in semi-shade and in a sheltered position in fertile, well-drained soil. Pinch out growing shoots of young plants to encourage a bushy habit. Propagate by seed sown under glass in early spring.
T. fournieri illus. p.285.

TORREYA (Cephalotaxaceae). See CONIFERS.
T. californica illus. p.81.

Tovara virginiana 'Painter's Palette'. See *Polygonum virginianum* 'Painter's Palette'.

TOWNSENDIA (Compositae)
Genus of evergreen, short-lived perennials and biennials, grown for their daisy-like flower heads. Suits alpine houses as dislikes winter wet. Fully hardy. Needs sun and moist soil. Propagate by seed in autumn.
T. grandiflora illus. p.331.
T. parryi. Evergreen, basal-rosetted, short-lived perennial. H 7–15cm (3–6in), S 5cm (2in). In late spring produces daisy-like, lavender or violet-blue flower heads, with bright yellow centres, above spoon-shaped leaves.

Toxicodendron succedaneum. See *Rhus trichocarpa*.
Toxicodendron verniciflua. See *Rhus verniciflua*.

TRACHELIUM, syn. DIOSPHAERA (Campanulaceae)
Genus of small perennials, useful for rock gardens and mixed borders. Some are good in alpine houses. Flowers of half-hardy species are ideal for cutting. Fully to half hardy, but protect fully-hardy species under glass in winter as they resent damp conditions. Grow in a sunny, sheltered position and in fertile, very well-drained soil (*T. asperuloides* prefers lime-rich soil). Propagate by seed in early or mid-spring or by softwood cuttings in spring.
T. asperuloides illus. p.331.
⚘ **T. caeruleum** illus. p.282.

TRACHELOSPERMUM (Apocynaceae)
Genus of evergreen, woody-stemmed, twining climbers with stems that exude milky sap when cut. Frost hardy. Grow in any well-drained soil and in sun or semi-shade. Propagate by seed in spring, by layering in summer or by semi-ripe cuttings in late summer or autumn.
⚘ **T. asiaticum**. Evergreen, woody-stemmed, much-branched, twining climber. H to 6m (20ft). Oval, glossy, dark green leaves are 2.5cm (1in) long. In summer bears scented, tubular, cream flowers, with expanded mouths, that age to yellow. Pairs of long, slender pods, 12–22cm (5–9in) long, contain silky seeds.
⚘ **T. jasminoides** illus. p.167.

TRACHYCARPUS (Palmae)
Genus of evergreen, summer-flowering palms, grown for their habit, foliage and flowers. Frost hardy. Requires full sun and does best in a position sheltered from strong, cold winds, especially when young. Needs fertile, well-drained soil. Propagate by seed in autumn or spring.
⚘ **T. fortunei** illus. p.58.

TRADESCANTIA

TRACHYMENE (Umbelliferae)
Genus of summer-flowering annuals. Half hardy. Grow in a sunny, sheltered position and in fertile, well-drained soil. Support with sticks. Propagate by seed sown under glass in early spring.
T. coerulea, syn. *Didiscus coeruleus* (Blue lace flower). Moderately fast-growing, upright, branching annual. H 45cm (18in), S 20cm (8in). Has deeply divided, pale green leaves. Spherical heads, to 5cm (2in) wide, of tiny, blue flowers are produced in summer. Flowers are excellent for cutting.

TRADESCANTIA (Commelinaceae)
Spiderwort
Genus of perennials, some of which are evergreen, grown for their flowers or ornamental foliage. Fully hardy to frost tender, min. 10–15°F (50–59°F). Grow in fertile, moist to dry soil and in sun or partial shade. Cut back or repropagate trailing species when they become straggly. Propagate hardy species by division, frost-tender species by tip cuttings in spring, summer or autumn.
T. albiflora. See *T. fluminensis*.
T. blossfeldiana. See *T. cerinthoides*.
T. cerinthoides, syn. *T. blossfeldiana*. Evergreen, creeping perennial. H 5cm (2in), S indefinite. Frost tender. Narrowly oval, fleshy, stem-clasping leaves, to 10cm (4in) long, are glossy, dark green above, purple with long, white hairs below. Intermittently has clusters of tiny, pink flowers, with white centres, surrounded by 2 leaf-like bracts. Leaves of ⚘ 'Variegata' have longitudinal, cream stripes.
T. fluminensis, syn. *T. albiflora* (Wandering Jew). Evergreen perennial with trailing, rooting stems. H 5cm (2in), S to 60cm (2ft) or more. Frost tender. Oval, fleshy leaves, 4cm (1½in) long, that clasp the stem, are glossy and green above, sometimes tinged purple below. Intermittently has clusters of tiny, white flowers enclosed in 2 leaf-like bracts. 'Albovittata' and ⚘ 'Variegata' illus. p.264.
⚘ **T. 'J.C. Weguelin'**. Clump-forming perennial. H to 60cm (2ft), S 45cm (1½ft). Fully hardy. Narrowly lance-shaped, fleshy, green leaves, 15–30cm (6–12in) long, clasp stems. In summer has clusters of 3-petalled, lavender-blue flowers, 2.5cm (1in) or more wide, surrounded by 2 leaf-like bracts.
T. navicularis. See *Callisia navicularis*.
⚘ **T. 'Osprey'** illus. p.240.
T. pallida, syn. *Setcreasea purpurea*.
⚘ **'Purple Heart'** illus. p.267.
T. pexata. See *T. sillamontana*.
⚘ **T. 'Purple Dome'** illus. p.247.
T. purpusii. See *T. zebrina* 'Purpusii'.
⚘ **T. sillamontana**, syn. *T. pexata*, *T. velutina*, illus. p.267.
T. spathacea, syn. *Rhoeo discolor*, *R. spathacea* (Boat lily, Moses-in-the-cradle). Evergreen, clump-forming

perennial. H 50cm (20in), S 25cm (10in). Frost tender. Short stem bears a rosette of lance-shaped, fleshy leaves, to 30cm (12in) long, glossy green above, purple below. Tiny, white flowers, enclosed in boat-shaped, leaf-like bracts, are produced year-round.
♛ **'Vittata'** has leaves striped longitudinally with pale yellow.
T. velutina. See *T. sillamontana.*
♛ *T. zebrina*, syn. *Zebrina pendula*, illus. p.265. ♛ **'Purpusii'** (syn. *T. purpusii*) is a strong-growing, evergreen, trailing or mat-forming perennial. H 10cm (4in), S indefinite. Frost tender. Has elliptic, purple-tinged, bluish-green leaves and tiny, shallowly cup-shaped, pink flowers.
♛ **'Quadricolor'** has leaves striped green, pink, red and white.

TRAPA (Trapaceae)
Genus of deciduous, perennial and annual, floating water plants, grown for their foliage and flowers. Frost hardy to frost tender, min. 5°C (41°F). Requires sun. Propagate in spring from seed gathered in autumn and stored in water or damp moss.
T. natans illus. p.389.

Trichocerus bridgesii. See *Echinopsis bridgesii.*
Trichocerus candicans. See *Echinopsis candicans.*
Trichocerus spachianus. See *Echinopsis spachianus.*

TRICHODIADEMA (Aizoaceae)
Genus of bushy, perennial succulents with woody or tuberous roots and cylindrical to semi-cylindrical leaves. Frost tender, min. 5°C (41°F). Needs sun and well-drained soil. Propagate by seed or stem cuttings in spring or summer.
T. densum. Tufted, perennial succulent. H 10cm (4in), S 20cm (8in). Cylindrical, pale green leaves are each 1–2cm ($1/2–3/4$in) long and tipped with clusters of white bristles. Roots and prostrate, green stem are both fleshy and form caudex. Stem tip carries daisy-like, cerise flowers, 3cm ($1 1/4$in) across, in summer.
T. mirabile illus. p.405.

TRICHOSANTHES (Cucurbitaceae)
Genus of annual and evergreen, perennial, tendril climbers, grown for their fruits and overall appearance. Frost tender, min. 15–18°C (59–64°F). Needs full sun or partial shade and humus-rich soil. Water freely in growing season, less in cool weather. Provide support. Propagate by seed in spring at not less than 21°C (70°F).
T. anguina. See *T. cucumerina* var. *anguina.*
T. cucumerina var. *anguina*, syn. *T. anguina* (Snake gourd). Erect to spreading, annual, tendril climber. H 3–5m (10–15ft). Has mid- to pale green leaves, to 20cm (8in) long, that are broadly oval to almost triangular and sometimes shallowly 3- to 5-lobed. In summer bears 5-petalled, white flowers, 2.5–5cm (1–2in) across, with heavily fringed petals; the females are solitary, the males in racemes. Cylindrical fruits, 60cm (2ft) or rarely to 2m (6ft) long, often twisted or coiled, are green-and-white striped and ripen to dull orange.
Tricuspidaria lanceolata. See *Crinodendron hookerianum.*

TRICYRTIS (Liliaceae)
Toad lily
Genus of late summer- and autumn-flowering, rhizomatous perennials. Fully hardy. Grows in sun or, in warm areas, in partial shade. Needs humus-rich, moist soil. Propagate by division in spring or by seed in autumn.
♛ *T. formosana*, syn. *T. stolonifera*, illus. p.223.
T. hirta. Upright, rhizomatous perennial. H 30cm–1m (1–3ft), S 45cm ($1 1/2$ft). In late summer and early autumn, clusters of large, open bell-shaped, white-spotted, purple flowers appear from axils of uppermost leaves. Leaves are narrowly oval, hairy and dark green and clasp stems. var. *alba* illus. p.258.
T. macrantha. Upright, rhizomatous perennial. H and S 60cm (24in). In early autumn bears loose sheaves of open bell-shaped, deep primrose-yellow flowers, spotted with light chocolate, at tips of arching stems. Small, oval leaves are dark green.
T. stolonifera. See *T. formosana.*

TRIFOLIUM (Leguminosae)
Clover
Genus of annuals, biennials and perennials, some of which are semi-evergreen, with rounded, usually 3-lobed leaves and heads of pea-like flowers. Some species are useful in rock gardens or on banks, others in agriculture. Many are invasive. Fully to frost hardy. Needs sun and well-drained soil. Propagate by division in spring or by seed in autumn. Self seeds readily.
T. repens **'Purpurascens'** illus. p.336.

TRILLIUM (Liliaceae)
Trinity flower, Wood lily
Genus of perennials with petals, sepals and leaves that are all borne in whorls of 3. Is excellent for woodland gardens. Forms of *T. cuneatum* and *T. sessile* are often grown as *T. chloropetalum.* Fully hardy. Enjoys full or partial shade and fertile, moist, preferably neutral to acid soil. Propagate by division after foliage has died down in summer or by seed in autumn.
T. cernuum f. *album* illus. p.231.
♛ *T. chloropetalum* illus. p.232.
♛ *T. erectum* illus. p.233.
♛ *T. grandiflorum* illus. p.232.
♛ **'Flore Pleno'** is a clump-forming perennial. H 38cm (15in), S 30cm (12in). Large, double, pure white flowers, borne singly in spring, turn pink with age. Has large, broadly oval, dark green leaves.
T. nivale (Dwarf white wood lily, Snow trillium). Early spring-flowering, rhizomatous perennial. H 7cm (3in), S 10cm (4in). Whorls of 3 oval leaves emerge at same time as outward-facing, slightly nodding, white flowers, each with 3 narrowly oval petals. Thrives in an alpine house or trough. Is difficult to grow.
T. ovatum illus. p.232.
♛ *T. rivale* illus. p.313.
T. sessile illus. p.233.
T. undulatum (Painted trillium, Painted wood lily). Clump-forming perennial. H 10–20cm (4–8in), S 15–20cm (6–8in). Open funnel-shaped flowers have red-bordered, green sepals and 3 white or pink petals, each with a carmine stripe at base. They are borne singly in spring, well above broadly oval, basal, blue-green leaves.

TRIPETALEIA (Ericaceae)
Genus of one species of deciduous shrub, grown for its flowers; is now often included in *Elliottia.* Fully hardy. Needs semi-shade and moist, peaty, neutral to acid soil. Propagate by softwood cuttings in summer or by seed in autumn.
T. paniculata. Deciduous, upright shrub. H and S 1.5m (5ft). Bears upright panicles of pink-tinged, white flowers, each with 3 (sometimes 4 or 5) narrow petals, from mid-summer to early autumn. Lance-shaped, dark green leaves persist well into autumn.

TRIPTERYGIUM (Celastraceae)
Genus of deciduous, twining or scrambling climbers, grown for their foliage and fruits. Frost hardy. Grow in any fertile, well-drained soil and in full sun or light shade. Water freely while in full growth, less in low temperatures. Provide stems with support. Thin out crowded stems in winter or early spring. Propagate by seed when ripe or in spring or by semi-ripe cuttings in summer.
T. regelii. Deciduous, slender-stemmed, twining or scrambling climber. H 10m (30ft). Leaves are oval and usually rich green. In late summer produces clusters, 20–25cm (8–10in) long, of small, off-white flowers, followed by 3-winged, pale green fruits.

TRISTANIA (Myrtaceae)
Genus of evergreen trees and shrubs, grown for their overall appearance when mature and for shade. Is related to *Eucalyptus.* Half hardy to frost tender, min. 3–5°C (37–41°F). Prefers fertile, well-drained soil and full light. Other than shaping plants in winter, pruning is seldom necessary. Propagate by seed in spring or by semi-ripe cuttings in summer.
T. conferta, syn. *Lophostemon confertus* (Brisbane box, Brush-box tree). Fast-growing, evergreen, round-headed tree. H and S 15–40m (50–130ft). Frost tender. Produces lance-shaped, leathery, lustrous leaves. In spring bears white flowers with prominent, feathery stamen bundles. **'Variegata'** has cream- or white-patterned leaves.

TRITELEIA (Liliaceae)
Genus of late spring- and early summer-flowering corms with wiry stems carrying *Allium*-like umbels of funnel-shaped flowers. Long, narrow leaves usually die away by flowering time. Frost hardy. Needs an open but sheltered, sunny situation and well-drained soil that dries out to some extent in summer. Dies down in mid- to late summer until winter or spring; plant during dormancy in early autumn. Propagate by seed or offsets in autumn.
T. hyacinthina, syn. *Brodiaea hyacinthina*, illus. p.363.
T. ixioides, syn. *Brodiaea ixioides.* Early summer-flowering corm. H to 50cm (20in), S 8–10cm (3–4in). Bears semi-erect, basal leaves. Stem produces a loose umbel, to 12cm (5in) across, of yellow flowers; petals each have a purple stripe.
T. laxa, syn. *Brodiaea laxa*, illus. p.366.
T. peduncularis, syn. *Brodiaea peduncularis.* Early summer-flowering corm. H 10–40cm (4–16in), S 10–15cm (4–6in). Has semi-erect, basal leaves. Stem produces a loose umbel, to 35cm (14in) across, of white flowers, each 1.5–3cm ($5/8–1 1/4$in) long, faintly tinged blue.

TRITONIA (Iridaceae)
Genus of corms, with flattish fans of sword-shaped, erect leaves, grown for their spikes of colourful flowers. Frost to half hardy. Needs a sunny, sheltered site and well-drained soil. Plant corms in autumn (*T. rubrolucens* in spring). Dry off once leaves start dying back in summer (winter for *T. rubrolucens*). Propagate by seed in autumn or by offsets at replanting time.
T. crocata, syn. *T. hyalina.* Spring-flowering corm. H 15–35cm (6–14in), S 5–8cm (2–3in). Half hardy. Has erect, basal leaves. Wiry stems each bear a loose spike of up to 10 widely cup-shaped, orange or pink flowers, 4–5cm ($1 1/2–2$in) across, with transparent margins.
T. disticha subsp. *rubrolucens*, syn. *T. rosea*, *T. rubrolucens*, *Crocosmia rosea*, illus. p.364.
T. hyalina. See *T. crocata.*
T. rosea. See *T. disticha* subsp. *rubrolucens.*
T. rubrolucens. See *T. disticha* subsp. *rubrolucens.*

TROCHOCARPA (Epacridaceae)
Genus of evergreen shrubs, grown for their nodding flower spikes. Frost hardy to frost tender, min. 7°C (45°F). Needs sun and moist but well-drained, peaty, sandy soil. Propagate by semi-ripe cuttings in summer.
T. thymifolia. Slow-growing, evergreen, erect shrub. H 30cm (12in), S to 20cm (8in). Frost hardy. Stems are covered in minute, thyme-like leaves. Carries 4cm ($1 1/2$in) long spikes of tiny, bell-shaped, pink flowers in summer-autumn. Suits an alpine house.

TROCHODENDRON (Trochodendraceae)
Genus of one species of evergreen tree, grown for its foliage and flowers. Frost hardy, but needs shelter from strong, cold winds. Tolerates sun or shade and requires moist but well-drained soil; dislikes very dry or very shallow, chalky soil. Propagate by semi-ripe cuttings in summer or by seed in autumn.
T. aralioides illus. p.58.

TROLLIUS (Ranunculaceae)
Globeflower
Genus of spring- or summer-flowering perennials that thrive beside pools and streams. Fully hardy. Tolerates sun or shade. Does best in moist soil. Propagate by division in early autumn or by seed in summer or autumn.
T. 'Alabaster' illus. p.238.
T. 'Earliest of All'. Clump-forming perennial. H 60cm (24in), S 45cm (18in). Globular, butter-yellow flowers are borne singly in spring, above rounded, deeply divided, mid-green leaves.
T. europaeus illus. p.239. 'Canary Bird' is a clump-forming perennial. H 60cm (24in), S 45cm (18in). In spring bears globular, canary-yellow flowers above rounded, deeply divided, mid-green leaves.
♛ *T.* 'Goldquelle'. Clump-forming perennial. H 60cm (24in), S 45cm (18in). Large, globular, rich orange flowers are borne singly in spring, above rounded, deeply divided, mid-green leaves.
♛ *T.* 'Orange Princess'. Clump-forming perennial. H 75cm (30in), S 45cm (18in). Globular, orange-gold flowers are borne singly in spring, above rounded, deeply divided, mid-green leaves.
T. pumilus illus. p.321.
T. yunnanensis. Clump-forming perennial. H 60cm (2ft), S 30cm (1ft). Has broadly oval leaves with 3–5 deep lobes. Produces buttercup-like, bright yellow flowers in late spring or summer.

TROPAEOLUM (Tropaeolaceae)
Nasturtium
Genus of annuals, perennials and herbaceous, twining climbers, grown for their brightly coloured flowers. Fully hardy to frost tender, min. 5°C (41°F). Most species prefer sun and well-drained soil. Propagate by seed, tubers or basal stem cuttings in spring. Aphids and caterpillars of cabbage white butterfly and its relatives may cause problems.
T. azureum. Herbaceous, leaf-stalk climber with small tubers. H to 1.2m (4ft). Frost tender. Leaves, to 5cm (2in) across, have 5 narrow lobes. Small, purple-blue flowers, with notched petals, open in late summer.
T. canariense. See *T. peregrinum.*
T. majus (Garden nasturtium, Indian cress). 'Alaska' illus. p.291. 'Empress of India' is a fast-growing, bushy annual. H 20cm (8in), S 30cm (12in). Fully hardy. Has rounded, mid-green leaves. Trumpet-shaped, spurred, dark crimson flowers, 5cm (2in) wide, are borne from early summer to early autumn. Gleam Series has a semi-trailing habit and double flowers in single colours or in a mixture that includes scarlet, yellow and orange. Jewel Series illus. p.293. Flowers of 'Peach Melba', H to 30cm (12in), are pale yellow, blotched with scarlet.
♛ Whirlybird Series, H to 30cm (12in), has single flowers in a mixture or in single colours.
T. peregrinum, syn. *T. canariense* (Canary creeper). Herbaceous, leaf-stalk climber. H to 2m (6ft). Frost tender. Grey-green leaves have 5 broad lobes. Small, bright yellow flowers, the 2 upper petals much larger and fringed, are borne from summer until first frosts. In cool areas is best grown as an annual.
T. polyphyllum illus. 257.
♛ *T. speciosum* illus. p.170.
♛ *T. tricolorum* illus. p.165.
T. tuberosum illus. p.178. ♛ 'Ken Aslet' illus. p.176.

TSUGA (Pinaceae). See CONIFERS.
T. canadensis illus. p.81. 'Aurea' (illus. p.85) is a broadly conical conifer, often with several stems. H 5m (15ft) or more, S 2–3m (6–10ft). Fully hardy. Shoots are grey with spirally arranged, needle-like, flattened leaves, golden-yellow when young, ageing to green in second year, those along top inverted to show silver bands. Has ovoid, light brown cones. 'Bennett', H 1–2m (3–6ft), S 2m (6ft), is a compact, dwarf form with arching branches and a nest-shaped, central depression. ♛ f. *pendula* has weeping branches that may be trained to create a domed mound, H and S 3–5m (10–15ft), or left to spread at ground level, H 50cm (20in), S 2–5m (6–15ft).
T. caroliniana (Carolina hemlock). Conifer with a conical or ovoid crown. H 10–15m (30–50ft), S 5–8m (15–25ft). Fully hardy. Red-brown shoots produce spirally set, needle-like, flattened, glossy, dark green leaves. Bears ovoid, green cones, ripening to brown.
T. diversifolia (Japanese hemlock, Northern Japanese hemlock). Conifer with a broad, dense crown. H 10–15m (30–50ft), S 8–12m (25–40ft). Fully hardy. Has orange shoots and needle-like, flattened, glossy, deep green leaves, banded with white beneath, that are spirally set. Ovoid cones are dark brown.
♛ *T. heterophylla* (Western hemlock). Vigorous, conical conifer with drooping branchlets. H 20–30m (70–100ft), S 8–10m (25–30ft). Fully hardy. Grey shoots bear spirally set, needle-like, flattened, dark green leaves with silvery bands beneath. Bears ovoid, pale green cones that ripen to dark brown.
T. mertensiana (Mountain hemlock). Narrowly conical conifer with short, horizontal branches. H 8–15m (25–50ft), S 3–6m (10–20ft). Fully hardy. Red-brown shoots bear needle-like, flattened, glaucous blue-green or grey-green leaves, spirally arranged. Cones are cylindrical and yellow-green to purple, ripening to dark brown.
T. sieboldii (Japanese hemlock, Southern Japanese hemlock). Broadly conical conifer. H 15m (50ft), S 8–10m (25–30ft). Fully hardy. Has glossy, buff shoots that bear needle-like, flattened, lustrous, dark green leaves, set spirally. Cones are ovoid and dark brown.

TSUSIOPHYLLUM (Ericaceae)
Genus of one species of semi-evergreen shrub, grown for its flowers. Is similar to *Rhododendron* and is suitable for rock gardens and peat beds. Frost hardy. Requires shade and well-drained, peaty, sandy soil. Propagate by softwood cuttings in spring or early summer or by seed in autumn or spring.
T. tanakae. Semi-evergreen, spreading shrub. H 15cm (6in) or more, S 25cm (10in). Twiggy, branched stems bear tiny, narrowly oval, hairy leaves. In early summer produces small, tubular, white or pinkish-white flowers at stem tips.

TUBERARIA (Cistaceae)
Genus of annuals. Fully hardy. Grow in sun and in any very well-drained soil. Propagate by seed sown outdoors in spring.
T. guttata, syn. *Helianthemum guttatum.* Moderately fast-growing, upright, branching annual. H and S 10–30cm (4–12in). Has lance-shaped, hairy, mid-green leaves and, in summer, yellow flowers, sometimes red-spotted at base of petals, that look like small, single roses.

TULBAGHIA (Liliaceae)
Genus of semi-evergreen perennials. Frost to half hardy. Needs full sun and well-drained soil. Propagate by division or seed in spring.
T. natalensis. Semi-evergreen, clump-forming perennial. H 12cm (5in), S 10cm (4in). Frost hardy. In mid-summer, umbels of delicately fragrant, tubular, yellow-centred, white flowers, with spreading petal lobes, open above fine, grass-like, mid-green foliage.
T. violacea illus. p.247.

TULIPA (Liliaceae)
Tulip
Genus of mainly spring-flowering bulbs, grown for their bright, upward-facing flowers. Each bulb has a few linear to lance-shaped, green or grey-green leaves, which are borne on the stem. Flowers have 6 usually pointed petals (botanically known as perianth segments) and 6 stamens; unless otherwise stated below, they are borne singly. Each plant has a spread of up to 20cm (8in). All are fully hardy unless otherwise stated. Requires a sunny position with well-drained soil and appreciates a summer baking; in cool, wet areas, bulbs may be lifted, when the leaves have died down, and stored in a dry place for replanting in autumn. Propagate by division of bulbs in autumn or for species by seed in spring or autumn.
Horticulturally, tulips are grouped into the following divisions:

Div.1 Single early – has cup-shaped, single flowers, often opening wide in sun, from early to mid-spring.
Div.2 Double early – double flowers open wide, appear in early and mid-spring and are long-lasting.
Div.3 Triumph – sturdy stems bear rather conical, single flowers, opening to a more rounded shape, in mid- and late spring.
Div.4 Darwin hybrids – large, single flowers of variable shape are borne on strong stems from mid- to late spring.
Div.5 Single late – single flowers, variable in shape but often ovoid or squarish and usually with pointed petals, appear in late spring and very early summer.
Div.6 Lily-flowered – strong stems bear narrow-waisted, single flowers, with long, pointed petals often reflexed at tips, in late spring.
Div.7 Fringed – flowers are similar to those in Div.6, but have fringed petals.
Div.8 Viridiflora – variably shaped, single flowers, with partly greenish petals, are borne in late spring.
Div.9 Rembrandt – comprises mostly very old cultivars, similar to Div.6, but has colours 'broken' into striped or feathered patterns owing to virus. Flowers in late spring.
Div.10 Parrot – large, single flowers of variable shape, with petals frilled or fringed and usually twisted, appear in late spring.
Div.11 Double late (peony-flowered) – double flowers are usually bowl-shaped and appear in late spring.
Div.12 Kaufmanniana hybrids – single flowers are usually bicoloured, open flat in sun and are borne in early spring. Leaves are usually mottled or striped.
Div.13 Fosteriana hybrids – has large, single flowers that open wide in sun from early to mid-spring. Leaves are often mottled or striped.
Div.14 Greigii hybrids – large, single flowers appear in mid- and late spring. Leaves are generally wavy-edged, always mottled or striped.
Div.15 Miscellaneous – a miscellaneous category of other species and their varieties and hybrids. Flowers appear in spring and early summer.

T. acuminata (Horned tulip; illus. p.355), Div.15. Mid-spring-flowering bulb. H 30–45cm (12–18in). Flowers are 7–13cm (3–5in) long, with long-pointed, tapered, pale red or yellow petals, often tinged with red or green outside.
T. 'Ad Rem' (illus. p.355), Div.4. Mid- to late spring-flowering bulb. H 60cm (2ft). Flowers are scarlet with black bases and yellow edges. Anthers are yellow.
T. aitchisonii. See *T. clusiana.*
♛ *T.* 'Ancilla', Div.12. Early spring-flowering bulb. H 15cm (6in). Flowers are pink and reddish outside, white inside, each with a central, red ring.
T. 'Angélique' (illus. p.354), Div.11. Late spring-flowering bulb. H 40cm (16in). Delicately scented, double, pale pink flowers deepen with age. Each petal has paler streaks and a lighter edge. Is good for bedding.
♛ *T.* 'Apeldoorn's Elite' (illus. p.355), Div.4. Mid- to late spring-flowering bulb. H 60cm (2ft). Has buttercup-yellow flowers feathered with cherry-red and with yellowish-green bases.
T. 'Apricot Beauty' (illus. p.356), Div.1. Early spring-flowering bulb. H 40cm (16in). Flowers are salmon-pink faintly tinged with red.
T. 'Artist' (illus. p.356), Div.8. Late spring-flowering bulb. H 45cm (18in). Flowers are salmon-pink and purple outside, sometimes marked with green, and deep salmon-pink and green inside.
T. 'Attila' (illus. p.355), Div.3. Mid-spring-flowering bulb. H 40cm (16in).

Strong stems carry long-lasting, pink flowers. Is good for bedding.

℗ *T. aucheriana*, Div.15. Early spring-flowering bulb. H to 20cm (8in). Has grey-green leaves. Flowers, 2–5cm (3/4–2in) long, each tapered at the base with oval petals, are pink with yellow centres.

T. bakeri. See *T. saxatilis*.

T. 'Balalaika' (illus. p.355), Div.5. Late spring-flowering bulb. H 50cm (20in). Bright red flowers each have a yellow base and black stamens.

℗ *T.* 'Ballade' (illus. p.354), Div.6. Late spring-flowering bulb. H 50cm (20in). Reddish-magenta flowers with a white-edged, yellow base have long petals that are edged with white.

℗ *T. batalinii*, syn. *T. linifolia* Batalinii Group (illus. p.356), Div.15. Early spring-flowering bulb. H 10–30cm (4–12in). Is often included under *T. linifolia*. Leaves are grey-green. Flowers, 2–6cm (3/4–2 1/2in) long, have broadly oval petals and are bowl-shaped at the base. Pale yellow petals are darker yellow or brown at bases inside. Several cultivars are hybrids between *T. batalinii* and *T. linifolia*. These include 'Apricot Jewel' with flowers that are orange-red outside, yellow inside; 'Bright Gem', which has yellow flowers flushed with orange; and 'Bronze Charm', which bears yellow flowers with bronze feathering.

T. 'Bellona' (illus. p.356), Div.1. Early spring-flowering bulb. H 30cm (12in). Fragrant flowers are deep golden-yellow. Is good for bedding and forcing.

T. biflora, syn. *T. polychroma* (illus. p.354), Div.15. Early spring-flowering bulb. H 5–10cm (2–4in). Has grey-green leaves. Stem bears 1–5 fragrant, yellow-centred, white flowers, 1.5–3.5cm (5/8–1 1/2in) long and tapered at the bases. Narrowly oval petals are flushed outside with greenish-grey or greenish-pink. Suits a rock garden.

T. 'Bing Crosby' (illus. p.355), Div.3. Mid- to late spring-flowering bulb. H 50cm (20in). Has glowing, scarlet flowers.

T. 'Bird of Paradise' (illus. p.355), Div.10. Late spring-flowering bulb. H 45cm (18in). Has cardinal-red flowers that are edged with orange and have bright yellow bases. Anthers are purple.

T. 'Blue Parrot' (illus. p.356), Div.10. Late spring-flowering bulb. H 60cm (2ft). Very large, bright violet flowers, sometimes bronze outside, are borne on strong stems.

T. 'Burgundy Lace', Div.7. Late spring-flowering bulb. H 60cm (2ft). Flowers are wine-red, each petal with a fringed edge.

T. 'Candela' (illus. p.356), Div.13. Early to mid-spring-flowering bulb. H 30cm (12in). Large flowers are yellow, with black anthers, and long-lasting.

T. 'Cape Cod' (illus. p.356), Div.14. Mid- to late spring-flowering bulb. H 45cm (18in). Grey-green leaves have reddish stripes. Yellowish-bronze flowers each have a black-and-red base; petals are edged with yellow outside.

T. 'Carnaval de Nice' (illus. p.354), Div.11. Late spring-flowering bulb. H 40cm (16in). Double flowers are white feathered with deep red.

T. 'China Pink' (illus. p.354), Div.6. Late spring-flowering bulb. H 55cm (22in). Flowers are pink, each with a white base, and have slightly reflexed petals.

T. 'Chopin', Div.12. Early spring-flowering bulb. H 20cm (8in). Has brown-mottled, grey-green leaves. Lemon-yellow flowers have black bases.

T. 'Clara Butt', Div.5. Late spring-flowering bulb. H 60cm (2ft). Flowers are salmon-pink. Is particularly good for bedding.

T. clusiana, syn. *T. aitchisonii* (Lady tulip; illus. p.355), Div.15. Mid-spring-flowering bulb. H to 30cm (12in). Has grey-green leaves. Each stem bears 1 or 2 flowers, 2–6.5cm (3/4–2 1/2in) long, that are bowl-shaped at the base. Narrowly oval, white petals are purple or crimson at base inside, striped deep pink outside. Stamens are purple. Flowers of ℗ var. *chrysantha* (illus. p.356) are yellow, flushed red or brown outside, with yellow stamens. var. *stellata* has white flowers with yellow bases and yellow stamens.

T. dasystemon of gardens. See *T. tarda*.

T. 'Dawnglow', Div.4. Mid- to late-spring flowering bulb. H 60cm (2ft). Pale apricot flowers are flushed with deep pink outside and are deep yellow inside. Has purple anthers.

T. 'Diana' (illus. p.354), Div.1. Early spring-flowering bulb. H 28cm (11in). Large, pure white flowers are carried on strong stems.

T. 'Dillenburg' (illus. p.356), Div.5. Late spring-flowering bulb. H 65cm (26in). Flowers are brick-orange and are good for bedding.

T. 'Don Quichotte' (illus. p.354), Div.3. Mid-spring-flowering bulb. H 40cm (16in). Purple-pink flowers are long-lasting.

T. 'Dreamboat' (illus. p.356), Div.14. Mid- to late spring-flowering bulb. H 25cm (10in). Has brown-striped, grey-green leaves. Urn-shaped, red-tinged, amber-yellow flowers have greenish-bronze bases with red blotches.

T. 'Dreaming Maid' (illus. p.356), Div.3. Mid- to late spring-flowering bulb. H 55cm (22in). Flowers have violet petals edged with white.

T. 'Dreamland' (illus. p.355), Div.5. Late spring-flowering bulb. H 60cm (2ft). Flowers are red with white bases and yellow anthers.

T. eichleri. See *T. undulatifolia*.

T. 'Estella Rijnveld' (illus. p.355), Div.10. Late spring-flowering bulb. H 60cm (2ft). Large flowers are red, streaked with white and green.

T. 'Fancy Frills' (illus. p.354), Div.10. Late spring-flowering bulb. H 50cm (20in). Fringed, ivory-white petals are striped and edged with pink outside; inside, base is rose-pink. Anthers are pale yellow.

T. fosteriana, Div.15. Early spring-flowering bulb. H 20–45cm (8–18in). Has a downy stem and grey-green leaves, downy above. Flowers, 4.5–10cm (1 3/4–4in) long, are bowl-shaped at the base with narrowly oval, bright red petals, and each has a purplish-black centre inside, ringed with yellow.

T. 'Fringed Beauty' (illus. p.355), Div.7. Early to mid-spring-flowering bulb. H 32cm (13in). Fringed petals are bright red with yellow edges. Is excellent for forcing.

T. 'Fringed Elegance', Div.7. Late spring-flowering bulb. H 50cm (20in). Has pale yellow flowers dotted with pink outside; inside, bases have bronze-green blotches. Each petal has a yellow fringe. Anthers are purple.

T. 'Gala Beauty', Div.9. Late spring-flowering bulb. H 60cm (2ft). Yellow flowers are streaked with crimson.

T. 'Garden Party' (illus. p.355), Div.3. Mid- to late spring-flowering bulb. H 40–45cm (16–18in). Has white flowers; petals are edged with deep pink outside and streaked with deep pink inside.

T. 'Glück' (illus. p.355), Div.12. Early spring-flowering bulb. H 15cm (6in). Has reddish-brown-mottled, grey-green leaves. Petals are red, edged with yellow, outside and yellow inside, each with a darker base.

T. 'Golden Apeldoorn', Div.4. Mid- to late spring-flowering bulb. H 50–60cm (20–24in). Golden-yellow flowers each have a black base and black stamens.

T. 'Golden Artist' (illus. p.356), Div.8. Late spring-flowering bulb. H 45cm (18in). Flowers are bright golden-yellow.

T. 'Gordon Cooper' (illus. p.355), Div.4. Mid- to late spring-flowering bulb. H 60cm (2ft). Petals are deep pink outside, edged with red; inside they are red with blue-and-yellow bases. Has black anthers.

T. 'Greenland'. See *T.* 'Groenland'.

T. greigii, Div.15. Early spring-flowering bulb. H 20–45cm (8–18in). Has downy stems. Leaves are streaked or mottled with red or purple. Cup-shaped flowers, 3–10cm (1 1/4–4in) long, with broadly oval, red or yellow petals, have yellow-ringed, black centres.

T. 'Greuze' (illus. p.356), Div.5. Late spring-flowering bulb. H 65cm (26in). Flowers are dark violet-purple and are good for bedding.

T. 'Groenland' (illus. p.354), Div.8. Late spring-flowering bulb. H 50cm (20in). Flowers have green petals that are edged with rose-pink. Is a good bedding tulip.

T. hageri (illus. p.355), Div.15. Mid-spring-flowering bulb. H 10–30cm (4–12in). Frost hardy. Stem has 1–4 flowers, 3–6cm (1 1/2–2 1/2in) long, tapered at the base and with oval, dull red petals tinged with green outside.

T. 'Heart's Delight', Div.12. Early spring-flowering bulb. H 20–25cm (8–10in). Green leaves are striped with red-brown. Petals are deep pinkish-red outside, edged with pale pink, and are pale pink inside with red-blotched, yellow bases.

T. 'Hollywood', Div.8. Late spring-flowering bulb. H 30cm (12in). Red flowers, tinged and streaked with green, have yellow bases. Is good for bedding.

T. humilis (illus. p.356), Div.15. Early spring-flowering bulb. A variable species, often considered to include *T. aucheriana*, *T. pulchella* and *T. violacea*. H to 20cm (8in). Has grey-green leaves. Stem bears usually 1, sometimes 2 or 3, pinkish-magenta flowers, 2–5cm (3/4–2in) long, tapered at the base and with a yellow centre inside. Petals are oval. Is suitable for a rock garden.

T. 'Jack Laan', Div.9. Late spring-flowering bulb. H 60cm (2ft). Purple flowers are shaded with brown and feathered with white and yellow.

T. 'Juan' (illus. p.355), Div.13. Early to mid-spring-flowering bulb. H 35cm (14in). Flowers are deep orange overlaid with scarlet. Leaves are marked with reddish-brown.

T. kaufmanniana (Water lily tulip; illus. p.356), Div.15. Early spring-flowering bulb. H 10–35cm (4–14in). Leaves are grey-green. Stem has 1–5 often scented flowers, 3–10cm (1 1/2–4in) long and bowl-shaped at the base. Narrowly oval petals are usually either cream or yellow, flushed with pink or grey-green outside; centres are often a different colour. Pink, orange or red forms occasionally occur.

℗ *T.* 'Keizerskroon' (illus. p.355), Div.1. Early spring-flowering bulb. H 35cm (14in). Flowers have crimson-scarlet petals, with broad, bright yellow margins. Is a good, reliable bedding tulip.

℗ *T.* 'Kingsblood' (illus. p.355), Div.5. Late spring-flowering bulb. H 60cm (2ft). Cherry-red flowers are edged with scarlet.

℗ *T. linifolia* (illus. p.355), Div.15. Early spring-flowering bulb. H 10–30cm (4–12in). A variable species, often considered to include *T. batalinii* and *T. maximowiczii*. Has grey-green leaves. Red flowers, 2–6cm (3/4–2in) long, are bowl-shaped at the base and, inside, have blackish-purple centres that are usually ringed with cream or yellow. Petals are broadly oval. **Batalinii Group** see *T. batalinii*.

T. 'Lustige Witwe'. See *T.* 'Merry Widow'.

T. 'Mme Lefèbre', syn. *T.* 'Red Emperor' (illus. p.355), Div.13. Early to mid-spring-flowering bulb. H 35–40cm (14–16in). Produces very large, brilliant red flowers.

T. 'Maja' (illus. p.356), Div.7. Late spring-flowering bulb. H 50cm (20in). Egg-shaped, pale yellow flowers, with fringed petals, are bronze-yellow at the base. Anthers are yellow.

T. 'Margot Fonteyn' (illus. p.355), Div.3. Mid- to late spring-flowering bulb. H 40–45cm (16–18in). Flowers have yellow-edged, bright red petals, each with a yellow base inside. Anthers are black.

T. marjolettii (illus. p.355), Div.15. Mid-spring-flowering bulb. H 40–50cm (16–20in). Flowers, 4–6cm (1 1/2–2 1/2in) long and bowl-shaped at the base, have broadly oval, creamy-white petals, edged and marked with deep pink.

T. maximowiczii, Div.15. Early spring-flowering bulb. H 10–30cm (4–12in). Has grey-green leaves. Bright red flowers, 2–6cm (3/4–2 1/2in) long, with

broadly oval petals, have white-bordered, black centres and are bowl-shaped at the base.
T. 'Menton' (illus. p.354), Div.5. Late spring-flowering bulb. H 60cm (2ft). Flowers have rose-pink petals that are edged with light orange, and bright yellow and white bases. Anthers are yellow.
T. 'Merry Widow' (syn. T. 'Lustige Witwe'; illus. p.355), Div.3. Mid- to late spring-flowering bulb. Flowers have deep glowing red petals edged with white.
♀ T. 'Monte Carlo', Div.2. Early spring-flowering bulb. H 40cm (16in). Has double, yellow flowers with sparse, red streaks.
T. 'New Design' (illus. p.354), Div.3. Mid-spring-flowering bulb. H 40cm (16in). Flowers have yellow petals that fade to pinkish-white and are edged with red outside and marked with apricot inside. Leaves have pinkish-white margins.
T. 'Orange Emperor', Div.13. Early to mid-spring-flowering bulb. H 40cm (16in). Flowers are bright orange, each with a yellow base inside, and have black anthers.
T. 'Orange Triumph', Div.11. Late spring-flowering bulb. H 50cm (20in). Has double, soft orange-red flowers flushed with brown; each petal has a yellow margin.
♀ T. 'Oranje Nassau' (illus. p.355), Div.2. Early spring-flowering bulb. H 25–30cm (10–12in). Has double, blood-red flowers flushed fiery orange-red. Is good for forcing.
T. 'Oratorio', Div.14. Mid- to late spring-flowering bulb. H 20cm (8in). Has reddish-brown-mottled, grey-green leaves. Broadly urn-shaped flowers are rose-pink outside, apricot-pink inside with black bases.
T. orphanidea (illus. p.356), Div.15. Mid-spring-flowering bulb. H 10–30cm (4–12in). Frost hardy. Green leaves often have reddish margins. Stem has 1–4 flowers, 3–6cm (1 1/4–2 1/2in) long and tapered at the base. Oval petals are orange-brown, tinged outside with green and often purple.
T. 'Page Polka' (illus. p.354), Div.3. Mid-spring-flowering bulb. Large, deep red flowers have white bases and are striped with white. Anthers are yellow.
T. 'Palestrina', Div.5. Late spring-flowering bulb. H 45cm (18in). Petals of large, salmon-pink flowers are green outside.
T. 'Peach Blossom', Div.2. Early spring-flowering bulb. H 25–30cm (10–12in). Produces double, silvery-pink flowers flushed with deep pink.
T. 'Peer Gynt' (illus. p.355), Div.3. Mid- to late spring-flowering bulb. H 50cm (20in). Flowers have fuchsia-red petals with purple edges and white bases spotted with yellow. Anthers are purplish-grey.
♀ T. 'Plaisir' (illus. p.355), Div.14. Mid- to late spring-flowering bulb. H 15–20cm (6–8in). Grey-green leaves are mottled with red-brown. Broadly urn-shaped, deep pinkish-red flowers have petals edged with pale yellow and with black-and-yellow bases.
T. polychroma. See T. biflora.

♀ T. praestans 'Fusilier', Div.15. Early spring-flowering bulb. H 10–45cm (4–18in). Has a minutely downy stem and downy, grey-green leaves. Stem bears 3–5 flowers that are 5.5–6.5cm (2 1/4–2 1/2in) long and bowl-shaped at the base. Oval petals are orange-scarlet. 'Unicum' (illus. p.355) has leaves that are edged with pale yellow. Flowers have bright red petals with yellow bases and blue-black anthers. 'Van Tubergen's Variety' (illus. p.356) produces 2–5 flowers per stem that are often yellow at the base; it increases very freely.
♀ T. 'Prinses Irene' (illus. p.356), Div.1. Early spring-flowering bulb. H 30–35cm (12–14cm). Produces orange flowers streaked with purple.
T. pulchella, Div.15. Early spring-flowering bulb. H to 20cm (8in). Has grey-green leaves. Flowers, 2–5cm (3/4–2in) long, are tapered at the base, have oval, purple petals and yellow or bluish-black centres inside. Is useful for a rock garden.
T. 'Purissima', syn. T. 'White Emperor' (illus. p.354), Div.13. Early to mid-spring-flowering bulb. H 35–40cm (14–16in). Flowers are pure white.
T. 'Queen of Night' (illus. p.356), Div.5. Late spring-flowering bulb. H 60cm (2ft). The darkest of all tulips, has long-lasting, very dark maroon-black flowers on sturdy stems. Is useful for bedding.
T. 'Red Emperor'. See T. 'Mme Lefèbre'.
T. 'Red Parrot' (illus. p.355), Div.10. Late spring-flowering bulb. H 60cm (2ft). Large, raspberry-red flowers are carried on strong stems.
♀ T. 'Red Riding Hood' (illus. p.355), Div.14. Late spring-flowering bulb. H 20cm (8in). Produces vivid, black-based, scarlet flowers amid spreading, dark green leaves that are mottled brownish-purple.
T. saxatilis, syn. T. bakeri (illus. p.354), Div.15. Early spring-flowering bulb. H 15–45cm (6–18in). Frost hardy. Has shiny, green leaves. Stem produces 1–4 scented flowers, 4–5.5cm (1 1/2–2 1/4in) long and tapered at the base. Oval, pink to lilac petals are yellow at the base inside.
T. 'Shakespeare' (illus. p.356), Div.12. Early spring-flowering bulb. H 12–15cm (5–6in). Petals are deep red, edged with salmon, outside and salmon, flushed red with a yellow base, inside.
♀ T. sprengeri (illus. p.355), Div.15. Late spring- and early summer-flowering bulb. H 30–45cm (12–18in). Flowers are 4.5–6.5cm (1 1/4–2 1/2in) long and tapered at the base. Has narrowly oval, orange-red petals, the outer 3 with buff-yellow backs. Is the latest flowering tulip. Increases very rapidly.
♀ T. 'Spring Green' (illus. p.354), Div.8. Late spring-flowering bulb. H 35–38cm (14–15in). Has white flowers feathered with green. Anthers are pale green.
T. sylvestris (illus. p.356), Div.15. Early spring-flowering bulb. H 10–45cm (4–18in). Yellow flowers,

usually borne singly, are 3.5–6.5cm (1 1/2–2 1/2in) long and tapered at the base. Narrowly oval petals are often tinged with green outside.
♀ T. tarda, syn. T. dasystemon of gardens (illus. p.356), Div.15. Early spring-flowering bulb. H to 15cm (6in). Has glossy, green leaves. Flowers, 4–6 per stem, are 3–4cm (1 1/4–1 1/2in) long and tapered at the base. Oval, white petals have yellow lower halves inside and are tinged with green and sometimes red outside. Suits a rock garden or raised bed.
♀ T. 'Toronto', Div.14. Mid- to late spring-flowering bulb. H 30cm (12in). Has mottled leaves and 2 or 3 long-lasting flowers per stem. Open, broadly cup-shaped flowers have pointed, bright red petals each with a brownish-green-yellow base inside. Anthers are bronze.
♀ T. turkestanica (illus. p.354), Div.15. Early spring-flowering bulb. H 10–30cm (4–12in). Has a hairy stem and grey-green leaves. Unpleasant-smelling flowers, up to 12 per stem, are 1.5–3.5cm (5/8–1 1/2in) long and tapered at the base. Oval, white petals are flushed green or pink outside; flowers have yellow or orange centres inside.
T. 'Uncle Tom' (illus. p.355). Late spring-flowering bulb. H 50cm (20in). Double flowers are maroon-red.
T. undulatifolia, syn. T. eichleri (illus. p.355), Div.15. Early to mid-spring-flowering bulb. H 15–50cm (6–20in). Has a downy stem and grey-green leaves. Flowers, 3–8cm (1 1/4–3in) long, are bowl-shaped at the base. Narrowly oval, red or orange-red petals each have a pale red or buff back and a yellow-bordered, dark green or black blotch at the base inside.
T. 'Union Jack' (illus. p.355), Div.5. Late spring-flowering bulb. H 60cm (2ft). Ivory-white petals, marked with deep pinkish-red 'flames', have blue-edged, white bases.
♀ T. urumiensis (illus. p.356), Div.15. Early spring-flowering bulb. H 10–20cm (4–8in). Stem is mostly below soil level. Leaves are green or greyish-green. Bears 1 or 2 flowers, each 4cm (1 1/2in) long and tapered at the base. Narrowly oval, yellow petals are flushed mauve or red-brown outside. Is useful for a rock garden.
T. violacea (illus. p.356), Div.15. Early spring-flowering bulb. H to 20cm (8in). Has grey-green leaves. Violet-pink flowers, 2–5cm (3/4–2in) long, are tapered at the base and have yellow or bluish-black centres inside. Petals are oval. Suits a rock garden or raised bed.
T. 'West Point' (illus. p.356), Div.6. Late spring-flowering bulb. H 50cm (20in). Primrose-yellow flowers have long-pointed, recurved petals.
T. 'White Dream' (illus. p.354), Div.3. Mid- to late spring-flowering bulb. H 40–45cm (16–18in). Flowers are white with yellow anthers.
T. 'White Emperor'. See T. 'Purissima'.
T. 'White Parrot' (illus. p.354), Div.10. Late spring-flowering bulb. H 55cm (22in). Large flowers have ruffled, white petals flecked green near the base. Is good for cutting.

T. 'White Triumphator' (illus. p.354), Div.6. Late spring-flowering bulb. H 65–70cm (26–28in). White flowers have elegantly reflexed petals.
T. whittallii (illus. p.356), Div.15. Mid-spring-flowering bulb. H 30–35cm (12–14in). Frost hardy. Stem has 1–4 flowers, 3–6cm (1 1/4–2 1/2in) long and tapered at the base. Oval petals are bright brownish-orange.
T. 'Yokohama' (illus. p.356), Div.1. Early to mid-spring-flowering bulb. H 35cm (14in). Pointed flowers are deep yellow.

Tunica saxifraga. See *Petrorhagia saxifraga*.

TURRAEA (Meliaceae)
Genus of evergreen trees and shrubs, grown for their flowers and foliage. Frost tender, min. 12–15°C (54–9°F). Prefers full sun. Needs fertile, well-drained soil. Water freely in full growth, less at other times. Young plants may need growing point removed to promote branching. Prune after flowering if necessary. Propagate by seed in spring or by semi-ripe cuttings in summer.
T. obtusifolia illus. p.142.

TWEEDIA (Asdepiadaceae)
Genus of herbaceous, twining climbers; only one species is in general cultivatation. Frost tender, min 5°C (41°F). In cool climates may be grown as an annual. Grow in sun and in well-drained soil. Pinch out tips of shoots to encourage branching. Propagate by seed in spring.
♀ T. caerulea, syn. Oxypetalum caeruleum, illus. p.175.

TYLECODON (Crassulaceae)
Genus of deciduous, bushy, winter-growing, succulent shrubs with very swollen stems. Frost tender, min. 7°C (45°F). Likes a sunny site with very well-drained soil. Propagate by seed or stem cuttings in summer.
T. paniculatus, syn. Cotyledon paniculatus (Butter tree). Deciduous, bushy, succulent shrub. H and S 2m (6ft). Swollen stem and branches have papery, yellow coverings. Leaves are oblong to oval, fleshy and bright green. In summer bears clusters of tubular, green-striped, red flowers at stem tips.
T. reticulatus, syn. Cotyledon reticulatus, illus. p.401.
T. wallichii, syn. Cotyledon wallichii. Deciduous, bushy, succulent shrub. H and S 30cm (1ft). Has 3cm (1 1/4in) thick stems with cylindrical, grooved-topped, green leaves at tips. After leaf fall, stems are neatly covered in raised leaf bases. Produces tubular, yellow-green flowers, 2cm (1/2in) long, in autumn.

TYPHA (Typhaceae)
Genus of deciduous, perennial, marginal water plants, grown for their decorative, cylindrical seed heads. Fully hardy. Grows in sun or shade. Propagate in spring by seed or division.
T. latifolia illus. p.388.
T. minima illus. p.389.

U

Ugni molinae. See *Myrtus ugni.*

ULEX (Leguminosae)
Genus of leafless, or almost leafless, shrubs that appear evergreen as a result of their year-round, green shoots and spines. Is grown for its flowers in spring. Fully hardy. Needs full sun, and prefers poor, well-drained, acid soil. Trim each year after flowering to maintain compact habit. Straggly, old plants may be cut back hard in spring. Propagate by seed in autumn.
U. europaeus illus. p.127. ♥ **'Flore Pleno'** is a leafless, or almost leafless, bushy shrub. H 1m (3ft), S 1.2m (4ft). In spring, bears masses of fragrant, pea-like, double, yellow flowers on leafless, dark green shoots.

ULMUS (Ulmaceae)
Elm
Genus of deciduous or, rarely, semi-evergreen trees and shrubs, often large and stately, grown for their foliage and habit. Inconspicuous flowers appear in spring. Fully hardy. Requires full sun and fertile, well-drained soil. Propagate by softwood cuttings in summer or by seed or suckers in autumn. Is susceptible to Dutch elm disease, which is quickly fatal, although *U. parvifolia* and *U. pumila* appear more resistant than other species and hybrids.
U. americana (American white elm, White elm). Deciduous, spreading tree. H and S 30m (100ft). Has grey bark and drooping branchlets. Large, oval, dark green leaves are sharply toothed and rough-textured.
U. angustifolia (Goodyer's elm). var. *cornubiensis* (Cornish elm) is a deciduous tree, conical when young and with a vase-shaped head when mature. H 30m (100ft), S 15m (50ft). Oval, toothed, glossy, bright green leaves turn yellow in autumn.
U. **'Camperdownii'** illus. p.66.
U. carpinifolia (Smooth-leaved elm). Deciduous, spreading tree with arching branches and drooping shoots. H 30m (100ft), S 20m (70ft). Small, oval, toothed, glossy, bright green leaves turn yellow in autumn. **'Sarniensis Aurea'** see *U.* 'Dicksonii'.
U. **'Dicksonii'**, syn. *U. carpinifolia* 'Sarniensis Aurea', *U.* 'Wheatleyi Aurea', illus. p.55.
U. glabra (Wych elm). Deciduous, spreading tree. H 30m (100ft), S 25m (80ft). Has broadly oval, toothed, very rough, dark green leaves, often slightly lobed at tips. From mid- to late spring bears clusters of winged, green fruits on bare branches. **'Exoniensis'** (Exeter elm), H 15m (50ft), S 5m (15ft), is narrow with upright branches when young, later becoming more spreading.
U. **'Hollandica'** (Dutch elm). Vigorous, deciduous, spreading tree with a short trunk and upright branches. H 30m (100ft), S 25m (80ft). Has oval, toothed, glossy, dark green leaves. Is very susceptible to Dutch elm disease.
U. **'Jacqueline Hillier'**. Slow-growing, deciduous, bushy shrub, suitable for hedging. H and S 2m (6ft). Small, oval, dark green leaves, rough-textured and sharply toothed, form 2 rows on each shoot; they persist into winter.
U. parvifolia (Chinese elm). Deciduous or semi-evergreen, rounded tree. H and S 15m (50ft). Small, oval, glossy, dark green leaves last well into winter or, in mild areas, may persist until fresh growth appears.
U. procera (English elm). Vigorous, deciduous, spreading tree with a bushy, dense, dome-shaped head. H 35m (120ft), S 15m (50ft). Broadly oval, toothed, rough, dark green leaves turn yellow in autumn.
U. pumila (Siberian elm). Deciduous, spreading, sometimes shrubby tree. H 15m (50ft), S 12m (40ft). Has oval, toothed, dark green leaves. Has some resistance to Dutch elm disease, but seedlings may be susceptible in hot summers.
U. **'Sarniensis'** (Jersey elm, Wheatley elm). Deciduous, conical, dense tree with upright branches. H 30m (100ft), S 10m (30ft). Leaves are small, broadly oval, toothed, glossy and mid-green.
U. **'Vegeta'** (Huntingdon elm). Vigorous, deciduous, spreading tree with upright, central branches and drooping, outer shoots. H 35m (120ft), S 25m (80ft). Leaves are broadly oval, toothed, glossy and dark green.
U. **'Wheatleyi Aurea'**. See *U.* 'Dicksonii'.

UMBELLULARIA (Lauraceae)
Headache tree
Genus of evergreen, spring-flowering trees, grown for their aromatic foliage. Frost hardy, but requires shelter from strong, cold winds when young. Needs sun and fertile, moist but well-drained soil. Propagate by seed in autumn.
U. californica illus. p.48.

Urceolina peruviana. See *Stenomesson miniata.*

URGINEA (Liliaceae)
Genus of late summer- or early autumn-flowering bulbs, growing on or near soil surface, with spear-shaped flower spikes up to 1.5m (5ft) high. Frost to half hardy. Needs a sunny site and well-drained soil that dries out while bulbs are dormant in summer. Plant in mid- to late summer. Water until leaves die down. Propagate by seed in autumn or by offsets in late summer.
U. maritima (Crusaders' spears, Sea onion, Sea squill). Late summer- or early autumn-flowering bulb. H 1.5m (5ft), S 30–45cm (1–1 1/2ft). Half hardy. Broadly sword-shaped, erect, basal leaves appear in autumn after a long spike of star-shaped, white flowers, each 1–1.5cm (1/2–5/8in) across, has developed.

URSINIA (Compositae)
Genus of annuals, evergreen perennials and sub-shrubs, grown mainly for their flower heads usually in summer, a few species for their foliage. Half hardy to frost tender, min. 5–7°C (41–5°F). Needs full light and well-drained soil. Water potted plants moderately, less when not in full growth. Requires good ventilation if grown under glass. Propagate by seed or greenwood cuttings in spring. Aphids are sometimes troublesome.
U. anthemoides illus. p.290.
U. chrysanthemoides. Evergreen, bushy perennial. H and S 60cm (2ft) or more. Frost tender. Narrowly oval, feathery, strongly scented, green leaves are 5cm (2in) long. Has small, long-stalked, daisy-like, yellow flower heads, sometimes coppery below, in summer.
U. sericea. Evergreen, bushy sub-shrub. H and S 25–45cm (10–18in). Frost tender. Leaves are cut into an elegant filigree of very slender, silver-haired segments. Daisy-like, yellow flower heads, 4cm (1 1/2in) across, open in summer. Is mainly grown for its foliage.

UTRICULARIA (Lentibulariaceae)
Genus of deciduous or evergreen, perennial, carnivorous water plants with bladder-like, modified leaves that trap and digest insects. Most species in cultivation are free-floating. Frost hardy to frost tender, min. 7°C (45°F). Some species are suitable only for tropical aquariums; those grown in outdoor pools require full sun. Thin out plants that are overcrowded or become laden with algae. Propagate by division of floating foliage in spring or summer.
U. exoleta. See *U. gibba.*
U. gibba, syn. *U. exoleta.* Deciduous, perennial, free-floating water plant. S 15cm (6in). Frost tender. Slender stems carry finely divided, mid-green leaves on which small bladders develop. Pouched, bright yellow flowers are borne in summer. Is evergreen in very warm water; suitable only for a tropical aquarium.
U. vulgaris. Deciduous, perennial, free-floating water plant. S 30cm (12in). Frost hardy. Much-divided, bronze-green leaves, studded with small bladders, are produced on slender stems. Bears pouched, bright yellow flowers in summer. May be grown in a pool or cold-water aquarium.

UVULARIA (Liliaceae)
Genus of spring-flowering perennials that thrive in moist woodlands. Fully hardy. Requires semi-shade and prefers peaty soil. Propagate in early spring, before flowering, by division.
♥ *U. grandiflora* illus. p.238.
U. perfoliata. Clump-forming perennial. H 45cm (18in), S 30cm (12in). In spring, clusters of pendent, bell-shaped, pale yellow flowers with twisted petals appear on numerous slender stems above stem-clasping, narrowly oval, mid-green leaves.

V

VACCINIUM (Ericaceae)
Genus of deciduous or evergreen sub-shrubs, shrubs and trees, grown for their foliage, autumn colour (on deciduous species), flowers and fruits, often edible. Fully to frost hardy. Requires sun or semi-shade and moist but well-drained, peaty or sandy, acid soil. Propagate by semi-ripe cuttings in summer or by seed in autumn.
V. angustifolium. Deciduous shrub, usually grown in the dense, bushy form var. *laevifolium* illus. p.142.
V. arctostaphylos (Caucasian whortleberry). Deciduous, upright shrub. H 3m (10ft), S 2m (6ft). Fully hardy. Has red-brown young shoots and oval, dark green leaves that turn red and purple in autumn. Carries bell-shaped, red-tinged, white flowers in spreading racemes in early summer, then spherical, purplish-black fruits.
⚇ *V. corymbosum* (Highbush blueberry) illus. p.130. **'Pioneer'** illus. p.143.
⚇ *V. glaucoalbum* illus. p.146.
V. myrtillus (Bilberry, Whortleberry). Deciduous, usually prostrate shrub. H 15cm (6in) or more, S 30cm (12in) or more. Fully hardy. Has small, heart-shaped, leathery, bright green leaves. Pendent, bell-shaped, pale pink flowers in early summer are followed by edible, round, blue-black fruits.
V. nummularia. Evergreen, prostrate shrub. H 10cm (4in), S 20cm (8in). Frost hardy. Slender stems, covered in red-brown bristles, bear oval, wrinkled, bright green leaves with red-brown bristles at margins. Has small racemes of bell-shaped, white to deep pink flowers at stem tips in early summer, followed by small, round, black fruits. Suits a rock garden or peat bed. Needs semi-shade. May also be propagated by division in spring.
V. parvifolium illus. p.143.
V. vitis-idaea. Vigorous, evergreen, prostrate shrub, spreading by underground runners. H 2–25cm ($^3/_4$–10in), S indefinite. Fully hardy. Forms hummocks of oval, hard, leathery leaves. Bell-shaped, white to pink flowers are borne in nodding racemes from early summer to autumn, followed by bright red fruits in autumn-winter. May also be propagated by division in spring. var. *minus* (syn. *V.v.-i.* 'Minus') illus. p.314.

VALERIANA (Valerianaceae)
Valerian
Genus of summer-flowering perennials that are suitable for borders and rock gardens. Fully hardy. Requires sun and well-drained soil. Propagate by division in autumn, *V. officinalis* by seed in spring.
V. officinalis illus. p.205.
V. phu 'Aurea' illus. p.238.

VALLEA (Elaeocarpaceae)
Genus of one species of evergreen shrub, grown for its overall appearance. Half hardy, but best at 3–5°C (37–41°F) to prevent foliage being damaged by cold. Prefers full sun and humus-rich, well-drained soil. Water potted plants freely when in full growth, moderately at other times. Untidy growth may be cut out in early spring. Propagate by seed in spring or by semi-ripe cuttings in summer. Red spider mite may be a nuisance.
V. stipularis. Evergreen, erect, then loose and spreading shrub. H and S 2–5m (6–15ft). Leaves are lance-shaped to rounded and lobed, deep green above, grey beneath. Small, cup-shaped flowers, each with 5 deep pink petals that have 3 lobes, are borne in small, terminal and lateral clusters in spring-summer.

VALLISNERIA (Hydrocharitaceae)
Genus of evergreen, perennial, submerged water plants, grown for their foliage. Is suitable for pools and aquariums. Frost tender, min. 5°C (41°F). Requires sun or semi-shade and deep, clear water. Remove fading foliage, and thin overcrowded plants as required. Propagate by division in spring or summer.
V. gigantea. Vigorous, evergreen, perennial, submerged water plant. S indefinite. Quickly forms colonies of long, strap-shaped, mid-green leaves. Insignificant, greenish flowers are produced year-round.
V. spiralis (Eel grass, Tape grass). Vigorous, evergreen, perennial, submerged water plant. S indefinite. Rapidly forms a mass of long, strap-shaped, mid-green leaves, but on a smaller scale than *V. gigantea*. Bears insignificant, greenish flowers year-round.

Vallota speciosa. See *Cyrtanthus purpureus*.

VANCOUVERIA (Berberidaceae)
Genus of perennials, some of which are evergreen, suitable for ground cover. Fully hardy. Prefers cool, partially shaded positions and moist, peaty soil. Propagate by division in spring.
V. chrysantha. Evergreen, sprawling perennial. H 30cm (12in), S indefinite. Oval, dark green leaves borne on flower stems are divided into rounded diamond-shaped leaflets with thickened, undulating margins. Loose sprays of small, bell-shaped, yellow flowers are borne in spring.
V. hexandra illus. p.295.

VANDA. See ORCHIDS.
V. Rothschildiana (illus. p.262). Evergreen, epiphytic orchid for a cool or intermediate greenhouse. H 60cm (24in). Sprays of dark-veined, violet-blue flowers, 10cm (4in) across, are borne twice a year in varying seasons. Has narrowly oval, rigid leaves, 10–12cm (4–5in) long. Grow in a hanging basket and provide good light in summer.

Vellozia elegans. See *Talbotia elegans*.

VELTHEIMIA (Liliaceae)
Genus of winter-flowering bulbs with dense spikes of pendent, tubular flowers and rosettes of basal leaves. Frost tender, min. 10°C (50°F). Requires good light, to keep foliage compact and to develop flower colours fully, and well-drained soil. Plant in autumn with tips above soil surface. Reduce watering in summer. Propagate by seed or offsets in autumn.
⚇ *V. bracteata*, syn. *V. undulata*, *V. viridifolia*, illus. p.369.
⚇ *V. capensis*, syn. *V. glauca*. Winter-flowering bulb. H 30–45cm (12–18in), S 20–30cm (8–12in). Has a basal rosette of lance-shaped leaves, usually with very wavy edges. Stem produces a dense spike of pink or red flowers, each 2–3cm ($^3/_4$–1$^1/_4$in) long.
V. glauca. See *V. capensis*.
V. undulata. See *V. bracteata*.
V. viridifolia. See *V. bracteata*.

x VENIDIO-ARCTOTIS. See ARCTOTIS x HYBRIDA.

VERATRUM (Liliaceae)
Genus of perennials, ideal for woodland gardens. Fully hardy. Requires semi-shade and fertile, moist soil. Propagate by division or seed in autumn.
V. album (White false hellebore). Clump-forming perennial. H 2m (6ft), S 60cm (2ft). Produces stems, leafy at base, that bear dense, terminal panicles of saucer-shaped, yellowish-white flowers in summer. Leaves are oval, pleated and dark green.
⚇ *V. nigrum* illus. p.192.

VERBASCUM (Scrophulariaceae)
Mullein
Genus of mainly summer-flowering perennials, some of which are semi-evergreen or evergreen, and evergreen biennials and shrubs. Fully to frost hardy. Tolerates shade, but prefers an open, sunny site and well-drained soil. Propagate species by seed in spring or late summer or by root cuttings in winter, selected forms by root cuttings only. Some species self seed freely.
⚇ *V. bombyciferum*. Evergreen, erect biennial. H 1.2–2m (4–6ft), S 60cm (2ft). Fully hardy. Oval leaves and stems are covered with silver hairs. Has upright racemes densely set with 5-lobed, yellow flowers in summer.
V. chaixii. Erect perennial, covered with silvery hairs. H 1m (3ft), S 60cm (2ft). Fully hardy. Produces slender spires of 5-lobed, yellow, sometimes white flowers, with purple stamens, in summer. Has oval, toothed, rough, nettle-like leaves.
V. 'Cotswold Queen'. Short-lived, rosette-forming perennial. H 1–1.2m (3–4ft), S 30–60cm (1–2ft). Fully hardy. Throughout summer, branched racemes of 5-lobed, apricot-buff flowers are borne on stems that arise from oval, mid-green leaves.
V. densiflorum, syn. *V. thapsiforme*. Fairly slow-growing, semi-evergreen, upright perennial. H 1.2–1.5m (4–5ft), S 60cm (2ft). Fully hardy. Has a rosette of large, oval, crinkled, hairy, mid-green leaves. Hairy stems each produce a bold spike of flattish, 5-lobed, yellow flowers in summer.
⚇ *V. dumulosum* illus. p.307.
⚇ *V.* 'Gainsborough' illus. p.218.
⚇ *V.* 'Letitia' illus. p.306.
V. lychnitis (White mullein). Slow-growing, evergreen, upright, branching biennial. H 60cm–1m (2–3ft), S 60cm (2ft). Fully hardy. Has lance-shaped, dark grey-green leaves. Flattish, 5-lobed, white flowers are borne on branching stems in summer.
V. nigrum illus. p.220.
V. olympicum illus. p.193.
⚇ *V.* 'Pink Domino'. Short-lived, rosette-forming perennial. H 1.2m (4ft), S 30–60cm (1–2ft). Fully hardy. Produces branched racemes of 5-lobed, rose-pink flowers throughout summer above oval, mid-green leaves.
V. thapsiforme. See *V. densiflorum*.

VERBENA (Verbenaceae)
Genus of summer- and autumn-flowering biennials and perennials, some of which are semi-evergreen. Frost hardy to frost tender, min. 1°C (34°F). Prefers sun and well-drained soil. Propagate by stem cuttings in late summer or autumn or by seed in autumn or spring.
V. alpina of gardens. See *V. tenera* var. *maonettii*.
V. bonariensis. See *V. patagonica*.
V. chamaedrioides. See *V. peruviana*.
V. chamaedryfolia. See *V. peruviana*.
V. x *hybrida*. 'Amethyst' is a fairly slow-growing, bushy perennial, grown as an annual. H 20–30cm (8–12in), S 30cm (12in). Half hardy. Has oval, serrated, mid- to deep green leaves. Clusters of small, tubular, lobed, blue flowers, with white eyes, appear in summer and early autumn. **'Defiance'** illus. p.280. **Derby Series**. H 20cm (8in), is very compact and has a wide colour range, including red, pink, blue, mauve and white. Flowers of **'Mme du Barry'** are carmine-crimson. **'Showtime'** illus. p.278. **'Springtime'**, H 20cm (8in), is a spreading cultivar, in a mixture of bright colours.
V. patagonica, syn. *V. bonariensis*, illus. p.192.
V. peruviana, syn. *V. chamaedrioides*, *V. chamaedryfolia*. Semi-evergreen, prostrate perennial. H to 8cm (3in), S 1m (3ft). Frost tender. Heads of small, tubular, brilliant scarlet flowers, with spreading petal lobes, are produced from early summer to early autumn. Oval, toothed leaves are mid-green. Prefers not too rich, dry soil.

VERONICA

♥ *V. rigida*, syn. *V. venosa*, illus. p.248.
♥ *V.* 'Sissinghurst' illus. p.246.
V. tenera var. *maonettii*, syn. *V. alpina* of gardens. Spreading perennial with a slightly woody base. H 8cm (3in), S 15cm (6in). Half hardy. Has oblong to oval leaves, deeply cut into linear, toothed, mid-green segments, and, in summer, terminal clusters of small, tubular, reddish-violet flowers, with white-edged lobes.
V. venosa. See *V. rigida*.

VERONICA (Scrophulariaceae)
Genus of perennials and sub-shrubs, some of which are semi-evergreen or evergreen, grown for their usually blue flowers. Fully to frost hardy. Needs sun and well-drained soil. Propagate by division in spring or autumn, by softwood or semi-ripe cuttings in summer or by seed in autumn.
V. austriaca. Mat-forming or upright perennial. H and S 25–50cm (10–20in). Fully hardy. Short, dense or lax racemes of small, saucer-shaped, bright blue flowers appear in early summer. Leaves are very variable: from broadly oval to narrowly oblong, and from entire to deeply cut and fern-like. Suits a rock garden or bank. subsp. *teucrium* (syn. *V. teucrium*) illus. p.306.
♥ subsp. *teucrium* 'Royal Blue' has deep royal-blue flowers. Propagate by division in spring or by softwood cuttings in summer.
♥ *V. cinerea*. Spreading, much-branched, woody-based perennial. H 15cm (6in), S 30cm (12in). Fully hardy. Has small, linear, occasionally oval, hairy, silvery-white leaves. Trailing flower stems bear saucer-shaped, deep blue to purplish-blue flowers, with white eyes, in early summer. Suits a sunny rock garden.
V. exaltata. Erect, elegant perennial. H 1.2m (4ft), S 30cm (1ft). Fully hardy. Has tall racemes of star-shaped, light blue flowers from mid- to late summer on stems clothed in narrowly oval, toothed, mid-green leaves.
V. fruticans (Rock speedwell). Deciduous, upright to procumbent sub-shrub. H 15cm (6in), S 30cm (12in). Fully hardy. Leaves are oval and green. Spikes of saucer-shaped, bright blue flowers, each with a red eye, are borne in summer. Suits a rock garden.
♥ *V. gentianoides* illus. p.251.
V. incana. See *V. spicata* subsp. *incana*.
V. longifolia. Clump-forming perennial. H 1–1.2m (3–4ft), S 30cm (1ft) or more. Fully hardy. In summer, long, terminal racemes of star-shaped, bright blue flowers are produced on stems clothed with whorls of narrowly oval to lance-shaped, toothed, mid-green leaves. 'Romiley Purple' illus. p.215.
V. pectinata. Dense, mat-forming perennial that is sometimes semi-erect. H and S 20cm (8in). Fully hardy. Has small, narrowly oval, hairy leaves and bears loose sprays of saucer-shaped, soft blue to blue-violet flowers in summer. Is good for a rock garden or bank. 'Rosea', H 8cm (3in), has rose-lilac flowers.
V. perfoliata. See *Parahebe perfoliata*.

♥ *V. prostrata* (Prostrate speedwell) illus. p.305. 'Kapitan' and 'Trehane' illus. p.305. 'Spode Blue' is a dense, mat-forming perennial. H to 30cm (12in), S indefinite. Fully hardy. Upright spikes of small, saucer-shaped, china-blue flowers appear in early summer. Leaves are narrowly oval and toothed.
V. spicata (Spiked speedwell). Clump-forming perennial. H 30–60cm (12–24in), S 45cm (18in). Fully hardy. Spikes of small, saucer-shaped, bright blue flowers are borne in summer above narrowly oval, toothed, mid-green leaves. subsp. *incana* (syn. *V. incana*) has star-shaped, clear blue flowers and linear to lance-shaped leaves. Plant is densely covered in silver hairs.
V. teucrium. See *V. austriaca* subsp. *teucrium*.
V. virginica. See *Veronicastrum virginicum*. f. *alba*, see *Veronicastrum virginicum* f. *alba*.

VERONICASTRUM (Scrophulariaceae)
Genus of evergreen perennials grown for their elegant, pale blue flowers. Fully hardy. Grow in sun and moist soil. Propagate by division in spring or autumn, by softwood or semi-ripe cuttings in summer or by seed in autumn.
V. virginicum, syn. *veronica virginia*. Upright perennial. H 1.2m (4ft), S 45cm (1½ft). Fully hardy. In late summer, racemes of small, star-shaped, purple-blue or pink flowers crown stems clothed with whorls of narrowly lance-shaped, dark green leaves. f. *alba* (syn. *veronica virginia* f. *alba*) illus. p.205.

VESTIA (Solanaceae)
Genus of one species of evergreen shrub, grown for its flowers and foliage. Frost hardy, but in cold areas is often cut to ground level and is best grown against a south-facing wall. Requires sun and fertile, well-drained soil. Propagate by semi-ripe cuttings in summer or by seed in autumn or spring.
♥ *V. foetida*, syn. *V. lycioides*. Evergreen, upright shrub. H 2m (6ft), S 1.5m (5ft). Has pendent, tubular, pale yellow flowers from mid-spring to mid-summer. Oblong, glossy, dark green leaves have an unpleasant scent.
V. lycioides. See *V. foetida*.

VIBURNUM (Caprifoliaceae)
Genus of deciduous, semi-evergreen or evergreen shrubs and trees, grown for their foliage, autumn colour (in many deciduous species), flowers and, often, fruits. Fruiting is generally most prolific when several plants of different clones are planted together. Fully to frost hardy. Grow in sun or semi-shade and in deep, fertile, not too dry soil. To thin out overgrown plants cut out some older shoots after flowering. Propagate by cuttings (softwood for deciduous species, semi-ripe for evergreens) in summer or by seed in autumn.
V. acerifolium illus. p.131.
V. betulifolium illus. p.118.
V. bitchiuense. Deciduous, bushy shrub. H and S 2.5m (8ft). Fully hardy. Has oval, dark green leaves, rounded heads of fragrant, tubular, pale pink flowers, from mid- to late spring, and egg-shaped, flattened, black fruits.
♥ *V.* x *bodnantense*. 'Dawn' illus. p.120. ♥ 'Deben' is a deciduous, upright shrub. H 3m (10ft), S 2m (6ft). Fully hardy. Oval, toothed, dark green leaves are bronze when young. Clusters of fragrant, tubular, white flowers, tinted with pale pink, open during mild periods from late autumn to early spring.
V. x *burkwoodii*. Semi-evergreen, bushy, open shrub. H and S 2.5m (8ft). Fully hardy. Rounded heads of fragrant, tubular, pink, then white flowers are borne amid oval, glossy, dark green leaves from mid- to late spring. ♥ 'Anne Russell', H and S 1.5m (5ft), is deciduous and has very fragrant, white flowers. ♥ 'Park Farm Hybrid' bears very fragrant, white flowers, slightly pink in bud, and older leaves turn bright red in autumn.
♥ *V.* x *carlcephalum* illus. p.87.
♥ *V. carlesii* illus. p.124. 'Diana' is a deciduous, bushy, dense shrub. H and S 2m (6ft). Fully hardy. Broadly oval leaves are bronze when young and turn purple-red in autumn. From mid- to late spring bears rounded heads of red buds that open to very fragrant, tubular, pink flowers fading to white.
♥ *V. cinnamomifolium*. Evergreen, bushy or tree-like shrub. H and S 5m (15ft). Frost hardy. Has large, oval, mid-green leaves, each with 3 prominent veins. Produces broad clusters of small, star-shaped, white flowers in early summer, then egg-shaped, blue fruits.
♥ *V. davidii* illus. p.145.
V. dilatatum. Deciduous, upright shrub. H 3m (10ft), S 2m (6ft). Fully hardy. Oval, sharply toothed, dark green leaves sometimes redden in autumn. Flat heads of small, star-shaped, white flowers in late spring and early summer are succeeded by showy, egg-shaped, bright red fruits. 'Catskill' illus. p.109.
♥ *V. farreri*, syn. *V. fragrans*, illus. p.117. 'Candidissimum' is a deciduous, upright shrub. H 3m (10ft), S 2m (6ft). Fully hardy. Oval, toothed, dark green leaves are pale green when young. Produces clusters of fragrant, tubular, pure white flowers in late autumn and during mild periods in winter and early spring.
V. foetens illus. p.119.
V. fragrans. See *V. farreri*.
V. grandiflorum. Deciduous, upright, open shrub. H and S 2m (6ft). Fully hardy. Stiff branches bear oblong, dark green leaves that become deep purple in autumn. Dense clusters of fragrant, tubular, white-and-pink flowers open from deep pink buds from mid-winter to early spring.
♥ *V.* x *juddii* illus. p.124.
V. lantana (Wayfaring tree). Vigorous, deciduous, upright shrub. H 5m (15ft), S 4m (12ft). Fully hardy. Has broadly oval, grey-green leaves that redden in autumn, flattened heads of small, 5-lobed, white flowers in late spring and early summer, and egg-shaped, red fruits that ripen to black.

V. lentago (Sheepberry). Vigorous, deciduous, upright shrub. H 4m (12ft), S 3m (10ft). Fully hardy. Oval, glossy, dark green leaves turn red and purple in autumn. Produces flattened heads of small, fragrant, tubular, star-shaped, white flowers in late spring and early summer, then egg-shaped, blue-black fruits.
V. odoratissimum (Sweet viburnum). Evergreen, bushy shrub. H and S 5m (15ft). Frost hardy. Clusters of small, fragrant, star-shaped, white flowers, borne amid oval, leathery, glossy, dark green leaves in late spring, are followed by egg-shaped, red fruits that ripen to black.
V. opulus (Guelder rose). Vigorous, deciduous, bushy shrub. H and S 4m (12ft). Fully hardy. Bears broadly oval, lobed, deep green leaves that redden in autumn and, in late spring and early summer, flattened, lace-cap-like heads of white flowers. Produces large bunches of spherical, bright red fruits.
♥ 'Compactum' illus. p.142.
♥ 'Xanthocarpum' has yellow fruits and mid-green leaves that become yellow in autumn.
V. plicatum, syn. *V.p.* f. *plicatum* (Japanese snowball tree). Deciduous, bushy, spreading shrub. H 3m (10ft), S 4m (12ft). Fully hardy. Leaves are oval, toothed, deeply veined and dark green, turning reddish-purple in autumn. Dense, rounded heads of large, sterile, flattish, white flowers are borne along branches in late spring and early summer. ♥ 'Mariesii' illus. p.86. 'Nanum Semperflorens' (syn. *V.p.* 'Watanabei', *V. watanabei*), H 2m (6ft), S 1.5m (5ft), is slow-growing, conical and dense, and produces small flower heads more or less continuously from late spring until early autumn.
♥ 'Pink Beauty' illus. p.100.
f. *tomentosum* has tiered branches, flattish, lace-cap-like flower heads and red fruits that ripen to black.
'Watanabei' see *V. p.* 'Nanum Semperflorens'.
V. x *pragense*. See *V.* 'Pragense'.
♥ *V.* 'Pragense', syn. *V.* x *pragense*, illus. p.109.
V. rhytidophyllum illus. p.88.
V. sargentii. Deciduous, bushy shrub. H and S 3m (10ft). Fully hardy. Maple-like, mid-green foliage often changes to yellow or red in autumn. Broad, flattish, lacecap-like heads of white flowers in late spring are followed by spherical, bright red fruits.
♥ 'Onondaga', S 2m (6ft), has bronze-red, young leaves, becoming deep green, then bronze-red again in autumn. Flower buds are pink.
V. sieboldii. Deciduous, rounded, dense shrub. H 4m (12ft), S 6m (20ft). Fully hardy. Has large, oblong to oval, glossy, bright green leaves, rounded heads of tubular, creamy-white flowers, in late spring, and egg-shaped, red-stalked, red fruits that ripen to black.
V. tinus (Laurustinus) illus. p.119.
♥ 'Eve Price' is an evergreen, bushy, very compact shrub. H and S 3m (10ft). Frost hardy. Flattened heads of small, star-shaped, white flowers are freely borne from deep pink buds amid oval, dark green leaves in winter-spring and are followed by ovoid, blue fruits.

♚ 'Gwenllian' has pale pink flowers and fruits very freely.
V. watanabei. See V. plicatum 'Nanum Semperflorens'.

VIGNA (Leguminosae)
Genus of evergreen, annual and perennial, erect or scrambling and twining climbers, grown mainly as crop plants for their leaves, pods and seeds. Frost tender, min. 13–15°C (55–9°F). Provide humus-rich, well-drained soil and full light. Water freely when in full growth, sparingly at other times. Stems require support. Thin crowded stems or cut back hard in spring. Propagate by seed in autumn or spring.
V. caracalla, syn. Phaseolus caracalla (Snail flower). Fast-growing, evergreen, perennial, twining climber. H 3–5m (10–15ft). Leaves comprise 3 oval leaflets. From summer to early autumn carries pea-like, purple-marked, cream flowers that turn orange-yellow.

VINCA (Apocynaceae)
Periwinkle
Genus of evergreen, trailing sub-shrubs and perennials, grown for their foliage and flowers. Flowers are tubular with 5 spreading lobes. Fully to frost hardy. Is useful for ground cover in shade, but flowers more freely given some sun. Grows in any soil that is not too dry. Propagate by semi-ripe cuttings in summer or by division from autumn to spring.
V. difformis. Evergreen, prostrate sub-shrub. H 30cm (12in), S indefinite. Frost hardy. Slender, trailing stems bear oval, glossy, dark green leaves. Erect flower stems produce pale blue flowers in late autumn and early winter.
V. major (Greater periwinkle, Quater). Evergreen, prostrate, arching sub-shrub. H 45cm (18in), S indefinite. Fully hardy. Leaves are broadly oval, glossy and dark green. Large, bright blue flowers are produced from late spring to early autumn. subsp. hirsuta has leaves, leaf stalks and calyces edged with long hairs. ♚ 'Variegata' illus. p.145.
V. minor (Lesser periwinkle) illus. p.146. 'Alba Variegata' is an evergreen, prostrate sub-shrub. H 15cm (6in), S indefinite. Fully hardy. Forms extensive mats of small, oval, glossy, dark green leaves, edged with pale yellow, above which white flowers are carried from mid-spring to early summer, then intermittently into autumn. 'Bowles' White' bears large, white flowers that are pinkish-white in bud. ♚ 'Gertrude Jekyll' is of dense growth and produces a profusion of small, white flowers. Flowers of ♚ 'La Grave' are large and lavender-blue.
V. rosea. See Catharanthus roseus.

VIOLA (Violaceae)
Violet
Genus of annuals, perennials, some of which are semi-evergreen, and deciduous sub-shrubs, grown for their distinctive flowers. Annuals are suitable as summer bedding, perennials and sub-shrubs are good in rock gardens, screes and alpine houses. Fully to half hardy. Grow in sun or shade and well-drained but moisture-retentive soil unless otherwise stated; a few species prefer acid soil. Propagate annuals by seed sown according to flowering season, perennials and sub-shrubs by softwood cuttings in spring unless otherwise stated. Species may also be propagated by seed in spring or autumn.
V. aetolica illus. p.319.
V. 'Azure Blue'. See V. x wittrockiana.
V. 'Baby Lucia'. See V. x wittrockiana.
V. biflora (Twin-flowered violet). Creeping, rhizomatous perennial. H 5–15cm (2–6in), S 15cm (6in). Fully hardy. Flat-faced, deep lemon-yellow flowers, veined dark brown, are borne singly or in pairs on upright stems in summer. Leaves are kidney-shaped and mid-green. Needs shade. May also be propagated by division.
V. calcarata illus. p.316.
V. cazorlensis. Tufted, woody-based perennial. H to 5cm (2in), S to 8cm (3in). Frost hardy. Has small, linear to lance-shaped leaves and in late spring carries small, flat-faced, long-spurred, deep pink flowers, singly on short stems. Suits an alpine house. Is difficult to grow.
V. cenisia. Spreading perennial with runners. H 7cm (3in), S 10cm (4in). Fully hardy. Small, flat-faced, bright violet flowers, each with a deep purple line radiating from the centre, are produced on very short stems in summer. Has a deep tap root and tiny, heart-shaped or oblong, dark green leaves. Suits a scree. Propagate by division in spring.
V., Clear Crystals Series. See V. x wittrockiana.
♚ V. cornuta illus. p.297.
V., Crystal Bowl Series. See V. x wittrockiana.
V. cucullata. See V. obliqua.
V. elatior. Upright, little-branched perennial. H 20–30cm (8–12in), S 15cm (6in). Fully hardy. Leaves are broadly lance-shaped and toothed. Produces flat-faced, pale blue flowers, with white centres, in early summer. Prefers semi-shade and moist soil. Propagate in spring by division.
V., Floral Dance Series. See V. x wittrockiana.
V. glabella. Clump-forming perennial with a scaly, horizontal rootstock. H 10cm (4in), S 20cm (8in). Fully hardy. Flat-faced, bright yellow flowers, with purplish veins on lower petals, open in late spring above heart-shaped, toothed, bright green leaves. Needs shade. Propagate by division in spring.
V. gracilis. Mat-forming perennial. H 12cm (5in), S 15cm (6in) or more. Fully hardy. Flat-faced, yellow-centred, violet-blue or sometimes yellow flowers are produced in summer. Has small, dissected leaves with linear or oblong segments. Needs sun.
♚ V. 'Haslemere' illus. p.329.
V. hederacea, syn. V. reniformis, Erpetion reniforme (Australian violet, Ivy-leaved violet). Evergreen, creeping, mat-forming perennial. H 2.5–5cm (1–2in), S indefinite. Half hardy. Has tiny, rounded leaves and bears purple or white flowers, with a squashed appearance, on short stems in summer. Is useful in an alpine house. Prefers semi-shade. Propagate by division in spring.
♚ V. 'Huntercombe Purple' illus. p.330.
V., Icequeen Series. See V. x wittrockiana.
V., Imperial Series, 'Orange Prince'. See V. x wittrockiana.
V., Imperial Series, 'Sky Blue'. See V. x wittrockiana.
V. 'Irish Molly'. Evergreen, clump-forming, short-lived perennial. H 10cm (4in), S 15–20cm (6–8in). Fully hardy. Has broadly oval, dissected leaves and, in summer, a succession of flat-faced, old-gold flowers with brown centres. Flowers itself to death. Needs sun.
♚ V. 'Jackanapes' illus. p.320.
♚ V., Joker Series (summer-flowering) illus. p.284.
V. labradorica 'Purpurea' illus. p.317.
V. 'Love Duet'. See V. x wittrockiana.
V. lutea (Mountain pansy). Mat-forming, rhizomatous perennial. H 10cm (4in), S 15cm (6in). Fully hardy. Has small, oval to lance-shaped leaves. Flat-faced, yellow, violet or bicoloured flowers appear in spring-summer.
V. 'Majestic Giants'. See V. x wittrockiana.
V. obliqua, syn. V. cucullata. Variable, spreading perennial with fleshy rhizomes. H 5cm (2in), S 10–15cm (4–6in). Fully hardy. Has kidney-shaped, toothed, mid-green leaves. In late spring produces flat-faced, blue-violet, sometimes white or pale blue flowers. Propagate in spring by division. Self seeds freely.
V. odorata (Sweet violet). Semi-evergreen, spreading, rhizomatous perennial. H 7cm (3in), S 15cm (6in) or more. Fully hardy. Leaves are heart-shaped and toothed. Long stems each carry a fragrant, flat-faced, violet or white flower from late winter to early spring. Is useful in a wild garden. Self seeds prolifically. May also be propagated by division.
V. palmata. Spreading perennial. H 10cm (4in), S 15cm (6in). Fully hardy. Has short-stemmed, flat-faced, pale violet flowers in late spring and deeply dissected leaves. Prefers dry, well-drained soil. Self seeds readily.
V. pedata illus. p.317. var. bicolor is a clump-forming perennial with a thick rootstock. H 5cm (2in), S 8cm (3in). Fully hardy. Flat-faced, velvety-purple or white flowers are borne singly on slender stems in late spring and early summer. Leaves are finely divided into 5–7 or more, narrow, toothed segments. Suits an alpine house. May be difficult to grow; needs peaty, sandy soil.
V. 'Queen of the Planets'. See V. x wittrockiana.
V. 'Redwing'. See V. x wittrockiana.
V. reniformis. See V. hederacea.
V. 'Roggli Giants'. See V. x wittrockiana.
V. 'Scarlet Clan'. See V. x wittrockiana.
V. 'Silver Princess'. See V. x wittrockiana.
V. 'Super Chalon Giants'. See V. x wittrockiana.
V. tricolor illus. p.317. 'Bowles' Black' illus. p.318.
V., Universal Series. See V. x wittrockiana.
V. x wittrockiana (Pansy). Group of slow- to moderately fast-growing, mainly bushy perennials, usually grown as annuals or biennials. H 15–20cm (6–8in), S 20cm (8in). Fully hardy. Has oval, often serrated, mid-green leaves. Flattish, 5-petalled flowers, 2.5–10cm (1–4in) across, in a very wide colour range, appear throughout summer or in winter-spring. The following are among those available:
'Azure Blue' (spring-flowering) illus. p.285.
'Baby Lucia' (summer-flowering) has small, deep blue flowers.
Clear Crystals Series (summer-flowering) is in a wide range of clear colours (yellow, illus. p.289).
Crystal Bowl Series (summer-flowering) is in a range of colours (yellow, illus. p.289).
Floral Dance Series (winter-flowering) has a wide range of colours (mixed, illus. p.281; white, illus. p.271).
Icequeen Series (winter- and early spring-flowering) has a range of colours (yellow, illus. p.289).
Imperial Series, 'Orange Prince' (summer-flowering) has orange flowers with black blotches.
Imperial Series, 'Sky Blue' (summer-flowering) has sky-blue flowers, each with a deeper blotch.
'Love Duet' (summer-flowering) has cream or white flowers, each with a deep pink blotch.
'Majestic Giants' (summer-flowering) has large flowers in a wide colour range.
'Queen of the Planets' (summer-flowering) has very large, multicoloured flowers.
'Redwing' (summer-flowering) illus. p.290.
'Roggli Giants' (summer- to autumn-flowering) illus. p.279.
'Scarlet Clan' (summer-flowering) illus. p.291.
'Silver Princess' (summer-flowering) has white flowers, each with a deep pink blotch.
'Super Chalon Giants' (summer- to autumn-flowering) illus. p.288.
♚ Universal Series (winter- to spring-flowering) has a range of separate colours (apricot, illus. p.291) or a mixture of colours.

VIRGILIA (Leguminosae)
Genus of short-lived, evergreen shrubs and trees, grown for their flowers in spring-summer. Frost tender, min. 5°C (41°F). Prefers well-drained soil and full light. Water pot-grown plants freely when in full growth, less at other times. Pruning is usually not required. Propagate in spring by seed, ideally soaked in warm water for 24 hours before sowing.
V. capensis. See V. oroboides.
V. oroboides, syn. V. capensis. Fast-growing, evergreen, rounded shrub or tree. H and S 6–10m (20–30ft). Has

leaves of 11–21 oblong leaflets. Racemes of fragrant, pea-like, bright mauve-pink flowers, sometimes pink, crimson or white, are produced in late spring and summer, usually in profusion.

Viscaria alpina. See *Lychnis alpina.*
Viscaria elegans. See *Silene coeli-rosa.*

VITALIANA (Primulaceae)
Genus of one species of evergreen, spring-flowering perennial, grown for its flowers. Is often included in *Douglasia* and is useful for rock gardens, screes and alpine houses. Fully hardy. Requires sun and moist but well-drained soil. Propagate by softwood cuttings in summer or by seed in autumn.
V. primuliflora, syn. *Douglasia vitaliana*, illus. p.320.

VITEX (Verbenaceae)
Genus of evergreen or deciduous trees and shrubs, grown for their flowers. Cultivated species are frost hardy, but in cold areas grow against a south- or west-facing wall. Needs full sun and well-drained soil. Propagate by semi-ripe cuttings in summer or by seed in autumn or spring.
V. agnus-castus (Chaste tree). Deciduous, spreading, open, aromatic shrub. H and S 2.5m (8ft). Upright panicles of fragrant, tubular, violet-blue flowers appear in early and mid-autumn. Dark green leaves are each divided into 5 or 7 long, narrowly lance-shaped leaflets.
V. negundo. Deciduous, bushy shrub. H and S 3m (10ft). Mid-green leaves are each composed of 3–7 narrowly oval, sharply toothed leaflets. Produces loose panicles of small, tubular, violet-blue flowers from late summer to early autumn.

VITIS (Vitaceae)
Vine
Genus of deciduous, woody-stemmed, tendril climbers, grown for their foliage and fruits (grapes), which are produced in bunches. Fully to half hardy. Prefers fertile, well-drained, chalky soil and sun or semi-shade. Produces the best fruits and autumn leaf-colour when planted in a warm situation. Propagate by hardwood cuttings in late autumn.
V. aconitifolia. See *Ampelopsis aconitifolia.*
V. amurensis (Amur grape). Vigorous, deciduous, woody-stemmed, tendril climber. H 6m (20ft). Fully hardy. Dark green, 3- or 5-lobed leaves, 12–30cm (5–12in) long, turn red and purple in autumn. Has inconspicuous flowers in summer, followed by tiny, black fruits in late summer and autumn.
♛ **V. 'Brant'.** Deciduous, woody-stemmed, tendril climber. H to 7m (22ft) or more. Fully hardy. Has lobed, toothed, bright green leaves, 10–22cm (4–9in) long, that except for the veins, turn brown-red in autumn. Inconspicuous flowers in summer are followed by green or purple fruits.
♛ **V. coignetiae** illus. p.178.
V. davidii. Deciduous, woody-stemmed, tendril climber; young stems have short prickles. H to 8m (25ft) or more. Half hardy. Heart-shaped leaves, 10–25cm (4–10in) long, are blue- or grey-green beneath, turning scarlet in autumn. Insignificant, greenish flowers are borne in summer, followed by small, black fruits.
V. henryana. See *Parthenocissus henryana.*
V. heterophylla. See *Ampelopsis brevipedunculata* var. *maximowiczii.*
V. quinquefolia. See *Parthenocissus quinquefolia.*
V. striata. See *Cissus striata.*
V. thompsonii. See *Parthenocissus thompsonii.*
V. vinifera (Grape vine).
♛ **'Purpurea'** illus. p.178.

VRIESEA (Bromeliaceae)
Genus of evergreen, rosette-forming, epiphytic perennials, grown for their flowers and overall appearance. Frost tender, min. 15°C (59°F). Needs semi-shade and a rooting medium of equal parts humus-rich soil and either sphagnum moss or bark or plastic chips used for orchid culture. Using soft water, water moderately when in growth, sparingly at other times, and from mid-spring to mid-autumn keep rosette centres filled with water. Propagate by offsets or seed in spring.
V. fenestralis. Evergreen, epiphytic perennial with dense, funnel-shaped rosettes. H and S 30–40cm (12–16in). Pale green leaves, with dark lines and cross-bands, are very broadly strap-shaped and arching or rolled under at tips. In summer, flat racemes of tubular, yellowish-green flowers, with green bracts, are carried above the foliage.
♛ **V. fosteriana.** Evergreen, epiphytic perennial with dense, funnel-shaped rosettes. H and S 60cm (24in) or more. Has broadly strap-shaped, arching, yellowish- to deep green leaves, cross-banded with reddish-brown, particularly beneath. In summer-autumn, flat spikes of tubular, pale yellow or greenish-yellow flowers, with brownish-red tips, are produced well above foliage.
V. hieroglyphica (King of the bromeliads). Evergreen, epiphytic perennial with dense, funnel-shaped rosettes. H and S 60cm–1m (2–3ft). Produces very broadly strap-shaped, arching, yellowish-green leaves, cross-banded and chequered with dark brownish-green. In summer bears panicles of tubular, yellow flowers well above leaves.
V. platynema. Evergreen, basal-rosetted, epiphytic perennial. H and S 60cm (24in). Broadly strap-shaped, mid- to light green leaves, with purple tips, form dense rosettes. In summer produces flat racemes of tubular, green-and-yellow flowers with red or yellow bracts.
♛ **V. psittacina.** Evergreen, spreading, basal-rosetted, epiphytic perennial. H and S 40–60cm (16–24in). Produces dense rosettes of strap-shaped, arching, pale green leaves. Flat spikes of tubular, yellow flowers with green tips, emerging from red-and-yellow or red-and-green bracts, are borne above foliage in summer-autumn.
♛ **V. splendens** (Flaming sword; illus. p.229). Evergreen, basal-rosetted, epiphytic perennial. H and S 30cm (12in). Strap-shaped, arching, olive-green leaves, with purple to reddish-brown cross-bands, form dense rosettes. Flat, sword-shaped racemes of tubular, yellow flowers, between bright red bracts, are borne in summer-autumn.

x VUYLSTEKEARA. See ORCHIDS.
x V. Cambria 'Lensing's Favorite' (illus. p.261). Evergreen, epiphytic orchid for a cool greenhouse. H 23cm (9in). Produces long sprays of wine-red flowers, 10cm (4in) across, heavily marked with white; flowering season varies. Has narrowly oval leaves, 10–15cm (4–6in) long. Needs shade in summer.

W

WACHENDORFIA (Haemodoraceae)
Genus of summer-flowering perennials with deep roots, to guard against frost. Half hardy. Requires full sun and moist soil. Propagate by division in spring or by seed in autumn or spring.
W. thyrsiflora. Clump-forming perennial. H 1.5–2m (5–6ft), S 45cm (1 1/2ft). Shallowly cup-shaped, yellow to orange flowers are produced in dense panicles in early summer. Mid-green leaves are narrowly sword-shaped, pleated and rather coarse.

WAHLENBERGIA (Campanulaceae)
Genus of summer-flowering annuals, biennials and short-lived perennials, grown for their bell-shaped flowers. Is useful for alpine houses. Frost hardy. Needs a sheltered site, partial shade and well-drained, peaty, sandy soil. Propagate by seed in autumn.
W. albomarginata (New Zealand bluebell). Basal-rosetted, rhizomatous perennial. H and S 15cm (6in) or more. Slender stems each carry a bell-shaped, clear blue flower that opens flat in summer. Has narrowly elliptic to oval, mid-green leaves in tufts. Is good in a rock garden.
W. congesta. Mat-forming, creeping, rhizomatous perennial. H 7cm (3in), S 10cm (4in). Has small, rounded or spoon-shaped, mid-green leaves and, in summer, bell-shaped, lavender-blue or white flowers held singly on wiry stems.

WALDSTEINIA (Rosaceae)
Genus of semi-evergreen, creeping perennials with runners. Makes good ground cover. Fully hardy. Needs sun and well-drained soil. Propagate by division in early spring.
W. fragarioides. Vigorous, semi-evergreen, mat-forming perennial. H 10–15cm (4–6in), S indefinite. Has 3-parted, hairy leaves and, in late spring and early summer, sparse clusters of small, saucer-shaped, yellow flowers.
W. ternata, syn. *W. trifolia*, illus. p.333.
W. trifolia. See *W. ternata*.

WASHINGTONIA (Palmae)
Genus of evergreen palms, grown for their stately appearance. Frost tender, min. 10°C (50°F). Grows in fertile, well-drained soil and in full sun. Water containerized specimens freely in summer, moderately at other times. Remove skirt of persistent, dead leaves as they are a fire risk. Propagate by seed in spring at not less than 24°C (75°F). Red spider mite may be a nuisance.
♀ *W. filifera* (Desert fan palm). Fast-growing, evergreen palm. H and S to 25m (80ft). Has fan-shaped, long-stalked, grey-green leaves, each lobe with a filamentous tip. Long-stalked clusters of tiny, creamy-white flowers are borne in summer and berry-like, black fruits in winter.
W. robusta illus. p.47.

WATSONIA (Iridaceae)
Genus of clump-forming corms, *Gladiolus*-like in overall appearance, although flowers are more tubular. Half hardy. Requires an open, sunny position and light, well-drained soil. Plant in autumn, 10–15cm (4–6in) deep; protect with bracken, loose peat or similar during first winter, if frost is expected. Feed with slow-acting fertilizer, such as bonemeal, in summer. Corms are best left undisturbed to form clumps. Propagate by seed in autumn.
W. beatricis. See *W. pillansii*.
W. borbonica, syn. *W. pyramidata* illus. p.344.
W. fourcadei. Clump-forming, summer-flowering corm. H to 1.5m (5ft), S 30–45cm (1–1 1/2ft). Sword-shaped, erect leaves are mostly basal. Has a dense spike of tubular, salmon-red flowers, each 8–9cm (3–3 1/2in) long and with 6 lobes.
W. meriana. Clump-forming, summer-flowering corm. H to 1m (3ft), S 30–45cm (1–1 1/2ft). Has sword-shaped, erect leaves both on stem and at base. Stem carries a loose spike of tubular, pinkish-red flowers, each 5–6cm (2–2 1/2in) long and with 6 spreading lobes.
W. pillansii, syn. *W. beatricis*, illus. p.344.
W. pyramidata. See *W. borbonica*.

Wattakaka sinensis. See *Dregea sinensis*.

WEIGELA (Caprifoliaceae)
Genus of deciduous shrubs, grown for their showy, funnel-shaped flowers. Fully hardy. Prefers sunny, fertile soil. To maintain vigour, prune out a few older branches to ground level, after flowering each year. Straggly, old plants may be pruned hard in spring (though this will lose one season's flowers). Propagate by softwood cuttings in summer.
W. 'Bristol Ruby'. Vigorous, deciduous, upright shrub. H 2.5m (8ft), S 2m (6ft). Deep red flowers open from darker buds amid oval, toothed, mid-green leaves in late spring and early summer.
W. 'Candida'. Deciduous, bushy shrub. H and S 2.5m (8ft). Pure white flowers appear in late spring and early summer. Leaves are oval, toothed and bright green.
W. 'Eva Rathke'. Deciduous, upright, dense shrub. H and S 1.5m (5ft). Has oval, toothed, dark green leaves. Broad-mouthed, crimson flowers open from darker buds from late spring to early summer.
W. florida. Deciduous, arching shrub. H and S 2.5m (8ft). Bears deep pink flowers, pale pink to white inside, in late spring and early summer. Oval, toothed leaves are mid-green.
♀ *'Foliis Purpureis'* illus. p.133.
♀ *'Variegata'* illus. p.130.
W. 'Looymansii Aurea'. Weak-growing, deciduous, upright shrub. H 1.5m (5ft), S 1m (3ft). From late spring to early summer produces pale pink flowers amid oval, toothed, narrowly red-rimmed, golden-yellow leaves. Protect from hot sun.
W. middendorffiana illus. p.138.
W. praecox. Deciduous, upright shrub. H 2.5m (8ft), S 2m (6ft). In late spring produces fragrant, pink flowers, marked inside with yellow. Leaves are bright green, oval and toothed.
♀ *'Variegata'* has leaves with broad, creamy-white margins.

Weingartia neocumingii. See *Rebutia neocumingii*.

WEINMANNIA (Cunoniaceae)
Genus of evergreen trees and shrubs, grown for their foliage, flowers and overall appearance. Frost tender, min. 5–7°C (41–5°F). Needs partial shade or full light and humus-rich, well-drained but not dry soil, ideally neutral to acid. Water potted plants freely when in full growth, moderately at other times. Pruning is tolerated if needed. Propagate by seed in spring or by semi-ripe cuttings in summer.
W. trichosperma. Evergreen, ovoid to round-headed tree. H 12m (40ft) or more, S 8–10m (25–30ft). Glossy, rich green leaves have 9–19 oval, boldly toothed leaflets borne on a winged midrib. Spikes of tiny, fragrant, white flowers, with pink stamens, are produced in early summer.

WELDENIA (Commelinaceae)
Genus of one species of summer-flowering, tuberous perennial, grown for its flowers. Half hardy. Needs sun and gritty, well-drained soil. Keep dry from autumn until growth restarts in late winter. Is suitable for alpine houses. Propagate by root cuttings in winter or by division in early spring.
W. candida illus. p.310.

WELWITSCHIA (Welwitschiaceae)
Genus of one species of evergreen, desert-growing perennial with a deep tap root. Has only 2 leaves, which lie on the ground and grow continuously from the base for up to 100 years. Frost tender, min. 10°C (50°F). Requires sun and sharply drained soil. Needs desert conditions: may succeed in a mixture of stone chippings and leaf mould, in a length of drainpipe to take its long tap root. Propagate by seed when ripe.
W. bainesii. See *W. mirabilis*.
W. mirabilis, syn. *W. bainesii*, illus. p.268.

WESTRINGIA (Labiatae)
Genus of evergreen shrubs, grown for their flowers and overall appearance. Frost tender, min. 5–7°C (41–5°F). Requires full light and well-drained soil. Water containerized specimens moderately, less when not in full growth. Propagate by seed in spring or by semi-ripe cuttings in late summer.
♀ *W. fruticosa*, syn. *W. rosmariniformis*, illus. p.128.
W. rosmariniformis. See *W. fruticosa*.

WIGANDIA (Hydrophyllaceae)
Genus of evergreen perennials and shrubs, grown for their flowers and foliage. Frost tender, min. 7–10°C (45–50°F). Needs full light and moist but well-drained soil. Water potted plants freely when in full growth, moderately at other times. Cut down flowered stems in spring to prevent plants becoming straggly. Propagate by seed or softwood cuttings in spring. Whitefly is sometimes troublesome.
W. caracasana. Evergreen, erect, sparsely branched shrub. H 2–3m (6–10ft), S 1–2m (3–6ft). Has oval, wavy-edged, toothed, deep green leaves, 45cm (18in) long and white-haired beneath. Carries 5-petalled, violet-purple flowers in large, terminal clusters from spring to autumn. Is often grown annually from seed purely for its handsome leaves.

Wigginsia vorwerkiana. See *Parodia vorwerkiana*.

Wilcoxia albiflora. See *Echinocereus leucanthus*.
Wilcoxia schmollii. See *Echinocereus schmollii*.

x **WILSONARA**. See ORCHIDS.
x *W. Hambühren Stern 'Cheam'* (illus. p.261). Evergreen, epiphytic orchid for a cool greenhouse. H 23cm (9in). Bears spikes of deep reddish-brown flowers, 9cm (3 1/2in) across, each with a yellow lip; flowering season varies. Narrowly oval leaves are 10cm (4in) long. Requires shade in summer.

WISTERIA (Leguminosae)
Genus of deciduous, woody-stemmed, twining climbers, grown for their spectacular flowers and suitable for walls and pergolas and for growing against buildings and trees. Fully to frost hardy. Grows in sun and in fertile, well-drained soil. Prune after flowering and again in late winter. Propagate by bench grafting in winter or by seed in autumn or spring. Plants produced from seed may not flower until some years old and often have poor flowers.
W. brachybotrys f. *alba*. See *W. venusta*.
W. chinensis. See *W. sinensis*.
W. floribunda (Japanese wisteria).
♀ *'Alba'* illus. p.167.
♀ *'Macrobotrys'* is a deciduous, woody-stemmed, twining climber. H to 9m (28ft). Fully hardy. Leaves,

25–35cm (10–14in) long, have 11–19 oval leaflets. Scented, pea-like, lilac flowers, flushed darker, are carried in racemes, 1–1.2m (3–4ft) long, in early summer, and may produce oblong, velvety pods in late summer and autumn.
W. × *formosa* illus. p.175.
♚ *W. sinensis*, syn. *W. chinensis*, illus. p.175. ♚ **'Alba'** illus. p.167. **'Black Dragon'** is a deciduous, woody-stemmed, twining climber. H to 9m (28ft). Fully hardy. Leaves, 25–35cm (10–14in) long, have 11–19 oval leaflets. Scented, pea-like, double, deep purple flowers are carried in racemes, 12–25cm (5–10in) long, in early summer, sometimes sparsely again in autumn. Oblong, velvety pods, 7–15cm (3–6in) long, appear in late summer and autumn. **'Prolific'**, H to 30m (100ft), is vigorous, with masses of single, lilac or pale violet flowers in longer racemes.
W. venusta, syn. *W. brachybotrys* f. *alba* (Silky wisteria). Deciduous, woody-stemmed, twining climber. H to 9m (28ft) or more. Fully hardy. Leaves are 20–35cm (8–14in) long, with 9–13 oval leaflets. Has 10–15cm (4–6in) long racemes of scented, pea-like, white flowers, each with a yellow blotch at base of upper petal, in early summer; sometimes flowers again sparsely in autumn. **'Alba Plena'** has double, white flowers.

WOLFFIA (Lemnaceae)
Duckweed
Genus of semi-evergreen, perennial, floating water plants, grown for their curiosity value as the smallest-known flowering plants. Is ideal for cold-water aquariums. Half hardy. Needs a sunny position. Remove excess plantlets as required. Propagate by redistribution of plantlets as required.
W. arrhiza (Least duckweed). Semi-evergreen, perennial, floating water plant. S 1mm ($^1/_{32}$in). Leaves are rounded and mid-green. Insignificant, greenish flowers appear year-round.

WOODSIA (Polypodiaceae)
Genus of deciduous ferns, suitable for rock gardens and alpine houses. Fully hardy. Tolerates sun or semi-shade. May be difficult to cultivate: soil must provide constant moisture and also be quick-draining, and crowns of plants must sit above soil to avoid rotting. Propagate by division in early spring.
W. alpina (Alpine woodsia). Deciduous fern. H and S 15cm (6in). Produces dense tufts of lance-shaped, mid-green fronds with oval pinnae.
W. ilvensis (Rusty woodsia). Deciduous fern. H and S 15cm (6in). Broadly lance-shaped, much-divided fronds, with oblong, wavy-toothed pinnae, are mid-green.

WOODWARDIA (Polypodiaceae)
Genus of evergreen or deciduous ferns. Half hardy. Prefers semi-shade and fibrous, moist, peaty soil. Remove faded fronds regularly. Propagate by division in spring.
♚ *W. radicans* (Chain fern). Vigorous, evergreen, spreading fern. H 1.2m (4ft), S 60cm (2ft). Large, broadly lance-shaped, coarsely divided, arching fronds, with narrowly oval pinnae, are mid-green.
W. virginica (American chain fern, Virginian chain fern). Deciduous, creeping fern. H and S 45cm (18in). Has broadly lance-shaped, divided, olive-green fronds with oblong to oval pinnae.

WORSLEYA (Amaryllidaceae)
Blue amaryllis
Genus of one species of evergreen, winter-flowering bulb, with a neck up to 75cm (2$^1/_2$ft) high crowned by a tuft of leaves and a 20–30cm (8–12in) leafless flower stem. Frost tender, min. 15°C (59°F). Needs full sun and well-drained soil, or compost mixed with osmunda fibre, perlite or bark chips and some leaf mould. Do not allow soil to dry out at any stage. Propagate by seed in spring.
W. procera. See *W. rayneri.*
W. rayneri, syn. *W. procera*, *Hippeastrum procerum*. Evergreen, winter-flowering bulb. H 1–1.2m (3–4ft), S 45–60cm (1$^1/_2$–2ft). Bears long, strap-shaped, strongly curved leaves and up to 14 funnel-shaped, lilac-blue flowers, 15cm (6in) long, with wavy-edged petals.

WULFENIA (Scrophulariaceae)
Genus of evergreen, summer-flowering perennials with rough-textured leaves. Is suitable for alpine houses as dislikes winter wet. Fully hardy. Requires full sun and well-drained soil. Propagate by division in spring or by seed in autumn.
W. amherstiana illus. p.304.
W. carinthiaca. Evergreen, basal-rosetted perennial. H and S 25cm (10in). Produces oblong to oval, toothed, dark green leaves that are hairy beneath. Top quarter of flower stem is covered in a dense spike of small, tubular, violet-blue flowers in summer.

X Y

XANTHOCERAS (Sapindaceae)
Genus of one species of deciduous, spring- to summer-flowering shrub or tree, grown for its foliage and flowers. Fully hardy. Requires sun and fertile, well-drained soil. Does best in areas with hot summers. Propagate by seed in autumn or by root cuttings or suckers in late winter. Is susceptible to coral spot fungus.
♇ *X. sorbifolium* illus. p.88.

XANTHORHIZA (Ranunculaceae)
Genus of one species of deciduous, spring-flowering shrub, grown for its foliage and flowers. Fully hardy. Prefers shade or semi-shade and moist soil. Propagate by division in autumn.
X. apiifolia. See *X. simplicissima.*
X. simplicissima, syn. *X. apiifolia* (Yellow-root). Deciduous, upright shrub that spreads by underground stems. H 60cm (2ft), S 1.5m (5ft). Bright green leaves, each consisting of usually 5 oval to lance-shaped, sharply toothed leaflets, turn bronze or purple in autumn. Produces nodding panicles of tiny, star-shaped, purple flowers from early to mid-spring as foliage emerges.

XANTHORRHOEA (Xanthorrhoeaceae)
Blackboy, Grass tree
Genus of evergreen, long-lived perennials, grown mainly as foliage plants. Frost tender, min. 10°C (50°F). Grow in full sun, in well-drained soil and in a fairly dry atmosphere. Propagate by basal offsets or seed in spring.
X. australis. Evergreen perennial with a stout, dark trunk. H 60cm–1.2m (2–4ft), S 1.2–1.5m (4–5ft). Very narrow, arching, flattened, silvery-green leaves, 60cm (2ft) or more long, 2mm (1/10in) wide, spread from top of trunk. In summer may have small, fragrant, 6-petalled, white flowers carried in dense, candle-like spikes, 60cm (2ft) or more long, on stems of similar length.

XANTHOSOMA (Araceae)
Genus of perennials, with underground tubers or thick stems above ground, grown mainly for their attractive foliage. Many species are cultivated in the tropics for their edible, starchy tubers. Frost tender, min. 15°C (59°F). Grow in partial shade, in rich, moist soil and in a moist atmosphere. Propagate by division or stem cuttings in spring or summer.
X. sagittifolium illus. p.230.
X. violaceum. Stemless perennial with large, underground tubers and leaves rising from ground level. H and S 1.2m (4ft). Purplish leaf stalks, to over 60cm (2ft) long, carry broadly arrow-shaped leaf blades, 70cm (28in) long, dark green with purple midribs and veins. Intermittently bears greenish-purple spathes, yellower within, surrounding a brownish spadix.

XERANTHEMUM (Compositae)
Immortelle
Genus of summer-flowering annuals. Half hardy. Grow in sun and in fertile, very well-drained soil. Propagate by seed sown outdoors in spring.
X. annuum. Fairly fast-growing, upright annual with branching flower heads. H 60cm (2ft), S 45cm (1 1/2ft). Has lance-shaped, silvery leaves. Daisy-like, papery, purple flower heads are produced in summer. Double forms are available in shades of pink, mauve, purple or white (illus. p.277).

XERONEMA (Liliaceae)
Genus of evergreen, robust, tufted perennials, with short, creeping rootstocks, grown for their flowers. Frost tender, min. 10°C (50°F). Grow in sun or partial shade and in humus-rich, well-drained soil. Propagate by seed or division in spring.
X. callistemon. Evergreen, iris-like, clump-forming perennial. H 60cm–1m (2–3ft), S indefinite. Erect, folded leaves, 60cm–1m (2–3ft) long, are very narrow and hard-textured. In summer, one-sided racemes, 15–30cm (6–12in) long, are crowded with short-stalked, 6-petalled, red flowers, to 3cm (1 1/4in) wide.

XEROPHYLLUM (Liliaceae)
Genus of elegant, summer-flowering, rhizomatous perennials. Frost hardy. Prefers full sun and moist, peaty soil. May be difficult to cultivate. Propagate by seed in autumn.
X. tenax. Clump-forming perennial. H 1–1.2m (3–4ft), S 30–60cm (1–2ft). Star-shaped, white flowers, with violet anthers, are borne in dense, terminal racemes in summer. Basal leaves are linear and mid-green.

YUCCA, syn. HESPEROYUCCA (Agavaceae)
Genus of evergreen shrubs and trees, grown for the architectural value of their bold, sword-shaped, clustered leaves and showy panicles of usually white flowers. Makes excellent container-grown plants. Fully hardy to frost tender, min. 7°C (45°F). Needs full sun and well-drained soil. Water potted specimens moderately, less when not in full growth. Remove spent flowering stems. Propagate in spring: frost-tender species by seed or suckers, hardier species by root cuttings or division.
Y. aloifolia illus. p.123.
♇ *Y. filamentosa* (Adam's needle). Evergreen, basal-rosetted shrub. H 2m (6ft), S 1.5m (5ft). Fully hardy. From mid- to late summer, tall panicles of pendulous, tulip-shaped, white flowers are produced, rising from low tufts of sword-shaped, deep green leaves edged with white threads.
Y. filifera **'Ivory'.** See *Y. flaccida* 'Ivory'.
♇ *Y. flaccida* **'Ivory',** syn. *Y. filifera* 'Ivory', illus. p.130.
♇ *Y. gloriosa* (Spanish dagger) illus. p.107. **'Nobilis'** is an evergreen shrub. H and S 2m (6ft). Frost hardy. Stout, usually unbranched stem is crowned with a large tuft of long, sword-shaped, sharply pointed, blue-green leaves, the outer ones semi-pendent. Pendulous, tulip-shaped, red-backed, white flowers in long, erect panicles are produced from mid-summer to early autumn.
Y. parviflora. See *Hesperaloe parviflora.*
Y. whipplei illus. p.130.

YUSHANIA (Bambusoideae).
See GRASSES, BAMBOOS, RUSHES and SEDGES.
Y. anceps, syn. *Arundinaria anceps, A. jaunsarensis, Sinarundinaria jaunsarensis,* illus. p.184.

Z

ZANTEDESCHIA (Araceae)
Genus of summer-flowering, tuberous perennials, usually evergreen in a warm climate, grown for their erect, funnel-shaped spathes, each enclosing a club-shaped spadix. Frost hardy to frost tender, min. 10°C (50°F). Needs full sun or partial shade and well-drained soil. Z. aethiopica also grows in 15–30cm (6–12in) of water as a marginal water plant. Propagate by offsets in winter.
Z. aethiopica (Arum lily).
 ‘**Crowborough**’ illus. p.341.
♀ ‘**Green Goddess**’ illus. p.341.
Z. albomaculata, syn. Z. melanoleuca. Summer-flowering, tuberous perennial. H 30–40cm (12–16in), S 30cm (12in). Frost tender. Has arrow-shaped, semi-erect, basal leaves with transparent spots. Produces a yellow spadix inside a white spathe, 12–20cm (5–8in) long, shading to green at the base with a deep purple blotch inside.
♀ **Z. elliottiana** illus. p.347.
Z. melanoleuca. See Z. albo-maculata.
♀ **Z. rehmannii** (Pink arum). Summer-flowering, tuberous perennial. H 40cm (16in), S 30cm (12in). Frost tender. Arrow-shaped, semi-erect, basal, green leaves are generally unmarked. Flower stem bears a yellow spadix surrounded by a reddish-pink spathe, 7–8cm (3in) long and narrowly tubular at the base.

ZANTHOXYLUM (Rutaceae)
Genus of deciduous or evergreen, spiny shrubs and trees, grown for their fruits, aromatic foliage and habit. Fully to frost hardy. Grows in sun or semi-shade and in fertile soil. Propagate by seed in autumn or by root cuttings in late winter.
Z. piperitum illus. p.115.
Z. simulans illus. p.118.

ZAUSCHNERIA (Onagraceae)
Genus of sub-shrubby perennials, grown for their mass of flowers. Frost hardy. Needs sun and well-drained soil. Propagate by seed or division in spring or by side-shoot cuttings in summer.
Z. californica, syn. Epilobium canum. Clump-forming, woody-based perennial. H and S 45cm (18in). Produces terminal clusters of tubular, bright scarlet flowers on slender stems in late summer and early autumn. Bears lance-shaped, rich green leaves.
♀ ‘**Glasnevin**’ illus. p.303.
Z. cana, syn. Epilobium canum. Clump-forming perennial. H 30cm (12in), S 45cm (18in). Leaves are linear and grey. Sprays of fuchsia-like, brilliant scarlet flowers appear from late summer to early autumn.

ZEA (Gramineae), Indian corn, Maize. See GRASSES, BAMBOOS, RUSHES and SEDGES.
Z. mays (Ornamental maize, Sweet corn). ‘**Gigantea Quadricolor**’ is a fairly fast-growing, upright annual. H 1–2m (3–6ft), S 60cm (2ft). Half hardy. Has lance-shaped leaves, 60cm (2ft) long, variegated with white, pale yellow and pink. Feathery, silky flower heads, 15cm (6in) long, are borne on long stems in mid-summer, followed by large, cylindrical, green-sheathed, yellow seed heads, known as cobs.
‘**Gracillima Variegata**’ illus. p.272. Cobs of ‘**Japonica Multicolor**’ contain yellow, red, black and orange seeds.

Zebrina pendula. See Tradescantia zebrina.

ZELKOVA (Ulmaceae)
Genus of deciduous trees, grown for their foliage and habit and best planted as isolated specimens. Produces insignificant flowers in spring. Fully hardy, but prefers some shelter. Does best in full sun and requires deep, fertile, moist but well-drained soil. Propagate by seed in autumn.
Z. abelicea, syn. Z. cretica. Deciduous, bushy-headed, spreading tree. H 5m (15ft), S 7m (22ft). Has small, oval, prominently toothed, glossy, dark green leaves.
Z. carpinifolia, syn. Z. crenata (Caucasian elm). Deciduous tree with a short, stout trunk from which many upright branches arise to make an oval, dense crown. H 30m (100ft), S 25m (80ft). Oval, sharply toothed, dark green leaves turn orange-brown in autumn.
Z. cretana. See Z. carpinifolia.
Z. cretica. See Z. abelicea.
♀ **Z. serrata** illus. p.45.

ZENOBIA (Ericaceae)
Genus of one species of deciduous or semi-evergreen, summer-flowering shrub, grown for its flowers. Fully hardy. Requires semi-shade and moist, peaty, acid soil. Prune out older, weaker shoots after flowering to maintain vigour. Propagate by semi-ripe cuttings in summer.
Z. pulverulenta illus. p.108.

ZEPHYRANTHES (Amaryllidaceae)
Rain lily, Windflower
Genus of clump-forming bulbs with an erect, crocus-like flower on each stem. Frost to half hardy. Needs a sheltered, sunny site and open, well-drained but moist soil. Pot-grown bulbs need a dryish, warm period after foliage dies down in summer, followed by copious amounts of water to stimulate flowering. Propagate by seed in autumn or spring.
Z. atamasco (Atamasco lily). Clump-forming, early summer-flowering bulb. H 15–25cm (6–10in), S 8–10cm (3–4in). Half hardy. Has very narrow, grass-like, semi-erect, basal leaves. Stems each bear a widely funnel-shaped, purple-tinged, white flower, 10cm (4in) wide.
Z. candida illus. p.381.
Z. carinata. See Z. grandiflora.
Z. citrina. Clump-forming, autumn-flowering bulb. H 10–15cm (4–6in), S 5–8cm (2–3in). Half hardy. Has rush-like, erect, basal, green leaves. Stems produce funnel-shaped, bright yellow flowers opening to 4–5cm (1½–2in) wide.
Z. grandiflora, syn. Z. carinata, Z. rosea, illus. p.368.
Z. rosea. See Z. grandiflora.

ZIGADENUS (Liliaceae)
Genus of summer-flowering bulbs with spikes of star-shaped, 6-petalled flowers. Frost hardy. Needs sun or partial shade and well-drained soil. Water copiously in spring and summer, when in growth; less at other times. Is dormant in winter. Propagate by division in early spring or by seed in autumn or spring.
Z. elegans. Clump-forming, summer-flowering bulb. H 30–50cm (12–20in), S 10–15cm (4–6in). Bears long, narrowly strap-shaped, semi-erect, basal leaves. Stem produces a spike of greenish-white flowers, each 1.5–2cm (⅝–¾in) wide with a yellowish-green nectary near the base of each petal.
Z. fremontii illus. p.364.

ZINNIA (Compositae)
Genus of annuals with large, dahlia-like flower heads that are excellent for cutting. Half hardy. Grow in sun and in fertile, well-drained soil. Dead-head regularly. Propagate by seed sown under glass in early spring.
Z. elegans. Moderately fast-growing, upright, sturdy annual. H 60–75cm (2–2½ft), S 30cm (1ft). Oval to lance-shaped leaves are pale or mid-green. Dahlia-like, purple flower heads, over 5cm (2in) wide, appear in summer and early autumn. Elegans group hybrids are available in shades of yellow, red, pink, purple, cream or white.
 ‘**Belvedere Dwarfs**’ illus. p.290.
 Big Top Series, H 60cm (2ft), has large, double flower heads like cactus dahlias, in a range of colours.
 ‘**Border Beauty Rose**’, H 45cm (1½ft), has double, deep pink flower heads.
 Burpee Hybrids illus. p.293.
 ‘**Envy**’ illus. p.287.
 Fantastic Series, H 20cm (8in), has double flower heads in a range of colours.
 Peppermint Stick Series, H 60cm (2ft), has flower heads like pompon dahlias that are striped, blotched and stippled in many colours.
 Peter Pan Series, H 20cm (8in), has double flower heads in a wide colour range.
 Pulcino Series, H 30cm (1ft), has double flower heads in a mixture of clear colours.
 Ruffles Series (mixed) illus. p.281, (scarlet) illus. p.281.
 Sunshine Series, H 60cm (2ft), has double flower heads like pompon dahlias in a wide range of colours.
 Thumbelina Series illus. p.274.
Z. ‘Persian Carpet’. Moderately fast-growing, upright, bushy annual. H 38cm (15in), S 30cm (12in). Has lance-shaped, hairy, pale green leaves. Small, dahlia-like, double flower heads, over 2.5cm (1in) wide, are produced in summer in a range of colours.

ZIZANIA (Gramineae). See GRASSES, BAMBOOS, RUSHES and SEDGES.
Z. aquatica (Canada wild rice). Annual, grass-like, marginal water plant. H 3m (10ft), S 45cm (18in). Half hardy. Has grass-like, mid-green leaves and, in summer, grass-like, pale green flowers. Rice-like seeds attract waterfowl. Requires sun and is suitable for up to 23cm (9in) depth of water. Propagate from seed stored damp and sown in spring.

Zygocactus truncatus. See Schlumbergera truncata.

ZYGOPETALUM. See ORCHIDS.
Z. mackaii. See Z. mackayi.
Z. mackayi, syn. Z. mackaii (illus. p.262). Evergreen, epiphytic orchid for a cool or intermediate greenhouse. H 30cm (12in). In autumn carries long sprays of fragrant, brown-blotched, green flowers, 8cm (3in) across, with reddish-indigo-veined, white lips. Leaves are narrowly oval, ribbed and 30cm (12in) long. Requires semi-shade in summer.
Z. Perrenoudii (illus. p.262). Evergreen, epiphytic orchid for a cool or intermediate greenhouse. H 30cm (12in). Spikes of fragrant, violet-purple-lipped, dark brown flowers, 8cm (3in) across, are borne in winter. Has narrowly oval, ribbed leaves, 30cm (12in) long. Requires semi-shade in summer.

Glossary of Terms

Terms printed in italics refer to other glossary entries.

Acid [of soil]. With a *pH* value of less than 7; see also *alkaline* and *neutral*.
Adventitious [of roots]. Arising directly from a stem or leaf.
Aerial root. See *root*.
Air-layering. A method of propagation by which a portion of stem is induced to root by enclosing it in a suitable medium such as damp moss and securing it with plastic sheeting; roots will form if the moss is kept moist.
Alkaline [of soil]. With a *pH* value of more than 7; some plants will not tolerate alkaline soils and must be grown in *neutral* or *acid* soil.
Alpine house. An unheated greenhouse, used for the cultivation of mainly alpine and bulbous plants, that provides greater ventilation and usually more light than a conventional greenhouse.
Alternate [of leaves]. Borne singly at each *node*, on either side of a stem.
Annual. A plant that completes its life cycle, from germination through to flowering and then death, in one growing season.
Anther. The part of a *stamen* that produces pollen; it is usually borne on a *filament*.
Apex. The tip or growing point of an organ such as a leaf or shoot.
Areole. A modified, cushion-like *tubercle*, peculiar to the family Cactaceae, that bears hairs, spines, leaves, side-branches or flowers.
Asclepiad. A member of the family Asclepiadaceae, e.g. *Asclepias*, *Hoya*, *Stephanotis*.
Auricle. An ear-like lobe such as is sometimes found at the base of a leaf.
Awn. A stiff, bristle-like projection commonly found on grass seeds and *spikelets*.
Axil. The angle between a leaf and stem where an axillary bud develops.
Bedding plant. A plant that is mass-planted to provide a temporary display.
Biennial. A plant that flowers and dies in the second season after germination, producing only stems, roots and leaves in the first season.
Blade. The flattened and often broad part of a leaf.
Bloom. 1. A flower or blossom. 2. A fine, waxy, whitish or bluish-white coating on stems, leaves or fruits.
Bog garden. An area where the soil is kept permanently damp but not waterlogged.
Bole. The trunk of a *tree* from ground level to the first major branch.
Bolt. To produce flowers and seed prematurely, particularly in the case of vegetables such as lettuce and beetroot.
Bonsai. A method of producing dwarf trees or shrubs by special techniques that include pruning roots, pinching out shoots, removing growth buds and training branches and stems.
Bract. A modified leaf at the base of a flower or flower cluster. Bracts may resemble normal leaves or be reduced and scale-like in appearance; they are often large and brightly coloured.
Bud. A rudimentary or condensed shoot containing embryonic leaves or flowers.
Bulb. A storage organ consisting mainly of fleshy scales and swollen, modified leaf-bases on a much reduced stem. Bulbs usually, but not always, grow underground.
Bulbil. A small, *bulb*-like organ, often borne in a leaf *axil*, occasionally in a *flower head*; it may be used for propagation.
Bulblet. A small *bulb* produced at the base of a mature one.
Bur. 1. A prickly or spiny *fruit*, or aggregate of fruits. 2. A woody outgrowth on the stems of certain trees.
Cactus (pl. cacti). A member of the family Cactaceae, often *succulent* and spiny.
Calyx (pl. calyces). The outer part of a flower, usually small and green but sometimes showy and brightly coloured, that encloses the petals in bud and is formed from the *sepals*.
Capsule. A dry *fruit* that splits open when ripe to release its seeds.
Carpel. The female portion of a flower, or part of it, consisting of an *ovary*, *stigma* and *style*.
Catkin. A flower cluster, normally pendulous. Flowers lack petals, are often stalkless, surrounded by scale-like *bracts*, and are usually unisexual.
Caudex (pl. caudices). The stem base of a woody plant such as a *palm* or tree fern.
Cladode. A stem, often flattened, with the function and appearance of a leaf.
Claw. The narrow, basal portion of petals in some genera, e.g. *Dianthus*.
Climber. A plant that climbs using other plants or objects as a support: a **leaf-stalk** climber by coiling its leaf stalks around supports; a **root** climber by producing aerial, supporting roots; a **self-clinging** climber by means of suckering pads; a **tendril** climber by coiling its tendrils; a **twining** climber by coiling stems. **Scandent**, **scrambling** and **trailing climbers** produce long stems that grow over plants or other supports; they attach themselves only loosely if at all.
Clone. A group of genetically identical plants, propagated vegetatively.
Compound. Made up of several or many parts, e.g. a leaf divided into 2 or more *leaflets*.
Cone. The clustered flowers or woody, seed-bearing structures of a conifer.
Coppice. To cut back to near ground level each year in order to produce vigorous shoots for ornamental use, as is usual with *Cornus* and *Eucalyptus*.
Cordon. A trained plant restricted in growth to one main stem, occasionally 2–4 stems.
Corm. A *bulb*-like, underground storage organ consisting mainly of a swollen stem base and often surrounded by a papery tunic.
Cormlet. A small *corm* arising at the base of a mature one.
Corolla. The part of a flower formed by the petals.
Corona (crown). A petal-like outgrowth sometimes borne on the *corolla*, e.g. the trumpet or cup of a *Narcissus*.
Corymb. A racemose flower cluster in which the inner flower stalks are shorter than the outer, resulting in a rounded or flat-topped head.
Cotyledon. See *seed leaf*.
Creeper. A plant that grows close to the ground, usually rooting as it spreads.
Crisped. Minutely wavy-edged.
Crown. 1. The part of the plant at or just below the soil surface from which new shoots are produced and to which they die back in autumn. 2. The upper, branched part of a tree above the *bole*. 3. A *corona*.
Culm. The usually hollow stem of a grass or bamboo.
Cutting. A section of a plant that is removed and used for propagation. The various types of cutting are: **basal** – taken from the base of a plant (usually *herbaceous*) as it begins to produce growth in spring; **greenwood** – made from the tip of young growth; **hardwood** – mature wood taken at the end of the growing season; **leaf** – a detached leaf or part of a leaf; **root** – part of a semi-mature or mature root; **semi-ripe** – half-ripened wood taken during the growing season; **softwood** – young growth taken at the beginning of the growing season; **stem** – a greenwood, hardwood, semi-ripe or softwood cutting; **tip** – a greenwood cutting.
Cyme. A flower cluster in which each growing point terminates in a flower.
Dead-head. To remove spent flower heads so as to promote further growth or flowering, prevent seeding or improve appearance.
Deciduous. Losing its leaves annually at the end of the growing season; **semi-deciduous** plants lose only some leaves.
Decumbent. Growing close to the ground but ascending at the tips.
Dentate. With toothed margins.
Die-back. Death of the tips of shoots due to frost or disease.
Dioecious. Bearing male and female flowers on separate plants.
Disbud. To remove surplus buds to promote larger flowers or fruits.
Disc floret, disc flower. A small and often individually inconspicuous, usually tubular flower, one of many that comprise the central portion of a composite flower head such as a daisy.
Division. A method of propagation by which a clump is divided into several parts during dormancy.
Elliptic [of leaves]. Broadening in the centre and narrowing towards each end.
Entire [of leaves]. With untoothed margins.
Epiphyte. A plant which in nature grows on the surface of another without being parasitic.
Evergreen. Retaining its leaves at the end of the growing season although losing some older leaves regularly throughout the year; **semi-evergreen** plants retain only some leaves or lose older leaves only when the new growth is produced.
F1 hybrid. The first generation derived from crossing 2 distinct plants, usually when the parents are pure bred lines and the offspring are vigorous. Seed from F1 hybrids does not come *true* to type.
Fall. An outer *perianth segment* of an iris which projects outwards or downwards from the inner segments.
Fan palm. A *palm* with *palmate* rather than *pinnate* leaves.
Farina. A powdery, white, sometimes yellowish deposit naturally occurring on some leaves and flowers.
Fibrous root. A fine, young root, usually one of many.
Filament. The stalk of an *anther*.
Floret. A single flower in a head of many flowers.
Flower. The basic flower forms are: **single**, with one row of usually 4–6 petals; **semi-double**, with more petals, usually in 2 rows; **double**, with many petals in several rows and few or no stamens; **fully double**, usually rounded in shape, with densely packed petals and with the stamens obscured.
Flower head. A mass of small flowers or florets that together appear as one flower, e.g. a daisy.
Force. To induce artificially the early production of growth, flowers or *fruits*.
Frond. The leaf-like organ of a fern. Some ferns produce both barren and fertile fronds, the fertile fronds bearing *spores*.
Fruit. The structure in plants that bears one or more ripe seeds, e.g. a berry or nut.
Glabrous. Not hairy.
Glaucous. Bluish-white, bluish-green or bluish-grey.
Globose. Spherical.
Glochid. One of the barbed bristles or hairs, usually small, borne on a cactus *areole*.
Grafting. A method of propagation by which an artificial union is made between different parts of individual plants; usually the *shoot* (scion) of one is grafted onto the *rootstock* (stock) of another.
Heel. The small portion of old wood that is retained at the base of a cutting when it is removed from the stem.
Herbaceous. Dying down at the end of the growing season.
Hose-in-hose [of flowers]. With one *corolla* borne inside another, forming a double or semi-double *flower*.
Inflorescence. A cluster of flowers with a distinct arrangement, e.g. *corymb*, *cyme*, *panicle*, *raceme*, *spike*, *umbel*.
Insectivorous plant. A plant that traps and digests insects and other small animals to supplement its nutrient intake.
Key. A winged seed such as is produced by the sycamore (*Acer pseudoplatanus*).
Lateral. A side growth that arises from the side of a shoot or root.
Layering. A method of propagation by which a stem is induced to root by being pegged down into the soil while

GLOSSARY OF TERMS

it is still attached to the parent plant. See also *air-layering*.
Leaflet. The subdivision of a compound leaf.
Lenticel. A small, usually corky area on a stem or other part of a plant which acts as a breathing pore.
Lime. Compounds of calcium; the amount of lime in soil determines whether it is *alkaline*, *neutral* or *acid*.
Linear [of leaves]. Very narrow with parallel sides.
Lip. A lobe comprising 2 or more flat or sometimes pouched *perianth segments*.
Loam. Well-structured, fertile soil that is moisture-retentive but free-draining.
Marginal water plant. A plant that grows partially submerged in shallow water or in moist soil at the edge of a pond.
Midrib. The main, central vein of a leaf or the central stalk to which the *leaflets* of a *pinnate* leaf are attached.
Monocarpic. Flowering and fruiting only once before dying; such plants may take several years to reach flowering size.
Mulch. A layer of organic matter applied to the soil over or around a plant to conserve moisture, protect the roots from frost, reduce the growth of weeds and enrich the soil.
Naturalize. To establish and grow as if in the wild.
Nectar. A sweet, sugary liquid secreted by the **nectary** – glandular tissue usually found in the flower but sometimes found on the leaves or stems.
Neutral [of soil]. With a *pH* value of 7, the point at which soil is neither *acid* nor *alkaline*.
Node. The point on a stem from which a leaf or leaves arise.
Offset. A small plant that arises by natural vegetative reproduction, usually at the base of the mother plant.
Opposite [of leaves]. Borne 2 to each *node*, one opposite the other.
Ovary. The part of the female portion of the flower, containing embryonic seeds, that will eventually form the *fruit*.
Palm. An evergreen *tree* or *shrub*-like plant, normally single-stemmed, with *palmate* or *pinnate* leaves usually in terminal rosettes; strictly a member of the family Palmae.
Palmate. Lobed in the fashion of a hand, strictly with 5 lobes arising from the same point.
Pan. A shallow, free-draining pot in which alpine plants or bulbs are grown.
Panicle. A branched *raceme*.
Papilla (pl. papillae). A minute protuberance or gland-like structure.
Pea-like [of flowers]. Of the same structure as a pea flower.
Peat bed. A specially constructed area, edged with peat blocks and containing moisture-retentive, acidic, peaty soil.
Pedicel. The stalk of an individual flower.
Peduncle. The stalk of a flower cluster.
Peltate [of leaves]. Shield-shaped, with the stalk inserted towards or at the centre of the blade and not at the margin.
Perennial. Living for at least 3 seasons. In this book the term when used as a noun, and unless qualified, denotes an *herbaceous* perennial. A woody-based perennial dies down only partially leaving a woody stem at the base.
Perianth. The outer parts of the flower comprising the *calyx* and the *corolla*. The term is often used when the calyx and the corolla are very similar in form.
Perianth segment. One portion of the *perianth*, resembling a *petal* and sometimes known as a tepal.
Petal. One portion of the often showy and coloured part of the *corolla*. In some families, e.g. Liliaceae, the *perianth segments* are petal-like and referred to horticulturally as petals.
Petaloid. Like a petal.
Petiole. The stalk of a leaf.
pH. The scale by which the acidity or alkalinity of soil is measured. See also *acid*, *alkaline*, *neutral*.
Phyllode. A flattened leaf stalk, which functions as and resembles a leaf.
Pinch out. To remove the growing tips of a plant to induce the production of side-shoots.
Pinna (pl. pinnae). The primary division of a *pinnate* leaf. The fertile pinnae of ferns produce *spores*, vegetative pinnae do not.
Pinnate [of leaves]. Compound, with *leaflets* arranged on opposite sides of a central stalk.
Pistil. The female part of a flower comprising the *ovary*, *stigma* and *style*.
Pollard [of a tree]. To cut back to its main branches in order to restrict growth.
Pollination. The transfer of pollen from the *anthers* to the *stigma* of the same or different flowers, resulting in the fertilization of the embryonic seeds in the *ovary*.
Procumbent. Prostrate, creeping along the ground.
Raceme. An unbranched flower cluster with several or many stalked flowers borne singly along a main axis, the youngest at the apex.
Ray floret, ray flower. One of the flowers, usually with strap-shaped petals, that together form the outer ring of flowers in a composite *flower head* such as a daisy.
Ray petal. The petal or fused petals, often showy, of a ray *floret*.
Recurved. Curved backwards.
Reflexed. Bent sharply backwards.
Revert. To return to its original state, as when a plain green leaf is produced on a variegated plant.
Rhizome. An underground, creeping stem that acts as a storage organ and bears leafy shoots.
Root. The part of a plant, normally underground, that functions as anchorage and through which water and nutrients are absorbed. An **aerial root** emerges from the stem at some distance above the soil level.
Rootball. The roots and accompanying soil or compost visible when a plant is lifted.
Rootstock. A well-rooted plant onto which a scion is grafted; see *grafting*.
Rosette. A group of leaves radiating from approximately the same point, often borne at ground level at the base of a very short stem.
Runner. A horizontally spreading, usually slender stem that forms roots at each node, often confused with *stolon*.
Scale. 1. A reduced or modified leaf. 2. Part of a conifer *cone*.
Scandent. See *climber*.
Scarify. To scar the coat of a seed by abrasion in order to speed water intake and hence germination.
Scion. See *grafting*.
Scree. An area composed of a deep layer of stone chippings mixed with a small amount of loam. It provides extremely sharp drainage for plants that resent moisture at their base.
Seed head. Any usually dry *fruit* that contains ripe seeds.
Seed leaf (cotyledon). The first leaf, pair of leaves or occasionally group of leaves produced by a seed as it germinates. In some plants they remain below ground.
Self seed. To produce seedlings around the parent plant.
Sepal. Part of a *calyx*, usually insignificant but sometimes showy.
Series. The name applied to a group of similar but not identical plants, usually annuals, linked by one or more common features.
Sessile. Without a stalk.
Sheath. A cylindrical structure that surrounds or encircles, partially or fully, another plant organ such as a stem.
Shoot. The aerial part of a plant which bears leaves. A **side-shoot** arises from the side of a main shoot.
Shrub. A plant with *woody stems*, usually well-branched from or near the base.
Shy-flowering. Reluctant to flower; producing few flowers.
Simple [of leaves]. Not divided into leaflets.
Soft-stemmed. The opposite of *woody-stemmed*.
Spadix (pl. spadices). A *spike*-like flower cluster that is usually fleshy and bears numerous small flowers. Spadices are characteristic of the family Araceae, e.g. *Arum*.
Spathe. A large *bract*, or sometimes 2, frequently coloured and showy, that surrounds a *spadix* (as in *Arum*) or an individual flower bud (as in *Narcissus*).
Sphagnum. Mosses common to bogs; their moisture-retentive character makes them ideal components of some growing media. They are used particularly for orchid cultivation.
Spike. A racemose flower cluster with several or many unstalked flowers borne along a common axis.
Spikelet. 1. The flowering unit of grasses comprising one or several flowers with basal *bracts*. 2. A small *spike*, part of a branched flower cluster.
Spore. The minute reproductive structure of flowerless plants, e.g. ferns, fungi and mosses.
Sporangium (pl. sporangia). A body that produces *spores*.
Sport. A mutation, caused by an accidental or induced change in the genetic make-up of a plant, which gives rise to a shoot with different characteristics to those of the parent plant.
Spur. 1. A hollow projection from a petal, often producing *nectar*. 2. A short stem bearing a group of flower buds such as is found on fruit trees.
Spur back. To cut back side-shoots to within 2 or 3 buds of the main shoot.
Stamen. The *anther* and *filament*.
Standard. 1. A *tree* or *shrub* with a clear length of bare stem below the first branches. Certain shrubs, e.g. roses and fuchsias, may be trained to form standards. 2. One of the 3 inner and often erect *perianth segments* of the iris flower. 3. The larger, usually upright back petal of a flower in the family Leguminosae, e.g. *Lathyrus*.
Stapeliad. A member of the genus *Stapelia* and closely related genera of the family Asclepiadaceae.
Stem segment. A portion of a jointed stem between 2 *nodes*, most frequently occurring in cacti.
Sterile. Infertile, not bearing *spores*, pollen, seeds etc.
Stigma. The part of the female portion of the flower, borne at the tip of the *style*, that receives pollen.
Stipule. A small scale- or leaf-like appendage, usually one of a pair, mostly borne at a *node* or below a leaf stalk.
Stock. See *rootstock*.
Stolon. A horizontally spreading or arching stem, usually above ground, which roots at its tip to produce a new plant.
Stop. To remove certain growing points of a plant so as to control growth or the size and number of flowers.
Stratify. To break the dormancy of some seeds by exposing them to a period of cold.
Style. The part of the flower on which the *stigma* is borne.
Sub-globose. Almost spherical.
Sub-shrub. A plant that is woody at the base although the terminal shoots die back in winter.
Succulent. A plant with thick, fleshy leaves and/or stems, in this book evergreen unless otherwise stated.
Sucker. A shoot that arises from below ground level, directly from the *root* or *rootstock*.
Summer-deciduous. Losing its leaves naturally in summer.
Taproot. The main, downward-growing root of a plant; it is also applied generally to any strong, downward-growing root.
Tendril. A thread-like structure, used to provide support; see also *climber*.
Tooth. A small, marginal, often pointed lobe on a leaf, *calyx* or *corolla*.
Tepal. See *perianth segment*.
Tree. A woody plant usually having a well-defined trunk or stem with a head of branches above.
Trifoliate. With 3 leaves; loosely, with 3 *leaflets*.
Trifoliolate. With 3 *leaflets*.
True [of seedlings]. Retaining the distinctive characteristics of the parent when raised from seed.
Truss. A compact cluster of flowers, often large and showy, e.g. those of pelargoniums and rhododendrons.
Tuber. A thickened, usually underground, storage organ derived from a stem or root.
Tubercle. A small, rounded protuberance; see also *areole*.
Turion. 1. A bud on a *rhizome*. 2. A fleshy, overwintering bud found on certain water plants.
Umbel. A usually flat-topped or rounded flower cluster in which the individual flower stalks arise from a central point. In a compound umbel each primary stalk ends in an umbel.
Upright [of habit]. With vertical or semi-vertical main branches.
Water bud. See *turion*.
Whorl. The arrangement of 3 or more organs arising from the same point.
Winged [of seeds or fruits]. Having a marginal flange or membrane.
Woody-stemmed. With a stem composed of woody fibres and therefore persistent, as opposed to soft-stemmed and *herbaceous*. A **semi-woody stem** contains some softer tissue and may be only partially persistent.
x The sign used to denote a hybrid plant derived from the crossing of 2 or more botanically distinct plants.
+ The sign used to denote a graft hybrid; see *grafting*.

Index of Common Names

A

Aaron's beard. See *Hypericum calycinum*, p.140.
Abele. See *Populus alba*, p.38.
Adam's needle. See *Yucca filamentosa*.
Aeroplane propeller. See *Crassula perfoliate* var. *falcata*, p.400.
African daisy. See *Arctotis stoechadifolia*; *Dimorphotheca*.
African fountain grass. See *Pennisetum setaceum*.
African hemp. See *Sparmannia africana*, p.88.
African lily. See *Agapanthus africanus*.
African marigold. See *Tagetes erecta*.
African milkbush. See *Synadenium grantii*.
African red alder. See *Cunonia capensis*.
African tulip tree. See *Spathodea campanulata*, p.45.
African violet. See *Saintpaulia*.
Albany bottlebrush. See *Callistemon speciosus*.
Alder. See *Alnus*.
Aleppo pine. See *Pinus halepensis*, p.81.
Alexandra palm. See *Archontophoenix alexandrae*, p.46.
Alexandrian laurel. See *Danäe racemosa*.
Algerian iris. See *Iris unguicularis*.
Algerian oak. See *Quercus canariensis*, p.40.
Algerian winter iris. See *Iris unguicularis*.
Alleghany vine. See *Adlumia fungosa*.
Almond. See *Prunus dulcis*.
Alpine anemone. See *Pulsatilla alpina*, p.294.
Alpine avens. See *Geum montanum*.
Alpine azalea. See *Loiseleuria procumbens*, p.325.
Alpine buttercup. See *Ranunculus alpestris*, p.311.
Alpine catchfly. See *Lychnis alpina*.
Alpine cinquefoil. See *Potentilla crantzii*.
Alpine coltsfoot. See *Homogyne alpina*.
Alpine columbine. See *Aquilegia alpina*, p.296.
Alpine fleabane. See *Erigeron alpinus*, p.303.
Alpine forget-me-not. See *Myosotis alpestris*, p.318.
Alpine heath. See *Erica carnea*.
Alpine lady's mantle. See *Alchemilla alpina*.
Alpine mouse-ear. See *Cerastium alpinum*.
Alpine penny-cress. See *Thlaspi alpinum*.
Alpine pink. See *Dianthus alpinus*, p.326.
Alpine poppy. See *Papaver burseri* (*P. alpinum* group).
Alpine snowbell. See *Soldanella alpina*, p.316.
Alpine thistle. See *Carlina acaulis*, p.323.
Alpine toadflax. See *Linaria alpina*.
Alpine totara. See *Podocarpus nivalis*, p.84.
Alpine woodsia. See *Woodsia alpina*.
Alum root. See *Heuchera*.
Aluminium plant. See *Pilea cadierei*, p.264.
American arbor-vitae. See *Thuja occidentalis*.
American arrowhead. See *Sagittaria latifolia*, p.386.
American aspen. See *Populus tremuloides*.
American beech. See *Fagus grandifolia*.
American bittersweet. See *Celastrus scandens*.
American chain fern. See *Woodwardia virginica*.
American chestnut. See *Castanea dentata*.
American elder. See *Sambucus canadensis*.
American holly. See *Ilex opaca*, p.72.
American hop hornbeam. See *Ostrya virginiana*, p.51.
American hornbeam. See *Carpinus caroliniana*.
American lime. See *Tilia americana*.
American lotus. See *Nelumbo lutea*.
American mountain ash. See *Sorbus americana*.
American spatterdock. See *Nuphar advena*.
American wall fern. See *Polypodium virginianum*.
American white elm. See *Ulmus americana*.
American white oak. See *Quercus alba*, p.45.
Amur cork tree. See *Phellodendron amurense*.
Amur grape. See *Vitis amurensis*.
Amur maple. See *Acer ginnala*, p.69.
Amur silver grass. See *Miscanthus floridulus*.
Anceps bamboo. See *Yushania anceps*, p.184.
Angelica. See *Angelica archangelica*, p.192.
Angel's fishing rod. See *Dierama*.
Angel's tears. See *Billbergia* x *windii*; *Narcissus triandrus*, p.361.
Angels' trumpets. See *Datura*.
Angels' wings. See *Caladium* x *hortulanum*.
Angelwing begonia. See *Begonia coccinea*.
Angled Solomon's seal. See *Polygonatum odoratum*.
Annual phlox. See *Phlox drummondii*.
Antarctic beech. See *Nothofagus antarctica*.
Apache plume. See *Fallugia paradoxa*, p.106.
Apennine anemone. See *Anemone apennina*.
Apple mint. See *Mentha suaveolens*.
Apple of Peru. See *Nicandra physalodes*.
Arctic birch. See *Betula nana*, p.297.
Arizona ash. See *Fraxinus velutina*, p.54.
Arizona cypress. See *Cupressus glabra*.
Armand pine. See *Pinus armandii*.
Armenian oak. See *Quercus pontica*.
Arolla pine. See *Pinus cembra*, p.80.
Arrow bamboo. See *Pseudosasa japonica*, p.184.
Arrowhead. See *Sagittaria*.
Arum lily. See *Zantedeschia aethiopica*.
Asarabacca. See *Asarum europaeum*, p.338.
Ash. See *Fraxinus*.
Ash-leaved maple. See *Acer negundo*.
Aspen. See *Populus tremula*.
Asphodel. See *Asphodelus aestivus*.
Atamasco lily. See *Zephyranthes atamasco*.
Atlas cedar. See *Cedrus atlantica*.
August lily. See *Hosta plantaginea*, p.253.
Australian banyan. See *Ficus macrophylla*.
Australian brush cherry. See *Syzygium paniculatum*, p.55.
Australian cabbage palm. See *Livistona australis*.
Australian firewheel tree. See *Stenocarpus sinuatus*.
Australian heath. See *Epacris impressa*, p.125.
Australian pea. See *Lablab purpureus*, p.171.
Australian rosemary. See *Westringia fruticosa*, p.128.
Australian sarsparilla. See *Hardenbergia violacea*.
Australian sassafras. See *Atherosperma moschatum*.
Australian tree fern. See *Cyathea australis*, p.74; *Dicksonia antarctica*, p.186.
Australian violet. See *Viola hederacea*.
Austrian pine. See *Pinus nigra* var. *nigra*, p.78.
Autograph tree. See *Clusia major*.
Autumn crocus. See *Colchicum autumnale*, p.381; *Crocus nudiflorus*, p.372.
Autumn snowdrop. See *Galanthus reginae-olgae*.
Autumn snowflake. See *Leucojum autumnale*, p.381.
Avens. See *Geum*.
Aztec lily. See *Sprekelia formosissima*, p.357.
Aztec marigold. See *Tagetes erecta*.

B

Baby blue-eyes. See *Nemophila menziesii*, p.285.
Baby rubber plant. See *Peperomia clusiifolia*.
Baby's tears. See *Soleirolia*.
Baby's toes. See *Fenestraria aurantiaca* f. *rhopalophylla*, p.415.
Bald cypress. See *Taxodium distichum*, p.78.
Baldmoney. See *Meum athamanticum*.
Balearic box. See *Buxus balearica*, p.123.
Balkan blue grass. See *Sesleria heufleriana*.
Balloon flower. See *Platycodon*.
Balloon vine. See *Cardiospermum halicacabum*.
Balm of Gilead. See *Populus* x *candicans*.
Balsam. See *Impatiens balsamina*, p.272.
Balsam fir. See *Abies balsamea*.
Balsam poplar. See *Populus balsamifera*.
Bamboo palm. See *Rhapis excelsa*, p.122.
Banana. See *Musa*.
Banana passion fruit. See *Passiflora antioquiensis*.
Baneberry. See *Actaea*.
Banksian rose. See *Rosa banksiae*.
Banyan. See *Ficus benghalensis*, p.46.
Baobab. See *Adansonia*.
Barbados gooseberry. See *Pereskia aculeata*, p.393.
Barbados pride. See *Caesalpinia pulcherrima*.
Barbed-wire plant. See *Tylecodon reticulatus*, p.401.
Barberry. See *Berberis*.
Barberton daisy. See *Gerbera jamesonii*, p.266.
Barrel cactus. See *Ferocactus*.
Barrenwort. See *Epimedium alpinum*.
Bartram's oak. See *Quercus* x *heterophylla*, p.56.
Basket grass. See *Oplismenus hirtellus*.
Basswood. See *Tilia americana*.
Bastard balm. See *Melittis*.
Bat flower. See *Tacca chantrieri*.
Bay laurel. See *Laurus nobilis*.
Bay tree. See *Laurus*.
Bay willow. See *Salix pentandra*.
Bayonet plant. See *Aciphylla squarrosa*, p.230.
Beach pine. See *Pinus contorta*, p.81.
Bead plant. See *Nertera granadensis*, p.335.
Bead tree. See *Melia azedarach*, p.50.
Bear grass. See *Dasylirion*.
Bearded bellflower. See *Campanula barbata*, p.304.
Bear's breeches. See *Acanthus*.
Beauty bush. See *Kolkwitzia amabilis*.
Bedstraw. See *Galium*.
Bee balm. See *Monarda didyma*.
Beech. See *Fagus*.
Beech fern. See *Phegopteris connectilis*.
Beefsteak plant. See *Iresine herbstii*.
Bell heather. See *Erica cinerea*.
Belladonna lily. See *Amaryllis belladonna*, p.352.
Bellflower. See *Campanula*.
Bell-flowered cherry. See *Prunus campanulata*.
Bells of Ireland. See *Moluccella laevis*, p.287.
Bellwort. See *Uvularia grandiflora*, p.238.
Benjamin. See *Lindera benzoin*, p.101.
Bentham's cornel. See *Cornus capitata*.
Bergamot. See *Monarda*.
Berlin poplar. See *Populus* x *berolinensis*.
Bermuda lily. See *Lilium longiflorum*, p.348.
Berry bladder fern. See *Cystopteris bulbifera*.
Besom heath. See *Erica scoparia*.
Betony. See *Stachys officinalis*.
Bhutan pine. See *Pinus wallichiana*, p.77.
Big tree. See *Sequoiadendron giganteum*, p.76.
Big-cone pine. See *Pinus coulteri*, p.76.
Bilberry. See *Vaccinium myrtillus*.
Biota. See *Thuja orientalis*.
Birch. See *Betula*.
Bird cherry. See *Prunus padus*, p.50.
Bird-catcher tree. See *Pisonia umbellifera*.
Bird-of-paradise flower. See *Strelitzia*.
Bird's-eye primrose. See *Primula farinosa*, p.236.
Bird's-foot ivy. See *Hedera helix* 'Pedata', p.181.
Bird's-foot violet. See *Viola pedata*, p.317.
Bird's-nest bromeliad. See *Nidularium innocentii*.
Bird's-nest fern. See *Asplenium nidus*, p.189.
Birthroot. See *Trillium erectum*, p.233.
Birthwort. See *Aristolochia*.
Bishop pine. See *Pinus muricata*, p.77.
Bishop's cap. See *Astrophytum myriostigma*, p.402.
Bishop's mitre. See *Astrophytum myriostigma*, p.402.
Bishop's weed. See *Aegopodium*.
Bistort. See *Polygonum bistorta*.
Biting stonecrop. See *Sedum acre*, p.332.
Bitter cress. See *Cardamine*.
Bitter root. See *Lewisia rediviva*, pp.321 and 327.
Bitternut. See *Carya cordiformis*.
Bitternut hickory. See *Carya cordiformis*.
Black alder. See *Alnus glutinosa*.
Black bamboo. See *Phyllostachys nigra*.
Black bean. See *Kennedia nigricans*.
Black bean tree. See *Castanospermum*.
Black cherry. See *Prunus serotina*, p.39.
Black chokeberry. See *Aronia melanocarpa*, p.106.

INDEX OF COMMON NAMES – C

Black cottonwood. See *Populus trichocarpa*.
Black false hellebore. See *Veratrum nigrum*, p.192.
Black gum. See *Nyssa sylvatica*, p.45.
Black huckleberry. See *Gaylussacia baccata*.
Black Jack oak. See *Quercus marilandica*, p.54.
Black mulberry. See *Morus nigra*.
Black oak. See *Quercus velutina*.
Black pine. See *Pinus jeffreyi*, p.77; *Pinus nigra*.
Black poplar. See *Populus nigra*.
Black sarana. See *Fritillaria camschatcensis*, p.357.
Black spruce. See *Picea mariana*.
Black tree fern. See *Cyathea medullaris*.
Black walnut. See *Juglans nigra*, p.41.
Black willow. See *Salix* 'Melanostachys'.
Blackberry. See *Rubus*.
Blackboy. See *Xanthorrhoea*.
Black-eyed Susan. See *Rudbeckia fulgida*; *Thunbergia alata*, p.176.
Blackthorn. See *Prunus spinosa*.
Bladder cherry. See *Physalis alkekengi*.
Bladder nut. See *Staphylea*.
Bladder senna. See *Colutea arborescens*, p.116.
Blanket flower. See *Gaillardia*.
Bleeding heart. See *Dicentra spectabilis*, p.209.
Blessed Mary's thistle. See *Silybum marianum*, p.274.
Blood flower. See *Asclepias curassavica*; *Scadoxus multiflorus* subsp. *katherinae*, p.345.
Blood leaf. See *Iresine lindenii*.
Blood lily. See *Haemanthus coccineus*, p.379.
Bloodroot. See *Sanguinaria canadensis*, p.311.
Bloody cranesbill. See *Geranium sanguineum*, p.302.
Blue amaryllis. See *Worsleya*.
Blue Atlas cedar. See *Cedrus atlantica* f. *glauca*, p.75.
Blue candle. See *Myrtillocactus geometrizans*, p.392.
Blue cupidone. See *Catananche*.
Blue Douglas fir. See *Pseudotsuga menziesii* subsp. *glauca*, p.76.
Blue fescue. See *Festuca glauca*.
Blue flag. See *Iris versicolor*, p.198.
Blue grama. See *Bouteloua gracilis*, p.183.
Blue gum. See *Eucalyptus globulus*.
Blue holly. See *Ilex* x *meserveae*.
Blue lace flower. See *Trachymene coerulea*.
Blue marguerite. See *Felicia amelloides*.
Blue oat grass. See *Helictotrichon sempervirens*, p.183.
Blue passion flower. See *Passiflora caerulea*, p.174.
Blue poppy. See *Meconopsis*.
Blue potato bush. See *Solanum rantonnetii*.
Blue star. See *Amsonia*.
Blue trumpet vine. See *Thunbergia grandiflora*.
Bluebell. See *Hyacinthoides*.
Bluebell creeper. See *Sollya*.
Blueberry ash. See *Elaeocarpus cynaneus*.
Bluebottle. See *Centaurea cyanus*.
Blue-flowered torch. See *Tillandsia lindenii*, p.229.
Blushing bromeliad. See *Neoregelia carolinae*; *Nidularium fulgens*.
Blushing philodendron. See *Philodendron erubescens*.
Bo. See *Ficus religiosa*.
Boat lily. See *Tradescantia spathacea*.
Bog arum. See *Calla palustris*, p.386.
Bog bean. See *Menyanthes trifoliata*, p.386.
Bog pimpernel. See *Anagallis tenella*.
Bog sage. See *Salvia uliginosa*.
Bonin Isles juniper. See *Juniperus procumbens*, p.84.
Borage. See *Borago*.
Bosnian pine. See *Pinus heldreichii* var. *leucodermis*, p.78.
Boston ivy. See *Parthenocissus tricuspidata*, p.178.
Bottle plant. See *Hatiora salicornioides*, p.404.
Bottlebrush. See *Callistemon*.
Bottlebrush buckeye. See *Aesculus parviflora*, p.89.
Bower vine. See *Pandorea jasminoides*, p.168.
Bowles' golden sedge. See *Carex elata* 'Aurea', p.185.
Box. See *Buxus*.
Box elder. See *Acer negundo*.
Box-leaved holly. See *Ilex crenata*.
Bracelet honey myrtle. See *Melaleuca armillaris*.
Bramble. See *Rubus*.
Branched bur reed. See *Sparganium erectum*, p.389.
Brandy bottle. See *Nuphar lutea*, p.391.
Brass buttons. See *Cotula coronopifolia*.
Brazilian firecracker. See *Manettia luteo-rubra*, p.166.
Breath of heaven. See *Diosma ericoides*.

Brewer's spruce. See *Picea breweriana*, p.79.
Bridal bouquet. See *Porana paniculata*.
Bridal wreath. See *Francoa appendiculata*; *Spiraea* 'Arguta'; *Spiraea* x *vanhouttei*, p.124.
Brisbane box. See *Tristania conferta*.
Bristle club-rush. See *Scirpus setaceus*.
Bristle-cone pine. See *Pinus aristata*, p.82.
Bristle-pointed iris. See *Iris setosa*, p.199.
Brittle bladder fern. See *Cystopteris fragilis*.
Broad beech fern. See *Phegopteris hexagonoptera*.
Broad buckler fern. See *Dryopteris dilatata*.
Broadleaf. See *Griselinia littoralis*.
Broad-leaved lime. See *Tilia platyphyllos*.
Brompton stock. See *Matthiola incana*.
Bronvaux medlar. See + *Crataegomespilus dardarii*.
Broom. See *Cytisus*; *Genista*.
Brush-box tree. See *Tristania conferta*.
Buckbean. See *Menyanthes trifoliata*, p.386.
Buckeye. See *Aesculus*.
Buckthorn. See *Rhamnus*.
Buffalo berry. See *Shepherdia argentea*.
Buffalo currant. See *Ribes odoratum*.
Bugbane. See *Cimicifuga*.
Bull bay. See *Magnolia grandiflora*.
Bullock's heart. See *Annona reticulata*.
Bunch-flowered daffodil. See *Narcissus tazetta*.
Bunnies' ears. See *Stachys byzantina*, p.268.
Bunny ears. See *Opuntia microdasys*.
Bur oak. See *Quercus macrocarpa*, p.54.
Bur reed. See *Sparganium*.
Burnet. See *Sanguisorba*.
Burnet rose. See *Rosa pimpinellifolia*.
Burning bush. See *Dictamnus albus*, p.204; *Kochia scoparia* f. *trichophylla*, p.287.
Burro's tail. See *Sedum morganianum*.
Bush groundsel. See *Baccharis halimifolia*.
Bush violet. See *Browallia speciosa*, p.230.
Busy lizzie. See *Impatiens* cvs.
Butcher's broom. See *Ruscus aculeatus*.
Butter tree. See *Tylecodon paniculatus*.
Buttercup. See *Ranunculus*.
Butterfly bush. See *Buddleja davidii*.
Butterfly flower. See *Schizanthus*.
Butterfly orchid. See *Psychopsis papilio*, p.263.
Butterfly weed. See *Asclepias tuberosa*, p.222.
Butternut. See *Juglans cinerea*.
Button-fern. See *Pellaea rotundifolia*.

C

Calamondin. See x *Citrofortunella microcarpa*, p.121.
Calico bush. See *Kalmia latifolia*, p.111.
Calico flower. See *Aristolochia littoralis*, p.171.
California allspice. See *Calycanthus occidentalis*, p.112.
California bluebell. See *Phacelia campanularia*, p.286.
California buckeye. See *Aesculus californica*, p.59.
California nutmeg. See *Torreya californica*, p.81.
Californian laurel. See *Umbellularia californica*, p.48.
Californian live oak. See *Quercus agrifolia*, p.59.
Californian pepper-tree. See *Schinus molle*.
Californian poppy. See *Romneya*.
Callery pear. See *Pyrus calleryana*.
Campernelle jonquil. See *Narcissus* x *odorus*.
Campion. See *Silene*.
Canada hemlock. See *Tsuga canadensis*, p.81.
Canada lily. See *Lilium canadense*.
Canada moonseed. See *Menispermum canadense*.
Canada wild rice. See *Zizania aquatica*.
Canadian burnet. See *Sanguisorba canadensis*, p.190.
Canadian columbine. See *Aquilegia canadensis*.
Canadian poplar. See *Populus* x *canadensis*.
Canary creeper. See *Tropaeolum peregrinum*.
Canary Island bellflower. See *Canarina canariensis*, p.179.
Canary Island date palm. See *Phoenix canariensis*.
Canary Island ivy. See *Hedera canariensis*.
Canary-bird bush. See *Crotalaria agatiflora*, p.93.
Candle plant. See *Senecio articulatus*.
Candytuft. See *Iberis*.
Canoe birch. See *Betula papyrifera*, p.46.
Canterbury bell. See *Campanula medium*.

Capa de oro. See *Solandra maxima*, p.166.
Cape blue water lily. See *Nymphaea capensis*.
Cape chestnut. See *Calodendrum capense*.
Cape dandelion. See *Arctotheca calendula*, p.258.
Cape gooseberry. See *Physalis peruviana*.
Cape grape. See *Rhoicissus capensis*.
Cape honeysuckle. See *Tecomaria capensis*.
Cape jasmine. See *Gardenia augusta*.
Cape leadwort. See *Plumbago auriculata*, p.175.
Cape marigold. See *Dimorphotheca*.
Cape myrtle. See *Myrsine africana*.
Cape pondweed. See *Aponogeton distachyos*, p.387.
Cape primrose. See *Streptocarpus rexii*.
Cape sundew. See *Drosera capensis*, p.269.
Cappadocian maple. See *Acer cappadocicum*.
Caraway thyme. See *Thymus herba-barona*, p.329.
Cardinal climber. See *Ipomoea* x *multifida*.
Cardinal flower. See *Lobelia cardinalis*.
Cardinal's guard. See *Pachystachys coccinea*.
Cardoon. See *Cynara cardunculus*, p.192.
Caricature plant. See *Graptophyllum pictum*.
Carnation. See *Dianthus*.
Carolina allspice. See *Calycanthus floridus*.
Carolina hemlock. See *Tsuga caroliniana*.
Carolina jasmine. See *Gelsemium sempervirens*.
Caspian locust. See *Gleditsia caspica*.
Cast-iron plant. See *Aspidistra elatior*.
Castor-oil plant. See *Ricinus communis*, p.287.
Catalina ironwood. See *Lyonothamnus floribundus*.
Catchfly. See *Silene*.
Catmint. See *Nepeta*.
Cat's claw. See *Macfadyena unguis-cati*, p.176.
Cat's ears. See *Antennaria*; *Calochortus*.
Cat's valerian. See *Valeriana officinalis*, p.205.
Cat's whiskers. See *Tacca chantrieri*.
Caucasian elm. See *Zelkova carpinifolia*.
Caucasian fir. See *Abies nordmanniana* 'Golden Spreader', p.85.
Caucasian lime. See *Tilia* 'Euchlora'.
Caucasian oak. See *Quercus macranthera*, p.40.
Caucasian spruce. See *Picea orientalis*.
Caucasian whortleberry. See *Vaccinium arctostaphylos*.
Caucasian wing nut. See *Pterocarya fraxinifolia*.
Cedar. See *Cedrus*.
Cedar of Goa. See *Cupressus lusitanica*.
Cedar of Lebanon. See *Cedrus libani*, p.77.
Celandine. See *Chelidonium*.
Celandine crocus. See *Crocus korolkowii*.
Century plant. See *Agave americana*.
Chain cactus. See *Rhipsalis paradoxa*.
Chain fern. See *Woodwardia radicans*.
Chamomile. See *Anthemis*; *Chamaemelum nobile*.
Channelled heath. See *Erica canaliculata*, p.148.
Chaste tree. See *Vitex agnus-castus*.
Chatham Island forget-me-not. See *Myosotidium*.
Cheddar pink. See *Dianthus gratianopolitanus*, p.325.
Cherimoya. See *Annona*.
Cherry. See *Prunus*.
Cherry laurel. See *Prunus laurocerasus*.
Cherry plum. See *Prunus cerasifera*.
Chestnut. See *Castanea*.
Chestnut vine. See *Tetrastigma voinierianum*, p.180.
Chicory. See *Cichorium*.
Chile pine. See *Araucaria araucana*, p.77.
Chilean bamboo. See *Chusquea culeou*, p.184.
Chilean bellflower. See *Lapageria rosea*, p.170.
Chilean blue crocus. See *Tecophilaea cyanocrocus*, p.377.
Chilean firebush. See *Embothrium coccineum*, p.67.
Chilean glory flower. See *Eccremocarpus scaber*, p.177.
Chilean hazel. See *Gevuina avellana*.
Chilean incense cedar. See *Austrocedrus chilensis*, p.80.
Chilean jasmine. See *Mandevilla laxa*.
Chilean laurel. See *Laurelia sempervirens*.
Chilean wine palm. See *Jubaea chilensis*, p.59.
Chimney bellflower. See *Campanula pyramidalis*.
China aster. See *Callistephus*.
Chincherinchee. See *Ornithogalum thyrsoides*, p.364.
Chinese anisatum. See *Illicium anisatum*.
Chinese anise. See *Illicium anisatum*.
Chinese arbor-vitae. See *Thuja orientalis*.
Chinese box thorn. See *Lycium barbarum*.
Chinese elm. See *Ulmus parvifolia*.

INDEX OF COMMON NAMES – D

Chinese evergreen. See *Aglaonema*.
Chinese fan palm. See *Livistona chinensis*, p.59.
Chinese fir. See *Cunninghamia lanceolata*, p.80.
Chinese fountain grass. See *Pennisetum alopecuroides*.
Chinese fountain palm. See *Livistona chinensis*, p.59.
Chinese fringe tree. See *Chionanthus retusus*.
Chinese gooseberry. See *Actinidia chinensis*.
Chinese hat plant. See *Holmskioldia sanguinea*.
Chinese horse-chestnut. See *Aesculus chinensis*, p.38.
Chinese juniper. See *Juniperus chinensis*.
Chinese lantern. See *Physalis*.
Chinese necklace poplar. See *Populus lasiocarpa*.
Chinese parasol tree. See *Firmiana simplex*, p.42.
Chinese persimmon. See *Diospyros kaki*.
Chinese privet. See *Ligustrum lucidum*.
Chinese tallow tree. See *Sapium sebiferum*.
Chinese thuja. See *Thuja orientalis*.
Chinese trumpet creeper. See *Campsis grandiflora*.
Chinese trumpet vine. See *Campsis grandiflora*.
Chinese tulip tree. See *Liriodendron chinense*.
Chinese walnut. See *Juglans cathayensis*.
Chinese wing nut. See *Pterocarya stenoptera*.
Chinese wisteria. See *Wisteria sinensis*, p.175.
Chinese witch hazel. See *Hamamelis mollis*.
Chinese-lantern lily. See *Sandersonia aurantiaca*, p.367.
Chives. See *Allium schoenoprasum*, p.380.
Chocolate cosmos. See *Cosmos atrosanguineus*, p.213.
Chocolate vine. See *Akebia quinata*, p.166.
Chokeberry. See *Aronia*.
Christmas begonia. See *Begonia* x *cheimantha* 'Gloire de Lorraine'.
Christmas berry. See *Heteromeles arbutifolia*.
Christmas box. See *Sarcococca*.
Christmas cactus. See *Schlumbergera* 'Buckleyi'.
Christmas fern. See *Polystichum acrostichoides*.
Christmas rose. See *Helleborus*.
Christ's thorn. See *Paliurus spina-christi*, p.92.
Chusan palm. See *Trachycarpus fortunei*, p.58.
Cider gum. See *Eucalyptus gunnii*, p.46.
Cigar flower. See *Cuphea ignea*, p.142.
Cineraria. See *Senecio* x *hybridus*; *Senecio* 'Spring Glory', p.284.
Cinnamon fern. See *Osmunda cinnamomea*.
Claret ash. See *Fraxinus oxycarpa* 'Raywood'.
Climbing butcher's broom. See *Semele androgyna*.
Climbing dahlia. See *Hidalgoa*.
Climbing fumitory. See *Adlumia fungosa*.
Climbing hydrangea. See *Hydrangea petiolaris*, p.168.
Cloth-of-gold crocus. See *Crocus angustifolius*.
Cloud grass. See *Aichryson* x *domesticum* 'Variegatum', p.413.
Clover. See *Trifolium*.
Club-rush. See *Scirpus*.
Cluster pine. See *Pinus pinaster*, p.77.
Clustered ivy. See *Hedera helix* 'Conglomerata'.
Coast redwood. See *Sequoia sempervirens*.
Coastal redwood. See *Sequoia sempervirens*.
Cobnut. See *Corylus avellana*.
Cobweb houseleek. See *Sempervivum arachnoideum*, p.337.
Cock's foot. See *Dactylis glomerata*.
Cockspur coral-tree. See *Erythrina crista-galli*, p.113.
Cockspur thorn. See *Crataegus crus-galli*.
Colombian ball cactus. See *Wigginsia vorwerkiana*, p.415.
Colorado spruce. See *Picea pungens*.
Columbine. See *Aquilegia*.
Comfrey. See *Symphytum*.
Common alder. See *Alnus glutinosa*.
Common arrowhead. See *Sagittaria sagittifolia*.
Common ash. See *Fraxinus excelsior*.
Common beech. See *Fagus sylvatica*, p.42.
Common box. See *Buxus sempervirens*.
Common broom. See *Cytisus scoparius*.
Common camassia. See *Camassia quamash*.
Common camellia. See *Camellia japonica*.
Common elder. See *Sambucus nigra*.
Common English ivy. See *Hedera helix*.
Common gardenia. See *Gardenia augusta*.
Common German flag. See *Iris germanica*.
Common hackberry. See *Celtis occidentalis*.
Common hawthorn. See *Crataegus monogyna*.
Common holly. See *Ilex aquifolium*, p.72.

Common honeysuckle. See *Lonicera periclymenum*.
Common hop. See *Humulus lupulus*.
Common hornbeam. See *Carpinus betulus*.
Common houseleek. See *Sempervivum tectorum*, p.337.
Common jasmine. See *Jasminum officinale*.
Common juniper. See *Juniperus communis*.
Common laburnum. See *Laburnum anagyroides*.
Common lime. See *Tilia* x *vulgaris*.
Common morning glory. See *Ipomoea purpurea*.
Common myrtle. See *Myrtus communis*, p.100.
Common oak. See *Quercus robur*.
Common passion flower. See *Passiflora caerulea*, p.174.
Common pear. See *Pyrus communis*.
Common pitcher plant. See *Sarracenia purpurea*.
Common polypody. See *Polypodium vulgare*, p.189.
Common quaking grass. See *Briza media*.
Common rue. See *Ruta graveolens*.
Common snowdrop. See *Galanthus nivalis*.
Common spruce. See *Picea abies*, p.78.
Common stag's-horn fern. See *Platycerium bifurcatum*, p.186.
Common stonecrop. See *Sedum acre*, p.332.
Common valerian. See *Valeriana officinalis*, p.205.
Compass plant. See *Silphium laciniatum*.
Coneflower. See *Echinacea*; *Rudbeckia*.
Confederate jasmine. See *Trachelospermum jasminoides*, p.167.
Confederate rose. See *Hibiscus mutabilis*.
Cootamundra wattle. See *Acacia baileyana*, p.71.
Copey. See *Clusia major*.
Copihue. See *Lapageria rosea*, p.170.
Copper beech. See *Fagus sylvatica* f. *purpurea*, p.39.
Copperleaf. See *Acalypha wilkesiana*, p.113.
Coquito. See *Jubaea chilensis*, p.59.
Coral berry. See *Aechmea fulgens*.
Coral bush. See *Templetonia retusa*.
Coral cactus. See *Rhipsalis cereuscula*, p.397.
Coral drops. See *Bessera elegans*.
Coral gem. See *Lotus berthelotii*, p.247.
Coral honeysuckle. See *Lonicera sempervirens*, p.170.
Coral pea. See *Hardenbergia violacea*.
Coral plant. See *Berberidopsis corallina*, p.171; *Russelia equisetiformis*, p.214.
Coral vine. See *Antigonon*.
Coral-bark maple. See *Acer palmatum* 'Senkaki', p.94.
Coralberry. See *Ardisia crenata*, p.120; *Symphoricarpos orbiculatus*.
Cork oak. See *Quercus suber*, p.47.
Corkscrew rush. See *Juncus effusus* f. *spiralis*, p.184.
Corn cockle. See *Agrostemma*.
Corn plant. See *Dracaena fragrans*.
Corn poppy. See *Papaver rhoeas*.
Cornelian cherry. See *Cornus mas*.
Cornflower. See *Centaurea cyanus*.
Cornish elm. See *Ulmus angustifolia* var. *cornubiensis*.
Cornish golden elm. See *Ulmus* 'Dicksonii', p.55.
Cornish heath. See *Erica vagans*.
Corsican heath. See *Erica terminalis*.
Corsican mint. See *Mentha requienii*.
Corsican pine. See *Pinus nigra* var. *maritima*.
Cotton ball. See *Espostoa lanata*, p.393.
Cotton lavender. See *Santolina chamaecyparissus*.
Cotton rose. See *Hibiscus mutabilis*.
Cotton thistle. See *Onopordum acanthium*, p.274.
Cottonwood. See *Populus deltoides*.
Coulter pine. See *Pinus coulteri*, p.76.
Cowslip. See *Primula veris*, p.237.
Cow's-tail pine. See *Cephalotaxus harringtonia*.
Crab apple. See *Malus*.
Crab cactus. See *Schlumbergera truncata*, p.409.
Crack willow. See *Salix fragilis*.
Cradle orchid. See *Anguloa clowesii*.
Cranesbill. See *Geranium*.
Crape myrtle. See *Lagerstroemia indica*, p.65.
Cream cups. See *Platystemon californicus*, p.287.
Creeping blue blossom. See *Ceanothus thyrsiflorus* var. *repens*, p.138.
Creeping bluets. See *Hedyotis michauxii*, p.331.
Creeping Charlie. See *Pilea nummulariifolia*, p.269.
Creeping dogwood. See *Cornus canadensis*, p.322.
Creeping fig. See *Ficus pumila*.

Creeping Jenny. See *Lysimachia nummularia* 'Aurea', p.334.
Creeping juniper. See *Juniperus horizontalis*.
Creeping phlox. See *Phlox stolonifera*.
Creeping soft grass. See *Holcus mollis*.
Creeping willow. See *Salix repens*, p.126.
Creeping zinnia. See *Sanvitalia procumbens*, p.288.
Cretan brake. See *Pteris cretica*, p.187.
Cretan dittany. See *Origanum dictamnus*.
Cricket-bat willow. See *Salix alba* 'Caerulea'.
Crimean lime. See *Tilia* 'Euchlora'.
Crimson glory vine. See *Vitis coignetiae*, p.178.
Cross vine. See *Bignonia capreolata*.
Cross-leaved heath. See *Erica tetralix*.
Croton. See *Codiaeum variegatum*, p.147.
Crown imperial. See *Fritillaria imperialis*, p.340.
Crown of thorns. See *Euphorbia milii*, p.125.
Cruel plant. See *Araujia sericofera*, p.167.
Crusaders' spears. See *Urginea maritima*.
Crystal anthurium. See *Anthurium crystallinum*, p.228.
Cuban royal palm. See *Roystonea regia*.
Cuckoo flower. See *Cardamine pratensis*.
Cuckoo pint. See *Arum*.
Cucumber tree. See *Magnolia acuminata*.
Cup-and-saucer vine. See *Cobaea scandens*, p.174.
Cupid peperomia. See *Peperomia scandens*.
Curled pondweed. See *Potamogeton crispus*, p.388.
Currant. See *Ribes*.
Curry plant. See *Helichrysum italicum*.
Custard apple. See *Annona*.
Cypress. See *Cupressus*.
Cypress vine. See *Ipomoea quamoclit*, p.170.
Cyprus turpentine. See *Pistacia terebinthus*.

D

Daffodil. See *Narcissus*.
Dahurian juniper. See *Juniperus davurica*.
Daimio oak. See *Quercus dentata*.
Daisy. See *Bellis*.
Daisy bush. See *Olearia*.
Dalmatian iris. See *Iris pallida*.
Dalmatian laburnum. See *Petteria ramentacea*.
Dalmatian toadflax. See *Linaria genistifolia* var. *dalmatica*.
Dame's violet. See *Hesperis matronalis*, p.203.
Dancing-doll orchid. See *Oncidium flexuosum*, p.263.
Dandelion. See *Crepis*.
Darling pea. See *Swainsona galegifolia*.
Darwin's barberry. See *Berberis darwinii*, p.87.
Date plum. See *Diospyros lotus*.
David pine. See *Pinus armandii*.
David's peach. See *Prunus davidiana*.
Dawn redwood. See *Metasequoia glyptostroboides*, p.76.
Day flower. See *Commelina coelestis*, p.286.
Daylily. See *Hemerocallis*.
Deadnettle. See *Lamium*.
Delavay fir. See *Abies delavayi*.
Delta maidenhair. See *Adiantum raddianum*.
Deodar. See *Cedrus deodara*.
Deptford pink. See *Dianthus armeria*.
Desert fan palm. See *Washingtonia filifera*.
Desert privet. See *Peperomia magnoliifolia*.
Desert rose. See *Adenium*.
Devil's club. See *Oplopanax horridus*.
Devil's fig. See *Argemone mexicana*, p.287.
Dickson's golden elm. See *Ulmus* 'Dicksonii', p.55.
Digger speedwell. See *Parahebe perfoliata*, p.250.
Disk-leaved hebe. See *Hebe pinguifolia*.
Dittany. See *Origanum*.
Dog fennel. See *Anthemis*.
Dog's-tooth violet. See *Erythronium dens-canis*, p.374.
Dogwood. See *Cornus*.
Doll's eyes. See *Actaea alba*, p.223.
Donkey-tail. See *Sedum morganianum*.
Dorset heath. See *Erica ciliaris*.
Double common snowdrop. See *Galanthus nivalis* 'Flore Pleno', p.383.
Double daisy. See *Bellis perennis*.
Double meadow buttercup. See *Ranunculus acris* 'Flore Pleno', p.256.

INDEX OF COMMON NAMES – E

Double sea campion. See *Silene uniflora* 'Flore Pleno'.
Double soapwort. See *Saponaria officinalis* 'Rubra Plena'.
Douglas fir. See *Pseudotsuga menziesii*.
Dove tree. See *Davidia involucrata*, p.51.
Downy cherry. See *Prunus tomentosa*.
Dragon arum. See *Dracunculus vulgaris*, p.345.
Dragon tree. See *Dracaena draco*, p.74.
Dragon's head. See *Dracocephalum*.
Dragon's mouth. See *Horminum pyrenaicum*.
Dragon's-claw willow. See *Salix matsudana* 'Tortuosa', p.59.
Drooping juniper. See *Juniperus recurva*.
Drooping star-of-Bethlehem. See *Ornithogalum nutans*.
Dropwort. See *Filipendula vulgaris*.
Drumstick primula. See *Primula denticulata*, p.237.
Drunkard's dream. See *Hatiora salicornioides*, p.404.
Duck potato. See *Sagittaria latifolia*, p.386.
Duckweed. See *Wolffia*.
Duke of Argyll's tea-tree. See *Lycium barbarum*.
Dumb cane. See *Dieffenbachia*.
Dumpling cactus. See *Lophophora williamsii*, p.407.
Durmast oak. See *Quercus petraea*.
Dusky coral pea. See *Kennedia rubicunda*, p.165.
Dutch crocus. See *Crocus vernus*, p.372.
Dutch elm. See *Ulmus* 'Hollandica'.
Dutchman's breeches. See *Dicentra cucullaria*, p.310; *Dicentra spectabilis*, p.209.
Dwarf bearded iris. See *Iris pumila*.
Dwarf fan palm. See *Chamaerops humilis*, p.146.
Dwarf mountain palm. See *Chamaedorea elegans*, p.122.
Dwarf palmetto. See *Sabal minor*, p.146.
Dwarf pine. See *Pinus mugo*.
Dwarf pomegranate. See *Punica granatum* var. *nana*, p.303.
Dwarf Siberian pine. See *Pinus pumila*.
Dwarf sumach. See *Rhus copallina*.
Dwarf white wood lily. See *Trillium nivale*.
Dwarf white-stripe bamboo. See *Pleioblastus variegatus*, p.182.
Dwarf willow. See *Salix herbacea*.
Dyers' greenweed. See *Genista tinctoria*, p.127.

E

East Indian arrowroot. See *Tacca leontopetaloides*.
Easter cactus. See *Rhipsalidopsis gaertneri*, p.411.
Easter lily. See *Lilium longiflorum*, p.348.
Eastern cottonwood. See *Populus deltoides*.
Eastern hemlock. See *Tsuga canadensis*, p.81.
Eastern redbud. See *Cercis canadensis*.
Eastern white cedar. See *Thuja occidentalis*.
Eastern white pine. See *Pinus strobus*, p.76.
Eau-de-Cologne mint. See *Mentha* x *piperita* 'Citrata'.
Edelweiss. See *Leontopodium*.
Edible prickly pear. See *Opuntia ficus-indica*.
Eel grass. See *Vallisneria spiralis*.
Eglantine. See *Rosa eglanteria*, p.151.
Egyptian star. See *Pentas lanceolata*, p.133.
Elder. See *Sambucus*.
Elephant apple. See *Dillenia indica*.
Elephant bush. See *Portulacaria afra*, p.123.
Elephant's ears. See *Hedera colchica* 'Dentata', p.181; *Philodendron domesticum*.
Elephant's foot. See *Beaucarnea recurvata*, p.74; *Dioscorea elephantipes*, p.403.
Elm. See *Ulmus*.
Emerald ripple. See *Peperomia caperata*, p.264.
Engelmann spruce. See *Picea engelmannii*, p.79.
English bluebell. See *Hyacinthoides non-scripta*, p.358.
English elm. See *Ulmus procera*.
English iris. See *Iris latifolia*, p.198.
Epaulette tree. See *Pterostyrax hispida*.
Etruscan honeysuckle. See *Lonicera etrusca*.
European fan palm. See *Chamaerops humilis*, p.146.
European larch. See *Larix decidua*.
European white lime. See *Tilia tomentosa*.
Evening primrose. See *Oenothera*.
Everlasting flower. See *Helichrysum bracteatum*.
Everlasting pea. See *Lathyrus grandiflorus*, p.169; *Lathyrus latifolius*, p.171; *Lathyrus sylvestris*.
Exeter elm. See *Ulmus glabra* 'Exoniensis'.
Eyelash begonia. See *Begonia bowerae*, p.259.

F

Fair maids of France. See *Saxifraga granulata*, p.294.
Fairy bells. See *Disporum*.
Fairy duster. See *Calliandra eriophylla*, p.142.
Fairy foxglove. See *Erinus*.
Fairy lantern. See *Calochortus*.
Fairy moss. See *Azolla caroliniana*, p.388.
Fairy thimbles. See *Campanula cochleariifolia*, p.331.
False acacia. See *Robinia pseudoacacia*.
False African violet. See *Streptocarpus saxorum*, p.248.
False anemone. See *Anemonopsis*.
False aralia. See *Schefflera elegantissima*, p.97.
False cypress. See *Chamaecyparis*.
False heather. See *Cuphea hyssopifolia*, p.128.
False indigo. See *Baptisia australis*, p.216.
False jasmine. See *Gelsemium sempervirens*.
False oat grass. See *Arrhenatherum elatius*.
False spikenard. See *Smilacina racemosa*, p.197.
Fanwort. See *Cabomba caroliniana*.
Fat pork tree. See *Clusia major*.
Feather grape hyacinth. See *Muscari comosum* 'Plumosum', p.376.
Feather-top. See *Pennisetum villosum*, p.183.
Fennel. See *Foeniculum vulgare*.
Fern-leaf aralia. See *Polyscias filicifolia*, p.122.
Fetterbush. See *Pieris floribunda*, p.97.
Feverfew. See *Tanacetum parthenium*, p.271.
Fiddle-leaf fig. See *Ficus lyrata*.
Field poppy. See *Papaver rhoeas*.
Figwort. See *Scrophularia*.
Filbert. See *Corylus maxima*.
Finger-leaved ivy. See *Hedera helix* 'Digitata', p.181.
Fire lily. See *Lilium bulbiferum*.
Firecracker vine. See *Manettia cordifolia*.
Firethorn. See *Pyracantha*.
Fish grass. See *Cabomba caroliniana*.
Fishtail fern. See *Cyrtomium falcatum*, p.187.
Fishbone cactus. See *Epiphyllum anguliger*, p.397.
Fishpole bamboo. See *Phyllostachys aurea*.
Five fingers. See *Pseudopanax arboreus*; *Syngonium auritum*.
Five-leaved ivy. See *Parthenocissus quinquefolia*.
Five-spot baby. See *Nemophila maculata*, p.271.
Flaky juniper. See *Juniperus squamata*.
Flame azalea. See *Rhododendron calendulaceum*.
Flame coral tree. See *Erythrina americana*.
Flame creeper. See *Tropaeolum speciosum*, p.170.
Flame flower. See *Pyrostegia venusta*, p.178.
Flame nasturtium. See *Tropaeolum speciosum*, p.170.
Flame-of-the-forest. See *Spathodea campanulata*, p.45.
Flame vine. See *Pyrostegia venusta*, p.178.
Flame violet. See *Episcia cupreata*, p.265.
Flaming Katy. See *Kalanchoe blossfeldiana*.
Flaming sword. See *Vriesea splendens*, p.229.
Flamingo flower. See *Anthurium scherzerianum*, p.266.
Flannel flower. See *Fremontodendron*.
Flax lily. See *Dianella*.
Fleabane. See *Erigeron*.
Floating water plantain. See *Luronium natans*.
Floss flower. See *Ageratum*.
Floss silk tree. See *Chorisia speciosa*, p.44.
Flower-of-the-hour. See *Hibiscus trionum*, p.272.
Flowering banana. See *Musa ornata*, p.196.
Flowering currant. See *Ribes sanguineum*.
Flowering dogwood. See *Cornus florida*.
Flowering gum. See *Eucalyptus ficifolia*.
Flowering quince. See *Chaenomeles*.
Flowering raspberry. See *Rubus odoratus*.
Flowering rush. See *Butomus umbellatus*, p.388.
Flowering tobacco. See *Nicotiana sylvestris*, p.190.
Fly honeysuckle. See *Lonicera xylosteum*, p.110.
Foam of May. See *Spiraea* 'Arguta'.
Foamflower. See *Tiarella*.
Forget-me-not. See *Myosotis*.
Fountain flower. See *Ceropegia sandersoniae*.
Four o'clock flower. See *Mirabilis*.
Foxglove. See *Digitalis*.
Foxglove tree. See *Paulownia tomentosa*, p.51.
Foxtail barley. See *Hordeum jubatum*, p.183.
Foxtail fern. See *Asparagus densiflorus* 'Myersii', p.231.
Foxtail lily. See *Eremurus*.
Foxtail millet. See *Setaria italica*.
Fragrant olive. See *Osmanthus fragrans*.
Fragrant snowbell. See *Styrax obassia*.
Frangipani. See *Plumeria*.
Freckle face. See *Hypoestes phyllostachya*, p.228.
French honeysuckle. See *Hedysarum coronarium*, p.214.
French lavender. See *Lavandula dentata*; *Lavandula stoechas*, p.137.
French marigold. See *Tagetes patula*.
Friendship plant. See *Pilea involucrata*.
Friendship tree. See *Crassula ovata*, p.393.
Fringe tree. See *Chionanthus virginicus*, p.89.
Fringecups. See *Tellima grandiflora*.
Fringed water lily. See *Nymphoides peltata*, p.391.
Frogbit. See *Hydrocharis morsus-ranae*, p.386.
Fuchsia begonia. See *Begonia fuchsioides*.
Fuchsia-flowered currant. See *Ribes speciosum*.
Fuji cherry. See *Prunus incisa*, p.60.
Full-moon maple. See *Acer japonicum*.
Fulvous daylily. See *Hemerocallis fulva*.
Funkia. See *Hosta*.

G

Galingale. See *Cyperus longus*.
Gandergoose. See *Orchis morio*, p.262.
Garden loosestrife. See *Lysimachia punctata*, p.219.
Garden nasturtium. See *Tropaeolum majus*.
Gardener's garters. See *Phalaris arundinacea* 'Picta', p.182.
Garland flower. See *Hedychium*.
Gay feathers. See *Liatris*.
Gean. See *Prunus avium*, p.45.
Gentian. See *Gentiana*.
Geraldton waxflower. See *Chamelaucium uncinatum*, pp.119 and 120.
Geranium. See *Pelargonium*.
German ivy. See *Senecio mikanioides*.
Ghost tree. See *Davidia involucrata*, p.51.
Giant Burmese honeysuckle. See *Lonicera hildebrandiana*.
Giant buttercup. See *Ranunculus lyallii*.
Giant cowslip. See *Primula florindae*, p.237.
Giant elephant's ear. See *Alocasia macrorrhiza*.
Giant fennel. See *Ferula*.
Giant fir. See *Abies grandis*, p.78.
Giant granadilla. See *Passiflora quadrangularis*, p.175.
Giant holly fern. See *Polystichum munitum*, p.186.
Giant larkspur. See *Consolida ambigua*, Imperial Series, p.284.
Giant lily. See *Cardiocrinum giganteum*, p.342.
Giant pineapple flower. See *Eucomis pallidiflora*, p.341.
Giant pineapple lily. See *Eucomis pallidiflora*, p.341.
Giant redwood. See *Sequoiadendron giganteum*, p.76.
Giant reed. See *Arundo donax*.
Giant scabious. See *Cephalaria gigantea*.
Giant Spaniard. See *Aciphylla scott-thomsonii*.
Giant wood fern. See *Dryopteris goldiana*.
Ginger lily. See *Hedychium*.
Gingham golf ball. See *Euphorbia obesa*, p.413.
Gippsland fountain palm. See *Livistona australis*.
Gladwin. See *Iris foetidissima*.
Gland bellflower. See *Adenophora*.
Glastonbury thorn. See *Crataegus monogyna* 'Biflora'.
Globe amaranth. See *Gomphrena globosa*, p.283.
Globe lily. See *Calochortus albus*, p.353.
Globe thistle. See *Echinops*.
Globeflower. See *Trollius*.
Glory bush. See *Tibouchina urvilleana*, p.92.
Glory lily. See *Gloriosa superba*.
Glory vine. See *Eccremocarpus scaber*, p.177.

INDEX OF COMMON NAMES – J

Glory-of-the-snow. See *Chionodoxa*.
Glory-of-the-sun. See *Leucocoryne ixioides*, p.358.
Gloxinia. See *Sinningia speciosa*.
Goat willow. See *Salix caprea*.
Goat's beard. See *Aruncus dioicus*, p.190.
Goat's rue. See *Galega*.
Gold dust. See *Aurinia saxatilis*, p.298.
Golden arum lily. See *Zantedeschia elliottiana*, p.347.
Golden ball cactus. See *Notocactus leninghausii*, p.404.
Golden bamboo. See *Phyllostachys aurea*.
Golden barrel cactus. See *Echinocactus grusonii*, p.396.
Golden chain. See *Laburnum anagyroides*.
Golden club. See *Orontium aquaticum*, p.391.
Golden elder. See *Sambucus nigra* 'Aurea'.
Golden fairy lantern. See *Calochortus amabilis*, p.378.
Golden flax. See *Linum flavum*.
Golden foxtail. See *Alopecurus pratensis* 'Aureovariegatus', p.185.
Golden globe tulip. See *Calochortus amabilis*, p.378.
Golden larch. See *Pseudolarix amabilis*, p.81.
Golden oak of Cyprus. See *Quercus alnifolia*.
Golden oats. See *Stipa gigantea*, p.183.
Golden rod. See *Solidago*.
Golden shower. See *Cassia fistula*; *Pyrostegia venusta*, p.178.
Golden Spaniard. See *Aciphylla aurea*, p.202.
Golden spider lily. See *Lycoris aurea*.
Golden top. See *Lamarckia aurea*.
Golden trumpet. See *Allemanda cathartica*.
Golden trumpet tree. See *Tabebuia chrysotricha*, p.71.
Golden weeping willow. See *Salix* 'Chrysocoma', p.48.
Golden willow. See *Salix alba* var. *vitellina*, p.48.
Golden wonder. See *Cassia didymobotrya*, p.117.
Golden-chalice vine. See *Solandra maxima*, p.166.
Golden-eyed grass. See *Sisyrinchium californicum*.
Golden-feather palm. See *Chrysalidocarpus lutescens*, p.74.
Golden-groove bamboo. See *Phyllostachys aureosulcata*.
Golden-rain tree. See *Koelreuteria paniculata*, p.66.
Golden-rayed lily of Japan. See *Lilium auratum*.
Goldfish plant. See *Columnea gloriosa*.
Goldilocks. See *Aster linosyris*, p.226.
Good-luck plant. See *Cordyline fruticosa*.
Goodyer's elm. See *Ulmus angustifolia*.
Gorgon's head. See *Euphorbia gorgonis*.
Gorse. See *Ulex europaeus*, p.127.
Gout weed. See *Aegopodium*.
Grand fir. See *Abies grandis*, p.78.
Granite bottlebrush. See *Melaleuca elliptica*, p.112.
Granny's bonnets. See *Aquilegia vulgaris*.
Grape hyacinth. See *Muscari*.
Grape ivy. See *Cissus rhombifolia*, p.180.
Grape vine. See *Vitis vinifera*.
Grass of Parnassus. See *Parnassia palustris*, p.298.
Grass tree. See *Xanthorrhoea*.
Grass widow. See *Sisyrinchium douglasii*.
Grass-leaved daylily. See *Hemerocallis minor*.
Great burnet. See *Sanguisorba officinalis*.
Great Solomon's seal. See *Polygonatum biflorum*.
Great white cherry. See *Prunus* 'Taihaku', p.60.
Great yellow gentian. See *Gentiana lutea*, p.219.
Greater celandine. See *Chelidonium*.
Greater periwinkle. See *Vinca major*.
Greater pond sedge. See *Carex riparia*.
Greater quaking grass. See *Briza maxima*.
Greater woodrush. See *Luzula sylvatica*.
Grecian strawberry tree. See *Arbutus andrachne*.
Greek fir. See *Abies cephalonica*.
Green ash. See *Fraxinus pennsylvanica*.
Green earth star. See *Cryptanthus acaulis*.
Green hellebore. See *Helleborus viridis*, p.269.
Green-veined orchid. See *Orchis morio*, p.262.
Grey alder. See *Alnus incana*, p.40.
Grey poplar. See *Populus* x *canescens*, p.39.
Ground elder. See *Aegopodium*.
Ground ivy. See *Glechoma hederacea*.
Guelder rose. See *Viburnum opulus*.
Guernsey lily. See *Nerine sarniensis*, p.368.
Gum tree. See *Eucalyptus*.
Gutta-percha tree. See *Eucommia ulmoides*.

H

Hackberry. See *Celtis*.
Hairy brome grass. See *Bromus ramosus*.
Hard fern. See *Blechnum spicant*.
Hard shield fern. See *Polystichum aculeatum*.
Hardy age. See *Eupatorium rugosum*.
Harebell poppy. See *Meconopsis quintuplinervia*, p.235.
Hare's-foot fern. See *Davallia canariensis*.
Hare's-tail grass. See *Lagurus ovatus*, p.182.
Harlequin flower. See *Sparaxis*.
Hart's-tongue fern. See *Asplenium scolopendrium*, p.189.
Hawkweed. See *Hieracium*.
Hawthorn. See *Crataegus*.
Hawthorn maple. See *Acer crataegifolium*.
Hazel. See *Corylus*.
Headache tree. See *Umbellularia*.
Heart leaf. See *Philodendron cordatum*; *Philodendron scandens*, p.180.
Heart of flame. See *Bromelia balansae*, p.229.
Heart pea. See *Cardiospermum halicacabum*.
Heart seed. See *Cardiospermum halicacabum*.
Heart vine. See *Ceropegia linearis* subsp. *woodii*, p.398.
Hearts-and-honey vine. See *Ipomoea* x *multifida*.
Heartsease. See *Viola tricolor*, p.317.
Heath banksia. See *Banksia ericifolia*.
Heavenly bamboo. See *Nandina domestica*.
Hedge bamboo. See *Bambusa multiplex*, p.183.
Hedgehog broom. See *Erinacea anthyllis*, p.296.
Hedgehog holly. See *Ilex aquifolium* 'Ferox'.
Hedgehog rose. See *Rosa rugosa*, p.153.
Helmet flower. See *Aconitum napellus*.
Hen-and-chicken fern. See *Asplenium bulbiferum*.
Herald's trumpet. See *Beaumontia grandiflora*, p.165.
Herb Paris. See *Paris*.
Heropito. See *Pseudowintera axillaris*.
Herringbone plant. See *Maranta leuconeura* 'Erythroneura', p.267.
Hers's maple. See *Acer davidii* var. *grosseri*.
Hiba. See *Thujopsis dolabrata*.
Hickory. See *Carya*.
Higan cherry. See *Prunus subhirtella*.
Highbush blueberry. See *Vaccinium corymbosum*, p.130.
Hill cherry. See *Prunus serrulata* var. *spontanea*, p.50.
Himalayan birch. See *Betula utilis*.
Himalayan box. See *Buxus wallichiana*.
Himalayan holly. See *Ilex dipyrena*.
Himalayan honeysuckle. See *Leycesteria formosa*.
Himalayan hound's tongue. See *Cynoglossum nervosum*, p.218.
Himalayan lilac. See *Syringa emodi*.
Himalayan May apple. See *Podophyllum hexandrum*, p.232.
Himalayan pine. See *Pinus wallichiana*, p.77.
Himalayan weeping juniper. See *Juniperus recurva*.
Hinoki cypress. See *Chamaecyparis obtusa*.
Hoary willow. See *Salix elaeagnos*.
Holford pine. See *Pinus* x *holfordiana*, p.75.
Holly. See *Ilex*.
Holly fern. See *Cyrtomium falcatum*, p.187.
Holly flame pea. See *Chorizema ilicifolium*, p.127.
Hollyhock. See *Alcea*.
Holm oak. See *Quercus ilex*.
Holy flax. See *Santolina rosmarinifolia*.
Honesty. See *Lunaria*.
Honey locust. See *Gleditsia triacanthos*.
Honeybush. See *Melianthus major*.
Honeysuckle. See *Lonicera*.
Hoop-petticoat daffodil. See *Narcissus bulbocodium* subsp. *bulbocodium*.
Hop. See *Humulus*.
Hop hornbeam. See *Ostrya carpinifolia*.
Hop tree. See *Ptelea trifoliata*.
Hornbeam. See *Carpinus*.
Hornbeam maple. See *Acer carpinifolium*, p.66.
Horned holly. See *Ilex cornuta*.
Horned poppy. See *Glaucium*.
Horned tulip. See *Tulipa acuminata*, p.355.
Horned violet. See *Viola cornuta*, p.297.
Hornwort. See *Ceratophyllum demersum*.
Horse-chestnut. See *Aesculus*.
Horseshoe vetch. See *Hippocrepis comosa*, p.334.
Hottentot fig. See *Carpobrotus edulis*.
Hot-water plant. See *Achimenes*.
Hound's tongue. See *Cynoglossum*.
Houseleek. See *Sempervivum*.
Huckleberry. See *Gaylussacia*.
Humble plant. See *Mimosa pudica*, p.146.
Hunangemoho grass. See *Chionochloa conspicua*.
Hungarian oak. See *Quercus frainetto*, p.43.
Huntingdon elm. See *Ulmus* 'Vegeta'.
Huntsman's cup. See *Sarracenia purpurea*.
Hupeh crab. See *Malus hupehensis*, p.48.
Hupeh rowan. See *Sorbus hupehensis*.
Hyacinth. See *Hyacinthus*.
Hyacinth bean. See *Lablab purpureus*, p.171.
Hyssop. See *Hyssopus officinalis*, p.138.

I

Iceland poppy. See *Papaver nudicaule*.
Ice-plant. See *Dorotheanthus bellidiformis*; *Sedum spectabile*.
Illawarra flame tree. See *Brachychiton acerifolius*, p.39.
Illawarra palm. See *Archontophoenix cunninghamiana*.
Immortelle. See *Helichrysum bracteatum*; *Xeranthemum*.
Incense cedar. See *Calocedrus decurrens*, p.80.
Incense plant. See *Calomeria amaranthoides*, p.282.
Incense rose. See *Rosa primula*, p.154.
India rubber tree. See *Ficus elastica*.
Indian almond. See *Terminalia catappa*.
Indian bean tree. See *Catalpa bignonioides*, p.52.
Indian corn. See *Zea*.
Indian cress. See *Tropaeolum majus*.
Indian currant. See *Symphoricarpos orbiculatus*.
Indian fig. See *Opuntia ficus-indica*.
Indian ginger. See *Alpinia calcarata*.
Indian hawthorn. See *Rhaphiolepis indica*.
Indian horse-chestnut. See *Aesculus indica*.
Indian laburnum. See *Cassia fistula*.
Indian pink. See *Dianthus chinensis*.
Indian plum. See *Oemleria cerasiformis*.
Inkberry. See *Ilex glabra*.
Interrupted fern. See *Osmunda claytoniana*.
Irish ivy. See *Hedera hibernica*, p.181.
Iron cross begonia. See *Begonia masoniana*, p.259.
Ironwood. See *Ostrya virginiana*, p.51.
Italian alder. See *Alnus cordata*, p.40.
Italian buckthorn. See *Rhamnus alaternus*.
Italian cypress. See *Cupressus sempervirens*, p.79.
Italian maple. See *Acer opalus*.
Italian millet. See *Setaria italica*.
Ivy. See *Hedera*.
Ivy of Uruguay. See *Cissus striata*.
Ivy peperomia. See *Peperomia griseoargentea*.
Ivy-leaved toadflax. See *Cymbalaria muralis*.
Ivy-leaved violet. See *Viola hederacea*.

J

Jack pine. See *Pinus banksiana*, p.81.
Jack-in-the-pulpit. See *Arisaema triphyllum*, p.366.
Jacobean lily. See *Sprekelia formosissima*, p.357.
Jacob's coat. See *Acalypha wilkesiana*, p.113.
Jacob's ladder. See *Polemonium*.
Jade tree. See *Crassula ovata*, p.393.
Jade vine. See *Strongylodon macrobotrys*, p.166.
Japan pepper. See *Zanthoxylum piperitum*, p.115.
Japanese anemone. See *Anemone* x *hybrida* 'September Charm', p.195.
Japanese angelica tree. See *Aralia elata*.
Japanese apricot. See *Prunus mume*.
Japanese aralia. See *Fatsia japonica*.
Japanese arrowhead. See *Sagittaria sagittifolia* 'Flore Pleno'.

633

INDEX OF COMMON NAMES – K

Japanese banana. See *Musa basjoo*, p.197.
Japanese big-leaf magnolia. See *Magnolia hypoleuca*, p.49.
Japanese bitter orange. See *Poncirus trifoliata*.
Japanese black pine. See *Pinus thunbergii*, p.80.
Japanese cedar. See *Cryptomeria japonica*.
Japanese climbing fern. See *Lygodium japonicum*.
Japanese flag. See *Iris ensata*.
Japanese hemlock. See *Tsuga diversifolia*; *Tsuga sieboldii*.
Japanese holly. See *Ilex crenata*.
Japanese honeysuckle. See *Lonicera japonica*.
Japanese horse-chestnut. See *Aesculus turbinata*.
Japanese hydrangea vine. See *Schizophragma hydrangeoides*.
Japanese ivy. See *Hedera rhombea*; *Parthenocissus tricuspidata*, p.178.
Japanese larch. See *Larix kaempferi*.
Japanese maple. See *Acer japonicum*; *Acer palmatum* 'Atropurpureum'.
Japanese pittosporum. See *Pittosporum tobira*.
Japanese privet. See *Ligustrum japonicum*.
Japanese quince. See *Chaenomeles japonica*.
Japanese red pine. See *Pinus densiflora*.
Japanese roof iris. See *Iris tectorum*, p.198.
Japanese rose. See *Rosa rugosa*, p.153.
Japanese sago palm. See *Cycas revoluta*, p.122.
Japanese shield fern. See *Dryopteris erythrosora*.
Japanese snowball tree. See *Viburnum plicatum*.
Japanese spindle. See *Euonymus japonicus*.
Japanese umbrella pine. See *Sciadopitys verticillata*, p.80.
Japanese walnut. See *Juglans ailantifolia*.
Japanese white pine. See *Pinus parviflora*, p.79.
Japanese wisteria. See *Wisteria floribunda*.
Japanese witch hazel. See *Hamamelis japonica*.
Japanese yew. See *Taxus cuspidata*, p.83.
Japonica. See *Chaenomeles*.
Jasmine. See *Jasminum*.
Jeffrey pine. See *Pinus jeffreyi*, p.77.
Jelly palm. See *Butia capitata*, p.74.
Jersey elm. See *Ulmus* 'Sarniensis'.
Jerusalem cherry. See *Solanum pseudocapsicum*.
Jerusalem cross. See *Lychnis chalcedonica*, p.222.
Jerusalem sage. See *Phlomis fruticosa*, p.140.
Jerusalem thorn. See *Paliurus spina-christi*, p.92; *Parkinsonia aculeata*.
Jessamine. See *Jasminum officinale*.
Jesuit's nut. See *Trapa natans*, p.389.
Job's tears. See *Coix lacryma-jobi*, p.184.
Joe Pye weed. See *Eupatorium purpureum*, p.195.
Jonquil. See *Narcissus* 'Jumblie', p.362.
Josephine's lily. See *Brunsvigia josephinae*.
Judas tree. See *Cercis*.
Juneberry. See *Amelanchier*.
Juniper. See *Juniperus*.

K

Kaffir fig. See *Carpobrotus edulis*.
Kaffir lily. See *Schizostylis*.
Kaki. See *Diospyros kaki*.
Kangaroo paw. See *Anigozanthos*.
Kangaroo vine. See *Cissus antarctica*, p.180.
Kansas gay feather. See *Liatris pycnostachya*.
Kapok. See *Ceiba pentandra*.
Karo. See *Pittosporum crassifolium*.
Kashmir cypress. See *Cupressus cashmeriana*, p.75.
Katsura. See *Cercidiphyllum japonicum*, p.45.
Kenilworth ivy. See *Cymbalaria muralis*.
Kentucky coffee tree. See *Gymnocladus dioica*.
Kermes oak. See *Quercus coccifera*.
Kilmarnock willow. See *Salix caprea* 'Kilmarnock'.
King anthurium. See *Anthurium veitchii*.
King of the bromeliads. See *Vriesea hieroglyphica*.
King palm. See *Archontophoenix*.
King protea. See *Protea cynaroides*, p.131.
King William pine. See *Athrotaxis selaginoides*.
Kingcup. See *Caltha palustris*, p.391.
Kingfisher daisy. See *Felicia bergeriana*, p.286.
King's crown. See *Justicia carnea*, p.133.
King's spear. See *Eremurus*.
Kiwi fruit. See *Actinidia chinensis*.

Knapweed. See *Centaurea*.
Knotweed. See *Polygonum*.
Korean fir. See *Abies koreana*, p.83.
Korean mountain ash. See *Sorbus alnifolia*.
Korean thuja. See *Thuja koraiensis*.
Kowhai. See *Sophora*.
Kudzu vine. See *Pueraria lobata*.
Kurrajong. See *Brachychiton populneus*.
Kusamaki. See *Podocarpus macrophyllus*.

L

Lablab. See *Lablab purpureus*, p.171.
Labrador tea. See *Ledum groenlandicum*, p.124.
Lace aloe. See *Aloe aristata*, p.416.
Lace cactus. See *Mammillaria elongata*, p.406.
Lace flower. See *Episcia dianthiflora*, p.263.
Lace-bark. See *Hoheria populnea*.
Lace-bark pine. See *Pinus bungeana*, p.82.
Lacquered wine-cup. See *Aechmea* 'Foster's Favorite', p.229.
Ladder fern. See *Nephrolepis cordifolia*.
Lad's love. See *Artemisia abrotanum*, p.146.
Lady fern. See *Athyrium filix-femina*.
Lady tulip. See *Tulipa clusiana*, p.355.
Lady-of-the-night. See *Brassavola nodosa*, p.260.
Lady's eardrops. See *Fuchsia magellanica* var. *molinae*, p.134.
Lady's mantle. See *Alchemilla*.
Lady's slipper orchid. See *Cypripedium calceolus*, p.262.
Lady's smock. See *Cardamine pratensis*.
Lamb's tongue. See *Stachys byzantina*, p.268.
Lamb's-tail cactus. See *Echinocereus schmollii*, p.398.
Lampshade poppy. See *Meconopsis integrifolia*, p.256.
Lancewood. See *Pseudopanax crassifolius*.
Lantern tree. See *Crinodendron hookerianum*, p.112.
Large self-heal. See *Prunella grandiflora*, p.329.
Large yellow restharrow. See *Ononis natrix*, p.307.
Large-leaved lime. See *Tilia platyphyllos*.
Larkspur. See *Consolida*.
Late Dutch honeysuckle. See *Lonicera periclymenum* 'Serotina'.
Laurel. See *Laurelia*; *Laurus*; *Prunus laurocerasus*; *Prunus lusitanica*.
Laurustinus. See *Viburnum tinus*, p.119.
Lavender. See *Lavandula*.
Lavender cotton. See *Santolina chamaecyparissus*.
Lawson cypress. See *Chamaecyparis lawsoniana*.
Lead plant. See *Amorpha canescens*.
Least bur reed. See *Sparganium minimus*.
Least duckweed. See *Wolffia arrhiza*.
Least snowbell. See *Soldanella minima*.
Least willow. See *Salix herbacea*.
Leatherleaf. See *Chamaedaphne calyculata*.
Leatherleaf sedge. See *Carex buchananii*.
Leatherwood. See *Cyrilla racemiflora*.
Lemon verbena. See *Aloysia triphylla*, p.113.
Lemon vine. See *Pereskia aculeata*, p.393.
Lent lily. See *Narcissus pseudonarcissus*, p.361.
Lenten rose. See *Helleborus*.
Leopard lily. See *Dieffenbachia*; *Lilium pardalinum*, p.349.
Leopard's bane. See *Doronicum*.
Lesser celandine. See *Ranunculus ficaria*.
Lesser periwinkle. See *Vinca minor*, p.146.
Leyland cypress. See x *Cupressocyparis leylandii*, p.75.
Lijiang spruce. See *Picea likiangensis*.
Lilac. See *Syringa*.
Lily. See *Lilium*.
Lily tree. See *Magnolia denudata*, p.49.
Lily-of-the-valley. See *Convallaria*.
Lily-of-the-valley tree. See *Clethra arborea*.
Lilyturf. See *Liriope*.
Lime. See *Tilia*.
Linden. See *Tilia*.
Ling. See *Calluna vulgaris*.
Lion's ear. See *Leonotis leonurus*, p.119.
Lipstick plant. See *Aeschynanthus pulcher*.
Liquorice. See *Glycyrrhiza*.
Liquorice fern. See *Polypodium glycyrrhiza*, p.186.

Little walnut. See *Juglans microcarpa*, p.66.
Living rock. See *Ariocarpus*; *Pleiospilos bolusii*, p.415.
Living stones. See *Lithops*.
Livingstone daisy. See *Dorotheanthus bellidiformis*.
Lizard's tail. See *Saururus cernuus*, p.387.
Lobel's maple. See *Acer lobelii*, p.40.
Loblolly bay. See *Gordonia lasianthus*.
Lobster cactus. See *Schlumbergera truncata*, p.409.
Lobster claws. See *Heliconia*.
Locust. See *Robinia pseudoacacia*.
Lodgepole pine. See *Pinus contorta* var. *latifolia*, p.80.
Lollipop plant. See *Pachystachys lutea*, p.127.
Lombardy poplar. See *Populus nigra* 'Italica', p.40.
London plane. See *Platanus* x *hispanica*, p.41.
London pride. See *Saxifraga* x *urbium*.
Loosestrife. See *Lysimachia*.
Loquat. See *Eriobotrya japonica*.
Lord Anson's blue pea. See *Lathyrus nervosus*.
Lords and ladies. See *Arum*.
Lorraine begonia. See *Begonia* x *cheimantha* 'Gloire de Lorraine'.
Love-in-a-mist. See *Nigella damascena*.
Love-lies-bleeding. See *Amaranthus caudatus*, p.278.
Low-bush blueberry. See *Vaccinium angustifolium* var. *laevifolium*, p.142.
Lucombe oak. See *Quercus* x *hispanica* 'Lucombeana', p.47.
Lungwort. See *Pulmonaria*.
Lupin. See *Lupinus*.
Lyme grass. See *Leymus arenarius*.

M

Macadamia nut. See *Macadamia integrifolia*, p.47.
Mace sedge. See *Carex grayi*.
Macedonian pine. See *Pinus peuce*, p.75.
Mackay's heath. See *Erica mackaiana*.
Madagascar dragon tree. See *Dracaena marginata*.
Madagascar jasmine. See *Stephanotis floribunda*, p.165.
Madeira vine. See *Anredera*.
Madonna lily. See *Lilium candidum*, p.348.
Madroña. See *Arbutus menziesii*.
Madroñe. See *Arbutus menziesii*.
Mahoe. See *Melicytus ramiflorus*; *Thespesia populnea*.
Maiden pink. See *Dianthus deltoides*.
Maidenhair fern. See *Adiantum capillus-veneris*.
Maidenhair spleenwort. See *Asplenium trichomanes*, p.187.
Maidenhair tree. See *Ginkgo biloba*, p.77.
Maiten. See *Maytenus boaria*.
Maize. See *Zea*.
Majorcan peony. See *Paeonia cambessedesii*, p.201.
Malay ginger. See *Costus speciosus*.
Male fern. See *Dryopteris filix-mas*, p.186.
Mallow. See *Malva*.
Maltese cross. See *Lychnis chalcedonica*, p.222.
Mamaku. See *Cyathea medullaris*.
Mandarin's hat plant. See *Holmskioldia sanguinea*.
Mandrake. See *Mandragora*.
Manipur lily. See *Lilium mackliniae*, p.348.
Manna ash. See *Fraxinus ornus*, p.50.
Manna gum. See *Eucalyptus viminalis*.
Manuka. See *Leptospermum scoparium*.
Manzanita. See *Arctostaphylos*.
Maple. See *Acer*.
Mapleleaf begonia. See *Begonia dregei*; *Begonia* x *weltoniensis*, p.259.
Marguerite. See *Chrysanthemum frutescens*, p.204.
Marigold. See *Calendula*.
Mariposa tulip. See *Calochortus*.
Maritime pine. See *Pinus pinaster*, p.77.
Marmalade bush. See *Streptosolen jamesonii*, p.180.
Marsh buckler fern. See *Thelypteris palustris*, p.188.
Marsh fern. See *Thelypteris palustris*, p.188.
Marsh marigold. See *Caltha palustris*, p.391.
Martagon lily. See *Lilium martagon*, p.348.
Marvel of Peru. See *Mirabilis*.
Mask flower. See *Alonsoa warscewiczii*, p.281.
Masterwort. See *Astrantia*.

INDEX OF COMMON NAMES – P

Mastic tree. See *Pistacia lentiscus*.
May. See *Crataegus laevigata*.
May apple. See *Podophyllum peltatum*.
May lily. See *Maianthemum*.
Mayflower. See *Epigaea repens*.
Meadow buttercup. See *Ranunculus acris*.
Meadow cranesbill. See *Geranium pratense*.
Meadow foam. See *Limnanthes douglasii*, p.288.
Meadow lily. See *Lilium canadense*.
Meadow rue. See *Thalictrum*.
Meadow saffron. See *Colchicum autumnale*, p.381.
Meadow saxifrage. See *Saxifraga granulata*, p.294.
Meadowsweet. See *Filipendula*.
Mediterranean cypress. See *Cupressus sempervirens*, p.79.
Medlar. See *Mespilus*.
Melon cactus. See *Melocactus intortus*, p.408.
Merry-bells. See *Uvularia grandiflora*, p.238.
Mescal button. See *Lophophora williamsii*, p.413.
Metake. See *Pseudosasa japonica*, p.184.
Metal-leaf begonia. See *Begonia metallica*, p.259.
Mexican blood flower. See *Distictis buccinatoria*, p.165.
Mexican bush sage. See *Salvia leucantha*.
Mexican cypress. See *Cupressus lusitanica*.
Mexican firecracker. See *Echeveria setosa*.
Mexican flame vine. See *Senecio confusus*, p.177.
Mexican foxglove. See *Tetranema roseum*, p.267.
Mexican giant hyssop. See *Agastache*.
Mexican hat plant. See *Kalanchoe daigremontiana*, p.398.
Mexican orange blossom. See *Choisya ternata*, p.97.
Mexican palo verde. See *Parkinsonia aculeata*.
Mexican stone pine. See *Pinus cembroides*, p.83.
Mexican sunflower. See *Tithonia rotundifolia*.
Mexican tulip poppy. See *Hunnemannia fumariifolia*.
Mexican violet. See *Tetranema roseum*, p.267.
Mezereon. See *Daphne mezereum*, p.144.
Michaelmas daisy. See *Aster*.
Mickey-mouse plant. See *Ochna serrulata*.
Mignonette. See *Reseda*.
Mignonette vine. See *Anredera*.
Mile-a-minute plant. See *Polygonum aubertii*; *Polygonum baldschuanicum*, p.177.
Milkweed. See *Euphorbia*.
Mimicry plant. See *Pleiospilos bolusii*, p.415.
Mimosa. See *Acacia*.
Mind-your-own-business. See *Soleirolia*.
Miniature date palm. See *Phoenix roebelenii*.
Miniature grape ivy. See *Cissus striata*.
Mint. See *Mentha*.
Mint bush. See *Elsholtzia stauntonii*, p.143; *Prostanthera*.
Mirbeck's oak. See *Quercus canariensis*, p.40.
Missouri flag. See *Iris missouriensis*, p.198.
Mist flower. See *Eupatorium rugosum*.
Mistletoe cactus. See *Rhipsalis*.
Mistletoe fig. See *Ficus deltoidea*, p.122.
Moccasin flower. See *Cypripedium acaule*, p.260.
Mock orange. See *Philadelphus coronarius*; *Pittosporum tobira*.
Monarch birch. See *Betula maximowicziana*.
Monarch-of-the-East. See *Sauromatum venosum*, p.357.
Money tree. See *Crassula ovata*, p.393.
Moneywort. See *Lysimachia nummularia* 'Aurea', p.334.
Mongolian lime. See *Tilia mongolica*.
Monkey musk. See *Mimulus*.
Monkey puzzle. See *Araucaria araucana*, p.77.
Monkshood. See *Aconitum*.
Montbretia. See *Crocosmia*.
Monterey ceanothus. See *Ceanothus rigidus*.
Monterey cypress. See *Cupressus macrocarpa*.
Monterey pine. See *Pinus radiata*, p.78.
Montezuma pine. See *Pinus montezumae*, p.76.
Montpelier maple. See *Acer monspessulanum*.
Moon flower. See *Ipomoea alba*.
Moon trefoil. See *Medicago arborea*.
Moonlight holly. See *Ilex aquifolium* 'Flavescens'.
Moonseed. See *Menispermum*.
Moonstones. See *Pachyphytum oviferum*, p.411.
Moreton Bay chestnut. See *Castanospermum*.
Moreton Bay fig. See *Ficus macrophylla*.
Morinda spruce. See *Picea smithiana*.
Morning glory. See *Ipomoea hederacea*, p.174.

Moroccan broom. See *Cytisus battandieri*, p.93.
Moses-in-the-cradle. See *Tradescantia spathacea*.
Mosquito grass. See *Bouteloua gracilis*, p.183.
Moss campion. See *Silene acaulis*, p.313.
Mother of thousands. See *Saxifraga stolonifera*; *Soleirolia*.
Mother spleenwort. See *Asplenium bulbiferum*.
Mother-in-law's seat. See *Echinocactus grusonii*, p.396.
Mother-in-law's tongue. See *Sansevieria trifasciata*.
Mother-of-pearl plant. See *Graptopetalum paraguayense*, p.414.
Mount Etna broom. See *Genista aetnensis*, p.67.
Mount Morgan wattle. See *Acacia podalyriifolia*, p.105.
Mountain ash. See *Sorbus aucuparia*, p.55.
Mountain avens. See *Dryas*.
Mountain buckler fern. See *Oreopteris limbosperma*.
Mountain dogwood. See *Cornus nuttallii*, p.50.
Mountain fern. See *Oreopteris limbosperma*.
Mountain fetterbush. See *Pieris floribunda*, p.97.
Mountain flax. See *Phormium cookianum*.
Mountain gum. See *Eucalyptus dalrympleana*, p.46.
Mountain hemlock. See *Tsuga mertensiana*.
Mountain pansy. See *Viola lutea*.
Mountain pepper. See *Drimys lanceolata*.
Mountain pine. See *Pinus mugo*.
Mountain pride. See *Penstemon newberryi*.
Mountain sow thistle. See *Cicerbita alpina*.
Mountain spruce. See *Picea engelmannii*, p.79.
Mountain tassel. See *Soldanella montana*.
Mountain willow. See *Salix arbuscula*.
Mountain wood fern. See *Oreopteris limbosperma*.
Mourning iris. See *Iris susiana*.
Mourning widow. See *Geranium phaeum*, p.202.
Mouse plant. See *Arisarum proboscideum*.
Moutan. See *Paeonia suffruticosa*.
Mulberry. See *Morus*.
Mullein. See *Verbascum*.
Muriel bamboo. See *Fargesia spathacea*.
Murray red gum. See *Eucalyptus camaldulensis*.
Musk. See *Mimulus moschatus*.
Musk willow. See *Salix aegyptiaca*.
Myrobalan. See *Prunus cerasifera*.
Myrtle. See *Myrtus*.
Myrtle flag. See *Acorus calamus* 'Variegatus', p.387.

N

Naked coral tree. See *Erythrina americana*.
Nankeen lily. See *Lilium* x *testaceum*.
Narihira bamboo. See *Semiarundinaria fastuosa*, p.184.
Narrow buckler fern. See *Dryopteris carthusiana*.
Narrow-leaved ash. See *Fraxinus angustifolia*.
Nasturtium. See *Tropaeolum*.
Natal grass. See *Rhynchelytrum repens*.
Natal ivy. See *Senecio macroglossus*.
Natal plum. See *Carissa macrocarpa*.
Native Australian frangipani. See *Hymenosporum flavum*.
Necklace poplar. See *Populus deltoides*.
Needle spike-rush. See *Eleocharis acicularis*.
Nepalese ivy. See *Hedera nepalensis*.
Nettle tree. See *Celtis*.
Nettle-leaved bellflower. See *Campanula trachelium*, p.216.
Net-veined willow. See *Salix reticulata*, p.318.
New Zealand bluebell. See *Wahlenbergia albomarginata*.
New Zealand cabbage palm. See *Cordyline australis*.
New Zealand Christmas tree. See *Metrosideros excelsa*, p.57; *Metrosideros robusta*.
New Zealand edelweiss. See *Leucogenes*.
New Zealand flax. See *Phormium*.
New Zealand honeysuckle. See *Knightia excelsa*.
New Zealand satin flower. See *Libertia grandiflora*, p.203.
New Zealand tea-tree. See *Leptospermum scoparium*.
Night-blooming cereus. See *Hylocereus undatus*.
Nikko fir. See *Abies homolepis*.
Nikko maple. See *Acer maximowiczianum*.
Ninebark. See *Physocarpus opulifolius*.

Nirre. See *Nothofagus antarctica*.
Noble fir. See *Abies procera*.
Nodding catchfly. See *Silene pendula*.
Nootka cypress. See *Chamaecyparis nootkatensis*.
Norfolk Island hibiscus. See *Lagunaria patersonii*.
Norfolk Island pine. See *Araucaria heterophylla*.
North Island edelweiss. See *Leucogenes leontopodium*.
Northern bungalow palm. See *Archontophoenix alexandrae*, p.46.
Northern Japanese hemlock. See *Tsuga diversifolia*.
Northern maidenhair fern. See *Adiantum pedatum*, p.188.
Northern pitch pine. See *Pinus rigida*, p.80.
Norway maple. See *Acer platanoides*.
Norway spruce. See *Picea abies*, p.78.

O

Oak. See *Quercus*.
Oak-leaved hydrangea. See *Hydrangea quercifolia*, p.114.
Obedient plant. See *Physostegia*.
Oconee bells. See *Shortia galacifolia*, p.312.
Ohio buckeye. See *Aesculus glabra*.
Old man. See *Artemisia abrotanum*, p.146.
Old man of the Andes. See *Borzicactus celsianus*, p.398; *Borzicactus trollii*.
Old man's beard. See *Clematis*.
Old-lady cactus. See *Mammillaria hahniana*, p.401.
Old-man cactus. See *Cephalocereus senilis*, p.396.
Old-witch grass. See *Panicum capillare*, p.184.
Old-woman cactus. See *Mammillaria hahniana*, p.401.
Oleander. See *Nerium oleander*, p.90.
Oleaster. See *Elaeagnus angustifolia*, p.92.
Olive. See *Olea europaea*.
Onion. See *Allium*.
Ontario poplar. See *Populus* x *candicans*.
Opium poppy. See *Papaver somniferum*.
Orange jasmine. See *Murraya paniculata*.
Orange lily. See *Lilium bulbiferum*.
Orchard grass. See *Dactylis glomerata*.
Orchid cactus. See *Epiphyllum*.
Oregon grape. See *Mahonia aquifolium*, p.127.
Oregon maple. See *Acer macrophyllum*, p.38.
Oregon oak. See *Quercus garryana*, p.54.
Organ-pipe cactus. See *Lemaireocereus marginatus*, p.393.
Oriental beech. See *Fagus orientalis*.
Oriental bittersweet. See *Celastrus orbiculatus*.
Oriental plane. See *Platanus orientalis*.
Oriental poppy. See *Papaver orientale*.
Oriental spruce. See *Picea orientalis*.
Oriental sweet gum. See *Liquidambar orientalis*.
Oriental white oak. See *Quercus aliena*.
Ornamental cabbage. See *Brassica oleracea* forms, p.276.
Ornamental maize. See *Zea mays*.
Ornamental pepper. See *Capsicum annuum*.
Ornamental yam. See *Dioscorea discolor*, p.179.
Osage orange. See *Maclura pomifera*.
Oso berry. See *Oemleria cerasiformis*.
Ostrich fern. See *Matteuccia struthiopteris*, p.188.
Ostrich-feather fern. See *Matteuccia struthiopteris*, p.188.
Ovens wattle. See *Acacia pravissima*, p.71.
Owl-eyes. See *Huernia zebrina*.
Oxlip. See *Primula elatior*, p.237.

P

Pacific dogwood. See *Cornus nuttallii*, p.50.
Pacific fir. See *Abies amabilis*.
Paddy's pride. See *Hedera colchica* 'Sulphur Heart', p.181.
Pagoda tree. See *Sophora japonica*.
Paintbrush. See *Haemanthus albiflos*.
Painted fern. See *Athyrium nipponicum*, p.189.
Painted net-leaf. See *Fittonia verschaffeltii*, p.266.
Painted trillium. See *Trillium undulatum*.

INDEX OF COMMON NAMES – Q

Painted wood lily. See *Trillium undulatum*.
Pampas grass. See *Cortaderia selloana*.
Panda plant. See *Kalanchoe tomentosa*, p.403.
Pansy. See *Viola wittrockiana*.
Pansy orchid. See *Miltoniopsis*.
Panther lily. See *Lilium pardalinum*, p.349.
Paper birch. See *Betula papyrifera*, p.46.
Paper mulberry. See *Broussonetia papyrifera*, p.53.
Paper reed. See *Cyperus papyrus*, p.183.
Paper-bark maple. See *Acer griseum*, p.74.
Paper-bark tree. See *Melaleuca viridiforma* var. *rubriflora*.
Papyrus. See *Cyperus papyrus*, p.183.
Para para. See *Pisonia umbellifera*.
Parachute plant. See *Ceropegia sandersoniae*.
Paradise palm. See *Howea forsteriana*.
Parlour palm. See *Chamaedorea elegans*, p.122.
Parrot leaf. See *Alternanthera ficoidea*.
Parrot's bill. See *Clianthus puniceus*, p.165.
Parrot's feather. See *Myriophyllum aquaticum*, p.388.
Parrot's flower. See *Heliconia psittacorum*, p.193.
Parrot's plantain. See *Heliconia psittacorum*, p.193.
Parsley fern. See *Cryptogramma crispa*, p.189.
Partridge berry. See *Mitchella repens*.
Partridge-breasted aloe. See *Aloe variegata*, p.400.
Pasque flower. See *Pulsatilla vulgaris*, p.296.
Passion flower. See *Passiflora*.
Patagonian cypress. See *Fitzroya cupressoides*, p.80.
Pawpaw. See *Asimina triloba*.
Peace lily. See *Spathiphyllum wallisii*, p.264.
Peach. See *Prunus persica*.
Peacock flower. See *Tigridia pavonia*, p.367.
Peacock plant. See *Calathea makoyana*, p.268.
Peanut cactus. See *Lobivia silvestrii*, p.411.
Pear. See *Pyrus*.
Pearl berry. See *Margyricarpus pinnatus*.
Pearl everlasting. See *Anaphalis*.
Pedunculate oak. See *Quercus robur*.
Peepul. See *Ficus religiosa*.
Pelican flower. See *Aristolochia grandiflora*.
Pencil cedar. See *Juniperus virginiana*.
Pendent silver lime. See *Tilia* 'Petiolaris', p.42.
Pendulous sedge. See *Carex pendula*, p.185.
Peony. See *Paeonia*.
Peppermint. See *Mentha* x *piperita*.
Peppermint geranium. See *Pelargonium tomentosum*, p.211.
Peppermint tree. See *Agonis flexuosa*, p.64.
Pepper-tree. See *Pseudowintera axillaris*.
Père David's maple. See *Acer davidii*.
Perennial pea. See *Lathyrus latifolius*, p.171; *Lathyrus sylvestris*.
Periwinkle. See *Vinca*.
Perry's weeping silver. See *Ilex aquifolium* 'Argentea Marginata Pendula', p.72.
Persian buttercup. See *Ranunculus asiaticus*, pp.365 and 367.
Persian everlasting pea. See *Lathyrus rotundifolius*.
Persian ironwood. See *Parrotia persica*, p.56.
Persian ivy. See *Hedera colchica*.
Persian lilac. See *Melia azedarach*, p.50; *Syringa* x *persica*, p.90.
Persian stone cress. See *Aethionema grandiflorum*, p.300.
Persian violet. See *Exacum affine*, p.283.
Persimmon. See *Diospyros kaki*.
Peruvian daffodil. See *Hymenocallis narcissiflora*, p.364.
Peruvian mastic tree. See *Schinus molle*.
Peruvian old-man cactus. See *Espostoa lanata*, p.393.
Peruvian pepper-tree. See *Schinus molle*.
Pestle parsnip. See *Lomatium nudicaule*.
Peyote. See *Lophophora*.
Pheasant grass. See *Stipa arundinacea*.
Pheasant's eye. See *Narcissus poeticus* var. *recurvus*, p.361.
Philippine violet. See *Barleria cristata*.
Piccabeen palm. See *Archontophoenix cunninghamiana*.
Pick-a-back plant. See *Tolmiea menziesii*.
Pickerel weed. See *Pontederia cordata*, p.388.
Pigeon berry. See *Duranta erecta*, p.120.
Pignut. See *Carya glabra*.
Pignut hickory. See *Carya glabra*.
Pin cherry. See *Prunus pensylvanica*.
Pin oak. See *Quercus palustris*, p.43.

Pincushion cactus. See *Mammillaria*.
Pine. See *Pinus*.
Pineapple broom. See *Cytisus battandieri*, p.93.
Pineapple flower. See *Eucomis*.
Pineapple guava. See *Feijoa sellowiana*, p.113.
Pine-mat manzanita. See *Arctostaphylos nevadensis*.
Pink. See *Dianthus*.
Pink arum. See *Zantedeschia rehmannii*.
Pink broom. See *Notospartium carmichaeliae*.
Pink dandelion. See *Crepis incana*.
Pink snowball. See *Dombeya* x *cayeuxii*, p.62.
Pink trumpet tree. See *Tabebuia rosea*.
Pinwheel. See *Aeonium haworthii*, p.402.
Pinyon. See *Pinus cembroides*, p.83.
Pitch apple. See *Clusia major*.
Pitcher plant. See *Nepenthes*; *Sarracenia*.
Plane. See *Platanus*.
Plantain lily. See *Hosta*.
Plovers' eggs. See *Adromischus festivus*.
Plum yew. See *Cephalotaxus harringtonia*; *Prumnopitys andinus*.
Plume poppy. See *Macleaya*.
Plum-fruited yew. See *Prumnopitys andinus*.
Plush plant. See *Echeveria pulvinata*, p.400.
Poached-egg flower. See *Limnanthes douglasii*, p.288.
Pocket handkerchief tree. See *Davidia involucrata*, p.51.
Poet's daffodil. See *Narcissus poeticus*.
Poet's narcissus. See *Narcissus poeticus*.
Poinsettia. See *Euphorbia pulcherrima*, p.120.
Polka-dot plant. See *Hypoestes phyllostachya*, p.228.
Polyanthus. See *Primula*.
Polyanthus daffodil. See *Narcissus tazetta*.
Polypody. See *Polypodium vulgare*, p.189.
Pomegranate. See *Punica*.
Pontine oak. See *Quercus pontica*.
Pony-tail. See *Beaucarnea recurvata*, p.74.
Poor man's orchid. See *Schizanthus*.
Poplar. See *Populus*.
Poppy. See *Papaver*.
Port Jackson fig. See *Ficus rubiginosa*.
Portia oil nut. See *Thespesia populnea*.
Portugal laurel. See *Prunus lusitanica*.
Portuguese heath. See *Erica lusitanica*.
Pot marigold. See *Calendula officinalis*.
Potato creeper. See *Solanum seaforthianum*.
Potato vine. See *Solanum jasminoides*.
Powder-puff cactus. See *Mammillaria bocasana*, p.407.
Prayer plant. See *Maranta leuconeura*.
Prickly Moses. See *Acacia verticillata*.
Prickly pear. See *Opuntia*.
Prickly poppy. See *Argemone mexicana*, p.287.
Prickly shield fern. See *Polystichum aculeatum*.
Pride of Bolivia. See *Tipuana tipu*.
Pride of India. See *Koelreuteria paniculata*, p.66; *Lagerstroemia speciosa*.
Primrose. See *Primula vulgaris*, p.237.
Primrose jasmine. See *Jasminum mesnyi*, p.167.
Prince Albert's yew. See *Saxegothaea conspicua*.
Prince's feather. See *Amaranthus hypochondriacus*, p.282.
Princess tree. See *Paulownia tomentosa*, p.51.
Privet. See *Ligustrum*.
Prophet flower. See *Arnebia pulchra*.
Prostrate coleus. See *Plectranthus oertendahlii*.
Prostrate speedwell. See *Veronica prostrata*, p.305.
Pudding pipe-tree. See *Cassia fistula*.
Puka. See *Meryta sinclairii*, p.74.
Pukapuka. See *Brachyglottis repanda*, p.97.
Pukanui. See *Meryta sinclairii*, p.74.
Purple anise. See *Illicium floridanum*.
Purple beech. See *Fagus sylvatica* f. *purpurea*, p.39.
Purple broom. See *Cytisus purpureus*.
Purple crab. See *Malus* x *purpurea*.
Purple loosestrife. See *Lythrum*.
Purple mountain saxifrage. See *Saxifraga oppositifolia*, p.314.
Purple osier. See *Salix purpurea*.
Purple rock brake. See *Pellaea atropurpurea*.
Purple toadflax. See *Linaria purpurea*.
Purple-leaved ivy. See *Hedera helix* 'Atropurpurea', p.181.
Purple-stemmed cliff brake. See *Pellaea atropurpurea*.

Pussy ears. See *Cyanotis somaliensis*, p.267; *Kalanchoe tomentosa*, p.403.
Pussy willow. See *Salix caprea*.
Pygmy date palm. See *Phoenix roebelenii*.
Pyramidal bugle. See *Ajuga pyramidalis*.
Pyrethrum. See *Tanacetum coccineum*.

Q

Quaking aspen. See *Populus tremuloides*.
Quaking grass. See *Briza*.
Quamash. See *Camassia quamash*.
Quassia. See *Picrasma ailanthoides*, p.71.
Quater. See *Vinca major*.
Queen Anne's double daffodil. See *Narcissus* 'Eystettensis'.
Queen Anne's jonquil. See *Narcissus jonquilla* 'Flore Pleno'.
Queen palm. See *Arecastrum*.
Queencup. See *Clintonia uniflora*.
Queen-of-the-night. See *Hylocereus undatus*; *Selenicereus grandiflorus*, p.392.
Queen's crape myrtle. See *Lagerstroemia speciosa*.
Queen's tears. See *Billbergia nutans*, p.229.
Queensland nut. See *Macadamia integrifolia*, p.47.
Queensland pyramidal tree. See *Lagunaria patersonii*.
Queensland silver wattle. See *Acacia podalyriifolia*.
Queensland umbrella tree. See *Schefflera actinophylla*, p.58.
Quince. See *Cydonia oblonga*.

R

Rabbit tracks. See *Maranta leuconeura* var. *kerchoviana*, p.268.
Rain daisy. See *Dimorphotheca pluvialis*, p.271.
Rain lily. See *Zephyranthes*.
Rainbow star. See *Cryptanthus bromelioides*.
Raisin-tree. See *Hovenia dulcis*, p.53.
Rangiora. See *Brachyglottis repanda*, p.97.
Rangoon creeper. See *Quisqualis indica*, p.171.
Rata. See *Metrosideros excelsa*, p.57; *Metrosideros robusta*.
Rat's-tail cactus. See *Aporocactus flagelliformis*, p.399.
Rauli. See *Nothofagus procera*, p.41.
Red ash. See *Fraxinus pennsylvanica*.
Red baneberry. See *Actaea rubra*.
Red buckeye. See *Aesculus pavia*.
Red chokeberry. See *Aronia arbutifolia*, p.100.
Red horse-chestnut. See *Aesculus* x *carnea*.
Red maple. See *Acer rubrum*, p.44.
Red morning glory. See *Ipomoea coccinea*.
Red mountain spinach. See *Atriplex hortensis* 'Rubra'.
Red oak. See *Quercus rubra*, p.43.
Red orach. See *Atriplex hortensis* 'Rubra'.
Red orchid cactus. See *Nopalxochia ackermannii*.
Red passion flower. See *Passiflora coccinea*, p.165; *Passiflora racemosa*.
Red pineapple. See *Ananas bracteatus*.
Red spider lily. See *Lycoris radiata*, p.364.
Red spike. See *Cephalophyllum alstonii*, p.410.
Red valerian. See *Centranthus ruber*, p.209.
Red-and-green kangaroo paw. See *Anigozanthos manglesii*, p.214.
Red-barked dogwood. See *Cornus alba*.
Red-berried elder. See *Sambucus racemosa*.
Redbird flower. See *Pedilanthus tithymaloides* 'Variegata', p.395.
Redbud. See *Cercis*.
Red-hot cat's tail. See *Acalypha hispida*.
Red-hot poker. See *Kniphofia*.
Red-ink plant. See *Phytolacca americana*.
Redwood. See *Sequoia sempervirens*.
Reflexed stonecrop. See *Sedum reflexum*, p.307.
Regal lily. See *Lilium regale*, p.348.
Resurrection plant. See *Selaginella lepidophylla*.
Rewa rewa. See *Knightia excelsa*.
Rex begonia vine. See *Cissus discolor*.
Rhubarb. See *Rheum*.

INDEX OF COMMON NAMES – S

Ribbon gum. See *Eucalyptus viminalis*.
Ribbon plant. See *Dracaena sanderiana*, p.121.
Ribbonwood. See *Hoheria sexstylosa*.
Rice-paper plant. See *Tetrapanax papyrifer*, p.96.
Rienga lily. See *Arthropodium cirrhatum*.
River red gum. See *Eucalyptus camaldulensis*.
Roast-beef plant. See *Iris foetidissima*.
Roble. See *Nothofagus obliqua*, p.42.
Rock lily. See *Arthropodium cirrhatum*.
Rock rose. See *Cistus*; *Helianthemum*.
Rock speedwell. See *Veronica fruticans*.
Rocky Mountain juniper. See *Juniperus scopulorum*.
Roman wormwood. See *Artemisia pontica*, p.254.
Roof houseleek. See *Sempervivum tectorum*, p.337.
Rosa mundi. See *Rosa gallica* 'Versicolor', p.153.
Rosary vine. See *Ceropegia linearis* subsp. *woodii*, p.398.
Rose. See *Rosa*.
Rose acacia. See *Robinia hispida*, p.111.
Rose cactus. See *Pereskia grandifolia*, p.394.
Rose of Jericho. See *Selaginella lepidophylla*.
Rose of Sharon. See *Hypericum calycinum*, p.140.
Rose periwinkle. See *Catharanthus roseus*, p.130.
Rose pincushion. See *Mammillaria zeilmanniana*, p.409.
Rosebud cherry. See *Prunus subhirtella*.
Rosemary. See *Rosmarinus officinalis*, p.137.
Roseroot. See *Sedum rosea*.
Rouen lilac. See *Syringa* x *chinensis*.
Rough-barked Mexican pine. See *Pinus montezumae*, p.76.
Round-headed club-rush. See *Scirpoides holoschoenus*.
Round-leaved mint-bush. See *Prostanthera rotundifolia*, p.115.
Round-leaved wintergreen. See *Pyrola rotundifolia*.
Rowan. See *Sorbus aucuparia*, p.55.
Royal agave. See *Agave victoriae-reginae*, p.397.
Royal fern. See *Osmunda regalis*, p.188.
Royal jasmine. See *Jasminum grandiflorum*.
Royal paintbrush. See *Scadoxus puniceus*.
Royal palm. See *Roystonea*.
Royal red bugler. See *Aeschynanthus pulcher*.
Rubber plant. See *Ficus elastica*.
Rubber vine. See *Cryptostegia grandiflora*.
Ruby grass. See *Rhynchelytrum repens*.
Rue. See *Ruta*.
Russian comfrey. See *Symphytum* x *uplandicum*.
Russian vine. See *Polygonum aubertii*; *Polygonum baldschuanicum*, p.177.
Rusty woodsia. See *Woodsia ilvensis*.
Rusty-back fern. See *Ceterach officinarum*, p.187.
Rusty-leaved fig. See *Ficus rubiginosa*.

S

Sacred bamboo. See *Nandina domestica*.
Sacred fig tree. See *Ficus religiosa*.
Sacred lotus. See *Nelumbo nucifera*, p.387.
Saffron crocus. See *Crocus sativus*.
Sage. See *Salvia*.
Saguaro. See *Carnegiea gigantea*, p.392.
St Augustine grass. See *Stenotaphrum secundatum*.
St Bernard's lily. See *Anthericum liliago*, p.240.
St Bruno's lily. See *Paradisea liliastrum*.
St Catherine's lace. See *Eriogonum giganteum*, p.109.
St Dabeoc's heath. See *Daboecia cantabrica*.
Sand phlox. See *Phlox bifida*, p.329.
Sandwort. See *Arenaria*.
Sapphire berry. See *Symplocos paniculata*, p.108.
Sargent cherry. See *Prunus sargentii*, p.61.
Sargent's rowan. See *Sorbus sargentiana*.
Satin poppy. See *Meconopsis napaulensis*, p.191.
Sausage tree. See *Kigelia africana*.
Savin. See *Juniperus sabina*.
Saw palmetto. See *Serenoa repens*.
Sawara cypress. See *Chamaecyparis pisifera*.
Sawfly orchid. See *Ophrys tenthredinifera*, p.261.
Sawtooth oak. See *Quercus acutissima*.
Saxifrage. See *Saxifraga*.
Scabious. See *Knautia arvensis*; *Scabiosa*.
Scarlet ball cactus. See *Parodia haselbergii*, p.404.
Scarlet banana. See *Musa uranoscopus*.

Scarlet fritillary. See *Fritillaria recurva*, p.340.
Scarlet oak. See *Quercus coccinea*, p.44.
Scarlet plume. See *Euphorbia fulgens*.
Scarlet trompetilla. See *Bouvardia ternifolia*, p.143.
Scarlet trumpet honeysuckle. See *Lonicera* x *brownii*.
Scarlet turkscap lily. See *Lilium chalcedonicum*, p.349.
Scented paper-bark. See *Melaleuca squarrosa*.
Scotch heather. See *Calluna vulgaris*.
Scotch laburnum. See *Laburnum alpinum*, p.67.
Scotch rose. See *Rosa pimpinellifolia*.
Scotch thistle. See *Onopordum acanthium*, p.274.
Scots pine. See *Pinus sylvestris*.
Screw pine. See *Pandanus*.
Scrub palmetto. See *Serenoa repens*.
Scrub pine. See *Pinus virginiana*, p.81.
Scurvy grass. See *Oxalis enneaphylla*.
Sea buckthorn. See *Hippophäe rhamnoides*, p.94.
Sea campion. See *Silene uniflora* 'Flore Peno'.
Sea daffodil. See *Pancratium maritimum*.
Sea holly. See *Eryngium*.
Sea kale. See *Crambe maritima*, p.240.
Sea lavender. See *Limonium*.
Sea lily. See *Pancratium maritimum*.
Sea onion. See *Urginea maritima*.
Sea pink. See *Armeria maritima*.
Sea squill. See *Urginea maritima*.
Sea urchin. See *Astrophytum asterias*.
Self-heal. See *Prunella*.
Sensitive fern. See *Onoclea sensibilis*, p.188.
Sensitive plant. See *Mimosa pudica*, p.146.
Sentry palm. See *Howea forsteriana*.
Serbian spruce. See *Picea omorika*, p.77.
Service tree of Fontainebleau. See *Sorbus latifolia*.
Serviceberry. See *Amelanchier*.
Sessile oak. See *Quercus petraea*.
Shadbush. See *Amelanchier*.
Shag-bark hickory. See *Carya ovata*, p.43.
Shallon. See *Gaultheria shallon*, p.132.
Shamrock pea. See *Parochetus communis*, p.332.
Shasta daisy. See *Chrysanthemum* x *superbum*.
Sheep laurel. See *Kalmia angustifolia*.
Sheepberry. See *Viburnum lentago*.
Sheep's bit. See *Jasione laevis*.
Shell flower. See *Alpinia zerumbet*, p.191; *Moluccella laevis*, p.287.
Shell ginger. See *Alpinia zerumbet*, p.191.
Shield ivy. See *Hedera helix* 'Deltoidea', p.181.
Shingle oak. See *Quercus imbricaria*.
Shingle plant. See *Monstera acuminata*.
Shining pondweed. See *Potamogeton lucens*.
Shoofly. See *Nicandra physalodes*.
Shooting stars. See *Dodecatheon*.
Shore juniper. See *Juniperus conferta*.
Shore pine. See *Pinus contorta*, p.81.
Showy lady's slipper orchid. See *Cypripedium reginae*, p.260.
Shrimp plant. See *Justicia brandegeana*, p.136.
Shrubby germander. See *Teucrium fruticans*.
Shrubby hare's ear. See *Bupleurum fruticosum*, p.116.
Shrubby restharrow. See *Ononis fruticosa*, p.301.
Siberian bugloss. See *Brunnera macrophylla*.
Siberian crab. See *Malus baccata*.
Siberian elm. See *Ulmus pumila*.
Siberian flag. See *Iris sibirica*.
Siberian melic. See *Melica altissima*.
Siberian squill. See *Scilla siberica*.
Siberian wallflower. See *Erysimum allionii*.
Silk cotton tree. See *Ceiba pentandra*.
Silk tree. See *Albizia julibrissin*, p.64.
Silk vine. See *Periploca graeca*.
Silk weed. See *Asclepias*.
Silk-tassel bush. See *Garrya elliptica*, p.95.
Silky oak. See *Grevillea robusta*.
Silky wisteria. See *Wisteria venusta*.
Silver ball cactus. See *Parodia scopa*.
Silver beech. See *Nothofagus menziesii*.
Silver bell. See *Halesia*.
Silver birch. See *Betula pendula*.
Silver cassia. See *Cassia artemisioides*.
Silver chain. See *Dendrochilum glumaceum*.
Silver dollar cactus. See *Astrophytum asterias*.
Silver fir. See *Abies*.
Silver heart. See *Peperomia marmorata*, p.268.
Silver hedgehog holly. See *Ilex aquifolium* 'Ferox Argentea'.

Silver inch plant. See *Tradescantia zebrina*, p.265.
Silver jade plant. See *Crassula arborescens*, p.394.
Silver lime. See *Tilia tomentosa*.
Silver maple. See *Acer saccharinum*.
Silver morning glory. See *Argyreia splendens*.
Silver net-leaf. See *Fittonia verschaffeltii* var. *argyroneura*, p.264.
Silver torch. See *Cleistocactus strausii*, p.395.
Silver tree. See *Leucadendron argenteum*, p.74.
Silver vase plant. See *Aechmea fasciata*, p.229.
Silver vine. See *Actinidia polygama*; *Epipremnum pictum* 'Argyraeus', p.179.
Silver wattle. See *Acacia dealbata*, p.57.
Silver willow. See *Salix alba* var. *sericea*.
Silver-leaf peperomia. See *Peperomia griseoargentea*.
Silver-leaved geranium. See *Pelargonium* 'Flower of Spring', p.211.
Silver-margined holly. See *Ilex aquifolium* 'Argentea Marginata', p.72.
Sitka spruce. See *Picea sitchensis*.
Skullcap. See *Scutellaria*.
Sky plant. See *Tillandsia ionantha*.
Skyflower. See *Duranta erecta*, p.120.
Sleepy mallow. See *Malvaviscus arboreus*, p.90.
Slender lady palm. See *Rhapis excelsa*, p.122.
Slipper orchid. See *Cypripedium*.
Sloe. See *Prunus spinosa*.
Small-leaved box. See *Buxus microphylla*.
Small-leaved lime. See *Tilia cordata*.
Smoke tree. See *Cotinus coggygria*.
Smooth cypress. See *Cupressus glabra*.
Smooth sumach. See *Rhus glabra*, p.113.
Smooth winterberry. See *Ilex laevigata*.
Smooth-leaved elm. See *Ulmus carpinifolia*.
Snail flower. See *Vigna caracalla*.
Snake bush. See *Duvernoia adhatodoides*.
Snake gourd. See *Trichosanthes curcumina* var. *anguina*.
Snake-bark maple. See *Acer capillipes*, p.56; *Acer davidii*; *Acer davidii* var. *grosseri*; *Acer pensylvanicum*, p.58; *Acer rufinerve*, p.56.
Snake's-head fritillary. See *Fritillaria meleagris*, p.357.
Snapdragon. See *Antirrhinum*.
Sneezeweed. See *Helenium*.
Snow brake. See *Pteris ensiformis*.
Snow bush. See *Breynia nivosa*, p.145.
Snow creeper. See *Porana paniculata*.
Snow gum. See *Eucalyptus niphophila*, p.58.
Snow poppy. See *Eomecon chionantha*.
Snow trillium. See *Trillium nivale*.
Snowball pincushion. See *Mammillaria candida*.
Snowbell. See *Soldanella*.
Snowberry. See *Symphoricarpos albus*.
Snowbush rose. See *Rosa* 'Dupontii', p.151.
Snowdrop. See *Galanthus*.
Snowdrop tree. See *Halesia*.
Snowdrop windflower. See *Anemone sylvestris*, p.232.
Snowflake. See *Leucojum*.
Snow-in-summer. See *Cerastium tomentosum*, p.309; *Euphorbia marginata*, p.271.
Snow-on-the-mountain. See *Euphorbia marginata*, p.271.
Snowy woodrush. See *Luzula nivea*, p.183.
Soapwort. See *Saponaria*.
Soft shield fern. See *Polystichum setiferum*.
Solomon's seal. See *Polygonatum*.
Sorrel tree. See *Oxydendrum arboreum*, p.52.
South American air plant. See *Kalanchoe fedtschenkoi*.
South Sea arrowroot. See *Tacca leontopetaloides*.
Southern beech. See *Nothofagus*.
Southern Japanese hemlock. See *Tsuga sieboldii*.
Southernwood. See *Artemisia abrotanum*, p.146.
Spade leaf. See *Philodendron domesticum*.
Spanish bayonet. See *Yucca aloifolia*, p.123.
Spanish bluebell. See *Hyacinthoides hispanica*, p.358.
Spanish broom. See *Spartium junceum*, p.117.
Spanish chestnut. See *Castanea sativa*.
Spanish dagger. See *Yucca gloriosa*, p.107.
Spanish gorse. See *Genista hispanica*, p.140.
Spanish heath. See *Erica australis*.
Spanish jasmine. See *Jasminum grandiflorum*.
Spanish moss. See *Tillandsia usneoides*, p.229.
Spanish tree heath. See *Erica australis*.
Spice bush. See *Lindera benzoin*, p.101.

637

INDEX OF COMMON NAMES – T

Spiceberry. See *Ardisia crenata*, p.120.
Spider fern. See *Pteris multifida*.
Spider flower. See *Cleome*.
Spider orchid. See *Ophrys sphegodes*.
Spider plant. See *Anthericum*; *Chlorophytum comosum*.
Spiderwort. See *Tradescantia*.
Spignel. See *Meum athamanticum*.
Spike heath. See *Bruckenthalia spiculifolia*.
Spiked speedwell. See *Veronica spicata*.
Spindle tree. See *Euonymus europaeus*.
Spinning gum. See *Eucalyptus perriniana*, p.74.
Spring beauty. See *Claytonia virginica*.
Spring bell. See *Sisyrinchium douglasii*.
Spring crocus. See *Crocus vernus*, p.372.
Spring gentian. See *Gentiana verna*, p.318.
Spring snowflake. See *Leucojum vernum*, p.370.
Spruce. See *Picea*.
Spurge. See *Euphorbia*.
Spurge laurel. See *Daphne laureola*.
Squawroot. See *Trillium erectum*, p.233.
Squirrel's-foot fern. See *Davallia mariesii*.
Squirrel tail grass. See *Hordeum jubatum*, p.183.
Staff tree. See *Celastrus scandens*.
Staff vine. See *Celastrus orbiculatus*.
Stag's-horn fern. See *Platycerium*.
Stag's-horn sumach. See *Rhus hirta*.
Stanford manzanita. See *Arctostaphylos stanfordiana*.
Star daisy. See *Lindheimera texana*, p.288.
Star flower. See *Stapelia orbea*, p.412.
Star ipomoea. See *Ipomoea coccinea*.
Star jasmine. See *Trachelospermum jasminoides*, p.167.
Star magnolia. See *Magnolia stellata*, p.49.
Star-cluster. See *Pentas lanceolata*, p.133.
Star-of-Bethlehem. See *Ornithogalum*.
Star-of-Bethlehem orchid. See *Angraecum sesquipedale*, p.260.
Statice. See *Psylliostachys*.
Stemless gentian. See *Gentiana acaulis*, p.318.
Stinking hellebore. See *Helleborus foetidus*, p.269.
Stinking iris. See *Iris foetidissima*.
Stock. See *Matthiola*.
Stone pine. See *Pinus pinea*, p.83.
Stone plant. See *Lithops*.
Stonecrop. See *Sedum*.
Strap cactus. See *Epiphyllum*.
Strawberry cactus. See *Mammillaria prolifera*.
Strawberry geranium. See *Saxifraga stolonifera* 'Tricolor'.
Strawberry tomato. See *Physalis peruviana*.
Strawberry tree. See *Arbutus unedo*, p.67.
Strawflower. See *Helichrysum bracteatum*.
String-of-beads. See *Senecio rowleyanus*, p.398.
String-of-hearts. See *Ceropegia linearis* subsp. *woodii*, p.411.
Striped squill. See *Puschkinia scilloides*, p.376.
Striped torch. See *Guzmania monostachia*, p.229.
Subalpine fir. See *Abies lasiocarpa*.
Sugar bush. See *Protea repens*.
Sugar maple. See *Acer saccharum*.
Sugared-almond plum. See *Pachyphytum oviferum*, p.411.
Sumach. See *Rhus*.
Summer cypress. See *Kochia scoparia* f. *trichophylla*, p.287.
Summer holly. See *Arctostaphylos diversifolia*.
Summer hyacinth. See *Galtonia candicans*, p.341.
Summer snowflake. See *Leucojum aestivum*, p.340.
Sun plant. See *Portulaca grandiflora*.
Sundew. See *Drosera*.
Sunflower. See *Helianthus*.
Sunrise horse-chestnut. See *Aesculus* x *neglecta*.
Swamp cypress. See *Taxodium distichum*, p.78.
Swamp lily. See *Lilium superbum*; *Saururus cernuus*, p.387.
Swamp pink. See *Helonias bullata*.
Swan flower. See *Aristolochia grandiflora*.
Swan River daisy. See *Brachycome iberidifolia*, p.286.
Swedish birch. See *Betula pendula* 'Dalecarlica', p.47.
Swedish ivy. See *Plectranthus australis*; *Plectranthus oertendahlii*.
Swedish whitebeam. See *Sorbus intermedia*.
Sweet alyssum. See *Lobularia maritima*.
Sweet bay. See *Laurus nobilis*; *Magnolia virginiana*.

Sweet box. See *Sarcococca*.
Sweet briar. See *Rosa eglanteria*, p.151.
Sweet buckeye. See *Aesculus flava*, p.56.
Sweet chestnut. See *Castanea sativa*.
Sweet Cicely. See *Myrrhis*.
Sweet corn. See *Zea mays*.
Sweet flag. See *Acorus calamus* 'Variegatus', p.387.
Sweet gum. See *Liquidambar styraciflua*, p.44.
Sweet pea. See *Lathyrus odoratus*.
Sweet pepper-bush. See *Clethra alnifolia*.
Sweet rocket. See *Hesperis matronalis*, p.203.
Sweet scabious. See *Scabiosa atropurpurea*.
Sweet sop. See *Annona*.
Sweet sultan. See *Centaurea moschata*, p.287.
Sweet viburnum. See *Viburnum odoratissimum*.
Sweet violet. See *Viola odorata*.
Sweet William. See *Dianthus barbatus*.
Sweetheart ivy. See *Hedera helix* 'Deltoidea', p.181.
Swiss mountain pine. See *Pinus mugo*.
Swiss-cheese plant. See *Monstera deliciosa*, p.180.
Sword fern. See *Nephrolepis cordifolia*; *Nephrolepis exaltata*, p.188.
Sycamore. See *Acer pseudoplatanus*.
Sydney golden wattle. See *Acacia longifolia*.
Syrian juniper. See *Juniperus drupacea*.
Szechuan birch. See *Betula szechuanica*.

T

Tacamahac. See *Populus balsamifera*.
Tail flower. See *Anthurium andraeanum*, p.228.
Taiwan cherry. See *Prunus campanulata*.
Taiwan spruce. See *Picea morrisonicola*, p.81.
Tall melic. See *Melica altissima*.
Tamarind. See *Tamarindus indica*.
Tamarisk. See *Tamarix*.
Tanbark oak. See *Lithocarpus densiflorus*.
Tansy-leaved thorn. See *Crataegus tanacetifolia*.
Tape grass. See *Vallisneria spiralis*.
Tarajo holly. See *Ilex latifolia*.
Taro. See *Alocasia macrorrhiza*; *Colocasia esculenta*, p.388.
Tasman celery pine. See *Phyllocladus aspleniifolius*.
Tasmanian blue gum. See *Eucalyptus globulus*.
Tasmanian podocarp. See *Prumnopitys alpinus*.
Tasmanian sassafras. See *Atherosperma moschatum*.
Tasmanian snow gum. See *Eucalyptus coccifera*, p.46.
Tasmanian waratah. See *Telopea truncata*, p.101.
Tassel flower. See *Amaranthus caudatus*, p.278; *Emilia coccinae*, p.293.
Tassel grape hyacinth. See *Muscari comosum*.
Tassel maidenhair. See *Adiantum raddianum* 'Grandiceps'.
Tawny daylily. See *Hemerocallis fulva*.
Teddy-bear vine. See *Cyanotis kewensis*.
Temple bells. See *Smithiantha cinnabarina*.
Temple juniper. See *Juniperus rigida*.
Tenby daffodil. See *Narcissus obvallaris*.
Terebinth tree. See *Pistacia terebinthus*.
Texan walnut. See *Juglans microcarpa*, p.66.
Thatch-leaf palm. See *Howea forsteriana*.
Thimbleberry. See *Rubus odoratus*.
Thistle. See *Carlina*.
Thorn. See *Crataegus*.
Thornless rose. See *Rosa* 'Zéphirine Drouhin', p.163.
Thread agave. See *Agave filifera*, p.401.
Thread palm. See *Washingtonia robusta*, p.47.
Three birds toadflax. See *Linaria triornithophora*, p.215.
Thrift. See *Armeria maritima*.
Throatwort. See *Trachelium caeruleum*, p.282.
Thyme. See *Thymus*.
Ti tree. See *Cordyline fruticosa*.
Tickseed. See *Coreopsis*.
Tidy tips. See *Layia platyglossa*.
Tiger flower. See *Tigridia pavonia*, p.367.
Tiger lily. See *Lilium lancifolium*.
Tiger-jaws. See *Faucaria tigrina*, p.415.
Timber bamboo. See *Phyllostachys bambusoides*, p.184.
Tingiringi gum. See *Eucalyptus glaucescens*.
Tipa tree. See *Tipuana tipu*.
Tipu tree. See *Tipuana tipu*.

Toad lily. See *Tricyrtis*.
Toadflax. See *Linaria*.
Toadshade. See *Trillium sessile*, p.233.
Toothwort. See *Lathraea clandestina*, p.235.
Torch cactus. See *Echinopsis spachianus*, p.393.
Torch lily. See *Kniphofia*.
Torch plant. See *Aloe aristata*, p.416.
Toyon. See *Heteromeles arbutifolia*.
Trailing arbutus. See *Epigaea repens*.
Trailing azalea. See *Loiseleuria procumbens*, p.325.
Traveller's joy. See *Clematis*.
Traveller's tree. See *Ravenala madagascariensis*.
Tree cotoneaster. See *Cotoneaster frigidus*.
Tree fuchsia. See *Fuchsia arborescens*, p.134.
Tree germander. See *Teucrium fruticans*.
Tree heath. See *Erica arborea*.
Tree ivy. See x *Fatshedera lizei*, p.122.
Tree lupin. See *Lupinus arboreus*, p.139.
Tree mallow. See *Lavatera*.
Tree medick. See *Medicago arborea*.
Tree of heaven. See *Ailanthus altissima*.
Tree purslane. See *Atriplex halimus*.
Tree tomato. See *Cyphomandra crassicaulis*, p.95.
Trident maple. See *Acer buergerianum*.
Trigger plant. See *Stylidium graminifolium*.
Trinity flower. See *Trillium*.
Tropical almond. See *Terminalia catappa*.
Trumpet creeper. See *Campsis radicans*.
Trumpet flower. See *Bignonia capreolata*.
Trumpet gentian. See *Gentiana clusii*.
Trumpet honeysuckle. See *Campsis radicans*.
Trumpet vine. See *Campsis radicans*.
Trumpets. See *Sarracenia flava*, p.254.
Tuberose. See *Polianthes tuberosa*.
Tufted hair grass. See *Deschampsia caespitosa*.
Tufted sedge. See *Carex elata*.
Tulip. See *Tulipa*.
Tulip tree. See *Liriodendron tulipifera*, p.39.
Tumbling Ted. See *Saponaria ocymoides*, p.326.
Tunic flower. See *Petrorhagia saxifraga*, p.323.
Tupelo. See *Nyssa*.
Turkey oak. See *Quercus cerris*.
Turkish hazel. See *Corylus colurna*.
Turk's cap. See *Melocactus*.
Turkscap lily. See *Lilium superbum*.
Turtle-head. See *Chelone*.
Twin flower. See *Linnaea*.
Twin-flowered violet. See *Viola biflora*.

U

Ulmo. See *Eucryphia cordifolia*.
Umbrella leaf. See *Diphylleia cymosa*.
Umbrella pine. See *Pinus pinea*, p.83.
Umbrella plant. See *Peltiphyllum*.
Unicorn plant. See *Martynia annua*, p.272.
Urn plant. See *Aechmea fasciata*, p.229.

V

Valerian. See *Valeriana*.
Van Volxem's maple. See *Acer velutinum* var. *vanvolxemii*.
Variegated apple mint. See *Mentha suaveolens* 'Variegata', p.240.
Variegated Bishop's weed. See *Aegopodium podagraria* 'Variegatum', p.240.
Variegated creeping soft grass. See *Holcus mollis* 'Albovariegatus', p.182.
Variegated gout weed. See *Aegopodium podagraria* 'Variegatum', p.240.
Variegated ground ivy. See *Glechoma hederacea* 'Variegata', p.264.
Variegated iris. See *Iris variegata*, p.199.
Variegated Leyland cypress. See x *Cupressocyparis leylandii* 'Harlequin', p.78.
Variegated purple moor grass. See *Molinia caerulea* subsp. *caerulea* 'Variegata'.
Varnish tree. See *Rhus verniciflua*.
Veitch fir. See *Abies veitchii*, p.76.

Veitch's screw pine. See *Pandanus veitchii*, p.145.
Velvet plant. See *Gynura aurantiaca*, p.180.
Venetian sumach. See *Cotinus coggygria*.
Venus flytrap. See *Dionaea muscipula*, p.269.
Venus's navelwort. See *Omphalodes linifolia*, p.270.
Vetch. See *Hippocrepis*.
Victorian box. See *Pittosporum undulatum*.
Vine. See *Vitis*.
Vine lilac. See *Hardenbergia violacea*.
Vine maple. See *Acer circinatum*.
Violet. See *Viola*.
Violet cress. See *Ionopsidium acaule*.
Violet willow. See *Salix daphnoides*, p.48.
Virginia creeper. See *Parthenocissus quinquefolia*.
Virginia pine. See *Pinus virginiana*, p.81.
Virginian bird cherry. See *Prunus virginiana*.
Virginian chain fern. See *Woodwardia virginica*.
Virginian pokeweed. See *Phytolacca americana*.
Virginian stock. See *Malcolmia maritima*, p.275.
Virginian witch hazel. See *Hamamelis virginiana*, p.95.
Virgin's palm. See *Dioon edule*.
Voodoo lily. See *Sauromatum venosum*, p.357.
Voss's laburnum. See *Laburnum* x *watereri*.

W

Wake-robin. See *Trillium grandiflorum*, p.232; *Trillium sessile*, p.233.
Walking fern. See *Camptosorus rhizophyllus*.
Wall flag. See *Iris tectorum*, p.198.
Wallflower. See *Cheiranthus*.
Wall-spray. See *Cotoneaster horizontalis*, p.142.
Walnut. See *Juglans*.
Wandering Jew. See *Tradescantia fluminensis*.
Wandflower. See *Dierama*.
Waratah. See *Telopea speciosissima*, p.112.
Warminster broom. See *Cytisus* x *praecox*, p.126.
Washington grass. See *Cabomba caroliniana*.
Washington thorn. See *Crataegus phaenopyrum*.
Water chestnut. See *Trapa natans*, p.389.
Water dragon. See *Saururus cernuus*, p.387.
Water fern. See *Azolla caroliniana*, p.388; *Ceratopteris thalictroides*.
Water figwort. See *Scrophularia auriculata* 'Variegata'.
Water forget-me-not. See *Myosotis scorpioides*.
Water fringe. See *Nymphoides peltata*, p.391.
Water hawthorn. See *Aponogeton distachyos*, p.387.
Water hyacinth. See *Eichhornia crassipes*, p.388.
Water lettuce. See *Pistia stratiotes*, p.389.
Water lily. See *Nymphaea*.
Water lily tulip. See *Tulipa kaufmanniana*, p.356.
Water moss. See *Fontinalis antipyretica*.
Water oak. See *Quercus nigra*, p.42.
Water plantain. See *Alisma plantago-aquatica*, p.386.
Water poppy. See *Hydrocleys nymphoides*, p.389.
Water soldier. See *Stratiotes aloides*, p.387.
Water violet. See *Hottonia palustris*, p.387.
Waterer's gold holly. See *Ilex aquifolium* 'Watereriana', p.73.
Watermelon begonia. See *Pellionia repens*, p.267.
Watermelon plant. See *Peperomia argyreia*.
Wax flower. See *Stephanotis floribunda*, p.165.
Wax plant. See *Hoya carnosa*, p.168.
Wax privet. See *Peperomia glabella*, p.269.
Wax tree. See *Rhus succedanea*.
Wax vine. See *Senecio macroglossus*.
Wayfaring tree. See *Viburnum lantana*.
Wedding-cake tree. See *Cornus controversa* 'Variegata', p.63.
Weeping aspen. See *Populus tremula* 'Pendula', p.53.
Weeping beech. See *Fagus sylvatica* f. *pendula*, p.40.
Weeping fig. See *Ficus benjamina*.
Weeping willow. See *Salix babylonica*.
Wellingtonia. See *Sequoiadendron giganteum*, p.76.
Welsh poppy. See *Meconopsis cambrica*, p.239.

West Himalayan birch. See *Betula utilis* var. *jacquemontii*, p.57.
West Himalayan spruce. See *Picea smithiana*.
West Indian jasmine. See *Plumeria alba*.
West Indian tree fern. See *Cyathea arborea*.
Western balsam poplar. See *Populus trichocarpa*.
Western hemlock. See *Tsuga heterophylla*.
Western prickly Moses. See *Acacia pulchella*, p.127.
Western red cedar. See *Thuja plicata*.
Western tea-myrtle. See *Melaleuca nesophila*, p.115.
Western yellow pine. See *Pinus ponderosa*, p.77.
Weymouth pine. See *Pinus strobus*, p.76.
Wheatley elm. See *Ulmus* 'Sarniensis'.
White ash. See *Fraxinus americana*.
White asphodel. See *Asphodelus albus*, p.203.
White baneberry. See *Actaea alba*, p.223.
White Chinese birch. See *Betula albosinensis*, p.48.
White cypress. See *Chamaecyparis thyoides*, p.81.
White elm. See *Ulmus americana*.
White evening primrose. See *Oenothera speciosa*.
White false hellebore. See *Veratrum album*.
White fir. See *Abies concolor*.
White ginger lily. See *Hedychium coronarium*.
White mugwort. See *Artemisia lactiflora*, p.191.
White mulberry. See *Morus alba*.
White mullein. See *Verbascum lychnitis*.
White poplar. See *Populus alba*, p.38.
White rosebay. See *Epilobium angustifolium* f. *album*, p.190.
White sails. See *Spathiphyllum wallisii*, p.264.
White Sally. See *Eucalyptus pauciflora*, p.58.
White snakeroot. See *Eupatorium rugosum*.
White spruce. See *Picea glauca*.
White trumpet lily. See *Lilium longiflorum*, p.348.
White water lily. See *Nymphaea alba*, p.390.
White willow. See *Salix alba*.
White-backed hosta. See *Hosta hypoleuca*.
Whitebeam. See *Sorbus aria*.
Whiteywood. See *Melicytus ramiflorus*.
Whorl flower. See *Morina*.
Whorled Solomon's seal. See *Polygonatum verticillatum*.
Whorled water milfoil. See *Myriophyllum verticillatum*, p.389.
Whortleberry. See *Vaccinium myrtillus*.
Widow iris. See *Hermodactylus tuberosus*, p.358.
Wild buckwheat. See *Eriogonum*.
Wild cherry. See *Prunus avium*, p.45.
Wild coffee. See *Polyscias guilfoylei*.
Wild daffodil. See *Narcissus pseudonarcissus*, p.361.
Wild ginger. See *Asarum*.
Wild iris. See *Iris versicolor*, p.198.
Wild Irishman. See *Discaria toumatou*.
Wild lily-of-the-valley. See *Pyrola rotundifolia*.
Wild marjoram. See *Origanum vulgare*.
Wild pansy. See *Viola tricolor*, p.317.
Wild pineapple. See *Ananas bracteatus*.
Wild rum cherry. See *Prunus serotina*, p.39.
Wild yellow lily. See *Lilium canadense*.
Willow. See *Salix*.
Willow gentian. See *Gentiana asclepiadea*, p.227.
Willow herb. See *Epilobium*.
Willow moss. See *Fontinalis antipyretica*.
Willow myrtle. See *Agonis*.
Willow oak. See *Quercus phellos*, p.45.
Willow-leaved magnolia. See *Magnolia salicifolia*, p.49.
Willow-leaved sunflower. See *Helianthus salicifolius*.
Windflower. See *Anemone*; *Zephyranthes*.
Windmill palm. See *Trachycarpus fortunei*, p.58.
Winecups. See *Babiana rubrocyanea*, p.375.
Wing nut. See *Pterocarya*.
Winged spindle. See *Euonymus alatus*, p.118.
Winter aconite. See *Eranthis hyemalis*, p.385.
Winter cherry. See *Cardiospermum halicacabum*; *Physalis alkekengi*; *Solanum capsicastrum*.
Winter cress. See *Barbarea vulgaris*.
Winter heath. See *Erica carnea*.
Winter heliotrope. See *Petasites fragrans*.
Winter iris. See *Iris unguicularis*.
Winter jasmine. See *Jasminum nudiflorum*, p.121.
Winter savory. See *Satureja montana*.
Winterberry. See *Ilex verticillata*, p.72.
Wintergreen. See *Pyrola*.

Winter's bark. See *Drimys winteri*, p.52.
Wintersweet. See *Acokanthera oblongifolia*, p.120; *Chimonanthus praecox*.
Wire-netting bush. See *Corokia cotoneaster*, p.122.
Wishbone flower. See *Torenia fournieri*, p.285.
Witch alder. See *Fothergilla gardenii*.
Witch hazel. See *Hamamelis*.
Woad. See *Isatis tinctoria*.
Wolf's bane. See *Aconitum*.
Wonga-wonga vine. See *Pandorea pandorana*.
Wood anemone. See *Anemone nemorosa*.
Wood lily. See *Trillium*.
Wood millet. See *Milium effusum*.
Wood rose. See *Merremia tuberosa*.
Wood sorrel. See *Oxalis acetosella*.
Wood spurge. See *Euphorbia amygdaloides*.
Woodbine. See *Lonicera periclymenum*.
Woodruff. See *Galium odoratum*, p.239.
Woodrush. See *Luzula*.
Woolly netbush. See *Calothamnus villosus*.
Woolly willow. See *Salix lanata*, p.126.
Wormwood. See *Artemisia*.
Wormwood cassia. See *Cassia artemisioides*.
Wych elm. See *Ulmus glabra*.

Y

Yarrow. See *Achillea millefolium*.
Yatay palm. See *Butia*.
Yellow asphodel. See *Asphodeline lutea*, p.202.
Yellow banksia. See *Rosa banksiae* 'Lutea', p.164.
Yellow bells. See *Tecoma stans*, p.70.
Yellow birch. See *Betula alleghaniensis*.
Yellow buckeye. See *Aesculus flava*, p.56.
Yellow cosmos. See *Cosmos*, Bright Lights Series.
Yellow elder. See *Tecoma stans*, p.70.
Yellow flag. See *Iris pseudacorus*, p.199.
Yellow flax. See *Linum flavum*; *Reinwardtia indica*, p.140.
Yellow foxglove. See *Digitalis grandiflora*.
Yellow fritillary. See *Fritillaria pudica*, p.378.
Yellow haw. See *Crataegus flava*, p.64.
Yellow jasmine. See *Jasminum humile*, p.116.
Yellow kangaroo paw. See *Anigozanthos flavidus*, p.218.
Yellow lady's slipper orchid. See *Cypripedium calceolus*, p.262.
Yellow mariposa. See *Calochortus luteus*, p.359.
Yellow morning glory. See *Merremia tuberosa*.
Yellow musk. See *Mimulus luteus*, p.256.
Yellow oleander. See *Thevetia peruviana*, p.67.
Yellow ox-eye. See *Buphthalmum salicifolium*, p.256.
Yellow palm. See *Chrysalidocarpus lutescens*, p.74.
Yellow parilla. See *Menispermum canadense*.
Yellow pitcher plant. See *Sarracenia flava*, p.254.
Yellow pond lily. See *Nuphar advena*.
Yellow rocket. See *Barbarea vulgaris*.
Yellow scabious. See *Cephalaria gigantea*.
Yellow skunk cabbage. See *Lysichiton americanus*, p.391.
Yellow turkscap lily. See *Lilium pyrenaicum*.
Yellow water lily. See *Nuphar lutea*, p.391.
Yellow whitlow grass. See *Draba aizoides*.
Yellow wood. See *Cladrastis lutea*, p.56.
Yellow-root. See *Xanthorhiza simplicissima*.
Yesterday-today-and-tomorrow. See *Brunfelsia pauciflora*.
Yew. See *Taxus baccata*.
Yoshino cherry. See *Prunus* x *yedoensis*, p.61.
Young's weeping birch. See *Betula pendula* 'Youngii', p.66.
Youth-on-age. See *Tolmiea menziesii*.
Yulan. See *Magnolia denudata*, p.49.

Z

Zebra plant. See *Aphelandra squarrosa*; *Calathea zebrina*, p.230.
Zigzag bamboo. See *Phyllostachys flexuosa*, p.184.

Acknowledgments

1	4	7	10
2	5	8	11
3	6	9	12

The publishers are grateful to the following for permission to reproduce illustrative material. Most of the photographs are found in *The Plant Catalogue* and are referenced by two numbers: the page number is given first, followed by the specific number or numbers of the photograph(s) separated by hyphens. The photograph number is determined by the position of the photograph's caption, according to the page grid shown alongside. The same numbering principle has been used for photographs in the Feature Boxes, with numbers from 1 to 42; illustrations elsewhere in the book are numbered from top to bottom of the page.

Alpine Garden Society Slide Library 286/5, 296/4-9, 302/3, 304/11, 310/3, 312/10, 321/9, 323/12, 326/6, 327/5, 331/5, 336/3, 340/10, 344/3, 372/13, 373/2-26, 379/1
Jacques Amand Ltd./John Amand 200/5-25, 253/42, 375/24
A-Z Botanical Collection 114/26, 145/9, 162/6, 164/10, 178/11, 198/35, 206/35, 223/7, 246/7, 295/6, 322/2, 348/31, 387/2-10, 388/6
Gillian Beckett 89/1, 93/2-3, 102/21, 110/5, 115/1, 120/5, 124/4, 125/7, 127/12, 138/12, 141/1, 144/11, 179/6, 187/4, 189/10, 212/5, 214/4, 215/4, 218/12, 220/1, 228/1, 237/12, 252/38, 254/5, 256/3, 272/1, 296/1, 302/11, 304/12, 308/2, 309/1, 311/1, 315/11, 316/1, 321/1, 322/8-9, 325/2, 326/1, 329/10, 331/2, 333/3, 337/1, 346/12, 349/5-33, 356/3-6, 357/3-6, 358/10, 359/3, 361/21, 362/5, 365/3, 367/2-6, 375/37, 376/2-3, 377/8, 378/1, 379/7, 383/9-10, 384/3-7, 386/6, 388/7-9
Kenneth A. Beckett 51/4, 166/8, 314/2, 316/4
Biofotos/Heather Angel 388/12
Patrick Booth/John Thirkell 194/26-37-38
Ann & Roger Bowden 252/25-27, 253/5
Christopher Brickell 63/4, 74/1, 101/2-4, 102/17, 103/4-25, 104/6, 105/1, 112/12, 117/4, 118/1-5, 135/22, 166/1
Pat Brindley 106/4, 168/12, 170/7, 175/4, 189/4, 199/5, 201/18, 206/26, 212/38, 259/19, 272/2, 273/4-7-9, 274/11, 276/7-11-12, 278/9, 280/3-12, 281/4, 283/7, 284/3-4, 285/9, 288/2-8, 289/6, 293/3, 343/5, 354/23, 355/8-41, 356/13-17-27-31, 373/7-29-37, 388/8
British Iris Society 199/2-14-20
Brinsley Burbridge 169/12
Ray Cobb 373/9
Eric Crichton 39/5, 44/4, 45/2, 67/1, 71/3, 73/10, 113/3, 127/9, 138/4, 147/11, 168/10, 175/1, 189/9, 195/3, 198/7-15-29, 202/1, 207/1, 219/1-10, 246/10, 272/10, 276/6, 290/7, 297/9, 302/10, 303/1, 305/9, 306/3, 317/1, 323/3, 327/1, 330/9, 331/3, 335/3, 340/7, 343/19, 350/36, 351/10, 354/17, 363/11, 364/1, 373/31, 390/31
Philip Damp 350/36, 351/6-40
Alan M. Edwards 361/35, 372/17
Raymond Evison 172/27-28-33, 173/13-14-16-25-35-41
Valerie Finnis 302/7
Ron & Christine Foord 236/11
Maureen Foster 199/13
John Glover 101/3, 104/27, 128/8, 206/42, 213/9, 2501,385/3
Derek Gould 86/11, 94/11, 104/4, 105/4, 114/40, 120/6, 123/6, 127/10, 140/2, 166/5, 168/7, 178/6, 197/7, 200/20, 204/4, 205/3-6-8, 212/33, 213/1, 217/12, 222/9-10, 232/12, 236/34, 241/6, 243/8, 254/3, 286/11, 293/6, 295/2, 307/7, 334/7, 338/4, 341/7, 356/11, 359/8, 371/3, 387/7
Diana Grenfell 221/6-10, 252/5-7-20-24, 253/19-35-40
Jerry Harpur 2 (facing title page)
Terry Hewitt 392/9, 408/2, 410/2, 416/1
Muriel Hodgeman 301/11, 307/10, 331/12, 333/1
Hortico 362/15
International Flower Bulb Centre 375/35
Mike Ireland 315/2, 322/3, 325/7, 373/4-38
Andrew Lawson 348/11-16, 349/3-30
Sidney Linnegar 198/4 (G.E. Cassidy)-11 (R. Henley)-31 (R. Henley) -37 (G.E. Cassidy), 199/3-34 (G.E. Cassidy) -40 (G.E. Cassidy)
Brian Mathew 346/3, 358/11, 363/1-4, 365/9, 367/1-5, 368/3, 370/9, 378/11
S & O Mathews 121/7
Oxford Scientific Films/Fredrik Ehrenstrom 389/11
Photos Horticultural 195/12, 198/10, 200/35, 206/18-21, 241/3, 281/1, 287/5, 349/11, 368/2
Collection & Photo Riviere (France, 26 Drôme) 200/18, 201/21-36
Royal Botanic Gardens, Kew 12/1
Royal Horticultural Society, Lindley Library: 10: Curtis's Botanical Magazine (CBM), lxxv (1849), T.4458: 11/1: CBM, cxvix (1923), T.9004; 11/2: R.J. Thornton: A New Illustration of the Sexual System of Carolus Von Linnaeus (1807); 12/2: CBM, cliv (1928), T.9241; 13: Edwards's Botanical Register, xx (1835), T.1686; 37: CBM, cl (1924), T.9036; 417: CBM, vii (1794), T.252.
A.D. Schilling 57/10, 64/11, 102/4, 136/8, 139/9
Arthur Smith 343/26-38-42
Harry Smith Collection 39/6, 43/3-12, 44/1, 45/4-10, 46/7-10, 49/7-22-36, 51/4-11 (inset), 55/3, 56/6-8, 57/9, 63/6, 65/1-6, 66/3-12, 67/2-6-7-9, 69/8, 71/4-6, 76/1, 78/3-5, 80/9, 82/8-12, 83/6, 85/33, 86/7, 87/1-6, 88/11, 89/4, 91/11, 93/1, 94/1-7, 95/6, 96/9, 99/31-35, 100/5-11, 102/20-22-27-32-35, 103/16, 104/16, 105/2, 111/4, 112/7, 113/5, 115/9, 116/7, 117/2-5-12, 118/10, 119/2, 120/1, 121/11, 124/7, 125/3-6, 128/12, 130/6, 134/25-39, 135/17, 138/10, 140/10, 141/3, 142/6, 143/5-6-12, 147/7, 162/8, 164/4-5-6, 165/7, 166/10, 167/3, 168/1-9, 169/3-9, 170/1-5-6, 171/2-3-7, 174/2-3-4, 177/2-12, 178/8-10, 179/2-3, 180/12, 181/26, 190/3, 192/9, 194/19-30-35-40, 195/9, 196/1-5, 199/4-23, 201/1-11-14-37, 206/13-19-29, 208/11, 212/28-39-40, 218/6, 221/16, 227/9, 230/1, 232/8, 233/7, 234/1-4, 235/8, 237/1-4-18, 239/12, 242/6, 246/1-9, 247/12, 250/4, 251/6, 252/9-31-35, 253/1, 257/6-10, 259/19, 271/10-12, 272/4, 273/8, 278/12, 279/3, 281/7-11, 282/4, 283/4-5-6-8, 284/12, 286/3-4-6-7, 290/10, 291/1-3, 292/9, 293/1-8-9, 295/1-5, 297/2-10, 299/9, 300/6, 304/9, 305/6, 308/1-4-7, 309/4, 312/5-12, 313/9, 314/11, 316/10, 317/9-11, 319/6-7-8-10, 320/7, 321/2, 323/5-9, 328/12, 329/5, 331/1-7, 334/8-9, 335/8-12, 336/4, 339/6, 341/4, 342/2, 343/11-16-28-33-35, 344/1-2, 345/12, 346/5, 347/6-12, 348/12-15-17-18-19, 349/10-12-19-40, 349/352/1-6, 353/6, 355/32, 356/5-38, 358/3-6, 359/6, 364/7, 365/6, 366/9-11, 368/9, 369/1-11-12, 370/7, 371/12, 372/16-28-30-42, 373/8-16-34,374/10,376/1-4-11, 378/9, 379/3, 380/4-8-10, 381/11, 382/1-4, 383/4, 384/2-11, 385/4-7-10, 387/3-4-12, 389/7-10, 390/9, 403/7,411/10
Sir Peter Smithers 200/28, 201/10-15-17
Thompson & Morgan 280/2, 283/9, 287/6
Unwins Seeds Ltd. 212/7, 274/12, 278/6, 280/10-11, 283/10-11, 284/1, 288/1, 292/3-6-9
Jack Wemyss-Cooke 237/13-14-35
John Wright 134/3, 135/15

Picture research: Andrew Brown, Susan Mennel
Picture research for the revised edition: Anna Lord

Abbreviations

C	centigrade	mm	millimetres
cm	centimetre(s)	Mme	Madame
cv(s)	cultivar(s)	min.	minimum
Dr	Doctor	p(p).	page(s)
f.	forma	pl.	plural
F	Fahrenheit	S	spread
ft	foot, feet	subsp.	subspecies
H	height (or length of trailing stems)	subspp.	subspecies (pl.)
illus.	illustrated	St	Saint
in	inch(es)	syn.	synonym(s)
m	metre(s)	var.	varietas

The publishers would like to thank all those who generously assisted the photographers and provided plants for photography, in particular the curators, directors and staff of the following organizations and those private individuals listed below. Special thanks are due to those at the Royal Botanic Gardens, Kew, and the Royal Horticultural Society's Garden, Wisley, for their invaluable assistance and support.

African Violet Centre, Terrington St Clement, Norfolk; Ken Akers, Great Saling, Essex; Jacques Amand Ltd, Clamphill, Middx; Anmore Exotics, Havant, Hants; David Austin Roses, Albrighton, Shrops; Avon Bulbs, Bradford-on-Avon, Wilts; Ayletts Nurseries, St Albans, Herts; Steven Bailey Ltd, Sway, Hants; Bill Baker, Tidmarsh, Berks; Batsford Arboretum, Moreton-in-Marsh, Glos; Booker Seeds, Sleaford, Lincs; Rupert Bowlby, Reigate, Surrey; Bressingham Gardens, Diss, Norfolk; Roy Brooks, Newent, Glos; British Orchid Growers' Association; Broadleigh Gardens, Somerset; Burford House Gardens, Tenbury Wells, Shrops; Cambridge Bulbs, Newton, Cambs; Nola Carr, Sydney, Australia; Beth Chatto Gardens, Colchester, Essex; Chelsea Physic Garden, London; Colegrave Seeds, Banbury, Oxon; County Park Nurseries, Hornchurch, Essex; Jill Cowley, Chelmsford, Essex; Mrs Anne Dexter, Oxford; Edrom Nurseries, Coldingham, Berwicks; Dr Jack Elliott, Ashford, Kent; Joe Elliott, Broadwell, Glos; Erdigg (National Trust), Clwyd, Wales; Fibrex Nurseries, Pebworth, Warwicks; Fisk's Clematis Nursery, Westleton, Suffolk; Mr & Mrs Thomas Gibson, Westwell, Oxon; Glasgow Botanic Garden, Glasgow; 'Glazenwood', Braintree, Essex.

R. Harkness & Co. Ltd, Hitchin, Herts; Harry Hay, Lower Kingswood, Surrey; Hazeldene Nurseries, East Farleigh, Kent; Hidcote Manor (National Trust), Chipping Camden, Glos; Hillier Gardens and Arboretum, Romsey, Hants; Hillier Nurseries (Winchester) Ltd, Romsey, Hants; Holly Gate Cactus Nursery, Ashington, Sussex; Hopleys Plants, Much Hadham, Herts; Huntingdon Botanical Gardens, San Marino, California; W.E.Th. Ingwersen Ltd, East Grinstead, Sussex; Clive Innes, Ashington, Sussex; Kelways Nurseries, Langport, Somerset; Kiftsgate Court Gardens, Chipping Camden, Glos; Lechlade Fuchsia Centre, Lechlade, Glos; The Living Desert, Palm Desert, California; Robin Loder, Leonardslee, Sussex; Los Angeles State and County Arboreta and Botanical Gardens, Los Angeles, California; Lotusland Foundation, Santa Barbara, California; McBeans Orchids, Lewes, Sussex; Merrist Wood Agricultural College, Worplesdon, Surrey; Mrs J.F. Phillips, Westwell, Oxon; Mr & Mrs Richard Purdon, Ramsden, Oxon; Ramparts Nurseries, Colchester, Essex; Ratcliffe Orchids, Didcot, Oxon; Mrs Joyce Robinson, Denmans, Fontwell, Sussex; Peter Q. Rose, Castle Cary, Somerset; Royal Botanic Garden, Edinburgh; Royal Botanic Gardens, Kew, Surrey; Royal Botanic Gardens, Sydney, Australia; Royal National Rose Society, St Albans, Herts; Royal Horticultural Society's Garden, Wisley, Surrey.

Santa Barbara Botanic Garden, Santa Barbara, California; Savill Garden, Windsor, Berks; Mr & Mrs K. Schoenenberger, Shipton-under-Wychwood, Oxon; Mrs Martin Simmons, Burghclere, Berks; Dr James Smart, Barnstaple, Devon; Arthur Smith, Wigston, Leics; P.J. Smith, Ashington, Sussex; Springfields Gardens, Spalding, Lincs; Staite & Sons, Evesham, Hereford and Worcs; Stapeley Water Gardens, Nantwich, Cheshire; Strybing Arboreta Society of Golden Gate Park, San Francisco, California; David Stuart, Dunbar, East Lothian; Suffolk Herbs, Sudbury, Suffolk; University Botanic Garden, Cambridge; University of British Columbia Botanical Garden, Vancouver; University of California Arboretum, Davis, California; University of California Arboretum, Santa Cruz, California; University of California Botanical Garden, Berkeley, California; University of California Botanical Gardens, Los Angeles, California; University of Reading Botanic Garden, Reading, Berks; Unwins Seeds Ltd, Histon, Cambridge; Jack Vass, Haywards Heath, Sussex; Rosemary Verey, Barnsley, Glos; Vesutor Air Plants, Ashington, Sussex; Wakehurst Place (Royal Botanic Gardens, Kew), Ardingly, Sussex; Primrose Warburg, Oxford; Waterperry Gardens, Wheatley, Oxon; Westonbirt Arboretum, Westonbirt, Glos; Woolman's Nurseries, Dorridge, West Midlands; Wyld Court Orchids, Newbury, Berks; Eric Young Orchid Foundation, Jersey, Channel Islands.